英汉船舶及海洋工程技术大辞典

(第2版)

AN ENGLISH-CHINESE DICTIONARY OF MARINE AND OCEAN ENGINEERING

顾　问　曾恒一
主　编　梁启康
副主编　王　磊　史本宁　林德辉　黄恒祥

联合组编单位
中国船舶及海洋工程设计研究院
上海交通大学海洋工程国家重点实验室
上海市船舶与海洋工程学会
国家深海技术试验大型科学仪器中心
上海市海洋工程科普基地
上海研途船舶海事技术有限公司

哈尔滨工程大学出版社
Harbin Engineering University Press

内 容 简 介

本辞典编写是一项总结和提升船舶工业科技水平的基础性工作,在包含传统、基本的船舶及海洋工程技术词汇的基础上,最大限度收录国际海事界最新、最优和最具代表性的词目,能最大限度满足我国船舶与海洋工程行业的发展需要。

本辞典大量引用近十年来出现的新船型、设备、工艺、方法以及国际公约、规则、规范、标准和法规的新变化,使读者可以方便地了解其含义和要求。创新了英汉船舶辞典的编排模式,对英语词目除中文释义外,还增加技术性的解释;对涉及国际公约、规则、规范、标准和法规的词目还进行中英(双语)解释,以便读者正确理解。

本辞典适用于船舶与海洋工程领域工程技术人员和高等院校的师生以及船舶及海洋工程爱好者查阅。

图书在版编目(CIP)数据

英汉船舶及海洋工程技术大辞典/梁启康主编. —哈尔滨:哈尔滨工程大学出版社,2021.4
ISBN 978-7-5661-2827-0

Ⅰ. ①英… Ⅱ. ①梁… Ⅲ. ①船舶工程—对照词典—英、汉②海洋工程—对照词典—英、汉 Ⅳ. ①U66-61 ②P75-61

中国版本图书馆 CIP 数据核字(2021)第 064774 号

选题策划	史大伟 薛 力
责任编辑	薛 力 王 宇 田雨虹

出版发行	哈尔滨工程大学出版社
社　　址	哈尔滨市南岗区东大直街 124 号
邮政编码	150001
发行电话	0451-82519328
传　　真	0451-82519699
经　　销	新华书店
印　　刷	吉林省吉广国际广告股份有限公司
开　　本	787 mm×1092 mm　1/16
印　　张	76.75
字　　数	2012 千字
版　　次	2021 年 4 月第 1 版
印　　次	2021 年 4 月第 1 次印刷
定　　价	500.00 元

http://www.hrbeupress.com
E-mail:heupress@hrbeu.edu.cn

出　　品	船海书局
网　　址	www.ship-press.com
告 读 者	如发现本书有印装质量问题请与船海书局发行部联系。
服务热线	400 867 0886
版权所有	侵权必究

第2版《英汉船舶及海洋工程技术大辞典》
编辑出版委员会

顾　问　曾恒一

主　任　邢文华

执行副主任（以姓氏笔画为序）

马胜远　孔　斌　占金锋　卢　霖　史本宁　冯学宝　严　俊　严浙平
杨建民　汪学锋　张宏军　林　枫　柳存根　候晓明　翁震平

副主任（以姓氏笔画为序）

王　征　王　朋　王永庆　王君农　孔　冰　叶志浩　史大伟　宁伏龙
朱　刚　朱　凌　刘文斌　孙树政　李　雄　李文华　李孝堂　李沂滨
杨　松　张越雷　陈　锋　欧金生　秦炳军　聂志斌　高霄鹏　黄　焱
黄　蔚　黄永仲　谌志新　隋江华　董国祥　樊兴龙

委　员（以姓氏笔画为序）

于霄雷　王　宇　王　林　王　磊　王广东　王凤生　王东辉　王孟军
王俊杰　王海荣　王朝杰　亢峻星　甘浪雄　古　彪　可　伟　叶　聪
叶邦全　兰志华　邢　瑜　吉雨冠　刘　昆　刘齐辉　刘承敏　刘厚德
刘剑军　孙宏亮　孙维生　芮　旻　李　俊　李世远　李存军　李英超
杨立军　吴　彬　吴开明　吴文伟　吴贻欣　沈　权　宋　磊　宋丹戎
张　健　张　萍　张太佶　张世联　张立华　张阿漫　陈　松　陈宝起
林德辉　周　龙　周长江　郑　宇　夏大男　奚崇德　黄恒祥　黄领才
梁圣童　梁启康　彭　涛　蒋志勇　韩　龙　傅　冰　薛　力　魏建平

秘书长　周长江

第 2 版《英汉船舶及海洋工程技术大辞典》编辑工作组

组　　长　　史大伟

副组长　　梁启康　　王　磊　　史本宁　　周长江

翻译人员

梁云芳　　魏建平　　彭　涛　　王海荣　　刘剑军　　江齐锋　　舒　展　　李辉绸

参编人员

蒋勇刚　　牟金磊　　严　谨　　李　威　　张大勇　　王　慧　　沈　权　　孙宏亮
周　龙　　傅　冰　　王俊杰　　张　萍　　樊兴龙　　王君农　　孔　冰

专业审校人员

段雪琼　　解洲水　　汪　骥　　桂洪斌　　史本宁　　王　朋　　杨　松　　李文华
吉雨冠　　刘承敏　　宋丹戎　　王东辉　　陈　锋　　张越雷　　欧金生

出版策划

薛　力　　左芳琪

前　　言

　　船舶工业是国家战略性产业之一,是国家经济技术、国防实力、工业化和信息化水平的主要标志之一。我国从自行设计、建造第一艘万吨级"东风号"货船到能建造包括豪华邮船在内的国内外船东所需的所有船型,船舶工业实现了跨越式发展成为世界第一造船大国。但是,在技术创新、环境保护、可靠性、人才培养、生产管理、效率、高附加值等诸多方面与世界造船强国还有一定的差距。

　　进入21世纪,船舶设计的理念已经发生了很大变化。新船型、新技术、新方法、新材料、新工艺、新产品大量涌现,特别是海洋工程和游艇市场异军突起,已成为我国船舶工业新的技术、经济增长点和产品结构调整方向。军品出口市场的活跃、对船舶的水上安全航行、环境保护、节能减排和船员健康的要求越来越高,对生态保护和经济可持续发展关注度也越来越高,对船舶经济性、高效率等也提出了更高的标准。这些变化促使相关的国际海事组织、国际船东协会和国际船级社协会等连续出台了许多新公约、规范、标准和法规。

　　为了应对国内外船舶市场的需求,应对国际海事组织造船新规则、新标准的变化;同时,随着我国船舶行业向国际化、市场化、商业化的深度发展,商业利益和行业先进技术与经验推广之间的矛盾日渐显现,行业的发展迫切需要高水平的出版物传播和积累先进造船技术与经验。自2008年开始,我们组织了多位长期从事船舶与海洋工程科研、设计方面的资深专家编写了第1版《英汉船舶与海洋工程技术大辞典》。多年来,这本辞典已为广大船舶行业专业读者所接受和认可,成为大家在工作和学习中的帮手。

　　这次,本辞典在原有的基础上进行了修订和补充。其再版的编撰原则是:在包含传统、基本的船舶及海洋工程技术词汇的基础上,最大限度收录最新、最优和最具代表性的词目。此次共新增词目600余条。

　　本辞典具有以下特点:

　　1. 创新性

　　摆脱了过去英汉船舶辞典的传统模式,对英语词目除中文释义外,还增加技术性的解释;对涉及国际公约、规则、规范、标准和法规的词目还进行中英(双语)解释,以便读者正确理解。

　　2. 前瞻性

　　从最新文本和资料中收录词目(截至2020年10月),包括:① 最新的设计理念(如EEDI)、设计路线和方法、检验方法和标准;② 最新的标准,特别是国际公约、规则、规范、标准和法规(包括即将生效的);③ 国内外最新、最先进的船舶设计技术、工艺、装备、产品和材料;④ 实用性强的相关附录。

　　3. 系统性

　　从专业角度涵盖了船舶总体、性能、结构、舾装、轮机、电气和计算机等;从船型角度涵盖了民用船舶、海洋工程、游艇、高性能船舶和军用舰船等;从现代造船技术角度涵盖了检验、材料、焊接、工艺和涂装等;从航运角度涵盖了苏伊士运河、巴拿马运河、美国水域航运规则;从绿色船舶和环境保护角度涵盖了防污染公约、拆船公约和新涂层标准等。

　　4. 实用性

　　一是大量引用近十年来出现的新船型、设备、工艺、方法以及国际公约、规则、规范、标准和法规的新变化,使读者可以方便地了解其含义和要求。二是编排时,采用三种方式:① 对传统和基本的词目采用英译中的方式;② 对新出现的专业性强的词目,除英译中外,还补充中文解释;③ 对引自国际公约、规则、规范、标准和法规的词目提供中英双语解释。

　　作为一本实用的工具书,本辞典主要面向下列读者群:① 船舶与海洋工程研究、设计的科技工作者;② 船舶与海洋工程建造部门的工程技术人员和管理人员；③船舶与海洋工程设备供应商；④船舶与海洋工程检验部门和验船师；⑤船舶航运及海洋工程装备操作者；⑥国防装备部门工程技术人员；⑦高等院校船舶及海洋工程专业师生；⑧船舶及海洋工程爱好者等。

船舶及海洋工程涉及的专业范围广,发展日新月异,加之编者学识水平的限制,辞典在收录准确性和实时性方面恐有不完善之处,期望广大读者提出宝贵意见。

编　者
2021年3月于上海

凡　　例

一、本辞典分为前后两部分，前半部分为英汉部分，后半部分为索引部分。

二、本辞典正文按照英文字母顺序排列，不考虑字母的大小写，数字及希腊字母另列，专用符号（空格、圆点、连字符、斜杠等）不参加排序。

三、本辞典中文索引按中文汉语拼音字母顺序排列，同一个词汇有不同的英文释义的归为同一个词条。

四、前半部分中词汇的译名用粗体，一个英文词汇有多个译名的用"/"隔开。

五、本辞典的词条包括下列5种形式：

①单词，如 Cavity 空泡；

②复合词，如 After body 后体；

③缩略语，如 DFD（Dual – fuel diesel）双燃料柴油机；

④缩写符号如 Air clr. = Air cooler 空气冷却器；

⑤专业上应用较多的常用词组视为一个词条，如 Electronic chart 电子海图。

六、圆括号（　　）内的内容表示：

①括号内的字母、音节、单词和汉字可以省略。

例如：Crafer's bottom(SWATH) → Crafer's bottom

压缩天然气(CNG)运输船 → 压缩天然气运输船；

②英文词条后括号内的内容表示可以替换某一词或字的缩略语、单词、词组，如 Deep sea vehicle(Deep diving submersible)深潜器；

③对于词汇的译名尚不足以理解其含义的少量词条在其后的圆括号中添加简要注释，以做进一步的说明，如 Deepest operating water line(fishing vessel)最深作业水线（渔船）。

七、每页的书眉左右两侧分别有两个单词（词组），左侧的单词为本页的第一个单词（词组），右侧的单词为本页的最后一页单词（词组）。

八、当名词和对应的形容词同为词目时，一般选名词词条，以不重复为度。

九、英文词汇除必须是复数外，一般用单数表示。

目　　录

A	1
B	65
C	104
D	203
E	266
F	314
G	367
H	402
I	442
J	486
K	488
L	489
M	530
N	584
O	614
P	650
Q	710
R	713
S	755
T	880
U	929
V	940
W	958
X	986
Y	987
Z	988
以数字起首的词条	989
附录1　设计图纸上的缩写符号	991
附录2　国际公约和规则专属名词解释	1000
附录3　不适用 IBC 规则的货物清单	1012
附录4　危险化学品清单	1014
附录5　国际组织及其规则、标准中与船舶有关缩写词英汉对照	1025
中文索引	1051
后记	1215

A

A-60 class division　A-60 级耐火分隔　系指由符合一定要求的舱壁和甲板所构成的分隔，使用经认可的不可燃性材料隔热，经 60 s 的燃烧试验，保持结构的完整。并在规定的时间内，其背火一面的平均温度较原温度升高不超过 140 ℃，且在包括任何接头在内的任何一点上的温度较原温度升高不超过 180 ℃。

A-60 standard　A-60 标准　表示一适当扶强的、用钢或其他等效的材料建造的防火结构，它能在整整 1 h 的标准耐火试验期间内防止烟和火焰通过。它们应用认可的不燃性材料隔热，从而使得在 60 min 时间内，其背火一面的平均温度较原温度增高不超过 139 ℃，且在包括任何接头在内的任何一点处的温度，较原温度增高不超过 180 ℃。A-60 standard means a fire-resisting construction of steel or other equivalent material, which is suitably stiffened and so constructed as to be capable of preventing the passage of smoke and flame for the complete period of the one-hour standard fire test. It is to be insulated with approved non-combustible materials, so that the average temperature on the unexposed side will not rise by more than 139 ℃ above the original temperature, nor will the temperature at any one point, including any joint, rise more than 180 ℃ above the original temperature within 60 minutes.

A&R winch　收放绞车

A. C. (alternating current)　交流电　系指：(1)一种周期性的电流，其平均值在一个周期内为零。(2)一种在有规则地重复出现的时间间隔时，发生反向且交替出现正值和负值的电流。

A. C. magnetic starter　交流磁力启动器　系指用于笼型电动机全电压从电网直接启动的控制电器。磁力启动器主要由电源开关、接触器、热继电器组成，并配以控制变压器、按钮、指示灯、电流表、计时器等辅助元件。启动器具有失压和过载保护功能。为了配合某些机械简单的自动控制的需要，有的启动器还带有手动-自动转换控制开关。

A-bracket　人字架　有两根支撑的艉轴架。

A class divisions　A 级分隔　系指由符合下列标准的适当的舱壁以及由甲板所组成的分隔：(1)它们用钢或其他等效的材料制成；(2)它们有适当的防挠材加强；(3)在下列时间之内露出火焰时，未露出火焰一侧的平均温度不可比初始温度高 140 ℃ 且不可再上升。通过得到认可的非可燃性材料放热并使在包括任何接头在内的任何一点的温度较初始温度升高不超过 180 ℃；(4)它们的构造应在 1h 的标准耐火试验至结束时能防止烟及火焰通过；(5)船级社已要求按《耐火试验程序规则》对原型舱壁或甲板进行一次试验，以确保满足上述完整性和温升的要求。见表 A-1。A class divisions are those divisions formed by bulkheads and decks which comply with the following criteria: (1) they are constructed of steel or other equivalent material; (2) they are suitably stiffened; (3) they are insulated with approved non-combustible materials such that the average temperature of the unexposed side will not rise more than 140 ℃ above the original temperature, nor will the temperature, at any one point, including any joint, rise more than 180 ℃ above the original temperature, within the time listed above; (4) they are constructed as to be capable of preventing the passage of smoke and flame to the end of the one-hour standard fire test; (5) a Society has required a test of a prototype bulkhead or deck in accordance with the Fire Test Procedures Code to ensure that it meets the requirements above for integrity and temperature rise. Table A-1 shows A class division.

表 A-1　A 级分隔
Table A-1　A class divisions

"A-60"级(class "A-60")	60 min
"A-30"级(class "A-30")	30 min
"A-15"级(class "A-15")	15 min
"A-0"级(class "A-0")	0 min

A class fire proof door　甲级防火门　符合甲级防火标准，装设在甲级防火舱壁上的门。

A class inflammable gas　A 级易燃气体　对美国国家消防委员会(NFPA)船上气体危害控制标准而言，系指雷氏蒸发气压力 $\geqslant 9.6 \times 10^3$ Pa(14 lbf/in^2)的易燃气体。

A frame　A 型吊架　系指由两根直立和一根水平的杆件构成并可绕其根部水平销轴转动，外形似 A 字母或倒 U 字母的框架结构。

A or B class standard fire test　A/B 级标准耐火试验　系指《耐火试验规则》中规定的 A/B 级耐火试验。A or B class standard fire test means the A or B class test specified in the Fire Test Procedures Code.

A sound level　A 声级　系指 A 计权网络在计权后

的声级。其频率响应曲线为 40 phon 等响曲线经整理后倒置。

A standard ship's station A 标准船站(卫星通信) 系指从 1982 年起开始进行全球性通信业务,能进行电话、电报、传真、数据以及高速数据的通信系统。

Abandon dive 放弃潜水 在潜水母船上将潜水员立即召回到潜水钟,并将潜水钟带到水面,与指示潜水控制器的红色警戒的要求等同。

Abandon ship drill 弃船演习 系指按下列程序举行的演习。其包括:(1)先使用相关要求的报警系统,然后通过公共广播或其他通信系统宣布进行演习,将乘客和船员召集至集合站,并确保他们知道弃船的命令;(2)向集合站报到,并准备执行应变部署表所述的任务;(3)查看乘客和船员穿着是否合适;(4)查看是否正确地穿好救生衣;(5)在完成任何必要的降落准备工作后,至少降下一艘救生艇;(6)启动并操作救生艇发动机;(7)操作降落救生筏所用的吊筏架;(8)模拟搜救几位被困于客舱中的乘客;(9)介绍无线电救生设备的使用方法。Abandon ship drill means a drill takes place on following procedures:(1) summoning passengers and crew to muster stations with the alarm required by relative regulation followed by drill announcement on the public address or other communication system and ensuring that they are aware of the order to abandon ship;(2) reporting to stations and preparing for the duties described in the muster list;(3) checking that passengers and crew are suitably dressed;(4) checking that lifejackets are correctly donned;(5) lowering at least one lifeboat after any necessary preparation for launching;(6) starting and operating the lifeboat engine;(7) operating davits used for launching liferafts;(8) a mock search and rescue of passengers trapped in their staterooms;(9) giving instructions of radio life-saving appliances.

Abandoning system 弃船系统

Abandonment 弃船 系指船上所有人员被迫离开遇难而且救助无效的船舶的行动。一经决定弃船,应做好妥善安排,并向船舶主管单位发报,人员要有秩序撤离(如先旅客,后船员,船长最后离船),并指定专人带下各种证书、记录文件、国旗及必要的贵重物品等。离船后继续发出求救信号,并观察船舶的有关结果(下沉、爆炸或燃烧等)后离去。弃船命令一般由船长发布。

Abandonment clause 弃船条款

Abandonment suit 抛弃服 系指设计成在被迫入水的危急状态下可迅速穿戴的救生服。Abandonment suit means the immersion suit designed to permit rapid donning in the event of an imminent unintended immersion in water.

Abnormal actions 异常的作用力 系指比预期和正常作用力大的作用力,例如规则波或非规则波。Abnormal actions are those actions larger than expected or normal actions, for example, rogue or freak waves.

Abnormal operating conditions 不正常工作状态 系指在内部技术系统故障而需要船上的备用系统工作,或在不规则状态期间发生故障,或在值班人员变得不适合执行其任务,且尚未由其他合格人员替代时的状态。 Abnormal operating conditions means that when internal technical system fails to operation of back-up systems on the ship is required or when they occur during an irregular operating condition, or when the officer of the watch becomes unfit to perform his duties and has not yet been replaced by another qualified officer.

Abrasive 磨料

Abrasive inclusions 磨料嵌入物

Absolute maximum craft speed 绝对最大船舶速度 处于地效航行状态的船舶,如果航速超过某值,船舶的空气动力稳定性就不能得到保证。达到该值的航速称为绝对最大船舶速度。超过该速度还会损及船舶的可控性。 Absolute maximum craft speed is the speed beyond which craft aerodynamic stability in ground effect cannot be assured. Craft controllability may also be injured beyond this speed.

Absolute probability judgment 绝对概率判断方法 系指当研究的情况不存在相关数据时,通过专家判断对人为失误概率直接进行数值估算的一系列方法,如:德尔菲法、义群体法、成对比较法。

Absolute quota 绝对配额 系指某些商品进口数量或金额达到进口额度后,便不准继续进口的配额管理。

Absolute transmissibility 绝对传递率 系指设有隔振器时,传递给基础的激励力与刚性连接的激励力之比。

Absorbent wall brick 吸声墙砖

Absorber 阻尼器(缓冲器,减震器) 机器或机械在运行中可能会产生引起有害振动的力,能全部或部分吸收这种力的机外设备称为阻尼器、缓冲器或减震器。

Absorption area 吸声量 材料式结构的吸声量定义为:$A = \alpha s$ (m^2) 式中,α—吸声系数;s—吸声材料或吸声结构的吸声面积(m^2)。

Absorption of electromagnetic radiation (AER) 电磁辐射吸收率 系指使用手机时,可能被人体吸收的电磁辐射量。韩国标准 1.6 W/kg,国际标准 2 W/kg。

Absorption-type refrigerating unit 吸收式制冷机组 系指不使用压缩机,不采用氯氟烃而使制冷剂在一系列热交换器中进行循环流动实现制冷的机组。

AC/DC motor　交直流两用电动机　系指一种串励或补偿串励电动机。该电动机设计成用直流电流或用频率不大于 60 Hz、电压近乎同一有效值的单相交流电流工作时，其转速和输出功率近乎相同。

Accelerated cooling　加速冷却（AcC）　系指最终轧制后，冷却小于 Ar3 温度区域时，将冷却速度控制在大于空气冷却速度，为了改进轧制出来的钢材的力学性能的工艺。Accelerated cooling processing (AcC) is a process, which aims to improve mechanical properties by controlled cooling with rates higher than air cooling in the range of Ar3 temperature or below. However, direct quenching is excluded from accelerated cooling.

Accelerated hot corrosion test　热腐蚀试验　系指为检查燃气轮机叶片在高温下耐腐蚀性所做的长期台架负荷试验。

Accelerating ducted propeller (Kort nozzle)　加速导管推进器　系指水进入导管后在叶轮处流速增高，静水压力降低，有利于增加推力的导管推进器。

Accelerating section　加速段　系指船模试验水池中船模拖曳速度从零增至预期试验速度的水池段。

Acceleration and parallel processing technology of AMD (APP)　AMD 公司的加速并行处理技术

Acceleration force derivatives　水动力线加速度导数　作用于船舶的水动力分力对线加速度和角加速度的偏导数。如 $X_u = \partial X/\partial u$ 等。

Acceleration level　加速度级　系指加速度与参考加速度之比的、以 10 为底的对数乘以 20。单位为分贝（dB）。加速度级的计算公式为：$L_a = 20\lg(a/a_0)$　式中，L_a—加速度级，dB；a—加速度有效值，m/s²；a_0—参考加速度，m/s²。参考加速度一般取 1×10^{-8} m/s²。

Acceleration moment derivatives　水动力矩线加速度导数　作用于船舶的水动力分力矩对线加速度和角加速度的偏导数。如 $K_p = \partial K/\partial p$ 等。

Acceleration of gravity　重力加速度　等于 9.81 m/s²。Acceleration of gravity is to be equal to 9.81 m/s².

Acceleration parameter a_0　加速度参数 a_0　取为 $a_0 = f_p(1.58 - 0.47/C_B)[(2.4/\sqrt{L}) + 34/L - 600 L^2]$。Acceleration parameter is taken equal to: $a_0 = f_p(1.58 - 0.47/C_B)[(2.4/\sqrt{L}) + 34/L - 600 L^2]$.

Acceleration test　加速性试验　系指为检查燃气轮机从空负荷升到各规定工况的时间和加、减运转速度时可靠性所做的试验。

Acceptable arrangements of means for safe passage of crew　船员安全通道的可接受的装置　系指下列装置：(1) 一条尽可能靠近干舷甲板的照明和通风条件良好的甲板下通道（净开口至少宽 0.8 m，高 2 m），该通道连接和通达各有关处所。(2) 在上层建筑甲板面或以上的船舶中心线处或尽实际可能地靠近船舶中心线处的一座结构坚固的固定步桥，用于提供一个至少宽 0.6 m 且表面防滑的连续平台，在其全长范围内两侧装设栏杆。栏杆应至少高 1 m，并按有关要求设三个开档，其间应设置挡脚板。(3) 一条固定走道，宽度至少 0.6 m 设在干舷甲板平面上，并由两排栏杆和间距不大于 3 m 的撑竿组成。栏杆的横挡数和间距按有关要求。在"B"型船舶上，可同意将高度不小于 0.6 m 的舱口围板作为走道的一侧，但在舱口之间应设有两排栏杆。(4) 一根直径不小于 10 mm 的钢丝安全绳，由间距不大于 10 m 的撑柱支持，或一根附设在舱口围板上并在舱口之间延续的有支撑的单根扶手或钢丝绳。(5) 一座固定步桥：①位于上层建筑甲板面或以上；②位于船舶中心线处或尽实际可能靠近船舶中心线处；③位于不至于妨碍容易穿过甲板工作区域处；④提供一个至少 1 m 的连续平台；⑤由防火和防滑材料构成；⑥在其全长范围内两侧装设栏杆，栏杆应至少高 1 m，开档应符合有关要求，并由间距不大于 1.5 m 的撑竿支持；⑦每侧设置挡脚板；⑧有开口通往甲板，如适合，配有梯子，开口间距应不大于 40 m；⑨如果所横穿的露天甲板的长度超过 70 m，在步桥处应设置间距不超过 45 m 的遮蔽设施。每个这种遮蔽设施应至少能容纳 1 人，且其结构应能在前部、左舷或右舷提供风雨密防护。(6) 设在船舶干舷甲板面中心线处或尽实际可能靠近船舶中心线处的固定走道，其技术规格和 (2) 对固定步桥所列的一样，但挡脚板除外。在核准载运散装液货的"B"型船舶上，当舱口围板和所设舱口盖的高度相加不小于 1 m 时，舱口围板可接受成为走道的一侧，但舱口之间应装设两排栏杆。Acceptable arrangements of means for safe passage of crew mean following equipments: (1) A well lighted and ventilated under-deck passageway (with a clear opening of at least 0.8 m wide and 2 m high), connecting and providing access to the locations in question. (2) A permanent and efficiently constructed gangway, fitted at or above the level of the superstructure deck, on or as near as possible to the centre line of the ship, providing a continuous platform at least 0.6 m in width and a non-slip surface and with guard rails extending on each side throughout its length. Guard rails shall be at least 1 m high with three courses and constructed as relative requirement. A foot-stop shall be provided. (3) A permanent walkway at least 0.6 m in width, fitted at freeboard deck level and consisting of two rows of guard rails with stanchions spaced not more than 3 m. The number of courses of rails and their spacing shall be in accordance with relevant requirement. For type 'B' ships, hatchway coamings not less than 0.6 m in height may be accepted as one side of the walkway, provided that two rows of guard rails are fitted between the hatchways. (4) A wire rope lifeline not less than 10 mm in diameter, supported by stanchions not more than 10 m apart, or a single hand rail or wire rope attached to hatch coamings, continued and supported between hatchways. (5) A permanent gangway that is: ①located at or above the level of the superstructure deck; ②located on or as near as possible to the centre line of the ship; ③located so as not to hinder easy access across the working areas of the deck; ④providing a continuous platform at least 1 m in width; ⑤constructed of fire resistant and non-slip material; ⑥fitted with guard rails extending on each side throughout its length; guard rails shall be at least 1 m high with courses in accordance with relative requirement and supported by stanchions spaced not more than 1.5 m apart; ⑦provided with a foot-stop on each side; ⑧having openings, with ladders where appropriate, to and from the deck. Openings shall not be more than 40 m apart; ⑨having shelters set in way of the gangway at intervals not exceeding 45 m if the length of the exposed deck to be traversed exceeds 70 m. Every such shelter shall be capable of accommodating at least one person and be constructed to afford weather protection on the forward, port and starboard sides. (6) A permanent walkway located at the freeboard deck level, on or as near as possible to the centre line of the ship, having the same specifications as those for a permanent gangway listed in (2) except for foot-stops. On type 'B' ships (certified for the carriage of liquids in bulk) with a combined height of hatch coaming and fitted hatch cover not less than 1 m in height, the hatchway coamings may be accepted as one side of the walkway, provided that two rows of guard rails are fitted between the hatchways.

Acceptable risk 可接受的风险

Acceptance 承兑 系指远期汇票的付款人对远期汇票表示承担到期付款责任的行为。

Acceptance 接受/验收 系指受盘人接到发盘人的发盘或发盘人接到受盘人的还盘,同意对方提出的条件,愿意与对方达成交易,订立合同的一种表示。

Acceptance criteria 接受准则 系指用于表述可以为问题活动所接受的风险等级准则(限于高等级风险的表述)。

Access 通道 包括出口。 Access includes egress.

Access for maintenance and repair 维修通道 系指设在机、炉舱内供操纵、维护和检修各种机械设备用的通道。 Access for maintenance and repair means an access provided in machinery and boiler spaces for the purpose of control, maintenance, inspection and repair of various machinery and equipment.

Access hatch-cover 出入舱口盖 系指装于由露天甲板或顶棚通入下层舱室的舱口上,便于启闭以供人经常出入的舱口盖。

Access to spaces in the cargo area 进入货物区域内各处所的通道 系指进入货物区域内的隔离舱、压载舱、液货舱和其他处所的通道,并应直接通到开敞甲板,和能确保对上述舱室的全面检查。进入双层底处所的通道可以通过货泵舱、泵舱、深隔离舱、管隧或类似舱室,但其通风方面必须予以考虑。 Access to spaces in the cargo area means an access to cofferdams, ballast tanks, cargo tanks and other spaces in the cargo area shall be direct from the open deck and such to ensure their complete inspection. Access to double-bottom spaces may be through a cargo pump-room, pump-room, deep cofferdam, pipe tunnel or similar compartments, subject to consideration of ventilation aspects.

Access to spaces in the cargo area of oil tankers 进入油船货物区域中处所的通道 系指供安全检查人员进行检查作业时,能直接从开敞甲板进入隔离空舱、压载舱、货油舱及货物区域的其他处所的安全通道。在设置安全通道时应注意:(1)允许从货油泵舱、泵舱、深隔离空舱、管隧及类似舱室进入双层底处所,但应对通风加以考虑;(2)如通过水平开口、舱口或人孔出入上述处所,则这些开口的尺寸应足以保证配戴自储式呼吸装置及保护设备的人员上、下梯子不受阻碍,而且这些开口的有效尺寸应足以能将负伤人员从该处所底部提升上来。最小的有效开口应不小于600 mm×600 mm;(3)如通过可提供贯穿处所长度和宽度的通道的垂直开口或人孔而出入上述处所,则其最小有效开口应不小于600 mm×800 mm,除非提供格栅或其他踏板,否则其应位于从底壳板量起不超过600 mm的高度处;(4)对载重量小于5 000 t的油船,在特殊情况下,如能证实通过较小尺寸的开口或转移伤员的能力可使主管机关满意,主管机关可允许较小尺寸的开口。

Access to structures 接近结构方式 (1)对全面检查,系指提供设施使验船师能用安全和可行的方法检查船体结构。(2)对近观检查,系指提供验船师可接受的下列一个或多个接近措施:①固定脚手架和通往结构的通道;②临时脚手架和通往结构的通道;③升降机和可移动台架;④便携式梯子;⑤其他等效设施。 Access to structures is: (1) For overall survey, means should be provided to enable the surveyor to examine the structure in a safe and practical way. (2) For close-up survey, one or more of the following means for access, acceptable to the surveyor, should be provided: ①permanent staging and passages through structures; ②temporary staging and passages through structures; ③lifts and moveable platforms; ④portable ladders; ⑤other equivalent means.

Access to the ship 船舶通道 系指外部人员能够合法或非法登船的可能的通道。Access to the ship means the possible means for persons outside the ship to board either legally or illegally.

Accessible 可达性 系指无须拆除船艇上结构的永久性部件（使用或不使用工具）就可到达并进行检查、拆除或维护的能力。Accessible means one is capable of being reached for inspection, removal or maintenance without removal of a permanent part of the craft structure, with or without the use of tools.

Accessible (as applied to equipment) 可接近（用于设备） 适用于能被人偶然触及或人员靠近到安全距离以内的物体或设备。本定义适用于无适当防护或绝缘的物体。

Accessible (as applied to wiring methods) 可接近（用于敷线的方法） 系指无遮蔽的。

Accessory 附具 系指与装置的敷线及用电设备有关的、除灯具之外的任何设备,如开关、熔断器、插头、插座、灯座或接线盒。Accessory is any device, other than a luminaire associated with the wiring and current-using appliances of an installation; for example, a switch, a fuse, a plug, a socket-outlet, a lamp-holder, or a junction box.

Accident 事故 系指非故意事件,包括死亡、伤害,船舶沉没或破损,其他财产损失或损毁,环境的毁坏。Accident means an unintended event involving fatality, injury, ship loss or damage, other property loss damage, or environmental damage.

Accident 意外事故 系指行为人的行为虽然在客观上造成了损害,但不是出于行为人的故意或者过失,而是出于不能抗拒或者不能预见的原因引起的。意外事故,法律不认为是犯罪,尽管行为人的行为已造成损害结果,但行为人在主观上既无故意也无过失,如果意外事故认定为犯罪,则堕入"客观归罪",有悖于主客观相统一的刑事责任原则的刑法要求。

Accident category 事故类型 系指根据事故属性如火灾、碰撞、搁浅等在统计表中所列的事故名称。Accident category means a designation of accidents in statistical tables according to their nature, e.g. fire, collision, grounding, etc.

Accident causality 事故原因

Accident conditions 事故状态/事故工况 系指由于环境条件构成威胁对船舶或其人员的状态。Accident conditions are situations where threats against the ship or the personnel on board arise due to the environmental conditions.

Accident meeting rule at sea 海上意外相遇规则 《海上意外相遇规则》是由澳大利亚、新西兰等国首先提出,在2000年西太平洋海军论坛会议上正式公布。2014年4月22日,由中国海军承办的第14届西太平洋海军论坛年会上获得通过。当一国海军舰艇和飞机与其他国家海军舰机偶然或不期而遇时,应采取哪些安全措施和手段减少相互干扰和不确定性,方便进行通信的规则。其阐述的绝不是相互打个招呼的问题,而是一个技术性、操作性和实用性都很强的安全规范,对于海上舰机安全航行、操作和防止与他国舰机发生碰撞具有很强的实际指导意义,只要各国海军信守诺言,严格执行操作规程就能够有效维护地区海上安全与稳定。海上意外相遇规则是自愿遵守的规定,不具有法律的约束力,仅适用于海军舰艇和飞机"偶然或意外"相遇的情况下,不适用于一国的领海内。但美国海军官员在此前表示,他们希望西太平洋海军论坛的所有成员在所有地区都遵守该准则,包括在南中国海和东中国海等有争议的海域。

Accident scenario 事故场景 系指触发事件到最终阶段之一的事件序列。Accident scenario means a sequence of events from the initiating event to one of the final stages.

Accidental actions 意外事件的作用力 系指在诸如倾覆、碰撞、搁浅、失火和爆炸等事故的情况下所产生的作用力。Actions applied in the event of accidents such as capsizing, collisions, grounding, fire, explosions, and so on.

Accidental effect 意外效应 系指意外事件的结果,表述为热流、冲击力或能量、加速度等,是用于安全评估的基础。

Accidental event 意外事件 系指可能引起人命、健康丧失或者环境、财产损失的事件或串联事件。

Accidental limit state 事故极限状态 事故极限状态认为任何一个货舱的进水不致扩展到其他舱室,包括:(1)船体梁的最大承受能力;(2)双层底结构的最大承受能力;(3)舱壁结构的最大承受能力。任意一个货舱内某一构件的单一失效事故,应考虑对整个加筋的板格作极限强度评估。Accidental limit state is considered as the flooding of any one cargo could hold uater without progression of the flooding to the other compartments and includes: (1) the maximum load-carrying capacity of hull girder; (2) the maximum load-carrying capacity of double bottom structure; (3) the maximum load-carrying capacity of bulkhead structure. Accidental single failure of one structural member of any one cargo hold is considered in the assessment of the ultimate strength of the entire stiffened panel.

Accidental loads 意外载荷 系指包括因船舶意外事故或操纵失灵的后果而产生的载荷。规范所涉及的意外载荷是由于舱室进水而导致增加的舱室压力。 Accidental loads include loads that result as a consequence of an accident or operational mishandling of the ship. The accidental loads covered by the Rules are increased tank pressures due to flooding of compartments.

Accidental oil outflow performance 意外泄油性能 目的是为了提供在碰撞或搁浅事故中防止油污染的足够保护而提出的假定泄油量的计算方法。适用于2010年1月1日或以后交船的船舶。

Accidentally damaged structure 偶发破损事故

Accidents 事故 系指可能导致人命丧失、人体伤害、环境破坏或财产和经济利益损失的无法控制的事件。 Accidents mean uncontrolled events that may entail the loss of human life, personal injuries, environmental damage or the loss of assets and financial interests.

Accommodation 居住处所 系指任何公共处所。如大厅、饭厅、餐厅、休息室、走道、厕所、住室、医疗室、电影院、娱乐休闲室、无烹调设备的配膳间及向乘客和船员开放的类似处所。 Accommodation means any public space such as a hall, dining room, mess room, lounge, corridor, lavatory cabin, office, hospital, cinema, game and hobby room, pantry that contains no cooking appliances and a similar space open to the passengers and crew.

Accommodation and service space 起居、服务处所 系指SOLAS公约的第Ⅱ-2/3.10,3.12及3.22定义的居住处所、服务处所和控制站。 Accommodation and service spaces mean the accommodation spaces, service spaces and control stations as defined in SOLAS, regulation Ⅱ-2/3.10,3.12, 3.22.

Accommodation areas 生活区 系指人们居住、休息和娱乐场所。 Accommodation areas are those used for living, rest and recreation.

Accommodation chamber on deck 甲板居住舱 系指创造饱和潜水员在高气压曝露条件下的生活环境的，一个坚固的钢质壳体，且端部设有舱门，可供潜水员进出及与潜水钟对接的舱室。舱内配有环控系统、卫生装置生活必需的管线、照明灯以及空调器等。双舱式甲板居住舱配备与潜水钟对接的过渡通道和确保该舱与潜水钟对接时起密封作用的卡箍。其由主舱和过渡舱（TUP）组成。主舱和过渡舱之间设有舱门。两个舱可以分别加以不同的压力，潜水员通过过渡舱进入潜水钟内可以有效保证居住舱内潜水员的安全。

Accommodation deck 居住甲板/起居甲板 系指主要布置船员起居舱室的甲板。 Accommodation deck is a deck used primarily for the accommodation of the crew.

Accommodation ladder 舷梯 系指舷侧的一具便携式梯子，提供人员从小艇或码头登船时使用。 Accommodation ladder means: a portable set of steps on a ship's side for people boarding from small boats or from a pier.

Accommodation ladder winch 舷梯绞车 系指吊放舷边舷梯的绞车。

Accommodation location 居住舱位置

Accommodation of quota 限额的融通

Accommodation outfit 舱室设备 系指配置在船舶舱室内的船用家具、厨房设备、卫生设备、医疗设备与舱室属具等的统称。

Accommodation outfitting 绞滩滑车 系指设置于内河船前部两舷，专供引绞滩绳的固定导向滑车。

Accommodation space 居住舱室 系指：（1）围蔽在其六个面上的坚固分隔，提供船上人员使用；（2）公共处所、走道、厕所、客舱、办公室、船员舱室、病室、游戏和娱乐舱室，不带炊具的配餐间和其他类似处所。 Accommodation space means: (1) any space, enclosed on all six sides by solid divisions, provided for the use of persons on-board; (2) those spaces used for public spaces, corridors, lavatories, cabins, offices, crew quarters, hospitals, game and hobby rooms, pantries containing no cooking appliances and similar spaces.

Accommodation spaces 起居处所 系指用作公共处所、走廊、盥洗室、住室、办公室、医务室、放映室、游戏室、娱乐室、理发室、无烹调设备的配膳室以及类似处所。 Accommodation spaces are those spaces used for public spaces, corridors, lavatories, cabins, offices, hospitals, cinemas, game and hobby rooms, barber shops, pantries containing no cooking appliances and similar spaces.

Accommodation spaces of greater fire risk 具有较大失火危险的起居处所 系指：（1）设有未限制失火危险的家具和陈设的公共处所，且其甲板面积等于或大于50 m^2；（2）理发室和美容室；（3）桑拿房；（4）小卖部。 Accommodation spaces of greater fire risk are: (1) public spaces containing furniture and furnishings without restricted fire risk and having a deck area of 50 m^2 or more; (2) barber shops and beauty parlours; (3) saunas; (4) sale shops.

Accommodation spaces of minor fire risk 具有较小失火危险的起居处所　系指：(1)设有限制失火危险的家具和陈设的居住处所；(2)设有限制失火危险的家具和陈设的办公室和诊疗室；(3)设有限制失火危险的家具和陈设的公共处所，且其甲板面积小于 50 m²。 Accommodation spaces of minor fire risk are: (1) cabins containing furniture and furnishings with restricted fire risk. (2) offices and dispensaries containing furniture and furnishings with restricted fire risk; (3) public spaces containing furniture and furnishings with restricted fire risk and having a deck area of less than 50 m².

Accommodation spaces of moderate fire risk 具有中等失火危险的起居处所　系指：(1)如同具有较小失火危险的起居处所，但其内设有未限制失火危险的家具和陈设；(2)设有限制失火危险的家具和陈设的公共处所，且其甲板面积等于或大于 50 m²；(3)起居处所内面积小于 4 m² 的独立小间及小储物间(不储存易燃液体)；(4)电影放映室和影片储藏室；(5)厨房(无明火者)；(6)清洁用具储藏室(不存放易燃液体)；(7)实验室(不存放易燃液体)；(8)药房；(9)小干燥间(面积等于或小于 4 m²)；(10)贵重物品保管室；(11)手术室。 Accommodation spaces of moderate fire risk are: (1) those spaces as thesame as accommodation spaces of minor fire risk but containing furniture and furnishings without restricted fire risk; (2) those public spaces containing furniture and furnishings with restricted fire risk and having a deck area of 50 m² or more; (3) isolated lockers and small store-rooms in accommodation spaces having areas less than 4 m² (in which flammable liquids are not stowed); (4) those motion picture projection and film stowage rooms; (5) those diet kitchens (having no open flame); (6) cleaning gear lockers (in which flammable liquids are not stowed); (7) those laboratories (in which flammable liquids are not stowed); (8) pharmacies; (9) those small drying rooms (having a deck area of 4 m² or less); (10) specie-rooms; (11) the operating rooms.

Accommodation tower 居住塔楼

Accommodation unit 居住平台　系指主要用于人员居住的生活平台。 Accommodation unit means a unit mainly used for crew habitation.

Accommodation vessel 生活平台　系指以提供和改善施工人员的生活条件为主要目的的海上施工辅助平台。一些生活平台配备了起重设施，也称其为起重生活平台。目前有 4 种船体结构形式的生活平台——半潜式、单体船形、双体船形和圆筒形。

Accommodation 舱室　系指用作公共场所、走廊、厕所、住舱、办公室、医疗室、电影院、游戏和娱乐室、理发店、无烹饪设备的配膳室和类似的处所。 Those spaces used for public spaces, corridors, lavatories, cabins, offices, hospitals, cinemas, game and hobby rooms, barber shops, pantries without cooking appliances and similar spaces.

Accompanying ecological research 附属生态研究　为了进一步认识海上风场在建造和营运时对海洋环境产生的影响所使用的分析方法。也用于开发气候和生态友好的海上风电。

Accumulation tank 贮存舱　系指用作保存从海水分离并回收的浮油的液舱。 Accumulation tank is a tank intended for the retention of oil removed and separated from sea water.

Accumulator 蓄能器　系指为每一气缸设置的小型压力容器，它被液压油送至燃油喷射装置所附的执行机构或排放阀驱动装置。 Accumulator is a small pressure vessel provided for each cylinder which provides hydraulic oil to the actuator attached to the fuel injection device or the exhaust valve driving gear.

Accurate Position Indicator (API) 精确位置指示器

Acid cleaning 酸洗　系指在适当温度下将酸性溶液，有时并加入阻蚀剂，在锅炉内循环或静泡，以除去金属表面的水垢或氧化物的方法。

Acid storage 酸液储存装置

Acknowledge 收妥结汇　又称收妥付款，系指国内议付行收到出口公司的各种单据后，首先进行审核，审核无误后将单据寄交开证银行(如有偿付银行则将单据寄交偿付银行)，开证银行审核无误后，立即付款或授权偿付银行对国内议付银行付款。国内议付银行收到开证银行(或偿付银行)将货款拨入议付银行账户的贷记通知书后，立即将货款结付给出口企业。

Acknowledge 应答　系指对接收到的警报或呼叫的手动响应。 Acknowledge means manual response to an alert or a call.

Acoustic alarm 声响报警器　系指用声音引起有关人员注意的报警器。

Acoustic booth 隔声间　系指在噪声强烈的处所内建造的有良好隔声性能的小舱室，以供工作人员在其中操作或观察、控制车间内各部分工作使用。良好的隔声间，能使其中的工作人员免受听力损失，改善精神状态，获得舒适的工作条件，从而提高劳动生产率。如在船上机舱里设置的集控室就是一种隔声间。

Acoustic current meter 声学海流计　系指利用声波在水中传播过程的某些效应取得海流值的测流仪器。

Acoustic enclosure 隔声罩　系指用来阻隔机器向

外辐射噪声的罩子,可以与机器的外壳结合在一起,也可以是离开机器的单独的罩。通常具有隔声、吸声、阻尼、隔振和通风、消声等功能。其通常由金属框架、隔声件(包括隔声门、观察窗、隔声板)、弹性元件(包括弹性支撑和隔声管)和冷却系统组成。

Acoustic filter　消声器　系指一种可以阻碍或减弱声音向外传播,而允许气流顺利通过的降噪或消音设备。

Acoustic hood　隔声罩　系指一种将声源封闭起来的罩形壳体结构。以减小声源向周围环境的声辐射,而同时又不妨碍声源设备的正常工作。

Acoustic reflection measurement　声反射测量　系指测量介质的声反射系数。一般用"脉冲管"法进行测量。

Acoustic releaser　声释放器　系指对应声指令脉冲信号,使水下仪器设备或其指示器与锚脱离,以便回收的装置。

Acoustic scattering measurement　声散射测量　测量声散射系数及其在不同方向的分布。

Acoustic transponder　声应答器　系指对应声询问脉冲信号发出声回答脉冲信号的水下装置。

Acoustolith tile　吸声贴砖

Acquired radar target　捕获的雷达目标　系指自动或手动捕获启动雷达跟踪。当数据达到稳定状况时,显示矢量和先前位置。 Acquired radar target means the automatic or manual acquisition initiates radar tracking. Vectors and past positions are displayed when data has achieved a steady state condition

Acquisition　捕获(捉)　系指选择要求跟踪程序及开始跟踪那些目标船。 Acquisition means the selection of those target ships requiring a tracking procedure and the initiation of their tracking.

Acquisition of a radar target　雷达目标的捕获　系指捕获目标并开始跟踪的过程。 Acquisition of a radar target is a process of acquiring a target and initiating its tracking process of acquiring a target and initiating its tracking.

Acquisition/activation zone　捕获/激活区　系指由操作人员建立的一个区域,进入该区域时,系统应自动捕获雷达目标并激活报告的 AIS 目标。 Acquisition/activation zone means zone set up by the operator in which the system should automatically acquire radar targets and activate reported AIS targets when entering the zone.

Act　法令　系指经修正的防止船舶造成污染的法令(美国法规第 33 篇 1901-1911)。 Act means the Act to Prevent Pollution from Ships, as amended (33 U. S. C. 1901-1900).

Action　作用力　系指直接施加到结构上或间接作用使结构产生强迫变形或加速度的外载荷,例如建造公差、装配和温度变化能产生强迫变形。该术语也称"载荷"。 Action is the external load applied to the structure (direct action) or an imposed deformation or acceleration (indirect action); for example, an imposed deformation can be caused by fabrication tolerances, settlement, or temperature charge. The term is also called "load".

Action effect　作用效应　系指在整个结构或结构构件上产生的效应,例如产生内力、弯矩、变形、应力。该术语也称"载荷效应"。 Action effect means the effect of actions on a global structure or structural component, for example, internal force, moment, deformation, stress. The term is also called "load effect".

Actionable subsidy　可申诉的补贴　系指政府通过直接转让资金,放弃财政收入,提供货物或服务和各种收入支持和价格支持对某些特定企业提供特殊补贴。这种特殊补贴实际上就是指一国政府实施有选择的、有差别或带有歧视性的补贴。如果这种特殊补贴造成其他缔约方国内有关工业的重大损害时,该国可诉诸争端解决机制加以解决。

Activated AIS target　被激活的自动识别系统(AIS)目标　系指一个静止目标被自动或手动激活的目标,用以显示附加的图表显示信息。目标通过"被激活的目标"符号显示,包括:(1)矢量(COG/SOG);(2)船首向;(3)ROT 或旋转方向指示(如有)以指示航向改变。 Activated AIS target is a target representing the automatic or manual activation of a sleeping target for the display of additional graphically presented information. The target is displayed by an "activated target" symbol including: (1) a vector (COG/SOG); (2) the heading; (3) ROT or direction of turn indication (if available) to indicate initiated course changes.

Activated anti-rolling tank stabilization system　主动水舱式减摇装置　系指由控制系统操纵泵水设备来完成舱内液位变化的水舱式减摇装置。

Activated fin　主动减摇鳍(可控的减摇鳍)　能用船上的动力对其进行控制的机械式鳍称为主动减摇鳍或可控减摇鳍。

Activated fin stabilizers　主动式减摇鳍　系指类似于自动控制的水平舵,以船的横摇角速度(主要信号)、角度及角加速度叠加而成的信号控制鳍转动,使鳍产生阻止横摇的力矩,以减小横摇幅度的装置。

Activated maneuvering device　主动转向装置　自身能发出推力并使船舶获得回转力的转向装置。

Activation of an AIS target　自动识别系统目标的

激活 系指为显示附加图示和字母数字信息而对一个静止自动识别系统(AIS)目标的激活。 Activation of an AIS target means an activation of a sleeping AIS target for the display of additional graphical and alphanumerical information.

Active antiroll fin 可控减摇鳍(主动式减摇鳍) 能用船上的动力对其进行控制而使船舶达到减摇目的的机械式鳍称为主动减摇鳍或可控减摇鳍。

Active anti-rolling tank 主动式减摇水舱 系指当船舶横摇时,采用机械装置将一舷的水输送到另一舷,利用左右水舱水体的重力差产生抵抗船舶横摇的力矩来达到船舶减摇目的的减摇水舱。

Active bow fin 可控艏(减纵摇)鳍 系指能用船上的动力对其进行控制的位于船首的机械式鳍。

Active electronically scanned array radar(AESAR)有源相控阵雷达 系指采用有源相控阵技术的雷达。有源相控阵雷达的每个辐射器都配装有一个发射/接收组件,每一个组件都能自己产生、接收电磁波,因此在频宽、信号处理和冗余度设计上都比无源相控阵雷达具有较大的优势。正因为如此,也使得有源相控阵雷达的造价昂贵,工程化难度加大。但有源相控阵雷达在功能上有独特优点,大有取代无源相控阵雷达的趋势。见图A-1。

图 A-1 有源相控阵雷达
Figure A-1 Active electronically scanned array radar(AESAR)

Active electronically scanned array(AESA) 有源电子扫描阵/自动电子扫描相控阵 通常也称为有源相控阵技术,终于在机载雷达上取得了成功的应用。AESA雷达可以使F-15C战斗机在21世纪仍然保持其空战优势。AESA的成功应用是对传统机载雷达的一次革命,它极大地扩展了雷达的应用领域和提高了雷达的工作性能,进而提高和丰富了作战飞机执行任务的能力和作战模式。1998年4月,诺·格公司已交付第一套APG-77雷达硬件和软件给波音飞机公司F-22航空电子综合实验室,对F-22的航空电子设备进行系统综合测试和鉴定试验。

Active failure 主动失效 系指对于设备的操作或监控回路有实时影响的所有失效。 Active failure concerns all failures which have an immediate effect either on the operation of the installations or on the monitoring circuits.

Active fin 可控鳍(主动式鳍) 系指能用船上的动力对其进行控制的机械式鳍。

Active fin stabilizer 主动式减摇鳍装置 系指能用船上的动力对其进行控制而使船舶在摇摆中达到稳定目的的装置。

Active fire fighting system 主动式灭火系统

Active power filter(APF) 有源滤波器 系指一种用于动态抑制谐波、补偿无功的新型电力电子装置,它能够对大小和频率都变化的谐波以及变化的无功进行补偿。通过使用有源滤波器可以提高通信系统及配电系统的稳定性,延长通信设备及电力设备的使用寿命,并且使配电系统更符合谐波环境的设计规范。有源滤波器是用电流互感器采集直流线路上的电流,经采样,将所得的电流信号进行谐波分离算法的处理,得到谐波参考信号,作为的调制信号,与三角波相比,从而得到开关信号,用此开关信号去控制单相桥,根据技术的原理,将上、下桥臂的开关信号反接,就可得到与线上谐波信号大小相等、方向相反的谐波电流,将线上的谐波电流抵消掉。

Active quota arrangement 主动配额管理 系指国家为保证出口符合国民经济计划的要求,对部分重要出口商品实行的一种出口配额管理。出口配额可以通过直接分配的方式分配,也可以通过招标等方式分配。

Active rise control 主动风险控制 系指通过安全设备或操作员的动作来控制风险。

Active roll-resisting fin 可控减摇鳍(主动式减摇鳍) 系指能用船上的动力对其进行控制而使船舶达到减摇效果的机械式鳍。

Active rudder 主动舵 为提高船的转向性能或用以推船缓行,在舵叶后部装有电动小螺旋桨或导管推进器的舵。

Active substance 活性物质 系指一种物质或生物,包括病毒和真菌,其对有害水生物和病原体具有一般的或特定的作用或抑制作用。 Active substance means a substance or organism, including a virus or a fungus that has a general or specific action on or against harmful aquatic organisms and pathogens.

Active system 主动系统 对计算机系统而言,系指采用传感器读取以及输入液舱内的货物容量等数据取代手工输入的计算机系统。 For the computer system, active system means a computer system which replaces the manual entry with sensors reading and entering the contents of tanks, etc.

Active vibration isolation 积极隔振

Activities in the area 区域内活动 系指勘探和开发区域的资源的一切活动。 Activities in the area mean all activities of exploration for, and exploitation of, the resources of the area.

Actual carrier 实际承运人 系指受承运人委托从事货物运输或部分货物运输的任何人,包括受托从事此项工作的任何其他人。 Actual carrier means any person to whom the performance of the carriage of the goods, or of part of carriage, has entrusted by the carrier, and includes any other person to whom such performance has been entrusted.

Actual cycle 实际循环

Actual rudder angle 实际舵角 系指舵叶环绕舵杆相对于零舵角位置的实际角位移。

Actual tare 实际皮重 系指包装材料的质量。

Actual total loss 实际全损 系指货物全部灭失或全部变质而不再有任何商业价值。推定全损系指货物遭受风险后受损,尽管未达实际全损的程度,但实际全损已不可避免,或者为避免实际全损所支付的费用和继续将货物运抵目的地的费用之和超过了保险价值。推定全损需经保险人核查后认定。

Actual value 实际价格 系指"在进口国立法确定的某一时间和地点,在正常贸易过程中充分竞争的条件下,某一商品或相同商品出售或兜售的价格"。

Actual working hours 实动工时 系指在生产活动中实际耗用的工时。

Actuating device/Actuator 执行机构 系指接受来自电子控制器(ECU)的控制指令后执行动作,实现对柴油机各相关部件操作的机构。执行机构主要由电磁阀、步进电机、电液装置及管路等组成。执行机构驱动的动力分电力和液压两种。电力驱动由步进电机操作,一般用在驱动不大和实时性要求不太严格的场合,如电子调速器、小型涡轮增压器可变喷嘴的控制、冷却剂润滑系统可调阀门的控制等。液力驱动由电磁阀控制的液压力进行操作,如用在取消凸轮轴以传输动力的喷油系统、进排气系统等。 Actuating device means a actuator which receives control orders from electronic control unit (ECU) to operate relevant components of diesel engine by driving system. Actuator is mainly composed of electromagnetic valve, step motors, electro-hydrautic equipment and pipes. Driving power of actuator includes electric power and hydrautic power. Electric drive is operated by step motor and generally used in cases with less driving power and less strict real-time requirements, such as electronic speed governor, control of variable nozzle of small type turbocharger, control of adjustable valve of cooling and lubricating system, etc. Hydraulic drive is operated by hydraulic power controlled by electromagnetic valve, and is used in oil injection system and air intake and discharge system which transmit power without camshaft.

Actuator disc 鼓动器 系指推进器动量理论中假设的能使流体自由通过并使压力突然增加,因而使流体加速的理想机构,即理想推进器。

Acute aquatic toxicity 急性水毒性

Acute mammalian toxicity 哺乳动物急性中毒 包括:(1)急性吸入中毒;(2)急性皮肤中毒;(3)误食急性中毒。 Acute mammalian toxicity includes: (1) acutely toxic by inhalation; (2) acutely toxic in contact with skin; (3) acutely toxic if swallowed.

Acute release 急性释放 通常缘于偶发事件或意外事故而以排出、排放或曝露形式突然释放。

Acutely toxic by inhalation 急性吸入中毒 见表A-2。

表 A-2 急性吸入中毒
Table A-2 Acutely toxic by inhalation

吸入毒性(LC_{50})	Inhalation toxicity(LC_{50})
危险程度 Hazard level	mg/1/4h
高 High	≤0.5
较高 Moderately high	>0.5 ~ ≤2
中等 Moderate	≥2 ≤10
轻微 Slight	≥10 ~ ≤20
没有 Negligible	≥20

Acutely toxic in contact with skin 急性皮肤中毒 见表 A-3。

表 A-3　急性皮肤中毒
Table A-3　Acutely toxic in contact with skin

皮肤接触中毒(LD_{50})	Dermal toxicity(LD_{50})
危险程度 Hazard Level	mg/kg
高 High	≤50
较高 Moderately high	≥50 ~ ≤200
中等 Moderate	≥200 ~ ≤1 000
轻微 Slight	≥1 000 ~ ≤2 000
没有 Negligible	≥2 000

Acutely toxic in contact with skin　误食急性中毒　见表 A-4。

表 A-4　误食急性中毒
Table A-4　Acutely toxic in contact with skin

口服毒性(LD_{50})	Oral toxicity(LD_{50})
危险程度 Hazard Level	mg/kg
高 High	≤5
较高 Moderately high	≥5 ~ ≤50
中等 Moderate	≥50 ~ ≤300
轻微 Slight	≥300 ~ ≤2 000
没有 Negligible	≥2 000

Acute-phase-protein(APP)　急性期蛋白　系指由于细菌、病毒、或寄生虫等感染，机械刺激或温度创伤、局部缺血性坏死或恶性增生等原因引起机体一系列的早期、高度复杂反应的产物。APP 在感染或炎症后的恢复中起重要作用。绝大多数 APP 由肝细胞合成，也有一些由其他细胞，如：单核细胞、成纤维细胞和脂肪细胞产生。与细胞免疫和体液免疫的特异性相比，急性期的变化是非特异的。

Adapter　接合器(适配器,转接器)

Adapter module　过渡舱　系指供潜水员进出的通道，有多个舱门，可与潜水钟或外界相通。舱内通常装有气压吹除式抽水马桶、盥洗池、淋浴喷头等设备。

Added mass of entrained water　附加质量　系指：(1)为便于计算物体在理想流体中运动时所产生的惯性附加力和惯性附加力矩，将物体视作在真空中运动而另外给物体的质量所附加的代替流体动力作用的适当的量；(2)当物体作非定常运动时，其周围的流体的质点，由于附加的当地加速度而相应产生的附加惯性力作用并附加于物体的具有质量因次量。

Added moment of inertia　附加惯性矩　系指：(1)为便于计算物体在理想流体中运动时所产生的惯性附加力和惯性附加力矩，将物体视作在真空中运动而另外给物体的惯性矩所附加的代替流体动力作用的适当的量；(2)当物体作非定常运动时，其周围的流体的质点，由于附加的当地加速度而相应产生的附加惯性力作用并附加于物体的具有惯性矩因次量。

Added products of inertia　附加惯性积　系指船舶运动时，其周围的流体介质使其总的有效惯性积比较船的实体的惯性积增加的当量部分。

Added value　附加值　系指由综合航行系统(INS)提供的各设备性能标准要求以外的功能和信息。Added value means the functionality and information, which are provided by the integrated navigation system (INS), in addition to the requirements of the performance standard for the individual equipment.

Added weight method　增加质量法　系指把进入船舱的水当作船上增加的液体载荷。以计算船舶进水后浮态和稳性的方法。

Additional audit　附加审核　系指在公司的管理体系发生重大的变化，船旗国政府或船级社特别要求等情况下进行的审核。

Additional award　补充裁决

Additional ballast water　额外压载水　系指在天气情况非常恶劣的少数航次，油船船长认为必须在货油舱中加装额外的压载水才能保证船舶安全时，或者由于油船的具体营运特性必须加装超额压载水时，这种加装的压载水称为额外压载水。

Additional bridge functions　附加桥楼功能　系指

船舶航行时在桥楼执行的,但不属于桥楼主要功能的功能。和不是航行值班驾驶人员负责执行的这类功能,例如:延伸的通信功能;压载水和货物操作监视和控制功能;对机械的监视和控制;船舶系统的监视和控制。Additional bridge functions mean those functions related to ship operations which shall be carried out on the bridge in addition to primary functions, and whether or not the officer in charge of the navigational watch(OOW) is responsible for the allocated tasks. Examples of such functions are: extended communication functions; monitoring and control of ballasting and cargo operations; monitoring and control of machinery; monitoring and control of domestic systems.

Additional notation 附加标志 系指是船舶不同特点的分级表述,包括船舶类型、货物特性、特殊任务、特殊的特征、航区、航线限制以及其他含义的1个或1组标志。通常附加标志加注在入级符号之后。

Additional repairs with add cost 附加工程 属于工程项目合同范围以内的"新增工程",是合同项目所必需的工程,缺少了这些工程,合同项目即不能发挥预期的作用。

Additional resistance 附加阻力 主要系指风和浪对船体增加的额外阻力,其他因海水温度、盐度、舵角和漂移等引起的额外阻力,由于其影响较小,且修正方法可操作性不强,故可忽略不计。

Additional survey 附加检验 (1)对 400 总吨(GT)及以上的船舶以及所有固定和移动钻井平台和其他平台而言,系指在按有关规定的任何重大修理或换新后,或在按有关规定的检查结果进行修理后应根据情况进行全面或部分的检验。该检验应确保已有效进行了必要的修理或换新,确保这种修理或换新所用的材料和工艺在各方面均属合格,并确保该船在各方面均符合有关要求。(2)对每艘 150GT 及以上的油船和 400GT 及以上的其他船舶而言,附加检验系指在有关规定的检查结果进行修理后或在任何重大修理或换新后,根据情况进行的全面或部分检验。该检验应确保已有效进行了必要的修理或换新,确保这种修理和换新所用的材料和工艺在各方面均属合格,并确保船舶在各方面均符合国际防止船舶造成污染公约附则Ⅰ的要求。 Additional survey:(1) For the ships of 400 Gross Tonnage and above and fixed and floating drilling rig and other platforms, additional survey means a survey either general or partial, according to the circumstances, shall be made whenever any important repairs or renewals are made as relevant requirement or after a repair resulting from investigations of relevant requirement. The survey shall ensure that the necessary repairs or renewals have been effectively made in all respects satisfactory and that the ship complies in all respects with the relevant requirements. (2) For the oil tankers of 150 Gross Tonnage and above, and every other ship of 400 Gross Tonnage and above, additional survey either general or partial, according to the circumstances, shall be made after a repair resulting from investigations prescribed in relevant requirement, or whenever any important repairs or renewals are made. The survey shall ensure that the necessary repairs or renewals have been effectively made, that the material and workmanship of such repairs or renewals are in all respects satisfactory and that the ship complies in all respects with the requirements of MARPOL Annex Ⅰ.

Additional survey for the International Certificate of Inventory of Hazardous Materials 《国际有害物质清单证书》附加检验 系指应船东请求,在结构、设备、系统、配件、布置和材料经过改变、更换或重大维修后,可根据情况进行的总体或局部的附加检验。该检验应确保船舶经过任何此类改变、更换或重大维修后仍符合 SRC 的要求,且对有害物质清单 第Ⅰ部分进行了修正(如必要),检验合格后对 ICIHM 进行签署。 Additional survey means a survey either general or partial, according to the circumstances, may be made at the request of the ship-owner after a change, replacement, or significant repair of the structure, equipment, systems, fittings, arrangements and material. The survey shall ensure that any such change, replacement, or significant repair has been made in the way that the ship continues to comply with the requirements of this Convention, and that Part I of the Inventory is amended as necessary. ICIHM will be signed after qualified test.

Additional survey on WIG craft 地效翼船的附加检验 系指如有特别情况发生时进行的检验。 Additional survey on WIG craft is a survey tested when the occasion arises.

Additional risk 附加险 附加险是相对于主险(基本险)而言的,顾名思义系指附加在主险合同下的附加合同。它不可以单独投保,要购买附加险必须先购买主险。一般来说,附加险所交的保险费比较少,但它的存在是以主险存在为前提的,不能脱离主险,构成一个比较全面的险种。比如,一般个人人寿保险可以附加意外伤害保险和医疗保险;普通家庭财产保险可以附加盗窃保险等。也有部分公司的险种既可以作为附加险购买,也可以作为主险单独投保。

Address book 通信录 系指方便个人或企业联系时的一种简单的、实用的记事载体。目前,通信录分为纸质通信录、电子通信录、手机通信录和网络通信录。你可以在个人电脑、掌上电脑、移动电话等任何联网设备上录入你的联系人的手机/电话号码、Email、QQ、MSN、通信地址等通信录信息，或对以前的信息进行分组、管理和更新，在你的许可下，该联系人可以看到他所在组内的其他联系人信息，从而实现通信录共享，如果该联系人更新自己的联系信息，随即你的通信录会自动更新，实现同步通信录，并留下旧版本的通信录信息。

Adherence 坚持/依附

Adhesive failure 附着力

Adiabatic efficiency(isentropic efficiency) 绝热系数 系指表示实际工作过程偏离同压比下理论绝热过程能量转换情况的指标。对压气机是理论绝热压缩功(温升)与实际压缩功(温升)之比，对涡轮则是实际膨胀功(焓降)与理论绝热膨胀功(焓降)之比。

Adiabatic system 绝热系统 系指其全部或部分由绝热材料所填充的处所，它可以是或不是屏壁间处所。

Adjacent areas 邻近区域 系指并非被保护区域的，在固定式水基的局部使用灭火系统起作用时曝露于直接喷淋中的区域或水可能扩散的其他区域。 Adjacent areas are those areas, other than protected areas, exposed to direct spray or other areas where water may extend when a fixed water-based local application firefighting system is activated.

Adjacent cabin 相邻舱室 系指与所论舱室有共同船体结构(舱壁或者甲板)的舱室。

Adjacent channel modulator 邻频调制器

Adjacent space 相邻处所 系指:(1)在各方向，包括所有接触点、角、对角线、甲板、舱顶和隔舱壁，与一个处所邻接的处所。(2)紧邻 CA 区域，由通有管子、电缆、风道、门、中间甲板等的水密舱壁或甲板分隔的围蔽处所。 Adjacent space:(1) those spaces bordering a space in all directions, including all points of contact, corners, diagonals, decks, tank tops and bulkheads. (2) an enclosed space adjoining a CA zone separated by watertight bulkheads or decks penetrated by pipes, cables, ducts, doors, tween deck, etc.

Adjustable(varying) speed motor 调速电动机 系指其转速在较大范围内可逐渐变化，但一经调定后实际上就不受负荷影响的电动机。诸如为获得较宽的调速范围而设计的、磁场电阻可进行调节的直流并励电动机。

Adjustable bolted propeller(ABP) 可变桨叶角螺旋桨 这种螺旋桨的桨叶与桨毂系采用螺栓与螺母的拧紧连接，但与 CPP 不同的是桨毂内无任何执行机构，航行中桨叶不能转动，仅能在今后船舶坞修时可将桨叶转动一个螺孔角度后重新拧紧，以获得一个新螺距而使效率不致进一步下降。

Adjustable change-speed motor 可调变速电动机 系指转速可逐渐调节，但一旦在确定的负载下调定后，该转速在很大程度上将随负载改变而变化的电动机。诸如通过控制磁场来调速的直流复励电动机或用变阻器调速的绕线转子式感应电动机。

Adjustable delivery pump 可调排量泵

Adjustable vane(variable stator blade) 可转静叶 系指为防止压气机在非设计工况发生喘振以及为改变涡轮部分负荷时的工作特性而设计的可在运行中改变安装角的静叶片。

Adjustable-pitch propeller(controllable-pitch propeller) 可调螺距螺旋桨 系指可通过毂内机构转动各叶，调节螺距以适应各种工作情况的螺旋桨。

Adjuster 调停者 协助各缔约方消除分歧，解决他们之间所发生的贸易争端。

Adjusting thrust bearing 自位式推力轴承 系指具有自动调节位置的平衡垫块的推力轴承。

Adjusting valve 调节阀(调整阀)

Administered price 垄断价格 系指国际垄断组织利用其经济实力和对市场的控制力决定的，在一定程度上加以操纵的一种旨在保证其最大利润量的市场价格。它有买方垄断价格和卖方垄断价格两种形式。垄断组织在国际间采用垄断价格是有条件的。它们是:某一部门和竞争的公司数量、产品的价格需求弹性、替代弹性的大小以及国际经济和政治形势等。

Administration 主管当局/主管机关 系指:(1)船舶在其管辖下进行营运的国家政府。对于悬挂某一国家国旗的船舶，主管机关即指该国政府，对在沿海国行使自然资源勘探和开发主权的海岸附近水域从事海床和底土勘探和开发的固定或浮动平台，主管机关即指相关沿海国政府;(2)船旗国政府;(3)对《2009年香港国际安全与环境无害化拆船公约》而言，系指经一缔约国指定的在特定地理区域或专业领域内负责按该公约规定的该缔约国管辖范围内作业的拆船厂相关事宜的一个或多个政府当局。 Administration means:(1)the Government of the State under whose authority the ship is operated. With respect to a ship entitled to fly a flag of a State, the Administration is the Government of that State. With respect to fixed or floating platforms engaged in exploration and exploitation of the seabed and subsoil thereof adjacent to the coast over which the coastal State exercises sovereign rights for the purposes of exploration and exploitation of their natural resources, the Administration is the Government of the coastal State concerned;(2) the Government of the State whose flag the ship is entitled to fly;(3)For the Hong Kong International Convention for the Safe and Environmentally

Sound Recycling of Ships, 2009, Administration means the Government of the State whose flag the ship is entitled to fly, or under whose authority it is operated.

Administration of the coastal State 沿岸国主管机关 系指平台作业水域所在国的安全管理机关。

Administration of the flag State 船旗国主管机关 系指平台登记国的海事当局。Administration of the flag State means the maritime authority of the country in which the unit is registered.

Administration(container) 主管机关(集装箱) 系指有权批准集装箱的缔约国政府。Administration means the Government of a Contracting Party under whose authority containers are approved.

Administrative fee 管理费

Administrator 主管人员 系指美国环境保护署的主管人员。Administrator means the Administrator of the United States Environmental Protection Agency.

Admiralty anchor 海军锚 属于有杆锚,结构简单。抓力大,主要用于小型船舶。

Admiralty board margin 海军裕度 系指一个满足船东在设计和建造阶段对舰船或设备进行修改的余量。Admiralty board margin is an allowance to cater for modifications made by the owner to the vessel or equipment during the design and build stages.

Admiralty coefficient(admiralty constant) 海军系数 系指用于约略估计船舶功率或比较型似船舶快速性能的一种系数;$C = \triangle^{2/3} V^3 / P$,式中,$\triangle$—排水量;$V$—航速;$P$—任一种功率。

Adult 成人 系指年龄至少为18岁的人员。Adult means a person who is at least 18 years of age.

Advalorem duties 从价税 系指是以进出口商品的价格为标准计征一定比率的关税。其税率表现为商品价格的百分率。从价税的计算公式如下:从价税额=商品总价×从价税率。从价税额与商品价格有直接关系。它与商品价格的涨落成正比关系,其税额随着商品价格的变动而变动,所以它的保护作用与价格有着密切的关系。如在价格下跌的情况下,其税率不变,从价税额相应减少,因而保护关税作用也有所下降。其完税价格标准,大体上可概括为以下3种:(1)以成本、保险费加运费价格(CIF)作为征税价格标准。(2)以载运港船上交货价格(FOB)作为征税价格标准。(3)以法定价格作为征税价格标准。

Advance 纵距 系指船舶自发出操舵令位置到艏向改变离初始航向90°位置,沿初始航线方向在船中点量取的距离。Advance is the distance travelled in the direction of the original course by the mid-ship point of a ship from the position at which the rudder order is given to the position at which the heading has changed 90° from the original course.

Advance angle of a blade element 叶元体进角 系指叶元体相对于附近未受螺旋桨扰动水的运动方向与其处沿圆周方向间的夹角。$\beta = \arctan V_A / 2\pi nr$,式中,$V_A$—轴向速度,即进速;$2\pi nr$—周向速度。

Advance charge 预付款 系指是用来描述买方在未得到任何有价值的作为回报的东西之前向供应商所做的支付,预付款为供应之前向供应商提供了工作资金。

Advance coefficient 进速系数 系指螺旋桨每转进程对其直径的比值 $J = V_A / nD$,式中,V_A—进速;n—螺旋桨转速;D—螺旋桨直径。

Advance per revolution 每转进程 系指螺旋桨或其叶元体每转一周相对附近水沿轴线前进的距离,即转速 n 除进速 V_A 之商。

Advance ratio of propeller 进速比 系指螺旋桨每转进程对螺旋桨盘面周长的比值 $\lambda = V_A / \pi nD$,式中,V_A—进速;n—螺旋桨转速;D—螺旋桨直径。

Advance price 增价拍卖 也称买方叫价拍卖。这是最常用的一种拍卖方式。拍卖时,由拍卖人(Auctioner)提出一批货物,宣布预定的最低价格,估价后由竞买者(Bidder)相继叫价,竞相加价,有时规定每次加价的金额额度,直到拍卖人认为无人再出更高的人。

Advanced 进距 系指船舶回转试验中,自转舵瞬间的船重心沿初始直线航线方向至艏向改变90°瞬间的船重心间的纵向距离。

Advanced electronic guidance information system(AEGIS)/Airborne early-warning ground

Advanced gun system(AGS) 先进舰炮系统 主要由火炮、隐身炮塔、供弹系统、自动化弹库、随动系统、电气控制系统和弹药等部分组成。其塔长约6.4 m(不含身管)、宽约7.6m、高约4.0 m、全重约95 t(不包括弹库质量)。是一种斜式发射的常规155 mm舰炮,其炮管长度为9.61 m,俯仰范围为-5°~+70°,初始射速为10~12发/min。

Advanced hypersonic weapon 超高音速武器 该技术拥有国:美国、中国。见图A-2。

Advanced micro devices(AMD) [美]超微电脑股份有限公司 AMD公司专门为计算机、通信和消费电子行业设计和制造各种创新的微处理器(CPU、GPU、APU、主板芯片组、电视卡芯片等)、闪存和低功率处理器解决方案,AMD致力于技术用户——从企业、政府机构到个人消费者——提供基于标准的、以客户为中心的解决方案。AMD是目前业内唯一一个可以提供高性能CPU、高性能独立显卡GPU、主板芯片组三大组件的半导体公

图 A-2 超高音速武器
Figure A-2 Advanced hypersonic weapon

司，AMD 提出 3A 平台的新标志，在笔记本领域有"AMD Vision"标志的就表示该电脑采用 3A 构建方案。2012 年 9 月 18 日，AMD 宣布，CFO 托马斯·赛菲特（Thomas Seifert）将会离职寻找其他机会。

Advanced quiet propulsion system 安静型先进推进系统

Advanced shipment billing notices（ASBN） 反舰弹道导弹 它实际上是普通弹道导弹的"改进版"，普通弹道导弹射程远，可达数千千米，但由于最后阶段速度太快（可达 10 倍以上音速），难以控制，故只能打击固定目标。为了突破这个局限，苏联在赫鲁晓夫时代曾秘密研制过具有末端制导能力，可打击移动目标的弹道导弹。由于此类弹道导弹主要针对移动在大洋中的航母战斗群，故称为"反舰弹道导弹"，也叫"航母杀手"。我国"东风 21D"弹道导弹即为一种"反舰弹道导弹"。

Advanced surface missile system（ASMS） 先进水面导弹系统

Adventure ocean 海上猎奇（豪华邮轮）

Adviser 顾问

Advising bank 通知银行 简称通知行，它系指受开证银行的委托，将信用证转交出口人的银行。它一般是出口人所在地银行，有时也可能是外地的银行。如香港某银行指定其在深圳的联行、分行或代理行为限制议付行，由其直接仰挂号信邮寄信用证正本给受益人在北海的某进出口公司。这深圳的银行就是外地的通知行。《跟单信用证统一惯例》第 500 号版本第 7 条 A 款"信用证可经另一家银行（'通知行'）通知受益人，而通知行无须承担责任，如通知行决定通知信用证，它应合理审慎地核检所通知信用证的表面真实性。如通知行决定不通知信用证，它必须不延误地告知开证行。B 款"如通知行不能确定信用证的表面真实性，它必须不延误地告知从收到该指示的银行"，说明它不能确定该信用证

的真实性。如通知行仍决定通知该信用证，则必须告知受益人，它不能核对信用证的真实性。"

Advisory 预警报 DP 操作状态，冗余设备故障。主要用在钻井船上。

Aegis combat system 宙斯盾战斗系统 系指美国海军现役最重要的整合式水面舰艇作战系统。20 世纪 60 年代末，美国海军认识到自己在各种环境中的反应时间，火力，运作妥善率都不足以应付苏联大量反舰导弹对水面作战系统的饱和攻击威胁。对此美国海军提出一个"先进水面导弹系统"（ASMS/the Advanced Surface Missile System）的提案，经过不断发展，1969 年 12 月改名为空中预警与地面整合系统（Advanced Electronic Guidance Information System/Airborne Early-warning Ground Integrated System），英文缩写刚好是希腊神话中宙斯之盾（AEGIS），所以也译为"宙斯盾"系统。

Aegis destroyer /Aegis warship 神盾舰 是一种驱逐舰。最早安装了"宙斯盾"系统的美国海军阿利伯克级驱逐舰和提康得罗加级巡洋舰都被称为"神盾舰"。这里的盾并不系指军舰舰身如盾牌般坚韧。而系指在中国海军 052C 型驱逐舰上安装了"宙斯盾"相控阵防空雷达。安装了"宙斯盾"相控阵雷达后，军舰可以侦测四面八方的有威胁的飞行器，无论是导弹还是飞机。由此，可以防卫"盾舰"所在区域的安全。仿佛构成了一面看不见的"盾牌"。在阿利伯克级基础上建造的日本"金刚"和"爱宕"、韩国的"KDX"驱逐舰等都被称为"宙斯盾舰"。而中国海军拥有与宙斯盾舰功能类似的 052C 型导弹驱逐舰以后，军迷也将 052C 称为"中华神盾舰"。总而言之，拥有相控阵防空雷达，并能进行中远程防空的军舰，一般被称为"神盾舰"。世界上拥有"神盾舰"的国家并不多，而舰型更少。目前被称为"神盾舰"的只有美国海军"提康得罗加"级导弹巡洋舰和"阿利伯克"级导弹驱逐舰、日本"爱宕"级导弹驱逐舰和"金刚"级导弹驱逐舰、韩国"KDX"导弹驱逐舰、中国 052C 型导弹驱逐舰这类大型水面舰艇，除此之外，英国 45 型导弹驱逐舰、法意合作生产的"地平线"导弹驱逐舰、德国 F-124 型导弹护卫舰、俄国 22350 型导弹护卫舰以及印度的"金奈"号导弹驱逐舰。这些在美中日韩之外的国家的"盾舰"要么是"神盾"的扫面范围不够大，要么是防空武器的射程不够远，不能与"伯克"级等同日而语，而日本和韩国的"神盾舰"只是在"伯克"级的基础上增加改进建造的，这样看来，实际上世界上拥有最强神盾舰的国家只有中国和美国。在国际驱逐舰的竞争中，美国除了不断升级"阿利伯克"级驱逐舰以外，还进行了 DDG-1000 万吨大驱的研制。而中国也紧跟美国步伐，在 052C 的研制基础上，研发 052D 和 055 型导弹驱逐舰。未来的"神盾"舰的竞争上，中美已经把其他国家甩开了。该技术拥有

国:中国、美国(日本、英国、法国、韩国是买美国的)。见图A-3。

图 A-3　神盾舰
Figure A-3　Aegis destroyer/Aegis warship

Aerodynamic control surfaces　艉翼　为保证全垫升气垫船航态的航向稳定性和操纵性而设的由垂直艉翼和水平艉翼所组成的艉部空气翼。

Aerodynamic noise　空气动力噪声　系指气流内部运动或与物体互相作用所产生的噪声。气流受空气动力扰动产生局部的压力脉动,并以波的形式通过周围的空气向外传播而形成噪声。飞机空气动力噪声在飞机全部噪声中占主要部分,它来自许多不同的声源。

Aerodynamic rudder(directional rudder)　空气方向舵　系指垂直艉翼中用以提供垫升航态时水平面内操纵性的可转动部分。

Aero-engine　航空发动机　系指为航空器提供飞行所需动力的发动机。作为飞机的心脏,被誉为"工业之花",它直接影响飞机的性能、可靠性及经济性,是一个国家科技、工业和国防实力的重要体现。目前,世界上能够独立研制高性能航空发动机的国家只有美、英、法、俄和乌克兰等少数几个国家,技术门槛很高。中国国防科工局表示,要结合"国防科技工业2025"和国防科技工业军民融合"十三五"规划的编制,推动我国装备升级。在"中国制造2025"战略的推动下,各行业都在积极承接并制定本行业规划。此次国防科工局表态,首次确认了军工领域正在编制"国防科技工业2025"。该技术拥有国:美国(惠普)、英国(罗尔斯)、俄罗斯、法国(斯奈克玛)、中国(中国航空工业)和乌克兰。见图A-4。

图 A-4　航空发动机
Figure A-4　Aero-engine

Aero-foam fire extinguishing system　蒸汽泡沫灭火系统　系指将空气和水混合到泡沫液内形成空气泡沫液灭火剂的灭火系统。

Aerofoil　翼　泛指运动时能提供升力的叶或物体。

Aerofoil section　机翼切面　系指导边或随边与随边具有翘度的不对称流线型切面。

Aerogel　气凝胶　系指世界上已知密度最低的人造发泡物质。气凝胶借由临界干燥法将凝胶里的液体成分抽出。这种方法会令液体缓慢地被脱出,但不至于使凝胶里的固体结构因为伴随的毛细作用被挤压破碎。最早由美国科学工作者Kistler在1931年制得。气凝胶的种类很多,有硅系、碳系、硫系、金属氧化物系、金属系等。气凝胶拥有诸多优异特性,其中热学方面特性最受人们重视,也是最具产业化价值的,被誉为超级隔热材料,其隔热保温上的应用形式,主要为气凝胶毡、气凝胶板等。

Aerographical ascent(aerographical sounding)　高空气象观测　为获取不同高度大气中空气温度、湿度、气压和风等气象要素,而进行的直接(如飞机)或间接(探空仪或雷达)的气象观测。

Aerosol　气溶胶　系指由浓缩的气溶胶或弥散的气溶胶组成的不会造成对臭氧层破坏的灭火剂。
Aerosol is the non-ozone depleting fire-extinguishing medium consisting of either condensed aerosol or dispersed aerosol.

Aerospace plane　空天飞机　系指既能航空又能航天的新型飞行器。它像普通飞机一样起飞,以高超音速在大气层内飞行,在30~100 km高空的飞行速度为12~25倍音速,并直接加速进入地球轨道,成为航天飞行器,返回大气层后,像飞机一样在机场着陆。在此之前,航空和航天是两个不同的技术领域,由飞机和航天飞行器分别在大气层内、外活动,航空运输系统是重复使用的,航天运载系统一般是不能重复使用的。而空天飞机能够达到完全重复使用和大幅度降低航天运输费用的目的。该技术拥有国:美国、中国。见图A-5。

图 A-5　空天飞机
Figure A-5　Aerospace plane

AFRA maximum ship 阿芙拉型船舶 系指载重量(DWT)80 000～110 000 t 的船舶。

Aframax tanker 阿芙拉型油船 系指载重量在 8～10万吨级的油船。该型船设计吃水一般控制在 12.20 m,设计载重量不超过 8 万载重吨,此船舶可以停泊在大部分北美港口,并可获得最佳经济性。该型船最大载重量通过调整结构吃水获得。一般又被称为"运费型船"或"美国油船"。

AFS Code 船底防污染公约(AFS 公约) 系指 2001 年国际海事组织(IMO)通过的"国际控制有害船底防污系统公约"。 AFS Code means the International Convention on the Control of Harmful Anti-fouling System on Ships adopted by IMO in 2001.

AFS statement 船底防污系统声明 系指适用于长度24 m 以上但小于 400GT 的国际航行船舶,由船舶所有人或其授权代理所签署,表明船上的防污系统符合"船底防污系统公约"要求的证明文件。 AFS statement means a documentary evidence to show the anti-fouling system on ships is in compliance with the requirements of AFS Code, endorsed by the ship-owner or agent authorized by the ship-owner for the ship of more than 24 m in length and less than 400 gross tonnage engaged on international voyages.

Aft and forward space 船首和船尾处所

Aft bulkhead 后端壁 系指上层建筑的后壁。

Aft draft 艉吃水 相当于艉垂线处的型吃水,或龙骨吃水,或距船底某一基准点的吃水。

Aft end(AE) 艉端 系指规范船长 L 的尾端。 Aft end(AE) is rear section of the rule length L.

Aft end region structure 后端区域结构 应认为包含船舯部 $0.4L$ 区域之后的所有结构。 Aft end region structure is considered to include all structure aft of $0.4L$ the midship region.

Aft limit 后部界限

Aft part 艉部 包括:(1)艉尖舱舱壁以后的结构;(2)艉尖舱舱壁之后的区域。 Aft part includes:(1) the structures located aft of the aft peak bulkhead;(2) the area aft of the aft peak bulkhead.

Aft peak bulkhead 艉尖舱舱壁 系指船尾向前第一道主水密舱壁。 Aft peak bulkhead is the first main watertight bulkhead forward of the stern.

Aft peak tank 艉尖舱 艉尖舱舱壁之后的船尾狭窄部分的舱室。 Aft peak tank is the compartment in the narrow part of the stern aft of the aft peak bulkhead.

Aft perpendicular AP 艉垂线(AP) 系指从 FP 量起规范船长 L 后端处的垂线。 Aft perpendicular is the perpendicular at the aft end of the rule length, L, measured from the FP.

Aft region 艉部区 系指舯部区后边界至船尾的范围。 Aft region extends from the aft boundary of the mid-ship region to the stern.

Aft region of ice belt 冰带艉部区域 系指冰带舯部区域的后限界至艉柱间的区域。 Aft region is the region from the aft boundary of ice belt to the stern.

Aft seal 艉封门 侧壁气垫船在艉部船底用以维持气垫的封闭装置。

Aft terminal 后端点(艉端) 系指:(1)船舶分舱长度的最后一点;(2)分舱长度的艉界限。 Aft terminal is:(1) the last terminal of the subdivision length;(2) the aft limit of the subdivision length.

After anchor light 后锚灯 系指设置在船舶尾部的锚灯。

After body 后体 系指中横剖面以后的船体部分。

After burning 后燃 气缸内气体在膨胀行程中燃烧的现象。

After condenser 后冷凝器 在喷射抽气冷凝器组中,用以冷凝第二级抽气器蒸汽工质的冷凝器。

After engine arrangement 艉机布置 系指主机安置在船舶尾部的布置形式。

After freeboard perpendicular(AP_{LL}) 干舷艉垂线 取自长度 L_{LL} 的后端。 After freeboard perpendicular is to be taken at the aft end of the length L_{LL}.

After mast 后桅杆 系指设置在前桅以后装设桅灯的一根桅杆。

After peak bulkhead 艉尖舱舱壁 该舱壁应水密延伸至干舷甲板,其可以在舱壁甲板以下具有台阶但船舶分舱的安全等级不会因此而降低。 After peak bulkhead are to be made watertight up to the freeboard deck. The after peak bulkhead may, however, be stepped below the bulkhead deck, provided the degree of safety of the ship as regards subdivision is not thereby diminished.

After perpendicular(AP) 艉垂线(AP) 为设计水线与舵柱后缘交点处的垂线。对无舵柱的舰船,其艉垂线 AP 为该水线与舵杆中心线或舵板剖面中心线交点处的垂线。 After perpendicular AP is the perpendicular at the intersection of the waterline at the design draught with the after side of the rudder post or to the centre of the rudder stock for vessels without a rudder post or to the intersection with the transom profile on the centerline.

After shoulder 后肩 系指去流段紧邻平行中体的部分。

Aftercondenser 后冷凝室 在冷凝系统中产生第

二次冷凝功能的腔室称为后冷凝室。

Aftercooled diesel 后冷式柴油机

Aftercooler 后冷却器 系指在冷却系统中产生第二次冷却功能的冷却器称为后冷却器。

Afterpeak 艉尖区 艉尖舱舱壁之后的高度符合有关规定的区域。 Afterpeak is defined as the area aft of after peak bulkhead, up to the height according to related requirements.

Aftfoot 后踵 系指艉柱与龙骨连接的部分。

Aft-middle type ship 中后机型船 系指主机舱位于船长中部偏后的船。

Aft-terminal 后端点 系指船舶分舱长度的最后一点。

Age hardening 人工时效硬化处理 系指为了提高合金的硬度,将合金置于较高温度下,使其过饱和固溶体析出溶质原子。 Age hardening is the increasing the hardness of an alloy by a relatively low-temperature heat treatment that causes precipitation of components or phases of the alloy from the supersaturated solid solution.

Age of ship 船龄 系指:(1)在船舶从其建造完成年份算起至今所过去的年限。(2)从安放龙骨之日,或处于类似建造阶段,或改造为其他类型船舶之日算起的时间。

Aged structure 老化的结构 系指在防腐和疲劳断裂方面的能力可能相对下降的现有结构。 Aged structure is an existing structure that may suffer age-related degradation such as corrosion and fatigue cracks.

Agency 代理 系指许多国家商人在从事进出口业务中习惯采用的一种贸易方式。在国际市场存在着名目繁多的代理商。其中包括采购、销售、运输、保险、广告等多方面的代理商。

Agency agreement 代理协议 系指规定委托人和代理人权利与义务的协议。代理协议的主要内容包括:(1)协议双方当事人;(2)指定的代理商品;(3)指定的代理地区;(4)授予代理的权利;(5)协议有效期和中止条款;(6)代理人的佣金条款。

Agency fee 代理费 即代理(服务)费,多指在办理法律及知识产权事务时,由当事人(被代理人)支付给代理人(律师事务所,知识产权代理公司)的,就代理人向被代理人提供双方约定的专业代理服务而支付的费用,即劳务报酬。

Agent 代理人 系指任何自然人和法人,其代表船东、租船人或船舶经营者或货主提供航运服务,包括给成为海上安全调查对象的船舶处理安排工作。 Agent means any person, natural or legal, engaged on behalf the owner, charterer or operator of a ship, or the owner of the cargo, in providing shipping services, including managing arrangements for the ship being the subject of a marine safety investigation.

Aggregate dredger(AD) 骨料疏浚船 系指专门用于以采集并筛选海底砂砾、砾石和卵石等材料装舱,然后利用船上的拉斗、抓斗或斗轮等卸料设备将物料从泥舱输送到舷边的料斗,再经过传送带将其输送到岸上,以供建筑使用的一种耙吸式挖泥船。

Aggregate limit 总限额 即协定商品自动出口的总额数。

Aggregated alert 组合警报 该警报指示存在多个警报。 Aggregated alert indicates the existence of multiple individual alerts.

Aggregation 组合 将单独的警报组合成一个警报(一个警报代表多个单独警报),例如,驾驶室推进系统即将减速或停车的警报。 Aggregation means combination of individual alerts to provide one alert (one alert represents many individual alerts), e. g. imminent slowdown or shutdown or shutdown of the propulsion system alarm at the navigation bridge.

Aggregation of marine cargo 海运货物集中运输

Aging of lubricating oil 滑油老化 系指滑油使用日久变质,使滑油效果减弱的现象。

Agitation dredger 搅动挖泥船

Agreement 协议书 有广义和狭义之分。广义的协议书系指社会集团或个人处理各种社会关系、事务时常用的"契约"类文书,包括合同、议定书、条约、公约、联合宣言、联合声明、条据等。狭义的协议书系指国家、政党、企业、团体或个人就某一问题经过谈判或共同协商,取得一致意见后,订立的一种具有经济或其他关系的契约性文书。协议书是应用写作的重要组成部分。协议书是社会生活中,协作的双方或数方,为保障各自的合法权益,经双方或数方共同协商达成一致意见后,签订的书面材料。协议书是契约文书的一种。是当事人双方(或多方)为了解决或预防纠纷,或确立某种法律关系,实现一定的共同利益、愿望,经过协商而达成一致后,签署的具有法律效力的记录性应用文。

Agreement of consignment 寄售协议 系指寄售人和代销人之间就有关权利、义务及有关寄售条件和具体做法而签订的书面协议。销售商委托代理商销售货物的协议。按照协议,销售商通过邮政或公共承运人把货物寄(运)送给代理商进行异地销售。代理商是销售商的代理人,负有推销和妥善保管货物的义务,售货款以及未售完的货物均属销售商所有,代理商只提取佣金。

Agreement quotas 协议配额 又称双边配额,是在贸易中进行的协议确定的配额。

Agreement voluntary export quotas 协定自动出口配额制

Agricultural vessel 农用船 各种农业用船的统称。

Aground sweeping 拖底扫海测量 系指扫海具底索全部着底的扫海测量。

AH = anchor handling 起、抛锚

Ahead and astern 倒顺车 系指主机转速不变，改变轴系转向使其正转或倒转的措施。

Ahead stage 正车级 系指使船舶获得前进动力的汽轮机级。

Ahead turbine 正车透平 在透平推进装置中使船舶做前进运动的透平缸称为正车透平。

AHTS = anchor handling, tug supply 起、抛锚，拖带，供应

Aim point 瞄准点 又称上线点。是每根测线起始点前的一个点。它位于该测线的延长线上。瞄准点至起始点之间的距离由操作员按需要确定。驶向瞄准点的过程中，显示器不断显示瞬时船位到瞄准点的距离和方法。

Air (pressure) testing 空气试验（渗漏试验） 系指通过空气压力差和渗漏探测方法来验证密性的试验，包括液舱空气试验和接缝空气试验，诸如压缩空气试验和抽真空试验。

Air bearing 空气轴承 系指利用空气作为轴承上下体之间膜体者谓之空气轴承。

Air bottle 空气瓶 系指储存一定压力空气的钢瓶称为空气瓶。

Air change ratio 换气次数 一个舱室的通风量与该舱室一般以空舱计算的容积的比值。

Air circulating rates 风量倍数 系指吹风冷却的冷藏货舱内循环的总风量与空舱容积的比值。

Air compressor 空气压缩机 将大气空气进行压缩形成一定压力空气的机器称为空气压缩机。压缩比一般在2以上。

Air compressor lubricating tank 压缩机滑油柜 系指贮存空气压缩机使用的滑油的舱柜。

Air conditioner 空调装置（空调器） （1）对局部环境内空气的温度、湿度乃至新风比进行调节的设备称为空调装置或空调器；（2）主要由通风机、表面式空气冷却器、散热器、空气过滤器、加湿器、挡水板、水盘、进风室、出风室等组成，对空气进行降温、去湿等处理的机械设备。

Air conditioning 空调系统（空调器）

Air conditioning plant 空气调节装置 对局部环境内空气的温度、湿度乃至新风比进行调节的装置称为空气调节装置。

Air conditioning pump 空调装置（用）泵

Air conditioning unit 空调装置 对局部环境内空气的温度、湿度乃至新风比进行调节的装置称为空调装置。

Air container[1] 空气瓶罐 系指不与艇体或甲板结构连为一体的，由刚性材料制成的容器。Air container is the container made of stiff material, and is not integral with the hull or deck structure.

Air container[2] 气罐（空气瓶） 储存一定压力空气的钢瓶称为空气瓶。

Air content in water 水中空气含量 系指水中所含溶解和未溶解的空气量。以百万分数计。

Air control valve 空气调节阀

Air cooled compressor (air flow cooled compressor) 风冷式压缩机 系指汽缸或冷却器均由空气冷却的压缩机。

Air cooled diesel 空气冷却式柴油机（风冷式柴油机）

Air cooler 空气冷却器（冷风机） 由通风机、冷却盘管、水盘、进风室、出风室等组成，使空气进行冷却的热交换器称为空气冷却器。

Air cooling 吹风冷却 热空气经过冷风机中的盘管得到冷却，然后将冷风送入冷藏舱、冷藏库内以冷却货物，而升温的热空气又被吸回冷风机循环使用的冷却方式。

Air cooling engine 气冷式发动机

Air cooling zone 空气冷却区 系指使冷凝器中的空气在被抽出冷凝器前先行冷却的区域。

Air curtain 气幕 由一薄层射流空气形成的，具有一定封闭或屏障作用，用以维持其内外气压差的气流屏幕。

Air cushion 气垫 系指由气垫船的底部、支承表面及围裙（或侧壁及首尾门）所封闭的，其压力高于大气的空气层。

Air cushion craft 全垫升气垫船 系指能借助气垫支承其全部重量的高速船（是借助柔性围裙保持气垫，并借助气垫支承其全部重量的一种气垫船）。Air cushion craft is a high speed craft wholly supported by air cushion.

Air cushion craft (hovercraft) 气垫船 系指船舶不论在静止或运动时，其全部重力或大部重力能被连续产生的气垫所支承的船舶。其利用气垫的原理将船托起，使船的航行能适应各种多变的航道情况，如浅水、沼泽、沙滩等。气垫的作用减少了船体与水的直接接触，因而气垫船一般具有很高的速度。气垫船按气垫的形式分为两种：全垫升气垫船和侧壁式气垫船。

Air cushion vehicle(ACV/hovering craft) 气垫船　系指利用高于大气压的空气在船底与支承表面间形成气垫,使全部或部分船体脱离支承表面的船舶。气垫船分为两大分支:(1)全垫升气垫船(hovercraft, surface effect vehicle),就是船体全部由气垫支承而和水(地)表面完全脱离的船舶,采用空气推进器推进,具有水陆两栖能力。其又分为周边射流气垫船、多气室气垫船和增压式气垫船。(2)侧壁式气垫船(side wall surface effect vehicle),也称表面效应船(SES)。船的两舷刚性侧壁插入水中,艏部和艉部仍有柔性围裙(气封装置),用以封闭气垫。船航行时,由于刚性侧壁仍在水中,只能在水面运行,所以没有两栖性,但节省不少气垫功率,又可以向大型化发展。这类气垫船用水中推进器推进。

Air defence missile 防空导弹　系指由地面、舰船或者潜艇发射,拦截空中目标的导弹,西方也称之为对空导弹。由于大多数空中目标速度高、机动性大,故防空导弹绝大多数为轴对称布局的有翼导弹;动力装置多采用固体火箭发动机,也可以采用液体火箭发动机、冲压式空气喷气发动机和火箭冲压发动机;从20世纪40年代初德国开始研究到目前经历近60个春秋,世界上的防空导弹已研制了三代,目前还在发展第四代。据不完全统计,已研制的型号达120余种,其中装备部队90多种,正在研制的有20多种。我国的防空导弹都有远程防空导弹:红旗2、红旗9、S300、FT-2000;中程:红旗61、KS-1、飞蠓80;近程/肩扛式:红缨5、前卫1、前卫2、前卫3和飞弩6等。

Air defense identification zone(ADIZ) 防空识别区　系指濒海国家或地区基于海空防安全需要,在面向海洋方向上空划定的特定空域。这个空域一般划设在领空外,有时也会覆盖部分领空,特别是远离本土的岛屿。在该空域内,划设国一般可要求他国(方)航空器提供飞行计划、无线电、应答机和标志等方式,以便及时进行识别。若不遵守规定,可能会被拒绝入境,甚至遭到拦截、迫降。这样做的目的是在不明航空器进入领空前,对其性质提前做出判定,以赢得处理时间,有效保证国防安全。防空识别区不是领空,而是在一国领空之外所划定的空域范围,以便留出预警时间,保卫国家空防安全。因此,设立防空识别区并不意味着领空范围的扩大,而是更加有效地保卫国家的领空的安全。目前,美国、加拿大、澳大利亚、缅甸、韩国、古巴、芬兰、希腊、冰岛、意大利、日本、利比亚、阿曼、巴拿马、菲律宾、德国、泰国、土耳其、印度、越南等国分别建立了防空识别区。中国于2013年11月23日划定东海防空识别区。

Air detector(air finder) 探空仪　又称"无线电气象仪",系指探测不同高度气象要素(气温、气压、湿度、风等)的仪器。

Air distributor 空气分配器　空气管路中将空气按气目的分配进入相应管路的设备或管路附件称为空气分配器。

Air diving 空气潜水　系指在潜水作业中,以压缩空气为呼吸介质,其作业深度超过60~70 m的潜水。

Air draught H_a 净空吃水 H_a　净空吃水 H_a应取空艇状态时的漂浮平面到艇的结构或桅的最高点之间测得的垂直距离。注意:制造厂商无须在艇主手册中说明,艇主需考虑桅顶灯和可能安装的天线。 Air draught H_a shall be measured as the vertical distance between the flotation plane in the light craft condition and the highest point of the craft's structure or mast. It is noted that the manufacturer is free to state in the owner's manual that the boat owner is to make an allowance for a masthead light and possible fitting of aerial(s).

Air dynamic noise 空气动力性噪声　系指由气体的振动而产生的噪声。例如气垫船上的升力风机的噪声,推进用的空气螺旋桨噪声,飞机上的喷气式发动机喷发出来的噪声,排气管喷出废气的噪声等。

Air ejector 抽气喷射器(抽气器)　系指:(1)利用蒸汽或压力水作为工质从冷凝器内抽除非凝结气体的喷射器;(2)为保持冷凝器真空度而将其内部空气不断抽出的装置。

Air ejector condenser 抽气冷却器(喷射抽气冷凝器组)　系指:(1)用以冷凝抽气喷射器排放的工作蒸汽以及由其所抽出的蒸汽和空气的冷却器;(2)为保持凝器内的真空度抽除其中空气并冷凝抽气器蒸汽工质的两级抽气成套设备。由抽气器及中、后冷凝器等组成。

Air engine 航空发动机(空气发动机)

Air express 急件传递　系指目前航空公司运输中最快捷的方式。它由专门经营此项业务部门和航空公司合作,以最迅速的方式传送急件。如急需药品、图纸资料、货样单据以及文件合同等。

Air extracting pump 抽气泵

Air extractor 抽气器　用于将混入系统中的空气抽除出去的器具称为抽气器。

Air fed hydrofoil 供气水翼　系指用人工方法使非凝结气体通进水翼,并控制气量的大小,以使升力发生变化的水翼。

Air filled cavity 空气泡　系指充满空气的泡。如全充气螺旋桨叶背的空泡。

Air filter 空气过滤器　系指空气在其中曲折流动时,能过滤尘埃杂质的设备。

Air filter silencer 空气滤清消音器　系指在内燃机进气管或增压器压气机的空气进口处,设置的滤清空气,并能减少噪声的装置。

Air filtration unit 空气滤清装置（空气过滤设备）
用于将混入空气中的杂质过滤出来而使该空气达到使用要求的设备称为空气滤清装置或空气过滤设备。

Air gap 气隙　系指平台主体升到作业位置时，主体结构最低构件下沿与最大设计波高的波峰之间的净空距离（见图 A-6）。气隙 C 为 C_1 和 1.2m 之小者：$C_1 = (H_1 + H_2 + H_3)$（m）；式中，H_1—天文潮高，m；H_2—风暴潮高，m；H_3—最大设计波浪在基准水面上的高度，m。Air gap refers to the clearance between the lowest edge of hull bottom plating of the unit and the crest of maximum design wave (Figure A-6.). Air gap C is to be the smaller one of C_1 and 1.2 m: $C_1 = (H_1 + H_2 + H_3)$ (m); where: H_1—elevation of astronomical tide, in m; H_2—elevation of storm tide, in m; H_3—elevation of crest of maximum design wave to cardinal water level, in m.

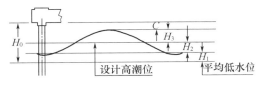

图 A-6　气隙
Figure A-6　Air gap

Air grill 通风棚　系指一般用于居住舱室或公共场所，装在通风管出口处或进风口的格栅，主要有直槽式、横槽式、线网式等。

Air gun 空气枪　曾称"空气脉冲发生器"，控制高压空气的充入和释放，从而对海水产生脉冲振动的一种活阀式机械装置。它由上、中、下气室、活塞组和电磁阀等组成。

Air gun array 组合空气枪　系指按不同容量、不同间距编组排列的空气枪群。

Air gun assembly controller 组合气枪控制器　系指用来控制和调节组合气枪同步激发的仪器。用微处理机反馈补偿时，精度可达 0.1 ms。

Air gun source 空气枪振源　系指利用高压空气充入空气枪，由电磁阀控制空气枪中的高压空气，在海水中瞬时释放，产生弹性波的振动源。

Air gun tiring controller 空气枪激发控制器　系指用来控制和调节空气枪阵列发火延迟时间，使之产生更一致、更强的宽频带脉冲特征波形的控制装置。

Air heater 空气加热器　系指用于将空气加热的热交换器称为空气加热器。

Air hole 透气孔　系指构件上为使空气能自由流通而开的孔。

Air induction valve 进气阀

Air injection diesel 空气喷射式柴油机　利用一定压力的空气将燃油以雾状喷入气缸进行燃烧做功的柴油机称为空气喷射式柴油机。

Air inlet unit 进气装置　系指装于压气机进气缸前用以吸入空气并具有稳流、稳压、除水、除盐和消声作用的装置。

Air jet diffuser 集散式布风器　系指可以任意转动的，一端为喷口，另一端呈扩散形的顶式风头。

Air lift dredger（ALD）/Air pneumatic dredger（APD） 气力提升挖泥船　又称气力输送挖泥船。是一种采用气力作用原理进行水下沉积物清除的疏浚机具。普遍用于气力清淤作业。

Air lock（vapor lock） 气塞　系指化油器或燃油喷射装置内有燃油蒸发或空气时阻塞燃油供应的现象。

Air noise of equipments 设备的空气噪声　系指机械动力设备运转时直接向空气中传播的噪声。

Air noise in ship's space 船舶舱室的空气噪声（船上噪声）　系指船舶中的声源辐射到舱室空气中的噪声。

Air oil separator 油-气分离器　将空气中的油分分离出去的设备称为油-气分离器。

Air operated valve 气动阀

Air outlet 送风头　通常用于机炉舱重点部位的通风，装在通风口出口处的风头。主要形式有转动喇叭式、播风器等。

Air outlet valve 排气阀

Air pipe 空气管　系指油舱和水舱的通气管路。

Air pipe closing appliance 空气管关闭装置　液舱的空气管在通向大气的端部所装的能使空气管予以关闭的机构为空气管关闭装置。

Air pipe heads 空气管头　系指设置在露天甲板上的空气管道接口，向干舷甲板或船楼甲板上部延长的空气管头。Air pipe heads installed on the exposed decks are those extending above the freeboard deck or the upper part of erectim decks.

Air pipe heads installed on the exposed decks 露天甲板上的空气管头　系指延伸至干舷甲板或上层建筑甲板以上的空气管头。Air pipe heads installed on the exposed decks are those extending above the freeboard deck or superstructure decks.

Air placed and dissipative acoustic filter 喷注耗散型消音器

Air plant 压缩空气装置　用压缩机将空气压缩至一定压力的设备称为压缩空气装置。

Air pollution index（API） 空气污染指数　系指反映空气污染物，如最常见是氮氧化物、悬浮粒子（来自汽车)和二氧化硫(来自工厂)水平的一个指数。空气污染指数也就是将常规监测的几种空气污染物浓度简化成为单一的概念性指数值形式,并分级表征空气污染程度和空气质量状况,适合于表示城市的短期空气质量状况和变化的趋势。空气污染的污染物有:烟尘、总悬浮颗粒物、可吸入悬浮颗粒物(浮尘)、二氧化氮、二氧化硫、一氧化碳、臭氧、挥发性有机化合物等。

Air post parcel receipt 航空邮包收据

Air pre-heater(air heater) 空气预热器 系指用回收烟气热量或用蒸汽加热锅炉燃烧用空气的装置。

Air pressure oil atomizer 空气雾化喷油器 系指利用压缩空气高速射流雾化燃油的喷油器。

Air pump 空气泵

Air purification system 空气净化系统 用于将空气中的杂质去除的系统称为空气净化系统。

Air purification unit 空气净化设备 用于将空气中的杂质去除的设备称为空气净化设备。

Air purifier 空气滤器 用于将空气中的杂质过滤掉的设备称为空气滤器。

Air quality index(AQI) 空气质量指数 系指一种评价大气环境质量状况简单而直观的指标。通过报告每日空气质量的参数,描述了空气清洁或者污染的程度,以及对健康的影响。计算空气质量指数通过五个主要污染标准:地面臭氧,颗粒物污染(也称颗粒物),一氧化碳,二氧化硫,二氧化氮。AQI 共分6级,从1级优,2级良,3级轻度污染,4级中度污染,直至5级重度污染,6级严重污染。当 PM2.5 日均值浓度达到 150 $\mu g/m^3$ 时,AQI 即达到 200;当 PM2.5 日均浓度达到 250 $\mu g/m^3$ 时,AQI 即达 300;PM2.5 日均浓度达到 500 $\mu g/m^3$ 时,对应的 AQI 指数达到 500。指数越大、级别越高说明污染的情况越严重,对人体的健康危害也就越大。

Air quantity 通风量 系指单位时间内输送或抽除的舱室空气量。

Air reactive substances 空气反应物质 系指与空气发生反应并造成潜在危险的货物,如形成的氧化物会导致爆炸反应。 Air reactive substances are products which react with air to cause a potentially hazardous situation, e.g. the formation of peroxides which may cause an explosive reaction.

Air receiver 空气瓶 储存一定压力空气的钢瓶称为空气瓶。

Air refreshing arrangement 空气更新设备 用换气的方式使空气保持一定清新度的设备称为空气更新设备。

Air register 调风器 系指油燃烧器中向炉内配送风并造成良好油风混合物以建立起稳定燃烧的部件。

Air release valve 放气阀

Air resistance(wind resistance) 空气阻力 系指航行中船舶其水上部分与空气的相对运动产生的阻力。其中包括由于风的作用产生的阻力。空气阻力由摩擦阻力和黏性阻力两部分组成。由于空气黏性小,摩擦阻力仅占很小部分,故船舶的空气阻力主要是黏性阻力。它与船舶水上部分的外形、尺寸以及船舶与空气相对运动的速度大小和方向有关。

Air resorber 空泡重溶装置 系指使空泡中所含空气有充分时间并在足够压力下重行溶解于水中的装置。

Air revitalization unit fan 再生通风机 系指空气再生装置用的通风机。

Air screen method 气幕法 系指利用压缩空气在水中产生气泡,气流使水面隆起,形成气幕,围住溢油,以防止溢油扩散的方法。其优点是不妨碍船舶航行,火灾时也有效。缺点是只宜用于狭窄、平静的水域。

Air screw 空气螺旋桨 系指气垫船、快艇等使用的在空气中工作的螺旋桨。

Air space 领空 系指主权国家领土和领海上空的空气空间,是国家领土的组成部分。一国领空具有完全排他性的国家主权,也就是说外国航空器在未经某一国允许的情况下,不得进入这一国家的领空。

Air starting valve 空气启动阀

Air strainer 空气滤清器(空气初滤器) 用于将空气中的杂质初步过滤出来而使该空气达到某种使用要求的设备称为空气初滤器。

Air suction filter 进气过滤器 系指进气装置中去除进气中水分、盐雾和杂质的装置。

Air tank 空气舱柜 系指与艇体或甲板结构连为一体的,由艇体结构材料制成的舱柜。 Air tank is the tank made of hull construction material, and is integral with hull or deck structure.

Air test 气密试验 系指用于检查舱室气密性的试验。其方法是:在舱内充气,再在舱外的焊缝上涂肥皂水,发现鼓肥皂泡的地方即为漏点。其优点在于可以随时发现缺陷,随时修补。

Air transport 航空运输 是一种现代化的运输方式,其有运输速度快,货物质量轻,且不受地面条件限制等优点。因此,它最适宜运送急需物资、鲜活商品、精密仪器和贵重物品。

Air transportation all risks 航空运输一切险 系指对航空运输途中所遭受的意外损失相对应的保险。保险公司赔偿包括被保险货物在运输途中遭受雷电、火灾、爆炸或由于飞机遭受恶劣气候或其他危难事故而被抛弃,或者由于飞机遭受碰撞、倾覆、坠落或失踪等自然灾害和意外事故所造成的全部或部分损失；还包括被保险货物由于一般外来原因所造成的全部或部分损失。

Air transportation cargo insurance 航空运输货物保险

Air transportation risks 航空运输险 系指对被保货物在运输途中遭受雷击、火灾、爆炸，或由于飞机遭受恶劣气候或其他危难事故而被抛弃，由于飞机遭受碰撞、倾覆、坠落等事故所造成的全部或部分损失。

Air tube cone clutch 锥形高弹性离合器 系指以压缩空气进行操纵，利用圆锥面接合并设有盘型高弹性橡胶元件的离合器。

Air tube disc clutch 盘型高弹性离合器 系指以压缩空气充入气胎进行操纵，利用圆盘面接合的离合器。

Air turbine pump 空气涡轮泵

Air valve 空气阀

Air vent and overflow pipe(air and overflow pipe) 空气兼溢流管 系指用以逸出空气，又能在液舱注满时将液体溢流的管路。

Air-augmented water-jet propulsion 加气喷水推进 系指将压缩空气加入水中形成极小气泡，利用空气膨胀的能量以增大水柱动量的一种双态喷射推进。

Air-blast atomizing nozzle(air blast atomizer) 空气雾化喷油嘴 系指燃油自中心部分的涡硫室喷出，而具有相当压力的空气在外缘和油流旋流相反的方向喷出，从而提高雾化质量并增加头部区域中的空气量以改善燃烧的喷油嘴。

Airborne early warning airplane(AEW) 空中预警机 系指为了克服雷达受到地球曲度限制的低高度目标搜索距离，同时减轻地形的干扰，将整套雷达系统放置在飞机上，自空中搜索各类空中、海上或者是陆上目标。空中预警机是集侦察、通信、指挥、控制于一体的作战飞机。

Airborne early-warning ground integrated system(AEGIS)[1] 宙斯盾 系指希腊神话中宙斯之盾。

Airborne early-warning ground integrated system(AEGIS)[2] 空中预警与地面整合系统 又称"宙斯盾"战斗系统。

Airborne high power generator power supply system and power supply system 飞机用大功率发电机及供电系统 该技术拥有国：美国、俄罗斯、中国。见图A-7。

Airborne noise in cabin 舱室空气噪声 系指主要来自舱室内主机、辅机的噪声。

Airborne warning and control system(AWACS) 机载空中预警和控制系统 系指利用飞机和以监视雷

图 A-7 飞机用大功率发电机及供电系统
Figure A-7 Airbone high power generator supply system and power supply system

达为主的机载电子设备探测目标和执行指挥使命的系统。这种系统用于防空和战术指挥。能与地面系统配合或独立完成战略和战术的预警、指挥和控制任务，还可担负交通管制和空中监视等非军事使命。空中预警和指挥系统能在数百公里的范围内监视上、下和周围的情况，凭借飞机的高度和速度而免受敌方攻击。它由空中监视雷达与自动计算、数据处理和通信设备构成防御中心，并能随时转移到局势需要的空域，提供有效的空中监视、指挥和控制防空武器等防御或攻击任务。空中预警和控制系统于20世纪60年代开始研制，70年代投入使用。它由作为载体的飞行器、天线罩和电子系统三大部分组成。空中预警和控制系统的电子系统分为监视雷达、数据处理、数据显示、控制、导航、通信和敌我识别等6个分系统。在敌我识别系统中使用一部专门的计算机处理回波，并识别结果送至数据处理系统。通信系统由数字式数据网和高频、超高频话音通信设备组成。导航系统由惯性导航装置、奥米加接收机和多普勒导航装置组成。正常工作时采用奥米加-多普勒-惯导组合导航方式。数据处理系统根据来自其他各系统的关于目标的报告实现多目标跟踪、控制通信系统发送和接收信息、执行操纵人员人工干预指定的计算和武器控制等任务。数据显示和控制系统由显示数据控制器接口和一个数据显示多用途控制盒组成，用来显示操纵人员要求的数据、表格等信息。

Airborne early warning aircraft(AEWA) 预警和控制系统飞机 简称预警机，是一种装有远距搜索雷达、数据处理、敌我识别以及通信导航、指挥控制、电子对抗等完善电子设备，集预警、指挥、控制、通信和情报于一体，用于搜索、监视与跟踪空中和海上目标并指挥、引导己方飞机遂行作战任务的作战支援飞机。目前，我国2000预警机是全世界最先进、领先美国E-3C预警机的一代预警机。

Air-cell combustion chamber 空气室燃烧室 系

指利用压缩过程中积蓄于空气室内的高压空气,随着活塞的下降以高压气流冲入主燃烧室与喷入的燃料进行混合形成涡流以达到较完全燃烧的燃烧室。

Air-conditioning apparatus dew point 空调机露点 系指空气调节器冷却盘管的外表面温度。

Air-conditioning unit room 空调室 系指供安置空调设备及其控制设备的舱室。

Air-cooled internal diesel engine 风冷内燃机 系指用空气冷却的内燃机。

Air-cooled refrigerated containers(ACRC) 风冷式冷藏集装箱

Aircraft 航空器 系指依靠空气的反作用力浮游在空中的任何机械,但任何靠贴近地球表面的空气反作用力的机械除外。 Aircraft means any machine that can derive support in the atmosphere from the reactions of the air other than the reactions of the air against the earth's surface.

Aircraft carrier battle group/Aero-plane carrier battle group 航空母舰战斗群 系指以大型航母为核心,巡洋舰、驱逐舰、护卫舰、核潜艇、支援舰为辅助,以舰载航空兵为主要作战手段,组成航空母舰战斗群,互为策应,互为依托,在火力和装甲层层的保护下,在有效地抵挡了外来威胁情况下,巡洋舰、驱逐舰、核潜艇上的对地攻击导弹将充分发挥战斗力,以导弹为先锋,撕碎敌人的对空防御和警戒,为舰载攻击机的空袭开路。集海军航空兵、水面舰艇和潜艇为一体,是空中、水面和水下作战力量高度联合的海空一体化机动作战部队。具有灵活机动,综合作战能力强,威慑效果好等特点,可以在远离军事基地的广阔海洋上实施全天候、大范围、高强度的连续作战。

Aircraft carrier/Aero-plane carrier 航空母舰 系指以舰载飞机或直升机为主要武器,并作为其海上活动基地的大型水面舰船。一般多以装载固定翼飞机为主,并装备有导弹、火炮等武器,电子设备完善,能执行多种战斗任务,是舰队或特混分舰队的核心军舰。现代航空母舰的满载排水量达4~10万吨,航速为20~35kn,携带飞机数量不等,大型者可装载100余架。推进动力装置多为汽轮机,有核动力和常规动力之分。主要航空设施有飞行甲板、弹射器、阻拦装置、机库、飞机升降装置、助航镜以及航空指挥、导航等设备。为保证有足够的起落滑行跑道和待命场所,上层建筑常设在舷侧(右舷),形成岛式建筑。飞行甲板为斜角型,可使飞机的起飞和着落作业同时进行而互不干扰。现代航空母舰按其承担的任务可分为攻击型航空母舰、护航型航空母舰和反潜型航空母舰,按推进动力装置可分为常规动力航空母舰和核动力航空母舰;按搭载机型可分为固定翼飞机航空母舰、直升机航空母舰和垂直/短距离起落航空母舰等。见图A-8。

图A-8 航空母舰
Figure A-8 Aircraft carrier/Aero-plane carrier

Aircraft mode 飞机状态 对地效翼船操作模式而言,系指C类地效翼船在ICAO规则规定的飞机最小安全高度以上飞行。 For the WIG craft operational modes, aircraft mode denotes the flight of a WIG craft of type C above the minimal safe altitude for an aircraft prescribed by ICAO regulations.

Air-cushion vehicle(ACV) 气垫船 系指船舶不论在静止或运动时,其全部重量或大部分重量能被连续产生的气垫所支承的船舶。 Air-cushion vehicle(ACV) is a craft that the whole or a significant part of its weight can be supported, whether at rest or in motion, by a continuously generated cushion of air dependent for its effectiveness.

Air-drop 空投 系指从飞行器上用降落伞或者其他有效减速装置投放物品或人员。

Airflow noise 气流噪声 系指气流的起伏运动或气动力产生的噪声。常见的气流噪声有喷气噪声、边棱声、卡门涡旋声、受激涡旋声、螺旋桨噪声、风扇声等。

Air-fuel ratio 空燃比(空气燃料比) 系指按质量计算的,进入发动机气缸内的空气与燃料的质量的比值。

Air-jet propelled boat 喷气推进船 系指依靠向后喷气的反作用力推进的船舶。

Air-launched launch vehicle 空射运载火箭 见图A-9。

Air-launched rocket system 空射运载火箭系统

Air-launched rocket technology/Ballistic missile technology 空射运载火箭/弹道导弹技术 技术拥有国:中国("开拓者")、美国("飞鸟座")、俄罗斯("Barge hauler")。

Airless injection diesel 无气喷射式柴油机

Airless injection engine 无气喷射发动机 不用压力空气而将燃油压缩达到以雾状喷入气缸进行燃烧做功的柴油机称为无气喷射式柴油机。

Air-lock 气闸 系指其应设有两扇间距不小于1.2m,但也不必大于2.5m的钢质气密门的设施,气密

图 A-9 空射运载火箭
Figure A-9 Air-launched launch vehicle

门应为自闭式,且没有任何门背钩装置。 Air-lock has two gastight steel doors which are self-closing without any hold-back arrangement, and the distance between which is not to be less than 1.2 m and not necessarily more than 2.5 m.

Airlock 气闸室 系指一种双门舱室,它的两扇门处于无毒区域和核、生、化与放射性污染危险物或洗消站之间,可以将无毒区域与有毒区域隔开。气闸室通常使用纯洁空气进行清洁,以避免人员跨区域行动时将沾染物带入无毒区。 Airlock is a compartment with two doors between the toxic free area and the source of the CBRN hazard or cleansing station. Airlocks are normally purged with clean air to allow personnel to pass from one area to another without contaminants entering the toxic free area.

Air-MAX ship 气泡船 系指一种突破传统船体设计的新船型。其创新在于 Air-MAX 船的船体部分被移除,形成一个庞大的空腔由压缩机压入空气以减少船体与水的接触面,从而提高航速;同时该船型还应用了新型球鼻艏。船模试验证明,这两项新技术的应用可使船舶降低约 30% 的燃油消耗量。

Airplane system 飞机式水翼系统 系指 65% 或更多的水翼面积分布于艇的重心之间的水翼系统。

Airport passenger processing system(APP) 机场旅客处理系统

Airscrew ship(aerial propeller vessel) 空气螺旋桨船 系指用空气螺旋桨推进的船舶。

Air-temperature automatic fire alarm(automatic thermal type fire alarm) 测温式失火自动报警器 系指依靠测量失火区舱室温度的变化情况作为失火信号源的自动报警器。

Air-water separator 气-水分离器 系指将水分从气体中分离出来的器具。

Airway bill 航空运单 系指承运人与托运人之间签订的运输合同,也是承运人或其代理人签发的货物收据。航空运单还可作为核收运费的依据和海关查验放行的基本单据。但航空运单不是代表航空公司的提货通知单。

Air-way system for diver 潜水用的中压空气系统 一般系指用于潜水深度不超过 60~70 m 时向潜水员供应呼吸用空气的气源系统。

AIS 自动识别系统 AIS means automatic identification system.

AIS target 自动识别系统(AIS)目标 系指自动识别系统(AIS)生成的目标。 AIS target is a target generated from an AIS message.

AIS-SART 搜救 AIS 应答器 系指能发射显示遇险单元位置、静态和安全信息的信息装置。所发射的信息应能与现有的 AIS 装置相容。发射的信息应能被 AIS-SART 接收范围内的辅助单元识别和显示,并能从 AIS 装置中清晰地分辨出 AIS-SART。 AIS-SART means a messages device which is capable of transmitting messages that indicate the position, static and safety information of a unit in distress. The transmitted messages should be compatible with existing AIS installations. The transmitted messages should be recognized and displayed by assisting units in the reception range of AIS-SART, and clearly distinguish the AIS-SART from an AIS installation.

Alarm[1] 报警 系指:(1)高度优先的报警。这状态需要立即引起注意并采取行动,以保持船舶的安全航行和操作。下列警报应列为报警:①机器报警;;②操舵装置报警;③控制系统故障报警;④舱底水报警;⑤进水探测预报警;⑥轮机员报警;;⑦人员报警;⑧驾驶室航行值班报警系统;⑨探火报警;⑩固定式局部使用灭火系统启动报警;⑪报警指示管理或探测系统发生故障或其失电的报警;⑫货物报警;⑬气体探测报警;⑭动力驱动的水密门故障报警;⑮"经修订的综合航行系统(INS)性能标准";[(MSC.252(83)决议附录5)]所规定的与航行有关的报警;⑯对特种船舶(如高速船,除上述定义的报警外,附加报警器也可列为报警;(2)当被监控的机电设备或系统超出预定参数范围时所发出的听觉和视觉信号。 Alarm means:(1) An alarm which has a high priority of an alert, when a condition vequires immediate attention and action, to maintain the safe navigation and operation of the ship. The following alerts are classified as alarms:①machinery alarm;②steering gear alarm;③control system fault alarm;④bilge alarm;⑤water ingress detection pre-alarm;⑥engineers' alarm;;⑦personnel alarm;⑧bridge navigation watch alarm system(BNWAS);⑨fire detection alarm;⑩fixed local application fire-extinguishing system activation alarm;⑪alarms indicating faults in alert management or detection systems or loss of their power supplies;⑫cargo alarm;⑬ gas detection alarm;⑭ power-operated watertight door fault alarm;⑮navigation related alarms as specified in the Revised Performance Standard for Integrated Navigation System(INS) [resolution MSC. 252(83), appendix 5.];⑯for special ships(e. g. high-speed craft), additional alerts may be classified as alarms in addition to the ones defined above;(2)visual and audible signal of a predetermined out-of-limits parameter for the monitored machinery or system.

Alarm^2 报警器(报警装置,报警显示屏) 能发出听觉和/或视觉或异常色调信号提示有关人员已有某项指标超标的器具或设备称为报警器或报警装置。

Alarm^3 报警装置 系指通过听觉装置或听觉和视觉装置通告需注意的状况的报警器或报警系统。 Alarm means an alarm or alarm system which is announced by audible means, or audible and visual means.

Alarm devices 报警设施 系指能使操作者立即识别定位系统的任何失效的视听信号。 Alarm devices mean those visual and audible signals enabling the operator to immediately identify any failure of the positioning system.

Alarm indicator 报警指示器 系指一种当出现一个或多个故障时,为了引起操作人员注意,而发出视觉和听觉报警的指示器。 Alarm indicator is an indicator which gives a visible and/or audible warning upon the appearance of one or more faults to advise the operator that his attention is required.

Alarm system 报警系统 系指用以指示任何需要注意故障的系统。此报警系统能:(1)在主机控制室或推进装置控制位置发出听觉报警,并能在适当位置以视觉方式显示每一独立报警功能;(2)通过选择开关与轮机员公用舱室和每一个轮机员居住舱室相连,以确保至少与这些舱室的其中一个相连。主管机关可允许采用等效装置;(3)对要求值班驾驶员采取行动或加以注意的任何情况,在驾驶室启动听觉和视觉报警;(4)应尽实际可能按故障安全原理设计;(5)如果报警在一个限定时间内未能就地引起注意,启动一个从发动机控制室或操纵平台(视具体情况而定)进行操作的轮机员报警装置。报警系统应持续得到供电,并应在失去正常供电的情况下自动转换成备用电源供电。报警系统的正常供电失效时应报警指示。报警系统应能同时提示一个以上故障,且任一报警的接受不应妨碍其他报警。在主机控制室或推进装置控制位置对任何报警的接受,应在表明存在该报警状态的各个位置显示。报警应保持到其被接受,各个报警的视觉显示应保持到故障被排除,此时报警系统应自动复位到正常运行状态。 Alarm system means a system which shall be indicating any fault requiring attention and shall:(1)be capable of sounding an audible alarm in the main machinery control room or at the propulsion machinery control position, and indicate visually each separate alarm function at a suitable position;(2)have a connection to the engineers' public rooms and to each of the engineers' cabins through a selector switch, to ensure connection to at least one of those cabins. Administrations may permit equivalent arrangements;(3)activate an audible and visual alarm on the navigation bridge for any situation which requires action by or attention of the officer on watch;(4)be designed on the fail-to-safety principle;(5)activate the engineers' alarm which shall be operated from the engine control room or at the manoeuvring platform as appropriate, if an alarm function has not received attention locally within a limited time. The alarm system shall be continuously powered and shall have an automatic change-over to a stand-by power supply in case of loss of normal power supply. Failure of the normal power supply of the alarm system shall be indicated by an alarm. The alarm system shall be able to indicate at the same time more than one fault and the acceptance of any alarm shall not inhibit another alarm. Acceptance in the main machinery control room or at the propulsion machinery control position of any alarm condition shall be indicated at the positions where it was shown. Alarms shall be maintained until they are accepted and the visual indications of individual alarms shall remain until the fault has been corrected, when the alarm system shall automatically reset to the normal operating condition.

Alarm transfer system 报警传送系统 除非 IMO 另有规定,否则在用户定义的时间之后发生未应答报警时,桥楼导航值班报警系统(BNWAS)的功能将启动"紧急呼叫"。 Functionality of the bridge navigational watch alarm system(BNWAS) actuates the "emergency call" in case of an unacknowledged alarm after a time defined by the user unless otherwise specified by IMO.

Alarm unit(alarm device) 报警装置 系指报告滑油或冷却水的温度和压力不符合规定值的装置。

Albatross ramjet/Aerofoil boat 冲翼艇 也称作气翼艇,是一种依靠艇与水面之间压缩"气垫"离开水面而又贴近水面飞行的水上交通工具。

Alerts 警报 系指:(1)通知和要求引起注意的异常状况。警报分为 4 个优先级:紧急报警、报警、警告和提醒。警报提供关于某一规定的状态发生改变的信息以及关于如何以规定的方式将该事件通知系统和操作员的信息。(2)对需要关注的异常情况和状况发出报警、警告和提醒。 Alerts are:(1)announcing abnormal situations and conditions requiring attention. Alerts are divided in four priorities: emergency alarms, alarms, warnings and cautions. An alert provides information about a defined state change in connection with information about how to announce this event in a defined way to the system and the operator. (2) announcing abnormal situations and conditions requiring attention. Alerts are divided in three priorities: alarms, warnings and cautions.

Alert announcements 警报发布 系指警报的听觉和视觉显示。 Alert announcements are visual and acoustical presentation of alerts.

Alert history list 警报历史清单 系指可访问的以前警报清单。 Alert history list is the accessible list of past alerts.

Alert management 警报管理 系指对驾驶台监测、处理、发布和显示警报的统一管理的概念。 Alert management is a concept for the harmonized regulation of the monitoring, handling, distribution and presentation of alerts on the bridge.

Alert phase 警戒阶段 系指对船舶及船上人员的安全感到忧虑的情况。 Alert phase means a situation wherein apprehension exits as to the safety of a vessel and of the persons on board.

Alerts given warning of immediate personnel hazard 警告对人员有紧迫危害的报警 包括:(1)灭火剂排放前的报警——警告即向某一处所施放灭火剂的报警;(2)动力驱动滑动水密门关闭报警——SOLAS 公约第Ⅱ-1/15.7.1.6 条所要求的报警,警告动力滑动水密门关闭。 Alerts given warning of immediate personnel hazard includes: (1) fire-extinguishing ore-discharge alarm—an alarm warning of the imminent release of fire-extinguishing medium into a space; (2) power-operated sliding watertight door closing alarm—an alarm required by SOLAS regulation Ⅱ-1/15.7.1.6, warning of the closing of a power-operated sliding watertight door.

Alfalfa 苜蓿 系指干苜蓿的衍生物质。以粗粉、丸粒形状装运。该货物为不燃物或失火风险低,无特殊危害。 Alfalfa means a material derived from dried alfalfa grass. Shipped in the form of meal, pellets etc. This cargo is non-combustible or has a low fire-risk. No special hazards.

Aligned indication 连续缺陷显示 系指在线像中存在的 3 个以上的,缺陷指示的最末端间距为 2 mm 以下的缺陷显示。 Aligned indication means three or more indications in a line, separated by 2 mm or less edge-to-edge.

Alkaline cleaning 碱洗 系指在适当温度下将碱性溶液在锅炉内进行循环,以除去黏附于金属表面的水垢和油脂的方法。

Alkaline corrosion 苛性腐蚀 系指锅炉中金属材料与碱性溶液接触,在高温下应力集中的区域发生的晶间腐蚀。

Alkalinity 碱度 系指在每公升水中以 OH、CO_3、HCO_3、PO_4 等一类能因离解或水解而致 OH 浓度增加的物质的阴离子毫克当量的总和表示的该类物质的总含量。

Alkylates-fuel 烷基化燃料

All oily mixtures 所有油性混合物

All purpose fuel rate 全船燃料消耗率 系指主机每单位输出功率每小时的全船燃料的消耗率。

All risks 一切险/保全险 是海上运输保险的基本种类之一,是保险人对保险标的物遭受特殊附加险以外的其他原因造成的损失均负赔偿责任的一种保险,包括平安险、水渍险和一般附加险。

All ships 所有船舶 系指:(1)在 2009 年 1 月 1 日或以前或以后建造的船舶;(2)所有船或艇,而不论其类型和用途。 All ships mean: (1) ships constructed before, on or after 1 January 2009; (2) any ship, vessel or craft irrespective of type and purpose.

All time saved 节省的全部时间

All total 总数

All up weight (AUW) 总质量 系指在特定应用需要考虑遇到的最大值,包括飞机、人员、燃料和有效载荷的最大质量:(1)对于直升机,总质量是在任何时候飞机、人员、燃料和有效载荷的最大质量;(2)对于固定翼飞机,总质量是飞机、人员、燃料和有效载荷的最大质量;对于折叠翼飞机的起飞,燃料质量的最大值应减去运输到离开位置的燃料质量;(3)对于固定翼飞机的降落,总质量是如上所述,除了燃料质量的最大值减去最短的可能飞行的消耗。 All up weight (AUW) is the maximum weight that will be considered for the specific application it includes the maximum weight of aircraft, personnel, fuel and payload: (1) For helicopters, the AUW is to be taken as the maximum weight of aircraft, personnel, fuel and payload at all times; (2) For manoeuvring of fixed wing aircraft, the AUW is to be taken as the maximum weight of aircraft, personnel, fuel and payload. For take off of fixed wing aircraft, the fuel weight is to be the maximum minus the fuel required to transit to the take off position; (3) For the landing of fixed wing aircraft, the AUW is to be as above except that the fuel weight is to be the maximum minus that consumed by the possible shortest flight.

All-caps 全部大写的

All-direction harbor tug 全回转拖船 系指装备

全回转螺旋推进器，具有极强的机动性，可以在原地360°回转的拖船。

All-direction propeller（Z-propeller） 全向推进器 简称Z推。通过斜齿轮系统能绕竖轴作360°转动，用以推进和操纵船舶的螺旋桨或导管推进器。

All-mountain amphibious chariot 两栖全地形战车 系指不受道路条件限制，可在普通车辆难以机动的地形上行驶的越野车辆。这类车往往具有宽大的轮胎，以增加与地面的接触面积，从而产生更大的摩擦力并降低车辆对地面的压强，使其容易行驶于沙滩、泥路、石路、河床、山间林道以及溪流。见图A-10。

图 A-11 全回转推进系统
Figure A-21 Azipod Allrotary system

图 A-10 两栖全地形战车
Figure A-10 All-mountain amphibious chariot

Allocation and transfer price 调拨价格 又称为转移价格，系指跨国公司为了在国际经营业务中最大限度地减轻税负，逃避东道国的外汇管制，以及为了扶植幼小的子公司等目的，在公司内部进行交易时采用的价格；该价格一般不受国际市场供求关系的影响，由公司上层管理者制定。

Allowable minimum angle 允许的最小角度 系指在安全工作载荷下的起重机系统允许操作的起重机吊杆与水平面的夹角。 Allowable minimum angle is the angle to horizontal of a derrick boom at which the derrick system is permitted to operate under the safe working load, and expressed in degrees(°).

Alloy 合金 系指由金属与另一种（或几种）金属或非金属所组成的具有金属通性的物质。一般通过熔合成均匀液体和凝固而得。根据组成元素的数目，可分为二元合金、三元合金和多元合金。

Allrotary propulsion system 全回转推进系统 这是一种悬挂在船体外的电力推进装置，电动机转速可以调节，而驱动的螺旋桨为定螺距，该推进装置可以其立轴为中心在360°范围内旋转，因而能在船舶的任何方向获得推力。采用这种推进装置的船舶无需设置舵、尾侧推等设备。见图A-11。

All-round flashing yellow light 黄色闪光灯 系指专供气垫船在非排水状态操作时用的显黄色的环照闪光灯。

All-round light 环照灯 系指在360°的水平弧度显示不间断灯光的号灯。 All-round light means a light showing an unbroken light over an arc of the horizon of 360°.

All-voltage starter 全电压启动器 系指不采用电阻或自耦变压器降压，而将电动机接至电源的一种启动器。该启动器可由一个手动操作开关或一个主控开关所组成。这种开关可使一个电磁操作的接触器通电。

All-weather aircraft（AWA） 全天候飞机 系指能够在昼夜和复杂气象条件下都能完成飞行任务的飞机。由于机载电子设备日趋完备，特别是有了各种性能良好的机载雷达，多数民用飞机和军用飞机都属于全天候飞机。

Alongside towing 绑拖 系指拖船舷靠舷地绑结被拖船或浮物一起行驶的一种拖带作业方式。

Alongside transfer 横向传送 系指补给船与接受船并列航行时通过索具横向传递物资或人员的方法。

ALP（articulated loading platform） 铰接式载油平台 也称铰接式载油柱（ALC）。 ALP means articulated loading platform, also termed as "articulated loading column"（ALC）.

Alterations and modifications of a major character 重大改装 系指任何影响到船舶分舱程度的结构改装，如对货舱做这种改装，则应证实按改装后船舶计算的 A/R 之比不小于按改装前船舶计算的 A/R 之比。但是，在改装前船舶的 A/R 之比等于或大于1的情况下，则仅需改装后船舶计算的 A 值不小于 R。 Alterations and modifications of a major character means, in the context of cargo ship subdivision and stability, any modification to the construction which affects the level of subdivision of that ship. Where a cargo ship is subject to such modification, it shall be demonstrated that the A/R ratio calculated for the ship after such modifications is not less than the A/R ratio calculated for the ship before the modification. However, in those cases where the ship's A/R ratio before modification is equal to or greater than unity, it is only necessary that the ship after modification has an A value which is not less than R, calculated for the modified ship.

Alternate Management System 替代的管理系统(AMS) 系指经外国主管机构依据国际海事组织的国际压载水管理公约提出的标准认可的并满足所有美国法律的适用要求的，且替代压载水调配的压载水管理系统。Alternate Management System (AMS) means a ballast water management system approved by a foreign administration pursuant to the standards set forth in the International Maritime Organization's International BWM Convention, and meets all applicable requirements of U.S. law, and is used in lieu of ballast water exchange.

Alternating flow 交变流 系指波浪场中的水质点往复运动。

Alternative communication system 替代通信系统 系指可供个人或船舶在没有海上甚高频/中频公共通信服务的区域使用的通信系统。Alternative communication system means a communication system available to individuals or to ships in areas where no maritime VHF/MF public correspondence services are available.

Alternative design and arrangements 替代设计和布置 系指偏离规范的任何指令性要求，但以满足所需相关章节目标和功能要求有可被船级社接受的安全和防止污染措施。这一术语包括一系列措施，包括可替代的舰上创新或独特设计基础上的结构和系统，以及安装于可替代布置或配置中的传统舰船结构和系统。根据偏差的不同性质和程度，通过特许权或其他设计理由证明的方式，这是可以接受的。Alternative design and arrangements mean safety and pollution prevention measures which deviate from any prescriptive requirement(s) of these rules, but are acceptable to LR to satisfy the objective(s) and the functional requirements of the relevant chapter. The term includes a wide range of measures, including alternative shipboard structures and systems based on novel or unique designs, as well as traditional shipboard structures and systems that are installed in alternative arrangements or configurations. Depending on the nature and extent of the deviation, it will be accepted by way of either a concession or alternative design justification.

Alternative duties 选择税 系指对于一种进口商品同时订有从价税和从量税两种税率，在征税时选择其税率较高的一种征税。

Alternative marine power(AMP) 船舶中压岸电系统 系指船舶在港停泊时停止船内发电机原动机和锅炉的使用而从岸上获得必要的电力的系统。其目的在于满足绿色环保港口的要求，减少船舶在港期间有毒有害气体和颗粒的排放。通常使用中压岸电系统的船舶有豪华邮轮、滚装船、集装箱船、液货船、LNG 运输船等。

Alternative means 替代方法 系指不采用按"原油油船货油舱保护涂层性能标准"[MSC.288(87)决议]涂装的保护涂层的方法。Alternative means is a means that is not a utilization of protective coating applied according to the performance standard for protective coating for cargo oil tanks of crude oil tankers [resolution MSC. 288(87)].

Alternative propulsion system 交替推进系统 系指一个适用于维持船舶在处于有下列限制的运行状态的系统。在主推进系统失效或者其他一些船东的需要，诸如在航行中主推进器或发电系统的一些重要部件进行维修的情况下，该推进系统既可用来允许船舶到达第一个合适的港口或避难的地方，也可避开严重的境况，此时可对航行、安全、货物的保存和可居住条件作最低的服务。交替推进系统也包括下列有关系统：(1)将扭矩转换成推力的设备；(2)操作所需的辅助系统；(3)控制、监测和安全系统。Alternative propulsion system is an arrangement of machinery suitable to maintain the ship in operating condition in case of loss of the main propulsion system. The alternative propulsion system may be used either to allow the ship to reach the first suitable port or place of refuge, or to escape from severe environment, allowing minimum services for navigation, safety, preservation of cargo and habitability. The alternative propulsion system also includes the following associated systems: (1) the equipment intended to convert the torque into thrust; (2) the auxiliary systems necessary for operation; (3) the control, monitoring and safety systems.

Alternative system 替代系统(涂层) 系指所有的根据所有类型船舶专用海水压载舱和散货船双舷侧处所保护涂层性能标准表1涂装的非环氧基涂层系统。Alternative system mean all systems that are not an epoxy-based system applied according to Table 1 of performance standard for protective coatings for dedicated seawater ballast tanks in all types of ships and double-side skin spaces of bulk carrier.

Alternatively-fired exhaust gas boiler 废气燃油交替式锅炉 系指自带燃油燃烧设备，并可交替使用燃油与废气的锅炉。

Alumina 氧化铝 系指一种白色无嗅细粉末，水分极少或无水分。不溶于有机液体。含水量 0% ~ 5%

的物质。氧化铝在受潮状态下不可泵送。该货物不溶于水,氧化铝粉尘的腐蚀性和穿透性很强,刺激眼睛和黏膜。该货物为不燃物或失火风险低。 Alumina is a fine, white odourless powder with little or no moisture. It is insoluble in organic liquids. Moisture content: is 0% to 5%. If it is wet, alumina is unpumpable. This cargo is insoluble in water. Alumina dust is very abrasive and penetrating. It is irritating to eyes and mucous membranes. This cargo is non-combustible or has a low fire-risk.

Alumina calcined 煅烧氧化铝 系指颜色呈淡灰至暗灰。不含水分的物质。该货物不溶于水。该货物为不燃物或失火风险低,无特殊危害。 Alumina calcined is light to dark grey in colour. It has no moisture content. This cargo is insoluble in water. This cargo is non-combustible or has a low fire-risk. It has no special hazards.

Alumina silica 硅石氧化铝 白色,由氧化铝和硅石晶体组成。含水量低(1%~5%)。结块60%。粗粒粉末40%。该货物不溶于水。该货物为不燃物或失火风险低,无特殊危害。 Alumina silica is white in colour, it consists of alumina and silica crystals. It has low moisture content (1% to 5%). It has 60% lumps. and 40% coarse grained powder. This cargo is insoluble in water. This cargo is non-combustible or has a low fire-risk. No special hazards.

Alumina silica, pellets 硅石氧化铝丸粒 白色至灰白色。不含水分。该货物为不燃物或失火风险低,无特殊危害。 Alumina silica, pellets is white to off-white in colour. It has no moisture content. This cargo is insoluble in water. This cargo is non-combustible or has a low fire-risk. No special hazards.

Aluminium alloy 铝合金 以铝为基的合金总称。主要合金元素有铜、硅、镁、锌、锰,次要合金元素有镍、铁、钛、铬、锂等。铝合金密度低,但强度比较高,接近或超过优质钢,塑性好,可加工成各种型材,具有优良的导电性、导热性和抗蚀性,工业上广泛使用,使用量仅次于钢。

Aluminium boat 铝合金艇 系指以铝合金作为艇体结构材料的船舶。

Aluminium ferrosilicon powder 硅铁铝粉末 细粉或砖状物。遇水可能释放氢,即可能在空气中形成一种爆炸性混合物的易燃气体。杂质在类似条件下可能产生磷化氢和砷化氢,两者均为剧毒气体。该货物为不燃物或失火风险低。 Aluminium ferrosilicon powder is fine powder or briquettes. In contact with water it may release hydrogen, flammable gas of explosive mixture may form in the air. Impurities may, under similar conditions, produce phosphine and arsine, which are highly toxic gases. This cargo is non-combustible or has a low fire-risk.

Aluminium smelting by-products or aluminium remelting by-products 铝熔炼副产品或铝再熔炼副产品 是铝制造过程中产生的废弃物。灰色或黑色粉末或团块,含有一些金属杂质。该名称涵盖各种不同的废弃物质,包括但不限于:铝渣、铝盐渣、铝渣沫、废阴极、废槽料。遇水可能产生热量并可能释放易燃和有毒气体:如氢、氨和乙炔。该货物为不燃物或失火风险低。不大可能着火,但易燃气体爆炸后可能随之起火并难以扑灭。在港口内可考虑向舱内灌水,但宜充分考虑船舶稳性。 Aluminium smelting by-products or aluminium remelting by-products are wastes from the aluminium manufacturing process. They are grey or black powder or lumps with some metallic inclusions. The term encompasses various different waste materials, which include but are not limited to: aluminium dross, aluminium salt slags, aluminium skimmings, spent cathodes, spent potliner. In contact with water it may cause heating with possible evolution of flammable and toxic gases such as hydrogen, ammonia and acetylene. This cargo is non-combustible or has a low fire-risk. Fire is unlikely but may follow an explosion of flammable gas and will be difficult to extinguish. In port, flooding may be considered, but due consideration should be given to stability.

Aluminum foil 铝箔 系指厚度小于0.20 mm、横断面呈矩形且均一的压延铝制品。

Aluminum foil paper 铝箔纸 作为一种工业制造原辅材料,产品主要应用于包装防护,生活用品,建筑等。于1932年开始量试成功的铝箔纸是一种新型的工业材料。将更多地取代单一性材料,广泛应用于各种软包装。随着经济发展,中国必将成为全球最主要的包装需方市场。铝箔包装发展前景十分广泛。

Always afloat 永远漂浮

Ambient pressure 周围压力 系指物体周围未受扰动的压力。

Ambient sea-noise measurement 海洋环境噪声测量 系指测量由于波浪、潮汐、风雨、海啸、湍流、热骚动、生物群、水体变化、船舶航行、混响及海洋工程等所产生的总噪声功率谱及其取向性。

Amendment and supplement of the repairing contract 船舶修理/修正补充合同

American Arbitration Association (AAA) 美国仲裁协会

American foreign trade definitions 美国对外贸易定义 系指由美国等9个商业团体制定的定义。该定义中所解释的贸易术语共有6种:分别为 Ex-Point of Origin, FOB, FAS, C&F, CIF 和 Ex Dock。

American National Standard Institute (ANSI) 美

American On Line (AOL) [美]

国标准化协会 系指非营利性质的民间标准化团体。但它实际上已成为国家标准化中心，各界标准化活动都围绕着它进行。通过它，使政府有关系统和民间系统相互配合，起到了联邦政府和民间标准化系统之间的桥梁作用。其协调与指导全国标准化活动，给标准制定、研究和使用单位以帮助，提供国内外标准化情报。它又起着行政管理机关的作用。

American On Line(AOL) [美] 在线服务公司
简称美国在线。系指总部设在弗吉尼亚州维也纳的一种在线信息服务公司，可提供电子邮件、新闻、教育和娱乐服务，并支持对因特网访问。美国在线服务(AOL)公司是美国最大因特网服务提供商之一。于1998年收购著名即时通信软件ICQ，网络浏览器Netscape，媒体播放器Winamp，但却有人士认为AOL是软件厂商的"百慕大"。2000年美国在线服务(AOL)公司与媒体巨人华纳时代公司合并，旨在扩展品牌内容服务以及通信服务的大众市场，合并后的公司形成了一个通信和媒体大公司，这个大公司拥有因特网最大用户群体，并有娱乐、出版和有线电视领域的广泛基础，并且在宽带接入方面有真广大的客户群，曾经是世界第一大ISP。近来由于业绩不佳，曾多次传出被微软、谷歌收购，与雅虎合并等负面新闻。目前已经拥有AOL中文(Chinese.aol.com)、AOL中国(cn.aol.com)、AOL香港(www.aol.hk)、AOL台湾(www.aol.tw)等多个处于测试期的中文门户网站。另有媒体报道AOL入主中国失败，于2009年3月关闭了其位于北京的研发中心。

American Petroleum Institute(API) 美国石油学会 建于1919年，是美国第一家国家级的商业协会，也是全世界范围内最早、最成功的制定标准的商会之一，是一家提供美国石油消耗及库存水平重要的每周数据的美国石油业机构。

Amicable settlement 和解 系指平息纷争，重归于好。今法律上指当事人约定互相让步，不经法院以终止争执或防止争执发生。和解一经成立，当事人不得任意反悔要求撤销。和解成立后，当事人所争执的权利即归确定，所抛弃的权利随即消失。诉讼外的和解法国1806年颁布的《民事诉讼法典》第127条规定，在诉讼进行的整个过程中，当事人双方可以自动地或在法庭建议下进行和解。破产程序进行中的和解系指具备破产原因的债务人，为了避免破产清算，而与债权人团体达成以让步方式了结债务的协议，协议经法院认可后生效的法律程序。

Amicable settlement agreement 和解协议

Amidships 船舯 系指：(1)是船长 L 的中点；(2)取为规范船长 L 中点；(3)从艏垂线量起的船长的中点。 Amidships is: (1) the middle of the length (L); (2) the middle of the rule length, L; (3) the middle of L, measured from F.P.

Amidships section 舯剖面 系指应取在自艏柱前缘量起的规范规定的船长 L 中点横截面。 Amidships section be taken as the middle of the rule length, L, measuring from the forward side of the stem.

Amidships(fishing vessel) 船舯(渔船) 系指船长 L 的中点处。 Amidships is the mid-length of L.

Amidships-engined ship 中机型船 系指主机舱位于船长中部的船舶。

Ammonium mitrate based fertilizer(UN 2067) 硝酸铵基化肥(UN 2067) 晶体、颗粒或丸粒。全部或部分溶于水。吸湿。硝酸铵基化肥为均匀混合物，硝酸铵为其主要成分，其组分限度如下：(1)硝酸铵含量不少于90%，以碳计算的可燃/有机物质含量总计不多于0.2%，附加物质(如有)为无机物并对硝酸铵呈惰性；(2)硝酸铵含量少于90%但多于70%，并含有其他无机物质，或硝酸铵含量多于80%但少于90%并混有碳酸钙和/或白云岩以及含量总计不多于0.4%的以碳计算的可燃/有机物质；(3)含有硝酸铵和硫酸铵混合物的硝酸铵基化肥，其硝酸铵含量多于45%但少于70%，以碳计算的可燃/有机物质含量总计不多于0.4%，从而使硝酸铵和硫酸铵成分的百分数比之和超过70%。助燃。这些物质如受到污染(例如被燃油污染)或处于严密封闭状态，则其载运船舶在发生重大火灾时可能有爆炸的风险。邻近的引爆也可能引起爆炸的风险。如遇到强热会发生分解，货物处所内和甲板上有出现有毒和助燃的烟气和气体的风险。化肥粉尘可能会刺激眼睛和黏膜。该货物吸湿，受潮会结块。 Ammonium nitrate based fertilizer (UN 2067) is crystals, granules or prills. It is wholly or partly soluble in water. It is hygroscopic. Ammonium nitrate-based fertilizers classified as UN 2067 are uniform mixtures containing ammonium nitrate as the main ingredient within the following composition limits: (1) not less than 90% ammonium nitrate with not more than 0.2% total combustible/organic material calculated as carbon and with added matter, if any, which is inorganic and inert towards ammonium nitrate; (2) less than 90% but more than 70% ammonium nitrate with other inorganic materials or more than 80% but less than 90% ammonium nitrate mixed with calcium carbonate and/or dolomite and not more than 0.4% total combustible/organic material calculated as carbon; (3) ammonium nitrate-based fertilizers containing mixtures of ammonium nitrate and ammonium sulphate with more than 45% but less than 70% ammonium nitrate and not more than 0.4% total combustible organic material calculated as carbon such that the sum of the percentage compositions of ammonium nitrate and ammonium sulphate exceeded by 70%. It is combustion supportive. in case of a ship carrying these substances in case of may involve a risk of explosion in the event of contamination (e. g. , by fuel oil) or strong confinement. An adjacent detonation may involve a risk of explosion. If heated strongly decomposes, risk of toxic fumes and gases which support combustion, in the cargo space and on deck. Fertilizer dust might be irritating to skin and mucous membranes. This cargo is hygroscopic and will cake if wet.

Ammonium nitrate 硝酸铵　白色晶体、丸粒或颗粒。全部或部分溶于水。助燃。吸湿。可燃物质含量不超过 0.2%，包括任何以碳计算的有机物质，而任何其他附加物质不计在内。这些物质如受到污染（例如被燃油污染）或处于严密封闭状态，则其载运船舶在发生重大火灾时可能有爆炸的风险。邻近的引爆也可能引起爆炸的风险。该货物如遇到强热会发生分解，放出有毒气体和助燃气体。硝酸铵粉尘可能会刺激眼睛和黏膜。该货物吸湿，受潮会结块。　Ammonium nitrate is white crystals, prills or granules. It is wholly or partly soluble in water. It is combustion supportive. It is hygroscopic. It has no more than 0.2% total combustible material, including any organic substance, calculated as carbon to the exclusion of any other added substance. In case of a ship carrying these materials in case of may involve a risk of explosion in the event of contamination (e. g. , by fuel oil) or strong confinement. An adjacent detonation may also involve a risk of explosion. If heated strongly, this cargo decomposes, giving off toxic gases and gases which support combustion. Ammonium nitrate dust might be irritating to skin and mucous membranes. This cargo is hygroscopic and will cake if wet.

Ammonium nitrate based fertilizer (non-hazardous) 硝酸铵基化肥（无危害）　干燥时无黏性的晶体、颗粒或丸粒。全部或部分溶于水。在本细目所述状况下运输的硝酸铵基化肥为均匀混合物，硝酸铵为其主要成分，其组分限度如下：（1）硝酸铵含量不多于 70%，含有其他无机物质；（2）硝酸铵含量不多于 80%，并混有碳酸钙和/或白云岩以及含量总计不多于 0.4% 的以碳计算的可燃/有机物质；（3）含有硝酸铵和硫酸铵混合物的氮类硝酸铵基化肥，其硝酸铵含量不多于 45%，以碳计算的可燃/有机物质含量总计不多于 0.4%；（4）硝酸铵基化肥与氮、磷酸盐或钾碱的均匀混合物，其中硝酸铵含量不多于 70%，以碳计算的可燃/有机物质含量总计不多于 0.4%，或硝酸铵含量不多于 45%，可燃物质含量不受限制。该货物为不燃物或失火风险低。该货物虽归入无危害类别，但在遇到强热时会发生分解并放出有毒气体，作用与 UN 2071 所属第 9 类硝酸铵基化肥相同。分解反应速度较慢，但如货物遇到强热，货物处所内和甲板上将有产生有毒烟气的风险。化肥粉尘可能会刺激眼睛和黏膜。该货物吸湿，受潮会结块。　Ammonium nitrate based fertilizer (non-hazardous) is non-cohesive crystals, granules or prills when it is dry. It is wholly or partly soluble in water. Ammonium nitrate based fertilizers transported in conditions mentioned in this schedule are uniform mixtures containing ammonium nitrate as the main ingredient within the following composition limits: (1) not more than 70% ammonium nitrate with other inorganic materials; (2) not more than 80% ammonium nitrate mixed with calcium carbonate and/or dolomite and not more than 0.4% total combustible organic material calculated as carbon; (3) nitrogen type ammonium nitrate based fertilizers containing mixtures of ammonium nitrate and ammonium sulphate with not more than 45% ammonium nitrate and not more than 0.4% total combustible organic material calculated as carbon; (4) uniform ammonium nitrate based fertilizer mixtures of the nitrogen, phosphate or potash, containing not more than 70% ammonium nitrate and not more than 0.4% total combustible organic material calculated as carbon or with not more than 45% ammonium nitrate and unrestricted combustible material. This cargo is non-combustible or with a low fire-risk. Even though this cargo is classified as non-hazardous, it will behave the same way as the ammonium nitrate based fertilizers classified in Class 9 under UN 2071 when heated strongly, by decomposing and giving off toxic gases. The speed of the decomposition reaction is lower, but there will be a risk of toxic fumes in the cargo space and on deck if the cargo is strongly heated. Fertilizer dust might be irritating to skin and mucous membranes. This cargo is hygroscopic and will cake if wet.

Ammonium nitrate based fertilizer (UN 2071) 硝酸铵基化肥（UN 2071）　通常为颗粒。全部或部分溶于水。吸湿。UN 2071 类的硝酸铵基化肥为硝酸铵基化肥与氮、磷酸盐或钾碱的均匀混合物，其中硝酸铵含量不多于 70%，以碳计算的可燃/有机物质含量总计不多于 0.4%，或硝酸铵含量不多于 45%，可燃物质含量不受限制。这些混合物受热后可能会自续分解。这种反应的温度能达到 500 ℃。分解一旦发生，可能遍及其余部分，并产生有毒气体。这些混合物都无爆炸危害。化肥粉尘可能会刺激眼睛和黏膜。该货物吸湿，受潮会结块。

Ammonium nitrate based fertilizer (UN 2071) is usually granules. It is wholly or partly soluble in water. It is hygroscopic. Ammonium nitrate-based fertilizers classified as UN 2071 are uniform ammonium nitrate based fertilizer mixtures of the nitrogen, phosphate or potash, containing not more than 70% ammonium nitrate and not more than 0.4% total combustible organic material calculated as carbon or with not more than 45% ammonium nitrate and unrestricted combustible material. These mixtures may be subject to self-sustaining decomposition if heated. The temperature in such a reaction can reach 500 ℃. Decomposition, once initiated, may spread throughout the remainder, producing gases which are toxic. None of these mixtures is subject to the explosion hazard. Fertilizer dust might be irritating to skin and mucous membranes. This cargo is hygroscopic and will cake if wet.

Ammonium polyphosphate(APP) 聚磷酸铵 又称多聚磷酸铵或缩聚磷酸铵,1965 年美国孟山都公司首先开发成功。聚磷酸铵无臭无味,不产生腐蚀气体,吸湿性小,热稳定性高,是一种性能优良的非卤阻燃剂。聚磷酸铵具有膨胀阻燃功能,故更有利于降烟和抗滴落,虽然为固体,但易于在多元醇中分散。聚磷酸铵的含磷量高达 30% ~ 32%,含氮为 14% ~ 16%。聚磷酸铵为白色结晶或无定形微细粉末。APP 广泛应用于膨胀型防火涂料、聚乙烯、聚丙烯、聚氨酯、环氧树脂、橡胶制品、纤维板及干粉灭火剂等,是一种使用安全的高效磷系非卤消烟阻燃剂。

Ammonium sulphate 硫酸铵 灰褐色至白色晶体。溶于水。自由流动。吸收水分。含水量 0.04% ~ 0.5%。具有氨的气味。其质量会自然损失。粉尘可能会刺激皮肤和眼睛。吞入有害。该货物虽归入无危害类别,但如货物处ับ潮湿,可能对骨架、舷侧壳板、舱壁等造成严重腐蚀。该货物为不燃物或失火风险低。
Ammonium sulphate is brownish grey to white crystals. It is soluble in water and free flowing. It absorbs moisture. Moisture content is 0.04% to 0.5%. It has Ammonia odour. It is subject to natural loss in weight. Dust may cause skin and eye irritation. It is harmful if swallowed. Even though this cargo is classified as non-hazardous, it may cause heavy corrosion of framing, side shell, bulkhead, etc., if sweating of cargo space occurs. This cargo is non-combustible or has a low fire-risk.

Ammunition lobby 弹药转运间 系指为了将弹药舱传递来的炮弹转运到随炮扬弹机上或输弹槽上,而设置在火炮下方的围闭空间。

Ammunition room ventilation 弹药舱通风 系指为使弹药舱温度不致太高而进行的自然或机械通风。

Amortization 摊销 系指包括所有折旧,但不包括对资本使用的补偿。Amortization includes all depreciations, but it does not include compensation for use of capital.

Amortization time 分期偿还时间

Amphibian mode 两栖状态 对地效翼船操作模式而言,系指两栖地效翼船的特殊短期状态,此时船舶主要由静力气垫支承且在水面以上的表面上缓慢移动。
For the WIG craft operational modes, amphibian mode is the special short-term mode of amphibian WIG craft when it is mainly supported by a static air cushion and moves slowly above a surface other than water.

Amphibious air-cushion vehicle 全垫升气垫船 系指能借助气垫支承其全部重力的高速船。

Amphibious assault ship(AAS) 两栖攻击舰/直升机登陆运输舰 在 20 世纪 50 年代美军诞生了登陆战的"垂直包围"理论,它要求登陆兵从登陆舰甲板登上直升机,飞越敌方防御阵地,在其后降落并投入战斗,这样可避开敌反登陆作战的防御阵地,并加快登陆速度。两栖攻击舰便是在这种作战思想指导下产生的新舰种。

Amphibious craft 两栖船 系指既可在水上航行,又可在陆上行驶的船舶。如全垫升气垫船、冲翼艇、螺杆艇等。

Amphibious dredger 两栖挖泥船 系指备有挖掘机械,可依靠安装在艏、艉端的 3 或 4 根带有轮子的液压机械腿自行移动,适宜于在浅水区或陆上进行作业的挖泥船。

Amphibious research ship 两栖调查船 系指用于沿海浅滩及沼泽地带进行地质、水文、资源调查的船舶。该型船舶借助柱体状船体上的螺纹板旋转产生的推力航行于水中或行走于沼泽地和陆地。因船小,续航力低,故一般不备有固定的调查设备和专用实验室。

Amphibious warships 两栖战舰 亦称登陆舰艇,它是一种用于运载登陆部队、武器装备、物资车辆、直升机等进行登陆作战的舰艇,出现于第二次世界大战中,并于 20 世纪 50 年代以后大力发展。该技术拥有国:中国(071 型、081 型)、美国(圣安东尼奥级)、英国(海洋级/不列颠级)、法国(西北风级)、荷兰(鹿特丹级)、日本(大隅级)、韩国(LPX)。见图 A-12。

Amplifier 放大器 系指音响系统中最基本的设备,俗称功放,其任务是把信号源的微弱电信号进行放大以驱动扬声器发出声音。

Amplitude differences 幅度差

Amplitude of heaving 垂荡幅值 系指船舶垂摇时,在每个循环中其重心瞬时位置离原平衡位置间的最大垂向距离。

Amplitude of pitching 纵摇幅值 系指船舶纵摇时,每个循环中的最大纵摇角。

A 两栖战舰之一
A One of Amphibious warships

B 两栖战舰之二
B Another one of Amphibious warships

图 A-12 两栖战舰
Figure A-12 Amphibious warships

Amplitude of roll 横摇幅值 系指船舶横摇时，每个循环中的最大横摇角。

Amplitude of yaw 艏摇幅值 系指船舶艏摇时，每个循环中的最大艏摇角。

Amyloid precursor protein（APP） 淀粉样前体蛋白

Anadromous species 溯河产卵物种 系指在淡水环境中产卵/繁殖，但至少有一部分成年物种的生活是在海洋环境中度过的。Anadromous species mean that the species spawn/reproduce in freshwater environments, but spend at least part of their adult life in a marine environment.

Analog set-top box 模拟机顶盒

Analog signals 模拟信号 系指信息参数在给定范围内表现为连续的信号。或在一段连续的时间间隔内，其代表信息的特征量可以在任意瞬间呈现为任意数值的信号。模拟信号系指用连续变化的物理量所表达的信息，如温度、湿度、压力、长度、电流、电压等，人们通常又把模拟信号称为连续信号，它在一定的时间范围内可以有无限多个不同的取值。优点：分辨率较高，高达 24 位；转换速率高，高于积分型和压频变换型 ADC；价格低；内部利用高倍频过采样技术，实现了数字滤波，降低了对传感器信号进行滤波的要求。

Analog-digital converter（ADC） 模数转换器 系指把经过与标准量（或参考量）比较处理后的模拟量转换成以二进制数值表示的离散信号的转换器，简称 ADC 或 A/D 转换器。

Analysis certificate 分析证书 系指涉及的项目是蛋白质、脂肪等项目，还有重金属、药残等安全性的项目的证书。

Analytical expression 解析表达法

Analytical model 解析模型

ANCC system 全球统一标识系统 国际上称为 EAN·UCC 系统。中国物品编码中心根据国际物品编码协会制定的 EAN·UCC 系统规则和我国国情，研究制定并负责在我国推广应用的一套全球统一的产品与服务标识系统。

Anchor 锚 系指一端与锚链相连，放入海床使船舶定位的装置；设计成当船舶在风和水流影响下要飘移而拉动它时，能咬住海底；通常由铸件或重型铸件制成。Anchor means a device which is attached to anchor chain at one end and lowered into the sea bed to hold a ship in position; it is designed to grip the bottom when it is dragged by the ship trying to float away under the influence of wind and current; usually made of heavy casting or casting.

Anchor（handling）gear 锚泊机械 系指起、抛锚的机械。如起锚机、起锚绞盘等。

Anchor（ship's anchor） 锚 具有特定形状，在抛入水中后能埋进底土，提供抓力而又易于从中起出，用以通过锚链或锚缆，将船舶或其他浮物系留于水域的专用器具。

Anchor arm 锚臂 系指自锚冠向外伸出的臂状部分。

Anchor arrangement 锚泊设备 船上锚、锚缆、锚链等锚具及其收放设备的总称。

Anchor aweigh lounge 悬锚舞厅（豪华邮轮）

Anchor bed 锚床 系指装在上甲板船首两舷边处，供导引锚链和存放锚用的敞露的槽状结构物。

Anchor boat 起锚船 系指在甲板上设有起锚设备，专为工程船移锚、起锚和抛锚及做其他用途的船舶。

Anchor buoy 锚浮标 系指为便于收锚和检查是否走锚而系于锚上用以标志锚在水中位置的浮体。

Anchor capstan 起锚系缆绞盘 系指立式的起抛锚及系缆用的机械。

Anchor chain 锚链 一种专供用作锚缆的重型锁环链条。其单位长度质量比锚缆或纤维缆大得多。可为钢缆的 4～5 倍，故吸收动载荷的能力强且与锚一起形成的锚泊抓持力也大。这是船舶在风浪中锚泊所必需的。

Anchor crown 锚冠 系指锚头的顶部。

Anchor davit 吊锚杆 系指可供吊锚用的船上木质吊柱。常用的有弧形和悬臂式两种。

Anchor fluke 锚爪 系指由锚臂延伸的用以产生锚抓力的楔状或三角状钩爪。

Anchor gill net 定刺网 系指用锚定位的刺网。

Anchor handing towing supply vessel(AHTS) 起抛锚/拖带/供应三用工作船

Anchor handling boat 抛锚艇 系指在甲板上设有起锚设备，专为工程船移锚、起锚和抛锚及作其他用途的船艇。见图A-13。

图 A-13　抛锚艇
Figure A-13　Anchor handling boat

Anchor head 锚头 系指连接于锚杆下端，包括锚爪、锚臂和锚冠，提供锚抓力的部分。

Anchor holding power-to-weight ratio 锚抓重比 系指锚抓力与锚重的比值。

Anchor lights(riding light) 锚灯 系指设置在船舶中线面上规定处，用以表明船舶停泊信号而增设的白环照灯。

Anchor moored positioning drilling unit 系泊定位式钻井装置

Anchor mooring and towing arrangement 系船设备 船舶停泊和拖带用的设备。通常系指锚泊设备、系缆设备和拖带用具。

Anchor pea 锚爪尖 系指锚爪的尖端部分。

Anchor rack 锚架 系指伸出船外，供收藏锚用的支架。

Anchor recess(anchor pocket) 锚穴 系指外板上用以收藏锚的龛状结构。

Anchor ring 锚环 系指穿在锚杆上端的圆环。现在都用直形或圆形锚卸扣代替。

Anchor rope 锚缆 系指连接锚和船的缆绳。

Anchor shackle 锚卸扣 系指穿在锚杆上端的卸扣。

Anchor shank 锚杆 系指锚的躯干部分。

Anchor stock 锚横杆 系指装在锚杆上端，并垂直于锚爪平面，用以阻止锚翻身，保证锚爪尖着地并啮入底土的杆件。

Anchor stopper(anchor lashings) 掣锚链条 系指其上配有脱钩和松紧螺旋扣，在船舶航行时将锚收紧用的链条。

Anchor swivel 锚链转环 系指为防止锚链过度绞扭而装在锚链两端的可转动的专用环。

Anchor windlass 起锚机 系指船舶上的一种大型甲板机械，用来收、放锚和锚链。起锚机通常安装在船舶艏、艉部主甲板上，供舰船起锚，抛锚和系缆时用。起锚机通常与绞车配合使用。

Anchored continuous observation 定点连续观测 系指在调查海区布设若干有代表性的观测站，按一定时间间隔取正点资料进行的观测。

Anchored hourly observation 正点观测 一般系指水文连续观测。按一定的时间间隔取正点资料进行的观测。

Anchor-handling tugs 带缆船/起锚拖船 系指在甲板上设有起锚设备，专为工程船移位、起锚和抛锚及作其他用途的船舶。

Anchoring and mooring equipment 系泊设备 系指将船系泊于码头、浮筒、船坞或邻船用的设备，主要包括缆索、带缆桩、导缆器、系缆绞车和系缆机械。常用的缆索有白棕缆、钢缆、合成纤维缆。带缆桩用于系结缆索，有柱式（双柱、独柱、直式、斜式）带缆桩和十字（单十字、双十字）带缆桩等。导缆器用于引导缆索通过并变换方向或限制其导出位置，并可减少缆索与船体间的摩擦，主要有滚轮导缆器、导缆钳、导向滚轮或滚柱、导缆孔等。系缆卷车用于收卷和保存缆索。系缆机械用于收紧缆索，有时用起锚机械兼作系缆机械。系泊设备布置于船的艏、艉和舷边，且多左右对称。现代船舶特别是大型船舶如集装箱船、油船、散货船等，大都能在数小时内装卸万吨以上货物，以致船舶吃水迅速变化，易使缆索过分张弛而出现险情，因此这类船舶多装有自动系缆机械，能根据缆索张力变化自动收放缆索，以保证船舶和码头的安全。

Anchoring/warping winch 起锚系缆机 系指用来使船舶安全地停泊于水面或系泊于码头或浮筒的，短期工作或断续周期工作的甲板机械。起锚系缆机械的形式根据船舶类型及布置有多种多样，如卧式起锚机（带系缆滚筒）、立式起锚绞盘、立式系缆系缆绞盘及卧式绞缆机等。

Anderson power-supply product(APP) 安德森电源产品

Android 安卓软件 是一种基于Linux的自由及开放源代码的操作系统，主要使用于移动设备，如智能手机和平板电脑，由Google公司和开放手机联盟领导及开发。尚未有统一中文名称,中国大陆地区较多人使用"安卓"或"安致"。Android 操作系统最初由 Andy Rubin 开发,主要支持手机。2005 年 7 月由 Google 收购注资。2007 年 11 月,Google 与 84 家硬件制造商、软件开发商及电信营运商组建开放手机联盟共同研发改良 Android 系统。随后 Google 以 Apache 开源许可证的授权方式,发布了 Android 的源代码。第一部 Android 智能手机发布于 2008 年 10 月。Android 逐渐扩展到平板电脑及其他领域上,如电视、数码相机、游戏机等。2013 年的第 4 季度,Android 平台手机的全球市场份额已经达到 78.1%。2013 年 09 月 24 日谷歌开发的操作系统 Android 在迎来了 5 岁生日,全世界采用这款系统的设备数量已经达到 10 亿台。2014 第一季度 Android 平台已占所有移动广告流量来源的 42.8%,首度超越 iOS。但运营收入不及 iOS。

Anemometer and anemoscope 船舶气象仪 用于船上测量瞬时风速、100 s 的平均风速、风向、空气温度、湿球温度的仪器,并用查表方法得出空气的相对湿度。

Angle between cargo runners 吊货索间夹角 系指定位双杆操作时,吊货索间的夹角。

Angle of attack 攻角 系指未受扰动流体的流向与所指的面或线间的夹角。

Angle of attack 锚爪袭角 系指锚爪嵌入土中后和底土表面的夹角。

Angle of attack at zero-lift 零升力攻角 系指翼元体零升力线与弦线之间的夹角。

Angle of bucket ladder 斗桥倾角 系指指不同挖深时斗桥侧面中心线与水平面的夹角。

Angle of divergent wave system 散波角 系指由散波所形成的扇形母线与船体中线面间的水平夹角。

Angle of encounter 遭遇浪角 系指船舶前进方向与波浪方向之间的水平夹角。右舷为正。

Angle of glide chute 溜泥槽倾角 系指溜泥槽与船体基平面之间的夹角。

Angle of heel (angle of list) 横倾角 系指船舶横倾时的水线平面与正浮的水线平面之间的夹角。

Angle of maximum righting level 最大复原力臂角 系指静稳性曲线图上最大复原力臂所对应的横倾角。

Angle of pitching 纵摇角 系指船舶纵摇时,其瞬时位置与原平衡位置间的夹角。

Angle of repose 静止角 系指非黏性(即自由流动)颗粒物质的最大斜角。它是在水平面和这类物质的锥形斜面之间量取的角度。 Angle of repose is the maximum slope angle of non-cohesive(i. e. free-flowing) granular material. It is measured as the angle between a horizontal plane and the cone slope of such material.

Angle of roll 横摇角 系指船舶摇摆时,其瞬时位置与原平衡位置间的夹角。

Angle of shafting declivity 轴系倾角 系指轴系中线与船体基线面之间的夹角。

Angle of two position lines 位置线交角 系指过定位线的两条位置线之间的夹角,它是衡量各种定位方法精度的一个量。位置线交角为直角时定位误差最小,交角越小(或接近 180°时)定位误差越大。

Angle of vanishing stability 稳性消失角 系指静稳性曲线图上超过最大复原力臂值后,复原力臂为零时所对应的横倾角。

Angle of vanishing stability φ_V 稳性消失角 φ_V 系指在相应的装载状态下,横向稳性恢复力矩为零时假定无偏移载荷和所有可能的下沉进口开口为水密时测定的最接近正浮(并非正浮)的横倾角。注意 1:如果艇具有非快速泄水的凹体,则这些凹体的下沉进水角应为 φ_V,除非在确定 φ_V 时已充分地考虑了这些凹体的影响。注意 2:稳性消失角以度表示。 Angle of vanishing stability means the angle of heel nearest to the upright (other than upright) in the appropriate loading condition at which the transverse stability righting moment is zero; It is assumed that there is no offset load, and that all potential down-flooding openings are assumed to be watertight. Note 1: Where a boat has recesses which are not quick-draining, φ_V is to be taken as the down-flooding angle to these recesses, unless such recesses are fully accounted for in determining φ_V. Note 2: Angle of vanishing stability is expressed in degrees.

Angle of velocity vector 航速角 系指船舶按指定航线航行时,从地球真北线、磁北线或罗北线北段等基准方向到其重心处瞬时航速矢量之间的水平夹角。顺时针为正。

Angle of yaw 艏摇角 系指船舶首摇时,其中线面瞬时位置与原平面位置间的水平夹角。

Angle valve 角阀

Angled deck 斜角甲板 系指在飞行甲板上,从后端起向左舷画出的与该甲板中心线成一定夹角,供舰载机着舰也可以起飞用的跑道区。

Angling 垂钓

Angling rate 钓获量 系指使用延绳钓时,每次起钓的鱼类上钓数与所放出的钓钩总数的百分比。

Angular acceleration components 角加速度分量 系指船舶运动时,其角加速度分别相对于运动坐标系各轴的分量,符号按右手法则。

Angular frequency 角频率 在物理学(特别是力学和电子工程)中,角频率 ω 有时也叫作角速度标量,是对旋转快慢的度量,它是角速度矢量的标量。角频率的国际单位是弧度每秒。频率是描述物体振动快慢的物理量,所以角频率也是描述物体振动快慢的物理量。频率、角频率和周期的关系为 $\omega = 2\pi f = 2\pi/T$。

Angular velocity 角速度 系指单位时间内转动的弧度 $\omega = 2\pi n$,式中,n—转速。

Angular velocity components 角速度分量 系指船舶运动时,其角速度分别相对于运动坐标系各轴的分量,符号按右手法则。

Angular velocity of blade of cycloidal propeller 平旋推进器叶角速 系指平旋推进器叶对叶元中心的角速度。

Animal carcasses 动物尸体 系指船上作为货物载运且在航行中死亡或被实施安乐死的任何动物的躯体。 Animal carcasses means the bodies of any animals that are carried on board as cargo and that die or are euthanized during the voyage.

Animal fiber 动物纤维 系指由动物的毛发或分泌液形成的纤维。最主要的品种是各种动物毛和蚕丝。主要成分是由一系列氨基酸经肽键结合成链状结构的蛋白质。

Animal/vegetable oil tanker 动物/植物油船 系指可装运食用油的船舶,如棕榈油船等。

Annealing colour 退火色斑

Annex 附则 系指经国际海事组织以1978年议定书修正并经1997年议定书修订的1973年国际防止船舶造成污染公约(MARPOL)的附则Ⅵ,这些修正案按本公约第16条的规定予以通过并生效。 Annex means Annex VI to the International Convention for the Prevention of Pollution from Ships, 1973 (MARPOL), as modified by the Protocol of 1978 relating thereto (MARPOL 73/78), and modified by the Protocol of 1997, amended by the organization, provided that such amendments are adopted and brought into force in accordance with the provisions of article 16 of the present Convention.

Anniversary date 周年日期 系指:(1)与有关证书失效日对应的每年的该月该日;(2)从入级检验或者定期检验结束日开始到下一次定期检验指定日期的期间每年接受定期检验年月日。 Anniversary date means: (1) the day and the month of each year which will correspond to the date of expiry of the relevant certificate; (2) the day and the month of each year which will correspond to the due date of the next special survey from the completion date of the initial classification survey or of the special survey.

Anniversary date 年度审核日期 系指每年的,与DOC和SMC的失效日期相对应的一个日期。

Annual audit 年度审核 系指每年一次的审核。

Annual earning capacity 年盈利能力

Annual fatality rate 年度死亡率 系指将频率和死亡结合成为的一种便利的一维的衡量社会风险的方法,也即潜在生命损失(PLL)。

Annual repair 年度修理 系指每年有计划地对船舶进行的维修和养护工作。

Annual survey(AS) 年度检验 对每艘150 GT及以上的油船和400 GT及以上的其他船舶而言,年度检验系指在证书的每个周年日之前或之后3个月内进行的检验。包括对初次检验所述的结构、设备、系统、附件、布置和材料的总体检查,以确保其已按有关规定进行保养,并确保其继续满足船舶预定的运营要求。该年度检验应在按上述附则第7条或第8条所签发的证书上予以签署。 For every oil tanker of 150 gross tonnage and above, and every ship of 400 gross tonnage and above, annual survey means the survey proceeded within three months before or after each anniversary date of the certificate, includes a general inspection of the structure, equipment, system, fitting, arrangements and material referred in initial survey to ensure that they have been maintained in according with relevant requirement and that they remain satisfactory for the service for which the ship is intended. Such annual surveys shall be endorsed on the certificate issued under regulation 7 or 8 of the above annex.

Annular blowoff preventer/Annular blowout preventer 环形防喷器 通常装有闸板式防喷器的大型闸门,运作时会在管柱和井筒之间形成一个密封的环形空间,在井内设有管柱的情况下,也能单独完成封井,但是不能反复使用,并且不允许长期关井使用。

Annular combustion chamber 环形燃烧室 系指在内、外壳之间的环腔内配置一个环形火焰筒所组成的燃烧室。

Annular gear 内齿轮 系指内圈带齿的齿轮。

Annulus vent 环形通气孔 系指为及时稀释并扩散油气,防止油蒸发气浓度增大到燃、爆极限,罐顶与罐壁周围开设有通气孔。孔口之间的环向间距应不大于10 m,每个油罐至少应开设4个;总的开孔面积要求每米油罐直径在0.06 m^2以上。通气孔出入口安装有金属丝网罩。

Anode 阳极 系指在腐蚀性电池中,直流电流经其流入电解液的电极。 Anode means an electrode through which direct current passes into electrolyte in the corrosive batteries.

Anodes 阳极 系指化学电池中,能使电解质发生氧化反应的电极称为阳极。此外,在电子管中用来接收或加速从阴极发射的电子的电极也叫阳极。阳极氧化系指以某种金属（主要是铝）制作为阳极，在适宜的电解液中进行电解，使制件表面形成无机氧化物薄膜的过程。阳极氧化后的铝或其合金，提高了其硬度和耐磨性，可达 250 ~ 500 kg/mm^2，良好的耐热性，硬质阳极氧化膜熔点高达 2 320 K，优良的绝缘性，耐击穿电压高达 2 000 V，增强了抗腐蚀性能，在 ω = 0.03 NaCl 盐雾中经几千小时不腐蚀。

Anstayed mast　无支索桅　系指支索桅支张仅由甲板支承的桅杆。

Antarctic areas (waters)　南极区域（水域）　系指南纬 60° 以南的区域。 Antarctic areas (waters) mean the sea south of 60° south latitude.

Antarctic research ship　南极调查船　系指从事南极海域科学考察，协助在南极洲建立科学考察站并提供后勤补给、营救等服务的综合性调查船。船型为破冰型。为提高船舶的可靠性，通常用双螺旋桨或三螺旋桨推进。船上设有海洋气象、冰象、水文化学、生物、地质等实验室。并设有直升机机库和直升机平台。有足够的货舱以存放供科学考察站人员用的物资和燃料，并备有特制的机动雪车和雪橇作为南极陆上的交通工具。

Antarctica　南极　系指南纬 60° 以南的区域。 Antarctica means the area south of 60 degrees south latitude.

Antenna height　天线高度　系指船载天线相对于参考椭球体的高度，也就是天线海拔高度和大地水准面高度之和。

Antenna staff　天线杆　系指安装在甲板或桅杆和起重柱的上端，专供支张无线电天线用的竖立杆件。

Antenna unit　天线单元

Anti aircraft artillery　防空火炮

Anti-sliding stability　抗滑移稳性　系指在坐底工况时，平台在相应工况的水平载荷作用下具有足够的抵抗水平滑动的能力。对海床土质较差的海域，其滑动面应选为沉垫或下壳体与土壤的交界面或轮廓面，不考虑地基深层滑动。平台的抗滑移稳性应满足下述要求：RH/FH ≥ KH，式中，RH—抗滑力，kN。包括土壤的黏聚力、摩擦力、被动土压力、抗滑装置产生的抗滑力；FH—滑移力，kN。包括作用在平台上所有的水平力；KH—抗滑安全系数，正常作业工况时应不小于 1.4，自存工况时应不小于 1.2。 Anti-sliding stability means that when resting on the seabed, an adequate ability to withstand horizontal sliding haven for the unit under horizontal loadings as appropriate. Where the soil of seabed is poor, the sliding area is to be taken as the contact area or profile of the mat or lower hull with the soil, without considering the sliding in the deep bottom soil. The anti-sliding stability of the unit is to comply with the following: RH/FH ≥ KH, where: RH—anti-sliding force, in kN, including adhesion and friction of the soil, passive soil pressure, anti-sliding force generated by anti-sliding device; FH—sliding force, in kN, including all horizontal forces acting on the unit; KH—anti-sliding safety factor, is not less than 1.4 for normal operating conditions and is not less than 1.2 for survival condition.

Anti-ballistic missiles　反弹道导弹　系指一种旨在拦截弹道导弹的导弹。它是国家战略防御系统的重要组成部分。弹道导弹能够依照弹道飞行轨迹投射核弹头、化学弹头、生物武器弹头或常规弹头。历史上只有两个反弹道导弹系统投入过正式使用，它们是美国的卫兵系统和俄国的 A-35 反弹道导弹系统。该技术拥有国：美国、俄罗斯、中国。见图 A-14。

图 A-14　反弹道导弹
Figure A-14　Anti-ballistic missiles

Anticipatory letter of credit　预支信用证　其特点是进口商先付款，出口商后交货的贸易方式，是进口商给予出口商的一种优惠、融通资金的便利。凡欲采用预支款的信用证，买卖双方于谈判时，出口商须向进口商提出支付条款，预支款额和方法列明在信用证内。经进口商同意后，进口商填写并签署开立信用证申请书中予以明示。

Anti-corrosion　防腐蚀　系指抗酸类物质对零部件的作用，或系指保护零部件免受腐蚀的性质。

Anti-corrosion system　防腐系统　通常可考虑全硬保护涂层。全硬保护涂层通常系指环氧树脂或等同物。除软涂层和半硬涂层以外的其他涂层系统只要根据制造厂的规定应用和维护，可以考虑作为替代品接受。

Anti-corrosive oil tank　防锈剂柜　系指贮存柴油机闭式冷却水系统中淡水回路的防锈剂的专用舱柜。

Anti-dumping duties　反倾销税　系指对实行商品倾销的进口商品所征收的一种进口附加税。进口商品以低于正常价值的价格进行倾销，并对进口国的同类产品造成重大损害是构成征收反倾销税的重要条件。反倾销税的税额一般以倾销差价征收,其目的在于抵制商品倾销,保护本国工业和市场。关税和贸易总协定第6条对倾销的规定,主要有:(1)用倾销手段将一国产品以低于正常的价格挤入另一国贸易时,如因此对某一缔约国领土内已建立的某项工业造成重大损害或产生重大威胁;或者对某一国内工业的新建产生严重阻碍,这种倾销应受到谴责。(2)缔约国为了抵消或防止倾销,可以对倾销的产品征收数量不超过这一产品的倾斜差额的反倾销税。(3)"正常价格"系指相同产品在出口国用于国内消费时在正常情况下的可比价格。如果没有这种国内价格,则是相同产品在正常贸易情况下向第三国出口的最高可比价格;和产品原产国的生产成本加上合理的推销费用和利润。(4)不得因抵消倾销或出口补贴,而同时对它既征收反倾销税又征收反补贴税。(5)为了稳定初级产品价格而建立的制度,即使它有时会使出口商品的售价低于相同产品在国内市场销售的可比价格,也不应认为造成了重大损害。

Anti-dumping laws 反倾销法 系指由一国立法机关制定,由国家行政机关保证执行,为规范进口产品价格秩序,保护国内相关产业,要求进口产品相关者必须遵守的行为规则。

Anti-explosion bulkhead stuffing box 防爆填料函 系指货油泵与原动机连接轴处设有冷却润滑设施,在工作中不致发生火花或过热现象的舱壁填料函。

Anti-exposure suit 抗曝露服 系指设计成供救助艇艇员和海上撤离系统人员使用的防护服。 Anti-exposure suit is a protective suit designed for rescue boat crews and marine evacuation system parties.

Anti-extrusion layer 抗挤压层

Anti-foaming 防沫 系指防止在蒸发液面上聚集泡沫的措施。

Anti-fouling coating 防污涂层

Anti-fouling coating system 防污底涂层系统 系指船上使用的用于控制或防止不利水生生物附着的所有成分的涂层、表面处理(包括底漆、封闭漆、黏结剂、防腐蚀和防污底涂层)或其他表面处理的组合。 Anti-fouling coating system means the combination of all component coatings, surface treatments (including primer, sealer, binder, anti-corrosive and anti-fouling coatings) or other surface treatments, used on a ship to control or prevent attachment of unwanted aquatic organisms.

Anti-fouling compose and system 防污底化合物和系统 系指在应用或解释拆管规则时,现行的《2001年国际控制船舶有害防污底系统公约》(AFS公约)附则I中所规定的防污底化合物和系统。(1)所有船舶不得施涂含有有机锡化合物作为杀生物剂的防污底系统或任何其他AFS公约禁止施涂或使用的防污底系统。(2)所有新船或船上的新装置不得施涂或采用不符合AFS公约规定的防污底化合物或系统。 Anti-fouling compounds and systems regulated under Annex I to the International Convention on the Control of Harmful Anti-fouling Systems on Ships, 2001 (AFS Convention) in force at the time of application or interpretation of this Annex. (1) No ship may apply anti-fouling systems containing organotin compounds as a biocide or any other anti-fouling system whose application or use is prohibited by the AFS Convention; (2) No new ships or new installations on ships shall apply or employ anti-fouling compounds or systems in a manner inconsistent with the AFS Convention.

Antifouling paints 防污油漆

Anti-fouling system (AFS) 防污底系统 (1)对2010年船舶防污底系统检验和发证指南而言,系指船上使用的用于控制或防止不利生物附着的涂层、油漆、表面处理、表面或装置;(2)不含作为生物杀灭剂的有机化合物的船舶防污底系统。(3)船上使用的用于控制或防止不利生物附着的涂层、油漆、表面处理、表面或装置。 Anti-fouling system: (1) For the 2010 guidelines for survey and certification of anti-fouling systems on ships, anti-fouling system means a coating, paint, surface treatment, surface, or device that is used on a ship to control or prevent attachment of unwanted organisms; (2) a system which dose not contain any organic compound acting as biocide. (3) a coating, paint, surface treatment, surface, or device that is used on a ship to control or prevent attachment of unwanted organisms.

Antifreeze system 防冻系统 防冻系统:(1)是一个湿管喷水器灭火系统,系统内含防冻液并与供水管相连接。喷水器被火灾的热量激发打开,防冻液就随水立刻排出。(2)将自动喷水器接至管内充满防冻液并与供水源相接的管路上的湿管式喷水系统。一旦喷水器感受着火引起的温升而开启时,管路首先排出防冻液,接着被水流注满并投入灭火工作。 Antifreeze system: (1) a wet pipe sprinkler system employing automatic sprinklers attached to a piping system containing an antifreeze solution and connected to a water supply. The antifreeze solution is discharged, followed by water, immediately upon operation of sprinklers opened by heat from a fire; (2) a wet pipe system employing automatic nozzles or sprinklers attached to a piping system containing an antifreeze solution and connected to a water supply. The antifreeze solution is discharged, followed by water, immediately upon operation of nozzles or sprinklers opened by heat from a fire.

Anti-guided missile system on passenger plane 客机反导系统 系指在客机上采用释放诱饵、干扰等"软手段"，肩扛防御式地空导弹攻击的系统。该系统由红外告警系统、控制系统和红外诱饵弹3部分组成，可在合适的距离向导弹来袭的方向投放红外诱饵弹，欺骗来袭导弹，使其偏离客机。

Anti-icing equipment (anti-icer) 防冰装置 系指在压气机进口处预防空气中水分结冰的装置。

Anti-impact gear 防撞装置 系指为防止进出坞船舶与坞墙尾端剧烈碰撞，而在坞墙尾端边缘设置的缓冲装置。

Anti-magnetic steel 防磁钢

Anti-missile interceptor 反导拦截弹

Antimony ore and residue 锑矿石和残留物 铅灰色矿物。表面会变黑。该货物为不燃物或失火风险低。如遇火会释放危险的锑和氧化硫烟气。Antimony ore and residue is lead grey mineral, subject to black tarnish. This cargo is non-combustible or has a low fire-risk. In contact with fire, antimong and sulful oxide smoke wil bereleased.

Anti-overturning stability 抗倾覆稳性 系指在坐底工况时，平台在相应工况的环境载荷作用下应具有足够的抗倾覆能力，平台的抗倾稳性应满足下述要求：$M_k/M_q \geqslant K_q$，式中，M_k——平台坐底时的抗倾力矩，kN·m；M_q——平台坐底时的倾覆力矩，kN·m；K_q——抗倾安全系数，见表A-5。

表 A-5 抗倾安全系数
Table A-5 Anti-overturning safety factor

工况 Condition	自升式平台 Self-elevating unit	坐底式平台 Unit resting on seabed
正常作业 Normal operation	1.5	1.6
自存 Survival	1.3	1.4

Anti-overturning stability means that when resting on the seabed, the unit has adequate ability to withstand the overturning moment of the combined environmental forces as appropriately as it can. The anti-overturning stability of the unit is to comply with the following: $M_k/M_q \geqslant K_q$, where: M_k—anti-overturning moment of the unit when resting on the seabed, in kN·m; M_q—heeling moment of the unit when resting on the seabed, in kN·m; K_q—anti-overturning safety factor. The anti-overturning safety factor is shown in Table A-5.

Anti-roll pump 减摇泵 系指用于减摇水舱系统的泵。

Anti-rolling device 抗摇设备（减横摇装置） 能防止或对抗船舶在航行中横摇的设备或装置。

Anti-rolling equipment 防摇装置 能防止或对抗船舶在航行中横摇的设备或装置。

Anti-rolling system (anti-rolling device) 减摇装置 系指旨在减小船的运动振幅，提高船的作业能力的装置。

Anti-rolling tank 减摇水舱 系指由两舷对称的水舱、连通水道及控制阀等组成的，通过水舱内的水往复运动产生抵抗船舶横摇力矩达到船舶减摇目的的水舱。减摇水舱可分为主动式减摇水舱和被动式减摇水舱两种。

Anti-satellite technologies 反卫星技术 系指从地面、空中或外层空间攻击敌方卫星的军事技术。即击落对方卫星的能力，策反对方卫星的能力、强行搞乱对方卫星轨道的能力。反卫星技术几乎是跟卫星技术本身同步发展起来的。各太空大国的太空计划基本上都包含着破与立两个方面，保全自己，算计敌人。该系统包括一枚装备常规弹头的导弹，其基本机制是：在敌方卫星的地球轨道上升到达发射阵地上空时，将反卫星导弹发射进入与目标卫星接近的轨道；在一至两个轨道的距离上，这枚重1 400 kg的拦截弹头将在弹上雷达的引导下实施机动，"俯冲"向目标卫星，并在一公里左右的距离上引爆，通过弹头的预制破片摧毁目标。该技术拥有国：美国、俄罗斯、中国。见图A-15。

图 A-15 反卫星技术
Figure A-15 Anti-satellite technologies

Anti-scouring device 防冲刷装置 系指为防止海流、潮汐在桩腿、桩靴或沉垫与海底泥地接触处的周围形成严重涡流，引起泥沙淘空或冲刷现象而设计安设的装置。

Anti-ship missile 反舰导弹 系指从舰艇、岸上或飞机上发射,攻击水面舰船的导弹。世界上最早的反舰导弹是德国于二战末期研制的 Hs-292 反舰导弹,在 1944 年末投入实战并击沉多艘盟军运输船。反舰导弹常采用半穿甲爆破型战斗部,固体火箭发动机为动力装置;采用自主式制导、自控飞行,当导弹进入目标区,导引头自动搜索、捕捉和攻击目标。反舰导弹多次用于现代战争,在现代海战中发挥了重要作用。反舰导弹发展到近代,已经可以从多种型态的载具上使用,包括从各类飞行器上发射的空射型,由地面发射的陆射型,由水面舰艇使用的舰射型以及自潜艇发射的潜射型。许多导弹在经过少许改装之后就可以在不同的载具上使用,不必另外发展专用衍生型。反舰导弹技术拥有国:中国、美国、法国、挪威、日本、俄罗斯、瑞典、意大利、印度(与俄罗斯合作)。见图 A-16。

图 A-16　反舰导弹
Figure A-16　Anti-ship missile

Anti-singing edge　抗谐鸣边　系指为了消除谐鸣现象将螺旋桨叶随边做成适当形状的部分。

Anti-spray enclosure　防溅外壳　系指其开孔应做到使其偏离垂直位置不大于 100° 的任意角度,以使得直线地掉落在其上或向其来的液滴或固体颗粒不会直接进入其中或不会在撞到其表面和沿该表面抖动的情况下进入的外壳。

Anti-subsidy duty　反补贴税　又称抵消税或补偿税。是对直接或间接地接受奖金或补贴的外国商品进口所征的一种进口附加税。进口商品在生产、制造、加工、买卖、输出过程中直接或间接地接受奖金或补贴,并使进口国生产同类产品遭受重大损害是构成征收反补贴税的重要条件。反补贴税的税额一般按"补贴数额"征收。其目的在于增加进口商品的成本,抵消出口国对该商品所做补贴的鼓励作用。

Antitank guided missile　反坦克导弹　系指用于击毁坦克和其他装甲目标的导弹。20 世纪 50 年代中期由法国率先投入使用,继而在众多国家掀起研制高潮。其发展经历了三代,到现在已经成为最有效的反坦克武器。反坦克导弹主要由战斗部、动力装置、弹上制导装置和弹体组成。反坦克导弹的发展趋势是"发射后不用管"、全天候作战能力、自动目标识别以及较强的抗干扰能力等。反坦克导弹导引体制的另一发展趋势,是由单模向多模发展,如红外/毫米波、激光/红外成像、双色红外等。

Anti-vibration device　减振装置(阻尼器)　系指防止或减轻振动有害后果的装置。

Anti-vibration gear　防振装置　系指能防止或减轻振动有害后果的机构。

Anti-vibrator　减振器　系指能防止或减轻振动有害后果的器具。

Anti-wear layer　抗磨层

APHIS = Animal and Plant Health Inspection Service of the U.S, Department of Agriculture　美国农业部动植物检疫服务中心

Apostal Inlands National Lakeshore　阿尔波斯特群岛的国家湖滨区　系指其位置处在国家公园管理局管理的苏必略湖、玛德林岛边或附近的,并包括威斯康星州贝菲尔德半岛海岸线,从沿岸线的西南的北纬 46°57′19.7″,西经 090°52′51.0″的一点至北纬 46°53′56.4″,西经 091°3′3.1″一点的陆地。 Apostal Inlands National Lakeshore is the site on or near Lake Superior and administered by the National Park Service, and Madeling Island, and including the Wisconsin shoreline, or land at 46°57′19.7″N, 090°52′51.0″W, southwest along the shoreline to a point of land at 46°53′56.4″N, 091°3′3.1″W.

Apparel/Deck equipment/Marine gear/Marine store/Tackle　船具　系指船舶舱面用具、甲板装备、船用装置、船用物料和滑车等的总称。

Apparent advance coefficient　表观进速系数　系指用船速计量的进速系数。 $J_v = V/n\ D$,式中,V—船速;n—螺旋桨转速;D—螺旋桨直径。

Apparent amplitude of roll　表观横摇幅值　系指在不规则横摇时,某一周期中的横摇幅值。

Apparent projection of blade section　叶切面表观投影　螺旋桨图中,展平的叶切面投影于叶面节线上所得的线段在相应半径的水平线上的投影。

Apparent slip ratio　表观螺距比　系指螺旋桨螺距与每转一周船的进程之差对螺距的比值。$S_A = [P - (V/n)]/P = 1 - V/nP$,式中,$V$—船速;$n$—螺旋桨转速;$P$—螺距。

Apparent wave amplitude　波幅　波峰或波谷到水平面之间的垂向距离。

Apparent wave height　表观波高　系指不规则波相邻波峰与波谷之间的垂向距离。

Apparent wave length　表观波长　系指不规则波在其前进方向上,两个相邻向上跨零点之间的水平

距离。

Apparent wave period　表观波浪周期　系指在某一空间固定点上不规则地出现两个相邻波峰或波谷所经历的时间间隔。

Appear procedure　二审程序

Appendage resistance　附体阻力　系指船舶设计水线以下的附体,如:舭龙骨、舵、轴包套、轴、轴支架、减摇鳍、导流罩、螺旋桨、船体开口等,这些突出船体表面外的附体在航行时产生的相对于裸船体阻力的增值。一般把船体表面上的开孔或凹槽的阻力也计入附体阻力中。立龙骨的阻力可按其湿面积计入船体摩擦阻力内。运转着的螺旋桨是推进装置主要部分,不计入附体阻力内。

Appendage scale effect factor　附体尺度效应增值　系指实船附体阻力系数与船模附体阻力系数的比值。

Appendages　附体　系指水线以下突出于船体型表面以外的物体。包括轴包套、艉轴架、轴、舵、舭龙骨、矩形龙骨、减摇鳍、导流罩、水翼等,但不包括外板。

Appendix/Annex　附录　置于参考文献之前,文字、图形、表格、数学公式与前述相同。图形编号如附图1。

Apple macintosh operating system(OS X)　苹果操作系统　系指苹果电脑所采用的操作系统,与微软的Windows操作系统以及开源的Linux等为最常用的操作系统。苹果操作系统是全球领先的操作系统。基于坚如磐石的UNIX基础,设计简单直观,让处处创新的Mac安全易用,高度兼容,出类拔萃。Mac OS X以简单易用和稳定可靠著称。

Appliance　设备(装置,仪表)

Applicable national law　所适用的国内法　系指根据核动力船舶的核动力船舶经营人责任公约具有管辖权的法院所在国的法律,包括该国内法中有关法律冲突的任何规则。　Applicable national law means the national law of the court having jurisdiction under the convention on the liability of operators of nuclear ships including any rules of such national law relating to conflict of laws.

Applicable Water of Alaska　阿拉斯加的适用水域　系指在亚历山大群岛水域,以及在阿拉斯加州之内和在卡奇玛克湾的国家港湾搜索预备队管辖范围内的美国可航行水域。Applicable Water of Alaska means the wastes of the Alexander Archipelago and the navigable waters of Alaska and within the Kachemak Bay National Estuarine Research Reserve.

Applicant[1]　开证申请人　系指根据商务合同的规定向银行(开证行)申请开立信用证的人,即是进口商。信用证的申请人包括名称和地址等内容,必须完整、清楚。

Applicant[2]　申请方　系指:(1)申请船级社产品检验的组织。申请方可以是一家产品制造厂商、代理商、产品设计者等;(2)与成员国或主管机关共同研发压载水管理系统(BWMS)并在研发压载水管理系统(BWMS)时拟对某一特定活生物质使用初始基本认可的任何生产厂商或研发者。　Applicant means: (1) an organization applying for an inspection of products by a society. An applicant may be a manufacturer, an agency or a designer institute; (2) any manufacturer or developer working with the Member State or Administration in the development of a BWMS that intends to use the original Basic Approval for a certain Active Substance in the development of the BWMS.

Applicant[3]　申请人　系指按有关规定,向船级社提交检验申请的海底管道系统的作业者或其代理人。

Application for arbitration　仲裁申请书　其内容包括:(1)申请人和被申请人的名称和住所(如有邮政编码、电话、电传、传真、电报号码或其他电子通信方式,也应写明);(2)申请人所依据的仲裁协议;(3)案情和争议的要点;(4)申请人的请求及所依据的事实和证据。仲裁申请书应由申请人及/或申请人授权的代理人签名及/或盖章。

Application for evidence preservation　证据保全申请书

Application for labour arbitration　劳动仲裁申请书　系指劳动争议一方或双方当事人向劳动仲裁机关,就劳动争议事项提出仲裁请求的法律文书,也是劳动仲裁机关立案的依据和凭证。劳动争议当事人认为自己的权利受到侵害,需要向仲裁机关提出申诉,要求劳动仲裁机关予以维护时,就应提供劳动仲裁申请书。劳动仲裁的时效为一年,需注意在时效内申请。

Application of port state's contingency strategies　港口国当局的应急策略　系指由于恶劣天气及海况或船上设备故障、船舶安全因素等,船舶不能够采用压载水交换措施,满足港口国的港口提出的应急策略和要求,船舶应到指定地点排放最少量的必要压载水或接受其他应急措施。　Application of port State' contingency strategies mean those contingency strategies that when the ballast water exchange measures are not practicable due to deteriorating weather and sea conditions, or shipboard equipment failure and ship safety factors, the ship should only discharge the minimum essential amount of ballast water in designed areas according to port State' contingency strategies and requirements or accept other contingency procedures.

Application platform　雪鲤鱼平台　系指由上海雪鲤鱼计算机科技有限公司开发的一种手机应用平台。

（国产机/有一大部分是用的这种平台）也就是 STEP 平台。该平台是一个基于 MTK、ADI 以及展讯等通用手机内置平台上的应用开发引擎。这些主流内置平台原本是种封闭的平台,通过安装 STEP 平台,可以实现通过外部存储加载外部应用,使用户手机平台拥有智能手机的扩展功能。平台附带有合理的计费方案,使各合作可以通过软件商店获得合理收益。

Application program 应用程序 系指为完成某项或多项特定工作的计算机程序,它运行在用户模式,可以和用户进行交互,具有可视的用户界面。应用程序通常又被分为两部分:图形用户接口(GUI)和引擎(Engine)它与应用软件的概念不同。应用软件指使用的目的分类,可以是单一程序或其他从属组件的集合,例如 Microsoft Office、Open Office。应用程序指单一可执行文件或单一程序,例如 Word、Photoshop。日常中可不将两者仔细区分。一般视程序为软件的一个组成部分。

Application program for mobile device 移动设备应用程序

Application program interface(API) 应用程序界面 被定义为应用程序可用以与计算机操作系统交换信息和命令的标准集。一个标准的应用程序界面为用户或软件开发商提供一个通用编程环境,以编写可交互运行于不同厂商计算机的应用程序。API 不是产品,而是战略,所有操作系统与网络操作系统都有 API。在网络环境中不同机器的 API 兼容是必要的,否则程序对其所驻留的机器将是不兼容的。

Application software 应用软件 系指一个专门针对计算机基本系统结构来执行任务并由基本软件支持的软件。 Application software is a software performing tasks specific to the actual configuration of the computer-based system and supported by the basic software.

Application softwares 应用软件 系指用户可以使用的各种程序设计语言,以及用各种程序设计语言编制的应用程序的集合,分为应用软件包和用户程序。应用软件包是利用计算机解决某类问题而设计的程序的集合,供多用户使用。计算机软件分为系统软件和应用软件两大类。应用软件是为满足用户不同领域、不同问题的应用需求而提供的那部分软件。它可以拓宽计算机系统的应用领域,放大硬件的功能。

Application specific integrated circuit(ASIC) 专用集成电路 在集成电路界被认为是一种为专门目的而设计的集成电路。ASIC 的设计方法和手段经历了几十年的发展演变,从最初的全手工设计已经发展到现在先进的可以全自动实现的过程。在集成电路界 ASIC 被认为是一种为专门目的而设计的集成电路。系指应特定用户要求和特定电子系统的需要而设计、制造的集成电路。ASIC 的特点是面向特定用户的需求,ASIC 在批量生产时与通用集成电路相比具有体积更小、功耗更低、可靠性提高、性能提高、保密性增强、成本降低等优点。

Applicator 施放装置(喷射器,喷射枪) 该名词原意为敷药的器具或涂药器,船舶上通常系指喷射灭火剂的器具,例如泡沫枪。

Applied end moment 吊杆附加弯矩 系指由于吊杆初挠度以及外力偏心作用,使吊杆受到的弯曲力矩。

Appointed defense 指定辩护

Appraisal of ship values 船价鉴定

Appraisal survey 鉴定检验 作为船级社的一种商业性的技术服务,有其非常丰富的内容,服务的范围也从船舶扩大到其他工业领域。鉴定的标准也是多种多样的,可以是国家标准、行业标准、企业标准、交易双方的合同条款等等。但对于船舶来讲,较常见的鉴定检验主要有:(1)保险鉴定——受保险公司的委托,代替保险公司进行的有关船舶保险方面的鉴定;(2)船况鉴定——是船舶买卖、租赁的需要。主要有:①买船前检验;②租赁契约前后检验;(3)耐航鉴定——对损伤船舶从事故发生地到预定修理地航行时的适航性鉴定,或对船舶进行超航区航行时的适航性检验;(4)损伤鉴定——对船舶损伤或机器故障原因的鉴定;(5)船价鉴定——多用于船舶被强制拍卖的场合;(6)载重量鉴定——主要是确认干舷标志的位置,从而算出对应的载货重量。

Appraiser 鉴定人

Approach(APP) 方法

Approach speed 回转初速 系指回转试验中,转舵阶段开始前瞬时船舶沿直线航行的稳定速度。

Appropriate action 相应的行动 系指收到通知的缔约国将所述港口不能处理某些消耗臭氧物质(ODS)和/或废气清除残余物传达给受其控制的船舶所采取的行动,和为使替代方法管理或处理这些物质船舶需要采取的行动。替代方法可包括在停靠受影响的港口之前或之后安排收集这些物质。如在停靠港口之后进行收集,应确保船上有充足的贮存容量。Appropriate action means those actions taken by informed Parties to communicate to ships under their control that the advised ports cannot handle certain ODS and/or exhaust gas cleaning residues and those actions ships will need to take necessary to manage or process those substances in an alternative manner. Such alternatives could include arranging for collection before or after visiting the affected port, and in the latter case, adequate storage on board should be ensured for those substances.

Appropriate authority 有关当局 系指经偷渡者离船的港口所在国政府授权，在该港内设置的，按照国际偷渡公约规定接受和处理偷渡者的机构或个人。 Appropriate authority means the body or person at the port of disembarkation authorized by the Government of the State in which that port is situated to receive and deal with stowaways in accordance with the provisions of the International Convention relating to stowaways.

Appropriate authority 主管机关 系指要求遵守其规定的政府机构。

Appropriate portfolio of up to date paper charts (APC) 适当的最新纸质海图卷(APC) 系指一套纸质海图，以一定的比例显示足够详细的地形、水深、航行危险、航标、绘制在海图上的航线划定措施，向航海人员提供全部航行环境的信息。APC应提供适当的预测能力。沿海国会提供满足该海图卷要求的海图详细情况，这些海图详细情况包括在由IHO维护的全球数据库中。确定APC内容时应考虑本数据库中的详细情况。 Appropriate portfolio of up to date paper charts (APC) means a suite of paper charts of a scale to show sufficient detail of topography, depths, navigational hazards, aids to navigation, charted routes, and routing measures to provide the mariner with information on the overall navigational environment. The APC should provide adequate look-ahead capability. Coastal States will provide details of the charts which meet the requirement of this portfolio, and these details are included in a worldwide database maintained by the IHO. Consideration should be given to the details contained in this database when determining the content of the APC.

Approval 认可 系指船级社对与入级有关的文件、程序或其他项目检查并接受，仅验证其与有关规范要求或其他参考要求的符合性。 Approval means the examination and acceptance by the society of documents, procedures or other items related to classification, verifying solely their compliance with the relevant rules requirements, or other references where requested.

Approval 认可(集装箱) 系指主管机关做出的决定，即某种定型设计或某个集装箱在国际集装箱安全公约条款范围是安全的。 Approval means the decision by an Administration that a design type or a container is safe within the terms of the present Convention.

Approval authority 审批机关 系指负责审批的组织的通用术语。审批机关包括船旗国管理机构、船级社和其他能够审批船舶设计和船舶系统设计的组织。

Approval expiry date 认可生效日期 系指证明其满足国际海上人命安全公约防火安全要求的后续认可有效的最后日期。 Approval expiry date means the last date on which the subsequent approval is valid as proof of meeting the fire safety requirements of the Convention.

Approval matrix 审批矩阵图

Approval of welder 焊工的认可 系指船级社对从事熔焊钢和非铁素体金属的焊工的认可。 Approval of welder means the Society's approval of welder for fusion welding of steel and non-ferrous metals.

Approval process 审批程序

Approval team 审批小组 对于特定的申请，审批机关在多数情况下都会成立一个审批小组。根据相应的申请，审批小组由来自船旗国管理机构、船级社和其他被认可的组织的代表组成。

Approval work 审批工作

Approved 批准 系指图纸资料或文件已审核，符合船级社规范的要求。船级社对图纸资料的批准仅包含船级社规范要求的项目，而不涉及船级社规范不要求的项目。若船级社同时承担法定检验，则船级社的"批准"还应包括有关法定规则要求的项目。经审查认为符合规定的图纸资料，应在已批准的图纸资料上，盖"批准"章。批准的条件和限制意见，可写在图纸资料上；也可在退审的信函中陈述，但应在图纸资料上注明。同时批准的图纸应注明有效期。

Approved 认可 系指经主管机关认可。 Approved means approved by the administration.

Approved method 认可方法 对防污公约附则Ⅵ防止船舶造成空气污染规则而言，系指应用于特定发动机或一系列发动机，确保其符合第13.7条所述的适用NO_x极限的方法。 For the MARPOL Annex Ⅵ Regulations for the Prevention of Air Pollution from Ships, approved method is a method for a particular engine, or a range of engines, which, when applied to the engine, will ensure that the engine complies with the applicable NO_x limit as detailed in regulation 13.7.

Approved method file 认可方法案卷 对防污公约附则Ⅵ防止船舶造成空气污染规则而言，系指描述认可方法及其检验方式的文件。 For the MARPOL Annex Ⅵ Regulations for the Prevention of Air Pollution from Ships, Annex Ⅵ, approved method file is a document which describes an approved method and its means of survey.

Approved/accepted 认可/验收(材料) 系指材料应具备合格证书、验收报告或船级社颁发的无异议的信函。 Approved/accepted refers to materials which hold a valid Certificate, Statement of Acceptance, or Letter of Non-Objection issued by the society.

Apron plate 老鹰板 系指船首端舷墙顶部的连

接平板。

Aquatic nuisance species task force (ANSTF) 水生物物种特遣队 系指强制执行 1990 年防止和控制外来水生物污染法令 (NANPCA) 的水生物物种特遣队。

Aquatic plants 水生植物

Aquatic product carrier (fish carrier) 水产品运输船 系指专门用于运输水产品的船舶。其特点是船舶不具有制冷装置，水产品冷藏方式采用物理冷媒，如冰。是在货舱内结构表面敷设隔热层的船舶。 Aquatic product carrier means a carrier dedicated to transporting aquatic products, characterized by using physical media such as ice, instead of refrigerating plant, for cold storage of aquatic products, with insulation layers fitted on inner surface of holds.

Aramid fiber 聚芳酰胺纤维 系指分子结构中芳酰胺链节占 85% 以上的纤维。

Arbitral subject matter/Arbitral tribunal 仲裁标的

Arbitral tribunal 仲裁庭 仲裁委员会在决定受理案件后，并不直接对案件进行仲裁，而是组成仲裁庭，由仲裁庭行使仲裁权。

Arbitral institution 仲裁机构 系指通过仲裁方式，解决双方民事争议，做出仲裁裁决的机构。分为国内仲裁机构和国际仲裁机构，后者又分为全国性的仲裁机构和国际性或地域性的仲裁机构。此外，按仲裁机构的设置情况，国际上进行仲裁的机构有三种：一种是常设仲裁机构，一种是临时仲裁机构，还有一种是专业性仲裁机构。常设仲裁机构有国际性的或区域性的，有全国性的，还有附设在特定行业内的专业性仲裁机构。它们都有一套机构和人员，负责组织和管理有关仲裁事务，可为仲裁的进行提供各种方便。所以大多数仲裁案件都被提交在常设仲裁机构进行审理。著名的常设仲裁机构有：国际商会仲裁院 (International Chamber of Commerce Court of Arbitration)、英国伦敦仲裁院、瑞士苏黎世商会仲裁院、日本国际商事仲裁协会、美国仲裁协会、瑞典斯德哥尔摩商会仲裁院、中国国际贸易促进委员会对外经济贸易仲裁委员会等。

Arbitration 仲裁 作为解决海事海商争议的一种方式，普遍地为国际航运所采用。当事人之所以选择仲裁，是因为仲裁具有许多独特的优势：(1) 充分自治。仲裁的本质就是当事人意思自治。选择仲裁方式，当事人可享有最大限度的自主权，包括自主选择仲裁机构、仲裁员、仲裁地点、仲裁所适用的法律、仲裁所使用的仲裁规则以及语言。(2) 程序简便。仲裁实行"一裁终局"制度，没有上诉或再审程序，裁决自做出之日起即发生法律效力，对双方当事人均有约束力。因而简化了程序，缩短了审理期限，提高了争议解决的效率。(3) 信息保密。仲裁实行不公开审理制度。未经当事人同意，第三人不可旁听案件审理，审理结果不公布于媒体。(4) 费用低廉。仲裁实行"一裁终局"制度，没有二审或再审程序，为当事人节省了时间和费用。(5) 易于执行。1968 年联合国在纽约通过的《关于承认和执行外国仲裁裁决公约》为国际社会提供了一项普遍接受的、简便的承认及执行外国仲裁裁决的制度。根据该公约的规定，缔约国的仲裁裁决能直接申请在 140 个缔约国法院得以强制执行。

Arbitration agent 仲裁代理人

Arbitration agreement/Clauses of arbitration 仲裁协议 系指当事人在合同、提单、运单或援引的文件中订明的仲裁条款或者以其他方式达成的提交仲裁的书面协议。仲裁协议独立存在，合同的变更、解除、终止、失效和无效以及存在与否，均不影响仲裁协议的效力。仲裁协议有两种形式：一种是在争议发生之前订立的，它通常作为合同中的一项仲裁条款出现；另一种是在争议之后订立的，它是把已经发生的争议提交给仲裁的协议。这两种形式的仲裁协议，其法律效力是相同的。

Arbitration clause 仲裁条款 又称公断，系指合同双方在争议发生之前或发生之后，签订书面协议，自愿将争议提交双方所同意的第三者予以裁决 (Award)，以解决争议的一种方式。由于仲裁是依照法律所允许的仲裁程度裁定争端，因而促裁决具有法律约束力，当事人双方必须遵照执行。

Arbitration commission 仲裁委员会 系指常设性仲裁机构，一般在我国的省、直辖市、自治区人民政府所在地的市设立，也可以根据需要在其他设区的市设立，不按行政区划层层设立。仲裁委员会由市的人民政府组织有关部门和商会统一组建，并应经省、自治区、直辖市的司法行政部门登记。仲裁委员会由主任 1 人，副主任 2 至 4 人和委员 7 至 11 人组成。仲裁委员会的主任、副主任和委员由法律、经济贸易专家和有实际工作经验的人员担任。仲裁委员会的组成人员中法律、经济贸易专家不得少于 2/3。

Arbitration document 仲裁文件

Arbitration fee schedule 仲裁费用表

Arbitration institute of Stockholm chamber of commerce 瑞典斯德哥尔摩商会仲裁院

Arbitration law of the People's Republic of China 中华人民共和国仲裁法

Arbitration Law 仲裁法

Arbitration procedure 仲裁程序/仲裁流程 系指国际商事仲裁中，从案件仲裁提起直至裁决做出的整个过程所应遵循的程序和规则,仲裁程序通常由当事人双方选择的仲裁规则加以确定。包括立案受理;交换申请书、答辩书、证据及其他书面材料;选择仲裁员组成仲裁庭;开庭审理(双方可协议书面审理);同意调解——进入调解程序调解成功——制作调解书/调解不成,或不同意调解——仲裁庭做出裁决并制作裁决书。

Arbitration rules　仲裁规则　系指规范仲裁进行的具体程序及此程序中相应的仲裁法律关系的规则。仲裁规则不同于仲裁法,它可以由仲裁机构制定,有些内容还允许当事人自行约定。因此,仲裁规则是任意性较强的行为规范。但是仲裁规则不得违反仲裁法中的强制性规定。仲裁规则应依据仲裁法和民事诉讼法的有关规定加以制定。根据我国仲裁法规定,我国仲裁委员会仲裁规则的制定分为两种情况:1.国内仲裁委员会的仲裁规则,由中国仲裁委员会统一制定,在中国仲裁协会制定仲裁规则之前,各种委员会可以按照仲裁法和民事诉讼法的有关规定仲裁暂行规则;2.涉外仲裁委员会的仲裁规则则由中国国际商会制定。

Arbitrator　仲裁员　系指对海事、海商、物流、保险以及法律等方面具有专门知识和实际经验的中外人士。

Arbitrator's declaration　仲裁声明书

Arbitrators' special remuneration　仲裁员办理案件的特殊报酬

Arc strikes　触发电弧

Archetype　原则

Archipelagic sea lanes passage　群岛海道通过　系指按照联合国海洋法公约规定,专为在公海或专属经济区的一部分和公海或专属经济区的另一部分之间不停地、迅速和无障碍地过境的目的,行使正常方式的航行和飞越的权利。Archipelagic sea lanes passage means the exercise in accordance with the United Nations Convention on the law of the sea, of the rights of navigation and over-flight in the normal mode solely for the purpose of continuous, expeditious and unobstructed transit between one part of the high seas or an exclusive economic zone and another part of the high seas or an exclusive economic zone.

Archipelagic State　群岛国　系指全部由一个或多个群岛构成的国家,并可包括其他岛屿。Archipelagic State means a State constituted wholly by one or more archipelagos and may include other islands.

Archipelago　群岛　系指一群岛屿,包括若干岛屿的若干部分,相连的水域和其他自然地形彼此密切相关,以致这种岛屿、水域和其他自然地形在本质上构成一个地理、经济、政治的实体,或在历史上已被视为这种实体。Archipelago means a group of islands, including parts of islands, interconnecting waters and other natural features which are so closely interrelated such islands, waters and other natural features from an intrinsic geographical, economic and political entity, or which historically have been regarded as such.

Archives　档案(国际海事卫星组织)　包括国际海事卫星组织所拥有或掌管的一切手稿、信件、文件、照片、影片、光记录和磁记录、数据记录、图表和计算机程序。 Archives includes all manuscripts, correspondence, documents, photographs, films, optical and magnetic recordings, data recordings, graphic representations and computer programmes, belonging to or held by INMARSAT.

Arcing time(of a pole or fuse)　(一极或熔断器的)燃弧时间　系指从开关电器一极或熔断器电弧产生的瞬间起,至该极和熔断器中电弧最终熄火之间的时间间隔。

Arcing time(of multi pole switching device)　(多极开关电器的)燃弧时间　系指从第一个电弧产生的瞬时起,到所有极电弧最终熄火的瞬间止的时间间隔。

Arctic ice-covered waters　北极冰覆盖区域　系指如下两个水域;(1)北面从格陵兰岛南端—格陵兰岛南岸至Kape Hoppe—恒向线至北纬67°03′9,西经026°33′4恒向线至SØrkapp, Jan Mayen,然后沿Jan Mayen南岸至Bjɇrnɇya—大圆线从Bjɇrnɇya岛至Cap Kanin Nos—沿亚洲大陆北岸由东至白令海峡—从白令海峡向西至北纬60°直到Il'pyrskiy,并沿北纬60°向东至并包括Etolin海峡—沿北美大陆北岸至北纬60°—再向东至格陵兰岛的南端;(2)海水冰集量覆盖面为该水域中1/10或以上,对船舶结构造成危险。 Arctic ice-covered waters means those waters which are both:(1) located north of a line from the southern tip of Greenland and thence by the southern shore of Greenland to Kape Hoppe and thence by a rhumb line to latitude 67°03′9 N, longitude 026°33′4 W and thence by a rhumb line to SØrkapp, Jan Mayen and by the southern shore of Jan Mayen to the Island of Bjɇrnɇya, and thence by a great circle line from the Island of Bjɇrnɇya to Cap Kanin Nos and thence by the northern shore of the Asian Continent eastward to the Bering Strait and thence from the Bering Strait westward to latitude 60° North as far as Il'pyrskiy and following the 60° North parallel eastward as far as and including Etolin Strait and thence by the northern shore of the North American continent as far south as latitude 60° North and thence eastward to the southern tip of Greenland;(2) in which sea ice concentrations of 1/10 coverage or greater are present and pose a structural risk to ships.

Arctic vessel(polar ship)　极区船　适宜于在北冰洋或南极圈内海区航行的船舶。

Arctic waters 北极水域 系指位于下述连线以北的水域：从北纬58°00′0，西经042°00′0 延伸至北纬64°37′0，西经035°27′0 的连线，再经一恒向线延伸至北纬67°03′9，西经026°33′4，再经一恒向线延伸至 S¢rkapp，Jan Mayen 并经由 Jan Mayen 南岸延伸至 Bj¢rn¢ya岛，再经一圆线从 Bj¢rn¢ya岛延伸至 Cap Kanin Nos，再经由亚洲大陆北岸向东延伸至白令海峡，再从白令海峡向西延伸至北纬60°直到 Il's pyrskiy，并沿北纬60°向东延伸至并包括 Etolin 海峡，再经由北美大陆向南延伸至北纬60°，再向东沿北纬60°平行线延伸至西经56°37′1，再延伸至北纬58°00′0，西经042°00′0。Arctic waters means those waters which are located north of a line extending from latitude 58°00′0 N, longitude 042°00′0 W to latitude 64°37′0 N, longitude 035°27′0 W and thence by a rhumb line to latitude 67°03′9 N, longitude 026°33′4 W and thence by a rhumb line to S¢rkapp, Jan Mayen and by the southern shore of Jan Mayen to the Island of Bj¢rn¢ya and thence by a great circle line from the Island of Bj¢rn¢ya to Cap Kanin Nos and thence by the northern shore of the Asian continent eastward to the Bering Strait and thence from the Bering Strait westward to latitude 60°N as far as Il's pyrskiy and following the 60° North parallel eastuard as far as and including Etolin strait and thence North American continent as far south as latitude 60°N and thence eastward along parallel of latitude 60°N, to longitude 56°37′1 W and thence to the latitude 58°00′0 N, longitude 042°00′0 W.

Area 区域 系指国家管辖范围以外的海床、洋底及其底土。Area means the sea-bed, and subsoil thereof, beyond the limits of national jurisdiction.

Area of platform 平台面积 系指钻井平台上甲板的总面积。

Area of rudder 舵面积 系指舵叶的侧投影面积。当部分舵叶露出水面时，则指设计水线以下的舵面积。

Area to be avoided 避航区 系指特殊敏感海域。Area to be avoided means the particularly sensitive sea area (PSSA).

Area under cut-up 船体空缺面积 船首或船尾下部实体的截空部分在中线面上的投影面积。

Area velocity (AV) value 面积速度值（AV） 系指通过催化剂块的废气流量（m^3/h）与选择性催化还原系统催化器中催化剂块的总活性表面的比值。因此，AV值的单位是 m/h。废气流体积系指在 0 ℃ 和 101.3kPa 定义的体积。Area velocity (AV) value means a value of the exhaust gas flow rate passing through the catalyst blocks (m^3/h) per volume of the catalyst block(s) in the selective catalytic reduction (SCR) chamber (m^3). Therefore, unit of AV value is (m/h). The exhaust gas flow volume is the volume defined at 0 ℃ and 101.3kPa.

Area-adjustable small water-plane area twin hull 可调面积小水线面双体船 系指一种根据不同载重工况的需要可以调整水线面面积的小水线面双体船。该船型的特点：(1)受海浪激动力小，横摇阻力大，自摇周期长并可避免谐摇，可控减摇效果显著，耐波性能好；(2)双体船航速快，航向稳定性好；(3)配备两对螺旋桨，相互间距大，操纵性好，船可原地旋转，回避碰撞，离靠码头方便；(4)甲板宽敞，可设置直升机平台。该船型适合在风浪条件恶劣的海峡航行。

Areas for which special conditions exist 特殊条件的区域 系指电子海图显示和信息系统（ECDIS）按照有关规定要求发现并提供报警或指示的区域：(1)分道通航区；(2)沿岸通航区；(3)限制区域；(4)警戒区域；(5)近海生产区域；(6)避航区；(7)用户定义的避航区；(8)军事演习区域；(9)水上飞机降落区域；(10)潜水艇过道；(11)渔场和水产养殖场；(12) PASS（特殊敏感海域）。Areas for which special conditions exist mean the areas which ECDIS should detect and alarm indication according to relevant requirement: (1) traffic separation zone; (2) inshore traffic zone; (3) restricted area; (4) caution area; (5) offshore production area; (6) areas to be avoided; (7) user defined areas to be avoided; (8) military practice area; (9) seaplane landing area; (10) submarine transit lane; (11) marine farm/aquaculture; (12) PSSA (particularly sensitive sea area).

Argument 辩论

Arithmetic and logic unit (ALU) 运算器 系指计算机中执行各种算术和逻辑运算操作的部件。运算器的基本操作包括加、减、乘、除四则运算，与、或、非、异或等逻辑操作，以及移位、比较和传送等操作，亦称算术逻辑部件（ALU）。计算机运行时，运算器的操作和操作种类由控制器决定。运算器处理的数据来自存储器；处理后的结果数据通常送回存储器，或暂时寄存在运算器中。

Arm hook 吊臂叉头 系指吊艇臂上端承载浮动滑车的叉状构件。

Armed robbery against ships 对船舶的武装抢劫 系指下列行为中的任何行为：(1)在国家内陆水域、群岛水域和领海范围内，为私人目的针对船舶或船舶上的人或财物实施"海盗行为"以外的任何非法的暴力和扣留行为，或其威胁；(2)教唆或故意便利上述行为的任何行为。Armed robbery against ships means any of the following acts: (1) any illegal act of violence or detention or any act of depredation, or threat thereof, other than an act of piracy, committed for private ends and directed against a ship or against persons or property on board such ship, within a State's internal waters, archipelagic waters and territorial sea; (2) any act of inciting or of intentionally facilitating an act described above.

Army Corps of Engineers(ACOE)　［美］陆军工程兵部队

Aromatic oil(excluding vegetable oil)　芳烃油类（不包括植物油）

ARPA　自动雷达标绘仪　ARPA is automatic radar plotting aid.

Arrangement of shipboard weapons　武备布置　系指根据战术技术任务书的要求，为充分发挥舰艇上各种武器装备的作战威力，而对其所做的全面、统一的规划和布置。

Arrest　扣押　系指通过司法程序滞留船舶，以保全海事请求，但不包括执行或满足某项判决中船舶的扣留。Arrest means the detention of a ship by judicial process to secure a maritime claim, but does not include the seizure of a ship in execution or satisfaction of a judgment.

Arrival ballast　到港压载　系指清洁压载水。Arrival ballast means clean ballast water.

Arrival condition　到达工况　系指消耗备品10%的工况。Arrival condition means a condition with 10% of consumables.

Arsine　砷化氢　是一种无色有毒气体，有类似大蒜的味道。砷化氢对神经系统和血液有毒。症状的出现通常滞后(有时1天左右)。起初症状不明显。症状：(1)感觉不适、呼吸困难、剧烈头痛、眩晕、阵发昏厥、恶心、呕吐和胃部紊乱；(2)在严重的情况下，呕吐加剧，黏膜可能带有浅蓝色，尿液变成深色并含血。经过1天左右出现严重贫血和黄疸。人在500 ppm① 浓度下曝露几分钟后会死亡，而在250 ppm 浓度下曝露30 min 后有生命危险。在6.25~15.5 ppm 浓度下，曝露30 min~60 min后有危险。0.05 ppm 浓度是人可以长期曝露和极限阈值。Arsine is a toxic, colourless gas with a garlic like odour. Arsine is a nerve and blood poison. There is generally a delay before the onset of symptoms (sometimes a day or so). These are at first indefinite. Symptoms: (1) Feeling of malaise, difficulty in breathing, severe headache, giddiness, fainting fits, nausea, vomiting and gastric disturbances. (2) In severe cases, vomiting may be pronounced, the mucous membranes may have a bluish discolouration and urine is dark and bloodstained. After a day or so there is severe anaemia and jaundice. A concentration of 500 ppm is lethal to humans after exposure of a few minutes, while concentrations of 250 ppm are dangerous to life after 30 minutes exposure. Concentrations of 6.25 to 15.5 ppm are dangerous after exposure of 30 to 60 minutes. A concentration of 0.05 ppm is the threshold long limit to which a person may be exposed.

Arsonist test　点火试验　系指在点火之前30 s，将1 L 白酒洒在一下床铺及靠背上，引燃物应放在下床铺枕块的中前部(朝向门外)的试验。Arsonist test means a test in which fire arranged by spreading 1 L of white spirits evenly over one lower bunk bed and backrest 30 s prior to ignition. The igniter should be located in the lower bunk bed at the front(towards door) centerline of the pillow.

Article Numbering Center of China（ANCC）　中国物品编码中心　系指统一组织、协调、管理我国商品条码、物品编码与自动识别技术的专门机构，隶属于国家质量监督检验检疫总局，1988年成立，1991年4月代表我国加入国际物品编码协会(GS1)，负责推广国际通用的、开放的、跨行业的全球统一编码标识系统和供应链管理标准，向社会提供公共服务平台和标准化解决方案。中国物品编码中心在全国设有46个分支机构，形成了覆盖全国的集编码管理、技术研发、标准制定、应用推广以及技术服务为一体的工作体系。物品编码与自动识别技术已广泛应用于零售、制造、物流、电子商务、移动商务、电子政务、医疗卫生、产品质量追溯、图书音像等国民经济和社会发展的诸多领域。

Articles　商品

Articulated buoyant tower　铰接式浮动塔　这种结构形式是依靠作用于接近水面的浮力提供必要的复原稳性。因为它往往有较大的水平位移，所以铰接式浮动结构可以在海底附近配备一个支枢。

Articulated column/Articulated tower　铰接柱　系指用铰接式接头连接到海底重力基础或桩基础上，在近水面处设有浮力舱或浮筒的管状结构或桁架结构。

Articulated connection PB combination-barge　铰接式连接顶推船驳船组合体—驳船　系指由顶推船和一艘驳船组成的船队。顶推船通过首部机械装置锁紧在驳船尾部凹槽内，顶推船与驳船之间仅有一纵摇的自由度，营运时保持联结状态，脱开后，两船可独立停泊或作业。驳船为组合体的组成部分。Articulated connection PB combination-barge means a combination of a pusher tug and a barge wherein the pusher tug is secured in the barge notch by mechanical means, allowing pitch between the tug

———————
① ppm 为百分比浓度，1 ppm = 0.0001%。

and the barge in only one degree of freedom. The two vessels act as a single unit in a seaway and when disconnected from each other, both may moor or operate independently. The barge is a component part of the combination.

Articulated connection PB combination-pusher 铰接连接顶推船驳船组合体—推船 系指由顶推船和一艘驳船组成的船队。顶推船通过首部机械装置锁紧在驳船尾部凹槽内，顶推船与驳船之间仅有一纵摇的自由度，营运时保持连接状态，脱开后，两船可独立停泊或作业。顶推船为组合体的组成部分。 Articulated connection PB combination-pusher means a combination of a pusher tug and a barge where in the pusher tug is secured in the barge notch by mechanical means, allowing pitch between the tug and the barge in only one degree of freedom. The two vessels act as a single unit in a seaway and when disconnected from each other, both may moor or operate independently. The pusher is a component part of the combination.

Articulated loading tower system(ALT) 铰接式装卸塔系统 系指其上部与一个刚性桁架连在一起形成塔柱，桁架下部用方向接头与海底基座连接，为保持稳定，桁架四周设有压载物，浮体上部为一旋转平台，并设有收放油管和油船首缆的绞车系统。用于环境较严酷的海域。

Articulated tug-barge combination（巴拿马运河） 推-驳组合体 系指由一艘推船和一艘前置的非舱式驳船刚性组合而成的组合体。组合体意味着一艘推船借助于"机械措施"刚性地与一艘被推驳船连接，使该组合体作为一个整体经受波浪和涌浪，从而可认为是一艘单一的动力驱动船。"机械措施"不包括绳索、缆索、钢索或链条。巴拿马运河认可的推-驳组合体还应符合巴拿马运河管理局所有有关通航的规范和规则，并能在相当尺度的船舶能够操作的所有条件下进行操作。 For Panama Canal, articulated tug-barge combination means a pushing vessel and a non-tank barge pushed ahead rigidly connects to each other to form a composite unit. A composite unit means the pushing vessel rigidly is connected by "mechanical means" to a barge being pushed so they react to the sea and swell as one vessel and is considered as a single power-driven vessel. "Mechanical means" does not include lines, hawsers, wires or chains. To be considered as an ITB at the Panama Canal, such vessels must meet all current ACP regulations and requirements for transit and be able to operate in all conditions under which a ship of equivalent size can operate.

Articulated tug-tank combination（巴拿马运河） 推-舱组合体 系指组合体由一艘推船和一艘前置的舱式驳船刚性组合而成的组合体。该组合体必须满足对推-驳组合体所有的规定和规则。 For the purpose of Panama Canal, articulated tug-tank combination means a pushing vessel and a tank barge pushed ahead rigidly connects to each other to form a composite unit. The composite unit must meet all specifications and requirements set forth for an integrated tug-barge combination.

Articulated wheel system 可升降转轮 全垫升气垫船用以在陆上驾驶时防止侧漂并提高登坡能力，在水面狭窄航道或拥挤港口航行时提高回转性能的，不用时可升降水面或地面兼作舵用的可升降轮子。

Artificial accelerated aging test of natural marine envirnment for ship and marine coating 涂料的海洋环境模拟加速试验 系指包括下列内容的试验:(1)盐雾试验;(2)人工加速老化(紫外、氙灯)试验;(3)高低温试验;(4)湿热试验;(5)海水喷淋试验;(6)防污漆动态模拟试验;(7)水线漆划水试验;(8)管道流动水;(9)压载舱涂料波浪舱试验;(10)压载舱涂料冷凝舱试验;(11)压载舱交叉试验;(12)油船货油舱涂层气密柜试验;(13)油船货油舱涂层浸泡试验;(14)油船货油舱耐蚀钢上甲板状态模拟试验(气氛试验);(15)油船货油舱耐蚀钢内底状态模拟试验(浸泡试验)等。 Artificial accelerated aging test of natural marine envirnment for ship and marine coating mean those tests included:(1) salt spray test;(2) artificial accelerated aging test (UV、Xenon radiation);(3) high/low temperature test;(4) heat and humidity test;(5) seawater sprinkling test;(6) dynamic test of antifouling paint;(7) water-thrashing test for boottoping paint;(8) pipeline flowing water;(9) PSPC test for WBT;(10) condensation test for WBT;(11) PSPCcrossover test for WBT;(12) gas-tight test for PSPC-COT;(13) immersion test for PSPC-COT;(14) upper deck condition test for corrosion resistant steel of COT;(15) inner bottom condition test for corrosion resistant steel of COT, etc.

Artificial biological cornea 人工生物角膜

Artificial island 人工岛 系指为了在浅水区进行海上油气开发，采用沉箱结构、泥沙吹填等方法建成的岛式油气生产基地。

Artificial neural network(ANN) 人工神经网络

Artificial rubber 人造橡胶

Artificial sea-grass 人工海草

Artificial ventilation 人工(机械)通风 通过人工设施(例如风机)，且适用于一般区域的空气的移动以及新鲜空气的更换。 Artificial ventilation means the movement of air and its replacement with fresh air by artificial means(for example fans) and applied to a general area.

Artimis "月女神"主动声呐系统 用于测量船舶位置的无线电系统。系统运行使用微波频率,从一个固定站位测量船舶的距离和方位,通常安装在平台上。

$A_s(\text{cm}^2)$ A_s 扶强材或主要支撑构件连同宽度 s 带板的净横剖面积(cm^2)。 A_s means the net sectional area of the stiffener or the primary supporting member, which are with attached plating of width $s(\text{cm}^2)$.

As low as reasonable practicable area 最低合理可行范围 最低合理可行范围是介于可忽略线和不可容忍线之间的风险水平,对于低于可忽略线或高于不可容忍线的风险水平,要运用成本效益分析来鉴定成本效果的风险控制方案。

As low as reasonable practicable (ALARP) 最低合理可行 最低合理可行原则是在风险评估过程中适用的一种原则。系指对于介于可忽视线和不可容忍线之间的事故,必须减少其风险水平,除非为降低风险而继续投资是不合理的。

As rolled (AR) 轧制 系指将钢材在高温下轧制后冷却在空气中的轧制工艺。通常轧制和最终轧制是在大于正火温度和奥氏体再结晶温度区域内进行。在该工艺中生产出来的钢材的强度及韧性低于轧制后经热处理的钢材或经改进的工艺而生产出来的钢材。
This procedure involves the rolling of steel at high temperature followed by air cooling. The rolling and finishing temperatures are typically in the austenite recrystallization region and above the normalizing temperature. The strength and toughness properties of steel produced by this process are generally less than that of steel heat treated after rolling or steel produced by advanced processes.

As rolled (AR) condition 轧制状态 标准轧制的目的是仅使制品达到最终的尺寸的过程。 As rolled (AR) condition is standard rolling arming only to give the product to its final dimension.

Asbestos 石棉 系指含有石棉的材料。对于所有的船舶,应禁止新装含有石棉的材料。 Asbestos mean those materials containing asbestos. For all ships, new installation of materials which contain asbestos shall be prohibited.

As-built thickness $t_{\text{as-built}}$ 建造厚度 为新建阶段中提供的实际厚度。包括船东的腐蚀损耗额外裕度 $t_{\text{voluntary-addition}}$(如有)。 As-built thickness $t_{\text{as-built}}$ is the actual thickness (mm), provided at the new-building stage, including $t_{\text{voluntary-addition}}$, if any.

$A_{sh}(\text{cm}^2)$ A_{sh} 扶强材或主要支撑构件净剪切横剖面积(cm^2)。 A_{sh} means the net shear sectional area (cm^2), of the stiffener or the primary supporting member.

Ash chute 出灰管 燃煤的烟管锅炉或焚烧炉上用于出灰的一段有一定坡度的管道。

Ash discharging gear 出灰装置 燃煤的烟管锅炉或焚烧炉上用于出灰的一种简单的机械装置。

Ash ejector 冲灰器 以前用于将残灰冲出舷外或冲入海中的喷射器。

Ash-ejector cock 冲灰旋塞

Asia Pacific Economic Cooperation (APEC) 亚太经济与合作组织 简称亚太经合组织,成立于1989年,是亚洲-太平洋地区级别最高、影响最大的区域性经济组织。该组织为推动区域贸易投资自由化,加强成员间经济技术合作等方面发挥了不可替代的作用。是亚太区内各地区之间促进经济成长、合作、贸易、投资的论坛。1991年11月,中国以主权国家身份,中国台北和香港(1997年7月1日起改为"中国香港")以地区经济名义正式加入亚太经合组织。2014年11月5日,APEC在北京怀柔雁栖湖举办,本届主题为共建面向未来的亚太伙伴关系。

Asian Development Bank (ADB) 亚洲开发银行 简称"亚行",是亚洲、太平洋地区的区域性政府间国际金融机构。它不是联合国下属机构,但它是联合国亚洲及太平洋经济社会委员会(联合国亚太经社会)赞助建立的机构,同联合国及其区域和专门机构有密切的联系。总部设在马尼拉。

Asian Infrastructure Investment Bank (AIIB) 亚洲基础设施投资银行 简称"亚投行"。是由中国国家主席习近平倡议成立亚洲基础设施投资银行以促进本地区互联互通建设和经济一体化。成立"亚投行"旨在为亚洲大型基础设施项目融资,并满足日益增长的需求,以实现更包容和更均衡的国际金融秩序。尽管其重心在于亚洲基础设施发展,亚投行为具有先进技术的发达国家提供了充裕的贸易和投资机遇。经济基础设施建设促进一个国家的经济活动,例如道路、高速公路、公路、铁路、航空港和海港等建设。交通基础设施对一些国家非常重要。中国已经帮助一些国家修建了许多新桥梁和道路。中国继续更新其技术基础结构的需求为国际公司提供了机会,尤其是在供水技术、医疗设备和高科技工业设备等领域。这个跨国公司新任命的首席财务官正在领导他的员工建立新的公司财务基础结构。这个基础结构计划旨在促进生产力,并提高中小企业的竞争力。

Askania marine gravimeter 阿斯卡尼亚海洋重力仪 系指利用弹簧扭矩与重力矩保持平衡的原理制成的摆杆型海洋重力仪。

Aspect ratio 展弦比 叶片高度与弦长的比值。

Aspect ratio of rudder 舵展弦比 系指舵高与舵宽之间的比值。矩形舵：$\lambda = h/b$；非矩形舵：$\lambda = h^2/A_R$。

Asphalt carrier 沥青运输船 系指运载沥青的船舶。

Asphalt carrier(independent tank/integral tank) 设有独立液货舱/整体液货舱的石油沥青船(独立液货舱/整体液货舱) 系指设有独立液货舱或整体液货舱，专运沥青的船舶。 Asphalt carrier means a carrier fitted with independent or integral cargo tanks, dedicated to carrying asphalt.

Asphalt damping materials 沥青型阻尼材料 系指以沥青为主，并配入大量无机填料混合制成的，需要时再加入适量的塑料、树脂和橡胶等。

Asphalt glass fiber felt 沥青玻璃纤维毡

Asphalt solutions 沥青溶液

As-rolled(AR) 轧后(AR)状态 系指：(1)产品的供应可以在轧制工序后不进行任何热处理，轧制温度减少所造成颗粒大小的变化不受到严格控制；(2)将钢材在高温下轧制后冷却于空气中的轧制工艺。通常轧制和最终轧制是在大于正火温度和奥氏体再结晶温度区域内进行。在该工艺中生产出来的钢材的强度及韧性低于轧制后经热处理的钢材或经改进的工艺而生产出来的钢材。 As-rolled(AR) refers to: (1) High temperature treatment will not be performed after rolling of steel. The change of particle size due to temperature variation is not rigidly controlled; (2) this procedure involves the rolling of steel at high temperature followed by air cooling. The rolling and finishing temperatures are typically in the austenite recrystallization region and above the normalizing temperature. The strength and toughness properties of steel produced by this process are generally lower than those of steel heat treated after rolling or steel produced by advanced processes.

Assassin's mace/Maces 撒手锏 美国五角大楼将其定义为使一个处于劣势的军事集团与处于优势的强国对抗中占上风的技术。即在战争关键时刻，如果某一种武器能起到"一击制敌"的作用，那么它就可以被称为"撒手锏"。

Assemble 管理

Assembled crankshaft 组装式曲轴 非整体制造，而将曲臂、曲柄等部件组合装配成的曲轴称为组装式曲轴。

Assembling search engine 集合式搜索引擎 如howsou.com在2007年底推出的引擎。该引擎类似meta搜索引擎，但区别在于不是同时调用多个引擎进行搜索，而是由用户从提供的多个引擎当中选择，搜索用户需要的内容，因此叫它"集合式"搜索引擎更确切些。集合式搜索引擎的特点是可以集合众多搜索引擎的特点，对比搜索，更能准确地找到目标内容。

Assembly line for panels 平面分段流水线 系指一种典型的混合型装配流水线。随着现代造船模式的深入发展，平面分段流水线被越来越多地用于大中型船舶制造企业的生产中。

Assembly of hull sections 船体分段吊装 整个船体由分段组合而成，要进行安装就位时，就要采用分段吊装的方法，使其成形。见图A-17。

Assembly station 集合站(登乘站) 系指船舶在紧急情况下，能够使乘客集中并给予指令以及必要时准备弃船的地方。乘客处所可以用作集合站；只要这些处所能容纳所有乘客接受指令，并准备好弃船。 Assembly station is an area where passengers can be gathered in the event of an emergency, given instructions and prepared to abandon the craft, if necessary. The passenger spaces may serve as assembly stations if all passengers can be instructed there and prepared to abandon the craft.

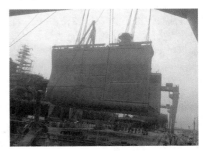

图 A-17 船体分段吊装
Figure A-17 Assembly of hull sections

Assessment of shipyard 造船厂评估 对于首次申请建造入级船舶的造船厂或首次建造入级的新船型的造船厂而言，系指验船师对造船厂的生产能力，包括生产场所、设施及造船厂的质量保证体系、施工人员的总体资质、分包方等各方面以及对即将建造船舶的适用性和有效性进行评估。 Assessment of shipyard: For a shipyard applying for building a ship or a novel ship classed with a society for the first time, the surveyor assesses the production capacity of the shipyard, covering locations and facilities of production, quality assurance system, competence of building personnel in general and subcontractors, and the applicability and effectiveness of the ship to be built.

Assigned freeboard 勘定干舷 系指从舯部甲板

线上缘向下垂直量至有关载重线上缘的距离。 Assigned freeboard is the distance measured vertical downwards amidships from the upper edge of the deck line to the upper edge of the related load line.

Assignee 受让人 系指接收"东西"那一方，A 有个东西，给了 B，那么 A 就是让与人，B 就是受让人。法律用语，相对出让人。出让人将自己的全部或者部分权利通过合同或者协议、赠予的方式转让给受让人。

Assignment 转让

Assistant engineer officer 助理轮机员 系指正在接受成为轮机员的培训并将由国家法律或法规指派为轮机员的人。 Assistant engineer officer means a person become an engineer officer is being trained and will be designatedk.

Assistant judge 助理审判员 系指由各级人民法院按照需要任免的协助审判员进行工作的人员。助理审判员，根据法官法规定助理审判员由本院院长任命，可以代行审判员职务。助理审判员享有和法官一样的权利与义务，助理审判员本身就是法官。助理审判员被法院任命后，按照《最高人民法院关于助理审判员可否作为合议庭成员并担任审判长问题的批复》，可以担任合议庭组成人员，并经过由院长或庭长指定也可以担任合议庭的审判长。助理审判员：三级法官至五级法官。

Assistant procurator 助理检察官 是检察官梦想的开始。成为助理检察员意味着成了庄严的检察官。利用法律武器将犯罪分子绳之以法。其工作任务是协助主办检察官办理相关检察业务，完成目标考核任务。协助主办检察官的工作包括:(1)协助主办检察官办理民事行政申诉案件;(2)协助主办检察官办理职务犯罪案件;(3)协助主办检察官对案件做出是否立案的决定;(4)协助主办检察官对立案案件进行审查、调阅审判卷宗、调查等工作,在规定期限审查终结,(5)协助制作《案件审结报告》《提请抗诉决定书》等法律文书;(6)协助主办检察官出席再审法庭;(7)协助主办检察官向人大报送重大民事行政案件;(8)负责本主办室各类法律文书的报审、送达;(9)协助主办检察官开展案件材料的移送和归档工作;(10)在办案过程中提出自己的意见,协助主办检察官,提高案件质量等。

Assisted craft 受援船 系指设计为已向船旗国和港口国做了具体说明的航线上营运的任一地效翼客船，并使其确信一旦在该航线任何地点出事而需撤离，有很大把握能在下述最短时间内将所有乘客和船员安全救出:(1)在最坏预计条件下,为保持救生艇筏中的人员免于因曝露而挨冻的时间;(2)与该航线所处的环境条件和地理特点相称的时间;(3)4 h。 Assisted craft is any passenger WIG craft designed for operation on a route where it has been demonstrated satisfaction to the flag and port States that there is a high probability that in the event of an evacuation at any point of the route all passengers and crew can be rescued safely within the least of:(1)the time to prevent persons in survival craft from exposure causing hypothermia in the worst intended conditions;(2)the time appropriate with respect to environmental conditions and geographical features of the route;(3)4 hours.

Associated apparatus 关联设备 系指电器设备电路或电路部件不必都是本质安全型的,但它包括能有效与电器设备连接的本质安全的安全的电路。

Associated piping 相关的管系 系指从货舱吸入点到岸接头用于卸货的管系,包括与卸货管路公开连接的船舶所有管系、泵和过滤器。 Associated piping means the pipeline from the suction point in a cargo tank to the shore connection used for unloading the cargo, it includes all ship's pipings, pumps and filters which are in open connection with the cargo unloading line.

Associated protective measure 相关保护措施 系指在国际海上组织权限范围内,为了保护危险区域而调节国际海上活动的规范或标准。 Associated protective measure means an international rule or standard that within the purview of IMO and by which international maritime activities are regulated for protection of the area at risk.

Associated resistance test 自航前阻力试验 自航试验前用带附体并装有螺旋桨轴的船模所进行的阻力试验。

Associated target 关联目标 (1)如果捕获的雷达目标和自动识别系统(AIS)报告目标有相似的符号联合编码要求的参数(例如位置、航向和速度),则这些目标被认为是同一目标并成为关联目标;(2)同时代表雷达跟踪目标和 AIS 目标的一个目标。雷达跟踪目标和 AIS 目标具有类似的参数(例如位置、航向、航速)并符合有关联合编码的要求。 Associated target means:(1)If an acquired radar target and an AIS reported target have similar parameters(e. g. position, course, speed)complying with an association algorithm, they are considered to be the same target, thus become an associated target;(2)a target simultaneously representing a tracked radar target and AIS target having similar parameters(e. g. position, course, speed)complying with an association algorithm.

Association of European Shipbuilders and Shiprepairers 西欧造船家协会

Association of Southeast Asian Nations(ASEAN) 东南亚国家联盟 简称东盟,也叫东南亚国家协会(东协)、亚细安组织(亚细安)及东南亚合作组织(东合),

是集合东南亚区域国家的一个政府性国际组织。预定 2015 年成立"东盟共同体"。东盟成立初期,基于冷战背景,主要任务之一为防止区域内共产主义势力扩张,合作侧重在军事安全与政治中立,冷战结束后各国政经情势趋稳。1997 年爆发亚洲金融风暴之后,东盟各国汲取教训,开始转向加强区域内经济、环保等领域的合作,并积极与区域外国家或组织展开对话与合作。东盟最知名的特点,就是在谈判协商时采取"东盟模式"(The ASEAN Way,或称亚洲方式),也就是对成员国内政、领土和主权采取不干涉的原则。

Assurance factor 安全系数 系指为避免事故发生的群体支付的意愿。

Astern gas turbine 倒车燃气轮机 系指在倒车航行时工作的燃气轮机。

Astern power(backing power) 倒车功率 主机倒车时发出的最大功率。

Astern power(performance)test 倒车性能试验 系指检查可逆转内燃机在规定的倒车工况下,能否连续运转一定时间的试验。

Astern power(backing power) 倒车功率 主机倒车时所发出的最大功率。

Astern stage 倒车级 使船舶获得后退动力的汽轮机级。

Astern turbine 倒车透平 系指在透平推进装置中使船舶作后退运动的透平缸。

Astern-exhaust chest sprayer 倒车排气室喷雾器 为防止汽轮机倒车时排气室温度过高而向排气室和冷凝器喉部喷水冷却的装置。

ASTM = American Society for Testing and Materials 美国材料与试验协会 其前身是国际材料试验协会(International Association for Testing Materials, IATM)。19 世纪 80 年代,为解决采购商与供货商在购销工业材料过程中产生的意见分歧,有人提出建立技术委员会制度,由技术委员会组织各方面的代表参加技术座谈会,讨论解决有关材料规范、试验程序等方面的争议问题。IATM 首次会议于 1882 年在欧洲召开,会上组成了工作委员会。

Asymmetric digital subscriber line(ADSL) 非对称数字用户环路 是一种新的数据传输方式。它因为上行和下行带宽不对称,因此称为非对称数字用户线环路。它采用频分复用技术把普通的电话线分成了电话、上行和下行 3 个相对独立的信道,从而避免了相互之间的干扰。即使边打电话边上网,也不会发生上网速率和通话质量下降的情况。通常 ADSL 在不影响正常电话通话的情况下可以提供最高 3.5 Mb/s 的上行速度和最高 24 Mb/s 的下行速度。

Asymptotically stable motion 渐近稳定运动 系指在受扰动完全消除后,经过一定的动荡仍能恢复其扰动前原平衡状态的船舶运动。

Asynchronous motor 异步电机 系指运行时的转速与其所接系统的频率不成正比例的一种电机。

Asynchronous transfer mode(ATM) 异步传输模式 是实现 B-ISDN 的业务的核心技术之一。ATM 是以信元为基础的一种分组交换和复用技术。它是一种为了多种业务设计的通用的面向连接的传输模式。它适用于局域网和广域网,它具有高速数据传输率和支持许多种类型如声音、数据、传真、实时视频、CD 质量音频和图像的通信。ATM 采用面向连接的传输方式,将数据分割成固定长度的信元,通过虚连接进行交换。ATM 集交换、复用、传输为一体,在复用上采用的是异步时分复用方式,通过信息的首部或标头来区分不同信道。

Asynchronous transmission 异步传输 系指将比特分成小组进行传送,小组可以是 8 位的 1 个字符或更长。发送方可以在任何时刻发送这些比特组,而接收方从不知道它们会在什么时候到达。一个常见的例子是计算机键盘与主机的通信。按下一个字母键、数字键或特殊字符键,就发送一个 8 比特位的 ASCII 代码。键盘可以在任何时刻发送代码,这取决于用户的输入速度,内部的硬件必须能够在任何时刻接收一个键入的字符。ATM 能够比较理想地实现各种 QoS,既能够支持有连接的业务,又能支持无连接的业务。是宽带 ISDN(B-ISDN)技术的典范。ATM 为一种交换技术,在传送资料时,先将数位资料切割成多个固定长度的封包,之后利用光纤或 DS1/DS3 传送。到达目的地后,再重新组合。ATM 网络可同时将声音、影像及资料整合在一起。针对各种资讯型态,提供最佳的传输环境。

At definite depth 定深扫海测量 系指扫海具的底索在深度基准面下保持一定深度的扫海测量。

Atactic polypropylene(APP/PP) 无规聚丙烯 是一种热塑性树脂。根据分子结构的不同,有无规聚丙烯、等规聚丙烯和间规聚丙烯 3 种。工业生产的等规聚丙烯为无色、无臭、无味的固体。相对密度 0.90~0.91。耐热性高,使用温度范围——30~140℃。韧性和耐化学腐蚀性都很好。缺点是耐低温冲击性差,较易老化,是一种通用塑料。

Athwartship thruster 侧推器 使船舶产生横向运动的推力器称为侧推器。见图 A-18。

Athwartships bow propeller 艏横向推力器 装于船舶艏部的侧推器亦称艏横向推力器。

ATM Wireless Access Communication System

图 A-18 侧推器
Figure A-18 Athwartship thruster

(AWACS) ATM 无线接入通信系统 第三代移动通信系统(IMT-2000),在第二代移动通信技术基础上进一步改进的以宽带 CDMA 技术为主,并能同时提供话音和数据业务的移动通信系统亦即未来移动通信系统,是一代有能力彻底解决第一、二代移动通信系统主要弊端的最先进的移动通信系统。第三代移动通信系统一个突出特色就是,要在未来移动通信系统中实现个人终端用户能够在全球范围内的任何时间、任何地点,与任何人,用任意方式、高质量地完成任何信息之间的移动通信与传输。可见,第三代移动通信十分重视个人在通信系统中的自主因素,突出了个人在通信系统中的主要地位,所以又称为未来个人通信系统。

Atmosphere noise 空气噪声

Atmospheric condenser (barometric condenser) 大气冷凝器 系指回收蒸汽辅机排气或废气锅炉多余蒸汽等的无真空度的冷凝装置。

Atmospheric drain(collecting) tank 大气泄水柜 聚集生活杂用、舱柜加热蒸汽等凝水的开式水柜。

Atomic battery 核电池 又称"放射性同位素电池"。它主要通过半导体换能器将同位素在衰变过程中不断放出的具有热能的射线转变为电能。而热源只提供热能。核电池体积小,硬币大小就可以具有普通化学电池上百万倍的效能。

Atomic spectrum analysis 原子光谱分析 系指通过测定润滑油中金属及非金属元素的种类、浓度及变化趋势,判断机械摩擦副的磨损状况及润滑油状态的方法。Atomic spectrum analysis is a method which is capable of judging the wear condition of machinery friction pairs and of lubricating oil by detecting the kinds, concentration and variation trend of metallic and non-metallic elements in the lubricating oil.

Atomizer 雾化器 将射出或喷出的液体达到雾化状态的器具称为雾化器。

Atriums 天井 系指在单一的主竖区内跨越三层或以上开敞甲板的公共处所。 Atriums are public spaces within a single main vertical zone spanning three or more open decks.

Attached cavities 附着空泡 与物体有明显接触线的空泡区。

Attached laboratory (auxiliary laboratory) 附属实验室 又称辅助实验室。为配合实验室进行试验、观测、记录和分析等科学研究工作而设置的相应辅助性工作舱室。

Attached pump 机带泵 由原动机直接或经齿轮或皮带驱动的并直接装于原动机机身上的泵称为机带泵。

Attaching organism research 附着生物调查 对海中物体表面上附着生活的动物、植物和微生物所进行的专题调查。内容一般有种类组成、附着量、季节变化与环境之间的关系。

Attachment 附件

Attachment of boiler and pressure vessel 锅炉及压力容器的附件 (1)直接与锅炉、锅炉附属装置及压力容器相连的法兰、竖管及隔片;(2)直接与锅炉、锅炉附属装置及压力容器相连的阀门。 Attachment of boiler and pressure vessel means the following:(1)Flanges, stand pipes and distance pieces attached directly to boilers, equipment subject to a boiler, and pressure vessels;(2)Valves attached directly to boilers, equipment subject to a boiler, and pressure vessels.

Attained EEDI 获得的船舶能效设计指数 系指单艘船按有关要求获得的 EEDI 值。可按下式求得:Attained EEDI ≤ Required EEDI = $\left(1 - \dfrac{X}{100}\right) \times$ Reference line value 式中,Attained EEDI—根据计算公式得到的 EEDI 值;Required EEDI—所要求的 EEDI 标准值;X—某阶段的折减系数;Reference line value—参考线值。EEDI 折减系数及强制执行时间见表 A-6。

表 A-6 EEDI 折减系数及强制执行时间表

船型	尺寸	阶段 0 2013.1.1~ 2014.12.31	阶段 1 2015.1.1~ 2019.12.31	阶段 2 2020.1.1~ 2024.12.31	阶段 3 2025.1.1 以后
散货船	20 000 DWT 以上	0	10	20	30
散货船	10 000~20 000 DWT	—	0~10*	0~20*	0~30*
气体运输船	10 000 DWT 以上	0	10	20	30
气体运输船	2 000~10 000 DWT	—	0~10*	0~20*	0~30*
油船	20 000 DWT 以上	0	10	20	30
油船	4 000~20 000 DWT	—	0~10*	0~20*	0~30*
集装箱船	15 000 DWT 以上	0	10	20	30
集装箱船	3 000~15 000 DWT	—	0~10*	0~20*	0~30*
常规货船	15 000 DWT 以上	0	10	15	30
常规货船	3 000~15 000 DWT	—	0~10*	0~15*	0~30*
冷藏船	5 000 DWT 以上	0	10	15	30
冷藏船	3 000~5 000 DWT	—	0~10*	0~15*	0~30*
兼装船	20 000 DWT 以上	0	10	20	30
兼装船	4 000~20 000 DWT	—	0~10*	0~20*	0~30*

* 该折减系数值是根据船舶载重量与折减系数的相对比例进行线性插值求得。

Attempted stowaway 偷渡未遂者 对防止偷渡者进入和寻求成功解决偷渡案件责任分配指南而言,系指未经船长或船东或任何其他负责人员同意,藏匿于船上或之后装载于船上的货物中,于船舶离港前在船上被发现的人员。 For Guidelines on the prevention of stowaway incidents and the allocation of responsibilities to seek the successful resolution of stowaway cases, attempted stowaway means a person who is hided secretly on a ship, or in cargo which is subsequently loaded on the ship, without the consent of the ship-owner or the master or any other responsible person, and who is found on board the ship before it has departed from the port.

Attenuator 衰减器 系指一种提供衰减的电子元器件,广泛地应用于电子设备中,它的主要用途是:(1)调整电路中信号的大小;(2)在比较法测量电路中,可用来直读被测网络的衰减值;(3)改善阻抗匹配,若某些电路要求有一个比较稳定的负载阻抗时,则可在此电路与实际负载阻抗之间插入一个衰减器,能够缓冲阻抗的变化。衰减器的工作频带是指在给定频率范围内使用衰减器,衰减器才能达到指标值。衰减器是一种能量消耗元件,功率消耗后变成热量。横向位移型光衰减器是一种比较传统的方法,由于横向位移参数的数量级均在微米级,所以一般不用来制作可变衰减器,仅用于固定衰减器的制作中,并采用熔接或黏接法,到目前仍有较大的市场,其优点在于回波损耗高,一般都大于 60 dB。

Attorney 律师所代理人

A-type hanging bracker and towing windlass aft 船艉 A 型吊架及拖缆绞车 系指一种拖缆机与门架吊机是组合式设备。属于海上装备中的起重及收放设备,主要安装在各种海洋工程船舶的艉部,用于水下设施如海缆埋设机、水下机器人的吊放、拖曳和回收,是海缆敷设、打捞系统的重要配套设备。其由 A 型吊架、伸缩吊具、摩擦绞车、储缆绞车、液压泵站、机旁控制台、遥控操作盒等组成。具有索具收放、滚轮吊放、吊架伸缩、横移、摆动、琐销插放、自动排绳、恒张力控制、紧急弃绳功能,可在 4 级海况、100 m 水深条件以及布缆船动力定位模式下进行精确作业。

A-type independent tank A 型独立液舱 此类液舱为平面重力式结构,设计蒸发气压力不得超过 0.07 MPa 须在大气压和接近大气压下采用全冷方式运输液货。

A-type stem　A型艏柱　系指边缘尖锐的艏柱。
A-type stem means a sharp edged stem.

Auction　拍卖　是一种具有悠久历史的交易方式，在今天的国际贸易中仍被使用。拍卖是由经营拍卖业务的拍卖行接受货主的委托，在规定的时间和场所，按照一定的章程和规则，以公开叫价的方法，把货物卖给出价最高的买主的一种贸易方式。

Auctioneer　拍卖人　系指依照《中华人民共和国拍卖法》和《中华人民共和国公司法》设立的从事拍卖活动的企业法人。

Audibility sound　可听声　系指能引起听觉的声波。一般简称声或声音。可听声的频率范围大致为20 Hz ~ 20 kHz。

Audible indication　听觉显示　系指在发出信号的场所能够察觉的声响信号。　Audible indication means an audible signal that is detectable at the location where it is signaled.

Audio frequency sound　音频声　系指人耳能听到的频率范围为 $f = 20 \sim 20\,000$ Hz 内的声音。

Audio guide　声讯服务

Audiovisual material　视听资料　系指以录音磁带、录像带、电影胶片或电子计算机相关设备存储的作为证明案件事实的音响、活动影像和图形，称称为"视听资料"。视听资料，既能使文字记载的文献再现，又能脱离文字形式而直接记录各种声音与图像；既能反映静态的书面文献，又有充分发挥其动态的特殊效果，具有以声传情、形象逼真、声像并茂的特色；尤其是可以运用放大或缩小、加速或减慢、剪辑合成等手法，其作用为一般传统印刷型出版物所无法比拟。视听资料，还便于运用现代通信技术及时地迅速传播。

Audit　审核/审计　系指确定质量活动及其相关结果是否符合计划安排，以及这些安排是否有效实施，并适合于达到预定的目标的系统和独立的检查。　Audit means a systematic and independent examination to determine whether quality activities and related results comply with planned arrangements and whether these arrangements are implemented effectively and is suitable to achieve the stated objectives.

Audit checklist　审核检查表　系指在指定范围内规定应进行审核项目的清单。　Audit checklist means a listing of specific items within a given area that are to be audited.

Audit for issuing interim DOC　签署临时DOC的审核　系指由审核机构派员到被审核的船舶管理公司进行的审核，并签署临时的DOC证书。

Audit of the manufacturing management system　产品制造管理体系审核　系指对制造厂商的产品质量保证和控制体系进行评价，以验证产品质量符合持续满足其规定的质量水平和规范要求的能力。根据制造厂商的产品质量保证和控制体系的类别，形式认可可分为如下两个模式：(1) 形式认可B：制造厂商具备适宜的产品生产和测试设备，并建立有效的质量控制制度。(2) 形式认可A：除满足形式认可B的要求外，制造厂商应建立和实施一个至少符合ISO 9000标准或等效标准的质量管理管理体系并具有船级社批准的按规范要求进行检验和试验程序。　Audit of the manufacturing management system means that the quality assurance and control system of the manufacturer is assessed to verify his ability to meet the specified level of product quality and a society's rules consistently. Depending on the types of manufacturers' quality assurance and control systems, the type approval is divided into the following modes：(1) Type approval B：The manufacturer has appropriate production and test equipment and has established an effective quality control system；(2) Type approval A：In addition to complying with the requirements for type approval B, the manufacturer should establish and implement a quality management system complying at least with ISO 9000 or another equivalent standard and has inspection and test procedures approved by a society according to its rules.

Audit of the manufacturing process　产品制造过程审核　系指对制造厂商的产品制造过程进行评价，以验证产品的制造工艺和检验计划适合于制造厂商规定的质量控制水平，并满足规范的要求。　Audit of the manufacturing process means a manufacturing process of the manufacturer is assessed to verify that the production technology and inspection scheme are suitable for the quality control level specified by the manufacturer and comply with the rules.

Audit plan　审核计划

Audit program　审核程序

Auditer　被审核者

Auditor　审核员　系指负责审核工作的人员。

Auditory impedance　声阻抗　系指媒质中声的吸收，等于界面声压与通过该面的声通量(质点流速或体速度乘以面积)之比。

Auditory threshold　听阈　系指人耳刚好能感觉到其存在的声音时的声压，听阈对于不同频率的声波是不相同的。人耳对1 000 Hz的声音感觉最灵敏，其听阈声压为 $P_0 = 2 \times 10^{-5}$ Pa(称为基准声压)。

Auditory threshold sound pressure　听阈声压　系

指人耳听到的最弱的声压(2×10^{-5} Pa)。

Augment of resistance 阻力增额 船在一定航速时由于螺旋桨作用引起的船舶阻力增加，即螺旋桨推力与船的总阻力之差。

Augmented stabilization 增稳 系指自稳和强制相结合的一种稳定。Augmented stabilization is a combination of self-stabilization and forced stabilization.

Aurora observation 极光观测 系指研究极地上空极光各种现象与地磁变化和太阳活动的关系。

Australia ladder 澳大利亚货舱梯 澳大利亚港口当局对船舶要求在其货舱内设置的梯子。

Australian Maritime Safety Authority (AMSA) 澳大利亚海事安全局 系指澳大利亚负责海事安全管理的机构。Australian Maritime Safety Authority (AMSA) means the Australian Marine Safety Authority.

AUT-0 机器处所周期性无人值班 主推进装置由驾驶室控制站遥控、机器处所包括机舱集控站(室)周期无人值班。在无人值班周期自动化系统应能保证下列机电设备连续正常运行：(1)主推进装置，包括主机(如柴油机、主汽轮机、主燃气轮机和电力推进装置等)、减振器(如有时)、传动装置(如离合器、减速齿轮箱等)以及推进器(如可调螺距螺旋桨等)；(2)为主推进装置服务的重要机械；(3)主、辅锅炉；(4)电站；(5)其他机电设备，如空气压缩机、舱底水系统(包括油水分离器)，与遥控有关的阀、燃油系统以及必要进行自动化监控的其他机电设备。

Author of the invention 发明人 系指发明创造的实质性特点做出创造性贡献的人。发明人应当使用本人真实姓名，不得使用笔名或者其他非正式的姓名。申请人改正请求书中所填写的发明人姓名的，应当提交补正书、当事人的声明及相应的证明文件。不公布姓名的请求提出之后，经审查认为符合规定的，专利局在专利公报、专利申请单行本、专利单行本以及专利证书中均不公布其姓名，并在相应位置注明"请求不公布姓名"字样，发明人也不得再请求重新公布其姓名。

Author patent 发明专利 是专利的一个种类。中国专利法规定可以获得专利保护的发明创造有发明、实用新型和外观设计3种，其中发明专利是最主要的一种。发明是发明人运用自然规律而提出解决某一特定问题的技术方案。所以中国专利法实施细则中指出"专利法所称的发明系指对产品、方法或其改进所提出的新的技术方案"。发明人只有将这种技术方案向专利局提出申请，并且通过一系列严格的审查，特别是新颖性、创造性和实用性的审查。申报发明专利的风险是比较大的，一定要做好充分的思想准备(发明专利授权率在40%~50%)。

Auto-correlation function 自相关函数 一个随机变量在时间 t 时的瞬时值与在时间 $(t+\tau)$ 时的瞬时值乘积的时间平均数。其表达式为 $R(\tau) = \lim(1/T) \times \int_0^T \xi(i)\xi(t+\tau)\mathrm{d}t$，$T \to \infty$。

Auto pilot 自动操舵仪 又称自适应舵，简称自舵。是根据指令信号自动完成操纵舵机，以使船舶能够保持在预定航线上稳定航行的设备。它通过电罗经不断地把传送来的船舶实际航向与给定的航向及舵角大小相比较，以获得控制信号来控制舵机适当转舵，使船舶自动保持在给定的航向上航行。自动保持给定航向有两个含义：一种是出航前将整个航线拟定后存入微机中，船舶航行时计算机按照航线控制自动舵自动改变航向，完成整个航程。另一种是航向的改变随时由驾驶人员给定，在没有新的航向信息输入之前，自动舵将保持船舶在原来给定的航向上航行，又称航向自动舵。自动操舵有两种工况：(1)自动稳定航向；(2)改变航向。自动操舵仪与 ECDIS 相结合，可实现航迹控制、在航路点(WP)处自动转向，在偏离航迹时，自动控制船舶回到设定的航迹。自动操舵仪已经历数代变化：第一代是机械式主动舵；第二代是比例-微分-积分控制自动舵；第三代是自适应自动舵；第四代是航迹舵(智能自动舵)。

Auto side blocks 自动边墩 系指放置在船坞的内侧、能自动调节高度的坞墩。

Automated pronunciation instructor (API) 自动发音教学机

Automatic 自动 在 GTAW 方法中，系指焊炬及分开的焊丝均由全机械控制和操作的工艺。In the GTAW process, automatic refers to the fully mechanized control and application of both torch and separate filler wire.

Automatic boiler control system 锅炉自动控制装置 保证锅炉在各种负荷工况下维持给定参数，经济而可靠运行的自动控制装置。

Automatic combustion regulating system 燃烧自动调节装置 为使锅炉在各种负荷工况下维持在规定蒸汽参数下运行，并保证以最适当的比例向锅炉供给必需的燃料和空气量而设置的自动调节装置。

Automatic control 自动控制 系指：(1)是一种不需人直接或间接介入，对预定状态进行操作的控制；(2)具有自动调节特性，无需操作人员参与即可按预定指令对设备进行操作的控制。Automatic control is: (1) the control of an operation without direct or indirect human intervention, in response to the occurrence of predetermined conditions; (2) self-regulating control carrying out ordered instruction to operate the machinery without action by an operator.

Automatic control functions 自动控制功能 包括舵向、和/或航速自动控制的功能,或其他航行相关自动控制功能。 Automatic control functions include these functions, such as automatic heading, and/or track and/or speed control or other navigation related automatic control functions.

Automatic control mode(DPS) 自动控制方式(动力定位系统) 系指包括位置和艏向控制,能独立地选择位置和艏向设置点的控制方式。 Automatic control mode(DPS) means a control mode which includes control of position and heading, set-points for control of position and heading is to be independently selectable.

Automatic control system 自动控制系统 系指无需操作员输入就能保持船舶的艏向和/或飞高的系统。 Automatic control system is a system which enables the craft's heading and/or altitude to be maintained without operators input.

Automatic expansion valve 恒压膨胀阀(自动膨胀阀) 制冷装置中使制冷剂维持恒压的膨胀阀称为恒压膨胀阀或自动膨胀阀。

Automatic feed water regulating system 给水自动调节装置 系指自动调节给水流量的装置。

Automatic fire damper 自动挡火闸 在失火时能自动关闭通道防止火焰蔓延和扩散的钢制闸板。

Automatic fire-extinguishing sprinkler system 自动喷水灭火系统 在失火时能根据升高的温度自动开启喷头使管道内的压力水喷出进行灭火的系统。

Automatic ignition system 自动点火系统

Automatic isolating arrangement 自动隔断装置

Automatic level control system 水位自控系统 系指锅炉或容器内的水位能利用给水泵等附属设备使其保持在上下水位限值之间的系统。

Automatic load shedding 自动卸载 系指在任何一台发电机过载时自动地将非重要负载卸除。根据发电机组的过载能力,此种卸载可分为一级或多级进行。 Automatic load shedding means when any one of the generators is overload, the excess non-essential load is disconnected automatically. This load shedding may be carried out in one or more stages according to the overload ability of the generating set.

Automatic lubricators 自动滑油注油器 系指内燃机的气缸与活塞之间所需的滑油能随活塞的运动自动同步注入的滑油加油器。

Automatic mooring winch 自动系泊绞车 系指在一定缆绳长度范围内,能自动调整系泊缆绳张力的绞车。见图 A-19。

图 A-19 自动系泊绞车
Figure A-19 Automatic mooring winch

Automatic pilot 自动舵 系指根据航向信号操纵方向舵,使船舶自动保持在给定航向上航行的操舵设备。为适应船的性能和气候条件,自动舵有:天气、舵角比例、压舵、反作用舵等调节环节。自动舵的参数要在试验中调整和确定。

Automatic power management system(DPS) 自动功率管理系统(动力定位系统) 系指能使电机随负荷的变动而启动或停止的系统。当没有足够的功率启动大功率的负载时,应阻止大功率设备的启动,并按要求启动备用发电机,然后再启动所需的负载。 Automatic power management system(DPS) means a system which can perform load dependent starting of additional generators, and can also conduct load dependent stop of running generators. The system is to block starting of large consumers when there is not adequate running generator capacity, and to start up back-up generators as required, and hence to permit required consumers start to proceed.

Automatic racking system 自动送钻杆辊道 钻井系统在能自动运送和水平排放钻杆的装置。

Automatic radar plotting aid(ARPA) 自动雷达标绘仪 系指能够使观测者自动获取信息,并能自动处理和标绘多处物标的系统,并通过自动标绘周围目标的方位、航向及速度来达到避碰目的。它不仅能够提供连续、准确和迅速的航行形势估计,而且能显示船舶周围的事态,确保船舶安全航行。该系统采用计算机技术控制整个系统,与导航雷达配套作为显示器和分显示器。它具有多批目标录取及自动跟踪,显示目标运动的轨迹,对危险目标自动发出告警信号,并能自动校准、自动调节和内部故障自检及指示。

Automatic shutdown system 自动关闭系统

Automatic smoke type fire alarm[1] 测烟式失火自动报警器 系指依靠测量失火区发烟程度作为失火信号源的自动报警器。

Automatic smoke type fire alarm[2] 感烟式自动失火报警器

Automatic sprinkler 自动洒水器

Automatic sprinkler fire extinguishing system 自动喷水灭火系统 在可能发生火灾的处所,为防止发生火灾而设置的一种达到规定温度后,喷水器自动喷水的灭火系统。包括压力水柜、自动启动的供水泵、失火报警设备、自动喷水器等。

Automatic sprinkler system 自动喷水器系统

Automatic sprinkler, fire detection and fire alarm systems 自动喷水器、探火及失火报警系统

Automatic sprinkling system 自动喷水灭火系统

Automatic starter 自动启动器 系指使其完成动作的作用是自动施加的一种启动器。

Automatic start-up equipment 自动启动装置

Automatic telephone 自动电话 系指用于各舱室日常工作和事务直接的通信联络的电话。

Automatic teller machine (ATM) 自动柜员机 系指银行在不同地点设置一种小型机器,利用一张信用卡大小的胶卡上的磁带记录客户的基本户口资料(通常就是银行卡),让客户可以透过机器进行提款、存款、转账等银行柜台服务,现在大多数客户都把这种自助机器称为提款机。

Automatic temperature control system 温度自控系统

Automatic temperature regulating equipment 自动温度调节装置

Automatic tensioner 自动张紧器

Automatic thermal type fire alarm 感温式自动失火报警器

Automatic thruster control (DPS) 推力器自动控制(动力定位系统) 系指由计算机系统组成,包括一台或多台带有处理装置、输入/输出设备和存储器的计算机。Automatic thruster control (DPS) consists of a computer system, comprising one and more computers with processing units, input/output devices, and memory.

Automatic towing winch (constant tension winch) 自动拖缆机 系指具有自动调整或维持给定的拖缆张力装置的拖缆机。

Automatic watch keeping equipment 自动监测装置

Automatic welding 自动焊接 系指诸如埋弧焊或气-电立焊等,焊接和焊接材料的供应均为自动操作。Automatic welding is used to describe processes in which the weld is made automatically by a welder using a continuously fed electrode wire such as submerged arc welding or electro-slag welding, etc.

Automatically controlled valve 自动控制阀

Automatically controller door 自动控制门 系指利用光电感应等装置,当人与物接近和离开时能自动启闭的门。

Automatically inflated PFD 自动气胀式个人漂浮装置 系指由浸水引起气体膨胀的,使用者在浸水时不必进行任何动作的个人漂浮装置。Automatically inflated PFD means that in which inflation occurs as a result of immersion without the user carrying out any action at the time of immersion.

Automation systems 自动化系统 包含有控制系统和监测系统的系统。Automation systems are those systems including control systems and monitoring systems.

Automobile carrier 汽车运输船 系指专运商品汽车的船舶。见图 A-20。

图 A-20 汽车运输船
Figure A-20 Automobile carrier

Automobile ferry 汽车渡船 系指专运汽车的渡船。

Automotive gasdine 车用汽油

Autonomous and remotely operated underwater vehicle (ARV) 遥控自主水下机器人 自带能源并通过光纤与母船联系,其综合了 ROV 和 AUV 的优点,可在复杂环境下实现自主、遥控和监控三种模式下的中等范围搜索,定点观测以及水下轻作业的机器人。在遥控模式下,水面人员可通过微细光纤操控 ARV 进行作业,此时的 ARV 相当于一台传统的轻作业型 ROV;在预编程工作模式下,ARV 不需要携带微细光缆,可以进行自主作业;而在监控模式下,水面人员可通过微细光缆对预编程 ARV 的作业任务或过程进行人工干预。

Autonomous quotas 自主配额 又称单方面配额,是由进口国家完全自主地,单方面强制规定在一定时期内从某个国家或地区进口某种商品的配额。这种配额不需征求输出国家的同意。

Autonomous underwater vehicle (AUV) 无缆直治

水下机器人　不设脐带缆，自带电源，依靠其本身的自治能力进行作业的机器人。与ROV相比AUV具有活动范围大、潜水深度，不怕电缆纠缠，不需要复杂水面支持系统等特点。在军用方面，被称作海军的"力量倍增器"。可作为潜艇自卫和进攻的手段，也可作为潜艇远距离航行时的通信中继站，提高潜艇的生存能力同时还能作为反水雷工具，深入敌水雷区绘制雷区图，以引导己方舰艇安全通过。但是这对智能控制、导航定位、决策规划、目标探测、水下通讯等方面提出了更高的要求。

Autonomy　自治

Auto-pilot　自动操舵控制　俗称自动舵。是一个自动调节系统，其特点是根据船舶航向的变化自动控制舵机偏舵，以保持某一航向的航行。自动舵主要用于现代化的远洋航行的船舶或长途海运船舶和高速快艇上。

Autopilot system　罗经自动操舵仪　系指能自动控制舵机，以保持船舶按规定航向航行的设备。又称自动操舵装置。它是在通常的操舵装置上加装自动控制部分而成。其工作原理是：根据罗经显示的船舶航向和规定的航向比较后所得的航向误差信号，即偏航信号，控制舵机转动舵，并产生合适的偏舵角，使船在舵的作用下，转向规定的航向。自动操舵仪具有自动操舵和手动操舵两种工作方式。船舶在大海中直线航行时，采用自动操舵方式，可减轻舵工劳动强度和提高航向保持的精度，从而相应缩短航行时间和节省能源；船舶在能见度不良或进出港时，采用手动操舵方式，具有灵活、机动的特点。见图A-21。

图A-21　罗经自动操舵仪
Figure A-21　Autopilot system

Autotransformer starter　自耦变压器式启动器　系指具有一台自耦变压器，在启动电动机时能提供已降低电压的一种启动器。这种启动器具有所需要的换接机构，通常称为补偿器或自动启动器。

Auxiliary air compressor　辅空压机

Auxiliary apparatus　辅助设备

Auxiliary boiler room　辅助锅炉舱

Auxiliary boiler/Donkey boiler　辅锅炉　系指向辅助机械、设备蒸汽和为生活杂用供蒸汽或热水的锅炉。

Auxiliary boilers for essential services　重要用途辅锅炉　系指：(1)为船舶安全航行服务的辅机供应蒸汽的辅助锅炉；(2)锅炉产生的蒸汽供给对平台在海上安全作业起重要作用的辅助机械，但不供给主推进机械的锅炉，如蒸汽用来加热高黏度的燃油。 Auxiliary boilers for essential services are: (1) those for the purpose of supplying steam to auxiliary machineries as to ensure safe navigation of ships; (2) boilers supplying steam for auxiliary machineries essential for the safety or the operation of the unit at sea, but not for main propulsion engines, such as for heating high-viscosity fuel oil.

Auxiliary boilers for essential services (auxiliary boilers, exhaust boilers, economizers and steam-heated steam generators)　重要用途的所有其他锅炉(辅锅炉、废气锅炉、经济器和蒸汽加热蒸汽发生器)　系指锅炉产生的蒸汽供给船舶在海上安全作业起重要作用的辅助机械，如蒸汽供给加热为柴油机工作的高黏度的燃油，但不供给主推进机械的锅炉。 Auxiliary boilers for essential services (auxiliary boilers, exhaust boilers, economizers and steam-heated steam generators) are boilers supplying steam for auxiliary machineries essential for the safety or the operation of the ship at sea, such as for heating high-viscosity fuel used for operation of diesel engines, but not for main propulsion engines.

Auxiliary boilers for nonessential services　非重要用途辅助锅炉　系指产生的蒸汽不是供给船舶或平台在海上安全作业所必需的锅炉，如生活锅炉。 Auxiliary boilers for nonessential services are boilers supplying steam not exclusively for the safety or the operation of the ship or unit at sea, such as domestic boilers.

Auxiliary bucket ladder　辅助斗桥　复式斗桥中位于上面的一个较短小的起辅助作用的斗桥。

Auxiliary condenser　辅冷凝器　冷凝辅汽轮机或蒸汽辅机排气的冷凝器。

Auxiliary condenser circulating pump　辅冷凝器循环泵　系指抽送蒸汽动力装置中辅冷凝器循环冷却水的泵。

Auxiliary control device　辅助控制装置　对73/78防污公约附则Ⅵ而言，系指船用柴油机上安装的用于保护柴油机和/或其辅助设备不受可导致其损坏或故障的操作条件的影响或有助于柴油机启动的系统、功能或控制策略。辅助控制装置也可以是业已证明为非抑制装置的策略或措施。 For the MARPOL73/78 Annex Ⅵ, auxiliary control device means a system, function, or control strategy installed on a marine diesel engine that is used to protect the engine and/or its ancillary equipment against operating conditions that could result in damage or failure, or that is used to facilitate the starting of the engine. An auxiliary control device may also be a strategy or measure that has been satisfactorily demonstrated not to be a defeat device.

Auxiliary control station 辅助控制站 系指除主控制站(中央控制站/室)和就地控制站以外的其他控制站。 Auxiliary control station means control stations other than the above-mentioned main control station (centralized control station/room) and local control station.

Auxiliary device 辅助装置(附加设备)

Auxiliary diesel 辅柴油机

Auxiliary engine 辅机(辅助机械) 带动辅机用的发动机,通常系指发电机组。

Auxiliary engine room 辅机舱 通常系指发电机舱。

Auxiliary engine/Auxiliary machinery 辅机/辅助机械 系指:(1)主推进机械以外的机械,在船舶运行时它们在使用中,例如辅柴油机、涡轮发电机、液压马达和液压泵、压缩机、锅炉通风机及齿轮泵。(2)船上除主机、主锅炉以外所有机械设备的总称。大多是由工作机械与原动机两部分联合组成形式。原动机可用电动机、汽轮机、蒸汽机、液压马达等。有些小型或应急用的辅机也可用人力驱动。

Auxiliary equipment 辅助装置(附加设备)

Auxiliary equipment of air line 辅助通风管路

Auxiliary feed pipe line 辅给水管路 蒸汽动力装置中,当主锅炉给水管路损坏或发生故障时应急代替用的给水管路。

Auxiliary gas turbine 辅燃气轮机 系指驱动各种辅机的燃气轮机。

Auxiliary generator 辅助发电机 即非主发电机,又非应急发电机,则谓之辅助发电机。

Auxiliary hook load 辅钩起重量 系指起重机副钩的额定起重能力(一般为主钩起重量的25%)。

Auxiliary machinery 辅机(辅助机械) 系指在船舶运行时在使用中的,主推进机械以外的机械,例如辅柴油机、涡轮发电机、液压马达和液压泵、压缩机、锅炉通风机及齿轮泵。 Auxiliary machinery is machinery other than main propulsion machinery that is in service when the ship is in normal service, e.g. auxiliary diesel engine, turbo-generators, hydraulic motors and pumps, compressors, boiler ventilation fans, gear pumps.

Auxiliary machinery compartment 辅机舱 系指设置辅助机械的舱室。

Auxiliary machinery spaces 辅机处所 系指设有驱动发电机的输出功率为 110 kW 及以下的内燃机、水喷淋器、消防泵、舱底泵等、加油站、总功率超过 800 kW 的配电板的诸处所;类似处所;以及处所的围壁通道。 Auxiliary machinery spaces are spaces containing internal combustion engines of power output up to 110 kW driving generators, sprinkler, drencher or fire pumps, bilge pumps, etc. oil filling stations, switchboards of aggregate capacity exceeding 800 kW, similar spaces and trunks to such spaces.

Auxiliary machinery spaces having little or no fire risk 无失火危险或失火危险较小的辅机处所 系指设置冷藏、稳定、通风和空调机械、总功率 800 kW 及以下配电板的诸处所、类似处所及通往这些处所的围壁通道。 Auxiliary machinery spaces having little or no fire risk are spaces such as refrigerating, stabilizing, ventilation and air conditioning machinery, switchboards of aggregate capacity 800 kW or less, similar spaces and trunks to such spaces.

Auxiliary machinery spaces, cargo spaces, cargo and other oil tanks and other similar spaces of moderate fire risk 具有中等失火危险的辅机处所、货物处所、货油舱和其他油舱以及其他类似处所 系指:(1)货油舱;(2)货舱、货舱围壁通道及舱口;(3)冷藏舱;(4)燃油舱(设在无机器的单独处所内);(5)允许储存可燃物的轴隧和管隧;(6)其内设置具有压力润滑系统的机器或允许储藏可燃物品的机器辅机处所;(7)燃油加油站;(8)设有浸油式电力变压器(10 kVA 以上)的处所;(9)设有由涡轮机及往复式蒸汽机驱动的辅助发电机、由输出功率为 110 kW 及以下的小内燃机驱动的发电机、喷水器、水幕喷头泵或消防泵、舱底泵等处所;(10)用于上述处所的封闭围阱。 Auxiliary machinery spaces, cargo spaces, cargo and other oil tanks and other similar spaces of moderate fire risk are:(1) cargo oil tanks;(2) cargo holds; trunkways and hatchways;(3) refrigerated chambers;(4) oil fuel tanks(which are installed in a separate space with no machinery);(5) shaft alleys and pipe tunnels allowing storage of combustibles;(6) auxiliary machinery spaces which contain machinery having a pressure lubrication system or where storage of combustibles is permitted.(7) oil fuel filling stations;(8) spaces containing oil-filled electrical transformers(above 10 kVA).(9) spaces containing turbine and reciprocating steam engine driven auxiliary generators and small internal combustion engines of power output up to 110 kW driving generators, sprinkler, drencher or fire pumps, bilge pumps, etc.;(10)closed trunks serving the spaces listed above.

Auxiliary means 辅助工具(辅助设备)

Auxiliary output gear (accessory output gear) 辅助输出装置 通常从船用主减速齿轮箱中分出传动轴以驱动发电机、水泵等辅机的齿轮传动设备。

Auxiliary propeller 辅助推进器

Auxiliary propelling/maneuvering unit 辅助推进/操纵装置 系指非航行用途的,仅用作局部调整作业船位使用的装置。 Auxiliary propelling/maneuvering unit means an arrangements which are intended not for navigation purposes, but only for locally adjusting operation position of the ship.

Auxiliary propulsion machinery 辅助推进装置 (辅助推进机械) 系指主推进机械以外的推进机械,往往起冗余的作用。

Auxiliary propulsion system 辅助推进系统 系指主推进系统以外的推进系统,往往起冗余的作用。

Auxiliary pump(donkey pump) 辅泵 系指作为待用和起辅助作用的泵。如给排水泵、辅冷却水泵等。

Auxiliary room 辅机舱 系指专供安置辅机用的舱室。

Auxiliary steam pipeline 辅蒸汽管系 锅炉蒸汽输送到辅机、甲板机械以及供其他用途等的管路及附件。

Auxiliary steam stop valve 辅蒸汽阀 系指装在锅炉的上锅筒、过热器、辅蒸汽出口处的蒸汽截止阀。

Auxiliary steam turbine 辅汽轮机 系指带动电站发电机以及水泵、风机等船用辅机的汽轮机。

Auxiliary steam turbine load turn-over test 辅汽轮机转移负荷试验 为考核蒸汽动力装置在应急情况下两台汽轮辅机相互转移负荷的工作可靠性所做的试验。

Auxiliary steering gear 辅助操舵装置 系指:(1)如主操舵装置失效时操纵船舶所必需的设备。其不属于主操舵装置的任何部分,但不包括舵柄、舵扇或作同样用途的部件。(2)在主操舵装置失效时为操纵平台而使舵运动的设备。 Auxiliary steering gear are:(1) the equipment other than any part of the main steering gear necessary to steer the ship in the event of failure of the main steering gear but does not include the tiller, quadrant or components serving the same purpose. (2) the equipment which provids movement of the rudder for the purpose of steering the unit in the event of failure of the main steering gear.

Auxiliary system 辅助系统 系指推进机械、推进器、转向系统和发电机组正常工作所必需的支持系统,如燃油系统、滑油系统、冷却水系统、压缩空气系统和液压系统等。

Auxiliary thrust bearing 辅推力轴承 轴系中,通常用以承受非工作状态时产生的轴向力的轴承。

Auxiliary turbine exhaust steam feed heating 辅机排气预热给水 系指用背压式辅汽轮机的排气来预热锅炉给水的预热形式。

Availability 可用性 系统或设备的可用性系指在某一时间点其不处于故障状态的可用性。 Availability is the probability of a system or equipment when it is not in a failed state at a point in time.

Availability 有效性 系指某系统或某设备规定的瞬间或一段时间内完成了其所需的功能的能力。

Availability A 可用度 系指实际使用时间对预期的海上使用时间的比值。可用度可用下列公式计算: A = 平均无故障时间/(平均无故障时间 + 平均修理所需时间)。 Availability is the ratio of actual service time to expected service time at sea. Availability A may be calculated from the following formula: $A = MTTF/(MTTF + MTTR)$.

Availability of a communication system 通信系统的可用性 系指系统可被用来进行访问和通信的时间的百分比,即:$A = [(预定运行时间 - 停工时间)/预定运行时间] \times 100\%$。 Availability of a communication system is defined as the percentage of time in which the system is available for access to and communications through the system, i. e. $A = \{[(\text{scheduled operating time}) - (\text{downtime})]/(\text{scheduled operating time})\} \times 100\%$.

Average 海损 系指船舶和货物等在海上运输中遭遇自然灾害、意外事故或其他特殊情况,为了解除共同危险而采取合理措施所引起的特殊损失(即牺牲)和合理的额外费用。海上运输中,由于自然灾害或意外事故引起的船舶或货物的任何损失,如船舶因触礁、搁浅、碰撞、沉没、火灾、风灾、爆炸等造成船舶或货物的物质损失及费用损失等,均属海损。

Average circumferential wake 沿周平均伴流 在螺旋桨盘面上某一半径处的平均标称伴流。

Average freight rate assessment 平均运费指数

Average individual risk 平均个体风险率 其值表示每个人在船上可能遭遇致命风险。

Average sheer, mean sheer 平均舷弧 系指艏舷弧与艉舷弧的平均值。

Average sound pressure 平均声压

Average speed for heaving anchor 起锚平均速度 锚泊机械在起锚深度下,自锚破土后起锚的平均

Average tare 平均皮重 又称标准皮重(Standard tare),系指对于那些材料和规格整齐划一的包装,抽取若干件进行衡量,所得的平均值即为平均皮重。

Average value 算术平均值 系指一日昼夜平均值。Average value means a average during one day and night.

Aviation fire control system 航空火力控制系统 系指由控制飞机火力的方向、密度、时机和持续时间的机载设备构成的系统。航空火力控制系统的基本功能为:引导飞机到达目标区和沿最佳航线接近目标;搜索、识别、跟踪目标;测量目标和载机的运动参数,进行火力控制计算;控制武器的发射方式、数量和装定引信;对需要载机制导的武器进行发射后的制导。

Aviation gasoline 航空汽油

Award 授标 系指中标。

Award 仲裁裁决

Awareness time 知道时间/觉察期 系指乘客知道弃船处境并采取行动的时间。它计自乘客听到最初的通知(例如警报)至乘客已经接受这一处境并开始向集合站行进为止。Awareness time is the time it takes for passengers to process and react to the situation. This time begins upon initial notification(e. g. alarm) of an emergency and ends when the passenger has accepted the situation and begins to move towards an assembly station.

Away from 远离 系指将不相容物质有效隔离,使其在发生事故时不会发生危险的相互作用,但只要最小水平隔距的垂直投影达到 3 m,仍可在同一货舱或舱室内或在甲板上载运。Away from means incompatible materials are effectively segregated so that they cannot interact dangerously in the event of an accident but they can still be carried in the same hold or compartment or on deck provided a minimum horizontal separation of 3 m, projected vertically.

Awning 天幕 系指张设在船舶敞露部位,供遮挡阳光、雨雪用的帐幕。

Awning bar 天幕压条 系指在固定天幕上,用以压盖玻璃钢瓦楞板板边的条状配件。

Awning curtains 天幕帘 系指在天幕边缘向下垂挂的布帘。

Awning hook 天幕钩 系指在固定天幕上,用以紧固玻璃钢瓦楞板的钩状配件。

Awning rope 天幕张索 系指穿绕在天幕柱顶端,用以支张天幕的绳索。

Awning stanchion 天幕柱 系指供张设天幕用的支柱。

Awning stop 天幕系索 系指用以把天幕系结在天幕张索或天幕桁架上的细绳。

Awning store 天幕贮藏室 系指船上供贮存天幕的舱室。

Awning stretcher and rafter 天幕桁架 系指架设在天幕柱上,用以支承和系结天幕的纵、横向杆件。

a_x 任何一点的 x 方向加速度(m/s^2) a_x 由下式得出:$a_x = C_{xg}g\sin\Phi + C_{xs}a_{surge} + C_{xp}a_{pitch\, x}$,式中,$a_{surge}$——纵荡引起的纵向加速度($m/s^2$),由下式得出:$a_{surge} = 0.2 \times a_0 g$;$a_{pitch\, x}$——纵摇引起的纵向加速度,由下式得出:$a_{pitch\, x} = \Phi(180/\pi)(2\pi/T_P)^2 R$。At any point, accelerations along x directions(m/s^2), are given by $a_x = C_{xg}g \times \sin\Phi + C_{xs}a_{surge} + C_{xp}a_{pitch\, x}$, where: a_{surge}——longitudinal acceleration due to surge (m/s^2) is given by: $a_{surge} = 0.2 a_0 \times g$; $a_{pitch\, x}$——longitudinal acceleration due to pitch (m/s^2), are given by $a_{pitch\, x} = \Phi(180/\pi)(2\pi/T_P)^2 R$.

Axial fan 轴流式风机 系指气流主要沿轴向流动的风机。

Axial flow marine jet unit 船用轴流泵喷水推进装置 系指采用轴流泵作为喷水推进动力的装置称为轴流泵喷水推进装置。

Axial flow pump(propeller pump) 轴流泵 系指叶轮呈推进器形状,液体在泵内沿轴向流动的泵。

Axial inflow factor 轴向进流因数 系指螺旋桨叶元体处轴向感生速度对其进速的比值。

Axial plunger pump 轴向柱塞泵

Axial thrust failure protective device 轴向位移保护装置 系指当汽轮机的轴向位移达到极限值时,发出信号,通过综合开关,如油遮断器使速关阀关闭,汽轮机紧急停车的装置。

Axial velocity 轴向速度 系指叶元体相对于附近水沿轴向的速度。对未受扰动的水等于进速。

Axial velocity induced at propeller 轴向感生速度 由于螺旋桨的作用,在螺旋桨叶的半径处使水产生的与前进方向相反的速度。

Axial vibration 轴向振动

Axial vibration of shafting(longitudinal vibration) 轴系纵向振动 轴系沿轴向的振动。

Axial-flow compressor(axial compressor) 轴流式压气机 空气或其他气体在压缩过程中沿轴向流动的压气机。

Axial-flow steam turbine 轴流式汽轮机 系指蒸汽在膨胀过程中沿轴向流动的汽轮机。

Axial-flow turbine(axial turbine) 轴流式涡轮 高温燃气或其他气体在膨胀过程中沿轴向流动的压

a_y 任何一点的 y 方向加速度(m/s^2) a_y 由下式得出:$a_y = C_{yg}g\sin\theta + C_{ys}a_{sway} + C_{yr}a_{roll\ y}$,式中,$a_{sway}$——横荡引起的横向加速度($m/s^2$),由下式得出:$a_{sway} = 0.3a_0 \times g$,$a_{roll\ y}$——横摇引起的横向加速度($m/s^2$),由下式得出:$a_{roll\ y} = \theta(\pi/180)(2\pi/T_R)^2R$。 At any point, accelerations along y directions(m/s^2) are given by:$a_y = C_{yg}g \times \sin\theta + C_{ys}a_{sway} + C_{yr}a_{roll\ y}$, where:$a_{sway}$—transverse acceleration due to sway(m/s^2) are given by:$a_{sway} = 0.3a_0g$,$a_{roll\ y}$—transverse acceleration due to roll(m/s^2) are given by:$a_{roll\ y} = \theta(\pi/180)(2\pi/T_R)^2R$.

a_z 任何一点的 z 方向加速度(m/s^2) a_z 由下式得出:$a_z = C_{zh}a_{heave} + C_{zr}a_{roll\ z} + C_{zp}a_{pitch\ z}$,式中,$a_{heave}$——垂荡引起的垂向加速度($m/s^2$),由下式得出:$a_{heave} = a_0 \times g$;$a_{roll\ z}$——横摇引起的垂向加速度($m/s^2$),由下式得出:$a_{roll\ z} = \theta(\pi/180)(2\pi/T_R)^2y$;$a_{pitch\ z}$——纵摇引起的垂向加速度($m/s^2$),由下式得出:$a_{pitch\ z} = \Phi(\pi/180)(2\pi/T_P)^2|(x-0.45L)|$,式中:$|(x-0.45L)|$ 应取不小于 $0.2L$;$R = z - \min(D/4 + T_{LC}/2, D/2)$。 At any point, accelerations along z directions(m/s^2) are given by:$a_z = C_{zh}a_{heave} + C_{zr}a_{roll\ z} + C_{zp}a_{pitch\ z}$, where:$a_{heave}$—vertical acceleration due to heave(m/s^2) is given by:$a_{heave} = a_0g$,$a_{roll\ z}$—vertical acceleration due to roll(m/s^2), is given by:$a_{roll\ z} = \theta(\pi/180)(2\pi/T_R)^2y$;$a_{pitch\ z}$—vertical acceleration due to pitch(m/s^2), is given by:$a_{pitch\ z} = \Phi(\pi/180)(2\pi/T_P)^2 \times |(x-0.45L)|$; where:$(x-0.45L)$ is to be taken not less than $0.2L$;$R = z - \min(D/4 + T_{LC}/2, D/2)$.

Azimuth indicator 方位指示器 系指能指示推力方向的仪器。

Azimuth of borehole inclination 井斜方位角

Azimuth propulsion arrangement 全方位推进装置 推进装置的推力方向能在360°方位随意变动者称为全方位推进装置。

Azimuth propulsion system 方位推进系统 系指由下列子系统构成的系统:(1)操纵装置;(2)轴承;(3)船体支架;(4)系统中的舵部件;(5)吊舱,当为吊舱系统时吊舱中装有电机。当船舶以最大航速航行时,方位推进系统在每舷的最大定向角度应由设计者提出。该最大角度在每舷一般应小于35°。一般而言,船级社可考虑方位推进系统在操纵时的定向角度大于该角度,但定向值以及相关航速值应提交船级社审批。 Azimuth propulsion system is constituted by the following sub-systems:(1) the steering unit;(2) the bearing;(3) the hull supports;(4) the rudder part of the system;(5) the pod, which contains the electric motor in the case of a podded propulsion system. The maximum angle at which the azimuth propulsion system can be oriented on each side when the ship navigates at its maximum speed is to be specified by the Designer. Such maximum angle is generally to be less than 35° on each side. In general, orientations greater than this maximum angle may be considered by the Society for azimuth propulsion systems during manoeuvres, provided that the orientation values together with the relevant speed values are submitted to the Society for approval.

Azimuth thruster[1] 全回转推力器 系指具有在360°方向旋转的能力,以在任何方向发出推力的推力器。
Azimuth thruster is a thruster which has the capability to rotate through 360° in order to develop thrust in any direction.

Azimuthing nozzled thruster 全回转导管推力器
全回转导管推力器系指推力器位于导管内的全回转推力器。

Azimuthing propeller[2] 全回转推进器(Z推,L推) 推进器的推力方向能在360°方位随意变动者称为全回转推进器或Z推或L推。

Azimuthing thruster 全向推力器(方位推力器,可转向推力器) 推力器的推力方向能在360°方位随意变动者称为全向推力器或方位推力器或可转向推力器。

Azipod propulsion unit 吊舱式推进装置 将整套全方位推进装置吊装在船外而实现对船舶进行推进及操纵的装置称为吊舱式推进装置。见图A-22。

图 A-22 吊舱式推进装置
Figure A-22 Azipod propulsion unit

B

B class divisions B 级分隔 系指由符合下列标准的适当的舱壁、甲板、天井或内置板所组成的分隔:(1)它们用认可的不燃材料制成,且"B"级分隔建造和装配中所用的一切材料均为不燃材料,但并不排除可燃装饰板的使用,只要这些材料符合有关的其他相应要求;(2)在下列时间之内露出火焰时,未露出火焰一侧的平均温度不可比试验初始温度高 140 ℃ 且不可再上升。保留放热值并使包括任何接头在内的任何一点的温度较试验初始温度升高不超过 225 ℃,见表 B-1;(3)它们的构造应在标准耐火试验最初的 0.5 h 结束时,能防止火焰通过;(4)船级社已要求按《耐火试验程序规则》对原型分隔进行一次试验,以确保满足上述完整性和温升的要求。 B class divisions are those divisions formed by bulkheads, decks, ceilings or linings which comply with the following criteria:(1)they are constructed of approved non-combustible materials and all materials used in the construction and erection of "B" class divisions are non-combustible, with the exception that combustible veneers may be permitted provided they meet other appropriate requirements of this chapter;(2)they have an insulation value such that the average temperature of the unexposed side will not rise more than 140 ℃ compared with the original temperature, nor will the temperature at any one point, including any joint, rise more than 225 ℃ compared with the original temperature, within the time listed above. Table B-1 shows B class divisions.(3)they are constructed as to be capable of preventing the passage of flame to the end of the first half hour of the standard fire test;and (4) a Society has required a test of a prototype division in accordance with the Fire Test Procedures Code to ensure that it meets the above requirements for integrity and temperature rise.

表 B-1 B 级分隔
Table B-1 B class divisions

"B-15"(class "B-15")	15 min
"B-0"(class "B-0")	0 min

B class fire proof door B 级防火门 符合 B 级防火标准,装设在乙级防火舱壁上的门。

B class inflammable gas B 级易燃气体 对美国国家消防委员会(NFPA)船上气体危害控制标准而言,系指雷氏蒸发气压力小于 9.6×10^3 Pa(14 lbf/in²) 大于 5.9×10^3 Pa(8.5 lbf/in²) 的易燃气体。

B sound level B 声级 系指 B 计权网络在计权后的声级。其频率响应曲线为 70 phon 等响曲线经整理后倒置。

B standard ship's station B 标准船站(卫星通信) 它是 A 标准船站的替代系统,1993 年开始投入使用。它是一个全新的全球话音和数字通信系统。主要提供数字传输方式的电话、数据、传真和群呼等业务,主要供通信业务量大的船舶使用。例如豪华邮轮、海上钻井平台、豪华游艇以及大型远洋船舶上应用。

B. D. G. inject comp. 燃气喷射压缩机

B. O. P stack recess storage 防喷器凹形区 系指浮式钻井装置上,为专门存放水下防喷器组而设计的其上装有桥式吊车的区域。

B1 ice class notation B1 冰级标志 对船舶及需要破冰船辅助的船舶而言,系指按下述规定授予的冰级标志:船舶的结构、主机功率及其他特性能够确保船舶在严重冰况下具有正常航行的能力,但在需要时应有破冰船的辅助。 For the ships and icebreaker-assisted ships, B1 ice class notation is an ice class notation assigned as following:ships with such structure, engine power and other properties are normally capable of navigating in difficult ice conditions with assistance of icebreakers when necessary.

B1* ice class notation B1* 冰级标志 对船舶及需要破冰船辅助的船舶而言,系指按下述规定授予的冰级标志:船舶的结构、主机功率及其他特性能够确保船舶在严重冰况下具有正常航行的能力,且不需要破冰船的辅助。 For the ships and icebreaker-assisted ships, B1* ice class notation is an ice class notation assigned as following:ships with such structure, engine power and other properties are normally capable of navigating in difficult ice conditions without assistance of icebreakers.

B2 ice class notation B2 冰级标志 对船舶及需要破冰船辅助的船舶而言,系指按下述规定授予的冰级标志:船舶的结构、主机功率及其他特性能够确保船舶在中等冰况下具有正常航行的能力,但在需要时应有破冰船的辅助。 For the ships and icebreaker-assisted ships, B2 ice class notation is an ice class notation assigned as following:ships with such structure, engine power and other properties are normally capable of navigating in moderate ice conditions with assistance of icebreakers when necessary.

B3 ice class notation　B3 冰级标志　对船舶及需要破冰船辅助的船舶而言,系指按下述规定授予的冰级标志:船舶的结构、主机功率及其他特性能够确保船舶在轻度冰况下具有正常航行的能力,但在需要时应有破冰船的辅助。　For ships and icebreaker-assisted ships, B3 ice class notation is an ice class notation assigned as following: ships with such structure, engine power and other properties are normally capable of navigating in light ice conditions with assistance of icebreakers when necessary.

Babcock and Wilcox boiler(B & W boiler)　小型联箱式锅炉

Babcock and Wilcox feed water regulator　B&W型给水调节器

Back balance anti-vibrator　减振器(平衡块减振器)

Back cavitation　背空泡　系指螺旋桨叶背产生的空泡。一般由叶梢开始向内发展成银色片形,也可发展于叶背最大厚度略后处。

Back coupling　后联轴节

Back guide　倒车导板

Back hoe dredger　铲斗式挖泥船(反铲)　系指用装于艏部甲板的全液压折臂反铲挖掘机挖掘水底泥沙、石块等的挖泥船。

Back hoe dredger (BHD)/Dipper dredger　铲斗式挖泥船　系指具有大多由陆用挖掘机派生的挖掘机构,能将全部功率集中在一个铲斗上进行特硬挖掘和适合其他挖掘船所不能胜任的场合的船舶。

Back of blade　叶背　系指推船前进时,螺旋桨叶产生吸力的一面,即向船前方的一面。

Back pressure　背压　系指工质在完成做功离开发动机处的压力。

Back pressure regulator　背压调节器

Back pressure regulator (evaporator pressure regulator)　蒸发压力调节阀　系指使自动控制盘管或蒸发器中压力保持稳定的阀。

Back pressure test　背压试验　系指测定内燃机在一定工况下排气背压变化或被动时各主要性能参数变化情况的试验。

Back pressure trip　背压脱扣器

Back pressure valve　止回阀(背压阀)

Back-scattered scanning　背散射扫描　系指一种X射线扫描技术。它能够透过被遮挡的物体来扫描隐藏的物品,例如通过人体的衣服产生高分辨率的人体图像,炸药与毒品类的物质对X射线的散射能力很强,采集到的散射信号也就是图像的灰度比较高,所以,背散射X射线检查设备能发现人体携带的炸药等危险品。同时,金属物品,如枪支和匕首等,对X射线的散射能力很低,采集到的散射信号就低,在人体散射背景下,人体携带的这些危险武器也极易被发现。所以说X射线背散射人体检查设备不仅能完成金属探测门探测金属武器的功能,还能探测炸药以及爆炸装置。常常有人担心背散射扫描危害人体健康,研究表明由于背散射人体检查设备采用了点扫描成像原理,每一时刻只有一个X射线的光点照射被检查的人员,大大减少了色斑的单次检查剂量。研究表明,一次背散射扫描和一次2 h的航空旅程所接受到的X射线辐射相当。美国约翰·霍普金斯大学应用物理实验室和联邦食药局对背散射扫描仪安全性进行了评估,多项数据显示该设备的辐射剂量符合美国法律规定的卫生和安全标准。背散射扫描安检设备多年来被广泛应用于监狱、钻石开采、海关搜索、医疗探测。近年来又被测试出可以用于机场、火车站等公共场所的安检。近年来,背散射扫描仪在美国机场已被越来越多地配置和使用,而且,有取代金属探测门的趋势。这种安检设备也逐步被广大旅客所接受。我国也拥有自主知识产权的背散射扫描技术,公安部第一研究所自主研制的人体背散射安全检查设备样机已经在首都机场实行试用。

Back to back credit　对背信用证　又称转开信用证,系指受益人要求原证的通知行或其他银行以原证为基础,另开一张内容相似的新信用证。对背信用证开证银行只能根据不可撤销信用证来开立。

Background noise　背景噪声　系指在发生、检查、测量或记录的系统中与信号存在与否无关的一切干扰。

Background-noise level　背景噪声级　系指测量系统(水听器和电子仪器)接收到的除潜艇螺旋桨噪声以外的一切干扰噪声级。

Backing propeller test　反转试验　系指将螺旋桨模型反转进行的试验。

Backslash　反斜杆

Back-pressure steam turbine　背压式汽轮机　系指蒸汽在汽轮机中只膨胀到高于大气压力,其排气被用于加热给水和其他用途的汽轮机。

Back-up control systems　备用控制系统　系指在主控制系统损坏或失效后,维持高速船安全运转所必需的控制设备组成。　Back-up control systems comprise all equipment that are necessary to maintain control of essential functions required for the craft's safe operation when the main control systems have failed or malfunctioned.

Back-up DP　DP备份　一种DP控制系统。与主DP控制系统实体上完全分离的,在主系统完全故障事件中可用。

Back-up navigator　后备驾驶人员　系指由船长指

定,在必要时被呼叫来协助或代替值班驾驶人员的驾驶人员。 Back-up navigator is a navigational officer who has been designated by the ship's master to be on call to assist or replace the officer of the watch when required.

Back-up protection 后备保护 系指由于最接近故障点的保护电器有故障或缺乏能力,或者其他保护电器有故障,以致不能及时清除系统故障的情况下起作用的保护设备或系统。 Back-up protection is the protection equipment or system which is intended to operate when a system fault is not cleared in due time because of failure or lack of ability of a protective device closest to the fault to operate or because of failure of other protective device.

Back-up recovery system 备用回收系统
Back-up system 备用系统
Back-up 备份
Baffle-plate column 折流板式塔
Baffler 挡板(导流板,消声器)
Bagasse dietary fiber 蔗渣膳食纤维
Bagasse fiber board 甘蔗纤维板
Bagasse fiber/PVC composite 蔗渣纤维 P/VC 复合材料

Baidu search engine 百度搜索引擎 于1999年底在美国硅谷由李彦宏和徐勇创建。致力于向人们提供"简单、可依赖"的信息获取方式。"百度"二字源于中国宋朝词人辛弃疾的《青玉案·元夕》诗句:"众里寻他千百度",象征着百度对中文信息检索技术的执着追求。目前是国内最大的商业化全文搜索引擎。

Bailey feed water regulator "贝来"给水调节器
Bailment 委托
Bait hold 饵料舱 钓鱼船上装载鱼饵的舱。
Balance of trade 贸易差额 系指一定时期内一国出口总额与进口总额增加的差额。用以表明一国对外贸易的收支状态。当出口总额超过进口总额时,称为贸易顺差,或贸易出超;当进口总额超过出口总额时,称为贸易逆差,或贸易入差。通常贸易顺差以正数表示,贸易逆差以负数表示。如出口总额与进口总额相等,则称为贸易平衡。

Balance pressure reducing valve 平衡减压阀
Balanced modulator 平衡调制器
Balanced rudder 平衡舵 系指舵杆轴线位于舵叶导边后面一定的距离。以使舵压力中心接近舵杆轴线而减少转舵扭矩的舵。
Balanced slide valve 平衡滑阀
Balanced ventilation system 平衡式通风系统 系指由强力通风和诱导或自然通风相结合的通风系统,其使货舱中的压力状态接近于大气压。 Balanced ventilation system means a ventilation system consisting of a combination of forced draught and induced or natural draught, to produce a pressure condition in the hold space approximately equal to atmospheric pressure.

Balancer 平衡器(平衡装置,稳定器)
Balancing holds 平衡孔 为减少冲动式汽轮机的轴向推力而均匀布置在叶轮上的圆孔。
Balancing valve 平衡阀
Bale capacity 包装货物舱容
Bale cargo capacity(bale capacity, bale measure, bale cable) 包装舱容 包括舱口围板在内,量自内底板之顶面、甲板纵骨下缘、肋骨或舷侧纵桁内缘、舱壁骨架的自由翼缘,或量至货舱护条的表面,但扣除舱内支柱、通风筒等所占空间后而得出的船舶各货舱的总容积或其中任一货舱的单舱容积。

Ball float valve 浮球阀
Ball governor 飞球调速器
Ball journal 球轴颈
Ball thrust bearing 球形止推轴承
Ball valve 球阀
Ballast 压载 系指专用以改变船舶的质量和重心位置的任何固体物或液体物。

Ballast branch pipe(branch ballast line, ballast pipe) 压载水支管 系指自压载水总管接至各压载水舱的管路。

Ballast draught 压载吃水 系指符合结构强度要求的最小设计压载吃水(m)。对于包括出港和到港状态的装载手册中的任何压载工况,最小设计压载吃水不必大于最小压载吃水,且从船中型基线量起。 Ballast draught is the minimum design ballast draught, in m, for the ships whose strength requirements of scantlings are met. The minimum design ballast draught is not greater than the minimum ballast draught, measured from the moulded base line at amidships, for any ballast loading conditions including both departure and arrival conditions in the loading manual.

Ballast oil separator 油舱压载水油分离装置 为避免造成污染,用以分离含油压载水,使剩油留在舱内的装置。

Ballast operating system 压载操纵系统
Ballast piping 压载水管系 系指用于注入、排出和调拨压载水舱的压载水的管系。
Ballast pump 压载泵 系指用于压载水舱注水或排水的泵。
Ballast system 压载水系统 系指用水作为压载,可对船的吃水、纵倾、横倾进行调整的管路系统。
Ballast main(water ballast main) 压载水总管

压载水管系中自压载水泵至各压载水支管或压载水支管分配阀箱的管路。

Ballast tank 压载水舱 系指：(1)用于存储压载水的舱室；(2)船上用于装载压载水的各种液舱或货舱，不管此液舱或货舱是否为指定供此用途者；(3)双层底舱加上双舷侧边舱加上双甲板舱（取适用者），即使它们是分开的，也是如此；(4)所有供压载水用的处所及通往这些处所的围壁通道；(5)也包括压载水和货油兼用舱，但不包括"1973/1978 国际防止船舶造成污染公约"第13(3)条所指可能装载压载水的货油舱。 Ballast tanks are: (1) a compartment used for the storage of ballast water; (2) any tank or hold on a vessel used for carrying ballast water, whether or not the tank or hold was designed for that purpose; (3) double bottom tank plus double side tank plus double deck tank, as applicable, even if these tanks are separate; (4) all spaces used for ballast water and trunks and spaces to such; (5) dual-purpose tanks for ballast and cargo oil, but cargo oil tanks which may carry water ballast according to Regulation 13(3), of MARPOL 73/78 not induded.

Ballast tank of a bulk carrier 散货船压载舱 系指单独用于海水压载的液舱，或适用时，对可用于装货和海水压载的处所，当发现其显著腐蚀时，将视为压载舱。 Ballast tank of a bulk carrier is a tank which is used solely for salt water ballast, or, where applicable, a space which is used for both cargo and ballast will be treated as a ballast tank when substantial corrosion has been found in that space.

Ballast tank of a double skin bulk carrier 双壳散货船压载舱 系指单独用于海水压载的液舱，或适用时对可用于装货和海水压载的处所，当发现其显著腐蚀时，将视为压载舱。即使两舷的边舱与顶边舱或底边舱相连，也应被认为是一个独立舱。 Ballast tank of a double skin bulk carrier is used solely for salt water ballast, or, where applicable, a space which is used for both cargo and ballast will be treated as a ballast tank when substantial corrosion has been found in that space. A double side tank is considered as a separate tank even if it is in connection to either the topside tank or hopper side tank.

Ballast valve 压载水舱通海阀

Ballast water 压载水 (1)传统定义（仅就压载水在船上的功能而言）：压载水系指为控制船舶纵倾、横倾、吃水、稳性或应力而在船上摄入的舷外水；(2)对《2004年压载水公约》而言（除功能外还涉及对环境的保护）：压载水系指为控制船舶纵倾、横倾、吃水、稳性或应力而在船上摄入的舷外水及其悬浮物。 Ballast water means: (1) For its function, ballast water means water taken on board of a ship to control trim, list, draught, sta-bility or stresses of the ship. (2) For the Ballast Water Management Convention, ballast water means water with its suspended solids taken on board of a ship to control trim, list, draught, stability or stresses of the ship.

Ballast water capacity 压载水容量 就"压载水公约"而言，系指船上用于承载、装填或排放压载水的任何液舱、处所或舱室，包括被设计成允许承载压载水的任何多用途液舱、处所或舱室的总体积容量。 For the Ballast Water Management Convention, ballast water capacity means the total volume capacity of any tanks, spaces or compartments on a ship used for carrying, loading or discharging ballast water, including any multi-use tank, space or compartment designed to allow carriage of ballast water.

Ballast water discharge 压载水排放 系指将压载水向船外排放。 Ballast water discharge means the ballast water is discharged overboard.

Ballast water exchange[1] 压载水交换 系指沿海附近的（包括港口和海湾）有机物释放到深海，或者深海的释放到沿海水域一般不会继续生存。在深海进行压载水交换，例如通过继续的排空和灌注压载舱或者促进水通过压载舱流动，因此能够减少有害水生物和病原体传播的可能性。 Ballast water exchange means a change that near coastal (including ports and estuares) organisms are released in deep ocean, and deep ocean organisms are released in coastal waters, and the organisms do not generally survive. Exchange of ballast water to deep ocean areas by sequentially pumping out and refilling ballast tanks or allowing water to flow through the tanks, offers a means of limiting the probability of transferring harmful aquatic organisms and pathogens.

Ballast water exchange 2(BWE) 压载水置换

Ballast water exchange standard (D-1) 压载水更换标准(D-1) 系指包括下列内容的标准：(1)船舶按国际船舶压载水和沉积物控制和管理公约第D-1条的要求进行压载水更换，其压载水容积更换率应至少为95%；(2)对于使用泵入-排出方法交换压载水的船舶，泵入-排出3倍于每一压载水舱容积应视为达到相应的标准。泵入-排出少于压载舱容积3倍，如船舶能证明达到了至少95%容积的更换，则也可被接受。 Ballast water exchange standard (D-1) means a standard includes the followings: (1) ships performing ballast water exchange in accordance with regulation D-1 of the Convention shall do so with an efficiency of at least 95% volumetric exchange of ballast water. (2) for ships exchanging ballast water by the pumping-through method, pumping through three times the volume of each ballast water tank shall be considered to meet the standard. Pumping through less than three times the volume may be accepted provided the ship can demonstrate that at least 95 percent volumetric exchange is met.

Ballast water management(BWM)　压载水管理
(1)就2004年《压载水公约》而言,系指旨在消除、无害处理、防止摄入或排放压载水和沉积物中的有害水生物和病原体的机械、物理、化学和生物的单一或综合方法;(2)用一种机械的、物理的、化学的或其他过程去杀灭、消除或破坏繁殖在压载水和沉积物中的有害或潜在有害的水生物和病原体。　Ballast water management means:(1) For the Ballast Water Management Convention, ballast water management means mechanical, physical, chemical, and biological processes, either singularly or in combination, to remove, render harmless treatment, or avoid the uptake or discharge of harmful aquatic organisms and pathogens within ballast water and sediments;(2)Mechanical, physical, chemical, and biological processes, either singularly or in combination, to remove, render harmless treatment, or avoid the uptake or discharge of harmful aquatic organisms and pathogens within ballast water and sediments.

Ballast Water Management Convention　压载水公约　系指"2004年船舶压载水和沉积物控制和管理国际公约(The International Convention for the Control and Management of Ships' Ballast Water and Sediments, 2004)"的简称。

Ballast water management plan　压载水管理计划
系指:(1)根据2004年国际船舶压载水和沉积物控制和管理公约B-1的规定描述在单艘船舶上压载水管理的文件;(2)为减少有害水生物和病原体传播的对船舶压载水提供安全有效和环境保护措施的管理计划。　Ballast water management plan is:(1) the document referred in Regulation B-1 of the International Convention for the Control and Management of Ships' Ballast Water and Sediments, 2004, describing the ballast water management processes and procedures on board of individual ships;(2) a plan specific to each ship providing safe, effective and environmentally procedures regarding ballast water management and control operations on board for ships to minimize the spread of harmful aquatic organisms and pathogens.

Ballast water management system(BWMS)　压载水管理系统　系指对压载水进行处理使其达到或高于有关规定的压载水性能标准的任何系统。压载水管理系统包括压载水处理设备、所有相关控制设备、监测设备以及取样设备。　Ballast water management system (BWMS) means any system which processes ballast water such that it meets or exceeds the ballast water performance standard in relevant requirement. The BWMS includes ballast water treatment equipment, all associated control equipment, monitoring equipment and sampling facilities.

Ballast water performance standard (D-2)　压载水更换标准(D-2)　系指包括下列内容的标准:(1)每立方米中最小尺寸大于或等于50 μm的可生存物少于10个;(2)每毫升中最小尺寸小于50 μm但大于或等于10 μm的可生存物少于10个;(3)指示微生物的排放不应超过:①有毒霍乱弧菌(O1和O139)——少于每100毫升1个菌落形成单位(cfu)或小于每一克(湿重)浮游动物样品1个cfu;②大肠杆菌——少于每100毫升250个cfu;和③肠道球菌——少于每100毫升100个cfu。　Ballast water performance standard (D-2) means a standard includes the followings:(1) less than 10 viable organisms per cubic meter greater than or equal to 50μm in minimum dimension;(2) less than 10 viable organisms per milliliter less than 50μm in minimum dimension and greater than or equal to 10μm in minimum dimension;(3) discharge of the indicator microbes shall not exceed:①Toxicogenic vibrio cholerae (O1 and O139) with less than 1 colony forming unit (cfu) per 100 milliliters or less than 1 cfu per 1 gramme (wet weight) zooplankton samples;②escherichia coli less than 250 cfu per 100 milliliters;③Intestinal enterococci less than 100 cfu per 100 milliliters.

Ballast water system　压载水处理系统　系指采用电解方法对压载水进行处理的装置。其核心技术是:在船上直接电解海水产生次氯酸钠对压载水进行消毒。其工作原理为:在船舶航行中进行压载时,首先通过自动反冲洗过滤器来去除较大的浮游生物和固体颗粒物,随后引入一定量海水进入电解槽产生次氯酸钠,并将次氯酸钠重新注回主管路,随着压载水进入压载舱进行消毒杀菌。当压载水被排放时,在其中加入中和剂,以中和剩余的氧化剂。将确保所排放的压载水不会对压载水接收环境产生危害。见图B-1。

图 B-1　压载水处理系统
Figure B-1　Ballast water system

Ballast water tanks 压载水舱 系指所有供压载水用的处所及通往这些处所的围壁通道。 Ballast water tanks mean all spaces used for ballast water and trunks to such spaces.

Ballast water treatment 压载水处理 对"压载水公约"而言,系指"压载水管理"范围之内的一种方法,系指为了使船上的压载水达到允许排放的标准,必须使压载水通过这样的设备,使其通过机械、物理、化学和生物的单一或综合过程而达到"压载水性能标准"(D-2 标准)的指标的方法。

Ballast water treatment equipment 压载水处理设备 系指采用单独或合并的机械、物理、化学或生物处理方法以清除、无害处置或避免摄入或排放压载水和沉积物中的有害水生物和病原体的设备。航行期间,压载水处理设备可在压载水摄入或排放时工作,或同时工作。 Ballast water treatment equipment means equipment which mechanically, physically, chemically, or biologically processes, either singularly or in combination, to remove, render harmless treatment, or avoid the uptake or discharge of harmful aquatic organisms and pathogens within ballast water and sediments. Ballast water treatment equipment may operate at the moment of uptake or discharge of ballast water, during the voyage, or at a combination of these events.

Ballast water working group(BWWG) 压载水工作小组

Ballast waterline 压载水线 系指最小艏、艉吃水的连线。

Ballast/solid cargo hold 压载兼用舱 系指油船运航时,邻接货油舱作为专用压载舱且在矿砂船或散装货船运航时作为固体货物荷载舱的处所。 Ballast/solid cargo hold is a compartment which is used as an exclusive tank for ballast adjacent to a cargo oil tank when the ship is not in the dry cargo mode and which is used as a solid cargo stowing hold when the ship is in the dry cargo mode.

Ballast-draft speed 压载航速 系指船在压载航行时的航速。

Ballistic missile 弹道导弹 是一种导弹,通常没有翼,在烧完燃料后只能保持预定的航向,不可改变,其后的航向由弹道学法则支配。为了覆盖广大的距离,弹道导弹必须发射很高,进入空中或太空。

Ballistic Missile Defense Organization under United States Department of Defense(BMDO) 美国国防部弹道导弹防御局 1993 年在比尔·克林顿总统领导下的政府机构,它的名字被更改为弹道导弹防御局(BMDO),它的重点从国家导弹防御系统被转移到区域导弹防御,从全球性的范围到更多地方的覆盖面。

Ballistic missile technology on railway maneuvering 铁路机动弹道导弹技术 该技术拥有国:中国、俄罗斯。见图 B-2。

图 B-2 铁路机动弹道导弹技术
Figure B-2 Ballistic missile technology on railway maneuvering

Baltic Sea area 波罗的海区域 系指波罗的海本身以及波的尼亚湾、芬兰湾和波罗的海入口,以斯卡格拉克海峡中斯卡晏角处的北纬 57°44.8′为界。 Baltic Sea area means the Baltic Sea proper with the Gulf of Bothnia and the Gulf of Finland and the entrance to the Baltic Sea bounded by the parallel of the Skaw in the Skagerrak at 57°44.8′N.

BAM = Bridge alert management 桥楼警报管理 它是一个整体管理的概念,用于处理和统一桥楼上的警报。 It is a concept of overall management, handling and harmonized presentation of alerts on the bridge.

Bamboo raft 竹筏 系指用一组竹竿并排连接而成的简易水上运载工具。

BAMS = bridge alert management system 桥楼警报管理系统 一个协调警报的优先级,分类,处理,分配和显示的系统,使桥楼团队能够充分重视船舶的安全操作,并立即确定需要采取行动维持船舶安全航行的任何警报情况。 A system that harmonizes the priority, classification, handling, distribution and presentation of alerts, to enable the bridge team to devote full attention to the safe operation of the ship and to immediately identify any alert situation requiring action to maintain the safe operation of the ship.

Band frequency spectrum analyser 宽频带频谱分析仪

Band plate 带板 系指在船体结构的强度或稳定性计算中,被认为同骨材一起工作的与骨材毗连的一部分板材。

Band sound pressure level of propeller noise 螺旋桨噪声频带声压级 系指带宽内螺旋桨噪声的声压级。

Bank distance parameter 岸距参数 系指船长与

船舶中心线离岸距离的比值。

Bank draft 银行汇票 系指汇款人将款项交存当地出票银行,由出票银行签发的,由其在见票时,按照实际结算金额无条件支付给收款人或持票人的票据。银行汇票有使用灵活,票随人到,兑现性强等特点,适用于先收款后发货或钱货两清的商品交易。单位和个人各种款项结算,均可使用银行汇票。

Bank guarantee 银行保函 又称银行保证书、银行信用保证书或简称保证书。银行作为保证人向受益人开立的保证文件。银行保证被保证人未向受益人尽到某项义务时,则由银行承担保函中所规定的付款责任。保函内容根据具体交易的不同而多种多样;在形式上无一定的格式;对有关方面的权利和义务的规定、处理手续等未形成一定的惯例。遇有不同的解释时,只能就其文件本身内容所述来做具体解释。其主要内容根据国际商会第458号出版物《UGD458》规定:有关当事人(名称与地址);开立保函的依据;担保金额和金额递减条款;要求付款的条件。

Bank Rakyat Indonesia(BRI) 印度尼西亚人民银行 是印度尼西亚主要的国有商业银行之一,成立于1895年,已经有100多年的历史。BRI的业务主要分为三个部分:商业金融、小额信贷金融、公司和国际金融。BRI的村行建立在乡镇并高度自治,村行经理拥有贷款决定权。村行对自然村派出工作站,工作站负责吸收储蓄和回收贷款,但是不发放贷款。BRI的村行系统的核心业务是按照商业化原则运行的小额信贷,其员工激励计划不以贷款户的增加为基数,而是以赢利为基础。BRI的存款和贷款产品都围绕客户需要,其显著的特征是简单易用并实行标准化管理,业务操作高度透明。印度尼西亚人民银行通过其超过4 000家分行为3 000万零售客户提供小规模、小额借款和贷款。

Bank suction 近岸吸力 船舶在近岸航行时,由于近岸一侧水流速度增大和压力下降而产生的使船向岸边横移的吸力。

Bar keel 方龙骨 系指矩形断面的龙骨。

Barcelona Mobile World Congress 巴塞罗那世界移动通信大会 系指全球通信行业顶级的展会。

Barcode 条形码/条码 系指将宽度不等的多个黑条和空白,按照一定的编码规则排列,用以表达一组信息的图形标识符。常见的条形码是由反射率相差很大的黑条(简称条)和白条(简称空)排成的平行线图案。条形码可以标出物品的生产国、制造厂家、商品名称、生产日期、图书分类号、邮件起止地点、类别、日期等许多信息,因而在商品流通、图书管理、邮政管理、银行系统等许多领域都得到了广泛的应用。

Barcode printer 条形码打印机

Bare hull, naked hull 裸船体 系指船体型表面所围蔽的船底或船模的赤裸船体。

Bare mobile phone 裸机 系指没有配置操作系统和其他软件的电子计算机,也指没有加入通信网的手机、寻呼机。即只有硬件部分,还未安装任何软件系统的电脑叫裸机。为什么销售商介绍不使用更精确的"非合约机"而使用模棱两可的"裸机"来介绍商品,而常常有消费者购买之后大呼上当,其实就是销售商利用消费者认知中的误区,误导消费者导致的。特别是网购,不拆包装很难分辨手机是不是定制机,而拆了包装,销售商往往以拆包不退为借口,拒绝退货。另一种解释,也就是整个手机包装内只含有一个手机,其他配件都没有(比如充电器、数据线),这样也称之为裸机。

Bareboat charter 光船租船 是海洋运输的一种租船方式。系指租期内船舶所有人只提供船舶的一种类似财产租赁的租船形式。

Bareboat charter party 光船租赁合同 系指船舶出租人向承租人提供不配备或配备船员的船舶,在约定的时期内由承租人占有、使用和营运,并向出租人支付租金的合同。在法律性质上,光船租赁合同是一种船舶租赁合同。光船租赁合同的标准合同格式中,极为突出的波罗的海国际航运公会(BIMCO)制定的标准光船租赁合同(BARECON),BARECON 有 A、B 两种格式,A 格式用于一般的船舶经营租赁;B 格式用于通过抵押融资的新建船舶的租赁。A、B 格式中均有船舶租购方面的规定,供当事人自由采用。

Bareboat charter/Demise charter 光船租赁 系指船舶出租人提供一艘不包括船员在内的船舶出租给船舶承租人使用一段时间,并由承租人支付租金的一种租船方式。

Bare-hull resistance 裸船体阻力 系指实船或船模的裸船体所引起的阻力。

Bargain 议价

Bargainee 买主

Bargainor 卖主

Barge 驳船 系指:(1)不装推进装置、具有重型结构的丰满的平底船舶。(2)一般不具备主推进设备的船舶。(3)有人或无人的非机动驳船,包括:①货舱内装载一般干货的驳船;②货舱内装载散装液体货的驳船;③货舱内装载一般干货,且适于装在大船上的船载驳;④在甲板上装载不易腐烂的货物而专门设计的箱形驳。 Barge means: (1) a flat-bottomed vessel of full body and heavy construction without installed means of propulsion. (2) a ship generally not provided with main propulsion; (3) a manned or unmanned non-self-propelled barge including: ① barges carrying general dry cargo in cargo holds;

②barges carrying liquid cargo in bulk in cargo holds;
③shipboard barges carrying general dry cargo in cargo holds and suitable for regular carriage on board a large ship;
④specially designed pontoons for the carriage of nonperishable cargo on deck.

Barge carrier 载驳船 系指具有较大的甲板面积，专载运载驳和重型设备的船舶。 Barge carrier is a barge dedicated to cargo barge and heavy cargo carriers with large deck area.

Barge unloading dredger 吹泥船 属于挖泥船范围，但它不具备对水下土层挖掘的能力，只有对疏浚泥浆进行吸入和吹出的功能，是一种简单的吹扬式船舶，故属于吸扬式挖泥船的类型。它基本上装备吸扬式挖泥船的一些设备，如吸泥头、吸泥管、泥泵和排泥管等。它依靠泥泵的吸、排能力，将泥驳载运来的疏浚泥沙，经稀释后以泥浆的形式吹送上岸，或用以进行其他的吹填工程。所以，吹泥船是机械式非自航挖泥船进行疏浚吹填和输泥上岸施工作业中的配套船舶之一。

Barge unloading suction dredger 吹泥船 用泥泵抽吸泥驳中被高压水冲成的泥浆并吹送至填泥区的船。

Barge warping winch 移驳绞车 系指挖泥船上，用以移动泥驳的绞车。

Barge-type unit 驳船式平台 系指无推进装置的水面式平台。

Barite storage tank 重晶石舱 系指贮存为加重泥浆比重所需要的重晶石的舱室。

Barratry 船员不法行为 系指船员非法走私、偷窃、聚众赌博和吸毒等行为。

Barred-speed range 主机转速禁区 一般系指为防止轴系扭转或横振时应力超过持续运转许可值而避免使用的主机转速范围。

Barring 盘车 系指在汽轮机启动前和停车后，为了减少或消除由于本体上、下部受热或冷却的速度不同造成的热弯曲而间断地或连续地低速转动汽轮机转子的操作。

Barring gear interlocking device 盘车连锁装置 系指当盘车装置为脱开时，使速关阀不能开启的连锁装置。

Barring motor 盘车电动机

Barter 易货贸易 系指在换货的基础上，把等值的出口货物和进口货物直接结合起来的贸易方式。传统的易货贸易，一般是买卖双方各以等值的货物进行交换，不涉及货币的支付，也没有第三者介入，易货双方签订一份包括相互交换抵偿货物的合同，把有关事项加以确定。在国际贸易中，使用较多的是通过对开信用证的方式进行易货，即由交易双方先签订易货合同，规定各自的出口商品均按约定价格以信用证方式付款。根据协定规定，任何一方的进口或出口，由双方政府的指定银行将货值记账，在一定时期内互相抵冲结算，其差额没有的规定结转下一年度。易货贸易在实际做法上比较灵活，如在交货时间上，可以进口与出口同时成交，也可以有先有后等。

Bar-type stopper 闸刀式制链器

Base component 基料

Base earth station 基地地球站

Base plane 基平面 根据不同船型，通过舭龙骨与中站面的交点或船体型表面最低点，并平行于设计水线的平面。

Base plate 底板（钢夹层板） 系指钢夹层板中用于与船体主要结构连接的一侧的钢板。

Base port 基地港 系指在营运手册中规定的专门港口，并备有：（1）任何时候都能与在港口或海上的高速船保持连续的无线电通信的设施；（2）能取得相应地区的可靠天气预报，并及时发送到所有营运中的高速船的手段；（3）能为 A 型船提供适当的救助设备的渠道；（4）以适当设备为高速船提供维修服务的渠道。

Base station 基站 即公用移动通信基站是无线电台站的一种形式，系指在一定的无线电覆盖区中，通过移动通信交换中心，与移动电话终端之间进行信息传递的无线电收发信电台。移动通信基站的建设是我国移动通信运营商投资的重要部分，移动通信基站的建设一般都是围绕覆盖面、通话质量、投资效益、建设难易、维护方便等要素进行。随着移动通信网络业务向数据化、分组化方向发展，移动通信基站的发展趋势也必然是宽带化、大覆盖面建设及 IP 化。一个基站的选择，需从性能、配套、兼容性及使用要求等各方面综合考虑，其中特别注意的是基站设备必须与移动交换中心相兼容或配套，这样才能取得较好的通信效果。基站子系统主要包括两类设备：基站收信/发信台（BTS）和基站控制器（BSC）。

Base station controller(BSC) 基站控制器 是基站收信/发信台和移动交换中心之间的连接点，也为基站收信/发信台与操作维修中心之间交换信息提供接口。一个基站控制器通常控制几个基站收发台，其主要功能是进行无线信道管理、实施呼叫和通信链路的建立和拆除，并为本控制区内移动台的跨区切换进行控制等。目前国内主要有 GSM 和 CDMA 两类基站。

Base transceiver station(BTS) 基站收/发台

Base transmit station（BTS） 基准传送站

Base weather station 基地气象台

Basel Convention on the Control of Trans-boundary Movements of Hazardous Wastes and Their Disposal

(Basel Convention) 《控制危险废料越境转移及其处置巴塞尔公约》 简称《巴塞尔公约》。1989 年 3 月 22 日在联合国环境规划署于瑞士巴塞尔召开的世界环境保护会议上通过,1992 年 5 月正式生效。1995 年 9 月 22 日在日内瓦通过了《巴塞尔公约》的修正案。已有 100 多个国家签署了这项公约,中国于 1990 年 3 月 22 日在该公约上签字。该公约旨在遏止越境转移危险废料,特别是向发展中国家出口和转移危险废料。公约明确规定:如确有必要越境转移废料,出口危险废料的国家必须事先向进口国和有关国家通报废料的数量和性质;越境转移废料时,出口国必须持有进口国政府书面批准书。公约明确,危险废料系指国际上普遍认为具有爆炸性、易燃性、腐蚀性、化学反应性、急性毒性、慢性毒性、生态毒性和传染性等特性中的一种或几种特性的生产性垃圾和生活性垃圾,其中生产性垃圾包括金属废料。

Base-line 基线 无线电导航系统中主台和副台之间的最短连线。不分主、副台的无线电导航系统,两发射台之间的最短连线亦称"基线"。

Baseline (fishing vessel) 基线(渔船) 系指在船中与龙骨线相交的水平线。 Baseline is the horizontal line intersecting at amidships the keel line.

Basic insulation 基本绝缘 系指用于防止电击而提供的基本保护所加在带电部件上的绝缘。基本绝缘不必包括专用于特殊用途的绝缘。 Basic insulation means the insulation applied to live parts to provide basic protection against electric shock. Basic insulation does not necessarily include insulation used exclusively for functional purposes.

Basic insulation 主绝缘 提供防电击的主要保护的带电部件的绝缘。注意:主绝缘不一定包括仅用于功能目的的绝缘。 Basic insulation is the insulation applied to live parts to provide basic protection against electric shock. Note: basic insulation does not necessarily include insulation used exclusively for functional purposes.

Basic rate interface (BRI) 基本速率接口 系指由两个 64 kb/s 组成的 B 通道和 1 个 16 kb/s 的 D 通道。因此,它可以提供速度为 128 Kbps 数据传输服务。

Basic-speed of adjustable speed motor 调速电动机的基本转速 系指该电动机在额定负载、额定电压一经定额中规定的温升下所获得的最低转速。

Batch of containers 集装箱堆码 一个堆码的集装箱或者简称为一个堆码,系指同一根风管和同一台风机所服务的一组集装箱。 Batch of containers is a batch of containers, or simply, is a set of containers served by the same duct and the same air cooler.

Batch testing 批量试验 系指对于绑扎用的杆、配件及系固装置,在每 50 件(不足 50 件仍按 50 件计)中应抽取一个试件,并对其进行验证负荷的试验。验证负荷为其安全工作负荷的 1.5 倍。对于绑扎装置用的链或钢丝绳,在每 50 件(不足 50 件仍按 50 件计)中应抽取一个试件,并对其进行破断试验。 For rod lashings, fittings and securing devices, batch testing means a testing in one sample from every fifty pieces, or from each batch less than fifty pieces. and the sample is proof loaded 1.5 time the safe working load for which the item is intended. For chain or wire rope lashings, one sample from every fifty pieces, or from each batch less than fifty pieces, is tested for breaking.

Bathyscaph support vessel 深潜器母船

Batten bar (battening iron, hatch batten) 封舱塞条 系指围固在舱口四周防止盖布松动的金属扁条。

Battery capacity 蓄电池容量 系指充满电的蓄电池用一定的电流放电至规定放电终止电压的放电量。通常采用如下两种表示方法:(1)安时容量 = 放电电流 × 放电时间;(2)瓦时容量 = 安时容量 × 平均放电电压。

Battery room 蓄电池室 系指供安置蓄电池的舱室。

Battery room exhaust blower 蓄电池舱通风机 用以排除蓄电池室内的有害气体的通风机。

Battery room ventilation 蓄电池舱通风 为使蓄电池室内温度不致太高,并排除充电时所产生的氢气以防止发生爆炸事故而进行的自然通风或机械通风。

Bauschinger effect Bauschinger 效应 系指初期屈服和对从拉伸或压缩状态卸载和反向加载的应力-应变响应曲线的范围。 Bauschiager effect means early yielding and rounding of a stress-strain response on unloading and reverse loading from a tensile or compressive stress stage.

Bay[1] 海湾 系指明显的水曲区域,其凹入程度和曲口宽度的比例使其具有被陆地环抱的水域,而不仅为海岸的弯曲。但是,除非水曲的面积等于或大于一横越曲口所划的直线作为直径的半圆形面积,否则不应视为海湾。 Bay is a well marked indentation whose penetration is in such proportion to the width of the mouth as to contain landlocked waters and constitute more than a mere curvature of the coast. An indentation shall not, however, be regarded as a bay unless its area is as large as, or larger than, that of the semi-circle whose diameter is a line drawn across the mouth of that indentation.

Bay[2]　**排位（集装箱）**　系指集装箱堆垛的纵向位置标识。一个集装箱横向堆垛，与舱口和舱口盖相关联，包含多垛（或排）集装箱。

Bay[3]　**跨**　系指相邻横框架或横舱壁之间的区域。
Bay means the area between adjacent transverse frames or transverse bulkhead.

Bayesian network　**贝叶斯网络**　一种概率图解模型（统计模型的一种），通过又向求循环图，呈现了一系列随机变量和随机变量间的互为条件的关系。例如，贝叶斯网络能够呈现出疾病和症状之间的概率关系。如果症状已知，利用该网络可以计算不同疾病出现的概率。

Bayesian weighting　**贝氏计权**

Bayonet nut connector/British naval connector (BNC)　**卡扣配合型连接器接头**　系指一种用于同轴电缆的连接器。

Bayonet type quick-disconnect coupling　**卡口式快速接头**

BC-A　**散装货船协调附加标志 BC-A**　授予设计装载货物密度为 $1.0 \ t/m^3$ 及以上的，最大吃水工况有指定空货舱组，装载工况中包括 BC-B 的要求，船长 150 m 及以上的干散货船的协调附加标志。

BC-A bulk carriers　**BC-A 型散货船**　系指船长为 150 m 及以上的，装载货物密度为 $1.0 \ t/m^3$ 及以上干散货物，且最大吃水工况中有指定空货舱组，并且装载工况中包括 BC-B 要求的散货船。BC-A bulk carriers means bulk carriers with the length of 150 m and above, and it can carry dry bulk cargoes with cargo density $1.0 \ t/m^3$ and above, with specified holds empty at maximum draught in addition to BC-B conditions.

BC-B　**散装货船协调附加标志 BC-B**　授予设计装载货物密度为 $1.0 \ t/m^3$ 及以上的，所有舱装货，装载工况中包括 BC-C 的要求、船长 150 m 及以上的干散货船的协调附加标志。

BC-B bulk carriers　**BC-B 型散货船**　系指船长为 150m 及以上的，装载货物密度为 $1.0 \ t/m^3$ 及以上干散货物，且所有货舱装货，并且装载工况中包括 BC-C 的要求的散货船。BC-B bulk carriers means bulk carriers with the length of 150 m and above, and it can carry dry bulk cargoes of cargo density of $1.0 \ t/m^3$ and above with all cargo holds loaded in addition to BC-C conditions.

BC-C　**散装货船协调附加标志 BC-C**　授予设计装载货物密度小于 $1.0 \ t/m^3$ 的，船长 150 m 及以上的干散货船的协调附加标志。

BC-C bulk carrier　**BC-C 型散货船**　系指船长为 150 m 及以上的，装载货物密度小于 $1.0 \ t/m^3$ 干散货物的散货船。BC-C bulk carrier means bulk carriers with the length of 150 m and above, and it can carry dry bulk cargoes of cargo density less than $1.0 \ t/m^3$.

Beacon　**信标**　系指在海床或结构物上的声学脉冲发生装置。以有规律的间隔时间重复其信号，用于确定船舶位置的水声位置基准系统（HPR）。

Beacon light　**航标灯**　系指为保证船舶在夜间安全航行而安装在某些航标上的一类交通指示灯。

Beacon light boat　**航标灯船**　系指布设在航道与其附近的暗礁、浅滩、岩石处，装有作为航标使用，有发光设备的专用船舶。

Beam at waterline B_{WL}　**水线宽度 B_{WL}**　水线宽度 B_{WL} 应在平行于艇中线面的两垂向平面之间进行测量，在指定装载状态下测量艇体表面与水平面的交线之间的最大距离。对多体艇，水线宽度应按每一单体艇分别确定。Beam at waterline B_{WL} shall be measured as the distance between two vertical planes parallel to the center-plane of the craft, as the maximum distance between the intersection of the hull surface and the flotation plane for a specific loading condition. For multi-hulls, the beam at waterline shall be established for each hull individually.

Beam element　**梁单元**　系指线单元，具有轴向、扭转和双向剪切和弯曲刚度，且沿单元长度其特性不变。
Beam element is line element with axial, torsional and bi-directional shear and bending stiffness and with constant properties along the length of the element.

Beam knee　**梁肘板**　系指连接横梁与肋骨的肘板。

Beam of the hull B_H　**艇体宽度 B_H**　艇体宽度 B_H 应在通过艇体最外侧的永久性固定部件的平行于艇中线面的两垂向平面之间进行测量。艇体宽度包括艇的所有结构或组成部件，诸如艇体的延伸部分，艇体/甲板连接件及舷墙。艇体宽度不包括能以不被损坏的方式及不影响艇完整性而拆卸的可拆部件，例如延伸到艇舷外的橡胶护舷材、碰垫、护栏和支柱以及其他类似设备。但艇体宽度包括艇在静止或航行时起静水力或动力支承作用的艇体可拆部件。对多体艇，艇体宽度应相应地按每一单独艇体确定。单体艇测量见图 B-3，多体艇测量见图 B-4。Beam of the hull B_H shall be measured as the distance between two vertical planes parallel to the center-plane of the craft, between the outermost permanently fixed parts of the hull. The beam of the hull includes all structural or integral parts of the craft such as extensions of the hull, hull/deck joints and bulwarks. The beam of the hull excludes removable parts that can be detached in a non-destructive manner and without affecting the integrity of the craft, e. g. rubbing strakes, fenders, guardrails and stanchions extending beyond the craft's side, and other similar equipment. The beam of the hull does not exclude detachable parts of the hull, which act as hydrostatic or dynamic support when the craft is at rest or underway. For multi-hulls, the beam of the hull shall be established accordingly for each individual hull. See Figure B-3 for mono-hull measurements and Figure B-4 for multihull measurements.

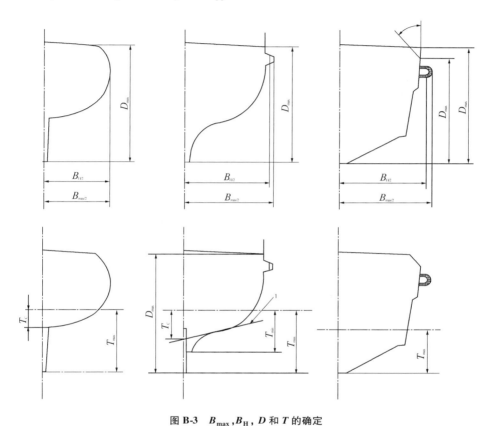

图 B-3　B_{max}，B_H，D 和 T 的确定

Figure B-3　Determination of B_{max}，B_H，D and T

说明(Key)：

1　切线　Tangent

注意：D_{max}的较高位置取决于在船体/甲板交线与实际甲板之间的倾角。如 α≥45°，则采用较低位置，如 α<45°，则采用较高位置。

Note：The upper position of D_{max} depends on the inclination between the hull/deck intersection and the actual deck. Where α≥45°, the lower position is adopted, where α<45°, the upper position is adopted.

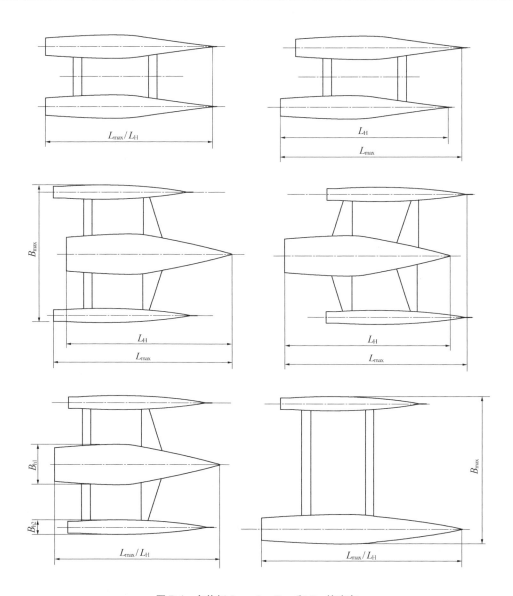

图 B-4　多体艇 L_{max}，L_H，B_{max} 和 B_H 的确定

Figure B-4　Determination of L_{max}, L_H, B_{max} and B_H for multihulls

Beam pin（pin of hatch beam）　舱口梁销　系指舱口梁端部插入承座后，用以止锁的金属销。

Beam sea　横浪　系指沿正方向或负方向行进的波浪。Beam sea are those waves propagating in the positive or negative direction，

Bearer B/L　不计名提单　系指提货单上的收货人栏内，没有指明特定收货人、承运人交货，只凭单，不凭人，采用这种提单风险大，故其在国际贸易中也很少采用。

Bearing burnt　轴承烧损　系指由于滑油供应失常

或其他原因,使轴承产生高温而导致部分轴承熔化甚至与轴胶结的现象。

Bearing clearance 轴承间隙 系指轴承内径和轴的外径之差。

Bearing ratio 承载比 系指甲板模块质量/壳体结构质量之比。

Bearing span of shafting 轴承跨距 系指轴系中,相邻两个轴承间的距离。

Bearing-only tracking (BOT) on under water target 水下目标纯方位角跟踪 系指根据目标本身发出的声波辐射以及无源声呐实时测得的目标相对方位信息,实时估算目标运动状态(航速、航向、相对距离等)的跟踪技术。

Bearth recess 卧龛 小船上,为充分利用狭小空间而设置的前边装有通风拉门的柜形卧铺。

Beaufort number 蒲氏风级 系指用于表示风的强度。风速与蒲氏风级的对应关系见表B-2。Beaufort Number is a measure of wind strength. The wind speed corresponding to each Beaufort number is shown in Table B-2.

表 B-2 蒲氏风级数据
Table B-2 Beaufort number data

蒲氏风级 Beaufort number	定义 definition	风速范围 Wind speed range (kn)	平均风速 Mean wind speed (m/s)	可能的平均浪高 Probable mean wave height (m)	海面征状 Descriptive sea criterion
0	无风 Calm	0~1	<0.5	0.0	海面如镜。Sea is like a mirror.
1	软风 Light air	1~3	0.5~1.7	0.1	鱼鳞状涟漪。Ripples with the appearance of scale are formed, but there is no foam crests.
2	轻风 Light breeze	4~6	1.8~3.3	0.2	小波,波长尚短,但波形显著,波峰呈玻璃外观,未破碎。There are small wavelets, still short but more pronounced, crests have a glassy appearance and do not break.
3	微风 Gentle breeze	7~10	3.4~5.4	0.6	较大小波,波峰开始破碎,出现玻璃色浪花,间或有稀疏白浪。There are large waves, crests begin to break, foam of glassy appears perhaps scattered white horse appear.
4	和风 Moderate breeze	11~16	5.5~8.4	1.0	小浪,波长逐渐变长,白浪较频。There are small waves, and they become longer and longer, fairly freguent white horses.
5	劲风 Fresh breeze	17~21	8.5~11.0	2.0	中浪,具有较显著的长波形状,白流很多(偶有飞沫)。There are moderate waves, more pronounced long form and many white horses are formed (some are sprays).

续表 B-2

蒲氏风级 Beaufort number	定义 definition	风速范围 Wind speed range（kn）	平均风速 Mean wind speed（m/s）	可能的平均浪高 Probable mean wave height（m）	海面征状 Descriptive sea criterion
6	强风 Strong breeze	22~27	11.1~14.1	3.0	大浪开始形成，带着白色浪花到处皆是（可能有些飞沫）。Large waves begin to form, the white foam crests are more extensive everywhere (probably some are sprays).
7	疾风 Near gate	28~33	14.2~17.2	4.0	浪头高耸，开化碎浪白沫开始随风吹成条纹成串飞溅。Waves heap up and white foam from breaking waves begins to be blown in streaks along the direction of the wind.
8	大风 Gale	34~40	17.3~20.8	5.5	较长的中高浪，波峰边缘开始破碎成为浪花，白沫明显地随风成串飞溅。There is greater and moderately high waves, edges of crests begin to break into spindrift. The foam is blown into well-marked streaks along the direction of the wind.
9	烈风 Strong gale	41~47	20.9~14.4	7.0	狂浪，沿风向出现密集的白流化条纹波峰开始翻滚，飞沫可能影响能见度。There is high waves. Dense streaks of foam are blown along the direction of the wind. Crests of waves begin to topple and roll over. Spray may affect visibility.
10	狂风 Storm	48~55	24.5~28.5	9.0	狂涛，波峰长而翻卷，大片白沫密集地随风成串飞溅，整个海面成白色，海面翻滚动荡汹涌澎湃，影响能见度。There is very high waves with long over hanging crests. The resulting foam in great patches is blown in dense white streaks along the direction of wind. On the whole the surface of the sea takes a white appearance. Tumbling of sea becomes heavy and is shock like. Visibility affected.

续表 B-2

蒲氏风级 Beaufort number	定义 definition	风速范围 Wind speed range (kn)	平均风速 Mean wind speed (m/s)	可能的平均浪高 Probable mean wave height (m)	海面征状 Descriptive sea criterion
11	暴风 Violent storm	56~63	28.6~32.6	11.5	异常狂涛(中小型船有时可能被遮蔽在浪后看不见)沿风向海面白沫弥漫,所有波峰边缘都被吹成泡沫。能见度受影响。Exceptionally high waves. (Small and medium-sized ships might be lost for a time to view behind the waves.). The sea is completely covered with long white patched of foam lying along the direction of the wind. Every where the edges of the wave crests are blown into froth. Visibility is affected.
12	飓风 Hurricane	>64	>32.7	14.0	空中充满白沫和飞溅水雾,风吹飞沫使海面白茫茫一片,能见度严重地受影响。The air is filled with foam and spray. The sea is completely white with driving spray. Visibility is seriously affected.

Water reactive substances 遇水反应物质 这些物质分为以下 3 类,见表 B-3。 These are classified into three groups shown in table B-3.

表 B-3 遇水反应物质
Table B-3 Water reactive substances

遇水反应指数(WRI) Water reactive index (WRI)	定义 Definition
2	接触水后,产生有毒、易燃或腐蚀性气体或气雾的化学品。 Any chemical which is in contact with water, may produce a toxic, flammable or corrosive gas or aerosol.
1	接触水后,发热或产生无毒、不可燃无腐蚀性气体的化学品。 Any chemical which is in contact with water, may generate heat or produce a non-toxic, non-flammable or non corrosive gas.
0	接触水后,不产生上述 1 类或 2 类反应的化学品。 Any chemical which is in contact with water, would not undergo a reaction to justify value 1 or 2.

Bech reversed spiral test 逆螺线操纵试验　系指调整舵角使船舶维持各指定的定常转艏角速度以检验航向稳定性和确定航向不稳定船舶在小舵角下的不稳定回线区的一种试验。

Becket 绳扣　系指装在滑车上，供套扣绳索用的附件。

Bed with drawers 柜床　系指在床铺与地板间装有柜子或抽屉的床。

Behavior managers 行为管理者　系指负责秘书处工作、管理预算、处理与缔约有关的行政事务的经理。

Behind ship tests propeller 螺旋桨船后试验　系指螺旋桨在受到船体干扰的水流中进行的试验。常用的方法是船模自航试验，也可在非自航的船模后或模拟的伴流中做试验。

BeiDou navigation satellite system (BDS) 北斗卫星导航系统　"北斗导航系统"是"北斗卫星导航系统"的全称。中国斗卫星导航系统是中国自行研制的全球卫星导航系统。是继美国全球定位系统(GPS)、俄罗斯格洛纳斯卫星导航系统(GLONASS)之后第三个成熟的卫星导航系统。北斗卫星导航系统(BDS)和美国GPS、俄罗斯GLONASS、欧盟GALILEO是联合国卫星导航委员会已认定的供应商。北斗卫星导航系统由空间段、地面段和用户段三部分组成，可在全球范围内全天候、全天时为各类用户提供高精度、高可靠定位、导航、授时服务，并具有短报文通信能力，已经初步具备区域导航、定位和授时能力。定位精度10 m，测速精度0.2 m/s，授时精度10 ns。2012年12月27日，北斗系统空间信号接口控制文件正式版1.0公布，北斗导航业务正式对亚太地区提供无源定位、导航、授时服务。2013年12月27日，北斗卫星导航系统正式提供区域服务一周年新闻发布会在国务院新闻办公室新闻发布厅召开，正式发布了《北斗系统公开服务性能规范(1.0版)》和《北斗系统空间信号接口控制文件(2.0版)》两个系统文件。2014年11月23日，国际海事组织海上安全委员会审议通过了对北斗卫星导航系统认可的航行安全通函，这标志着北斗卫星导航系统正式成为全球无线电导航系统的组成部分，取得面向海事应用的国际合法地位。中国的卫星导航系统已获得国际海事组织的认可。最近，通过差分仪试验并获得成功，使其精度从10 m提升到1 m配合地基增强系统精度将达到厘米级，并拥有短信功能。

Being closed manually 可手动关闭　系指分隔两侧依靠故障安全型电器开关或气动式释放装置(弹簧加载式等)，远程操纵关闭防火阀或用机械方式关闭。但是，截面面积不到0.075 m² 的导管或配管，通过A级分隔时，不需安装防火风闸。　Being closed manually means closing by mechanical means of release or by remote operation of the fire damper by means of a fail-safe electrical switch or pneumatic release (spring-loaded, etc.) on both sides of the division. However, ducts or pipes with free sectional area of 0.075 m² or less do not need be fitted with fire damper at their passage through Class "A" division.

Bell[1] 号钟　系指供船上作为音响信号器具使用的铜钟。

Bell[2] 潜水钟　系指在潜水作业时将从甲板居住舱进入其内的潜水员运送至潜水作业地点，并能兼作潜水员加、减压舱室用的装置。潜水钟壳体呈椭圆形，顶端配有主吊缆的吊孔，外周设防碰撞保护环形圈。在回收系统故障，不能回收潜水钟的紧急情况下，可以借助解脱装置作为应急手段，使其上浮至水面。

Bell controlling system 潜水钟控制系统　系指对潜水钟内供气、热水供应、照明、环境(压力、氧气、二氧化碳浓度)、通信、吊放操作等实施监视和控制的系统。

Bell lifting and launching system 潜水钟吊放系统　系指在额定的安全工作负荷下，完成潜水钟下放和回收的装置。该系统由吊架、滑车、绞车、脐带、液压滑车、升沉补偿绞车钢缆、恒张力绞车和液压动力站和导向压载等组成。

Bell umbilical 潜水钟主脐带　潜水钟、水下居住舱及系缆可潜器从水面工作船或补给浮标获得电能、气体、热水和联络信号等的一束管线，是饱和潜水系统的生命线。其基本组成包括：供气软管、测深软管、抗压排气管、热水软管、电源电缆、通信/钟内摄像电缆、钟外摄像电缆等。此外，在管缆间填充有聚酯绳，以增加脐带耐压和抗压性。软管端部用磷青铜压环固定在螺旋芯子上。脐带外包有标准紧网状单纤维聚丙烯织物保护层。

Belt tightener 皮带张紧器

Bending moment curve 弯矩曲线　系指船体或构件弯曲时，其横剖面内的弯矩沿船长或构件长度方向变化的曲线。

Bending ratio 弯曲比率　对导缆装置而言，系指导缆装置的支承表面直径与短拖索的直径之比。　For fairleads, bending ratio means towing pennant bearing surface diameter to towing pennant diameter.

Bending stiffener 限弯器

Bending wave 弯曲波　系指在点、线力驱动下，或入射声波的激励下，板或棒作弯曲运动并向周围空间辐射的声波。由于弯曲波的传播速度与频率有关，因此任一复杂波形将随传播距离而改变其形状。

Beneficiary 受益人 系指人身保险合同中由被保险人或者投保人指定的享有保险金请求权的人,投保人、被保险人可以为受益人。如果投保人或被保险人未指定受益人,则他的法定继承人即为受益人。一般见于人身保险合同。这种单据的名称因所证明事项不同而略异,可能是寄单证明、寄样证明(船样、样卡和码样等)、取样证明、证明货物产地、品质、商标、包装和标签情况、电抄形式的装运通知、证明产品生产过程、证明商业已检验、环保人权方面的证明(非童工、非狱工制造)等。

Benign area 良好海况区域 系指不受热带风暴或运动低气压影响的区域。然而,这些区域受热带风暴或运动低气压影响时除外,如西南季风时的北印度洋或东北季风时的南中国海区域。良好海况区域的气象条件:风速 15 m/s;有义波高 2 m。

Benthic trawl winch 底栖拖网绞车 系指为收放大中型生物网具的工作绞车。该绞车也用于底质采样。

Benthic trawling gear 底栖拖网装置 系指保证底栖拖网作业的一种专门装置。由网具、网位仪、底栖拖网绞车、缓冲器及起重吊杆等组成。

Benthos 底栖生物 系指生活在海底的固着或爬行的水动植物群(包括蜗牛、虾、蠕虫和螃蟹)。

Benthos trawl 底栖生物拖网 系指利用底栖生物网具在海底进行航行拖曳并采集底栖生物样品的一种作业。常用网具有:阿拖网、桁拖网、双刃拖网、雪橇网、板式拖网。

Benthos trawl (dredge) 底栖生物拖网 系指用于定性采集深海底栖生物样品的袋形网具。

Bern carbon cycle model 伯尔尼碳循环模型 系指设计用来研究人为的二氧化碳排放量与大气中的二氧化碳浓度之间关系,以及前者与地球的辐射平衡中产生表面温度信号的瞬间响应扰动之间关系的模型。伯尔尼碳循环模型于 1996 年首次被 IPCC SAR(Secord assessment report)定义为 CO_2 情况分析的标准和计算全球变暖潜值的标准,并于 2001 年出版了 Bern 碳循环模式的修正版。

Berth charter 泊位装货租船合同

Berth ladder 床梯 系指架设于上、下铺之间的垂直梯子。

Berth recess 卧室 小船上,为充分利用狭小空间而设置的前边装有通风拉门的柜形卧铺。

b_f (mm) 普通扶强材或主要支撑构件的面板宽度(mm)。 b_f is the face plate width (mm) of ordinary stiffener or primary supporting component.

Bi-colored light 双色舷灯 系指设置在船舶中线面上规定处,并在水平面上 225°内,即自艏向至左、右舷 112.5°,显示左红、右绿不间断光的舷灯。

Bid bond/Lend bond SLC 投标备用信用证 系指对投标申请人中标后执行合同的责任和义务进行担保。

Bidder 竞买人/竞买者 系指参加竞购拍卖标的的公民、法人或者其他组织。

Biennial survey 两年度检验 系指两年进行一次的检验。

Bifurcation buckling 分叉屈曲 系指描述使原先单一解决方法分叉成两个可能解决方法点的一个数学术语。Bifurcation buckling means a mathematical term that describes the point at which a previously unique solution bifurcates into two possible solutions.

Bilateral contract 双务合同 系指当事人双方互负对待给付义务的合同,买卖合同是双务合同的典型。双务合同中的抗辩权是在合同履行过程中产生的,在符合法定条件时,当事人一方得以对抗另一方的履行请求权,起到暂时拒绝履行己方义务的作用。现实生活中的合同大多数为双务合同,如买卖、互易、租赁、承揽等。

Bilge 舭部 系指船底和船舷之间的连接部分。

Bilge alarm 舱底报警 系指示舱底高水位异常的报警。Bilge alarm means an alarm which indicates an abnormally high level of bilge water.

Bilge and ballast oily water separator 舱底水和压载水的油水分离器

Bilge and ballast system 舱底和压载水系统 舱底水及压载水系统的总称。

Bilge area 舱底范围 系指机舱底层花钢板和机舱底部之间的空间。Bilge area is the space between the solid engine-room floor plates and the bottom of the engine-room.

Bilge automatic discharging device 自动舱底水排出装置 待舱底水积聚到一定水位即能自动启动舱底水泵及有关阀件将舱底水排除而在排除后又能自动停止的一套自动控制排水装置。

Bilge brackets 舭肘板 系指肋板与肋骨之间连接用的肘板。Bilge brackets means a brackets used for connecting floors to frames.

Bilge branch pipe line (bilge pipe) 舱底水支管 由舱底水总管或舱底水集合阀箱分接到各水密分舱污水沟或污水阱处的管路。

Bilge diagonal 舭斜剖线 系指通过舭部的斜剖线。

Bilge grab rail 龙骨扶手 救生艇底上,供艇翻覆时落水人员把握用的开有间断长孔的舭龙骨。

Bilge high level alarm　舱底水高位报警器　即舱底水的高水位报警装置。

Bilge injection vavle　舣部吸水阀

Bilge keel　舭龙骨　一块沿舭部垂直设置在船壳上以减少横摇运动的板材。Bilge keel is a piece of plate set perpendicular to a ship's shell along the bilges to reduce the rolling motion.

Bilge level alarm　舱底水液位报警装置　系指舱底水高水位报警装置。

Bilge level detection and alarm system　舱底水水位探测与报警系统

Bilge level monitoring device　舱底水位监测装置

Bilge main　舱底水总管　系指自舱底水泵接至各舱底水支管或支管集合阀箱的管路。

Bilge monitoring device　舱底水监视装置　系指安装在泵舱内的监视舱底水位的装置。

Bilge oily water separator　舱底水油水分离器　利用重力、吸附及离心原理将舱底水中所含的油分分离出来的设备。

Bilge piping　舱底水管路　系指排除舱底积水和海损时排除破舱进水的管系。按照国际防止船舶造成污染公约的规定和要求，管系中需设置油水分离器，舱底水必须在达到允许含油量的标准后才能排放。

Bilge plating　舭部板　系指在船底外壳和舷侧外壳间的曲线型板的区域。如下所取：从船底处舭部圆弧下缘的开始处到舷侧外壳的舭部圆弧上缘终端处，或基线/局部中心线升高以上 0.2D 处两者的较小位置。Bilge plating means an area of curved plating between the bottom shell and side shell. It is taken as follows: from the start of the curvature at the lower turn of bilge on the bottom to the lesser of, the end of curvature at the upper turn of the bilge on the side shell or 0.2D above the baseline/local centerline elevation.

Bilge primary tank　舱底水预处理柜　对 MEPC.1/Circ.642 中"舱底水综合处理系统指南"而言，舱底水预处理柜系指为对含油舱底水进行分离的一个预处理单元。For the purposes of MEPC. 1/CIRC. 642 "Guideline for Inteqrated Bilge water Treatment System", the bilge water pretreament tank means a pretreament Unit for the Separation of oil.

Bilge pump　舱底泵　一般具有自吸能力，用于抽吸舱底水并将其排至舷外的水泵，属于保船的设备。

Bilge pump numeral　舱底泵数　用于衡量一艘客船是否需要增设一台舱底泵的计算值，计算公式如下：当 P_1 大于 P 时，舱底泵数 $= 72 \times [(M + 2P_1)/(V + P_1 - P)]$；在其他情况下，舱底泵数 $= 72 \times [(M + 2P)/V]$；式中：$L=$ 船长，m；$M=$ 机器处所的容积，m³，它位于舱壁甲板以下，并加上机器处所前方或后方位于内底以上的如何固定燃油舱的容积；$P=$ 舱壁甲板以下的乘客处所和船员处所的总容积，m³，它为乘客和船员提供居住和使用的处所，但不包括行李、物料、食品和邮件舱；$V=$ 舱壁甲板以下的船舶总容积，m³，$P_1 = KN$；式中：$N-$核准该船搭载的乘客数；$K-$ 0.056 L。但是，如 KN 的数值大于 P 与舱壁甲板以上的实际乘客处所总容积之和，则 P_1 应取上述之和，或 KN 值的 2/3，取两者中的较大者。

Bilge pumping system　舱底水泵送装置　系指在所有实际工况下均能抽除及排干任何水密舱室的水的装置。但固定用于装载淡水、压载水、燃油或液货并设有其他有效排水设备的处所除外。在冷藏舱应设置有效的排水装置。Bilge pumping system is a system which is capable of pumping and draining from any watertight compartment other than a space permanently appropriated for the carriage of fresh water, water ballast, oil fuel or liquid cargo and for which other efficient means of pumping are provided, under all practical conditions. Efficient means shall be provided for draining water from refrigerated holds.

Bilge radius　舭部半径　舭部呈圆弧形时，最大横剖面上舭部的圆弧半径。

Bilge sludge box (bilge mud box, rose box)　舱底水过滤箱　为滤去机炉舱和轴隧内舱底水中混有的棉纱、布条油污的污物，而在舱底水吸入管路中设置的滤箱。

Bilge strake　舭列板　系指在舭部板的较低的列板。Bilge strake means a lower strake of bilge plating.

Bilge suction strum plate　舱底水吸入口滤网　系指设置在舱底污水吸入口处的滤网。

Bilge suction valve　舱底水吸入阀　系指吸入舱底水的阀。

Bilge system　舱底水系统　系指将舱底污水排出舷外的系统。

Bilge tangential point　舭部切点　定义为一条斜线与舭部的切点，该斜线与 LCG 处的水平面成为 50°角。见图 B-5。As Figure B-5 shows, bilge tangential point is defined as the tangential point of the bilge with an oblique line sloped at 50° to the horizontal at the LCG.

Bilge tank　污水舱　系指供贮存污水的舱室。

Bilge water　舱底水　系指船舶舱底部的积水。

Bilge water disposal　污水处理　为使污水达到排水某一水体或再次使用的水质要求对其进行净化的过

程。污水处理被广泛应用于建筑、农业、交通、能源、石化、环保、城市景观、医疗、餐饮等各个领域,也越来越多地走进寻常百姓的日常生活。

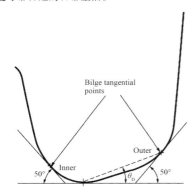

图 B-5 舷部切点的定义

Figure B-5　Definition of bilge tangential point

注意:(1) Bilge tangential point 舷部切点
　　　(2) outer 外切点
　　　(3) inner 内切点

Bilge water system　油污水收集和处理系统　系指收集和处理船舶或海上平台甲板以上各机器和各舱室处所的滴漏油、冲洗水等舱底水的系统。其主要由油水分离器、分离水舱、污油舱、舱底水舱及各种阀门、连接件等组成。

Bilge water tank　污水舱　系指专门设计用来储存含油污水的舱柜。

Bilge well　污水井　系指用以聚集和排出舱底水的井状结构。

Bilge-pumping arrangement　舱底排水设备

Bilge-pumping plant　舱底排水装置

Bill　票证　我国计划经济时期极具时代特色的票证,经历了40多年的风风雨雨,终于在20世纪90年代逐步退出了经济舞台,完成了历史使命。中国的票证历史是一部凝重浑厚的中华民族创业史,是一部华夏子孙与贫穷、饥饿的抗争史,是囊括中国农业、商业、工业、服务业的发展史,是中国计划经济这段历史的真实写照和证明。

Bill of exchange　汇票　系指一个人向另一个人签发的要求受票人在见票时或在将来固定的时间或可以确定的时间向某人或其指定人或持票人无条件地支付一定金额的书面支付命令。

Bill of lading　提单　系指用以证明海上货物运输合同和货物由承运人接收或装船,以及承运人保证据以交付货物的单证。单证关于货物应按记名人的指示交付,或者按指示交付,或者向提单持有人交付的规定,构成此种保证。

Bill of landing　海运提单　系指承运人收到承运货物后签发给出口商的证明文件,它是交接货物、处理索赔与理赔以及向银行结算货款或进行议付的重要单据。

Binary code　二进制代码　二进制就是当到2后就进一位。比如 0 + 1 = 1,1 + 1 = 10,10 + 1 = 11,11 + 1 = 100……特征是只有0和1。一串代码只有0和1构成就是二进制代码。

Binary notation　二进制计数器

Bioaccumulation tests　生物体内积累测试　系指通过使用测得达到海洋(或淡水)生物的生物聚集系数对潜在的生物体内积累进行评估的试验。如果这些测试不适用,或 $\log P_{ow} < 3$,可采用量化的结构活性关系[(Q)SAR]模式估算生物聚集系数(BCF)值。 Bioaccumulation tests mean those tests in which the assessment of (potential for) bioaccumulation should use measured bio concentration factors in marine (or freshwater) organisms. Where these tests are not applicable, or if $\log P_{ow} < 3$, Bio Concentration Factor (BCF) values may be estimated using (Quantitative) Structure-Activity Relationship [(Q)SAR] models.

Bioavailability (biomagnifications, bio concentration) 生物药效率(生物放大作用,生物集结)

Bio-computer　生物计算机　主要原材料是生物工程技术产生的蛋白质分子,并以此作为生物芯片,利用有机化合物存储数据。信息以波的形式传播,当波沿着蛋白质分子链传播时,会引起蛋白质分子链中单键、双键结构顺序的变化。运算速度要比当今最新一代计算机快10万倍,它具有很强的抗电磁干扰能力,并能彻底消除电路间的干扰。能量消耗仅相当于普通计算机的十亿分之一,且具有巨大的存储能力。生物计算机具有生物体的一些特点,如能发挥生物本身的调节机能,自动修复芯片上发生的故障,还能模仿人脑的机制等。

Biodegradable　生物降解功能　系指诸如塑料等化工产品在大自然中能随时间的推移而逐渐分解的功能。

Biodiesel　生物柴油　系指利用有机生物或其废料提炼出来的具有热值能进行燃烧的类油液体。

Bio-fouling　生物污底(污垢)　系指在浸入或曝露于水生环境的表面和结构上的水生生物(例如微生物、植物和动物)的积聚。生物污底可包括大型污底和微型污底(垢)。 Bio-fouling means the accumulation of aquatic organisms such as micro-organisms, plants, and animals on surfaces and structures immersed in or exposed to the aquatic environment. Bio-fouling can include micro-fouling and macro-fouling.

Bio-fouling management plan　生物污垢管理计划　系指为生物污垢管理提供有效程序的计划。生物污垢管理计划可以是单独文件,或者部分或全部纳入船舶操作和程序手册和/或计划维护保养系统。该计划包括:(1)2011 年为最大程度减少入侵生物种转移的船舶生物污垢控制和管理指南的相关部分;(2)防污底系统和使用操作方法或处理的细节(包括不利地区);(3)易受生物污垢影响的船体位置、防污底系统的计划检查、修理、维护保养和换新的日程表;(4)适合于选择的防污底系统和操作方法的推荐工作条件的细节;(5)与船员安全相关的细节,包括使用的防污底系统的细节;和(6)为验证生物污垢计划簿中记录的任何处理要求的文件的细节。 Bio-fouling management plan means a plan provided effective procedures for bio-fouling management. The bio-fouling management plan may be a stand-alone document, or integrated in part or fully, into the existing ship' operational and procedural manuals and/or planned maintenance system. The bio-fouling management plan includes: (1) relevant parts of 2011 guidelines for the control and management of ship' bio-fouling to minimize the transfer of invasive aquatic species; (2) details of the anti-fouling systems and operational practices or treatments used, including those for niche areas; (3) schedules of planned inspections, repairs, maintenance and renewal of hull locations susceptible to bio-fouling, and anti-fouling system; (4) details relevant for the working condition for anti-fouling system and operation method; (5) details relevant for the safety of the crew, including details on use of the anti-fouling system(s); (6) details of the documentation required to verify any treatments in the Bio-fouling Record Book.

Bio-fuel blends　生物燃油混合物　系指生物燃油中的货品与石油混合后形成的混合物。 Bio-fuel blends are mixtures resulting from the blending of those bio-fuel products with petroleum oil.

Bio-fuels　生物燃油　系指 IBC 规则第 17 章或第 18 章和 MEPC.2 通函/三方协议中所列的乙醇、脂肪酸甲酯、植物油(甘油-脂酯)和闪点为 60 ℃或低于或高于 60 ℃直链和带有支链的烷烃。Bio-fuels are ethyl alcohol, fatty acid methyl esters (FAME), vegetable oils (triglycerides) and alkanes (C-10-C36), linear and branched with a flashpoint of either 60 ℃ or less or more than 60 ℃, as identified in chapter 17 and 18 of the IBC Code or the MEPC.2/Circular/tripartite agreements.

Bio-geographic region　生物地理区域　系指由地理和生物特性界定大的自热区域,区内的动物和植物物种有高度相似性。无明显和绝对界限,只有或多或少清晰可见的过带。 Bio-geographic region is a large natural region defined by physiographic and biologic characteristics within which the animal and plant species show a high degree of similarity. There are no sharp and absolute boundaries but rather more or less clearly expressed transition zones.

Biological research institute(BRI)　生物研究所

Biological sample chamber　生物样品舱　用于保存海洋调查所采集的生物固定样品(用福尔马林、酒精等浸泡固定)的舱室,该室要通风降温。

Biota　生物群

Bipod mast　人字桅　系指通常兼作桅用的由两根斜柱构成的呈人字形的起重柱。

Bit　比特　是一个包含很多原子的系统,由一个或多个连续参数如电压描述。

Bitwise operation　逐位计算

Bitt(mooring bitt)　带缆桩　系指固定在甲板上用以系缚缆绳的桩柱。

Bituminous mineral wool felt　沥青矿棉毡

Black basket　黑色鱼篮　系指供小型渔船用的黑色提篮状的号型。

Black box　黑匣子　学名飞行数据记录仪(Flight data recorder),记录飞机飞行期间的详细资料。现代商用飞机一般安装两个黑匣子,分别是"驾驶舱语音记录器"和"飞行数据记录器"。"驾驶舱语音记录器"用来记录驾驶舱和客舱的飞行员、飞行员之间及客舱内乘客的讲话录音及各种可听见的声响。"飞行数据记录器"则记录飞行数据,如:水平速度、垂直速度、加速情况和磁角等飞行数据。黑匣子通常为橙红色,长约 10 cm,宽与高约为 20 cm 的长方形金属盒子,每个质量为 20~30 kg。两个黑匣子一般安装在飞机尾部,因为机尾在空难事故中通常保存相对完整。黑匣子携带水下信息识别器,一旦落入水中受到浸泡,自动激活电池供电系统,不间断地向外发射频率为 37.5 千赫的电磁波,供搜寻用专门接收器确认其方位。法律规定,黑匣子入水后电池必须保证发射信号至少 30 天。一些专家指出,在 30 天之后,依据黑匣子的电池电力不同,还能发出 15 天或更长时间的噪声。

Black conical shape　黑锥号型　系指圆锥体的高度等于其底部直径的黑色圆锥体号型。

Black cylinder shape　黑柱号型　系指黑色圆柱体状的号型。

Black diamond shape　黑菱号型　系指正倒两个

Black globular shape (black ball) 黑球　系指黑色球体状或由两正交的黑色圆片构成的号型。

Black Sea area 黑海区域　系指黑海本身,与地中海以北纬41°为界。

Black water 黑水　系指 MARPOL73/78 附则Ⅳ所定义的"生活污水",并见之于 ISO 文件及某些国家的法规(例如德国的"SBG")中。

Blackout 失电(动力定位系统)　推力器或 DP 控制系统的全部主电源丧失。电源丧失妨碍 DP 控制系统操作称为 DP 失电。

Blackout 中断　系指主要和辅助的机械装置,包括主电源都不运行,但仍然可得到使它们投入运行的各项服务(例如压缩空气、从蓄电池来的启动电流等)。Blackout means the main and auxiliary machinery installations, including the main power supply, are out of operation but the services for putting them back into operation (e. g. compressed air, starting current from batteries etc.) are available.

Blackout period 停电期　系指主发电机和应急发电机停止电力供应的时期。　Blackout period is the period suffering loss of electric power from the main and emergency generating plants.

Blackout situation 船舶失电　系指主、辅助机械,包括主电源不能工作,但启动它们的设备(如压缩空气、启动蓄电池等)仍然可用的状态。　Blackout situation is a situation main and auxiliary machinery, including main power supply, are out of operation, but the services for putting them back into operation (e. g. compressed air, starting current from batteries, etc.) are available.

Blade 叶片　由叶根和具有特定曲线型面的叶杆组成,用以改变工况、工质压力和流向从而实现能量转换的通流部分的主要元件。

Blade area ratio 叶面比　系指螺旋桨展开或伸张面积对盘面积的比值。

Blade corrosion 叶片侵蚀　系指湿蒸汽区域工作的汽轮机叶片由于水滴冲击所引起的叶片材料表面的损伤。

Blade developed area 叶片展开面积　系指在一个平面上叶片展开的面积。　Blade developed area is the area of the blade surface expanded in one plane.

Blade element 叶元体　系指假定的组成螺旋桨叶的共轴线弧形薄片。

Blade element theory of screw propeller 螺旋桨叶元体理论　假定螺旋桨叶是由许多叶元体所组成,叶的推力与转矩是各叶元体推力与转矩之和,但未考虑螺旋桨对水感生速度作用的理论。

Blade face reference line 叶面参考线　在螺旋桨图上据已量出各半径处导边和随边的线。通常即叶面的母线,也可用连接各半径处切面弦长中点的线,连接切面最大厚度处的线或其他适宜的线作为参考线。

Blade failure load F_{ex}　叶片失效载荷 F_{ex}　系指塑性弯曲导致叶片损失的极限叶片载荷。该作用力可导致叶片完全失效,即在根部区域产生了塑性弯曲。该作用力作用于 $0.8R$ 处。转力力臂为 $0.8R$ 处的导边或随边至桨叶自转轴线距离的 $2/3$,取较大者。叶片失效载荷用于确定叶片螺栓、螺距控制机构、螺旋桨轴和推力轴承的尺寸。目的是保证整个螺旋桨的叶片失效不会导致其他部件的损坏。　Blade failure load F_{ex} means ultimate blade load resulting from blade loss through plastic bending. The force may cause total failure of the blade so that plastic hinge is caused to the root area. The force is acting on $0.8R$. Spindle arm is taken as $2/3$ of the distance between the axis of blade rotation and the leading/trailing edge (whichever is the greater) at the $0.8R$ radius. Blade failure load is used to determin the size of the blade bolts, pitch control mechanism, propeller shaft, propeller shaft bearing and thrust bearing. The objective is to guarantee that total propeller blade failure should not cause damage to other components.

Blade handle 叶柄　系指通至毂内用以转动可调螺距螺旋桨叶的柄。

Blade root 叶根　系指:(1)使叶片固定在压气机或涡轮的轮盘和机匣上的具有一定尺寸和形状的叶片根部。(2)螺旋桨叶连接于毂的部分。

Blade thickness on axial line 轴线上叶厚　在螺旋桨图中将表示各半径处最大叶厚的线延长到毂内在轴线上得到的假想厚度。最大叶厚不按直线规律变化者,此厚度只能大略估计。

Blade thickness ratio 叶厚比　系指螺旋桨轴线上叶厚对其直径的比值。

Blade tip 叶梢　系指螺旋桨距轴线最远处。

Blade tip clearance 梢隙　系指螺旋桨在旋转中叶梢与船体间的最小距离或导管推进器叶轮与导管壁间的最大间隙。

Blade tip thickness 梢厚　系指螺旋桨叶梢部的最大厚度。在叶梢附近背削薄者,取假定未削薄的厚度作为厚度。

Blade vibration 叶片振动

Blade wheel(rotor wheel) 叶轮　系指装上动叶后的轮盘。

Blast engine(blower) 鼓风机(通风机、增压器、压

气机) 压力较高的风机,一般锅炉鼓风用。

Blast furnace 鼓风炉(高炉)

Blast sweep 扫掠式喷射

Blasting and profile 喷射处理和粗糙度

Blasting treatment 抛(喷)射处理 系指利用高速磨料的冲击作用,清理和粗化基底表面的工艺过程。

Blasting treatment means a technological process to clean and rough the ground surface by the impacting action of high speed abradant.

Bleed air rate 抽气量 系指单位时间内从蒸馏装置冷凝器中抽除的空气量。

Bleed steam system (extraction steam pipe line) 抽气管系 从汽轮机中间级抽出蒸汽输送到给水预热器或其他热交换器的管路及附件。

Bleeding steam turbine (extraction steam turbine) 抽气式汽轮机 从中间级中抽出一定压力的一部分蒸汽供辅汽轮机,抽气器工作或其他用途的汽轮机。

Blending on board 船上混合 系指将两种货品混合成一种货品(经混合的混合物),并仅表现为与任何其他化学处理截然不同的物理混合。此类混合作业应仅在船舶位于港口限制范围内时进行。 Blending on board describes mixing two products to get one single product (a blended mixture) which reflects only physical mixing distinct from any chemical processing. Such mixing operations should only be undertaken whilst the ship is within port limits.

Blending stocks 调和油料

Blending system 混砂系统

Blind flange 封口法兰

Blind plate 盲板

Blind sector 盲区 系指视野中的障碍物,在其两侧有明显的扇区。 Blind sector means an obstruction in a field of vision with a clear sector on both sides.

Blisters 起泡

Block (pulley block) 滑车 由滑轮、车壳、车轴、滑车箍及滑车接头等组成,供变换作用力方向,起省力或减少绳索摩擦等作用的简单起重器。

Block assembly 分段组装

Block coefficient C_b 方形系数 C_b 系指:(1)由下式确定:$C_b = \nabla/LBd_1$,式中,∇——对于金属船壳的船舶是船舶的型排水体积,不包括附体;对于其他材料船壳的船舶是量到船体外表面的排水体积;两者均取自 d_1 处的型吃水;d_1——最小型深的85%;(2)计算多体船的方形系数时,应取用全宽(B)而不是单个船体的宽度。 Block coefficient (C_b):(1) it is given by:$C_b = \nabla/LBd_1$ where:∇ is the volume of the moulded displacement of the ship, excluding bossing, in a ship with a metal shell, and is the volume of displacement to the outer surface of the hull in a ship with a shell of any other material, both are taken at a moulded draught of d_1, and where:d_1 is 85% of the least moulded depth;(2) when calculating the block coefficient of a multi-hull craft, the full breadth (B) is to be used instead of the breadth of a single hull.

Block coefficient C_B (barge) 方形系数 C_B(驳船) $C_B = D/1.025LBT$。对于与推船刚性连接的驳船,C_B 应按驳船/推船的组合体进行计算。 Block coefficient $C_B = D/1.025LBT$. For barge rigidly connected to a push-tug, C_B is to be calculated for the combination barge/push-tug.

Block connection 滑车接头 系指滑车上,供悬吊的部件。

Block for signal flag 旗索滑车 系指安装在旗杆顶端、信号张索或桅横桁等处供导引旗索和升降号旗及号型用的单饼滑车。

Block loading condition 块状装载工况

Block section 块截面积 系指根据催化剂块的外部尺寸计算的横截面积(m^2)。 Block section means the cross-sectional area (m^2) of the catalyst block based on the outer dimensions.

Block with becket 带绳扣滑车 系指配有绳扣的滑车。

Blockage correction 阻塞修正 考虑到阻塞效应而作的修正。如将有限剖面船模试验水池中所做的水动力试验结果,变换为无限水域中或另一剖面水池中的数值而作的修正。

Blockage effect 阻塞效应 系指物体在受池壁和池底或筒壁限制的流体中运动时,较其在无限流体中运动时所引起的流场和受力情况等的差异。

Blockage factor 堵塞系数 系指在多级轴流压气机设计中,用调整平均轴内速度的方法考虑环面附面层的存在和发展的影响时所采取的修正系数。

Blockage ratio 阻塞比 系指物体浸水部分横剖面积与水池水面下横剖面积的比值,或物体的横剖面积与水筒(风筒)测试段横剖面积的比值。

Blocking 闭锁 系指能避免意外开启或关闭的动作。

Blocking valve 阻流阀(阻断阀)

Blogger "博客" 系指书写网络日志(Blog)的人。通常也指谷歌公司免费为"网络博客"提供的发布平台。

Blogger Logo Weblog (Blog) 网络日志 其精髓是以个人的视角,以整个互联网为视野,精选和记录精彩内容。

Blow down (blow-off) 排污 系指为使锅水维持

规定质量标准而自锅炉中排出一部分盐碱浓度的锅水和泥渣的锅水的过程。

Blow down valve(blow off valve) 排污阀 系指锅炉排污或泄水的阀。

Blow off preventer（BOP） 防喷器 系指用于试油、修井、完井等作业过程中关闭井口，防止井喷事故发生，将全封闭和半封闭两种功能合为一体，具有结构简单、易操作、耐压高等特点，是油田常用的防止井喷的安全密封井口装置。

Blow-down pump 排盐泵 用于抽除蒸发器内盐水的泵。

Blow-down valve 底部吹泄阀(排污阀)

Blower scavenging 扫气泵扫气 系指利用扫气泵压缩空气进行的扫气。

Blower turbine 鼓风机透平 系指以驱动鼓风机的透平机。

Blowing wool 喷吹玻璃棉

Blow-off 放气 根据运行要求在压缩过程中或压气机出口处向外界放出部分空气或其他气体的过程。

Blow-out limit 吹熄极限 系指保持燃烧室稳定工作的空气燃料比的上下极限值。

Blue peter 开船旗 系指升起该旗，即表示该船即将起航。

Blue visual signal 蓝色的视觉信号 系指根据需要给予特定的含义的指导/信息信号。例如：电动机准备启动；空载发电机准备合闸；停转电动机加热电路接通。Blue visual signal means an indication for which specific meaning is assigned according to the need in the case considered. For example: motor begins to start, unloading generator begins to switch on, heating circuit of stopping motor is connected.

Bluetooth file transfer "蓝牙"传输 系指比较普遍使用的传输方式。对距离的限制没有红外大，一般是 10 m 左右，而且可以传输过程中随便移动。缺点是比较容易中病毒。

Bluetooth headset "蓝牙"耳机 就是将"蓝牙"技术应用在免持耳机上，让使用者可以免除恼人电线的牵绊，自在地以各种方式轻松通话。自从问世以来，"蓝牙"耳机便成了行动商务族提升效率的好工具。"蓝牙"是一种低成本大容量的短距离无线通信规范。"蓝牙"笔记本电脑，就是具有"蓝牙"无线通信功能的笔记本电脑。"蓝牙"这个名字还有一段传奇故事。公元 10 世纪，北欧诸侯争霸，丹麦国王挺身而出，在其不懈的努力下，血腥的战争被制止了，各方都坐到了谈判桌前。通过沟通，诸侯们冰释前嫌，成为朋友。由于丹麦国王酷爱吃蓝梅，以致于牙齿都被染成了蓝色，人称"蓝牙国王"，所以，"蓝牙"也就成了沟通的代名词。一千多年后的今天，当新的无线通信规范出台时，人们又用"蓝牙"来为其命名。1995 年，爱立信公司最先提出"蓝牙"概念。"蓝牙"规范采用微波频段工作，传输速率每秒 1 M 字节，最大传输距离 10 m，通过增加发射功率可达到 100 m。"蓝牙"技术是全球开放的，在全球范围内具有很好的兼容性，全世界可以通过低成本的无形蓝牙网连成一体。

Blue-tooth technology "蓝牙"技术 系指一种尖端的开放式无线通信标准，能够在短距离范围内无线连接桌上型电脑与笔记本电脑、便携设备、PDA、移动电话、拍照手机、打印机、数码相机、耳麦、键盘甚至是电脑鼠标。"蓝牙"无线技术使用了全球通用的频带（2.4 GHz），以确保能在世界各地通行无阻。简言之，"蓝牙"技术让各种数码设备之间能够无线沟通，让散落各种连线的桌面成为历史。有了整合在 Mac OS X 中的"蓝牙"无线技术，你就可以轻松连接你的 Apple 电脑和基于 Palm 操作系统的便携设备、移动电话以及其他外围设备——在 9 m (30 ft) 距离之内以无线方式彼此连接。

Blunt bow 丰满型船首 系指用板材加工制造的艏柱和外板在俯视图上与中心线成 30°或 30°以上角度的船首。Blunt bow means a bow in which any part of the shell forms an angle of 30° or more to the centerline.

BNWAS = bridge navigational watch alarm system 桥楼驾驶值班报警系统

Board and lodging expenses 食宿费

Board of directors 董事会

Board spectrophotmeter 船用分光光度计 系指适用于船载的分光光度计。

Boat and boat handling gear 艇装置 系指船上设置的艇和吊艇装置及其附件的总称。

Boat boom 吊艇杆 吊艇架中承吊小艇的杆状或杆状组合构件。

Boat chock 艇座 船上存放小艇时直接支承艇体的座子。

Boat cover 艇罩 小艇存放时，用以遮盖艇体的防护罩。

Boat davit 吊艇架 由吊艇臂或吊艇杆、吊艇架座架及各种联动机构组成的用以吊放小艇的专门装置。

Boat deck 艇甲板 放置救生艇或工作艇的一层甲板。

Boat engine 救生艇用机器 安装在救生艇上用以驱动的发动机。

Boat fail 吊艇索 直接承担吊艇时荷重的绳索。

Boat grip gear 固艇索具 艇存放时可利用松紧

螺旋扣收紧,卸艇时可利用滑脱钩迅速解脱的,固定救生艇的专用索具。

Boat handling gear 吊艇装置 船上吊卸救生艇、工作艇的专门设备的总称。包括吊艇架、起艇机、滑车、索具和其他属具。见图 B-6。

图 B-6 吊艇装置
Figure B-6 Boat handling gear

Boat mast 艇桅 设置于小艇上供张帆用的活动桅。

Boat plug 艇底塞 设置在艇底,用以释放艇内积水的专门塞子。

Boat rudder 艇舵 设置于小艇上,一般为可折并与艇系连的舵。

Boat sail 艇帆 设置于小艇上的颜色鲜明的帆。

Boat winch 起艇机 起吊船舶上小艇的绞车。

Boat/craft 艇 某些较小船的惯称。在军用船中,通常指排水量在 500 t 以下的船,唯有潜艇,不论其吨位大小均称为艇。

Boatmain's chair 座板 供高空、舷外作业用的吊在舷外的活动座板。

Boatswain's tool 帆缆用具 为帆缆作业所配备的各种工具、属具及其备品的统称。

Body coordinate system 运动坐标系 一般以船舶重心位置为原点,固定于船体的直角坐标系。符号按右手法则。

Body lines 横剖线 系指平行于中站面平面与船体型表面的交线。

Body plan 横剖线图 系指型线图中,各站处的横向垂直平面与船体型表面的交线在正视方向的投影图。

BOEMRE 美国海洋能源管理局

Boiler 锅炉(蒸汽发生器) 系指:(1)用来接收燃油或燃气燃烧产生的热量使蒸汽或水的温度在 120 ℃ 以上的有关一个或以上,接着管路系统的受火压力容器和附属的管系。任何与锅炉直接相连的设备,例如节能器、过热器、安全阀,如果不能用隔离阀与蒸汽发生器隔离的设备,则认为是锅炉的一部分。与锅炉相连的在隔离阀以上的部分管系看作是锅炉的上游部分,而在隔离阀下面的部分看作锅炉的下游部分;(2)通过火焰、燃气、其他高温气体发生蒸汽或热水装置,包括其附属装置。 Boiler is:(1)One or more fired pressure vessels and associated piping systems used for generating steam or hot water using fuel oil or gas at a temperature above 120 ℃. Any equipment directly connected to the boiler, such as economizers, super-heaters and safety valves, is considered as part of the boiler, if it is not separated from the steam generator by means of any isolating valve. Piping connected to the boiler before the isolating valve is considered as upstream the part and piping connected to the boiler after the isolating valve is considered as downstream part of the boiler;(2)plant which generates steam or hot water by means of flame, combustion gas or other hot gases, including the equipment subject to a boiler.

Boiler alarm 锅炉报警器

Boiler and thermal oil heater survey 锅炉和热油加热器检验

Boiler bearer 锅炉基座 系指装设锅炉用的基座。

Boiler blow-out pipe line 锅炉排污管路 系指排除锅炉水面浮油和底部污垢的管路。

Boiler burner 锅炉燃烧器 系指把燃油和空气以一定方式引入炉膛进行迅速、稳定、完全燃烧的设备。

Boiler check valve 锅炉止回阀

Boiler compound tank 锅炉药剂柜 贮放锅水处理用的药剂的舱柜。

Boiler efficiency 锅炉热效率 系指向锅炉供入的总热量中转为被锅炉工质吸收的热量的百分数。

Boiler evaporative test(boiler performance test) 锅炉热工试验 测量锅炉在承受各种负荷时的各项主要热工参数并确定其效率的试验。

Boiler feed pump 锅炉给水泵 将给水输入锅炉的泵。

Boiler feed water 锅炉给水 锅炉运行时供入锅炉的水。

Boiler feed water compound pump 锅炉给水复合泵 向锅炉给水注入药剂,以降低炉水硬度的泵。

Boiler forced-circulation pump 锅炉水强制循环泵 在强制循环锅炉中驱使炉水循环的泵。

Boiler forced-draft fan 锅炉鼓风机 供给锅炉燃料燃烧所需空气的鼓风机。

Boiler fronts 锅炉前部 系指无论锅炉设计如何，都解释为锅炉点火处所。 Boiler fronts should be interpreted as the boiler burner location irrespective of the boiler design.

Boiler fuel oil daily tank 锅炉燃油日用柜 存放锅炉日常使用的燃油的舱柜。

Boiler fuel oil heater 锅炉燃油加热器 将锅炉燃烧的重油加热到适当黏度以保证良好雾化的加热器。

Boiler fuel oil pump (boiler fuel oil shaft pump) 锅炉燃油泵 将燃油升压至所需压力,输至锅炉燃烧器喷油嘴的泵。

Boiler fuel oil settling tank 锅炉燃油澄清柜 使锅炉燃油中杂质、水分沉淀的油柜。

Boiler fuel oil system 锅炉燃油系统 系指燃油从燃油舱经日用油柜输送至锅炉的机械设备、管路和附件的总称。

Boiler fuel oil tank 锅炉燃油柜 存放辅锅炉使用的燃油的舱柜。

Boiler fully-automatic control 锅炉全自动控制 锅炉的启动和停止都用手动以及建立火焰后的控制都采用自动控制,可实现无人值班的控制方式。一般包括:程序控制、压力和温度控制、水位控制及安全、极限保护等。

Boiler furnace 锅炉炉膛

Boiler heat balance 锅炉热平衡 输入锅炉和自锅炉输出的热量之间的平衡。

Boiler heating surface 锅炉热交换表面 系指可以通过它向加热介质传热的表面,可在和火或者热气体接触的一侧进行测定。 Heating surface is the area of the part of the boiler through which the heat is supplied to the medium, on the side exposed to fire or hot gases.

Boiler ignition light oil pump 锅炉点火泵 系指抽送供锅炉点火用的轻油的泵。

Boiler ignition oil tank 锅炉点火油柜 存放锅炉升气点火用的轻柴油的舱柜。

Boiler induced-draft fan 锅炉引风机 抽除锅炉中油气的风机。

Boiler maneuvering test 锅炉机动性试验 检查并测定锅炉冷态升压、升降负荷及变工况所需时间的试验。

Boiler maximum continuous rating 锅炉最大蒸发量 锅炉长期连续可靠工作的最大蒸发量。

Boiler overload capacity 锅炉超负荷蒸发量 经过试验确定的锅炉最大可能的蒸发量。

Boiler program control (boiler sequence control) 锅炉程序控制 系指对锅炉启动、预扫气、自动点火后扫气及停炉等操作按一定程序进行的控制。

Boiler rated evaporating capacity 锅炉额定蒸发量 锅炉全负荷工作的蒸发量。

Boiler room 锅炉舱 供安置锅炉及其附属设备的舱室。

Boiler room casing 锅炉舱棚 自锅炉舱通至露天的甲板的围井及顶盖。

Boiler saddle (boiler stool) 锅炉底座 系指使锅炉固定在船体基座上的座架。

Boiler secondary air blower 锅炉二次鼓风机 供应锅炉二次风的鼓风机。

Boiler semi-automatic control 锅炉半自动控制 锅炉的启动和停止都用手动,在建立火焰以后才转入自动控制,有时也包括一些安全和极限保护装置在内的控制方式。

Boiler survey (BS) 锅炉检验 每两年对锅炉进行一次的检验。主要检验项目:(1)各腔体、阀、塞的开放检查;(2)各安全报警装置的效用试验;(3)安全阀的开启试验及设定压力确认等。

Boiler unit 锅炉装置

Boiler water 锅水 系指锅炉运行时锅炉内部正在不断蒸发中的水。

Boiler water circulating system 锅炉水循环系统 系指强制循环锅炉内锅水的循环系统。由锅炉循环泵、管路和附件等组成。

Boiler water circulation 锅炉水循环 依靠水及汽水混合物的比重差或用水泵压头形成的锅水的循环。

Boiler water density 锅水浓度 系指锅水中有机杂质、盐、碱等的总含量。

Boiler water hardness 锅水硬度 系指以每公斤锅水含钙、镁离子的毫克当量浓度表示的水中钙、镁盐的浓度。

Boiler 锅炉 蒸汽锅炉的简称。利用燃料燃烧放出的热量,将水加热成具有一定压力和温度的蒸汽或热水的设备。

Boiling evaporation 沸腾蒸发 系指水加热至汽化温度时转化为蒸汽的相态变化。

Boiling off gas (BOG) propulsion system 蒸发气体推进系统

Boiling out 煮炉 将浓度很高的碱性溶液在锅炉中加热使之沸腾,以除去油脂等杂物的操作过程。

Boiling point 沸点 系指在大气气压为 101.3 kPa (1 013 mbar)时液体沸腾的温度(是在标准室内进行蒸馏而不发生分解时测得)。

Boiling point elevation 沸点升高 当水中含有溶解盐类时,在同一蒸发压力下,其沸腾温度高出纯水沸

Bollard pull 系柱拖力　系指拖船系柱拖力证书上证明的连续系柱拖力。系柱拖力通常是拖船在静水（蒲氏风力小于 3 级，即风速不超过 5 m/s，流不超过 0.5 m/s）条件下，主推进装置连续额定输出功率，航速为零时的拖力。

Bollards 双柱带缆桩　系指具有两个柱头的带缆桩。

Bolt pin 叉头横销　系指连接吊杆叉头与吊杆转枢用的带有螺母的销轴。

Bond 跨接　非带电部件之间的连接，用于确保电气连接的连续性，或使部件之间的电位相等。 Bond is a connection of non-current-carrying parts to ensure continuity of electrical connection, or to equalize the potential between parts.

Bonded area 保税区　系指经由国务院批准建立的、海关总署实施监管的特殊经济区域，同时也是我国目前拥有最大开放度和自由度的经济区域。它是海关设置的或经海关批准注册的，受海关监督的特定地区和仓库，外国商品存入保税区内，可以暂时不缴纳进口税；如再出口，不缴纳出口税；如要运进国内市场，则需缴纳进口税。

Bonded exhibition 保税陈列场　系指经海关批准在一定期限内用于陈列外国货物进行展览的保税场所。

Bonded shed 保税货棚　系指经海关批准，由私营企业设置的用于装卸、搬运或暂时储存进口货物的场所。与保税货棚相关联的是指定的保税区，两者之间的区别是：指定保税区是公营的，而保税货棚是私营的。由于保税货棚是经海关批准的，因此必须缴纳规定的批准手续费，储存的外国货物如有丢失须缴纳关税。

Bonded factory (BF) 保税工厂　系指经海关批准的，并在海关监管之下，用免税进口的原材料、零配件进行加工、制造外销商品，可以对外国货物进行加工、制造、分类以及检修等保税业务活动的场所。凡经国家批准有进口经营权的生产企业（包括外商投资企业），均可向企业所在地主管海关申请建立保税工厂。

Bonded warehouse 保税仓库　是保税制度中应用最广泛的一种形式，系指经海关批准设立的专门存放保税货物及其他未办结海关手续货物的仓库。例如：龙口港公用型保税油库和保税堆场、江门市"日新日盈"公用型保税仓库。保税仓库经理人应于货物入库后即在上述报关单上签收，其中一份留存保税仓库，作为入库的主要凭证，一份交回海关存查。保税仓库验收合格后，经海关注册登记并核发《中华人民共和国海关保税仓库注册登记证书》，方可投入运营。保税仓库与一般仓库最不同的特点是，保税仓库及所有的货物受海关的监督管理，非经海关批准，货物不得入库和出库。

Bonjean's curves 邦戎曲线　系指在各站点上，以吃水为纵坐标，相应的横剖面面积为横坐标所绘制的一组积分曲线。其包括横剖面面积曲线和横剖面面积矩曲线。

Book a whole dock 包坞　系指预定占用整个船坞。

Book rack 书架　固定在舱壁上，用以存放书刊的小型搁架。

Boolean algebra 布尔运算

Boom fittings 吊杆零部件　除起重柱、吊杆、绞车以外构成吊杆装置的全部零部件的统称。分可卸的和不可卸的两类。

Boom holder band 吊杆抱合箍　设置在起重柱上部或桅肩处，用以紧固非工作状态的吊杆的一种可开闭的抱合杆。

Boom length 吊杆长度　吊杆叉头横销中心至吊杆头端吊货眼板中心的吊杆轴线长度，即为吊杆的计算长度。

Boom topping angle 吊杆仰角　系指吊杆工作时，其轴线与船体基平面间的夹角。

Booster gas turbine 加速燃气轮机　系指在高航速航行或应急机动时开动的燃气轮机。

Booster gas turbine engine 加速燃气轮机

Booster installation (booster engine unit) 加速机组　系指在多机组的推进装置中，加速航行时使用的机组。

Booster pump (boosting pump) 增压泵　产生一定压力，为另一泵建立吸入条件的泵。例如装设在锅炉给水泵前的增压泵。

BOP = blow off preventer 防喷器

BOP stack trolley 防喷器叉车

Boring machine 镗床　系指主要利用镗刀对工件已有的预制孔进行镗削的机床。通常，镗刀旋转为主运动，镗刀或工件的移动为进给运动。它主要用于加工高精度孔或一次定位完成多个孔的精加工，此外还可以从事与孔精加工有关的其他加工面的加工。使用不同的刀具和附件还可进行钻削、铣削、切削的加工精度和表面质量要高于钻床。镗床是大型箱体零件加工的主要设备。可加工螺纹及外圆和端面等。

Borrowing standby L/C 借款备用信用证

Boss 桨毂　系指螺旋桨毂，是桨叶所附的中心部分，桨轴端在此穿过。见图 B-7。 As Fig. B-7 shows, boss means a boss of propeller, it is the central part to which propeller blades are attached and through which the shaft end passes. .

图 B-7　桨毂
Figure B-7　Boss

Boss plate　轴包板　系指包裹轴毂或艉轴管的外板。

Bossing (shaft bossing, shell bossing)　轴包套/轴包架　系指凸出于船体型表面，沿艉轴方向，全部或局部围封舷侧艉轴的船体鳍状部分。

Bottom blow (down blow)　下排污　系指从锅炉下联箱或载水部件的最低处排污点排掉带泥渣的锅水的过程。

Bottom blow valve　底部吹泄阀

Bottom brass (propeller)　下半轴承　系指推进器轴轴承的下半衬套。

Bottom ceiling　舱底木铺板　系指货舱内从一舷至另一舷铺设密排舱底的木铺板。Bottom ceiling is a tight bottom ceiling fitted from board to board in holds.

Bottom center girder (vertical keel)　中底桁　系指双层底内中线面处的纵桁。

Bottom construction　舱底结构　系指船体的基础，是保证船体总纵强度、横向强度和船底局部强度的重要结构。舱底结构有双层底结构和单层底结构两种类型。按骨架的排列方式可分为横骨架式结构和纵骨架式结构两种形式。

Bottom dead center　下止点　系指活塞在气缸内与曲轴中心线距离达最小值的位置。

Bottom door　底开门　系指开设在船体底部，供抛投物件或进行某种作业而特别设置的门。

Bottom drag net　底拖网　系指拖扫于水域底层的拖网。

Bottom frame　船底肋骨　系指船底部分的肋骨。

Bottom girder　底纵桁　系指双层底内的纵桁。

Bottom log　海底计程仪　系指放置在船底部的，计量船舶航程的航海仪器，也是推算航迹的基本工具之一。有些计程仪也指示速度。计程仪改正率通常在船速校验场实测求得。此法是利用1个14 s或28 s的沙漏计时，另以一块木板连接绳索1根，在绳索上等距打结，两个结之间称为1节。20世纪50年代出现的多普勒计程仪和20世纪70年代制成的声相关计程仪，在一定水深内可以直接测量船舶相对于水底的航速和航程，使计程仪发展到一个新的水平。这种计程仪工作性能较可靠，但线性差，低速误差大，不能测后退速度，机械结构复杂，使用不便，逐渐被淘汰。

Bottom long line　底延绳钓　系指沉于海底的延绳钓。

Bottom long liner　底拖钓船

Bottom longitudinal　船底纵骨　系指船底板上的纵骨。

Bottom plate　船底板/舵叶底板　系指：(1) 船体两侧舭列板之间的底部外板。(2) 舵叶底面的封板。

Bottom plug　船底塞　救生艇或工作艇底供泄放积水用的孔座和塞。

Bottom shell　船底外板　系指主要组成船体外壳和平底的外壳板材，包括平板龙骨。Bottom shell means shell envelope plating forming the predominantly flat bottom portion of the shell envelope including the keel plate.

Bottom side girder　旁底桁　系指双层底内中线面两侧的纵桁。

Bottom strengthened for operating a ground　坐底作业船底加强　系指挖泥船坐底作业加强。Bottom strengthened for operating a ground means bottom of dredgers strengthened for operating aground.

Bottom suction in salvaging　触底吸力　拖开搁浅船和捞起坐底船或沉船时，由于船体与海底之间构成真空而产生的使船向岸边横移的吸力。

Bottom suction in shallow water　浅水底吸力　系指船舶在浅水中航行时，由于船体与水底间的局部压降而产生的使船身下沉的吸力。

Bottom supporting drilling platform　定柱沉垫式钻井平台　系指由沉垫、立柱、上船体组成，工作时向沉垫的压载舱注水，沉垫坐落海底，平台留在水上进行工作，工作水深一般在20 m以下的非自升式钻井平台。

Bottom tank　双层底舱　系指利用双层底的空间构成的液舱。

Bottom zone　舱底区域　对散货船而言，包括底边舱斜板上端水平线或内底板(如无底边舱)以下对船体梁强度起作用的所有下列项目：(1) 龙骨板；(2) 船底板；(3) 舭板；(4) 船底桁材；(5) 内底板；(6) 底边舱斜板;(7)舷侧外板;(8)与上述板材连接的纵骨。 For a up to the upper level of the hopper sloping plating or up to the inner bottom plating if there is no hopper tank bulk carrier, the bottom zone includes the following items contributing to the hull girder strength:(1)keel plate;(2)bottom plating;(3)bilge plating;(4)bottom girders;(5)inner bottom plating;(6)hopper tank sloping plating;(7)side shell plating;(8)longitudinal stiffeners connected to the above mentioned plating.

Bottom-side tank 底边舱 散货船上,位于货舱底部两侧角隅处的船舱。

Bottom-up 自上而下法 系指基于设备类型的计算海运温室气体排放量的方法。计算中也使用各类燃料的燃料消耗数据,但需要在各模式下分类采用检验较大特殊性的国家特定的排放因子(即:按设备类型等采用不同的排放因子),包括按照船型(如远洋船和艇)、燃料类型(如燃料油)、发动机类型(如柴油发电机)等。计算公式如下:$E = FC_{ij}EF_{ij}$,式中,E 为海运温室气体分类的温室气体排放量(t);FC_{ij} 为不同类型设备、不同类型燃料消耗量;EF_{ij} 为不同类型设备、不同类型燃料的分类排放因子。

Bound vortex 附涡 按螺旋桨环流理论,就所在位置而言,系指附着于螺旋桨叶产生升力的涡线部分。

Boundary layer 边界层 系指流体流经固体或固体在流线中运动时,靠近固体表面明显受到黏性影响的流体层。

Boundary layer separation 边界层分离 系指流体流过固体时,由于方向或固体表面曲度突然改变和流纵向压力梯度变化而导致的流体主体从固体面分离的现象。

Boundary layer thickness 边界层厚度 边界层外缘至固体表面的垂直距离,其外缘处流速为理想流体流速的99%。

Boundary strip 边界条 对钢夹层板而言,系指与底板直接连接形成灌注空间的条形钢板。

Boussinesq number 布新尼斯克数 系指用于分析浅水狭航道中水流现象和船舶运动的一个无量纲参数。$B_n = V/(gR_n)^{1/2}$,式中,V—船速;g—重力加速度;R—水力半径。

Bow 艏 系指船舶首端的结构布置和外形。Bow means structural arrangement and form of the forward end of the ship.

Bow and aft 艏与艉 非自航起重船船首与船尾的称谓,因其所处的状态不同而不同,在调遣时一般船舶的规定相同,即将前进方向的船体的前端称为艏,相应另一端称为艉。在调遣以外的其他状态,则将安装有起重臂的一端称为艏,相应另一端称为艉。

Bow door 艏门 通常系指供车辆等进出或登陆用的开设在船首端的门。

Bow line 艏缆 当船舶带缆时,从船首部向前引出的缆绳。

Bow reference height H_b 船首基准高度 H_b 定义如下:(1)对于长度小于 250 m 的船舶:$H_b = 0.056L_L \times [1-(L_L/500)][1.36/(C_{bL}+0.68)]$;(2)对于长度 250 m 及以上的船舶:$H_b = 7[1.36/(C_{bL}+0.68)]$,式中:$L_L$—载重线长;$C_{bL}$—载重线方形系数。 Bow reference height H_b is defined as:(1)For ships less than 250 m in length:$H_b = 0.056L_L[1-(L_L/500)][1.36/(C_{bL}+0.68)]$;(2)For ships 250 m or greater in length:$H_b = 7[1.36/(C_{bL}+0.68)]$.

Bow region 艏部区(冰区加强) 系指从艏柱至与船壳前端线(该位置处的水线与中心线平行)平行向后 0.04L 处之间的部分。(1)对于 B1*和 B1 冰级,超过前端线的水平距离不必大于 6 m;(2)对于 B2、B3 冰级长度不必大于 5 m。 Bow region is the region from the stem to a line parallel to and 0.04L aft of the forward borderline of the part of the hull where the waterlines is parallel to the centerline. (1)For ice class B1* and B1 the overlap over the borderline does not need to exceed 6 m,(2)For ice class B2 and B3, this overlap over the borderline does not need to exceed 5 m.

Bow rudder 艏舵 系指设置在船首以改善船舶倒车时操纵性能的舵。

Bow screw 艏螺旋桨 系指破冰船,渡船等装在船首的螺旋桨。

Bow sea 艏斜浪 在左、右舷 15°～75°之间方向内与船首遭遇的波浪。

Bow shackle 圆形卸扣 系指 U 形体的弯曲部分向外鼓出,且其直径较开口为大的呈圆环状的卸扣。

Bow sheave assembly 艏滑轮架 安装在布缆船首端突出部分上,在布缆时供引导电缆及工作人员操作、指挥用的滑轮和架子。

Bow thruster 艏推力器(艏转向装置)

Bow transom plate 艏封板 方型船首端外板。

Bow transverse thrust unit 艏侧推器(船首横向推力器)

Bow up display 船首向上显示 系指船舶首向位于图像最上端的非稳定显示。

Bower anchor 艏锚 系指安置在船首的锚。 Bower anchor means an anchor carried at the bow of the ship.

Bowl 分离筒 装固于轴上,其内装有分离盘组的圆筒状零件。

Box cooler 箱式冷却器 这种冷却器的冷却盘管直接浸没在机舱的海水箱内,船舶航行时利用海水的相对运动流速直接对冷却盘管中的液体进行冷却。

Box cross deck structure 箱型横向结构 系指货舱横舱壁顶部的横向箱型桁材结构。Box cross deck structure means the transverse box girder structure at the top of transverse bulkheads of container holds.

Box keel 箱形龙骨 系指在船体中心线上,纵向布置于开底泥舱下部的泥舱前后舱壁之间的一种箱形组合结构。Box keel is a built-up box-shaped structure on the centerline, arranged longitudinally at the lower portion of, and between the fore and aft end bulkheads of the hopper or hoppers.

b_p(mm) b_p 扶强材或主要支撑构件带板的宽度(mm)。用于屈服强度校核。b_p is the width, in m, of the plating attached to the stiffener or the primary supporting member, used for the yielding check.

Bracing members 撑杆 系指在柱稳式或坐底式平台中,将平台各主结构(即上壳体、立柱和下壳体)连接成一个结构整体的圆管状或其他形状的连接构件。

Bracket 肘板 系指用以增加两个构件结合部强度的外加构件。Bracket is an extra structural component used to increase the strength of a joint between two structural members.

Bracket floor(open floor) 框架肋板 双层底内,由船底肋骨、内底横骨和肘板等组成的框架。

Bracket notation 中括号

Bracket plate of foundation 基座隔板 系指基座桁侧面的肘板。

Bracket toe 肘板趾端 系指锥形肘板的窄端。Bracket toe means a narrow end of a tapered bracket.

Brain trust (think tank/policy institute) 智库 俗称"智囊团",但其含义已超越"智囊团",可以指专门组织。(1)智囊团是一组就政治、社会或经济问题提供建议和想法的专家组。(2)智库是一个组织、机构、公司或群体,在诸如社会政策、政治策略、经济、科学或技术问题、工业或商业政策以及军事咨询等领域进行研究并提供支持;(3)智库是民主、科学决策的一个重要保障;(4)智库是一个受雇在军事、政治或社会领域解决复杂问题或预测并规划未来发展的研究机构或组织。

Brainstorming 发挥想像/头脑风暴 通过发挥想象去发现解决问题的可能方案和潜在的改进的机会。发挥想象是一种技巧,它可以发现一些创造性的思想从而形成并澄清一系列的思想、问题和争议。

Brake power 制动功率 对"73/78防污公约"附则Ⅵ而言,系指在曲轴或其等效设备处测得的测量功率,为了在试验台上运转该发动机仅设有必要的标准辅助设备。For the "MARPOL73/78" Annex Ⅵ, brake power is the observed power measured at the crankshaft or its equivalent, as the engine being equipped only with the standard auxiliaries necessary for its operation on the test bed.

Brake thermal efficiency 有效热效率 系指转变为曲轴输出功的热量与产生此功所消耗的全部燃料热量的比值。

Branch circuit 分支电路 系指敷线系统中延伸到最终保护该电路的过电流装置以外的那部分电路。

Brass/bronze 黄铜/青铜 系指由铜和锌所组成的合金。如果只是由铜、锌组成的黄铜就叫作普通黄铜。如果是由两种以上的元素组成的多种合金就称为特殊黄铜。如由铅、锡、锰、镍、铅、铁、硅等组成的铜合金。黄铜有较强的耐磨性能。特殊黄铜又叫特种黄铜,它强度高,硬度大,耐化学腐蚀性强,还有切削加工的机械性能也较突出。由黄铜所拉成的无缝铜管,质软,耐磨性能强。黄铜无缝管可用于热交换器和冷凝器、低温管路、海底运输管、制造板料、条材、棒材、管材,铸造零件等。含铜在62%~68%,塑性强,制造耐压设备等。

Brass bush/shaft liner 轴套 系指推进器轴或艉轴上的铜套。

Brazil, Russia, India, China and South Africa (BRICS) 金砖国家 2001年,美国高盛公司首席经济师吉姆·奥尼尔(Jim O'Neill)首次提出"金砖四国"这一概念,特指新兴市场投资代表。"金砖四国"(BRIC)引用了巴西(Brazil)、俄罗斯(Russia)、印度(India)和中国(China)的英文首字母。由于该词与英语单词的砖(Brick)类似,因此被称为"金砖四国"。2008年~2009年,相关国家举行系列会谈和建立峰会机制,拓展为国际政治实体。2010年南非(South Africa)加入后,其英文单词变为"BRICS",并改称为"金砖国家"。金砖国家的标志是五国国旗的代表颜色做条状围成的圆形,象征着"金砖国家"的合作,团结。2015年1月1日,俄罗斯开始担任金砖国家机制轮值主席国。俄总统普京表示,将利用主席国身份进一步提高金砖国家在世界范围的影响力。

Breadth[1] 船宽 系指船舶的最大宽度。对金属壳体的船舶,其宽度是在船长中点处量到两舷的肋骨型线;对其他任何材料壳体的船舶,其宽度在船长中点处量到船体外面。Breadth is the maximum breadth of the ship, measured amidships to the moulded line of the frame in a ship with a metal shell and to the outer surface of the hull in a ship with a shell of any other material.

Breadth[2] 船宽 B（内河船） 对钢质船舶系指不包括船壳板在内的船体最大宽度；对纤维增强塑料船舶系指船体两侧外表面之间的最大宽度。但舷伸甲板或护舷材等不计入船宽。

Breadth（SWATH） 船宽（小水线面双体船） 系指刚体水密船体的最大宽度，不包括水密水线处及以下的附体。Breadth means breadth of the broadest part of the moulded watertight envelope of the rigid hull, excluding appendages at or below the design waterline in the displacement mode.

Breadth B 船宽 B 通常为舰船型深范围内的最大型宽，或按有关章节中的定义。对于具有非常规横剖面的舰船，其船宽应予以特殊考虑。 Breadth B is generally the greatest moulded breadth, in m, throughout the depth of the ship or as defined in appropriate Chapters. For vessels of unusual cross section the breadth will be specially considered.

Breadth B of a bulk carrier 散货船宽度 系指现行"国际载重线公约"所定义的宽度。 Breadth B of a bulk carrier means the breadth defined in the International Convention on Load Lines in force.

Breadth B（special purpose ship） 船宽（特种用途船舶） 系指船舶的最大宽度，金属船壳的船舶是在船中部量至肋骨型线。其他材料的船舶，在船中部量至船壳外表面。船宽应以米计算。 Breadth B means the maximum breadth of the ship, measured amidships to the moulded line of the frame in a ship with a metal shell and to the outer surface of the hull in a ship with a shell of any other material. The breadth (B) should be measured in meters.

Breadth B_S 船宽 B_S 就 MARPOL 而言，系指在最深载重线吃水处 d_B 处或下面的船舶最大的型宽，m。 For MARPOL, breadth B_S is the greatest moulded breadth of the ship, in m, at or below the deepest load line d_S.

Breadth draft ratio 宽度吃水比 一般系指设计水线宽度与设计吃水之比。

Breadth for freeboard B_f 干舷船宽 B_f 系指在 L_f 的中央处，从一舷肋材的外缘量至另一舷肋材外缘的最大水平距离(m)。 Breadth of ship for freeboard B_f is the maximum horizontal distance in meters from the outside of a frame to the outside of another frame measured at the middle of L_f.

Breadth $_{MLD}$ 型宽（中部） 系指船舶的最大宽度。对金属壳板的船舶，其宽度是在船长中点处量到两舷的肋骨型线；对其他任何材料壳板的船舶，其宽度是在船长中点处量到船体外面。

Breadth moulded（moulded beam） 船宽 通常系指型宽。型宽是在外板内缘测定的，也称为理论宽度或计算宽度。通过宽度系指通过运河水闸的宽度，还要加上外板的板厚以及可能有的护舷材的厚度。

Breadth of lower hull（SWATH） 下潜体宽（小水线面双体船） 系指单个下潜体的最大型宽。Breadth of lower hull means the maximum moulded breadth of one lower hull.

Breadth of waterline B_{WL}（m） 水线宽 B_{WL} 系指船舶静浮于水面时，沿设计水线量得的最大宽度。对于多体船（如双体船、穿浪型双体船、水面效应船等）系指设计水线处各片体最大型宽之和。

Breadth of yacht B（m） 艇宽 对游艇而言，系指在艇的最宽处，由一舷的肋骨外缘量至另一舷的肋骨外缘之间的水平距离；对纤维增强塑料艇为艇体两侧外表面之间的最大宽度，但不包括护舷材等突出物。

Breadth overall submerged 浸体宽 设计水线下，船体水下部分型表面之间垂直于中线面的最大水平距离。

Breadth（fishing vessel）B 船宽 B（渔船） 系指船舶的最大宽度。对于金属船壳的船舶是在船中处由两舷肋骨型线间量得，对于其他任何材料船舶，在船中处由船体外壳面间量得。 Breadth B for fishing vessel is the maximum breadth of the vessel, measured amidships to the moulded line of the frame in a vessel with a metal shell and to the outer surface of the hull in a vessel with a shell of any other material.

Breadth（floating dock） 坞宽 B_D（m） 系指由浮船坞最大宽度处的肋骨外缘之间量得的宽度。其中内坞墙宽系指坞墙内壁板（内坞墙板）面向中纵线一侧外缘之间量得的型宽；净内坞宽系指内坞墙宽减去内坞墙上悬挑结构最大伸出长度之后的净宽度。

Break bulk 开舱 系指船舶到达装卸站时，打开货舱盖，准备卸货的作业。

Break rating of circuit breaker（rated break current） 断路器的断开定额（额定断开电流） 系指断路器在规定工作电压时，能断开的最大的有效值电流。此时断路器在规定的工作制下，且正常频率恢复电压等于规定的工作电压。

Break time 分断时间 系指从开关电器的断开瞬间（或熔断器的弧前时间）开始时起，到燃弧时间结束瞬间止的时间间隔。注意：①为了防止产生歧义，本处称"全分断时间"。②对于熔断器，另有"熔断时间 operating time"术语，其含义为"弧前时间和燃弧时间之和"。实质上与术语"分断时间"是一致的。

Break water 防波堤 为阻断波浪的冲击力，围护港池,维持水面平稳以保护港口免受坏天气影响,以便船舶安全停泊和作业而修建的水中建筑物。防波堤还可起到防止港池淤积和波浪冲蚀岸线的作用。它是人工掩护的沿海港口的重要组成部分。一般规定港内的容许波高在 0.5～1.0 m 之间,具体按水域的不同部位、船舶的不同类型与吨位的需要确定。防波堤常由一、二道与岸连接的突堤或不连接的岛堤组成,或由突堤和岛堤共同组成。防波堤掩护的水域常有一个或几个入口供船舶进出。

Breakage of packing risk 包装破裂险 投保平安险和水渍险的基础上加保此险,保险人负责赔偿承保的货物在运输过程中因搬运或装卸不慎造成包装破裂所引起的损失,以及因继续运输安全的需要修补或调换包装所支出的费用。

Breaker 断路器 系指一种能接通、承载以及分断正常电路条件下的电流,也能在规定的非正常电路(例如短路)下接通、承载一定时间和分断电流的一种机械开关电器。

Breaking current(of a switching device or a fuse)(开关电器的或熔断器的)分断电流 系指在规定的使用和性能条件下,开关电器或熔断器在规定电压下能分断的预期分断电流值。

Breaking load(BL) 破断载荷 系指证书上证明的拖曳索具的最小破断载荷。

Breaking test load 锚链破断试验载荷 系指锚链的最小破断试验载荷。 Breaking test load of the anchor chain cables is the minimum breaking test load.

Breakwater 挡浪板 系指露天甲板上的倾斜加强板结构,用以挡开涌上船舶的水流。Breakwater means an inclined and stiffened plate structure on a weather deck to break and deflect the flow of water coming over the bow.

Breast hook 艏肘板 系指在艏柱处连接左右舷构件的三角形平板肘板。Breast hook means a triangular plate bracket joining port and starboard side structural members at the stem.

Breast line 横缆 系指船舶带缆时横向引出的船舶缆绳。

Breadth of hatchboard 舱盖板宽度 系指舱盖板在较短的一边的外形尺寸。

Breadth of hatchcover(hatchcover width) 舱盖宽度 整个大舱盖沿船的横向的外形尺寸。

Breadth of ladder 梯宽 系指梯架内间的水平距离。

Breadth of lifeboat 救生艇宽 系指艇舯横剖面上在舷顶处两舷壳板外表面间的水平距离。

Breather valve 呼吸阀

Breathing apparatus 呼吸器 系指一具自给式压缩空气呼吸器。其筒内储气量应至少为 1 200 L,或一具其他类型的自给式呼吸器,其可供使用的时间至少为 30 min。呼吸器所用的空铅瓶应能互换。 Breathing apparatus means a self-contained compressed air-operated breathing apparatus in which the volume of air contains shall be at least 1 200 L, or other self-contained breathing apparatus which shall be capable of functioning for at least 30 min. All air cylinders for breathing apparatus shall be interchangeable.

Breathing valve 呼吸阀(通气阀) 系指将液舱内的压力维持在一定范围内的一种阀件,系利用舱内外的压差自动进行排气或吸气。

Breeches buoy 兜带救生圈 在其下方牢固连着兜带,可承载人员以备吊送过舷或靠引绳拉动从水面引渡的一种救生圈。

BRICS bank 金砖银行 系指中国国家主席习近平、巴西总统罗塞夫、俄罗斯总统普京、印度总理莫迪、南非总统祖马在 2014 年 7 月 15 日举行的金砖国家领导人第六次会晤上决定,成立金砖国家开发银行,其总部设在中国上海。该银行初始核定资本为 1 000 亿美元,初始认缴资本为 500 亿美元并由各创始成员国均摊。银行资金将主要用于支持金砖国家和其他新兴经济体及发展中国家的基础设施建设。首任理事会主席将由俄罗斯提名、首任董事会主席将由巴西提名、首任行长将由印度提名。同时,该银行非洲区域中心将设在南非。

Bridge 桥楼 系指:(1)进行船舶导航和操纵的场地,包括驾驶室及两翼平台;(2)升高的上层建筑,在前方和两侧有清晰的视野,在此驾驶船舶;(3)不延伸到艏垂线或艉垂线的上层建筑。见图 B-8。Bridge means:(1)the area where the navigation and control of the ship is exercised, including the wheelhouse and bridge wings;(2)an elevated superstructure having a clear view forward and on each side, and from which a ship is steered;(3)a superstructure which does not extend to either the forward or after perpendicular. As is shown in Fig. B-8.

Bridge alert management(BAM) 警报管理 系指驾驶室警报监控、处理、分类和显示的协调一致管理的概念。 Bridge alert management (BAM) means the overall concept for management, handling and harmonized presentation of alerts on the bridge.

Bridge control 驾驶室遥控 系指由驾驶室控制站遥控主、辅机的操纵方式。

Bridge control console 驾驶室集控台 系指位于驾驶室内设置的集中控制台。

图 B-8 桥楼
Figure B-8 Bridge

Bridge control devices 桥楼控制设备 系指一个对主推进装置的遥控控制设备，或者是对一个在驾驶桥楼或船桥上主控制站的调距螺旋桨的推进装置的遥控控制设备。 Bridge control devices are remote control devices for the main propulsion machinery or controllable pitch propellers provided on a navigation bridge or a main control station on bridge.

Bridge control station（BCS） 驾驶室控制站 系指设在驾驶室内对推进装置及其他设备进行监控的控制站。 Bridge control station means a control station monitoring the propelling plant and other equipment in bridge.

Bridge crane 桥式起重机 系指桥架在高架轨道上运行的一种桥架型起重机，又称天车。桥式起重机的桥架沿铺设在两侧高架上的轨道纵向运行，起重小车沿铺设在桥架上的轨道横向运行，构成一个矩形的工作范围，就可以充分利用桥架下面的空间吊运物料，不受地面设备阻碍的影响。这种起重机广泛用在室内外仓库、厂房、码头和露天贮料场等处。桥式起重机可分为普通桥式起重机、简易梁桥式起重机和冶金专用桥式起重机3种。

Bridge crane and heavy equipment carrier 大件运输船 系指具有较大的甲板面积，专运桥式吊机和重型设备的船舶。 Bridge crane and heavy equipment carrier is a carrier with large deck area, and dedicated bridge crane and heavy equipment carriers with heavy equipment.

Bridge deck 驾驶甲板/桥楼甲板 系指位于桥楼顶部，布置有驾驶室的一层甲板。

Bridge navigation watch alarm system（BNWAS） 驾驶室航行值班报警系统 系指MSC.128（75）决议所要求的第2级和第3级远距离听觉报警。其监视桥楼操作情况是否正常，值班人员是否正常，如有异常情况，则向其他航海人员、公共场所发出报警。

Bridge system 桥楼系统 系指用于执行桥楼各项功能的整个系统。其包括桥楼操作人员、技术系统、人机界面和操作程序。 Bridge system mean a total system for the performance of bridge functions, comprising bridge personnel, technical systems, man and machine interface and operation procedures.

Bridge wings 桥楼翼台 系指从驾驶室的两侧到船舷扩张出的船桥的一部分。 Bridge wings are parts of the bridge on both sides of the ship to which wheelhouse extendes from the ships side.

Bridge 桥楼 系指进行船舶导航和操纵的场地。其包括驾驶室及其两翼平台。 Bridge is the area from which the navigation and control of the ship are exercised, comprising the wheelhouse and the bridge wings.

Bridge-to-bridge communications 驾驶室对驾驶室的通信 系指在船舶通常驾驶位置进行的船舶之间的安全通信。 Bridge-to-bridge communications means safety communications between ships from the position where the ships are normally navigated.

Bridle 龙须缆（链） 系指用于大型被拖物，为保持被拖物拖航时的航向稳定性，从被拖物两侧的拖力点（拖力眼板或拖桩）引出缆或摩擦链至三角板的连接缆。

Bridle apex 龙须缆/链顶点 系指龙须缆/链顶点与短缆起点的连接装置，如三角板、拖曳环、或卸扣。使用单龙须链/缆的拖索装置中，龙须缆/链常采用拖曳环或卸扣同短缆链接。

Brine 盐水 系指：（1）是由工业盐溶液组成的一种载冷剂。盐水通常作为冷媒，在间接冷却系统中冷却冷藏室，盐水一词也代表其他冷媒，例如乙二醇溶液；（2）一种次要制冷剂的一般用语，次要制冷剂是被主要制冷剂所冷却并且作为热媒去冷却货物的。 Brine is: (1) a refrigerant constituted by a solution of industrial salts, which is normally used to cool the chambers in the indirect cooling systems, as secondary refrigerant. In general, the word brine is also used to cover other types of secondary refrigerants, for instance, refrigerants based on glycol; (2) a general term for the secondary refrigerants which is cooled by the primary refrigerants and which is a thermal medium to cool the cargo.

Brine balance tank 盐水膨胀箱

Brine ejector 排盐喷射器 系指利用压力水为工

质,抽除蒸发器内盐水的喷射器。

Brine head tank 盐水膨胀箱

Brine heat loss 排盐热损失 系指单位排盐带走的热量与单位给水带入的热量之差。

Brine pump 盐水泵 系指抽送盐水的泵。

Brine rate (blow-down rate) 排盐量 系指单位时间内从淡化装置中排出的盐水量。

Brine valve 盐水阀

Brine water dissolving tank 熔盐水箱 系指融制间接冷却系统中盐水溶液的容器。

British national overseas passport (BNOP) 英国海外公民护照

British Ship Research Association (BSRA) 英国船舶研究协会

British Standard Institute (BSI) 英国标准协会 是一家全球领先的商业标准服务机构,成立于1901年。并于1929年获得英国皇家特许,成为世界上第一个国家标准机构,同时,作为国际标准化组织(ISO)创始成员之一。BSI 创立了全球最值得信赖和得到广泛认可的 ISO 系列管理体系。而与 BSI 联合推出云安全国际认证的云安全联盟(CSA),是一家非营利性组织,遵循 BSI 标准,获得世界的广泛认可。

British textile agreement (BTA) [英] 纺织品协定

Broaching 横甩 船舶随浪航行由波峰落入波谷时,由于艏摇角过大而引起使船舶于极短时间内转到横浪位置的现象。

Broadband Global Area Network (DGAN) 全球海事宽带通信网络 系指一套能覆盖全球的兼容第三代通信技术(3G)的系统。该网络能向用户提供帮助 IP 数据服务和流量 IP 数据服务,其功能类似传统的语音和移动综合业务数字网(ISDN)数据服务。该服务可提供与互联网和内部局域网的连接,并可通过卫星终端进行视频点播、视频会议、传真、电子邮件、语音和宽带可达 492 kb/s 的虚拟专用网络。

Broadband noise 宽频带噪声

Broadcaster's table 广播桌 系指专供安置广播器材和供广播员工作用的桌子。

Broadside guide bollard (wing bollard) 自由导向滑车 安装在围网渔船舷边的立式可转动弧形铁架顶部,在捕捞操作中,用以调节钢索方向的滑车。

Broken 破碎 系指易碎物品遭受碰压造成破裂、碎块。

Broken stowage (breakage) 亏损舱容 系指货物积载中,货物之间或货物与货舱之间所不能利用的货舱容积。

Brokerage 经纪人佣金 系指经纪人完成了为委托人提供订约机会或者充当订约介绍人的中介活动后,由委托人支付给他的劳动报酬。是一种合法的劳务费用。

Brokerage clause 经纪人佣金条款 系指合同中对经纪人佣金的具体规定的条款。

Broom clean 打扫干净 系指这样的一个状况。船舶的甲板显示采取措施防止和限制任何的干散装货物残余物的可见的浓缩物,以至于任何残留的散装干货物残余物仅仅由灰尘、粉末或单独和随机的、尺寸不超过 1 英寸的片状物组成。Broom clean means a condition in which the vessel's deck shows that measures has been taken to prevent or eliminate any visible concentration of bulk dry cargo residues, so that any remaining bulk dry cargo residues consist only of dust, powder, or isolated and random pieces, none of which exceeds 1 inch in diameter.

Brushless exciter 无刷励磁机 系指一种交流(旋转电枢型)励磁机,其输出经由半导体器件整流,给电机提供励磁。该半导体器件安装在交流励磁机电枢上并与之一起旋转。

Brushless synchronous motor 无刷同步电机 系指一种具有无刷励磁机的同步电机。励磁机的旋转电枢和半导体器件与主电机的磁场在一根公共轴上。该型电机及其励磁机没有集流环、换向器或点刷。

B-type spherical independent tank B 型球形独立液舱 此类液舱可以是平面形结构,也可以是压力容器结构,其设计中采用模型试验,与 A 型相比具有更精确的应力分析,包括疲劳寿命和裂纹扩展的分析等,所以此类液舱具有更可靠的安全性。

Bubble cavitation 泡状空泡 通常系指在叶背上剖面最大厚度处产生的空泡,呈泡沫状。这时叶剖面的攻角较小,导缘未出现负压峰,压力最低处大致在最大叶厚附近,由此所产生的空泡因前后压力变化比较缓和,单个空泡的成长清晰可见,成长到相当长度后,被水带向下游时尺寸减小。

Bubble jet 气泡技术 系指通过加热喷嘴,使墨水产生气泡,喷到打印介质上的喷墨技术。

Bubbles 气泡

Bucket 链斗 系指链斗式挖泥船上,用于挖泥的斗状工具。

Bucket (ladder/elevator) dredger (BD) 链斗式挖泥船 系指具有链斗机构的借助斗链绕其上下斗轮连续地运转的形式实施挖泥作业的挖泥船。它能适应各种土质,广泛应用于采金、采砂、采盐、采蚝壳、采锡等开采船。

Bucket chain 斗链 系指安装在斗桥上的带有链

斗的履带式装置。

Bucket chain pitch 斗链节距 系指前后两个斗链的斗销中心线间的直线距离。

Bucket chain speed 斗链转速 系指以链斗在泥阱上每分钟倒泥数计算的斗链运转速度。

Bucket chain tensioner 斗链张紧装置 系指用以调节斗桥上支承点的跨距和斗链张力的装置。

Bucket dredger 链斗挖泥船 系指装有一套链斗挖掘机构，由斗链带动链斗连续运转进行作业的挖泥船。

Bucket ladder 斗桥 系指以钢板和型钢焊成，其上等距离分布有导链滚轮，在上端设有一横轴装于斗塔斜撑上，下部有起吊支点，在下顶端还设有下导轮的斗链托架。

Bucket ladder gantry 斗桥吊架 用以支撑斗桥上升和下放的门式构架。

Bucket ladder hoisting gear 斗桥起落装置 由斗桥下部横轴、斗桥吊架、拉杆、横担、滑轮组、钢缆、斗桥起落绞车等组成的，用以操纵斗桥起落的装置。

Bucket thermometer 表层温度表 一种安装在金属外壳内的普通水银温度表。水银泡外围装有绝热的贮水筒。使用时，放入海水中表层感温数分钟后，再提取上来读数。

Bucket wheel dredger 链斗式挖泥船 系指具有链斗挖泥设备的挖泥船。

Bucket wheel suction dredger (BWSD) 斗轮式挖泥船 系指具有斗轮机构的挖泥船。

Buckhoe & Dipper dredge (BD/DD) 铲斗式挖泥船 系指一种采用单斗作业的挖泥船。铲斗式挖泥船有正铲挖泥船和反铲挖泥船之分。

Buckhoe dredge (BD) 正铲挖泥船 系指具有铲斗挖泥设备的，作业时，船随作业过程向正前方行进的铲斗式挖泥船。

Bucking strength 屈曲强度 其规定了结构稳定性的要求，以防止构件承受压应力和剪切力时发生屈曲或倾侧。Bucking strength gives the requirements for structural stability to prevent bucking and tripping of structural elements when subjected to compressive and shear stresses.

Buckle arrestor 止屈器 系指沿管路以一定的间隔放置的环形加强筋。其用途是阻止屈曲的蔓延和限制对两个相邻的止屈器之间距离管子的损害。Buckle arrestor means circumferential stiffener placed at regular intervals along a pipeline with the purpose of arresting a propagating buckle and limiting the damage to the distance between two adjacent arrestors (usually several hundred feet apart).

Buckler 锚链筒盖 系指盖在锚链筒甲板开口处，其上有供锚链穿过的开口，用以阻挡海水从锚链筒中大量涌上甲板的专用盖子。

Buckling 屈曲 为一专用术语：(1) 通常用以描述面内受压和/或受剪情况下的结构强度。屈曲强度或能力可视情况考虑内力重新分布的影响；(2) 从刚性和结构坚固的形状转变为刚性小的形状和经常是处于危险状态的过程。弹性屈曲系指结构屈曲，而材料是刚性的线弹性。塑性屈曲系指材料在屈曲时部分或全部塑化。Buckling is used as a generic term to describe the strength of structures, (1) generally under in-plane compressions and/or shear. The buckling strength or capacity can take into account the internal redistribution of loads depending on the situation; (2) The process of switching from a stiff and structurally sound configuration to one means less stiff is often dangerous. Elastic buckling means that the structure buckles while the material is stiff linearly elastic. Plastic buckling implies that when buckling happens, the material is either partially or fully plasticized.

Buckling capacity accepting local elastic plate buckling with load redistribution (Method 1) 允许载荷重新分布的屈曲能力(方法1) 此方法定义了屈曲能力的上限值。为表征板格不致产生显著永久残余变形所能承受最大承载载荷，且为板格的有效极限承载能力。该屈曲能力取为导致加筋板格中的任一处开始出现膜屈服应力的载荷。为此在计算时应考虑结构中的载荷重新分布。该载荷重新分布为板件，如加筋之间板材发生弹性屈曲的结果。对细长结构，用该方法计算能力典型比理想弹性屈曲应力(最小特征值)要高。Buckling capacity accepting local elastic plate buckling with load redistribution is referred to as Method 1. The buckling capacity is the load that results in the first occurrence of membrane yield stress anywhere in the stiffened panel. Buckling capacity based on this principle gives a lower bound estimate of ultimate capacity, or the maximum load the panel can carry without suffering major permanent set. Method 1 buckling capacity assessment utilizes the positive elastic post-buckling effect for plates and accounts for load redistribution between the structural components, such as between plating and stiffeners. For slender structures the capacity calculated using this method is typically higher than the ideal elastic buckling stress (minimum Eigen-value).

Buckling capacity does not accept load redistribution between structural components (Method 2) 不允许载荷重新分布的屈曲能力(方法2) 此方法定义了屈

曲能力的下限值。对细长结构按理想弹性屈曲应力定义(最小特征值截止)。对于理想弹性屈曲强度高的短粗形板格,膜屈服应力将开始出现在任一载荷重新分布出现之前,并由此得到与上限值相同的屈曲强度,与方法 1 相同。在计算该屈曲强度时,不考虑内力重型分布,因此,该方法比方法 1 给出的高限量值更为保守,以确保板格不会有大的弹性变形而导致面内刚度降低。Buckling capacity does not accept load redistribution between structural components is referred to as Method 2. It gives the minimum value of buckling capacity. Method 2 buckling capacity normally equals the same stress as Method 1 for stocky panels, while it is the ideal elastic buckling stress (minimum Eigen-value cut-off) for slender panels. By applying the ideal elastic buckling stress limitation, large elastic deflections and reduced in-plane stiffness will be avoided at higher buckling utilization levels.

Buckling failure mode 屈曲失效模式 系指屈曲失效的一种特定模式。具有开敞形状的加筋板格的典型失效模式如下:(1)板屈曲;(2)加强筋扭转屈曲;(3)加强筋腹板屈曲;(4)加强筋侧向屈曲。 Buckling failure mode refers to a specific pattern of buckling failure. Typical failure modes of stiffened panels with open profiles are:(1) plate buckling;(2) torsional stiffener buckling;(3) stiffener web plate buckling;(4) lateral stiffener buckling.

Buddy line 互助索 系指一段可以绑扎或用其他方法固定在其他人或此人的个人漂浮装置或其他物体上,以保持使用者在那个人或物体附近,从而便于确定位置和易于救援的绳索。 Buddy line is a length of cord which can be tied or otherwise fixed to another person or to that person's PFD or other objects, so as to keep a user in the vicinity of that person or object with a view to confirm location and thus make rescue easier.

Budget 预算 系指管理的主要职能——计划和控制的工具;是将远期战略目标和近期实施过程相联系的桥梁。

Buffering coefficient 缓冲系数

Build margin 建造裕度 系指一个由舰船建造者可能需要做出的不可预料的改变的余量。 Build margin is an allowance for unforeseen changes that may need to be made by the builder of the vessel.

Builder 建造厂 系指船舶修造厂。

Building related illness(BRI) 大楼并发症 系指与室内悬浮粒子及污染物有关的病症。

Building research institute(BRI) 建筑研究所

Build-up crankshaft 组合曲轴 非整体制造的,而将曲臂、曲柄等几个部件组合装配而成的曲轴。

Built-up pillar 组合支柱 系指用组合型材料制成的支柱。

Built-in sheave 嵌入滑轮 通常系指重型吊杆上嵌在吊杆上部杆体内供导引吊货索用的滑车。

Built-up propeller 组装螺旋桨 系指由几个铸件组成的螺旋桨。通常,组装螺旋桨分别由叶片铸件通过一系列螺栓和双头螺栓和轴毂连接在一起。 Built-up propeller is a propeller cast in more than one piece. In general, built up propellers have the blades cast separately and fixed to the hub by a system of bolts and studs.

Bulb profile 舱壁 系将船舶内部分为各个舱室的结构分隔壁。 Bulkhead means a structural partition wall sub-dividing the interior of the ship into compartments.

Bulb profile 球扁钢 系指一种主要应用于造船和造桥领域的中型材,其中船用球扁钢是造船用辅助中型材。近年来,随着造船业的迅猛发展,船用球扁钢需求旺盛。一般较大的船舶和正规的船舶在设计时主船体大多选用船用球扁钢,采用与相连板材相同厚度与材质的球扁钢作骨材(也称筋骨或龙骨)。

Bulb section 球臌型剖面 船首或船尾部在设计水线以下局部呈球状形臌出形式的横剖面。

Bulbous bow(bulb bow) 球鼻艏 系指设计水线以下艏部前部呈球形的船首部。见图 B-9。

图 B-9 球鼻艏
Figure B-9 Bulbous bow(bulb bow)

Bulb-type rudder 导流罩舵(整流帽舵) 系指舵叶中部位于螺旋桨轴线处装有回转体状的导流体,具有整流作用的舵。

Bulk cargo 散货 系指由粒子、颗粒或其他类似碎片或聚合体组成的,除液体或气体以外的任何材料,它无须任何形式的容器而可以直接装载至船上货物处所。

Bulk cargo container 散货集装箱 系指适用于装运谷物类农产品及粉状类矿产品的集装箱。

Bulk carrier 散货船 系指:(1)通常具有单甲板,在货物区域具有顶边舱和底边舱的构造,且主要从事运输散装干货的船舶,包括兼用船等船型;(2)主要用于运输散装干货的船舶,在装货处所通常具有单甲板、顶边舱和底边舱,货舱边界为舷侧壳板;(3)SOLAS 第Ⅻ章第2条定义的主要用于运输散装干货的船舶,包括矿砂船等船型,但不包括兼装船;(4)通常为在货物长度区域具有单层甲板、双层底、顶边舱和斜边舱及单舷侧结构的自推进海船,主要用于载运散装干货,且包括诸如矿砂船的船型。也应包括双舷侧散货船,但另有规定者除外。见图 B-10。 Bulk carrier is: (1) a ship which is constructed generally with single deck, topside tanks and hopper side tanks in cargo spaces, and is intended primarily to carry dry cargo in bulk. Combination carriers are included. (2) a ship carrying mainly dry cargo in bulk, normally constructed with single deck, topside tanks and hopper tanks in cargo spaces, cargo holds bounded by side shell; (3) a ship which is intended primarily to carry dry cargo in bulk, including such types as ore carrier as defined in SOLAS chapter Ⅻ, regulation 1, but excluding combination carrier; (4) a sea going self-propelled ship which is constructed generally with single deck, double bottom, topside tanks and hopper side tanks and with single side skin construction in the cargo length area, and is intended primarily to carry dry cargo in bulk and includes ship types such as ore carriers. It shall also include double skin bulk carriers except those specified. One bulk carrier is shown Fig. B-10.

图 B-10 散货船
Figure B-10 One bulk carrier

Bulk carrier bulkhead and double bottom strength standards 散货船舱壁和双层底强度标准 系指 1974 年国际海上人命安全公约,缔约国政府大会在 1997 年 11 月 27 日通过的决议 4"对最前两个货舱之间垂向槽形水密横舱壁尺寸和最前部货舱许可装载的评估标准",该标准可能经 IMO 组织修正,但这类修正案应按 SOLAS 公约第Ⅷ条有关的除第 1 章外适用的附则修正程序的规定予以提高、生效和实施。 Bulk carrier bulkhead and double bottom strength standards means "Standards for the evaluation of scantlings of the transverse watertight vertically corrugated bulkhead between the two foremost cargo holds and for the evaluation of allowable hold loading of the foremost cargo hold" adopted by Resolution 4 at the Conference of Contracting Governments to the International Convention for the Safety of Life at Sea, 1974, on 27 November 1997, which may be amended by the Organization, provided that such amendments are adopted, brought into force and take effect in accordance with the provisions of article Ⅷ of the present Convention concerning the amendment procedures applicable to the Annex other than Chapter 1.

Bulk carrier of double-side skin construction 双舷侧结构的散货船 系指主要用于运输散装干货的散货船。该船所有货舱边界均为双舷侧结构,2000 年 1 月 1 日以前建造的散货船,该双舷侧结构宽度等于或大于 760 mm;2000 年 1 月 1 日或以后,但在 2006 年 7 月 1 日以前建造的散货船,该双舷侧结构宽度等于或大于 1 000 mm;该宽距按垂直于舷侧壳板量取。 Bulk carrier of double-side skin construction means a bulk carrier which is intended primarily to carry dry cargo in bulk, in which all cargo holds are bounded by a double-side skin, the width of which is not less than 760 mm in bulk carriers constructed before 1 January 2000 and not less than 1,000 mm in bulk carriers constructed on or after 1 January 2000 but before 1 July 2006, the distance being measured perpendicular to the side shell.

Bulk carrier of single-side skin construction 单舷侧结构的散货船 系指主要用于运输散装干货的船舶,该型船:(1) 货舱任何边界均为舷侧壳板;或(2) 一个或多个货舱边界为双舷侧结构,2000 年 1 月 1 日以前建造的散货船,该双舷侧结构宽度小于 760 mm;2000 年 1 月 1 日或以后,但在 2006 年 7 月 1 日以前建造的散货船,该双舷侧结构宽度小于 1 000 mm 该宽距按垂直于舷侧壳板量取。此型船包括货舱任何边界均为舷侧壳板的兼装船。 Bulk carrier of single-side skin construction means a ship which is intended primarily to carry dry cargo in bulk, in which: (1) any part of a cargo hold is bounded by the side shell; or (2) one or more cargo holds are bounded by a double-side skin, the width of which is less than 760 mm in bulk carriers constructed before 1 January 2000 and less than 1 000 mm in bulk carriers constructed on or after 1 January 2000 but before 1 July 2006, the distance being measured perpendicular to the side shell. Such ships include combination carriers in which any part of a cargo hold is bounded by the side shell.

Bulk carrier with double-hull 双壳散货船 系指通常采用单层甲板,在货舱区有顶边舱和底边舱,主要拟散装运输干货而建造的船舶，包括指定运输矿砂或散装固体货物的兼用船，并且所有的货舱均为双舷侧船壳（不考虑舷舱的宽度）。对于附加了纵舱壁的兼用货船，也可视为双壳散装货船。

Bulk carrier with large opening 大舱口散货船 系指货舱开口特别大，有的达到船宽的 80% 以上，适合于装载各种货物的船舶。

Bulk carriers constructed 建造的散货船 系指安放龙骨或处于类似建造阶段的散货船。 Bulk carriers constructed means bulk carriers the keels of which are laid or which are at a similar stage of construction.

Bulk cement tank 散装水泥舱 系指贮存散装水泥的船舱。

Bulk Chemical Code 散装化学品规则 系指由 IMO 组织海上环境保护委员会以 MEPC. 20（22）决议通过的并经 IMO 组织修正的"散装运输危险化学品船舶构造和设备规则"。 Bulk Chemical Code means the Code for the Construction and Equipment of Ships carrying Dangerous Chemicals in Bulk adopted by the Marine Environment Protection Committee of IMO by resolution MEPC. 20 (22), which is amended by IMO.

Bulk density 散装货物密度 系指单位体积内固体、空气及水分的质量。它包括货物的含水量及货物内充满空气或水分的气孔。密度应用 kg/m³ 来表示。 Bulk density is the weight of solids, air and water per unit volume. It includes the moisture content of the cargo and the voids whether filled with air or water. The density should be expressed in kilograms per cubic meter (kg/m³).

Bulk dry cargo residues 散装干货物残余物 系指无害和无毒的散装运输干货物的残余物，不计及其颗粒尺寸的，包括石灰岩和其他精选的矿石、铁矿石、煤炭、盐和水泥。但它不包括已知其具有毒性或有害的任何物质的残余物，诸如镍、铜、锌、铅或在法律或条约规定中列为有害的材料。Bulk dry cargo residues means non-hazardous and non-toxic residues of dry cargo carried in bulk, regardless of particle size, including limestone and other clean stone, iron ore, coal, salt, and cement. It does not include residues of any substances known to be toxic or hazardous, such as nickel, copper, zinc, lead, or materials classified as hazardous in provisions of law or treaty.

Bulk grain carrier 散装谷物运输船 系指运散装谷物的船舶。

Bulk mud tank 散装泥浆舱 系指贮存坩子土的船舱。

Bulk solid cargo 固体散货 系指除液体或气体以外的由粒子、颗粒或较大块状物质组成的任何货物，成分通常一致，并直接装入货舱处所而无需任何中间围护形式。

Bulk/sulphuric acid carrier 散货/硫酸运输船

Bulkhead 舱壁 系指将船舶内部分为各个舱室的结构分隔壁。 Bulkhead is a structural partition wall sub-dividing the interior of the ship into compartments.

Bulkhead construction 舱壁结构 舱壁将船体内部分割成多个舱室，并且有提高船体强度、控制火灾蔓延、增强船舶抗沉性能等功能。按舱壁的适用性能，可分为水密舱壁、油密舱壁及耐火舱壁等。舱壁的结构形式分为平面舱壁和槽型舱壁两种。

Bulkhead deck 舱壁甲板 系指：（1）连接到水密横舱壁和船壳板的最上层连续甲板；（2）除艏舱、艉舱壁以外的水密横舱壁所到达形成有效结构的最高层甲板。 Bulkhead deck is: (1) the uppermost continuous deck to which transverse watertight bulkheads and shell are carried; (2) the highest deck to which the watertight transverse bulkheads, except both peak bulkheads, extend and are made effective.

Bulkhead deck 舱壁甲板 （1）参照 SOLAS 公约第Ⅱ-Ⅰ/2.5 条，舱壁甲板是除两尖舱舱壁外，各水密横舱壁有效延伸至的最上层甲板；（2）横向水密舱壁所到达的最高一层甲板。 (1) Ref. SOLAS Reg. Ⅱ-1/2.5, bulkhead deck is the uppermost deck to which the transverse watertight bulkheads, except both peak bulkheads, extend and are made effective; (2) the uppermost deck up to which the transverse watertight bulkheads are carried.

Bulkhead deck in a passenger ship 客船的舱壁甲板 系指水密主舱壁和水密船壳在分舱长度 L_s 范围内任何一点所达到的最高一层甲板，以及在 SOLAS 公约的第Ⅱ-Ⅰ章第 8 条和 B-2 部分所定义各种破损情况下进水的任何阶段乘客和船员撤船时不会被水阻挡的最低一层甲板。舱壁甲板可为阶梯形甲板。货船的干舷甲板可视为舱壁甲板。 Bulkhead deck in a passenger ship means the uppermost deck at any point in the subdivision length L_s to which the main bulkheads and the ship's shell are carried, watertight and the lowermost deck from which passenger and crew evacuation will not be impeded by water in any stage of flooding for damage cases defined in regulation 8 and in part B-2 of chapter Ⅱ-Ⅰ of SOLAS. The bulkhead deck may be a stepped deck. In a cargo ship the freeboard deck may be taken as the bulkhead deck.

Bulkhead insulation 舱室绝缘 系指船舶舱室绝热、吸声/隔声和防火方面的材料和工艺系统。

Bulkhead isolation valve 货油舱壁隔离阀

Bulkhead plate 舱壁板 系指构成舱壁表面的

板材。

Bulkhead recess 舱壁龛　系指舱壁前的一部分凹入而形成的龛状结构。

Bulkhead stool 舱壁凳座　系指槽形舱壁的上座或下座。Bulkhead stool means a lower or upper base of a corrugated bulkhead.

Bulkhead structure 舱壁结构　系指横向或纵向的舱壁板及其扶强材和桁材的总称。Bulkhead structure is defined as transverse or longitudinal bulkhead plating with stiffeners and girders.

Bulkhead stuffing box 舱壁隔填料函　设置在轴系所穿过的隔舱壁处,在轴承旋转的情况下,借以保证该处水密的密封装置。

Bulkhead valve 舱壁阀　系指货油舱舱壁的隔离阀。

Bulkhead ventilation gate 隔舱通风闸阀

Bulkhead-mounted socket 扶手支座　系指装在板壁上,支承扶手用的座子。

Bull rope 曳纲　系指拖网捕鱼时,连接在船与网之间的绳索。

Bull trawler 对拖网渔船　系指由两艘船共同拖带一顶拖网,进行捕捞作业的拖网渔船。

Bulwark 舷墙　系指露天甲板四周紧接船舷上缘的垂直板材。Bulwark is the vertical plating immediately above the upper edge of the ship's side surrounding the exposed deck(s).

Bulwark rail 舷墙顶舷材　系指加强舷墙上缘的型材。

Bumper 缆绳缓冲器　为调查船在定点作业或航行拖曳作业时缓和外界(风、浪)突加的冲击负荷用,以保护有关设备及缆绳免遭损坏的一种装置。在调查船上多用于深水抛锚作业、深水底栖作业或大负荷的地质绞车上。

Bundesarnt für See Schiffahrt(BSH) (德国)海事局　总部设在汉堡,德国联邦运输部下属的,负责对挂德国旗新船的安全、事故预防和防污染进行法定检验的机构之一。由其主管的最重要的一套德国法规为"海船安全法规"(Safety Regulation for Seagoing Ships/SSV)。该局不进行例行的图纸审批及检验工作。

Bundled cable 成束敷设　系指从机器处所及具有较大火危险处所中出来的 5 根或多于 5 根,以及在其他处所中多于 10 根电缆紧靠敷设在电缆槽中者。Bundled cable means 5 or more cables from the machinery spaces and spaces of high fire risk, and more than 10 cables from the other spaces are closely laid in the cable trunking.

Bunker 燃料舱　系指存储船舶机器所用燃料的舱室。Bunker is a compartment for the storage of fuel oil used by the ship's machinery.

Bunker clause 燃料条款

Bunker tax(emission trading) 燃料税(碳排放税)

Bunkering facility 加油设备

Buoy dues 浮筒损

Buoy tender 航标船　又称"布标船"。在航道与其附近的暗礁、浅滩、岩石处进行航标布设、巡检、补给、修理、维护的船舶。有内河和沿海之分。艉甲板上设有起吊航标用的起重机和绞盘,甲板下设航标储存舱,甲板室内设的航标修理室。有的航标船兼作航道测量及海洋水文调查用。其其有良好的操纵性和低速航行性能。

Buoy tour & inspection ship 航标巡检船　系指对航标进行巡检、补给、修理、维护的船舶。

Buoyage winch 浮标系统绞车　为收放浮标系统绳索及调查仪器进行海流、水文等昼夜连续观测的绞车。

Buoyancy 浮性　系指船舶于各种载重情况下,保持一定浮态的性能。

Buoyancy curve 浮力曲线　系指表示船舶在某一状态下,浮力沿船长分布状况的曲线。

Buoyancy loads 浮力载荷　系指船舶浮力。Buoyancy loads are buoyancy of the ship.

Buoyancy tube 浮胎　一般分上下两层围设于气胀式救生筏周缘以提供浮力的软胎。

Buoyant apparatus 救生浮具　系指设计用以支持在水中的一定数量人员,并在构造上能保持本身形状及性能的漂浮设备(救生艇、救生筏、救生圈、救生衣等除外)。Buoyant apparatus means flotation equipment (other than lifeboats, liferafts, lifebuoys and lifejackets) designed to support a specified number of persons who are in the water, and with such construction that it can retain its shape and properties.

Buoyant oar 浮桨　系指配置于救生筏、浮具上供划水用的能自浮的桨。

Buoyant rescue quoit 施救浮索　系指穿有色彩鲜明的浮环,抛投水面供救援用的绳索。

Buoyant raft 浮筏　系指将多台机械设备分别弹性或刚性地安装在其上的中间基座,即筏架。一个完整的浮筏系统通常由被隔离的设备(组)、隔振器组(第一层隔振器组,第二层隔振器组)、筏架、支撑基座、管路挠性接管、限位器等组成。

Burean of Political Military affairs(PM) [美]政治军事局　其使命:致力于整合军事与外交,为共同应

对安全挑战，锻造强有力的国家伙伴。该机构成为国务院与国防部的重要纽带，在国际安全领域、安全救援、军事行动、防御战略计划、国防贸易对等领域，政治军事局提供积极的政策牵引。美国政治军事局的成立是美国实现军事、外交以及经济、科技的对外运用的力量再整合，同时，也是现实需求与政治目标之间的战略平衡。

Burner boom 燃烧臂

Burton rig 通索索具 系指在补给船与接收船之间，由两条联动的传送干货的绳索所组成的索具。

Bury barge 埋管船 系指备有一套水底埋管机，供在水域埋设石油或天然气输送管道用的船舶。

Bushel 蒲式耳 是一个计量单位。它是一种定量容器，好像我国古时的斗、升等计量容器。1蒲式耳在英国等18加仑，相当于36.268升(公制)。在美国，1蒲式耳相当于35.238升(公制)。1Bushel油料或谷类的质量各异。即使同一种油料或谷物也因不同品种或产地实际换算也有些差别。1英制蒲式耳(1.0321美制蒲式耳)合36.3677升。

Business 经营 一般系指"商业"的意思，还有交易、商业、营业、行业的意思，另外，还有职责、本分、分内事、权利的意思。在不同的语境中，可以有不同的含义。

Business people 商务人员 系指由熟悉交易惯例、价格谈判条件、了解交易行情，有经验的业务员或厂长、经理担任。

Business process automation 业务流程自动化

Business tax 营业税 是针对在中国境内提供应税劳务，转让无形资产或销售不动产的单位和个人，就其所取得的营业额征收的一种税。营业税属于流转税制中的一个主要税种。2011年11月17日，财政部、国家税务总局正式公布营业税改征增值税试点方案。2015年5月，营改增的最后3个行业建安房地产、金融保险、生活服务业的营改增方案将推出，不排除分行业实施的可能性。其中，建安房地产的增值税税率暂定为11%，金融保险、生活服务业为6%。这意味着，进入2015年下半年后，中国或将全面告别营业税。

Business visa 商务签证 主要系指有关人员因公务或者个人原因去目的地国家从事投资、贸易、会议、展览、劳务等方面事务进行的实地考察或洽谈。这类人员进入目的地国时需持商务访问签证。

Butt 对接缝 系指板材短边的连接缝。

Butterfly valve 蝶形阀

Butterworth hatch 清洗口 系指货油舱用高压高温热水进行洗舱的清洗口。

Butterworth opening 巴氏货油舱清洗开口

Butterworth pump (tank cleaning pump) 洗舱泵(油船洗舱泵) 系指油船上抽送清洁油舱用水的泵。

Buttocks 纵剖线 系指平行于中线面平面与船体型表面的交线。

Buy boat 收鲜船 系指设有保鲜设备，专门去渔场收运新鲜渔获物的船。

Buyaney tank 打捞浮筒 系指打捞沉船、沉物时，提供浮力用的圆柱形水密筒。

Buyer credits 买方信贷 系指出口国银行直接向外国的进口厂商或进口方银行提供的贷款。其附带条件就是贷款必须用于购买债权国的商品，因而起到了促进商品出口的作用，这就是所谓的约束性贷款。买方信贷不仅使进口厂商可以较快地得到贷款和减少风险，而且使进口厂商对货价以外的费用比较了解，便于其与出口厂商讨价还价。出口方银行直接向买方(进口商、进口国政府机构或银行)提供贷款，使国外进口商得以即期支付本国出口商货款的一种融资方式。买方信贷涉及的当事人常有4类：出口商、出口国银行、进口商、进口国银行。

Buying insurance 投保 系指凡是按CIF价格成交的合同，卖方在装船前，须及时向保险公司办理投保手续，填制投保单的行为。

B_{WL} line B_{WL}线 系指船首和船尾处最小吃水的连线。B_{WL} line means the line between by the minimum draughts fore and aft.

By mutual consent 双方同意

By-pass damper 烟气调节挡板 系指布置在烟道中，可改变其开度大小旁通烟气，以调节过热蒸汽温度的挡板。

By-pass governing 旁通调节 以使调节级前面或后面压力较高的蒸汽绕过低速级组的办法来改变蒸汽流量，从而改变蒸汽轮机功率的调节方式。分内旁通和外旁通两种。

By-pass steam dump 旁路排气 系指没有进汽轮机做功而直接引入冷凝器的一部分蒸汽。

By-pass valve 旁通阀 为实现旁通调节而设置在旁通气道中的阀。

B-δ design charts (Tayloe's design charts) B-δ式设计图谱 以收到功率系数B_p或推功率系数B_u为横坐标，螺距比为纵坐标，并绘有直径系数$δ$和敞水效率等值线的螺旋桨设计图谱。

C

C 系指在燃油舱允装率为98%时船舶的总燃油量(包括小燃油舱在内)，m³。 *C is the ship's total volume of oil fuel including that of the small oil fuel tanks, in cubic meters, at 98% tank filling.*

C class combustible liquid　C级易燃气体　对美国国家消防委员会(NFPA)船上气体危害控制标准而言，系指雷氏蒸发气压力≤5.9×10^3 Pa(8.5 lbf/in²)的易燃气体。

C class divisions　C级分隔　系指用认可的不燃材料制成的分隔，不必满足防止烟和火焰通过以及限制温升的要求。允许使用可燃装饰板，只要这些材料满足相应要求。 *C class divisions are divisions constructed of approved non-combustible materials. They need meet neither requirements relative to the passage of smoke and flame nor limitations relative to the temperature rise. Combustible veneers are permitted provided they meet the relative requirements.*

C disk　C盘　系指电脑硬盘主分区之一，一般用于储存或安装系统使用。针对安装在本地硬盘的单操作系统来说，是默认的本地系统启动硬盘。大部分C盘内文件主要由Documents and Settings、Windows、Program Files等系统文件夹组成，Program Files文件夹一般都是安装软件的默认位置，但是也是病毒的位置，所以要对C盘进行严密保护。C盘对于本地硬盘的单操作系统来说，是极其重要的，所以平时存放数据尽量不要放C盘。默认在C盘需要移动出来的个人目录是:C:\Documents and Settings\你的登录账号\Documents。

C sound level　C声级　系指C计权网络在计权后的声级。其频率响应曲线为100 phon等响曲线经整理后倒置。

C standard ship's station　C标准船站(卫星通信)　是从1991年初正式投入商业使用。该站是可以提供数据和电传通信的通信系统。

C zone　C区域(螺旋桨)　系指经受低的工作载荷和螺旋桨相对薄的区域。在这种情况下，采用焊补方法进行修补可能是安全的。

CA zone　CA区域　系指在气密外壳之内包含的一个或多个货物舱。 *CA zone means one or more cargo chambers enclosed in an air-tight envelope.*

Cabin　客舱　系指设有铺位或卧铺，专供船上旅客休息用的舱室。

Cabin balcony　客舱阳台　系指单个客船的居住者专用的且从该客舱可直接进入的开敞甲板的处所。 *Cabin balcony is an open deck space which is provided for the exclusive use of the occupants of a single cabin.*

Cabin doors and windows　舱室门窗　系指设有密性要求的舱室开口上的盖闭部件及其附属件。

Cabin hardware　船用小五金　系指供舱室设备上使用，一般都具有防腐要求的铰链、插销、拉手、防撞门钩、移门滑轮、拉帘轨、门钩、窗钩、风暴钩以及各种门锁、抽屉锁的小五金物品的统称。

Cabin ladder　舱室梯　系指设置在生活舱室内的梯子。

Cabin luggage　自带行李　系指旅客在其客舱内的行李，或其他由其携带、保管或控制的行李。 *Cabin luggage means luggage which the passenger has in his cabin or is other wise in his possession, custody or control.*

Cabin outfit　舱室属具　舱室设备中的各种舱室门窗、舱室梯、搁架和船用小五金的统称。

Cabin wall ventilator(cabin fan)　舱室通风机　系指舱室通风换气用的通风机。

Cam angle　凸轮轴转角　凸轮配气机构的凸轮轴旋转的角度。

Cable　锚缆　与锚连接的索或链。 *Cable is a rope or chain attached to the anchor.*

Cable buoy　布缆浮筒　系指从船上向岸上布设电缆时，供在浅水区支乘和运送电缆登陆用的浮筒。

Cable burying machine　电缆埋设机　系指在浅海区布缆时可放至海底由布缆船上的绞车牵引，依靠机器的自重把电缆埋入海底泥沙中的机械设备。

Cable charges　电报费

Cable cutter　切线器　系指用以捞获电缆并靠拖索拉力能自动切断损坏电缆的器具。

Cable data communication　有线数据通信

Cable layer　布缆船　系指设有布缆机等专用设备，在海上布设和维修水底电缆的船舶。布缆船是专门用于敷设海底电缆的海洋工程船舶。由于海底电缆的铺设，使人们能远隔重洋进行通话，互相间的声音就像市内电话一样清晰。建设海底电缆是为了进行有线通信。有线通信具有容量大，距离远，安全可靠，抗干扰能力强等特点。对于没有陆路相通的国家、地区之间，需要在海底敷设通信电缆。布缆船是"海上架线兵"，不仅担负敷设海底电缆，沟通大洋彼岸通信的重任，还担负维修海底电缆，保证海上通信畅通的任务。见图C-1。

图 C-1 布缆船
Figure C-1 Cable layer

Cable management system 电缆管理系统 典型的电缆管理系统是由电缆绞车、电缆长度或张力自动控制设备和相关仪表组成。船舶通过电缆管理系统收放岸电电缆,与岸上电源进行连接。

Cable releaser 弃锚器 系指装在锚链末端用以在紧急情况下迅速脱弃锚链的专用器具。

Cable tank 电缆舱 系指在布缆船上,供储存电缆用的舱室。

Cable tension indicator 缆绳测力计 系指指示绳索张力的仪表。

Cable winch 声呐电缆绞车 为船舶航行中拖曳或收放电缆及拖带式换能器(包括发射及接收部分)进行海底地形、地貌地磁的测量,海底沉碍物等探测用的绞车。

Cable-laying ship 电缆敷设船 系指用于敷设海底电缆的工作船舶。

Cables 成束电缆 系指两根或多根电缆敷设在单独的管道、电缆槽或电缆通道内,或者是未加封闭且相互间不可分开的一束电缆。 Cables are said to be bunched when two or more are laid within a conduit, trunking or duct, or, if not enclosed, are not separated from each other.

Cage mast (lattice mast) 桁架桅 系指由型钢或管材构成的桁架式的桅。

Caisson 桩腿沉箱 系指装在桁架形桩腿底部,用以加大支撑面积的箱型结构。

Caisson dock 沉箱浮坞

Caisson type floating dock 整体式浮船坞 系指底部浮箱与两舷坞墙为连续,且不可分离的浮船坞。

Caissons or footings 桩靴 系指与自升式平台单个桩腿底部相连的独立水密结构。 Caissons or footings are separate watertight structures connected to the bottom of each leg of a self-elevating unit.

Calcium nitrate (UN 1454) 硝酸钙 白色易潮解固体,溶于水。不可燃物质。如遇火将大大加剧可燃物质的燃烧。虽不可燃,但其与可燃物质的混合物容易点燃并可能猛烈燃烧。该货物吸湿,受潮会结块。吞入该货物有害。 Calcium nitrate (UN 1454) is white deliquescent solid soluble in water. It is non-combustible materials. If involved in a fire, it will greatly intensify the burning of combustible materials. Although it is non-combustible, mixtures with combustible material are easily ignited and may burn fiercely. This cargo is hygroscopic and will cake if wet. This cargo is harmful if swallowed.

Calcium nitrate fertilizer 硝酸钙化肥 颗粒,主要由复盐(硝酸钠和硝酸铵)构成,氮含量总计不超过15.5%,水含量至少12%。无特殊危害。该货物为不燃物或失火风险低。 Calcium nitrate fertilizer is granule consisting mainly of double salt (calcium nitrate and ammonium nitrate) and containing not more than 15.5% total nitrogen and at least 12% water. It has no special hazards. This cargo is non-combustible or has a low fire-risk.

Calculated flow of persons (p/s) 计算的人流 系指每单位时间通过脱险通道特定点的人数。其按下式求得:$F_c = F_s W_c$。 Calculated flow of persons (p/s) is the predicted number of persons passing a particular point in an escape route per unit time. It is obtained from: $F_c = F_s W_c$.

Calculation of stability 稳性计算

Calculation wind speed V_W 计算风速 V_W 计算用的稳态风速的中间值或平均值。注意:计算风速以m/s表示。 Calculation wind speed means mean or average steady wind speed to be used for calculations. Note that calculation wind speed is expressed in meters per second.

Calculations of cathodic protection design 阴极保护设计计算书 是对需进行阴极保护的部位(船体结构水线以下部位、船舶的液货舱和压载水舱内部)进行保护设计的计算内容。阴极保护设计计算书应包括下列内容:(1)阴极保护方式的选择;(2)被保护部位(区域)的面积;(3)保护电位范围、保护对流电流密度和保护电流总量;(4)当采用牺牲阳极保护时,牺牲阳极材料、大小、形式、质量、电容量、发生电流、使用寿命和设计使用数量;(5)当采用外加电流保护时,外加电流保护装置的型号、辅助阳极的型号、数量和质量参比电极型号和螺旋桨及舵接地装置型号。 Calculations of cathodic protection design includes the calculated content of protection design necessary for the cathodically protected positions (positions under hull structural waterline, the internal of ship's cargo tanks and ballast tanks). The calculations of cathodic protection design includes the following: (1) selection of cathodic protection methods; (2) area of the protected position (area);(3) protective potential range, protective current density and protective current capacity;(4) sacrificial anodic material, size, type, weight, capacity, generating current, service life and designed service number if the sacrificial anodic protection is used;(5) type of external current protection unit, type of auxiliary anode, number and weight, type of reference electrode and type of earthing devices for propeller and rudder if the external current protection is used.

Call 呼叫 系指一个人向另一个人或一组人发出的关于进行联络、提供援助和/或采取行动的请求,即发出信号和表明这种请求的全过程。 Call is the request for contact, assistance and/or action from an individual to another person or group of persons, i. e. the complete procedure of signaling and indicating this request.

Call sign 呼号 系指从事无线电通信的法定代码。是从事无线电操作人员或电台,在无线电通信时使用的电台站代号,有标准的字母解释法发音朗读,中国的业余无线电台呼号以字母 B 开头,第二位表示电台的性质或等级,Y 表示集体台,A 表示个人一级台,D 表示个人二级台,G、H 表示个人三～五级台,第三位数字 0～9 表示电台站所在的分区,分别表示中国大陆地区的 10 个区域,再后面接 2～3 个英文字母代码。中国的业余无线电台被划分为 10 个区,其中北京是第 1 区,福建、江西、浙江三省是第 5 区,因此 BG5VIP 这个电台在这三省中的某一省。

CALM 悬链线锚腿系泊 CALM is a catenary anchor leg mooring.

Calm water service restriction 平静水域营运限制 系指对高速船及小船营运限制。限制授予航行于距岸不超过 5 n mile 的水域,满载并以其营运航速航行,航程不超过 2 h,并限制在风力不超过 6 级(蒲氏风级),且目测波高不超过 1.0 m 的海况的船舶的附加标志。

CAM = central alert management 中央警报管理 用于管理 CAM-HMI 上警报显示的功能,CAM-HMI 与导航系统和传感器之间警报状态的通信。(该功能可以集中或部分集中在子系统中并通过标准化的警报相关通信互连。) It is used for the management of the display of alerts on the CAM-HMI, the communication of alert states among CAM-HMI and navigational systems and sensors. [The function(s) may be centralized or partly centralized in subsystems and interconnected via a standardized alert-related communication].

Cam clearance 凸轮间隙 冷态柴油机在气门杆或其气门盖端面与相应的摇臂或其延伸件端面接触时挺杆滚轮与凸轮圆柱部分挺杆滚轮与凸轮圆柱部分的间隙。

Cam shaft 凸轮轴 系指具有凸轮的轴。

Camber[1] 拱度 系指翼切面的拱线与其头尾线间的最大距离。

Camber[2] 梁拱 系指露天甲板从两舷向船舶中线升高。 Camber means upward rise of the weather deck from both sides towards the centerline of the ship.

Camber correction factor 拱度修正因数 系指叶切面实效拱度对按形状确定的拱度的比值。

Camber curve 梁拱线 系指甲板具有梁拱时,各肋骨号处的横向垂直平面与甲板型表面的交线。无特别注明时,一般系指最大横剖面处者。

Camber line 拱线 系指连接叶切面各站厚度中点的曲线。

Camber ratio 拱度比 系指翼切面的拱度与其弦长的比值。

Camera phone 拍照手机 系可照相/录像的手机。

CAM-HMI-central alert management-human machine interface 中央警报管理-人机界面 用于显示和处理桥楼上报警的 BAMS 接口。 It is the BAMS interface for display and handling of alerts on the bridge.

CAM-HMI 用于显示和处理桥楼上报警的人机界面。 It is human machine interface for presentation and handling of alerts on the bridge.

CAMS-central alert management system 中央警报管理系统 统一的系统,用于监视、处理、分发和显示用于导航的设备和系统在桥楼上发出的警报。CAM 可以是独立系统或 INS 的一部分。CAM 也可以是桥楼警报管理系统的一部分,用于监视、处理、分发和显示所有必须在驾驶桥楼显示的强制性报警。 It is a unified system for the monitoring, handling, distribution and display of alerts on the bridge from equipment and systems used for navigation. The CAM can be a stand-alone system or part of an INS. The CAM can be part of a bridge alert management system for the monitoring, handling, distribution and display of all mandatory alarms on the navigating bridge.

Camshaft lubricating oil pump 凸轮轴滑油泵 系指将润滑油抽送至柴油机凸轮轴所用的泵。

Camshaft lubricating oil tank 凸轮轴滑油循环柜 在柴油机的凸轮轴独立润滑系统中,用以汇集凸轮轴润滑油供再循环使用的舱柜。

Camshaft tensioner 凸轮轴张紧器

Canada-euro Free Trade Agreement(CETA) 加拿大-欧盟自由贸易协定 系指加拿大-欧盟的贸易实体间所签订的具有法律约束力的契约,其目的在于促进经济一体化,消除贸易壁垒(例如关税、贸易配额和优先级别),允许货物与服务在国家间自由流动。

Canard system 鸭式水翼系统 65%或更多的水翼面积分布于艇的重心之后,较小的前翼往往作为控制面或稳定面的水翼系统。

Cancel an offer 撤销报价

Cancelled repairing items 减账工程 系指在履行合同过程中,因合同双方同意放弃的工程。

Cancelling date 解约日 系指:(1)租方可以接受装船的最晚交货日期。(2)船东有权选择解除合同的日期。

Cannabis 大麻 是最普通的违法毒品。它可有3种形式:(1)草本(大麻叶和花),有绿色、黄色的或棕色的草药材料,质地有粗有细,取决于样本的等级,外观与干的带刺的荨麻干草相似。有茎、杆和嫩枝,还带有小白籽。这种物质有浓烈的湿土气味和淡淡的腐烂植物气味。抽吸时有明显的辛辣"篝火"气味。这种气味在没有通风的环境中会久留不散。(2)大麻脂,呈米色,还有深褐色或黑色(偶尔也有淡黄色或浅绿色),通常是厚片或小块,但偶尔也有粉末状或模压成形。质地上有些粘连性。如果是厚片或模压块中,在质量上通常为 500 g 或 1 kg,尺寸分别为 130 mm × 100 mm × 25 mm(5 in × 4 in × 1 in)或 260 mm × 200 mm × 25 mm(10 in × 8 in × 1 in)。这种厚片通常是用聚乙烯或亚麻布包起来的。这种物质可以模压成不同的形状,如鞋底、小珠、雕像等。(3)大麻油,呈深绿色,还有黑色,偶尔也有金黄色,是具有黏性的油类液体,与草本大麻有相似的气味,但更强烈。通常是装在 5 kg 或 1 gal 玻璃或金属容器内进行运输,但有时也可能是更小的容器。大麻油可溶解聚乙烯或塑料。Cannabis is the most common illicit drug. It can be found in three forms:(1)herbal (marijuana)—this is found as a green, yellow or brown herbal material, it is rough or fine in texture depending on the grade of the sample and has similar appearance to dried stinging nettles. Stalks, stems and twigs may be present as well white seeds. The substance smells like spicy damp earth and mild rotting vegetation. There is a noticeably acrid "bonfire" smell when being smoked. The smell will linger in a non-ventilated environment;(2)resin—this appears as beige to dark brown or black (occasionally with a yellowish or greenish tinge)and is normally found as slabs or small chunks, although occasionally in powdered form or moulded shapes. It is slightly sticky in texture. If it is in slabs or moulded blocks, these are normally 0.5 or 1 kg in weight with dimensions 130 mm × 100 mm × 25 mm(5 in × 4 in × 1 in)or 260 mm × 200 mm × 25mm(10 in × 8 in × 1 in)respectively. The slabs will usually be wrapped in polythene or linen. The substance can be moulded into various shapes such as soles of shoes, beads, carved heads, etc;(3)oil—this appears as a dark green to block, occasionally golden, viscous oily liquid and has smell similar to herbal cannabis, but stronger. It is normally transported in glass or metal containers of 5 liter or 1 gallon though they may sometimes be smaller. Cannabis oil dissolves polythene or plastic.

Canoe body draught T_C 单艇体吃水 T_C 单艇体吃水 T_C 应在单艇体的最低点处量至单艇体与艇中心线的交点。在龙骨形状难以与艇体分离的情况下,单艇体吃水应通过船体表面与中线面坡度最小的切线的交点来测量。Canoe body draught T_C shall be measured between the intersection of the canoe body with the centerline of the craft at the lowest point of the canoe body. In cases where the keel form cannot be easily separated from that of the hull, the canoe body draught shall be determined by the intersection of the least steep tangent to the hull surface with the centerline plane.

Canon 佳能公司 是全球领先的生产影像与信息产品的综合集团。佳能总部位于日本东京,并在美洲、欧洲、亚洲及日本设有 4 大区域性销售总部,自 1937 年成立以来,经过多年不懈的努力,佳能已将自己的业务全球化并扩展到各个领域。目前,佳能的产品系列共分布于三大领域:个人产品、办公设备和工业设备,主要产品包括照相机及镜头、数码相机、打印机、复印机、传真机、扫描仪、广播设备、医疗器材及半导体生产设备等。

CANREP system CANREP 系统 系指加纳利群岛船舶强制报告系统。CANREP system means the mandatory ship reporting system for the Canary Islands.

Cant frame 斜肋骨 系指不在船体横剖面内而与之成一定角度的肋骨。

Cantilever 吊架 系指一种由三角形悬伸臂架构成的,供装设起重索具用的舱面吊重设备。

Cantilever jack-up drilling unit 悬臂自升式钻井平台 系指井架及钻台安装在悬臂结构上,且能沿轨道滑移到平台甲板以外一定距离进行钻井的自升式钻井平台。

Cantilevers 悬臂梁 系指从舷边延伸至其所支持的舱口甲板纵桁的甲板强横梁。Cantilevers are the deck transverses extending from ships side to the hatch side deck girders supported by them.

Canvas awning 帆布天幕 系指可收折的帆布做幕棚。

Canvas gear 帆布具 系指船上使用的帆布制品。包括舱口盖布、天幕、围帘、艇罩、机具罩套和帆布通风筒等。

Canvas ventilator　帆布通风筒　系指临时张设的用帆布制成的通风筒。

Canvas wind-sail　帆布通风筒　系指用帆布制的通风筒。

Capability plot　性能曲线图表　对各种不同方向的风、浪、流的特殊条件的理论上的船舶性能极坐标曲线图表。这些曲线图表可由不同的推力器组合来确定，并应按照IMCAM-DP性能曲线图表技术说明书来制订。

Capacitance　电容(电容量)　系指在导体和介质组成的系统中，当两个导体之间存在电位差时，可储藏电量的一种特性。

Capacitor (electrical condenser)　电容器　系指以将电容引入电路为主要目的的一种器件。根据其介质，电容器通常分为空气电容器、云母电容器、纸质电容器等。

Capacity[1]　载运能力　对能效设计指数而言，定义为：(1)对于散货船、液货船、气体运输船、滚装货船、杂货船、冷藏货物运输船和兼装船，载重吨应用作载运能力；(2)对于客船和滚装客船，《1969年国际吨位丈量公约》附则Ⅰ第3条所述总吨应用作载运能力；(3)对于集装箱船，载重吨(DWT)的70%应用作载运能力。集装箱船的EEDI值计算如下：①attained EEDI 按照EEDI公式计算，将70%的载重吨用作载运能力进行计算；②2012年新船达到的能效设计指数(EEDI)计算方法指南中用于计算基准线的估计指数值使用70%的载重吨计算如下：

$$\text{估计指数值} = 3.1144 \times \frac{190 \sum_{i=1}^{NME} P_{MEi} + 215 P_{AE}}{70\% \text{DWT} \times V_{ref}}$$

③MARPOL附则Ⅵ第21条的表2中对集装箱船的参数 a 和 c 通过按100%载重吨标出估计指数值来确定，即确定 $a = 174.22$ 和 $c = 0.201$；④新集装箱船的 required EEDI 使用100%载重吨计算如下：required EEDI = $(1 - X/100) \times a \times 100\%$ 载重吨-c，式中：X—关于新集装箱船的适用阶段和尺度的 MARPOL 附则Ⅵ第21条表1的折减系数(百分比)。 For the EEDI, capacity is defined as follows: (1) For bulk carriers, tankers, gas tankers, Ro-Ro cargo ships, general cargo ships, refrigerated cargo carrier and combination carriers, deadweight should be used as capacity. (2) For passenger ships and Ro-Ro passenger ships, gross tonnage in accordance with the International Convention of Tonnage Measurement of Ships 1969, Annex I, regulation 3 should be used as capacity. (3) For containerships, 70% of the deadweight (DWT) should be used as capacity. EEDI values for containerships are calculated as follows: ①attained EEDI is calculated in accordance with the EEDI formula using 70% deadweight for capacity; ②estimated index value in the Guidelines for calculation of the reference line is calculated using 70% deadweight as: Estimated Index Value = $3.1144 \times [(190 \sum_{i=1}^{NME} P_{MEi} + 215 \times P_{AE})/70\% \text{DWT} \times V_{ref}]$; ③parameters a and c for containerships in Table 2 of Regulation 21 of MARPOL Annex VI are determined by plotting the estimated index value against 100 percent deadweight i.e. $a = 174.22$ and $c = 0.201$ were determined; ④required EEDI for a new containership is calculated using 100 percent deadweight as: required EEDI = $(1 - X/100) \times a \times 100\%$ deadweight-c, Where X is the reduction factor (in percentage) in accordance with Table 1 in Regulation 21 of MARPOL Annex VI relating to the applicable phase and size of new containership.

Capacity[2]　装载量　定义为：(1)对于散货船、液货船、气体运输船、集装箱船、滚装货船和普通货船，载重吨应作为装载量；(2)对于客船和滚装客船，按照1969年国际吨位丈量公约附则Ⅰ第3条的总吨位应用作装载量；(3)对于集装箱船，装载量参数应根据载重吨的65%确定。 Capacity is defined as follows: (1) for dry cargo carriers, tankers, gas tankers, containerships, Ro-Ro cargo ships and general cargo ships, deadweight should be used as capacity; (2) for passenger ships and Ro-Ro passenger ships, gross tonnage in accordance with the International Convention of Tonnage Measurement of ships 1969, Annex I, regulation 3 should be used as capacity; (3) for containerships, the capacity parameter should be established as 65% of the deadweight.

Capacity at standard condition　供气量　系指压缩机在单位时间内排出的，相应于规定状态的干燥气体容积值。

Capacity curve　容积曲线　通常系指全船各液舱、煤舱和货舱在不同载重量时，舱容及其几何形状中心位置随载重量变化的曲线。

Capacity plan　容积图　绘有各货舱和液舱的布置并标明各舱容积及其几何形状中心位置的图纸。

Cape of Good Hope route　好望角航线　被称为"西方海上生命线"，西欧进口石油的近80%、战略原料的70%、粮食的25%都要通过这里运输。随着苏伊士运河的开通，人类更多地选择经济、快捷又安全的苏伊士运河，避开因受强烈西风带影响而风大浪急的好望角。由于造船工业和技术进步，以及货运量大增，大型船舶日益增多，苏伊士运河渐渐不能完全满足现代海运要求，使得好望角航线的地位重新增强。

Cape of good hope-type bulk carrier 好望角型散货船 系指能绕好望角航行的,载重量为 150 000 ~ 200 000 DWT 的大型散货船。

Cape size ship 好望角型船舶 系指载重量为 150 000 ~ 200 000 DWT 的船舶。

Cape size type carriers 好望角型船 系指在远洋航行中可以通过好望角或者南美洲海角最恶劣天气,载重在 10 万吨以上的干散货船。

Capillarity constant 毛细常数 单位长度的表面张力。

Capillary wave 表面张力波 一般由风速不到半节的微风产生,其表面张力比重力所引起的作用更大的波。

Capital 大写的(字母)

Capsizing level (upsetting level) 倾覆力臂 使船舶倾覆的最小外力矩臂。

Capsizing moment (upsetting moment) 倾覆力矩 使船舶倾覆的最小外力矩。

Capstan 绞盘 系指具有垂直安装的绞缆筒,在动力驱动下能卷绕但不储存绳索的机械,也指转动轴线与甲板垂直的绞车,是车辆、船舶的自我保护及牵引装置,可在雪地、沼泽、沙漠、海滩、泥泞山路等恶劣环境中进行自救和施救,并可在其他条件下,进行清障、拖拉物品、安装设施等作业,是军警、石油、水文、环保、林业、交通、公安、边防、消防及其他户外运动不可缺少的安全装置。主要用于越野汽车、农用汽车、ATV 全地形车、游艇、消防救援车、道路清障车以及其他专用汽车、特种车辆。

Captain of the port 港区司令官(COTP) 系指被委派出任纽约州布法罗港海上检验区的司令官或按有关规定的纽约州纽约港的港区司令官的海岸警卫队官员及港区司令官授权代表。Captain of the port means the Coast Guard officer designated as COTP of either the Buffalo, NY, Marine Inspection Zone and Captain of the Port Zone or the New York, NY, Captain of the Port Zone described in Part 3 of this chapter or an official designated by the COTP.

Captain's room 船长室 系指供船长办公和住宿用的舱室。

Captive model test 约束模操纵性试验 在船模试验水池中采用一定设备,强制船模作规定的运动,以测定作用于船模上水动力和力矩的试验。

Car/passenger ferry 车客渡船 系指一种同时装载汽车和旅客的船舶。车客渡船多用于海湾、海峡及沿海岛屿间作穿梭航运,并且多为定期班船。见图 C-2。

图 C-2 车客渡船
Figure C-2 Car/passenger ferry

Car carrier 车辆运输船 系指在设计和制造用于运输商品轮式车辆的船舶。见图 C-3。

图 C-3 车辆运输船
Figure C-3 Car carrier

Car PC 车载电脑 系指专门针对汽车特殊运行环境及电器电路特点开发的具有抗高温、抗尘、抗震功能并能与汽车电子电路相融合的专用汽车信息化产品,是一种高度集成化的车用多媒体娱乐信息中心。能实现所有家用电脑功能,支持车内上网、影音娱乐、卫星定位、语音导航、游戏、电话等功能,同时也能实现可视倒车、故障检测等特定功能。其主要功能包括车载全能多媒体娱乐,GPS 卫星导航,对汽车信息和故障专业诊断,移动性的办公与行业应用。

Carbamic acid foamed plastics 氨基甲酸泡沫橡胶

Carbon capture storage (CCS) 二氧化碳捕捉和储运

Carbon equivalent 碳当量 系指根据桶样分析,由下式 $C_{eq} = C + M_n/6 + (C_r + M_o + V)/5 + (Ni + Cu)/15$ 确定的系数(%)。Carbon equivalent (C_{eq}) is determined from the ladle analysis in accordance with the following equation: $C_{eq} = C + M_n/6 + (C_r + M_o + V)/5 + (Ni + Cu)/15 (\%)$.

Carbon ring packing (carbon ring gland) 碳环式气封 系指由若干段石墨环组成的气封。

Carbonic acid gas compressor (carbon dioxide compressor) 二氧化碳压缩机 系指舰艇中将舱室中经净化装置脱析的二氧化碳气体排出舷外的压缩机。

Carborundum 碳化硅 一种坚硬的碳和硅的黑

色晶体状化合物。无嗅,无含水量。吸入后有轻微毒害。该货物为不燃物或失火风险低。 Carborundum is a hard black crystalline compound of carbon and silicon. It is odourless. It has no moisture. It is slightly toxic by inhalation. This cargo is non-combustible or has a low fire-risk.

Card room 纸牌室(豪华邮轮)

Cargo 货物 包括:(1)①船上的贮藏品、食品、设备和燃料;②邮件;③旅客行李;④修船或货物处所的附件用的材料或用于紧固货物的材料和设备;⑤当使用起重机或吊杆装置装载货物或转运货物时,机械装卸设施和运输设备。(2)①除邮件、船用物料、船舶备件、船舶设备、船员物品和旅客携带的行李以外的任何货物、制品、商品以及船上运载的任何种类的物品;②物品、器皿、商品和用集装箱载运的各种物件。 Cargo includes: (1)① ship's stores, provisions, equipment and fuel; ②mails; ③passengers' baggage; ④material for the repair of a ship or for the fitting of a cargo space or material and equipment used for securing cargo; ⑤mechanical stowing appliances and transport equipment when carried as cargo or being handled by means of cranes or derricks; (2)①any goods, wares, merchandise, and articles of every kind whatsoever carried on a ship, other than mail, ship's stores, ship's spare parts, ship's equipment, crew's effects and passengers' accompanied baggage; ②any goods, wares, merchandise and articles of every kind whatsoever carried in the containers.

Cargo (oil) pump 货油泵 系指油船上抽送货油的泵,有时亦兼作压载泵。

Cargo alarm 货物报警 系指指示货物或货物的保障或安全系统异常的报警。 Cargo alarm means an alarm which indicated abnormal conditions originating in cargo, or in systems for the preservation or safety of cargo.

Cargo area 装货区域 系指船上包括液货舱、污液货舱、液货泵舱,包括相邻于液货舱或污液舱的泵舱、隔离舱、压载舱或留空处所,以及这些处所上方的整个长度和宽度的甲板区域。如货舱处所内设有独立的液舱时,则位于最后一个货舱处所后端或位于最前一个货舱处所前端的隔离舱、压载舱或空舱不算在装货区域内。 Cargo area is that part of the ship which includes cargo tanks, slop tanks, cargo pump rooms including pump rooms, cofferdams, ballast or void spaces adjacent to cargo tanks or slop tanks, and also deck areas throughout the entire length and breadth over the above-mentioned spaces. If independent tanks are installed in hold spaces, cofferdams, ballast or void spaces at the after end of the aftermost hold space or at the forward end of the forward most hold space are excluded from the cargo area.

Cargo area or cargo length area 货物区域或货物长度区域 系指包含所有货舱及其邻近区域,包括燃油舱、隔离舱、压载舱和空舱的船舶部分。对于油船和化学品船,货物区域系指包括液货舱、污油舱、液货/压载泵室、隔离舱、压载舱和邻接液货舱的空舱以及上述处所以上的船舶全长和全宽部分甲板区域的船舶部分。 Cargo area or cargo length area is that part of the ship which includes all cargo holds and adjacent areas including fuel tanks, cofferdams, ballast tanks and void spaces. For oil tankers and chemical tankers, the cargo area is that part of the ship which includes cargo tanks, slop tanks and cargo/ballast pump-rooms, cofferdams, ballast tanks, and void spaces adjacent to cargo tanks, and also deck areas throughout the entire length and breadth over the above mentioned spaces.

Cargo associated wastes 与货物相关联的废弃物 系指在船上堆装和装卸货物用后成为废弃物的所有材料。与货物相关联的废弃物包括但不限于垫舱物料、衬材、托板、衬板和包装材料、胶合板、纸、告示牌、钢丝和钢质的捆扎条。 Cargo associated wastes mean all materials which have become wastes as a result of the use of a ship for cargo stowage and handling on board. Cargo training manual include, but are not limited to, dunnage, shoring, pallets, lining, and packing materials, plywood, paper, cardboard, wire and steel strapping.

Cargo block 起货滑车 系指装设于吊杆头端处,用以吊货的滑车。

Cargo boat 小型货船 系指船长20 m以下的货船。 Cargo boat is a boat less than 20 m in length.

Cargo capacity 舱容 系指全船各货舱和液舱的总容积或任何某一货舱(或液舱)的单舱容积。

Cargo capacity plan 舱容图 系指绘有上甲板以下全船各个货舱和液舱及贮藏室等位置,并标明其容积和形心位置的图。有纵剖面图、甲板及舱底平面图、各舱容积一览表及载重量标尺,同时注明船舶的主尺寸、总吨位、净吨位、吊杆的起重能力和工作半径等。它在船舶建成后由造船厂提供给船东,供用船部门使用,便于货物装载和做必要的营运计算。

Cargo carriage contracts 货运合同 即货物运输合同,系指当事人为完成一定数量的货运任务,约定承运人使用约定的运输工具,在约定的时间内,将托运人的货物运送到约定地点交由收货人收货并收取一定运费而明确相互权利义务的协议。

Cargo carrying area 载货区域 系指船上包括货物围护系统部分和液货泵舱,压缩机舱以及上述区域上方覆盖整个船长和船宽的甲板区域。最后容纳舱柜空间后端和最前容纳舱柜空间前端,如设有隔离空舱、压载舱或空置处所,则不属于载货区域。

Cargo chain 吊货短链 系指在轻型吊杆装置中,为使空钩易于下降而装于吊货钩上方的一段短链。

Cargo compressor 液货压缩机 系指升高货物系统的蒸汽压力的设备。其用于输送货物、再液化、扫舱等工序。是液化气船必备的设备之一。分为离心式、活塞式和螺杆式等类型。

Cargo containment system 货物围护系统 对散装运输液化气体船而言,系指用于围护货物的装置,包括所设的主屏壁和次屏壁以及附属的绝热层和屏壁间处所,必要时还包括用于支持这些构件的邻接结构。如果次屏壁是船体结构的一部分,则它可以是货舱处所的边界。

Cargo control room 货物控制室 对散装运输液化气体船而言,系指用于控制货物装卸作业,并符合有关要求的处所。

Cargo control stations 货物控制站 系指控制货物操作的处所。 Cargo control stations means a space controlling cargo handling.

Cargo craft 货船 系指:(1)客船以外的任何船舶,这类船的任意一舱破损后,其他未破损处所的主要功能和安全系统仍能维持正常状态。(2)对地效翼船而言,系指客船以外的任何地效翼船。 Cargo craft is:(1)any high-speed craft other than passenger craft,which is capable of maintaining the main functions and safety systems of unaffected spaces after damage in any one of the compartments on board. (2)for the of WIG, cargo craft is any WIG craft other than a passenger craft.

Cargo deadweight 载货量 系指载重量中允许装载货物的最大值。

Cargo deck 装货甲板 系指装货区域内的开敞甲板,并构成货舱顶部,或上面设有液货舱,液货舱舱口,液货舱清洗舱口,液货舱测量开口和检查孔以及泵、阀和其他为装卸所需的装置和附件。 Cargo deck means an open deck within the cargo area, which forms the upper crown of a cargo tank;or above which, cargo tanks, tank hatches, tank cleaning hatches, tank gauging openings and inspection holds as well as pumps, valves and other appliances and fittings required for loading and discharging are fitted.

Cargo evaporator 液货蒸发器 系指在液化气船上将液货加热蒸发补充蒸发气压力的设备。常用于卸货、再液化作业中,以保持货舱内的蒸发气压力。

Cargo ferry 货物渡船 系指在规定航线运航的船舶,通过车辆门(指作为船身结构一部分的船艏门、

艉门、舷侧门和坡道)装载和运输车辆。 Cargo ferry means that a ship which operates in regular route and carries vehicles through the vehicle doors (bow doors , inner doors , side doors , and ramp form part of the hull structures).

Cargo fittings 货物附属装置 系指鹅颈式支架、顶部支架、起重机吊杆头部配件、起重机根部接线片、支索夹板、监视器配件等。它们都是永久固定安装的结构件或者货物装配用的船体结构。 Cargo fittings include goose neck brackets , topping brackets , fittings at the derrick boom head , derrick heel lugs , guy cleats , eye fittings , etc. which are permanently fitted to the structural members or the hull structure for the purpose of cargo handling.

Cargo gear 起货装置 系指在装卸货物时与使用吊杆或起重机有关的装备部件。它们是:(1)不与吊杆或起重机铆接、焊接或永久固定者;(2)设计成可从吊杆或起重机上卸下者,且包括各种钢索,纤维索、吊索、网、夹头、抓头、可拆卸的零部件、电磁起吊装置、真空起吊装置、专制的货物转运系统或自卸系统,但不包括运输设备或包装。 Cargo gear means an article of equipment used in loading or unloading cargo with a crane or derrick, that:(1)is not riveted, welded or otherwise permanently attached to the crane or derrick; (2)is designed to be detachable from the crane or derrick, and includes any wire rope, fiber rope, sling, net, clamp, grab, loose gear, magnetic lifting device, vacuum lifting device, patent cargo handling system or self unloading system but does not include transport equipment or packaging.

Cargo gear coefficient 起货设备利用系数 系指起货设备实际装卸能力与理论装卸能力的比值。

Cargo gears 装卸设备 系指用于装载和卸载货物和其他物品的起重机系统、起重机、货物升降装置和其他附件,不包含货物坡道,但包括以上装置驱动系统安装和货物装配。 Cargo gears are derrick systems, cranes, cargo lifts and other machinery used for the loading and unloading of cargo and other articles except cargo ramps, and include their installations of driving systems and cargo fittings.

Cargo handling appliances 起货机械 从形式上分为:吊杆式(derrick system)、吊车(crane)、电梯(lift)和坡道(ramp)等4类。

Cargo handling appliances survey 起货设备检验 系指对起货设备进行下列的检查:(1)对吊杆型的起货设备进行年检;(2)对甲板吊杆型的起货设备进行年度彻底检验;(3)对吊杆型的起货设备,每4年要进行1次包括鹅颈头拆检在内的,对系统内部各部件的彻底检验。

Cargo handling appliances 货物装卸设备 系指升降设备和传送设备。 Cargo handling appliances are lifting appliances and loose gear.

Cargo handling by conveyor system 自卸货系统 系指装备货物传送设备，能自装或卸货功能的系统。 Cargo handling by conveyer system means a system fitted with belt driving conveyors for cargo handling, which is capable of self-loading and unloading.

Cargo handling by deck cranes system 起重机装卸 用船上配置的起重机装卸货物的一种吊装装卸方法。

Cargo handling gear 起货设备 船上装卸货物的装置的总称。主要有吊杆装置、起重机及其他装卸机械等设备。

Cargo handling productivity 装卸能力 系指全船的起货设备或某一套起货设备每小时装货或卸货的吨数。

Cargo handling rate 装卸率 系指每日装卸货物的速度。

Cargo handling with grabs 抓斗装卸 用抓斗攫取散货进行装卸的一种吊装装卸方法。

Cargo hatch 货舱口 供装卸货物用的舱口。

Cargo hatch trunk 货舱围井 系指连接上下货舱口使之与其他部分隔开的围井。

Cargo hold 货舱 系指兼用舱、压载兼用舱及固体货物专用舱的总称。 Cargo hold is a general term for solid cargo/oil hold, ballast/solid cargo hold and exclusive solid cargo hold.

Cargo hold bulkhead 货舱舱壁 系指货舱的边界舱壁。 Cargo hold bulkhead is a boundary bulkhead for cargo hold.

Cargo hold dehumidification 货舱空气干燥 当舱外空气温度、湿度较高不能进行通风时，进行封闭循环。同时加入部分经过干燥处理的空气，以保持舱内干燥的技术。

Cargo hold ladder 货舱梯 系指设置在货舱内的梯。

Cargo hold ventilation 货舱通风 为使货舱内温度不致太高而进行的自然通风或机械通风。

Cargo hook 吊货钩 系指钩尖具有遮蔽件的吊货用钩子。

Cargo hook gear (hook assembly) 吊钩装置 由吊货钩、转环、吊钩卸扣、压重和三角眼板等零件所组成，用以钩取货物的装置。

Cargo hook swivel 吊钩转环 为使吊货钩可以自由转动和吊货索具免受扭转而装于吊货钩上部的转环。

Cargo HSC 高速货船 系指载货的高速船。 Cargo HSC is a high speed cargo craft carrying cargoes.

Cargo information 货物资料 系指：(1)托运人在装货前及早向船长或其代表提供关于该货物的相应资料。以便能实施为此种货物的正确积载和安全运输可能是必需的预防措施。此类资料应在货物装船前以书面形式和相应的运输单证予以确认。货物资料应包括：①对于杂货和货物单元运输的货物，应有货物的一般说明、货物或货物单元的毛重和货物的任何特性。同时，应提供IMO组织A.714(17)决议通过并可能经修正的《货物积载和系固安全操作规则》第1.9节所要求的货物资料。对第1.9节的任何此种修正案应经SOLAS公约第Ⅷ条有关适用于除第1章外的附则修正程序的规定予以通过、生效和实施；②对于散装货物，应有关于货物积载因数、平舱方法、移动的可能性(包括稳定角，如适用)以及任何其他有关特性的资料。对于浓缩物或可流态化的其他货物还应有关于货物含水量及其可运含水极限证书的资料；③对于未按第Ⅶ/1.1条所定义的IMDG规则的规定分类，但具有可能造成潜在危险的化学特性的散装货物，除上述各项要求的资料外，还应有关于其化学特性的资料。在货物单元装船前，托运人应确保这类货物单元的毛重与运输单证中表明的毛重一致；(2)托运人在装货前及早向船长或其代表通知关于该货物的适当资料，以便能实施为此种货物的适当积载因数和安全运输可能是必需的预防措施。此类资料应在货物装船前以书面形式和适当的运输单证予以确认。货物资料包括：(1)对于杂货和以货物单元运输的货物，应有货物的一般说明、货物或货物单元的毛重和货物的任何有关特征。(2)对于固体货物：①BCSN（国际海运固体货物规则列出该货物时），除BCSN外还可使用辅助名称；②货物组别(A和B，A、B或C)；③货物的IMO类别(如适用)；④货物以字母UN开头的联合国编号(如适用)；⑤交运货物的总量；⑥积载因数；⑦平舱的需要和平舱程序(必要时)；⑧移动的可能性，包括静止角(如适用)；⑨对精矿或其他可流态化货物，以证书形式提供的关于货物的含水量及其适运水分极限的补充信息；⑩形式潮湿底层的可能性；⑪货物可能产生的有毒或易燃气体(如适用)；⑫货物的易燃性、毒性、腐蚀性以及耗氧倾向(如适用)；⑬货物的自热特性，以及平舱的需要(如适用)；⑭与水接触后散发的易燃气体的特性(如适用)；⑮放射特性(如适用)；⑯国家主管当局要求的任何其他信息。 Cargo information means: (1) the shipper shall provide appropriate information on the cargo sufficiently to the master or his representative in advance of loading to enable the precautions which may be necessary for proper stowage and safe carriage of the cargo to be put into effect. Such information shall be confirmed in writing and by appropriate shipping documents prior to loading the cargo on the ship: ① For general cargo, and cargo carried in cargo units, a general description of the cargo, the gross mass of the cargo or of the cargo units, and any relevant special properties of the cargo shall be prouided. The cargo information required in sub-chapter 1.9 of the Code of Safe Practice for Cargo Stowage and Securing, adopted by IMO Organization by resolution A.714(17), as may be amended, shall be provided. Any such amendment to sub-chapter 1.9 shall be adopted, brought into force and take effect in accordance with the provisions of article Ⅷ of SOLAS Convention concerning the amendment procedures applicable to the Annex other than chapter I. ②For bulk cargo, information on the stowage factor of the cargo, the trimming procedures, likelihood of shifting including angle of repose, if applicable, and any other relevant special properties shall be provided. For concentrate or other cargo which may liquefy, additional information shall be provided on the moisture content of the cargo and its transportable moisture limit shall be provided. ③For bulk cargo which is not classified in accordance with the provisions of the IMDG Code, as defined in regulation VII/1.1, but has chemical properties that may create a potential hazard, in addition to the information required by the preceding subparagraphs, information on its chemical properties shall be provided. Prior to loading cargo units on board ships, the shipper shall ensure that the gross mass of such units is in accordance with the gross mass declared on the shipping documents; (2) Appropriate information on the cargo provided by the shipper, to the master or his representative sufficiently in advance of loading to enable the precautions which may be necessary for proper stowage and safe carriage of the cargo to be put into effect. Such information shall be confirmed in writing and by appropriate shipping documents prior to loading the cargo on the ship: (1) For general cargo, and cargo carried in cargo units, a general description of the cargo, the gross mass of the cargo or of the cargo units, and any relevant special properties of the cargo shall be provided; (2) For solid bulk cargo, information as following: ① the BCSN when the cargo is listed in the International Maritime Solid Bulk Cargoes Code. Secondary names may be used in addition to the BCSN; ②the cargo group (A and B, A, B or C); ③the IMO Class of the cargo, if applicable; ④the UN number preceded by letters UN for the cargo, if applicable; ⑤the total quantity of the cargo offered; ⑥the stowage factor; ⑦the need for trimming and the trimming procedures, as necessary; ⑧the likelihood of shifting, including angle of repose, if applicable; ⑨additional information in the form of a certificate on the moisture content of the cargo and its transportable moisture limit in the case of concentrate or other cargo which may liquefy; ⑩likelihood of formation of a wet base; ⑪toxic or flammable gases which may be generated by cargo, if applicable; ⑫flammability, toxicity, corrosiveness and propensity to oxygen depletion of the cargo, if applicable; ⑬self-heating properties of the cargo, and the need for trimming, if applicable; ⑭properties on emission of flammable gases in contact with water, if applicable; ⑮radioactive properties, if applicable; and ⑯any other information required by national authorities shall be provided.

Cargo introducing additional fire hazards 能引起额外失火危险的货物 系指在37.8℃时蒸发气绝对压力大于1.013 bar的货物。载运此类货物的船舶应符合《国际散装化学品规则》第15.14条的要求。若船舶在限制时间内航行于限制区域，有关主管机关可根据《国际散装化学品规则》第15.14.3条免除对制冷系统的要求。 Cargo introducing additional fire hazards is a liquid cargo with a vapour pressure greater than 1.013 bar absolute at 37.8 ℃. Ships carrying such substances shall comply with paragraph 15.14 of the International Bulk Chemical Code. When ships operate in restricted areas and at restricted times, the Administration concerned may agree to waive the requirements for refrigeration systems in accordance with paragraph 15.14.3 of the International Bulk Chemical Code.

Cargo length area 货物长度区域 系指所有货舱或邻近区域，包括燃油舱、隔离舱、压载舱或空舱。 Cargo length area is that part of the ship which includes all cargo holds and adjacent areas including fuel tanks, cofferdams, ballast tanks and void spaces.

Cargo lifts 货物提升设备 系指被指定为包含货物和装卸货物的设备。 Cargo lifts are the installations designed to contain the cargo in their structure for loading and unloading the cargo.

Cargo net 吊货网兜 由钢索或纤维索编成并附有吊攀，用以堆集货物后可供起吊的索网。

Cargo notation 货种标志 表明船舶的设计、改造或布置适于载运一种或多种特殊货物的标志，例如硫酸。具有一种或多种特殊货种标志的船舶并不因此而妨碍其载运适合于该船的其他货物。 Cargo notation is a notation indicating that the ship has been designed, modified or arranged to carry one or more particular cargoes, e.g. sulphuric acid. Ships with one or more particular cargo notations are not thereby excluded from carrying other cargoes for which they are suitable.

Cargo oil 货油　系指是对油船可以装运的液体货物的总称。　Cargo oil will be used as a collective term for liquid cargoes which may be carried by oil carrier.

Cargo oil deck pipe lines 甲板货油管系　系指布置在油船甲板上的货油装卸管系。

Cargo oil heating system 货油加热系统　系指为适于泵运，将货油加热或保温的系统。

Cargo oil hose（cargo hose） 货油软管　油船装卸货油时与岸上连接的软管。

Cargo oil pump room pipe lines 货油泵舱管系　系指布置在货油泵舱内的货油装卸管系。

Cargo oil stripping pump 货油清舱泵　系指船上抽送油舱中残余剩油的泵。

Cargo oil suction 吸油口　系指货油舱管系上用以将货油吸出的专用吸口。分高位吸口和低位吸口两种。

Cargo oil suction heating coils 吸油口加热盘管　货油加热系统中，为节省加热蒸发气和缩短加热时间而设置于油舱吸口附近对邻近货油进行局部快速加热的盘管。

Cargo oil tank 货油舱　系指兼用舱、兼用油舱及污油舱的总称。　Cargo oil tank is a general term for solid cargo/oil hold, oil/ballast tank and slop tank.

Cargo oil tank breathing system（breathing valve, pressure and vacuum relief valve） 货油舱呼吸阀　系指设置在透气管路上，用以调节油舱压力和防止油气大量外逸的一种特种阀。

Cargo oil tank gas pressure indicator 货油舱气压指示器　封闭式货油装卸系统中测量油舱压力的仪器仪表。

Cargo oil tank pipe lines 货油舱管系　货油装卸系统中布置在货油舱内部的管路。由总管、支管和高、低位吸口等组成。

Cargo oil tank stripping system 货油舱清舱系统　系指为尽量吸净货油舱剩油所设的专用系统。包括清舱管路和清舱泵等。

Cargo oil valve（cargo valve） 货油阀　系指货油管路上使用的阀。

Cargo operations 货物操作　系指在船上和海上设施之间转移或接收一般混合货物或液体货物有关的作业，包括控制和监视本船和起货设备。仅用于货物操作的船舶被称为平台供应船（见 PSV）。　Cargo operations mean related to transferring or receiving general mixed cargo or liquid cargo between ship and offshore installation, including control and monitoring of own ship and cargo gear. Ships only designed for cargo operation is named platform supply vessel（PSV）.

Cargo permeability 货物渗透率　系指货物堆内的空隙与货物所占容积之比。

Cargo plan/stowage plan 配载图　系指表示船上各舱室货物装载位置的计划图。

Cargo pump room 货油泵舱　系指：（1）为包括泵及其附件，用来装载授予营运标志的船舶的货物的处所；（2）装有货泵的处所及通往这些处所的出入口和围壁通道；（3）装有供装卸货品用的泵及其属具的处所。　Cargo pump rooms are：（1）a space containing pumps and their accessories for the handling of products covered by the service notation granted to the ship；（2）spaces containing cargo pumps and entrances and trunks to such spaces；（3）a space containing pumps and their accessories for the handling of the products.

Cargo pump room ventilation 油泵舱通风　为不断排除油船油泵舱内油气，并补充新鲜空气使舱内油气浓度不致升高而进行的自然通风或机械通风。

Cargo purchase 吊货索具　系指钩住或抓住货物以及传递起货绞车的拉力而提升和下降货物的成套索具。一般包括吊钩装置、滑车。吊货索以及卸扣、眼板、套环等连接部件。

Cargo purchase eye 吊货眼板　系指位于吊杆头端处，专供连接吊货索具或同时连接吊货索具和千斤索用的眼板。

Cargo ramps 货物坡道　系指安装在船壳上或在船上提供用来安排进入汽车或载货汽车的设备，配备有机械设备用来打开、关闭或旋转该设备。　Cargo ramps are the installation mounted on the shell or provided in the ship, and arranged to permit passage of vehicles as cargo or vehicles loaded with cargo on themselves and having mechanism enabling its opening and closing or turning.

Cargo receipt 承运货物收据　系指在特定运输方式下所使用的一种运输单据，它既是承运人出具的货物收据，也是承运人与托运人签订的运输合同。

Cargo Record Book 货物记录簿　系指 MARPOL 73/78 附则 II 所适用的每艘船舶均应备有一份符合该附则附录 IV 规定格式的货物记录簿。不论其是作为船舶正式航海日记的组成部分，还是另外形式均可。　Cargo Record Book means that every ship to which Annex II of MARPOL 73/78 applies shall be provided with a Cargo Record Book, whether as part of the ship's official log-book or otherwise, in the form specified in appendix IV to the Annex.

Cargo residues 货物残余物　系指 MARPOL 公约

其他附则未涵盖、且在装载或卸载后仍然留在甲板上或货舱内的任何货物的残余物,包括装载和卸载的多余货物或溢出物,无论其处于潮湿或干燥条件下或是夹带在洗涤水中,但不包括进行清扫后在甲板上的残留的货物灰尘或船舶外表面上的灰尘。 Cargo residues means the remnants of any cargo which are not covered by other Annexes to the MARPOL Convention and which remain on the deck or in holds following loading or unloading, including loading and unloading excess or spillage, whether they are in wet or dry condition or entrained in wash water but it does not include cargo dust remaining on the deck after sweeping or dust on the external surface of the ship.

Cargo runner（cargo fail） 吊货索 系指吊货索具中用以直接承吊货物的钢索。

Cargo runner guide 护索环 装设在吊杆中段下侧,用以防止空载吊货索过度悬垂的U型杆件。一般有带滚轮和不带滚轮两种形式。

Cargo Securing Manual 货物系固手册 系指在整个航程中,除散装固体和液体货物以外的所有货物、货物单元和货物运输单元,应按主管机关认可的"货物系固手册"进行装载、积载和系固。对于具有第Ⅱ-2/3.41条定义的滚装处所的船舶,应在离开泊位之前按"货物系固手册"完成所有这些货物、货物单元和货物运输单元的系固。所有运输除散装固体和液体货物以外的货物的各类船舶均应要求配备"货物系固手册",手册的编制标准应至少等效于IMO组织制定的指南。 Cargo Securing Manual means that all cargoes, other than solid and liquid bulk cargoes, cargo units and cargo transport units, shall be loaded, stowed and secured throughout the voyage in accordance with the Cargo Securing Manual approved by the Administration. In ships with Ro-Ro spaces, as defined in regulation Ⅱ-2/3.41, all securing of such cargoes, cargo units and cargo transport units, in accordance with the Cargo Securing Manual, shall be completed before the ship leaves the berth. The Cargo Securing Manual is required on all types of ships engaged in the carriage of all cargoes other than solid and liquid bulk cargoes, which shall be drawn up to a standard at least equivalent to the guidelines developed by the Organization.

Cargo service spaces 理货（货物）服务处所 系指:(1)货物区域内的工作间、物料间以及面积为2 m²以上的储存货物装卸设备的处所;(2)位于货油区域内为货油装卸设备服务的工作,贮藏室和超过2 m²的贮藏处所。 Cargo service spaces are:(1)spaces within the cargo area used for workshops, lockers and store-rooms of more than 2 m² in area, used for cargo-handling equipment;(2)spaces within the cargo oil area used for workshops, lockers and store-rooms of more than 2 m² in area, intended for cargo oil handling equipment.

Cargo shackle 吊货卸扣 当作吊货钩用的卸扣。

Cargo Ship Safety Certificate 货船安全证书 系指经检验并符合经1988年修订的1974年 SOLAS第Ⅱ-1章、第Ⅱ-2章、第Ⅲ章、第Ⅳ章和第Ⅴ章以及其他有关要求的货船,签发的货船安全证书,以代替上述的货船构造安全证书、货船设备安全证书和货船无线电安全证书。 Cargo Ship Safety Certificate means a certificate issued after a survey to a cargo ship which complies with the relevant requirements of chapters Ⅱ-1, Ⅱ-2, Ⅲ, Ⅳ and Ⅴ and other relevant requirements of SOLAS 1974 modified by the 1988 SOLAS Protocol, as an alternative to the above cargo ship safety certificates. A Record of Equipment for the Cargo Ship Safety Certificate (Form C) shall be permanently attached.

Cargo ship safety construction certificate 货船构造安全证书 系指:(1)由SOLAS公约缔约国政府授权的个人或组织,按经1988年议定书修订的1974年国际海上人命安全公约的规定对货船构造签发的安全证书;(2)对500GT及以上的船舶经检验、满足1974年SOLAS第Ⅰ/10条所述关于货船检验要求,并符合除关于灭火设备和防火控制图等要求以外的第Ⅱ-1章和第Ⅱ-2章适用的要求签发的货船构造安全证书。 Cargo ship safety construction certificate means:(1) a Safety Certificate for cargo Ship's Construction Issued by a person or an organization authorized under the authority of the Government and under the provisions of the International Convention for the Safety of Life at sea. 1974, as modified by the Protocol of 1988 relating thereto;(2) a certificate issued after a survey to a cargo ship of 500 gross tonnage and above which satisfies the requirements for cargo ships on survey and regulation I/10 of SOLAS 1974, and complies with the applicable requirements of chapters Ⅱ-1 and Ⅱ-2, except for those relating to fire-extinguishing appliances and fire control plans.

Cargo ship safety equipment certificate 货船设备安全证书 系指:(1)对500GT及以上的船舶经检验满足1974年SOLAS第Ⅱ-1章、第Ⅱ-2章和第Ⅲ章用的有关以及任何其他有关要求,签发的货船设备安全证书,签发的货船设备安全证书,应永久性附有一份货船设备安全证书的设备记录簿;(2)由SOLAS公约缔约国政府授权的个人或组织,按经1988年议定书修订的1974年国际海上人命安全公约的规定对货船设备签发的安全证书。 Cargo ship safety equipment certificate means:

(1) a certificate issued after a survey to a cargo ship of 500 gross tonnage and above which complies with the relevant requirements of chapters Ⅱ-1, Ⅱ-2 and Ⅲ and any other relevant requirements of SOLAS 1974. A Record of Equipment for the Cargo Ship Safety Equipment Certificate shall be permanently attached; (2) a Safety Certificate for cargo Ship's Equipment Issued by a person or an organization authorized under the authority of the Government and under the provisions of the International Convention for the Safety of Life at sea, 1974, as modified by the Protocol of 1988 relating thereto.

Cargo Ship Safety Radio Certificate 货船无线电安全证书 系指：(1)500GT 及以上的船舶所安装的无线电装置，包括用于救生设备上的无线电装置，经检验符合1974年SOLAS第Ⅲ章和第Ⅳ章以及任何其他有关要求签发的货船无线电安全证书，签发的货船无线电安全证书，应永久性附有一份货船无线电安全证书的设备记录簿；(2)由 SOLAS 公约缔约国政府授权的个人或组织，按经1988年议定书修订的1974年国际海上人命安全公约的规定对货船无线电签发的安全证书。 Cargo Ship Safety Radio Certificate means: (1) a certificate issued after a survey to a cargo ship of 500 gross tonnage and above, and the ship is fitted with a radio installation, including those used in life-saving appliances, which complies with the requirements of chapters Ⅲ and Ⅳ and any other relevant requirements of SOLAS 1974. A Record of Equipment for the Cargo Ship Safety Radio Certificate shall be permanently attached; (2) a Safety Certificate for cargo Ship's Radio Issued by a person or an organization authorized under the authority of the Government and under the provisions of the International Convention for the Safety of Life at sea, 1974, as modified by the Protocol of 1988 relating thereto.

Cargo ship (freight vessel) 货船 系指客船、军用舰艇和运兵船、非机动船、原始工艺制造的木船、渔船或海上移动式钻井平台以外的任何船舶。 Cargo ship is any ship which is not a passenger ship, a ship of war and troopship, a ship which is not propelled by mechanical means, a wooden ship of primitive build, a fishing vessel or a mobile offshore drilling unit.

Cargo slewing guy 货钩牵索 系指在重型吊杆装置操作时，系结在吊货钩处，用以控制被吊货物位置的钢索。

Cargo space 货物处所 系指：(1)船上拟用于装载货物的处所，它包括通往该处所的围井和舱口；(2)用作装载货物的处所、货油舱、装载其他液体货物的液货舱和通往此种处所的围壁通道；(3)净吨位计算中所包括的载货处所，系指适宜于运载可由船上卸下货物的围蔽处所，而且这些处所已经列入总吨位计算之内。上述载货处所应在易于看到的地方用字母CC（货舱）作永久性标志，字母的高度应不小于100 mm，以便查核；(4)所有特种处所以外的装货处所和装货的滚装处所及通往这些处所的围壁通道。 Cargo space means: (1) a space in a ship intended for the carriage of cargo, including any trunk-way and hatchway to that space; (2) those spaces used for cargo, cargo oil tanks, tanks for other liquid cargo and trunks to such spaces; (3) enclosed spaces included in the computation of net tonnage that are appropriate for the transport of cargo which is to be discharged from the ship, provided that such spaces have been included in the computation of gross tonnage. Such cargo spaces shall be certified by permanent marking with the letters CC (cargo compartment) to be so positioned that they are readily visible and are not less than 100 mm (4 inches) in height; (4) all spaces other than special category spaces and Ro-Ro spaces used for cargo and trunks to such spaces.

Cargo space construction of liquefied gas carrier 液化气船货舱结构 在IGC规则中将液化气船的货舱分为：(1)整体式货舱结构(integral tank)，其主要特点为货舱作为整个船体的一部分，货舱结构件与船体受力相互影响。此类货舱的设计蒸气压力小于0.025 MPa，用于装载沸点大于 -10 ℃的货物。(2)薄膜式货舱结构(membrane tank)，其主要特点为货舱内表面覆盖一层薄膜，热绝缘层介于薄膜与船体之间。薄膜不能独立承受货物的重力，需要通过绝缘层由船体结构支持，称为非自己支持型。此类货舱的设计蒸气压力小于0.025 MPa，薄膜厚度小于10 mm。(3)半薄膜式货舱结构(semi-membrane tank)，其主要特点为液舱空载时是自己支持的，在载重状态下为非独立自己支持，其热绝热层介于薄膜与船体之间。此类货舱的设计蒸气压力小于0.025 MPa。(4)独立式货舱结构(independent tank)，其主要特点为货舱为自己独立支持型。其货舱与船体结构相对独立分开，货舱的结构件不承受船体强度。此类货舱含A,B,C三种形式。(5)内部绝缘液舱(internal insulation type)，其主要特点为货舱为非自己支持型，热绝热层与货物直接接触。目前大多数液化气船的货舱结构为第(2)和第(4)种。

Cargo spaces 货物处所 (1)对高速船而言，系指除特种处所和滚装处所以外的所有装货处所和通往这些处所的围壁通道；(2)对非高速船而言，系指包括滚装处所、特种处所和开式甲板处所的货物处所。 (1) For high-speed craft, it means all spaces other than special category spaces and Ro-Ro spaces used for cargo and trunks to

Cargo sweat 货物结露 系指货舱内货物温度低于舱内空气露点温度时,货物表面产生的凝水现象。

Cargo tank 液货舱 系指用于装运液体货物的容器。 Cargo tank means container used to carry liquid cargo.

Cargo tank bulkhead[1] 货油舱区域 系指船舶这一部位,包括货油舱和货/污油水舱以及相邻区域(包括压载水舱、燃油舱、隔离舱、空舱),并且还包括上述处所之上的船舶部位全长和全宽范围内的甲板区域。它包括防撞舱壁和货油区域后端的横舱壁。 Cargo tank bulkhead means a part of the ship that consists of cargo tanks and cargo/slop tanks and adjacent areas including ballast tanks, fuel tanks, cofferdams, void spaces and also deck areas throughout the entire length and breadth of the part of the ship over the mentioned spaces. It includes the collision bulkhead and the transverse bulkhead at the aft end of the cargo block.

Cargo tank bulkhead[2] 货油舱舱壁 系指分隔货油舱的限界舱壁。 Cargo tank bulkhead means a boundary bulkhead separating cargo tanks.

Cargo tank cleaning system 液货舱清洗系统 对油船而言,系指货油舱清洗系统。对 20 000 载重吨及以上的原油船,则应使用原油洗舱的货油舱清洗系统—原油洗舱系统。

Cargo tank pressure relief valve 液货舱压力释放阀 液化气船货舱内设置的,保证货舱压力在正常的范围内安全装置。该系统由压力释放阀、透气桅及与其连接的排气管组成。

Cargo tank stirpping procedures 液货船扫舱程序 系指每一液货舱扫舱期间要遵循的程序。其应包括下列内容:(1)扫舱系统的作业;(2)横倾和纵倾的要求;(3)管路泄放和清扫或吹除布置(如适用);(4)水试验扫舱时间。 Cargo tank stirpping procedures mean those procedures followed during the stripping of each cargo tank. The procedures shall include the following: (1) operation of stripping system; (2) list and trim requirements; (3) line draining and stripping or blowing arrangements if applicable and (4) duration of the stripping time of the water test.

Cargo tank vapour piping system (cargo tank gas piping system) 货油舱透气管系 为调节货油舱内正负压力而专门设置的经由呼吸阀与大气相连通的管系。

Cargo tanks 货油舱 系指所有用于装载液体货物的处所及通往这些处所的围壁通道。 Cargo tanks mean all spaces used for liquid cargo and trunks to such spaces.

Cargo winch gear 起货机 系指船上装卸货物的装置的总称。主要有吊杆装置、起重机及其他装卸机械等设备。

Cargo zone 货物区域 对散装运输液化气体船而言,系指船上设有货舱围护系统、货泵舱和压缩机的部分,并包括在上述处所上方的船上该部分的整个长度和宽度范围内的甲板区域。对于在最后一个货舱处后面或最前一个货舱处前面所设的隔离舱、压载舱或留空处所,不应算作货物区域。

Cargoes which may liquefy 可流态货物 系指含有一定比例细微颗粒和一定数量水分的货物。这类货物装运时如含水量超过其适运水分极限,就有可能流态化。 Cargoes which may liquefy mean cargoes which contain a certain proportion of fine particles and a certain amount of moisture. They may liquefy if shipped with moisture content in excess of their transportable moisture limit.

Cargohook swivel 吊钩转环 为使吊货钩可以自由转动和吊货索具免受扭曲而装于吊货钩上部的转环。

Cargo-pumping system (cargo oil pumping system) 货油装卸系统 油船上专门用以完成装卸货油任务,由货油舱管系、货油泵舱管系和甲板货油管系等组成的系统。

Caribbean Region 泛加勒比海区域 系指墨西哥湾和加勒比海本区域,包括其中的海湾和海区以及以下边界组成的大西洋的一部分;在北纬30°自佛罗里达向东至西经77°30′,然后连一条恒向线至北纬20°与西经59°的交叉点,然后再连一恒向线至北纬7°20′与西经50°的交叉点,然后再连一恒向线沿西南方向至法属圭亚那的东部边界。 Wider Caribbean Region means the Gulf of Mexico and Caribbean Sea proper including the bays and seas therein and that portion of the Atlantic Ocean within the boundary constituted by the 30° N parallel from Florida eastward to 77°30′ W meridian, thence a rhumb line to the intersection of 20° N parallel and 59° W meridian, thence a rhumb line to the intersection of 7°20′ N parallel and 50° W meridian, thence a rhumb line drawn south-westerly to the eastern boundary of French Guiana.

Caribou island and Southwest Bank Protection Area 北美驯鹿岛(卡里布岛)和西南浅滩保护区 系指从最北的一点开始,并顺时针方向连接下列坐标的恒向线所包围的区域:北纬47°30.0′,西经085°50.0′;北纬47°34.2′,西经085°38.5′;北纬47°04.8′,西经085°48.0′;北纬47°05.7′,西经085°58.0′;北纬47°18.1′,西经086°05.0′。 Caribou island and Southwest Bank Protection Area means the area enclosed by rhumb lines connecting the following co-ordinates, beginning on the northernmost point and proceeding clockwise:47°30.0′N, 085°50.0′W;47°34.2′N, 085°38.5′W;47°04.8′N, 085°48.0′W;47°05.7′N, 085°58.0′W;47°18.1′N, 086°05.0′W.

Caribrite 校准

Carlings 短梁 系指用于补足正常加强结构布置的加强构件。 Carlings means a stiffening member used to supplement the regular stiffening arrangement.

Carpenter's store 木工间 系指供从事木工作业和贮存木工工具用的舱室。

Carpenter's stores 木工用具 船上木工专用的工具、属具及其备品的统称。

Carriage and insurance paid to named place of destination(CIP) 运费、保险费付至指定目的地 系指卖方向其指定的承运人交货,但卖方还必须支付将货物运至目的地的运费,亦即买方承担卖方交货之后的一切风险和额外费用。但是,按照CIP术语,卖方还须办理买方货物在运输途中灭失或损坏风险的保险。

Carriage by sea 海上运输 系指从在装船时有害有毒物质进入船舶设备的任何部分之时起至在卸船时它们不再存在与船舶设备的任何部分时止的期间。如未使用任何船舶设备,则该期间分别起止于有害有毒物质越过船舷之时。 Carriage by sea means the period from the time when the hazardous and noxious substances enter any part of ship's equipment, on loading to the time they cease to be present in any part of the ship's equipment, on discharge. If no ship's equipment is used, the period begins and ends respectively when the hazardous noxious substances cross the ship's rail.

Carriage contract at sea 海上货物运输合同 系指承运人收取运据以将货物从一个港口运往另一个港口的合同。对于既涉及海上运输又涉及某些其他运输方式的合同而言,只有在其涉及海上运输时,才应视为联合国海上货物运输公约所指的海上货物运输合同。

Carriage of goods 货物运输 包括货物自装上船之时起至货物卸离船之时止的一段时间。 Carriage of goods covers the period from the time when the goods are loaded on to the time they are discharged from the ship.

Carriage of refrigerated containers in holds(CRC) 舱内载运冷藏集装箱 将货物冷藏集装箱放置在货舱内。

Carriage paid to—named place of destination (CPT) 运费付至指定目的地 按此术语成交,货价的构成因素中包括从装运港至约定目的地港的通常运费和约定保险费,故卖方除具有与CFR术语的相同的义务外,还就为买方办理货运保险,交支付保险费,按一般国际贸易惯例,卖方投保的保险金额应按CIF价加成10%。(Carriage paid to)运费付至指定目的港 CPT 是贸易术语,是 Carriage paid to named place of destination 的缩写,即"运费付至指定目的地"。它系指卖方向其指定的承运人交货,支付将货物运至目的地的运费,办理出口清关手续。亦即买方承担交货之后一切风险和其他费用。

Carriage paid to...(CPT) 运费付至……(……指定目的地) 系指卖方向其指定的承运人交货,但卖方还必须支付将货物运至目的地的运费。即买方承担交货之后一切风险和其他费用。"承运人"系指任何人,在运输合同中,承诺通过铁路、公路、空运、海运、内河运输或上述运输的联合方式履行运输或由他人履行运输。如果还使用接运的承运人将货物运至约定目的地,则风险自货交给第一承运人时转移。CPT 术语要求卖方办理出口清关手续。根据《INCOTERMS 2000》该术语可适用于各种运输方式,包括多式联运。

Carriage return 回车

Carrier 承运人 系指:(1)包括与托运人签订运输合同的船舶所有人或承租人。(2)由其本人或以其名义与托运人签订海上货运运输合同的任何人。 Carrier includes are:(1)the owner or the charterer who enters into a contract of carriage with a shipper;(2)any person by whom or in whose name a contract of carriage of goods by sea has been concluded with a shipper.

Carrier for fresh fish 活鲜鱼运输船 系指设计和制造用于运输活鲜鱼的船舶。 Carrier for fresh fish means a ship designed and constructed for the carriage of fresh fish.

Carrier ring (holding ring, diaphragm housing ring) 隔板套 外缘装在冲动式汽轮机气缸槽内,内缘可装两级以上隔板的中间支承元件。

Carrier rocket 运载火箭 是航天运载工具的一种,是将有效载荷按照预定的速度和方向送入太空的火箭。一般情况下,运载火箭将有效载荷送入轨道。它是由多级火箭组成的航天运输工具。用途是把人造地球卫星、载人飞船、空间站、空间探测器等有效载荷送入预定轨道。是在导弹的基础上发展的,一般由 2～4 级组成。每一级都包括箭体结构、推进系统和飞行控制系统。末级有仪器舱,内装制导与控制系统、遥测系统和发射场安全系统。级与级之间依靠级间段连接。有效载荷装在仪器舱的上面,外面套有整流罩。1980 年 5 月 18 日,中国第一枚运载火箭发射成功。

Carrier-bone vertical launching system(CVLS) 舰载垂直发射系统 是一种用在潜艇和某些水面舰船上的导弹发射系统,最早产生于弹道导弹的发射系统。

舰载垂直发射系统技术拥有国：中国（170/177）、美国（Mk41）、俄罗斯（3k95）、以色列（巴拉克-1）、英国（海狼）、法国（海响尾蛇 VT-1）、南非（Umkhonto）。见图C-4。

图 C-4 舰载垂直发射系统
Figure C-4 Carrier-bone launching system

Carrier-borne aircraft 舰载机 以航空母舰或其他军舰为基地的海军飞机。主要用于攻击空中、水面、水下和地面目标，并遂行预警、侦察、巡逻、护航、布雷、扫雷和垂直登陆等任务，是海军航空兵的主要作战手段之一，也是海洋战场上夺取和保持制空权、制海权的重要力量。

Carry over 蒸汽带水 系指锅炉运行时，锅筒蒸发引起管中的饱和蒸汽携带锅水的现象。

Carrying capacity of lifeboat 救生艇乘员定额 根据救生艇立方容积核算，并经实艇搭载试验后实际核定的单艘救生艇允许搭载的人数。

Cascade effect（gap effect） 叶栅作用 引起螺旋桨每叶与其单独在开阔水域中运动时水动力性能差异的螺旋桨各叶间的相互影响。

Case of retrial 再审案件 再审是为纠正已经发生法律效力的错误判决、裁定，依照审判监督程序，对案件重新进行的审理。法院对已经审理终结的案件，依照再审程序对案件的再行审理，其目的是纠正已经发生法律效力但确属错误的判决或裁定。再审的特点是：（1）提起再审的主体必须是最高人民法院和上级人民法院；最高人民检察院和上级人民检察院或本院院长；（2）提起再审的客体是已经发生法律效力的第一审或第二审案件的判决或裁定；（3）提起再审的时间是判决或裁定生效以后，没有截止时间限制。再审是一项重要的诉讼程序制度，也是各国刑事诉讼法和民事诉讼法的重要组成部分。纵观各国的刑事诉讼法和民事诉讼法，对再审制度的规定大致可分为两类：一类是规定审判监督程序，即法定的机关和公职人员，基于法律赋予的审判监督权，对有错误的已经发生法律效力的裁判，提起再行审判。因为审判监督程序是以审判监督权为基础的，因此，一般对提起的期限不做强制性规定，对提起再审的条件和理由等也只作原则性规定。另一类是基于当事人诉权的再审，即当事人不服已经生效的裁判，向审法院提起再审之诉，再审法院对案件再行审理。各国一般对再审的条件和理由、再审的范围以及提起再审的期限都做了具体的规定。

Case-by case approval 逐个产品认可 系指对仅使用于某一特定船舶的产品的认可。逐个产品认可仅对适用的特定船舶有效。 Case-by case approval means an approval by which a product is approved for installation on board of a specific ship without using a type approval certificate. The case-by-case approval is only valid for the specific ship.

Cash with order 随订单付款

Cashier's check 银行本票 在银行开立存款账户的持票人向开户银行提示付款时，应在银行本票背面"持票人向银行提示付款签章"处签章，签章须与预留银行签章一致。

Casing pipe 套管 系指套在另一部件上的管子。断面为环形，用于保护其他设备的各种材质的圆管，多为铁、钢质。套管是放入裸眼的大直径管，并且需要使用水泥固定。油井设计者须设计套管能够承受各种外力，如碰撞、爆裂、拉力等，以及化学侵蚀。

Casing[1] 舱棚 系指任何处所的防护盖或舱壁。 Casing means a covering or bulkhead around or about any space for protection.

Casing[2] 炉板 包覆整个锅炉起气密和保温作用的金属壳体。

Casing[3] 套管 系指用作油井衬层的管状结构。 Casing is a tubular structure used for lining oil wells.

Casing（cylinder, shell） 气缸/机匣 安装静叶、静叶环、隔板套、隔板、气封等固定部件并覆盖和支承汽轮机转子使与外界隔开的壳体。

Casing head（natural） 套管头

Casino royal 皇家赌场（豪华邮轮）

Cast shop 铸造车间 系指从事铸件生产、加工的工作场所。

Cast steel chain 铸钢锚链 系指以钢铸成的锚链。

Casting defect 铸造缺陷

Castor beans or castor meal or castor pomace or castor flake（UN 2969） 蓖麻籽或蓖麻粉或蓖麻油渣或蓖麻片 系指榨过油的蓖麻籽。含有一种强烈的变应原，某些人吸入粉尘或皮肤与碎蓖麻籽货品接触会严重刺激皮肤、眼睛和黏膜。摄入也有毒害。 Castor beans or castor meal or castor pomace or castor flake (UN 2969) are the beans from which castor oil is obtained. It contains a powerful allergen. Inhalation of dust or by skin contact with crushed bean products, can give rise to severe irritation of the skin, eyes, and mucous membranes in some persons. It will be toxic by ingestion.

Casualty 灾害 系指对人的生命、健康或环境的严重破坏,诸如导致人命损失或人身严重伤害的事件。

Cat tackle 吊锚索具 用于吊放锚的滑车和绳索的总称。

Catadromous species 下海产卵物种 系指在海洋环境中产卵/繁殖,但至少有一部分成年生活是在淡水环境中度过的物种。 Catadromous species mean that some species spawn/reproduce in marine environments, but they spend at least part of their adult life in a freshwater environment.

Catalyst block 催化剂块 系指废气通过,其内表面含有减少废气中 NO_x 的催化剂成分的一定尺寸的块体。 Catalyst block means a block of certain dimension through which exhaust gas passes and which contains catalyst composition on its inside surface to reduce NO_x from exhaust gas.

Catalytic agent 催化剂

Catamaran HSC 双体高速船 系指具有两个相互平行的船体,其上部用强力构架联成一个整体的高速船。 Catamaran HSC is a high speed craft in which upper parts of two parallel hulls being connected by strength framing.

Catamaran research ship 双体调查船 船体下部为两个单体,上部用强力构架连成一体,能在海上从事有关海洋调查研究的船舶。它具有大的舱容和开阔的甲板操作面积,艉部有两套螺旋桨和舵,操纵性能好,船舶的回转圈较小,比单体船具有较大的初稳性。

Catamaran(twin-hull ship) 双体船 系指具有两个相互平行的船体,其上部用强力构架联成一个整体的船舶。 Catamaran is a craft in which upper parts of two parallel hulls being connected by strength framing.

Catamaran-foil 水翼双体船 一种以双体船为主并辅以水翼的船舶。即在通常的尖舭型双体船两片体间加装水翼,采用尖舭片体是为了便于安装水翼。加水翼后不仅可以降低双体船的阻力而且使其在波浪中的运动和垂向加速度均有明显减小。

Catastrophic effect 灾难性影响(灾难性后果) 系指导致损失船舶和/或多人死亡的故障状况的影响。 Catastrophic effect means the effect of failure conditions that leads to a loss of the craft and/or multiple fatalities.

Catcher boat 捕捞渔船

Categories of structural members of the unit 平台结构构件分类 系指根据构件所承受的载荷、应力水平及模式、关键载荷传递和应力集中以及失效后果,所有平台结构构件可分为:次要构件、主要构件和特殊构件等。 Categories of structural members of the unit means that all structural members of the unit may be grouped into the following structural categories, such as according to applied loading, stress level and associated stress pattern, critical load transfer points and stress concentrations, and consequence of failure, such as.

Category 1 oil tanker 第1类油船 就 MARPOL 而言,系指不符合附则Ⅰ第1.28.4条所定义的在1982年6月1日以后交付的油船的要求,且装载量为20 000 t 及以上装运原油、燃油、重柴油或润滑油作为货物的油船,以及载重量为30 000 t 及以上载运除上述油类以外的其他油类的油船。 For MARPOL, category 1 oil tanker means an oil tanker of 20,000 t deadweight and above carrying crude oil, fuel oil, heavy diesel oil or lubricating oil as cargo, and of 30,000 t deadweight and above carrying oil other than the above, which does not comply with the requirements for oil tankers delivered after 1 June 1982, as defined in regulation 1.28.4 of the Annex Ⅰ, MARPOL.

Category 1-1 vessel 1-1类船舶 系指其甲板具有特大开口,有必要对垂直和水平方向上的合成弯矩和扭矩予以特殊考虑的船舶。 Category 1-1 vessel means a vessel having large opening on the deck and combined stress of bending moment and torsional moment in vertical and horizontal directions has to be considered.

Category 1-2 vessel 1-2类船舶 系指非均匀装载货物和压载的船舶。 Category 1-2 vessel means a vessel having non-homogeneous cargo and ballast loading.

Category 1-3 vessel 1-3类船舶 系指危险化学品船或液化气散装船。 Category 1-3 vessel means a chemical tankers or ships carrying liquified gases in bulk.

Category 2 oil tanker 第2类油船 就 MARPOL 而言,系指符合附则Ⅰ等1.28.4条所定义的在1982年6月1日以后交付的油船的要求,且装载量为20 000 t 及以上装运原油、燃油、重柴油或润滑油作为货物的油船,以及载重量为30 000 t 及以上载运除上述油类以外的其他油类的油船。 For MARPOL, Category 2 oil tanker means an oil tanker of 20,000 t deadweight and above carrying crude oil, fuel oil, heavy diesel oil or lubricating oil as cargo, and of 30,000 t deadweight and above carrying oil other than the above, which complies with the requirements for oil tankers delivered after 1 June 1982, as defined in regulation 1.28.4 of the Annex Ⅰ.

Category 2 vessel　**Ⅱ类船舶**　系指货物和压载的装载分布变化可能性较小，具有一定装载状态的船舶，例如：(1)不标记最大载重线的船舶；(2)不载货的船舶；(3)车辆运输船；(4)装货均匀的船舶。　Category 2 vessel means a vessel having homogeneous cargo and ballast loading, such as: (1) vessel having no loadline mark; (2) vessel not carrying out cargoes; (3) cargo vehicle carrier; (4) vessel having homogeneous cargo loading.

Category 3 oil tanker　**第3类油船**　就MARPOL而言，系指载重量为5 000t及以上但低于上述第1类油船和第2类油船规定载重量的油船。　For MARPOL, Category 3 oil tanker means an oil tanker of 5,000 t deadweight and above but less deadweight than the specified in Category 1 oil tanker or Category 2 oil tanker.

Category Ⅳ ships　**第Ⅳ类船舶**　就装载状态而言，系指布置上，货物和压载的分布变化可能性很小的船舶，以及装载手册给出充分指导性资料的定期船舶，另外还有那些不属于第Ⅱ类和第Ⅲ类的船舶　For loading condition, Category Ⅳ ships mean those ships with arrangement giving small possibilities for variation in the distribution of cargo and ballast, and ships on regular and fixed trading pattern where the loading manual gives sufficient guidance, and in addition to those exceptions given under Category Ⅱ and Ⅲ.

Category A alerts　**A类警报**　系指：(1)发出警报时，必须在直接指定为生成警报的任务指南站显示图像信息，作为评估警报相关条件的决策支持。(2)在该警报状态下，直接具有产生警报功能的任务的图表信息需用作评估警报相关状况的决策支持。例如：①碰撞危险；和②搁浅危险。如果不能在HMI应答A类警报，该事实应清楚地向用户指明。　Category A alerts means: (1) alerts where graphical information is displayed at the task station directly assigned to the function that generating the alert is necessary, used as decision support for the evaluation the alert related condition; (2) alerts in which graphical information is displayed at the task station directly assigned to the function that generating the alert is necessary, used as decision support for the evaluation of the alert-related condition, e.g.: ①danger of collision; ②danger of grounding. When category A alerts cannot be acknowledged at a HMI, this fact should be clearly indicated to the user.

Category A craft　**A类客船**　系指满足下列条件的客船：(1)船舶在其规定的营运航线的任何地点出事，有很大把握能在以下三者中的最短时间内将船上所有乘客和船员救出：①救生艇筏内的人员因受冻以至伤亡的时间；②与该航线所处的环境条件和地理特点相适应的时间；③4 h；(2)载客不超过450人。　Category A craft is any high-speed passenger craft: (1) operating on a route where it has been demonstrated to the satisfaction of the flag and port States that there is a high probability that, in the event of an evacuation at any point of the route, all passengers and crew can be rescued safely within the least of: ①the time to prevent persons in survival craft from exposure causing hypothermia in the worst intended conditions; ②the time appropriate with respect to environmental conditions and geographical features of the route; ③4 hours; (2) carrying not more than 450 passengers.

Category A machinery spaces　**A类机械场所**　是指下述空间及场所，这些场所内装有：(1)作为主推进装置用的内燃机；(2)除了作为主推进装置用的内燃机以外的内燃机，其输出总功率超过375kW；或者(3)任何燃油锅炉或燃油装置。　Machinery spaces of Category A are those spaces and trunks to such spaces which contain: (1) internal combustion machinery used for main propulsion; (2) internal combustion machinery used for purposes other than main propulsion where such machinery has a total power output of not less than 375 kW in the aggregate; (3) any oil-fired boiler or oil fuel unit.

Category A training　**A类培训**　对气体燃料船舶上的培训而言，系指对负责基本安全的船员的基础培训。(1)A类培训旨在为负责基本安全的船员提供对作为燃料的气体、液体和压缩空气的技术属性、爆炸限值、着火源、风险降低措施，以及在正常操作和紧急情况下必须遵守的规则和程序的一个基本了解；(2)一般基础培训是基于假定船员以前对气体、气体发动机和气体系统一无所知的前提下制定的。培训教师中应有一人或多人为用气设备或气体系统的供应厂商，和对所用的气体和船上安装的用气设备有深入了解的气体专家；(3)培训应包括与气体、相关系统以及液体和压缩气体操作时个人防护有关的理论培训和实践操作培训。熄灭气体火灾的实际操作也应为培训的一部分，并应由经认可的安全中心进行。　For the training on gas-fuelled ships, Category A training means a basic training for crew who is responsible ofr the basic safety. (1) The goal of Category A training should provide the crew with a basic understanding of the technical properties, gas, explosion limits, ignition sources, risk reducing and consequence reducing measures of gas, liquid and compressed gas used as afuel, and the rules and procedures that must be followed during normal operation and in emergency situations; (2) The general basic training is based on the assumption that the crew do not have any prior knowledge of gas, gas engines and gas systems. The instructors should include one or more technicals form gas equipment or gas systems suppliers, and other specialists with in-depth knowledge of the gas and the technical gas systems that are installed on board; (3) The training should consist of both theoretical and exercises that involve gas and the relevant systems, as well as personal protection while handling liquid and compressed gas. Extinguishing of gas fires should be traind practically, and should take place at an approved safety centre.

Category A yacht　A 类游艇　系指航行于距岸不超过 20 n mile(中国台湾海峡及类似海域距岸不超过 10 n mile)的海上航行的游艇。

Category B alerts　B 类警报　系指在该警报状况下，除能在中央警报管理－人机界面(CAM-HMI)显示的信息外，不需要用于决策支持的附加信息。　Category B alerts are alerts where no additional information for decision support is necessary besides the information which can be presented at the central alert management human machine interface (CAM-HMI).

Category B craft　B 类客船　系指 A 类客船以外的客船，这类船的机械和安全系统的设置应保证：一旦某一舱发生破损且舱内的主要机械和安全系统失效，该船仍能保持安全航行的能力　Category B craft is any high-speed passenger craft other than a Category A craft, with machinery and safety systems arranged such that, in the event of any essential machinery and safety systems in any one compartment being disabled, the craft retains the ability to navigate safely.

Category B training　B 类培训　对气体燃料船舶上的培训而言，系指对驾驶员的补充培训。培训内容：(1)驾驶员应接受超出一般基础培训范围以外的气体相关培训。应从技术上对驾驶员划分 B 类和 C 类培训。应由公司的培训经理或船长决定甲板操作需要哪些培训。(2)参与气体燃料充装以及驱气操作，或进行与气体发动机或用气装置有关工作的普通船员应参加所有或部分 B/C 类培训。公司和船长负责根据对这类船员的工作要求/船上责任范围的评估安排这些培训。(3)补充培训教师应与 A 类培训教师的要求相同。(4)应回顾船上所有与其他有关的系统。船舶的维护手册、供气系统手册和爆炸危险处所和区域的电气设备手册应用作该部分培训的基础。(5)作为 SMS 体系的一部分由公司和船上高级管理层定期检查。应强调风险分析，并在培训期间，应将所有进行的风险分析资料提供给参加课程的学员。(6)如由船舶自身的船员对用气设备进行技术维护，则针对此类工作的培训应形成文件。(7)在船舶投入营运之前，船长和轮机长应对船上负责基本安全的船员给出最终的鉴定。该鉴定文件仅适用于与其他相关的培训，并应同时经船长/轮机长和参加课程的学员签字。与气体相关的培训鉴定文件可纳入船舶总培训大纲，但应清楚说明哪些是气体相关培训，哪些是相关培训。(8)与气体系统相关的培训要求应每年至少评估一次，评估方式应与船上其他培训要求的评估相同。应定期对培训计划进行评估。　For the training on gas-fuelled ships, Category B training means a supplementary training for deck officers. the training includes: (1) deck officers should have on gas training beyond the general basic training. Category B and Category C training should be divided technically between deck and engineer officers. The company's training manager or the master should determine what comes under deck operations; (2) those ordinary crew members who are to participate in the actual bunkering work, as well as gas purging, or are to perform work on gas engines or gas installations, etc., should participate in all or parts of the training for Category B/C. The company and the master are responsible for arranging such training based on an evaluation of the concerned crew member's job instructions/area of responsibility on board; (3) the instructors used for such supplementary training should be the same as outlined for Category A; (4) all gas-related systems on board should be reviewed, the ship's maintenance manual, gas supply system manual and manual for electrical equipment in explosion hazardous spaces and zones should be used as a basic for this part of the training; (5) this regulation should be regularly reviewed by the company and onboard senior management team as part of the SMS system. Risk analysis should be emphasized, and any risk analysis and subanalyses performed should be available to course participants during training; (6) if the ship's own crew will be performing technical maintenance of gas equipment, the training for this type of work should be documented; (7) the master and the chief engineer officer should give the basic safety crew on board their final clearance prior to shipping business. The clearance document should only apply to gas-related training, and it should be signed by both the master/chief engineer office and the course participant. The clearance document for gas-related training may be integrated into the ship's general training programme, but it should be clearly evident about what is regarded as gas-related training and what is regarded as other training; (8) the training requirements related to the gas system should be evaluated in the same manner as other training requirements on board at least once a year.

The training plan should be evaluated at regular intervals.

Category B yacht B类游艇 系指航行于下列水域的游艇:(1)沿海海岸与岛屿,岛屿与岛屿围成的遮蔽条件较好、波浪较小的海域。在该海域内岛屿之间、岛屿与海岸不超过 10 n mile；或在距岸不超过 10 n mile 的水域,并限制在风级不超过 6 级(蒲氏风级)且目测波高不超过 2 m 的海况下航行的游艇。(2)内河 A 级航区。

Category C alerts C类警报 系指不能在驾驶室应答但要求关于警报状况和处理信息的警报,例如来自发动机的某些警报。Category C alerts are alerts that cannot be acknowledged on the bridge but whose information is required about the status and treatment of the alerts, e. g. certain alerts from the engine.

Category C training C类培训 对气体燃料船舶上的培训而言,系指对轮机员的补充培训。应从技术上对轮机员划分 B 类和 C 类培训。培训内容:(1)应由公司的培训经理或船长决定轮机需要哪些培训。(2)参与气体燃料充装以及驱气操作,或进行与气体发动机或用气装置有关工作的普通船员应参加所有或部分 B/C 类培训。公司和船长负责根据对该类船员的工作要求/船上责任范围的评估安排这些培训。(3)补充培训教师应与 A 类培训教师的要求相同。(4)应回顾船上所有与其他有关的系统。船舶的维护手册、供气系统手册和爆炸危险处所和区域的电气设备手册应用作这部分培训的基础。(5)作为 SMS 体系的一部分由公司和船上高级管理层定期检查。应强调风险分析,并在培训期间,应将所有进行的风险分析资料提供给参加课程的学员。(6)如由船舶自身的船员对用气设备进行技术维护,则针对此类工作的培训应形成文件。(7)在船舶投入营运之前,船长和轮机长应对船上负责基本安全的船员给出最终的鉴定。该鉴定文件仅适用于与其他相关的培训,并应同时经船长/轮机长和参加课程的学员签字。与气体相关的培训鉴定文件可纳入船舶总培训大纲,但应清楚说明哪些是气体相关培训,哪些是相关培训。(8)与气体系统相关的培训要求应每年至少评估一次,评估方式应与船上其他培训要求的评估相同。应定期对培训计划进行评估。 For the training on gas-fuelled ships, Category C training means a supplementary training for engineer officers. The training includes: (1) engineer officers should have training on gas beyond the general basic training. Category B and Category C training should be divided technically between deck and engineer officers. The company's training manager or the master should determine what comes under deck operations; (2) those ordinary crew members who are to participate in the actual bunkering work, as well as gas purging, or are to perform work on gas engines or gas installa-tions, etc. , should participate in all or parts of the training for category B/C. The company and the master are responsible for arranging such training based on an evaluation of the concerned crew member's job instructions/area of responsibility on board; (3) the instructors used for such supplementary training should be the same as outlined for Category A; (4) all gas-related systems on board should be reviewed, the ship's maintenance manual, gas supply system manual and manual for electrical equipment in explosion hazardous spaces and zones should be used as a basic for this part of the training; (5) this regulation should be regularly reviewed by the company and onboard senior management team as part of the SMS system. Risk analysis should be emphasized, and any risk analysis and sub-analyses performed should be available to course participants during training; (6) if the ship's own crew will be performing technical maintenance of gas equipment, the training for this type of work should be documented; (7) The master and the chief engineer officer should give the basic safety crew on board their final clearance prior to shipping business. The clearance document should only apply to gas-related training, and it should be signed by both the master/chief engineer office and the course participant. The clearance document for gas-related training may be integrated into the ship's general training programme, but it should be clearly evident about what is regarded as gas-related training and what is regarded as other training; (8) the training requirements related to the gas system should be evaluated in the same manner as other training requirements on board at least once a year. The training plan should be evaluated at regular intervals.

Category C yacht C类游艇 系指航行于下列水域的游艇:(1)距岸不超过 5 n mile 的水域,并限制在风级不超过 6 级(蒲氏风级)且目测波高不超过 1 m 的海况下航行的游艇。(2)内河 B 级航区的游艇。

Category D yacht D类游艇 系指航行于内河 C 级的游艇。

Category I ships I类船舶 系指:(1)有甲板大开口的船舶,需考虑垂向和水平船体梁弯曲和扭转的合成应力及侧向载荷;(2)非均匀装载的船舶,货物和/或压载不均匀分布,除长度小于 120 m 的船舶之外,若设计时考虑到货物或压载不均匀分布,此类船舶属于 II 类;(3)具有化学品船或液化气船营运标志的船舶;(4)对 1998 年 7 月 1 日或以后签约建造的船长为 65 m 及以上的海船而言,系指:①具有甲板大开口($b_L/B_M \geq 0.6$;或 $l_L/l_M \geq 0.7$, 式中:b_L—舱口宽度;多排舱口时,b_L—各个舱口宽度的总和；l—舱口长度；B_M—在舱口长度中点处量取的甲板宽度；l_M—舱口每一端的横向甲板板条中心之间的距离。如除所考虑的舱口之外，再没有舱口时，l_M将予特别考虑。）的船舶，必须考虑由于船体梁的垂直弯曲和水平弯曲、扭转和侧向载荷引起的合成应力；②化学品船和液化气船；③船长大于 120 m 其货物和/或压载可为非均匀分布。但船长小于 120 m，设计时已考虑到货物或压载非均匀分布的船舶属于第 II 类。 Category I ships are：（1）ships with large deck openings where combined stresses due to vertical and horizontal hull girder bending and torsional and lateral loads need to be considered；（2）ships liable to carry non-homogeneous loadings, where the cargo and/or ballast may be unevenly distributed; exception is made for ships less than 120 m in length, when their design takes into account uneven distribution of cargo or ballast：such ships belong to Category II；（3）ships having the service notation of chemical tanker ESP or liquefied gas carrier；（4）For all classed seagoing ships of 65 m and above and in length which are contracted for construction on or after 1st July 1998, means： ①ships with large deck openings ($b_L/B_M \geq 0.6$；or $l_L/l_M \geq 0.7$, where：b_L—breadth of hatchway, in case of hatchways, b_L is the sum of the individual hatchway-breadths；l_L—length of hatchway；B_M—breadth of deck measured at the mid length of hatchway；l_M—distance between centers of transverse deck strips at each end of hatchway. Where there is no further hatchway beyond the one under consideration, l_M will be specially consideration), combined stresses due to vertical and horizontal hull girder bending and torsional and lateral loads have to be considered. ②chemical tankers and gas carriers；③ships more than 120 m in length, where the cargo and/or ballast may be unevenly distributed；But ships less than 120 m in length, when their design takes into account uneven distribution of cargo or ballast, belong to Category II.

Category II ships II 类船舶 系指：（1）船舶布置使货物或压载分布变化较小的船舶；（2）具有规则而固定营运模式的船舶，此时装载手册可提供足够的指导；（3）I 类船舶中例外的情况；（4）对 1998 年 7 月 1 日或以后签约建造的船长为 65 m 及以上的海船而言，系指：①船上的布置使得货物或压载分布变动不大的船舶（例如客船），②在装载手册中给出足够指导，且以有规律和固定方式营运的船舶；③以及第 I 类以外的船舶。 Category II ships are：（1）ships whose arrangement provides small possibilities for variation in the distribution of cargo and ballast；（2）ships on a regular and fixed trading pattern where the loading manual gives sufficient guidance；（3）the exception given under Category I；（4）for ships with length 65 m and above constructed on or after July 1 1998, mean：①those ships with arrangement giving small possibilities for variation in the distribution of cargo and ballast (e. g. passenger vessels)；②ships on regular and fixed trading patterns where the loading manual gives sufficient guidance；③in addition those exceptions given under Category I ships.

Category of liquid 液体类别 系指 MARPOL 73/78 附则 I 所涵盖的液体物质的类别，细分如下：（1）原油、（2）燃油和残油，包括船舶燃油；（3）未经加工的馏分油、液压油和润滑油；（4）汽油，包括船舶燃油；（5）煤油；（6）石脑油和凝析油；（7）汽油混合原料；（8）汽油和酒精；（9）沥青溶液。 Category of liquid means the category of liquid substances covered by Annex I of MARPOL 76/78, which is subdivided as following：（1）crude oils；（2）fuel and residual oils, including ship's bunkers；（3）unfinished distillates, hydraulic oils and lubricating oils；（4）gas oils, including ship's bunkers；（5）kerosenes；（6）naphthas and condensates；（7）gasoline blending stocks；（8）gasoline and spirits；（9）asphalt solutions.

Category X noxious liquid substances X 类有毒液体物质 就 MARPOL 附则 II 而言，这类有毒液体物质如从洗舱或排除压载的作业中排放入海，将被认为会对海洋资源或人类健康产生重大危害，因而应严禁向海洋环境排放该类物质。 For the regulations of the Annex II, MARPOL, Category X noxious liquid substances, if discharged into the sea from tank cleaning or de-ballasting operations, are deemed to present a major hazard to either marine resources or human health, and, therefore, the prohibition of the discharge into the marine environment is justified.

Category Y noxious liquid substances Y 类有毒液体物质 就 MARPOL 附则 II 而言，这类有毒液体物质如从洗舱或排除压载的作业中排放入海，将被认为会对海洋资源或人类健康产生危害，或对海上的休憩环境或其他合法利用造成损害，因而对排放入海的该类物质的质和量应采取严格的限制措施。 For the regulations of the Annex II, MARPOL, Category Y noxious liquid substances, if discharged into the sea from tank cleaning or de-ballasting operations, are deemed to present a hazard to either marine resources or human health or cause harm to amenities or other legitimate uses of the sea, therefore a limitation on the quality and quantity of the discharge into the marine environment is justified.

Category Z noxious liquid substances Z 类有毒液体物质 就 MARPOL 附则 II 而言，这类有毒液体物质如从洗舱或排除压载的作业中排放入海，将被认为会对海洋资源或人类健康产生较小的危害，因而对排放入海的该类物质应采取较严格的限制措施。 For the regulations of the Annex Ⅱ, MARPOL, Category Z noxious liquid substances, if discharged into the sea from tank cleaning or de-ballasting operations, are deemed to present a minor hazard to either marine resources or human health, and therefore less stringent restrictions on the quality and quantity of the discharge into the marine environment is justified.

Category Ⅲ ships 第Ⅲ类船舶 就装载状态而言，系指化学品船和液化气船。长度小于 65 m，在其设计时考虑货物和压载可不均匀分布的船舶属于第Ⅳ类船舶。 For loading condition, Category Ⅲ ships mean chemical tankers and gas carriers. Ships of less than 65 m in length, when their design takes into account uneven distribution of cargo or ballast, belong to Category Ⅳ.

Catenary anchor leg mooring buoy (CALM buoy) 单点系泊的浮筒

Catenary anchor leg mooring system (CALM) 悬链锚腿系泊系统 系指以一个大直径的浮筒为主体，由不小于 4 套锚（或锚桩）及锚链呈辐射状系住，浮筒顶部为一个装有大滚珠轴承的转盘，可作 360°回转，转盘上设有输油管系、阀门及缆绳装置，运油船用船缆与系缆装置连接，水下输油管从浮筒下面进入浮筒内与输油转换接头连接，浮筒上部转盘处的输油管则连到油船上，与油船的管系连接的系统。

Catenary mooring (CALM) 悬链式系泊

Catenary riser 悬链线立管/柔性立管 系指一段自由悬浮的管子。该管子将海底上的管路或油井与到海面上的设施连接，在连接过程中该管子已形成悬链线。 Catenary riser is a freely suspended pipe that connects a pipeline or a well on the sea floor to a facility above the surface of the sea, and is a catenary shape.

Catering to the register 注册登记

Cathodic protection 阴极保护 系指通过电化学方式，降低被保护金属腐蚀电位，使之成为腐蚀电池中的阴极，从而获得防腐蚀效果的保护方法。阴极保护可采用牺牲阳极保护法和外加电流保护法两种形式，外加电流保护法不适用于船舶结构内部浸水表面的防腐。 Cathodic protection means a protective method to reduce corrosive potential of the protected metal by electrochemical method and make it to be the cathode in corrosive batteries and obtain anticorrosion effect. The ship's cathodic protection may be two types, which are sacrificial anodic protection and external current protection. The external current protection does not apply to the corrosion protection for the submersed surface of structural interior.

Cathodic protection system 阴极保护系统 系指采用牺牲阳极或外加电流的阴极保护系统，也可采用两者联合的系统。 Cathodic protection system is a system of sacrificial anodic protection or external current protection or both.

Cattle carrier 牲畜船 系指专运牲畜的船舶。见图 C-5。 Cattle carrier is a dedicated cattle ship, which is shown in Fig. C-5.

图 C-5 牲畜船
Figure C-5 Cattle carrier

Catwalk 船员步桥 系指允许人员安全接近驾驶室前壁窗户的驾驶室外的装置。 Catwalk is an arrangement outside the wheelhouse allowing a person gets safe access to window along the front bulkhead(s).

Catwalk machine 猫道机

Causal factor 起因 系指行动、疏忽、事件或情况，如其不存在则:(1)海难或海上事故不会发生;(2)与海难或海上事故相关的不利影响可能不会发生或者如此严重;(3)与(1)或(2)中的结果相关的其他行为、疏忽、事件或情况可能不会发生。 Causal factor means actions, omissions, events or conditions, without which: (1) the marine casualty or marine incident would not have occurred; (2) adverse consequences associated with the marine casualty or marine incident would probably not have occurred or have been as serious; (3) another action, omission, event or condition, associated with an outcome in (1) or (2) would probably not have occurred.

Cause analysis 因果分析 系指确定导致顶事件发生的潜在情形组合的过程。

Caustic carrier 苛性钠运输船

Caution 提醒（当心） 系指优先度最低的警报。提醒表明情况还不足以发出报警或警告，但仍需对该状况或给定信息予以超出常规的关注。 Caution is an alert with the lowest priority. Caution brings an awareness of a condition which does not warrant an alarm or warning condition, but still requires attention out of the ordinary consideration of the situation or given information.

Cavitating propeller 空泡螺旋桨 系指桨叶表面产生空泡的螺旋桨。

Cavitation 空泡现象 系指一部分压力降至水饱和蒸气压以下时产生的气泡,这些气泡是由蒸气和某些溶解于水中的气体组成的。空泡产生于发声的振动板的周围,或产生与单位面积升力大的水翼(如水翼或船的螺旋桨)的周围。也就是说,这些物体的表面压力剧烈下降,最终下降到所在温度的饱和蒸气压以下,这时先生成微小的气泡,然后气泡吸收去气体而逐渐增大。随着空泡产生,气泡流向下游,一般说,空泡周围的压力会再度上升,而空泡趋向消失。然而,由于空泡是突然产生的,冲过来的水力会使气泡破裂,并且又激烈地相互碰撞和相互排斥,气泡在造成的冲击状态中反复地被破坏和产生。这种现象会引起周围的流体产生剧烈的压力变化,发出噪声。在那里物体表面除受到冲击引起损伤之外,还会导致阻力显著地增大。为此,在设计水中运转的机械时,不要设计成会产生空泡的形状,或者设计成即使产生空泡,也不至于对该机械带来不良影响的形状。

Cavitation criteria 空泡标准 系指衡量能否避免空泡或产生空泡程度的准则。对不同类形式螺旋桨需用不同的标准,以曲线图或公式表达。

Cavitation erosion 空泡剥蚀 系指空泡进入压力较高的区内,在物体表面上迅速坍毁时产生的极大压力对物体所造成的损伤。

Cavitation inception (onset of cavitation) 空泡开始 由于周围压力稍减或流速稍增,空泡核暴胀开始形成空泡的情况。

Cavitation number (based on vapour pressure) 空泡数 根据水气压力衡量空泡现象的一种无因次参数。$\sigma_V = (P - P_V)/q$,式中,P—绝对压力;P_V—水气压力;q—水的动压力。

Cavitation number based on actual cavity pressure 实压力空泡数 根据空泡实际压力衡量空泡现象的一种无因次参数。$\sigma = (P - P_c)/q$,式中,P—绝对周围静压力;P_c—空泡压力;q—水的动压力。

Cavitation tunnel (varying-pressure water tunnel) 空泡试验水筒 系指主要用于测试产生空泡现象的螺旋桨模型或其他物体性能的,能调节压力的循环水筒。

Cavity 空泡 水中压力低至或接近于其温度的水气压力时产生的气泡。其中一般也含有小量原混于或溶于水中的空气或其他气体。

Cavity 空腔 对钢夹层板而言,系指由底板、顶板和边界围成的空间。

Cavity pressure 空泡压力 系指定常或准定常空泡内的实际压力。均等于空泡的所含水汽及其他气体压力之和。

CCTV = closed circuit television 闭路电视

CD-ROM 只读光盘 系指一种能够存储大量数据的外部存储媒体,一张压缩光盘的直径大约是4.5英寸,1/8英寸厚,能容纳约660兆字节的数据。CD-ROM盘都是用一张母盘压制而成,然后封装到聚碳酸酯的保护外壳里。

DT cable winch CDT 绞车 为收放电缆及测量仪器,在船舶航行中进行平面视察温盐深水文要素用的绞车。

Ceiling type induction unit 顶式诱导器 装于平顶下,冷风或热风从出风口的四周沿平顶吹出的诱导器。

Cell 信元 系指专用于 ATM 网络,原点到目的结点传输的是信元,信元是一种特殊的数据结构,不同于普通网络传输的帧或包,因为帧和包是变长的,而 ATM 的信元是定长的,非常小的,长度只有 53 个字节,其中 5 个字节是信元头,48 个字节是信息段。信息段中可以是各类业务的用户数据,在信元头中包含各种控制信息。在信元中包括 CRC 校验,其生成公式为 $X^8 + X^2 + X + 1$,校验和仅对信元头进行校验。

Cell guides 集装箱导轨架/格栅 系指由垂直角钢制成的刚性系固系统,在集装箱的长度和宽度方向留有间隙,为集装箱堆垛提供定位和水平约束用。

Cell spar 蜂巢形柱体式平台/多柱型立柱式平台

Cell Truss Spar platform 多桁架立柱式平台 第三代 Spar 平台。其采用多个小直径深吃水圆柱体组合在一起代替原来的单个大直径圆柱构成平台主体的 Spar 平台。

Cellar deck 泥浆净化平台 系指供安装泥浆净化设备用的平台。

Cellular 蜂窝电话 为了能容纳大量的用户,可以把一个地理区域划分成许多小区。不是采用单个大功率的发射器,而是每一个小区由一个小功率的基站(Base station)来提供服务。由于这些基站能够影响的范围比较有限,因此同一频谱可以在一个远离一定距离的另一个小区中再次使用。频率的再使用以及越区切换(Handoff)的操作方式合在一起就构成了小区(蜂窝)概念的主体。

Cellular construction 分隔结构 系指用两个间距紧密的界限和利用内部隔板构成数个小舱室的结构布置。Cellular construction means a structural arrangement where there are two closely spaced boundaries and internal diaphragm plates arranged in such a manner to create small compartments.

Cement 水泥　是一种细研粉末,含有空气或受到极大扰动时实际上几乎成为流体,因而形成的静止角很小。该货品在装载完成后几乎立刻脱气,货品变得密实而稳定。如船舶不是专门设计的水泥运输船或岸上设备没有安装专门的粉尘控制装置,则水泥粉尘在装载和卸货期间会是一个问题。含有空气时会造成流动。该货物为不燃物或失火风险低。　Cement is a finely ground powder which becomes almost fluid in nature when aerated or significantly disturbed, thereby creating a very minimal angle of repose. After loading, almost complete de-aeration occurs immediately and the product settles into a stable mass. Cement dust can be a major concern during loading and discharge if the vessel is not specially designed as a cement carrier or shore equipment is not fitted with special dust control equipment. It may shift when aerated. This cargo is non-combustible or has a low fire-risk.

Cement box　堵漏盒　船舶破损时,用以盖住不平整破洞口的钢质四方盒。

Cement carrier　散装水泥运输船　系指设计和制造用于运输散装水泥的船舶。　Cement carrier is a ship designed and constructed for the carriage of cement in bulk.

Cement clinker　水泥熔块　水泥由石灰石和黏土烧结而成。在此燃烧过程中产生的粗糙渣块随后压碎成细粉末产生出水泥。粗糙渣块称为熔块并以此种形态装运,以避免载运水泥粉末的困难。无特殊危害。该货物为不燃物或失火风险低。　Cement clinker is formed by burning limestone with clay. This burning produces rough cinder lumps that are later crushed to a fine powder to produce cement. The rough cinder lumps are called clinkers and are shipped in this form to avoid the difficulties of carrying cement powder. It has no special hazards. This cargo is non-combustible or has a low fire-risk.

Cement tank　水泥舱　系指贮存固井用水泥的舱室。

Cement-mixed tank　水泥搅拌舱　调配固井水泥用的舱室。

Center blade　中间舵叶　设置多叶舵的船舶,系指位于中间的舵叶。

Center height of upper tumbler　上导轮中心高　系指上导轮中心线离设计水线的高度。

Center keelson　中内龙骨　单层船底结构中,中线面处的纵桁。

Center moored drilling ship /**Turret moored drilling ship**　中心系泊定位钻井船　系指将定位用系泊缆系到船体中央的转筒下方,船体可绕转筒旋转,使作用在船体上的风浪流外力减到最小,以改善定位与运动性能,从而提高钻井效率的系泊定位钻井船。

Center of buoyancy　浮心　浮力的作用点。即型排水体积的中心。

Center of floatation　漂心　船舶水线面的面积中心。

Center of gravity　重心　系指船舶各部分重力的合力的作用点。

Center of rudder pressure　舵压力中心　系指舵压力合力的作用点。

Center of turning circle　回转中心　系指定常阶段回转圈的中心。

Center shafting　中轴系　在船中线面的轴系。

Center(- line) plane [**central longitudinal plane, longitudinal middle(- line) plane**]　中线面　系指将船体分为左右两个对称部分,并垂直于基平面的纵向平面。

Centerline bulkhead　中纵舱壁　位于中线面上的舱壁。

Centerline girder (cellular construction)　中纵桁材　位于船舶中纵线处的纵向构件。　Centerline girder is a longitudinal member located on the centerline of the ship.

Centerline rudder　中舵　在设置多舵的船上,设置在船体中线面上的舵。

Center-mooring drilling ship　中心锚泊钻井船　在船体的中心有一垂直的回转筒,其上部装有钻机、井架和锚机,下部引出对称反射状锚链抛锚定位,工作时船体可随风、流绕筒回转,以保持船首迎风迎流而减少摇摆的钻井船。

Centimetre wave　厘米波

Central air conditioner　集中式空气调节器　系指供一个区域内许多舱室合用的空气调节器。

Central air conditioning system　集中式空气调节系统　部分舱外新鲜空气和部分舱室内循环空气两者混合,或全部是舱外新鲜空气经集中式空气调节器处理后由通风机经通风管分别送至各舱室的空气调节系统。

Central alert management (CAM)　中央警报管理　中央警报管理-人机界面(CAM-NMI)警报显示,中央警报管理-人机界面(CAM-NMI)和航行系统以及传感器之间警报状态通信的管理功能。功能可集中或在子系统部分集中并通过标准化警报相关通信互相连接。　Central alert management (CAM) means the functions for the management of the presentation of alert states between CAM-NMI and navigation systems and sensors. The functions may be centralized or partly centralized in subsystems and interconnected via a standardized alert-related communication.

Central alert management human machine interface (CAM-NMI) 中央警报管理-人机界面

Central control of machinery spaces (MCC) 机舱集控站(室) 系指有人值班对机电设备进行监控的处所。 Central control of machinery spaces (MCC) means a central control space constantly attended by watch-keepers.

Central control room 中央控制室 系指用于对饱和潜水设备和各分系统进行集中操纵、控制和监护的舱室。

Central control station 集中控制站 系指：(1)具有下列集中控制和显示功能的控制站：①固定式探火和失火报警系统；②自动喷水器、探火和失火报警系统；③防火门位置指示；④防火门锁闭；⑤水密门位置指示；⑥水密门锁闭；⑦风机；⑧通用/失火报警；⑨包括电话在内的通信系统；⑩公共广播系统的扩音器。(2)是船舶控制站的一种，是为推进机械、发电装置、船舶主推进器中的基本辅助设备（也称为"基本辅助设备"）或其他被船级社认为必需的辅助装置的所必需并充分的系统，是一个为了安装机械的中央监控系统的特殊的房间，通常在这里操控船舶的主推进机械。 Central control station is: (1) a control station in which the following control and indicator functions are centralized: ①fixed fire detection and fire alarm systems; ② automatic sprinkler, fire detection and fire alarm systems; ③fire door indicator panels; ④fire door closure; ⑤watertight door indicator panels; ⑥watertight door closures; ⑦ventilation fans; ⑧general/fire alarms; ⑨ communication systems including telephones; ⑩microphones to public address systems. (2) Centralized control station is one of the control stations of a ship which has necessary and sufficient systems to control main propulsion machinery, generating sets, auxiliary machinery essential for main propulsion of the ship (hereinafter referred to as "essential auxiliary machinery") and other auxiliaries considered necessary by a Society and a room specially provided for the purpose of installing centralized monitoring and control systems for machinery, from which main propulsion machinery is normally controlled.

Central part 船舯部 包括位于防撞舱壁之后与艉尖机舱舱壁之间的结构。 Central part includes the structures located between the collision bulkhead and the after peak bulkhead.

Central processing unit^1 (CPU) 微处理器 系指对测量值进行计算、处理的单片机。经过处理后的信号可以显示瞬时声级、最大声级、统计声级、等效连续声级、噪声曝露级等各种评价值。

Central processing unit hour (CPUH) 中央处理机时间 即反映CPU全速工作时完成该进程所花费的时间。可以反映CPU性能的一个指标。

Central processing unit^2 (CPU) 中央处理器 系指一块超大型的集成电路，是一台计算机的运算核心和控制核心，也是电子计算机的主要设备之一。主要包括运算器(ALU/Arithmetic and logic unit)和控制器(CU/Control unit)两大部件。此外，还包括若干个寄存器和高速缓冲存储器及实现它们之间联系的数据、控制及状态的总线。它与内部存储器和输入/输出设备合称为电子计算机三大核心部件。其功能主要是解释计算机指令以及处理计算机软件中的数据。计算机的性能在很大程度上由CPU的性能所决定，而CPU的性能主要体现在其运行程序的速度上。

Central tank 中间舱 系指纵向舱壁间的任何舱柜。 Central tank means any tank inboard of a longitudinal bulkhead.

Central tower 斗塔 设置在链斗挖泥船甲板中部，在其上部设有传动斗链的装置和上导轮座，用以支持斗桥的刚性构架。

Centralised plants for oxyacetylene welding 集中式氧乙炔焊接设备 是一个固定式设备，它由气瓶室、分配站和分配管，总数超过4个的氧气瓶、乙炔瓶组成。 Centralised plant for oxyacetylene welding is a fixed plant consisting of a gas bottle room, distribution stations and distribution piping, where the total number of acetylene and oxygen bottles exceeds 4.

Centralized control console (centralized control board) 主/辅机集中操纵台 系指设置在机舱或控制室内的主、辅机集中操纵台。

Centralized control station (room) of engine room 机舱集控站(室) 系指机舱内集中布置自动化设备的所有监控设施的控制站(室)。 Centralized control station (room) of engine room means a control station (room) in which all monitoring means for automated equipment in engine room are concentrated.

Centralized monitoring and control station on bridge 桥楼上的集中监测控制站 系指在桥楼上具备机械的集中监测和控制系统，并通常在控制主推进机械的船舶的驾驶桥楼上。 Centralized monitoring and control station on bridge is a navigation bridge of a ship which has centralized monitoring and control systems for machinery on the bridge and from which main propulsion machinery is normally controlled.

Centralized operation cargo oil pumping system

集中操纵货油装卸系统　整艘油船的装卸工作和油船装卸过程中需要配合动作的机械设备仪表,均可在集中控制室中进行监控的系统。

Centralized transport package　**集合运输包装**　或称成组化运输包装,系指在单件运输包装基础上,为了适应运输、装卸工作现代化的要求,将若干件单件运输包装组合成一件大包装。

Centre line screw　**中线螺旋桨**　系指轴线在船中线面内的螺旋桨。

Centre of screw propeller　**螺旋桨中点**　系指螺旋桨基准线与螺旋桨轴线的交点。

Centre well　**中央井**

Centric reduction gear　**同心式减速齿轮箱**　轴系中线与发动机轴轴线重合的减速齿轮箱。

Centrifugal (oil) separator (centrifuge)　**油分离机**　系指利用分离筒的高速旋转,使混合物中具有不同比重的油、水或机械杂质在离心力场作用下,获得不同的离心力以达到分离目的的设备。

Centrifugal compressor　**离心式压缩机**　利用叶轮的旋转作用,压缩气体在叶轮内作径向离心流动,而后再进入导向器的一种压缩机。

Centrifugal fan　**离心式风机**　系指气流主要沿叶轮径向流动的风机。

Centrifugal pump　**离心泵**　系指利用叶轮旋转时的离心力作用,提高液体的压力能以获得扬程的泵。

Centrifugal refrigerating compressor　**离心式制冷压缩机**　一般用于大能量空气调节系统,主要由工作轮、叶片、扩散器、轴、轴承和油泵等组成,用离心作用使气体得到压缩的制冷压缩机。

Centripetal turbine (inward-flow turbine)　**向心式涡轮**　高温燃气或其他气体在膨胀过程中是从外向内作径向流动并沿径向出口的涡轮。

Centrum baleony　**中央露天舞台(豪华邮轮)**

Certificate[1]　**证书**　系指由主管机关颁发或经主管机关授权颁发或主管机关所认可的一种有效文件(不论其名称如何),该文件许可其持有人担任该文件所指定的或国家法规所规定的职务。　Certificate means a valid document, whatever the name it may be known, issued by or under the authority of the Administration or recognized by the Administration authorizing the holder to serve as stated in this document or as authorized by national regulations.

Certificate[2]　**签证**

Certificate for cargo ships　**货船安全证书**　对货船而言,系指由 SOLAS 公约缔约国政府授权的个人或组织,按经 1988 年议定书修订的 1974 年国际海上人命安全公约的规定签发的安全证书。　For cargo ships, Safety Certificate for cargo Ships means a Safety Certificate issued by a person or an organization authorized under the authority of the Government of under the provisions of the International Convention for the Safety of Life at sea, 1974, modified by the Protocol of 1988 relating thereto.

Certificate of export subsidy system　**出口奖励证制**　政府对出口商出口某种商品以后发给一种奖励证,持有该证可以进口一定数量的外国商品,或将该证在市场中自由转让出售,从中获利。

Certificate of fitness　**适装证书**　系指由国家政府或船级社代表政府签发的证书,证明该船的构造和设备符合散装危险化学品规则或散装液化气体规则或类似国家认可的散装货物条款或达到最低的有效的标准。　Certificate of fitness means a certificate issued by a national government, or a society on behalf of government, certifying that the construction and equipment of the ship are in accordance with the code for dangerous chemical in bulk or the code for liquefied gases in bulk or to similar recognized national provisions in bulk or to the least effective standards.

Certificate of fitness for offshore support vessels　**近海供应船适装证书**　系指当近海供应船装运所规定的货物时,所持有的一份适装证书。该证书应按"关于在近海供应船上运输和装卸有限数量的散装危险和有毒液体物质的指南"予以签发。如果某一近海供应船仅载运有毒液体物质,则可对其签发经适当批注的国际防止散装运输有毒液体物质污染证书,以代替上述的适装证书。　Certificate of fitness for offshore support vessels means that when carrying such cargoes, a Certificate of Fitness carried for offshore support vessels, is issued under the "Guidelines for the Transport and Handling of Limited Amounts of Hazardous and Noxious Liquid Substances in Bulk on Offshore Support Vessels". If an offshore support vessel carries only noxious liquid substances, a suitably endorsed International Pollution Prevention Certificate for the Carriage of Noxious Liquid Substances in Bulk may be issued instead of the above Certificate of Fitness.

Certificate of Fitness for the Carriage of Dangerous Chemicals in Bulk　**散装运输危险化学品适装证书**　系指对经初次检验或定期检验之后,符合 BCH 规则的有关要求的国际航行化学品液货船,签发的散装运输危险化学品适装证书。该证书的标准格式见 BCH 规则的附录。按 MARPOL 73/78 附则 II 的规定,对于 1986 年 7 月 1 日以前建造的化学品液货船,该规则是强制性的。　Certificate of Fitness for the Carriage of Dangerous Chemicals in Bulk means a certificate called a Certificate of Fitness for the

Carriage of Dangerous Chemicals in Bulk, the model form of which is set out in the appendix to the Bulk Chemical Code. It should be issued after an initial or periodical survey to a chemical tanker engaged in international voyages which complies with the relevant requirements of the Code. The Code is mandatory under Annex II of MARPOL 73/78 for chemical tankers constructed before 1 July 1986.

Certificate of fitness for the carriage of INF cargo 国际装运 INF 货物适装证书　系指对装运 INF 货物的船舶经检验并符合"1974 年国际海上人命安全公约"任何适用要求和"国际船舶装运密封装辐射性核燃料、钚和强放射性废料规则"的要求后,由主管机关或其认可的组织签发的证书。

Certificate of fitness for the carriage of liquefied gases in bulk 散装运输液化气体适装证书　系指对经初次检验或定期检验后,符合气体运输船舶规则有关要求的气体运输船,签发的散装运输液化气体适装证书。该证书的标准格式见 GC 规则附录。 Certificate of fitness for the carriage of liquefied gases in bulk means a certificate called a Certificate of Fitness for the Carriage of Liquefied Gases in Bulk, the model form of which is set out in the appendix to the Gas Carrier Code, it should be issued after an initial or periodical survey to a gas carrier which complies with the relevant requirements of the Code.

Certificate of inspection 检验合格证书

Certificate of insurance or other financial security in respect of civil liability for oil pollution damage 关于油污损害民事赔偿责任的保险和其他财务保证的证书　系指对每艘载运 2 000 t 以上散装货油的油船,均应给其签发一份证明其保险或其他财务保证的证书。该证书应在确定该船已符合 1992 年 CLC 公约第Ⅶ条 1 的要求以后,由该船登记国的有关当局签发或证明。对于在缔约国登记的船舶,该证书应由船舶登记国的有关当局签发;对于不在缔约国登记的船舶,该证书可以由任何缔约国有关当局签发或验证。 Certificate of insurance or other financial security in respect of civil liability for oil pollution damage means a certificate attesting that insurance or other financial security is in force in accordance with the provisions of the 1992 CLC Convention. It shall be issued to each ship carrying more than 2,000 t of oil in bulk as cargo after the appropriate authority of a Contracting State has determined that the requirements of article Ⅶ, paragraph 1, of the Convention have been complied with. With respect to a ship registered in a Contracting State, such certificate shall be issued by the appropriate authority of the State of the ship's registry; with respect to a ship not registered in a Con-

tracting State, it may be issued or certified by the appropriate authority of any Contracting State.

Certificate of origin 产地证明书　即"原产地证明书",是出口商应进口商要求而提供的由公证机构或政府或出口商出具的证明货物原产地或制造地的一种证明文件。产地证书是贸易关系人交接货物、结算货款、索赔理赔、进口国通关验收、征收关税的有效凭证,它还是出口国享受配额待遇、进口国对不同出口国实行不同贸易政策的凭证。

Certificated 具有了证书的　系指持有适当的证书。 Certificated means properly holding a certificate.

Certificated person 持证人员　系指持有主管机关按照现行的《国际海员培训、发证和值班标准公约》要求,授权颁发的或承认有效的精通救生艇、筏业务证书的人员;或持有非该公约缔约国的主管机关为公约证书同一目的而签发或承认的证书的人员。 Certificated person is a person who holds a certificate of proficiency in survival craft issued under the authority, or recognized as valid by, the Administration in accordance with the requirements of the International Convention on Standards of Training, Certification and watch-keeping for Seafarers, in force; or a person who holds a certificate issued or recognized by the Administration of a State not a Party to that Convention for the same purpose as the convention certificate.

Certificates for masters, officers or ratings 船长、高级船员和普通船员等级证书　系指主管机关对那些满足工作、年龄、健康、培训等各方面要求并满意的人员,按照 1978 年海员培训、发证和值班标准公约附则的规定经考试合格后,签发的船长、高级船员和普通船员等级证书。证书的格式列于 STCW 规则第 A-1/2 节。证书的正本必须保存在该持证者工作的船上。 Certificates for masters, officers or ratings means Certificates for masters, officers or ratings issued to those candidates who, to the satisfaction of the Administration, meet the requirements for service, age, medical fitness, training, qualifications and examinations in accordance with the provisions of the STCW Code annexed to the Convention on Standards of Training, Certification and Watch-keeping for Seafarers, 1978. Formats of certificates are given in section A-I/2 of the STCW Code. Certificates must be kept available in their original form on board the ships on which the holder is serving.

Certification[1] 证书　是获得书面证据过程,证明其结构、各项设备、或其他装置已按规定的方式完成计、制造、安装或维护。"规定的方式"通常系指在规则中形成文件由管理部门诸如主管当局作为命令或规范予以颁布。"书面证据"可由公认的专业团体或代理人授权的证明者颁发的任何形式的证书或其他文件。 Certification is the process of obtaining written evidence that a structure, item of equipment, or other arrangement has been designed, constructed, installed, or maintained in a prescribed manner. The "prescribed manner" is normally documented in the regulations that are defined as orders or rules issued by a regulatory regime, such as a governmental authority. The "written evidence" may be in the form of any certificate or other document issued by the certifier who is authorized to do so by the appropriate professional organization or agency.

Certification^2 认证 系指对所提交文件的审核和船舶在建造及试航时的检验。

Certification of the inclining test weights 倾斜试验重物证书 系指在试验重物上所标明质量的证明。应使用经认可的标准质量来鉴定试验重物。称重应在尽量接近倾斜试验时进行,以确保所测质量的准确。Certification of the inclining test weights is the verification of the weight marked on a test weight. Test weights should be certified using a certificated scale. The weighing should be performed close enough in time to the inclining test to ensure the measured weight is accurate.

Certification survey 发证检验 系指按照委托方规定的要求进行验证,以确认合同规定的各项已满足要求。这些要求一般为公认的规范及标准、业界标准和/或有关导则。检验完成时,船级社将签发有关的噪声和提供/签署有关的检验文件。

Certified safe type 合格安全型 系指由公认的机构按公认的标准核准为安全型的电气设备。对电气设备的核准应与甲烷气体的类别和组别相对应。 Certified safe type means electrical equipment that is certified safe by a recognized organization based on a recognized standard. The equipment standard is correspond to the category and group for methane gas.

Certified safe-type equipment 合格安全型设备 系指已由国家的主管机关或其他有关主管机关进行为证实其在爆炸性气体环境中使用时该设备对爆炸危险之安全性所必需的形式鉴定和试验之类型的电气设备。 Certified safe-type equipment means the electrical equipment of a type for which a national or other appropriate authority has carried out the type verifications and tests necessary to certify the safety of the equipment with regard to explosion hazard when used in an explosive gas atmosphere.

Certified value 核准值 系指废气滤清(EGC)装置核准的生产厂商股东规定的排放极限。Certified value means that emission limit specified by the manufacturer that the EGC unit is certified.

CFR = Code of Federal Regulation/Federal Regulation [美]联邦政府行政法规汇编/美国联邦政府法规

Chafing plate at well side 桥挡板板 为避免斗桥在作业时擦损其槽侧板而装在该侧板上的防护板条。

Chain 锚链 系指由金属圆环或链环构成,用以拉住锚或拴紧木材货等。 Chain means a connected metal rings or links used for holding anchor, fastening timber cargoes, etc.

Chain drum 舵链卷筒 系指带有导槽,用以卷动操舵链或操舵索的卷筒。

Chain for rigging 索具链 作索具用的链条。

Chain link 锚链环 组成锚链的各种单环的统称。有普通、加大、末端及有档、无档之分。

Chain locker 锚链舱 通常系指在船舶艏端,用于存放锚链的舱室。 Chain locker means a compartment usually at the forward end of a ship which is used to store the anchor chain.

Chain pipe 锚链管 供锚链穿入或穿出锚链舱的一段管子。 Chain locker is a section of pipe through which the anchor chain enters or leaves the chain locker.

Chain stopper(chain cable compressor) 制链器 系指在抛锚时以及在将锚固定于锚链筒内遮蔽位置时拉住链缆的装置,可减轻锚机负荷。 Chain stopper means a device for securing the chain cable when riding at anchor as well as securing the anchor in the housed position in the hawse. pipe, there by relieving the strain on the windlass.

Chain stripper 分链器 防止锚链卡在锚链轮上连续转动,以便于保护起抛锚的起锚机械部件。

Chain tensioner 锚链张紧器/链条张紧器

Chain wheel for hawse pipe(wildcat) 导链滚轮 为使锚链顺利收放并避免与锚链筒口摩擦而设置在锚链筒与止链器之间的导链轮。

Chain wheel type skylight controlling gear 链轮式天窗传动装置 利用链条链轮启闭天窗盖的天窗传动装置。

Chain wire diameter(of common link) 链径 系指以普通锚链的圆钢断面直径表示的锚链公称直径。

Chair barrel steering gear(chain rod steering gear) 舵链传动操舵装置 以操舵链和操舵拉杆为传动器件的人力操舵装置。

Chair table 海图桌 专供存放海图和进行海图作业用的桌子。

Chalk test 压痕试验 用于检查风雨密门、舱口盖、舷窗等带密封胶条的开口处密性的试验。

Challenge 对…有异议

Challenge against an arbitrator 对仲裁员回避的书面请求

Chamber winch 浮筒绞车 作为收、放浮筒进行海流（表层流）观测工作的绞车。

Chamotte 耐火黏土 经焙烧的黏土。灰色。以细碎石形态运输。用于炼锌和制造耐火砖（筑路碎石）。多粉尘。无特殊危害。该货物为不燃物或失火风险低。

Chamotte are the burned clay. It is grey, shipped in the form of fine crushed stone. Used by Zinc smelters and in manufacture of firebrick (road metal). It is dusty. It has no special hazards. This cargo is non-combustible or has a low fire-risk.

Champagne bar 香槟酒吧（豪华邮轮）

Change of heading test 改向试验 选择一定范围内的几种航速和舵角改变艏向10°、20°、30°，并沿原来艏向方向及沿操舵到开始进入稳定新艏向时瞬时船舶重心位置之间距离的试验。

Change of trim 纵倾调整 改变船舶重心和浮心的相对位置，对其纵倾状态所进行的调整。

Change-speed motor 变速电动机 系指转速随负载变化的，一般当负载增加时，转速降低的电动机。诸如串励电动机或推斥电动机。

Channel ship 海峡船 系指往返于海峡两岸的船舶。

Characteristic curves of screw propeller 螺旋桨特性曲线 以螺旋桨进速系数为横坐标，以推力系数、转矩系数和敞水效率为纵坐标的一组性能特性曲线。

Characteristic ratio (Parson's number) 特性比 用来大致判断多级汽轮机设计完善程度的比值。

Charcoal 木炭 系指在尽可能少接触空气的情况下经高温燃烧后的木材。该货物粉尘极多，质量轻。所吸收水分能达到自身质量的18%~70%。黑色粉末或颗粒。颗粒自燃。遇水后颗粒自热。易使货物处所缺氧。不得装载超过55℃的热木炭筛屑。Charcoal is the wood burnt at a high temperature with as little exposure to air as possible. It is very dusty and light. It can absorb moisture to about 18% to 70% of its weight. It is black powder or granules. It may be ignited spontaneously. Contact with water may cause self-heating. It is liable to cause oxygen depletion in the cargo space. Hot charcoal screenings in excess of 55 ℃ should not be loaded.

Charge forward 费用先付

Charge-coupled device (CCD) 电荷耦合元件 可以称为CCD图像传感器，也叫图像控制器。CCD是一种半导体器件，能够把光学影像转化为数字信号。CCD上植入的微小光敏物质称作像素（Pixel）。一块CCD上包含的像素数越多，其提供的画面分辨率也就越高。CCD的作用就像胶片一样，但它是把光信号转换成电荷信号。CCD上有许多排列整齐的光电二极管，能感应光线，并将光信号转变成电信号，经外部采样放大及模数转换电路转换成数字图像信号。

Charging by breathe 呼吸充电 巴西一名设计师设计了一款用呼吸为手机充电的新型装置——AIBE面罩，里面装有一个小型的风力涡轮机，能够把呼吸产生的风能转化成电能。

Charging by friction 摩擦充电 这利用的是静电原理。一片指甲大小的纳米材料通过摩擦就能产生8毫瓦的电量，足够让心脏起搏器运行起来了。摩擦25 mm²的材料足以点亮600盏LED灯。

Charging by voice 声音充电 整个装置就像"三明治"，特定化合物把氧化锌纳米管线夹在当中。当外界声音传至表面时，声波振动会导致氧化锌纳米管线压缩和伸展，从而产生微量的电压。目前，这种装置能将100 dB左右的音量转换成50 mV的电压。

Charging by WiFi WiFi 充电 美国华盛顿大学正在开发"WIFI 充电系统"。这项技术可以通过WIFI信号为8.5 m内的设备充电。传感器接收射频信号中的电能，并将其转化为直流电。

Charging efficiency 充气效率 系指每一循环充入气缸内的新鲜空气或可燃混合气在标准环境状态下的容积与气缸工作容积之比。

Charging to furniture 家具充电 宜家公司推出一系列支持无线充电的家居用品。最吸引人关注的是支持无线充电的台灯、书桌、床头柜和落地灯等。这些家具自带无线充电模块，用户买回去连上电源就可以使用。

Charpy V-notch (CVN) impact test 夏比V型缺口冲击试验 系指测量材料断裂韧性的通用试验。
Charpy V-notch (CVN) impact test means the common test for measuring the fracture toughness of a material.

Chart room 海图室 设有海图桌和必要的航海仪器，用以进行海图作业和保管海图、航海日志以及有关航海资料的舱室。

Charter of concession 特许证

Charter party 租船合同 系指采用船舶运输方式的出租人与承租人之间关于租赁船舶所签订的一种海上运输合同。其规定承租人以一定的条件向船舶所有人租用一定的船舶或一定的舱位以运输货物，并就双方的权利和义务、责任与豁免等各项以条款形式加以规定，用以明确双方的经济、法律关系。租船合同分为定程租船合同和定期租船合同,简称定程租约(Voyage charter party)和定期租约(Time charter party)。

Charter period　租期　系指承租人使用船舶的期限,始于出租人交船,可以用日、月或年来表示。承租人须在租期届满前还船。

Chartered carrier transport　包机运输　系指航空公司按照约定的条件和费率,将整架飞机租给一个或若干个包机人(包机人系指发货人或航空货运代理公司),从一个或几个航空站装运货物至指定目的地。包机运输适合于大宗货物运输,费率低于班机,但运送时间则比班机要长些。

Charterer　租船者　系指购买船舶运输服务的买方,他是船东的顾客。 Charterer is a buyer of vessel transportation services who is a vessel owner's customer.

Chartering and booking shipping space　租船订舱

Chartering shipping　租船运输　又称不定船期运输。是相对于班轮运输而言的另一种船舶营运方式。它与班轮运费不同,没有预定的船期表、航线和停靠港口也不固定,须依据船舶所有人与承租人双方签订的租船合同安排船舶就航的航线。

Chatroom　聊天室/私人聊天室　是一个网上空间,为了保证谈话的焦点,聊天室有一定的谈话主题。聊天室可以建立在即时通信软件(如 MSN Messenger、QQ、Anychat)、P2P 软件、万维网(如 Halapo、Meebo)等基础上,万维网方式更为普通和种类繁多,交谈的手段不局限于文本,更包括语音、视频。通常聊天室是按照房间或频道为单位的,在同一房间或频道的网人可以实时地广播和阅读公开消息。一般情况下,与其他网络论坛、即时通信不同的是,聊天室不保存聊天记录。

Checklist　检查清单

Checklist analysis(in hazard)　(危害的)清单分析　一种建立在经验基础上的方法。它使用一份含有条目或过程步骤的书面清单,用于鉴定设备、系统或操作中已知的危害类型、设计缺陷和潜在事故场景。

Chemical　化学品　系指任何固态、液态或气态的化合物、混合物或溶体,其特性(包括或不包括易燃性)有危害性者,或化合物的特性可能在热态作业或冷态作业时显示其危害性者。

Chemical barge　化学品驳　系指船舱内装载化学品的驳船。 Chemical barge is a barge carried chemicals in holds.

Chemical carrier　化学品船　系指散装运输危险化学品的船舶。有专用化学品船,如硫酸运输船等;有可同时装运多种化学品的多功能化学品船。见图 C-6。

Chemical coagulant　化学凝聚剂　系指能在水面扩散并压缩油膜,使其面积大大缩小,以防止浮油扩散

图 C-6　化学品船
Figure C-6　Chemical carrier

的化学物质。因其布散迅速,对煤油、柴油、重油的扩散约束有效,常与围油栅一起联合使用。

Chemical de-sealing　药剂除垢　系指将药剂注入蒸发器的容水腔内,借药剂与换热面上的垢发生化学反应以去除污垢的除垢方式。

Chemical storage tank　化学剂贮存舱　系指贮存处理泥浆所需要的化学药剂的舱室。

Chemical tanker　化学品液货船　系指其构造适用于散装运输《国际散装运输危险化学品船舶构造和设备规则》(IBC 规则)第 17 章所列任何易燃液体货品的液货船。又称化学品船。类似于油船,设有货物围护系统,被设计用来专运《散装运输危险化学品船舶构造与设备规范》中所列的货品,如浓硫酸、盐酸、溶化硫、过氧化氢等液体货品、动植物油、高温熔融物(硫黄等)、载重量 40 000 t 左右,货舱可达 40 多个的船舶。对于 1986 年 7 月 1 日及以后建造并符合 IBC 规则的船舶,根据载运化学品类别加以下标志:(1)1 型载运对环境或人身安全有非常严重危险的化学品,需要采取最严格措施来预防泄漏的货物。其货舱形式包括为整体液舱和独立液舱;(2)2 型载运对环境或人身安全有相当严重危险的化学品,需要采取较严格措施来预防泄漏的货物。其货舱形式包括为整体液舱和独立液舱;(3)3 型装载对环境或人身安全有足够严重危险的化学品,需要采取适当的措施来预防泄漏的货物。其货舱形式包括为整体液舱和独立液舱;对于 1986 年 7 月 1 日以前建造并符合 BC 规则的船舶,上述的船舶 1 型/2 型/3 型分布由Ⅰ型/Ⅱ型/Ⅲ型替代。其按用途可分为专用于运输硫酸、过氧化氢溶液、磷酸等化学品的专用化学品船,如苛性钠运输船、沥青运输船和设有独立液货舱或整体液货舱专运石油沥青运输船以及用来运输包括石油产品在内的多种货物的多用途化学品船,如糖蜜/化学品运输船、散货/硫酸运输船等。

Chemical/oil tanker　化学品/油液货船　系指既可装运化学品亦可装运石油产品的船舶。　Chemical/

oil tanker means that a tanker is capable of carrying both chemicals and oil products.

Chemical/product oil tanker 化学品/成品油船 系指可装运成品油和低级化学品的船舶。

Chemistry pumping 化学注入 是最简单的修井作业,不需要向井内放置硬件设施,通常仅需将化学药剂注入管线与采油树的压井翼阀连接并将化学药剂注入井内则可。

Chenxi shipyard (Guangzhou) Company Limited 中船澄西远航船舶(广州)越秀公司——华南海上改装修理"桥头堡" 简称澄西广州,是中国船舶工业集团有限公司在华南地区的海洋工程改装修理的重要基地,公司毗邻香港,拥有便利的地理位置和深水码头条件,适合承修各类海工业务。2013 年 10 月,公司历时 26 个月完成了目前世界上海上浮式生产储油船(FPSO)改装,历史上工程量、工程最复杂、技术最先进的项目"伊利亚贝拉"号改装,将一艘超大型原油船(VLCC)改装为集生产、处理、储存和卸载为一体的 FPSO。其采用多点系泊的形式,具备自航能力,改装完成后 25 年不需进坞修理。

Cheque 支票 是以银行为付款人的即期汇票,即存款人对银行的无条件支付一定金额的委托或命令。

Chequered plate 花钢板 系指装于机炉舱底层、辅机周围及主要作业场所的防滑钢板。

Chief engineer officer 轮机长 系指负责船舶机械推进职能的高级船员。Chief engineer officer means the senior engineer officer, who is responsible for the mechanical propulsion of the ship.

Chief information officer (CIO) 首席信息官/首席资讯官 通常负责对企业内部信息系统和信息资源规划和整合的高级行政管理人员。CIO 通常归公司执行主管(CEO)、运作主管(COO)或财务主管(CFO)领导。CIO 是一个比较新的职位,随着商业领域多极化的竞争与发展,越来越多的企业开始将信息(Innovation)这一概念作为企业的持续发展的动力和竞争优势,CIO 将成为未来企业最为重要的职位领导人之一。顺应此潮流,国外已经开始有各种各样 Chief Innovation Officer 培训,其中最有名的是 Langdon Morris,他出版了一本名为《Leading Innovation Workbook》,并在世界各地进行巡回演讲授课,颇受好评。

Chief marine surveyor 海事总验船师 对澳大利亚海事安全局海事指令而言,系指澳大利亚海事安全局的验船部门经理,或由该验船部门经理委派的负责特定目的的,具有适当资格的人员。For the maritime order of the Australian Maritime Safety Authority (AMSA), chief marine surveyor means the manager (survey operations) in AMSA, or in respect of any particular purpose under this part, a suitably qualified person authorized by the Manager (survey operations) for that purpose.

Chief mate 大副 系指级别仅低于船长,并在船长不能工作时由其指挥船舶的驾驶员。Chief mate means the deck officer whose rank next to the master and upon whom the command of the ship will fall in the event of the incapacity of the master.

Chilled cargo 冷却货物 系指经预冷到冻结温度以上的,在 $-2\ ℃ \sim +12\ ℃$ 舱内温度下装运的货物。

China academy of engineering physics (CAEP) 中国工程物理研究院 创建于 1958 年,在国家计划中单列的我国唯一的核武器研制、生产基地。是以发展国防尖端科学技术为主的集理论、实验、设计、生产为一体的综合性研究院。最早位于北京;随后主要工程和生产部门迁入青海(现青海海燕);1970 年,主要工程和生产部门迁往四川,分布在四川北部的山区中。其科研基地主体坐落在四川省绵阳市,占地 4 000 多亩,建筑面积 150 多万平方米,是一座设施齐全、文明美丽的现代化科学城。在北京、上海、深圳、成都等地设有科研分支机构或办事机构。

China Classification Society (CCS) 中国船级社 (1)中国船级社是由中国有关法律授权的、经法律登记注册的、从事船舶入级服务与法定服务等的专业技术机构/组织;(2)中国船级社主要承担国内外船舶、海上设施、集装箱及其相关工业产品的入级服务、鉴证检验、公证检验和经中国政府、外国(地区)政府主管机关授权,执行法定服务等具体业务,以及经有关主管机关核准的其他业务。(1) China Classification Society (CCS) is a specialized technical organization, authorized under the relevant laws of China and registered in accordance with the laws for providing classification services and statutory services for ships and offshore installations. (2) It mainly undertakes classification services, certification surveys and surveys relating to notarial matters for ships, offshore installations, containers and the related industrial products both at home and abroad, and performs specific services such as statutory services, etc., on behalf of the Chinese Government and the governments of foreign countries or regions when so authorized, and other service approved by the relevant administrations.

China Council for the Promotion of International Trade 中国国际贸易促进委员 成立于 1952 年 5 月,是由中国经济贸易界有代表性的人士、企业和团体组成的全国民间对外经贸组织。负责指导、协调中国贸促会各地方分会、行业分会、支会和各级国际商会的工作;负责对各分支机构及会员的服务及培训工作等。1986 年，中国国际贸易促进委员会以国家委员会名义申请加入国际商会，经过 8 年多的谈判和努力，1994 年 11 月，国际商会第 168 次理事会正式通过决议，同意中国加入国际商会并组建国际商会中国国家委员会。

China Foreign Economic Trade Arbitration Commission 中国对外经济贸易仲裁委员会

China International Capital Corporation Limited (CICC) 中国国际金融有限公司 简称中金公司，成立于 1995 年，是由国内外著名金融机构和公司基于战略合作关系共同投资组建的中国第一家中外合资投资银行，注册资本为 1.25 亿美元。

China International Marine Containers (CIMC) 中国国际海运集装箱(集团)股份有限公司 简称中集集团，创立于 1980 年 1 月，最初由香港招商局和丹麦宝隆洋行合资组建，是中国最早的集装箱专业生产厂商和最早的中外合资企业之一。中集集团于 1982 年 9 月 22 日正式投产，1987 年改组为中远、招商局、宝隆洋行的三方合资企业，1993 年改组为公众股份公司，1994 年在深圳证券交易所上市，1995 年起以集团架构开始运作。集团致力于为现代化交通运输提供装备和服务，主要经营集装箱、道路运输车辆、罐式储运设备、机场设备制造和销售服务。截至 2012 年 6 月，中集集团总资产 652.32 亿元，净资产 207.43 亿元，流通市值 127 亿，2011 年销售额 641.25 亿元，净利润 36.91 亿元。在中国以及北美、欧洲、亚洲、澳洲等国家和地区拥有 150 余家全资及控股子公司，员工超过 6.4 万人，初步形成跨国公司运营格局。

China Maritime Arbitration Commission (CMAC) 中国海事仲裁委员会 原名中国国际贸易促进委员会海事仲裁委员会，是仲裁机构，它以仲裁的方式，独立、公正地解决海事、海商、物流以及其他契约性争议，以保护当事人的合法权益，促进国际国内经济贸易和物流的发展。仲裁委员会受理下列争议案件：(1) 租船合同、多式联运合同或者提单、运单等运输单证所涉及的海上货物运输、水上货物运输、旅客运输争议。(2) 船舶、其他海上移动式装置的买卖、建造、修理、租赁、融资、拖带、碰撞、救助、打捞，或集装箱的买卖、建造、租赁、融资等业务所发生的争议；(3) 海上保险，共同海损及船舶保赔业务所发生的争议；(4) 船上物料及燃料供应、担保争议、船舶代理、船员劳务、港口作业所发生的争议；(5) 海洋资源开发利用、海洋环境污染所发生的争议；(6) 货运代理、无船承运、公路、铁路、航空运输、集装箱的运输、拼箱和拆箱、快递、仓储、加工、配送、仓储分拨、物流信息管理、运输工具、搬运装卸工具、仓储设施、物流中心、配送中心的建造、买卖或租赁、物流方案设计与咨询，与物流有关的保险，与物流有关的侵权争议，以及其他与物流有关的争议；(7) 渔业生产、捕捞等所发生的争议；(8) 双方当事人协议仲裁的其他争议。

China Merchants Group Ltd 招商局集团有限公司 创立于 1872 年 12 月 26 日，1873 年 1 月 17 日在上海正式开业，是中国民族工商业的先驱，被誉为"中国民族企业百年历程缩影"。招商局集团现为国家驻港大型企业集团，香港四大中资企业之一，总部设于香港。主要经营活动分布于香港、内地、东南亚等地区。2015 年 12 月 29 日，中国外运长航集团整体并入招商局集团，成为其全资子企业。

China Mobile Communication Corporation (CMCC) 中国移动通信集团公司 简称中国移动通信或中国移动，是 2000 年 4 月成立的中国国有重要骨干企业，注册资本为 518 亿人民币，截至 2008 年 9 月 30 日，资产规模超过 8 000 亿人民币。

China Network Corporation Ltd 中国网络通信有限公司

China Shipbuilding Standard Contact 中国船舶建造标准合同 目前我国造船业通用的造船合同，是原中国船舶工业总公司时期制定的，业内称为 CSTC 版本。该合同在内容上吸取了欧洲、日本、韩国、美国造船合同的长处，并结合我国造船业的实际情况，比较合理地兼顾了船东和造船厂的相关权益，至今仍然受到了船东和船厂的广泛认可。实践证明《中国船舶建造标准合同》仍是合理且适用的造船合同，体现了造船厂和船东双方的利益。但在目前国际社会日益提高环保标准以及船东不断提出节能降耗的要求的双重背景下，该版本合同尚未能完全适应形势发展的需要，再加上新标准造船合同(NEWBUILDCON)等版本合同推介步伐的加快，因此，有必要对其进行修改及完善，以期更好地保障双方当事方的合法权益，降低海事纠纷发生概率，同时又与时俱进，满足来自造船厂及船东的相关要求。

China Unicom 中国联合通信有限公司 简称中国联通，是经国务院批准于 1994 年 7 月 19 日成立的我国唯一一家综合性电信运营企业，经营范围包括移动通信(GSM 和 CDMA)、电信增值、国内国际长途电话、批准范围内的本地电话、数据通信及互联网、IP 电话等业务。中国联通在全国 31 个省、自治区、直辖市设立了分支机构，是中央直接管理的国有重要骨干企业。

China waters 中国水域 系指中华人民共和国领海、沿海港口、内水以及中国政府管辖的一切水域。

China-ASEAN Free Trade Area (CAFTA) 中国-东盟自贸区 是中国和东盟双方各自与外部建设的第一个自贸区，目标是构建一个拥有 18 亿消费者的庞大统一市场。该自贸区从中国提议到双方《货物贸易协议》

的达成,仅用了短短4年时间,这充分反映了中国与东盟增进合作、共同发展的迫切愿望。CAFTA 建设,具有十分积极的现实意义和深远的历史意义。一方面,我们已经看到,双方的贸易、投资等合作呈现了良好的发展势头;另一方面,我们也应意识到,CAFTA 建设是一项庞大的系统工程。《中国—东盟全面经济合作框架协议货物贸易协议》实施后,必将出现许多新情况、新变化、新问题,这就特别需要双方积极探讨和应对。中国和东南亚各国资源禀赋各具优势,产业结构各有特点,互补性强,合作潜力巨大。近些年来在 CAFTA 这一区域,中国与东盟国家本着平等互利的方针,在多领域的交流与合作正呈活跃之势,逐渐形成了宽领域、多层次、广支点、官民并举的良好的区域合作新局面。中国与东盟在半年前也就是第八次中国-东盟领导人会议上,已签署了《落实中国与东盟面向和平与繁荣的战略伙伴关系联合宣言的行动计划》,作为 2005~2010 年间全面深化和拓展双方关系和互利合作的"总体计划",这一总体计划实施必将极大推进 CAFTA 建设。随着 CAFTA 建设的全面实施,中国与东盟及其成员国必须也能够加大合作力度,丰富合作内涵,拓宽合作领域,创新合作方式。中国与东盟在 CAFTA 建设的同时,合力推动整个东亚地区的经济一体化,则符合我们共同的经济利益和经济增长要求。中国与东盟及其成员国的合作,已成为一个古老的亚洲正在走向复兴、一个崭新的亚洲正在蓬勃崛起的重要力量。

China's Shipbuilding Industry Systems Engineering Institute/Ocean Electronic Science and Technology Ltd. 中国船舶工业系统工程研究院/海洋电子科技有限公司——海洋防务装备的"推动者" 中国船舶工业系统工程研究院和海洋电子科技有限公司隶属于中国船舶工业集团有限公司,是面向海军装备体系,以系统集成为主要业务领域,覆盖"海军装备体系研究和顶层规划、系统综合集成、系统核心设备研制"三个层次,涵盖海军综合电子综合信息系统、舰艇作战系统、电子武器系统、舰载航空系统、舰船平台系统等五大领域的骨干军工单位。中国船舶工业系统工程研究院完成了 40 余型舰艇的千余套系统和 1500 多台套设备的交付,覆盖了专项工程一、二、三期多项任务;共获得科技进步奖 339 项,其中国家级奖 20 项、省部级奖 204 项,曾被中共中央、国务院、中央军委授予"高技术武器装备发展建设工程重大贡献奖"。2013 年,中国船舶工业系统工程研究院紧紧抓住国家建设海洋强国的战略新机遇,以"智慧海洋"为旗帜,以成立中船电子科技有限公司为契机,切实贯彻实施"做强军工、做大产业"的双轮驱动战略目标。中船电科作为产业平台,利用市场机制探索军民两用技术和民品市场,致力发展全球航海运营服务、海洋卫星应用、船舶自动化、航海导航、民用水声电子、舰船电力推进、通用航空、装备健康管理、船舶信息化等系统解决方案和核心产品以及计算机系统集成与服务、基础软件和应用技术开发与技术服务等领域。

Chinese Academy of Social Sciences 中国社会科学院 是中共中央直接领导、国务院直属的国家哲学社会科学研究的最高学术机构和综合研究中心,正部级事业单位,其前身是成立于 1955 年的中国科学院哲学社会科学部。时任中国科学院院长的郭沫若兼任哲学社会科学部主任。1977 年 5 月 7 日,经中央批准,在中国科学院哲学社会科学部的基础上,正式成立中国社会科学院。党中央对中国社会科学院提出的三大定位是:马克思主义的坚强阵地、我国哲学社会科学研究的最高殿堂、党中央、国务院重要的思想库和智囊团。中国社会科学院拥有文学哲学部、社会政法学部、历史学部、经济学部、国际研究学部、马克思主义研究学部等 6 大学部,40 个研究院所,10 个职能部门,包括中国社会科学院研究生院在内的 8 个直属机构,2 个直属公司,180 余个非实体研究中心,主管全国性学术社团 105 个,并代管中国地方志指导小组办公室。全院有二、三级学科近 300 个,其中国家重点学科 120 个。全院在职总人数 4 200 余人,科研业务人员 3 200 余人,其中高级专业人员 1 676 名,学部委员 61 人,荣誉学部委员 133 人。研究生院有在校生 3 100 余人。在美国宾夕法尼亚大学发布的《2014 全球智库报告》中,中国社会科学院以第 20 名的成绩跻身"全球智库 50 强",并蝉联"亚洲最高智库"。2015 年 1 月,中办、国办印发《关于加强中国特色新型智库建设的意见》指出,"发挥中国社会科学院作为国家级综合性高端智库的优势,使其成为具有国际影响力的世界知名智库"。

Chip 芯片 又称微电路(Microcircuit)、微芯片(Microchip)、集成电路(Integrated circuit/IC)。在电子学中是一种把电路(主要包括半导体设备,也包括被动组件等)小型化的方式,并通常制造在半导体晶圆表面上。前述将电路制造在半导体芯片表面上的集成电路又称薄膜(Thin-film)集成电路。另有一种厚膜(Thick-film)混成集成电路(Hybrid integrated circuit)是由独立半导体设备和被动组件,集成到衬底或线路板所构成的小型化电路。

Chlorination technology 氯化技术

Chlorinity titration box 氯度滴定箱 系指专供海水滴定用的仪器箱。它包括整套滴定仪器和电磁搅拌器、照明灯具及指示剂等。

Chock 导缆钳 通常装在舷边设有栏杆的区域和舷墙顶面上的钳状导缆器。有开式、闭式、直式、斜式以

Chocks fairlead rollers 导缆滚轮 系指通常装在舷边设有栏杆的区域和舷墙顶面上的钳状导缆器中的滚轮。

Choking 阻塞 亚音速压气机内部通道中某一截面上的相对流动达到声速后空气或其他气体流量为最大，不能再有增加的现象。

Choking limit (choking line) 阻塞边界 由压气机特性曲线上所示的在各种转速下各阻塞开始点连接而成的边界线。

Chopped rubber and plastic insulation 橡胶和塑料绝缘碎料 塑料和橡胶绝缘材料，清洁，不含其他物质，呈颗粒状。无特殊危害。该货物为不燃物或失火风险低。Chopped rubber and plastic insulation is a plastic and rubber insulation material, and it is clean and free from other materials, it is in granular form. It has no special hazards. This cargo is non-combustible or has a low fire-risk.

Chopper 斩波器 系指一种把恒定直流电压变换成负载所需的可调直流电压的装置。它通过周期性的快速通断可关断器件，而把恒定直流电压斩成一系列的脉冲电压，通过控制这一下脉冲的占空比来实现输出电压平均值的调节。它可以调阻、调磁和调压，也能使电动机实现再生制动，把电流反馈电源。

Chord length 弦长 翼形体平行于其运动方向的尺度或螺旋桨叶切面沿其面节线的长度。

Christmas tree 信号灯杆 安装在甲板或桅上，带有数根横杆，供装设通信信号灯用的竖杆。

Christmas tree/Xmas tree 采油树 是安装在井口（包括井口的延伸装置，如海洋石油中的采油立管）的油气井必备装置，是一种用于控制生产，并为钢丝、电缆、连续油管等修井作业提供条件的装置。由于采油树结构形状酷似圣诞树，故其英文名称为 Christmas tree，或 Xmas tree。

Chrome pellets 铬丸 丸粒。含水量最多2%。无特殊危害。该货物为不燃物或失火风险低。Chrome pellets are pellets. Moisture: up to 2% maximum. It has no special hazards. This cargo is non-combustible or has a low fire-risk.

Chromite ore 铬铁矿石 精矿或块状，呈深灰色。吸入粉尘后有毒害。该货物为不燃物或失火风险低。Chromite ore is concentrates or lumpy. It is dark grey. It is toxic by dust inhalation. This cargo is non-combustible or has a low fire-risk.

Chronic aquatic toxicity 慢性水毒性

Chronic release 慢性释放 以排出、排放或曝露形式连续或持续释放。

CIMC Raffies offshore Co. Ltd 烟台中集来福士海洋工程有限公司 其前身是1977年建成的烟台造船厂。1994年，烟台造船厂与新加坡造船私人有限公司合资成立了烟台普泰造船有限公司，该合资公司于1997年正式更名为烟台莱佛士船业有限公司。1996年烟台普泰造船有限公司与胜利油田实业集团公司、新加坡泰山烟台造船私人有限公司投资成立了烟台泰山造船有限公司；2001年烟台泰山造船有限公司更名为烟台来福士海洋工程有限公司；2010年烟台来福士海洋工程有限公司更名为烟台中集来福士海洋工程有限公司。经过十多年的发展，烟台中集来福士现已成为国际领先的船舶及海工建造企业。烟台中集来福士致力于为客户提供技术领先、安全高质的产品，业务范围涵盖海工及船舶的建造、维修、改造等。烟台中集来福士是国内最大的半潜式钻井平台制造基地之一，也是目前国内唯一一个拥有自升平台、半潜平台、海工特种船舶系列产品线的海工企业。

Circle net 围网 网片垂直置于水中，并可通过收绞穿于底环中的括网，使网具下端封闭，以围捕中、上层鱼类为主的长带形网具。

Circle-hyperbolic system 圆-双曲线系统 利用圆和双曲线位置线交点来定位的系统。该系统的优点是适当配置圆心台的位置可以在主要工作区得到较好的位置线交角，并且几何精确度高。

Circuit-breaker (mechanical) （机械式）断路器 系指能接通、承载以及分断正常电路条件下的电流，也能在规定的非正常电路（例如短路）下接通、承载一定时间和分断电流的一种机械开关电器。

Circuit-breaker of Category A A类断路器 系指在短路状态下相对于其他串联在负载侧的短路电流保护装置并不特别指定选择性保护的断路器，即在短路状态下不带短延时选择性保护的断路器。Circuit-breaker of Category A means a circuit-breaker not specifically intended for selectivity under short-circuit conditions with respect to other short-circuit protective devices in series on the load side, i. e. without an intentional short-time delay provided for selectivity under short-circuit conditions.

Circuit-breaker of category B B类断路器 系指在短路状态下相对于其他串联在负载侧的短路电流保护装置为特别指定选择性保护的断路器，即在短路状态下带有短延时选择性保护的断路器。Circuit-breaker of category B means a circuit-breaker specifically intended for selectivity under short-circuit conditions with respect to other short-circuit protective devices in series on the load side, i. e. with an intentional short-time delay (which may be adjustable) provided for selectivity under short-circuit condi-

Circular error probability 概率误差圆 目前较常用的一种定位误差表示法。以平均位置为中心包含50%定位点的圆为概率误差圆。

Circular frequency of encounter 遭遇圆频率 船舶在波浪中运动时，其遭遇周期用圆周运动表达时的角速度。$\omega_E = 2\pi/T_E$。

Circulars 通函 是推荐性的文件，也是一种统一解释。船舶设计、建造中主要采用海上安全委员会（Maritime Safety Committee，MSC）和海上环境保护委员会（Maritime Environment Committee，MEPC）的通函。

Circulating current 环流 在流场中沿封闭曲线，其线积分不等于零的速度。

Circulating lubricating oil tank 循环滑油舱 供贮存主机所需循环滑油用的舱柜。

Circulating water channel 循环水槽 水能作循环流动的船模试验设施。

Circulating water ejector 水循环喷射器 利用蒸汽为工质引射并加热在蒸发器内再循环的加热水的设备。

Circulating water ratio 循环水倍率 再循环闪发式蒸馏装置内盐水循环量与蒸馏水产量的比值。

Circulating water system 循环水系统 供应冷凝器、冷却器及轴系等冷却水的机械设备、管路和附件。

Circulation 环量 在流场中沿一封闭曲线积分 $\Gamma = \int V d_s$，式中，d_s—曲线的一小段；V—沿此段曲线的速度分量。

Circulation ratio 循环倍率 锅炉或锅炉某一水循环回路中循环水量和生成的蒸汽量的质量比。

Circulation ratio (cooling water ratio) 冷却倍率 流经冷凝器的冷却水质量与被冷却的蒸汽质量的比值。

Circulation theory of screw propeller (vortex theory) 螺旋桨环流理论 将螺旋桨叶看作是一扭曲的翼，应用流体动力学的环流理论来分析螺旋桨推进作用的理论。

Circumferential inflow factor 周向进流因数 螺旋桨叶元体处周向感生速度对其周向线速度的比值。

Circumferential velocity 周向速度 叶元体环绕轴线转动的线速度。周向速度 $= 2\pi n r$，式中，r—叶切面所在的半径；n—螺旋桨转速。

Circumferential velocity induced of propeller 周向感生速度 由于螺旋桨的作用，在螺旋桨叶的各半径处使水产生的与其转向相同的周向速度。

Circumferentially varying pitch 周向变螺距 描出螺旋桨叶面的母线在不同角的位置的前进率不同者。月牙切面和对称流线型切面的螺距由其面节线方向决定，不看作是周向变螺距。

Citadel 堡垒（密闭区） 系指船体和上层建筑的气密外壳。堡垒内部应设有多个相互连通的舱室，这些舱室都处于同一个气密边界中。堡垒内部还应设有必要的独立系统，以提供一个不含任何核、生、化与放射性污染的无毒区域。大型舰艇应设置子堡垒或多个堡垒。

Citadel is the gastight envelope of the hull and superstructure. It consists of a group of interconnecting compartments enclosed by a gas-tight boundary with the independent systems necessary to provide a toxic free area free from any CBRN hazard. Large ships may have subcitadels or more than one citadel.

Civil ship/merchant ship 民用船舶 各类非军用船舶的统称。

CKYH alliance CKYH联盟 由中远集运、川崎汽船、阳明海运和韩进海运4家班轮公司组成。2012年4月CKYH联盟宣布同长荣海运进行舱位互换合作，主要合作航线为亚欧航线与亚洲-地中海航线。目前，该联盟的运力规模为217.3万TEU，占世界集装箱船队总运力的12.3%。

Clad or cladding structure 覆盖层或覆盖结构 系指芯材和新的钢顶板按照有关要求覆盖到原有钢板上的结构。

Clad steel 复合钢

Claim indemnity 索赔

Claim letter 索赔书

Claim/Claim for damages 索赔 系指投保人或被保险人在发生保险事故、遭受财产损失或人身伤亡以后，要求保险人履行赔偿或给付保险金义务的行为。保险索赔是被保险人获得实际的保险保障和实现其保险权益的具体体现。需要注意的是，索赔作为被保险人一项权利是有时效限制的，保险种类不同，其时效也有所不同。根据《保险法》规定，人寿保险的索赔时效为5年，除人寿保险以外的其他保险索赔时效为2年。

Claimant 请求人 系指提出存在对其有利的海事请求的人。 Claimant means a person who alleges that a maritime claim exists in his favour.

Claimant/Plaintiff 仲裁申诉人

Claims settlement 理赔

Claims statement 索赔清单

Clamp coupling 夹壳联轴器 沿轴向剖分为两半，用螺栓将其连接在两轴的端部，依靠夹紧产生的摩擦力传递推力与扭矩的刚性联轴器。

Clamshell grab 双腭抓斗 由两瓣斗体组成，形如蚌壳的抓斗。

Clarified oil 澄清油

Clarifier 分杂机构 分离油中含有的大量机械杂质和微量水分的机构。

Clark sampler for dissolved organic constituents 克拉克溶解有机物采水器 克拉克设计的专门用于采集供有机物分析的水样,容量为 15 L 的采水器。

Clash and breakage risk 碰损、破碎险 投保平安险和水渍险的基础上加保此险,保险人负责赔偿承保的金属、木材等货物因震动、颠簸、碰撞、挤压而造成货物本身的损失,或易碎性货物在运输途中由于装卸野蛮、粗鲁,运输工具的颠震所造成货物本身的破裂、断碎的损失。

Clashing 碰损 主要系指金属及其制品在运输途中因受震动、受挤压而造成变形等损失。

Class 1 INF ship INF 1 级船舶 系指所装 INF 货物,验证其总放射性强度小于 4 000 TBq 的船舶。Class 1 INF ship means ships which are certified to carry INF cargo with an aggregate activity less than 4 000 TBq.

Class 1 pressure vessel (PV-1) 1 级压力容器 (PV-1) 系指:(1)设计压力超过 0.35 MPa 蒸汽发生器。(2)贮存温度 38 ℃ 时具有蒸汽压力不小于 0.2 MPa 的易燃高压气体的容器。但当压力容器的容量为 0.5 m³ 或以下时,其材料、结构和焊接应满足 2 级压力容器的规定。(3)壳板厚度超过 38 mm 或设计压力超过 4 MPa 的各种压力容器,以及最高工作温度超过 350 ℃ 的压力容器。但即使壳板厚度超过 38 mm 或设计压力超过 4 MPa,只要其在大气温度下仅承受液压或水的压力,则归属 PV-2 级。(4)贮存氨或其他毒性气体等的压力容器。Class 1 pressure vessel (PV-1) means:(1) steam generators whose design pressure exceeds 0.35 MPa;(2) Pressure vessels in which inflammable high pressure gas having the vapour pressure not less than 0.2 MPa at 38 ℃ is contained. However, the requirements for "PV-2" may be applied to the pressure vessels with the capacity of 0.5 m³ or under with respect to their materials, construction and welding. (3) Pressure vessels whose shell plates exceed 38 mm in thickness, and/or whose design pressures exceed 4 MPa, and/or whose maximum working temperatures exceed 350 ℃. However, the pressure vessels in which the shell plates exceed 38 mm in thickness and/or the design pressure exceed 4 MPa are classified as "PV-2", provided that they are subject to hydraulic pressure or water pressure at the atmospheric temperature. (4) Pressure vessels contain ammonia or toxic gases.

Class 2 INF ship INF 2 级船舶 系指验证其总放射性强度小于 2×10^6 TBq 的辐射性核燃料和强放射性废料的船舶和验证其装运的钚其总放射性强度小于 2×10^6 TBq 的船舶。 Class 2 INF ship means ships which are certified to carry irradiated nuclear fuel or high-level radioactive wastes with an aggregate activity less than 2×10^6 TBq and ships which are certified to carry plutonium with an aggregate activity less than 2×10^6 TBq.

Class 2 pressure vessel (PV-2) 2 级压力容器 (PV-2) 系指:(1)设计压力不超过 0.35 MPa 的蒸汽发生器;(2)壳板厚度超过 16 mm 或设计压力超过 1 MPa,或最高工作温度超过 150 ℃ 的压力容器。 Class 2 pressure vessel (PV-2) means:(1) steam generators whose design pressure do not exceed 0.35 MPa. ;(2) pressure vessels whose shell plate exceed 16 mm in thickness, and/or whose design pressure exceed 1 MPa, and/or whose maximum working temperature exceed 150 ℃.

Class 3 INF ship INF 3 级船舶 系指验证其装运的辐射性核燃料和强放射性废料的船舶和验证其装运的钚最大总放射性强度不受限制的船舶。 Class 3 INF ship means ships are certified to carry irradiated nuclear fuel or high-level radioactive wastes and ships which are certified to carry plutonium with no restriction of the maximum aggregate activity of the materials.

Class 3 pressure vessel (PV-3) 3 级压力容器 (PV-3) 系指 1,2 级以外的压力容器。 Class 3 pressure vessel (PV-3) means vessels are not included in Class 1 and 2.

Class 4.1 flammable solids 第 4.1 类易燃固体 系指具有易被火花和火焰等外部火源点燃、易于燃烧、受摩擦时易引起燃烧或会助燃等特性的物质。 Class 4.1 flammable solids means these materials have the properties of being easily ignited by external source such as sparks and flames and of being readily combustible or of being liable to cause or contribute to fire through friction.

Class 4.2 substances liable to spontaneous combustion 第 4.2 类易自燃物质 系指具有易自热并自燃的共同特性的物质。 Class 4.2 substances liable to spontaneous combustion means these materials have the common property of being liable to heat spontaneously and to ignite.

Class 4.3 substances which, in contact water emit flammable gases 第 4.3 类 遇水产生可燃气体的物质 系指具有遇水产生可燃气体的共同特性。在某些情况下,这些气体是易于自燃的物质。 Class 4.3 substances which, in contact water emit flammable gases means these materials have the common property, when in contact with water, of evolving flammable gases. In some cases these gases are liable to spontaneous ignition.

Class 5.1 oxidizing substances (agents) 第 5.1 类具有氧化性的物质(氧化剂) 系指尽管本类物质本身不一定可燃,但与其他物质接触时,其产生的氧气或发生的类似反应会增加燃烧的危险和烈度的物质。 Class 5.1 oxidizing substances (agents) means these materials, although by the are not necessarily combustible by they may, either by yielding oxygen or by similar processes, increase the risk and intensity of fire in other materials with which they come into contact.

Class 6.1 toxic substances 第 6.1 类有毒物质 系指如被吞咽、被吸入或与皮肤接触,则易于造成死亡或产生严重损伤或危害人的健康的物质。 Class 6.1 toxic substances means these materials are liable either to cause death or serious injury or to harm human health if swallowed or inhaled, or by skin contact.

Class 6.2 infectious substances 第 6.2 类 感染性物质 系指含有能引启动物或人体发病的活体微生物或毒素的物质。 Class 6.2 infectious substances means these materials contained viable micro-organisms or their toxins which are known or suspected to cause disease in animals or humans.

Class 7 radioactive materials 第 7 类放射性物质 系指能释放出大量射线,其放射性比度大于 70 kBq/kg(0.002 μCi/g)的物质。 Class 7 radioactive materials mean these materials spontaneously emit a significant radiation. Their specific activity is greater than 70 kBq/kg (0.002 μCi/g).

Class 8 corrosives 第 8 类腐蚀性物质 系指具有在原来形态下在某种程度上严重损伤活体组织的共同特性的物质。 Class 8 corrosives mean these materials had their original state the common property of being able more or less severely to damage living tissue.

Class 9 miscellaneous dangerous substances and articles 第 9 类其他危险物质和物品 系指具有第 1 类~第 8 类以外危险的物质。 Class 9 miscellaneous dangerous substances and articles mean these materials present a hazard not covered by the Class 1 to Class 8.

Class A fire divisions A 级防火分隔 系指那些由舱壁和甲板组成的分舱,它们应符合国际海事组织(IMO)海上安全委员会(MSC)第 61(67)号决议"国际耐火试验程序规则"附录 1 第 3 篇的规定。 Class A fire divisions are those divisions formed by bulkheads and decks which comply with the requirements of IMO resolution MSC 61 (67) Fire test procedures code, Annex 1, Part 3.

Class A-0 divisions A-0 级防火分隔 系指在 0 min 内满足该隔壁或甲板包有船级社认可的隔热材料,可使非向火一侧的平均温度升高在规定的时间内不大于 140 ℃,且包括隔热材料节点在内的任何一点的温升也不大于 180 ℃ 的 A 级防火分隔要求的隔壁或甲板。

Class A-15 divisions A-15 级防火分隔 系指在 15 min 内满足该隔壁或甲板包有船级社认可的隔热材料,可使非向火一侧的平均温度升高在规定的时间内不大于 140 ℃,且包括隔热材料节点在内的任何一点的温升也不大于 180 ℃ 的 A 级防火分隔要求的隔壁或甲板。

Class A-30 divisions A-30 级防火分隔 系指在 30 min 内满足该隔壁或甲板包有船级社认可的隔热材料,可使非向火一侧的平均温度升高在规定的时间内不大于 140 ℃,且包括隔热材料节点在内的任何一点的温升也不大于 180 ℃ 的 A 级防火分隔要求的隔壁或甲板。

Class A-60 divisions A-60 级防火分隔 系指在 60 min 内满足该隔壁或甲板包有船级社认可的隔热材料,可使非向火一侧的平均温度升高在规定的时间内不大于 140 ℃,且包括隔热材料节点在内的任何一点的温升也不大于 180 ℃ 的 A 级防火分隔要求的隔壁或甲板。

Class B-0 divisions B-0 级防火分隔 系指在 0 min 内满足该隔壁或甲板包有船级社认可的隔热材料,可使非向火一侧的平均温度升高在规定的时间内不大于 140 ℃,且包括隔热材料节点在内的任何一点的温升也不大于 225 ℃ 的 B 级防火分隔要求的隔壁、甲板、天花板及衬垫。

Class B-15 divisions B-15 级防火分隔 系指在 15 min 内满足该隔壁或甲板包有船级社认可的隔热材料,可使非向火一侧的平均温度升高在规定的时间内不大于 140 ℃,且包括隔热材料节点在内的任何一点的温升也不大于 225 ℃ 的 B 级防火分隔要求的隔壁、甲板、天花板及衬垫。。

Class maintenance survey 维持船级的检验

Class of dynamic positioning equipment 动力定位设备等级 系指衡量动力定位能力的可靠性程度。每个设备等级对应的最恶劣故障模式如下:(1)设备等级 1——单故障时定位能力丧失。(2)设备等级 2——单故障时定位能力不丧失。(3)设备等级 3——单故障时定位能力不丧失,且单个故障包括任一水密舱室失火或进水造成的全部设备损失。所以对于设备等级 2 和设备等级 3,要求动力定位系统具备冗余性。冗余性对设备等级 2 系指所有的活动部件,对设备等级 3 冗余性扩展到所有的部件,并且要求隔舱布置。

Class survey 入级检验 系指船舶所有人自愿接受的,由船级社进行的对其所拥有的入级船舶的检验。船级社根据船舶的用途、技术状态和航行区域按照船级社的规则和规范对船舶进行检验。船级社对不同的船舶都授予规定的主船级和附加船级符号,并将新入级的船舶登载在本船级社定期出版的"船名录"中向全世界公布,以表明此船舶经船级社检验,其性能和安全性符合船级社规范的要求,并由船级社负责向该船核发船级证书。只有具有船级证书的船舶,才能取得保险,投入运营。入级检验分为:(1)入级检验;(2)维持船级的检验;(3)鉴定检验。其中入级检验又分为:新建船入级检验和船舶转级检验;维持船级的检验又分为:年度检验、中间检验、特别检验、临时检验、坞内检验、水下检验、螺旋桨轴和艉轴管检验、锅炉和热油加热器检验和起货设备检验等。鉴定检验是船级社开展的一种技术性服务。常见的鉴定检验有:保险检验;船舶状况检验;损伤检验;载重量鉴定等。船级检验发放的证书有:(1)船体船级证书;(2)轮机船级证书。

Class suspense 中止船级 在船级指定项目到期或过期,到期或过期的检验的项目没有安排检验的情况下,船级就会被中止。

Classed ship 入级船舶 系指船级社根据其规范签发入级证书的船舶。 Classed ship is a ship to which a classification certificate is issued by a classification society in accordance with its rules.

Classes of piping systems 管系的等级 其定义见表 C-1。但以下系统不包括在该表中:(1)油船、液化气船和化学品船的货品的管系;(2)冷藏装置的液体管系。Classes of piping systems are defined in following table. The following systems are not covered by Table C-1:(1)cargo piping for oil tankers, gas tankers and chemical tankers,(2)fluids piping systems are for refrigerating plants.

表 C-1 管系等级
Table C-1 Class of piping systems

管系输送的介质 Media conveyed by the piping system	Ⅰ级 Class Ⅰ	Ⅱ级 Class Ⅱ①④	Ⅲ级 Class Ⅲ⑦
有毒介质 Toxic media	无特殊安全措施③ without special safeguards	不适用 not applicable	不适用 not applicable
腐蚀性介质 Corrosive media	无特殊安全措施③ without special safeguards	无特殊安全措施③ without special safeguards	不适用 not applicable
腐蚀性介质 Corrosive media	无特殊安全措施③ without special safeguards	无特殊安全措施③ without special safeguards	不适用 not applicable
易燃物:(1)加热超过闪点;(2)闪点小于 60 ℃ 和液化气 Flammable media:(1)heated above flashpoint, or(2)having flashpoint < 60 ℃ and liquefied gas	无特殊安全措施③ without special safeguards	无特殊安全措施③ without special safeguards	不适用 not applicable
氧乙炔气 Oxyacetylene	和 p 无关 irrespective of p	不适用 not applicable	不适用 not applicable
蒸汽 Steam	$p > 1,6$ 或 $T > 300$	其他 ② other	$P \leqslant 0,7$ 和 $T \leqslant 170$

续表 C-1

管系输送的介质 Media conveyed by the piping system	Ⅰ级　Class Ⅰ	Ⅱ级　Class Ⅱ ①④	Ⅲ级⑦　Class Ⅲ
热油　Thermal oil	$p > 1,6$ 或 $T > 300$	其他② other	$P \leqslant 0,7$ 和 $T \leqslant 150$
燃油⑧　Fuel oil 润滑油　Lubricating oil 易燃液压油 ⑤　Flammable hydraulic oil	$p > 1,6$ 或 $T > 150$	其他② other	$P \leqslant 0,7$ 和 $T \leqslant 60$
其他介质⑤⑥　Other media	$p > 4$ 或 $T > 300$	其他② other	$P \leqslant 1,6$ 和 $T \leqslant 200$

注意：①Valves under static pressure on oil fuel tanks or lubricating oil tanks belong to Class Ⅱ. 在燃油舱中承受静压力的阀属于Ⅱ级。
②Pressure and temperature conditions other than those required for Class Ⅰ and Class Ⅲ. 不在Ⅰ、Ⅲ级中规定的压力、温度。
③Safeguards for reducing leakage possibility and limiting its consequences: e. g. pipes led in positions where leakage of internal fluids will not cause a potential hazard or damage to surrounding areas which may include the use of pipe ducts, shielding, screening etc. 降低泄漏可能性和限制其的安全措施，例如通到内部液体泄漏部位的管子不会对周围区域引起潜在危险或损害，这包括使用导管、护套、屏壁等。
④Valves and fittings fitted on the ship side and collision bulkhead belong to Class Ⅱ. 安装于船舷和防撞舱壁的阀和附件属于Ⅱ级。
⑤Steering gear hydraulic piping system belongs to Class Ⅰ irrespective of p and T. 操舵液压管系属于Ⅰ级，与压力或温度无关。
⑥Including water, air, gases, non-flammable hydraulic oil. 包括水、空气、气体和非易燃液压油。
⑦The open ended pipes, irrespecitve of T, generally belong to Class Ⅲ (as drains, overflows, vents, exhaust gas lines, boiler escape pipes, etc). 端部敞开的管子与T无关，一般属于Ⅲ级(如泄漏、溢流、透气、排气、锅炉逸气管等)。
⑧Design pressure for fuel oil systems is to be determined in accordance with relevant requirement. 燃油系统的设计压力是根据有关要求确定的。
Note 1: p—— Design pressure, in MPa. 注意1：P—设计压力，单位 MPa。
Note 2: T—— Design temperature, in ℃. 注意2：T—设计温度，单位 ℃。
Note 3: Flammable media generally include the flammable liquids as oil fuel, lubricating oil, thermal oil and flammable hydraulic oil.
注意3：易燃介质一般包括易燃液体，如燃油、润滑油热油和易燃液压油。

Classic spar　经典柱体式平台/传统型立柱式平台
Classics bit (Cbit)　经典比特
Classification[1]　船级　系指表示船舶技术状态的一种指标。在国际航运界，凡注册总吨 100 GT 以上的海运船舶，必须在某船级社或船舶检验机构监督之下进行监造。在船舶开始建造之前，船舶各部分的规格须经船级社或船舶检验机构批准。每艘船建造完毕，由船级社或船舶检验局对船体、船上机器设备、吃水标志等项目和性能进行鉴定，发给船级证书。证书有效期一般为 4 年，期满后须予以重新鉴定。船舶入级可保证船舶航行安全，有利于国家对船舶进行技术监督，便于租船人和托运人选择适当的船舶，以满足进出口货物运输的需要，便于保险公司决定船、货的保险费用。

Classification[2]　入级　这是船级社按照选定船级社已生效的规范和规则对设计和建造进行监督的一种形式。入级过程可包括对按有关入级规范提交的技术图纸进行审查和批准，对制造、加工、组装或部件的安装或完工的项目按照已批准的图纸和资料进行物理性确认，并按照规范要求进行后续的试验和颁发相应的文件以证明其与该船级社要求的符合度。　Classification is a form of design and construction oversight carried out by a classification society in accordance with the published rules and guidelines of the selected classification society. The classification process may consist of the review and approval of technical submissions in accordance with relevant class rules, physical confirmation of manufacture, fabrication, assembly or installation of components or finished item in accordance with approved drawings and dates, and subsequent testing as required by the rules and associated issuance of documents attesting to the degree of compliance with the requirements of the classifi-

Classification activities 入级活动 系指：(1)研究和制订规范及指导性文件等；(2)对图纸、计算书、说明书和其他技术文件进行审查，以确认满足规范有关要求；(3)对有关制造厂商和服务厂商进行认可或认证；(4)对与船舶或平台级有关的项目进行检验和试验，以确认其材料、尺度、构造和布置与批准的图纸、计算书、说明书和其他技术文件相符，且工艺和安装等在各方面都令人满意，满足船舶或平台预定用途；(5)向船级委员会建议授予船舶或平台入级符号和附加标志，签发有关证书和必要的文件；(6)对授予船舶或平台入级后的营运船舶或平台进行保持船舶或平台级的检验，确认船舶或平台保持良好技术状况；(7)承办鉴证检验和公证检验；(8)出版船舶录和产品录。 Classification activities mean: (1) To research and formulate rules and guidance notes; (2) To review plans, calculations, specifications and other technical documents so as to ascertain that they meet the relevant requirements of the Rules; (3) To carry out approval or certification of relevant manufacturers or service suppliers; (4) To carry out surveys and tests for the items related to ship/unit's class so as to ascertain that the ship/unit's material, scantlings, construction and arrangements are in compliance with the approved plans, calculations, specifications and other technical documents, and the workmanship and installation are satisfactory in all aspects to ensure that the ship/unit is fit for the service for which it is intended; (5) To recommend to the Class Committee that the ship/unit's characters of classification and class notation(s) be assigned and issue the relevant certificates and necessary documents; (6) To carry out surveys for ship/units in service for class maintenance after classes have been assigned to ship/units so as to confirm that the ship/units are in good working condition; (7) To undertake certification surveys and surveys relating to notarial matters. (8) To publish Register of Ships and Lists of Approved Marine Products.

Classification certificate 船级证书 系指船级社根据船舶入级规则的规定，对符合入级条件的船舶所签发的证书。分船体入级证书和轮机入级证书。证明入级船舶的船体、设备、轮机、电气和消防的部分均符合船级社所颁布的现行船舶制造规范的规定，或符合船级社认可的等效的技术要求。船级证书的有效期一般为4年，在此期间，应遵照船舶入级规则的规定，进行保持船级的各种检验。

Classification condition 船级条件 系指须限期处理的特定措施、修理和检验等实施要求，以保持船级。

Classification of computer system 计算机(自动控制)系统的分类 根据单一故障可能引起损害的程度，计算机(自动控制)系统可分为Ⅰ类、Ⅱ类和Ⅲ类(见表C-2)。该损害是直接由某一事件引发的，而非间接产生的损害。

表 C-2 计算机(自动控制)系统分类
Table C-2 Classification of computer system

类型	影响	系统功能	举例
Ⅰ类	这些系统的故障不会对人员的安全,船舶的安全以及环境产生危害	监视功能和日常管理功能	维修保养支持系统；日常信息处理
Ⅱ类	这些系统的故障最终会对人员的安全,船舶的安全以及环境产生危害	监视和报警功能；对保持船舶处于正常运营和起居状况所必要的控制功能	监视和报警装置；液柜容量测量系统；辅机控制系统；主推进装置遥控系统；探火和灭火系统；舱底水系统；调速器
Ⅲ类	这些系统的故障即刻会对人员的安全,船舶的安全以及环境产生危害	保持船舶推进和操舵的控制功能；安全功能	机械保护系统或设备；燃烧器控制系统；内燃机的电子喷油器；推进和操舵控制系统；发电机同步单元

Classification of ice strengthening 冰区加强的冰级 根据船体结构的加强和推力程度,冰区加强分为5个冰级,船级附加标志如下:(1)IA Super级—其结构、功率和其他功能可保证在没有破冰船支援的情况下也可在遇冰情况下正常航行的船舶;(2)IA 级—其结构、功率和其他功能在必要时,得到破冰船的支援后,可在险冰情况下正常航行的船舶;(3)IB级—其结构、功率和其他功能在要时,得到破冰船的支援后,可在中等冰情况下正常航行的船舶;(4)IC 级—其结构、功率和其他功能在必要时,得到破冰船的支援后,可在微冰情况下正常航行的船舶;(5)ID 级—具有钢质船体结构,其结构适合航行于开阔海域,且可依靠本船推进装置在微冰海域航行的船舶。 Strengthening for navigation in ice is classified into the following 5 classes dependent on the degree of reinforcement and engine output of the ship: (1) IA Super—ships with such structure, engine output and other properties that are normally capable of navigating in difficult ice conditions without the assistance of icebreakers; (2) IA—ships with such structure, engine output and other properties that are capable of navigating in difficult ice conditions, with the assistance of icebreakers when necessary; (3) IB—ships with such structure, engine output and other properties that are capable of navigating in moderate ice conditions, with the assistance of icebreakers when necessary; (4) IC—ships with such structure, engine output and other properties that are capable of navigating in light ice conditions, with the assistance of icebreakers when necessary; (5) ID—ships that have a steel hull and that are structurally fit for navigation in the open sea and that, are capable of navigating in very light ice conditions with their own propulsion machinery.

Classification of ship 船级 是船级社根据其规范或标准对某一特殊用途或营运的设施在结构上和机械性能上合适性的一种表述。

Classification of WIG craft 地效翼船分类 地效翼船可划分为以下三类:"A"型艇:系指无地面效应,不能飞行的艇。"B"型艇:系指能在地效应区外作短时且增加高度有限的,以越过某一船舶、障碍物飞行或具有其他用途的艇。此类"高飞"的最大高度应小于国际民航组织(ICAO)规定的最小安全飞行高度。"C"型艇:系指能从地面起飞,在超出 ICAO 规定的最小安全飞行高度巡航的艇。一般,动力气垫船和动力气垫地效翼船属于"A"型艇,一般的地效翼船与动力增升地效翼船属于"B"型艇,水上飞机和飞艇属于"C"型艇。 Classification of WIG craft can be subdivided into three types as follows: "A" craft is not capable of operation without the ground effect. "B" craft is capable to increase its altitude limited in time and magnitude outside influence of the ground effect in order to over fly a ship, an obstacle or for other purpose. The maximal height of such an "over flight" should be less than the minimal safe altitude of an aircraft prescribed by ICAO. "C" craft is capable to take-off from the ground and cruise at an altitude that exceeds the minimal safety altitude of an aircraft prescribed by ICAO. In general, DACC and DACWIG belong to "A" type, classic WIG and PAR-WIG belong to "B" type, and seaplanes and flying boats belong to "C" type.

Classification rules 入级规范 系指内容完整的规定,包括入级条件与范围、与之相配套的技术要求,旨在控制安全与质量达到适当水平,并得到广泛的认同。 Classification rules are such provisions that have entire content comprising conditions and scope of classification and supporting technical requirements. The aim of the rules is to ensure the safety and quality is controlled to an appropriate level, and is generally acknowledged.

Classification Societies 船级社 (1)船级社是从事船舶与海上设施入级服务的独立、公正的组织。船级社与船舶和海上设施的设计、建造、买卖、营运、管理、保养、维修、融资、保险、租赁组织之间,没有任何商业关系;(2)船级社致力于船舶与海上设施安全和环境保护,通过技术支持、符合性确认和研究开发,对海上安全和入级规范制定做出独特的贡献。船级社按其颁布的入级规范,为客户提供入级服务以及法定服务和其他服务;(3)船级社提供船舶、造船、海上开发、相关工业产品制造业、保险、金融以及其他有关业界普遍接受和认可的合理标准——入级规范,并依照此规范在船舶设计中进行审图,在建造中和建造后进行检验,以确认船舶符合入级规范的要求,并独立签发入级证书;(4)船级社接受船旗国政府的授权,按照船旗国政府的要求进行法定服务,以确认船舶满足国际公约和/或船旗国有关法规的要求,并签发法定证书。 (1) Classification societies are independent and impartial organizations that undertake classification services for ships and offshore installations. Classification societies have no commercial interests related to design, building, ownership, operation, management, maintenance or repairs, financing, insurance or chartering of ships and offshore installations; (2) Classification societies work for the safety of ships and offshore installations and environmental protection, and make a unique contribution to maritime safety and the development of classification rules through technical support, compliance verification and research and development. Classification societies provide classification services, statutory services and other services for clients in accordance with the classification rules published by them;(3) Classification societies furnish reasonable standards—the classification rules, which are generally accepted and recognized, on ships, shipbuilding, marine exploitation and related manufacturing industries as well as insurance, financing and other related sectors, carry out plan approval in ship design and surveys during and after construction so as to ascertain that ships are in compliance with the requirements of the classification rules, and issue classification certificates independently, in accordance with such rules;(4) When authorized by the Government of the flag State, classification societies carry out statutory services in accordance with the requirements of the Government of the flag State with a view to ascertaining the ships is compliance with the requirements of international conventions and/or relevant regulations of the flag State, and issue statutory certificates.

Classification survey 入级检验 系指船东自愿申请船舶接受某船级社的检验。船级社与船东、造船厂和设计单位的关系属于民间技术服务性质，船级社对不同的船舶都授予规定的主船级符号和附加船级标志，并将新入级的船舶登记在本船级社定期出版的"船舶录"中向全世界公布，以表明该船舶经船级社检验，其性能和安全性符合船级社规范要求，并由船级社负责向该船核发船级证书。只要具有船级证书的船舶，才能取得保险，投入营运。由于海上航行风险大，故船东往往要求保险商提供保险，而保险商则根据船级确定保险费用，因此可认为入级检验是为了"保险"。

Classification survey during construction 新造船入级检验 船级社从图纸审查开始到试航交船为止全过程地按船级社规范对新造船进行的检验。满足规范要求的船舶被授予相应船级符号，以区别从其他船级社转级过来的船舶。

Classification survey of ships not built under survey 船舶转级检验 船舶在运营过程中将其船舶从一个船级转到另一个船级的检验。这种转级通常在 IACS 成员船级社之间进行。

Classified ships 入级船舶 系指船级社根据其规范签发入级证书的船舶。

Clay 黏土 通常呈淡灰至深灰色，由10%软块和90%的软粒构成。该物质通常潮湿，但触摸并无湿感。含水量可达25%。无特殊危害。该货物为不燃物或失火风险低。 Clay is usually light to dark grey and comprises 10% soft lumps and 90% soft grains. The material is usually moist but not wet to the touch. Moisture is up to 25%. It has no special hazards. This cargo is non-combustible or has a low fire-risk.

Clean 洁净 系指环境保护。 Clean means environmental protection.

Clean (permanent water) ballast pump 专用清洁压载泵 油船上抽送专用清洁压载水的泵。

Clean B/L 清洁提单 系指货物在装船时"表面状况良好"，船公司在提单上未加注任何有关货物受损或包装不良等批注的提单。

Clean ballast[1] 清洁压载 系指舱中的压载在舱内最后装置的油已清洁到这种程度，即从舱中泄放出的流液在晴天从船上固定排入平静而洁净的水域后，水面上或邻近海岸线不会产生可见的油迹，或在水面下或邻近海岸线上发生油污或乳液堆积。如果为洁净压载水的泄放是通过船级社认可的油泄放监控系统，则由此系统提供的证据表明泄放的流液的含油成分不超过 15 ppm 的有效性是能确定其压载是干净的，尽管还存在可见油迹。 Clean ballast means the ballast in a tank in which oil is last carried, has been so cleaned that if the effluent are discharged from a ship into clean calm water on a clear day in the same way, it would not produce visible traces of oil on the surface of the water or on adjoining shorelines or cause a sludge or emulsion to be deposited beneath the surface of the water or upon adjoining shorelines. If the ballast is discharged through an oil discharge monitoring and control system approved by the Society, evidence based on such a system shows that the oil content of the effluent does not exceed 15 parts per million, thus, it is determinative that the ballast is clean, notwithstanding the presence of visible traces.

Clean ballast[2] 清洁压载水 （1）对"73/78 防污公约"附则Ⅰ而言，系指这样一个舱内的压载水，该舱自从上次装油后，已清洗到如此程度，以致倘若在晴天从一静态船舶将该舱中的排出物排入清洁而平静的水中，不会在水面或邻近的岸线上产生明显的痕迹，或形成油泥或乳化物沉积于水面以下或邻近的岸线上。如果压载水是通过经主管机关认可的排油监控系统排出的，而根据这一系统的测定查明该排出物的含油量不超过15ppm，则尽管有明显的痕迹，仍应确定该压载水是清洁的；(2) 对"73/78 防污公约"附则Ⅱ而言，清洁压载水系指装入一个舱内的压载水，该舱自从上次用于装载含有 X，Y 或 Z 类物质的货物以来，已予彻底清洗，所产生的残余物也已按该公约附则Ⅱ的相应要求全部排空。(1) For "MARPOL73/78" AnnexⅠ, clean ballast means the ballast in a tank in which, oil is last carried, has been so cleaned that if effluent are discharged from a ship into clean calm water on a clear day in the same way, it would not produce visible traces of oil on the surface of the water or on adjoining shorelines or cause a sludge or emulsion to be deposited beneath the surface of the water or upon adjoining shorelines. If the ballast is discharged through an oil discharge monitoring and control system approved by the Administration, evidence based on such a system shows that the oil content of the effluent does not exceed 15 parts per million, thus, it is determinative that the ballast is clean, notwithstanding the presence of visible traces; (2) For "MARPOL73/78" Annex II, clean ballast means ballast water carried in a tank which, since it is last used to carry a cargo containing s substance in category X, Y or Z, has been thoroughly cleaned and the residues resulting from there have been discharged and the tank emptied in accordance with the appropriate requirements of the Annex.

Clean ballast tank（CBT） 清洁压载水舱　系指这样的压载舱，对油船而言，即该舱自从上次装完油后，其中的废液在晴天从静态的船舶排入清洁而平静的水中时，也不会在水面或附近的海岸产生明显的痕迹、或形成油泥或乳化物沉淀积于水面下或邻近的岸线上。如果压载水是通过经认可的排油监控系统排出的，且根据该系统的检测证明排出液中的含油量不超过 15×10^{-6} 时，即使出现明显的痕迹，仍认为该压载水是清洁的。对化学品船而言，即该舱自从上次装载有毒液体物质已于彻底地清洗，所产生的残余物已按要求全部排空。

Clean bill 光票　系指不随附任何商业单据（如货运等相关单据），而在国外付款的外币票据。光票的流通全凭出票人、付款人或背书人的信用。在国际结算中，一般仅限于贸易从属费用、货款尾数、佣金等费用的支付。光票一般用于偿债、赠予、接济、留学支出等。由于光票签章不一定能鉴定，必须寄送国外代收银行才可收到票款，因此目前许多银行外汇部门皆有办理光票托收（clean collection）业务。

Clean credit 光票信用证　系指不附单据、受益人可以凭开立收据或汇票分批或一次在通知行领取款项的信用证。在贸易中它可以起到预先支取货款的作用。

Clean drains 清洁泄放水　系指诸如使用海水、淡水、蒸汽、空调等设备中漏泄出来以及冷凝出来的内部泄放水，这些水通常未被油污染。 Clean drains mean internal drains resulting from the leakage of and condensate from equipment by use of seawater, fresh water, steam, air conditioning, etc., which are normally not contaminated by oil.

Clean water holding tank 清洁水储存柜　系指储存从滤油设备来的工作水的舱柜。 Clean water holding tank means tanks which hold processed water from the oil filtering equipment.

Cleaning additives 清洁添加剂　系指为方便货舱清洗而在水中加入少量的添加剂（洗涤产品）。

Cleaning agents 清洁剂　系指洗舱用的非水清洗介质。

Cleaning appliances 清洁用具　保持环境清洁卫生用的各种用具的统称。如吸尘器、绞拖把器、打蜡电刷、挂式痰盂等。

Cleaning interval 清洗周期　海水淡化装置在两次清洗之间运行的时间。

Cleanliness factor 清洁系数　减少表面式冷凝器传热系数时考虑冷却水管表面清洁程度而取用的修正系数。

Cleansing station 洗消站　系指适当布置并配备适当设备的一组舱室,沾染核、生、化与放射性物质的人员与物资可以在这里进行洗消。 Cleansing station is a group of compartments suitably arranged and equipped whereby CBRN decontamination of personnel and materials can take place.

Clear grounds that the ship is not in compliance 船舶不符合要求的明显理由　系指考虑了 ISPS 规则 B 部分中的导则，有证据或可靠信息表明船舶保安体系和任何相关的保安设备不符合 SOLAS 第 XI-2 章或 ISPS 规则 A 部分的要求。该证据或可靠信息可以从正式授权的官员在验证按 ISPS 规则 A 部分的要求签发的船舶国际保安证书或临时国际保安证书时，根据其专业判断或观察结果得出或从其他来源得到。即使船上备有有效的证书，正式授权的官员仍可根据其专业判断有明显理由确信船舶不符合要求（ISPS 规则 B/4.32）。 Clear grounds that the ship is not in compliance means evidence or reliable information shows that the security system and any associated security equipment of the ship does not correspond with the requirements of SOLAS Chapter XI-2 or Part A of the ISPS Code, taking the guidance given in Part B of the ISPS Code into account Such evidence or reliable information may arise from the duly authorized officer's professional judgment or observations gained while verifying the ship's International Ship Security Certificate or Interim International Ship Security Certificate issued in accordance with Part A of the ISPS Code or from other sources. Even a valid certificate is on board the ship, the duly authorized officers may still have clear grounds for believing that the ship is not in compliance based on their professional judgment (ISPS Code paragraph B/4.32).

Clear sight of light 透光尺寸　系指船用窗户能透

Clear size of opening 通孔尺寸 舱口、人孔和门等可供人和物通过的净尺寸。

Clear waters 开敞水域 有足够水深,能使风生波正常发展的水域。 Clear waters are the waters having sufficient depth to permit the normal development of wind generated waves.

Clear width of corridors and stairways 走廊和楼梯的净宽 系指扣除栏杆后的宽度。 Clear width is measured off the handrail(s) for corridors and stairways.

Clear width of door 门的净宽 系指门全开状态下的实际通过宽度。 Clear width of door means the actual passage width of a door in its fully open position.

Clear width of escape routes 脱险通道的净宽 系指:(1)走廊,梯道扣除栏杆后的宽度;(2)门处于全开位置时实际通过宽度;(3)公共场所内固定座位与过道空隙;(4)公共场所内一排固定座位(当无人占用时)最大向内凸出部分间空隙。 Clear width of escape route is:(1) corridors and stairways measured off the handrail(s);(2) the actual passage width of a door in its fully open position;(3) the space between the fixed seats for aisles in public spaces;(4) the space between the most intruding portions of the seats (when unoccupied) in a row of seats in public space.

Clearance hole 通焊孔 为使焊缝通过而在构件上开的孔。

Clearance volume (compression space) 压缩容积 压缩终了时被压缩气体所占的容积。

Clear-view screen 雨雪扫除器 用以去除船窗玻璃上雨雪的专用设备。

Cleat 系索拴(系索耳,羊角) (1)系索环和羊角的统称;(2)用圆钢制成的供系拴绳索用的羊角索具配件。

Client 客户 船级社与之签订合同并承担工作的组织。 Client means the organization with whom the society sings a contract and who is going to undertake work. contracting it to undertake work.

Clinker system 鱼鳞式 板的两边搭接相邻两板异端的排列连接方式。

Clinometer 倾斜仪 系指测量物体随时间的倾斜变化及铅垂线随时间变化的仪器。仪器中感应倾变量的检测器是摆,有铅垂摆、水平摆、交叉摆及水准器、连通管等多种形式。它们都可分为摆基座和摆体两部分,一旦摆基座出现倾斜,或铅垂线发生变化,就会引起摆体的角位移,仪器的量测系统就将这一角位移检测和记录下来。

Clip 夹扣 舱盖、门、窗等处,用以压紧填料保证密性的压紧器。

Clipper bow (fiddle bow, cut water bow, knee bow, overhanging bow) 飞剪型艏 设计水线以上具有较大悬伸部,艏柱侧影呈凹形曲线形式的艏部。

Clo value (Immersion suit) Clo 值(救生服) 用于表示各种服装组件相对热绝缘值的单位。一个 clo 等于 $0.155 K \cdot m^2 \cdot W^{-1}$。 Clo value means the unit to express the relative thermal insulation values of various clothing assemblies. One clo is equal to (Immersion suit) $0.155 K \cdot m^2 \cdot W^{-1}$.

Close connected bucker chain 连续斗链 前后两个斗链与斗销直接铰接的斗链。

Close-coupled rudder 并联舵 处在同一螺旋桨尾流中的两个并联同步转动的舵。

Closed container 封闭集装箱 系指其由永久性结构完全封闭起来的集装箱。如果其开口能满意地密封,以阻止火花的进入,则具有小尺寸通风开口的集装箱可视为封闭集装箱。 Closed container means a container that totally encloses its contents by permanent structures.

Closed cooling water system 闭式冷却水系统 主、辅柴油机用封闭循环的淡水冷却,淡水再由舷外水冷却的冷却水系统。

Closed cycle 闭式循环 工质与大气隔绝,在密封系统中连续反复压缩、加热、膨胀和冷却的燃气轮机循环。

Closed cycle gas turbine plant 闭式循环燃气轮机装置 按闭式循环工作的燃气轮机动力装置。

Closed feed water system 闭式给水系统 凝水从冷凝器出来送到锅炉的过程中均不与空气接触的给水系统。

Closed gauging device[1] 闭式测量装置 系指与液舱大气隔离的装置,且保持舱内货液不泄漏,它可以:(1)穿舱,如浮子式系统、电探头、磁探头或保护玻璃管;(2)不穿舱,如超声波装置或雷达装置。 Closed gauging device means a device which is separated from the tank atmosphere and keeps tank contents from being released. It may: (1) penetrate the tank, such as float-type systems, electric probe, magnetic probe or protected sight glass; (2) not penetrate the tank, such as ultrasonic or radar device.

Closed gauging device[2] 封闭式液位测量装置 系指将此装置伸入液货舱内,成为封闭系统的一部分,而且能防止舱内货物溢出,例如浮筒式系统、电子探头、磁性探头和带有防护装置的观察器,也可采用不用穿过液货舱壳板而与液货舱无关的间接式装置,如货物称重装置和管式流量计等。 Closed gauging device means a device which penetrates the tank, but is a part of a closed system and keeps tank contents from being released. Examples are the float-type systems, electronic probe, magnetic probe and protected sight-glass. Alternatively, an indirect device which does not penetrate the tank shell and is independent of the tank. Examples are weighing of cargo, pipe flow meter.

Closed inspection 近观检查 系指包括目视检查以及使用设备,如活动扶梯(必要的地方)和工具才能识别明显缺陷(如螺栓松动)的检查。

Closed Ro-Ro spaces 闭式滚装处所 系指既不是开式滚装处所,也不是露天甲板的滚装处所。 Closed Ro-Ro spaces are Ro-Ro spaces which are neither open Ro-Ro spaces nor weather decks.

Closed session 不公开场合

Closed socket 闭式索节 具有环状连接端的索节。

Closed stokehold draft 闭式炉舱通风 鼓风机向密闭的炉舱送入锅炉燃烧用的空气,炉舱内压力高于大气压的一种锅炉强力通风的形式。

Closed vehicle space 封闭车辆区域 系指风雨密围蔽的车辆区域,而不是开放的车辆区域。 Closed vehicle space means closed space with weather-tight other than open vehicle space.

Closed-type cavitation tunnel 封闭式空泡试验水筒 水完全封闭在筒内进行循环的空泡试验水筒。

Close-in fueling rig 近距离加油索具 由悬吊在吊杆上的鞍座索和软管等组成的传送油料的索具。船间距离一般为 18~24 m。

Closest approach 最接近点 系指在一次卫星通过中,卫星距观察者距离最近的一点。卫星通过最接近点时称为卫星过顶。在卫星过顶瞬间,卫星与观察者的连线垂直于卫星的轨迹线,因此该瞬间的多普勒频移为零。

Close-up survey 近观检查 系指通常在验船师的手可以触及的距离之内对船体构造附属物的状态通过肉眼进行的精密检验。 Close-up survey is a survey that the details of structural components are within the close visual inspection range of the surveyor, i. e. normally within reach of hand.

Closing alarms for refrigerated spaces 冷藏处所关闭报警 系指如遇冷藏货舱和伙食冷藏库等冷藏处所的门不能从其内部开启,设置在能从该处所内部触发误关报警,并将其传送至通常有人位置的设施。 Closing alarms for refrigerated spaces mean a device capable of activating within the spaces and transmitting to the spaces where personnel are normally present, when the doors to refrigerated spaces such as refrigerated holds and refrigerating food chamber cannot be opened from the interior.

Closing appliances and stopping devices of ventilation 通风的关闭和停止装置 系指在所有通风系统的主要进口和出口处能从被通风处所的外部予以关闭的装置。关闭装置操作位置应易于到达,有明显的永久性标志,且应指示出关闭装置是处在开启位置还是处在关闭位置。 Closing appliances and stopping devices of ventilation is a means which shall be capable of being closed from outside the spaces being ventilated in main inlets and outlets of all ventilation systems. The means of closing shall be easily accessible as well as prominently and permanently marked and shall indicate whether the shut-off is open or closed.

Closing meeting 审核结束会议 系指由审核机构、被审核单位的相关人员参加的,旨在说明审核结果的会议。

Cloud "云" 系指在线数据储存的地点。

Cloud cavitation 云状空泡 片状或沫状空泡后的水流不再顺物体切面轮廓前进时,在整个漩涡尾流区形成的云雾状空泡。对剥蚀而言,云状空泡是最危险的一种空泡类型,成群的气泡发生崩溃比大量气泡周期性的生灭而造成的剥蚀强度大得多。这种剥蚀使叶片表面上产生凹坑,随之在叶片端部产生变形(随缘弯曲)。

Cloud computing "云计算" 系指基于互联网的相关服务的增加、使用和交付模式,通常涉及通过互联网来提供动态易扩展且经常是虚拟化的资源。云是网络、互联网的一种比喻说法。过去在图中往往用云来表示电信网,后来也用来表示互联网和底层基础设施的抽象。因此,"云计算"甚至可以让你体验每秒10万亿次的运算能力,拥有这么强大的计算能力可以模拟核爆炸、预测气候变化和市场发展趋势。用户通过计算机、笔记本电脑、手机等方式接入数据中心,按自己的需求进行运算。对"云计算"的定义有多种说法。对于到底什么是"云计算",至少可以找到100种解释。现阶段广为接受的是美国国家标准与技术研究院(NIST)定义:"云计算"是一种按使用量付费的模式,这种模式提供可用的、便捷的、按需的网络访问,进入可配置的计算资源共享池(资源包括网络、服务器、存储、应用软件、服务),这些资源能够被快速提供,只需投入很少的管理工作,或与服务供应商进行很少的交互。全球首台"云计算机""紫云1000"在中国问世,这标志着中国在"云计算"核心技术领域取得了重大突破。

Cluster　群呼　系指高级别的功能组，例如航行、自动化。Cluster is the group of high level functions, e. g., navigation, automation.

Cluster of pores　集结的孔穴（集结孔隙）　应包括集结区域内全部孔穴的面积并按其稍大的两个区域：包含所有孔穴的包络线或直径与焊缝尺寸相匹配的一个圆圈的百分比进行计算。许可的孔穴集结区应是局部化的。The entire pore area within a cluster should be included and calculated as a percentage from, the larger of the two areas: envelope curve encompassing all the pore or a circle with a diameter that matches that of the side of the weld. The permitted pore area should be localized.

Clutch　离合器　在主机与轴系、主机与减速齿轮箱、发动机与电机等之间根据运行需要，使主、从动轴接合或脱离的传动组件。

CM (construction monitoring)　船舶结构监控　系指对船舶的结构关键位置结构精度控制，包括对中、装配、边缘处理以及工艺标准符合批准的计划的船长150 m及以上的油船、散货船和集装箱船，根据船东申请，可授予该标志。

CMA ship　CMA船　系指一种其主推进器和基本辅助机械的中央监控符合有关规定的并已经注册的船舶。CMA ship is the ship of which centralized monitoring and control system for main propulsion and essential auxiliary machinery comply with the relative requirements and is registered.

CNG carrier　压缩天然气（CNG）运输船　系指载运压缩天然气的液化气体船。CNG carrier means a ship carrying compressed natural gas.

CO_2 carrier　二氧化碳运输船　系指专门运输压缩二氧化碳的液化气船。

CO_2 emissions per unit load　单位货运量的CO_2排放量

CO_2 fire extinguishing system　二氧化碳灭火系统　系指采用二氧化碳作灭火剂，由二氧化碳站室、二氧化碳贮气瓶、二氧化碳瓶头阀以及二氧化碳管路、施放报警装置等组成的成套灭火系统。

CO_2 releasing alarm　二氧化碳施放报警器　采用二氧化碳灭火时在灭火剂发送前和发送过程中能自动发出警报的报警器。

Coagulant　凝聚剂　系指含高分子混凝剂和助凝剂，能破坏水中的胶体粒子稳定性，降低彼此间的排斥力，从而形成细小的絮体，达到固液分离，净化水质目的的化合物。用于金属加工废水，肉类加工废水、含磷废水、含氟废水、造纸废水、煤气洗涤水的处理。

Coal　煤　（沥青质的及无烟煤）是一种由非晶体形碳和碳氢化合物组成的天然固体可燃物质。煤可能产生易燃气体，可能自热，可能大大减少氧气浓度；可能腐蚀金属结构。如大部分为碎煤且其中小于5 mm者占75%，则可能流态化。Coal (bituminous and anthracite) is a natural, solid, combustible material consisting of amorphous carbon and hydrocarbons. Coal may create flammable atmospheres, may heat spontaneously, may deplete the oxygen concentration, may corrode metal structures. It can liquefy if predominantly fine less than 5 mm accounts for 75%.

Coal hole cover　煤舱盖　设置于日用煤舱上的舱口盖。

Coal slurry　煤泥　是细颗粒煤和水的一种混合物。煤泥在海上运输期间易流态化。煤若干细颗粒会自燃，但在正常条件下不大会。该货物为不燃物或失火风险低。Coal slurry is a mixture of fine particles of coal and water. Coal slurry is liable to liquefy during sea transport. Spontaneous combustion is possible if the coal dries out but is unlikely under normal conditions. This cargo is non-combustible or has a low fire-risk.

Coal supply boat　供煤船　系指专门供应煤炭的供应船。

Coaming　舱口围板　系指舱口或天窗的垂直限界结构。Coaming means a vertical boundary structure of a hatch or skylight.

Coarse chopped tyres　轮胎粗碎块　旧轮胎剁碎或切碎的粗块。装运前如未适当老化且交运尺寸若小于有关规定的尺寸，受含油类留物污染后可能缓慢老化。该货物为不燃物或失火风险低。Coarse chopped tyres are chopped or shredded fragments of used tyres in coarse size. It may self-heat slowly if contaminated by oily residual, if not properly aged before shipment and if offered to the shipment in smaller size than relevant requirement size. This cargo is non-combustible or has a low fire-risk.

Coast earth station (CES)　海岸地面站

Coast Guard　海岸警卫队　即水上警察，但它的体制和职能则是警察和武警这两支武装部队所无法比拟的，堪称一个准军事化的部队。它主要负责一个国家的所有海岸线上的警戒、巡逻、执法等任务，是一支军事化的综合执法队伍。它有利于整合海上战略资源，并给外交上和军事留下回旋余地。其职能覆盖了相当于中国当今海军、公安边防武警、海监、海事、渔政、海关、环境保护等部门的部分业务。

Coast watching unit　海岸值守单位　系指为对沿海地区船舶安全保持值守的固定或流动的陆上单位。Coast watching unit means a land unit, stationary or mobile, designed to maintain a watch on the safety of vessels in coastal areas.

Coastal engineering　海岸工程　主要包括海岸防护工程、围海工程、海港工程、河口治理工程、海上疏浚工程、沿海渔业设施工程、环境保护设施工程。

Coastal fishing ship　沿岸渔船　系指在近海、沿海作业,大多数为垂线间长 30 m 以内的小型渔船。

Coastal Service area　沿海区域　系指从海岸开始 20 海里(1n mile = 1 852 m)以内的水域。 Coastal service area means water area within 20 Nautical miles (1 Nautical mile = 1 852 m) of the shore.

Coastal service restriction　沿海航区营运限制　系指对高速船营运限制。授予航行于距岸不超过 20 n mile 的水域,且船舶在其经营的航线上,满载并以其营运航速至庇护地的航行时间满足如下规定的船舶的附加标志:(1)客船不超过 4 h;(2)货船不超过 8h。

Coastal State　沿海国/沿岸国家　系指:(1)在其领土范围内(包括领海)发生海难或海上事故的国家。(2)对平台的钻井作业行使行政管理国家的政府。Coastal State means: (1) a State in whose territory, including territorial sea, a marine casualty or marine incident occurs. (2) the Government of the State exercising administrative control over the drilling operations of the unit.

Coastal waters　近海海区　系指距海岸距离小于相当于以相应船速航行30min 的水域。而在此海区的另一侧船舶可以相应船速在任何方向至少航行 30 min 的相当距离。 Coastal waters mean waters that encompass navigation along a coast at a distance less than the equivalence of 30 minutes of sailing with the relevant ship speed. The other side of the course line allows freedom of course setting in any direction for a distance equivalent to at least 30 minutes of sailing with the relevant speed.

Coastal Zone Management Act (CZMA)　[美]海岸带管理法　1972 年 10 月 27 日,美国国会颁布了该法,从而使海岸带综合管理(Integrated Coastal Zone Management/ICZM)作为一种正式的政府活动首先得到实施,也标志着海岸带管理实行了法治。

Coastal zone survey　海岸带调查　对海岸进行的多科学综合调查。范围为潮上带并向陆延伸约10 km、潮间带和潮下带向海一般延至 10～15 m 等深线处。内容有海洋水文、陆地水文、地质、地貌、海水化学、生物和环境污染、海滨沼泽、土壤和土地利用等。

Coaster　沿海船　航行于沿海各港口之间的船舶。

Coating　涂料

Coating conditions for steel ships　钢船涂层状况　定义如下:良好——系指仅有微小点状锈斑,所考虑部位受影响面积小于 20% 的情况;尚可——系指在扶强材边缘和焊接接缝处有局部脱落和/或轻度锈蚀影响到所考虑面积的 20% 或以上的情况;差——系指涂层普遍脱落,影响到所考虑面积的 20% 或以上,或有硬质锈皮,影响到所考虑面积的 10% 或以上的情况。见表 C-3。
Coating conditions for steel ships are defined as follows: GOOD means the condition with only minor spot rusting affecting not more than 20% of areas under consideration. FAIR means the condition with local breakdown at edges of stiffeners and weld connections and/or light rusting affecting 20% or more of areas under consideration. POOR means the condition with general breakdown of coating affecting 20% or more of areas under consideration or hard scale affecting 10% or more of the area under consideration. Asis shown in Table. C-3.

表 C-3　涂层状态
Table C-3　Coating condition

良好 GOOD	仅有微小点状锈斑 Condition with only minor spot rusting
尚好 FAIR	在扶强材和焊接接缝边缘处有局部脱落和/或轻度锈蚀,影响所考虑面积的 20% 或以上的情况,但比"差"情况要少 Condition with local breakdown of coating at edges of stiffeners and weld connections and/or light rusting affecting over 20% or more of areas under consideration, but less than as defined for POOR condition
差 POOR	涂层普遍脱落,影响所考虑面积的 20% 或以上,或有硬质锈皮,影响所考虑面积的 10% 或以上的情况 Condition with general breakdown of coating over affecting 20% or more of areas or hard scale affecting 10% or more of areas under consideration

Coating prequalification test 涂层合格预试验

Coating specification 涂层技术规格书（涂层技术条件） 系指涂层系统的规格，包括涂层系统的类型、钢板处理、表面处理、表面清洁度、环境条件、涂装程序、检查和验收标准。 Coating specification means the specification of coating systems which include the type of coating systems, steel preparation, surface preparation, surface cleanliness, environmental conditions, application procedure, inspection and acceptance criteria.

Coating standard 涂层标准 本标准基于这样的技术条件和要求，即为使涂层达到15年的目标使用寿命，这是从最初的涂装开始，涂层系统维持良好状态的持续时间。涂层的实际使用寿命是变化的，取决于很多的变化因素，包括使用中遇到的真实条件。 This standard is based on specifications and requirements which intend to provide a target useful coating life of 15 years, which is considered to be the time period, from initial application, over which the coating system is intended to remain in "GOOD" condition. The actual useful life will vary, depending on numerous variables including actual conditions encountered in service.

Coating technical file (CTF) 涂层技术文件 系指用于船舶专用压载舱和双舷侧处所的涂层体系的技术条件，造船厂和船东的涂装记录，涂层系统选择的详细标准，工作说明书，检查、维护和修补报告均应形成文件列入涂层技术文件。涂层技术文件应经主管机关审查。 Coating technical file means a file which includes specification of the coating system applied to the dedicated seawater ballast tanks and double side skin spaces, record of the shipyard and ship-owner's coating work, detailed criteria for coating selection, Job specifications, inspection, maintenance and repair shall be documented in the coating technical file (CTF), and the coating technical file shall be reviewed by the Administration.

Coating type 涂层类型

Coaxial cable 同轴电缆 是由内外两层同心圆柱体构成，在这两根导体之间用绝缘体隔开。内导体多为实心导线，外导体是一根空心导电管或金属编织网，在外导体外面有一层绝缘保护层，在内外导体之间可以填充实心介质材料火绝缘支架，起到支撑和绝缘的作用。由于外导体通常接地，因此能够起到很好的屏蔽作用。随着光纤的广泛应用，远距离传输信号的干线线路多采用光纤替代同轴电缆，在有线电视广播（CATV: Cable Television）中还广泛地采用同轴电缆为用户提供电视信号，另外在很多程控电话交换机中PCM群路信号仍然采用同轴电缆传输信号，同轴电缆也作为通信设备内部中频和射频部分经常使用传输的介质，如连接无线通信收发设备和天线之间的馈线。

Coaxial contra-rotating propellers 对转螺旋桨 在同轴线的内外两轴上所装转向相反的一对螺旋桨。

Coca leaf 古柯叶 呈椭圆形叶子，颜色为褐绿或红色等，外观上与大的月桂树的叶子相似，通常是干的，没有气味。 Coca leaf appears as an elliptical leaf, it is greenish brown to red in colour, it is similar to large bay leaves in appearance, and it is usually dried. It is odourless.

Coca paste 古柯膏 呈白色至灰白色或奶油色油泥状物质，其强烈的化学品气味很像亚麻籽油。 Coca paste appears as a white to off-white or creamy coloured putty-like substance. It has a strong chemical odour, rather like linseed oil.

Cocaine 可卡因 是从安第斯古柯灌木的叶子中衍生的，有很强的刺激性，与安非他明相似。主要产于南美的北半部分，特别是哥伦比亚和委内瑞拉，可卡因的利润对其经济有着重要的影响。制造者面临的主要问题是将这些物质运输至消费区域。呈蓬松的白色结晶粉末像雪一样闪光，但偶尔作为无色的溶剂运输。没有气味。 Cocaine is derived from the leaves of the Andean coca shrub and has powerful stimulant properties similar to those of amphetamine. It is produced mainly in the northern half of South America, especially in Colombia and Venezuela, where cocaine profits have a major influence on the economy. The main problem the producers face is transporting the substance to consumption areas. Cocaine appears as a fluffy white erystalline powder which glistens like snow, though occasionally transported as a colourless solution. It is odourless.

Co-cumulative spectrum 共积谱 表示从∞频率到某一指定频率能量分布累积成的曲线。

CODAG 柴油机与燃气轮机联合动力装置

Code for dangerous chemical in bulk 散装危险化学品规则 系指由国际海事组织颁布的，并即时修订的运输散装危险化学品船舶构造和设备规则。 Code for dangerous chemical in bulk means the code for the construction and equipment rules of ships carrying dangerous chemical in bulk, as amended, published by IMO.

Code for liquefied gases in bulk 散装液化气体规则 系指由国际海事组织（IMO）颁布的并即时修订的散装运输液化气体船舶构造和设备规则。 Code for liquefied gases in bulk means the code for the construction and equipment rules of ships carrying liquefied gases in bulk, as amended, published by IMO.

Code for solid bulk cargoes 固体散装货物规则 系指固体散装货物安全操作规则。 Code for solid bulk cargoes means the code of safe practice for solid bulk cargoes.

Code of Federal Regulations（CFR） ［美］联邦政府法规 是美国联邦政府各行政部门和机构所出版的一部全面的永久性规则的汇编。所有这些官方的规则都一起汇集在 CFR 中，其所包括的大约 180 余卷本（Volume）中。CFR 共分 50 篇（Title），每一篇分成若干章（Chapter）、每一章（或每一分章）又分为若干节（Part）、每一节又可分为若干条（Section），条是 CFR 的基本单元。与船舶设计、建造和营运等有关的 USCG 的规则的要求列入其中。其相应的规定适用于挂美国旗的商船和在美国可航水域中航行的外国船。

Code on noise levels on board ships 舱室噪声级规则 系指国际海事组织（IMO）在 1981 年 11 月 19 日在第 12 届大会上通过的 A.468（Ⅻ）决议《船上噪声级规则》。并作为推荐性要求纳入 SOLAS 公约第Ⅱ-1/36 条。该决议注意到船上高噪声会影响船员的健康及损坏船舶安全，为此，规定船上允许的噪声级，以保护船员健康及确保船舶安全使用。该规则分为 7 章，分别为总则、测量、测量设备、最大允许声压级、噪声曝露限值、起居处所之间的隔声和护耳及警告标志。

Codeine 可待因 通常为白色的药片或药丸。 Codeine is usually found as white tablets or pills.

Codes 规则 系指：(1)自愿地由各船旗国作为其国家规则，一旦采用，对该国船舶即具强制性；(2)涉及较广泛的领域，如移动式近海装置构造和设备规则（MODU）、国际散装运输危险化学品船构造和设备规则（IBC）、国际散装运输液化气体船舶构造和设备规则（IGC）等；(3)可用来代替决议，某些规则是以决议形式通过的，例如，A.468（Ⅻ）"船上噪声级规则"，既是 IMO 规则，又是 IMO 大会决议；(4)一些规则因对某一公约的修正而已被纳入了该公约，对于已作为公约之一部分的规则，具有与公约相同的性质和特点，例如国际海上人命安全公约（SOLAS）1983 年修正案规定，把 IBC 和 IGC 看作是该公约的一部分，因此这两项规则具有与 SOLAS 相同的强制性。

Coding room 译电工作室 供译电员从事密电码翻译工作用的机要舱室。

Coefficient corresponding to the probability level f_p 与概率水平对应的系数 f_p 取 $f_p = 1.0$，对应 10^{-8} 概率水平的强度评估；$f_p = 0.5$，对应 10^{-4} 概率水平的强度评估。 f_p means a coefficient corresponding to the probability level, taken equal to: $f_p = 1.0$ for strength assessments corresponding to the probability level of 10^{-8}; $f_p = 0.5$ for strength assessments corresponding to the probability level of 10^{-4}.

Coefficient of appendage resistance 附体阻力系数 以附体阻力对动压力与湿表面乘积的比值，表示附体阻力特征的一个无量纲的参数。$C_{AP} = R_{AP}/[(1/2)\rho V^2 S]$，式中，$R_{AP}$—附体阻力；$V$—船速；$S$—湿面积；$\rho$—流体质量密度。

Coefficient of dynamic viscosity 动力黏性系数 表示流体内部抵抗切变能力的系数。对于单向切变流，$\mu = \tau/[dU/dy]$，式中，τ—切应力；dU/dy—切变。

Coefficient of equivalence 等值系数

Coefficient of frictional resistance 摩擦阻力系数 以摩擦阻力对动压力与湿表面乘积的比值，表示阻力特征的一个无量纲的参数。$C_F = R_F/[(1/2)\rho V^2 S]$，式中，$R_F$—摩擦阻力；$V$—船速；$S$—湿面积；$\rho$—流体质量密度。

Coefficient of frictional resistance of plank 平板摩擦阻力系数 以平板阻力对动压力与湿表面乘积的比值，表示平板阻力特征的一个无量纲的参数。$C_{FO} = R_{FO}/[(1/2)\rho V^2 S]$，式中，$R_{FO}$—平板阻力；$V$—船速；$S$—湿面积；$\rho$—流体质量密度。

Coefficient of kinematic capillarity 运动毛细常数 水的毛细常数对质量密度的比值。$X = \sigma/\rho$，式中，σ—水的毛细常数；ρ—水的质量密度。

Coefficient of kinematic viscosity 运动黏性系数 流体的动力黏性系数与其质量密度的比值。$\nu = \mu/\rho$，式中，μ—流体动力黏性系数；ρ—流体质量密度。

Coefficient of lift（lift coefficient） 升力系数 表示升力特征的一个无量纲的系数：$C_L = L/0.5\rho U^2 S$，式中，L—升力；S—特征面积；U—流速；ρ—流体质量密度。

Coefficient of reserve（factor of reserve） 贮备系数 离合器最大扭矩与发动机标定扭矩的比值。$K = M_{max}/M_{标定}$，式中，M_{max}—离合器最大扭矩；$M_{标定}$—发动机标定扭矩。

Coefficient of roughness resistance 粗糙度阻力系数 以粗糙度阻力对动压力与湿表面乘积的比值，表示粗糙度阻力特征的一个无量纲的参数。$C_{AR} = R_{AR}/[(1/2)\rho V^2 S]$，式中，$R_{AR}$—粗糙度阻力；$V$—船速；$S$—湿面积；$\rho$—流体质量密度。

Coefficient of submerged lateral area 水下侧面积系数 船舶水下侧面积对相应水线长度和平均吃水乘积的比值。

Coefficient of total resistance of model 船模总阻力系数 以船模试验拖索上测得的总阻力对动压力与湿表面积乘积的比值，表示船模总阻力特征的一个无量纲系数：$C_{Tm} = R_{Tm}/[(1/2)\rho V^2 S_m]$，式中，$R_{Tm}$—船模总阻力；$S_m$—船模的湿面积；$V$—船速；$\rho$—水的质量密度。

Coefficient of total resistance of ship 实船总阻力系数 以计算的或实测的总阻力与湿表面积乘积的比值，表示实船总阻力特征的一个无量纲系数：$C_{Ts} = R_{Ts}/[(1/2)\times\rho V^2 S_s]$，式中，$R_{Ts}$—实船总阻力；$S_s$—实船的湿面积；$V$—船速；$\rho$—水的质量密度。

Coefficient of viscous pressure resistance 黏压阻力系数 以黏性阻力对动压力与湿表面乘积的比值，表示黏性阻力特征的一个无量纲的参数。$C_{PV} = R_{PV}/[(1/2)\rho V^2 S]$，式中，$R_{PV}$—黏压阻力；$V$—船速；$S$—湿面积；$\rho$—流体质量密度。

Coefficient of viscous resistance 黏性阻力系数 以黏性阻力对动压力与湿表面积的比值，表示黏性阻力特征的一个无量纲的参数。$C_V = R_V/[(1/2)\rho V^2 S]$，式中，$R_V$—黏性阻力；$V$—船速；$S$—湿面积；$\rho$—流体质量密度。

Coefficient of wave making resistance 兴波阻力系数 以兴波阻力对动压力与湿表面乘积的比值，表示兴波阻力特征的一个无量纲的参数。$C_W = R_W/[(1/2)\rho V^2 S]$，式中，$R_W$—兴波阻力；$V$—船速；$S$—湿面积；$\rho$—流体质量密度。

Coefficient of wetted surface 湿面积系数 表示湿面积大小的无量纲系数：$C_S = S/(\nabla L)^{1/2}$，式中，∇—型排水体积；L—船长。

Coefficients of form 船型系数 用以表示船体线型肥瘦特征的各种无因次系数的统称。

Coefficients of hydrodynamic force components 水动力分量系数 一般系指水动力分量与$(1/2)\rho L^2 U^2$ 的比值的无量纲系数；其表达式如下：纵向分力系数 $X/[(1/2)\rho L^2 U^2]$；横向分力系数 $Y/[(1/2)\rho L^2 U^2]$；垂向分力系数 $Z/[(1/2)\rho L^2 U^2]$；升力系数 $L/[(1/2)\times\rho L^2 U^2]$；阻力系数 $D/[(1/2)\rho L^2 U^2]$；正交力系数 $C/[(1/2)\rho L^2 U^2]$，式中，ρ—水的密度；L—运动物体的特征长度；U—运动物体的速度。

Coefficients of hydrodynamic moment components 水动力矩分量系数 一般系指水动力矩分量与$(1/2)\rho L^2 U^2$ 的比值的无量纲系数；其表达式如下：横倾力矩系数 $K/[(1/2)\rho L^2 U^2]$；纵倾力矩系数 $M/[(1/2)\times\rho L^2 U^2]$；力矩系数 $N/[(1/2)\rho L^2 U^2]$，式中，ρ—水的密度；L—运动物体的特征长度；U—运动物体的速度。

Cofferdam 围堰 系指在水利工程建设中，为建造永久性水利设施，修建的临时性围护结构。其作用是防止水和土进入建筑物的修建位置，以便在围堰内排水、开挖基坑、修筑建筑物。一般主要用于水工建筑中，除作为正式建筑物的一部分外，围堰一般在用完后拆除。围堰高度高于施工期内可能出现的最高水位。对海上风电场建设而言，围堰是在海中建造的一个充满空气的密闭空间，工作人员可以在里面进行工作。因此围堰是像安装到海底的盒子，而且在正常情况下海水被抽空，海底曝露于空气中，那么就有可能作业时不使用呼吸器。

Cofferdam 隔离舱/隔离空舱 系指：（1）两相邻的钢质舱壁或甲板之间的隔离空间。其可为空舱或压载舱。下列处所也可作隔离舱，燃油舱以及液货泵舱和与机器处所、通道和起居处所无直接连通的泵舱。隔离舱舱壁的净间距不得小于600 mm；（2）其设置是为了使每一侧的舱室没有共同的界面，隔离舱可以垂直或水平设置。通常隔离舱应适当通风，并应有足够大的尺寸，以便可以进入检查、维护和安全撤离；（3）两个邻接的钢制舱壁或甲板间的隔离区域。两舱壁或甲板间的最小距离，应对其细致地检查以及确保可以安全地通行。为避免在拐角处发生事故，可在拐角处焊接警示牌；（4）两道舱壁或甲板之间的处所，主要设计为防止油从一个舱室渗漏至另一舱室。 Control stations are：（1）the isolating space between two adjacent steel bulkheads or decks. This space may be a void space or a ballast space. The following spaces may also serve as cofferdams：oil fuel tanks as well as cargo pump rooms and pump rooms not having direct connection to the machinery space, passage ways and accommodation spaces. The clear spacing of cofferdam bulkheads is not to be less than 600 mm；（2）an empty space arranged so that compartments on each side have no common boundary；a cofferdam may be located vertically or horizontally. As a rule, a cofferdam is to be properly ventilated and of sufficient size to allow proper inspection, maintenance and safe evacuation；（3）an isolating space between two adjacent steel bulkheads or decks. The minimum distance between the two bulkheads or decks should be sufficient for safe access and inspection. In order to meet the single failure principle, in the particular case when a corner-to-corner situation occurs, this principle may be met by welding a diagonal plate across the corner；（4）those spaces between two bulkheads or decks primarily designed as a safeguard against leakage of oil from one compartment to another.

Cofferdams, voids, etc. 隔离舱、空舱等 系指两个相邻舱室的舱壁之间的空处所。 Cofferdams and voids are those empty spaces between two bulkheads separating two adjacent compartments.

COG = Course over ground 对地航向 船舶的航向是相对于地球表面而言的。Ship's course measured relatively to the earth surface.

COGES 蒸汽轮机与燃气轮机联合动力推进

Cohesive failure 内聚力

Cohesive material 黏性物质　系指除非黏性物质以外的物质。 Cohesive material means materials other than non-cohesive materials.

Coil cooling 盘管冷却　载冷剂或制冷剂在冷藏舱、冷藏库的盘管内吸热使舱库内温度降低的冷却方式。

Coil tube ship 盘管船　系指设有闭式盘管系统，使用热油为加热介质的液货船。

Coiled tubing(CT) 连续油管　系指用于直接向井底注入化学药剂，如：循环或化学清洗。当井筒偏差过大无法依靠重力下放工具或无法使用钢丝牵引器时，钢丝作业的任务也由连续油管来完成。

Coin certificate 兑换券

Coincidence effect 吻合效应　系指当一定频率的声波以某一角度入射到壁板上，如果正好和声波激发的壁板的弯曲波发生吻合，壁板弯曲波振幅就最大，因而壁板的另一面发射较强的声波，这时壁板隔声量降至最小，对于薄板可以认为几乎失去隔声能力，这种壁板的运动和空气中声波运动高度耦合的现象。

Coke 焦炭　灰色块状，可能含有微粒(炭屑)。无特殊危害。该货物为不燃物或失火风险低。 Coke is a grey lumps which may contain fines (breeze). It has no special hazards. This cargo is non-combustible or has a low fire-risk.

Coke breeze 焦屑　灰色粉末。焦屑如含水量足够高则易于流动。该货物为不燃物或失火风险低。 Coke breeze is greyish powder. Coke breeze is liable to flow if it has sufficiently high moisture content. This cargo is non-combustible or has a low fire-risk.

Cold blow-off operation 冷吹运行　燃气轮机机组在启动不成或假启动之后，所进行的以空气吹除残余机积油的操作过程。

Cold forming 冷成形　系指对板进行弯曲、折边或卷边处理。对板作冷处理时，最小平均弯曲半径应不小于3t (t 为板的总厚度)。为防止产生裂缝，火焰切割毛刺或剪切毛刺应在冷处理前除去。在冷成形后，应对所有构件，尤其是弯曲端部(板边)进行裂缝检查。除可忽略不计的边缘裂缝外，所有存在裂缝的构件均不得使用。不允许进行补焊。 Cold forming are carried out bending, flanging, beading for plates. For cold forming of plates, the minimum average bending radius is to be not less than 3t (t = gross plate thickness). In order to prevent cracking, flame cutting flash or sheering burrs shall be removed before cold forming. After cold forming, all structural components and, in particular, the ends of bends (plate edges)

are to be examined for cracks. Except in cases where edge cracks are negligible, all cracked components are to be rejected. Repair welding is not permissible.

Cold service systems 冷却装置　系指冷冻装置和空调机用冷却水管。例如：比大气及海水温度低的装置。 Cold service systems means refrigeration systems and chilled water piping for air-conditioning systems, i.e. systems with temperature below ambient air and sea water.

Cold shuts 冷塞

Cold standby system 冷备用系统　系指一种带有人工交换或人工替换的或非操作性程序的双套系统。这个双套系统在 10 min 之内动作，能够完成对具有相同性能的主系统的操作。 Cold standby system is a duplicated system with a manual commutation or manual replacement of cards. The duplicated system is able to achieve the operation of the main system with identical performance, and be operational within 10 minutes.

Cold starting 冷态启动　汽轮机在其汽缸温度接近环境温度状态下的启动。

Colemanite 硬硼酸钙石　一种天然水合硼酸钙呈微粒至块状。淡灰色，类似黏土。含水量约7%。无特殊危害。该货物为不燃物或失火风险低。 Colemanite is a natural hydrated calcium borate. It is fine to lumps, it has light grey appearance similar to clay. Its moisture is approximately 7%. It has no special hazards. This cargo is non-combustible or has a low fire-risk.

Collapse 破坏　系指在不需要附加外力的情况下，有负表面结构通常出现大的变形(破坏)。 A structure with negative surface usually undergoes large deformations (collapses) without additional external effort.

Collapse pressure 坍毁压力　空泡进入较高压力区后迅速缩小的最后阶段所产生的极大压力。

Collapsed slip joint 伸缩接头　是泵、阀门，管道等设备与管道连接的新产品，通过螺栓把它们连接起来，使其成为整体，并有一定的位移量，这样就可以在安装维修时，根据现场安装尺寸进行调整，在工作时，可以把轴向推力传还至整个管道系统。这样不仅提高工作效率，而且对泵、阀等管道设备起到一定保护作用。

Collapsible mast 可倒桅杆　可绕其根部的支承座转动而放倒的桅杆。

Collapsible railing 活动栏杆　可拆卸或倒下的栏杆。

Collapsing security condition 破坏保安状况　对 ISPS 规则而言，系指由于非法行动对船舶、财产、设施以及人员造成的后果。

Collar bearing(collar thrust bearing) 环形推力

轴承　承压面是一个连续环带的推力组成轴承。

Collar plate　补板　系指用以部分或完全闭合纵向加强筋穿过横向腹板所开洞口的补板。 Collar plate means a patch used to, partly or completely, close a hole cut for a longitudinal stiffener passing through a transverse web.

Collection　托收　系指由卖方开立汇票，委托出口地银行通过其在国外的分行或代理行，向买方收取货款或劳务费用的一种结算方式。

Colliding icebreaking　"冲撞式"破冰法　系指破冰船船艏部位吃水浅，会轻而易举地冲到冰面上去，船体就会压下面厚厚的冰层压为碎块。然后破冰船倒退一段距离，再开足马力冲上去，把船下的冰层压碎的方法。

Collier / coal carrier　运煤船　专运煤炭的船舶。

Collision avoidance　避碰　系指探测和标绘其他船舶和物体以避免碰撞的导航任务。 Collision avoidance means the navigational task of detecting and plotting other ships and objects to avoid collisions.

Collision avoidance functions　避碰功能　系指探测其他船舶和运动目标并标出其位置，确定并执行航向和航速改变，以避免碰撞。 Collision avoidance functions means functions of detecting and plotting of other ships and moving objects, determination and execution of course and speed deviations to avoid collision.

Collision bulkhead　防撞舱壁　系指在船舶艏部直通至工作甲板的水密舱壁。它应具备下述条件：(1)此船舶应与艏垂线有一定距离；①长度为 45 m 以上的船舶，不小于 0.05 L，且不大于 0.08 L；②长度 45 m 以下的船舶，除主管机关许可外，不小于 0.05 L 且不大于 0.05 L + 1.35 m；③不得小于 2.0 m；(2)如船体水下部分向艏垂线的前部延伸，例如球鼻艏，则按(1)项所规定的距离；(3)假如在(1)项的规定范围内，舱壁可以是台阶式或者凹形。 Collision bulkhead is a watertight bulkhead up to the working deck in the forepart of the vessel which meets the following conditions：(1) The bulkhead shall be located at a distance from the forward perpendicular：①not less than 0.05L and not more than 0.08L for vessels of 45 m and over in length；②not less than 0.05L and not more than 0.05L plus 1.35 m for vessels of less than 45 m in length except those as may be allowed by the Administration；③in no case, less than 2.0 m. (2) The submerged part of the hull extends to the front of the bow perpendicular, such as the bulbous bow. at the distance specified in subparagraph (1). (3) The bulkhead may have steps or recesses provided they are within the limits prescribed in subparagraph (1).

Collision bulkhead　升降口　系指船舶甲板通往下面处所的风雨密入口。 Companionway means a weather-tight entrance leading from a ship's deck to spaces below.

Collision mat　堵漏席　船舶破损时，用以堵塞破洞的防水席垫。

Coloration　显色性　系指光源射到物体上所呈现的色彩，在视觉上存在一定的失真度。这种失真度用"显色性"表示。与标准光的显色越是一致，显色性就越高，显色性可用显色指标表示。显色指数在 90 ~ 100 范围内，显色物体的颜色可达到正确可靠的程度。

Colour meter　水色计　系指目测海水水色的参比仪器。由 22 只无色玻璃管，内装 21 种不同验色的标准色级溶液组成。海水水色由 1/2 透明度深度处的水色与标准色级相比较得出，以标准色级号码表示。

Colour table　色表　系指为观察光源本身时所接受到的颜色印象。

Colour vertical projection fish finder　彩色鱼探仪　系指用来探测鱼群的水声设备。是渔用声呐的一种。

COLREG　国际海上避碰规则　系指经修正的"1972 年国际海上避碰规则"，包括其附录。 COLREG means the International Regulations for Preventing at sea, 1972, as amended, including their annexes.

Column (column-stabilized)　立柱(稳柱)　半潜式钻井装置中，用于连接下浮体和上平台，并能对整个装置起稳定作用的柱式结构物。

Column footings　立柱浮箱　半潜式钻井装置中，位于立柱之下的并能对整个装置起沉浮和稳定作用的箱型结构物。

Column stabilized semi-submersible drilling unit　半潜式钻井平台　系指由上部结构、立柱与浮垫或桩靴等组成，大部分浮体深埋于水面以下，依靠立柱的小水线面保证漂浮稳性，并在波浪中具有良好运动性能的移动式钻井平台。见图 C-7。

图 C-7　半潜式钻井平台
Figure C-7　Column stabilized semi-submersible drilling unit

Column stabilized semisubmersible production unit　柱稳、半潜式生产平台　系指在海上进行油气生产的

柱稳、半潜式平台。

Columns 立柱 系指柱稳式平台或坐底式平台连接上壳体和下壳体或柱靴的柱形结构物。 Columns are the tubular structural members of a column-stabilized or submersible unit, with which the upper hull and lower hulls or footings are connected into a integral structure.

Column-stabilized drilling unit 柱稳式钻井平台 系指依靠稳柱的浮力在装置的所有浮态作业模式或提升模式或下降模式时保持浮动和稳性的装置。其下船体或桩靴可以安装在稳柱的底部。 Column-stabilized drilling unit means a unit that depends upon the buoyancy of columns for flotation and stability for all floating modes of operation, or in the raising or lowering of the unit. Lower hulls or footings may be provided at the bottom of the columns.

Column-stabilized unit（semi-submersible unit） 柱稳式平台(半潜式平台) 系指用立柱或沉箱将上壳体连接到下壳体或柱靴上的平台。漂浮作业时下壳体或柱靴潜入水中,部分立柱露出海面,为半潜状态;坐底作业时下壳体或柱靴坐落在海底上,部分立柱露出海面,为坐底状态。 Column-stabilized unit (semi-submersible unit) means a unit with the upper hull connected to the lower hull or footings by columns or caissons. Drilling operations may be carried out in the floating condition with lower hull or footings submerged and some columns above sea surface, in which condition the unit is described as a semi-submersible, or when the unit is supported by the sea bed with lower hull or footings thereon and with some columns above sea surface, in which condition the unit is described as a submersible.

Combat information center 情报中心 系指将海上敌我作战动态等各种情报通过电子应用设备加以汇集并进行综合分析,为指挥员作战提供情报的场所。

Combination carrier 兼装船 系指:(1)所有用于装运散装油类和固体货物的船舶的总称;除污油舱内残留的含油混合物外,但这些货物不同时装运;(2)载运散货类或交替载运散装固体货物的液货舱。其设计类似于散货船包括矿砂船,但装设了管系、泵和惰性气体装置以便能够装卸指定处所油类货物;(3)设计用于既可载运100%载重量的散装液货也可装载100%散装干货的船舶。 Combination carrier is:(1) ship for carrying oil and soild cargo in bulk. But they can not be shipped at the same time except for for the residuals of oil mixture in the slop tanks.(2) the tanker carrying oil or alternatively solid cargoes in bulk, it is similar to bulk carriers (including ore carriers) in design, but fitted with piping, pumps and inert gas systems for loading and unloading of oil in specified spaces;(3) ship designed to load 100% deadweight with both liquid and dry cargo in bulk.

Combination fishing vessel 多种作业渔船 能从事两种或两种以上捕捞作业的渔船。

Combination piston 组合活塞 由几部分组合而成的活塞。

Combination purseanchor winch 上纲起锚联合绞机 在围网渔船起锚机中部增加一绳索卷筒,用以绞收围网引纲的绞机。

Combination system of framing 混合骨架式 主船体中段横剖面内,部分采用横骨架式、部分采用纵骨架式的船体骨架形式。

Combine engine drive 并车 由两台或两台以上主机拖动一个轴系的措施。

Combine engine drive gear 并车传动装置 通常是由联轴器或耦合器与多输入轴的减速齿轮箱组成的并车的传动设备。

Combined cargo/ballast tank 货油/压载兼用舱 系指作为船舶操作常规部分,用于运载货油或压载水的液舱,并将按压载舱处理。对于由 MARPOL 附则Ⅰ第18.3条规定,仅在例外情况下可以装载压载水的货油舱应按货油舱处理。 Combined cargo/ballast tank is a tank which is used for the carriage of cargo or ballast water as a routine part of the vessel's operation and will be treated as a ballast tank. Cargo tanks in which water ballast might be carried only in exceptional cases listed in MARPOL Ⅰ/18 (3) are to be treated as cargo tanks.

Combined control system for UUVs 多艘无人水下航行器联合控制系统

Combined cycle 联合循环 汽轮机和燃气轮机在热力上实现联合的循环。

Combined diesel and/or gas turbine (propulsion) plant 柴-燃联合装置 由柴油机和燃气轮机组成的交替使用或同时使用的联合推进装置。

Combined framed ship 混合骨架船 主船体中段,部分采用纵骨架式,部分采用横骨架式的船舶。

Combined gas turbine and/or gas turbine (propulsion) plant 全燃联合装置 由多台燃气轮机组成的交替使用或同时使用的联合推进装置。

Combined machinery (combined propulsion) plant 联合装置 由燃气轮机与其他发动机或多台燃气轮机组成的,在热力上实现联合循环或无热力联系的联合推进装置。有多机共轴、多机多桨等机型。

Combined navigation 组合导航系统 又称综合导航(integrated navigation)。是一套能将船舶上各种单一导航设备的信息(如电罗经的航向、计程仪的速度、

GPS 和罗兰 C 的位置、时间等)采入计算机,并经过综合处理得到本船的最佳的导航参数,从而实施对一些导航、航海设备的自动化控制。它是一种具备丰富导航服务功能的系统。船舶组合导航系统既是导航信息处理中心,可提供精确可靠的导航参数和各种导航辅助决策,又是导航、航海设备的控制中心,可对挂接的设备进行集中控制、故障检测等。其主要功能:(1)导航功能——这是组合导航系统的基本功能,包括确定船舶位置和运动参数、航路点导航、航线导航、导航报警等。组合导航系统输出的导航参数经过综合优化处理,精度比单一的导航设备高、可靠性好,因为系统的输入信息来源多,具有"冗余性",即使某一设备发生故障,系统仍有可靠的导航数据输出,这对航海船舶尤其重要。(2)避碰功能——组合导航系统可从导航雷达、红外线激光测距仪等装置来获得船的距离方位信息;或根据友邻船舶通报的船位、航速、航向、结合本船的运动参数,计算最近相遇距离,判断碰撞危险程度,并可做出避碰决策。(3)自动驾驶功能——如将组合导航系统与自动操舵仪连接起来,组合导航系统可提供自动舵调节参数,发出修正航向指令。当船舶偏离航线时,组合导航系统控制自动操舵仪使船舶返回航线,实现自动航迹保护。(4)对显示、记录、绘图设备的控制功能——综合导航系统可绘制航迹,自动形成航海目标,有的还配有电子海图。(5)故障处理功能——判断各信息源数据的正确性并进行报警等处理。

Combined steam and/or gas turbine (propulsion) plant 蒸-燃联合装置 系指由汽轮机和燃气轮机组成的交替使用或同时使用的联合推进装置。

Combined steam engine and exhaust turbine installation 蒸汽机-乏汽轮机联合装置 系指由蒸汽往复机和利用其排气为工质的乏汽轮机,通过液力耦合器、齿轮减速器作用于同一轴来驱动螺旋桨的联合装置。或由蒸汽机驱动螺旋桨而以乏汽轮机来驱动发电机或压气机的联合装置。

Combined stresses 组合应力 系指由组合载荷引起的应力,其载荷包括适用静载荷与相应的设计环境载荷的组合,并包括由加速度和倾斜引起的载荷,所对应的工况称为组合工况。 Combined stresses mean stresses due to combined loadings, where the applicable static loads are combined with relevant environmental loadings, including acceleration and heeling forces. The corresponding condition is called combined loading condition.

Combined transport 多式联运 系指至少以两种不同的运输方式,将货物在一国内接管的地点,运至另一国内指定交付的目的地的货物运输。 Combined transport means the carriage of goods by at least two different modes of transport, from a place at which the goods are taken in charge situated in one country to a place designated for delivery situated in a different country.

Combined transport document (CT document) 多式联运单证 系指证明履行货物多式联运和/或实现履行货物联运合同的一种单证,并在正面载有这样的标题:"根据多式联运单证统一规则签发的可转让的多式联运单证"或"根据联运单证统一规则签发的不可转让的多式联运单证"。 Combined transport document (CT document) means a document evidencing a contract for the performance and/or procurement of performance of combined transport of goods and bearing on its face either the heading "Negotiable combined transport document issued subject to United Rule for a Combined transport document" or the heading "non-negotiable combined transport document subject to United Rule for a Combined transport document".

Combined transport documents 联合运输单据

Combined transport documents (CTD) 多式联运单据 系指证明国际多式联运合同成立及证明多式联运经营人接管货物,并负责按照多式联运合同条款交付货物的单据。

Combined transport operator (CTO) 多式联运经营人 系指签发多式联运单证的人(包括任何法人、公司或法律实体)。如果国内法律规定,如果此人须经授权或得到许可后才有权签发多式联运单证,则多式联运经营人仅指这种经授权或得到许可的人。 Combined transport operator (CTO) means a person (including any corporation, company or legal entity) issuing a combined transport document. Where a national law requires a person to be authorized or licensed before being entitled to issue a combined transport document, then combined transport operator can only refer to a person so authorized or licensed.

Combined type metallic damping materials 复合型阻尼金属板材 系指在 2 块钢板或铝板之间夹有一层非常薄的高分子黏性材料构成的板材。也称夹心钢板或夹心铝板,金属板弯曲振动时,通过高分子黏性材料的剪切变形,发挥其阻尼作用。

Combined type starter 组合启动器 系指由多个不同类型和规格的启动器组合而成的控制设备。

Combined type valve 组合阀 系指吸、排气阀组合在一起的气阀。

Combustible gas detecting and alarm system 可燃气体探测报警系统 系指在船上存在可燃气体的处所,用来探测该处的可燃气体浓度,并在浓度超过爆炸上限时发出报警和发出相应的信号、动作,如增强通风、停止油泵或压缩机工作等,以保证安全运行的系统。

Combustible gas indicators 可燃气体探测仪 系指用来测定可燃蒸发气浓度的手提式测量仪器，并配有足够的备件。 Combustible gas indicator is an portable instrument used for measuring flammable vapour concentrations, together with a sufficient set of spares.

Combustible ice 可燃冰 学名甲烷水合物。它是甲烷气体和水分子形成的笼状结晶，如将两者分离，就能获得天然气。可燃冰是一种潜在的能源，储量很大，理论上来说，它储存的天然气比世界上现有的传统天然气的储量还要大。利用可燃冰的诀窍就是挖掘深埋在海底的资源。

Combustible ice research ship 可燃冰调查船 从事可燃冰调查的一种配置较完善的综合地质地球物理调查船。是一艘全天候综合调查船，不仅可以在深海水域以及恶劣海况下开展活动，还集合了勘探系统、地质系统以及水温系统，具有物理勘探、生态监测、水温监测等方面的功能。与国外仅具备1~2项勘探功能相比，它具备综合考察能力，往往一趟就能彻底摸清相关海域的整体情况以及水温、海流等具体细节，而且在解决动力定位与多波束相冲突方面也取得较好的效果。

Combustible liquid 可燃液体 系指闪点（开杯试验）≥26.6 ℃（80 °F）的液体。

Combustible material 可燃性材料 系指除不燃性材料以外的所有物质。 Combustible material is any material other than a non-combustible material.

Combustible range 燃烧范围 系指与空气中含氧量所对应的、有可能发生燃烧或爆炸的可燃性气体的浓度范围，又称爆炸范围。

Combustion 燃烧 系指氧化剂（通常是氧气或空气）与可氧化的材料迅速产生化学反应，足以产生辐射作用，即光和热的过程。 Combustion means a rapid chemical process that involves reaction of an oxidizer (usually oxygen or air) with an oxidizable material, sufficient to produce radiation effects, that is, heat and light.

Combustion 燃烧 可燃性材料和氧气发生反应并放出光和热的快速过程。其按燃烧的模式分为：发烟燃烧（flame combustion）、表面燃烧（surface combustion）、扩散燃烧（diffuse combustion）、一般燃烧（ordinary combustion）和预燃烧（pre-mixed combustion）5种。

Combustion chamber 燃烧室 (1)供燃料和空气混合燃烧的空间（内燃机）；(2)供燃料和空气混合并连续燃烧最终产生高温燃气的部件。(3)火焰筒外侧的空气流向与火焰筒中燃气流向相反的燃烧室。

Combustion efficiency 燃烧效率 燃料在燃烧室中燃烧实际放出的热量与理论上完全燃烧时放出热量的比值。

Combustion gas detecting device 可燃气体探测器 系指探测处所内可燃气体浓度的检测装置。其探头安装在泵舱、机舱等气体危险区域，而显示及报警装置设置在货物控制室、驾驶室内。可燃气体探测器有固定式（fixed combustion gas detecting device）和便携式（portable combustion gas detecting device）两种。

Combustion noise 燃烧噪声 系指燃料（包括固态、液态和气态燃料）在燃烧过程中发出的噪声。它不仅与燃烧过程的气体流动力和火焰的能量交换有关，还与燃烧过程的发声机制有关。燃烧系统发生的噪声又受到燃料喷嘴、燃烧室的几何形状等因素的影响，因此它的机制非常复杂。关于产生燃烧噪声的机制的研究工作还限于某些燃料混合气的自由火焰方面。

Combustion outer casing 燃烧室外壳 燃烧室外壁具有支撑火焰筒内部零件及承受力的作用等的筒形或环形结构。

Combustor (combustion chamber) 燃烧室 系指：(1)供燃料和空气混合燃烧的空间（内燃机）；(2)供燃料和空气混合并连续燃烧最终产生高温燃气的部件（燃气轮机）。

Comma 逗号

Command and control goal 1[1] 指挥和控制目标1（救生和人员撤离） 每艘舰船应配备设备和人员，以保持所有救生和人员撤离位置的指挥：(1)每艘舰船应装有满足SOLAS公约第3章B部分第6.4.2条要求的通用紧急警报装置。警报应在所有开敞的甲板以及每一个舱室可以听见；(2)每艘舰船应装有满足SOLAS公约第3章B部分第6.5条要求的扩音系统；(3)紧急情况指示装置应满足SOLAS公约第3章B部分第8条和第9条的要求；(4)人员配置和救生设备监督应满足SOLAS公约第3章B部分第10条的要求；(5)紧急和训练演习应满足SOLAS公约第3章B部分第19条的要求。 Every ship is to be equipped and manned so that command of all life-saving and evacuation situations can be maintained: (1) Every ship is to be fitted with a general emergency alarm in accordance with SOLAS Chapter Ⅲ, Part B, Regulation 6.4.2 as applicable. The alarm is to be audible on all open decks and every compartment. (2) Every ship is to be fitted with a public address system in accordance with SOLAS Chapter Ⅲ, Part B, Regulation 6.5. (3) Emergency situation instructions are to be in accordance with SOLAS Chapter Ⅲ, Part B, Regulations 8 and 9. (4) The manning and supervision of life saving appliances are to be in accordance with SOLAS Chapter Ⅲ, Part B, Regulation 10. (5) Emergency and training drills are to be in accordance with SOLAS Part B, Chapter Ⅲ, Regulation 19.

Command and control goal 1[2]　指挥和控制目标 1（消防控制）　应设置消防控制站以便对所有可进行灭火的着火点进行统一指挥。　Command and control goal 1 means that fire control stations are to be provided so that there will be a central point of command in all fire situations where fire-fighting may be expected.

Command and control goal 2[1]　指挥和控制目标 2（救生和撤离）　在中央司令部与战略救生和撤离点之间应有可靠语音通信手段:每艘舰船应提供双向 VHF 无线电话设备。　Command and control goal 2 means that reliable means of speech communication are to be provided between the central point of command and strategic life-saving and evacuation stations:Two-way VHF radiotelephone apparatus is to be provided on every ship.

Command and control goal 2[2]　指挥和控制目标 2（消防控制）　应在消防控制站和确认的火灾危险区域之间设置固定的双通道语音通话设备。　Command and control goal 2 means that fixed means of two-way speech communication are to be provided between the fire-fighting control station and identified fire risk areas.

Command and control objective[1]　指挥和控制目标（消防控制）　应提供适当的方法来确保任何积极的防火控制方法能被安全、有效地结合在一起。　Command and control objective means that suitable means are to be provided to ensure any active fire control measures can be safely and effectively orchestrated.

Command and control objective[2]　指挥和控制目标（救生和人员撤离）　提供适宜的方式以确保救生和撤离活动能够安全和有效地进行。　Command and control objective means that suitable means are to be provided to ensure that life-saving and evacuation operations can be safely and effectively orchestrated.

Command telephone for navigating　航行指挥电话　系指用于航行驾驶和操纵各作战部位之间进行指挥和通信联络的电话。

Command,Control,Communication,Computer,Intelligence,Surveillance and Reconnaissance（C4ISR）　美军情报监视与侦察系统　C4ISR 是军事术语,意为自动化指挥系统,它是现代军事指挥系统,C4 代表指挥,控制,通信,计算机,4 个字的英文开头字母均为"C",所以称"C4","I"代表情报;"S"代表电子监听;"R"代表侦察,7 个子系统的英语单词的第一个字母的缩写,即指挥 Command、控制 Control、通信 Communication、计算机 Computer、情报 Intelligence、监视 Surveillance、侦察 Reconnaissance。C4ISR 就是美国开发的一个通信联络系统。科索沃战争是美军第一次大规模实战运用 C4ISR 系统。该系统包括"通信"与"电脑"两大子系统。

Commandant　司令官　系指(美国)海岸警卫队司令官或其授权代表。Commandant means the Commandant of the Coast Guard or an authorized representative(U.S.).

Commanding view　指挥视野　系指无可能干扰驾驶人员执行其主要任务的障碍物的视野,至少包括安全执行避碰功能所需的视野。　Commanding view means a view without obstructions, which could interfere with the navigator's ability to perform his main tasks, at least covering the field of vision required for safe performance of collision avoidance functions.

Commencement of repairs　开工　系指船舶铺龙骨或修船工程点火施工的时刻。

Commercial bill　商业汇票　系指由债权人向债务人发出的支付命令书,命令他在约定的期限支付一定款项给第三人或持票人。

Commercial invoice　商业发票　系指出口方向进口方开列发货价目清单,是买卖双方记账的依据,也是进出口报关交税的总说明。商业发票是一笔业务的全面反映,内容包括商品的名称、规格、价格、数量、金额、包装等,同时也是进口商办理进口报关不可缺少的文件,因此商业发票是全套出口单据的核心,在单据制作过程中,其余单据均需参照商业发票缮制。

Commercial paper　商业本票　又叫一般本票。货币市场主要的交易工具就是商业本票,早期为确保债权,主要流通的票券是有实质交易基础的"交易性商业本票",也就是俗称的 CP1,不过有交易作基础的商业本票,金额往往不整齐、常有零头,造成交易上的不便,因此在时间推进的发展结构中,逐渐产生变革,借由银行保证程序,发行的"融资性商业本票"兴起,也就是 CP2,逐渐在整数金额方便交易的优势,以及银行保证的信用作后盾下,跃升市场交易的主流,目前融资性商业本票交易约占货币市场交易的九成以上。

Commercial towing　商业拖航　系指非救助拖航(not in nature of salvage)和非应急拖航(non-emergency towing)。

Commercial vessel　商船　(1)有自航能力的船舶,但军舰、军用船舶和其他公务船除外。(2)在五大湖的美国水域装载、卸载干散货物的商船。或在五大湖任何区域运输散装货物和营运的美国船舶。但该术语并不包括非自航的驳船,除非它是拖船与驳船的组合体。Commercial vessel means:(1)a self-propelled vessel other than a naval, military or other public vessel.(2)a commercial vessel loading, unloading, or discharging bulk dry cargo in the U.S. waters of the Great Lakes, or a U.S. commercial vessel transporting bulk dry cargo and operating anywhere on the Great Lakes, but the term does not include a non-self-propelled barge unless it is part of an integrated tug and barge unit.

Comminuter 粉碎机 (1) 对"73/78 防污公约"附则Ⅳ而言，系指用于将生活污水中的杂物进行粉碎的机械。(2) 对"73/78 防污公约"附则Ⅴ而言，系指用于将船上垃圾进行粉碎的机械。

Commission 佣金 系指在国际商业活动中，代办处理人或经纪人为委托人服务而收取的报酬，或者是中间商、代办处理商在介绍交易成交后而取得的收入。

Commission agency 佣金代理 又称一般代理。系指在同一代理地区、时间及期限内，同时有几个代理人代表委托人行为的代理人。

Committee 委员会 系指 IMO 组织的海上环境保护委员会。Committee means the Marine Environment Protection Committee of the Organization.

Committee of experts on the transport of dangerous goods 联合国危险品运输专家委员会

Commodity tax 商品税 国际上通称为"商品和劳务税"，是以商品为征税对象的一类税的总称，主要包括增值税、消费税、营业税、关税 4 个税种。

Commom rail pipe (accumulator) 共轨管(蓄能器) 系指电控柴油机的燃油或液压油系统中，用来提供高压燃油或润滑油的压力容器。

Common accumulator 共用蓄能器(共轨管) 系指为所有气缸共用的压力容器，其用于提供液压油或加压燃油。Common accumulator is a pressure vessel common to all cylinders for providing hydraulic oil or pressurized fuel oil.

Common antenna system (CAS) 共用天线系统 系指用于接收岸上的无线电和电视广播信号，并分配至船上的居住舱室、公共场所的系统。它由全向船用电视、广播接收天线、电视信号控制箱、串行信号输出终端盘等组成。

Common but differentiated responsibilities (CBDR) 共同但有区别的责任

Common cause 共同原因 系指影响到几个要素的事件，这些要素在其他方面被视为独立的或多余的。Common cause means an occurrence that affects several elements which are otherwise considered independent or redundant.

Common cause failure 共因失效 系指可能导致多个组件、子系统和系统同时失败的条件。例如，这些条件可以同时突破多层保护。导致共因失效的原因可能有：环境(如火灾、洪灾)、设计缺陷、制造误差、检测、维护、操作失误等。

Common failure mode 公共故障模式 一种影响重要设备的两种或两种以上相似项目的故障。

Common link 普通链环 组成锚链各链节的基本锚链环。

Common loading hold 共同装载舱 系指具有一层或多层中间甲板，且当货物超过中间甲板时，中间甲板的开口没有有效的舱密装置来关闭的装货处所。

Common market/Single market 共同市场 系指 2 个或 2 个以上的国家之间通过达成某种协议，不仅要实现共同市场的目标，还要在共同市场的基础上，实现成员国经济政策的协调。共同市场有：南方共同市场，或称"南锥共同市场"(Southern Common Market)；中美洲共同市场[Central American Common Market (CACM)]；加勒比共同体，Caribbean Community，旧称"加勒比共同体与共同市场"，"Caribbean Community and Common Market"；欧洲共同市场；两岸共同市场；白种人共同市场(Caucasian Common Market)。

Common rail system 共轨系统(共管系统) 其主要结构特点是柴油机的各缸的燃油喷射都由 1 个蓄压容器(或 1 压力油管)提供，使柴油机各缸燃油喷射均匀，并能通过电控系统调节燃油喷射压力、喷射正时和喷射模式等。共轨中的介质有燃油和润滑油两类：共轨系统按压力高低区分有中压及高压两类。凡共轨中的燃油压力已达喷油压力的，称高压共轨系统；凡共轨中的燃油压力较低尚需通过增压泵才能达到喷油压力的，称中压共轨系统。有的柴油机中也兼有高压共轨及中压共轨，高压喷油，中压排气。Common rail system means a system in which its main features are that fuel oil injection of each cylinder of diesel engine is provided by a pressure chamber (or a pressure oil pipe) to ensure even fuel oil injected into each cylinder, and the fuel oil injection pressure, time and mode can be adjusted by electronic control system. Fuel oil or lubricating oil are used as fluid media in common rail. According to the operating pressure, common rail system is divided into medium pressure and high pressure. High pressure common rail system means that the fuel oil pressure in the common rail has reached fuel injection pressure. If the fuel oil pressure in common rail is lower and fuel injection pressure will be achieved by pressure pump, it is called medium pressure common rail system. Some diesel engines have both high pressure common rail system (fuel oil injection) and medium pressure common rail system (exhaust).

Common stator electric generator 共定子电动发电机 系指一种具有 1 个磁场和 2 个电枢或 1 个含有几个独立绕组的电枢，并兼有电动机和发电机两种作用的变流机。

Common structural rules(CSR) 共同结构规范 系指国际船级社协会通过的"散货船共同结构规范"和"双壳油船共同结构规范"。 Common structural rules means the "Common Structural Rules for Bulk Carriers" and the "Common Structural Rules for Double Hull Oil Tankers" adopted by IACS.

Commutator room 变流机室 供安置各种变流器用的舱室。

Communication branch circuit 通信分支电路 系指在船内或船上,用于声、光信号已经由一处到另一处进行信息交流的电路。

Communication engineering 通信工程 是电子工程的一个重要分支,同时也是其中一个基础学科。

Communication light on topmast 桅顶通信灯 安装在桅顶的通信灯。

Communication searchlight 通信探照灯 固定安装在船上,按编码符号,用可见光与船外进行远距离通信联络的探照灯。见图 C-8。

图 C-8 通信探照灯
Figure C-8 Communication searchlight

Communication system(DPS) 通信系统(动力定位系统) 系指在 DP 控制站和下列位置之间设有一个双向的通信设施:(1)驾驶室;(2)主机控制室;(3)有关操作控制站。 Communication system (DPS) is a two way communication device fitted between DP control center and following locations:(1) wheelhouse;(2) main engine control room;(3) relevant operating control room.

Communication technology 通信技术 又称通信工程(也作信息工程、电信工程,旧称远距离通信工程、弱电工程)是电子工程的重要分支,同时也是其中一个基础学科。该学科关注的是通信过程中的信息传输和信号处理的原理和应用。通信工程研究的是,以电磁波、声波或光波的形式把信息通过电脉冲,从发送端(信源)传输到一个或多个接收端(信宿)。接收端能否正确辨认信息,取决于传输中的损耗高低。信号处理是通信工程中一个重要环节,其包括过滤、编码和解码等。专业课程包括计算机网络基础、电路基础、通信系统原理、交换技术、无线技术、计算机通信网、通信电子线路、数字电子技术、光纤通信等。

Communication unit 通信单元

Communications in an emergency 救生用通信 系指在救生艇、筏集合和登乘地点与主控制站和/或驾驶台(如设有)之间,设有能传送命令的相互通信系统。该通信系统可由可携式设备或固定安装的设备组成,并应在主电源失电情况下仍能工作。 Communications in an emergency means an intercommunication system provided between muster and embarkation stations for survival craft and main control station and/or if fitted, navigating bridge, which enables commands to be transmitted. The communication system may comprise portable or permanently installed equipment, and is to be operable in the case of a failure of the main power supply.

Commuter(crew boat) 交通船 为港区内船舶或海上作业平台运送工作人员的专用船舶。此类船舶的航程离岸基地较近,船上一般不设炉灶、生活舱室及其他生活设施。

Compact 合同

Compact disc read-only memory(CD ROM) 光驱 系指只读光盘驱动器。

Compact flash CF 卡 最初是一种用于便携式电子设备的数据存储设备。作为一种存储设备,它革命性地采用了闪存技术,于 1994 年首次由 SanDisk 公司生产并制定了相关规范。当前,它的物理格式已经被多种设备所采用。从外形上 CF 卡可以分为两种:CF I 型卡以及稍厚一些的 CF II 型卡。从速度上它可以分为 CF 卡、高速 CF 卡(CF + /CF 2.0 规范),更快速的 CF 3.0 标准也即将在 2005 年被采用。CF II 型卡槽主要用于微型硬盘等一些其他的设备。

Compact semi-submersible platform(CSSP) 紧凑型半潜式平台

Compacting hammer barge 打夯船 系指专供打夯用的船舶。设有吊夯架、绞车等。重力锤由绞车提起,根据需要的高度自由降落,利用其下落时的夯击力夯实地基。

Companion 舱口围罩 系指出入舱口上用以遮挡

Companion ladder 舱室梯　系指设置在生活舱室内的梯。

Companionway 升降口　系指风雨密性的甲板结构，保护通向干舷甲板以下处所的通道开口或进入围蔽上层建筑处所的通道开口。Companionway is defined as a weathertight deck structure, protecting an access opening leading below the freeboard deck, or into a space within an enclosed superstructure.

Companionways cover 舱棚出入口盖　系指设置在小型船舶的半升高舱棚或棚壁与棚顶连接处出入口上的舱口盖。

Company 公司（ISM 规则）　对 2010 年船舶防污底系统检验和发证指南而言，系指船舶所有人或任何其他组织或个人，例如管理者或光船承租人，其已从船舶所有人处接受船舶营运的责任，并在接受责任时，已同意承担国际安全管理（ISM）规则规定的所有责任和义务。For guidelines for survey and certification of antifouling systems on ships in 2010, company means the owner of the ship or any other organization or person such as the manager or the bareboat charterer, who has assumed the responsibility for the operation of the ship from the owner of the ship and who, on assuming such responsibility, has agreed to take over all duties and responsibility imposed by the International Safety Management (ISM) Code.

Company 公司（ISPS 规则）　系指按 SOLAS 第Ⅸ章第 1 条所定义如下任何一个管理涉及船舶保安体系的船舶的公司：(1) 与船舶所有人以管理合同或光船承租合同方式，承担了船舶的操作、维护、船员配备等责任的一个独立的组织或个人；(2) 构成组织的一部分并承担管理职能的部门，在此情况下，该部门（或几个部门）负责船舶操作、维护和船员配备等试验管理活动。对于仅履行部分管理活动的组织不符合"公司"的定义；(3) 船舶经营人、船舶管理人、光船租船人或已承担船舶营运责任并在承担此种责任时，同意承担 SOLAS 第Ⅸ章要求的所有责任和义务的任何其他组织和个人。

Company audit 公司审核　系指审核机构对船舶管理公司的审核。

Company audit on site 公司现场审核　系指由审核机构派员到被审核的船舶管理公司进行的审核。

Company auditor 公司审核员　公司内部的审核员。

Company Energy Efficiency Management Plan (CEEMP) 公司能效管理计划

Company security officer (CSO) 公司保安员　系指由公司指定的人员，负责确保船舶保安评估得以展开、"船舶保安计划"得以制订、提交批准，而后得以实施和维持，并与港口设备保安员和船舶保安员进行联络。Company security officer means the person designated by the Company for ensuring that a ship security assessment is carried out, and that a ship security plan is developed, submitted for approval, implemented and maintained, and cso acts as a liaison with port facility security officers and the ship security officer.

Compaq 康柏电脑公司　是由罗德·肯尼恩（Rod Canion）、吉米·哈里斯（Jim Harris）和比利·默顿（Bill Murto）三位来自德州仪器公司的高级经理于 1982 年 2 月，分别投资 1 000 万美元共同创建的。2002 年康柏公司被惠普公司收购。

Comparable characteristics, cargoes, and operations 可比较的特征、货物和营运　系指类似船舶的设计、尺度、船龄、船员的补给物、货物、营运航线。甲板和货舱的结构和固定的货物传输设备的配置。Comparable characteristics, cargoes, and operations means similar vessel design, size, age, crew, complement, cargoes, operational routes, deck and hold configuration, and fixed cargo transfer equipment configuration.

Compartment 舱室　(1) 一般系指船体内部，由纵、横舱壁或其他构件分隔成，供船上人员工作、生活或安置、存放、装载各种设备、物品等用的空间；(2) 由船外板、甲板、和舱壁形成的一部分船体。原则上应是水密的。

Compass deck 罗经甲板　系指安置标准磁罗经的一层甲板。

Compass-WALCS 三维波浪载荷计算软件系统　其主要功能，是解决船舶与海洋工程结构弹性效应与流场的耦合作用以及海洋平台与柔性构件的耦合问题，给出作用于船体或海洋平台的波浪载荷响应结果。该软件系统包括船舶与海洋工程三维波浪载荷计算基本模块、线性水弹性分析模块、非线性水弹性分析模块、有航速船舶运动与载荷计算模块以及浮式海洋平台与柔性附连构件的耦合分析模块。

Compatibility 兼容性　系指有点透明度的互配，该透明度对系统实体之间的连接足以支持一可接受的服务等级。

Compatibility with slops 污液相容性

Compelling 引人入胜的

Compensating water system 补重系统　对排水量和稳性要求比较严格的船舶，为避免大量燃油消耗之后影响船舶的纵横倾吃水而设置以舷外水压载以调整纵横倾状态的补偿压载水系统。

Compensation 补偿金　系指当一个公共部门必须征购土地用于建设时，应当给予土地原来所有者的补偿。

Compensation for expansion 膨胀补偿 系指承受胀缩或其他应力的管子,应采取管子弯曲或膨胀接头等必要的补偿措施。油管和消防水管的膨胀补偿装置和法兰垫片应由不燃材料制成。管路中所使用的膨胀接头应为认可的型号,与膨胀接头毗连的管子应适当校直和固定,必要时,波纹管型膨胀接头需加以防护,以防机械损伤。铺设在平台间步桥上的管子应设有补偿位移措施。 Compensation for expansion means a suitable provision for compensation is to be made for all pipes subject to expansion, contraction or other strain, such as bends, loops, or expansion joints as required. Expansion appliances and flange washers on oil and fire service pipes are to be made of non-combustible materials. Where expansion pieces are fitted in piping, they are to be of an approved type. The adjoining pipes are to be suitably aligned and anchored. Where necessary, expansion pieces of bellows type are to be protected against mechanical damage Suitable provision for compensation of shifting is to be made for pipes laid on any gangway between platforms.

Compensation trade 补偿贸易 又称产品返销,系指交易的一方在对方提供信用的基础上,进口设备技术,然后以该设备技术所生产的产品,分期抵付进口设备技术的价款及利息。

Competent authority 适任机构(耐火试验) 系指由主管机关授权执行"国际耐火试验程序应用规则"要求的功能的组织。 Competent authority means an organization authorized by the Administration to perform functions required by The International Code for Application of Fire Test Procedures.

Competent Authority 主管当局 系指:(1)与国际海运固体散装货物规则相关的任何目的而指定或以其他方式认可的在任何国家的管理机构或机关;(2)经某一缔约国指定的在特定地理区域或专业领域内负责按2009年香港国际安全与环境无害化拆船公约规定的该缔约国管辖范围内作业的拆船厂相关事宜的有关或多个政府当局。 Competent Authority means:(1) any national regulatory body or authority designated or otherwise recognized as such for any purpose in connection with this Code;(2) A governmental authority or authorities designated by a Party as responsible, within specified geographical area(s) or area(s) of expertise, for duties related to Ship Recycling Facilities operating within the jurisdiction of that Party as specified in Hong Kong International Convention for the safe and environmentally sound cycling of ships, 2009.

Competent person 适任人员[1] 对2009年香港国际安全与环境无害化拆船公约而言,系指具备适当的资质、经培训和具有足够的知识、经验和技能来开展具体工作的人员。具体来说,适任人员可以是受过培训的工人或管理人员,其能够识别和评估拆船厂中的职业危险、风险和员工是否曝露于潜在有害物质或不安全的条件下,并能制定必要的保护和预防措施来消除或减少这些危险、风险或曝露情况。主管机关可制定指定此类人员的适当标准,并可确定应授予其的职责。 For Hong Kong International Convention for the Safe and Environmentally Sound Recycling of Ships, 2009, competent person means a person with suitable qualifications, training and sufficient knowledge, experience and skill, for the performance of the specific work. Specifically, a competent person may be a trained worker or a managerial employee capable of recognizing and evaluating occupational hazards, risks, and employee exposure to potentially hazardous materials or unsafe conditions in a ship recycling facility, and who is capable of specifying the necessary protection and precautions to be taken to eliminate or reduce those hazards, risks, or exposures. The Competent Authority may define appropriate criteria for the designation of such persons and may determine the duties to be assigned to them.

Competent person 责任人员[2] 系指有权进入某一围蔽处所和具有对紧接着各种程序足够知识的人员能对已存在或要出现的危险情况作出判断。 Competent person means a person with sufficient theoretical knowledge and practical experience to make an informed assessment of the likelihood of a dangerous atmosphere present or subsequently arising in the space.

Competent person 主管人员[3] 对澳大利亚海事安全局海事指令而言,系指具有实践经验和理论知识,以及相关经验的人员,这足以能使该类人员发现和评估可能影响该设备预定功能的任何缺陷和任何薄弱环节。

For the maritime order of the Australian Maritime Safety Authority (AMSA), competent person means a person having practical and theoretical knowledge and relevant experience, sufficient to detect and evaluate any defects and any weaknesses that may affect the intended performance of the equipment.

Competition 竞争

Competitive bids/Open tendering 公开招标 系指招标人在公开媒介上以招标公告的方式邀请不特定的法人或其他组织参与投标,并向符合条件的投标人中择优选择中标人的一种招标方式。相对于公开招标,称之为邀请招标。按照招标人和投标人参与程度,可将公开招标过程粗略划分成招标准备阶段、招标投标阶段和决标成交阶段。

Compiler 编译器　系指将便于人们编写、阅读、维护的高级计算机语言翻译为计算机能识别、运行的低级机器语言的程序。编译器将源程序（Source program）作为输入，翻译产生使用目标语言（Target language）的等价程序。

Complementary metal oxide semiconductor（CMOS） 互补性氧化金属半导体芯片　在计算机领域，CMOS 通常系指保存计算机基本启动信息（如日期、时间、启动设置等）的芯片。有时人们会把 CMOS 和 BIOS 混称，其实 CMOS 是主板上的一块可读写的 RAM 芯片，是用来保存 BIOS 的硬件配置和用户对某些参数的设定。CMOS 可由主板的电池供电，即使系统失电，信息也不会丢失。

Complete economic integration 完全经济一体化　系指两个或两个以上的国家在现有生产力发展水平和国际分工的基础上，由政府间通过协商缔结条约，建立多国的经济联盟。在这个多国经济联盟的区域内，商品、资本和劳务能够自由流动，不存在任何贸易壁垒，并拥有一个统一的机构，来监督条约的执行和实施共同的政策及措施。

Complete inspection/Complete survey 全面检查　系指用目测检查，必要时辅之以其他方法，并尽可能仔细进行，以使对所检查的部件得出安全可靠的结论。为此目的，必要时应将部件或机件拆开检查。

Complete tank 完整的液舱　系指包括所有液舱的边界和内部结构，以及液舱甲板上的外部结构的构件。Complete tank mean structural members including all tank boundaries and internal structure, and external structure on deck.

Complete transverse web frame ring 完整的横向环状框架　系指包括相邻的结构构件的构件。Complete transverse web frame ring means a structural member including adjacent structural members.

Complex cycle 复杂循环　系指具有回热、中间冷却剂中间加热等工作过程的复杂形式的燃气轮机循环。

Complex cycle gas turbine plant 复杂循环燃气轮机装置　系指按复杂循环工作的燃气轮机动力装置。

Complex deemulsifier 复合型破乳剂

Complex impedance sound eliminator 阻抗复合式消音器　系指由低、中频消音效果较好的抗性消音器和中、高频消音效果较好的阻性消音器组合而成的消音器。

Compliance declaration of a ship's energy efficiency management plan 船舶能效管理计划符合声明　系指根据经修订的 MARPOL 公约附则 VI 第 22 条要求，对船舶能效管理计划进行符合性核查合格后签发的符合声明。该符合声明不代表任何主管机关签发，仅为满足 MARPOL 附则 VI 对 SEEMP 文件符合性验证之目的，从 SEEMP 验证实施程序方面提出的控制要求。

Compliance officer 法律人员　系指律师或学习经济、法律专业知识的人员。通常由特聘律师、企业法律顾问或熟悉有关法律规定的人员担任。

Compliant lower 柔性塔

Compliant pile tower（CPT） 顺应式桩承塔

Compliant platform /Compliant tower（CT） 顺应式平台　系指一种具有很大柔性的坐底式结构。

Compliant production system 顺应式生产系统　系指以顺应式结构支承海上油气处理装置的生产系统。

Compliant structure 顺应式结构/顺应式平台　系指利用拉索、张力腿、万向接头等构件，对结构物在外载荷作用下产生的 6 个自由度的运动加以某种限制与约束，以满足定位与运动要求的半固定式结构。

Compliant Tower 柔性塔架　系指瘦长型的，相当匀称的框架式塔架。在其顶部支撑甲板并用桩固定到海底。其瘦长型的设计使其更柔性，因此它比其他固定式平台更易于偏转，但具有比台风波浪的周期长得多的周期。Compliant Tower is a slender, nearly uniform frame tower that supports a deck at the top and is fixed to the seabed with piles. Its slender design makes it more compliant, and thus it is easier to deflect than other fixed platforms, but with periods that are much longer than those of hurricane sea waves.

Compliant vertical accessing riser（CVAR） 垂直通路顺应式立管

Component 部件　系指：（1）用材料制成产品或系统的零件/构件。（2）影响 NO_x 排放特性的那些互换性部件，由其设计/部件号标识。Component mean：（1）parts/members of a product or system formed from material.（2）those interchangeable parts which influence the NO_x emissions performance, identified by their design/parts number.

Component video connector 色差端子　又称分量接口，是把类比视频中的明度、彩度、同步脉冲分解开来各自传送的端子。

Composite oil-exhaust gas fired boiler 废气燃油组合式锅炉　系指将两个独立使用的燃油锅炉和废气锅炉组合在一个壳体里的锅炉。

Composite armor 复合装甲　系指由两层以上不同性能的防护材料组成的非均质坦克装甲，一般来说，是由一种或者几种物理性能不同的材料，按照一定的层次比例复合而成，依靠各个层次之间物理性能的差异来干扰来袭弹丸（射流）的穿透，消耗其能量，并最终达到阻止弹丸（射流）穿透的目的。这种装甲分为金属与金属复合装甲、金属与非金属复合装甲以及间隔装甲三种,它们均具有较强的综合防护性能。该技术拥有国:英国(酋长)、美国(M1)、法国(Leclerc)、日本(90式)、中国(99式)、韩国(K1)、俄罗斯(T-90)、以色列(梅卡瓦)。见图C-9。

图C-9　复合装甲
Figure C-9　Composite armor

Composite damping structure　复合阻尼结构
Composite material　复合材料
Composite ship　混合结构船　系指外板和骨架分别以不同材料构成的船舶。
Composite slab　叠合板
Composite video　合成视频信号　是将全部信号打包成一个整体进行传送。
Composite video port　复合视频端子　又称AV端子,是几乎所有的电视机、影碟机类产品都用的接口,是目前最普遍的一种视频接口。其用于NTSC、PAL、SECAM等电视制式中的常见端子。AV端通常是黄色的RCA端子,另外配合两条红色与白色的RCA端子传送音信。欧洲的电视机通常以SCART端子取代RCA端子,不过SCART的设计上可以载送画质比YUV更好的RGB信号,故也被用来连接显示器、电视游乐器或DVD播放机。在专业应用当中,也有使用BNC端子以求获得更佳信号品质。在20世纪80年代早期,当时的个人电脑与电视游乐器通常也输出复合视信,使用者必须使用一台RF调变器来将其载波导向至电视的特定频道,在北美常为第3或第4频道(66~72 MHz),欧洲为36频道,日本日规第1或第2频道(90~102 MHz);台湾则因当时仅开放1~7CH(即美国规则VHF 7~13CH,但7~12CH被台视、中视、华视使用中),故仅剩第13频道(210~216 MHz)可用。
Composition of foreign trade　对外贸易货物结构　系指一定时期内一国进出口贸易中各类货物的构成,即某大类或某种货物出口贸易与整个进出口贸易额之比,以份额表示。其可以反映出一国的经济发展水平、产业结构状况和第三产业发展水平等。
Composition of international trade　国际贸易货物结构　系指一定时期内各大类货物或某种货物在整个国际贸易中的构成,即各大类货物或某种货物贸易额与整个世界进出口贸易额相比,以比重表示。其可以反映出世界的经济发展水平、产业结构状况和第三产业发展水平等。
Compound (expansion) engine　双胀式蒸汽机　系指蒸汽依次在高压及低压缸内进行二次膨胀,将蒸汽热能转换成机械功的蒸汽往复机。
Compound compressor　联合式压缩机　一种以离心式压缩机或回转式压缩机作为增压级的活塞式压缩机。
Compound generator　复励发电机　系指一种具有两个独立磁场绕组的直流发电机。其中提供主要励磁的一个绕组与电枢电路并联,而只提供部分励磁的另一个绕组则与电枢电路串联,此串联绕组提供励磁所占的比例为:要采用均压线才能实现满意的并联运行。
Compound motor　复励电动机　系指具有两个独立磁场绕组的直流电动机。一个绕组与电枢电路并联连接,通常产生主要的磁场;另一个绕组则与电枢电路串联连接。就转速和转矩而言,其特性介于并励和串励电动机之间。
Compound = supercharged diesel engine　复合式增压柴油机　废气涡轮增压柴油机中,废气涡轮与柴油机曲轴用机械方式连接的柴油机。
Comprehensive export processing zone　综合性出口加工区　即在区内可以经营多种出口加工工业。
Comprehensive laboratory (universal laboratory)　综合实验室　又称通用实验室,安装通用和专用仪器及实验设备,能从事多项目或多学科的试验、观测、记录和分析用的科学研究实验室。
Comprehensive research ship　综合调查船　用于对海洋进行多种学科海洋调查研究的船舶。船上设有海洋水文、海洋物理、海洋化学、地球物理、海洋地质地貌、海洋生物、海洋气候等实验室及相应的仪器设备,并能同时进行多科目的调查研究工作。见图C-10。
Comprehensiveness principle　综合性原则　是在风险评估和提出建议时考虑范围的价值,包括经济、环境、社会和文化价值的原则。　Comprehensiveness principle is a principle that the full range of values, including economic, environment, social and cultures, are considered when assessing risks and making recommendation.
Compressed air　压缩空气
Compressed air testing for fillet welding　压缩空气

图 C-10　综合调查船
Figure C-10　Comprehensive research ship

填角焊试验(渗漏试验)　系指适用于填角焊的带有渗漏指示方法的T型接头填角焊空气试验。

Compressed natural gas（CNG）　压缩天然气　CNG means compressed natural gas.

Compression chamber　加压舱　通过注入压缩气体,首先提高舱压,创造高气压环境条件,而后逐步减压,可供科学研究,模拟潜水,潜水减压及医疗救治等目的使用的一种钢质承压容器。

Compression pressure　压缩压力　压缩行程终了时气缸内的压力。

Compression ratio　压缩比　系指气缸总容积和压缩容积之比,为理论值。

Compressor　压气机　系指:(1)使空气或其他气体压缩以供给燃气轮机燃烧室等高压气流的叶轮式工作机械;(2)是将气态制冷剂压缩成液态供循环使用的装置。

Compressor and turbine equilibrium running line　压气机涡轮共同工作线　表示在压气机特性曲线上的由增压涡轮与压气机功率平衡的工作点连接而成的共同工作曲线。

Compressor performance characteristic diagram (curve)　压气机特性曲线　表示压气机各特性参数之间关系的曲线。

Compressor surging test　压气机喘振边界测定　为测定压气机在不使用防喘装置时的喘振裕度所做的试验。

Compressor turbine (charging turbine, gas generator turbine, compressor-driving turbine)　增压涡轮　所产生的动力全部用来驱动压气机,而其排气侧进入动力涡轮继续膨胀的燃气发生器中的涡轮。

Compulsory license　强制许可证

Computed tare　约定皮重　系指买卖双方事先协商确定的皮重,不必一一衡量。计算皮重究竟采取哪一种方法,在磋商交易和签订合同时,应明确加以规定。

Computer　计算机　系指用来储存和处理数据,进行计算或执行控制的一种可编程的电子装置。它包括基本的计算单元、输入和输出设备、控制单元以及程序和数据存储器。它可以是单机或由几个内部相连的单元构成,并且应包括由主机和小型机或微型机构成的可编程电子系统(PES)。Computer means a programmable electronic device for storing and processing data, making calculations, or performing control. It comprises basic compute elements, input and output devices, control units, as well as programs and data memories. It may consist of a stand-alone unit or may consist of several interconnected units and include any programmable electronic system (PES) which is made up of mainframe, minicomputer or microcomputer.

Computer aided business administrate　(CABA)　计算机辅助经营管理

Computer aided design（CAD）　计算机辅助设计　系指利用计算机及其图形设备帮助设计人员进行设计工作。在工程和产品设计中,计算机可以帮助设计人员担负计算、信息存储和制图等项工作。CAD 能够减轻设计人员的劳动,缩短设计周期和提高设计质量。在设计中通常要用计算机对不同方案进行大量的计算、分析和比较,以决定最优方案;各种设计信息,不论是数字的、文字的或图形的,都能存放在计算机的内存或外存里,并能快速地检索;设计人员通常用草图开始设计,将草图变为工作图的繁重工作可以交给计算机完成;利用计算机可以进行与图形的编辑、放大、缩小、平移和旋转等有关的图形数据加工工作。

Computer aided design calculating　计算机辅助设计计算

Computer aided engineering（CAE）　计算机辅助工程　CAE 技术的提出就是要把工程(生产)的各个环节有机地组织起来,其关键就是将有关的信息集成,使其产生并存在于工程(产品)的整个生命周期。因此,CAE 系统是一个包括了相关人员、技术、经营管理及信息流和物流的有机集成,且优化运行的复杂的系统。随着计算机技术及应用的迅速发展,特别是大规模、超大规模集成电路和微型计算机的出现,使计算机图形学(Computer Graphics,CG)、计算机辅助设计(Computer Aided Design/CAD)与计算机辅助制造(Computer Aided Manufacturing/CAM)等新技术得以十分迅猛的发展。CAD、CAM 已经在电子、造船、航空、航天、机械、建筑、汽车等各个领域中得到了广泛的应用,成为最具有生产潜力的工具,展示了光明的前景,取得了巨大的经济效益。计算机技术的迅速发展还推动了现代企业管理的发展,企业管理借助于管理信息系统的。

Computer aided manufacturing （CAM） 计算机辅助制造　系指在机械制造业中,利用电子数字计算机通过各种数值控制机床和设备,自动完成离散产品的加工、装配、检测和包装等制造过程。简称 CAM。计算机辅助制造系统的组成可以分为硬件和软件两方面:硬件方面有数控机床、加工中心、输送装置、装卸装置、存储装置、检测装置、计算机等,软件方面有数据库、计算机辅助工艺过程设计、计算机辅助数控程序编制、计算机辅助工装设计、计算机辅助作业计划编制与调度、计算机辅助质量控制等。

Computer aided process planning （CAPP） 计算机辅助工艺过程设计

Computer aided solving （CAS） 计算机辅助计算

Computer aided solving system 计算机辅助计算系统

Computer auxiliary analysis and calculation 计算机辅助分析与计算

Computer graphics(CG) 计算机图形学　其主要研究内容就是研究如何在计算机中表示图形以及利用计算机进行图形的计算、处理和显示的相关原理与算法。图形通常由点、线、面、体等几何元素和灰度、色彩、线型、线宽等非几何属性组成。

Computer integrated manufacture system（CIMS） 计算机集成制造系统

Computer mouse 电脑鼠标

Computer net 计算机网络　系指计算机技术与通信技术相互渗透的一门新兴技术,是用通信线路将分散在不同地点并具有独立功能的多台计算机互相连接,按网络协议进行数据通信,实现资源共享的信息系统。计算机网络按其分布范围的大小,分为广域网和局域网。广域网分布范围大,但信道传输速率较低;局域网的范围小,它的信道传输速率高,可达 20Mb/s,网络拓扑结构比较简单,常用的有总线型、星型或环型。计算机网络的主要功能是:共享硬件资源,在计算机网络中的用户可以共享网络中的硬件设施,如高性能的主机、绘图机、打印机等;共享软件资源,如大型软件包、数据库等,实现数据传输、交换,集中管理,提高可靠性,可用性进行分布处理。

Computer network 计算机网络　系指将地理位置不同的具有独立功能的多台计算机及其外部设备,通过通信线路连接起来,在网络操作系统,网络管理软件及网络通信协议的管理和协调下,实现资源共享和信息传递的计算机系统。它极大扩大了计算机系统的功能。

Computer numerical control(CNC) 数控技术　简称数控。即采用数字控制的方法对某一工作过程实现自动控制的技术。熟悉的有 CAD,MILL9 等。

Computer room 电子计算机室　系指安装电子计算机,并对海域调查资料进行数据处理、运算以及对海洋调查仪器进行自动控制的工作舱室。该舱室应防振、防尘、隔音、恒温、恒湿。

Computer software 计算软件　系指在审图、建造中和建造后检验等方面发挥重要作用的计算软件系统。其中包括船舶性能、结构计算与评估、轴系振动与强度、短路电流等计算软件。

Computer storage 计算机存储

Computer system 计算机系统　系指:(1)由1台或多台计算机、有关软件、外围设备和接口组成的系统。能够提供控制、报警、安全、监视4种功能;(2)由1台或多台计算机组成的系统,配备软件、外围设备和接口、计算机网络及其协议。　Computer system means:(1) a system of one or more computers, associated with software, peripherals and interfaces, having functions of control, alarm, safety and monitoring; (2) one or more computers, associated with software, peripherals and interfaces, and a computer network with its protocol.

Computer system（DP system） 计算机系统(动力定位系统)　系指由一台或多台计算机组成的系统,配备软件、外围设备和接口,计算机网络及其协议。Computer system means a system consisting of one or more computers, including software, peripherals and interfaces, and a computer network with its protocols.

Computer Technology 计算机技术　系指计算机领域中所运用的技术方法和技术手段。计算机技术的内容非常广泛,可粗分为计算机系统技术、计算机器件技术、计算机部件技术和计算机组装技术等几个方面。计算机技术具有明显的综合特性,它与电子工程、应用物理、机械工程、现代通信技术和数学等紧密结合,发展很快。第一台通用电子计算机 ENIAC 就是以当时雷达脉冲技术、核物理电子计数技术、通信技术等为基础的。

Computer virus 计算机病毒　系指编制或者在计算机程序中插入的"破坏计算机功能或者毁坏数据,影响计算机使用,并能自我复制的一组计算机指令或者程序代码"。本词条为消岐义词条。常说的病毒有2种,本词条介绍的是计算机病毒,要了解生物方面的病毒,请参看另一词条"病毒"。编制者在计算机程序中插入的破坏计算机功能或者破坏数据,影响计算机使用并且能够自我复制的一组计算机指令或者程序代码被称为计算机病毒。具有非授权可执行性、隐蔽性、破坏性、传染性、可触发性。

Computer-aided calculation & analysis 计算机辅助计算与分析

Computer-based education （CBE） 计算机辅助教

育　系指以计算机为主要媒介所进行的教育活动。也就是使用计算机来帮助教师教学,帮助学生学习,帮助教师管理教学活动和组织教学等。

Computer-based system　**基于计算机的系统**　系指一个或者多个计算机的系统,该系统联合了软件、外围设备和接口及其具备协议的计算机网络。 Computer-based system is a system of one or more computers, associated with software, peripherals and interfaces, and the computer network with its protocol.

Comsumption-based digital camera　**消费型数码相机**

Concealed spaces or inaccessible spaces　**隐蔽处所或不能接近的处所**　系指如顶棚内、内装和外板间以及双重板之间等空间。 Concealed spaces or inaccessible spaces is, for instance, the spaces in the rear side of ceilings, spaces between lining and shell plating, spaces in double-plated bulkheads and other similar spaces.

Concentrate circulating pump　**泡沫液循环泵**　泡沫灭火系统中将泡沫液自储存舱吸出并使泡沫液与水混合成泡沫混合液的泵。

Concentrate pump　**泡沫原液泵**　系指泡沫灭火系统中用作抽送泡沫原液的泵。

Concentrated venting system　**油舱集中换气系统**　系指将各舱中排气管连接起来集中到一个排气桅排放的系统。各舱中的排气管都必须连接压力真空阀,以保证舱内的压力控制在 +0.021 ~ +0.041 MPa 及 -0.003 ~ -0.007 MPa 范围内。

Concentrates　**精矿**　系指通过富集或精选过程,利用物理或化学方法从天然矿石中分离并去除不需要的成分而获得的物质。 Concentrates means materials obtained from a natural ore by a process of enrichment or beneficiation by physical or chemical separation and removal of unwanted constituents.

Concentration ratio　**浓度比**　系指盐水浓度与给水盐度的比值,亦即给水量与排盐量的比值。

Concept development　**概念开发**　即设计要素输入(技术设计),是对产品主要技术参数、功能、特性等限制条件的描述。

Concept development for Lean product　**精益产品概念开发**　就是要求研发人员对各种概念进行全面的综合分析比较,尽量做到在无余量地满足顾客需要的前提下,选择合适的产品概念,以降低产品成本,增加产品的利润。

Conceptual design　**方案报价设计/概念设计**　系指:(1)设计部门根据设计技术任务书的要求进行论证、各方案的比较,并逐项落实任务书的要求的设计。该设计阶段要确定船舶的主要参数,如主尺度、船型、载重量、吨位、总布置格局、主要机电设备的功率及通信导航设备等,根据该方案设计,造船厂能预估该船的造价。(2)设计师有意识地针对大到如宇宙天体、地球环境等。小到产品功能产品需求等的元素进行深入分析、提炼、浓缩而成的一种可以统领全局,有序的、可组织的有目标的设计活动。它的设计过程表现为一个由粗到精,由模糊到清晰,由抽象到具体的不断进化的过程。

Conciliate/Conciliation　**调解**　系指发生争议后,双方协商不成则可邀请第三方居间调停。是处理企业劳动争议的基本办法或途径之一。事实上,调解可以贯穿着整个劳动争议的解决过程。它既指在企业劳动争议进入仲裁或诉讼以后由仲裁委员会或法院所做的调解工作,也指企业调解委员会对企业劳动争议所做的调解活动。这里所说的调解指的是后者。企业调解委员会所做的调解活动主要系指,调解委员会在接受争议双方当事人调解申请后,首先要查清事实、明确责任,在此基础上根据有关法律和集体合同或劳动合同的规定,通过自己的说服、诱导,最终促使双方当事人在相互让步的前提下自愿达成解决劳动争议的协议。

Conciliation　**调解愿望**

Conciliation of execution　**执行和解**　系指在法院执行过程中,双方当事人经过自愿协商,达成协议,结束执行程序的活动。和解的内容,可以是一方自愿放弃一部分或全部权利,也可以是一方满足另一方的要求,还可以是双方都做一些让步。和解虽然发生在双方当事人之间,是双方自己的事,但也要符合一定的条件,即这种和解必须基于双方当事对双方当事人达成的和解协议,人民法院执行员应当将协议内容记入笔录中,由双方签名或盖章。

Conclusion of execution　**执行终结**　系指人民法院在执行过程中,由于出现了某种特殊情况,使执行程序无法或无须继续进行,从而结束执行程序

Conclusion of investigation　**侦查终结**

Concrete gravity platform　**混凝土重力式平台**　系指由钢筋混凝土建造,具有钻井、采油、储油等多种功能的大型重力式平台。

Concrete ship　**水泥船**　用水泥、砂、石、钢筋或钢丝网作为船体结构材料的船舶。

Condensate depression (condensate under-cooling)　**过冷度**　冷凝器喉部进口蒸汽与热井凝结水温度之差。

Condensate inspection tank (condensate observation tank)　**蒸馏水检验柜**　用以检验由海水淡化装置制造出来的蒸馏水质量的专用柜。

Condensate inspection tank (condensate observation)　**凝水观察柜**　用以检验凝水含油情况的专用柜。

Condensate pump 凝水泵 抽送各种冷凝装置中冷凝水的泵。

Condensate re-circulating pipe line 凝水再循环管路 在低负荷时为保证中、后冷凝器有足够冷却水量而在凝水管路进入除氧器前将一部分凝水回流到冷却器实行再循环的管路。

Condensate under-cooling 凝水过冷 凝水温度低于蒸汽饱和温度的状态。

Condensate = feed water system 凝水-给水系统 将主、辅冷凝器的凝水抽出经加热除氧，再由给水泵送入锅炉的机械设备、管路及附件。

Condensation chamber test 冷凝舱试验

Condenser 冷凝器 使汽轮机的排汽，以及其他方面排来的蒸发气在其中冷凝并建立和保持高度真空的一种热交换器。

Condenser anti-rolling sliding support 冷凝器防摇滑动支座 用于限制冷凝器在船舶横倾时的横向位移，并保证冷凝器沿汽轮机轴向膨胀的支座。

Condenser characteristic curve 冷凝器特性曲线 表示冷凝器真空度与冷却水进口温度、冷凝器气量和冷却水流量等特性参数之间关系的曲线。

Condenser circulating water pass/number of passes 冷凝器流程 冷却水流经冷凝器冷却水管的次数。

Condenser cooling surface 冷凝器冷却面积 表面式冷凝器中冷却水管的有效冷却面积。

Condenser duty(condenser heat load) 冷凝器热负荷 单位时间内排入冷凝器中的蒸汽传给冷却水的热量。

Condenser end cover 冷凝器封头 构成冷凝器壳体两端的皿形盖板。

Condenser expansion joint 冷凝器补偿器 用于补偿冷却水管和冷凝器壳体之间由于温度差以及材料线膨胀系数不同而产生的热变形差的部件。

Condenser flow 冷凝器通道 冷却水引入冷凝器的近路，有单通道和双通道之分。

Condenser inlet 冷凝器喉部 冷凝器接受汽轮机排汽的进口部分。

Condenser leakage test 冷凝器密封性试验 为检查冷凝器的真空密封性和冷却水侧密封性所做的试验。

Condenser microphone 电容传声器 其原理与动圈传声器的原理有所不同，它不是利用电磁感应的原理，而是利用电容容量变化而拾音的原理。

Condenser shell 冷凝器壳体 包容冷凝器冷却表面的外围薄壁筒体结构。

Condenser spring support 冷凝器弹簧支座 承托冷凝器质量，并保证冷凝器可靠连接和自由膨胀的支座。

Condenser tube 冷却水管 表面式冷凝器中用以通过冷却水并组成冷凝器冷却表面的管子。

Condenser vacuum 冷凝器真空度 大气压力与冷凝器喉部测点的静压之差占大气压力的百分数。

Condensing pressure 冷凝压力 在一定温度下，载冷剂由气态转变为液态时的压力。

Condensing steam turbine 冷凝式汽轮机 蒸汽在汽轮机中一直膨胀到低于大气压力，然后排入冷凝器的汽轮机。

Condition Assessment Scheme（CAS）Statement of Compliance（CAS Final Report and Review Records） 状态评估计划(CAS)符合证明（CAS 最终报告和审核记录） 系指对每艘按状态评估计划(CAS)[经修正的MEPC.94(46)决议]的要求已进行检验，并符合其要求的油船，由主管机关签发符合证明。此外，经主管机关审核已签发符合证明的 CAS 最终报告的副本以及相关审核记录的副本，应随同符合证明一起放置在船上。Condition Assessment Scheme（CAS）Statement of Compliance,（CAS Final Report and Review Records）means a Statement of Compliance issued by the Administration to every oil tanker which has been surveyed in accordance with the requirements of the Condition Assessment Scheme（CAS）（resolution MEPC.94（46），as amended）and found to be in compliance with these requirements. In addition, a copy of the CAS Final Report which is reviewed by the Administration for the issue of the Statement of Compliance and a copy of the relevant Review Record shall be placed on board to accompany the Statement of Compliance.

Condition inspection（anti-corrosion） 状态(施工)检验(防腐) 系指：(1)船舶在营运过程期间对其结构的防腐涂层和装置的有效性进行评定的检验；(2)防腐工程施工过程中，为保证施工质量而进行的检验。

Condition inspection（anti-corrosion）means：(1) the inspection for evaluating the effectiveness of the structural anti-corrosive coating and units during the ship's operation；(2) the inspection carried out during the anti-corrosive construction in order to ensure the quality.

Condition monitoring 状态监控 系指用于监控系统状态以预先采取措施防止和消除潜在故障的程序控制的诊断技术。Condition monitoring is the scheduled diagnostic technologies used to monitor the system condition to anticipate and defect a potential failure.

Condition monitoring equipment 状态监控设备 系指利用状态监测技术，如振动信号、滑油分析、冲击脉冲分析、温度测量及气缸内部探测等方法，对设备定期进行监测(监测的频度应按设备制造厂商说明书的规定),由监测得到的数据来分析确定设备是否需要进行维修保养,这种采用状态监测技术来分析判别运行状态的设备,称为状态监控设备。 Condition monitoring equipment means that the monitoring is to be carried out periodically by the equipment (the frequency of monitoring is to be in accordance with the specifications of manufacturers) by means of condition monitoring techniques, e. g. analyzing vibration signal, lubricating oil and impact impulses, measuring temperature and internal detection of cylinder. The monitored data are analyzed to determine whether repair or maintenance is necessary. Such equipment, by which the operation condition is analyzed and judged by applying condition monitoring techniques, is called condition monitoring equipment.

Condition monitoring system 状态监控系统 系指包括油品理化分析、原子光谱分析、铁谱分析等的三类方法。 Condition monitoring system means the three methods, i. e. physical and chemical analysis of oils, atomicspectrum analysis and ferrographic analysis.

Condition survey 船况鉴定

Conditioned weight 公量 有些商品,如棉花、羊毛、生丝等有比较强的吸湿性,所含的水分受客观环境的影响较大,其质量也就很不稳定。为了准确计算这类商品的质量,国际上通常采用按公量计算,其计算方法是以商品的干净重(即烘去商品水分后的质量)加上国际公定回潮率与干净重的乘积所得出的质量,即为公量。

Conditions of Class(CC) 船级条件 系指对在规定的期限内(如必要时应立即实施)应完成规定作业(例如修理、调整、补强和检验)的要求。 Conditions of Class to be carried out means that specified operations (e. g, repairs, adjustments, reinforcements or surveys) are to be carried out within a specified time limit (immediately if necessary).

Conductivity sensor 电导率传感器 系指一种测量海水电导率的敏感器件。主要有电极式和感应式两种。

Conductor[1] 隔水导管

Conductor[2] 导体 系指善于传导电流的物质。

Conductors 钻模 系指供钻井进行钻探、完井和生产用的,部分打入海底的管道。 Conductors are tubes partially driven into the sea floor through which wells are drilled, completed and produced.

Cone clutch 锥形离合器 系指利用圆锥面接合的摩擦离合器。

Conference center 会议中心(豪华邮轮)

Conference on Interaction and Confidence Building Measures in Asia(CICA) 亚洲相互协作与信任措施会议 简称亚信,1992年10月5日哈萨克斯坦总统纳扎尔巴耶夫首次在第47届联合国大会上倡议成立唯一的一个覆盖亚洲所有地区的重要国际会议,其成员国与观察员覆盖亚洲超过90%的面积和人口。从1992年起,它经历了以下4个阶段:1992～1995年的第一阶段——专家磋商阶段;1995～1998年的第二阶段——建立论坛阶段;1998～2000年的第三阶段——机制推动阶段和2000～2005年欧亚互动阶段与欧安组织建立合作关系,并逐渐向美洲、非洲、大洋洲等扩展,最终形成全球性安全合作组织。其秘书处设在阿拉木图。如今,"亚信"拥有26个成员国:中国、阿富汗、阿塞拜疆、埃及、印度、伊朗、以色列、哈萨克斯坦、吉尔吉斯斯坦、蒙古、巴基斯坦、巴勒斯坦、俄罗斯、塔吉克斯坦、土耳其、乌兹别克斯坦、泰国、韩国、约旦、阿联酋、越南、伊拉克、巴林、柬埔寨、孟加拉国、缅甸和卡塔尔。11个观察员:印度尼西亚、马来西亚、美国、乌克兰、日本、菲律宾、斯里兰卡;以及联合国、欧洲安全与合作组织(欧安组织)、阿拉伯国家联盟和突厥语国家议会大会。其特长是安全合作,在以下5大领域落实信任措施:(1)应对新威胁与新挑战——反三股势力、反毒品、走私、非法交易、非法移民、洗钱、跨国犯罪、非法销售武器,防范经济和网络犯罪;执法合作,应对突发性流行疾病。(2)经济——交通联通,运输效率和安全;能源安全合作;旅游合作、签证便利化;金融合作;中小企业合作;投资合作;体系和信息技术合作;建立共享的经贸数据库。(3)生态——分享环境政策领域信息;分享可持续发展领域最佳实践的信息;在早期预警等灾害管理体系方面开展合作;建立共同的环保方案;对可能危害邻国的生态或人为灾难进行通报。(4)人文——促进不同文明、文化和宗教对接;媒体合作;文化活动;通过科技、教育、体育机构、非政府组织加强国民交流;组织考古探险发掘共同文化遗产;促进对基本权利和自由的尊重。(5)军事-政治——军方权威机构以及军事院校代表互访;军方相互邀请参加对方国家举办的国家日和文化体育活动;军方顶层人士分享简历;军方分享有关军控、裁军、外空等领域信息。

Confession to justice 自首

Confidence 可信度 系指某一样本排除采样误差以百分数表示的概率。75%的可信度意味着结果可能出现采样误差(采样和试验的随机波动)的概率是25%。 Confidence is the probability, in %, that a given observation is not the result of sampling error. A confidence level of 75% means that there is a probability of 25%, that sampling error (chance fluctuations of sampling and testing) cowld occur.

Confidence limits 置信限 对随机取样可容许的、能包含其平均值的概率即置信系数一般在90%以上的、误差范围的上下限。

Configuration available 可用配置 系指在每个工作站所规定的和可实施的操作。 Configuration available is operation(s) allocated and available at each workstation.

Configuration data 配置数据 描述了船舶的设备、其在船上安装以及与VDR的关系。存储和回放软件使用该数据来存储数据记录并在回放时将数据记录转换为有助于事故调查的信息。 Configuration data describes the vessel's equipment, its installation on the vessel and its relation to the VDR. The storage and playback software uses this data to store the data record and to convert the data record into information that assists casualty investigation during playback.

Configuration in use 在用配置 系指每个工作站在用的操作和任务。 Configuration in use is operation(s) and task(s) currently in use at each workstation.

Configuration of complete system 整个系统配置 系指业已安装的综合桥楼系统所有操作功能。 Configuration of complete system means all operational functions of the integrated bridge system as installed.

Confined space 限制处所 系指由以下特征之一确定的处所:进出开口受限,自然通风不良或设计为非连续工作场所。 Confined space means a space identified by one of the following characteristics: limited openings for entry and exit, un-favourable natural ventilation or not designed for continuous worker occupancy.

Confirm 确认

Confirmation 确认书 是合同的简化形式,对于讲格、索赔、仲裁、不可抗力等一般条目都不会列入,使用第一人称语气。

Confirmatory audit 确认性审核 系指对附加标志船舶机械计划保养系统(PMS)的有效性进行确认。在船舶机械计划保养系统(PMS)检验时,应在每年的年度/中间/特别检验时按有关规定进行年度审核。检查合格后,验船师应在入级证书的签证栏目进行相应的签署。 Confirmatory audit means confirmation of validity of the class notation PMS. For the survey of PMS, an annual audit is to be made in accordance with the relative requirements when the annual/intermediate/special survey is carried out every year. Upon satisfactory confirmation, the endorsement column of classification certificate is to be signed by the Surveyor accordingly.

Confirmed letter of credit 保兑信用证 系指在国际贸易结算中较为常见的一种信用证。这类信用证与普通信用证最大的区别在于有两个银行承诺向受益人付款。

Conformity 合格 系指满足某项要求。 Conformity means fulfillment of a requirement.

Connecting bridge 天桥 系指架空在上甲板之上,沟通分立的上层建筑的通道。

Connecting plug 连接插头 通过插入插座中,可以使其所接软线的导线与永久接于插座的导线之间接通的器件。

Connecting rod 连杆 连接曲柄销与活塞销或十字头销的零部件。

Connecting shackle 连接卸扣 专用于连接锚链各链节的直形卸扣。

Connecting traverse 吊钩梁 一般用于重型吊杆装置,供连接吊货钩与两套吊货索滑车用的承载杆件。

Connection by hard bracket 用刚性肘板固定 系指双层底板之间或相邻平面内类似的加强筋与肘板之间连接处的端部固定或等效的固定情况。 Connection by hard bracket is a connection by bracket to the double bottoms or to the adjacent members, such as longitudinals or stiffeners in line, or a connection equivalent to the connections mentioned above.

Connection by soft brackets 用柔性肘板固定 系指与横梁等交叉构件及肘板间的端部固定。 Connection by soft brackets is a connection by bracket to the transverse members such as beams or equivalent thereto.

Connection point 连接点 对风电系统而言,系指将海上风电馈入电网的陆上发电站。

Connection-type A of vertical stiffeners 垂直加强筋的A型连接 是通过肘板连接到纵向构件或相邻的结构上,在加强肋方向上有相同的或更大的截面。 Connection-type A of vertical stiffeners is a connection by bracket to the longitudinal members or to the adjacent members, in line with the stiffeners, of the same or large sections.

Connection-type B of vertical stiffeners 垂直加强筋的B型连接 是通过肘板连接到横向构件例如横梁或其他上述的等效连接的支架进行的连接。 Connection-type B of vertical stiffeners is a connection by bracket to the transverse members such as beams, or other connections equivalent to the connections mentioned above.

Connectivity 连同性 系指完整的数据链并显示有效数据。 Connectivity is a complete data link and the display of valid data.

Conning information display 驾驶信息显示 系指通过屏幕显示信息的系统。其能清晰地提供从传感器输入的有关导航和操纵的信息，以及当自动导航系统与操舵和推进系统相连时，由后者提供的所有相关的和即将执行的指令。 Conning information display means a screen-based information system that clearly displays information relevant to navigation and maneuvring from sensor inputs, as well as all corresponding and upcoming orders given by an automatic navigation system to steering and propulsion systems if connected.

Conning position 操舵位置 系指控制船舶正车或倒车操作装置所处的位置。 Conning position means the stations in which the ship's control devices for ahead or astern operations are located.

Conning position 驾驶指挥位置 系指：(1)驾驶室内具有指挥视野的位置，该处为监视和指挥船舶运动时由驾驶员所用；(2)在指挥、操纵和控制船舶时由驾驶员使用的、具有宽阔视野的场所。(3)船舶操舵控制或正车或倒车操作装置所在位置。 Conning position is: (1) the place in the wheelhouse with a commanding view and which is used by navigators when monitoring and directing the ship's maneuvring; (2) a place on the bridge with a commanding view and which is used by navigators when commanding, maneuvering and controlling a ship; (3) the stations in which the ship's steering control and devices for ahead or astern operations are located.

Conning station or position 驾驶台或驾驶位 系指驾驶室内具有指挥视野的位置，可提供驾驶必需的信息，以供驾驶人员在监视和引导船舶运动时使用。 Conning station or position means a place in the wheelhouse with a commanding view providing the necessary information for conning, and which is used by navigators when monitoring and directing the ship's movements.

Consecutive voyages 定期租船 又称"期租船"。系指船舶所有人按照租船合同约定，将特定的船舶，在约定的期限内，交给承租人使用的一种租船方式。

Consequence 后果 系指事故产生的影响，诸如伤害、死亡、环境破坏和财产损失。 Consequence means those effects of an accidental event, such as injuries, fatalities, environmental damage, and property damage.

Consequence evaluation 后果评价 系指对缘于事故如火灾和爆炸载荷的物理效应的评估。

Consignee 收货人 系指有权提取货物的人。 Consignee means the person entitled to take delivery of the goods.

Consignment 寄售 系指是把货物放在客户那里，客户要用的时候才付款的库存，一般系指卖方把货物存放在买方所属仓库，消耗后结账。

Consignment 托运货物 系指托运人送交运输的固体散货。 Consignment means solid bulk cargoes presented by a shipper for transport.

Consignor 发货人 系指其本人，或以其名义，或其代表同多式联运经营人订立多式联运合同的任何人，或指其本人，或以其名义，或其代表将货物实际交给多式联运经营人的任何人。 Consignor means any person by whom or in whose name or on whose behalf a multimodal transport contract has been conducted with the multimodal transport operator, or any person by whom or in whose name or on whose behalf the goods are actually delivered to the multimodal transport operator in relation to the multimodal transport contract.

Consistency principle 透明性原则/一致性原则 系指：(1)支持风险评估建议的行动的理由和证据，以及不确定区域（及其对这些建议的可能后果），均有明确的文字记载供决策者使用的原则。(2)风险评估使用共同的步骤和方法达到统一的高效能的原则。 Consistency principle is: (1) a principle that the reasoning and evidence supporting the action recommended by risk assessments, and areas of uncertainty (and their possible consequences to those recommendations), are clearly documented and made available to decision-makers. (2) principle that risk assessments achieve a uniform high level of performance using a common process and methodology.

Consistent common reference point (CCRP) 统一共同基准点 系指本船的一个位置，所有水平测量，例如目标距离、方位相对航向、相对航速、相遇最近点（CPA）或至相遇最近点的时间（TCPA），均参照此位置，一般为驾驶台的指挥位置。 Consistent common reference point (CCRP) is a location of the ship, to which all horizontal measurements such as target range, relative course, relative speed, closest point of approach (CPA) or time to closest point of approach (TCPA) are referenced, and is typically the conning position of the bridge.

Consistent common reference system (CCRS) 统一共同基准系统 INS的一个子系统，用于搜集、处理、存储、监测和分配数据和信息，向INS的子系统和后续功能，以及其他连接的设备（如有时）提供统一和强制参考。 Consistent common reference system (CCRS) is a sub-system or function of an INS for acquisition, processing, storage, surveillance and distribution of data and information providing identical and obligatory reference to sub-systems and subsequent functions within an INS and to other connected equipment, if available.

Console for electric propulsion 电力推进操纵站（台） 它集中了电力推进装置（推进电动机）的监视、调节和控制设备（也包括发电机组的遥控启动和停车设备），通过它可方便地对电力推进系统进行操纵及控制。它可装于机舱外一个或数个地点分别地进行操纵。

Constant pitch 等螺距 螺旋桨叶面各处螺距皆相同者。

Constant pressure cycle(diesel cycle) 等压循环 由绝热压缩、等压加热、绝热膨胀和等容放热等4个过程组成的理想热力循环。

Constant pressure supercharging 等压增压 利用较大直径的排气管使排气通向涡轮时的速度和压力比较均匀的废气涡轮增压方式。

Constant temperature and humidity room 恒温恒湿室 用于海洋生物的培养和其他要求恒温恒湿条件的设备使用，具有可控温度和湿度的舱室。

Constant tension winch 恒张力绞车 系指一种能保持张力恒定的绞车。如张力一旦在外界风、浪、流影响下超出预先设定的数值范围时，能自动放出或收进一定长度的绳索，使张力维持恒定的绞车。见图C-11。

图 C-11 恒张力绞车
Figure C-11 Constant tension winch

Constant volume cycle（Otto cycle） 等容循环 系指由绝热压缩、等容加热、绝热膨胀和等容放热等4个过程组成的理想热力循环。

Constant wear suit 常穿服 设计成用于在水面上或近水面处活动时日常穿戴以防偶然落水的救生服，但允许穿戴者身体活动到使之可从事活动而无过分累赘的程度。 Constant wear suit means the immersion suit designed to be routinely worn for activities on or near water in anticipation of accidental immersion in water, but permitting physical activity by the wearer to such an extent that actions may be undertaken without undue encumbrance.

Constant-speed motor 恒速电动机 系指一种正常运行时转速恒定或实际上恒定的电动机。诸如同步电动机、小滑差感应电动机或普通的直流并励电动机。

Constant-torque resistor 恒定转矩电阻器 系指一种用于电动机电枢或转子电路内使其电流在整个转速范围内实际上处于恒定的电阻器。

Constrained damping layer 约束阻尼层

Constrained layer damping configuration 约束阻尼层结构

Constrained layer damping structure 约束阻尼结构 黏弹性阻尼结构

Constraint 刚性固定 系指加强筋用肘板刚性连接至其他构件上，或加强筋穿过支承它的桁材的端部连接方式。 Constraint means an end attachment manner in which the stiffeners are rigidly connected to other members by means of brackets or by running throughout supporting girders.

Construct equipment 施工装备 系指在深水油气田开发过程中，执行海上油气田施工建设和生产维护的平台或船舶，主要包括起重铺管船、水下建设船、潜水支持船、修井船、增产作业船、半潜运输船、生活（船）平台和生产支持船等。

Constructed in respect of a ship 船舶的建造 系指下述建造阶段：（1）安放龙骨；（2）可辨认出某一具体船舶建造开始；（3）该船业已开始的装配量至少为50 t，或为全部结构材料估算质量的1%，两者中取较小者。 Constructed in respect of a ship means a stage of construction：(1) the keel is laid；(2) construction is identifiable for the be ginning a specific ship；(3) assembly of the ship has commenced and comprises at least 50 t or 1% of the estimated mass of all structural material, whichever less is chosen.

Constructed in respect to a vessel 与船舶建造有关的建造阶段 系指——（1）船舶安放龙骨；（2）可以认为某一具体船舶建造开始；（3）该船业已开始的装配量至少为50吨，或为所有结构材料估算质量的1%，以较小者为准。（4）该船经过重大改建。 Constructed in respect to a vessel means a stage of construction：(1) the keel of a vessel is laid；(2) constructing is identifiable for the beginning of the specific vessel；(3) assembly of the vessel has commenced and comprises at least 50 tons or 1 percent of the estimated mass of all structural material, whichever is chosen；(4) the vessel undergoes a major conversion.

Construction 建造 系指：（1）安放龙骨；（2）处于下述相应建造阶段：①可以认定某一具体船舶建造开始；或②该船业已开始的装配量至少为50 t 或为所有结

构材料估算质量的 1%，以较小者为准。Construction means：(1) the keel of a ship is laid；(2) it is at the follouing stage of construction：①construction is identifiable for the beginning of a specific ship；②assembly of the ship has commenced and comprises at least 50 t or 1% of the estimated mass of all structural material, whichever less is chosen.

Construction booklet 建造说明书 系指一份经船级社批准的说明书。该说明书应包括能表明不同等级及力学性能结构材料的适用位置和范围的图纸、主要结构所采用的焊接工艺，以及包括修理和改装须知在内的任何其他有关的结构资料，结构材料包括钢、铝合金和其他材料。Construction booklet means a copy of booklet approved by a Society, containing a set of plans showing the exact location and extent of application of different grades and mechanical properties of structural materials (steel, aluminum alloy or other materials), together with welding procedures employed for primary structure and any other relevant construction information including instructions regarding repairs or modifications.

Construction noise 结构噪声

Construction plan 基本结构图 表示全船船体主要结构的图纸。一般包括船体纵剖面图和各甲板及船底的平面图。

Construction profile & deck plan 外板展开图 系指在肋位上按肋骨的型线周长展开的外板布置图。因船体外形具有双曲度，按几何方法难以精确展开，故只能按肋位横向展开。一般船体是左右对称的，故仅画出其中的一舷。图中表明外板及其骨架的布置、外板的尺寸和厚度、边接缝、端接缝的位置和连接方式、分段的划分线的位置、外板开口的位置和加强覆板的形状和尺度，以及在各有冰区加强时冰区带的范围等。

Constructive total loss 推定全损 系指实际全损已不可避免，或受损货物残值，如果加上施救、整理、修复、续运至目的地的费用之和超过其抵达目的地的价值时，视为已经全损。推定全损与"实际全损"相对称。保险标的受损后并未完全丧失，是可以修复或可以收回的，但所花的费用将超过获救后保险标的的价值，因此得不偿失。在这种情况下，保险公司放弃努力，给予被保险人以保险金额的全部赔偿即为推定全损。在下列情况下为推定全损：由于实际全损似乎无法避免，或为避免实际全损所支付的费用将超过被保险财产的价值而将被保险财产委付；当被保险人因承保的危险（peril insured against）丧失对被保险财产的占有，而无法恢复占有或意图恢复占有的费用太高时；或当对被保险财产的修复费用太高时。参见1906年（英国）海上保险法（Marine Insurance Act, 1906）第60条。在我国，船舶发生保险事故后，认为实际全损已经不可避免，或者为避免发生实际全损所需支付的费用超过保险价值的，为推定全损；货物发生保险事故后，认为实际全损已经不可避免，或者为避免发生实际全损所需支付的费用与继续将货物运抵目的地的费用之和超过保险价值的，为推定全损。

Constructor 建造者 是负责进行下列任意一项或所有各项工作的任何个人或组织：建造、装配、检查、试验、装卸、运输和安装。

Consultant 顾问 系指通过培训和实践经验持有专业证书并在所述领域中具有专长的任何个人。

Consulting final opinion of the plaintiff and defendant 征询原、被告最后意见

Container 集装箱 系指一种运输设备：(1) 具有耐久性，因而其相应的强度足能适合于重复使用；(2) 经专门设计，便于以一种或多种运输方式运输货物，而无须中途重新装卸；(3) 为了系固和/或便于装卸，设有角配件；(4) 4个外底角所围蔽的面积应为下列两者之一：①至少为 14 m^2(150 ft^2)；②如装有顶角配件，则至少为 7 m^2(75 ft^2)。集装箱一词既不包括车辆，也不包括包装，但是，集装箱在底盘车上运输时，则连同底盘车包括在内。见图 C-12。常用集装箱外形尺寸及质量见表 C-4。Container means an article of transport equipment：(1) of a permanent character and accordingly strong enough to be suitable for repeat use；(2) specially designed to facilitate the transport of goods by one or more modes of transport without intermediate reloading；(3) designed to be secured and/or readily handled, having corner fittings for these purposes；(4) of a size such that the area enclosed by the four outer bottom corners is either：①at least 14 m^2 (150 ft^2) or ②at least 7 m^2 (75 ft^2), if it is fitted with top corner fittings. The term container includes neither vehicles or packaging, however, chassis on which the containers are carried are included. A container can be see in Fig. C-12. External dimension and weight of common container can be seen in Table C-4.

图 C-12　集装箱
Figure C-12　Container

表 C-4　常用集装箱外形尺寸及质量
Table C-4　External dimension and weight of common container

名称 Name	长度 Length/(ft/mm)	高度 Height/(ft/mm)	宽度 Breadth/(ft/mm)	质量 Weight/kg
53′	53′/16 154	$9'6\frac{1}{2}''$/2 098	8′6″/2 591	30 480
49′ ISO（草案）	49′/14 935	9′6″/2 896 8′6″/2 591	$8'6\frac{5}{32}''$/2 595	30 480
24′ ISO（草案）	$24'2\frac{1}{2}''$/7 430	9′6″/2 896 8′6″/2 591	$8'6\frac{5}{32}''$/2 595	30 480
48′	48′/14 631	$9'6\frac{1}{2}''$/2 908	8′6″	30 480
45′	45′/16 716	9′6″/2 896 9′/2 743	8′/2 438	32 500 30 480
43′	43′/13 107	8′6″/2 591	8′/2438	30 480
40′ ISO	40′/12 192	9′6″/2 896 9′/2 743 8′6″/2 591 8′/2 438	8′/2 438	30 480
20′ ISO	$19'10\frac{1}{2}''$/6 050	8′6″/2 591 8′/2 438	8′/2 438	24 000
20′	$19'10\frac{1}{2}''$/6 050	9′6″/2 896 9′/2 743	8′/2 438	24 000
40′ EURO（欧洲）	40′/12 192	9′6″/2 896 9′/2 743 8′6″/2 591	/2 500	30 480
40′BELLLINE	40′/12 192	9′6″/2 896 9′/2 743 8′6″/2 591	/2 500	30 480
35′ SEALAND（海陆公司）	35′/10 659	8′6″/2 591	8′/2 438	30 000

续表 C-4

名称 Name	长度 Length/(ft/mm)	高度 Height/(ft/mm)	宽度 Breadth/(ft/mm)	质量 Weight/kg
30′ISO	29′11$\frac{1}{4}$″/9125	8′6″/2 591 8′/2 438	8′/2 438	25 400
24′MATSON	24′/7 315	8′6″/2 591 8′/2 438	8′/2 438	

Container B/L 集装箱提单 系指采用集装箱运输货物而由承运人签发给托运人的提单。

Container block 集装箱块 系指一定数量的集装箱箱堆垛之间通过双头堆锥和/或桥接件连接。

Container cell 集装箱架 系指各个集装箱的位置。其通常处于垂直的箱格导轨的架中,且通常被位于集装箱上方或下方的横桁所围住。 Container cell means the position of an individual container. It is usually within a set of vertical cell guides and is normally enclosed by transverse stringers located above and below the container.

Container corner fittings 集装箱角附件

Container electrical power supply 集装箱电源 系指专门用于向所有各冷藏集装箱供电和保持通风系统各风机电动机供电的电源。 Container electrical power supply means the generated power supply which is dedicated to supply electricity for the refrigerated containers and the fan motors of ventilation system.

Container plug-in point 集装箱插头点 系指位于甲板上每一合适的集装箱位置,以及位于甲板下每一集装箱架位置的符合 ISO 1496-2:1996 附录 L 要求的插座。 Container plug-in point means an electrical socket located at each applicable container's location on deck and each cell's location below deck being in accordance with Annex L of ISO 1496-2:1996.

Container securing arrangement 集装箱系固件/集装箱紧固设备 对集装箱应用一种或几种装置的组合进行系固的装置。如角锁紧装置、绑扎装置、箱格导轨、撑柱、单压撑柱或其他等效的支撑结构等。

Container securing system 集装箱系固系统 系指以下任一方式或和不同方式组合对集装箱进行系固,且以其他方式系固集装箱时,应是船级社认为恰当的方式:(1)集装箱锁定设备;(2)由拉杆、钢丝或锁链绑固;(3)由支肋、支撑柱或其他同等构件固定;(4)箱格导轨。 Container securing system means that containers are to be secured by one, or a combination, of the following systems. other methods used to seaure containers are to be in accordance with the requirements deemed as appropriate by a society:(1)securing by container locking devices;(2)securing by rod, wire and chain lashing;(3)securing by buttresses, shores or equivalent structural restraint;(4)securing by cell guides.

Container security fittings 集装箱系固装置 包括箱体的定位、固定、连接和绑扎等,适用于船舱内、甲板上及集装箱专用运输车辆等。

Container ship 集装箱船 系指专门设计用于在货舱内和甲板上载运集装箱的船舶。 Container ship means a ship designed exclusively for the carriage of containers in holds and on deck.

Container stack 集装箱堆垛 系指单独的垂直集装箱堆,可通过扭锁、扭锁加绑件或导轨进行系固。

Container transport 集装箱运输 系指把货物装进集装箱,然后再以集装箱作为货物单元装船运输或装车运输的运输方式。

Container vessel 集装箱船 系指具有双层底、双壳、舱顶设抗扭箱(或其他等效的单层壳舷侧结构)、甲板开口大的结构形式,在货舱内和甲板上专门装载标准和非标准集装箱的专用船舶。在货舱内设有格栅结构,以便垂直堆放集装箱,及防止因船的摇摆使集装箱倾倒。一般货舱可堆放 4~9 层集装箱。其装卸方式是垂直方向进行的。现代集装箱船依靠港口的起货设备进行装卸,船上不设起货设备。少数集装箱船自备起货机。集装箱船为"尾机型船",机舱和上层建筑均设在船尾部位。船体的线型比较削瘦,航速较高,一般船首为球鼻艏型。集装箱船一般设有长的艏楼,艏楼甲板上一般不堆放集装箱。现代的集装箱船都设计为双层船壳。另外,为了便于调整集装箱船的纵倾和稳性,需要配备大量的压载水,双层船壳对此也很有利。舷侧的双层壳内通常设有压载水舱或燃油舱。集装箱船还设置专用的集装箱绑扎设备。从 1956 年第一艘改装集装箱船诞生到 10 000 TEU 集装箱船出现,再到 20 000 TEU 集装箱船进入研究阶段，航运业用了50多年的时间完成了集装箱运输工具的历史性换代升级。20世纪70年代末，国际标准化组织制订了统一的集装箱规格标准，从此集装箱运输进入高速发展时代，而集装箱船大型化的速度更是令人眩晕。70年代航运公司开始订造装箱量在1 000 TEU以上的第二代集装箱船；80年代3 000 TEU集装箱船开始登台亮相，船宽设定为32.2 m被称为巴拿马型集装箱船；1988年美国总统轮船公司订造了4 340 TUE集装箱船"杜鲁门总统"号，为超巴拿马型集装箱船；90年代中期，船宽超过40 m装箱量为5 000～6 000 TEU集装箱船投入营运；1996年7 000 TEU级集装箱船"Regina"号问世。进入21世纪铁行渣华定造的6 674 TEU集装箱船"南安普顿"号和马士基定造的7 660 TEU集装箱船"马士基君主"号先后交付；2006年6月现代重工正式开工建造世界首艘10 000 TEU集装箱船；2007年三星重工研发出16 000 TEU集装箱船；2010年18 000 TEU集装箱船概念诞生，STX造船集团提出22 000TEU集装箱船设计方案。集装箱船装箱量以惊人的速度在增加。见图C-13。

图C-13　集装箱船
Figure C-13　Container vessel

Containerized cargo system　**集装装卸**　系指以集装箱或集装用的驳船作为货物装卸单元进行装卸的方式。

Container-type laboratory（mobile laboratory）**集装箱式实验室**　又名活动实验室。一般都为专业实验所用，室内按专业需要，装有各种专业仪器设备和水电等接口，可随时预装到海洋调查船上工作及随时从船上卸下，以减少船舶停港时间并充分利用甲板使用面积的一种工作舱室。

Containment objective　**抑制目标**　每艘舰船的布置应使其尽快限制火、烟和有毒燃烧产物蔓延至整个火源所在处所。Containment objective means that every ship is to be arranged, to control the spread of fire, smoke and toxic by-products to the space of origin as quidely as possible.

Containment of fire　**火灾的限制**　系指将失火遏制在火源处所内。为此，应满足下列功能要求：(1)应通过耐热和结构性限界将船舶分成若干区；(2)限界内的隔热应充分考虑到处所及其相邻处所的失火危险；(3)在开口和贯穿件处应保持分隔的耐火完整性。Containment of fire is to contain a fire in the space of origin. For this purpose, the following functional requirements shall be met: (1) the ship shall be subdivided by thermal and structural boundaries; (2) thermal insulation of boundaries shall have due regard for the fire risk of the space and adjacent spaces; (3) the fire-resistant integrity of the divisions shall be maintained at openings and penetrations

Continental method of model self-propulsion test　**定转速船模自航试验**　系指保持螺旋桨转速不变，而改变船模航速的自航试验。

Continental shelf of a coastal State　**沿海国的大陆架**　系指包括其领海以外依其陆地领土的全部自然延伸，扩展到大陆边缘外的海底区域的海床和底土，如果从测量领海宽度的基线量起到大陆边外缘的距离不到200海里，则扩展到200海里的距离。Continental shelf of a coastal State comprises the sea-bed and subsoil of the submarine areas that extend beyond its territorial sea throughout the natural prolongation of its land territory to the outer edge of the continental margin, or to a distance of 200 nautical miles from the baseline from which the breadth of the territorial sea is measured where the outer edge of the continental margin does not extend up to that distance.

Contingency planning　**应急规划**　系指对处理紧急情况，包括实际制定应急行动而提供设施、培训及钻探的规则。

Continuity of service　**供电连续性**　系指在某电路发生故障期间以及故障之后，非故障电路的供电能始终得以保证。Continuity of service is the condition that during and after a fault in a circuit, the supply to the healthy circuits is permanently ensured.

Continuous "B" class ceilings or linings　**连续"B"级天花板或衬板**　系指终于"A"或"B"级分隔的"B"级天花板或衬板。Continuous "B" class ceilings or linings are those "B" class ceilings or linings which terminate at "A"or"B"class division.

Continuous feeding　**连续进料**　系指当焚烧炉在正常操作情况下，燃烧室工作温度在850 ℃和1 200 ℃之间时，无需人工辅助将废料送入燃烧室的过程。Continuous feeding is defined as the process whereby waste is fed into a combustion chamber without human assistance while the incinerator is in normal operating conditions with the combustion chamber operative temperature between

850 ℃ and 1 200 ℃.

Continuous hull survey(CHS) 船体循环检验 除液货船和散货船以外的船舶,将特别检验的项目均匀分配在5年内轮流检查,以代替特别检验时需作内部检验和试验项目的检验。

Continuous icebreaking "连续式"破冰法 系指依靠螺旋桨的力量和船舶把冰层劈开撞碎的方法。

Continuous improvement principle 持续改进原则 系指任何风险模型均应定期评核和更新以计及理解上的提高的原则。 Continuous improvement principle is a principle that any risk model should be periodically reviewed and update in account for improved understanding.

Continuous machinery survey (CMS) 轮机循环检验 以5年为周期,连续、分别地对机器设备进行的检验,即将特别检验的项目均匀分配在5年内轮流检查,以代替特别检验时需作内部检验和试验项目的检验。

Continuous member 连续构件 系指构件相交处保持连续未被隔断的构件。

Continuous plankton recorder 浮游生物连续采集器 系指航行中连续采集浮游生物定量和定性样品的仪器。用于调查浮游生物的连续分布。

Continuous rating 持续功率 系指主机在现行有关标准规定的环境条件下,允许长期持续运转的输出功率。

Continuous sludging centrifuge 连续去污油分离机 系指机器在工作过程中分离出来的残渣不断移向喷嘴然后由经常流出的水将残渣排去的去污油分离机。

Continuous speed 持续转速 相应于持续功率的转速。

Continuous stratification profiler 连续地层剖面仪 船舶在航行中连续发射声波并接收海底反射波信号,用以给出所测得的海底地层剖面的仪器。

Continuous survey 连续检验 又称循环检验,分为:(1)船体循环检验,即除液货船和散货船以外的船舶,将特别检验的项目均匀分配在5年内轮流检查,以代替特别检验时需作内部检验和试验项目的检验。(2)轮机循环检验,即以5年为周期,连续、分别地对机器设备进行的检验,即将特别检验的项目均匀分配在5年内轮流检查,以代替特别检验时需作内部检验和试验项目的检验。

Continuous synopsis record 连续概要记录 系指旨在就其中记录的信息在船上提供一份船舶历史记录。其应至少包括以下信息:(1)该船船旗国国名;(2)该船在该国注册的日期;(3)船舶识别号;(4)该船船名;(5)该船的船籍港;(6)注册船东姓名及其注册地址;(7)注册船东识别号;(8)注册光船租赁人姓名及其注册地址(如适用);(9)公司的名称,其注册地址及其开展安全管理活动的地址;(10)公司识别号;(11)该船所入级的所有船级社的名称;(12)向经营该船的公司签发的ISM规则规定的"符合证明"(或"临时符合证明")的主管机关或缔约国政府或认可组织名称,如果进行审核并据此发证的机构不是同一机构,还要有审核机构的名称;(13)向该船签发的ISM规则规定的"安全管理证书"(或"临时安全管理证书")的主管机关或缔约国政府或认可组织名称,如果进行审核并据此发证的机构不是同一机构,还要有审核机构的名称;(14)向该船签发的ISPS规则A部分规定的"国际船舶保安证书"(或"临时国际船舶保安证书")的主管机关或缔约国政府或认可组织名称,如果进行审核并据此发证的机构不是同一机构,还要有审核机构的名称;(15)该船终止在该国注册的日期。 Continuous synopsis record is intended to provide an on-board record of the history of the ship with respect to the information recorded therein. it shall include, at least, the following information:(1) the name of the State whose flag the ship is entitled to fly;(2) the date on which the ship was registered with that State;(3) the ship's identification number in accordance with regulation;(4) the name of the ship;(5) the port at which the ship is registered;(6) the name of the registered owner(s) and their registered address(es);(7) the registered owner identification number;(8) the name of the registered bareboat charterer(s) and their registered address(es), if applicable;(9) the name of the Company, as defined in regulation IX/1, its registered address and the address(es) from which it carries out the safety-management activities;(10) the Company identification number;(11) the name of all classification society(ies) with which the ship is classed;(12) the name of the Administration or of the Contracting Government or of the recognized organization which has issued the Document of Compliance (or the Interim Document of Compliance), specified in the ISM Code as defined in Regulation IX/1, to the Company operating the ship and the name of the body which has carried out the audit on the basis of which the Document was issued, if the Document is issued by a different organization form that it is auditted;(13) the name of the Administration or of the Contracting Government or of the recognized organization that has issued the Safety Management Certificate (or the Interim Safety Management Certificate), specified in the ISM Code as defined in Regulation IX/1, to the ship and the name of the body which has carried out the audit on the basis of which the Certificate was issued, if the Certificate is issued by a different organization form that it is auditted;(14) the name of the Administration or of the Contracting Government or of the recognized security organization that has issued the International Ship Security Certificate (or the Interim International Ship Security Certificate), specified in Part A of the ISPS Code as defined in regulation XI-2/1, to the ship and the name of the body which has carried out the audit on the basis of which the Certificate was issued, if the Certificate is issued by a different organization form that it is auditted; (15) the date on which the ship ceased to be registered with that State.

Continuous system for hull survey 船体循环检验系统 系指特别检验的替代检验系统,适用于除普通干货船、油船、散货船和兼用船及化学品船以外的船舶。 Continuous system for hull survey is an alternative survey system for special survey and is applicable to ships other than general dry cargo ships, oil tankers, bulk carriers, combination carriers and chemical tankers.

Continuous tank cover 连续凸形液舱盖

Continuous ventilation 持续通风 系指一直运行的通风。 Continuous ventilation means ventilation that is operating at all times.

Continuous watch 连续值班 系指有关的无线电值班不会中断,除非当船舶接收能力由于自身通信被削弱和阻塞时,或当设备处于维护或检查时而引起短暂间隔。 Continuous watch means that the radio watch concerned shall not be interrupted, unless there are brief intervals when the ship's receiving capability is impaired or blocked by its own communications or when the facilities are under periodical maintenance or checks.

Continuous working system 连续工作制 系指需要在实质上恒定的负载下无限长时间运行的一种使用要求。

Continuously manned central control station 持续有人操作的中心控制站 系指船舶在正常营运期间,总有1名负责的船员持续操作的控制站。 Continuously manned central control station is a central control station which is continuously operated by a responsible member of the crew, while the craft is in normal service.

Continuously manned control station 连续有人控制站 系指船舶在正常航行时由值班船员连续人工操作的控制站。 Continuously manned control station is a control station which is continuously operated by a responsible member of the crew while the craft is in normal service.

Contra propeller (reaction propeller) 反应推进器 固定在螺旋桨后,一般作叶轮形,吸收尾流中旋转动能,以提高对船推力的装置。

Contra propulsion bulb 扩流毂锥 为限制尾流收缩并增加推力,而装在螺旋桨毂后的大直径锥形体。

Contra rotating propeller (CRP) 对转螺旋桨

Contract 合同 系指船级社和客户之间的专项协议。它规定了客户所要求的服务范围,且涉及:(1) 无论是新造或营运中船舶的入级;(2) 代表国家海事主管机关执行的法定检验工作;(3) 船用设备和材料。 Contract is the specific agreement between the Society and the client. It defines the extent of services requested by the client, and is concerned with: (1) the classification of ships, both new-buildings and in operation; (2) statutory work carried out on behalf of National Maritime Authorities; (3) equipment and materials for ships.

Contract design 合同设计 系指较完整体现一艘准备建造的新船的技术状态,并提供合同商务谈判的合同文本附件的设计阶段。合同设计的主要内容包括:(1) 详细的船、机、电的技术规格说明书;(2) 全船总布置图(包括各类房间布置图);(3) 船舶结构的中横剖面图;(4) 主要机电设备的供应厂商表;(5) 机舱布置图;(6) 电力负荷估算书。一般在合同设计中需解决的关键技术问题有:(1) 航速——与此有关的为主机功率、船舶线型、船的空船质量、载重量、分舱、螺旋桨、上层建筑及船舶水下部分的附体(如侧推装置、人字架、边龙筋等)船的质量、重心等;(2) 稳性——按初步的质量、重心的估算,预估全船的质量和重心的位置,以保证船舶建造完成后能满足任务书上规定的抗风浪的要求;(3) 载重量——不同种类的船舶,其载重量的含义不同:货船系指装货的吨位,油船系指装油的吨位,集装箱船系指装载一定箱重的载箱量。

Contract of affreightment 包运合同 又称数量合同。是承运人在规定的时间内,分批将一定数量的货物运至约定港口,而由托运人支付运费的合同。

Contract of carriage 运输合同 系指任何全部或部分经由海上运输货物的协议。 Contract of carriage means any agreement to carry goods wholly or partly by sea.

Contract of international goods sales 国际货物买卖合同 系指营业地处于不同国家的当事人之间所订立的,由一方提供货物并转移所有权,另一方支付价款的协议。国际货物买卖合同是国际贸易交易中最为重要的一种合同,是各国经营进出口业务的企业开展货物交易最基本的手段。

Contract review 合同检查 对质量管理体系而言,系指制造厂商应建立并执行在验收前及验收后进行合同检查的各种程序,以保证:(1) 合同的要求被充分地用文件加以确定;(2) 与原始询价和投标中规定有所不同的要求均已被解决;(3) 制造厂商有能力满足并保证符合规定要求的能力。 For the quality management system, contract review means that the manufacturer is to establish and implement procedures for conducting a contract review prior to and after acceptance to ensure that: (1) the requirements of the contract are adequately defined and documented; (2) any requirements differing from those specified in the original enquiry/tender are resolved; and (3) the manufacturer has the capability to meet and verify compliance to the specified requirements.

Contracted for construction 签订建造合同 系指预期船东和造船厂之间签订船舶建造合同的日期。有关"签订建造合同"日期的进一步详细说明，见 IACS 程序要求（PR）No. 29。Contracted for construction means the date on which the contract to build the ship is signed between the prospective owner and the shipbuilder. For further details regarding the date of "contracted for construction", refer to IACS Procedural Requirement (PR) No. 29.

Contracting Government 缔约国 一般系指"1974 年海上人命安全公约"和"国际船舶和港口设施保安规则"的缔约国，包括船旗国、港口国、沿岸国。Contracting Government mean generally the contracting Government in 1974 SOLAS Convention and the ISPS Code, including flag State, port State, coastal State.

Contractor 接触器 系指用来实现电动机主电路通断和改变其接线状态的电动机控制电路中的主要器件。

Contracts for construction project 建设工程合同 系指承包人进行工程建设，发包人支付价款的合同。通常包括建设工程勘察、设计、施工合同。在传统民法上，建设工程合同属承揽合同之一种，德国、日本、法国及中国台湾地区民法均将对建设工程合同的规定纳入承揽合同中。

Contractual license 契约许可证

Contributing cargo 摊款货物 系指作为货物由海上运输到一当事国境内的港口或码头并卸于该当事国的任何有害有毒物质。在从最初装船况港口或码头至最后目的地港口或码头的运输过程中直接地或通过港口或码头从一船全部或部分地转到另一船的转口货物，仅应在最后目的地接收时，被视为摊款货物。Contributing cargo means any hazardous and noxious substances which are carried by sea as cargo to a port or terminal in the territory of a State Party and discharged in that State. Cargo in transit which is transferred directly, or through a port or terminal, from one ship to another, either wholly or in part, in the course of carriage from the port or terminal, from one ship to another, either wholly or in part, in the course of carriage from the port or terminal of original loading to the port or terminal of final destination shall be considered as contributing cargo only in respect of receipt at the final destination.

Contributing Government 分摊国政府 系指遵照有关规定承担义务分摊冰区巡逻服务费用的缔约国政府。 Contributing Government means a Contracting Government undertaking to contribute to the costs of the ice patrol service pursuant to relative Rules.

Control buoy 可控浮筒

Control by magnetic starter 磁力启动器控制 是一种适用于交流单速电动机的用可逆磁力启动器控制电动机的启动、停止和换向，并能对电动机进行过载和失压保护的控制方式。它适用于一般性系泊或移船绞车。

Control circuit 控制电路 系指传送控制设备或系统的电信号使控制器动作的电路。但此种电路不包含主电力电路。

Control equipment 控制设备 系指：(1) 要求为确保原型压载水处理技术正常工作而安装的设备；(2) 操作和控制压载水处理设备要求的安装的设备。 Control equipment refers to: (1) the installed equipment required for proper functioning of the prototype ballast water treatment technology; (2) to the installed equipment required to operate and control the ballast water treatment equipment.

Control equipment for electric propulsion 电力推进控制装置 主要指各种推进电气回路的转换和控制装置。它主要包括发电机主电路转换及控制，发电机励磁回路及推进电动机的信号及保护回路的转换及控制，操纵台选择的转换等设备。它一般位于机舱。

Control for cargo tanks 液货舱的环境控制 液货舱的环境控制通常有以下 4 种不同方式：(1) 惰化法：用不助燃且不与货物反应的气体或蒸发气充入液货舱以及其管系和 IBC 规则的第 15 章有规定的液货舱周围空间，以维持状态；(2) 隔绝法：用能使货物与空气隔绝的液体、气体或蒸发气充入液货舱以及其管系来维持状态；(3) 干燥法：用大气压力下露点为 -40 ℃ 或更低的干燥气体或蒸发气充入液货舱及其管系来维持状态；(4) 通风法：进行强制通风或自然通风。 There are four different types of control for cargo tanks, as follows: (1) Inerting: by filling the cargo tank and associated piping systems and, the spaces surrounding the cargo tanks, specified in Chapter 15 of IBC Code, with a gas or vapour which will not support combustion and which will not react with the cargo, and maintaining that condition; (2) Padding: by filling the cargo tank and associated piping systems with a liquid, gas or vapour which separates the cargo from the air, to maintain that condition;(3)Drying: by filling the cargo tank and associated piping systems with moisture free gas or vapour with a dewpoint of －40 ℃ or below at atmospheric pressure, to maintain that condition;(4)Ventilation: forced or natural.

Control method for hull damage 控制船体受损程度方法 有以下两种方法:(1)进行残余强度分析,确保舰船受损后仍能保持足够的整体强度。有些部位可能还有必要进行撞击分析。(2)确保局部结构可以抵抗威胁或者限制损害。可以通过加固和加强某些区域的局部结构来达到此目的。 For the hull, the effect of the threat can be limited in two ways:(1)ensuring that there is adequate global strength following damage using a residual strength analysis. Where appropriate, a whipping analysis may also be necessary;(2)ensuring that local structure can resist the threat or limit the damage. Individual items of structure can be hardened or strengthened in certain areas to achieve that.

Control monitoring plan(CMP) 船体建造监控计划 系指针对业已识别的船体结构关键位置,规定专门的建造质量标准和控制程序,是船舶建造质量计划的补充。

Control of hazards 危险的控制 为预防升级而限制危险事件的范围和/或持续时间。

Controls of hazardous materials 有害物质的控制。详见表 C-5。

表 C-5 有害物质的控制
Table C-5 Controls of hazardous materials

有害物质 Hazardous material	定义 Definitions	控制措施 Control measures
石棉 Asbestos	含有石棉的材料 Materials containing asbestos	对于所有船舶,应禁止新装含有石棉的材料。 For all ships, new installation of materials which contain asbestos shall be prohibited.
消耗臭氧物质 Ozone-depleting substances	消耗臭氧物质系指在应用或解释本附则时有效的 1987 年消耗臭氧层物质蒙特利尔议定书第 1 条第 4 款中定义的并在该议定书附件 A,B,C 或 E 中所列的受控物质。在船上可能有的消耗臭氧物质包括但不限于下列各项: Halon 1211　溴氯二氟甲烷 Halon 1301　溴三氟甲烷 Halon 2402　1,2-二溴化物-1,1,2,2-四氟乙烷（亦称作 Halon114B2） CFC-11　三氯氟甲烷 CFC-12　二氯二氟甲烷 CFC-113　1,1,2-三氯-1,2,2-三氟乙烷 CFC-114　1,2-二氯-1,1,2,2-四氟乙烷 CFC-115　氯五氟乙烷 Ozone-depleting substances means controlled substances defined in Paragraph 4 of Article 1 of the Montreal Protocol on Substances that Deplete the Ozone Layer, 1987, listed in Annexes A,B,C or E to the Protocol in force at the time of application or interpretation of this Annex.	除 2020 年 1 月 1 日前允许含有氢化氯氟烃（HCFC）的新装置以外,所有船上应禁止使用含有消耗臭氧物质的新装置。 New installations which contain ozone-depleting substances shall be prohibited on all ships, except that new installations containing hydrochlorofluorocarbons (HCFCs) are permitted until 1 January 2020.

续表 C-5

有害物质 Hazardous material	定义 Definitions	控制措施 Control measures
消耗臭氧物质 Ozone-depleting substances	Ozone-depleting substances that may be found on board ship include, but are not limited to: Halon 1211 Bromochlorodifluoromethane Halon 1301 Bromotrifluoromethane Halon 2402 1,2-Dibromo-1,1,2,2-tetrafluoro-ethane (also known as Halon 114B2) CFC-11 Trichlorofluoromethane CFC-12 Dichlorodifluoromethane CFC-113 1,1,2-Trichloro-1,2,2-trifluoroeth-aneCFC-114 1,2-Dichloro-1,1,2,2-tetrafluoro-ethane CFC-115 Chloropentafluoroethane	
多氯联苯 Polychlorinated biphenyls (PCB)	多氯联苯系指联苯分子(2个苯环被一个碳—碳键连在一起)上的氢原子被可多至10个氯原子取代而形成的芳香族化合物。 Polychlorinated biphenyls means aromatic compounds formed in such a manner that the hydrogen atoms on the biphenyl molecule (two benzene rings bonded together by a single carbon-carbon bond) may be replaced by up to ten chlorine atoms.	对于所有船舶,应禁止新装含有多氯联苯的材料。 For all ships, new installation of materials which contain Polychlorinated biphenyls shall be prohibited.
防污底化合物和系统 Anti-fouling compounds and systems	在应用或解释2009年香港国际安全与环境无害化拆船公约附则时现行的《2001年国际控制船舶有害防污底系统公约》(AFS公约)附则 I 中所规定的防污底化合物和系统。 Anti-fouling compounds and systems regulated under Annex I to the International Convention on the control of harmful anti-fouling systems on Ships, 2001 (AFS Convention) in force at the time of application or interpretation of this Annex to the Hong Kong International Convention for the safe and environmentally sound recycling of ships, 2009.	(1)所有船舶不得施涂含有机锡化合物作为杀生物剂的防污底系统或任何其他AFS公约禁止施涂或使用的防污底系统。 (2)所有新船或船上的新装置不得施涂或采用不符合AFS公约规定的防污底化合物或系统。 (1) No ship may apply anti-fouling systems containing organotin compounds as a biocide or any other anti-fouling system whose application or use is prohibited by the AFS Convention. (2) No new ships or new installations on ships shall apply or employ anti-fouling compounds or systems in a manner inconsistent with the AFS Convention.

Control panels 控制面板 系指操作动力定位系统在定位场所和测量处安装的控制面板。Control panels comprise centrally and locally situated panels for operating the dynamic positioning system.

Control room 操纵室 系指执行船舶航行和控制的封闭区域。

Control section for monitoring and control system 监控系统的控制部分 监控系统的控制部分包括：(1)处理机,它接收排放物含油量；排放物流速和航速信号,并将这些数值换算成每海里的排油量升和排油总量；(2)提供报警和向舷外排放控制命令信号的设备；(3)提供自动数据记录的记录设备；(4)展示目前操作数据的数据显示器；(5)在监控系统发生故障时使用的越控系统；(6)提供信号给启动连锁以防止在监控系统完全运作前排放任何排放物的设备。Control section for monitoring and control system comprises: (1) a processor, which accepts signals of oil content in the effluent, the effluent flow rate and the ship's speed and computes these values into liters of oil discharged per nautical mile and the total quantity of oil discharged; (2) equipment to provide alarms and command signals to the overboard discharge control; (3) a recording device to provide a record of data; (4) a data display to exhibit the current operational data; (5) a manual override system to be used in the event of failure of the monitoring system; (6) means to provide signals to the starting interlock to prevent the discharge of any effluent before the monitoring system is fully operational.

Control stage (regulating stage, governing stage) 调节级 喷嘴调节时,其进气度随负荷大小而变化的汽轮机级。

Control stations 控制站 系指：(1)①设有应急电源和应急照明电源的处所；②驾驶室和海图室；③设有船舶无线电设备的处所；④消防控制站；⑤位于推进装置处所外面的推进装置控制室；⑥设有集中失火报警设备的处所；⑦设有集中应急公共广播系统站和设备的处所；(2)具有监视功能且能够对机电设备实施控制的处所。控制站主要有下列4类：①机舱集控站(室)；②驾驶室控制站；③就地控制站；④其他控制站。 Control stations are: (1) ① spaces containing emergency sources of power and lighting; ② wheelhouse and chartroom; ③ spaces containing the ship's radio equipment; ④ fire control stations; ⑤ control room for propulsion machinery when located outside the propulsion machinery space; ⑥ spaces containing centralized fire alarm equipment; ⑦ spaces containing centralized emergency public broadcast station and equipment; (2) spaces fitted with monitoring means capable of controlling the machinery and electrical installations. They are mainly divided into four categories as flowing: ①centralized control stations (rooms) of engine room (CCS); ②bridge control station (BCS); ③local control station (LCS); ④other control stations.

Control stations 控制站(平台) 系指平台无线电设备或主要航行设备或应急电源所在的处所,火警指示器或失火控制设备或动力定位控制系统集中的处所,以及用于不同场所的灭火系统所在的处所。对于柱稳式平台,压载水集中控制站即是"控制站"。 Control stations are those spaces in which the unit's radio or main navigating equipment or the emergency source of power is located or where the fire recording or fire control equipment or the dynamic positioning control system is centralized or where a fire-extinguishing system serving various locations is situated. For column-stabilized units, a centralized ballast control station is a "control station".

Control surface 控制面 控制船舶运动方向的作用面。

Control system[1] 控制系统 系指一个将预期的动作施加到一个设备上,从而来达到指定目的的系统。Control system is a system by which an intentional action is exerted on an apparatus to attain given purposes.

Control system[2] 控制设备 系指自动或手动控制船舶位置时所需要的硬件和软件。 Control system means all central hardware and software necessary to dynamically position the vessel by automatic or manually control.

Control system fault alarm 控制系统故障报警 系指指示自动控制系统或遥控系统故障的报警,例如驾驶室推进控制故障报警。 Control system fault alarm means an alarm which indicates a failure of an automatic or remote control system, e.g. the navigation bridge propulsion control failure alarm.

Control system on MGO viscosity 低硫油(MGO)黏度控制系统 主要通过控制船用低硫油温度进而改变其黏度,从而使其符合现有柴油机的使用要求的装置。其由制冷系统、冷冻水系统和油循环系统三部分组成。制冷系统采用了对臭氧层不会起破坏作用的R404a制冷剂；冷却水系统采用闭式循环,并选用30%的乙二醇环保型水溶液作为介质,可最大限度降低热损失。该装置采用模糊控制技术,可实时监控设备压力、温度和流量,从而使压缩机、水泵和三通阀处于最佳工作状态。同时该装置具有运行状态数据存储功能,便于对系统运行状态进行统计和检查。

Control transformer 控制变压器 系指用于控制系统的变压器。如提供多种电压的电源变压器,用于阻抗匹配的输入/输出变压器和具有足够电感量的脉冲变压器等。

Control unit (CU) 控制装置/控制器 系指一种接收下列自动信号的装置：(1)排放物的含油量,10^{-6}；(2)排放率,m^3/h；(3)船舶速度,kn；(4)船舶位置——经度和纬度；(5)日期和时间(GMT)；(6)舷外排放控的状况。 Control unit is a device which receives automatic signals of: (1) oil content of the effluent in ppm; (2) flow rate of discharge in m³/hour; (3) ship's speed in knots; (4) ship's position-latitude and longitude; (5) date and time (GMT); (6) status of the overboard discharge control.

Control valve 控制阀 系指控制液压油传输以驱动执行机构的部件,按其属名为开-关控制电磁阀、比例控制阀或变量控制阀等。 Control valve is a component to control the delivery of hydraulic oil to drive the actuator, with a generic name of on-off-controlled solenoid valve, proportional-controlled valve or variable-controlled valve, etc.

Controllable active anti-rolling tank 可控主动式减摇水舱 采用机械装置将一舷的水输送到另一舷,并通过控制左右两水舱间的控制阀来调节水往复运动的速度,实现可控制减摇目的的主动式减摇水舱。

Controllable constrained damping layer 可控约束阻尼层

Controllable passive anti-rolling tank 可控被动式减摇水舱 通过水舱顶部连通道或两边舱顶部安装气阀,或在水舱底部连通道安装可调节的栅板,用少量能量控制阀门或栅板的开关,实现对水舱中液体流动的控制,使水舱能够在更宽的频率范围内实现可控制减摇目的的被动式减摇水舱。通常可控被动式减摇水舱采用"开关式"的控制策略,即通过阀门或栅板的开启和关闭将水舱内的液体保持在船舶向上运动的一侧边舱内,从而使水舱中液体产生的力矩用于减小船舶横摇运动。

Controllable passive tank stabilization system 可控被动水舱式减摇装置 流道面积可以控制的被动水舱式减摇装置。

Controllable pitch propellers 可调螺距螺旋桨 为组装螺旋桨,它还包括在桨毂中安装的旋转叶片的机构,为的是在不同运行工况下有可能调控螺旋桨螺距。 Controllable pitch propellers are built-up propellers which include in the hub a mechanism to rotate the blades in order to have the possibility of controlling the propeller pitch in different service conditions.

Controlled disconnection 可控制脱开 以有计划的可控方式将连接着的船舶/平台装置,包括双船操作及其实体分隔的所有实体性连接解脱。

Controlled hydrofoil 可控水翼 为保护艇的起飞性能、稳定性及在波浪中的航行性能等,能用人工或自控方法进行控制的水翼。如攻角可以控制的水翼、襟翼、供气水翼等。

Controlled rolling [**normalizing rolling, CR (NR)**] 控制轧制[正火轧制,CR(NR)] 系指将最终轧制温度控制于正火热处理区域之内的轧制工艺。采用此轧制工艺生产出来的钢材的力学性能等同于进行通常正火热处理的钢材。 Controlled rolling is a rolling procedure in which the final deformation is carried out in the normalizing temperature range, resulting in a material condition generally equivalent to that obtained by normalizing.

Controller (DPS) 控制器(动力定位系统) 系指船舶实现动力定位所必需的一切集中控制的硬件和软件。控制器一般应由一台或几台计算机组成。 Controller (DPS) means all concentrated control hardware and software necessary to supply DP of the vessel. The controller is generally composed of one or more computers.

Controller and measuring system (DPS) 控制器和测量系统(动力定位系统) 系指由下列设备组成的系统:(1)计算机系统;(2)推力器手动控制;(3)推力器的联合操纵杆控制;(4)推力器的自动控制;(5)位置参照系统;(6)传感器系统;(7)显示和报警系统;(8)通信系统。 Controller and measuring system (DPS) means a system comprised by the following equipment: (1) computer system; (2) manual thruster controls; (3) joystick thruster controls; (4) aytomatpc thruster controls; (5) position reference systems; (6) sensor systems; (7) display and alarm system; (8) communication system.

Controls-fixed stability 锁定控制面稳定性 方向舵或升降舵能保持固定舵角时,船舶能保持一定航向或潜深的特性。

Control 控制 要么执行行动要么执行命令。

Convention 公约 系指:(1)2009年香港国际安全与环境无害化拆船公约;(2)国际压载水和沉积物控制和管理公约;(3)经修正的1974年国际海上人命安全公约等。 Convention means: (1) the Hong Kong International Convention for the Safe and Environmentally Sound Recycling of Ships, 2009; (2) the International Convention for the Control and Management of Ships' Ballast Water and Sediments; (3) the International Convention for the Safety of Life at Sea, 1974, as amended.

Convention for Limiting the Manufacture and Regulating the Distribution of Narcotic Drugs 限制麻醉药品制造和管制麻醉药品运销公约 系指1931年日内瓦"限制麻醉药品制造和管制麻醉药品运销公约"。 Convention for Limiting the Manufacture and Regulating the Distribution of Narcotic Drugs means 1931, Geneva, Convention for limiting the manufacture and regulating the distribution of narcotic drugs.

Convention for the Suppression of Illicit Traffic in Dangerous Drugs 取缔非法贩运危险毒品公约 系指

1936 年日内瓦取缔非法贩运危险毒品公约。 Convention for the Suppression of Illicit Traffic in Dangerous Drugs means the Convention for the suppression of Illicit traffic in dangerous drugs, 1936, Geneva.

Convention on Psychotropic Substance 精神药物公约 系指1971年维也纳精神药物公约。Convention on Psychotropic Substance means the pollutant listed in Chapter 40 Rules 401.16 of Laus and regulations of Federal government of the United States.

Convention ship 公约船舶 系指持有按国际有关公约规定签发国际证书的船舶。Convention ship is a ship holding an international certificate issued according to a relevant international convention.

Conventional diving 常规潜水 潜水员曝露在水下高压环境中，作业时间不长，机体各组织尚未完全被吸入气中的中性气体所饱和的潜水。

Conventional pollutants 常规污染物 系指美国联邦政府法规第40篇第401.16条中所列的污染物。Pollutants listed in Chapter 40 Rules 401.16 of Lans and regulations of Fedesal government of the United States.

Conventional stopping 常规停船 船舶在航行中，使螺旋桨执行一般倒车的停船方式。

Conventional submarine 常规潜艇 系指采用柴油机作为动力源，一边航行一边带动发电机给电池充电。由于柴油机工作需要大量的氧气，因此只有在水面状态、半潜状态和通气管状态航行时才能充电。常规潜艇技术拥有国：中国（029/041）、德国（209/214）、法国（鲉鱼/阿格斯塔91B）、俄罗斯（K基洛）、日本（苍龙）、瑞典、荷兰。见图C-14。

图 C-14 常规潜艇
Figure C-14 Conventional submarine

Conventional tension leg platform (CTLP) 传统张力腿平台/第一代张力腿平台 是张力腿平台家族中排水量最大的一种结构形式，因此，具有较大的油气处理能力，大多数传统张力腿平台还配备了钻井模块，可自行完成所管理干树井的钻井作业，是钻、采、修井及生产（油气处理）的一体化平台，也可作为井口平台（但不具有油气处理能力）使用。

Conventional torpedo submarine 常规鱼雷潜艇 系指以鱼雷作为攻击敌方武器的柴油机动力的潜艇。

Convergence angle of shafting 轴系内倾角 内斜轴系中线与船体中线面间的夹角。

Conversion engineering/Remould engineering 改装工程 系指对现有船舶进行重大改造的工程。改造的内容有使用任务的变更（如货船改成客船）、船体接长或进行技术更新（如更换主机）等。

Convex deck 凸形甲板 系指在船体的货舱区域内且高于上甲板的连续露天甲板。

Conveyor 输送机 一种在水平或坡度不大的方向连续输送货物的机械。

Cookie 小型文本文件 系指某些网站为了辨别用户身份，进行Session跟踪而储存在用户本地终端上的数据（通常经过加密）。定义于RFC2109（已废除）。为网景公司的前雇员 Lou Montulli 在 1993 年 3 月所发明。Cookie是由服务器端生成，发送给User-Agent（一般是浏览器），浏览器会将cookie的key/value保存到某个目录下的文本文件内，下次请求同一网站时就发送该cookie给服务器（前提是浏览器设置为启用cookie）。Cookie名称和值可以由服务器端规定自己定义，对于JSP而言也可以直接写入 jsessionid，这样服务器可以知道该用户是否合法用户以及是否需要重新登录等，服务器可以设置或读取cookies中包含信息，借此维护用户跟服务器会话中的状态。

Cooking appliances room 烹饪炊具室 包括以下设备。（1）每台超过5 kW功率的自动咖啡机、吐司炉、洗碟机、微波炉、热水炉和类似设备；（2）每台最大5 kW功率的料理电热板及食物保温电热板。 Cooking appliances room may contain the following devices. (1) Coffee automats, toasters, dish washers, microwave ovens, water boilers and similar appliances, each of which has a power of more than 5 kW. (2) Electrically heated cooking plates and hot plates for keeping food warm, each of which has a maximum power of 5 kW.

Cooking oil 食用油 系指用来或拟用来预制或烹饪食物的可食用的任何类型油或动物脂肪，但不包括这些油预制的食物本身。 Cooking oil means any type of edible oil or animal fat used or intended to be used for the preparation or cooking of food, but does not include the food itself that is prepared using these oils.

Cooking utensile 炊具 各种直接用于烹饪饭菜的锅、勺、铲、笼等器具的统称。

Cooling blade 冷却叶片 系指从结构上采取冷却措施以降低金属表面温度及温度梯度的涡轮叶片。有对流冷却、气膜冷却、发散冷却、撞击冷却等不同冷却方式。

Cooling fan　冷却风扇　空调装置中输送冷却空气的风扇。

Cooling lagging　Insulation plate　冷却罩壳　为减少燃气轮机对舱室的热辐射,在机匣外表面敷设的带有绝热材料的外罩。

Cooling pump（cooling medium pump）　载冷剂泵　在间接冷却系统中抽送载冷剂如盐水、三氧乙烯等的泵。

Cooling sea water control system　海水冷却控制系统

Cooling spray descaling　冷淋除垢　将蒸发器放于盐水并通入蒸汽加热,随后突然用冷水对换热面淋洗,使换热面上的污垢碎裂脱落的除垢方式。

Cooling state operating　冷态作业　系指船舶在建造、改建、修理或拆船作业中无加热、无明火或会产生火花的作业。

Cooling system for MGO　低硫燃油冷却系统　依据降温原理,将燃油的温度降低到17 ℃左右使其黏度增加到2厘斯以满足现有船用柴油机和供油系统的要求。其由1台船用冷水机组、1套冷媒水泵模块和1套水油换热系统组成。其特点是采用PLC控制系统来实现冷水机组、水泵和水油换热系统的自动控制,以确保工况稳定。该冷却系统采用模块化设计,冷水机组可靠性强,能根据负荷的变化,实现机组的多级能量调节,精确控制水油换热器的出油温度。

Cooling water pressure　冷却水压力　通常在进水总管上测得的进入内燃机的冷却水的压力。

Cooling water ratio　冷却水倍率　蒸馏装置中冷凝器中冷却水量与蒸馏水产量的比值。

Cooling water system　冷却水系统　以水为冷却介质的冷却系统。由水泵、冷却器、管路及附件等组成。

Cooling water temperature　冷却水温度　进出内燃机的冷却水温度。

Coordinate measuring machine（CMM）　坐标测量机　是最有代表性的坐标测量仪器。最早的坐标测量机是1个仅仅配备XYZ三轴数显的三维设备。对于谁最先发明坐标测量机一直存在争议。1950年代末,第一台坐标测量设备可能是由意大利DEA（Digital Electronic Automation Spa）公司发明的门式结构并配置刚性测量头的测量机。几个月后,苏格兰的Ferrantimetrology（现在的IMS）发明了悬臂测量机,配置数显设备和刚性测头。1959年,Ferranti发明了第一台真正意义上的坐标测量机（Direct computer assist,DCC）。这台坐标测量机的发明可以标志着坐标测量机正式走上历史舞台。同样来自英国的LK Tool公司随后发明了第一台桥式坐标测量机,并在过去的数年中成为坐标测量机的标准构架。随着越来越多制造厂商的加入并对测量机进行改造,悬臂测量机、桥式测量机、龙门测量机、水平臂测量机、门式测量机、活动平台测量机、固定桥式测量机和关节臂测量机都成为坐标测量机的通用构架。见图C-15。

图 C-15　坐标测量机
Figure C-15　Coordinate measuring machine（CMM）

Co-ordinate surface vessel　海面搜寻协调船　指定在特定搜寻区域内对海面搜救工作进行协调的非救助单位。 Co-ordinate surface vessel means a vessel, other than a rescue unit, designed to conduct co-ordinate surface search and rescue operations within a specified search area.

Copper alloy　铜合金　系指以纯铜为基体加入一种或几种其他元素所构成的合金。纯铜呈紫红色,又称紫铜。纯铜密度为8.96,熔点为1 083 ℃,具有优良的导电性、导热性、延展性和耐蚀性。主要用于制作发电机、母线、电缆、开关装置、变压器等电工器材和热交换器、管道、太阳能加热装置的平板集热器等导热器材。常用的铜合金分为黄铜、青铜、白铜3大类。

Copper alloy pipe fitting series　铜合金管件　系指以铜合金作为材料制成的管件。

Copper granules　铜粒　球形小颗粒。含铜75%,另有铅、锡、锌及少量其他杂质。含水量约1.5%。干燥时呈淡灰色,潮湿时呈深绿色。无嗅。无特殊危害。该货物为不燃物或失火风险低。 Copper granules are sphere shaped pebbles. It has 75% of copper with lead, tin, zinc, traces of others. Moisture content is 1.5% approximately. It is light grey colour when dry, dark green when wet. Odourless, It has no special hazards. This cargo is non-combustible or has a low fire-risk.

Copper inclusions 铜夹杂物

Copper matte 冰铜　天然黑色铜矿石。成分为75%铜和25%杂质。小的金属圆石或丸粒。无嗅。无特殊危害。该货物为不燃物或失火风险低。 Copper matte is a crude black copper ore. It is composed of 75% copper and 25% impurities. It is small metallic round stones or pellets and odourless. It has no special hazards. This cargo is non-combustible or has a low fire-risk.

Copra (dry) UN 1363 椰肉(干燥)　干燥的椰仁，具有刺鼻的陈腐气味，颗粒会沾污其他货物。易自热和自燃，特别在遇水时。易使货物处所缺氧。 Copra (dry) UN 1363 is dried kernels of coconuts with a penetrating rancid odour which may taint other cargoes. It is liable to heat and ignite spontaneously especially when in contact with water. It is liable to cause oxygen depletion in the cargo space.

Copy 复制件　系指与原件内容相同的复制品。

Copying mechanism 复印机属模拟方式　系指只能如实进行文献的复印。今后复印机将向数字式复印机方向发展，使图像的存储、传输以及编辑排版(图像合成、信息追加或删除、局部放大或缩小、改错)等成为可能。它可以通过接口与计算机、文字处理机和其他微处理机相连，成为地区网络的重要组成部分。多功能化、彩色化、廉价和小型化、高速仍然是重要的发展方向。

CORA wind-finding and radiosonde system 导航测风系统　用于收集高空气象资料、大气压力、温度、湿度、风速和风向及其位置等。系统中配有NOVA计算机。

Core material 芯材　对钢夹层板而言，系指通过灌注封闭于底板和顶板之间的直接与钢表面接触的一层高分子材料或高分子增强材料。

Corn Laws 谷物法　或称"玉米法案"。系指英国1672年制定的限制谷物进口的法律，并在1815年通过新的谷物法提高对农业的保护力度。

Corner fittings 角件　是一种固定装置，典型为铸件，为集装箱搬运和系固提供连接用的标准开口和接触面。为集装箱端部框架结构的组成部分，通常符合ISO 1161标准的规定。类似的装置可能布置在距离端部框架结构有一定距离的位置还有内部角柱处，如45ft集装箱的40ft位置的角件。

Corner fittings 角配件　系指为了装卸、堆码和/或紧固的目的而在集装箱顶部和/或底部上安装的一种表面有孔的支撑配件。Corner fittings mean an arrangement of fittings with apertures and which are installed at the top and/or bottom of a container for the purposes of handling, stacking and/or securing.

Corner post 角柱　是集装箱加强的垂直结构，位于集装箱端部的角件之间，用于承受起吊、堆装及系固时的压力和拉力。一些集装箱在距离端部一定距离的还有内部角柱，如40 ft的位置。

Corporation 公司　系指圣劳伦斯航道开发公司。 Corporation means the Saint Lawrence Seaway Development Corporation.

Correction layer 修正水层　在大倾角稳性计算中，使船舶作等体积倾斜时，假定的倾斜水线与实际的等体积倾斜水线间的水层。

Correction of rate of revolution 转速修正　将船模自航试验结果换算至实船的有关数值时就伴流的尺度效应对螺旋桨转速所做的修正。

Corrective action 纠正措施　对质量管理体系而言，系指制造厂商应建立和维持规定的步骤来复查不合格的性质和不合格材料的处理办法。这是为了：(1)提出生产监控和施工程序以及记录分析等以查明和消除不合格材料产生的潜在原因；(2)继续分析所给予的通融办法和材料报废或返工的原因以及需要采取的纠正措施；(3)对用户申诉的分析；(4)关于接受不合格材料问题，与供应商或分包商采取适宜的行动；(5)保证采取的纠正措施是有效的。 For the quality management system, corrective action means that the manufacturer is to establish and maintain documented procedures for the review of non-conforming materials and their disposition. These should provide for: (1) monitoring of process and work operations and analysis of records to detect and eliminate potential causes of non-conforming materials; (2) continuing analyzing concessions granted and material scrapped or reworked to determine causes and the corrective action required; (3) an analysis of customer complaints; (4) the initiation of appropriate action with suppliers or sub-contractors with regard to receipt of non-conforming material; (5) an assurance that corrective actions are effective.

Corrective action requests 改正措施请求书　系指在公司内部审核时，内审员要求被审核者提交的文件。

Corrective measures 改进措施　系指激活工程或管理和操作程序以降低产生故障的可能性。Corrective measures mean that engineering or administrative and operational procedures activated to reduce the likelihood of a failure.

Correlation coefficient 相关函数　系指：(1)衡量2个随机变量 ξ_1 和 ξ_2 间相关性的一个数字特征。其表达式为 $P = COV(\xi_1, \xi_2)/(D\xi_1 \cdot D\xi_2)^{1/2}$；$P = 0$ 时，ξ_1 与 ξ_2 不相关；$P = 1$ 时，$P \leq 1$；(2)表示随机过程间线性

相关的一种函数。有自相关函数与互相关函数两种。在无特别注明时,通常系指自相关函数。

Correspondence 信函

Correspondent banks 代理银行 系指为了促进资金转移,一些银行会为某些在相关地区没有分行的银行提供定期服务。这些提供服务的银行被称为代理银行。

Corresponding speed 相应速度 在某种相似条件下,如傅汝德数相同,船模和实物间相对应的速度。

Corridor test 走廊试验 系指火源贴近走廊墙壁且在一个喷嘴下方,或火源贴近走廊墙壁且在两个喷嘴之间的试验。在第一个喷嘴动作后火灾试验应进行10 min,任何余火应人工扑灭。 Corridor test means a test in which fire source is located against the wall of the corridor under one nozzle, or fire source is located against the wall of the corridor between two nozzles. The fire tests should be conducted for 10 min after the activation of the first nozzle, and any remaining fire should be put out.

Corridors 梯道 系指内部梯道、电梯和自动扶梯以及上述梯道等的围闭;至于仅在一层甲板围闭的梯道,应作为没有被防火门隔开的处所一部分。 Corridors mean interior stairways, lifts, and escalators and enclosures thereto. However, a stairway which is enclosed by one layer of deck is regarded as part of the space from which it is not separated by a fire door.

Corridors 走廊 系指旅客及船员使用的走廊和门厅。 Corridors are corridors and lobbies for passenger and crew.

Corrosion 腐蚀 系指金属与环境之间发生物理-化学作用,使其性能发生变化,并导致金属、环境或由它们组成的体系的功能受到损伤。 Corrosion means physical-chemical action taking place between metal and environment, which makes its properties change and causes the function damages of metal, environment or the system consisting of them.

Corrosion addition t_c 腐蚀增量 系指为保证船舶营运寿命,应在结构强度计算所要求的净尺寸上增加足够的增量。腐蚀增量应根据内部和外部结构的使用情况和对腐蚀介质接触情况确定,如水、货物或腐蚀性空气,另外,还应考虑防腐系统,如涂层、阴极保护或采用替代措施。 Corrosion addition is to be added to the net scantling required by structural strength calculations in order to make sure there is adequate operating life for the ship. The corrosion addition is to be assigned in accordance with the use and exposure of internal and external structure to corrosive agents, such as water, cargo or corrosive atmosphere, in addition to the corrosion prevention systems, e.g. coating, cathodic protection or by alternative means.

Corrosion prevention system 防腐系统/防腐措施 通常可考虑全面硬化保护涂层。全面硬化保护涂层通常系指环氧树脂或同等物。除软涂层或半硬涂层以外的其他涂层系统只要根据制造厂商的规定应用或维护,可以考虑作替代品接受。这里所说的软化涂层指,以一般(植物性或石油)油或羊毛脂(lanolin: sheep wool grease)为基础材料,碰撞或小的机械性冲击都能使之脱落的软质的涂层;半硬化涂层指,即使碰撞或在上面反复行走依然很坚固,干燥或固化后依然具有柔性的涂层。涂层状态区分如下:(1)良好:没有锈迹或者只有很小锈迹的状态;(2)普通:系指在扶强材边缘和焊缝的连接处涂层有局部脱落或者所检验的区域中有超过20%或更大的范围轻度锈蚀,但小于定义"差"的程度;(3)差:检验部位20%以上出现脱落或者10%以上有严重腐蚀的状态。 Corrosion prevention system normally adopts a full hard protective coating. A full hard protective coating is usually to be epoxy coating or equivalent. Other coating systems except thd soft coating and semihard coating may be considered acceptable as alternatives provided that they are applied and maintained in compliance with the manufacturer's specifications. Where soft coating means a coating that takes oil (vegetable or petroleum) or lanolin (sheep wool grease) as its basic materiol, and is usaally uorn off by low mechanical impact or touch. Semi-hard coating means a coating that dries or converts in such a way that it stays flexible through collision or walk. Coating condition is defined as follows: (1) GOOD condition with only minor spot rusting; (2) FAIR condition with local breakdown at edges of stiffeners and weld connections and/or light rusting over 20% or more of areas under consideration, but less than that defined for POOR condition; (3) POOR condition with general breakdown of coating over 20% or more, or hard scale at 10% or more, of areas under consideration.

Corrosion resistant steel 耐腐蚀钢 系指在货舱的内底或内顶的腐蚀性能经过试验和认可,除满足船用材料、结构强度和构造方面的其他相关要求外,还满足《原油油船货油舱保护涂层性能标准》要求的钢材。 Corrosion resistant steel is steel whose corrosion resistance performance in the bottom or top of the internal cargo oil tank is tested and approved to satisfy the requirements in the Performance Standard for Protective Coating for Cargo Oil Tanks of Crude Oil Tankers in addition to other relevant requirements for ship material, structures strength and construction.

Corrosion resistant steel standard 耐腐蚀钢的标准 该标准基于的规格和要求拟在达到25年的目标使

用寿命,这视为从最初的使用开始,钢材的厚度的折减小于许用厚度的折减,并且货油舱内保持水密完整性的持续时间。实际使用寿命会有变化,取决于许多可变因素,包括在使用中遇到的实际情况。 Corrosion resistant steel standard is based on specifications and requirements which intend to provide a target useful life of 25 years, which is considered to be the time period, from initial application, over which the thickness diminution of the steel is intended to be less than the diminution allowance and watertight integrity is intended to be maintained in cargo oil tanks. The actual useful life will vary, depending on numerous variables including actual conditions encountered in service.

Corrosion test 腐蚀试验 系指确定腐蚀凹坑和裂缝的试验。 Corrosion test means a test determining the corrosion pitting and crevice.

Corrosion-resisting materials for cathode protection 阴极保护用防腐材料

Corrosive substances (Class 8 dangerous materials) 腐蚀品(第 8 类危险货物) 系指通过化学作用会对所接触的活体组织造成严重损害,或实质性损害甚至毁坏其他货物或运输工具的任何物质。 Corrosive substances mean the materials in this class which, by chemical action, will cause severe damage when in contact with living tissue or will cause materially damage, or even destroy, other goods or the means of transport.

Corrosive to skin 腐蚀皮肤 见表 C-6。

表 C-6 腐蚀皮肤
Table C-6 Corrosive to skin

危险程度 Hazard Level	使皮肤完全坏死的接触时间 Exposure time to cause full thickness necrosis of skin	观察时间 Observation time
严重腐蚀皮肤 Severely corrosive to skin	≤3min	≤1h
高度腐蚀皮肤 Highly corrosive to skin	≥3min ~ ≤1h	≤14 days
轻微腐蚀皮肤 Moderately corrosive to skin	≥1h ~ ≤4h	≤14 days

Corrugated bulkhead 槽形舱壁 系指以槽形式样布置的板材构成的舱壁。 Corrugated bulkhead means a bulkhead comprised of plating arranged in a corrugated fashion.

Corrugated hatch covers 波形舱盖 由金属或玻璃钢制成的波纹形剖面舱盖板组成的拼装舱盖。

Cosine wave 余弦波 与正弦波呈 90° 相位差的波。其表达式为 $\xi = \xi_A \cos(kx - \omega t)$。

Cosmic rays observation 宇宙射线观测 连续观测穿过大气进入地球底层高能粒子的来源、数量和强度的变化以及能量范围在空间的分布,为研究太阳活动的规律提供资料。

Cosmic rays observation room 宇宙射线观测室 使用超多段中子监测器和塑料闪烁介子望远镜。连续记录穿过大气层进入地球表面层高能粒子的来源、数量和强度的变化,以及能量范围在空间的分布。为研究太阳活动的规律提供资料的工作舱室。

Co-spectrum 同向谱 互谱中,表示 2 个随机函数中同向位的组成分量间的相对能量大小的实数部分。其表达式为: $C_{1,2}(\omega) = (1/2\pi) \int_{-\infty}^{+\infty} R_{1,2}(\tau) \cos \omega \tau \, d\tau$。

Cost 价款/成本 系指合同的买方对合同的卖方履行合同,交付货物所应支付的货币为表现形式的价金。

Cost and freight (CFR) 成本加运费 系指卖方必须在合同规定的装运期内,在装运港将货物交至运往指定目的港的船上,负担货物越过船舷为止的一切费用和货物灭失或损坏的风险,并负责租船或订舱,支付抵达目的港的正常运费。

Cost and freight—named port of destination (CFR) 成本加运费(……指定目的港) 系指卖方必须支付把货物运至指定的目的港所需的开支和运费,但从货物交至船上甲板后,货物的风险、灭失或损坏以及发生事故后造成的额外开支,在货物越过指定港的船舷后,就由卖方转向买方负担。另外要求卖方办理货物的出口结关手续。本术语适用于海运或内河运输。

Cost benefit analysis 费用/效益分析(成本效益分析) 一种合理的系统性的框架,该框架用可比较的衡量单位,来评价各种降低风险方案的优缺点。

Cost effectiveness 成本/效益分析 系指以特定的临床治疗目的(生理参数、功能状态、增寿年等)为衡量指标,计算不同方案或疗法的每单位治疗效果所用的成本。其结果不用货币单位来表示,而通常使用健康结果或临床治疗指标,如抢救患者数、治愈率、延长的生命年、血压降低值等指标的变化来表示。

Cost for remanent repairs 预留成本 即不可预见

的费用。又称为预备费,系指在工程投资概(估)算中,预留的为支付施工中可能发生的、比预期的更为不利的水文、天气、地质及其他社会、经济条件而需增加的费用,一般以总投资的某一百分数计。

Cost insurance and freight—named port of destination(CIF) 指定目的港的成本加保险费加运费 按此术语成交,货价的构成因素中包括从装运港至约定目的地港的通常运费和约定的保险费,故卖方除具有与CFR术语的相同的义务外,还有为买方办理货运保险,交支付保险费,按一般国际贸易惯例,卖方投保的保险金额应按CIF价加成10%。如买卖双方未约定具体险别,则卖方只需取得最低限底的保险险别,如买方要求加保战争保险,在保险费由买方负担的前提下,卖方应予加保,卖方投保时,如能办到,必须以合同货币投保。

Cost/benefit evaluation 成本/收益评价 成本与收益的定量评估和比较。在收益就是降低安全性或环境危害的现实情况下系指安全措施或环境保护措施。

Costa propulsion bulb and rudder thrust fin composite device(CPB-RTF) 舵球鳍 一种装于舵叶两侧桨轴中心线延长线上,可以回收一部分螺旋桨尾流中损失的旋转能量,并将其转化为推力的节能装置。

Counter 艉伸部 突出于设计水线后端位于艉垂线以后的艉悬伸部。

Counter 缆绳计数器 系指绞车收放缆绳时指示其长度的仪器。

Counter weight 吊货索压重 在轻型吊杆装置中,为使空钩易于下降而装于吊货索上方的金属重块。

Counterbalanced window 平衡窗 依靠人力提拉,并以弹簧或重锤等构件平衡其重力使其定位于任何高度的窗。

Counter-offer 还盘 受盘人在接到发盘后,不能完全同意发盘的内容,为了进一步磋商交易,对发盘提出修改意见用口头或书面形式表示出来,就构成还盘。

Counter-rotating bearing 反向转动轴承 通常用于对转螺旋桨轴系中,在一个旋转中心上,转向相反,外轴承是内圈成为内轴承外圈的两个轴承的组合体。

Counter-rotating propellers(CRP) 反向旋转螺旋桨 固定在螺旋桨后,一般作叶轮形,吸收尾流中旋转动能以提高对船推力的装置。

Counter-weight tensioning device 重锤张力装置 系指用重锤保持绳索恒张力的装置。

Countra propeller(reaction propeller) 反应螺旋桨 固定在螺旋桨后,一般作叶轮形,吸收尾流中旋转动能以提高对船推力的装置。

Country quotas 国别配额 或称"地区配额",是进口配额制的一种,是在总配额内按国别或地区分配给固定的配额,超过规定的配额便不准进口。一般来说,国别配额可以分为自主配额和协议配额两种。为了区分来自不同国家和地区的商品,在进口商品时进口商必须提交原产地证明书。实行国别配额可以使进口国家根据其与有关国家或地区的政治、经济关系分配给予不同的额度。这是在总配额中按国别和地区分配配额。不同国家和地区如超过所规定的配额,就不准进口。

Coupling 耦合 几种方式的运动同时出现又互相影响的现象。

Coupling(shaft coupling) 联轴器 系指轴系中使两段轴互相连接的组件。见图C-16。

图 C-16 联轴器
Figure C-16 Coupling(shaft coupling)

Coupling and damper 联轴节 又名联轴器。用来连接不同机构中的两根轴(主动轴和从动轴)使之共同旋转以传递扭矩的机械零件。在高速重载的动力传动中,有些联轴器还有缓冲、减振和提高轴系动态性能的作用。联轴器由两半部分组成,分别与主动轴和从动轴连接。一般动力机大都借助于联轴器与工作机相连接。

Coupling efficiency(hydraulic coupling efficiency) 耦合器效率 耦合器从动轴功率与主动轴功率的比值。$n = n_2/n_1$;式中,n_1—泵轮转速;n_2—涡轮转速。

Coupling loss factor 耦合损耗因子 系指耦合子系统在连接处振动能量的传输损耗的百分比。

Courage, Strength, Sincerity and Competence(CSSC) 勇气、力量、忠诚、能力

Courier services 快递服务

Course 航向 系指船舶航行的方向。它既表示航线或航迹线的方向,如计划航线和航迹,又表示船航行中船首所指的方向,如真航向、磁航向、罗航向等。

Course change lag 转艏滞后 在Z形操纵试验中,从回复到正舵的时刻至最大转艏角一瞬间的时间间隔。

Course change quality number 转舵指数 用以判别操舵效应的,操舵后船舶移动一个船长时,每单位舵角的艏向改变值。

Course changing quality 转舵性 系指船舶应能转舵的性能。以转舵角速度表示。

Course of great circle 大圆航线 把地球当作一个球体,通过地面上任意两点和地心作一平面,该平面与地球表面的交线称大圆。两点间的大圆弧线是两点在地面上的最短距离,沿这一段大圆弧线航线即为大圆航线。一般远洋航海均采用大圆航线航行,故又称经济航线。

Course over ground(COG) 对地航向(COG) 系指从船上测量的,以自真北向的角度单位表示的船舶相对陆地的运动方向。Course over ground (COG) means a direction of the ship's movement relative to the earth, measured on board the ship, expressed in angular units from true north.

Course through water(CTW) 对水航向(CTW) 系指船舶对水运动的方向,通过穿越船舶的子午线与对水船舶运动方向的角度定义,以自真北的角度单位表示。Course through water (CTW) means a direction of the ship's movement through the water, defined by the angle between the meridian through its position and the direction of the ship's movement through the water, expressed in angular units from true north.

Course up display 航向向上显示 系指采用电罗经输入或等效方法,且选择时船舶航向位于图像的最上端的方位角稳定的显示。Course up display means an azimuth stabilized presentation which uses the gyro input or equivalent method and the ship's course is uppermost on the presentation at the time of selection course in relation to a pre-planned route and the waters.

Course-keeping ability 航向保持能力 系指衡量船舶在舵或船舶向没有过度摆动情况下,按预定航向保持直线航线的能力。Course-keeping ability is a measure of the ability of the steered ship to maintain a straight path in a predetermined course direction without excessive oscillations of rudder or heading.

Court of law 法院 系指一个司法活动中心。

Court president 法院院长

Covariance 协方差 系指两个随机变量分别与其期望值之差的乘积的统计平均值。其表达式为 $COV(\xi_1, \xi_2) = M(\xi_1 - M\xi_1)(\xi_2 - M\xi_2)$。

Cover keep lever 支撑器 系指用以支撑并定位开启后的天窗盖或舱盖的部件。

Cover of liquid tank 液货舱罩 对散装液化气运输船而言,系指用于保护突出于甲板以上的货物围护系统免受损坏的结构或用来不在甲板的连续性和完整性的防护。

Cover uppermost 顶篷 救生筏上为保护乘员免受因曝露所引起的伤害而覆盖和围蔽在筏顶和四周的蓬盖,在气胀救生筏中,它是用气柱支撑的软篷,并兼收集雨水的装置。

Coverage area of satellite system 卫星系统的覆盖区域 系指某一地理区域,在此区域内卫星系统根据有关要求在船对岸和岸对船方向上提供可用性,并在此区域内具有连续报警能力。这应与 SOLAS 公约定义的海区相对比来描述,即,A4 海区系指 A1,A2 和 A3 海区以外的海区;A3 海区系指 A1,A2 海区以外,由具有连续报警能力 INMARSAT 同步卫星所覆盖的海区;A2 海区系指由一个具有连续报警能力的中频(MF)海岸电台的无线电话所覆盖的区域;A1 海区系指至少由一个具有连续报警能力的甚高频(VHF)海岸电台的无线电话所覆盖的区域。Coverage area of satellite system is the geographical area within which the satellite system provides an availability in accordance with the relative criteria in the ship-to-shore and shore-to-ship directions, and within which continuous alerting is available. This should be described in relation to any of the sea area as defined in the SOLAS Convention, i. e. Sea Area A4 is an area outside sea areas A1, A2 and A3; Sea A3 is within the coverage of an Inmarsat geostationary satellite in which continuous alerting is available, excluding Sea Area A1 and A2; Sea Area A2 is within the radiotelephone coverage of at least one MF coast station in which continuous DSC, alerting is available; and SEA Area A1 is within the radiotelephone coverage of at least one VHF coast station in which continuous DSC, alerting is available.

Covered barge(deck barge) 甲板驳 系指不设货舱,货物堆在甲板上的货驳。见图 C-17。

图 C-17 甲板驳
Figure C-17 Covered barge(deck barge)

COW 原油洗舱 系 Crude oil washing 的缩写。

Cowl head ventilator 烟斗式通风帽　呈烟斗状的通风帽。

CPA 预期相遇的最近点　是在航向和船速不变时,计算出的与目标船相遇的最近距离。CPA is the closest point of approach, i. e. the shortest target ship-own ship calculated distance in the case of no change in course and speed data.

CPA/TCPA 相遇最近点/相遇最近点时间　系指与相遇最近点的距离以及至相遇最近点时间。由本船操作人员设定界限。CPA/TCPA is a closest point of approach/time to the closest point of approach: distance to the closest point of approach (CPA) and time to the closest point of approach (TCPA). Limits are set by the operator related to own ship.

C_P-C_{Th} type design chart C_P-C_{Th}式设计图谱　系指以进速系数为横坐标,功率载荷系数 C_P 或推力载荷系数 C_{Th} 为纵坐标,并绘有敞水效率和螺距比等值线的螺旋桨设计图谱。

Crabbing 蟹行　气垫船在侧风中前进或在前进并回转时,由于侧漂较大。呈现向艏外舷斜行的状态。

Crack 强力可卡因　作为"时髦"毒品出现于20世纪80年代初期,最初是在美国,现在也传播到其他国家使用。其制造方式是将可卡因氢氧化物与小苏打或氨水和/或安非他明粉末混合在一起,然后浸入水形成糊状,并加热烘干。烘干后将强力可卡因碎成小片。Crack emerged as the "in" drug in the early 1980s, initially in the United States. Its use has now spread to other countries. It is produced by mixing cocaine hydrochloride with baking soda or ammonia and/or amphetamine powder. Water is then added to form a paste which is heated and dried. After drying, the crack is broken into small pieces.

Crack arrestor 止裂器　系指沿管路以一定的间隔放置的环形加强筋,其用途是阻止裂缝的蔓延。Crack arrestor is a circumferential stiffener placed at regular intervals along a pipeline with the purpose of arresting a running crack.

Crack case 曲轴箱　系指供曲轴在其中运转的空间。

Cracked 裂化瓦斯油

Cracks 裂纹　系指各种裂纹,但 $(h \times l < 1 \text{ mm}^2)$ 的微型裂纹除外。Cracks are all type of cracks except micro cracks $(h \times l < 1 \text{ mm}^2)$.

Craft constructed 建造的高速船　系指:(1)安放龙骨或处于类似阶段的高速船;(2)无论何时建造的高速货船改装成高速客船,从开始改装之日应视作建造高速船。Craft constructed means: (1) craft the keels of which are laid or which is at similar stage of construction; (2) a cargo craft, whenever built, converted to a passenger craft shall be treated as a passenger craft constructed on the date on which such a conversion commences.

Craft identification number (CIN) 艇的识别号(CIN)

Craft's bottom (SWATH) 船底(小水线面双体船)　系指下潜体最宽处的以下部分船体,通常指椭圆形或圆柱体的下半体。Craft's bottom means the part under the maximum breadth of the lower hulls, generally indicating the lower part of the oval or cylinder.

Crane[1] 吊车　包括甲板吊(deck crane)、吊柱(davit)和行车(traveling crane)三类。

Crane[2] 起重机　包括吊杆起重机,但不包括机器处所内的起重机。Crane includes a derrick crane but excludes cranes in a machinery space.

Crane ship 起重船　属于工程船的一种,主要承担海上结构物的起吊工作。其功能:(1)导管架的辅助下水、安装就位;(2)平台上钻井、采油、处理、生活模块的安装;(3)沉船、海底结构物的起重打捞;(4)跨海大桥桥面的吊装,海底隧道沉管的吊放等。起重船可分为:(1)按起重系统的载体形式分为船式起重系统、浮式起重船、半潜式起重船和single-lift起重系统;(2)按船型可分为驳船型、机动船型和半潜船型;(3)按起重形式可分为固定扒杆式起重船、全回转式起重船和举力式起重船;(4)按用途可分为用于海上吊装、拆卸(平台、大件、结构物等)的大型起重船(大部分装设回转起重机),用于铺设海底油气管的起重铺管船,其上装有大起重量回转式起重机。用于大型水上工程吊装的起重船,其上多数装设固定臂架式起重机。见图C-18。

图 C-18　起重船
Figure C-18　Crane ship

Cranes 起重机或吊机　系指手动臂起重机、龙门起重机、高架起重机和提升同一货物吊柱等,并且能够完成货物装卸、快速定位和独立或同时完成水平移动货物的操作。Cranes cover jib cranes, gantry cranes, overhead cranes and hoists, cargo davits, etc. and they are capable of performing the works of cargo loading and unloading, quick positioning, slewing and/or horizontal movement simultaneously or separately.

Crank 曲柄 系指由两个曲柄臂和一个曲柄销组成的曲轴部分。

Crank arm 曲柄臂 系指主轴颈与曲柄销之间的连接件。

Crank pin 曲柄销 系指曲轴与连杆大端相连接的轴颈。

Crankcase explosion 曲轴箱爆炸 曲轴箱中的空气和油雾相混合遇到高温或火花导致着火或爆炸的现象。

Crankcase explosion proof door 曲轴箱防爆门 当曲轴箱内气压高于一定值时,能自动开启的阀门。

Crankcase scavenging 曲轴箱扫气 利用活塞往复运动,在曲轴箱内压缩空气进行的扫气。

Crankcase vent pipe 曲柄箱透气管路 自柴油机曲柄箱引出至上层甲板的油气透气管。

Crankcase ventilation installation 曲轴箱透气装置 导出曲轴箱中油气的装置。

Crankshaft 曲轴 通过连杆,使活塞的往复运动变为回转运动,并由其端部输出功率的轴。

Crankshaft deflection 曲轴臂距差 在整机状态下曲轴回转一周曲轴梢两侧间距离的变动值。

Crash-stop 紧急停船 船舶在航行中,使螺旋桨尽可能短时间内由正车转为倒车的一种停船方式。

Crater 弧坑 弧焊时由于断弧或收弧不当,在焊道末端处形成的低洼部分。由于弧坑低于焊道表面,且弧坑中常伴有裂纹和气孔等缺陷,因而该处焊缝性能严重削弱。

Credit 货款/信贷

Credit payable with a bank 银行付款信用证 是付款信用证的一种。汇票付款人为银行的一种信用证。这种信用证由银行直接对出口商提交的汇票承担付款的责任。汇票付款的银行,可以是进口地的开证行,也可以是其他地点的付款行。

Credit sale 赊销 是信用销售的俗称。赊销是以信用为基础的销售,卖方与买方签订购货协议后,卖方让买方取走货物,而买方按照协议在规定日期付款或分期付款形式付清货款的过程。赊销使商品的让渡和商品价值的实现在时间上分离开来,使货币由流通手段转变为支付手段。它实质上是提供信用的一种形式。赊销商品使卖者成为债权人,买者成为债务人,这种债务关系是在商品买卖过程中产生的。由于赊销赖以实现

的信用基础客观上存在不确定性和多变性,增大了赊销的风险,从而使企业往往处于两难的境地。所以要加强对赊销商品和应收账款的管理,重视赊销风险的防范。

Creditors' meeting 债权人会议 是全体债权人参加破产程序进行权利自治的临时机构,其权利范围和行使方式均由法律直接规定,主要是决议职能和监督职能。债权人会议是人民法院审理企业破产案件中一个重要的环节,是"实现债权人破产程序参与权的机构",享有听取报告权、选任常设的监督机构权、决定营业的继续和停止、指示破产财产的管理方法等职权。

Creep speed 最小绳速 系指绞车卷筒在用单排缆绳承受额定卷筒拉力时,能保持的最低平均绳速。

Crescent section 月牙切面 叶面与叶背皆为左右对称曲线,两端较尖,叶面凹入的切面。

Crescent type davit 镰刀型艇架 吊艇臂形如镰刀,小艇收置时坐设其中,放艇时摇动手柄,使伸缩机构的螺杆和套筒伸长,以吊艇出舷的摇倒式艇架。

Crew[1] 船员/乘员 系指:(1)船上所有为船舶航行及为保养船舶、机器、系统和推进与安全航行重要装置而配备的人员或为船上其他人员提供服务的人员;(2)船长和在船上以任何职位从事或参加该船业务工作的所有人员。(3)某一航次在船上为船舶工作或服务尽职的,在船员花名册上列有其名字的任何实际雇佣的人员。 Crew means: (1) all persons carried on board the ship to provide navigation and maintenance of the ship, its machinery, systems and arrangements essential for propulsion and safe navigation or to provide services for other persons on board; (2) the skipper and all persons employed or engaged in any capacity on board a vessel or persons engaged in the business of that vessel; (3) all persons actually employed for duties on board during a voyage in the working or service of a ship and included in the crew list.

Crew[2] 舰员(舰艇) 在舰上具有操作角色的所有人员。包括负责舰船及其机械和武器/飞机的航行和维修的人员。以上所指定的海上操作训练为目的海军学员通常也在舰员的定义范围内。 Crew means all personnel on board the ship for its operational role. This includes personnel for navigation and maintenance of the ship, its machinery and weapons/aircraft systems. Naval trainees on board for the purpose of training in naval operations identified in the previous sentence are also within the scope of the definition of crew.

Crew accommodations 船员起居舱室 系指用于船员的处所,包括船员舱室、医务室、办公室、盥洗室、休息室及类似的处所。 Crew accommodations are those spaces allocated for the use of the crew, and include cabins, sick bays, offices, lavatories, lounges and similar spaces.

Crew and embarked personnel spaces 舰员起居室 系指供舰上全体人员使用的区域，包括以下各项：(1)起居住所（例如：住舱、走廊、办公室、餐厅、娱乐室等）；(2)工作舱室；(3)驾驶室。 Crew and embarked personnel spaces are defined as all areas intended for crew and embarked personnel use, and include the following: (1) accommodation spaces (e.g. cabins, corridors, offices, mess rooms, recreation rooms); (2) work spaces; (3) navigation spaces.

Crew spaces 船员处所 仅供船员使用的所有区域，包括以下各项：(1)起居处所（例如船员舱室、办公室、餐室、休闲室）；(2)工作处所；(3)驾驶处所。 Crew spaces mean all areas intended for crew use only, and include the following: (1) accommodation spaces (e.g. cabins, offices, mess rooms, recreation rooms); (2) working spaces; (3) navigation spaces.

Crew's effects 船员物品 系指属于船员并携带在船上的、可能包括货币在内的衣服、日用品及其他物品。 Crew's effects mean clothing, items in everyday use and other articles which may include currency, belonging to the crew carried on the ship.

Crew's life jacker 工作救生衣 系指供船员工作时穿着，以备落水自救的一种救生衣。

Crew's limit(CL) 乘员定额(CL) 在评定设计类别时所用的最大乘员数（每位乘员的质量按 75 kg 计）。 Crew's limit means the maximum number of crew (with a mass of 75 kg each) used when assessing the design category.

Crew's room 船员室 供船员住宿用的舱室。

Criteria of seakeeping qualities 耐波性指标 常以完成任务的时间的百分比表示，有两种形式：(1)耐波性指标 1(SPI-1) 称为任务有效率（或作业时间百分比），系指船舶在规定的装载及环境条件下，能够完成作业的时间百分比，即：SPI-1 =（波浪中能够完成作业的时间/静水中能够完成作业的时间）×100%；耐波性指标一亦可用船舶在海浪中不能作业的时间（即误工期）计算，即为误工率(d_1)，误工率与 SPI-1 存在下列关系：$d_1 = 1 - $(SPI-1)；(2)耐波性指标 2(SPI-2) 称为航行时间指标——系指船舶在同样污底及装载情况下静水中航行两地（或多地）所需的时间(t_s)对其给定海况或季节航行该地实际所需时间 t_w 的比值，即：SPI-2 = t_s/t_w；航行时间指标等价于期望航速比率，用 V_s 及 V_w 分别表示航行于静水的设计航速及航行于海浪中的平均航速，则 SPI-2 可以表示为：SPI-2 = V_w/V_s。 耐波性指标取决于船舶任务、环境条件、船舶响应及耐波性标准（要素及其数值）4 个因素。这些因素必需全面考虑，不可或缺。

Criterion of service numeral 业务标准数 由限界线下旅客处所与机器处所相对体积、旅客人数和船舶长度所决定的用以查取一定长度客船分舱因数的数值。

Critical areas 关键区域 系指易受腐蚀、屈曲和/或疲劳开裂的处所。这应在图纸认可阶段确定。 Critical areas are locations where corrosion, buckling and/or fatigue cracking easily occur. These will be identified at the plan approval stage.

Critical cavitation number 临界空泡数 在某一流动系统中开始发生空泡时的空泡数。

Critical compartment 重要舱室(舰船) 是一个包含装备或人员的作战部位，没有这些装备或人员，作战生存能力的重要功能将失去。这些功能包括作战能力、操纵或通信。重要舱室的典型代表是海图室、作战室、指挥部位、舰船操纵室和主要通信办公室。依据舰船布置和设计，其他舱室也可视为重要舱室。可以通过避免单点失效和集中保护的方法来简化对重要舱室的保护需求。可以通过易损性分析来鉴别易受攻击的需要被保护的重要舱室和必要的设备或系统。 Critical compartment is a battle station, which contains equipment or personnel without whom functions critical to combat survivability would be lost. These functions include the ability to fight, manoeuvre or communicate. Critical compartments are typically the chart room, operations room, conning position, ship's control room and main communications office. Other compartments may be considered to be critical depending on ship's layout and design. The need for protecting critical compartments can be reduced by avoiding single point failure nodes and by concentrating and protecting those which cannot be avoided. A vulnerability analysis can be used to identify vulnerable critical compartments and the essential pieces of equipment or systems that are required to be protected.

Critical condition of operation 临界区域 船舶在航行中，其摇荡运动与波浪遭遇周期接近于发出谐振的速率范围。

Critical design conditions 临界设计工况 系指为设计目的而选择的限制性特定条件，高速船在排水状态下应维持这些条件。该条件应比最坏预计工况更恶劣，是通过合适的界限为船舶在残存情况下提供足够的安全性。 Critical design conditions mean the limiting specified conditions, chosen for design purposes, with which the craft shall keep in displacement mode. Such conditions shall be more severe than the "worst intended conditions" performance is provided by a suitable margin in the survival condition.

Critical failure　临界故障　系指规定的功能明显降低或丧失,并认为是不可能在船上修复的故障。Critical failure is a failure which involvs loss or significant deterioration of required function, and which is considered not repairable on board.

Critical human factors　关键人为因素　系指人的行为在风险控制中起重要作用,人的行为的失败直接造成事故和使事故持续发展。如果明确了关键的人为因素,在风险控制测算中人的行为或关键任务应该有明确定义。

Critical pipe and cable runs　重要管路和电缆线路　是关系到重要部件生存性的管线。其包括单独的管线或集中的区域。例如桅杆上的管线包括所有水上传感器的波导管和信号电缆。Critical pipe and cable runs are routes in which critical components run. They cover individual routes or concentrated areas. An example is a run containing wave guides and signal cables for all the above water sensors on the mast.

Critical position　关键位置　系指关键区域内高应力或易发生裂纹、屈曲和变形等结构破坏的位置。

Critical pressure of cavitation　空泡临界压力　在某一流动系统中开始发生空泡的外界压力。

Critical reliability　临界可靠性　系指某一设备或某一系统在规定的时间段内,设备或系统不发生临界故障的概率。Critical reliability of a component or a system is the probability that a critical failure will not occur to the component or system during a specified time interval.

Critical Reynolds number of experiment　试验临界雷诺数　进行船模试验时,为测得稳定可靠的量值所必需具有的,边界层中的水由层流变为湍流的最低雷诺数。

Critical speed in shallow water　浅水临界航速　浅水中,接近于重力加速度 g 与水深 h 的乘积之平方根 $\sqrt{gh}$ 的船速。

Critical speed of flow condition (or regime)　流态临界速度　层流开始转变为湍流时的速度。

Critical speed of shafting　轴系临界转速　轴系的回转转速与其对应固有频率相等时的转速。

Critical structural areas　临界结构区域　系指通过计算确定需要进行监控的区域/从该船舶或类似船舶或姊妹船(适用时)的营运历史中确定的容易发生影响船舶结构整体性的破裂、屈曲或腐蚀的区域。Critical structural areas are locations required monitoring which have been identified from calculations of or locations where cracking, buckling or corrosion from the service history of the subject ship or from similar or sister ships (if available) to is easy to occur and which would impair the structural integrity of the ship and which have been identified.

Critical system　关键系统　系指需连续运转的系统支持设备。对钻井装置而言,该系统是阻止事故的进一步发展或转移、人员保护、环境保护和装置保护所必需的。

Critical torque　临界转矩　系指电动机在施以额定频率的额定电压,且转速不出现骤然下跌时产生的最大转矩。

Critical velocity of cavitation　空泡临界速度　在一定外界压力下,开始发生空泡时的自由流速。

Cross beam　架空横梁　系指在甲板开口线内,横跨于泥舱纵舱壁或泥舱舱口围板之间的一种组合梁结构。Cross beam is a built-up beam spanning transversely the longitudinal bulkheads or hatch coaming of the hopper within the lines of deck openings.

Cross bitt　十字缆桩　系指具有十字形桩头的带缆桩。有单十字的和双十字的带缆桩两种。

Cross curves of stability　形状稳性臂曲线　系指一组对应于一定横倾角的形状稳性臂和排水体积的关系曲线。

Cross curves of stability　稳性十字曲线

Cross deck　横向甲板/横跨甲板　系指在舱口围板间和舷内侧的主甲板横向区域。Cross deck means the transverse area of main deck which is located inboard and between hatch coamings.

Cross force　正交力　作用于船体运动物体的流体动力与升力和阻力组成的平面相正交的分量。

Cross head type diesel engine　十字头式柴油机　用十字头滑板机构引导活塞运动的柴油机。

Cross license　交叉许可证

Cross over pipe　联通管　多缸汽轮机中用于连接相邻气缸并带有膨胀接头的蒸汽管道。

Cross scavenging (port-to-port scavenging)　横流扫气　由气缸一端的进气口向气缸盖方向流动,然后折回至与进气口对向的排气口排出的扫气形成。

Cross section　横剖面　系指包括所有纵向构件,如板以及在甲板、舷侧外板、船底板、内底板、纵舱壁(如适用时,还包括底边舱斜板和顶边舱底板)上的纵骨和纵桁。对横骨架式船,横剖面包括邻接的骨架及其在横剖面处的端部连接。(1)散货船横剖面:系指包括所有纵向构件,如板和在甲板、舷侧外板、船底板、内底板、底边舱斜板以及纵舱壁和顶边舱底板上的纵骨和纵桁。(2)双壳散货船横剖面:系指包括所有纵向构件,如板和在甲板、舷侧外板、船底板、内底板及底边舱斜板,以及内侧板、顶边舱斜底板和纵舱壁上的纵骨和纵桁。

Cross spectrum 互谱 系指表达 2 个随机函数相互间关系的频率的复函数。其表达式为：$S_{1,2}(\omega) = (1/2\pi)\int_{-\infty}^{+\infty} R_{1,2}(\tau) e^{-i\omega\tau} d\tau = C_{1,2}(\omega) + iQ_{1,2}(\omega)$。

Cross tie 边舱撑材 连接纵舱壁垂直桁与强肋骨用的横向或斜向构件。

Cross ties 撑材 系指连接到纵向舱壁且用以抵抗静水力载荷和水动力载荷的大型横向结构构件。Cross ties means a large transverse structural members joining longitudinal bulkheads and used to support them against hydrostatic and hydrodynamic loads.

Cross-compound engine 双流双胀蒸汽机 由 2 组高压和低压缸组成的成对二次膨胀做功的蒸汽往复机。其配气机构有提阀式和滑阀式两种形式。

Cross-compound gas turbine 差排式燃气轮机 低压压气机与高压涡轮相连接,高压压气机与低压涡轮轴相连接的燃气轮机。

Cross-correlation function 互相关函数 一个随机变量在时间 t 时的瞬时值与另一个随机变量在时间 $(t+\tau)$ 时的瞬时值乘积的时间平均数。其表达式为 $R_{1,2}(\tau) = \lim(1/T)\int_0^T \xi_1(i)\xi_2(t+\tau)dt$,式中：$T \to \infty$。

Cross-deck（SWATH） 连接桥结构(小水线面双体船) 系指连接左右两片体的甲板结构。

Crosshead 十字头 系指用以连接活塞与连杆,形如"十"字的部件。

Crosshead guide（guide plate） 导板 供滑块沿其导面滑动的部件。

Crotch 桨叉 划桨时支撑桨的架子。

Crow's nest 望鱼台 设置在围网渔船或捕鲸船桅杆上部,用以瞭望和侦察鱼群的装有围栏的平台。

Crowding mechanism 推压机构 安装在铲斗吊杆上,由动力传动,用以推压铲斗柄的齿轮装置。

Crown mounted compensator 天车补偿器

Crows nest 瞭望台 设置在桅上,供瞭望用的平台或遮蔽良好的小室。

Crude oil 原油 系指：(1)未经提炼加工的碳氢混合物。它是出自地下的液态石油,范围从很轻的(含汽油量高的)到很重的(含渣油量高的)。含硫原油含硫量高,但脱硫原油含硫量低而且是更有价值。(2)地下自然形成的液态碳氢化合物,不论是否是提炼到适合运输的状态并包括：①从部分蒸馏物已提取的原油；②从部分蒸馏物已添加的原油。 Crude oil are：(1) hydrocarbon mixtures that have not been processed in a refinery. It is a liquid petroleum coming out of the ground, ranging from very light (gasoline) to very heavy (high in residual oils). Sour crude has a high sulfur content, but sweet crude has a low sulfur and is more valuable. (2) any oil occured naturally in the earth whether or not refined to be suitable for transportation and includes：①crude oil from which certain distillate fractions have been removed；②crude oil to which certain distillate fractions may have been added.

Crude oil burner 试油燃烧器 钻井船或其他钻井装置(包括固定钻井平台在内)的平台上的油井,在试油时为防止海底地下原油及其他喷出物对海洋造成污染而将所有喷出物完全燃尽的一种设备。

Crude oil storage areas 原油储存区 系指原油贮存设备所在区域。Crude oil storage areas are areas where crude oil storage means are located.

Crude oil tank area 原油舱区 系指包括边压载舱、原油舱、原油舱区两端或原油舱间的隔离空舱,对平台的露天甲板上设置储油罐的储油平台,原油舱区系指储油罐。Crude oil tank area means an areas inculds side ballast tanks, crude oil tanks, cofferdams at both ends of crude oil tank area or between crude oil tanks. For storage units with oil storage tanks fitted on weather deck, crude oil tank area means oil storage tanks.

Crude oil tanker 原油油船 系指从事原油运输业务的,但禁止装运成品油的油船。见图 C-19。Crude oil tanker means an oil tanker engaged in carrying crude oil, but it is prohibited from carrying produce oil. Crude oil tank is shown in Fig. C-19.

图 C-19 原油油船
Figure C-19 Crude oil tanker

Crude oil washing equipment 原油洗舱设备 原油洗舱装置中的关键,亦即原油洗舱喷嘴组。

Crude oil washing installation 原油洗舱装置 系原油洗舱系统的重要组成之一,连同附属设备及布置即构成原油洗舱系统。

Crude oil washing operation and equipment manual（COW Manual） 原油洗舱操作与设备手册(COW 手册) 系指每艘使用原油洗舱系统的油船,均应备有一份详细说明该系统及设备并规定操作程序的"操作与设备手册"。该手册应使主管机关满意,并应包括73/78防污染公约附则Ⅰ第13B条(2)所述的技术条件中规定的全部资料。 Crude oil washing operation and equipment manual (COW Manual) mean that every oil tanker operating with crude oil washing systems shall be provided with an Operations and Equipment Manual detailing the system and equipment and specifying operational procedures. Such a Manual shall be to the satisfaction of the Administration and shall contain all the information set out in the specifications referred to in Paragraph 2 of Regulation 13B of Annex I of MARPOL 73/78.

Crude oil washing(COW) 原油洗舱 系指利用所载运的原油对油舱进行清洗的洗舱方法。

Crude oil washing system (COW) 原油洗舱系统 系指:(1)20 000DWT以上的油船要求安装的、对货油舱进行清洗的装置;(2)利用所载运的原油对油舱进行清洗的洗舱系统。

Crude oil/product carrier 原油/成品油油船 系指从事原油或成品油运输业务的油船。 Crude oil tanker means an oil tanker engaged in carrying crude oil or product oil.

Cruise missile(CM) 巡航导弹 是一种用动力推进,以机翼来产生升力的导弹。其大多数动力源是喷射发动机。简单来说巡航导弹就是飞行炸弹。它们可以携带常规弹头或核弹头,射程可达数百英里。近代的巡航导弹可以以超音速或亚音速飞行。巡航导弹,是导弹的一种。即主要以巡航状态在稠密大气层内飞行的导弹,旧称飞航式导弹。巡航状态系指导弹在火箭助推器加速后,主发动机的推力与阻力平衡,弹翼的升力与重力平衡,以近于恒速、等高度飞行的状态。在这种状态下,单位航程的耗油量最少。其飞行弹道通常由起飞爬升段、巡航(水平飞行)段和俯冲段组成。它依靠喷气发动机的推力和弹翼的气动升力。巡航导弹与无人驾驶飞机的不同之处,在于巡航导弹不担任侦察任务,弹头整合为系统的一部分,而且最后会在攻击中损失。这是多数反导系统难以拦截的导弹。巡航导弹可以用来从低空突破敌人防空网络,它们具有非常高的精确度和非常优异的机动性能,并能用来从任一方向攻击目标。通过陆基雷达进行探测非常困难,因为巡航导弹采用贴地飞行方法和隐身技术,并且防御方的探测还受到地球曲率的影响。陆地和海水的混乱也是降低探测和鉴别巡航导弹可能性的一个重要因素。技术拥有国:中国、美国、俄罗斯。见图C-20。

Cruise ship 旅游船 系指在国际航行中运载参加团体活动并在船上食宿的旅客的船舶,为了在一个或几个不同港口作预定的短期游览,在航中一般:(1)不上与

图 C-20 巡航导弹
Figure C-20 Cruise missile

其他旅客;(2)不装卸任何货物。 Cruise ship means a ship on an international voyage carrying passengers participating in a group programme and accommodated aboard, for the purpose of making scheduled temporary tourist at one or more different ports, and which during the voyage does not normally: (1) embark or disembark any other passengers; (2) load or discharge any cargoes.

Cruise vessel 大型邮轮 系指美国法规第46篇第2101(22)条中所定义的客船。此术语不包括由联邦政府所使用的美国船舶或由州政府拥有和使用的船舶。见图C-21。 Cruise Vessel means a passenger vessel as defined in section 2101(22) of Title 46. Vessels owned and operated by the are excluded. the Federal Government government of a State. It is shown in Fig. C-21.

图 C-21 大型邮轮
Figure C-21 Cruise vessel

Cruiser 巡洋舰 系指在排水量上仅次于航空母舰、在海战中起骨干作用的较大型的水面战斗舰艇,可以长时间巡航在海上,并以机动性为主要特性。

Cruiser stern 巡洋舰型艉部 具有近似锥形平顺曲面的艉伸部,其水平剖面则略似半卵形的船尾部。

Cruising engine unit 巡航机组 在多机组的推进装置中,巡航时使用的机组。

Cruising gas turbine 巡航燃气轮机 在舰船巡航航速下提供舰船推进动力的燃气轮机。

Cruising power 巡航功率 主机在舰艇巡航速度下的长期运转功率。

Cruising speed 巡航航速 舰船在执行巡航任务时所规定的航速。

Cruising steam turbine 巡航汽轮机 系指在巡航工况下投入运行而在全速工况下不工作的汽轮机。

Cruising turbine stages 巡航级组 设置在经济级组后面于巡航工况下投入运行而在全速工况下不工作的低速级组。

Crush condition 阻塞条件 系指脱险通道或处所最大允许乘客密度,其值定为 3.5 人/m²。 Crush condition is the maximum allowable density of persons in escape routes or spaces, fixed at 3.5persons/m².

Cryolite 冰晶石 钠和铝的一种氟化物,用于生产铝和用作陶瓷釉料。灰色丸粒。长时期接触可能严重损害皮肤和神经系统。该货物为不燃物或失火风险低。 Cryolite is a fluoride of sodium and aluminium used in the production of aluminium and for ceramic glazes. Grey pellets. Prolonged contact may cause serious damage It is of skin and nervous system. This cargo is non-combustible or has a low fire-risk.

Cryptogenic species 来源不明的物种 系指不知起源的物种,即不能证明是当地的或引入区域的物种。 Cryptogenic species mean that they are of unknown origin, i. e. species that are not demonstrated to be native or to be introduced to a region.

Crystal diode 晶体二极管 系指固态电子器件中的半导体两端器件。这些器件主要的特征是具有非线性的电流-电压特性。此后随着半导体材料和工艺技术的发展,利用不同的半导体材料、掺杂分布、几何结构,研制出结构种类繁多、功能用途各异的多种晶体二极管。制造材料有锗、硅及化合物半导体。晶体二极管可用来产生、控制、接收、变换、放大信号和进行能量转换等。

CSSC Huangpu Wenchong Shipbuilding Company Limited 中船黄埔文冲船舶有限公司 ——海洋维权利剑的"锻造者" 简称黄埔文冲,是中国船舶工业集团有限公司旗下的大型骨干造船企业和国家核心军工生产企业,是国内军用舰艇、特种工程船舶和海洋工程船舶的主要建造基地,是中国疏浚工程船和支线集装箱船的最大生产基地。黄埔文冲先后为海军建造了 20 多型 200 多艘主战舰艇,装备质量经受了从浅蓝到深蓝的重大考验,为政府部门建造了多种型号的海关缉私船、海巡船、海监船、救助船、渔政船,在维护国家海洋权益的过程中发挥了重要作用;批量建造的 14 000 kW、8 000 kW 系列海洋救助船作为我国海上救助的主力船型,在承担国家重大海上值班执勤、"神舟"系列飞船发射保障、马航 MH370 搜救等急难险重的海上救助任务中发挥了重要作用,充分体现了我国政府"以人为本"的执政概念。海洋工程装备是黄埔文冲的战略发展方向之一,已成功建造钻井平台、生活模块、多用途工作船、半潜船、铺管船、深水工程勘察船、多功能水下工程主持船、大型勘察船等海洋工程及保障船舶,使黄埔文冲在海洋工程领域的实力得到迅速提升。

CT-Spar 立柱式平台

C-type cylindrical independent tank C 型圆柱形的独立液舱 此型液舱符合压力容器的设计标准。一般为圆筒形或球形。设计蒸发气压力大于 0.2 MPa。一般用于全冷式或半冷半压式液化气船。为了提高船舶装货容积和利用率,有的船舶采用双联圆筒形液货舱。

Cubic capacity of lifeboat 救生艇立方容积 据以核算救生艇乘员定额的救生艇上容纳乘员的空间立方米数。

Cumulative spectrum 累积谱 表示从零频率到某一指定频率能量分布累积成的曲线。

Cup board (teacup rack) 茶杯架 系指固定在舱壁上,供安放茶杯用的架子。

Curly braces 大括号

Current 流 系指:(1)通过固定位置的水流——更准确地表述为欧拉流。拉格朗日流是跟踪水分子运动测得的流。流通常量其速度和方向,测量通常通过潮流和残流进行分析的。(2)对设计环境而言,设计流速应取风生流的流速与潮流流速之和。在计算时,潮流流速可以假定在水深范围内为常数,而风生流的流速可取为在海平面处的流速值为最大,海平面以下 50 m 处流速值为零,其间呈线性递增。 Current is: (1) a flow of water past a fixed location-more precisely described as an Eulerian current. A Lagrangian current is measured by following the movement of a water particle. Current are usually measured by speed and direction, measurements are usually analyzed in terms of the tidal current and residual currents. (2) For the design environment, the design current speed is to be taken as the sum of wind-induced and tidal current velocities. For calculation purposes, tidal current velocity can be assumed constant over water depth, and wind-induced current velocity can be taken to reduce linearly from its maximum value at the surface to zero at 50 m below sea level.

Current amplifier 电流放大器

Current meter 海流计 系指测量海流的流速和流向的仪器总称海流计。(1)应用活动的惯性元部件的海流计有:①转子式海流计,如印刷海流计;②旋转式海流计,如厄克曼海流计等。(2)无活动的惯性元、部件的海流计:如电磁海流计、声学海流计等。

Current observation 海流观测 利用测流仪器对海水流动的速度和方向进行的观测。

Current repair 小修 营运期中的船舶按规定周期结合定期检验而进行的短期计划性修理。其目的在于消除船体与机械设备产生的过度磨损,以保证船舶能安全营运至下一个计划修理年度。小修时结合坞修对船体结构、推进装置、舵装置、主辅机、锅炉和其他设备以及工程船的专用设备等机械重点检查和修理。小修船舶的修理间隔期,由航运部门根据航区、营运条件及设备状况而定。

Current setting (of a over-current overload relay or release) (过电流或过载继电器或脱扣器的)电流整定值 系指与继电器或脱扣器的动作特性有关,且用来确定继电器和脱扣器动作的主电路电流值。注意:继电器或脱扣器可有一个以上的电流整定值,整定值可采用可调的刻度盘、可更换的加热器等方式确定。

Current setting range (of a over-current overload relay or release) (过电流或过载继电器或脱扣器的)电流整定值范围 系指可调整的继电器或脱扣器电流整定值的最大值与最小值之间的范围。

Current speed (flow velocity) 流速 系指任何方向流的水平速度。速度变化遍及整个水柱。平均深度处的流速是整个水柱的平均流速或在指定深度处的流速。 Current speed is the horizontal speed of the current at any direction. The speed varies throughout the water column. Depth-averaged current speed is the speed of the current averaged throughout the water column or current of a specified depth is.

Curtain box 窗帘匣 供遮蔽拉帘轨用的长形匣子,明式的安装于天花板下,暗式的安装于天花板内。

Curtain plate 檐板 甲板室至甲板自由边缘的围边板条。

Curve of areas of water-planes 水线面面积曲线 系指船舶水线面面积与吃水的关系曲线。

Curve of block coefficients 方形系数曲线 方形系数与吃水的关系曲线。

Curve of declining angles 减摇曲线 表示船舶在静水中受阻尼横摇时,其摆幅随摇摆序列数或时间经历而衰减的曲线。

Curve of dynamical stability 动稳性曲线 船舶于某一装载情况下,复原力矩所做的功与横倾角的关系曲线,即静稳性曲线的积分曲线。

Curve of extinction 衰减曲线 以船舶在静水中有阻尼横摇时的相邻横摇幅值的平均值作横坐标,并以相邻摆幅的递减量为纵坐标所绘成的曲线。

Curve of floatation 浮心曲线 系指浮心轨迹在横剖面或中线面上的投影曲线。

Curve of limiting positions of center of gravity 极限重心垂向坐标曲线 在船舶符合稳性规范要求的条件下,计算并画出的以各排水量为横坐标,船舶重心距基线最大距离允许值为纵坐标的关系曲线。

Curve of longitudinal centers of buoyancy 浮心纵向坐标曲线 浮心纵向坐标与吃水的关系曲线。

Curve of longitudinal centers of floatation 漂心纵向坐标曲线 漂心纵向坐标与吃水的关系曲线。

Curve of mid-ship section coefficients 中剖面系数曲线 中剖面系数与吃水的关系曲线。

Curve of molded volumes 型排水体积曲线 船舶型排水体积与吃水的关系曲线。

Curve of moment to change trim one centimeter 每厘米纵倾力矩曲线 船舶纵倾 1 cm 所需的力矩与吃水的关系曲线。

Curve of prismatic coefficients 棱形系数曲线 棱形系数与吃水的关系曲线。

Curve of tons per centimeter of immersion 每厘米吃水吨数曲线 船舶吃水平行改变 1 cm 所引起的排水量的变化吨数与吃水的关系曲线。

Curve of vertical centers of buoyancy 浮心垂向坐标曲线 浮心垂向坐标与吃水的关系曲线。

Curve of vertical positions of transverse metacenter 横稳心垂向坐标曲线 横稳心垂向坐标与吃水的关系曲线。

Curve of volumes of total displacements 总排水体积曲线 船舶总排水体积与吃水的关系曲线。

Curve of water-plane coefficients 水线面系数曲线 水线面系数与吃水的关系曲线。

Curved tread 弓形梯步 为适应舷梯斜度变化和便于行走而将其横截面作成弧形的固定舷梯上的梯步。

Curves of hopper capacity 泥舱容量曲线 表示泥舱的不同舱深与其相应容积的关系曲线。

Curves of sectional area and their static moment 横剖面面积及其净力矩曲线 在各站站上,以吃水为纵坐标,相应的横剖面面积,以及该面积对基平面的净力矩为横坐标所绘制的三组积分曲线。

Cushion area 气垫面积 对全垫升气垫船而言,系指在气垫稳定状态下,围裙下缘周边所包围的面积;对侧壁气垫船而言,系指在两侧壁内水线和艏艉封口下缘所包围的面积。

Cushion beam 气垫宽 对全垫升气垫船而言,系指在水平垫升航态下,两舷围裙下缘之间的平均距离;对侧壁气垫船,指两侧壁内水线之间的平均宽度。

Cushion length 气垫长 气垫面积除以气垫宽的商。

Cushion system 气垫系统 由进气口、风机、气道及气囊、围裙等所组成,用以产生并维持气垫的一系列装置的总称。

Cushion thrust 气垫推力 气垫船在艉纵倾垫升航行时,由于艉部喷口及围裙逸出的空气流量大于艏部空气逸出流量而产生的推力。

Cushion-horne 垫升航态 系指气垫船的重力基本上由气垫所支承时的航行状态。

Cushion-lifting and/or planning mode 垫升和/或滑行模式(地效翼船) 系指船总重由气垫升力(作用于机翼上)和水动升力(作用于主船体和侧浮舟上)支撑的航行模式。动力气垫船和动力气垫地效翼船的大部分升力由气垫静升力提供,比动力增升地效翼船的比例高得多。由于气垫系统设计不同,地效翼船可能在该运行模式(如动力增升地效翼船),或在第一过渡模式(如动力气垫船和动力气垫地效翼船)经历气垫阻力峰速。当船继续加速时,机翼升力逐渐增大,直到船体由机翼升力和机翼下的动力气垫支撑。对于动力气垫船和动力气垫地效翼船,垫态时能够达到中性纵倾非常重要,因此船滑行时的气垫面积中心应该在船重心之下。一旦动力气垫船或动力气垫地效翼船加速到峰速以上,阻力随着推进器对气垫供气的减少而减小,从而能够持续加速,直至由机翼气动力支撑船体的第二过渡阶段。动力增升地效翼船必须通过比气垫阻力峰更尖的滑行阻力峰。动力增升地效翼船的第二过渡一般与船体的起飞密切相关。艏喷管或艏推进器导管位置以及主翼下气流提供升力的比例对于船从主船体起飞到稳定巡航过渡期间的稳性非常重要。 Cushion-lifting and/or planning mode means a mode in which the total craft weight is supported by air cushion lift (on the wings) and hydrodynamic lift (on main hull and side buoys). Most of the supporting lift for DACC and DACWIG is supplied by static air cushion lift, occuping a much higher proportion than in the case of PARWIG. Depending on the design of the air cushion system, the air cushion drag hump speed may be experienced in this mode (PARWIG) or in the first transitional mode (DACWIG, DACC). As the craft accelerates, lift on the wings gradually increases, offloading the hull planes until the craft is supported on the wing lift and dynamic air cushion under the wings. It may be noted that for DACC and DACWIG, it is important that neutral trim is achievable while hovering, so the cushion centre of area needs to be under the centre of gravity of the craft in plane. Once a DACC or DACWIG has accelerated above the hump speed, the drag reduces and acceleration can continue while cushion feed from thrusters is reduced until the lifting wing supports the craft on the dynamic air feed at the second transition. A PARWIG has to transit the drag hump of its planing hull, which is a sharper peak than that of an air cushion. The second transition is generally closely linked to hull lift-off for a PARWIG. The position of the bow jet nozzles or bow-thruster ducts and so the proportion of lift provided by forced air feed under the main wings is significant in determining the stability of the craft through the transition from main hull lift-off to steady cruise.

Custom 海关 系指设在边境上的国家行政管理机构,是执行本国进出口政策、法令和规章的重要工具。其任务是根据国家进出口政策、法令和规章对进出口货物、货币、金银、行李、邮件、运输工具等实行监督管理、征收关税、查禁走私、临时保管统管货物和统计进出口商品等。海关还有权对不符合国家规定的进出口货物不予放行、罚款直至没收和销毁。

Custom duties 关税 系指进出口商品经过一国关境,由政府所设置的海关向进出口商征收的税收。关税征收是通过海关执行的。关税是国家财政收入的一个重要组成部分。它与其他税收一样,具有强制性、无偿性和预定性。关税的主要特点:(1)关税是一种间接税。其主要征收对象是进出口货物,其税赋是由进出口商先行垫付,而后把它作为成本的一部分计入商品的价格、转嫁给最终消费者承担,因而关税属于间接税。(2)关税的税收主体和客体是进出口商和进出口货物。在税法中,征税涉及税收主体与客体。税收主体,也称课税主体,系指在法律上负担纳税的自然人和法人,也称纳税人。税收客体,也称课税客体或课税对象。关税的税收主体是本国的进出口商。当商品进出国境或边境时,进出口商根据海关规定向当地海关缴纳关税,他们是税收主体,是纳税人。关税的客体是进出口商品。根据海关税法和有关部门规定,海关对各种进出口商品依据不同的税目和税率征收关税。(3)关税可以起到调节一国进出口贸易的作用。许多国家通过制定和调整关税税率来调节进出口贸易。在出口方面,通过低税、免税来鼓励商品出口;在进口方面,通过税率的高低、减税来调节商品进口。对于国内不能生产或生产不足的商品,制定较低税率或免税以鼓励进口;对于国内能大量生产或非必需品的进口,则制定较高税率,以限制进口或达到禁止进口的目的。(4)关税是一国对外贸易政策的重要手段。关税体现着一国的对外贸易政策。关税税率的高低直接影响着一国进出口贸易,影响一国同其他国家经济贸易关系的发展,从而影响着一国经济贸易的发展。

Customary tare 习惯皮重 有些比较标准化、规格化的包装,其质量也为市场所公认,因此不必逐件重复过秤,而以习惯上公认的包装质量计算,称为习惯皮重。

Customs boat 海关艇 系指专门用于接送海关检查人员的船艇。

Customs cooperation council nomenclature (CCCN) 海关合作理事会税则目录 是海关合作理事会编制的商品分类目录。为了减少各国在海关税则商品分类上的矛盾，欧洲关税同盟研究小组于1952年12月拟定了"关税税则商品分类公约"（Convention on Nomenclature for the Classification of Goods in Customs Tariff），并设立了海关合作理事会。制定了"海关合作理事会税则目录"，因该税则目录是在布鲁塞尔制定的，故又称"布鲁塞尔税则目录（Brussels Tariff nomenclature, ABTN）"除美国、加拿大外，已有100多个国家和地区采用。海关合作理事会税则目录的商品分类的划分原则，是以商品的自然属性为主，结合加工程度等来划分的。它把全部商品共分为21类（Section）、99章（Chapter）、1015项税目号（Heading No.）。1~24章（前4类）为农畜产品，25~29章为工业制成品。

Customs declaration 报关 系指出口货物的发货人或其代理人、出境运输工具的负责人向海关申报、交验规定的单据和证件，请求办理货物或运输工具出境手续等。

Customs duty 关税 是进出口货物经过一国关境时，由政府所设的海关向进出口商征收的税收。

Customs office of destination 目的地海关 系指结束过境作业地的任何海关。 Customs office of destination means any customs office at which a customs transit operation is terminated.

Customs quota 关税配额 系指征收关税与进口配额相结合的一种限制进口的措施。它实行时要预先规定有关商品在一定时期内的关税配额，在配额以内进口的商品，给予低税或免税待遇（一般为优惠税率），对超过配额的进口商品征收较高的关税（一般为普通税率），或者征收进口附加税或罚款。

Customs tariff 海关税则 又称关税税则，是一国对进出口商品计征关税的规章和对进出口的应税与免税商品加以系统分类的一览表。海关凭此征收关税，是关税政策的具体体现。海关税则一般包括两个部分：一部分是海关课征关税的规章条例和说明；另一部分是关税税率表、税则号列（Tariff No. 或 Heading No. 或 Tariff item），简称号列、货物分类目录（Description of goods）、税率（Rate of duty）。

Customs transit document 海关过境单证 系指载有海关过境作业所需数据和资料记录的表格。 Customs transit document means a form containing the record of data entries and information required for the customs transit operation.

Customs transit procedure 海关过境手续 系指在海关管制下将货物从一处海关运到另一处海关的海关手续。 Customs transit procedure means the customs procedure under which goods are transported under customs control from one customs office to another.

Customs union 关税同盟 系指两个或两个以上国家缔结协定，建立统一的关境，在统一关境内缔约国相互间减让或取消关税，对从关境以外的国家或地区的商品进口则实行共同的关税税率和外贸政策。关税同盟从欧洲开始，是经济一体化的组织形式之一。对内实行减免关税和贸易限制，商品自由流动；对外实行统一的关税和对外贸易政策。关税同盟有两种经济效应，静态效应和动态效应。

Cutlery 餐具 系指供进餐用的器具。

Cut-out cylinder test 停缸试验 检验柴油机在持续功率运转工况下按规定停止部分气缸供油时能否继续运转一定时间的试验。

Cut-out turbo-supercharger test 停增压器试验 检验非机械连接增压柴油机的增压器部分或全部停止工作后柴油机能否继续运转一定时间的试验。

Cutter 铰刀 绞吸挖泥船上，装在铰刀轴下端的绞切泥土的工具。

Cutter drive 铰刀驱动装置 带动铰刀的动力装置。

Cutter ladder 铰刀轴 由几段圆钢或圆管以法兰盘或套筒联轴器连接而成。安装在铰刀架上。上端连接传动齿轮。下端装有铰刀的轴。

Cutter ladder gantry 铰刀吊架 系指用以吊放绞刀架的固定钢架。

Cutter suction dredger (CSD) 绞吸式挖泥船 系指具有铰刀等挖泥设备的挖泥船。见图 C-22。 Cutter suction is a dredger fitted with cutter head and other dredging equipment. it can be seen in Fig. C-22.

Cutter wheel dredger 斗轮式挖泥船 系指具有斗轮挖泥设备的挖泥船。 Cutter wheel dredger is a dredger fitted with cutter wheel dredging equipment.

Cutting machine 切割机 是一种切削物料，使其形状达到目的状态的工具。切割机主要使用外部能源来破坏原有物料的形状。随着现代机械加工业的发展，对切割的质量、精度要求的不断提高，对提高生产效率、降低生产成本，具有高智能化的自动切割功能的要求也在提升。数控切割机的发展必须要适应现代机械加工业发展的要求。切割机分为火焰切割机、等离子切割机、激光切割机、水下切割机等。激光切割机为效率最高，切割精度最高，切割厚度一般较小。等离子切割机切割速度也很快，切割面有一定的斜度。火焰切割机针

图 C-22 绞吸式挖泥船
Figure C-22 Cutter suction dredger(CSD)

对厚度较大的碳钢材质。

Cycle 周 系指:(1)在一个时间间隔出现的一个周期量的、完整系列的量值。(2)一个交变电流的一组完整的正值和负值。

Cyclic frequency variation 频率周期性变化 在正常运行期间频率的周期性偏差,例如可能由有规律的重复载荷引起的。频率周期性变化 = $\frac{\pm(f_{max}-f_{min})\times 100}{2f_{nominal}}\%$。Cyclic frequency variation is a periodic deviation in frequency during normal operation, such as, it might be caused by regularly repeated loading.

Cyclic loading 循环载荷 波浪作用产生的载荷,包括惯性载荷。Cyclic loading is loading due to wave action, including inertia loads.

Cyclic voltage variation 电压周期性变化 诸如可能由规律性的重复负载引起的额定电压的周期性偏差(均方根值最大~最小)。电压周期性变化 = $\frac{\pm(u_{max}-u_{min})\times 100}{2u_{nominal}}\%$ Cyclic voltage variation is a periodic voltage deviation (max. to min. r.m.s values) of the nominal voltage, such as, it might be caused by regularly repeated loading.

Cycloidal propeller (vertical axis propeller) 平旋推进器 在能旋转的圆盘下装有若干能转动的直叶伸入水中,可将推力指向任何方向的一类推进器。

Cyclone separation method 气旋分离法

Cylinder block 气缸体 支撑汽缸盖和固定汽缸套等零件的部件。

Cylinder bore (bore size) 缸径 系指气缸的内径。

Cylinder clearance 气缸间隙 汽缸套内径和活塞裙部外径之差。

Cylinder control unit(CCU) 气缸控制单元

Cylinder cowling 圆柱形通风帽 呈圆柱状的通风帽。

Cylinder head 气缸盖 系指固定在气缸顶部,与气缸、活塞组成工作室的内燃机部件。

Cylinder head cover 气缸盖罩 系指气缸盖上部使各种阀机构等密封的罩子。

Cylinder housing(LPG system) 气罐箱(LPG 系统) 预定只用于储存一个或多个 LPG 罐、压力调节器和安全装置,位于小艇外部,且所有泄漏都流向舷外的通风包壳。 Cylinder housing means the ventilated enclosure intended solely for storage of one or more LPG cylinders, pressure regulators and safety devices, and located on the exterior of the craft, where leakage would flow overboard.

Cylinder liner 汽缸套 镶在气缸体内圆柱面上,与缸盖、活塞共同组成工作室供活塞在其中作往复运动的圆形衬套。

Cylinder locker(LPG system) 气罐柜(LPG 系统) 预定用于储存一个或多个 LPG 罐的在艉舱中或凹入小艇内的具有舷外泄水孔的蒸气密包壳。 Cylinder locker means the vapour tight enclosure with an overboard drain intended solely for storage of one or more LPG cylinders in a cockpit or recessed into the craft.

Cylinder lubricating system (CLS) 气缸润滑系统

Cylinder oil measuring tank 气缸油计量柜 重型低速柴油机用以计量气缸油日常消耗量的舱柜。

Cylinder oil storage tank 气缸油贮存柜 贮存重型低速油机使用的气缸油的舱柜。

Cylinder power equalizing test 各缸工作均匀性试验 内燃机在持续运转工况下,测定各缸参数,如压缩压力、最大爆发压力、平均指示压力、排气温度等是否均匀的试验。

Cylinder transfer pump 汽缸油泵 将汽缸油从大油柜抽送至计量柜的泵。

Cylinder volume 气缸总容积 气缸工作容积与压缩容积之和。

Cylindrical chain locker 筒形锚链舱 系指能自动盘放锚链呈圆筒形状的锚链舱。

Cylindrical drilling platform 圆筒形钻井平台

Cylindrical FDPSO 圆筒形浮式钻井生产储油船 系指在圆筒形 FPSO 和圆筒形钻井平台的基础上组合而成的,具有钻井功能的平台。

Cylindrical moon pool 圆柱形月池

Cylindrical surface sound wave 柱面声波 系指波阵面为同轴圆柱面的声波。

D

D 柴油机推进

D class combustible liquid D 级可燃液体 对美国国家消防委员会（NFPA）船上气体危害控制标准而言，系指闪点低于 65.5 ℃（150 ℉）高于 26.6 ℃（80 ℉）的可燃液体。

D disk D 盘 系指电脑硬盘主分区之一，属于硬盘驱动器。

D or D-value D 或 D 值 系指直升机当旋翼旋转时从主旋翼的翼尖轨迹平面最前端至尾旋翼的翼尖轨迹平面或直升机结构的最后端所量得的最大尺寸。 D or D-value means the largest dimension measured from the most forward position of the main rotor tip path plane to the most rearward position of the tail rotor path plane or helicopter structure, when rotor(s) are turning.

D sound level D 声级 系指 D 计权网络在计权后的声级。其频率响应曲线为 40 noy 等响曲线经整理后倒置。用于对航空噪声测量与评价。

D. C.（direct-current） 直流电 系指一种单一方向的、实际上无脉动的电流，其数值变化为零，或小至可以忽略不计。

D. C. commutated motor 直流换向电机 系指含有一个直流电源激励的或由永久磁铁形成的磁场，一个电枢以及一个与其连接的换向器的电机。直流换向电机的具体形式有直流发电机、电动机、同步交流机、升压机、均压机和共定子电动发电机。

D. C. component of the short-circuit current 短路电流直流分量 系指突然短路发生后，短路电路中电流的一个组成部分，所有基波和谐波均不计算在内。 D. C. component of the short-circuit current means the component of current in the circuit immediately after it has been suddenly short-circuited, all components of fundamental and harmonics being excluded.

D. C. pressure equalizing device 直流均压机 系指由两台或两台以上相似直流电机（通常具有并励或复励励磁）组成的电机装置。这些直流电机彼此直接耦合联结、串联连接在一个多线配电系统的外导线上，用以使该系统中间导线上的电位保持不变，此中间导线接在这两台或多台直流电机之间的连接点上。

D. C. electrical telegraph gong 直流电动传令钟 系指利用直流同步传信原理的传令钟。它由发送器、接收同步机和连接导线组成。

D + G 柴油机发电 + 燃气轮机动力推进装置

DACC or GEM craft 动力气垫船或地面效应器船 系指运行于非常靠近地面、强地效区的船舶。 DACC or GEM craft means a craft operated very close to the ground in the strong surface effect region.

DACWIG craft 动力气垫型地效翼船 系指运行于地效区内具有较大飞高，并且在低速时用装在船艏的推进器以及机翼端板而在机翼下方生成气垫而非增强升力的船舶。见图 D-1。 DACWIG craft means a craft operated at a larger flying height in the surface effect region and that use air from bow-mounted propulsors and wing endplates to create an air cushion under the wings at low speed rather than just enhanced lift. It is seen in Fig. D-1.

图 D-1　动力气垫型地效翼船
Figure D-1　DACWIG craft

Daily (service) tank 日用油柜 存放供零星使用的油的舱柜。

Daily sewage treatment 生活污水处理 其核心技术为活性污泥法和生物膜法，对活性污泥法（或生物膜法）的改进及发展形成了各种不同的生活污水处理工艺，传统的活性污泥法处理工艺在中小型生活污水处理已较少使用。

Daily sewage treatment unit 生活污水处理装置 系指利用污水处理技术实施污水处理的装置。

Damage 损伤（结构） 系指一个结构如果原来形状发生某种有损于今后功能的变化，即使没有立即丧失功能，也认为这个结构受了损伤。损伤包括结构构件失去稳定性而产生的永久性变形或出现裂纹。在载重的情况下，结构虽能承受它的设计载荷，但是给结构留下了隐患，对发挥功能带来了不利影响。

Damage 损害 系指：（1）由有害有毒物质造成的、在运输这些物质的船舶上或船舶外的人身伤亡;（2）由有害有毒物质造成的、在运输这些物质的船舶上或船舶外的财产的灭失或损坏;（3）由有害有毒物质造成的环境污染所致的灭失或损害，但对不包括环境损害所致的利润损失在内的环境损害,应限于实际采取或将要采取的合理恢复措施的费用;(4)预防措施的费用和补救措施造成的新的灭失或损害。 Damage means:(1) loss of life or personal injury on board or outside the ship carrying the hazardous and noxious substances, which is caused by those substances;(2) loss or damage of property on hoard or outside the ship carrying the hazardous and noxious substances which is caused by those substances;(3) loss or damage by contamination of the environment caused by the hazardous and noxious substances, provided that compensation for impairment of the environment other than loss of profit from such impairment shall be limited to costs of reasonable measures of reinstatement actually undertaken or to be undertaken;(4) the costs of preventive measures and further loss or damage caused by preventive measures.

Damage control booklet 破损控制手册 在客船和货船上永久性固定展示的,并向船上高级船员提供的,清晰地标示各层甲板及货舱的水密舱室界限、界限上的开口及其关闭装置和控制位置,以及扶正由于进水产生的横倾的装置的控制图或小册子。

Damage control bulkhead 损伤控制舱壁

Damage control deck 破损控制甲板 是为帮助通信和恢复后续破损而设置的延伸艏部和艉部通道的最低甲板。其通常高于水密完整性垂直限界线最低处,准确位置取决于有关分舱和水密完整性标准。 Damage control deck is the lowest deck on which continuous fore and aft access is provided to aid communications and recovery following damage. It is normally above the lowest vertical limit of watertight integrity, the exact location is determined by the relevant sub-division and watertight integrity standard.

Damage control information 稳性控制资料 系指在驾驶室永久展示或随时可用的控制图。用于指导船上负责的高级船员。图上应清晰显示每层甲板及货舱的水密舱室界限面,上面的开口及其关闭装置和任何控制位置,以及扶正由于进水产生的横倾装置。此外,还应给船上高级船员提供包括上述资料的小册子。在客船上允许在航行中保持开启的水密门应清晰记载于船舶的稳性资料内。收入资料中的一般预防措施应包括主管机关认为在船舶正常营运时为保持水密完整性所需的设备、条件和操作程序清单。应收入资料的特殊预防措施应包括主管机关认为对船舶、乘客和船员的生存至关重要的各种事项(即关闭装置、货物系固和听觉报警等)。对适用 SOLAS 公约第 Ⅱ-Ⅰ 章 B-1 部分破损稳性要求的船舶,破损稳性资料应为船长提供一种简单易懂的方式评估船舶在涉及一个或一组舱室的所有破损情况下的残存能力。 Damage control information means control information permanently exhibited and readily available on the navigation bridge, for the guidance of the officer in charge of the ship, plans showing clearly for each deck and cargo hold's the boundaries of the watertight compartments, the openings therein with the means of closure and position of any controls thereof, and the arrangements for the correction of any list due to flooding. In addition, booklets containing the aforementioned information shall be made available to the officers of the ship./ Watertight doors in passenger ships permitted to remain open during navigation shall be clearly indicated in the ship's stability information. General precautions to be included shall consist of a listing of equipment, conditions, and operational procedures, considered by the Administration to be necessary to maintain watertight integrity under normal ship operations. Specific precautions to be included shall consist of a listing of elements (i.e. closures, security of cargo, sounding of alarms, etc.) considered by the Administration to be vital to the survival of the ship, passengers and crew. In case of ships to which damage stability requirements of part B-1 of chapterⅡ-Ⅰ, SOLAS apply, damage stability information shall provide the master a simple and easily understandable way of assessing the ship's survivability in all damage cases involving a compartment or group of compartments.

Damage control plan and damage control booklet 破损控制图和破损控制手册 系指旨在为船上高级船员提供有关船舶水密舱室以及维护舱室边界和保持分隔有效性的准确信息,以便在船舶破损情况下,能给予合适的补救措施以避免通过开口进一步进水,并采取有效措施以快速减轻损失,可能的话,使船舶已损失的稳性得到恢复的图纸和手册。 Damage control plan and damage control booklet mean a plan and booklet which are intended to provide ship's officers with clear information on the ship's watertight compartmentation and equipment related to maintain the boundaries and effectiveness of the compartmentation, so that, in the event of damage to the ship, such as, flooding, proper precautions can be taken to prevent progressive flooding through opening therein and effective action can be taken quickly to mitigate and, where possible, recover the ship's loss of stability.

Damage control plans and booklets 破舱控制图和手册 系指在客船和货船上应有永久展示的控制图。该图应清晰标明各层甲板及货舱的水密舱室限界,限界上的开口及其关闭装置和控制位置,以及扶正由于进水产生的横倾的装置,应向船上高级船员提供包含上述资料的小册子。 Damage control plans and booklets mean that on passenger and cargo ships, plans permanently exhibited showing clearly for each deck and cargo hold's boundaries of the watertight compartments, the openings therein with the means of closure and position of any controls thereof, and the arrangements for the correction of any list due to flooding. Booklets containing the aforementioned information shall be made available to the officers of the ship.

Damage method by high flow rate 高流速损伤法

Damage radii 损伤半径 系指水上攻击情况造成损伤范围的度量。假设舱室爆炸，爆炸半径显示爆炸的范围。假设位置和损伤半径都依据弹头的特性。一般，半径应垂直于所标位置，否则拆除大量的材料或对剖面惯性造成很大的影响。损伤半径可由下式推出：$r = fz\ W^{1/3}$(m)，式中，fz—根据船型的换算距离；W—TNT 当量，kg。一旦确定损伤半径，从残余强度计算出发，解决损伤的最简单方法是移出损伤范围内的所有结构和通过剩余强度计算接触损伤半径。 Damage radius is a measure of the extent of the damage caused by specific above water attack scenarios. Where the assumption of a detonation mid-compartment is shown and the extent of damage is indicated by the extent of the damage radii. Assumptions about position and extent of damage radii are dependent on warhead characteristics. In general, the radii is to be vertically positioned or removing a large amount of material has the greatest effect on the sectional inertia. Damage radii can be determined from: $r = fz\ W^{1/3}$ (m), where: fz—scaled distance dependent on ship type; W—equivalent weight of TNT, in kg. Once a damage radius has been determined, the simplest method of accommodating damage is to remove all structure within and touching the damage radii from the residual strength calculation.

Damage stability 破舱稳性 系指船舶由于出现事故发生进水的破损情况下的稳性。也称为"船舶破损稳性"。Damage stability is the stability of a vessel in damaged condition due to accidental flooding, and also termed "damaged vessel stability".

Damage stability of a unit 平台的破舱稳性 系指平台破舱后，依靠其自身倾斜后的复原力矩，在规定的外加风压作用下仍能保持不再继续进水的能力。Damage stability of a unit is the ability of a damaged unit, under the action of a specified wind pressure, to keep itself from continuous flooding, by means of the righting moment induced by the inclination of that damaged unit.

Damage survey 损伤鉴定

Damaged load 海损载荷 系指由于海损事故引起的作用在船体结构上的载荷。

Damaged stability 破舱稳性 船舶破损进水后的稳性。

Damp coating 阻尼敷层 系指使用内损耗、内摩擦大的材料，如沥青、软橡胶以及其他高分子涂料用于减弱结构振动与增大结构噪声传递损耗的涂层。

Damped structures 阻尼结构 系指将阻尼材料与构件结合成一体以消耗振动能量的结构。

Damped vibration 阻尼振动 由于阻尼作用，振动系统所具有的能量在振动过程中不断减少，振幅也随时间增加而减少的振动。

Damper 防火风闸 为防止烟、火在诸防火分隔间流窜，而安装在截面积超过 $0.075m^2$ 的风道中的、可开闭的翻板。

Damper winding 阻尼绕组 是一种永久性短路的绕组，由埋置在同步电机极靴内的导线所组成。这些导线在极的端部处连接在一起，但不一定在极与极之间加以连接。当用于凸极电机时，这种绕组有时不一定穿过极靴，但在两极尖之间的极间空间内得到支撑的导电条。

Damping constrained plate 阻尼约束板

Damping force 阻尼力 系指使物体运动衰减的流体动力。

Damping loss factor 内损耗因子 系指子系统在单位频率(每振动一次)内单位时间损耗能量与平均储存能量之比。

Damping material 阻尼材料 系指将固体机械振动能转换为热能而耗散的材料。主要用于振动和噪声控制。其性能可根据其耗散振动能的能力来衡量。评价阻尼大小的标准是阻尼系数。

Damping moment 阻尼力矩 使物体运动衰减的流体动力矩。

Damping paint 阻尼涂料 由高分子树脂加入适量的填料以及辅助材料配制而成，是一种可涂敷在各种金属板状结构表面上，具有减振、绝热和一定密封性能的特种涂料。可以广泛用于飞机、船舶、车辆和各种机械设备的减振。

Dan boat 航标工作船 供在航道上管理和维修航标用的船舶。见图 D-2。

图 D-2　航标工作船
Figure D-2　Dan boat

Danforth anchor 丹福氏锚 属于有杆锚,为一种大抓力锚。主要用于海上工程船舶。见图 D-3。

图 D-3 丹福氏锚
Figure D-3 Danforth anchor

Danger messages 危险通报 系指每艘船舶的船长如遇到危险冰块、危险漂浮物,或其他任何对航行的直接危险,或热带风暴,或遇到伴随强风的低于冰点的气温致使上层建筑严重积聚冰块,或未曾收到暴风警报而遇到蒲福风级 10 级或 10 级以上的风力时,均有责任自行采取一切措施将此信息通知附近各船及主管当局的信息。发送这种信息的形式不受限制,可用明语(最好用英文)或按《国际信号规则》发送。各缔约国政府应采取所有必要的步骤,确保其在获悉上述的任何危险的情报时,迅速通知有关各方并通报其他相关国家的政府。向有关船舶发送的上述危险通报,不收费用。上述发送的一切无线电通报应冠以安全信号,并按 SOLAS 公约第Ⅳ/2 条定义的《无线电规则》的程序办理。在危险通报内要求有下列信息:(1)冰块、漂浮物及其他对航行的直接危险:①所观察到的冰块、漂浮物或危险的种类;②最后观察到的冰块、漂浮物或危险的位置;③最后观察到的危险的时间和日期(协调世界时);(2)热带气旋(风暴):①遭遇热带气旋的报告书。这项义务应从广义理解,每当船长有充分理由确信附近正在形成或存在热带气旋时,即须发送信号。②观察的时间、日期(协调世界时)和船舶的位置:③在通报内应尽可能包括下列信息:(a)气压,最好是修正过的气压(注明其为 millibars, mm 或 in 以及是否已经修正);(b)气压趋势(过去 3 h 内气压的变化);(c)真风向;(d)风力(蒲氏风级);(e)海况(小浪、中浪、大浪、巨浪)(f)涌级(低、中、巨)及涌浪来自真方向。涌的周期或长度(短、中、长)也具有重要性;(g)船舶真航向及航速。 Danger messages means when the master of every ship encounters such conditions as dangerous ice, a dangerous derelict, or any other direct danger to navigation, or a tropical storm, or encounters sub-freezing air temperatures associated with gale force winds causing severe ice accretion on superstructures, or winds of force 10 or above on the Beaufort scale for which no storm warning has been received, is bound to report the message by all means at his disposal to ships in the vicinity, and also to the competent authorities. The form in which the message is sent is not obligatory. It may be transmitted either in plain language (preferably English) or by means of the International Code of Signals. Each Contracting Government will take all steps necessary to ensure that when intelligence of any of the dangers specified above is received, it will promptly bring the knowledge of those concerned and communicated to other related Governments. The transmission of messages respecting the dangers specified above is free of cost to the ships concerned. All radio messages issued as specified above shall be preceded by the safety signal, using the procedure as prescribed by the Radio Regulations as defined in regulation Ⅳ/2, SOLAS. The following information is required in danger messages: (1) ice, derelicts and other direct dangers to navigation: ①the kind of ice, derelict or danger observed. ②the position of the ice, derelict or danger when last observed. ③the time and date (Universal Coordinated Time) when the danger was last observed; (2) Tropical cyclones (storms): ① a statement that a tropical cyclone has been encountered. This obligation should be interpreted in a broad sense, and information is transmitted whenever the master has good reason to believe that a tropical cyclone is developing or exists in the neighbourhood. ②time, date (Universal Coordinated Time) and position of ship observed. ③the following information should be included in the message if possible: (a) barometric pressure, preferably corrected (stating in millibars, millimetres, or inches, and whether corrected or uncorrected); (b) barometric tendency (the change in barometric pressure during the past three hours); (c) true wind direction; (d) wind force (Beaufort scale); (e) state of the sea (smooth, moderate, rough, high); (f) swell (slight, moderate, heavy) and the true direction from which it comes. Period or length of swell (short, average, long) would also be of value; (g) true course and speed of ship.

Dangerous atmosphere 危险气体环境 系指会使工作人员曝露于死亡、失去能力、损害自救能力(如无须协助从处所中逃离)、受伤或急性疾病危险中的气体环境。 Dangerous atmosphere means an atmosphere that may lead exposure to the risk of death, incapacitation, impairment of ability to self-rescue (i. e. to escape unaided from a space), injury or acute illness to workers.

Dangerous cargo[1] 危险货物 系指:(1)任何易爆、易燃、具有放射性或者对人体或环境有害的货品；(2)拟进行运输或储存，并列入即时修订的"国际海运危险货物规则"（IMDG 规则，简称《国际危规》）中的任何包装或散装货物；(3)未列入"IMDG 规则"，但符合即时修订的散装危险化学品、散装液化气体和散装固体货物规则要求装运的任何散装货物。 Dangerous cargo means: (1) any material which is explosive, flammable, radioactive or toxic to humans or the environment. (2) any substance whether packaged or in bulk, intended for carriage or storage and having properties in the classes listed in the IMDG Code as amended form time to time. (3) any substance shipped in bulk not coming within the IMDG Code classes but is subject to the requirements of the Codes for the dangerous chemical in bulk, liquefied gases in bulk and solid bulk as amended from time to time.

Dangerous cargo[2] 危险货物（巴拿马运河） 系指任何易爆、易燃、具有放射性或者对人体或环境有害的货品。 For Panama Canal, dangerous cargo means any material which is explosive, flammable, radioactive or toxic to humans or the environment.

Dangerous cargo in bulk 散装危险货物 系指装载于船舶构造部分中的某一液舱或货物处所或永久固定在船上的液舱中且无任何中间形式围护的危险货物。在苏伊士运河规则中，按货物的危险程度分为：（1）高度危险性散装危险货物，如 ①A 级品石油（闪点低于 23 ℃）超过 3 000 kg；②液化易燃气体；③散装危险化学品；④未清除 A 级品石油气；⑤C 级品 + A 级品蒸发气；⑥未消除油气的液化易燃气；⑦危险废物。（2）中度危险性散装危险货物，如 ①B 级品石油（闪点 23 ~ 66 ℃）超过 3 000 kg；②未清除 B 级品石油气；③C 级品 + B 级品蒸发气；（3）低度散装危险货物，如：①C 级品石油（闪点高于 66 ℃）；②消除了爆炸气体、船舶运输 B 级品石油气不超过 3 000 kg 和 3 个组别危险货物不超过 9 000 kg。

Dangerous chemical in bulk carrier 散装危险化学品运输船 系指装载运输危险化学品的任何船舶。该类船舶应符合"1974/1978 国际海上人命安全公约"的标准，还必须是按照国际海事组织即时修订或达到最低有效标准的关于运输散装危险化学品船舶构造和设备的规则进行建造的船舶，并必须在属于国际船级社协会（IACS）会员之一的船级社入级，而且仍然受该船级社的监督。 Dangerous chemical in bulk carrier means any vessel that transports dangerous chemical in bulk. Which shall comply with standards of SOLAS 74/78 and must be constructed according to IMO code for the construction and equipment of ships carrying dangerous chemical in bulk, as amended from time to time or to standards at least effective, and must be classified in one of the recognized classification societies belonging to the IACS and still under its supervision.

Dangerous chemicals 危险化学品 系指散装运输国际散装运输危险化学品船舶构造和设备规则（IBC 规则）第 17 章中货品安全标准所规定的会引起安全危害的液体化学品。 Dangerous chemicals means any liquid chemicals designated as presenting a safety hazard, based on Chapter 17 of the safety criteria for assigning products of the construction and equipment of ships carrying dangerous chemicals in bulk (IBC) Code.

Dangerous coating 危险涂料 系指由于油漆和刷漆的，具有可燃、播焰或有毒性质的材料。或在作业过程中可产生有毒或爆炸性气体、气味或粉尘的材料。

Dangerous goods 危险货物 系指包括国际海事组织在 IMDG 规定中所涉及的物质。 Dangerous goods are those goods referred to in the IMDG Code of IMO.

Dangerous goods in solid form in bulk 固体散装危险货物 系指除液体或气体以外，由粒子、颗粒或较大块状物质组成的并在 IMDG 规则中列明的任何物质，成分通常一致，并直接装入船舶的货物处所而无须任何中间围护形式，包括装入载驳船上的驳船内的此类物质。 Dangerous goods in solid form in bulk means any material, other than liquid or gas, consisting of a combination of particles, granules or any larger pieces of material, generally uniform in composition, which is covered by the IMDG Code and is loaded directly into the cargo spaces of a ship without any intermediate form of containment, and includes such materials loaded in a barge on a barge-carrying ship.

Dangerous goods incident 危险货物事故 系指发生包装危险货物从船上落入海中灭失或可能灭失的事故。

Dangerous goods manifest or stowage plan 危险货物舱单或配载图 系指每一艘运送包装危险货物的船舶具有的一份特别清单或舱单，或每一艘载运散装固体危险货物的船舶具有的一份特别清单或舱单。按 IMDG 规则规定的分类，列出船上危险货物及其位置。标明所有危险货物的类别并表明其在船上位置的详细配载图可用来代替上述特别清单或舱单。船舶驶离前应备有一份这些单证的副本，以供港口国当局指定的人员或组织使用。 Dangerous goods manifest or stowage plan means a special list or manifest setting forth for each ship carrying dangerous goods in packaged form, in accordance with the classification set out in the IMDG Code, the dangerous goods on board and the location thereof. Or a list or manifest setting forth the dangerous goods on board and the location thereof for each ship carrying dangerous goods in solid form in bulk. A detailed stowage plan, which identifies the class and sets out the location of all dangerous goods on board, may be used in place of such a special list or manifest. A copy of one of those documents shall be made available before departure to the person or organization designated by the State Port authority.

Dangerous spaces[1] 危险处所 包括：(1)贮存舱；(2)隔离舱和与贮存舱毗邻或直接位于贮存舱上面的围蔽或局部围蔽处所；(3)设置用于输送回收油的泵舱；(4)位于贮存舱下面的双层底或管弄；(5)遍及与贮存舱毗邻的泵舱或隔离舱，由没有用气密甲板与这种舱分隔的围蔽或局部围蔽处所；(6)设置用于装卸回收油的管路、阀或其他设备的围蔽或局部围蔽处所；(7)距离回收装置、贮存舱的舱口或任何其他开口和未设置在泵舱中用于装卸回收油的泵的 3m 范围之内的位于露天甲板上的区域或局部围蔽处所；(8)在遍及贮存舱的露天甲板以上高度 2.4m 之内，对长度大于 50 m 的船来说，上述区域应超出贮存舱的前后端延伸 3m；(9)放置可能含有回收油残余的移动式泵及其附带软管和其他设备用的储藏室。但是，如备有能提供每小时至少换气 20 次的机械通风并具有维持这种通风效能的性能，则上述的(2)、(5)和(6)中规定的处所可被认为是安全处所。 Dangerous spaces include: (1) accumulation tanks; (2) cofferdams and enclosed or partially enclosed spaces adjacent to or immediately above accumulation tanks; (3) spaces containing pumps for the handling of oil collected; (4) double bottoms or duct keels located under accumulation tanks; (5) enclosed or partially enclosed spaces over pump rooms or cofferdams adjacent to accumulation tanks, which are not separated from such spaces by means of a gas-tight deck; (6) enclosed or partially enclosed spaces containing piping, valves or other equipment for the handling of oil removed; (7) zones or partially enclosed spaces on the weather deck, within a range of 3 m from equipment for oil removal, hatches or any other openings in accumulation tanks, and any pump not fitted in a pump room for the handling of oil removed; (8) zones of the weather deck over accumulation tanks within a height of 2.4 m above the deck. For ships whose length $L > 50$ m, the above zone is to be extended 3 m beyond the fore and aft end of the tanks; (9) storerooms for floating pumps and associated hoses and other equipment which may similarly contain residues of oil removed. Spaces defined in (2), (5) and (6) above may, however, be considered safe spaces which is fitted with forced ventilation capable of giving at least 20 air changes per hour and having such characteristics as to maintain the effectiveness of such ventilation.

Dangerous spaces[2] 危险区域 系指下列区域或场所，这些场所放置易燃或易爆物质，或这些物质有泄漏易燃或易爆气体或蒸发气的危险性。这些物质按照气体爆炸氛围的生成程度和持续时间来分类：(1)区域"0"(zone 0)：气体爆炸氛围持续或长时间存在的区域；(2)区域"1"(zone 1)：气体爆炸氛围在正常作业状态下偶尔出现的区域；(3)区域"2"(zone 2)：气体爆炸氛围在正常的作业状态下不发生，仅在异常作业状态下产生的区域(出现频度非常稀少且时间很短)。 Dangerous spaces are the following areas or spaces where flammable or explosive substances are placed and where it is likely to arise flammable or explosive gases or vapours from these substances and they are classified according to generation frequency and period of life of explosive gas atmosphere: (1) zone 0: area in which an explosive gas atmosphere is present continuously or is present for long periods; (2) zone 1: area in which an explosive gas atmosphere is sometimes likely to occur under normal operation; (3) zone 2: area in which an explosive gas atmosphere is not likely to occur under normal operation or it is likely to occur under abnormal operation with a very low frequency and a short period.

Dangerous target 危险目标 系指预计 CPA 和 TCPA 违反操作人员设定值的目标。各自的目标以"危险目标"符号标示。 Dangerous target means a target with a predicted CPA and TCPA that violates values preset by the operator. The respective target is marked by a "dangerous target" symbol.

Dan-layer [buoy (age) vessel] 布标船 设有起放航标用的起重机和绞盘等设备，主要用于在航道上布设和更换航标用的船舶。

Dark room 暗室 用于冲洗、印刷和放大感光胶片的工作室。

DARPA 美国国防部高级研究计划局

Data carrier 信息载体 系指设计用以携带信息登记记录的工具。 Data carrier means the medium designed to carry records of data entries.

Data circuit-terminating equipment (DCE) 数据通信设备 它在 DTE 和传输线路之间提供信号变换和编码功能，并负责建立、保持和释放链路的连接，由开放软件基金开发。

Data communication 数据通信 系指通信技术和计算机技术相结合而产生的一种新的通信方式。要在两地间传输信息必须有传输信道，根据传输媒体的不同，有有线数据通信与无线数据通信之分。但它们都是通过传输信道将数据终端与计算机联结起来,而使不同地点的数据终端实现软、硬件和信息资源的共享。移动通信系统从20世纪80年代诞生以来,到2020年将大体经过5代的发展历程,而且2010年,将从第3代过渡到第4代(4G)。目前已进入第5代(5G)。

Data communication 资料通信 系指透过电脑及通信线路将各类资料从传送端传送至接收端,以进行信息传递与交换的过程。

Data definition language (DDL) 数据库模式定义语言 系指用于描述数据库中要存储的现实世界实体的语言。一个数据库模式包含该数据库中所有实体的描述定义。DDL描述的模式,必须由计算机软件进行编译,转换为便于计算机存储、查询和操纵的格式,完成这个转换工作的程序称为模式编译器。数据字典和数据库内部结构信息是创建该模式所对应的数据库的依据,根据这些信息创建每个数据库对应的逻辑结构;对数据库数据的访问、查询也根据模式信息决定数据存取的方式和类型,以及数据之间的关系和对数据的完整性约束。

Data definition language (DDL) 数据定义语言 系指SQL语言集中负责数据结构定义与数据库对象定义的语言,由CREATE、ALTER与DROP三个语法所组成,最早是由Codasyl(Conference on Data Systems Languages)数据模型开始,现在被纳入SQL指令中作为其中一个子集。目前大多数的DBMS都支持对数据库对象的DDL操作,部分数据库(如Postgre SQL)可把DDL放在交易指令中,也就是它可以被撤回(Rollback)。较新版本的DBMS会加入DDL专用的触发程序,让数据库管理员可以追踪来自DDL的修改。

Data field 数据字段

Data line transfer 数据线传输 现在的手机一般都带有U盘功能,所以这种传输方式用得比较多,速度快,就是一般的USB 3.0的数据接口。

Data logging 巡检 系指为确认有关规定的建造流程中适用过程、活动和相关文件持续地符合船级社和法定要求所进行的独立和非预定的验证活动。

Data manipulation language (DML) 数据操作语言 系指用来存储和保护所有已授权的被确认版本介质配置项,由CMDB(ITIL配置管理)联邦提出。它们存储经过质检的主拷贝版本。这个库可以有一个或多个软件库或存放区来存放开发、测试和实时存储文件。它们包含组织所有软件的主拷贝、购买软件的副本及受控文件的电子版。DML包含物理的拷贝存储,DML是发布管理的基础。配置管理负责控制组织接收到的所有IT组件并需确保这些组件被记录在系统中。硬件可在其已订购或已交付时进行记录,而软件则通常在其被纳入DML时进行记录。DML分成交互型DML和嵌入型DML两类。

Data on environmental fate and effect under aerobic and anaerobic conditions 需氧和厌氧情况下的环境后果和影响资料 系指包括下列项目的资料:(1)退化方式(有生命和无生命的);(2)生物体内积累、分配数、辛醇/水系数;(3)在相关介质(压载水、海水和淡水)内的主要代谢物的留存和鉴定;(4)与生物质的反应;(5)对野生植物和海底生物栖息地的潜在物理影响;(6)海洋食物中潜在的残留物;(7)任何已知的相互影响。Data on environmental fate and effect under aerobic and anaerobic conditions means data including: (1) modes of degradation (biotic; abiotic); (2) bioaccumulation, partition coefficient, octanol/water coefficient; (3) persistence and identification of the main metabolites in the relevant media (ballast water, marine and fresh waters); (4) reaction with organic matter; (5) potential physical effects on wildlife & benthic habitats; (6) potential residues in seafood; (7) any known interactive effects.

Data on mammalian toxicity 哺乳动物的毒性资料 系指包括下列项目的资料:(1)急性毒性;(2)对皮肤和眼睛的影响;(3)慢性和长期的毒性;(4)发展和再生的毒性;(5)致癌性;(6)诱变性。Data on mammalian toxicity means data including: (1) acute toxicity; (2) effects on skin and eyes; (3) chronic and long-term toxicity; (4) developmental and reproductive toxicity; (5) carcinogenicity; (6) mutagenicity.

Data retrieval 数据检索 系指将经过选择、整理和评价(鉴定)的数据存入某种载体中,并根据用户需要从某种数据集合中检索出能回答问题的准确数据过程或技术。按查询问题的要求,分为简单检索(即单一因素的检索)和综合检索(即综合条件检索)。

Data room 资料室 存放图书、图表和有关海洋调查资料,供查阅和复制资料的工作舱室。该舱室可兼图书馆。

Data sink 数据宿

Data source 数据源 系指数据的来源,即提供某种所需要数据的器件或原始媒体。可以是文件,数据库等。在数据源中存储了所有建立数据库连接的信息。就像通过指定文件名称可以在文件系统中找到文件一样,通过提供正确的数据源名称,你可以找到相应的数据库连接。用户DSN允许单个用户在单个计算机上访问数据库,系统DSN允许在某个计算机上的多个用户访问数据库。C3P0是一个开放源代码的JDBC数据源实现项目,它在lib目录中与Hibernate一起发布,实现了JDBC3和JDBC2扩展规范说明的Connection和Statement池。

Data terminal 数据终端 即计算机显示终端,是计算机系统的输入、输出设备。计算机显示终端伴随主机时代的集中处理模式而产生，并随着计算技术的发展而不断发展。迄今为止，计算技术经历了主机时代、PC 时代和网络计算时代这 3 个发展时期，终端与计算技术发展的 3 个阶段相适应，应用也经历了字符哑终端、图形终端和网络终端这 3 个形态。

Data transmission 数据传输 系指依照适当的规程，经过一条或多条链路，在数据源和数据宿之间传送数据的过程。也表示借助信道上的信号将数据从一处送往另一处的操作。串行传输是构成字符的二进制代码在一条信道上以位(码元)为单位，按时间顺序逐位传输的方式。对于模拟传输信道，DCE 的发送部分就是调制器，它将二进制数字信号变换成模拟信号，使发送信号的频谱与传输信道的频带相匹配，以便数据信号能在传输信道中有效和可靠地传送。

Data transmission by induction radio 感应无线数据通信

Database 数据库 系指按照数据结构来组织、存储和管理数据的仓库。它产生于距今 50 年前，随着信息技术和市场的发展，特别是 20 世纪 90 年代以后，数据管理不再仅仅是存储和管理数据，而转变成用户所需要的各种数据管理的方式。数据库有很多种类型，从最简单的存储有各种数据的表格到能够进行海量数据存储的大型数据库系统都在各个方面得到了广泛的应用。

Database management system (DBMS) 数据库管理系统 系指一种操纵和管理数据库的大型软件，用于建立、使用和维护数据库。它对数据库进行统一的管理和控制，以保证数据库的安全性和完整性。用户通过 DBMS 访问数据库中的数据，数据库管理员也通过 dbms 进行数据库的维护工作。它可使多个应用程序和用户用不同的方法在同时或不同时刻去建立，修改和询问数据库。大部分 DBMS 提供数据定义语言(Data definition language/DDL)和数据操作语言(Data manipulation language/DML)，供用户定义数据库的模式结构与权限约束，实现对数据的追加、删除等操作。数据库管理系统是数据库系统的核心，是管理数据库的软件。数据库管理系统就是实现把用户意义下抽象的逻辑数据处理，转换成为计算机中具体的物理数据处理的软件。有了数据库管理系统，用户就可以在抽象意义下处理数据，而不必顾及这些数据在计算机中的布局和物理位置。

Data-logging system (data logger) 巡回检测装置 一种能对每个测试点的温度、压力等参数自动依次逐点进行测量、监控的装置。

Data-set for active substances and preparations 活性物质和配制品的数据集 系指活性物质和配制品(包括其成分)的性质或作用方面的如下资料:(1)对水生植物、无脊椎动物、鱼类和其他生物群，包括敏感的和有代表性的生物产生的影响的资料:①急性水毒性;②慢性水毒性;③内分泌干扰;④沉积物毒性;⑤生物药效率/生物放大作用/生物集结;⑥食物链/种群数量影响。(2)哺乳动物的毒性资料:①急性毒性;②对皮肤和眼睛的影响;③慢性和长期的影响;④发展和再生的毒性;⑤致癌性;⑥诱变性。(3)需氧和厌氧情况下的环境后果和影响资料:①退化方式(有生命和无生命的);②生物体内积累、分配系数、辛醇/水系数;③在相关介质(压载水、海水和淡水)内的主要代谢的留存和鉴定;④与生物质的反应;⑤对野生植物和海底生物栖息地的潜在物理影响;⑥海洋食物中潜在的残留物;⑦任何已知的相互影响。(4)活性物质和配制品以及经处理的压载水(如适用)的物理和化学性质;①熔点;②沸点;③可燃性;④密度(相对密度);⑤蒸发气压力，蒸发气密度;⑥水溶性/电离常数(pKa);⑦氧化作用/还原势;⑧对普通船舶结构材料或设备的腐蚀性;⑨自燃温度;⑩其他已知的相关物理和化学危害。 Data-set for active substances and preparations means the data on the properties or actions of the active substances and preparation including any of its components as follows: (1) data on effects on aquatic plants, invertebrates, fish, and other biota, including sensitive and representative organisms: ①acute aquatic toxicity; ② chronic aquatic toxicity; ③ endocrine disruption; ④ sediment toxicity; ⑤ bioavailability/biomagnifications/bioconcentration; ⑥ food web/population effects. (2) Data on mammalian toxicity: ①acute toxicity; ②effects on skin and eyes; ③chronic and long-term toxicity; ④developmental and reproductive toxicity; ⑤carcinogenicity; ⑥mutagenicity. (3) data on environmental fate and effect under aerobic and anaerobic conditions: ①modes of degradation (biotic; abiotic); ②bioaccumulation, partition coefficient, octanol/water coefficient; ③persistence and identification of the main metabolites in the relevant media (ballast water, marine and fresh waters); ④ reaction with organic matter; ⑤ potential physical effects on wildlife & benthic habitats; ⑥ potential residues in seafood; ⑦any known interactive effects. (4) physical and chemical properties for the active substances and preparations and the treated ballast water, if applicable: ①melting point; ②boiling point; ③flammability; ④density (relative density); ⑤ vapour pressure, vapour density; ⑥ water solubility/dissociation constant (pKa); ⑦oxidation/reduction potential; ⑧corrosivity to the materials or equipment of normal ship construction; ⑨ autoignition temperature; ⑩other known relevant physical or chemical hazards.

Date of build 建造日期 对于新造船舶，建造日期系指新船建造检验过程全部结束时的年份和月份,如果在建造检验完成与投入实际营运间隔相当长的时间,则投入营运的日期也可以特别注明。如果船舶进行改装,已确定的船舶建造日期保持不变。如果船舶主要部分完全替换或增加(如艏分段、艉分段、主货舱分段)则按照以下方式处理:(1)船舶的每一个主要部分的建造日期证明在入级证书上;(2)检验要求应基于船舶的每一主要部分的建造日期。　For a new building, the date of build is the year and month when the construction and survey process is completed. Where there is a substantial delay between the completion of the construction and survey process and the ship commencing active service, the date of commissioning may also be specified. If modifications are carried out, the date of build remains assigned to the ship. Where a complete replacement or addition of a major portion of the ship (e.g. forward section, after section, main cargo section) is involved, the following applies: (1) the date of build associated with each major portion of the ship is indicated on the Classification Certificate; (2) survey requirements are based on the date of build associated with each major portion of the ship.

Date of contract for construction[1]　建造合同的日期　定义如下:(1)船舶"建造合同"的日期系指未来船东和船厂之间签订船舶建造合同的日期。该日期一般由申请授予新船船级的建造合同方向船级社声明。(2)系列姐妹船,包括最终行使选择权的特定可选的船舶的"建造合同"的日期,系指未来船东和船厂之间签订建造系列船合同的日期。就本款而言,姐妹船是按相同的入级批准图纸建造的船舶。如系列船建造合同签订后1年之内行使续建选择权,则该可选续建的船舶将被认为相同的系列姐妹船的一部分;(3)如随后对建造合同进行修改,以及包括增建船舶或附加选择权,这类船的建造合同的日期,系指未来船东和船厂之间签订该合同修改的日期。该合同修改应被认为新合同,并应符合上述(1)和(2)款要求。Date of contract for construction is defined as follows: (1) the date of "contract for construction" of a ship is the date on which the contract to build the ship signed between the prospective owner and the shipbuilder. This date is normally to be declared to a Society by the Party applying for the assignment of class to a new-building. (2) The date of "contract for construction" of a series of sister ships, including specified optional ships for which the option is ultimately exercised, is the date on which the contract to build the series is signed between the prospective owner and the shipbuilder. For this subparagraph, sister ships are ships built under the same approved plans for classification purposes. The optional ships will be considered part of the same series of sister ships if the option is exercised within 1 year after the contract to build the series is signed. (3) If a contract for construction is later amended to include additional ships or additional options, the date of contract for construction for such ships is the date on which the amendment to the contract is signed between the prospective owner and the shipbuilder. The amendment to the contract is to be considered as a new contract to which above (1) and (2) apply.

Date of contract for construction[2]　建造签约日　系指:(1)船舶所有者和船舶建造者在建造合同上签字的日期。检验申请者在申请时要向船级社通报所有包括建造合同签约日和合同上所有船舶建造的号码(即船号);(2)系列姊妹船,包括最终行驶选择权的特定可选的船舶的合同日期是指船舶所有者和船舶建造者之间在系列姊妹船的合同上最终签字的日期;(3)姐妹船系列系指在单一的建造合同下在同一建造厂根据对其他船业经认可的图纸进行建造的船舶,船级社认为是同类型或者类似的船舶。但姐妹船系列中因下列原因可更改先前的设计:①当这些变更不会对船级相关事项产生影响,或 ②应符合船级条件时,则其应符合船舶所有者和船舶制造厂商就变更所签订的,带有有效日期的船级要求。如果没有关于变更的合约,则应符合由原船级提交的,具有有效期限的船级要求,以使变更得到认可。如果加船合同在姐妹船系列建造合约签字之日起1年内签字生效,则该船视为系列姊妹船;(4)但是如果在建造合同书上进行修改,添加船舶或者添加别的事项时,建造签约日改为改定合同日。这种合同的变更在因(1)～(3)号的作用被看作新的合同;(5)一旦出现因要更改船型而改变合约的情况,船型变更后的船舶建造签约日期即为船舶所有者和船舶制造者改订的建造合约书或新的建造合约书上签订的日期。　(1) Date of contract for construction of a vessel is the date on which the contract to build the vessel is signed between the prospective owner and the shipbuilder, this date and the construction numbers (i.e. hull numbers) of all the vessels included in the contract are to be declared to the Society by the Party applying for the assignment of class to a new-building; (2) The date of contract for construction of a series of vessels, including specified optional vessels for which the option is ultimately exercised, is the date on which the contract to build the series is signed between the prospective owner and the shipbuilder; (3) Vessels in application to (2) built under a single contract for construction are considered a series of vessels if they are built under the same approved plans for classification purpose by the same shipbuilder, and considered as same or similar by the Society. However, vessels of a series may have design alterations from the original design provided: ① such alterations do not affect matters related to classification, or ② if the alterations are subject to classification requirements, these alterations are to comply contract between the prospective owner and the shipbuilder or, in the absence of the alteration contract, comply with the classification requirements in effect on the date on which the alterations are submitted to the Society for approval. The optional vessels will be considered part of the same series of vessels if the option is exercised within 1 year after the contract to build the series is signed. (4) If a contract for construction is later amended to include additional vessels or additional options, the date of contract for construction for such vessels is the date on which the amendment to the contract, is signed between the prospective owner and the shipbuilder. The amendment to the contract is to be considered as a new contract to which above (1) to (3) apply. (5) If a contract for construction is amended to change the ship type, the date of contract for construction of this modified vessel, or vessels, is the date on which revised contract or new contract is signed between the owner and the shipbuilder.

Date of delivery 交船日期 系指建造完工的船舶,在系泊试验和航行试验合格后,由造船厂向船东移交的日期。

Date of initial classification for new buildings 新造船舶初始入级日期 通常新造船舶初始入级日期与建造日期一致。 As a general rule, for new buildings the date of initial classification coincides with the date of build.

Date of keel laid 铺龙骨日期 系指船舶正式开工的日期。

Date of ship's significant alteration contract 船舶重大改装合同日期 系指船东与造船厂之间签订现有船舶重大改装合同的日期。该日期应由申请授予改装船舶船级的重大改装合同方向船级社声明。当无改建合同时,改建设计审图申请日期代替重大改装合同日期。

Date on which a ship is built 船舶建造日期 系指在此日期,该船舶业已完成的装配量不小于 50 t,或为该船全部结构材料估计总质量的 1%,取小者。 Date on which a ship is built means the date on which not less than 50 t or 1% of the proposed total mass of the structural material of the ship, whichever is the less, has been assembled.

Datum 基准面 系指水密甲板或非水密甲板覆盖一个风雨密结构而组成的等效结构。该结构为保持风雨具有足够强度保持风雨密完整性并设有风雨密关闭装置。 Datum means a watertight deck or equivalent structure of a non-watertight deck covered by a weathertight structure of adequate strength to maintain the weathertight integrity and fitted with weathertight closing appliances.

Davit 吊柱 一种上部弯曲悬伸成钩状,其顶端装有起重吊具,能旋转的用以起吊重物的吊杆柱。

Davit arm 吊艇臂 吊艇架中用于吊载小艇的组合箱型构件。

Davit frame 吊艇架座架 吊艇架中用于支持吊艇臂或吊艇杆并承受其反力的组合构件。

Davit releasing device 放艇联动装置 重力式艇架中用以解开固艇索具同时释放吊艇臂的放艇组合机构。

Davit span 横张索 联结吊艇架上的两吊艇臂或吊艇杆顶端并起张紧作用的绳索及其索具。

Davit-launched type liferaft 可吊放救生筏 在满载乘员后,可由吊卸装置放到水面上的救生筏。

Daylight 白昼 系指日出后 1 h ~ 日落后 1 h。 Daylight means one hour before sunrise until one hour after sunset.

D-E propulsion 柴油机-发电机电力推进

Dead freight 空舱费 又称亏舱费,系指租船人向船东或租船经纪人在租船人所装载货物少于租约约定时支付的赔偿金数额。空舱费通常按全运费率支付,但如果运输费包括装卸费,则要减去装卸费。

Dead light (blind cover) 风暴盖 安装在舷窗上,在风暴中玻璃碎裂后可保证舱内不致进水的盖子。

Dead nip stopper 凸轮止链器 用摇手柄转动偏心凸轮而使闸块升降以将锚链止住或松开的止链器。

Dead reckoning 船位推算法 从一已知点开始,根据船舶的航向、航速、航行时间和流速流向推算出下一点的船位。常用的船位推算法是利用电(磁)罗经、计程仪、船钟等人工推算。多普勒声呐、惯性导航也是用测量速度(加速度)对时间的积分和航向数据来实现导航的。这是一种自动船位推算法,在航海与海上测量中广泛使用这种方法。

Dead ship condition[1] 瘫船状态(平台) 系指包括动力源的整个平台动力装置停止工作,而且使主推进装置运转和恢复主动力源的辅助用途的压缩空气和启动用蓄电池等不起作用。 Dead ship condition is understood to mean that the entire machinery installations, including the power supply, and the auxiliary services such as compressed air, and storage battery, etc., for bringing the main propulsion into operation are out of operation for the reason that the main power supply are not available.

Dead ship condition[2] 瘫船状态 系指：(1)因主电源失效而导致主推进器、锅炉及辅机不能工作；(2)假设为恢复主推进力而不能使用为启动推进装置、主电源装置及其他重要辅机而储存的能源，但假设用以启动应急发电机的程序可一直使用。(3)主推进装置、锅炉和辅机已停止运行，且在恢复推进的过程中，假定已没有储能可用于启动和运行推进装置、主发电机及其他必需的辅助设备的一种状态。 Dead ship condition means a condition under which：(1)the main propulsion plant, boilers and auxiliaries are not in operation due to the loss of the main source of electrical power；(2)in order to restore the propulsion, no stored energy for starting the propulsion plant, the main source of electrical power and other essential auxiliary machinery is assumed to be available. It is assumed that means are available to start the emergency generator at all times. (3) a condition under which the main propulsion plant, boilers and auxiliaries are not in operation and in order to restore the propulsion, no stored energy for starting and operating the propulsion plant, the main source of electrical power and other essential auxiliaries is assumed to be available.

Deadlights 窗盖 系指装设在窗和舷窗内侧的盖子。Deadlights mean those covers fitted to the inside of windows and side scuttles.

Deadrise〔rise of floor, rise of bottom〕 舭部升高 在最大横剖面上，自船底斜升线与舷侧切线的交点量至基平面的垂直距离。

Deadrise angle β 舭部升高角 β 系指在特定位置的横剖面上测得的艇底与水平线之间的夹角(°)。应按图 D-4 中所示的要求进行测量。Deadrise angle β is the angle of the bottom from the horizontal measured athwartship, at a specific position, in degrees. The measurement shall be taken as indicated in Figure D-4.

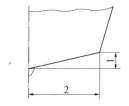

（a）直线形底部
(a) Straight bottom
注意：忽略阶梯和其他突出物形
Steps and other protrusions are ignored

（b）凹底加龙骨
(b) Concave bottom with keel
注意：从龙骨交点至舭缘线之间测量舭部升高
Deadrise is measured between keel intersection and chine

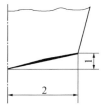

（c）Convex bottom 凸底
注意：在 $b_{H/2}$ 的 1/4 与 3/4 点之间测量舭部升高
NOTE: Deadrise is measured between 1/4 and 3/4 of $b_{H/2}$.

（d）Concave bottom with wing 带舭舱的凹底
注意：在中心线和舭舱外侧之间测量舭部升高
NOTE: Deadrise is measured between centreline and outer wing end.

说明(Key)：1 高度 height
2 宽度 width

图 D-4 舭部升高的确定
Figure D-4 Determination of deadrise

Deadweight 载重量 系指船舶在密度为 1.025 kg/m^3 的海水中,吃水相应于所勘划的夏季干舷时,排水量与该船空船排水量之差,以吨计。 Deadweight is the difference in tonnes between the displacement of a ship in water of a specific density of 1.025 kg/m^3 at the draught corresponding to that of the assigned freeboard in summer with lightweight of the ship.

Deadweight displacement ratio 载重系数 系指民用船舶的载重量与其设计排水量的比值。

Deadweight load 自重载荷(起重设备) 系指不包括在起升载荷中的起重设备部件的质量。

Deadweight scale[displacement scale] 载重线标尺 按船舶空船重力(或质量)至满载排水量之间的平均吃水高度,标有与其相对应的排水量、载重量、每厘米吃水吨数以及附有载重线标志的图表。

Deadweight survey 载重量鉴定

Deadweight tonnage DW 载重吨位 DW 系指满载排水量与轻载排水量之差(t)。 Deadweight DW is the difference in tons between full load displacement and light weight.

Deadwood 艉鳍 位于船尾底部中线面上或其两侧,供坐坞和提高航向稳定性用的船体鳍状部分结构。

Deaerator (deaerating feed water heater) 脱氧器 闭式给水系统中,用主机撇气或辅机排气来直接加热给水以排除溶解于水中的气体的设备。

Decca navigation system 台卡导航系统 "台卡"系英文 Decca 的译音。一种甚低频连续波相位双曲线导航系统。它是使用持续电波的相位差测定两点的距离差。用于海上近程高精度定位。该系统具有设备体积小、质量轻、可临时布设、工作区域较小的特点,在海道测量中应用广泛。

Decelerating ducted propeller (pump jet) 减速导管推进器 水进入导管后在叶轮处流速降低,静压力增高,有利于延缓产生空泡和减轻噪声的导管推进器。

Decelerating section 减速段 船模试验水池中船模拖曳速度从预期试验速度减至零的水池段。

Decentralized refrigerating plant 分散式制冷装置 系指每个集装箱与各个独立制冷单元在船上连接,以产生和分配冷量的装置。 Decentralized refrigerating plant is a decentralized refrigerating plant in which each container is connected on board to a separate unit for cold production and distribution.

Decimetre wave 分米波

Decision support system for masters 船长决策支持系统 系指所有客船在驾驶室配置的一个处理紧急情况的决策支持系统。 Decision support system for masters means that in all passenger ships, a decision support system for emergency management shall be provided on the navigation bridge.

Decision support system for masters of passenger ships 客船船长决策支持系统 系指在所有客船驾驶室中设置的,至少由 1 个或几个印制的应急计划构成的系统。所有可预计的紧急状态均应在应急计划中标明,包括但不限于下列各类主要的紧急情况:(1)火灾;(2)船舶破损;(3)污染;(4)威胁到船舶安全及乘客和船员保安的非法行为;(5)人员事故;(6)与货物相关的事故;(7)对其他船舶的应急救援。应急计划中所建立的应急程序,应向船长提供用以处理各种组合紧急状态的决策支持方案。应急计划应有统一的格式并易于使用,如果适用,为客船航行稳性而计算的实际装载工况应用于破损控制。除印制的应急计划外,主管机关也可接受在驾驶室使用以计算机为基础的决策支持系统。该系统能提供应急计划中包括的所有的信息、程序、检查清单等,能针对可预防的紧急情况提出拟采取的建议措施的清单。 Decision support system for masters of passenger ships means a decision support system provided on the navigation bridge for emergency management in all passenger ships. The system shall, as a minimum, consist of a printed emergency plan or plans. All foreseeable emergency situations shall be identified in the emergency plan or plans, including, but not limited to, the following main groups of emergencies: (1) fire; (2) damage to ship; (3) pollution; (4) unlawful acts threatening the safety of the ship and the security of its passengers and crew; (5) personnel accidents; (6) cargo-related accidents; (7) emergency assistance to other ships. The emergency procedures established in the emergency plan or plans shall provide decision support to masters for handling any combination of emergency situations. The emergency plan or plans shall have a uniform structure and be easy to use. Where applicable, the actual loading condition as calculated for the passenger ship's voyage stability shall be used for damage control purposes. In addition to the printed emergency plan or plans, the Administration may also accept the use of a computer-based decision support system on the navigational bridge which provides all the information contained in the emergency plan or plans, procedures, checklists, etc., which is able to present a list of recommended actions to be carried out in foreseeable emergencies.

Deck 甲板 限定舱室上部或上部界限的水平构件。 Deck is a horizontal structure element that defines the upper or lower boundary of a compartment.

Deck and sanitary water overboard discharge hole (scuppers and sanitary overboard discharges) 甲板水舷外排水口 甲板疏水管路向舷外排水的出口。

Deck and sanitary water piping system (deck scupper pipes and sanitary discharges) 甲板排水及疏水系统 甲板排水和污水疏水系统的总称。

Deck cargo barge 甲板驳 系指不设货舱、全部货物堆装在甲板上的驳船。

Deck cargo ship 甲板货船 系指不设货舱、全部货物堆装在甲板上的货船。

Deck chock (close chock) 甲板导缆孔 专指装在甲板上的闭孔状导缆器。有带折边的和无折边的两种。

Deck construction 甲板结构 一般系指船舶的上甲板结构。它承担船舶纵总弯曲时的最大拉、压力,以及货物、设备重力及波浪冲击力,通常称为强力甲板。舱底结构可分为横骨架式和纵骨架式两种形式。货船在甲板上开有货舱口,它是装卸货物的通道又具有防护的作用,通常对甲板上货舱口区域结构进行局部加强处理。甲板结构分为横骨架式和纵骨架式两种。

Deck craft 甲板艇 系指从艇舯至艇艉范围具有风雨密的连续露天甲板的游艇。

Deck crane 起重机/甲板吊 是将吊杆和绞车组合在一个机组内,并能在起吊货物后,直接操纵旋转机构进行旋转,随意改变吊货钩落点位置的货物装卸设备。起重机按动力源分为电动式和液压式两种。按吊臂类型分为单臂吊(single crane)、双臂吊(twin crane)和双座吊(tandem crane)3 种。见图 D-5。

图 D-5 起重机/甲板吊
Figure D-5 Deck crane

Deck equipment (deck sundries) 甲板用具 船舶甲板部门所配属的帆缆、消防、堵漏、木工等用具及其备品的统称。

Deck equipment and fittings 舱面属具 配备在船舶舱面上的各种辅助设备的统称。包括甲板梯、舷梯、栏杆、天幕、属具和甲板用具等。

Deck girder 甲板纵桁 甲板骨架中的纵桁。

Deck guide bollard 落地导向滑车 安装在渔船甲板面上,在起放网操作过程中,用以定向传导钢索的滑车。

Deck habitat chamber 甲板居住舱 是供潜水员加压,饱和深度停留和减压使用的主要舱室。舱内有卧室和起居室,设有供氧、二氧化碳吸收、温湿度控制等装置。

Deck house 甲板室 系指:(1)强力甲板以上,其侧壁距离船壳板向内大于 4% 型宽(B)的设有甲板的结构;(2)干舷甲板或上层建筑甲板上不延伸到船舶两舷的结构。 Deck house is:(1) defined as a decked structure, above the strength deck, with the side plating being inboard of the shell plating by more than 4% of the ship's breadth B;(2) a structure on the freeboard or superstructure deck not extending from side to side of the ship.

Deck house casing 甲板室围壁 甲板室四壁的统称。

Deck house top 甲板室甲板 各层甲板室顶部的甲板。

Deck ladder 甲板梯 供人员在船内上、下用的工作梯子的统称。

Deck light (deck glass) 甲板窗 水密地嵌固于甲板上,用来为下面舱室透光的窗。

Deck line 甲板线 (1)甲板边线和甲板中线的统称;(2)按海船载重线规范规定的式样和位置,水平勘绘在干舷甲板处船中两舷,为勘定载重线标志而设的基准线。

Deck line at center (deck center line) 甲板中线 甲板型表面与中线面的交线。

Deck line at side (deck side line) 甲板边线 甲板型表面的边缘线,对于甲板舷边为圆弧形的金属船体,则至横梁上缘延伸线与肋骨外缘延伸线之交点的连接线。

Deck longitudinal 甲板纵骨 甲板板下的纵骨。

Deck machinery 甲板机械 系指安装在露天甲板上的机械设备。对客船和货船而言有下列几种:(1)按用途分——锚泊机械装置的锚机(windlass)、系泊机械装置的系泊绞车(mooring winch)、用于货物装卸的起货机(cargo winch/cargo gear)、用于船舶操纵的舵机(steering gear)、用于货物、食品冷藏的冷藏设备(refrigerating equipment)、用于船员、引水员、乘客上、下的舷梯(accommodation ladder)、用于人、车上、下的吊桥(ramp way)等。(2)按动力源分——多用于油船、液化气船、化学品船等的气体危险区域的蒸汽驱动系统(steam power system);低振动、易遥控的交、直流电动系统(electric system);经济、易遥控的电动—液压式(electric-hydraulic type)等。(3)按操纵压力分——25~30 kg/cm² 的低压式(lower pressure type)、30~70 kg/cm² 的中压式(medium pressure type)和超过120 kg/cm² 的高压式(high pressure type)。按应用最为广泛的电动-液压式甲板机械。(3)按速度的控制方法分：①阀控机构(valve control system)——油泵和阀都是固定型的，且在系统中装有流量控制阀(flow control system)和方向控制阀(direction control valve)；②泵控型(pump control system)——油泵和马达的容量是各不相同的。根据油泵和马达的构造又分为——轴向活塞式油泵(axial piston type oil pump)、径向活塞式油泵(redial piston type oil pump)、齿轮式油泵(gear type oil pump)、滑动叶片式油泵(sliding vane type oil pump)。

Deck officer 甲板部高级船员 系指符合海员培训、发证和值班标准国际公约第Ⅱ章规定的合格的高级船员。 Deck officer means an officer qualified in accordance with the provisions of chapter Ⅱ of the International Convention on Standards of Training, Certification and Watch-Keeping for seafarers.

Deck operated isolating valve 甲板操纵隔离阀 能在油船甲板上操纵其启闭的隔离阀。

Deck safety harness and safety line 甲板安全索具和安全索 为防止使用者落水，或当他掉入水中后为防止其与船或岸分离，容许把使用者牢固地系于船或岸上的一个强支撑点的装置。 Deck safety harness and safety line are the device that allows a user to be securely attached to a strong point on a vessel or on shore, so as to prevent him from falling into the water, or, if he does fall into the water, to prevent him from being separated from the vessel or shore.

Deck socket for rail stanchion (deck socket for stanchion) 栏杆插座 供活动栏杆柱插入固定用的底座或眼板。

Deck sprinkler system (deck sprinkling system) 甲板洒水系统 为避免油船在夏季航行太阳暴晒所引起的油气外逸和降低油舱温升而设置的冷却主甲板的洒水系统。

Deck structure 甲板结构 系指有加强筋、桁材和支撑支柱的甲板板材。 Deck structure means deck plating with stiffeners, girders and supporting pillars.

Deck transverse 甲板强横梁 系指包括相邻的甲板结构构件的构件。 Deck transverse means a structural member including adjacent deck structural members.

Deck wash pump (flushing pump) 甲板冲洗泵 抽送冲洗甲板用水的泵。

Deck washing piping system (wash deck pipe) 甲板冲洗系统 由总用泵或消防泵等供水至甲板各部作冲洗或消防用的系统。

Deck water discharge pipe (deck scupper pipe) 甲板排水管 用以疏排甲板上的积水、冲洗水和消防水的管路。

Deck water discharge valve (deck draining valve) 甲板排水阀 安装在甲板排水管路上，为防止舷外水倒灌的阀。

Deck wetness (shipping of water) 上浪 船舶在波浪中剧烈摇荡时波浪涌上甲板的现象。

Deck zone 甲板区域 对散货船而言，包括顶边舱的水平列板或基线以上 0.9D(如无顶边舱)所对应的水平面之上对船体梁强度起作用的所有下列项目：(1)强力甲板板；(2)甲板边板；(3)舷顶列板；(4)舷侧外板；(5)顶边舱斜板，包括水平和垂直列板；(6)与上述板材连接的纵骨 For bulk carriers, the deck zone includes all the following items contributing to the hull girder strength above the horizontal strake of the topside tank or above the level corresponding to 0.9D of the base line if there is no topside tank：(1)strength deck plating；(2)deck stringer；(3)sheer strake；(4)side shell plating；(5)top side tank sloped plating, including horizontal and vertical strakes；(6)longitudinal stiffeners connected to the above mentioned plating.

Deck-end roller (guide roller) 导向滚柱 装在甲板端部，带有柱状滚筒，用于导引缆绳转向的导缆器。

Deckhouse 甲板室 (1)是位于干舷甲板或以上的除上层建筑外的甲板建筑物；(2)在强力甲板以上其侧壁板距外板内侧大于4%宽度(B)的带甲板的结构；(3)位于干舷甲板上或以上的不属于上层建筑的甲板建筑物。 Deckhouse are：(1)a decked structure other than a superstructure, located on the freeboard deck or above；(2)a decked structure above the strength deck with the side plating being more than 4% of the breadth B inboard of the shell plating；(3)a decked structure other than a superstructure, located on the freeboard deck or above.

Declaration of conformity (DOC) 符合声明

Declaration of security 保安声明 系指船舶与作为其界面活动对象的港口设施或其他船舶之间达成谅解的书面记录，规定各自实行的保安措施。 Declaration of security means that written agreement reached between a ship and either a port or another ship with which they interfaces specifying the security measures that each will implement.

Decline meter 定位倾斜仪

Decommissioning 退役 系指在使用年限末期封闭船舶(平台)结构和拆除有害材料的过程。 Decommissioning means the process of shutting down a structure and removing hazardous materials at the end of its service life.

Decontamination room 洗消室 供船上人员冲洗被沾染有害物质和更换被沾染的衣着用的舱室。

Decoupled 不挂钩/解耦

Dedicated Clean Ballast Tank Operation Manual 专用清洁压载舱操作手册 系指每艘按73/78防污公约附则Ⅰ第13(10)条规定采用专用清洁压载舱的油船，均应备有一份详细说明该系统并规定操作程序的专用清洁压载舱操作手册。该手册应使主管机关满意，并应包括73/78防污公约附则Ⅰ第13A条(2)所述的技术条件中规定的全部资料。 Dedicated Clean Ballast Tank Operation Manual means that every oil tanker operating with dedicated clean ballast tanks in accordance with the provisions of regulation 13(10) of Annex I of MARPOL 73/78 shall be provided with a Dedicated Clean Ballast Tank Operation Manual detailing the system and specifying operational procedures. Such a manual shall be to the satisfaction of the Administration and shall contain all the information set out in the Specifications referred to in Paragraph 2 of Regulation 13A of Annex I of MARPOL 73/78.

Dedicated reserve power source 专用的备用电源 系指专用于VDR的带有合适的自动充电装置的蓄电池，其容量须符合，专用的备用电源，应使VDR在2h内连续记录驾驶室声音的要求。 Dedicated reserve power source means a battery, with suitable automatic charging arrangements, dedicated solely to the VDR, of sufficient capacity to operate as required, that the VDR should continue to record bridge audio from the dedicated reserve power source for a period of 2 hours.

Dedicated RO-RO ship 专用滚装船 系指专用于装运轿车、拖车等的滚装船。

Dedicated slop tank 专用污液舱

Deducted space 减除处所 系指吨位丈量中，包括在总吨位中，但不包括在净吨位中的各处所。

Deep diving system 深潜系统 又称"下潜式加压舱系统"。一种以下潜式加压舱为基地，潜水员可出舱到深海潜水作业的饱和潜水系统。其主要由下潜式加压舱、甲板加压舱、总操纵室、能源操纵室、生命支持系统、甲板吊放装置等组成。

Deep draft multi-spar(DDMS) 深吃水多用途柱体式平台

Deep draught 满载吃水 指舰船在各项工作准备完毕且全船定员、储备、燃料、淡水和有效载荷时的排水量所对应的吃水。 Deep draught is measured at a displacement such that the ship is in all respects complete, and is fully loaded with full complement, stores, fuel, water and payload.

Deep mixing ship 深层软地基固化船

Deep sea anchor 深水锚 适用于船舶在深水区停泊的备用锚。

Deep sea anchor gear 深水抛锚装置 由锚、锚链或锚索、锚绞车及其他属具所组成的在深水海域起、抛锚装置。

Deep sea anchor winch 深水抛锚绞车 船舶深水作业时起锚用的绞车。

Deep sea rescue vehicle 深潜救生艇 系指配置在援潜救生船上，装有能与失事潜艇的救生钟平台对接的特殊装置，专用于从失事潜艇中营救艇员集体脱险出水的小型潜水艇。

Deep sea trawl winch 深水底栖拖网绞车 用于深水底质取样和底栖生物拖网调查的绞车。

Deep sea vehicle (deep diving submersible) 深潜器 通常由一个充满轻液体的船形浮筒与可携带2~3人的耐压球所构成，下潜深度很大，排水量超过100 t的深海潜水船。主要用于海底地貌的观察和标本的采集、搜集沉船和打捞沉船等海底物体的工具。其外形像船，它的外壳是用高强度钢板制成的耐压球体，球体的下端有观察窗和机械手等操作装置。深潜器装有螺旋桨推进器，在深海中有一定的活动能力，深潜器自身通过一条缆绳与母船连接，通过缆绳向深潜器供电、供气和通信联络。

Deep sea voyage 深水航行 系指船舶航行在距岸200 n mile 以外并且水深超过500 m的海域。 Deep sea voyage means a voyage in which a ship navigates in waters 500 m or more in depth and 200 n mile or more away from shore.

Deep sea/pelagic fishing ship 深海/远洋渔船 系指无限航区、海域作业，垂线间长50m以上的大型渔船。见图 D-6。

图 D-6 深海/远洋渔船
Figure D-6 Deep sea/pelagic fishing ship

Deep submergence rescue vehicle 深潜救生艇 配

置在援潜救生船上，装有能与失事潜艇的救生钟平台对口衔接的特殊装置，专用于从失事潜艇中营救艇员集体脱险出水的小型潜水艇。

Deep tank 深舱 (1)在两层甲板间或外板/内底和以上或更高甲板间延伸的任何液舱;(2)用于装载水、燃油或其他液体的舱，其在货舱或甲板间构成船体的一部分。装载油的深舱如必要表示可明示为"深油舱(deep oil tank)";(3)双层底以外的压载舱、船用水舱、货油舱(例如植物油舱)及按闭杯试验法闪点不低于60 ℃的燃油舱等。 Deep tank is:(1)any tank which extends between two decks or the shell/inner bottom and the deck above or higher;(2)a tank used for carriage of water, fuel oil and other liquids, forming a part of the hull between containers or decks. The deep tanks used for carriage of oil are designed as "deep oil tanks", if necessary;(3)a tank other than the double bottom tanks, refer to ballast tanks, water tanks for ship's use, tanks carrying oil as cargo (e.g. vegetable oil), and oil fuel tanks carrying the oil having a flash point not lower than 60 ℃ (closed cup test).

Deep tank bulkheads 深舱舱壁 系指用于压载舱、日用水舱和燃油舱等的舱壁结构。 Deep tank bulkheads are the bulkhead structures for ballast tanks, daily service tanks and oil fuel tanks.

Deepest operating waterline (fishing vessel) 最深作业水线(渔船) 系指允许的最大容许营运吃水的水线。 Deepest operating waterline is the waterline related to the maximum permissible operating draught.

Deepest subdivision draught d_s 最深分舱水线(分舱载重线) 系指相应于船舶夏季载重线吃水的水线。 Deepest subdivision draught is the waterline which corresponds to the summer load line draught of the ship.

Deepest subdivision load line 最深分舱载重线 系指相应于勘定的船舶夏季吃水的分舱载重线。 Deepest subdivision load line is the subdivision load line which corresponds to the summer draught assigned to the ship.

Deeply submerged hydrofoil 深浸式水翼 翼航时水翼浸深大于弦长，水翼升力基本上不随浸深而变化，因此必须设置专门的控制装置才能保持艇的飞高及纵、横稳定性的水翼。

Deep-water anchor gear 深水抛锚装置 供船舶在水深大于150 m的海域锚泊的一种专门装置。通常由深水锚绞车、锚缆、导缆器、缓冲器、涂油器、导向滑轮及锚等组成。

Deep-water exploration vessel 深水勘察船 主要用于模拟工程物探调查作业、单电缆二维高精度数字地震调查作业、工程地质钻探作业、海底表层采样、大型海洋工程起吊作业、起降直升机等的工程船舶。

Deep-water offshore engineering 深海工程 包括无人深潜的深潜水器和遥控的海底采矿设施等建设工程。

Deep-water tumbler platform (DIP) 深水不倒翁平台 系指一种可采用干式采油树的新型深水浮式平台。DIP平台系统由甲板、浮体结构、系泊系统和立管系统组成。其浮体结构由主体结构和伸缩结构组成，它可视为在传统的深吃水环形浮箱半潜式平台基础上，通过伸缩立柱连接下部浮箱(lower tier pontoon/LTP);LTP采用的是正多边形结构，其下部设有与其连接为一体的垂荡板，垂荡板中间开孔，使油气生产立管及钻井从中间穿过。系泊系统采用传统锚泊方式，采用链-缆-链三段组合方式。

Deep-well pump 深井泵 液化气船普遍采用的货油泵。

Default 违约 是违反合同规定的行为。

Defeat device 抑制装置 对"73/78防污公约"附则Ⅵ而言，系指为激活、调整、推迟或阻碍激活排放控制系统的任何部件或功能而对操作参数(如:发动机速度、温度、进气压力或任何其他参数)进行测量、检测或响应的装置，从而在正常操作遇到的工况下降低排放控制系统的有效性，但在适用的排放发证试验程序中大量使用该装置者除外。 For the "MARPOL73/78" Annex Ⅵ, defeat device means a device which measures, senses, or responds to operating variables (e.g., engine speed, temperature, intake pressure or any other parameter) for the purpose of activating, modulating, delaying or deactivating the operation of any component or the function of the emission control system/such that the effectiveness of the emission control system is reduced under conditions encountered during normal operation, unless the use of such a device is substantially included in the applied emission certification test procedures.

Defect 缺陷 可以采用机加工(machining)、磨削(grinding)，或火焰气割(flame scarfing)，或火焰气刨(flame gouging)，或采用焊接的方法(welding)加以修补。

Defect type 缺陷种类 包括:(1)裂纹;(2)纵向裂纹;(3)横向裂纹;(4)末端弧坑裂纹;(5)孔穴;(6)延伸孔(气孔);(7)蛀孔;(8)弧坑管(末端弧坑穴);(9)夹渣;(10)金属夹杂物;(11)侧壁未熔合;(12)内层焊道未熔合;(13)根部未熔合;(14)未焊透(未完全焊透);(15)连续的咬边;(16)间断的咬边;(17)坡口收缩，根部坡口收缩;(18)对接焊过大的增强高;(19)角接焊过大的凸出;(20)过大的根部增强高;(21)边缘未对准;(22)焊穿;(23)不完全的熔敷坡口;(24)根部凹陷;(25)不良再引弧。 Defect type includes:(1)crack;(2)

longitudinal crack;(3) transverse crack;(4) end crater crack;(5) hole;(6) elongated cavity (gas pocket);(7) worm hole;(8) crater pipe (end crater cavity);(9) slag inclusion;(10) metallic inclusion;(11) lack of side-wall fusion;(12) lack of inter-run fusion;(13) lack of root fusion;(14) lack of penetration (incomplete penetration);(15) under cut, continuous,(16) undercut intermittent;(17) shrinkage groove, groove in the root;(18) excessive weld reinforcement (butt weld);(19) excessive convexity (fillet weld);(20) excessive root reinforcement;(21) misalignment of edges;(22) burn-through;(23) incompletely fillet groove;(24) root concavity;(25) poor restart.

Defense 答辩

Defense intelligence agency (DIA) 美国国防情报局 是美国情报界重要成员,同时也是国防部对外军事情报的主要生产者和管理者,负责对国家决策者、军事人员等情报用户提供服务的机构。

Defensive rope 拦阻索 用于吸收着舰飞机的动能,缩短其滑行距离的装置,是航母飞行系统的核心部件之一。一般来说,拦阻索的直径为 35 mm,以油麻绳为核心,围绕其以高强度钢丝编织而成,油麻绳浸润有润滑油可以润滑钢丝。

Defferred payment/Delay in payment 延期付款 系指大部分货款在交货后若干年内分期摊付。具体做法是:合同签订后,买方预付的一小部分货款做定金,也有合同规定按工程进度或交货进度支付一部分货款,其余大部分货款凭远期信用证支付,货物所有权一般在交货时转移。

Defined situations of hazard and accident 危险与事故状态定义 系指基于对事故大小、风险增加等意外事件的危险情况分析,对事故发生时可通过应急反应处理的一系列可能事件的选择。

Defined situations of hazard and accident 已定义危险与事故状态 根据作业的设计意外事件、涉及风险暂时增加的危险和意外情况以及范围较小的意外事件,作业中的应急准备应该能够处理的一些可能事件。

Definite time-delay relay or release 定时限过电流继电器或脱扣器 系指经一定延时后动作的过电流继电器或脱扣器,其延时动作时间可以调整,但不经受过电流值的影响。

Deflecter 导向机构 系指当用作舵的机构,其将从喷嘴喷出来的水导向左舷或右舷。Deflecter is the device serving as a rudder by leading the water injected from the nozzle either to port or to starboard.

Deflection of floating dock 浮船坞挠度 浮船坞受自重和外力作用后产生的纵向弯曲度。

Deflection plate type 折流板式

Deformation loads 变形载荷 由热载荷和残余应力所造成。其包括热载荷和因建造产生的变形。 Deformation loads are caused by thermal loads and residual stresses. They are grouped into thermal loads and deformations due to construction.

Defrost 融霜 用冷水、热排气、热盐水和电热等方法除去盘管表面霜层的过程。

Defrost receiver 融霜贮液器 融霜时贮存从盘管出来的液态制冷剂的容器。

Degraded condition 功能弱化状态 系指由于故障造成的系统功能的降低。 Degraded condition means reduction in system functionality resulting from failure.

Degree of admission 进气度 部分进气级中喷嘴所占的弧段长度与整个圆周之比。

Degree of expansion 胀管度 冷却水管在管板孔中胀开的程度。

Degree of hazard and safety label 危险程度和安全标记 在表 D-1 中规定了危险的程度和相应的安全标记。 Degree of hazard and corresponding safety label, as defined in Table D-1.

表 D-1 危险程度和相应的安全标记
Table D-1 Degree of hazard and corresponding safety labels

危 险 DANGER	表示存在极其实质性的危险,若不采取适当的补救措施,将导致高死亡率或无可挽回的伤害 Denotes that an extreme intrinsic hazard exists which would result in high probability of death or irreparable injury if proper precautions are not taken
警 告 WARNING	表示存在危险,若不采取适当的补救措施,可能导致伤害或死亡 Denotes that a hazard exists which can result in injury or death if proper precautions are not taken
注 意 CAUTION	表示对安全的提示,或者是对可能导致人员伤害或者损坏艇或部件的不安全的提示 Denotes a reminder of safety practices or directs attention to unsafe practices which could result in personal injury or damage to the craft or components

Degree of protection of enclosures 外壳防护等级

表示防护等级的标志,由特征字母 IP 及随后的两位数字("特征数字")构成,特征数字与对应的条件见表 D-2 和表 D-3。 Degree of protection of enclosures means that designation to indicate the degree of protection, consisting of the characteristic letters IP followed by two numerals (the "characteristic numerals") indicating conformity with the conditions stated in table D-2 and table D-3 below.

表 D-2 第一位特征数字表示的防护异物的等级
Table D-2 Degree of unusual body-protected indicated by the first characteristic numeral

第一位 特征数字	保 护 等 级	
	简述	定义
0	无防护	—
1	防护大于 50 mm 的固体物	人体某一大面积部分,如手(但不防护故意接近),直径超过 50 mm 的固体物
2	防护大于 12 mm 的固体物	手指或长度不超过 80 mm 的类似物体,直径超过 12 mm 的固体物
3	防护大于 2.5 mm 的固体物	直径或厚度超过 2.5 mm 的工具、线材等,直径超过 2.5 mm 的固体物
4	防护大于 1 mm 的固体物	厚度大于 1 mm 的线材或带材,直径超过 1 mm 的固体物
5	防尘	不能完全防止灰尘进入,但进入的灰尘数量不足以影响设备的良好运行
6	尘密	灰尘不能进入

表 D-3 第二位特征数字表示的防护水的等级
Table D-3 Degree of water-protected indicated by the second characteristic numeral

第二位 数字特征	保 护 等 级	
	简述	定义
1	无防护	—
2	防护垂直滴落的水滴	垂直滴落的水滴应无有害的影响
3	在外壳倾斜不超过 15 ℃ 时防护垂直滴落的水滴	当外壳偏离其法线位置倾斜不超过 15 ℃ 的任一角度时,垂直滴水应无有害影响
4	防淋	偏离垂线不超过 60 ℃ 的任一角度的淋水应无有害影响
5	防溅	从任何方向向外壳溅水应无有害影响
6	防喷	用喷嘴从任何方向向外壳喷水应无有害影响
7	防护强力喷水	强力喷嘴从任何方向对外壳喷水无有害的影响
8	防护短时浸水	当外壳在标准的压力及时间条件下短时浸入水中时,进入的水的数量应不可能引起有害影响。

续表 D-3

第二位数字特征	保 护 等 级	
	简述	定义
9	防护持续浸水	当外壳在制造厂和用户同意的条件下,但不比7更严酷的条件下持续浸入水中时,进入的水的数量应不可能引起有害影响

对表 D-2 和表 D-3 的注意:
(1)对于接近危险部分的防护等级,通过附加和/或补充字母表示,见 IEC 60529。
(2)对于旋转电机的防护等级,见 IEC60034—5。旋转电机—第5篇:旋转电机外壳防护分类(IP 代号)。
(3)第2个特征数字6,也包括对汹涌海浪的防护。

Degree of reaction 反动度 在转子中转换的能量与在整个级中转换的能量之比。

Delaminated protecting coating 可剥离防护涂料 系指利用涂层本身的机械隔阻作用、涂层中缓蚀剂及抗菌添加剂的联合作用,有效防止空气和水分子等渗入到物件表面,达到防腐、防锈、防酸碱、防擦伤及局部防火等效果,并具有临时保护作用的涂料。该涂料具备好的耐火焰、耐焊接飞溅、耐打磨火花、防污(粉尘、油污、污水)和耐磨等特性。其最大的特点就是可剥离性。它的涂膜除了具有普通涂膜一般特性,还具有一定的韧性和强度,并对其保护基面有合适的附着力与良好的可剥离性,能大面积地从基材上剥离。

Delay in payment 延期支付

Deliver goods 交货

Delivered at frontier-named place(DAF) 指定地点的边境交货 系指当卖方在边境的指定的地点和具体交货点,在毗邻国家海关边界前,将仍处于交货的运输工具上尚未卸下的货物交给买方处置,办妥货物出口清关手续但尚未办理进口清关手续时,即完成交货。

Delivered duty paid-named place of destination(DEQ) 完税后交货到指定目的港 系指卖方在指定的目的港码头将货物交给买方处置,不办理进口清关手续,即完成交货。卖方应承担将货物运至指定的目的港并卸至码头的一切风险和费用。DEQ 和 DIS 都是在目的港交货的术语,在如何安排途中的运输、保险以及做好货物的交接手续等问题上,两者的注意事项也基本相同。鉴于世界各国在使用 DEQ 术语时,对于由谁负责办理进口手续的问题,做法并不完全统一,因此,使用 DEQ 术语时,必须加以注意。《90 通则》规定,办理货物进口报关的责任、费用和风险由卖方承担。《2000 通则》中还强调指出,DEQ 术语仅仅适用于水上运输和多式联运方式。

Delivered Ex ship-named port of destination(DES) 指定目的港的船上交货 系指卖方负责租订运输工具,在规定的时间内将已清关货物运抵指定的目的港,在目的港船上交货并承担交货前的费用、风险的贸易术语。DES 不同于象征性交货的 CIF,它是实际交货,因此,卖方是为自己的利益办理运输和保险事宜。系指在指定的目的港,货物在船上交给买方处置,但不办理货物进口清关手续,卖方即完成交货。卖方必须承担货物运至指定的目的港卸货前的一切风险和费用。只有当货物经由海运或内河运输或多式联运在目的港船上交货时,才能使用该术语。如果当事各方希望卖方负担卸货的风险和费用,则应使用 DEQ 术语。

Delivered power 收到功率 螺旋桨收到的机械功率。装在船上或船模上的螺旋桨收到的功率为:$P_D = 2\pi n Q$,式中,n —转速;Q—螺旋桨转矩。有时对敞水中螺旋桨收到功率写为 $2\pi n Q_0$,其中:Q_0—敞水中转矩。

Delivery 交付 系指:(1)将货物交给收货人,(2)按照多式联运合同或者交付地法律或特殊贸易习惯,将货物置于收货人的支配之下;(3)根据交付地适用的法律或规定,将货物交给的当局或第三方。 Delivery means:(1)the handing over of the goods to the consignee;(2)the placing of the goods at disposal of the consignee in accordance with the multimodal transport contract or with the law or usage of the particular trade applicable at the place of delivery;(3)the handing over of the goods to an authority or other third party to whom, pursuant to the law or regulations applicable at the place of delivery, the goods must be banded over.

Delivery and acceptance tests 交接船试验 鉴定该船舶在建造完工后或作重大的修理后是否满足技术说明书规定的各项性能要求进行的试验。

Delivery order 提货单 又称小提单。收货人凭正本提单或副本提单随同有效的担保向承运人或其代理人换取的,可向港口装卸部门提取货物的凭证。发放小提单时应做到:(1)正本提单为合法持有人所持有。(2)提单上的非清洁批注应转上小提单。(3)当发生溢短短卸情况时,收货人有权向承运人或其代理获得相应的签证。(4)运费未付的,应在收货人付清运费及有关费用后,方可放小提单。

Delta internal combustion engine ▽型内燃机 三列对置活塞气缸排列成▽形,在三角形的三个角上的三根曲轴用齿轮连接到一根功率输出轴的内燃机。

Deluge system 雨淋系统 系指将处于开启状态的喷水器通过一个阀门与供水源相连接的喷水系统,管路上的阀门由与喷水器处于同一区域内的探测系统操纵开启。当阀门开启时,水流流入管路系统并通过装设在其上的所有喷水器将水排出。 Deluge system means a system employing open nozzles or sprinklers attached to a piping system connected to a water supply through a valve that opened by the operation of a detection system installed in the same area as the nozzles or sprinklers. When this valve opens, water flows into the piping system and discharges from all nozzles or sprinklers attached thereto.

Demand draft 票汇 系指银行开出汇票交汇款人自行邮寄给收款人或自行携带到指定付款行取款的业务。

Demand factor 需用系数 系指"总安装负荷功率"与连续负荷及间断负荷之和的"实际负荷功率"之比。该系等于负荷使用系数、负荷连续使用系数及负荷间断使用系数的乘积。

Demulsifier 破乳剂 系指能破坏乳浊液使其中的分散相凝聚析出的物质。破乳剂是一种表面活性物质,它能使乳化状的液体结构破坏,以达到乳化液中各相分离开来的目的。SP型破乳剂的主要组分为聚氧乙烯聚氧丙烯十八醇醚,理论结构式为 R(PO)$_x$(EO)$_y$(PO)$_z$H,式中,EO—聚氧乙烯;PO—聚氧丙烯;R—脂肪醇;x、y、z—聚合度。多支链的特点决定了 AP 型破乳剂具有较高的润湿性能和渗透性能,当原油乳状液破乳时,AP 型破乳剂的分子能迅速地渗透到油水界面膜上,比 SP 型破乳剂分子的直立式单分子膜排列占有更多的表面积,因而用量少,破乳效果明显。

Demurrage 滞期费 系指在约定的允许装卸时间内未能将货物装卸完,致使船舶在港内停泊时间延长,给船方造成经济损失,则延迟期间的损失,应按约定每天若干金额补偿给船方的费用。

Demurrage clause 滞期条款

Demurrage days 滞期期限

Denial of service(DoS) 拒绝服务/DOS 攻击 系指故意的攻击网络协议实现的缺陷或直接通过野蛮手段残忍地耗尽被攻击对象的资源,目的是让目标计算机或网络无法提供正常的服务或资源访问,使目标系统服务系统停止响应甚至崩溃,而在此攻击中并不包括侵入目标服务器或目标网络设备。这些服务资源包括网络带宽,文件系统空间容量,开放的进程或者允许的连接。这种攻击会导致资源的匮乏,无论计算机的处理速度多快、内存容量多大、网络带宽的速度多快都无法避免这种攻击带来的后果。攻击现象 DoS 攻击,是网络攻击最常见的一种。

Density 密度 系指某一货品的质量与其容积之比值(以 kg/m³ 表示)。该定义适用于液体、气体及蒸发气。 Density is the ratio of the mass to the volume of a product, expressed in terms of kilograms per cubic meter. This applies to liquids, gases and vapours.

Density of persons in an escape route 脱险通道人员密度 脱险通道内人员密度为人员数(p)除以根据设计安排有人员的处所有关的可利用的脱险通道面积,以 p/m^2 表示。 Density of persons in an escape route is the number of persons (p) divided by the available escape route area pertinent to the space where the persons are originally located, expressed in (p/m^2).

Density ρ_c(t/m^3) 所载散装干货的密度 ρ_c Density ρ_c is density of the dry bulk cargo carried.

Density ρ_L(t/m^3) 所载液体的密度 ρ_L Density ρ_L is density of the liquid carried.

Deoiler 凝水除油器 汽轮机动力装置中,为除去蒸汽辅机排汽和油舱加热蒸汽的凝水中油分而设置的器具。

Deoxidation treatment biotechnology 脱氧处理生物技术

Departure ballast 离港压载 系指除到港压载以外的压载。 Departure ballast means ballast other than arrival ballast.

Departure condition 离港工况 系指燃油舱不小于 95% 满,其他消耗备品 100% 满的工况。 Departure condition means a condition with bunker tanks not less than 95% full and other consumables 100%.

Dependent rise control 依赖性风险控制 系指风险控制的措施能够影响风险贡献树中的其他因素。

Deprecated interfaces 过时的接口

Depression 凹痕

Depressurized towing tank 减压船模试验水池 建在可降低空气压力的气密室内,主要用于模拟实船流场和空泡现象等的船模试验水池。

Depth 型深 系指:(1)在舯横剖面上自型基线量到最上层连续甲板横梁上缘的垂直距离。对于具有立龙骨的船舶型基线取自船底板上表面与立龙骨的交线；(2)是在船中横剖面上由型基线量至最上层连续甲板船舷处甲板横梁上缘的垂直距离。有圆弧形舷弧的船舶，D 应量至型甲板线的延长部分；(3)在船长 L 中点沿船舷从基线到最上层连续甲板横梁上缘处的垂直距离。在有效上层建筑处，为确定船体结构尺寸，型深应量至上层建筑甲板；(4)在船长 L 中点外从平板龙骨上缘量至船侧干舷侧甲板横梁上缘的垂直距离(m)。如水密舱壁延伸至干舷甲板以上的某一甲板，并在船舶录中记为有效舱壁情况下，则型深应量至该舱壁甲板。 Depth (D) is: (1) the distance, in m, measured vertically on the mid-ship transverse section, from the moulded base line to the top of the deck beam at side on the uppermost continuous deck. In the case of a ship with a solid bar keel, the moulded base line is to be taken at the intersection between the upper face of the bottom plating with the solid bar keel; (2) the distance, in m, measured vertically on the mid-ship transverse section, from the moulded base line to the top of the deck beam at side on the upper-most continuous deck. On vessels with a rounded gunwale, D is to be measured to the continuation of the moulded deck line; (3) the vertical distance, at the middle of the length L, from the base line to top of the deck beam at side on the uppermost continuous deck. In way of effective superstructures the depth is to be measured up to the superstructure deck for determining the ship's scantlings; (4) the vertical distance in meters at the middle of L measured from the top of keel to the top of the freeboard deck beam at side. Where watertight bulkheads extend to a deck above the freeboard deck and are to be registered as effective to that deck, D is the vertical distance to that bulkhead deck.

Depth 型深（内河船）D 对钢质船舶系指在船长中点处沿舷侧自平板龙骨上表面量至干舷甲板下表面的垂直距离。对甲板转角为圆弧形的船舶，应由平板龙骨上表面量至干舷甲板下表面的延伸线与舷侧板内缘延伸线的交点；对纤维增强塑料船舶系指船长中点处沿船侧自船底板外表面至干舷甲板上表面之间的垂直距离。

Depth（floating dock） 坞深 D_D（m） 系指在外坞墙板内缘位置上从坞底板上表面量至顶甲板下表面的垂直距离。

Depth charge magazine 深弹舱 供储存深水炸弹用的舱室。

Depth charge rack room 深弹滚架舱 供安置深水炸弹投掷架的舱室。

Depth controller 定深器 又称水鸟。是控制和调整海洋地震电缆工作深度的装置。它由受微电机控制的双翼、蓄电池、信号接收部分组成。

Depth for freeboard D 计算型深 系指：(1)船舯处型深加干舷甲板边板的厚度。如果露天干舷甲板有覆盖物，则加上 $[T(L-S)]/L$，式中，T 是甲板开口外露天覆盖层平均厚度；S 是上层建筑的总长；(2)对于圆弧形舷弧半径大于宽度(B)的4%或上部舷侧为特殊形状的船舶，计算型深 D 系取自一中央截面的计算型深，此截面的两舷上侧垂直并具有同样的梁拱，且上部截面面积等于实际的中央截面的上部截面面积。 (1) Depth for freeboard D is the moulded depth amidships, plus the thickness of the freeboard deck stringer plate, where fitted, plus $[T(L-S)]/L$, if the exposed freeboard deck is sheathed, where: T is the mean thickness of the exposed sheathing clear of deck openings, and S is the total length of superstructures; (2) the depth for freeboard D in a ship having a rounded gunwale with a radius greater than 4 percent of the breadth B or having topsides of unusual form is the depth for freeboard of a ship having a mid-ship section with vertical topsides and with the same round of beam and area of topside section equal to that provided by the actual mid-ship section.

Depth for strength computation Ds 强度计算用型深 Ds 系指在 L 中点外，对上层建筑甲板为强力甲板的地方，从龙骨上缘量至上层建筑甲板，对无上层建筑的地方，量至船侧干舷甲板横梁上缘的垂直距离(m)。如该甲板不延伸至 L 的中点，则型深量自某一假想甲板线，该假想甲板线延伸至 L 中点并平行于强力甲板。 Depth of ship for strength computation D is the vertical distance (m) from the top of keel to the top of beam at side of the superstructure deck at the middle of L, for the part where the superstructure deck is strength deck, or the freeboard deck for other parts. Where the deck does not cover midship, the depth is to be measured at the imaginary deck line which is extended to the middle of L along the strength deck line.

Depth of hatch-board 舱盖板主梁深度 每块舱盖板跨度中点处的高度。

Depth of hatch-cover 舱盖深度 舱盖两侧边缘的外形高度。

Depth of hull 平台型深 钻井平台基线至上甲板之间的高度。

Depth of ladder 梯步深度 两个相邻梯步的同侧边缘的水平距离。

Depth of lifeboat 救生艇型深 在艇中横剖面上，从艇壳板内缘的最低点量至舷侧外板水密部分的上缘的垂直高度。但在计算艇的立方容积时，所用的艇深最大不得超过艇宽的45%。

Depth of mat　沉垫型深　沉垫由底至顶的高度。

Depth of water　水深　系指海图深度。Depth of water means the charted depth.

Depth of yacht $D(m)$　型深(游艇)　系指在艇长 L 中点处，沿舷侧从平板龙骨上缘量至甲板(甲板艇)横梁上缘或舷侧板顶端(敞开艇)的垂直距离；对纤维增强塑料艇，由平板龙骨下表面量至干舷甲板(甲板艇)上缘或舷侧板顶端(敞开艇)的操作距离。

Depth sounder　回声测深仪　(1)测量超声波信号自发射经水底反射至接收的时间间隔，用以确定水深的一种水声仪器。其用途有：船舶在不明的海域或水道航行时，用测量水深确保船舶航行安全；在能见度不良或其他导航仪器失效时，用测量水深来辨认船位；对海域的水深进行精密测量，提供确保船舶安全航行的水深资料；(2)是以回声方式探测船舶下方水深和水中障碍物(如沉船、暗礁或浅滩)并把探测结果记录或显示出来的仪器。

Depth temperature meter winch　深温计绞车　为收、放深度温度计(BT绞车)测量水深、水温进行水文观测工作的绞车。

Depth(SWATH)　型深 D (小水线面双体船)　系指片体的下潜体纵中剖面处的最低点量至干舷甲板边线的垂直距离。

Derrick　吊杆装置　包括各种支承结构和定位设施，诸如桅、起重柱、稳索、吊耳、鹅颈通气管、吊环螺栓、顶牵索、牵索、辅助索和绞车。Derrick includes the supporting structure and positioning devices, such as mast, king post, Sampson post, stay, lugs, goosenecks, eyebolts, topping lift, guys, preventers and winches.

Derrick boom　吊杆　在吊杆装置中，用以支承吊货滑车的支撑杆件。

Derrick crane　吊杆起重机　系指带有操作绞车和两个横移滑车的吊杆，其设计使该吊杆在吊重时可以变幅和旋转。Derrick crane means a derrick fitted with operating winches and two span tackles of such design that the derrick can be topped and slewed while hoisting a load.

Derrick hand　吊杆箍　位于吊杆头端处，供连接吊货索具、千斤索具和牵索索具用的金属箍。

Derrick head span block　上千斤滑车　位于吊杆头端处的千斤滑车。

Derrick heel eye　吊杆叉头　焊装于吊杆筒体根部，用以与吊杆座连接的叉头状眼板。

Derrick mast　起重桅　位于船体中线面上兼作起重柱用的桅。

Derrick mobile structure　井架移动机构　专设置在井架底部，用以驱动井架和移动底座使之达到钻井作业要求的移动机构。

Derrick post　起重柱　用以支索吊杆装置其上端可系固千斤索具的柱子。

Derrick rest　吊杆托架　吊杆不工作时，用以支托和紧固吊杆的支架。

Derrick rig　吊杆装置　由吊杆、起重柱、起货索具以及绞车等部件所组成的一种起货设备。

Derrick rig system　吊杆装卸　用吊杆装置进行的吊装装卸。主要分单杆和双杆两种。

Derrick rigging　起货索具　起货设备中所有索具的总称。在吊杆装置中则是吊货索具、千斤索具、牵索索具的统称。

Derrick system　吊杆式　分为吊杆起重机式和摆动吊杆式两种。

Derrick systems　悬臂起重机设备　系指通过旋吊货物的方法从悬臂起重机吊杆到悬臂起重机支柱卸载货物的设备。包括以下(1)、(2)和(3)所指定的设备：(1)顶部滑车的末端被固定后，在悬臂起重机顶端处两根钢缆被分别独立地缠绕在绞盘上将吊杆水平转动，简称为旋转悬臂起重机设备；(2)两个悬臂起重机吊杆在预先确定的位置成对地在左舷和右舷处被固定，为了装卸货物两个悬臂起重机的货物下降并连接在一起。简称为组合-滑车悬臂起重机系统。(3)货物下降能够被缓缓放松或被举起，而且当货物被悬吊起来的时候能够单个或同时摆动和回转悬臂起重机吊杆。简称为桅杆起重机系统。Derrick systems are installations for handling cargo by suspending the cargo from the top of the derrick boom fitted to derrick post or mast, including those specified below: (1) the end of topping lift being fixed, two guy ropes fitted at the top of the derrick boom are wound by independent winches respectively to swing the boom horizontally (hereinafter referred to as "swinging derrick system"); (2) two derrick booms, on port and starboard sides, in pair are fixed at predetermined positions. Two derricks for cargo falls are connected to load or unload the cargo (hereinafter referred to as "union-purchase derrick system"); (3) the cargo fall can be paid out or heaved in and luffing and slewing of derrick boom can be carried out singly or simultaneously while the cargo is suspended (hereinafter referred to as "derrick crane system").

Derrick table　吊杆台　连接在起重桅下部两侧，用以安装及支承吊杆座的小平台或横向构件。

Descaling　除垢　对附着在传热面上的污垢的清除。

Descending and ascending stability of a unit　平台

的沉浮稳性 系指平台通过压、卸载而引起的自身浮态的变化,以平稳而又有控制地由漂浮状态过渡到坐底状态,或由坐底状态过渡到漂浮状态的能力。 Descending and ascending stability of a unit is the ability of that unit to move smoothly in a controlled manner from floating condition to resting condition or vice versa, by changing its buoyancy through ballasting and deballasting.

Description of imperfections 未熔合 如果有几段未熔合 $h_1, h_2, h_3, \cdots$,则其总和为: $\sum h = h_1 + h_2 + h_3 + \cdots$。 Where there are several instances of lack of fusion $h_1, h_2, h_3, \cdots$, the sum is $\sum h = h_1 + h_2 + h_3 + \cdots$

Design 设计 (1)对质量管理体系而言,系指制造厂商应制定并执行与从事设计工作水平相适应的设计管理制度。文件规定的设计程序如下:①指明制造厂商执行设计的机构包括对该部门的指示,保证设计前的准备工作和随后的校核工作,有序和顺利地进行;②对所有设计的变化和改进,做出验证、归档和认可的规定;③对于不完善的,不明确的或矛盾的要求,提出处理的方法;④明确设计所需的资料,例如数据的来源,优先采用的标准件或材料以及设计资料,提出选用资料的程序,并交制造厂商审查其适用性;(2)所有描述产品的性能、安装和制造工艺的相关图纸、文件和计算报告。 (1) For the quality management system, design means that the manufacturer to establish and maintain a design control system appropriate to the level of undertaken design. Documented design procedures are to be established which include: ①identify the design practices of the manufacturer's organization including departmental instructions to ensure the orderly and controlled preparation of design and subsequent verification; ②make provisions for the identification, documentation and/appropriate approval of all design changes and modifications; ③prescribe methods for resolving incomplete, ambiguous or conflicting requirements; ④identify design inputs such as sources of data, preferred standard parts or materials and design information and provide procedures of their selection and submit fot the manufacturer to review the adequacy; (2) all relevant drawings, documents and calculation reports describing the performance, installation and manufacturing technologies of products.

Design accidental event (dimensioning accidental event) 设计意外事件 为满足已定义的风险接受准则而用于海上设施布局、定尺寸和使用以及全部作业基础的意外事件。

Design accidental load (dimensioning accidental load) 设计意外载荷 功能或系统在所需时间区间内为满足已定义的风险接受准则而应承受的最大意外载荷。

Design and manufacturing technologies for computer CPU 计算机CPU设计制造技术 该技术拥有国:美国(inter、AMD)、中国(龙芯、申威、飞腾)、中国台湾(威盛)。见图D-7。

图 D-7 计算机 CPU 设计制造技术
Figure D-7 Design and manufacturing technology for computer CPU

Design approval 设计认可 系指船级社准予设计在特定条件下适用于规定用途的认定过程,一般包括图纸审查和原型/形式试验(如适用时)。 Design approval means the process whereby permission is granted by a society for the design to be used for a stated purpose under specific conditions, generally comprising drawing approval and prototype/type test(if applicable).

Design breaking load 设计破断载荷 部件的设计破断载荷是由原型试验确定。设计破断载荷不应超过试验在发生破坏前最后记录的载荷数据。

Design cases of position mooring system 定位系泊系统的设计工况 系指下述设计工况:(1)完整作业工况——在规定的作业环境条件下,系泊平台能进行预定作业而不使平均偏移及锚索张力超过规定值。作业工况视平台的用途可分为钻井作业工况和/或生产作业工况;(2)完整自存工况——在规定的自存环境条件下,系泊平台的最大偏移及锚索张力不超过规定值;(3)破损作业工况——系泊系统中任一根锚索失效时的作业工况;(4)破损自存工况——系泊系统中任一根锚索失效时的自存工况。 Design cases of position mooring system mean the following design cases: (1) intact operating case—The moored unit is in an intended operating mode with mean offset and anchor line tension not exceeding specified limits, subject to specified operating environmental conditions. The operating mode may be drilling operation and/or production operation, depending on the purpose of the unit; (2) Intact survival case—maximum offset and anchor line tension of the moored unit do not exceed specified limits, subject to specified operating environmental conditions; (3) Damaged operating case—working condition under failure of any single anchor line in the mooring system; (4) Damaged survival case—survival condition under failure of any single anchor line in the mooring system.

Design category 设计类别 可适用于对小艇进行评定的海浪和风的状态的描述。设计类别定义的汇总见表D-4。Design category is the description of the sea and wind conditions for which a boat is assessed to be suitable. Summary of design category definitions is seen in Table D-4.

表 D-4 设计类别定义的汇总
Table D-4 Summary of design category definitions

设计类别 Design category	A	B	C	D
波高小于或等于 Wave height up to	有义波高7 m approx. 7 m significant	有义波高4 m 4 m significant	有义波高2 m 2 m significant	有义波高0.5 m 0.5 m significant
典型的蒲氏风级 Typical Beaufort wind force	≤10	≤8	≤6	≤4
计算风速/(m/s) Calculation wind speed /(m/s)	28	21	17	13

Design category A(category for "ocean" sailing) 设计类别A("远洋"航行的类别) 系指设计用于在7m及以下的有义波高和蒲氏风级为10级或以下航行，且在更恶劣的海况下能生存的船艇。在远程航线上可能遇到这些海况，例如在横渡远洋，或在对风浪无遮蔽的几百海里的近海。假定风力为28m/s的狂风。Design category A is considered to be designed to operate for waves of up to 7 m significant height and winds of Beaufort force 10 or less, and to survive in more severe conditions. Such conditions may be encountered on extended voyages, for example voyages across oceans, or inshore when unsheltered from the wind and waves for several hundred nautical miles. Winds are assumed to gust to 28 m/s.

Design category B(category for "offshore" sailing) 设计类别B("近海"航行的类别) 系指设计用于在4 m及以下的有义波高和蒲氏风级为8级或以下航行的船艇。这些海况可在远距离的近海航行或在不可能总是立即地得到遮蔽的沿海中遇到。这样的海况也可在出现足够大的波高的内陆海中遇到。假定风力为21 m/s的烈风。Design category B is considered to be designed for waves of up to 4 m significant height and a wind of Beaufort force 8 or less. Such conditions may be encountered on offshore voyages of sufficient length or on coasts where shelter may not always be immediately available. These conditions may also be experienced on inland seas of sufficient size for the wave height to be generated. Winds are assumed to gust to 21 m/s.

Design category C(category for "inshore" sailing) 设计类别C("沿海"航行的类别) 系指设计用于在2 m及以下的有义波高和典型稳态风力为蒲氏6级或以下航行的船艇。这样的海况可在通海的内陆水域、河口海湾和中等气候条件的沿海水域中遇到。假定风力为17 m/s的大风。Design category C is considered to be designed for waves of up to 2 m significant height and a typical steady wind force of Beaufort force 6 or less. Such conditions may be encountered on exposed inland waters, in estuaries, and in coastal waters in moderate weather conditions. Winds are assumed to gust to 17 m/s.

Design category D (category for "sheltered waters") 设计类别D("遮蔽水域"中航行的类别) 系指设计用于在偶发波高为0.5 m和典型稳态风力为蒲氏4级或以下航行的船艇。这样的海况可在遮蔽的内陆水域和在良好气候的沿海水域中遇到。假定风力为13 m/s的强风。Design category D is considered to be designed for occasional waves of 0.5 m height and a typical steady wind force of Beaufort force 4 or less. Such conditions may be encountered on sheltered inland waters, and in coastal waters in fine weather. Winds are assumed to gust to 13 m/s.

Design condition 设计工况 系指证实设计在一定时间间隔内未超过相关的极限状态的一组物理状态。Design condition is a set of physical situations occurring over a certain time interval for which the design needs to dem-

Design conditions 设计条件(温度) 系指集装箱的最低设计内部温度以及货舱的设计最高温度。 Design conditions mean the lowest internal container design temperature and the maximum hold space design temperature.

Design current velocity 设计流速 系指在平台作业海区范围内可能出现的最大流速值,包括潮流流速、风暴涌流速和风成流流速。应考虑作业海区流速的垂向分布。在波浪存在时,应对无波浪时的流速垂向分布进行修正,以使瞬时波面处的流速保持不变。当 $z < h_0$ 时, $V = V_t + V_s + V_w [(h_0 - z)/h_0]$, m/s, 当 $z > h_0$ 时, $V = V_t + V_s (m/s)$;式中,各字母的含义见图 D-3。 Design current velocity is the maximum possible current velocity in the sea area where the unit is to operate. The current velocity is to include components of tidal current, storm surge current and wind driven current. Consideration is to be given to the vertical distribution of current velocity in the sea area where the unit is to operate. In the presence of waves, the current velocity profile in still water is to be modified such that the current velocity at the instantaneous free surface is a constant. $V = V_t + V_s + V_w [(h_0 - z)/h_0] (m/s)$, for $z \leq h_0$; $V = V_t + V_s (m/s)$; for $z > h_0$, where: those letters are defined in Figure D-3.

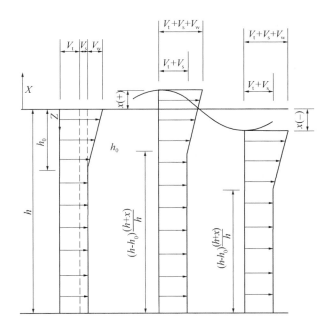

图 D-8 设计流速
Figure D-8 Design current velocity

图中: V—设计海流流速 design current velocity; V_t—潮流流速 tidal current velocity; V_s—风暴涌流速 storm surge current velocity; V_w—风成流流速 wind driven current velocity; h_0—风成流的参考水深 reference depth for wind driven current; z—水质点在静水面以下的垂直距离 vertical distance to water particle below still water level; h—静水水深 still water depth.

Design draft 设计吃水(小水线面双体船) 系指船舶静浮于水面时,沿设计水线在船中处量得的刚性水密船体的型吃水。

Design draught 设计吃水 可由舰船在满载状态加上其他指定边界条件的水线确定。在特定环境下,舰船的操作或船东规格书中可要求使用较高的水线。 Design draught may be determined from the waterline when the vessel is in a deep condition plus any specified margins.

In special circumstances, the operation of the vessel or the owner's specification may require that a higher waterline be used.

Design elevation capability 设计举升能力 系指在正常使用状态下,浮船坞举升的最大船舶重力。

Design force T 设计拖力 T 相当于由船东规定的拖索的拉力(或如拖索拉力不明则采用系柱拉力)。可以通过系柱拉力试验来验证设计拖力。 Design force T corresponds to the towrope pull (or the bollard pull, if the towrope pull is not defined) stipulated by the owner. The design force may be verified by a bollard pull test.

Design GM value 设计初稳性高度 用于集装箱绑扎计算的 GM 值由设计方确定,并在集装箱系固手册中明确。至少应包含最小值和最大值两个值。最小值由不小于装载最多集装箱数量工况时的 GM 值,最大值应不小于所有预期装载集装箱工况时的最大值。在初步设计时,设计初稳性高度的最小值可取为 $0.025B$,最大值可取为 $0.075B$,其中 B 为船舶的宽度。

Design hazard 设计危险 系指在设计阶段确定的工程危险性描述。 Design hazard means an engineering description of a hazard which is identified at the design stage.

Design load[1] 设计载荷(钻井船) 通常包括环境载荷、作业载荷和部件自重等。(1)环境载荷主要考虑下列几类:船体运动(升沉、纵摇、横摇、漂移等)、风载荷、温度和湿度、冰雨雪等。船体运动载荷对钻井系统的影响主要体现在船体倾角及加速度上,在船体主尺度确定后可以根据不同工况预报出风浪流等环境条件下对应的船体倾角和船体某些典型位置的运动加速度等浮体特性数据;风载荷对钻井系统的影响主要考虑处于船体外部、甲板以上的系统和设备(如:井架及井架内设备、管子处理设备等),船体或舱室之内的设备(如:泥浆泵、配浆设备等)不受风载荷的影响;温度和湿度影响钻井系统的各个系统,对材料的选取和共用系统的设计影响较大,温度和湿度的设定要与钻井船作业区域相适应;若钻井船作业于冰、雪区域,各个工况的冰雪载荷需要明确定义。(2)作业载荷主要是大钩载荷、转盘载荷、隔水管张紧载荷等。

Design load[2] 设计载荷(kN/m²) 系指:(1)船体局部强度计算中,构件单位面积上承受正压力的设计值;(2)为计算船体结构强度而取的作用载荷。(3)使用载荷乘以安全系数所得的载荷。安全系数根据使用载荷的可靠程度选定,一般取 1.5~2.0。

Design load combinations 设计载荷组合 由许多不同类型的载荷组成 诸如:营运载荷、环境载荷、意外载荷和变形载荷。 Design load combinations are composed of many different types of loads, such as operational loads, environmental loads, accidental loads and deformation loads.

Design loading conditions 设计装载工况 对于散货船而言,设计装载工况包括:(1)BC-C——最大吃水时的均装载工况,系指所有货舱货物密度相同,100%装满至舱口,所有压载舱为空舱;(2)BC-B——按 BC-C 的要求,加上最大吃水时的均匀装载工况,货物密度为 3.0 t/m³,所有货舱装货且装载比值相同(货物质量/货物舱体积),所有压载舱为空舱。如船舶在该设计工况中货物密度小于 3.0 t/m³,则货物密度用设计允许的最大货物密度,并应在协调附加标志 后面的最大货物密度附加标志注明:(maximum cargo density x. y t/m³);(3)BC-A——按 BC-B 的要求,加上最大吃水时,有指定的空舱组,所有装货舱中货物密度为 3.0 t/m³,且装载比值相同(货物质量/货物舱体积),所有压载舱为空舱。但是,该设计工况的指定空舱组应在协调附加标志后面的附加标志注明:(Holds Nos. … may be empty)。如船舶在该设计装载工况中货物密度小于 3.0 t/m³,则货物密度用设计允许的最大货物密度,并应在协调附加标志后面注明:(Holds Nos. … may be empty, with max cargo density… t/m³)。 For bulk carriers, design loading conditions includes:(1)BC-C—homogeneous cargo loaded condition where the cargo density corresponds to all cargo holds, including hatchways, being 100% full at maximum draught with all ballast tanks empty;(2)BC-B—as required for BC-C, plus: homogeneous cargo loaded condition with cargo density 3.0 t/m³, and the same filling rate (cargo mass/hold cubic capacity) in all cargo holds at maximum draught with all ballast tanks empty. In cases where the cargo density applied for this design loading condition is less than 3.0 t/m³, the maximum density of the cargo that the vessel is allowed to carry is to be indicated with the additional notation (max cargo density… t/m³);(3)BC-A—as required for BC-B, plus: at least one cargo loaded condition with specified holds empty, with cargo density 3.0 t/m³, and the same filling rate (cargo mass/hold cubic capacity) in all loaded cargo holds at maximum draught with all ballast tanks empty. The combination of specified empty holds shall be indicated with the annotation (Holds Nos. … may be empty). In such cases where the design cargo density applied is less than 3.0 t/m³, the maximum density of the cargo that the vessel is allowed to carry shall be indicated within the annotation, e. g. (Holds Nos. … may be empty, with max cargo density… t/m³).

Design loading of the unit 平台设计载荷 系指对于每种设计工况所考虑的静载荷和静载荷与环境载荷相组合的载荷。例如:(1)在正常作业工况时考虑:①静载荷包括平台重力、所有固定装置、供应品和压载重力以及作业载荷;②环境载荷取操作手册中正常作业允许的最大风、波浪、海流要素或载荷以及海床支承力。(2)在迁移工况时考虑:①静载荷包括迁移时平台重力、所有固定装置、供应品和压载重力以及浮力;②环境载荷:(a)油田内迁移:风速不小于 36 m/s(70 kn),波浪载荷按油田内迁移环境条件确定波浪要素,且与海流按最不利情况进行组合;(b)远洋迁移:风速不小于 51.5 m/s (100 kn),波浪载荷按最大迁移环境条件确定波浪要素,且与海流载荷按最不利情况进行组合;③应考虑平台倾斜和运动加速度以及拖缆拉力的影响。(3)在升降工况时考虑:①静载荷为升降平台主体或桩腿时的平台重力、固定装置、供应品和压载重力等;②环境载荷取操作手册中规定的允许升降平台主体或桩腿时的最大风、浪和海流要素或载荷。(4)在自存工况时考虑:①静载荷为适应自存状态的平台重力、固定装置、供应品和压载重力等;②环境载荷取操作手册中规定的平台自存时的最大风暴条件。 Design loading of the unit, for each design condition, are static loading and combined loading of static loading and environmental loading taken into account: (1) for normal operating condition: ①static loading includes the weight of the unit, the weight of all permanent installations and supplies, ballast weight and operational loads; ② environmental loading consists of the maximum wind, wave and current elements or loading and seabed supporting capabilities as specified for normal operations in the operating manual. (2) for transit condition: ①static loading includes the weight of the unit, the weight of all permanent installations and supplies, ballast weight and buoyancy during transit; ② environmental loading: (a) field transit condition: wind velocity is to be not less than 36 m/s (70 kn), elements of wave loading are to be determined according to environmental conditions for field transit and combined with current loading under the most unfavorable condition; (b) ocean transit condition: wind velocity is to be not less than 51.5 m/s (100 kn), elements of wave loading are to be determined according to extreme environmental conditions for field transit and combined with current loading under the most unfavorable condition; ③the effects of the inclination and acceleration of the unit and towing forces are to be taken into account. (3) for jacking condition: ①static loading includes the weight of the unit, the weight of all permanent installations and supplies, ballast weight during elevating or lowering the hull or legs; ②environmental loading consists of the maximum wind, wave and current elements or loading as specified for elevating or lowering the hull or legs in the op-erating manual. (4) for survival condition: ①static loading includes the weight of the unit, the weight of all permanent installations and supplies, ballast weight for achieving survival; ②environmental loading is to comply with the most severe storm conditions as specified for survival in the operating manual.

Design margin 设计裕度 是一个用于设计时重力估算误差的余量。 Design margin is an allowance for uncertainties used in the estimation of weight for design purposes.

Design operating draught 设计营运吃水 系指设计者规定且用于结构尺度计算的吃水。 Design operating draught is draught specified by the designer and is to be used to derive the appropriate structural scantling.

Design patent/Industrial design 外观设计专利 系指对产品的形状、图案、色彩或者其结合所做出的富有美感并适用于工业上应用的新设计。外观设计系指工业品的外观设计,也就是工业品的式样。外观设计专利办理可由咨询公司进行办理。该公司可以是一家专为中小微企业做商务服务的一站式 O2O 平台。

Design pressure[1] 设计压力(稳性) 系指完整稳性和破损稳性计算所假定的各个水密结构或设备按设计所应承受的静水压力。 For stability calculations, design pressure means the hydrostatic pressure which each watertight structure or appliance is to withstand and is assumed in the intact and damage stability calculations.

Design pressure[2] 设计压力(设备) 系指:(1)工作条件下可能遇到的最大工作压力或系统中任何安全阀或压力释放装置的最高设定压力(如装有这种装置);(2)用于确定零部件尺寸进行计算用的压力和零部件的许用的最大工作压力。但该设计压力应不小于任何安全阀的最大设定压力。 Design pressure means: (1) the maximum working pressure which is expected under operation conditions or the highest set pressure of any safety valve or pressure relief device on the system, if fitted; (2) a pressure used in the calculations made to determine the scantlings of such component and is the maximum permissible working pressure to the component. However, the design pressure is to be not less than the highest set pressure of any relief valve.

Design pressure[3] 设计压力(零部件) 系指用于确定零部件尺寸进行计算用的压力和零部件的许用的最大工作压力。但该设计压力应不小于任何安全阀的最大设定压力。 Design pressure is a pressure used in the calculations to determine the scantlings of such component and is the maximum permissible working pressure to the component. However, the design pressure is to be not less than the highest set pressure of any relief valve.

Design pressure [4] (pipeline and piping)　设计压力（管道和管系）　系指:(1)管系的设计压力由制造厂商考虑并决定系统部件的构件尺寸。它应不小于管系最大工作压力或安全阀和释放阀中最高设置压力,取其中的大者;(2)锅炉给水系统的设计压力不小于锅炉设计压力的1.25倍或给水系统中最大压力取其中的大者;(3)在减压阀前的(高压侧)蒸汽管系设计压力应不小于锅炉或过热器安全阀的设计压力;(4)在减压阀低压侧且没有安全阀门的管系设计压力不小于减压阀高压侧的最大压力;(5)泵或压缩机出口处管系的设计压力,对于容积式泵不小于安全阀的设计压力或对于离心泵不小于特性(压头-排量)曲线的最高值,取其中的大者。
(1) The design pressure of a piping system is the pressure considered by the manufacturer to determine the scantling of the system components. It is to be taken not less than the maximum working pressure expected in this system or the highest setting pressure of any safety valve or relief device, whichever is the greater; (2) The design pressure of a boiler feed system is to be not less than 1.25 times the design pressure of the boiler or the maximum pressure expected in the feed piping, whichever is the greater; (3) The design pressure of steam piping located upstream of pressure reducing valves (high pressure side) is to be not less than the setting pressure of the boiler or superheater safety valves; (4) The design pressure of a piping system located on the low pressure side of a pressure reducing valve not is to be less than the maximum pressure on the high pressure side of the pressure reducing valve; (5) The design pressure of a piping system located on the delivery side of a pump or a compressor is to be not less than the setting pressure of the safety valve for displacement pumps or the maximum pressure resulting from the operating (head-capacity) curve for centrifugal pumps, whichever is the greater.

Design pressure for piping　管系的设计压力　系指工作条件下可能遇到的最大工作压力或系统中任何安全阀或压力释放装置的最高设定压力(如装有这种装置)。　Design pressure for piping means the maximum working pressure which is expected under operation conditions or the highest set pressure of any safety valve or pressure relief device on the system, if fitted.

Design pressure[5] (boiler)　设计压力(锅炉)　设计压力系指锅炉和压力容器的最大许用工作压力,锅炉和压力容器的强度计算应以设计压力为依据,并应不小于任一个安全阀的最高设定压力。经济器的设计压力应为经济器内介质的最大工作压力。　Design pressure is the maximum permissible working pressure for boilers and pressure vessels. Strength calculations for boilers and pressure vessels are to be based on the design pressure and to be not less than the highest set pressure of safety valve. The design pressure of economizers is the maximum working pressure of the medium in the economizer.

Design pressure[6] (piping and component)　设计压力(管系和零部件)　(1)就管系而言,系指管内介质的最高工作压力,并应不小于下述压力:①对装有泄压阀或过压保护装置的管系,其设计压力应取泄压阀或过压保护装置的设定压力。但对与锅炉连接的蒸汽管系或与压力容器连接的管系,其设计压力应取锅炉或压力容器的设计压力;②对泵的排出口的管路,其设计压力应取当泵在额定转速运行并关闭释放阀时泵的释放压力。但对装有泄压阀或过压保护装置的泵,其设计压力应取其设定压力;③对于锅炉给水泵和锅炉给水止回阀之间的给水管,设计压力应取锅炉设计压力1.25倍的压力或关闭给水泵释放阀时泵的释放压力中较大的压力;④对于装有减压阀的管路,其低压侧如未装有泄压阀,设计压力则应取有泄压阀高压侧的设计压力;⑤对于锅炉吹泄管,其设计压力应取锅炉设计压力1.25倍的压力;⑥对于制冷装置的冷媒管,系指第9篇第1章102.5节中所定的压力;⑦燃油管应符合下列要求:(a)当工作压力不大于0.7 MPa,工作温度不高于60 ℃时,取0.3 MPa或最高工作压力中较大者;(b)当工作压力不大于0.7 MPa,工作温度超过60 ℃时,取0.3 MPa或最高工作压力中的较大者;(c)当工作压力超过0.7 MPa,工作温度不高于60 ℃时,取最高工作压力;(d)当工作压力超过0.7 MPa,工作温度超过60 ℃时,取1.4 MPa或最高工作压力中的较大者;⑧如无法适用本项规定时,取船级社认可的压力。(2)就零部件而言,系指用于确定零部件尺寸进行计算用的压力和零部件的许用的最大工作压力。但该设计压力应不小于任何安全阀的最大设定压力。
(1) For piping, design pressure is the maximum working pressure of a medium inside pipes and is to be not less than the following pressures given in ① to ⑧ : ①For piping fitted with a relief valve or other over-pressure protective device, the design pressure is based on the set pressure of the relief valve or over-pressure protective device. However, for steam piping connected to the boiler or piping fitted to the pressure vessel, it is based on the design pressure of the boiler or pressure vessel; ②For piping on the discharge side of the pumps, the design pressure is based on the delivery pressure of the pump with the valve on the discharge side closed while the pump is running at rated speed. However, for pumps

having a relief valve or a over-pressure protective device, the pressure based on its set pressure; ③For feed water piping on the discharge from the feed water pumps to feed water check valves, the design pressure is the 1.25 times the design pressure of the boiler or the pump pressure against a shut valve, whichever is the greater; ④For piping with relief valves if there is no pressure reducing valve on the low pressure side, it is based on the design pressure on the high-pressure side with pressure reducing valve; ⑤For boiler blow-off piping, the design pressure is 1.25 times the design pressure of the boiler; ⑥For refrigerating machinery piping, the pressure is prescribed in Pt 9, Ch 1, 102.5.; ⑦For pipes containing fuel oil, the design pressure is to comply with the following requirements: (a) Where the working pressure is not more than 0.7 MPa and the working temperature is not more than 60 ℃: 0.3 MPa or maximum. working pressure, whichever is greater; (b) Where the working pressure is not more than 0.7 MPa and the working temperature exceeds 60 ℃: 0.3 MPa or maximum. working pressure, whichever is greater; (c) Where the working pressure exceeds 0.7 MPa and the working temperature is not more than 60 ℃: the design pressure is maximum working pressure; (d) Where the working pressure exceeds 0.7 MPa and the working temperature exceeds 60 ℃: 1.4MPa or maximum working pressure, whichever is greater; ⑧Where it is impracticable to adopt the above values, the design pressure is to be specially considered by a Society in each case. (2) For the component, design pressure is a pressure used in the calculations to determine the scantling of such component and is the maximum permissible working pressure to the component. However, the design pressure is to be not less than the highest set pressure of any relief valve.

Design pressure (pressure vessels) 设计压力（压力容器） 设计压力系指锅炉和压力容器的最大许用工作压力，锅炉和压力容器的强度计算应以设计压力为依据，并应不小于任一安全阀的最高设定压力。经济器的设计压力应为经济器内介质的最大工作压力。 Design pressure (pressure vessels) is the maximum permissible working pressure for boilers and pressure vessels. Strength calculations for boilers and pressure vessels are to be based on the design pressure and not to be less than the highest set pressure of safety valve. The design pressure of economizers is the maximum working pressure of the medium in the economizer.

Design pressure[7] 设计压力（制冷剂） 系指其最高的设计压力，但是，设计压力不能小于表 D-5 中所给定的数值。 For refrigerants, design pressure mean the maximum design pressure. However, design pressure are to be not less than the values specified in Table D-5.

表 D-5 最低设计压力（制冷剂）
Table D-5 The Lowest Design Pressure

制冷剂的种类 refrigerants	高压侧 high Pressure Side ①(MPa)	低压侧 low pressure side ②(MPa)
R22	1.9	1.5
R134a	1.4	1.1
R404A	2.5	2.0
R407C	2.4	1.9
R410A	3.3	2.6
R507A	2.5	2.0
R717	2.3	1.8

注意：(1)高压侧——从压缩机出口侧到膨胀阀的压力部分。 high pressure side: The pressure part from the compressor delivery side to the expansion valve;
(2)低压侧——从膨胀阀到压缩机气门的压力部分。如果引入一个多级压缩系统，那么高压侧包括从低级的出料侧到高级吸引侧。 low pressure side: The pressure part from the expansion valve to the compressor suction valve. In case where a multistage compression system is adopted, the pressure part from the lower-stage delivery side to the higher-stage suction side is to be included.

Design scenario 应急预案 系指一整套条件和在系统的生存周期中可能发生事件的合理预计。这些条件和预期用于确定设计危险。 Design scenario means a set of conditions and incidents which may be reasonably expected to occur during the life of a system. These conditions and incidences are to be used in identifying the design hazards.

Design service life 设计使用寿命 系指设定用作预定用途的结构物经预期维护但无须进行实质性修理的周期。 Design service life means an assumed period for which a structure is to be used for its intended purpose with anticipated maintenance, but without substantial repair being necessary.

Design situation 设计状态 系指证实设计值在一定时间间隔内不超过有关极限状态的、代表真实状态的一组物理状态。 Design situation is a set of physical conditions representing real conditions during a certain time interval and the conditions will demonstrate that relevant limit states are not exceeded.

Design specifications 设计任务书 系指指导设计工作的最基本的文件。其主要包括下列内容:(1)船舶的用途——主要说明该船主要装什么货物、载重量或载客量多少;(2)船舶航行的海区——船舶适应什么航区(内河、沿海或远洋)始点港口和终点港口、主尺度有何限制;(3)船级和满足的规范、规则,必需入某一船级社的船级;(4)动力装置——确定主机、发电机的形式和功率有什么特殊要求、辅机的选型、自动化的程度;(5)船的航速、续航力(船在海上不补充燃料需航行的天数)、自持力(船在海上航行和停留的总的时间);(6)船体结构设计的标准、材料的要求(是否采用合金钢);(7)船舶设备——装卸设备:对杂货船、散货船等货船,主要为这些设备的类型、负荷大小、舷外跨距及传动方式等;对液货船为货油泵的类型、能量大小及扬程等;甲板机械:系指舵机、锚机、绞缆机、救生艇卷扬机等装置的类型和技术要求;对消防救生设备、冷藏通风设备的要求等;(8)船舶定员和舱室布置——船舶定员一般取决于船舶吨位大小和自动化程度,并与各国的传统习惯、劳动法规、工会要求和船员的素质有关,客船和客货船还要考虑旅客人数和舱室等级等;舱室布置是对船上各类人员居住舱室、公共处所(包括会议室、餐厅、休息室等)的要求;(9)通信和导航设备——按船舶航行区域的不同配置不同的助航设备,包括雷达、电罗经、卫星导航装置、卫星通信设备、避碰设备、自动操舵设备以及内部通信设备等。

Design speed 设计航速 船舶设计时,通过计算要求达到的航速。

Design still water bending moment M_{SW} (kN – m) 设计静水弯矩 M_{SW} 系指所考虑船体横剖面处的设计静水弯矩:(1)中拱工况:$M_{SW} = M_{SW,H}$;(2)中垂工况:$M_{SW} = M_{SW,S}$。M_{SW} means design still water bending moment, in kN – m, at the hull transverse section considered:(1)$MSW = M_{SW,H}$ in hogging conditions;(2)$M_{SW} = M_{SW,S}$ in sagging conditions.

Design still water shear force Q_{SW} 设计静水剪力 Q_{SW} 系指所考虑船体横剖面处的设计静水剪力。Q_{SW} means design still water shear force, in kN, at the hull transverse section considered.

Design stress 设计应力(起重设备) 系指起重设备在安全工作负荷作用下,起重设备部件允许承受的最大应力。及应考虑起重设备在环境载荷作用下,同时受到侧向载荷和与风载荷。

Design temperature 设计温度 (1)系选择钢级所依准的参考温度。露天结构的设计温度为航行区域的最低每日平均大气温度的统计平均温度。此温度被认为与航行区域中的最低月间温度的统计平均温度比较低2 ℃。局限于《夏季》航海时,最低月间温度的平均值可与相关月份的较暖 0.5 个月进行比较。对应的最低温度应考虑为比设计温度低 20 ℃的温度;(2)应取对船舶航行海域的一年最低日平均气温的平均值。对于季节性限制航行的船舶,可取相应航行期间内的最低温度作为设计温度。 Design temperature is:(1)a reference temperature used as a criterion for the selection of steel grades. The design temperature for external structures is defined as the lowest mean daily average air temperature in the area of operation. This temperature is considered to be 2 ℃ lower compared with the lowest monthly mean temperature in the area of operation. If operation is restricted to "summer" navigation, the lowest monthly mean temperature comparison may only be applied to the warmer half of the month in question. The corresponding extreme low temperature is generally considered to be 20 ℃ lower than the design temperature;(2)to be taken as the lowest mean daily average air temperature in the area of operation. For seasonally restricted service, the low value within the period of operation applies.

Design temperature for piping 管系的设计温度 系指管内流体的最高温度,但不低于 50 ℃。Design temperature means the maximum temperature of the internal fluid, but in no case is it to be less than 50 ℃.

Design temperature t_D 设计温度 t_D 系指营运区域的昼夜平均(average)空气温度的最低中间(mean)值。其中:(1)中间值(mean)——系指至少 20 年以上观察周期统计的平均值;(2)平均值(average)——系指每昼夜的平均值;(3)最低值(lowest)——系指一年中的最低值。对于季节性的有限航区,设计温度应采用在航行期间内最低的期望值。 Design temperature t_D means the lowest mean daily average air temperature in the area of operation, in which:(1)mean value means statistical mean over an observation period of at least 20 years;(2)average value means average during one day and night;(3)lowest value means lowest during/year. For seasonally restricted service, the lowest expected value within the period of operation applies.

Design temperature(vessel) 设计温度(容器) 系指在运行工况下部件的实际温度。该温度由制造厂商指定,应考虑运行时的各种温度变化的影响。除非制造厂商和船级社按具体情况专门商定者外,设计温度不能小于表 D-6 给出的温度。 Design temperature is the actual metal temperature of the applicable part under the expected operating conditions. This temperature is to be stated by the manufacturer and is to take in account of the effect of any temperature fluctuations which may occur during the service. The design temperature is to be not less than the temperatures stated in Table D-6, unless the specially agreed case between the manufacturer and the Society.

表 D-6　最小设计温度表
Table D-6　Minimum design temperature

容器类型 Type of vessel	最小设计温度 Minimum design temperature
压力容器和锅炉,不受高温气体加热的,或有绝缘保护良好的压力容器的压力部件和锅炉 Pressure parts of pressure vessels and boilers not heated by hot gases or adequately protected by insulation	内部液体的最高温度 Maximum temperature of the internal fluid
被高温气体加热的压力容器 Pressure vessel heated by hot gases	超过内部流体温度 25 ℃ 25 ℃ in excess of the temperature of the internal fluid
主要承受传递热量的锅炉水管 Water tubes of boilers mainly subjected to convection heat	超过饱和蒸汽温度 25 ℃ 25 ℃ in excess of the temperature of the saturated steam
主要承受辐射热量的锅炉水管 Water tubes of boilers mainly subjected to radiant heat	超过饱和蒸汽温度 50 ℃ 50 ℃ in excess of the temperature of the saturated steam
主要承受传递热量的锅炉过热水管 Superheater tubes of boilers mainly subjected to convection heat	超过饱和蒸汽温度 35 ℃ 35 ℃ in excess of the temperature of the saturated steam
主要承受辐射热量的锅炉过热水管 Superheater tubes of boilers mainly subjected to radiant heat	超过饱和蒸汽温度 50 ℃ 50 ℃ in excess of the temperature of the saturated steam
节能器管 Economiser tubes	超过内部流体温度 35 ℃ 35 ℃ in excess of the temperature of the internal fluid
湿背锅炉燃烧室用的那种类型的燃烧室 For combustion chambers of the type used in wet-back boilers	超过内部流体温度 50 ℃ 50 ℃ in excess of the temperature of the internal fluid
对干背锅炉的火炉、炉膛以及后管板和其他承受相似热传递比率的压力部件 For furnaces, fire-boxes, rear tube plates of dry-back boilers and other pressure parts subjected to similar rate of heat transfer	超过内部流体温度 90 ℃ 90 ℃ in excess of the temperature of the internal fluid

Design trim　设计纵倾　当艇正浮,且乘员、贮藏品和设备都位于由设计者或制造者所指定的位置时船艇的纵向姿态(角)。Design trim is the longitudinal attitude of a boat when upright, with crew, stores and equipment in the positions designated by the designer or builder.

Design value　设计值　系指在设计检查程序中使用的偏离典型值的数值,该值考虑了与典型值相关的不确定性。Design value is a value derived from the representative value for use in the design checking procedure, which accounts for the uncertainties associated with the representative value.

Design voltage　设计电压　系指为使灯泡发出额定坎德拉的光强度或额定瓦特的功率(如未规定额定坎德拉)所必需的输入电压。

Design waterline 设计水线 系指船舶在无升力或推进机械不开动的排水状态(即静浮于水面)时,其最大营运质量或满载排水量所对应的水线。 Design waterline means the waterline corresponding to the maximum operational weight of the craft with no lift or no propulsion machinery active.

Design wave height H_{dw} 设计波高 H_{dw} 运行区域一百年观察到的最大波高的平均值,取为 H_{dw} = 1.67 Hs (m)。 Design wave height H_{dw} means the average of the highest observed wave heights for the service area with in 100 years. It is to be taken as: H_{dw} = 1.67 Hs (m)。

Design wave period range T_{drange} 设计波浪周期范围 取为设计波浪周期加上两倍的跨零周期的标准偏差,至设计波浪周期减去两倍的跨零周期的标准偏差, T_{drange} 为 $T_{dw} - 2T_{dsd} \sim T_{dw} + 2T_{dsd}$ (s)。 Design wave period range T_{drange} is to be taken as the design wave period plus and minus 2 standard deviations of the zero crossing period, i.e. T_{drange} is $T_{dw} - 2Td_{sd}$ to $T_{dw} + 2T_{dsd}$ (s)。

Design wave period T_{dw} 设计波浪周期 T_{dw} 在运行区域所有的海况下,跨零周期的平均值: $T_{dw} = T_z$,(s)。 Design wave period T_{dw} means the average zero crossing period of all sea states in the service area: $T_{dw} = T_z$,(s)。

Designated Authority 指定当局 系指一个设在政府内的负责从港口设施的角度确保实施涉及港口设施保安和船/港界面活动规定的机构或行政机关。 Designated Authority means the organization(s) or the administration(s) identified, within the Contracting Government, as responsible for ensuring the implementation of pertaining to port facility security and ship/port interface, from the point of view of the port facility.

Designated bonded area 指定保税区 是为了在港口或国际机场简便、迅速地办理报关手续,为外国货物提供装卸、搬运或暂时储存的场所。

Designated danger areas 指定危险区 指非专用的军火存储舱室或空间,但这些地方偶尔会放置军火。 Designated danger areas are compartments and spaces not fitted out specifically for the stowage of munitions, but where munitions are occasionally present.

Designated person (DP) 指定人员 可以直接接触公司的最高经营决策层的,拥有对各船的安全航行及防污染状态进行监督责任,并能在必要时为船上提供岸基支援的人员。

Designated place(s) of safety 指定安全地点 是要在设计中明确,指定安全地点必须是舰船上的地方,以此可以合理预计将作为逃生后的撤离平台使用。 Designated place(s) of safety is/are to be declared in the design disclosure. These are to be places on board the vessel which may be reasonably expected to be used as platforms for evacuation following escape.

Designed displacement 设计排水量 新船设计中作为基准和预期达到的某一装载情况的排水量。对于民用船舶通常系指设计预期的满载排水量;对于军用舰船则是设计预期的正常排水量。

Designed draft 设计吃水 基平面与设计水线之间的垂直的距离。

Designed maximum load draught 最大设计满载吃水 系指在船长中点处从龙骨上缘量至设计最大载重线的垂直距离。

Designed maximum load line 最大设计满载载重线 系指与满载状态对应的水线。

Designed pitch 设计螺距 根据可调螺距螺旋桨主要工作情况,如自由航行、拖曳别船的要求、所设定的螺距。

Designed waterline 设计水线 船舶在预期设计状态自由正浮于静水上时,船体型表面与水面的交线,亦即对应于设计排水量的水线。

Designed waterline breadth 设计水线宽 设计水线平面处船体型表面之间垂直于中线面的最大水平距离。

Designed waterline length 设计水线长 设计水线平面与船体型表面艏艉端交点之间的水平距离。

Designer 设计者(机构) 提供设计和建造图纸的团体。 Designer is the organization which provides the design and constructional plans.

Desinent cavitation (cavitation disappearance) 空泡消失 由于周围压力稍增或流速稍减,部分空泡变为无空泡的情况。

Desktop computer 台式计算机 又称桌上型计算机,其优点是耐用,以及价格实惠,与笔记本电脑相比,相同价格前提下配置较好,散热性较好,配件若损坏更换价格相对便宜,缺点就是:笨重,耗电量大。计算机(Computer)是一种利用电子学原理根据一系列指令来对数据进行处理的机器。计算机可以分为两部分:软件系统和硬件系统。第一台计算机 ENIAC 于 1946 年 2 月 15 日宣告诞生。

Desuperheated steam 减热蒸汽 过热蒸汽通过减热器后形成的低过热度蒸汽。

Desuperheater 减温器 为调节锅炉过热蒸汽温度或为获取辅机用汽而用来降低过热蒸汽温度的装置。

Detachable gear (parts) 可卸零部件 系指非永

久性附连于起重设备上的零部件,如链条、三角眼板、吊钩、滑车、卸扣、转环、钢索索节、有节定位索和松紧螺旋扣等。吊梁、吊架、吊框与类似设备亦称为可卸零部件。

Detachable link 连接锚链(连接链环) 连接锚链各链节的可拆链环。分锻造和铸造两种。

Detachable rail stanchion 活动栏杆柱 可拆卸或倒下的栏杆柱。

Detail design (production design) 施工设计(生产设计) 系指根据船级社和船东已审查同意的技术设计(详细设计)的图纸和资料进行的设计。目的是根据承造厂的工艺设备条件绘制图纸或计算机的软盘,提供船厂生产。

Detailed inspection 逐项检查 系指包括近观检查以及只有打开外壳和/或(必要时)采用工具和检测设备才能识别明显缺陷(如接线端子松动)的检查。

Detection 探测 系指救助艇或救生艇筏位置的测定。Detection is the determination of the location of survivors or survival craft.

Detection and alarm 探测和报警 系指探测火源处的火灾,并规定发出安全撤离和采取灭火行动的警报。为此,应满足下列功能要求:(1)固定式探火和失火报警系统装置应适合于处所的性质,潜在的火势增大和潜在的烟气产生;(2)应有效地设置手动报警按钮,以确保随时可使用的报警通知方式;(3)消防巡逻应能作为一种有效方式来探测和确定火灾位置以及向驾驶室和船上消防队发出警报。Detection and alarm is to detect a fire in the space of origin and to provide for alarm for safe escape and fire-fighting activity. For this purpose, the following functional requirements shall be met:(1) fixed fire detection and fire alarm system installations shall be suitable for the nature of the space, fire growth potential and potential generation of smoke and gases;(2) manually operated call points shall be placed effectively to ensure a readily accessible means of notification;(3) fire patrols shall provide an effective means of detecting and locating fires and alerting the navigation bridge and fire teams.

Detection and alarm system for combustible gas 可燃气体检测与报警系统 系指当碳氢化合物和类似产品的气体浓度超过碳氢化合物和空气混合物爆炸最低极限30%时,能在驾驶室、开敞甲板或其他合适部位发出听觉和视觉报警的系统。

Detection radar 探测雷达

Detection system for fiber optic hydrophone 光纤水听器探测系统 该系统为基于大规模光纤水听器阵列的新一代水声探测系统,可广泛应用于海军水声装备、水下安防装备、海洋石油钻探和海洋地质调查装备的开发和应用。该系统由光纤水听器阵列、信号传输光缆及信息处理系统组成,若加上一套收放绞车系统,即可用于拖曳探测。该系统的工作原理是通过光纤水听器阵列,将感应到的水下噪声信号转换成光信号,然后通过传输光缆将光信号传送到信息处理系统,信息处理系统将光信号转化为电信号并将其转化的数据信号后再调出水声信号。利用阵列水声信号处理技术对水声信号进行分析,即可获得水下噪声目标特性。与传统压电式水听器相比,光纤水听器具有诸多优势:(1)灵敏度高、频响特性好,由于其自噪声很低,因此可检测到的最小信号比传统压电式水听器要高2~3个数量级;(2)动态范围大,其动态范围在12~140 dB,比压电式水听器的80~90 dB要广,这使得光纤水听器既可以提出弱信号,也可以提出强信号;(3)抗电磁干扰与信号串扰能力强,因其信号传感与传输均以光为载体,几百兆赫以下的电磁干扰影响非常小,各通道信号串扰也很小;(4)采用频分、波分及时分等技术进行多路复用,光纤传输损耗小,适于远距离传输与组阵 及制造大规模水声探测阵列;(5)信号传感与传输一体化,光纤水密性要求低,系统可靠性高。此外,该系统的探测缆和传输缆均为光缆,质量轻、体积小,系统容易收放。

Detection, alarm and control system in engine room 机舱检测、报警和控制系统 是用来检测、控制机电设备的工作状态和参数,如压力、温度、电流、电压、功率、频率、转速、流量、液位、绝缘电阻、电动机的启动和停止、阀的开和关等,并进行指示、报警、控制和记录的系统。当机电设备发生故障、工作参数越限时,发出报警指示。

Detector 探测器 系指一种通过热、烟或其他燃烧产物、火焰或任何这些因素而动作的仪器。 Detector is an instrument operated by heat, smoke or other products of combustion, flame, or any combination of these factors.

Detention 扣留

Determination 确定 系指主管机关决定是否签发、修正、中止、吊销或换新授权拆船文件的过程。Determination means the process by which the Competent Authority(ies) decides whether to issue, amend, suspend, withdraw or renew a DASR.

Deterministic method of ship motions 船舶运动确定式处理法 利用冲浪响应函数等概念,按不规则波面对时间的实际记录确定船舶运动历程的一种数据处理方法。

Detonation 爆燃 燃烧室内火焰以过高速度传布的现象。

Detroit River International Wildlife Refuge 底特律河国际野生动物保护区 系指底特律河的美国水域,

其边界从胭脂河的西南出口处的密歇根沿岸延伸至美国、加拿大边界向南的北纬41°54.0′，西经083°06.0′，并顺时针方向连接下列各点的区域：北纬43°02.0′，西经083°08.0′；北纬41°54.0′，西经083°08.0′；北纬41°50.0′，西经083°16.0′；北纬41°44′53″，西经083°22.0′，北纬41°44′18″，西经083°27.0′。Detroit River International Wildlife Refuge means the U. S. waters of the Detroit River bounded by the area extending from the Michigan shore at the southern outlet of the Rouge River to 41°54.0′N, 083°06.0′W, along the U. S., Canada boundary southward and clockwise connecting point: 43°02.0′N, 083°08.0′W; 41°54.0′N, 083°08.0′W; 41°50.0′N, 083°16.0′W; 41°44′53″N, 083°22.0′W; 41°44′18″N, 083°27.0′W

Developed area 展开面积 螺旋桨各叶片展开轮廓内面积之总和。

Developed area ratio 展开盘面比 系指全部叶片展开面积与螺旋桨外径和桨毂直径之间环形面积之比。

Developed area ratio is the ratio of the total blade developed area to the area of the ring included between the propeller diameter and the hub diameter.

Developed outline 展开轮廓 将螺旋桨叶曲面近似地展开放在垂直于轴线平面上的外形。

Device for directional control 方向控制装置/系统 系指：（1）能使船舶的艏向和航向在主要工况和航速下能最大可能地得到有效的控制，而无须在所有航速下和证书核定的所有工况中借助不合适的人力操纵的装置。（2）包括任何操舵装置或装置组群、任何机械联动装置和所有动力或人力操纵装置、控制器和驱动系统。(1) Device for directional control means a device which is of adequate strength and suitable design to enable the craft's heading and direction of travel to be effectively controlled to the maximum extent possible in the prevailing conditions and craft speed without undue physical effort at all speeds and in all conditions for which the craft is to be certificated. (2) Directional control system includes any steering device or devices, any mechanical linkages and all power or manual devices, controls and actuating systems.

Devil's claw 掣链钩 其上配有松紧螺旋扣，收锚时，能钩住锚链环，使锚贴紧船体的双爪U形钩。

Devit-launched type liferaft 可吊放救生筏 在满载乘员后，可由吊卸装置放到水面上的救生筏。

Dew point 露点 为空气被所含潮气饱和时温度。

Dew point is the temperature at which air is saturated with moisture.

Dewatering system 舱室排水系统 系指散货船上，用于排放和泵吸位于防撞舱壁前方的压载舱的压载水，和任何部位延伸至艏货舱前方的干舱（容积不超过船舶最大排水量0.1%的干舱和锚链舱除外）中的舱底水排放系统。Dewatering system means on bulk carriers, the means for drainage and pumping ballast tanks forward of the collision bulkhead, and bilges of dry spaces, any part of which extends forward of the foremost cargo hold (excluding dry spaces the volume of which does not exceed 0.1% of the ship's maximum displacement volume and the chain locker).

DEWI 德国风能研究所 位于德国威廉港，是全球风电专业技术领先者，可为风电机组制造商、部件供货商、项目业主、贷款方、投资方等客户提供风电技术服务。该研究所一部分职责是在FINO 1研究站上测量风速、协调即将开始的RAVE部分研究项目。

DFT 干膜厚度 DFT is dry film thickness.

DGPS = Differential global positioning system 差分全球定位系统 全球定位系统加上由一个或多个已知的固定位置接收器提供的差分修正，以增加位置固定的精度。

DHS = Department of Homeland Security （美国）国土安全部

Diagonal 斜剖线 斜交于基平面和中线面，但垂直于中站面的斜平面与船体型表面的交线。

Diameter D 直径 系指以英寸（in）为单位的指定用外径作为管子的尺寸。标准直径的范围从0.405~80 in（英寸）。直径2.375~12.75 in也能用小数来表示（因此"8 in"管子，其实际直径为8.625 in）。从14 in及以上，管子的尺寸以2 in的增幅递增。这些尺寸是世界通用的，因此在管子制造厂的模具都应符合这些标准。Diameter D means the size of the pipe is specified by the outside diameter in units of inches. The standard diameters range from 0.405 to 80 in. Diameters from 2.375 to 12.75 in can be fractional (thus an "8-inch" pipe has actual diameter of 8.625 in). From 14 inch and higher, the sizes go up by 2-inch increments. These sizes are accepted worldwide, and thus forming tools at pipe mills follow these standards.

Diameter of leg 桩腿直径 圆柱桩腿的外径。

Diameter of propeller 螺旋桨直径 螺旋桨半径的两倍。

Diammonium phosphate (DAP) 磷酸二铵 无嗅白色晶体或粉末。视来源而决定是否粉末状。吸湿。无特殊危害。该货物为不燃物或失火风险低。该货物吸湿，在货物处所内如受潮可能硬化。Diammonium phosphate (DAP) is odourless white crystals or powder. Depending on source it can be dusty. It is hygroscopic. It has no special hazards. This cargo is non-combustible or has a low fire-risk. This cargo is hygroscopic and may harden in the cargo space under humid conditions.

Diamorphine 药物海洛因 是对吗啡进一步蒸馏的产物。外观上通常与扑面用的香粉相似，可能略微有些粗糙，呈奶油色或浅褐色等。通常没有气味，但可能有淡淡的醋味。这种物质可以药丸、胶囊或针剂形式作为商品生产。与吗啡相比，更为上瘾者喜爱，因为它能更快引起更加强烈的快感。 Diamorphine is a further distillation of morphine. It is generally similar to face powder in appearance, It is perhaps slightly coarse. It is cream to light brown in colour. It is generally odourless but may have a faint vinegary smell. The substance may be commercially produced in pill, capsule or ampoule form. It is more popular with addicts than morphine since it gives a quicker and more intense "high".

Diaphone and signal control device 雾笛和信号控制装置 系指用于发出雾航信号，也可用于发出通用紧急报警信号的装置。

Diaphragm 隔板 装在冲动式汽轮机气缸或隔板套中，其上沿圆周分别嵌有喷嘴、静叶和气封，用以保持汽轮机各级压力差的两个半圆形板。

Diaphragm gland 隔板气封 减少级前蒸汽经隔板和转子间的间隙向隔板和叶轮间的空间泄漏的气封。

Diaphragm type compressor 膜式压缩机 通过液压驱动膜片来压缩气体的压缩机。

Die press enclosure type circuit breaker 模压外壳式断路器 系指作为一个整体单元组装在用绝缘材料支承的外壳内的断路器。其过电流和脱扣设施为热式、电磁式或两者的组合。

Diesel direct drive 柴油机直接传动 由柴油机直接带动轴系的传动方式。

Diesel engine 柴油机 用柴油或燃油作燃料的内燃机。

Diesel engine lube oil condition monitoring (ECM) 柴油机滑油状态监控 对柴油机零部件用的润滑油进行各种测试分析，掌握滑油分析结果及其他性能参数等情况，以决定是否拆检。

Diesel oil 柴油

Diesel oil daily tank 柴油日用柜 存放柴油机或辅锅炉日常使用的燃油的柜。

Diesel oil separator 柴油分离机 一般用于分离轻柴油中的杂质和水分的油分离机。

Diesel oil storage tank 柴油澄清柜 使柴油中杂质、水分沉淀的柜。

Diesel power plant 柴油机动力装置 以柴油机为主机的动力装置。

Diesel ship 柴油机船 以柴油机作为主机的船舶。

Diesel = LNG dual-fuel engine 柴油-液化天然气双燃料发动机 其技术改造原理是在不改变发动机原有结构的前提下，加装一套天然气喷射控制装置，辅以电子调节系统和燃料燃气供给系统，将船舶原有的柴油发动机转化为以少量柴油为引燃燃料，天然气为主燃燃料的双燃料发动机。

Diesel-compressor 柴油压缩机 在结构上与柴油机组成一个不可分割的整体的往复式压缩机。

Diesel-compressor aggregate 柴油压缩机组 由压缩机和单独的柴油机组成的机组。

Diesel-gas turbine ship 柴油机-燃气轮机船 以柴油机和燃气轮机作为主机的船舶。

Dieudonne direct spiral test 螺线操纵试验 从左舷或右舷约20°分阶段变化舵角，并在每阶段维持一定舵角达到稳定转首角速度为止的一种检验船舶航向稳定性的试验。

Different modes of transport 不同的运输方式 系指以两种或两种以上的运输方式，例如海上运输、内河运输、航空运输、铁路或公路运输。 Different modes of transport mean the transport of goods by two or more modes of transport, such as transport by sea, inland waterway, air, rail or road.

Differential gear (differential gear arrangement) 差动式减速齿轮箱 由运动度为2的差动式行星齿轮系所构成的减速齿轮箱。

Differential piston (differential motion piston) 级差活塞 呈阶梯状的活塞。

Differentiated compliance anchoring system (DI-CAS) 差异化顺应式系泊系统

Diffuse sound field 扩散声场 系指当封闭空间内被激发起足够多的简正方式时，由于不同方式有各种特定的传播方向，因而使达到某点的声波包括了各种可能的入射方向的声场。

Diffuser 扩压器 为降低气体速度使一部分动能转换成静压的通道。

Diffuser (flame holder) 稳焰器 位于喷油器头部，为使油雾蒸气着燃并维持火焰稳定的锥形罩或风轮。

Diffusion factor 扩压因子 从叶型表面速度分布及附面层分离的角度出发，以进出口气流速度及叶栅稠度之间的一定关系表示压气机叶栅负荷的一个设计参数。

Digital camera 数码照相机 简称数码相机，是一种利用电子传感器把光学影像转换成电子数据的照相机。与普通照相机在胶卷上靠溴化银的化学变化来记录图像的原理不同,数码相机的传感器是一种光感应式的电荷耦合器件(CCD)或互补性氧化金属半导体(CMOS)。在图像传输到电脑以前,通常会先储存在数码存储设备中。通常是使用闪存;而软磁盘与可重复擦写光盘(CD-RW)已很少用于数码相机设备。

Digital display working group(DDWG) 数字显示工作组

Digital electronic war system(DEWS) 数字式电子战系统 美国推出 F-15SE"沉默鹰",及一套能够与之协调运行的新一代数字式电子战系统,有利于战机的隐身化;它是采用先进的 AESA 机载有源相控阵雷达,进一步减少雷达反射面。

Digital Equipment Corporation(DEC) [美]数字设备公司

Digital image 数字图像 是以二维数字组的形式所表示的图像,其数字单元为像元,数字图像的恰当应用通常需要数字图像与观察到的现象之间关系的知识,也就是几何和光度学或者传感器校准,数字图像处理领域就是研究它们的变换算法。数字图像,又称数码图像或数位图像,是二维图像用有限数字数值像素的表示。数字图像可以许多不同的输入设备和技术生成,例如数码相机、扫描仪、坐标测量机、地震仪分析仪(seismographic profiling)、机载雷达(airborne radar)等等,也可以从任意的非图像数据合成得到,例如数学函数或者三维几何模型,三维几何模型是计算机图形学的一个主要分支。

Digital image processing 数字图像处理 系指通过计算机对图像进行去除噪声、增强、复原、分割、提取特征等处理的方法和技术。数字图像处理的产生和迅速发展主要受 3 个因素的影响:(1)计算机的发展;(2)数学的发展(特别是离散数学理论的创立和完善);(3)广泛的农牧业、林业、环境、军事、工业和医学等方面的应用需求的增长。

Digital indicator 数字指示器

Digital recorder 数字录音机

Digital seismic recording system 数字地震仪 将来自压力检波器的地震信号经过放大滤波,以数字形式记录在磁带上的仪器。

Digital selective calling(DSC) 数字选择性呼叫 系指使用数码使一无线电台与另一电台或一组电台建立联系和传递信息,并符合国际无线咨询委员会(ITU-R)有关建议的一种技术。 Digital selective calling(DSC) means a technique using digital codes, which enables a radio station to establish contact with, and transfer information to, another station or group of stations, and complying with the relevant recommendations of the International Telecommunication Union Radio-communication Sector(ITU-R)。

Digital single lens reflex camera(DSLR) 数码单镜反光相机 简称数码单反相机,是一种以数码方式记录成像的照相机。属于数码静态相机(digital still camera,DSC)与单反相机(SLR)的交集。单反系指单镜头反光,即 SLR(single lens reflex),这是当今最流行的取景系统,大多数 35 mm 照相机都采用这种取景器。在这种系统中,反光镜和棱镜的独到设计使得摄影者可以从取景器中直接观察到通过镜头的影像。因此,可以准确地看见胶片即将"看见"的相同影像。该系统的心脏是一块活动的反光镜,它呈 45°角安放在胶片平面的前面。进入镜头的光线由反光镜向上反射到一块毛玻璃上。早期的 SLR 照相机必须以腰平的方式把握照相机并俯视毛玻璃取景。毛玻璃上的影像虽然是正立的,但左右是颠倒的。为了校正这个缺陷,在平式 SLR 照相机在毛玻璃的上方安装了一个五棱镜。这种棱镜将光线多次反射改变光路,将其影像送至目镜,这时的影像就是上下正立且左右校正的了。

Digital still camera(DSC) 数码静态相机

Digital television(DTV) 数字电视 系指播出、传输、接收等环节中全面采用数字信号的电视系统。对该系统所有的信号传播都是通过由 0、1 数字串所构成的数字流来传播的电视类型。使用数字电视信号或其信号损失小,接收效果好。数字电视系统可以传送多种业务,如高清晰度电视、标准清晰度电视、智能型电视及数字业务等等。数字电视机顶盒的工作过程:数字电视机顶盒通过网络接口模块选择频道,并进行解调和信道解码处理,输出 MPEG-2 多节目传输流数据,送给解复用器,解复用器从 MPEG-2 传输流数据中抽出一个节目的已打包的视音频基本流(PES)数据,包括视频 PES,音频 PES 和辅助数据 PES,解复用器中包含一个扰引擎,可在传输流层和 PES 层对加扰的数据进行解扰,解复用器输出的是已解扰的视音频 PES。

Digital video disc(DVD) 数字化视频光盘

Digital visual interface(DVI) DVI 端子/数字视频接口 它是 1999 年由 Silicon Image、Intel(英特尔)、Compaq(康柏)、IBM、HP(惠普)、NEC、Fujitsu(富士通)等公司共同组成 DDWG(digital display working group,数字显示工作组)推出的接口标准。是一种高速传输数字信号的技术。有 DVI-A、DVI-D 和 DVI-I 3 种不同的接口形式。DVI-D 只有数字接口,DVI-I 有数字和模拟接口,目前应用主要以 DVI-D 为主。DVI 接口有 3 种类型 5 种规格,端子接口尺寸为 39.5 mm×15.13 mm。它是以 Silicon Image 公司的 PanalLink 接口技术为基础,基于 TMDS(transition minimized differential signaling,最小化传

输差分信号)电子协议作为基本电气连接。TMDS 是一种微分信号机制,可以将像素数据编码,并通过串行连接传递。显卡产生的数字信号由发送器按照 TMDS 协议编码后通过 TMDS 通道发送给接收器,经过解码送给数字显示设备。一个 DVI 显示系统包括一个传送器和一个接收器。传送器是信号的来源,可以内建在显卡芯片中,也可以以附加芯片的形式出现在显卡 PCB 上;而接收器则是显示器上的一块电路,它可以接受数字信号,将其解码并传递到数字显示电路中,通过这两者,显卡发出的信号成为显示器上的图像。目前的 DVI 接口分为两种,一个是 DVI-D 接口,只能接收数字信号,接口上只有 3 排 8 列共 24 个针脚,其中右上角的一个针脚为空,不兼容模拟信号。

Digital set-top box 数字机顶盒 是一种多媒体终端,有类似于家用电脑的硬件体系结构和专用的实时操作系统及应用软件。数字机顶盒工作在有线电视网络状态下,有线电视网采用模拟传输,因此必须对数字信号进行调制和解调才能在模拟信道传输,调制解调器是系统关键的组成部分,在技术上类似电话调制解调器的原理,但采用了更高的调制方法,下行多采用 64QAM 或 256QAM,在 DVB-C(digital video broadcast by cable)和 DAVIC 中采用 64QAM 作为标准调制方法,以 Motorola 的 MC92305QAM 解调芯片为例,在 7M 模拟带宽上采用 64QAM 调制的数字信号速率可达 42Mbit/S,上行采用两种方式,一种是采用电话线作为上行信道,另一种是采用双向 HFC 网的上行通道,采用 HFC 网时采用 QPSK 作为调制方案。

Digitizer plotter 数字化器绘图机

Diluted 稀释的 系指可燃气体在其与空气的混合物中的浓度低于其爆炸下限 50% 。 Diluted means a flammable gas or mixture is defined as diluted when its concentration in air is less than 50% of its lower explosive limit.

Dilution(Q_d) 稀释物(Q_d) 系指在流入液体取样点之后和流入液体流量测量装置之后引入污水处理装置的稀释水、灰水、处理水和/或海水。 Dilution is dilution water, grey water, process water, and/or seawater introduced to the sewage treatment plant after the influent sample point and after the influent flow measurement device.

Dilution air(mixing air, tertiary air) 掺混空气 系指进入掺混区的空气。

Dilution method 稀释法 系指替换的压载水从用于装载压载水的压载水舱顶部注入并同时以相同流速从底部排出的过程。舱内水位在压载水置换作业全过程中保持不变。 Dilution method is a process by which replacement ballast water is filled through the top of the ballast tank intended for the carriage of ballast water with simultaneous discharge from the bottom at the same flow rate, maintaining a constant level in the tank through out the ballast exchange operation.

Dilution zone(mixing zone) 掺混区 燃烧室中用大量空气掺混燃气以获得所需的涡轮前进口稳定并使稳定分布均匀的区域。

Dimension 维度

Determinate accidental load 确定的事故载荷 系指为了符合所定义的风险接受准则,即使结构在遭遇最严重的事故载荷时,仍应在一定时间内保持系统稳定的性能。对于某些类型的事故,可能难以定义事故载荷,比如船舶由于空舱进水导致完整浮性的丧失,在这种情况下,应针对特殊情况做出具体的规定。

Diode 二极管 又称晶体二极管,另外,还有早期的真空电子二极管;它是一种能够单向传导电流的电子器件。在半导体二极管内部有一个 P-N 结两个引线端子,这种电子器件按照外加电压的方向,具备单向电流的传导性。一般来讲,晶体二极管是一个由 P 型半导体和 N 型半导体烧结形成的 P-N 结界面。在其界面的两侧形成空间电荷层,构成自建电场。当外加电压等于零时,由于 P-N 结两边载流子的浓度差引起扩散电流和由自建电场引起的漂移电流相等而处于电平衡状态,这也是常态下的二极管特性。

Dip net 抄网 在网口框架上装有手柄或吊索,用以从大网中捞取渔获物的截头圆锥状小网。

Dipper 铲斗 前壁斗刃上镶有 4～6 个锰钢齿、后壁系有供操纵起落用的钢索,底部做成能卸泥的活门板,装在铲斗柄前端用以挖掘石块和硬质泥沙的挖泥工具。

Dipper boom 铲斗吊杆 借吊索紧固于铲斗支架上,底部支承于旋转平台,中部设有一套推压机构的刚性吊杆。

Dipper dredger(DD) 反铲式挖泥船 系指具有铲斗挖泥设备的,作业时船随作业过程后退的铲斗式挖泥船。

Dipper frame 铲斗支架 铲斗挖泥船上,用以支持铲斗吊杆的刚性构架。

Dipper machine 铲斗机 由铲斗、铲斗柄、铲斗吊杆、铲斗支架、推压机构和机棚等组成的一种船用挖掘机械。

Dipper stick 铲斗柄 端部接铲斗,底面装有齿条,穿插于铲斗吊杆中部,依靠推压机构做旋转和推压动作的刚性杆件。

Dipper type dredger 铲斗式挖泥船 系指具有铲斗挖泥设备的船舶。 Dipper dredger is a dredger fitted dipper.

Dipping 艏沉 船舶纵向下水过程中,发生艏落时,因动力作用使船首自静浮状态艏吃水继续下沉的现象。

Direct air cooling system 直接式风冷系统 是通过空气冷却器风机循环冷却空气进行制冷的系统。Direct air cooling system is the system by which the refrigeration is obtained by circulation of air refrigerated by an air cooler.

Direct air-flow valve 直流阀 介质流动方向基本不变的气阀。

Direct B/L 直达提单 系指船舶中途不经过换船而直接驶往目的港卸货所签发的提单,凡规定不准转船者,必须使用这种直达提单。

Direct bilge suctions 舱底水直通支管 不经总管,而单独直接与舱底水泵连接的舱底水管。

Direct cooling system 直接冷却系统 是由安装在冷藏室天花板或墙壁上的冷却盘管内的制冷剂经直接蒸发进行制冷的系统。Direct cooling system is the system by which the refrigeration is obtained by direct expansion of the refrigerant in coils fitted on the walls or ceilings of the refrigerated chambers.

Direct expansion 直接蒸发 制冷剂直接在盘管内蒸发吸热的冷却方式。

Direct expansion air cooler 直接蒸发式空气冷却器 管内通入制冷剂直接蒸发吸热的表面式空气冷却器。

Direct loading pipe line (Direct filling line) 直接装注油管 不通过油泵和泵舱管系而由甲板岸接头直接接入货油舱系统的注入接管。

Direct product offsets 直接产品补偿 即双方在协议中约定,由设备供应方向设备进口方承购买一定数量或金额的由该设备供应方直接生产出来的产品。

Direct radiative forcing 直接辐射强迫(正强迫) 系指气体通过吸收和发射产生直接辐射强迫。

Direct reading current meter 直读式海流计 一种由转子或旋桨测流、舡舵测向并直接读出测量值的仪器。有的附有深度指示器,以确定仪器测量时所处的位置。

Direct reduced iron (A) briquettes (hot-moulded) 直接还原砖状块铁(A)(热铸) 是由增密处理工艺产生的一种灰色金属物质,用铸模制成的砖块。在此过程中,直接还原铁进料的成形温度高于650 ℃,密度大于5 000 kg/m^3。微粒和小颗粒(不足6.35 mm)含量按质量计不超过5%。该物质在散装状态下装卸后,温度预计可能因自热而暂时升高约30 ℃。该物质遇水(特别是盐水)后可能缓慢释放氢。氢是易燃气体,按体积计大于4%浓度与空气混合后能形成爆炸性混合物。易使货物处所缺氧。该货物为不燃物或失火风险低。Direct reduced iron (A) briquettes (hot-moulded) is a metallic grey material, moulded in a briquette form, emanating from a densification process whereby the direct reduced iron (DRI) feed material is moulded at a temperature greater than 650 ℃ and has a density greater than 5 000 kg/m^3. Fines and small particles (under 6.35 mm) shall not exceed 5% by weight. Temporary increase in temperature of about 30 ℃ due to self-heating may be expected after material handling in bulk. The material may slowly evolve hydrogen when contacting with water (notably saline water). Hydrogen is flammable gas that can form an explosive mixture when mixed with air in concentration above 4% by volume. It is liable to cause oxygen depletion in cargo spaces. This cargo is non-combustible or has a low fire-risk.

Direct reduced iron (B) lumps, pellets (cold-moulded briquettes) 直接还原铁(B) 是一种多孔黑色/灰色金属物质,由氧化铁在低于铁的熔点温度下还原(除氧)而成。冷铸块形状为成形温度低于630 ℃或密度小于5 000 kg/m^{3}的砖块块。大小不足6.35 mm的微粒和小颗粒含量按质量计不超过5%。该物质在散装状态下装卸后,温度预计可能因自热而暂时升高约30 ℃。运输期间有热量过大、火灾和爆炸的风险。该货物与空气和淡水或海水发生反应而产生热量和氢气。而氢是易燃气体,按体积计大于4%浓度与空气混合后能形成爆炸性混合物。该货物的反应作用视矿源、还原过程和温度及后续老化工艺而定。货物发热可能产生的极高温度足以引燃货物。微粒的增加也可能导致自热、自燃和爆炸。货物处所和封闭处所可能缺氧。Direct reduced iron (DRI)(B) is highly porous, black/grey metallic material formed by the reduction (removal of oxygen) of iron oxide at temperatures below the fusion point of iron. Cold-moulded briquettes are defined as those which have been moulded at a temperature less than 650 ℃ or which have a density less than 5 000 kg/m^3. Fines and small particles under 6.35 mm in size shall not exceed 5% by weight. Temporary increase in temperature of about 30 ℃ due to self-heating may be expected after material handling in bulk. There is a risk of overheating, fire and explosion during transport. This cargo produces heat and hydrogen when reacting with air and with fresh water or seawater to. Hydrogen is flammable gas that can form an explosive mixture when mixed with air in concentrations above 4% by volume. The reactivity of this cargo depends upon the origin of the ore, the process and temperature of reduction, and the subsequent ageing proce-

Direct reduced iron（C）(by-product fines) 直接还原铁(C) 是一种有孔的黑色/灰色金属物质,是 DRI (A)和/或 DRI(B)制造和处理过程中产生的一种副产品。DRI(C)的密度小于 5 000 kg/m³。该物质在散装状态下装卸后,温度预计可能因自热而暂时升高约30 ℃。运输期间有热量过大、火灾和爆炸的风险。该货物与空气和淡水或海水发生反应而产生热量和氢。氢是易燃气体,按体积计以大于 4% 浓度与空气混合后能形成爆炸性混合物。货物发热可能产生的极高温度足以导致自热、自燃和爆炸。货物处所和封闭处所可能缺氧。这些处所的易燃气体也可能增加。进入货物处所和相邻封闭处所时应采取一切预防措施。由于该类别所能容纳的物质的性质,该货物的反应作用极难评估。因此,任何时候均宜假定最恶劣的情况。 Direct reduced iron (DRI)(C) is porous, black/grey metallic material generated as a by-product of the manufacturing and handling processes of DRI (A) and/or DRI (B). The density of DRI (C) is less than 5,000 kg/m³. Temporary increase in temperature of about 30 ℃ due to self-heating may be expected after material handling in bulk. There is a risk of overheating, fire and explosion during transport. This cargo hydrogen and heat produce when reacting with air and fresh water or seawater to. Hydrogen is a flammable gas that can form an explosive mixture when mixed with air in concentrations above 4% by volume. Cargo heating may generate very high temperatures that are sufficient to lead self-heating, auto-ignition and explosion. Oxygen in cargo spaces and in enclosed adjacent spaces may be depleted. Flammable gas may also build up in these spaces. All precautions shall be taken when entering cargo and enclosed adjacent spaces. The reactivity of this cargo is extremely difficult to assess due to the nature of the material that can be included in the category. A worst case scenario should therefore be assumed at all times.

Direct spiral test 正螺线试验 (1)是回转圆操纵试验。该试验可通过增加回转操纵中的舵角对各稳定状态偏航率/舵角值进行测试。应留有足够的时间使船舶达到稳定的偏航率以避免失真的不稳定指示;(2)是一系列为得到对应舵角的稳定回转率的有序回转试验。 Direct spiral test is: (1) a turning circle manoeuvre test through which various steady state yaw rate/rudder angle values are measured by making incremental rudder changes. Adequate time must be allowed for the ship to reach a steady yaw rate so that false indications of stability are avoided; (2) an orderly sequence of turning circle tests to obtain a steady turning rate versus rudder angle relation.

Direct subsidy 直接补贴 是一国政府对本国或本行业的出口商在出口某项商品时,直接给予出口商的现金补贴,主要来自财政拨款。

Direct trade 直接贸易 系指货物消费国、生产国直接买卖货物的行为。货物从生产国直接卖给消费国,对生产国而言,是直接出口;对消费国而言,是直接进口。

Direct transit trade 直接过境贸易 系指外国货物到港后,在海关的监督下,从一个港口通过国内航线运到另一个港口,而后离境的活动。

Direct-contact condensation 直接接触冷凝 蒸汽和冷却水直接接触无传热面换热而使蒸汽凝结成水的冷凝方式。

Direct-coupled steam turbine 直接传动式汽轮机 不经变速直接驱动发电机、水泵等工作机械的汽轮机。

Direct-injection combustion chamber 直接喷射燃烧室 由活塞顶气缸壁及气缸盖底所组成的燃烧室。

Direction of foreign trade 对外贸易地理方向 又称对外贸易地区分布或国别结构,系指一定时期内各个国家或区域集团在一国对外贸易中所占有的地位,通常以它们在该国进出口总额或进口总额、出口总额中的比重来表示。对外贸易地理方向指明一国出口商品的去向和进口商品的来源,从而反映一国与其他国家或区域集团之间经济贸易联系的程度。一国的对外贸易地理方向通常受经济互补性、国际分工的形式与贸易政策的影响。

Direction of well deflection（DWD） 井斜方位 系指井眼轴线的切线在水平投影面上的方向。以正北方向线为始边顺时针转至该水平投影线之间所夹的角度来表示。

Directional control system 方向控制系统 系指包括所有用于驾驶船舶的主要或辅助设备的系统。该系统包括所有相关的电源、连接装置、控制装置和启动系统。 Directional control system means a system includs any device or devices intended either as a primary or auxiliary means for steering the ship. The directional control system includes all associated power sources, linkages, controls and actuating systems.

Directional infra-red jamming technology 定向红外干扰技术 系指利用精确指向来袭导弹的定向激光干扰导弹的导引装置,使其失效或发生错误进而使导弹脱靶的技术。

Directional spectral density 方向谱密度 其在任

何区间的积分代表该区间内一切组成分波能量的波浪频率和波向的函数。

Directional stability 方向稳定性 船舶在水平面或垂直面内运动受扰动偏离平衡状态,当扰动完全消除后仍能循原有方向运动的性能。

Directional, attitude and altitude control system 方向、姿态和飞高控制系统 系指借助下列装置控制方向、姿态和飞高的系统:空气舵或水舵、气翼、襟翼、可转向螺旋桨或喷射器、偏航控制孔或侧推器、差动推进器、船舶的可变形状或其垫升系统部件,或这些装置的组合。对地效翼船而言,方向、姿态和飞高控制系统包括任何推进、垫升或操舵装置、任何机械联动装置和所有动力和人力装置、控制装置和启动装置。 Directional, attitude and altitude control system means a system which may be achieved by means of air or water rudders, foils, flaps, propellers or jets which may be steerable, yaw control ports or side thrusters, differential propulsive thrust, variable geometry of the craft or its lift-system components or by a combination of these devices. For the WIG craft, directional, attitude and altitude control system includes any propulsion, lift or steering devices, any mechanical linkages and all power or manual devices, controls and actuating systems.

Direction-finding system 测向系统 在船上用测量其相对于2个已知地面台的方位角来定位的系统。由2个方位角便可得到2条位置线,两者之交点即为船位。侧向方法可以是在船上用方向性接收天线测出全向发射台的方位,也可以是地面台以某一固定的或旋转的天线方向性图发射,供船舶测定用。这种系统常用近岸定位。

Directory index 目录索引

Direct-printing telegraphy 直接印字电报 系指符合国际无线咨询委员会(ITU-R)有关建议案的自动电报技术。 Direct-printing telegraphy means automated telegraphy techniques which comply with the relevant recommendations of the International Telecommunication Union Radio-communication Sector (ITU-R).

Dirty ballast 污压载水 对"73/78防污公约"附则Ⅰ而言,系指已被油类污染的压载水。

Disability adjusted life years (DALY)/Quality adjusted life years (QALY) 失能调整生命年/质量调整生命年 DALY的基本内容是一年内完全健康的寿命预期值为1,非完全健康的寿命预期值小于1。与DALY不同,QALY将一年内完全健康的寿命预期值视为0,一年内非完全健康的寿命预期值小于0。

Disabled nozzle test 喷嘴失去作用试验 系指舱室内的喷嘴应不起作用。火源布置在下床铺上并点燃放在枕块中前部(朝向门外)的引燃物。如果舱室内的喷嘴与走廊上的喷嘴相连,则一个故障会对它们共同产生影响,所以所有舱室内和走廊上连起来使用的喷嘴都应不起作用的。 Disabled nozzle test means that the nozzle(s) in the cabin should be disabled. Source of fire is arranged in one lower bunk bed and ignited with the igniter located at the front (towards door) in the corridor, such that a malfunction would affect them all, all cabin and corridor nozzles linked should be disabled.

Disbursements 船舶使用费

Disc 轮盘 冲动式汽轮机中装配动叶用的旋转圆盘体。

Disc area 盘面积 螺旋桨盘的面积。

Disc clutch 圆盘离合器 利用圆盘端面接合的摩擦离合器。

Discharge 排放 按照MARPOL的定义,对有毒物质或含有这种物质的废液而言,排放系指不论由于何种原因所造成的船舶排放。包括任何的逸出、处理、漏泄、溢出、泵出、冒出或排空。排放不包括下列情况——(1)1972年11月13日在伦敦签订的《防止倾倒废弃物和其他物质污染海洋公约》所指的倾倒;(2)由于对海底矿产资源的勘探、开发及与之相关联的近海加工处理所直接引起的油类或油性混合物的排放;(3)为了减少或控制污染的合法科学研究而进行的有害物质的排放。 Discharge as defined by MARPOL in relation to harmful substances or effluent containing such substances means any release however caused from a ship, and includes any escape, disposal, spilling, leaking, pumping, emitting, or emptying. It does not include: (1) the dumping specified in Dumping of Wastes and Other Matter, done in London on 13 November 1972; (2) the release of harmful substances directly arising from the exploration, exploitation and associated offshore processing of seabed mineral resources; (3) the release of harmful substances for purposes of legitimate scientific research relating to pollution abatement or control.

Discharge 卸货 系指将货物从船上卸到码头/岸上/驳船等处的作业方式。

Discharge manifold 排放汇集管 这是"73/78防污公约"附则Ⅰ中要求的每艘油船必须设置的一根管路,用以连接船外的接收设备以便排放船上的污压载水或污油水。

Discharge pressure 排气压力 压缩机排出口处的气体压力。

Discharge rate of battery 蓄电池放电率 系指相对于蓄电池容量的放电电流大小,通常用放电至终止电压时,可维持放电电流的时间表示。

Discharge recording device of residues 残留物排放记录仪

Discharge to special reception facilities 排放到专用接收装置 系指港口国设有为压载水和沉淀物提供专用的接收装置。压载水则可排放到该接收装置,并应关注有关港口国的专用接收装置的接口的要求。 Discharge to special reception facilities means that if special reception facilities for ballast water and sediments are provided by a port State, they should, where appropriate, be utilized, and the requirements of related port State for connection of the reception facilities are to be taken into account.

Discharges 排放管 系指通过船舶两舷的任何管路,用以输送舱底水、循环水、排放水等。 Discharges mean any piping leading through the ship's sides for conveying bilge water, circulating water, drains, etc.

Discharging berth 卸货泊位 系指实施卸货作业所占用的码头泊位。

Dis-connectable mooring 可分离系泊

Dis-connectable turret 可分离转塔

Discontinuous working system 间断工作制 系指需要在下述负载下作周期性交替运行的一种使用要求:(1)负载和空载;(2)负载和休止;(3)负载、空载和休止。这种交替的时间间隔是明确规定的。

Discontinuance of execution 执行中止 系指执行过程中,因为某种特殊情况的发生而使执行程序暂时停止,待这种情况消失后,再行恢复执行程序的,称为执行中止。依据法律规定,申请人认为可以延期执行的,即可以向人民法院提出申请中止执行,法院应当裁定许可。

Discount 贴现 系指远期汇票承兑后,尚未到期,由银行或贴现公司从票面金额中扣除按一定贴现率计算的贴现息后,将余款付给持票人的行为。

Discount 折扣 是卖方按原价给予买方一定的百分比减让,即在价格上给予适当的优惠。国际贸易中,凡价格中包括佣金的称含佣价,不包括佣金、不含折扣的称为净价。

Dishonour 拒付 (1)持票人提示汇票要求承兑时,遭到拒绝而不获承兑(dishonour by non-acceptance),或持票人提示汇票要求付款时,遭到拒绝而不获付款(dishonour by non-payment),均称退票(dishonour),也称拒付。(2)除了拒绝承兑和拒绝付款以外,付款人避免不见、死亡或宣告破产,以上导致付款事实上已不可能时,也称为拒付。

Dishwater 洗碗水 系指人工或自动清洗碗碟和烹饪用具的液态残余物,碗碟和烹饪用具已经预先清洁的程度使黏附在其上的任何食物颗粒通常不致妨碍自动洗碗机的运行。 Dishwater means the liquid residue from the manual or automatic washing of dishes and cooking utensile which have been ore-cleaned to the extent that any food particles adhering to them would not normally interfere with the operation of automatic dishwashers.

Disinfection inspection certificate 消毒检验证书 是证明出口动物性产品包括猪鬃、皮张、马尾、山羊毛、羽毛和羽绒制品等已经过消毒处理的证书。

Disk 磁盘 计算机的外部存储器中也采用了类似磁带的装置,比较常用的一种叫磁盘。

Disk acoustic filter 盘式消音器

Disk coupling 盘式联轴节

Disk drives 磁盘驱动 作为工作站数据的主要载体,磁盘的可靠性和可用性直接关系着工作站整体性能和可靠性。工作站产品应用的都是目前比较先进的技术,这包括标准Ultra160 SCSI 磁盘驱动、10 K 或 15 K r/min高转速磁盘以及未来的Ultra320 控制器。

Disk operating system(DOS) 操作系统 系指"磁盘操作系统"。DOS 系统主要是一种面向磁盘的系统软件。

Disk valve 碟状阀 阀片呈碟状的气阀。

Dismantlement 拆除

Dismissal of the case 撤销案件 系指侦查机关对立案侦查的案件,发现具有某种法定情形,或者经过侦查否定了原来的立案根据,所采取的诉讼行为。撤销案件,公安机关应当报经上级领导审查批准,然后写出撤销案件报告。"填写案由时应当同时写明犯罪嫌疑人的姓名,如"××抢劫"案;填写原因应当根据当地法律规定,属于何种情形,如因"没有犯罪事实",或"犯罪已过追诉时效期限的";法律条款应针对撤销案件的具体原因,填写第 15 条或第 130 条。

Dispatch days 速遣日数

Dispatch money 速遣费 系指约定的装卸时间和装卸率,提前完成装卸任务,使船方节省了船舶在港口的费用开支,船方将其获得的利益一部分给租船人作为奖励的费用。

Displacement[1] 移位 钻井船或其他钻井装置在钻井作业时因受风浪、潮流等作用而产生的井位偏离。

Displacement[2] 排水量(艇) 由艇(包括所有附件)所排开的水的质量。注意:排水量以 kg 或 t 表示。 Displacement means the mass of water displaced by the craft, including all appendages. Note: Displacement is expressed in kilograms or tonnes.

Displacement △ 排水量(船舶) 系指吃水为平均夏季型吃水时,在海水(密度 1.025 t/m³)中的型排水量。 Displacement △ means moulded displacement in t, in salt water (density is 1.025 t/m³) on draught T.

Displacement and other curves 静水力曲线 是表示船舶在静水中任何平浮吃水时的各项静水力要素数值的曲线。其包括：水线面积、排水体积、排水量、漂心纵向位置、浮心纵向位置、浮心垂向位置、横稳心垂向位置、纵稳心垂向位置、每厘米纵倾力矩、每厘米吃水排水量吨数、每厘米艉纵倾排水量变化、水线面系数、舯剖面系数、方形系数和棱形系数等曲线。

Displacement craft 排水量艇 系指其重力主要由浮力支持的艇。

Displacement margin 贮备排水量 系指在新船设计阶段中，为了弥补排水量估算不足，以及其他因素对于船舶静力计算可能产生的影响，而预先计入设计排水量中的一项备用量。

Displacement method 位移法 系指从有限元计算得到基本板格（EPP）的屈曲应力和边缘应力比的方法。 Displacement method is a method to obtain the buckling stresses and edge stress ratios for elementary plate panels (EPP) from a finite element calculation.

Displacement mode 排水模式 系指不论船舶是静止还是运动时，若船舶的重力全部或绝大部分是由静水力来支承的一种状态。当 Γ 小于 3 时，这种模式与泰勒公式一起普遍地适用于小艇。但是，一些 Γ 小于 3 并针对滑行而设计的小艇，应对其在非排水模式的运行予以考虑。 Displacement mode means the regime, whether at rest or in motion, where the weight of the ship is fully or predominantly supported by hydrostatic forces. Typically this applies to craft with a Taylor Quotient Γ less than 3. However, some craft are designed to plane with Γ less than 3 and these should be considered as operating in the non-displacement mode.

Displacement mode 排水状态 对地效翼船操作模式而言，系指船舶不论在静止或运动时，其全部或大部分重力由水静力支承的一种状态。 For operational modes of the WIG crafts, displacement mode means the regime, whether at rest or in motion, where the weight of the craft is fully or predominantly supported by hydrostatic forces.

Displacement ship 排水型船 航行于水面，其重力全部靠水的浮力支承的船。

Displacement thickness of boundary layer 边界层排挤厚度 边界层外侧流线相对于理想流体流过时沿物体表面的垂向偏移距离。对于平面流：$\delta^* = \int_0^\infty (1 - U/U_0) dy$，式中，$U_0$—边界层外的流速；$U$—边界层内的流速。

Displacement volume V_D 排水体积 V_D 与排水量对应的艇所排开的水的体积。注意 1：如计算排水体积时所用水的密度不是 1 025 kg/m³ 的盐水的密度，则需规定计算排水体积所用的水的密度。注意 2：排水体积以 m³ 表示。 Displacement volume is the volume of water displaced by the craft that corresponds to the displacement mass. Note 1: Where the density of water used to calculate the volume of displacement is not salt water at a density of 1 025 kg/m³, the density of water used to calculate the volume of displacement is specified. Note 2: Displacement volume is expressed in cubic meters.

Displacement/Displacement tonnage 排水量 是用来表示船舶尺度大小的重要指标，系指船装满货物后排开水的重力，也就是船满载后受到水的浮力。根据物体漂浮的条件，即可得出下列公式：排水量（浮力）= 船自身的重力 + 满载时货物的重力。排水量通常用吨位来表示，所谓排水量吨位是船舶在水中所排开水的吨数。排水量可分为轻载排水量、标准排水量、正常排水量、满载排水量、超载排水量。

Display 显示器 系指利用其可为驾驶员提供包括常用仪表在内的视觉信息的装置。在其屏幕上可观察到图像、场景或数据的图示。 Display is the means by which a device presents visual information to the navigator, including conventional instrumentation. An observable illustration of an image, scene of data shall be observed on its screen.

Display base 基本显示 系指列于下表的海图内容，不能从显示中消除。基本显示并非用于为安全航行提供足够信息。永久保留在 ECDIS 显示器上的基本显示包括：(1) 海岸线（高水位）；(2) 本船的安全轮廓线；(3) 安全轮廓线所定的安全水域里的单独危险物；其水下深度小于安全轮廓线的水下深度；(4) 安全轮廓线所定的安全水域里的单独危险物，例如固定结构、船舶上方的电线等；(5) 比例、范围和指北针；(6) 深度和高度单位；(7) 显示模式。

Display base[1] 显示库 不能从 ECDIS 显示器上删除的信息的级别，由所有地理区域和所有情况下在任何时候都要求的信息组成，但并不打算足以用于安全航行。 Display base means a level of information which cannot be removed from the ECDIS display, consisting of information which is required at all times in all geographic areas and all circumstances. It is not intended to be sufficient for safe navigation.

Display modes[2] 显示模式 系指本船的位置保持固定且所有目标相对本船移动的一种显示。 Display modes mean the display on which the position of own ship remains fixed, and all targets move relative to own ship.

Display orientation　显示方向　包括北朝上显示、航向向上显示和船艏向上显示。Display orientation includes north up display, course up display and bow up display.

Display resolution　显示分辨率　是显示器在显示图像时的分辨率，分辨率是用点来衡量的，显示器上这个"点"就是指像素（Pixel）。显示分辨率的数值系指整个显示器所有可视面积上水平像素和垂直像素的数量。

Display station　显示器控制台　作为计算机的一个终端，主要用于为操作员提供操作控制和人机对话的手段。当操作员在键盘上向计算机发出一条键盘命令，管理程序接收命令后进行分析判断，若命令无误，且在机器状态允许时便执行该条命令，并在 CRT 监视器上打印出回答信息。操作员可随机地使用显示控制台，计算机通过中断系统接收来自显示控制台的命令或数据；在此间隙中计算机仍执行原来的程序。一般显示控制台有三个相对独立的部分组成；键盘、CRT 监视器和微处理机。

Display system　显示系统　系指提供视觉信息的电子系统。显示系统按照不同的应用，采用一种或多种、一台或多台显示设备、提供单人或成组人所需的视觉信息，接收来自不同电子设备或系统的信号，一般需要配备适当的输入装置以便实现人－机联系和必要的记录设备供以后查用。

Display　显示　系指在屏幕上可观察到的图像、场景或数据的图示。Display is an observable illusration of an image, scene or data on a screen.

Displeasing sound　不愉快声　就是使人难受，感到突然发生的声音，或给人以厌恶、不喜欢的声音。如锯齿的声音、小孩的啼哭声。

Disputes　争议　系指合同当事人中的一方认为对方不履行合同义务或履行合同义务但不符合约定的条件而引起的纠纷。

Distance along course　沿航线距离　瞬时船位到测线起始点之间沿测线方向的距离。船舶超过起始点为正，在起始点之前为负。

Distance below deck head　上部甲板至下端的距离　系指自板材至下方的距离。Distance below deck head means the distance below the plating.

Distance cross course　正横航线距离　瞬时船位偏离测线的正横距离。船舶偏测线右边为正，偏测线左边为负。

Distance piece in boilers and pressure vessels　锅炉及压力容器的隔片　保持与锅炉或压力容器直接相连的法兰或竖管以及与锅炉或压力容器直接相连的阀门或各种计量器之间距离的隔片。Distance piece in boilers and pressure vessels is a piece used for keeping the distance between flange or stand pipe attached directly to boilers and pressure vessels, and valve attached directly to boilers and pressure vessels, or gauges.

Distance sailed　航行距离　系指在所考虑的航次或时间段的实际航行距离（单位 n mile）（甲板航海日志数据）。Distance sailed means the actual distance sailed in nautical mile (deck log-book data) for the voyage or period considered.

Distance to the end along course　到航线终点距离　瞬时船位到测线终点之间沿测线方向的距离。船舶在终点之前为正，超过终点为负。

Distillate　蒸馏水　用蒸馏方法得到的成品水。

Distillate fuel　馏分油　系指输送到船上并用于燃烧的燃油，其在 40 ℃时的运动黏度小于或等于 11.00 cs（mm^2/s）。Distillate fuel means the fuel oil for combustion purposes delivered to and used on board ships with a kinematic viscosity at 40 ℃ lower than or equal to 11.00 centistokes (mm^2/s).

Distillate heat loss　蒸馏水带热损失　单位蒸馏水带走的热量与单位给水带入的热量之差。

Distillate teat tank　蒸馏水检验柜　用以检验海水淡化装置制造出来的蒸馏水的质量的专用柜。

Distillates　馏分油

Distillation methods　蒸馏法　利用水在一定的压力、温度条件下发生相变的特性，使海水蒸发，随后冷却蒸汽使之凝结成淡水的舰船上最常用的海水淡化方法。

Distillation plant　蒸馏装置　按蒸馏法制造淡水的装置。

Distilled water tank　蒸馏水柜　贮存海水淡化装置制造出来的蒸馏水的柜。

Distiller　蒸馏器　通常由蒸发器或闪发室、汽水分离器和冷凝器等组合而成的蒸馏装置的主体。

Distilling plant　蒸馏设备　船用的海水淡化装置。其几乎都是蒸发式。又分为真空式（vacuum pressure type）和大气压式（atmospheric pressure type）2 种。由于后者在使用时要求将海水加热到 100 ℃以上，而在热效率及水垢的清除方面存在弊端，所以较少被采用。真空式又分为潜式（submerged tube type）和闪发式（flash type）2 种。这 2 种方法都是利用主机冷却淡水系统的余热进行工作的。闪发式的优点是不易结水垢，且不大受船舶摇晃的影响，但缺点是其产淡水量受海水温度影响较大。

Distilling ship　海水淡化船　设有海水淡化设备，专为舰艇、岛屿或哨所等应急供应饮用水的船舶。

Distress phase　遇险阶段　有理由确信船舶或人员有严重和紧急危险而需要立即救援的情况。Distress phase means a situation wherein there is a reasonable certainty that a vessel or a person is threatened by grave and imminent danger and requires immediate assistance.

Distress situations 遇险情况 推进和/或操舵能力缺失,或者由于其他原因使船舶不能航行(弃船状况前的情况)。 Loss of propulsion and/or steering, or when the ship is not seaworthy due to other reasons (situation prior to abandoning ship).

Distress, urgency and safety alarms 危难、紧迫和安全警报 系指某一呼叫者处于危难中或有紧迫信息须发送的警报。 Distress, urgency and safety alarms mean those alarms which indicate that a caller is in distress or has an urgent message to transmit.

Distribution board 分配电板 系指:(1)用于控制和分配电能至最后分路的开关设备和控制设备组件;(2)用来对最后分路进行控制和配电的一个或多个过电流保护装置的组合装置。见图 D-9。 Distribution board is:(1) an assembly arranged for the control and distribution of electrical power to final sub-circuits;(2) an assembly of one or more over-current protective devices for the control and distribution of electrical power to final sub-circuits. It can be seen in Fig. D-9.

图 D-9 分配电板
Figure D-9 Distribution board

Distribution center 配电中心 系指通常由接于汇流排的过载自动保护装置组成的设备所布置之处所。其主要功能是分配供电以及对于馈线、副馈线或分支电路,或对于这三者的任意组合进行控制和保护。

Distribution indexes for noise control 噪声控制指标分配 系指考虑噪声的传递与衰减确定各噪声源对目标舱室噪声级的贡献量,并针对影响目标舱室噪声级的主要噪声源或减振降噪措施的声学要求。包括:(1)复合岩棉板随频率变化的隔声量、吸声系数、阻尼损耗因子;(2)敷设各种材料涂层的甲板随频率变化的隔

声量、吸声系数、阻尼损耗因子;(3)消声器的消声量坐标;(4)各主要噪声源设备在陆上台架弹性安装的机脚振动指标限值和空气辐射噪声指标限值(10 Hz~8 kHz);(5)各主要噪声源设备的隔振装置隔振效果指标;(6)其他机械设备降噪措施的振动和噪声指标限值。

Distribution mode of electric power 配电方式 系指主配电板和区配电板、分配电板及其与用电设备之间的电缆和电线的连接和布设方式。通常应根据各用电设备的具体要求,并考虑整个电力系统的可靠性、灵活性、经济性及操作管理方便等因素,选择合理的方式。目前,采用的方式有:(1)馈线式——系指各区配电板、分配电板和重要负载分别由各自馈线电路直接从主配电板获得供电的输配电的方式,也称干馈混合式。其主要优点是:①可降低电缆数量;②保护装置数量少,便于维护;③主要供电开关均在一个控制场所,便于集中控制;④增加用电设备比较方便;⑤与环路式相比,设备价格便宜。(2)环路式——系指干线形成闭合环路,所有用电设备均从该环路获得供电的输配电方式。其主要特点是:①与馈线式相比,可靠性高;②可降低电压波动和功率损耗,但其维护复杂,价格较昂贵。此外,还有干线式、棋盘式、两舷供电式和混合式等。

Distribution stations 分配站 是有足够防护面积的空间,它配备截止阀、压力调节装置、压力计、止回阀和(连在焊枪上)氧气、乙炔软管。 Distribution stations are adequately protected areas or cabinets equipped with stop valves, pressure regulating devices, pressure gauges, non-return valves and oxygen as well as acetylene hose connections for the welding torch.

Ditching 迫降 气垫船由垫升航态突然丧失垫升力而转变为排水航行的状态。

Diurnal anchored continuous observation 周日连续观测 船舶到达观测站抛锚后,按一定的时间间隔,持续一昼夜的观测。

Diver training vessel 潜水训练船

Diver's boat /Diving vessel 潜水工作驳/船 系备有加压舱、潜水用器材和配气设备等,保障潜水员进行潜水作业用的驳/船。

Diver's ladder 潜水梯 供潜水员从潜水工作船上下潜和从水下登船用的软梯。

Divergence angle of shaft 轴系外斜角 外斜轴系中线与船体中线面间的夹角。

Divergence damping 发散衰减 系指声源辐射的声波在传播过程中,波阵面随距离的增加而增大,声能扩散,因而声强或声压随距离的增加而衰减。也称距离衰减。

Divergent waves 散波 船舶航行时,从艏、艉或突肩处产生的由两舷向外并向后倾斜扩张的波系。

Diverging shafting 内斜轴系 轴系中线从船首端到船尾端是渐近中线面的轴系。

Diverter flex joint 分流器柔性接头

Diving apparatus room 潜水器材舱 存放潜水器材用的舱室。

Diving bell 潜水钟 是一种无动力单人潜水运载器。由于早期的潜水器是由一个底部开口的容器,外形与钟相似,故得此名。现代潜水钟大多数已改为全封闭结构,外形也有很大改变,但仍沿用旧名。潜水钟内的空间里都充满新鲜的空气,以保证潜水员的生存。它是潜水员从甲板居住舱到海底作业,工作结束后回到甲板居住舱的运载设备,呈圆柱形或球形,可承受内压和外压,一般可容纳 3 名潜水员,潜水钟底部有一通道供潜水员出入。潜水钟以钟脐带同水面的潜水工作母船连接。

Diving bell handling system 潜水钟吊放系统 是保障潜水钟吊放及与居住舱对接的设备。

Diving cabin 潜水工作间 设有供气、配气及有关仪器、仪表等设备,供指挥和控制潜水作业用的工作舱室。

Diving operation 潜水作业 系指人在水下环境中从事的各种活动。

Diving support boat 潜水支持船 它是水下作业的水面操作平台。船上通常设有饱和潜水系统,包括潜水钟、减压舱和高压救生艇等。大多数潜水支持船都配有 ROV,包括观察级 ROV 和作业级 ROV。

Diving support vessel(DSV) 潜水作业母船 又称潜水支援船。能提供或搭载潜水设备所需的各种配套设备,并能确保潜水设备、生命支持系统和辅助机械设备正常运行的潜水作业专用的船舶。见图 D-10。

图 D-10 潜水作业母船
Figure D-10 Diving support vessel(DSV)

Diving system 潜水系统 系指海上移动式钻井平台上安全进行潜水作业所必需的装置和设备。Diving system is the plant and equipment necessary for the safe conduct of diving operations from a mobile offshore drilling unit.

Diving System Safety Certificate 潜水系统安全证书 系指经检验或检查后,对符合潜水系统安全规则要求的潜水系统,由主管机关或其正式授权的任何个人或机构签发的证书。在任何情况下,主管机关均应对该证书承担全部责任。 Diving System Safety Certificate means a certificate issued either by the Administration or any person or organization duly authorized by it after survey or inspection to a diving system which complies with the requirements of the Code of Safety for Diving Systems. In every case, the Administration should assume full responsibility for the certificate.

Diving vessel/divers boat 潜水工作船 备有加压舱、潜水用器材和配气设备等,保障潜水员进行潜水作业用的船舶。

Division board for berths 铺间挡板 相邻两铺位之间的分隔板。

DMT DMT 迷幻剂之一,外形为小的黑籽或磨细的黑色/褐色粉末。没有气味。 DMT is one of hallucinogens, which comes either as small black seeds, or as a finely ground black/brown powder. It is odourless.

DMZ 国际标准组织(ISO)制定的标准"ISO 8217:2010"(船用燃油标准——直馏油)中新增加的一档燃油规格,它的运动黏度为 3.000cSt(40 ℃),制定此项标准的主要原因是为了配合船上柴油机使用低硫燃油的要求。

Dock landing ship(DLS) 船坞登陆舰 就是可承载两栖登陆艇、两栖坦克和气垫船的战舰。其船舱是半吃水状态以方便两栖登陆艇、两栖坦克和气垫船的进出,就像船坞一样。其作战方式主要以承载为主,将参与两栖攻击的两栖登陆艇、两栖坦克或气垫船送至距离海岸线最佳的距离,由于船坞登陆舰一般都比较大在万吨以上,因此可作为海上两栖攻击临时基地,为滩头补充弹药和给养。船坞登陆舰上的武器一般以防空武器为主,必要时也可以对滩头进行射击。

Dock moving equipment 移坞装置 在水上移动浮船坞的装置。

Dock receipt/receiving note 码头收据 又称场站收据、港站收据。系指一般由发货人或其代理人根据公司已制定的格式填制,并跟随货物一齐运至集装箱码头堆场,由接受货物的人在收据上签字后交还给发货人,证明托运的货物已到的一种单据。

Dock sharing 并坞 系指合用一个船坞。

Dock tonnage dues 码头捐

Docking 靠泊 系指操纵船舶靠拢码头,并控制系泊作业。 Docking means a manoeuvring the ship alongside a berth and controlling the mooring operations.

Docking bracket 坞墩肘板 系指在双层底内为进坞而局部加强船底结构的肘板。 Docking bracket means a bracket located in the double bottom to locally strengthen the bottom structure for the purpose of docking.

Docking experiment 抬船试验 浮船坞在搁置有船舶时进行的下沉起浮试验。

Docking keel 坞龙骨 在大型船舶的船底外部两侧，专为船舶入坞坐墩而设的纵向箱形构件。

Docking operations 靠泊作业 系指操纵船舶靠拢码头，并控制系泊作业。 Docking operations mean manoeuvring the ship alongside a berth and supervising the mooring operations.

Docking plan 进坞平面图 是说明所有设计过程中做出的任何或全部假定，包括（但不限于）坞墩的布置、进坞时允许的最大装载量以及每个坞墩上的相应载荷的图纸。 Docking plan is a plan indicated any and all assumptions made during the design, including but not limited to, the arrangement of docking blocks, the maximum permissible loading during docking and the corresponding load at each block.

Docking strength 坐坞强度 船舶进坞置于坞墩上时，船体结构承受相应载荷的能力。

Docking survey（DS） 坞内检验 每两年半在坞内或船排上进行一次的检验。主要有下列检验项目：(1)水下部分船体外板、舵的检查；(2)舵间隙的测量；(3)螺旋桨的着色检查；(4)测量艉管轴间隙或艉轴下沉量；(5)通海阀的开放检查；(6)锚链直径的测量。坞检可有条件地使船舶不进坞，由潜水员在水下对外板、螺旋桨、通海阀进行探摸，验船师在水面上通过水下摄像装置观看的水下检测来代替。

Docking workstation 靠泊作业工作站/进坞工作站 系指：(1)一个配备有用于船舶进坞所必需装置的处所；(2)桥楼翼台上，在靠泊、锁定航道、引航员的转移等过程中可以操纵船舶的工作站。 Docking workstation is: (1) a place equipped with necessary means for docking the craft; (2) a workstation in the bridge wings, from which the ship can be operated during berthing, lock passage, pilot transfer, etc.

Doctrine of discretional evaluation of evidence 自由心证制度 系指"证据之证明力，通常不以法律加以拘束，听任裁判官之自由裁量"。自由心证是以证据的存在为前提，而不是以单纯的"自由"心证而认定事实。所谓允许心证者，不是证据的能力即何种资料具有证据的资格，而是证据的证明力即证据的价值。证据证明力的有无及大小，允许法官自由判断、选择、舍去，而无法律形式上的约束，以利于发现案件的实体真实。法官根据法庭审理过程中形成的内心信念自由裁断证据证明力的大小。因为，证明力只能由法官根据长期裁判活动中形成的经验，结合个案的具体情况做出判断，而无法由一条一般性的规则事先规定。

Document 证件 系指：(1)证明设计、产品、服务和过程符合规定要求的正式文件；(2)带有信息登记项的信息载体。 Document means: (1) a formal document showing compliance of a design, product, service or process with specified requirements; (2) a data carrier with data entries.

Document assess 文件评审 系指以评价公司已建立船舶安全管理体系，并符合 ISM 规则的要求以及有关规范要求的评审。

Document of authorization for the carriage of grain 谷物装运的批准文件 系指对每艘按照国际散装谷物安全装运规则装载的船舶，由主管机关或其认可的组织或代表缔约国政府签发的一份批准文件。该文件应随同或包括在所提供的谷物装载手册之内，以使船长能满足该规则的稳性要求。 Document of authorization for the carriage of grain means a document of authorization issued for every ship loaded in accordance with the regulations of the International Code for the Safe Carriage of Grain in Bulk either by the Administration or an organization recognized by it or by a Contracting Government on behalf of the Administration. The document shall accompany or be incorporated into the grain loading manual to enable the master to meet the stability requirements of the Code.

Document of Authorization to conduct Ship Recycling（DASR） 拆船厂授权书

Document of compliance（DOC）[1] DOC 证书 由船旗国的主管机构或经船旗国政府授权的有 ISM 审核资格的组织签发给符合 ISM 规则要求船舶管理公司的证书。该组织大多数是船级社，也有一些海运业的咨询公司从事这项工作。DOC 证书有长期、短期和临时之分。

Document of compliance（DOC）[2] 符合证明 系指符合 ISM 规则要求的每一公司签发的符合证明。船上应存有一份该证明的副本。 Document of compliance means a document of compliance issued to every company which complies with the requirements of the ISM Code. A copy of the document shall be kept on board.

Document of compliance with the special requirements 符合船舶载运危险货物特别要求的文件 系指一份证明符合"1974年国际海上人命安全公约"第Ⅱ-2/54.3条中有关构造和设备要求的适当的文件。

Document of compliance with the special requirements for ships carrying dangerous goods 载运危险货物船舶特殊要求的符合证明 系指主管机关为船舶提供一份适当的证明。作为其构造和设备符合1974年SOLAS公约第Ⅱ-2章第19条要求的证据。除固体散装货物外,对于被确定为第6.2和7类货物以及数量有限的危险货物,不要求危险货物证书。 Document of compliance with the special requirements for ships carrying dangerous goods means an appropriate document provided to the ship by the Administration, as evidence of compliance of construction and equipment with the requirements of regulation Ⅱ-2/19 of SOLAS 1974. Certification for dangerous goods, except solid dangerous goods in bulk, is not required for those cargoes specified in Class 6.2 and 7 and dangerous goods in limited quantities.

Documentary bill 跟单汇票 又称信用汇票、押汇汇票,是需要附带提单、仓单、保险单、装箱单、商业发票等单据,才能进行付款的汇票,属于有价证券的范畴。商业汇票多为跟单汇票,在国际贸易中经常使用。跟单汇票与光票相对应,系指附带有商业单据的汇票。银行汇票一般为光票,商业汇票一般为跟单汇票。

Documentary bills 押汇 又称买单结汇,系指议付行在审单无误的情况下,按信用证条款买入受益人(外贸公司)的汇票和单据,从票面金额中扣除从议付日到估计收到票款之日的利息,将余款按议付日外汇牌价折成人民币,拨给外贸公司。议付行向受益人垫付资金买入跟单汇票后,即成为汇票持有人,可凭票向付款行索取票款。银行做出口押汇,是为了对外贸公司提供资金融通,利于外贸公司的资金周转。商业银行为进口商开立信用保证文件的这一过程,称为进口押汇。

Documentary credit 跟单信用证 是银行用以保证买方或进口方有支付能力的凭证。在国际贸易活动中,买卖双方可能互不信任,买方担心预付款后,卖方不按合同要求发货;卖方也担心在发货或提交货运单据后买方不付款。因此需要两家银行作为买卖双方的保证人,代为收款交单,以银行信用代替商业信用。银行在这一活动中所使用的工具就是信用证。可见,信用证是银行有条件保证付款的证书,成为国际贸易活动中常见的结算方式。按照这种结算方式的一般规定,买方先将货款交存银行,由银行开立信用证,通知异地卖方开户银行转告卖方,卖方按合同和信用证规定的条款发货,银行代买方付款。

Documentary evidence 书面审理

Documentation 文件(文件化) 系指描述设计、过程、产品或服务的所有必需的相关图纸、文件和计算报告等。 Documentation means all necessary written information regarding design processes, products or services and includes relevantplans, papers and calculation reports.

Documentation and change control 文件及其修改的控制 对质量管理体系而言,系指制造厂商应编制和管理有关本方案要求的所有文件。这种管理将保证做到:(1)文件在使用前,由授权人员检查和认可其适当性,并包括认可和修改情况的指示;(2)所有对文件的修改都应以书面形式,保证发至各有关部门,并防止使用不适用的文件;(3)应从所有发放或使用部门迅速收回已作废的文件;(4)文件经多次修改后,应发放新的版本。 For the quality management system, documentation and change control means that the manufacturer is to establish and maintain control of all documentation that relates to the requirements of this scheme. This control shall ensure that: (1) documents are reviewed and approved for adequacy by authorized personnel prior to use, are uniquely identified and include indication of approval and revision status; (2) all changes to documentation are in writing and are processed in a manner that will ensure their availability at the appropriate location and preclude the use of non-applicable documents; (3) provision is made for the prompt removal of obsolete documentation from all points of issue or use; (4) documents are to be re-issued after a practical number of changes have been issued.

Dolomite 白云石 是一种浅黄色/褐色矿石,非常坚固和密实。有时会将一种由钙和镁的氧化物组成的物质(白云石生石灰)错误地称为"白云石"。无特殊危害。该货物为不燃物或失火风险低。 Dolomite is a light yellow/brown coloured mineral stone which is very hard and compact. Dolomite may sometimes, incorrectly, be used to describe a material consisting of the oxides of calcium and magnesium (dolomite quicklime). It has no special hazards. This cargo is non-combustible or has a low fire-risk.

Documents 单证 系指货物运输的凭证。

Documents against acceptance(D/A) 承兑交单 系指出口人的交单以进口人在汇票上承兑为条件。即出口人在装运货物后开具远期汇票,连同商业单据,通过银行向进口人提示,进口人承兑汇票后,代收银行即将商业单据交给进口人,在汇票到期时,方履行付款义务。所谓"承兑"就是汇票付款人(进口方)在代收银行提示远期汇票时,对汇票的认可行为,付款人于汇票到期日凭票付款。承兑交单方式只适用于远期汇票的托收。

Documents against payment(D/P) 付款交单 是经济贸易交易中付款方式的一种。是出口人的交单以进口人的付款为条件,即出口人将汇票连同货运单据交给银行托收,指示银行只有在进口人付清货款时,才能交出货运单据。即期交单（D/P Sight）系指出口方开具即期汇票，由代收行向进口方提示，进口方见票后即须付款，货款付清时，进口方取得货运单据。远期汇票的付款日期又有"见票后××天付款""提单日后××天付款"和出票日后××天付款"3种规定方法。

Documents against payment at sight (D/P at sight) 即期付款交单 系指（1）出口方开具即期汇票，通过代收银行向进口方提示、进口方见票后必须立即付清货款才能领取货运单据的付款交单方式。远期付款交单是卖方给予买方的资金融通，融通时间的长短取决于汇票的付款期限，通常有两种规定期限的方式：一种是付款日期和到货日期基本一致。买方必须请求代收行同意其凭信托收据（T/R）借取货运单据，以便先行提货。（2）表示出口人按合同规定装货后，填写托收委托书，开出即期汇票，连同全套货运单据送交银行代收货款。

Documents preparation for bank negotiation 制单结汇 系指制单结汇收款。出口货物装运以后，出口公司应按照信用证的规定，正确缮制各种单据，在信用证规定的有效期内递交银行办理议付结汇手续。

Dog 压紧器 压紧装置中直接压紧舱盖板的器件。

Dog type cable clench 闩式弃链器 一种以横闩扣住锚链的弃链器。

Dogging device 舱盖压紧装置 压紧舱盖板四周的填料，以达到密性要求的专用装置。一般有电动或手动、液压、气胀等形式。

Dogging wedge 压紧楔 利用斜面压紧舱盖板密封填料的专用楔块。

Domain name service (DNS) 域名服务 系指一个协议，它提供主机和域名的一个Internet范围内的数据库。例如，DNS用于查找主机名写作 microsoft.com 的IP地址。域名地址的使用为用户上网提供了极大的方便，但域名地址不能直接用于TCP/IP协议的路由选择。当用户输入域名地址上网时，必须通过域名服务器进行域名解析。域名解析，就是把 Internet 上主机的域名地址解析成 IP 地址，或者主机的 IP 地址解析成域名地址的过程。这项工作由域名服务器来完成，也称域名服务，是 Internet 的一项核心。

Domestic projector 家用投影机 是各大投影机厂家，专门针对在家庭里面使用来看电影的家庭影院投影机。如今家用投影机和普通商务投影机的最大区别在于分辨率，目前大部分主流家用投影机的分辨率已经达到了1080P全高清。家用投影机主要是追求对比度，层次感，真实感。一般的特性主要为低亮度流明，在800~1300流明左右（商用机都是高流明的，所以人们看到超过1500流明以上亮度会影响画面的对比度和细腻感），对比度按照趋势来看，入门级家用投影机的对比度都在

10 000∶1以上，主流的分辨率都是1 920×1 080的，简称1080P的家用投影机，也有720P的，不过已逐步停产。

Domestic railway transport 国内铁路运输 系指仅在本国范围内按《国内铁路货物运输规程》的规定办理的货物运输。

Domestic wastes 家居生活污水（生活废弃物） 系指：（1）在船上居住处所中生成的各种类型的生活污水，但残羹剩饭除外；（2）在 MARPOL 公约其他附则未涵盖的在船上起居处所产生的所有类型的废弃物。生活废弃物中不包括灰水。 Domestic wastes mean: (1) all type of wastes generated in the living spaces on board a ship, except victual wastes; (2) all type of wastes not covered by other Annexes to MARPOL that are generated in the accommodation spaces on board the ship. Domestic waste does not include grey water.

Domestic water system 生活用水系统 船上食用和洗涤用的冷热水系统。

Donor port 供体港 系指船舶装载压载水上的港口或地点。 Donor port is a port or location where the ballast water is taken onboard.

Door buffer 门顶缓冲器 安设在门上方，关门时起缓冲作用的装置。

Door closed mode for emergency conditions 应急状态关闭门模式 系指在应急情况下使用的，能自动关闭任何一扇开启着的水密门，同时可就地开启任何门，而在脱开其就地控制机构后能将其中的再关闭的模式。 Door closed mode for emergency conditions means a mode which is to automatically close any watertight that is open and permit doors to be opened locally and to automatically reclose the doors upon release the local control mechanism in emergency conditions.

Door frame 门框 与门板紧贴在门开口周沿的框架。

Door plate 门板 盖闭门开口的板件。

Door sill 门槛 门框最下边至甲板间的结构。对于露天甲板或侧壁甲板以上的门，门槛高度应符合有关规定。

Door stopper 防撞门钩 对门起防撞和固定作用的专用钩。

Doors used while at sea 在海上使用的门 系指滑动水密门。该门能从驾驶室遥控关闭，也能从该舱壁的每一侧就地操纵。在控制位置应装设显示门是开启或关闭的指示器，并且在门关闭时发出听觉报警。在主动力失灵时，动力、控制和指示器应能操作。特别应注意减少控制系统失灵的影响。每一扇动力操纵的滑动水密门应有一个独立的手动机械操纵装置。该装置应能

从该门的任何一侧用手开启和关闭该门。Doors used while at sea means sliding watertight doors capable of being remotely closed from the bridge and are also to be operable locally from each side of the bulkhead. Indicators are to be provided at the control position showing whether the doors are open or closed, and an audible alarm is to be provided at the door closure. The power, control and indicators are to be operable in the event of main power failure. Particular attention is to be paid to minimize the effect of control system failure. Each power-operated sliding watertight door is to be provided with an individual hand-operated mechanism. It should be possible to open and close the door by hand at the door itself from both sides.

Doppler count 多普勒计数 卫星电波的发射频率和观察者测得的接收频率是变化的,这两个频率之差叫作多普勒频移。它反映了卫星和观察者之间的距离变化。在一定的采样时间内把多普勒频移积累起来,即采用计数的办法得到采样时间内的总相位周数称为多普勒计数,又叫多普勒积算值。

Doppler current meter 多普勒海流计 利用声学多普勒原理测量海流的仪器。仪器向海水中发射声波,测量随海水流动的悬浮粒子所散射声波的多普勒频移,得到流速值。

Doppler sonar 多普勒声呐 利用声波的多普勒原理制成的精密测速和计程的仪器。它是一种脉冲多普勒声呐系统,利用4个晶体组合向海底前、后、左、右方向发出4束声波。每个晶体具有发送、接收两种功能。因此从4个声波束中提取多普勒信息,再计算出航速和航程。多普勒声呐的测速精度为 0.1% ~ 0.5%。多普勒声呐的主要特点是:(1)它所测得的航速和航程是相对于海底的称为"绝对航速"和"绝对航程"。一般多普勒声呐的工作水深约 460 m 之内,此时多普勒声呐工作在跟踪海底方式。超过工作水深时,多普勒声呐工作在跟踪水团的方式,测得的航速和航程是相对于海水的称为"相对速度"和"相对航程"。但精度较低。(2)它可测定浅水和低航速航行时的航速。测速的最浅深度允许在 0.3 ~ 0.5 m。可测的最低航速为 0.01 kn (0.005 m/s)。见图 D-11。

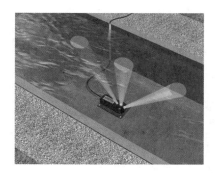

图 D-11 多普勒声呐
Figure D-11 Doppler sonar

Double bearing rudder 双支承舵 舵叶上、下均设有支承件的舵。

Double block and bleed valve 连锁气体阀 系指安装在每台气体燃料发动机的燃料供应管路上的一套3个的自动阀。Double block and bleed valve means a set of three automatic valve located at the fuel supply to each of the gas engine.

Double block conical shape 黑双锥号型 正、倒两个黑锥号型共用一个顶端所构成的号型。

Double bottom 双层底 船底、内底及两者之间的船底结构与空间的统称。

Double bottom compartment 双层底舱 双底船上,位于内底板以下的舱。

Double bottom ship 双底船 设有内底的船舶。

Double bottom structure 双层底结构 系指为内底顶部以上具有肋板的外板和内底板及其以下的其他构件的总称。注意:底边舱斜板应视为纵舱壁。Double bottom structure is defined as shell plating with stiffeners below the top of the inner bottom and outer elements below and including the inner bottom plating. Note that stopping hopper tank top side is to be regarded as longitudinal bulkhead.

Double classed ship 双船级船舶 系指某一船舶在两家船级社入级,且每家船级社均按照各自的规范和检验时间,独立完成所有的检验。Double classed ship is one which is classed by two Societies and each Society works as if it is the only Society classing the ship, and does all surveys in accordance with its own requirements and schedule.

Double decked ship 双甲板船 设有两层甲板的船舶。

Double factorial terms of trade 双因素贸易条件 双因素贸易条件下不仅考虑到出口商品劳动生产率的变化,而且考虑到进口商品劳动生产率的变化。其计算方法为:$D = (P_X/P_M) \times (Z_X/Z_M) \times 100\%$ 其中:D—双因素贸易条件;P_X—出口商品价格指数;P_M—进口商品价格指数;Z_X—出口商品劳动生产率指数;Z_M—进口商品劳动生产率指数。其结果大于1时,表明该国贸易条件得到改善;小于1时,则表明该国的贸易条件恶化。

Double-fluked anchor 两爪锚 锚头下端装有

固定或可拆的横档,并具有两个固定的钩爪的锚。

Double hook (Ramshorn hook) 山字钩 用于重型吊杆装置呈"山"字形的双钩吊货钩(拉姆斯霍恩吊货钩)。

Double house planer 龙门刨床 主要用于刨削大型工件,也可在工作台上装夹多个零件同时加工,是工业的母机。龙门刨床因有一个由顶梁和立柱组成的龙门式框架结构而得名,工作台带着工件通过龙门框架作直线往复运动,多用于加工大平面(尤其是长而窄的平面),也用来加工沟槽或同时加工数个中小零件的平面。

Double hull craft (catamaran) 高速双体船 系指具有 2 个片体的高速双体船。片体具有瘦长的特点,减少了水对船的阻力;另外 2 个片体的距离较大,双体船具有较宽的船身,有很好的稳性,航行安全。在与水翼船和气垫船的竞争中具有相当的优势。

Double hull oil tanker 双壳油船 系指:(1)为散装运输油品而建造的船舶,其货油舱由与整个货油区域等长的双壳船体保护,有双层侧壁和双层底处所的油船;(2)主要用于运输散装油类的船舶。其货油舱受双层船壳保护,该双层船壳覆盖货物区域的全长;包括用于装载压载水的双侧边舱和双层底处所或空舱。 Double hull oil tanker is:(1)a ship which is constructed primarily for the carriage of oil in bulk, which has the cargo tanks and cargo area protected by a double hull which extends for the entire length of the cargo area, consisting of double sides for the carriage of water ballast and double bottom spaces or void spaces;(2)a ship which is constructed primarily for the carriage of oil in bulk, which have the cargo tanks protected by a double hull which extends for the entire length of the cargo area, consisting of double sides for the carriage of water ballast and double bottom spaces for void spaces.

Double insulation 双重绝缘 系指由基本绝缘和辅助绝缘两者组成的绝缘。 Double insulation means the insulation comprising both basic insulation and supplementary insulation.

Double leaves door 两截门 常用于厨房,分上下两截,可根据通风或起分隔作用的不同要求来任意开闭全部或其中一截的门。

Double model 叠合船模 用 2 只相同船模的水下部分,倒正叠合而成的一个船模。

Double plate bulkhead 双板舱壁 由 2 层舱壁板及其间的骨架组成的舱壁。

Double plate rudder 覆板舵 舵叶由舵板覆合于构架 2 面而成的舵。

Double pressure boiler (double evaporation boiler, dual circulation boiler) 双路循环锅炉 用于油船的,

其汽、水循环有 2 个相互独立的回路,在第一回路中所产生的汽和热水被用作第二回路的热源,船上所需的蒸汽则全部由第二回路供应的锅炉。

Double shield riser 双屏立管

Double skin bulk carrier 双舷侧散货船 系指通常为在货物长度区域具有单层甲板、双层底、顶边舱和斜边舱,以及具有双舷侧结构(不论边舱的宽度如何)的自行推进海船,主要用于载运散装干货,且包括如矿砂船之类的船型。 Double skin bulk carrier is a sea going self-propelled ship which is constructed generally with single deck, double bottom, topside tanks and hopper side tanks and with double side skin construction in the cargo length area (regardless of the width of the wing space) and is intended primarily to carry dry cargo in bulk and includes such types as ore carriers.

Double skin member 双壳构件 双壳构件定义所指的构件,系由理想化的梁组成腹板,其顶折边和底折边由附板构成。 Double skin member is defined as a structural member where the idealized beam comprises webs, with top and bottom flanges formed by attached plating.

Double stage compression refrigerating system 双级压缩制冷系统 制冷剂经过低压级和高压级制冷压缩机后进入冷凝器放热液化的制冷系统。

Double tube sheet 双管板 由 2 层板叠置而成的管板。

Doubled up swinging boom 摆动双杆操作 当被装卸的货物重量超过一根单杆的安全工作负荷时,利用 2 根吊杆同时起吊,而且 2 根单杆随同货物同时摆向舷外或船内的操作方法。

Double-flow steam turbine 双流式汽轮机 蒸汽由气缸中间进入,向两端流动或由气缸两端进入,向中间流动的汽轮机。

Double-fluked anchor 双爪锚 锚头下端装有固定或可拆卸的横杆,并具有 2 个固定的钩爪的锚。

Double-hull tank 双壳液舱 系指包括双层底舱和舷侧液舱,即便这些液舱是独立的。

Double-hydrofoil craft 双水翼艇 在艏、艉部各装有水翼的水翼艇。

Double-pass boiler 双烟道锅炉 具有双侧排烟烟道的锅炉。

Double-plate clutch 双盘式离合器 有两个摩擦盘的离合器。

Double-purchase winch (double-drum winch) 双卷筒绞车 具有两个卷筒的绞车。

Doubler 覆板 系指一小板块用以附连到一大块要求在确定位置进行加强的板,通常在加强筋的连接

处。 Doubler means a small piece of plate which is attached to a larger area of plate and it requires strengthening in that location, usually at the attachment point of a stiffener.

Double-reduction gear (double-gear, double-reduction gear) 二级减速齿轮箱 通过两对齿轮的啮合作用,实现二次减速的减速齿轮箱。

Double-shaft reduction gear 双轴并列减速齿轮箱 由汽轮机高、低压气缸拖动一个轴系的减速齿轮箱。

Double-side skin 双舷侧 系指船舶每侧均为由舷侧壳板与纵舱壁组成的构造形式,该纵舱壁连接双层底和甲板。底边舱和顶边舱(如设有)可为双舷侧结构的组成部分。 Double-side skin means a configuration where each ship side is constructed by the side shell and a longitudinal bulkhead connecting the double bottom and the deck. Hopper side tanks and top-side tanks may, where fitted, be integral parts of the double-side skin configuration.

Doublet 偶极子 轴距趋向于零,力矩不变的一对源与汇(由源至汇为偶极子的轴,其轴距与源强乘积为其力矩)。

Down comer (downcast, down-take pipe) 下降管 锅水在其中作下降运动的水管。

Down flooding angle 进水角 是在船体、上层建筑或甲板室的开孔处进水时的最小横倾角。这些开孔在进水和允许注水时,不能保持风雨密。除了舰船主管机关指定的稳性和分舱规范要求进水角大于 40°外,最小进水角 θ_{df} 应为 40°。 Down flooding angle is the least angle of heel of openings in the hull, superstructure or deckhouses, which cannot be closed weathertight, immerse and allow flooding to occur. The minimum required angle of down flooding, θ_{df}, is to be taken as 40° except when a higher minimum angle is required by the stability and subdivision standard specified by the Naval Authority.

Down payment 入门费 是引进方在签订合同之后一段时间内,在支付提成费之前,按合同规定的一笔预先约定好的固定金额,有时也叫作定金。

Down-cast ventilator 进风帽 供吸入空气用的通风帽。

Down-flooding 向下进水 系指通过因要符合完整稳性或破损稳性标准而不能视具体情况关闭成水密或风雨密的开口或因操作原因需要保持敞开的开口,造成平台的浮体结构任何内部的任何进水。 Down-flooding means any flooding of the interior of any part of the buoyant structure of a unit through openings which cannot be closed watertight or weathertight, as appropriate, in order to meet the intact or damage stability criteria, or which are required for operational reasons to be left open.

Down-flooding angle curves 进水角曲线 系指进水角和排水体积(或排水量)的关系曲线。

Down-flooding point 向下进水点 系指:(1)可让水通过水密/风雨密结构上的任何尺寸开口(如敞开的窗),但不包括按相应水密/风雨密标准始终保持关闭,仅在应急情况下需用于人员出入或用于操作移动式舱底潜水泵的开口(例如,与其所安装处的结构具有类似强度和风雨密完整性的非开启窗)。(2)当船在完整或破损状态下,横倾至越过平衡角的一个角度时,使构成储备浮力的处所可能发生进水的任一开口。 Down-flooding point means: (1) any opening, irrespective of size, that would permit passage of water through a water/weathertight structure (e.g. opening window), but excludes any opening kept closed to an appropriate standard of water/weather-tightness at all times other than when required for access or for operation of portable submersible bilge pumps in an emergency (e.g. non-opening windows of similar strength and weather-tight integrity to the structure in which they are installed); (2) any opening through which flooding of the spaces which comprise the reserve buoyancy could take place while the craft is in the intact or damage condition and heels to an angle past the angle of equilibrium.

Downstream industry 下游产业

Dozer dredger 钢扒船

DP accident 动力系统事故 自动控制丧失、位置丧失或已经导致或必将导致红色警戒的事故。

DP class notation DP 入级附加标志 船级社,例如 DNV 使用的入级附加标志,对 DP 船舶应基于 IMO 的设备等级原则。

DP control location DP 控制位置 在 DP 船舶或平台装置上永久有人的位置,DP 操作员在该位置可监控 DP 系统的性能,DP 操作员在该位置可与 DP 系统配合工作,必要时进行干预。

DP control station 动力定位控制站 系指进行动力定位操作和控制的工作站。相关的指示器、报警器、控制板和通信系统应安装在该控制站上。 DP control station is a station used for DP operating and controlling. The relevant indicators, alarms, control panel and communication system are to be fitted at the control station.

DP control system 动力定位控制系统 即动力定位船舶所必需的所有的控制元件和/或系统、硬件和软件。由下列组成:(1)计算机系统和控制器;(2)传感器系统;(3)显示系统(操作面板)/自动驾驶仪;(4)位置参照系统;(5)相关的电缆和电缆布线。 DP control system means all control components and systems, hardware and software necessary to dynamically position the vessel.

The DP control system consists of the following: (1) computer system and controller; (2) sensor system; (3) display system (operator panels)/autopilot; (4) position reference system; (5) associated cabling and cable routing.

DP downtime　DP 停工期　位置保持不稳定,冗余丧失既不保证红色警戒,也不保证黄色警戒,不过,对于勘探、矫正、试验等,丧失了把握性就必然使其从操作状态退回到顺风顺流移动状态。

DP hazard observation　DP 危险观察　设定一种已经识别的环境状态,该状态已潜在地逐步升级到近乎失去状态或更为严重的状态。

DP near-miss　DP 近乎失去　对 DP 性能、可靠性或冗余已发生的不利影响,但不会逐步升级到"DP 事故""不希望事故"或"停工期"。

DP ship　动力定位船舶　系指仅用推力器的推力保持其自身位置(固定位置或预先确定的航迹)的船舶。

DP system(DPS)　动力定位系统　系指在风、浪、流的干扰下,不借助锚泊系统,利用自身的推力器系统使海上浮动装置保持一定的位置和航向,或按预定的轨迹运行的闭环控制系统。其工作原理是:根据位置参照系统测得的信息与 DP 传感器系统测得环境的信息,经滤波后得到估算值,根据估算值,然后经推力器分配模块计算后,发出对各推力器的指令。其包括下列分系统:(1)动力系统;(2)推力器系统;(3)动力定位控制系统和测量系统。

DP thruster　动力定位推力器

DP transducer　动力定位传感器

DP undesired event　DP 不希望事故　位置丧失或其他意外的/不可控的和已导致或必将导致黄色警戒的事件。

DP unit　动力定位装置

DP UPS　动力定位不间断供电电源系统

DP vessel　DP 船舶　具有动力定位功能的船舶。

DP = Dynamic positioning　动力定位　系指用与一个或多个位置基准有关的推力器来自动控制船舶的位置和首向。也可用于中间的已产生的动态定位。

DP-1 system　动力定位系统 1　系指安装有动力定位系统的船舶可在规定的环境条件下,自动保持船舶的位置和艏向,同时还应设有独立的集中于手动船位控制和自动艏向控制。　Dynamic positioning system 1 means the vessel with dynamic positioning system can keep the position and heading of the vessel under the specified environmental conditions. And at the same time, independent, concentrated manual control of vessels' position and automatic heading control is to be fitted.

DP-2 system　动力定位系统 2　系指安装有动力定位系统的船舶在出现单个故障(不包括一个舱室或几个舱室的损失)后,可在规定的环境条件下,在规定的作业范围内自动保持船舶的位置和艏向。　Dynamic positioning system 2 means the vessel with dynamic positioning system can automatically keep the position and heading of the vessel when single failure (excluding loss of a cabin or cabins) occurs under the specified environmental conditions and in specified operating fields.

DP-3 system　动力定位系统 3　系指安装有动力定位系统的船舶在出现任一故障(包括由于失火或进水造成一个舱室的完全损失)后,可在规定的环境条件下,在规定的作业范围内自动保持船舶的位置和艏向。　Dynamic positioning system 3 means the vessel with dynamic positioning system can automatically keep the position and heading of the vessel when any failure (including entire loss of a cabin caused by fire or flood) occurs under the specified environmental conditions and in specified operating fields.

DP-1 vessel　DP-1 船舶　安装有动力定位系统的船舶,可在规定的环境条件下,自动保持船舶的位置和艏向,同时还应设有独立的联合操纵杆系统。

DP-2 vessel　DP-2 船舶　安装有动力定位系统的船舶,在出现单个故障(不包括一个舱室或几个舱室的损失)后,可在规定的环境条件下,在规定的作业范围内自动保持船舶的位置和艏向。

DP-3 vessel　DP-3 船舶　安装有动力定位系统的船舶,在出现单个故障(包括由于失火或进水造成一个舱室的完全损失)后,可在规定的环境条件下,在规定的作业范围内自动保持船舶的位置和艏向。

DP-control system　动力定位控制系统　系指船舶动力定位所必需的所有控制单元和系统、硬件和软件。动力定位控制系统的组成包括:(1)计算机系统/手柄操作系统、传感器系统;(2)显示系统(操作板);(3)自动导航;(4)坐标参照系统;(5)相关的电缆敷设及线路选择。　DP-control system means all control components and systems, hardware and software necessary to dynamically poison the vessel. The DP-control system consists of the following: (1) computer system/joystick system, sensor system; (2) display system (operator panels); (3) autopilot; (4) poison reference system; (5) associated cabling and cable routing.

DPO = dynamic positioning system operator　动力定位系统操作人员

Draft　吃水　系指:(1)在船中部从龙骨线至相应水线之间的垂直距离;(2)船营运时夏季载重线吃水(m),量自船舯型基线。注意:这可能小于最大的许用夏季载重吃水;(3)平均夏季型吃水;(4)在船长 L 中点处从基线到夏季载重线干舷标志处的垂直距离。对于具有木材载重线的船舶，吃水应量至木材载重水线干舷标志处；(5)由龙骨上缘量起的夏季吃水。

Draft 吃水(内河船) d(m) 对钢质船舶系指在船长中点处沿船侧自平板龙骨上表面量至满载水线的垂直距离；对纤维增强塑料船舶系指船长中点处沿船侧自船底板外表面量至满载水线的垂直距离。

Draft award 裁决书草案

Draft depth ratio 吃水型深比 一般系指设计吃水与型深之比。

Draft mark 吃水标志 置于船舶艏、艉及中部两舷，用以分别标明相当于艏垂线、艉垂线及中横剖面处龙骨吃水值或距船底某一基准点之吃水值的数字和线条标志。每个数字的底缘系该数字所指的水线位置。见图 D-12。

图 D-12 吃水标志
Figure D-12 Draft mark

Drafting 绘图机 是一种自动化绘图的设备。可使计算机的数据以图形的形式输出。其笔可在 x、y 两个方向自由移动，并可放下或抬起，从而在平面上绘出图形。其品种很多，常见的有滚筒式、带台式、平台式等各种笔式绘图机。有的一台绘图机有两支绘图笔、有的 3 支或 4 支绘图笔。绘图面积也有各种尺寸。能在一幅图上绘出多种宽度的线条和多种颜色，具有较高的速度和精度。另外，还有静电绘图机、数字化器绘图机以及彩色绘图机系统、人机对话式绘图系统等。

Drag and energy reduction technology 减阻降耗技术 (1)附体节能技术——该技术主要是一些水动力节能附加装置，如鳍、导流管等，以起到改善艉部流场，使船舶降低黏性阻力的作用，还可以提高螺旋桨的推进效率。(2)空气润滑技术——当在船舶外壳设置空气润滑系统时，船舶湿表面摩擦阻力可减少 40%，节能可达 10%。

Drag arm 耙臂 边挖泥船上，由上部吸管、底部吸管、中间管、软管、活络接头等组成的其前端装有耙头的臂架。

Drag due to surface slope 江面坡度阻力 船在有坡度的江、河面逆水航行时，船体重力沿江、河面的分力。

Drag net (dredge net) 拖网 一种拖曳于水域底层或中、下层，用以捕捞鱼、虾等的袋形网具。

Drag rope winch (quarter rope winch) 围网引钢绞机 围网作业中用于绞收围网上纲的绞机。

Drag-head 耙头 在耙吸挖泥船上，由吸头、泥耙下唇、耙齿、格栅等部件组成，作业时贴于水底斜拖的挖泥工具。

Drag-head ladder 耙头架 艏耙或中耙挖泥船上，装在挖泥管前端，用以承托和固定吸泥管，其下端连接耙头，并可绕上端支承点起落的刚性构架。

Drag-lift ratio 升阻比 系指水翼形体阻力对升力的比值。

Dragline dredger (DD) 拉铲式挖泥船

Drain cooler 泄水冷却器 为回收泄水的热量用凝水或给水来冷却泄水的热交换器。

Drain hole 流水孔 构件上为使水或其他液体能自由流通而开设的孔。

Drainage collecting tank 泄水柜 聚集从泄水管路来的凝水的柜。

Drainage pipe line 泄水管路 为防止汽轮机启动时，凝水损坏叶片及发生管路水击现象等而设置的泄放汽轮机及蒸汽管路、附件等的凝结水的管路。

Drainage piping 污水疏水管系 用以疏泄各种洗涤污水包括配膳间、厨房、茶水间以及舱室、走廊的积水等的管路。

Drainage pump 疏水泵 排去船上积水的泵。

Draught (SWATH) 吃水(小水线面双体船) 系指船舶静浮于水面时，沿设计水线量得刚体水密船体的型吃水。 Draught means the moulded draught of the rigid waterline hull at mid-ship measured along the design waterline with no lift or propulsion machinery active.

Draught d 吃水 系指：(1)在船艏部从龙骨线至相应水线间的垂直距离；(2)船营运时夏季载重线吃水(m)，量自船舯型基线。注意：这可能小于最大许用夏季载重吃水；(3)平均夏季型吃水；(4)在船长 L 中点处从基线到夏季载重线干舷标志处的垂直距离。对于具有木材载重线的船舶，吃水应量至木材载重水线干舷标志处；(5)由龙骨上缘量起的夏季吃水；(6)船舶静浮于水面时，在船长 L 中点处，由基线量至设计水线的垂直距离。 Draught d is: (1) the vertical distance from the keel line at mid-length to the waterline in question; (2) the summer load line draught for the ship in operation, measured from the moulded base line at amidships. Note this may be

Draught d(m) less than the maximum permissible summer load waterline draught; (3) mean moulded summer draught in m; (4) the vertical distance at the middle of the length L from base line to freeboard marking for summer load waterline. For ships with timber load line the draught is to be measured up to the freeboard mark for timber load waterline; (5) the summer draught, in meters, measured from top of keel; (6) the vertical distance measured at the middle of the length L from top of base line to the summer load waterline, in the displacement mode with no lift or propulsion machinery active.

Draught d(m) 吃水 d(平台) d 为从基线量至勘划的载重线的垂直距离。平台结构和机械设备的某些构件或部件可以伸展到基线以下。 Draught d(m) is the vertical distance measured from the moulded baseline to the assigned load line. Certain components or parts of a unit's structure, machinery or equipment may extend below the moulded baseline.

Draught T 吃水 T(艇) 吃水 T 应取艇满载备用品状态时其水线与水下艇体规定点之间测得的垂直距离。 Draught T shall be measured as the vertical distance between the waterline and a specific point of the underwater body, when the vessel is in the fully loaded ready-for-use condition.

Draught T_F 艏吃水 系指在艏柱处的吃水。 Draught T_F is the draught at the forward perpendiculars.

Drawing room 绘图室 用于海洋资料的分析、整理和绘制图表的工作室。

Drawing-in factor 穿管系数 系指电缆外径截面积的总和与管子或管道或电缆槽内截面积之比。 Drawing-in factor means a ratio of the sum of the cross-sectional areas of the cables to the internal cross-sectional area of the pipe.

Dredge net 捞网

Dredger 采矿船 设有采矿、选矿和脱水设备，专供开采水底矿石或矿砂的船舶。见图 D-13。

图 D-13 采矿船
Figure D-13 Dredger

Dredger 挖泥船 系指开底式挖泥船、泥驳、开底泥驳和能自航及非自航的采用各种常规挖泥方式的类似船舶（例如链斗式挖泥船、吸扬式挖泥船、抓斗式挖泥船等）。 Dredger means hopper dredgers, barges, hopper barges and similar vessels which may be self-propelled and non-self-propelled and which are designed for all common dredging methods (e.g. bucket dredgers, suction dredgers, grab dredgers, etc.).

Dredger capacity 挖泥船生产力 挖泥船在单位时间内所挖出的土方量。

Dredging pipeline float 排泥管浮筒 用以承托水上排泥管的2个独立而又桡性相连的浮筒。

Dredging pump 泥泵 挖泥船上抽送泥浆的泵。

Dredging within coastal area 在沿海海域作业 系指工程船舶在2类航区内作业。

Dredging within greater coastal area 在近海海域作业 系指工程船舶在1类航区内作业。

Dredging within R1 挖泥船作业限定海域R1 系指挖泥船限制在1类航区内作业。

Dredging within R2 挖泥船作业限定海域R2 系指挖泥船限制在2类航区内作业。

Dredging within R3 挖泥船作业限定海域R3 系指挖泥船限制在3类航区内作业。

Dredging within sheltered water area 在遮蔽海域作业 系指工程船舶在3类航区内作业。

Drier 干燥器 通常用于氟利昂制冷系统，以吸收液态制冷剂内所含水分，内装有干燥剂的圆筒。

Drift 横漂 船舶受风浪、水流等外扰动力所引起的横向漂移。

Drift angle 漂角 从船的艏向至其重心处瞬时航速矢量间的水平夹角。顺时针为正。

Drift bottle (current bottle) 漂流瓶 投入海中测量海流的可在水中漂浮的瓶子。内装有卡片；注明投放地点和日期，并要求拾到者在卡片上填明发现此瓶的地点和日期，寄还投放者。

Drifting 侧漂 气垫船遭遇横风或受侧风外力干扰，或在回转时由离心力而引起的侧向漂移。

Drill 演习 系指一个涉及船舶或港口设施一个功能要素的全面训练事件，并测试通信、协调、可获得资源和响应。

Drill and practice 演练 系指为保持船舶处于一种高水平的保安准备状态而开展的至少涉及船舶保安计划的一部分的一次训练事件。

Drill demolition ship (DDS) 钻孔爆破船 又称水下炸礁船。是在对于具有水下岩基航道、运河、港区、码头等实施开挖及拓深、拓宽的改扩建工程中，用于进行水下钻孔爆破的工程船舶。

Drill pipe 钻杆 系指钻孔工具中连接钻头、用以传递动力的杆件。钻柱通常的组成部分有：钻头、钻铤、钻杆、稳定器、专用接头及方钻杆。

Drill string compensator（DSC） 钻柱补偿器

Drilling area 钻井区 系指包括钻台区、泥浆循环区和处理区。Drilling area includes drill floor area, mud circulation and treatment area.

Drilling barge 钻井驳船 系指：(1) 设有钻井设备，一般在浅水域进行钻井作业的驳船。(2) 在海上能独立进行钻井作业的非机动钻井船。

Drilling condition 钻井工况 系指船位能够维持，并且所有与其作业相关的活动可以开展的工况。包括下列工况：(1) 钻杆/套管处理及排放——在海上一定风速及船体摇摆情况下吊机一种常规作业；(2) 隔水管处理及排放——在海上一定风速及船体摇摆情况下吊机一种常规作业；(3) 海水表层钻进——开式大排量钻进，轻载提升，旋转作业；(4) 防喷器组下放及回收——重载搬运，重载提升，保证隔水管顺利通过转盘开口；(5) 泥浆配置——散料输送作业，泥浆池内循环作业；(6) 泥浆灌注隔水管——泥浆大循环，提升作业，旋转作业；(7) 水基泥浆钻进——泥浆常规循环，提升作业，旋转作业；(8) 油基泥浆钻进——泥浆常规循环，提升作业，旋转作业；(9) 起下钻——泥浆小循环，提升钻杆入井眼或者出井眼；(10) 下放套管——泥浆不循环，提升套管柱入井眼；(11) 固井——泥浆不循环，提升作业，水泥入井作业；(12) 试油——泥浆不循环，提升作业，开启保护水幕。

Drilling deck 钻井甲板 用于安装井架、钻机和钻井附属设备的平台甲板。

Drilling draft 钻井吃水 浮式钻井装置在钻井作业时的吃水。

Drilling fluid（mud）processing areas 钻井液（泥浆）处理区 系指钻井液处理和贮存区域。Drilling fluid（mud）processing areas are areas where drilling fluids are processed and stored.

Drilling installation 钻井装置 系指进行钻井作业所需的全部设备和系统的总称。见图 D-14。

Drilling load 满载钻井作业负荷 钻井船或钻井平台在钻井作业时所允许承受的最大负荷。

Drilling organism research 钻孔生物调查 对能穿凿海洋中岩石、贝壳、珊瑚礁和非金属材料等的钻孔生物所做的调查。钻井生物大部分是动物、少数是海藻。

Drilling platform 钻井平台 系指其上安装钻井装置从事海上钻井、完井、修井或测试作业的固定式或移动式结构物或其他海上结构物。

图 D-14 钻井装置
Figure D-14 Drilling installation

Drilling positioning（column locating） 钻井定位 用各种定位系统来确定浮式钻井装置在预定海域内的井位的作业。

Drilling rigs 钻井平台 系指主要用于钻探油井的海上结构物。随着人类对油气资源开发利用的深化，油气勘探开发从陆地转入海上。因此，钻井工程作业也必须在浩瀚的海上进行。在海上进行油气钻井施工时，几百吨重的钻机要有足够的支撑和放置的空间，同时还要有钻井人员生活居住的地方，海上石油钻井平台就担负起了这一重任。由于海上气候的多变、海上风浪和海底暗流的破坏，海上钻井装置的稳定性和安全性更显重要。见图 D-15。

图 D-15 钻井平台
Figure D-15 Drilling rigs

Drilling riser 钻井立管/钻井隔水导管

Drilling ship 采油船/钻井船 (1) 设有钻井设备能在水上钻井并移位的船舶。初期多用旧船或驳船改装,其后有专门设计制造的。它与半潜式钻井平台一样,钻井时漂浮在水上,适于深水作业,但需有相应的定位设施。它在波浪上的运动幅度比半潜式钻井平台大,但动力定位所需的功率较小。通常为了减少船体摇荡对钻井工作的影响,多将井架设在船的中央。一般为单体船,但也有采用双体船。多具有自航能力。无自航能力的称为钻井驳;(2) 能在海上进行石油钻探作业的专用船舶。一般能自航,并有单体、双体、三体等不同类型;(3) 设有钻探架、钻探机、定位绞车以及采样、化验等设备,为配合水上建筑进行水下地质钻探用的船舶。见图 D-16。

图 D-16 采油船/钻井船
Figure D-16 Drilling ship

Drilling slot 钻井凹槽

Drilling system 钻井系统

Drilling tender 钻井供应船/钻井辅助船 系指为钻井平台供应管子、泥浆、水泥、燃料、备品等物资,同时也提供人员居住舱室的船舶。

Drilling unit 钻井平台 系指能为勘探和开采海底下资源从事钻井作业的平台。 Drilling unit means a unit capable of engaging in drilling operation for the exploration for or exploitation of resources beneath the sea bed.

Drilling vessel 钻井船 是利用单船体、双船体、三船体或驳船的船体作为钻井工作平台的一种海上移动式钻井装置。与钻井平台相比,其钻井更具有灵活性。但钻井船,特别是动力定位钻井船造价高。钻井船到达井位后先抛锚定位或动力定位,钻井时和半潜式平台一样,整个装置处于漂浮状态,在风浪的作用下,船体也会上下浮动、前后、左右摇摆及在海面上漂移,因此需要安装升沉补偿装置、减摇装置,或采用动力定位等多种措施来保证船体在需要的范围内进行钻井作业。目前钻井船的钻井深度主要有 4 个系列,分别是 20 000 ft (6 096 m)、25 000 ft (7 620 m)、30 000 ft (9 144 m) 和 35 000 ft (10 668 m)。由于多数钻井船作业深度大于 1 500 m,因此采用动力定位约占 79%;锚泊定位占 21%。钻井船是漂浮于水面上的作业平台,通常适合在各种水深条件下进行钻探作业,但对船的定位要求很高,多采用多锚定位方式。

Drilling water tank 钻井水舱 用于贮存为配制泥浆、固井水泥和冷却机械设备等用水的淡水舱。

Drilling well 井口 系指钻探/生产用的平台开口。 Drilling well is an opening in the deck and bottom for drilling/production operation.

Drinking water boiler 沸水器 将饮用水用蒸汽或电加热成沸水的器具。

Drinking water cooler 冷饮用水机 带有制冷装置的冷饮用水设备。

Drinking water ozone disinfector 食用水臭氧消毒器 为防止远洋船舶所带淡水贮存时间久长水质变坏而对食用水用臭氧进行消毒的器具。由臭氧发生器等主要部件组成。

Drinking water pump (portable water pump) 饮用水泵 抽送饮用水的泵。

Drinking water system 食用水 船上专供食用的水。

Drip pan 水盘 专为积聚和排除空气在通过翅片管时所析出凝结水的盘。

Dripping height 艏沉深度 船舶纵向下水过程中,发生艏落后的瞬间,船首继续下沉至最大值时的艏吃水与静浮状态艏吃水之差。

Dripping oil range 滴油灶 采用滴油方式供燃油的燃油炉灶。

Drip-protected enclosure 防滴外壳 系指其开孔应做得以偏离垂直位置小于 15°的任意角度,使掉落在其上的液滴或固体颗粒不能进入该外壳;或如果的确进入该外壳,则不会妨碍所封装设备有效运行或使该设备损坏的外壳。

Drive endless chain 传动循环链 带动滚翻式舱盖板滚移的闭合循环链条。

Drive gear (driving gear, driving toothed gear) 主动齿轮 一对相啮合齿轮中按功率传动方向而言相对靠近发动机的齿轮,在减速齿轮系中它通常是小齿轮。

Drive ratio (gear ratio, transmitting ratio, velocity ratio) 传动比 单级或多级齿轮传动装置中输入轴转速与输出轴转速的比值。

Driven gear 被动齿轮 一对相啮合齿轮中按功率传动方向而言相对远离发动机的齿轮,在减速齿轮系中它通常是大齿轮。

Droneware "蒜靶"软件　系指面向宿主在线发送垃圾邮件或载性网站的资料。

Drop chute 泥井　两侧通溜泥槽,用以承接从链斗倒出的泥沙的围井。

Drop net 绳网　悬吊在救生浮内缘,以绳结成,用以吊挂格栅的网。

Drop weight tear test（DWTT） 落锤撕裂试验　系指测量大型剖面材料断裂韧性的试验方法。Drop weight tear test（DWTT）is the test method for determining the fracture toughness of heavy sections of materials.

Dropping 艏落　船舶纵向下水过程中,当前支架脱离滑道末端后的瞬间船的艏吃水小于静浮状态艏吃水时所发生的船首快速下落的现象。

Dropping height 艏落高度　船舶纵向下水过程中,当前支架脱离滑道末端时,船的艏吃水与静浮状态艏吃水之差。

Dropping speed motor unit 降速电动机组　系指具有一附соединená之机械装置的电动机,借以获得与电动机转速不同的转速。降速电动机组通常是为获得低于电动机转速的转速而设计的,但是也可以制造得使其获得高于电动机转速的转速。

Dropwise condensation 滴状冷凝　蒸汽在冷却壁面上冷凝时,凝水结成液滴状的冷凝方式。

Drug 毒品　系指当活的机体摄入时可能改变机体的一个或多个功能的某种物质。其中有些既可以自由得到,又为社会所接受。例如:(1)社会上接受和可以自由得到的物质:咖啡因、烟草(尽管越来越不为社会接受)、酒(在多个国家中);(2)社会上不接受但可自由得到的物质:胶水、甲基化酒精、汽油、溶剂、洗涤液;(3)社会上接受和可以自由得到的药物:阿司匹林、扑热息痛、维生素片;(4)社会上接受和受控的药品:巴比妥酸盐、安定、安定(利眠宁)和许多处方药品;(5)社会上不接受和受控的药品或物质:大麻、LSD、可卡因、吗啡、海洛因、安非他明、鸦片。每一类别中的许多物质都具有某种毒品依赖性的危险,但最后一项物质的这种危险性最大。虽然后者的某些物质在严格的医疗监督下可以使用,但毒品依赖性仍可在短期内发生。如果滥用这些毒品(即在不受控的情况下使用),很快就能成瘾。 Drug means any substance that, when taken into the living organism, may modify one or more of its functions, some of which are both freely available and socially acceptable. Examples: (1) socially acceptable and freely available substances: caffeine, tobacco (although becoming increasingly less socially acceptable), alcohol (in most countries); (2) socially unacceptable and freely available substances: glue, methylated spirit, petrol, solvents, cleaning fluids; (3) socially acceptable and freely available pharmaceuticals: aspirin, paracetamol, vitamin tablets; (4) socially acceptable and controlled pharmaceuticals: barbiturates, valium, diazepam (librium), and numerous other prescription drugs; (5) socially unacceptable and controlled pharmaceuticals or substances: cannabis, LSD, cocaine, morphine, heroin, amphetamines, opium. Most of the substance in each category carry some risk of drug dependence, but those in the last category carry by far the greatest. Although some of the latter substance may be used under strictly controlled medical supervision, total dependence can still occur within a short period of time. When these drugs are abused (i.e. used in uncontrolled circumstance) addiction can result very rapidly.

Drum 轮鼓　反动式汽轮机中安装动叶的鼓筒形旋转体。

Drum（barrel） 卷筒　两端具有凸缘用以收放和储存绳索的圆筒体。对调查船绞车为缩小卷筒外形,绳索均采取多层卷绕在其上。

Drum head（end drum） 封头　锅筒、联箱两端呈半球形、椭圆形或平板形的封板。

Drum internals（drum internal fittings） 锅内设备　各种置于锅筒内的设备,如汽水挡板、水下孔板、分离和汽水洗气设备以及表面排污管、集气管、给水内管、加药管和减温器等的统称。

Drum load（hauling load, rated load） 卷筒额定拉力　当卷筒绕上单排缆绳,绞车以额定绳速收绞时在卷筒缆索出口处测得的缆绳最大拉力。

Drum type cable laying machine 鼓轮式布缆机　安装在布缆船船舱内,供布放或捞起电缆用的鼓轮或绞缆机。

Drum type controller 鼓形控制器　系指利用鼓形开关作为主要换接元件的一种电控制器。鼓形控制器通常由一个鼓形开关和有关电阻器组成。

Dry cargo residue（or DCR）management plan 干货物残余物管理计划

Dry cargo ship 干货船　以运输干燥货物为主的船舶。是散货船和杂货船的统称。

Dry clutch 干式离合器　摩擦接合面不使用润滑冷却剂的离合器。

Dry compression chamber 干加压舱　舱内不能注水的加压舱。

Dry dock/Graving dock 干船坞　系指将水抽掉,使船舶在此进行出水检查、修理的封闭的船池。待出售的船舶通常在干船坞让有意向的买主查看。见图D-17。

Dry docking 坞(排)修　系指船舶入坞进行检查

图 D-17　干船坞
Figure D-17　Dry dock/Graving dock

和修理。

Dry docking period　坞修期　系指船舶入坞进行检查和修理的周期。

Dry laboratory　干实验室　适用于对湿度要求较低的海洋调查仪器和实验设备安装与使用的工作室。该室不设置供排水系统,可供各学科共用或专用。如计算机室、重力仪室等。

Dry pipe system　干管系统　系指将自动喷水器接至内部充满带压空气或氮气的管路上的喷水系统。当喷水器开启后,由于管内气体被释放,在另一侧水的压力作用下,称之为干管阀的阀门被打开,水流随之进入自开启的自动喷水器排出。　Dry pipe system means a system employing automatic nozzles or sprinklers attached to a piping system containing air or nitrogen under pressure, the release of which (as from the opening of a nozzles or sprinklers) permits the water pressure to open a valve known as a dry pipe valve. The water then flows into the piping system and out of the opened sprinklers.

Dry powder foam fire system　干粉灭火系统　以高压惰性气体作为动力,将干粉灭火剂吸出并喷成粉雾以扑救失火的一种灭火系统。

Dry preservation　干保养　锅炉长期停用时,先烘干锅炉,然后在锅筒内加入干燥剂并密封全部锅炉阀门,以保护锅炉不受氧化腐蚀的措施。

Dry pump room　干泵舱　浮船坞下沉时,舱内不允许进水的泵舱。

Dry submarine launch for nuclear bomb　潜射核弹干发射　即潜艇用压缩空气直接将导弹推出水面,导弹在距离海面一定高度时点燃发动机的发射方式。例如中国几十年前开发成功的第一种潜射弹道导弹—巨浪-1 所采用的发射方式。

Dry suit (Immersion suit)　干式服(救生服)　设计成在落水时阻止水进入的衣服。Dry suit is the garment designed to preclude the entry of water upon immersion.

Dry sump lubrication　干底润滑　系指柴油机的机外设有滑油循环柜,曲柄箱不兼作油池的润滑形式。

Dry tree　干式采油树

Dry type evaporator　干式蒸发器　卧式圆筒状,内装有很多金属管和挡板,制冷剂在管内蒸发、吸热,载冷剂在管外流动冷却的热交换器。

Dry-dock survey　坞内检验(平台)　系指在任何 5 年期限内至少进行 2 次的检验。在所有情况下,任何 2 次坞内检验的间隔期不得超过 36 个月。　Dry-dock survey means there is a minimum of two dry-dock surveys during any five-year period. In all cases the intervals between any two such surveys should not exceed 36 months.

Drying method　干燥法　对于液货舱的环境控制,系指将干燥气体或在大气压力下其露点为 -40 ℃ 或更低的蒸发气充入液货舱及相关管系并维持这种状态。For environmental control for cargo tanks, drying method means a method by filling the cargo tank and associated piping systems with moisture free gas or vapour with a dew-point of -40 ℃ or below at atmospheric pressure, to maintain that condition.

DSV = Deep submarine vessel　潜水母船　部署潜水员的船舶。

Dual casing riser　双屏立管

Dual channel satellite navigator　双频道卫星接收机　能同时接收卫星发射的两个相关频率(如 150 MHz 和 400 MHz)的双锁向接收机。双频道卫星接收机能较好地消除电离层折射对多普勒频移的影响。因此定位精度较高,均方根误差约 32 m,单频道卫星接收机的均方根误差约 52 m。

Dual class ship　双重船级船舶　系指某一船舶在两家船级社入级,且两家船级社有书面协议,工作共享,检验互认,检验信息和检验报告全面交换。　Dual class ship is one which is classed by two Societies between which there are written agreement, sharing of work, reciprocal recognition of surveys carried out by each of the Societies on behalf of the other Society and full exchange of information on the class status and survey reports.

Dual cycle (mixed cycle)　混合循环　等压和等容循环的组合循环。

Dual fuel diesel engine　油气两用双燃料内燃机　是气体发动机的一种。它能以 LNG 作为燃料,但又与纯气体发动机不同,还能以柴油(轻质燃油和重油)为燃料,并且可在工作过程中实现不同燃料之间的自由转换。其优点:(1) 由于 LNG 属于清洁能源,以其为燃料的船用发动机废气排放量大大降低,环保性能好,能满足 IMO 目前最严格的 TierⅢ 排放要求，可使船舶在排放控制区内自由航行而不必缴纳排放税；(2) 不用为船用发动机安装废气处理系统，能够降低购置成本和运行成本；(3) 在排放控制区外，船用发动机燃料可在气体燃料和燃油之间自由切换，选择使用成本最低的燃料，可避免燃料价格波动对船舶营运成本产生较大的影响；(4) 与纯气体发动机相比，双燃料内燃机的安装成本更低，前者需要安装 2 个及以上大小相同的燃料储存罐，占有空间较大，挤占了货舱的空间，而后者只需要安装 1 个同样大小的燃料储存罐，经济性能相对较高；(5) 双燃料内燃机的安全性能高，即使发生气体泄漏，仍可使用柴油作为燃料，且不影响船舶的动力和速度。见图 D-18。

图 D-18　油气两用双燃料内燃机
Figure D-18　Dual fuel diesel engine

Dual fuel diesel engines（DFD engines）　双燃料柴油机　系指可以使用油燃料和高压甲烷气燃料喷射的发动机（即装有气体运输船发动机并使用气化气体作为燃料的发动机）。其操作模式：(1) 应为双燃料型，采用引燃油点火，且可快速切换至燃油模式；(2) 启动时，只能使用油燃料；(3) 运行状态不稳定和/或机动运行时间，一般只能使用燃油运行；(4) 气体燃料供应切断时，应能以燃油持续运行。Dual fuel diesel engines mean an engine which may utilize both oil fuel and high pressure methane gas fuel injection (i.e. engines fitted on gas carriers and employing boil-off gas as fuel). Operation mode of DFD engines: (1) are to be of the dual-fuel type employing pilot fuel ignition and to be capable of immediate change-over to oil fuel only; (2) only oil fuel is to be used when starting the engine; (3) only oil fuel is, in principle, to be used when the operation of an engine is unstable, and/or during manoeuvring and port operations; (4) in case of shut-off of the gas fuel supply, the engines are to be capable of continuous operations by oil fuel only.

Dual tandem gear　二级串联减速齿轮箱　功率分支的二级串联式减速齿轮箱。

Dual-activity drilling system　双作业钻机
Dual-derrick drilling system　双井架钻机
Dual-duct air-conditioning system　双风管空气调节系统　部分舱外新鲜空气和部分舱内循环空气两者混合后，分别由 2 个集中式空气调节器降温去湿和加热加湿，然后经过 2 根平行的冷风管和暖风管送至各舱室的混合器内，再根据舱内温度调节器的控制，使冷空气和暖空气在混合器内自动或手动按比例混合后再送入舱室内的空气调节系统。

Dual-fuel engines　双燃料发动机　系指：(1) 可同时燃烧天然气和燃料或仅依靠燃料油或气体运转的发动机；(2) 可以使用燃油和高压甲烷气体作为燃料的柴油机。Dual-fuel engines mean: (1) engines that can burn natural gas and fuel oil simultaneously or operate on oil fuel or gas only; (2) a diesel engine utilizing oil fuel and high pressure methane gas as fuel.

Dual-fuel engines compartment　双燃料发动机舱室　系指位于机舱内，用于安装双燃料发动机的独立机器处所。

Dual-purpose carrier/oil and bulk carrier/oil and ore carrier　两用散货船　在构成一个航次的各阶段中，可分别载运石油、粮食或煤、矿砂中任何两种货物的散货船。

Duck（shroud）　导管　导管推进器中控制水流的喷管形外罩。

Duck control system　气道控制系统　设于气道内，可于驾驶室控制气道气流分布的一套设备。

Duck propeller　导管螺旋桨
Duckpeller harbour tug boat　全回转港作拖船　系指利用由螺旋桨式叶轮与控制水流的喷管形外罩共同组成的机构作为推进装置的拖船。见图 D-19。

图 D-19　全回转港作拖船
Figure D-19　Duckpeller harbour tug boat

Duct keel　箱形龙骨　系指箱形板材制成的龙骨，延伸至货油舱的长度。用以容纳通向船首的压载水管路和其他管路。如无此龙骨，则这些管路即须穿越货油舱。

Duct keel means a keel built of plates in box form extending the length of the cargo tank. It is used to house ballast piping and other piping leading forward, without which otherwise the piping would have to run through the cargo tanks.

Duct keel 箱型中底桁 在船底中线面附近,由两道底纵桁与内底板、船底板、骨架等组成的箱形结构。

Ducted propeller 导管螺旋桨 系指装在导管中的螺旋桨。 Ducted propeller is a propeller installed in a nozzle.

Ducted propeller(shrouded propeller) 导管推进器 由螺旋桨式叶轮与控制水流的喷管形外罩共同组成的推进机构。

Ductile cast iron 可锻铸铁 系指由一定化学成分的铁液浇注成白口坯件,再经退火而成的铸铁。它与灰口铸铁相比,可锻铸铁有较好的强度和塑性,特别是低温冲击性能较好,耐磨性和减振性优于普通碳素钢。

Ductile material 延展性材料 系指其延伸率超过12%的材料。 Ductile material is a material having an elongation over 12%.

Ductility 韧性 系指材料在外载荷作用下抵抗开裂和裂缝扩展的能力,也就是材料在断裂前经历的弹塑性变形过程中吸收能量的能力,是强度和塑性的综合体现。测量材料能够承受塑性变形范围的指标。通常在2英寸测量长度(包括颈部)拉伸试验以应变的百分数表示。

Duly authorized officer 正式授权的官员 系指由缔约国政府正式授权的按 SOLAS 第Ⅺ-2/9 条的规定采取控制和符合性措施的官员。 Duly authorized officer means an official of the Contracting Government duly authorized by that Government to carry out control and compliance measures in accordance with the provisions of SOLAS regulation Ⅺ-2/9

Dumb terminal 哑终端 即功能有限的计算机终端,本身几乎没有处理能力,只被用作文字编辑或向中央计算机提出需求。

Dummy piston 平衡活塞 形成反向蒸汽压差以平衡多级汽轮机的轴向推力的装置。

Dump door (hopper door) 泥门 系指供泥舱卸泥用的活动门。

Dump door closing system 泥门启闭装置 开闭泥门用的操纵装置。

Dumping 倾倒 系指:(1)从船舶、飞机、平台或其他海上人造建筑物上有意地向海上倾倒任何废弃物及其他物质;(2)有意地向海上倾弃任何船舶、飞机、平台或其他在海上的人造建筑物。但倾倒不包括:①从船舶、飞机、平台或其他海上人造建筑物及其设备的正常操作所引起的或偶尔产生的废物及其他物质的倾弃,而不是由船舶、飞机、平台或其他海上人造建筑物运送的或专门用于在海上倾弃这类物质的船舶、飞机、平台或其他海上人造建筑物上为处理这类废物及其他物质而产生的物质的倾弃;②在不违背防止倾倒废物及其他物质污染海洋的公约目的的条件下,不是为了倾弃目的而将物质留存在海中。(3)由于海底矿物资源的调查、开发及与调查和开发有关的海上加工所直接产生的或有关的废物及其他物质的处理,不受防止倾倒废物及其他物质污染海洋的公约的限制。 Dumping means: (1) any deliberate disposal at sea of waters or other matter from vessels, aircrafts, platforms or other men-made structures at sea; (2) any deliberate disposal at sea of vessels, aircrafts, platforms or other man-made structures at sea. Dumping does not include: ①the disposal at sea of waters or other matter incidental to, or derived from the normal operations of vessels, aircrafts, platforms, or other man-made structures at sea and their equipment, other than wastes or other matter transported by or to vessels, aircraft, platforms or other man-made structures at sea, operating for the purpose of disposal of such matter or derived from the treatment of such wastes or other matter on such vessels, aircraft, platform of structures;②placement of matter for a purpose other than the mere disposal thereof, provided that such placement is not contrary to the aims of the Convention on the prevention of marine pollution by dumping of wastes and other matter;(3) the disposal of wastes or other matter directly arising from, or related to the exploration, exploitation and associated offshore processing of sea-bed mineral resources will not be covered by the provision of the Convention on the prevention of marine pollution by dumping of wastes and other matter.

Dumping 商品倾销 系指出口商以低于正常价格的出口价格,集中地或持续大量地向国外抛售商品。这是资本主义国家常用的行之已久的扩大出口的有力措施。商品倾销通常由私人大企业进行,但随着国家垄断资本主义的发展,一些国家设立专门机构直接对外进行商品倾销。

Dunnage 垫舱物料 系指用于舱内堆物的垫衬物料。

Duplex DP 双重 DP 具有完全冗余的 DP 控制系统,包括在两套 DP 控制系统之间能自动平稳转换。

Duplex nozzle 双油路喷油嘴 具有主、辅油路的喷油嘴。

Duplex stainless steel(DSS) 双相不锈钢 系指在不锈钢中既有奥氏体又有铁素体组织的钢种,其中铁素体与奥氏体各约占 50%，一般较少相的含量最少也需要达到 30% 的不锈钢。在含 C 较低的情况下，Cr 含量在 18%~28%，Ni 含量在 3%~10%。有些钢还含有 Mo、Cu、Nb、Ti、N 等合金元素。较常用的奥氏体不锈钢具有更高的强度和耐腐蚀性度。而且，其含镍量低，用于制造货舱时体现的经济性显著。双相不锈钢从 20 世纪 40 年代在美国诞生以来，已经发展到第三代。它的主要特点是屈服强度可达 400~550 MPa，是普通不锈钢的两倍，因此可以节约用材，降低设备制造成本。在抗腐蚀方面，特别是介质环境比较恶劣（如海水，氯离子含量较高）的条件下，双相不锈钢的抗点蚀、缝隙腐蚀、应力腐蚀及腐蚀疲劳性能明显优于普通的奥氏体不锈钢，可以与高合金奥氏体不锈钢媲美。

Duplication 复制品

Duplication survey 重复测量 在所布置磁力测量的测线内，安排 30‰~50‰ 重复检查测量。用以计算磁力测量的观测精度。

Duration of extreme storm 极限风暴持续时间 假定极限风暴持续时间为 3h。It is to be assumed that extreme storm events are to persist for three hours.

Duration of sea state 海况持续时间 假定该量级海况持续时间为 12h。Duration of sea state is to be assumed that the duration of sea states of this magnitude is 12 hours.

Duration of voyage 航行期间 系指航次开始船舶离港到航次结束船舶到港之间的时间。Duration of voyage means the interval between the time when the ship leaves the port at which the voyage commences and the time when she arrives at the port where the voyage terminates.

Dust 灰尘 为经过处理的待涂表面上的松散微粒物质，由喷砂清理或其他表面处理工艺产生，或因环境作用造成。Dust is loose particulate matter present on a surface prepared for painting, arising from blast-cleaning or other surface preparation processes, or resulting from the action of the environment.

Dustpan suction dredger 冲吸式挖泥船/吸盘式挖泥船 用高压水将水底泥土冲成泥浆，并通过安装在艏部或船侧的吸头进行吸泥的吸扬式挖泥船。

Dust-protected enclosure 防尘外壳 系指制造成或防护得使其内部可能出现的任何积尘都不会妨碍所封装设备的有效运行或使该设备损坏的外壳。

Dust-tight enclosure 尘密外壳 系指制成使灰尘不能进入其封装设备的外壳。

Dutch auction 减价拍卖 又称荷兰式拍卖，是拍卖价格从高到低的一种拍卖形式。

Duty pump (service pump) 值班泵 工作中的泵。

Duty stations 工作场所 系指配置有主要的导航设备、船舶的无线电设备或应急电源，或者集中了火警设备或消防设备的舱室。这类各种场所包括厨房、主配膳室、洗衣间、储藏室（除独立的配膳室和小船室外）、邮件室、贵重物品保管室、集中控制室、除机舱备件间外的工作室以及类似这样的舱室。

DVB-S（ETS300421） 数字卫星广播系统标准 卫星传输具有覆盖面广、节目容量大等特点。数据流的调制采用四相相移键控调制（QPSK）方式，工作频率为 11/12 GHz。在使用 MPEG-2MP@ML 格式时，若用户端达到 CCIR601 演播室质量，码率为 9 Mb/s；达到 PAL 质量，码率为 5 Mb/s。一个 54 MHz 转发器传送速率可达 68 Mb/s，可用于多套节目的复用。DVB-S 标准几乎为所有的卫星广播数字电视系统所采用。我国也选用了 DVB-S 标准。

DVD player 数位光碟播放机 即播放数位光碟的设备。使用方式大多数皆需要连接到电视机上。

Dynamic action 动载荷 导致结构或结构部件加速度幅值足以引起需要特别考虑的载荷。Dynamic action means the action that induces acceleration of a structure or a structural component of a magnitude sufficient to require specific consideration.

Dynamic air cushion 动力气垫 系指：(1)机翼在强地效区内移动时，在机翼和水面或一些其他表面之间生成的高压区。(2)由于机翼在水面或其他某种表面的气动效应区域内运动而在机翼和该表面之间产生的高压区域。动力气垫可以由两种不同的方式产生：①船体几何形状可以设计成空气进入一个槽形开口并能维持，只有通过位于除船艏以外的船底下的小缝隙泄漏。这是 19 世纪 60 年代的气泡生成概念，大多被称为地面效应器。②空气以比船的前进速度更高的速度流入机翼下以增强升力。这可以通过放下襟翼、设置翼梢端板以及调整机翼几何形状来实现。Dynamic air cushion is：(1) a high-pressure region originating between the airfoil and a water surface or some other surface as the airfoil moves within the zone of enhanced aerodynamic effect; (2) a high pressure region originating between the airfoil and a water surface or some other surface as the airfoil moves within the zone of the aerodynamic effect of this surface. A dynamic air cushion can be created in two different ways: ①the craft hull geometry can be shaped so that air enters an opening and is retained except for air released through a small gap under the craft at the bow. This is the captured air bubble concept of the 1960s mostly called the ground effect machine. ②air is blown under a wing at a much higher speed than the craft's forward speed, to enhance lift. This may be enhanced by lowering wing flaps and wing tip fences and adjusteing wing geometry.

Dynamic amplification coefficient　动力放大系数
Dynamic factor KV　动载系数 KV　系指作用于齿侧的最大动力载荷与最大外加载荷（Ft·KA·Kγ）之比值。用以考虑大小齿轮互相啮合振动而引起的内部附加载荷的影响。　Dynamic factor KV means a ratio between the maximum load which dynamically acts on the tooth and flanks and the maximum externally applied load (Ft·KA·Kγ) accounts for internally generated dynamic loads due to vibrations of pinion and wheel against each other.

Dynamic frequency spectrum analyser　动态频谱分析仪

Dynamic load coefficient　动载系数(起重设备)　系指在起重设备工作时，考虑所有动载效应的一个系数。此系数乘以起升系数后，代表包括所有动载效应作用于系统上的载荷。

Dynamic loads　动力载荷　系指持续时间比诱导波浪周期载荷短得多的载荷。　Dynamic loads are those that have a duration much shorter than the period of the wave loads.

Dynamic loads and dynamic tank pressures due to ship accelerations　船舶运动加速度引起的动载荷和液舱晃动压力　是由于船舶运动加速度引起的动载荷和液舱晃动压力。　Dynamic loads and dynamic tank pressures due to ship accelerations are dynamic loads and liquid sloshing pressure in tanks due to ship accelerations.

Dynamic positioning system-1（DP-1）　1级动力定位系统　系指安装有动力定位系统的船舶，可在规定的环境条件下，自动保持船舶的位置和艏向，同时还应设有独立的集中手动船位控制和自动艏向控制。　Dynamic poisoning system-1 (DP-1) means vessels with dynamic poisoning system can keep the position and heading of the vessels under the specified environmental conditions. And at the same time, independent, concentrated manual control of vessels' position and automatic heading control is to be fitted.

Dynamic positioning system-2（DP-2）　2级动力定位系统　系指安装有动力定位系统的船舶，在出现单个故障(不包括一个舱室或几个舱室的损失)后，可在规定的环境条件下，在规定的作业范围内自动保持船舶的位置和艏向。　Dynamic poisoning system-2 (DP-2) means vessels with dynamic poisoning system can automatically keep the position and heading of the vessels when single failure (excluding loss of a cabin or cabins) occurs under the specified environmental conditions and in specified operating

fields.

Dynamic positioning system-3（DP-3）　3级动力定位系统　系指安装有动力定位系统的船舶，在出现任一故障(包括由于失火或进水造成一个舱室的完全损失)后，可在规定的环境条件下，在规定的作业范围内自动保持船舶的位置和艏向。　Dynamic poisoning system-3 (DP-3) means vessels with dynamic poisoning system can automatically keep the position and heading of the vessels when any failure (including entire loss of a cabin caused by fire or flood) occurs under the specified environmental conditions and in specified operating fields.

Dynamic positioning　动力定位　系指：(1)借助自动和/或手动控制的水动力系统，使船舶在其作业时，能够在规定的作业范围和环境条件下保持其船位和艏向；(2)主要由自动控制船上推力器的系统组成的定位技术。其产生适当的推力矢量以抵消由风、浪、流诱导均衡和缓慢变化的载荷。　Dynamic poisoning means：(1) a hydrodynamic system with automatic and/or manual control capacity of maintaining the heading and position of the ship during operation within specified operating limits and environmental conditions；(2) a station-keeping technique primarily consisting of a system of automatically controlled onboard thrusters, which generate appropriate thrust vectors to counteract the mean and slowly varying actions induced by wind, wave, and current.

Dynamic positioning control station　动力定位控制站　系指进行动力定位操作和控制的 DP 控制站，相关的指示器、报警器、控制板和通信系统应安装在该控制站内。　Dynamic positioning control station means a control station used for operating and controlling of dynamic position and the relevant indicators, alarms, control panel and communication system are to be fitted on the control station.

Dynamic positioning control system（DPCS）　动力定位控制系统　是一种闭环控制系统。是动力定位系统三个组成部分之一。其功能是不断检测船舶的实际位置与目标位置的偏差，再根据外界风、浪、流等外界挠动力的影响计算出使船舶恢复到目标位置所需推力的大小，并对船上各推力器进行推力分配，从而使船尽可能地保持在要求的位置上。早期动力定位中采用 PID 控制器，分别对船舶在海平面内纵摇、横荡和艏摇3个自由度上的运动实施控制，同时为了避免响应高频运动，还采用滤波器以剔除偏差信号中的高频成分。目前应用比较广泛的是第二代动力定位系统，该系统将卡尔曼(Kalman)滤波器引入动力定位的控制中。卡尔曼(Kalman)滤波器的作用主要是测量所得到的船舶综合运动的位置信息，估计出其低频运动状态并将之反馈形成针对船舶低频运动的线性随机最优控制。此外卡尔曼(Kalman)滤波器,使得取样和修正在同一周期内完成,因而解决了控制中存在的由于滤波而导致的相位滞后问题。近年来出现第三代动力定位系统采用了智能控制的理论和方法。智能控制建立在模糊控制理论基础上,智能控制不依赖于对象的精确数学模型,抗干扰能力强、响应速度快、鲁棒性好。智能控制中加入自调节功能,可在系统的运动过程中根据外干扰条件自动对控制策略进行调整。

Dynamic positioning controller 动力定位的控制器 系指船舶实现动力定位所必需的一切集中控制的硬件和软件。控制器一般应由一台或几台计算机组成。 Dynamic positioning controller means all concentrated control hardware and software necessary to supply DP of the vessel. The controller is generally composed of one or more computers.

Dynamic positioning drilling ship 动力定位钻井船 系指依靠自动控制的动力定位装置使船体保持所需位置以进行钻井作业的船舶。

Dynamic positioning drilling unit 动力定位式钻井装置 系指依靠自动控制的动力定位装置使平台保持所需位置以进行钻井作业的平台。

Dynamic positioning system 动力定位系统 系指使动力定位船舶实现动力定位所必需的一整套系统,包括下列分系统:(1)动力系统;(2)推力器系统;(3)动力定位控制系统和测量系统;(4)独立的操纵杆系统。 Dynamic positioning system means the complete installation necessary for dynamically positioning a vessel, comprising the sub-systems:(1)power system;(2)thruster system;(3)DP-control system and measuring system;(4)independent joystick.

Dynamic positioning thruster system 动力定位的推进器系统 系指用于动力定位的各推进器及其控制装置,包括:(1)具有驱动设备和必要的附属系统(包括管路)的推进器;(2)在动力定位系统控制下的主推进器和舵;(3)推进器电子控制设备;(4)手动推进器控制设备;(5)相关的电缆和电缆布线。

Dynamic positioning vessel 动力定位船舶 系指仅用推进器的推力自动保持其自身位置(固定的位置或预先确定的航迹)和艏向的船舶。 Dynamic positioning vessel means a vessel which automatically maintains its position (fixed location or predetermined track) and heading, exclusively by means of thruster force.

Dynamic pressure[1] 动压力 流体流动所产生的压力。其值为:$q = 1/2 \rho U^2$,式中,ρ—流体质量密度;U—流速。

Dynamic pressure[2] 动压力 系指由于水流冲击或爆炸事件而产生的压力,其使作用在爆炸区域的结构物上的压力—时间过程得以提升,例如由于砰击或晃动产生的压力载荷。 Dynamic pressure is a pressure due to water impact or explosive events, which give rise to a pressure-time history acting on the exposed area of a structure, for example, pressure loads arising from slamming or sloshing.

Dynamic stability 动稳性 系指在倾侧力矩突然加于船上,船舶的角速度有明显变化的倾斜时的稳性。

Dynamic stability control(DSC) 动态稳定控制系统

Dynamic stress variations 动态应力变量 系指应力范围 S 或者应力幅度 σ。 Dynamic stress variations are referred to as either stress range, S, or stress amplitude, σ.

Dynamical bending moment 动弯矩 船舶在波浪中运动时,由于波浪水动力、惯性力等的作用而产生的弯矩。

Dynamical heeling angle 动横倾角 船舶在动态外力作用下,外力矩所做的功等于复原力矩所消耗的功时产生的横倾角。

Dynamical similarity 动力相似 几何相似的物体,当运动相似时,对应点上作用的同类力方向相同,大小成一比例的相似性。

Dynamical stability curves 动稳性曲线 系指船舶横倾角与稳性力臂的关系曲线。

Dynamical stability level 动稳性臂 船舶在外力作用下倾斜时,重心与浮心间垂向距离相对于正浮状态时的变化值。

Dynamical upsetting angle 动倾覆角 倾覆力矩所对应的动横倾角。

Dynamically positioned vessel(DP-vessel) 动力定位船舶 是仅靠其推力器(包括轴系)就能自动定位(固定位置或预定航迹)的船舶或平台;动力定位系统包括所有实现该目的所必需的手段。 Dynamically positioned vessel (DP-vessel) means a unit or a vessel which automatically maintains a fixed position or follows a preset track exclusively by the action of its thrusters (including shaft lines), the dynamic positioning system (DP-system) comprises all means necessary for this purpose.

Dynamically supported craft construction and equipment certificate 动力支承船舶构造和设备证书 系指对动力支承船舶在按动力支承船舶安全规则 1.5.1 (a)的规定进行检验后签发的证书。 Dynamically supported craft construction and equipment certificate means a Certificate issued after survey carried out in accordance with

Paragraph 1.5.1(a) of the Code of Safety for Dynamically Supported Craft.

Dynamically-positioned drilling ship 动力定位钻井船 依靠配置在船上的动力定位系统进行定位的钻井船。

Dynamometer 缆绳测力计 为各类绞车收放绳索过程中测量绳索上拉力的设备。

Dynamometer calibrating equipment 测力仪校验设备 校验螺旋桨测力仪精密度并测定其内部损耗的设备。

E

E class combustible liquid E级可燃液体 对美国国家消防委员会(NFPA)船上气体危害控制标准而言,系指闪点≥65.5 ℃(150 °F)的可燃液体。

Eagle strike 18 "鹰击18" 是中国新型导弹之一,它无须命中目标就可以摧毁其60%电子系统,将其列入全球七大反舰巡航导弹之首,认为只需1枚就足以使一艘"宙斯盾"战舰失去作战能力,即使在距敌舰50 m处爆炸,其强大的反辐射功能也可以将敌方60%的舰载电子系统摧毁。见图E-1。

图 E-1 "鹰击18"
Figure E-1 Eagle strike 18

Ear nut 蝶形螺母
Ear protector 耳塞(耳罩)

Early implementation of the technical standards of the Hong Kong International Convention for the Safe and Environmentally Sound Recycling of Ships, 2009 尽早实施2009年香港国际安全与环境无害化拆船公约中的技术标准 系指2009年5月在香港举行的拆船公约外交大会上还通过了第5号决议——"尽早实施2009年香港国际安全与环境无害化拆船公约中的技术标准",该决议敦促IMO各成员国在公约生效前,在自愿基础上尽早实施公约附则"安全与环境无害化拆船规则"中的技术标准,并邀请工业界给以配合。随着社会对环保呼声日益高涨,不排除一些区域或国家提前强制实施拆船公约标准的可能性,为避免一些不必要的麻烦,建议各方熟悉相关要求,建立符合要求的体系和文件,并及早做好提前实施的准备。

Early production system (EPS) 早期生产系统 系指利用已有的少数勘探井、试油井和快速改装完成的采油设施先行生产,提前获得经济收益,有时也结合进行延长测试,以期进一步探明油藏,为最终制定油田开发方案提供有价值的油层资料的海上油气生产系统或生产与测试系统。

Early warning airplane 预警机 全称预警和控制系统飞机,是一种装有远距搜索雷达、数据处理、敌我识别以及通信导航、指挥控制、电子对抗等完善电子设备,集预警、指挥、控制、通信和情报于一体,用于搜索、监视与跟踪空中和海上目标并指挥、引导己方飞机遂行作战任务的作战支援飞机。该技术拥有国:美国(E-3A"哨兵")、以色列(费尔康)、俄罗斯(A-50"中坚")、英国(Mk3"猎迷")、中国(空警-2000/空警-200)、瑞典/荷兰(萨伯-2000)。见图E-2。

图 E-2 预警机
Figure E-2 Early warning airplane

Early-warning radar (EWR) 预警雷达 属于一种远距离搜索雷达。主要用来发现远、中、近程弹道导弹,测定其瞬间位置、速度、发射点和弹着点等关键参数,为最高军事机关提供导弹预警情报。

Earnings before interest and tax (EBIT) 息税前利润 系指不扣除利息也不扣除所得税的利润,也就是在不考虑利息的情况下在交所得税前的利润,也可以称为息前税前利润。息税前利润,顾名思义,系指支付利息和所得税之前的利润。税前利润就是在交税前的利润总和,如果在追求利润的时候不考虑利息,那么计算出来的就是息税前利润(息前税前利润)。无论企业营业利润多少,债务利息和优先股的股利都是固定不变的。当息税前利润增大时,每一元盈余所负担的固定财务费用就会相对减少,这能给普通股股东带来更多的盈余。

Earth station 地球站 通称卫星地球站,是由天线分系统、发射放大分系统、接收放大分系统、地面通信设备分系统、终端分系统和通信控制分系统以及电源分系统组成。分为固定式地球站、可搬运地球站、便携式地球站、移动地球站以及手持式卫星移动终端。

Earth 地 船舶或近海装置的全部金属结构或船体的整体。(注意:在美国和加拿大,"地"用"ground"而不用"earth")。 Earth is the general mass of the metal structure or hull of the ship or unit. (Note: In the U.S.A. and Canada "ground" is used instead of "earth").

Earthed 接地 (1)以确保在任何时候均能即时释放电能而不发生危险的与船舶或近海装置的金属结构或船体的整体的连接;(2)连接到船体总的接地块以确保总是立即释放电能而不致造成危险的接地线。 Earthed means: (1) connected to the general mass of the structure or hull of the ship or unit in such a manner as will ensure at all times an immediate discharge of electrical energy without danger; (2) earth connection to the general mass of the hull of the ship in such a manner as will ensure at all times an immediate discharge of electrical energy without danger.

Earthing device for portable instrument 活动仪器的接地装置 用于将活动或临时安装于实验室的电子仪器的外壳接地(船壳)的一种设施。

Earthing factor 接地系数 系指相对地电压与相对相的电压之比,在(1/sqrt 3)和 1 之间取值。 Earthing factor is defined as the ratio between the phase to earth voltage of the health phase and the phase to phase voltage. This factor may vary between (1/sqrt 3) and 1.

Easily accessible 易于接近 系指离操作位置 5 m 的距离以内。 Easily accessible means distance within 5 m from working position.

Easily attacked area 易受攻击区域 系指需要适当控制的场所,可包括:(1)甲板储藏室;(2)货物机械存藏室;(3)氧气和乙炔气瓶储存场所。 Easily attacked area means the location where appropriate control is needed. These areas include: (1) deck storage room; (2) cargo machinery storage room; (3) storage location for oxygen and acetylene.

Easily readable 易于阅读 为了使工作站上的信息易于阅读,所有相关的指示器和显示器都位于从操作位置看的向前 180°视野区域内。指示器和显示器的放置方向应垂直于从操作位置看到的导航仪的视线,如果信息应由位于多个工作站的人员使用,则放置在平均值(角度)处。 For information to become easily readable at the workstation all relevant indicators and displays are located within the forward 180° view sector seen from the operating position. The indicators and displays are placed with its front perpendicular to the navigator's line of sight seen from the operating position, or lacated within a mean value (angle) if the information shall be used by personnel located at more than one workstation.

E-business 电子商务 系指使用 Web 技术帮助企业精简流程,增进生产力,提高效率。使公司易于沟通合作伙伴、供货商和客户,连接后端数据系统,并以安全的方式进行商业事项处理。

Eccentric 偏心轮(偏心的,离心的)
Eccentric angle 偏心角
Eccentric arm 吊杆偏心距/偏心半径 作用在吊杆头端部的合力的作用线偏离吊杆轴线的距离。
Eccentric deviation 偏心度
Eccentric impact 偏心冲击力
Eccentric link 偏心轮机构
Eccentric motion 偏心运动
Eccentric point of cycloidal propeller 平旋推进器偏心点 决定平旋推进器推力方向和大小的点。在外摆线推进器是各叶垂线相会的点,正摆线推进器是顺叶水平直线相会的点。
Eccentric pulley 偏心滑轮
Eccentric radius 偏心半径
Eccentric reduction gear 偏位式减速齿轮箱 轴系中线与浮动件轴的轴线平行但不重合的减速齿轮箱。
Eccentric rod 偏心杆(蒸汽机)
Eccentric roller 偏心滚轮
Eccentric roller jacking device (eccentric roller) 偏心起升装置 装设在可转的偏心轮轴上,可利用偏心距抬升舱盖板,然后借助他力使之移动的滚轮装置。
Eccentric sheave 偏心环(偏心滑轮)
Eccentric shoe 偏心杆滑块
Eccentric strap 偏心环
Eccentric wheel 偏心轮
Eccentricity 偏心距(偏心率,偏心度)

ECDIS = Electronic chart display and information system 电子海图显示和信息系统 系指导航信息系统。其具有足够的备用装置,通过显示从电子航行海图系统(SENC)选取的信息与《国际海上人命安全公约》(SOLAS)第 V 章/第 19 条所要求的新海图相符合,并满足经海安会 MSC.99（73）决议修正的 SOLAS 第 V 章中的海图输送要求。 A navigation information system, with adequate back-up arrangements, can be accepted as complying with the up-to-date chart required by regulation 19 of SOLAS Chapter V, and be accepted as meeting the chart carriage requirements of SOLAS Chapter V, as amended by Res. MSC.99(73), by displaying selected information from a system electronic nautical chart (SENC)

ECDIS display base 电子海图显示和信息系统显示库 系指不能从电子海图显示和信息系统（ECDIS）显示器上删除的信息的级别，由所有地理区域和所有情况下在任何时候都要求的信息组成。但并不打算足以用于安全航行。 ECDIS display base means the level of information which cannot be removed from the ECDIS display, consisting of information which is required at all times in all geographic areas and all circumstances. It is not intended to be sufficient for safe navigation.

ECDIS standard display 电子海图显示和信息系统标准显示器（ECDIS 标准显示器） 系指当初次显示在电子海图显示和信息系统上时应显示的信息级别。用于航线制定或航线监控的信息级别可由航海者按其需求进行修改。 ECDIS standard display means the level of information that should be shown when a chart is first displayed on ECDIS. The level of the information it provides for route planning or route monitoring may be modified by the mariner according to the mariner's needs.

Echo sounder/Echo sounding device 回声测深仪 通过测量声脉冲到海底返回的时间间隔来确定海水深度的仪器。有浅海和深海 2 种，后者只用于海洋调查船。

Echo sounding device 回声测深仪 通过测量声波自发射经海底反射至接收的时间间隔推算水深的设备。

Eco ship 环保型船 其最突出的技术特点是采用的 LNG 高压直喷式发动机，将液化天然气（LNG）作为动力燃料，通过高压的方式将 LNG 直接喷入发动机。其优点是可以提高燃料的燃烧效率。同时，使用该型发动机可将 CO_2 和 NO_x 等有害气体及污染物的排放量减少 30%。采用的 LNG 高压直喷式发动机是仿效欧洲国家汽车发动机的研究成果，以柴油为燃料，采用高压直喷燃料技术，能大大提高燃料的燃烧效率从而降低有害气体的排放量。此外，环保型船还利用安装在甲板上的太阳能电池作为动力，利用风能的风筝船帆提供辅助性推力，采用电动机和柴油机相结合的"柴-电混合动力"（也称"双动力"）推进方式，以及有纯粹电力推进、燃料电池推进、核能推进等方式的新型船舶。

Economic integration 经济一体化 理论上系指加入同盟的国家有区别地减少或消除贸易壁垒的商业政策。即世界经济一体化，指世界各国经济之间彼此相互开放，形成相互联系、相互依赖的有机体。在这个多国经济联盟的区域内，商品、资本和劳务能够自由流动，不存在任何贸易壁垒，并拥有一个统一的机构，来监督条约的执行和实施共同的政策及措施。

Economic life of hull 船体经济寿命 系指从经济角度分析船体的最佳使用时间。

Economic turbine stages 经济级组 设置在调节级后面于经济工况下投入运行而在巡航工况以上的航速时不工作的低速级组。

Economic union 经济同盟 系指不但成员国之间废除贸易壁垒，统一对外贸易政策，允许生产要素的自由流动。

Economical dredging depth 采用挖深 挖泥船经常使用的高效率的挖深。

Economizer 经济器 置于锅炉的低烟温区烟道中用来回收烟气热量以加热给水的装置。

Economy entity 经济体 系指对某个区域的经济组成进行统称和划分，经济系指社会生产关系的总和。系指人们在物质资料生产过程中结成的，与一定的社会生产力相适应的生产关系的总和或社会经济制度，是政治、法律、哲学、宗教、文学、艺术等上层建筑赖以建立起来的基础。

Economy-based safety policy and objective 基于安全方针和目标的绩效 系指与公司海上安全和保安风险控制有关的，船舶保安管理体系的可测量的结果。 Economy-based safety policy and objective means the measurable effect of ship security management system related to the maritime safety and security risk control of the company.

Edelweiss dining room 罗密欧与朱丽叶餐厅（豪华邮轮）

Edge corrosion 边缘腐蚀 是板材、扶强材、主要支撑构件和开孔的自由边缘的局部厚度减少。 Edge corrosion is defined as local corrosion at the free edges of plates, stiffeners, primary support members and around openings.

Edge grinding 边缘打磨 系指二次表面处理前对边缘的处理。 Edge grinding is the treatment of edge before secondary surface preparation.

Edge rounded 倒角

Edge stress 边缘应力 板构件的边缘应力包括板的公称应力和边缘的应力增加影响，应采用有限元方法进行结构分析以求得边缘应力。 Edge stress on plate members includes the nominal stress of the plates and stress-raising effects on the edge of plates. It is calculated by using the finite element method.

EDI 电子数据交换 系指通过电信传输进行贸易数据交换。 EDI means electronic data interchange, i. e. the interchange of trade data effected by tele-transmission.

Editable field 可编辑字段

EDP = Emergency disconnect programme 应急脱开程序

Eductor 喷射器(排泄器)

EEDI 新船能效设计指数 2009 年 IMO 基于"强制性新船二氧化碳设计指数"引入的船舶减排新概念。该指数进一步强调船舶节能减排增效的目标，对船舶设计、生产工艺技术、配套设备、新能源技术应用提出了更高的要求。针对新船能效设计指数提出了船型优化、螺旋桨优化、推进系统标准优化、配套设备优化等25项可行的应对技术措施。这意味着节能减排将成为未来船舶的强制性要求。目前国内外船型研发机构和造船企业已开始通过各种途径开发新船型，以达到节能减排的目的:(1)优化线型，光滑和流线型的船型可改善船体受阻面，从而减少阻力;(2)应用高强度钢，减轻空船质量;(3)优化推进装置。应用节能推进装置是最常用的方法，此外螺旋桨优化也是研发的重点，如采用混合推进方式，在船尾安装前后2个螺旋桨，后方螺旋桨旋转时，可以吸收前方螺旋桨的能源，使推进效率大幅提高;(4)是采用新型涂料减阻，如新型船舶仿生涂层，能够在船体表面形成空气膜，使船与水摩擦造成能耗降低从而大量节省燃料。

EEOP 船舶高效率航海软件 可对气候、海潮、海流、风力等影响航海的多项海上环境要素进行综合分析，并将分析结果显示在航海模拟仪的屏幕上，为航海人员提供船舶最佳航线、最佳航行速度和发动机的最佳工作状态等信息的软件。

Effect 结果/后果 是一个事件的后果所引起的局面。 Effect is a situation arising as a result of an occurrence.

Effect of centrifugal force of propeller race 尾流离心作用 螺旋桨尾流旋转产生的离心力对推力及尾流形状的影响。

Effective angle of attack 实效攻角 计入感生速度的流向与叶元体零升力线间的夹角。

Effective aspect ratio of rudder 舵实效展弦比 与安装在船体上的舵，有等效水动力性能的敞水舵的展弦比。

Effective camber 实效拱度 按叶曲面形状所确定的拱度与感生拱度之差。

Effective clearing of the ship 有效脱离船舶 系指自由降落救生艇在自由降落后，不用其发动机而脱离船舶的能力。 Effective clearing of the ship is the ability of the free-fall lifeboat to move away from the ship after free-fall launching without using its engine.

Effective compression ratio 有效压缩比 压缩行程在进排气门或完全关闭的瞬时，工质所占的容积和压缩容积之比。

Effective earthing system 有效接地系统 系指中性点直接接地或经一低值阻抗接地的系统。通常，该系统的零序电抗与正序电抗的比值 $x_0/x_1 \leq 3$ ；零序电阻与正序电抗的比值 $R_0/x_1 \leq 1$ ，该系统又称大接地电流系统。中性点直接接地是有效接地系统之一。

Effective efficiency 有效效率 汽轮机的实际有效功与理想可用功之比。

Effective gusset plates 有效扣板 系指:(1)与符合有关规定的厚度、材料性能和焊接接头的卸货板组合在一起;(2)其高度不小于面板的宽度的1/2;(3)与下墩舱侧板对齐;(4)与槽形和下墩舱顶板焊接;(5)其厚度和性能不小于面板。 Effective gusset plates are those which:(1) are in combination with shedder plates having thickness, material properties and welded connections according to relevant requirement;(2) have a height not less than half of the flange width;(3) are fitted in line with the lower stool side plating;(4) are welded to the lower stool plate, corrugations and shedder plates;(5) have thickness and material properties not less than those required for the flanges.

Effective horse power 有效功率(有效马力)

Effective longitudinal bulkhead 有效纵舱壁 系指从船底延伸到甲板并且通过其前后的横舱壁与船舷相连接的舱壁。 Effective longitudinal bulkhead is a bulkhead extending from bottom to deck and which is connected to the ship's side by transverse bulkheads both forward and aft.

Effective pitch of cycloidal propeller 平旋推进器节距 平旋推进器在无滑脱状态每转一周的进程。

Effective plate flange area 有效的带板面积 系指有效带板宽度内板的横剖面面积，可包括有效带板内连续扶强材的面积。 Effective plate flange area means the cross sectional area of plating within the effective flange width. Areas of continuous stiffeners within the effective flange may be included.

Effective power 有效功率 船舶以一定的航速航行时，单纯为克服船舶阻力所需的功率。 $P_E = RV$ ，式中，R—船舶阻力；V—航速。

Effective rudder angle 实效舵角 由于船型和螺

旋桨尾流以及船体运动等影响，改变舵叶水流方向的实际生效的舵角。

Effective satellite fix 卫星有效定位 指符合标准的卫星所确定的船位。

Effective sectional area of strength deck 强力甲板的有效剖面积 系指船的每一舷在船中 $0.5 L$ 长度范围内延伸的钢板、甲板纵骨和纵桁等的剖面积。 Effective sectional area of strength deck is the sectional areas, on each side of the ship's centre line, of steel plating, longitudinal beams, longitudinal girders, etc., extending for $0.5 L$ amidships.

Effective shedder plates 有效卸货板 系指：(1)无折边；(2)与槽形和下墩舱顶板焊接；(3)至少安装成 $45°$ 倾斜角，且其下端边缘与下墩舱的侧板对齐；(4)其厚度不小于槽形面板厚度的 75%；(5)材料性能不小于面板的材料性能。 Effective shedder plates are those which: (1) are not knuckled; (2) are welded to the corrugations and the lower stool top plate; (3) are fitted with a minimum slope of $45°$, their lower edge being in line with the lower stool side plating; (4) have thickness not less than 75% of that required for the corrugation flanges; (5) have material properties not less than those required for the flanges.

Effective sound pressure 有效声压 是在一段时间内瞬时声压的均方根值。这段时间应为周期的整数倍或长到不影响计算结果的程度。

Effective superstructure 有效上层建筑 系指延伸到船中 $0.4L$ 区域内，且其长度超过 $0.15L$ 的上层建筑，其侧壁板应视作外板，其甲板应视作强力甲板。 Effective superstructure is a superstructure which extends into the range of $0.4L$ amidships and the length of which exceeds $0.15L$. Their side plating is to be treated as shell plating and their deck as strength deck.

Effective thrust 有效推力 就螺旋桨而言，相当于用于克服船的总阻力部分。

Effective wake 实效伴流 螺旋桨在船模后和敞水中分别运转，在同转速时，其推力或转矩也相等时，船模航速与螺旋桨进速之比。

Effective waterplane area 剩余水线面面积 船舶破损进水后，水线面未浸水部分的面积。

Effective wave slope 有效波倾 对船舶整个水下体积起等效作用的波倾。

Effective width of escape routes 脱险通道的有效宽度 为了身体横向摆动和保持平衡，人员通过脱险通道时，应与墙壁和/或其他固定物体(例如栏杆和固定椅子等)保持一定的间隙，其有效宽度系指脱险通道总宽度减去总间隙所得值。 In order to accommodate lateral body sway and balance, persons moving through the escape routes maintain a clearance from walls and/or other fixed items (e.g. handrail, fixed seats, etc.). The effective width of any portion of an escape route is the clear width of that portion reduced by the sum of the clearness.

Effective wood fire retardant(EWR) 高效木材阻燃剂

Effectiveness analysis of safety and emergency preparedness measures 安全与应急准备措施的有效性分析 记录安全与应急准备功能要求满足情况的分析。

Effectiveness principle 有效性原则 是风险评估精确衡量风险以达到适当保护所必需的风险程度的原则。 Effectiveness principle is a principle that risk assessments accurately measure the risks to the extent necessary to achieve an appropriate level of protection.

Effectiveness test 效用试验

Effects of finite number of blades 有限的叶数影响 仅有少数叶的实际螺旋桨与无限多叶的理想螺旋桨间的水动力性能差异。

Effervescing steel 沸腾钢

Efficiency 效率（高效）

Efficiency of blade element 叶元体效率 某半径处叶元体发出的推力功率对其吸收的转矩功率的比值。

Efficiency of ideal propeller 理想推进器效率 根据动量理论理想推进器所能达到的效率极限。这种理想推进器除了留在尾流中的动能外别无其他损失。

Effluent intensity indication for CO₂ 二氧化碳排放强度指标 系指营运船舶单位运输周转量 CO_2 排放量。

Effluent(Q_e) 排出物 系指由污水处理装置所产生的经处理的水。 Effluent means the treated wastewater produced by the sewage treatment plant.

Effuser 喷管

EGC 接收机 是单频道接收机并有专门的信息处理器。

EGC record book EGC 记录簿 系指与监测系统正常工作或其废气滤清装置使用操作参数、部件调节、维护和使用的记录簿。 EGC record book means a record book of the EGC unit in-service or operating parameters, component adjustments, maintenance and service records.

EGCS residues 废气滤清系统(EGCS)残余物 系指水处理过程的产物。该残余物可通过使用不同的处理技术形成并从水中去除，其含有硫酸盐、灰、积碳、金属和碳氢化合物，并可能特别含有亚硫酸盐($CaSO_x$)、还可能含有其他金属亚硫酸盐(NaSO$_x$ 和 KSO$_x$)和金属氧化物,包括钒(V)、镍(Ni)、镁(Mg)铝(Al)铁(Fe)和硅(Si)。 Exhaust gas cleaning system (EGCS) residues are products from the water process. The residues can be formed and removed from the water with different treatment techniques. Such residues contain sulphates, ash/soot, metals and hydrocarbons removed from the water. Specifically it may contain sulphate salts (CaSO$_x$) and may also include other metal sulphates (NaSO$_x$ and KSO$_x$) and metal oxides and including Vanadium (V), Nickel (Ni), Magnesium (Mg), Aluminium (Al), Iron (Fe), and Silicon (Si).

EGCS-SOx technical manual (ETM) 废气(SO$_x$)滤清系统技术手册 系指至少包括下列内容的技术手册:(1)装置的标识(生产厂商/型号/类型、序列号和其他必要的细节)包括对装置和要求的辅助系统的描述;(2)装置的证书上规定的工作限定值范围,其应至少包括:①最大和最小(如适用)废气质量流量;②应安装废气(SO$_x$)滤清系统装置的燃油燃烧装置的功率、类型和其他相关参数。对于锅炉,应给出在 100% 载荷时的最大空气/燃油比率。对于柴油机,应说明发动机是 2 冲程还是 4 冲程;③最大和最小洗涤水的流速、进口压力和最小进水口碱性(pH);④废气(SO$_x$)滤清系统装置工作时废气进口温度范围以及最大废气出口温度;⑤燃油燃烧装置以 MCR 或 80% 的额定功率(取其中的合适者)工作时,废气压差范围和废气进口最大压力;⑥提供足够中和剂所需的盐度或淡水成分;⑦为确保最大排放值不超过 6.0 g SO$_x$/kW·h,与之相关的废气(SO$_x$)滤清系统装置设计和操作方面的其他因素。(3)为确保装置的最大排放值不超过 6.0 g SO$_x$/kW·h 所必需的适用于废气(SO$_x$)滤清系统装置或相关设备的任何要求或限定。(4)维修保养、服务或调整要求以使废气(SO$_x$)滤清系统装置能保持最大排放值不超过 6.0 g SO$_x$/kW·h。(5)废气(SO$_x$)滤清系统装置的检验方法以确保保持其性能且装置按要求使用。(6)通过洗涤水特征中的范围性能变化。(7)洗涤水系统的设计要求;(8)SECA(SO$_x$ emission control area 硫氧化物排放控制区)符合证书。废气(SO$_x$)滤清系统技术手册应经主管机关认可。 EGCS-SO$_x$ technical manual (ETM) is a manual, as a minimum, containing the following information: (1) the identification of the unit (manufacturer/model/type, serial number and other details as necessary), including a description of the unit and any required ancillary systems; (2) the operating limits, or range of operating values, for which the unit is certified. These should, as a minimum, include: ①maximum and, if applicable, minimum mass flow rate of exhaust gas; ②the power, type and other relative parameters of the fuel oil combustion unit for which the EGCS-SO$_x$ unit is to be fitted. In the cases of boilers, the maximum air/fuel ratio at 100% load should also be given. In the cases of diesel engines, whether the engine is of 2 or 4 stoke cycle should also be given; ③maximum and minimum wash water flow rate, inlet pressures and minimum inlet water alkalinity (pH); ④exhaust gas inlet temperature ranges and maximum exhaust gas outlet temperature with the EGCS-SO$_x$ unit in operation; ⑤exhaust gas differential pressure range and the maximum exhaust gas inlet pressure with the fuel oil combustion unit operating at MCR or 80% of power rating, whichever is appropriate; ⑥salinity levels or fresh water elements necessary to provide adequate neutralizing agents; ⑦other factors concerning the design and operation of the EGCS-SO$_x$ unit relative to achieving a maximum emission value no higher than 6.0 g SO$_x$/kW·h; (3) any requirements or restrictions applicable to the EGCS-SO$_x$ unit or associated equipment necessary to enable the unit to achieve a maximum emission value no higher than 6.0 g SO$_x$/kW·h; (4) maintenance, service or adjustment requirements in order that the EGCS-SO$_x$ unit EGCS-SO$_x$ unit can continue to achieve a maximum emission value no higher than 6.0 g SO$_x$/kW·h; (5) the means by which the EGCS-SO$_x$ unit is to be surveyed to ensure that its performance is maintained and that the unit is used as required; (6) through range performance variation in wash-water characteristics; (7) design requirement of the wash water system; (8) the SECA (SO$_x$ emission control area) and EGCS-SO$_x$ technical manual (ETM) should be approved by the Administration.

Egyptian environmental protection act No. 4 1994 (EEPA) 埃及环境保护法规(第 4 号,1994 年)

EIAPP Certificate EIAPP 证书 对"73/78 防污公约"附则 VI 而言,系指与 NO$_x$ 排放有关的发动机国际防止空气污染证书。 EIAPP Certificate is the Engine International Air Pollution Prevention Certificate which relates to NO$_x$ emissions.

Ejection 喷出(排出)

Ejector 喷嘴(喷射器,喷吸器,弹射器)

Ejector pump 喷流泵(射流泵,抽气泵) 利用小股流体的高速喷射作用提升及运送大量流体的泵。

Ekman current meter 厄克曼海流计 测量流速和流向的旋转式海流计。仪器有两个旋桨:轻桨用于测弱流,重桨用于测强流。

Elastic 弹性的

Elastic coupling 弹性联轴节

Elastic deformation 弹性变形

Elastic element 弹性元件

Elastic vibration 弹性振动

Elastic wave 弹性波 应力波的一种,扰动或外力作用引起的应力和应变在弹性介质中传递的形式。弹性波理论已经比较成熟,广泛应用于地震、地质勘探、采矿、材料的无损探伤、工程结构的抗震抗爆、岩土动力学等方面。

Elasticity 弹性(恢复性,伸缩性)

Elasticity sandwich plate 弹性夹心板

Elbow 弯管(弯头)

Elbow acoustic filter 弯头消音器

Elbow connection 弯接头

Elbow of capture union 弯头套环/弯头联管节

Elbow union joint 弯管连接头

Elbow union 弯头套管

Elbow ventilator 弯管通风筒(鹅颈式通风筒)

Electric (cooking) range 电灶 船上用以烧煮、烘烤、炒炸各种食物的电热炉灶。

Electric appliance branch circuit 电气器具分支电路 系指向一个或多个用以连接电气器具的引出线端供电的电路。这种电路上没有永久性连接的、不属于电气器具一部分的照明灯具。

Electric auxiliary 电动辅机

Electric cargo winch 电动起货机 电力驱动的起货绞车。

Electric controlled diesel engine 电控柴油机 系指具有电控系统的柴油机,即是通过电子控制器等控制装置,柔性调节各系统参数,进一步提高各系统与柴油机匹配的灵活性及适应能力,实现最佳的运行,使性能得到优化的柴油机。Electric controlled diesel engine means diesel engine with electronic control system, i. e. diesel engine whose system parameters can be adjusted flexibly by control equipment such as ECU to further improve the flexibility and adaptability between individual system and system and diesel engine, and to optimize the performance.

Electric controller 电控制器 系指以某种预定方式对输送到其所接设备的电力进行控制的器件或器件组合。

Electric controlling unit(ECU) 电子控制器 对各种来自传感器的信息进行处理,并将结果输出,驱动执行机构动作。电子控制器主要由微电脑、接口电路、驱动电路及必要的软件组成。Electric controlling unit (ECU) deals with information from sensor, outputs the result and drives actuator. ECU is mainly composed of microcomputer, interface circuit, drive circuit and necessary software.

Electric coupling 电耦合(电磁联轴节)

Electric defrost 电热融霜 利用电加热方式将盘管表面霜层融化的融霜方式。

Electric drive 电力传动装置

Electric frying-pan 电炒锅 船上用以煎炒各种食物的电锅。

Electric generating set 发电机组

Electric heater 电暖器 供船上取暖用的电热电器。

Electric heating 电暖 采用电能转换成热能的取暖方式。

Electric heating belt 电加热带

Electric heating device 电热器

Electric heating element 电热元件

Electric heating unit 电热元件

Electric igniter 电火花点火器

Electric ignition 电点火

Electric immersion heater 水电加热器 以电能加热蒸馏水或盐水产生蒸汽用于启动或补偿压气式蒸馏装置热损失的设备。

Electric lighter 电点火器

Electric propulsion 电力推进

Electric propulsion plant 电力推进装置

Electric propulsion system 电力推进系统 由电动机直接带动船舶推进器(或螺旋桨)的推进系统。按原动机类型可分为柴油机电力推进系统,汽轮机电力推进系统,燃气轮机电力推进系统,原子能装置电力推进系统;按电流种类可分为直流、交流和交直流电力推进系统;按装置的功能可分为独立电力推进装置,联合电力推进装置,辅助电力推进装置,主动舵电力推进装置和特殊电力推进装置。见图 E-3。

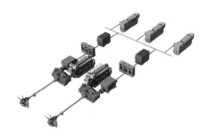

图 E-3 电力推进系统

Figure E-3　Electric propulsion system

Electric pump 电动泵

Electric radiator 电取暖器

Electric resistance welded pipe (ERW) 电阻焊接管 系指采用对接焊焊成带纵向焊缝的管子。 Electric resistance welded pipe (ERW) is a pipe with a longitudinal weld formed by butt welding.

Electric shop 电工车间　系指从事电工器材安装、调试、贮存和检修的场所。

Electric space heating appliance 电暖器

Electric steering control system 电力操舵系统　用电力机械传动方式操纵舵机的系统。

Electric steering gear 电动舵机　电力驱动的舵机。

Electric teapot 电茶壶　用以煮沸饮用水的茶壶状电炊具。

Electric towing winch 电动拖缆机　电力驱动的拖缆机。

Electric/steam rice cooker 电气两用饭锅　既可利用蒸汽,也可利用电加热的煮饭锅。

Electrical anchor capstan 电动起锚绞盘　电力驱动的起锚绞盘。

Electrical equipment with type of protection "n" "n"防护型电气设备　在正常运行时,该型设备不会引燃周围易爆性气体,且不可能发生引燃的故障。 Electrical equipment with type of protection "n" is the electrical equipment such that, in normal operation, it is not capable of igniting a surrounding explosive gas atmosphere and a fault capable of causing ignition is not likely to occur.

Electrical installations 电气设备　系指这样的设备:(1)对为船舶正常操作和居住条件所必需的电气辅助设备保证供电,而不必求助于应急电源;(2)在各种应急情况下,能保证对安全所必需的电气设备供电;(3)能确保乘客、船员和船舶的安全,免受电气事故的危害。 Electrical installations are such installations that: (1) all electrical auxiliary services necessary for maintaining the craft in normal operation and habitable conditions will be ensured without recourse to the emergency source of electrical power; (2) electrical services essential for safety will be ensured under various emergency conditions; (3) the safety of passengers, crew and craft from electrical hazards will be ensured.

Electrical installations 电气设备(动力定位系统)　系指动力定位系统的能量所需要的所有装置。 Electrical installations comprise all units necessary to supply the dynamic poisoning system with power.

Electrical motor generator unit 电动发动机组　是由一台或多台电动机在机械上与一台或多台发电机相连接而组成的一种电机。

Electrical network 电力网　又称电网。系指介于船舶电源和用电设备之间的传输、分配电能的全部配电装置和网络的总称。根据供电电源的不同、负载的性质和用途不同,船舶电力网可分为:(1)主电网——由主电源通过主配电板供电的电网;(2)应急电网——由应急发电机通过应急配电板供电,或由蓄电池通过蓄电池充放电板供电的网络;(3)临时应急电网——由蓄电池通过蓄电池充放电板用以传输、分配临时应急电能的网络;(4)一次网络——由主配电板直接向区配电板、分电板和发展供电的网络,也称一次系统或一次动力系统;(5)二次网络——由区配电板和分配电板向负载供电的网络,也称二次系统或二次电力系统;(6)动力网络——船舶电网中向动力设备供电的网络;(7)照明网络——船舶电网中向照明设备、电风扇及小容量电热设备等供电的网络;(8)弱电网络——船舶电网中向各导航、通信及无线电设备等供电的网络。

Electrical propulsion system 电力推进系统

Electrical prospecting 电法勘探　是以岩/矿石电学性质的差异为基础,通过观测和研究与这些差异有关的电场或电磁场在空间和时间上的分布特点和变化规律,查明地下地质构造和寻找有用矿产的物探方法。

Electrical services 电器用途　区分为重要用途和居住便利用途。 Electrical services are classified into essential services and services for habitability.

Electrical windlass 电动起锚机　电力驱动的起锚机。

Electrically welded chain 电焊锚链　用圆钢弯制并电焊而成的锚链。

Electrically welded stud link chain 有挡焊接锚链

Electrically-propelled ship 电力推进船　利用本船电源或蓄电池驱动的电动机作为主机的船舶。

Electrician's store 电工间　供贮存和检修电工器材的房间。

Electrode type salinometer 电极式盐度计　利用电极式电导率传感器测量海水盐度的仪器。由2~6个电极传感器、温度和压力补偿电器、显示和记录装置组成。

Electro-dialysis methods 电渗析法　利用相同排列的有选择性的阳、阴离子渗透膜,在直流电场作用下,使离子通过渗透膜进入相邻的隔室而使海水得以淡化的方法。

Electro-discharge treatment method (EDTM) 放电处理法

Electro-fin stabilizing system 电动减摇鳍装置　减摇鳍的运动执行机构全部为电力拖动的减摇鳍装置。

Electro-hydraulic steering gear 电动液压操舵装置(电动液压舵机)

Electrolysis-based ship ballast water management system 电解法船舶压载水管理系统

Electro-magnetic radiation（ER） 电磁辐射 又称电子烟雾,是由空间共同移送的电能量和磁能量所组成,而该能量是由电荷移动所产生；例如,正在发射信号的射频天线所发出的移动电荷,便会产生电磁能量。电磁"频谱"包括形形色色的电磁辐射,从极低频的电磁辐射至极高频的电磁辐射。两者之间还有无线电波、微波、红外线、可见光和紫外光等。电磁频谱中射频部分的一般定义,系指频率约由 3 千赫至 300 吉赫的辐射。而有些电磁辐射对人体也有一定的影响。

Electromagnetic clutch 电磁离合器 系指借助于电磁力传递转矩的一种装置。在这种装置中,驱动部件与从动部件之间无机械转矩联系。滑差式电磁离合器具有在一个旋转部件上的、由直流电流激励的磁极,而通常是双鼠笼型的电枢绕组,则在另一个旋转部件上。

Electromagnetic compatibility 电磁兼容性 系指采取适当的措施,以减小由于电磁能量所产生的干扰,从而保证所有电气设备和电子设备在船舶电磁环境中能正常工作。 Electromagnetic compatibility means that appropriate measures are to be taken to reduce the interference due to electromagnetic energy, so that all electrical and electronic equipment can operate normally in a ship's electro-magnetic environment.

Electromagnetic compatibility of electrical equipment and electronic equipment 电气和电子设备的电磁兼容 系指设备具有抗电磁干扰和抑制电磁干扰的能力。以保证设备在船舶电磁环境中正常工作。船舶电磁兼容应实现：(1)抑制干扰源的干扰电平或场强；(2)保证接收机一定的防干扰能力；(3)无线电接收设备与干扰源按必需的要求布置在船上；(4)载有干扰的电缆与对干扰灵敏的电缆进行屏蔽；(5)在船舶电缆传输干扰途径中,采取增加衰减干扰的措施；(6)在船舶舾装设备中采取防干扰措施。

Electromagnetic compatibility of ship's overall 船舶总体电磁兼容 系指船舶设计或建造应满足电磁兼容性要求。也就是说,在天线布局、舱室布置、设备安装、供电配电和电缆选择及其敷设等方面,严格按照电磁兼容技术要求实施,提供良好的电磁环境,保证电气设备和电子设备正常工作。

Electromagnetic current meter 人工磁场电磁海流计 一种以海水作导体通过一个人工磁场产生电动势来测定海流的仪器。

Electromagnetic disturbance 电磁骚扰 系指任何可能引起具、设备或系统性能降低,或者对有生命或无生命物质产生损害作用的电磁现象。

Electromagnetic emission 电磁发射 系指从源向外发出电磁能的现象。

Electromagnetic interference 电磁干扰 系指电磁骚扰引起的设备、传输通道或系统性能的下降。

Electromagnetic radiation 电磁辐射 系指能量以电磁波形式由源发射到空间或以电磁波形式在空间传播的现象。

Electromagnetic valve 电磁阀

Electromagnetism noise 电磁性噪声 是由于磁场脉动、伸缩引起周围空气运动而产生的噪声。

Electron beam ion trap（EBIT） 电子束离子阱 电子束从电子枪中被引出后,被漂移管和电子枪之间的高电压差加速,与之同时,电子束被位于漂移管段的超导亥姆霍兹线圈产生的强磁场约束,电子束在漂移管段时半径达到最小。在漂移管中与低电荷态离子碰撞,使离子的电子不断被剥离,电荷态不断地升高,直至平衡。通过选择电子束的能量和密度可以选择离子的电荷态。电子经过漂移管后被减速,最后被电子接收部件收集。

Electron diode 电子二极管 系指只具有一个阴极与一个阳极(板极)的电子管。它是靠被灯丝加热的阴极发射电子导电的。具有单向导电性能,即阳极电位高于阴极时,阴极发射的电子在电场的作用下,向阳极运动形成电子流。而阴极电压比阳极高时,电子所受到的电场力是将电子拉回阴极的,不能产生电流,一般用于整流与检波,有真空与充气(充有惰性气体)两种电子二极管。

Electronic anti-suppress 电子反压制 系指在敌人进行电子压制情况下,维持我方电磁波的正常使用。

Electronic bathy thermograph（EBT） 电子式测温计 简称 EBT。一种测量海水温度随深度变化的仪器。由传感器将温度和压力电信号用电缆传到船上,由表头指示和 X－Y 指示记录仪记录。

Electronic chart display and information system（ECDIS） 电子海图显示和信息系统 系指一种有足够后备布置的能视为符合并经修正的 1974 年 SOLAS 公约第Ⅴ章第 19 条和第Ⅴ章第 27 条要求的最新海图的航行信息系统。可有选择地显示系统航海图(SENC)信息及航行传感器的位置信息来帮助航海人员计划航线和监控航线,如有要求,还可显示其他关于航行的信息。Electronic chart display and information system (ECDIS) means a navigation information system which with adequate back-up arrangements can be accepted as complying with the up-to-date chart required by regulation Ⅴ/19 and Ⅴ/27 of the 1974 SOLAS Convention, as amended, by displaying selected information from a system electronic navigation chart (SENC) with positional information from navigation sensors to assist the mariner to route planning and route monitoring and if required, additional navigation related information can be displayed.

Electronic control system 电子控制系统

Electronic control system of diesel engine 柴油机电控系统 系指根据所控对象、方法和参数的不同,柴油机的电子控制系统有各种方案和形式,最主要的是电子控制燃油系统。它控制喷油压力、喷油时间、喷油量及喷油模式等参数。能灵活调节柴油机各系统参数的电子控制系统,简称柴油机电控系统。柴油机电子控制系统是由传感器、电子控制器(ECU)及执行机构组成。Electronic control system of diesel engine means that according to different objects, methods and parameters, electronic control system of diesel engine has different schemes and forms, of which the most important is electronic control fuel oil system. It controls parameters such as fuel injection pressure, time, amount and mode. The electronic control system which can flexibly adjust parameters of each system of diesel engine is called electronic control system of diesel engine. Electronic control system of diesel engine is composed of sensors, electronic control units (ECU) and actuators.

Electronic filter 电子滤波器

Electronic listening device 电子监听装置

Electronic mail system 电子邮件系统 系指由用户代理 MUA(Mail user agent)以及邮件传输代理 MTA(Mail transfer agent),邮件投递代理 MDA(Mail delivery agent)组成的系统。

Electronic mail(E-mail) 电子邮件 标志:@,也被大家昵称为"伊妹儿",又称电子信箱、电子邮政。它是一种用电子手段提供信息交换的通信方式。是 Internet 应用最广的服务:通过网络的电子邮件系统,用户可以非常低廉的价格(不管发送到哪里,都只需负担电话费和网费即可),以非常快速的方式(几秒钟之内可以发送到世界上任何你指定的目的地),与世界上任何一个角落的网络用户联系,这些电子邮件可以是文字、图像、声音等各种方式。同时,用户可以得到大量免费的新闻、专题邮件,并实现轻松的信息搜索。

Electronic monitoring 电子监听

Electronic monitoring system(data) 电子监控系统(数据) 系指用于检查计算机系统中所记载的交易手段,如贸易数据日志或跟踪审核。Electronic monitoring system means the device by which a computer system can be examined for the transactions that it records, such as a trade data log or an audit trail.

Electronic nautical chart(ENC) 电子航行海图 系指由政府的主管机关,或政府授权的航道测量机构或其他相关政府机构发布,其内容、结构和格式都已标准化,且符合 IHO 标准的与 ECDIS 一起使用的数据库。ENC 包含安全航行所需的所有海图信息,并可包含纸质海图上没有但可视为安全航行所需的补充信息(例如航行指南)。Electronic nautical chart(ENC)means a database comforming to IHO standards, standardized as the content, structure and format, issued for use with ECDIS by or on the authority of a government, or a authorized Hydrographic office or other relevant government institution. The ENC contains all the chart information necessary for safe navigation and may contain supplementary information that is not contained in the paper chart (e.g. sailing directions), but may be considered necessary for safe navigation.

Electronic photographing technology 电子照相技术

Electronic position fixing system(EPFS) 电子定位系统

Electronic storage 电子储藏 系指电子数据的临时、中期或永久性储藏,包括此数据的替代或原始储藏。Electronic storage means any temporary, intermediate or permanent storage of electronic data, including the primary and back-up storage of such data.

Electronic support 电子支援 系指侦察搜索、截收、标定、记录、分析敌方电磁波参数,以支持己方作战。

Electronic suppress 电子压制 系指对敌方使用的电磁波的装备和手段进行压制或破坏的行为。

Electronic visa 电子签证 系指把传统的纸质签证"电子化",以电子文档的形式将护照持有人签证上的所有信息存储在签证签发的系统中,电子签证办理成功后,游客将签证打印,附上译文即可成行。"电子签证"大致可以分为两类。一类是不需要递交相关材料到对应国家的大使馆和领事馆,只需在官网上填写简单信息,上传相关信息后,直接支付签证费用便可得到电子签证。另一类依然要求申请人递交相关材料到对应国家的大使馆和领事馆。目前在华发放电子签证的国家主要有新加坡、澳大利亚、柬埔寨、斯里兰卡、土耳其、卡塔尔、印度等。

Electronically Scanned Array radar(ESA radar) 电子扫描阵列雷达 系指一类通过改变天线表面阵列所发出波束的合成方式,来改变波束扫描方向的雷达。这种设计有别于机械扫描的雷达天线。

Electronics 电子学 是科学与技术中的一门述及主要靠电子在真空、气体或半导体中移动而导电的各种器件的分支学科。

Electrostatic plotter 静电绘图机

Elementary plate panel(EPP) 基本板格 是扶强材之间板材未被加强的最小部分。Elementary plate panel(EPP)is the smallest unstiffened part of plating between stiffeners.

Elevating room 升降机构室 系指桩腿周围安设升降机构的舱室。

Elevation deck 举升甲板 系指承载货物并在装卸货物作业过程中被水淹没的开敞露天甲板。

Elevator 提升机/水平尾舵 (1)一种在垂直方向或较大的倾斜方向连续输送货物的机械。(2)水平尾翼中用以控制垫升航态纵倾的可转动部分。

Elevator (cargo lift) 升降机 设置在船上供在各层甲板间提升和下降货物的装置。

Elevator alarm 升降机报警 系指在具有内部控制的升降机内设置的,能从其轿厢内操作的应急听觉呼叫器。这一报警信号应传送至通常有人的位置。 Elevator alarm means an emergency audible device provided in elevator having central control and capable of being operated in its car. The alarm is to be transmitted to the positions where personnel are normally present.

Elevator barge 封底泥驳 设有封底泥舱,依靠吹泥船将泥沙输送至填泥区的泥驳。

Elimination of quantitative restrictions 一般禁止数量限制原则

Ell fitting 弯管接头

Elliptical blade 椭圆叶 展开或伸张轮廓为部分椭圆形的螺旋桨叶。

Elliptical stern 椭圆型艉 具有较短的艉伸部,折角线以上呈椭圆体向上扩大,折角线以下与船体底部线型相连接的船尾部。

EM adjustable asynchronous motor 电磁调速异步电动机 是一种交流无级调速电动机。它由交流三相异步电动机、涡流离合器(又称电磁转差离合器)和测速发电机组装而成。电磁调速异步电动机的调速是通过电磁转差离合器的作用来实现的。改变离合器的励磁便可调节离合器的输出转速或转矩。通常与控机器配合后组成一套交流调速驱动装置,装有测速负反馈的自动调节系统,能在比较宽广的范围内进行平滑的无级调速。

E-M marine speed log 船用电磁计程仪 是现代船舶上的一种新型航海计程仪器,是根据电磁感应原理来测量船舶行程的。

Embarkation 登乘 系指旅客和船员从救生甲板登上救生艇/救生筏。

Embarkation deck 登乘甲板 位于艇甲板之上或装置吊艇筏架处,在船舶遇难时,供船上人员集聚和登乘救生艇或可吊救生筏的甲板区域。

Embarkation ladder 登乘梯 系指设置在救生艇筏登乘站以供安全登上降落下水后的救生艇筏用的梯子。 Embarkation ladder is the ladder provided at survival craft embarkation stations to permit safe access to survival craft after launching.

Embarkation station 登乘站 系指登乘救生艇筏的地方。登乘站如有足够的场地并能安全进行各种集合行动则可以兼作集合地点。 Embarkation station is the place at which a survival craft is boarded. An embarkation station may also serve as an assembly station, provided there is sufficient room, and the assembly station activities can safely take place there.

Embarkation time and launching time 登乘时间和降落时间 系指登乘时间和降落时间之和。它被定义为船撤离船上所有人员所需的时间。 Embarkation time and launching time, the sum of which defines the time required to provide for abandonment by the total number of persons on board.

Embarkation time E and launching time L 登乘时间及下水时间 此两时间之和决定了船上所有人员准备弃船所需要的时间。 Embarkation time E and launching time L, the sum of which defines the time required to provide for abandonment by the total number of persons on board.

Embarked personnel 乘员 除了舰员以外的额外人员,在舰上扩展为了特殊任务或目的,或军事目的。这些人员可包括额外的舰船系统专业维修人员、正常试验中的技术人员、飞行员和执行任务的军事人员,这种任务可以是在海军的控制下任何海军的相关活动,包括试验、训练、人道主义援助和军事活动。 Embarked personnel mean additional personnel other than crew, embarked on the ship for a specific task or purpose or for military purposes. Such personnel may include additional specialised maintenance personnel for ship systems, technicians on trials during normal operation, aircraft crew, and military personnel on a mission which may be any naval related activity under naval command including trials, training, humanitarian aid and military activities.

Embedded temperature detector 埋入式测温计 系指制作在电机内、用于测量温度的电阻式温度计或热电偶。

Emerged wedge volume 出水楔形体积 系指船舶相邻两等体积倾斜水线面所夹的露出水面的楔形体积。

Emergence fire pump 应急消防泵 具有独立动力源能快速启动并有自吸能力的应急时使用的消防泵。

Emergence rudder 应急舵 当船舶遭遇失舵事件后,利用应急器材临时装配而成的舵。

Emergence steering gear 应急操舵装置 在主、辅助操舵装置发生故障时，能应急投入使用的操舵装置。

Emergency 应急（事变，紧急关头，非常时期）

Emergency access 应急通道 其布置应允许人员在舰船上以控制损害和灭火为目的的移动。 Emergency access arrangements are to allow for movement of personnel within the ship for the purposes of damage control and fire-fighting.

Emergency access objective 应急通道目标 为了损害管制和灭火目的及演习，舰船布置能使人员配备必要设备进入所有区域。应急通道布置应不利于火灾、海水、烟雾和其他有毒气体扩散到指定安全地点。 Every ship is to be arranged so that personnel can access all areas with necessary equipment, for damage control and fire-fighting purposes and exercises. Provisions for emergency access are to be arranged such that they do not contribute to the spread of fire, flood, smoke or other toxic gases to the designated places of safety.

Emergency air bank 应急空气瓶组

Emergency air blower 应急空气鼓风机

Emergency air compressor 应急用空气压缩机（应急空压机） 供船舶应急情况下使用的压缩机。

Emergency alarm 紧急报警 系指优先级最高的警报。是指示存在对人的生命或对船舶及其机器的紧迫危险时应立即采取措施的报警。下列警报应列为紧急报警：(1)通用紧急报警；(2)失火报警；(3)进水探测主报警；(4)警告对人员有紧迫危害的报警：包括①灭火剂排放前的报警；②动力驱动滑动水密门关闭报警；(5)对特种船舶（如高速船），除上述定义的紧急报警外，附加报警器也可列为紧急报警。 Emergency alarm is the highest priority of an alert. Emergency alarm means an alarm which indicates that immediate danger to human life or to the ship and its machinery exists and immediate action should be taken. The following alerts are classified as emergency alarms：(1)general emergency alarm；(2)fire alarm；(3)water-ingress detection main alarm；(4)those alerts given warning of immediate personnel hazard, including ①fire-extinguishing ore-discharge alarm；②power-operated sliding watertight door closing alarm；(5)For special ships(e. g. high-speed craft) additional alarms may be classified as emergency alarms in addition to the ones defined above.

Emergency ammonia drainer 应急泄氨器 在紧急情况下，能迅速将贮液器、中间冷却器和蒸发器等的氨液与消防水一起泄入海水中的容器。

Emergency ballast system 应急压载系统

Emergency bilge pumping system 应急舱底排水系统

Emergency bilge suction 应急舱底水吸口

Emergency blower 应急鼓风机 供船舶应急情况下使用的鼓风机。

Emergency call 应急呼叫 该功能可立即启动 BNWAS 第二阶段警报，然后立即启动第三阶段远程声响警报。 Emergency call is a function that immediately initiates BNWAS second stage alert and subsequently third, stage remote audible alarms.

Emergency cargo shot-off system 应急切断系统 在装卸作业中发生意外时及时切断货物管路的系统。

Emergency cold starting test 冷态应急启动试验 为检查冷态的船用主、辅汽轮机的应急启动可靠性和在最短时间内承受规定负荷的能力所做的试验。

Emergency condition 紧急状态（应急状态） 系指：(1)由于主电源发生故障以致正常操作和居住条件所需的设施，均处于工作失常状态；(2)正常运行和适于居住状态所需的设备由于主电源故障而不能工作时的一种状态。(3)由于严重和急迫的危险构成对船舶和/或其人员威胁的状态。 Emergency condition is：(1) a condition under which any services needed for normal operational and habitable conditions are not in working order due to failure of the main source of electrical power；(2) a condition in which equipment required for normal operation and habitability is inoperable due to a failure of the main power supply；(3) a situations that the ship and/or the personnel on board are threatened by a serious and imminent danger.

Emergency consumer 应急设备 系指在主电源失电后，须由应急电源供电的设备。 Emergency consumer is a mandatory consumer which, after breakdown of the main energy supply, must be fed by the emergency energy supply.

Emergency cut-off device (emergency cut-off gear) 应急切断装置

Emergency dewatering system 应急排水系统 用以排除船舶在紧急情况下由于破舱所进的水或大量集聚的灭火用水的系统。该系统与舱底水系统分开。

Emergency disconnection 应急脱开 非计划中的快速关闭和解脱所有实体连接，使船舶/平台与装置能得以分隔。

Emergency engine 应急发动机（备用发动机）

Emergency escape 应急脱险（紧急逃生）

Emergency escape breathing devices (EEBD) 紧急逃生呼吸装置 系指提供空气或氧气的装置，仅用于从有危险气体的舱室逃生的目的，并且应为认可型的装置。紧急逃生呼吸装置不得用于救火、进入缺氧空舱或液货舱，也不得供消防员穿着使用。在这些场合，应采用特别适合这种目的的自给式呼吸器。紧急逃生呼吸装置应至少提供 10 min 的持续使用时间。紧急逃生呼吸装置应包括一具合适的头罩或全面罩用于在逃生期间为眼睛、鼻和嘴提供保护。头罩和面罩应用防火焰材料制成并应包括一扇清洁明亮的观察窗。暂时不使用的紧急逃生呼吸装置,应能佩戴在身上而使双手保持自由。简要的使用说明和示意图应清晰地打印在紧急逃生呼吸装置紧急逃生呼吸装置上。佩戴的程序应既快又简单以使能在最短的时间内从危险气体的环境中获得安全保护。 Emergency escape breathing devices (EEBD) is a supplied air or oxygen device only used for escape from a compartment that has a hazardous atmosphere and shall be of an approved type. EEBD shall not be used for fighting fires, entering oxygen deficient voids or tanks, or worn by firefighters. In these events, a self-contained breathing apparatus, which is specifically suited for such applications, shall be used. The EEBD shall have a service duration of at least 10 min. The EEBD shall include a hood or full face piece, as appropriate, to protect the eyes, nose and mouth during escape. Hoods and face pieces shall be constructed of flame resistant materials and include a clear window for viewing. An inactivated EEBD shall be capable of being carried hands-free. Brief instructions or diagrams clearly illustrating their use shall be clearly printed on the EEBD. The donning procedures shall be quick and easy to allow for situations where there is little time to seek safety from a hazardous atmosphere.

Emergency escape trunk 应急逃生通道 为机舱工作人员提供应急逃生用的通道。其暴露在机舱中的围壁要求为 A-60 级的防火分隔,通道上设置的门要求是 A-60 级的防火自闭式的,通道在甲板上的开孔的风雨密装置要求能双向开启的。

Emergency fire extinguishing appliance 应急灭火装置

Emergency fire pump 应急消防泵

Emergency fresh water tank 应急淡水柜 设置于水线以上,贮存应急饮用水的舱柜。

Emergency generating set 应急发电机组

Emergency generation system 应急发电系统

Emergency generator 应急发电机 系指船舶应急发电站中的发电机。

Emergency generator engine fuel oil tank 应急柴油发电机柴油柜 贮存应急柴油发电机组使用的轻柴油的柜。

Emergency generator room 应急发电机室 供单独安置应急发电机及其控制设备的舱室。

Emergency hot starting test 热态应急启动试验 为检查处于不同热状态的船用主、辅汽轮机的应急启动可靠性和操作程序的合理性所做的试验。

Emergency instruction (station bill instruction) 应变程序 附带在应变部署表内的,对清单中的内容进行解析及说明的文件。

Emergency lifeboat 应急救生艇 又称值勤救生艇。在船舶起航后,处于准备卸放状态,以备有人落水后及时营救的救生艇。

Emergency light (personal flotation device) 应急灯(个人漂浮装置) 为增加确定使用者位置之机会的发光装置。 Emergency light is the device which emits light so as to increase the chances of a user being located.

Emergency maneuvering quality 应急操纵性 船舶应付紧急情况的操纵性能。常用满舵回转试验和紧急停车试验测定。

Emergency over-speed trip 应急超速保护器

Emergency phase 紧急阶段 根据具体情况而定的船舶和船上人员安全存在不确定性、警戒或遇险阶段的总称。 Emergency phase means a generic term meaning, as the case may be, uncertainty, alerts or distress to the safety of a vessel and the persons on board.

Emergency position indication radio beacon (EPIRB) 紧急无线电示位标 是一种能发射无线电信标的设备。其用途是利用声波发射的射频信号表示自己的存在状态及其位置,以期待搜救其接收装置找到它。在船舶航行遇险必须弃船时,人工启动紧急无线电的示位标开关,发射示位信号,以示请求营救。

Emergency power station 应急电站

Emergency power supply (EPS) 应急电源(应急动力源,应急供电)

Emergency power-off (EPO) 紧急断电

Emergency preparedness 应急准备 如果发生了危险或意外情况,为保护人员及环境资源和财产,在应急管理组织下计划实施的技术、作业及组织措施。

Emergency preparedness analysis 应急准备分析 分析包括确定已定义的危险和事故状态,包括计算意外事件、确定应急准备的功能要求以及应急准备措施。

Emergency preparedness organization 应急准备组织 组织的计划、成立、培训和训练的目的是为了处理发生的危险和意外情况。

Emergency pump 应急泵

Emergency response service (SERS) 应急响应服务 系指根据船东选择,预先建立船舶稳性与结构强度有关的电子数据库,当船舶遭遇海上碰撞、搁浅、溢油等事故后,提供快速的稳性、强度与溢油量评估,提出应急处理建议,协助船舶脱离危险。 Emergency response service (SERS) means at the option of owner, an electronic database for stability and structural strength of a ship will be established in advance so that a quick assessment of stability, structural strength and spillage may be made and emergency measures recommended in case of collision at sea, grounding, oil spillage, etc., assisting the ship in getting out of danger.

Emergency shutdown 应急关断/事故停车 系指:(1)可燃气体非正常大量漏泄的情况下,通过人工有选择地关断非防爆的机、电设备和通风;(2)在失火后,通过人工关断为火灾继续提供燃料的设施(如燃油管路、可燃气管路、油柜出油口、油泵和可燃气体压缩机)以及关闭为火灾继续提供助燃剂的风机和围蔽开口;(3)在弃平台/船舶时,人工启动整个平台/船舶的关断。 Emergency shutdown means:(1) manually and selectively shutting down non-explosion-proof machinery, electrical installations and ventilation in the event of abnormally excessive leakage of flammable gas;(2) in the event of fire, manually shutting down any facilities supplying fuel to the fire (e.g. oil lines, flammable gas lines, outlets of oil tanks, oil pumps and flammable gas compressors) and closing any fans and enclosed openings contributing to the fire;(3) manually actuating shutdown of the whole unit/ship when abandoning it.

Emergency shutdown device 紧急停止装置/紧急关断装置 系指独立于自动化系统,在紧急情况下用人工停止机电设备运行的装置。如主柴油机紧急停车按钮、锅炉紧急停炉按钮、用电设备电源紧急切断按钮等。 Emergency shutdown device means a device independent of any automated system and intended for manual activation in an emergency to stop the operation of machinery and electrical installations, e.g. emergency stop button of main engines, emergency cutoff button of boiler fuel oil supply and emergency cutoff button of electric power supply to consumers.

Emergency shut-down mechanism/emergency trip mechanism 紧急停车装置 使内燃机在紧急情况下迅速停止运转的装置。

Emergency shutdown system 应急关断系统 系指执行安全功能的安全相关系统。

Emergency situations 应急状况 当事件由于严重威胁船舶的安全,完整性或安全而影响正常运营条件和优先事项时。 Emergency situations mean incidents affect regular operating conditions and priorities due to grave threats against the ship's safety, integrity or security.

Emergency source of electrical power 应急电源 系指:(1)当主电源发生故障时,拟向应急配电板供电的一种电源;(2)在主电源发生故障的情况下,用于为各种必要的用途供电的电源。 Emergency source of electrical power is:(1) a source of electrical power, intended to supply the emergency switchboard in the event of failure of the supply from the main source of electrical power;(2) a source of electrical power intended to supply the necessary services in the event of failure of the main source of electrical power.

Emergency steering gear 应急操舵索具 各种操舵装置损坏后,应急使用于操舵的滑车组与绳索的统称。

Emergency stop oil pump and fan (Emergency strip) 应急停止 又称应急脱扣。系指在火灾发生时要迅速关闭火灾发生区域内的油泵、净油机、油加热器、冷却器以及通风换气装置。应急脱扣的功能在主配电板上得以实现,通常在配电板上用红色或黄色来标出要应急脱扣的开关。应急脱扣的按钮通常设置在机舱棚的出口处、消防控制中心(fire control station)或驾驶台内。

Emergency switchboard 应急配电板 系指:(1)在主电源供电系统发生故障的情况下,由应急电源或临时应急电源直接供电,并将电能分配给应急用途的配电板;(2)正常情况下由主配电板供电,而在主电源供电系统发生故障的情况下,由应急电源或临时应急电源直接供电,并分配和控制电能至各种应急设备的开关设备和控制设备组件。 Emergency switchboard is:(1) a switchboard which, in the event of failure of the main electrical, power supply system, is directly supplied by the emergency source of electrical power or the transitional source of emergency power and is intended to distribute electrical energy to the emergency services;(2) a switchboard which, in the normal conditions is supplied by the main switchboard but in the event of failure of the main electrical power supply system, is directly supplied by the emergency source of electrical power or the transitional source of emergency power and is intended to distribute and control the electrical energy to the assembly of switch gears and control gears of the ship's emergency services.

Emergency towing arrangement 应急拖带装置 系指供 20 000 载重吨及以上的液货船,包括油船、化学品液货船和液化气体船,在应急情况下拖离危险区时的所用的拖带设备。应急拖带装置应经船级社认可,并应符合下列规定:(1)艏部应急拖带装置应预先装配好,并能在泊港状态下不超过 15 min 内投入使用;(2)艏部短拖索的回收装置应设计成在失去动力和不利环境下,能由 1 个人进行手工操作;回收装置应予以保护,以防不利的天气和其他情况;(3)艏部应急拖带装置应设计成至少用 1 个适当定位的导向滚轮将短拖索紧固到防擦装置上,以便拖索的连接;(4)艏部应急拖带装置应能在泊港状态下不超过1 h内投入使用;(5)符合艉部应急拖带装置规定的艏部应急拖带装置可被接受;(6)所有应急拖带装置均应有明显的标志,以便在黑暗中和能见度差的情况下,也能安全和有效地使用。 Emergency towing arrangement means an arrangement used for liquid cargo carriers of 20 000 tons deadweight and above, including oil tankers, chemical liquid carriers and liquefied gas carriers, for being towed out of dangerous areas in emergency cases with the arrangements needed. The emergency towing arrangements are subject to approval by a Society and in accordance with the following requirements: (1) The aft emergency towing arrangements are to be pre-rigged and capable of being deployed in a controlled manner in harbour condition in not more than 15 min; (2) The pick-up gear for aft towing pennant is to be designed at least for manual operation by one person taking into account the absence of power and the potential of adverse environmental conditions in an unfavorable condition. The pick-up gear is to be properly protected against the weather and other adverse conditions; (3) The forward emergency towing arrangement is to be designed at least with a means of securing a towline to the chafing gear using a suitably positioned pedestal roller to facilitate connection of the towing pennant; (4) The forward emergency towing arrangement is to be capable of being deployed in harbour conditions in not more than 1 h; (5) Forward emergency towing arrangements that comply with the requirements for aft emergency towing arrangements may be accepted; (6) All emergency towing arrangements are to be clearly marked to facilitate safe and effective use even in darkness and under condition with poor visibility.

Emergency towing arrangements on tankers 液货船应急拖带装置 系指:载重量不小于20 000 t的每艘液货船在其艏艉两端设置的应急拖带装置。对于2002年2月1日或以后建造的液货船:(1)该装置应能始终在被拖船上主动力失效时迅速展开并易于与拖船连接。至少一台应急拖带装置应预先设置或待命状态用于迅速展开;(2)考虑到船舶大小和载重量以及恶劣天气下预期的力作用,艏艉两端的应急拖带装置的设计与建造以及原型试验应经主管机关根据IMO组织制定的指南予以批准。对于2002年7月1日以前建造的液货船/应急拖带装置的设计与建造应经主管机关根据IMO组织制定的指南予以批准。 Emergency towing arrangements shall be fitted at both ends on board every tanker of not less than 20 000 tonnes deadweight. For liquid cargo tankers constructed on or after 1 July 2002: (1) the arrangements shall, at all times, be capable of rapid deployment in the absence of main power on the ship to be towed and easily connected to the towing ship. At least one of the emergency towing arrangements shall be pre-rigged ready for rapid deployment; (2) emergency towing arrangements at both ends shall be of adequate strength taking into account the size and deadweight of the ship, and the expected forces during bad weather conditions. The design and construction and prototype testing of emergency towing arrangements at both ends shall be approved by the Administration, based on the Guidelines developed by IMO. For tankers constructed before 1 July, 2002, the design and construction of emergency towing arrangements shall be approved by the Administration, based on the Guidelines developed by IMO.

Emergency trip 应急停机(应急脱扣)

Emergency tripping device 应急脱扣装置(应急断开设备,应急跳闸装置)

Emergency unit 应急装置

Emergency valve 应急阀

Emergency vent 应急风口

Emission 排放(辐射,发射,放射) 对"73/78防污公约"附则Ⅵ而言,系指从船舶上向大气或海上排放受1973年国际防止船舶造成污染公约(MARPOL)的附则Ⅵ控制的任何物质。 For "MARPOL73/78" Annex Ⅵ, emission means any release of substances, subject to control by Annex VI to the International Convention for the Prevention of Pollution from Ships, 1973 (MARPOL), from ships into the atmosphere or sea.

Emission control 排放控制

Emission control areas 排放控制区 对MARPOL附则Ⅵ而言,排放控制区应为:(1)北美区域,系指MARPOL附则Ⅵ的附录Ⅶ中坐标所述区域;(2)美国加勒比海区域,系指MARPOL附则Ⅵ的附录Ⅶ中坐标所述区域;(3)由IMO根据MARPOL附则Ⅵ的附录Ⅲ中设定的标准和程序所指定的任何其他海域,包括任何港口区域。 For Annex Ⅵ to MARPOL, emission control areas shall be: (1) the North American area, which means the area described by the coordinates provided in Appendix Ⅶ to the Annex Ⅵ to MARPOL; (2) the United States Caribbean Sea area, which means the area described by the coordinates provided in Appendix Ⅶ to the Annex Ⅵ to MARPOL; (3) any other sea area, including any port area, designated by the Organization in accordance with the criteria and procedures set forth in Appendix Ⅲ to the Annex Ⅵ to MARPOL.

Employer 雇主 对2009年香港国际安全与环境无害化拆船公约而言,系指雇佣一个或多个工人从事拆船的自然人或法人。 For Hong Kong International Convention for the Safe and Environmentally Sound Recycling of Ships, 2009, employer means a natural person or a legal person that employs one or more workers engaged in ship recycling.

Emptying outlet 排空

Emulsified oil (emulsion oil) 乳化油

Emulsify (emulsion) 乳化(乳状)

En round 航行途中　系指船舶在海上，包括偏离最短直线航道的航行，就实际可行的航行目的而言，其领海排放将遍及与一个合理可行的大的海域范围。En round means that the ship underway at sea on a course or courses, including deviation from the shortest direct route, which as far as practicable for navigational purpose, will cause any discharge to be spread over as great an area of the sea as is reasonable and practicable.

Encapsulation "m" (electrical equipment) "m"密封型(电气设备)　系指使因火花或受热而可能点燃爆炸性气体环境的部件密封在复合物中，以使此爆炸性气体不可能被点燃的保护形式。 Encapsulation "m" (electrical equipment) means a type of protection, the parts which could ignite an explosive atmosphere by either sparking or heating are enclosed in a compound in such a way that this explosive atmosphere cannot be ignited.

Enclosed and open superstructure 围蔽和敞开的上层建筑　上层建筑为：(1)围蔽的。则①由符合一个要求的艏端壁、侧壁和后壁所围成；②所有前端、侧向和后端开口配有有效的风雨密关闭装置。(2)敞开的，则不是围蔽的。 Superstructure may be: (1) enclosed, where: ①it is enclosed by front, side and aft bulkheads complying with the relevant requirements;②all front, side and aft openings are fitted with efficient weather-tight means of closing. (2) open, where it is not enclosed.

Enclosed container 密封集装箱　系指适用于装运普通件杂货的集装箱。

Enclosed external-ventilating type motor 封闭式外通风型电机　系指具有允许风冷空气进入和排出的开孔，而风冷空气是靠其外部的、并非为电机一部分的设施进行环流的、且在其他方面是完全封闭的电机。这些开孔设置成可以使进风管道和出风管道与其接通。

Enclosed flexible accommodation ladder 封闭式伸缩舷梯

Enclosed or semi-enclosed sea 闭海或半闭海　系指2个或2个以上国家所环绕并由一个狭窄的出口连接到另一个海或洋，或全部或主要由两个或两个以上沿海国的领海和专属经济区构成的海湾、海盆或海域。 Enclosed or semi-enclosed sea means a gulf, basin, or sea surrounded by two or more States and connected to another sea or the ocean by a narrow outlet or consisting entirely or primarily of the territorial seas and exclusive economic zones of two or more coastal States.

Enclosed sanitary system (closed sanitary system) 封闭式粪便水系统　将粪便水先泄放到粪便柜内并经消毒处理后再用粪便泵排出船外的闭式管路系统。

Enclosed self-ventilating type motor 封闭式自通风型电机　系指具有允许风冷空气进入和排出的开孔，而风冷空气是靠与其组装在一起的设施进行环流，且在其他方面是完全封闭的电机。这些开孔设置成可以使进风管道和出风管道或管子与其接通。如使用这种管道或管子，必须具有足够的截面积，且应布置得能向电机提供规定容量的空气，否则，通风量是不足的。

Enclosed space 封闭处所　系指具有下列某一特征的处所：(1)进出口受限制的；(2)不便于进行自然通风的；(3)不是为工作人员连续作业而设计的。还包括(但不限于)货物处所、双层底、燃油舱、压载水舱、泵舱、压缩机舱、隔离舱、空的处所、箱型龙骨、障碍物之间的处所、机舱曲轴箱和污水舱；(4)由舱壁和甲板所围蔽的处所，可以有可关闭的门、窗或其他开口；(5)在没有机械通风的情况下，通风受到限制且任何爆炸性环境不能被自然驱散的处所；(6)由船壳、固定的或可移动的隔板或舱壁、甲板或盖板所围成的所有处所，但永久的或可移动的大篷除外。无论是甲板上有间断处，或船壳上有开口，或甲板上有开口，或某一处所的盖板上有开口，或某一处的隔板或舱壁上有开口，永久一面未设隔板或舱壁的处所，都不妨碍将这些处所计入围蔽处所之内。 Enclosed space means a space, which has any of the following characteristics: (1) limited openings for entry and exit; (2) unfavourable natural ventilation; (3) is not designed for continuous worker occupancy, and includes, but is not limited to, cargo spaces, double bottoms, fuel tanks, ballast tanks, pump-rooms, compressor rooms, cofferdams, void spaces, duct keels, inter-barrier spaces, engine crankcases and sewage tanks; (4) a space sheltered by bulkhead and deck, with doors, windows or other opening switch that may be opened or closed; (5) any space within which, in the absence of artificial ventilation, the ventilation will be limited and any explosive atmosphere will not be dispersed naturally; (6) all those spaces which are bounded by the ship's hull, by fixed or portable partitions or bulkheads, by decks or coverings other than permanent or movable awnings. Neither break in a deck, nor any opening in the ship's hull, in a deck or in a covering of a space, or in the partitions or bulkheads of a space, nor the absence of a partition or bulkhead, shall preclude a space from being included in the enclosed space.

Enclosed spaces 围蔽处所(游艇) 系指由地板、舱壁和/或甲板为界限的处所，可以有门或窗。Enclosed spaces are spaces delineated by floors, bulkheads and/or decks which may have doors or windows.

Enclosed superstructure 封闭的上层建筑 (1)是一种具备下列设施的上层建筑：①结构坚固的封闭舱壁；②这些舱壁的出入开口(如有)设有符合有关要求的门；③上层建筑侧壁或端部的所有其他开口设有有效的风雨密关闭装置。桥楼或艉楼不应视为封闭的，除非当端壁开口封闭时，有通道供给船员随时自全通的最上层露天甲板或更高甲板上的任何一处用其他方式前往这些上层建筑内的机器处所和其他工作处所。(2)前或后舱壁设有风雨密门和关闭装置的上层建筑。(3)①符合有关要求的前、侧和后围壁封闭的上层建筑；②其所有前、侧和后开口设有有效的风雨密关闭装置。 Enclosed superstructure is: (1) a superstructure with: ①enclosing bulkheads of efficient construction; ②access openings, if any, in these bulkheads fitted with doors complying with the relevant requirements; ③all other openings in sides or ends of the superstructure fitted with efficient weather-tight means of closing. A bridge or poop shall not be regarded as enclosed unless access is provided for the crew starting from any point on the uppermost complete exposed deck or higher to reach machinery and other working spaces inside these superstructures by alternative means which are available at all times when bulkhead openings are closed; (2) a superstructure with bulkheads forward and/or aft fitted with weather tight doors and closing appliances; (3) a superstructure enclosed, where: ①it is enclosed by front, side and aft bulkheads complying with relative requirements; ②all front, side and aft openings are fitted with efficient weather-tight means of closing.

Enclosure 外壳 用于确保电气设备之防护的围绕电气设备之带电部件的所有外壁，包括门、盖、电缆入口、杆、支柱和轴。 Enclosure is used for all the walls which surround the live parts of electrical apparatus including doors, covers, cable entries, roads, spindles and shafts, ensuring the protection of the electrical apparatus.

Encountered waves 遭遇角 按不同舷角划分的各种与船舶相遭遇的波浪的统称。按不同舷角划分为顶浪、顺浪、横浪、艏斜浪、艉斜浪等。

End cover 端盖

End cover gasket 端盖垫料

End crater 端部弧坑

End crater crack 末端弧坑裂纹 是一种不允许存在的焊缝缺陷。以明显的线性形式出现的末端弧坑空穴可认为是末端弧坑裂纹，例如在陶瓷衬垫上进行单面焊时出现的末端裂纹痕。非线性形成的弧坑空穴则视为孔穴。 End crater crack is one of the defects that are not permitted. End crater cavities which occur for example, during single side welding on ceramic backings are regarded as end crater cracks if they appear as distinct linear indications. Non-linear indications count as pores.

End effect 末端后果 系指设备或子系统故障对输出(系统功能)的影响。末端后果应按以下类型进行评估和划分严重程度：(1)灾难性的；(2)危险的；(3)重大的；(4)轻微的。 End effect means the impact of an equipment or sub-system failure on the system output (system function). End effects shall be evaluated and their severity is classified in accordance with the following categories: (1) catastrophic; (2) hazardous; (3) major; (4) minor.

End face 端面

End fitting 端部接头

End link(end-link) 末端链环 当锚链两端用连接卸扣相连时，附加在锚链或链节末端的，其链径比普通链环大1.2倍的加强的锚链环。

End of blade 叶端 侧斜螺旋桨叶轮廓的尖端。

End or side-rolling hatch-board 侧移式舱盖 舱盖板向舱口两侧或两端平移启闭的滚移式舱盖。

End to end scenarios 端到端方案

End to end tests 端到端测试

End panel 末端盖板 滚翻式舱盖中，距收藏一端最近的一块舱盖板。

End parts of ship 船端部 系指距船每端$0.1L$的部分。

End plate 端板

End shackle 链端卸扣 锚链两端处，用以与锚或锚链舱内固定眼板相连的加大型直形卸扣。

Endangered areas of propeller 螺旋桨的危险区域 为建立螺旋桨桨叶中缺陷的影响与试验范围对应的关系和避免修补后出现疲劳断裂的危险，根据其承受工作载荷和危险程度而划分的区域。其分为A，B，C三个区域。其中A区域系指经受最大工作载荷并要求进行最充分试验的区域。通常这一区域内是在桨叶最厚处并对焊接的热膨胀产生最大阻力，从而导致修补的焊缝内和周围区域存在最大的残余应力。高的残余应力经常会导致在后续运行时出现疲劳裂纹，所以焊接部位必须经受热处理以降低应力。通常不允许在A区域施焊。

Endangered Species Act 濒危物种法 是美国联邦保护野生动物的主要法律。

Endless screw 蜗杆

Endorsement 背书 系指在票据上签名,票据所规定之权利由背书人转让给被背书人。票据的收款人或持有人在转让票据时,在票据背面签名或书写文句的手续。背书的人就会对这张支票负某种程度、类似担保的偿还责任,之后就引申为担保、保证的意思。即为自己的事情或为自己说的话作担保、保证。

Endurance 续航力(持续时间) 船舶一次装足燃料在平常海面,以规定航速航行时所能达到的最大距离。

Energy consumption intensity indication 能耗强度指标 系指营运船舶单位运输周转量能耗。

Energy efficiency 能效 系指能源利用效率,即得到的结果与所使用的能源之间的关系。

Energy efficiency data 能效数据 系指计算能耗、能效和 CO_2 排放的所有相关数据。

Energy efficiency design index[1](EEDI) 能效设计指数 是衡量船舶 CO_2 效能的有关指标。Energy efficiency design index (EEDI) is a measure of CO_2 efficiency of the ships.

Energy efficiency design index[2](EEDI) 新船能效设计指数 它系指在船舶设计时每单位船舶运输所创造的社会效益(benefit for society,载货量)而产生的环境成本(environmental cost, CO_2 的排放量),它不考虑船舶运营情况,只考虑船舶设计采取的提高效能的措施。对 EEDI 起决定作用的主要参数有:航速、船舶装载量或总吨位、为达到该航速而需要的安装功率。为了降低 EEDI 可采取下列措施:(1)选用功率小一点的主机,适当降低航速,从安全角度出发适当留有余度,以保证操纵性。采用合适的推进方式,提高螺旋桨的推进效率;(2)增加运力;(3)增加装载量,把船造得稍微大一点,通过结构优化、采用轻质材料(复合材料)和其他措施,也可采用少量或无压载水设计,使空船质量轻一点;(4)船型优化,如主尺度优化,艏部线型优化,艉部流场改善,调整船体、螺旋桨和舵之间的间隙以提高舵和螺旋桨的效率;(5)装载条件的优化,如充分利用装载手册航速与吃水、纵倾的关系,通过 CFD 找出尚有多少余度,且余度在哪里;(6)船体抛光打磨,在船体表面采用特别光滑、低摩擦系数的涂料,对螺旋桨进行涂装和打磨,采用艉部节能装置;(7)采用 LNG 代替柴油,或采用双燃料,甚至采用极其安全的核动力;(8)在动力系统中加大废气回收力度;(9)采用燃料电池作为发电机的电力推进方式;(10)采用其他节能技术,如太阳能、风帆、空气润滑等。

Energy efficiency factor 能效因素 系指在船舶运输/作业服务中,影响船舶能源利用效率和 CO_2 排放的因素。船舶能效因素包括但不限于:(1)航线设计;(2)航速管理;(3)气象航线;(4)主/辅机、锅炉的燃油和润滑油消耗;(5)燃油和润滑油管理;(6)船体保养;(7)压载航行;(8)吃水和吃水差;(9)船舶载货量;(10)船舶参数的保持;(11)推进系统;(12)螺旋桨状况;(13)焚烧炉和惰气发生器的燃油消耗;(14)舵和舵向控制系统;(15)装卸货操作;(16)废热回收;(17)隔热、通风与保暖;(18)燃料类型;(19)港口岸上供电的使用;(20)节能新技术;(21)趁潮航行;(22)内河船编队方式;(23)调度管理。

Energy efficiency index 能效指标 系指由能效目标产生的,为实现能效目标所需规定的具体要求。它们可适用于整个公司或其局部(包括船舶)。能效指标由能效目标产生,是对能效目标的细化、分解及实现能效目标的具体要求。能效指标应具有可量化的特点,由一定的参数指示。重要的能效因素对量化和测量能效具有决定性作用,因此,能效指标同样要以重要的能效因素为基础。能效指标有多种指标可以测量船舶能效,如船舶能效营运指数(EEOI)、千吨海里(kt-n mile)油耗、主机油耗、每海里(n mile)油耗、年度总油耗、单位产值油耗等,船舶可以用一个或几个指标作为测量船舶能效的指标工具。

Energy efficiency management plan (SEEMP) 能效管理计划

Energy efficiency management system (EEMS) 能效管理体系 系指公司管理体系的一部分,用于建立能效方针、目标和管理能效因素,并实现这些方针和目标的一系列相互关联要素的集合,包括公司结构、职责、惯例、程序、过程和资源。

Energy efficiency object 能效目标 系指公司所要实现的降低船舶单耗、提高能源利用效率的总体要求。能效目标应符合当前的国家、国际法律法规设定的要求,满足能效方针的要求,以重要的能效因素为基础,以可测量的绩效参数衡量。

Energy efficiency operational index (EEOI) 能效营运指数 系指为船舶单位运输作业所排放的 CO_2 量,即消耗燃油所排放的 CO_2 与货物/人的数量和运输距离乘积和的比值,用来衡量阶段时期内船舶营运能效的高低。计算如下:

$$\text{Average } EEOI = \frac{\sum_i \sum_j (FC_{ij} \times C_{Fj})}{\sum_i (m_{\text{cargo},j} \times D_i)}$$

式中,j—为燃油类;i—为航程数;FC_{ij} 为在航程 i 中燃油 j 的消耗量;C_{Fj}—为燃油 j 的燃油量与 CO_2 量转换系数;$m_{\text{cargo},j}$—货物为客船所载货物(t)或所做的功(TEU 或乘客数量)或 GT；D—为对应于所载货物或所做的功的距离(n mile)。

Energy efficiency operational indicator 能效营运指数 系指船舶单位运输作业艘排放的 CO_2 量。指数 $= M_{CO_2}/$航次。 Energy efficiency operational indicator is the ratio of mass of CO_2 (M) emitted per unit of transport work. Indication = $M_{CO_2}/$Transport work.

Energy efficiency police 能效方针 系指由公司最高管理者正式发布的船舶能效管理的宗旨和方向。能效方针的内容应反映最高管理者对遵守法律法规要求、充分利用能源、提高能源效率和持续改进的一项承诺。能效方针是建立能效目标和指标的基础，应有文件规定，内容应当清晰明确，使得执行人员能够理解，并应对方针进行定期评审与修订，以反映不断变化的内、外部条件和信息。公司制定的能效方针可能包括以下方面：(1)适用于船舶运输/作业服务的特点，并与公司已有的其他管理体系方针相协调；(2)包含对降低船舶单耗、提高能效并持续改进的承诺；(3)包含对遵守与船舶能效管理适用的国际公约、法律法规、标准及其他要求的承诺；(4)为制定和评价能效目标、指标提供框架；(5)形成文件，使全体员工能充分理解并实施；(6)可为相关方所获取。

Energy efficiency reference 能效基准 系指公司针对船舶/船队历年能效状况，确定的用于比较能源利用效率及 CO_2 排放的基础水平，也是船舶/船队能效水平纵向比较的参考点。

Energy efficiency staff 能效标杆 系指公司参照船舶同类可比活动所确定的能源利用效率和 CO_2 排放的水平。能效标杆是根据行业或不同公司同类船舶的能效水平制定，把其作为本公司船队/船舶能效管理目标或基准考核点，是船舶/船队能效水平横向比较的参考点。能效标杆的确立相对复杂，在船舶能效管理计划中可以灵活运用。

Energy source 震源 海洋地震勘探中，产生弹性波的震动源。

Energy spectrum 能谱 一般表示海浪或船舶运动等各组成分能量按频率的相对分布曲线。

Energy thickness of boundary layer 边界层能量厚度 表明由于边界层引起的能量传递损失的一个参数。对于平面流：$\theta^* = \int_0^\infty U/U_0 (1 - U^2/U_0^2) \, dy$，式中，$U_0$—边界层外的流速；$U$—边界层内的流速。

Energy-saving and environment protection technology for marine diesel engine 船用柴油机节能环保技术 系指提高船用柴油机节能、环保性能的方法。其中包括改善柴油机燃料燃烧过程、使用新型替代燃料以及采用尾气处理系统等。(1)改善柴油机燃料燃烧过程是在不降低船用柴油机功率的前提下，提高船用柴油机节能、环保性能最根本的方式。该技术包括：①涡轮增压技术；②共轨燃油喷射技术；③米勒循环技术等。通过增加气缸空气量、精确控制喷油时间和喷油速率来控制燃烧过程，提高燃料的燃烧效率，从而实现节约燃油和降低排放的目的。(2)使用新型替代燃料降低有害物质含量。从当前技术水平来看，液化天然气、液化石油气以及生物燃料等燃料都可以作为船用柴油机的替代燃料。该措施包括：①气体燃料发动机，其具有高效率、低污染的良好性能。该发动机还采用了稀薄燃烧技术，大幅度降低了废气的排放量，与现存的柴油发动机相比，二氧化碳的排放量降低 20% 以上，氮氧化物的排放量削减 50 ppm，削减幅度高达 97%，达到当前世界柴油机的最低排放水平；②使用液化石油气和船用柴油的双燃料柴油机，使其在保持原发动机高效率的同时，能够大幅减少烟煤和氮氧化物的排放量，环保性能好。这种双燃料柴油机，在不影响发动机基本设计和降低发动机高效燃烧过程的前提下，能够控制进入发动机增压气流的液化石油气数量并能在选定的低工况范围内保证双燃料柴油机运行的可靠性。在该双燃料柴油机中，液化石油气替代了大约 20% 的普通液体燃料；液化石油气经过冷却器后混入增压空气中。液化石油气双燃料发动机具有如下优点：将天然气-空气混合物的浓度保持在爆炸水平以下，在气门重叠期进气阀和出气阀同时打开的情况下，仅有少量的液化石油气会排放到空气中，能够限制发动机工作时产生的振动。但该发动机仅在低工况下才能使用液化石油气燃料。在该发动机中，天然气控制装置发挥着关键作用。该装置既能控制进入发动机中的天然气的数量，也能与船舶监控系统进行可靠的信息传递，从而决定该发动机何时使用液化石油气燃料。同时丁烷、丁烯、丙烷、丙烯、乙烷、乙烯等 7 种液化天然气燃料可在该发动机中应用；③生物燃料发动机，将生物燃料作为低速柴油机燃料和/或替代性燃料，以减少 CO_2 的排放量。此外，在生物柴油的应用方面，目前已开发出可燃烧混合型生物柴油的柴油机，既能使用柴油，又能以由 30% 生物柴油和 70% 普通矿物柴油组成的 B30 混合型生物柴油作为燃料。(3)尾气处理系统是降低硫氧化物排放量的有效方法。船用硫氧化物洗涤器(SO_x scrubber)的应用，能够去除发动机尾气中 99% 的硫氧化物；同时氮氧化物的去除率也可得到 3%~7%，烟尘的去除率可达 30%~60%。

Energy-saving propulsion system 节能推进系统
(1)高效特殊螺旋桨 大侧斜螺旋桨——增大螺旋桨侧斜度是减少噪声的常用技术。大侧斜螺旋桨，其叶片能以渐进的方式穿过不同的尾流场(尤其在弧顶),从而可以改进叶片的空化模式。K-3 螺旋桨——通过优化设计,其推进效率比传统的螺旋桨高出 2%～3%;(2)燃料乳化技术 若燃料中的水分比例增加,NO_x 排放量就会降低,通常来说,此时燃料消耗量会增加,不过,一定范围内的燃料乳化也有可能降低燃油消耗。

Enforceability 易于执行
Enforcement 执行
Engagement 接合
Engaging and disengaging gear 离合装置
Engaging lever 开动杆
Engine 发动机(引擎)
Engine aft ship (stern-engined ship) 艉机型船 主机舱位于艉部的船舶。
Engine and boiler grating 机炉舱格栅 系指机炉舱空间架设的便于机炉舱通风及设备维修的防滑通道。
Engine and boiler room 机炉舱 供安置主机、锅炉及其附属设备的船舱。
Engine and boiler room ventilation 机炉舱通风 为使机炉舱内操作地点温度不致太高而进行的自然通风或机械通风。
Engine base 机座
Engine bearer 机座
Engine bearing bed 发动机座
Engine bed (engine seat, engine bearer, engine bed plate) 机座 安装在船体基座上,用以承受内燃机重力和作用力的部件。
Engine block 机体 汽缸体与曲轴箱铸或焊成一体的零件。
Engine column 发动机机架
Engine compartment 机舱
Engine control room 机舱集控室 系指设置在机舱内对船舶的机械设备进行操纵和控制的处所。
Engine control station 机器控制站
Engine counter 发动机转速表 见图 E-4。
Engine crew 轮机班
Engine cutout 停车
Engine cylinder 发动机气缸
Engine department 轮机部
Engine efficiencyv 机器效率
Engine foundation 机座(发动机底座)
Engine frame 机架(发动机机架) 船用低速重型柴油机在汽缸体与机座之间的 A 字形的构件。
Engine framework 发动机机座
Engine hatch 机舱口
Engine International Air Pollution Prevention (EIA-

图 E-4　发动机转速表
Figure E-4　Engine counter

PP) Certificate 柴油机国际防止空气污染证书 系指与 NO_x 排放有关的柴油机国际防止空气污染证书(简称 EIAPP 证书)。
Engine log 轮机日记
Engine margin 柴油机功率裕度 系指造船合同中规定的满载服务航速所需的柴油机功率还留有一定的裕量(10%～15%)。
Engine man 机匠
Engine noise 发动机噪声
Engine oil tank 机油柜
Engine on outside of boat 舷外机 又称艉挂机、船外机,顾名思义发动机在船舷之外,悬挂在艉板之上,最顶部为发动机,发动机曲轴连接立轴,然后横轴最后输出给螺旋桨,这种机器在转向时,整个发动机都随同转向装置左右摆动,由于螺旋桨直接随驱动装置转向所以灵活性极强、整机安装简单、携带方便,广泛应用于小型游艇、充气动力艇、冲锋舟及各种钓鱼船。多用于游艇等小体积船型,如江河上面的捕鱼船、橡皮艇等,常见于江、湖渔业密集地区。一般是水冷发动机,分为二冲程或四冲程汽油机船外机。见图 E-5。
Engine opening 机舱开口
Engine operation 机器操作
Engine order telegraph 对机舱传令钟(对机舱车钟)
Engine output P 主机输出功率 系指主机能够传达到螺旋桨的最大持续输出功率。如主机的输出功率受技术条件或受适用规范的限制时,则此时主机输出功率系指其限制后的输出。 Engine output P is the maximum output the propulsion machinery can continuously delivers to the propeller(s). If the output of the machinery is restricted by technical means or by any regulations applicable to the ship, P shall be taken as the restricted output.
Engine parameter record book 柴油机参数记录簿

图 E-5 舷外机
Figure E-5　Engine on outside of boat

系指用于柴油机参数检验方法的、记录可能影响柴油机 NO_x 排放的所有参数变化包括构件和柴油机设定值的文件。

Engine performance　发动机特性

Engine power　主机功率　主机功率系指主机连接法兰上发出的最大的连续功率,以 kW 为单位。

Engine removal hatch　装卸发动机舱口(机舱顶可拆盖)

Engine room　机舱　对海上风电场而言,系指安装在塔架顶部的发动机室。

Engine room (E/R)　机舱　系指安装推进机械和发电机械的处所。 Engine room is the spaces containing propulsion machinery and machinery for generation of electrical power.

Engine room annunciator　机舱传令钟(机舱传令指示器)

Engine room arrangement　机舱布置　对机舱内部的主机及其有关设备、装置和系统所做的合理布置。

Engine room automation　机舱自动化

Engine room bulkhead　机舱舱壁　系指直接位于机舱前部或后部的横舱壁。 Engine room bulkhead means a transverse bulkhead either directly lolated forward or aft of the engine room.

Engine room casing　机舱棚　自机舱通至露天甲板的围井及顶盖。

Engine room central operating console　机舱集控台

Engine room centralized control room　机舱集控室

Engine room control cabin　机舱控制室

Engine room control station　机舱操纵台

Engine room crane　机舱行车(机舱起重吊车)　机舱检修用的起重设备。

Engine room emergency bilge suction valve　机舱应急舱底水阀　安装在机舱应急舱底水管上,用以排除由于各种事故所进的水的截止止回阀。

Engine room emergency direct bilge suction (emergency bilge drainage)　机舱应急舱底水管　机舱内用以专门排除由于各种事故所进的水的舱底水管。

Engine room fan　机舱通风机　机舱通风换气用的通风机。

Engine room fan room　机舱通风机间

Engine room flat　机舱平台

Engine room floor　机舱底层地板

Engine room hatch　机舱舱口

Engine room ladder　机舱梯

Engine room log　机舱日志

Engine room log book　机舱日志

Engine room log-book desk　机舱记录台　系指机舱内记录和存放轮机日志的台子。

Engine room noise level　机舱噪声级/机舱噪声电平

Engine room noise　机舱噪声

Engine room plan　机舱图

Engine room platform　机舱平台

Engine room skylight　机舱天窗

Engine room structure　机舱结构

Engine room tank　机舱油柜

Engine room telegraph　机舱传令钟(机舱车钟)

Engine room ventilator room　机舱通风机间

Engine seating girder　机座纵桁

Engine shipping hatch　装卸发动机舱口(机舱顶可拆盖)

Engine space　机器处所(机舱)

Engine starter　发动机的启动器

Engine starting gear　发动机启动器

Engine store room　机舱储藏室

Engine support　发动机底座

Engine system fitted with selective catalytic reduction (SCR)　装有选择性催化还原系统发动机　系指由船用柴油机、选择性催化还原系统催化器和还原剂喷射系指组成的一个系统。如设有 NO_x 减少性能的控制装置,也将其认为该系统的一部分。 Engine system fitted with selective catalytic reduction means a system consisting of a marine diesel engine, an SCR chamber and a redu ctant injection system. When a control device with NO_x reducing performance is provided, it is also regarded as a part of the system.

Engine telegraph (ET) 机舱传令钟(车钟)

Engine telegraph receiver 机舱传令钟接收器(车钟接收器)

Engine telegraph transmitter 机舱传令钟发送器

Engine trouble 发动机故障

Engine type 柴油机类型 通常,柴油机形式是以下列特性参数进行定义:(1)气缸直径;(2)活塞冲程;(3)喷油方式(直接或间接喷油);(4)燃油种类(液体、气体或双燃料);(5)工作循环(4冲程或2冲程);(6)燃气交换(自然吸出或增压);(7)在相应转速下的单缸最大连续功率和/或相应于上述最大连续功率的制动平均有效应力;(8)增压方式(脉冲系统或等压系统);(9)增压空气冷却系统(带或不带中间冷却器、级数等);(10)气缸排列(直列或V型)。 Engine type generally means that the type of an engine defined by the following characteristics: (1) the cylinder diameter; (2) the piston stroke; (3) the method of injection (direct or indirect injection); (4) the kind of fuel (liquid, gaseous or dual-fuel); (5) the working cycle (4-stroke, 2-stroke); (6) the gas exchange (naturally aspirated or supercharged); (7) the maximum continuous power per cylinder at the corresponding speed and/or brake mean effective pressure corresponding to the above-mentioned maximum continuous power; (8) the method of pressure charging (pulsating system or constant pressure system); (9) the charging air cooling system (with or without intercooler, number of stages, etc.); (10) cylinder arrangement (in-line or V-type).

Engine warning steam pipe line 暖机蒸汽管系 供汽轮机启动前暖机用的蒸汽管路及附件。

Engine-driven supercharging 机械增压 用机械方法驱动增压器进行的增压。

Engineer 工程师(机械师,轮机员)

Engineer department 轮机部门

Engineer in charge 主管工程师

Engineer officer 轮机员 系指:(1)轮机部合格的高级船员;(2)符合海员培训、发证和值班标准国际公约第Ⅱ章规定的合格的高级船员。 Engineer officer means: (1) a qualified officer in the engine department; (2) an officer qualified in accordance with the provisions of Chapter Ⅱ of the International Convention on standards of training, certification and watch-keeping for seafarers.

Engineer on duty 当值轮机员

Engineer surveyor 轮机验船师

Engineer workshop 轮机人员工作间

Engineer's accommodation 轮机员居住舱室

Engineer's alarm system 轮机员报警系统 又称轮机员值班呼叫系统。是在机舱集控室向轮机员居住区发出报警呼叫信号的船内通信装置。在机舱集控室设有呼叫开关、指示灯,在轮机员居住区设有视觉和听觉报警器。

Engineer's cabin 轮机员室

Engineer's call bell 轮机员呼叫铃(轮机员信号铃)

Engineer's call push button 轮机员呼叫铃按钮

Engineering 工程(工程的)

Engineering analysis on fire safety operation 消防安全操作工程分析 对消防安全操作而言,系指根据IMO组织制定的《消防安全替代设计和标准指南》编写并提交主管机关的报告。其应至少包括下列要素:(1)确定有关船型和处所。(2)判定船舶或处所不符合的规定要求。(3)判定有关船舶或处所的失火和爆炸危险,包括:①判定可能的着火源;②判定各有关处所火势增大的可能性;③判定各有关处所烟气和有毒物的可能性;④判定火灾、烟气和有毒物从有关处所向其他处所蔓延的可能性。(4)确定规定要求对有关船舶或处所提出的消防安全标准:①性能标准应基于消防安全目标和功能要求;②性能标准所规定的安全度应不低于应用规定要求所达到的安全度;③性能标准应可量化并具备可测量性。(5)替代设计和布置的细节描述,包括列出设计时采用的假设,以及所建议的任何操作限制或条件;(6)表明替代设计和布置符合所要求的安全性能标准的技术论据。 For fire safety operation, engineering analysis on fire safety operation means a report prepared and submitted to the Administration, based on *The Guidelines on Alternative Design and Arrangements for Fire Safety* developed by IMO Organization, and shall include, as a minimum, the following elements: (1) determination of type of the ship and space(s) concerned; (2) identification of prescriptive requirement(s) with which the ship or the space(s) will not comply; (3) identification of the fire and explosion hazards of the ship or the space(s) concerned, including: ①identification of the possible ignition sources; ②identification of the fire growth potential of each space concerned; ③identification of the smoke and toxic effluent generation potential of each space concerned; ④identification of the potential for the spread of fire, smoke or of toxic effluents from the space(s) concerned to other spaces. (4) determination of the required fire safety performance criteria for the ship or the space(s) concerned which is addthe purpose of ressed by the prescriptive requirement (s) in particular: ① performance criteria shall be based on the fire safety objectives; ② performance criteria shall provide a degree of safety not less than that achieved by using the prescriptive requirements; ③ performance criteria shall be quantifiable and measurable; (5) detailed description of the alternative design and arrangements, including a list of the assumptions used in the design and any proposed operational restrictions or conditions; (6) technical justification demonstrating that the alternative design and arrangements meet the required fire safety performance criteria.

Engineering calculation　工程计算
Engineering design　工程设计
Engineering drawing　工程图
Engineering log　工程日志
Engineering machinery　工程机械　泛指工程船舶的特种机械。工程船舶以工程类别可分为港口工程、航道工程、海洋工程、航保工程、水利工程、宇宙工程等。各种工程船舶中，特种机械性能要求也各有不同。主要的特种机械有：(1)疏浚机械——如铰刀、链斗、抓斗、铲斗、泥泵等。(2)特种起重机械——如大质量起重机、宽调速起重机、专用起吊机械(耙头架、铰刀架、斗桥架、定位桩等)。(3)特种系缆——如移船绞车(横移、定位等)、布缆机、自动拖缆机等。(4)钻探机械——如钻机、泥浆泵等。(5)特种推进——如低速推进、主动舵推进、侧向推进等。
Engineering plastics　工程塑料
Engineering reliability　工程可靠性
Engineering risk control　工程风险控制　涉及在设计中所包含的安全设备(固有设备或外置设备)。当缺乏某安全设备而造成不可容忍风险时，这些安全设备仍能保持安全状态。
Engineering specifications　技术规格(技术要求)
Engineering works　机械工程
Engineers' alarm　轮机员报警装置　系指一个从发动机控制室或操纵平台(视具体情况而定)进行操作的轮机员报警装置，且报警信号应能在轮机员居住舱室清晰地听到。 Engineers' alarm means an alarm which shall be operated from the engine control room or at the manoeuvring platform as appropriate, and shall be clearly audible in the engineers' accommodation.
Engine base　机座(机架)
Engine room　机舱
Engine-room auxiliary machine　机舱辅机　装设于船舱内，除基本机械以外，为主机及其他系统服务的机械。

Engine-room manned　有人机舱
Engine-room platform　机舱平台
Engineer alarm　轮机员呼叫装置(机舱警报)
Englers viscosity　恩氏黏度
English Channel　英吉利海峡
English method of model self-propulsion tests (model self-propulsion overload tests)　定航速模自航试验　保持船模航速不变，而改变螺旋桨转速的自航试验。
English standard thread　英制标准螺丝
Enhanced group call (EGC)　增强群呼系统　是第二代 INMARSAT，是以 C 标准站为基础的报文广播业务，主要用于陆地向船舶发送信息，即把地面上的信息通过岸站、卫星、以报文或数据流的形式送到具有 EGC 接收机的船站上。
Enhanced survey program (ESP)　加强检验程序　油船、油/散货船、油/散/矿砂运输船、化学品船、散装货船等应经受的检验程序。
Enhanced survey programme　检验强化制度　系指在货物区域内的货舱/油舱，泵舱，隔离舱，管隧道，货物间的空隙空间，和所有的压载舱等船体构造和管道装置要与有关强化检验的方法相一致。 Enhanced survey programme means an enhanced survey method applied for hull structure and piping systems and for cargo holds/tanks, pump rooms, cofferdams, pipe tunnels, void spaces within the cargo area and all ballast tanks in accordance with the relevant requirement.
Enhanced survey report file　加强检验报告卷宗　系指散货船和油船具有的一份符合 A.744(18)决议《散货船和油船检验期间加强检验程序指南》附件 A 和附件 B 中 6.2 和 6.3 规定的检验报告对散货船和油船卷宗和支持文件。 Enhanced survey report file means that bulk carriers and oil tankers shall have a survey report file and supporting documents complying with Paragraphs 6.2 and 6.3 of Annex A and Annex B of resolution A.744 (18): guidelines on the enhanced programme of inspections during surveys of bulk carriers and oil tankers.
Enhanced survivability　增强残存能力　系指单壳散货船在进水状态下的总纵强度水密槽型横舱壁强度及货舱许用装载量的标准。 Enhanced survivability means the criteria for longitudinal strength of single hull bulk carriers and strength of their corrugated transverse watertight bulkheads and permissible loading of their holds under flooded conditions.
Enlarged link　加大链环　当锚链两端用连接卸扣相连接时，附加在链节两端，连接于普通锚链与末端锚链之间,其链径比普通链环大 1.1 倍的加强的锚链环。

Enquiry 询盘/询价 是交易一方欲购买或出售某种商品,向另一方发出的探询购买该商品及有关条件的一种表示,也称询价、探盘。

Ensemble 总体 就统计而言,系指随机函数所有可能出现的函数曲线的集合。

Ensign staff 艉旗杆 设置在船艉部的旗杆。

Entanglement bit (Ebit) 纠缠比特

Enter 引入(进港,进刀,参加)

Enter into and get out of a dock 进出坞/上、下排

Enter visa 入境签证

Enthalpy 焓

Enthalpy entropy diagram 焓熵图(*i-s* 图)

Entrained droplet 夹带盐水 逸出液面的蒸汽中携带的盐水滴。

Entrance[1] 进流段 设计水线下,最大横剖面或平行中体以前的船体部分。

Entrance[2] 入口(开始,进口,入港手续)

Entrance angle 进水角 船舶横向倾斜至水开始由开口进入船内时的横倾角。

Entrepot trade 转口贸易 也称中转贸易。系指货物消费国与货物生产国通过第三国进行的贸易活动。对生产国是间接出口;对消费国是间接进口,而对第三而言,便是转口贸易。转口贸易的货物可以直接从生产国运送到消费国。

Entropy 熵

Entry 进入 系指人员通过开口进入处所。进入包括在该处所内的后续工作活动并在一旦进入者身体的任何部分突破进入该处所的开口平面即被视为已经发生。 Entry means the person enters the premises through an opening. Entry includes ensuing work activities in that space and is considered to have occurred as soon as any part of the entrant's body breaks the plane of an opening into the space.

Entry locker 过渡舱 加压舱结构中,供人员调压进出的部分。

Environment 环境 系指包括风、浪和流等状况。冰载荷可不予考虑。Environment means a status including wind, wave and current. Ice load may not be considered.

Environment condition 环境条件

Environment controlling unit 环境控制单元 对甲板居住舱内环境温度和湿度进行控制的装置。通常要求环境温度控制在 30~32 ℃,变化范围不超过 1 ℃;环境相对湿度控制在 50%~70%,变化范围不超过 ±10%。

Environment hazard 危害周围环境

Environment management system (EMS) 环境管理体系

Environment pollution 环境污染

Environment protection 环境保护

Environment protection boat 环保工作船 系指从事水域环境保护的特种工程船舶。主要用于海上石油勘探开发工作,包括勘探测试过程中的废液回收及海上溢油防治工作,并具备海洋工程船舶辅助作业功能,满足对浮油回收作业船以及对油田守护船、近海供应船及消防船的有关要求和规定,具有 DP-1 动力定位能力的海洋工程船舶。此类船舶一般为小型船舶,排水量仅几百吨甚至不足百吨,但其技术含量较高,均有特殊工作机构。

Environment temperature 环境温度

Environmental abnormality 环境异常

Environmental addiction 环境毒瘾 系指当上瘾者习惯于特定生活方式时就会产生的毒瘾。与鸦片或大麻服用者共享的社交聚会或会聚场所,对环境毒瘾有帮助的作用,并为上瘾者或"推销者"提供了机会。如果毒品在特定场所流传,上瘾者就有了固定的来源,"推销者"就有了固定的市场。艾滋病病毒对世界上许多地方的影响越来越大,这为减少毒品的滥用提供了新的动力,因为传播传染的主要渠道之一就是毒品服用者共同使用受污染的皮下注射针头。对于毒品服用者来说,在社会上没有区别,也没有阶层。在各行各业和各个社会阶层都可能发现毒品服用者。上瘾者的体质特征取决于所使用的毒品的类型以及自上一次吸毒以来消逝的时间。毒品服用者一般会逐渐具有对自己的习惯说假话和保密的能力。船员可能不会注意到在其同事中有人服用毒品。在封闭的群体中,如在船舶的船员中,对团队的忠诚可能有强烈的亲和力,这可能使人不愿意相信某个同事会做坏事。毒品滥用者和吸毒者懂得这点,如果被引起怀疑,他们会利用这点。 Environmental addiction means an addiction which can occur, when the addict becomes accustomed to a particular lifestyle. Social meetings or meeting places, not just of opium or cannabis users, have been conductive to environmental addiction and provide opportunities for both addicts and "pushers". If drugs circulate in particular places, the addict has a permanent source and the "pushers" will have a constant market. The increasing incidence of the AIDS virus in many parts of the world has given new impetus to reduce drug abuse, one of the main conduits for spreading infection is the use of contaminated hypodermic needles that have been shared. There are no social divisions or classes of drug users. They may be found in all walks of life and at all social level. The physical characteristics of drug addicts depend on the type of drug used and the time that has elapsed since the last dose. The drug user generally develops an ability to lie about his habit and keep it secret. Crew members may not notice the drug user among in their colleagues. In a closed community, such as in a ship's crew, there may be a strong bond of group loyalty which may result in an unwillingness for a person to believe the worst about a colleague. Drug abusers and drug traffickers are aware of this and will, if suspicious are aroused, take advantage of this.

Environmental compliance records 环境符合性记录 包括污水和灰水排放记录簿、所有排放报告、所有排放取样试验报告，以及必须保存的任何其他记录。Environmental compliance records includes the sewage and graywater discharge record book, all discharge reports, all discharge sampling test reports, as well as any other records that must be kept.

Environmental conditions 环境条件 一般包括：(1)海浪；(2)风；(3)水流；(4)潮汐和暴风涌浪；(5)空气和海水温度；(6)冰和雪；(7)海生物；(8)地震；(9)海冰。根据特定的安装现场，还可以要求调查其他一些现象，如海啸、海底滑坡、假潮、空气和水的异常成分、空气湿度、盐度、冰漂移、海中冰山冰冲刷等。标准气象和海况条件下，被拖物所要求的系柱拖力应确保拖带航向稳定，一般以作用在同一方向的下列气象与海况环境条件进行衡定：风速 20 m/s；有义波高 5 m；流速 0.5 m/s。

Environmental conditions for electric equipment operating 电气设备工作的环境条件 系指在下列环境条件下所有电气设备能进行正常工作的条件：(1)环境空气温度；(2)初级冷却水温度；(3)倾斜和摇摆；(4)船舶正常营运中产生的振动和冲击；(5)潮湿空气、盐雾、油雾和霉菌。Environmental conditions for electric equipment operating means the following conditions under which all electric equipment are to operate normally: (1) ambient temperature; (2) primary cooling temperature; (3) an inclination of ship's from normal; (4) the vibration and shock likely to arise under normal service of ships; (5) moisture, sea air, oil vapour and mould.

Environmental control 环境控制 系指在运输 IBC 规则的第 17 章中的货物时，要求对货舱中的蒸发气空间(vapor space)及货舱的周围进行的控制。环境控制的方法有：(1)惰性化(deactivation)——往货舱及相关的管系中充入惰性的、不和货物起化学反应的气体，例如锅炉燃烧后的废气经涤气机处理后使用；(2)充填(padding)——往货舱及相关的管系中充入液体、气体或蒸汽将货物与空气隔开；(3)干燥(drying)——往货舱及相

关的管系中充入不含水蒸气的气体或露点低于零下 40 ℃ 的蒸汽；(4)通风(ventilation)——包括强制通风和自然通风。

Environmental control system 环境控制系统
Environmental effect 环境效应
Environmental engineer 环境工程师
Environmental factors 环境因素
Environmental force 环境力
Environmental impact assessment (EIA) 环境影响评估 系指对规划和建设项目实施后可能造成的环境影响进行分析、预测和评估，提出预防或者减轻不良环境影响的对策和措施。其目的是为了实施可持续发展战略，预防因建设项目实施后对环境造成不良影响，促进经济、社会和环境的协调发展。它是强化环境管理的有效手段，对确定积极发展方向和保护环境等一系列重大决策都有重要作用。审核可能对环境造成危害的项目的程序。在决定发放项目许可证时必须考虑环境影响评估(EIA)报告。

Environmental load 环境载荷(风、浪、流作用)
Environmental load condition 环境载荷条件
Environmental loading 环境载荷 环境载荷是指直接或间接由环境作用引起的载荷，包括由环境载荷引起的所有外力，如系泊力、运动惯性力、液舱晃荡力等。一般由下列载荷组成：(1)风载荷；(2)波浪载荷；(3)海流载荷。如果业主/设计者认为需要，则地震、海床承载能力、温度、污底、冰/雪等对载荷的影响也应考虑。如可能，设计环境条件应根据可靠及足够的实测资料由统计分析确定，自存工况设计环境条件的重现期建议不小于 50 年。环境载荷除按规范给出的方法外，还可采用其他公认的方法进行计算，必要时应通过数学模拟计算或物理模型试验来确定。在操作手册中应注明每种工况的设计限制条件。Environmental loading is the loading which is coused due directly or indirectly to environmental actions, including all external forces which are responses to environmental loading, e. g. mooring forces, inertia forces, sloshing forces. Environmental loading comprise generally: (1) wind loading; (2) wave loading; (3) current loading. Where deemed necessary by the owner/designer, the effects of earthquake, sea bed supporting capabilities, temperature, fouling, ice/snow, etc. are also to be taken into account. Where possible, the design environmental criteria determining the loads on the unit and its individual elements are to be based upon significant statistical information and should have a return period (period of recurrence) of at least 50 years for survival conditions. In addition to the methods given en above, the environmental loading may be calculated by

any other recognized method and where necessary, they are to be determined by mathematical simulations or physical model tests. The unit's limiting design criteria for each mode of operation are to be included in the operating manual.

Environmental monitoring installation 环境监控装置

Environmental monitoring ship (environmental protection ship) 环境监测船 亦称环境保护船,是根据国家的《海洋环境保护法》对所管辖海域实行海洋环境保护和进行污染监测的船舶。船上设有水文、化学、生物、地质等实验室和海洋环境调查、监测等专用仪器设备,可提取海水、生物、地质的试样进行化验。经常密切注视海上油区污染情况。

Environmental protection 环境保护

Environmental Protection Agency (EPA) [美]联邦环境保护局

Environmental protection dredger (EPD) 环保挖泥船 属于专门水域环境生态保护的一类工程船舶。其确保在不造成水体二次污染的前提下,清除和处置污染底泥的挖泥船。

Environmental protection equipment 环保设备

Environmental protection material 环保材料

Environmental protection notation 环境保护附加标志

Environmental resource 环境资源 包括生物群体和栖息地,定义为:生物群体(stock)——在一定时间一定地理区域内出现的一群生物个体或单独在一定地理区域内繁殖的某一物种的所有个体。栖息地(habitat)——几种生物出现并相互影响的一个有限区域,比如海滩。

Environmental safety 环境安全 可能造成损害的意外溢出所导致的环境安全。

Environmental severity factor 环境严重性因素 系指在极端或疲劳载荷方面,对相对于不受约束的作业环境进行长期或预期的环境严重性的测量。 Environmental severity factor is a severity measure of historical or intended environment relative to the unrestricted service environment in terms of extreme or fatigue actions.

Environmental standard 环境标准

Environmental test 环境试验

Environmental transit condition 环境转移条件

Environmental vibration 环境振动

Environmentally sound method 环境无害化方法 系指用于防止引起或者用于控制水生有害物质蔓延,以使其对生态系统的结构和功能的有害影响减至最小,对无预定目标的有机体系和生态体系的有害影响减至最小,且强调综合的有害物质控制技术和非化学措施的方法、计划、行动和规划。 Environmentally sound method means methods, efforts, actions, or programs, either to prevent introductions or to control infestations of aquatic nuisance species, that will minimize adverse impacts to the structure and function of an ecosystem, minimize adverse effects on non-target organisms and ecosystems, and that emphasize integrated pest management techniques and non-chemical measures.

Environment-protecting ship 环保工作船 系指从事水域环境保护的特种工程船舶。大致有下述几种:(1)水面清扫船;(2)浮油回收船及油污水处理船;(3)环保挖泥船;(4)多用途清污船。此类船舶一般为小型船舶,排水量仅几百吨甚至不足百吨,但其技术含量较高,均有特殊工作机构。

EPA = Environmental Protection Agency 美国环境保护委员会/环境保护局

EPCI 采购、设计、建造和安装调试

EPFS = electronic position fixing system 电子定位系统

Epicyclic reduction gear unit 行星减速齿轮装置

Epicyclic train 行星齿轮

Epicyclic unit 外摆线齿轮装置

Epoxide number 环氧值

Epoxy 环氧树脂

Epoxy adhesive 环氧黏合剂/环氧胶黏剂

Epoxy coating 环氧树脂涂层

Epoxy film 环氧膜

Epoxy foam 环氧泡沫塑料

Epoxy glue 环氧胶

Epoxy paint 环氧树脂涂料

Epoxy primer 环氧底漆(环氧涂料)

Epoxy putty 环氧泥子

Epoxy resin 环氧树脂 是泛指分子中含有2个或2个以上环氧基团的有机化合物,除个别外,它们的相对分子质量都不高。环氧树脂的分子结构是以分子链中含有活泼的环氧基团为其特征,环氧基团可以位于分子链的末端、中间或成环状结构。由于分子结构中含有活泼的环氧基团,使它们可与多种类型的固化剂发生交联反应而形成不溶的具有三向网状结构的高聚物。凡分子结构中含有环氧基团的高分子化合物统称为环氧树脂。固化后的环氧树脂具有良好的物理、化学性能,它对金属和非金属材料的表面具有优异的黏接强度,介电性能好,变定收缩率小,制品尺寸稳定性好,硬度高,柔韧性较好,对碱及大部分溶剂稳定,因而广泛应用于国防、国民经济各部门,作浇注、浸渍、层压料、黏接剂、

涂料等用途。

Epoxy resin chock 现场浇注的环氧树脂定位垫
Epoxy resin-based paint 环氧树脂涂料
Epoxy tar anti-corrosive paint 环氧沥青防锈涂料
Epoxy value 环氧值
Epoxy zinc-rich primer 环氧富锌底漆
Epoxy-based system 环氧基体系
Epoxy-bonded 环氧树脂黏合的
Epoxy-screed deck coating 环氧树脂裂片甲板涂料

Equal loudness curve 等响曲线 系指描述等响条件下声压级与声波频率的关系曲线称为等响曲线,是重要的听觉特征之一。即在不同频率下的纯音需要达到何种声压级,才能获得对听者来说一致的听觉响度。图中每条曲线上对应于不同频率的声压级是不相同的,但人耳感觉到的响应却是一样,每条曲线上注有一个数字,为响度单位,由等响曲线族可以得知,当音量较小时,人耳对高低音感觉不足,而音量较大时,高低音感觉充分,人对1 000~4 000 Hz之间声音最为敏感。A计权网络特性曲线对应于倒置的40方等响曲线,B计权网络曲线对应于倒置的70方等响曲线,C计权网络曲线对应于倒置的100方等响曲线。

Equal loudness curve 等噪度曲线 类似于等响曲线,只是它以复合音为基础,以吵闹程度为目标效应,在同一的声呐值曲线上噪度相同。

Equalization arrangement 平衡装置
Equalizing compartment 横倾平衡舱
Equalizing pump 平衡泵
Equalizing valve 平衡阀
Equilibrated expansion joint 平衡膨胀连接
Equilibrium 平衡
Equilibrium slide valve 平衡滑阀
Equipment and outfits for an unit 平台的舾装设备 系指舵设备、临时锚泊设备及拖曳设备。Equipment and outfits for an unit means rudder, temporary anchoring and towing equipment.
Equipment class 设备等级 在IMO MSC 第654通函中描述的,按它们最坏状况的故障模式定义DP船舶的设备能力分级。
Equipment compatibility 设备互换性
Equipment component list 设备零件明细表
Equipment number 船具数(舾装数) 船舶建造规范规定的系船设备中表征锚、锚链、绳索等应配备的数量和尺寸的标准数。
Equipment of lifeboat 救生艇属具 配置于救生艇上,供救生艇操纵、维修和维护艇上乘员生命以及发出求救信号进行联络等所用的全部附件的总称。
Equipment of liferaft 救生筏属具 指配置于救生筏上,供救生筏操纵、维修和维护艇上乘员生命以及发出求救信号进行联络等所用的全部附件的总称。包括工具、器材、食物、药品。
Equipment of lifesaving apparatus 救生属具 附属于救生筏、浮具上或进行救生操作时所需的全部附件的总称。
Equipment of secondary lifesaving appliances 辅助救生用具附件 为操作使用辅助救生用具所必需的全部附属件的总称。
Equipment subject to a boiler 锅炉附件 (1)法兰、立管和直接附到锅炉或压力容器上的隔片;(2)直接附到锅炉和压力容器上的阀。Equipment subject to a boiler means the following: (1) flanges, stand pipes and distance pieces attached directly to boiler and pressure vessel. (2) valves attached directly to boilers and pressure vessels.
Equipment under test (EUT) 受试设备 系指为形式认可试验而指定的设备样机,包括使该设备功能完整化的任何辅助部件和系统,例如制冷、加热和机械减振器等。
Equipotential line 等势线 在势流场中速度势为一常数的线。
Equipotential plan 等势面 在势流场中速度势为一常数的面。
Equivalent 当量(等价,等值,同意义的)
Equivalent/bending rigidity 等效/弯曲刚度
Equivalent continuous sound pressure level ($L_{peq,t}$) 等效连续声压级 在测量时段 T 内的连续稳态声的声压,具有与随时间变化的噪声相同的均方声压,则这一连续稳态声的声压级就为该测量时段 T 内随时间而变化的等效声压级,单位dB,由下式给出:

$$L_{peq,t} \equiv 10\lg\left[(1/t_2 - t_1)\int_{t_1}^{t_2}\frac{P2(t)}{P0}dt\right](dB),$$

其中:$(t_2 - t_1)$ 即为时段 T,在开始时间 t_1 和结束时间 t_2 的时段内取均值。

Equivalent design wave (EDWs) 等效设计波 系指所产生的响应值与对结构构件起主要作用载荷分量的长期响应值相当的规则波,被设定为等效设计波,由以下组成:(1)在迎浪时垂向波浪弯矩达到最大时的规则波(EDW"H");(2)在随浪时垂向波浪弯矩达到最大时的规则波(EDW"F");(3)横摇运动达到最大时的规则波(EDW"R");(4)水线处水动压力达到-最大时的规则波(EDW"P")。Equivalent design wave (EDWs) means regular waves that generate response values equivalent to the long-term response values of the load components considered being predominant to the structural members are set as equivalent design waves (EDWs). They consist of: (1) regular waves when the vertical wave bending moment becomes maximum in head sea (EDW "H"); (2) regular waves when the vertical wave bending moment becomes maximum in following sea (EDW "F"); (3) regular waves when the roll motion becomes maximum (EDW "R"); (4) regular waves when the hydrodynamic pressure at the waterline becomes maximum (EDW "P").

Equivalent document　等效证明文件　系指本身不以船级社名义出具的，但经船级社盖章和船级社验船师签署的用于证明产品按船级社要求经过检验并合格的证书、报告等文件。Equivalent document means certificate, report, etc, issued not in the name of a Society, but stamped by the Society and endorsed by the Surveyor, showing that the products have been satisfactorily inspected according to a Society requirements.

Equivalent fixed gas fire-extinguishing systems　等效固定式气体灭火系统

Equivalent generator　等效发电机　系指为计算短路电流，将运行中的各台发电机和各台电动机综合成一台等效发电机，该等效发电机馈送的短路电流等效于各台发电机和各台电动机馈送的短路电流之和。 Equivalent generator means that in order to calculate short-circuit current, each individual generator and motor in service must be combined to form an equivalent generator, the short-circuit current fed through the equivalent generator is to be equivalent to the sum of the short-circuit current fed through each individual generator and motor.

Equivalent material　等效材料　系指由于本身性能或者用隔热物保护而在经过标准耐火试验后，在结构性能和完整性上与钢具有同等性能的任何不燃材料（例如有适当隔热性能的铝合金）。 Equivalent material means any non-combustible material which, by itself or due to insulation provided, has structural and integrity properties equivalent to steel after the applicable exposure to the standard fire test (e.g. aluminium alloy with appropriate insulation).

Equivalent means　等效设施

Equivalent motor　等效电动机　系指为简化短路电流的计算，将运行中除大电动机以外的各台电动机综合成一台等效电动机，该等效电动机馈送的短路电流等效于上述各台电动机馈送的短路电流之和。 Equivalent motor means that, in order to simplify the calculation of short-circuit current, each individual motor in service, other than large motors, must be combined to form an equivalent motor, the short-circuit current fed through the equivalent motor is to be equivalent to the sum of the short-circuit current fed through the above-mentioned motors.

Equivalent of heat　热当量

Equivalent plank　相当平板　与船体等湿面积和等长度的水力光滑平板。

Equivalent pressure water-spraying fire-extinguishing system　等效压力水雾灭火系统

Equivalent sand diameter roughness　相当砂径粗糙度　用产生相当摩擦阻力的黏附在平板上均匀砂粒的直径表示的粗糙度。

Equivalent sprinkler systems　等效喷水器系统

Equivalent static load coefficient　等效静载荷系数

Equivalent water-based fire-extinguishing systems　等效水基灭火系统

Equivalent water-mist fire-extinguishing system　等效细水雾灭火系统

Equivolume inclination axis　等体积倾斜轴线　船舶两相邻等体积倾斜水线间的交线。

Equivolume inclinations　等体积倾斜　船舶排水体积保持不变时的倾斜。

Equivolume inclined waterlines　等体积倾斜水线　船舶作等体积倾斜时的水线。

Erbium doped fiber amplifier (EDFA)　掺铒光纤放大器　系指在信号通过的纤芯中掺入了铒离子Er^{3+}的光信号放大器。它是1985年英国南安普顿大学首先研制成功的光放大器，它是光纤通信中最伟大的发明之一。掺铒光纤是在石英光纤中掺入了少量的稀土元素铒（Er）离子的光纤，它是掺铒光纤放大器的核心。从20世纪80年代后期开始，掺铒光纤放大器的研究工作不断取得重大的突破。WDM 技术、极大地增加了光纤通信的容量。成为当前光纤通信中应用最广的光放大器件。

ERBL　带指示器的电子方位线　系指与距离指示器一起用于测量自本船起或2个物体之间的距离和方位。 ERBL is electronic bearing line carrying a marker, which is combined with the range marker, used to measure range and bearing from own ship or between two objects.

Erection　合拢

Ergodicity　各态历经性　求取平稳随机过程的数字特征时，可用时间平均代替其总体平均的特征。

Ergonomics　人机工程学（人类工程学）　系指：(1)把人的因素应用于分析、设计设备、工作和工作环境；(2)隐含在工作场所和设备中的分析和设计中人的因素的体现。 Ergonomics means: (1) the application of the human factor in the analysis and design of equipment, work and working environment;(2) application of the human factors implication in the analysis and design of the workplace and equipment.

Erosion　腐蚀(侵蚀)

Erosion corrosion　侵蚀腐蚀

Erosion damage　剥蚀损伤(剥蚀损坏)

Erosion intensity　剥蚀强度(侵蚀烈度,冲刷强度)

Erosion number　腐蚀值

Erosion pit　剥蚀坑

Erosion resistance　抗腐能力

Error　错误　是一种由于操作人员或维修人员的不正确行为造成的事件。　Error is an occurrence arising as a result of incorrect action by the operating crew or maintenance personnel.

Error code　错误代码　表示网页不存在或已被删除。

Error ellipse　误差椭圆　表示随机定位误差的一种方法。当两条位置线的误差统计分布按正态分布时,定位点等概率分布密度是轨迹是以平均位置为中心的一族椭圆,即等概率误差椭圆。

Error limit　误差范围

Error producing condition　失误产生条件　系指对人的行为产生负面影响的因素。例如,因为使用人因失误评估与消除技术而增加失误的数量级、频率和可能性(与行为形成因子概念类似)。

Escalation　事态加剧　危险事件或系列事件的后果加大。

Escalation factor　事态加剧因素　因失去控制及缓解或恢复能力而导致风险加大的情况。

Escape　逃生/脱险(排泄,应急出口)　到达舰船上指定安全地点的人员转移(这可能是协调运送或个人的行动。因为这主要是关注人员在舰船上的流动,当然也包括通常的通道)。　Escape means the movement of personnel to a designated place of safety on board. (This may be coordinated movement or the action of individuals. Since this is mainly concerned with the flow of personnel through the ship, it may also include normal access).

Escape and rescue trunk　脱险围井(救生通道)

Escape and survival equipment　逃生救生设备(应急逃生设备)

Escape cover　安全口盖(逃口盖)　装设在供舱内人员在紧急情况下离开轴隧或机、炉舱的安全口上,能迅速开启的舱口盖。

Escape exit　脱险口(逃生出口)

Escape goal 1　逃生目标1(脱险)　所有逃生路线应指出有效和显而易见的通道进入指定安全地点。(1)一般规定,所有舰船上提供的逃生路线应满足 SOLAS 公约第Ⅱ章第2节第13.1条~13.4条的要求;(2)一般规定,所有舰船从机舱处所的逃生路线应满足 SOLAS 公约第Ⅱ章第2节第13.4条的要求;(3)一般规定,滚装空间逃生路线应满足 SOLAS 公约第Ⅱ章第2节第13.5条~第13.7条要求。　Escape goal 1 means that all escape routes are to provide effective and obvious access to designated places of safety. (1) In general, the provision of escape routes on all ships is to be in accordance with SOLAS Chapter Ⅱ-2, Regulation 13.1 to 13.4. (2) In general, escape routes from machinery spaces on all ships are to be in accordance with SOLAS Chapter Ⅱ-2, Regulation 13.4; (3) In general, escape route arrangements for RO-RO spaces are to be in accordance with SOLAS Chapter Ⅱ-2, Regulation 13.5 to 13.7.

Escape goal 2　逃生目标2(脱险)　所有逃生路线应易于到达。　Escape goal 2 means that all escape routes are to be readily.

Escape goal 3　逃生目标3(脱险)　所有逃生路线无不当风险。　Escape goal 3 means that all escape routes are to be free.

Escape goal 4　逃生目标4(脱险)　逃生路线的尺寸应适于可预见的紧急情况下预期人员的流动。一般规定,逃生路线的外形尺寸和设计应满足 SOLAS 公约第Ⅱ章第2节第13条的要求。　Escape goal 4 means that physical dimensions of escape routes are to be suitable for the anticipated flow of personnel during all foreseeable emergency conditions. In general, the dimensions and design of escape routes are to be in accordance with SOLAS Chapter Ⅱ-2, Regulation 13.

Escape goal 5　逃生目标5(脱险)　人员在逃生的同时,应足以免受火灾、烟雾和有害气体的伤害。　Escape goal 5 means that personnel are to be adequately protected from fire, smoke and hazardous vapours while escaping.

Escape goal 6　逃生目标6(脱险)　单独舱室的逃生布置应适合舱室及其预期的使用者目的。　Escape goal 6 means that the escape arrangements for individual compartments are to be suitable for the purpose of the compartment and its intended occupants.

Escape hatch　应急舱口/逃生舱口　在危急情况下,供人员脱离危险区的舱口。

Escape objective　逃生目标(脱险)　每艘舰船都应布置成在可预见的紧急情况下,所有处所的人员都能以安全有效的方式逃生到指定的安全地点。　Escape objective means that every ship is to be so arranged that all spaces have a means of safe and effective escape for personnel to a designated place of safety, during anticipated emergency situations.

Escape pipe 逸气管(放气管)

Escape quotes 转义符号

Escape route 逃生通道(脱险通道)

Escape scuttle 应急通孔/脱险口(安全口) 开设在舱室门下部,以备在无法开门时的应急情况下使用的,易于打开的出口。

Escape trunk 应急围井/应急出口(救生口) 在紧急情况下供人员脱险用的围井。

Escape valve 逸出阀

Escape way 逃生通道 平台上特别指定的通道路线,从危险区域到集合区、救生艇站或庇护区。

Escapement 应急出口(擒纵机构)

Escaping steam 逸出蒸汽

Escort 护航 系指与另一艘船舶同时行驶,具备有优异破冰能力的船舶。 Escort means any ship with superior icebreaking capability in transit with another ship.

Escorted operation 护航作业 系指通过护航的介入而便利另一艘船舶移动的作业。 Escorted operation means any operation that a ship's movement is facilitated through the intervention of an escort.

ESD 紧急关闭 ESD means emergency shutdown.

ESD1 应急关闭和脱开 1

ESD2 应急关闭和脱开 2

Essential auxiliaries 重要辅机 系指有重要用途的辅机,与船舶推进、人命、船舶的安全以及船舶的用途等相关的辅机。 Essential auxiliaries are the auxiliary machinery for important use and are those for propulsion of ships and those for safety of lives and ships or those for facilities in relation to the purpose of ships.

Essential auxiliary boiler 重要辅助锅炉 主锅炉以外的锅炉,系指为与诸如发电机、船舶的推进、人命和船舶的安全或船舶的用途等相关辅机的运行提供蒸汽的锅炉。 Essential auxiliary boiler means the auxiliary boiler other than the main boiler, which is used to supply steam for operating generators, auxiliary machinery in relation to the propulsion of ships, safety of lives and ships or the purposes of ships.

Essential components 重要部件 包括泵、热交换器、阀门执行机构和经形式认可的电气元件。 Essential components include pumps, heat exchangers, valve actuators, and electrical components approved.

Essential equipment 重要设备 系指推进、操舵和船舶安全所必需的设备,以及具有特殊附加标志船舶上的特殊设备。包括:(1)主要设备:系指为保持推进和操舵需连续运转的设备。例如:①操舵装置;②调距桨装置;③为主、辅柴油机服务的鼓风机、燃油供给泵、喷油嘴冷却泵、滑油泵和冷却水泵,以及推进用涡轮机所必需的相应设备;④为向主要设备供汽的辅锅炉服务的和在蒸汽轮机船上为蒸汽装置服务的强力鼓风机、给水泵、循环水泵、真空泵、冷凝水泵以及油燃烧装置;⑤单独作推进/操舵用的方位推进器连同其滑油泵和冷却水泵;⑥用于电力推进装置的电气设备连同其滑油泵和冷却水泵;⑦向上述 ①～⑥ 设备供电的发电机及有关电源;⑧向上述①～⑥设备提供动力的液压泵;⑨重油黏度控制设备;⑩消防泵和其他灭火剂泵;⑪航行灯、航行设备和信号设备;⑫船内安全通信设备;⑬照明系统;⑭以上①～⑬所列设备的控制、监视和安全设备/系统。(2)次重要设备:系指为保持推进和操舵不必连续运转的设备,以及为保持船舶安全必需的设备。例如:①锚机;②燃油输送泵和燃油处理设备;③滑油输送泵和滑油处理设备;④重油预热设备;⑤启动空气和控制空气压缩机;⑥舱底、压载和平衡泵;⑦机舱和炉舱通风机;⑧保持危险区域处于安全状态必需的设备;⑨探火与失火报警系统;⑩水密关闭设备;⑪为降低环境空气温度的制冷设备;⑫向上述①至⑪设备供电的发电机和有关电源;⑬向上述 ①～⑪设备提供动力的液压泵;⑭货物围护系统的控制、监视和安全系统;⑮上述①至⑪设备的控制、监视和安全设备/系统。(3)具有附加标志船舶上的特殊设备可作为重要设备。 Essential equipment is the equipment necessary for the propulsion, maneuverability, navigation and safety of the ship and the type-specific equipment on ships with special class notation, including: (1) primary essential equipment which needs to be in continuous operation for maintaining the ship's maneuverability with regard to propulsion and steering: ①steering gear; ②controllable pitch propeller installation; ③charging air blowers, fuel feeder pumps, fuel booster pumps, lubricating oil pumps and fresh cooling water pumps for main and auxiliary engines and turbines, so far as required for propulsion; ④forced draught fans, feed water pumps, water circulating pumps, vacuum pump, condensate pumps and oil burning installation for auxiliary steam boilers for the operation of primary essential equipment and for steam installations on steam turbine ships; ⑤azimuth thrusters which are the sole means for propulsion/steering with lubricating oil pumps, cooling water pumps; ⑥electrical equipment for electric propulsion plant with lubricating oil pumps and cooling water pumps; ⑦electric generators and associated power sources supplying the equipment mentioned in ① to ⑥ above; ⑧ hydraulic pumps supplying the equipment mentioned in ①to ⑥above;

⑨viscosity control equipment for heavy oil;⑩fire pumps and other extinguishing agent pumps;⑪navigation lights, aids and signals;⑫internal safety communication equipment;⑬lighting system;⑭control, monitoring and safety devices/systems for equipment mentioned in ①to ⑬above. (2) secondary essential equipment which do not necessarily need to be in continuous operation for maintaining for the ship's maneuverability and steering, but which are necessary for maintaining the ship's safety, e. g. :①windlasses;②fuel oil transfer pumps and fuel oil treatment equipment;③lubricating oil transfer pumps and lubricating oil treatment equipment;④pre-heaters for heavy fuel oil;⑤starting air and control air compressors;⑥bilge, ballast and heeling pumps;⑦ventilating fans for engine and boiler rooms;⑧equipment necessary for maintaining the safety in dangerous spaces;⑨fire detection and alarm system;⑩watertight closing appliances;⑪refrigerating equipment for lowering ambient air temperature;⑫electric generators and associated power sources supplying the equipment mentioned in ① to ⑪ above;⑬hydraulic pumps supplying the equipment mentioned in ①to ⑪above;⑭control, monitoring and safety systems for cargo containment systems;⑮control, monitoring and safety devices/systems for equipment mentioned in ① to ⑪ above;(3)type-specific equipment on ships with a special class notation may be classified as essential equipment.

Essential functions 主要功能（关键功能） 系指进行相关操作所需要的必不可少的功能。相对于水域、交通和气象条件、关于确定、实施维持船舶的安全航向、速度和位置的相关功能。这些功能包括但不限于:(1)规划航线;(2)航行;(3)避碰;(4)操纵;(5)进坞;(6)监视内部安全系统;(7)在桥楼操作和遇险情况下,与安全有关的对外通信和内部通信。 Essential functions are the indispensable functions to be available as required for the relevant operational use. Those functions relate to determination, execution and maintenance of safe course, speed and position of the ship in relation to the waters, traffic and weather conditions. Such functions include but are not limited to:(1) route planning;(2) navigation;(3) collision avoidance;(4) manoeuvring;(5) docking;(6) monitoring of internal safety systems;(7) external and internal communication related to safety in bridge operation and distress situations.

Essential information 主要信息（关键信息） 系指那种监视主要功能所必需的信息。Essential information means that information which is necessary for the monitoring of essential functions.

Essential machinery 重要机械
Essential part 主要部件
Essential piping systems 重要的管路系统 系指用于在机动类和船型类中舰船的推进和安全的系统,具体如下:(1)透气及溢流装置;(2)测深装置;(3)舱底及疏排水系统;(4)压载系统;(5)燃油系统;(6)燃气系统;(7)润滑油系统;(8)热油系统;(9)液压油系统:①操舵装置;②可调螺距螺旋桨;③推进和/或动力定位推进单元;④锚泊机械;⑤船首、船尾、舷侧和内部的水密门、电控阀系统等;(10)机械淡水冷却系统;(11)海水冷却系统;(12)主机启动压缩空气控制和报警系统;(13)蒸汽及冷凝系统;(14)排气及废气系统;(15)阀门及通风闸的遥控操作控制系统。 Essential piping systems are those systems installed for the propulsion and safety of the ship within the Mobility category and Ship Type category and include the following:(1) air and overflow arrangements;(2) sounding arrangements;(3) bilge and dewatering systems;(4) ballast systems;(5) oil fuel systems;(6) gas fuel systems;(7) lubricating oil systems;(8) thermal oil systems;(9) hydraulic oil systems:①steering gears;②controllable pitch propellers;③thrust units for propulsion and/or dynamic positioning;④windlass machinery;⑤watertight bow, stern, side and internal doors;⑥valve control systems, etc.;(10) fresh water cooling systems for machinery;(11) sea water cooling systems;(12) compressed air systems for starting engines, control and alarms;(13) steam and condensate systems;(14) exhaust and flue gas systems;(15) control systems for remote operation of valves and ventilation flaps.

Essential power supply 重要用途供电
Essential pressure vessel 重要用途的压力容器 对主机、重要的辅助锅炉、船舶的推进和人命的安全及船舶的安全至关重要的辅机,用于与船舶用途相关设备的辅机等直接相关的压力容器。Essential pressure vessel is a pressure vessel having relevance to main engines, essential auxiliary boilers, auxiliary machinery having relevance to propulsion, safety of lives and ships, and service of ships.

Essential service 主要营运 系指规范涉及的一种营运需求,海上的船舶必须能操纵或承担其营运中的有关作用及其保证生命的安全。 Essential service is intended to mean a service necessary for a ship to proceed at sea, be steered or manoeuvred, or undertake activities connected with its operation, for the safety of life, as far as class is concerned.

Essential services 重要设备（舰船） 系指机动类和船型类舰船中对舰船推进和安全所必需的设备,包括下列设备:(1)燃油发动机用空压机;(2)空气泵;(3)自动喷水系统;(4)压载水泵;(5)舱底和排水系统泵;(6)循环和冷却水泵;(7)通信系统;(8)冷凝器循环水泵;(9)电力推进设备;(10)燃油发动机电力启动系统;(11)抽吸泵;(12)锅炉强力通风用风机;(13)给水泵;(14)探火和报警系统;(15)燃油阀冷却泵;(16)对可调螺距螺旋桨以及此处所列重要设备提供液压的液压泵,否则这些重要设备应直接由电力拖动;(17)滑油泵;(18)舰船上通常有人员进出和使用的场所的照明系统;(19)生产和清洁水设备;(20)海军主管当局要求的导航设备;(21)海军主管当局要求的航行灯和特殊用途灯;(22)燃油泵和燃油燃烧装置;(23)油分离器;(24)消防系统用泵;(25)扫气风机;(26)操舵装置;(27)动力定位用推力器;(28)遥控操纵的阀;(29)机舱和锅炉舱的通风机;(30)水密门、边门和电气操纵的门;(31)关闭装置;(32)锚机;(33)为上述设备供电的电源和供电系统;(34)若适用,应规定按舰船类型设置的重要军事系统和损管装置为重要设备。Essential services are those necessary for the propulsion and safety of the ship within the Mobility category and Ship Type category and include the following:(1) air compressors for oil engines;(2) air pumps;(3) automatic sprinkler systems;(4) ballast pumps;(5) bilge and dewatering system pumps;(6) circulating and cooling water pumps;(7) communication systems;(8) condenser circulating pumps;(9) electric propulsion equipment;(10) electric starting systems for oil engines;(11) extraction pumps;(12) fans for forced draught to boilers;(13) feed water pumps;(14) fire detection and alarm systems;(15) fuel valve cooling pumps;(16) hydraulic pumps for controllable pitch propellers and those serving essential services here listed that would otherwise be directly electrically driven;(17) lubricating oil pumps;(18) lighting systems for those parts of the ship normally accessible to and used by personnel;(19) equipment for producing and cleaning water;(20) navigational aids where required by the Naval Authority;(21) navigation lights and special purpose lights where required by the Naval Authority;(22) oil fuel pumps and oil fuel burning units;(23) oil separators;(24) pumps for fire extinguishing systems;(25) scavenge blowers;(26) steering gear;(27) thruster for dynamic positioning;(28) valves which are required to be remotely operated;(29) ventilating fans for engine and boiler rooms;(30) watertight doors, shell doors and other electrical operated doors;(31) closing appliances;(32) windlasses;(33) power sources and supply systems for supplying the above services;(34) Essential military systems and damage control arrangements relating to the ship type are to be stated where applicable.

Essential services 重要用途 系指舰船的推进,操舵和安全有关的重要用途。它分为第一重要用途和第二重要用途。其定义和举例如下。(1)第一重要用途系指连续运行所必需的用途。这样才能维持推进力和操舵性能,使用第一重要用途的装置如下:①操舵器;②可变纵摆(螺距)螺旋桨用泵;③为了推进而使用在主/辅进气机和汽轮机上的定期空压机,燃油阀冷却泵,滑油泵和冷却水泵;④汽轮机船的蒸汽装置用以及第一重要用途供电的装置中使用蒸汽的船舶,其辅助锅炉用强制通风机、给水泵、循环水泵、真空水泵以及饮水泵;⑤带有向蒸汽涡轮船舶的蒸汽装置喷燃泵和第一重要用途供电装置的船舶,其辅助锅炉喷燃装置;⑥推进/操舵的唯一工具,即旋转推进装置(包括润滑油和冷却水泵);⑦带有滑油泵和冷却水泵的电力推进装置的电气设备;⑧向第一重要用途装置供电的发电机和相关电源;⑨向第一重要用途装置供油压的油压泵;⑩重油黏度控制装置;⑪使用第一重要用途装置的控制,检查和安全装置/系统。(2)第二重要用途虽然在维持船舶安全方面需要,但在保持推进力和操舵性能而进行的连续运转这一方面不需要的用途。使用第二重要用途的装置如下:①锚机;②燃料油输送泵和燃料油处理装置;③润滑油输送泵和润滑油处理装置;④重油预热器;⑤启动和控制用空气压缩机;⑥舱底,镇流器和侧倾泵;⑦灭火泵和其他固定灭火器泵;⑧机舱和锅炉室通风扇;⑨确认能够使危险区域保持安全的必要用途;⑩航海灯,航海用具和航海信号;⑪船内通信装置;⑫火灾探测和火灾警报装置;⑬照明装置;⑭水密封闭装置的电气设备;⑮向第一重要用途装备供电的发电机和相关电源;⑯向第一重要用途装备供给油压的油压泵;⑰货物围护装置所使用的控制,检测和安全系统;⑱使用第二重要用途装置的控制,检测和安全系统。Essential services are those services essential for propulsion and steering, and safety of the ship, which are made up of "primary essential services" and "secondary essential services". Definitions and examples of such services are given in (1) and (2) below:(1) Primary essential services are those services which need to be in continuous operation to maintain propulsion and steering. Examples of equipment for primary essential services are as follows:① steering gears;② pumps for controllable pitch propellers;③ scavenging air blower, fuel oil supply pumps, fuel valve cooling pumps, lubricating oil pumps and cooling water pumps for main and auxiliary engines and turbines necessary for propulsion;④ forced draught fans, feed water pumps, water circulating pumps, vacuum pumps and condensate pumps for steam plants on steam turbine ships, and also for auxiliary boilers on ships where steam is used for equipment supplying primary essential services; ⑤ oil burning installations for steam plants on steam turbine ships and for auxiliary boilers where steam is used for equipment supplying primary essential services; ⑥ azimuth thrusters which are the sole means for propulsion/steering with lubricating oil pumps, cooling water pumps; ⑦ electrical equipment for electric propulsion plant with lubricating oil pumps and cooling water pumps; ⑧ electric generators and associated power sources supplying equipment for the primary essential service; ⑨ hydraulic pumps supplying equipment for the primary essential service; ⑩ viscosity control equipment for heavy fuel oil; and ⑪ control, monitoring and safety devices/systems for equipment to primary essential services. (2) secondary essential services are those services which need not necessarily be in continuous operation to maintain propulsion and steering but which are necessary for maintaining the vessel's safety. Examples of equipment for secondary essential services are as follows: ① windlass; ② fuel oil transfer pumps and fuel oil treatment equipment; ③ lubrication oil transfer pumps and lubrication oil treatment equipment; ④ pre-heaters for heavy fuel oil; ⑤ starting air and control air compressors; ⑥ bilge, ballast and heeling pumps; ⑦ fire pumps and other fire extinguishing medium pumps; ⑧ ventilating fans for engine and boiler rooms; ⑨ services considered necessary to maintain dangerous spaces in a safe condition; ⑩ navigation lights, aids and signals; ⑪ internal safety communication equipment; ⑫ fire detection and alarm system; ⑬ lighting system; ⑭ electrical equipment for watertight closing appliances; ⑮ electric generators and associated power sources supplying equipment for the secondary essential service; ⑯ hydraulic pumps supplying equipment for the secondary essential service; ⑰ control, monitoring and safety systems for cargo containment systems; ⑱ control, monitoring and safety devices/systems for equipment to secondary essential services.

Essential services (ship or mobile offshore unit) 重要设备(船舶或移动式近海装置) 为船舶或移动式近海装置的航行、操舵或机动,或为人身安全,或为船舶或近海装置的特性(如特殊用途)所必需的设备。 Essential services mean services essential for the navigation, steering or manoeuvring of the ship or mobile offshore unit, or for the safety of human life, or for special characteristics of the ship or unit (for example, special services).

Essential services(ship) 重要设备(船舶) 对船舶推进和安全所必需的设备,诸如下列设备:燃油发动机用空压机;空气泵;自动喷水系统;压载水泵;舱底泵;循环和冷却水泵;通信系统;冷凝器循环水泵;电力推进设备;燃油发动机电力启动系统,抽吸泵,锅炉强力通风用风机;给水泵;探火和报警系统;燃油阀冷却系统;对可调螺距螺旋桨以及为此处所列重要设备提供液体压力的液压泵,否则这些重要设备应直接由电力拖动;滑油泵,惰性气体风机和洗涤器以及甲板密封泵;船上通常有船员和旅客进出和使用的场所的照明系统;法定规则要求的导航设备;法定规则要求的航行灯和特殊用途灯;燃油泵和燃油燃烧装置;分油机;消防系统用泵;扫气泵;操舵装置;动力定位用推力器;需要遥控操纵的阀;机舱和锅炉舱的通风机;水密门、舷门和电气操纵的其他关闭装置;锚机;为上述设备供电的电源和供电系统。 Essential services(ship) mean those necessary for the propulsion and safety of the ship, such as the following: air compressors for oil engines; air pumps; automatic sprinkler systems; ballast pumps; bilge pumps; circulating and cooling water pumps; communication systems; condenser circulating pumps; electric propulsion equipment; electric starting systems for oil engines; extraction pumps; fans for forced draught to boilers; feed water pumps; fire detection and alarm systems; fuel valve cooling pumps; hydraulic pumps for controllable pitch propellers and those serving essential services here listed that would otherwise be directly electrically-driven; lubricating oil pumps; inert gas fans and scrubber and deck seal pumps; lighting systems for those parts of the ship normally accessible to and used by personnel and passengers; navigational aids where required by Statutory Regulations; navigation lights and special purpose lights where required by Statutory Regulations; oil fuel pumps and oil fuel burning units; oil separators; pumps for fire-extinguishing systems; scavenge blowers; steering gear; thrusters for dynamic positioning; valves which are required to be remotely operated; ventilating fans for engine and boiler rooms; watertight doors, shell doors and other electrical operated closing appliances; windlasses; power sources and supply systems for supplying the above services.

Establishment of emergency preparedness 确立应急准备 基于风险及应急准备分析,涉及计划及实施适当应急准备措施的系统化过程。

Estimate horse power 估计功率(估计马力)

Estimate time of arrival (ETA) 预计到达时间

Estimated repair cost 预估费用/暂定价

Estimated working hours 估价工时

Estimation 估价

Estuarine survey 河口调查 对上起潮区界,下至口外深水线或水下三角洲外缘的河口范围进行的科学调查。内容有河口水文、地质、地貌、沉淀物、化学、生物和环境污染等。

Ethane carrier　乙烯运输船　系指在 -103 ℃超低温条件下运输乙烯的船舶。 Ethane carrier is a ship used for the carriage of ethane under -103 ℃ (super low temperature).

Ethernet　以太网　系指由 Xerox 公司创建并由 Xerox、Intel 和 DEC 公司联合开发的基带局域网规范，是当今现有局域网采用的最通用的通信协议标准。以太网络使用 CSMA/CD (载波监听多路访问及冲突检测) 技术，并以 10 m/s 的速率运行在多种类型的电缆上。以太网与 IEEE802.3 系列标准相类似。包括标准的以太网 (10 Mbit/s)、快速以太网 (100 Mbit/s) 和 10 G (10 Gbit/s) 以太网。它们都符合 IEEE802.3。

Ethoxyline resin　环氧树脂

ETM"scheme A"　ETM"方案 A"　系指方案 A 的 (废气滤清) $EGC-SO_x$ 技术手册。 ETM "scheme A" means the $EGC-SO_x$ technical manual for scheme A.

ETM"scheme B"　ETM"方案 B"　系指方案 B 的 (废气滤清) $EGC-SO_x$ 技术手册。 ETM "scheme B" means the $EGC-SO_x$ technical manual for scheme B.

EU2005/33/EC　欧盟法令的编号　这是一个关于在欧盟港口内船上使用燃油含硫量的新规定，自 2010 年 1 月 1 日起生效，具体要求是：凡在欧盟港口停泊 (包括锚泊、系浮筒、码头靠泊) 超过 2 h 的船舶不得使用硫含量超过 0.1% 的燃油 (该要求不适用于停掉所有机器而使用岸电的船舶)。 EU2005/33/EC takes effect from 1 January 2010, requiring that ships at berth in Europe Union ports for more than 2 hours shall not use marine fuels with a sulphur content exceeding 0.1% by mass except that ships switch off all engines and use shore side electricity while at berth in ports.

European Article Number Association (EANA)　欧洲物品编码协会

European Article Number (EAN)　EAN 码　是国际物品编码协会制定的一种商品用条码，通用于全世界。EAN 码符号有标准版 (EAN-13) 和缩短版 (EAN-8) 两种标准版表示 13 位数字，又称为 EAN13 码，缩短版表示 8 位数字，又称 EAN8 码。两种条码的最后一位为校验位，由前面的 12 位或 7 位数字计算得出。两种版本的编码方式可参考国标 GB-12094-1998。

European Atomic Energy Community (EAEC)　欧洲原子能联营

European Bank for Reconstruction and Development (EBRD)　欧洲复兴开发银行　成立于 1991 年，建立欧洲复兴开发银行的设想是由法国总统密特朗于 1989 年 10 月首先提出来的，于 1991 年 4 月 14 日正式开业，总部设在伦敦。主要任务是帮助欧洲战后重建和复兴。该行的作用是帮助和支持东欧、中欧国家向市场经济转化。2015 年 10 月，中国已正式申请加入"欧洲复兴开发银行"。欧洲复兴开发银行 (EBRD) 董事总经理兼代理首席经济学家兰克斯日前表示，欧洲复兴开发银行董事会将会积极考虑中国加入该行的申请。欧洲复兴开发银行股东将在 12 月中旬就中国加入事宜做出最后决定。不过，银行董事会预计会在 11 月中旬向银行股东建议同意接受中国加入该行的申请。有关专家据此认为，中国加入欧洲复兴开发银行的申请有望在 2015 年 12 月获得批准。2015 年 12 月 14 日，欧洲复兴开发银行理事会通过接受中国加入该行的决议。在履行国内相关法律程序后，我国正式成为该行成员。

European Coal and Steel Community (ECSC)　欧洲煤钢联营

European Commission　欧洲委员会　简称欧委会，于 1949 年 5 月 5 日在伦敦成立，原为西欧 10 个国家组成的政治性组织，现已扩大到整个欧洲范围，共有 46 个成员国，5 个部长委员会观察员国 (梵蒂冈、加拿大、美国、日本和墨西哥) 以及 3 个议会观察员国 (加拿大、墨西哥和以色列)。其宗旨是保护欧洲人权、议会民主和权利的优先性；在欧洲范围内达成协议以协调各国社会和法律行为；促进实现欧洲文化的统一性。欧委会通过审议各成员国共同关心的除防务以外的其他重大问题，推动各成员国政府签订公约和协议以及向成员国政府提出建议等方式，谋求在政治、经济、社会、人权、科技和文化等领域采取统一行动，并经常对重大国际问题发表看法。

European Court of Justice　欧洲法院　是欧洲联盟法院的简称，根据 1951 年《巴黎条约》设立，设于卢森堡，是欧洲联盟的最高司法机关。主要审理以成员国为当事人的违反欧盟法律的案件。

European Environment Agency　欧盟环境署

European Free Trade Association (EFTA)　欧洲自由贸易联盟　又称"小自由贸易区"。1960 年 1 月 4 日，奥地利、丹麦、挪威、葡萄牙、瑞典、瑞士和英国在斯德哥尔摩签订《建立欧洲自由贸易联盟公约》，即《斯德哥尔摩公约》。该公约经各国议会批准后于同年 5 月 3 日生效，欧洲自由贸易联盟正式成立，简称欧贸联，总部设在日内瓦。

European Investment Bank (EIB)　欧洲投资银行　是欧洲经济共同体成员国合资经营的金融机构。根据 1957 年《建立欧洲经济共同体条约》(《罗马条约》) 的规定，于 1958 年 1 月 1 日成立，1959 年正式开业。总行设在卢森堡。欧洲投资银行对欧洲原子能联营的筹资活动也提供帮助,它受该联营的委托,负责审查其成员国的借款申请书,并在贷款确定后,负责管理贷款的使用。欧洲投资银行的分支机构有:设在罗马的意大利分部和设在伦敦的联络处,它们主要为在意大利和英国兴建的工程项目提供信贷。

European Monitoring Centre for Drugs and Drug Addiction(EMCDDA) 欧洲毒品和毒瘾监管中心

European Parliament 欧洲议会 其前身是1952年成立的欧洲煤钢共同体议会,当时由法国、联邦德国、意大利、荷兰、比利时、卢森堡6个成员国的78名议员组成,1962年改称"欧洲议会",它是欧盟三大机构(欧盟理事会、欧盟委员会、欧洲议会)之一,为欧盟的立法、监督和咨询机构。2014年3月13日,欧洲议会称,若俄罗斯不从克里米亚撤军,则将考虑与俄中止外交关系。2015年6月2日晚欧洲议会决定限制俄驻欧盟代表自由进入欧洲议会,未来限制禁令将扩大到所有俄杜马议员。

European Union(EU) 欧洲联盟 简称欧盟,总部设在比利时首都布鲁塞尔,是由欧洲共同体(European Community,又称欧洲共同市场,简称欧共体)发展而来的,初始成员国有6个,分别为法国、联邦德国、意大利、比利时、荷兰以及卢森堡。该联盟现拥有27个会员国,正式官方语言有24种。1991年12月,欧洲共同体马斯特里赫特首脑会议通过《欧洲联盟条约》,通称《马斯特里赫特条约》(简称《马约》)。1993年11月1日,《马约》正式生效,欧盟正式诞生。欧洲理事会主席为范龙佩。欧盟委员会主席为巴罗佐。欧洲议会议长为马丁·舒尔茨。欧洲联盟的条约经过多次修订,目前欧洲联盟的运作方式是依照《里斯本条约》。政治上所有成员国均为民主国家(2008年《经济学人》民主状态调查),军事上绝大多数欧洲联盟成员国为北大西洋公约组织成员。2016年6月24日,英国退出欧洲联盟。

Euryhaline species 广盐性物种 系指盐度耐受范围大的物种。 Euryhaline species mean those species are able to tolerate a wide range of salinity.

Eurythermal species 广温性物种 系指温度耐受范围大的物种。 Eurythermal species mean those species which are able to tolerate a wide range of temperatures.

Eutectic alloy 低熔点合金(易熔合金)

Evacuation 撤离 在庇护区通过专用撤离方法放弃海上设施。通常主要考虑紧急撤离,因为预防性撤离不需要太多的撤离资源。

Evacuation analysis 撤离分析 系指在设计过程的早期对脱险通道进行的评估。这种分析应用于确定并尽可能消除在弃船过程中由于乘客和船员沿脱险通道正常移动,包括可能有船员需沿这些通道朝着与乘客相反的方向移动时可能造成的拥挤。此外,这种分析还应应用于证实逃生布置具有充分的灵活性以适应可能由于事故而引起某些脱险通道、集合站、登乘站或救生艇筏不能使用的情况。 Evacuation analysis means an evaluating on escape routes at a early design process. The analysis shall be used to identify and eliminate, as far as practicable, congestion which may develop during an abandonment, due to normal movement of passengers and crew along escape routes, including the possibility that crew may need to move along these routes in a direction opposite the movement of passengers. In addition, the analysis shall be used to demonstrate that escape arrangements are sufficiently flexible to provide for the possibility that certain escape routes, assembly stations, embarkation stations or survival craft may not be available as a result of a casualty.

Evacuation goal 1 撤离目标1(救生和救助) 救生设备在舰船上应以这样的方式布置,这种方式就是在紧急情况下易于接近和易于部署。(1)通常,救生衣应按照SOLAS公约第Ⅲ章B部分7.2条要求来提供;(2)救生设备和撤离装备的布置应按SOLAS公约第Ⅲ章B部分第11,12,13,15和16条;(3)战备状态的布置应按SOLAS公约第Ⅲ章B部分第20条。 Evacuation goal 1 means that life-saving equipment is to be arranged on the vessel in such a manner that it is easily accessible and readily deployed in case of emergency. (1)In general, the provision of life jackets is to be in accordance with SOLAS Chapter Ⅲ, Part B, Regulation 7.2. (2) Arrangements for life-saving and evacuation equipment are to be in accordance with SOLAS Chapter Ⅲ, Part B, Regulations 11, 12, 13, 15 and 16, as applicable. (3) Arrangements for operational readiness are to be in accordance with SOLAS, Part B, Chapter Ⅲ, Regulation 20.

Evacuation goal 2 撤离目标2(救生和救助) 舰船救生艇筏的总容量应足以确保所有工作人员能够在可预见的紧急撤离条件下撤离。救生艇筏及救助艇应按SOLAS公约第Ⅲ章B部分第31条提供。 Evacuation goal 2 means that the total capacity of the ship's survival craft is to be sufficient to ensure that all personnel can be evacuated during foreseeable emergency conditions. The provision of survival craft and rescue boats is to be in accordance with SOLAS Part B, Chapter Ⅲ, Regulation 31.

Evacuation goal 3 撤离目标3(救生和救助) 所有救生和救助设备应经过船级社、海军主管当局和国家主管部门的形式认可。(1)通常,所有救生和救助设备原型应进行测试,以确保其满足国际救生设备规则或其他船级社和海军主管当局接受的准则;(2)所有救生和救助设备应进行生产试验,以确保其生产的标准与认可的原型相同。 Evacuation goal 3 means that all life-saving and rescue equipment is to be of an approved type acceptable to the Society, the Naval Authority and National Administration where applicable. (1) In general, all life-saving and rescue equipment prototypes are to be tested to confirm that they comply with the International Life Saving Appliance Code or other standard acceptable to the Society and the Naval Authority. (2) All life-saving and rescue equipment is to subject to production tests to ensure that they are constructed to the same standard as the approved prototype.

Evacuation goal 4 撤离目标4(救生和救助) 提供布置来确保部署的救生艇、筏可以从损坏的舰船转移到安全处,直到所有人员被救助。 Evacuation goal 4 means that provision is made to ensure that the deployed survival craft can be moved to safety location from a damaged vessel until such time all personnel can be rescued.

Evacuation goal 5 撤离目标5(救生和救助) 提供布置来确保无行为能力的人可撤离到安全地点。 Evacuation goal 5 means that provision is made for incapacitated people to be evacuated to safety location.

Evacuation procedures 撤离程序 包括如下内容:(1)船长发出应急通知;(2)与基地港联系;(3)穿着救生衣;(4)救生艇筏和应急站人员就位;(5)关闭机器和燃油供应管路;(6)发出撤离命令;(7)降落救生艇筏、海上脱险系统和救助艇;(8)救生艇筏系拢航行;(9)监视乘客;(10)乘客在监视下有秩序地撤离;(11)船员检查所有乘客已全部离船;(12)船员撤离;(13)救生艇筏脱离大船;(14)救助艇(如有的话)集结救生艇筏。 Evacuation procedures shall include: (1) the emergency announcement made by the master; (2) contact with base port; (3) the donning of jackets; (4) manning of survival craft and emergency stations; (5) the shutting down of machinery and oil fuel supply lines; (6) the order to evacuate; (7) the deployment of survival craft and marine escape systems and rescue boats; (8) the bowsing of survival craft; (9) the supervision of passengers; (10) the orderly evacuation of passengers under supervision; (11) crew checking that all passengers have left the craft; (12) the evacuation of crew; (13) releasing the survival craft from the craft; (14) the concentrating of survival craft by the rescue boat, where provided.

Evacuation pump 真空泵 抽吸舱、柜或管路内气体的泵。

Evacuation stations and external escape routes 撤离站和外部脱险通道 系指:(1)救生艇筏存放区;(2)作为救生艇和救生筏登乘与降落站的开敞甲板处所和围蔽游步甲板处所;(3)内部和外部集合站;(4)用作脱险通道的外部通道和开敞甲板;(5)最轻航行水线之上的舷侧,位于救生艇筏和撤离滑道的登乘区域下方且相邻的上层建筑和甲板室舷侧。 Evacuation stations and external escape routes are: (1) survival craft stowage area; (2) open deck spaces and enclosed promenades forming lifeboat and liferaft embarkation and lowering stations; (3) assembly stations, internal and external; (4) external stairs and open decks used for escape routes; (5) the ship's side to the waterline in the lightest seagoing condition, superstructure and deckhouse sides situated below and adjacent to the liferaft and evacuation slide embarkation areas.

Evacuation time 撤离时间 系经过验证的时间,系指对应于乘客和船员总数的一定数量未经培训人员接受撤离指令后从船上撤离的时间。撤离时间T应不超过7 min 40 s,或如果结构防火时间小于30 min,则不应超过:$(T-7)/3$ min。 Evacuation time is the demonstrated time taken for a number of untrained people to escape from the craft following the order to evacuate corresponding to the total number of passengers and crew. The evacuation time should not exceed 7 min and 40 s or, where the structural fire protection time T is less than 30 min, a time of: $(T-7)/3$ min.

Evaluation 评估 系指对某一形式的救生艇释放和回收系统进行设计评审和性能试验。 Evaluation is a design review and performance test of lifeboat release and retrieval system.

Evaluation system 评价体系

Evaporate 蒸发

Evaporating gas pressure 蒸发气压力 系指按规定温度下液体上方饱和蒸发气的平衡压力(绝对压力),以MPa计。

Evaporating point 汽化点

Evaporating pressure 蒸发压力 在一定温度下,载冷剂由液态转变为气态时的压力。

Evaporation 汽化(蒸发)

Evaporation duct 大气波道 系指捕获雷达能量以贴近海面传播的低波道(空气密度的改变)。波道可增强或降低雷达目标探测距离。 Evaporation duct is a low lying duct (a change in air density) that traps the radar energy so that it propagates close to the sea surface. Ducting may enhance or reduce radar target detection ranges.

Evaporation rate of heating surface 受热面蒸发率 蒸发受热面包括经济器在内按单位面积小时计算产生的蒸发气量。

Evaporation surface load 镜面负荷 蒸发气量即

体积流量与蒸发液面面积的比值。

Evaporative 汽化的(蒸发的)

Evaporative boiling 汽化沸腾

Evaporative heat 蒸发热

Evaporator 蒸发器

Evaporator coil 蒸发盘管

Evaporator room 蒸发器舱

Even mode of descending and ascending 均匀沉浮方式 系指是在整个沉浮过程中平台没有或仅出现微不足道、因而可以忽略不计的倾斜的沉浮方式。Even mode of descending and ascending means a mode in which no or very slight inclination will be observed during the descending and ascending of the unit, thus the effect of the inclination can be neglected.

Event 偶然事件 是一种由船舶外部原因(例如,被波浪)产生的事件。Event is an occurrence which has its origin outside the craft (e.g. waves).

Event tree analysis 事件树分析 通过图表研究事故、失败或不理想的事件的发展及升级的方法。该图表从最初事件开始,分支部分是控制或缓和措施产生的影响,直到鉴定出最后结果。这些方法成功的可能性或频率反映出每种结果的可能性。

Evidence list 证据目录清单

Evidence to be preserved 证据保全 系指人民法院在起诉前或在对证据进行调查前,依据申请人的申请或当事人的请求,以及依职权对可能灭失或今后难以取得的证据,予以固定和保存的行为。人民法院解决民事纠纷,认定案件事实是必不可少的环节,认定案件事实必须依靠证据。由于民事案件的事实是过去发生的事实,民事案件的起诉、审理直至判决需要一个过程,所需的证据可能会由于没有及时地收集到而因人为或客观的原因灭失或难以取得,于是为了维护当事人的合法权益,为了人民法院公正审理民事案件,需要采取一定的措施,将可能灭失或以后难以取得的证据固定或保存下来,因此,有必要建立这样一套保全证据的制度。

Evidence 证据/凭证

EVTMS 增强的船舶交通管理系统 按照巴拿马运河管理局和国际安全要求设置和管理船舶交通的系统。

EWEA 欧洲风能协会

Ex work-named place(EXW) 指定地点的工厂交货 该术语是卖方承担责任最小的术语。买方必须承担在卖方所在地受领货物的全部费用和风险。但是,若双方希望在起运时卖方负责装载货物并承担装载货物的全部费用和风险时,则须在销售合同中明确写明。在买方不能直接或间接地办理出口手续时,不应使用该术语,而应使用 FCA,如果卖方同意装载货物并承担费用和风险的话。

Examination 检查(检测,验证)

Examine and verify 审核 系指通过代表性样本、验证船舶保安体系已被有效地实施、验证船舶保安计划规定的所有保安设备是否符合适用的要求并适于预期的服务。

Exceptional circumstances 例外情况 系指下列一种或多种情况:(1)无法获得坞内设备;(2)无法获得修理设备;(3)无法获得所需材料、设备或备件;(4)由于避免恶劣天气情况而导致的延期。Exceptional circumstances mean one or more of the following cases: (1) unavailability of dry-docking facilities; (2) unavailability of repair of facilities; (3) unavailability of essential materials, equipment or spare parts; (4) delays incurred by action taken to avoid severe weather conditions.

Exceptional control flow 异常控制流

Exceptions 例外 (1)对 MARPOL 附则Ⅰ而言,油类或油性混合物排放的例外,系指下列情况:①将油类或油性混合物排放入海,系为保障船舶安全或救护海上人命所必需者;②将油类或油性混合物排放入海,系由于船舶或其设备损坏而导致;(a)但须在发生损坏或发现排放后,为防止排放或使排放减少至最低限度,已采取了一切合理的补救措施;(b)但是,如果船东或船长是故意造成损坏,或轻率行事而又知道可能会招致损坏,则不在此列;③将经主管机关批准的含油物质排放入海,用以对抗特定污染事故,以使污染损害减少至最低限度。但任何这种排放,均应经拟进行排放所在地区的管辖国政府批准。(2)对 MARPOL 附则Ⅱ而言,有害液体物质或含有这种物质的混合物排放的例外,系指:①此排放系为保障船舶安全或救护海上人命所必需者;或 ②由于船舶或其设备损坏而导致;(a)但须在发生损坏或发现排放后,为防止排放或使排放减少至最低限度,已采取了一切合理的补救措施;(b)但是,如果船东或船长是故意造成损坏,或轻率行事而又知道可能会招致损坏,则不在此列;或③此排放系经主管机关批准用以对抗特定污染事故,以使污染损害减少至最低限度。但任何这种排放,均应经拟进行排放所在地区的管辖国政府批准。(3)对 MARPOL 附则Ⅲ而言,包装形式装运的有害物质入海的例外,系指为保障船舶安全或救护海上人命所必需者。(4)对 MARPOL 附则Ⅳ而言,船舶生活污水排放的例外,系指:①从船上排放生活污水,系为保障船舶安全或救护海上人命所必需者;②由于船舶或其设备损坏而导致排放生活污水,且在发生损坏前后已采取了一切合理的补救措施防止排放或使排放减少至最低限度。(5)对 MARPOL 附则Ⅴ而言,船舶上处理垃圾的例外，系指：①系为保障船舶安全或救护海上人命所必需者；或②由于船舶或其设备损坏而导致垃圾泄漏，且在发生泄漏前后已采取了一切合理的补救措施防止泄漏或使泄漏减少至最低限度；③合成渔网意外落失，且已采取了一切合理的补救措施防止这种落失。(6)对MARPOL 附则Ⅵ而言，向大气或海洋释放受该附则控制的任何物质的例外，系指：①任何为保障船舶安全或救护海上人命所必需的排放；②任何因船舶或其设备遭到损坏的排放：(a)但须在发生损坏或发现排放后，为防止排放或使排放减少至最低限度，已采取了一切合理的补救措施；(b)但是，如果船东或船长是故意造成损坏，或轻率行事而又知道可能会招致损坏，则不在此列。 Exceptions mean：(1)for Annex Ⅰ to MARPOL, exceptions for discharging into the sea of oil or oily mixture, such discharge mean：① it is necessary for the purpose of securing the safety of a ship or saving lives at sea；② discharge resulting from damage of a ship or its equipment：(a) provided that all reasonable precautions have been taken after the occurrence of the damage or discovery of the discharge for the purpose of preventing or minimizing the discharge；(b) Except if the owner or the master acts either with intent to cause damage, or recklessly and with knowledge that damage would probably occur；③ the discharge into the sea of substances containing oil, approved by the Adiministration, when being used for purpose of combating specific pollution incidents in order to minimize the damage from pollution. Any such discharge shall be subject to the approval of any Government in whose jurisdiction it is contemplated the discharge will occur. (2) for Annex Ⅱ to MARPOL, exceptions for discharging into the sea of noxious liquid substances or mixtures containing such substances, such discharge mean：① it is necessary for the purpose of securing the safety of a ship or saving lives at sea；or ② discharge resulting from damage of a ship or its equipment：(a) provided that all reasonable precautions have been taken after the occurrence of the damage or discovery of the discharge for the purpose of preventing or minimizing the discharge；(b) Except if the owner or the master acts either with intent to cause damage, or recklessly and with knowledge that damage would probably occur；③ it is approved by the Adiministration, when being used for purpose of combating specific pollution incidents in order to minimize the damage from pollution. Any such discharge shall be subject to the approval of any Government in whose jurisdiction it is contemplated the discharge will occur. (3) for Annex Ⅲ to MARPOL, exceptions for jettisoning of harmful substances carried in packaged form, such jettison means that it is necessary for the purpose of securing the safety of the ship or saving lives at sea. (4) for Annex Ⅳ to MARPOL, exceptions for the discharging of sewage from a ship, such discharge mean：① it is necessary for the purpose of securing the safety of a ship and those on board or saving lives at sea；② the discharge of sewage resulting from damage to a ship or its equipment if all reasonable precaution have been taken before and after the occurrence of the damage for the purpose of preventing or minimizing the discharge. (5) for Annex Ⅴ to MARPOL, exceptions for the disposal of garbage from a ship, such disposal means：① it is necessary for the purpose of securing the safety of a ship and those on board or saving lives at sea；② the escape of garbage resulting from damage to a ship or its equipment provided all reasonable precautions have been taken before and after the occurrence of the damage, for the purpose of preventing or minimizing the escape；③ the accidental loss of synthetic fishing nets, provided that all reasonable precautions have been taken to prevent such loss. (6) for Annex Ⅵ to MARPOL, exceptions for releasing of substances, subject to control by this Annex, from ships into the atmosphere or sea, such releasing mean：① any emission necessary for the purpose of securing the safety of a ship or saving lives at sea；② any emission resulting from damage to a ship or its equipment：(a) provided that all reasonable precautions have been taken after the occurrence of the damage or discovery of the emission for the purpose of preventing or minimizing the emission；(b) except if the owner or the master acts either with, intent to cause damage, or recklessly and with knowledge that damage would probably occur.

Excess air coefficient 过量空气系数

Excess air ratio (excess air factor) 过量空气系数 燃料燃烧时超过需要的实际空气量和所需空气量与理论空气量的比值。

Excess hatch 出入舱口 供人员进出的舱口。

Excessive convexity 过大的凸出 见图 E-6。

Excessive corrosion 过度腐蚀 系指损耗超出有关结构的可接受限度。 Excessive corrosion is wastage in excess of acceptable limits for the structure concerned.

Excessive foot (root) reinforcement 过大的根部增强高 见图 E-7。

Excessive sound 过响声 系指超过一定强度标准的噪声。往往使人突然受惊，产生恐惧感，或使人耳超负荷，非常难受或振破耳膜而造成耳聋，并可危害人体健康。

Excessive weld reinforcement 过大的焊缝增强高

Excessive weld thickness（fillet weld）

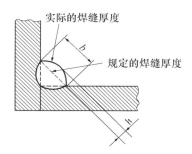

图 E-6　过大的凸出
Figure E-6　Excessive convexity

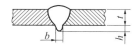

图 E-7　过大的根部增强高
Figure E-7　Excessive foot（root）reinforcement

（过大的增强高）　要求平顺地过渡。见图 E-8。

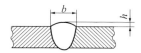

图 E-8　过大的焊缝增强高
Figure E-8　Excessive weld reinforcement

Excessive weld thickness（fillet weld）　过大的焊缝厚度（角接焊缝）　大多数应用场合，超过规定尺寸的焊缝厚度但不构成拒收的原因。见图 E-9。　For many applications a weld thickness in excess of the specified size does not represent grounds for rejection. It is seen in Fig. E-9.

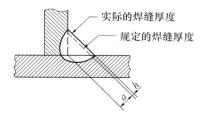

图 E-9　过大的焊缝厚度
Figure E-9　Excessive weld thickness

Excessively unequal leg lengths of fillet welds　过大的不等边臂长的角接焊缝　假设对不对称的角接焊缝未做明确的规定。见图 E-10。　It is assumed that an asymmetric fillet weld has not been expressly prescribed. It is seen in Fig. E-10.

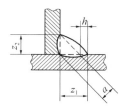

图 E-10　过大的不等边臂长的角接焊缝
Figure E-10　Excessively unequal leg lengths of fillet welds

Exchange　调配　系指使用下列之一种方法来更换压载舱内的压载水：(1)流通交换，系指在液舱底部通过泵入海洋中部的海水冲刷出压载水和连续地使该液舱的压载水从舱顶溢出直至整个舱内的水交换过 3 次为止，以减小残留在该液舱内水生有机物的数量至最低程度；(2)抽空/再注入调配，系指在港口、港湾或领海上泵出压载水直至该舱被抽空，然后重新用海洋中部的海水注入舱内；如果这样做是安全的话，船长/驾驶员应泵出几乎 100% 压载水。　Exchange means to replace the water in a ballast tank using one of the following methods：(1) flow through exchange means to flush out ballast water by pumping in mid-ocean water at the bottom of the tank and continuously overflowing the tank from the top until three full volumes of water has been changed to minimize the number of original organisms remaining in the tank；(2) empty/refill exchange means to pump ballast water taking place in ports, harbour or territorial waters until the tank is empty, then refill it with mid-ocean waters, master/operators should pump out as close to 100 percent of the ballast waters as is to do so.

Exchange dividend　外汇分红　系指政府允许出口厂商从其所得的外汇中提取一定百分比的外汇用于进口，鼓励其出口积极性的措施。

Exchange of a cylinder　换气气缸

Exchange of heat　热交换

Exchange over valve　交换阀

Exchange rate　汇率

Exchange ratio of ballast water　压载水交换率　系指通过注入或流通法加入的压载水总容积与压载水舱或货舱容积的比值。　Exchange ratio of ballast water means a ratio of total volume of ballast water, added by refilling or flowing through over capacity of ballast tank or hold.

Exchange dumping 外汇倾销 利用本国货币对外贬值的机会，向外倾销商品和争夺市场的行为称为外汇倾销。这是因为本国货币贬值后，出口商品用外国货币表示价格降低，提高了该国商品在国际市场上的竞争力，有利于扩大出口；而因本国货币贬值，进口商品的价格上涨，削弱了进口商品的竞争力，限制了进口。外汇倾销需要一定的条件，主要是本国货币对外贬值速度要快于对内贬值以及对方不进行报复等。

Exchangeability 互换性

Exchanger 交换器（转换装置）

Exciter 励磁机 系指为电机的励磁提供全部或部分磁场电流的电源装置。

Exciting force 扰动力/激振力 风浪、水流等作用在物体上的外力。

Exciting moment 扰动力矩/（激动力矩） 风浪、水流等作用在物体上的外力矩。

Excluded liability 除外责任 系指保单列明的不负赔偿责任的范围。除外责任可以以列举式的方式在保单中列举除外事项，也可以以不列举方式明确除外责任，即凡未列入承保范围的灾害事故均为除外责任。任何保险商品都不是万能的，总有一些不能赔付的责任内容。这是因为，保险公司都是经营性的，以盈利为目的，要考虑承担的风险的问题。保险公司要对风险认真考察，避免承保风险过大，造成保险公司的亏损。对应地，保险公司对风险进行筛选，把一些发生的可能性较大，损失较多的风险列为除外责任。

Excluded spacesV 免除处所 虽然对围蔽处所有所规定，但下列各项(1)~(5)所述处所仍应称作免除处所，不计入围蔽处所容积之内。但符合以下3个条件之一者，应作为围蔽处所：①设有框架或其他设施保护货物和物料的处所；②开口上设有某种封闭设备；③具有能使开口封闭的建筑物。(1)①甲板上建筑物内某一处所，它面对着高度为全甲板间的端部开口，且开口上沿板的高度不超过其邻近甲板横梁的高度 25 mm（1 英寸），如开口的宽度等于或大于该开口处甲板宽度的90%，则从实际端部开口起，至等于开口处甲板宽度的一半距离绘一条与开口线或面相平行的线，这个处所可不计入围蔽处所之内。②如该处所的宽度由于任何布置上的原因，不包括由于该船外壳板的收敛，使其宽度小于开口处甲板宽度的90%，则从开口线起，至船体横向宽度等于开口处甲板宽度的90%处绘一与开口平行的线，这个处所可不计入围蔽处所之内。③如果 2 个处所由一间隔区分开，而且间隔区除了舷墙和栏杆外是完全开敞的，则可按(1)①或(1)②的规定将其中一个或 2 个处所免除计量，但如果栏杆处所之间的间隔距离小于间隔区甲板最小宽度的1/2，就不适用这种免除。(2)在架空露天甲板下的处所，并开敞的两侧与船体除了必要的支柱外并无其他连接。在这种处所可以设置栏杆、舷墙及舷边上沿板，或在船边安设支柱，但栏杆顶或舷墙顶舷边上缘板的距离应不小于0.75 m(2.5 英尺)，或不小于该处所高度的1/3，以较大者为准。(3)伸展到两舷的建筑物内的处所，其两侧的相对开口的高度不小于0.75 m(2.5 英尺)，或不小于建筑物高度的1/3，以较大者为准。如果这种建筑物只在一侧有开口，则从围闭处所中免除计量的处所仅限于从开口向内最多伸到该开口处甲板宽度的1/2。(4)建筑物内，直接位于其顶甲板上无覆盖的开口之下的某一处所，如果这种开口是露天的，则从围闭处所中免除计量的处所仅限于此开口区域。(5)由建筑物界限舱壁形成的某一壁龛，这种壁龛是露天的，其开口高度为甲板间的全高度，无封闭设备，而且壁龛内宽度不大于其入口处宽度，同时从入口伸至内壁的深度不大于入口处宽度的两倍。 Notwithstanding the relevant provisions, the spaces referred to in subparagraphs (1)~(5) inclusive of this paragraph shall be called excluded spaces and shall not be included in the volume of enclosed spaces, except that any such space which fulfil at least one of the following three conditions shall be treated as an enclosed space: ① the space is fitted with shelves or other means for securing cargo or stores; ② the openings are fitted with any means of closure; ③ the construction provides any possibility of such openings being closed. (1)① A space within an erection opposite an end opening extending from deck to deck except for a curtain plate of a depth not exceeding by more than 25 mm (one inch) the depth of the adjoining deck beams, such opening has a breadth equal to or greater than 90% of the breadth of the deck at the line of the opening of the space. This provision shall be applied so as to exclude from the enclosed spaces only the space between the actual end opening and a line drawn parallel to the line or face of the opening at a distance from the opening equal to one half of the width of the deck at the line of the opening; ② If the width of the space dees not include convergence of the outside plating for the reason of arrangement, and become less than 90% of the breadth of the deck, only the space between the line of the opening and a parallel line drawn through the point where the athwartships width of the space becomes equal to, or less than, 90% of the breadth of the deck shall be excluded from the volume of enclosed spaces; ③ Where an interval which is completely open except for that bulwarks or open rails separates any two spaces, the exclusion of one or both spaces is permitted under sub-paragraphs (1)① and/or (1)②, such exclusion shall not apply if the separation between the two spaces is less than the least half breadth of the deck in way of the separation. (2) A space under an overhead deck covering open to the sea and weather, having no connexion on the exposed sides with the body of the ship other than the stanchions necessary for its support. In such a space, open rails or a bulwark and curtain plate or stanchions may be fitted at the ship's side, provided that the distance between the top of the rails or the bulwark and the curtain plate is not less than 0.75 m (2.5 feet) or one-third of the height of the space, whichever is the greater. (3) A space in a side-to-side erection directly in way of opposite side openings not less in height than 0.75 m (2.5 feet) or one-third of the height of the erection, whichever is the greater. If the opening in such an erection is provided on one side only, the space to be excluded from the volume of enclosed spaces shall be limited inboard from the opening to a maximum of one-half of the breadth of the deck in way of the opening. (4) A space in an erection immediately below an uncovered opening in the deck overhead, provided that such an opening is exposed to the weather and the space excluded from enclosed spaces is limited to the area of the opening. (5) A recess in the boundary bulkhead of an erection which is exposed to the weather and the opening of which extends from deck to deck without means of closing, provided that the interior width is not greater than the width at the entrance and its extension into the erection is not greater than twice the width of its entrance.

Exclusive agency 独家代理 系指在指定地区内,由其单独代表委托人行为的代理人。

Exclusive ballast tank 专用压载舱 系指油船运营时邻接货油舱的处所和矿砂船、散装货船或油船用于某种用途运营时,作为专用压载舱使用的舱。 Exclusive ballast tank is a compartment which is adjacent to a cargo oil tank when the ship is not in the dry cargo mode and which is used as an exclusive tank for ballast even when the ship is in or not in the dry cargo mode.

Exclusive chemical carrier 专用化学品船

Exclusive economic zone (EEZ) 专属经济区 系指领海以外并邻接领海的有关区域,受联合国海洋法公约规定的特定法律制度的限制,在这个制度下,沿海国的权利和管辖权以及其他国家的权利和自由均受联合国海洋法公约有关规定的支配。 Exclusive economic zone is an area beyond and adjacent to the territorial sea, subject to the specific legal regime established by the United Nations Convention on law of the sea, under which the right and jurisdiction of the coastal State and the rights and freedoms of other States are governed by the relevant regulations of the United Nations Convention on law of the sea.

Exclusive jurisdiction 专属管辖 系指法律强制规定某类案件只能由特定法院管辖,其他法院无权管辖,也不允许当事人协议变更管辖。与其他法定管辖相比,专属管辖具有优先性、排他性与强制性。专属管辖是法院管辖独有的制度,仲裁没有专属管辖。

Exclusive license 独占许可证

Exclusive sales 包销 系指出口人通过协议把某一种商品或某一类商品在每一个地区和期限内的经营权单独给予某个客户或公司的贸易方式。

Exclusive solid cargo hold 固体货物专用舱 系指油船运航时,邻接货舱的空舱或者用矿砂船或散装货船运营时作为固体货物装载舱的处所。 Exclusive solid cargo hold is a compartment which is used as a void space adjacent to a cargo oil tank when the ship is not in the dry cargo mode and which is used as a solid cargo stowing hold when the ship is in the dry cargo mode.

Excrement container 粪便贮容器

Excursion vessel (tourist ship) 旅游船 专供旅游者旅行游览用的船舶。

Execute heading angle 换舵艏向角 在 Z 形操纵试验中,在换反舵瞬时的转艏角。一般用 $\psi = 10°$。

Executed rudder angle 执行舵角 操舵系统的实际输入舵角。

Execution dead time 倒车时滞 从开始执行倒车操纵的瞬时到出现负推力或螺旋桨反转的瞬时之间所经的时间间隔。

Execution of arrest 执行逮捕

Executive tribunal 执行庭

Executor 执行员

Exempt and emergency action principle 豁免与紧急行动原则 系指总协定有免除承担总协定某项义务(捐税、劳役等)和争取保障措施的规定。

Exempted space 免除处所 指吨位丈量中,不包括在总吨位中的各处所。

Exemption Certificate 免除证书 系指根据并按 1974 年 SOLAS 公约的规定,对船舶准予免除某项规定后,除货船构造安全证书、货船设备安全证书、货船无线电安全证书或货船安全证书之外,向该船签发的免除证书。 Exemption Certificate means that when an exemption is granted to a ship under and in accordance with the provisions of SOLAS 1974, an Exemption Certificate issued to a ship in addition to the Cargo Ship Safety Construction certificates, Cargo Ship Safety Equipment Certificate and Cargo Ship Safety Radio Certificate.

Exemption of tax/duty-free 免税 系指按照税法规定不征收销项税额,同时进项税额不可抵扣应该转出,是对某些纳税人或征税对象给予鼓励、扶持或照顾的特殊规定,也是世界各国及各个税种普遍采用的一种税收优惠方式。免税是把税收的严肃性和必要的灵活性有机结合起来制定的措施。免税在经济生活中,税收立法确定的税收制度,包括设立税种、设计税目和税率,是根据社会经济发展的一般情况群体的平均承受能力来制定的,适应了普遍性、一般性的要求。

Exhaust (gas) temperature 排气温度 排气支管进口处的排气温度。

Exhaust advance 提前排气

Exhaust cam 排气凸轮

Exhaust casing 排气蜗壳

Exhaust chamber 排气室

Exhaust chest 排气室 为均匀排放汽轮机末级出口的排气进入冷凝器或其他部件而设置的气室。

Exhaust collector 排气总管(废气集合器)

Exhaust curve 排气线(示功图)

Exhaust cycle 排气循环

Exhaust door 排气门

Exhaust fan 排气风扇(抽风机) 将舱室内的污浊气体排至舷外大气中的通风机。

Exhaust fume 排烟(废气)

Exhaust gas 废气(排气,排出)

Exhaust gas (SO_x) cleaning system (EGCS-SOx) 废气(SO_x)滤清系统

Exhaust gas analysis 废气分析

Exhaust gas boilers 废气锅炉 系指:(1)能直接提供蒸汽的废气热交换器或强制循环废气热交换器,对于后者,如不能达到直接提供蒸汽,而能用其本身带有的不用燃烧的蒸汽贮存器来提供蒸汽;(2)仅利用柴油机的废气产生蒸汽或热水的装置,具备独立的汽腔或热井,此处系指具备蒸汽或热水出口。 Exhaust gas boilers are:(1) exhaust gas heat exchangers or forced-circulation exhaust gas heat exchangers. The latter, if not capable of directly supplying steam, is capable of supplying steam by means of non-burning-type steam reservoir;(2) a boiler which generates steam or hot water solely by exhaust gas from internal combustion engine and has a steam space or a hot well and has an outer of steam or hot water.

Exhaust gas change-over valve 废气(锅炉)调节阀(废气转换阀)

Exhaust gas cleaning system (EGCS) 废气滤清系统

Exhaust gas economizer 废气锅炉 船舶航行时,将主机的排气通入在主机烟道中特设的锅炉中,利用主机排气的热能,把锅炉中的水加热成饱和蒸汽,替代辅助锅炉,提高了燃料燃烧效率。

Exhaust gas economizer 炉水废气预热器 仅利用柴油机的废气产生蒸汽或热水并供给其他锅炉,不具备独立汽腔或热井的装置。 Exhaust gas economizer is the equipment which generates steam or hot water solely by exhaust gas from internal combustion engine and supplies to other boilers without any steam space or hot well.

Exhaust gas filtering system 废气滤清系统 将废气中的NO_x及SO_x等过滤去除的系统。

Exhaust gas heat exchanger 废气热交换器

Exhaust gas pipe 排气管

Exhaust gas recirculating technology (EGR) 排气再循环技术 为内燃机在燃烧后将排出气体的一部分离出并导入排气侧,使其再度燃烧的技术。其主要目的为减少排出气体中的氮氧化物与分担部分负荷时提高燃料消耗率。

Exhaust gas recirculating system (EGRS) 废气再循环系统 系指把发动机排出的部分废气回送到进气歧管,并与新鲜混合气一起再次进入气缸。由于废气中含有大量的CO_2等多原子气体,而CO_2等气体不能燃烧却由于其比热容高而吸收大量的热,使气缸中混合气的最高燃烧温度降低,从而减少了NO_x的生成量。

Exhaust gas temperature after turbine 涡轮后废气温度 废气离开涡轮时的温度。

Exhaust gas temperature before turbine 涡轮前排气温度 排气在涡轮喷嘴前的温度。

Exhaust gas turbine 废气轮机(废气透平)

Exhaust gas turbo supercharger 废气涡轮增压器

Exhaust gas turbocharger 废气涡轮增压器

Exhaust heat boiler 废热锅炉

Exhaust heat recovery unit 废热回收装置

Exhaust heating economizer 废气加热经济器 系指强制循环废气热交换器,其不能直接提供蒸汽,而通过燃油锅炉作为蒸汽储存器来提供蒸汽。

Exhaust lap (蒸汽机)排气余面

Exhaust line 排气管路

Exhaust manifold 排气歧管(排气阀箱)

Exhaust method 排空法 该方法是逐舱把在河口或近海水域中注入的压载水全部排出,直到把压载水排空为止,然后用深海海水加满,即用新鲜的深海水更换大量初始压载水。该方法的优点是压载水几乎完全有效地置换，泵和管系的工作负荷只需中等程度地增加；缺点是操作比较复杂，需要比较复杂的压载计划以避免超出船舶强度、稳性和纵倾的许可限制，同时需考虑动载荷的影响，包括未满舱的晃荡等(这对大部分船舶意味着只能在平静的海上进行压载水交换)。

Exhaust passage 排气管(排气道)

Exhaust pipe 排气管(排出管) 排出废气的管道。

Exhaust port 排气口 控制废气排出气缸的开口。

Exhaust pressure 排气压力(排出压力)

Exhaust silencer 乏气孔

Exhaust steam 排气(乏气)

Exhaust steam pipe line 排气管系 由主、辅汽轮机排到冷凝器或其他热交换器的排气管路及附件。

Exhaust stroke 排气冲程

Exhaust system 排气系统

Exhaust temperature 排气温度

Exhaust trunk 排气总管

Exhaust turbine 废气轮机(废气透平)

Exhaust turbine supercharger 废气涡轮增压器

Exhaust unit 排气装置 装于涡轮排气缸后，用以将废气排出并具有降温、消音等作用的装置。

Exhaust valve 排气门(排气阀) 控制废气排出气缸的阀门。

Exhaust ventilation 排气通风

Exhaust ventilator 排气通风筒(抽风机)

Exhaust whistle 汽笛

Exhaust-driven turbo-supercharger 废气透平增压器 实际上是一种空气压缩机，通过压缩空气来增加进气量。它是利用发动机排出的废气惯性冲力来推动涡轮室内的涡轮，涡轮又带动同轴的叶轮，叶轮压送由空气滤清器管道送来的空气，使之增压进入气缸。当发动机转速增快，废气排出速度与涡轮转速也同步增快，叶轮就压缩更多的空气进入气缸，空气的压力和密度增大可以燃烧更多的燃料，相应增加燃料量和调整发动机的转速，就可以增加发动机的输出功率。

Exhauster 排气器(抽风机)

Exhaust-gas charge over valve 废气锅炉烟气调节阀 调节进入锅炉的柴油机排气量的阀。

Exhaust-gas turbo-charger lubricating oil gravity tank 废气涡轮增压器的滑油重力柜 在废气涡轮增压器滑油系统中，布置在一定高度靠重力将滑油供给废气涡轮增压器轴承滑油的油柜。

Exhaust-gas turbo-charger lubricating oil sump tank 废气涡轮增压器的滑油循环柜 在废气涡轮增压器滑油系统中，用以汇集废气涡轮增压器润滑后的滑油供再循环使用的柜。

Exhausting (exhaustion) 排气

Existing bulk 现有散货船 一般系指按照有关的新规范生效前的规范批准图纸建造的散货船。 Existing bulk carrier is generally a bulk carrier which has been constructed in compliance with plan approved in accordance with the rules effective before the new relevant rules come into effect.

Existing container 现有集装箱 系指不属于新集装箱的集装箱。 Existing container means a container which is not a new one.

Existing engine 现有发动机 对"73/78 防污公约"附则Ⅵ的 NO_x 技术规则而言，系指受第13.7条约束的发动机。 For Annex Ⅵ of "MARPOL73/78" existing Engine is an engine which is subject to Regulation 13.7 of the Code.

Existing installation 现有设备 系指非新设备的任何设备。 Existing installation means any installation which is not a new one.

Existing lifeboat release and retrieval system 现有的救生艇释放和回收系统 系指未按经MSC.320(89)决议修正的LAS规则的第Ⅳ章4.4.7.6认可的救生艇释放和回收系统。 Existing lifeboat release and retrieval system is a lifeboat release and retrieval system that has not been approved in accordance with Paragraph 4.4.7.6 of Chapter Ⅳ of the LAS Code, as amended by resolution MSC.320(89).

Existing passenger ship 现有客船 系指非新客船的客船。 Existing passenger ship is a passenger ship which is not a new one.

Existing ship 现有船舶 系指：(1)并非新船的船舶；(2)包括1975年1月30日前开工建造的任何船舶。 Existing ship means (1) a ship that is not a new one; (2) any vessel, the construction of which was initiated before January 30, 1975.

Existing special trade passenger ship 现有特种业务客船 系指非新的特种业务客船业务的特种业务客船。 Existing special trade passenger ship means a special trade passenger ship which is not a new special one.

Existing standard 现行标准 系指自签订船舶建造合同或重大改装合同之日起适用于被改装船舶的标准。

Existing unit 现有平台 系指非新平台的平台。 Existing unit means a unit which is not a new one.

Existing vessel (fishing vessel) 现有船舶(渔船)

系指非新造的渔船。 Existing vessel is a fishing vessel which is not a new one.

Exit 出口(太平门,排气管)

Exit casing/exhaust casing 排气机匣 将燃气或其他气体自涡轮导出的通道。

Exit end 出口端

Exit pipe 排气管(排出管)

Exotherm temperature 散热温度

Exotherm value 散热值

Expanded area 伸张面积(展开面积) 螺旋桨各叶伸张轮廓内面积之和。

Expanded area ratio 伸张面比 螺旋桨伸张面积对盘面积的比值。

Expanded blade area 桨叶展开面积

Expanded outline 伸张轮廓 在螺旋桨正视图中将叶切面在母线以前和以后的部分弦长各标记于相应半径处的水平线上,连接各弦长端点所构成的轮廓。

Expanded perlite 膨胀珍珠岩 一种天然酸性玻璃质火山熔岩非金属矿产,包括珍珠岩、松脂岩和黑曜岩,三者只是结晶水含量不同。由于在 $1\,000 \sim 1\,300\ ℃$ 高温条件下其体积迅速膨胀 $4 \sim 30$ 倍,故称称为膨胀珍珠岩。一般要求膨胀倍数 $>7 \sim 10$ 倍(黑曜岩 >3 倍,可用),二氧化硅 70% 左右。均为露天开采。不用选矿,只需破碎,筛分即可。

Expander 扩张器(扩管器,辘)

Expanding 膨胀(扩张,胀开)

Expanding drill 扩孔钻

Expanding nozzle 扩张型喷嘴

Expanding test 扩口试验

Expanding valve 膨胀阀

Expansibility 膨胀性(延伸性)

Expansion 膨胀(扩大,展开)

Expansion arrangement 膨胀装置

Expansion crack 膨胀裂纹(膨胀开裂)

Expansion curve (示功图)膨胀曲线

Expansion factor 膨胀因数(膨胀系数)

Expansion joint 伸缩接头(膨胀接头) 系指:(1)用于吸收热膨胀的可伸缩的管路连接件;(2)为避免结构内产生过大的应力而设置的可以伸缩的接头。

Expansion link 伸缩杆

Expansion ratio 膨胀比 涡轮排气绝对压力(一般系指总压力)与进气压力的比值。

Expansion stroke 膨胀冲程

Expansion tank 膨胀箱(柜)

Expansion thermometer 膨胀温度计

Expansion valve 膨胀阀

Expendable bathy thermograph (XBT) 投弃式测温计 简称XBT。一种测量海水温度随深度变化的仪器。由感温探头、发射器和记录仪组成。

Expense 费用

Experience 经验 对质量管理体系而言,系指制造厂商应能证明工厂对于申请认可产品类型的复杂性和施工工艺已具备与此相符的经验,且工厂的产品一直符合高标准。 For the quality management system, experience means that the manufacturer is to demonstrate that the firm has experience consistent with technology and complexity of the product type for which approval is sought and that the firm's products have been of a consistently high standard.

Expert 专家(国际海事卫星组织) 系指职员以外的被指定的为国际海事卫星组织或以该组织名义完成某种特殊使命,其费用由国际海事卫星组织支付的人员。 Expert means a person other than a staff member appointed to carry out a specific task for or on behalf of INMARSAT and at its expense.

Expert system 专家系统 是一个人们用专业知识编辑的信息去解决问题的智能性知识基本系统。 Expert system is an intelligent knowledge-based system that is designed to solve a problem with information that has been compiled using some form of human expertise.

Explicit 清楚的/清晰的

Explosive 爆炸物

Exploitation of patent 专利实施

Exploration 勘探

Explosion 爆炸 系指:(1)燃烧无法控制而导致的爆燃事件;(2)引起爆炸能量的释放。 Explosion means:(1) a deflagration event of uncontrolled combustion;(2) a release of energy that causes a blast.

Explosion gas atmosphere 爆炸性气体环境 在环境条件下,气体、蒸发气或雾状的易燃物质与空气的混合物,在此混合物中,点燃以后,其燃烧扩大至所有未燃烧的混合物。 Explosion gas atmosphere means a mixture with air, under atmosphere conditions, of flammable substances in the form of gas, vapour or mist, in which, after ignition, combustion spreads throughout the unconsumed mixture.

Explosion pressure relief 爆炸压力释放 系指为防止容器或围蔽处所的爆炸压力超过该容器或处所的最大设计超压值而采取的将超压通过指定开口释放的措施。 Explosion pressure relief means measures provided to prevent the explosion pressure in a container or an enclosed space exceeding the maximum over-pressure the container or space is designed for, by releasing the over-pressure through designated openings.

Explosion protected enclosure 防爆外壳 其机械的完整性对于其合格性或接受其在危险区域中使用十分重要,且应仔细检查的外壳。 Explosion protected enclosure means an enclosure, the mechanical integrity of which is considered essential for, and is examined in detail for, its certification acceptance for use in a hazardous area.

Explosion protection 防爆

Explosion relief device 防爆释放装置

Explosion relief valve 防爆(安全)阀

Explosion-proof door 防爆门

Explosion-proof fan 防爆式风机 其结构在使用时不会引起周围环境中爆炸性混合物爆炸的风机。

Explosive limit 爆炸限值

Explosive vessel 爆炸品船舶 系指载运政府或商业爆炸品的船舶。如爆炸品符合美国的"危险货物条例"和"国际海运危险货物规则"第Ⅰ类第 1.1～1.5 分类的定义者则应视为爆炸品船舶。 Explosive vessel means a vessel carrying explosives, either under Government or commercial. A vessel defined in the Dangerous cargo Act of the United States and in the International Maritime Dangerous Goods Code, Class 1 Divisions 1.1 to 1.5 inclusive, shall be deemed to be an explosive vessel.

Explosive/flammability limits/range 易爆性/可燃性极限/范围 系指在给定的试验装置中,对燃料-氧化剂混合物施以一个足够强大的点火源,使其足以能形成燃烧的条件。 Explosive/flammability limits/range are the conditions defining the state of fuel-oxidant mixture, under which, when given an adequately strong external ignition source, the mixture is only just capable of producing flammability in a given test apparatus.

Explosive-protected enclosure 防爆外壳 系指设计和制成能承受在其内部可能出现的某种规定的气体或蒸发气的爆炸,还能防止由于外壳内部可能出现的火花、闪烁或所规定的气体或蒸发气的爆炸而点燃其周围所规定的气体或蒸发气的外壳。

Explosives 炸药 系指用于进攻、防御、训练或非作战行动,含有能产生爆炸、推进、燃烧或烟火效应的特别设计成分的所有武器、导弹或装备。 Explosives are all weapons, missiles or stores containing substances especially designed to produce an explosive, propulsive, incendiary or pyrotechnic effect for use in conjunction with offensive, defensive, training, or non-operational purposes.

Export 出口(输出)

Export commodity 出口商品

Export credit 出口信贷 是一种国际信贷方式,它是一国政府为支持和扩大本国大型设备等产品的出口,增强国际竞争力,对出口产品给予利息补贴、提供出口信用保险及信贷担保,鼓励本国的银行或非银行金融机构对本国的出口商或外国的进口商(或其银行)提供利率较低的贷款,以解决本国出口商资金周转的困难,或满足国外进口商对本国出口商支付货款需要的一种国际信贷方式。出口信贷名称的由来就是因为这种贷款由出口方提供,并且以推动出口为目的。出口信贷在国际贸易中出口信贷是垄断资本争夺市场、扩大出口的一种手段。

Export credit guarantee system 出口信贷国家担保制 是一国政府设立专门机构,对本国出口商和商业银行向国外进口商或银行提供的延期付款商业信用或银行信贷进行担保,当国外债务人不能按期付款时,由这个专门机构按承保金额给予补偿。这是国家用承担出口风险的方法,鼓励扩大商品出口和争夺海外市场的一种措施。

Export credit insurance 出口信用保险 是国家为了推动本国的出口贸易,保障出口企业的收汇安全而制定的一项由国家财政提供保险准备金的非营利性的政策性保险业务。出口信用保险诞生于19世纪末的欧洲,最早在英国和德国等地萌芽。

Export duty / export tax 出口税 系指对出口货物和物品征收的关税。征收出口税的目的主要是:(1)增加财政收入;(2)限制重要的原材料大量输出,保证国内供应;(3)提高以使用该国原材料为主的国外加工品的生产成本,削弱其竞争能力;(4)反对跨国公司在发展中国家低价收购初级产品。例如,1975 年 1 月,几内亚对铝矾土及其副产品征收特别出口税。

Export line 输出线 系指在平台或平台与岸上设施之间输送已加工的油和天然气的管道。 Export line is a pipeline that transports processed oil and gas between platforms or between a platform and a shore facility.

Export permit 出口许可证 在国际贸易中,根据一国出口商品管制的法令规定,由有关当局签发的准许出口的证件。出口许可证制是一国对外出口货物实行管制的一项措施。一般而言,某些国家对国内生产所需的原料,半制成品以及国内供不应求的一些紧俏物资和商品实行出口许可证制。通过签许可证进行控制,限制出口或禁止出口,以满足国内市场和消费者的需要,保护民族经济。此外,某些不能复制,再生的古董文物也是各国保护对象,严禁出口;根据国际通行准则,鸦片等毒品或各种淫秽品也禁止出口。

Export price index 出口商品价格指数

Export processing zone 出口加工区 是国家划定或开辟的专门制造、加工、装配出口商品的特殊工业区,经济特区的形式之一，常享受减免各种地方征税的优惠。出口加工区一般选在经济相对发达、交通运输和对外贸易方便、劳动力资源充足、城市发展基础较好的地区，多设于沿海港口或国家边境附近。世界上第一个出口加工区为1956年建于爱尔兰的香农国际机场。中国台湾高雄在20世纪60年代建立出口加工区。以后，一些国家也效法设置。中国在20世纪80年代实行改革开放政策后，沿海一些城市开始兴建出口加工区。全世界约有40个国家建立170多个出口加工区，其中马来西亚22个，菲律宾16个，印尼9个，中国台湾地区有高雄、楠梓、台中等3个，绝大部分都取得显著效果。

Export quota 出口配额 又称自愿出口限制是进口国的政府或一个工业同出口国的政府或相竞争的工业安排的、对后者所出口的一种或一种以上产品的数量加以限制的措施。

Export subsidy 出口补贴 又称出口津贴，是一国政府为了降低出口商品的价格，增加其在国际市场的竞争力，在出口某商品时给予出口商的现金补贴或财政上的优惠待遇。

Export tax 出口税 系指出口商品国家的海关对本国产品输往国外时，对出口商品征收的关税。鉴于出口税的征收将提高本国商品在国外市场的销售价格，从而削弱本国商品的竞争力，因而，大多数国家都不再征收。只有一些发展中国家出于增加财政收入，或保证本国的生产或市场供应的考虑，继续对其全部或部分出口商品征收出口税。例如，瑞典、挪威对于木材出口征税，以保护其纸浆和制纸工业。

Exposed location single buoy mooring system (ELSBM) 旷海单浮筒系泊系统 系指其浮筒上大下小，且制成内外壳，浮筒上部设有转盘，转盘上部有动力设备、系泊绞车和输油软管绞车。转盘顶部为直升机平台，浮筒底部有固定压载和压载舱以降低重心，使浮筒保持稳定，浮筒用多点锚泊固定的系统。

Exposed spring type safety valve 弹簧外露式安全阀

Exposed zones 曝露区（露天区域） 系指与船舷距离等于或小于0.04 B处设置的上层建筑或甲板室的边界。 Exposed zones are the boundaries of superstructures or deckhouses set in from the ship's side at a distance equal to or less than 0.04 B.

Exposure test to marine environment for ship and marine 船舶涂料海洋环境曝露试验 系指包括下列内容的试验：(1)沿海大气曝晒；(2)海面大气曝晒；(3)平台海水飞溅；(4)动态海水飞溅；(5)海水潮差；(6)海水全浸；(7)海泥；(8)海滨舱储；(9)海洋长期试验等。 Exposure test to marine environment for ship and marine mean those tests include：(1) exposure to inshore weathering；(2) exposure to marine weathering；(3) seawater splashing on platform；(4) dynamic seawater splashing；(5) seawater tidal range；(6) full seawater immersion；(7) sea mud；(8) seashore-storage；(9) marine test for long-scale sample；etc..

Express airmail 航空特快专递
Express order 汇票
Extend 延伸(扩展)
Extended alarm 延伸报警
Extended alarming function 延伸报警功能
Extended hazardous areas 扩大危险区 对载运闪点低于23℃有包装的易燃液体和易燃气体船舶和只载运散装固体危险货物及MHB的船舶而言，扩大危险区系指在正常工作状态下不大可能出现爆炸性环境，即使出现也仅仅是偶然的和短时间的区域或处所。例如：(1)以自闭式气密门（水密门可以认为是气密门）与①闭式装货处所和闭式滚装装货处所；②危险区域的通风管道（如设有）分隔，并有自然通风的处所；(2)气阻装置（如有）连接到①闭式装货处所和闭式滚装装货处所；②危险区域的通风管道（如设有）；③离危险区任何排风口周围1.5 m范围内的露天甲板区域，或者露天甲板上的半围蔽处所；④有开口直接通向以上①和②所列处所的围蔽或半围蔽处所，并无防止易燃气体进入该处所的适当措施相毗邻的气闸（如设有）的里面；(3)在离危险区任何排风口周围1.5 m范围内的露天甲板区域，或者露天甲板上的半围蔽处所之外1.5 m范围的露天或半围蔽处所。 For the ship carrying inflammable liquids with a flashpoint below 23 ℃ in packaged form and the ship carrying only dangerous solid goods in bulk and MHB, extended hazardous areas mean those zones or spaces in which an explosive atmosphere is likely to occur in normal operation, if it does, is likely to occur only infrequently and will exist for a short period only. Such as：(1) spaces which may be separated by self-closed gas-tight doors (watertight doors can be regarded as the gas-tight) doors from ① enclosed cargo spaces or enclosed RO-RO cargo spaces；② ventilation ducts (if any) for hazardous areas and have natural ventilation；(2) space inside the air-lock (if any) adjoining to ①enclosed cargo spaces or enclosed RO-RO cargo spaces；② ventilation ducts (if any) for hazardous areas；③weather deck areas, or semi-enclosed space so on weather deck round 1.5 m from any draft outlet of hazardous areas；④enclosed or semi-enclosed spaces with direct openings to ①or ②mentioned above, without appropriate measures to prevent inflammable dust from entering；(3) weather deck areas, or semi-enclosed

spaces on weather deck 1.5 m outside weather deck areas, or semi-enclosed space so on weather deck round 1.5 m from any draft outlet of hazardous areas.

Extended protected waters service　扩大的保护水域航区　系指在保护水域并包括在"适宜气候"条件下超出保护水域某一短距离（通常少于 15 n mile）的航区，例如："扩大的拉各斯港保护水域航区"。 Extended protected waters service mean the service in protected waters and also for short distances (generally less than 15 nautical miles) beyond protected waters in 'reasonable weather', e.g. "Extended Protected Waters Service from the Port of Lagos".

Extended tension leg platform (ETLP)　延伸式张力腿平台

Extension clause　扩展条款　出口货物到香港（包括九龙）或是澳门存仓火险责任扩展条款。这是中保财产保险公司所开办的一种特别附加险。它对于被保险货物自内地出口运抵香港（包括九龙）或澳门，卸离运输工具，直接存放于保险单载明的过户银行所指定的仓库期间发生火灾所受的损失，承担赔偿责任。该附加险是一种保障过户银行权益的险种。因为，货物通过银行办理押汇，在货主未向银行归还贷款前，货物的权益属于银行，所以，在该保险单上必须注明过户给放款银行。相应地，货物在此期间到达目的港的，收货人无法提货，必须存入过户银行指定的仓库。从而，保险单附加该险条款的，保险人承担火灾责任。该附加险的保险期限，自被保险货物运入过户银行指定的仓库之时起，至过户银行解除货物权益之时，或者运输责任终止时起满 30 天时止。若被保险人在保险期限届满前向保险人书面申请延期的，在加缴所需保险费后可以继续延长。

Extension shaft　中介轴　连接动力涡轮输出轴与减速齿轮箱输入轴的中间传动轴。

Extensive corrosion　普遍腐蚀　系指参考区域中含有超过 70% 或更大范围的硬质和/或松脱的锈块腐蚀状态，包括点腐蚀，且伴随厚度减薄的证据。

Exterior passageway　外通道　船上敞露的通道。

External audit　外部审核　系指由公司以外的审核机构进行的审核。

External characteristic curve　外特性曲线　表示内燃机在各种转速下的最大功率和扭矩及其他主要参数的曲线。

External combustion gas turbine plant　外燃式燃气轮机装置　工作介质不在燃烧室中与燃料掺混，而是通过热交换器加热后进入涡轮做功的燃气轮机装置。

External criteria for the assessment of flag State performance　船旗国履约评估外部标准　系指参考那些能够被作为船旗国评估指示器的信息，特别是港口国控制数据及海难事故数据。下列标准为船旗国履约指示器，但这些指示器与船旗国主管机关的组织没有直接的关系。如将下列标准（1）至（5）作为指示器，应考虑船旗国的船舶总数及其为履约国的国际文件：(1) 按国际海难数据库要求上报的事故、海难和事件数目；(2) 导致船上人员受伤，3 天不能上班的事故数目；(3) 船舶操作过程中，船上人员死亡数目；(4) 丢失的船舶数目；(5) 按 MARPOL 73/78 报告标准（包括对严重事故的衡量标准），对海洋产生污染的事故数目；(6) 船舶被港口国按港口国控制程序滞留的数目；(7) 向 IMO 提交强制性文件所要求的资料。　External criteria for the assessment of flag State performance refers to information, in particular port State control data and casualty accident data, which may also be taken as indicators of the way in which a flag State is performing. The following are indicators of the way in which the flag State is performing but do not relate directly to the organization of the flag State's Administration. When the criteria listed in (1) to (5) are used as indicators, the overall number of ships flying its flag must be considered in proportion, subject to international documents to which the State is a Party: (1) number of accidents, casualties and incidents reported to IMO in terms of the requirements of the international casualty database; (2) number of accidents involving personal injuries leading to absence from duty of 3 days or more on board ships flying the flag of the State concerned; (3) Number of lives lost on its ships resulting from the operation of ships flying its flag; (4) Number of ships lost; (5) Number of incidents of pollution to the sea according to MARPOL 73/78 reporting standards including a measure of the seriousness of the incidents; (6) number of ships detained by other States under port State control procedures; (7) information of communication to IMO required in mandatory documents.

External current cathodic protection　外加电流阴极保护　系指由外部电源提供保护电流的阴极保护。　External current cathodic protection means the cathodic protection by which protective current is supplied by external electrical source.

External diameter　外径

External dimensions　外形尺寸

External environment noise　外环境噪声　系指所有建筑物外部的噪声。不管离建筑物的距离有多远，包括陆地和海洋、河川的上空，统称外环境噪声。如建筑物附近马路上的汽车声及其喇叭声、铁路车辆的隆隆声及汽笛声、港口上船舶的汽笛声、空中飞机的轰鸣声、人们户外走路的脚步声及讲话声,建筑工地上的施工声、水泥搅拌机的机械声,挖土机的敲击声,鞭炮声等都属于外环境噪声。

External examination 外部检验

External gear 外齿轮 外圈带齿的齿轮。

External opening 外部开口 在构成舱室的外板、露天甲板或舱壁上的开口。

External processing and assembling business 对外加工装配业务 是一种委托加工的方式,外商将原材料等运交我方,并未发生所有权的转移,我方企业只是作为受托人按照外商的要求,将原材料加工成为成品。在加工过程中,我方付出了劳务,获取的加工费是劳动报酬,因此,可说对外加工装配属于劳务贸易的一种形式,是以商品为载体的劳务出口。

External safety operations 外部安全操作 在紧急情况下协助其他人员。

External safety related messages 外部安全相关信息 系指通过SOLAS第Ⅴ章所列设备和/或 NAVTEX 从船外接收到的有关安全航行的数据。External safety related messages means the data received from outside of the ship concerning the safety of navigation, through equipment listed in SOLAS Chapter Ⅴ and/or NAVTEX.

External structure with respect to design temperature 与设计温度有关的外部结构 定义为加强筋距壳板、露天甲板、露天舷侧和上层建筑/甲板室的露天两侧/两端 0.5 m 的板结构。External structure with respect to design temperature is defined as the plating with stiffening to distance of 0.5 m from the shell plating, exposed decks and exposed sides and ends of superstructure and deckhouses.

External turret 外转塔

External ventilating type motor 外通风型电机 系指其风冷空气是由其外部的单独风机和鼓风机供给的电机。

Extinguishant 灭火剂

Extinguisher 灭火器(熄灭器)

Extra ballast operation for heavy equipment carrier 大件运输船超压载作业 系指:(1)船舶为了在港内装/卸成套大件和项目设备进行压载作业,使船舶吃水达到不超过作业吃水标志的作业状态;(2)船舶在达到(1)压载状态下进行限定条件下的港内移船作业,如过桥等。Extra ballast operation for heavy equipment carrier means: (1) an operational condition for a ship with the draft not exceeding operational draft mark during the ballasting operation in order to load/unload the complete sets of equipment and project facilities at port; (2) ship shifting at port with restrictions such as passing through bridge, etc., under the ballast condition of (1).

Extra feed 冷凝器真空补给水

Extra, reasonable and actual expenses 额外的、合理的实际开支

Extracting pump 抽吸泵

Extracting regulating valve 撤气调节阀

Extraction 抽吸

Extraction display 抽取显示器

Extraction opening 抽气口

Extraction pump 抽水泵

Extraction test 抽取试验(抽提试验)

Extraction type 抽吸式

Extraction valve 抽气阀

Extractor 分离器(抽出器、脱模器)

Extra-fine glass wool 超细玻璃棉 超细玻璃棉绝热材料是以石英砂、长石、硅酸钠、硼酸等为主要原料。经过高温熔化制得小于 2 μm 的纤维棉状,再添加热固型树脂黏合剂加压高温定型制造出各种形状、规格的板、毡、管材制品。其表面还可以粘贴铝箔或 PVC 薄膜。该产品具有容重轻、导热系数小、吸收系数大、阻燃性能好。可广泛用于热力设备、空调恒温、冷热管道、烘箱烘房、冷藏保鲜及建筑物的保温、隔热、隔音等方面。

Extraneous 外附的(外来的,无关的)

Extraneous risks 外来风险 系指由于自然灾害和意外事故以外的其他外来原因造成的风险,但不包括货物的自然损耗和本质缺陷。依风险的性质可分为一般外来风险和特殊外来风险两类:(1)一般外来风险系指由于一般外来原因所造成的风险,主要包括:偷窃、渗漏、短量、碰损、钩损、生锈、雨淋、受热受潮等;(2)特殊外来风险系指由于军事、政治、国家政策法令和行政措施等以及其他特殊外来原因,如战争、罢工、交货不到、被拒绝进口或没收等。

Extraordinary 非常的(格外的)

Extrapolation 外推(外插)

Extrapolation method 外插法

Extreme action 极端载荷 系指在其设计服务寿命期间施加到结构物上的最大载荷。Extreme action is the maximum action applied to a structure during its service life design.

Extreme breadth 最大宽度 系指船舶包括外板和永久性固定突出物如护舷材、水翼等在内,垂直于中线面的最大水平距离。

extreme ice condition 最严重冰况

Extreme length 最大长度 系指船舶最前端至最后端之间包括外板和两端永久性固定突出物如顶推装

置等在内的水平距离。

Extreme oil outflow parameter 极端泄油参数 极端泄油在所有情况下的泄油量以由小到大顺序排列后计算,即 0.9~1.0 之间累积概率的泄油量与其各自概率的乘积之和。得到的值再乘以 10。极端泄油参数以货油舱注油至 98% 舱容时全部货油量的分数来表示。Extreme oil outflow parameter is calculated as follows: after the volumes of all outflow cases have been arranged in ascending order the sum of the outflow volumes with cumulative probability between 0.9 ~ 1.0, is multiplied by their respective probability. The value so derived is multiplied by 10. The extreme oil outflow parameter is expressed as a fraction of the total cargo oil capacity at 98% tank filling.

Extreme value 极值 系指一个的海洋气象学变量的估算值,其随着规定的重现期而变化。其他定义也有可能性。 Extreme value is an estimate of the value of a metocean variable with a stated return period. Other definitions are also possible.

Extreme wave period range T_{xrange} 极限波浪周期范围 T_{xrange} 取为极限设计波浪周期加减 1.5 倍标准偏差: $T_{xrange} = T_{xw} - 1.5\ T_{dsd} \sim T_{xw} + 1.5\ T_{dsd}(s)$。 Extreme wave period range (T_{xrange}) is to be taken as the extreme design wave period plus and minus 1.5 standard deviations: T_{xrange} is $T_{xw} - 1.5\ T_{dsd}$ to $T_{xw} + 1.5\ T_{dsd}(s)$.

Extremely improbable (Probability of occurrence) 极不可能的概率 对事故的概率而言,系指故障情况不大可能发生在一种类型的所有船舶的整个使用期限中(最坏 10^{-9})。 For the probability of occurrences, extremely improbable means the failure conditions would be unlikely to arise in the entire operational life of all craft of one type (at worst 10^{-9}).

Extremely remote (probability of occurrences) 极少可能的概率 对事故的概率而言,系指故障情况在考虑到一种类型的所有船舶的整个所有期限时不大可能发生,但仍然认为是可能的(最坏 10^{-7})。 For the probability of occurrences, extremely remote means the failure conditions are unlikely to occur when considering the total operational life of all craft of one type, but neverthe less have to be considered as being possible (at worst 10^{-7}).

Extruded polystyrene foam sheet 挤塑式聚苯乙烯泡沫塑料板

Eye brow (wriggle) 眉毛板 在门窗上方用以导流壁上雨水的零件。

Eye guard 护目罩

Eye plate 眼板(系缆钮,吊环板) 焊固于甲板、舷墙或其他构件上,供连接钩、环、卸扣、滑车等用的带眼孔的金属件。见图 E-11。

图 E-11 眼板(系缆钮,吊环板)
Figure E-11 Eye plate

Eye screw shackle 环眼螺栓卸扣
Eye shield 护目镜
Probability of dangerous failure (PFD_{avg}) (平均要求)危险失效概率

F

Fclass divisions F 级分隔 系指由舱壁、甲板、天花板或衬板所组成的分隔,并应符合下列要求:(1)它们的构造应在最初半小时的标准耐火试验至结束时,能防止火焰通过;(2)它们应具有这样的隔热值:使在最初半小时的标准耐火试验至结束时,其背火的一面的平均温度,较原温度升高不超过 139 ℃,其在包括任何接头在内的任何一点的温度较原温度升高不超过 225 ℃。主管机关可要求将原型分隔进行一次试验,以保证满足上述完整性和温升的要求。 F class divisions are those divisions formed by bulkheads, decks, ceilings or linings which comply with the following: (1) they shall be so constructed as to be capable of preventing the passage of flame to the end of the first one half hour of the standard fire test; (2) they shall have an insulation value such that the average temperature of the unexposed side will not rise more than 139 ℃ compared with the original temperature, nor will the temperature at any one point, including any point, rise more than 225 ℃ compared with the original temperature, up to the end of the first one half hour of the standard fire test. The Administration may require a test of a prototype division to ensure that it meets the above requirement for integrity and temperature rise.

Fstandard ship's station　F 标准船站（卫星通信）　系指能提供高性能的船对岸和岸对船高速通信业务的通信系统。

F.P　闪点　系指石油闪点，并必须采用开杯试验法或任何其他接近同等准确度的闭杯试验法来确定。　F.P means point for petroleum and must be ascertained by open cup test or any other closed cup test of an equal degree of accuracy.

Fabrication standard　建造标准　系指包括确立以下规定项目的范围和公差限制内容的标准：（1）切边——切边的斜度和切边粗糙度；（2）折边纵骨和肘板以及组合型材——翼板宽度和腹板高度、翼板与腹板之间的角度，以及翼板平面或面板顶部的平直度；（3）支柱——各层甲板之间平直度和圆柱结构直径；（4）肘板和加强筋——防倾肘板和加强筋自由边缘线的线型；（5）骨材组件——次要面板和骨材的端部细节；（6）板材组件——对平面和曲面分段的尺寸（长和宽）、分段扭曲度和方正度，以及内部构件与板材面的偏差；（7）立体分段装配——对平面和曲面立体分段，除板材组件标准以外的上下板材之间扭曲偏差；（8）特殊装配——上下舵钮间距、螺旋桨毂后缘和艉尖舱壁间距、艉框架扭曲、舵与舵轴中心线偏差、舵板扭曲、主机座顶板平面度、宽度和长度。螺旋桨毂的镗孔和艉框架。艉鳍或舵托，以及舵、舵销和承舵轴的装配和对中，应在船舶尾部的主要部分焊接完成之后进行。最终装配之前，应检验舵拴、舵杆和舵轴之间的圆锥接触面；（9）板对接——板材对接对齐；（10）十字形接头——在十字形接头的中线和底线测量对齐情况；（11）内部构件对齐——T形纵骨翼板的对齐、板格扶强材的对齐、T形接头和搭接接头的间隙，以及分段组装和合拢接头处连续骨材上切口与扇形孔的间距；（12）龙骨和船底观察——船舶全长以及相邻两舱壁间距的偏差、船舶前体和后体翘起以及船中肋板升高；（13）尺寸——垂线间长、船中型宽和型深以及螺旋桨毂后缘和主机的距离；（14）骨材面板的光顺度——外板肋骨、液舱顶、舱壁、上甲板、上层建筑甲板、甲板室甲板和舷墙板的骨材之间的平整度；（15）骨材处板的光顺度——在肋骨处量取的外板、液舱顶、舱壁、强力甲板板和其他结构的平整度。Fabrication standard is a standard included information, to establish the range and the tolerance limits, for the items specified as follows: (1) cutting edge—the slope of the cut edge and the roughness of the cut edges; (2) flanged longitudinals and brackets and built-up sections—the breadth of flange and depth of web, angle between flange and web, and straightness in plane of flange or at the top of face plate; (3) pillars—the straightness between decks, and cylindrical structure diameter; (4) brackets and small stiffeners—the distortion at the free edge line of tripping brackets and small stiffeners; (5) sub-assembly stiffeners—details of snipe end of secondary face plates and stiffeners; (6) plate assembly—for flat and curved blocks the dimensions (length and breadth), distortion and squareness, and the deviation of interior members from the plate; (7) cubic assembly—in addition to the criteria for plate assembly, twisting deviation between upper and lower plates, for flat and curved cubic blocks; (8) special assembly—the distance between upper and lower gudgeons, distance between aft edge of propeller boss and aft peak bulkhead, twist of stern frame assembly, deviation of rudder from shaft centerline, twist of rudder plate, and flatness, breadth and length of top plate of main engine bed. The final boring out of the propeller boss and stern frame, skeg or solepiece, and the fit-up and alignment of the rudder, pintles and axles, are to be carried out after completing the major part of the welding of the aft part of the ship. The contacts between the conical surfaces of pintles, rudder stocks and rudder axles are to be checked before the final mounting; (9) butt joints in plating—alignment of butt joint in plating; (10) cruciform joints—alignment measured on the median line and measured on the heel line of cruciform joints; (11) alignment of interior members—alignments of flange of T longitudinals, alignment of panel stiffeners, gaps in T joints and lap joints, and distance between scallop and cut outs for continuous stiffeners in assembly and in erection joints; (12) keel and bottom sighting—deflections for whole length of the ship, and for the distance between two adjacent bulkheads, cocking-up of fore body and of aft body, and rise of floor amidships; (13) dimensions—dimensions of length between perpendiculars, moulded breadth and depth at midship, and length between aft edge of propeller boss and main engine; (14) fairness of plating between frames—deflections between frames of shell, tank top, bulkhead, upper deck, superstructure deck, deck house deck and wall plating; (15) fairness of plating in way of frames—deflections of shell, tank top, bulkhead, strength deck plating and other structures members.

Face　面（叶面，压力面）

Face cavitation　面空泡　螺旋桨叶面产生的空泡。一般当叶切面攻角为较大负值时发生在导边的附近。

Face of blade 叶面 推船前进时，螺旋桨叶产生压力的，通常是向船后方的一面。

Face piece 面罩 系指设计成通过适当方式使之固定就位并把眼睛、鼻和嘴全部罩住的面套。 Face piece means a face covering that is designed to form a complete seal around the eyes, nose and mouth which is secured in position by a suitable means.

Face pitch 面螺距 螺旋桨叶切面展开后其面部为直线或大部分为直线者，其面部螺旋线的螺距。普通形式螺旋叶的面螺距简称螺距。

Face pitch line 面节线 表示螺旋桨某一半径处叶面方向的线，即将其处螺旋线伸开的直线。月牙切面和对称流线型切面的面节线是平行于头尾线与叶面相切的线。

Face plate 面板 系指一段加强构件经由腹板附于板材的部分，通常与板材表面平行。Face plate means a section of a stiffening member attached to the plate via a web and is usually parallel to the plated surface.

Face plate of foundation 基座面板 基座顶部的平板。

Face plate width b_f (mm) 面板宽度 b_f 系指普通扶强材或主要支撑构件的面板宽度（mm）。Face plate width b_f means a face plate width (mm), of ordinary stiffener or primary supporting member, as the case may be.

Facilitate 帮助

Facilities 设备 对质量管理体系而言，系指要求制造厂商具备足够的符合设计、发展和生产水平运行的设备和设施。 For the quality management system, facilities mean that the manufacturer is required to have adequate equipment and facilities for those operations appropriate to the level of design, development and manufacture being undertaken.

Factor 要素 系指产品技术和特征的总称。对船舶产品而言，包括船型、主尺度、载重量、舱室布置、主机型号、航速及产品质量、制造成本等。

Factor of subdivision 分舱因数 由船长和船舶的业务性质等而定的，用以决定许可舱长的因数。其值不大于1.0。

Factor of variable cross section 吊杆变断面系数 两端呈锥体或阶梯状断面变化的吊杆，按吊杆中段等断面计算其临界轴向压力时所采用的系数。

Factor verification 要素验证 是对产品目标规格的模拟实验，以验证将要开发的产品各项指标能否达到目标要求，或要进行如何修正。

Factorial load 因素载荷 对起重设备而言，系指设计起重设备时应考虑的载荷，此载荷可用下式表示：因素载荷 = 起升载荷 × 作业系数 × 动载系数。

Fail safe 故障安全 是一种设计特性，在该项设计中规定的故障模式是以与船舶安全原则为主。 Fail safe is a design property of an item in which the specified failure mode is predominantly in a safe direction with regard to the safety of the ship, as a primary concern.

Fail-safe principle 故障安全原则 系指当一个元件或一个系统出现故障或误动作时，该元件或系统的输出能自动处于预定的安全状态而不使故障扩大。 Fail-safe principle means that upon failure or malfunction of a component or system, the output automatically reverts to a predetermined design state of least critical consequence.

Failure 故障 系指船舶的一个或几个部件失效或工作不正常的事故，例如失控。故障包括：(1) 个别故障；(2) 与一个系统有关的独立故障；(3) 涉及一个以上系统的独立故障，包括：①任何已出现但未被探测到的故障；②有理由预计将跟随正在处理中的故障之后，还会发生的进一步的故障；(4) 共同原因的故障（由相同原因引起一个以上部件或系统的失效）；(5) 设备或系统产生下列一种或两种后果的事件：①设备或系统丧失功能；②执行功能的能力恶化到使船舶、人员或环境的安全度明显降低；(6) 系统或系统的一部分失效或失灵。

Failure is an occurrence in which a part, or parts of the craft fail or malfunction, e. g. out of control. A failure includes: (1) a single failure; (2) independent failures related to a system; (3) independent failures involving more than one system, taking into account: ①any undetected failure that is already present; ②such further failures as would be reasonably expected to follow the failure under consideration; (4) common cause failure (failure of more than one component or system due to the same cause); (5) an event to a component or a system causing one or both of the following effects: ①function loss of a component or a system; ②deterioration of functional capability to such an extent that the safety of the ship, personnel or environment is significantly reduced; (6) a loss of function or a malfunction of a system or part of a system.

Failure analysis 故障分析 (1) 对一个单项的合乎逻辑、系统的检查，包括其图表或公式，以确定和分析潜在及实际故障的概率、原因和后果；(2) 旨在证明系统具备故障-安全功能。针对所安装部件对故障的影响及其结果的评估。 Failure analysis means: (1) the logical, systematic examination of an item, including its diagrams or formulas, to identify and analyze the probability, causes and consequences of potential and real failures; (2) an analysis which aims to demonstrate that the system has a fail-to-safe functionality. The failure effects and their consequences are assessed for the installed components.

Failure condition 故障情况　系指由一个或多个故障引起的对船舶和船上人员造成影响的情况,并计及不利的营运和环境条件。故障情况按其影响的严重性分级。　Failure condition is a condition with an effect on the craft and its occupants caused by one or more failures, taking into account relevant adverse operational or environmental conditions. Failure condition is classified according to the severity of its effects.

Failure criterion 故障标准　系指在成功和故障的极限状态下所考虑的阈值。　Failure criterion means the threshold value considered in a limit state that separates success and failure.

Failure effect 故障影响(故障后果)　系指船舶、系统或物件层面的故障情况的后果。故障影响分为以下级别:(1)轻微影响;(2)重大影响;(3)危险影响;(4)灾难性影响;(5)危险。　Failure effect is the consequence of a failure condition at craft, system or item level. Failure effects are categorized as follows: (1) minor effect; (2) major effect; (3) catastrophic effect; (4) hazardous effect; (5) hazard.

Failure mode 失效模式　失效发生的观察模式或方法。

Failure mode and effect analysis(FMEA) 疲劳模式和结果分析　(1)是一种疲劳分析的方法学。该方法用在设计中假定每一个疲劳模式和结果。(2)对船舶及其构成系统的故障特征进行实用而现实的书面评估。其主要目的是提供一个综合的、系统的和书面的调查。该调查建立了对船舶推进、操纵和发电系统,以及任何其他船东要求系统的重要故障条件,并且评估它们对于船舶和船上人员安全的重要性。　Failure mode and effect analysis(FMEA) is: (1) a failure analysis methodology used during design to postulate every failure mode and the corresponding effect or consequences. (2) a practical, realistic and documented assessment of the failure characteristics of the ship and its component system. The primary objective of failure modes and effect analysts(FMEA) is to provide a comprehensive, systematic and documented investigation which establishes the important failure conditions of the ship propulsion, steering and power generation systems, as well as any other system required by the Owner, and assesses their significance with regard to the safety or the ship and its occupants.

Failure modes effect and criticality analysis(FMECA) 故障模式和影响概要　系指提供设备潜在的故障模式的系统检查。其旨在确认原因,分析对系统运行的影响,量化发生概率(故障率)并确认纠正措施,即修改设计。　FMECA provides a systematic examination of potential failure modes of an equipment. It seeks to identify causes, analyze effects on system operation, quantify occurrence probabilities(failure rates), and identify corrective actions, i.e. design modifications.

Failure severity 故障严重性　分为:(1)灾难性——全部功能丧失、爆炸、丧生⇒必须更改设计;(2)重大——部分功能丧失或恶化⇒应研究更改设计的可能;(3)轻微——对功能的影响很小⇒可不要求更改设计。　Failure severity includes: (1) catastrophic—loss of complete function, explosion, loss of lives(Design change is to be compulsory); (2) major—loss or deterioration of part of function(Possible design change is to be investigated); (3) minor—negligible affect on function(Design change may be not required).

Failure to delivery risk 交货不到险　系指货物装船后6个月之内未运到目的地交货,无论何种原因,保险公司按全损赔付。交货不到险承保的损失主要是由于政治上的因素而导致货主无法按时收到货物,例如由于中途某国对货物扣押而导致交货不到的损失。卖方应承担将货物运至指定的目的地的一切风险和费用,不包括在需要办理海关手续时在目的地国进口应交纳的任何"税费"(包括办理海关手续的责任和风险,以及交纳手续费、关税、税款和其他费用)。

Failures 故障　系指预料的或要求可能会出现的不能确保运行的配件或装置的损失,还有性能不良和脱离正常的运转状态。　Failures are deviations from normal operating conditions such as loss or malfunction of a component or system such that it cannot perform an intended or required function.

FAIR coating condition 尚好的涂层状况　系指在扶强材边缘和焊缝的连接处涂层有局部脱落和/或所检验的区域中有超过20%或更大范围的轻度锈蚀,但小于定义"差"的程度的状态。　FAIR coating condition is a condition with local breakdown at edges of stiffeners and weld connections and/or light rusting over 20% or more of areas under consideration, but less than as defined for POOR condition.

Fair trade 互惠贸易

Fairing(nose cone) 整流罩　与进气机匣一起组成环形通道的轴对称静止体。

Fairlead 导缆器　供引导缆索通过并变换方向或限制其导出位置而使缆索免受擦损的器具。主要有导缆钳、滚轮导缆器、滚柱导缆器、导向滚轮、导向滚柱、导缆孔以及转动导缆孔。

Fairlead with horizontal roller 滚柱导缆器 设置在船舷,由直立和水平滚柱组成的可导引来自任何方向缆绳的导缆器。有四滚柱、六滚柱等类型。

Fairly 公平地

Fall pipe vessel 落管船 又称落管抛石船。是海上油气及风电工程等基础设施领域新崛起的一种作业船舶。通常其具有多种作业功能,通过船上专门设置的卸料装置及定位系统向海底投放石料,以期保护海底埋设的各类管缆和电源线设备不受到损坏。部分落管船还能实施海底挖沟、电缆管线布设以及回填作业。见图F-1。

图 F-1　落管船
Figure F-1　Fall pipe vessel

Falling film evaporator 降膜蒸发器 由一束竖立的管组成加热管组,水在管内向下流动形成薄膜蒸发的蒸发器。

False accepted rate(FAR) 误识率 系指将其他人员误识为预识别人员的概率。

False bottom 假底 安装在船模试验水池中,可以上下移动,用以调节水深的活动池底。

False rejection rate(FRR) 拒识率 系指将预识别人员误识为其他人员的概率。

Fan 风扇 出口处全压低于10 mm水柱的无机壳的风机。

Fan chamber 鼓风机室

Fan characteristic resister 风扇特性电阻器 系指一种用于电动机的电枢或转子电路内使其电流与电动机转速近似成正比的电阻器。

Fan dial 扇面形刻度盘

Fan engine 通风机(打风机)

Fan flat 鼓风机平台

Fan noise 风机噪声

Fan propulsion gas turbine 风扇推进燃气轮机 在气垫船中驱动空气螺旋桨以获得推进动力的燃气轮机。

Fan room 通风机舱(通风机室) 供安置通风机及其控制设备的舱室。

Fan trunk 空气动力管(压力通风总管)

Fan ventilator 叶片式通风机

Fan-shaped blade 扇形叶 外形大致如折扇的螺旋桨叶。

Fast attack catamaran(FAC) 双体近岸快速攻击艇

Fast attack craft(FAC) 快速攻击艇 是现代海军近岸防御的重要舰型,在海岸防御、对台作战、专署经济区巡逻和海上反恐作战中,都可能发挥非常重要的作用。

Fast automatic shuttle transfer system 快速自动往复传送系统 在补给船与接收船之间进行快速自动往复传送物资的系统。

Fast craft 快艇 是小型高速船的总称。主要用途为执勤、水上救援、娱乐、体育等。快艇也根据其用途分为巡逻艇、救生艇、游艇、赛艇等。见图F-2。

图 F-2　快艇
Figure F-2　Fast craft

Fast infrared transfer 高速红外线 系指传道输送速度在每一秒1或是4 mbits的红外线。

Fasten 固定(上紧,握紧)

Fastener 紧固件(接合件,扣闩)

Fastening 紧固(紧固件)

Fastening device 紧固装置 系指通过防止围绕铰链的转动来保持门关闭的装置。

Fatal accident rate 致命事故发生率 其值表示一组人员每一亿小时的死亡数。

Fatigue limit state 疲劳极限状态 疲劳极限状态与因周期性载荷而造成的损坏的可能性有关。 Fatigue limit state relates to the possibility of failure due to cyclic loads.

FATO 进场和起飞区 为完成接近飞行的最后阶段而进行悬停或着陆和开始起飞而规定的区域。它是供操作1级直升机的进场和起飞区,还应包括可以利用的被废弃的起飞区域。

Faucet　旋塞(开关,龙头,套管)
Faucet joint　(管端的)弯筒结合
Fault　缺陷(故障,损伤,断层)
Fault check(fault detection)　探伤(故障检验)
Fault detector(fault finder)　故障探测器(探伤器)
Fault free analysis　故障树分析　归纳定量分析技术,用于确定失效和事故的原因及量化其概率。
Fault tolerance　故障承受性　系指系统对故障的承受程度。尽管发生了有限数量的故障,但该系统还能确保系统主要的功能继续发挥作用。 Fault tolerance means an extent to which a system tolerates faults, i. e., the system is capable of continuing its main functions despite of a limited number of faults.
Fault tolerant　容错　系指在故障存在的情况下计算机系统不失效,仍然能够正常工作的特性。例如在双机容错系统中,一台机器出现问题时,另一台机器可以取而代之,从而保证系统的正常运行。在早期计算机硬件不是特别可靠的情况下,这种情形比较常见。现在的硬件虽然较之从前稳定可靠得多,但是对于那些不允许出错的系统,硬件容错仍然是十分重要的途径。计算机系统的容错性通常可以从系统的可靠性、可用性、可测性等几个方面来衡量。可靠性对于火箭发射之类关键性应用领域来说尤为重要。而对于通用计算机来说,一个重要的指标就是系统的可用性。可用性系指在一年的时间中确保系统不失效的时间比率。可测性在容错系统的设计过程中也是一个非常重要的指标,如果无法对某个系统进行测试,又如何能保证其不出问题呢?此外还有 MTBF(故障间的平均时间),即当系统正常运行后能坚持多长时间不失效。MTTR(故障修理的平均时间),系指系统要清除故障所需的时间。MTTR 的大小直接影响着系统的可用性,而 MTBF 则反映了系统的可靠性。
Fault tree analysis(FTA)　故障树分析　用于初步系统安全评估(PSSA)过程中确定导致功能危险评估(FHA)确认的不良的顶级事件的原因。它是导致顶级事件的各事件,或更经常是组合事件的图示。它通过以下方式提供不同分析方法之间的联系:(1)将功能危险评估(FHA)确认的危险或灾难性故障情况用作顶级事件;(2)产生可能必须在故障模式和影响概要(FMECA)中进一步分析的基本事件;(3)说明基本事件的组合任何导致故障模式和影响概要(FMES)和区域危险分析(ZHA)推导的故障模式;(4)量化基本或中间事件的故障率预估;(5)推导基本事件的允许故障率。 FTA is an analysis employed in the PSSA process to determine the causes leading to undesirable top events identified in the FHA. It is a graphical representation of events, or more often combinations of events, that contribute to the top event. It provides the link between the different analysis methods described in the present section by: (1) using failure conditions identified as hazardous or catastrophic in the FHA as top event; (2) generating basic events that may have to be further analyzed in an FMECA; (3) demonstrating how combinations of basic events lead to failure modes derived by FMES and ZHA; (4) quantifying failure rate budgets for basic or intermediate events; (5) deriving permissible failure rates for basic events.

Fax machine　传真机　是应用扫描和光电变换技术,把文件、图表、照片等静止图像转换成电信号,传送到接收端,以记录形式进行复制的通信设备。
Feasibility　可行性
Feathering float　摆动蹼　明轮上所装可稍为摆动的蹼板。
Feathering tread　活动梯步　装在舷梯上,能随船的不同斜度而保持梯步表面始终水平的能动的梯步。
Fecal coliform bacteria　粪便中的大肠杆菌　系指与温血动物大肠有关联的有机物,通常是在粪便物中出现,并可能引起人类疾病潜在危险的有机物。 Fecal coliform bacteria are those organisms associated with the intestine of warm-blooded animals that are commonly used to indicate the presence of fecal material and the potential presence of organisms capable of causing human disease.
Feces water　粪便水
Federal advisory council(FAC)　[美]联邦咨询委员会
Federal Aviation Administration(FAA)　[美]联邦航空管理局　隶属于美国运输部,其职责为负责民用航空安全、联邦航空机构的行为。在1958年成为联邦航空机构,并于1967年成为交通部的下属单位。其主要任务包括:为促进民航安全管理;鼓励和发展民用航空,包括航空新技术;开发和经营空中交通管制、导航系统的民用和军用飞机;研发体系和民用航空领空;制定和实施控制飞机噪声和其他节目环境影响因素;美国商业空间运输管理等。
Federal aviation commission(FAC)　[美]联邦航空委员会
Federal Food, Drug and Cosmetic Act　[美]联邦食品、药品及化妆品法
Federal Maritime and Hydrographic Agency(FMHA/BSH)　德国联邦海事与水文局
Federal Trade Commission(FTC)　[美]联邦贸易委员会

Federal water pollution control act (FWPCA) [美国]联邦政府水污染控制法令

Feed 给水(供电,供给)

Feed(er) line 馈线 起始于主配电中心,并向一个或几个二次配电中心;或一个或几个分支电路配电中心,或向这两类配电中心的任意组合进行供电的一组导线。但发电机和配电板之间的汇流排连接电路,包括主、应急配电板之间的汇流排连接电路不作为馈线考虑。

Feed appliance 给水设备

Feed check valve 给水止回阀

Feed hopper 装料斗

Feed pipe 给水管

Feed pressure 供水压力

Feed pump 给水泵

Feed ratio 给水倍率 给水量与蒸馏水产量的比值。

Feed system 馈电系统(供给系统)

Feed water 给水 海水淡化装置的原料水。

Feed water filter (feed filter) 给水滤器 装设在锅炉给水泵之后的给水管路上,去除给水中油分和杂质的过滤器。

Feed water heater 给水加热器 加热给水用的换热器。

Feed water heater (feed pre-heater) 给水预热器 利用主机撤气或辅机排气加热锅炉给水的热交换器。

Feed water jet pump 给水喷射泵

Feed water supply 给水供应

Feed water tank 锅炉水舱 供贮存锅炉用水的舱。

Feed water test 给水分析

Feed water treatment 给水处理 通过过滤、除气、软化或蒸馏等过程以除去水中有害杂质的给水水质处理。

Feed water valve (feed valve) 给水阀 向锅炉供水的阀。

Feeder 给水器(进料器)

Feeder ship 支线船 又称喂给船。用以从停泊在港口或锚地的大型船舶中,通过转驳运输方式将货物运往内地的船舶。

Feeding 给水(进料,进刀)

Feeler 测隙器(量隙规,塞尺)

Feeler gauge (feeler knife) 量隙规(塞尺)

Feeler pin 探针

Felspar lump 长石块 由硅酸铝及钾、钠、钙和钡组成的状物质。白色或淡红色。无特殊危害。该货物为不燃物或失火风险低。 Felspar lump is a crystalline minerals consisting of silicates of aluminium with potassium sodium, calcium and barium. It is white or reddish in colour. It has no special hazards. This cargo is non-combustible or has a low fire-risk.

Felt 毛毡 工业常用工具,采用羊毛制成,利用加工黏合而成。主要特征有富有弹性,可作为防振、密封、衬垫和弹性钢丝针布底毡的材料。应用于各种产业机械—防振、含油润滑、耐摩擦等行业。毛毡的伸缩性能比较好,可达到规定的长度可用于皮革滚动带、造纸吸浆带。毛毡具有保湿性且具有弹性可制成汽车门窗密封条,中央门窗密封条。书画毡,以优质纯羊毛或纤维为原料,书画皆宜,毡面平整均匀,洁白柔软,富有弹性,可衬托宣纸,使书画爱好者运笔自如,手感舒适。在作品墨多时,不会跑墨,有托墨的作用,又因羊毛可吸烘水分,又具有吸墨的作用,大大体现了脱水保墨的良好性能,特别是绘制国画,能够妙笔生花,书画毡经济实惠,等级齐全,方便耐用,是文房四宝必需的辅助材料。

Felt fiber 毡纤维

Felt gasket 毡垫片

Felt packing 毡填料

Felt washer 毡垫圈

Female fittings 内螺纹管接头配件(阴螺纹管接头配件)

Female mold 阴模(下模)

Female rotor 凹形转子

Female thread 阴螺纹

Fence 隔栅(围墙,雷达警戒网)

Fender 护舷材 装设在舷侧满载水线以上,保护舷侧外板的构件。

Fender 碰垫 船舶靠离码头或他船时,为保护船舷和增加缓冲能力而挂在船舷的固定或活动的衬垫。

Fender (pudding fender) 防撞碰垫 固定设置在拖船首、尾端的护船软碰垫。

Ferries 渡船 系指:(1)为载运乘客,且不设卧铺和/或车辆往返于海峡两岸或岛屿间作定班期营运而专门设计的船舶;(2)具有全通甲板结构,适用于短途、定班期车辆和/或乘客摆渡特点的船舶。见图 F-3。 Ferries are: (1) the ships specially designed for the carriage of passengers (without sleeping berths) and /or vehicles engaged on regular voyages between two sides of straits or islands; (2) a ship having a continuous deck and fit for short-distance and regular ferrying of vehicles and/or passengers. It can be seen in Fig. F-3.

Ferritic steel pressure pipe 铁素体受压钢管

Ferro cast (FC) 铸铁

图 F-3 渡船
Figure F-3 Ferries

Ferro-alloy 铁合金 是铁与一种或几种元素组成的中间合金，主要用于钢铁冶炼。铁合金在钢铁工业中一般还把所有炼钢用的中间合金，不论含铁与否（如硅钙合金），都称为"铁合金"。习惯上还把某些纯金属添加剂及氧化物添加剂也包括在内。铁合金一般用作：脱氧剂：在炼钢过程中脱除钢水中的氧，某些铁合金还可脱除钢中的其他杂质如硫、氮等。合金添加剂：按钢种成分要求，添加合金元素到钢内以改善钢的性能。孕育剂：在铸铁浇铸前加进铁水中，改善铸件的结晶组织。此外，还用作以金属热还原法生产其他铁合金和有色金属的还原剂；有色合金的合金添加剂；还少量用于化学工业和其他工业。

Ferro-cement ship 钢丝网水泥船 用钢丝网水泥作为船体结构的船舶。

Ferrochrome 铬铁白金 铁与铬混合物。极重的货物。无特殊危害。该货物为不燃物或失火风险低。Ferrochrome is a raw material of iron mixed with chrome. It is extremely heavy. It has no special hazards. This cargo is non-combustible or has a low fire-risk.

Ferrochrome, exothermic 放热铬铁白金 铁与铬混合物。极重的货物。无特殊危害。该货物为不燃物或失火风险低。Ferrochrome, exothermic is an alloy of iron and chromium. It is extremely heavy. It has no special hazards. This cargo is non-combustible or has a low fire-risk.

Ferro-graphic analysis 铁谱分析 系指通过对润滑油中磨损颗粒的形貌分析（形状、表面纹理、边缘、颜色等）及浓度分析，进一步判断机械摩擦副的磨损状态，确定磨损机理及磨损部位，从而为早期预报故障及保养、维修决策提供依据的方法。Ferro-graphic analysis is a method It has no analyzing the appearance（shape, surface texture, edges, colors, etc.）and concentration of the wear particles in the lubricating oil, further judging the wear condition of machinery friction pairs and determining the mechanism and positions of wear so as to provide the basis for early failure forecasts and maintenance decisions.

Ferromagnetic metal 铁磁金属

Ferromanganese 锰铁（合金） 是铁与锰混合原料。无特殊危害。该货物为不燃物或失火风险低。Ferromanganese is a raw material of iron mixed with manganese. It has no special hazards. This cargo is non-combustible or has a low fire-risk.

Ferromanganese steel 锰钢

Ferronickel 镍铁合金 一种铁与镍的合金。无特殊危害。该货物为不燃物或失火风险低。Ferronickel is an alloy of iron and nickel. It has no special hazards. This cargo is non-combustible or has a low fire-risk.

Ferro-phosphorus（including briquettes） 磷铁合金（包括砖形块） 是铁与磷的一种合金。用于钢铁工业。该货物为不燃物或失火风险低。Ferro-phosphorus（including briquettes）is an alloy of iron and phosphorus used in the steel industry. This cargo is non-combustible or has a low fire-risk.

Ferrosilicon 硅铁合金（硅钢） 是硅铁中加入钙、稀土配置的合金，亦称镁合金球化剂，是一种良好的孕育剂，其机械强度大，脱氧、脱硫的效果较强。是生产球化剂、蠕化剂、孕育剂使用的轻稀土镁硅铁合金，也用于在生产钢、铁中作添加剂、合金剂。硅铁、稀土矿、金属镁是生产稀土镁硅铁合金的主要原料。稀土镁硅铁合金的生产是在矿热炉中进行，耗电量大，也可以使用中频炉生产。

Ferrosilicon（UN 1408）with 30% or more but less than 90% silicon（including briquettes） 硅铁（硅含量30%以上，但小于90%，包括砖形块） 是一种极重的货物。遇水分可能释放氢，可能在空气中形成爆炸性混合物的易燃气体，在类似条件下并可能产生磷化氢和砷化氢，两者均为极毒气体。该货物为不燃物或失火风险低。Ferrosilicon（UN 1408）is an extremely heavy cargo. In contact with moisture or water it may evolve hydrogen, a flammable gas which may form explosive mixtures with air and may, under similar circumstances, produce phosphine and arsine, which are highly toxic gases. This cargo is non-combustible or has a low fire-risk.

Ferrous material 铁质材料

Ferrous metal 铁类金属（黑色金属）

Ferrous metal borings, shavings, turnings or cuttings（UN 2793）（in a form liable to self-heating） 黑色金属钻屑、削屑、旋屑或切屑（呈自然状态） 金属钻屑通常潮湿，或沾染不饱和切削油，含油碎布和其他可燃

物质。 Ferrous metal borings, shavings, turnings or cuttings are metal drillings that are usually wet or contaminated with such materials as unsaturated cutting oil, oily rags and other combustible material.

Ferrovanadium steel 钒铁合金

Fertilizer capable of self-sustaining decomposition 能发生自续分解的化肥 系指:在其中局部开始的分解将扩散至其全部的化肥。提交运输的化肥的这一性质可放在试验槽中进行测定。试验中,将拟交付运输的化肥盛入水平试验槽中,使分解从局部开始,移去热源后,测出其分解的扩散速度。 Fertilizer capable of self-sustaining decomposition: it is defined as one whose decomposition initiated in a localized area will spread throughout the mass. The tendency of a fertilizer offered for transport to undergo this type of decomposition can be determined by means of trough test. In this test localized decomposition is initiated in a bed of fertilizer contained in a horizontally mounted trough. The amount of propagation, after removal of the initiating heat source, of decomposition through the mass is measured.

Fertilizers without nitrates (non-hazardous) 不含硝酸盐的化肥 呈粉末状和颗粒状,浅绿色、褐色或米色。无嗅。含水量很低(0%~1%)。无特殊危害。该货物为不燃物或失火风险低。该货物吸湿,受潮会结块。 Fertilizers without nitrates (non-hazardous) is powder and granular. It is greenish, brown or beige in colour. odourless. It has very low moisture content (0% to 1%). No special hazards. This cargo is non-combustible or has a low fire-risk. This cargo is hygroscopic and will cake if it gets wet.

Fetch 风程/风区长度 系指风在吹达舰船之前越过开敞水域的距离。 Fetch is the extent of clear water across which a wind has blown before reaching the ship.

Ffire fighting system test 灭火系统效能试验

Fiber board 纤维板 系指由木质纤维素纤维交织成型并利用其固有胶黏性能制成的人造板。制造过程中可以施加胶黏剂和/或添加剂。纤维板又名密度板,是以木质纤维或其他植物素纤维为原料,施加脲醛树脂或其他适用的胶黏剂制成的人造板。制造过程中可以施加胶黏剂和/或添加剂。纤维板具有材质均匀、纵横强度差小、不易开裂等优点,用途广泛。

Fiber glass 玻璃纤维/玻璃棉/纤维玻璃 是一种性能优异的无机非金属材料。成分为二氧化硅、氧化铝、氧化钙、氧化硼、氧化镁、氧化钠等。它是以玻璃球或废旧玻璃为原料经高温熔制、拉丝、络纱、织布等工艺。最后形成各类产品,玻璃纤维单丝的直径从几个微米到二十几个微米,相当于一根头发丝的1/20~1/5,每束纤维原丝都有数百根甚至上千根单丝组成,通常作为复合材料中的增强材料、电绝缘材料和绝热保温材料、电路基板等,广泛应用于国民经济各个领域。

Fiber material 纤维材料 电工领域用的纤维材料有天然纤维(包括植物纤维和动物纤维)、无机纤维(如石棉、玻璃纤维)和合成纤维(如聚酯纤维、聚芳酰胺纤维等)三大类。它具有如下优点:①在超导磁体线圈中,能使冷却剂浸透所有的截面,增加传热面积;②保证浸渍漆或包封胶直接与超导纤维及复合层接触。

Fiber mooring line 合成纤维缆 用尼龙、锦纶等合成纤维制成的缆绳。

Fiber reinforced plastics (FRP) 玻璃钢 亦称作纤维强化塑料(GFRP)。一般系指用玻璃纤维增强不饱和聚酯、环氧树脂与酚醛树脂基体。以玻璃纤维或其制品作增强材料的增强塑料,称为玻璃纤维增强塑料,或称为玻璃钢,注意与钢化玻璃区别开来。由于所使用的树脂品种不同,因此有聚酯玻璃钢、环氧玻璃钢、酚醛玻璃钢之分。质轻而硬,不导电,性能稳定,机械强度高,回收利用少,耐腐蚀。可以代替钢材制造机器零件和汽车、船舶外壳等。玻璃钢别名玻璃纤维增强塑料,即纤维增强复合塑料。根据采用的纤维不同分为玻璃纤维增强复合塑料(GFRP)、碳纤维增强复合塑料(CFRP)、硼纤维增强复合塑料等。它是以玻璃纤维及其制品(玻璃布、带、毡、纱等)作为增强材料,以合成树脂作基体材料的一种复合材料。纤维增强复合材料是由增强纤维和基体组成。纤维(或晶须)的直径很小,一般在 10 μm 以下,缺陷较少又较小,断裂应变约为30‰以内,是脆性材料,易损伤、断裂和受到腐蚀。基体相对于纤维来说,强度、模量都要低很多,但可以经受住大的应变,往往具有黏弹性和弹塑性,是韧性材料。

Fiber rope 纤维索 系指仅由天然或人造纤维制成的绳索,它包括由天然或人造纤维制成的绳索或扁平索。 Fiber rope means a rope constructed of natural or synthetic fiber only and includes rope or flat-woven webbing constructed of natural or synthetic fibers.

Fiber rope block 绳索滑车 穿绕纤维绳的滑车。

Fiberglass 玻璃纤维

Fiberglass reinforced plastic boat 纤维增强塑料艇 以玻璃纤维增强塑料作为艇体结构基本材料的船艇。

Fiberglass-reinforced plastic 玻璃纤维增强塑料(玻璃钢)

Fiber-reinforced material 纤维增强材料

Fiber-reinforced plastic 纤维增强塑料

Fibrous glass 玻璃纤维(玻璃丝)

Fiddley 锅炉舱棚(安装围栏的舱口)

Field of vision 视野 系指:(1)在驾驶桥楼内部可以瞭望的角度;(2)从驾驶桥楼的某一位置可看到景物的角度范围。 Field of vision is:(1) an angular range that can be observed from a position on the ship's bridge;(2) an angular size of a scene that can be observed from a position on the ship's bridge.

Field Terminal platform 油区终端平台 系指作为油气开发区平台群中的终端站,通过单点系泊装置或其他形式的输油、装油设施将油气开发区生产和储存的原油输送出去的平台。

Fighter 战斗机 系指主要用于保护我方运用空投以及摧毁敌人使用空权之能力的军用飞机。其特点是飞行性能优良、机动灵活、火力强大。现代的先进战斗机多配备各种搜索、瞄准火控设备,能全天候攻击所有空中目标。该技术拥有国:美国(F15/16/18)、俄罗斯(SU-27系列)、法国(幻影2000)、中国(J10/AB)、英德意联合(阵风)、日本(算半个)。见图F-4。

图 F-4 战斗机
Figure F-4 Fighter

Filing procedure 立案流程 包括:(1)申请材料——准备仲裁的申请材料;(2)审核——立案检查申请材料;审查是否有仲裁协议或者合同中的仲裁条款;(3)缴费——受理案件,领取缴费通知至财务室缴费或发送预缴费通知,银行划账;(4)立案受理——交给具体办案秘书。

Filled compartment trimmed 平舱的满载舱 系指经装载及平舱后的任何货物处所。散装谷物达到其可能的最高水平面。 Filled compartment trimmed refers to any cargo space in which, after loading and trimming, the bulk grain is at its highest possible level.

Filled compartment untrimmed 未平舱的满载舱 系指在舱口处散装谷物已灌注至可能最大程度,但在舱内四周围未进行平舱的货物处所。 Filled compartment untrimmed refers to a cargo space which is filled to the maximum extent possible in way of the hatch opening but which has not been trimmed outside the periphery of the hatch opening.

Filler 装填物(装填器,熔接料,填充丝,浇铸设备)
Fillet 内圆角(角焊缝,焊脚,填角料)
Fillet cavitation 叶根圆角空泡
Fillet of the journal 轴颈圆角
Filling and sounding pipe 注入兼测量管 既作注入管,同时又可利用其对舱内液位进行测量的管道。
Filling pipe(filling line) 注入管 向油舱、水舱等注入液体的管路。
Filling block 垫块 对钢夹层板而言,系指固定在底板上的方形和圆形的钢或高分子材料,用于保持底板和顶板间的距离。
Filling piece(locking piece) 锁口件 最后装入叶轮或鼓轮以某种特殊方法与之固定并对整列叶片起锁紧作用的隔离件。
Filling pump 灌注泵
Filling sieve 注入滤网
Filling tank 补给油柜
Filling up 充满
Filling valve 注入阀
Fillister 凹槽
Fill-up 填塞(加注)
Film 薄膜(涂膜,效片)
Film coefficient of heat transfer 膜传热系数
Film recording thermometer 照相温度表 一种用照相方法定时记录水温的装置。由玻璃水银温度计、照相机和日期编码装置等组成。取样时间和计时由定时机构控制。
Film thickness distribution 膜厚分布 系指所有膜厚检测点涂层干膜厚度的数学分布状态。 Film thickness distribution means a mathematical distribution mode of dry film thickness of all film thickness measuring points for coating.
Filmwise boiling 膜状沸腾 在传热温差低于 23 ℃ ~27 ℃时,水在传热面上产生无数小泡泡,泡泡彼此连成一层气膜的沸腾现象。
Filmwise condensation 膜状冷凝 蒸汽在冷却壁面上冷凝时,结成完整液膜的冷凝方式。
Filter unit 过滤元件
Filter 滤波器 系指一种从声音信号频谱中除去非选择的频率,从而调整音色,或者说它是用来消除或减少音频信号中某些不必要频率的装置或电路。
Filtering system 过滤系统
Filtration method 过滤法 就是把溶于液体的固态物质跟液体分离的一种方法。
Fin 鳍板(减摇鳍,翼片,散热片)

Fin control unit 减摇鳍控制器
Fin filter 薄片滤器
Fin housing and extending gear 减摇鳍收放装置 使鳍放出舷外或收入鳍箱的装置。
Fin stabilizer 减摇鳍 （1）一种特别有效的减摇装置。一般安装在客船和一些对稳性有特别要求的船上。减摇鳍在舯部相当于水翼，在船舶有航速时，它产生由流体动力组成的抵抗船舶横摇的力矩。其具有一套自动操纵系统。可分为不可收放式减摇鳍和折叠收放式减摇鳍2种；（2）在一定航速下，控制外伸与两舷的鳍的攻角，以产生减摇力矩的装置。见图F-5。

图 F-5 减摇鳍
Figure F-5 Fin stabilizer

Fin tilting gear 减摇鳍转鳍机构
Fin tilting machinery 水平舵转舵机构（减摇鳍传动机构）
Fin tip propeller 端翼螺旋桨
Fin tube 肋片管（翅片管）
Final approach and take-off area(FATO) 最终进场/起飞区域(FATO) 系指一限定区域，直升机要在该区域之上完成悬停或降落的进场动作最后阶段和开始进行起飞动作。 Final approach and take-off area(FATO) is a defined area over which the final phase of the approach manoeuvre to hover or landing of the helicopter is intended to be completed and from which the take-off manoeuvre is intended to be commenced.

Final award 最终裁决/最后裁决书
Final blade 末叶片 最后装入叶轮或鼓轮以某种特殊方法之固定并对整列叶片起锁紧作用的叶片。
Final decision 一裁终局 仲裁实行"一裁终局"制度，即仲裁做出后即发生法律效力，没有上诉或再审程序。
Final document review 文件终审 系指初审后的体系文件，在按照审核机构要求进行修改后，再次提交审核机构的审核。

Final inspection 最终检查 对质量管理体系而言，系指制造厂商应对证实产品符合规定要求所必需的所有检查和试验工作都已完成。最终检查和试验的程序应保证：（1）在说明书、质量计划或其他文件中所规定的活动都已完成；（2）所有在早期阶段应进行的各种检查试验都已完成，且其数据是可以被接受的；（3）在说明书、质量计划或其他文件规定的活动没有完成前，产品不应发放。除非是验船师已同意放行的产品。 For the quality management system, final inspection means that the manufacturer is to perform all inspections and tests on the finished product necessary to complete the evidence of conformance to the specified requirements. The procedures for final inspection and test are to ensure that：（1）all activities defined in the specification, quality plan or other documented procedure have been completed；（2）all inspections and tests that should have been conducted at earlier stages have been completed and that the data is acceptable；and（3）no product is to be dispatched until all the activities defined in the specifications, quality plan or other documented procedure have been completed, unless products have been released with the permission of the Surveyors.

Final recording medium 最终记录介质 系指记录数据的硬件装置，通过采用适当的设备，访问其中任何一个，便能获取到数据并回放。固定记录介质、自浮记录介质和长期记录介质组合在一起，即为最终记录介质。 Final recording medium means the items of hardware on which the data is recorded such that access to any one of them would enable the data to be recovered and played back by use of suitable equipment. The combination of a fixed recording medium and float-free recording medium and long-term recording medium, together, is recognized as the final recording medium.

Final stage 最终状态
Final sub-circuit 最后分路 系指延伸在分配电板最后需要的过电流保护装置范围以外的那部分的分线路系统。 Final sub-circuit is that portion of a wiring system extending beyond the final required over-current protective device of a board.

Final survey for the International Certificate of Inventory of Hazardous Materials 《国际有害物质清单证书》最终检验 系指该检验在船舶退役前和拆船开工前进行。该检验应验证：（1）参照IMO组织制定的指南第5.4条要求的有害物质清单符合2009年香港国际安全与环境无害化拆船公约的要求；（2）第9条所要求的拆船计划正确反映了按第5.4条要求的有害物质清单所包含的信息，并包括建立、保持和监控进入安全和热做安全条件的有关信息;(3)拆除船舶的拆船厂持有按 2009 年香港国际安全与环境无害化拆船公约规定的有效授权书。 Final survey for the International Certificate of Inventory of Hazardous Materials means a survey started before the ship being taken out of service and before the recycling of the ship. This survey shall verify:(1) that the Inventory of Hazardous Materials as required by Regulation 5.4 is in accordance with the requirements of the Hong Kong International Convention for the Safe and Environmentally Sound Recycling of Ships, 2009, taking into account the guidelines developed by the Organization;(2) that the Ship Recycling Plan, as required by Regulation 9, properly reflects the information contained in the Inventory of Hazardous Materials as required by Regulation 5.4 and contains information concerning the establishment, maintenance and monitoring of Safe-for-entry and Safe-for-hot work conditions;(3) that the Ship Recycling Facility(ies) where the ship is to be recycled holds a valid authorization in accordance with the Hong Kong International Convention for the Safe and Environmentally Sound Recycling of Ships, 2009.

Final testing 最终试验 系指检验的一种方式。旨在为接受产品所进行的所有在产品检验证书中记载的试验。 Final testing, as an inspection method, means all acceptance tests recorded in the products certificate.

Final working repairs 扫尾工程 系指完成最后剩下的工作,使其结束的工程。

Finance staff 财务人员 由熟悉成本情况、支付方式及金融知识,具有较强的财务核算能力的财务会计人员担任。

Financial SLC 融资备用信用证 主要是对申请人应履行的付款责任进行担保。广泛用于国际信贷融资安排。

Finder 探测器(瞄准装置,选择器,寻线机)

Finding 发现 系指观察项或不合格项。 Finding means an observation or a non-conformity item.

Fine alignment 精对中

Fine file 细锉

Fine filter 细滤器

Fine for violation of the contract 违约金 系指债权人或债务人完全不履行或不适当履行债务时,必须按约定给付对方的一定数额的金钱。

Fine gravity measurement 引力精密测量 罗俊院士团队算出世界最精确万有引力常数,其引力实验室也被外国专家称为"世界的引力中心"。

Finish machining(finished machining) 精加工(精机加工)

Finish with engine 完车(本机使用完毕)

Finished plan 完工文件(竣工图) 系指船舶建造完工后把与原设计有出入的实际情况反映到图纸文件中的阶段。一般完工文件还包括如何使用这些资料,例如装载手册。

Finned 翅片式的(有安定面的,装鳍的)

Finned radiator tubular 翅片管式散热器

Finnish-Swedish ice class rules 芬兰-瑞典冰级规则

Fin-shaft 鳍轴 减摇鳍与转轴机构的连接轴。

Fin-tilting gear 转鳍机构 用来旋转鳍,使其与船进速形成所需水动力攻角的机构。

Fire 失火 系指易燃蒸发气或气体与氧化剂(通常是氧或空气)合成,通过光、热和火焰逸出来表征的燃烧过程。 Fire is a combustible vapor or gas combining with oxidizer(usually oxygen or air) in a combustion process manifested by the evolution of light, heat, and flame.

Fire alarm 失火报警 系指发生火灾时召集船员的警报。 Fire alarm means an alarm to summon the crew in the case of fire.

Fire alarm device 火警装置

Fire alarm equipment 火灾报警器

Fire alarm signal 火警信号

Fire alarm signal systems in passenger ships 客船上失火报警信号系统 系指客船在海上或港口的所有时间内(非营运时除外)为船员配置或配备的设备,以保证负责船员能立即接到任何初始失火报警。 Fire alarm signal systems in passenger ships is a means manned or equipped for the crew in passenger ships at all times when at sea, or in port(except when out of service), to ensure that any initial fire alarm is immediately received by a responsible member of the crew.

Fire alarm system 失火报警系统

Fire alarm system panel 失火报警系统板

Fire and by-product containment goal 1 火和燃烧产物抑制目标 1 为了使火、烟和有毒产物的蔓延限制在限定区域内,在火灾发生时仍没有弃船的情况下使舰船尽可能地接近正常工作状态,在舰船内应设置足够的限制边界:(1)抑火布置应符合 SOLAS 第Ⅱ-2 章 C 部分第 9 条的要求;(2)给处所补充空气和可燃液体的控制应符合 SOLAS 第Ⅱ-2 章 B 部分 5.2 条和 C 部分第 8.2 条的要求;(3)关于关闭设备,设计用于搭载人数超过 50 人的舰船,除了在机器处所和控制站不设置一直通向开敞甲板的出口外,应设置有带控制的机械通风,以便所有风扇能被从两处地点关闭,该两处地点间应尽可能地相隔较远距离;(4)所有灭火系统的控制应设置在一个控制地点或设置在尽可能少的地点,以满足舰船主管当局的要求;(5)防火材料的使用应符合 SOLAS 第Ⅱ-2 章 B 部分第 5.3 条的要求;(6)对于控制烟蔓延的布置应符合 SOLAS 公约第Ⅱ-2 章 C 部分第 8 条的要求。 Fire and by-product containment goal 1 means that adequate containment boundaries are to be fitted within the ship in order that spread of fire, smoke and toxic by-products is limited to a predetermined area, allowing the vessel operate as normally as possible in the event of fire, without evacuation. (1) Arrangements for the containment of fire are to be in accordance with SOLAS Chapter Ⅱ-2, Part C, Regulation 9; (2) Control of air supply and flammable liquid to the space is to be in accordance with SOLAS Chapter Ⅱ-2, Part B, Regulation 5.2 and Part C, Regulation 8.2; (3) With regard to closing devices, vessels designed to carry in excess of 50 embarked personnel, power ventilation, except in machinery spaces and control stations not having an exit to an open deck, is to be fitted with controls so that all fans can be stopped from either of two positions which should be located as far apart from each other as is practicable; (4) All controls for the fire-extinguishing system shall be situated in one control position or in as few positions as possible to the satisfaction of the Naval Authority; (5) The use of fire protection materials is to be in accordance with SOLAS Chapter Ⅱ-2, Part B, Regulation 5.3; (6) The arrangements for the control of smoke spread are to be in accordance with SOLAS Chapter Ⅱ-2 Part C, Regulation 8.

Fire and by-product containment objective 火和燃烧产物的抑制目标 每艘舰船的布置应使其尽快地限制火、烟和有毒燃烧产物蔓延至整个火源所在处所。 Fire and by-product containment objective means that every ship is to be arranged, so far as is practicable, to limit the spread of fire, smoke and toxic by-products to the space of origin.

Fire and insurance code 防火险规则

Fire and rescue bill 消防救生部署表

Fire appliances 消防设备(灭火器具)

Fire bar 炉条

Fire bell 救火钟

Fire blanket 消防毯 供消防用的用耐火材料制成或浸渍的专用毯。

Fire boat(fire fighting ship) 消防船 对发生火灾的船舶提供灭火和救援的专用船舶。或用于扑灭船舶或港口岸边火灾的船舶。又分为第 1 类消防船——具有扑灭初期火灾能力消防设备的船舶、第 2 类消防船——具有扑灭大火能力消防设备的船舶和第 3 类消防船——具有扑灭大火和油类火灾能力消防设备的船舶。此类船一般设有高压水炮、水/空气泡沫两用炮、干粉炮等和大功率消防泵,以及装载灭火专用化学品的舱柜。船上配备水幕自救喷射系统,能有效地降低热辐射对其自身的影响。见图 F-6。

图 F-6 消防船
Figure F-6 Fire boat(fire fighting ship)

Fire box 烟箱

Fire box frame 烟箱架

Fire brick 耐火砖

Fire bridge (锅炉)坝(矮墙)

Fire bucket 消防水桶 供消防用的手提水桶。

Fire chamber 燃烧室

Fire clay 耐火黏土/耐火泥 又称无序高岭石,是以高岭石为主要矿物成分的天然硅酸铝质材料,由硅酸盐质岩石风化作用形成,是最基本、最常用的耐火原料。工业上使用的耐火黏土主要分为硬质黏土和软质黏土两类。

Fire clothes 防火服 供消防人员穿着的隔热耐火的专用服装。

Fire coat 氧化皮

Fire cock 救火旋塞(救火龙头)

Fire control 火势控制 系指:(1)通过预先湿润邻近可燃材料和控制天花板气体温度以防止结构损坏来限制火势发展;(2)通过水的分布限制火势的规模,其目的是降低火焰的热释放量,并预湿邻近的易燃物,同时控制天花板气体温度,以避免结构性损坏。 Fire control means:(1) the limitation of the growth of the fire by pre-wetting adjacent combustibles and controlling ceiling gas temperatures to prevent structural damage; (2) to limit the spread of a fire by distribution of water so as to decrease the heat release rate and pre-wet adjacent combustibles, while controlling ceiling gas temperatures to avoid structural damage.

Fire control plans 防火控制图 系指永久展示的、向高级船员提供指导的总布置图。该图清晰标明每层甲板的控制站、"A"级分隔围蔽的各防火区域、"B"级分隔围蔽的各防火区域、连同探火和失火报警系统、喷水器装置、灭火设备和各舱室、甲板等的出入通道以及通风系统的细节：包括风机控制位置、挡火闸的位置和服务于每一区域的通风机识别号码的细节。作为替代，经主管机关同意，上述细节可合并成册，每位高级船员人手一册，而在船上易于到达的位置应有一份副本可供随时取用。控制图和小册子应保持更新，任何改动应尽快予以记录。该防火控制图和小册子的说明文字应以主管机关要求的一种或几种语言写成。如果该语言既不是英文也不是法文，则应包括其中一种语言的译文。一套防火控制图或附有防火控制图小册子的副本应永久存放于甲板室外标有明显标志的风雨密盒中，供岸上消防人员使用。 Fire control plans means a general arrangement plans permanently exhibited for the guidance of the ship's officers, showing clearly for each deck the control stations, the various fire sections enclosed by "A" class divisions, the sections enclosed by "B" class divisions together with particulars of the fire detection and fire alarm systems, the sprinkler installation, the fire-extinguishing appliances, means of access to different compartments, decks, etc., and the ventilating system including particulars of the fan control positions, the position of dampers and identification numbers of the ventilating fans serving each section. Alternatively, at the discretion of the Administration, the aforementioned details may be set out in a booklet, a copy of which shall be supplied to each officer, and one copy shall at all times be available on board in an accessible position. Plans and booklets shall be kept up to date; any alterations thereto shall be recorded as soon as possible. Description in such plans and booklets shall be in the language or languages required by the Administration. If the language is neither English nor French, a translation into one of those languages shall be included. A duplicate set of fire control plans or a booklet containing such plans shall be permanently stored in a prominently marked weathertight enclosure outside the deckhouse for the assistance of shore-side fire-fighting personnel.

Fire control stairs 应急消防梯

fire control stations 消防控制站 系指火警指示器或防火控制设备集中的处所。 Fire control stations are those spaces where the fire recording or fire control equipment is centralized.

Fire control system 火力控制系统 系指控制火炮、火炮群或导弹发射器瞄准和射击的整套设备。火力控制系统常用于地面和舰上火炮、防空火炮、轰炸机的防御火炮以及船上和飞机上的火箭、导弹的控制。广义的火力控制系统还包括指挥截击机的飞机、导弹的地面引导站、弹道导弹防御系统中的地面系统。简单的火力控制系统主要由敏感元件、计算机和定位伺服机构组成。常用的火力控制系统有防空系统、航空火力控制系统、舰载火力控制系统、反坦克导弹控制系统、反导弹防御火力控制系统等。见图 F-7。

图 F-7 火力控制系统
Figure F-7 Fire control system

Fire damper 防火风闸

Fire dampers automatically 自动防火风闸 系指保险丝型风闸或船级社认可的同类部件。Fire dampers automatically are to be the fuse type dampers or those considered to be equivalent by a Society.

Fire detecting arrangement 失火探测装置（探火装置）

Fire detecting device 探火装置（失火探测器）

Fire detecting system 失火探测系统（探火系统）

Fire detecting system alarm 探火系统报警 系指由手动或自动探火装置发出的报警。

Fire detection 火警探测 采用电气或物理等方法对船上货舱及居住处所等舱室发生的火警的探察。

Fire detection 烟火探测（探火）

Fire detection alarm 探火报警 系指提醒船上安全中心、连续有人值班的集中控制站、驾驶室或主火灾控制站或其他处所的船员发现了灾情的报警。 Fire detection alarm means an alarm to alert the crew in the onboard safety centre, the continuously manned central control station, the navigation bridge or main fire control station or elsewhere that a fire has been detected.

Fire detection and alarm device 探火与失火报警装置

Fire detection and fire alarm system 探火与失火报警系统

Fire detection goal 1 火灾探测目标 1 每艘舰船应装备能够有效地探测热/火焰和烟的系统，系统在火灾预期可能发生的情况下应当准确动作:（1）一般，探测和报警系统的布置应符合 SOLAS 第 Ⅱ-2 章 C 部分 7.2

条的要求;(2)固定式火灾探测和报警系统的试验应符合 SOLAS 第Ⅱ-2 章 C 部分 7.3 条的要求;(3)机器处所的保护应符合 SOLAS 第Ⅱ-2 章 C 部分 7.4 条的要求;(4)起居处所、服务处所和控制站的保护应符合 SOLAS 第Ⅱ-2 章 C 部分 7.5 条的要求;(5)手动操纵呼叫点的设置应符合 SOLAS 第Ⅱ-2 章 C 部分 7.7 条的要求;(6)防火巡逻的安排应符合 SOLAS 第Ⅱ-2 章 C 部分 7.8 条的要求;(7)火灾报警信号应符合 SOLAS 第Ⅱ-2 章 C 部分 7.9 条的要求。 Fire detection goal 1 means that every ship is to be equipped with effective heat/flame and smoke detection systems that will function correctly in the environment in which a fire may be reasonably expected to occur. (1) In general, the arrangements for detection and alarm are to be in accordance with SOLAS Chapter Ⅱ-2, Part C, Regulation 7.2; (2) The testing of fixed fire detection and alarm systems are to be in accordance with SOLAS Chapter Ⅱ-2, Part C Regulation 7.3; (3) the protection of machinery spaces is to be in accordance with SOLAS Chapter Ⅱ-2, Part C, Regulation 7.4; (4) the protection of accommodation, service spaces and control stations is to be in accordance with SOLAS Chapter Ⅱ-2, Part C, Regulation 7.5; (5) the provision of manually operated call points is to be in accordance with SOLAS Chapter Ⅱ-2, Part C, Regulation 7.7; (6) fire patrols are to be arranged in accordance with SOLAS Chapter Ⅱ-2, Part C, Regulation 7.8; (7) fire-alarm signaling is to be in accordance with SOLAS Chapter Ⅱ-2, Part C, Regulation 7.9.

Fire detection goal 2　火灾探测目标 2　系统应设置成可以探测到一处已被探测和扑灭火的复燃。 Fire detection goal 2 means that systems are to be arranged so as to detect the re-ignition of a detected and extinguished fire.

Fire detection objective　火灾探测目标　每艘舰船的设计和装备应能使其尽快探测到任何潜在的火灾或爆炸危险。 Fire detection objective means that every ship is to be designed and equipped, as far as is practicable, to detect any potentially hazardous fire or explosion.

Fire detection system　失火探测系统

Fire detector　火灾探测器(火警探测器)

Fire door holding and release system　防火门的吸持和释放系统

Fire door of self-closing type　自闭式防火门　在相邻舱室失火后能自行关闭的防火门。

Fire door(fire gate)　防火门　装设在船舶防火舱壁上,关闭后能隔阻火灾蔓延的门。

Fire drills　消防演习　系指根据船型和货物类型可能发生的各种紧急情况下举行的演习。每次消防演习应包括:(1)到集合点集中报到,并准备执行应变部署表中所述的任务;(2)启动一个消防泵,要求至少射出两股水柱,以表明该系统是处于正常的工作状态;(3)检查消防员装备和其他个人救助设备;(4)检查有关的通信设备;(5)检查演习区域内的水密门、防火门和挡火风闸以及通风系统主要进出口的工作情况;(6)检查供随后弃船用的必要装置。演习中使用过的设备应立即恢复到完好的操作状态;演习中发现的任何故障和缺陷应尽快予以消除。 Fire drills means drills taking place in the various emergencies that may occur depending on the type of ships and the cargo. Each fire drill shall include: (1) reporting to stations and preparing for the duties described in the muster list; (2) starting a fire pump, using at least the two required jets of water to show that the system is in proper working order; (3) checking fireman's outfit and other personal rescue equipment; (4) checking relevant communication equipment; (5) checking the operation of watertight doors, fire doors, fire dampers and main inlets and outlets of ventilation systems in the drill area; (6) checking the necessary arrangements for subsequent abandoning of the ship. The equipment used during drills shall immediately be brought back to its fully operational condition and any faults and defects discovered during the drills shall be remedied as soon as possible.

Fire endurance　耐火性　系指:(1)管子遇到火灾时,在预先设定的时间内保持其强度与完整性(能够起到其应起的作用)的能力;(2)在曝露于火中一段预定的时间后,仍可完成其预定功能,例如保持其强度和完整性的能力。 Fire endurance means: (1) the capability for piping to maintain its strength and integrity(i. e. capable of performing its intended function) for some predetermined period of time while exposed to fire; (2) the capability for the piping system to perform its intended function, i. e. maintain its strength and integrity, for some predicted period of time while exposed to fire.

Fire endurance for piping　管系的耐火性　系指管子遇到火灾时,在一预先设定的时间内保持其强度与完整性(能够起到其应起的作用)的能力。 Fire endurance for piping means the capability for piping to maintain its strength and integrity(i. e. capable of performing its intended function) for some predetermined period of time while exposed to fire.

Fire escape　太平门(安全门,火警应急出口)

Fire extinction　灭火　是一个将火源释放出来的热量减少,并采用直接手段和足够的灭火剂将所有火焰及炽热物件的暗火全部灭除的过程。 Fire extinction is a reduction of the heat release from the fire and a total elimination of all flames and glowing parts by means of direct and sufficient application of extinguishing media.

Fire extinction arrangement 灭火装置
Fire extinguisher 灭火器
Fire extinguishing agent applicating alarm 灭火剂施放报警 系指对将充满 CO_2 或其他灭火剂的处所,如机舱、泵舱,在施放之前和施放过程中发出的紧急报警。
Fire extinguishing agent container 灭火剂容器
Fire extinguishing appliance 灭火装置(消防设备)
Fire extinguishing arrangement 消防设备(灭火装置)
Fire extinguishing equipment 灭火装置(消防设备)
Fire extinguishing goal 1 灭火目标1 舰船上的布置应能通过使用适合火灾特性的灭火介质将被探测到的火灾扑灭:(1)除非特别规定,在下述要求中灭火布置应符合 SOLAS 公约第Ⅱ-2 章 C 部分 10 条的要求;(2)当高压海水系统用于灭火时,系统应符合第 2 卷第 7 部分第 5 章第 10 节的要求;(3)水灭火系统应具有等同于高压海水系统的能力,在第 2 篇第 7 部分第 5 章 10.2.3 中对高压海水系统所要求具有的能力进行了规定;(4)消火栓的数量和位置应满足:至少来自不同消火栓的 2 股水流能到达舰船上被探测到火的任何部位,其中一股仅靠 1 根消防水带的长度。上述消火栓应布置位于接近被保护处所出入口处。用于预期火情边界冷却的消火栓也可以使用;(5)考虑到消火栓的压力,水灭火系统应具有在邻近的消火栓上输送所需流量的能力,且所有消火栓应满足下列的压力规定:①NS1 和 NS2 型舰船和所有总排水量不小于 4 000 t 的舰船为 4 bar;②NS3 型舰船为 3 bar。压力不能超过那些已经完成设计系统的压力,或者不得超过可有效控制消防水带所限制的压力;(6)关于消防泵的设置,总排水量不小于 4 000 t 的舰船应安装至少 3 台消防泵,总排水量小于 4 000 t 的舰船应至少安装 2 台消防泵。如果这些泵的布置在单一火灾时会使其全部失效,需要准备一个应急消防泵;(7)关于消防水带,每个消火栓应至少配有 1 条满足长度要求的消防水带。这些消防水带能单独地用于消防和设备试验;(8)关于喷水系统,NS1 和 NS2 型舰船,和那些设计用来运载超过 50 名人员的舰船,应在所有控制站、起居和服务处所安装一套符合舰船主管当局要求的自动喷水、火灾探测和火灾报警系统。如果喷水系统可能会对重要设备造成损坏,允许用一种舰船主管当局认可的固定式消防系统来代替。在空间狭小或者没有火灾隐患的处所,不需要安装一套这样的系统;(9)在 NS3 型舰船和设计用来搭载少于 50 名人员的舰船上,应安装一套符合舰船主管当局要求的自动喷水、火灾探测和报警系统来保护控制站。应布置一个固定火灾探测系统和报警装置来探测走廊、楼梯、起居处所内的逃生和出入通道中的烟雾;(10)关于消防员装备,消防员装备应达到舰船主管当局所要求的标准。作为最小配置,应符合国际消防安全规则(FSS)的要求;(11)每艘舰船都应携带足够的消防员装备,以满足由舰船主管当局同意的每个消防队所要求的数量。每艘舰船应至少携带 2 套消防员装备。沿甲板纵向每 80 m 设置两套消防员装备,并且此外每个主竖区设置两套消防员装备。在每个呼吸器具旁应储存一个烟雾施放器。在每个防火区域应能直接获取消防员装备和烟雾施放器;(12)任何以水作为灭火介质的舱室,均应能够在灭火后将水排走。 Fire extinguishing goal 1 means that arrangements on board are to be such that all detected fires can be extinguished using a medium which is suitable for the nature of the fire. (1) Unless otherwise given, in the following paragraphs, arrangements for the extinction of fire are to be in accordance with SOLAS Chapter Ⅱ-2, Part C, Regulation 10;(2) Where high pressure sea-water systems are used for fire-fighting purposes, they are to be in accordance with the requirements of Vol. 2, Pt 7, Ch 5, 10;(3) The water fire-fighting system is to have a capacity, equal to the requirements of HPSW systems as outlined in Vol. 2, Pt 7, Ch 5, 10.2.3;(4) The number and position of hydrants are to be such that at least two jets of water not emanating from the same hydrant, one of which shall be from a single length of hose, may reach any part of the ship where a fire can be reasonably expected. Such hydrant is to be located near the access to protected spaces. Hydrants for boundary cooling for the expected fire scenarios are also to be made available;(5) With regard to pressure at hydrants, the water fire-fighting system is to be capable of delivering the required quantity of water at adjacent hydrants with the following pressures being available at all hydrants: ①4 bar for NS1 and NS2 ships and all ships of 4 000 gross tonnage or greater;②3 bar for NS3 ships. The pressure is not to exceed that for which the system has been designed, or that above which the fire hose cannot be demonstrated as being controllable;(6) With regard to the provision of fire pumps, ships of 4 000 gross tonnage and above are to be fitted with at least three fire pumps, and ships of less than 4 000 gross tonnage are to be fitted with at least two. If the pumps are arranged such that a single fire will put all pumps out of action, an additional emergency fire pump will be required;(7) With regard to the requirements for fire hoses, at least one fire hose of the required length is to be permanently

available at each required hydrant. These are to be for the sole use of fire-fighting and testing the equipment;(8)With regard to sprinkler systems, NS1 and NS2 ships, and those designed to carry in excess of 50 embarked personnel, are to be fitted with an automatic sprinkler, fire detection and fire alarm system of a type acceptable to the Naval Authority in all control stations, accommodation and service spaces. Alternatively, if a water sprinkler system may cause damage to essential equipment, a fixed fire-fighting system acceptable to the Naval Authority is to be used. Spaces where there is little or no risk of fire does not need to be fitted with such a system;(9)On NS3 ships and vessels designed to carry less than 50 embarked personnel, an automatic sprinkler, fire detection and fire alarm system of a type acceptable to the Naval Authority is to be installed to protect control stations. A fixed fire detection system and alarm is to be arranged to provide smoke detection in corridors, stairways, escape and access routes within accommodation spaces;(10)With regard to fire-fighters' outfits, the fire-fighters' outfit is to be to a standard acceptable to the Naval Authority. As a minimum they are to be in accordance with the Fire Safety Systems Code(FSS Code);(11)Each ship is to carry sufficient fire-fighters' outfits for the number required for each fire party as agreed with the Naval Authority. Each ship is to carry at least two firefighters' outfits. Two fire-fighters' outfits are to be provided for every 80 m of longitudinal deck space and in addition, two in every vertical zone. One water fog applicator is to be stored adjacent to each set of breathing apparatus. The fire-fighters' outfits and fog applicators are to be readily accessible in each fire zone;(12)Where water is used for fire-fighting purposes in any compartment, these compartments are to be provided with arrangements for removing the water after fire extinguishing.

Fire extinguishing goal 2   灭火目标 2   在灭火系统防护区域内,灭火系统应具有扑灭任何预测到可能发生的火灾的足够能力。关于水的利用方面,至少一股有效的水柱能够迅速地从所在处所内的任何消火栓上获得。一台自动启动的消防泵应能保证水的连续供应。

Fire extinguishing goal 2 means that the extinguishing system sufficient capacity to extinguish any fire which may has reasonably has expected to occur within the jurisdiction of that extinguishing system. With regard to the availability of water, at least one effective jet of water is to be immediately available from any hydrant in an interior space. The continued supply of that water is to be ensured by the automatic starting of a fire pump.

Fire extinguishing installation   消防设备(灭火装置)

Fire extinguishing medium   灭火剂

Fire extinguishing objective   灭火目标   每艘舰船的装备应使其能够快速地将探测到的火灾安全、有效地扑灭。   Fire extinguishing objective means that every ship is to be equipped, so far as is practicable, so that all detected fires can be safely and effectively extinguished.

Fire extinguishing plant   灭火装置

Fire extinguishing system   灭火系统(消防系统)

Fire fighter outfit   消防员装备

Fire fighting   灭火/消防   系指抑制并将火灾迅速地扑灭在火源处。为此,应满足下列功能要求:(1)应安装固定式灭火系统,并充分考虑到受保护处所的潜在火势增大;(2)灭火器材应随时可用。   Fire fighting is to suppress and swiftly extinguish a fire in the space of origin. For this purpose, the following functional requirements shall be met:(1)fixed fire-extinguishing systems shall be installed and the fire growth potential of the protected spaces shall be fully considered;(2) fire-extinguishing appliances shall be readily available.

Fire fighting appliances   消防设备(灭火器具)

Fire fighting equipment   消防用具/消防设备   船上供消防用的各种器具的统称。

Fire fighting gun   消防炮   固定安装在消防船上的炮式灭火剂喷射器。见图 F-8。

图 F-8   消防炮
Figure F-8   Fire fighting gun

Fire fighting medium   灭火剂(灭火介质)

Fire fighting ship 1   第 1 类消防船   系指具有扑灭初期火灾能力的消防设备的船舶。   Fire fighting ship 1 means a ship used for early stage fire fighting.

Fire fighting ship 2   第 2 类消防船   系指具有扑灭大火能力的消防设备的船舶。   Fire fighting ship 2 means a ship used for large fire fighting.

Fire fighting ship 3　第 3 类消防船　系指具有扑灭大火和油类火灾能力的消防设备的船舶。　Fire fighting ship 3 means a ship used for large or oil fire fighting.

Fire fighting system　灭火系统(消防系统)

Fire fighting tug　消防拖船　系指兼作对海上或近岸失火舰船进行灭火施救的拖船。

Fire flaps　防火闸

Fire functional requirements　消防功能要求　系指为了达到消防安全目标,应体现下列功能:(1)用耐热与结构限界面,将船舶划分为若干主竖区和水平区;(2)用耐热与结构限界面,将起居处所与船舶其他处所隔开;(3)限制可燃材料的使用;(4)探知火源区域内的任何火灾;(5)遏制和扑灭火源区域内的任何失火;(6)保护脱险通道和消防通道;(7)灭火设备的随时可用性;(8)将易燃货物蒸发气着火的可能性降至最低。Fire functional requirements means that in order to achieve the fire safety objectives, the following functional requirements are embodied: (1) division of the ship into main vertical and horizontal zones by thermal and structural boundaries; (2) separation of accommodation spaces from the remainder of the ship by thermal and structural boundaries; (3) restricted use of combustible materials; (4) detection of any fire in the zone of origin; (5) containment and extinction of any fire in the space of origin; (6) protection of means of escape and access for fire-fighting; (7) ready availability of fire-extinguishing appliances; and (8) minimization of possibility of ignition of flammable cargo vapour.

Fire gong　火警锣(火警钟)

Fire gun　灭火枪

Fire hazard　失火危险(易燃性)

Fire hazardous　易燃的

Fire hook　消防火钩　供消防人员使用的长柄铁钩。

Fire hose box　消防水带箱　存放消防水带的专用箱。

Fire hose canvas　帆布救火水龙

Fire hoses　消防水带　系指由主管机关认可的由耐腐蚀材料制成,并具备足够的长度供水柱喷射到可能需要使用消防水带的任何处所。每条消防水带应配有一支水枪和必要的接头,并且与其必要的配件和工具一起,应存放在供水消防栓或接头附近的明显位置,以备随时取用。　Fire hoses mean hoses made of non-perishable material approved by the Administration and shall be sufficient in length to project a jet of water to any of the spaces in which they may be required to be used. Each hose shall be provided with a nozzle and the necessary couplings, together with any necessary fittings and tools, be kept ready for use in conspicuous positions near the water service hydrants or connections.

Fire indicator board　火警指示板

Fire insulation　耐火绝缘

Fire integrity　耐火完整性

Fire lamp　火警指示灯

Fire life line　耐火救生绳

Fire locker　消防员装备储存箱

Fire main　消防总管

Fire main pipe　消防总管

Fire main system　消防水总管系统

Fire mask　防火面罩

Fire nozzles　救火水龙喷嘴

Fire passage　消防通道

Fire patrol　消防巡逻

Fire patrol system　消防巡逻制度

Fire plug　消防栓(消火栓)

Fire point　着火点

Fire prevention　防火

Fire prevention equipment　防火设备

Fire prevention goal 1　防火目标 1　舰船内的火源应适当地保持在一个尽可能小的数量。通常,火源设备的布置和可燃性物质应满足 SOLAS 公约 B 篇第Ⅱ-2 章 4.4 条的要求。　Fire prevention goal 1 means that sources of ignition within the ship are to be kept to a number that is as low as reasonably practicable. In general, the arrangements for items of ignition sources and ignitability are to be in accordance with SOLAS Chapter Ⅱ-2, Part B, Regulation 4.4.

Fire prevention goal 2　防火目标 2　应认识到在火源位置,易燃和潜在爆炸危险材料的使用是受限制和控制的:(1)燃油、滑油和其他可燃油液的布置应符合 SOLAS 公约第Ⅱ-2 章 B 部分 4.2 条的要求;(2)舰船内气体燃料的布置应符合 SOLAS 公约第Ⅱ-2 章 B 部分 4.3 条的要求。　Fire prevention goal 2 means that the use of combustible and potentially explosive materials is to be restricted and controlled, taking due cognizance of the locality of ignition sources. (1) Arrangements for oil fuel, lubrication oil and other flammable oils are to be in accordance with SOLAS Chapter Ⅱ-2, Part B, Regulation 4.2; (2) Arrangements for gaseous fuel for domestic purposes are to be in accordance with SOLAS Chapter Ⅱ-2, Part B, Regulation 4.3.

Fire prevention goal 3　防火目标 3　与直升机设备有关的火灾危险应尽可能合理地降至最低。通常,直升机设备的布置应符合 SOLAS 公约第 Ⅱ-2 章 18 条的要求。 Fire prevention goal 3 means that the fire hazards associated with helicopter facilities are to be as low as reasonably practicable. In general, the arrangements for helicopter facilities are to be in accordance with SOLAS Chapter Ⅱ-2, Regulation 18.

Fire prevention goal 4　防火目标 4　与危险货物运送有关的火灾危险应尽可能合理地降至最低。通常，危险货物运送的布置应符合 SOLAS 公约第 Ⅱ-2 章 19 条的要求。 Fire prevention goal 4 means that the fire hazards associated with the carriage of dangerous goods are to be as low as reasonably practicable. In general, the arrangements for the carriage of dangerous goods are to be in accordance with SOLAS Chapter Ⅱ-2, Regulation 19.

Fire prevention goal 5　防火目标 5　与车辆运送、特种和滚装处所有关的火灾危险应尽可能合理地降至最低。通常，车辆运送、特种和滚装处所的布置应符合 SOLAS 第 Ⅱ-2 章 20 条的要求。 Fire prevention goal 5 means that the fire hazards associated with the carriage of vehicles, special category and RO-RO spaces are to be as low as reasonably practicable. In general, the arrangements for the carriage of vehicles, special category and RO-RO spaces are to be in accordance with SOLAS Chapter Ⅱ-2, Regulation 20.

Fire prevention objective　火灾预防目标　每艘舰船的设计和装备,应使其在执行相应民事和军事任务时能够防止火灾或爆炸的发生。 Fire prevention objective means that every ship is to be designed and equipped so as to prevent the occurrence of fire or explosion, taking due account of its civil and military operational role.

Fire proof rock wool-panel　复合岩棉板(耐火)　系指以玄武岩为材料，经过高温融熔加工成的人工无机纤维，具有质量轻、导热系数小、吸热、不燃的特点，初始研制在建筑中是常见的应用类型，多用于工业建筑。应符合《建筑绝热材料的应用类型和基本要求》的规定。1981 年 6 月量试成功岩棉板是一种新型的保温、隔燃、吸声材料。

Fire protecting arrangement　消防设备布置(消防装置)

Fire protection　耐火分隔　系指在一定时间内能把火势控制在一定空间内,阻止其蔓延扩大的一系列分隔设施。

Fire protection device　防火设备(消防设备)

Fire protection equipment　防火设备(消防设备)

Fire protection facility　消防设施

Fire protection rules　消防规范

Fire protection system　防火系统

Fire pumps　消防泵　卫生泵、压载泵或通用泵均可接受作为消防泵,但其通常不得用于抽送油类,且其如偶尔用于驳运或泵送燃油,应装设合适的转换装置。 Fire pumps mean that sanitary, ballast, bilge or general service pumps may be accepted as fire pumps, provided that they are not normally used for pumping oil and that if they are subject to occasional duty for the transfer or pumping of oil fuel, suitable change-over arrangements are fitted.

Fire quarters　消防站

Fire resistance　耐火性

Fire resistance gate(fire resisting door)　耐火门

Fire resistant insulation　耐火绝缘

Fire resistant paint　防火漆

Fire resisting cable　耐火电缆

Fire resisting closing lifeboat　封闭式耐火救生艇　配置在油船或海上钻井平台装置上,能供火灾时搭载人员后迅速驶离着火海区的救生艇。

Fire resisting equipment　防火设备

Fire resisting insulation material　耐火绝热材料

Fire resisting material　耐火材料

Fire resisting paint　防火漆　是由成膜剂,阻燃剂,发泡剂等多种材料制造而成的一种阻燃涂料;由于家居中大量使用木材、布料等易燃材料,所以,舱室防火已经是一个值得注意的问题。

Fire resisting wood　耐火木材

Fire retardant　阻火的(滞燃的,滞火剂)

Fire retardant linoleum　耐火漆布

Fire retardant paint　耐火漆

Fire retardant polyester resin　耐火聚酯树脂

Fire retardant resin　阻燃型树脂

Fire retardant tile　耐火砖　简称火砖。具有一定形状和尺寸的耐火材料。按制备工艺方法划分可分为烧成砖、非烧成砖、电熔砖(熔铸砖)、耐火隔热砖;按形状和尺寸可分为标准型砖、普通砖、特异型砖等。可用作建筑窑炉和各种热工设备的高温建筑材料和结构材料,并在高温下能经受各种物理化学变化和机械作用。

Fire room　锅炉舱

Fire room flat　锅炉舱平台

Fire safety　消防　包括探火、防火和灭火。 Fire safety includes fire detection, fire protection and fire extinction.

Fire safety appliances　防火器材

Fire safety objectives　消防安全目标　是:(1)防止火灾和爆炸的发生;(2)减少火灾造成的生命危险;

(3)减少火灾对船舶、船上货物和环境破坏的危险;(4)将火灾和爆炸抑制、控制和扑灭在火源舱室内;(5)为乘客和船员提供充分和随时可用的脱险通道。 Fire safety objectives are to: (1) prevent the occurrence of fire and explosion; (2) reduce the risk to life caused by fire; (3) reduce the risk of damage caused by fire to the ship, its cargo and the environment; (4) contain, control and suppress fire and explosion in the compartment of origin; (5) provide adequate and readily accessible means of escape for passengers and crew.

Fire safety operation booklets   消防安全操作手册
系指包含与消防安全有关的船舶安全操作和货物装卸安全操作的必要信息和须知的小册子。该手册应包括关于船员在船舶装卸货物时和航行时对船舶总体消防安全所负责任方面的信息,还应对装卸一般货物时需采取的消防安全预防措施进行解释。对于装运危险货物和易燃散货的船舶,消防安全车操作手册还应相应提及"固体散装货物安全操作规则""国际散装化学品规则""国际气体运输船规则"和"国际海运危险货物规则"中有关消防和紧急货物装卸的须知。应在每一船员餐厅和娱乐室或在每一船员居住舱室内配备一本消防安全操作手册。消防安全操作手册应以船上的工作语言写成。消防安全操作手册可与培训手册合并。 Fire safety operation booklets mean booklets contained the necessary information and instructions for the safe operation of the ship and cargo handling operations in relation to fire safety. The booklet shall include information concerning the crew's responsibilities for the general fire safety of the ship underway loading and discharging cargo. Necessary fire safety precautions for handling general cargoes shall be explained. For ships carrying dangerous goods and flammable bulk cargoes, the fire safety operational booklet shall also provide reference to the pertinent fire-fighting and emergency cargo handling instructions contained in the Code of Safe Practice for Solid Bulk Cargoes, the International Bulk Chemical Code, the International Gas Carrier Code and the International Maritime Dangerous Goods Code, as appropriate. The fire safety operational booklet shall be provided in each crew mess room and recreation room or in each crew cabin. The fire safety operational booklet shall be written in the working language of the ship. The fire safety operational booklet may be combined with the training manuals.

Fire safety system   消防安全系统
Fire Safety Systems Code   《消防安全系统规则》
系指 IMO 的海上安全委员会以 MSC.98(73)决议通过的《国际消防安全系统规则》。此规则可由 IMO 修正,但此种修正应根据 SOLAS 公约第Ⅷ条关于适用于附则(除第Ⅰ章外)的修正程序的规定予以通过、生效和实施。 Fire Safety Systems Code means the International Code for Fire Safety Systems as adopted by the Maritime Safety Committee of the Organization by Resolution MSC.98(73), as may be amended by IMO, provided that such amendments are adopted, brought into force and take effect in accordance with the provisions of article Ⅷ of the present Convention concerning the amendment procedures applicable to the annex other than Chapter 1 thereof.

Fire safety training manual   消防安全培训手册
系指用船上的工作语言写成的,并在每一船员餐厅和娱乐室或在每一船员居住舱室内配备一份的培训手册。培训手册应包括 SOLAS 公约第Ⅱ-1 章第 15.2.3.4. 条所要求的须知和资料。这些资料的部分可以用视听辅助教材形式提供,用以代替手册。 Fire safety training manual means a training manual written in the working language of the ship and provided in each crew mess room and recreation room or in each crew cabin. The manual shall contain the instructions and information required in SoLAS Part Ⅱ-1 Regulation 15.2.3.4. Part of such information may be provided in the form of audio-visual aids in lieu of the manual.

Fire sand box   灭火沙箱
Fire screen   (锅炉炉门)防火筛(挡火屏)
Fire service pipe   消防管
Fire service pump   消防泵
Fire signal   火警信号
Fire slice   长柄火铲
Fire smothering system   窒息灭火系统
Fire smothering unit room   窒息灭火装置室
Fire source   火源 系指起火的可燃材料以及覆盖在墙面和天花板内的可燃物。 Fire source is defined as the combustible material that can start a fire and the combustible material covering walls and ceiling.

Fire sprinkler   喷洒灭火器
Fire sprinkling system   喷洒灭火系统
Fire stopper   灭火器
Fire strip   挡火条
Fire suppression   火的抑制(火灾抑制)   系指:(1)降低火燃烧产生的热量,通过对火的控制及其从起火处向外扩散以及减少过火面积;(2)迅速降低火焰热释放量并通过用足量的水直接覆盖在燃烧物表面,以防止其重新产生火灾。 Fire suppression means: (1) a reduction in heat output from the fire and control of the fire to restrict its spread from its seat and reduce the flame area;

(2) sharply reducing the heat release rate of a fire and preventing its regrowth by means of a direct and sufficient application of water through the fire plume to the burning fuel surface.

Fire suppression system(FSS) 灭火系统 分为两大类,(1)多个灭火器通过连接物相互关联形成一个整体灭火;(2)单个灭火装置其本身就是一个灭火系统。

Fire tender 消防船

Fire test 耐火试验

Fire test procedure(FTP) 耐火试验程序

Fire Test Procedures(FTP) Code "耐火试验程序规则" 系指 IMO 的海上安全委员会以 MSC.61(67)决议通过的"国际耐火试验程序应用规则"。此规则可由该组织修正,但此种修正应根据 SOLAS 公约的第Ⅷ条关于适用于附则(除第Ⅰ章外)的修正程序的规定予以通过、生效和实施。 Fire Test Procedures Code means the International Code for Application of Fire Test Procedures as adopted by the Maritime Safety Committee of the Organization by Resolution MSC.61(67), as may be amended by the Organization, provided that such amendments are adopted, brought into force and take effect in accordance with the provisions of article Ⅷ of the SOLAS Convention concerning the amendment procedures applicable to the annex other than chapter Ⅰ thereof.

Fire tube boiler(fire tube boiling) 火管锅炉

Fire up (锅炉)生火

Fire valve 消防阀(消火栓)

Fire vessel 消防船(救火船)

Fire warning 火警报警

Fire welding 锻接

Fire zones 防火区 系指用防火分隔将船体、上层建筑和甲板室划分成的区域。防火分隔是以限制火焰蔓延为目的的那些安装和/或保护。 Fire zones are those sections which the hull, superstructure and deckhouses are divided into by fire resistant divisions. Fire resistant divisions are those which are installed and/or protected for the purpose of restricting the spread of fire.

Fire-control radar 火控雷达 也称照射火控雷达。是用于给导弹或者火炮提供目标参数并引导武器系统进行攻击的雷达。一般火控雷达都与探测雷达和火力控制系统配合使用。为提高系统的整体工作效率,火控雷达的火力控制功能一般都需要通过辅助的计算机实现,因此一般在综合武器平台如战斗机、军舰、防空导弹系统上,火控雷达才会得到应用。在一些低强度摩擦中,使用火控雷达照射对方,这种自卫行为,是带有警告性质的。

Fired boiling 燃油锅炉

Fired pressure vessel 受火压力容器 系指完全或者部分与燃烧气体或者燃烧器的火焰接触的压力容器。 Fired pressure vessel is a pressure vessel which is completely or partially exposed to fire from burners or combustion gases.

Fire-director room(weapon control room) 指挥仪室 供安置射击指挥仪并设置相应操作部位的舱室。

Fire-extinguishing 灭火剂 分为采用喷射(jet)、喷淋(spraying)、喷雾(fine spraying)方式的水(water);为提高抑制效果而在水中加入碳酸钙等药剂的强化液(water with chemical additive)、气体(gas)、泡沫(foam)、粉末(powder)和砂(sand)等。

Fire-fighter's outfit 消防员装备(消防员装具) 系指一套为个人配备的一具呼吸器和一根救生绳。个人配备包括防护服、消防手套、消防靴、消防头盔、安全灯(手提灯)和太平斧组成的装备。 Fire-fighter's outfit is a set of personal equipment, including a breathing apparatus and a lifeline. Personal equipment is to consist of protective clothing, gloves, boots, a helmet, an electric safety lamp (hand lantern) and an axe.

Fireman 生火(锅炉工)

Fireman's axe 消防斧 系指供消防专用的长柄斧。

Fireman's outfits 消防员装备 其组成应如下所列:.(1)防护服,其材料应能保护皮肤不受火焰的热辐射,并不受蒸汽的灼伤和烫伤。衣服的外表面应是防水的;(2)长筒靴和手套应由橡胶或其他电绝缘材料制成;(3)一顶能有效地防护撞击的硬头盔;(4)一盏认可型的电安全灯(手提灯),其最短照明时间应为 3 h;(5)一把具有绝缘手柄的太平斧;(6)一具储压式呼吸器,其可供使用的时限应至少达 30 min,且自由空气的容积至少为 1 200 升。每具呼吸器至少应配置一套备用的装满空气的空气瓶;(7)每具呼吸器应配有一根足够长度与强度的耐火救生绳,此绳用弹簧卡钩系在呼吸器背带上,或系在一条独立的腰带上,以便在使用救生绳时防止与呼吸器脱开。 The composition of a fireman's outfit is to be as follows:(1)protective clothing, whose material to protect the skin from heat radiating from the fire and from burns and scalding by steam. The outer surface is to be water-resistant;(2)boots and gloves of rubber or other electrically non-conducting material;(3)a rigid helmet providing effective protection against impact;(4)an electric safety lamp(hand lantern)of an approved type with a minimum operating period of three hours;(5)an axe having an insulated handle;(6)a self-contained breathing apparatus, which is to be capable of functioning for a period of at least 30 minutes and having a capacity of at least 1 200 litres of free air. Spare and fully charged air bottles are to be provided at the rate of at least one set per required apparatus; (7) for each breathing apparatus, a fireproof lifeline of sufficient length and strength is to be provided capable of being attached by means of a snaphook to the harness of the apparatus or to a separate belt, in order to prevent the breathing apparatus becoming detached when the life-line is operated.

Fire-proof 防火的(耐火的,不燃的)

Fire-proof alloy 耐火合金

Fire-proof arrangement 防火装置

Fire-proof bulkhead 防火舱壁(耐火舱壁)

Fire-resisting closing lifeboat 封闭式耐火救生艇 配置在油船或海上石油钻井装置上,供火灾时搭载人员尽快离开着火区的全封闭式机动救生艇。

Fire-resisting divisions 耐火分隔 系指由符合以下规定的舱壁和甲板组成的分隔:(1)它们应由符合有关要求的具有隔热或阻燃性质的不燃或阻燃材料制成;(2)它们应有适当的加强;(3)它们的构造应在相应的防火时间范围内,能防止烟及火焰通过;(4)需要时,它们应在相应的防火时间范围内,仍具有承受载荷的能力;(5)它们应有这样的温度特性,即在相应的防火时间范围内,背火面的平均温度较初始温度的温升不大于140℃,而且包括任何接头在内的任一点的温升不超过180℃;(6)应按耐火试验程序规则的要求对原型舱壁和甲板进行一次试验,以确保满足上述要求。 Fire-resisting divisions are those divisions formed by bulkheads and decks which comply with the following: (1) they shall be constructed of non-combustible or fire-restricting materials whose insulation or inherent fire-resisting properties satisfy the relevant requirements; (2) they shall be suitably stiffened; (3) they shall be so constructed as to be capable of preventing the passage of smoke and flame up to the end of the appropriate fire protection time; (4) where required they shall maintain load-carrying capabilities up to the end of the appropriate fire protection time; (5) they shall have thermal properties such that the average temperature on the unexposed side will not rise more than 140 ℃ above the original temperature, nor will the temperature, at any one point, including any joint, rise more than 180 ℃ above the original temperature during the appropriate fire protection time; (6) a test for a prototype bulkhead or deck in accordance with the Fire Test Procedures Code shall be required to ensure that it meets the above requirements.

Fire-restricting materials 阻燃材料 系指能符合国际耐火试验程序规则(FTP 规则)的材料。 Fire-restricting materials are those materials which have properties complying with the Fire Test Procedures(FTP) Code.

Fire-smothering unit room 灭火装置室 系指供集中安置化学灭火装置的房间。

Fire-tested penetration device 经耐火试验的贯穿装置

Fire-tube boiler 火管锅炉 烟气在置于容水空间内的烟管内流动的锅炉。

Firewall 防火墙 是一个或一组系统,用来在两个或多个网络间加强访问控制,限制入侵者进入,从而起到安全保护的作用。

Fire-water-tube boiler 烟-水管锅炉 在卧式烟管锅炉的烟室中布置水管束和水管的锅炉。

Firing 点火(燃烧)

Firing floor 炉底(炉舱工作台)

Firing order (ignition order) 点火次序 内燃机各气缸轮流点火的次序。

Firing test 点火试验

Firing tools 生火工具

Firing-up pump 点火泵

Firm offer 实盘 系指发盘人/发价人对接受人所提出的,是一项内容完整、明确、肯定的交易条件,一旦送达受盘人(即接受人或称受发价人)之后,则对发盘人产生拘束力,发盘人在实盘规定的有效期内不得将其撤销或加以变更。表明发盘人有肯定订立合同的意图。如果受盘人在有效期内无条件地接受,就可以达成交易,成为对买卖双方都有约束力的合同。实盘必须同时具备3个条件才能成立:内容必须是完整和明确的;内容必须是肯定的,无保留条件的;实盘必须规定有效期限。其内容包括货物品名、品质规格、包装、数量、价格等。

Firsov's diagram 费尔索夫图谱 以舯吃水为横坐标,艉吃水为纵坐标,所绘制的等排水体积和等浮心纵向坐标2组曲线的图谱。用以求得大纵倾下船舶的排水量和浮心纵向位置。

First filter 粗滤器

First island chain 第一岛链 系指冷战期间,美国为阻止共产势力扩张,而在亚洲东部岛屿协助建立"民主"政权,同时提供军力以及其他各种形式上的援助。其地理位置是北起日本群岛、琉球群岛、中接台湾岛,南至菲律宾、大巽他群岛组成的链形岛屿带。

First order divided difference filter 一阶差分滤波器

First overshoot angle 第一超越角 系指Z形操纵试验中紧接着第二次操舵增加的偏离值。First overshoot angle is the additional heading deviation experienced in the zig-zag test following the second execute.

First pickling 初次酸洗

First principle procedures 直接计算 当船级社规范有专门要求，或者采用新颖的结构形式；或者结构的布置、船舶尺度超出船级社规范规定时，而采用适用的通用程序或经船级社认可的程序进行的构件尺寸、强度校核等的计算。直接计算所考虑的装载工况应包括船舶营运中最为严重的装载工况。

First-aid repair 抢修 系指事故发生后或因时间紧而动员人们抓紧时间高速度地修理。

Fish(in bulk) 鱼(散装) 系指冷冻后散装载运的鱼。散装载运的鱼可能流态化。该货物为不燃物或失火风险低。 Fish(in bulk) is fish carried in bulk after freezing. Fish carried in bulk may liquefy. This cargo is non-combustible or has a low fire-risk.

Fish box 鱼箱 系指盛装渔获物的无盖箱子。

Fish carrier 水产品运输船

Fish catch(fishing game) 渔获物 在水中捕获的水产品的总称。主要指鱼、虾、蟹、软体动物及贝类等。

Fish factory ship 鱼类加工船 又称水产品加工船，专门在渔场将渔获物就地加工成成品或半成品的船舶。

Fish finder 鱼探仪 利用鱼体对声波的反射搜索海中鱼群的设备。

Fish hatch cover 鱼舱盖 设置于渔船的鱼舱上，具有保温性能的舱口盖。

Fish hold 鱼舱 渔业船上装载渔获物的舱室。

Fish luring light boat, lurker 诱鱼灯船 通常与围网渔船配合作业，其上设有专用的诱鱼灯，用以在水上或水下诱集鱼群的船舶。

Fish pond(deck pond) 拦鱼池 为清洗和处理渔获物，在甲板上以活动的木格栅围成的用以临时存放渔获物的区域。

Fish processing room 鱼品加工间 船上把渔获物加工成鱼品的工作间，如鱼粉间、鱼油间、鱼片间等。

Fisheries administration ship 渔政船 各种指导、监督渔业生产，以及负责渔场救助工作的船舶的总称。包括渔业指导船、渔业监督船、渔业救助船等。见图F-9。

Fisheries auxiliary vessel 渔业辅助船 在渔业生产中，各种从事加工、补给、运输、医疗等辅助工作的船舶的总称。

Fisheries guidance ship 渔业指导船 装有探鱼仪、渔具、测量水文气象的仪器和通信设备等，在海上对渔场船队实行指导和指挥渔业生产的船舶。

Fisheries inspection boat 渔业监督船 对渔船在

图 F-9 渔政船
Figure F-9 Fisheries administration ship

渔业生产中执行国家有关规定和国际渔业协定情况进行监督的船舶。

Fisheries rescue ship 渔业救助船 在渔场上担负人员医疗和救助工作的船舶。

Fisheries tender 渔业补给船 向渔船供应生活用品和渔需品的船舶。

Fisheries vessel 渔船 各种直接从事水生物捕捞生产的船舶的统称，包括渔政船、渔船、渔业辅助船等。

Fishery Dispute Resolution Center of the Arbitration Commission 仲裁委员会的渔业争议解决中心

Fishery inspection vessel 渔业检查船

Fishery mother ship 渔业基地船 在渔船队中，能较长时间地支持整个船队进行捕捞生产，处理全部渔获物，供给生活用品和渔需品，提供人员休息、医疗和船舶修理等条件的船舶。

Fishery patrol 渔业监视指导船

Fishery products processing 鱼品加工 为保持渔获物所采用的各种加工方法，包括腌制、干制、制罐、制粉、炼油的总称。

Fishery research 渔业调查船

Fishery ship 渔业船舶 系指与渔业生产活动有关的捕捞、加工、运输、养殖、采集和科研等船舶的总称。通常将其中直接利用网具、渔具等捕捞鱼类或其他水生物的船舶称为捕捞生产渔船；其余的称为渔业辅助船。

Fishery supervision boat 渔业监督船

Fishery training 渔业实习船

Fishery vessel 渔业船 从事渔业生产以及为渔业生产服务的船舶。如渔船、拖网渔船、捕捞渔船、捕鲸船、水产品加工船、水产品运输船、渔业训练/调查船、渔业检查船、渔业监督船(fishery supervision boat)等。

Fish-factory ship 鱼类加工船 系指专用于加工鱼类的船舶。 Fish-factory ship means a ship specialized for fish processing.

Fishing auxiliary vessel 渔业辅助船 在渔业生产中，各种从事加工、补给、运输、医疗等辅助工作的船的统称。

Fishing boat carrier 母子式渔船 一艘母船装载若干艘子船到渔场，并向子船供给油、水及各种备品，而子船则脱离母船专门从事捕捞作业，并把渔获物交给母船处理的组合式渔船。

Fishing deck 渔捞甲板 渔船上供捕捞作业用的甲板。

Fishing equipment 捕捞设备 渔船捕捞生产过程中，所用各种简单装置的统称。

Fishing gear 渔具 系指可放置于水上或水中或海底拟用于捕捞或后续的捕捞或控制或采收海洋或淡水有机物的任何物理装置或其组件或各种工具的组合。Fishing gear means any physical device or part thereof or combination of items that may be placed on or in the water or on the sea-bed with the intended purpose of capturing, or controlling for subsequent capture or harvesting, marine or fresh water organisms.

Fishing gear hold 渔具舱 渔业船上装载渔具和渔具材料的舱室。

Fishing motor sailer 机帆渔船 备有风帆装置或形似帆船的机动渔船。

Fishing sailer 风帆渔船 配置桅杆和帆，依靠风力推进的渔船。

Fishing ship 渔船 系指用于捕捞鱼类或其他海洋生物资源等的船舶。渔船的种类很多，按捕鱼的区域分为远洋渔船、近海渔船和内河渔船。按捕捞方式分为网式渔船（包括拖网、围网、刺网渔船等）、钓式渔船、特种渔船等。按船体材料分为木质、钢质、玻璃钢质、铝合金及各种混合结构渔船。渔船的主要特征如下：(1) 多数捕捞船的船型较小，为适应在风浪中连续航行和作业的需要，要求渔船具有较好的稳性、耐波性和适航性；(2) 渔船在作业期间载重量变化很大；(3) 船用设备要求结构性能可靠，坚固耐用，维修方便；(4) 主机功率较大航速较快；(5) 除配置一般船用设备外，还需配备捕捞设备、保鲜和加工设备、助渔和导航设备等，鱼舱要求隔热设施性能好。见图F-10。

Fishing training/research ship 渔业训练/调查船

Fishing vessel 渔船 系指：(1) 具有捕鱼设备的船舶；(2) 用于捕捞鱼类、鲸鱼、海豹、海象或其他海洋生物资源的船舶。Fishing vessel means: (1) a vessel provided with fishing equipment; (2) a vessel used for catching fish, whales, seals, walrus or other living resources of the sea.

Fishing yield 渔获量 渔获物的数量。

Fishmeal (fish-scrap) (stabilized UN 2216) (Antioxidant treated) 加稳定剂的鱼粉（鱼渣）（经抗氧处

图 F-10 渔船
Figure F-10 Fishing ship

理) 系指将含有脂肪的鱼加热和烘干而制成的褐色至褐绿色物质。含水量按质量计大于5%但不超过12%。强烈气味可能影响其他货物。脂肪含量按质量计不大于15%。易自燃，除非脂肪含量低或经有效的抗氧处理。易使货物处所缺氧。 Fishmeal (fish-scrap), stabilized is a brown to greenish-brown material obtained through heating and drying of oily fish. Moisture content: greater than 5% but not exceeding 12%, by mass. Strong odour may affect other cargoes. Fat content is not more than 15%, by mass. It is liable to heat spontaneously unless it has low fat content or it is effectively anti-oxidant treated. It is liable to cause oxygen depletion in cargo space.

Fit 装配（配合）

Fit tolerance 配合公差

Fitter 装配工（钳工）

Fitting tolerance 装配公差（配合公差）

Fittings 附件（塑料管） 系指用塑料制成的弯头、肘形弯管、组装分支管部件等。 Fittings mean bends, elbows, fabricated branch pieces etc. made of plastic materials.

Fit-up 装置（装配，供给设备）

Fit-up gap 装配间隙

Five-dimensional space 五维空间 系指海陆空天电磁。是空间一个集合，最基本的元素是点，点的集合是线面体，线面体运动产生了三维体，三维体的运动产生了时间，以此类推。这一个说法，也就是人类给四维的最好说法。简单地说，五维就是由于四维运动产生，假设四维空间可以对折，那么对折后的那部分所谓的无，就会由于四维的运动而给填补，那样大家也许会说，这样并不能影响时间的运动，也就是没对四维造成改变，不能是四维运动。不是那样的，时间就是由三维运动产生，既然这样不就是三维的改变，变的让时间需要变短，那样不就成了五维，也就是说那个轴就是速度。三维线，四维面，五维扇，六维1/4 球，七维1/2 球。

Five-layer plywood 五夹板 又叫多层板和五合板,层数不同叫法不同,其性能优劣主要看原料。

Fix 固定(安装,确定)

Fixation 固定(锁紧)

Fixed(locked) propeller test(fixed propeller test) (螺旋桨)锁定试验 将螺旋桨或其模型制住不转以测定其阻力的试验。

Fixed aerosol fire-extinguishing system 固定式气溶胶灭火系统

Fixed ballast 固定压载

Fixed blade 固定桨叶

Fixed blade(stator blade, stationary blade) 静叶片 装配在隔板、机匣等固定部件上的叶片。

Fixed blade propeller 固定桨叶螺旋桨

Fixed blade ring 静叶环 外缘装在反动式汽轮机气缸槽内,内缘可装两级以上隔板的中间支承元件。

Fixed blading 固定叶片

Fixed bow fin 固定艏鳍(固定船首减纵摇鳍)

Fixed carbon dioxide fire-extinguishing system 固定式二氧化碳灭火系统

Fixed coordinate system 固定坐标系 固定于地球的直角坐标系。符号按右手法则。

Fixed deck foam fire-extinguishing system 固定式甲板泡沫灭火系统

Fixed deck foam system 固定式甲板泡沫(灭火)系统

Fixed drilling platform 固定式钻井平台

Fixed emergency fire pumps 固定式应急消防泵

Fixed fire detection and fire alarm system 固定式探火和失火报警系统

Fixed fire extinguishing system 固定灭火系统 固定安装在船舶上的灭火系统。

Fixed fire-extinguishing arrangement 固定式灭火装置

Fixed fire-extinguishing system 固定式灭火系统

Fixed fire-fighting system 固定式消防系统

Fixed float(radial float) 定蹼 明轮上所装固定的蹼板。

Fixed foam fire-extinguishing system 固定式泡沫灭火系统

Fixed foam monitor system 固定式泡沫炮系统

Fixed frame(outside rim) 窗座 与船体紧固,用以安装窗框或窗玻璃的座子。

Fixed gas fire-extinguishing system 固定式气体灭火系统

Fixed gear(parts) 固定部件(不可卸零部件) 铆焊于吊杆、起重柱以及船体结构上的吊杆装置零部件。如眼板、吊货杆叉头、吊货杆承座包括转轴、吊杆箍环和嵌入滑轮等。

Fixed guide vanes 固定导流片 为提高推进效能,装在螺旋桨前或后的固定导流叶片。

Fixed high-expansion foam fire-extinguishing system 固定式高倍泡沫灭火系统

Fixed hydrocarbon gas detection systems 固定式碳氢化合物气体探测系统

Fixed inert gas fire-extinguishing system 固定式惰性气体灭火系统

Fixed installation type fish-pump 固定鱼泵 固定安装于渔船或码头,吸送鱼水混合物的泵。

Fixed local application fire-extinguishing system 固定式局部使用灭火系统

Fixed local application fire-extinguishing system activation alarm 固定式局部使用灭火系统启动报警 系指通过指示已启动的部分,提醒船员该系统已排放的报警。 Fixed local application fire-extinguishing system activation alarm: an alarm to alert the crew that the system has been discharged, with indication of the section activated.

Fixed local water based fire-extinguishing arrangement 固定式局部水基灭火装置

Fixed low-expansion foam fire-extinguishing system 固定式低倍泡沫灭火系统

Fixed network 固定网

Fixed or floating platforms 固定或浮动平台 系指位于海上从事海底矿产资源的勘探、开发或相关联的近海加工的固定或浮动结构物。 Fixed or floating platforms means fixed or floating structures located at sea which are engaged in exploration, exploitation or associated offshore processing of sea-bed mineral resources.

Fixed oxygen analyzer 固定式氧气分析仪

Fixed phase-shift 固定相移 系指以相位周为单位的基线长的小数与仪器的相位延迟两部分之总和。无线电双曲线相位系统工作过程中,副台是在收到主台发射的信号后保持原相位转发的,因此由船台测得的相位差中,包括基线所对应的相位周整数和小数两部分。部分可在连测中消去,剩下小数部分与副台从接收主台到发射信号,船台从接收信号到终端显示所存在的相位延迟均属固定相移,测量中应消去固定相移。

Fixed pitch propeller 定螺距螺旋桨 系指螺距固定的螺旋桨。

Fixed platform 固定式平台 系指设有工作甲板的桁架式管状结构。该平台是以桩腿固定到海底。

Fixed platform is a tubular truss structure with a working deck. The platform is secured to the seabed with piles.

Fixed pressure water-spraying and water-mist fire-extinguishing system　固定式压力水雾和细水雾灭火系统

Fixed pressure water-spraying fire-extinguishing system　固定式压力水雾灭火系统

Fixed pressure water-spraying system　水雾灭火系统　设置于燃油锅炉及柴油机舱内在火灾发生时通过高压水雾化喷嘴，喷出水雾灭火的消防系统。

Fixed price　定价

Fixed price　固定价格　系指各个成员之间达成协议，彼此同意以相同的价格出售其产品，以消除各成员之间在产品售价方面竞争的一种做法。企业之间之所以通过协议固定产品的价格，其主要目的是为了消灭彼此之间的竞争，从而达到维护自身利益的目的。

Fixed production rig　固定式采油平台　又称生产平台，用于汇集各个井的原油管系，是一种下端固定于海底的平台。形式很多，常见的有：导管架平台、重力式平台、拉索塔平台和张力腿平台等。

Fixed production system　固定式生产系统　系指以固定式结构支承海上油气处理装置的生产系统。

Fixed propeller　整叶桨(定距螺旋桨)

Fixed railing　固定栏杆　直接固定在船体上，既不可拆卸又不可倒下的栏杆。

Fixed recording medium　固定记录介质　系指最终记录介质的一部分，具有保护以防止在失火、冲击、贯穿和海底长期存在后损坏。它可从沉船的甲板处回收，并有指示其位置的设施。 Fixed recording medium means a part of the final recording medium which is protected against fire, shock, penetration and a prolonged period on the ocean floor. It is expected to be recovered from the deck of the ship that has sunk. It has means of indicating location.

Fixed station observation　大面调查　在调查海区布设若干测站，于一定时间内断面测站上进行观测。

Fixed type floating crane　定机式起重船　起重吊杆为固定式，其舷外跨距不能变幅的起重船。

Fixed water-based local application fire-fighting system(FWBLAFFS)　固定式水基局部使用灭火系统

Fixed water-mist fire-extinguishing system　固定式细水雾灭火系统

Fixed-supported　固定支承的

Fixed-wing aeroplane　固定翼飞机　简称定翼机，常被再简称为飞机，系指由动力装置产生前进的推力或拉力，由机身的固定机翼产生升力，在大气层内飞行的重于空气的航空器。它是固定翼航空器的一种，也是最常见的一种，另一种固定翼航空器是滑翔机。飞机按照其使用的发动机类型又可被分为喷气式飞机和螺旋桨飞机。

Fixing screw　定位螺钉

Flag chest　旗箱　专供存放国际信号旗、国旗和公司旗等旗帜用，并按号旗所表示的字母分格的箱柜。

Flag hook　旗钩　用以将号旗或号型与旗索连接的扣钩。

Flag line　旗索　悬挂号旗及号型的纤维绳。

Flag of convenience　方便旗　系指不论船东的国籍(也许因为某些原因或吨位费)如何都可以进行船舶登记的主管国家。 Flag of convenience is a country that admits the vessel registration regardless of the owner's nationality(perhaps for some reasons or tonnage frees).

Flag staff　旗杆　船上供悬挂旗帜用的各种杆件的统称。

Flag State　船旗国　系指船舶被授权悬挂其国旗的国家。 Flag State means a State whose flag a ship flies and is entitled to fly.

Flake　氧化皮

Flame　火焰(火舌,燃烧)

Flame annealing　火焰退火

Flame arrester　火焰制止器(防焰器)　即根据规定的功能标准防止火焰通过的装置。它的火焰制止成分是根据火焰熄灭原理。 Flame arrester is a device to prevent the passage of flame in accordance with a specified performance standard. Its flame arresting element is based on the principle of quenching.

Flame bending of pipe　火焰弯管

Flame brazing　火焰钎焊(气焊)

Flame cleaning　火焰清理(火焰除锈)

Flame cleaning torch　火焰清净炬

Flame colour test　火焰试验(焰色试验)

Flame cutting　气割

Flame detector　火焰探测器

Flame failure　火焰故障

Flame failure protective device　熄火保护装置　系指当燃烧室熄火时 能自动切断燃油供给的装置。 Flame failure protective device means a device capable of shutting off the fuel supply automatically in the event of flame failure in the combustion chambers.

Flame failure safeguard　熄火保护　为防止燃烧过程中因炉膛突然熄火导致事故而采取的安全保护措施。

Flame furnace　火焰炉(反射炉)

Flame lighter　点火器

Flame locator(flame detector) 火焰探测器 用来检验燃气轮机运行时燃烧室内是否着火以及负荷突然下降时火焰是否熄火等情况的器具。

Flame resistance 耐燃性

Flame resisting 耐火的(防火的)

Flame retardant 阻燃剂

Flame retarding 滞燃

Flame retarding materials 滞燃材料 系指具有不会传播火焰,且其延续燃烧时间不大于相应燃烧试验中所规定值的耐火特性的材料。

Flame screen 防火网 系指根据规定的功能标准利用金属网防止无限制的火焰通过的装置。Flame screen is a device utilizing wire mesh to prevent the passage of unconfined flames in accordance with a specified performance standard.

Flame speed 火焰速度 系指火焰沿着管系或其他系统传播的速度。 Flame speed is the speed at which a flame propagates along a pipe or other system.

Flame supervision device(LPG system) 火焰监控装置(LPG 系统) 具有传感元件,能因火焰出现或消失而动作,使 LPG 气源至燃烧器的进气口打开或关闭的装置。 Flame supervision device is the device that has a sensing element, activated by presence or absence of a flame, which causes the inlet of the LPG supply to a burner to be opened or closed.

Flame temperature 火焰温度(着火温度)

Flame tube(combustion liner, cans) 火焰筒 燃油喷入其中与空气混合并连续燃烧产生高温燃气的燃烧室内层筒形或环形结构。

Flame tube air film cooling 火焰筒气膜冷却 空气流经大量分布在火焰筒壁面上的缝隙和小孔,而在火焰筒内壁形成空气薄膜使炽热燃气不直接与火焰筒壁面接触而使壁面温度下降的一种冷却方式。

Flameproof enclosure(electrical equipment) "d"隔爆型(电气设备) 其外壳能承受已进入其内部之易燃混合物的内部爆炸而不足以使其损坏,且不会通过此外壳的任何接缝或结构开口,而引起由一种或多种已指定的气体或蒸发气所构成的外部爆炸性大气的点燃的电气设备的防护形式。注意:IEC 60079-1 规定了使用此防护方法之设备的结构特征和试验要求。 Flameproof enclosure(electrical equipment) is a type of protection of electrical apparatus in which the enclosure will withstand an internal explosion of a flammable mixture which has penetrated into the interior, without suffering damage and without causing ignition, of an external explosive atmosphere consisting of one or more of the gases or vapours for which it is designed through any joints or structural openings in the enclosure. Note: IEC 60079-1 specifies the constructional features and test requirements for apparatus using this method of protection.

Flammability limit 可燃性极限 系指在给定的试验装置中,对燃料氧化剂混合物施以一个足够强大的着火源后,使其正好能产生燃烧的条件。

Flammability point 燃点

Flammable cargoes 易燃货物 符合以下标准的为易燃货物(见表 F-1)。 A cargo is defined as flammable according to the following criteria(Table F-1).

表 F-1 易燃货物
Table F-1 Flammable cargoes

IBC 规则规定 IBC Code descriptor	闪点(℃) Flash point ℃
高度易燃 Highly flammable	<23
易燃 Flammable	≤60,但≥23

Flammable gas 可燃气

Flammable gas or vapour 易燃的气体或蒸发气 当其与空气按一定比例混合时将形成爆炸性气体环境的气体或蒸发气。 Flammable gas or vapour means gas or vapour which, when mixed with air in certain proportions, will form an explosive gas atmosphere.

Flammable liquid 易燃液体(可燃液体) 在任何可预见的运行条件下可能产生易燃蒸发气或雾的液体。 Flammable liquid means a liquid capable of producing a flammable vapour or mist under any foreseeable operating conditions.

Flammable material 易燃材料 构成易燃气体、蒸发气、液体和/或雾的材料。 Flammable material means a material consisting of flammable gas, vapour, liquid and/or mist.

Flammable mist 易燃的雾 弥散在空气中,使之形成爆炸性气体环境的易燃液体的微滴。 Flammable mist means droplets of flammable liquid, dispersed in air, so as to form explosive atmosphere.

Flammable oils 易燃油 包括燃油、滑油、热油以及液压油。 Flammable oils include fuel oils, lubricating oils, thermal oils and hydraulic oils.

Flammable solids(Class 4.1 dangerous materials)

易燃固体(第4.1类危险货物) 系指易燃固体和通过摩擦可能引起火灾的固体。 Flammable solids mean the materials in this class are readily combustible solids and solids which may cause fire through friction.

Flammable(inflammable) 易燃
Flange[1] 法兰(凸缘,缘板,折边)
Flange[2] 翼板 系指一段加强构件,一般连接到腹板,但有时由腹板弯曲形成,通常与板材表面平行。 Flange means a section of a stiffening member, typically attached to the web, but is sometimes formed by bending the web over. It is usually parallel to the plated surface.
Flange[3] 折边 将板的边缘折弯而形成的窄条。
Flange bearing 法兰轴承
Flange connection 法兰连接
Flange connection bolt 法兰连接螺栓
Flange fillet 法兰圆角
Flange gland 填料箱凸缘
Flange joint 法兰接合
Flange spreader 法兰撑开器(拆开法兰用)
Flange stress 法兰应力(凸缘应力)
Flange tap 法兰式接头
Flanged 折边的(用法兰连接的,带凸缘的)
Flanged coupling 可拆联轴器 系指安装在轴端可以装拆并能传递推力和扭矩的法兰形刚性联轴器。
Flange-mounted alternator 法兰连接交流发电机
Flank of tooth 齿面
Flanking propeller 舷侧推进器
Flanking rudder 倒车舵 设置于螺旋桨前面推进器轴两侧,主要用以改进船舶倒车操纵性的舵。
Flap 襟翼 系指组成水翼或气翼的完整部分或延伸部分的一个部件。用以调节该翼的水动或气动升力。 Flap means an element formed as an integrated part of, or an extension of, a foil, used to adjust the hydrodynamic or aerodynamic lift of the foil.
Flap angle 襟舵角 襟翼舵的襟叶与主舵叶的相对转动角度。
Flap type wave generator 摇板式造波机 在船模试验水池中,用下端铰接、上端摆前后摆动的摇板制造人工波浪的设备。
Flap valve 片阀(舌阀,瓣阀,止回阀)
Flapper-nozzle 挡板喷嘴
Flap-type rudder(articulated rudder) 襟叶舵 可较大幅度提高舵力的舵。舵叶后部的局部,作成一个可独立地由随动机构控制的襟翼的舵。
Flare 外倾(外飘) 系指设计水线以上船长中部舷侧表面向外倾斜。其值以最大横剖面上自横梁上缘线与龙骨外缘线之交点至中线面的水平距离与设计水线半宽值之差表示。在无特别注明时,一般系指上甲板处者。
Flare back 回火
Flare platform 火炬平台 系指将油气处理过程中分离出的少量天然气引至火炬塔顶放空燃烧的专用平台。
Flare point 闪点
Flare tower 火炬塔
Flash 毛边/毛刺
Flash 闪光 是爆炸的产物,爆炸会产生过度焰、压力波和电磁波。 Flash is a product of an explosion producing transient flame and associated pressure wave and the electromagnetic wave.
Flash chamber 闪发室 系指闪发蒸发海水的腔室。
Flash evaporation 闪发蒸发 热海水进入低于其饱和压力的容器时,其中一小部分骤然转化为蒸汽的相态变化。
Flash light control box 闪光(通信)灯控制箱 安装在驾驶室内部对闪光灯进行集中控制并标明闪光灯工作情况的装置。
Flash point 闪点 系指:(1)液体能放出足够蒸发气与直接在液体表面上或在规定的条件下在点火源上使用的船内的空气形成可点燃混合物的最低温度;(2)货品释放出的易燃蒸发气足以点燃时的摄氏温度。国际散装运输危险化学品船构造和设备规则(IBC规则)所列数值是用认可的闪点装置按"闭杯试验"测定的。 Flash point is:(1) the minimum temperature of a liquid at which it gives off sufficient vapors to form an ignitable mixture with air immediately above the surface of the liquid or within the vessel used on the application of an ignition source under specified conditions;(2) the temperature in degrees Celsius at which a product will give off enough flammable vapour to be ignited. Values given in the Code are those for a "closed-cup test" determined by an approved flashpoint apparatus.
Flash point test 闪点试验
Flash point tester 闪点测定器
Flashback 逆火 限制火焰通过装置传播。 Flashback is the transmission of a flame through a device.
Flashed feed stocks 闪发原料油
Flashing light 闪光灯 系指每隔一定时间以每分钟频率120闪或120闪以上闪次的闪光的号灯。 Flashing light means a light flashing with a frequency of 120 flashes or more per minute at regular intervals.
Flashing turn light 转向闪光灯 用以表明船舶转向、调头信号,而于船舶驾驶室及室顶左、右侧规定处增设的红、绿各一盏的闪光灯。

Flat bar 扁钢 系指仅由一个腹板构成的加强筋。 Flat bar means a stiffener comprising only of a web.

Flat blade 平板叶 大致为平板形的螺旋桨叶。

Flat end 压力容器的平封头

Flat face 平面

Flat file 扁锉

Flat slide valve 平滑阀

Flat type glass water level indicator 平板式玻璃水位表

Flat-bottomed ship 平底船 舭部升高较小或等于零的船舶。

Flattened fully-loading hold 经平舱的满载舱 系指在任何货物处所内对散装谷物的装载和平整,应使甲板和舱口盖下方的所有空间装满到可能的最大限度的货舱。

Flay 抨击 系指船体某处出水后,由于相对运动而迅速浸没于水中,导致船体与波浪之间发生剧烈相互作用的一种现象。抨击由波浪引起,船体与波浪之间的垂向相对运动将直接决定抨击是否发生以及抨击的剧烈程度。抨击可分为3种:底部抨击、艏部外飘抨击和艉部抨击。

Fleet broad band(FBB) 宽带F船站

Flettner rotor propeller(Flettner propeller) 旋筒推进器 由立于甲板面,使空气发生环流的旋动圆筒构成的风力推进器。遇风时利用所产生的纵向分力推船前进。

Flex lay 柔性铺管

Flexibility sealing arrangement 挠性密封装置 用于机械舱盖,关舱后在舱口围板变形的情况下仍能保证密性的具有挠性的封舱装置。

Flexible coupling(elastic coupling) 弹性联轴器/弹性联轴节 有扭转弹性组件的联轴器。

Flexible damping coupling 阻尼挠性联轴节 是一种挠性联轴节,其具有第一轮毂和第二轮毂,每一个轮毂都具有凸缘并连接到传动轴。每个螺旋弹簧支撑在弹簧孔中,以使螺旋弹簧的相邻端面相接触并且彼此串联。

Flexible hose assembly 挠性软管组件 系指通常带有管端安装配件的金属或非金属短软管。Flexible hose assembly means short length of metallic or non-metallic hose normally with prefabricated end fittings ready for installation.

Flexible hoses 挠性软管 通常系指带有为安装而预制的端部配件的金属或非金属的短软管。挠性软管可用于固定管路系统和机械部件之间的固定连接,亦可用于便携式设备和管路之间的临时连接。 Flexible hoses mean short length of metallic or non-metallic hoses normally with prefabricated end fittings ready for installation. Flexible hoses may be used for a permanent connection between a fixed piping system and items of machinery, as well as for temporary connection between portable equipment and pipes.

Flexible installing 弹性安装

Flexible manufacturing system(FMS) 柔性制造系统 系指以数控机床、自动机器、工业机器人、自动输送系统和电子计算机等为主体,按系统工程的原理把它们组织并运行起来,使制造过程高度自动化的一种系统。

Flexible member 柔性构件 由于可能失稳或其他原因使抵抗总纵弯曲能力减弱的纵向构件。

Flexible pipe 柔性管 系指用聚合物隔开的,绞合钢丝层状结构的管子。以使管子具有相对小的弯曲刚度。 Flexible pipe is a pipe with layered construction of steel strands separated by polymeric layers which results in relatively small bending stiffness.

Flexible shaft 挠性轴 其一次临界转速低于额定转速的汽轮机轴。

Flexible shafting 挠性轴系 临界转速小于额定转速的轴系。

Flex-lay vessel 柔性铺管船

Flight deck 飞行甲板 舰船上供舰载机起飞、着舰和待机用的一层甲板。

Flight of an inclined ladder 斜梯的梯段 系指斜撑架的实际长度,垂直扶梯则指平台之间的距离。 Flight of an inclined ladder means the actual stringer length of an inclined ladder. For vertical ladders, it is the distance between the platforms.

Flight trim 飞行纵倾 系指船舶在无人操作情况下,借助船舶的控制翼面而使飞机保持飞行姿态和方向。 Flight trim means the condition of the craft whereby control surface settings are such that attitude and direction the craft is maintained without significant operator input.

Flight-trim angle 飞行纵倾角 飞行纵倾船在受到外力干扰后恢复的角度。 Flight-trim angle is an angle to which a flight-trimmed craft will return after being disturbed by an external force.

Flight-trim speed 飞行纵倾速度 飞行纵倾船在受到外力干扰后恢复的速度。 Flight-trim speed is a speed to which a flight-trimmed craft will return after being disturbed by an external force.

Flitter bow ship 竖立式调查船 一种无推进装置

FLNG 浮式液化天然气

Float(paddle) 蹼板 装在明轮上用以泼水向后使船前进的平面或弧形板。

Float free survival craft 自行浮起式救生艇筏 系指一种带有装置和备品以允许其脱离沉船,并能自动漂浮在海面上的艇筏。 Float free survival craft is craft whose installations and stowage are intended to permit them to separate from a sinking vessel and float to the surface automatically.

Float production, storage and offloading system (FPSO) 浮式生产储油船 该装备为由锚系到海底而进行定位的大型油船型驳船结构。其通常与开口平台或海底采油系统组成一个完整的采油、原油处理、储油和卸油系统。FPSO 具有投资少、建造周期短、迁移方便、可连续生产、重复使用等优点、适应水深范围广泛,具有风标效应,被广泛应用于环境恶劣的海域。其主要由系泊系统、船体部分、油生产设备、舾卸载系统等部分组成。随着海洋工程装备的不断发展,新型 FPSO 不断涌现,并朝着深水和超大型化的方向发展。见图 F-11。

图 F-11 浮式生产储油船
Figure F-11 Float production, storage and offloading system(FPSO)

Float valve 浮阀(浮子阀)

Float-free launching 自由飘浮下水 系指救生艇筏从下沉中的船舶自动脱开并立即可用的下水方法。 Float-free launching is that method of launching whereby a survival craft is automatically released from a sinking craft and is ready for use.

Float-free recording medium 自浮记录介质 系指最终记录介质的一部分,其应在船舶沉没后自浮于水面,并有指示其位置的设施。 Float-free recording medium means a part of the final recording medium which should float-free after a sinking. It has means of indicating location.

Floating accelerometer 重力式测波仪 应用装在随波单锚系浮体内的加速度计测量海浪中水质点沿重力方向的加速度来测量波浪的仪器。

Floating block 浮动滑车 重力式艇架中直接承吊小艇的动滑车。

Floating body 浮体 抬船甲板下,产生浮力的整体或组合式的箱型结构体。

Floating booster station 接力泵船 为增加吸扬式挖泥船的排泥距离,而串联在水上远程排泥管线上,装有1~2台泥泵的方箱形驳船。

Floating concrete mix ship 混凝土搅拌船 系指在水上从事混凝土搅拌并在施工场地直接进行混凝土浇筑的驳船。 Floating concrete mix ship means a barge mixing concrete on the water and depositing concrete directly at the construction site.

Floating condition 浮态 船舶在静水中的平衡状态。主要系指船的吃水、纵倾角和横倾角。

Floating crane 起重船 系指甲板上有起重设备,专供水上作业起吊重物的船舶。 Floating crane is ship fitted with hoisting appliances on deck, dedicated to hoisting operations on water.

Floating crane(FC) 起重船/浮吊 系指漂浮在水上的,在甲板上有起重设备,专供水上作业起吊重物的船舶。一般分成两大类,一类是起重臂能够360°回转的,另一类是吊臂固定在船上的一个方向,整个船靠拖船拖带转向,或是靠锚向各个方向抛锚,通过牵拉不同方向的锚链,而实施重物回转的。前者的结构和机械构造非常复杂,而且起重能力也比较小。起重船,又称浮吊船,用于水上起重、吊装作业、一般为非自航,也有自航。起重船上装有吊机。作业频繁的起重船通常为自航式,其起重机可旋转,当吊重特大件时,可用两个起重船合并作业。见图 F-12。

图 F-12 起重船/浮吊
Figure F-12 Floating crane(FC)

Floating crane(lifting within coastal service) 沿海航区内作业的起重船

Floating crane(lifting within greater coastal serv-

ice) 近海航区内作业的起重船

Floating crane(lifting within harbor) 港口水域内作业的起重船

Floating crane(lifting within sheltered water service) 遮蔽航区内作业的起重船

Floating discharging pipeline 水上排泥管 排泥管线中,位于水面上的排泥管。

Floating dock 浮船坞 简称浮坞,是一种用于修、造船用的,在海上驳船的大型漂浮设备工程船舶。它不仅可用于修、造船舶,还可用于打捞沉船,运送深水船舶通过浅水的航道等。它可承担万吨级油船、散货船、大型全集装箱船和其他大型海上工程建筑物的坞修工程,也可用于大型船舶和海洋工程的对接、改建和制造。2012 年 12 月,广船国际向中国海军交付世界第一艘自航浮船坞。2016 年 2 月 8 日,中国第一艘自航浮船坞"华船一号"问世,标志着大型战舰修理由岸基定点保障向远海机动保障实现突破——自航浮船坞:开启大型舰船海上坞修新纪元。

Floating docks 浮船坞 系指具有底部浮箱,两舷为坞墙,供抬起船舶进行修理的船舶。Floating docks means a ship fitted with buoyant boxes at bottom, both sides constructed as bulwarks, for repairing hoisted ships.

Floating drilling production storage and offloading (FDPSO) 浮式钻井生产储油船 系指具有钻井功能的半潜式生产平台。

Floating drilling rig 浮式钻井装置 在海上用锚泊或动力定位的,浮在海面上进行钻井作业的装置。通常分为半潜式钻井装置、钻井船、钻井驳船等。

Floating drilling unit 浮式钻井装置

Floating floor 浮动地板 系指在钢质甲板上敷设由岩棉、陶瓷棉、阻尼材料、浮动敷料、地板等各种材料构成的结构。

Floating gill net 流刺网 可在水中漂移的刺网。

Floating grab dredge 抓斗式挖泥船 系指利用吊在旋转式吊机上的抓斗的下放、闭合、提升和张开来抓取和抛卸被挖物的船舶。其船型一般为单抓斗机非自航方箱型船体,也有多抓斗(2 个或 3 个)的自备泥舱和自航式船舶。

Floating interval of floating dock 升浮时间 浮船坞从最大沉深吃水升至工作吃水所需的时间。

Floating liquefied natural gas(FLNG) 浮式液化天然气生产储卸装置

Floating liquefied natural gas(FLNG) production vessel 液化天然气的生产船 系指主要用于天然气的开发生产的船舶。

Floating liquefied natural gas unit 浮式液化天然气装置

Floating living quarter 住宿船 设有各种生活设施,主要供水域工程的施工人员休息和住宿的船舶。

Floating long line 浮延绳钓 浮于海面的延绳钓。

Floating on even keel, zero trim 正浮 船舶无横倾和纵倾时的浮态。

Floating part 浮动件 在行星齿轮传动中起运动误差补偿作用,从而使负荷均匀分配的零件。

Floating pile driver 打桩船 用于水上打桩作业,船体为钢箱型结构,专为水上工程打桩用的船舶。在甲板的端部装有打桩架,可俯后仰以适应施打斜桩的需要。打桩船为非自航船,用推(拖)船牵引就位。打桩船广泛应用于桥梁、码头、水利工程施工。打桩船的建造过程包括:钢板切割,分段制作,船台阶段,设备调试阶段和试航阶段。打桩船配备打桩锤,桩架,附属设备。一般,打桩船配有工作间或维修间,可以对简单的问题(例如焊接)进行修复。与国内其他打桩船相比,"海威951"打桩船有三大特点:(1)该船集打桩和吊装为一体,一次最大抬吊重 240 吨,还可以做 −27°～+25°的变幅;(2)可插打 18.5°的斜桩,其桩基施工的仰俯角度在国内为最大;(3)95 m 高的桩架可进行 3 次变幅折叠,变幅后的桩架距水面高度仅有 25.4 m,可以抵达长江中下游的任何施工地点。见图 F-13。

图 F-13 打桩船
Figure F-13 Floating pile driver

Floating pile presser 压桩船 装有由绞车、钢绳、滑轮组、大压梁等组成的压桩设备,在开动绞车收紧钢缆后可使压梁产生成倍压力,将桩柱无振动地压入软地基中的打桩船。

Floating pipeline 浮管 外部包扎有闭孔泡沫橡胶并覆以保护层,两端配以球节或法兰,能自浮于水上的输泥管。

Floating production storage and offloading(FPSO) 浮式生产储油/卸油船 是对海上开采的石油进行油

气分离、处理含油污水、动力发电、供热、原油产品的储存、集人员居住与生产指挥系统于一体的综合性大型海上石油生产基地,俨如是一座"海上石油加工厂"。其主要由系泊系统、载体系统、生产工艺系统、卸载外输系统及配套系统所组成。涵盖了数十个系统,作为集油气生产、储存及外输功能于一身的浮式生产储油/卸船具有高风险、高技术、高附加值、高投入、高回报的综合性海洋工程特点,故被称为"油田心脏",与其他形式石油生产平台相比,浮式生产储油/卸油船具有抗风浪能力强、适应水深范围广、储/卸油能力大及可以转移、重复使用等优点,广泛适用于远离海岸的深海、浅海海域及边际油田的开发。目前已成为海上油气田开发的主流生产方式。

Floating production storage unit (FPSU)/Oil production and storage unit 浮式生产储油装置 系指以船或驳船作为支承结构,具有油气处理及原油储存功能的浮式装置。见图F-14。

图 F-14 浮式生产储油装置
Figure F-114 Floating production storage unit (FPSU)/Oil production and storage unit

Floating production system (FPS) 浮式生产系统
系指以系泊缆和多个锚固定在井位上的半潜式浮动平台。 Floating production system (FPS) is a semi-submersible floating platform secured in place with mooring lines and anchors.

Floating sand piler 砂桩打桩船 将空心桩打入水下软地层里,而后用大型喷砂泵将砂注入桩孔以形成砂桩的打桩船。

Floating storage re-gasifieation unit (FSRU) 浮式再液化气储存装置

Floating structure 浮动结构物 系指其全部重力由浮力支承的结构物。全部重力包括空船重力、系泊系统重力、任何立管预张力和作业重力。 Floating structure is a structure where the full weight is supported by buoyancy. The full weight includes lightship weight, mooring system weight, any riser pretension, and operating weight.

Floating switch 浮子开关(浮动开关)
Floating welder 电焊工作船

Floating winch station for warping 绞滩船
Floatover engineering 浮托工程
Flood pipe 浸水管
Flood protection materials 堵漏用具 船舶破损时,用以堵塞漏洞的各种应急器材的统称。
Floodable length 可浸长度 船上某点的可浸长度,系指沿船长方向上以该点为中心的舱在规定的分舱载重线和渗透率情况下破损进水后,船舶不致淹过限界线的最大允许舱长。
Floodable length curve 可浸长度曲线 可浸长度沿船长各点的分布曲线图。
Flooded tank (surge tank) 液体分离器 贮存低压液态制冷剂供盘管作重力循环使用并分离随盘管回气中带来的液滴的容器。
Flooded waterline 破舱水线 船舶破损进水后的水线。
Flooding 进水 系指水通过不能关闭成完整稳性和破舱稳性标准要求为风雨密或水密的开口,或由于作业要求在任何风浪条件下保持开敞的开口,流入平台浮力结构之内。
Flooding angle 进水角
Flooding calculation 进水计算
Flooding of fish-holds 鱼舱进水 系指渔捞作业期间,不能迅速封闭舱盖的敞开舱口,从而可能从其他方向鱼舱连续进水。 Flooding of fish-holds means progressive flooding of fish-holds which could occur through hatches which remain open during operations and which cannot rapidly be closed.
Flooding pipe line 溢流管路
Flooding probability 浸水概率
Flooding valve 浸水阀
Floor 肋板 系指:(1)船底的横桁材;(2)船底横向构件。 Floor is: (1) a bottom transverse girder; (2) a bottom transverse member.
Floor area of liferaft 救生筏底面积 用以核定救生筏乘员定额的救生筏内底的面积。
Floor boring machine 落地镗床 系指工件安置在落地工作台上,立柱沿床身纵向或横向运动。用于加工大型工件的镗床。主要用镗刀对工件已有的预制孔进行镗削的机床。通常,镗刀旋转为主运动,镗刀或工件的移动为进给运动。它主要用于加工高精度孔或一次定位完成多个孔的精加工,此外还可以从事与孔精加工有关的其他加工面的加工。使用不同的刀具和附件还可进行钻削、铣削、切削的加工精度和表面质量要高于钻床。镗床是大型箱体零件加工的主要设备。可加工螺纹及外圆和端面等。

Floor line 船底斜升线 在最大横剖面上,自船底斜升线与设计水线半宽处所作垂线的交点量至基平面的垂直距离。

Floppy drive 软驱

Flora-fauna-habitat protection area(FFH area) 动植物栖息保护区域 根据欧盟动植物栖息保护指令设立的保护区域,旨在保护大自然动植物及其栖息地。也包括对鸟类栖息地的保护。目的是将全欧洲的动物保护区汇总(2000年自然保护规划)。该区域是由德国联邦州提议,经欧盟委员会分析后通过申请注册。2008年末,全欧洲22 945个区域进行了汇总。陆地表面积达到661 503 km²(占欧洲陆地面积的13.3%)。海洋面积达92 893 km²作为欧洲共同利益区域。德国拥有4 675个区域;陆地面积达54 343 km²(9.9%为陆地面积),海洋面积达19 134 km²。这些区域几乎包括了德国北海沿线的整个浅滩海。

Flotation element 浮性器材 为船艇提供浮力,且因此而影响其浮性特征的器材。 Flotation element is the element which provides buoyancy to the boat and thus influences its flotation characteristics.

Flotsam cleaner boat 水面清扫船 系指在游人如积的风景名胜地,担负着景区水面的保洁和美容作用的船舶。

Flotsam cleaner boat /rubbish recovery ship 清扫船 供清扫水面杂物用的船舶。

Flow 流(流通,流程)

Flow ability 通流能力

Flow area 通流面积(流截面,流通截面)

Flow coefficient 流量系数 气流轴向速度与叶片圆周速度的比值。

Flow graph 流程图(流图)

Flow meter 流量计(流量表)

Flow moisture point 流动水分点 系指在按规定的方法试验时,使物质的代表性样品产生流动状态的含水量百分数(按湿重计)。 Flow moisture point means the percentage of moisture content(wet mass basis)at which a flow state of representative sample of the material develops under the prescribed method of test.

Flow noise 流动噪声

Flow passage deposit 流道结垢 盐分、灰渣和油垢等杂质在压气机和涡轮通道内,特别是叶片上的沉积现象。

Flow path(steam path, flow passage) 通流部分 从速关阀到冷凝器喉部蒸汽流经的通道和元件的总称。

Flow regime 流态 黏性流体在任何区域中流动的状态(主要指层流、过度流、湍流和分离流)。

Flow state 流动状态 系指大量颗粒物质被液体浸湿到一定程度时,在振动、撞击或船舶运动等主要外力影响下丧失内部抗剪强度并呈现液体的状态。 Flow state means a state occurring when a mass of granular material is saturated with liquid to an extent that, under the influence of prevailing external forces such as vibration, impaction or ships motion, it loses its internal shear strength and behaves as a liquid.

Flow times 流动时间 系指N个人通过出口系统的一点所需的总时间。 Flow times is the total time needed for N persons to move past a point in the egress system, and is calculated as: $t_F = N/F_c$.

Flow velocity 流速

Flow-line 流线段

Flowline 流送管 系指从总管或靠近油井顶部的拖运器将油井采集的液体输送到首次加工后阶段的加工装置(例如平台)的一段管路。 Flowline is a pipeline that transports well fluids from a manifold or a sled near the wellhead to the first downstream process component (such as a platform).

Flow-through method 溢流法 系指将替换的压载水泵入用于装载压载水的压载水舱过程,水从溢流口或其他装置流出。当采用该方法时,在深海由泵向已满的压载水舱注水,让水溢流,至少应与以3倍该舱容积的水量流经该舱。 Flow-through method is a process by which replacement ballast water is pumped into a ballast tank intended for the carriage of ballast water, allowing water to flow through overflow mouth or other arrangements. When this method is used, fully filled ballast water tanks are simultaneously filled and discharged by pumping water while the ship is at deep-ocean, and at least the water of three times the tank volume should be pumping through the tank.

Flue 烟道(烟管,烟囱)

Flue gas 炉膛烟气

Flue gas detector 烟气检测器(检烟器)

Flue gas extinguisher system 烟气灭火系统

Flue gas isolating valve 烟气隔离阀

Flue gases water scrubber 烟气洗涤器 烟气防爆系统中用以将烟气中的烟灰含硫量洗涤干净,并将烟气温度冷却的专用设备。

Flue pass 烟道 烟气通至烟囱的通道。

Flue-extracted type fire alarm system 抽烟式火警报警系统 系指利用火灾发生之前所出现的烟雾,引起探测器动作产生的自动报警。它由抽烟式火警报警器、集烟器和抽风机箱等组成。

Fluid 流体(液体,介质)

Fluid coupling 液力联轴节
Fluid fire extinguisher 液式灭火器
Fluid transfer system 液体输送系统　系指将FPSO的成品油(通过海底管线和单点系泊自身的管系装置)输送到穿梭油船的系统。　Fluid transfer system is a system to transport the produced oil of FPSO (via the subsea pipeline and through the single point mooring's own piping arrangement) to the shuttle tanker.
Fluke angle 锚爪折角　系指锚爪与锚杆之间的夹角。对无杆锚则指锚爪转动后的最大夹角。
Flume tank (被动)减摇水舱
Flume type anti-rolling tank 自由液面减摇水舱　对称设置于船舶两舷，有深槽横向连通，利用横摇时自由液面起减摇作用的封闭式水舱。
Fluorescent crack detection 荧光探伤
Fluorescent inspection 荧光检验(荧光检测)
Fluorescent lamp 荧光灯　是一种管状发光体，两端设有钨丝电极，管内充有低压汞蒸气及少量惰性气体，管子内壁涂有荧光粉。当灯丝通电预热后，起辉器断开镇流器，产生反电势使灯管两电极间被击穿放电，汞原子在电离过程中被激发出紫外线照射到荧光粉即发出可见光。
Fluoro rubber 氟橡胶　系指主链或侧链的碳原子上含有氟原子的合成高分子弹性体。最早的氟橡胶为1948年美国DuPont公司试制出的聚-2-氟代-1.3-丁二烯及其与苯乙烯、丙烯等的共聚体，但性能并不比氯丁橡胶、丁橡胶突出，而且价格昂贵，没有实际工业价值。20世纪50年代后期，美国Thiokol公司开发了一种低温性好，耐强氧化剂(N_2O_4)的二元亚硝基氟橡胶，氟橡胶开始进入实际工业应用。此后，随着技术进步，各种新型氟橡胶不断开发出来。氟橡胶具有稳定性、耐高温性、耐老化性等特征，主要应用于现代航空、导弹、火箭等尖端技术及汽车、造船、化学等工业领域。
Fluorometer 荧光计　测量海水荧光特性的仪器。由装有紫外线灯的散射仪与两个单色仪构成。散射仪测量荧光强度，选择在粒子散射小的角度90°附近记录，单色仪分别用以确定激励波长和测量荧光辐射波长。
Fluorspar 萤石　黄色、绿色或紫色晶体。粗粉粒。该物质如装运时含水量超过适合运输水分极限，可能流态化。吸入其粉尘后有害并有刺激性。　Fluorspar is a yellow, green or purple crystal. It is coarse dust. This material may liquefy if shipped at moisture content excess of their transportable moisture limit. It is harmful and irritating after dust inhalation.
Flush deck ship 平甲板船　系指干舷甲板上没有上层建筑的船舶。　Flush deck ship is one which has no superstructure on the freeboard deck.
Flush manhole cover 平置人孔盖(齐平人孔盖)　用于周缘仅设一圈扁平座板的开口上，并依靠双头螺栓紧固的人孔盖。
Flush system 平接式　板与板对接，板面齐平的排列连接方式。
Flush valve 冲洗阀
Flush water 冲洗水　系指用于把污水或其他废物从厕所或小便池运送到处理系统的传输介质。　Flush water means the transport medium used to carry sewage or other wastes from toilets or urinals to the treatment system.
Flushing equipment 冲洗设备
Flushing pump 冲洗泵
Flutter 颤振
Flutter calculation 颤振计算(抖振计算)
Flutter failure 颤振破坏
Flutter frequency 颤振频率
Flutter of hydrofoil 水翼颤振　水翼在惯性力、水动力和结构弹性力的综合作用下，于临界航行速度时产生的一种自激振荡现象(常见的有水翼弯扭颤振、水翼弯曲、襟翼扑动颤振等)。
Flutter test 颤振试验
Flux 流(通量，焊剂)
Flux density 通量密度
Fly ash 飘尘　是燃烧煤和燃油的火力发电厂产生的轻细碎且多粉尘的细粉末状残留物。不得与黄铁矿的烧渣混淆。含有空气时可能移动。该货物为不燃物或失火风险低。　Fly ash is the light, finely divided dusty fine powder residue from coal and oil fired power stations. Avoid confusion with calcined pyrites. It may shift when aerated. This cargo is non-combustible or has a low fire-risk.
Fly ash carrier 烟灰运输船　系指专门设计和制造用于运输烟灰的船舶。　Fly ash carrier is a ship designed and constructed for the carriage of fly ash.
Fly block 动滑车
Flying bridge 浮船坞飞桥　位于坞墙端，连接左右坞墙顶甲板，作为通用的可启闭的连接桥。
Flying mode beyond the GEZ as an airplane 在地面效应区外飞行(像飞机一样)模式　系指对于设计为可在地效区外飞行的地效翼船，在该模式下将完全升起机翼襟翼以减小阻力，使船能够加速到最大速度。对于设计为可短时爬高的船，这将是一个"跃升模式(Jump mode)"。　Flying mode beyond the GEZ as an airplane means a mode in which a WIG designed to fly above the GEZ will lift the wing rear flap completely to reduce drag and allow accelera-

Flying mode in the ground effect zone(GEZ)　地效区(GEZ)内飞行模式　起飞后,地效翼船的总重由作用在机翼、船体和侧浮舟上的气动升力支撑的飞行模式。飞行模式时,需调节主翼随边的襟翼以平衡所需飞行高度下的动力气垫(在此速度下不是静态气垫)或地效升力。此外,动力气垫船或动力气垫地效翼船艏推进器后的导流叶片或射流导叶片需向上转动以获得更高的推力恢复系数。对于设计巡航速度很高的动力增升地效翼船,应减小艏推进器功率,使主升力翼完全以地效运行。升力的作用中心应始终维持在重心之前。 Flying mode in the ground effect zone(GEZ) means a mode in which the WIG's total weight will be supported by the aerodynamic lift acting on wings, hull and side buoys after take-off. In this flying mode, the flap at the trailing edge of the main wing will be adjusted so as to balance out the dynamic air cushion(not static air cushion at this speed) or wing-in-ground effect lift at the desired flight elevation. In addition, the guide vanes or air-jet deflectors aft of the bow thrusters will be rotated upwards to obtain higher propulsion recovery coefficient on DACC or DACWIG. PARWIG designed to cruise at significantly higher speed will reduce the power on the bow thrusters, allowing the main lifting wing to operate purely as a wing-in-ground effect. The centre of effort for lift forces should remain just forward of the centre of gravity.

Fly-over mode　飞行状态　对地效翼船操作模式而言,系指在有限时间内增加 B 类地效翼船和 C 类地效翼船的飞行高度,该高度超过地效的垂直范围,但不超过 ICAO 条款规定的飞机最小安全高度(150 m)。 For operational modes of WIG craft, fly-over mode denotes increase of the flying altitude for WIG craft of types B and C within a limited period, which exceeds the vertical extent of the ground effect but does not exceed the minimal safe altitude for an aircraft prescribed by ICAO provisions(150 m).

Flywheel　飞轮

Flywheel governor　飞轮式调速器(蒸汽机用)

FMEA　故障模式与分析

FMEA report　FMEA 报告　系指一份完备的文件,其应对船舶、船舶的系统及其功能、建议的操作和故障模式、原因及后果和环境条件进行充分的阐述,且均不必借助于不在报告之内的其他图纸和文件而能够理解。如需要,该文件应包括分析的假设和系统框图,报告应包含结论的摘要,以及系统故障分析和设备故障分析中的每一个分析系统的说明。如需要,还应列出所有可能的故障及故障概率,在每一种所分析操作模式中对每一个系统的纠正措施或操作限制。该报告应包含有试验程序、所参考的所有其他试验报告和 FMEA 试验。 FMEA report means a self-contained document with a full description of the craft, its systems and their functions and the proposed operation and environmental conditions for the failure modes, causes and effects to be understood without any need to refer to other plans and documents not in the report. The analysis assumptions and system block diagrams shall be included, where appropriate. The report shall contain a summary of conclusions and recommendations for each of the systems analyzed in the system failure analysis and the equipment failure analysis. It shall also list all probable failures and their probability of failure, where applicable, the corrective actions or operational restrictions for each system in each of the operational modes under analysis. The report shall contain the test programme, reference of any other test reports and the FMEA trials.

F-N curve　F-N 曲线　用于表示社会风险。它是由表征每船年发生一次死亡 N 人及以上事故的累积频率(f)分布的纵坐标和表征一次事故的后果(N 人死亡)的横坐标构成的坐标图。

Foam　泡沫　系指泡沫溶液通过泡沫发生器并与空气混合而生成的灭火剂。 Foam is the extinguishing medium produced when foam solution passes through a foam generator and is mixed with air.

Foam absorption material　泡沫尖劈材料

Foam appliance(foam applicator)　泡沫喷枪

Foam cavitation(burbling cavitation)　沫状空泡　大量形似泡沫,在进入较高压力区时坍塌消失的空泡。

Foam concentrate　泡沫浓缩剂(泡沫灭火剂原液)　系指泡沫灭火剂原液按一定浓度和水生成的溶液。 Foam concentrate is the liquid which, when mixed with water with appropriate concentration forms a foam solution.

Foam concentrate tank　泡沫浓缩剂柜(泡沫液柜)

Foam container　泡沫容器

Foam extinguisher　泡沫灭火器

Foam extinguishing system　泡沫灭火系统

Foam fire extinguisher　泡沫灭火器

Foam fire-extinguishing mediums　泡沫灭火剂　分为:(1)利用泡沫原液的基剂(蛋白质加水分解后生成物或合成界面活性剂)和海水混合,通过机械搅拌产生的、用于固定式泡沫灭火器的空气泡沫。按其膨胀率(expansion rate)不同又分为膨胀 12 倍以下为低膨胀泡沫(low-expansion foam)和膨胀 1 000 倍以下的为高膨胀泡沫(high-expansion foam);(2)通过碳酸钠水溶液与硫酸铝水溶液的化学反应而产生的二氧化碳泡沫的、用于便携式泡沫灭火器的化学泡沫。按其是否对酒精类火灾有效又分为:①对油火灾有效、对酒精类火灾无效的、其泡沫原液为由在加水分解蛋白质(如牛血等)中添加一价铁盐构成的普通泡沫(ordinary type foam);②对油火灾无效但对酒精类火灾有效的,其泡沫原液为由在加水分解蛋白质中添加含氟表面活性剂构成的耐酒精泡沫(alcohol-resistant type foam);(3)既对油火灾有效,又对酒精类火灾有效的两用型泡沫(dual-purpose type foam)。

Foam forming liquid 泡沫生成液

Foam generating equipment 泡沫发生设备

Foam generator 泡沫发生器 系指泡沫溶液通过其与空气混合形成泡沫,并直接排放至被保护处所的排放设备或装置。通常由一个喷嘴或一组喷嘴和一个外壳组成。外壳一般由穿孔的钢/不锈钢板制成盒子状并将喷嘴围住。 Foam generators are discharge devices or assemblies through which foam solution is aerated to form foam that is discharged directly into the protected space, typically consisting of a nozzle or set of nozzles and a casing. The casing is typically made of perforated steel/stainless steel plates shaped into a box that encloses the nozzle(s).

Foam insulating material 泡沫保温材料

Foam making branch pipe 泡沫发生支管

Foam mixing rate 泡沫混合率 系指泡沫灭火剂原液与水混合形成泡沫溶液的百分比。Foam mixing rate is the percentage of foam concentrate mixed with water forming the foam solution.

Foam mixing unit 泡沫混合器

Foam monitor 泡沫炮

Foam rubber 泡沫橡胶 是一种海绵状,有很多小孔的橡胶;是用生橡胶加起泡剂或用浓缩胶乳边搅拌边鼓入空气,再经硫化制成。特点:质轻、柔软、有弹性,能隔热、隔音、耐油、耐化学药品。

Foam smothering system 泡沫灭火系统

Foam solution 泡沫溶液 系指泡沫灭火剂原液和水生成的溶液。 Foam solution is solution of foam concentrate and water.

Foam sprinkler system 泡沫灭火喷淋系统

Foam suppressor 泡沫灭火器

Foamed buoyant material 泡沫浮力块 由吸水率极小的硬质泡沫塑料作成,充填在救生载具、浮具内,用以提供内部浮力的块状物。

Foamed cement 泡沫水泥

Foamed glass 泡沫玻璃 最早是由美国彼得堡康宁公司发明的,是由碎玻璃、发泡剂、改性添加剂和发泡促进剂等,经过细粉碎和均匀混合后,再经过高温熔化、发泡、退火而制成的无机非金属玻璃材料。它是由大量直径为1~2 mm的均匀气泡结构组成。泡沫玻璃,不论用于屋面或外墙都是一种优良的保温隔热材料。泡沫玻璃用于屋面的保温隔热层时,由于其不吸水、不吸湿、用水泥砂浆黏铺后,水分不可能进入,所以不会造成防水层的起鼓破坏现象,不需作排气孔,也不存在长久使用后性能下降的问题。

Foamed material 泡沫材料

Foamed plastics 泡沫塑料 是由大量气体微孔分散于固体塑料中而形成的一类高分子材料,具有质轻、隔热、吸音、减振等特性,且介电性能优于基体树脂,用途很广。几乎各种塑料均可作成泡沫塑料,发泡成型已成为塑料加工中一个重要领域。泡沫塑料中加入$CaCO_3$后能提高其强度、耐热性,减小线胀系数、收缩率。泡沫塑料有硬质、软质2种。硬质聚氯乙烯低发泡板材、管材或异型材则用挤出法成型,发泡剂在机筒中分解,物料离开机头时,压力降到常压,溶入气体即膨胀发泡,如果发泡过程与冷却定型过程配合得当,就可得到结构泡沫制品。

Foamed rubber 泡沫橡胶 是一种海绵状,有很多小孔的橡胶;是用生橡胶加起泡剂或用浓缩胶乳边搅拌边鼓入空气,再经硫化制成。特点:质轻、柔软、有弹性,能隔热、隔音、耐油、耐化学药品。

FOB liner term 班轮条件下的船上交货 系指装船费用按照班轮的做法办理,即卖方不负担装船的费用。

FOB stowed 理舱费在内的船上交货 系指卖方负责将货物装入船舱,并承担理舱费在内的装船费用。

FOB trimmed 平舱费在内的船上交货 系指卖方负责将货物装入船舱,并承担平舱费在内的装船费用。

FOB under tackle 吊钩下交货的船上交货 系指卖方承担的费用截止到买方指定船舶的吊钩所之处,包括装运港驳船费在内,有关装船的各项费用一概由买方负担。

Fog 雾 系指较浓的雾。

Fog horn 雾角 供在雾天时使用的非动力号角。

Fog system 喷雾系统

Fog test 喷雾试验

Fog whistle 雾笛

Foil 水翼 系指高速船航行时产生流体动升力的一块翼状板或三维结构物。 Foil means a profiled plate or three dimensional construction at which hydrodynamic lift is generated when the craft is under way.

Folding flat 活动桌面板 一边以铰链连接于舱壁上,靠活动撑脚支撑,不用时可翻转紧贴于舱壁的桌面板。

Folding hatch cover 折叠式舱口盖 由成对的互相绞接在一起的盖板组成的舱口盖。可分为液压折叠式舱口盖和钢索曳式折叠式舱口盖。

Folding retractable fin stabilizer 折叠收放式减摇

鳍可收入船体内的减摇鳍。其采用十字头结构来保证各自独立的鳍收放和鳍角转动。

Folding step board 活动踏脚板 供登床用的可放的踏脚板。

Folding stretcher 船用担架 船上运送伤病员用的担架。

Following sea 随浪 系指正 x 方向行进的波浪。Following sea is waves propagating in positive x-direction.

Food wastes 食品废弃物 对"73/78 防污公约"附则Ⅴ而言,系指任何变质或未变质的食物,包括水果、蔬菜、乳制品、家畜、肉制品和船上产生的食品的碎屑。For 73/78 MARPOL Annex Ⅴ, food wastes means any spoiled or unspoiled food substances and includes fruits, vegetable, daily products, poultry, meat products and food scraps generated aboard ship.

Foot grating 格子板(地格栅,踏脚格栅)

Foot step 踏步 系指直接装设于船体各部或桅、柱等处,供人员踏登的凹穴或由方钢或圆钢制成的突出物。

Foot valve 脚底止回阀

Foot-grating(foot platform) 踏脚格栅 悬吊在救生浮绳网下,用以支承人员的格栅。

Footing column stabilized semi-submersible drilling unit 柱靴柱稳半潜式钻井平台 系指其每一立柱下端设一柱靴作为主要浮体的柱稳、半潜式钻井平台。

Footing jack-up drilling unit 桩靴自升式钻井平台 系指在每一桩腿的下端设一面积较大的箱体以增大海底支承面积的自升式钻井平台。

Footings 柱靴 系指与柱稳式平台单个立柱相连接的独立浮体。 Footings are separate buoyant structures connected to the bottom of each column of a column-stabilized unit.

Forbidden zone 禁区

Forbidden zone of speed 速度禁区

Force air circulation 强制空气循环

Force and moment detector(force and moment sensor) 力和力矩传感器

Force circulation boiler 强力循环锅炉

Force draught 强力通风

Force fitting 压入配合

Force lubrication 压力润滑法

Force majeure 不可抗力 系指船舶损坏;由于港口当局对人员入境或行动的限制导致的船级社验船师意外地无法登船;由于非正常的持续的恶劣天气、罢工或国内动乱造成的船舶在港口意外的拖延或无法卸货;战争或其他不可抗拒的外力。 Force majeure means damage to the vessel; unforeseen inability of the Surveyors to attend the vessel due to the governmental restrictions on right of access or movement of personnel; unforeseeable delays in port or inability to discharge cargo due to unusually lengthy periods of severe weather, strikes, civil strife; acts of war, or other cases of force majeure.

Force of buoyancy, buoyant force 浮力 作用于船舶浸水外表面上的静水压力垂向分力的合力,其值等于排水量。

Force transducer 测力传感器

Force-circulation 强制循环 是借助外力实现循环,一般为介质原料在加热后密度比重变化不大,或自身动力流不能实现循环。自然循环是借助介质自身由于温度差引起的比重差来完成介质循环,一般为比重密度受影响比较大。另外,由于温度的提升容易形成一些不必要因素的,如接胶时,需采用强制循环。

Forced circulation boiling 强制循环锅炉

Forced circulation exhaust gas heat exchanger 强制循环废气热交换器

Forced circulation pump 强制循环泵

Forced draft 强力通风

Forced draft fan 强力鼓风机

Forced draught trunk 强力通风通道

Forced frequency 强制频率(强制振动频率)

Forced oil circulating lubrication system 强制循环式润滑系统

Forced perturbation motion 强制扰动运动 在外力持续扰动下的船舶运动。

Forced rolling 强制横摇 船舶在周期性外力作用下所产生的横摇。

Forced stabilization of the craft 强制稳定 系指依靠下列手段使船舶达到稳定:(1)自控系统;(2)手控辅助系统;(3)自控及手控辅助相结合的联合系统。 Forced stabilization of the craft is stabilization achieved by: (1)an automatic control system; (2)a manually assisted control system; (3)a combined system incorporating elements of both automatic and manually assisted control systems.

Forced ventilation 强制通风

Forced vibration 强迫振动 系指振动系统在周期性外力(强迫力)的持续作用下的振动。

Fore and aft end part 艏、艉部 系指距船各端 $0.1L$ 的部分。 Fore and aft end part means the part covering $0.1L$ from the fore and aft end of the ship.

Fore body 前体 中横剖面以前的船体部分。

Fore draft 艏吃水 相当于艏垂线处的型吃水,或龙骨吃水,或距船底某一基准点吃水。

Fore end 艏端 系指:(1)在测定船长 L 时船艏侧的起点;(2)规范船长 L 的艏端,是艏柱前端至夏季载重水线的垂线。 Fore end is:(1) the start point of forward side, where measuring the length of ship; (2) the perpendicular to the summer load waterline at the forward side of the stem.

Fore end region structure 艏端区域结构 应认为包含船舯部的 $0.4L$ 区域之前的结构。 Fore end region structure is considered to include structure forward of the midship $0.4L$ region.

Fore mast(main mast) 前桅杆 船舶具有 2 个以上桅杆时,指设置在船前部的一根桅杆。

Fore panel 艏端盖板 滚翻式舱盖中,距收藏一端最远的一块舱盖板。

Fore part 艏部 包括位于防撞舱壁之前的结构,如:(1)艏尖舱结构;(2)艏柱。另外包括:艏部平底加强和艏部外飘加强。 Fore part includes the structures located forward of the collision bulkhead, i. e. : (1) the fore peak structures; (2) the stems. In addition, it includes: the reinforcements of the flat bottom of the forward area and the reinforcements of the bow flare area.

Fore peak 艏尖区域(艏尖舱) 系指防撞舱壁向前的船舶区域。 Fore peak means an area of the ship forward of the collision bulkhead.

Fore peak deck 艏尖甲板 系指一段短的升高甲板,从船艏向后延伸。 Fore peak deck means a short raised deck extending aft from the bow of the ship.

Forecastle 艏楼 (1)参照经修正的国际载重线公约[MSC.143(77)决议第3(10,g)条],艏楼是从艏垂线向后延伸至艉垂线前方一点的上层建筑。艏楼起点可以是艏垂线前方的一点;(2)艏部的短上层建筑。(1)Ref. ILLC, as amended [Resolution MSC. 143 (77) Reg. 3 (10, g)], forecastle is a superstructure which extends from the forward perpendicular aft to a point which is forward of the after perpendicular. The forecastle may originate from a point forward of the forward perpendicular; (2) Forecastle means a short superstructure situated at the bow.

Forecastle deck 艏楼甲板 艏楼顶部的甲板。

Forefoot 前踵 艏柱与龙骨连接部分。

Forefoot region 1 艏踵区1 在主冰带区以下,从艏柱(或设有球鼻艏时为球鼻艏前端)开始,延伸至水平龙骨线和倾斜艏柱线交点处向后 5 档肋距的地方。Forefoot region 1 is the area below the main ice belt zone extending from the stem, or the fore end of the bulb where a bulbous bow is fitted, to a position five frame spaces aft of the point of intersection between the level keel line and the raked stem.

Forefoot region 2 艏踵区2 在主冰带区以下,从艏踵区 1 的后边界延伸至艏部区的后边界,包括舷侧板和船底板在内的区域。 Forefoot region 2 is the area below the main ice belt extending from the aft boundary of Forefoot Region 1 to the aft boundary of the forward region and encompasses both side and bottom shell plating.

Foreign Economic and Trade Arbitration Commission [中]对外经济贸易仲裁委员会

Foreign trade 对外贸易 系指一国或地区同别国或地区进行货物和服务交换的活动,从一个国家来看这种交换活动称为对外贸易。广义的对外贸易包括服务贸易,狭义的对外贸易只包含货物贸易的内容。

Foreign trade coefficient 对外贸易依存度 又称为对外贸易系数(传统的对外贸易系数),系指一国的进出口总额占该国国民生产总值或国内生产总值的比重。其中,进口总额占 GNP 或 GDP 的比重称为进口依存度,出口总额占 GNP 或 GDP 的比重称为出口依存度。对外贸易依存度反映一国对国际市场的依赖程度,是衡量一国对外开放程度的重要指标。

Foreign trade policy 对外贸易政策 是对各国在特定时期对进出口贸易进行管理的原则、方针和措施手段的总称。其所包含的基本因素包括:(1)政策主体。系指政策行为者,即政策制定者和实施者,一般来说系指各国政府。(2)政策客体或政策对象。就是贸易政策规范、指导、调整的贸易活动和从事贸易活动的企业、机构和个人。(3)政策目标。贸易政策行为是有目的的行动。贸易政策的内容首先是在一定政策目标的指导下确定的,政策目标是政策内容制定的依据。(4)政策内容。即为实现既定的政策目标、实施政策内容所采用的对外贸易管理措施,如关税、非关税等。各国制定对外贸易政策的目的在于:(1)保护本国市场;(2)扩大本国产品的国外市场;(3)优化产业结构;(4)积累发展基金;(5)维护和发展同其他国家和地区的政治经济关系;(6)其他。各国对外贸易政策一般由下述内容构成:①总贸易政策。其中包括货物进口与服务获取总政策和货物出口与服务提供总政策。它是从整个国民经济出发,在一个较长的时期内实行的贸易政策。②商品和服务贸易政策。根据总贸易政策、国内经济结构与市场供求状况针对不同的商品分别制定。③国别、地区贸易政策。它是根据总贸易政策,同别国或地区的政治、经济关系分别具体制定的。

Foreign water 舷外水

Foreign-related case 涉外案件

Foreign-trade zones 外贸保税区

Forelock 开口销(扁销,销住)

Forelock pin 开尾销(锚杆销)

Forelock shackle 开口销卸扣

Foremast light 前桅灯 设置在船舶前桅上的桅灯。

Forepeak 艏尖舱 防撞舱壁之前的高度符合有关规定的区域。 Forepeak is defined as the area forward of collision bulkhead, up to the height according to related requirements.

Fore-propeller hydrodynamic fin sector 桨前扇形整流鳍 一种置于桨前,由 2~3 片机翼形剖面的支臂构成的节能装置。其一端固定于支架,外端可加装扇形环与船体连成一体以增加强度。也可与普通导管的导缘部分连接。该装置特别适用于导管桨推进的船舶。

Forest product carrier 木材制品运输船 系指专门设计和制造用于运输木材制品的船舶。 Forest product carrier is a ship designed and constructed for the carriage of forest product.

Forewarmer 预热器

Forged piece 锻件

Forged steel 锻钢

Forging shop 锻工车间 系指从事锻造工件生产的场所。

Form coefficient 形状系数 船体黏压阻力系数对同面积平板摩擦阻力系数的比值。$k = C_{pv}/C_{FO}$ 或 $(C_V - C_{FO})/C_{FO}$,式中,C_V—黏性阻力系数;C_{pv}—黏压阻力系数;C_{FO}—平板摩擦阻力系数。

Form effect 形状效应 由于船体形状不同于相当平板面引起的对摩擦阻力的影响。

Form factor 形状因数 船体黏性阻力系数对相当平板摩擦阻力系数的比值。$r = C_V/C_{pv} F = 1 + k$,式中,k—形状系数。

Form resistance 形状阻力 由于边界层排挤作用与分离现象所引起的阻力。

Form stability level cross-curves 形状稳性力臂曲线 系指一组表示在各不同横倾角时排水体积(或排水量)与形状稳性力臂关系的曲线。

Form the nearest land 距最近陆地 系指按照国际法则划定领土所属领海的基线,但下述情况除外:在澳大利亚东北海面"距最近陆地"系指澳大利亚海岸下述各点的连线而言——自南纬 11°00′东经 142°08′的一点起,至南纬 10°35′东经 141°55′的一点,然后至南纬 10°00′东经 142°00′的一点,然后至南纬 13°00′东经 144°00′,然后至南纬 18°00′东经 147°00′,然后至南纬 21°00′东经 153°00′的一点,然后至澳大利亚海岸南纬 24°42′东经 153°15′的一点所划的一条连线。 Form the nearest means from the baseline from which the territorial sea of the territory in question is established in accordance with international law, except that "form the nearest" off the north eastern coast of Australia shall mean from a line drawn from a point on the coast of Australia in—latitude 11°00 south, longitude 142°08′ east to a point in latitude 10°35′ south, longitude 141°55′ east, thence to a point- latitude 10°00′ south, longitude 142°00′ east, thence to a point latitude 13°00′ south, longitude 144°00′ east, thence to a point latitude 18°00′ south, longitude 147°00′ east, thence to a point-latitude 21°00′ south, longitude 153°00′ east, thence to a point on the coast of Australia in latitude 24°42′ south, longitude 153°15′ east.

Formal contract 正式合同 是带有"合同"字样的法律契约。合同的文字要清楚、经济责任要明确,并对双方有约束性,签订手续完备。

Formal safety assessment 综合安全评估 是一种结构化的系统方法,在规范制定中应用这一方法,目的是安全而综合的考虑影响安全的诸方因素,通过风险评估、费用和效益评估,提出合理的并能有效地控制风险的规范要求,从而不断改进和提高规范的水平。

Forschungs plattform in Nord and Ostsee(Research Platform on North and Baltic Seas) (FINO) 北海和波罗的海研究站 至今已建立了 3 个研究站,FINO 1 位于阿尔法·文图斯海上风电试验场边缘,FINO 2 位于波罗的海,大约在 Rugen 岛北 30 km 处,FINO 3 位于北海,约在 Sylt 岛西 80 km 处。这些研究站的主要任务是测量风速和不同高度的风向。此外,他们也收集所有的气象信息,如有关海洋和生态的数据。

Forsters anchor dredge 福斯特锚式拖网 用于采集砂质或硬质海底底栖生物的网具。由漏斗、叉形拖臂和网衣所组成。可深入底质达 25 mm。

Forty foot equivalent units (FEU) 40 英尺集装箱

Forward air controller (FAC) 空军前进引导员

Forward and after perpendiculars 艏、艉垂线 系指取自船长 L 的前后两端,艏垂线应与计量长度水线上的艏柱前缘相重合。 Forward and after perpendiculars shall be taken at the forward and after ends of the length L. The forward perpendicular shall coincide with the foreside of the stem on the waterline on which the length is measured.

Forward engine room 前机舱

Forward freeboard perpendicular FP_{LL} 干舷艏垂线 是取自长度 L_{LL} 的前端,并与在计量长度 L_{LL} 的水线上的艏柱前边相重合。 Forward freeboard perpendicular is to be taken at the forward end of the length L_{LL} and is to coincide with the foreside of the stem on the waterline on

Forward perpendicular FP 艏垂线 系指夏季载重水线与船首前缘交点处的垂线。对具有非常规船艏形状,其艏垂线的位置应作特别考虑。 Forward perpendicular is the perpendicular at the intersection of the summer load waterline with the fore side of the stem. For ships with unconventional stem curvature, the position of FP will be specially considered.

Forward region 艏部区的范围 系指从艏柱至舷侧平直部分的前端边界线之间的区域。对冰级1A Super和1A,此距离等于 $0.04 L_R$ 或 5 m 中的较大者;对于冰级 1B 和 1C,等于 $0.02 L_R$ 或 2 m 中的较大者。舷侧平直部分无明显的前端界线时,则艏部区的后边界可取为艏垂线后的 $0.4 L_R$ 处(见图F-15)。 Forward region extends from the stem to aft of the forward borderline of the flat side of the hull by a distance equal to the greater of $0.04 L_R$ or 5 m for Ice Classes 1 A super and 1A, or the greater of $0.02 L_R$ or 2 m for Ice Classes 1B and 1C. Where no clear forward borderline of the flat side of the hull is discernible, the aft boundary of the forward region is to be taken $0.4 L_R$ aft of the forward perpendicular. It is seens in Fig. F-15.

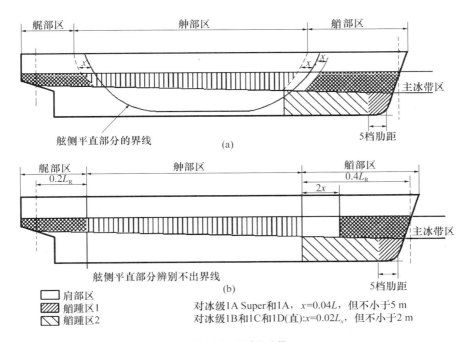

图 F-15 区域和冰带
Figure F-15 Region and ice belt

Forward region of ice belt 冰带艏部区域 系指从艏柱向后至舷侧平直部分前限界线之后 $0.04 L$ 处之间的区域。对 IA Super 冰级和 IA 冰级,与前限界线重叠的距离不必大于 6 m,对 IB、IC 和 ID 冰级,不必大于 5 m。 Forward region of ice be it: from the stem to a line parallel to and $0.04 L$ aft of the forward borderline of the part or the hull where the waterlines run parallel to the centerlines. For ice classes IA Super and IA the overlap over the borderlines does not need to exceed 6 m, and for ice classes IB, IC and ID this overlap does not need to exceed 5 m.

Forward seal 艏封门 侧壁气垫船在艏部船底用以维持气垫的通常采用囊指或囊筒围裙构成的封闭装置。

Forward shoulder 前肩 系指进流段紧邻平行中体的部分。

Forward terminal 前端点(艏端) 船舶分舱长度的最前一点(分舱长度的艏界限)。 Forward terminal is the forward limit of the subdivision length.

Foul gas 不凝性气体(惰性气体)

Fouling 污底(弄脏) 船体水下部分表面生锈及

滋生海藻、贝类等的现象。

Fouling by marine organism 海生物污底

Fouling resistance 污底阻力 船舶长期浸在水中,随时间推移,由于油漆失效,船壳板锈蚀和水生物的黏附使船体表面粗糙度增大,从而增大了摩擦阻力,即形成了污底阻力。污底阻力的大小与航区、气候、水文、油漆质量、出坞时间、航速及停泊时间等许多因素有关。

Foundation(seating) 基座 用以装设船舶设备的座子。

Foundation structure 基础结构 系指海上风能转换系统的桩基,由钢或混凝土构成。在阿尔法-文图斯试验场中使用了两种钢制桩基。三脚式桩基和导管架式桩基。最简单的桩基由打入海底的钢管构成(单桩)。然而,这只适合于浅水域中,因而阿尔法-文图斯试验场的建造并未采用这种桩基。

Foundry shop 铸造车间 系指从事铸造工件生产的场所。

Fountain 喷水器(中心注管)

Four way join box 四路联箱

Four-cycle 四冲程循环

Four-cycle period 四冲程循环周期

Four-cylinder triple expansion engine 四冲程三胀式蒸汽机

Four-stroke engine 四冲程发动机

Fourth generation fighter 第四代战机 系指在20世纪70年代陆续开始服役的、吸收第三代战斗机设计与使用上的经验,加上诸多空中冲突与演习显示出来的问题和需求,融合之后成为冷战结束前后最主要的角色。美国曾很长时间称呼这类战机为"第三代战机",不少中文媒体也延续美国分代法第三代战机的称呼。由于苏联传统分类和美国2009年后分类方式两者已统一,所以以上就是唯一的国际第四代战斗机标准。该技术拥有国:美国(F22/F35)、俄罗斯(T50)、中国(歼-20/歼-31)。见图F-16。

图 F-16 第四代战机
FigureF-16 Fourth generation fighter

f_p 与概率水平对应的系数 取 $f_p = 1.0$,对应 10^{-8} 概率水平的强度评估率;取 $f_p = 0.5$,对应 10^{-4} 概率水平的强度评估率。 Coefficient corresponding to the probability level, taken equal to: $f_p = 1.0$, for strength assessments corresponding to the probability level of 10^{-8}; $f_p = 0.5$, for strength assessments corresponding to the probability level of 10^{-4}.

FPSO 浮式生产、储油和卸油船 通常系指一艘系泊于海底或使用内或外转台系泊系统,允许直接从海底装置生产、储存和卸载加工过液体的船舶。 FPSO is a floating production, storage, and offloading vessel. A marine vessel usually moors to the seabed or uses an internal or external turret system, allowing direct production, storage, and offloading of process of fluids from subsea installations.

FPSO with single point mooring system(SPM) 单点转塔系泊式浮式生产储/卸油船(FPSO) 系指在强台风中不解脱的FPSO。其采用船舶深海停泊技术,永久系泊在油田上作业。不仅可以在浪高20 m、风速50 m/s强台风环境中保持岿然不动,同时还配有公共系统、安全控制系统两套三级处理的原油生产加工系统等多个系统,并可作为操作员生活基地。

FPU 浮式生产装置 FPU means a floating production unit.

Fraction 分数

Fracture detection procedures 破裂探测方法 系指下列方法:(1)射线照相设备;(2)超声波设备;(3)磁粉探伤设备;(4)染色渗透液;(5)其他等效方式。 Fracture detection procedures mean the following procedures:(1) radiography equipment;(2) ultrasonic equipment;(3) magnetic particle equipment;(4) dye penetrant;(5) other equivalent means.

Fracture toughness 断裂韧性 系指衡量材料阻止裂缝蔓延的能力。 Fracture toughness means the measure of material resistance to crack propagation.

Frame 肋骨 舷侧和船底部分,用以支承外板和保持船形的横向和纵向骨材。

Frame cooling bender 肋骨冷弯机

Frame size 壳架等级电流 表示一组断路器特性的术语,其结构尺寸对几个电流额定值者相同。壳架等级电流以相应于该组电流额定值的最大值表示。在一壳架等级中,宽度可随极数而不同。注意:此定义不包含尺寸标准化。

Frame spacing a 肋骨间距 a 系指相邻肋骨理论线间的距离。 Frame spacing a will be measured from moulding edge to moulding edge of a frame.

Frame type container 框架集装箱 系指便于装运活的牲畜用的集装箱。

Framing 骨架 船体内支承外板、甲板板、内底板等纵横骨架的统称。

Frapping lines 靠舷索 为使救生艇在放至登乘甲板时,可靠依船舷以便人员登乘而设置在救生艇前、后艇吊钩处供收放用的索具。

Fraunhofer Institute for Wind Energy and Energy System Technology 弗劳霍夫风能及能源系统技术研究所 成立于 2009 年 1 月,由位于卡塞尔的太阳能供应技术研究所(ISET)和位于不莱梅的弗劳霍夫风能和海事技术中心(Fraunhofer CWNI)合并而成。在不莱梅港和卡塞尔均设有办事处。

Free alongside-named port of shipment(FAS) 指定装运港的船边交货 系指卖方在指定的装运港码头或驳船上把货物交至船边,从这时起买方须承担货物灭失或损坏的全部费用和风险,另外买方须办理出口结关手续。以前版本要求买方安排办理出口手续。

Free carrier(FCA) 货交承运人(指定地点) 系指卖方只要将货物在指定的地点交给买方指定的承运人,并办理了出口清关手续,即完成交货。FCA 是国际贸易术语之一,《2000 年国际贸易术语解释通则》对其规定如下:需要说明的是,交货地点的选择对于在该地点装货和卸货的义务会产生影响。若卖方在其所在地交货,则卖方应负责装货,若卖方在任何其他地点交货,卖方不负责装货。该术语可用于各种运输方式,包括多式联运。若买方指定承运人以外的人领取货物,则当卖方将货物交给此人时,即视为已履行了交货义务。

Free carrier-named place(FCA) 指定地点的货交承运人 系指卖方应负责将其移交的货物,办理出关后,在指定的地点交付给买方指定的承运人照管。

Free damping layer 自由阻尼层

Free fall totally-enclosed lift-boat 抛落式救生艇 系指安放在船舶艉部,使用时,利用救生艇自身的重力,自行滑落的全封闭救生艇。

Free field 自由声场 是一种理想的声场。在均匀、各向同性的介质中,边界影响可以忽略不计的声波传播场所。例如宽阔的空间或消音室中,在自由声场中声音的传播不受阻挡,不受干扰,辐射特性不变。

Free flooding 通海的(自由进水的)

Free flooding opening 自由流孔

Free for all links 免费链接列表 这类网站一般只简单地滚动排列链接条目,其中有一部分有简单的分类目录,不过其规模比起 Yahoo 目录索引要小得多。

Free from particular average(FPA) 平安险 这一名称在我国保险行业中沿用甚久,其英文原意指单独海损不负责赔偿。根据国际保险界对单独海损的解释,它系指保险标的物在海上运输途中遭受保险范围内的风险直接造成的船舶或货物的灭失或损害。因此,平安险的原来保障范围只赔全部损失。但在长期实践的过程中对平安险的责任范围进行了补充和修订,当前平安险的责任范围已经超出只赔全损的限制。其中,水渍险的责任范围除了包括上列"平安险"的各项责任外,还负责被保险货物在运输过程中由于恶劣气候、雷电、海啸、地震、洪水等自然灾害所造成的部分损失。

Free of all average 全损赔偿 系指船舶在运营过程中,所载的货物全部受到损坏或灭失后,保险公司根据全损险的规定向货主进行的赔偿。

Free of charge(FOC) 免费

Free of turn 不论到港顺序

Free on board-named port of shipment(FOB) 指定目的港的装运港船上交货 系指卖方负责在合同规定的日期或期间内,在指定的装运港把货物装到买方指定的船上,卖方即完成交货。承担货物越过船舷以前的费用和风险的贸易术语。而买方必须从该点起承担货物灭失或损坏的一切风险。

Free on board-named port of destination(FOB) 指定目的港的船上交货 系指当货物在指定的装运港越过船舷,卖方即完成交货。这意味着买方必须从该点起承当货物灭失或损坏的一切风险。船上交货分为装运港船上交货和目的港船上交货。FOB 术语要求卖方办理货物出口清关手续。该术语仅适用于海运或内河运输。如当事各方无意越过船舷交货,则应使用 FCA(货交承运人)术语。

Free perimeter 自由边界区 也称自由贸易区。一般设在本国的一个省或几个省的边境地区。对于在区内使用的生产设备、原材料和消费品可以免税或减税进口。如从区内转运到本国其他地区出售,则需照章纳税。外国货物可在区内储存、展览、混合、包装、加工和制造等活动。

Free perturbation motion 自由扰动运动 船舶受外力扰动引起的运动在扰动完全消除后,所遗留的运动。

Free port 自由港

Free rolling 自由横摇 船舶在静水中当产生初始横摇的外力消失后,自由摇摆的横摇。

Free running 空车运行

Free running model test 自由自航模操纵性试验 在水池中,用自由自航船模进行的操纵性试验。

Free running speed 自由航速 船舶不拖带它物时的航速。

Free ship 中立国船(战时) 系指战时中立国拥

Free span 自由张紧索 系指在 2 个海底斜面之间自由悬浮的部分管路。 Free span is the section of pipeline that is freely suspended between two seabed shoulders.

Free standing fuel tank 独立架设燃油柜

Free standing hybrid riser(FSHR) 自由站立式组合立管

Free stern 优先靠码头 系指不受装载时间限制，不受协议装货日期限制随时随地直接靠泊码头。

Free surface 自由液面 船舶倾斜时，舱柜内能自由变动的液面。

Free surface correction 自由液面修正 指由于自由液面的影响而对船舶稳性所做的修正。

Free trade 自由贸易 系指政府不采用关税、配额或其他形式来干预国际贸易的政策。自由贸易(Free trade)系指国家取消对进出口贸易的限制和障碍，取消本国进出口商品各种优待和特权，对进出口商品不加干涉和限制，使商品自由进出口，在国内市场上自由竞争的贸易政策。它是"保护贸易"的对称。这并不意味着完全放弃对进出口贸易的管理和关税制度，而是根据外贸法规即有关贸易条约与协定，使国内外产品在市场上处于平等地位，展开自由竞争与交易，在关税制度上，只是不采用保护关税，但为了增加财政收入，仍可征收财政关税。自由贸易的静态利益包括两方面。

Free trade policy 自由贸易政策 是自由放任经济政策的一个重要组成部分。国家取消对进出口贸易和服务贸易的限制和障碍，取消对本国进出口贸易和服务贸易的各种特权和优待，使商品自由进出口，服务贸易自由经营，也就是说国家对贸易活动不加或少加干预，任凭商品、服务和有关要素在国内外市场公平、自由的竞争。

Free trade zone 自由贸易区 又称为对外贸易区(Foreign trade zone)或免税贸易区(Tax-free trade zone)，系指 2 个或 2 个以上的国家通过达成某种协定或条约取消进出口商品的全部或大部分关税和与关税具有同等效力的其他措施，并且允许港内或区内进行商品的自由储存、展览、加工和制造等业务活动，以促进地区经济和对外贸易发展的一个区域。一般建立在一个港口的港区或邻近港口的地区，它实际上是采取自由港政策的关税隔离区。2016 世界自由贸易大会于 2016 年 11 月 5 日在中国澳门召开，主题为《自由无界的世界贸易》。

Free vehicle respirometer 活动式海底生物呼吸测量器 可在海底现场测量海底栖生物群落代谢率的一种仪器装置。并可采集海底栖生物以及拍摄动物照片等。

Free vibration 自由振动 系指船体受外力干扰，在该力去除后所发生的振动。

Free vortex 自由涡 就所起作用而言，仅在流体中感生流场但不对物体产生升力的涡线部分。

Free voyage or excursion 收费航行或游览 系指:(1)除(2)和(3)中的规定外，不付费且无人提供食物和服务;(2)从事航行或游览业务的船东，可以收费或对任何人提供食物或服务;(3)从事航行或游览业务的船东仅可收取: ①为与该船操作或在船上搭载任何人员有关的或在航行或游览期间在船上使用或提供的食品和服务的费用。②作为对在航行或游览期间操作该船发生的直接花费的补偿。 Free voyage or excursion means a voyage or excursion in respect of which;(1) subject to paragraphs(2) and(3), no money is paid, and no goods or services are provided, by any person;(2) the owner of the vessel engaged in the voyage or excursion may pay money, or provide goods and services, to any person; and(3) the owner of the vessel engaged in the voyage or excursion may only receive;①money for, or in connection with, the operation of the vessel or the carrying of any person in the vessel as a contribution to the direct expenses of the operation of the vessel incurred during the voyage or excursion;②goods and services which are used or provided on the vessel during the voyage or excursion.

Free water effect 自由液面影响

Freeboard 干舷 系指勘定的载重线和干舷甲板之间的距离。 Freeboard assigned means the distance between the assigned load line and freeboard deck.

Freeboard deck 干舷甲板 (1)干舷甲板通常系指最上层露天连续甲板，其上的所有的露天开口设有永久性关闭装置;其下在船侧处的所有开口设有永久性水密关闭装置;(2)下层甲板作为干舷甲板:由船东选择并经主管机关批准，可将下一层甲板指定为干舷甲板，但该甲板至少在机器处所与艏、艉尖舱舱壁之间应是全通的和永久性的前后连续甲板，并且横向也是连续的。①当该上层甲板为阶梯形时，甲板最低线及其平行于甲板上部的延长部分取为干舷甲板;②当一下层甲板设计为干舷甲板时，就干舷的核定条件和计算而言，该干舷甲板以上的船体部分作为上层建筑处理。干舷是从这层甲板算起;③当下层甲板设计为干舷甲板时，在货舱范围内，这种干舷甲板的结构最低限度应在船侧和通至上甲板的每一水密舱壁处设有适当的框架结构桁材。这些桁材的宽度应适合于方便安装，并应考虑船舶的结构和操作情况。桁材的任何布置也应能满足结构上的要求;(3)不连续干舷甲板，阶梯形干舷甲板;①如果干舷甲板的凹槽延伸到两舷侧且长度超过 1 m，则该露天甲板的最低线及其平行于甲板上部的延伸部分取为干舷甲板;②如果干舷甲板的凹槽未延伸到两舷侧,则甲板上部取为干舷甲板;③如果露天甲板以下的一层甲板指定为干舷甲板且其设有未从一舷侧延伸至另一舷侧的凹槽,只要露天甲板上的所有开口设有风雨密关闭装置,则该凹槽可以不计;④应充分考虑露天凹槽的排水系统和自由液面对稳性的影响。⑤①～④项规定不适用于挖泥船、开底泥驳或设有大开口舱的其他类似船舶,对这类船舶的每一种均需单独考虑。 Freeboard deck means: (1) normally the uppermost complete deck exposed to weather and sea, which has permanent means of closing for all openings in the weather part thereof, and below which all openings in the sides of the ship are fitted with permanent means of watertight closing; (2) depending on the option of the owner and subject to the approval of the Administration, a lower deck may be designed as the freeboard deck provided that it is a complete and permanent deck continuous in a fore and aft direction at least between the machinery spaces and peak bulkheads and continuous athwartships: ①when this lower deck stepped the lower line of the deck and the continuation of that line parallel to the upper part of the deck is taken as the freeboard deck; ②when a lower deck is designated as the freeboard deck, that part of the hull which extends above the freeboard deck is treated as a superstructure concerning the application of the conditions of assignment and the calculation of freeboard. It is from this deck that the freeboard is calculated; ③when a lower deck is designated as the freeboard deck, such a minimum shall consist of suitably framed stringers at the ship sides and transversely fitted at each watertight bulkhead which extend to the upper deck, within cargo spaces. The width of these stringers shall not be less than that can be conveniently fitted having regard to the structure and the operation of the ship. Any arrangement of stringers shall be such that structural requirement can also be met; (3) discontinuous freeboard deck, stepped freeboard deck: ①where a recess in the freeboard deck extends to the sides of the ship and is in excess of 1 m in length, the lower line of the exposed deck and the continuation of that line parallel to the upper part of the deck is taken as the freeboard deck; ②where a recsee in the freeboard deck does not extend to the sides of the ship, the upper part of the deck is taken as the freeboard deck; ③recesses not extending from side to side in a deck below the exposed deck, designated as the freeboard deck, may be disregarded, provided all openings in the weather deck are fitted with weathertight closing appliances; ④due regard shall be given to the drainage of exposed recesses and to free surface effects on stability; ⑤provisions of subparagraphs ①～④ are not intended to apply to dredgers, hopper barges or other similar types of ships with large open holds, where each case requires individual consideration.

Freeboard deck of open-top container ships 敞口集装箱船的干舷甲板 对国际航行敞口集装箱船,系指对"1966 国际载重线公约"附则Ⅰ第1章和第2章而言,视舱口围板顶部设置舱口盖按1966LL 公约确定的干舷甲板。 Freeboard deck of open-top container ships engaged on international voyages, for Chapter I and II of Annex I of the International Convention on Load Lines, 1966, is the freeboard deck according to the LL, 1966 as if hatch covers are fitted on top of the hatch coamings.

Freeboard F 干舷 F 干舷 F 应取在规定的纵向位置处甲板舷弧线与各种特定的装载条件下漂浮面之间测得的距离。 Freeboard F shall be measured as the distance between the sheerline at the defined lengthwise location and the flotation plane in any specified loading condition.

Freeboard length 干舷船长 系指:(1)参照经修正的载重线公约 | MSC.143(77)决议第 3(1,a)条 | 干舷船长 L_{LL} 系指自龙骨顶部向上量至最小型深85%处水线从艏柱前缘量至舵杆中心的距离。L_{LL} 应不小于该水线总长的96%;(2)参照经修正的载重线公约 | MSC.143(77)决议第 3(1,e)条 | 如在最小型深85%处水线以上的艏柱外形为凹入的,则总长的前端和艏柱前边均应从(该水线以上的)艏柱外形最后一点垂直投影在该水线的点量起。 Freeboard length means: (1) Ref. ILLC, as amended(Resolution MSC.143(77) Reg. 3(1,a)). The freeboard length L_{LL} is the distance, in m, on the waterline at 85% of the least moulded depth from the top of the keel, measured from the forward side of the stem to the centre of the rudder stock. L_{LL} is to be not less than 96% of the extreme length on the same waterline. (2) Ref. ILLC, as amended(Resolution MSC.143(77) Reg. 3(1,c)). Where the stem contour is concave above the water-line at 85% of the least moulded depth, both the forward end of the extreme length and the forward side of the stem are to be taken at the vertical projection to that waterline of the aftermost point of the stem contour(above that waterline)

Freeboard of lifeboat 救生艇干舷 救生艇载足额定乘员和规定属具时的干舷。

Freeboard of liferaft 救生筏干舷 救生筏满载额定乘员后自水面量至上浮胎或刚性浮体上缘的干舷。

Freeboard on fresh water 淡水干舷 船舶在相对

密度为 1.000 g/cm³ 的淡水中时，各季节干舷应从各季节相应的夏季干舷减去 $\triangle /40T(\text{cm})$，其中：$\triangle$ 为夏季载重水线时的海水排水量 t，T 为夏季载重水线处在海水中每 1 cm 浸水的吨数，t/cm。

Freeboard, aft F_A 艉干舷 F_A 艉干舷 F_A 应取在规定的纵向位置处甲板舷弧线与各种特定的装载条件下漂浮面之间测得的距离，在甲板舷弧线的最后点处测量。 Aft freeboard F_A shall be measured as the distance between the sheerline at the defined lengthwise location and the flotation plane in any specified loading condition, at the aftermost point of the sheerline.

Freeboard, forward F_F 艏干舷 F_F 艏干舷 F_F 应取在规定的纵向位置处甲板舷弧线与各种特定的装载条件下漂浮面之间测得的距离，在甲板舷弧线/舷侧处甲板线的最前点测定。 Forward freeboard F_F shall be measured as the distance between the sheerline at the defined lengthwise location and the flotation plane in any specified loading condition, at the most forward point of the sheerline/deck at the side.

Freeboard, mid-ship F_M 舯干舷 F_M 舯干舷 F_M 应取在规定的纵向位置处甲板舷弧线与各种特定的装载条件下漂浮面之间测得的距离，在 1/2 船体长度处测量。 Mid-ship freeboard F_M shall be measured as the distance between the sheerline at the defined lengthwise location and the flotation plane in any specified loading condition, at half-length of the hull.

Free-fall acceleration 自由降落加速度 系指在自由降落救生艇降放期间，登乘者所经受的速度变化率。 Free-fall acceleration is the rate of change of velocity experienced by the occupants during launching of a free-fall lifeboat.

Free-fall certification height 自由降落的核准高度 系指救生艇被批准的最大降落高度，即从静水表面量至救生艇在降放状态时救生艇外形的最低点。 Free-fall certification height is the greatest launching height for which the lifeboat is to be approved, measured from the still water surface to the lowest point on the lifeboat when the lifeboat is in the launch configuration.

Free-fall launching 自由降落下水 系指载足全部乘员和属具的救生艇筏在船上脱开并在没有任何约束载荷的情况下，任其下降到水面的降落方法。 Free-fall launching is that method of launching a survival craft whereby the craft with its complement of persons and equipment on board is released and allowed to fall into the sea without any restraining apparatus.

Free-fall lifeboat 自由抛落式救生艇 系指采用自由抛落的方式进行释放的救生艇。在救生艇抛落前，人员提前进入艇内，并启动发动机怠速运转，救生艇落水后可以迅速离开母船。该方式以其安全和及时的特点，为大型船舶和海洋平台提供了有效的安全疏散通道。见图 F-17。

图 F-17 自由抛落式救生艇
Figure F-17 Free-fall lifeboat

Free-form deformation approach (FFD) 自由变形方法

Freeing 排除（免除）

Freeing arrangement 排水口

Freeing pipe 排水管（逸气管，油舱的油气溢出管）

Freeing port 排水舷口 系指舷墙上的开口，可让从舷侧进入甲板的水自由流向舷外。 Freeing port means an opening in the bulwarks to allow water shipped on deck to run freely overboard.

Freeze to 冻凝（冷藏）

Freeze up 冻住（水塞）

Freezer 冻结器（间）（冻结设备）

Freezer stern trawler 冷冻艉拖网渔船 系指将渔获物储藏在船上的冷库中的艉滑道拖网渔船。

Freezing 制冷（冻结，凝固，冻融）

Freezing and thawing test 冻融试验（冻结解冻试验）

Freezing capacity 冻结能力

Freezing mixture 冷冻剂

Freezing plant 冷冻装置

Freezing point 冰点（凝固点）

Freezing rate 冻结速度

Freezing room 冷冻舱

Freezing shrinkage 冷凝收缩

Freezing test 冻结试验（抗冻性试验）

Freight barge / cargo barge 货驳 专运货物的驳船。有干货驳、矿砂驳、煤驳、猪驳、军火驳、石驳等。

Freight ferry 货物渡船 专运货物的渡船。

Freight ferry is a ship dedicated to carry freight.

Freight forward 目的港支付运费 系指货物到达目的港后,托运人向承运人支付的费用。

Freight in full 全包运费

Freight prepaid 运费预付 系指货物在起运前,托运人向承运人预付的部分或全部的费用。

Freight steamer 蒸汽机货船 系指以蒸汽机作为主机的货船。

Freon 氟利昂（制冷剂）

Freon refrigeration 氟利昂制冷

Frequency 频率（频度） 系指表示单位时间（例如,年）预期出现的次数或诸如出现事故的范围。该术语也称为"可靠性"。 Frequency is a score indicating the expected number of occurrences per unit time (e. g. , year) or an event such as an accident. It is also termed "likelihood".

Frequency alias analyser 频率分析器/谐波分析器

Frequency conversion power 变频电源 经过交-直-交变换的逆变电源。其主要功用是将现有交流电网电源变换成用户所需的频率。

Frequency of encounter 遭遇频率 船舶在波浪中运动时,单位时间内按不变的方向经过船体上一定点的波峰或波谷的数量。

Frequency response function of pitching 纵摇频率响应函数 在规则波中,系指船舶纵摇幅值与波浪幅值的比值。而在不规则波中则一般系指船舶纵摇响应频谱密度对波浪频谱密度比值的平方根。$V_{\theta\xi} = [S_\theta(\omega)/S_\xi(\omega)]^{1/2}$。

Frequency response function of yaw 艏摇频率响应函数 在规则波中,系指船舶首摇幅值与波浪幅值的比值。而在不规则波中则一般系指船舶首摇响应频谱密度对波浪频谱密度比值的平方根。$V_{\psi\xi} = [S_\psi(\omega)/S_\xi(\omega)]^{1/2}$。

Frequency selective damping material 选频阻尼材料

Frequency selective material (FSM) 频率选择材料

Frequency selective surface (FSS) 频率选择表面 是由大量无源谐振单元组成的单屏或多屏周期性阵列结构,由周期性排列的金属贴片单元或在金属屏上周期性排列的孔径单元构成。这种表面可以在单元谐振频率附近呈现全反射（贴片型）或全传输特性（孔径型）。分别称为带阻或带通型 FSSC3。目前 FSS 的应用十分广泛,也可用于反射面天线的负反射器以实现频率复用,提高天线的利用率;可用于波极化器、分波束仪和激光器的"腔体镜",以提高激光器的泵吸功率;还可用于隐身技术,如雷达天线罩以降低天线系统的雷达截面（RCS）。

Frequency spectrum 频谱 是表示声音的频率成分与声能量分布的关系,就是频率的分布曲线,表示出声音的频率特性。复杂振荡分解为振幅不同和频率不同的谐振荡,这些谐振荡的幅值按频率排列的图形叫作频谱。广泛应用在声学、光学和无线电技术等方面。频谱利用率为每小区每 MHz 支持的多少对用户同时打电话;而对于数据业务来讲,定义为每小区每 MHz 支持的最大传输速率。在这里,小区的频率复用系数（f）非常重要:f 越低,则意味着每小区可选的频率自由度越大。在 CDMA 系统中,每个小区都可以重复使用同一频带（$f=1$）。在一个小区内对每个移动台的总干扰是同区内其他移动台干扰加上所有邻区内移动台干扰之和。

Frequency spectrum analyzer 频谱分析仪 是将声级计和滤波器组合在一个机壳内的,用于研究电信号频谱结构的仪器。用于信号失真度、调制度、谱纯度、频率稳定度和交调失真等信号参数的测量,可用以测量放大器和滤波器等电路系统的某些参数,是一种多用途的电子测量仪器。它又可称为频域示波器、跟踪示波器、分析示波器、谐波分析器、频率特性分析仪或傅里叶分析仪等。现代频谱分析仪能以模拟方式或数字方式显示分析结果,能分析 1Hz 以下的甚低频到亚毫米波段的全部无线电频段的电信号。仪器内部若采用数字电路和微处理器,具有存储和运算功能;配置标准接口,就容易构成自动测试系统。

Frequency tolerance 频率偏差 在正常运行状态下,偏离额定频率的最大值,但不包括瞬态和周期性频率变化。（注意:频率偏差是稳态偏差,并包含由负载和调速器特性引起的变化,也包括由环境条件引起的变化。） Frequency tolerance is the maximum departure from nominal frequency during normal operation conditions excluding transient and cyclic frequency variations. (NOTE: Frequency tolerance is a steady state tolerance and includes variations caused by loads and governor characteristics. It also includes variations due to environmental conditions.)

Frequency transient 频率瞬变 频率的突然变化,在扰动开始后,频率变化超出频率偏差限值,且在规定的恢复时间之内（时间范围为 s）返回并保持在这些偏差之内。 Frequency transient is the sudden change in frequency which goes beyond the frequency tolerance limits and returns to and remains within these limits within a specified recovery time after initiation of the disturbance (time range: seconds).

Frequency transient recovery time 频率瞬变恢复

时间 从频率超出额定偏差到频率恢复并保持在额定偏差限值内所经过的时间。 Frequency transient recovery time is the time elapsed from exceeding the normal tolerance until the frequency recovers and remains within the frequency tolerance limits.

Frequent 经常的概率 系指故障情况的概率大于 10^{-3} 的故障情况。 Frequent means the fault conditions having a probability greater than 10^{-3}.

Frequent (Probability of occurrences) 频繁的(事件的概率) 是一艘特定的船舶在营运生命期内,经常可能发生的(事件)。 Frequent means that an incident is likely to occur often during the operational life of a particular ship.

Fresh and live fish vessel 鲜活鱼运输船 系指用于运输鲜活鱼的船舶。 Fresh and live fish vessel is a ship used for the carriage of fresh and live fish.

Fresh and rain water damage 淡水雨淋 系指由于淡水、雨水或融雪而导致货物水残的损失。

Fresh water 淡水

Fresh water and rain damage risk (FWRD) 淡水雨淋险 投保平安险和水渍险的基础上加保此险,保险人负责赔偿承保货物在运输途中遭受雨水、淡水以及雪融水浸淋造成的损失,包括船上淡水舱、水管漏水以及舱汗所造成的货物损失。不过,保险人承担赔偿责任,要求被保险人必须在知道发生损失后的10天内申请检验,并要以外包装痕迹或其他证明为依据。

Fresh water circuit 淡水管路

Fresh water cooler 淡水冷却器

Fresh water cooling pump 淡水冷却泵

Fresh water distilling unit 蒸馏装置(制淡水装置)

Fresh water filter 淡水滤器

Fresh water generator 造水机

Fresh water pressure tank 淡水压力柜

Fresh water pump 淡水泵

Fresh water tank 淡水舱(淡水柜) 供贮存淡水的舱。

Fresh water transfer pump 淡水驳运泵

Freshen the ballast 压载舱换水

Friction clutch 摩擦式离合器

Friction correction curves for screw dynamometer 测力仪摩擦修正曲线 按不同转速和载荷,修正测力仪内部摩擦损失用的曲线。

Friction correction factor 摩擦修正因数 通过实验得出的叶色体升力系数对理论计算值的比值。

Friction damping 摩擦阻尼 在机械物理学中,系统的能量的减小——阻尼振动不都是因"阻力"引起的,就机械振动而言,一种是因摩擦阻力生热,使系统的机械能减小,转化为内能,这种阻尼叫摩擦阻尼;摩擦的需要稳定的时间! 指针万用表表针稳定住的时间! 现有利用摩擦阻尼生产出一种摩擦阻尼器,摩擦阻尼器是典型的耗能元件,多用于振动能量的衰减。

Friction data (frictional coefficient) 摩擦系数

Friction gear 摩擦轮传动装置

Friction hardening 摩擦硬化

Friction head 摩擦头

Friction loss 摩擦损失

Friction of motion 动摩擦

Friction of rest 静摩擦

Friction of rolling 滚动摩擦

Friction pair 摩擦副

Friction reducing polymer 降阻剂 加入水中以降低摩擦阻力的添加剂。

Friction roller 摩擦滚筒

Frictional force 摩擦力

Frictional resistance (friction drag) 摩擦阻力 系指船舶航行时,由于水有黏性,在船体周围形成边界层,水与船体之间产生切向摩擦力,它在航向分量的合力即为摩擦阻力。

Frictional wake 摩擦伴流 船前进时,由于水与船体摩擦作用使水在螺旋桨处产生的伴流。

Front bulkhead (deckhouse front) 前端壁 上层建筑的前壁。

Front header 前联箱

Front tube plate (锅炉的)前管板

Frost formation 结霜

Frost valve 防冻阀

Froth applicator unit 灭火泡沫发生器

Froth fire extinguisher 泡沫灭火器

Froth suppressor 泡沫灭火器

Frother 起泡剂(泡沫发生器)

Froude number F_n 傅汝德数 F_n 傅汝德数为一个无量纲参数,它是兴波阻力的主要要素,它规定了最大的位移速度。它的定义为: $F_a = \dfrac{V_m}{\sqrt{gL_{WL}}}$;式中,g—重力加速度,取 9.81m/s^2 ;L_{WL}—水线长(m);V_m—适当的航速(m/s)。 Froude number is a non-dimensional parameter which is the primary constituent part of the wave making resistance and which dictates the maximum displacement speed. It is defined as: $F_a = \dfrac{V_m}{\sqrt{gL_{WL}}}$;where: g is the acceleration due to gravity and is taken to be 9.81 m/s²; L_{WL} is waterline length in m; V_m is the appropriate ship speed in m/s.

Froude wake factor 傅氏伴流因数 船速对螺旋桨进速的比值。$1-\omega_F = T - V_A/V$，式中，V—船速；V_A—螺旋桨进速。

Froude wake fraction 傅氏伴流分数 船速与螺旋桨进速之差对螺旋桨进速的比值。$\omega_F = (V - V_A)/V$，式中，V—船速；V_A—螺旋桨进速。

Frozen products insulated container (thermally container) 保温集装箱 系指适用于装运冷藏货物或怕冰坏货物的集装箱。

Fruit carrier 运水果船 系指运输水果的冷藏船舶。Fruit carrier is a refrigerated fruit ship used for the carriage of fruit.

FSA 综合安全评估 系指开发用以协助风险评估和简化主动防范风险管理的综合配置和系统的方法学。FSA is a formal safety assessment. A formal, structured, and systematic methodology, developed to assist risk assessment and to facilitate proactive risk management.

FSO 浮式储油和卸油船 系指类似于FPSO，但不能提供生产能力的浮动储油和卸油船。它确实能够储存油和气体产品并将其卸载到水下管线或穿梭油船上。FSO is a floating, storage, and offloading vessel. Similar to FPSO, but no production capability is provided. It stores oil and gas products that can be offloaded to pipelines or shuttle tanker.

FSS Code FSS 规则 系指 IMO 组织海上安全委员会 MSC.98(73)决议通过并经修正的"国际消防安全系统规则"。FSS Code means the International Code for Fire Safety Systems, adopted by the Maritime Safety Committee of the Organization by resolution MSC.98(73), as amended.

FSU 浮式储油装置 系指浮动的储油装置。它是浮式装置的统称，包括FPSOs和FSOs。FSU is a floating storage unit. It is generic term for floating installations, including FPSOs and FSOs.

FSUR 浮式储存-再液化船

FTP Code FTP 规则 系指 IMO 组织海上安全委员会 MSC.61(67)决议通过并经修正的"国际耐火试验程序规则"。FTP Code means the International Code for Application of Fire Test Procedures, adopted by the Maritime Safety Committee of the Organization by resolution MSC.61(67), as amended.

Fuel 燃料(燃油)

Fuel battery 燃料电池 是直接或间接地使燃料和氧化剂进行化学反应，把燃料的化学能转变为电能的化学电池。它与热机发电不同，其电化学转换过程不受卡诺循环限制，其热效率可达90%，大大高于现有的一切热机。燃料电池的单位容积和单位质量的能量输出大(可达70 kW·h/m³)。它与通常的电池不同，只要连续供应燃料就能连续产生电能。

Fuel cell 燃料电池(燃料舱，燃料箱)

Fuel cell battery 燃料电池组

Fuel cell electric propulsion apparatus 燃料电池电力推进装置

Fuel cell propulsion system 燃料电池推进系统

Fuel changeover operation 燃油转换操作

Fuel compartment 燃料舱

Fuel component 燃料的成分

Fuel consumption 燃料消耗量

Fuel consumption trial 燃料消耗量试验

Fuel control unit 燃油控制器

Fuel economy trial 经济工况试验

Fuel gas 燃料气体(气体燃料)

Fuel gravity tank 重力燃油柜

Fuel injection 燃油喷射

Fuel injection nozzle 喷油嘴

Fuel injection pipe 高压油管(喷油管)

Fuel injection pump 燃油喷射泵

Fuel injection valve cooling pump 喷油嘴冷却泵

Fuel injector 喷油器(油头)

Fuel knock 燃料爆震

Fuel level indicator 油位指示器

Fuel line 燃油管路

Fuel loading 燃油装载

Fuel oil 燃油 (1)对"73/78防污公约"附则Ⅰ第20条而言，系指重蒸馏物或原油的残油或性质相当于IMO所接受的规格，拟用于产生热量或电力的燃料的此类物质的混合物；(2)对"73/78防污公约"附则Ⅵ而言，系指为了船舶推进或运转而交付船上的用于燃烧的任何燃料，包括蒸馏燃油和残余燃油。(1)For the purposes of the Regulation 20 of the "MARPOL 73/78" Annex Ⅰ, fuel oil means heavy distillates or residues from crude oil or blends of such materials equivalent to the specification acceptable to the IMO; intended for use as a fuel for the production of heat or electric power; (2)For "MARPOL 73/78" Annex Ⅵ, fuel oil means any fuel delivered to and intended for combustion purposes for propulsion or operation on board a ship, including distillate and residual fuels.

Fuel oil analysis 燃油分析

Fuel oil atomization 燃油雾化

Fuel oil combustion unit(fuel oil combustion installation) 燃油燃烧装置 系指除船上焚烧炉以外的任何发动机、锅炉、燃气轮机或其他燃油燃烧的设备。

Fuel oil daily tank 燃油日用柜

Fuel oil deep tank 燃油深舱

Fuel oil double wall piping 双层壁燃油管系

Fuel oil drainage tank 燃油泄油柜

Fuel oil filter 燃油滤器

Fuel oil gravity tank 重力燃油柜

Fuel oil leakage tank 燃油泄油柜

Fuel oil management procedure 燃油管理程序

Fuel oil no.1(kerosene) 1号燃料油(煤油)

Fuel oil no.1-D 1-D号燃料油

Fuel oil no.2 2号燃料油

Fuel oil no.2-D 2-D号燃料油

Fuel oil no.4 4号燃料油

Fuel oil no.5 5号燃料油

Fuel oil no.6 6号燃料油

Fuel oil pipe 燃油管

Fuel oil precipitation 燃油沉淀

Fuel oil pulverizer valve 燃油喷雾阀

Fuel oil pump 燃油泵

Fuel oil separator 燃油分油器

Fuel oil service tank 日用燃油柜

Fuel oil settling tank 澄油箱(燃油沉淀柜)

Fuel oil sludge tank 燃油油渣柜

Fuel oil storage tank 燃油贮存柜

Fuel oil sump tank 污油舱

Fuel oil system 燃油系统

Fuel oil tank protection 燃油舱保护

Fuel oil tanks(fuel tank) 燃油舱 系指所有用于装载液体燃油(不包括货油)的处所及通往这些处所的围壁通道。 Fuel oil tanks mean all spaces used for fuel oil(excluding cargo tanks) and trunks to such spaces.

Fuel oil technology for environmental protection 环保燃料技术 (1)液化天然气:LNG燃料能降低20～25%的二氧化碳(CO_2)排放、80～90%的NO_x排放,基本消除SO_x和颗粒物排放,因此,LNG动力推进系统将成为今后绿色航运的重要选择之一。(2)生物柴油燃料:从供应能力、价格及品质等方面考虑,生物柴油燃料作为船用主要燃料使用还存在一定困难,但使用生物柴油燃料的主机与使用传统重油的柴油机在推进船舶航行的原理上完全相同,所以生物柴油比较适用于现有船舶。此外,生物柴油燃料可自由调整与重油的混合比例,从而使船舶温室气体排放在一定程度上可控。因此,这是一种值得期待的替代燃料。(3)燃料电池技术:燃料电池最开始主要用于常规动力潜艇领域,其可使用多种燃料,例如纯氢气、天然气、甲烷、液态碳氢燃料及煤气等,具有良好的环保性能。近年来,随着该项技术的不断发展以及船舶节能减排要求的日趋严格,燃料电池开始进入商船领域,且越来越受到重视。

Fuel oil transmission operation 燃油传输作业

Fuel oil valve 燃油阀

Fuel oil-sump tank 污油舱 供贮存污油的舱。

Fuel pipe 燃油管

Fuel piping 燃油管系

Fuel pressure pump 燃油增压泵

Fuel pump 燃油泵

Fuel quality setting(FQS) 燃油质量设定

Fuel resin 燃料树脂

Fuel settling tank 燃油沉淀柜

Fuel space 燃料舱(燃油舱)

Fuel supply 供油

Fuel supply pump 燃料供应泵

Fuel supply unit 供油单元 是现代燃烧重油柴油机必需配备的辅助装置,一般有供油泵及增压泵组成一套模块化的装置。

Fuel tank protection 燃油舱保护

Fuel tank removal hatch 燃油柜取出舱口

Fuel transfer pump 燃油驳运泵

Fuel valve 燃油阀

Fuel/water supply vessel 油水供给船 对不能进港的大型船舶或对航行途中的船舶补充燃料和淡水的船舶。船上设有储存油水的专用液舱和供给泵系统。

Fujitsu Kabushiki-gaisha 富士通株式会社 简称富士通,是一家日本公司,专门制作半导体、电脑(超级电脑、个人电脑、服务器)、通信装置及服务,总部位于东京。1935年"富士通信机制造"成立,1954年,富士通研制出日本第一台电脑FACOM 100,1967年,公司的名字正式改为缩写Fujitsū(富士通)。

Full 满的(充分的,肥满线型)

Full admission turbine 全(周)进气透平

Full ahead 全速正车

Full ahead test 全速正车试验

Full and down 完全满载 即满舱满载。

Full annealing 完全退火 在大于A_3点(亚共析钢)或A_1点(过共析钢)的温度下加热,在该温度中保持充分的时间后,极其缓慢地将其冷却软化的退火。 Full annealing means an annealing treatment in which the steel is austenitized by heating to a temperature above the upper critical temperature(A_3 or A_{cm}) and then cooling slowly to room temperature.

Full astern　全速倒车

Full astern stopping test　全速倒车停船惯性试验　是测量船从执行全速倒车指令到船停止在水中的轨迹的试验。　Full astern stopping test determines the track of a ship from the time an order for full astern is given until the ship stops in water.

Full astern test　全速倒车试验

Full cargo condition　满载状态

Full container vessel　全集装箱船（巴拿马运河）　用来载运装有货物的标准尺寸大型集装箱的商船。　Full container vessel means a merchant ship to carry large containers of standard size packed with cargo.

Full developed sea　充分发展的风浪　系指风浪发展到一定尺度时，停止发展的风浪。

Full high-definition（FHD）　全高清　是针对平板电视的屏幕分辨率而言的，必须同时满足2个条件：（1）屏幕的物理分辨率为符合1 920×1 080 p；（2）电视内部电路可以处理并输出1 080 p格式的画面信号。显示屏整体物理分辨率要达到1 920×1 080 p，也就是水平方向的分辨率要达到1 920个像素，垂直分辨率要达到1 080条扫描线。全高清系指屏幕固有物理特性，不因信号画面改变而改变，无论播放的是有线电视、DVD影碟还是PC上的高清视频，全高清电视始终保持着1 920×1 080 p的物理分辨率。符合全高清标准的液晶电视，面板附近会以标签或作为电视机设计的一部分标示"FULL HD"。

Full load design draught　满载设计吃水 T_{full}　系指满载设计吃水 T_{full}（m），装载手册中的均匀满载工况下从船舯型基线量起的出港时的吃水。　Full load draught T_{full} in m, is the draught at departure given for the homogeneous full load condition in the loading manual, measured from the moulded base line at amidships. This draught is also known as the full load design draught.

Full load displacement（loaded displacement）　满载排水量　船舶空载排水量或空船质量加上全额可变质量时的排水量。对于民用船舶系指空船质量加上载重量时的排水量；对于军用船舰则是标准排水量加上全部燃料、滑油和备用锅炉水时的排水量。

Full load displacement（SWATH）　满载排水量（小水线面双体船）　系指船舶满载出港状态静浮时的营运排水量，通常等于最大的营运质量。　Full load displacement means the weight of sea water displaced by a craft with no lift or propulsion machinery active under a full loaded departure condition, usually equal to the maximum operational weight.

Full load displacement（WIG）　满载排水量（地效翼船）　系指船上所有按规定配备的船员、乘客、设备、货物、备品、附件及索具等都装备齐全，并装满燃油、滑油、淡水、食品和供应品，处于立即可以起航状态时所排开水的质量。

Full load displacement△　满载排水量△　系指与满载状态对应的型排水量（包括外板等附加物），以吨（t）计。　Full load displacement △ is the displacement (including shell plating and appendages, etc.) in tons at the summer load line.

Full load draught T_{full}　满载吃水　装载手册中的均匀满载工况下，从船舯型基线量起的、出港时的吃水。该吃水也通常被称为满载设计吃水。　Full load draught in metres, is the draught at departure given for the homogeneous full load condition in the loading manual, measured from the moulded base line at amidships. This draught is also known as the full load design draught.

Full power　全功率

Full power trial　全功率试航（全功率试车）

Full pressure　全压［满压（锅炉）］

Full redundant　全冗余　是用来描述一个由两个（相同或不相同的）执行同一功能并同时工作的独立系统组成的自动化系统。　Full redundant is used to describe an automation system comprising two (identical or nonidentical) independent systems which perform the same function and operate simultaneously.

Full speed　全速

Full speed operational condition　全速运转状态（全负荷运转状态）

Full speed trial　全速试航

Full superstructure　全上层建筑　系指最低限度自首垂线延伸到艉垂线的上层建筑。　Full superstructure is a superstructure which, as a minimum, extends from the forward to the after perpendicular.

Full-formed ships　肥大船型

Full-load displacement△（游艇）　满载排水量　对游艇而言，系指艇上所有按规定配备的设备、货物、备品、附件及索具都装备齐全，并装满燃油、滑油、淡水、食品和供应品，额定乘员全部上艇，游艇处于满载使用状态时所排开水的质量。

Full-load draft d（m）　满载吃水　对游艇而言，系指满载排水量静浮于水面时，在艇长 L 中点处由平板龙骨上缘（对纤维增强塑料船为龙骨下表面）量到满载水线的垂直距离。

Full-tern DOC　长期DOC证书　由审核机构总部签发的，有效期为5年的长期证书。

Full-text index　全文索引　是目前搜索引擎的关键技术。试想在1M大小的文件中搜索一个词，可能需要几秒,在 100M 的文件中可能需要几十秒,如果在更大的文件中搜索那么就需要更大的系统开销,这样的开销是不现实的。

Fully decked boat 全甲板艇 其舷弧线区域的水平投影由下列各项的任何组合所构成的艇:水密甲板和上层建筑,和/或符合 ISO 11812 要求的快速泄水凹体,和/或符合 ISO 11812 要求的合计容积小于 $L_H B_H F_M/40$ 的水密凹体。所有关闭装置按 ISO 12216 要求为水密。

Fully decked boat means the boat in which the horizontal projection of the sheerline area comprises any combination of watertight deck and superstructure, and/or quick-draining recesses complying with ISO 11812, and/or watertight recesses complying with ISO 11812 with a combined volume of less than $L_H B_H F_M/40$, all closing appliances being watertight in accordance with ISO 12216.

Fully developed cavitation (super-cavitation) 全空泡 物体表面,例如螺旋桨叶背,为充分发展的空泡所笼罩并延伸到物体以后的情况。

Fully developed sea 充分发展风浪 在一定风速下,风区和风时都足够长,风浪要素达到稳定状态时的风浪。

Fully magnetic controller 全磁控制器 系指其全部基本功能是依靠由电磁操作的器件来完成的一种电控制器。

Fully planning mode or non-displacement mode 全滑行模式或非排水模式 非排水模式系指船舶的重力完全或绝大部分由非静水力来支承的一种模式。当 Γ 大于 3 时,这种模式与泰勒公式一起普遍地适用于小艇。但是,一些 Γ 大于 3 且并非针对滑行而设计的小艇,应对其在排水模式的运行予以考虑,除非将它们归类为高速艇。 Non-displacement mode means the normal operational regime of a ship when non-hydrostatic forces substantially or predominantly support the weight of the ship. Typically this applies to craft with a Taylor Quotient, Γ, greater than 3. However, some craft are designed not to plane with Γ greater than 3 and these should be considered as operating in the displacement mode unless they are classified as a high speed craft.

Fully pressurized gas tanker 全压式液化气船 在常温下,将气体加压至液化压力,把液化气储存在高压容器中进行运输。货舱设计压力一般达到 1.75 MPa。这种运输方式,船体和操作技术都比较简单,但容器质量大,船舶的容积利用率低。此类船舶主要从事液化石油气(LPG)和氨等的短途运输。由于受压力容器材料和制造条件的限制,单罐最大容积为 1 500 m³ 左右,船的装货总容积最大约为 6 000 m³,均为小型船舶。

Fully refrigerated gas tanker 全冷式液化气船 在很低的温度下,将气体冷却至液态,如液化天然气船。设计最低温度可以达到 -161 ℃,货舱内的压力为 0.07 MPa。在如此低温度以下进行运输,货舱通常用垫块支撑,并与船体活动连接,允许膨胀和收缩。船上设有绝热材料和再液化装置。此类船舶适合于大量运输液化气体,但是设计和制造技术要求很高。

Fully revolving crane vessel 全回转起重船 系指装备全回转起重机,并能 360°回转的起重船。

Fully submerged foil 全浸深水翼 系指在翼航状态下无割划水面产生升举力构件的水翼。 Fully submerged foil means a foil having no lift components piercing the surface of the water in the foil-borne mode.

Fully submerged hydrofoil 全浸式水翼 通常指翼航时全部浸入水中的深浸式水翼。

Fully-cavitating propeller (super-cavitating propeller) 全空泡螺旋桨 叶背为充分发展的空泡所笼罩的螺旋桨。

Fully-mechanized welding processes 全自动焊接工艺 如:(1)埋弧焊;(2)熔弧和药芯焊丝金属极的电弧焊;(3)多极埋弧焊;(4)单面焊;(5)填角焊和双面填角焊;(6)带或不带易熔焊丝导向喷嘴电极的电渣焊;(7)气体保护金属极电弧焊;(8)电气焊等。 Fully-mechanized welding processes mean: (1) submerged arc welding; (2) fus-arc and flux-cored wire metal-arc welding; (3) multiple-electrode submerged arc welding; (4) single-side welding; (5) fillet and double-fillet welding; (6) electro-slag welding with and without fusible wire guide nozzle electrode; (7) gas-shielded metal-arc welding; (8) electro-gas welding, etc.

Fully-ventilated propeller 全充气螺旋桨 在叶背输入空气,其气泡延伸到叶边以后的螺旋桨。

Fumigation 熏蒸 主要是应部分进口国海关要求,对木制品以及有木质包装的货物进行的下药熏蒸杀虫的行为,合格之后由商检局盖章确认并出具熏蒸证。

Fumigation certificate 熏蒸证书 是用于证明粮谷、油籽、豆类、皮草等商品,以及包装用木材与植物性填充物等,已经过灭虫的证书。

Function[1] 功能 系统预期目的/任务的一种表现。

Function[2] 职能 系指"海员培训、发证和值班标准国际公约"指明的船舶操作、海上人命安全或保护海洋环境所需的一组任务、职责和责任。 Function means a group of tasks, duties and responsibilities, as specified in the International Convention on Standards of Training, Certi-

Function for alarm indication and acknowledgement 报警显示和报警应答

除满足有关要求外,报警和信息的显示还应满足下列要求:(1)报警和信息显示应能根据其功能特点,采用字母数字形式或者图形或图表的形式进行显示;(2)如除其他信息和图形外单独显示报警,则应提供报警列表显示以便操作人员调用查看;(3)如报警可以分区显示,则可以提供可调用的分区报警显示;(4)对字母数字的显示、列表和图形显示应能迅速调用。其最大调用时间应满足下列要求:①优先级1和2内的所有报警和报警列表显示:<2 s;②较低级别的列表和图形显示:<30 s;(5)每一报警应以视觉信号进行显示。该视觉报警还应能被消除,同时不会阻碍新的报警的产生。故障未消除之前,视觉报警应维持其视觉信号显示,同时,应答前后的报警信号应能明显区别,如采用闪光变为平光等。当故障消除后,报警通道应自动恢复到正常工作状态;(6)如对计算机的操作跨越几个工作站,则应显示哪一个工作站处于工作状态;(7)如采用符号来表示报警和信息内容,则应有对这些符号的解释性列表清单,并在整个计算机系统中均应统一;(8)如报警采用彩色显示器显示,按照上述(5)中要求的报警显示状态应明显区别于无关的颜色;(9)应对屏幕(CRT)采取一定的措施(如进行反光处理或加装滤光器等),以确保在光照条件下清楚地显示报警;(10)在驾驶室应能在夜晚光线条件下对显示器的亮度进行调节。如使用彩色显示器,不允许颜色发生失真;(11)在所有的光照条件下,字母数字显示和图形显示应在距离显示1.5 m处都清晰可辨;(12)如报警是以符号的形式在荧光屏或条形显示器[由发光二极管(LED)组成]上显示的,则应另备一套显示设备。如彩色荧光屏、单色荧光屏、条形显示器、信号灯以及LED等。 Function for alarm indication and acknowledgement means that in addition to the relative requirements, indication of alarms and messages is to comply with the following requirements: (1) alarms and messages are to be indicated alphanumerically or in the form of graphics or diagrams, according to their functional characteristics; (2) where alarm is to be separately indicated in addition to other messages and graphics, display of an alarm list is to be available for the operator; (3) where alarms may be indicated in sections, displays of section alarms may be provided for call; (4) alphanumeric, tabular and graphic displays are to be capable of being promptly called. The maximum call time is to comply with the following: ①displays of all alarms and an alarm list at priority levels 1 and 2: <2 s; ②tabular and graphic displays at low level: <30 s; (5) each alarm is to be indicated by a visual signal, which is to be capable of being extinguished without interfering with new alarming. The visual alarm signal is to be maintained until the fault is removed. And the alarm signals before and after the acknowledgement of an alarm are to be clearly distinguishable from each other, e.g. from flashing to a steady light. After the fault has been removed, the alarm channel is to automatically restore its normal working condition; (6) where computer operations are covered by several workstations, it is to be indicated which station is in working condition; (7) where alarms and messages are indicated by symbols, a unified list of explanations of these symbols is to be available throughout the computer system; (8) if alarm messages are displayed on color monitors, the distinctions in the alarm status required by (e) are to be clearly distinguishable from irrelevant colors; (9) certain measures (e.g. reflectorization, or fitting of an optical filter) are to be taken to the screen (SRT) to ensure clear indication of alarms in lighting condition; (10) it is to be possible to adjust the brightness of monitors on the navigation bridge at night. Falsification of colors is not allowed for color monitors in use; (11) alphanumeric and graphic displays are to be clearly recognizable at a distance of 1.5 m in all lighting conditions; (12) where alarms are indicated as symbols on a screen or bar display [composed of a light-emitting diode (LED)], an additional display unit is to be provided, e.g. a color screen, a monochrome screen, a bar display, a signal lamp or LED.

Function for alarming 报警功能

在报警触发时应在2 s时间内显示出报警通道,报警功能应优先于计算机其他操作动作,包括故障搜寻程序。 Function for alarming means that the alarm channel is to be indicated in 2 seconds from initiation of an alarm, the alarm function is to have priority over other operational actions of the computer, including fault detection program.

Function for locking of alarms 报警闭锁功能

在设备正常停机和启动中,出现偏离报警整定值时,应能切断报警线路;如采用手动报警闭锁时,则应用视觉信号清楚地显示。 Function for locking of alarms means that alarm link is to be cut off in case of deviation from alarm setting at normal stopping and starting of the equipment; manual locking of alarms is to be indicated by a clear visual signal.

Function for monitoring and modifying set values of alarming 监视和报警整定值修改功能

系指监视和报警整定值的修改,仅限于指定人员借助钥匙等工具,或使用专用代码等方法才能进行。在建造后检验时，应特别注意重要功能报警整定值的修改是否合理。 Function for monitoring and modifying set values of alarming means that this is to be limited to designated persons with keys or special codes, etc. During surveys after construction, particular attention is to be given to reasonableness of any modification of alarming setting for essential functions.

Function for self-checking and monitoring 自检和监视功能 即能对系统自身的故障自动进行检测和报警或显示，当系统内部出现故障时（如电源故障、传感器故障等）应能发出报警并显示，且应与非内部故障的报警和显示有明显的区别。如只设置手动的自检操作设施，应经船级社确认是否满足上述功能要求。此外，还能自动、周期性地监视程序的执行和信息的传递，如必要，还能对数据传递阻断进行报警。 Function for self-checking and monitoring means that faults of the system are automatically tested and alarmed or indicated by itself. The system is able to give alarm and indicate any internal fault (e.g. fault of power supply, sensors) of the system in a way clearly distinguishable from the alarm and indication for faults other than internal ones. Where only manual operating means of self-checking are provided, compliance with the above functional requirements is to be confirmed by Society. In addition, program execution and data transfer are to be monitored automatically and periodically; where necessary, data transfer interruption is to be alarmed.

Function test 功能试验 系指为证实设备符合其设备标准（技术条件）要求的运行功能的试验。

Functional block diagram 功能方块图

Functional hazard assessment（FHA） 功能危险评估 系指在开发阶段初期进行的评估。它应清楚地确认与船舶功能有关的故障情况并对其进行分级。这些故障情况分级确立了安全目标。 Functional hazard assessment（FHA）is the assessment conducted at the beginning of the development cycle. It should clearly identify and classify failure conditions associated with the craft's functions. These failure condition classifications establish the safety objectives.

Functional reliability test 功能可靠性试验

Functional requirements 功能性要求 通常系指舰船和舰船上系统/设备必须提供什么功能，以满足规范的安全目标。 Functional requirements explain, in general terms, what function the ship and shipboard systems/equipment must provide in order to meet the safety objectives of these Rules.

Functional requirements for a safety management system（SMS） 安全管理体系（SMS）的基本要求 包括下列主要要求：(1)安全和环境保护方针；(2)保障船舶安全营运及环境保护、符合有关国际公约及船旗国政府法规的须知和程序；(3)明确岸上人员和船上人员的权限和相互间的通信联络方式；(4)按本规则规定报告事故及不合格的程序；(5)应急情况的防备及处理程序；(6)内审及管理评审程序。 Functional requirements for a safety management system（SMS）include the following functional requirements：(1) a safety and environmental protection policy；(2) instructions and procedures to ensure safe operation of ships and protection of the environment in compliance with relevant international and flag State legislation；(3) defined levels of authority and lines of communication between, and amongst, shore and shipboard personnel；(4) procedures for reporting accidents and non-conformities with the provisions of this Code；(5) procedures to prepare for and respond to emergency situations；(6) procedures for internal audits and management reviews.

Functional requirements to safety and emergency preparedness 安全与应急准备的功能要求 针对安全及应急准备措施有效性的可能验证要求，应确保在设计和作业中满足安全目标、风险接受准则、管理机构的最低要求以及已制定的标准。

Functional sequence diagram 功能顺序图

Functional test 功能试验

Functionality 功能度 系指完成预计功能的能力。功能的实施通常涉及显示系统和仪表。 Functionality means the ability to perform an intended function. The performance of a function normally involves display system and instrumentation.

Fundamental frequency 基频 系指基音的频率，它决定整个音的音高。电机中的基频系指电机在额定扭矩时的频率，又叫基带（Base band）。基频是复杂声音中最低且通常情况下最强的频率，大多数通常被认为是声音的基础音调（fundamental）：复合振动或波形（如声波）的谐波成分，它具有最低频率，且通常具有最大振幅——亦称"基谐波"在大部分手机中，基频都是最昂贵的半导体元件，随着TFT-LCD显示屏价格的下降，基频正在成为手机中占成本比例最高的元件。

Funnel damper 烟囱的调节风门

Funnel draught 自然通风

Funnel flue 烟道

Furnace 炉（炉膛,反应堆,熔炼）

Furnace oil 锅炉重油

Furnace pipe 烟管

Furnace wall 炉墙

Fuse 熔断器 是船舶电力系统中常用的过电流保护电器之一,也是一种机械开关电器,当电流超过规定值足够长的时间,通过熔断一个或几个特殊设计的和成比例的熔体分断此电流,由此断开其所接入电路的电器。熔断器由形成完整电器的所有部件组成。

Fuse-element 熔体 系指在熔断器动作时熔化的熔断器部件,熔体可以由几个并联的熔体组成。

Fuse-link 熔断体 系指带有熔体的熔断器部件,在熔断器熔断后可以更换。

Fusion line(FL) 熔敷线

G

G6 alliance G6 联盟 目前 G6 联盟的成员为赫伯罗特、日本邮船、东方海外、美国总统轮船、现代商船、商船三井等6家班轮公司。该联盟成立于2012年3月起在亚欧航线上展开合作,并于2013年5月进一步在远东——美东航线上深入合作。目前G6联盟的运力规模为318万TEU,占世界集装箱船队总运力的18%。

Gage 规(表,计,轨距)

Gage cock 试水位旋塞

Gale ballast tank 风暴压载舱 系指因考虑安全的原因允许装载压载水的货油舱。Gale ballast tank is a cargo tank that is allowed to load ballast water for reasons of safety.

Galileo 欧洲卫星导航系统 欧洲卫星导航系统设计成完全的民用系统,在公共操纵下操作。欧洲卫星导航系统(Galileo)包含在3个圆形轨道中的30个中地球轨道(MEO)卫星。每个轨道倾斜56°,并包括9个运行卫星加上1个备用卫星。这种几何形状确保至少6个卫星面向全球用户,其位置精度衰减因子(PDOP)≤3.5。Galileo 发射10个导航信号和1个搜救(SAR)信号。SAR信号在为应急服务保留的波段(1 544～1 545MHz)上广播,而10个导航信号在分配给无线电导航卫星服务(RNSS)的波段上提供:(1)4个信号占据频率范围1 164～1 215 MHz(E5a～E5b);(2)3 个信号占据频率范围1 260～1 300 MHz(E6);(3)3 个信号占据频率范围1 559～1 591 MHz(E2,L1,E1)。每个频率携带2个信号:第一个是跟踪信号即所谓的导频信号,无数据但增加接收机处的跟踪强度,而另一频率携带航行数据信息。Galileo 为海事界提供2种不同的使用服务:①Galileo 开放服务提供定位、导航和授时服务,对直接使用者免费。开放服务可在1个(L1),2个(L1 和 E5a 或 L1 和 E5b)或3个(L1,E5a 和 E5b)频率上使用;②Galileo生命安全服务可在1个(L1 或 E5b)或2个(L1 和 E5b)频率上使用。L1 和 E5b 频率均各自携带包括完整性信息的航行数据信息。E5a 频率不包括完整性数据。Galileo is the European satellite navigation system. Galileo is designed as a wholly civil system, operated under public control. Galileo comprises 30 medium earth orbit (MEO) satellites in 3 circular orbits. Each orbit has an inclination of 56° and contains 9 operational satellites plus one operational spare. This geometry ensures that a minimum of 6 satellites are in view to users world-wide with a position dilution of precision(PDOP) ≤ 3.5. Galileo transmits 10 navigation signals and 1 search and rescue(SAR) signal. The SAR signal is broadcast in one of the frequency bands reserved for the emergency services(1 544～1 545 MHz) whereas the 10 navigation signals are provided in the radio-navigation satellite service(RNSS) allocated bands:(1)4 signals occupy the frequency range 1 164～1 215 MHz(E5a～E5b);(2)3 signals occupy the frequency range 1 260～1 300 MHz(E6);(3)3 signals occupy the frequency range 1 559～1 591 MHz(E2, L1, E1). Each frequency carries two signals: the first is a tracking signal-the so-called pilot signal-that contains no data but increases the tracking robustness at the receiver whereas the other carries a navigation data message. Galileo provides two different services of use for the maritime community:①the Galileo open service provides positioning, navigation and timing services, free of charges for direct users. The open service can be used on one(L1), two(L1 and E5a or L1 and E5b) or three(L1, E5a and E5b) frequencies;②The Galileo safety of life service can be used on one(L1 or E5b) or two(L1 and E5b) frequencies. Each of the L1 and E5b frequencies carries a navigation data massage that includes integrity information. The E5a frequency does not include integrity data.

Galileo positioning system(GPS) 伽利略定位系统 是欧盟一个正在建造中的卫星定位系统。有"欧洲版GPS"之称,也是继美国现有的"全球定位系统(GPS)"、俄罗斯的"GLONASS"、中国北斗卫星导航系统之后第四个卫星导航系统。

Gallery 格栅平台(通道,走廊)

Galley baking oven 厨用烘箱 装有加热设备,用以烘烤食品的炉箱。

Galley equipment 厨房设备 船上炊事工作所需的各种器械和用具的统称。

Galley machine 厨房机械 为炊事工作服务的各种专用机械。如绞肉机、揉面机、洗碗机、厨房升降机等。

Galley range 船用炉灶 能适应船上环境条件的专用炉灶。包括滴油灶、汽化油灶、电灶、燃气灶、燃煤灶等。

Galleys 厨房 系指设有加热表面外露烹饪设备的或每台功率大于 5 kW 的烹饪设备或食品加热器具的围蔽处所。 Galleys are those enclosed spaces containing cooking facilities with exposed heating surfaces, or which have any cooking or food heating appliances, each having a power of more than 5 kW.

Gallow(trawl gallow) 网板架 安装在单拖网渔船艉部或舷侧前后,供吊挂网板用的铁架。

Galvanized steel 镀锌钢

Galvanizing 镀锌(电镀)

Game console 电视游乐器 具有互动性的视听娱乐设备。

Gangboard 跳板 船舶靠泊时,向舷外搭搁,供船舶与岸或其他船间交通用的板件。

Gangway 舷梯 悬挂于舷外水面上供人员登、离船用的活动斜梯。常设于两舷侧处不被外板封闭的最低一层甲板上,不用时可收存于舷旁,使其不突出舷外。

Gangway 步桥 系指上层建筑之间,例如艏楼和桥楼之间或桥楼和艉楼之间的升高走道。 Gangway means a raised walkway between superstructures, such as, between the forecastle and bridge, or between the bridge and poop.

Gangway bridge(wharf ladder) 舷桥 船舶靠泊时,临时向舷外搭搁,两侧有支索桁架和栏杆等,供船舶与岸或其他船间交通用的活动板桥。

Gangway platform 舷梯平台 设置于舷梯端部或中部的平台。

Gantry 龙门型吊架 由一根水平主梁支承在两根支柱上所组成形似门字的刚性构件。作为重型绞车钢索导向滑车的支承或拉曳网具的支点。

Gantry crane 高架门座起重机 是可移动的起重装置(简称转动部分)装在高架门形座架上的一种臂架型起重机。门形座架的 4 条腿构成 4 个"门洞",可供铁路车辆和其他车辆通过。见图 G-1。

Gap 间隙(余隙,双体船的船体间的距离)

Gap of propeller blades 叶间隙 螺旋桨一定半径处相邻两叶面节线的垂直距离。$G_Z = (2\pi\sin\Phi)/Z$,式中,r—叶切面所在的半径;Φ—螺距角;Z—叶数。

Gap ratio of blade elements 叶元体间隙比 螺旋桨一定半径处两叶元体的周向间距对其弦长的比值。

图 G-1 高架门座起重机
Figure G-1 Gantry crane

叶元体间隙比 $C = 2(F/Z)$,式中,F—叶元体所在的半径;Z—叶数;C—叶元体间隙比弦长。

Garbage 垃圾(73/78 防污公约) (1)对"73/78 防污公约"附则 V 而言,垃圾系指产生于船舶正常营运期间并需要持续或定期处理的各种食品废弃物、生活废弃物和作业废弃物、所有塑料制品、货物残余、焚烧炉灰渣、食用油、渔具和动物尸体,但 MARPOL 其他附则中所规定或列出的物质除外。垃圾不包括在航行期间进行捕鱼活动或水产养殖活动获得的鲜鱼及其各部分,该水产养殖活动涉及将鱼包括贝类运至养殖设施内放置,以及从该类设施内将收获的鱼包括贝类运至岸上供加工;(2)就垃圾记录簿(或船舶的正式航海日记)而言,垃圾分类如下:①塑料;②食品废弃物;③生活废弃物;④食用油;⑤焚烧炉灰渣;⑥作业废弃物;⑦货物残余;⑧动物尸体;⑨渔具。 (1) For the "MARPOL73/78" Annex V, garbage means all kinds of food wastes, domestic wasters and operational wasters, all plastics, cargo residues, incinerator ashes, cooking oil, fishing gear, and animal carcasses generated during the normal operation of the ship and liable to be disposed of continuously or periodically except those substances which are define or listed in other Annexes to MARPOL. Garbage does not include fresh fish and parts thereof generated as a result of fishing activies undertaken during the voyage, or as a result of aquaculture activies which involve the transport of fish including shellfish for placement in the aquaculture facility and the transport of harvested fish including shellfish from such facilities to shore for processing;(2)for the garbage record book(or ship's official log-book), garbage is to be grouped into categories as following:①plastic material;②food wastes;③domestic wasters;④ cooking oil;⑤ incinerator ashes;⑥ operational wastes;⑦cargo residues;⑧animal carcass(es);⑨fishing gear.

Garbage boat / sewage boat 垃圾船 系指为港内船舶卸运垃圾的船舶。

Garbage chute 垃圾滑槽

Garbage disposal system 垃圾处理系统

Garbage disposal unit(GDU) 垃圾处理装置

Garbage management plan 垃圾管理计划 "73/78 防污公约"附则 V 规定,凡 400 总吨及以上的船舶和核准载运 15 名或以上人员的船舶,均应备有一份船员必须遵守的垃圾管理计划。该计划应就收集、储藏、加工忽然处理垃圾及船上设备使用等提供书面程序,还应指定负责执行该计划的人员。该计划应按 MEPC 制定的"垃圾管理计划编制指南"编制,并用船员的工作语言书写。 As "MARPOL 73/78" Annex V prescribes, every ship of 400 gross tonnage and above and every ship which is certified to carry 15 persons or more shall carry a Garbage Management Plan which the crew shall follow. This plan shall provided written procedures for collecting, storing, processing and disposing of garbage, including the use of the equipment on board. It shall also designate the person in charge of carrying out the plan. Such a plan shall be in accordance with the guidelines for the development of garbage management plans developed by MEPC, and written in the working language of the crew.

Garbage pollution 垃圾污染

Garbage record book 垃圾记录簿 这是《73/78 防污公约》附则 V 的强制规定,凡 400 总吨及以上的船舶和核准载运 15 名或以上人员、航行于其他缔约国政府管辖权范围内的港口或近海装卸站的船舶,以及从事海底矿产勘探和开发的固定和移动平台,均应备有一份《垃圾记录簿》。《垃圾记录簿》应记录每次排放作业或完成的焚烧作业。 It is mandatory provision in "MARPOL 73/78" Annex V: every ship of 400 gross tonnage and above and every ship which is certified to carry 15 persons or more engaged in voyages to ports or offshore terminals under the jurisdiction of other Parties to the Convention and every fixed and floating platform engaged in exploration and exploration of the seabed shall be provided with a Garbage Record Book. Each discharge operation, or completed incineration, shall be recorded in the Garbage Record Book.

Garbage removal 清除垃圾 系指在船舶进、出坞前后打扫、清理船上废弃物的作业。

Garbage shoot 垃圾滑槽

Garboard strake 龙骨翼板 邻接龙骨的左、右各一列船底板。

Gas 气体(煤气,毒气) (1)对氨冷冻装置而言,气体系指用作制冷剂的氨气;(2)温度在 37.8 ℃时蒸发压力超过 2.8 bar(1 bar = 10^5 Pa)绝对压力的液体;(3)延缓鲜品代谢过程的合适的气态混合物。 (1)For refrigerating machinery using ammonia as refrigerant, gas means ammonia gas used as the refrigerant;(2)a fluid having a vapour pressure exceeding 2.8 bar absolute at a temperature of 37.8 ℃;(3)a suitable gaseous mixture to retard the metabolic process of fresh products.

Gas accommodation of liquid tank 液货舱气室 对散装运输液化气船而言,系指液货舱向上延伸部分。如货物围护系统位于甲板以下时,液货舱气室应伸出露天甲板或液货舱罩之上。

Gas analysis instrument 烟气分析器

Gas annealing 气体退火

Gas arc welding 气体保护电弧焊

Gas barge 气体运输驳船 系指舱内装载液化气体的驳船。 Gas barge means a barge carried liquefied gas in holds.

Gas bottle rooms 气瓶室 是容纳乙炔和氧气瓶的房间,分配集管、止回阀、减压装置、到分配站的供应管出口也安装在室内。 Gas bottle room is a room containing acetylene and oxygen bottles, where distribution headers, non-return and stop valves, pressure reducing devices and outlets of supply lines to distribution stations are also installed.

Gas brazing 气焰铜焊(气焰硬焊)

Gas carburizing 气体渗碳(充式渗碳)

Gas carrier 气体运输船 系指经建造或改建用于散装运输下述规则之一(视何者适用而定)所列任何液化气体或其他货品的货船;(1)经海安会 MSC.5(48)决议通过的,并可能由 IMO 组织修正的《国际散装运输液化气体船舶构造和设备规则》(以下简称《国际气体运输船规则》)第 19 章;(2)经 IMO 组织 A.328(Ⅸ)决议通过的,并已经或可能由 IMO 组织修正的《散装运输液化气体船舶构造和设备规则》(以下简称《气体运输船规则》)第 XIX 章。 Gas carrier is a cargo ship constructed or adapted and used for the carriage in bulk of any liquefied gas or other products listed in either(which ever is applicable):(1)Chapter 19 of the International Code for the Construction and Equipment of Ships Carrying Liquefied Gases in Bulk adopted by the Maritime Safety Committee by resolution MSC.5(48), hereinafter referred to as "the International Gas Carrier Code", as may be amended by IMO;(2) Chapter XIX of the Code for the Construction and Equipment of Ships Carrying Liquefied Gases in Bulk adopted by the Organization by Resolution A.328(IX), hereinafter referred to as "the Gas Carrier Code", as has been or may be amended by IMO.

Gas cavity 气孔

Gas chamber 燃气室

Gas cleaning 气体净化

Gas collector (combustion header, turbine entry duct) 燃气收集器 (1)从火焰筒出口到高压涡轮进口的过渡导管;(2)自由活塞燃气轮机装置中,用以收集各发生器燃气并减少燃气压力波动的情况。

Gas combustion unit (GCU) 燃气燃烧装置

Gas constant 气体常数

Gas content 气体含量

Gas control room 燃气系统控制室

Gas corrosion 气体腐蚀

Gas cutter (gas cutting torch, gas torch) 气割炬(气焊炬)

Gas cutting 气割

Gas dangerous space 气体危险处所 系指下列处所:(1)载货区域内没有按认可的方式保证其大气在任何时间保持气体安全状态进行布置或配备的处所;(2)载货区域外的围蔽处所。通过该处所的管道内可能包含液货或其蒸发气,或该处所为这类运输管道的终点。设有认可的设备,以防止液货的蒸发气逸入该处所大气中的除外;(3)货物围护系统和货运管系:①货物装于需要次防壁的货物围护系统的容纳舱柜空间;②货物装于不需要次防壁的货物围护系统的容纳舱柜空间;(4)由单层气密钢限界面从(3)所述的容纳舱柜空间分隔出的处所;(5)液货泵舱和液货压缩机舱;(6)距离散货舱出口、气体或蒸发气出口、液货管道法兰、运货阀门或液货泵舱和液货压缩机舱的进出口和通风开口3 m(9.84 ft)之内的露天甲板上的半围蔽处所;(7)载货区域之上的露天甲板和露天甲板上距载货区域前后3 m (9.84 ft),到露天甲板以上2.4 m(7.88 ft);(8)距货物围护系统曝露在大气中的外表面2.4 m(7.88 ft)内的处所;(9)设有载货管系的围蔽处所或半围蔽处所;(10)液货软管舱室;(11)直接有开口通向气体危险处所或区域的围蔽处所或半围蔽处所。

Gas dangerous space 危险气体处所

Gas dangerous zone 气体危险区域 系指可燃气体或爆炸性气体或蒸发气易聚集至危险浓度的区域。气体危险区域可分为下列2类:(1)0类危险区域——系指可燃气体或爆炸性气体或蒸发气与空气混合物连续或长时间存在的区域;如:①回收油贮存舱;②属于回收油贮护系统的管路和容器的内部。(2)1类危险区域——系指可燃气体或爆炸性气体或蒸发气与空气混合物在正常作业中可能出现的区域;如:①与回收油贮存舱相邻接的隔离舱或其他处所;②属于回收油输送系统的管路法兰、阀、软管、泵和其他设备所在的围蔽或半围蔽处所;③距分离器、回收油输送系统的软管和阀、回收油贮存舱开口和泵舱或隔离舱等1类危险处所的开口3 m以内的开敞甲板处所,包括半围蔽处所;④回收油贮存舱上的开敞甲板区域及其前后各加3 m,高度为2.4 m的空间;⑤位于回收油贮存舱外面,回收油管路(中间管段或管路终端)所在的围蔽处所,如设有符合有关规定的通风,则可除外;⑥能直接从1类危险区域进入(无气闸)或有开口通向1类危险区域的围蔽或半围蔽处所,如设有符合有关规定的通风,则可除外。

Gas density 燃气密度

Gas detecting device 有毒气体探测装置 为了检测货舱加热蒸汽管中是否有因管系破损而混入有毒气体的装置。一般有毒气体探测装置安装在该管系的末端。毒性的探测采用与特定的试剂进行化学反应的方法,但由于化学品船所运输的货物种类繁多,毒性也各有差异,故需要储存多种化学试剂。

Gas detection alarm 气体探测报警 系指指示发现气体的报警。 Gas detection alarm means an alarm which indicates that gas has been detected.

Gas detection and alarm system 气体探测与报警系统

Gas detection equipment 气体探测设备

Gas detector 气体检漏器(气体分析仪,气体检定器)

Gas diode 充气二极管 充气管是管内充有气体或蒸发气的电子管,又名离子管。在充气管内,电子在电极间运动时与气体原子和分子碰撞,产生电离现象,运动较慢的正离子抵消电子的负空间电荷作用,使管子具有电流大、内阻低的特点。充气管内的放电形式取决于管子结构、气体种类、气压和外电路参数。人们利用各种放电形式的不同特性制出一系列不同性能的充气管。

Gas driven blower 燃气鼓风机

Gas engine 煤气机 用煤气或天然气作燃料的内燃机。

Gas equation 气体方程

Gas expulsion system 气体驱逐装置 对氨冷冻装置而言,气体驱逐装置系指迅速地把气体从舱室内排出的装置。其包括通风装置,气体吸收装置,水的筛选装置,气体吸收水箱等。 For refrigerating machinery using ammonia as refrigerant, gas expulsion system means the system excludes gas quickly from a compartment, and consists of ventilation system, gas absorption system, water screening system, gas absorption water tanks, etc.

Gas flue 烟道

Gas freeing plant 除气装置

Gas fuel technology 气体燃料技术 使用的是一种优质、高效、清洁的燃料,其着火的温度较低,火焰传播速度快,燃烧非常容易和简单,很容易实现自动输气、混合、燃烧过程。气体燃料主要指液化天然气(LNG)和液化石油气(LPG)。

Gas generator 气体发生器

Gas generator(gas producer, charging set, gasifier) 燃气发生器 产生驱动动力涡轮用的高温燃气的装置。

Gas hazardous space 气体危险处所 系指:(1)在货物区域内,未装置或配备认可设备的处所,因而不能确保该处所内的空气在任何时候均处于安全状态的处所;(2)货物区域以外有含有液体或气体货物的任何管系通过或终止的围蔽处所,但装有认可的装置能防止货物蒸发气逸入该处所内空气中的处所除外;(3)货物围护系统和货物管系;(4)①要求设置次屏壁的货物围护系统的货舱处所;(4)②不要求设置次屏壁的货物围护系统的货舱处所;(5)以单层钢质气密周界与(4)①所述货舱处所相隔离的处所;(6)货泵舱和货物压缩机舱;(7)在开敞甲板上或开敞甲板上的半围蔽处所内,离液货舱出口、气体或蒸发气出口、货物管法兰或货物阀门,或离开货泵舱或货物压缩机舱的入口或通风口3 m范围内的区域;(8)在货物区域内的开敞甲板上和在开敞甲板上货物区域前后3 m内直至露天甲板上2.4 m高度范围内的处所;(9)距该货物完好系统露天表面2.4 m货物内的区域;(10)内部含有货物管路的围蔽或半围蔽处所。但需要蒸发气体作为燃料并符合有关要求设有符合有关要求的气体探测设备的处所,应不认为其是气体危险处所;(11)储存货物软管的舱室;(12)其开口直接通向气体危险处所的围蔽处所或半围蔽处所。

Gas heater 燃气加热器

Gas holder(gas tank) 气罐 对游艇而言,系指艇上用于储存液化石油气(LPG)的专用钢瓶。

Gas holder(gas tank) space 气罐处所 对游艇而言,系指艇上用于存放气罐的处所。

Gas hood 烟罩(排气罩)

Gas inclusion 气体夹杂物 包括孔隙、集结的孔隙和孔隙。集结区内全部孔隙的面积应包括:包含所有孔隙的包络线或直径与焊缝宽度尺寸相匹配的一个圆圈。许可的孔隙集结区应是局部化的。 Gas inclusion encompasses porosity, clusters and pores. The entire pore area within a cluster should be included: an envelope curve that encompasses all the pores or a circle with a diameter that matches that of the width of the weld. The permitted pore area should be localized.

Gas injection diesel engine(GIDE) propulsion system 喷射气体柴油机推进系统

Gas jet propulsion 喷气推进

Gas knock 气体爆震

Gas laser 气体激光器

Gas leakage 漏气

Gas leakage test 漏气试验

Gas lock 气封

Gas mask 防毒面具

Gas metal arc welding 金属极气体保护电弧 系指以电弧作为热源、利用气体保护熔池的焊接方法。气体的作用主要是保护熔化金属不受空气中氧、氮、氢等有害元素和水分的影响,但它同时对电弧的稳定性、熔滴过渡形式和熔池的活动性有一定影响。因此,采用不同的气体会产生不同的冶金反应和工艺效果。气体保护电弧焊的主要特点是电弧可见,熔池较小,易于实现机械化和自动化,生产率高。20世纪70年代迅速发展的焊接机器人主要就是用于电阻点焊和气体保护电弧焊。气体保护电弧焊适用于钢铁、铝和钛等金属的焊接,广泛应用于汽车、船舶、锅炉、管道和压力容器等产品的制造,特别是其中要求质量较高或全位置焊接的场合。气体保护电弧焊,按电极类型可分为钨极惰性气体保护焊和熔化极气体保护焊。

Gas metal arc welding(gas metal arc process) 熔化极气体保护焊(金属极气体保护电弧焊)

Gas mixture 燃气混合物

Gas mixture metal arc welding 金属极混合气体保护电弧焊

Gas nitrided alloy steel 气体渗氮合金钢

Gas nitrided quenched and tempered steel 气体渗氮淬火回火钢

Gas nozzle 气焊喷口

Gas oil 汽油(瓦斯油)

Gas path 气道(烟气通路)

Gas pickling 气体侵蚀

Gas pipe 燃气管

Gas planning 气刨

Gas pocket(gas porosity) 气孔

Gas power 燃气功率 自由活塞燃气发生器出口

的燃气,绝热膨胀至大气压力时所能发出的功率。

Gas pressure 气体压力

Gas pressure regulating valve 气体压力调节阀

Gas pressure welding 气压焊(加压气焊)

Gas producer 煤气发生炉

Gas proof 气密的

Gas protection 毒气防护

Gas purging 气体净化　对氨冷冻装置而言,气体净化系指在从冷凝器中排出不冷凝的气体。 For refrigerating machinery using ammonia as refrigerant, gas purging means the discharge of non-condensing gases from the condenser.

Gas quenching 气冷淬火

Gas re-heater 烟气加热式再热器　在再热循环汽轮机装置中用锅炉烟气在汽轮机内做过部分功的蒸汽的热交换器。

Gas safe space 气体安全处所　系指气体危险处所以外的处所。

Gas scrubbing unit 气体净化装置

Gas separator 燃气分离器

Gas shield 气体防护

Gas storage cylinder 储气罐

Gas system 气体系统　系指控制氧气和/或二氧化碳含量的系统。 Gas system means a system which controls the levels of oxygen and/or carbon dioxide.

Gas temperature 气体温度

Gas tight 气密

Gas tightness 气密性　不透气的性能。

Gas tightness test 气密性试验

Gas torch tip 气焊炬喷嘴

Gas turbine(gas turbine engine) 燃气轮机　空气或气体由压气机压缩后到燃烧室与喷入的燃料混合燃烧或由加热器加热,然后进入涡轮中连续膨胀,使其部分热能转换成功最后由转子输出机械功的一种叶轮式动力机械。

Gas turbine alternator 燃气轮机交流发电机组

Gas turbine characteristic line 燃气轮机工作线　由压气机特性曲线图上表示燃气轮机与推进器负荷协调的工作的曲线。

Gas turbine driven alternator 燃气轮机驱动交流发电机组

Gas turbine module 燃气轮机组装体　为提高可靠性和利用率以节省安装工时和运行维修费用而设计的一种通常包括燃气发生器、动力涡轮、进排气装置、所有附属设备、抗震支座以及具有隔声、隔热、抗冲击、抗污染能力的外壳等组件,小功率机组还包括减速齿轮箱的组装式标准燃气轮机组。

Gas turbine power plant 燃气轮机动力装置　以燃气轮机作为主机的船舶动力装置。

Gas turbine room 燃气轮机舱

Gas turbine ship 燃气轮机船　以燃气轮机作为主机的船舶。

Gas turbo-alternator 燃气轮机交流发电机组

Gas venting mast 通气桅(杆)

Gas welding apparatus(gas welding equipment) 气焊设备

Gas welding flux 气焊焊剂

Gas welding rod 气焊条

Gaseous 气态的

Gaseous cavity 含气空泡　泡中主要是其周围水中放出的所含气体的空泡。此类空泡发展较慢,可在大于或小于水汽压力下产生。

Gaseous cavity 含气空泡

Gaseous fuel 气态燃料

Gaseous hydrocarbon 气态碳氢化合物

Gas-freeing 除气　系指向舱内送入新鲜空气以排除有毒、可燃及惰性气体,并使舱内空气的体积含氧量增到21%。 Gas-freeing means sending fresh air into a tank to remove toxic, flammable or inert gas so as to increase the oxygen content to 21% by volume.

Gas-hazardous zone 气体危险区域　系指可燃体或爆炸性气体或蒸发气易聚集至危险浓度的区域。气体危险区域可分为2类:0类气体危险区域和1类气体危险区域。 Gas-hazardous zone means an area in which flammable or explosive gas or vapor is liable to accumulate to a dangerous concentration. The gas-hazardous zones may be divided into two categories: hazardous zone of categrory 0; hazardous zone of categrory 1.

Gasification 汽化

Gasifier 燃气发生炉

Gasify 使汽化

Gas-jet propulsion 喷气推进　用气体作为推进剂的喷射推进。

Gasket 垫圈(垫片,垫密片,填料)

Gasket joint 垫圈(垫片)

Gasket lip 垫料框

Gasket material 垫衬材料

Gasoline 汽油

Gasoline blending stocks 汽油调和料类

Gasoline compartment 汽油舱

Gasoline engine 汽油机　用汽油作燃料的内燃机。

Gasoline engine ship 汽油机船 以汽油机作为主机的船舶。

Gasoline tank 汽油箱

Gasolines 汽油类

Gas-safe space 气体安全区域 系指因碳氢化合物气体的流入而造成存在着火性或有毒性危险的处所。

Gas-safe space is a space in which the entry of hydrocarbon gases would produce hazards with regard to flammability or toxicity.

Gas-shielded arc welding 气体保护电弧焊

Gas-shielded gravity welding set 气体保护重力焊设备

Gas-shielded type electrode 造气涂料焊条(气体保护型焊条)

Gas-shielded welding 气体保护焊 系指利用气体作为电弧介质并保护电弧和焊接区的电弧焊称作气体保护电弧焊,简称气体保护焊。

Gassing 充气

Gas-tight 气密/气密的 用于防止任何显著数量的易燃气体或蒸发气进入相邻区域的物理屏障的特性。

Gas-tight means the attribute of a physical barrier which prevents any significant quantity of flammable gas or vapour from entering an adjoining area.

Gastight door 气密门 系指一种设计为在正常大气条件下能阻止气体通过的配合紧密的实心门。

Gastight door is a solid, close-fitting door designed to resist the passage of gas under normal atmospheric conditions.

Gas-tight seam 气密缝

Gastightness 气密性 不透气的性能。

Gas-turbine power plant 燃气轮机动力装置 以燃气轮机作为主机的动力装置。

Gate valve 闸阀

Gate valve for dredging pipeline 泥管闸阀 安装在吸、排泥管线上的闸阀。

Gateway 网关 系指本地主机将IP包转发到其他网络时所经过的IP地址。网关可以是本地网络适配器的IP地址或者是同一网段内的路由器的IP地址。

Gathering system 集输系统 系指汇集各油井的井液并输送给油气处理装置的系统。

Gauge 计(表,测量,满载吃水)

Gauge annunciator 表式示号器

Gauge cock (test cock) 验水阀 当水位计玻璃模糊或破损失效时,用以检查水位的阀组。

Gauge hatch 测量液位口(量油口)

Gauge lamp 表灯

Gauge length 计算长度(标准距离)

Gauge mark 基准标志(标距标记)

Gauge pressure 表压力

Gauge terminal 应变片接线板

Gauged thickness 实测厚度

Gauged thickness t_{gauged} 一个项目的测量厚度 即在船舶定期的营运检验过程中,在同一项目上所进行不同测量值的项目平均厚度(mm)。 Gauged thickness is the thickness in mm, on one item, i.e. average thickness on one item using the various measurements taken on this same item during periodical ship's in service surveys.

Gauging 计量(校准)

Gauging devices 测量装置 液货舱应设有下列形式之一的测量装置:(1)开式装置:利用液货舱的开口进行测量,可以将测量仪表放置于货物或其蒸发气之中。如空挡测量孔就是一例。(2)限制式装置:此装置伸入液货舱,使用时允许少量货物蒸发气或液体逸入大气。不使用时,这种装置是完全封闭的。其设计应确保在打开这种装置时不致使舱内货物(液体或气雾)发生危险的外溢;(3)闭式装置:此装置伸入液货舱,成为封闭系统的一部分,且能防止舱内货物逸出。例如浮筒式系统、电子探测器、磁性探测器和带有防护的观察装置等;也可采用不穿过液货舱壳板而与液货舱无关的间接式装置,如货物称重装置和管式流量计等。 Cargo tanks shall be fitted with one of the following types of gauging devices:(1) open device: it makes use of an opening in the tanks and may expose the gauger to the cargo or its vapour. An example of this is the ullage opening;(2) restricted device: which penetrates the tank and which, when in use, permits a small quantity of cargo vapour or liquid to be exposed to the atmosphere. When not in use, the device is completely closed. The design shall ensure that no dangerous escape of tank contents (liquid or spray) can take place when opening the device;(3) closed device: which penetrates the tank, but which is part of a closed system and keeps tank contents from being released. Examples are the float-type systems, electronic probe, magnetic probe and protected sight-glass. Alternatively, an indirect device which does not penetrate the tank shell and which is independent of the tank may be used. Examples are weighing of cargo, pipe flow meter, and soon.

Gauging methodology 测量方法

Gauging system 计量系统

Gauss 高斯

Gauze 金属丝网

GBS 目标型船舶建造

Gear 装置(机构,齿轮,滑车,器具,开动,啮合)

Gear box　齿轮箱体　支撑并保护轮系的外壳。
Gear case(gear casing)　齿轮箱
Gear counter　成对齿轮
Gear coupling(tooth coupling)　齿形联轴器　通过齿轮与齿套的配合以传递扭矩的联轴器。
Gear drive　成对转动
Gear engagement　啮合
Gear plane　齿轮平面
Gear pump　齿轮泵　利用齿轮副旋转时，其啮合容积的变化，完成吸入和压出过程，以抽送液体的泵。分为内齿轮泵和外齿轮泵两种。
Gear quadrant　扇形齿轮
Gear quadrant steering gear　齿扇式舵机　系指利用扇齿机构带动舵杆的舵机。
Gear rack　齿条
Gear ratio　齿轮比
Gear reduction unit　齿轮减速装置(减速器)
Gear room　属具间
Gear rotary pump　齿轮回转泵
Gear shaft　齿轮轴
Gear teeth　轮齿
Gear train　齿轮组(齿轮系)
Gear tumbler　齿轮换向器
Gear unit　齿轮装置
Gear wheel　齿轮
Gear wheel(bull gear, main gear wheel)　大齿轮　一对相啮合齿轮中尺寸较大的齿轮。
Gear wheel pump　齿轮泵
Geared capstan　齿动传动的手摇绞盘
Geared diesel　齿动传动式柴油机
Geared diesel drive　柴油机齿轮传动　由柴油机通过减速齿轮箱带动轴系的传动方式。
Geared door　齿动水密门
Geared engine　齿轮传动柴油机
Geared steam turbine　齿轮传动式汽轮机　通过齿轮减速驱动螺旋桨、发电机、水泵等工作机械的汽轮机。
Geared turbine unit　齿轮传动涡轮机组
Geared turbo-generator　齿轮传动涡轮发电机组
Gearing　齿轮装置(传动装置，啮合)
Gearing chain　传动链
Gearing compartment　减速齿轮舱(齿轮传动舱)
Gearing down　脱开齿轮传动机构
Gearing face　齿轮面
Gearing oil sprayers　连动喷油器
Gearing room　变速齿轮舱

Gearing up　接入齿轮传动装置
Gearing worm　涡轮传动装置
Geislinger elastic coupling　盖斯林格联轴节　即"簧片式联轴节"
Gelatin　凝胶　又称冻胶。溶胶或溶液中的胶体粒子或高分子在一定条件下互相连接，形成空间网状结构，结构空隙中充满了作为分散介质的液体(在干凝胶中也可以是气体，干凝胶也称为气凝胶)，这样一种特殊的分散体系称作凝胶。没有流动性。内部常含有大量液体。例如血凝胶、琼脂的含水量都可达99%以上。可分为弹性凝胶和脆性凝胶。弹性凝胶失去分散介质后，体积显著缩小，而当重新吸收分散介质时，体积又重新膨胀，例如明胶等。脆性凝胶失去或重新吸收分散介质时，形状和体积都不改变，例如硅胶等。
Gelation　胶凝作用　系指由溶液或溶胶形成凝胶的过程。
General acceptance list of repairs　工程总验收单
General additional risk　一般附加险　承保一般外来原因引起的货物损失，亦称普通附加险，它们包括在一切险之中。若投保了一切险，就无须另行加保。若投保了平安险或水渍险，则由被保险人根据货物特性和运输条件选择一种或几种附加险，经与保险人协议加保。
General agency　总代理　系指委托人在指定地区的全权代表，他有权代表委托人从事一般商务活动和某些非商务性事务，并可以将有权指定地区内细分地区，委托下放给区域代理，从事一般商务活动和某些非商务性事务。总代理应区分于独家代理和一般代理，也是事务的主要负责人。总代理经过委托人的同意，可以发展、产生下层代理，下层代理行使代理权限内的权利，并接受上层的管理。按有关规定合同双方同意卖方与委托方签订转让技术价格条款及条件时，委托方的义务应根据相应规定支付佣金，同时总代理人按照相应规定有权收取佣金，届时不得以任何借口延迟，应即时支付。
General Agreement on Tariffs and Trade(GATT)　关税与贸易总协定　是由美国等23个国家于1947年在日内瓦签订，并从1948年1月1日正式生效的一项多边国际条约。其宗旨是通过削减关税和其他贸易壁垒，削除国际贸易中的差别待遇，促进国际贸易自由化，以充分利用世界资源，扩大商品的生产与流通。关贸总协定于1947年10月30日在日内瓦签订，并于1948年1月1日开始临时适用。应当注意的是，由于未能达到GATT规定的生效条件，作为多边国际协定的GATT从未正式生效，而是一直通过《临时适用议定书》的形式产生临时适用的效力。

General Agreement on Trade in Service(GATS) 服务贸易总协定 是世界贸易组织管辖的一项多边贸易协议。《服务贸易总协定》由三大部分组成：一是协定条款本身，又称为框架协定，二是部门协议，三是各成员的市场准入承诺单。《服务贸易总协定》适用于各成员采取的影响服务贸易的各项政策措施，包括中央政府、地区或地方政府和当局及其授权行使权力的非政府机构所采取的政策措施。

General assembly 总装配

General assembly drawing(GAD) 总装图

General average 共同海损 系指在同一海上航程中，当船舶、货物和其他财产遭遇共同危险时，为了共同安全，有意地、合理地采取措施所直接造成的特殊牺牲、支付的特殊费用，由各受益方按比例分摊的法律制度。只有那些确实属于共同海损的损失才由获益各方分摊，因此共同海损的成立应具备一定的条件，即海上危险必须是共同的、真实的；共同海损的措施必须是有意的、合理的、有效的；共同海损的损失必须是特殊的、异常的，并由共损措施直接造成。共同海损损失应由船、货（包括不同的货主）各方共同负担。所采取的共同海损措施称共同海损行为。

General aviation 通用航空 系指使用民用航空器从事公共航空运输以外的民用航空活动，包括从事工业、农业、林业、渔业和建筑业的作业飞行以及医疗卫生、抢险救灾、气象探测、海洋监测、科学实验、教育训练、文化体育等方面的飞行活动。通用航空具有机动灵活、快捷高效等特点，对于国民生产具有重大的意义。

General cargo ship 普通货船 系指设有多层甲板或单层甲板主要用于装运普通货物的船舶。该定义不包括专用干货船，其不属于普通散货船基线计算范围，即牲畜运输船、载驳母船、重货运输船、游艇运输船和燃料运输船。General cargo ship means a ship with a multi-deck or single deck hull designed primarily for the carriage of general cargo. This definition excludes specialized dry cargo ships, which are not included in the calculation of reference lines for general cargo ships, namely livestock carrier, barge carrier, heavy load carrier, yacht carrier, nuclear fuel carrier.

General corrosion 全面腐蚀/平均腐蚀 系指：(1)使强力构件厚度全面均匀地减薄的耗蚀，也称为"均匀腐蚀"。(2)一定面积范围内材料厚度均匀减少。General corrosion means: (1) a corrosion wastage in which the thickness of structural members idealized to be uniformly reduced. Also it is called "uniform corrosion"; (2) It is defined as areas where general uniform reduction of material thickness is found over an extensive area.

General corrosion 普遍腐蚀 系指参考区域中含有超过70%或更大范围的硬质和/或松脱的锈块腐蚀状态，包括点腐蚀，且伴随厚度减薄的证据。

General design 总体设计 是为了设计预览和初始分析阶段而开发的设计。总体设计是一种考虑了总布置、主要系统和组件等的高层次的设计。

General dry cargo ship 普通干货船 系指以载运干货为主，也可装运桶装液货的船舶。但不包括散货船、集装箱船、滚装货船、冷藏货船、水泥运输船、牲畜运输船、坞式甲板船、从事木材制品运输船和从事碎木运输船。General dry cargo ship is a ship mainly carrying dry cargo, and also liquid cargo, excluding bulk carriers, container ships, mainly Ro-Ro cargo ships, refrigerated cargo ships, cement carriers, livestock carriers, dock/deck ships, forest product carriers and wood chip carriers.

General dry cargo ship with double side skin 双舷侧普通干货船 系指整个货舱区域长度和至上甲板的整个货舱高度范围内设置双舷侧、以载运干货为主，也可载运桶装液货的船舶。

General emergency alarm 通用紧急报警 系指在发生紧急情况时向船上所有人员发出召集乘客和船员到集合站进行集合的报警。General emergency alarm means an alarm given in the case of an emergency to all persons on board summoning passengers and crew to assembly stations.

General emergency alarm system 通用应急报警系统 系指由驾驶室操作，在紧急情况下向全船人员发出的、召唤全船人员至集合部位的紧急报警的系统。

General extraneous risks 一般外来风险 系指由一般外来原因所造成的风险称为一般外来风险。一般外来风险：由一般外来原因所造成的风险称为一般外来风险，如，偷盗、破碎、雨淋、受潮、受热、发霉、串味、玷污、短量、渗漏、钩损、锈损等。保险人认定的一般外来原因包括：偷盗、提货不着、淡水雨淋、短量、混杂、玷污、渗漏、碰撞破损、串味异味、受潮受热、钩损、包装破裂、锈损等原因。

General harmful substances distribution on board ship 常见有害物质在船舶上的分布 见表G-1。

表 G-1　常见有害物质在船舶上的分布
Table G-1　General harmful substances distribution on board ship

1. 石棉 Asbestos

结构和/或设备 Structure and/or equipment	部件 Component
螺旋桨轴 Propeller shaft	低压液压管子法兰填料 Packing with low pressure hydraulic piping flange
螺旋桨轴 Propeller shaft	外壳填料 Packing with casing
螺旋桨轴 Propeller shaft	离合器 Clutch
螺旋桨轴 Propeller shaft	制动衬片 Brake lining
螺旋桨轴 Propeller shaft	合成艉轴管 Synthetic stern tubes
柴油机 Diesel engine	管子法兰填料 Packing with piping flange
柴油机 Diesel engine	燃料管护层隔热材料 Lagging material for fuel pipe
柴油机 Diesel engine	排气管护层隔热材料/排气管填料 Lagging material for exhaust pipe /Exh. pipe packing
柴油机 Diesel engine	涡轮增压器护层隔热材料 Lagging material for the turbocharger
涡轮发动机/蒸汽涡轮 Turbine engine/steam turbine	蒸汽管路、排气管路和泄水管路的管子和阀门的法兰填料 Packing with flange of piping and valve for steam line, exhaust line and drain line
涡轮发动机/蒸汽涡轮 Turbine engine/steam turbine	蒸汽管路、排气管路和泄水管路的管子和阀门护层隔热材料 Lagging material for piping and valve of steam line, exhaust line and drain line
锅炉 Boiler	燃烧室隔热 Insulation in combustion chambers

续表 G-1

结构和/或设备 Structure and/or equipment	部件 Component
锅炉 Boiler	锅炉外壳覆层和绝缘 Boiler claddings casings and insulation
锅炉 Boiler	外壳门填料 Packing for casing door
锅炉 Boiler	排气管护层隔热材料 Lagging material for exhaust pipe
锅炉 Boiler	耐火砖和炉衬 Fire bricks and furnace linings
锅炉 Boiler	人孔垫片 Gasket for manhole
锅炉 Boiler	手孔垫片 Gasket for hand hole
锅炉 Boiler	吹灰器和其他孔的气体保护填料 Gas shield packing for soot blower and other holes
锅炉 Boiler	蒸汽管路、排气管路、燃料管路和泄水管路的管子和阀门的法兰填料 Packing with flange of piping and valve for steam line, exhaust line, fuel line and drain line
锅炉 Boiler	蒸汽管路、排气管路、燃料管路和泄水管路的管子和阀门护层隔热材料 Lagging material for piping and valve of steam line, exhaust line, fuel line and drain line
废气经济器 Exhaust gas economizer	外壳门填料 Packing for casing door
废气经济器 Exhaust gas economizer	手孔填料 Packing with hand holes
废气经济器 Exhaust gas economizer	吹灰器气体保护填料 Gas shield packing for soot blower
废气经济器 Exhaust gas economizer	蒸汽管路、排气管路、燃料管路和泄水管路的管子和阀门的法兰填料 Packing with flange of piping and valve for steam line, exhaust line, fuel line and drain line

续表 G-1

结构和/或设备 Structure and/or equipment	部件 Component
废气经济器 Exhaust gas economizer	蒸汽管路、排气管路、燃料管路和泄水管路的管子和阀门护层隔热材料 Lagging material for piping and valve of steam line, exhaust line, fuel line and drain line
焚烧炉 Incinerator	外壳门填料 Packing for casing door
焚烧炉 Incinerator	人孔填料 Packing with manholes
焚烧炉 Incinerator	手孔填料 Packing with hand holes
焚烧炉 Incinerator	排气管护层隔热材料 Lagging material for exhaust pipe
辅机(泵、压缩机、净油器、起重机、起锚机、舵机、绞车、制动器轴、起货设备、辅机、分离器、液压系统) Auxiliary machinery (pump, compressor, oil purifier, crane, windlass, steering gear, winch, shaft brake, cargo gear, auxiliary engine separators, hydraulic systems)	外壳门和阀门填料 Packing for casing door and valve
辅机(泵、压缩机、净油器、起重机、起锚机、舵机、绞车、制动器轴、起货设备、辅机、分离器、液压系统) Auxiliary machinery (pump, compressor, oil purifier, crane, windlass, steering gear, winch, shaft brake, cargo gear, auxiliary engine separators, hydraulic systems)	压盖填料 Gland packing
辅机(泵、压缩机、净油器、起重机、起锚机、舵机、绞车、制动器轴、起货设备、辅机、分离器、液压系统) Auxiliary machinery (pump, compressor, oil purifier, crane, windlass, steering gear, winch, shaft brake, cargo gear, auxiliary engine separators, hydraulic systems)	制动系统的耐磨材料如制动衬片 Friction material for brakes (brake lining)
热交换器 Heat exchanger/heaters	外壳填料 Packing with casing
热交换器 Heat exchanger/heaters	阀门压盖填料 Gland packing for valve

续表 G-1

结构和/或设备 Structure and/or equipment	部件 Component
热交换器 Heat exchanger/heaters	护层隔热材料和绝缘 Lagging material and insulation
阀件 Valve	阀门填料/阀门压盖填料,管子法兰薄板填料,阀帽 Valve packing/Gland packing with valve, sheet packing with piping flange, bonnet
阀件 Valve	高压和/或高温法兰垫片 Gasket with flange of high pressure and/or high temperature
管线、导管 Pipe, duct	护层隔热材料和绝缘,管线压盖填料 Lagging material and insulation, gland packing for piping
液舱(燃料舱、热水舱、冷凝器),其他设备(燃料过滤器、润滑油过滤器) Tank(fuel tank, hot water tank, condenser), other equipments(fuel strainer, lubricant oil strainer)	护层隔热材料和绝缘 Lagging material and insulation
电气设备 Electric equipment	隔热材料,断路器和熔断器,断路器电弧隔板,电缆材料/绝缘(特别是有编织物护套的电缆) Insulation material, electrical equipment such as circuit breakers and fuses, circuit breaker arc chutes, electrical cable materials/insulation (particularly cables with cloth like sheathes)
空间中的石棉 Airborne asbestos	墙壁、天花板 Wall, ceiling
居住舱室区域、厨房和餐厅的天花板、地板和墙壁 Ceiling, floor and wall in accommodation area, galleys and messes	天花板、天花板敷料、地板和墙壁 Ceiling, ceiling covering, floor, wall
防火分隔,如起居间、机舱、烟囱、服务处所、防火控制室/货物控制室、驾驶台、储藏室 Fire Insulation(accommodation, engine room, funnel and uptakes, auxiliary and service spaces, control spaces such as fire control spaces/cargo control spaces, navigation spaces, lockers, etc.)	门(防火门的填料、构件和隔热)、地板、板材、贯穿件(特别是防火舱壁的电缆和管线)、舱壁、防火屏蔽、密封胶绳门、喷涂保温 Doors(packing, construction and insulation of the fire door), floors, boards, penetrations(particularly cables and pipes in fire bulkheads), bulkheads, fireshields and fireproofing, rope door sealants, sprayed on insulation
惰性气体系统 Inert gas system	外壳填料等 Packing for casing, etc.

续表 G-1

结构和/或设备 Structure and/or equipment	部件 Component
空调系统 Air-conditioning system	管子和挠性连接的薄板填料和护层隔热材料,HVAC 导管（用于供暖,通风和空调设备的导管） Sheet packing, lagging material for piping and flexible joint, HVAC ducts (Ducts are used in heating, ventilation, and air conditioning)
其他 Miscellaneous	隔热材料,如"A-60"级分隔隔热材料 Thermal insulating materials, i.e. Class "A-60" insulation materials 甲板敷料 Deck covering 绳索 Ropes/cords 火屏蔽/防火装置 Fire shields/fire proofing 厨房设备 Galley equipment 电气舱壁贯穿填料 Electrical bulkhead penetration packing 制动衬片 Brake linings 蒸汽/水/通风口法兰垫片 Steam/water/vent flange gaskets 高温设备的护层隔热和绝缘材料,特别是全船其他设备管线和高温管道、排烟管、服务处所蒸汽管、高温燃油/水/液体管的护层隔热、垫片、压盖 Thermal laggings and insulation for high temperature applications, special pipes and high temperature conduits uptakes, exhausts, steam pipes in service spaces, high temperature fuel/oil/water/other fluid laggings, gaskets, glands 油漆（有隔热要求部位,如主机外壳） Paints (temperature insulation intention, i.e. paints for M. E. casing) 黏合剂/胶水/胶黏剂/密封剂/填料（填充剂） Adhesives/glues/mastics/sealants/fillers 瓷砖/地砖/甲板衬垫物 Tiles/floor tiles/deck underlay 隔音 Sound damping/sound insulation 石膏（包括装饰线条） plaster (including decorative mouldings) 塑胶如模压塑料产品 Plastics (Moulded plastic products)

续表 G-1

结构和/或设备 Structure and/or equipment	部件 Component
其他 Miscellaneous	泥子如密封泥子 Putty (Sealing putty) 轴,如螺旋桨轴密封,螺旋桨轴轴承 Shaft (seal for propeller shaft, bearing part for propeller shaft) 衬垫 Underlays 吊架(挂钩) Hangars 衬垫 Inserts 管吊架衬垫 Pipe hanger inserts 填充物 Padding 接头 Joints 铺装材料 Surfacing materials 焊接帘 Welding curtain 焊接设备如焊接车间护罩/燃烧罩 Welding equipment (Weld shop protectors/burn covers) 消防设备,如消防毯/服/手套、工作服、防热毯 Firefighting equipment (Fire-fighting blankets /clothing/ gloves, overalls, heat protective blankets) 混凝土压载块 Concrete ballast 被动消防系统用混凝土 Concrete laid for passive fire protection 屏蔽 Shielding 织物 Textiles

2. 多氯联苯(PCB)

设备 Equipment	设备部件 Component of equipment
变压器 Transformer	绝缘油 Insulating oil

续表 G-1

设备 Equipment	设备部件 Component of equipment
冷凝器 Condenser	绝缘油 Insulating oil
燃料加热器 Fuel heater	加热介质 Heating medium
电缆 Electric cable	覆盖、绝缘胶带 Covering, insulating tape
润滑油 Lubricating oil	
热油 Heat oil	温度计、传感器、指示器 Thermometers, sensors, indicators
橡胶/毛毡垫片 Rubber/felt gaskets	
橡胶软管 Rubber hose	
泡沫塑料隔热 Plastic foam insulation	
隔热材料 Thermal insulating materials	
电压调节器 Voltage regulators	
开关/自动开关/轴衬 Switch/recloser/bushing	
电磁铁 Electromagnets	
黏合剂/纸带 Adhesives/tapes	
机械表面污染 Surface contamination of machinery	
油基涂料 Oil-based paint	

续表 G-1

设备 Equipment	设备部件 Component of equipment
捻缝 Caulking	
橡胶隔离架 Rubber isolation mounts	
管吊架 Pipe hangers	
灯用镇流器（荧光灯设备中的部件） Light ballasts (component within fluorescent light fixtures)	
增塑剂 Plasticizers	
船底上隔板下的毛毡 Felt under septum plates on top of hull bottom	

3. 消耗臭氧物质（ODS）

物质 Materials	设备部件 Component of equipment
CFC（R11，R12）	冰箱制冷剂 Refrigerant for refrigerators
CFC	尿烷构成的材料 Urethane formed material
CFC	LNG 船隔热的起泡剂 Blowing agent for insulation of LNG carriers
卤素灭火剂 Halons	灭火剂 Extinguishing agent
其他完全卤化的 CFC Other fully halogenated CFCs	船上使用的可能性低 The possibility of usage in ships is low
四氯化碳 Carbon tetrachloride	船上使用的可能性低 The possibility of usage in ships is low
1,1,1-三氯乙烷（甲基氯仿） 1,1,1-Trichloroethane (Methyl chloroform)	船上使用的可能性低 The possibility of usage in ships is low

续表 G-1

物质 Materials	设备部件 Component of equipment
HCFC(R22, R141b)	制冷机的制冷剂(可能使用至2020年) Refrigerant for refrigerating machine (It is possible to use it until 2020)
HBFC	船上使用的可能性低 The possibility of usage in ships is low
甲基溴 Methyl bromide	船上使用的可能性低 The possibility of usage in ships is low

4. 其他有害物质

物质 Materials	设备部件 Component of equipment
镉和镉化合物 Cadmium and cadmium compounds	镍镉电池、电镀膜、轴承 Nickel-cadmium battery, plating film, bearing
六价铬化合物 Hexavalent chromium compounds	电镀膜 Plating film
汞和汞化合物 Mercury and mercury compounds	荧光灯、汞电灯、汞电池、液体电平开关、电罗经、温度计、测量工具、锰电池、压力传感器、灯具、电气开关、火灾探测器 Fluorescent light, mercury lamp, mercury cell, liquid-level switch, gyro compass, thermometer, measuring tool, manganese cell, pressure sensors, light fittings, electrical switches, fire detectors
铅和铅化合物 Lead and lead compounds	铅酸蓄电池、防腐底漆、焊料(几乎所有电气装置含有焊料)、涂料、防腐涂层、电缆绝缘、铅压载、发电机 Lead-acid storage battery, corrosion-resistant primer, solder (almost all electric appliances contain solder), paints, preservative coatings, cable insulation, lead ballast, generators
多溴化联(二)苯(PBB) Polybrominated biphenyls (PBBs)	不易燃塑料 Non-flammable plastics
多溴二苯醚(PBDE) Polybrominated diphenyl ethers (PBDE)	不易燃塑料 Non-flammable plastics
多氯化联萘 Polychlorinated naphthalenes	涂料、润滑油 Paint, lubricating oil

续表 G-1

物质 Materials	设备部件 Component of equipment
放射性物质 Radioactive substances	荧光涂料、离子型烟探测器、水准仪 Fluorescent paint, ionic type smoke detector, level gauge
某些短链氯化石蜡 Certain shortchain chlorinated paraffins	不易燃塑料 Non-flammable plastics

General inspection 一般检验

General machinery 通用机械

General machinery areas 通用机械区 系指为油、气勘探、生产和安全服务的机械设备（如，发电机和锅炉）所在区域。 General machinery areas are areas where machinery installations serving exploration, production of oil and gas and safety(e. g. generators and boilers) are located.

General offshore anchor 一般性能海洋工程用锚 系指适用于砂或硬泥底质海洋工程用的有（横）杆的转爪锚。如丹福斯锚、LWT 锚、斯达托锚等。其具有很好的稳定性、抓力比船用锚的大得多，常将其归类为"大抓力锚"。其中将 LWT 锚、斯达托锚设计成折角可以改变，以适应软泥底质。在 20 世纪的 60 年代～70 年代设计的海洋工程装置如钻井（驳）船、半潜式钻井平台、自升式平台上一般性能海洋工程用锚的应用十分广泛。

General permit 一般许可证 系指根据有关规定事先发给的许可证。 General permit means permission granted in advance and in accordance with relevant requirements.

General purpose casing 一般用途外壳 系指主要用于防护偶然接触和轻微的非直接溅水，但既非防滴、也非防溅的外壳。

General purpose manipulator 万能机械手（通用机械手）

General radio communications 一般无线电通信 系指通过无线电进行的除遇险和安全信息以外的业务和信息通信。然而业务和公共信息通信可包括与船舶安全航行有关的其他无线电通信，例如船舶移动、医疗指导、天气报告等。 General radio communications means operational and public correspondence traffic, other than distress, urgency and safety messages, conducted by radio. However, operational and public correspondence traffic could include other radio-communications related to the safe navigation of the ship such as ship movement, medical advice, weather reports, etc.

General service pump 总用泵 作为日常杂用的泵，也常作为其他专用泵的备用泵。

General trade system 总贸易体系 又称一般贸易体系。系指贸易国家进行对外贸易统计所采用的统计制度之一。它是以货物通过国境作为进出口的标准。据此，所有进入本国国境的货物一律计入进口；所有离开本国国境的货物一律计入出口。

General tug 常规拖船 系指以拖带驳船（队）运输货物为主要用途的拖船。航区有近海、沿海、锚地至河口、江河联运等。风浪较缓和的近海、沿海、常规拖船也可采用顶推形式。

Generalized Processor Sharing (GPS) 处理器分享

Generalized system of preferences(GSP) 普遍优惠制 是发展中国家在联合国贸易与发展会议上进行长期斗争，在 1968 年通过建立普惠制决议之后取得的。该协议规定，发达国家承诺对从发展中国家和地区输入的商品，特别是制成品和半制成品，给予普遍的、非歧视的和非互惠的关税优惠待遇，这种税称作普惠税。普惠制的主要原则是普遍的、非歧视的和非互惠的。所谓普惠的，系指发达国家应对发展中国家或地区出口的制成品或半制成品给予普遍的优惠待遇。所谓非歧视的，系指应使所有发展中国家和地区都不受歧视，无例外享受普惠制的待遇。所谓非互惠的，系指发达国家应单方面给予发展中国家和地区关税优惠，而不要求发展中国家和地区提供反向的优惠。普惠制的目的：增加发展中国家或地区的外汇收入，促进发展中国家或地区工业化；加速发展中国家或地区的经济增长率。世界上，现在有给惠国家实行 16 个普惠制方案。它们是欧盟 15 国、日本、新西兰、挪威、瑞士、加拿大、瑞典、奥地利、澳大利亚、美国、捷克、斯洛伐克、保加利亚、匈牙利、波兰和独联体，其中欧盟 15 个成员国执行一个共同的方案。接受普惠制关税优惠的发展中国家或地区达到 170 多个。

Generating boat/power supply boat 供电船 系

指装有小功率发电机组,在农村流动供电的小型船舶。

Generating set (generator set) 发电机组（生成集）

Generating set mooring trial 发电机组系泊试验

Generating set sea trial 发电机组航行试验

Generating ship 发电船/电站船　系指装有整套发电和输配电设备,专门担负流动供应电能任务的船舶。

Generating tube bank (generating tube nest) 蒸发管束　借对流传热加热或蒸发锅水的管束。

Generator 发电机　是一种将机械功率转变为电功率的机器。

Generator line 母线　通过螺旋桨中点,绕轴线转动并沿轴线移动扫描出叶面或部分叶面的线。在正视图上与螺旋桨基准线重合,在侧视图上是与基准线间夹角为纵斜角的直线。

Generator sets 发电机组　系指船上的电站。

Generic model 通用模型　系指对所有研究中的船舶或领域均为通用的一套功能。 Generic model means a set of functions common to all ships or areas under consideration.

Genetic technology 基因技术　基因由人体细胞核内的 DNA（脱氧核糖核酸）组成,变幻莫测的基因排序决定了人类的遗传变异特性。人类基因组研究是一项生命科学的基础性研究。有科学家把基因组图谱看成是路路图,或化学中的元素周期表;也有科学家把基因组图谱比作字典。但不论是从哪个角度去阐释,破解人类自身基因密码,以促进人以及相应的染色体位置被破译后,将成为医学和生物制药产业知识和技术创新的源泉。中国首次完成人类单个卵细胞高精度基因组测序,可有力提高试管婴儿活产率。

Geographically disadvantaged States 地理不利国　系指其地理条件使其依赖于开发同一分区域或区域内的其他国家专属经济区的生物资源,以供应足够的鱼类来满足其人民或部分人民的营养需要的沿海国,包括闭海或半闭海沿岸国在内,以及不能主张自己的专属经济区的沿海国。 Geographically disadvantaged States means coastal States, including States bordering enclosed or semi-enclosed seas, whose geographical situation makes them dependent upon the exploitation of the living resources of the exclusive economic zones of other States in the sub-region or region for adequate supplies of fish for the nutritional purposes of their populations or parts thereof, and coastal States which can claim no exclusive economic zones of their own.

Geological laboratory 海洋地质实验室　对地质样品进行现场观测、处理、分析以及对声呐和浅地层地震仪器所获得的资料进行分析研究的工作室。该室一般设有分析间和仪器间。

Geological prospecting 地质勘查　是根据地质知识,利用罗盘、铁锤等简单工具,通过直接观察和研究裸露在地面上的地层、岩石等资料,了解沉积地层及其构造特征,收集所有地质资料,以便查明油气生成和聚集的有利地带和分布规律,达到找油气田的目的。

Geological research ship 地质调查船　用于对海洋地形地貌、地质构造、矿藏资源调查研究的船舶。船上设有地质地貌实验室、重力实验室、磁力实验室、泥样储存室。备有重力仪、测深仪、地貌仪、拖曳式磁力仪、浅地层剖面仪以及表层采泥器、重力采泥器。甲板上安置着各种不同深度的地质绞车。有些还设有地震实验室,用于人工激发弹性波的方法来获取海底地质资料。

Geological sample chamber 地质样品舱　用于保存海洋调查所采集的地质样品（海底表层样品、柱状样品、拖网样品）的舱室。该室贮存的样品应低温冷藏保存。

Geological winch 地质绞车　为收、放地质采样仪器进行水文观测工作的绞车。

Geomagnetic electro kinetograph (GEK) 地磁场电磁海流计　一种根据导体在地球磁场中运动产生的感应电动势测定流速的仪器。

Geomagnetic research ship 地磁测量船　系指用于测量海洋地球磁场异常分布的船舶。船上设有船用核子旋进式的磁力测量仪、测深仪、地貌仪以及重力采泥器和地质绞车。

Geometric 几何平均数　系指 n 个数乘积的第 n 次根。 Geometric means the nth root of the product of n numbers.

Geometric similarity 几何相似　形状相似,但大小不同的相似性。

Geometrical product specifications (GPS) 产品几何技术规范　是一套有关工件几何学的规范。它覆盖了尺寸、尺度、几何公差和表面几何性能的标准。新一代 GPS 系统是国际上近几年才提出的、正在研究与发展中的、引领世界制造业前进方向的、基础性的新型国际标准体系,是一套新的产品开发与制造工程的工具。

Geophysical prospecting 地球物理勘探　简称物探。它是根据地质学和物理学的原理,通过不同的物理仪器观察地面上各种物理现象,推断地下地质情况,达到找油的目的。根据物理方法的不同,物探包括重力勘探、磁法勘探、电法勘探和地震勘探。

Geophysical research ship 地球物理勘探船　用于对海洋石油、天然气等矿物资源勘探研究的船舶。设有地震、磁力、重力、资料处理设备、勘探工具和相应的实验室。通过重力、地震、磁力等仪器进行勘查，用计算机进行整理，亦可将数据通过卫星发射到基地进行整理。船上设有重力仪器、地球物理实验室、计算机以及地震、磁力探查等仪器室。见图 G-2。

图 G-2　地球物理勘探船
Figure G-2　Geophysical research ship

Geosims(geometrically similar model)　几何相似的船模　形状完全相同，而大小不同的船模。

Geothermal heat flow survey system　地热流仪　处理海底沉积物的热导率和温度梯度参数的装置，经计算得出热流值。

German Offshore Test Field and Infrastructure GMBH & Co. KG(DOTI)　[德国]海上试验风场与基础设施公司　系指 2006 年 7 月由意昂公司、EWE 和大瀑布公司 3 家能源公司共同出资组成 DOTI，以有效地共同开展海上活动。

GILI coefficient　基尼系数　是意大利经济学家基尼(CORRADO GILI, 1884~1965)于 1912 年提出的，在国际上用来综合考察居民内部收入分配差异状况的一个重要分析指标。即判断收入分配公平程度的指标。它是一个比值，数值在 0~1 之间。基尼系数的数值越低，表明财富在社会成员之间的分配越均匀。一般发达国家的基尼系数在 0.24 到 0.36 之间。一般在高于 0.5 的情况，则可视为严重的贫富不均。

Gill net　刺网　网片垂直置于水中，使鱼缠于网衣上或刺入网目中的捕鱼网具。

Gill net hauler　刺网起网机　系指绞收刺网网列的机械。用以直接绞网的；有滑轮式起网机、滚柱式起网机；用以绞钢带的，有夹爪式起网机，夹板式起网机。

Gill netter　刺网渔船　将刺网的网片垂直放直在水中，使鱼缠于网上或刺入网目中，以此进行捕捞作业的渔船。

Gillnet capstan　刺网绞盘　用以绞收刺网张纲的一种直立式卷筒。

Girder　桁材　系指主要支撑结构构件的通用术语。Girder means a collective term for primary supporting structural members.

Girder(stringer)　纵桁　船底、舷侧、甲板等处，一般用于支承横向骨材的较大尺寸的纵向构件。

Girder of foundation　基座桁材　基座结构中的桁材。

Gland　填料函(压盖, 密封装置)

Gland box　填料函

Gland bush　压盖衬套

Gland leakage　轴封泄漏

Gland leak-off condenser (gland seal condenser)　气封冷凝器　冷凝汽轮机气封漏气的冷凝器。

Gland leak-off steam condenser　气封抽气器　在轴封的外部气室中建立并保持一定的压力，同时把气封蒸汽和工作蒸汽冷凝成水的装置。

Gland seal　压盖密封(封闭装置)

Gland seal leakage　轴封泄漏

Gland steam pressure regulator　气封压力调节器　使气封平衡箱中的压力自动保持在一定范围内的调节器。

Gland steam receiver (gland steam bottle, gland steam collector)　气封平衡器　汽轮机气封系统中用以保持恒定气封蒸汽压力的箱柜。

Glandless pump　无填料函泵　泵与电机封闭在气密外壳内，泵叶轮直接安装在电机轴上，或通过磁性与之连接成一体，轴不伸出壳体外，故不与空气接触，因此不需要填料函，具有无漏泄特点的泵。通常有：(1)磁性连接式；(2)屏蔽式；(3)湿定子式等 3 种形式。

Glass cotton felt　玻璃棉毡　该产品为玻璃棉施加黏合剂，加温固化成型的毡状材料。其容重比板材轻，有良好的回弹性，价格便宜，施工方便。这种材料在施工中还可根据需要任意剪裁，主要用于建筑的室内，消声系统、交通工具，制冷设备，家用电器的减振、吸声、降噪处理，效果十分理想。有铝箔贴面的玻璃棉毡，还具有较强的抗热辐射能力，是高温车间、控制室、机房内壁、隔间及平顶极好的内衬材料。离心玻璃棉内部纤维蓬松交错，存在大量微小的孔隙，是典型的多孔性吸声材料，具有良好的吸声特性。

Glass fiber　玻璃纤维　是一种性能优异的无机非金属材料。其成分为二氧化硅、氧化铝、氧化钙、氧化硼、氧化镁、氧化钠等。它是以玻璃球或废旧玻璃为原料经高温熔制、拉丝、络纱、织布等工艺。最后形成各类产品，玻璃纤维单丝的直径从几个微米到 20 微米，相当于一根头发丝的 1/20~1/5，每束纤维原丝都有数百根甚至上千根单丝组成，通常作为复合材料中的增强材料，电绝缘材料，绝热保温材料和电路基板等，广泛应用于国民经济各个领域。

Glass fiber wool 纤维玻璃棉

Glass gage 玻璃水位计

Glass microfibers 微纤维玻璃棉

Glass wool 玻璃棉 属于玻璃纤维中的一个类别,是一种人造无机纤维。玻璃棉是将熔融玻璃纤维化,形成棉状的材料,化学成分属玻璃类,是一种无机质纤维,具有成型好、体积密度小、热导率低、保温绝热、吸音性能好、耐腐蚀、化学性能稳定。

Glass-holder frame(glass rim) 窗框 窗结构中供装设玻璃的框架。

Glide chute 溜泥槽 安装在斗塔两侧,其上段和斗塔固定连接,下段可起落转动的倾斜式卸泥槽。

Global corrosion 整体腐蚀 被定义为较大面积,如主要支持构件和船体梁的总平均腐蚀。是新建船舶评审和营运船评估的基础。 Global corrosion is defined as the overall average corrosion of larger areas such as primary support members and the hull girder. It is used as a basis for the new building review and is to be confirmed during operation of the vessel.

Global dynamic stress components(primary stresses) 总体动态应力分量(主要应力) 系指船体梁的垂向波浪弯曲应力 σ_v 和船体梁的水平波浪弯曲应力 σ_h。 Global dynamic stress components(primary stresses) are vertical wave hull girder bending stress, σ_v, and horizontal wave hull girder bending stress, σ_h.

Global maritime distress and safety system(GMDSS) identities 全球海上遇险和安全系统(GMDSS)标识 可由船舶设备发送,并用于识别船舶的海上移动业务识别码、船舶呼号、IMMARSAT 识别码和系列号识别码。 Global maritime distress and safety system(GMDSS) identities mean those maritime mobile services identity, the ship's call sign, IMMARSAT identities and serial number identity which may be transmitted by the ship's equipment and used to identify the ship.

Global navigation satellite system (GLONASS) 全球卫星导航系统/俄罗斯格洛纳斯卫星导航系统 格洛纳斯卫星导航系统作用类似于美国的 GPS、欧洲的伽利略卫星定位系统和中国的北斗卫星导航系统。该系统最早开发于苏联时期,后由俄罗斯继续该计划。俄罗斯 1993 年开始独自建立本国的全球卫星导航系统。该系统于 2007 年开始运营,当时只开放俄罗斯境内卫星定位及导航服务。到 2009 年,其服务范围已经拓展到全球。该系统主要服务内容包括确定陆地、海上及空中目标的坐标及运动速度信息等。"格洛纳斯"导航系统目前在轨运行的卫星已达 30 颗,俄航天部门计划 2014 年内再发射 3 颗。2014 年 11 月 11 日报道,"俄罗斯航天系统"公司公告说,该公司准备好在中国部署俄罗斯卫星导航系统格洛纳斯(GLONASS)差分校正和监测系统站(SDCM),安装工作将在 12 月份开始。

Global positioning system(GPS) 全球定位系统 又称导航星系统(navstar system)。由空间部分、地面控制部分和用户设备三部分组成:(1)空间部分——GPS 的全部卫星均位于地面 20 200 km 高度的圆形轨道上,轨道周期为 12 h,共布置 24 颗卫星,其中 21 颗工作卫星,3 颗备用卫星。平均布置在 6 个轨道平面内。这种配置可以使全球任何地方的用户在地平线 9.5°以上可以"看到"4 颗卫星,而且在地平线上则至少可以观察到 5 颗卫星。卫星上的主要设备是具有长期稳定的、产生精度时间的铯原子钟以及连续反射无线电导航定位信息的发射机;(2)地面控制部分——GPS 地面控制部分由主控站(MCS)、地面控制站(GCS)和监控站(MCS)组成。其地面站的位置已由精确测量获得。作用是对导航卫星进行连续监控。每个监控站设置带有铯原子钟的 GPS 接收机和具有半球覆盖的天线。天线自动跟踪视野中的所有卫星,并接收来自卫星的各种信息。这些信息,包括环境数据在内一起送往主控站。主控站对全部信息进行处理,并单项地将有关信息发给注入站以向每颗卫星注入导航信息和其他控制信息;(3)用户设备部分——主要是 GPS 导航仪。一般由天线、接收机、数据处理机和 I/O 装置组成。GPS 导航仪根据接收到的导航卫星信号。选择最佳星座,并由导航电文与伪距测量方程解算出用户位置。该系统的主要特点:一是实现全球连续定位;二是提供精确的时间标准;三是速度数据和坐标位置。GPS 起始于 1958 年美国军方的一个项目,1964 年投入使用。20 世纪 70 年代,美国陆海空三军联合研制了新一代卫星定位系统 GPS 。主要目的是为陆海空三大领域提供实时、全天候和全球性的导航服务,并用于情报搜集、核爆监测和应急通信等一些军事目的,经过 20 余年的研究实验,耗资 300 亿美元,到 1994 年,全球覆盖率高达 98% 的 24 颗 GPS 卫星座已布设完成。

Global quotas 全球配额 属于世界范围的绝对配额,对于来自任何国家和地区的商品一律适用。系指主管当局通常按进口商的申请先后或过去某一时期的实际进口额批给一定的额度,直至总配额发放完为止,超过总配额就不准进口。

Global radio navigation system 全球无线电导航系统 是利用无线电技术对飞机、船舶或其他运动载体进行导航和定位的系统。无线电导航技术的基本要素是测角和测距,因此可以组成测角-测角、测距-测距、测角-测距、测距差(双曲线)等多种形式的系统。

Global system of mobile communication(GSM)　全球移动通信系统　是当前应用最为广泛的移动电话标准。全球超过 200 个国家和地区超过 10 亿人正在使用 GSM 电话。GSM 标准的无处不在使得在移动电话运营商之间签署"漫游协定"后用户的国际漫游变得很平常。GSM 较之它以前的标准最大的不同是它的信令和语音信道都是数字式的，因此 GSM 被看作是第二代(2G)移动电话系统。这说明数字通信从很早就已经构建到系统中。GSM 是一个当前由 3GPP 开发的开放标准。

Global warming potential(GWP)　全球变暖潜能值　是一个表示某一种温室气体能够捕获得到空气中热量相对值。一定质量的温室气体所捕获得到的热量相对于同样质量 CO_2 所捕获得到的热量之比。即将特定气体和相同质量 CO_2 比较之下，造成全球变暖化的相对能力，以衡量温室气体对全球暖化的影响。

Global warming/Climate change　全球气候变暖　系指地球-大气系统平均温度较长时期的升高现象。由于大气中 CO_2 等多种温室气体增加，地球和低层大气放射的长波辐射被更多的吸收，地球吸收的太阳辐射多于它放射并离开大气顶的长波辐射而导致全球气候变暖。

Globe valve　球形阀

Globoid worm　球形蜗杆

GM(m)　GM　初稳性高度，m。　GM means the metacentric height, in m.

GMEC　全球海事环境大会

Goal new ship build standard(GBS)　目标型新船建造标准　系指 IMO 组织在国际公约的框架内规定的国际航行船舶必须满足的基本结构建造标准。海上安全委员会设有 GBS 通信组，在"等效或替代设计批准导则"适用范围问题上，因各方意见分歧较大，该导则被定为非强制性文件，具体如何应用将由船旗国主管机关自行决定(但不超出 IMO 强制性文件规定的范围)。

Goal post　门形柱　由两根立柱和一横向构件所构成的呈门形的起重柱。

Goal-based ship construction standard for bulk and oil tanker　国际散货船和油船目标型船舶建造标准　系指海上安全委员会以 MSC.287(87)决议通过并可经 IMO 组织修正的"国际散货船和油船目标型船舶建造标准"，但这类修正案应按 SOLAS 第Ⅷ条关于除第一章以外适用的附则修正程序的规定予以通过、生效和实施。 Goal-based ship construction standard for bulk and oil tanker means the International goal-based ship construction standard for bulk and oil tanker, adopted by the Maritime Safety Committee by Resolution MSC. 287(87), as may be amended by the IMO organization, provided that such a-mendments are adopted, brought into force and take effect in accordance with the provisions of Article Ⅷ of the SOLAS convention concerning the amendment procedures applicable to the annex other than Chapter 1 thereof.

Goalpost mast　门形桅杆　由两根立柱和一根常设有顶桅的横向构件所组成，呈门形的桅杆。

Gold dredger　采金船　系指专供在水中采集金矿砂用的采矿船。

Gong　号锣　供在雾天时使用的一种铜锣状音响信号器具。

GOOD condition　良好状况　系指国际海事组织 A.744(18)决议中为评估液货船压载舱涂层而定义的有少量点锈的状况。 GOOD condition is the condition with minor spot rusting as defined in IMO resolution A.744(18) for assessing the ballast tank coatings for tankers.

Goods　货物　(1)包括活动物，如果货物用集装箱、货盘或类似装运工具集运，或者货物带有包装，而此种装运工具或包装由托运人提供，则货物应包括此种装运工具或包装；(2)任何财产，包括活动物，也包括非由多式联运经营人提供的集装箱、货盘或类似的装载或包装工具，不论它们将要或已经装在舱面或舱内。 (1)Goods includes live animals, where the goods are consolidated in a container, pallet or similar article of transport or where they are packed, goods includes such article of transport or packaging it supplied by the shipper; (2)Goods mean any property including live animals as well as containers, pallets or similar articles of transport or packaging not supplied by the multimodal transport operator, irrespective of whether such property is to be or is carried on or under deck.

Goods carried by the conference　公会承认的货物　系指公会会员航运公司按照公会协议运输的货物。 Goods carried by the conference mean the cargo transported by shipping lines members of conference in accordance with the conference agreement.

Goods rejected　退货

Goods trade　货物贸易　就是传统意义上的狭义的贸易，系指有形货物的交换活动。货物贸易的媒介是商业资本。按照 1950 年版(经 1960 年和 1974 年分别修正)的《联合国国际贸易标准分类》，货物贸易共分为 10 大类、63 章、233 组、786 个分组和 1 941 个申报项目。这 10 大类商品分别为：食品及主要供食用的活动物(0)；饮料及烟类(1)；燃料以外的非食用粗原料(2)；矿物燃料、润滑油及有关原料(3)、动植物油脂及油脂(4)；未列名化学品及有关产品(5)；主要按原料分类的制成品(6)；机械及运输设备(7)；杂项制品(8)；没有分类的其他商品(9)。在国际贸易统计中:一般把 0 到 4 类商品称为初级产品;把 5 到 8 类商品称为制成品。

Google search engine 谷歌搜索引擎 是由两位斯坦福大学的博士 LARRY PAGE 和 SERGEY BRIN 在 1998 年创立的,几年间发展为目前规模最大的搜索引擎。目前,每天需要处理 2 亿次搜索请求,数据库存有 30 亿个 WEB 文件。其提供常规搜索和高级搜索两种功能。信息条目数量。多种语言。

Gooseneck 吊杆转枢 其上端与吊杆叉头相连接的吊杆座中的直立转轴。

Gooseneck bearing 转枢座 用以支承吊杆转枢的座子。

Gooseneck bracket 吊杆座 由吊杆转枢及转枢座两个主要零件所组成。用以支承吊杆并使吊杆可左右摆动及上下转动的支座。

Gooseneck connection 鹅颈管接头(S 形管接头,鹅颈连接法,S 形弯管连接法)

Governing 调整(控制,操纵)

Governing(performance) test 调速性能试验 测量调速器与内燃机配合运转中各项性能,如调速特性、突变负荷、转速波动率、最低工作稳定转速等的试验。

Governing device test 调车机构试验 检查主汽轮机的调车机构,如加速关阀、倒顺车操纵阀等,在各工况下的密封性和操作灵活性的试验。

Governing performance test 调速性能试验

Governing screw 调节螺丝

Governing system 调节系统

Governing test 调速试验

Governing valve 调节阀 用改变蒸汽流量和参数以控制汽轮机功率的阀。

Government Communications Head Quarters (GCHQ) [英]政府通信总部 是从事秘密通信电子监听的机构,也是该国应对国际网络战的前沿阵地,与军情五处、军情六处,并列为英国三大情报机构。

Government behavior 政府行为 是一种行政行为。政府行为涉及国家社会经济发展的方方面面。即当事人在订立合同以后,政府当局颁布新政策、法律和行政措施而导致合同不能履行。例如:订立合同以后,由于政府颁布禁运的法律,使合同不能履行。

Governor 调速器(调节阀,平衡器,稳定器)

Governor(speed regulator) 调速器 当转速偏离给定值时能发出信号,通过调速系统使转速恢复到给定值的机构。

Governor gear 调节器传动装置(调节装置,调速装置)

Governor lever 调节杆

Governor test 调速器试验(突卸负荷试验)

GPS receiver 全球卫星定位系统(GPS)卫星导航仪 利用全球卫星定位系统(GPS)信号进行定位和导航的接收设备。

GPS relative 相对 GPS 差分和相对定位系统(如通常标为 DGPS,但用于双 DGPS 的同样原理可应用于双 DARPS 或双 GPS)。

Grab 抓斗 用以攫取泥沙、石子或各种散货的抓挖工具。

Grab dredger(GD) 抓斗式挖泥船 系指具有一台或多台抓斗机挖泥设备的挖泥船。 Grab dredger is a dredger fitted with one or more grab machines.

Grab machine 抓斗机 操纵和控制抓斗进行挖掘工作的起重机械。

Grab type dredger 抓斗挖泥船 系指具有一台或多台抓斗机挖泥设备的挖泥船。

Grab-X 抓斗装卸结构加强 系指对货舱内底板、底边舱斜板最下列板和横舱壁的底凳板具有重大 X 吨重的抓斗装卸货的结构加强的附加标志。 Grab-X means an additional notation of the strengthening for hold inner bottom plating, tank sloping plate and transverse lower stool plating for holds loading/unloading designed for loading/unloading by grabs having a maximum by grabs weight up to X tons.

Grade 品质(等级,度)

Grade 1 manganese bronze(Cu1) 1 级锰青铜

Grade 2 Ni-manganese brorize(Cu2) 2 级镍锰青铜

Grade 3 Ni-manganese brorize(Cu3) 3 级镍锰青铜

Grade 4 Mn-aluminium brorize(Cu4) 4 级锰铝青铜

Grain 谷物 包括小麦、玉蜀黍(苞米)、燕麦、稞麦、大麦、大米、豆类、种子以及加工后在自然状态下具有类似特点的制成品。 Grain covers wheat, maize (corn), oats, rye, barley, rice, pulses, seeds and processed forms thereof, whose behavior(u)r is similar to that of grin in its natural state.

Grain capacity(grain cargo capacity, grain cubic) 散装舱容(谷物舱容) 包括舱口围板在内,量自内底板之顶面、舱壁板表面,甲板和外板之内面,但扣除舱内骨架、支柱、货舱护条、通风筒等所占空间后而得出的船舶各货舱的总容积或其中任一货舱的单舱容积。

Grain fittings 散装谷物积载设备

Grain size 颗粒度

Grain/bale 散货/包装 系指货物的类型,散装货

Granulate tyre rubber 颗粒状轮胎橡胶 切碎的橡胶轮胎材料,经过处理,不含其他物质。无特殊危害。该货物为不燃物或失火风险低。 Granulate tyre rubber is a fragmented rubber tyre material cleaned and free from other materials. It has no special hazards. This cargo is non-combustible or has a low fire-risk.

Granulated slag 颗粒状炉渣 炼钢厂产生的残留物,呈肮脏灰色块状。铁含量0.5%。无特殊危害。该货物为不燃物或失火风险低。 Granulated slag is a residue from blast furnaces of steelworks with a dirty grey, it has lumpy appearance. Iron: 0.5%. It has no special hazards. This cargo is non-combustible or has a low fire-risk.

Graphene 石墨烯 作为目前发现的最薄、强度最大、导电导热性能最佳的一种新型纳米材料,被誉为"新材料之王"。它是从石墨材料中剥离出来,由碳原子组成只有原子厚度的二维晶体。而肯塔基大学(University of Kentucky)现在从理论上验证了一种一层原子厚度的二维晶体存在,只不过它由硅、硼、氮三种原子构成,类似石墨烯以六边形机构排列成单原子层状二维晶体。科学家从理论模型上推测这种材料同样具有薄、强韧性,而且这种材料由于具备硅原子,可以制备成半导体材料广泛应用于消费电子产品中。

Graphics terminal 图形终端

Grapnel 抓缆钩 检修海底电缆时捞取电缆用的钩。

Grapnel anchor 四爪锚 具有四个固定钩爪的锚。

Grate 格栅(炉排)

Grate area 炉排面积

Grate bar 炉栅

Grate shaker 炉箅摇动器

Grated hatch 格栅舱口(栅形舱盖)

Grating 格栅(格子板)

Grating cover 格子盖(格栅盖)

Grating deck 格栅甲板

Grating frame 格栅框架

Grating hatch 格栅舱口(栅形舱盖)

Gratings 格栅 系指机舱内各种小平台上的钢质铺板。

Gravimeter winch 重力仪绞车 作为收、放海底重力仪进行海洋重力测量工作的绞车。

Gravimetric foundation 重力式基础 系指非常重的位移结构。通常由混凝土制成。重力式基础利用作用于地面的垂直压力,矗立在海底。基础的直径通常为15~25 m,所有的力和弯矩通过基础底座传递。一般情况下,重力式基础用于半坚硬的均匀的海底和水深较浅处。

Gravitation 重力(万有引力)

Gravitation tank 重力水柜 安装于船舶高处,依靠重力,水流往各用水处的水柜。

Gravitational prospecting 重力勘探 根据重力学的原理,通过在地面、水面或在海底测得的重力值相对变化,来研究地质构造和勘探矿藏。

Gravitational tank 重力液舱 系指蒸汽压力不大于70 kPa的液舱。

Gravitational wave 重力波 在重力作用下,产生振荡,消耗能量所形成的波。

Gravity 重力

Gravity acceleration g 重力加速度 取值为9.81 m/s²。 Gravity acceleration is taken equal to 9.81 m/s².

Gravity anchor 重力锚 即重块锚。由混凝土块或钢块、碎金属或其他高密度材料制成。裙板依靠其自重穿透,设计的提升能力取决于沉没的锚重,其承受水平负荷的能力是锚与底质之间的摩擦力以及锚下底质的剪切强度函数。重力锚适用于小型泊系统。

Gravity balanced hatch cover 重力平衡舱盖 利用舱盖自重,以减少翻启用力的舱口盖。

Gravity chute 甲板排水槽(排水管,斜槽)

Gravity circulation 重力循环

Gravity corer 重力取样管 利用重物下沉的冲击力插入海底取管状样品的装置。

Gravity fed self-unloading bulk carrier 重力输送直卸式散货船 系指在货舱底部设有重力输送系统的船舶。利用可以打开或关闭的货门将货物送到传送带上。传送带在货舱底以下沿船舶艏艉方向运转,货物从该处通过传送系统送至甲板,用能伸到岸上并配有传送带的自卸式吊杆卸到岸上。这不适用于设有起重机和抓斗等卸载系统的船舶。 Gravity fed self-unloading bulk carrier means a vessel that has gravity fed systems from the bottom of cargo holds, using gates that may be opened or closed to feed the cargo onto conveyor belts. Such belts run in fore and aft direction underneath the holds, from there the cargo is carried by means of conveyor systems to the deck and discharged onto shore with a self-unloading boom that can extend over the shore and has a conveyor belt. This is not applicable for the vessels with unloading systems such as cranes and grabs.

Gravity forced-feed oiling system 重力式滑油系统 (1)利用滑油重力柜将润滑油输送到主汽轮机各润滑点的滑油系统;(2)由压力式和重力式两种形式组成的

主汽轮机滑油系统。

Gravity line survey 重力路线测量　在调查区域内，为控制地质的构造特征，按规定的调查路线所进行的重力测量。

Gravity loadings 重力载荷　系指在静水条件下由平台重力、使用及作业引起的载荷。重力载荷一般由下列载荷组成：(1)平台重力；(2)作业载荷；(3)甲板载荷；(4)露天结构上积聚的冰/雪载荷(适用时)。Gravity loadings are these loadings which exist due to the unit's weight, use and treatment in still water conditions. Gravity loadings comprise generally: (1) weight of the unit; (2) gravitational loadings due to operation; (3) deck loadings; (4) loadings of ice/snow accumulated on exposed structures.

Gravity oil system 重力润滑系统

Gravity piston type corer 重力活塞式取样管　应用快速下降的冲击力和活塞的吸力取海底管状样品的装置。用于粗粒沉积物地区的采样。

Gravity platform 重力式平台　系指大型混凝土加强平台。其主要依靠自身重力放置在井位上，一般用于水深小于 1 000 ft(300 m)的大型油田。Gravity platform means the large reinforced concrete platform, which is held in place mainly by their own weight. They are used in large field in water depths less than about 1 000 ft(300 m).

Gravity prospecting 重力勘探　是以牛顿万有引力定律为基础，利用探测对象与其周围岩/矿石之间的密度差异引起的地表的重力加速度值的变化，通过观测和研究重力场的变化规律，查明地质构造、寻找矿藏及探测物的一种物探方法。

Gravity survey 重力测量　对不同海区的重力加速度值的测量。

Gravity tank 重力液货舱　系指液货舱顶部设计压力不大于 0.07 MPa(表压)的液货舱。重力液货舱可以是独立液货舱或整体液货舱。重力液货舱的建造和试验应按照公认的标准,考虑载运货物的温度和相对密度。Gravity tank means a tank having a design pressure not greater than 0.07 MPa gauge at the top of the tank. A gravity tank may be independent or integral. A gravity tank shall be constructed and tested according to recognized standards, taking account of the temperature of carriage and relative density of the cargo.

Gravity type boat davit 重力式艇架　在操作时,依靠重力作用将艇吊出舷的艇架。

Gravity-type lubrication system 重力式润滑系统

Gray water 灰水　系指：(1)从洗碗机、洗衣机、洗脸盆和洗澡盆排出的排出物,但不包括从厕所、小便池、医务室或货物处所排出的排出物；(2)仅从厨房、洗碗机、浴缸和洗衣机排出的废水。此术语不包括其他废水或废液。Gray water means: (1) drainage from dishwasher, shower, laundry, bath and washbasin drains, not including drainage form toilets, urinals, hospitals and cargo spaces; (2) only galley, dishwasher, bath, and laundry waste water. The term does not include other wastes or liquid waste.

Gray water control(GWC) 灰水的控制　系指对船上所设的洗衣房、浴室、厨房、住舱房的排出废水的控制。Gray water control(GWC) means control of drainage from the laundry, bathrooms, gallies, and accommodations.

Grease 油脂(涂油,润滑)
Grease cup 牛油杯
Grease gun 牛油枪
Grease lubrication 滑脂润滑法
Grease lubricator 牛油杯(滑油杯)
Grease nipple 牛脂填料
Grease oil 滑脂油
Grease packing 油脂填料
Grease seal 油脂密封
Grease trap 注油头
Greaser 油脂杯(牛油杯)
Greasy 被涂油的

Great lakes 大湖　系指：(1)美国北部的五大湖和从罗斯尔斯角(Cap des Rosiers)至安提科斯蒂岛(Anticosti Island)两点之间连一条航向线以西,以及西经63°的安提科斯蒂岛北侧的圣·劳伦斯河。(2)安大略湖、伊利湖、休伦湖(包括圣·克劳尔湖)、密歇根湖、苏必利尔湖和与其相连的水道(苏圣·玛丽河、圣·克劳尔河、底德律河。尼亚加拉河与加拿大相邻的圣·劳伦斯运河)并包括上述湖泊及与其相连的水系流域范围内的所有主干道。Great lakes means: (1) the Great Lakes of North America and the St·Lawrence River west or a rhumb line drawn from Cap des Rosiers to West Point Anticosti Island and on the north side of Anticosti Island, the meridian of longitude 63 degrees west. (2) Lake Ontario, Lake Erie, Lake Huron(including Lake Saint Clair), Lake Michigan, Lake superior, and the connecting channels (Saint Mary's River, Saint Clair River, Detroit River, Niagara River, and Saint Lawrence River to the Canadian border), and includes all other bodies of water within the drainage basin of such lakes and connecting channels.

Great Lakes region 五大湖区　系指位于加拿大与美国之间交界处的湖泊区域。即美国中西部和美国中北部的五大湖区及密西西比河上游河谷周围的地区。

Great Portage National Monument 大河谷国家纪念碑 系指在北纬41°57.521′,西经089°41.245′纪念碑所在地的一点到北纬47°57.888′,西经089°40.725′纪念碑所在地东北角的,(美国)国家公园管理局管理的苏必略湖边或邻近的区域。Great Portage National Monument means the site on or near Lake Superior administered by the National Park Service, from the southwest corner of the monument point to land at 47°57.521′N, 089°41.245′W, to the northeast corner of the monument point of land 47°57.888′N, 089°40.725′W.

Greater coastal service restriction 近海航区营运限制 系指航行于距岸不超过 200 n mile 的水域,且船舶在其经营的航线上,满载并以其营运航速航行至庇护地:对客船不超过 4 h;对货船不超过 8 h。如上述某些水域的海况较为恶劣,则船级社可视其情况对上述距离提出更严格的要求。如船旗国主管机关或其所在营运区的海岸主管机关对该水域有特定距离的规定时,则应根据该主管机关的规定执行。

Great-Lake bulk carrier 大湖型散货船 系指能进入北美五大湖航运的船舶,其尺度受五大湖圣劳伦斯航道的限制,吃水不得超过 7.92 m。见图 G-3。

图 G-3　大湖型散货船
Figure G-3　Great-Lake bulk carrier

Green all-round light 绿环照灯 显示绿色光的环照灯。

Green construction 绿色建造

Green design 绿色设计

Green environment protection technology 绿色环保技术 不同的绿色环保技术在船舶上的应用效果可用船舶生态矩阵(EPM)来表达。船舶生态矩阵(EPM)能表现不同绿色环保解决方案对船舶建造成本的影响,也能表现不同绿色环保解决方案对船舶营运成本的影响的复杂程度。例如在采用替代燃料中,用船用柴油(MDO)或低硫油(MGO)替代重油和利用液化天然气(LNG)替代重油都可减少排放,但成本明显增加。在船上配备航路优先系统可利用实时伴流、波浪和天气预报进行航路优化,进而最大限度地利用自然界的能量,降低燃油消耗和废气排放。另外纵倾优化也是一个不错的低成本解决方案,既能降低船舶营运成本又能降低废气排放。利用DTA软件,根据计算流体力学(CFD)系列计算和船模试验验证可得到船舶各种营运吃水,不同航速下的最佳纵倾,取得了明显的节能效果。(1)绿色环保技术有:①减少风阻;②合理设计航速;③采用电喷主机;④采用废热回收装置;⑤利用清洁能源,如风能、太阳能、核动力以及清洁能源驱动的燃料电池、清洁能源驱动的柴油机、燃气轮机低排放柴油机和压载水处理装置。(2)在船舶建造方面:①推行三维建模和数字化设计;②推行"5S"管理和清洁生产、优化作业流程;③采用符合环保标准的钢板表面处理设备,最大限度地减少船舶建造过程中能源浪费和排放污染,提高生产过程中资源利用率;④采用丙烷和天然气代替乙炔作为切割用气;⑤在生产过程中推广节水技术,使用较高标准的个人防护用具;⑥采用高效低排放的焊接技术提高钢材等材料的利用率等。(3)在船型设计方面,基于目标型船舶设计标准进行船型研发,采用风险评估手段控制技术风险以提高船舶的燃油效率和货物运输效率为目标,不断优化船舶线型、船舶布置、推进器和舵体等。(4)在系统设计上,重点开展能源的再利用,推进装置功率管理、低硫油应用,电网低损概念,岸电系统以及环境保护技术研究,采用成熟可靠和成本可控的新技术和新配置,提高船型的能效水平。在设计中充分考虑营运保养方面的要求,如货物安全、防火安全、装卸安全等,并同时注重操作便利与维护成本控制。

Green lighting 绿色照明 是节约能源、保护环境、有益于提高人们生产、工作、学习效率和生活质量、保护身心健康的照明。

Green operation 绿色营运

Green passport for recycling(GPR) 船舶绿色护照 记录下列内容的、符合 IMO《拆船公约》定义的护照:(1)船舶资料;(2)船舶潜在有害材料清单,包括船上已知每种材料分布区域和大约数量或体积,潜在有害材料包括船舶结构和设备中潜在有害材料、船上操作产生的废料及备品中潜在有害材料。上述资料的任何变动/更换情况应记入绿色护照;新旧记录连同变动历史资料应同时保存在船上。

Green port power supply technology 港口船舶岸电供电技术 系指船舶靠港期间,停用船舶上的发电机电源供电,由(港区)码头的岸电通过电缆对船上的电器设备供电的技术。这对节能、减排、建设绿色港口、解决船上发电对港口水域污染都有非常积极的意义。

Green ship 绿色船舶 系指:(1)通过采用先进技术,把"使用功能和性能要求"与"节约资源和保护环境要求"紧密结合,在船舶设计、建造、使用、拆解的全寿命周期内,节省资源和能源,减少消除环境污染,保障生产者和使用者健康安全、友好舒适的新技术船舶。(2)通过应用诸多绿色技术，在船舶建造及其交付后的整个寿命周期内，达到安全、环境友好、效能最优和职业健康目的，并在船舶退役后船体和组件都能拆解再利用，使其对环境的影响最小化的船舶。简而言之，绿色船舶就是对环保有贡献并有环保竞争力的船舶。

Green ship rule 绿色船舶规范 是中国船级社在对近年来国际船舶技术标准发展、船舶绿色特性和绿色技术跟踪研究的基础上编制而成的，旨在促进造船业、航运业和相关制造业优化升级，促进航运业对新建船舶和现有船舶采取有效的技术和管理措施，在安全的前提下实现船舶低消耗、低排放、低污染和环境舒适的目标。绿色船舶规范基于国际公约的最新发展成果以及船舶工业和航运业的最新发展成果，对船舶能效、环境保护、船员工作环境舒适度等方面的功能要求进行了规定，并首次界定了"绿色船舶"的概念，对造船界转变船舶设计和建造理念，提升整体竞争力，占领国际市场具有积极的指导和促进作用。绿色船舶规范还强调打造绿色船舶的目标涉及环境保护目标、能效目标、和规则环境目标三个方面。其中环境保护目标是减少船舶对海洋、陆地、大气环境造成污染或破坏；能效目标是减少船舶营运所产生的二氧化碳排放量，提高船舶能效水平；工作环境目标则是改善船员工作和居住条件，降低船员劳动强度。该规范适用于申请中国船级社绿色船舶附加标志的船舶。申请船舶应满足中国船级社"钢质海船入级规范"和船旗国主管机关的相应要求，提交申请后，经中国船级社审图和检验，并通过综合评定，确认符合"绿色船舶规范"要求的船舶，将获得绿色船舶相应的附加标志。

Green ship Ⅰ Ⅰ级绿色船舶 系指在环境、能效（包括设计能效和营运能效）、工作环境三个方面的绿色要素满足绿色船舶Ⅰ级所有的适用要求的船舶。

Green ship Ⅱ Ⅱ级绿色船舶 系指在环境、能效（包括设计能效和营运能效）、工作环境三个方面的绿色要素满足绿色船舶Ⅱ级的所有适用要求的船舶。

Green ship Ⅲ Ⅲ级绿色船舶 系指在环境、能效（包括设计能效和营运能效）、工作环境三个方面的绿色要素满足绿色船舶Ⅲ级所有适用要求的船舶。

Green side light (starboard light) 绿舷灯 设置在船舶右舷侧，在船首至左舷112.5°内显示绿色光的舷灯。

Green status 绿色状态 正常操作状态，有足够的DP设备在线使用，以符合公开宣称的安全工作极限内所要求的性能。

Green stern light 绿艉灯 船舶通过基尔运河时专用的显示绿色光的艉灯。

Green visual signal 绿色的视觉信号 系指安全状态的指示。例如：机械正常运转；液体正常循环；压力、温度和电流等在限定值以内。 Green visual signal means a indication of normal operating conditions. For example, normal operation of machinery, normal circulation of liquids, pressure, temperature and current, etc is within the limited value.

Green warning condition of DP 动力定位绿色警戒状态 系指正常操作状态。足够的设备在线，以符合在所标明的安全感在极限内要求的性能。

Green water 甲板上浪 系指船舶在正常营运情况下，除飞溅上船的以外，打上甲板的海水。Green water is sea water other than spray shipped aboard the ship under normal operating conditions.

Greenhouse gas emission from international shipping by fuel burning 国际海运燃料排放 系指从事国际水运的所有船舶因燃料燃烧而排放的温室气体。该运输可发生于海上、内陆湖泊或水道和沿海水域，包括从出发国到目的地国全程的排放，但不包括渔船的排放。海运温室气体分类主要包括二氧化碳（CO_2）、甲烷（CH_4）、氧化亚氮（N_2O）、氢氟碳化合物（HFCs）、全氟碳（PECs）、六氟化硫（SF_6）以及其他相关物质。来自船舶的温室气体排放归类为以下4类：(1)船舶废气排放；(2)货物的排放；(3)制冷剂的排放；(4)其他排放。其中船舶废气排放主要涉及由船舶主发动机、辅助发动机和锅炉排放的废气；货物的排放系指与所载韵味的货物有关的排放；包括各种货物的排放和泄漏，冷藏集装箱和卡车的制冷剂泄漏，运输液体货物时挥发性化合物（甲烷和非甲烷挥发性有机化合物）的挥发释放等；制冷剂排放主要是用于冷藏/冷冻货物以及空调的制冷剂在制冷过程和空调设备的维修过程中发生的泄漏，从而排放到大气中。

Greenhouse gases (GHG) 温室气体

Greenwich Mean Time (GMT) 格林尼治标准时间

Grey cast iron 灰铸铁 系指具有片状石墨的铸铁，因断裂时断口呈暗灰色，故称为灰铸铁。主要成分是铁、碳、硅、锰、硫、磷，是应用最广的铸铁，其产量占铸铁总产量80%以上。根据石墨的形态，灰铸铁可分为：普通灰铸铁，石墨呈片状；球墨铸铁，石墨呈球状；可锻铸铁，石墨成团絮状；蠕墨铸铁，石墨呈蠕虫状。

Grey cast iron specimen 灰铸铁试样

Grey flake iron casting 片状灰铸铁件

Grey iron casting 灰口铸铁件（灰铁铸件，生铁铸件）

Grey water 灰水 系指：(1)洗碗碟水、厨房洗涤槽水、淋浴水、洗衣水、洗澡水以及洗脸水等排出的水，其不包括从厕所、小便池、医务室和MARPOL附则 IV 第1.3条所定义的动物处所排出的水，且其不包括从货物处所排出的水。(2)对ISO15749-1：2004(E)而言，灰水是除生活污水以外的需要处理的废水。 Grey water is：(1) drainage from dishwater, galley sink, shower, laundry, bath and washbasin drains and it does not include drainage from toilets, urinals, hospitals, and animal spaces, as defined in Regulation 1.3 of MARPOL Annex IV and does not include drainage from cargo spaces. (2) For ISO15749-1：2004(E), grey water is drainage to be disposed of, excluding sanitary sewage.

Grey water control(GWC) 灰水控制

Grey water holding tank (grey water retention tank) 灰水储存舱

Grey water treatment system 灰水处理系统

Grid 格子线 系指在型线图中，三个相互垂直的投影面上，由基线、水线、站线、纵剖线、中线、半宽边框线组成矩形格子的水平线和垂直线。

Grid mode 格网方式 是确定测线的一种方法。测线由一族等距离的平行线组成。

Gridiron expansion valve 栅形膨胀阀

Gridiron valve 栅形阀

Grill 格栅(格子,百叶栅)

Grill flooring 格栅铺板

Grillage 交叉梁系(格架,格栅,板架)

Grind 磨(磨光,磨碎)

Grind a valve 研阀

Grinder 磨碎机(磨床,砂轮机,磨工) 对"73/78防污公约"附则 V 而言，系指用于将船上垃圾进行磨碎的机械。

Grinding 磨削(打磨,研磨法)

Grinding compound 磨料(磨剂)

Grinding crack 磨痕(磨削裂纹)

Grinding machine 耐磨机

Grinding oil 润磨油

Grinding out 磨孔

Grinding test 磨损试验

Grinding wheel spectacle 磨轮保护镜片

Grindstone 磨石

Grip 夹扣(夹具,桨柄)

Grip hand 手柄

Grip holder 夹头

Grip nut 扣紧螺母

Grip type 夹扣型

Gripping pliers 夹管钳

Gripping unit 压紧装置

Grit 砂砾(颗粒,砂粒,金属屑)

Grit blast 喷丸设备(抛丸设备,喷丸处理)

Grit blasting 喷丸除锈/喷丸处理(喷砂处理) 系指利用喷丸，在封闭的车间内对钢板进行除锈的作业。

Groove angle 坡口角度

Groove corrosion 沟槽腐蚀 是扶强材与扶强材、扶强材与板材的焊缝连接处典型的局部厚度减少。Groove corrosion is typically local material loss adjacent to weld joints along abutting stiffeners and at stiffener or plate butts or seams.

Groove cutting 开坡口

Groove cutting machine 开槽机(开沟机)

Groove depth 坡口深度

Groove preparation 坡口加工

Groove radius 坡口半径

Groove weld joint 坡口焊接接头

Grooved insulator 环槽绝缘体

Grooved pulley 槽轮

Grooved rail 槽形轨道

Grooved weld 开槽焊缝(坡口焊缝)

Grooved welding 开坡口焊(开槽焊)

Grooves 沟槽(开槽,压槽)

Grooving 开槽 系指：(1)开坡口；(2)焊缝边缘准备；(3)为布置挖泥设备，在挖泥船船体上的凹进部分或阱。

Grooving chesel 开槽凿

Grooving machine 开槽机(开沟机)

Gross cost of averting a fatality(GCAF) 总灾难转移费用 一种成本测算方式，即风险控制的附加成本与为避免领域死亡所降低的风险的比例。GCAF = 附加成本/降低的风险的危险因素并加以排序，对排序后的危险因素依次确定减小发生频率和产生后果的措施。

Gross domestic product(GDP) 国内生产总值 系指在一定时期内(一个季度或一年)，一个国家或地区的经济中所生产出的全部最终产品和劳务的价值。国内生产总值是国民经济核算的核心指标，在衡量一个国家或地区经济状况和发展水平亦有相当重要性。2015年4月15日，中国国家统计局公布的数据显示，一季度中国GDP增速放缓至7%，为2009年第一季度以来最低。2015年7月15日，2015年中国经济半年报正式揭晓，2015年上半年中国GDP生产总值达296 868亿元，同比增长7.0%。2015年9月7日，统计局发布《国家统计局关于2014年国内生产总值(GDP)初步核实的公告》，2014年GDP增速为7.3%，下调0.1个百分点。

Gross shipping mass m_G　总装运质量 m_G　总装运质量 m_G 系有关规定的净装运质量加上装运材料,诸如垫架、支架、紧固材料及覆盖物等,该质量以吨(t)或千克(kg)表示。　Gross shipping mass m_G is the net shipping mass, as defined by relevant provisions, plus shipping materials such as cradles, supports, fastening material and covers. The mass is expressed in tons, t, or kilograms, kg.

Gross thickness offered $t_{\text{gross-offered}}$　总提供厚度　为新建阶段中提供的总厚度。是从建造厚度减去自愿增加厚度后得到的,按下式计算:$t_{\text{gross-offered}} = t_{\text{as-built}} - t_{\text{voluntary-addition}}$。　Gross thickness offered $t_{\text{gross-offered}}$ is the gross thickness provided at the newbuilding stage, which is obtained by deducting the thickness for voluntary addition from the as-built thickness, as follows: $t_{\text{gross-offered}} = t_{\text{as-built}} - t_{\text{voluntary-addition}}$.

Gross thickness required $t_{\text{gross-required}}$　总要求厚度　为要求净厚度加上 t_C 所得到的总(全)厚度(mm)。其不小于腐蚀增量 t_C 与净要求厚度相加所得总厚度。按下式计算:$t_{\text{gross-required}} = t_{\text{net-required}} + t_C$。　Gross thickness required is the gross (full) thickness, in mm, obtained by adding t_C to the net thickness required. Gross thickness required $t_{\text{gross-required}}$ is not less than the gross thickness which is obtained by adding the corrosion addition t_C, as follows: $t_{\text{gross-required}} = t_{\text{net-required}} + t_C$.

Gross ton　总吨

Gross tonnage　总吨位　系指按"1969 年国际吨位丈量公约"附则 1 或任何后继公约中的吨位丈量规定计算的总吨位。　Gross tonnage means the gross tonnage calculated in accordance with the tonnage measurement regulations contained in Annex 1 to the International Convention on Tonnage Measurement of Ships, 1969, or any successor convention.

Gross tonnage GT of a ship　舰船总吨位 GT　按照规范,通过下列公式确定:$GT = K_1 V$;式中,V 为舰船所有封闭空间总体积,单位为 m^3,其中包括转动炮塔,雷达拱形结构,桅杆等;$K_1 = 0.2 + 0.02 \lg V$。　Gross tonnage GT of a ship is to be determined for the purposes of these Rules, by the following formula:$GT = K_1 V$;where:V = total volume of all enclosed spaces in the ship, in m^3 and includes gun turrets, radar domes, masts, etc. $K_1 = 0.2 + 0.02 \lg V$.

Gross volume of a space　舱室总容积　系指各甲板之间和船侧肋骨、护条、衬板表面之间量得的容积。　Gross volume of a space is the volume measured between the decks and between the face of the frames, sparrings or linings at the ship's side.

Gross weight　毛重　系指货物连同其包装的质量。引证解释货物连同包装材料或牲畜家禽连同皮毛在内的质量。与"净重"相对。国际上有按实际皮重(Actual tare)、平均皮重(Average tare)、习惯皮重(Customary tare)、按约定皮重(Computed tare)、等计算皮重的方法,究竟采用哪一种计算方法来求得净重,应根据商品的性质、所使用的包装的特点、合同数量的多寡以及交易习惯,由双方当事人事先约定并列入合同,以免事后引起争议。平均度重:在包装物比较划一的情况下,可从全部商品中抽取一定件数的包装物,加以称重,求出平均每件包装物的质量。

Gross-compound steam turbine　多轴并列式汽轮机　汽轮机转子相互并列,其轴经齿轮减速器驱动同一螺旋桨的主汽轮机。

Ground　地　系指金属船体的整体或专门设置的金属接地板。

Ground　基底　系指底材的表面,此表面或无覆盖层或有覆盖层。　Ground means the surface of base material with or without coat.

Ground effect　地面效应　系指:(1)当机翼贴近地面或水面(通常飞行高度不超过机翼弦长)掠海飞行时作用在机翼上增加的垫升力。升力增加是由于地面和机翼表面之间空气减速使机翼下表面压力增加而产生的。这可以通过放下襟翼、在翼梢下方安装短板以及调节机翼平面几何形状等来实现;(2)接近表面的机翼达到升力增加和诱导阻力减少的现象。这种现象的程度取决于船舶的设计,但一般发生在小于翼的平均弦长的高度处。　Ground effect is:(1)the enhanced lift force acting on a wing that is travelling close to the ground or water surface, the height is commonly less than one wing chord. The enhanced lift is generated by the greater pressure increase on the undersurface of the wing due to higher deceleration of the air trapped between the ground and wing surfaces. This can be enhanced by lowering the wing flaps, installing fences below the wing tips and adjusting the wing plan geometry;(2)a phenomenon of increase of a lift force and reduction of inductive resistance of a wing approaching a surface. The extent of this phenomenon depends on the design of the craft but generally occurs at an altitude less than the mean chord length of the wing.

Ground effect craft　地效翼船　是一种利用机翼在地面效应下的气动升力将船体托抬出水面,可以在地面效应区内作长时间稳定飞行的新型超高速船舶。

Ground effect mode　地效状态　对地效翼船的操作模式而言,系指在地效区中驾驶地效翼船的主要稳定状态。　For operational modes of WIG craft, ground effect mode is the main steady state operational mode of flying the WIG craft in ground effect.

Ground effect zone 地面效应作用区(地面效应区) 系指存在地面效应影响的空间。

Ground power block 落地式动力滑车 安装在渔船舷边或尾部甲板上的动力滑车。

Ground reference plane 接地参考平面 系指一块导电平面,其电位用作公共参考电位。

Ground stabilization mode 地面稳定模式 系指航速和航向信息系参照地面并采用地面轨迹输入数据或EPFS作为参照的显示模式。 Ground stabilization mode means a display mode in which speed and course information are referred to the ground, using ground track input data, or EPFS as reference.

Ground swell 海涌 系指周期超过4.5 s的波浪。

Ground tackle 锚具 锚和锚链或锚缆及其配件的统称。

Ground-based interceptor(GBI) 地基拦截弹

Grounding device 接地装置

Group A cargo A组货物 系指装运时如含水量超过其适运水分极限,就有可能流态化的货物。 Group A cargo consists of cargoes which may liquefy if shipped at a moisture content in excess of their transportable moisture limit.

Group alarm 组合报警 系指被监控的机电设备或系统处于任何非正常状况下发出的任一报警,通过一个报警通道所送出的公用报警。 Group alarm means a common alarm activated by any abnormal conditions of the monitored machinery or system.

Group B cargo B组货物 系指具有化学危害,会使船舶产生危险情况的货物。 Group B cargo consists of cargoes which produce a chemical hazard that could give rise to a dangerous situation on a ship.

Group C cargo C组货物 系指既不易流体化(A组),也不具有化学危害(B组)的货物。 Group C cargo consists of cargoes which are neither liable to liquefy(Group A) nor to produce chemical hazards(Group B).

Group of compartments 舱室组 由2个或2个以上相邻舱室组成的部分船体。

Group of LR Society 英国劳氏船级社集团 包括劳氏船级社、它的分支机构及子公司以及它的官员、董事、雇员、代表和代理人中的任何一员,个别的或集体的。

Group of twenty finance ministers and central bank governors(G20) 20国集团 于1999年9月25日由8国集团(G8)的财长在华盛顿宣布成立,属于布雷顿森林体系框架内非正式对话的一种机制,由原8国集团以及其余12个重要经济体组成。宗旨是为推动已工业化的发达国家和新兴市场国家之间就实质性问题进行开放及有建设性的讨论和研究,以寻求合作并促进国际金融稳定和经济的持续增长,按照以往惯例,国际货币基金组织与世界银行列席该组织的会议。20国集团成员涵盖面广,代表性强,该集团的GDP占全球经济的90%,贸易额占全球的80%,人口约为40亿。因此已取代G8成为全球经济合作的主要论坛。2016年9月4日至5日G20领导人峰会在中国杭州举办。2016年G20首次财长和央行行长会议2016年2月27日在上海开幕,这是中国接任2016年G20主席国后召开的首个高级别会议。20国集团建立最初由美国等8个工业化国家的财政部长于1999年6月在德国科隆提出的,目的是防止类似亚洲金融风暴的重演,让有关国家就国际经济、货币政策举行非正式对话,以利于国际金融和货币体系的稳定。20国集团会议当时只是由各国财长和各国中央银行行长参加,自2008年由美国引发的全球金融危机使得金融体系成为全球的焦点,开始举行20国集团首脑会议,扩大各个国家的发言权,这取代之前的8国首脑会议或20国集团财长会议。中国经济网专门开设了"G20财经要闻精粹"专栏,每日报道G20各国财经要闻。20国集团的成员包括:美国、日本、德国、法国、英国、意大利、加拿大、俄罗斯、欧盟、澳大利亚、中国、南非、阿根廷、巴西、印度、印度尼西亚、墨西哥、沙特阿拉伯、土耳其、韩国。20国集团是布雷顿森林体系框架内非正式对话的一种机制,旨在推动国际金融体制改革,为有关实质问题的讨论和协商奠定广泛基础,以寻求合作并促进世界经济的稳定和持续增长。

Group quota 组限额 即按不同类别的商品分为若干组,分别规定不同的额数。

Group survival kit(GSK) 团体救生用具 团体救生用具的内容在表G-2中列举。A sample of the contents of the group survival kit is listed in the Table G-2.

表 G-2 团体救生用具
Table G-2 Group survival kit

属具 Equipment	数量 Quantity
团体用属具 Group equipment	

表 G-2

属具 Equipment	数量 Quantity
帐篷 Tents	6 人 1 顶　1 per 6 person
气垫 Air mattresses	2 人 1 个　1 per 2 person
睡袋（真空包装）Sleeping bags（vacuum packed）	2 人 1 个　1 per 2 person
炉灶 Stove	每个帐篷 1 个　1 per tent
炉灶燃料 Stove fuel	每人 0.5 升　0.5 litres per person
燃料罐 Fuel paste	每个炉灶 2 罐　2 tubes per stove
火柴 Matches	每个帐篷 2 盒　2 boxes per tent
平底锅（带密封盖）Pan（with sealing lid）	每个炉灶 1 个　1 per stove
强化健康饮用水 Fortified health drinks	每人 5 包　5 packets per person
手电筒 Flashlights	每个帐篷 1 个　1 per tent
蜡烛和蜡烛座 Candles and holders	每个帐篷 5 个　5 per tent
雪铲 Snow shovel	每个帐篷 1 把　1 per tent
雪锯、雪刀 Snow saw and snow knife	每个帐篷 1 把　1 per tent
防水油布 Tarpaulin	每个帐篷 1 块　1 per tent
脚保护——靴子（真空包装）Foot protection—booties	每个帐篷 1 双　1 per tent
备用个人用具 Spare personal equipment	每个团体救生工具箱 1 袋，可视为有关规定的 110% 的部分。（1 set per GSK container, which may be considered as part of the 110% as relevant requirement.）
SGK 箱 GSK container	1 箱
头盔（真空包装）Head protection（vacuum packed）	1 个
颈、面护具（真空包装）Neck and face protection（vacuum packed）	1 个
手保护——连指手套（真空包装）Hand protection—Mitts（vacuum packed）	1 双　1 pair

表 G-2

属具　Equipment	数量　Quantity
手保护——手套(真空包装) Hand protection—Gloves(vacuum packed)	1 双　1 pair
脚保护——短袜(真空包装) Foot protection—Socks(vacuum packed)	1 双　1 pair
脚保护——靴子(真空包装) Foot protection—Boots(vacuum packed)	1 双　1 pair
保温服(真空包装) Insulated suit(vacuum packed)	1 件
保暖内衣(真空包装) Thermal underwear(vacuum packed)	1 套　1 pair
手取暖器　Handwarners	1 套　1 set
太阳眼镜　Sunglasses	1 副
哨子　Whistle	1 个
饮水杯 Drinking mug	1 个

Grouping　编组(分组)　是一个通用的术语,系指:(1)警报操纵台上各个警报器的布置或指示操纵台上各个指示器的布置,例如,操舵装置警报器布置在驾驶室航行和操纵工作站,或门指示器布置在驾驶室安全工作站的水密门位置指示台上;(2)对警报按其功能或优先级安排。　Grouping is a generic term meaning:(1) the arrangement of individual alerts on alert panels or individual indicators on indicating panels, e. g. steering gear alerts at the workstation for navigating and manoeuvring on a navigation bridge, or door indicators on a watertight door position indicating panel at the workstation for safety on the navigation bridge;(2) arrangement of alerts in terms of their function or priority.

Growth margin　增加裕度　是一个对舰船整个寿命周期中,未来能控制或不能控制的重力增加的余量。Growth margin is an allowance for future controlled and uncontrolled weight growth during the life of the ship.

Guangzhou Shipyard International Company Limited　广船国际股份有限公司——海洋装备制造企业上市的"先驱"　1993 年 6 月 7 日,广州广船国际股份有限公司由成立于 1954 年的广州造船厂改制设立。同年在香港和上海上市,总股本为人民币 4.98 亿元,是中国第一家造船上市公司,中国船舶工业集团公司旗下华南地区重要的现代化核心造船企业,广东省 50 家重点装备制造企业和中国制造业 500 强之一,中国最大的灵便型液货船制造厂商,国家高技能人才培育示范基地,国家高新技术企业。广船国际以造船(民品及特种军辅船)为核心业务,涉及大型钢结构、船舶轴系、舵系的加工,船舶内装,防腐涂装、船舶劳务、机电产品和软件开发等,并已成功地进入滚装船、客滚船、半潜船等高技术、高附加值船舶市场。在海洋工程领域,广船国际近 5 年来供交付 4 艘 VLCC、4 艘半潜船,56 艘化学品船。在成功实现"中国第一,世界第三灵便型液货船设计和制造企业"的目标之后,面对内外部环境的巨大变化,公司于 2013 年重新确定战略目标,致力于成为全球海洋与重型装备市场技术领先、服务卓越的遗留船舶企业。

Guarantee letter　保函　系指由托运人出具的用以担保承运人签发某种提单而产生的一切法律后果的一种担保文件。

Guarantee period　保修期　是制造厂商对其产品性能担保的正向目标,保修期越长,意味着厂家负责的保修职守越重。

Guarantee repair　保修　系指产品提交用户,在保修期内提供免费修理、有偿更换零部件的服务。

Guarantee risk　保障的风险　系指保险人即保险公司承担保险的风险。

Guardian 监护者 他可以最大限度地向各缔约方施加其影响,要求他们遵守总协定,但无指令权。

Gudgeon 舵钮 系指中间有一孔能容纳舵销的垫块,位于艉柱上,对舵起支撑作用并使其可摆动。 Gudgeon means a block with a hole in the centre to receive the pintle of a rudder; located on the stern post, it supports and allows the rudder to swing.

Gudgeon pin 活塞销(杆销,十字头销,轴头销)

Guests relation desk 客服中心(豪华邮轮)

Guidance note 指导性意见 是建议,其对于授予船级并非为强制性的,而是船级社依据一般经验建议遵守的内容,因此可以由客户决定是否采用这些指导性意见。 Guidance note is advice which is not mandatory for assignment of class, but with which the Society, in right of general experience, advises compliance, hence, it is for the client to decide whether to apply the note or not.

Guide bearing 导向轴承(辅助轴承)

Guide blade 导叶(导向叶片) 主要只起改变蒸汽流向作用的速度级的第二列静叶。

Guide block 导向滑车(导块)

Guide bracket 导向支架

Guide gear 导向装置

Guide plate(baffle plate, impingement) 导流挡板 在表面式冷凝器中使蒸汽或水依一定方向流动的板。

Guide valve 导向阀

Guide vane 导流器 设置在进气装置、进气机匣及排气机匣中用来引导气流的加肋叶片、曲线环状形扩压器的总称。

Guide vane 导翼(导流叶片)

Guide vane(guide blade) 导向叶片 用来引导气流使具有一定流向的一种叶片。分进口导向叶片和出口导向叶片两种。

Guided missile destroyer 导弹驱逐舰 是以舰对舰导弹为主要武器对海上目标实施打击,兼有防空、反潜、护航等任务的多用途的水面攻击型战舰。一般排水量在3 000~7 000 吨。其主要作战任务是为大型舰队和运输船队护航。见图 G-4。

图 G-4 导弹驱逐舰
Figure G-4 Guided missile destroyer

Guided missile frigate 导弹护卫舰 是护卫舰的一种,系指装有导弹的护卫舰。它主要用于反潜和防空护航,以及侦察、警戒巡逻、布雷、支援陆军和保障陆军濒海作战等,在现代海军编队中,护卫舰是在吨位和火力上仅次于驱逐舰的水面作战舰船。两次世界大战促使护卫舰迅速发展,20 世纪 70 年代以后,护卫舰开始装备导弹和直升机,出现了导弹护卫舰和第一艘隐形护卫舰。见图 G-5。

图 G-5 导弹护卫舰
Figure G-5 Guided missile frigate

Guidelines 指南 系指对现有规范中没有包括的内容,或规范中有原则要求,但需进一步细化的内容,或需增加具体可操作性的内容,或新型船舶、设备、系统。 Guidelines includes those not covered in present rules, or the principled requirements therein which need to be further defined in details, or where specific applicability of the rules is needed, or for novel ships or equipment or systems.

Guidelines for ballast water management and development of ballast water management plan 压载水管理指南和制定压载水管理计划指南

Guidelines for Inspection of Ships 船舶检查指南

Guidelines for preparation of ship's ballast water management plan 船舶压载水管理计划编制指南

Guidelines for Safe and Environmentally Sound Ship Recycling 安全和环境无害化拆船指南

Guidelines for Survey and Certification 检验和发证指南

Guidelines for surveys for pollution prevention of garbage from ship's 防止船舶垃圾污染检验指南

Guidelines for the Authorization of Ship Recycling Facilities 拆船厂批准指南

Guidelines for the Designation of Special Area and Identification of Particularly Sensitive Sea Area(Guidelines) 指定特殊区域和确定特别敏感海区导则("导则") 系指大会在 1991 年以 A.720(17)决议通过的"导则"及其修正案。该"导则"最初目的是帮助国际海事组织及成员国政府指定、管理和保护特别敏感海区。 Guidelines for the Designation of Special Area and Identification of Particularly Sensitive Sea Area(Guidelines) means the Guidelines adopted by Assembly Resolution A.720(17) in 1991, as amended, which are primarily intended to assist IMO and Member Governments to identifying, managing, and protecting sensitive sea areas.

Guidelines for the Development of the Ship Recycling Plan 拆船计划编制指南

Guidelines for the prevention and suppression of the smuggling, psychotropic substances and precursor chemicals on ships engaged in internatonal maritime traffic 防止和制止从事国际海上运输船舶走私毒品、精神药物和前体化学品指南

Guidelines for vessel with dynamic positioning systems 配备动力定位系统船舶的指南 系指国际海事组织在1994年举行的IMO第63届海安会通过的MSC/Circ 645"Guidelines for vessel with dynamic positioning systems"指南。其目的是对动力定位系统的设计要求必须配备的设备、操作要求、试验程序和文件等做出规定,以减少动力定位作业给人员、船舶及水下作业和海洋工程施工带来的风险。由于动力定位系统最核心的问题是可靠性,级别高的动力定位系统必配有冗余设备,以保证在船舶控制系统和部分设备出现故障后不受影响而正常工作,所以IMO在其制定和出台的规范中将动力定位系统分为3个级别,对每个级别的动力定位系统设备冗余度都做出了具体的规定,并分别授予每个级别不同的附加标志。其中,符合Class I 要求的动力定位系统不需要冗余设计;符合Class II 要求的动力定位控制系统应至少配置两套独立的计算机系统,以防其中一套系统的一般设备和流程失效导致全部系统停止运行;符合Class III 要求的动力定位控制系统,则需要在Class II 的基础之上,另外安装一套备份的控制系统,如果有计算机失效或未能进入准备状态,能自动报警。

Guidelines on the prevention of stowaway incidents and the allocation of responsibilities to seek the successful resolution of stowaway cases 防止偷渡者进入和寻求成功解决偷渡案件责任分配指南

Gulf cooperation council(GCC) 海湾阿拉伯国家合作委员会 原名海湾合作委员会,简称海合会,1981年5月在阿联酋阿布扎比成立。是一个包括阿拉伯海湾地区的6个国家(即海湾六国)在内的国际组织(政府间国际组织)和贸易集团,其目标主要针对经济和社会方面。

Gulf of Aden areas 亚丁湾区域 系指红海和阿拉伯海之间的亚丁湾部分,西以拉斯西尼(北纬12°40.4′,东经43°19.6′)和胡森穆拉保(北纬12°40.4′,东经46°30.2′)之间的恒向线为界,东以拉斯阿西尔(北纬11°50′,东经51°16.69′)拉斯法尔塔克(北纬15°35′,东经52°13.8′)之间的恒向线为界。 Gulf of Aden areas mean the part of the Gulf of Aden between the Red Sea and the Arabian Sea bounded to the west by the rhumb line between Ras, st Ane(12°40.4′N,43°19.6′E) and Husn Murad(12°40.4′N,46°30.2′E) andto the eastby the rhumb line between Ras Asir(11°50′N,51°16.69′E) and the Ras Fartak(15°35′N,52°13.8′E).

Gulf Oil Corporation Limited 阿拉伯海湾石油公司

Gulfs area 海湾区域 系指位于拉斯尔哈得(北纬22°30′,东经59°48′)和拉斯阿尔法斯特(北纬25°04′,东经61°25′)之间的恒向线西北的海域。 Gulfs area means the sea area located north-west of the rhumb line between Rasal Hadd(22°30′N,59°48′E) and Ras al Fasteh(25°04′N,61°25′E).

Gun 枪(喷射器,注射器,焊枪)

Gun brass(gun bronze) 炮铜

Gun foundation 火炮基座 装设火炮运动基座。

Gun metal 炮铜(锡锌青铜)

Gun platform 火炮平台 装设火炮的平台。

Gun steel 炮钢 制造不同类型大炮的炮身、炮尾和炮闩等主要结构件用钢。加农炮、榴弹炮的炮身是主要部件。在连续射击作战中,炮管承受高膛压和高温烧蚀作用,要求钢管应具有高强度、高韧性和耐烧性能。通常使用中碳CrNiMo钢,如PCrNi1Mo和PCrNi3Mo钢,比例极限强度达550~850 MPa。薄壁迫击炮管要求质量轻,一般使用更高强度的合金钢,钢的比例极限强度要求在1 200MPa以上。

Gun tackle 双饼滑车组

Gun type burner 总装式燃烧器 将喷油器、调风器、风机、油泵、电动机、有关附件及自动控制装置等部件都总装成一体的燃烧器。

Gun welder 焊枪(焊接钳)

Gunwale 舷缘 系指船舷上缘。 Gunwale means upper edge of the ship's sides.

Gusset 角板(封槽板) 系指通常用以在两个构件的强度接合部对力起分布作用的板。 Gusset means a triangular plate, usually fitted to distribute forces at a strength connection between two structural members.

Gutter-way 甲板水沟 甲板上的流水沟。

Guy(slewing guy) 牵索 牵索索具中用以摆动吊杆或调整吊杆偏角的绳索。

Guy block 牵索滑车 用于牵索索具中的滑车。

Guy eye 牵索眼板 位于吊杆头端处,供连接牵索索具用的眼板。

Guy post 牵索柱 设置在舷侧供系固牵索索具用的短柱。

Guy tackle(rigging guy tackle) 牵索索具 牵住吊杆于某一工作位置和使吊杆改变偏角的成套索具。

Guyed tower 拉索(顺应式)塔 系指由支承在海底处的一个装置基础(或一个大型定位罐)上的细长塔及锚泊在海底的若干条拉索所组成的结构物。

GWC = Gray water control 灰水控制 系指船上所设的洗衣房、浴室、厨房、住舱房排出的污水按规定得以控制,并且设置了符合规定容积的灰水集污舱,高液位报警器符合规定能力的污水处理系统的要求。

Gypsum 石膏 一种天然的水合硫酸钙。不溶于水。装载时呈细粉末状,聚集成块。平均含水量为1%~2%。无特殊危害。该货物为不燃物或失火风险低。

Gypsum is a natural hydrated calcium sulphate. It is insoluble in water. It is loaded as a fine powder that aggregates into lumps. Average moisture content is 1% to 2%. It has no special hazards. This cargo is non-combustible or has a low fire-risk.

Gyro navigation system 陀螺导航系统 根据陀螺仪的特性和地球自转作用构成的导航仪器。航向指示器、摆式陀螺罗经、电控罗经、平台罗经以及和高精度、直主式的惯性导航系统等均属此类仪器。其能连续地给出本船舶航向或方位,平台罗经和惯性导航系统,还能提供船舶的纵横摇角,惯性导航系统和具有短期惯导作用的平台罗经更能确定船舶的位置。

Gyro-compass 陀螺罗经/电罗经 利用陀螺特性,能自动地找北跟踪地理子午线的航海仪器。其中利用重力矩作为修正力矩的称摆式罗经;利用电磁力矩作为修正力矩的称电控罗经;它为船舶提供航向基准和航向传输系统,航向精度为0.2°~0.5°。陀螺罗经通电后一般需4 h才能稳定并投入工作,在综合导航系统中罗经作为航向传感器使用。在高纬区(纬度75°以上)陀螺罗经不能正常工作,需改用方位仪。其分类如下:(1)按陀螺罗经灵敏部具有的转子个数划分,可分为单转子陀螺罗经和双转子陀螺罗经;(2)按结构和工作原理划分:可分为安修茨系列、斯伯利系列和勃朗系列;(3)按罗经施加力矩的形式划分:可分为机械摆式和电磁控制式。见图G-6。

图 G-6 陀螺罗经/电罗经
Figure G-6 Gyro-compass

Gyro-compass room 陀螺罗经室 供安置陀螺罗经的主罗经及其附属设备的房间。

Gyroscopic stabilizer 陀螺式减摇装置 利用动力陀螺产生陀螺力矩作为减摇力矩的船舶减摇装置。

Gyro-stabilized platform 陀螺稳定平台 使被稳定对象能在相互垂直的两根轴组成的平面内(一般指地理水平面,也可以是空间给定的平面)保持稳定的陀螺装置。水平精度1~4,惯性系统中的陀螺稳定平台可达秒级。

H

H class divisions H级分隔 是由符合下列要求的舱壁与甲板组成的分隔:(1)它们应以钢或其他等效的材料制造;(2)它们应有适当的防挠加强;(3)它们的构造应在2 h的标准耐火试验至结束时能防止烟及火焰通过;(4)它们应用认可的不燃材料隔热,使在下列时间内,其背火一面的平均温度,较原始温度增高不超过140 ℃,且在包括任何接头在内的任何一点的温度较原始温度增高不超过180 ℃。见表H-1。 H class divisions are those divisions formed by bulkheads and decks which comply with the following criteria:(1) they are constructed of steel or other equivalent material;(2) they are suitably stiffened;(3) they are so constructed as to be capable of preventing the passage of smoke and flame to the end of the two-hour standard fire test;(4) they are insulated with approved non-combustible materials such that the average temperature of the unexposed side will not rise more than 140 ℃ above the original temperature, nor will the temperature, at any one point, including any joint, rise more than

180 ℃ above the original temperature, within the time listed above. It can be seen in table H-1.

表 H-1　H 级分隔
Table H-1　H class divisions

H-120 级(class "H-120")	120 min
H-60 级(class "H-60")	60 min
H-0 级(class "H-0")	0 min

H class standard fire test　H 级标准耐火试验　系指按 ISO 834 耐火试验中规定的 H 级耐火试验。H class standard fire test means the H class test specified in ISO 834: Fire-resistance tests.

Habitat　栖息地

Habitual residence　惯常住所

Half angle of entrance　半进流角　设计水线前端处的切线与中线面最近的夹角。

Half beam　半梁　舷侧至舱口边之间的横梁。

Half breadth plan　半宽水线图　系指型线图中，对称于中线面一舷的各水线、甲板边线及舷墙线在俯视方向的投影图。

Half depth girder　半高底纵桁　为局部加强而加设的较低的底纵桁。

Half free sound field　半自由声场　系指既不是完全自由的声场，也不是完全的混响声场的声场。

Half tine grab　半齿抓斗　斗刃上加装短齿，适宜于抓泥沙的双腭抓斗。

Half-free space sound field　半自由空间声场　系指一半辐射空间受到限制的自由声场。

Half-siding　龙骨水平宽度　中线面与龙骨折角线之间的水平距离。

Hall's anchor　霍尔锚　较常用的一种无杆锚。因无横杆，便于收藏锚链孔中，选用为艏锚。但抓力较小，常用增大锚质量来增大锚的抓力。见图 H-1。

Hallucinogens　迷幻剂　包括麦角酸二乙胺、酶斯卡灵、二甲4-羟色胺/二甲4-羟色胺磷酸、DMT、蟾毒色胺及合成药等。Hallucinogens include lysergic acid diethylamide, mescaline, psilocin/psilobycin, DMT, bufotenine and synthetics, etc..

Halocarbon(Halon)　卤代烃

Halogen compound fire extinguishing system　卤化物灭火系统　采用"1211"，"1202"，"1301"等卤化物液体作为灭火剂的灭火系统。

Hammer forging　锤锻

Hammer guide frame　龙口　打桩架上，用以引导

图 H-1　霍尔锚
Figure H-1　Hall's anchor

打桩锤和桩的导轨。

Hammock　吊床　一边用铰链联结，另一边悬吊，或全部悬吊的床。

Hand anchor capstan　人力起锚绞盘　人力转动的起锚绞盘。

Hand control　手动调节器(手动调节，手动控制)

Hand extinguisher　手提式灭火器

Hand fire alarm device　手动火警器

Hand fire pump　手摇救火泵

Hand flag　手旗　一副带有执手棒的"O"字母旗，作为旗语用的号旗。

Hand operated　手操纵的

Hand operation　手动操作(人工操作)

Hand pump(manual pump)　手动泵　由人力驱动的小型泵。

Hand rail(handhold)　扶手(栏杆,把柄)　装设在围壁、舷墙或梯子上，供人员扶靠用的硬木、钢管或其他材料制成的长条杆件。

Hand rail bracket　扶手支架(栏杆支架)

Hand rail stanchion　栏杆柱

Hand regulation　人工调节

Hand release　手动释放装置(手动投掷装置)

Hand reset　人工复原(人工重调)

Hand reversing gear　手动倒顺车机

Hand trip gear　手动停车装置　系指在应急情况下能人工迅速切断燃料供应的装置。Hand trip gear means a device capable of shutting of the fuel supply in an emergency.

Hand turning gear 手动转轴机

Hand vice 手钳

Hand wheel type skylight controlling gear 手轮式天窗传动装置

Hand windlass 人力起锚机 人力转动的起锚机。

Hand-cleaning centrifuge 人工去污油分离机 在停机的条件下,用人力方法,将分离筒内积聚的残渣在油分离机。

Hand-grasp area 伸手可及区域

Handhold stanchion 扶手支柱

Hand-hydraulic steering gear 人力液压操舵装置 通过液压系统传动的人力操舵装置。

Handle 柄(把手,搬运,处理)

Handle controlling gear for sliding door 水密滑门手动装置 用人力启闭水密滑门的传动装置。

Handling appliance(equipment) 装卸设备(搬运设备)

Handling facilities 操纵设备(航运设备)

Handling facility 操纵灵活性

Handling of fish catch 渔获物处理 渔船上各种对渔获物处理方法,包括渔获物保鲜、过船和鱼品加工的统称。

Handling of the ship 船舶的管理 系指船舶、机器装置和设备方面应适当地配备人员和适当地管理。这也包括货物的装运、压载和燃料的分配,以及在恶劣气候下掌握航速和航行,应观察与该船的使用有关的船级条件。 Handling of the ship means that the ship, machinery installations and equipment are to be adequately manned and competently handled. This also includes stowage of cargo, distribution of ballast and bunkers, and speed and navigation under heavy weather. Class conditions regarding the use of the ship are to be observed.

Handling, storage, and delivery 处理、贮存、发送 对质量管理体系而言,系指:(1)制造厂商应制定和维持一套制度,以识别、保存、分类和处理各种材料,从材料进厂直到整个生产过程,这一制度应该包括防止滥用、误用、损伤或变质的处理方法;(2)对于暂时不用的材料,应该提供一个用以隔离和保护这些材料的贮藏处所或房屋。为了防止材料变质,在早期就应该进行定期评估;(3)制造厂商应该做出安排,保护其产品在运输中的质量,制造厂商应尽可能保证产品能安全到达接收地点并随时可以识别。 For quality management system, handling, storage, and delivery means: (1) the manufacturer is to establish and maintain a system for the identification, preservation, segregation and handling of all material from the time of receipt through the entire production process. The system is to include methods of handling that prevent abuse, misuse, damage or deterioration; (2) secure storage areas or rooms are to be provided to isolate and protect material pending use. To detect deterioration, at an early stage, the condition of material is to be periodically assessed; (3) manufacturer is to arrange for the protection of the quality of his product during transit. The manufacturer is to ensure, so far as it is practicable, the safe arrival and ready identification of the product at destination.

Handoff 跨区切换 系指在目前的无线移动网络中,严重的带宽限制已经不能满足人们越来越多的需求,迫使设计者将服务区域分为能够重复使用无线频谱的蜂窝,而且蜂窝还有越来越小的趋势。当移动台从一个蜂窝移动到另一个蜂窝时,为保持通话的连续性,呼叫需要从一个基站切换到另一个基站。蜂窝越小,切换的平均次数将越多。通信系统的不同业务对切换的要求不一样,切换时要区别对待,这样也加大切换技术的复杂性。

Hand-operated 人力操纵的(人工驱动的,手动的)

Hand-operated door(hand-operated gate) 手动启闭门

Hand-operated mechanism 手动机械装置

Hand-power steering gear 人力操舵装置 人力驱动的操舵装置。

Hand-wheel type skylight controlling gear 手轮式天窗传动装置 利用操纵手轮和传动部件启闭天窗盖的天窗传动装置。

Handy lifting hook 手拉葫芦

Handy maximum/Handy size carriers 灵便型船舶 系指 DWT 20 000 ~ 50 000 t 的散货船或油船。

HANDYMAX oil tanker 灵便型油船(HANDYMAX) 系指载重量为 40 000 ~ 50 000 t 的油船。

Hangar deck 机库甲板 设置机库并用来检修直升机用的甲板。

Hanging spittoon 挂式痰盂 挂在床边,供旅客晕船时呕吐用的盛器。

Harbor 港口 多指海港港口。即具有水陆联运设备和条件,供船舶安全进出和停泊的运输枢纽。是水陆交通的集结点和枢纽,工农业产品和外贸进出口物资的集散地,船舶停泊、装卸货物、上下旅客、补充给养的场所。由于港口是联系内陆腹地和海洋运输(国际航空运输)的一个天然界面,因此,人们也把港口作为国际物流的一个特殊结点。

Harbor and emergency generator room 停泊和应

急发电机间

Harbor dues 海损 船舶和货物等在海上运输中遭遇自然灾害、意外事故或其他特殊情况，为了解除共同危险而采取合理措施所引起的特殊损失（即牺牲）和合理的额外费用。海上运输中，由于自然灾害或意外事故引起的船舶或货物的任何损失，如船舶因触礁、搁浅、碰撞、沉没、火灾、风灾、爆炸等造成船舶或货物的物质损失及费用损失等，均属海损。

Harbor generator 停泊发电机

Harbor pump 停泊（用）泵

Harbor tug 港作拖船 系指在港区内协助大型船舶，如拖带客船、货船、工程船等进出港口、靠离码头、移泊和拖曳港驳等作业用的拖船。修造船厂用于协助船舶进出坞的拖船亦属于此类。

Harbour boat 港务船 各种专门从事港务工作的船的统称。

Harbour research ship 港湾调查船 用于对河口、港湾及沿岸进行水文、地质、地形测量的小型调查船。船上设有海洋水文、海洋化学、海洋物理等实验室，由于港湾调查船长度较小，船上仅有很小的试验室进行记录测试，大量的整理工作需在岸上进行。

Harbour survey 港湾调查 对沿岸港湾水域进行的科学调查。有关港湾水文、地质、地貌、沉淀物、化学、生物和环境污染等。

Hard chine 尖舭部 呈尖角的舭部。

Hard coating 硬涂层 系指在固化过程中，发生化学变化的涂层或非化学变化，在空气中干燥的涂层。硬涂层可用于维护目的。类型可以是无机的也可以是有机的。 Hard coating is a coating that chemically converts during its curing process or a non-convertible air drying coating which may be used for maintenance purpose. It can be either inorganic or organic.

Hard corners 硬角 系指船体横剖面上刚度较大的构成单元主要按弹塑性失效模式损坏（材料屈服）。这些单元一般由两块不共面的板组成。舭部、舷顶列板—甲板边板单元，纵桁—甲板的连接、大桁材的腹板—面板的连接都是典型的硬角单元。 Hard corners are sturdier elements composing the hull girder transverse section, which collapse mainly according to an elasto-plastic mode of failure(material yielding). These elements are generally constituted of two plates not lying in the same plane. Bilge, sheer strake-deck stringer elements, girder-deck connections and face plate-web connections on large girders are typical hard corners.

Hard disk drive(HDD) 硬盘驱动器 简称硬盘，是电脑上使用坚硬的旋转盘片为基础的非易失性（Non-volatile）存储设备，由一个或者多个铝制或者玻璃制的碟片组成。电脑中常说的电脑硬盘 C 盘，D 盘为磁盘分区都属于硬盘驱动器。这些碟片外覆盖有铁磁性材料。绝大多数硬盘都是固定硬盘，被永久性地密封固定在硬盘驱动器中。不过，现在可移动硬盘越来越普及，种类也越来越多。它在平整的磁性表面存储和检索数字数据。信息通过离磁性表面很近的写头，由电磁流来改变极性方式被电磁流写到磁盘上。由于其体积小、容量大、速度快、使用方便，已成为 PC 的标准配置。硬盘有固态硬盘（SSD 盘，新式硬盘）、机械硬盘（HDD，传统硬盘）、混合硬盘（HHD，一块基于传统机械硬盘诞生出来的新硬盘）。SSD 采用闪存颗粒来存储，HDD 采用磁性碟片来存储，混合硬盘（HHD：Hybrid hard disk）是把磁性硬盘和闪存集成到一起的一种硬盘。绝大多数硬盘都是固定硬盘，被永久性地密封固定在硬盘驱动器中。目前硬盘一般常见的磁盘容量为 80G、128G、160G、256G、320G、500G、750G、1TB、2TB 等等。硬盘按体积大小可分为 3.5 寸、2.5 寸、1.8 寸等；按转数可分为 5400 r/min/7200 r/min/10000 r/min 等；按接口可分为 PATA、SATA、SCSI 等。PATA、SATA 一般为桌面级应用，容量大，价格相对较低，适合家用；而 SCSI 一般为服务器、工作站等高端应用，容量相对较小，价格较贵，但是性能较好，稳定性也较高。

Hard disk Silencer 硬盘消音器 是一款为硬盘运行降低噪声的绿色免费软件。它通过设置硬盘寄存器设置磁头移动及磁头停靠来在几乎不影响性能的前提下显著降低硬盘的噪声。

Hard scrape 重铲

Hard tank 硬舱 系指在柱体式平台结构中，提供平台所需浮力结构部分。又称浮力舱。

Harden 硬化（凝固，淬火）

Harden quenching 硬淬

Hardenability test 淬硬性试验

Hardened and tempered steel 调质钢 是机械制造业中应用十分广泛的重要材料之一。所谓调质钢，一般系指含碳量在 0.3% ~ 0.6% 的中碳钢。一般用这类钢制作的零件要求具有很好的综合力学性能，即在保持较高的强度的同时又具有很好的塑性和韧性，人们往往使用调制处理来达到这个目的，所以人们习惯上就把这一类钢称作调质钢。各类机器上的结构零件大量采用调质钢，是结构钢中使用最广泛的一类钢。

Hardened area(hardened zone) 淬硬区（硬化区，硬化层）

Hardened steel 淬硬钢（硬化钢）

Hardener component 固化剂

Hardness index 硬度指数

Hardness limit 硬度极限

Hardness measurement 硬度测量

Hardness number 硬度数

Hardness ratio factor 工作硬化系数

Hardness resin 固体树脂

Hardness scale 硬度标准(硬度计)

Hardness survey 硬度测算

Hardness test 硬度试验

Hardness tester 硬度测试器

Hardness testing-machine 硬度试验仪(回跳硬度计,硬度试验机)

Hardover angle 满舵舵角 满舵时,舵叶偏离正舵位的角度。

Hard-over angle 最大转舵角 按舵机性能可以转动舵叶偏离正舵位的最大角度。

Hardover rudder stopper 舵叶舵角限位器 设置在舵叶和舵柱上部,用以防止舵叶转动角度过大的装置。其限制舵角应比甲板舵角限位器所限的舵角稍大。

Hardware 硬件 是计算机系统中各种设备的总称。计算机的硬件应包括5个基本部分,即运算器、控制器、存储器、输入设备和输出设备。

Hardwood 硬木

Harmful aquatic organism 有害水生物

Harmful aquatic organisms and pathogens 有害水生物和病原体 (1)就"压载水公约"而言,系指如被引入海洋,包括河口,或引入淡水水道则可能危害环境、人体健康、财产或资源、损害生物多样性或妨碍此种区域的其他合法利用的有害水生物和病原体;(2)水生物或病原体,如果被扩散到海洋、海湾或淡水河道,将产生有害于人体健康,损害生存资源和水中生命,毁坏身体的变化以及影响到海洋的其他合法利用。 (1) For Ballast Water Management Convention, harmful aquatic organisms and pathogens means aquatic organisms or pathogens which, if introduced into the sea including estuaries, or into fresh water courses, may create hazards to the environment, human health, property or resources, impair biological diversity or interfere with other legitimate uses of such areas; (2) aquatic organisms or pathogens which, if introduced into the sea including estuaries, or into fresh water courses, may create hazards to the environment, human health, property or resources, impair biological diversity or interfere with other legitimate uses of such areas.

Harmful damage 有害损伤

Harmful defect 有害缺陷

Harmful gas and evaporation gas 有毒气体和蒸发气 主要系指:(1)一氧化碳;(2)原油、汽油和苯的蒸发气;(3)若干种消毒、清洁和油漆稀释液的蒸发气;(4)由装运的物品或其配料逸散出的气体和蒸发气;(5)甲烷。这些物质污染室内空气,如果空气污染的浓度超过了允许的限度,就应考虑对健康的危害,例如当污脏的压载水与植物、动物和油类的残余物混合时就会产生甲烷。

Harmful liquid substance 有毒液体物质 系指在"IBC"规则第17章或第18章的污染栏中所指明的列为X、Y或Z类的任何物质。

Harmful substance 有害物质/有毒物质 系指:(1)任何进入海洋后易于危害人类健康、伤害生物资源和海洋生物、损害休息环境或妨碍对海洋其他合法利用的物质,并包括受国际防止船舶造成污染公约控制的任何物质;(2)对"73/78防污染公约"附则Ⅲ而言,系指那些在《国际海运危险货物规则》(IMDG规则)确定为海洋污染物的物质,对以前曾载运过有害物质的空容器,除已采取足够的预防措施确保其已无危害海洋环境的残余物外,应视其本身为有害物质。符合下列任何一种识别标志的物质均为有害物质:①急性1-96hrLC$_{50}$(对鱼类)≤1mg/L 和/或 48hrEC$_{50}$(对甲壳动物)≤1mg/L 和/或 72 or 96ErC$_{50}$(对海藻或其他水生植物)≤1mg/L;②慢性1-96hrLC50(对鱼类)≤1mg/L 和/或 48hrEC$_{50}$(对甲壳动物)≤1mg/L 和/或 72 or 96ErC$_{50}$(对海藻或其他水生植物)≤1mg/L,且该物质不能很快降解和/或 log K_{OW} ≥ 4(除非经实验确定 BCF < 500);③慢性2-96hrLC$_{50}$(对鱼类)≥1≤10mg/L 和/或 48hrEC$_{50}$(对甲壳动物)≥1≤10mg/L 和/或 48hrEC$_{50}$(对甲壳动物)≥1≤10mg/L 和/或 72 or 96ErC$_{50}$(对海藻或其他水生植物)≤1mg/L,且该物质不能很快降解和/或 log K_{OW} ≥4(除非经实验确定 BCF < 500),除非慢性毒性 NOEC > 1mg/L。

Harmful substance means:(1) any substance which, if introduced into the sea, is liable to create hazards to human health, to harm living resources and marine life, to damage amenities or to interfere with other legitimate uses of the sea, and includes any substance subject to control by the present Convention;(2) For "MARPOL73/78" Annex Ⅲ, "harmful substances" are those substances which are identified as marine pollution in the International Maritime Dangerous Goods Code(IMDG Code), emply packagings which have been used previously for the caggiage of harmful substances shall themselves be treated as harmful substances unless adequate precautions have been tanken to ensure that they contain no residue that is harmful to the marine environment. Substances identified by any one of the following criteria are harmful substances: ① Acute 1-96hrLC$_{50}$(for fish)≤1mg/L and/or

$48hrEC_{50}$ (for crustaces) $\leqslant 1mg/L$ and/or 72 or $96ErC_{50}$ (for algae or other aquatic plants) $\leqslant 1mg/L$; ② Chronic 1-$96hrLC_{50}$ (for fish) $\leqslant 1mg/L$ and/or $48hrEC_{50}$ (for crustaces) $\leqslant 1mg/L$ and/or 72 or $96ErC_{50}$ (for algae or other aquatic plants) $\leqslant 1mg/L$ and the substance is not rapidly degradable and/or the log $K_{OW} \geqslant 4$ (unless the experimentally determined BCF <500); ③Chronic 2-$96hrLC_{50}$ (for fish) $\geqslant 1 \leqslant 10mg/L$ and/or $48hrEC_{50}$ (for crustaces) $\geqslant 1 \leqslant 10mg/L$ and/or 72 or $96ErC_{50}$ (for algae or other aquatic plants) $\geqslant 1 \leqslant 10mg/L$ and the substance is not rapidly degradable and/or the log $K_{OW} \geqslant 4$ (unless the experimentally determined BCF <500); unless the chronic toxicity NOEC >1mg/L.

Harmful substances in packaged form 包装形式的有害物质 系指那些在《国际海运危险货物规则》(IMDG 规则)中第1(b)分节中确定为海洋污染物的物质。

Harmful substances in packaged form referred to in Subparagraph 1(b) of this article means substances which are identified as marine pollutants in the International Maritime Dangerous Goods Code (IMDG Code).

Harmful to the marine environment in relation to the discharge of 对海洋环境有害并与排放有关的(货物残余物) (1)货物残余物系指按联合国化学品全球标记和协调制度(UN GHS)标准分类的、满足下列特性的散装固体物质的残余物:(ⅰ)1类急性水生物毒性残余物;和/或(ⅱ)1类或2类慢性水生物毒性残余物;和/或(ⅲ)1A或1B类不能迅速降解的和具有生物体内积累度高的致癌性残余物;和/或(ⅳ)1A或1B类不能迅速降解的和具有生物体内积累度高的诱变残余物;和/或(ⅴ)1A或1B类不能迅速降解的和具有生物体内积累度高的再生毒性残余物;和/或(ⅵ)1类重复暴露的特定目标器官毒性的和具有生物体内积累度高的残余物;和/或(ⅶ)含有或由合成聚合物、橡胶、塑料或塑料给料颗粒(这包括可切碎的、磨碎的、斩碎或浸渍的或类似的材料)组成的散装固体货物。(2)清洁剂或添加剂系指下列的清洁剂或添加剂:(ⅰ)符合MARPOL附则Ⅲ中标准的"有毒物质";(ⅱ)任何含有已知的能诱变致癌或毒性的合成物。对海洋环境有害定义的注:1.这些标准是基于联合国化学品全球标记和协调制度(UN GHS)标准2011年修正案第4版。对于特殊的产品(例如金属和无机金属合成物),联合国化学品全球标记和协调制度附则8和10中的指南是正确解释该标准和分类的依据,且应予遵守。2.这些是按致癌性、诱变性、再生毒性或重复曝露的特定目标器官毒性 对口腔有害的、皮肤有害的或无曝露途径的技术规格的有害程度来分类的产品。

Harmful to the marine environment in relation to the discharge of:(1) Cargo residues means residues of solid bulk substances which are classified according to the criteria of the United National Globally Harmonized System for Classification and Labeling of Chemicals (UN GHS) meeting the following parameters;(ⅰ) Acute Aquatic Toxicity Category 1, and/or;(ⅱ) Chronic Aquatic Toxicity Category 1, and/or;(ⅲ) Carcinogenicity Category 1A or 1B combined with not being bioaccumulation; and/or (ⅳ) Mutagenicity Category 1A or 1B combined with not being rapidly degradable and having high bioaccumulation; and/or(ⅴ) Reproductive Toxicity Category 1A or 1B combined with not being rapidly degradable and having high bioaccumulation; and/or(ⅵ) Specific Target Organ Toxicity Repeated Exposure Category 1 combined with not being rapidly degradable and having high bioaccumulation; and/or(ⅶ) Solid bulk cargoes containing or consisting of synthetic polymers, rubber, plastics, or plastic feedstock pellets (this includes materials that are shredded, milled, chopped, or macerated or similar materials.) (2) Cleaning agents or additives means a cleaning agent or additive that is:(ⅰ) A "harmful substance" in accordance with the criteria in MARPOL Annex Ⅲ, and/or (ⅱ) Contains any components which are known to be carcinogenic, mutagenic or reprotoxic. Notes to definition of harmful to the marine environment:1. These criteria are based on UN GHS fourth revised edition (2011). For specific products (e.g. metals, and inorganic metal compounds) guidance available in UN GHS, annexes 9 and 10 is essential for proper interpretation of the criteria and classification and should be followed. 2. These are products with a hazard statement classification for Careinogenicity, Mutagenicity, Reproductive Toxicity, or Specific "Target Organ Toxicity Repeated Exposure for oral hazards, dermal hazards or without specification of the exposure route."

Harmonic 谐波(谐波的,谐函数)
Harmonic cancellation 谐波消除
Harmonic excitation 谐波激励(谐波励磁)
Harmonic frequency 谐频(谐波频率) 系指周期波中频率为基频频率整数倍的频率分量,也可称为谐音。
Harmonic motion 谐和运动(简谐运动)
Harmonic stress function 应力调和函数
Harmonic vibration 谐振动(谐波振动)
Harmonization 谐和(谐波),调整)
Harness 硬度 衡量材料对利用球形或角锥形压头进行压痕试验的承受能力(洛氏硬度、维氏角锥硬度、布氏硬度)。硬度大概与材料的屈服应力相关,因此可采用局部测量屈服应力的非破坏性方法。 Harness is

the measure of the resistance of a material to indention by a spherical or pyramidal indenter(Rockwell, Vickers, Brinell scales). Hardness can be approximately related to the material yield stress and thus is a non-destructive method of measuring yield stress locally.

Harpoon gun platform 鲸炮台 在捕鲸船首部,专为安装捕鲸炮而设置的加强台座。

Hashtag 主题标签 系指发"微博"和推特时标注的主题。

Hatch batten cleat(coaming cleat, batten cleat) 封舱楔耳 供封舱楔楔入的金属耳板。

Hatch battening arrangement 封舱装置 大舱盖上,用以保证舱口密性的所有部件的总称。

Hatch beam 舱口梁 泛指架设于舱口上的活动梁,尤指沿舱口横向架设的活动梁。

Hatch beam carrier 舱口梁承座 设置在舱口围板上,直接支承舱口梁的托座。

Hatch coaming 舱口围板 系指与船用甲板垂直的围绕在舱口周围的钢板,其作用是防止水进入货舱并减少甲板工作人员从敞开的舱口摔下去的危险性。

Hatch cover 舱口盖 货舱的关闭设备。它具有保证舱内货物安全、船体水密的作用。舱口盖上有时还堆放一些货物,因此它还具有抵抗货物压力的能力。可分为:移动式舱口盖(portable hatch cover)、机械开启式舱口盖(mechanical hatch cover)、移动式浮筒舱舱口盖(portable pontoon hatch cover)、折叠型舱口盖(fold type hatch cover)、侧移型舱口盖(sliding type hatch cover)、单拉折叠型折叠(single-pull type hatch cover)、小舱口盖(small access hatch cover)以及油密舱口盖(oil-tight hatch cover)等。

Hatch cover(handling)winch 舱口盖绞车 启闭船舶货舱舱口盖的绞车。

Hatch cover controlling gear(hatch cover lifting and locking gear) 舱盖启闭装置 机械舱盖中,用以启闭舱盖板的机械设备及其附件的总称。

Hatch cover gasket 舱盖填料 填在舱盖板周缘,使舱盖板和邻贴部件之间达到密性要求的填料。

Hatch cover jacking device 舱盖起升装置 机械舱盖中,将装设滚轮的舱盖板抬高,以便使舱盖启闭滚移的专门机构。

Hatch cover roll 舱盖滚轮 装在舱盖板上的滚轮。

Hatch end beam 舱口端横梁 舱口前后端的强横梁。

Hatch ladder 舱口梯 设置于船上出入舱口处的梯。

Hatch side girder 舱口纵桁 沿舱口边设置的纵桁。

Hatch ways 舱口 系指:(1)船舶甲板上的开口,一般为矩形,作为向下通入舱室的出入口。(2)船舶甲板上的用于装卸货物时进入货物处所的开孔。 Hatch ways mean:(1)a openings, generally rectangular, in a ship's deck affording access into the compartment below. (2)an aperture in a deck of a ship providing access to a cargo space for loading or unloading.

Hatchboard 舱盖板 大舱盖中用以盖闭舱口的板件或组合构件的统称。

Hatchcover roll 舱盖滚轮 装在舱盖板上的滚轮。

Having low flame-spread characteristics 具有缓慢的火焰扩散的特性 系指《制造方法及类型认证等相关标准》所规定的通过下列试验合格的可燃性材料:(1)火焰传播性试验;(2)发烟性试验;(3)有毒性气体试验。 Having low flame-spread characteristics means those combustible materials which passed the following tests in accordance with the requirements specified in the Guidance for Approval of Manufacturing Process and Type Approval, Etc.(1)Flame propagation test;(2)Smoking test;(3)Toxic gas test.

Hawse pipe 锚链筒 锚索或锚缆通过的钢管,在船首位于艏柱任一侧,也称为锚链管。 Hawse pipe is a steel pipe through which the hawser or cable of anchor passes, located in the ship's bow on either side of the stem, also known as spurling pipe.

Hawser store 帆缆舱 船上供贮存帆布、缆绳等的房间。

Hawsers 缆索 系指船用缆绳。缆索用得最普遍的材料是聚酯、聚乙烯、尼龙(聚酰胺)。在3种材料中,聚酯的抗疲劳性能最好,当浸在水中时它不会降低强度。但是其弹性最差,而且它不像尼龙具有浮力。聚乙烯仅仅是具有自浮能力并不需要除去浮子情况下进行检查的材料,但是它在3种纤维中耐磨性能最弱和最差的。聚乙烯的弹性比尼龙差,而且曝露在阳光下其强度会降级,但是它使缆索容易进入绞车,所以它经常用作系浮筒索。尼龙具有高强度、良好的耐磨性以及良好的弹性,但是其抗疲劳性能比其他两种纤维差,而且当处于湿态时其破断强度会降低。尼龙不具有浮性,如果要求能浮起则需要浮箱或浮子。 Hawsers are marine ropes. The most common materials used for hawsers are polyester, polypropylene, and nylon(polyamide). Polyester offers the best fatigue properties of the three materials, and it does not lose strength when submerged in water. But it has the lowest elastic properties, and it is not buoyant as with nylon. Polypropylene is the only material that is self-floating and can be inspected without having floaters removed, but it is the weakest and least abrasion resistant of the three fibers. Polypropylene is less elastic than nylon and degrades in strength when exposed to sunlight, but it makes winching-in of the hawser easy so that it is often used for mooring pick-up ropes. Nylon offers high strength and good abrasion resistance together with good elastic properties, but it is less fatigue resistance than the other two fibers, and its breaking strength can be reduced when it is wet. Nylon is not buoyant and needs buoyancy tanks or floaters if it is required to float.

Hazard 危险 系指具有引起潜在的伤害、死亡、环境破坏和/或财产损失的事件。根据 IMO 的规定，对人类生命、健康、财产或环境具有潜在危险的事件举例如下：(1) 对船舶造成的外部危险—风暴、闪电、低的可见度、海图上未标明的水下物体、其他船舶、战争和罢工；(2) 对船上造成的危险—①在舱室区域：可燃的家具、储存室内储存的清洁剂、在厨房设备中的油/脂肪；②在甲板区域：货物、由于油漆/油/黄油/水造成的光滑甲板、舱口盖和电气设备连接件；③在机器处所：电缆、主机的燃油和柴油、锅炉、燃油管系和阀、含油舱底水和冷冻剂；④点火源：电器设备、热表面、从热工产生的或烟囱废气、海上在甲板上工作、货物作业、液舱检验和船上修理。 Hazard is an event with the potential to cause injuries, fatalities, environmental damage, and/or property damage. According to IMO, examples of an event with potential to threaten human life, health, property, or the environment are as follows: (1) hazards external to the vessel: storms, lightning, poor visibility, uncharted submerged objects, other ships, war, and sabotage; (2) hazards on board a vessel: ①in accommodation areas—combustible furnishings, cleaning material in stores, and oil/fat in galley equipment; ②in deck areas—cargo, slippery deck due to paint/oils/grease/water, hatch covers, and electrical connections; ③in machinery spaces—cabling, fuel and diesel oil for engines, boilers, fuel oil piping and valves, oily bilge, and refrigerants; ④ sources of ignition—naked flame, electrical appliances, hot surface, sparks from hot work or funnel exhaust, and deck and engine room machinery.

Hazard and operability studies 危险和可操作性分析 识别可能导致严重后果的系统偏差及产生原因；确定减少这类偏差可能发生频率和产生后果的措施。

Hazard identification 危险确定/危险识别 是在设计阶段所有危险分类确定的过程。 Hazard identification is the process whereby all hazards identified at the design stage are catalogued.

Hazard rating 危害级别 对"73/78 防污染公约"附则Ⅲ而言，系指标示于海运包装有害物质上的标志或标签牌上的表明该物质对海洋污染危害程度的符号。

Hazard situation 危险状态 一种实际的具有潜在威胁人的生命、健康、财产或环境的状态。

Hazardous and noxious chemical 危险化学品或有毒化学品 系指对人员、船舶及环境(水、海洋、空气)产生危害性的化学品。

Hazardous area 0 0 类危险区域 系指持续存在或长时间存在爆炸性气体环境或闪点低于 60 ℃ 的易燃气体的区域。该区域包括：气体柜的内部、用于气体柜压力释放或其他透气系统的任何管路、内含其他的管路和设备。 Hazardous area 0 is an area in which an explosive gas atmosphere or a flammable gas with a flashpoint below 60 ℃ is present continuously or is present for long periods. This zone includes: the interiors of gas tanks, any pipework of pressing-relief or other venting systems for gas tanks, pipes and equipment containing gas.

Hazardous area 1 1 类危险区域 系指在正常存在情况下，可能出现爆炸性气体环境或闪点低于 60 ℃ 的易燃气体的区域。该区域包括：(1) 燃料储舱室；(2) 气体压缩机舱，按有关要求设有通风装置；(3) 距离任何其他柜出口、气体或蒸发气出口、燃料总管阀门、其他气体阀气体管法兰、气体燃料泵舱出口和为让温度变化产生的少量气体或蒸发气混合物流动而设置的气体柜压力释放口 3 m 以内的开敞甲板上的区域或甲板上的半围蔽处所；(4) 距离气体压缩机舱和泵舱入口、气体燃料泵和压缩机舱通风进口以及通往 1 类区处所的其他开口 1.5 m 的开敞甲板上的区域或甲板上的半围蔽处所；(5) 开敞甲板上的包括其他燃料总管阀门的防溢挡板以内及挡板向外 3 m，并不高于甲板以上 2.4 m 的处所；(6) 气体管路所在的围蔽或半围蔽处所，例如其他管路周围的管道、半围蔽的燃料充装站；(7) 在正常操作情况下，ESD 保护的机器处所视为非危险区域，但当出现其他泄漏时，该处所变为 1 类区域。 Hazardous area is an area in which an explosive gas atmosphere or a flammable gas with a flashpoint below 60 ℃ is likely to occur in normal operation. This zone includes: (1) tank room; (2) gas compressor room arranged with ventilation according to relevant requirement; (3) areas on open deck, or semi-enclosed spaces on deck, within 3 m of any gas tank outlet, gas or vapour outlet, bunker manifold valve, other gas valve, gas pipe flange, gas pump room ventilation outlets and gas tank openings for pressure release provided to permit the flow of small volumes of gas or vapour mixtures caused by thermal variation; (4) areas on open deck or semi-enclosed spaces on deck, within 1.5 m of gas compressor and pump room entrances, gas pump and compressor room ventilation inlets and other openings into zone 1 spaces; (5) areas on the open deck within spillage coamings surrounding gas bunker manifold valves and 3 m beyond these, up to a height of 2.4 m above the deck; (6) enclosed or semi-enclosed spaces in which pipes containing gas are located, e.g., ducts around gas pipes, semi-enclosed bunkering stations; (7) the ESD-protected machinery spaces is considered as non-hazardous area during normal operation, but changes to zone 1 in the event of gas leakage.

Hazardous area 2 2 类危险区域 系指在正常存在情况下，不大可能出现爆炸性气体环境或闪点低于 60 ℃ 的易燃气体的区域。即使出现，也可能仅偶然发生并且存在的时间短。该区域包括：距离 1 类的开敞或半围蔽处所 1.5 m 以内的区域。 Hazardous area 2 is an area in which an explosive gas atmosphere or a flammable gas with a flashpoint below 60 ℃ is not likely to occur in normal operation and if it does occur, is likely to do so only infrequently and will exist for a short period only. This zone includes: areas within 1.5 m surrounding open or semi-enclosed spaces of zone 1.

Hazardous area classification 危险区域分级 是一种用来对可能出现爆炸性气体环境的区域进行分析和分类的方法。分级的目的是为了选择能够在这些区域内安全操作的电气设备。 Hazardous area classification is a method of analyzing and classifying the areas where explosive gas atmospheres may occur. The object of the classification is to allow the selection of electrical apparatus able to be operated safely in these areas.

Hazardous areas 危险区域 对载运危险货物船舶而言，系指在正常工作状态下可能出现爆炸性环境的区域或处所。例如：(1) 装运有包装的 1 类爆炸品的下列处所或区域属于危险区：①闭式装货处所和闭式或敞开式滚装装货处所；②固定安装的容器（例如：弹药箱）；(2) 载运只产生爆炸性粉尘环境的散装固体危险货物的下列处所或区域属于危险区：①闭式装货处所；②危险区域的通风管道（如设有）；③由井口直接通向 ①或②所列处所的围蔽或半围蔽处所，并无防止易燃粉尘进入该处所的适当措施；(3) 载运闪点低于 23 ℃ 有包装的易燃液体和易燃气体的下列区域或处所：①闭式装货处所和闭式滚装装货处所；②危险区域的通风管道（如设有）；③离危险区任何排风口周围 1.5 m 范围内的露天甲板区域，或者露天甲板上的半围蔽处所；④有开口直接通向以上①和②所列处所的围蔽或半围蔽处所，并无防止易燃气体进入该处所的适当措施；(4) 只载运散装固体危险货物和 MHB 的下列处所或区域属于危险区：①闭式装货处所和闭式滚装装货处所；②危险区域的通风管道（如设有）；③离危险区任何排风口周围 1.5 m 范围内的露天甲板区域，或者露天甲板上的半围蔽处所；④有开口直接通向以上①和②所列处所的围蔽或半围蔽处所，并无防止易燃气体进入该处所的适当措施。 For ship carrying dangerous goods, hazardous areas mean those zones and spaces in which an explosive atmosphere is likely to occur in normal operation. Such as: (1) the following spaces or areas carried explosive substances in packaged form, conforming to class 1: ①enclosed cargo spaces, and enclosed or open Ro-Ro cargo spaces; ②permanently fixed containers (e.g. magazines); (2) the following spaces or areas carried solid dangerous goods in bulk which may develop explosive dust only: ①enclosed cargo spaces; ②ventilation ducts (if any) for hazardous areas; ③enclosed or semi-enclosed spaces with direct openings to ①or ②mentioned above, without appropriate measures to prevent inflammable dust from entering in; (3) the following spaces or areas carried inflammable liquids with a flashpoint below 23 ℃ in packaged form: ①enclosed cargo spaces or enclosed Ro-Ro cargo spaces; ②ventilation ducts (if any) for hazardous areas; ③weather deck areas, or semi-enclosed space so on weather deck round 1.5 m from any draft outlet of hazardous areas; ④enclosed or semi-enclosed spaces with direct openings to ①or ②mentioned above, without appropriate measures to prevent inflammable dust from entering in; (4) the following spaces or areas carried only dangerous solid goods in bulk and MHB: ①enclosed cargo spaces or enclosed Ro-Ro cargo spaces; ②ventilation ducts (if any) for hazardous areas; ③weather deck areas, or semi-enclosed space so on weather deck round 1.5 m from any draft outlet of hazardous areas; ④enclosed or semi-enclosed spaces with direct openings to ①or ②mentioned above, without appropriate measures to prevent inflammable dust from entering in.

Hazardous areas 危险区域（平台） 就机电设备而言，系指危险区域中的 0 类危险区、危险区域中的 1 类危险区、危险区域中的 2 类危险区。 For machinery and electrical installations, hazardous areas are hazardous areas zone 0, hazardous areas zone 1 and hazardous areas zone 2.

Hazardous areas 危险区域（电气设备） 对电气设备而言，系指存在爆炸性气体、或预料由于存在蒸发气体、可燃粉尘或易爆物品以致特别需要对结构、安装和电气器具的使用特别预先注意的区域。危险区域的定级是基于爆炸性气体发生的频度和持续时间。有爆炸性气体的危险区域按下列规定定级：(1) 0 区域：爆炸性气体连续或长期存在的区域;(2)1 区域:爆炸性气体在正常工作时可能出现的区域;(3)2 区域:在正常工作情况下不可能出现爆炸性气体,以及如果出现,可能为不频繁并且仅短期存在的区域。　For electrical apparatus, hazardous areas are those areas in which an explosive atmosphere is or may be expected to be present in quantities such as to require special precautions for the construction, installation and use of electrical apparatus. Hazardous areas are classified in zones based upon the frequency and the duration of the occurrence of explosive atmosphere. Hazardous areas for explosive gas atmosphere are classified in the following zones:(1) zone 0: an area in which an explosive gas atmosphere is present continuously or is present for long periods;(2) zone 1: an area in which an explosive gas atmosphere is likely to occur in normal operation;(3) zone 2: an area in which an explosive gas atmosphere is not likely to occur in normal operation and if it does occur, is likely to do only infrequently and will exist for a short period only.

Hazardous areas of category 0　0 类危险区(平台)　对平台而言,系指:(1)在正常工作条件下持续和长期存在爆炸性气体环境的区域。(2)易燃气体或蒸发气的引燃浓度连续或长期存在的危险区。划分为 0 类危险区的处所如下:(1)井液循系统中从井口至除气排出管终端之间的内部空间;(2)油、气、水处理系统中从采油树至油、气、水处理终端一切含有烃类物质的内部空间;(3)原油储存容器及外输系统的内部空间;(4)其他一切运送、储存、处理天然气、原油或闪点低于 60 ℃ 油类产品系统的内部空间。　For the unit, hazardous areas of category 0 means:(1) a zone in which an explosive gas-air mixture is continuously present or present for long periods in normal operating conditions. (2) a hazardous zone in which ignitable concentrations of flammable gases or vapours are continuously present or present for long periods. The spaces classified as hazardous areas of category 0 are as follows:(1) The internal spaces of the mud circulating system between the well and the final degassing discharge;(2) All internal spaces containing hydrocarbon in oil, gas and water processing systems between the Christmas tree and processing terminals;(3) The internal spaces of crude oil storage vessels and outlet system;(4) The internal spaces of all other systems conveying, storing and processing natural gas, crude oil or oil products having a flash point below 60 ℃.

Hazardous areas of category 1　1 类危险区(平台)　对平台而言,系指:(1)在正常工作条件下可能出现爆炸性气体环境的区域。(2)在正常作业中可能出现易燃气体或蒸发气的引燃浓度的危险区。划分为 1 类危险区的处所如下:(1)钻井液系统中,从井口至最终除气口之间的一段 3 m 以内的区域。如井液循环系统在围蔽的处所内,则整个围蔽处所划为 1 类区;(2)在钻井阶段围蔽的钻井架以内的区域;(3)采油树周围和下方的半围蔽、有遮挡且通风不良的地方;(4)油、气、水处理系统中以及原油储存系统中任何泄放口、放气口周围半径为 3 m 以内的区域;(5)原油储存罐的透气装置出口及其他一切天然气的冷空放的周围半径为 3 m 的区域;(6)闪点不低于 60 ℃ 的燃料油柜的内部空间;(7)内含 1 类释放源且通风合格的任何围蔽处所。　For the unit, hazardous areas of category 1 means:(1) a zone in which an explosive gas-air mixture is likely to occur in normal operating conditions. (2) a hazardous zone in which ignitable concentrations of flammable gases or vapours are likely to occur in normal operation. The spaces classified as hazardous areas of category 1 are as follows:(1) The areas 3 m beyond the mud circulating system between the well and the final degassing discharge. Where the mud circulating system is located in an enclosed space, the complete enclosed space is to be classified as hazardous areas of category 1;(2) Areas to be enclosed in the drilling derrick during drilling;(3) Semi-enclosed, obstructed and poorly ventilated spaces around and below the christmas tree;(4) Areas within a radius of 3 m from any drainage and degassing discharge in oil, gas and water processing systems and crude oil storage system;(5) Areas within a radius of 3 m from venting outlets of crude oil storage tanks and all other cool discharges of natural gas;(6) Internal spaces of tanks used for storage of fuel oil having a flash point of not less than 60 ℃;(7) Any enclosed spaces containing a release source of category 1 and satisfactorily ventilated.

Hazardous areas of category 2　2 类危险区(平台)　对平台而言,系指:(1)在正常工作条件下不大可能出现爆炸性气体环境,即使出现也只是短时间存在的区域。(2)不大可能出现易燃气体或蒸发气的引燃浓度,或这种混合气体即使出现也只会在短时间存在的危险区。划为 2 类危险区的处所如下:(1)包含有从最终除气排出口至坑池处井液泵吸入接管间的井液循环系统的围蔽处所;(2)在钻井架限界以内、钻台以上 3 m 高度内的露天部位;(3)钻台下面邻接于钻台和钻井架限界或邻接于易积聚气体的任何围壁范围内的半围蔽部位;(4)在钻台下面一个可能的释放源(如钻井喇叭口管的顶部)周围 3 m 以内的露天部位;(5)除 0 类危险区及 1 类危险区规定之外的整个油、气、水处理系统所在的区域,并包括以油、气、水处理系统中的任何设备及管路为界向外再延伸 3 m 的区域;(6)原油储存区域并包括以管路和储油罐为界再向外延伸 3 m 的区域;(7)其他一切运送、储存、处理天然气、原油或闪点低于 60 ℃ 油类的系统中的管道及设备周围 3 m 以内的区域;(8)天然气或原油燃料管的通风导管内以及使用天然气或原油做燃料的燃烧设备所在的罩壳内;(9)天然气冷放空口以及原油储存罐的透气口周围,从 1 类危险区之外再向外延伸半径为 7 m 的区域;(10)内含 2 类释放源且通风合格的任何围蔽处所;(11)油漆间含有原油软管的围蔽处所;(12)气锁间;(13)储存乙炔瓶的围蔽处所,如乙炔瓶存放在开敞区,则瓶头阀周围 3 m 以内的空间。 For the unit, hazardous areas of category 2 means:(1) a zone in which an explosive gas-air mixture is not likely to occur, and if it occurs, it will exist only for a short time. (2) a hazardous zone in which ignitable concentrations of flammable gases or vapours are not likely to occur, or in which such a mixture, if it does occur, will only exist for a short time. The spaces classified as hazardous areas of category 2 are as follows:(1) Enclosed spaces which contain open sections of the mud circulating system from the final degassing discharge to the mud pump suction connection at the mud pit;(2) Outdoors locations within the boundaries of the drilling derrick up to a height of 3 m above the drill floor;(3) Semi-enclosed locations below and contiguous with the drill floor and to the boundaries of the derrick or to the extent of any enclosure which is liable to trap gases;(4) Outdoor locations below the drill floor and within a radius of 3 m from a possible source of release gas such as the top of a drilling nipple;(5) The areas occupied by the entire oil, gas and water processing systems, other than those specified in Hazardous areas of category 0 and Hazardous areas of category 1, including the areas 3 m beyond the boundaries consisting of any equipment and pipelines in the oil, gas and water processing systems;(6) The crude oil storage area including the areas 3 m beyond the boundaries consisting of pipelines and oil storage tanks;(7) The areas 3 m beyond the pipelines and equipment of all other systems conveying, storing and processing natural gas, crude oil or oil products having a flash point below 60 ℃;(8) Internal spaces of ventilation ducts of natural gas or crude oil fuel pipes or internal spaces of housings of burning equipment fuelled by natural gas or crude oil;(9) Areas within a radius of 7 m from hazardous areas of category 1 around cool discharges of natural gas and vents of crude oil storage tanks;(10) Any enclosed spaces containing a release source of category 2 and satisfactorily ventilated;(11) The enclosed space containing crude oil hoses in paint lockers;(12) Air lock spaces;(13) Enclosed spaces used for storage of acetylene cylinders and where such cylinders are stored in an open area, areas 3 m beyond the cylinder head valve.

Hazardous areas zone 0 危险区域中的 0 类危险区(平台) 系指容纳未脱气的活性钻井泥浆、闭杯闪点低于 60 ℃ 的油或易燃气体和蒸发气及其产出的油和气的封闭舱柜和管道的内部空间,油/气/空气的混合物在此连续或长期存在的区域。 Hazardous areas zone 0 are the internal spaces of closed tanks and piping for containing active non-degassed drilling mud, oil that has a closed-cup flashpoint below 60 ℃ or flammable gas and vapour, as well as produced oil and gas in which an oil/gas/air mixture is continuously present or present for long periods.

Hazardous areas zone 1 危险区域中的 1 类危险区(平台) 系指:(1)装有泥浆循环系统任何部分的围蔽处所,而该部分位于油(气)井并与最终除气排出口之间并有开口通入该围蔽处所;(2)位于转台以下并含有一个可能释放源(例如隔水套管的喇叭口上端)的围蔽处所或半围蔽处所;(3)位于转台以下一个可能的释放源(例如隔水套管的喇叭口上端)1.5 m 半径以内的露天部位;(4)位于转台上而没有以全实地板与位于转台以下并含有一个可能释放源(例如隔水套管的喇叭口上端)的围蔽处所或半围蔽处所隔开的围蔽处所;(5)在露天或半围蔽部位内,除(2)中所规定者外,距通向 1 中所述泥浆系统所属设备的任何开口的界限、1 类危险区处所的任何通风出口的界限,或 1 类危险区处所任何出入口的界限 1.5 m 以内的区域;(6)在本应属 2 类危险区,布置成气体不易散开部位的坑池、导管或类似结构。 Hazardous areas zone 1 are:(1) enclosed spaces containing any part of the mud circulating system that has an opening into the spaces and is between the well and the final degassing discharge;(2) enclosed spaces or semi-enclosed locations that are below the drill floor and contain a possible source of release such as the top of a drilling nipple;(3) outdoor locations below the drill floor and within a radius of 1.5 m from a possible source of release such as the top of a drilling nipple;(4) enclosed spaces that are on the drill floor and which are not separated by a solid floor from the spaces in paragraph 2;(5) in outdoor or semi-enclosed locations, except as provided for in paragraph 2, the area within 1.5 m from the boundaries of any openings to equipment which is part of the mud system as specified in paragraph 1, any ventilation outlets of zone 1 spaces, or any access to zone 1 spaces;(6) pits, ducts or similar structures in locations which would otherwise be zone 2 but which are so arranged that dispersion of gas may not occur.

Hazardous areas zone 2 危险区域中的 2 类危险区

(平台)系指:(1)在最终除气排出口至泥浆池处的泥浆泵吸入接头之间容纳泥浆循环系统敞露部分的围蔽处所;(2)在钻井架限界以内钻台以上3 m高度内的露天部分;(3)从下方邻接于钻台井和井架限界或任何易于积聚气体的围蔽范围相比邻的半围蔽处所;(4)钻台以下在位于转台以下一个可能的释放源(例如隔水套管的喇叭口上端)1.5 m半径以内的露天部位所规定的1类危险区和位于转台以下并含有一个可能释放源(例如隔水套管的喇叭口上端)的围蔽处所或半围蔽处所所规定的半围蔽处所以外1.5 m的区域;(5)在露天或半围蔽部位内,除危险区域中的1类危险区2中所规定者外,距通向危险区域中的1类危险区1中所述泥浆系统所属设备的任何开口的界限、1类危险区处所的任何通风出口的限界,或1类危险区处所任何出入口的限界1.5 m以内的区域;(6)距2类危险区处所任何通风出口或出入口的限界1.5 m以内的露天区域;(7)半围蔽井架在其钻台运输围蔽范围内或钻台以上3 m高度范围内,取大者;(8)1类危险区和非危险区域之间的气闸。 Hazardous areas zone 2 are:(1)enclosed spaces which contain open sections of the mud circulating system from the final degassing discharge to the mud pump suction connection at the mud pit;(2)Outdoor locations within the boundaries of the drilling derrick up to a height of 3 m above the drill floor;(3)Semi-enclosed locations below and contiguous to the drill floor and to the boundaries of the derrick or to the extent of any enclosure which is liable to trap gases;(4)in outdoor locations below the drill floor, within a radius of 1.5 m area beyond the zone 1 area as outdoor locations below the drill floor and within a radius of 1.5 m from a possible source of release such as the top of a drilling nipple;(5)the areas 1.5 m beyond the zone 1 areas specified in outdoor or semi-enclosed locations, except as provided for in paragraph 2, the area within 1.5 m from the boundaries of any openings to equipment which is part of the mud system as specified in paragraph 1, any ventilation outlets of zone 1 spaces, or any access to zone 1 spaces and beyond the semi-enclosed locations specified in enclosed spaces or semi-enclosed locations that are below the drill floor and contain a possible source of release such as the top of a drilling nipple;(6)Outdoor areas within 1.5 m of the boundaries of any ventilation outlet from or access to a zone 2 space;(7)Semi-enclosed derricks to the extent of their enclosure above the drill floor or to a height of 3 m above the drill floor, whichever is greater;(8) air locks between a zone 1 and a non-hazardous area.

Hazardous atmosphere 危险气体 系指能直接对人命或健康造成损害的任何气体。 Hazardous atmosphere means any atmosphere that is immediately dangerous to life or health.

Hazardous cargo vessel 灾害性危险品船舶 灾害性危险品船舶系指下列船舶:(1)散装运输闪点61 ℃以下的燃料、汽油、原油或其他易燃液体的液货船,包括先前运输过闪点在61 ℃以下货物但未除气的液货船;(2)运输压缩过的液化气、散装酸或液化化学品的液货船:①压缩、液化或压力下的气体(IMO 第2 类)超过50 t;②闪点61 ℃以下的易燃液体(IMO 第3 类)超过50 t;③易燃固体、易自燃的物质或受潮后逸出易燃气体的物质(IMO 第4 类)超过50 t;④起氧化作用的物质或有机过氧化物(IMO 第5 类)超过50 t;⑤任何数量的有毒(毒性)物质和传染性物质(IMO 第6 类);⑥任何数量的放射性物质(IMO 第7 类);⑦腐蚀性物质(IMO 第8 类)超过50 t;⑧任何数量的散装的车削、钻孔、切割或剪切加工后的金属,其装载或运输温度超过65.5 ℃的;⑨任何数量的烟熏谷物(烟熏所用的化学品对人体生命有害);⑩任何数量的直接还原铁(DRI)。 Hazardous cargo vessel means a vessel shall be deemed to be a hazardous cargo vessel in the following cases:(1)A tanker carrying fuel oil, gasoline, crude oil or other flammable liquids in bulk, having a flashpoint below 61 ℃, including a tanker that is not gas free where its previous cargo had a flashpoint below 61 ℃;(2)a tanker carrying compressed liquefied gases, bulk acids or liquefied chemical:①In excess of 50 t of gases, compressed, liquefied or dissolved under pressure(IMO Class 2);②in excess of 50 t of flammable liquids, having a flashpoint below 61 ℃(IMO Class 3);③in excess of 50 t of flammable solids, spontaneously combustible material or substances emitting combustible gasses when wet(IMO Class 4);④in excess of 50 t of oxidizing substances or organic peroxides(IMO Class 5);⑤any quantity of poisonous(toxic) substances and infectious substances(IMO Class 6);⑥any quantity of radioactive substances(IMO Class 7);⑦in excess of 50 t of corrosive substances(IMO Class 8);⑧any quantity of metal turnings, borings, cuttings, or shavings in bulk having a temperature on loading or in transit in excess of 65.5 ℃, ⑨any quantity of grain that is under fumigation, where the chemical being used is hazardous to human life, ⑩any quantity of direct reduced iron(DRI).

Hazardous effect 危险结果 是产生下列情况的结果:(1)危险地增加了船员的操作责任以及他们履行职责时他们的难度,以致使他们不能够期望去应付它们,并将可能要求外界帮助;(2)危险地降低处理特性;(3)危险地降低船舶的强度;(4)船员处于生存的边缘条件,或受到伤害;(5)必需要外界的援救。 Hazardous effect is an effect which produces: (1) a dangerous increase in the operational duties of the crew or in their difficulty in performing their duties of such magnitude that they cannot reasonably be expected to cope with them and will probably require outside assistance; (2) dangerous degradation of handling characteristics; (3) dangerous degradation of the strength of the ship; (4) marginal conditions for, or injury to, occupants; (5) an essential need for outside rescue operations.

Hazardous effect 危险影响　系指故障情况的影响使船舶或船员对付不利营运状况的能力减少到危险的程度。例如极大地减少安全裕度和营运能力，由于实地遇险状况或较高工作量而不能依靠飞行乘务员来正确或充分完成其任务，或对相对少量的船上人员造成严重或致命的伤害。 Hazardous effect means the effect of failure conditions that reduces the capability of the craft or the ability of the crew to cope with adverse operating conditions to the extent that there would be, for example, a large reduction in safety margins or functional capabilities, physical distress or higher workload such that the flight crew cannot be relied upon to perform their tasks accurately or completely, or serious or fatal injuries to a relatively small number of occupants.

Hazardous gas emission controls 有毒有害气体排放控制　MEPC 修订了 MARPOL 附则Ⅵ中关于船用柴油机的定义，要求以气体为燃料的发动机也要符合 MARPOL 附则Ⅵ第 13 条关于加油站氮氧化物的排放标准。批准了"不需要满足 Tier Ⅲ标准的非相同的柴油机替代柴油机的导则"，对在 2016 年 1 月 1 日或以后由不相同的柴油机替代的船用柴油机，如果不能满足 Tier Ⅲ标准，则可以满足 Tier Ⅱ标准，并同意将 Tier Ⅲ标准实施时间由 2016 年 1 月 1 日推迟 5 年。

Hazardous material 有害物质/有害材料　系指危害人类健康和/或环境造成危害的任何材料或物质。 Hazardous material means any material or substance which is liable to create hazards to human health and/or the environment.

Hazardous spaces of larger fire risk 较大失火危险处所　系指下列处所:(1)SOLAS 第Ⅱ-2 章第 3.30 所定义的机器处所;(2)装有燃油处理设备或其他易燃物质的处所;(3)厨房和装有烹调设备的配膳间;(4)装有烘干设备的洗衣房;(5)载客超过 36 人客船上，SOLAS 第Ⅱ-2 章第 9.2.2.条第(8)、(12)和(14)所定义的处所。

Hazardous wastes 危险废物　系指按巴塞尔(Basel)公约控制危险废物越界调迁规定中具有危险特性的废物。 Hazardous wastes means wastes having hazardous characteristics according to Basel Convention on the control of trans-boundary movements of hazardous wastes.

Hazardous zone of category 0 0 类危险区域(浮油回收船)　系指可燃气体或爆炸性气体或蒸发气与空气混合物连续或长时间存在的区域。在浮油回收船上，下列区域或处所属于 0 类危险区域:(1)回收油储存舱。(2)属于回收油围护系统的管路和容器内部。 Hazardous zone of category 0 means an area in which the mixture of flammable or explosive gas or vapor with air exists continuously or for long time. The following zones or spaces on oil recovery ships are to be regarded as gas-hazardous zones of category 0: (1) the interiors of recovered oil tanks; (2) the interiors of piping system and containers of the containment system for recovered oil.

Hazardous zone of category 1 1 类危险区域(浮油回收船)　系指可燃气体或爆炸性气体或蒸发气与空气混合物在正常作业中可能出现的区域。在浮油回收船上，下列区域或处所属于 1 类危险区域:(1)与回收油储存舱相邻接的隔离舱或其他处所。(2)属于回收油输送系统的管路、阀、软管、泵和其他设备所在的围蔽或半围蔽处所。(3)距分离器、回收油输送系统的软管和阀、回收油储存舱开口和泵舱或隔离舱等 1 类危险处所的开口 3 m 以内的开敞甲板处所，包括半围蔽处所。(4)回收油储存舱上的开敞甲板区域及其前后各加 3 m，高度为 2.4 m的空间。(5)位于回收油储存舱外面，回收油管路(中间管段或管路终端)所在的围蔽处所，如设有符合有关规定的通风，则可除外。(6)能直接从 1 类危险区域进入(无气闸)或有开口通向 1 类危险区域的围蔽或半围蔽处所如设有符合有关规定的通风，则可除外。 Hazardous zone of category 1 means an area in which the mixture of flammable or explosive gas or vapor with air may occur during normal operation. The following zones or spaces on oil recovery ships are to be regarded as gas-hazardous zones of category 1: (1) cofferdams or other spaces adjacent to any recovered oil tank; (2) enclosed or semi-enclosed spaces in which pipe flanges, valves, hoses, pumps and other equipment for handling of recovered oil are located; (3) spaces including semi-enclosed spaces on open deck within 3 m radius of the opening of gas-hazardous zones of category 1, such as separator, hoses and valves which belong to oil recovered transfer system, a recovered oil tank, pump room or cofferdam; (4) spaces on open deck zone within 3 m radius, a height 2.4 m above oil recovered tank; (5) enclosed spaces outside oil recovered tank, in which oil recovered piping (intermediate spool or terminal) are located; (6) enclosed or semi-enclosed spaces which can be entered directly from zones of category 1 (without air lock) or which have openings into zones of category 1 may be excluded, provided that ventilation is fitted as relevant provisions.

Hazardous zone or space 危险区域或处所 系指可由于任何下列情况的存在所引起:(1)处所或舱柜,包含:①闪点(闭杯试验)不超过60 ℃的可燃液体;②闪点超过60 ℃,由环境条件加热或升温至不超过其闪点15 ℃以内的可燃液体;③可燃气体;(2)管系或设备,包含(1)所规定的液体,并有凸缘接头或填料函或其他开口,在正常工作情况下可能发生漏液;(3)包含诸如煤或谷类等固体,易于释放可燃气体和/或可燃粉尘的处所;(4)包含包装形式危险货物(在 IMDG 规则中规定的货物)的处所;(5)与化学过程有关的管系或设备(诸如蓄电池充电或电氯化处理),产生可燃气体作副产品,并有开口在正常工作情况下可能从此逸出气体;或(6)包含(1)中没有规定的可燃液体的管系或相当设备,有凸缘接头、填料函或其他开口,在正常工作情况下可能从此泄漏以烟雾或细喷液形式的泄漏液体。 Hazardous zone or space may arise from the presence of any of the following: (1) spaces or tanks containing either: ①flammable liquid having a flashpoint (closed-cup test) not exceeding 60 ℃; ② flammable liquid having a flashpoint exceeding 60 ℃, heated or raised by ambient conditions to a temperature within 15 ℃ of its flashpoint; or ③flammable gas; (2) piping systems or equipment containing fluid defined by (1) and having flanged joints or glands or other openings through which leakage of fluid may occur under normal operating conditions; (3) spaces containing solids, such as coal or grain, liable to release flammable gas and/or combustible dust; (4) spaces containing dangerous goods in packaged form, of the following Classes as defined in the IMDG Code; (5) piping systems or equipment associated with processes (such as electro-chlorination) generating flammable gas as a by-product and having openings from which the gas may escape under normal operating conditions; or (6) piping systems or equivalent containing flammable liquids not defined by (1) having flanged joints, glands or other openings through which leakage of fluid in the form of a mist or fine spray may occur under normal operating conditions.

Hazardous zones and spaces 危险区域和处所(浮油回收作业) 下列区域或处所,在浮油回收作业期间和完成后,被认为是危险的,直至验证为气体安全为止:(1)拟用于储存回收油舱柜的内部;(2)拟用于装卸回收油管系的内部;(3)从拟用于回收油油舱的内部,由单舱壁、单甲板或其他舱室界面隔开的处所,或者其舱壁紧靠回收油舱舱壁之上或之下并与油舱壁成直线的处所,但采用符合有关要求的斜拉板或采用满足有关要求的装置进行保护的处所除外;(4)罩盖含有回收油或被回收油污染的管系或其他设备,且在正常作业条件下,可能在其法兰接头、填料函或其他开口处出现油液渗漏的处所;(5)开敞甲板上回收油舱的通风出口,或在正常作业条件下允许开启的检验舱口,在其3 m半径之内的区域;(6)在回收油舱的任何抽样或测深点1.5 m半径之内的区域;(7)在开敞甲板上,任何含有回收油或被回收油污染的法兰接头、填料函或其他设备,且在正常作业条件下可能出现渗漏者,在其1.5 m半径之内的区域;(8)在汇集回收油漏油用的任何围油栏或围油屏的界定范围并外延1.5 m,高度至1.5 m的开敞甲板上的区域;(9)在开敞甲板上,通向(3)或(4)规定处所的任何开口1.5 m半径之内的区域;(10)跨越舱顶曝露在露天的所有回收油舱,至整个船宽再加上最前和最后舱舱壁前后各3 m,高度至甲板上方0.45 m或至任何防浪板高度的开敞甲板上的区域;(11)超过(5)至(9)规定范围延伸1.5 m的开敞甲板上的区域;(12)有直接开口进入上述所鉴别的危险区域或处所的任何围蔽或半围蔽的处所。 The following zones or spaces are regarded as hazardous during and on completion of oil recovery operations, until proven gas-safe: (1) the interiors of tanks intended for the storage of recovered oil; (2) the interiors of piping systems intended for the handling of recovered oil; (3) spaces separated by a single bulkhead, deck or other tank boundary, from the interior of a tank intended for recovered oil, or having a bulkhead immediately above or below and in line with a bulkhead of a tank intended for recovered oil, unless protected by a diagonal plate in accordance with relevant requirements or the arrangements comply with the relevant requirements; (4) spaces housing piping systems or other equipment containing or contaminated with recovered oil and having flanged joints or glands or other openings from which leakage of fluid may occur under normal operating conditions; (5) zones on open deck within a 3 metre radius of the ventilation outlets, or inspection hatches permitted to be opened under normal operating conditions, of tanks intended for recovered oil; (6) zones on open deck within a 1.5 m radius of any sampling or sounding point of a tank intended for recovered oil; (7) zones on open deck within a 1.5 m radius of any flanged joints, glands or other parts of any equipment containing or contaminated with recovered oil from which leakage may occur under normal operating conditions; (8) zones on open deck within the confines of, and extending 1.5 m beyond, any bund or barrier intended to contain a spillage of recovered oil, up to a height of 1.5 m; (9) zones on open deck within a 1.5 m radius of any opening into a space described by (3) or (4); (10) zones on open deck over all tanks intended for recovered oil, where the tops of the tanks are exposed to the weather, to the full width of the ship plus 3 m fore and aft of the forwardmost and aftmost tank bulkhead, up to a height of 0.45 m above the deck or to the height of any bulwarks; (11) zones on open deck extending 1.5 m beyond those defined by (5) to (9); (12) Any enclosed or semi-enclosed space having a direct opening into a hazardous zone or space identified above.

Hazards of products 货品的危险性 包括:(1)由化学品的闪点、易爆性/易燃性的限制/范围和自燃温度所确定的火灾危险性。(2)由下述情况确定的健康危险性:①在液体状态下,对皮肤的刺激作用;②急性毒性作用,确定时要考虑到以下数值:口服致死剂量 LD_{50}(口服):系指口服时,导致50%的接受试验者死亡的剂量;皮肤致死剂量 LD_{50}(皮肤):系指作用于皮肤时,导致50%的接受试验者死亡的剂量;致死浓度 LC_{50}(吸入):系指吸入时,导致50%的受试验者死亡的浓度。③其他如致癌及敏感的健康危害作用。(3)由与下列物质的反应性确定的反应危险性:①水;②空气;③其他化学品;④化学品本身(例如:聚合作用)。(4)由下述情况确定的海洋污染危险性:①生物积聚性;②缺乏生物易降解性;③对水中有机体的急性毒性作用;④对水中有机体的慢性毒性作用;⑤对人类健康的长期影响及⑥引起货物漂浮或下沉的物理特性并因此造成对海洋生物的负面影响。 Hazards of products include: (1) fire hazard, defined by flashpoint, explosive/flammability limits/range and auto-ignition temperature of the chemical. (2) health hazard, defined by: ① corrosive effects on the skin in the liquid state; ② acute toxic effect, taking into account values of: LD_{50} (oral): a dose, which is lethal to 50% of the test subjects when administered orally; LD_{50} (dermal): a dose, which is lethal to 50% of the test subjects when administered to the skin; LC_{50} (inhalation): the concentration which is lethal by inhalation to 50% of the test subjects; ③ other health effects such as carcinogenicity and sensitization. (3) reactivity hazard, defined by reactivity: ① with water; ② with air; ③ with other products; ④ of the product itself (e.g. polymerization); (4) marine pollution hazard, as defined by: ① bioaccumulation; ② lack of ready bio-degradability; ③ acute toxicity to aquatic organisms; ④ chronic toxicity to aquatic organisms; ⑤ long term human health effects; and ⑥ physical properties resulting in the product floating or sinking and so adversely affecting marine life.

Haze 霾(阴霾、灰霾) 系指混有烟、尘埃或蒸汽的雾。

HAZID 危险识别 HAZID is a hazardous identification.

HAZOP 危险操作 HAZOP is a hazardous operation.

HCS = heading control system 艏向控制系统

Head sea 迎浪 系指负 x 方向行进的波浪。Head sea is waves propagating in the negative x-direction.

Head up display 船首向上显示 系指本船首向位于图像最上端的非稳定显示。 Head up display means an un-stabilized presentation in which own ship's heading is uppermost on the presentation.

Header 联箱(集管,磁头,镦锻机) 与锅炉上升管、下降管或过热管群相连接,汽水或汽水混合物在其中汇集、分配和混合的集管。

Heading 艏向 船舶的艏部所指的方向,以距正北的角位移表示。 Heading is a direction in which the bow of a ship is pointing expressed as an angular displacement from north.

Heading and/or track control systems 艏向和/或航迹控制系统 系指在高密度航运区域能见度受限制的条件下以及其他危险的航行情况下,能立即确立人工操舵的系统。从自动操舵转换为人工操舵,以及从人工操舵转换为自动操舵,应由1名负责的驾驶员操作或在其监督下进行操作。

Heading angle 艏向角 从基准方向,如地球真北线、磁北线或罗北线到艏向间的水平夹角。顺时针为正。

Heading sensitivity 艏向敏感性 船舶在水道近岸航行时,由于螺旋桨在船尾使局部水流加速,靠岸侧水流的补充受到限制而产生吸力,致使船首有离岸转向的特征。

Headline anchoring winch 艏锚绞车 在用锚定位的工程船上,设置在艏端,用以收放艏锚索使船身前移的绞车。见图 H-2。

图 H-2 首艏锚绞车
Figure H-2 Headline anchoring winch

Headquarters Party 总部国(国际海事卫星组织) 系指国际海事卫星组织在其领土上建立该组织总部之缔约国。 Headquarters Party means the Party to the International maritime satellite organization Convention in whose territory INMARSAT has established its headquarters.

Head-reach 停船冲程 停车或倒车后,船舶沿原航向惯性前移的最大距离。

Headroom 吊钩净高 吊货钩处于最高位置时,距舷墙或舱口围板上缘的净高度。在定位双杆操作时,系指吊货索间夹角为120°情况下,两吊货索连接点距舷墙或舱口围板上缘的高度。

Headroom 净空高度 净空高度应取在客舱/舱室地板的顶层与在指定位置处的甲板横梁或天花板的下缘(取低值)之间测得的垂直距离。制造厂商无须说明在其他位置,例如上层固定床铺的净空高度。 Headroom shall be measured as the vertical distance between the top of the cabin/compartment floor and the underside of the deck beam or deck head (whichever is lower) at a designated position. The manufacturer is free to state the headroom in other locations, e. g. above bunks.

Headset "耳麦" 是耳机与麦克风的整合体,它不同于普通的耳机。普通耳机往往是立体声的,而"耳麦"多是单声道的,同时,"耳麦"有普通耳机所没有的麦克风。"耳麦"分为"无线耳麦"和"有线耳麦"。这两种"耳麦"有各自的特点。

Health certificate 健康证书 是证明可供人类食用的出口动物产品、食品等经过卫生检验或检疫合格的证书。

Health, safety and environmental(HSE) 健康/安全和环境 也称环境/健康和安全(EH&S)。该集合名词用于评价众多大企业采用的不同环境和工作安全管理体系。

Hearing 审理

Heartcut distillate oil 窄馏分油

Heat absorption capacity 吸热量

Heat affected zone(heat-affected zone)(HAZ) 热影响区 系指与在焊接时通过加热修补的焊缝邻近的部分母体金属。焊接时热影响区处于AC3以上的区域,由于这类钢的淬硬倾向较大,故焊后得到淬火组织(马氏体)。母材被加热到AC1~AC3温度之间的热影响区,在快速加热条件下,铁素体很少溶入奥氏体,而珠光体、贝氏体、索氏体等转变为奥氏体。原铁素体保持不变,并有不同程度的长大,最后形成马氏体-铁素体的组织,故称不完全淬火区。对于含碳高、合金元素较多、淬硬倾向较大的钢种,还出现淬火组织马氏体,降低塑性和韧性,因而易于产生裂纹。

Heat aging 热时效

Heat and moisture resistance test 抗热抗潮试验

Heat balance 热平衡

Heat balance by direct method 正平衡 锅炉热效率试验时,利用锅炉工质吸收热量和输入锅炉热量之比直接计算锅炉效率的方法。

Heat balance by indirect method 反平衡 锅炉热效率试验时,根据各项热损失间接推算锅炉效率的方法。

Heat balance calculation 热平衡计算 配合热线图进行的从燃料燃烧到功率输出过程中有关热交换、热能分布、损失等热量平衡情况及热效率的计算。

Heat balance diagram (steam and feed flow diagram) 热线图 表示蒸汽动力装置汽水工作循环、组成及流程,并注有工质在流程中的压力、温度、流量等热力参数以及热能分布情况的基本原理图。

Heat balance test 热平衡试验 对动力装置工作范围内各稳定工况所做的全面热工测量试验。

Heat capacity 热容量

Heat carrier 载热体(传热介质)

Heat coil 热线圈

Heat consumption 耗热量

Heat consumption(specific heat consumption) 热耗率 汽轮机单位功率、单位时间的热消耗量。

Heat content 焓(热含量)

Heat corrosion 热腐蚀

Heat crack 热裂纹/热裂 系指焊接过程中,焊缝和热影响区金属冷却到固相线附近的高温区间产生的焊接裂纹。在焊接热影响区或多层焊的前一焊道上,因焊接热循环的作用致使塑性陡降,在拉伸应力下沿二次结晶界形成的热裂纹。其裂纹敏感温度区域略低于再结晶温度。多数发生在奥氏体钢和合金及少数高强度钢的焊接接头中。其裂纹产生条件有些类同于多边化裂纹,但其裂纹形成机制和裂纹形态却各不相同。

Heat cure 热固化(热塑化,热硫化)

Heat cycle 热循环

Heat deflection temperature 热挠曲温度

Heat deformation temperature(heat distortion temperature) 热变形温度

Heat detector 感温探测器

Heat dissipation capacity 散热量

Heat drop 焓降

Heat efficiency 热效率

Heat elimination equipment 散热设备 系指在自由空气条件和试验用标准大气条件规定的大气压力

（86kPa～106kPa）条件下,在温度稳定后测得的表面最热点温度与环境温度之差大于5k的受试设备。

Heat emission　放热(热量发射)

Heat endurance　耐热性(耐热度)

Heat energy　热能

Heat engine　热机

Heat exchanger　热交换器　是用一种液体加热或者冷却另一种流体的压力容器。除非另有说明,通常热交换器由一系列相邻的室组成。两种液体分别在不同的室内流动。一个或多个室可由管束组成。 Heat exchanger is an unfired pressure vessel used to heat or cool a fluid with another fluid. In general heat exchangers are composed of a number of adjacent chambers, the two fluids flowing separately in adjacent chambers. One or more chambers may consist of bundles of tubes.

Heat fatigue　热疲劳　产生热疲劳的环境是有一些金属材料构件,在高温环境中受力,特别容易产生的疲劳。

Heat flux　热流量　系指在热流方向单位标称面积的热传输率。它是由辐射、传导和对流传热量的总和。Heat flux is the rate of heat transfer per unit area normal to the direction of heat-flow. It is the total heat transmitted by radiation, conduction, and convection.

Heat fusion　熔化热

Heat hardening　加热硬化(加热淬火)

Heat input　热输入　系指单位长度焊缝输入的能量E,按下式确定：$E = [U(V) \times I(A) \times$ 焊接时间$(min) \times 6]/[$焊缝长度$(mm) \times 100](kJ/mm)$。 Heat input means energy input per unit length of weld E applied during welding shall be determined by the following formula: $E = [U(\text{volts}) \times I(A) \times \text{welding time}(min) \times 6]/[\text{length of seam}(mm) \times 100](kJ/mm)$.

Heat insulating　隔热的(绝热的)

Heat insulating material/Thermal insulation material　隔热材料　能阻滞热流传递的材料,又称热绝缘材料。隔热材料分为多孔材料、热反射材料和真空材料三类。传统绝缘材料,如玻璃纤维、石棉、岩棉、硅酸盐等,新型绝热材料,如气凝胶毡、真空板等。隔热材料的物质构成不同,其物理热性能也就不同;隔热机理存有区别,其导热性能或导热系数也就各有差异。隔热材料中,大部分热量是从孔隙中的气体传导的。隔热材料的比热对于计算绝热结构在冷却与加热时所需要冷量(或热量)有关。

Heat insulation　隔热(热绝缘)

Heat insulation test　热绝缘试验

Heat interchange　热交换

Heat loss due to combustibles in refuse　机械不完全燃烧热损失　由于燃料在炉膛中燃烧后,尚有未燃尽的固体可燃成分而造成的化学热量损失。

Heat loss due to exhaust gas　排烟热损失　烟气离开锅炉时,以物理显热形式表现的带走的热量损失。

Heat loss due to radiation　散热损失　由锅炉外表面的周围环境散热而造成的热量损失。

Heat loss due to unburned gas　化学不完全燃烧热损失　由于燃料在炉膛中燃烧时形成的可燃气体未在炉膛中燃烧完全即排出炉外而造成的化学热量损失。

Heat of combustion　燃烧热

Heat power engineering　热力工程

Heat pressing condition　热压条件

Heat producer　热量发生器

Heat property　热性能

Heat radiation　热辐射

Heat reducer　减热器

Heat release rate in furnace　炉膛容积热负荷　按炉膛单位容积计算每小时投入的燃料热能的数量。

Heat resisting　耐热的(抗热的,难熔的)

Heat resisting alloy　耐热合金　又称高温合金,它对于在高温条件下的工业部门和应用技术,有着重大的意义。能在 >700 ℃高温下工作的金属通称耐热合金,"耐热"系指其在高温下能保持足够的强度和良好的抗氧化性。

Heat resisting alloy steel　耐热合金钢

Heat resisting cast iron　耐热铸铁

Heat resisting material　耐热材料

Heat resisting steel　耐热钢　在高温下具有较高的强度和良好的化学稳定性的合金钢。它包括抗氧化钢(或称高温不起皮钢)和热强钢两类。抗氧化钢一般要求较好的化学稳定性,但承受的载荷较低。热强钢则要求较高的高温强度和相应的抗氧化性。耐热钢常用于制造锅炉、汽轮机、动力机械、工业炉和航空、石油化工等工业部门中在高温下工作的零部件。这些部件除要求高温强度和抗高温氧化腐蚀外,根据用途不同还要求有足够的韧性、良好的可加工性和焊接性,以及一定的组织稳定性。此外,还开发出一些新的低铬镍抗氧化钢种。

Heat resistor　热源

Heat retaining　保热的(保温的,贮热的)

Heat run test　热运行试验

Heat sensitive material　易熔材料

Heat sensitivity　热敏性(热灵敏度)

Heat sensor　热传感器

Heat sinking　散热的

Heat strain (immersion suit) 热应变(救生服) 由于不可能被温度调节所完全补偿的持续的热应力所引起的人体温度的增加,或热效应器官为适应能引起其他非热的调节系统状态持续变化的热应力而产生的活化作用。 Heat strain means that the increase of body temperature induced by sustained heat stress which cannot be fully compensated by temperature regulation, or activation of thermo-effect or activities in response to heat stress which cause sustained changes in the state of other, non-thermal, regulatory systems.

Heat test 耐热试验(加热试验)
Heat transfer 热传导
Heat transfer coefficient 传热系数
Heat treatment condition 热处理状态
Heat treatment for hydrogen removal 脱氢热处理
Heat treatment furnace 热处理炉
Heat treatment temperature 热处理温度
Heat treatment with a laser beam 激光束热处理
Heat unit 热量单位
Heat up 加热(加温)
Heat up and sweat 受热受潮 系指由于气温的骤然变化或者船上的通风设备失灵,使船舱内的水汽凝结,引起发潮发热导致货物的损失。
Heat value 热值
Heat water circulating pump 热水循环泵
Heat working 热加工
Heated tank 加热液舱
Heater 加热器(热源,加热工) 通过某种高温热源间接加热气体最终产生高温气体的部件。
Heater unit 加热器部件
Heat-expansion-prevention separator 防止热膨胀隔板
Heating 加热(供暖)
Heating apparatus 加热设备(加热器)
Heating arrangement 加热装置
Heating coil 加热盘管
Heating element 加热元件
Heating furnace 加热炉
Heating jacket 热套
Heating period 加热周期
Heating pipe 加热管
Heating plant 加热装置(暖气装置)
Heating power 发热量
Heating steam 加热蒸汽 作为蒸馏海水热源用的蒸汽。
Heating surface 受热面 一侧与水和蒸汽、空气等受热工质接触,另一侧与火焰和烟气等加热质接触,实现两侧间热交换的传热体的壁面。
Heating surface area on a boiler 锅炉的传热面积 一侧接触燃烧气体,另一侧接触水,按照燃烧气体一侧计算的面积;若无另外规定,不包括过热器、再热器、经济器及废气经济器等的传热面积。 Heating surface area on a boiler is the areas calculated on the combustion gas side surface.
Heating time 加热时间
Heating torch 加热喷灯
Heating unit 加热元件
Heating value 热值
Heating warpage test 加热弯曲试验
Heating water ratio 加热水倍率 以水为加热介质的蒸馏装置中,加热水量与蒸馏水产量的比值。
Heating zone 加热区通
Heating ventilation and air conditioning system (HVACS) 取暖、风和空调系统
Heat-proof 防热的(耐热)
Heat-resistant paint 耐热漆
Heat-sensitive 热敏的
Heat-sensitive paint 热敏涂料
Heat-treatable 可热处理的
Heat-treated steel 热处理钢
Heat-treatment furnace 热处理炉 是对金属工件进行各种金属热处理的工业炉的统称。分为罩式炉、辊底式炉、链式炉、牵引式热处理炉、钢丝铅淬火炉。
Heave compensator 垂荡补偿器 浮式钻井装置在钻井过程中为克服由于波浪作用引起船体相对于钻井井口的垂荡运动而采取的补偿装置。
Heave up depth 起锚深度 锚泊机械配用规定的锚及锚链时,能进行正常锚泊工作的最大深度。
Heaving 正垂荡 船舶沿垂向 z 轴的上下升沉运动。 Heaving is the ship's linear motion along the z axis.
Heaving line 撇缆绳 船舶靠岸系泊时引送系缆用的,一端是编织的包有重物的撇缆头的绳索。
Heavy ballast condition 重压载工况 系指压载(无货物)工况为:(1)压载舱可为满舱、部分压载或空舱。压载舱如为部分压载应满足相关条件;(2)至少一个在海上用于装载压载水的货舱应为满舱;(3)螺旋桨浸深 I/D 应至少为60%,其中:I—螺旋桨中线至水线距离;D—螺旋桨直径;(4)船舶应尾倾且应不超过 $0.015L_{BP}$;(5)重压载工况下的船首型吃水应不小于 $0.03\ L_{BP}$ 或 8 m 两者中之小值。 Heavy ballast condition is a ballast (no cargo) condition where: (1) the ballast tanks may be full, partially full or empty. Where ballast tanks are

partially full, the relative conditions are to be complied with; (2) at least one cargo hold adapted for carriage of water ballast at sea is to be full; (3) the propeller immersion I/D is to be at least 60%, where: I = Distance from propeller centerline to the waterline; D = Propeller diameter; (4) the trim is to be by the stern and is not to exceed $0.015L_{BP}$; (5) the moulded forward draught in the heavy ballast condition is not to be less than the smaller of $0.03L_{BP}$ or 8 m.

Heavy derrick boom 重型吊杆 单杆操作时安全工作负荷大于 98kN 的吊杆装置和吊杆式起重机。

Heavy derrick stool 重吊杆座 通常装设在甲板或起货平台上,由吊杆转枢、上支承、下支承等零部件所组成。用以支承重型吊杆并使重型吊杆可左右摆动及俯仰的支座。

Heavy diesel oil 重柴油 对"73/78 防污公约"附则Ⅰ而言,系指除采用 IMO 所接受的方法试验时,在不超过 340 ℃温度下有 50%(体积)以上馏化的蒸馏物以外的柴油。 For "MARPOL73/78" Annex Ⅰ, heavy diesel oil means diesel oil other than those distillates of which more than 50 percent by volume distils at a temperature not exceeding 340 ℃ when tested by the method acceptable to the IMO.

Heavy equipment vessel 大件运输船 系指具有较大的甲板面积,专门用于在甲板上装/卸,并进行海上远程运输桥吊和重型设备等尺寸/质量相对很大的成套大件和项目设备的运输船舶。见图 H-3。 Heavy equipment vessel means a carrier having large deck areas and specially engaged in loading/unloading on decks and long-distance transporting the complete sets of heavy equipment facilities with the large dimension/weight such as bridge cranes and heavy equipment at sea. See Fig. H-3.

图 H-3 大件运输船
Figure H-3 Heavy equipment vessel

Heavy fuel oil (HFO) 重燃油 即 15 ℃时密度大于 900 kg/m³ 或 50 ℃时运动黏度大于 180 mm²/s 的燃油。 Heavy fuel oil means the fuel oil with a density at 15 ℃ of higher than 900 kg/m³, or a kinematic viscosity at 50 ℃ of higher than 180 mm²/s.

Heavy goods vehicle (HGV) 载重物车辆

Heavy grade oil (HGO) 重级别油/重质油 对"73/78 防污公约"附则Ⅰ而言,系指下列任何油类:(1)在 15 ℃时密度高于 900 kg/m³ 的原油;(2)除原油外,在 15 ℃时密度高于 900 kg/m³ 或在 50 ℃时运动黏度高于 180 mm²/s 的油类;(3)沥青、焦油及其乳化物。 For "MARPOL73/78" Annex Ⅰ, heavy grade oil means any of the following: (1) crude oil having a density at 15 ℃ higher than 900 kg/m³; (2) oils, other than crude oil, having either a density at 15 ℃ higher than 900 kg/m³ or a kinematic viscosity at 50 ℃ higher than 180 mm²/s; (3) bitumen, tar and their emulsions.

Heavy grain 重谷物 系指比重大的谷物。

Heavy intervention 重型修井作业 系指需采用钻井隔水管和防喷器进行的修井作业,因此,只能采用半潜式钻井平台或钻井船来完成。

Heavy lift derrick cargo winch 重吊起货机 配合重吊使用的起货机。

Heavy oil 重油

Heavy seamless steel pipe 加厚无缝钢管

Heavy side scuttle 重型舷窗 具有较好抗风浪性能,装设在邻近载重线的规定船舷范围内的舷窗。

Heel angle 横倾角

Heel block 吊杆座滑车 安装在吊杆座处供引导吊货索用的滑车。

Heel block holder 吊杆座滑车眼板 设置在吊杆座处供安装吊杆座滑车用的眼板。

Heel on turning 回转横倾角 船舶回转时所产生的,向回转中心一侧的内倾角或向另一侧的外倾角。

Heeling 倾斜/横倾 系指船舶在任何方向偏离垂直方向的倾斜。船舶作任何倾斜时出现横侧力矩,此时船上的重力与浮力不在一条线上并使该船偏离垂直位置。 Heeling means the inclination of the vessel from the upright in any direction. A heeling moment exists at any inclination of the vessel where the forces of weight and buoyancy are not aligned and act to move the vessel away from the upright position.

Heeling ballacing system 横倾平衡系统

Heeling mode of descending and ascending 倾斜沉浮方式 系指是在沉浮过程中,人为地使平台纵倾,因而在下沉时平台一端先接触地面,然后另一端再以接地端为支点慢慢下降直至平台坐底;反之,在起浮时也是先让一端绕着另一端转动,然后再让着地端抬起、调平平台的沉浮方式。 Heeling mode of descending and ascending means a mode in which the unit will pitch to a certain degree during the descending in a controlled manner, so one end of the unit touches the seabed first and then the other end lowers slowly with the first end as a pivoting point; alternatively, when the unit is ascending from the on-bottom condition, one end will first ascend with the other end as a pivoting point and then let the other end ascend to even the unit.

Heeling moment 横倾力矩 使船舶产生横向倾斜的外力矩。

Heeling tank 横倾水舱

Heeling water pump 横倾平衡水泵

Height above the hull 船体以上高度 系指最上层连续甲板以上的高度。Height above the hull means height above the uppermost continuous deck.

Height between base of pontoons and upper deck 下浮体底与上甲板间距 半潜式钻井装置的下浮体的下底与上甲板面之间的距离。

Height of a superstructure 上层建筑高度 H 系指在舷侧所量得的自最上层建筑甲板横梁上缘至干舷甲板横梁上缘的最小垂直高度(m)。Height of a superstructure is the least vertical height measured at side from the top of the superstructure deck beams to the top of the freeboard deck beams.

Height of a superstructure or other erection 上层建筑或其他建筑物的高度 系指沿船侧从上层建筑或其他建筑物的甲板横梁上缘量到工作甲板横梁上缘的最小垂直距离。Height of a superstructure or other erection is the least vertical distance measured at side from the top of the deck beams of a superstructure or an erection to the top of the working deck beams.

Height of a tank h (m) 液舱高度 系指取自舱底至舱顶的垂直距离。不包括任何小舱口。Height of a tank(m) is to be taken as the vertical distance from the bottom to the top of the tank, excluding any small hatchways.

Height of handrail 栏杆高 扶栏顶缘至甲板面的垂直距离。

Helical type hydraulic steering gear 螺旋式液压舵机 采用活塞与螺旋副组成的特殊形式液压缸带动舵杆的液压舵机。

Helicoid 螺旋面 与轴线成固定角度的直线沿轴线方向以等线速前进,同时以等角速绕轴线旋转所形成的曲面。

Helicopter deck 直升机甲板 系指船上专门建造的直升机降落区域,包括所有结构、消防设备和其他为直升机的安全操作所必需的设备。见图 H-4。Helicopter deck is a purpose-built helicopter landing area located on a ship including all structure, fire-fighting appliances and other equipment necessary for the safe operation of helicopters. See Fig. H-4.

图 H-4 直升机甲板
Figure H-4 Helicopter deck

Helicopter facilities 直升机设施 系指:(1)直升机起降场地和结构、存放、消防、供油等设施;(2)包括任何加油和机库设施在内的直升机甲板。Helicopter facilities mean:(1) the areas and structures for takeoff and landing of helicopters, and storage, fire protection and oil supply facilities for helicopters;(2) a helideck including any refuelling and hangar facilities.

Helicopter landing deck 直升机降落甲板 作为直升机正常起降场地和布置有关设施的甲板。

Helicopter port 直升机港 人工建筑物上拟全部或部分用于直升机降落、起飞和地面转移的机场或规定区域。

Helicopter transit suit(immersion suit) 直升机运输服(救生服) 直升机乘客所穿戴的常穿服。Helicopter transit suit is the constant wear suit worn by helicopter occupants.

Helideck 直升机甲板(平台) 系指在海上移动式钻井平台(MODU)上专门建造的直升机降落平台。Helideck is a purpose-built helicopter landing platform located on a mobile offshore drilling unit(MODU).

Helium gas compressor 氦气压缩机 压缩介质为氦气的压缩机。

Helium storage room 储氦室 系指储存供充灌测风及探空气球用的压缩氦气(气瓶压缩,压力箱压缩)的舱室。

Helium-oxygen diving(heliox diving) 氦氧潜水 在潜水作业中,以氦氧为呼吸介质,其作业深度超过 $60 \sim 70$ m 的潜水。

Helix 螺旋线 与轴线保持一定距离的点,沿轴线方向以等线速前进,同时以等角速绕轴线旋转的轨迹。

Helm indicator　舵角指示器　系指用于驾驶室与操舵站指示舵板转动角度,即舵板对船中剖面的相对位置的仪器。

Helm telegraph gong　舵角传令钟　系指在船上用来传送操舵命令和检测舵板转向转角的设备。

Helm/Rudder　舵　用以驾驶船舶的装置。普通型舵设有垂直尾鳍、能从左舷35°移至右舷35°;舵的特征由面积、长宽比和形状表示。见图H-5。

图 H-5　舵
Figure H-5　Helm/Rudder

Helmsman　舵工　系指船舶航行时操纵舵的人员。Helmsman means a person who steers the ship under way.

Hemispherical end　压力容器的半球形封头

Hemp plant (cannabis sativa)　大麻植物　是一种野生的灌木植物。遍及世界上多数热带和温带地区,特别是中东、北美洲西北部分,东南亚和墨西哥。实际上它在世界上任何地方都可以生长,但主要"商业"运输一般源于西印度群岛、非洲、土耳其、印度次大陆和泰国。最重要的活性成分集中在这种植物顶部的树脂中。从印度大麻提炼出的麻药就是从这种植物中刮出来的树脂压成块的树脂。Hemp plant (cannabis sativa), is a bushy plant which grows wild throughout most of the tropical and temperate regions of the world, especially in the Middle East, south-western North America, South East Asia and Mexico. It can be grown virtually anywhere in the world although the major "commercial" movements generally originate in the West Indies, Africa, Turkey, the Indian subcontinent and Thailand. The most important active ingredients are concentrated in the resin at the top of the plant. Hashish or "hash" is resin scraped from the plant and compressed into block.

He-O$_2$ air-way system　氦氧混合气系统　在超过60~70 m的深潜作业中,向潜水员供应呼吸用的氦、氧混合气体的系统。

Hermetically sealed refrigerating unit　全封闭式制冷压缩机

Hermetically seated refrigerating compressor unit　全封闭式制冷压缩机组　制冷压缩机、电动机装在同一个密封的壳体内的压缩机组。

Hertz (Hz)　赫兹　频率的单位,即每秒一周。

Hewlett Packard (HP)　惠普　是面向个人用户、大中小型企业和研究机构的全球技术解决方案提供商。惠普(HP)提供的产品涵盖了IT基础设施,个人计算及接入设备,全球服务,面向个人消费者、大中小型企业的打印和成像等领域。在截止至2007年10月31日的2007财年中,惠普(HP)的营业额达1043亿美元。HP在2007美国财富500强中名列第14位。

Hewlett-Packard Company (HP)　惠普公司　美国电子工业企业,世界IT巨头。成立于1954年。创业者是威廉·休利特和大卫·帕卡德。当时他们的"公司"只是设在一个汽车库里。这就是著名的"博士+车库=惠普"的出处。他们的最初产品是音频振荡器。在用集成电路芯片从事商业生产和制造消费品,特别是在制造电子计算机主机和微型计算机方面,取得了迅猛的发展。同时惠普公司是面向个人用户、大中小型企业和研究机构的全球技术解决方案提供商。

Hexagonal head wrench　六角扳手

Hexagonal nut　六角螺母

Hidden arc welding　埋弧焊　利用在焊剂层下燃烧的电弧进行焊接的方法。在焊接过程中,焊剂熔化产生的液态熔渣覆盖电弧和熔化金属,起保护、净化熔池、稳定电弧和渗入合金元素的作用。埋弧焊分为自动埋弧焊和半自动埋弧焊两种。近年来,虽然先后出现了许多种高效、优质的新焊接方法,但是,埋弧焊的应用领域依然未受任何影响。从各种熔焊方法的熔敷金属质量所占份额的角度来看,埋弧焊约占10%左右,且多年来一直变化不大。

Hidden repairs　隐蔽工程　(1)即在装修后被隐蔽起来,表面上无法看到的施工项目。根据装修工序,这些"隐蔽工程"都会被后一道工序所覆盖,所以很难检查其材料是否符合规格、施工是否规范。(2)隐蔽工程系指敷设在装饰表面内部的工程。家庭装修的隐蔽工程主要包括6个方面:给排水工程、电气管线工程、地板基层、护墙板基层、门窗套板基层、吊顶基层。(3)隐蔽工程系指地基、电气管线、供水供热管线等需要覆盖、掩盖的工程。还应包括:基础各分项工程:混凝土、钢筋、砖砌体等及其他各部位的钢筋分项:屋面工程的,找平层,保温层,隔热层,防水层等。

Hierarchy 层次

High and low pressure controller 高低压控制器 当高压侧压力超过额定值或低压侧压力低于额定值时,控制器能使制冷压缩机停止工作,以起保护和控制作用的压力控制装置。

High and low water alarm 高低水位报警器

High brass 高锌黄铜(硬质黄铜,优质黄铜)

High definition multimedia interface(HDMI) 高清晰度多媒体接口 2002年4月,日立、松下、飞利浦、索尼、汤姆逊、东芝和Silicon Image等7家公司联合组成HDMI组织。HDMI能高品质地传输未经压缩的高清视频和多声道音频数据,最高数据传输速度为5Gbps。同时无须在信号传送前进行数字/模拟或者模拟/数字转换,可以保证最高质量的影音信号传送。HDMI可搭配宽带数字内容保护(HDCP),以防止具有著作权的影音内容遭到未经授权的复制。HDMI所具备的额外空间可应用在日后升级的音视频格式中。

High definition television(HDTV) 高清晰度电视 是一种电视业务下的新型产品,高清晰度电视应是一个透明系统,一个正常视力的观众在距该系统显示屏高度的3倍距离上所看到的图像质量应具有观看原始景物或表演时所得到的印象。HDTV 技术源之于 DTV(Digital television)"数字电视"技术,HDTV 技术和 DTV 技术都是采用数字信号,而 HDTV 技术则属于 DTV 的最高标准,拥有最佳的视频、音频效果。HDTV 有3种显示格式,分别是:720P(1280×720P,非交错式)、1080 i(1920×1080i,交错式)、1080P(1920×1080p,非交错式)。HDTV 基本可以分为 MPEG2-TS、WMV-HD 和 H.264 这三种算法,不同的编码技术自然在压缩比和画质方面有着区别。HDTV 文件都比较大,即使是经过重新编码过后的 WMV 文件也非同小可。

High efficient welding electrod and flux-cored welding wire 高效焊条及药芯焊丝

High elastic rubber coupling 高弹性橡胶联轴节

High expansion foam 高膨胀泡沫(高倍泡沫)

High expansion foam-extinguishing system 高倍泡沫灭火系统

High frequency response(HFR) 高频响应/和频响应 是由二阶波浪力中的和频项或更高阶的波浪荷载引起的运动响应。

High frequency sound 高频声 系指频率大于1000Hz的声音。

High frequency vibration 高频振动

High frequency(HF) 高频 频率通常为200~300千赫。

High gas temperature protective device 燃气高温保护装置 系指当燃烧室出口处的燃气温度高于许用值时,能发出报警的装置。 High gas temperature protective device means a device capable of sending out alarm when the gas temperature at burner outlet exceeds the permissible value.

High liquid level 超高液位

High holding power anchor 大抓力锚 系指:(1)适合于船舶任何时候使用,且不需要预先调整或在海底处于特殊位置的锚。这种锚的抓力应至少为同等质量专利锚(标准的无杆海军锚)抓力的两倍。(2)亦即按船级社材料规范适用要求证明为抓力高于普通锚抓力的锚,不需在海底作预调整或在海底做特殊布置。大抓力锚如用作艏锚,每只锚质量应不小于相应普通无杆锚质量的75%。 High holding power anchor are:(1)anchors which are suitable for ship's use at any time and which do not require prior adjustment or special placement on the sea bed. The anchor is to have a holding power at least twice that of a patent anchor(admiralty standard stockless) of the same mass.(2)anchors for which a holding power higher than that of ordinary anchors has been proved according to the applicable requirements of the Society's Rules for Materials, do not require prior adjustment or special placement on the sea bottom. Where HHP anchors are used as bower anchors, the mass of each anchor is to be not less than 75% of that required for ordinary stockless anchors.

High level language 高级语言 由于汇编语言依赖于硬件体系,且助记符量大难记,于是人们又发明了更加易用的所谓高级语言。在这种语言下,其语法和结构更类似汉字或者普通英文,且由于远离对硬件的直接操作,使得一般人经过学习之后都可以编程。高级语言通常按其基本类型、代系、实现方式、应用范围等分类。

High lift rudder(articulated rudder, flap-type rudder) 襟翼舵 舵叶后部的局部,作成一个可独立地由随动机构控制的襟翼的舵。

High line 高架索 设置在补给船与接收船桅杆指定高度上具有恒张力的绳索。

High manganese steel(HMS) 高锰钢 即含锰量在10%以上的合金钢,是专供重工业使用的一种防磨钢材。高锰钢是典型的抗磨钢,铸态组织为奥氏体+碳化物。经1 000 ℃左右水淬处理后,其组织转变为单一的奥氏体或奥氏体加少量碳化物,韧性反而增加。极低温用的高锰钢有一定量的锰元素,经过有关热处理后,即使在极低温条件下也可以保持理想的性能。另外,它的强度比目前 LNG 储存舱罐等系统使用的铝合金钢、镍合金钢等材料的强度要大得多。

High order harmonic　高次谐波

High order vibration　高阶振动

High peak stress　高应力

High performance ship　高性能船舶　系指这样的一类船舶,其航行性能的某一个(或几个)方面与常规船相比具有明显的优越性。比如,具有快速性优势的高速双体船、滑行艇、水翼艇、气垫船和地效翼船;对耐波性有重大改进的小水线面双体船和深 V 型船;具有良好两栖性的全垫升气垫船和气垫平台等。

High pitch propeller　大螺距螺旋桨

High power　大功率的

High pressure air conditioning　高压空调系统

High pressure air system　高压空气系统

High pressure charging　高压增压

High pressure compressor　高压压缩机　排气压力 9 800 ~ 98 000 kPa 表压的压缩机。

High pressure cylinder　高压缸

High pressure fluorescent mereury lamp　荧光高压汞灯　系指在石英放电管内充有高压汞蒸发气和少量惰性气体,玻璃外壳内壁涂有荧光粉的灯具。

High pressure fuel delivery line　高压燃油管

High pressure fuel pump　高压油泵

High pressure indication meter　高压指示表　能承受并指示高压空气压力数值的指示器。

High pressure oil pump　高压油泵

High pressure pumping system　高压泵送系统

High pressure receiver　高压容器

High pressure safety valve　高压安全阀

High pressure sea-water (HPSW) systems　高压海水(HPSW)系统　应能为全舰(船)灭火、弹药库喷淋、预湿、舱底水和脱水排泄器以及应急冷却连续供给和分配通常压力不小于 7 bar 的海水。 High pressure seawater(HPSW) systems are to continuously supply and distribute sea-water at a pressure of generally not less than 7 bar throughout the ship to provide water for fire-fighting, magazine spraying, pre-wetting, bilge and dewatering eductors and emergency cooling.

High pressure side　高压侧　对制冷装置而言,系指压缩机排气侧至膨胀阀之间的受压部件。 For the purpose of refrigerating Plant, high pressure side means the pressed parts of the compressor between its exhausting side and expansion valve.

High pressure side (supply-pressure side) (LPG system)　高压端(供气压力端)(LPG 系统)　LPG 供气管直接暴露于液化石油气系统中压力调节器进口压力的部分。注意:在 20 ℃ 时,丙烷蒸发气压力为 0.7 MPa(7 bar),丁烷蒸发气压力为 0.175 MPa(1.75 bar)。 High pressure side is the part of the LPG supply line directly exposed to pressure at the inlet of a pressure regulator in a liquefied petroleum gas system. Note: Vapour pressure propane at 20 ℃ is 0.7 MPa, vapour pressure butane at 20 ℃ is 0.175 MPa.

High pressure sodium lamp　高压钠灯　是在放电管内除汞及惰性气体外,还加入过量钠的灯具。

High pressure washing　高压水清洁

High pressure water jet cleaning　高压水喷射清洗

High pressure-fuel gas system(HP-FGS)　高压气体燃料供应系统

High recovery thermocouple　高恢复系数热电偶

High resistance alloy　高电阻合金　用以制造电阻元件的合金导电材料。它主要不是用于传导电流,而是以其高电阻来限制或控制电路中的电流,如在电机、电位器、标准电阻器、电位差计、滑线电阻器中应用的电阻元件、电阻线等;制造反映应变、温度、磁场、压力等参数的传感元件用的电阻合金;温度补偿器、测温电阻器中的电阻元件。高电阻合金还可用于制造各种电阻加热设备中的发热元件。

High sea　公海　系指不包括在一国领海或内海的全部海域。公海对所有国家开放,任何国家不得有效地声称将公海的任何部分置于其主权之下。公海自由是在日内瓦公海公约和其他国际法规则所规定的条件下行使的。公海自由对沿海国家和非沿海国家而言,包括:(1)航行自由;(2)捕鱼自由;(3)铺设海底电缆和管道的自由;(4)公海上飞越自由。所有国家行使这些自由以及国际法的一般原则所承认的其他自由时,都应适当顾及其他国家行使公海自由的利益。 High sea means all parts of the sea that are not included in the territorial sea or in the internal waters of a State. The high sea being open to all nations, no State may validly purpose to subject any part of them to its sovereignty. Freedom of the high sea is exercised under the conditions laid down by these Articles of the Geneva Convention on the high seas and by the other rules of international law. It comprises, inter alia, both for coastal and non-coastal States:(1) freedom of navigation;(2) freedom of fishing;(3) freedom to lay submarine cables and pipelines;(4) freedom to fly over the high seas. These freedoms, and others which are recognized by the general principles of international law, shall be exercised by all states with reasonable regard to the interests of other States in their exercise of the freedom of the high seas.

High sea suction　高位海水门　由进水格栅、海水箱、通海阀组成的装于船侧上的舷外水吸入口。

High skewed propeller(highly skewed propeller) 大侧斜螺旋桨

High sound insulation panels 高隔声壁板

High speed cavitation labouratory 高速空泡实验室 该技术拥有国：美国、中国、俄罗斯。见图 H-6。

图 H-6　高速空泡实验室
Figure H-6　High speed cavitation labouratory

High speed diesel engine 高速柴油机 曲轴转速在 1 500 r/min 以上的柴油机。

High speed life boat 救生快艇 专供援救海上遇难人员或护送伤病员用的快艇。

High speed motor 高速内燃机

High speed towing tank 高速船模试验水池 试验高速船模的船模试验水池。

High speed underway replenishment 高速航行补给 补给船在 16 节以上航速航行时对其他舰船进行的补给。

High speed yacht 高速游艇 系指其满载排水量时的最大航速 V 同时满足下式的游艇：$V \geqslant 3.7\ \nabla^{0.1667}$ m/s；$V \geqslant 25$ kn，式中，∇—满载排水量 $\triangle$ 对应的排水体积，m^3。

High-strength(quenched and tempered fine-grained structural steel) 高强度(淬火加回火)细化晶粒结构钢

High strength brass 锰黄铜(高强度黄铜)

High strength low-alloy steel 高强度低合金钢

High temperature alarm 高温报警器

High temperature alarm for bearings 轴承高温报警装置 系指当轴承的滑油温度过度升高时，能发出警报的装置。 High temperature alarm for bearings means a device capable of sending out warning when the lubricate oil is going up excessively.

High velocity air conditioning 高速空调系统

High velocity vent 高速通风 系指由机械阀组成的根据注入阀的压力调节开口使流速不低于 30 m/s，阻止火焰通过的装置。 High velocity vent is a device to prevent the passage of flame consisting of a mechanical valve which adjusts the opening available for flow in accordance with the pressure at the inlet of the valve in such a way that the flow velocity cannot be less than 30 m/s.

High viscosity category B NLS 高黏度的 B 类有毒液体物质 系指 20 ℃时其黏度至少为 25 MPa·s 和卸货时其黏度至少为 25 MPa·s 的 B 类有毒液体物质。 High viscosity category B NLS means any category B NLS having a viscosity of at least 25 MPa·s at 20 ℃, and at least 25 MPa·s, at the time it is unloaded.

High viscosity category C NLS 高黏度的 C 类有毒液体物质 系指 20 ℃时其黏度至少为 60 MPa·s 和卸货时其黏度至少为 60MPa·s 的 C 类有毒液体物质。 High viscosity category C NLS means any category C NLS having a viscosity of at least 60 MPa·s at 20 ℃, and at least 60 MPa·s, at the time it is unloaded.

High viscosity fuel 高黏度燃油

High viscosity NLS 高黏度有毒液体物质 系指包括 20 ℃时其黏度至少为 25 MPa·s 和卸货时其黏度至少为 25 MPa·s 的 A 类有毒液体物质、高黏度的 B 类有毒液体物质和高黏度的 C 类有毒液体物质。 High viscosity NLS includes category A NLSs having a viscosity of at least 25 MPa·s at 20 ℃, and at least 25 MPa·s, at the time they are unloaded, high viscosity category B NLSs and High viscosity category C NLSs.

High voltage 高电压 系指导线之间超过交流 1 000 V 或直流 1 500 V 的电压。 High voltage is a voltage exceeding 1,000 V A.C. or 1,500 V D.C. between conductors.

High voltage electrical installations 交流高压电气装置 系指额定电压(相向电压)超过 1 kV 的交流三相电气装置。 High voltage electrical installations mean the A, C, three-phase systems with nominal voltage exceeding 1 kV.

High voltage shore power supply system 交流高压岸电系统 系指在船舶靠港期间向船舶供电的设备，包括船载装置和岸基装置。港口向船舶配电系统供电的电源额定电压(相间电压)为 1 kV 以上，15 kV 以下。

High water lever alarm 高水位报警器(高水位报警)

High(higher) calorific value 高发热值

High-bandwidth digital content protection(HDCP) 高带宽数字内容保护技术 随着 HDTV(高清电视)时代即将来临，为了适应高清电视的高带宽，出现了 HD-MI。HDMI 是一种高清数字接口标准，它可以提供很高的带宽，无损地传输数字视频和音频信号。为了保证 HDMI 或者 DVI 传输的高清晰信号不会被非法录制，就出现了 HDCP 技术。HDCP 技术规范由 Intel 领头完成，当用户进行非法复制时，该技术会进行干扰，降低复制出来的影像的质量，从而对内容进行保护。

High-definition (HD) 高清 系指高清晰度。

High-density solid bulk cargo 高密度固体散货 系指积载因数等于或小于 $0.56 \, m^3/t$ 的固体散货。High-density solid bulk cargo means a solid bulk cargo with a stowage factor of $0.56 \, m^3/t$ or less.

Higher bronze 铝铁镍锰高级青铜

Higher harmonic 高次谐波

Higher harmonic term 高谐波项

Higher high level 高高水位

Higher low level 高低水位

Higher order vibration 高阶振动

Higher order wave 高阶波

Higher strength hull structural steel 高强度船体结构钢 系指最小屈服强度较高的钢材。Higher strength hull structural steel means a steel having a higher specified minimum yield stress.

Higher strength steel factor k 高强度钢系数 系指用于在使用高强度船体结构钢情况下确定船体梁剖面模数的系数。k 值见表 H-2。Higher strength steel factor k means a factor used for the determination of hull girder section modulus strength hull, where higher strength hull structural steel is used, a higher strength steel factor, k is given in table H-2.

表 H-2 高强度钢系数 k 值
Table H-2 Higher strength steel factor k

规定最小屈服强度 N/mm^2	k
235	1.00
265	0.93
315	0.78
340	0.74
355	0.72
390	0.68

Higher tensile strength iron casting 高强度铸铁 是将灰口铸铁铁水经球化处理后获得，析出的石墨呈球状。它与钢相比，除塑性、韧性稍低外，其他性能均接近，是兼有钢和铸铁优点的优良材料，在机械工程上应用广泛。

Higher-performance offshore anchor 较高性能海洋工程用锚 系指其抓力比一般性能海洋工程用锚更大的有(横)杆的转爪锚。如布鲁斯铸钢(Bruce SS)锚和布鲁斯(Bruce)锚及 TS 锚(前两种均为固定爪锚、后者可改变锚爪折角)和阔鳍型德尔泰(Flipper Delita)锚(锚爪折角可改变)以及史蒂汶(Stevin)系列的史蒂夫狄格(Stevdig)锚、史蒂夫莫特(Stevmud)锚、史蒂夫菲克斯(Stevfix)锚、霍克(Hook)锚等(均为转抓锚)。除史蒂夫莫特锚专用于淤泥底质外，其余的均能适应多种底质。

Higher-strength hull structural steel 高强度船体结构钢

High-holding power anchor 马氏大抓力锚 一种大抓力锚。其特点是在锚臂上有固定的横杆。主要用于海洋工程船舶。见图 H-7。

图 H-7 马氏大抓力锚
Figure H-7 High-holding power anchor

High-level radioactive wastes 强放射性废料 系指由第一阶段提炼系统作业所产生的废液或随后提炼阶段在辐射性核燃料再加工装置中聚集的废液或由液体废料转换成的固体。High-level radioactive wastes means liquid wastes resulting from the operation of the first stage extraction system or the concentrated wastes from subsequent extraction stage, in a facility for reprocessing irradiated nuclear fuel, or solid into which such liquid wastes have been converted.

High-line transfer system 高架索传送系统 由架设在桅杆上承受负荷的高架索、牵引传送吊车的内、外牵索及绞车、作动筒、张力器等组成的传送系统。

Highline trolley 传送吊车 在高架索上由内外牵索牵引用以承受货物的吊车。

Highly skewed blade 大侧斜叶(大侧斜桨叶) 近梢的部分向正车旋转相反方向弯曲很大的螺旋桨叶。

Highly skewed propellers and very highly skewed

propellers 大侧斜螺旋桨和超大侧斜螺旋桨 有侧斜角超过25°的叶片的螺旋桨称为大侧斜螺旋桨;有侧斜角超过50°的叶片的螺旋桨称为超大侧斜螺旋桨。见图H-8。 Highly skewed propellers are propellers having blades with skew angle exceeding 25°. Very highly skewed propellers are propellers having blades with skew angle exceeding 50°. See Fig. H-8.

图 H-8 大侧斜螺旋桨和超大侧斜螺旋桨
Figure H-8 Highly skewed propellers and very highly skewed propellers

High-performance offshore anchor 高性能海洋工程用锚 系指均由钢板焊接而成的固定爪锚。如史蒂芙帕瑞斯(Stevpris)锚、史蒂芙夏克(Stevshark)锚、布鲁斯(Bruce)锚、FFTS锚等。该类锚的抓力远大于各种锚,甚至在软泥中也有很大的抓力,是当前最先进的拖starter埋置锚。

High-pressure compressor 高压压气机 压缩过程由几台压气机完成时,最后对空气或其他气体进行压缩的压气机。

High-pressure feed water heater 高压给水预热器 设置在给水泵排出管路中加热给水的热交换器。

High-pressure piping 高压管路(高压管系) 系指:(1)最大工作压力大于 10 bar(1 bar = 10^5 Pa) 的气体燃料管路;(2)在燃油压力泵或液压油压力泵的出口以下的管系。 High-pressure piping means: (1) gas fuel piping with maximum working pressure greater than 10 bar(1 bar = 10^5 Pa); (2) a piping in the downstream of the fuel oil pressure pump or hydraulic oil pressure pump.

High-pressure rescue chamber (HRC) 高压逃生舱 特殊情况下,供潜水员在高压状态下逃生和转运用的装置。其由一个传物舱和一个能快速解脱开的人行通道组成,并在舱外建造有一高强度的管形构架提供支撑和保护。高压逃生舱具备多名潜水员(如 8 ~ 10 名)的救生和逃生功能,并能用作正常情况下两名潜水员的加压和减压。高压逃生舱携带有氧气和混合气,能支持 72 h,携带的电池也要求维持 72 h。有适配的、安放在其他地方的生命支持控制系统(Fly away package),并能在 24 h 内到达高压逃生舱所在的位置。

High-pressure turbine 高压涡轮 高温燃气或其他气体首先在其中进行膨胀的涡轮。

High-pressure water pump 高压水泵(挖泥船) 系指挖泥船泥泵装置中的冲水泵以及为了提高泥浆浓度而在耙管吸口或铰刀头附近安装的冲水泵。

High-pressured water jet gear 高压水松土装置 装在吸扬式挖泥船上,用高压水冲松泥土,以提高挖泥效率的装置。

High-silica glass wool 高硅氧玻璃棉

High-speed craft 高速船 系指最大航速(m/s)等于或大于下列值的船:$3.7 \nabla^{0.1667}$,式中,∇ = 相应于设计水线的排水量,m^3。不包括在非排水状态下船体由地效应产生的气动升力完全支承在水面以上的船舶。 High-speed craft is a craft capable of a maximum speed, in m per second (m/s), equal to or exceeding: $3.7 \nabla^{0.1667}$, where: ∇ = volume of displacement corresponding to the design waterline (m^3), excluding craft the hull of which is supported completely clear above the water surface in non-displacement mode by aerodynamic forces generated by ground effect.

High-Speed Craft Code, 1994 (1994 HSC Code) 1994 年高速船规则 系指 IMO 组织海上安全委员会 MSC.36(63)决议通过并可能经 IMO 组织修正的"国际高速船安全规则",但该修正案应按 SOLAS 公约的第Ⅷ条有关适用于除第 1 章以外的附则修正程序的规定予以通过、生效和实施。 High-Speed Craft Code, 1994 (1994 HSC Code) means the International Code of Safety for High-Speed Craft adopted by the Maritime Safety Committee of IMO by resolution MSC.36(63), as may be amended by IMO, provided that such amendments are adopted, brought into force and take effect in accordance with the provisions of article Ⅷ of SOLAS Convention concerning the amendment procedures applicable to the annex other than chapter I.

High-Speed Craft Code, 2000 (2000 HSC Code) 2000 年高速船规则 系指 IMO 组织海上安全委员会 MSC.97(73)决议通过并可能经 IMO 组织修正的"国际高速船安全规则",但该修正案应按 SOLAS 第Ⅷ条有关适用于除第 1 章外的附则修正程序的规定予以通过、生效和实施。High-Speed Craft Code, 2000 (2000 HSC Code) means the International Code of Safety for High-Speed Craft, 2000, adopted by the Maritime Safety Committee of IMO by resolution MSC. 97(73), as may be amended by IMO, provided that such amendments are adopted, brought into force and take effect in accordance with the provisions of article Ⅷ of SOLAS Convention concerning the amendment procedures applicable to the annex other than chapter Ⅰ.

High-speed craft safety certificate 高速船安全证书 系指经初次检验或换证检验完成后,对全部符合高速船(HSC)规则要求的高速船签发的高速船安全证书。

High-speed craft safety certificate means a certificate called a High-Speed Craft Safety Certificate issued after completion of an initial or renewal survey to a craft which complies with the requirements of the 1994 HSC Code or the 2000 HSC Code, as appropriate.

High-speed engine 高速机 系指燃气轮机或高速柴油机。 High speed engine is the high-rotating-speed internal combustion engine or gas turbine.

High-supersonic weapon 高超音速武器 系指以高超音速飞行技术为基础,飞行速度超过5倍音速的武器。此类武器航程远,速度快,性能卓越,被军事专家称为继螺旋桨、喷气推进之后航空史上第三次革命性成果,将对未来的战争产生巨大的影响。

High-vacuum treatment equipment for welding flue gas and dust 高真空焊接烟尘治理设备 利用高压风机产生的巨大负压,在焊接烟尘生成初期,第一时间捕集尚未扩散的烟尘微粒,并使其经过高真空软管进入除尘器进行过滤和集中收集的装置。其由主体设备、管路系统和自动控制系统组成。

High-velocity induction air conditioning system 高速诱导空气调节系统 部分舱外新鲜空气和部分舱室内循环空气两者混合,或全部是舱外新鲜空气经集中式空气调节器处理后的一次风由高压高速通风机经通风管分别送至各舱室内的诱导器的空气调节系统。

High-viscosity substance 高黏度物质 对"73/78防污公约"附则Ⅱ而言,高黏度物质系指在卸货温度下黏度等于或高于50 MPa·s的X或Y类有毒液体物质。

For the purposes of the "MARPOL73/78" AnnexⅡ, high-viscosity substance means a noxious liquid substance in category X or Y with a viscosity equal to or greater 50 MPa·s at the unloading temperature.

High-voltage system 高压系统 系指额定电压大于1 kV但不超过15 kV,额定频率为50 Hz或60 Hz的交流系统,或在额定工作条件下最高瞬时电压超过1 500 V的直流系统。High-voltage system is a system operating with the rated voltage more than 1 kV but not exceeding 15kV and with rated frequencies of 50 Hz or 60 Hz or direct-current system where the maximum instantaneous value of the voltage under rated operating conditions exceed 1,500 V.

High voltage direct current transmission (HVDCT) 高压直流输电 系指直流电压在10万伏特到100万伏特范围内电力输送。发电机(如风力发电机)产生的三相交流电首先经过整流变成直流,然后在传输线末端再次转换成三相交流电。HVDC适用于远距离电力传输和转换大容量电力。一条三相交流电传输线对于阿尔法·文图斯海上风电试验场而言已经足够。然而在未来几年内,距离海岸100 km以上的风电场将由HVDC线路连接陆地的连接点。

Hinder 阻碍

Hinged cradle type davit 铰链型艇架 吊艇臂下端前后各设一副铰链,放艇时,铰链转动使吊艇臂外移和倾倒,从而吊艇伸出舷外的重力式艇架。

Hinged flap hatch covers 铰翻型舱盖 舱盖板一端装有铰链与舱口相连,可以翻转开启的铰链式舱盖。

Hinged folding hatch cover 折叠型舱盖 舱盖板分为多块,块间装设铰链,依靠翻转、折合以进行开启的铰链式舱盖。

Hinged hatch covers 铰链式舱盖 舱盖板依靠铰链,翻转折叠而开启的机械舱盖。

Hinged manhole cover 铰链式人孔盖 用盖板铰链和人孔的围板或座板连接的人孔盖。

Hinged rudder (multi-pintle rudder) 多支承舵 用三个以上的舵销将舵叶与舵柱连接的舵。

Hinging cradle type davit 铰链式艇架 吊艇臂下端前后各设一副铰链,放艇时,铰链转动使吊艇臂外移和倾倒,从而吊艇出舷的重力式艇架。

Hipped plate acoustic filter 折板式消音器

Hipped plate construction 折板式结构

Histogram 直方图 以观测值作横坐标并细分成大量区间,观测值落入各区间的频率则用相应的矩形面积表示的分布图。

HMI = human machine interface 人机界面

HNS Fund 有害有毒物质基金 系指根据国际海上运输有害有毒物质损害责任和赔偿公约第13条设立的国际有害有毒物质基金。 HNS Fund means the International Hazardous and Noxious Substances Fund established under article 13 of the International Convention on liability and compensation for damage in connection with the carriage of hazardous and noxious substances by sea.

Hogging 中拱 船体中部上拱,艏艉部下垂的弯曲状态。

Hoist trunk 升降通道 系指在住舱和作业区域的所有通道。其净空尺寸必须不受各种装置和设备的限制。在此区域的甲板不准有会导致绊倒的突出物。

Hoisting 变量提升

Hoisting apparatus 起重设备 系指安装于船上或海上设施上的吊杆装置、吊杆式起重机、起重机以及升降机和跳板，用以吊运或装运货物、设备、物品及人员等的设备。

Hoisting gear for cutter ladder 绞刀架起落装置 用以操纵绞刀架起落的装置。

Hoisting gear for drag-head ladder 耙头架起落装置 用以控制耙头架起落的装置。

Hoisting gear for glide chute 溜泥槽起落装置 用以起落溜泥槽活动段的装置。

Hoisting machinery 起重机械 是在货船、客货船及大多数工程船上用来吊放货物或工程机械等的甲板机械。

Hoisting winch 起落绞车 挖泥船上，用以起落绞刀架、耙臂、斗桥或吸泥管等的绞车。

Hold 货舱 对集装箱船而言，系指装有冷藏集装箱的封闭处所。这些集装箱通常被限定在集装箱格导轨之内。对于无舱口盖船，货舱系指舱口围板以下的处所。

Hold freeing ports 货舱排水舷口 设于货舱区域满载水线以上船体两侧贯通船壳内外装有截止止回阀的开口，即任何情况下，舱内水能通过该阀门溢出船外而舱外海水不能进入船内。Hold freeing ports are openings penetrating the hull on both sides of the ship above full load waterline within hold area, fitted with non-return valves through which water in a hold overflows outboard while external seawater will not enter the hold in any case.

Hold space 货舱处所 系指：(1)由船舶结构围蔽，且其内部设有独立的液货舱处所。(2)对集装箱船而言，系指装有冷藏集装箱的封闭处所。这些集装箱通常被限定在格栅导轨之内。对于无舱口盖船，货舱系指舱口围板以下的处所。 Hold space is (1) the space enclosed by the ship's structure in which an independent cargo tank is fitted. (2) For the purpose of container vessel, hold space means an enclosed space containing refrigerated containers. The containers are usually restrained within cell guides. For hatch coverless ships, hold space means the space below the hatch coamings.

Holder 持有人 系指享有根据国际海事组织电子提单规则一个条款所列权利并拥有有效密码的一方。 Holder means the party who is entitled to the rights described in relevant article of CMI Rule for electronic bills of lading by virtue of its possession of a valid private key.

Holding bolt 固定螺栓

Holding down bolt 底脚螺栓

Holding load(brake load) 制动载荷 系指锚链轮制动装置应承受的锚链上的最大静载荷。 Holding load is the maximum static load on the anchor chain cables which the cable lifter brake should withstand.

Holding moment 制动力矩 绞车制动器(或绞车)能产生保证载荷维持在静止状态下的力矩。该力矩并具有一定制动安全系数；其值的大小按工作类型不同而异。

Holding power 锚抓力 锚躺卧在水底所提供的抓住力。

Holding screw 埋头螺钉

Holding tank 集污舱 对 MARPOL73/78 附则 Ⅳ 而言，系指用于收集和储存生活污水的舱柜。 For "MARPOL73/78" Annex Ⅳ, holding tank means a tank used for the collection and storage of sewage.

Holland profile(HP) 荷兰型材 HP means Holland profile.

Hollow 中空(空心的,凹槽)

Hollow blade 空心叶片 为实现冷却或减少应力等目的而制成的具有内部空腔的涡轮叶片。

Hollow of wave making resistance curve 波阻谷点 由于船首与船尾横波的抵消作用，在兴波阻力曲线上出现的凹入点。

Hollow piston 空心活塞

Hollow plate 空心板 (1)用混凝土浇筑而成。将板的横截面做成空心的称为空心板。空心板较同跨径的实心板质量轻，运输安装方便，建筑高度又较同跨径的T梁小，因之小跨径桥梁中使用较多。其中间挖空形式有很多种。(2)也叫中空隔子板、万通板、瓦楞板、双壁板。是一种质量轻(空心结构)、无毒、无污染、防水、防震、抗老化、耐腐蚀、颜色丰富的新型材料，相比于纸板结构产品，中空板具有防潮、抗腐蚀等优势。相比于注塑产品，中空板具有防震、可灵活设计结构，不需开注塑模具等优势。同时，该板材可通过原料的控制灵活地加入防静电、导电母料等，生产出具有导电、防静电功能的塑料中空板板材，防静电板材表面电阻率可控制在 $10^6 \sim 10^{11}$ 之间。导静电板材表面电阻率可控制在 $10^3 \sim 10^5$ 之间。

Hollow screw 空心螺钉

Hollow shaft(tubular-shaft) 空心轴 整个轴段上有轴向通孔的轴。

Hollow structure 中空结构 系指舵、舵杆、艉鳍、铸件、桅杆、吊杆栏杆等船上附件具有封闭空间者。

Hollow wadding plate 中空填棉板

Hollowness 凹度(空心度)

Home computer 家用电脑　是一款电子产品,是供家庭上网,学习,听音乐,看电影用的电脑,主要用于家庭生活。

Homogeneous loading condition 均匀装载工况　对散货船而言,系指货物密度修正后两个最前部货舱最高与最低装载货物高度之比不超过 1∶20 的装载工况。　For bulk carrier, homogeneous loading condition means a loading condition in which the ratio between the highest and the lowest filling ratio, evaluated for the two foremost cargo holds, does not exceed 1∶20, to be corrected for different cargo densities.

Homogeneous material 均质材料　系指成分完全一致的不能经机械拆解为不同物质的材料,即指该材料原则上不能由机械作用,例如拆卸、切割、粉碎、打磨和研磨过程予以分离。　Homogeneous material means a material of uniform composition throughout that cannot be mechanically disjointed into different materials, meaning that the materials cannot, in principle, be separated by mechanical actions such as unscrewing, cutting, crushing, grinding and abrasive processes.

Homogenization 均质处理(均质化,均质化处理)

Homogenizer 均质器(均化器)

Honeycomb sandwich plate 蜂窝夹心板

Honeycomb fashion acoustic filter 蜂窝式消音器

Hong Kong International Convention for the Safe and Environmentally Sound Cycling of Ships, 2009 2009 年香港国际安全与环境无害化拆船公约

Honorary Chairman 名誉主席

Hood 头罩(盖,罩形舾楼)　系指其能全部覆盖头部、颈部,并且能覆盖肩膀部位的头套。　Hood means a head covering which completely covers the head, neck, and may cover portions of the shoulders.

Hood support 机罩(支架)

Hook 吊钩　是起重机械中最常见的一种吊具。吊钩常借助于滑轮组等部件悬挂在起升机构的钢丝绳上。吊钩按形状分为单钩和双钩;按制造方法分为锻造吊钩和叠片式吊钩。叠片式吊钩由数片切割成形的钢板铆接而成,个别板材出现裂纹时整个吊钩不会破坏,安全性较好,但自重较大,大多用在大起重量或吊运盛钢水桶的起重机上。吊钩在作业过程中常受冲击,须采用韧性好的优质碳素钢制造。吊钩成型需要一系列复杂的工艺,透热锻造是所有工艺中最重要的工艺。

Hook 索具钩　连接在滑车或绳索端头,用以吊挂、固定或牵引重物的钩状件。常用的有:正面钩、侧面钩、抱钩、旋转钩、牵索钩、弹簧钩、脱钩等。

Hook assembly 艇钩装置　系指附于救生艇,连接救生艇与艇索的装置。　Hook assembly is the mechanism, attached to the lifeboat, which connects the lifeboat to the lifeboat falls.

Hook cycle 钩吊周期　吊装装卸货物时完成一次循环所需的时间。

Hook damage 钩损　系指货物在装卸搬运的操作过程中,由于挂钩或用手钩不当而导致货物的损失。

Hook damage risk 钩损险　投保平安险和水渍险的基础上加保此险,保险人负责赔偿承保的货物(一般是袋装、箱装或捆装货物)在运输过程中使用手钩、吊钩装卸,致使包装破裂或直接钩破货物所造成的损失及其对包装进行修理或调换所支出的费用。如粮食包装袋因吊钩钩坏而造成粮食外漏的损失。

Hook locking part 艇钩锁定装置　系指艇钩装置中保持活动钩体处于闭锁位置直至被操作装置激发以释放艇钩的部件。该激发可通过艇钩装置的其他部件进行。　Hook locking part is the component(s) within a hook assembly which holds the movable hook component in the closed position until activated by the operating mechanism to release the hook. This activation may be performed through other components within the hook assembly.

Hook quick-release device 拖钩快速释放装置　是在拖钩受载时,能由设在驾驶室或尽可能近的地方的一个遥控装置操纵。要求在危险状况下,不必顾及横倾角和拖索方向而立即释放拖钩的装置。　Quick-release device is a device which is capable of being operated from a remote control device on the bridge, or as near as practicable, while the hook is under load. It is required that, in the case of a critical situation, the towline can be immediately released regardless of the angle of heel and the direction of the towline.

Hook shackle 吊钩卸扣　供吊货钩与吊钩装置的其他零件相连接用的,其横销带有螺纹且两端不凸出在外的卸扣。

Hook-bolt(patch bolt) 堵漏螺丝杆　用以固定和压紧堵漏板的螺杆夹紧器。有活动型、钩头型和 T 型等。

Hoop reinforcement 环向增强层

Hopper 开底泥舱　系指在挖泥船和泥驳上能打开舱底泥门,自动卸出挖掘物的泥舱。　Hopper means a mud hold designed to carry dredged spoil and arranged to be opened from the dredger or hopper barge to enable the spoil to be automatically discharged through the doors in the bottom of the ship.

Hopper barge 泥驳 系指专门用于输送泥浆的驳船。 Hopper barge is a barge dedicated to carrying mud.

Hopper dump barge 开底泥驳 系指能打开舱底活动门自动卸泥的泥驳。启、闭泥门的动力有电动、液压和手动。

Hopper plating 底边舱斜板 系指内底和内壳纵舱壁垂直部分之间舱室的斜面全长板材。 Hopper plating means plating running the length of a compartment sloping between the tank top and inner side shell.

Horizontal 横向的(水平的)

Horizontal accommodation 水平融通 系指在同一年度内组与组之间，项与项之间在一定百分率内互相使用的权限。

Horizontal axis 水平轴

Horizontal bracket 水平肘板 近似水平地装设的肘板。

Horizontal control surfaces 水平舵翼 舵翼中用以提供垫升航态时的纵倾稳定性并控制纵倾姿态的部分，包括水平安定面和水平舵。

Horizontal displacement compensator 水平位移补偿装置 浮式钻井装置在钻井过程中为克服由于波浪作用引起船体相对于钻井井口的水平位移而采取的补偿装置。

Horizontal engine 卧式发动机

Horizontal flow type 水平吹风式 冷风从舱壁送风管上各个出风口水平吹出，经过货堆后仍返回冷风机的吹风方式。

Horizontal girder 水平桁 支承舱壁扶强材或槽型舱壁板的水平构件。

Horizontal integration 水平一体化 它是由经济发展水平相同或接近的国家所形成的经济一体化组织。

Horizontal internal combustion engine 卧式内燃机 气缸中心线平行于基座安装面的内燃机。

Horizontal lay 水平铺管

Horizontal projection fish finder 水平探鱼仪 利用回声原理，探测渔船周围各方位鱼群的仪器。

Horizontal pump 卧式泵 系指回转泵的轴为卧轴，或往复泵的缸体为卧式的泵。

Horizontal return flue boiler("Scotch" boiler) 卧式烟管锅炉 烟气出炉膛后在水平布置的烟管内作多次来回流动，整体外形呈卧置的回焰式烟气锅炉。

Horizontal shell and tube evaporator 卧式壳管蒸发器 卧式圆筒状，两端焊有管板，管板上扩胀或焊有很多冷却管，制冷剂在管外蒸发，载冷剂在管内流动冷却的热交换器。

Horizontal sliding door 侧移式水密滑门 门板沿着水平方向移动开闭的水密滑门。

Horizontal stabilizing fin 水平安定面 水平舵翼中用以提供垫升航态时垂直面内稳定性的不能转动的部分。

Horizontal vibration 水平振动 系指船体在水平面上所产生的横向水平弯曲振动。

Horizontal warping winch 卧式绞缆机 用来绞收或放出缆索的动力机械。在船舶靠泊过程中使用。卧式绞缆机的卷筒中心线为水平位置。

Horizontal wave bending moment M_{WH}(kN-m) V 波浪水平弯矩 M_{WH} 系指所考虑船体横剖面处的波浪水平弯矩。 M_{WH} means horizontal wave bending moment(kN-m), at the hull transverse section considered.

Horizontal web 水平隔板 舵构架中与舵杆轴线垂直的构件。

Hormuz Strait 霍尔木兹海峡 位于亚洲西南部，介于伊朗与阿拉伯半岛之间，东接阿曼湾，西连海湾(伊朗人称之为波斯湾，阿拉伯人称之为阿拉伯湾)，呈人字形。由于它是海湾与印度洋之间的必经之地，霍尔木兹海峡素有"海湾咽喉"之称，具有十分重要的战略和航运地位。海湾沿岸产油国的石油绝大部分通过这一海峡输往西欧、澳大利亚、日本和美国等地，合计承担着西方石油消费国60%的供应量，西方国家把霍尔木兹海峡视为"生命线"。

Horn/Whistle 汽笛 系指能够发出规定笛声并符合国际海上避碰规则附录Ⅲ所载规格的任何声响信号器具。

Horseshoe thrust bearing 扇形块推力轴承 由许多扇形推力块组成轴承承压面的一种推力轴承。

Hose 软管(挠性管,消防水带)
Hose carrier 软管车(水龙带小车)
Hose connection 软管连头
Hose connection valve 软管连接阀
Hose coupling 软管接头
Hose davit 装油软管吊杆
Hose extension 软管加长节
Hose fire 救火水龙管
Hose rack 软管架(消防水龙管架)
Hose survey 冲水检验

Hose test(hose testing) 冲水试验 其进行是用来证明未做静水压或渗漏试验的结构项目和对船体的水密或风雨密完整性有作用的其他结构件的紧密性。这种试验应在一水管压力不小于 2.0 bar(2.0 kgf/cm²) 和最大距离为1.5 m 的条件下进行。喷嘴直径应不小于12 mm。冲水水柱应直接对准试验的焊缝或密封处。

Hose testing is carried out to demonstrate the tightness of

structural items not subjected to hydrostatic or leak testing, and other components which contribute to the watertight or weathertight integrity of the hull. This is to be carried out at a maximum distance of 1.5 m with a hose pressure not less than 2.0 bar(2.0 kgf/cm²). The nozzle diameter is not to be less than 12 mm. The jet is to be targeted directly onto the weld or seal being tested.

Hose(LPG system) 软管(LPG 系统) 柔性材料制成的管道。 Hose is the pipeline of flexible material.

Hospital ship/Hospital vessel 医院船 系指海上浮动医院。按照1949年《改善海上武装部队伤者、病者及遇船难者境遇之日内瓦公约》规定,医院船壳体的水线以上涂白色,在两舷和甲板上标有红十字(或红新月或红狮日)图案,悬挂本国国旗和红地白十字旗,在任何情况下不受攻击和捕拿。根据相关国际法规定,医院船是不可侵犯的,医院船有义务救助交战双方的伤员,交战各方均不得对其实施攻击或俘获,而应随时予以尊重和保护。大型医院船是现代海军的重要标志之一。目前,世界上共有美国、英国、加拿大、日本、中国等少数国家拥有具有远海医疗救护能力的医院船,这些医院船大多数由民船改装而成。其中,美国海军有两艘"仁慈"级医院船,主要在南美和东南亚国家定期进行医疗救援活动,并在发生大规模灾害之际提供紧急救援。英国有一艘私人医疗船"非洲爱心号",主要为世界上欠发达地区或战乱地区提供慈善医疗活动。我国也拥有此类船舶。见图 H-9。

图 H-9 医院船
Figure H-9 Hospital ship/Hospital vessel

Hot air heating 热空气加热(热空气供暖)
Hot air welding 热气焊
Hot and cold water pipelines 冷热水管路 船上供应生活用的冷水和热水管路。
Hot bending 热弯
Hot brine defrost 热盐水融霜 利用热盐水在盘管内循环将盘管表面霜层融化的融霜方式。
Hot cargo 热货 水果、蔬菜等未经预冷的货物。
Hot corrosion 热腐蚀

Hot dip galvanizing 热浸镀锌法
Hot dip plating 热浸涂镀 是将已清洗洁净的铁件,经由 Flux 的润湿作用,浸入锌浴中,使钢铁与熔融锌反应生成一合金化的皮膜。
Hot dipped galvanized steel wire 热浸镀锌钢丝
Hot dipping process 热浸法
Hot finished seamless steel pipe 热轧无缝钢管
Hot finished seamless tube 热轧无缝钢管
Hot galvanizing 热镀锌(热电镀)
Hot gas defrost 热排气融霜 利用制冷压缩机高温高压排气通入盘管,将盘管表面霜层融化的融霜方式。
Hot gas pass 热气通道
Hot oil quenching 热油淬火
Hot peening 热喷丸
Hot press 暖菜柜 装有加热设备,用于保暖和加温食品的柜子。
Hot setting glue 热固树胶
Hot setting resin 热凝树脂
Hot spot 热点 系指可能产生疲劳裂纹的部位。 Hot spot is the location where fatigue crack may initiate.
Hot spot stress 热点应力 是包括公称应力和构件结构的不连续所产生的应力的增加,不考虑焊道形状等局部缺口所产生的应力集中效应。在板结构处,集中应力沿板厚度方向呈直线分布,分为膜应力和板的弯曲应力。通常集中应力大于公称应力,但在距离结构的不连续点足够远的位置,集中应力和公称应力相同。计算集中应力时,采用有限元方法进行结构分析;在焊接结构中,为去除缺口效应,应采用离焊趾足够远处的应力,通过直线外推法求得集中应力。另外,由于在焊趾附近的应力分布随有限元大小和种类的不同影响非常大,因此在采用时应注意保持其一贯性。 Hot spot stress includes all stress-raising effects by a structural discontinuity. It does not include the stress peak effect caused by the local notch and the weld bead shape. The hot spot stress in plate structure is divided into the membrane stress and the bending stress. It is linearly distributed in thicknesswise. In general, hot spot stress is greater than the nominal stress, but hot spot stress is equal to nominal stress at the position far enough from the discontinuity of structures. For the calculation of the hot spot stress, the three dimensional finite element analysis is to be performed. Then, it can be determined by extrapolating maximum principal stresses outside the region affected by the weld geometry. The stress range near welding toe is to be used consistently depending on the effect by type and size of the finite element.

Hot stand-by(DPS) 热备用(动力定位系统) 系指动力定位系统的冗余单元和系统能立即投入运行。
Hot stand-by(DPS) means redundant components and system of dynamic positioning system are to be immediately available.

Hot standby system 热备用系统 是一个由两个(相同或不相同的)执行同一功能的独立系统组成的自动化系统,其中一个独立系统工作时,另一个带有转换开关的独立系统就作为备用。 Hot standby system is used to describe an automation system comprising two(identical or non-identical) independent systems which perform the same function, one of which is in operation while the other is on standby with an automatic change-over switch.

Hot starting 热态启动 汽轮机在停车后短时间内尚未充分冷却,温度还高于环境温度状态下的重新启动。

Hot state operating 热态作业 系指船舶在建造、改建、修理或拆船作业中的铆、焊、气割或类似的明火作业。除有船舶化学家的认可外,磨、钻、喷砂研磨或会产生火花的作业都应作为热态作业。

Hot water circulating pipe lines 热水循环管路 为防止生活用热水在管路内间歇地使用而自然冷却所设置的循环管路。由热水供应管与热水回流管两部分组成。

Hot water circulating pump 热水循环泵 生活用热水系统中,使热水通过热水器循环,以防止热水冷却的泵。

Hot water heating 热水供暖(供热设备)

Hot water heating system 热水供暖系统 采用热水作工质的舱室取暖系统。

Hot water jacket 热水套

Hot water return pipe lines 热水循环回流管 系指热水管路或热水供应管最远端,即冷端接回加热水柜的回流管路。

Hot water tank 水加热柜(热水柜) 热水管路中用以加热冷水的柜。

Hot water unit 热水加热器 系指由燃油、废气或电加热水的一种加热装置,其中被加热水的温度始终低于该液体在大气压下的沸点温度。

Hot well(condensate sump, cascade tank, feed water filter well) 热井 (1)表面式冷凝器底部贮存冷凝水的容器;(2)开式给水系统中用于过滤给水中油分和调节系统内水量的敞开式水柜。

Hot work 高温作业 系指无论在船上何处,需要使用电弧或气体焊接设备、切割喷枪设备或其他形式的火焰,以及加热或产生火花的工具的任何活动。 Hot work means any activity requiring the use of electric arc or gas welding equipment, cutting burner equipment or other forms of flame, as well as heating or spark-generating tools, regardless of where it is carried out on board a ship.

Hot-bulb diesel engine(semi-diesel engine) 热球柴油机 用气缸盖上热球的高温热量或其中的电热丝来点火燃烧的柴油机。

Hot-water boiler 热水锅炉 用以制取热水的锅炉。

Hot-water unit 热水单元 能为潜水员提供稳定热水供应的加热系统用的机组。该机组可使用淡水或海水作为加热工质。其由泵和控制装置组成。

Hot-work 热作 系指焊接或钢材面板切割。
Hot-work means the welding or steel flame cutting.

House "室" 这一术语包括了上层建筑和甲板室。最低层或第一层"室"通常系指直接位于计量规范型深 D 时所量到的这层甲板。第二层指位于最低层之上的那一层,依次类推。The term 'house' is used to include both superstructures and deckhouses. The lowest, or first tier of a house, is normally that which is directly situated on the deck to which, D, is measured. The second tier is the next tier above the lowest tier and so on.

House flag 公司旗 用以表示船舶所属航运单位或公司的号旗。

House flag staff 公司旗杆 设置在后桅上,供悬挂公司旗用的斜置式旗杆。

House ship 房船 一种类似于水上旅馆、游艇和陆地上房车的移动式水上船舶。可替代陆地上的房屋作为居住地。房船具有时尚的外形、精致的内部结构、"水上房屋"的功能和游艇的休闲式,适合于短期出差的外籍人员及其家庭租用或使用。也可作为豪华游艇使用。房船一般设有主人房、客房、客厅、书房以及厨房、卫生间、洗衣间、储藏室等。与普通游船相比,房船在内部布置方面具有更多的家居气息,特别是在家居配置方面突出了生活感、趣味感和舒适性。比如船上设有可以种花的大面积露天阳台,卫生间配备了带有可升降电视的浴缸和带有音响功能的淋浴系统等。房船可停靠港湾、码头的专用泊位或港池。

Hovercraft/air-cushion ship 气垫船 是一种以空气在船舶底部衬垫承托的交通工具。其气垫通常是由持续不断供应的高压气体形成。气垫除了在水上航行外,还可以在某些比较平滑的陆上地形行驶。气垫船是高速行驶船舶的一种,行驶时因为船体受气压自水面抬升而起,大幅降低船体的流体阻力,以致行驶速度比动力输出接近的一般船舶快上许多。很多气垫船的速度都可以超过 50 kn,但与另一种可以高速行驶的船舶水翼船不同,由于气垫船的升力并非依靠速度产生,因此即使以非常缓慢的速度行驶,也不影响其效率。见图 H-10。

图 H-10 气垫船
Figure H-10 Hovercraft/air-cushion ship

Hovercraft/surface effect vehicle 全垫升气垫船 船体硬结构全由气垫升,不与支承表面接触,能两栖运行的气垫船。

Hovering 垫升 气垫船在垫升系统工作并通过喷口向支承表面喷射空气,从而形成气垫并依靠气垫支承的状态。

Hovering craft twin hull 双体气垫船 是由侧壁式气垫船引申发展而成,它将侧壁式气垫船和高速双体船两者的优点结合起来。为了改善侧壁式气垫船的耐波性,开始是由薄侧壁变为厚侧壁,进而成为双体气垫船。它采用大气垫长宽比和容积型侧壁,使阻力峰值和中低速阻力值下降,既提高了装载量,又改善了经济性、耐波性和可靠性。气垫船由于在波浪中所受的冲击力相对空气作用力而言是很大的量值,加上气垫的波浪泵气作用等,其耐波性较差。表现为波浪中颠簸严重,失速明显。以船体静浮力替代部分气垫升力将会改善耐波性。双体船提供的静浮力可以达到 1/4~1/3 船重,耐波性要求高时可取大些。当对耐波性要求更高时可以将排水体积移至水下,成为小水线面气垫船。见图 H-11。

图 H-11 双体气垫船
Figure H-11 Hovering craft twin hull

HPR 任何一种水声位置基准
HSC 高速船 系指符合 SOLAS 中对高速船定义的船舶。 HSC is a high-speed craft(HSC)are vessels which comply with the definition in SOLAS for high speed craft.
HSRFRV 高超声速可重复使用飞行器
H-type internal combustion engine H 型内燃机 两列对置气缸内燃机并列,两根曲轴用齿轮连接到一根功率输出轴上的内燃机。
HUB 集线器 HUB 是一个多端口的转发器,当以 HUB 为中心设备时,网络中某条线路产生了故障,并不影响其他线路的工作。所以 HUB 在局域网中得到了广泛的应用。大多数的时候它用在星型与树型网络拓扑结构中,它以 RJ45 接口与各主机相连(也有 BNC 接口),HUB 按照不同的说法有很多种类。HUB 按照对输入信号的处理方式上,可以分为无源 HUB、有源 HUB、智能 HUB。集线器(HUB)属于数据通信系统中的基础设备,它和双绞线等传输介质一样,是一种不需任何软件支持或只需很少管理软件管理的硬件设备。它被广泛应用到各种场合。集线器工作在局域网(LAN)环境,应用于 OSI 参考模型第一层,因此又被称为物理层设备。
Hub(boss) 轮毂 (1)套接在舵轴端的螺旋桨筒形部分。(2)风能转换系统(风力发电机)的组成部件。该部件安装在主驱动轴的末端,上面装有一个叶轮。
Hub diameter 毂直径 毂表面在其与螺旋桨叶母线相交处的直径。
Hub diameter ratio 毂径比 毂直径与螺旋桨直径的比值。
Hub fairing 桨毂导流帽
Hub height 轮毂高度 风力轮毂中心与地面或平均海平面的距离。
Hub length 毂长 毂的长度。
Hub sealing 桨毂密封
Hub tip ratio 轮壳比
Hub vortex 毂涡
Hub vortex absorbed fins 消涡鳍 是在螺旋桨将军帽上增设与螺旋桨叶数相同的小叶片,以回收螺旋桨毂涡能量为目的的一种水动力节能装置。无论在新船还是在旧船上安装都可收到 2%~5% 的节能效果。
Hub vortex cavitation 毂涡空泡
Hudong-zhonghua shipyard(Group)Co. Ltd 沪东中华——海上超级"LNG 船的摇篮" 沪东中华造船(集团)有限责任公司是中国船舶工业集团有限公司旗下的骨干企业,是既建造民用船舶、军用舰船,又制造海洋工程和大型钢结构的特大型企业集团。年销售额达到 200 亿元人民币。沪东中华建造的国内第一艘 LNG

船和拥有完全自主知识产权的 8530TEU 集装箱船,填补了国内空白,提高了我国造船业的水平和国际地位,建造的 17 万吨级和 20.6 万吨散货船被誉为"绿色环保型散货船";建造的 32.9 万吨 VLCC 是世界上载重量最大、款式最新的超级油船。沪东中华具有雄厚的船舶开发、设计建造实力,具有 80 多年的丰富造船经验,先后为国内外船东建造过液化天然气(LNG)船、大中型集装箱船、化学品船、滚装船、油船、散货船、军舰和军辅船等军民用船舶共 3 000 多艘。

Hull 船体 又称"船壳"。不包括船内外任何设备、装置和系统等的船舶壳体。

Hull auxiliaries 船体设备 各种舵、系船、关闭、桅与信号、起货、拖曳等设备和舱面属具的总称。

Hull auxiliary 船用辅机

Hull-borne engine unit 排水航行机组 在水翼艇的推进装置中,供排水状态下航行时使用的机组。

Hull braced structure 船体支撑结构 系指上部或内部安装船用配件,并直接承受作用在船用配件上的力的部分船体结构。用于正常拖带、系泊操作的绞盘、绞车等的船体支撑结构也应符合有关规定。

Hull classification certificate 船体船级证书

Hull closures(closing appliance) 关闭设备 在船体结构的一切开口上的盖闭部件及其控制装置和附件的总称。

Hull construction monitoring plan(HCMP) 船体建造监控计划 系指:(1)造船厂针对用于保障船舶中被定义为关键位置的结构处所采用的加强的建造质量管理的质量标准和程序编制的文件。(2)船体建造监控计划应在建造开始前提交船级社审批。船体建造监控计划应由船级社现场验船师和审图验船师评估以确认该计划已充分反映该船的实际构造、结构分析和疲劳分析。批准后,现场验船师应采取措施确保船体建造监控计划能被相关人员充分理解和有效贯彻。(3)船体建造监控计划为由造船厂提供的质量计划的附加要求。(4)船体建造监控计划一经批准,造船厂应在船级社验船师的参与下确认处已采用的标准外,是否符合船体建造监控计划中的所有要求事项。(5)船体建造监控计划应包括下列信息:①每个结构关键位置的结构详图;②关键位置的结构许用误差详情;③标有所有关键位置许用公差的概要表;④所采用的对准确认方法;⑤从预装配到分段装配,直到船台合龙,各个装配阶段的质量控制概要;⑥所采用的质量保障程序概要;⑦校核结果的记录和报告方法;⑧纠正措施方法详细标准。(6)批准的船体建造监控计划副本应在船舶整个寿命期内以电子格式或复印件形式保存在船上,供对船舶营运有效性而校核设计阶段定义为关键位置处时使用。Hull construction monitoring plan (HCMP) means: (1) the hull construction monitoring plan (HCMP) is a document complied by the shipyard to provide a record of the enhanced quality standards and procedures employed by the Shipbuilder to ensure that an increased level of construction quality control is employed at those areas of the structure that have been identified as critical to the vessel. (2) the HCMP is submitted to Head Office of this Society's site Surveyor and Plan Approval Surveyor in order that the findings of practical construction, structural analysis and fatigue analysis are uniquely reflected in the plan. Once approval is given, this society's sits Surveyors maintain efficient contact between all interested Parties to ensure that the requirements of the HCMP are fully understood and are complied with. (3) the HCMP is supplemental to and does not replace the Quality Plan provided by the Shipbuilder. (4) on receipt of the approved HCMP, the Shipbuilder, in association with this Society's Surveyor, ensures that all of the requirements contained within the HCMP are met in addition to any shipbuilding standards used. (5) a typical HCMP is to contain the following information:①appropriate structural plans with the critical locations clearly marked;②details of appropriate construction tolerances including any "design offset" at the critical locations are to be included on the appropriate structural plans for approval. Summary table of all critical locations indicating tolerances applied;③a lignment verification methods used, i.e. offset marking. Outline of quality controls in place during block construction, per-erection and erection;④outline of Q = A procedures used methods for recording and reporting of inspection results;⑤details of standard remedial measures to be employed where required;⑥a summary of the quality assurance procedures used;⑦methods for recording and reporting the results of cheeks;⑧corrective action method detailed standard. (6) A copy of the approved HCMP is maintained on board either in electronic or hard copy format through-out the life of the vessel. The HCMP is to be used to focus survey on those areas of the structure identified during the design process as being critical to the operational effectiveness of the vessel.

Hull cooling(shell cooling) 船舷冷却 用舷外冷却管或船壳代替淡水冷却器的一种简化闭式冷却方式。

Hull efficiency 船身效率 伴流与推力减额对船舶推进的综合影响。$\eta_H = (1-i)/(1-\omega)$ 或 $1/(1+\alpha)(1-\omega)$,式中,$(1-i)$—推力减额因数;$(1-\omega)$—伴流因数;$(1+\alpha)$—阻力增额因数。

Hull girder loads 船体梁载荷 系指局部载荷作

用在整个船体作为梁考虑时产生的力和弯矩（包括静水、波浪和动力）。 Hull girder loads are (still water, wave and dynamic) forces and moments which result as effects of local loads acting on the ship as a whole and considered as a beam.

Hull girder ultimate bending moment capacity M_U 船体梁极限弯矩 M_U 定义为船体梁的最大弯曲能力，超过此弯矩船体梁将发生崩溃。船体梁的失效由纵向构件的屈曲、极限强度和屈服控制。 Hull girder ultimate bending moment capacity M_U is defined as the maximum bending capacity of the hull girder beyond which the hull will collapse. Hull girder failure is controlled by buckling, ultimate strength and yielding of longitudinal structural elements.

Hull life 船体寿命 可分为自然寿命、经济寿命及技术寿命。

Hull monitoring system 船体监控系统 是：（1）将船舶在航行过程中及船舶在停泊时装卸货物期间所经受的船体纵向总应力和运动的实时数据提供给船长和官员；（2）允许实时数据压缩到一组主要的统计结果；该组统计结果应周期性地刷新、显示和存贮在活动媒介中。考虑到船东以后开发时可能选用存贮的信息，例如在船舶开发中用作一个单元或作为对航海日志的补充。船体监控系统提供的信息应视为船长的辅助手段，它不应取代船长自己的判断能力和责任。 Hull monitoring system is a system which: (1) provides real-time data to the Master and officers of the ship on hull girder longitudinal stresses and vertical accelerations the ship experiences while navigating and during loading and unloading operations in harbour; (2) allows the real-time data to be condensed into a set of essential statistical results. The set is to be periodically updated, displayed and stored on a removable medium. Extra information may be added in view of later exploitation by the Owner, for instance as an element in the exploitation of the ship or as an addition to its logbook. The information provided by the Hull Monitoring System is to be considered as an aid to the Master. It does not replace his-own judgement or responsibility.

Hull number 船体号 系指船舶建造时，为了方便组织生产和管理，给予该船舶的施工代号。

Hull strength 船体强度 船体结构抵抗内外作用力的能力。

Hull structural noise 船体结构噪声 系指船体的主体结构及附体结构，如机器底座、装置或设备的支架、加强筋加固的板或板架等，受到外力的激励而振动所产生的噪声，包括空气噪声和水动力噪声。

Hull structural steel 普通强度船体结构钢 具有最小标称屈服强度为 235 N/m² 和抗拉强度为 400～520 N/m² 的船体结构钢。

Hull structure (ship structure) 船体结构 （1）包括所有内部及外部结构的船体；（2）上层建筑、甲板室及套管；（3）焊接的基座，如主机基座；（4）舱口围板、舷墙；（5）舱壁、甲板及外板处设置的所有焊接贯通部件；（6）通风管及舷外阀与甲板；舱壁及外板连接的所有设备——经修正的国际载重线公约（ILLC）1966所要求的所有项目；（7）焊接在外板、甲板及主要构件上的附着物，如起重机基座、系缆桩及系船柱，但对船体结构只考虑两者的相互作用。 Hull structure is as follows: (1) hull envelope including all internal and external structures; (2) superstructures, deckhouses and casings; (3) welded foundation e.g. main engine seatings; (4) hatch coamings, bulwarks; (5) all penetrations fitted and welded into bulkheads, deck and shell; (6) the fittings of all connections to decks, bulkheads and shell, such as air pipes and ship side valves—all ILLC 1966, as amended terms; (7) welded attachments to shell, decks and primary member e.g. crane pedestals, bitts and boards, but only as regards their interaction on the hull structure.

Hull structure of a self-elevating unit 自升式平台主体结构 系指自升式平台的上部平台结构，其上放置的各种设备和设施在作业时均处在海平面以上。 Hull structure of a self-elevating unit means the upper platform structure of the unit, the equipment and installations mounted on which are above sea level during operation.

Hull valve 舷侧阀（通海阀）

Hull ventilation 全船通风

Hull vibration 壳体振动（船体振动）

Hull vibration damping 船体振动阻尼

Hull-borne (displacement mode) 排水航态 系指地效翼船/气垫船不论在静止或运动时其重力主要由所排水的浮力所支承时的航行状态。

Hull-borne operation mode 船体排水航行模式（地效翼船） 系指低速时船的总重由主船体和侧浮舟的浮力支撑。船的力和纵倾平衡由重力和重心控制，由浮力、浮心和稳心位置平衡的航行模式。 Hull-borne operation mode means a mode at which the total weight of the craft is supported by the main hull buoyancy and side buoys at very slow speed. The balance of forces and craft trim are controlled by weight and centre of gravity, balanced by buoyancy, centre of buoyancy and position of the metacenter.

Human element 人为因素 是一个复杂的多维问题，其对海上安全和海上环境污染产生影响。涉及全部人员的行为,包括船员、岸上管理人员、主管机构、认可组织、造船厂、立法者和其他相关人员,要想有效解决人为因素问题,需要这些人的共同努力。

Human error 人为失误 相对于可接受的或要求的人的实施或操作的偏离,从而导致了不可接受的或不符合要求的结果的出现。

Human error assessment and reduction technique (HEART) 人为失误评估和减小技术 是由威廉姆斯于1985年针对不利于人的行为的人机工程学,任务和环境等因素提出的。该技术能量化每种因素对人的行为的独立影响,根据这些影响计算人为失误概率。HEART提供了防止人为失误的可补救的风险控制方案的具体方法。主要研究5种导致人为失误的原因:系统知识受损,回应时间不足,系统回锁不明确,操作人员重要判断失误和职责,健康,环境对警觉性的影响。

Human error probability 人为失误概率 已发生的人为失误占所有可能的任务失误的比例。

Human error recovery 人为失误恢复 系指在造成不理想结果之前,通过本人或其他人,使失误恢复的潜在可能性。

Human exposure scenario (HES) 人体曝露情景

Human factor 人为因素 包括人类科学和工业工程,还涉及人们及其工作行为和工作环境之间关系的优化技术。该学科涉及技术和组织系统的设计和操作,目的是对人工任务做出合理改变,通过人体工程学解决人为因素。

Human machine interface (HMI) 人机界面 系指:(1)与操作人员互动的系统部分。界面是一个集合手段,使用户与机器、装置和系统进行互动。界面提供输入手段,使用户能控制系统和输出,使系统能通知用户;(2)一个系统中与操作员互动的部分。界面是用户与机器、设备和系统互动的各种手段的聚集体。界面提供输入,允许用户控制系统和输出,允许系统通知用户。

Human machine interface (HMI) means:(1) the part of a system an operator interacts with. The interface is the aggregate of means by which the users interact with a machine, device, and system (the system). The interface provides means for input, allowing the users to control the system and output, allowing the system to inform the users; (2) the part of a system an operator interacts with. The interface is the aggregate of means by which the users interact with a machine, device, and system (the system). The interface provides means for input, allowing the users to control the system and output, allowing the system to inform the users.

Human reliability 人为可靠性 可分为两种情况:(1)(如有时间限制)在既定时间内正确无误按照系统规定实施行为的概率;(2)不实施降低系统性能的外部行为的概率。

Human reliability analysis 人因可靠性分析 此分析涉及定性/定量的方法,用来决定由特殊操作人员执行特殊任务时发生失误的可能性和潜在的后果。

Humid clutch 湿式离合器 摩擦接合面使用润滑冷却剂的离合器。

Humidifier 加湿器 将蒸汽或水雾均匀地喷在加热的空气中的加湿设备。

Humidity controller 湿度控制器

Humidity removing 干燥(去湿)

Humidity resistance 抗湿性

Hump 阻力峰 系指在水面航行时,由于气垫兴波阻力和侧壁或侧体兴波阻力在不同航速下的变化特征而呈现的峰值。

Hump of wave making resistance curve 波阻峰值 由于船首与船尾横波的叠加作用,在兴波阻力曲线上出现的凸出点。

Hump speed 阻力峰速度 系指水动阻力达到峰值时的速度,此时船有严重的飞溅,从而影响主机、艉推进器运行及驾驶舱的驾驶人员视线。此时主机功率高,而由于航速低,主机的冷却空气流量小。若在这个速度段运行时间过长,则可能由于冷却流量小导致主机气缸和滑油过热,因此驾驶人员应尽快越过此峰值速度。

Hump speed means a speed at which the hydrodynamic resistance reaches a peak, and with heavy spray acting on the craft so as to influence the main engine, bow thruster, and the pilot's field of view from the cockpit. The main engine power setting will be high, while for reciprocating engines cooling airflow is low due to the low speed. The low cooling can sometimes make main engine cylinders and lubrication oil overheat if operation is prolonged at this speed, so pilots make an effort to pass through hump speed as quickly as possible.

Humphree control system-5 HCS-5型船舶运动阻截系统 是配置了一些新的部件,能改善各种船舶运动状态的系统。

h_w(**mm**) 普通扶强材或主要支撑构件的腹板高度(mm) h_w means web height of ordinary stiffener or primary supporting member(mm).

Hybrid fiber-coaxial (HFC) 混合光纤同轴网络 是以光纤为骨干的网络,同轴电缆为分支网络的高带宽网络,传输速率可达20 Mb/s以上。目前国内的广电行业系统网络即采用HFC网络体系。数字电视以HFC为传输基础网络,其与传统有线电视结构基本一致,主要存储及传送的内容是MPEG-2流,采用IP OVER DWDM

技术,基于 DVD IP 光纤网传输。

Hybrid hard disk(HHD) 混合硬盘 是一个标准的硬盘驱动器上装有一个大的缓冲区或内存高速缓存不需要一个旋转的磁盘记录数据。

Hybrid propulsion system 混合推进系统

Hybrid riser tower(HRT) 组合立管塔

Hybrid type high performance marine vehicles 复合型高性能船 是以某一种高性能船为基础,吸收其他高性能船技术(指水翼技术和气垫技术)以改进其短处而改变了个别类型外,大多数复合型高性能船都是水翼技术或气垫技术在各类高性能船上的应用。复合型高性能船除各种形式的小水线面水翼船外,还有双体水翼船、水翼双体船、双体气垫船、双体穿浪船、水翼双体气垫船、水翼气垫小水线面船、水翼三体船、水翼穿浪船、气垫穿浪船、动力气垫地效翼船等等。其中由高速双体船与小水线面双体船复合而成的双体穿浪船,以及由侧壁式气垫船演变而来的双体船与气垫船复合而成的双体气垫船两种类型比较成熟,得到了较广泛应用。

Hybrid-type PFD 混合型个人漂浮装置 具有组合的浮力类型,也就是固有浮性材料和气胀式相混合的个人漂浮装置。 Hybrid-type is the PFD of combined buoyancy types, i.e. inherent and inflatable.

Hydraulic part shop 液压件车间 系指从事液压件生产、安装、调试和储存的场所。

Hydrant(fire hydrant) 消火栓(消防龙头) 系指固定水灭火管路与消防水带相接处的由截止阀、快速接头、保护盖等组成的连接件。

Hydrant valve 消防龙头

Hydraulic accumulation 液压蓄能器

Hydraulic actuated governor 液体传动调速器

Hydraulic actuator 液压执行器

Hydraulic anchor capstan 液压起锚绞盘 液力驱动的起锚绞盘。

Hydraulic bender 液压折弯机 包括支架、工作台和夹紧板,工作台置于支架上,工作台由底座和压板构成,底座通过铰链与夹紧板相连,底座由座壳、线圈和盖板组成,线圈置于座壳的凹陷内,凹陷顶部覆有盖板。使用时由导线对线圈通电,通电后对压板产生引力,从而实现对压板和底座之间薄板的夹持。由于采用了电磁力夹持,使得压板可以做成多种工件要求,而且可对有侧壁的工件进行加工。见图 H-12。

Hydraulic bending of pipe 液压弯管

Hydraulic berth carriage 液压船架小车

Hydraulic bolt tensioning jack 拉紧螺栓油压机

Hydraulic braking 水力制动

Hydraulic bronze 耐蚀青铜

图 H-12 液压弯折机
Figure H-12 Hydraulic bender

Hydraulic caisson crane 液压沉箱吊

Hydraulic capstan 液压绞盘

Hydraulic cargo winch 液压起货机 液力驱动的起货绞车。

Hydraulic circuit 液压管路

Hydraulic cleaning 水力清理

Hydraulic cleat 使滚装船门固定的液压夹扣

Hydraulic clutch 液力离合器

Hydraulic component 液压部件

Hydraulic control actuator 液压控制执行机构

Hydraulic coupling 液力联轴节

Hydraulic coupling(fluid coupling) 液力耦合器 由主动件泵轮和从动件涡轮组成,通过液压传递能量,主、从动轴向的转速比能自动无级变化,而扭矩始终相等的传动组件。

Hydraulic crane 液压起重机

Hydraulic cylinder 液压油缸

Hydraulic damper 液力减震器

Hydraulic derrick 液力吊杆装置 由液压机构直接改变吊杆偏角和吊杆仰角的吊杆装置。

Hydraulic device 液压装置

Hydraulic door 液压传动门

Hydraulic dredger(HD) 水力式挖泥船

Hydraulic drive 液力传动

Hydraulic dynamometer 液力测功器

Hydraulic elevation structure 液压升降机构 采用液压传动,使自升式钻井平台完成平台升降的机构。

Hydraulic elevation structure 液压升降结构

Hydraulic equipment 液压装置(液压设备)

Hydraulic explosion test 液压爆炸试验(液压胀裂试验)

Hydraulic fender 液压防碰设备

Hydraulic fitting 液压装配

Hydraulic fluid　液压流体
Hydraulic fluid reservoir　储液器
Hydraulic governing　液压调节
Hydraulic grab　液压抓斗　由液压控制其开闭的抓斗。
Hydraulic jet unit　喷水推进装置
Hydraulic joint　液压连接
Hydraulic jump　水跃
Hydraulic lift　液压升降机
Hydraulic loader　液压装载机
Hydraulic lock　液压锁(液压封闭,液阻塞)
Hydraulic machinery　液压机械
Hydraulic manipulator　液压机械手
Hydraulic cylinder unit(HCU)　液压气缸单元
Hydraulic motor　液压马达
Hydraulic oil storage tank　液压油贮存柜　贮存液压系统使用的工作油的柜。
Hydraulic oil sump tank　液压油循环柜　汇集液压系统工作油的回油供再循环使用的柜。
Hydraulic oil supply tank　液压油补给柜　贮存和补充液压系统内工作油的柜。
Hydraulic oil　液压油
Hydraulic oil outlet pressure gauge　液压油出口压力表
Hydraulic oil pressure pump　液压油压力泵　系指通过共用蓄能器向诸设备,例如燃油喷射装置、排放阀驱动装置或控制阀提供液压油的泵。 Hydraulic oil pressure pump is a pump to provide hydraulic oil for the equipment, e.g. fuel injection devices, exhaust valve driving gears or control valves, through the common accumulator.
Hydraulic operating system　液压操作系统
Hydraulic piping　液压管路
Hydraulic power　液压动力(水力,液力)
Hydraulic power capsule　液压动力舱
Hydraulic power operated system　液压动力操纵系统
Hydraulic power pack　液压泵站
Hydraulic power system(HPS)　液压动力系统(液压动力单元)
Hydraulic power transmitter　液压功率发送器
Hydraulic pressure　液体压力
Hydraulic pressure device　液压设备
Hydraulic pressure equipment　液压设备
Hydraulic pressure test　水压试验(液压试验)
Hydraulic pressure treatment　液压处理
Hydraulic propulsion　液力推动

Hydraulic propulsion unit　喷水推进装置
Hydraulic pump　液压泵
Hydraulic push hinge　液力顶推铰链　用液压油缸带动铰链启闭舱盖板的组合装置。
Hydraulic radius　水力半径　系指船舶在水道中航行时,水道的净横剖面积与水道总浸湿周长的比值。$R_n = (F_0 - A_x)/(U_k + U_s)$,式中,$F_0$—水道横剖面;$A_x$—船体浸水横剖面;$U_k$—水道横剖面浸湿周长;$U_s$—船体浸水横剖面浸湿周长。
Hydraulic ram tensioner　作动筒张力器　具有与蓄压器联动工作的作动筒,对承载绳索起缓冲作用的装置。
Hydraulic regulation　液压调节器
Hydraulic seal　水压密封(水封)
Hydraulic shock absorber　液力减振器
Hydraulic spring-type　平衡块减振器(液压弹簧式减振器)
Hydraulic steering control system　液压操舵系统　用液压传动方式操纵舵机的系统。
Hydraulic steering gear　液压舵机　液力驱动的舵机。
Hydraulic steering telemotor　液压操舵器　设置在驾驶室内部通过液压系统来控制舵机等的装置。
Hydraulic system　液压系统
Hydraulic telemotor　液压遥控传动装置
Hydraulic test　水压试验(液压试验)
Hydraulic towing hook　液压拖钩　利用液压操纵控制的拖钩。
Hydraulic towing winch　液压拖缆机　液力驱动的拖缆机。
Hydraulic transmission　液压传动
Hydraulic transmitter　液压发送器
Hydraulic tubing　液压管路
Hydraulic unit　液压机构
Hydraulic valve　液压阀
Hydraulic water seal　水压密封(水封)
Hydraulic windlass　液压起锚机　液力驱动的起锚机。
Hydraulically operated valve　液力作用阀
Hydraulically-operated cargo oil valve　液压操纵货油阀　用液压传动的遥控货油阀。
Hydraulically-operated isolating valve　液压操纵隔离阀　采用液压作为动力进行启闭的隔离阀。
Hydraulic-electric hatch cover　电动液压舱口盖　系指电动液压驱动的舱口盖。
Hydrautorque hinge　液力铰链　供启闭舱盖用的

依靠液力能自身转折的铰链。

Hydro survey 水路测量

Hydro survey ship 水路测量船

Hydro-biological research ship 生物调查船 用于对海洋生物的种群组成、数量分布和变化规律以及生物和环境调查研究的船舶。船上设有生物实验室、微生物实验室及同位素实验室等。船舶要求低速航行以满足底栖拖网和浮游生物调查的需要。

Hydrocarbon 碳氢化合物 系指仅含氢和碳的有机化合物。碳氢化合物最简单的例子是在正常温度下的煤气，但是，随着分子量的增加，它们转化成液态形式并最终转化为固态形式。碳氢化合物构成石油和天然气的主要组成部分。 Hydrocarbon is an organic compound containing only hydrogen and carbon. The simplest examples of hydrocarbon are gases at ordinary temperatures, but, with increasing molecular weight, they change to the liquid form and finally to the solid state. Hydrocarbons form the principal constituents of petroleum and natural gas.

Hydrocarbon gas 碳氢气体

Hydrochloric acid 盐酸

hydrochlorofluorocarbon(HCFC) 氢化氯氟烃

Hydrodynamic advance angle 水动力进角 在叶元体处水的实效合速与圆周方向之间的夹角。

Hydrodynamic electromagnetic propulsion 电磁喷水推进 水中混有磁铁细粒，由外部磁场使细粒加速，以水柱形态喷出产生推力的一种喷射推进。

Hydrodynamic force components 水动力分量 作用于运动物体重心的水动力分别沿运动坐标系各轴的分量。

Hydrodynamic force energy-saving device 水动力节能装置 包括伴流补偿导管、前置导叶、前置导管、消涡鳍、舵球、舵附加鳍、挠曲舵以及各种节能装置的组合装置。

Hydrodynamic moment components 水动力矩分量 作用于运动物体重心的水动力矩分别相对于运动坐标系各轴的分量。符号按右手法则。

Hydrodynamic moment transformer (hydraulic torque converter) 液力变矩器 由主动件泵轮、固定件导向器和从动件涡轮组成，通过液体传递能量，主、从动轴向的转速比和扭矩比均能自动无级变化的传动组件。

Hydrodynamic noise 水动力性噪声 是由水体的振动而出现水波动所产生的噪声。例如一般排水量船舶在航行时，破浪前进，使海水表面或河川水的表面振动而产生的拍击声，其尾部的水螺旋桨在旋转时，拍击水产生的噪声；又如潜艇潜航时，迫使周围的水体振动而产生的噪声，深水炸弹在水下爆炸引起强力冲击波而产生高强度的水下噪声。

Hydrodynamic partition plate 整流隔板 安装于螺旋桨前、船体尾部的中纵剖面的，剖面形状为平板或曲面隔板。可以起改善螺旋桨进流提高螺旋桨推进效率的节能装置。

Hydro-dynamically smooth 水力光滑 物体表面突出物高度或粗糙度小于层流底层厚度，即使该面的粗糙度再继续减小，其摩擦阻力也不得减小时的光滑度。

Hydrodynamics 水动力学 研究水的运动规律及水与物体相互作用的学科。

Hydro-elasticity 水弹性 具有弹性的物体在水动力作用下产生变形，为使水动力性能改变的特性。

Hydrofoil 水翼 在水中运动时能产生升力并用以支撑艇重的翼形结构物。

Hydrofoil catamaran(HYCAT) 双体水翼船 双体水翼船以水翼为主。船重的大部分由水翼的升力支承，船体只提供一小部分浮力支承这些水翼。与双体复合型比双体船有更优良的快速性，可以不必采用复杂的水翼自控技术也能达到较优良的耐波性能，且可以使船大型化，其速度大约在40~50 kn之间。

Hydrofoil craft 水翼船 系指:(1)非排水状态航行时，能被水翼产生的水动升力支承在水面以上的船舶;(2)通过悬挂于船体下的水翼在水中起到如飞机翼那样的作用，从而实现水面上运行的船艇。 Hydrofoil craft is:(1) a craft supported completely clear above water surface in non-displacement mode by hydrodynamic forces generated on foils;(2) a craft operated above the water surface by having foils suspended beneath the hull that act like an aircraft's wings in the water.

Hydrogen compressor 氢气压缩机 舰船中将从水电解槽中收集来的氢气排至舷外所用的压缩机。

Hydrogen cylinder 氢气瓶

Hydrogen making room 制氢室 为充灌测风及探空气球而制造氢气(通常分为电介制氢、化学制氢)的工作室。该室应采取防爆措施。

Hydrogen removal 除氢(脱氢)

Hydrogen storage room 贮氢室

Hydrogen sulphide 硫化氢 高度易燃(闪点 -82 ℃)，遇空气能形成爆炸性混合物，潮湿时有腐蚀性，必须远离者火源存放，有刺激性和窒息性，长期曝露极限值(LTEL)5 ppm，短期曝露极限值(STEL)10 ppm，浓度再高会致命，无气味，反复低浓度曝露能导致对气味的嗅觉丧失。 Hydrogen sulphide is highly flammable (flash point of -82 ℃), can form an explosive mixture with air, corrosive when wet, causes burns, has to be kept away from sources of ignition, irritant and asphyxiant, LTEL 5 ppm, STEL 10 ppm, higher concentrations can be fatel and have no odour. Repeated exposure to low concentrations can result in the sense of smell for the gas being diminished.

Hydrogen test　测氢试验　是确定焊缝金属中的可扩散氢含量的试验。　Hydrogen test is a test determined the diffusible hydrogen content of the weld metal.

Hydrogen treatment　去氢处理

Hydrographic services　水文服务　系指安排水文资料的收集和编制、出版、传播以及不断更新为安全航行所必需的所有航海资料的业务。其主要内容为：(1)确保尽可能按安全航行的要求进行水文勘测；(2)编制和公布海图、航行指南、灯塔表、潮汐表和其他航海出版物(如适用)以满足安全航行的需要；(3)向航海者颁布通告以使海图和航海出版物尽可能及时更新；(4)提供数据管理安排以支持这些服务。

Hydrographic survey ship　航道测量船　系指用于对航道深度、海流及标准海洋学参数测量的船舶。

Hydrographic survey ship　水文调查船　用于对海洋水文流速、流向、地质进行调查的船舶。船上设有水文、化学实验室。

Hydrographic winch　水文绞车　为收、放水文测量仪器进行水文观测工作的绞车。

Hydrographical survey　海道测量　又称水道测量、航道测量。其任务为编绘海图和海区资料，对水域进行测量和调查。

Hydro-jet　喷水推进

Hydro-jet propelled boat [water jet (propelled) boat]　喷水推进船　依靠向后喷水的反作用力推进的船舶。

Hydrokineter　炉水循环加速器

Hydrological davit　水文吊杆　一种海洋考察用吊杆。它为钢管或型钢结构。具有一定跨距及高度，可旋转。在进行水文调查时与水文绞车配合供挂放测量仪器用。

Hydrophone　水听器　又名接收换能器，用于接收水声信号的水传感器。等同于水声接收器。

Hydro-pneumatic testing　液-气试验(静水压气动试验)　系指静水压试验和空气试验的组合，它将水灌进液舱，然后施加一额外的空气压力，其状态应尽实际可能模拟液舱的实际载荷，但无论如何空气压力应不小于规定的值[0.015 MPa(0.15 kg/cm^2)]。进行这种试验时，应遵守有关规定的安全注意事项。　Hydro-pneumatic testing is a combination of hydrostatic and air testing, consisting of filling the tank with water and applying an additional air pressure. The conditions are to simulate, as far as practicable, the actual loading of the tank and in no case is the air pressure to be less than given value [0.015 MPa (0.15 kg/cm^2)]. When this is performed, the relevant safety precautions are to be followed.

Hydrosound cable winch　水声学电缆绞车　利用水声学原理进行海洋地形测量(水深)、海洋底质测量(浅地层剖面测量)、海洋地质调查(海洋地质构造和地壳构造)等用的绞车。

Hydrostatic contents gauge(HCG)　静压式舱容计

Hydrostatic curves　静水力曲线　表示船舶正浮状态时的浮性要素、初稳性要素和船型系数等与吃水之间关系的各曲线之总称。

Hydrostatic table　排水量数值图

Hydrostatic testing　静水压试验　通过注入规定水位产生的水压以验证液舱结构的密封性和结构设计的合适性的一种试验。水压试验是用于结构试验的一种常规方式，但如受到实际限制而不应进行或允许采用气密试验的情况除外。　Hydrostatic testing is a test to verify the structural adequacy of the design and the tightness of the tank's structure by means of water pressure, produced by filling water to the level given. Hydrostatic testing is the normal means for structural testing, with exception, where severe practical limitations prevent it or where air testing is permitted.

Hydrostatically balanced loading (HBL) operational manual　静压平衡装载(HBL)操作手册　系指每艘符合第13G(6)(b)条要求采用静压平衡装载的油船，均应按MEPC.64(36)决议的要求备有的一份操作手册。　Hydrostatically balanced loading (HBL) operational manual means that every oil tanker which, in compliance with regulation 13G(6)(b), operates with Hydrostatically Balanced Loading shall be provided with an operational manual in accordance with resolution MEPC.64(36).

Hygrometer　湿度器

Hygroscopic　吸湿的

Hygrostater　恒湿器

Hyperbolic navigation system　双曲线导航定位系统　利用双曲线的交点来实现导航定位的系统。该系统先后或同时对几组地面站进行距离差的测定，获得两条或两条以上的位置线，以确定船舶的位置。按测定距离差所用的无线电参数的不同，双曲线系统可分为：(1)测定两个脉冲信号之间的时间差得到距离差的称脉冲双曲线系统，如罗兰 A；(2)测定两个连续波信号之间的相位差得到距离差的称脉冲相位双曲线系统，如罗兰 C；(3)同时测定两个脉冲信号的时间差及其包内载频的相位差得到距离差的称脉冲相位双曲线系统，如罗兰 B 和罗兰 D。双曲线导航系统中，接收设备不必装备发射装置和高精度频标。用户数量不限，设备简单、便宜，故这种系统得到广泛应用。

Hyperbolic system 双曲线系统 利用两条双曲线位置线的交点来确定船位的系统。该系统先后或同时对两对地面站进行距离差的测定获得两条位置线，两者交点即为船舶的位置。一般双曲线系统最少需要三个地面台，该系统定位设备不需要发射台和高精度频标，而且传播路径上的误差在主要工作区域可抵消一部分，因此这种系统得到广泛应用。如罗兰 C，奥米茄等。

Hypersonic 高超音速 系指超过 5 倍音速的速度。5 倍音速又被称为 5 马赫，相当于海平面每小时 6 200 km 的速度或是 10 km 以上的高海拔 5 300 km/h 的速度。

Hypersonic glide vehicle (HGV) 高超声速滑翔飞行器

Hypersonic vehicle (HSV) 高超声速飞行器 系指飞行速度超过 5 倍音速的飞机、导弹、炮弹之类的有翼或无翼飞行器，是一种高超声速滑翔飞行器（HGV）。由进气道、压气机、燃烧室、涡轮和尾喷管组成。部分军用发动机的涡轮和尾喷管间还有加力燃烧室。这种飞行器通过弹道导弹作为推进系统发射升空，在大约 100 km 临近空间高度与火箭分离，进行俯冲机动滑翔飞行，飞行速度最高达到 10 马赫，大约相当于每小时 12 359 km。高超声速武器一直被国际防务界视为改变游戏规则的武器，因为它能在任何现有导弹防御系统做出有效反应之前命中并摧毁目标。一旦部署将极大推进我国战略和传统导弹实力。

Hypothermia (Immersion suit) 体温过低（救生服） 人体内部温度低于 35 ℃ 的状态。 Hypothermia is the condition where body core temperature is below 35 ℃.

Hypothetical outflow of oil 假定的泄油量 对"73/78 防污公约"附则 I 而言，系指为了提供在碰撞或搁浅事故中防止油污染的足够保护而提出的假定泄油量的计算方法。

I

IACO 国际民航组织

IAPP Certificate IAPP 证书 "国际防止空气污染证书"（International Air Pollution Prevention Certificate）的简称，当相关船舶经主管机关检验确认其符合 MARPOL73/78 附则 Ⅵ 的规定后由该主管机关授予的"国际防止空气污染证书"。

IATM = International Association for Testing Materials 国际材料试验协会

Ice belt 冰带 对在冰区航行的船舶而言，系指舷侧抗冰加强部分，分为下述 3 个区域：(1) 冰带艏部区——从艏柱向后至舷侧平直部分前端线之后 0.04 L 处之间的区域。对 B1* 和 B1 冰级，超过前端线的水平距离 x_1 不必大于 6 m，对 B2、B3 冰级则不必大于 5 m；(2) 冰带中部区——从冰带艏部区的后边界线向后至舷侧平直部分后端线之后 0.04 L 处之间的区域。对冰级 B1* 和 B1，超过后端线的水平距离 x_2 不必大于 6 m，对冰级 B2 和 B3，不必大于 5 m；(3) 冰带艉部区——从冰带中部区的后边界线至艉柱之间的区域。上述三个区域应在外板展开图上标明。当一个构件跨越两个区域时，应按照加强要求高的区域设计。 Ice belt—the part of sides which has to be reinforced is to be divided into three regions as follows: (1) Forward region—from the stem to a line parallel to and 0.04 L aft of the forward borderline of the part of the hull where the waterlines run parallel to the centerline. For Ice Classes B1 * and B1 the overlap x_1 over the borderline need not exceed 6 m, for Ice Classes B2 and B3 this overlap need not exceed 5 m; (2) Midship region—from the aft boundary of the forward region to a line parallel to and 0.04 L aft of the aft borderline of the part of the hull where the waterlines run parallel to the centerline. For Ice Classes B1 * and B1 the overlap x_2 over the borderline need not exceed 6 m, for Ice Classes B2 and B3 this overlap need not exceed 5 m; (3) Aft region—from the aft boundary of the midship region to the stern. The above-mentioned three regions are to be indicated on the shell expansion plan. Where a structural member is located in two regions, this member is to be designed according to the severer requirements of the two regions.

Ice belt area 冰带区 可能接触浮冰的水线附近的舷侧部分。

Ice breaker 破冰船 系指用于破碎水面冰层，开辟航道，保障舰船进出冰封港口、锚地，或引导舰船在冰区航行的勤务船，分为江河、湖泊、港湾或海洋破冰船。其推进系统多采用双轴和双轴以上的多螺旋桨装置，以柴油机为原动力的动力推进。第一艘在北极航行的破冰船是由英国建造的"叶尔马克"号。1957 年，苏联建造出第一艘核动力破冰船——"列宁"号。如果核动力破冰船带上10 kg铀，就相当于带上25 000吨标准煤，可以在远离港口的冰封海域常年作业。目前，破冰船一般采用两种常见的破冰方法：(1)"连续式"破冰法——依靠螺旋桨的力量和船舶把冰层劈开撞碎。(2)"冲撞式"破冰法——破冰船船艏部位吃水浅，会轻而易举地冲到冰面上，船体就会把下面厚厚的冰层压为碎块。然后破冰船倒退一段距离，再开足马力冲上去，把船下的冰层压碎。目前，破冰船主要聚焦4种船型，分别是LNG破冰船、斜破冰船、核动力破冰船和浅吃水破冰船。LNG破冰船采用柴油和LNG作为燃料，能有效降低NO_x和SO_x的排放量，并降低成本。除采用"双向作用"破冰法外，斜破冰船还可以利用船舶的一侧前进破冰，使破冰宽度达到50 m。2011年11月15日，中国国家海洋局宣布，中国拟自主建造第一艘极地考察破冰船。该船将可搭载直升机和水下机器人，船首、船尾均有破冰能力，已于2019年完成建造。见图I-1。

图 I-1　破冰船
Figure I-1　Ice breaker

Ice class B(navigation in floating ice conditions)　小块漂浮冰况区域航行　在严重冰况区航行时，船舶应根据有关规范的B级冰区加强的要求对其船首、舯部和船尾的满载吃水和压载吃水之间的结构进行加强。

Ice class B1(navigation in severe ice conditions)　严重冰况区域航行　在严重冰况区航行时，船舶应根据有关规范的B1级冰区加强的要求对其船艏、舯部和船艉的满载吃水和压载吃水之间的结构进行加强。

Ice class B1*(navigation in extreme ice conditions)　最严重冰况区域航行　在最严重冰况区航行时，船舶应根据有关规范的B1*级冰区加强的要求对其船艏、舯部和船艉的满载吃水和压载吃水之间的结构进行加强。

Ice class B2(navigation in intermediate ice conditions)　中等冰况区域航行　在中等冰况区航行时，船舶应根据有关规范的B2级冰区加强的要求对其船艏、舯部和船艉的满载吃水和压载吃水之间的结构进行加强。

Ice class B3(navigation in light ice conditions)　轻度冰况区域航行　在轻度冰况区航行时，船舶应根据有关规范的B3级冰区加强的要求对其船艏、舯部和船艉的满载吃水和压载吃水之间的结构进行加强。

Ice class draught　冰级吃水　应为夏季淡水中载重水线的吃水。如船舶具有木材载重线水线，则应为夏季淡水中木材载重水线的吃水。

Ice deep waterline　冰区满载水线　相当于夏季淡水载重线。若经特别请求且海军主管当局准许，冰区满载水线的规定可以与前述不同，但应相当于舰船在冰区航行时预期的最深吃水。　Ice deep waterline corresponds to the deep draught waterline. Where specially requested and where permitted by the Naval Authority, an ice deep waterline may be specified which differs from the foregoing, but corresponds to the deepest condition in which the ship is expected to navigate in ice.

Ice hatch　加冰孔　输送碎冰到鱼舱的开孔。

Ice hole cover　加冰孔盖　设置于冰藏保鲜鱼舱的加冰孔上的舱口盖。

Ice light waterline　冰区轻载水线　相当于在冰区航行时预期的最浅吃水。但是，建议船首端最小吃水不小于$T_f = (1.5 + 0.1\sqrt[3]{\Delta})h$, (m)，式中，$h$——名义冰层厚度，m，与所需要的冰级有关；$\Delta$——排水量。　Ice light waterline is that corresponding to the lightest condition in which the ship is expected to navigate in ice. However, it is recommended that the minimum draught at the fore end is not to be less than: $T_f = (1.5 + 0.1\sqrt[3]{\Delta})h$, (m), Where: h = the nominal ice thickness, in m, associated with the desired Ice Class; Δ = displacement.

Ice load　冰雪负荷　钻井船或其他钻井平台所允许考虑的冰雪沉积的量。

Ice navigator　冰区驾驶员　系指除按STCW公约的要求取得资格外，还经特别训练或以其他方式取得资格能在冰覆盖水域指挥船舶移动的任何人员。　Ice navigator means any individual who, in addition to being qualified under the STCW Convention, is specially trained and otherwise qualified to direct the movement of a ship in ice-covered waters.

Ice patrol service　冰区巡逻服务　系指在冰季期间，为穿越北大西洋内冰山区的船舶提供海上人命安全、航行安全和航行效率以及海洋环境保护等的巡逻警戒服务。

Ice season　冰季　系指每年2月15日至7月1日这段时期。　Ice season means the annual period between February 15 and July 1.

Ice strengthening 冰区加强　对航行冰区的船舶所做的局部结构加强。

Ice torque 冰块扭矩　螺旋桨叶片与冰块碰撞时所产生的冲击载荷。

Iceberg and floating ice observation room 冰山和浮冰观察室　利用观测仪器,对冰山的类型、数量、高度、差度、移动方向和分布情况,以及浮冰的类型、范围、堆积程度和覆盖密度进行观测和分析的工作室。该室应位于船的首部。

Icebreaker 破冰船　系指能实现护航或破冰功能的船舶,其动力和规模可在冰区覆盖水域内进行超常规作业。Icebreaker means any ship whose operational profile may include escort or ice management functions, whose powering and dimensions allow it to undertake aggressive operations in ice-covered waters.

Icebreaker bow 破冰型艏　设计水线以下艏柱侧影与基平面成45°左右或更小夹角形式的船首部。

Ice-breaking 具有破冰能力　系指授予具有航行冰区的加强要求,且航行于当年结冰区域,具有独立破冰能力的非破冰专用船的附加标志。

Ice-breaking tug 破冰拖船　系指兼作对冰封航道进行破冰的拖船。

Ice-covered water 冰覆盖水域　系指当地冰况对船舶结构构成危险境地的水域。Ice-covered water means polar waters where local ice conditions present a structural risk to a ship.

Ichthyology laboratory 鱼类实验室　用于鱼类样品观测、处理、分析;并对获取的鱼群信号进行分析、研究其形态、分类、生理生态特征、种群组成、鱼类分布(鱼群回流和渔场分布)数量变动与环境之间关系的工作室。

ICLL Convention 国际载重线公约　首次颁布的国际载重线公约于1930年正式实施,其依据为储备浮力原理。虽然随后被认为是干舷亦应确保有适当的稳性并避免因超载而使船体产生过度应力。因此,对船舶可能载货的吃水限制,对其安全性有显著贡献。除了外部气密和水密完整性之外,以干舷形式给定的此类限制构成了该公约的主要目标。与其他公约一样 ICLL 经国际海事组织的连续多年的增补修订。The first ICLL Convention (International Convention on Load Lines), adopted in 1960, was based on the principle of reserve buoyancy, although it was recognized then that the freeboard should also ensure adequate stability and avoid excessive stress on the ship's hull as a result of overloading. Thus, limitations on the draught to which a ship may be loaded make a significant contribution to her safety. These limits are given in the form of freeboard, which constitute, besides external weathertight and watertight integrity, the main objective of this Convention. As other conventions, it was continually updated at IMO by amendments through the years.

ICS = Network based integration of navigation system　基于网络的,组合导航系统

IDAS　潜射交互防御与进攻武器系统

Ideal efficiency of screw propeller 螺旋桨理想效率　螺旋桨在理想流体中的推进效率。

Ideal fluid 理想流体　一种无黏性的假想流体。

Identification of a particularly sensitive sea area 指定一个特别敏感海区　系指由于某一特定海区公认的生态学的、社会经济的或科学的及其易受到确定的国际海事活动的损害(伤害或环境危害)脆弱性的特点,由某一提出成员国政府根据《指定特殊区域和确定特别敏感海区导则》要求建议对该海区需要采取相关的保护措施并由国际海事组织确定。Identification of a particularly sensitive sea area means a determination by IMO that a proposing Member Government, in accordance with the guidelines, has established a need for Associated Protective Measures for a particular sea area because of the sea's recognized ecological, socioeconomic, or scientific characteristics and its vulnerability to damage (that is, injury or environmental harm) by identified international maritime activities.

Identification survey 鉴定检验　系指按照委托方规定的要求在各有关方同意的范围内,通过检查和试验进行符合性校核,以确认合同规定的各项已满足要求。这些要求一般为公认的规范及标准、业界标准和/或有关导则。检查完成时,船级社将签发有关的证书和提供有关的检验文件。

Identification system (AIS) target 自动识别系统目标　系指自动识别系统产生的目标,如被激活目标、失踪目标、被选目标和静止目标等。Automatic identification system (AIS) target is a target generated from an AIS message. See activated target, lost target, selected target and sleeping target.

Identify speed 识别速度

Idle capacity 空转功率

Idle gear (idling gear, idler gear) 惰轮　置于减速齿轮箱轮系中不改变传动比,而仅起改变转动方向或调整中心距作用的中间齿轮。

Idle pulley (idling gear) 惰轮(空转轮)

Idle time (race time) 惰转时间　从汽轮机停止进气时起直至到机组完全静止所需的时间。

Idler 空转轮(惰轮,中介齿轮)

Idling 空转

Idling speed 空载速度(惰速)

IEC = International Electro-technical Committee 国际电工委员会

IEEE = Institute of Electrical and Electronics Engineers 电气和电子工程师协会

IGC Code 系指经修正的"国际散装运输液化气体船舶构造和设备规则"。 IGC Code means the International Code for the construction and equipment of ships carrying liquefied gases in bulk, as amended.

Ignitability 可燃性(易燃性)

Ignite 着火(点燃)

Igniter 引燃物 系指用于引燃火源的物品。 Igniter means a device used to ignite the fire source.

Igniting 点火(发火,起爆) 燃气轮机启动时利用点火器使喷入火焰筒中的燃油着火的过程。

Ignition 点火 系指通过输入能量开始燃烧的过程。当物质温度升高到其分子能同时与氧化剂发生反应点时出现点火并随后出现燃烧现象。 Ignition means the process of starting a combustion process through the input of energy. Ignition occurs when the temperature of substance is raised to the point at which its molecules react simultaneously with an oxidizer and combustion occurs.

Ignition 引火源

Ignition advance 发火提前

Ignition delay 发火滞后时间

Ignition lag 发火滞后

Ignition order 发火次序

Ignition point 着火点

Ignition temperature (of an explosion gas atmosphere) (爆炸性气体环境的)点火温度 系指在按 IEC 60079-4 的规定条件下,受热表面点燃以气体或蒸发气与空气混合物形式出现的易燃材料的最低温度。 Ignition temperature(of an explosion gas atmosphere) means the lowest temperature of a heated surface at which, under specific conditions according to IEC 600798-4, the ignition of a flammable material in the form of a gas or vapour in mixture with air will occur.

Ignition test 点火试验

Ignition timer 点火定时器

Ignition timing 点火定时

Ignition unit 点火装置

Ignitor(ignition device) 点火器(发火器,点火剂) 燃气轮机启动时利用电火花、电弧、焰炬等使喷入火焰筒中的燃油着火的元件。

IGS 惰性气体系统 系指在任何时候使用惰性气体以保持货油舱内的大气避免燃烧的系统。 Inert gas system(IGS) means a system used inert gas to keep atmosphere within cargo tanks form burning at any time.

Illegal act 非法行为 系指针对船舶、海上井架、人员、货物以及公开设施保安所实施的犯罪行为。 Illegal act means a criminal act to the ship, offshore derrick, persons, cargoes and port facilities.

Illuminance 照度 系指单位面积上接收的光通量。单位为勒克斯(lx)。

Illumination 照明 系指高于工作表面 1m 处的水平面上的照明。 Illumination means illumination in the horizontal plane at a height of one meter above a working surface.

Ilmenite clay 钛铁矿土 极重的黑色黏土。有磨蚀性。可能多粉尘。从钛铁矿土中能提取钛、硅酸盐和氧化铁。含水量 10%~20%。该物质如装运时含水量超过其适运水分极限(TML),可能流态化。该货物为不燃物或失火风险低。 Ilmenite clay is very heavy black clay. It is abrasive. It may be dusty. Titanium, silicate and iron oxides are obtained from ilmenite clay. Moisture content: 10% to 20%. The material may liquefy if shipped at moisture content in excess of its transportable moisture limit (TML). This cargo is non-combustible or has a low fire-risk.

Ilmenite sand 钛铁矿砂 极重的黑砂。有磨蚀性。可能多粉尘。从钛铁矿土中能提取钛、独居石和锌。该货物 C 组中的含水量为 1%~2%。如含水量超过 2%,该货物归入 A 组。C 组中的该货物无特殊危害。A 组中的该货物如装运时含水量超过其适运水分极限(TML),可能流态化。该货物为不燃物或失火风险低。 Ilmenite sand can be categorized as Group A or C. it is very heavy black sand. Abrasiveness. It is may be dusty. Titanium, monazite and zinc ore are obtained from ilmenite sand. The moisture content of this cargo in Group C is 1% to 2%. When moisture content is above 2%, this cargo is to be categorized in Group A. This cargo in Group C has no special hazards. This cargo in Group A may liquefy if shipped at moisture content in excess of its TML. This cargo is non-combustible or has a low fire-risk.

Image resolution 图像分辨率 系指图像中存储的信息量,是每英寸图像内有多少个像素点,分辨率的单位为 PPI(Pixels per inch),除图像分辨率这种叫法外,也可以叫作图像大小、图像尺寸、像素尺寸和记录分辨率。对于计算机的显示系统来说,一幅图像的 PPI 值是没有意义的,起作用的是这幅图像所包含的总的像素数,也就是上述的另一种分辨率表示方法:水平方向像素数×垂直方向的的像素数。在大多数印刷方式中,都使用 CMYK（品红、青、黄、黑）四色油墨来表现丰富多彩的色彩，但印刷表现色彩的方式和电视、照片不一样，它使用一种半色调点的处理方法来表现图像的连续色调变化，不像后两者能够直接表现出连续色调的变化。

Imaging 成像 就是生物样本的造影技术，依照样本尺度大小可以概分为组织造影与细胞分子的显微技术。这些大致都需要光学技术配合生物样本的特性发展，少数会使用光以外的波动性质，例如核磁共振、超音波等等。成像系统使得网络用户可以从中央图像存储系统中存储和调用图像文档。成像是文档处理和工作流应用程序（管理文档在组织机构内传送的方式）的组成部分。

IMDG Code IMDG 规则 系指 IMO 组织海上安全委员会 MSC.122(75)决议通过并可能经 IMO 组织修正的"国际海运危险货物（IMDG）规则"，但该修正案应按 SOLAS 公约第Ⅷ条有关适用于除第 1 章外的附则修正程序的规定予以通过、生效和实施。 IMDG Code means the International Maritime Dangerous Goods(IMDG) Code adopted by the Maritime Safety Committee of IMO by resolution MSC. 122 (75), as may be amended by IMO, provided that such amendments are adopted, brought into force and take effect in accordance with the provisions of article Ⅷ of SOLAS Convention concerning the amendment procedures applicable to the Annex other than chapter I.

Immersed clo value(Immersion suit) 浸水 clo 值（救生服） 在服装组件浸水且经受静水压力时测取的 clo 值。 Immersed clo value is the clo value measured when a clothing assembly is immersed and subjected to the effect of hydrostatic compression.

Immersed transom beam 方艉浸宽 方艉端面在设计水线处的宽度。

Immersed transom draft 方艉浸深 方艉端面最低点至设计水线之间的垂直距离。

Immersed wedge volume 入水楔形体积 系指船舶相邻两等体积倾斜水线面所夹的没入水中的楔形体积。

Immersion of propeller axis 螺旋桨浸深 螺旋桨中点在平静水面以下的深度。

Immersion suit 救生服 系指：(1)减少在冷水中穿着该服的人员体热损失的保护服。(2)保护穿戴者避免意外落水时受冷的衣服。 Immersion suit are: (1) a protective suit which reduces the body heat-loss of a person wearing it in cold water. (2) a suit designed to protect the wearer from the cooling effects of unintended immersion in water.

Immunity 豁免 系指权利人的法律地位免受法律的约束。

IMO Maritime Safety Committee(MSC) IMO 海事安全委员会

IMO requirements IMO 要求 系指 IMO 公约、规则、决议、法规、推荐、指南、通函和有关的 ISO 标准和 IEC 标准。 IMO requirements are the IMO Conventions, Regulations, Resolutions, Codes, Recommendations, Guidelines, Circulars and related ISO and IEC standards.

IMO Subcommittee on Safety of Navigation(NAV) IMO 航海安全委员会

Impact 冲击 系指在很短的时间周期通常由于意外的或非正常事件产生的作用，例如，跌落的物体、碰撞。 Impact is an action(s) due to accidental or abnormal event usually during a very short period of time, for example, dropped objects, collisions.

Impact energy absorber 缓冲器（冲击能吸收器）

Impact test 冲击试验

Impact testing 冲击试验

Impact testing machine 冲击试验机

Impeller 泵轮（叶轮，工作轮，钻子） 耦合器中与主动轴连接并起着泵作用的工作轮。

Impeller 叶轮 系指设有叶片的旋转组合体，其给水以能量。 Impeller is a rotating assembly provided with blades to give energy to the water.

Impeller pump 叶轮泵

Imperfect combustion 不完全燃烧

Imperfections in welded joints in steel 钢焊接接头的缺陷

Implead 控告 系指机关、团体、企事业单位和个人向司法机关揭露违法犯罪事实或犯罪嫌疑人，要求依法予以惩处的行为。控告一般是由遭受犯罪行为直接侵害的被害人或其近亲属提出，主要是基于维护自身权益而要求追究被控告人刑事责任。控告是公民享有的重要权利和同违法犯罪行为做斗争的重要手段，也是刑事案件立案材料的主要来源。公民的控告权受到国家《宪法》和其他法律的保护。

Impleader 控告人 系指向司法机关揭露违法犯罪事实或犯罪嫌疑人的机关、团体、企事业单位和个人。

Implementation survey 实施检验 系指对申请船舶机械计划保养系统（PMS）检验的船舶，在其 1 年试运行期后进行的首次船舶机械计划保养系统（PMS）确认性检验。 Implementation survey means the first confirmatory survey of PMS for the ship which applies for survey of PMS after one year of its trial.

Implosion 内爆 空泡进入压力较高的区内，其中水汽突然凝结，空泡迅速坍毁的现象。

Import/export duties and taxes　进出口关税及其他税　系指关税及所有其他税项、费用或对货物进出口或与其有关而征收的其他款项,但不包括金额大致相当于所提供服务的成本费用和款项。　Import/export duties and taxes means customs duties and all other duties, fees or other charges which are collected on or in connection with the import/export of goods, but not including fees and charge which are limited in amount to the approximate cost of services rendered.

Important indication　重要标示　系指需要特别注意的显示信息的操作状态的标本,例如低完整性或无效消息。　Important indication means a marking of an operational status of displayed information which needs special attention, e.g. information with low integrity or invoilid information.

Important system　重要系统　系指确保装置可靠运行,并维持装置在操作限制之内的系统支持设备。

Important telephone system　重要电话系统　系指下列处所之间设置的,能保证在船舶各种工况下通话清晰的电话系统:(1)驾驶室-机器控制室;(2)驾驶室-舵机舱内操舵装置控制位置;(3)驾驶室-无线电室(如不用电话就能通话,则不需设此电话);(4)驾驶室-螺旋桨通常控制位置,(如设有应急主机传令钟系统,则不需设此电话);(5)驾驶室-螺旋桨其他控制位置(如这些位置上设有主机传令钟的复示器,则不需设此电话);(6)机器控制室-螺旋桨其他控制位置(如这些位置上设有主机传令钟的复示器,则不需设此电话)。　Important telephone system means telephone system provided in the following spaces,(1)bridge-engine control room;(2)bridge-steering gear control position in steering gear compartment;(3)bridge-radio room, not required if repeater of main engine telephone system;(4)bridge-normal control position of propeller, not required if emergency main engine telephone system is provided;(5)bridge-other control positions of propeller, not required if repeater of main engine telephone is provided in these positions;(6)engine control room-other control positions of propeller, not required if repeater of main engine telephone is provided in these positions. The system is to be such as to ensure fully satisfactory vocal intercommunication under all working conditions.

Impression　压痕

Impulse　冲击[冲击,冲量(自动化)]

Impulse blade　冲动式叶片

Impulse turbine　冲动式涡轮机

Impulse type steam turbine (action steam turbine)　冲动式汽轮机　蒸汽的膨胀过程名义上是全部在静叶中完成,动叶的作用只是使高速蒸汽的动能转换成机械功的汽轮机。

Impulses theory　冲量理论　假设推进器动量理论中的鼓动器不但能使流体沿轴向加速,也能使其沿周向加速以较为适应螺旋桨实际情况的一种理论。

Impulsive action　激振载荷　系指动载荷或作用持续周期(t)很短的冲击载荷。激振载荷通常能够用两个参数(冲击力和作用时间)来表征。当 $t \leqslant 0.3t$ 时,动载荷的问题需要根据激振范围加以处理。　Impulsive action is a dynamic action or impact action with a very short duration(t) of action persistence, for example, an explosion compared to the natural period(t) of the exposed structure. The Impulsive action profile can usually be characterized by two parameters, an impact load and its duration time. A dynamic action problem needs to be dealt with in the impulsive domain when $t \leqslant 0.3t$.

Impulsive noise　脉冲噪声　持续时间少于1s的孤立事件,或重复率小于15次/s的连续事件中的一次的噪声。脉冲噪声的存在可由时间计权I和F测得的等效连续声压级之间的差值得以确定。假如差值大于2dB,就可假定脉冲噪声的存在。

Impurities(inclusion)　杂质

IMSBC Code　IMSBC规则　系指IMO组织海上安全委员会以MSC. 268(85)决议通过并可能经IMO组织修正的"国际海运固体散货规则",但这类修正案应按SOLAS公约第Ⅷ条规定的关于除第1章以外适用的附则修正程序予以通过、生效和实施。　IMSBC Code means the International Maritime Solid Bulk Cargoes(IMSBC)Code adopted by the Maritime Safety Committee of IMO Organization by resolution MSC. 268(85), as may be amended by IMO Organization, provided that such amendments are adopted, brought into force and take effect in accordance with the provisions of article Ⅷ of the present Convention concerning the amendment procedures applicable to the annex other than chapter I.

In and out system　内外搭接式　一列板两边均搭接在相邻两列板同一侧的排列连接方式。

In site　就地　对船舶废气清洗系统试验和检验而言,系指废气清洗系统的废气管内直接取样。

In situ　现场　系指特定的近海油田或地区。　In situ is a specific offshore field or location.

Inactive gas　惰性气体

Inboard(derrick) boom　舱口吊杆　定位双杆操作时,吊杆头端位于本船舱口上方的一根吊杆。

Incident　事故　系指:(1)具有同一起源的造成污染损害或形成此种损害的严重和紧迫威胁的任何一个

或一系列事件；(2)涉及实际或可能将有害物质或含有这种物质的废液排放入海的事件；(3)对于"国际船舶装运密封装辐射性核燃料、钚和强放射性废料规则"而言，系指任何包括容器完整性的损坏而引起或可能引起泄漏或可能的INF货物的泄漏，相同隐患的发生或连续发生。Incident means: (1) any occurrence, or series of occurrences having the same origin, which causes pollution damage or creates a grave and imminent threat of causing such damage; (2) an event involving the actual or probable discharge into the sea of a harmful substance, or effluents containing such a substance; (3) for International Code for the safe carriage of packaged irradiated nuclear fuel, plutonium and high-level radioactive wastes on board ships, incident means any occurrence or series of occurrences, including loss of container integrity, having the same origin which results or may result in a release, or probable cargo release of INF cargo.

Incidents involving harmful substances 涉及有害物质的事故 系指下列情况：(1)排放超过允许排放标准或无论何种原因有可能排放油类或有毒液体物质，包括为保障船舶安全或救护海上人命而进行的排放；或(2)排放或可能排放包装形式的有害物质，包括装在货运集装箱、可移动罐柜、公路和铁路槽罐车以及船载驳船中的有害物质；或(3)船长为15 m或以上的船舶发生的损坏、故障或失灵：①影响船舶安全——包括但不限于碰撞、搁浅、火灾、爆炸、结构失效、进水以及货物移动；或②导致影响船舶航行安全——包括但不限于操舵装置、推进装置、发电系统和船上主要导航设备的故障或失灵；或(4)船舶营运期间排放油类或有毒液体物质超过国际防止船舶造成污染公约允许的排放量或瞬间排放速率。Incidents involving harmful substances mean those incidents: (1) a discharge above the permitted level or probable discharge of oil or of noxious liquid substances for whatever reason including those for purpose of securing the safety of the ship or for saving life at sea; or (2) a discharge or probable discharge of harmful substances in packaged form, including those in freight containers, portable tanks, road and rail vehicles and shipborne barges; or (3) damage, failure or breakdown of a ship of 15 metres in length or above which: ① affects the safety of the ship, including but not limited to collision, grounding, fire, explosion, structural failure/flooding and cargo shifting; or ② results in impairment of the safety of navigation, including but not limited to, failure or breakdown of steering gear, propulsion plant, electrical generating system, and essential shipborne navigational aids; or (4) a discharge during the operation of the ship of oil or noxious liquid substances in excess of the quanlity or instantaneous rate permitted under the MARPOL.

Incinerator 焚烧炉 是一种焚烧固体垃圾(成分近似家用垃圾)和液体垃圾，这类垃圾是船舶营运中产生的(例如：生活垃圾、与货物有关的垃圾、维修保养垃圾、运行中产生的垃圾、货舱舱底残余物、渔业机械垃圾等)以及闪点大于60 ℃的渣油的船用设备。这种设备设计成可以利用焚烧产生的热量。见图I-2。Incinerator is a shipboard facility for incinerating solid garbage approximating in composition to household garbage and liquid garbage deriving from the operation of the ship (e. g. domestic garbage, cargo-associated garbage, maintenance garbage, operational garbage, cargo residue, and fishing gear), as well as for burning sludge with a flash point above 60 ℃. These facilities may be designed to use the heat energy produced. See Fig. I-2.

图 I-2 焚烧炉
Figure I-2 Incinerator

Incinerator arrangement 焚烧炉装置

Incinerator ashes 焚烧炉灰渣 系指用于焚烧垃圾的船上焚烧炉产生的灰和熔渣。Incinerator ashes mean ash and clinkers resulting from shipboard incinerators used for the incineration of garbage.

Inclination of ladder 梯斜度 斜梯梯架与船的基平面所形成的夹角。

Inclined ladder 斜梯 以固定的斜度设置的梯。

Inclined ramp 斜跳板 系指与船体中心线形成一定夹角设置的跳板。其夹角一般为30°~45°，由3段铰接而成，它使滚装船在一般沿岸码头也可停靠，大大增强了船舶使用的生命力。

Inclined rolling 带倾横摇 有初始横倾或遇强侧风时的船舶摇摆。

Inclined ship condition 船舶横浪状态 在此状态下，船舶遭遇波浪时产生在x-y平面和y-z平面内的船舶运动，即横荡、横摇和艏摇。 Inclined ship condition means a condition, in which the ship encounters waves which produce ship motions in the x-y and y-z planes, i.e. sway, roll, yaw and heave.

Inclining moment 倾斜力矩 使船舶产生倾斜的外力矩。

Inclining test 倾斜试验 通常系指以横向移动已知的系列重物，然后测量船舶平衡横倾角的变化结果。使用这一数据并运用造船基本原理来确定船舶的重心垂向位置(VCG 或 KG)。Inclining test is a procedure which involves moving a series of known weights, normally in the transverse direction, and then measuring the resulting change in the equilibrium heel angle of the ship. By using this information and applying basic naval architecture principles, the ship's vertical centre of gravity(VCG or KG) is determined Inclining test. The inclining test is a procedure which involves moving a series of known weights, normally in the transverse direction, and then measuring the resulting change in the equilibrium heel angle of the ship. By using this information and applying basic naval architecture principles, the ship's vertical centre of gravity(VCG or KG) is determined.

Inclusive resistance test 带附体阻力试验 在船模试验水池中拖曳带附体的船模，测出各种航速时阻力的试验。

Incombustibility 不可燃性

Incombustible material 不燃材料

Income tax 所得税 是以企业的生产经营所得和其他所得征收的一种税。

Incompatibility 不兼容

Incompatible materials 不相容物质 系指混合在一起会发生危险反应的物质。这些物质需满足9.3条规定和归入B组的各项货物的细目隔离要求。 Incompatible materials means materials that may react dangerously when mixed. They are subject to the segregation requirements of subsection 9.3 and the schedules for individual cargoes classified in Group B.

Incomplete penetration 未完全焊透 焊接过程中，接头根部未完全熔透的现象。未焊透是一种比较危险的缺陷。焊缝出现间断或突变部位，焊缝强度大大降低甚至引起裂缝。

Incompletely fillet groove 不完全的熔敷坡口 要求平顺地过渡。见图I-3。 Smooth transition is required. See Fig. I-3.

Incompressible fluid 不可压缩流体

Increase of R. F. M in waves 波浪中平均转速增

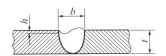

图 I-3　不完全的熔敷坡口
Figure I-3　Incompletely fillet groove

值 船舶在风浪中维持一定航速较在静水中对应航速所受平均转速的增量。

Increase production 增产作业 系指油气生产过程中为了确保油井的正常生产而采取的一种提高油层出油能力的措施,增产作业包括:酸化、压裂和防砂。

Increase production workboat 增产作业船 是海上油气田生产维护的主要装备之一，其可完成井下压裂和酸化等增产措施的作业。

Increased safety "e"(electrical equipment) "e"增安型(电气设备) 适用于在正常使用中不会产生电弧或火花的电气设备的防护形式，在这些设备中已采取附加的措施,增加其安全性,以防止过高的温度以及电弧和火花出现的可能性。注意:IEC 60079-7规定了使用此防护方法之设备的结构特征和试验要求。 Increased safety "e" (electrical equipment) means a type of protection applied to electrical apparatus that does not produce arcs or sparks in normal service, in which additional measures are applied so as to give increased security against the possibility of excessive temperatures and the occurrence of arcs and sparks. Note: IEC 60079-7 specifies the constructional features and test requirement for apparatus using this method of protection.

Increaser 异径接管(异径接头)

Increasing gear 增速齿轮

Increasing gear(gear-up) 增速齿轮箱 通过传动齿轮来实现增速并传递功率的齿轮传动组件。

Increasing oxygen device(increasing oxygen gear) 增氧装置

Increasing ratio 增速比 当输出轴转速大于输入轴转速时两者的比值。

Increment value duty 增值税 是以商品(含应税劳务)在流转过程中产生的增值额作为计税依据而征收的一种流转税。从计税原理上说，增值税是对商品生产、流通、劳务服务中多个环节的新增价值或商品的附加值征收的一种流转税。实行价外税,也就是由消费者负担,有增值才征税，没增值不征税。增值税是对销售货物或者提供加工、修理修配劳务以及进口货物的单位和个人就其实现的增值额征收的一个税种。增值税已经成为中国最主要的税种之一,增值税的收入占中国全部税收的 60% 以上,是最大的税种。增值税由国家税务局负责征收,税收收入中 75% 为中央财政收入,25% 为地方收入。进口环节的增值税由海关负责征收,税收收入全部为中央财政收入。

Independent[1] **独立性(电气系统)** 两个电气系统具有独立性系指,除了能源和馈电板外,当其他系统的任何部分出故障时,任何一个系统都能够继续工作。 Two electrical systems are considered independent when any one system may continue to operate with a failure of any part of the other system, except the power source and electrical feeder panels.

Independent[2] **独立性(蒸汽控制系统)** 系指:(1)例如一个管系或透气系统决不与另一系统连接,并且也没有任何设施与其他系统进行潜在的连接;(2)适用于两套系统,系指除动力源和馈电板外,当一套系统中的任何部分出现故障时,另一个系统都能够继续工作。 Independent means that:(1)a piping or venting system, for example, is in no way connected to another system and that there are no provisions available for the potential connection to other systems;(2) is suitable for two electrical systems which are considered independent when any one system may continue to operate with a failure of any part of the other system, except the the power source and electrical feeder panels.

Independent gear 独立装置

Independent hydraulic system 独立液压系统

Independent joystick 独立的联合操纵杆 系指独立于 DP 控制系统的单手柄控制器。

Independent joystick control system 独立的联合操纵杆控制系统 系指由推进器和舵(如适用时)等组成的综合控制系统。联合操纵杆系统能实现纵向推力、横向推力、转向力矩和这些推力分量的一切组合的控制。 Independent joystick control system is an integrated control system consisting of thrusters and rudders(if applicable). The joystick system is to enable the control of longitudinal and transverse thruster forces, turning moments, and all combinations of these thruster force components.

Independent magazine(explosives) **(爆炸品)独立库房(独立式仓库)** 系指容积为 3 m^3 及以上和非整体性可移动库房。独立库房应位于露天甲板上不受海浪直接冲击的位置。该位置应对从配餐间、泵房等处排出的热空气或危险气体设有足够的防护。应注意爆炸品遭受无线电波辐射时产生危险的可能性。独立库房应为水密金属结构。其内部应以达到 A-15 级标准的不燃材料进行绝缘。独立库房与船舶电气系统相连接的电源接头应为水密构造,并应设有说明库房的电源要求的标牌。独立库房上应设有标牌,标明其空载重量及爆炸品的最大允许装载量。独立库房应设有带防火网设施的有效自热通风。独立库房应有明显的标牌注明:(1)该处是危险库房;(2)明火或火焰不得靠近;(3)库房门应保持关闭;(4)进入库房前留下火柴、打火机;(5)不得吊装内存物。 Independent magazine is non-integral, portable magazine with a capacity of 3 m^3 or grater. Independent magazine should be located on a weather deck in a location protected from direct impact of the sea. The location should provide sufficient protection against warm air or hazardous vapours being emitted from galleys, pump-room, etc. Due regard should be paid to the possible risk of subjecting centain explosives to radio emissions. Independent magazine should be of weather-tight metal construction. The interior should be insulated with a non-combustible insulation providing an A-15 standard. The electrical terminals on independent magazine for connection to the ship's electrical system should be of watertight construction and should bear a label plate denoting the power requirement of the magazine. Independent magazine should be provided with efficient natural ventilation fitted with flame screen. Independent magazine should be clearly labeled indicating:(1) the space is a magazine;(2) open light and flame should be kept away;(3) the magazine door should be kept shut;(4) matches and lighters should be removed prior to entering;(5) not to lift with contents.

Independent magazines 独立弹药库 系指容量大于 3 m^3 的可移动非整体弹药库,此类弹药库的设计和建造应符合适用的整体弹药库要求。 Independent magazines are those that are non integral, portable magazines greater than 3 m^3 and the requirements for integral magazines are to be applied where applicable.

Independent opening 独立开口 系指在船上的横向/垂向开口间隔不小于 1 m 的开口。

Independent operation of a component 设备的独立工作 系指其功能和动力源均不依赖于主机。 Independent operation of a component is when the function of the component and the power supply of the component are independent of main engine.

Independent pump 独立泵

Independent risk control 独立性风险分析 系指控制风险的措施不影响其他因素。

Independent tank 独立液货舱 系指:(1)不与船体结构相连接或不是船体结构的组成部分的货物围护容器。建造和安装独立液货舱是为了在所有可能的时刻,能消除因相邻的船体结构的应力或移动对液货舱所造成的应力(或降至最小)。独立液货舱对船体的结构完整性不是必需的;(2)独自支撑的液舱。 Independent tank means:(1)a cargo-containment envelope, which is not contiguous with, or part of, the hull structure. An independent tank is built and installed so as to eliminate whenever possible(or in any event to minimize)its stressing as a result of stressing or motion of the adjacent hull structure. An independent tank is not essential to the structural completeness of the ship's hull;(2)a self supporting tank.

Independent ventilation system 独立通风系统
系指与别的通风系统完全不相连接,也没有允许连接至其他通风系统等设施的通风系统。 Independent ventilation system means a ventilation system that is in no way connected to another ventilation system and there is no provision available to allow connection to another ventilation system.

Indiana Dunes National lakeshore 印第安纳沙丘国家湖滨区 系指就在或邻近(美国)国家公园管理局管理的密歇根湖的地区,从沿岸线向东北纬41°42′59.4″,西经086°54′50.0″邻近印第安纳州陆地的一点到邻近印第安纳州密歇根市北纬41°37′08.8″,西经087°17′18.8″的区域。Indiana Dunes National lakeshore means the site on or near Lake Michigan, administered by the National Park Service, from a point of land near Gary, Indiana at 41°42′59.4″N,086°54′W eastward along the shoreline to 41°37′08.8″N, 087°17′18.8″W, near Michigan City Indiana.

Indicated efficiency 指示效率
Indicated horse power(i. h. p) 指示功率
Indicated instrument 指示仪表
Indicated power 指示功率(额定功率) 从示功图上求得的功率。
Indicated pressure 指示压力
Indicated thermal efficiency 指示热效率 转变为指示功的热量与产生此功所消耗的全部燃料热量的比值。
Indicated value 指示值
Indicated work 指示功
Indicating diagram 示功图 气缸内压力与活塞位置变化关系的图形。
Indicating diagram 指示图
Indicating display 指示器(指示设备)
Indicating element 指示元件
Indicating lamp 指示灯(信号灯)
Indicating liquid 显示液
Indicating panel 指示面板

Indicating plan 指示图
Indicating system 显示系统
Indicating type echo sounder 指示式回声测深仪
Indication 标示 系指显示正常信息和情况,不属于警报管理部分。 Indication means a display of regular information and conditions, not part of alert management.
Indication 显示 用于着色渗透方法检测螺旋桨缺陷的一项评定标准。分为非线性显示、线形显示和连续显示三种。
Indication circuit 指示电路 系指控制设备或系统的控制电路的一部分。其功能是将信息传送给指示被控设备状态的可见或声响器件。
Indication panel 指示面板
Indicator 指示器(显示器,计量表) 系指给出关于系统或设备状况信息的视觉指示。 Indicator means a visual indication giving information about the condition of a system or equipment.
Indicator cock 指示阀 供测绘示功图用的阀门。
Indicator diagram 示功图
Indicator dial 指示表刻度盘
Indicator gear 示功器转动装置
Indicator of fuse 熔断器的指示器 系指指示熔断器是否动作的部件。
Indicator valve 示功阀
Indicator 指示器
Indices of sea-keeping performances 船舶耐波性指标 衡量船舶耐波性的标准值。例如:纵摇幅值对波倾的比值;纵荡幅值对波高幅值的比值;垂向加速度对波高幅值的比值;相对船首运动幅值对波高幅值的比值;每小时内甲板上浪次数或砰击次数。
Indirect cooling 间接冷却 载冷剂在蒸发器内被冷却后由泵输送到盘管内流动吸热再返回蒸发器循环使用的冷却方式。
Indirect cooling air cooler 间接冷却式空气冷却器 管内由载冷剂循环吸热的表面式空气冷却器。
Indirect cooling system 间接冷却系统 是经初级制冷剂冷却的盐水或其他冷媒循环流经安装在冷藏室天花板或墙壁上的冷却盘管进行制冷的系统。Indirect cooling system is the system by which the refrigeration is obtained by brine or other secondary refrigerant, which is refrigerated by a primary refrigerant, circulated through pipe grids or coils fitted on the walls and ceilings of the refrigerated chambers.
Indirect radiative forcing 间接辐射强迫(负强迫) 系指由于潜艇上化学反应而影响其他温室气体或颗粒的吸收和发射辐射。

Indirect gauging device 非直接测量装置 系指确定液位的装置。如通过称重或管流量表。Indirect gauging device means a device which determines the level of liquid, for instance by means of weighting or pipe flow meter.

Indirection 间接

Individual alerts 单个警报 通知一个要求引起注意的异常状况的警报。Individual alerts are alerts announcing one abnormal situation and condition requiring attention.

Individual assembly time 个人集合时间 系指个人反应时间和集合时间的总和。Individual assembly time is the sum of the individual response time and the individual travel time.

Individual ballast system 专用压载系统 当油船上设有专用压载水舱时所配置的压载水管路和设备。

Individual display 单独显示 系指根据需要的其他信息显示：(1)现场测深；(2)水下电缆和管道；(3)所有单独危险物的详细情况；(4)助航设备的详细情况；(5)注意事项的内容；(6)ENC 版本的日期；(7)最近海图更新号；(8)磁偏角；(9)经纬线；(10)地名。Individual display means other displays on demand: (1) spot soundings; (2) submarine cables pipelines; (3) details of all isolated dangers; (4) details of aids to navigation; (5) contents of cautionary notes; (6) ENC edition date; (7) most recent chart update number; (8) magnetic variation; (9) graticule; (10) place names.

Individual protective equipment 个人防护设备 系指保护个人不受核、生、化与放射性物质毒害所需的服装与工具。通常包括防护服和防毒面具。Individual protective equipment is the personal clothing and equipment required to protect an individual from CBRN hazard. It normally consists of a protective suit and respirator.

Individual risk 个人风险 在某一特定位置长期生活的未采取任何防护措施的人员遭受特定危害的频率。一般来说，风险系指死亡风险，特定周期系指一年或者人的一生。在 FSA 中，个人风险系指在一个给定的地点，船舶事故给个人（船员或者船上的乘客）带来的死亡、受伤或者患病的风险，或者给第三方所有的财产带来的风险。

Individual submarine escape 潜艇艇员单人脱险 艇员穿着装具出水的潜艇艇员水下脱险。

Individual submarine escape breathing apparatus 潜艇艇员脱险呼吸器 在潜艇失事的情况下，供潜艇艇员在艇内和离艇出水时使用的呼吸器。

Individual testing 逐件试验 系指对每个配件、系固装置及绑扎用的杆均应按其相应的许用负荷逐件进行的试验。但对绑扎装置用的链或钢丝绳，在每批产品出厂前，从中抽取 1 个试件，并对其进行破断试验。Individual testing means a testing in which alternatively, every piece of fittings, securing devices and rod lashings is to be proof loaded to the safe working load of the item. In addition, one sample from every batch of chain or wire rope lashings is to be tested to breaking by the manufacturer prior to delivery.

Individual travel time 个人移动时间 系指个人从他/她的起始地至集合站的时间。Individual travel time is the time incurred by an individual in moving from his/her starting location to reach the assembly station.

Individually identifiable 可逐一识别 对国际消防安全系统规则（FSS 规则）而言，系指系统有能力识别已触发的探测器手动报警按钮的准确位置和类型，并且能将该设备信号与所有其他信号区分。For the International Code for fire safety systems (FSS Code), individually identifiable means a system with the capability to identify the exact location and type of detector or manually activated call point which has activated, and which can differentiate the signal of that device from all others.

Induced camber 感生拱度 由于沿螺旋桨叶切面水的感生速度分布作用，对水流感生的弯度。以水流方向曲线与切面头尾线间最大距离计量。

Induced resistance（induced drag） 感生阻力 由于涡紊等感应而产生的阻力。

Induced velocity 感生速度 由于螺旋桨或水翼作用使水产生的速度。

Inducer 导风轮

Induction air-conditioning system 感应空调系统

Induction electric motor 感应电动机 是一种交流电动机，其中一个部件（通常是定子）上的初级绕组与电源相连接，而另一个部件（通常是转子）上的多相次级绕组组成鼠笼型次级绕组则承载感应电流。

Induction fan 抽风（通风）机

Induction generator 感应发电机 是一种借助于一个外部的机械动力源驱动使其转速高于同步转速的感应电机。

Induction melting furnace 感应熔炼炉 系指采用 200～2 500 Hz 中频电源进行感应加热，功率范围为20～2 500 kW 的熔炼炉。

Induction motor 感应电机 一种异步交流电机，含有一个磁路，该磁路与两个电路或几组电路相交链，且这些电路彼此相对旋转。在这种电机中，功率是靠电磁感应从一个电路转移至另一电路的。感应电机的例子有感应发电机、感应电动机、某种形式的变频机和变相机。

Induction ratio 诱导比 诱导器的二次风量——经进风棚被引入诱导器的舱室空气量和一次风量——从诱导器喷嘴吹出的风量的比值。

Induction salinometer 感应式盐度计 利用电磁感应式电导率传感器测量海水电导率,以确定其盐度的仪器。

Induction squirrel-cage motor 鼠笼式感应电动机 一种次级电路系由适当地配置在次级铁芯槽内的鼠笼绕组所组成的电动机。

Induction unit 诱导器 使一次风——经过集中式空气调节器处理的空气,从喷嘴口高速吹出时形成真空,诱导一定比例的二次风——舱室空气,经进风棚送入混合室内与一次风混合,然后从出风棚吹入舱室的内部带有喷嘴和混合室的高速出风器。

Inductive type synchronous generator 感应子式同步发电机 系指磁场线圈的磁性位置相对于电枢导体为固定的一种发电机,其电动势由若干个磁性材料体的运动而产生。

Industrial Development Agency (IDA) 联合国工业开发署

Industrial felt 工业毛毡

Industrial machinery and components 生产机器和部件 系指用途与钻井作业有关的机器和部件。 Industrial machinery and components are the machinery and components which are used in connection with the drilling operation.

Industrial personal computer (IPC) 工控机 是一种加固的增强型个人计算机。工控机是一种采用总线结构,对生产过程及机电设备、工艺装备进行检测与控制的工具总称。工控机具有重要的计算机属性和特征,如具有计算机 CPU、硬盘、内存、外设及接口,并有操作系统、控制网络和协议、计算能力、友好的人机界面。工控机经常会在环境比较恶劣的环境下运行,对数据的安全性要求也更高,所以工控机通常会进行加固、防尘、防潮、防腐蚀、防辐射等特殊设计。

Industry 工业(产业,勤勉)

Industry standard 工业标准 系指国际规范(ISO等)或船舶制造国家认可的国内规范(KS, DIN, JMSA等)。 Industry standard means international standard (ISO etc.) or standards issued by national association (KS, DIN, JMSA etc.) which are recognized in the country where the ship is built.

Industry standard practices 工业标准化实践 系指在商船上或商船内,确保船上货物传输设备和干散装货物移动设备正确的安装、维护和营运的实践,船员在干散装货物最小化程序培训方面的和正确检验干散装货物堆积物最小化操作的实践。 Industry standard practices means practices that ensure the proper installation maintenance, and operation of shipboard cargo transfer and DCR removal equipment, proper crew training in DCR minimization procedures and cargo transfer operations and proper supervision or cargo transfer operations to minimize DCR accumulation on or in a commercial vessel.

Inert 惰性的(不活泼的)

Inert gas (I.G) 惰性气体 系指液货舱内的惰性气体。需进行检验以确信液货舱含氧浓度低于可燃极限,即可认为不存在爆炸气体。 Inert gas (I.G) means inert gas used in cargo tanks and must be checked to ensure that the oxygen concentration is below the flammable limit i. e. can be consisted as free from explosive gases.

Inert gas blower 烟气鼓风机 烟气防爆系统中将经过洗涤器后的烟气,自烟道抽出再输送到各油舱处的风机。

Inert gas generator 惰性气体发生器

Inert gas main 惰性气体总管

Inert gas plant 惰性气体装置

Inert gas shield 惰性气体保护

Inert-gas smothering system 惰性气体灭火及防火系统 利用含氧量较低或不含氧的惰性气体进行长期预防性充气防火和紧急充气窒息灭火的消防系统。

Inert gas smothering system 惰性气体窒息灭火系统

Inert gas system (IGS) 惰性气体系统 系指具有下列功能的系统:(1)降低每一个散货舱大气的含氧量,而使空的液货舱惰性化,以达到不能支持燃烧的水平;(2)在港内停泊和海上航行的任何时间,保持任一货舱内一部分大气的含氧量(以体积计)不超过8%,并保持正压;但当需要对液货舱除气时除外;(3)在正常作业中空气不得进入液货舱,但当需要对液货舱除气时除外;(4)驱除空的液货舱的碳氢化合物气体,使随后除气工作时不会在舱内产生可燃气体。 Inert gas system is a system provided following functions: (1) inerting empty cargo tanks by reducing the oxygen content of the atmosphere in each tank to a level at which combustion cannot be supported; (2) maintaining the atmosphere in any part of any cargo tank with an oxygen content not exceeding 8% by volume and at a positive pressure at all times in port and at sea except when it is necessary for such a tank to be gas-free; (3) eliminating the need for air to enter a tank during normal operations except when it is necessary for such a tank to be gas-free; and (4) purging empty cargo tanks of a hydrocarbon gas, so that subsequent gas-freeing operations will at no time create a flammable atmosphere within the tank.

Inert shielding gas 惰性保护气体

Inert state 惰性状态 系指由于充入惰性气体而使整个被惰化舱柜内气体的体积含氧量降低到8%或以下的状态。 Inert state means a state of an inerted tank, in any part of which the oxygen content of the atmosphere is reduced to 8% or less by volume.

Inerted 惰性化 系指：(1)这样的状态，在此状态下可燃气体与空气的混合物中氧气的体积含量不大于8%；(2)为了达到惰性状态而向被惰化舱柜送入惰性气体。 Inerted is：(1)the condition in which the oxygen content in a flammable gas/air mixture is 8% or less by volume；(2)sending inert gas into a tank to bring it into a inert state.

Inert-gas blower 惰性气体鼓风机 惰性气体防爆系统中将经过洗涤器后的惰性气体，自烟道抽出再输送到各油舱处的风机。

Inert-gas generator 独立式惰性气体发生器 专用以燃烧燃油直接产生含氧量很低的供消防用的惰性气体的装置。

Inert-gas prevent explosion system (inert-gas prevent explosion prevention system) 惰性气体防爆系统 油船上采用锅炉烟气、柴油机排气或专设的惰性气体发生器产生的惰性气体，经冷却洗涤后输送到货油舱内使油气与空气隔绝，进行预防性充气防火的一种惰性气体消防系统。

Inertia 惯性（惰性，惯量）

Inertia axis 惯性轴

Inertia balance 动平衡（惯性平衡）

Inertia coordinate system 惯性坐标系

Inertia force 惯性力

Inertia system 惯性系统

Inertia turning test 惯性回转试验 在较低航速下，操满舵使船舶作90°转舵后沿直线航行，并测定其纵距的一种试验。

Inertial navigation system (INS) 惯性导航系统 简称"惯导"，是一种不依赖于外部信息、也不向外部辐射能量的自备式（自主式）导航系统。它主要由惯性测量装置、计算机和控制显示器组成。该系统通过陀螺、加速度计测量船舶相对惯性空间的角速度或角位移和线加速度，并经过计算机的计算获得船舶的速度、位置等导航数据及其他信息。工作过程中不需要地面任何无线电导航设备，不受外界电磁波、光波或云层气候的干扰，也不向外发射电磁波，因而隐蔽性好。但由于导航精度随着系统工作时间的增加而降低，且部件成本高，维护复杂，因而使该系统不能被广泛应用。该系

统最适合于水下导航。其工作环境不仅包括空中、地面，还可以在水下。惯导的基本工作原理是以牛顿力学定律为基础，通过测量载体在惯性参考系的加速度，将其对时间进行积分，且把它变换到导航坐标系中，就能够得到在导航坐标系中的速度、偏航角和位置等信息。

Inertial-type stabilized gyro-compass 惯性式平台罗经 除输出船舶的方位、纵摇、横摇信息外还能输出船舶速度和船舶的准确位置。它有两种工作方式：平台罗经方式和惯性导航方式。一般船舶执行长期任务时，仪器工作于罗经状态；而在武器发射时需要精确的舰船导航参数或其他需要短期高精度时，平台则使用舒拉调谐实现指北，转入惯导工作，此时的精度与计程仪误差、水流、舰船偏航及操纵均无关，只与陀螺精度有关，因此减少了许多误差源，所以其精度高。其主要特点是：精度高（航向角度3.5角分，纵摇和横摇精度为0.6角分）对准时间短（小于30min）；质量轻（61kg）；体积小（92L）。

Inerting 惰性化 系指：(1)将氧化剂（通常是氧或空气）清除的过程，以防止通常是伴随除气而出现的燃烧或腐蚀扩展。(2)为了达到惰性状态而向被惰化的舱柜送入惰性气体。 Inerting is：(1) the process of removing an oxidizer (usually oxygen or air) to prevent a combustion process or corrosion progress from occurring normally accomplished by purging. (2) sending inert gas into a tank to bring it into a inert state.

Inerting method 惰化法 对于液货舱的环境控制，系指用不助燃也不与货物反应的气体或蒸发气充入液货舱和相关管系及液货舱周围的空间（若用于运输散装运输国际散装运输危险化学品船舶构造和设备规则（IBC规则）第15章有规定时），并维持这种状态。 For environmental control for cargo tanks, inerting method means a method by filling the cargo tank and associated piping systems and, where specified in Chapter 15 of the construction and equipment of ships carrying dangerous chemicals in bulk (IBC) Code, the spaces surrounding the cargo tanks, with a gas or vapour which will not support combustion and which will not react with the cargo, and maintaining that condition.

INF cargo INF 货物 系指作为按 IMDG 规则的表10,11,12 或 13 第7级货物运输密封装辐射性核燃料、钚和强放射性废料。 INF cargo means packaged irradiated nuclear fuel, plutonium and high-level radioactive wastes carried as cargo in accordance with class 7 of the IMDG Code, schedule 10, 11, 12/13.

INF Code 辐射性核燃料（INF）规则 系指 IMO 组织海上安全委员会 MSC.88(71)决议通过并可能经 IMO 组织修正的《国际船舶安全运输密封装辐射性核燃

料、钚和强放射性废料规则》，但该修正案应按 SOLAS 第 Ⅷ 条有关适用于除第 1 章外的附则修正程序的规定予以通过、生效和实施。 INF Code means the International Code for the Safe Carriage of Packaged Irradiated Nuclear Fuel, Plutonium and High-Level Radioactive Wastes on Board Ships, adopted by the Maritime Safety Committee of IMO by resolution MSC. 88 (71), as may be amended by IMO, provided that such amendments are adopted, brought into force and take effect in accordance with the provisions of article Ⅷ of SOLAS Convention concerning the amendment procedures applicable to the annex other than chapter I.

Infield cabling 内场布线 风电场内风能转换系统的电路连接。将 6 台风力发电机每台的电缆绕成一个环，然后作为海底电缆对阿尔法·文图斯海上风电试验场的变电站进行布线。

Infinite aspect ratio 无限展弦比 假想流体为二维流动时的无限大的展弦比。

Inflame 燃烧

Inflammability 易燃性

Inflammability cargo 易燃货物

Inflammable compressed air 易燃压缩气 系指任何易燃气体被压缩或液化以便于运输，且其具有雷氏蒸发气压力超过 2.76×10^5 Pa (40 lbf/in^2) 者。

Inflammable gas 易燃气体 系指具有闪点（闭杯试验）低于 26.6 ℃ (80 ℉) 和蒸发气压力在 26.6 ℃ (80 ℉) 时不超过绝对压力 275.8 kPa [2,068.6 mm 水柱 (40 lbf/in^2)] 的液体。

Inflammable gas detector 易燃性气体探测器

Inflammable liquid 易燃液体

Inflammable liquid product 易燃液体货品

Inflammable material 易燃材料

Inflammable point 着火点（燃点）

Inflammable product 易燃货品

Inflatable appliance 气胀式设备/充气式设备 系指依靠非刚性的充气室提供浮力，而在准备使用前通常保持不充气状态的设备。 Inflatable appliance is an appliance which depends upon non-rigid, gas-filled chambers for buoyancy and which is normally kept uninflated until ready for use.

Inflatable embarkation platform 充气登乘滑梯 船舶遇难时，供人员从船上滑降至水面救生载具的充气成形的滑架。

Inflatable lifejacket 充气救生衣 依靠充气提供浮力的一种救生衣。

Inflatable liferaft 气胀救生筏 具有气胀浮胎、软质水密蒙层的筏底和保护乘员的顶篷，备于特制容器中，使用时用拉绳或其他简单方法自动充气成形的救生筏。

Inflatable sealing rubber 气胀密封垫 用空心橡皮制成。靠空气充胀来保持密性的舱盖填料。

Inflated bag 充气袋囊 由柔性材料制成的袋囊，不与艇体或甲板连为一体，可靠近进行外观检查，且当小艇开始使用时，需要预先充气的。注意：在浸水时拟自动充气的袋囊（例如：置于桅顶，作为一种防止倾覆的手段）但不应视为浮性器材。 Inflated bag is the bag made of flexible material, not integral with hull or deck, accessible for visual inspection and intended always to be inflated when the boat is being used. Note：Bags intended to be inflated automatically when immersed (e.g. at the masthead as a means to prevent inversion) are not regarded as flotation elements.

Inflated lifeboat 气胀式救生筏 系指一种具有分隔的结构，坚固耐磨，永久充气的救生浮具。 Inflated lifeboat is a permanently inflated survival craft subdivided and of strong, abrasion resistant construction.

Inflow 进流 流体向静止物体的流动。

Influent (Q_i) 流入液体 (Q_i) 系指须由处理装置处理的包括污水、灰水或其他液体流的液体。 Influent (Q_i) means liquid containing sewage, grey water or other liquid streams, to be processed by the treatment plant.

Information 信息 对检验而言，特指船级社或进行法定检验的机构，在检查报告上注明的某检验已完成、某种设备已被废弃等的信息。

Information access 信息存取 系指计算机访问文档或数据集的方式。信息存取将所有信息的组织、检索活动及其先进的技术手段融合在一起。

Information and communications technology (ICT) 信息和通信技术 它是信息技术与通信技术相融合而形成的一个新的概念和新的技术领域。也是在线测试仪的简称。业内称为 MDA 测试。主要测试电路板的开短路、电阻、电容、电感、二极管、三极管、电晶体、IC 等元件。

Information channels 信道 是信号的传输媒质，可分为有线信道和无线信道两类。有线信道包括明线、对称电缆、同轴电缆及光缆等。无线信道有短波传播、短波电离层反射、超短波或微波视距中继、人造卫星中继以及各种散射信道等。如果把信道的范围扩大，它还可以包括有关的变换装置，比如：发送设备、接收设备、馈线与天线、调制器、解调器等，一般称这种扩大的信道为广义信道，而称前者为狭义信道。

Information collection 信息采集 系指根据特定的目标和要求，将分散蕴涵在不同时空域的有关信息，通过特定的手段和措施,采集和汇聚的过程。

Information network system(INS) 信息网络系统

Information processing 信息加工 系指对收集来的信息进行去伪存真、去粗取精、由表及里、由此及彼的加工过程。

Information processing system 信息处理系统 系指以计算机为基础的处理系统。由输入、输出和处理3部分组成,或者说由硬件(包括中央处理机、存储器、输入/输出设备等)、系统软件(包括操作系统、实用程序、数据库管理系统等)、应用程序和数据库所组成。

Information service business 信息服务业务 系指通过信息采集、开发、处理和信息平台的建设,通过固定网、移动网或因特网等公众通信网络直接向终端用户提供语音信息服务(声信服务)或在线信息和数据检索等信息服务的业务。

Information service platform 信息服务平台

Information Technology(IT) 信息技术 是主要用于管理和处理信息所采用的各种技术的总称。它主要是应用计算机科学和通信技术来设计、开发、安装和实施信息系统及应用软件。它也常被称为信息和通信技术(Information and communications technology/ICT)。主要包括传感技术、计算机技术和通信技术。

Information terminal company(ITC) 信息终端公司

Information transmission 信息传输 是从一端将命令或状态信息经信道传送到另一端,并被对方所接收。包括传送和接收。传输介质分有线和无线两种,有线为电话线或专用电缆;无线是利用电台、微波及卫星技术等。信息传输过程中不能改变信息,信息本身也并不能被传送或接收。必须有载体,如数据、语言、信号等方式,且传送方面和接收方面对于载体有共同解释。

Infrared antistealth technology 红外反隐身技术 是研究如何采用红外线技术使隐身措施的效果降低甚至失效的技术。

Infrared communicating light 红外线通信灯 供舰船上按编码符号用红外线进行隐蔽通信联络的灯具。

Infra-red radiation thermometer 红外辐射测温仪 通过测量海水表面或地表面自身的红外辐射,来测定其表面温度的仪器。由探测器、参考黑体、信号处理装置、记录装置等组成。

Infrared transfer(IR) 红外线传输 有一部分手机带红外功能,也比较普遍。但是传速度比较慢,而且对距离限制比较大,一般是30 cm以内。基本上红外线跟蓝牙同样都是无线传输,但是使用的频带不同。红外线的原理就是利用可视红光光谱之外的不可视光,就因为红外线也是光的一种,所以它同样具有光的特性,它无法穿越不透光的物体。并非因为我们看不到红外线,就表示它不存在,在我们生活的四周即充斥着红外线光,它可能是从电灯发出,也可能太阳光发出,使用者并不需要使用执照即可以使用红外线。例如,低速红外线(Slow IR)应用在电视的遥控器上已有相当长的一段时间了,其他如录像机、音响等遥控器也是;电视遥控器将特定的信号编码,然后透过红外线通信技术将编码送出(通常你可以看到遥控器的信号灯亮了一下),而设置在电视上的红外线接收器收到编码之后,将其进行译码而得到原来的信号;例如,电视端解得的信号为加大音量,则译码后即进行加大音量的动作。低速红外线系指其传输速率在每秒115.2 Kbits者而言,它适用于传送简短的信息、文字或是档案。有低速红外线也有高速红外线(Fast IR),系指传输速率在每秒1 Mbits 或是 4 Mbits 者而言,其他更高传输速率则仍在发展中。对于网络解决方案而言,高速红外线可以说是其基础,包括档案传输、局域网络连接甚至是多媒体传输。

Infrasound 次声 系指频率低于人类听觉可分辨的频率范围的声波。人耳可分辨的频率范围大致为20 Hz~20 000 Hz。

Infrastructure Planning Acceleration Act 《加快基础设施规划法案》 德国为加快规划大规模基础设施建设而通过的法律文件。于2000年12月17日生效,强制要求电网运营商在其平衡区域为海上风电场装备并网连接设备。

Ingress 进入

Ingress of air 进气管

Ingress of water 进水

Ingress pipe 导入管

Inhaul whip 内牵索 将传送吊车自接收船牵回补给船的绳索。

Inherent buoyancy 内部浮力 储备在救生艇内,当艇体破损进水时,仍能维持艇漂浮的浮力。

Inherent dynamic stability 固有动稳定性 系指直线航行时船舶保持动稳定,略有干扰之后,又会很快恢复保持到一个新的直线航向而无须任何舵修正。固有稳定度以及干扰程度和持续时间将取决初始船首向的偏航。 Inherent dynamic stability means that a ship is dynamically stable on a straight course if it, after a small disturbance, soon will settle on a new straight course without any corrective rudder. The resultant deviation from the original heading will depend on the degree of inherent stability and on the magnitude and duration of the disturbance.

Inherent risk control 固有风险控制 系指设计过程中,在最高概念级水平,做出限制潜在风险水平的选择。

Inherent stability 自稳性 水翼工作时不需控制

而自动保持飞高及纵、横稳定性的能力。

Inherent vice　内在缺陷　系指货物本身所固有的容易引起损坏的特性。这种特性在正常情况下亦可使货物发生质的变化而受损。例如新鲜水果会腐烂,煤炭会自燃等均属内在缺陷。

Inherently buoyant material　固有浮性材料　系指固有密度小于水的材料。 Inherently buoyant material means the material which is permanently less dense than water.

Inhibit　制止(禁止)

Inhibit behaviour　禁用工况　由于温度、振动和喘振裕度的限制而规定的燃气轮机组某些禁止使用的工况。

Initial audit　初始审核　系指对公司或船舶的最初的审核。

Initial classification survey　初次入级检验　系指对申请入级的船舶,在第1次授予其船级和颁发入级证书之前所进行的符合性检查。以确认其文件、结构、设备的设计、配置和技术状况以及管理等符合船级社入级规范、规则及船级社承认的其他技术要求。 Initial classification surveys means an initial classification survey of a ship requesting to be classed with the society is the examination of compliance of its documentation and of the design, configuration, technical condition and management of its structure and equipment with the society rules and regulations for classification and other technical requirements recognized by it, prior to assigning the society class and classification certificates to it for the first time.

Initial deflection of boom　吊杆初挠度　由于制造误差和自重引起的吊杆轴线的挠度。

Initial density of persons(D) in an escape route　初始人员密度　脱险通道中的初始人员密度系指人员数量除以人员在原来位置可用的脱险通道面积。 Initial density of persons in an escape route is the number of persons(p) divided by the available escape route area pertinent to the space where the persons are originally located and expressed in (p/m^2).

Initial mathcentre height　稳心高度　系指船舶重心和稳心之间的距离。这个距离很关键。因为如果距离太小,则船舶不稳,有慢摇的倾向,这样的船称作稳性过小(Tender);如果距离太大,则有快摇倾向,这样的船称作稳性过大(Stiff)。稳心高度以 GM. 表示,G 是重心,M 是稳心。

Initial meta-center　初稳心　船舶正浮时和小角度倾斜时的稳心。通常系指初横稳心。

Initial meta-center height　初稳心高　船舶正浮时和小角度倾斜时的稳心高度。通常系指初横稳心高。

Initial meta-center radius　初稳心半径　船舶正浮时和小角度倾斜时的稳心半径。通常系指初横稳心半径。

Initial set rudder angle　初始舵角　舵处于正舵位时,舵剖面纵向中心线与船中线面所成的夹角。

Initial stability　初稳性　船舶作小角度(一般为10°以内)横向倾斜时的稳性。

Initial survey[1]　初次检验　对每艘150GT及以上的油船和400GT及以上的其他船舶而言,初次检验系指在船舶投入营运前或首次签发国际防止船舶造成污染公约附则I第7条所要求的证书之前进行的检验。该检验应包括对上述附则所涉及船舶的结构、设备、系统、附件、布置和材料完全符合上述附则的适用要求。 For every oil tanker of 150 gross tonnage and above, and every other ship of 400 gross tonnage and above, initial survey means the survey proceeded before the ship is put in service or before the certificate required under regulation 7 of the MARPOL Annex I is issued for the first time, which shall include a complete survey of its structure, equipment, system, fitting, arrangements and material in so far as the ship is covered by the above Annex. The survey shall be as to ensure that the structure, equipment, system, fitting, arrangements and material fully comply with the applicable requirements of the above Annex.

Initial survey[2]　初次检验(MARPOL)　系指在船舶投入营运前或首次签发 MARPOL73/78 附则Ⅳ第5条所要求的证书之前进行的检验。该检验应包括对 MARPOL73/78 附则Ⅳ所涉及船舶的结构、设备系统、附件、布置及材料的全面检查。该检验应确保其结构、设备系统、附件、布置及材料完全符合 MARPOL73/78 附则Ⅳ的适用要求。 Initial survey is a survey before the ship is put in service or before the Certificate required under regulation 5 of the Annex Ⅳ to MARPOL 73/78 is issued for the first time, which shall include a complete survey of its structure, equipment, systems, fittings, arrangements and material in so far as the ship is covered by this Annex to MARPOL 73/78. This survey shall be such as to ensure that the structure, equipment, systems, fittings, arrangements and material fully comply with the applicable requirements of the Annex Ⅳ to MARPOL 73/78.

Initial survey[3]　初次检验(平台)　系指在平台投入营运之前或在首次签发证书之前进行的检验。 Initial survey is a survey before the unit is put in service or before the certificate is issued for the first time.

Initial survey for the International Certificate of Inventory of Hazardous Materials　《国际有害物质清单

证书》初次检验　系指在船舶投入营运前或在《国际有害物质清单证书》(ICIHM)签发前进行的检验。该检验应验证香港国际安全与环境无害化拆船公约附则第5条所要求的有害物质清单 第1部分符合 SRC 的要求,检验合格后,签发有效期最多不超过 5 年的 ICIHM。　Initial survey means a survey which before the ship is put in service, or before the International Certificate on Inventory of Hazardous Materials is issued. This survey shall verify that Part I of the Inventory required by regulation 5 is in accordance with the requirements of the Hong Kong International Convention for the Safe and Environmentally Sound Recycling of Ships.

Initial survey on WIG craft　地效翼船的初次检验　系指在船舶投入营运之前或首次签发地效翼船安全证书之前进行的检验。　Initial survey on WIG craft is a survey before the craft is put into service or before the WIG Craft Safety Certificate is issued for the first time.

Initial turning time　初转期　在 Z 形操纵试验中,从首次操舵的瞬时到第一次换操反舵瞬时之间的时间间隔。其无量纲值为 $A = t_a V/L$。

Initial turning/course-changing ability　初始回转/航向改变能力　系指用单位航程的船首偏航(P 数),或用船首偏航前的迹程(例如 Z 形操纵时显示的"到第二次操舵的时间"),由对应适度操舵的艏向变化定义的初始回转能力。　Initial turning/course-changing ability is the initial turning ability which is defined by the change-of-heading response to a moderate helm, in terms of heading deviation per unit distance sailed (the P number) or in terms of the distance covered before realizing a certain heading deviation (such as the "time to second execute" demonstrated when entering the zig-zag manoeuvre).

Initiating event　初始事件　系指可能导致确实出现危险的硬件失效、控制系统失灵、人为错误、极端天气或地球物理事件。　Initiating event means a event such as hardware failure, control system failure, human error, extreme weather, or geophysical event, which could lead to a hazard being realized.

Initiating event　触发事件　系指导致危险情形或事故的事件序列中的第一个。　Initiating event means the first of a sequence of events leading to a hazardous situation or accident.

Inject　注入(喷射)

Injection　喷射(注射,注入,铸入)

Injection advance　喷射提前

Injection advance angle　供油提前角　从燃油机构开始供油的瞬间到活塞至上止点时曲轴所回转的角度。

Injection angle　(燃油)喷射角(喷射持续角)　表示自喷射器开始向燃烧室内喷射的瞬时到喷射终了时间的曲轴回转角。

Injection equipment　喷射设备

Injection gear　喷射装置

Injection line　喷射管路　系指引导液体流入维护碳氢化合物生产系统的一段管路。　Injection line is a pipeline that directs fluids into a formation to support hydrocarbon production.

Injection lag　喷射滞后

Injection nozzle　喷嘴

Injection of fuel　燃油喷射

Injection oil pressure　喷油压力　喷油器正常喷射时喷油嘴内针阀开启时的瞬时燃油压力。

Injection pressure　喷射压力

Injection pump　喷射泵

Injection timing　喷射定时

Injector　喷射器(喷嘴喷油器)

Injector blowpipe　低压喷焊器

Injector cooling pump　喷油器冷却器

Injector head　喷油器头

Injector nozzle　喷油嘴

Injector pump　喷射泵(喷油泵,高压油泵)

Injure well-repair　损害修井　系指当井的产量在一定程度上有所降低时,对油管、井筒、射孔孔眼、储层孔隙和储集层裂缝的堵塞进行旁通或清除。

Ink-jet printer　喷墨打印机　按工作原理可分为固体喷墨和液体喷墨两种(目前又以后者更为常见),而液体喷墨方式又可分为气泡式(Canon 和 HP)与液体压电式(Epson)。气泡技术(Bubble jet),而 HP 采用的热感应式喷墨技术。

Inland push boat　内河推船　其艉部装有顶推设备和连接装置。推船和驳船组成顶推船队,进行内河顶推运输的船舶。见图 I-4。

图 I-4　内河推船
Figure I-4　Inland push boat

Inland vessel, river boat 内河船 以江、河、湖泊为其航区的船舶。

Inlet casing (air-intake casing) 进气机匣 将开启或其他气体均匀引入压气机的通道。

Inlet port (intake port, suction port) 进气口 控制新鲜空气或可燃混合气进入气缸内的开口。

Inlet valve (intake valve) 进气阀 控制新鲜空气或可燃混合气进入气缸内的阀门。

In-line engine 单排直立式发动机

Inmarsat 国际海事卫星组织 系指按1976年9月3日通过的国际海事卫星组织公约成立的组织。Inmarsat means the Organization established by the Convention on the International Maritime Satellite Organization (Inmarsat) adopted on 3 September 1976.

INMARSAT space segment 国际海事卫星组织的公务活动（空间段） 系指：(1)卫星以及支持这些卫星运行所需的跟踪遥测、指挥、控制、监视和相关设施与设备。这些卫星由INMARSAT拥有或租赁。 INMARSAT space segment means: the satellites, and tracking, telemetry, command, control, monitoring and related facilities and equipment required to support the operation of these satellites, which are owned or leased by INMARSAT.

INMASAT 国际海事卫星系统 这是一个海事移动卫星通信系统。它主要用于岸与船之间的通信，终端用户可通过卫星进行全球话音、用户电报和数据通信。其特点可进行电话、电报、话路数据、遇难安全通信、群呼(电报)，为岸-船和船-岸之间提供即时通信联络。通信卫星都是利用人造地球卫星作中继站来转发无线电信号，在两个或多个地面站之间进行通信。目前绝大多数卫星通信系统采用都采用同步轨道卫星，其上装有电子设备和电源，它能接收和放大地面站发出的信号然后变换频率发回地面。

Inner & outer tensile armour layer 内（外）抗拉铠装层

Inner bottom 内底 双层底的上面一层底板。

Inner bottom longitudinal 内底纵骨 内底板下的纵骨。

Inner bottom plate 内底板 构成双层底内表面的底板。

Inner case of stern tube 艉轴管衬套 艉轴管内，安装轴衬的套管。

Inner hull 内壳 系指构成船舶壳体第二层的最内部板材。 Inner hull means the innermost plating forming a second layer to the hull of the ship.

Inner passage of pontoon 浮船坞内通道 经过浮体内部，以沟通两侧坞墙间的通道。

Inner screw 内螺旋桨 船上装有两对边螺旋桨时，近中线面的一对螺旋桨。

Inner shafting (inboard shafting) 内侧轴系 四轴推进船舶中，靠近船中线面的一对轴系。

Inner side 内侧 系指双壳船舶限定内壳的纵舱壁。 Inner side is the longitudinal bulkhead which limits the inner hull for ships fitted with double hull.

Inner streak 内列板 两边均搭接在相邻列板内侧的列板。

Innocent passage (ship) 无害通过（船舶） 系指只要不损害沿海国的和平、良好秩序或安全的通过。 Innocent passage means that the passage is innocent so long as it is prejudicial to the peace, good order or security of the coastal State.

Inorganic fiber material 无机纤维材料 是以矿物质为原料制成的化学纤维吸声材料。主要品种有玻璃纤维、石英玻璃纤维、硼纤维、陶瓷纤维和金属纤维、玻璃丝、玻璃棉、岩棉和矿渣棉及其制品等。

In-process inspection 生产过程中的检查 对质量管理体系而言，系指制造厂商应该：(1)对于下一步工序不能检查的所有事项应在制造过程中进行检查；(2)根据规定的要求，进行检查试验和产品识别；(3)根据适用的情况，用生产过程监控和管理方法使产品能符合规定的要求；(4)在所要求的检查和试验没有完成和验证前，不发放产品；(5)清楚地标明不合格产品，以防止误用、装运或混入合格的产品中。 For quality management system, in-process inspection means that the manufacturer is to: (1) perform inspection during manufacture on all characteristics that cannot be inspected at a later stage; (2) inspect test and identify products in accordance with specified requirements; (3) establish product conformance to specified requirements by use of process monitoring and control methods where appropriate; (4) hold products until the required inspections and tests are completed and verified; (5) clearly identify non-conforming products to prevent unauthorized use, shipment, or mixing with conforming material.

Input power 输入功率 系指从外部输入到系统的能量。

Input/Output (I/O) 输入输出 系指产品输入输出视频信号的端口，比较常见的是S端子和复合视频端口。系指在计算机上输入输出数据的操作系统、程序或设备。实际上，有些设备只有输入功能，如键盘和鼠标；有些设备只有输出功能，如打印机；还有些设备具有输入输出两种功能，如硬盘、磁盘和可写性只读光盘(CD-ROM)。

INS = integrated navigation system 综合导航系统 组合导航系统，其至少执行以下任务：避碰，航线监视,从而为操作人员提供"附加价值"的信息,以计划、监视和安全地导航船舶的进程。 Integrated navigation system which performs at least the following tasks: collision avoidance, route monitoring thus providing "added value" for the operator to plan, monitor and safely navigate the progress of the ship.

Insert control system for electric propulsion 嵌入式电力推进控制系统 采用"基于嵌入式系统的船用发电柴油机组机旁控制"技术,通过按钮可实现电力推进装置启动、转向、调速,进行安全限制、应急预控、状态监测和故障报警。并能对突发的污染物进行定速、定点、定向控制清扫的系统。

Insertion loss(IL) 插入损失 系指未设置隔声结构时噪声源向周围辐射的声功率级为 Lw1,设置隔声结构后,噪声源透过隔声结构向周围辐射噪声的声功率级为 Lw2,那么,隔声结构的插入损失为:IL = Lw1 – Lw2。如果设置隔声结构前后,声源的方向性和室内声场分布的情况大致相同,插入损失 IL 也就是结构外给定的声压级差,即离声源一定距离外某测量点测得的隔声结构设置前的声压级 Lp1 和设置后的声压级 Lp2 之差值,即为 IL = Lp1 – Lp2。插入损失通常用于现场评定隔声罩、隔声屏障等的效果。

In-service inclining test 营运期间倾斜试验 系指为验证在实际装载工况下预先计算的许用初稳性高度 GM_C 和载重量重心而进行的倾斜试验。 In-service inclining test means an inclining test which is performed in order to verify the pre-calculated GM_C and the deadweight's centre of gravity of an actual loading condition.

Inside air foam system 内部空气泡沫系统 系指在泡沫灭火系统中泡沫发生器位于被保护处所内并从该处所吸取空气的固定式高倍泡沫灭火系统。使用内部空气的高倍泡沫灭火系统由泡沫发生器和泡沫灭火剂原液组成。 Inside air foam system is a fixed high-expansion foam fire-extinguishing system with foam generators located inside the protected space and drawing air from that space. A high-expansion foam system using inside air consists of both the foam generators and the foam concentrate.

In-side audit 现场审核 系指以评价公司和船舶的安全管理活动符合批准的船舶安全管理体系的评审。

Inside diameter 内径

Inside gauge 内径规(塞规)

Inside lap 内搭接(内搭接边、内余面)

Inside micrometer gage 内径千分卡

Inside thread 内螺纹

In-site colour meter 现场水色计 通过测量海面下两个选定光波长(450 nm 及 520 nm)的海底辐射强度之比来测定水色的比色指数的仪器。

In-site extraction sampler 现场萃取采水器 采用吸附原理设计的供现场连续定量萃取溶解有机组分用的采水器。

In-site salinometer 现场盐度计 用于现场测量海水盐度的仪器。有自容式和直读式两种。用于定点或航行测量。

In-site seawater oxygenometer 海水溶解氧现场测定仪 一种在现场测定海水溶解氧的仪器。由电源、记录器和装于水下探头中的氧传感器组成。氧传感器能将氧含量转换成电信号。

In-situ 原位 系指在废气流内直接取样。 In-situ means the sampling directly within an exhaust gas stream.

Inspected vessel 已检验的船舶 系指已按美国联邦政府法规第 46 篇第 1 章要求进行检验的船舶。Inspected vessel means any vessel that is required to be inspected under 46 CFR Ch.1.

Inspection 检查 系指通过目视、电子或其他方法探测和评估部件、结构、设备或装置损坏情况的活动。 Inspection is an activity to detect and evaluate deterioration in components, structure, equipment, or plants by visual, electronic, or other means.

Inspection 检验 (1)对澳大利亚海事安全局海事指令而言,系指经过仔细地目视检验,其包括(如有必要)进行拆卸,以便评估部件或零件的状况,是否存在变形、歪曲、损坏、磨耗、腐蚀或削弱其操作可靠性的任何其他缺陷;(2)由验船师在产品生产前和/或制造过程中和/或完工后对船级社规范要求的项目进行验证、检验和试验。 (1) for the maritime order of the Australian Maritime Safety Authority (AMSA), Inspection means a careful visual examination including, if necessary, dismantling, to assess the condition of the assembly or article for any deformation, distortion, damage, wear, corrosion or any other defect impairing its operational reliability;(2) the verification, examination and test carried out by the surveyor for items required by a society's rule before commencement and/or during process and/or after completion of production.

Inspection and test status 检查和试验情况 对质量管理体系而言,系指制造厂商应该建立并维持一套识别制度,用以保证能区分哪些是合格的,哪些是不合格的和哪些是没有检查过的项目。有关检查和试验步骤和记录应能区分发放合格产品的责任人。 For quality management system, inspection and test status mean that the manufacturer is to establish and maintain a system for the identification of inspection status of all material, components and assemblies by suitable means which distinguish between conforming, non-conforming and uninspected items. The relevant inspection and test procedures and records are to identify the authority responsible for the release of conforming products.

Inspection and verification for repairs　勘验工程
系指对待修的船舶进行实地探查的工程,以确定修理项目。

Inspection certificate of quality　品质检验证书
是出口商品交货结汇和进口商品结算、索赔的有效凭证;法定检验商品的证书,是进出口商品报关、输出输入的合法凭证。

Inspection certificate of quantity　数量检验证书
是证明出口商品出口数量的证单。

Inspection certificate of weight　质量检验证书　是证明进口商品的质量,如毛重、净重等的证书。

Inspection certificate on damaged cargo　验残检验证书

Inspection equipment　检查设备　对质量管理体系而言,系指制造厂商应负责提供、管理、校验和维护保养那些为证明材料及服务能符合规定要求所必需的,在生产管理系统所规定的检查、测量和试验所用的设备。

For quality management system, inspection equipment means that the manufacturer is to be responsible for providing, controlling, calibrating and maintaining the inspection, measuring and test equipment necessary to demonstrate the conformance of material and services to the specified requirements or used as part of the manufacturing control system.

Inspection machine　检测仪　应用十分广泛,主要应用于石化工业、电力工业、航空工业、造船业、造纸业、纺织业、冶金工业等。检测仪愈靠近泄漏点,则急流声会愈大,指示读数值会更高。功能:实时监测空气中香烟烟雾的含量,禁烟标志动态显示,丰富的语音提醒,灵活组网控制吸烟,贯彻法规控烟。该仪器操作简单,测试准确,显示部分采用液晶显示屏-使屏幕更加清晰明亮,示值清晰可见,分别可显示水分值、样品初值、终值、测定时间、温度初值,最终值等数据,并具有与计算机、打印机连接功能、符合GLP规范的打印数据输出。

Inspection, maintenance, repair(IMR)　检查、维护和修复

Installation process　总体安装方案

Installation stability　安装稳性

Installations　装置/设备　对"73/78防污公约"附则Ⅵ而言,系指与该附则Ⅵ第12条有关的在船上安装的系统、设备,包括手提式灭火器、绝缘体或其他材料,但不包括对以前安装的系统、设备、绝缘体或其他材料的修理或重新充注或者对手提灭火器的重新充注。For "MARPOL73/78" Annex Ⅵ, installations in relation to regulation 12 of this Annex means the installation of systems, equipment including portable fire-extinguishing units, insulation, or other material on a ship, but excludes the repair or recharge of previously installed systems, equipment, insulation, or other material, or the recharge of portable fire-extinguishing units.

Installed　安装　对"73/78防污公约"附则Ⅵ而言,系指安装或拟安装上船的船用柴油机,包括可移动式辅助船用柴油机,只要其加油、冷却或排气系统是船舶的组成部分。加油系统只有在永久附于船上时才可视为船舶的组成部分。该定义包括用于补充或增强船舶已装动力容量并拟成为船舶组成部分的船用柴油机。

For "MARPOL73/78" Annex Ⅵ, installed means a marine diesel engine that is or is intended to be fitted on a ship, including a portable auxiliary marine diesel engine, only if its fuelling, cooling, or exhaust system is an integral part of the ship. A fuelling system is considered integral to the ship only if it is permanently affixed to the ship. This definition includes a marine diesel engine that is used to supplement or augment the installed power capacity of the ship and is intended to be an integral part of the ship.

Instant messaging(IM)　即时通信　系指能够即时发送和接收互联网消息等的业务。1998年即时通信的功能日益丰富,逐渐集成了电子邮件、博客、音乐、电视、游戏和搜索等多种功能。即时通信不再是一个单纯的聊天工具,它已经发展成集交流、资信、娱乐、搜索、电子商务、办公协作和企业客户服务等为一体的综合化信息平台。随着移动互联网的发展,互联网即时通信也在向移动化扩张。微软、腾讯、AOL、Yahoo等重要即时通信供应商都提供通过手机接入互联网即时通信的业务,用户可以通过手机与其他已经安装了相应客户端软件的手机或电脑收发消息。

Instant messenger software(IMS)　即时通信软件
是一种可以让使用者在网络上建立某种私人聊天室(Chatroom)的实时通信服务。即时通信软件是一个终端连往一个即时通信网络的服务。即时通信不同于E-mail在于它的交谈是即时的。

Instantaneous availability　即刻可用性　救生设备本身及其存放状态允许救生设备在规定条件下,发生紧急情况时可立刻投入使用的性能。

Instantaneous rate of discharge of oil content　油量瞬间排放率　系指在任一瞬间每小时排油的升(L/h)数除以同一瞬间船速[节(kn)数]之值。Instantaneous rate of discharge of oil content means the rate of discharge of oil in liters per hour at any instant divided by the speed of the ship in knots at the same instant.

Instantaneous relay or release 瞬时继电器或脱扣器 系指无任何人为延迟动作的继电器或脱扣器。

Instantaneous sound pressure 瞬时声压 系指声波通过媒介质某一点时,某一瞬时的声压。

Institute cargo clause 协会货物保险条款 即1982年1月1日由英国伦敦保险人协会修改形成的新的协会的海运货物保险条款(Institute cargo clauses)的简称,即1995年11月1日开始使用的"伦敦海上保险人协会定期船舶保险条款"(Institute time clauses-hull)的简称。

Institute cargo clauses 货物保险条款

Institute for Intelligence and Special Operations (MOSSAD) 以色列情报和特殊使命局 简称摩萨德,由以色列军方于1948年建立,成立之初,摩萨德有多种名称,"以色列秘密情报局""中央协调局""中央情报与安全局"。以大胆、激进、诡秘称著于世,与美国中央情报局、英国军情六处、俄罗斯联邦安全局(克格勃)一起,并称为"世界四大情报组织"。自从成立以来,摩萨德进行了多次让世界震动的成功行动。它的成功,成为世界情报史上的传奇。2002年,梅尔·达甘出任摩萨德局长,并担任至今。

Institute for Scientific Information (ISI) [美]科学信息研究所

Institute for solar energy technology(ISET) 太阳能供应技术研究所 系指位于德国卡塞尔的,2009年1月与位于不莱梅的弗劳霍夫风能和海事技术中心(Fraunhofer CWNI)合并而成的弗劳恩霍夫风能及能源系统技术研究所。机构的精简是为了合作研究阿尔法斯文图斯项目。

Instruction on how to survive in the survival craft 海上获救须知 置于救生载具中用耐水纸印刷,供海上遇难人员参阅的有关获救知识的小册子。

Instructions for on board maintenance 维护保养手册 包括维护保养及修理的指南、维护保养的间隔、备品存放的位置、检查及维护的记录等内容的小册子。

Instrument 仪器 系指传感器或检测元件。Instrument is a sensor or monitoring element.

Instrument repairing room 仪器检修舱 用于海洋调查仪器和设备进行检修的工作舱室。

Instrument transformer 互感器(仪表用) 系指用于电压、电流测量的变压器。如电源互感器和电流互感器等。其主要特点是磁路不饱和,励磁电流小。

Instrumentation amplifier instrument 仪表放大器

Insubmersibility(floodability) 抗沉性 船舶破损进水后仍能保持一定的浮态和稳性的能力。

Insulated door(cold storage door) 冷藏门 用于冷藏库并带有绝热层的门。

Insulated gate bipolar transistor(IGBT) 高压绝缘栅双极晶体管

Insulated hatch cover 冷藏舱口盖 装于冷藏舱的敷贴绝热层的舱口盖。

Insulating layer 保温层

Insulation arrangement 隔热设施

Insulation lining 隔热层衬板

Insulation material 保温材料/绝缘材料 一般系指导热系数小于或等于0.2的材料。保温材料发展很快,在工业和建筑中采用良好的保温技术与材料,往往可以起到事半功倍的效果。建筑中每使用一吨矿物棉绝热制品,一年可节约一吨石油。工业设备和管道的保温,采用良好的绝热措施和材料气凝胶最早应用于美国国家航天局研制的太空服隔热衬里上。

Insulation plate 隔热罩 保护涡轮轮盘的后部、表面免受排气温度影响的遮板。

Insulation system 绝缘系统 系指一个具体模式的设备中绝缘材料的总称。此种绝缘系统的等级可以用文字、数字或其他符号标明。某一绝缘等级使用的材料具有适当的温度指标,且工作在设备标准规定的、高于所述环境温度的温度下,此温度是根据经验或已认可的试验数据确定的。该系统也可以包含任何等级的材料,只要经验和该设备的认可试验过程已经证明此种材料具有相等的预期寿命。

Insulator 绝缘体 系指不善于传导电流的物质。绝缘体又称为电介质引。它们的电阻率极高。绝缘体的定义:不容易导电的物体叫作绝缘体。绝缘体和导体,没有绝对的界限。绝缘体在某些条件下可以转化为导体。这里要注意导电的原因:无论固体还是液体,内部如果有能够自由移动的电子或者离子,那么它就可以导电。没有自由移动的电子,但在某些条件下,可以产生导电粒子,那么它也可以成为导体。

Insurance 保险 系指投保人根据合同约定,向保险人支付保险费,保险人对于合同约定的可能发生的事故因其发生所造成的财产损失承担赔偿保险金责任。

Insurance certificate 保险凭证 又称"小保单"。系指在保险凭证上不印保险条款,实际上是一种简化的保险单。保险凭证与保险单具有同等效力,凡是保险凭证上没有列明的,均以同类的保险单为准。保险凭证是保险人发给投保人以证明保险合同业已生效的另一文件形式,是一种简化了的保险。我国还有一种联合保险凭证,主要用于保险公司同外贸公司合作时附印在外贸公司的发票上,仅注明承保险别和保险金额,其他项目均以发票所列为准。

Insurance policy 保险单 是正式的保险单据,背面载有保险条款及双方当事人所约定的权利义务，是证实保险人与被保险人建立的保险合同。

Insurance standby LC 保险备用信用证 用于保险公司保险和分保业务的偿付。备用信用证还可以用于商品和金融期货交易的保证金以及股票和债券买空卖空交易的保证金等。

Intact stability 完整稳性 系指船舶在完整状态时的稳性，也称"船舶完整稳性"。Intact stability means the stability of a vessel in intact condition and also termed "intact vessel stability".

Intact stability booklet 完整稳性手册 系指所有客船不论其大小和所有船长为 24 m 及以上的货船，应在完工后进行倾斜试验，并确定它们的稳性要素。应向船长提供一本稳性手册，其中包括必需的资料，能使船长在各种装载状态下迅速而简便地得到准确的指导。对于散货船，散货船手册中所需的资料可包括在稳性手册内。 Intact stability booklet means that every passenger ship regardless of size and every cargo ship of 24m and over shall be inclined on completion and the elements of their stability determined. The master shall be supplied with a Stability Booklet containing such information as is necessary to enable him, by rapid and simple procedures, to obtain accurate guidance as to the ship under varying conditions of loading. For bulk carriers, the information required in a bulk carrier booklet may be contained in the stability booklet.

Intact stability of a floating unit 平台的完整稳性 系指漂浮着的平台依靠倾斜后其自身的复原力矩来抵抗外加倾覆力矩的能力。 Intact stability of a floating unit is the ability of that unit, when inclined, to withstand external overturning moments, by means of its righting moment.

Intake 进口

Intake advance 进气提前

Intake manifold 进口阀箱

Intake pipe 进气管（吸入管） 新鲜空气或可燃混合气进入气缸的管道。

Intake stroke 吸入冲程

Intake system 进气系统

Intake valve 进气阀

Integer 整数

Integral buckle arrestor 组合式屈曲限制器 系指以一定间隔焊接在管路上的一段稍厚的管环。其用途是限制新产生的屈曲扩散。 Integral buckle arrestor is a thicker section of pipe ring that is welded in a pipeline at chosen intervals with the purpose of arresting an incoming propagating buckle.

Integral magazines 整体弹药库 是被主船体结构部件包围的弹药库。有些指定军火的主要装备需要安全地进行长期存储，这种弹药库就是为了存放这些主要装备而特殊设计和建造的。 Integral magazines are those which are bounded by the elements of the main hull structure. They are specifically designed and constructed for the safe permanent stowage of the main outfit of designated munitions defined in the armaments requirement.

Integral magazines (explosives) （爆炸品）整体库房（整体式仓库） 系指与船舶构成一个整体的库房。整体库房的位置不得与起居处所相邻近，不得在起居处所之下，更不得与控制处所相邻近。整体库房也不得紧靠锅炉舱、机舱、厨房及具有火灾危险的其他处所。如有必要将库房置于上述这些区域邻近的处所，则应设有至少 0.6 m 的隔离舱将此两种处所隔开。此类隔离舱应设有通风装置，舱内不得用来存放物品。构成隔离舱的舱壁应为 A-15 级结构，而与 A 类机器处所相邻的舱壁应为 A-30 级结构。整体库房应从露天甲板进出，但在任何情况下不允许通过起居处所、锅炉舱、机舱、厨房及具有火灾危险的其他处所进出。整体库房应为永久性水密结构并由永久性 A-15 级分隔构成。如库房的邻近处所不存放可燃物质，则可允许用 A-0 级分隔。整体库房应设有带防火网设施或自然通风，足以使库房的温度能保持在 38 ℃ 以下。整体库房应设有喷水系统，其出水率为 $24L/(m^2 \cdot min)$。主管机关可接受其他等效装置。控制器上应清楚地标出其功能。整体库房应有明显的标牌注明：(1) 该处是危险库房；(2) 明火或火焰不得靠近；(3) 库房门应保持关闭；(4) 进入库房前留下火柴、打火机。 Integral magazines (explosives) mean those forming an integral part of the ship. Integral magazines should not be located in close proximity to and never below accommodation spaces and not in close proximity to control spaces. Integral magazines should not located adjacent to a boiler room, engine room, galley or other space presenting a fire hazard. If it is necessary to construct the magazine in proximity to these areas, a cofferdam of at least 0.6 m should be provided separating the two spaces. Such a cofferdam should be provided with ventilation and should not be used for stowage. One of bulkheads forming the cofferdam should be of A-15 construction unless there is adjacent machinery space of category A in which case A-30 is appropriate. Access to integral magazines should preferably be from the open deck, but in no case through following spaces: accommodation spaces, boiler room, engine room, galley or other space presenting a fire hazard. Integral magazines should be of permanent watertight construction and formed by permanent A-15 class divisions.

A-0 class divisions may be allowed if spaces adjacent to the magazine do not contain flammable products. Integral magazines should be provided with natural or mechanical ventilation fitted with flame screen sufficient to maintain the magazine temperature below 38 ℃. In integral magazines a sprinkler system should be installed with an application rate of 24 L/m^2 per minute. Equivalent means may be accepted by the Administration. The controls should be clearly marked as to their function. Integral magazines should be clearly labeled indicating:(1)the space is a magazine;(2)open light and flame should be kept away;(3)the magazine door should be kept shut;(4)matches and lighters should be removed prior to entering.

Integral shroud blade(shrouded blade) 带冠叶片 顶部带有叶冠的叶片。

Integral tank 整体液货舱 系指一个货物防护壳,它构成船体的一部分,以相同的方式受力且承受与相邻的船体结构相同的负荷。它是船体结构完整性所必需的。 Integral tank means a cargo containment envelope which forms part of the ship's hull and which may be stressed in the same manner and by the same loads which stress the contiguous hull structure and which is normally essential to the structural completeness of the ship's hull.

Integral tip shroud 叶冠 叶片顶部向外延伸的带有气封的冠状复块。

Integrated air-drop technology for paratrooper and car 伞兵人车一体式空投技术 该技术拥有国:俄罗斯、中国。见图I-5。

图 I-5 伞兵人车一体式空投技术
Figure I-5 Integrated air-drop technology for paratrooper and car

Integrated barge 分节驳 由舯艉两节端驳和夹在其间的若干节箱型驳相互连接而成的货驳。

Integrated bilge water treatment system(IBTS) 舱底水综合处理系统 MEPC 第 54 次会议认为"舱底水综合处理系统"能使机器处所所产生的含油舱底水量减到最少,在这个系统中对漏水和漏油予以分别处理,且处理过程变得最少和最简单,这是一个防止船舶机器处所油污染的彻底的解决办法。MEPC 第 54 次会议认为有必要将 IBTS 概念向船东及造船工作者进行推广。"舱底水综合处理系统"的具体内容可在 2008 年 11 月 12 日通过的 MEPC.1/Circ.642 通函"处理机器处所含油废弃物系统 2008 导则(含舱底水综合处理系统指南 IBTS)"中查到。

Integrated bridge system(SYS-IBS) 综合桥楼系统 系指内连系统的任意组合,以求能从工作站集中访问传感器信息,以便实行下述两种或两种以上的操作:(1)航程实施;(2)通信;(3)机器设备监视;(4)装载卸载监视和货品监视,包括客船的取暖通风和空调系统(HVAC)监视在内。Integrated bridge system(SYS-IBS) means any combination of systems which are interconnected in order to allow centralized access to sensor information from workstations to perform two or more of the following operations:(1)passage execution;(2)communications;(3)machinery monitoring;and(4)loading, discharging and cargo monitoring, including HVAC for passenger ships.

Integrated circuit(IC) 集成电路 我国台湾称为"积体电路",是一种小型化的电路(主要包括半导体设备,也包括被动元件),制造在半导体晶圆表面上。其成本低,性能高,普及度广,很多学者认为集成电路带来的数位革命是人类历史中最重要的事件之一。

Integrated floating dock 组合式浮船坞 其浮体和坞墙均由不同数量的箱形体拼装而成的浮船坞。

Integrated gate commutated thyristors(TGCT) 集成门极换流晶闸管

Integrated hull construction, outfitting and painting 壳舾涂一体化 船体(壳体)建造、舾装作业和涂装作业的综合集成。即在船体建造的同时,有机地结合进行机械舾装和涂装作业,实行同步或三位一体。

Integrated hull structure(IHS)design IHS 设计 对 LNG 船而言,系指采用连续凸形液舱舱盖甲板的设计,以代替传统的半球液舱盖的形式。采用此种设计可大幅度地增加传统的总纵强度。

Integrated lift and propulsion system 升推联动系统 用同一台动力机械通过齿轮、轴系驱动垫升风机及推进器的整套系统。

Integrated marine propulsion system 综合船用推进系统

Integrated monitoring system 集成监测系统

Integrated navigation and tactical plotting system(INTPS) 导航与绘图系统 它是以高速计算机为核心,通过接口设备和数据总线同时接收多种导航设备给出的导航信息。由于系统各部分通过总线联结,因此 INTPS 的系统配置比较简单,由用户根据需要在基本系统的基础上进行配置。系统可对罗兰 C、卫导、惯导、天文、地文等导航信息进行综合处理,给出高精度的定位信息。

Integrated navigation system(INS) 综合航行系统 是一个合成的航行系统。至少执行下述任务:避碰、航线监测,从而为操作人员计划、监测和安全导航船舶的进程提供附加值。综合航行系统(INS)可符合 SOLAS 第 V 章第 19 条的相关内容,并支持合理应用 SOLAS 第 V 章第 15 条。 Integrated navigation system (INS) is a composite navigation system which performs at least the following tasks: collision avoidance, route monitoring thus providing added value for the operator to plan, monitor and safely navigate the progress of the ship. The INS allows meeting the respective parts of SOLAS regulation V/19 and supports the proper application of SOLAS.

Integrated power system(IPS) 船舶综合电力系统 系指采用电力推进系统的船舶,以"电力导式"为指导思想,来研究船舶电能的产生、输送、变换、分配及利用电能,其利用 1 套发电机组(含原动机)产生共同的电能既用于推进,也用于其他非推进系统电气负荷的供电系统。采用综合电力系统的船舶可以根据推进和非推进电气负荷的需要,实时调整电能的分配。

Integrated power dependent liner cooling(IPDLC) 功率综合控制缸套冷却

Integrated satellite navigation system 综合卫星导航系统 系指以卫导为基础的综合导航系统。所谓综合导航系统是把载体上的某种导航设备组合在一个统一的系统内,成为一个有机整体。综合卫星导航系统从 1968 年开始用于远洋船舶,系统的基本设备有:陀螺罗经(作为航向传感器);多普勒声呐(作为航速传感器);卫星导航接收机(作为位置校准设备),小型电子计算机(作为系统的处理和控制中心);必要的输入输出设备(如电传打印机、分显示器、磁带机、绘图仪等)。根据需要也有把无线电导航设备组合在系统中的。在进行海洋调查、航道测量和石油勘探方面的任务时,还要把重力仪、磁力仪、测深仪和地震仪等组合在系统中,以实现对测量的控制,自动记录定位数据以及采集辅助的地球物理测量和地形测量的数据。

Integrated services digital network(ISDN) 综合业务数字网 是数字传输和数字交换综合而成的数字电话网。它能实现用户端的数字信号进网,并且能提供端到端的数字连接,从而可以用同一个网络承载各种话音和非话音业务。ISDN 基本速率接口包括两个能独立工作的 64 kbps 的 B 信道和一个 16 kbps 的 D 信道,选择 ISDN 2B+D 端口一个 B 信道上网,速度可达 64 kb/s,比一般电话拨号方式快 2.2 倍(若 Modem 的传输速率为 28.8 kb/s)。若两个 B 信道通过软件结合在一起使用时,通信速率则可达到 128 kb/s。

Integrated system 集成系统(计算机) 对计算机系统而言,系指:(1)基于传感器提供的输入来控制或触发动作的计算机系统;(2)一个有两个或两个以上具有独立功能的子系统组成的一个系统。该子系统是通过数据传输网络连接在一起,是从一个或多个工作站进行操作的。 For calculating system, integrated system means: (1) a calculating system which controls or initiates actions based on the sensor-supplied inputs; (2) a system consisting of two or more subsystem having independent functions connected by a data transmission network and operated from one or more workstations.

Integrated system 组合系统 系指由内部互联的多个计算机系统构成的才能集中处理信息、命令以及控制。如组合系统可以进行一个或多个下列操作:(1)航行控制(如:操舵、速度控制、交通监测、航行计划等);(2)通信联络(如:无线电话、电传等);(3)机器控制(如:电源管理、机器监测、燃油/滑油驳运等);(4)货物操作(如:货物监测、惰性气体产生、货物装卸等);(5)安全和保护(如:消防监测、消防泵控制、水密门等)。 Integrated system means a combination of computer systems which are interconnected in order to allow centralized access to sensor information and/or command/control, integrated systems may, for example, perform one or more of the following operations: (1) passage execution (e.g. steering, speed control, traffic surveillance, voyage planning); (2) communications (e.g. radiotelephone, radio telex); (3) machinery (e.g. power management, machinery monitoring, fuel oil/lubrication oil transfer); (4) cargo (e.g. cargo monitoring, inert gas generation, loading/discharging); (5) safety and security (e.g. fire detection, fire pump control, watertight doors).

Integrated tug and barge unit 拖船和驳船组合体 系指通过使用特殊设计特征的或特别设计的连接系统,能以传统的推进模式改善拖船和驳船操纵性的拖船-驳船组合体。 Integrated tug and barge unit means any the barge combination which through the use of special design features or a specially designed connection system, has increased sea keeping capabilities relative to a tug and barge in the conventional pushing mode.

Integrated search 综合检索 系指综合条件检索。

Integrating sound level meters 积分声级计 系指用于测量声曝露的声级计,又称声级曝露计。

Integrating-averaging sound level meters 积分平

均声级计　是一种直接显示某一测量时间内被测量噪声的时间平均声级，即等效连续声级（L_{eq}）的仪器。

Integrative underwater weapon system（IUWS）综合水下作战系统

Integrity　完整性　系指：（1）系统不用时，在指定时间内提供给用户精确、及时、完整、清楚的信息和警告的能力；（2）INS 向用户及时、完整和明确地提供达到规定精确度的信息的能力，以及系统慎用或不能使用时，在规定时间内向用户发出警报的能力。 Integrity is: (1) the ability of a system to provide users with accurate, timely, complete and unambiguous information and warnings within a specified time when the system is not in use; (2) the ability of the INS to provide the user with information within the specified accuracy in a timely, complete and unambiguous manner, and alerts within a specified time when the system should be used with caution or not at all.

Intel corporation（INTC）　英特尔公司　是世界上最大的半导体公司，也是第一家推出 X86 架构处理器的公司，总部位于美国加州圣克拉拉。由罗伯特·路易斯、高登·摩尔、安迪·葛洛夫，以集成电路之名（Integrated electronics）在 1968 年共同创办 Intel 公司，将高阶芯片设计能力与领导业界的制造能力结合在一起。

Intellectual property　知识产权　系指由人类的智慧活动而产生出来的成果，常附着于有形的媒体，其价值不在媒体本身，而在于人们透过媒体所表达出来概念的欣赏。

Intellectual property tribunal　知识产权庭

Intelligent character recognition（ICR）　利息备付率　又称利息覆盖率、利息覆盖倍数，衡量公司产生的税前利润能否支付当期利息的指标。公式为：借款偿还期内各年的息税前利润（EBIT）÷当年应付利息（PI）。利息备付率应分年计算，从付息资金来源的充裕性角度反映企业偿付债务利息的能力，表示企业使用息税前利润偿付利息的保证倍率。这基本上是一个风险提示指标，特别是在公司经历业绩低谷、自由现金流脆弱的时期更为关键，它可以说明公司是否还有能力支付利息以避免偿债风险，以及是否还有融资能力来扭转困境。显然，该比率低于 1 时，公司情况就已经很危急了，说明公司产生的利润连支付银行利息都不够。事实上，当该比率低于 1.5 时，就要引起投资者的警惕。根据我国企业历史数据统计分析，一般情况下，利息备付率不宜低于 2。该指标中的利息支出可以在财务报表附注中的财务费用明细中找到。

Intelligent diesel engine　智能型柴油机　一般系指对燃油喷油系统、进排气系统、油水冷却、气缸润滑、废气涡轮增压、振动平衡等各个系统实行综合电子控制，使系统参数全面优化，降低能耗、减少环境污染（低有害物质排放、低噪声、低振动），并能自动监控自身运行状况及故障，提高安全可靠性的柴油机。 Intelligent diesel engine generally means diesel engine that comprehensive electronic control is applied in individual system such as fuel oil injection system, air intake and exhaust system, oil or water cooling system, cylinder lubricating system, exhaust turbo-charging system and vibration balance system to fully optimize system parameters, lower consumption, decrease environmental pollution (low hazardous discharge, low noise, low vibration), automatically monitor operation conditions and failure, and improve safety and reliability.

Intelligent network processor（INP）　智能网络处理器

Intelligent network service（INS）　智能网络服务

Intent letter　意向书　其内容包括并购意向、非正式报价、保密义务和排他性等条款。

Inter panel　中间盖板　滚翻式舱盖中，位于首端和末端盖板之间的所有舱盖板的统称。

Interception missile　拦截弹

Interchange ability　交换性

Interchangeable　可互换的

Inter-condenser　中间冷凝器　在喷射抽气冷凝器组中，用以冷凝第一级抽气器蒸汽工质的冷凝器。

Interconnecting pipe　内部连接管

Interconnection（cross-ignition tube, cross over tube）　联焰管　在管型和环管燃烧室中使邻近各火焰筒互相连接以便在点火后传播火焰起均压作用的导管。

Intercooled air engine　中间冷却空气发动机

Intercooler　中间冷却器（内冷器）　在压缩过程中为减少压缩功而对空气或其他气体进行冷却的一种热交换器。

Inter-cooling　中间冷却（中冷）　为降低空气经过增压器后升高的温度而进行的冷却。

Intercostal　间断纵向构件　系指船舶肋板或肋骨之间的纵向非连续构件。 Intercostal means a longitudinal member between the floors or frames of a ship; it is non-continuous.

Inter-costal member　间断构件　系指构件相交处被连续构件隔断的构件。

Inter-crystalline corrosion test　晶间腐蚀倾向试验　系指表明钢材不易为过量的铬碳化合物的晶体析出而产生晶间腐蚀的试验。 Inter-crystalline corrosion test means a test demonstrated that the material is not susceptible to inter-granular corrosion resulting from grain boundary precipitation of chromium-rich carbides.

Inter-electrical signal device 船舶内部电气信号装置 系指用于向旅客、船员传送与安全有关的视觉信号和听觉信号指示的系统。

Interest payable(IP) 应付利息 系指企业按照合同约定应支付的利息,包括吸收存款,分期付款到期还本的长期借款,企业债券等应支付的利息。本科目可按存款人或债权人进行明细核算。应付利息与应计利息的区别:应付利息属于借款,应计利息属于企业存款。甲股份有限公司于2007年1月1日向银行借入一笔生产经营用 短期借款,共计120 000元,期限为9个月,年利率为8%。根据与银行签署的借款协议。本例中,1月至2月已经计提的利息为1 600元,应借记"应付利息"科目,3月份应当计提的利息为800元,应借记"财务费用"科目;实际支付利息2 400元。

Interested party 利益方 系指:(1)船级社以外的、对经入级或认证的船舶、产品、装置和系统拥有利益或有责任的一方(诸如船东及其代理人、船舶建造者、主机制造者或待检验零部件的供应商),其要求提供服务或其代理方要求提供服务。(2)海上安全调查国确定的对海上安全调查的结果有很大利益关系、权利或合法预期的组织或个人。 Interested party mean:(1)the party, other than the Society, having an interest in or responsibility for the ship, product, plant or system subject to classification or certification(such as the owner of the ship and his representatives, the ship builder, the engine builder or the supplier of parts to be tested)who requests the services or on whose behalf the services are requested.(2)an organization, or individual, who, as determined by the marine safety investigating State(s), has significant interests, right or legitimate expectations with respect to the outcome of a marine safety investigation.

Interface 接口 系指:(1)一个用作数据交换的传递设备。(例如:输入输出接口,通信接口);(2)能够进行信息交换的连接点。它可以包括:①输入/输出接口(供与传感器和执行机构的内部连接用);②人/机接口(如:显示单元、键盘、鼠标和专用控制器,以及供操作人员和计算机之间进行通信联络的仪器);③通信接口(使之能与其他计算机或外围设备进行串行通信和联网的装置)。 Interface is:(1)a transfer point at which information is exchanged.(examples:input/output interface; communications interface);(2)a transfer point at which information is exchanged. It comprises:①input/output interface(used for interconnection with sensors and actuators);②man/machine interface(e.g. visual display units, keyboards, tracker-balls, and dedicated controls and instruments used for communication between the operator and the computer);③communication interface(used to enable serial communications/networking with other computers or peripherals).

Interface technology 接口技术 系指可极大地提高硬盘的最大外部数据传输率的技术。现在普遍使用的UltraATA/66已大幅提高了E-IDE接口的性能,所谓Ultra DMA66系指一种由Intel及Quantum公司设计的同步DMA协议。

Interference power conducting suppression 电源传导干扰抑制 系指沿着导线传输的干扰,船上较大的传导干扰是电源传导干扰。各种用电设备从电网上获得电源,同时,用电设备产生的电磁干扰也通过供电电网传输给其他用电设备,因此,必须采取有效措施抑制电源传导干扰。

Interference rejection (对干扰的)抗扰度 系指器件、设备或系统面临电磁骚扰而不降低其工作性能的能力。

Intergovernmental panel on climate change(IPCC) 政府间气候变化专门委员会 是对全球范围内有关气候变化及其影响、气候变化减缓和适应参数的科学、技术、社会、经济方面的信息进行评估,并根据需求为《联合国气候变化框架公约》[The United Nations Framework Convention on Climate Change(UNFCCC)]实施提供科学技术咨询的国际性组织。IPCC不直接评估政策问题,但所评估的科学问题均与政策相关。IPCC已经于1990年、1995年、2001年和2007年相继完成了四次评估报告,报告已成为国际社会认识和了解气候变化问题的主要科学依据,对气候变化国际谈判产生了重要影响。

Interim DOC 临时证书 是发给那些没有安全管理体系运行经验公司的证书。

Interim guidelines on safety natural gas-fuelled engine installations in ship 船上天然气燃料发动机装置安全暂行指南

Interior arrangement 舱室布置 对舱室内部的设备、装置和系统所做的统一而又合理的布置。

Interior passageway 内通道 船上非敞露的通道。

Interleaved gear(nested reduction gear) 嵌入式减速齿轮箱 第一级大齿轮安置在第二级大齿轮两齿圈轴向空间之中,或第二级大齿轮安置在第一级大齿轮两齿圈轴向空间之中的减速齿轮箱。

Interlinkage 联动装置

Interlock 连锁(连锁装置)

Interlock device 连锁装置

Interlock protection 联锁保护

Interlock valve 连锁阀

Interlocked carcass 骨架层

Interlocked pressure armour　压力铠装层

Interlockeed gas valves (double block and bleed valve)　连锁气体阀　系指安装在每台双燃料发动机气体供应管路上的一套自动阀(3只)，其中两只串接在通向发动机的其他燃料管路上，第3只安装在处于两只串接阀之间的气体燃料管上，该透气管应通向露天的安全位置。

Interlocking frame　连锁架(互联机构)

Interlocking safety system　连锁安全系统

Interlocking switch　连锁开关

Interlocutory award　中间裁决　在仲裁程序中，如果仲裁庭认为争议中的一个或多个法律或事实问题已经审理清楚，为有利于继续审理而做出的部分裁决。

Intermediary trade/Re-export trade　转口贸易　又称中转贸易或再输出贸易，系指国际贸易中进转口贸易出口货物的买卖，不是在生产国与消费国之间直接进行，而是通过第三国转手进行的贸易。

Intermediate　中间的(中级的)

Intermediate bearer　中间支承

Intermediate bearing　中间轴承

Intermediate bulk container　中间散装集装箱　系指货物容器，其:(1)由刚性、半刚性或柔性材料或由这些材料组合在一起制成;(2)设计成可由物资装运设备起吊的整体或可分离的装置;(3)具有不小于 $0.25 m^3$ 和不大于 $3\ m^3$ 的容积。　Intermediate bulk container means a cargo receptacle: (1) constructed of rigid, semi-rigid or flexible material, or a combination of such materials; (2) designed to be lifted by materials handling equipment, by means of either integral or detachable device, (3) having a capacity, of not less that 0.25 cubic meters and not more than 3 m³.

Intermediate casing　中间机匣　连接燃气发生器的增压涡轮机机匣和动力涡轮机机匣的中间环形扩压器。

Intermediate frame　中间肋骨　为局部加强而增设在肋骨间距中点位置上的肋骨。

Intermediate People's Court　中级人民法院　我国一级审判机关，是我国法院体系的一个层级，其上级单位是高级人民法院。中级人民法院设刑事审判庭、民事审判庭、行政审判庭，根据需要可以设置其他审判庭。

Intermediate platform　舷梯中间平台　装于分节舷梯中间的舷梯平台。

Intermediate pressure　中间压力　除去第一级吸气压力和最后一级排气压力以外的各级吸气压力和排气压力的统称。

Intermediate pressure compressor　中压压缩机　排气压力 980~9 800 kPa 表压的压缩机。

Intermediate pressure cylinder　(蒸汽机)冲压缸

Intermediate rail　横栏　位于扶栏下方，并与之平行，起拦护作用的横向杆件。

Intermediate shaft　中间轴　系指主机和减速器之间，或主机和齿轮机座之间，或主机和万向连轴节之间的连接轴。

Intermediate shaft(line shaft, tunnel shaft)　中间轴　轴系中用以连接推力轴与艉轴或推进器轴的轴。

Intermediate shaft pedestal　中间轴承座

Intermediate starting valve　(蒸汽机)旁通阀

Intermediate stern tube　中艉轴管　多轴系船的侧轴上，艉轴管分三段组成时的中段部分。

Intermediate survey　中间检验　对400GT及以上的船舶以及所有固定和移动钻井平台和其他平台而言，系指在证书的第2个周年日之前或之后3个月内或第3个周年日之前或之后3个月内进行的检验，并应取代有关规定的其中一次年度检验。中间检验应确保设备及其布置完全符合有关要求，并处于良好的工作状态。该中间检验应在按有关规定签发的IAPP证书上予以签署。　For vessels of 400 gross tonnage and above and fixed and floating drilling rig and other platforms, intermediate survey means a survey within three months before or after the second anniversary date or within three months before or after the third anniversary date of the certificate which shall take the place of one of the annual surveys of relevant requirement. The intermediate survey shall be such as to ensure that the equipment and arrangements fully comply with the relevant requirement and are in good working order. Such intermediate surveys shall be endorsed on the IAPP certificate issued under relevant requirement.

Intermediate zone (secondary combustion zone)　过渡区　主燃区和掺混区之间使残余燃料在其中继续燃烧的过渡区域。

Intermittent or predatory dumping　间歇性或掠夺性倾销　是以低于国内价格甚至低于成本价格，在某一国外市场上倾销商品，在打垮了或摧毁了所有或大部分竞争对手，垄断了这个市场后，再提高价格。

Intermixture and contamination risk　混杂、沾污险　投保平安险和水渍险的基础上加保此险，保险人负责赔偿承保的货物在运输过程中因混进杂质或被沾污,影响货物质量所造成的损失。此外保险货物因为和其他物质接触而被沾污，例如布匹、纸张、食物、服装等被油类或带色的物质污染因而引起的经济损失。

Internal audit　内部审核　对质量管理体系而言，系指制造厂应组织内部审核以保证质量制度的持续执行。应建立质量审核计划，按照各项活动的重要性和情况确定审核，并根据以往结果进行调整。　For quality management system, internal audit means that the manufacturer is to conduct internal audits to ensure continued adherence to the system. An audit programme is to be established with audit frequencies scheduled on the basis of the status and importance of the activity and adjusted on the basis of previous results.

Internal combustion engine 内燃机

Internal combustion gas turbine plant 内燃式燃气轮机装置 燃料与空气在燃烧室中混合燃烧的燃气轮机装置。

Internal cone propeller 双锥筒推进器 船内两截锥筒中的水轮将水由船底吸入加以动能向后排出以起推进作用的一种喷水推进器。

Internal control 内控 为确保工作符合所有要求和规范而实施的所有管理措施。

Internal criteria for the assessment of flag State performance 船旗国履约评估内部标准 系指与船旗国主管机关的运作有直接关系,并对船旗国主管机关按有关文件履行其义务的有效性给出明确指示的标准。船旗国职责指南包含在大会决议 A.847(20)"帮助船旗国履行 IMO 文件的导则"中。联合国海洋法第 94 条也给出了缔约国的职责[第 1,2(1)款];联合国海洋法第 217 条给出了船旗国实施职责。根据国际文件,其为文件缔约国的船旗国有责任使其在国内实施,执行这些要求。 Internal criteria for the assessment of flag State performance are those which are directly relevant to the operation of the flag State as an Administration and are designed to give a clear indication of the effectiveness of a flag State Administration in fulfilling its obligations under the instruments. Guidance on flag State responsibilities is contained in Assembly resolution A.847(20) on Guidelines to assist flag States in the implementation of IMO instruments. Article 94 of UNCLOS also sets out the duties of the State Parties(Article 1,2(1)). Article 217 of UNCLOS is also relevant in detailing the enforcement responsibilities of flag States. Based on international instruments a flag State has responsibilities relating, in particular, to setting legal requirements to give national effect to the instruments to which it is a Party; enforcement of those requirement.

Internal deck 内甲板 系指在船体的货舱区域内,上甲板以上的构成船体内部第二层壳体的甲板。

Internal diameter 内径

Internal efficiency 内效率 汽轮机不计机械损失的内部有用能量与理想可用能量之比。

Internal examination 内部检验

Internal insulation tank 内部绝热液货舱

Internal noise in ground transportation 陆上交通工具内部的噪声 主要指汽车内,铁路车辆内的噪声。这类噪声直接影响到旅客的休息、睡眠等,也影响驾驶人员注意力不能集中而发生意外事故,也妨碍服务人员的正常工作。

Internal noise in land buildings 陆上建筑物内部噪声 系指处在陆地上的所有建筑物,如居住用的房屋,办公大楼、高楼大厦等建筑物,工矿企业的工厂、车间、工场等的固定建筑物,其内部的房间(如居民住家的卧室、书房、厨房、餐厅等,旅馆或宾馆的卧室、会客室或厅、办公室、娱乐室等)、走廊、梯道、平台等的噪声。

Internal noise in ship cabin 船舶内舱室的噪声 主要指船舶舱室内部的噪声。这类噪声对乘客和船员危害特别大。

Internal noise in trans atmospheric vehicle 空中交通工具内部的噪声 主要指飞机内部的噪声。飞机舱内的噪声直接影响到旅客旅途休息和服务人员的正常工作,也影响驾驶员注意力不能集中而发生意外事故。

Internal opening 内部开口 在露天舱室以外的舱室、在其甲板或舱壁上的开口。

Internal pressure sheath 内压护套层

Internal screw thread 内螺纹

Internal treatment 锅内处理 对运行中的锅炉,利用成垢物质在某些条件下会在锅水中形成泥渣而便于排放的特性来防止锅炉结垢、金属腐蚀的水处理办法。

Internal turret 内转塔

Internal wave(boundary wave) 界面波 在两种不同密度水界面上形成的波。

International aeronautical and maritime search and rescue(IAMSAR)manual 国际航空和海上搜救手册

International Air Pollution Prevention Certificate (IAPP Certificate) IAPP 证书 "国际防止空气污染证书"的简称,当相关船舶经主管机关检验确认其符合 MARPOL73/78 附则Ⅵ的规定后由该主管机关授予的"国际防止空气污染证书"。

International Anti-dumping Laws 国际反倾销法

International anti-fouling system certificate 国际控制有害船底防污系统证书 系指由政府主管机关或经其授权的任何人员或机构签发或签证的,表明船底防污系统符合"船底防污公约"要求的证书。 International anti-fouling system certificate means a documentary evidence to show the anti-fouling system on ships is in compliance with requirements of AFS Code, issued or endorsed by the Administration or any person or organization authorized by the Administration.

International Article Number Association(IANA)

国际物品编码协会

International Association of Classification Societies (IACS) 国际船级社协会 总部设在伦敦,由10个成员船级社、两个准成员船级社组成。其组织机构为:由主席及副主席、来自各成员船级社的高级管理代表组成的理事会。其下设有:(1)来自各成员船级社的高级管理代表组成的总政策组、(2)由常设秘书、技术官员和管理职员组成的常设秘书处、(3)由成员船级社代表组成的质量委员会。第三层机构为:(1)来自各成员船级社的高级管理代表组成的工作组;(2)负责外部事务追踪的ISO质量代表;(3)在质量委员会下设有成员秘书组成的审查委员会和外审小组。IACS成员包括:美国船级社、法国船级社、中国船级社、挪威-德国劳氏船级社、韩国船级社、英国劳氏船级社、日本海事协会、意大利船级社、俄罗斯船舶登记局、克罗地亚船级社、印度船级社《(Associate)Indian Register of Shipping/IRS》、和波兰船级社等。此外,非IACS成员有:希腊船级社和巴拿马船级社等。见表1-1。

表1-1 国际船级社协会成员

IACS 成员	IACS members	缩写
美国船级社	America Bureau of Shipping	ABS
法国船级社	Bureau veritas	BV
中国船级社	China Classification Society	CCS
挪威船级社-德国劳氏船级社	Det Norske Veritas-Germanischer Lloyd	DNV-GL
韩国船级社	Korean Register of Shipping	KR
英国劳氏船级社	Lloyd's Register of Shipping	LR
日本海事协会	Class NK Nippon kaiji kyokai	NK
意大利船级社	Registro Italiano Navale	RINA
俄罗斯船舶登记局	Russian Maritime Register of Shipping	RS
克罗地亚船级社	(Associate)Croatian Register of Shipping	CRS
印度船级社	(Associate)Indian Register of Shipping	IRS
波兰船级社	(Associate)Polski Regestr Statkow	PRS
非IACS成员	Non-IACS member	
希腊船级社	Hellenic Register of Shipping	HR
巴拿马船级社	Panama Bureau of Shipping	PBS

International Bank for Reconstruction and Development (IBRD) 国际复兴开发银行 通称"世界银行"(World Bank)。世界银行(WBG)是世界银行集团的俗称,"世界银行"这个名称一直是用于指国际复兴开发银行(IBRD)和国际开发协会(IDA)。它是根据1944年联合国国际货币金融会议通过的《国际复兴开发银行协定》建立的全球性国际金融组织。这些机构联合向发展中国家提供低息贷款、无息信贷和赠款。它是一个国际组织,其一开始的使命是帮助在第二次世界大战中被破坏的国家的重建。今天它的任务是资助国家克服穷困,各机构在减轻贫困和提高生活水平的使命中发挥独特的作用。

International biding competitive (IBC) 国际竞争性招标 是在世界范围内进行招标,国内外合格的投标商均可以投标。要求制作完整的英文标书,在国际上通过各种宣传媒介刊登招标公告。例如,世界银行对贷款项目货物及工程的采购规定了三个原则:(1)必须注意节约资金并提高效率,即经济有效;(2)要为世界银行的全部成员国提供平等的竞争机会,不歧视投标人;(3)有利于促进借款国本国的建筑业和制造业的发展。

International Bulk Chemical Code 国际散装化学品规则 系指由IMO组织海上环境保护委员会以MEPC.19(22)决议通过的并经IMO组织修正的"国际散装运输危险化学品船舶构造和设备规则"。 International Bulk Chemical Code means the International Code for the Construction and Equipment of Ships Carrying Dangerous Chemicals in Bulk adopted by the Marine Environment Protection Committee of the Organization by resolution MEPC.19(22), as amended by IMO.

International Bulk Chemical Code(IBC Code)
"国际散装危险化学品规则(IBC规则)" 系指 IMO 组织海上安全委员会 MSC.4(48)决议通过并可能经 IMO 组织修正的"国际散装运输危险化学品船舶构造和设备规则",但该修正案应按 SOLAS 第Ⅷ条有关适用于除第1章外的附则修正程序的规定予以通过、生效和实施。

International Bulk Chemical Code(IBC Code) means the International Code for the Construction and Equipment of Ships Carrying Dangerous Chemicals in Bulk adopted by the Maritime Safety Committee of IMO by resolution MSC. 4 (48), as may be amended by IMO, provided that such amendments are adopted, brought into force and take effect in accordance with the provisions of article Ⅷ of SOLAS Convention concerning the amendment procedures applicable to the annex other than chapter I.

International Business Machines(IBM) [美]国际商业机器公司

International business negotiation 国际商务谈判 是国际商务活动中不同的利益主体,为了达成某笔交易,而就交易的各项条件进行协商的过程。国际商务谈判,实际上是通过协商弥合分歧使各方利益目标趋于一致而最后达成协议的过程。在国际经济往来中,企业之间的洽谈协商活动不仅反映着企业与企业的关系,还体现了国家与国家的关系,相互间要求在尊重各自权利和国格的基础上,平等地进行贸易与经济合作事务。在谈判的信息资料方面,谈判者既有获取真实资料的权利,又有向对方提供真实资料的义务。

International cargo agreement transportation 国际货协运单 系指参加国际货协各国之间办理铁路联运时所用的运输单据。

International cargo insurance 国际货物保险 是以运输过程中的各种货物作为保险标的,被保险人(买方或卖方)向保险人(保险公司)按一定金额投保一定的险别,并缴纳保险费。保险人承保以后,如果保险标的在运输过程中发生的约定范围内的损失,应按照规定给予被保险人经济上补偿的一种财产保险。

International Center for Settlement of Investment Disputes(ICSID) 国际投资纠纷解决中心 是1966年10月14日根据1965年3月在世界银行赞助下于美国华盛顿签署的,1966年10月14日生效的《解决各国和其他国家国民之间投资争端的公约》(即1965年华盛顿公约)而建立的一个专门处理国际投资争议的国际性常设仲裁机构,它是复兴开发银行下属的一个独立机构。其宗旨在于:以调解和仲裁的方式,为解决其国同外国私人投资者之间投资争议提供便利。

International certificate of fitness for the carriage of dangerous chemicals in bulk 国际散装运输危险化学品的适装证书 系指对经初次检验或定期检验之后,对符合 IBC 规则有关要求的国际航行化学品液货船签发的散装运输危险化学品适装证书。该证书的标准格式见 IBC 规则的附录。按1974SOLAS 公约的第Ⅶ章和 MARPOL 73/78 附则Ⅱ的规定,对于1986年7月1日以前建造的化学品液货舱,该规则是强制性的。 International certificate of fitness for the carriage of dangerous chemicals in bulk means a certificate called an International certificate of fitness for the carriage of dangerous chemicals in bulk, the model form of which is set out in the appendix to the International Bulk Chemical Code, should be issued after an initial or periodical survey to a chemical tanker engaged in international voyages which complies with the relevant requirements of the Code. The Code is mandatory under both chapters Ⅶ of SOLAS 1974 and Annex II of MARPOL 73/78 for chemical tankers constructed on or after 1 July 1986.

International certificate of fitness for the carriage of INF cargo 国际运输密封装辐射性核燃料货物的适装证书 系指装运密封装辐射性核燃料货物的船舶除应符合 SOLAS 规则的任何适用要求外,还应符合国际船舶装运密封装辐射性核燃料、钚和强放射性废料规则的要求,并应经受检验和备有国际装运密封装辐射性核燃料货物适装证书。 International certificate of fitness for the carriage of INF cargo means that a ship carrying INF cargo shall comply with the requirements of the International Code for the safe carriage of packaged irradiated nuclear fuel, plutonium and high-level radioactive wastes on board ships(INF Code) in addition to any other applicable requirements of the SOLAS regulations and shall be surveyed and be provided with the International certificate of fitness for the carriage of INF cargo.

International certificate of fitness for the carriage of liquefied gases in bulk 国际散装运输液化气体的适装证书 系指对经初次检验或定期检验之后,对符合国际气体运输船舶规则有关要求的气体运输船签发的国际散装运输液化气体适装证书。该证书的标准格式见 IGC 规则附录。根据1974年 SOLAS 公约第Ⅶ章的规定,对1986年7月1日或以后建造的气体运输船,该规则是强制性的。 International certificate of fitness for the carriage of liquefied gases in bulk means a certificate called an International certificate of fitness for the carriage of liquefied gases in bulk, the model form of which is set out in the appendix to the International Gas Carrier Code should be issued after an initial or periodical survey to a gas carrier which complies with the relevant requirements The Code is mandatory under chapter Ⅶ of SOLAS 1974 for gas carriers constructed on or after 1 July 1986.

International Certificate on Inventory of Hazardous Materials(ICIHM) 国际有害物质清单证书

International Chamber of Commerce Court of Arbitration 国际商会仲裁院

International Chamber of Commerce(ICC) 国际商会 成立于1919年，发展至今已拥有来自130多个国家的成员公司和协会，是全球唯一的代表所有企业的权威代言机构。国际商会以贸易为促进和平、繁荣的强大力量，推行一种开放的国际贸易、投资体系和市场经济。由于国际商会的成员公司和协会本身从事于国际商业活动，因此它所制定用以规范国际商业合作的规章。

International closed market price 世界"封闭市场价格" 与"自由市场价格"不同，它是买卖双方在一定特殊关系下形成的价格。在这种情况下，商品在国际间的供求关系，一般对价格不会产生实质性的影响，其主要包括：调拨价格、垄断价格、区域性经济贸易集团内的价格和国际商品协定下的协定价格。

International code flag 国际信号旗 由26面拉丁字母旗、10面数字旗、3面代用旗和1面回答旗所组成的一整套国际通用的作为通信联络信号用的号旗。

International Code for fire safety systems(FSS Code) 国际消防安全系统规则(FSS规则)

International Code for the Safe Carriage of Packaged Irradiated Nuclear Fuel, Plutonium and High Level Radioactive Wastes on Board Ships 辐射性核燃料(INF)规则 系指由国际海事组织海上安全委员会以MSC.88(71)决议通过并可能经国际海事组织修正的"国际船舶载运密封装辐射性核燃料、钚和强放射性废料规则"。

International Code of Signals 国际信号规则

International commercial arbitration 国际商事仲裁 又称对外经济贸易及海事仲裁、涉外仲裁等，系指不同国家的公民、法人将他们在对外经济贸易及海事中所发生的争议，以书面的形式，自愿交由第三者进行评断和裁决。

International commercial arbitration association 国际商事仲裁协会

International commercial arbitration association of Japan 日本国际商事仲裁协会

International commodity agreement 国际商品协定 系指某种商品的主要生产出口国之间，或者主要生产国与主要进口国之间为了稳定或者操纵该种商品的世界市场价格，获得足够的垄断利润，保证世界范围内的供求基本平衡而签订的多边国际协议。

International comparison program 国际比较项目

International compliance certificate of energy efficiency 国际能效符合证明 系指船旗国主管机关接受MARPOL附则Ⅵ，经公司申请，按MARPOL附则Ⅵ船舶能效要求检验合格后签发的符合证明。

International Convention 国际公约 系指各国之间用书面签订，并受国际法规制约的国际协议。 International Convention means an international agreement concluded among States in written form and governed by international law.

International Convention for Prevention of Pollution from ships 73/78, as amended from time to time published by IMO(MARPOL) 1973/1978 年国际防止船舶造成污染公约，经常修正，并由国际海事组织出版。

International Convention for the Control and Management of Ship's Ballast Water and Sediments, 2004 2004年船舶压载水和沉积物控制和管理国际公约 这是IMO成员国经过多年协商和努力，终于在2004年制订的关于防止因压载水的异域随意排放而导致有害水生物和病原体的转移，以尽可能减少对环境、人体健康、财产和资源造成损伤或损害。该公约生效后所有船舶必须遵守，前期执行"压载水置换标准"，即"D-1标准"，数年后全部执行"压载水性能标准"，即"D-2标准"。

International Convention for the Prevention of Pollution from Ships, 1973, as modified by the Protocol of 1978 relating thereto(MARPOL 73/78) 73/78防污公约 系指经1978年议定书修订的1973年国际防止船舶造成污染公约的简称。公约沿革简述如下：由国际海事组织(IMO)在1973年10月8日至11月2日召开的国际海洋污染会议上通过了《1973年国际防止船舶造成污染公约》(简称1973防污公约)。议定书Ⅰ(关于涉及有害物质事故报告的规定)和议定书Ⅱ(仲裁)在同一会议上通过。随后，公约经IMO在1978年2月6日至17日召开的国际油船安全和防污染会议(TSPP会议)通过的1978年议定书的修订。经1978年议定书修订的本公约，称为"经1978年议定书修订的1973年国际防止船舶造成污染公约"，或简称《73/78防污公约》，涉及船舶造成污染各种成因的规则包括在本公约的5个附则内。本公约还经1997年议定书作了修订，增加了第6个附则。

International Convention for the Safety of Life at Sea 74/78, as amended from time to time published by IMO(SOLAS) 1974/1978 国际海上人命安全公约，经常修正，并由国际海事组织出版

International court of justice 国际法院 是联合国的司法裁决机构。其根据《国际法院规约》于1946年2月成立。院址在荷兰海牙的和平宫，亦称"海牙国际法

庭"。国际法院的主要功能是对各国所提交的案件做出仲裁,或在联合国大会及联合国安理会的请求下提供咨询性司法建议。它还可以审理涉嫌违反国际法的案件。它作为联合国 6 大机构之一,国际法院是唯一具有一般管辖权的普遍性国际法院。国际法院由 15 名法官组成。法官候选人需在联合国安理会和联合国大会分别获得绝对多数赞成票方能当选,每届任期 9 年,每 3 年改选 1/3,以保持工作连续性;全体法官以无记名投票方式推举院长,院长每届任期 3 年。国际法院受理的案件中,半数以上是领土和边界纠纷。

International criminal court 国际刑事法院 成立于 2002 年,具有国际法律人格,对灭绝种族罪、危害人类罪、战争罪、侵略罪等这些国际社会中的最严重犯罪具有管辖权,通常只追究个人的刑事责任,并且是在各个所属的法院不能自主审理的情况下才介入。国际刑事法院享有为行使其职能和实现其宗旨所必需的法律行为能力,可以在缔约国境内、在有特别协定的前提下在非缔约国境内,行使其职能和权力。

International criminal police organization (ICPO/ INTERPOL) 国际刑警组织

International Development Association (IDA) 国际开发协会 是根据 1960 年 9 月 24 日通过的《国际开发协会协定》成立,它是世界银行的附属机构之一,也是联合国专门机构之一。总部设在华盛顿。

International division of labor 国际分工 系指世界上各国(地区)之间的劳动分工,是国际贸易和各国地区经济联系的基础,它是社会生产力发展到一定阶段的产物,是社会分工超越国界的结果,是生产社会化向国际化发展的趋势。

International energy efficiency certificate(IEEC) 国际能效证书 系指根据经 MEPC. 203(62)决议修订的 MARPOL 附则 VI,为满足能效要求的船舶签发的法定证书。

International Finance Centre(IFC) 香港国际金融中心 简称国金,是香港作为世界级金融中心的著名地标,位于香港岛中环金融街 8 号,面向维多利亚港。由地铁公司(今港铁公司)及新鸿基地产、恒基兆业、香港中华煤气及中银香港属下新中地产所组成的 IFC Development Limited 开发、著名美籍建筑师 César Pelli 及香港建筑师严迅奇合作设计而成,总楼面面积达 436 000 km²。现为恒基集团和香港金融管理局的总部所在地。

International Finance Corporation(IFC) 国际金融公司 是世界银行下属机构之一。1956 年 7 月正式成立,总部也设在华盛顿。它虽是世界银行的附属机构,但其本身具有独立的法人地位。国际金融公司的人民币债券发行说明了中国资本市场的发展又往前迈进了重要的一步,它将进一步推动中国非政府债券市场的发展,从而增加私营企业的融资渠道。国际金融公司的宗旨主要是:配合世界银行的业务活动,向成员国特别是其中的发展中国家的重点私人企业提供无须政府担保的贷款或投资,鼓励国际私人资本流向发展中国家,以推动这些国家的私人企业的成长,促进其经济发展。

International free market price 世界"自由市场价格" 系指在国际间不受垄断或国家垄断力量干扰的条件下,由独立经营的买者和卖者之间进行交易的价格。国际供求关系是这种价格形成的客观基础。"自由市场"是由较多的买主和卖主集中在固定的地点,按一定的规则,在规定的时间进行的交易,尽管这种市场也不可避免会受到国际垄断和国家干预的影响。但是,由于商品价格在这里是通过买卖双方公开竞争而形成的,所以,它常常比较客观反映商品供求关系的变化。

International Gas Carrier Code(IGC Code) "国际气体运输船规则(IGC 规则)" 系指 IMO 组织海上安全委员会 MSC.5(48)决议通过并可能经 IMO 组织修正的《国际散装运输液化气体船舶构造和设备规则》,但该修正案应按 SOLAS 公约第Ⅷ章有关适用于除第 1 章外的附则修正程序的规定予以通过、生效和实施。 International Gas Carrier Code(IGC Code) means the International Code for the Construction and Equipment of Ships Carrying Liquefied Gases in Bulk as adopted by the Maritime Safety Committee of IMO by resolution MSC.5(48), as may be amended by IMO, provided that such amendments are adopted, brought into force and take effect in accordance with the provisions of article Ⅷ of SOLAS Convention concerning the amendment procedures applicable to the annex other than chapter I.

International Grain Code 国际谷物规则 系指 IMO 组织海上安全委员会 MSC.23(59)决议通过并可能经 IMO 组织修正的"国际散装谷物安全运输规则",但该修正案应按 SOLAS 公约第Ⅷ条有关适用于除第 1 章外的附则修正程序的规定予以通过、生效和实施。 International Grain Code means the International Code for the safe carriage of grain in bulk adopted by the Maritime Safety Committee of IMO by resolution MSC. 23(59) as may be amended by IMO, provided that such amendments are adopted, brought into force and take effect in accordance with the provisions of article Ⅷ of SOLAS Convention concerning the amendment procedures applicable to the annex other than chapter I.

International Labo(u) r Organization(ILO) 国际劳工组织 成立于 1919 年,其宗旨是改善劳动者的工作

条件和健康状况。1946 年成为联合国的一级咨询组织。目前,由 ILO 制订的与船舶有关的现行重要公约有 5 项,见表 I-2。

表 I-2　国际劳工组织公约
Table I-2　International Labo(u)r Organization Convention

序号 No.	正式标题 Official title	通过年份 Adopted
1	ILO 第 68 号船员膳食与餐桌公约 ILO68 Convention concerning food and catering for crews on board ship	1946 年
2	ILO 第 92 号船员舱室公约 ILO92 Convention concerning crew accommodation on board ship (revised 1949)	1949 年
3	ILO 第 133 号船员舱室(补充规定)公约 ILO133 Convention concerning crew accommodation on board Ship (Supplementary Provisions)	1970 年
4	ILO 第 147 号商船最低标准公约 ILO147 Convention concerning minimum standards in merchant ships	1976 年
5	ILO 第 152 号职业安全与健康(码头工作)公约 ILO152 Convention concerning occupational safety and health in dock work	1979 年

International Life-Saving Appliance(LSA) Code 国际救生设备(LSA)规则　系指 IMO 组织海上安全委员会 MSC.48(66) 决议通过的《国际救生设备(LSA)规则》。该规则可能经 IMO 组织修正。但该修正案应按 SOLAS 第Ⅲ章有关适用于除第 1 章外的附则修正程序的规定予以通过、生效和实施。　International Life-Saving Appliance(LSA)Code means the International Life-Saving Appliances(LSA)Code adopted by the Maritime Safety Committee of the Organization by resolution MSC.48(66), as it may be amended by the Organization, provided that such amendments are adopted, brought into force and take effect in accordance with the provisions of article Ⅷ of the present Convention concerning the amendment procedures applicable to the annex other than chapter I.

International load line exemption certificate　国际载重线免除证书　系指根据并按国际载重线公约或经 1988 年议定书修订的该公约条文第 6 条的规定,应为已准予免除的任何船舶签发的国际载重线免除证书。　International load line exemption certificate means an International Load Line Exemption Certificate issued to any ship to which an exemption has been granted under and in accordance with article 6 of the Load Line Convention or the Convention as modified by the 1988 LL Protocol, as appropriate.

International Management Code for the Safe Operation of Ships and Pollution Prevention(ISM)　国际航运安全及防止污染管理规则　该规则分为 A 和 B 两部分。A 部分是规则的实施,其中包括对公司的安全及环保方针、公司的责任和权利、船长的职责和权利、管理资源、应变程序、船舶的维护保养等的要求;B 部分是对检验和发证的要求。

International marine contractor association(IMCA)　国际海事承包商协会

International maritime activities　国际海事活动　系指在国际海上组织权限范围内,受国际规则和标准约束的船舶交通及其他船舶有关的运营活动。　International maritime activities mean vessel traffic and other vessel-based operations that are subject to regulation by international rules and standards within the purview of IMO.

International Maritime Dangerous Goods(IMDG) Code　国际海运危险货物输规则(IMDG 规则)　系指按国际海上人命安全公约第Ⅶ章定义的国际海运危险货物输规则。　IMDG Code means the International Maritime Dangerous Goods(IMDG)Code as defined in chapter Ⅶ of the Convention.

Intersecting member 交叉构件 船体板架中骨材数目较少的那个方向上的骨材。

International load line certificate 国际载重线证书 系指按国际载重线公约或1988年议定书修订的该公约的规定,经过检验并勘划了载重线标志以后,应根据1966年国际载重线公约的规定为每艘船舶签发的国际载重线证书。International load line certificate means an International Load Line Certificate issued under the provisions of the International convention on load lines, 1966, to every ship which has been surveyed and marked in accordance with the Convention or the Convention as modified by the 1988 LL Protocol, as appropriate.

International Maritime Organization(IMO) 国际海事组织 其前身是IMCO(International Maritime Coordinate Organization)它隶属于联合国的社会经济委员会,但其人员和经费是独立的。其总部设在伦敦。目前有158个成员国和两个准成员国。其功能在于颁布促进国际船舶航运、保证海事安全、防止海洋污染的条约和法规。在其海上安全委员会和海上环境保护分委员会下设有:散装液体和气体分委员会,船旗国履约委员会,航行安全分委员会,稳性、载重线和渔船安全分委员会,消防分委员会,船舶设计和设备分委员会,无线电通信与搜救分委员会,培训和值班标准分委员会,危险品、固体货物和集装箱运输分委员会。见表I-3。

表 I-3 国际海事组织委员会和分委员会
Table I-3 Committee and Sub-committee of International Maritime Organization

国际海事组织委员会和分委员会	Name of Committee and Sub-committee	缩写
海上安全委员会	Maritime Safety Committee	MSC
海上环境保护委员会	Marine Environment Protection Committee	MEPC
散装液体和气体分委员会	Sub-committee on Bulk Liquids and Gases	
船旗国履约委员会	Sub-committee on Flag State Implementation	
航行安全分委员会	Sub-committee on Safety of Navigation	
稳性、载重线和渔船安全分委员会	Sub-committee on Stability and Load Lines and Fishing Vessel Safety	
船舶设计和设备分委员会	Sub-committee on Ship Design and Equipment	
无线电通信与搜救分委员会	Sub-committee on Radio Communication and Search and Rescue	
消防分委员会	Sub-committee on Fire Protection	
培训和值班标准分委员会	Sub-committee on Standard of Training and Watch-keeping	
危险品、固体货物和集装箱运输分委员会	Sub-committee on Dangerous Goods Solid Cargos and Container	

International Maritime Organization(IMO) ballast water management guidelines 国际海事组织(IMO)压载水管理指南 系指《为减少有害水生物和病原体对船舶压载水控制和管理的指南》[1997年11月IMO通过的A.868(20)决议]。International Maritime Organization(IMO) ballast water management guidelines means the guidelines lines for the Control and Management of Ships Ballast Water Minimize the Transfer of Harmful Aquatic Organisms and Pathogens[IMO resolution 888(20), adopted November 1997].

International maritime organization(IMO) guideline 国际海事组织指南 系指实施(2012年3月2日IMO MEPC219(63)决议通过的)MARPOL附则V和经IMO批准或采用的与垃圾污染有关的其他指南。International maritime organization(IMO) guideline means the guidelines for the implementation of MAEMYMPOL Annex V (IMO) Resolution MEPC219(63) adopted March 2 2012 and other garbage pollution related guidance or adopted by the IMO.

International maritime satellite organization(IMSO) 国际海事卫星组织 是一个监督某些公共卫星安全以及通过海事卫星提供安全通信服务的国际组织。

一些相关服务包括搜索和营救协调通信、海上安全信息广播、航空移动卫星服务、全球海上遇险与安全系统协调及一般通信。该组织成立于1976年9月4日，当时在英国首都伦敦签署了《国际海事卫星组织公约》，主管单位为国际海事组织。这份公约1979年7月16日正式生效。1994年12月，国际海事卫星组织代表大会决定把该组织改名为国际移动卫星组织。1998年4月24日，《国际移动卫星组织公约》在伦敦签署，重建国际海事卫星组织，1999年4月15日继承国际移动卫星组织。此后，国际海事卫星组织空间部分和经营业务被移交英国一家私人公司——国际海事卫星组织公司，但同时保留了规模较小的国家组织机构，专门负责监督海事卫星履行遇险安全通信的责任。截至目前，国际海事卫星组织共拥有98个成员国，包括中国、美国、俄罗斯、英国、日本和澳大利亚等国。中国1979年以创始成员国身份加入该组织，并指定交通部北京船舶通信导航公司作为中国的（股东）签字者，承担有关海事卫星的一切经营和管理实务。

International maritime standards 国际海事标准 系指国际海事组织（IMO）/国际劳工组织（ILO）的公约、议定书、规则和决议。其目的是为了安全和防止污染，但不包括涉及政府间关系、法律和正规的条款和规定。International maritime standards mean IMO/ILO conventions, protocols, codes and resolutions, in so far as their purpose is safety and pollution prevention, excluding articles and regulations dealing with intergovernmental relations, legal and formal aspects.

International market price 国际市场价格 系指一种商品在国际贸易中被广泛承认的具有代表性的成交价格。亦称世界市场价格。国际市场价格是商品国际价值的货币表现。

International Mobile Satellite Organization（INMARSAT） 国际移动卫星组织

International Monetary Fund（IMF） 国际货币基金组织 是根据1944年7月在布雷顿森林会议签订的《国际货币基金协定》，于1945年12月27日在华盛顿成立的。与世界银行同时成立、并列为世界两大金融机构之一，其职责是监察货币汇率和各国贸易情况，提供技术和资金协助，确保全球金融制度运作正常。其总部设在华盛顿。我们常听到的"特别提款权"就是该组织于1969年创设的。国际货币基金组织于2015年4月6日早间发布了一份官方公报，在公报中，IMF主席拉加德确认，希腊方面已经同意在2015年4月9日如期偿还其应偿付的IMF到期借款。2015年11月30日，国际货币基金组织执董会决定将人民币纳入特别提款权（SDR）货币篮子。

International multimodal transport 国际多式联运 是在集装箱运输的基础上产生和发展起来的一种综合性连贯运输方式。它一般是以集装箱为媒介，把海陆空各种传统的单一运输方式有机地结合起来，组成一种国际间的连贯运输。

International multi-model transport document 国际多式联运单据 系指证明国际多式联运合同成立及证明多式联运经营人接管货物，并负责按照多式联运合同条款支付货物的单据。

International multi-modes transport 国际多式联运 系指按照多式联运合同，以至少两种不同的运输方式，由多式联运经营人将货物从某一国境内接管货物的地点运至另一国指定交付货物的地点。为履行单一方式运输合同而进行的该合同所规定的货物接收业务，不应视为国际多式联运。International multi-modes transport means the carriage of goods by sea at least two different modes of transport on the basis of a multi-modes transport contract from a place in one country at which the goods are taken in charge by the multimodal transport operator to a place designated for delivery situated in a different country. The operation of pick-up and delivery of goods carried out in the performance of a unimodal transport contract, as defined in such contract, shall not be considered as international multi-modal transport.

International Narcotics Control Board（INCB） 国际麻醉药品管制委员会

International NAVTEX service 国际奈伏斯泰业务（国际NAVTEX） 系指在518 kHz上，使用窄带直接印字电报手段用英语协调广播和自动接收海上安全信息。International NAVTEX service means the co-ordinated broadcast and automatic reception on 518 kHz of maritime safety information by means of narrow-band direct-printing telegraphy using the English language.

International Network Processors Conference（INPC） 国际网络处理器会议

International News Service（INS） 国际新闻社 由原人民日报社社长、中国杰出新闻记者、新闻学家范长江创办，抗日战争时期和第三次国内战争时期中国共产党领导的通讯社，简称'国新社'。1952年已更名为中国新闻社。

International oil pollution prevention certificate 国际防止油类污染证书 系指对航行于73/78防污公约其他缔约国管辖范围内的港口或近海装卸站的150总吨及以上任何油船以及400总吨及以上的任何其他船舶，经按73/78防污公约附则1第4条的规定检验以后，签发的国际防止油类污染证书。按适用情况，该证书应附有油船以外船舶构造和设备记录或油船构造和设备记录。International oil pollution prevention certificate means an International oil pollution prevention certificate issued, after survey in accordance with regulation 4 of Annex I of MARPOL 73/78, to any oil tanker of 150 gross tonnage and above and any other ship of 400 gross tonnage and above which is engaged in voyages to ports or offshore terminals under the jurisdiction of other Parties to MARPOL 73/78. The certificate is supplemented with a record of construction and equipment for ships other than oil tankers or a record of construction and equipment for oil tankers, as appropriate.

International Opium Conventions 《国际鸦片公约》 系指1912年《海牙公约》和1925年《日内瓦公约》。International Opium Conventions mean the Hague conventions, 1912, and Geneva conventions, 1925.

International pollution prevention certificate for the carriage of noxious liquid substances in bulk(NLS Certificate) 国际防止散装运输有毒液体物质污染证书(NLS证书) 系指对航行于MARPOL 73/78其他缔约国管辖范围内的港口和装卸站的散装运输有毒液体物质的任何船舶,经按MARPOL 73/78附则II第10条规定检验之后,签发国际防止散装运输有毒液体物质污染证书(NLS证书)。关于化学品液货船,分别根据BCH规则和IBC规则的规定签发的散装运输危险化学品适装证书和国际散装运输危险化学品适装证书,应与NLS证书具有同等的效力和得到同样的承认。International pollution prevention certificate for the carriage of noxious liquid substances in bulk(NLS Certificate) means an international pollution prevention certificate for the carriage of noxious liquid substances in bulk(NLS Certificate) shall be issued, after survey in accordance with the provisions of regulation 10 of Annex II of MARPOL 73/78, to any ship carrying noxious liquid substances in bulk and which is engaged in voyages to ports or terminals under the jurisdiction of other Parties to MARPOL 73/78. In respect of chemical tankers, the Certificate of fitness for the carriage of dangerous chemicals in bulk and the International certificate of fitness for the carriage of dangerous chemicals in bulk, issued under the provisions of the Bulk Chemical Code and International Bulk Chemical Code, respectively, shall have the same force and receive the same recognition as the NLS Certificate.

International practices 国际惯例

International production price 国际生产价格

International railway transport 国际铁路联运 系指使用一份统一的国际联运票证,由铁路部门负责经过两国或两国以上铁路的全程运送,并由一国铁路向另一国铁路移交货物时,不需要发货人和收货人参加。

International ready for recycling certificate(IRRC) 国际适合拆船证书

International roaming(IR) 国际移动电话漫游

International Rules for the Interpretation of Trade Terms(INCOTERMS) 国际贸易术语解释通则 是国际商会为统一各种贸易术语的不同解释于1936年制定的。其宗旨是为国际贸易中最普遍使用的贸易术语提供一套解释的国际规则,以避免因各国不同解释而出现的不确定性,或至少在相当程度上减少这种不确定性。INCOTERMS2010中将贸易术语划分为适用于各种运输的CIP、CPT、DAP、DAT、DDP、EXW、FCA和只适用于海运和内水运输的CFR、CIF、FAS、FOB,并将术语的适用范围扩大到国内贸易中,赋予电子单据与书面单据同样的效力,增加对出口国安检的义务分配,要求双方明确交货位置,将承运人定义为缔约承运人,这些都在很大程度上反映了国际货物贸易的实践要求,并进一步与《联合国国际货物销售合同公约》及《鹿特丹规则》衔接。

International Safety Management(ISM) Code "国际安全管理(ISM)规则" 系指IMO组织A.741(18)决议通过并可能经IMO组织修正的"国际船舶安全营运和防污染管理规则",但该修正案应按SOLAS公约的第Ⅷ条有关的适用于除第1章以外的附则修正程序的规定予以通过、生效和实施。International Safety Management(ISM) Code means the International Management Code for the Safe Operation of Ships and for Pollution Prevention adopted by IMO by resolution A.741(18), as may be amended by IMO, provided that such amendments are adopted, brought into force and take effect in accordance with the provisions of article Ⅷ of SOLAS Convention concerning the amendment procedures applicable to the annex other than chapter I.

International service trade 国际服务贸易 系指:(1)从某一参加方境内向任何其他参加方境内提供服务;(2)在某一参加方境内向任何其他参加方的服务消费者提供服务;(3)某一参加方在其他任何参加方境内提供服务的实体的介入而提供服务;和(4)某一参加方的自然人在其他任何参加方境内提供服务。服务贸易多元无形,不可储存的;服务提供与消费同时进行;其贸易额在各国海关统计中没有反映,部分体现在各国国际收支平衡表中。世界贸易组织列出服务行业包括以下12个部门:商业、通信、建筑、销售、教育、环境、金融、卫生、旅游、娱乐、运输、其他。

International sewage pollution prevention certificate 国际防止生活污水污染证书 系指对航行于SOLAS公约其他缔约国所管辖的港口或近海装卸站且要求符

合 73/78 防污染公约附则 IV 规定的船舶，在按照附则 IV 第 4 条的规定进行初次检验或换证检验后，予以签发的"国际防止生活污水污染证书"。 International sewage pollution prevention certificate means an International sewage pollution prevention certificate issued, after an initial or renewal survey in accordance with the provisions of regulation 4 of Annex IV of MARPOL 73/78, to any ship which is required to comply with the provisions of that Annex and is engaged in voyages to ports or offshore terminals under the jurisdiction of other Parties to the Convention.

International Ship and Port Facility Security (ISPS) Code "国际船舶和港口设施保安(ISPS)规则"
系指"1974 年国际海上人命安全公约"缔约国政府会议于 2002 年 12 月 12 日以第 2 号决议通过的"国际船舶保安和港口设施保安规则"，由 A 部分(其规定视为具有强制性)和 B 部分(其规定视为建议性)组成。该规则可能经 IMO 组织修正，但：(1)该规则 A 部分的修正案应按 SOLAS 公约的第Ⅷ条有关适用于除第 1 章以外的附则修正程序的规定予以通过、生效和实施；(2)该规则 B 部分的修正案应由海上安全委员会按其议事规则通过。 International Ship and Port Facility Security (ISPS) Code means the International Code for the Security of Ships and of Port Facilities consisting of Part A (the provisions of which shall be treated as mandatory) and part B (the provisions of which shall be treated as recommendatory), as adopted, on 12 December 2002, by resolution 2 of the Conference of Contracting Governments to the International Convention for the Safety of Life at Sea, 1974 as may be amended by IMO Organization, provided that: (1) amendments to part A of the Code are adopted, brought into force and take effect in accordance with article Ⅷ of the present Convention concerning the amendment procedures applicable to the annex other than chapter I; and (2) amendments to part B of the Code are adopted by the Maritime Safety Committee in accordance with its Rules of Procedure.

International ship security certificate (ISSC) or interim International ship security certificate 国际船舶保安证书 系指主管机关或其认可的组织为每艘船舶签发国际船舶保安证书(ISSC)，以验证船舶符合 SOLAS 公约第Ⅺ-2 章和 ISPS 规则 A 部分的海上保安规定。根据 ISPS 规则 A 部分第 19.1 节，可签发临时 ISSC。 International ship security certificate (ISSC) or interim International ship security certificate means an International ship security certificate (ISSC) issued to every ship by the Administration or an organization recognized by it to verify that the ship complies with the maritime security provisions of SOLAS chapter Ⅺ-2 and part A of the ISPS Code. An interim ISSC may be issued under the ISPS Code part A, section 19.1.

International shore connections 国际通岸接头
系指由钢或其他等效材料制成的并设计成能承受 1.0 N/mm^2 工作压力。其一端应为平面法兰，另一端则有永久附连的配合船上消火栓和消防水带的接口。国际通岸接头应与 1 只承受 1.0 N/mm^2 工作压力的任何材料的垫片及 4 只长度为 50 mm、直径为 16 mm 的螺栓，直径为 16 mm 的螺母和 8 只垫圈一起保存在船上。国际通岸接头标准尺寸如表 I-4 所示：

表 I-4 国际通岸接头
Table I-4 International shore connections

名称 Description	尺寸 Dimension
外径 Outside diameter	178 mm
内径 Inside diameter	64 mm
螺栓节圆直径 Bolt circle diameter	132 mm
法兰槽口 Slots in flange	直径为 19 mm 的孔 4 个，等距离分布，在上述螺栓节圆直径上，开槽口至法兰盘外缘 (4 holes, 19 mm in diameter space equidistantly on a bolt circle of the above diameter, slotted to the flange periphery)
法兰厚度 Flange thickness	至少为 14.5 mm (14.5 mm minimum)
螺栓及螺母 Bolts and nuts	4 副，每只直径 16 mm，长度 50 mm (4, each 16 mm diameter, 50 mm in length)

International shore connections mean these connections of steel or other equivalent material and shall be designed for 1.0 N/mm^2 services. The flange shall have a flat face on one side and, on the other side, it shall be permanently attached to a coupling that will fit the ship's hydrant and hose. The connection shall be kept aboard the ship together with a gasket of any material suitable for 1.0 N/mm² services, together with four bolts of 16 mm diameter and 50 mm in length, four 16 mm nuts, and eight washers.

International Tea Committee(ITC) 国际茶叶委员会

International tonnage 国际吨位

International tonnage certificate(1969) 国际吨位证书 系指按国际吨位公约确定了总吨位和净吨位之后,应为每艘船舶签发的国际吨位证书。 International tonnage certificate (1969) means an International tonnage certificate(1969)issued to every ship, the gross and net tonnage of which have been determined in accordance with SOLAS Convention.

International trade 国际贸易 系指世界各国之间货物和服务交换的活动,是各国之间分工的表现形式,反映了世界各国在经济上的相互依靠。

International Trade Administration Commission (ITAC) 南非国际贸易委员会

International trade by region 国际贸易地理方向 又称国际贸易地区分布,用以表明世界各国或各个国家贸易集团在国际贸易中所占有的地位。计算各国在国际贸易中的比重,即可以计算各国的进、出口额在世界进、出口总额中的比重,也可以计算各国的进出口总额在世界进出口总额中的比重。

International Trade Center(ITC) 国际贸易委员会

International Transmission Center(ITC) 国际传输中心

International transport 国际运输 系指:(1)位于两个国家领土上的起运和目的地之间的运输。而国际集装箱安全公约至少适用其中一国,两国间运输业务的一部分在一个适用国际集装箱安全公约的国家领土范围内进行时,国际集装箱安全公约也应适用;(2)在运输港站经营人接管货物时确定其起运地和目的地位于两个不同国家的任何货物运输。 International transport means:(1)Transport between points of departure and destination situated in the territory of two countries to at least one of which the present Convention applies. The International Convention for safe container Convention shall also apply when part of a transport operation between two countries takes place in the territory of a country to which the International Convention for safe container applies;(2)any carriage in which the place of departure and the place of destination are identified as being located in two different State when the goods are taken in charge by the operator.

International Transport Committee(ITC) 国际运输委员会

International Transport Convention Act 1983 (ITC1983) 1983 年国际运输公约法

International treaties 国际条约 系指国际法主体之间以国际法为准则而为确立其相互权利和义务而缔结的书面协议。国际条约包括一般性的条约和特别条约。国际条约的名称很多,有条约、公约、协定、协定书、宪章、签约和宣言等。除国家元首、政府首脑和外交部部长外,开始时通常须审查代表是否奉有谈判条约的全权。在国际组织范围内缔结国际公约,有时不经过签字这种传统程序,而是根据该组织的组织文件规定,由主管机关将公约拟定后,径送各国审议批准。

International tribunal for the law of the sea 国际海洋法法庭 是根据《联合国海洋法公约》设立的独立司法机关,旨在裁判因解释或实施《公约》所引起的争端。法庭总部设在德国汉堡。法庭管辖权包括根据《公约》及其《执行协定》提交法庭的所有争端,以及在赋予法庭管辖权的任何其他协定中已具体规定的所有事项。《公约》缔约国都可参加法庭,在某些情况下,除缔约国之外的实体(例如国际组织)也可参加。

International voyage 国际航行(国际航线) 系指:(1)悬挂某一国家国旗的船舶从或至另一国管辖的港口、船厂或海上中转站的航行;(2)由适用SOLAS 公约的一国驶往该国以外港口或与此相反的航行;(3)在特种业务指定的区域内由适用特种业务客船规则的一国港口驶往该国以外港口或与此相反的航行,为此,凡由国际海上人命安全公约缔约国政府对其国际关系负责或由联合国管理的每一领土,都视为单独国家;(4)在国际水域内的航行。 International voyage means:(1)a voyage by a ship entitled to fly the flag of one Stats to or from a port, shipyard, or offshore terminal under the jurisdiction of another State;(2)a voyage from a country to which the present Convention applies to a port outside such country, or conversely;(3)a voyage within the area prescribed in special trade from a port in a country to which the special trade passenger ship rules applies to a port outside such country or conversely. For this purpose every territory for the international relations of which a Contracting Government to the SOLAS Convention is responsible or for which the United Nations are administrating authority is regarded as a separate country;(4)voyages in international waters.

International wool secretariat(IWS) 国际羊毛局 成立于1937 年,是一个非牟利性机构。其宗旨是为各成员国的养羊人士建立羊毛制品在全球的长期需求。成员国中最大的羊毛出口国是澳大利亚、新西兰及南半球一些国家,他们出口的原毛占全球年成交量的80%左右。

Internet protocol(IP) 网际协议 是"网络之间互连的协议",简称"网协",也就是为计算机网络相互连接进行通信而设计的协议。在因特网中,它是能使连接到网上的所有计算机网络实现相互通信的一套规则,规定了计算机在因特网上进行通信时应当遵守的规则。任何厂家生产的计算机系统,只要遵守IP协议就可以与因特网互连互通。正是因为有了IP协议,因特网才得以迅速发展成为世界上最大的、开放的计算机通信网络。因此,IP协议也可以叫作"因特网协议"。

Internet+ 互联网+ 代表一种新的经济形态,即充分发挥互联网在生产要素配置中的优化和集成作用,将互联网的创新成果深度融合于经济社会各领域之中,提升实体经济的创新力和生产力,形成更广泛的以互联网为基础设施和实现工具的经济发展新形态。"互联网+"行动计划将重点促进以云计算、物联网、大数据为代表的新一代信息技术与现代制造业、生产性服务业等的融合创新,发展壮大新兴业态,打造新的产业增长点,为大众创业、万众创新提供环境,为产业智能化提供支撑,增强新的经济发展动力,促进国民经济提质增效升级。2015年3月5日十二届全国人大三次会议上,李克强总理在政府工作报告中首次提出"互联网+"行动计划。

Interpolate 插入/篡改
Interpreter 译员
Intersecting axes drive gear 斜交轴传动减速齿轮箱 各传动轴轴线斜交成一定角度的减速齿轮箱。
Intersecting member 交叉构件 船体板架中骨材数目较少的那个方向上的骨材。
Intersection algorithm 求交算法
Interstage cooling 级间冷却
Interstage packing 级间密封
Interstage water seal 级间水封 多级闪发蒸发器级与级之间只能通过水流,而不让蒸汽串通的浸水连通管或孔口。
Interval of ladder(interval of tread) 梯步间距 两个相邻梯步间的垂直距离。
Intervention riser system(IRS) 修井隔水管系统
Intrinsically-safe circuit"i" "i"本质安全电路 在所述的试验条件(包括在正常运行和规定的故障状态)下所产生的火花和任何热效应均不致引起点燃所指定的爆炸性大气环境的电路。注:IEC 60079-11规定了使用此防护方法之设备的结构特征和试验要求。 Intrinsically-safe circuit"i" means a circuit in which no spark or any thermal effect produced in the test conditions prescribed(which include normal operation and specified fault condition)is capable of causing ignition of a given explosive gas atmosphere. Note: IEC 60079-11 specifies the constructional features and test requirement for apparatus using this method of protection.

Invasive aquatic species 入侵水生物种 系指可能会对人类、动物和植物的生命,经济与文化活动和水生环境造成威胁的物种。 Invasive aquatic species means a species which may pose threats to human, animal and plant life, economic and cultural activities and the aquatic environment.

In-vehicle network(IVN) 车载网络 是汽车环境中的一种通信基础设施,用于在不同电子元件之间交换信息,以充分发挥汽车各种功能所需。

Inventory of Hazardous Materials(IHM) 有害物质清单 系指IMO海上环境保护委员会(MEPC)于2009年7月17日通过MEPC.179(59)决议《有害物质清单编制指南》(Guidelines for the Development of the Inventory of Hazardous Materials)。该指南提供了编制有害物质清单的建议,为相关利益方(例如造船厂、设备供应商、修船厂、船东和船舶管理公司)提供了切实可行和合理地编制清单的基本要求。本清单旨在提供船上存在的实际有害物质的船舶特定信息,以在拆船厂保护健康和安全,且防止环境污染。拆船厂将使用此信息,以决定如何管理有害物质清单中所列的物质。清单应包括三个部分:第Ⅰ部分:船舶结构或设备中包含的物质(Materials contained in ship structure or equipment);第Ⅱ部分:运行产生的废料(Operationally generated wastes);第Ⅲ部分:物料(Stores)。

Inverse time-delay relay or release 反时限过电流继电器或脱扣器 系指经一定延时动作的过电流继电器和遥控器,其延时动作时间与通过的过电流成反比。注意:上述继电器或脱扣器应设计成在较高电流时接近一个确定的最小值。

Invertebrates 无脊椎动物
Investigate for criminal responsibility 追究刑事责任
Investigation 审查 系指为确定其可追溯性和真实性,并确认过程持续地满足船级和法定要求所进行的文件检查活动。
Investigation stage 侦查阶段
Invitation of tender 招标、投标买卖 系指以公开竞价的方式买受人买受符合标的物质量要求的,最低应价者的标的的买卖方式。

Invitation of tender/Invitation of tender 招标 是一种商业行为。系指招标人(买方)在规定时间、地点、

发出招标公告或招标单,提出准备买进商品的品种、数量和一个买卖条件、邀请投标人(卖方)投标的行为。这也是一种有组织的交易方式。

Invitation to offer 邀请发盘 是国际贸易中的一个术语,系指交易的一方打算购买或出售某种商品,向对方询问买卖该项商品的有关交易条件,或者就该项交易提出带有保留条件的建议。

Invoice 发票 是一切单位和私人在购销商品、提供劳务事项以及处置其他筹划活动,所提供给对方的收付款的书面证明,它也是财务收支的依据。

Involute cam 渐开线凸轮

Involve use of materials that may be hazardous to health 涉及使用可能对健康造成危害的材料 包括:(1)二氧化硫——潮湿时有腐蚀性,吸入有毒,造成灼伤,对眼睛和呼吸系统有刺激;(2)硫化氢——高度易燃(闪点 -82 ℃),遇空气能形成爆炸性混合物,潮湿时有腐蚀性,造成灼伤,必须远离着火源存放,有刺激性和窒息性,长期曝露极限值(LTEL)5 ppm,短期曝露极限值(STEL)10 ppm,浓度再高会致命,无气味。反复曝露于低浓度硫化氢能导致对气体的嗅觉消失;(3)苯——高度易燃(闪点 -11 ℃),遇空气能形成爆炸性混合物,有毒,致癌,严重危及健康;(4)甲苯——高度易燃(闪点 4 ℃),遇空气能形成爆炸性混合物,有刺激性,严重危及健康,对生殖器官有毒性。 Involve use of materials that may be hazardous to health are as follows:(1) Sulphur Dioxide—corrosive when wet, toxic if inhaled, causes burns, and is an irritant to the eyes and respiratory system;(2) Hydrogen Sulphide—highly flammable (Flash point of -82 ℃), can form an explosive mixture with air, corrosive when wet, causes burns, has to be kept away from sources of ignition, irritant and asphyxiant, LTEL 5 ppm, STEL 10 ppm, higher concentrations can be fatal and have no odour. Repeated exposure to low concentrations can result in the sense of smell for the gas being diminished. (3) Benzene—highly flammable (Flash point of -11 ℃), can form an explosive mixture with air, toxic, carcinogenic, acute health risk. (4) Toluene—highly flammable (Flash point of 4 ℃), can form an explosive mixture with air, irritant, acute health risk, reprotoxin.

Involved human factors 相关人为因素 系指通过人的行为控制风险,但人的行为的失败本身并不能造成事故或导致事故持续发展。

Inward turning 内旋 推船向前时,边螺旋桨上部向船中线面的旋向。

Inward turning propeller 内旋螺旋桨

In-water cleaning 水下清洗 系指当船舶在水中时,使用物理方法去除其生物污底(垢)。In-water cleaning means the physical removal of bio-fouling from a ship while in the water.

In-water survey(IWS) 水下检验 系指船舶在漂浮状态下对其进行的检验。 In-water survey means a survey of the ship under floating condition.

Ionosphere observation room V 电离层观察室 为研究极地电离层吸收电磁波的规律,使用相对电离层吸收仪,测定和记录极地宇宙射电噪声水平的工作室。

Ionospheric refraction correction 电离层折射校正 电离层折射是卫星电波在穿过电离层时传播路径会产生弯曲的现象。多普勒频移积将会产生误差,影响到定位精度。采用双频道接收机时,把接收到的两个不同频率的信号进行比较,就能计算出电离层所折射误差,从而进行校正,以消除这一误差。经过电离层折射校正后的剩余误差小于 0.005 海里(10 m)不考虑电离层折射时的定位误差,夜间为 0.1 海里,白天为 0.3 海里。

IOPP certificate 国际防止油污染证书 系指国际海事组织国际防止油污染证书,证明该船已按照"1973/1978 国际防止船舶造成污染公约"的规则进行检验。IOPP certificate means the IMO International oil pollution prevention certificate, certifying that the ship has been surveyed in according to MARPOL 73/78.

I_p (cm^4) 普通扶强材或主要支撑构件与板材连接处的净极惯性矩(cm^4)。 I_p means the net polar moment of inertia, in cm^4, of ordinary stiffener or primary supporting member, about its connection to plating.

IP × × 系指电气设备的外壳防护等级的标志。由 IP 字母后面加两位数字组成:其中,IP 为特征字母;第 1 位数字见表 I-5;第 2 位数字见表 I-6。 IP × × means a designation of protective enclosures for electrical equipment. The designation to indicate the degree of protection consists of the characteristic letters IP followed by two numerals. in which IP is characteristic letters、1st characteristic numeral see Table I-5 and 2nd characteristic numeral see Table I-6.

表 I-5　第 1 位特征数字所代表的防护等级
Table I-5　Degree of protection indicated by the first characteristic numeral

第 1 位特征数字 1st characteristic numeral	简要说明 Brief description	防护等级 Degree of protection 定义 Definition
0	无防护 Non-protected	无专门的防护 No special protection
1	防护大于 50 mm 的固体 Protected against solid objects greater than 50 mm	人体大面积部分,如手(但对有意识的接触并无防护),直径超过 50 mm 的固体 Large surface of human body, e. g. a hand (but no protection against deliberate access), solid objects greater than 50 mm
2	防护大于 12 mm 的固体 Protected against solid objects greater than 12 mm	手指或类似物,长度不超过 80 mm,直径超过 12 mm 的固体 Fingers or similar objects not exceeding 80 mm, solid objects greater than 12 mm
3	防护大于 2.5 mm 的固体 Protected against solid objects greater than 2.5 mm	直径或厚度大于 2.5mm 的工具、电线等,直径超过 2.5 mm 固体 Tools, wires, etc. of diameter or thickness greater than 2.5 mm, solid objects greater than 2.5 mm
4	防护大于 1.0 mm 的固体 Protected against solid objects greater than 1.0 mm	厚度大于 1 mm 的线或片状物,直径不超过 1 mm 的固体 Wires or strips of thickness greater than 1 mm, solid objects not exceeding 1.0 mm
5	防尘 Dust-protected	并不能完全防止尘土进入,但进入量不能达到妨碍设备正常运转的程度 Ingress of dust is not totally prevented, but dust allowed to enter is not to interfere with normal operation of equipment
6	尘密 Dust-tight	无尘进入 No ingress of dust

表 I-6　第 2 位特征数字所代表的防护等级
Table I-6　Degree of protection indicated by the second characteristic numeral

第 2 位特征数字 2nd characteristic numeral	简要说明 Brief description	防护等级 Degree of protection 定义 Definition
0	无防护 Non-protected	无专门的防护 No special protection
1	防滴 Protected against dripping water	垂直滴水应无有害影响 Vertically dripping is to have no harmful effect
2	防 15°倾斜的水滴 Protected against dripping water when tilted water up to 15°	设备与垂直线成 15°时,滴水应无有害影响 Dripping water is to have no harmful effect when the enclosure is tilted to any angle up to 15° from its normal position
3	防淋水 Protected against spraying water	与垂直线成 60°时的淋水应无有害影响 Water falling as spray at an angle up to 60° from the vertical is to have no harmful effect
4	防溅 Protected against splashing water	任何方向溅水应无有害影响 Water splashed against the enclosure from any direction is to have no harmful effect
5	防冲水 Protected against water jets	任何方向冲水应无有害影响 Water projected by a nozzle against the enclosure from any direction is to have no harmful effect
6	防猛烈海浪 Protected against heavy seas	猛烈海浪或强烈冲水时进入机壳水量应无有害影响 Water from heavy seas or water projected in powerful jets is not to enter the enclosure in harmful quantities
7	防浸水 Protected against effects of immersion	浸沉在规定压力的水中经规定的时间后,进入水量应无有害影响 Ingress of water in a harmful quantity is not be possible when the enclosure is immersed in water under defined conditions of pressure and time
8	防潜水 Protected against submersion	能长期潜水,其技术条件由制造厂规定 注:通常设备应完全密封,但对某些类型设备,在不产生有害影响的前提下,可允许水进入设备中 The equipment is suitable for continuous submersion in water under conditions to be specified by manufacturer Note: normally, the equipment is to be hermetically sealed. For certain types of equipment, however, water is allowed to enter in such a manne that it produces no harmful effects

iPad　苹果平板电脑　是由苹果公司于 2010 年开始发布的平板电脑系列,定位介于苹果的智能手机 iPhone 和笔记本电脑产品之间,通体只有一个按键,(屏幕中有 4 个虚拟程序固定栏)与 iPhone 布局一样,提供浏览互联网、收发电子邮件、观看电子书、播放音频或视频、玩游戏等功能。由英国出生的设计主管乔纳森·伊夫(Jonathan Ive)(有些翻译为乔纳森·艾维)领导的团队设计的,这个圆滑、超薄的产品反映出了伊夫对德国天才设计师 Dieter Ram 的崇敬之情。iPad 在欧美称 Tablet PC。具备浏览网页、收发邮件、普通视频文件播放、音频文件播放、玩一些游戏等基本的多媒体功能。由于采用 ARM 架构,不能兼容普通 PC 台式机和笔记本的程序,可以通过安装由 Apple 官方提供的 iWork 套件进行办公,可以通过 iOS 第三方软件预览和编辑 Office 和 PDF 文件。

iPhone　苹果手机　是苹果公司旗下研发的智能手机系列,是苹果公司和 Cingular 电信公司推出的一款手机,它搭载苹果公司研发的 iOS 手机操作系统。第 1 代 iPhone 于 2007 年 1 月 9 日由当时苹果公司 CEO 史蒂夫·乔布斯发布,并在同年 6 月 29 日正式发售。2013 年 9 月 10 日,苹果公司在美国加州举行新产品发布会上,推出第 7 代产品 iPhone5C 及 iPhone5S。第 8 代的 iPhone 6 和 iPhone6 Plus 于 2014 年 9 月 10 日发布,中国大陆地区销售时间定为 10 月 17 日。

Iron ore　铁矿石　呈深灰至锈红色等各种颜色,铁含量从赤铁矿(高品位矿粒)到商用品级较低的菱铁矿各不同。含水量 0% ~ 16%。无特殊危害。该货物为不燃物或失火风险低。铁矿石货物可能影响磁罗经。　Iron ore is a ore varies in colour from dark grey to rusty red, varies in iron content from haematite, (high grade ore) to ironstone of the lower commercial ranges. Moisture content: 0% to 16%. Mineral Concentrates are different cargoes. It has no special hazards. This cargo is non-combustible or has a low fire-risk. Iron ore cargoes may affect magnetic compasses.

Iron ore pellets　铁矿石丸　大致呈球形团状,将铁矿石压碎成粉末而成。用黏土作为黏合剂将这种氧化铁制成球团,然后在窑中 1 315 ℃温度下用火烧硬。含水量 0% ~ 2%。无特殊危害。该货物为不燃物或失火风险低。　Iron ore pellets are approximately spherical lumps formed by crushing iron ore into a powder. This iron oxide is formed into pellets by using clay as a binder and then hardening by firing in kilns at 1,315 ℃. Moisture content: 0% to 2%. It has no special hazards. This cargo is non-combustible or has a low fire-risk.

Iron oxide, spent or iron sponge, spent(UN 1376)

废氧化铁或废海绵铁　粉末状物质。呈黑色、褐色、红色或黄色。气味强烈,可能污染其他货物。易自热和自燃,特别是沾染油或受潮时。有毒气体,可能产生硫化氢、二氧化碳和氰化钾。粉尘可能造成爆炸危害。易减少货物处所的氧气含量。　Iron oxide, spent or iron sponge, spent (UN 1376) is a powdery material, black, brown, red or yellow. Strong odour may taint other cargo. Liable to heat and ignite spontaneously, especially if contaminated with oil or moisture. Toxic gases: hydrogen sulphide, sulphur dioxide, and hydrogen cyanide may be produced. Dust may cause an explosion hazard. It is liable to reduce the oxygen in the cargo space.

Ironstone　磷铁矿　矿石。含水量 1% ~ 2%。无特殊危害。该货物为不燃物或失火风险低。　Ironstone is a ore. Moisture: 1% to 2%. It has no special hazards. This cargo is non-combustible or has a low fire-risk.

Irradiated nuclear fuel　辐射性核燃料　系指以使用过而仍含有保持自行持续核连锁反应的铀、钍和/或钚的同位素的材料。　Irradiated nuclear fuel means material containing uranium, thorium and/or plutonium isotopes, which has been used to maintain a self-sustaining nuclear chain reaction.

Irrational emission control strategy　不合理排放控制策略　对"73/78 防污公约"附则Ⅵ而言,系指当船舶在正常使用条件下营运时将排放控制系统的有效性降至低于适用的排放试验程序所预期的水平的任何策略或措施。　For the purposes of the "MARPOL73/78" Annex Ⅵ, irrational emission control strategy means any strategy or measure that, when the ship is operated under normal conditions of use, reduces the effectiveness of an emission control system to a level below that expected on the applicable emission test procedures.

Irregular operating conditions　不规则工作状态　系指外部条件给操作人员增添的过度工作负担。　Irregular operating conditions mean that when external conditions cause excessive operator workloads.

Irregular sea　不规则波　实际存在于海面上,波面随时间而变化,不能用简单函数式表达的波浪。

Irreversible engine　不可逆转式发动机

Irrevocable letter of credit　不可撤销信用证　系指证行一经开出、在有效期内未经受益人或议付行等有关当事人同意,不得随意修改或撤销的信用证;只要受益人按该证规定提供有关单据,开证行(或其指定的银行)保证付清货款。凡使用这种信用证,必须在该证上注明"不可撤销"(Irrevocable)的字样,并载有开证行保证付款的文句。按《跟单信用证统一惯例》(第 600 号出版物)第 3 条 C 款的规定:"信用证是不可撤销的,即使信用证中对此未做指示也是如此。"

Irrigating carrying vessel 排灌运输船 装有水泵机组,既可用于农田灌溉,又可将喷水口转向后方,借喷水推进方式进行水上运输的船舶。

Irrigating vessel 排灌船 装有水泵机组,专用于农田排灌的船舶。

Irrotational flow 无旋流 在流场中流体质点的平均旋转速度为零时,沿任何封闭曲线的速度线积分也为零的无旋性的流动。

I_S(cm^4) 扶强材或主要支撑构件连同宽度 s 船壳板对其与板材平行的中和轴的净惯性矩(cm^4) I_S means the net moment of inertia, in cm^4, of the stiffener or the primary supporting member, with attached shell plating of width s, about its neutral axis parallel to the plating.

IS Code IS 规则 系指经修正的"IMO 文件包括的所有类型船舶的完整稳性规则"。 IS Code means the Code on intact stability for all type of ships covered by IMO instruments, as amended.

ISA = Instrument Society of America 美国仪表协会

Island 岛屿 是四面环水并在高潮时高于水面的自然形成的陆地区域。 Island is a naturally formed area of land, surrounded by water, which is above water at high tide.

Isle Royale National Park 罗亚尔岛国家公园 系指就在或邻近国家公园管理局管理的苏必尔湖的地方,其边界包括罗亚尔岛及其周围的群岛,包括帕西奇岛(Passage)和海鸥岛(Gull)的海岸线 4.5 英里范围内的美国政府管辖的领土范围内的任何水下湿地。Isle Royale National Park means the site on or near Lake Superior, administered by the National Park Service, where the boundary includes, any submerged lands within the territorial jurisdiction of the United States within 4.5 mile of the of the shoreline of Isle Royale and the surrounding island, including Passage Island and Gull Island.

ISO 15016 航速修正法 系指国际标准组织制定的根据螺旋桨负荷变化修正功率、速度和转速的方法。

ISO freight container ISO 货运集装箱 其设计尺寸和额定质量满足 ISO 集装箱标准的集装箱。如:(1) ISO 1496-1 系列 1 货运集装箱(ISO 1496-1series 1 freight container);(2) ISO 668 外部尺寸和额定质量(ISO 668 external dimensions aqnd ratings);(3) ISO 1161 集装箱角件和规格(ISO 1161 corner fittings and specifications)。

ISO = International Standards Organization 国际标准化组织

Isolated openings 独立开口 系指在船的横向/垂向开口间隔不小于 1m 的开口。 Isolated openings are openings spaced not less than 1m apart in the ship's transverse/vertical direction.

Isolating valve 隔离阀(隔流阀,截止阀) 货油装卸管路在通过每一隔舱壁处设置的,需要时可把邻近的油舱隔开的阀。

Isolation booth 隔声间 系指在噪声强烈的车间内建造的,使用隔声结构组成的具有良好隔声性能的小房间,如机舱集控室,以供工作人员在其中操作或观察、控制车间内各部分工作之用。良好的隔声间,能使在其中的工作人员免受听力损害,改善精神状态,得到舒适的工作条件,从而提高劳动生产率。

Iso-potential connection 等电位连接 系指使导电部件之间电位基本相等的电气连接。

Isotherm follower 等温线跟踪仪 一种能自动跟踪测量等温层深度分布的装置。由热敏电阻传感器和对应于所需温度的电阻组成的平衡电桥,自动地控制绞车收放传感器,从而跟踪等温层,深度用压力传感器测量,最后记录出等温层深度的跟踪线。

Isothermal annealing 等温退火 在大于 A_3 点(亚共析钢)或 A_1 点(过共析钢)的温度下加热,并在小于 A_1 点的温度下急速冷却至较快地发生珠光体发生的温度,在保持该温度的情况下,将奥氏体发生为铁素体和碳化物,并在较短的时间内将其软化的退火,也称其为循环退火(Cycle annealing)。 Isothermal annealing means a process in which a ferrous alloy is produce a structure partly or wholly austenitic, and is then cooling to and held at a temperature that cause transformation of the austenite to a relatively soft ferrite-carbide aggregate.

ISPS Code ISPS 规则 系指 1974 年"国际海上人命安全公约"缔约国政府会议于 2002 年 12 月 12 日以第 2 号决议通过的"国际船舶和港口设施保安规则"。

Issuing Authority 证书签发机关 系指主管机关,代表主管机关的经认可的保安组织或签发证书的主管机关请求的缔约国政府。 Issuing Authority means the Administration, the recognized security organization who acting on behalf of the Administration, or the Contracting Government who at the request of the Administration, has issued the certificate.

ITU-R 国际电信联盟-无线电报 ITU-R means the International Telecommunication Union-radio sector.

ITU-T 国际电信联盟-通信组 ITU-T means the International Telecommunication Union-telecommunication sector.

Iurning upper platform 可转上平台 能旋转的舷梯上平台。

$I_w(cm^4)$ 普通扶强材或主要支撑构件与板材连接处的净剖面惯性矩(cm^4) I_w means the net sectional moment of inertia, in cm^4, of ordinary stiffener or primary supporting member about its connection to plating.

$I_y(m^4)$ 船体横剖面对其水平中和轴的惯性矩(m^4) I_y means the moment of inertia, in m^4, of the hull transverse section about its horizontal neutral axis.

$I_z(m^4)$ 船体横剖面对其垂向中和轴的惯性矩(m^4) I_z means the moment of inertia, in m^4, of the hull transverse section about its vertical neutral axis.

Inspection certificate of value 价值检验证书 是证明出口商品价值的证书,即由商检机构对卖方出具的发票开列的商品名称、数量、单价及总值进行核实后出具的(或在发票上加盖印章)证书。

J

Jack and pinion rack 齿轮齿条起重器

Jack screw 螺旋起重器(螺旋千斤顶,螺旋夹)

Jacket 导管架[井架,(蒸发气)套,(水)套] 系指一种井圈形框架结构。在其顶部支撑一层甲板并通过桩将其固定到海底上。 Jacket is a frustum-shaped frame structure that carries a deck at the top and is fixed to the seabed by piles.

Jacket cooling water pump 缸套冷却泵 抽送柴油机缸套冷却介质的泵。

Jacket drain valve 蒸发气套的泄水阀

Jacket space 护套空间(隔层空间)

Jacket water pump 缸套水泵

Jacketed protection system 套管保护系统

Jacketing 外套

Jacketing gear 盘车机

Jacking and fixation system 升降和锁紧系统

Jacking condition 升降工况 系指:(1)自升式平台升降桩腿,预压及升、降平台主体时的状态;或(2)自升式平台在下放桩腿和升起主体结构或下降主体结构和拔起桩腿时的状态。 Jacking condition refers to a condition wherein a unit is elevating or lowering the legs, preloading the legs and elevating or lowering the hull; or (2) a condition during which a self-elevating unit lowers its legs and raises its hull structure or lowers its hull structure and raises its legs.

Jacking conditions 平台升降海况 允许自升式钻井平台升降的海况。

Jacking engine 盘车机

Jacking pitch 升降节距 液压自升式钻井平台桩腿上相邻两齿块或开孔之间的距离。

Jacking speed 平台升降速度 自升式钻井平台的平台升降的速度。

Jack-over 盘车(转车)

Jack-up drilling platform 自升式钻井平台 系指具有活动桩腿且其主体能沿支撑于海底的桩腿升至海面以上预定高度进行作业的平台,它属于移动式平台。该平台结构分为钻井平台主体、桩腿和升降装置三部分。这种平台的角处安装桩腿。每根桩腿可各自相对平台上下升降。移动时将所有的桩腿升起,由拖船拖到井位后,将桩腿降下,插入海底固定。将平台升到一定高度,进行钻探作业。其优点是钻井作业平稳、效率高、造价相对较低;缺点是桩腿长度有限,致使工作水深受到限制。自升式钻井平台适合在大陆架浅水区作业。自升式钻井平台是近海石油勘探开发的主要装备。见图J-1。

图 J-1 自升式钻井平台
Figure J-1 Jack-up drilling platform

Jack-up load(normal for one leg) 单桩腿负荷 每根桩腿所允许承受的负荷。

Jack-up self-elevating drilling platform 桩腿自升式钻井平台 桩腿直接插入海底的自升式钻井平台。

Jam nut 锁紧螺母(防松螺母)

Jamming 紧压(机器等)

Jargon 行话/专业术语

Jaw clutch 爪形离合器(爪盘联轴节)

Jet 喷柱/喷水(射流,喷射器) 由喷管喷出的柱形流体。

Jet aircraft 喷气式飞机 系指使用喷气发动机依靠燃料燃烧时产生的气体向后高速喷射的反作用使飞机向前飞行的飞机。德国研制成功的世界上第一架喷气式飞机飞上天空。该技术拥有国:美国(波音)、俄罗斯(苏霍伊)、欧盟(空客)、法国(达索)、英国(BAE 系统公司)、中国(成飞)、加拿大(庞巴迪公司)、巴西(巴西航空工业公司)、日本。见图 J-2。

图 J-2 喷气式飞机
Figure J-2 Jet aircraft

Jet carburettor 喷射式汽化器

Jet condensation 喷射式冷却

Jet current 喷流(射流)

Jet efflux 射流

Jet fire 喷火 系指汹涌扩散的火焰,它使压力作用下连续释放的液体或气体在特定方向进行燃烧。Jet fire is a turbulent diffusion flame resulting from the combustion of a liquid or gas continuously released under pressure in a particular direction.

Jet fuels 喷气燃料类

Jet lift dredger 喷射挖泥船 装有喷射泥泵,当用高压水通过文氏管装置时,周围的水及泥土被吸起形成泥浆而后喷出的吸扬挖泥船。

Jet nozzle 喷管 喷射推进器的流体出口部分。

Jet propeller 喷射推进器 任何能对流体供给动能,利用其反作用使船前进,所有活动部分不露出船外的推进装置。

Jet propulsion 喷射推进 推进机构的活动部分不露出船外,利用喷出流体的向前反作用力使船舶前进的一种船舶推进方式。

JIS 日本工业型材标准 JIS is the Japanese industrial standard profile.

J-Lay J 式敷管法 系指管子以垂直(或接近垂直)方向离开敷管船并其悬浮段形成 J 形的一种管路安放方法。 J-Lay is a pipeline installation method where the pipe leaves the vessel in a vertical(or nearly vertical)orientation, and the suspended section acquires a "J" shape.

Jogging track 慢跑道(豪华邮轮)

Joggled planting 折曲式 一块板的板边折曲后与另一块板搭接的排列连接方式。

Joggling 啮合

Joining 连接(接合)

Joint 接头(接合,接缝,关节) 系指使用黏结剂黏合、层压、焊接等连接管路。 Joint means joining pipes by adhesive bonding, laminating, welding, etc.

Joint coupling 活节连接器

Joint probability 联合作用的可能性 系指当两个或以上的变量交叉作用时将导致结构做出响应,可能需要确定各种变量组合出现的可能性,即它们同时出现的可能性。 Joint probability means that when two or more variables interact to producing a response on a structure, it may be necessary to determine the probabilities with which various combinations of variables occur, that is, their joint probability of occurrence.

Jointer 接合件(接缝器,非标短截管子)

Joints 接合(塑料管) 包括所有管子组装装置和方法,例如黏结剂黏合、层压、焊接等。 Joints include all pipe assembling devices or methods, such as adhesive bonding, laminating, welding, etc.

Jolly boat 工作艇 (1)配备在大船上,主要供维修船体、人员交通或执行一定任务用的小艇;(2)某些执行一定任务的非船载小船。

Journal 轴颈(枢轴,航海日记)

Journal bearing 轴颈轴承

Journal friction 枢摩擦(轴颈摩擦)

Journal packing 轴颈填料(轴颈油封)

Joystick 单手柄控制器 用一个单手柄来控制纵荡、横荡和艏摇的定位设施。

JP-1(kerosene) JP-1(煤油)喷气燃料

JP-3 JP-3 喷气燃料

JP-4 JP-4 喷气燃料

JP-5(kerosene, heavy) JP-5(煤油,重质)喷气燃料

J-tube J 形管路 系指安装在平台上的一段 J 形管路,其从海底延伸到平台甲板上。拉伸管子穿过甲板以形成一根立管。 J-tube is a J-shaped tube installed on a platform extending from the sea floor to the platform deck through which a pipe is pulled to form a riser.

Jumper 跨接管 系指从油井顶部连接到拖运器或总管的一段预制短管。 Jumper is a short section of prefabricated tube that connects the wellhead to a sled or a manifold.

Junction pipe 连接管

K

Keel 龙骨 系指船舶主要构件或主骨架,沿船底中线纵向延伸。通常为在其壳内中纵线处由垂直板加强的平板。见图 K-1。 Keel means a main structural member or backbone of a ship running longitudinal along centerline of bottom, usually a flat plate stiffened by a vertical plate on its centerline inside the shell. See Fig. K-1.

图 K-1 龙骨
Figure K-1 Keel

Keel draft 龙骨吃水 龙骨线呈直线时,系指自龙骨底线或其延伸线沿垂直于基平面的方向量至任一水线的距离。龙骨线呈曲线或折角线时,则系自通过龙骨底下缘的最低点且平行于设计水线的平面沿垂直于基平面的方向量至任一水线的距离。无特别注明时,通常尤指在中横剖面处,按以上所述量至设计水线的距离。

Keel line 龙骨线 系指在船中穿过以下部位与龙骨斜面平行的线:(1)金属船壳船舶中心线或船壳外板内侧与龙骨交线(如有方龙骨延伸至该线之下)处的龙骨顶端;或(2)对木质或混合结构船舶,该距离自龙骨镶口下缘量起。当船中剖面下部为凹形时,或如设有厚的龙骨翼板,则该距离自船底平面向内延伸线与船中心线的交点量起。 Keel line is a line parallel to the slope of the keel passing amidships through:(1)the top of the keel at centerline or line of intersection of the inside of shell plating with the keel if a bar keel extends below that line, on a ship with a metal shell; or (2) in wood and composite ships, the distance is measured from the lower edge of the keel rabbet. When the form at the lower part of the mid-ship section is of a hollow character, or where thick garboards are fitted, the distance is measured from the point where the line of the flat of the bottom continued inward intersects the centerline amidships.

Keelson 内龙骨 单层船底结构中的纵桁。
Keeper 锁紧螺母(定位用附件,护板)
Keeping quality 保持性能
Kelvin degree(K) 开氏温标(绝对温度)
Kelvin scale 开氏温标(绝对温度度数)
Kerosene 煤油
Kerosene oil 煤油
Kerosene test 煤油渗透试验 用于检测焊缝的质量的试验。其操作方法为在焊缝的一侧涂煤油,另一侧涂白石灰粉,隔一定的时间后看涂白石灰粉的一侧有没有煤油渗透的痕迹。
Kettle 蒸汽锅(水壶)
Key 开关(按钮,扳手)
Key extractor 取键工具
Key fit 键配合
Key groove 键槽
Key inspection 重点检查
Key joint 键接
Key locker 钥匙箱
Key on 用键固定(接通)
Key operation 关键操作 系指那些因任何人为的过失或疏忽均可能导致人员非法登船、非法进入限制区域、威胁船舶及人员安全、非法行动等严重后果的所有操作活动。 Key operation means all the operating activities likely to cause serious consequence of illegal boarding, illegal entry into restricted areas, a threat to the safety of the ship and persons, illegal action due to any human fault or ignorance.
Key plan 关键图(概略原理图,解说图)
Key position 关键部位
Key principles 关键原则 系指确定风险评估的性质和性能的原则,包括:(1)有效原则;(2)透明原则;(3)一致原则;(4)完整原则;(5)风险管理原则;(6)预防原则;(7)基于科学原则;和(8)持续改进原则。 Key principles are those principles defined the nature and performance of risk assessment, including: (1) effectiveness principle;(2)transparency principle;(3)consistency principle;(4) comprehensiveness principle;(5) risk management principle;(6) precautionary principle;(7) science based principle; and (8) continuous improvement principle.
Key way 键槽

Keyboard 键盘(开关板,用键盘输入)

Keyed shaft taper 带键轴锥体

Keyhole 键孔

Keyless shaft taper 无键轴锥体

Keystrokes 按键

Kick 反横矩 回转试验中,过渡阶段内船舶重心离开初始直线航线向回转中心的反侧横移的最大距离。

Killed steel 镇静钢/脱氧钢 根据冶炼时脱氧程度的不同,钢可分为沸腾钢、半镇静钢和镇静钢。镇静钢为完全脱氧的钢,使得氧的质量分数不超过0.01%(一般常在0.002%~0.003%)。通常注成上大下小带保温帽的锭型,浇注时钢液镇静不沸腾。由于锭模上部有保温帽(在钢液凝固时作补充钢液用),这节帽头在轧制开坯后需切除,故钢的收缩率低,但组织致密,偏析小,质量均匀。优质钢和合金钢一般都是镇静钢。

Killing action 脱氧作用(镇静作用)

Killing agent 脱氧剂(镇静剂)

Kiln 干燥室(干燥装置,炉)

Kind of explosion-protected construction used for electrical equipment on board ships 船上电气设备所用防爆结构的类型 如下:(1)隔爆型(Ex-d);(2)本质安全型:①a类本质安全防爆构造(Ex-ia);②b类本质全防爆构造(Ex-ib);(3)增安型(Ex-e);(4)正压型(Ex-p);(5)充砂型(Ex-q);(6)油浸型(Ex-o);(7)密封型(Ex-m);和(8)防护型(Ex-n)。 Kind of explosion-protected construction used for electrical equipment on board ships is to be selected from the followings. (1) Flameproof type (Ex-d); (2) Intrinsically safe type: ①Category 'ia' intrinsically safe type (Ex-ia); ② Category 'ib' intrinsically safe type (Ex-ib); (3) Increased safety type (Ex-e); (4) Pressurized protected type (Ex-p); (5) Powder filling type (Ex-q); (6) Oil immersion type (Ex-o); (7) Encapsulation type (Ex-m); and (8) Non-sparking type (Ex-n).

Kinematical similarity 运动相似 实物与其模型的对应点的速度方向相同,大小成同一比例的相似性。

Kinetic viscosity 动力黏度

King bolt 主螺栓(中心主轴)

Kingstone valve 海水阀(通海阀)

Kirsten-Boeing propeller 正摆线推进器 叶的轨迹是一正摆线,推进器盘每转一周各叶也自转半周的平旋推进器。

K-J type design chart K-J式设计图谱 以进速系数 J 为横坐标,推力系数和转矩 K_Q 为纵坐标,并绘有敞水效率等值线的螺旋桨设计图谱。

Knee 肘板(管弯头,顶推架)

Knock 敲缸 由于燃烧异常,在燃烧室内出现如金属敲击之声的现象。

Knot(kn) 节 船舶航速的单位。1 kn相当于每小时1海里的航速(1.852 km/h)。

Knuckle 折边 系指构件的不连续处。 Knuckle means a discontinuity in a structural member.

Knuckle line 折角线 船体表面或船体结构件曲度突变呈折角形成的棱角线。

Knuckle line of keel 龙骨折角线 船底具有舭部升高,龙骨剖面形同碟盆的剖面,船体型剖面底部近中线面处的纵向折角线。

Knudsen pipette 克努森移液管 摩尔-克努森(Mohr-Knudsen)氯度滴定法规定使用的一种高精确度的移液管。

K_r(m) 横摇回转半径(m) K_r means the roll radius of gyration, in m.

Kyoto Protocol 京都议定书

L

L frame L型吊架 外形似倒"L"字母的构件。它具有固定式和活动式两种。固定式作为深海绞车的钢索导向滑车的支点;活动式除起支点作用外还借液压油缸或其他机械装置的作用,将海洋调查仪器,装备倒出船外或翻进船内。

L sound level L声级 系指未计权的声压级,其频率响应呈平直响应。用于飞机、气垫船噪声测量与评价。

Labor productivity 劳动生产率 系指劳动者在一定时期内创造的劳动成果与其相适应的劳动消耗量的比值。劳动生产率水平可以用同一劳动在单位时间内生产某种产品的数量来表示,单位时间内生产的产品数量越多,劳动生产率就越高;反之,则越低;也可以用生产单位产品所耗费的劳动时间来表示,生产单位产品所需要的劳动时间越少,劳动生产率就越高。具体劳动生产使用价值的能力或效率。反之,则越低。

Labor productivity index 劳动生产率指数

Laboratory(Lab.) 实验室 安装仪器和实验设备,从事试验、观测、记录和分析用的科学研究工作室。

Laboratory instrument system 实验室仪器系统 仪器的各组成部分之间的电缆线路系统。它包括仪器的各个分机、电源装置、控制箱和电缆线路等。

Laboratory recognized by the Administration 主管机关公认的实验室　系指有关主管机关接受的实验室。其他实验室根据实际情况由有关主管机关根据协议特别认可。　Laboratory recognized by the Administration means a testing laboratory which is acceptable to the Administration concerned. Other testing laboratories may be recognized on a case-by-case basis for specific approvals as agreed upon by the Administration concerned.

Laboratory signal socket box　实验室信号插座箱　用于转接信号电缆的插座箱。一般采用无线电电子设备用的多芯屏蔽插座。插座箱装于室外时应为防水型，用于室内时应为防滴型。

Laboratory telephone system　实验室电话系统　用于调查船各实验室之间或实验室内与各实验部位之间通信联络的电话系统。

Labour compensation　劳务补偿　系指用加工、装配、效力等劳务收取补偿进口设置装备技术、技能的价款。

Labour supplying country　劳工提供国　系指提供船员在悬挂另一国国旗的船舶上服务的国家。　Labour supplying country means a country which provides seafarers for service on a ship flying the flag of another country.

Labradorite　拉长石　一种石灰钠岩状长石。可能产生粉尘。无特殊危害。该货物为不燃物或失火风险低。　Labradorite is a lime-soda rock form of felspar. It may give off dust. It has no special hazards. This cargo is non-combustible or has a low fire-risk.

Labyrinth　迷宫式音箱

Labyrinth acoustic filter　迷宫式消音器

Labyrinth gland　迷宫式压盖(曲径式压盖)

Labyrinth packing　迷宫式密封

Labyrinth packing(labyrinth gland)　曲径式气封　利用一系列容积突然变化的微小空间来降低蒸汽压力，减少漏气的气封。

Labyrinth ring　曲径式密封圈

Labyrinth seal　曲径密封(迷宫式密封)

Labyrinth stuffing box　曲径填料箱

Lacing wire　拉筋　从中间部位将若干叶片连接起来的金属线。

Lack of fusion　未熔合　系指焊缝金属与母材金属，或焊缝金属之间未熔化结合在一起的缺陷。未熔合是一种面积缺陷，坡口未熔合和根部未熔合对承载截面积的减小都非常明显，应力集中也比较严重，其危害性仅次于裂纹。主要是焊接热输入太低，电弧指向偏斜，坡口侧壁有锈垢及污物，层间清渣不彻底等。适当加大焊接电流，正确地选择焊接工艺参数，注意坡口及层

间部位的清洁。

Ladder guide rollers　导链滚轮　安装在斗桥或副斗桥上面，用以支承和导引斗链运转的滚轮。

Ladder hoisting gear　吊梯装置　供收放舷梯用的吊杆、索具以及绞车等的总称。

Ladder overturning gear(ladder turning gear)　翻梯装置　收放舷梯过程中，用以将舷梯翻覆于船舷的装置。

Ladder type hydrofoil　阶梯式水翼　由两个或两个以上的翼片以一定的间隔叠置，每个翼片类似于一个梯级，利用翼片的出水、入水引起的升力变化而保持艇的飞高及纵、横稳定性的水翼。翼片有平翼、V 型翼或斜升翼等多种形式。

Ladder-type pier　阶梯式码头　系指呈阶梯形的码头。渡船与码头用跳板连接，供汽车上下。这种码头投资少，且可供其他船舶停靠，但较难克服横向流的作用，对潮差的适应性也较差。

Lag　隔热套(外罩，延迟)

Lagging　罩壳　为减少从汽轮机对舱室的热辐射，在气缸和冷凝器喉部的外表面敷设的带有绝热材料的外罩。

Lagging　滞后(绝热装置，绝缘层)

Lagging material　绝缘材料(隔热材料)

Laminar flow　层流　在低雷诺数黏性流中，流体层平滑地相对移动的流态。

Laminar sublayer　层流底层　在湍流边界层和固体面之间，一层很薄的层流层。

Laminate bearing　层压胶木轴承　以层压胶木作为轴衬材料的轴承。

Lamination　分层(叠层)

Lamp　灯　系指光源，包括白炽光源，发光二极管(LED)和其他非白炽光源。　Lamp means a source producing light, including incandescent sources, Light Emitting Diodes(LED) and other non-incandescent sources.

Lamp(navigation light)　灯具(航行灯)　包括灯泡、灯座、外罩、透镜和为使灯完成预定功能所必需的其他部件的整个航行灯组件。

Lamp base　灯座　系将灯泡支承在工作位置，且为该灯泡提供电连接的航行灯的部件。

Lamp bulb　灯泡　系指在通电时发光的航行灯的部件。

Lamp store　灯具贮藏室　船上供贮存灯具的房间。

Land based power box　岸电箱　系指安装在岸上的、供船舶停靠码头时接收电源的配电箱。

Land earth station　陆上地面站　系指固定卫星服

务中的地面站,或在某一特定地点或陆上某一特定区域内的移动卫星服务中为移动卫星服务提供馈送电路的地面站。 Land earth station means an earth station in the fixed satellite service or, in some cases, in the mobile-satellite service, located at a specified fixed point or within a specified area on land to provide a feeder link for the mobile-satellite service.

Land-based testing 陆基试验 系指为确认压载水管理系统满足国际船舶压载水和沉积物控制和管理公约第 D-2 条规定的标准,按压载水管理系统认可导则附件第 2 和第 3 部分的规定,在实验室、设备厂或中间试制工厂(包括系泊的试验驳船或试验船)进行的压载水管理系统的试验。 Land-based testing is a test of the BWMS carried out in a laboratory, equipment factory or pilot plant including a moored test barge or test ship, according to Parts 2 and Parts 3 of the annex to the Guidelines for approval of ballast water management systems, to confirm that the BWMS meets the standards set by regulation D-2 of the international Convention for the Control and Management of Ships' Ballast Water and Sediments, 2004.

Landing mode for carrier-borne aircraft 航母舰载机降落方式 (1)利用拦阻索——飞机降落时尾部放下拦阻钩钩住一根拦阻索,吸收着舰飞机的动能,缩短其滑行的距离。目前世界上大部分航母舰载机均采用这种方式降落;(2)垂直/短距离降落。

Landing speed range 着陆速度范围 操作人员在整个飞机着陆操作过程中得以保持对船舶控制的速度范围。 Landing speed range is a range of speeds that allow operator to maintain craft control throughout a landing manoeuvre.

Landing visa/visa on arrival 落地签证 系指申请人不需直接从所在国家取得前往国家的签证,而是持护照和该国有关机关发给的入境许可证明等抵达该国口岸后再签发签证。

Land-locked State 内陆国 系指没有海岸的国家。 Land-locked State means a State which has no seacoast.

Language interpretation 语言翻译
Language translation(LT) 数据处理 系指对大量的数据进行加工处理,如收集、存储、传送、分类、检测、排序、统计和输出等,从中筛选出有用信息。

Laptop computer/Laptop 膝上电脑 系指便携式个人电脑。

Large car transport carrier(LCTC) 大型汽车运输船

Large hatch covers(cargo hatch covers) 大舱盖 装于尺度较大的舱口上,一般由若干块舱盖板组成,并具有专门封舱装置的舱口盖。

Large landing ship 大型登陆舰 是一款现代军事海上登陆战最实用的武器装备,也叫作两栖舰艇,是为输送登陆兵及其武器装备,补给品登陆而专门制造的舰艇。登陆舰又称坦克登陆舰,它的排水量 600 ~ 10 000 吨,可载坦克几辆至几十辆和士兵数百名。它的续航能力一般为 200 ~ 6 000 km,航速 20 ~ 40 km/h,这就使登陆部队可从出发地直抵登陆点滩头,无须中途换乘,大大简化和强快了登陆过程,其登陆作战能力比登陆艇大为增强。德国还造过两艘被称为战列巡洋舰的战舰:"沙恩霍斯特"号、"格奈森瑙"号,主要被用作巡洋作战,而类似战列巡洋舰的特点,但它们的装甲比战列巡洋舰要强使用 11 英寸口径主炮。见图 L-1。

图 L-1 大型登陆舰
Figure L-1 Large landing ship

Large motor 大型电动机 系指任何额定输出功率大于 100 kW,或者大于系统中最大发电机额定功率的 25% 的电动机。 Large motor means any motor rated more than 100 kW or 25% the rated power of the maximum generator on the systems.

Large opening barge 舱口驳 设有较大货舱口的货驳。

Large openings 大开口 系指:(1)椭圆开口长超过 2.5 m 或宽超过 1.2 m;(2)圆直径超过 0.9 m。大开口和扇形孔(若采用挖孔焊),在计算船体梁横剖面时应从横剖面面积中扣除;(3)符合下述任一条件的甲板开口为大开口:①$b/B_1 \geq 0.7$;②$l_H/l_{BH} \geq 0.89$;③$b/B_1 > 0.6$ 和 $l_H/l_{BH} > 0.7$,式中,b——开口宽度(m),如有几个舱口并列,则 b 代表各开口宽度之和,即 $b = b_1 + b_2$,B_1——在开口长度中点处包括开口在内的甲板宽度(m);l_H——舱口长度(m);l_{BH}——每一舱口两端横向甲板条中心线之间的距离(m),如舱口前或后再无其他舱口时,则 l_{BH} 算到舱壁为止。 Large openings means:(1) elliptical openings exceeding 2.5 m in length or 1.2 m in breadth; (2) circular openings exceeding 0.9 m in diameter. Large openings and scallops, where scallop welding is applied, are always to be deducted from the sectional areas included in the hull girder transverse sections；(3) a deck opening is to be regarded as a large opening if any of the following conditions applies：①$b/B_1 \geq 0.7$；②$l_H/l_{BH} \geq 0.89$；③$b_1/B_1 > 0.6$ and $l_H/l_{BH} > 0.7$，where：b—breadth of the opening (m)．Where there are several hatches abreast，b is the sum of individual widths of these openings，i.e. $b = b_1 + b_2$，B_1—maximum breadth of deck (m)，including opening，measured at the mid-length of the opening；l_H—length of the hatch (m)；l_{BH}—distance between centerlines of the deck strips at each end of the opening, in m．Where there are no further hatches fore or aft the one under consideration，then l_{BH} is to be measured to the bulkhead．

Large quantities of oil fuel　大量燃油　对"73/78防污公约"附则Ⅰ而言，是针对因其营运和运输的具体特点而需要在海上待很长持续时间的船舶(大型渔船、远洋拖船等)而提出的。根据已考虑的情况，此类船舶将需要在其空燃油舱中加入压载水，已保持足够的稳性和安全航行条件。

Large volume sampler　大容量采水器　主要用于放射性同位素研究的，容量为50升以上的采水器。

Large-scale military transport aircraft　大型军用运输机　系指起飞质量超过100吨，装载量超过40吨的飞机，其具有快速运送大量兵员、武器装备和其他军用物资到作战前线的能力，并能进行空投空降，确保部队战略机动、战术投送的规模化、快捷性和突然性，在近年来多次现代战争中发挥了至关重要的作用。由于大型军用运输机用途广泛，因此还可作为空中预警机、空中加油机、电子干扰机、海上巡逻机。特种任务飞机等支援机型的改装基础平台，以及执行人道主义救援等任务。如运-20以66吨的有效载荷，跻身全球10大运力最强运输机之列，与美国C-17"环球霸王"和俄罗斯伊尔-76属同一级别。

Largest effective wave steepness　有效波陡最大值

Laser dynamic inspection machine　激光动态检测仪　系指采用激光技术实时检测的仪器。

Laser plotter　激光绘图机

Laser printer　激光打印机　系指脱胎于20世纪80年代末的激光照排技术，流行于20世纪90年代中期。它是将激光扫描技术和电子照相技术相结合的打印输出设备。其基本工作原理是由计算机传来的二进制数据信息，通过视频控制器转换成视频信号，再由视频接口/控制系统把视频信号转换为激光驱动信号，然后由激光扫描系统产生载有字符信息的激光束，最后由电子照相系统使激光束成像并转印到纸上。较其他打印设备，激光打印机有打印速度快、成像质量高等优点，但使用成本相对高昂。激光打印机按其打印输出速度可分为3类：即低速激光打印机(每分钟输出10～30页)；中速激光打印机(每分钟输出40～120页)；高速激光打印机(每分钟输出130～300页)。

Laser scanning techniques　激光扫描技术

Laser technologies (LT)　激光技术　是依据一定的原理，改变激光振荡或激光辐射的参数，使之适合于某一目的的技术。由于激光具有方向性好、亮度高、单色性好等特点而得到广泛应用。目前，我国深紫外全固态激光器通过验收，成为目前世界上唯一能够制造实用化、精密化深紫外全固态激光器的国家。该技术拥有国：俄罗斯、中国、美国。见图L-2。

图L-2　激光技术
Figure L-2　Laser technology

Laser power transmission　激光电能传输　系指利用激光方向性强，可以携带大量能量的优势，实现较远距离的输电。

Laser-based system　激光系统　定位和跟踪系统。利用反射的脉冲激光来测量距离和角度，如Fanbeam和Cyscan。

Lashing assembly　绑扎装置　由杆、钢丝绳或链制成的拉力元件、张紧装置以及绑扎点，用于系固集装箱堆垛的装置。

Lashing bridge　绑扎桥　甲板上的横向升高平台，舱口盖和甲板上的集装箱可以通过绑扎装置绑扎到绑扎桥上。其设置在集装箱船的上甲板，用于系固堆放在上甲板上的多层集装箱用的装置。

Lashing of container　集装箱绑架　系指由杆、钢丝绳或链制成的拉力元件、张紧装置以及绑扎点，用于系固集装箱堆垛的装置。

Lashing point　绑扎点　焊接在甲板、舱口盖、箱柱及绑扎桥上，用于绑扎装置与船体结构或舱口盖连接的装置，包括D形环、绑扎眼板等。

Latency　等待时间　系指实际与显示数据间的延迟。Latency means a delay between actual and presented data.

Latency　执行时间　系指事件和由此产生的信息之间的时间间隔，其中包括信息处理、传输和接收时间。

Latency is a time interval between an event and the resulting information, including time for processing, transmission and reception.

Lateral deviation 停船横矩 船舶在以定常航速沿直线前进中,从执行停车或倒车起到船舶完全停止移动止,在垂直于原航向上的最大横移距离。

Lateral force 横向力 作用于船体的流体动力正交于其中线面的分量。

Lateral loads in still water 静水中的侧向载荷 系指外部静水压力及货物和压载引起的内部静压力。Lateral loads in still water mean an external hydrostatic pressure and internal static pressure due to cargo and ballast.

Lateral loads in waves 波浪中的侧向载荷 系指外部水动压力以及货物和压载引起的内部惯性压力。Lateral loads in waves mean external hydrodynamic pressure and internal inertial pressure due to cargo and ballast.

Lateral position of propeller 螺旋桨横向位置 螺旋桨中点至船中线面的垂直距离。

Lateral thrust 侧推推力

Lateral thrust unit 侧推器(横向推力器)

Lateral thrust unit compartment 侧推装置舱

Lateral thruster 侧推器(侧推装置)

Lateral thruster 侧向推力器 系指安装在船舶的左右两舷,能产生侧向推力,用于船舶的机动操纵、靠离码头和船舶动力定位的装置。

Lateral thruster turning maneuver test 侧向推力器回转试验 船舶在正舵情况下,侧向推力器用全推力作左、右舷回转操纵达90°转艏角,而初速保持在0~8 kn内的试验。

Lateral thruster zig-zag maneuver test 侧向推力器Z形操纵试验 以3~6 kn的转向速度及零舵角达到10°时变换侧向推力器方向的Z形操纵试验。

Lateral thrusting bow propeller 艏部侧向推力器

Lateral vibration 横向振动(侧向振动)

Lateral vibration of shafting (transverse vibration of shafting) 轴系横向振动 轴系在与其中线交叉方向的抖动振动。

Latin American Free Trade Association (LAFTA) 拉丁美洲自由贸易协会

Lattice division (wall) 花格窗 开在相邻两空间的壁上。带有镂空图案,起装饰、间隔及通气作用的窗。

Lattice door 栅栏门 仅起分隔作用的栅栏形式的门。

Launch 汽艇/摩托艇 起源于19世纪末。是以汽油机、柴油机或涡轮喷气发动机等为动力的机动艇。作为在水上竞速的一种体育活动的工具。

Launch barge 下水驳 系指专用于海洋工程导管架结构的运输,在导管架下水作业时采用艉倾的方式将导管架滑入水中。

Launching 下水 将船舶从岸边移至水域的过程。

Launching appliance 降落装置 系指将满载额定人员及设备的救生艇从其搭载的位置处降落到水面上的装置。Launching appliance is a device capable of launching, from the embarkation position, a craft fully loaded with the number of persons it is permitted to carry and with its equipment.

Launching appliance or arrangement 降落设备或装置 系指将救生艇筏或救助艇从其位置安全地转移到水上的设施。Launching appliance or arrangement is a means of transferring a survival craft or rescue boat from its stowed position safely to the water.

Launching barge 下水驳 系指平底平甲板船型,甲板上设有导管架专用下水滑道及摇臂机构,把坐落在下水滑道上的导管架滑移到水中的驳船。

Launching by hand 人力抛投 系指救生浮具全靠人力抬动抛落水面的操作方式。

Launching condition 降落条件 在船上遇难紧急情况下,进行卸放救生艇操作所必须满足的船舶纵、横倾条件。

Launching curves 下水曲线 由下水计算结果所得出的各种曲线,通常包括:(1)浮力曲线;(2)下水重力曲线;(3)浮力对前支架和滑道末端的力矩曲线;(4)下水重力对前支架和滑道末端的力矩曲线等。

Launching ramp angle 降落滑道角度 系指船舶处于正浮时,救生艇滑道与水平面形成的角度。Launching ramp angle is the angle between the horizontal and the launch rail of the lifeboat in its launching position with the ship on even keel.

Launching ramp length 降落滑道长度 系指救生艇尾部至降落滑道下端的距离。Launching ramp length is the distance between the stern of the lifeboat and the lower end of the launching ramp.

Launching weight 下水重力 下水时船舶的总重力与滑板、下水架等重力的总和。

Laundry equipment 洗涤设备 各种清洗整理衣、被、床单等物品用的器具的统称。包括洗衣机、脱水机、烘衣机、洗涤池等。

Lawyer/solicitor 律师 系指具备一定资质的、从事法律工作者。

Lay time 装卸时间 系指允许完成装卸任务所约定的时间,它一般以天数或小时数来表示。

Lay tower 铺设塔

Laydays date 受载日　系指租方可以接受装船的最早装货日期。

Layer exchange technique/Layer-switching technologies 层交换技术　可以识别数据帧中的 MAC 地址信息，根据 MAC 地址进行转发，并将这些 MAC 地址与对应的端口，记录在自己内部的一个 MAC 地址表中。目前，第2层交换技术已经成熟。从硬件上看，第2层交换机的接口模块都是通过高速背板/总线（速率可高达几十 Gbps）交换数据的，第2层交换机一般都含有专门用于处理数据包转发的 ASIC（Application Specific Integrated Circuit）芯片，因此转发速度可以做到非常快。

Layer 布设船　在内河或海上布设航标、电缆或管道的船舶。

Layer-7 load balance 第7层负载均衡

Lay-up survey(LS) 搁置检验　对船舶在搁置阶段开始时所进行的检验。其旨在确认船舶安全状态、保养措施、搁置位置和系泊布置是否符合船级社已同意的搁置维护方案。搁置检验完成并认为满意后，在入级证书上签署并注明船舶已处于搁置期。

Lazy s-riser 缓S形立管

Lazy wave riser 惰性波纹立管　系指将悬链线立管的上段进行加工以形成局部的波纹，使悬链线下拐折点与平台群的运动分离。 Lazy wave riser means wave shaped the upper section of a catenary riser to form a local undulation that isolates the sagbend from motions of the host platform.

Lead auditor 主任审核员　具有足够的管理能力和经验，对审核的各种局面负有最终责任的审核员。

Lead block 导向滑车　导引和改变钢索走向用的滑车。

Lead craft 首制艇　系指按批准的设计文件制造的首艘游艇。

Lead nitrate(UN 1469) 硝酸铅　白色晶体。溶于水。硝酸作用于铅后的衍生物。吞入或吸入粉尘后有毒害。自身不可燃，但与其他可燃物质的混合物容易点燃并猛烈燃烧。 Lead nitrate(UN 1469) is a white crystal. It is soluble in water and derived from the action of nitrate acid on lead. Toxic if swallowed or dust inhaled. Not combustible by itself, but mixtures with combustible materials, are easily ignited and burn fiercely.

Lead ore 铅矿砂　重而软的灰色固体物质。有毒。与酸性物质接触会释放剧毒蒸发气。该货物为不燃物或失火风险低。 Lead ore is a heavy soft grey solid material. Toxic, with acids evolves highly toxic vapour. This cargo is non-combustible or has a low fire-risk.

Leading block 舵链导轮　舵链传动装置中用以引导舵链转向，其上带有凹槽的链轮。有卧式舵链导轮和立式舵链导轮两种。

Leading edge 导边　螺旋桨叶片的导边系指当螺旋桨旋转时，叶片开始进入水里的一边。 Leading edge of a propeller blade is the edge of the blade at side entering the water while the propeller rotates.

Leading-out terminal 引出线端　系指在敷线上的某一点，电流于此点输出供给用电设备。

Leading-out terminal for lighting 照明引出线端　系指用于直接连接一个灯座或一个照明灯具的引出线端。

Leading-out terminal for socket 插座引出线端　系指装有一个或多个插座的引出线端。

Leak 漏泄（漏损，漏泄电阻）

Leak detection 泄漏检查（漏电检查）

Leak detector 检漏器（漏电检测仪）

Leak free 密封的

Leak out 漏出

Leak proof 防漏

Leak test(air test) 漏风试验　用正压或负压风检查锅炉风道内外夹层和烟道密封性的试验。

Leak testing 渗漏试验（泄漏试验）　是进行空气或其他介质的试验，以证明结构的紧密性。（1）这种试验采用一种能有效显示的液体，例如肥皂水溶液，涂在要试验的焊缝或舾装贯穿件处进行。此时，液舱或舱室将承受至少为 0.15 bar（0.15 kgf/cm²）的空气压力。（2）建议将空气压力升至 0.2 bar（0.2 kgf/cm²），且在这个压力下保持约 1 h，以达到一个平衡状态，舱周围的人员应尽量减少，然后在检查前降至试验压力。应装设一U型管，将水灌至相当于试验压力的高度，以供验证并避免过压。U型管的截面应大于空气供应管的截面。此外，试验压力还应该用压力计或其他相当的系统予以验证。（3）应对液舱周界的所有填角焊缝与装配焊缝及所有舾装贯穿件上的焊缝，在涂保护涂层前进行渗漏试验。自动焊缝及药芯焊丝电弧焊（FCAW），建筑物连接的半自动对接焊缝，只要认真地目视检查显示焊缝形状呈连续均匀形状，没有经过修理，选择的无损试验的结果显示没有重大缺陷就不需要试验。（4）验船师考虑到造船厂的质量控制程序，也可要求在施加涂层前，对自动装配焊缝以及预装配手工或自动焊缝的选定位置进行渗漏试验。若免除这个要求，则可在施加了防护涂层后进行渗漏试验，条件是焊缝已经过仔细检查，并使验船师满意。 Leak testing is an air or other medium test carried out to demonstrate the tightness of the structure. (1)This is carried out by applying an efficient indicating liquid (e. g. soapy water solution), to the weld or outfitting penetration being tested, while the tank or compartment is subject to an air pressure of at least 0.15 bar (0.15 kgf/cm^2). (2) It is recommended that the air pressure be raised to 0.2 bar (0.2 kgf/cm^2) and kept at this level for about one hour to reach a stabilized state, with a minimum number of personnel in the vicinity, and then lowered to the test pressure prior to inspection. A U-tube filled with water to a height corresponding to the test pressure is to be fitted for verification and to avoid overpressure. The U-tube is to have a cross-section larger than that of the air supply pipe. In addition, the test pressure is to be verified by means of a pressure gauge, or alternative equivalent system. (3) Leak testing is to be carried out, prior to the application of a protective coating, on all fillet welds and erection welds on tank boundaries, and on all outfitting penetrations. Automatic and Flux Core Arc Welding (FCAW) semi-automatic butt welds of the erection joints need not be tested, provided that careful visual inspections show continuous uniform weld profile shape, free from repairs, and the results of selected NDE testing show no significant defects. (4) Selected locations of automatic erection welds and pre-erection manual or automatic welds may also be required to be tested before coating, at the discretion of the Surveyor, taking account of the quality control procedures of the shipyard. Where exempt from this requirement, leak testing may be carried out after the protective coating has been applied, provided that the welds have been carefully inspected to the satisfaction of the Surveyor.

Leakage 渗漏/漏泄（逸流，漏出量） 系指流质或半流质的物质因为容器的破漏引起的损失。

Leakage 逸流 从气垫向外界大气逸散的气流。

Leakage detecting system 漏液探测系统

Leakage detector 泄漏探测器

Leakage protective system 防漏系统（漏电防护装置）

Leakage risk 渗漏险 投保平安险和水渍险的基础上加保此险，保险人负责赔偿承保的流质、半流质和/或油类货物在运输途中因容器损坏而引起的渗漏损失，或用液体储藏的货物因液体渗漏而引起的腐烂变质造成的损失。如以流体装存的湿肠衣，因为流体渗漏而使肠衣发生腐烂、变质等损失，均由保险公司负责赔偿。

Leakage test 紧密性试验（泄漏试验，漏电试验）

Leakage tester 测漏器（探漏器）

Leakage trace 泄漏痕迹

Leakage-proof 防漏（避漏）

Leaking 泄漏 对于《国际船舶装运密封装辐射性核燃料、钚和强放射性废料规则》而言，系指INF货物由其盛装系统漏出或INF货物密封包装损坏。 For the purpose of International Code for the safe carriage of packaged irradiated nuclear fuel, plutonium and high-level radioactive wastes on board ships, leaking means the escape of INF cargo from the containment system or the loss of an INF cargo package.

Leaks from packing 填料泄漏

Lean production 精益生产方式 以最小的资源和投入，包括人力、设备、资金、材料、时间与空间，创造出尽可能多的价值，为顾客提供满意的产品与及时服务的一种生产方式。换言之，即为了满足市场和客户要求而获取更高利润的方法和途径。其旨在通过全员的激励和努力，优化生产组织结构、去掉一切无效（不增值）的生产过程和环节，通过减少生产过程中的一切浪费来缩短生产周期、降低成本、保证生产质量、提高产品利润。它以准时生产、成组技术和全面质量管理为支柱，并引入并行工程和整体优化概念，在时间和空间上合理配置和利用生产要求，发挥以人为核心的整体制造系统效益。

Lease 出租 系指将船舶租借给承租人的业务。

Leasing contract 租赁合同 系指出租人将租赁物交付承租人使用、收益，承租人支付租金的合同。交付租赁物的一方为出租人，接受租赁物的一方为承租人，被交付使用的财产即为租赁物，租金就是承租人向出租人交纳的使用租赁物的代价。

Least moulded depth D_F 最小型深 取为平板龙骨上缘至舷侧干舷甲板横梁上缘之间的垂直距离。在具有圆弧舷边的船上，此型深应量至甲板与舷侧外板型线的交点，这些型线的延伸就像舷边为角状设计。如干舷甲板呈梯级状，且其升高部分延伸于确定型深之处，则此型深应量至沿甲板较低部分参考线。 Least moulded depth taken as the vertical distance in m, from the top of the keel to the top of the freeboard deck beam at side. In ships having rounded gunwales, the moulded depth shall be measured to the point of intersection of the moulded lines of the deck and side shell plating, the lines extending as though the gunwale were of angular design. Where the freeboard deck is stepped and the raised part of the deck extends over the point at which the moulded depth is to be determined the moulded depth shall be measured to a line of reference extending from the lower part of the deck along a line parallel with the raised part.

Leave-joint (engaging and disengaging) 离合 主机与轴系的分离或接合。

Left hand engine　左旋发动机

Left-hand nut　左向螺母

Left-hand revolving engine unit　左转机组　正车时,由功率输出端向自由端看,做逆时针方向旋转的主机机组。

Left-hand turning　左旋　由船后向前看,螺旋桨推船向前时,为逆时针的旋向。

Left-handed propeller　左旋螺旋桨

Left-handed thread　左螺纹

Leg(spud)　桩腿　用于支持自升式钻井平台升离水面进行钻井的腿柱(一般有圆筒形、箱形和桁架形等)。

Leg wells　桩孔　系指自升式平台上桩腿围井处的平台开口。 Leg wells are openings in the deck in way of leg trunk on a self-elevating unit.

Legal net weight　法定净重　系指从商品总质量中扣除内外包装的法定质量后的质量。

Legal person　法人　系指按照法定程序设立,有一定的组织机构和独立财产,并能以自己的名义享有民事权利和承担民事义务的社会组织。

Legal weight　法定质量　系指货物的净重加上与货物直接接触的,可以连同货物零售的销售包装的质量。

Legitimate right　合法权益　系指符合法律规定的权力和利益。在我国,公民的合法利益包括宪法和法律所规定的政治权利、民主权利、人身权利、经济权利、教育权利等。

Legs　桩腿　是一种在自升式平台上借助电动机械、液压机械或电动与液压相结合的机械与平台主体结构作预定相对运动的柱形或桁架式结构,桩腿可插入海床并将平台主体结构抬出海面到一定高度。 Legs are tubular or truss-type structures of a self-elevating unit capable of relatively moving with the hull in a pre-set manner by means of electrical or hydraulic devices or both, and of penetrating into seabed, supporting the hull and enabling it torise to a predetermined elevation over the sea surface.

LEL　爆炸下限　LEL means the lower explosive limit.

Length[1]　船长　系指:(1)应取为量自龙骨板上缘的最小型深85%处总长的90%,或沿该水线从艏柱前边至舵杆中心线的长度,取大者;(2)对于无舵杆的船舶,长度 L 取为最小型深85%处总长的90%;(3)若在最小型深85%处水线以上的艏柱外形为凹的,则总长的最前端和艏柱前边都应在该水线以上的艏柱外形最后一点垂直投影在该水线上的点量起;(4)龙骨设计成倾斜的船舶,其计量本长度的水线应和最小型深的85%处的设计水线平行,该水线由绘一平行于船舶(包括呆木)的龙骨线并与干舷甲板型舷弧相切的切线得到。此最小型深为在切点处从龙骨板上缘量至干舷甲板舷侧处横梁上缘的垂直距离。 Length mean:(1)the length L shall be taken as 96% of the total length on a waterline at 85% of the least moulded depth measured from the top of the keel, or as the length from the fore side of the stem to the axis of the rudder stock on that waterline, if that be greater. (2) For ships without a rudder stock, the length L is to be taken as 90% of the waterline at 85% of the least moulded depth;(3) where the stem contour is concave above the waterline at 85% of the least moulded depth, both the forward terminal of the total length and the fore-side of the stem respectively shall be taken at the vertical projection to that waterline of the aftermost point of the stem contour (above that waterline);(4) In ships designed with a rake of keel the waterline on which this length is measured shall be parallel to the designed waterline at 85% of the least moulded depth D_{min} found by drawing a line parallel to the keel line of the vessel (including skeg) tangent to the moulded sheer line of the freeboard deck. The least moulded depth is the vertical distance measured from the top of the keel to the top of the freeboard deck beam at side at the point of tangency.

Length[2]　船长(内河船)L(m)　系指沿满载水线自艏柱前缘量至舵柱后缘的长度;无艏柱船舶的船长应自船体中纵剖面前缘与满载水线的交点量起;无舵柱船舶量至舵杆中心线;但均应不大于满载水线长度,亦不小于满载水线长度的96%。无舵船舶的船长取满载水线长度。

Length[3]　船长(非高速小水线面双体船)　系指沿夏季载重线,由最前端支柱前缘量至最后端支柱体尾缘的长度,并应计入下潜体从艏至艉的长度与刚性水密船体位于设计水线以下部分的总长(不包括设计水线处及以下的附体)之差的50%。

Length(SWATH)　船长(高速小水线面双体船)　系指船舶静浮于水面时,其刚体水密船体位于设计水线以下部分的总长,不包括水密水线处及以下的附体。 Length means the overall length of the underwater watertight envelope of the rigid hull, excluding appendages at or below the design waterline in the displacement mode with no lift or propulsion machinery active.

Length and breadth of a vessel　船长和船宽　系指总长和最大宽度。 Length and breadth of a vessel mean her length overall and greatest breadth.

Length at waterplane Lw(m)　水线面处船长 Lw

对小水线面双体船而言，系指船舶静浮于水面时，位于设计水线处量得的船体前后缘纵向之距离。

Length between perpendiculars L_{pp} 垂线间长 L_{pp} **(m)** (1) 为沿夏季载重水线由艏柱前缘量至舵柱后缘的长度，若无舵柱则为量至舵杆中心线的长度。对艉部布置独特的船舶，其 L_{pp} 将另行考虑。艏垂线 FP 为夏季载重水线与艏柱前缘交点处的垂线。艉垂线 AP 为夏季载重水线与舵柱后缘交点处的垂线。对无舵柱的船舶，其艉垂线 AP 为该水线与舵杆中心线交点处的垂线；(2) 是取自最深分舱载重[即分舱要求（适用的）所允许的最大吃水的相应水线]的两端的垂线间量得的船舶长度。 Length between perpendiculars (L_{pp}) is: (1) the distance, in m, on the summer load waterline from the fore side of the stem to the after side of the rudder post, or to the centre of the rudder stock if there is no rudder post. In ships with unusual stern arrangements the length, L_{pp}, will be specially considered. The forward perpendicular, FP, is the perpendicular at the intersection of the summer load waterline with the fore side of the stem. The after perpendicular, AP, is the perpendicular at the intersection of the summer load waterline with the after side of the rudder post. For ships without a rudder post, the AP is the perpendicular at the intersection of the waterline with the centreline of the rudder stock; (2) the length of the ship measured between perpendiculars taken at the extremities of the deepest subdivision load line, i.e. of the waterline which corresponds to the greatest draught permitted by the subdivision requirements which are applicable.

Length breadth ratio 长宽比 一般系指垂线间长与设计水线宽之比。

Length depth ratio 长深比 一般系指垂线间长与型深之比。

Length draft ratio 长度吃水比 一般系指垂线间长与吃水之比。

Length L 船长（地效翼船） 系指船舶在排水状态而且垫升和推进机械不工作时，设计水线处及以下的刚性船体水下水密外壳的总长，不包括附体。 Length L means the overall length of the underwater watertight envelope of the rigid hull, excluding appendages, at or below the design waterline in the displacement mode with no lift or propulsion machinery active.

Length of a bulk carrier 散货船的船长 系指现行《国际载重线公约》所定义的船长。 Length of a bulk carrier means the length as defined in the International Convention on Load Lines in force.

Length of a floating dock 浮船坞长 L_D(m) 系指由浮船坞最前面浮箱前端缘量至最后面浮箱后端缘的长度。对于不属于浮箱整体结构的端部平台以及天桥部分不包括在坞长的长度范围内。

Length of blade 叶长 螺旋桨半径与毂半径之差。

Length of brackets l_b(m) 肘板长度 l_b(m) l_b means the length, in m, of brackets.

Length of combination L_c 组合体长度 L_c 系指顶推船-驳船组合体视作单独船的长度。组合长度 L_c 应取为如下距离(m)：在驳船最小型深的 85% 水线处（如顶推船干舷甲板最低点位于或高于该水线），或在组合体设计水线处（如顶推船干舷甲板最低点低于该水线），从驳船的艏柱前缘量至顶推船的舵杆或艉柱的后缘，如无舵柱或艉柱，L_c 应量至顶推船舵杆的中心线。L_c 应取不小于上述组合体水线总长度的 96%，但也不必大于 97%。 Length of combination L_c means the length of the pusher tug-barge combination acting as a single unit. The combined length L_c is to be taken as the distance, in m, measured on a waterline at 85% of the least molded depth of the barge (where the lowest point of freeboard deck of the pusher tug is at or above this waterline) or on the design waterline of the combination (where the lowest point of freeboard deck of the pusher tug is below this waterline), from the fore side of the stem of the barge to the after side of the rudderpost or stern post of the pusher tug. Where there is no rudder post or stern post, L_c is to be measured to the centerline of the rudder stock of the pusher tug. L_c, however, is not to be taken less than 96% and need not be taken greater than 97% of the total waterline length of the combination measured at 85% of the least molded depth of the barge.

Length of crane boat 起重船的船长 系指在上甲板下缘，自船首端外板的外表面到船尾端外板的外表面的水平距离。

Length of entrance 进流段长 船体水下部分前端至平行中体前端或最大横剖面之间的水下距离。

Length of floating body 浮体长 系指浮体前端舱壁到后端舱壁之间的距离。

Length of hatchboard 舱盖板长度 (1) 每块舱盖板在较长的一边的外形尺寸。(2) 整个大舱盖沿船的纵向的外形尺寸。

Length of hull 平台长 钻井平台沿设计拖航方向上的最大长度。

Length of leg penetrating in sea-bed 桩腿入土深度 桩腿插入海底泥土中的深度。

Length of lifeboat 救生艇长 救生艇艏、艉间平

行于其基线的直线距离。通常用以下三个量度表达：(1)计算长度——自艇艏舷处壳板的内缘量至艉柱处壳板的内缘的直线距离。(2)丈量长度——一般与计算长度相同，但在用近似公式计算木质艇的容量时，系指自艇艏柱处壳板的外缘量至艉柱处壳板的外缘的直线距离。方艉量至艉封板的外缘。(3)最大长度——自艇一端的最外缘量至另一端最外缘的直线距离。

Length of mat 沉垫长 沉垫沿设计拖航方向上的长度。

Length of overall 总长

Length of panels(length of hatchcover section) 舱盖分块长度 沿船纵向分块布置的舱盖板的纵向外形尺寸。

Length of run 去流段长 船体水下部分后端至平行中体后端或最大横剖面之间的水下距离。

Length of ship for freeboard L_f 干舷船长 L_f 系指从平板龙骨上缘量至最小型深85%处水线上从艏柱前缘至艉端船体板后缘之船长的96%，或在此水线上从艏柱前缘至舵杆中心线的距离(m)，取大者。但如艏柱的轮廓线在最小型深85%的水线上方为凹形，则干舷船长的前端点应选在艏柱轮廓线最后一点在此水线处的垂直投影点上。在其上量取干舷的船长的水线应与载重线平行。 Length of ship for freeboard L_f is 96% of the length in m measured from the fore side of stem to the aft side of aft end shell plate on the waterline at 85% of the least moulded depth measured from the top of keel, or the length in m measured from the fore side of stem to the axis of rudder stock on that waterlines, whichever is the greater. However, where the stem contour is concave above the waterline at 85% of the least moulded depth, the forward terminal of this length is to be taken at the vertical projection to this waterline of the aftermost point of the stem contour. The waterline on which this length is measured is taken to be parallel to the load line.

Length of stroke 冲程长度

Length of superstructure 上层建筑长度 是位于船长 L 之内的上层建筑部分的平均长度。 Length of superstructure is the mean length of the part of the superstructure which lies within the length L.

Length of the arm 锚臂长度 锚臂长度是销中心(对于有锚头销的锚)或锚冠顶(对于其他种类的锚)至锚爪尖端之间的距离；如锚冠为凹型，则锚杆中心线和与锚臂顶平面的交点应认为是锚冠顶。 Length of the arm is the distance from the centre of the pin in case of anchors having the head pin and from the top of the crown in case of anchors of the other types to the tip of the flukes.

Where the crown is of concave form, the intersection of the centre line of the shank with the plane in contact with the top of the arms is considered as the top of the crown.

Length of the compartment l_H (m) 舱室长度 l_H (m) l_H means the length, in m, of the compartment.

Length of the hull L_H 艇体长度 L_H 艇体长度 L_H 应取在垂直于艇中线面的两垂向平面之间平行于基准水线和艇的中心线的距离，它的一个平面通过艇的最前端部件，另一个平面通过艇的最后端部件。该长度包括艇的所有结构和组成部件，诸如木质、塑料或金属艏柱或艉柱、舷墙与船体/甲板连接件。该长度不包括能以不被损坏的方式及不影响艇结构完整性而拆卸的可拆部件，例如帆桁、艇首斜撑帆杆、艇两端的操纵台、艏柱附件、舵、舷外挂机、舷外发动机及其安装支架和安装平台、潜水平台、登艇平台、橡胶护舷材及碰垫。但该长度包括当艇在静止或航行时起静水力或动力支承作用的艇体可拆部件。对于多体艇，应分别测量每一艇体的长度。应取各次分别测量之最长者为艇体长度 L_H。单体艇测量见图 L-3a，多体艇测量见图 L-3b。 Length of the hull L_H shall be measured parallel to the reference waterline and craft centerline as the distance between two vertical planes, perpendicular to the center-plane of the craft, one plane passing through the foremost part and the other through the aftermost part of the craft. This length includes all structural and integral parts of the craft, such as wooden, plastic or metal stems or sterns, bulwarks and hull/deck joints. This length excludes removable parts that can be detached in a non-destructive manner and without affecting the structural integrity of the craft, e. g. spars, bowsprits, pulpits at either end of the craft, stemhead fittings, rudders, outdrives, outboard motors and their mounting brackets and plates, diving platforms, boarding platforms, rubbing strakes and fenders. This length does not exclude detachable parts of the hull, which act as hydrostatic or dynamic support when the craft is at rest or underway. With multihull craft, the length of each hull shall be measured individually. The length of the hull L_H shall be taken as the longest of the individual measurements. See Figure L-3a for monohull measurements and Figure L-3b for multihull measurements.

Length of the member 构件长度 系指构件(包括端部肘板)的总长度。 Length of the member means the overall length of the member(including the end brackets).

Length of the tug ship L 拖船船长 系指垂线间纵向距离。船长应取为不小于夏季载重水线总长的96%，但不必大于其97%。对于与推船刚性连接的驳船,船长应为驳船首垂线和推船尾垂线之间的距离。
Length of the ship defined as longitudinal distance in m, between perpendiculars. L is not to be taken less than 96%, and need not to be taken greater than 97%, of the extreme length on the summer load waterline. For barge rigidly connected to a push-tug L is to be measured between FP of barge and AF of push-tug.

Length of the unit $L(m)$ 平台长度 $L(m)$ 系指:(1)自升式平台——L 为在 $0.85D$ 处,沿平台中纵剖面上艏艉壳板内缘之间的水平距离,但不考虑井口槽的影响;(2)柱稳式平台和坐底式平台——L 为平台在中纵剖面的上最大投影水平尺度;(3)水面式平台——L 为在 $0.85D$ 处水线总长的96%,或沿该水线由艏柱前缘量到舵杆中心线的长度中的较大者,具有倾斜龙骨的平台,其计量长度的水线应和设计水线平行。 Length of the unit $L(m)$ are:(1) Self-elevating units—L is the horizontal distance measured on the centerline of the unit between the inside surfaces of fore and aft shell plating at $0.85D$, without taking the well slot into account;(2) Column-stabilized units and submersible units—L is the extreme horizontal length of the unit on a projection of the centerline;(3) Surface-type units—L means 96% of the total length on a waterline at 85% of D, or the length from the foreside of the stem to the axis of the rudder stock on that waterline, if that be greater. In units designed with a rake of keel, the waterline on which this length is measured is to be parallel to the designed waterline.

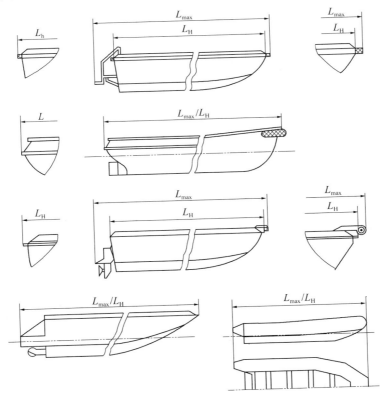

(a) 单体机动艇 L_{max} 和 L_H 的确定;

图 L-3 单体机动艇 L_{max} 和 L_H 的确定

Figure L-3 Determination of L_{max} and L_H, for monohull power boats

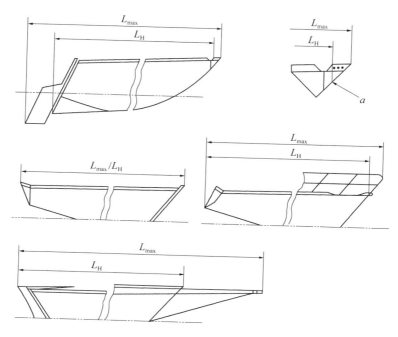

(b) 单体帆艇 L_{max} 和 L_H 的确定

图 L-3　单体艇 L_{max} 和 L_H 的确定
Figure L-3　Determination of L_{max} and L_H, for monohull sailing boats

Length of yacht L(m) 艇长　对游艇而言，系指在满载排水量状态下，静浮于水面时，其刚性水密艇体位于水线以下部分的艇体长度，但不包括水线处及以下的附体。

Length over keel blocks 布墩长　最后中墩与最前中墩的中心线之间的距离。

Length overall L_{oa} 总长 L_{oa}　平行于满载水线的自艏柱前缘至艉部或艉部封板后缘的长度，不包括护舷材和其他突出物。 Length overall L_{oa} is the distance, in m, measured parallel to the deep load waterline from the fore side of the stem to the after side of the stern or transom, excluding rubbing strakes and other projections.

Length overall submerged 浸体长　设计水线下，船体水下部分型表面前后端之间的水平距离。当艏部或艉部无水下伸出部分时则同设计水线长。

Length waterline 设计水线长 L_{WL}　沿设计水线由艏柱前缘量至船尾后缘或艉封板的长度。 Length waterline L_{WL} is the distance, in m, measured on a waterline at the design draught from the fore side of the stem to the after side of the stern or transom.

Length(fishing vessel) 船长(渔船)　应取由龙骨线起量至最小型深85%处水线总长的96%；或者是该水线上从艏柱前缘至舵杆中心线之间的长度，取大者。船舶设计为倾斜龙骨时，其计量长度的水线应与设计水线平行。 Length shall be taken as 96% of the total length on a waterline at 85% of the least moulded depth measured from the keel line, or as the length from the foreside of the stem to the axis of the rudder stock on that waterline, if that be greater. In vessels designed with rake of keel the waterline on which this length is measured shall be parallel to the designed waterline.

Lens-shaped section 梭形切面　叶面与叶背以及前后部分全对称的切面。

Letter meassage(LM) 书信电报

Letter of commitment 承诺信

Letter of credit(L/C) 信用证　系指开证行应申请人的要求并按其指示向第三方开立的载有一定金额的、在一定期限内的凭符合规定的单据付款的书面保证文件。

Letter of indemnity 赔偿保证书 系指由托运人向承运人出具的保函。

Letter of intent 意向书 其内容包括并购意向、非正式报价、保密义务和排他性等条款。

Level indicator 液位指示器

Level indicator system 液位指示系统

Level measuring system 液位测量系统

Level of form stability 形状稳性臂 船舶沿正浮状态的浮心至倾斜状态时浮力作用线的距离。

Level of stability by weights 重力稳性臂 船舶沿正浮状态的浮心至倾斜状态时重力作用线的距离。

Level of statical stability 静稳性臂 静稳性曲线上的纵坐标,其物理意义同复原力臂。

Level of vacuum 真空度

Level of wind pressure 风压倾覆力臂 风压倾覆力矩与船舶排水量的比值。

Level trim 水平纵倾

Level I vulnerability criteria for parametric roll 参数横摇第一层薄弱性标准

Level I vulnerability criteria for pure loss for stability 纯稳性丧失第一层薄弱性标准

Level I vulnerability criteria for surf-riding and broaching 骑浪和横甩第一层薄弱性标准

Level II vulnerability criteria for parametric roll 参数横摇第二层薄弱性标准

Level II vulnerability criteria for pure loss for stability 纯稳性丧失第二层薄弱性标准

Level II vulnerability criteria for surf-riding and broaching 骑浪和横甩第二层薄弱性标准

Leveling pipe 水平管

Lever chain stopper 闸刀止链器 在底座上装一活动的横向闩条,以止住锚链的止链器。

Liability for breach of contract 违约责任 也称为违反合同的民事责任,系指合同当事人因不履行合同义务或者履行合同义务不符合约定,而向对方承担的民事责任。违约责任与合同债务有密切联系。

Liability exemption 免责

Liaison offices 办事处 是仲裁委员会的宣传、咨询和联络机构,在仲裁委员会的统一领导下,从事海事仲裁的宣传、调研和咨询工作,协助仲裁委员会或其分会在当地安排开庭,但不从事仲裁案件的受理、收费和审理。

Library 图书馆(豪华邮轮)

License 许可证 是在银行服务基础上推出的一项业务模式,通过提供涵盖多品牌、多系列 IT 基础架构产品的 MA 服务为客户 IT 系统提供保障,其优势是突出的。

License agent 专利代理人 系指获得了专利代理人资格证书,持有专利代理人执业证并在专利代理机构专职从事专利代理工作的人员。专利代理人代码标准涉及的相关术语由中国国家知识产权局赋予规范注释并公之于众,中国国家知识产权局拥有对专利术语的唯一和最终解释权。专利代理人代码要始终坚持唯一性,确定一个代码只对应一个专利代理人的规则,每一个代码不随其拥有者的法律意义的消失而被他人使用。

License agreement 许可证协议

Licensee 引进方/受让方

Licensing 许可证贸易 系指跨国公司在获得一定收益的情况下,授权东道国企业使用某种工业产权或技术。

Licensing State 签发许可证的国家 系指经营核动力船舶或授权经营悬挂其国旗的核动力船舶的核动力船舶经营人责任公约缔约国。 Licensing State means the Contracting State of the convention on the liability of operators of nuclear ships which operates or which has authorized the operation of a nuclear ship under its flag.

Licensor 许可方

Life belt 救生带 可缠绕于腋下,在水中具有一定浮力以支持落水人员用的充气带子。

Life bench 救生凳 带有浮箱、救生索,在应急时可兼作救生浮具用的木质凳椅。

Life boat 救生艇 系指设置在船上的用于救生的小艇。

Life buoy 救生圈 供落水人员套于腋下,能在水中提供浮力的环状体。

Life float 救生浮 四周为浮体,内沿设绳网,下挂格栅,遇难人员可站浮其中的救生浮具。

Life jackers for children(lifejacket for children) 儿童救生衣 按儿童身材特制的救生衣。

Life line 放艇安全索 为保证放艇人员安全和降登已放落水面的救生艇用的,悬吊在横张索上,能放伸到水面的绳索。

Life line 救生绳(舷沿救生索) (1)用以在火场营救遇险人员的绳索;(2)沿救生艇、筏、浮具周缘布设,有的还带有浮子,可供落水人员把握和攀登用的索具。

Life quality index(LQI) 生活质量指数 系指在工作状态下,社会、健康、环境和经济等层面的生活质量指数。LQI 可以用来概括影响人们生活的重要问题,这些问题有助于公开讨论如何提高社会的生活质量。

Life support system 生命支持系统 又称环境控制系统。主要是控制甲板居住舱环境内的压力、温度、相对湿度、氧分压、二氧化碳浓度等条件,提供居住环境必需的生活条件的,用以保证潜水员正常生活和生命安全,在发生事故时能提供应急措施而使潜水员安全脱险的一个配套系统。其由配气供气系统、压力监控系统、温度和湿度监控系统、氧分压监控系统、CO_2 及有害气体成分监测装置、气体回收系统、供气及排污系统、电视监控系统、通信和信号呼叫系统、医学监护系统气体、循环净化装置等组成。

Lifeboat release and retrieval system　救生艇释放和回收系统　系指将救生艇与艇索连接并将其从艇索释放的措施,用于救生艇的降放、降落和回收。其包括艇钩装置和操作装置。Lifeboat release and retrieval system is the means by which the lifeboat is connected to, and released from, the lifeboat falls for lowering, launch and retrieval. It comprises the hook assembly and operating mechanism.

Life-combination suit　救生服　供船员穿着,具有浮力,有些还能保暖,以便在水面扶正气胀救生筏和协助遇难人员登乘的防水连衣裤。

Lifejacket　救生衣　系指能提供浮力以防止人员溺毙的衣服。救生衣还能支持一个无知觉的人员处于口鼻离水的位置,但救生衣实际上不具备保温能力。Lifejacket means a clothe provided buoyancy and prevent a person from drowning. They also support an unconscious person in a poison where the mouth and nose are clear of the water. However, lifejackets virtually offer no thermal protection.

Life-jacket locker　救生衣柜　标有明显标志,用以存放救生衣的柜子。

Life-jacket rack　救生衣架　存放救生衣的专用架子。

Lifeline pistol(lifeline- throwing gun)　抛绳枪　供发射抛射绳用的枪状器具。

Liferaft　救生筏　系指符合国际海上人命安全公约第Ⅲ章第15条和第16条规定的救生筏。Liferaft means a liferaft complying with either regulation 15 or regulation 16 of chapter Ⅲ of the SOLAS Convention.

Life-salvage ship　救生打捞船　系指装有打捞救生设备、从事打捞救生作业的船舶。

Life-saving appliances arrangement　救生设备布置图　明确地标出各种救生设备安放位置的图纸。

Life-Saving Appliances Code(LSA Code)　救生设备(LSA)规则　系指 IMO 组织海上安全委员会 MSC.48(66)决议通过并经修正的"国际救生设备规则"。LSA Code means the International Life-Saving Appliance Code, adopted by the Maritime Safety Committee of the Or-ganization by resolution MSC.48(66), as amended.

Lifesaving equipment　救生设备　船上人员在水域救助落水人员或本船遇难时供人员自救而设置在船上的专门设备及其附属件的统称。包括救生载具、救生浮具、辅助救生用具及其存放、收卸等辅助设备和救生属具等。

Life-throwing appliances　抛绳器具　配置在船上,主要供海上施救时将绳索抛射一定距离以达到确定目标的器具。

Lift　电梯　分为货梯(cargo lift)和车辆电梯(car elevator)两种。

Lift　升力　一般系指物体在流体中由于运动所产生的铅垂于运动方向的合力。

Lift boat　救生艇　设置在船上,主要用作船舶遇难时,供船上人员搭乘自救的小艇。

Lift boat engine　救生艇发动机

Lift by the stern, lifting　艉浮　船舶纵向下水过程中,浮力对前支架的力矩大于重力对前支架的力矩的瞬间所发生的艉部上浮的现象。

Lift devices　垫升装置　系指直接提高空气压力并主要是为气垫船提供垫升力的机器设备。Lift devices are those items of machinery which directly raise the pressure of the air and move it for the primary purpose of providing lifting force for an air-cushion vehicle.

Lift fan(hover fan)　垫升风机　气垫船中用以产生气垫和升力的风机。

Lift net　扳罾网

Lift on/lift off system　吊装装卸　一种将货物吊进和吊离船舶的装卸方式。

Lift power　垫升功率　气垫船垫升时,供给垫升系统的功率。

Lift power limitations　垫升动力限制　系指除排水状态以外的其他营运状态下对提供升力的机械和设备所做的各种限制。Lift power limitations are those limitations imposed upon machinery and components which provide lift in operational modes other than the displacement mode.

Lift theorem(Kutta-Joukowski theorem)　升力定律　阐述翼型体的升力与流体环量间关系的定理。$dL = \rho V \Gamma dx$,式中,dL—小段翼所受升力;ρ—流体质量密度;Γ—环量;dx—小段翼的翼展。

Lift-drag ratio　升阻比　翼形体升力对阻力的比值。

Lifting accommodation platform　起重生活平台　系指主要用于油气田日常生产维护作业并为施工人员提供生活起居条件的平台。由于对平稳和舒适性要求较高,起重生活平台主要采用半潜式结构。

Lifting appliances 升降设备 系指装卸设备和装卸坡道及其驱动和装卸配套安装设备。Lifting appliances are cargo gears and cargo ramps include their installations of driving systems and cargo fittings.

Lifting beam 吊梁(提升梁) （1）装卸重大件货物时系在货物上方,便于吊装的金属梁；(2)提升平台的,其截面呈矩形、环形或三角形等的梁。

Lifting capacity of floating dock 浮船坞举力 浮船坞处于正常工作状态时,所能抬起的最大船舶的排水量。

Lifting cargo deck 举升甲板 系指承载货物并在装卸货物作业过程中被水淹没的开敞露天甲板。Lifting cargo deck means the open weather deck which is used for the carriage of goods and is submerged during loading and unloading cargoes.

Lifting celler platform 井口升降甲板 井口处用于安装液压防喷器的升降平台。

Lifting eye bolt(rudder eye) 舵杆吊环 设置在舵杆头部端面上,供拆装舵时吊舵用的环首螺钉。

Lifting lime theory 升力线理论 用一束强度沿半径变化的附涡代替螺旋桨叶,产生的自由涡逸入尾流中的一种螺旋桨环流理论。

Lifting link(lifting controlling link) 起升联动杆 使移动舱盖的若干滚轮或嵌在舱口围板水平材缺口中的托板一起抬升的联合操纵杆。

Lifting load 提升负荷/起升载荷 系指安全工作载荷的总和。安全工作载荷是被吊起货物的最大自身质量加上一些附属设备的质量,例如吊钩、吊货滑车、起重钩、铲斗、起重横梁、扩张器等。除非被船级社认定为必须考虑的,除了当设备被提升的高度大于50m,否则用来下降货物的钢缆质量是不需要考虑的。 Lifting load is the sum of the safe working load defined as the maximum mass of cargoes themselves to be suspended and the mass of accessories such as hooks, cargo blocks, grabs, buckets, lifting beams, spreaders, etc. Unless otherwise deemed necessary by the Society, the mass of wire ropes used as cargo falls need not be taken into account except when the installation is designed for a lift of 50m or more.

Lifting loop(personal flotation device) 提升环(个人漂浮装置) 为便于手动打捞水中人员的装置。Lifting loop(personal flotation device) means the device which facilitates manual recovery of a person from water.

Lifting rudder 可升降舵 安装于船尾,能上下移动的舵。

Lifting surface theory 升力面理论 用分布于叶面或其投射平面上的附涡系统代替螺旋桨叶,产生的自由涡逸入尾流中的一种螺旋桨环流理论。

Lifting vortex 升力涡 就所起作用而言,对物体产生升力的涡线部分。

Lifting/pipe-laying ship 起重铺管船 系指主要承担海上大型结构件水面吊装的施工作业和海底管道与立管以及脐带缆的铺设安装的工程船舶或平台。主要用于平台上部组块的吊装、张力腿和立管安装、Spar平台安装、海底管道及悬链式立管以及脐带缆的铺设安装等作业。其结构形式有半潜式和船形结构两种形式。

Lift-off covers 拼装舱盖 由若干块舱板拼合盖覆,外加防水盖布和封舱压条、闩条、楔等组成的舱口盖的统称。

Lift-on/lift-off system 吊装装卸 一种将货物吊进吊离船舶的装卸方式。

Light boat 灯标船 装有发光灯等信号设备,停泊在航道重要地段,如航道转向处、宽广的河道口、大桥桥洞进口等处,作为航标使用的船舶或从事航标作业的船舶。

Light buoy crane 航标起重机 安装在航标船上,供起吊航标用的起重机。

Light buoy room 航标舱 布标船上供储存航标用的舱。

Light dead weight 轻吨

Light derrick boom 轻型吊杆 单杆操作时安全工作负荷等于和小于98 kN的吊杆装置和吊杆式起重机。

Light detection and ranging(LIDAR) 光达技术 系指用激光测量光速。通过粉尘和悬浮微粒来测量激光束的反向散射。该方法测得的范围为几百米之内。

Light displacement(floating dock) （浮船坞）轻载排水量 系指包括浮箱、坞墙、垫墩、起重机、所有机械和舾装设备、以及端封板/工作平台、浮船坞使用的淡水和燃油、调节压载水(如要求)和残余水等的质量。

Light displacement(light weight) 空载排水量 系指军用舰船装备齐全但无载重时的排水量。其中包括固定压载、备件、管系中的液体,以及液舱中不能吸出的剩余液体在内。但不包括人员、淡水、粮食、燃料、滑油、弹药、供应品、给水以及锅炉、冷凝器中的水。

Light displacement of floating dock 空坞排水量 装有全部机械起重机和设备,但不包括浮船坞用的淡水、燃油、调整压载水和剩余水的坞体排水量。

Light displacement/light weight 空船排水量 是船舶本身加上船员和必要的给养物品三者质量的总和,是船舶最小限度的质量轻载排水量,又称轻排水量、空船排水量。对于运输船舶来说是船舶没有装货物、旅客、燃料、淡水和供应品等时的排水量。

Light distribution curve 配光曲线 系指把光强用矢量表示,将各矢量的端点连成曲线,用以表示光强分布的状态(同一灯具使用同一种光源,其配光曲线与灯具功率大小无关)。

Light draft 浅吃水/空载吃水 系指船舶处于空载状态时的吃水。空载状态在实际营运中很少出现,主要用于设计计算。空载水线与满载水线之间的外板干湿交替,易受波浪冲蚀,应涂刷特种的涂料。

Light draft of floating dock 空坞吃水 浮船坞在无油水及压载水和剩余水的空载状况下,由基线至水面的垂直距离。

Light dues 灯塔费

Light emitting diode(LED) 发光二极管 是一种半导体固体发光器件,它是利用半导体固体芯片作为发光材料,光两端加上正向电压,半导体中的载流子发生复合引起光子发射而产生光。发光二极管(LED)可以直接发出红、黄、蓝、绿、青、橙、紫和白色的光。它由镓(Ga)与砷(As)、磷(P)、氮(N)、铟(In)的化合物制成的二极管,当电子与空穴复合时能辐射出可见光,因而可以用来制成发光二极管。在电路及仪器中作为指示灯,或者组成文字或数字显示。磷砷化镓二极管发红光,磷化镓二极管发绿光,碳化硅二极管发黄光,钢镓氮二极管发蓝光。当给发光二极管加上正向电压后,从P区注入N区的空穴和由N区注入P区的电子,在PN结附近数微米内分别与N区的电子和P区的空穴复合,产生自发辐射的荧光。因化学性质又分有机发光二极管OLED和无机发光二极管LED。

Light flux 光通量 系指单位时间内向空间辐射并引起人眼视觉的能量。单位为流明(lm)。

Light grain 轻谷物 系指比重较轻的谷物。

Light intensity 光强 系指光源在某方向单位立体角内辐射的光通量。单位为坎德拉(cd)。

Light intervention 轻型修井作业 系指采用钢丝绳、电缆或连续油管(Coiled tube)作业的方式在采油树和完井管柱中完成的修井作业。由于不抽出油管,因此无须安装隔水管,被称为无立管修井,可采用配备了无立管修井系统的轻型修井船完成。

Light line speed (no-load speed) 空钩绳索 绞车卷筒在用单排缆绳起吊空钩时,能保持的最大平均绳速。

Light oil 轻油

Light running magin 轻运行裕度 船舶设计时为达到相同航速而储备的转速裕度(约5%~7%)。

Light sector 光弧 号灯光源的光束在水平面上或垂直面向外散射所及的角度。

Light service draught d_1 轻载航行吃水 系指相应于最轻预计装载量和相关液舱容量的航行吃水,但应计入稳性和/或浸水可能需要的压载。客船应足额计入乘客和船员。Light service draught is the service draught corresponding to the lightest anticipated loading and associated tankage, including, however, such ballast as may be necessary for stability and/or immersion. Passenger ships should include the full complement of passengers and crew on board.

Light side scuttle 轻型舷窗 具有一般抗风浪性能,仅允许装设于距载重线较高的规定范围内的舷窗。

Light source 光源 系指产生可见光的辐射体。电气照明光源基本有以下几类:(1)热辐射光源,如白炽灯;(2)气体放电光源,如荧光灯、高压汞氙灯、高压钠灯等;(3)绿色光源,如LED发光体。

Light telegraph gong 灯光传令钟 系指利用指示灯系统传送原理的传令钟。灯光传令钟由发送器和接收器组成。发送器是一转换开关,用来接通信号指示灯电源;接收器是指装在彼此分隔、互不透光的壳体内的指示灯组,用于指示发送器送来的信号。

Light trap for plankton 浮游生物光诱捕装置 利用光亮诱捕浮游生物的装置。它由浮于水面的浮生物网架和罩在其内的防水光源组成。

Light weight (floating dock) 浮船坞的空坞质量 系指包括浮箱、坞墙、垫墩、起重机、所有的机械和舾装设备以及端封板/工作平台等的质量。

Light weight LW 空船排水量 LW 系指不包括货物、燃油、滑油、舱内压载水和淡水、储存物、乘务员及其携带物的船舶的排水量(t)。Light weight LW is the displacement in tons excluding cargoes, fuel oil, lubricating oil, ballast and fresh water in tanks, stored goods, crew and their properties.

Lightening hole 减轻孔 系指构件上用以减轻其质量的开孔。Lightening hole means a hole cut in a structural member to reduce its weight.

Lighter 点火装置 利用电弧、电火花或电热等方法点火的装置。

Lighterage 驳船费 系指从海港到内支线或者小港口的支线船费用。

Lighter/barge 驳船 简称"驳"。本身无动力装置,依靠拖船或拖船带动的平底船。上层建筑很简单,一般不设起货设备。按船型分有普通驳、推驳、分节驳等。按结构形式分有甲板驳、舱口驳、敞舱驳和半舱驳等。按用途分有客驳、货驳、液驳、油驳、石驳、集装箱驳等。带有动力装置的称为机动驳。驳船型肥宽,线型平直、吃水浅,造价低廉,特别适宜在内河浅海区域短途运送客货,有的也可用于港口和水上工程作业。随着海上运送的发展,出现了载驳货船、顶推驳运船队等运输方法,为驳船的应用开创了新的前景。见图 L-4。

图 L-4 驳船
Figure L-4 Lighter/barge

Lightering 过驳 一般系指大船停靠码头、浮筒、装卸平台,或大船在锚地用驳船或其他小船装卸货物。

Lightering operation 驳运作业 系指将液体货物从一艘船输送至另一艘辅助船上,或将固体货物/集装箱过驳到另一艘辅助船上。

Lightest sea going condition 最轻载航行状态 系指船舶处于平浮、无货物,剩有 10% 的备品和燃料的装载状态。 Lightest sea going condition is the loading condition with the ship on even keel, without cargo with 10% stores and fuel remaining and in the case of a passenger ship with the full number of passengers and crew and their luggage.

Lighting boat 灯船 系指作为航标使用,有发光设备的专用船舶。锚泊于离岸较远、水深浪大、难以建造灯塔的地点,如港口入口,用以引导航行或标示危险物或浅滩所在。多数灯船船身和灯架涂成红色,甲板上其他建筑涂白色,两舷标有白色的灯船名或锚泊地名。发光设备装在甲板高处;灯具有的安置在平衡架上,使灯光不受船体摆动影响,以接近于水平方向发射。灯光射程可达 10 余海里。见图 L-5。

图 L-5 灯船
Figure L-5 Lighting boat

Lighting branch circuit 照明分支电路 系指仅供电给照明引出线端的电路。照明分支电路也可给可移动式台扇或挂扇、小型加热器、1/4 马力(196.5W)和以下的电动机以及其他每个不超过 600W 的可移动式电器供电。

Lightship condition 空船状态(空船质量) 系指船舶没有装载船用消耗备品、物料、货物、船员和行李,以及除位于工作水平的机械和管系中的液体,除润滑剂和液压油以外,没有装载任何液体的状态。 Lightship condition is a ship complete in all respects, but without consumables, stores, cargo, crew and effects, and without any liquids on board except that machinery and piping fluids, such as lubricants and hydraulics, are at operating levels.

Lightship weight 空船质量 系指钢材质量加上舾装、机器和永久性设备的质量。 Lightship weight is steel weight and outfit, machinery and permanent equipment.

Light-telegraph 灯光传令钟 系指利用指示灯系统传送原理的传令钟。灯光传令钟由发送器和接收器组成。发送器是一转换开关,用来接通信号指示灯电源;接收器是一装在彼此分隔、互不透光的壳体内的指示灯组,用于指示发送器送来的信号。

Lightweight 空船排水量 系指:(1)船舶在没有货物、舱柜内无燃油、润滑油、压载水、淡水、锅炉给水,消耗物料,且无乘客、船员及其行李物品时的排水量,以 t 计。(2)不计货物、燃油、滑油、压载水、淡水和(锅炉)给水,消耗品,旅客和船员及行李的排水量,但包括管路中的液体。 Lightweight is:(1)the displacement of a ship in t without cargo, fuel, lubricating oil, ballast water, fresh water and feedwater in tanks, consumable stores, and passengers and crew and their effects. (2)the displacement, in t, without cargo, fuel, lubricating oil, ballast water, fresh water and feed water, consumable stores and passengers and crew and their effects, but including liquids in piping.

Lightweight 空船排水量(平台) 系指整个平台连同安装的机械、设备和舾装,包括固定压载、备件以及机械和管路中至正常工作水平面的各种液体的质量,但不包括贮存在船舱内的油、水、消耗品或可变载荷、贮存物品、船员和行李质量。 Lightweight is defined as the weight of the complete unit with all its permanently installed machinery, equipment and outfit, including permanent ballast, spare parts normally retained on board and liquids in machinery and piping to their normal working levels, but does not include liquids in storage or reserve supply tanks, items of consumable or variable loads, stores or crew and their effects.

Lightweight anchor 大抓力锚 在锚尖上装设有与锚爪处于同一平面的突出杆体,其锚爪可转动的锚。

Lightweight check 空船质量测定 系指对船上应予增、减或在倾斜试验时搬动位置的所有项目进行审核,以便该船舶能从观测状态调整至空船状态。每一项目的质量和纵向、横向、垂向位置应予准确记录下来。使用这些资料以及通过测量干舷或核对船舶吃水标志、船舶静力学数据和海水密度而确定倾斜试验时船舶的静浮水线,从而可得出空船排水量和重心纵向位置(LCG)。 Lightweight check is a procedure which involves auditing all items which are to be added, deducted or relocated on the ship at the time of the inclining test so that the observed condition of the ship can be adjusted to the lightship condition. The weight and longitudinal, transverse and vertical location of each item are to be accurately determined and recorded. The lightship displacement and longitudinal centre of gravity(LCG) can be obtained using this information, as well as the static waterline of the ship at the time of the inclining test as determined by measuring the freeboard or verified draught marks of the ship, the ship's hydrostatic data and the sea water density.

Lignum vitae bearing 铁梨木轴承 以铁梨木作为轴衬材料的轴承。

Lignum vital 铁梨木 又称铁力木。木质坚硬耐久,富含油脂的常绿大乔木。藤黄科。多产于中美洲及南美洲北岸一带、东南亚、南亚和我国云南、广西两省也有出产。是优良的建筑和家具用料,种子可榨油供工业用。用作海水润滑的船舶尾轴轴承材料已有100多年历史。其比重为 1.23~1.31,顺纹压缩强度 88×10^6 – 102.9×10^6 帕(900~1050 千克力/厘米^2),水润滑摩擦系数 0.08~0.09,允许耐压负重 17.6×10^4 帕(1.8 千克力/厘米^2)。由于系天然产,资源不多,作为船舶尾轴轴承材料现正试以橡胶、层压板、纤维缠绕增强塑料、布质层压板、铸型尼龙等替代。

Limber hole 流水孔 系指肋骨或板材上的小型排水孔,用以防止水或油积聚。 Limber hole means a small drain hole cut in a frame or plate to prevent water or oil from collecting.

Lime(unslaked) 石灰(生) 呈白色或灰白色的物质。生石灰与水结合形成氢氧化钠(熟石灰)。这一反应产生巨大的热量,足以引燃附近的可燃物质。对眼睛和黏膜有腐蚀性。该货物为不燃物或失火风险低。 Lime(unslaked) White or greyish-white material. Unslaked lime combines with water to form calcium hydroxide(hydrated lime)or magnesium hydroxide. This reaction develops a great deal of heat which may be sufficient to cause ignition of nearby combustible materials. This is not combustible or has a low fire-risk corrosive to eyes and mucous membranes. This cargo is non-combustible or has a low fire-risk.

Limestone 石灰石 颜色各异。石灰石呈乳黄、白、中度深灰(刚破碎时)等各种颜色。含水量最大为4%。无特殊危害。该货物为不燃物或失火风险低。 Limestone varies in colour from cream through white to medium dark grey(when freshly broken). Moisture: up to 4%. No special hazards. This cargo is non-combustible or has a low fire-risk.

Limit alarms 极限报警 位置偏移和艏向偏移的可选值。达到这点时,激活报警。

Limit load 极限载荷 系指在结构响应中最大的载荷,其通常指示破坏现象的出现。 Limit load means a load maximum in a structural response that usually indicates the onset of collapse.

Limit of elasticity 弹性限度

Limit of side projection of blade 叶侧的投影界限 螺旋桨绕轴旋转一周在若干角位置时在侧视图上各叶片侧投影的包络线。

Limit state 极限状态 系指超过它,结构不能再使用的状态。 Limit state is a state beyond which a structure is no longer operable.

Limitation liability for maritime claims 海事赔偿责任限制制度 是发生重大海损事故时,对事故负有责任的船舶所有人、救助人或其他人对海事赔偿请求人的赔偿请求依法申请限制在一定额度内的法律制度。这是海商法中特有的赔偿制度。

Limitation of actions 诉讼时效 系指民事权利受到侵害的权利人在法定的时效期间内不行使权利,当时效期间届满时,人民法院对权利人的权利不再进行保护的制度。在法律规定的诉讼时效期间内,权利人提出请求的,人民法院就强制义务人履行所承担的义务。而在法定的诉讼时效期间届满之后,权利人行使请求权的,人民法院就不再予以保护。值得注意的是,诉讼时效届满后,义务人虽可拒绝履行其义务,权利人请求权的行使仅发生障碍,权利本身及请求权并不消灭。当事人超过诉讼时效后起诉的,人民法院应当受理。受理后,如另一方当事人提出诉讼时效抗辩且查明无中止、中断、延长事由的,判决驳回其诉讼请求。如果另一方当事人未提出诉讼时效抗辩,则视为其自动放弃该权利,法院不得依照职权主动适用诉讼时效,应当受理支持其诉讼请求。

Limitation of arbitration 仲裁时效 系指权利人向仲裁机构请求保护其权利的法定期限,也即权利人在法定期限内没有行使权力,即丧失提请仲裁以保护其权益的权利。仲裁分为商事仲裁和劳动仲裁两个大类。《中华人民共和国仲裁法》第 74 条规定:法律对仲裁时效有规定的,适用该规定。法律对仲裁时效没有规定的,适用诉讼时效的规定。

Limitations in the use of oils as fuel 燃油的使用限制 燃油的使用应受到下列限制:(1)除另有许可外,不得使用闪点低于 60 ℃ 的燃油。(2)应急发电机可使用闪点不低于 43 ℃ 的燃油。(3)如符合下列条件,可以使用闪点低于 60 ℃ 但不低于 43 ℃ 的燃油;①除布置在双层底舱内的燃油柜外,其他燃油柜应位于 A 类机器处所外;②在燃油泵的吸管上设有油温测量装置;③燃油滤净器的进口侧和出口侧均设有截止阀和/或旋塞;④尽可能使用焊接结构的或圆锥型的或球型的管接头。(4)在货船上,可准许使用闪点低于 60 ℃ 的燃油。例如原油,但此种燃油不得储存在任何机器处所内,且整套装置应经主管机关认可。 Limitations in the use of oils as fuel mean following limitations shall apply to the use of oil as fuel:(1) except as otherwise permitted, no oil fuel with a flashpoint of less than 60 ℃ shall be used; (2) in emergency generators oil fuel with a flashpoint of not less than 43 ℃ may be used; (3) the use of oil fuel having a flashpoint of less than 60 ℃ but not less than 43 ℃ may be permitted (e. g. , for feeding the emergency fire pump's engines and the auxiliary machines which are not located in the machinery spaces of category A) subject to the following: ① fuel oil tanks except those arranged in double bottom compartments shall be located outside of machinery spaces of category A; ② provisions for the measurement of oil temperature are provided on the suction pipe of the oil fuel pump; ③ stop valves and/or cocks are provided on the inlet side and outlet side of the oil fuel strainers; and ④ pipe joints of welded construction or of circular cone type or spherical type union joint are applied as much as possible; and (4) in cargo ships the use of fuel having a lower flashpoint than 60 ℃, for example crude oil, may be permitted provided that such fuel is not stored in any machinery space and subject to the approval by the Administration of the complete installation.

Limited obstacle sector (LOS) 障碍限制区(LOS) 系指以向外延伸的扇形区,由 360°圆弧中除无障碍区以外的弧段形成,其中心为确定障碍区的参照点。在障碍限制区内的障碍物受到规定高度的限制。 Limited obstacle sector (LOS) is a sector extending outward which is formed by that portion of the 360° arc, excluding the obstacle-free sector, the centre of which is the reference point from which the obstacle-free sector is determined. Obstacles within the limited obstacle sector are limited to specified heights.

Limiter 限位器(限制器)

Limiting angle of turning 极限转向角 系指非传统的船舶推进和转向系统(诸如全方位推进器或喷水推进系统,但不仅限于此)用最大转向角来表示的一个操纵极限值,或根据制造厂商安全操作指南给出的等效值,且都应顾及船舶航速或螺旋桨扭矩/转速或其他限制,对于每艘船舶特定的非传统转向装置,"极限转向角"应由方向控制系统的制造厂商予以明示。

Line winder 盘线机 配合延绳钓起线机以盘绕整理钩线入笼的机械。

Limits for dimensions and draft of ship 船舶的尺度限制和吃水限制(巴拿马运河) (1)最大总长——①对商船或非商用船,准予经常通过的最大总长(包括球鼻艏在内)为 289.6 m(950 ft);但对客船和集装箱船则总长可放宽至 294.13 m(965 ft)。总长超过 274.32 m(900 ft)的船舶,不论是新建造的或是新改装的,在其首次通过本运河前应按规定加以检查,并事先审批其图纸。凡事先未获审批和(或)不符合本运河规定的船舶,将可能拒绝其通过运河。②对拖-驳组合体,准予经常通过的最大总长(包括拖船在内)为 274.32 m(900 ft)。对拖-驳组合体必须作为一个单元通过,由拖船提供推进动力。③对非机动船舶,准予经常通过的最大组合长度(包括伴随的拖船)为 259.1 m(850 ft)。伴随的拖船必须与非机动船贯串连锁。超过上述限制者也可能在逐个审查基础上获准作单次性通过。但须符合有关的规定,且事先得到通航作业处经理的同意。(2)最大船宽——①对商船或非商用船和拖-驳组合体,准予经常通过的最大船宽(在船壳板的外侧面之间测得)为 32.31 m(106 ft)。②在事先经通航作业处经理或其指定人批准的情况下,更宽的、但宽度不超过 32.61 m(107 ft)的商船(包括拖-驳组合体)可允许其作单次性通过,但其船体在热带淡水中的浸没最深点不超过 11.3 m(37ft)。③对非机动船舶(拖-驳组合体除外),准予通过的最大船宽为 30.5 m(100 ft)。超过上述限制者也可能在逐个审查的基础上获准作单次性通过,但须符合有关的规定,且事先得到通航作业处经理的同意。(3)最大总宽——①最大总宽超过最大船宽的船舶,如其图纸未事先审查获准,将不得通过本运河。凡事先未获准和(或)不符合本运河要求的船舶,将可能拒绝其通过运河。②如船舶所载货物距其最大宽度仅为 2.5 cm(1in)或以下,则应采取认可的措施(如防擦条),以便船舶在船闸内靠在闸室壁边时能对货物加以保护。不得因采取这种防护方法而超出 32.31 m(106 ft)的最大宽度。制订本款是由于大量集装箱船在设计中已将集装箱的堆载达到了最大船宽。如果船舶以横倾状态靠在闸室壁上，或者闸室壁或闸门上设有突出的防护物，就可能发生损坏。(4)突出物——①突出在船舶船体以外的任何东西(主锚除外)将被视为突出物,对此将有适用的规定和限制。②运河管理局对突出物(不论是永久性的还是暂时的)所遭受的损坏不承担任何责任。③具有突出物的船舶也可以获准通过,只要通航作业处经理认为此类突出物不会影响船舶的安全通过或者不会危及运河的结构物。不论属于何种情况,在允准船舶通过之前,均要求船舶的船长签署一份文件解除运河管理局对因突出物之故而发生事故时的责任,并在运河蒙受损失时向运河管理局做出赔偿。④具有永久性突出物的船舶必须在其驶往运河之前提供有关突出物的详尽资料(包括图纸)并申请通过。事先提供资料将使延误通过运河或被拒通过的可能性减少至最低程度。有关此项详细事宜请与通航作业处经理联系。⑤具有突出物的船舶,如果已超过上述(2)①款和(2)②款中对于最大总长和最大总宽的限制,也可能在逐个审查的基础上允许其通过运河。条件是:事先获得通航作业处经理的同意,确认突出物不会危及或影响船闸结构、运河设备和(或)运河作业,而且该船的船长必须签署一份文件解除运河管理局的责任[见(2)⑧款]。(5)最大高度——任何船舶在通过运河或进入巴尔博亚港(Balboa)时,不论潮水的情况如何,允许的最大高度为 57.91 m(190 ft),这一高度系从水面量到船舶的最高点。如事先得到通航作业处经理的同意,在逐个审查的基础上,在低水位通过时(巴尔博亚 MLWS),允许最大高度可增加到 62.5 m(205 ft)。最大高度限制起因是由于潮水或不可预测的水位上升(因海浪、涌浪和风暴潮等),以及巴尔博亚桥下悬挂的维护设备。(6)吃水——①巴拿马运河的最大允许通航吃水规定为 TFW(热带淡水)12.04 m(39 ft 6 in),相应于夏通湖水位为 24.84 m(81 ft 6 in)或以上。[夏通湖水的密度在 29.4 ℃(85 °F)时为 0.995 4 g/cm³]。这就为船舶在运河本体中航行时对一些关键地段保证了至少有 1.52 m(5ft)的安全通航裕度,并且当密拉夫劳湖水位为 6.61 m(54 ft 6 in),在通过米盖尔船闸的南面闸槛时,可以有 0.50 m(1ft 8 in)的净空。通过吃水超过 10.82 m(35 ft 6 in)的船舶,在其首次通过运河之前,船东、经营或代理人必须在船舶装货前不迟于两周内向通航作业处经理提交巴拿马运河航行海事规则(巴拿马运河航行海事规则第 52 条)所要求的全部资料,并申请允准最大通过吃水(在热带淡水中浸没的最深点)。该申请在退走时应盖有核准的允准最大通过吃水的印记。(7)潜在的季节性吃水限制——①如逢异常的干旱(一般为 10 年一遇),有必要发布吃水限制。②在雨季(5 月～12 月)时,作为运河上游蓄水池和巴拿马城市供水源的夏通湖和马登湖都是满的。在旱季(12 月～5 月)时,马登湖的蓄水量被用来使夏通湖能保持住一个合适的水位。遇上异常的干旱季节,马登湖将会干涸,夏通湖的水位也不断下降。当夏通湖水位低于 24.84 m (81.5 ft)时,就有必要降低允准的最大吃水值以保证安全航行裕度。吃水降低值将根据运河管理局的水文学家和气象学家所做的湖水水位和降雨量计算机辅助预报,按 15.24 cm(6 in)递减,并尽可能提前 3 周做出通报。在新的吃水限制发布时已按原先吃水限制装载货物的船舶将出于重要的安全考虑而不准其通过。船舶在新吃水限制发布后装载时可允许以不大于 15.24 cm (6 in)的裕度超过限制值,但可能要求船舶做适当的纵倾或部分卸载,以获得安全的通过吃水。

Line 线(管路,航线)

Line a bearing 浇铸轴承

Line fishing boat(liner) 钓鱼船 系指使用钓具进行捕捞作业的渔船。

Line interactive UPS unit 联机交换型 UPS 装置 系作为脱机 UPS 的一种,当输入电源脱离事先设定的电压和频率界限值时,把支电路换成被储备的能源电源的情况。 Line interactive UPS unit means a system where the bypass line switch to stored energy power when the input power goes outside the preset voltage and frequency limits.

Line of position 位置线 利用无线电导航台发射的电磁波的几个特性,通过测量时间、相位、幅度、频率确定出基本的导航参数、角度、距离、距离差、距离和等,从而建立船舶与电台的相对位置关系。某种几何参数相等的点之轨迹称为位置线。最常见的位置线有直线、圆、双曲线、椭圆。

Line of shafting 轴系中线 系指整个轴系的中心线。

Line pipe 管系(管路,总管)

Line printer 行式打印机 是针式打印机的一种,可用来打印大批量的工单、表单等,使用成本较低。

Line shaft 轴承

Line shafting 轴系(轴线)

Linear acceleration components 线加速度分量 船舶运动时,其重心相对于周围静水的线加速度分别沿运动坐标系各轴的分量。

Linear cable laying machine 履带式布缆机 安装在布缆船甲板上由同步运转的履带夹持着电缆进行作业的布缆机。

Linear expansion 线膨胀

Linear expansion factor 线膨胀系数

Linear indication 线形缺陷显示 系指长度为厚度 3 倍以上的缺陷显示。 Linear indication means an indication in which the length is at least three times the width.

Linear velocity(LV) 线性速度(LV) 系指通过催化剂块的废气流量(m^3/h)与废气流垂直方向的催化剂块截面积(m^2)的比值。因此，LV 值的单位是 m/h。废气流体积系指在 0 ℃ 和 101.3kPa 定义的体积。Linear velocity(LV) means a value of the exhaust gas flow rate passing through the catalyst blocks(m^3/h) per catalyst block's section(m^2) in a noemal direction of exhaust gas flow. Therefore, unit of LV value is(m/h). The exhaust gas flow volume is the volume defined at 0 ℃ and 101.3kPa.

Linear velocity components 线速度分量 船舶运动时，其重心相对于周围静水的线速度分别沿运动坐标系各轴的分量。

Liner 衬条 在铆接结构的搭接处，为使连续密合而衬垫在相连构件间的窄板条。

Liner conference or conference 班轮公会或公会 系指两个或两个以上使用船舶的运输商的团体，这些运输商在特定的地理范围内，在某一条或数条航线上提供运送货物的国际班轮性质服务，并在一项不论何种性质的协定或安排的范围内，按照整齐划一的或共同的运费率及任何其他有关提供班轮服务的协议条件而经营业务。 Liner conference or conference means a group of two or more vessel-operating carriers which provided international liner services for the carriage of cargo on a particular route or routes within specified geographical limits and which has an agreement or arrangement, whatever its nature, within the framework of which they operate under uniform or common freight rates and any other agreed conditions with respect to the provision of liner services.

Liner extrusion 内衬层

Liner freight 班轮运费 系指班轮承运人根据运输合同完成货物运输后，从托运人取得的报酬。

Liner shipping 班轮运输 系指船舶按照公开的时间表航行于固定的航线，即以预先设定的挂靠港口顺序，按照固定的频率运输航线上各港口间的货物。是航运公司提供的一种服务，货船定期在预定的公告的装卸港之间营运。运费按照该公司运价表的费率计算。具有"四固定"的特点，即是固定航线、固定港口、固定船期和相对固定的费率。船舶按固定的航线和预先公布的船期表在固定港口间运送旅客和货物的运输。班轮运输适合于货流稳定、货种多、批量小的杂货运输。旅客运输一般采用班轮运输。通常的做法是船公司在所经营的班轮航线的各挂靠港口及货源腹地通过自己的营业机构或船舶代理人与货主建立业务关系；通过报纸、杂志刊登船期表；通过与货主、无船承运人或货运代理人等签订货物运输服务合同或揽货协议来争取货源。

Lines plan 型线图 在三个相互垂直的投影面上，以船体型表面的剖切线、投影线和外廓表示船体外形的图纸或图形。

Line-throqing appliance 抛绳装置 配置在船上，主要供海上施救时将绳索抛射一定距离以达到确定目标的器具。

Line-throwing appliances and distress signals 抛绳器 遇难船艇向前来救助的船发射救生绳用的工具。有两种形式：(1)火箭筒式(rocket type)；(2)手枪式(pistol type)。前者在发射时几乎无后坐力，但发射时后方不可站人；后者后坐力较大，在打出火箭弹的同时点燃其引信，利用火箭弹的反作用力将救生绳抛到救助船艇上。在无风情况下，抛绳器的最小射程应不小于 230 m。

Line-throwing rocket 抛绳火箭 带引抛射绳的火箭。

Line-throwing rocket gun 抛绳火箭筒 发射抛绳火箭的专用火箭筒。

Lining of bearing 轴承衬

Lining with corrosion resistant lining 装设防腐材料 系指授予货物处所采取防腐蚀材料的化学品船的附加标志。

Link connection 链接和(连杆连接)

Link drive 连接传动装置

Link gear 连杆机构(联动装置)

Linked 联结 系指在操作中，用控制绳、悬索或供应联系缆与母船(可为岸边、水下或水面的船舶)相联结。 Linked means connected, while in operation, to an attendant ship(which may be on shore, submerged or afloat) by a restraining line, suspension cable or umbilical cord.

Linted cotton seed with not more than 9% moisture and not more than 20.5% oil 带棉绒的棉籽(含水量不大于9%，油含量不大于20.5%) 经机器脱棉大约 90%~98% 后，附有短棉纤维的棉籽。可能自热并使服务处所缺氧。 Linted cotton seed is a cottonseed with short cotton fibres adhering to the kernel after approximately 90%~98% of the cotton has been removed by machine. May self-heat and deplete oxygen in cargo space.

Liquefied gas 液化气 由在 37.8 ℃ 温度时蒸发气压力超过 2.8bar(绝对值)的气体通过压缩和/或冷却而形成的液体。 Liquefied gas means a liquid formed by pressurization and/or a gas having a vapour pressure exceeding 2.8 bar absolute at a temperature of 37.8 ℃.

Liquefied gas cargo 液化气体货物

Liquefied gas carrier 液化气船 系指其构造适用于散装运输《国际散装运输液化气体船舶构造和设备规则》(IGC 规则)第 19 章所列液化气体及其他易燃货品

的液货船。其设有货物围栏系统,将《散装运输液化气体船舶构造与设备规范》中所列的液化气体通过加压或冷却后变成液体或其他易燃液货进行运输的船舶。对于1986年7月1日及以后建造并符合IGC规则的船舶,根据所采取的防漏保护措施分别加以下标志:(1)1G型(type 1):采用最严格防漏保护措施的货物;(2)2G型(type 2):采取中等的防漏保护措施的货物;2PG型(type 2PG)适用 L≤150 m,采用相当严格防漏保护措施的货物,且释放阀最大调定值至少为0.7 MPa,设计温度为 -55 ℃或以上;(3)3G型(type 3):采取中等的防漏保护措施的货物。其按用途分为:运载液化石油气的液化石油气运输船(LPG carrier)、运载液化天然气的液化天然气运输船(LNG carrier)、运载压缩天然气的压缩天然气运输船(CNG carrier)。按气体液化方式分为:(1)全压式(fully pressurized gas carrier)——按压力容器设计货舱,主要用来进行LPG的短途运输,货舱为球状、单圆筒状或多筒状,货舱的上部突出在上甲板之上;(2)半冷半压式(semi-refrigerated and semi-pressurized gas carrier)——用来运输多种或特定货物,如用来运输多种货物,其货舱的设计压力、温度范围很宽。例如设计温度为 -45 ℃~-104 ℃,设计压力为 3~8 kg/cm²;如用来运输某种特定货物,则货舱的设计温度和设计压力局限在该种货物的运输要求值;(3)全冷式(fully refrigerated gas carrier)——大型LPG船、LNG船和LEG船所采用的方式,在IGC规则中将液化气船的货舱分为:(1)一体式货舱构造(integral tank),其特点:货舱作为船体结构的一部分,与邻接的船体构造受力相互影响;设计蒸发气压力≤0.025 MPa,用于装载沸点≥-10 ℃的货物;船体构造与一般的化学品船类同。(2)薄膜式货舱构造(membrane tank),其特点:非自己支持型,热绝热层介于薄内膜和船体内壳之间、设计蒸发气压力≤0.025 MPa,薄内膜厚度<10 mm,其安全性要通过模型试验来确认。薄膜式货舱构造又分为 ①TECHNIGAZ薄膜式货舱系统(TECHNIGAZ membrane system)、②气体运输薄膜型货舱系统(gas-transport membrane system)。(3)半薄膜式货舱构造(semi-membrane system),其特点:在载重状态下为非自己支持型,热绝热层介于薄内膜和船体内壳之间、设计蒸发气压力≤0.025 MPa。(4)独立式货舱构造(independent tank)其特点:自己支持型,货舱与船体构造相对独立,可分为:①A型——货舱为平板重力式,船体结构按深舱设计,设计蒸发气压力≤0.07 MPa。②B型——要求用精密的手段对应力、疲劳寿命、裂纹进展进行解析,要做模型试验。③C型——完全按压力容器的规则进行设计,被全压式的LNG船采用。(5)内部绝缘式货舱构造(internal insulation type)其特点:为非自己支持型,热绝热层与货物直接接触,需要进行模型试验

和构造解析,设计蒸发气压力≤0.07 MPa。内部绝缘式货舱又可分为 ①Ⅰ型——热绝热层具有初次防护功能;②Ⅱ型——热绝热层具有初次、二次防护功能。见图L-6。

图 L-6 液化气船
Figure L-6 Liquefied gas carrier

Liquefied gas carrier type （1G/Type 2G/Type 2PG/Type 3G） 液化气体船(1G/2G/2PG/3G 型) 系指设有货物围护系统,专运《散装运输液化气体船舶构造与设备规范中所列的液化气体或其他易燃液货的船舶。对1986年7月1日及之后建造并符合IGC规则的船舶,根据所采取的防漏保护措施分别加如下标志:(1)1G型/Type 1G:采用最严格防漏保护措施的货物。(2)2G型/Type 2G:采用中等防漏保护措施的货物。(3)2PG型/Type 2PG:适用 L≤150 m,采用相当严格防漏保护措施的货物,且释放阀的最大调定值至少为0.7MPa,设计温度为 -55 ℃或以上;(4)3G型/Type 3G:采用中等防漏保护措施的货物。对于1986年7月1日之前建造并符合GC规则的船舶,上述1G型/2G/2PG型/3G型分别由IG型/IIG型/IIPG/IIIG型替代。 Liquefied Gas Carrier Type （1G/Typ2G/Typ2PG/Typ3G）is a ship provided with cargo containment system, dedicated to carry liquefied gases or other flammable liquid cargos as listed in IGC Code Liquefied gas carriers constructed on or after 1 July 1986 and complying with IGC Code are to be assigned the following notations respectively, depending on the preventive measures to preclude the escape of cargo: (1) Type 1G: Maximum preventive measures are required to preclude the escape of cargo; (2) Type 2G: Significant preventive measures are required to preclude the escape of cargo; (3) Type 2PG: For gas carriers of 150 m in length or less and significant preventive measures are required to preclude the escape of cargo, with a MARVS of at least 7 bar gauge and a cargo containment system design temperature of -55 ℃ or above; (4) Type 3G: Moderate preventive measures are required to preclude the escape of cargo. Liquefied gas carriers constructed before 1 July 1986 and complying with GC Code are to be assigned the notations of Type IG/Type IIG/Type IIPG/Type IIIG respectively instead of Type 1G/Type 2G/Type 2PG/Type 3G.

Liquefied inflammable gas carrier　液化易燃气体运输船　系指运输液化易燃气体的任何船舶。该类船舶应符合"1974/1978 国际海上人命安全公约"的标准，还必须是按照国际海事组织即时修订或达到最低有效标准的关于运输散装液化易燃气体船舶构造和设备的规则进行建造的船舶，并必须在属于国际船级社协会（IACS）会员之一的船级社入级，而且仍然受该船级社的监督。 Liquefied inflammable gas carrier means any vessel which transport bulk liquefied inflammable gas. She shall comply with standards of SOLAS 74/78 and must be constructed according to IMO code for the construction and equipment of ships carrying liquefied inflammable gas in bulk, as amended from time to time or to standards at least as effective, and must be classified in one of the recognized classification societies belonging to the IACS and still under its supervision.

Liquefied natural gas(LNG)　液化天然气　天然气经压缩、冷却、液化而成，以液态形式储存在特定容器中。

Liquefied natural gas carrier　液化天然气船　系指在常压、-163 ℃下装运液化天然气的船舶。其载货总容积大多在 120 000 ~ 130 000 m³ 之间。液货舱需设绝热物和次屏壁。一般不设再液化装置。而将蒸发的气体用作燃料。其装有保持低温的设施和存贮液货的高压容器，专运液化天然气的船舶。

Liquefied petroleum gas carrier　液化石油气船　装有保持低温的设施和储存液货的高压容器，专运液化石油气的船舶。

Liquefied petroleum gas(LPG)　液化石油气（LPG）　在正常温度和压力条件下为气体，通过增加压力或降低温度使之保持液态的轻碳氢化合物的混合物。注意(1)：其基本成分是丙烷、丙烯、丁烷和丁烯。注意(2)：LPG 可由商用丁烷、商用丙烷或由这两者的混合物构成。 Liquefied petroleum gas means the mixture of light hydrocarbons, gaseous under conditions of normal temperature and pressure, and maintained in the liquid state by increase of pressure or lowering of temperature. Note(1): The principal components are propane, butanes or butenes. Note(2): LPG can be obtained as commercial butane, commercial propane or a mixture of the two.

Liquid　液体　对"IBC 规则"而言，系指那些在温度为 37.8 ℃时，其蒸发气压力不超过 0.28 MPa 绝对压力的液体。 For the purpose of "IBC Code", liquid Means a liquid with a vapour pressure not greater than 0.28 MPa absolute at 37.8 ℃.

Liquid bulk cargo carrier　液体散货船　系指其结构主要适用于载运液体货物的，专门从事运输大宗液体散装货物的船舶。按液体性质不同，此类船舶又分为油船（专门运输原油和成品油的船舶）、液化气船（专门运输液化石油气或液化天然气的船舶）和液体化学品船，此类船舶都为单甲板尾机型船。

Liquid carburizing　液体渗碳（充液渗碳）

Liquid cargo hold space　液货舱处所　系指由船体结构所围成且设有货物围护系统的处所。

Liquid cargo pump　液货泵　抽送液体货物的泵。

Liquid cargo ship　液货船　专运散装液态货物的船舶。

Liquid chemical tanker　液体化学品船　专运液体化学品的船舶。

Liquid chemical wastes　液体化学品废弃物　系指提供装运的，且其所含的或被污染的一种或多种成分是受用于散装运输的国际散装运输危险化学品船舶构造和设备规则（IBC 规则）约束的物质、溶液或混合物，同时认为它们已无直接用途，对其进行载运是为了能在除海上以外的地方对其进行倾倒、焚烧或其他方式的处理。 Liquid chemical wastes are substances, solutions or mixtures, offered for shipment, containing or contaminated with one or more constituents which are subject to the requirements of this Code and for which no direct use is envisaged but which are carried for dumping, incineration or other methods of disposal other than at sea.

Liquid fuel　液体燃料

Liquid level alarm device　液位报警装置　当液舱内的液位超过设定值时能发出警报的装置。其探头设置在各货舱内，而显示及报警装置设在货物控制室内。货舱液位报警装置主要有：高液位报警（high level alarm）、溢流报警（overflow alarm）、装载过度报警（overfilling alarm）等。

Liquid mud handling system　船舶泥浆系统　又称船舶泥浆输送系统，是一种新型货物输送系统。目前主要用于海上石油平台供应船（PSV、AHTS）等向钻井平台以及采油平台提供泥浆，或将平台上不用的泥浆回收（送到岸上）处理。

Liquid nitrogen　液氮

Liquid pump recirculation　液泵供液　由液泵将低压循环贮液器大量的液态制冷剂输送至盘管或蒸发器中蒸发吸热，未蒸发完的液体和气体一同在回入低压循环贮液器循环使用的冷却方式。

Liquid receiver　贮液器　贮存经冷凝器冷凝的液

态制冷剂以供制冷系统循环使用的容器。

Liquid substances 液体物质 对"73/78 防污公约"附则Ⅱ而言,系指在温度为37.8℃时,绝对蒸发气压力不超过0.28MPa的物质。 For the purposes of the "MARPOL73/78" Annex Ⅱ, liquid substances are those having a vapour pressure not exceeding 0.28MPa absolute at a temperature of 37.8 ℃.

Liquid tank 液舱 (1)贮存燃油、滑油、淡水、蒸馏水等各种船用液体的舱的统称。(2)对散装运输液化气船而言,系指被设计成货物的主要容器的液密壳体,包括不管其是否有绝热层或次屏壁或两者的所有这类容器。

Liquidation 液化 系指物质由气态转变为液态的过程,会对外界放热。实现液化有两种手段:(1)降低温度;(2)压缩体积。临界温度是气体能液化的最高温度。由于通常气体液化后体积会变成原来的几千分之一,便于贮藏和运输,所以现实中通常对一些气体(如氨气、天然气)进行液化处理,由于这两种气体临界点较高,所以在常温下加压就可以变成液体,而另外一些气体如氢、氮的临界点很低,在加压的同时必须进行深度冷却。

Liquidation ethylene gas ship(LEGS) 液化乙烯船 系指运载液化乙烯的船舶。

Liquid-suction heat exchange 气-液换热器 装在氟利昂回气管上与高压液管接触的,使液态制冷剂得到过冷,提高回气过热的热交换器。

Liquified natural gas loading unit 液化天然气装载装置 系指为船舶装载液化天然气的单点系泊装置。

Liqutd ink-jet printer 液体喷墨打印机

List/transverse inclination/heel 横倾 船舶具有横向倾斜的浮态,一般以横倾角表示。

List for original repairs 原始修理工程单

List of operational limitations 操作限制清单 系指适用SOLAS公约第Ⅰ章的客船,在船上保存一份该船所有操作限制清单。清单中应包括对SOLAS规则的任何免除、航区的限制、天气的限制、海况的限制、许用载荷的限制、纵摇限制、速度限制以及其他任何限制,不论这些限制是由主管机关强制规定还是在设计或建造阶段就已制订。 List of operational limitations means that passenger ships to which chapter I of SOLAS Convention applies shall keep on board a list of all limitations on the operation of the ship, including exemptions from any of the SOLAS regulations, restrictions in operating areas, weather restrictions, sea state restrictions, restrictions in permissible loads, trim, speed and any other limitations, whether imposed by the Administration or established during the design or the building stages.

List pump 横倾平衡泵 将水或油从左舷或右舷液舱抽送到右舷或左舷液舱以调节船舶横倾平衡的泵。

Lists of approved marine products 产品录 船级社对认可工厂和船舶产品,将其有关产品的名称及其主要性能要素和细节及其制造厂商的细资料汇编成册,供船舶设计单位、造船厂、船东、贸易商和出口商等提供信息的文件。

Lithium bromide 溴化锂

Lithium bromide absorption refrigerating unit 溴化锂吸收式制冷装置

Lithium bromide water absorption refrigerating plant 溴化锂-水吸收式制冷装置 由本体及蒸发器、吸收器、冷凝器、发生器、屏蔽泵和真空泵等组成,在真空容器内以水为制冷工质,借助于溴化锂溶液的蒸发吸收作用,以获得低温冷水的制冷装置。

Litigation 诉讼 即打官司。系指由司法部门按法律程序来解决双方的贸易争端。这通常是由于争议所涉及的金额较大,双方都不肯让步,或者一方缺乏解决问题的诚意,通过协商或调解难以达成协议,以致诉诸法律。

Live 带电的 当导体或电路对地之间存在电位差时,该导体或电路为带电的。 A conductor or circuit is live when a difference of potential exists between it and earth.

Live fish carrier 活鱼运输船 系指设有活鱼舱,采用循环水或换水方式,有些还备有增氧、净水和/或降温等装置,专用于运输活鱼的船舶。 Live fish carrier is a ship fitted with live fish holds, provided with water cycling or exchanging, in some cases provided with devices for increasing oxygen, purifying water and reducing temperature, dedicated to carry live fish.

Live steam 新蒸汽

Livestock carrier 运牲畜船 系指专门运载牲畜,舱内设置分隔围栏的船舶。 Livestock carrier is a ship dedicated cattle carrier, with holds divided into enclosures.

Living accommodation 生活舱室 船上居住舱室、厨房、餐室、盥洗室、浴室、俱乐部、厕所等与人员日常生活有关的各种舱室的统称。

Living accommodation ventilation 舱室通风 为使居住舱室、会议室、餐厅等舱室内温度不致太高而进行的自然通风或机械通风。

Living quarter 生活区

LNG 系指液化天然气。 LNG means liquefied natural gas.

LNG boil-off gas 液化天然气蒸发气体

LNG carrier 液化天然气(LNG)运输船 系指能

够载运液化天然气的船舶。 LNG carrier means a ship capable of carrying liquefied natural gas.

LNG fueled ship 液化天然气燃料推进船 系指以液化天然气作为发动机燃料的船舶,它是新一代环保绿色船型。见图 L-7。

图 L-7 液化天然气燃料推进船
Figure L-7 LNG fueled ship

LNG-FPSO 液化天然气-浮式生产储油/卸油船 为解决环境污染问题,提高浮式生产储油/卸油船的环保性能,采用液化天然气作为清洁能源的一种新概念浮式生产储油/卸油船。随着深海油气田开发的不断升温,海洋工程装备日益向专业化、功能复合化等新型概念的方向发展,LNG-FPSO 装置是开发海上边际气田及回收海上油田排入空气中油气的有效工具。

LNG-FSRU 带气化外输装置的液化天然气储存和再液化装置

LNG-PAC system 天然气储存系统 系指为装在船上的双燃料发动机储存液化天然气的装置。该系统由加气站(与岸上液化天然气加气站或液化天然气供气船连接)、液化天然气管道、液化天然气储存舱、相关的连接部件和阀门等组成。该系统符合国际海事组织(IMO)有关装船天然气发动机的标准。当双燃料发动机以天然气作为燃料时,船舶能够实现硫化物零排放,且其氮氧化物排放量与国际海事组织的 Tier-Ⅰ 排放标准相比减少了 80%;其二氧化碳排放量与以柴油机为动力的船舶相比减少了 20%。

Load 载荷 系指船体结构承受该船处于"临界设计条件"下作用在船体结构上的各种静力、动力和周期性激振力。其中静力如船上的人员、设备、车辆、行李和货物的重力、船体内液体静压力、船体外水浮力、气垫内压力、停放船体的支架支撑力、吊运船体的起吊力等;动力如船体在波浪上运动时产生的惯性力、波浪冲击力、气垫支撑力、水翼的水动升力、全垫升气垫船在陆上降落时地面的反冲力等;周期性激振力,主要指主机和螺旋桨周期性运转所产生的力。

Load area 负荷面积 系指单个车轮的轮印。当一组车轮印的间距小于单个车轮印的尺寸时,那么这一组车轮印也认为是负荷面积。 Load area is defined as the footprint area of an individual wheel or the area enclosing a group of wheels when the distance between footprints is less than the smaller dimension of the individual prints.

Load balance 负载平衡

Load calculation 负荷计算

Load carrying capacity 载重量(载货量,承载能力)

Load case 装载工况 系指船体结构承受船体梁和局部载荷组合的一种状况。 Load case is a state of the ship structures subjected to a combination of hull girder and local loads.

Load characteristic test 负荷特性试验 内燃机在持续转速下进行的测定其空负荷逐步增至最大负荷时的功率、燃油消耗量、排气温度等主要性能指标的试验。

Load coefficient 齿轮负荷系数 衡量轮齿表面负荷强度的一个指标。在平行传动中:$\kappa = (P/d)(i+1)/i$;对太阳轮与行星轮之间的啮合:$\kappa = (P/d)(i+1)/(i-1)$;对行星轮与内齿轮之间的啮合:$\kappa = (P/d) \times (i+1)/(2-i)$;式中,$P$—齿的周向负荷;$d$—小齿轮的节轮直径,mm;$i$—传动比$(i>1)$。

Load coefficient 载荷系数

Load curve 载荷曲线 船体或构件上载荷沿船长或构件长度分布的曲线。在总纵强度计算中是重力曲线和浮力曲线相应垂向坐标差值的曲线。

Load dependent cylinder lubrication (LDCL) 负载的气缸润滑

Load draught 满载吃水 d 系指:对需要标记满载吃水线的船舶,指在船长(L)中点处从平板龙骨外缘量至载重线的垂直距离(m)。 Load draught d is the vertical distance in m from the top of keel to the load line measured at the middle of L_f, in case of a ship which is required to be marked with load lines and at the middle of L.

Load factor in power prediction (power prediction factor) 功率估计载荷因数 由船模试验数据估计实船在试航或营运情况受到功率时所用的因数,与阻力、推进尺度效应、船体粗糙度、风浪情况和船型以及测试技术等有关:$(1+X) = \eta_D P_D/P_E$,式中,η_D—推进效率;P_D—受到功率;P_E—有效功率。

Load line 载重线 是船舶按其航行的区域、区带和季节期而定的载重水线。对需要标记载重线的船舶,系指其设计夏季满载吃水线,对不标记载重线的船舶,系指其设计最大吃水线。 Load line is a deep waterline determined in according to his navigation area, zone and seasonal period. It is the waterline corresponding to the designed summer load draught in case of a ship which is required to be marked with load lines and the waterline corresponding to the designed maximum draught in case of a ship which is not required to be marked with load lines.

Load line block coefficient C_{bL}　载重线方形系数 C_{bL}　在"国际载重线公约"中的定义如下：$C_{bL} = \nabla_L / L_L BT_L$，式中，$\nabla_L$——在型吃水 T_L 下的型排水体积，m³；L_L——载重线长度，m；B——型宽，m；T_L——量至最小型深 85% 处水线，m。 Load line block coefficient, is defined in the International Convention on Load Lines as follows: $C_{bL} = \nabla_L / L_L BT_L$, where: ∇_L—moulded displacement volume at the moulded draught, T_L, in m³; L_L—load line length, in m, B—moulded breadth, in m, T_L—the moulded draught measured to the waterline at 85% of the least moulded depth, in m.

Load line draught d_S　载重线吃水　就 MARPOL 而言，系指自船中处的型基线至相应于船舶核定夏季干舷的水线之间的垂直距离(m)。 For the purpose of MARPOL, load line draught(d_S) is the vertical distance, in m, from the moulded baseline at mid-length to the waterline corresponding to the summer freeboard to be assigned to the ship. Calculations pertaining to this regulation should be based on draught d_S.

Load line length L_L　载重线长 L_L(m)　系指"1966 年国际载重线公约"附则Ⅰ第 3 条定义的船长。应取为由龙骨上缘量至 85% 最小型深处的水线总长的 96%，但若由艏柱前缘至舵杆中心线的该水线长度大于前者，则取后一长度。对设计有倾斜龙骨的船舶，量取上述长度的水线应平行于设计水线。 Load line length L_L(m) is the ship's length as defined in Regulation 3 of Annex Ⅰ to the International Convention on Load Lines, 1966. Load line length L_L is to be taken as 96% of the total length on a waterline at 85% of the least moulded depth measured from the top of the keel, or as the length from the fore side of the stem to the axis of the rudder stock on that waterline, if that is greater. In ships designed with a rake of keel, the waterline on which this length is measured is to be parallel to the designed waterline. The length L_L is to be measured in m.

Load line length L_{LL}　干舷船长 L_{LL}　系指龙骨板上缘向上最小型深 85% 处水线沿艏柱前缘量至舵杆中心线的距离。L_{LL} 不应小于同样水线总长的 95%。如果在最小型深 85% 处以上艏柱的外形为凹入的，艏柱前缘取作前端，此处艏柱为船体在前端外表面的轮廓，并且在球鼻艏外不包括任何附件。船舶设计为倾斜龙骨时其计量长度的水线应和设计水线平行。 Load line length L_{LL} is the distance, in m, on the waterline at 85% of the least moulded depth from the top of the keel, measured from the forward side of the stem to the centre of the rudder stock. L_{LL} is to be not less than 96% of the total length on the same waterline. In ship design with a rake of keel, the waterline on this length is measured is parallel to the designed waterline at 85% of the least moulded depth D_{min} found by drawing a line parallel to the keel line of the ship (including skeg) tangent to the moulded sheer line of the freeboard deck.

Load line mark for passenger cargo ships　客货船的载重线标志　系指下列标志：热带淡水载重线(TF)、夏季淡水载重线(F)、热带载重线(T)、夏季载重线(S)、冬季载重线(W)、北大西洋冬季载重线(WNA)、热带淡水木材载重线(LTF)；夏季淡水木材载重线(LF)；热带木材载重线(LT)；夏季木材载重线(LS)；冬季木材载重线(LW)；北大西洋冬季木材载重线(LWNA)；客船分舱载重线(C1)；和交替运载客货分舱载重线(C2)。 Load Line Mark for Passenger Cargo Ships are as following: Tropical Fresh Water Load Line(TF); Summer Fresh Water Load Line(F); Tropical Load Line(T); Summer Load Line(S); Winter Load Line(W); Winter North Atlantic Load Line(WNA); Tropical Fresh Water Timber Load Line(LTF); Summer Fresh Water Timber Load Line(LF); Tropical Timber Load Line(LT); Summer Timber Load Line(LS); Winter Timber Load Line(LW); Winter North Atlantic Timber Load Line(LWNA); Passenger Ship Subdivision Load Line(C1); and Subdivision Load Line in Carrying Passengers/Cargoes(C2).

Load of elevating platform　平台升降负荷　自升式钻井平台在升降平台作业时所允许承受的最大负荷。

Load range　负荷范围　对船舶废气清洗系统试验和检验而言，系指柴油机最大额定功率或锅炉最大蒸发量。

Load sharing factor K_y　负荷分配系数 K_y　系指实际传动的最大负荷与平均分配负荷之比值。K_y 值其主要影响因素为分支的精度和柔度。是用以考虑多分支传递中负荷分配不均的影响(如双重排列、行星排列、双斜齿等)。 Load sharing factor K_y means ratio between the maximum load through an actual path and the evenly shared load. The factor mainly depends on accuracy and flexibility of the branches. it accounts for the maldistribution of load in multiple path transmissions(dual tandem, epicyclic, double helix, etc.).

Load shim 吊索垫片

Load shutdown test/governor test/speed variation test 突卸负荷试验 检查汽轮机或柴油机等机组在额定工况运转下突然卸去全部负荷,其调速系统性能,如瞬间最高转速、转速重新稳定时间和转速等的试验。

Load ting 吊索环

Load to value(LTV) 运载效率系数 LTV 概念来源于经济学领域的贷款风险评估和控制的指标—按揭成数(Loan to value, LOV),但它并不是风险评估指标,而是揭示了船舶运载量的运行成本之间的关系。可用于比较不同船型的运载效率。如何在不影响船舶航速的情况下,增加运载量,降低油耗,减少排放,同时降低成本,提高安全性实现环境保护已成为造船企业研究的新课题。采用 LTV 概念设计的船舶与常规设计的同型船舶相比,可保证相同航速的前提下,载重量可以增加,耗油量却下降,能以最小的油耗量实现载重量的最大化,同时还可以减少有害气体排放量,更加环保。

Load waterline length 满载水线长 满载水线平面与船体型表面艏艉端交点之间的水平距离。

Loaded displacement 满载排水量 △(t) 系指船舶满载出港状态静浮于水面时的排水量,通常等于最大营运质量。

Loaded speed 满载航速 船舶装载至满载吃水,主机在额定工况时的航速。

Loaded waterline 满载水线 船舶在满载状态自由正浮于静水上时,船体型表面与水面的交线,亦即对应于满载排水量的水线。

Loaded waterline 满载水线(内河船) 系指船舶在核定适航航区内所允许达到的最大载重水线。如果船舶适航于数级航区,满载水线系指核定的最高级别航区的最大载重水线。满载水线长度系指船舶的满载水线面在中纵剖面上的投影长度。

Loading[1] 负载 系指船舶用电设备所需的功率。在计算电力负载时,可根据使用情况将负载做如下分类:(1)第一类负载——连续使用的负载,(2)第二类负载——短时或重复短时使用的负载;(3)第三类负载——偶然短时使用的负载,以及按操作规程规定可以在电站尖峰负载实际外使用的负载。

Loading[2] 装载 系指把位于船外的货物运至船上或在船上不同地点转运货物,并包括相关的操作,诸如捆扎和紧固货物,以及在紧固装置中插入夹头和销子。Loading means conveying cargo from a location outside a ship to a location on board a ship, or transferring cargo between location on board a ship, and includes associated operations such as lashing and securing of cargo and inserting clamps and pins in securing devices.

Loading capacity of boat davit 吊艇架额定负荷 吊艇架所能承吊的最大的救生艇总重。

Loading computer D 装载仪 D 系指可用于破舱稳性的计算及校核的装载仪。 Loading computer D means a loading computer capable of calculating and checking damage stability.

Loading computer G 装载仪 G 系指可用于散装谷物稳性的计算及校核的装载仪。Loading computer G means a loading computer capable of calculating and checking stability of grain in bulk.

Loading computer I 装载仪 I 系指可用于完整稳性计算及校核的装载仪。 Loading computer I means a loading computer capable of calculating and checking intact stability.

Loading computer S 装载仪 S 系指可用于各种装载工况下船体确定的计算及校核的装载仪。 Loading computer S means a loading computer capable of calculating and checking hull strength under various loading conditions.

Loading computer system 装载计算机系统 是用以方便迅速地确定给定读出点的静水弯矩、剪力以及静水扭矩和侧向载荷(如适用时)的系统。这些读数在任何装载和压载状态下均应不超过规定的许用值。Loading computer system is a system by means of which it can be easily and quickly ascertained that, at specified read-out points, the still water bending moments, shear forces and the still water torsional and lateral loads, where applicable, in any load or ballast condition will not exceed the specified permissible values.

Loading computer system 装载仪系统 装载仪系统是一种数字的能够简单快速地辨别任何装载工况是否超出操作限制的系统。装载仪系统应由各船级社根据自己的规范进行批准。装载仪系统应能计算任何指定的装载工况,核实是否符合图表中给出的操作限制。并提供包括输入输出数据的图表。如果有任何一项操作限制未经检查,用户在使用装载仪系统时应得到适当的提示并且借助图表可以通过其他途径核实每一个项目。装载仪系统至少应核实满足下列项目:(1)吃水限制;(2)在指定位置/读数点报告静水弯矩和剪力。装载仪最终试验工况基于最后的装载手册给出的工况。试验工况应经批准,由装载仪系统计算出的各读数点处的剪力和弯矩应分别在装载手册中给出的 $0.02Q_{\text{sw-perm}}$ 和 $0.02M_{\text{sw-perm}}$ 范围之内。这里 $Q_{\text{sw-perm}}$ 和 $M_{\text{sw-perm}}$ 分别为指定的每个读数点处的许用剪力和许用弯矩。在接受装载仪系统之前,应向验船师演示说明计算机的各有关方面,包括但不限于以下:(1)确认使用船舶的最终数据;(2)确认所有读数点的相关限制是正确的;(3)在系统安装在船上后,其操作符合批准的试验工况;(4)船上具备批准的试验工况;(5)船上具备操作手册。 Loading computer system is a digital system and that can easily and quickly ascertain whether operational limitations are exceeded for any loading condition. The loading computer system is to be approved based on the Rules of the individual Classification Society. The loading computer system is to be capable of producing any specific loading condition and verify that these comply with all the operational limitations given in and provide plots including input and output. If any of the operational limitations are not checked, the user is to be properly informed when using the system, and by the plots provided, so that each such item is verified by other means. The loading computer system is as a minimum to verify that the following are satisfied:(1) draught limitations;(2) still water bending moments and shear forces are reported at the specified locations/read-out points. The final test conditions for the loading computer are to be based on conditions given in the final Loading Manual. The test conditions are subject to approval and the shear forces and bending moments calculated by the loading computer system, at each read out point, are to be within $0.02Q_{\text{sw-perm}}$ or $0.02M_{\text{sw-perm}}$ of the results given in the loading manual, where $Q_{\text{sw-perm}}$ and $M_{\text{sw-perm}}$ are the assigned permissible shear force and bending moment at each read out point respectively. Before a loading computer system is accepted, all relevant aspects of the computer, including but not limited to the following, are to be demonstrated to the Surveyor:(1) verification that the final data of the ship has been used;(2) verification that the relevant limits for all read-out points are correct;(3) that the operation of the system after installation onboard, is in accordance with the approved test conditions;(4) that the approved test conditions are available onboard;(5) that an operational manual is available onboard.

Loading condition 装载状态 系指一种船舱内所载装载重量的分布状态。 Loading condition is a distribution of weights carried in the ship spaces arranged for their storage.

Loading draft 满载吃水 系指船舶处于满载排水量状态时的吃水或夏季载重线吃水,集装箱船为70%载重量(DWT)对应的吃水。

Loading factor (pressure coefficient, temperature drop coefficient, temperature rise coefficient) 负荷系数 流经压气机或涡轮级以后,气流等熵总焓的变化与圆周速度的平方或其1/2的比值。

Loading guidance information 装载指导资料 系指符合 LLC66 第10(1)条的一种手段,其使船长在装载和压载时能使船舶处于安全状态而不超过许用应力。 Loading guidance information is a means in accordance with Regulation 10(1) of LLC 66, which enables the master to load and ballast the ship in a safe manner without exceeding the permissible stresses.

Loading instrument 装载仪 是一种模拟式或数字式仪器,通过它可容易地快速确定各规定读出点在任何装载或压载状况下的静水弯矩、剪力、静水扭矩和侧向压力(如有)确实不超过规定的允许值。对于具有下列营运标志之一:散货船、矿砂船或兼装船,且船长等于或大于150 m 的船舶装载仪除上述外,它也能按适用情况,确定:(1)作为每个货舱中部位置吃水函数的货舱处货物质量和双层底容量;(2)作为任何函数相邻两货舱平均吃水的这两个货舱的货物质量和双层底容量;(3)货舱进水状态下的静水弯矩和剪力不超过规定的允许值。其包括:(1)装载计算机(硬件)和(2)装载程序(软件)。 Loading instrument is an instrument which is either analog or digital and by means of which it can be easily and quickly ascertained that, at specified read-out points, the still water bending moments, shear forces and still water torsional moments and lateral loads, where applicable, in any load or ballast condition, do not exceed the specified permissible values. For ships with one of the service notations bulk carrier ESP, ore carrier ESP or combination carrier ESP, and equal to or greater than 150 m in length, the loading instrument is an approved digital system as relevant requirement. In addition to above, it is also to ascertain as applicable that:(1) the mass of cargo and double bottom contents in way of each hold as a function of the draught at mid-hold position;(2) the mass of cargo and double bottom contents of any two adjacent holds as a function of the mean draught in way of these holds;(3) the still water bending moment and shear forces in the hold flooded conditions, where required do not exceed the specified permissible values. Loading instrument is an approved analog or digital instrument consisting of(1) loading computer(hardware) and(2) loading program(software).

Loading manual 装载手册 装载手册是:(1)描述基于航行和港内/遮蔽水域营运进行设计和批准的装载工况;(2)描述静水弯矩和剪力的计算结果,以及如适用时,由于扭矩与侧向载荷的限制;(3)描述如表 L-1 给出的相关操作限制。 Loading manual is a document that:(1) describes the loading conditions on which the design and approval of the ship has been based for seagoing – and harbour/sheltered water operation;(2)describes the results of the calculations of still water bending moments, shear forces and where applicable, limitations due to torsional and lateral loads;(3)describes relevant operational limitations as given in table L-1.

表 L-1 设计参数
Table L-1 Design parameter

参数	Parameter
许用静水弯矩限制（航行和港内/遮蔽水域营运）	Permissible limits of still water bending moments(seagoing operation and harbour/sheltered water operation)
许用静水剪力限制（航行和港内/遮蔽水域营运）	Permissible limits of still water shear forces(seagoing operation and harbour/sheltered water operation)
结构吃水：T_{sc}	Scantling draught, T_{sc}
船中处的设计最小压载吃水：T_{bal}	Design minimum ballast draught at mid-ships, T_{bal}
首部双层底压载舱装满时的设计砰击吃水：$T_{FP\text{-}Full}$	Design slamming draught forward with forward double bottom ballast tanks filled, $T_{FP\text{-}full}$
首部双层底压载舱空舱时的设计砰击吃水：$T_{FP\text{-}mt}$	Design slamming draught forward with forward double bottom ballast tanks empty, $T_{FP\text{-}mt}$
允许的货物最大密度	Maximum allowable cargo density
装载手册中任何装载工况下的货物最大密度	Maximum cargo density in any loading condition in Loading Manual
包括任何限制的压载水交换操作说明	Description of the ballast exchange operations including any limitations
设计航速	Design speed

Loading manual 装载手册（货船） 是描述下列内容的文件：(1)作为船舶设计基础的装载工况，包括静水弯矩和剪力的允许值；(2)静水弯矩、剪力及扭矩和侧向压力界限（如适用）的计算结果；(3)允许的结构局部载荷（舱口盖、甲板、双层底等）。除上述外，对于具有下列营运标志之一：散货船、矿砂船或兼装船，且船长等于或大于150 m 的船舶装载手册还应描述：(1)对于营运标志为散货船的双壳船，货舱进水情况下静水弯矩和剪力的包络线和允许范围；(2)满载吃水时可为空舱的货舱或货舱组合。如满载吃水时没有货舱允许为空舱，则装载手册上要清楚地说明；(3)作为货舱中部吃水函数的最大允许和最低要求的质量（用于确定货舱的允许装载质量）；(4)作为两相邻货舱平均吃水函数的两个货舱的货物和双层底容量的最大允许和最低要求质量。平均吃水可为两个货舱中部吃水的平均值（用于确定货舱的允许装载质量）；(5)最大允许舱顶装载和非散货的货物性质说明；(6)甲板和舱口盖上最大允许装载。如船的甲板或舱口盖上未批准装载，则装载手册上应清楚地说明；(7)压载最大变化率并劝告应在可达到的压载变化率基础上与目的港商定装载计划。 Loading manual is a document which describes:(1)the loading conditions on which the design of the ship has been based, including permissible limits of still water bending moment and shear force;(2)the results of the calculations of still water bending moments, shear forces and, where applicable, limitations due to torsional and lateral loads;(3)the allowable local loading for the structure(hatch covers, decks, double bottom, etc.). In addition to above, for ships with one of the service notations bulk carrier ESP, ore carrier ESP or combination carrier ESP, and equal to or greater than 150 m in length, the loading manual is also to describe:(1)for cargo holds of single side skin construction of ships with the service feature BC-A or BC-B: the envelope results and permissible limits of still water bending moments and shear forces in the hold flooded condition;(2)for ships with the service feature BC-A: the cargo hold(s) or combination of cargo holds which might be empty at full draught;(3)hold mass curves for each single hold in the relevant loading conditions showing the maximum allowable and minimum required mass of cargo and double bottom contents of each hold as a function of the draught at mid-hold position (for determination of permissible mass in cargo holds); (4) hold mass curves for any two adjacent holds in the relevant loading conditions, showing the maximum allowable and minimum required mass of cargo and double bottom contents of any two adjacent holds as a function of the mean draught in way of these holds. This mean draught may be calculated by averaging the draught of the two midhold positions (for determination of permissible mass in cargo holds); (5) maximum allowable tank top loading together with specification of the nature of the cargo for cargoes other than bulk cargoes; (6) maximum allowable load on deck and hatch covers. If the ship is not approved to carry load on deck or hatch covers, this is to be clearly stated in the loading manual; (7) the maximum rate of ballast change together with the advice that a load plan is to be agreed with the terminal on the basis of the achievable rates of change of ballast.

Loading mooring storage(LMS) 储油系泊装油装置 系指具有储油和为处于系泊状态的穿梭油船装油等多种功能，且可在环境条件恶劣海域使用的单点系泊装置。

Loading rate 负载系数 系指某一期间内的负载平均需要功率与同一时期内的负载最大需要功率的比值。

Loading time 装货时间

Loadline 载重线 系指民用船舶按载重线公约或相关规则根据其航行的区带、区域和季节期，以载重线标志水平线段上边缘表示的满载水线。

Loadline mark 载重线标志 民用船为表明其载重线位置，用以检查装载状况不违背以核定的最小干舷，而按载重线公约或相关规则所规定的式样勘绘于船中两舷的标志。

Loan contract 借款合同 是当事人约定一方将一定种类和数额的货币所有权移转给他方，他方于一定期限内返还同种类同数额货币的合同。其中，提供货币的一方称贷款人，受领货币的一方称借款人。借款合同又称借贷合同。按合同的期限不同，可以分为定期借贷合同、不定期借贷合同、短期借贷合同、中期借贷合同、长期借贷合同。按合同的行业对象不同，可以分为工业借贷合同、商业借贷合同、农业借贷合同。借款合同是借款人向贷款人借款，到期返还借款并支付利息的合同。

Local area network(LAN) 局域网 系指一般用于小范围场所内的计算机信息传输网，通过该网能进行数据交换和传输，且能与互联网络的设备共用。 Local area network means a general purpose computer communication network, servicing a small geographical location, where exchange and transmission of data take place, and is in common use for the devices interconnected with the network.

Local control 就地控制 是：(1)一个机械设备的直接手动控制装置并且可以在靠近机械设备的地方进行操作，接收来自测量仪器和指示器等发出的信息；(2)位于机电设备近旁由操作人员对设备进行直接的人工操作的控制。 Local control is: (1) direct manual control of the machinery and equipment performed at or near their locations, receiving the necessary information from the measuring instruments, indicators and so on; (2) direct manual control by an operator of machinery through a device located on or adjacent to the controlled machinery.

Local control mechanism 就地控制机构

Local control mode for watertight doors 水密门就地控制模式 系指在正常情况下使用的，可不使用自动关闭设备即能对任何水密门就地开启或关闭的模式。 Local control mode for watertight doors means a mode which is to allow any watertight door to be locally opened and locally closed after use without automatic closure in normal conditions.

Local control station(LCS) 就地控制站 系指对机电设备旁或附近实施就地控制的控制站。 Local control station means the control station where machinery and electrical installations are locally controlled on or adjacent to them.

Local corrosion 局部腐蚀 系指：(1)点腐蚀、沟槽型腐蚀、边缘腐蚀、颈缩效应或其他局部性很大的腐蚀；(2)被定义为局部结构单元，如单块板和加强筋的均匀腐蚀。是新建船舶审评和营运船评估的基础。 Local corrosion is: (1) pitting corrosion, grooving, edge corrosion, necking effect or other corrosions of very local aspect; (2) defined as uniform corrosion of local structural elements, such as a single plate or stiffener. It is used as a basis for the new building review and are to be confirmed during operation of the vessel.

Local dynamic stress amplitudes 局部动应力幅值 系指外部波动压力或舱内动压力引起的总局部应力幅值 σ_{e-i}。 Local dynamic stress amplitudes are defined as the total local stress amplitude due to dynamic wave pressure loads or dynamic tank pressure loads, σ_{e-i}.

Local excess penetration 局部过大的熔深 系指个别焊缝根部过大的偏差。 Local excess penetration means individual excessive root deviations.

Local failure effects 局部故障后果 系指所考虑的以特定子系统或设备方面故障模式的结果。 Local failure effects mean failure effects on a specific sub-system or equipment under consideration.

Local frictional resistance 局部摩擦阻力 物体表面各处由于相应的雷诺数不同,所产生的不同摩擦阻力。

Local loading and unloading rate 管内装卸率

Local loads 局部载荷 系指直接作用在独立构件板格,普通骨材和主要支承构件上的压力和力。(1)静水局部载荷是由外部海水静水压力,以及船舶舱室内装载的重力引起的静压力和力所组成。(2)波浪局部载荷是由波浪引起的外部海水压力,以及船舶加速度作用于船舶所载重力引起的压力和力组成;(3)动力局部载荷是由冲击和晃动压力所组成。 Local loads are pressures and forces which are directly applied to the individual structural members: plating panels, ordinary stiffeners and primary supporting members. (1) Still water local loads are constituted by the hydrostatic external sea pressures and the static pressures and forces induced by the weights carried in the ship spaces. (2) Wave local loads are constituted by the external sea pressures due to waves and the inertial pressures and forces induced by the ship accelerations applied to the weights carried in the ship spaces. (3) Dynamic local loads are constituted by the impact and sloshing pressures.

Local pitch 局部螺距 螺旋桨叶面某半径处叶元体的螺距。

Local strength 局部强度 船体中某个构件或部分结构承受相应载荷的能力。

Local stress components 局部应力分量 系指由桁架系弯曲引起的二阶应力幅值 σ_2,由桁材支撑构件之间骨材弯曲引起的压力幅值 σ_{2a},由纵骨和横向肋骨之间非加筋板单元弯曲引起的二阶应力幅值 σ_3。 Local stress components are secondary stress resulting from bending of girder systems, σ_2, stress amplitude produced by bending of stiffeners between girder supports, σ_{2a}, and tertiary stress amplitude produced by bending of unstiffened plate elements between longitudinals and transverse frames, σ_3.

Local support members 局部支撑构件 局部支撑构件定义为局部加强构件,仅对单一板格的结构(例如甲板梁)完整性有影响。 Local support members are defined as local stiffening members which only influence the structural integrity of a single panel, e.g. deck beams.

Local vibration 局部振动 系指凡在某一局部范围内能感受到较大的振动。

Localization of buckling or collapse 屈曲或破坏的局部化 系指非弹性屈曲模式倾向于局部化,导致与其全部破坏,倒不如局部破坏。 Localization of buckling or collapse means the inelastic buckling modes tend to localize leading to local rather than global collapse.

Localized corrosion 局部腐蚀 系指在局部区域出现的一种腐蚀耗蚀,例如麻点或凹槽沟。 Localized corrosion is a type of corrosion wastage that occurs in local regions, for example, pitting or grooving.

Locating 定位(寻位) 对无线电通信而言,系指发现遇险的船舶、航空器、海上设施或人员。 For the purpose of radio communications, locating means the finding of the ships, craft, aircraft, units or persons in distress.

Locating pin 定位销

Locating signals 定位信号 是便于发现了处于困境或幸存位置的移动单元的无线电传输。这些信号包括那些由搜索机构、救生艇筏、自由漂浮 EPIRB(紧急定位无线电示位标)、卫星 EPIRB 以及由搜索-救援雷达应答器传播到协助单位的信号。 Locating signals are radio transmissions intended to facilitate the finding of a mobile unit in distress or the location of survivors. These signals include those transmitted by searching units, and those transmitted by the mobile unit in distress, by survival craft, by float-free EPIRBs (emergency position-indicating radio beacons) by satellite EPIRBs and by search and rescue radar transponders to assist the researching units.

Location prone to rapid wastage 快速易损耗区域 系指在区域内出现如下一种情况时即称为"快速易损耗区域":(1)长期存在舱底污水的区域;(2)贴近燃油舱一面的舱壁存在被加热状况。 Location prone to rapid wastage means one of the following cases among the location: (1) Area with standing bilges; (2) Bulkheads facing fuel oil tanks being heated.

Locator 定位器(漏电防护装置)
Lock 闸(闭锁装置,自动跟踪)
Lock chamber 闸门室(水闸)
Lock file 共享文件
Lock nut 调整螺母
Lock screw 锁紧螺钉(止动螺钉)
Lock-in lock-out submersible system (LILOSS) 闸式可潜器系统 又称"出潜式可潜器"。能把潜水员送到水下并可外出作业的系统。其由甲板加压舱、吊车、可潜器和生命支持系统等组成。潜水员先在可潜器的加压舱中加压,然后由吊车把可潜器吊放入海,可潜器把潜水员送到水下作业地点,潜水员外出作业完成后回到可潜器的加压舱里再转移到甲板加压舱中休息或减压。它不单用于常规潜水,而且适用于饱和潜水,在水下活动范围广,可直接在水下完成复杂的作业和检查工作。

Locking apparatus 锁紧装置

Locking arrangement 闭锁装置

Locking bar(hatch securing bar) 封舱压条 置于舱盖上,用以扣住防水盖布,以防被风吹倒,并压牢舱盖板的金属扁条。

Locking device 锁紧装置(舵) 系指将舵固定在任意位置的装置。该装置以及与船壳相连的底座应有坚固结构,以在规定的舵杆设计屈服扭矩下不超过所用材料的屈服点。船舶航速如超过 12 kn,则仅需对基于航速 V_0 = 12 kn 时的舵杆直径来计算设计屈服力矩。Locking device means a device kept the rudder fixed at any position. This device as well as the foundation in the ship's hull is to be of strong construction so that the yield point of the applied materials is not exceeded at the design yield moment of the rudder stock as specified. Where the ship's speed exceeds 12 knots, the design yield moment need only be calculated for a stock diameter based on a speed V_0 = 12 knots.

Locking device on door 门的锁紧装置 系指将紧固装置锁紧在关闭位置的一种装置。Locking device on door means a device that locks a securing device in the closed position.

Locking gear 锁定装置(锁紧装置)

Locking washer 锁紧垫圈

Locus of centers of buoyancy 浮心轨迹 船舶作不同倾角的等体积倾斜时,浮心移动所构成的空间曲线。

Locus of metacenter 稳心曲线 稳心的轨迹曲线。

Log 计程仪 重要的导航设备之一,由其测出船舶的速度并输出航程。计程仪的种类很多,最早的拖曳计程仪,现在用的计程仪主要有两大类:一类是相对计程仪。它是测量船舶相对海水的速度及航程;另一类是绝对计程仪,它测出船舶对海底(地球)的速度及航程。(1)电磁计程仪:电磁计程仪具有测速灵敏度高、可测出较低的航速、使用维护方便、工作稳定可靠等优点。此外其不受水的密度、黏度、温度等的影响,因而近来得到广泛使用。电磁计程仪根据电磁感应原理,将非电量的航速转变成电量的航速信号。所以电磁计程仪基本原理主要由电磁传感器的工作原理来体现。(2)多普勒计程仪:多普勒计程仪是绝对计程仪的一种。它是测量船舶对海底的速度。而不像水压、电磁计程仪是测量船舶相对海水速度,故称绝对计程仪。它的工作原理是应用多普勒效应。(3)声相关计程仪:第四代计程仪。它也是一种可选用的绝对计程仪。它是利用测量垂直方向的收、发信号并对回波信号幅度包络进行相关信息处理,测量船舶航速、累计航程并可测出水深的一种电航仪器。声相关计程仪是利用水声信息相关处理技术来实现其测速工作的。

Log raft 木筏 用一组树干并排连接而成的简易水上运载工具。

Log room 计程仪舱 供安置计程仪的舱室。

Logistics 物流 是集成现代运输、仓储管理、装卸搬运、包装、产品流通加工、配送、信息处理整合为一体的综合服务业务。为了满足客户的需要,对商品、信息以及其他资源从产地到消费地的高效、低成本流动和存储,进行规划、实施与控制。也就是说,从产地到消费地,对商品和服务的需求、供给和运输,这三者之间存在着一种最优关系。同时也要避免物流链中的任一环节,出现任何一种商品或服务供应过剩。

Logistics Dispute Resolution Center of the Arbitration Commission 仲裁委员会物流争议解决中心

London Court of International Arbitration(LCIA) 英国伦敦国际仲裁院 1892 年 11 月 23 日成立的伦敦仲裁会,1903 年 4 月 2 日改名为伦敦仲裁院,由伦敦城市和伦敦商会各派 12 名代表组成的联合委员会管理。1975 年伦敦仲裁院与女王特许仲裁员协会合并,并于 1978 年设立了由来自 30 多个国家的具有丰富经验的仲裁员组成的"伦敦国际仲裁员名单"。1981 年改名为伦敦国际仲裁院,这是国际上最早成立的常设仲裁机构,现由伦敦市、伦敦商会和女王特许协会三家共同组成的联合管理委员会管理,仲裁院的日常工作由女王特许协会负责,仲裁协会的会长兼任仲裁院的主席。伦敦国际仲裁院的职能是解决国际商事争议提供服务,它可以受理当事人依据仲裁协议提交的任何性质的国际争议。该仲裁院在组成仲裁庭方面确定了一项重要的原则,即在涉及不同国籍的双方当事人的商事争议中,独任仲裁员和首席仲裁员必须由 1 名中立国籍的人士担任。伦敦国际仲裁院于 1985 年 1 月 1 日起实行新的《伦敦国际仲裁院规则》,仲裁庭组成后,一般应当按照伦敦国际仲裁院的仲裁规则进行仲裁程序,但同时,该仲裁院也允许当事人约定按《联合国国际贸易委员会仲裁规则》规定的程序仲裁。它是目前英国最主要的国际商事仲裁机构,可以审理提交给它的任何性质的国际争议,尤其擅长国际海事案件的审理。由于其较高的仲裁质量,它在国际社会上享有很高的声望。

Long blade 长叶片 一般系指径高比 $D/L \leq 10$ 的叶片。

Long bossing 长轴包架 舷侧艉轴穿出船底处至螺旋桨前,全围封艉轴的轴包架。

Long bridge 长桥楼 长度大于 0.15 倍船长,且不小于其本身高度 6 倍的桥楼。

Long bridge ship 长桥楼船 通常系指其桥楼长

度大于 0.15 L,且不小于本身高度 6 倍的船舶。

Long deckhouse 长甲板室 为船中 0.4 L 区域内长度超过 0.2 L 或 12 m(取大者)的甲板室。长甲板室的强度应特别考虑。 Long deckhouse means a long deckhouse is a deckhouse the length of which within 0.4 L amidships exceeds 0.2 L or 12 m, whichever is the greater. The strength of a long deckhouse is to be specially considered.

Long deckhouse and short deckhouse 长甲板室及短甲板室 系指:(1)长度大于 0.15 L,且不小于其高度 6 倍的甲板室为长甲板室;(2)不符合长甲板室条件的为短甲板室。 Long deckhouse and short deckhouse mean:(1) a deckhouse having a length greater than 0.15 L and not less than 6 times its height is to be termed as a long deckhouse;(2) otherwise it is to be regarded as a short deckhouse.

Long forecastle 长艏楼 一般系指长度大于 0.25 倍船长的艏楼。

Long forecastle ship 长艏楼船 通常系指其艏楼长度大于 0.25 倍船长的船舶。

Long imperfection 长的缺陷 系指在任意的 100 mm 长的焊缝中,一个或多个缺陷的总长度大于 25 mm 或对于长度小于 100 mm 的焊缝,缺陷的长度最小为 25%焊缝长度。 Long imperfection is one or more imperfections of total length greater than 25 mm in any 100 mm length of the weld or a minimum of the weld length for a weld shorter than 100 mm.

Long line 延绳钓 以一根长的干线,每隔若干米敷设一短支线,支线的末端有带饵的钓钩,整根干线依靠浮子敷设于预定水层,用以捕捞鱼类的钓具。

Long line hauler 延绳钓起线机 利用数个动力滚轮对绳索的摩擦力绞收延绳钓干线和起上渔获物的机械。

Long liner 延绳钓渔船 使用延绳钓具,即以一根依靠浮子敷设于预定水层的干线,其上每隔一定距离敷设一短支线,支线末端装有带鱼饵的钓钩,以此进行捕捞作业钓渔船。

Long poop 长艉楼 一般系指长度大于 0.25 倍船长的艉楼。

Long poop ship 长艉楼船 通常系指其艉楼长度大于 0.25 倍船长的船舶。

Long range navigation 罗兰 C 导航系统(LORAN) 是一种双曲线无线电导航系统。它具有最长的持续记录期,并具有易于读取的接口,用于下载其存储的数据。又称远程导航。它由地面无线电发射台(地面台)和航行体上的接收机(用户设备)组成的无线电导航系统。载有用户设备的航行体,借助于测量来自地面站台的脉冲信号到达的时间差确定自己的当前位置。罗兰 C 系统采用脉冲相位双曲线导航体制,具有作用距离远、精度高,无多值、价格低、用途广、技术新以及可靠性高等特点。罗兰 C 导航定位的基本原理是:从解析几何可知,若某点到两定点的距离差为常值,则该点处于某条确定的双曲线上,如果还已知该点到另两点的距离差,则该点的位置便确定了。上述的定点,即罗兰 C 台链。罗兰 C 台链系指建在陆上的反射网,每一台链由一个主台(M)和至少两个副台(X,Y,Z 等)组成各台链以其特定的重复周期(GRI)发射罗兰 C 信号。全世界所有的罗兰 C 台链,其载波频率均为 100 kHz,但其重复周期是不同的,利用其特定的重复周期来代表这个台链。罗兰 C 定位,就是罗兰 C 接收机精确测量所选台链各副台信号与主台信号到达的时间差——时差(TD)每一组时间差对应海图上的一条位置线(LOP),两条位置线的交战即为所需的船位。罗兰 C 系统主要误差有:同步误差(范围在 ±0.05 ps ~ ±0.06 ps 之间)、接收测量误差(范围约在 0.01 ps ~ 0.1 ps 之间、当信噪比降低到 0 dB 以下时,测量误差至少增大一个数量级)、天线修正误差(范围在 1 ps ~ 1.5 ps 左右)、地波传递误差(需实时测量)等。而其精度通常是用"绝对精度"和"重复精度"来衡量的。绝对精度用来度量罗兰 C 系统确定地理位置的能力。随着在覆盖区位置的变化,罗兰 C 的绝对精度在 0.1 n mile ~ 2.5 n mile 之间变化。重复精度用来度量依靠罗兰 C 系统引导返回先前位置的能力。它系指罗兰船位(经纬度)在同一地点由一天到另一天的变化,它不包括系统性的偏差,重复精度的典型值为 0.01 n mile(依赖于在罗兰 C 覆盖区的位置和环境)。

Long shank blade root 长颈叶根 为使涡轮叶片减振及轮盘冷却在叶身与型线配合部分之间设置长颈以形成冷却通道的叶根。

Long stroke engine 长冲程柴油机

Long superstructure 长上层建筑 长度大于 0.15 L,且不小于其高度 6 倍的上层建筑。

Long superstructure and short superstructure 长上层建筑及短上层建筑 系指:(1)长度大于 0.15 L,且不小于其高度 6 倍的上层建筑为长上层建筑。(2)不符合长上层建筑条件的为短上层建筑。 Long superstructure and short superstructure mean:(1) a superstructure having a length greater than 0.15 L and not less than 6 times its height is to be termed as a long superstructure;(2) otherwise it is to be regarded as a short superstructure.

Long term evolution(LTE) 长期演进 是基于正交频分多址 Orthogonal frequency division multiple access (OFDMA)技术、由 3GPP 组织制定的全球通用标准,包

括 FDD 和 TDD 两种模式用于成对频谱和非成对频谱。LTE 标准中的 FDD 和 TD 两个模式间只存在较小的差异,相似度达 90%。FDD(频分双工)是该技术支持的两种双工模式之一,应用 FDD(频分双工)式的 LTE 即为 FDD-LTE。作为 LTE 的需求,TD 系统的演进与 FDD 系统的演进是同步进行的。绝大多数企业对 LTE 标准的贡献可等同用于 FDD 和 TD 模式。由于无线技术的差异、使用频段的不同以及各个厂家的利益等因素,FDD-LTE 的标准化与产业发展都领先于 TD-LTE。FDD-LTE 已成为当前世界上采用的国家及地区最广泛的,终端种类最丰富的一种 4G 标准。

Long ton 长吨 等于 1.01 605 公吨 = 1 016.046 kg = 2 240 磅 = 1.12 短吨。

Long-crested regular waves 长峰规则波 系指仅存在于主方向上且有着无穷长单向的不同间距的波峰彼此保持平行的二因次不规则波。

Long-crested waves 长峰浪 由同方向波列组成的二维波系。

Longitudinal 纵骨 纵骨架式结构中,用于支承外板、甲板板和内底板等布置得较密的纵向骨材。

Longitudinal axis 纵轴

Longitudinal bending 总纵弯曲 由于作用在船体上的重力、浮力、波浪水动力和惯性力等而引起的船体整体绕水平横轴的弯曲。

Longitudinal bulkhead 纵舱壁 沿船长方向设置的舱壁。

Longitudinal center of buoyancy(from AP) 纵向浮心(自艉垂线距离) 系指从基准点至船舶浮心的纵向距离。基准点通常在艉垂线处(前+/后-)。Longitudinal center of buoyancy(from AP) is longitudinal distance form reference point to centre of gravity, reference point usually at aft perpendicular(forward+/aft-).

Longitudinal center of floatation(from AP) 纵向漂心 系指从基准点至船舶漂心的纵向距离。基准点通常在艉垂线处(前+/后-)。Longitudinal center of floatation(from AP) is longitudinal distance form reference point to centre of gravity, reference point usually at aft perpendicular(forward+/aft-).

Longitudinal center of gravity(from AP) 纵向重心(自艉垂线距离) 系指从基准点至船舶重心的纵向距离。基准点通常在艉垂线处(前+/后-)。Longitudinal center of gravity(from AP) is longitudinal distance form reference point to centre of gravity, reference point usually at aft perpendicular(forward+/aft-).

Longitudinal centerline bulkhead 中纵舱壁 系指位于船舶中心线上的纵舱壁。Longitudinal centerline bulkhead means a longitudinal bulkhead located on the centerline of the ship.

Longitudinal cross beam 架空纵梁 系指在甲板开口线内,纵跨于泥舱前后端壁之间的一种组合梁结构。Longitudinal cross beam is a longitudinal built beam spanning longitudinally the fore and aft transverse bulkheads of the hopper within the lines of deck openings.

Longitudinal division wall 纵向隔壁 系指作为减少谷物移动所产生的不利横倾影响的一种装置。可用于经平舱的满载舱、未经平舱的满载舱和部分转载舱内。但其应符合下列条件:(1)谷密;(2)结构强度应符合要求,(3)在甲板间舱内应从甲板延伸到甲板,其他情况应从甲板或舱口盖的下面向下延伸一定的符合要求的距离。

Longitudinal hull girder structural members 纵向船体梁结构构件 对船体梁纵向强度有贡献的结构构件,包括甲板、舷侧板、内底板,如设置则还包括上端斜板的纵向舱壁、底边舱斜板、舭部板、纵舱壁、双层底桁材和边压载水舱的水平桁材。Longitudinal hull girder structural members are those structural members that contribute to the longitudinal strength of the hull girder, including: deck, side, bottom, inner bottom, inner hull longitudinal bulkheads including upper sloped plating where fitted, hopper, bilge plate, longitudinal bulkheads, double bottom girders and horizontal girders in wing ballast tanks.

Longitudinal inclination 纵倾 船舶相对于设计水线具有纵向倾斜的浮态。纵倾值一般以艏艉吃水差表示,或以该值除以设计水线长的无因次方式表示。对于具有龙骨设计斜度的船舶,则其值系指艉吃水减去艏吃水加上龙骨设计斜度来表示,或以该值除以设计水线长的无因次方式表示。

Longitudinal launching cradle 纵向下水船架 系指船舶纵向下水时,为了防止船舶埋艏,而在船舶首部加装的托架。

Longitudinal members on transverse section 横剖面上的纵向构件 包括所有纵向构件,诸如在所考虑横剖面上的甲板板、甲板纵骨和纵桁,外板、船底板、内底板和纵舱壁。Longitudinal members on transverse section includes all longitudinal members such as plating, longitudinal and girders at the deck, sides, bottom, inner bottom and longitudinal bulkheads.

Longitudinal metacenter 纵稳心 船舶纵向倾斜时的稳心。

Longitudinal metacenter height 纵稳心高 纵稳心在重心之上的高度。

Longitudinal metacenter radius 纵稳心半径 船

舶纵向倾斜时的稳心半径。

Longitudinal position of propeller 螺旋桨纵向位置 螺旋桨中点在艉垂线所在横剖面以前的距离。

Longitudinal slipway with self leveling side transition cradle 变坡滑道 系指船舶下水用的坡度可变化的滑道。

Longitudinal space of two legs 桩腿纵距 三条、四条桩腿等的自升式钻井平台平行于设计拖航方向两桩腿中心线之间的距离。

Longitudinal stability 纵稳性 船舶纵向倾斜时的稳性。

Longitudinal stay(tube plate stay) 纵向联杆 为防止管板变形而在两管板之间装设的一种支撑元件。

Longitudinal strength 总纵强度 船体结构抵抗总纵弯曲的能力。

Longitudinal strength member 纵向强力构件 沿船长方向设置的,参与抵抗总纵弯曲的构件。

Longitudinal system of framing 纵骨架式 纵向骨架较密、横向骨架较疏的骨架形式。

Longitudinal transfer 纵向传送 补给船与接受船串列航行时通过索具纵向传送物资或人员的方式。

Longitudinal vibration 纵向振动 系指船体横剖面绕其纵轴前-后往复的振动。

Longitudinal wave 纵向波 系指传播方向与介质质点的振动方向平行的波。又称纵波,L 波。

Longitudinally framed ship 纵骨架式船 船体结构中纵向骨材比横向骨材较密的船舶。

Longitudinally framed side structures 纵骨架式舷侧结构 系指由用垂直主要支撑构件支撑的纵向普通扶强材组成的舷侧结构。 Longitudinally framed side structures means a structures built with longitudinal ordinary stiffeners supported by vertical primary supporting members.

Long-range identification and tracking (LRIT) information 船舶远程识别和跟踪信息 系指下列信息:(1)船舶的识别标志;(2)船舶的位置(经度和纬度);(3)提供位置的日期和时间。 Long-range identification and tracking information as following:(1) the identity of the ship;(2) the position of the ship (latitude and longitude); and (3) the date and time of the position provided.

Long-range identification and tracking(LRIT) system 船舶远程识别和跟踪系统 系指提供对船舶的全球识别和跟踪的系统。船舶远程识别和跟踪系统由船载船舶远程识别和跟踪系统信息发送设备、通信服务提供方、应用服务提供方、船舶远程识别和跟踪数据中心组成,包括任何相关的船舶监控系统、船舶远程识别和跟踪数据记录。船舶远程识别和跟踪协调方代表所有缔约国政府对船舶远程识别和跟踪系统性能的某些方面进行审查或审核。 Long-range identification and tracking(LRIT) system means a system provided for the global identification and tracking of ships. The LRIT system consists of the shipborne LRIT information transmitting equipment, the Communication Service Provider(s), the Application Service Provider(s), the LRIT Data Centre(s), including any related Vessel Monitoring System(s), the LRIT Data Distribution Plan and the International LRIT Data Exchange. Certain aspects of the performance of the LRIT system are reviewed or audited by LRIT Coordinator acting on behalf of all Contracting Governments.

Long-run dumping 长期性倾销 这种倾销是无限期地、持续地以低于国内市场的价格在国外市场销售商品。20 世纪 70 年代以来,长期性倾销日益增多。其之所以能够存在和维持,一般来说必须具备 3 个条件:(1)出口商品生产企业在本国市场上有一定的垄断能力,在很大程度上可以决定价格的形成。(2)本国与外国的市场隔离,不存在倒买倒卖的可能性。(3)两国的需求价格弹性不同,出口国需求价格弹性低于进口国需求价格弹性。

Long-tackle block(fiddle block) 串列滑车 由两个不同直径滑轮同向串列组合而成的滑车。

Long-term credit 长期信贷 系指贷款期限为 5~10 年的信贷。

Long-term recording medium 长期记录介质 系指最终记录介质中永久安装的那一部分。它具有最长的持续记录期,并具有易于读取的接口,用于下载其存储的数据。 Long-term recording medium means a permanently installed part of the final recording medium. It provides the longest record duration and has a readily accessible interface for downloading the stored data.

Lookout 瞭望(监视) 系指通过看、听以及在当时环境和条件下可采用的一切可行的方法,全面评估所发生的情况和碰撞危险的行为。 Lookout means activity carried out by sight and hearing as well as by all available means appropriate in the prevailing circumstances and conditions so as to make a full appraisal of the situation and of the risk of collision.

Loop 回路 系指按顺序将各分区的探测器联结形成的,并接(输入和输出)至显示装置的电器回路。 Loop is an electrical circuit linking detectors of various sections in a sequence and connected(input and output) in the indicating unit(s).

Loop directions finder 环状天线测向器 测定无线电台的方向,或通过两次测量来确定电台或测向器本身位置的一种仪器。它的测向原理是以环状天线"8"字形方向性图为基础。

Loop scavenging 回流扫气 由气缸一端的进气口向气缸盖方向流动,然后折回至和进气口同一侧上面的排气口排出形成的扫气。

Loose gear 可拆卸零部件 系指通过其可将负载系留到起重机或吊杆装置上的装备件,它不成为该起重机或吊杆装置的不可分割的部件,但它包括任何种类的滑车、卸扣、吊钩、转环、连接板、环、起重滑车或起重滑车组、链条或检修用的重物和集运架梁的组成件、吊梁、吊架、扩张器、盘、管或除中间散集装装箱或运输设备以外的其他货物容器。 Loose gear means an item of equipment by means of which a load may be attached to a crane or derrick, which does not form an integral part of the crane or derrick, and includes any block, shackle, hook, swivel, connecting plate, ring, pallet bar, lifting beam, lifting frame, spreader, tray, tub or other cargo receptacle other than an intermediate bulk container or transport equipment.

Loose gears 传送装置 系指滑车、绳索、链、环、吊钩、缆绳、转环、夹具、起重钩、起重磁铁、扩张器等。它们都是用来将货物载荷传送到结构件上的可移动的零件。 Loose gears are blocks, ropes, chains, rings, hooks, shackles, swivels, clamps, grabs, lifting magnets, spreaders, etc. which are removable parts used for transmitting the loads of cargo to the structural members.

Loose smut of wheat wool 短纤维玻璃棉

Loran navigation system 罗兰导航系统 "罗兰"是远程导航(long range navigation)英文字母的缩写"Loran"的译音。旧译"劳兰"是一种双曲线无线电导航系统。

Loran-C navigation system 罗兰C导航系统 低频脉冲相位双曲线远程导航系统。精密时差测量靠载频比相来实现;无多值性,粗测靠包络测量来实现。

Loran-D navigation system 罗兰D导航系统 罗兰C和罗兰D在技术上基本相同,罗兰D的地面台可以搬迁,因而适应战术与测量定位。作用距离比罗兰C近。20世纪70年代以来英国研制的"脉冲8定位系统"是在罗兰D的基础上发展起来的,应用于500 n mile范围内的海洋测量定位及海上与空中导航。

Losing party 败诉方

Loss 亏损

Loss of or damage to luggage 行李丢失或损坏 系指包括在运输或本应运输行李的船舶到达后的合理时间内,未能将该行李交还旅客而引起的经济损失,但不包括劳资纠纷引起的延误。 Loss of or damage to luggage includes pecuniary loss resulting from the luggage not having been re-delivered to the passenger within a reasonable time after the arrival of the ship on which the luggage has been or should have been carried, but does not include delays resulting from labour disputes.

Loss of position 位置丧失 船舶和/或平台装置从计划或目标位置移动。

Lost AIS target 失踪的AIS目标 代表AIS目标在其数据丢失前最后有效方位的目标。该目标以"失踪AIS目标"符号表示。 Lost AIS target is a target representing the last valid position of an AIS target before the reception of its data was lost. The target is displayed by a "lost AIS target" symbol.

Lost buoyancy 损失浮力 船舶破损进水后失去的浮力。

Lost buoyancy method 损失浮力法 在全船浮力中扣除破损部分失去的浮力,而仍按原有排水量计算船舶浸水后的浮态和稳性要素的方法。

Lost target 失踪目标 代表目标在其数据丢失前最后有效方位的目标。该目标以"失踪目标"符号表示。 Lost target is a target representing the last valid position of a target before its data was lost. The target is displayed by a "lost target" symbol.

Lost tracked target 失踪的被跟踪的目标 由于信号微弱,失踪或被遮蔽而不再收到目标信息,目标以"失踪的被跟踪的雷达目标"符号显示。 Lost tracked target is the target information is no longer available due to poor, lost or obscured signals. The target is displayed by a "lost tracked radar target" symbol.

Lost waterplane area 损失水线面面积 船舶破损进水后,水线面浸水部分的面积。

Loudness 响度 又称音量。人耳感受到的声音强弱,它是人对声音大小的一个主观感觉量。响度的大小决定于声音接收处的波幅,就同一声源来说,波幅传播的愈远,响度愈小;当传播距离一定时,声源振幅愈大,响度愈大。响度的大小与声强密切相关,但响度随声强的变化不是简单的线性关系,而是接近于对数关系。当声音的频率、声波的波形改变时,人对响度大小的感觉也将发生变化。

Loudness level 响度级 系指根据听力正常的听音判断为等响的1 000Hz纯音(来自正前方的平面行波)的声音级。

Low carbon 低碳 以低能耗、低污染、低排放为基础的发展模式。其实质在于能源高效运用,开发清洁能源,追求绿色GDP。核心是能源技术创新、制度创新和人类生存发展观念的根本性改变。发展低碳经济,推进节能减排,建设资源节约型、环境友好型社会积极推进生态文明建设。加强节能、提高能效,大力发展可再生能源和核能增加森林碳汇、积极发展低碳经济和循环经济,提倡低碳生活。尽可能实现节能减排,减少温室气体特别是 CO_2 排放,低碳代表更健康、更自然、更安全,同时也是低成本,低能耗的生产生活方式。

Low carbon steel 低碳钢 为碳含量低于 0.25% 的碳素钢,因其强度低、硬度低而软,故又称软钢。它包括大部分普通碳素结构钢和一部分优质碳素结构钢,大多不经热处理用于工程结构件,有的经渗碳和其他热处理用于要求耐磨的机械零件。低碳钢退火处理后形成铁素体和少量珠光体,其强度和硬度较低,塑性和韧性较好。因此,其冷成形性良好,可采用卷边、折弯、冲压等方法进行冷成形。这种钢还具有良好的焊接性。含碳量从 0.10% ~ 0.30% 低碳钢易于接受各种加工如锻造、焊接和切削,常用于制造链条,铆钉、螺栓、轴等。

Low density material 低密度材料 主要将其注入艇内,以在灌水时增加浮力的密度小于 1 的材料。Low density material means the material with a specific gravity of less than 1, primarily incorporated into the boat to enhance the buoyancy when swamped.

Low flame-spread 低播焰性 系指所述表面能有效地限制火焰蔓延,应根据国际耐火试验程序规则(FTP规则)确定。 Low flame-spread means that the surface thus described will adequately restrict the spread of flame, this being determined in accordance with the Fire Test Procedures(FTP)Code.

Low frequency(LF) 低频 系指 20 Hz ~ 160 Hz 这一段频率。在人耳所能听到的声音中,低频是声音的基础。

Low frequency sound 低频声 系指频率小于 300 Hz 的声音。

Low-low water level alarm 过低水位报警

Low lubricating oil pressure protective device 低滑油压保护装置 系指在滑油压力低于许用值时,能自动切断燃油供应的装置。 Low lubricating oil pressure protective device means a device which is capable of shutting off the fuel supply automatically when lubricating oil pressure is lower than the specified value.

Low noise burner 低噪声烧嘴

Low noise circuit design 低噪声电路设计

Low noise design for ventilation pipeline 通风管路系统低噪声设计

Low noise propeller 低噪声推进器

Low pitch propeller 低螺距螺旋桨

Low pressure(LP) 低压

Low pressure air conditioning 低压空调系统

Low pressure CO_2 system 低压 CO_2(灭火)系统

Low pressure compressor 低压压缩机 排气压力 196 ~ 980 kPa 表压的压缩机。

Low pressure deliver system 低压泵送系统

Low pressure feed water heater 低压给水预热器 设置在给水泵吸入管路中加热给水的热交换器。

Low pressure piping 低压气体管路 系指最大工作压力不超过 1MPa 的气体燃料管路。

Low pressure receiver 低压循环贮液器 贮存低压液态制冷剂供液泵循环使用的容器。

Low pressure side 低压侧 对制冷装置而言,系指膨胀阀后到压缩机吸入阀之间的受压部件;但若装置的切换(如为了热融霜)可能使它们处于高压,则这些零部件均应按规定的高压侧压力进行设计和试验。 For the purpose of refrigerating Plant, low pressure side means the pressed parts from behind the expansion valve to suction valve of the compressor. Where the change-over of the plant (e. g. for defrosting) will put these parts under high pressure, they are to be designed and tested to the specified pressure of high pressure side.

Low pressure side(LPG system) 低压端(LPG系统) LPG 供气管的暴露于气体调节器调节压力的部分。 Low density material is the part of the LPG supply line exposed to the regulated pressure of the gas regulator.

Low pressure steam generator (steam generator, steam converter) 低压蒸汽发生器 用锅炉减压蒸汽或主机撤气作为热源供低压系统使用的蒸汽发生器。

Low pressure steam generator feed(water) pump 低压蒸汽发生器给水泵 将给水输入低压蒸汽发生器的泵。

Low sea suction 低位海水门 由进水格栅、海水箱、通海阀组成的装于靠近舭部或船底的舷外水吸入口。

Low speed diesel engine 低速柴油机 曲轴转速在 300 r/min 以下的柴油机。

Low speed maneuverability 低速操纵性 船舶在浮游生物、底栖拖网、地震勘探等多种调查作业时,要求船舶低速航行(约 5 kn 及以下)并保持一定的航向或保持一定的船位。可采用主动舵、可调螺旋桨、电力推进等来满足低速航行的要求。

Low speed propulsion 低速推进 调查船进行科研与考察工作,常需低速航行(约 5 kn 及以下),要求船舶低速航行时维持其低速所需相应的推进工况。低速推进一般可用电力推进、可调螺矩螺旋桨或主动舵等实现。

Low speed wind tunnel 低速风筒 利用空气作介

质,风速一般低于0.4马赫数,用于进行船舶模型试验的风筒。

Low suction valve 低水位吸入阀

Low sulphur fuel oil 低硫燃油 硫含量不超过0.1%(m/m)的船用直馏型燃油,EU(欧盟)2005/33/EC法令规定,自2010年1月1日起,凡在欧盟港口停泊(包括锚泊、系浮筒、码头靠泊)超过2 h的船舶不得使用硫含量超过0.1%(m/m)的燃油。

Low temperature alarm 低温报警

Low temperature annealing 低温退火 消除由于常温加工等导致的内部应力,为了使其软化或降低由于淬火导致的畸变而采用的热处理前的退火。其温度在450 ℃ ~ 600 ℃为适当。如在低于450 ℃的条件下加热将无法消除应力。 Low temperature annealing means an annealing treatment which is performed to eliminate internal stress or reduce quenching strain.

Low temperature corrosion 低温腐蚀 在金属表面温度低于露点的条件下,锅炉排烟中的二氧化硫和水蒸气所形成的硫酸蒸发气凝结成硫酸和亚硫酸,从而引起的金属腐蚀。

Low temperature liquid cargo 低温液货 系指沸点低于 -90 ℃(-130 ℉)的冷冻液化气。

Low water alarm 低水位报警器(低水位信号)

Low water level alarm 低水位报警器

Low yield stress 最低屈服强度

Low-draft full type bulk carriers 浅吃水肥大型散货船 系指方形系数大于0.75的大型散货船,因相对速度较低(傅汝德数约在0.20以下)故采用大方形系数对船体阻力性能的影响甚微,而在同样的载重量下,船体主尺度可以减小,从而降低船舶的造价、提高船舶的经济性。

Lower between deck 下层中间甲板 系指在上层中间甲板以下的甲板。 Lower between decks means the deck below the upper between decks.

Lower bunk bed test 下床铺试验 系指火源布置在一个下床铺上,并点燃放在枕块中前部(朝向门外)的引燃物的试验。 Lower bunk bed test means a testing which fire arranged in one lower bunk bed and ignited with the igniter located at the front(towards door)centerline of the pillow.

Lower cargo purchase block 下吊货滑车 吊货滑车组中位于吊钩座处的滑车。

Lower cross seat 下座板 在救生艇内底横向水平布设,供人员乘坐的板材。

Lower dead center(LDC) 下死点

Lower decks 下甲板 系指上层连续甲板以下第1层连续甲板为第2甲板,依次向下为第3甲板……,总称为下甲板。 Lower decks are continuous deck next below the uppermost continuous deck is to be named as 2nd deck, and so on. They are generally called lower decks.

Lower explosive limit(LEL) 爆炸下限(LEL) 系指在空气中易燃气体、蒸发气或雾的某一浓度,在此浓度以下,将不会形成爆炸性气体环境。 Lower explosive limit(LEL)means a concentration of flammable gas, vapour or mist in air, below which an explosive gas atmosphere will not be formed.

Lower flammable limit(LFL) 最低燃烧下限 系指可能引起燃烧的最低温度极限点。

Lower frequency response(LFS) 低频响应 也称为差频响应,是由风和二阶波浪力的差频成分(波漂力)引起的锚泊浮式结构的慢漂运动。

Lower hull(lower pontoon) 下浮体 半潜式钻井装置中,位于立柱之下,具有浮力的结构物。

Lower hull(SWATH) 下潜体(小水线面双体船) 系指连接在支柱体下面沉浸在水下的圆形或类似椭圆形的鱼雷结状结构。 Lower hull means round or oval torpedo-like structure positioned at the base of vertical struts submerged in water.

Lower hull breadth 下潜体宽度 对小水线面双体船而言,系指单个下潜体的最大宽度。

Lower hull length 下潜体长度 对小水线面双体船而言,系指单个下潜体从艏缘量至艉缘的水平长度。若一个片体中的下潜体只数为1个以上时,应作累加计入。

Lower hulls 下壳体 系指柱稳式或坐底式平台下部与若干立柱相连接的连续浮体。 Lower hulls are the continuous buoyant structures connected to the bottom of columns of a column-stabilized or submersible unit.

Lower ice waterline(LIWL) 低位冰区水线 系指船舶在冰区航行预定水线最低点的包络线。该水线可为折线。 Lowest waterline is to be the envelope of the lowest points of the waterline at which the ship is intended to operate in ice. The line may be a broken line.

Lower intervention package(LIP) 底部修井装置总成

Lower lubricator package(LLP) 底部防喷系统总成

Lower mast 桅柱 系指桅肩以下的桅体部分。

Lower pintle(bottom pintle) 下舵销 系指将舵叶与艉框底部或挂舵臂下部连接的舵销。

Lower platform 舷梯下平台 装于舷梯下端的舷梯平台。

Lower portions of the cargo holds and ballast tanks 下层部分货舱和压载舱　系指在空载水线以下的那部分。　Lower portions of the cargo holds and ballast tanks are considered to be the parts below light ballast water line.

Lower stock 下舵杆　系指平板舵中位于舵杆轴线处的杆件。

Lower support of bucket ladder 斗桥下支承　固定在斗桥下部腹板上,作为起吊斗桥的着力点的轴和轴套的总称。

Lower tank 底舱　系指在单底船内,位于船底部的舱。

Lower tumbler 下导轮　安装在斗桥下端专用承座上,当斗链运转时,起导向作用的比上导轮大的带轴从动导轮。

Lowermost steady speed test 最低稳定转速试验　在航行试验时,为检查主机在带动轴系时正车最低转速的稳定性所做的试验。

Lowest design temperature of steel 钢材的最低设计温度　系指构件钢材等级选择时的基准温度,并假定等于日平均大气温度的最低值,该值应根据平台预期作业海域的气象资料所确定。

Lowest mean daily temperature 最低日统计的平均温度　系指相应区域一年内日温度曲线图上的最低值。对受季节限制的营运,适用营运期内的最低值。Lowest mean daily temperature is the lowest value on the annual mean daily temperature curve for the area in question. For seasonally restricted service the lowest value within the time of operation applies.

Lowest monthly mean temperature 最低月统计的平均温度　系指相应年最冷月份的月统计平均温度。Lowest monthly mean temperature is the monthly mean temperature for the coldest month of the year.

Lowing gear of hatch cover 舱盖拖曳装置　机械舱盖中,拖曳舱盖板使其移动就位的装置。

Low-load tuning 低负荷调整　为执行 IMO 环保会关于"新船能效设计指数 EEDI"的要求,柴油机制造厂商采用了一种能使柴油机在低负荷时降低燃油消耗的调整方法。

Low-noise electronic design 低噪声电子设计

Low-noise optimization design/Optimized design for low noise 低噪声优化设计

Low-pressure compressor 低压压气机　压缩过程由几台压气机完成时,首先对空气或其他气体进行压缩的压气机。

Low-pressure turbine 低压涡轮　高温燃气或其他气体最后在其中进行膨胀的涡轮。

Low-tide elevation 低潮高地　系指在低潮时四面环水并高于水面,但在高潮时没入水中的自然形成的陆地。如果低潮高地全部或一部分与大陆或岛屿的距离不超过领海的宽度,该作为高地的低潮线可作为测算领海宽度的基线。如果低潮高地全部与大陆或岛屿的距离超过领海的宽度,则该高地没有其自己的领海。Low-tide elevation is a naturally-formed area of land which is surrounded by and above water at low-tide but submerged at high-tide. Where a low-tide elevation is situated wholly or partly at a distance not exceeding the breadth of the territorial sea from the mainland or an island, the low-water line on that elevation may be used as the baseline for measuring the breadth of the territorial sea. Where a low-tide elevation is wholly situated at a distance exceeding the breadth of the territorial sea from the mainland or an island, it has no territorial sea of its own.

Low-viscosity substance 低黏度物质　对"73/78 防污公约"附则 II 而言,系指非高黏度物质的有毒液体物质。　For the purposes of the "MARPOL73/78" Annex II, low-viscosity substance means a noxious liquid substance which is not a high-viscosity substance.

Low-voltage main busbar 低压主汇流排　系指在采用较高电压的主发电机时,通过变压器将其变为较低电压,直接(不再经变压器变压)对船舶正常操作和满足正常居住条件所需的动力辅助设备供电的汇流排。

Low-voltage system 低压系统　系指:(1)工作于额定频率为 50 Hz 或 60 Hz、导体之间最高电压不超过 1 000 V 的交流系统,或在额定工作条件下导体之间最高瞬时电压不超过 1 500 V 的直流系统;(2)额定电压大于 50 V 均方根值到 1 000 V 在内的交流系统和在额定工作条件下最高瞬时电压值在 50 V~1 500 V 之间的直流系统。　Low-voltage system is:(1) a system operating with the maximum rated voltage not exceeding 1,000 V inclusive and with rated frequencies of 50 Hz or 60 Hz, or direct-current system where the maximum instantaneous value of the voltage under rated operating conditions does not exceed 1,500 V;(2) those alternating current systems with rated voltages greater than 50 V r.m.s. up to 1000 V r.m.s. inclusive and direct current systems with a maximum instantaneous value of the voltage under rated operating conditions greater than 50 V up to 1 500 V inclusive.

LP cylinder (蒸汽轮机)低压缸

LPG 液化石油气　系指在常温和大气压下呈气态,通过增压和降温可使之保持液态的轻质碳氢化合物的混合物,其基本成分为丙烷、丙烯、丁烷、丁烯。它也可由商用丁烷、商用丙烷和两者混合物构成。

LPG carrier LPG 运输船　系指运载液化石油气的船舶。 LPG carrier means a ship carrying liquefied petroleum gas.

LPG fuel system 液化石油气燃油系统　系指以液化石油气为燃料的船舶的供油系统。 LPG fuel system means a oil supplying system in a ship using liquefied petroleum gas as fuel.

LPG system LPG 系统　预定用于贮存、供应、监测或控制燃气流至燃气器具(包括该器具)的罐、安全装置、压力调节器、连接器、阀、管路、短管、软管、附件和装置的组合所构成的系统。注意：罐为可替换件，且可不随船上的 LPG 系统一起供应。 LPG system means the system consisting of an arrangement of cylinder(s), safety device(s), pressure regulator(s), connection(s), valve(s), piping, tubing, hose, fitting(s) and devices intended to store, supply, monitor or control the flow of fuel gas up to and including the appliance. Note：The cylinders are replacement items and may not be supplied with the LPG system in the craft.

LPG tourist boat 液化石油气作燃料动力游艇　系指用液化石油气作燃料(LPG)(放入动力装置中)的游艇。 LPG Tourist Boat is a motorboats using liquefied petroleum(LPG) as fuel(in power plant).

LR certificate 英国劳氏船级社证书　该证书的形式是由英国劳氏船级社根据其规范的要求，对所进行试验和检查的结果认为满意而颁发的。 This type of certificate is issued by LR based on the results of testing and inspection being satisfactorily carried out in accordance with the requirements of these Rules.

LR type approval LR 形式认可　是一个公正的发证体系，它提供了独立的第三方形式认可证书，证明产品符合特定的标准或技术条件。它是根据对设计的审阅和形式试验，或者如试验并不合适，则需根据对设计的分析。 LR type approval is an impartial certification system that provides independent third-party Type Approval Certificates attesting to a product's conformity with specific standards or specifications. It is based on design review and type testing or where testing is not appropriate, a design analysis.

LR type approval system LR 形式认可体系　是对产品按技术条件、标准或规定进行评定，以校核其满足所述的要求和通过选定的试验，且验证其符合特定的性能要求的过程。此试验在原型或代表已制造的被认可产品的随机选取的产品上进行。此后，要求生产者采用质量管理程序和过程，以确保所提交的每一项产品都与已经形式认可者相一致。 LR type approval system is a process whereby a product is assessed in accordance with a specification, standard or code to check that it meets the stated requirements and through selective testing demonstrates compliance with specific performance requirements. The testing is carried out on a prototype or randomly selected product(s) which are representative of the manufactured product under approval. Thereafter, the producer is required to use Quality Control procedures and processes to ensure that each item delivered is in conformity with that which has been Type Approved.

LRIT Data User 船舶远程识别和跟踪用户　系指决定接收其有权接收的船舶远程识别和跟踪信息的缔约国政府或搜救(SAR)服务机构。 LRIT Data User means a Contracting Government or a search and rescue (SAR) service which opts to receive the LRIT information it is entitled to.

LRIT information LRIT 信息　系指第Ⅴ/19-1.5条规定的信息。 LRIT information means the information specified in regulation Ⅴ/19-1.5.

Lube oil 润滑油(滑油)
Lube oil piping 滑油管系
Lube oil tank 滑油舱(柜)
Lubricant 润滑剂(浸渍润滑剂)
Lubricant separator 滑油分离机　用于分离滑油中的杂质和水分的油分离机。
Lubricate 润滑
Lubricating 润滑
Lubricating oil 润滑油(滑油)
Lubricating oil analysis 滑油分析
Lubricating oil analysis information 滑油分析资料
Lubricating oil analysis record table 滑油分析记录表
Lubricating oil batch purification 滑油定期净化　定期将滑油系统内一定数量的滑油进行净化的滑油处理方式。
Lubricating oil by pass purification 滑油连续净化　将滑油系统内的滑油连续进行净化的滑油处理方式。
Lubricating oil comsumption record table 滑油消耗量记录表
Lubricating oil condition 滑油状态
Lubricating oil condition monitoring system 滑油状态监控系统
Lubricating oil consumption 滑油消耗率　单位时间内发动机单位输出功率的滑油消耗量。
Lubricating oil cooler 滑油冷却器　滑油系统中，

用水冷却滑油的热交换器。

Lubricating oil cooler for exhaust-gas turbo-charger 废气涡轮增压器滑油冷却器 冷却废气涡轮增压器滑油的热交换器。

Lubricating oil dilution 滑油稀释 在运行中燃油逐渐混入滑油,使滑油黏度降低的现象。

Lubricating oil drainage tank 污滑油柜 收集各种污滑油的柜。

Lubricating oil filling main 滑油注入总管 注入滑油的总管路。

Lubricating oil filter (lubricating oil strainer) 滑油滤清器 过滤润滑油杂质的滤器。

Lubricating oil gravity tank 滑油重力柜 在重力式滑油系统中,布置在一定高度的滑油柜。

Lubricating oil heater 滑油加热器 滑油系统中,供冷天启动柴油机时加热循环滑油的热交换器。

Lubricating oil low-pressure automatic shut-down mechanism 滑油低压自动停车装置 当内燃机滑油压力低于一定值时,能使燃油立即停止供应的安全装置。

Lubricating oil pressure 滑油压力 内燃机各润滑点滑油的压力。

Lubricating oil pump 滑油泵 抽送润滑油的泵。

Lubricating oil purification heater 滑油分离加热器 将进入油分离机前的滑油加热到适当黏度使之易于净化分离的加热器。

Lubricating oil separator 滑油分油器

Lubricating oil settling tank 滑油澄清柜 使润滑油中杂质释放沉淀的柜。

Lubricating oil sludge tank 滑油油渣柜 收集从滑油分离机分离出来的油渣的柜。

Lubricating oil storage tank 滑油贮存柜 贮存新润滑油的柜。

Lubricating oil sumo tank 滑油循环柜 在润滑系统中用以汇集主、辅柴油机底部润滑油供再循环使用的柜。

Lubricating oil sump tank 污滑油舱 系指供贮存污滑油的舱。

Lubricating oil system 滑油系统 将滑油输送到主、辅机、减速齿轮箱、轴承等进行润滑和冷却的机械设备、管路及附件的总称。

Lubricating oil tank (lubricating oil compartment) 滑油舱 供贮存滑油的舱柜。

Lubricating oil temperature 滑油温度 进出内燃机的冷却水滑油温度。

Lubricating oil test 润滑油试验

Lubricating oil test for tightness 润滑油密性试验

Lubricating oil transfer pump (lubricating oil pump) 滑油输送泵 把润滑油从一个舱(柜)输送到另一个舱(柜)的泵。

Lubricating oils and blending stocks 润滑油和调和油料

Lubricating pipe 润滑油管

Lubricating property 润滑性质

Lubrication 润滑(润滑法)

Lubrication groove 润滑脂

Lubrication ring 润滑环

Lubrication system 润滑系统

Lubrication trouble 润滑故障

Lubricator 涂油器(注油器) (1)对缆绳涂刷滑油的设备;(2)向气缸臂供应润滑油的装置。(3)为保护各类绞车上的绳索以防锈蚀并延长其使用寿命,对绳索进行去水干燥、涂油的一种设备。

Lubricator tubular (LUB) 防喷管

Lubricity 润滑性(含油性)

LUF system LUF 系统 系指集装箱成组运输系统。

Luff type floating crane 定机变幅式起重船 起重吊机为固定式,其舷外跨距能变幅的起重船。有工作变幅和非工作变幅两种。

Luffing arm type davit 倒臂型艇架 放艇时,吊艇臂仅绕其下端单销轴转动倾倒,从而吊艇出舷的重力式艇架。

Luffing type boat davit 摇倒式艇架 靠摇动手柄,转动吊艇架机构,使吊艇杆倒出而吊艇出舷的吊艇架。

Lug-connection 直接连接 系指构件的腹板和面板都直接连接到舱壁、甲板或内底板上,并在壁板的另一面也有构件的有效支撑。 Lug-connection is such a connection as both web and face bar of stiffener are effectively attached in the bulkhead plating, decks or inner bottoms which are strengthened by effective supporting members on the opposite side of plating.

Luggage 行李 系指承运人根据运输合同运输的任何物品或车辆,但不包括:(1)根据租船合同、提单或主要与货物运输有关的其他合同所运输的物品和车辆;(2)活动物。 Luggage means any article or vehicle carried by the carrier under a contract of carriage, excluding: (1) article and vehicle carried under a charter party, bill of lading or other contract primarily concerned with the carriage of goods, and (2) live animals.

Luggage room 行李舱 供存放旅客行李的舱室。

Luminance 亮度 系指发光体在给定方向上单位

投影面积中发的光强。单位为坎每平方米(cd/m^2)。

Luminescent efficiency 发光效率 系指光源所发射的光通量与所消耗的功率之比。单位为流明每瓦(lm/W)。

Lump sum charter 整笔运费租船

Lurching 横摇突倾 在横摇中突然出现的特大倾侧。

L_{WL} line L_{WL} 线 系指船艏、船舯和船艉最大吃水的连线(允许折线)。L_{WL} line means the line defined by the maximum draughts fore, amidships and aft(may be a broken-line).

Lysergic acid diethylamide(LSD) 麦角酸二乙胺(LSD) 迷幻剂之一,是一种合成的白色粉末,能制成粗糙的药丸或其他粗糙的形状。也有浸渍纸形状,尺寸如同邮票,经常带有神秘的符号或卡通人物或微型画。其纯净形式是灰白色的或无色的溶剂。没有气味。Lysergic acid diethylamide(LSD) is one of hallucinogens, it is found as impreguated papers the size of postage stamps, often with mystic signs or sheets of cartoon characters or miniature pictures. It is a pale or colourless solution in its pure form. It is odourless.

M

M standard ship's station M 标准船站(卫星通信) 是一种小型的数据通信船站。于 1993 年投入商业使用。它是可以提供电话、低速传真和低速数据业务移动终端。

M type single-pull hatch cover 斜置型舱盖 舱口开启后,舱盖板倾斜的舱口开启后,以减小舱盖板收藏位置的滚翻式舱盖。

Machine 机械

Machine component 机械构件(零件)

Machine finish margin 机械加工裕度

Machine forging 机械锻造

Machine riveting 机铆

Machined condition 机械加工后状态

Machined surface 机加工面

Machine-grooved type 机械槽型

Machinery 机械(机器)

Machinery alarm 机器报警 系指指示机器和电气设备故障或其他异常情况的报警。Machinery alarm means an alarm which indicates a malfunction or other abnormal condition of the machinery and electrical installations.

Machinery arrangement 机舱布置

Machinery bed 机座

Machinery casing 机舱棚(机舱口围阱)

Machinery chock 甲板机械下的木质垫块(机械座垫块)

Machinery chocking compound 机械填垫复合物

Machinery class notation 轮机入级附加标志

Machinery classification certificate 轮机船级证书

Machinery compartment 机舱

Machinery condition monitoring(MCM) 机械状态监测

Machinery control and automation 机舱自动化

Machinery control compartment 机器操纵室

Machinery damage 机损

Machinery foundations 机械设备底座

Machinery friction pair 机械摩擦副

Machinery installation 机械装置

Machinery items 机械零件

Machinery manufacturing 机械制造

Machinery noise 机械噪声

Machinery number 主机特征数(机号)

Machinery planned maintenance scheme(MPMS) 机械计划维修安排

Machinery plant 机械装置

Machinery record 机舱记录

Machinery room 机舱

Machinery space 机器处所 系指:(1)设有输出功率 110 kW 以上的内燃机、发电机、燃油装置、推进装置、主要电机的处所和类似的处所,以及通往这些处所的围壁通道;(2)介于一个处所的水密限界面之间,供安置主辅推进机械,包括主要供推进装置用的锅炉、发电机和电动机的各个处所。对于特殊布置的船舶,主管机关可以规定机器处所的范围;(3)所有 A 类机器处所和所有其他设有推进装置、锅炉、燃油装置、蒸汽机和内燃机、发电机和主要电动机、加油站、制冷机、防摇装置、通风机和空调机的处所以及类似处所,和通往这些处所的围壁通道。Machinery space are:(1) machinery spaces of category A and other spaces containing propulsion machinery, boilers, oil fuel units, steam and internal combustion engines, generators and major electrical machinery, oil filling stations, refrigerating, stabilizing, ventilation and air conditioning machinery, and similar spaces, and trunks to such spaces; (2) spaces between the watertight boundaries of a space containing the main and auxiliary propulsion machinery, including boilers, generators and electric motors primarily intended for propulsion. In the case of unusual arrangements, the Administration may define the limits of the machinery spaces; (3) all machinery spaces of category A and all other spaces containing propelling machinery, boilers, oil fuel units, steam and internal combustion engines, generators and major electrical machinery, oil filling stations, refrigerating, stabilizing, ventilation and air conditioning machinery, and similar spaces, and trunks to such spaces.

Machinery space be periodically unmanned 无人值班机器处所

Machinery space bulkhead 机舱舱壁(机器处所舱壁)

Machinery space periodically unattened(AUT-0) 机器处所周期无人值班 系指推进装置由驾驶室控制站遥控,机器处所包括机舱集控站(室)周期性无人值班。 Machinery space periodically unattened (AUT-0) means main propulsion machinery remotely controlled from bridge control space (BCS), machinery pace including central control space periodically unattented.

Machinery spaces and main galleys 机器处所和主厨房 系指：(1)主推进机舱(电力推进电动机舱除外)及锅炉舱；(2)极少或无失火危险的液舱、空舱及辅机处所和具有中等失火危险的辅机处所、货物处所、货油舱和其他油舱以及其他类似处所以外的设有内燃机或其他燃油、加热或泵送装置的辅机处所；(3)主厨房及其附属间；(4)上述处所的围阱及舱棚。 Machinery spaces and main galleys are: (1) Main propulsion machinery rooms (other than electric propulsion motor rooms) and boiler rooms. (2) auxiliary machinery spaces other than tanks, voids and auxiliary machinery spaces having little or no fire risk and auxiliary machinery spaces, cargo spaces, cargo and other oil tanks and other similar spaces of moderate fire risk which contain internal combustion machinery or other oil-burning, heating or pumping units. (3) main galleys and annexes. (4) trunks and casings to the spaces listed above.

Machinery spaces of category A A类机器处所 系指装有下列任何一个设备的处所和通往这些处所的围壁通道：(1)用作主推进的内燃机；(2)用作非主推进以外的合计总输出功率不小于375kW的内燃机；(3)燃油锅炉或燃料油装置或除锅炉之外的惰性气体发生器、焚烧炉等的油烟装置。 Machinery spaces of category A are those spaces and trunks to such spaces which contain either: (1) internal combustion machinery used for main propulsion; (2) internal combustion machinery used for purposes other than main propulsion where such machinery has in the aggregate a total power output of not less than 375kW; or (3) any oil-fired boiler or oil fuel unit, or any oil-fired equipment other than boilers, such as inert gas generators, incinerators, etc.

Machinery spaces(ships and offshore units) 机器处所(船舶及近海装置) 装有推进机械、锅炉、燃油装置、蒸汽机和内燃机、碳氢化合物处理设备、水处理设备、钻井及有关设备,发电机及主要电机,加油站、冷冻、减摇、通风和空调机械的处所,以及类似处所和通往这些处所的围壁通道。 Machinery spaces mean those spaces containing propelling machinery, boilers, oil fuel units, steam and internal combustion engines, hydrocarbon process equipment, water treatment and handling equipment, drilling and associated equipment, generators and major electrical machinery, oil filling stations, refrigerating, stabilizing, ventilation and air-conditioning machinery, and similar spaces and trunks to such spaces.

Machinery trial 机械试车
Machinery weight 机械质量
Machining 机加工(刨尽)
Machining accuracy 机加工精度
Machining alligatoring 加工龟裂
Machining crack 机加工开裂
Machining works 机加工工厂
Machinist 机工(机械师)
Macro-fouling 大型污底 系指人类肉眼清晰可见的大型多细胞生物,例如藤壶、管虫或藻类的叶子。 Macro-fouling means large, distinct multicellular organisms visible to the human eye such as barnacles, tubeworms, or fronds of algae.

MAFI truck MAFI拖车 系指一种低拖曳架多轴车辆。

Magazine 弹药舱/弹药库 供存放弹药的舱室。
Magazine 钻探管架(杂志)

Magazine boxes 弹药箱 系指容量不大于$3m^3$的非整体式可抛弃轻便弹药库。 Magazine boxes are non-integral, portable magazines with a capacity less than or equal to $3m^3$ and capable of being jettisoned overboard.

Magazine boxes(explosives) (爆炸品)箱形库房(箱式仓库) 系指容积小于$3m^3$非整体式可移动库房。箱形库房应位于露天甲板上不受海浪直接冲击的位置。该位置应对从配餐间、泵房等处排出的热空气或危险气体设有足够的防护。应注意爆炸品遭受无线电波辐射时产生危险的可能性。箱形库房应位于距甲板及任何甲板室至少为 0.1m 的露天甲板上,并且适合将内存物抛弃的位置。箱形库房应为水密金属结构,箱体及箱盖的厚度不得小于 3 mm。如箱体可能会暴露于阳光直射之下,应设有遮阳板。箱形库房应有明显的标牌注明:(1)此箱为危险箱形库房;(2)明火或火焰不得靠近;(3)此箱应保持关闭。 Magazine boxes are non-integral, portable magazine with a capacity of less 3 m³. Magazine boxes should be located on a weather deck at least 0.1 m from the deck and any deckhouse and in position suitable for jettisoning the contents. Magazine boxes should be of watertight metal construction having a body and lid thickness of no less than 3 mm. Where the box may be exposed to direct sun, sun shields should be provided. Magazine boxes should be clearly labeled indicating:(1) the container is a magazine box;(2) open lights and flame should be kept away;(3) the box should be kept shut.

Magazine lockers 弹药室 系指容积不大于 3 m³ 的弹药库,用于安全地储存爆炸性物品,而已经建成的其他弹药库设施不能存储这些物品。这种弹药库应该独立设置,其四周应保持开阔,不可有任何舱室与其相连。 Magazine lockers are magazines less than or equal to 3 m³, designed and constructed for the safe stowage of explosive stores for which in built magazine facilities have not been provided. They are to be free standing and surrounded by an air gap such that they do not have an adjacent compartment.

Magazines for stowing explosives on board special purpose ships 特殊用途船舶装运爆炸品的储存仓库 系指用于装运爆炸品的各类仓库。 Magazines for stowing explosives on board special purpose ships mean all categories of magazines used for carrying explosives.

Magdynamo 永磁发动机
Magnaflux 磁粉检测法(磁粉检测机,磁通量)
Magnaflux inspection 磁力检测(磁力检验)
Magnaflux test 磁性检测(磁性探伤)
Magnesia(deadburned) 氧化镁(烧僵) 制成砖形块。通常呈白色、褐色或灰色。大小、外形和装卸方式与碎石类似,干燥且多粉尘。烧僵的氧化镁是天然菱镁矿砂经高温焙烧后形成的非活性氧化镁,与水结合不会产生自热。无特殊危害。该货物为不燃物或失火风险低。 Magnesia(deadburned) is a material manufactured in briquette form and is usually white, brown or grey. It is very similar in size, appearance and handling to gravel and is dry and dusty. Deadburned magnesia is natural magnesite calcined at very high temperatures, which results in a non-reactive magnesium oxide, which does not hydrate or produce spontaneous heat. No special hazards. This cargo is non-combustible or has a low fire-risk.

Magnesia(unslaked) 氧化镁(未熟化) 与水结合形成氢氧化镁,体积膨胀并释放热量。可能将引燃温度低的物质引燃。与石灰(生的)相似,但活性较弱。对眼睛和黏膜有腐蚀性。该货物为不燃物或失火风险低。 Magnesia(unslaked) combines with water to form magnesium hydroxide with an expansion in volume and a release of heat. It may ignite materials with low ignition temperatures. Similar to LIME(UNSLAKED) but is less reactive. It is corrosive to eyes and mucous membranes. This cargo is non-combustible or has a low fire-risk.

Magnesite(natural) 天然菱镁矿石 呈白色至黄色。无特殊危害。该货物为不燃物或失火风险低。 Magnesite is white to yellow in colour. No special hazards. This cargo is non-combustible or has a low fire-risk.

Magnesium nitrate(UN 1474) 硝酸镁 白色晶体,溶于水。吸湿。自身虽不可燃,但其与可燃物质的混合物容易点燃并可能猛烈燃烧。该货物吸湿,受潮会结块。 Magnesium nitrate(UN 1474) is a white crystals, soluble in water. It is hygroscopic. Although non-combustible by itself, mixtures with combustible material are easily ignited and may burn fiercely. This cargo is hygroscopic and will cake if wet.

Magnet 磁铁(磁性)
Magnet crane 电磁吊车(电磁吸盘式起重机)
Magnet steel 磁性钢
Magnet valve 电磁阀
Magnetic 磁的
Magnetic ageing 磁老化
Magnetic alloy 磁性合金
Magnetic amplifier 磁放大器
Magnetic attraction 磁吸引(磁引力)
Magnetic auto steering gear 磁性自动操舵装置
Magnetic axis 磁轴
Magnetic azimuth 磁方位角
Magnetic brake 电磁闸
Magnetic clutch 电磁离合器
Magnetic compass 磁罗经 是一种精密完善的指南针,它是利用磁性指针在地球磁场的作用下自动指北特性制作而成。标准磁罗经和操舵磁罗经是船舶航行必需的航海设备。

Magnetic controller 磁力控制器 是一种供鼠笼型交流电动机频繁启动(或正/反向)、停止运行、且有过载、短路、断相和失压保护,可以遥控按钮操作,也可以作为用手动-自动转换控制的简单可靠的控制装置。按不同启动要求分成全电压直接启动，△-Y 降压启动和自耦变压器降压启动三种形式，每种形式还按电流等级分成若干规格。其中的控制仅适用于二位式开关量控制，通过压力控制器、温度控制器、液位控制器、行程控制器等器件的上、下来实现控制。

Magnetic coupling　电磁连轴节

Magnetic crack detection　磁力裂纹探测（磁粉检测）

Magnetic crack detector　磁力裂纹检测仪

Magnetic discharge welding　电磁焊

Magnetic door holder　门定位磁座

Magnetic drive　磁力传动（带电磁离合器的传动装置）

Magnetic fault detection　磁粉检测

Magnetic field　磁场

Magnetic field equalizer　磁场均衡器（补偿磁铁系统）

Magnetic field intensity　磁场强度

Magnetic flaw detector　磁粉检测仪（磁力检测仪）

Magnetic flux　磁通量（磁性焊剂，磁性熔剂）

Magnetic flux arc welding　磁性焊剂电弧焊

Magnetic flux gas-shielded arc welding　磁性焊剂气体保护电弧焊

Magnetic ignition　电磁点火

Magnetic induction　磁感应

Magnetic inspection　磁力检验（磁力检测法）

Magnetic iron　磁铁

Magnetic lubricating oil filter　滑油磁性滤清器　装有永久磁铁过滤润滑油中铁屑的过滤器。

Magnetic material　磁性材料　系指强磁性物质，是古老而用途十分广泛的功能材料，磁性材料早期被大比特公司根据用途分为：铁氧体、钕铁硼、钐钴磁体、铝镍钴磁铁、铁铬钴磁铁等五大类。而物质的磁性早在3 000年以前就被人们所认识和应用，例如中国古代用天然磁铁作为指南针。现代磁性材料已经广泛的用在我们的生活之中，例如将永磁材料用作马达，作为变压器中的铁心材料，作为存储器使用的磁光盘，计算机用磁记录软盘等。大比特公司资讯上说，磁性材料与信息化、自动化、机电一体化、国防、国民经济的方方面面紧密相关。而通常认为，磁性材料系指由过渡元素铁、钴、镍及其合金等能够直接或间接产生磁性的物质。磁性材料按磁化后去磁的难易可分为软磁性材料和硬磁性材料。磁化后容易去掉磁性的物质叫软磁性材料，不容易去磁的物质叫硬磁性材料。一般来讲软磁性材料剩磁较小，硬磁性材料剩磁较大。

Magnetic metal　磁性金属

Magnetic particle　磁粉

Magnetic particle(flux test)　磁粉探伤

Magnetic particle equipment　磁粉检测设备（磁粉探伤设备）

Magnetic particle examination　磁粉检测（磁粉检查）

Magnetic particle inspection(MPI)　磁粉检测（磁粉检验）

Magnetic particle method　磁粉检测法

Magnetic particle test　磁粉检测（磁粉探伤）

Magnetic prospecting　磁力勘探　曾称"磁法勘探"。应用磁力仪测量海区磁力异常现象；来勘探矿藏和研究地下地质构造。

Magnetic spark plug　电磁火花塞

Magnetic steel　磁性钢

Magnetic tape recording　磁带记录　信息以一连串单元的形式在磁带表面上储存起来。这样在磁带上记录了各种类型的数据。实际工作中采用了双磁带机。当一盘磁带记满后，磁带机会自动转到另一台磁带机，且互相有一段互相重叠以避免丢失数据。磁带记录的好处是便于计算中心进行后处理，且后处理所获得的成果精度要高于测量所获得的成果精度。

Magnetic tapes　磁带机　将二进位的信息记录在磁带上，并能把记录在磁带上的信息读出的装置。每个磁带机包括读写部分和磁带驱动部分。磁带机可用来记录各种数据，把它们保存起来供资料分析和处理成果时使用。磁带机也可用来向计算机输入各种应用程序。

Magnetic test　磁性检测

Magnetic trip　电磁式脱扣器

Magnetic valve　电磁阀

Magnetoelectric sensor　电磁探头

Magnetometer transducer　磁力仪探头　使磁力仪产生和接收核子旋进信号的敏感元件。

Magnetometer winch　磁力仪绞车　为收、放磁力仪探头、电缆进行海洋磁力测量工作用的绞车。

Magneto-fluid　磁流体　又称磁性液体、铁磁流体或磁液，是一种新型的功能材料，它既具有液体的流动性又具有固体磁性材料的磁性。是由直径为纳米量级（10纳米以下）的磁性固体颗粒、基载液（也叫媒体）以及界面活性剂三者混合而成的一种稳定的胶状液体。该流体在静态时无磁性吸引力，当外加磁场作用时，才表现出磁性，正因如此，它才在实际中有着广泛的应用，在理论上具有很高的学术价值。用纳米金属及合金粉末生产的磁流体性能优异，可广泛应用于各种苛刻条件的磁性流体密封、减振、医疗器械、声音调节、光显示、磁流体选矿等领域。

Magneto-fluid propulsion technologies for nuclear submarine 磁流体推进核潜艇技术 该技术拥有国:中国。见图 M-1。

图 M-1 磁流体推进核潜艇技术
Figure M-1 Magneto-fluid propulsion technology for nuclear submarine

Magnification factor 放大因数 在一定频率时输出幅值对输入幅值的比值。

Magnification factor for deflection 挠曲增大因数

Mail 邮件 系指邮政部门托运的和交给邮政部门的信件或其他物品。 Mail means the dispatches of correspondence and other objects tendered by intended for delivery to postal administrations.

Mail room 邮件舱(总管,主母线,主要的)

Main 总线(总管,主母线,主要的)

Main air bottle 主空气瓶

Main air compressor 主空压机

Main alarm level 主报警水位 系指货舱内传感器工作的较高水位。或除货舱以外的舱室内适用 SOLAS 第Ⅻ/12.1 条要求的唯一水位。 Main alarm level means the higher level at which the sensor(s) in the cargo hold space will operate. or the sole level in spaces other than cargo holds to which the requirements of SOLAS regulation Ⅻ/12.1 apply.

Main attachments 主要附件

Main ballast sea valve 主压载水舱通海阀

Main bearing 主轴承 支撑曲轴主轴颈的轴承。

Main bearing bolt 主轴承螺栓

Main bilge pipe 舱底水总管

Main bilge pumping system 主舱底排水系统

Main boiler 主锅炉 系指:(1)用于驱动推进用蒸汽机的锅炉;(2)锅炉产生的蒸汽供给主推进机械的锅炉。 Main boiler means:(1)the boiler used in moving the propulsion steam engine;(2) main boilers are boilers supplying steam for main propulsion engines.

Main bucket ladder 主斗桥 复式斗桥中位于下面的一个较长的能承受链斗导轮主力的斗桥。

Main bus-bar 主汇流排 系指由主发电机直接供电的汇流排。

Main cargo line(cargo oil transfer main pipe line) 货油总管 与各货油舱支管连接的总管。

Main circulating pump 主循环水泵

Main class symbol 主要船级标志 表示船舶在符合规范建造和维护方面要求的符合程度。对每一艘入级船舶都有的主要入级标志。 Main class symbol expresses the degree of compliance of the ship with the rule requirements as regards its construction and maintenance. There is one main class symbol, which is compulsory for every classed ship.

Main combustion zone(flame zone, primary combustion zone) 主燃烧区 燃烧室中空气燃油比接近化学当量值,因而燃烧连续进行且大部分燃料在其中燃烧完毕的区域。

Main condenser 主冷凝器 以冷凝主汽轮机排气为主的冷凝器。

Main condenser circulating pump 主冷凝器循环泵 抽送蒸汽动力装置中主冷凝器循环冷却水的泵。

Main conning position 主指挥站 系指主要由驾驶员指挥用的场所。 Main conning position is a conning position which is mainly used by navigators.

Main console 主控台

Main control board 主控制板

Main control station 主控台(主指挥所)

Main control station 主控制站 (1)是一个具备足够设备的并且足够控制主推进器机械的控制站。通常主推进器机械是通过这里进行控制的,在驾驶桥楼外面安装主控制设备;(2)系指所有自动控制设备的监视设施集中布置并在正常工作实施操作监控的处所。 Main control station is (1)a control station provided with equipment necessary and sufficient to control the main propulsion machinery, and from which the main propulsion machinery is normally controlled, of the ship which provides the main control equipment at the outside of the navigation bridge;(2) the space in which monitoring means for all automated installations are concentrated and operations of such installations are monitored under normal conditions.

Main control station on bridge 桥楼上的主控制站 是一个船舶在驾驶桥楼上提供主控制设备的导航桥楼并控制着船舶的主推进装置。 Main control station on bridge is a navigation bridge of the ship which provides main control equipment at the navigation bridge and that the main propulsion machinery is normally controlled there.

Main control system 主控制系统

Main control valve 主控制阀

Main dimensions 主尺度 船舶的主尺度系指船舶的主要参数。主尺度不仅用于在许多方面比较一艘船与另一艘船,而且用于确定港口费、码头预定、向各有关行政当局和租船人提供信息,也用于船舶建造和入级。主尺度包括船长、船宽、型深、吃水、干舷、排水量、载重量和主机功率等。

Main engine(M/E) 主机 一般系指船舶动力装置中用于船舶推进用的发动机。

Main engine control console 主机控制台(主机操纵台)

Main engine control room 主机操纵室 通常系指以机械或电气方式集中操纵主机及其附属设备的舱室。

Main engine driven 由主机驱动

Main engine foundation 主机机座

Main engine fuel oil heater 主机燃油加热器 将进入主柴油机前的燃油加热到适当黏度以保证良好雾化的加热器。

Main engine output 主机输出功率 系指主机在船上实际持续输出的最大功率。 Main engine output is the actual maximum continuous output of the ship's propulsion machinery.

Main engine remote control system 主机遥控系统

Main engine revolution measuring system 主机转速测量系统 系指为主机遥控,安全系统提供转速信号且供主机安全保护装置用的系统。

Main engine revolution meter 主机转速表 系指用于驾驶室、机舱和其他有关部位指示主机转速的仪器。

Main engine room 主机舱 供安置主机及其附属设备的舱室。

Main engine room alarm ringer system 主机舱警铃系统

Main engine seating 主机基座 装设主机用的基座。

Main engine trial 主机试车

Main engine-driven generating set 主机驱动发电机组

Main entrance 大堂(豪华邮轮)

Main equipment 主要设备

Main essential services 主要设备 系指为保持推进、操舵需连续运转的设备,例如:(1)操舵装置;(2)调距螺旋桨装置;(3)为主辅柴油机服务的鼓风机、燃油供给泵、喷油嘴冷却泵、滑油泵和冷却水泵,以及推进用涡轮机所必需的相关设备;(4)为向主要设备供气的辅助锅炉服务的和在蒸汽轮机船上为 蒸汽装置服务的强力鼓风机、给水泵、循环水泵、真空泵、冷凝水泵以及油燃烧装置;(5)单独作推进/操舵用的方位推进器连同其滑油泵和冷却水泵;(6)用于电力推进装置的电气设备连同其滑油泵和冷却水泵;(7)向上述(1)至(6)设备供电的发电机及有关电源;(8)向上述(1)~(6)设备提供动力的液压泵;(9)重油黏度控制设备;(10)消防泵和其他灭火剂泵;(11)航行灯、航行设备和信号设备;(12)船内安全通信设备;(13)照明系统;(14)以上(1)~(13)所列设备的控制、监控和安全设备/系统。 Main essential services is that which need to be in continuous operation for maintaining the ship's maneuverability with regard to propulsion and steering:(1) steering gear;(2) controllable pitch propeller installation;(3) charging air blowers, fuel feeder pumps, fuel booster pumps, lubricating oil pumps and fresh cooling water pumps for main and auxiliary engines and turbines, so far as required for propulsion;(4) forced draught fans, feed water pumps, water circulating pumps, vacuum pump, condensate pumps and oil burning instalation for auxilatry steam boilers for the operation of main essential equipment and for steam instalations on steam turbine ships;(5) azimuth thrusters which are the sole means for propulsion/steering with lubricating oil pumps, cooling water pumps;(6) electrical equipment for electric propulsion plant with lubricating oil pumps and cooling water pumps;(7) electric generators and associated power soures supplying the equipment mentioned in(1) to(6) above;(8) hydraulic pumps suppying the equipment mentioned in(1) to(6) above;(9) viscosity control equipment for heavy oil;(10) fire pumps and other extinguishing agant pumps;(11) navigation lights, aids and signals;(12) internal safety communication equipment;(13) lighting system;(14) control, monitoring and safety devices/systems for equipment mentioned in(1) to (13) above.

Main feed pipe line 主给水管路 蒸汽动力装置中,锅炉装置在正常情况下使用的给水管路。

Main frame 主肋骨 防撞舱壁与艉尖舱壁之间,最下层甲板以下的肋骨。

Main gas turbine(propulsion gas turbine) 主燃气轮机 作为船舶推进用的燃气轮机。

Main gas turbine set 主燃气轮机组 作为船舶推进用的燃气轮机组。

Main generating station 主电站(主发电站) 系指主电源所在的处所。 Main generating station is the space in which the main source of electrical power is situated.

Main girders　主桁　是连接桩腿围阱的强力结构,通常由底板、舷侧板、强力甲板、沿主桁长度方向的内侧壁或水密舱壁以及连接并支撑这些舱壁的桁材所组成。　Main girders are the strength structures connecting the leg wells and generally composed of bottom plate, side plates, strength deck, internal side walls or watertight bulkheads along the length of the main girders as well as girders connecting and supporting such bulkheads.

Main hook load　主钩起重量　起重机主钩的额定起重能力。

Main hull　主船体　船体的主要部分,系指强力甲板以下的船体。

Main ice belt zone　主冰带区　自冰区满载水线和冰区轻载水线垂直向上和向下延伸,延伸距离见表 M-1。对设有球鼻艏的舰船,主冰带区的垂向范围应适当增加。　Main ice belt zone extends vertically above and below the ice deep waterline and ice light waterline by the distances shown in Table M-1. For ships fitted with a bulbous bow, the vertical extent of the main ice belt zone will be suitably increased.

表 M-1　主冰带区的垂向范围
Table M-1　Vertical extent of the main ice belt zone

冰级	冰区满载水线以上(mm)	冰区轻载水线以下(mm)
1AS	600	750
1A	500	600
1B	400	500
1C	400	500

Main journal　主轴颈　曲轴搁置在主轴承上的轴颈。

Main line　总管(主要管路)

Main lubricating oil pipeline　主要的滑油管路

Main machinery　主机(主要机械)

Main machinery console　主机控制台(主机操纵台)

Main member of main structure　主体结构的主要构件(半潜式平台)　系指保持平台的整体完整性的重要构件,如立柱的外壳、沉垫和上平台的外壳、水平桁撑和斜撑的外壳等。

Main motor switchboard　主推进电动机配电板

Main opening　主要开口

Main operating mode of dynamic positioning system　动力定位系统的主要操作模式　包括:(1)自动模式——自动定位舯向控制;(2)操纵杆模式——手动定位,可选手动/自动舯向控制;(3)手动模式——单独控制每个推力器的螺距、转速、方向、启动和停止;(4)自动航迹模式——当作自动定位控制的变化方式,船舶在可编程的参照点上移动。

Main operation plate(MOP)　主操作板

Main piece　主隔板　复板舵的构架中位于舵杆轴线处,保证舵叶强度的构件。

Main pipe　总管(主管)

Main pool　游泳池(豪华邮轮)

Main propulsion engines and turbines　主推进发动机和涡轮机　系指那些直接驱动或通过机械轴系非直接驱动主推进的机器,它们也可能驱动发电机为辅助机械提供动力。辅助发动机和涡轮机系指那些与为辅助用途提供动力的发电机组连接在一起为电力主推进、电动机或两者联合装置的机器。　Main propulsion engines and turbines are defined as those which drive main propelling machinery directly or indirectly through mechanical shafting and which may also drive electrical generators to provide power for auxiliary services. Auxiliary engines and turbines are defined as those coupled to electrical generators which provide power for auxiliary services, for electrical main propulsion motors or a combination of both.

Main propulsion gas turbine　全工况燃气轮机　在舰船各种航速下都能工作的主燃气轮机。

Main propulsion machinery　主推进装置(主推进机械)

Main propulsion machinery space　主推进机械处所

Main propulsion plant　主推进装置

Main propulsion system　主推进系统

Main propulsion thruster　主推力器

Main pump　主泵　主要的和经常使用的泵。如主给水泵、主冷却泵等。

Main push-towing rope　主缆　顶推系统部件中,保持顶推船队联结,承受一定预张力及回航时拉力的缆绳。

Main rotor　主转子　由压气机和涡轮的转子组合而成的旋转共同体。

Main safety function　主安全功能　为确保人员不直接立刻暴露于危险并以有组织的方式到达安全区域而必须保持的安全功能,安全区域在设施上或可以通过可控撤离到达。

Main sea discharge　主海水排水孔

Main sea inlet　主海水进水孔

Main shaft 主轴 系指传递能量给叶轮的轴。Main shaft is a shaft that transmits power to the impeller blades.

Main shaft driven feed pump 轴带给水泵 由主机减速齿轮轴系驱动的给水泵。

Main shut-off valve (LPG system) 主截止阀（LPG 系统） 使整个 LPG 系统与供气压力端气源隔离的装置。Main shut-off valve means the device to isolate the entire LPG system from the high pressure side.

Main source of electrical power 主电源 系指向主配电板供电,并通过主配电板对于保持船舶处于正常操作和居住条件所必需的所有设备配电的电源。Main source of electrical power is a source intended to supply electrical power to the main switchboard for distribution to all services necessary for maintaining the ship in normal operational and habitable condition.

Main source of electrical power 主电源（平台） 系指为保持平台正常作业和居住条件所需要的一切设施供应电力的电源。Main source of electrical power is a source intended to supply electrical power for all services necessary for maintaining the unit in normal operational and habitable conditions.

Main starting valve 主启动阀

Main steam pipe 主蒸汽管

Main steam pipeline (main steam line) 主蒸汽管系 锅炉蒸汽输送到主机的管路及附件。

Main steam stop valve 主蒸汽阀 装在锅炉的上锅筒、过热器、再热器、主蒸汽出口处的蒸汽截止阀。

Main steam turbine (propulsion steam turbine) 主汽轮机组 作为船舶推进动力用的主汽轮机、主冷凝器及减速齿轮箱三者的总称。

Main steam turbine condition monitoring (TCM) 主汽轮机状态监测

Main steam turbine set 主汽轮机组 作为船舶推进动力用的主汽轮机、主冷凝器及齿轮减速器三者的总称。

Main steering gear 主操舵装置 系指在正常情况下,为操纵船舶而使舵产生动作所必需的机械、舵执行器、操舵动力设备（如设有）和附属设备以及对舵杆施加扭矩的装置（如舵柄或舵扇）。Main steering gear is the machinery, rudder actuators, steering gear, power units, if any, and ancillary equipment and the means of applying torque to the rudder stock (e. g. tiller or quadrant) necessary for effecting movement of the rudder for the purpose of steering the ship under normal service conditions.

Main steering position 主操舵部位 航行工作站中控制船舶航向的有关控制器和仪表所在的部分。Main steering position means that part of the navigation workstation where those controls and instrumentation relevant to controlling the ship's course are located.

Main stop valve (main throttle valve) 主阀［总阀,(汽轮机)速关阀］ 汽轮机进气总管上能快速关闭的蒸汽阀。

Main structure 主要结构（钢夹层板） 是支撑钢夹层板的构件,典型的有:(1)甲板结构——甲板强横梁和甲板纵桁;(2)舷侧结构——主肋骨和舷侧纵桁;(3)舱壁结构——垂直桁和水平桁;(4)单层底和双层底结构——实肋板和桁材。

Main structure member 主要骨材 系指纵桁、龙骨、强横梁、实肋板等船体结构中的主要构件。

Main switchboard 主配电板 系指由主电源直接供电、并分配和控制电能至船上各种设备的开关设备和控制设备的配电板。Main switchboard is a switchboard which is directly supplied by the main source of electrical power and is intended to distribute and control the electrical energy to the assembly of switch gears and control gears of the ship's services.

Main switchboard 主配电板（平台） 系指由主电源直接供电并将电能分配给平台上各种设施的配电板。Main switchboard is a switchboard directly supplied by the main source of electrical power and intended to distribute electrical energy to the unit's services.

Main tank valve 储存柜主阀 系指位于气体储存柜的气体出口上的遥控操纵的阀门,其位置应尽可能靠近储存柜出口位置。Main tank valve means a remote operated valve on the gas outlet from a gas storage tank, located as close to the tank outlet point as possible.

Main tank vent 主液舱透气阀

Main throttle valve 主节流阀

Main transmission gear 主传动装置

Main trunnion 主耳轴装置

Main valve 主阀 根据通过导阀的继动压力,自动控制盘管或蒸发器中压力或流量的阀。

Main vertical zones 主竖区 系指由"A"级分隔分成的船体、上层建筑和甲板室区段。其在任何一层甲板上的平均长度和宽度一般不超过 40 m。Main vertical zones are those sections into which the hull, superstructure and deckhouses are divided by "A" class divisions, the mean length and width of which on any deck does not in general exceed 40 m.

Main vibration 主振动

Main wheel 主操纵盘（主驾驶盘,主齿轮）

Main-engine stuffing box lubricating oil dirty tank 主机活塞杆填料函污滑油柜 聚集从主机活塞杆填料函等处漏出的污滑油的舱柜。

Maintain 保护(维护)

Maintain ability 可维护保养能力

Maintaining 保护/维护

Maintenance 维护 在构件的设计使用寿命期间所进行的全部活动包括更换、修理或调整,以使之仍然适合其用途。 Maintenance is the total set of activities performed during the design service life of a structure to enable it to remain fit for purpose, which includes replacement, repair, or adjustment.

Maintenance equipment 维护保养设备

Maintenance expense 维修费用

Maintenance facilities 维修设备

Maintenance inspection 维护检查(定期检验)

Maintenance instruction 维护保养说明书(维修说明书)

Maintenance level 维护等级

Maintenance load 维修工作量

Maintenance management system 维护管理计划 系指由计算机支持的,以及维护计划和历史记录的内容。 Maintenance management system means the computerized support, as well as the content, that is the maintenance plan and the history data.

Maintenance manipulator 维护控制设备

Maintenance manual 维护保养手册

Maintenance measure 维护措施

Maintenance of approval 认可保持

Maintenance of the ship 船舶的维护 系指船舶、机器装置和设备应按有关要求进行保养。 Maintenance of the ship means that the ship, machinery installations and equipment are to be maintained at a standard complying with the relative requirements.

Maintenance plan 维护保养计划

Maintenance prevention 安全措施(防护检修)

Maintenance record 维护保养记录

Maintenance route 维修程序(维修工作)

Maintenance routine 维修程序(维修工作,日常例行维护保养)

Maintenance scheme 维护方案

Maintenance standby time 维修准备时间

Maintenance waste 维修中生成的垃圾 系指在船舶维修和营运时收集的废弃物,包括但不限于烟灰、机械装置的沉积物、废油漆、散落在甲板上的垃圾、清理后的废弃物和清理用的碎布。 Maintenance waste means materials collected white maintaining and operating the ship, including, but not limited to, soot, machinery deposits, scraped paint, deck sweepings, wiping wastes and rags.

Maintenance work 维修工作

Maintenance, assembly and disassembly (MAD) 维护保养、装配与拆卸,维修拆装

Major accidents 重大事故 可能导致多人死亡(通常5人或更多)的事故,经常由油气泄漏或严重结构损坏所致。

Major alterations of units 平台重大改装 系指:(1)实质上改变了平台的主尺度和能力;(2)实质上改变了平台用途;(3)实质上影响了分舱因素。 Major alterations of units means that:(1)Dimensions and capacity of the unit have been substantially altered;(2)The purpose of the unit has been substantially changed;(3)The factor of subdivision has been substantially affected.

Major axis 长轴

Major conversion 重大改建 系指对现有船舶所做的下列改建——(1)该船的主尺度或装载量做了重大改变;或(2)该船船舶型的改变;或(3)根据船舶营运管辖国政府的意见,这种改变的目的实际上是为了延长船舶的使用年限;或(4)这种改建在其他方面已使该船舶变成一艘新船,以至于应遵守MARPOL中不适用于现有船舶的有关规定;(5)实质上改变了该船的能效,并且包括哪些能使该船超出MARPOL附则Ⅵ第21条所述适用Required EEDI 的改装。 Major conversion means a conversion of an existing ship;(1)That substantially alters the dimensions or carrying capacity of the ship;or(2)That changes the type of the ship;or(3)The intent of which, in the opinion of the government of the country under whose authority the ship is operating, is substantially to prolong its life;or(4)Which otherwise so alters the ship that if it were a new ship, it would become subject to relevant provisions of MARPOL not applicable to it as an existing ship;(5) which substantially alters the energy effceiency of the ship and includes any in regulation 21 of MARPOL Annex Ⅵ.

Major Conversion on diesel engine 柴油机的重大改装 系指2000年1月1日或以后对尚未按有关规定所述标准发证的柴油机的改变,即:(1)柴油机由其他船用柴油机代替或新增安装的柴油机;(2)对柴油机进行了经修订的"2008年NO_x技术规则"中定义的任何实质性改变;或(3)柴油机的持续额定功率与柴油机初始证书上的最大持续额定功率相比,增加超过10%。 Major Conversion on diesel engine means:a modification on or after 1 January 2000 of a marine diesel engine that has not already been certified to the standards set forth the relevant regulation where: (1) the engine is replaced by a marine diesel engine or an additional marine diesel engine is installed, (2) any substantial modification, as defined in the revised NO_x technical Code 2008, is made to the engine, (3) the maximum continuous rating of the engine is increased by more than 10% compared to the maximum continuous rating of the original certification of the engine.

Major defect　主要缺陷

Major effect　重大后果(重大影响)　系指:(1)产生下列情况的后果:①明显加重船员的工作任务,或增加其执行任务的困难,如果没有其他重大后果同时发生,该任务不应超出合格船员的能力;②操作性能明显降低;③明显地改变许可的工作条件,但不要求操作船员具有超出正常的技能仍具有可安全完成一个航程的能力。(2)故障情况的影响明显降低了船舶或船员对付不利营运状况的能力,例如明显减少安全裕度和营运能力、明显增加船员工作量或明显增加船员工作效率的状况,或引起船上人员的困难,可能包括受伤。 Major effect is: (1) an effect which produces: ①a significant increase in the operational duties of the crew or in their difficulty in performing their duties which by itself shall not be outside the capability of a competent crew provided that another major effect does not occur at the same time; or ②significant degradation in handling characteristics; or ③significant modification of the permissible operating conditions, but will not remove the capability to complete a safe journey without demanding more than normal skill on the part of the operating crew; (2) the effect of failure conditions that reduces that reduces the capability of the craft or the ability of the crew to cope with adverse operating conditions to the extent that there would be, for example, a significant reduction in safety margins or functional capabilities, a significant increase in crew work load or in conditions impairing crew efficiency, or discomfort to occupants, possibly including injuries.

Major modification on computer system　计算机系统的重大改装　系指船上的计算机系统除外围设备外,发生下列之一或组合的情况导致功能和安全性能发生实质性改变:(1)计算机硬件配置的改变;(2)软件升级;(3)网络(包括拓扑结构)的更改。 Major modification on computer system means one of the following cases or a combination thereof which will cause a substantial change to functions or safety features of an onboard computer system (excluding peripherals): (1) change of hardware configuration; (2) update of software; (3) alteration of network (including topology structure).

Major non-conformity　严重不合格　系指可标识的对人身或船舶安全构成严重威胁或对环境构成严重危险,要求立即采取纠正措施的偏差,并包括缺乏对ISM规则有效和系统的实施。任何这些情况被认为严重不合格。 Major non-conformity means an identifiable deviation that poses a serious threat to the safety of personnel or the ship or a serious risk to the environment that requires immediate corrective action or the lack of effective and systematic implementation of a requirement of ISM Code.

Major non-conformity(MNC)　主要不符合项　系指在审核中发现的,严重威胁船员生命及船舶安全,严重威胁海洋环境的与ISM规则的要求不一致的项目。这些项目需要立即进行整改。如公司的安全管理体系未能按照ISM规则的要求体系化的运行也将被视为一个主要不符合项。

Major overhaul　大修

Major parts　主要零件

Major principal stress　大主应力

Major repair　重大修理　对船体结构而言,系指影响结构完整性的修理。

Majority　大多数

Make　制造(组成)

Make certain　使……确信

Make for　使……成为(向去,做成)

Make off　突然离去(离岸,划线)

Make sure　使……明确(弄清楚)

Make up　造成(补充)

Makeshift material　代用材料

Make-up feed water　补给水　由于系统漏泄,锅炉排污,吹灰等引起给水系统内水量不足而要求补充的给水。

Make-up pump　补给泵

Make-up valve　补给阀

Make-up water　补给水(补加水)

Making capacity(of a switching device)　(开关电器的)接通能力　系指在规定的使用和性能条件下,开关电器在规定的电压下能接通的预期接通电流值。

Male　雄性的(阳性的)

Male adapter　外螺纹接合器

Male coupling　阳端接头

Male fitting　外螺纹管接头配件(阳螺纹管接头配件)

Male flange　阳凸面

Male mold　阳模

Male rotor　凸形转子

Male screw　外螺纹螺丝钉(阳螺纹螺丝钉)

Malleable cast iron(MCI) 可锻铸铁(韧性铸铁) 由一定化学成分的铁液浇注成白口坯件,再经退火而成的铸铁。它与灰口铸铁相比,可锻铸铁有较好的强度和塑性,特别是低温冲击性能较好,耐磨性和减振性优于普通碳素钢。

Malleable casting(malleable iron) 可锻铸铁件(马铁铸件,韧性铸件)

MAMDAT MDAT 的月算术平均值 MAMDAT means monthly average of MDAT.

Man rope(wire-rope guard rail) 舷梯扶索 位于舷梯两侧,供上下人员作扶手用的绳索。

Manage and process 管理或处理 系指按最佳操作对消耗臭氧物质(ODS)和/或废气清除残余物进行收集、储存、运输、处理和处置的措施,从而使其处于安全和利于环保的状态。 Manage and process mean action related to the collection, storage, transport, treatment and disposal of ODS and/or exhaust gas cleaning residues such that they are rendered in a safe and environmentally benign condition in accordance with best available practices.

Managed services 托管

Management level 管理级(人员) 系指与下列内容有关的责任级别:(1) 在海船上担任船长、大副、轮机长或大管轮,并(2) 确保正确履行指定职责范围内的所有职能。 Management level means the level of responsibility associated with:(1) serving as master, chief mate, chief engineer officer or second engineer officer on board a ship, and(2) ensuring that all functions within the designed area of responsibility are properly performed.

Management plan 垃圾管理计划

Management representative 管理者代表 对质量管理体系而言,系指制造厂商应指定一位管理者代表,这一代表最好不兼其他职务,该代表应具有在贯彻和持续执行质量体系中的授权和责任。 For the purpose of quality management system, management representative means that manufacturer is to appoint a management representative preferably independent of other functions, who is to have defined authority and responsibilities for the implementation and maintenance of the quality system.

Management review 管理检查 对质量管理体系而言,系指按照有关要求所建立的质量体系应在适当的时间间隔内由制造厂商进行系统的检查,以保证其持续有效性。这些管理制度检查的记录应保存,以供验船师所用。 For the purpose of quality management system, management review means that the quality system established in accordance with the relevant requirements is to be systematically reviewed at appropriate intervals by the manufacturer to ensure its continued effectiveness. Records of such management reviews are to be maintained and be made available to the surveyors.

Management system 安全管理体系 系指 ISM 规则中所述的国际安全管理体系。Safety management system means the international safety management system as Safety described in the ISM Code.

Managing and operating 管理和操作 就"北大西洋冰区巡逻、运作和费用规则"而言,系指冰区巡逻的保持,行政管理和运作,包括传播由此收到的信息。 Managing and operating means maintaining, administering and operating the Ice Patrol, including the dissemination of information received therefrom.

Mandatory prewash 强制预洗 对"73/78 防污公约"附则Ⅱ而言,系指必须执行的预洗。

Mandrel 芯轴(型芯,紧轴)

Mandrel for pipe bending 弯管用心轴

Mandrill 尖嘴锤

Maneuverability 操纵性 系指船舶用其控制装置来保持或改变其运动速率、姿态和方向的能力。船舶操纵性的优劣关系到航运安全及船舶的经济性和快速性。运输船舶操纵性的主要性能包括:(1) 足够的航向稳定性;(2) 中小舵角良好的应舵性能;(3) 符合要求的大舵角回转性能;(4) 适中的主机停车和主机逆转的停船性能。

Maneuverability 操纵性

Maneuvering characteristics curve 操纵性特征曲线 利用螺线或逆螺线操纵试验的数据按定常转艏角速度对舵角所标绘的曲线。

Maneuvering period 转舵阶段 回转试验中,船舶从开始执行转舵命令起到实现命令舵角为止的阶段。

Maneuvering surface of WIG craft 操纵面(地效翼船) 系指组成机翼的完整部分或延伸部分的一个部件,用以调节机翼的气动升力或水动升力。

Manganese brass 锰黄铜

Manganese bronze 锰青铜

Manganese content 锰含量

Manganese dioxide 二氧化锰

Manganese iron brass 锰铁黄铜

Manganese ore 锰矿砂(石) 呈黑色至棕黑色。是一种极重的货物。含水量最高为 15%。无特殊危害。该货物为不燃物或失火风险低。 Manganese ore is black to brownish black in colour. It is a very heavy cargo. Moisture content: up to 15%. It has to special hazards. This cargo is non-combustible or has a low fire-risk.

Manganese steel 锰钢

Manhole 人孔 系指甲板、液舱等的圆形或椭圆形开孔,用作通道。 Manhole means a round or oval hole cut in decks, tanks, etc., for the purpose of providing access.

Manhole cover 人孔盖 盖闭人孔的盖子及其固定附件的总称。

Manifold 总管/集管(分配阀箱) 系指将若干根来自油井的输入流送管连接到通往生产设施(如平台)的输出流送管的海底结构物。 Manifold is the seabed structure that connects several incoming flowlines from wells to an outgoing flowline that connects to a production facility such as a platform.

Manifold center deck 管汇甲板(集管中心甲板)
Manila rope 白棕缆 用白棕纤维制成的缆绳。
Manned 有人驾驶的(有人操纵的,有人管理的)
Manned space 有人处所
Manned space flight 载人航天 是人类驾驶和乘坐载人航天器在太空中从事各种探测、研究、试验、生产和军事应用的往返飞行活动。其目的在于突破地球大气的屏障和克服地球引力,把人类的活动范围从陆地、海洋和大气层扩展到太空,更广泛和更深入地认识整个宇宙,并充分利用太空和载人航天器的特殊环境进行各种研究和试验活动,开发太空极其丰富的资源。目前仅美、中、俄三国拥有自主载人航天能力。需要指出的是:欧盟、印度、日本可以列为"准载人航天能力国"。中国航天员见图 M-2。

图 M-2 中国航天员
Figure M-2 China astronauts

Manner 方式(态度)
Manning the propulsion machinery space 有人值班机器处所
Manoeuvering gear 操纵机构 控制内燃机启动、换向、调速等的联合装置。
Manoeuvring light 操纵号灯 当船舶在相遇中进行操纵时,用以补充号笛而装设在桅灯以上的闪光灯。
Manoeuvring shaft 分配轴(柴油机) 系指控制轴(涡轮机喷嘴)和倒车轴(齿轮传动装置)。
Manoeuvring 操纵 系指:(1)按要求使船舶的操舵系统和推进装置动作,以使船舶按预定的舷向、速度和/或方向航行;(2)按要求使船舶的操舵系统和推进装置动作,以使船舶按规定的航向、船位或航迹航行。 Manoeuvring means:(1)the operation of steering and propulsion machinery, as required to alter the ship's heading, speed and/or directional movement;(2)an operation of steering systems and propulsion machinery as required to move the ship into predetermined directions, positions or tracks.

Manograph 压力计
Manometer 压力表(液位压力计)
Manometer pressure 表压
Manometer vacumn 压力真空表
Manometric 压力计的(用压力计量的)
Man-over-board mode(MOB) 人员落水模式 系指当人员落水事故发生后,针对船舶操作和运动的显示模式(释放安全设备如救生圈和救生带,操纵船舶回转等)。Man-over-board mode(MOB)means the display mode for operations and actions of a ship after a Man-over-board accident happened(release of safety equipment, e. g. , life buoy and life belt, performance of a return manoeuvre etc.).

Manpowered boat 人力船 用摇橹、划桨、撑篙和拉纤等方法,依靠人力推进的船舶。

Manual 手工(焊接) 这一术语是用来描述一只手持有保护气体的钨极焊炬,而另一只手独立地加焊料的工艺。 Manual is used to describe the technique where the gas-shielded tungsten arc torch is held in one hand and the filler is added separately by the other hand.

Manual alarm 手动报警器
Manual bending of pipe 手工弯管
Manual call point 手动报警按钮
Manual control 手动控制 使用单独的推力器控制器或具有或没有自动舷向控制的单手柄控制器。
Manual control device 手动控制装置
Manual control mode(DPS) 手动控制方式(动力定位系统) 系指包括用单独的控制器来控制各个推力器的螺距/转速和方向,以及使用联合操纵杆进行组合推力遥控的控制方式。 Manual control mode(DPS)means a control mode which is to include control of thrusters by individual control device for pitch/speed and azimuth of each thruster, and an integrated remote thruster control by use of joystick.

Manual controller 手动控制器 系指其全部基本功能是依靠手动操作的器件来完成的一种电控制器。

Manual metal arc welding(SMAW) 手工金属电弧焊

Manual override function　手动越控功能
Manual starting　手动启动(手动启动)
Manual steering workstation　人工操舵工作站　可以由舵工操纵船舶的工作站。Manual steering worksta-tion is a workstation from which the ship can be steered by a helmsman.

Manual thruster controller　推进器手动控制器　系指各个推进器的手动操作的控制器,用以完成启动、停车、方位和螺距/速度控制(可不包括高压电动机的启动/停止)。Manual thruster control means a controller used in individual and separate manual operation thrusters for start, stop, azimuth and pitch/speed control (start/stop of high voltage motors may be excluded).

Manual welding　手工焊接　系指诸如手工电弧焊等,即用手工进行焊接操作,焊条也用手来供给。见图 M-3。Manual welding is used to describe processes in which the weld is made manually by a welder using a manu-ally fed electrode such as shield metal arc welding, etc. See Fig. M-3.

图 M-3　手工焊接
Figure M-3　Manual welding

Manual welding processes　手工焊接工艺　如垂直下行焊(vertical-down welding)、深熔焊(deep penetration welding)、带垫板的单面焊(single-side welding with back-ing)等。

Manually inflated PFD　手动气胀式个人漂浮装置　由使用者操作机械以引起气体膨胀的 PFD。Manu-ally inflated PFD is the PFD in which inflation is effected as a result of the user operating a mechanism.

Manually mechanically propelled lifeboat　人力机械推进救生艇　依靠人力通过机械装置传动推进器的救生艇。

Manually operated call point　手动火灾报警按钮
Manually operated clutch　手动离合器　利用人力通过机械部件实现离合的离合器。
Manually operated valve　手动阀

Manually-operated air compressor　手动空气压缩机　用人力驱动的独立空气压缩机。
Manufacture information system(MIS)　生产信息管理系统

Manufacturer　制造厂商　系指生产和/或装配最终产品,并对最终产品负有全部责任的组织。 Manu-facturer means an organization producing and/or assembling final products and fully responsible for such products.

Manufacturer　制造厂商(救生艇)　就现有救生艇释放和回收系统而言,系指:(1)原设备制造厂商;或(2)负责某一系列或形式的救生艇释放和回收系统制造厂商;或(3)当原设备制造厂商不存在或不能支持设备时,负责某一系列或形式的救生艇释放和回收系统的其他任何个人或实体。Manufacturer, with respect to existing lifeboat release and retrieval system, is: (1) the original e-quipment manufacturer; or(2) a manufacturer of lifeboat re-lease and retrieval system who has taken on the responsibility for a range or type of lifeboat release and retrieval system; or (3) any other person or entity which taken responsibility for a range or type of lifeboat release and retrieval system when the original manufacturer on longer exists or supports the e-quipment.

Manufacturer　制造厂商(船用卫生装置)　系指任何从事制造、装配或进口符合按联邦水域防污染法规第 312 条要求制定的标准和规则的船用卫生装置或船舶的法人。Manufacturer means any person engaged in manu-facturing, assembling, or importing of marine sanitation de-vices or of vessel subject to the standards and regulations promulgated under section 312 of the Federal Water Pollution Control Act.

Manufacturer's certificate validated by LR　英国劳氏船级社认可有效的制造商证书　这一证书是交付的产品根据有关规范的要求进行过检查和试验后由英国劳氏船级社认可有效者可以被接收的,在这一情况下,证书应包括以下说明:"我们在此地证明这些材料的制造是经认可的过程并经按照英国劳氏船级社规范进行过令人满意的试验"。Manufacturer's certificate vali-dated by LR is a manufacturer's certificate, validated by LR on the basis of inspection and testing carried out on the de-livered product in accordance with the requirements of these Rules may be accepted. In this case, the certificate will in-clude the following statement: "We hereby certify that the material has been made by an approved process and satisfacto-rily tested in accordance with the Rules of Lloyd's Register".

Manufacturer's certificate　制造厂商证书　由制造厂商颁发这种证书的形式应根据有关规范或适用的

国家或国际标准的要求试验和检查其结果是令人满意的。这种证书应由制造厂商授权的代表认可后生效,这一授权代表应独立于制造部门,这一证书应包括一个声明,说明这些产品是符合有关规划或适用的国家或国际标准的要求的。 This type of certificate is issued by the manufacturer based on the results of testing and inspection being satisfactorily carried out in accordance with the requirements of relevant Rules or the applicable National or International standard. The certificate is to be validated by the manufacturer's authorized representative, independent of the manufacturing department. The certificate will contain a declaration that the products are in compliance with the requirements of these Rules or the applicable National or International standard.

Manufacturer's certificate issued under the Materials Quality Scheme 制造厂商颁发的根据"材料质量方案"的证书 对于一个根据"材料质量方案"认可的制造厂商可以颁发由认可标记支持的制造厂商证书。这一证书应附有下列内容的申明:"本证书是根据英国劳氏船级社(操作部门)授权按照'材料质量方案'和证书号MSO(……)而颁发的"。 Where a manufacturer is approved according to the Materials Quality Scheme, they will issue manufacturer's certificates bearing the scheme mark. The certificates must also bear the following statement: "This certificate is issued under the arrangements authorized by Lloyd's Register(operating group) in accordance with the requirements of the Materials Quality Scheme and certificate number MQS(…)".

Manufacturer's document 制造厂商证明 系指由制造厂商独立行使职责所签发的事实陈述或证书,作为自检的结果。 Manufacturer's document means factual statements or certificates issued by the manufacturer as the result of independently exercising his inspection duty.

Manufacturing control 生产管理 对质量管理体系而言,系指制造厂商应保证对于直接影响产品质量的一切活动都是在有管理的情况下进行,包括:(1)编写施工说明书,如没有这种说明书,就可能影响产品而不能符合规定的要求,应该确定监控的方法和对产品性能的控制;(2)通过书面标准或代表性样品来订立工艺标准。 For the purpose of quality management system, manufacturing control means that the manufacturer is to ensure that those operations which directly affect quality are carried out under controlled conditions. These are to include the following:(1) Written work instructions wherever the absence of such instructions could adversely affect compliance with specified requirements. These should define the method of monitoring and control of product characteristics.(2) Established criteria for work manship through written standards or representative samples.

Manufacturing management system 产品制造管理体系 系指影响产品制造过程的一组要素,这包括过程输入的控制,过程控制因素(如人员能力、程序、设施和设备、培训等)、过程输出以及旨在实现持续改进的质量、过程和产品的测量。 Manufacturing management system means a group of elements affecting the manufacturing process, including process input control, process control factors(e.g. competence of personnel, procedures, facilities and equipment, training, etc.), process output, and measurement of quality, processes and products for continued improvement.

Manufacturing process 生产过程 系指制造产品所采取的步骤。 Manufacturing process is the steps that one takes to produce(manufacture) product.

Maps/navigation lines 地图/航线 操作人员定义的或制订的航线,用以表示航道、分航计划或航行重要区域的边界。 Maps/navigation lines are those lines defined or created lines by operator to indicate channels, Traffic Separation Schemes or borders of any area important for navigation.

Marble chips 大理石碎粒 干燥,多尘,白色至灰色碎块、颗粒和粉末,混有少量砾石和卵石。无特殊危害。该货物为不燃物或失火风险低。 Marble chips are dry, dusty, white to grey lumps, particles and powder mixed with a small amount of gravel and pebbles. No special hazards. This cargo is non-combustible or has a low fire-risk.

Margin line 限界线 船舶分舱的计算中,为检查船舶破损进水后的水线是否超过极限位置而作的距舱壁甲板上表面以下不少于76mm处并平行于甲板边线的一根限制线。

Margin plate 内底边板 内底外侧列板、内底边板(或桁材)在向下翻到舭部时,构成双层底的外部界限。 Margin plate is the outboard strake of the inner bottom and when turned down at the bilge the margin plate(or girder) forms the outer boundary of the double bottom.

Marine 海上的(船用的,海运业)

Marine accident 海损事故

Marine air conditioning 船舶空气调节 为改善船员和旅客的生活、工作条件,对船舶舱室内空气温度、相对湿度与洁净程度的调节。

Marine air conditioning 船用空调

Marine alloy 船用合金

Marine aluminium alloy 船用铝合金

Marine anchor 船用锚 系指适用于多种海底底质(砂、硬泥、软泥、淤泥以及这些底质的混合物)的无杆转爪锚。其抓力都不是很大。锚爪的折角一般在35°~45°的范围内。如霍尔锚、斯贝克锚等普通船用无杆锚的折角为42°,其最大抓力仅为其自重的4倍左右;AC-14型大抓力无杆锚的折角为35°,其最大的抓力可达到其自重的8倍以上,但是当拉力超过抓力额限时容易翻转而导致失效。该类锚的优点是抛起锚方便,可收藏在锚链筒内。

Marine atmospherc corrosion 海洋大气腐蚀

Marine atmosphere corrosion resisting steel 耐海洋大气腐蚀钢

Marine auxiliary machinery 船舶辅机 船上除主机、主锅炉以外所有机械设备的统称。

Marine benthos laboratory 海洋底栖生物实验室 用于底栖生物样品的处理、观察和分析以及对所获取的海底生物图像进行分析。研究其形态、分类、种群组成、数量分布以及与环境之间关系的工作室。底栖生物系指生活于水域底上和底内的动物、植物和微生物。

Marine biological laboratory 海洋生物实验室 用于海洋生物样品的观测、处理、分析和有控试验,以及对获取的生物信号(图像和声信号)进行分析。研究其形态、分类、种群组成、数量分布、生理生态特征、生物之间和生物与环境之间关系的工作室。

Marine biological research vessel 海洋生物调查船 用于对海洋生物的区系分布、生态、海洋生物分类和海洋初级生产力进行调查研究的海洋专业调查船。

Marine boiler 船用锅炉 通过吸收燃料释放的热量或利用废气余热等其他热源来加热水工质,并以规定压力和温度输出蒸汽或热水,供船舶推进或船上其他用途的设备。

Marine casualty 海难 系指造成与船舶操作直接有关的下列任一后果的一个或一连串事件:(1)人员死亡或严重受伤;(2)船上人员失踪;(3)船舶火灾、推测可能发生的火灾或弃船;(4)船舶材料损坏;(5)船舶搁浅或无法航行,或船舶碰撞;(6)会严重危及船舶、其他船舶或人员安全的船舶外部的船上基础设施的材料损坏;(7)一艘或数艘船舶损坏对环境造成严重破坏或可能对环境造成严重破坏。但是海难不包括有意对船舶、人员或环境安全造成危害的故意的行为或疏忽。见图 M-4。

Marine casualty means an event, or a sequence of events, that has resulted in any of the following which has occurred directly in connection with the operation of a ship:(1) the death of, or serious injury to, a person;(2) the loss of a person from a ship;(3) the loss, presumed loss or abandonment of a ship;(4) material damage to a ship;(5) the stranding of a ship, or the involvement of a ship in a collision;(6) material damage to marine infrastructure external to a ship, that could seriously endanger the safety of the ship, another sip or an individual;(7) severe damage to the environment, or the potential for severe damage to the environment, brought about by the damage of a ship or ships. However, a marine casualty does not include a deliberate act or omission, with the intention to cause harm to the safety of a ship, an individual or the environment. See Fig. M-4.

图 M-4 海难
Figure M-4 Marine casualty

Marine centrifugal separator 船用油分离机 符合船舶规范规定或船用技术条件要求的各种供船舶上使用的油分离机的统称。

Marine chemical laboratory 海洋化学实验室 用于对海水化学成分的测定及其方法的研究;海水中化学要素分布变化规律的研究、海水与边界之间化学要素交换过程的研究和分析用的工作室。

Marine compressor 船用压缩机 符合船舶规范规定或船用技术条件要求的各种船舶上使用的压缩机的统称。

Marine diesel engine 船用柴油机 系指1973年国际防止船舶造成污染公约(MARPOL)的附则Ⅵ第13条适用的以液体或双燃料运行的任何往复式内燃机,包括增压/复式系统(如适用)。Marine diesel engine means any reciprocating internal combustion engine operating on liquid or dual fuel, to which regulation 13 of Annex VI to the International Convention for the Prevention of Pollution from Ships, 1973 (MARPOL) applies, including booster/compound systems if applied.

Marine engineer 轮机工程师(轮机员,船舶机械工程师)

Marine environment 海洋环境

Marine environmental protection 海洋环境保护

Marine environmental protection Committee 海洋环境保护委员会 是 IMO 下属专门从事海洋环境保护的一个机构,简称 MEPC(环保会)。

Marine environmental research 海洋环境调查
对海洋水文、气象、物理、地质、化学和生物进行的调查。它又分为多学科综合性调查和单学科专题性调查。

Marine equipment 船用设备（海用设备）

Marine equipment certificate 船用设备证书

Marine evacuation system 海上撤离系统 系指将人员从船舶的登乘甲板迅速转移到漂浮的救生艇筏上的设备。 Marine evacuation system is an appliance for the rapid transfer of persons from the embarkation deck of a ship to a floating survival craft.

Marine facility 船用设施

Marine fishery resources research 海洋水产资源调查 对有捕捞价值的水产动、植物种类的数量分布、生活习性、群落结构、季节变化以及与环境条件关系的调查。

Marine fittings 船用配件 系指下列正常系泊船用的系缆桩与缆柱、系缆器、立式滚轮、导缆孔以及用于正常拖带船舶的类似部件，其他部件如绞盘、绞车等不包括在内。任何船用配件与船体支撑结构的焊接、螺栓或其他等效设施是船用配件的部分，应满足船用配件所适用的工业标准。

Marine fouling 海生物污底

Marine gas turbine（marine gas turbine） 船用燃气轮机 在参数、材料、结构及运行性等方面都满足船用要求，供船舶推进或带动发电机、水泵等船用辅机之用的燃气轮机，分主燃气轮机和辅燃气轮机两种。

Marine gas turbine unit 船用燃气轮机机组 由一台或几台燃气轮机及相应的传动部分和辅助系统组成的动力设备。

Marine gear 船具（船用装置，船用齿轮组）

Marine geological research vessel 海洋地质调查船 用于对海底地质、地貌和地球物理进行调查研究的海洋专业调查船。

Marine geological sampling 海洋地质取样 用取样器或潜水采集海底沉积物和岩石样品的作业。海水中的悬浮物和海洋上空大气微粒的取样亦属此范围。

Marine geology and geomorphology laboratory 海洋地质地貌实验室 利用测深仪、旁侧声呐、浅地层剖面仪、导航定位仪和水下摄影装置等观测和记录海底地形地貌概况。为进一步研究海底地质地貌提供资料的工作室。

Marine geophysical exploration laboratory 海洋物探实验室 应用各种地球物理方法对海洋声学、重力、地震、磁力、热流等项目的单项或多项地球物理场进行现场测量与研究的工作室。

Marine geophysical prospecting 海洋地球物理勘探 利用地震、压力和磁力的地球物理方法，对海底的地质构造进行调查研究和勘探。

Marine gravimeter 海洋重力仪 设置在船上进行连续重力观测用的仪器。主要由探头、记录装置和重力平台组成。

Marine gravimetric survey room 海洋重力测量室 又称重力室。安装海洋重力仪和其他附属设备，并对海洋重力场进行测量和资料分析的工作室。该室要接近船舶重心位置上。

Marine growth prevention system（MGPS） 海生物生长预防系统 系指在内部海水冷却系统和通海阀箱中用于防止生物污垢积聚的防污底系统，且其可包括采用阳极、喷射系统和电解。 Marine growth prevention system（MGPS）means an anti-fouling system used for the prevention of bio-fouling accumulation in internal seawater cooling systems and sea chests and can include the use of anodes, injection systems and electrolysis.

Marine hydrographic and chemical laboratory 海洋水文化学实验室 用于对海洋水文、化学等各种要素：温度、电导（电导率、电导比）、盐度、深度、溶解氧、酸碱度（即pH值）、浊度、声速、密度等进行综合观测和研究用的工作室。

Marine hydrographic laboratory 海洋水文实验室 又称物理海洋实验室（physical oceanography laboratory）。用于现场观测海洋各水文要素：水深、水色、透明度、温度、盐度、海流、波浪、潮汐、海冰、海发光等，并研究其时间分布、变化规律与机制以及探讨水文预报方法的工作室。

Marine hydrographic observation 海洋水文观测 对海洋水文要素时空分布进行的观测。一般包括：水深、水温、盐度、海流、海浪、透明度、水色、海发光、海冰及有关的气象资料。

Marine hydro-meteorological research vessel 海洋水文气象调查船 用于对海水温度、盐度、海流、内波和海浪等水文要素的分布和变化规律，以及海洋表面及其高空气象要素进行测定和研究的海洋专业调查船。

Marine incident 海上事故 系指除海难外直接与船舶营运有关的一个或一连串事件，这些事件危及或如不经纠正会危及船舶、人员或附近的安全。但是，海上事故不包括有意对船舶、人员或环境安全造成危害的故意的行为或疏忽。 Marine incident means an event, or sequence of events, other than a marine casualty, which has occurred directly in connection with the operation of a ship that endangered, or, if not corrected, would endanger the safety of a ship, its occupants or any other of the environment. However, a marine incident does not include a deliberate act or omission, with the intention to cause harm to the safety of a ship, an individual or the environment.

Marine installation 船用装置

Marine installation certificate 船用装置证书

Marine living resources research 海洋生物资源调查 对海洋生物的种类组成、数量分布、生活习性、群落结构、季节变化以及与环境条件关系的调查。

Marine loading computer 船用装载仪 以装载计算软件为核心,用于船舶安全装载分析计算系统。该系统通过船舶性能,强度和破舱稳性的实时分析计算,确保船舶高效装载货物和安全航行。

Marine log 船用计程仪 在航海中用来测定船舶航行速度和累计船舶航程的仪器。也是推算航迹的基本工具之一,并将船舶航速、航程的信息输出给所需的设备和部门进行数据处理和显示、记录。船用计程仪可分为两大类:(1)相对计程仪——所测定的船舶速度和累计的航程均指船舶相对于水而言,其有水压计程仪和电磁计程仪;(2)绝对计程仪——是直接测定船舶相对于地球的速度和航程,如多普勒计程仪和声相计程仪等。

Marine machinery 船用机械

Marine magnetic survey room 海洋磁力测量室 又称磁力室。安装海洋磁力仪和其他附属设备,并对海洋磁场进行测量和资料分析的工作室。

Marine mammal and birds observation room 海上哺乳动物和鸟类观察室 用于观测和研究极地海域和冰上的哺乳动物和鸟类的种类、数量、分布和生活习性的工作室。

Marine materials 船用材料(耐海洋环境材料) 系指普通强度船体结构钢、较高强度船体结构钢和类似等级的锻钢或铸钢、高强度(淬火加回火)细化晶粒结构钢、低温韧性钢、不锈钢及复合钢、铝合金及其他有色金属等。 Marine materials means normal strength hull structural steels, higher strength hull structural steels, and comparable grades of forged steels or steel casting, high strength (quenched and tempered) fine grained structural steels, steels tough at subzero temperatures, stainless and clad steels, aluminium alloys and other non-ferrous metals.

Marine meteorological laboratory 海洋气象实验室 进行海面气象观测,低空探测、高空探测、气象情报的收集及天气分析等海上气象综合观测的工作室。同时为本船海上作业和气象科研服务。该室应位于视野开阔的上甲板以上。

Marine microbiology laboratory 海洋微生物实验室 用于海洋微生物无菌采样具灭菌;样品的处理、培养、分离、观测、计数;生理生化特征的初步测定以及样品的保存。研究其形态分类和数量分布的工作室。微生物系指个体微小、形态结构比较简单的单细胞或近单细胞的生物。

Marine microorganism sampling 海洋微生物取样 利用微生物采水器和采泥器,对预定的水层和海底,分别采取水样和底质样品。在无菌操作的条件下,将采取的样品接种、培养分离和纯化。

Marine microwave ranging system 海用微波测距定位系统 利用微波频率测定船台到两个或两个以上岸台的距离进行定位的系统。其特点是定位精度比一般双曲线定位系统高。但作用距离较短,适合于沿岸测量定位时使用。

Marine mineral resource research 海洋矿藏资源调查 对特定海域的成矿地质条件、矿物种类、富脊的规律和蕴藏贮量的调查。

Marine nozzle 导流管 系指由内、外壳板、内部环形隔板与纵向筋板焊接而成的环形结构物。螺旋桨位于导流管的最小截面处,导流管的转动轴线通过螺旋桨圆盘面,桨叶端部与该处导流管内壁之间的间隙应尽可能地小,通常不超过螺旋桨直径的0.5%或1 cm。根据工作状态,导流管划分固定式和转动式两种:固定式导流管固定在船尾,通常将轴支架和舵中下支承组合在一起;转动式导流管可绕竖轴转动一般约转40°,并可兼作舵使用。采用转动导流管,可改变螺旋桨尾流方向,提高船舶的操纵性,特别是低速航行时的机动能力。见图M-5。

图 M-5 导流管
Figure M-5 Marine nozzle

Marine optical laboratory 海洋光学实验室 用于现场测量海洋中的光场和海水光学特性,并研究其变化,以及与海洋生物、水团等要素之间关系的工作室。

Marine physical laboratory 海洋物理实验室 用于对海洋的声、光、磁和热的物理特性进行测量以及研究其规律的综合性的工作室。在进行声学特性测量时需具备低噪声供电。

Marine physiology and ecology laboratory 海洋生物生理生态实验室 对获取的海洋生物样品进行培养、

观测、处理分析以及有控试验。研究其生理生态特征以及生物之间和环境之间关系的工作室。

Marine plankton laboratory 海洋浮游生物实验室 用于浮游生物样品的观测、处理和分析,以及初级生产力的测定。研究其形态、分类、种群组成、数量分布,以及与环境之间关系的工作室。浮游生物系指无游泳能力或具有一定的浮游能力但随波逐流的生物。

Marine pollutant 海洋污染物 系指在美国联邦政府法规第47篇172.101附则B中所列出的、以包装形式出现的有害物质。Marine pollutant means a harmful substance in packaged form, as it appears in Appendix B of 49 CFR 172.101.

Marine pollution monitoring laboratory 海洋污染监测实验室 用于对特定海区或污染区以不同的频率所获得的样品(气候、水样、生物和底质样品)进行处理并测定其中污染物质的组成和含量,掌握污染物质的分布和变化,为评价该海区的生态环境提供依据的工作室。该室在建造、使用和管理上,应考虑排除室内本底污染,在防污染上要采取相应的措施。

Marine pollution observation 海洋污染调查 对进入海洋的污染物所进行的调查。内容包括水质、底质、水文、气象、海洋生物生态、生物体内有害物质、污染源、污染途径、污染程度、分布状态、影响和危害等。

Marine power plant(ship's power plant) 船舶动力装置 为船舶推进和其他需要提供机械能、电能、热能的成套装置。

Marine primary productivity determination 海洋初级生产力测定 对含叶绿素的海洋植物,利用光能将无机物转换成有机物的初级生产能力所进行的调查。

Marine primary-secondary clock 船用子母钟 是船上的时间服务系统。它能同时提供当地时间、世界时间或某一特定地方的时间。该时间服务系统由一个母钟和若干个子钟组成,通过信号电缆连接起来。所有子钟都由母钟统一控制运行,除了标准子钟指示世界时间或某一特定时间外,其他子钟都保持时间同步。

Marine product 船用产品

Marine product certificate 船用产品证书

Marine products survey 船用产品检验 系指对船用产品进行的入级检验和法定检验。经检验合格后可颁发的证书有:工厂认可证书、形式认可证书和产品认可证书等。

Marine proton gradiometer 海洋质子梯度仪 测量海上地磁场及其梯度值的质子磁力仪系统。主要由主、从质子磁力仪和数据采集系统组成。

Marine proton magnetometer(marine proton precession magnetomter) 海洋质子磁力仪 又称核子旋进式磁力仪。通过测量氢质子在地磁场中的旋进频率来求出磁场绝对值的磁力仪。

Marine pump(ship's pump) 船用泵 符合船舶规范规定或船用技术条件要求的各种供船舶使用的泵的统称。

Marine reduction gear 船用减速齿轮箱 按一定的传动比将船用发动机的功率传递给推进器或发电机、水泵等辅机的减速齿轮箱,有主减速齿轮箱和辅减速齿轮箱之分。

Marine refrigerating plant 船舶制冷装置

Marine refrigeration 船舶冷藏 在冷藏货舱、鱼舱、冷藏库内创造一定的低温环境,以运输或储存需要低温保藏的货物的技术。

Marine safety investigation 海上安全调查 系指为了防止今后发生海难和海上事故的目的而进行的海难和海上事故调查和追查(无论各国对此如何称呼)。调查包括收集和分析证据,确定起因并在必要时提出安全建议。Marine safety investigation means an investigation or inquiry(however referred to by a State), into a marine casualty or marine incidents in the future. The investigation includes the collection of, and analysis of, evidence, the identification of causal factors and the making of safety recommendations as necessary.

Marine safety investigation authority 海上安全调查当局 系指负责按海难或海上事故安全调查国际标准和建议措施规则进行调查的国家当局。Marine safety investigation authority means an Authority in a State, responsible for conducting investigations in accordance with the Code of the International standard and recommended practices for a safety investigation into a marine casualty or marine incident(casualty investigation code).

Marine safety investigation report 海上安全调查报告 系指包括以下内容的报告:(1)概述海难或海上事故的基本事实并说明是否最终导致死亡或污染;(2)船旗国、船东、船舶经营者、安全管理证书中确定的船公司和船级社的识别标志(根据关于隐私的国内法而定);(3)如适当,相关船舶的主尺度和发动机情况以及船员、工作程序和其他事宜(例如船上规则工作时间)的描述;(4)海难或海上事故的详细记述;(5)对起因(包括任何机械、人为和组织因素)的分析和意见;(6)对海上安全调查结果的讨论,包括安全问题的确定以及海上安全调查的结论;(7)如适当,为防止今后发生海难和海上安全事故而提出的建议。Marine safety investigation report means a report that contains:(1) a summary outlining the basic facts of the marine casualty or marine incident and stating whether any deaths, injuries or pollution occurred as a result;(2) the identity of the flag State, owners, operators, the company as identified in the safety management certificates, and the classification society(subject to any national laws concerning privacy);(3) where relevant the details of the dimensions and engine of any ship involved, together with a description of the crew, work routine and other matters, such as time served on the ship;(4) a narrative detailing the circumstances of the marine casualty or marine incident;(5) analysis and comment on the causal factors including any mechanical, human, and organizational factors;(6) a discussion of the marine safety issues, and the marine safety investigation's conclusions; and(7) where appropriate, recommendations with a view to preventing future marine casualties and marine incidents.

Marine Safety Investigation State(s) 海上安全调查国家 系指船旗国或如适当,根据海难或海上事故安全调查国际标准和建议措施规则相互达成的协议,负责进行海上安全调查的一个或多个国家。 Marine Safety investigation State(s) means the flag State or, where relevant, the State or States that the responsible for the conduct of the marine safety investigation as mutually agree in accordance with the Code of the International standard and recommended practices for a safety investigation into a marine casualty or marine incident(casualty investigation code).

Marine safety record 海上安全记录 系指为海上安全调查设计的下列类型的记录:(1)为海上安全调查所做的所有声明;(2)船舶操作相关人员之间的所有通信;(3)所有关于海难或海上事故所涉及人员的医疗或私人信息;(4)海上安全调查期间获得的信息或证据材料分析的所有记录;(5)来自航行数据记录仪的信息。 Marine safety record means the following types of records collected for a marine safety investigation:(1) all statements for the purpose of a marine safety investigation;(2) all communications between persons pertaining to the operation of the ship;(3) all medical or private information regarding persons involved in the marine casualty or marine incident;(4) all records of information or analytical evidence data collected during a marine safety investigation;(5) information from the voyage data recorder.

Marine sanitation device 船用卫生装置 包括安装在船上,指定用来接收、留存、处理或排放污水和处理这些污水过程所需的设备。 Marine sanitation device and device includes any equipment for installation on board a vessel which is designed to receive, retain, treat, or discharge sewage, and any process to treat such sewage.

Marine seismic prospecting 海洋地震勘探 以人工激发地震波,经海洋地震电缆和数字地震仪,接收并记录来自海底及其下面介质的反射和折射地震波,通过研究地震波的运动学和动力学的特点,以便探索海区地质构造及寻找有用的矿藏。

Marine seismic streamer 海洋地震电缆 又称"漂浮电缆"。在海上地震勘探时,沉放在水下规定深度中接收和传递人工地震信号,并将各道压力检波器组与地震仪器连接的特种的多芯电缆。

Marine seismic streamer depth system 海洋地震电缆深度控制系统 通过装配在海洋地震电缆上的定深器,由船上遥控,保持电缆作业深度,并能指令电缆在水下0 m~30 m范围内上浮下潜的控制系统。

Marine seismic survey room 海洋地震测量室 又称地震室。利用数字地震仪、组合气枪控制器、海洋地震电缆控制器以及外围设备,接收和记录来自地下界面的反射和折射地震波的信号。为研究海底地质构造提供资料的工作室。

Marine steam turbine 船用汽轮机 在参数、材料、结构及运行性能等方面均满足船用要求,供船舶推进或带动辅机之用的汽轮机。

Marine surveyor 验船师 是船级社雇用的人员,其主要职责是在船舶建造中监督和检验海洋结构物及其构件,使其符合船级社的规范或船级社认为合适的其他一些标准。

Marine system(ship system) 船舶系统 船舶上除动力系统外,为船舶生活、卸载、安全所设的系统,包括:舱底和压载系统、甲板排水和疏水系统、冷藏与制冷系统、通风、空调系统、供暖和生活用水系统、灭火系统及注入、空气与测量管等。

Marine Technology Society(MTS) 海洋技术协会

Marine transmission gear(marine gearing) 船用传动装置 系指联轴器、离合器、减速齿轮箱、液力耦合器等主机与轴系间的传动设备或这些设备的组合。

Marine underwater room 海洋水声实验室 对海洋中的声场、声学参数、声在海水中的传播、散射、吸收、混响、噪声等要素进行现场测量,并运用声学方法进行水文要素的测量、分析和研究的工作室。该室必须附有低噪声电源设备。

Marine water 海水 系指盐度大于30psu的水。 Marine water mean the water with salinity higher than 30psu.

Marine wired interphone 船用有线对讲机 系指用于船舶操纵部位之间指挥通信用的设备。它由对讲主机和对讲分机及扬声器等组成。其功能如下:(1)主机可任选一分机,进行呼叫和对讲;(2)分机可向主机呼叫并进行对讲;(3)主机可向所有分机通播并进行对讲;(4)主机可外接50W号筒扬声器进行喊话。

Marine-type fan 船用风扇 符合船舶规范规定或船用技术条件要求的各种供船舶上使用的风机的统称。

Marinized aircraft gas turbine (aircraft-derived marine gas turbine) 喷水推进型船用燃气轮机 在喷水推进船中驱动高速水泵以获得推进动力的燃气轮机。

Maritime casualty 海上事故 系指船舶碰撞、搁浅或其他航行事故，或是在船上或船舶外部发生对船舶或货物造成物质损失或紧急威胁的事件。 Maritime casualty means a collision of ships, stranding or other incident of navigation, or other occurrence on board a ship or external to it resulting in material damage or imminent threat of material damage to a ship of cargo.

Maritime claim 海事请求 系指由于下列一个或多个原因引起的请求：(1)碰撞或其他情况下任何船舶所造成的损害；(2)由任何船舶所造成的或任何船舶营运有关而发生的人身伤亡；(3)海难救助；(4)不论采取租船合同与否，凡与任何船舶使用或租赁有关的协议；(5)不论采取租船合同或其他形式，凡与任何船舶上货物运输有关的协议；(6)包括行李在内的任何船舶所载货物的灭失或损坏；(7)共同海损；(8)船舶抵押贷款；(9)拖带；(10)引航；(11)在任何地方提供给船舶为其营运或维持所需的物品和材料；(12)任何船舶的建造、修理或装备，或船坞费用和款项；(13)船长、高级船员或一般船员的工资；(14)船长所支付的费用，包括托运人、承租人或代理人代表船舶或其所有人支付的费用；(15)有关任何船舶的权利或所有权的争议；(16)任何船舶的共有人之间对该船的所有权、占用、使用或收益的争议；(17)任何船舶的抵押权或质押权。 Maritime claim means a claim arising out of one or more of the following: (1) damage caused by any ship either in collision or otherwise; (2) loss of life or personal injury caused by ship or occurring in connection with the operation of any ship; (3) salvage; (4) agreement relating to the use or hire of any ship whether by charter party or otherwise; (5) agreement relating to the carriage of goods in any ship whether by charter party or otherwise; (6) loss of or damage to goods including baggage carried in any ship; (7) general average; (8) bottomry; (9) towage; (10) pilotage; (11) goods or materials wherever supplied to a ship for her operation or maintaince; (12) construction, repair or equipment of any ship or dock charges and dues; (13) wages of masters, officers, or crew; (14) master's disbursements, including disbursements made by shippers, charterers or agents on behalf of a ship or her owner; (15) disputes as to the title to or ownership of any ship; (16) disputes between co-owners of any ship as to the ownership, possession employment or earnings of that ship; (17) the mortgage or hypothecation of any ship.

Maritime Environment Protection Committee (MEPC) 海上环境保护委员会 国际海事组织(IMO)的五个主要的委员会之一，其主要职责是审议有关组织范围内有关防止和控制船舶所造成的污染问题，为大会通过的有关公约、规则及其修正案制订实施措施。此外，提供各国有关防止和控制船舶对海上污染的科学技术实用资料，提出有关建议和拟定指导原则；以及组织区域性合作和其他国际组织的合作，以防止和控制船舶对海上的污染。

Maritime Operation Department 海事作业部(巴拿马运河) 海事作业部是巴拿马运河管理局的一个职能机构，由海事作业主任领导，负责处理有关运河及其终端港的通航事宜。巴拿马运河管理局向航运界提供的各种服务均通过海事作业部进行。以下是海事作业部所属的一些单位：(1)丈量组(Admeasurement Unit)由丈量组经理领导负责。确定船舶在通过运河时具有准确的巴拿马运河吨位；审计和确定全集装箱船的 TEU 数(TTA)和其他船舶在甲板上载运的 TEU 数(NNT)；登船检查和消除疫情；船舶的一般检查；为船舶数据库(Ship Date Bank)搜集资料；并为通航及与之有关的各种服务出具收款单据。(2)验船师委员会(Board of Inspectors)由该验船师委员会主席领导，负责对发生在运河作业区、港口、运河锚地和与上述地区相邻区域内涉及运河管理局工作人员和/或设备的各种海事事故进行正式调查和处理，并负责对管理局的雇员签发海事方面的证件。(3)通航作业处(Transit Operations Division)由通航作业处经理领导，负责处理日常的海事管理和紧急海事事故，监督实施运河航行规则和条例，审批新的结构设计(包括导缆器和带缆桩、登船设施、驾驶室设计和能见度要求等)以确保到来待航的船舶已正确地装备。此外，该处还负责处理有关船舶安全方面的交通流量及其控制、船舶的材料状况及其检查、危险货物、运河的物理状况，以及对火灾或油料/化学品的外溢等应急情况。通航作业处经理的职责通过运河的港区值班船长执行；(4)交通管理组(Traffic Management Unit)由交通管理组经理领导，负责处理预计到达时间(ETA)信息，编制每天的通航计划，监督并协调所有在运河区操作区内船舶的移动，管理巴拿马运河通航预约系统。

Maritime Regulations for the Operation of the Panama Canal (MROPC) 巴拿马运河航行海事规则

Maritime Safety Committee (MSC) 海上安全委员会 国际海事组织(IMO)的五个主要的委员会之一，是IMO的最高技术机构，其主要职责是研究本组织范围内有关海上安全、助航设备、船舶建造和装备、船员配备、避碰规则、危险货物装卸、航道信息、航海日志、航行记录、救助救生、海上事故调查以及直接影响海上安全的任何其他事宜。MSC 每年召开一至两次会议。

Maritime safety information 海上安全信息 系指向船舶播放的航行和气象警报、气象预报和与安全有关的其他紧急信息。 Maritime safety information means navigational and meteorological warnings, meteorological forecasts and other urgent safety related messages broadcast to ships.

Maritime security auditor (MSA) 海事保安审核员

Mark of leg 桩腿标尺 自升式钻井平台桩腿上指示桩腿升高高度的标尺。

Marking of lifesaving apparatus 救生设备标志 救生载具、救生浮具和救生圈上注明所属船舶的船名、船籍港及救生设备本身编号等的标志。

MARPOL = the IMO Convention for the Prevention of Pollution by Ships. 国际防止船舶造成污染公约 系指经1978年议定书修正的1973年国际防止船舶造成污染公约。可向国际海事组织（英国伦敦 4 Alber Embankment, London, SE1, 7SR.）申购 MARPOL 的副本。MARPOL means the International Convention for the Prevention of Pollution from Ships, 1973, as modified by the Protocols of 1978 and 1997 relating to that Convention. A copy of MARPOL is available from the International Maritime Orhanizayion(4 Alber Embankment, London, SE1, 7SR, United Kingdom).

MARPOL Annex Ⅰ Regulations for the Prevention of Pollution by Oil 防污公约 附则Ⅰ"防止油类污染规则" 该附则于1983年10月2日生效并在"防污公约"缔约国之间替代早期生效并经1962年和1969年修正的"1954年国际防止海上油污公约"。附则Ⅰ的修正案已由MEPC通过并已生效。 The Annex entered into force on 2 October 1983 and, as between the Parties to MARPOL73/78, supersedes the International Convention for the prevention of pollution of the sea by oli, 1954, as amended in 1962 and 1969, which was then in force. A number of amendments to Annex Ⅰ have been adopted by the MEPC and have entered into force.

MARPOL Annex Ⅱ Regulations for the Control of Pollution by Noxious Liquid Substances in Bulk 防污公约 附则Ⅱ"控制散装有毒液体物质污染规则" 为便于该附则的实施，原版本在1985年经 MEPC.16(22)决议的修正，这次修正涉及泵吸、管路和控制要求。MEPC 在其22届会议上根据1978年议定书第11条还决定，"各缔约国应自1987年4月6日起受经修正的'73/78防污公约'附则Ⅱ各项规定的约束"。 To facilitate implementation of the Annex, the original text underwent amendments in 1985, by resolution MEPC.16(22), in respect of pumping, piping and control requirements. At its twenty-second session, the MEPC also decided that, in accordance with article of the 1978 Protocol,"Parties shall be bound by the provosions of Annex Ⅱ of MARPOL 73/78 as amended from 6 April 1987".

MARPOL Annex Ⅲ Regulations for the Prevention of Pollution by Harmful Substances Carried by Sea in Packaged Form 防污公约 附则Ⅲ"防止海运包装有害物质污染规则" 该附则于1992年7月1日生效。但早在该生效日期前，MEPC 与海上安全委员会（MSC）即一致同意将该附则通过 IMDG 规则予以实施。IMDG 规则包括由 MSC 编制的涉及海洋污染的修正案（修正案25-89），这些修正案自1991年1月1日起实施。 The Annex entered into force on 1 July 1992. However, long before this entry into force date, the MEPC, with the concurrence of the Maritime Safety Committee(MSC), agreed that the Annex should be implemented through the IMDG Code. The IMDG Code had amendments covering marine pollution prepared by the MSC(Amendment 25-89).

MARPOL Annex Ⅳ Regulations for the Prevention of Pollution by Sewage from Ships 防污公约 附则Ⅳ"防止船舶生活污水污染规则" 该附则于2003年9月27日生效。其后的修正案已由 MEPC 通过年已生效。

The Annex entered into force on 27 September 2003. Subsequent amendments have been adopted by the MEPC and have entered into force.

MARPOL Annex Ⅴ Regulations for the Prevention of Pollution by Garbage from Ships 防污公约 附则Ⅴ"防止船舶垃圾污染规则" 该附则于1988年12月31日生效。其后的修正案已由 MEPC 通过年已生效。

The Annex entered into force on 31 December 1988. Subsequent amendments have been adopted by the MEPC and have entered into force.

MARPOL Annex Ⅵ Regulations for the Prevention of Air Pollution from Ships 防污公约 附则Ⅵ"防止船舶造成空气污染规则" 由"防污公约"缔约国国际会议于1997年9月通过，是"经1978年议定书修订的1973年国际防止船舶造成污染公约"的附件。该附则于2005年5月19日生效。 The Annex is appended to the Protocol of 1997 to amend the International Convention for the Prevention of Pollution from ships, 1973, as modified by the Protocol of 1978 relating thereto, which was adopted by the International Convention of Parties to the MARPOL Conventionin September 1997. Annex Ⅵ entered into force on

19 May 2005.

MARPOL Convention 73/78 "1973/1978 国际防止船舶造成污染公约" 是防止船舶因营运或事故原因造成海洋环境污染的主要国际公约。它是国际海事组织分别于 1973 年和 1978 年正式通过的两个协定的合并公约，MARPOL 已经过连续多年的增补修订。MARPOL Convention (International Convention for the prevention of pollution from ships) is the main international convention covering the prevention of pollution of the marine environment by ships from operational or accidental causes. It is a combination of two treaties adopted at IMO in 1973 and 1978 respectively. MARPOL was continuously updated by amendments through the years.

MARPOL tanker MARPOL 油船 系指满足国际海事组织 MARPOL73/78 要求设置专用压载的单壳油船。即 MARPOL 前油船系指 MARPOL73/78 设置专用压载要求生效前建造的单壳油船。MARPOL tanker means a single skin tanker meeting the requirements of IMO MARPOL73/78 for segregated ballast. A pre-MAEPOL tanker is a single skin tanker build before the MARPOL 73/78 segregated ballast requirements.

MARPOL = Maritime Agreement Regarding Oil Pollution 国际防止船舶造成污染公约

Married hook 双杆吊货钩 可连接两根吊货索专供定位双杆操作时使用的吊钩。

Marsh screw amphibian 螺杆艇 在水域、泥地、沼泽地或冰雪地，依靠安装在艇底两侧的两个螺杆旋转而推进的船舶。

MARVS 释放阀最大的调定值 系指气体柜释放阀的最大许用设定值。MARVS means the maximum allowable relief valve setting of a gas tank.

Mass displacement of naked hull 裸船舶排水质量 实船或船模的裸船体水线以下部分排开水的质量；$\Delta = \rho \nabla$，式中，Δ——裸船体排水质量的通用符号；ρ——水的质量密度；∇——船舶的型排水体积。

Mass of solid body 质量（船舶实体） 不考虑周围介质影响，度量实体惯性大小的物理量，对于船舶实体用 $\Delta = \rho \nabla$，式中，ρ——水的质量密度；Δ——船舶的排水质量；∇——船舶的排水体积。

Mass production 批量生产 对主、辅用途的船用柴油机制造而言，批量生产定义为：(1) 按同一的程序，在严格控制材料和部件质量的条件下，大量生产；(2) 采用专门设计的夹具和自动化机床，使加工件达到接近互换性的公差范围，且定期检查进行确认；(3) 通过对取自仓库的零部件进行组装，只需很少或不需加工，并接受；(4) 单个柴油机按试验大纲进行的台架试验；(5) 随机选取台架试验后的数台柴油机，通过最终试验进行评估。应注意：凡用于上述机械设备的所有铸件、锻件和其他部件，亦应以类似方法进行制造和检查。用上述方法生产的机械设备，其技术要求应明确所有零部件的制造允许范围，制造厂商应证明产品总产量，可要求由检验机构进行确认。For the purpose of the construction of marine engines for main and auxiliary purpose, mass production may be defined as that machinery which is produced: (1) in quantity under strict quality control of material and parts according to an agreed programmed; (2) by the use of jigs and automatic machines designed to machine parts to close tolerances for interchanreability, and which are to be verified on a regular inspection basis; (3) by assembly with parts taken from stock and requiring little or no fitting of the parts and which is subject to; (4) bench tests carried out on individual engines on a program basis; (5) appraisal by final testing of engines selected at random after bench testing. It should be noted that all casings, forgings and other parts for use in the forgoing machinery are also to be produced by similar methods with appropriate inspection. The specification for machinery produced by the forgoing method must be define the limits of manufacturer of all component parts. The total production output is to be certified by the manufacturer and verified as may be required, by the inspecting authority.

Mass transport 质移 波浪中水质点就地作轨圆运动的同时，本身的净纯的微小水平位移。

Mast 桅 用以装设号灯、天线和悬挂号型、号旗或扬帆，有的还兼作起货设备用的立柱或组合构架。

Mast and rigging 桅设备 船上各种桅、索具、帆及其附件的总称。

Mast head span block 上千斤滑车 位于起重柱处的千斤滑车。

Mast house 桅室 设置在最上层连续甲板上，位于桅的周围或门型柱的两柱间的房间。

Mast ladder (spar ladder) 桅梯 设置在桅、柱或烟囱等处的梯。

Mast light (head light) 桅灯 设置在船舶桅杆规定高度处，并在水平面上 225° 内，即自艏向至左、右 112.5°，显示白色不间断光的灯具。

Mast power block (ceiling power block) 悬挂式动力滑车 悬挂在渔船起重吊杆上的动力滑车。

Mast rigging 桅索具 桅设备中各种支索、张索及其配件的总称。

Mast thwart 桅箍座板 救生艇上可支持桅的横座板。

Master 船长 系指指挥一艘船的人员。Master

means the person having command of a ship.

Master gas fuel valve 气体燃料总阀 系指在气体燃料发动机所在的机器处所的各发动机的气体供应管路上安装的自动控制阀,并应尽可能靠近气体加热器(如设有)。 Master gas fuel valve means an automatic valve in the gas supply line to each engine located outside the machinery space for gas-fuelled engines and as close to the gas heater(if fitted) as possible.

Master start button 主启动按钮

Master supervisory and alarm frame 主监视报警装置机架

Master switch 主控开关 系指使控制接触器、继电器或其他远距离操作器件动作的开关。

Master switch direct control 主令开关直接控制 是一种适用于11kW及以上的交流小型双速电动机用主令开关直接控制电动机的启动、停止(包括刹车)、变速和换向,并能对电动机及其有关电路实施过载和失压保护功能的控制方式。

Master's discretion for ship safety and security 船长对船舶安全和保安的决定权 系指:(1)船长按照其专业判断而做出或执行为维护船舶安全或保安所必需的决定,应不受公司、承租人或任何他人的约束。这包括拒绝人员(经确认的缔约国政府正式授权的人员除外)或其物品上船和拒绝装货,包括集装箱或其他封闭的货运单元;(2)如果按照船长的专业判断,在船舶操作中出现该船的安全和保安之间发生冲突的情况,船长应执行为维护船舶安全所必须的要求。在这种情况下,船长可以实施临时性保安措施并应随即通知主管机关;如情况适宜,还应随即通知该船所在或拟进入的港口所属缔约国政府。据此采取的任何此类临时性保安措施应尽最大可能相当于主要的保安等级。在识别这种情况后,主管机关应确保此类冲突得以解决并使其再次发生的可能性减至最低。 Master's discretion for ship safety and security mean that: (1) The master shall not be constrained by the Company, the charterer or any other person from taking or executing any decision which, in the professional judgement of the master, is necessary to maintain the safety and security of the ship. This includes denial of access to persons (except those identified as duly authorized by a Contracting Government) or their effects and refusal to load cargo, including containers or other closed cargo transport units; (2) If in the professional judgement of the master, a conflict between any safety and security requirements applicable to the ship arises during its operations, the master shall give effect to those requirements necessary to maintain the safety of the ship. In such cases, the master may implement temporary security measures and shall forthwith inform the Administration and, if appropriate, the Contracting Government in whose port the ship is operating or intends to enter. Any such temporary security measures under this regulation shall, to the highest possible degree, be commensurate with the prevailing security level. When such cases are identified, the Administration shall ensure that such conflicts are resolved and that the possibility of recurrence is minimized.

Masthead light 桅灯 系指安置在船的首、尾中心线上的白灯,在225°在水平弧内显示不间断的灯光,其装置要使灯光从船的正前方到每一舷正横后22.5°内显示。 Masthead light means a white light placed over the fore and aft centerline of the vessel showing an unbroken light over an arc of the horizon of 225° and so fixed as to show the light from right ahead to 22.5° abaft the beam on either side of the vessel.

Mat 沉垫 系指为降低自升式平台桩腿对地基的压力而把各桩腿底部连接起来的整体式水密箱形结构。 Mat is an integral watertight box-type structure supported by soft seabed and connected to the bottoms of legs of a self-elevating unit.

Mat self-elevating drilling platform 沉垫自升式钻井平台 桩腿均安装在同一个整体沉垫上,升降时不要预压定位的自升式钻井平台。

Mat skirt 沉垫裙板 为防止沉垫底面与海底泥面的相对位移而在沉垫下底面装设的纵向和横向的立板。

Material 材料 系指船舶建造中所采用的金属材料和非金属材料等。如普通强度船体结构钢(normal-strength hull structural steel)、高强度船体结构钢(higher-strength hull structural steel)、和类似等级的锻钢(forged steel)、铸钢件(steel castings)、高强度(出火加回火)细化晶粒结构钢[high-strength(quenched and tempered)fine-grained structural steel]、低温韧性钢(steels tough at subzero temperatures)、不锈钢(stainless steel)、和复合钢(clad steel)、铝合金(aluminum alloys)、有色金属(non-ferrous metals)和非金属材料(non-metallic material)等。

Material damage in relation to marine casualty 与海难相关的材料破损 系指:(1)破损——①对船上基础设施或支承结构的完整性、性能或操作特性有严重影响;②要求大修或更换一个或多个主要部件;(2)船上基础设施或船舶的毁坏。 Material damage in relation to marine casualty means:(1)damage that:①significantly affects the structural integrity, performance or operational characteristics of marine infrastructure or a ship; and ②requires major repair or replacement of a major component or component(s); (2) destruction of the marine infrastructure or ship.

Material damage safety　物资损失安全　根据生产延误及装备和结构重建的意外后果而涉及的设施、设施结构和设备安全。

Material declaration identification number　材料声明标识号

Material declaration(MD)　材料声明　系指在材料声明中规定,造船工业的供应商应标识和声明表中所列物质的存在是否高于规定的阈值水平。但是,该规定不适用于并非构成成品之一部分的化学品(chemicals which do not constitute a part of the finished product)。材料声明应至少包括:(1)声明日期;(2)材料声明标识号;(3)供应厂商名称;(4)产品名称(通用产品名称或生产厂商使用的名称);(5)产品编号(供生产厂商标识);(6)声明中所列物质是否以高于规定的阈值水平存在于产品中;以及(7)如果以高于阈值水平存在,应则列出每一构成物质的质量。

Material factor k　材料系数　系指用于确定一般强度钢和高强度钢尺寸的系数。该系数是最小屈服应力 R_{eH} 的函数。屈服强度大于 390 N/mm² 的钢材,船级社应逐个考虑。见表 M-2。　Material factor k means the coefficient for scantling of normal and higher strength steel. The material factor k is as a function of the minimum yield stress R_{eH} Steels with a yield stress greater than 390 N/mm² are considered by the Society on a case by case basis. See table M-2.

表 M-2　材料系数
Table M-2　Material factor k

最小屈服应力 (Minimum yield stress R_{eH}) (N/mm²)	材料系数 (Material factor k)
235	1.0
315	0.78
355	0.72
390	0.68

Materials found on board ships that the ship recycling facility should be prepared to handle　拆船设施应做好准备进行处理的船上可见材料 (有害材料清单第Ⅲ部分中所包括者)见表 M-3。

表 M-3　拆船设施应做好准备进行处理的船上可见材料(有害材料清单第Ⅲ部分中所包括者)
Table M-3　Materials found on board ships that the ship recycling facility should be prepared to handle(included in Part Ⅲ of the inventory of hazardous materials)

序号	Hazardous materials	有害材料
1	Kerosene	煤油
2	White spirit	石油溶剂
3	Lubricating oil	润滑油
4	Hydraulic oil	液压油
5	Anti-seize compounds	防黏剂
6	Fuel additive	燃料添加剂
7	Engine coolant additives	发动机冷却添加剂
8	Antifreeze fluids	防冻液
9	Boiler and feed water treatment and test reagents	锅炉和给水处理剂测试试剂
10	Deionizer-regenerating chemicals	脱离子再生化学剂
11	Evaporator dosing and descaling acids	蒸发器给料与除垢酸
12	Paint stabilizers/rust stabilizers	油漆稳定剂/锈稳定剂

续表 M-3

序号	Hazardous materials	有害材料
13	Solvents/thinners	溶剂/稀释剂
14	Paints	油漆
15	Chemical refrigerants	化学制冷剂
16	Battery electrolyte	电池电解质
17	Alcohol/methylated spirits	酒精/甲基酒精
18	Acetylene	乙炔
19	Propane	丙烷
20	Butane	丁烷
21	Oxygen	氧气
22	Carbon dioxide	二氧化碳
23	Perfluorocarbons(PFCs)	全氟化碳(PFCs)
24	Methane	甲烷
25	Hydrofluorocarbons(HFCs)	氢氟碳(HFCs)
26	Nitrous oxide(N_2O)	氮氧化物(N_2O)
27	Sulfur hexafluoride(SF6)	六氟化硫(SF6)
28	Bunkers, e.g. fuel oil	燃料,如燃油
29	Grease	脂
30	Perfluorocarbons(PFCs)	全氟化碳(PFCs)
31	Fuel gas	燃气
32	Batteries(including lead-acid batteries)	电池(包括铅酸电池)
33	Pesticides/insecticide sprays	杀虫剂/杀虫剂喷雾
34	Extinguishers	灭火器
35	Chemical cleaner (including electrical equipment cleaner, carbon remover)	化学清洁剂 (包括电气设备清洁剂、碳清除剂)
36	Detergent/bleacher(potentially a liquid)	洗涤剂/漂白剂(潜在液体)
37	Miscellaneous medicines	杂项药品
38	Fire-fighting clothing and personal protective equipment	消防服装和个人保护设备
39	Spare parts containing hazardous materials	含有有害材料的备件

Material resource plan(MRP) 物料资源规划
Materials exposed to cargo 曝露在货物中的材料 系指在正常作业状态下与货物(液体或蒸气)接触的那些组成系统、装卸货设备或装置的材料。 Materials exposed to cargo are those constituting systems, cargo appliances or arrangements which are in contact with (liquid or

vapour) cargo in normal operating conditions.

Materials handling equipment　物资装运设备　系指设计用于运送货物或在运送货物时使用的部件或这些部件的完整组件,其包括起货装置、起重机、吊杆、货物提升机、舷侧装载平台、机械装载设施和机械堆装设施。　Materials handling equipment means an article or an integrated assembly of articles designed to convey or for use in conveying cargo and includes cargo gear, a crane, derrick, cargo lift, side loading platform, mechanical loading appliance and mechanical stowing appliances.

Materials hazardous only in bulk(MHB)　仅在散装时有危害的物质(MHB)　系指散装载运时可能具有化学危害的物质,但在 IMDG 规则中归入危险货物类别的物质除外。　Materials hazardous only in bulk(MHB) mean those materials which may possess chemical hazards when transported in bulk other than materials classified as dangerous goods in the IMDG Code.

Materials safety data sheet(MSDS)　材料安全数据单

Materials that may be hazardous to health　可能对健康造成危害的材料　包括:(1)二氧化硫——潮湿时有腐蚀性,吸入有毒,造成灼伤,对眼睛和呼吸系统有刺激;(2)硫化氢——高度易燃(闪点 -82 ℃),遇空气能形成爆炸性混合物,潮湿时有腐蚀性,造成灼伤,必须远离着火源存放,有刺激性和窒息性,长期曝露极限值(LTEL)5×10^{-6},短期曝露极限值(STEL)10×10^{-6},浓度再高会致命,无气味。反复曝露于低浓度硫化氢能导致对气体的嗅觉消失;(3)苯——高度易燃(闪点 -11 ℃),遇空气能形成爆炸性混合物,有毒,致癌,严重危及健康;(4)甲苯——高度易燃(闪点 4 ℃),遇空气能形成爆炸性混合物,有刺激性,严重危及健康,对生殖器官有毒性。　Materials that may be hazardous to health are as follows:(1) Sulphur Dioxide—corrosive when wet, toxic if inhaled, causes burns, and is an irritant to the eyes and respiratory system;(2) Hydrogen Sulphide—highly flammable (Flash point of -82 ℃), can form an explosive mixture with air, corrosive when wet, causes burns, has to be kept away from sources of ignition, irritant and asphyxiant, LTEL 5×10^{-6}, STEL 10×10^{-6}, higher concentrations can be fatal and have no odour. Repeated exposure to low concentrations can result in the sense of smell for the gas being diminished. (3) Benzene—highly flammable(Flash point of -11 ℃), can form an explosive mixture with air, toxic, carcinogenic, acute health risk. (4) Toluene—highly flammable (Flash point of 4 ℃), can form an explosive mixture with air, irritant, acute health risk, reprotoxin.

Mathematical expectation　期望值　随机变量的统计平均值。其表达式为 $\mu = M\xi = \int_{-\infty}^{+\infty} x_p(x)\mathrm{d}x$。

Mathematical lines　数学型线　系指用数学公式表示的船体型线。

Matn steering gear　主操舵装置　系指在设有两套操舵装置的船舶上,其首先和经常使用的操舵装置。

Max. cargo density　最大货物密度　允许散装货船装载货物的最大密度。

Maximum advanced　最大纵距　回转试验中,从转舵瞬时的船舶重心位置,沿初始直线航线方向到重心运动轨迹最远处的纵向距离。

Maximum ahead(Maximum beam)　最大船宽　船壳板外表面之间的最大船体宽度。　Maximum breadth means a width of the hull between the outside surfaces of the shell plating.

Maximum ahead service speed　最大营运前进航速　系指船舶在最大航海吃水情况下,螺旋桨转速为最大值以及相应的主机为最大持续功率时保持海上营运的最大设计航速。　Maximum ahead service speed is the greatest speed which the ship is designed to maintain in service at sea at her deepest seagoing draught at maximum propeller RPM and corresponding engine MCR.

Maximum allowable response time　最大允许响应时间　对巴拿马运河而言,系指主机在收到传令钟指令后从停机到前进或从停机到后退过程所耗时的最大值。该数值必须在营运开始前进行测试。　For Panama Canal, maximum allowable response time is the maximum allowable amount of time that it takes the main propulsion to respond from stop to ahead or from stop to astern, after a telegraph order is received, which must be tested before the transit begins.

Maximum allowable start time　最大允许启动时间(巴拿马运河)　对巴拿马运河而言,系指启动船舶主机所需的最大允准时间,在通航开始前必须加以测定。　For Panama Canal, maximum allowable start time means the maximum allowable amount of time that it takes for a vessel engine to start, which must be tested before the transit begins.

Maximum allowable transfer rate　最大允许传送率　对巴拿马运河而言,系指船舶装载货物或压载所能达到的最大容量速率。　For Panama Canal, maximum allowable transfer rate is the maximum volumetric rate at which a ship may receive cargo or ballast.

Maximum allowable vertical center of gravity　最大许用重心高度

Maximum astern speed 最大倒车速度　系指船舶在最大航海吃水情况下用设计的最大倒车功率估算能达到的速度。 Maximum astern speed is the speed which it is estimated the ship can attain at the designed maximum astern power at the deepest sea-going draught.

Maximum authorized draft 允准的最大吃水（巴拿马运河）　对巴拿马运河而言，系指小于允许的最大通过吃水，或者由其载重线证书允准的最大热带淡水吃水的吃水值。 For the purpose of Panama Canal, maximum authorized draft means the lesser of the maximum authorized transit draft or the maximum tropical freshwater draft by Load Line Certificate.

Maximum authorized transit draft 允许的最大通过吃水（巴拿马运河）　对巴拿马运河而言，系指受夏通湖水位和运河的限制，在任何时间允许某一艘特定船舶在热带淡水中浸没的最深点。 For Panama Canal, maximum authorized transit draft means the deepest point of immersion in TFW of a particular vessel permitted at anytime, Gatun Lake level and Canal restrictions permitting.

Maximum beam 最大船宽（巴拿马运河）　对巴拿马运河而言，系指船壳板外表面之间的最大船体宽度。 For Panama Canal, maximum beam means the maximum breadth(width) of the hull between the outside surfaces of the shell plating.

Maximum beam B_{max} 最大宽度（艇）B_{max}　最大宽度 B_{max} 应在通过艇体最外侧部件的平行于艇中线面的两垂向平面之间进行测量。最大宽度包括艇的所有结构或组成部件，诸如艇体的延伸件、艇体/甲板连接件，延伸件诸如覆板、舷侧厚板、桅侧支索牵条、橡胶护舷材、固定碰垫等，以及延伸到艇舷外的船舷扶手。 Maximum beam B_{max} shall be measured between planes passing through the outermost parts of the craft. The maximum beam includes all structural or integral parts of the craft, such as extensions of the hull, hull/deck joints, extensions such as doublings, sheer planks, chain plates, rubbing strakes, permanent fenders and liferails extending beyond the craft's side.

Maximum blade width ratio 最大叶宽比　螺旋桨的最大叶宽对直径的比值。

Maximum breadth under waterline B_w (m) 水线下最大船宽　对小水线面双体船而言，系指船舶静浮于水面时，位于设计水线以下取的最大船宽。

Maximum capacity 最大容量（最大功率，最大生产率）

Maximum clearance 最大间隙

Maximum climbing slope 爬坡坡度　全垫升气垫船以一定航速在陆地上持续上坡能安全行驶的最大坡度。

Maximum combustion pressure (maximum explosive) 最大爆发压力　内燃机正常运转时循环中气缸内的最高瞬时压力。

Maximum continuous load 最大连续载荷

Maximum continuous output of engine 最大持续输出功率　(1)主推进装置在满载吃水情况下航行时，使主推进装置以外的机械能够安全连续工作的最大功率；(2)在规定的环境条件下，在额定最大转速以下以及在两次依次大修之间的时间间隔内柴油机能连续运行而发出的最大功率。功率、转速和两次依次大修之间的时间间隔应由制造厂商规定并经船级社同意。 Maximum continuous output of engine is：(1) the maximum output at which the engine can run safely and continuously in the running condition at full load draught, in the engine for propulsion. In the engine excluding main engine, the maximum continuous output of engine is the maximum output at which the engine can run safely and continuously in the intended condition；(2) the maximum power at ambient reference conditions which the engine is capable of delivering continuously, at nominal maximum speed, in the period of time between two consecutive overhauls. Power, speed and the period of time between two consecutive overhauls are to be stated by the Manufacturer and agreed by the Society.

Maximum continuous rating (MCR) 最大持续功率

Maximum depth D_{max} 最大型深 D_{max}　最大型深 D_{max} 应取在1/2的水线长度 L_{WL} 处的甲板舷弧线与龙骨的最低点之间测得的垂直距离。注意：对于传统的长龙骨的艇，该龙骨的坡度可导致增加艉吃水，其不在1/2的水线的长度或船体的长度处测量最大型深。 Maximum depth D_{max} shall be measured as the vertical distance between the sheerline at half-length of the waterline L_{WL} and the lowest point of the keel. Note：With traditional long-keeled craft, the slope of the keel may result in increased draught aft, which is not at half-length of the waterline or length of the hull.

Maximum design draught T_{sc} 最大设计吃水 T_{sc}　系指满足关于船舶尺度的强度要求下的最大设计吃水(m)。 Maximum design draught T_{sc}, in m, means a draught at which the strength requirements for the scantlings of the ship are met.

Maximum discharge distance 最大排泥距离　系指挖泥船在一定土质条件和泥浆浓度下，按一定的泥泵转速、功率、排泥管直径和排泥高度排泥时所能达到的最大距离。

Maximum discharge height 最大排泥高度 系指吸扬式挖泥船在设计计算最大排泥距离时所用的最大的排泥高度。

Maximum displacement 最大排水量 军用舰船装载超过全额的人员、淡水、粮食、弹药、燃料、滑油以及非常备武器等所允许的最大额外载重时的排水量。

Maximum diving depth 最大沉深 系指半潜作业状态下允许下潜到的最大吃水。

Maximum draught[1] 最大吃水 系指夏季淡水载重线的吃水。如船舶有木材载重线，系指夏季淡水木材载重线的吃水。 Maximum draught means the draught on the fresh water load line in summer. If the ship has a timber-load line, the fresh water timber load line in summer is to be used.

Maximum draught[2] 最大吃水（冰区航行） 系指船舶在冰区航行时，应不超过由满载水线界定的吃水和纵倾。 Maximum draught are the draught and trim limited by the UIWL are not to be exceeded when the ship is navigating in ice. The salinity of the sea water along the intended route is to be taken into account when loading the ship.

Maximum draught amidships 船舯最大吃水 船舯最大吃水是相应于夏季淡水载重线的吃水。如果船舶具有木材载重线，则取夏季淡水木材载重线。 Maximum draught amidships is the draught corresponding to the fresh water load line in summer. If the ship has a timber load line, the fresh water timber load line in summer is to be used.

Maximum draught T_{max} 最大吃水（艇）T_{max} 最大吃水 T_{max} 应测量至处于最低位置的水下艇体或附件（包括处于其最低点的中插板）。 Maximum draught T_{max} shall be measured to the lowest point of the underwater body or appendage, including centerboards, in their lowest position.

Maximum dredging depth 最大挖深 挖泥船所能挖掘的水面以下最大深度。

Maximum dredging width 最大挖宽 系指绞吸挖泥船的铰刀，在垂直于前移向的方向上所能摆动切削的最大宽度。

Maximum drilling water depth 最大工作水深 钻井船或其他钻井装置能进行钻井作业的最大水深。

Maximum free space(maximum air gap) 最大升船高度 自升式钻井平台底部升离水面的最大高度。

Maximum hauling and rendering speed 最大收放绳速度 当绳索卷绕至卷筒最外层时，原动机以最高转速收放绳能达到的最大线速度。

Maximum height of lift 最大起升高度 起重机在船舶正浮状态工作时，吊钩最高位置与最低位置间的垂直距离。

Maximum height under sailing condition 航行高度 起重船处于航行状态时，其结构最高点离水面的高度。

Maximum ice class draught amidships 船舯处最大冰级吃水 系指夏季淡水中载重水线的吃水。如船舶具有木材载重水线，则应采用夏季淡水中的木材载重水线。 Maximum ice class draught amidships is draught on the water load line in summer. If the ship has a timber load line the fresh water timber load line in summer is to be used.

Maximum length[1] 最大总长 船舶首、尾终端（包括球鼻艏和突出物）之间的距离（即总长 L_{OA}）。 Maximum length means a distance between the forward and after extremities of a vessel, including the bulbous bow and protrusions(also length overall L_{oa}).

Maximum length[2] 最大船长 对巴拿马运河而言，系指船体最前端和最后端之间的距离，包括球鼻艏和突出物（也称为总长 L_{OA}）。 For Panama Canal, maximum length is the distance between the forward and after extremities of a vessel, including the bulbous bow and protrusions(also length over-all-L_{OA}).

Maximum length L_{max} 最大长度（艇）L_{max} 最大长度 L_{max} 应取在垂直于艇中线面的两垂向平面之间平行于基准水线和艇的中心线的距离，其一个平面通过艇的最前端部件，另一平面通过艇的最后端部件。该长度包括艇的所有结构和组成部件，诸如木质、塑料或金属艏柱或艉柱，舷墙与艇体/甲板连接件。该长度包括通常为固定的部件，诸如固定的帆桁、艇首斜撑帆杆、艇两端的操纵台、艉柱附件、舵、舷外挂机支架、舷外挂机、喷水推进器及延伸到艉板外的各种推进装置，潜水和登艇平台，橡胶护舷材及固定式碰垫等。 Maximum length L_{max} shall be measured parallel to the reference waterline and craft centerline as the distance between two vertical planes, perpendicular to the center-plane of the craft, one plane passing through the foremost part and the other through the aftermost part of the craft. This length includes all structural and integral parts of the craft, such as wooden, plastic or metal stems or sterns, bulwarks and hull/deck joints. This length includes parts which are normally fixed, such as fixed spars, bowsprits, pulpits at either end of the craft, stemhead fittings, rudders, outboard motor brackets, outdrives, water-jets and any propulsion units extending beyond the transom, diving and boarding platforms, rubbing strakes and permanent fenders.

Maximum normal ground effect speed 最大正常地效速度 船舶以地效状态在已被证实的载荷和稳性范围内正常运行的最高航速。为保持合理预见的运行条件瞬间变化留有足够的安全裕度。 Maximum normal ground effect speed is the highest speed at which the craft is normally operated in ground effect through its proven load and stability range allowing sufficient safety margin for reasonable foreseeable transient variations in operating conditions.

Maximum offset 最大偏移 系指平均偏移加上经适当组合的平台波频和低频运动幅值。最大偏移应按下述两式计算，并取其中较大值：$X_{max} = X + X_{max}, l_f + X_{1/3}, w_f$；$X_{max} = X + X_{1/3}, l_f + X_{max}, w_f$，式中，$X_{max}$——最大偏移，m；$X$——平均偏移，m；$X_{max}$，$l_f$——平台低频运动最大单幅值，m；$X_{1/3}$，$l_f$——平台低频运动有义单幅值，m；$X_{max}$，$w_f$——平台波频运动最大单幅值，m；$X_{1/3}$，$w_f$——平台波频运动有义单幅值，m。 Maximum offset is the mean offset plus the amplitude of appropriately combined oscillatory motions and low frequency motions of the unit. The maximum offset is to be taken as the greater value obtained by the following formulas：$X_{max} = X + X_{max}, l_f + X_{1/3}, w_f$；$X_{max} = X + X_{1/3}, l_f + X_{max}, w_f$, where：$X_{max}$—maximum offset, in m；$X$ – mean offset, in m；X_{max}, l_f—maximum single amplitude of unit's low frequency motion, in m；$X_{1/3}$, l_f—significant single amplitude of unit's low frequency motion, in m；X_{max}, w_f—maximum single amplitude of unit's oscillatory motion, in m；$X_{1/3}$, w_f—significant single amplitude of unit's oscillatory motion, in m.

Maximum operating gross mass or rating 最大营运总质量或额定质量 系指集装箱和所载货物的最大允许总质量。 Maximum operating gross mass or rating means the maximum allowable sum of the mass of the container and its cargo.

Maximum operational Froude number 最大营运傅汝德数

Maximum operational weight 最大营运质量(t) 系指船舶在主管机关允许的装载状态营运时达到的最大总质量。 Maximum operational weight means the overall weight up to which operation in the intended mode is permitted by the Administration.

Maximum permissible payload 最大允许载货质量 系指最大营运总质量或预定质量与皮重之间的差数。 Maximum permissible payload means the difference between maximum operating gross weight or rating and tare weight.

Maximum relay 过载继电器 当过电流达到某一预定值时动作，使之断开电源的一种过电流继电器。过载继电器用来保护电动机或控制器，但并不一定能保护其自身。

Maximum safe speed 最大安全速度(地效翼船) 船舶持续显示出安全的稳定性特征的最大速度。该速度应不小于最大正常地效速度和绝对最大船舶速度的中间值。 Maximum safe speed is the maximum speed at which the craft will continue to demonstrate safe stability characteristics. This speed should be no less than midway between maximum normal ground effect speed and absolute maximum craft speed.

Maximum section, fullest section 最大横剖面 至横剖面中的最大剖面。当船体具有平行中体时，则系指中横剖面。

Maximum service speed V 最大服务航速 系指：(1)船舶在最深航行吃水时持续使用的设计最大航速；(2)船舶按其设计在营运中以最深航行吃水、螺旋桨最大转速(RPM)和发动机的相应最大持续功率(MCR)所保持的最大航速。 Maximum service speed in knots, mean：(1) defined as the greatest speed which the ship is designed to maintain in service at her deepest seagoing draught；(2) the greatest speed which the ship is designed to maintain in service at her deepest sea-going draught at the maximum propeller RPM and corresponding engine MCR (maximum continuous rating).

Maximum service weight 最大营运质量(t) 系指船舶在允许的装载状态营运时达到的最大总质量。

Maximum sinkage draft (floating dock) 最大沉浮吃水 d_{max}(m) 系指在平浮状态下，浮船坞允许下沉的最大吃水。

Maximum skew angle of a propeller blade 螺旋桨桨叶最大倾斜角 系指在桨叶的投影视图中，连接叶梢和轴中心点的直线与连接轴中心点和通过桨叶螺旋面的中点轨迹线相切直线之间的夹角。见图M-6。 Maximum skew angle of a propeller blade is defined as the angle, in projected view of the blade, between a line drawn through the blade geometric tip and the shaft centreline and a second line through the shaft centreline which acts as a tangent to the locus of the mid-points of the helical blade sections, see Figure M-6.

Maximum slewing radius 最大回转半径 系指在安全工作载荷下，动臂起重机允许操作的半径，单位是m。 Maximum slewing radius is the radius at which a jib crane is permitted to operate under the safe working load, and expressed in meters(m).

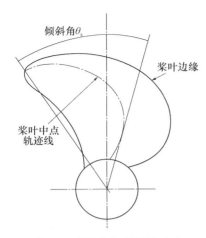

图 M-6　螺旋桨桨叶倾斜角定义
Figure M-6　Propeller skew angle definition

Maximum speed　最大航速　系指在满载排水量时以核定的最大持续推进功率在静水中航行能达到的航速。　Maximum speed is the speed achieved at the maximum continuous propulsion power for which the craft is certified at maximum operational weight and in smooth water.

Maximum steady state noise level　最大稳定噪声级

Maximum steam output　最大蒸汽输出　系指由锅炉或者蒸汽发生器在设计蒸汽工况下连续产生的最大的蒸汽产量。　Maximum steam output is the maximum quantity of steam than can be produced continuously by the boiler or steam generator operating under the design steam conditions.

Maximum submerged depth　最大沉深　系指半潜作业状态下允许下潜到的最大吃水。　Maximum submerged depth means the maximum draft to which the vessel is allowed to be submerged in semi-submersible operation condition.

Maximum submerged draft　最大沉浮吃水　浮船坞处于最大沉浮时，由基线至水面的垂直距离。

Maximum submerged freeboard　最大沉深干舷　浮船坞处于最大沉深时，坞墙顶甲板至水面间的垂直距离。

Maximum suface temperature　最高表面温度　系指电器设备在允许范围内最不利条件下运行时，能引起周围爆炸性气体环境点燃的任何部分或表面的最高温度。

Maximum sustained speed　最大持续船速　系指考虑由于船舶在规则波中航行时阻力增加所造成失速后的最大服务航速，但自愿降速不予考虑。　Maximum sustained speed is defined as the maximum service speed taking into account speed loss due to resistance increase in regular waves. Voluntary speed loss is not taken into consideration.

Maximum tension　最大张力　系指平均张力与经适当组合的波频及低频张力之和。最大张力应按下述两式计算，并取其中较大值：$T_{max} = T + T_{max,lf} + T_{1/3},w_f$；$T_{max} = T + T_{1/3,lf} + T_{max,w_f}$，式中，$T_{max}$—最大张力，kN；$T$—平均张力，kN；$T_{max,lf}$—最大低频张力，kN；$T_{1/3,lf}$—有义低频张力，kN；$T_{max,w_f}$—最大波频张力，kN；$T_{1/3,w_f}$—有义波频张力，kN。　Maximum tension means the sum of the mean tension and the appropriately combined oscillatory tension and low frequency tension. The maximum tension is to be taken as the greater value obtained by the following formulas：$T_{max} = T + T_{max,lf} + T_{1/3},w_f$，$T_{max} = T + T_{1/3,lf} + T_{max,w_f}$, where：$T_{max}$—maximum tension, in kN；$T$—mean tension, in kN；$T_{max,lf}$—maximum low frequency tension, in kN；$T_{1/3,lf}$—significant low frequency tension, in kN；T_{max,w_f}—maximum oscillatory tension, in kN；$T_{1/3,w_f}$—significant oscillatory tension, in kN.

Maximum transfer　最大横距　回转试验中，船舶重心自初始直线航线向回转圈中心侧横移的最大距离。

Maximum transfer rate　最大输送率　系指船舶可接受货物或压载所能达到的最大容积率。　Maximum transfer rate is the maximum volumetric rate at which a ship may receive cargo or ballast.

Maximum transver section coefficient　最大横剖面系数　最大横剖面浸水型面积与其相对应的水线宽和吃水的乘积之比。对于民用船舶，各水线处最大横剖面系数之计算，其中宽常用满载水线宽。

Maximum width　最大总宽　系指：(1) 船舶最宽点(包括突出物)之间的极端宽度；(2) 对巴拿马运河而言，系指船舶最宽点(包括突出物)之间的极端宽度。 (1) Extreme width means a total width of a vessel, including protrusions, at its widest point；(2) For the purpose of Panama Canal, maximum width means the extreme width of a vessel, including protrusions, at its widest point.

Maximum width of blade　最大叶宽　螺旋桨叶展开的或伸张轮廓的最大宽度。

Maximum working external pressure　最大工作外压　是管内真空与管子外部液体压头的总和。　Maximum working external pressure is a sum of the vacuum inside the pipe and a head of liquid acting on the outside of the pipe.

Maximum working pressure　最大工作压力　系指操舵装置按有关的规定操作时，系统中最大预期的压力。　Maximum working pressure means the maximum expected pressure in the system when the steering gear is operated to comply with relevant requirement.

MAX-Sterco　船舶多通道立体播放软件

MCC　机器处所集中控制　系指船舶设置机舱集控站(室)和就地控制站，并在机电设备正常运行时，机舱集控站(室)连续有人值班。　Central control of machinery spaces means that ship equipped with central control station(CCS) and local control station, CCS constantly attended by watch-keepers during the normal operating of machinery and electrical equipment.

MCR = maximum continuous rating　最大连续功率

MDHT　统计平均每日最高温度　MDHT mean daily high(or maximum) temperature.

MDO = marine diesel oil　船用柴油

MDRARP = Marine Debris Research Prevention and Reduction Act　搜索、预防和减少海上碎片公约

Meal carrier　运肉船　专运鲜肉的船舶。

Mean amplitude of roll　平均横摇幅值　船舶向一舷和随后向另一舷横摇幅值的平均值，对不规则横摇则是在一定时间间隔内横摇幅值的平均值。

Mean axial velocity at impeller plane　叶轮面平均流速　导管推进器叶轮处平面上水流的平均轴向速度。

Mean blade width　平均叶宽　螺旋桨叶长除以螺旋桨叶的展开或伸张面积之商。

Mean blade width ratio　平均叶宽比　螺旋桨的平均叶宽对其直径的比值。

Mean daily average temperature(MDAT)　统计平均的每日温度　系指以多年观察为准的对某个特定日历日平均温度的统计平均。　Mean daily average temperature(MDAT) is the statistical mean average temperature for a specific calendar day, based on a number of years of observations.

Mean draught T_M　平均吃水　系指(艏吃水 T_A + 艉吃水 T_F)/2，式中，T_A—艏吃水，T_F—艉吃水。　Mean draught T_M is defined as: $T_M = (T_A + T_F)/2$, where T_A—bow draugh, T_F—aft draugh.

Mean effective pressure　平均有效压力　假设的，在活塞工作行程作用在活塞上为完成每一循环有效功的不变压力。

Mean frictional resistance　平均摩擦阻力　局部摩擦阻力的平均值。

Mean increase in resistance in waves　波浪中阻力平均增值　船舶在波浪中航行比在静水中以相同速度航行时平均增加的阻力。

Mean indicated pressure　平均指示压力　假设的，在活塞工作行程作用在活塞上为完成每一循环指示的不变压力。

Mean of 1/10 highest waves　十一平均波高　将在不规则海面上定点或定时观测的波高数据，按大小序列，取其中大端的 1/10 区间的平均值，所代表的海上典型波浪上限的波高平均值。

Mean of 1/100 highest waves　百一平均波高　将在不规则海面上定点或定时观测的波高数据，按大小序列，取其中大端的 1/100 区间的平均值，所代表的海上最大波浪的统计平均波高。

Mean offset　平均偏移　系指在海流力、风力和波浪力联合作用下平台的位移。　Mean offset is the offset due to the combined effects of wind, current, and wave-drift forces.

Mean oil outflow parameter　平均泄油参数　平均泄油代表所有泄油量与其各自概率的乘积之和。平均泄油参数以货油舱注油至 98% 舱容时全部货油量的分数来表示。　Mean oil outflow parameter represents the sum of all outflow volumes multiplied by their respective probability. The mean oil outflow parameter is expressed as a fraction of the total cargo oil capacity at 98% tank filling.

Mean power increase in waves　波浪中平均功率增值　船舶在风浪中维持一定航速较在静水中对应航速所受平均功率的增量。

Mean resistance increase in waves　波浪中平均阻力增值　船舶在风浪中维持一定航速较在静水中对应航速所受平均阻力的增量。

Mean speed　平均速度　系指开始时有 3 节锚链进入水中并且是自由悬挂的状态下，回收两节锚链时的速度。　Mean speed means the speed for raising two lengths of cable chains when three lengths of cable chains are in the water with the anchor hanged free.

Mean tension　平均张力　系指与平台平均偏移相应的锚索张力。　Mean tension means the anchor line tension corresponding to the mean offset of the unit.

Mean thrust increase in waves　波浪中平均推力增值　船舶在风浪中维持一定航速较在静水中对应航速所受平均推力的增量。

Mean time necessary to repair(MTTR)　平均修理所需时间　系指从发生故障到重新恢复功能所需时间的平均值。　Mean time necessary to repair(MTTR) is the mean value of time from occurrence of failure to reestablishment of lost function.

Mean time to failure(MTTF) 平均无故障时间 系指直到发生故障时使用时间的平均值。Mean time to failure is the mean value of service time until failure occurs.

Mean torque increase in waves 波浪中平均转矩增值 船舶在风浪中维持一定航速较在静水中对应航速所受平均转矩的增量。

Mean value 统计平均值 系指一定观察期间内的统计平均值(最少 20 年)。Mean value means a statistical mean over observation period(at least 20 years).

Mean-diameter blade(length ratio) 径高比 系指螺旋桨的平均直径与叶片高度的比值。

Means of access 通道设施 系指能够确保安全进入船舶结构的设施，以便对结构进行全面检查、近观检查和厚度测量。通道设施应在船舶结构通道手册中予以说明。 Means of access are a means capable of ensuring safe access to the structures, in order to proceed overall and close-up inspection to ship structures and thickness measurements. The means of access are to be described in a Ship Structure Access Manual.

Means of access to cargo and other spaces 进入货舱和其他处所的通道 系指在船舶整个寿命期间内，在每一处所设置的通道，以供对船舶结构进行全面检查、近观检查和厚度测量。当正常装卸货物的操作容易损坏该固定通道，或在该处设置固定通道不切实际，作为代替，主管机关可允许设置移动式或便携式通道，只要固定、安装、悬挂和支撑便携式通道的装置构成船舶结构的固定部分。所有便携式设备应易于船上人员使用。所有通道的构造和材料及其与船舶结构连接的附件均应经船级社同意。 Means of access to cargo and other spaces means that each space is to be provided with means of access to enable, throughout the life of a ship, overall and close-up inspections and thickness measurements of the ship's structures. Where a permanent means of access may be susceptible to damage during normal cargo loading and unloading operations or where it is impracticable to fit permanent means of access, the Administration may allow, thereof, the provision of movable or portable means of access, provided that the means of attaching, rigging, suspending or supporting the portable means of access forms a permanent part of the ship's structure. All portable equipment are to be capable of being readily erected or deployed by ship's personnel. The construction and materials of all means of access and their attachment to the ship's structure are to be to the satisfaction of the Society.

Means of drainage 排水装置

Means of escape 脱险通道 系指一条使船上人员能够安全迅速撤向救生艇和救生筏登乘甲板的通道。为此应满足下列功能要求：(1)脱险通道应保持安全状态，无障碍物；(2)应提供其他必要的辅助逃生设施，确保其易于到达，标志清晰，设计能满足紧急情况需要。 Means of escape is a means escaped persons onboard safely and swiftly to the lifeboat and liferaft embarkation deck. For this purpose, the following functional requirements shall be met: (1) escape routes shall be maintained in a safe condition, clear of obstacles; and (2) additional aids for escape shall be provided as necessary to ensure accessibility, clear marking, and adequate design for emergency situations.

Means of escape plans 脱险通道图 系指表明下述内容的图纸：(1)在所有通常有人占据的处所内船员和乘客人数；(2)预计通过梯道并穿越门道、走廊及梯道平台逃生的船员的乘客人数；(3)集合站和救生艇筏登乘位置；(4)主要和次要脱险通道；(5)梯道、门、走廊及梯道平台区域的宽度。脱险通道图应附有确定梯道、门、走廊及梯道平台区域宽度的详细计算。 Means of escape plans is a plan provided indicating the following: (1) the number of the crew and passengers in all normally occupied spaces; (2) the number of crew and passengers expected to escape by stairway and through doorways, corridors and landings; (3) assembly stations and survival craft embarkation positions; (4) primary and secondary means of escape; (5) width of stairways, doors, corridors and landing areas. Means of escape plans shall be accompanied by detailed calculations for determining the width of escape stairways, doors, corridors and landing areas.

Means of feeding 给水方法

Means of going astern 后退措施 系指船舶应具有足够的后退能力，以保证在所有情况下均能正确控制船舶。 Means of going astern means what ship shall be provided with sufficient power for going astern to secure proper control of the ship in all normal circumstances.

Means of pumping 排水装置

Means of transport 运输工具 系指：(1)铁路车辆、海洋、湖泊和河川船舶以及公路车辆；(2)在当地情况需要时，搬运工人或驮兽。 Means of transport means: (1) railway rolling stock, sea, lake and river craft and road vehicles; (2) where local conditions so require, porters and pack animals.

Measured block coefficient C_{BF} 计量方形系数 系指按《1966 年国际载重线公约》定义的方形系数 $C_{BF} = \nabla / 1.025 L_F B T_F$。 Measured block coefficient means block coefficient defined in the International Convention of Load

Line 1966, $C_{BF} = \nabla / 1.025 L_F B T_F$.

Measured draught T_F　计量吃水　系指最小型深 D_F 的 85%。　Measured draught T_F means 85% of the least moulded draught.

Measured length L_F　计量船长　系指按《1966 年国际载重线公约》定义的船长。即量自平板龙骨上缘的最小型深的 85% 处水线总长的 96%，或为沿该水线从艏柱前缘至舵杆中心线的长度，取其大者。当船舶设计具有倾斜龙骨时，其计量船长应与设计水线平行。　Measured length L_F means a length of the ship as defined in the International Convention of Load Line 1966. The length shall be taken as 96% of the total length on a waterline at 85% of the least moulded depth measured from the top of the keel, or as the length from the fore side of the stem to the axis of the rudder stock on that waterline, if that be greater. In ships designed with a rake of keel the waterline on which this length is measured shall be parallel to the designed waterline.

Measurement of depth　水深测量　系指测量海面至海底的垂直距离。

Measurement of wake by thrust identity　等推力法　将螺旋桨模型在船模后和在敞水中的试验数据进行比较，以同转速、同推力时船模速度减去螺旋桨敞水进速以求得等推力实效伴流的计算方法。

Measurement of wake by torque identity　等转矩法　将螺旋桨模型在船模后和在敞水中的试验数据进行比较，以同转速、同转矩时船模速度减去螺旋桨敞水进速以求得等转矩实效伴流的计算方法。

Measuring section　测试段　系指：(1)船模试验水池中，拖曳速度稳定，适于进行测量的水池段；(2)空泡试验水筒或风筒内安装测试对象的筒段。

Measuring system　测量装置　系指为了参考船舶的位置及船首方位，而提供信息的以下参考装置和环境传感器的所有硬件及软件。(1)参考装置：①由下列测量手段组成的位置参考系统：声学装置；无线装置；雷达；惯性导航；卫星导航；绷紧线或者根据船舶的运行条件而使用的手段；②船首方位参考系统——旋转罗盘或与其等效的装置。(2)环境传感器：①为测量船舶倾斜及摇晃的垂直参考传感器；②风速仪与风向仪。　Measuring system comprise all hardware and software for the following reference system and environmental sensor to supply information and corrections necessary to give position and heading reference: (1) reference system: ①position reference system: position reference system is to incorporate suitable position measurement techniques which may be by means of the followings: acoustic device; radio; radar; inertial navigation; satellite navigation; taut wire; or other acceptable means depending on the service conditions for which the ship is intended. ②heading reference system—gyrocompass or equivalent means. (2) environmental sensor: ①vertical reference sensor to measure the pitch and roll of the ship; ②means to ascertained the wind-speed and direction acting on the ship.

Mechanic　力学(机械学)

Mechanical　机械的(机械能)

Mechanical air circulation system　机械空气循环系统

Mechanical and electrical service equipment　机电维护设备

Mechanical arm　机械手

Mechanical automatic control system　机械自动控制系统

Mechanical bathy　机械式探温计　简称探温器(BT)，一种测量海水温度随深度变化的仪器。在船舶抛锚或航行时用钢丝绳投放使用。

Mechanical bathy thermometer　机械式探温计

Mechanical bending of pipe　机械弯管

Mechanical characteristic　机械特性

Mechanical cleaning　机械清洗

Mechanical control actuator　机械控制执行结构

Mechanical damage　机械损伤

Mechanical desealing　机械除垢　利用机械工具如刮刀、木槌、钢丝刷、振荡器等去除附着在换热面上污垢的除垢方式。

Mechanical detail　机械(装置)[细节(部件)]

Mechanical drawing　机械画(工程画)

Mechanical drawing　机械制图

Mechanical efficiency　机械效率

Mechanical elevation structure　机械升降机构　采用齿轮、齿条或其他机械传动方式，使自升式钻井平台完成平台升降的机构。

Mechanical engineer　机械工程师

Mechanical equipment　机械设备

Mechanical equivalent of heat　热功当量

Mechanical guy derrick　动力牵索吊杆装置　借助两套配有专用动力绞车的牵索索具使吊杆连同货物能在舷内外来回摆动，进行货物装卸的吊杆装置。

Mechanical hatch covers　机械舱盖　用动力驱动其本身机械装置，能迅速进行启闭的各种形式舱口盖的统称。

Mechanical inclusion　机械夹杂物

Mechanical inspection　力学性能试验

Mechanical joint　机械连接(接线夹)

Mechanical linkage　机械联动装置
Mechanical loss　机械损失
Mechanical manipulator　机械手
Mechanical means　机械方法
Mechanical means of propulsion　机械推进装置
Mechanical mixture　机械混合物
Mechanical pilot hoists　引航员机械升降装置　系指一种供引航员登离船用的机械装置:(1)其应是主管机关认可的型号,并设计成像活动梯一样工作,供一人在船舷升降;或像平台一样工作,供一人或多人在船舷升降。其设计和构造应确保引航员能安全地登船和离船,包括从升降器到甲板和从甲板到升降器的安全通道。这种通道应有栏杆可靠保护的平台直接构成;(2)应设有有效的手动装置,以降下和送回所载人员,且在万一动力失效时随时可以使用;(3)升降器应牢固地固定在船体结构上,其固定不应仅依靠船舷栏杆而应在船舶的每一舷为可携式升降器提供适当和牢固的系固点;(4)如果在升降器处装有外护舷材,则这种外护舷材应该充分截短,以使升降器可以靠在船舷上工作;(5)引航员软梯应装在升降器附近,并可供立即使用,以便从升降器行程的任何位置上均可接近或使用,引航员软梯应能从其自身的登船处直达海面;(6)在升降器下降的船舷位置上应有标志;(7)可携式升降器应有适当的保护的贮存位置。天气冷时,为避免结冰危险,应在临近使用之前才将可携式升降器安装就位。

Mechanical polishing　机械抛光(法)
Mechanical protection　机械性保护
Mechanical purchase　链条滑车(链条绞辘,神仙葫芦)
Mechanical remote operating　机械遥控操纵
Mechanical sounding　机械测深
Mechanical steering control system　机械操舵系统　用机械传动方式操纵舵机的系统。
Mechanical steering gear　机械操舵装置
Mechanical stoker　机械加煤机
Mechanical stowing appliance　机械堆装设施　系指设计用于运送或移动货物的轮式或拖带的机械或车辆,它包括各种起吊卡车、跨运车、舷侧装载机、拖拉机、推土机、平板车、拖车和卡车。Mechanical stowing appliance means a wheeled or tracked machine of vehicle designed to convey or move cargo and includes any lift truck, straddle truck, sideloader, tractor, bulldozer, frontend loader, trailer and truck.
Mechanical switching device　机械开关电器　借助可分开的触头的动作闭合和打开一个或多个电路的开关电器。注意:任何机械开关电器可根据触头打开或闭合所处的介质(例如:空气、SF6、油)来命名。
Mechanical telegraph　机械式传令钟(机械车钟)
Mechanical timer　机械定时器
Mechanical treatment　机械处理(机械加工)
Mechanical ventilation　机械通风　系指通过动力产生的通风。Mechanical ventilation means power-generated ventilation.
Mechanical ventilation system　机械通风系统
Mechanical vibration　机械振动
Mechanical vibrograph　机械式示振仪
Mechanical wear　机械磨损
Mechanical whistle　动力号笛　由蒸汽、压缩空气或电力作为发声源的号笛。
Mechanical zero　机械零位
Mechanical-injection diesel engine　机械喷射柴油机　用高压喷油泵通过喷油器,使柴油在气缸内喷射雾化燃烧的柴油机。
Mechanically door　机械传动门
Mechanically operated cargo oil valve　机械操纵货油阀　用机械传动方式在甲板上用于操纵的货油阀。
Mechanically operated valve　机械开关的阀
Mechanically propelled lifeboat　机械推进救生艇
Mechanism　机械(机械装置,机械原理)
Mechanize　机械化
Mechanized derrick rig　改良型吊杆装置　配有专用的绞车或机械装置可进行回转和变幅操作的吊杆装置。
MED　欧洲船用设备指令　MED means the EU Marine Equipment Directive.
Media　介质
Medical apparatus　医疗设备　各种医疗器械和用具的统称。包括病床、诊疗床、手术台、担架、急救箱等。
Medical waste　医疗废弃物　系指消毒用过的废液、已感染的药剂、人类的血液和血制品、病理学的废液、针头、人体的部位、受污染的床单、外科的废弃物和可能受到污染的实验室废弃物,透析用过的废液和环境保护局(EPA)主管机关根据规则的要求指定的辅助医疗的废弃物。Medical waste means isolation wastes; infections agents, human blood and blood products pathological wastes, sharps, body parts, contaminated bedding, surgical wastes and potentially contaminated laboratory wastes, dialysis wastes, and such additional by the Administrator of the EPA by regulation.
Medicinal opium　医用鸦片　医用的或制成粉状的鸦片是在适中温度下干燥而成的鸦片并浓缩成了细粉,通常是淡褐色。它有鸦片特有的气味,但可用樟脑之类添加剂伪装。这种制品可用用于药物中,如果吗啡含量大于 0.1%，即被归入医用鸦片类。 Medicinal or powered opium is opium that has been dried at moderate temperatures and reduced to a fine powder, usually light brown in colour. It has characteristic smell of opium, though this may be disguised by additives such as camphor. The product cab be used in medicines, any of which are classed a medicinal opium if they have a morphine opium if they have a morphine content greater than 0.1%.

Mediterranean Sea area　地中海海区　系指地中海本身，包括其中的各个海湾和内海，与黑海以 41°N 纬线为界，西至直布罗陀海峡，以 005°36′W 经线为界。 Mediterranean Sea area means the Mediterranean Sea proper including the gulfs and seas therein with the boundary between the Mediterranean and the Black Sea constituted by the 41°N parallel and bounded to the west by the Straits of Gibraltar at the meridian of 005°36′W.

Medium frequency bending of pipe　中频弯管

Medium speed diesel engine　中速柴油机　曲轴转速在 300~1 000 r/min 的柴油机。

MEHT　每月最高温度值（记录至今）　MEHT means monthly extreme high temperature (ever recorded).

MELT　每月最低温度值（记录至今）　MELT means monthly extreme low temperature (ever recorded).

Member Governments　成员国政府　系指《国际海事组织公约》缔约国政府。 Member Governments mean those governments which are Contracting Parties to the Convention on the International Maritime Organization.

Membrane cargo tank　薄膜式液货舱　薄膜式货舱非由自身结构支持，薄膜为货物围护系统的主要屏蔽，其本身不能承受货物质量，需要通过隔热层由船体结构支持。此类货舱要有一个次屏壁，以保证货物围护系统的完整性。

Memoranda for owners (MO)　致船东的备忘录　系指给船东的信息，其可以是对操作手册中所包含的与入级有关内容的补充，规范要求的免除，说明小的破损对该船无大的影响的信息。致船东的备忘录并不要求进行修理或检验。 Memoranda for owners are information to the owners, and may be additional to the class related information contained in the Operating Manual, exemptions from Rule requirements, minor damage considered of no importance to the ship's safety, etc. Memoranda for owners will not require repairs or surveys to be carried out.

MEPC　环保会　海洋环境保护委员会（Marine Environmental Protection Committee）的简称　是 IMO 下属专事海洋环境保护的一个机构。

Merchant cargo ship　商用货船　系指载运商品的船舶。 Merchant cargo ship is a ship carrying merchant cargo.

Mescaline　酶斯卡灵　迷幻剂之一，外观上像黑色至褐色的纽扣，常常有白色菌丝，或像黑色磨碎的粉末。没有气味。 Mescaline is one of hallucinogens, which appears either as black to brown buttons with white, thready fungus often present or as a black ground powder. It is odourless.

Metacenter　稳心　船舶浮心曲线的曲率中心。

Metacenter height　稳心高　稳心在重心之上的高度。

Metacenter radius　稳心半径　船舶浮心曲线的曲率半径。

Metacentric diagram　稳心图　以浮心垂向坐标、横稳心垂向坐标为纵坐标，对应的吃水为横坐标所绘制的曲线图。

Metacentric height GM(m)　初稳性高度 $GM(m)$　GM means a metacentric height, in m.

Metacentric stability　静稳性　系指在倾侧力矩的作用是从零开始逐渐增加的，且船舶倾斜时的角速度很小，可以忽略的倾斜时的稳性。

Metal halid lamp　金属卤化物灯　简称金卤灯。是在放电管内除涂有惰性气体外，还会有一种或多种金属卤化物，如碘化钠、碘化铟等的灯具。

Metal sheathings　金属护套

Metal sulphide concentrates (see also mineral concentrates schedule)　金属硫化物精矿　是精炼矿石，通过清除绝大部分废料而使其具有价值的成分富集。粒度一般很小，但在生产时间较长的精矿内有时存在团粒。这类精矿中最常见的有：锌精矿、铅精矿、铜精矿和低品位中档精矿。某些含硫化物易于氧化并有自热趋势，同时会造成缺氧而散发有毒烟气。某些物质可能产生腐蚀问题。 Mineral sulphide concentrates are refined ores in which the valuable components have been enriched by eliminating the bulk of waste materials. Generally the particle size is small although agglomerates sometimes exist in concentrates which have not been freshly produced. The most common concentrates in this category are: zinc concentrates, lead concentrates, copper concentrates and low grade middling concentrates. Some sulphide concentrates are liable to oxidation and may have a tendency to self-heat, with associated oxygen depletion and emission of toxic fumes. Some materials may present corrosion problems.

Metal temperature　金属温度　金属温度系指受压构件工作时可能达到的实际温度。 Metal temperature is actual metal temperature expected under operating conditions for the pressure part concerned.

Meteorological information room 气象情报室
安装和使用各种无线电收信设备；无线电收信机、天气传真机、电传打字机、移频机等，以获取各地转发的气象资料，为天气预报提供依据的工作室。

Meteorological radar observation room 气象雷达观察室 应用气象雷达进行观测雨和云的各种变化，以及研究各种大气现象对雷达波散射的工作室。应附有暗室。

Meteorological research ship 气象观测船 用于对海洋进行气象观测研究、收集海上气象资料和进行分析研究工作的船舶。设备少的船舶仅能起到"气象站"的作用，即仅收集海上气象资料。设备齐全的可起到"气象中心"的作用，即可发布中、短期气象预报。船上设有探空系统、气象收信系统、资料处理和预报系统及气象发播系统。

Meteorological rocket(rocket sound) 气象火箭 又称"火箭测候器"，是将各种测量仪器发射至天空，用于探测高层大气性能的一种火箭。

Meteorological rocket room 气象火箭室 对气象火箭及探测器进行装配、基测、监视、控制发射和信号接收与分析的工作室。

Meteorological rocket spare parts storage room 气象火箭备品室 用于存放备用气象火箭及其配件的舱室。

Metering 测量(记录，调节)

Metering equipment 计量设备(调节设备)

Metering pump 计量泵 能直接计量输出流量的一种变量泵。

Meterological radar 气象雷达 气象上用来探测空中雨的变化，测定云层的高度和厚度，测定不同大气层里的风向、风速和其他气象要素的雷达。

Methane 甲烷

Methane number 甲烷值 表示点燃式发动机燃料抗爆性的一个约定数值，最低 80，如 $MN = 80$ 则柴油机功率将降低，点火及喷油须重新调整，而当 $MN = 60$ 时就不能使用。

Method 方法(规律，秩序)

Method IC IC 法 系指对起居处所、服务处所和控制站保护的方法之一。其安装和布置一个固定式探火和失火报警系统，以探测起居处所的所有走廊、梯道和脱险通道内的烟雾。在起居处所、服务处所内以不燃的"B"级或"C"级分隔作内部分隔舱壁。Method IC is one of methods of protection in accommodation area、service spaces and control stations. A fixed fire detection and fire alarm system shall be so installed and arranged as to provide smoke detection in all corridors, stairways and escape routes within accommodation spaces. The construction of internal divisional bulkheads is of non-combustible " B " or " C " class divisions.

Method IC(cargo ship) IC 法(货船) 系指安装和布置固定探火和失火报警系统，以在起居处所内的所有走廊、梯道和脱险通道内提供感烟式探测；CO_2 室不须用探火系统或喷淋系统进行保护。Method IC means a fixed fire detection and fire alarm system shall be so installed and arranged as to provide smoke detection in all corridors, stairways and escape routes within accommodation spaces. CO_2 rooms need not be protected by a fire detection system or a sprinkler system.

Method IC, IIC and IIIC IC 法，IIC 法和 IIIC 法 系指：(1)除装货处所内或服务处所的冷藏库外，隔热材料应为不燃的。用于冷却系统的隔热材料连同防潮层和黏合剂以及导管附件的隔热物不必为不燃材料，但其使用量应尽可能维持在最低数量，且其外露表面应具有低播焰性。(2)如在起居处所和服务处所内采用不可燃舱壁、衬板和天花板，则可覆以可燃的镶片，但其按所用厚度范围内的比热值应不超过 45 MJ/m^2。(3)任何由不燃舱壁、天花板和衬板围成的任何起居处所和服务处所内的可燃的贴面、嵌条、装饰板及镶片的总体积应不超过相当于各围壁和天花板组合面积上敷设 2.5 mm 镶片的体积。(4)围蔽在天花板、镶片或衬板背面的空隙应用紧密安装的其间隔不大于 14 m 的挡风条分隔。在垂直方向，这种空隙，包括那些在梯道、围壁通道等背面的空隙，应在每层甲板处封闭。 Method IC, IIC and IIIC mean that: (1) Except in cargo spaces or refrigerated compartments of service spaces, insulating materials shall be non-combustible. Vapour barriers and adhesives used in conjunction with insulation, as well as the insulation of pipe fittings, for cold service systems, need not be of non-combustible materials, but they shall be kept to the minimum quantity practicable and their exposed surfaces shall have low flame spread characteristics; (2) Where non-combustible bulkheads, linings and ceilings are fitted in accommodation and service spaces they may have a combustible veneer with a calorific value not exceeding 45 MJ/m^2 of the area for the thickness used; (3) The total volume of combustible facings, mouldings, decorations and veneers in any accommodation and service space bounded by non-combustible bulkheads, ceilings and linings shall not exceed a volume equivalent to a 2.5 mm veneer on the combined area of the walls and ceilings; (4) Air spaces enclosed behind ceilings, panellings, or linings, shall be divided by close-fitting draught stops spaces not more than 14 m apart. In the vertical direction, such air spaces, including those behind linings of stairways, trunks, etc, shall be closed at each deck.

Method IF　F 级第 I 法　对长度为 55 m 或 55 m 以上的渔船而言,系指施行于一切内部隔壁构造采用不燃材料的 B 级或 C 级分隔,而一般不具备探火或喷火器系统的起居处所和服务处所的防火安全措施。在起居处所、服务处所和控制站内的所有衬板、通风挡板、天花板和它们的结合处,均应采用不燃材料。此外,(1)除装货处所或服务处所的冷冻舱外,隔热材料应是不燃材料。制冷系统中与隔热物一同使用的阻气物和黏结剂以及管路附件的隔热材料不需要是不燃材料。但应视实际可能令其使用数量为最少,且其外露表面应具有令主管机关满意的能充分阻止火焰蔓延的性质。在可能渗入油类物质处,隔热物表面应不渗透油或油气;(2)凡安装在起居处所和服务处所的不燃舱壁、衬板和天花板采用的可燃饰面板除在走廊、梯道和控制台等处,其厚度不应超过 1.5 mm 外,在任何此类处所内,均可采用不超过 2 mm 的厚度;(3)封闭在天花板、镶板或衬板后面的空隙应以紧密安装的,间距不超过 14 m 的挡风条分隔;上述空隙:包括梯道、围壁梯道等衬板后面的空隙,在垂直方向上,应在各层甲板处加以封堵。　For fishing vessels of 55 m or above, method IF means one of methods of protection adopted in accommodation and service spaces, in which construction of all internal divisional bulkheads of non-combustible "B" or "C" class divisions generally without the installation of a detection or sprinkler system in the accommodation and services spaces. In accommodation and service spaces and control stations all linings draught stops, ceilings and their associated grounds shall be of non-combustible materials. In addition: (1) except in cargo spaces or refrigerated compartments of service spaces insulating materials shall be non-combustible. Vapour barriers and adhesives used in conjunction with insulation, as well as the insulation of pipe fittings, for cold service systems need not be of non-combustible material, but they shall be kept to the minimum quantity practicable and their exposed surfaces shall have qualities of resistance to the propagation of flame to the satisfaction of the Administration. In spaces where penetration of oil products is possible, the surface of insulation shall be impervious to oil or oil vapour; (2) where non-combustible bulkheads, linings and ceilings are fitted in accommodation and service spaces they may have a combustible veneer not exceeding 2.0 mm in thickness within any such space except corridors, stairway enclosures and control stations, where it shall not exceed 1.5 mm in thickness; (3) air spaces enclosed behind ceilings, panellings, or linings shall be divided by close-fitting draught stops spaced not more than 14 m apart. In the vertical direction, such spaces, including those behind linings of stairways, trunks, etc., shall be closed at each deck.

Method IIC　IIC 法　系指对起居处所、服务处所和控制站保护的方法之一。其应安装和布置一个符合《消防安全系统规则》和相关要求认可的自动喷水器、探火和失火报警系统,以保护起居处所和厨房及其他服务处所,但空舱和卫生处所等基本上无失火危险的处所除外。此外,还应安装和布置一个固定式探火和失火报警系统,以探测起居处所的所有走廊、梯道和脱险通道内的烟雾。一般对内部分隔舱壁的类型不予限制。　Method IIC is one of methods of protection in accommodation area、service spaces and control stations。An automatic sprinkler, fire detection and fire alarm system of an approved type complying with the relevant requirements of the Fire Safety Systems Code shall be so installed and arranged as to protect accommodation spaces, galleys and other service spaces, except spaces which afford no substantial fire risk such as void spaces, sanitary spaces, etc. In addition, a fixed fire detection and fire alarm system shall be so installed and arranged as to provide smoke detection in all corridors, stairways and escape routes within accommodation spaces. Generally with no restriction on the type of internal divisional bulkheads.

Method IIC and IIIC　IIC 法和 IIIC 法　系指在起居处所、服务处所和控制站使用的走廊及梯道环围内,天花板、衬板、风挡及其相关的衬挡均应为不燃材料。

Method IIC and IIIC mean that in corridors and stairway enclosures serving accommodation and service spaces and control stations, ceilings, linings, draught stops and their associated grounds shall be of non-combustible materials.

Method IIC(cargo ship)　IIC 法(货船)　系指安装符合有关要求的被认可的自动喷淋、探火和失火报警系统,以保护起居处所、厨房和其他服务处所,但基本没有失火危险的处所如空舱处所、卫生处所等除外。此外,还应安装和布置固定探火和失火报警系统,以在起居处所和所有走廊、梯道和脱险通道内提供感烟式探测。　Method IIC means an automatic sprinkler, fire detection and fire alarm system of an approved type complying with the relevant requirements of Chopter 4, Sec 14 shall be so installed and arranged as to protect accommodation spaces, galleys and other service spaces, except spaces which afford no substantial fire risk such as void spaces, sanitary spaces, etc. In addition, a fixed fire detection and fire alarm system shall be so installed and arranged as to provide smoke detection in all corridors, stairways and escape routes within accommodation spaces.

Method ⅡF F级第Ⅱ法 对长度为55 m 或55 m 以上的渔船而言,系指施行于一切可能产生火灾的处所,具有用以探火和灭火的自动喷水器和报警系统的设施,对内部分隔舱壁类型一般不加限制的起居处所和服务处所的防火安全措施。在供起居处所、服务处所和控制站使用的走廊和梯道围壁、衬板、通风挡板、天花板和它们的结合处,均应采用不燃材料。此外,(1)除装货处所或服务处所的冷冻舱外,隔热材料应是不燃材料。制冷系统中与隔热物一同使用的阻气物和黏结剂以及管路附件的隔热材料不需要是不燃材料。但应视实际可能令其使用数量为最少,且其外露表面应具有令主管机关满意的能充分阻止火焰蔓延的性质。在可能渗入油类物质处,隔热物表面应不渗透油或油气;(2)凡安装在起居处所和服务处所的不燃舱壁、衬板和天花板采用的可燃饰面板除在走廊、梯道和控制台等处,其厚度不应超过1.5 mm外,在任何此类处所内,均可采用不超过2 mm 的厚度;(3)封闭在天花板、镶板或衬板后面的空隙应以紧密安装的,间距不超过14 m 的挡风条分隔。上述空隙,包括梯道、围壁梯道等衬板后面的空隙,在垂直方向上,应在各层甲板处加以封堵。 For fishing vessels of 55 m or above, method ⅡF means one of methods of protection adopted in accommodation and service spaces, in which provided with fitting of an automatic sprinkler and fire alarm system for the detection and extinction of fire in all spaces in which fire might be expected to originate, generally with no restriction on the type of internal divisional bulkheads. In addition:(1) except in cargo spaces or refrigerated compartments of service spaces insulating materials shall be non-combustible. Vapour barriers and adhesives used in conjunction with insulation, as well as the insulation of pipe fittings, for cold service systems need not be of non-combustible material, but they shall be kept to the minimum quantity practicable and their exposed surfaces shall have qualities of resistance to the propagation of flame to the satisfaction of the Administration. In spaces where penetration of oil products is possible, the surface of insulation shall be impervious to oil or oil vapour;(2) where non-combustible bulkheads, linings and ceilings are fitted in accommodation and service spaces they may have a combustible veneer not exceeding 2.0 mm in thickness within any such space except corridors, stairway enclosures and control stations, where it shall not exceed 1.5 mm in thickness;(3) air spaces enclosed behind ceilings, panellings, or linings shall be divided by close-fitting draught stops spaced not more than 14 m apart. In the vertical direction, such spaces, including those behind linings of stairways, trunks, etc., shall be closed at each deck.

Method ⅢC ⅢC法 系指对起居处所、服务处所和控制站保护的方法之一。其应安装和布置一个固定式探火和失火报警系统,以探测所有起居处所和服务处所内的失火,以及各起居处所内所有走廊、梯道和脱险通道内的烟雾,但空舱处所、卫生处所等基本上无失火危险的处所除外。一般对内部分隔舱壁的类型不予限制。但无论在何种情况下任一居住处所,或用"A"级或"B"级分隔作为界面的各个处所的面积不得超过50 m²。但对于公共处所,主管机关可考虑增加这一面积。 Method ⅢC is one of methods of protection in accommodation area、service spaces and control stations. A fixed fire detection and fire alarm system shall be so installed and arranged as to detect the presence of fire in all accommodation spaces and service spaces, providing smoke detection in corridors, stairways and escape routes within accommodation spaces, except spaces which afford no substantial fire risk such as void spaces, sanitary spaces, etc. Generally with no restriction on the type of internal divisional bulkheads, except that in no case must the area of any accommodation space or spaces bounded by an "A" or "B" class divisions exceed 50 m². Consideration may be given by the Administration to increasing this area for public spaces.

Method ⅢC(cargo ship) ⅢC法(货船) 系指安装和布置固定探火和失火报警系统,以在起居处所和服务处所内能探知火灾的发生,但基本没有失火危险的处所如空舱处所、卫生处所等除外。CO_2 室不须用探火系统或喷淋系统进行保护。 Method ⅢC means a fixed fire detection and fire alarm system shall be so installed and arranged as to detect the presence of fire in all accommodation spaces and service spaces, providing smoke detection in corridors, stairways and escape routes within accommodation spaces, except spaces which afford no substantial fire risk such as void spaces, sanitary spaces, etc. In addition, a fixed fire detection and fire alarm system shall be so installed and arranged as to provide smoke detection in all corridors, stairways and escape routes within accommodation spaces. CO_2 rooms need not be protected by a fire detection system or a sprinkler system.

Method ⅢF F级第Ⅲ法 对长度为55 m 或55 m 以上的渔船而言,系指施行于一切可能产生火灾的处所的自动报警和探火系统的设施,一般对内部分隔舱壁类型不加限制的,除非由A级或B级分隔所接界的任何处所或起居处所的面积不超过 50 m² 的情况。但主管机关可为公共处所增大面积的限额。在供起居处所、服务处所和控制站使用的走廊和梯道围壁、衬板、通风挡板、天花板和它们的结合处，均应采用不燃材料。在供起居处所、服务处所和控制站使用的走廊和梯道围壁、衬板、通风挡板、天花板和它们的结合处，均应采用不燃材料。此外，(1)除装货处所或服务处所的冷冻舱外，隔热材料应是不燃材料。制冷系统中与隔热物一同使用的阻气物和黏结剂以及管路附件的隔热材料不需要是不燃材料。但应视实际可能令使用数量为最少，且其外露表面应具有令主管机关满意的能充分阻止火焰蔓延的性质。在可能渗入油类物质处，隔热物表面应不渗透油或油气；(2)凡安装在起居处所和服务处所的不燃舱壁、衬板和天花板采用的可燃饰面板除在走廊、梯道和控制台等处，其厚度不应超过 1.5 mm 外，在任何此类处所内，均可采用不超过 2 mm 的厚度；(3)封闭在天花板、镶板或衬板后面的空隙应以紧密安装的、间距不超过 14 m 的挡风条分隔。上述空隙，包括梯道、围壁梯道等衬板后面的空隙，在垂直方向上，应在各层甲板处加以封堵。

For fishing vessels of 55 m or above, method IIIF means one of methods of protection adopted in accommodation and service spaces: in which provided with fitting of an automatic fire alarm and detection system in all spaces in which a fire might be expected to originate, generally with no restriction on the type of internal divisional bulkheads, except that in no case shall the area of any accommodation space or spaces bounded by an "A" or "B" class division exceed 50 m². However, the Administration may increase this area for public spaces. In corridors and stairway enclosures serving accommodation and service spaces and control stations, ceilings, linings, draught stops and their associated grounds shall be of non-combustible materials. In addition: (1) except in cargo spaces or refrigerated compartments of service spaces insulating materials shall be non-combustible. Vapour barriers and adhesives used in conjunction with insulation, as well as the insulation of pipe fittings, for cold service systems need not be of non-combustible material, but they shall be kept to the minimum quantity practicable and their exposed surfaces shall have qualities of resistance to the propagation of flame to the satisfaction of the Administration. In spaces where penetration of oil products is possible, the surface of insulation shall be impervious to oil or oil vapour; (2) where non-combustible bulkheads, linings and ceilings are fitted in accommodation and service spaces they may have a combustible veneer not exceeding 2.0 mm in thickness within any such space except corridors, stairway enclosures and control stations, where it shall not exceed 1.5 mm in thickness; (3) air spaces enclosed behind ceilings, panellings, or linings shall be divided by close-fitting draught stops spaced not more than 14 m apart. In the vertical direction, such spaces, including those behind linings of stairways, trunks, etc., shall be closed at each deck.

Method of course and angle 航向-角定位法 按测量船航向和用六分仪测定水平角的定位方法。此种方法在海道测量定位中是精度最低的一种。一般情况下不使用。

Method of forward intersection 前方交会定位法 在两个岸上已知点上架设经纬仪，同时观测船舶的水平角，然后以作图或计算方法将测得位置标到图上的方法。这种方法比后方交会定位法和侧方交会定位法测量精度要高。大比例尺沿岸海道测量或海港工程常用该法对深度点进行定位。

Method of leading marks and angle 导标-角定位法 测量船舶沿导航方向和用六分仪向岸上两控制点测定单角的定位方法。这种方法用于近岸大比例测图，能使测深线顺直，间隔匀称。

Method of ranging 距离定位法 在船上同时观测岸上两个控制点距离的方法。一般使用船用微波测距仪对近海岸进行定位。对中、远海航区可用无线电测距方法进行定位。此时船上应配有高精度频标，且频标与岸台频标严格校准的情况下，才能保证足够定位精度。如罗兰 C、ARGO 等无线电定位系统均有测距的性能。

Method of resection by sextant 六分仪后方交会定位法 又称三标两角定位法。定位时，两观测者各持六分仪同时观测三个控制点的两个水平角，借助三杆分度仪或预先绘制的等角线网，把所测定的点标在图上。在海上近岸导航和测量中常用此法测定船位。

Methyl chloride system 氯甲烷制冷系统

Methyllene chloride 二氯甲烷

Metocean 海洋气象学 是气象学(meteorological)和海洋学(oceanographic)的缩写。 Metocean is an abbreviation of meteorological and oceanographic.

MF(medium frequency) 中频

MGO 船用低硫油 MGO means marine gasoline oil.

MHB 系指散装运输时会有危险，要求采取特殊措施的物质。 MHB means a substance required specific precautions when carried in bulk present sufficient hazards.

Michell type bearing 米歇尔式轴承

Micro-fouling 微型污底 系指包括细菌和硅藻及其分泌的黏性物质的微型生物。仅由微型污底组成的生物污底通常称为黏液层。 Micro-fouling means microscopic organisms including bacteria and diatoms and the slimy substances that they produce. Biofouling comprised of only microfouling is commonly referred to as a slime layer.

Micro-structural examination 微观结构试验 系指确定焊缝的铁素体/奥氏体比率的试验。 Micro-structural examination means a test determined the ferrite/austenite-ratio of the weld.

Mid-body region 舯部区(冰区加强) 是从艏部区的后边界线至船体后端线(该位置处的水线与中心线平行)平行向后 0.04 L 处之间的部分。对 B1* 和 B1 冰级超过后端线的水平距离不必大于 6 m,对 B2 和 B3 冰级,不必大于 5 m。 Mid-body region is the region from the aft boundary of the bow region to a line parallel to and 0.04 L aft of the aft borderline of the part of the hull where the waterlines run parallel to the centerline. For ice class B1* and B1 the overlap over the borderline need not exceed 6 m, for ice classes B2 and B3 this overlap need not exceed 5 m.

Middle-high speed diesel engine 中高速柴油机 曲轴转速在 1 000 ~ 1 500 r/min 的柴油机。

Mid-length 船长中点 系指:(1)船舶分舱长度的中点;(2)艏端之后 0.5L 距离处至水线的垂线。 Mid-length is:(1) the mid-point of the subdivision length of the ship;(2) the perpendicular to the waterline at a distance 0.5L aft of the fore end.

Mid-ship 船舯部 系指船舯 0.4L 部分。 Mid-ship means the part covering 0.4L amidships.

Mid-ship depth $D_{LWL/2}$ 艇舯型深 $D_{LWL/2}$ 艇舯型深 $D_{LWL/2}$ 应取在 1/2 的水线长度 L_{WL} 处的甲板舷弧线与同一位置龙骨的最低点之间测得的距离。 Mid-ship depth $D_{LWL/2}$ shall be measured at half-length of the waterline, L_{WL}, as the distance between the sheerline and the lowest point of the keel at the same position.

Mid-ship draught T_{LC}(m) 船舯吃水 系指所计及装载工况下的船舯吃水。 Mid-ship draught T_{LC} means midship draught in the considered loading condition.

Mid-ship part 船舯部分 系指舯端之后 0.4L 区域内的部分。 Mid-ship part of a ship is the part extending 0.4L amidships, unless otherwise specified.

Mid-ship perpendicular 舯垂线 系指艏、艉垂线距离一半处水线的垂线。 Mid-ship perpendicular is the perpendicular to the waterline at half the distance between forward and after perpendiculars.

Mid-ship region 舯部区 系指从艏部区的后边界线向后至船侧平直部分后端界线。对于冰级 1AS 和 1A,此距离等于 0.04LR 或 5 m 中之较大者;对冰级 1B 和冰级 1C 则等于 0.02LR 或 2 m 中之较大者。若舷侧平直部分无明显的后端界线,则舯部区的后边界取为艉垂线向首 0.2LR 处。 Mid-ship region extends from the aft boundary of the forward region to aft of the aft borderline of the flat side of the hull by a distance equal to the greater of 0.04LR or 5 m for Ice Classes 1AS and 1A or the greater of 0.02LR or 2 m for Ice Classes 1B and 1C. Where no clear aft borderline of the flat side of the hull is discernible, the aft boundary of the midship region is to be taken 0.2LR forward of the aft perpendicular.

Mid-ship region of ice belt 冰带舯部区域 系指从冰带舯部区域的后限界线向后至舷侧平直部分后限界线之后 0.04L 处之间的区域。对 IA Super 冰级和 IA 冰级,与后限界线重叠的距离不必大于 6 m,对 IB、IC 冰级,不必大于 5 m。 Mid-ship region: from the aft boundary of the forward region to a line parallel to and 0.04L aft of the aft borderline of the part of the hull where the waterlines run parallel to the centerline. For ice classes IA Super and IA the overlap over the borderline need not exceed 6 m, and for ice classes IB, IC this overlap need not exceed 5 m.

Mid-ship section(amid-ship section) 舯横剖面 系指位于垂线间长或设计水线中点处的横剖面。

Mid-ship section coefficient 舯横剖面系数

Mid-station 中站 系指位于垂线间长或设计水线中点处的站。

Mid-station plane 中站面 位于垂线间长或设计水线长中点处垂直于基平面和中线面的横向平面。

Mile 海里 1 海里为 1 852 m 或 6 080 ft。 A mile is 1 852 m or 6 080 ft.

Milwaukee Mid-Lake Special Protection Area 蜜尔沃基湖心特殊保护区 系指在最北部一点开始顺时针方向连接下列坐标点的恒向线所包围的区域:北纬 43°27.0′,西经 087°14.8′;北纬 43°21.2′,西经 087°03.3′;北纬 43°03.3′,西经 087°04.8′;北纬 42°57.5′,西经 087°21.0′;北纬 43°16.0′,西经 087°39.8′。 Milwaukee Mid-Lake Special Protection Area means the area enclosed by rhumb lines connecting the following co-ordinates, beginning on the northernmost point and proceeding clockwise: 43°27.0′N, 087°14.8′W; 43°21.2′N, 087°03.3′W; 43°03.3′N, 087°04.8′W; 42°57.5′N, 087°21.0′W; 43°16.0′N, 087°39.8′W.

Mineral concentrates 精矿 是精炼矿石,通过清除绝大部分废料而使其具有有价值的成分富集。包括:沉积铜、铜精矿、铁精矿、铁精矿(球团粒)、铁精矿(烧结料)、铅锌熔砂(混合)、中档铅锌、铅精矿、铅矿渣、铅银精矿、锰精矿、霞石正长石(矿石)、镍精砂、五水合物原矿、黄铁矿、黄铁矿灰(铁)、黄铁矿熔渣、银铅精矿、斯利格(铁矿砂)、锌铅熔砂(混合)、中档锌铅、锌精矿、锌烧结渣和锌淤渣。上述货物如装运时含水量超过其适运水分极限,可能流态化。这些货物会使遮盖舱底污水井的粗麻布或帆布腐烂。长期连续载运该货物可能对结构有破坏作用。 Mineral concentrates are refined ores in which valuable components have been enriched by eliminating the bulk of waste materials. Mineral concentrates includes: cement copper, copper concentrate, iron concentrate, iron concentrate (pellet feed); iron concentrate (sinter feed); lead and zing calcunes (mixed); lead and zing middlings, lead concentrate, lead ore residue, lead silver concentrate, manganese concentrate, nefelent syenite (mineral); nickel concentrate, pentahydrate crude, pyrites, pyritic ashes (iron); pyritic cinders, silver lead concentrate, slig (iron ore); zing and lead calcines (mixed); zing and lead middlings, zing concentraye, zing sinter, zing sludge. The above materials may liquefy if shipped at moisture content in excess of their transportable moisture limit (TML). This cargo will decompose burlap or canvas cloth covering bilge wells. Continuous carriage of this cargo may have detrimental structural effects over a long period of time.

Mineral oil 矿物油

Mineral spirit 矿物溶剂油

Mineral wool 矿渣棉

Minerals 矿物 系指从区域回收的资源。 Minerals mean those resources, when recovered from the area.

Minicaps bulk carrier 迷你好望角型散货船 是一型绿色环保、性能指标优越的散货船。其采用隐形球鼻艏(VS-BOW)线型,此型球鼻艏融入仿生学概念,能有效降低不同转载状态下船舶在波浪中的航行阻力。在船体设计方面,该船采用高干弦设计,有效地保证了舱容,降低了主甲板上浪概率。破舱稳性满足国际海上人命安全公约(SOLAS)的最新要求,并严格按照共同结构规范(CSR)的要求设计。在环保设计方面,该船型满足国际防止船舶污染公约(MARPOL)关于"燃油舱保护"、"污水处理装置"的要求、欧盟港口的超低硫油要求,船舶压载水管理要求、国际海事组织(IMO)关于主机氮氧化物的 Tier-Ⅱ标准等级新要求,是一型名副其实的、适合在拓宽后的巴拿马运河上航行的绿色船舶。

Minimization 最小化 系指最大实际可能地减少船上排放的任何干散货物残余物。 Minimization means the reduction, to the greatest extent practicable, of any bulk dry cargo residue discharge from the vessel.

Minimum ballast draught T_B (m) 最小压载吃水 系指在正常压载工况下船中最小压载吃水。 Minimum ballast draught means minimum ballast draught at midship in normal ballast condition.

Minimum bow height F_b 最小船首高度 系指在艏垂线处,自相应于核定夏季干舷和设计纵倾的水线,量至船侧露天甲板上边的垂直距离,该高度不小于:
$F_b = [6075(L_{LL}/100) - 1875(L_{LL}/100)^2 + 200(L_{LL}/100)^3] \times [2.08 + 0.609C_B - 1.603C_{wf} - 0.0129(L/T_1)]$,式中,$F_b$—计算的最小船首高度,(mm);$T_1$—计算型深 D_1 的 85% 处的吃水,(m);D_1—计算型深是船中处深加干舷甲板边板的厚度。对于圆弧形舷缘半径大于型宽(B)的 4% 或上部舷侧为特殊形状的船舶,计算型深取自一中央截面的计算型深。此截面两舷上侧垂直并具有同样的梁拱,且上部截面面积等于实际的中央截面的上部截面面积。C_{wf}——$L_{LL}/2$ 的前体水线面面积系数;$C_{wf} = A_{wf}/(L_{LL}/2)B$,式中,$A_{wf}$—吃水 T_1 处 $L_{LL}/2$ 的前体水线面面积,(m²)。对核定木材干舷的船舶,在应用上述公式时应采用夏季干舷(而不是木材夏季干舷)。 Minimum bow height F_b is defined as the vertical distance at the forward perpendicular between the waterline corresponding to the assigned summer freeboard and the designed trim and the top of the exposed deck at side, is to be not less than: $F_b = [6075(L_{LL}/100) - 1875(L_{LL}/100)^2 + 200(L_{LL}/100)^3] \times [2.08 + 0.609C_B - 1.603C_{wf} - 0.0129(L/T_1)]$, where: F_b—Calculated minimum bow height, (mm); T_1—Draught at 85% of the depth for freeboard D_1, (m); D_1—Depth for freeboard, is the moulded depth amidship plus the freeboard deck thickness at side. The depth for freeboard in a ship having a rounded gunwale with a radius greater than 4% of the breadth B or having topsides of unusual form is the depth for freeboard of a ship having a midship section with vertical topsides and with the same round of beam and area of topside section equal to that provided by the actual midship section. C_{wf}—Waterplane area coefficient forward of $L_{LL}/2$; $C_{wf} = A_{wf}/(L_{LL}/2)B$; A_{wf}—Waterplane area forward of $L_{LL}/2$ at draught T_1, (m²). For ships to which timber freeboards are assigned, the summer freeboard (and not the timber summer freeboard) is to be assumed when applying the formula above.

Minimum breaking strength 最小破断强度 系指经认证的锚链、钢丝绳及其他材料的破断强度。 Minimum breaking strength means the breaking strength of chain and for wire and other materials. It is the certified breaking strength.

Minimum clearance 最小间隙

Minimum comfortable condition of habitability 最

低舒适居住条件

Minimum design ballast draught T_{bal} 最小设计压载吃水 系指符合结构强度要求的最小设计压载吃水 (m)。对于包括出港和到港状态的装载手册中的任何压载工况，最小设计压载吃水不必大于包括压载水交换操作的压载工况的最小吃水，且从船中型基线处量起。Minimum design ballast draught T_{bal}, is the minimum design ballast draught, in m, at which the strength requirements for the scantlings of the ship are met. The minimum design ballast draught is not to be greater than the minimum ballast draught, measured from the moulded base line at amidships, for any ballast loading condition in the loading manual including both departure and arrival conditions.

Minimum design temperature of steel 材料的最低设计温度 钢材的最低设计温度系指构件钢材等级选择时的基准温度，并假定等于日平均大气温度的最低值。该值应根据平台预期作业海域的气象资料所确定。水下结构钢材的设计温度通常可取 0 ℃。除另有规定者外，内部结构的设计温度一般假定与邻接的外部结构的设计温度相同，但永久加热舱室内部结构的设计温度通常不必低于 0 ℃。Minimum design temperature of steel means the minimum design temperature of steel is a reference temperature used as a criterion for the selection of the grade of steel to be used in the structure and is to be assumed equal to the lowest of the average daily atmospheric temperatures, based on meteorologicaldata for any anticipated area of operation. A design temperature of 0 ℃ is generally acceptable for determining the steel grades for underwater structure. The internal structure of all units is normally assumed to have the same design temperature as the adjacent external structure unless defined otherwise. However, internal structures in way of permanently heated compartments need not normally be designed for temperatures lower than 0 ℃.

Minimum dimension 最小尺寸 对压载水取样导则而言，系指生物的最小尺寸，基于该生物身体的尺寸，不考虑脊椎、鞭毛或触须等的尺寸。最小尺寸应为"身体"的最小部分，即从各个角度观察的主要身体表面之间的最小尺寸。对于球形生物，最小尺寸约应为球的直径。对于菌落形成物种，应对个体进行测量，因为它是能复制该需求以在生存试验中进行测试的最小单位。For the purpose of guidelines for ballast water sampling, minimum dimension means the minimum dimension of an organism based upon the dimensions of that organism's body, ignoring e. g. , the size of spines, flagellae, or antenna. The minimum dimension should therefore be the smallest part of the "body", i. e. the smallest dimension between main body surfaces of an individual when looked at from all perspectives. For spherical shaped organisms, the minimum dimension should be the spherical diameter. For colony forming species, the individual should be measured as it is the smallest unit able to reproduce that needs to be tested in viability tests.

Minimum draught 最小吃水（冰区航行） 系指船舶在冰区航行时，应始终装载到不低于压载水线。位于压载水线以上为使船舶压载到该水线所需要的压载水舱必须装备防止压载水冻结的设备。Minimum draught that the ship is always to be loaded down to at least the LIWL when navigating in ice. Any ballast tank situated above the LIWL and needed to load down the ship to this waterline is to be equipped with devices to prevent the water from freezing.

Minimum draught T_{min} 最小吃水 T_{min} 最小吃水 T_{min} 应测量至船艇的最低点或不可收放附体的最低点，取较低者。所有可拆的水下部件均应处于其可能最高的位置。Minimum draught T_{min} shall be measured to the lowest point of the craft or non-retractable appendage, whichever is lower. All movable underwater parts shall be in their uppermost possible position.

Minimum dredging width 最小挖深 系指绞吸挖泥船的铰刀，在垂直于前移向的方向上保持正常作业的最小摆动切削宽度。

Minimum drilling water depth 最小工作水深 钻井船或其他钻井装置能进行钻井作业的最小水深。

Minimum forward draught 最小艏吃水 系指：(1)在确定压载水线时，应适当考虑船舶在压载状态时有一定程度的冰区航行能力。螺旋桨要求完全浸没，如果可能，全部处于冰层以下。最小艏吃水应不小于由下式给出的数值：$T_{AV} = (2 + 0.000\ 25 \triangle_1) h_0$，其中：$\triangle_1$—船中最大冰区级吃水时的排水量(t)，$h_0$—冰层厚度(m)，但吃水 T_{AV} 不必大于 $4h_0$；(2)应不小于下式确定的值，但不必大于 $4h_0$：$d_1 = (2 + 0.000\ 25\triangle) h_0$ (m)，式中，$\triangle$—最大吃水所对应的排水量(t)；h_0—名义冰厚(m)。Minimum forward draught means: (1) while determining the LIWL, due regard is to be paid to the need to ensure a reasonable degree of ice going capability in ballast. The propeller is to be fully submerged, if possible entirely below the ice. The minimum forward draught is to be at least equal to the value T_{AV}, in m, given by the following formula: $T_{AV} = (2 + 0.000\ 25 \triangle_1) h_0$, where: $\triangle_1$—Displacement of the ship(t), on the maximum ice class draught, h_0—Ice thickness(m). The draught T_{AV} need not, however, exceed $4h_0$;(2) not to be less than that obtained from the following formula but need not exceed $4h_0$: $d_1 = (2 + 0.00025)h_0(m)$, where: Δ—displacement of the ship(t), on the maximum ice class draught, h_0—nominal ice thickness(mm).

Minimum full ahead speed 最小全速前进船速(巴拿马运河) 系指巴拿马运河管理局为使船舶在标准时间内完成通航所规定的最小全速前进船速,其为 8 节(kn)。 For the purpose of Panama Canal, minimum full ahead speed means that the ACP has determined that the minimum full ahead speed required for vessels in order to complete transit in standard times is 8 knots.

Minimum hauling and rendering speed 最小收、放绳速度 当绳索卷绕至卷筒最外层时,原动机以最低转速收放绳能达到的最小线速度。

Minimum list of items for the inventory of hazardous materials 有害物质清单应列的最少项目 见表 M-4。

表 M-4 有害物质清单应列的最少项目
Table M-4 Minimum list of items for the inventory of hazardous materials

序号	应列的最少项目 Minimum list of items
1	AFS 公约附录 I 所列的任何有害物质 Any hazardous materials listed in Appendix I
2	镉和镉化合物 Cadmium and Cadmium compounds
3	六元铬和六元铬化合物 Hexavalent Chromium and Hexavalent Chromium compounds
4	铅和铅化合物 Lead and Lead compounds
5	汞和汞化合物 Mercury and Mercury compounds
6	多溴化联(二)苯(PBB) Polybrominated Biphenyl(PBBs)
1	AFS 公约附录 I 所列的任何有害物质 Any hazardous materials listed in Appendix I
7	多溴二苯醚(PBDE) Polybrominated Diphenyl Ethers(PBDEs)
8	多氯化联萘(超过 3 个氯原子) Polychlorinated Naphthalenes(more than 3 chlorine atoms)
9	放射性物质 Radioactive substances
10	某些短链氯化石蜡(烷类、C10-C13、氯基) Certain Shortchain Chlorinated Paraffins(Alkanes, C10-C13, chloro)

Minimum main engine output 主机最小功率

Minimum normal ground effect speed 最小正常地效速度 船舶以地效状态在已被证实的载荷和稳性范围内正常运行的最低航速。为合理预见的运行条件瞬间变化留有足够的安全裕度。 Minimum normal ground effect speed is the lowest speed at which craft can be operated in ground effect throughout proven load and stability range, allowing sufficient safety margin for reasonable foreseeable transient variations in operating conditions.

Minimum quantity of water to be used in a prewash 用于预洗的最小水量 预洗中所用的最小水量由舱内有毒液体物质的残余量、液货舱大小、货物性质、洗舱水排出物的许可浓度以及操作区域来确定。由下列公式计算最小水量:$Q = k(15r^{0.8}5r^{0.7}V/1000)$。式中,$Q =$ 要求的最小水量,m^3,$r =$ 每个液货舱的残余量,m^3。r 值应为实际扫舱效率试验中显示的值,但对于舱容为 500 m^3 及以上的液货舱,应取不低于 0.100 m^3;对于舱容为 1 000 m^3 及以下的液货舱,应取不低于 0.040 m^3;对于舱容在 1 000 m^3 和 500 m^3 之间的液货舱,在计算中允许使用的 r 的最小值由线性内插值法求得。对于 X 类物质,r 值应按《手册》根据扫舱试验予以确定,注意上述给出的较低限值,或取 0.9 m^3,$V =$ 舱容,m^3。$k =$ 具有下列值的系数:X 类,非固化的低黏度物质,$k = 1.2$;X 类,固化物质或高黏度物质,$k = 2.4$;Y 类,非固化的低黏度物质,$k = 0.5$;Y 类,固化物质或高黏度物质,$k = 1.0$。下表是当 k 系数取 1 时用公式计算所得,可作为方便参考。见表 M-5。 Minimum quantity of water to be used in a prewash is determined by the residual quantity of noxious liquid substance in the tank, the tank size, the cargo properties, the permitted concentration in any subsequent wash water effluent, and the area of operation. The minimum quantity is given by the following formula: $Q = k (15r^{0.8}5r^{0.7}V/1\,000)$. Where: Q = the required minimum quantity in m^3, r = the residual quantity per tank in m^3. The value of r shall be the value demonstrated in the actual stripping efficiency test, but shall not be taken lower than 0.100 m^3 for a tank volume of 500 m^3 and above and 0.040 m^3 for a tank volume of 100 m^3 and below. For tank sizes between 100 m^3 and 500 m^3 the minimum value of r allowed to be used in the calculations is obtained by linear interpolation. For Category X substances the value of r shall either be determined based on stripping tests according to the Manual, observing the lower limits as given above, or be taken to be 0.9 m^3, V = tank volume in m^3. k = a factor having values as follows: Category X, non-Solidifying, Low-Viscosity Substance, k = 1.2; Category X, Solidifying or High-Viscosity Substance, k = 2.4; Category Y, non-Solidifying, Low-Viscosity Substance k = 0.5; Category Y, Solidifying or High-Viscosity Substance k = 1.0. The table M-5 is calculated using the formula with a k factor of 1 and may be used as an easy reference.

表 M-5 用于预洗的最小水量
Table M-5 Minimum quantity of water to be used in a prewash

扫舱容量	舱容 m^3		
	100	500	3000
≤0.04	1.2	2.9	5.4
0.10	2.5	2.9	5.4
0.30	5.9	6.8	12.5
0.90	14.3	16.1	27.7

Minimum safe manning document 最低安全配员文件 系指 SOLAS 公约第 1 章所适用的每艘船舶，均应备有一份主管机关签发的适当的安全配员文件或实效文件，作为最低安全配员的证明。Minimum safe manning document means that every ship to which chapter I of SOLAS Convention applies shall be provided with an appropriate safe manning document or equivalent issued by the Administration as evidence of the minimum safe manning.

Minimum ship manoeuvring speed 最小船舶操纵船速 系指维持航向控制且符合船舶操纵特性的最小航速。Minimum ship manoeuvring speed is defined to be the minimum speed which maintains directional control and is consistent with the operating characteristics of the ship.

Minimum starting pressure test 最低启动压力试验 测定使汽轮机转子转动的最低蒸汽压力并检查汽轮机装配质量的试验。

Minimum steady speed 最低稳定工作转速 内燃机按推进特性能稳定工作运转的最低转速。

Minimum steady speed test 最低稳定工作转速试验 内燃机按推进特性曲线测定其最低稳定工作转速的试验。

Minimum variance updates 最小方差更新 用概率论中的最小方差原则估算最佳值来更新卫星定位位置。用这种方法更新船位时，若推算船位和卫星定位点之间的差值过大，就不进行最小方差更新计算，系统自动转入卫星更新。

Minor axis 短轴

Minor effect 轻微后果 系指：(1)可能由于故障、事件和差错造成的，可由操作船员迅速补救的后果，其包括：①稍微增加船员的工作任务，或稍微增加其执行任务的困难；②操作性能中等程度的降低；③轻微地改变许可的工作条件；(2)不明显减少船舶安全的故障情况的影响，它涉及完全在船员能力范围内的船员行为。例如，具有轻微影响的故障情况可包括安全裕度和营运能力的轻微减少，船员工作量的轻微增加，或船上人员的一些不方便。Minor effect is: (1) an effect which may arise from a failure, an event, or an error, which can be readily compensated for by the operating crew. It may involve: ①a small increase in the operational duties of the crew or in their difficulty in performing their duties; ②a moderate degradation in handling characteristics; ③slight modification of the permissible operating conditions; (2) the effect of failure conditions that does not significantly reduce craft safety, and which involve crew actions that are well within their capabilities. Failure conditions with a minor effect may include, for example, a slight reduction in safety margins or functional capabilities, a slight increase in crew workload, or some inconvenience to occupants.

Minus 负数(数号，减)

Minus exhaust lap 负排气余面

Misalignment of edges 边缘未对准 偏差允许的极限是相对于正确位置而言。除非另有规定，否则正确的位置是两构件的中心线重合。见图 M-7 和图 M-8。The deviation limits relate to the correct position. Unless otherwise specified, the connect position is when the centre lines coincide. See Figure M-7 and Figure M-8.

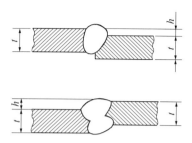

图 M-7　板和纵向焊缝
Figure M-7　Plates and longitudinal welds.
t 代表较小的板厚。t relates to the smaller thickness.

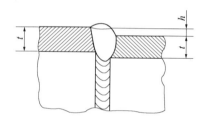

图 M-8　环形焊缝
Figure M-8　Circumferential welds

Miscellaneous dangerous substances and articles (Class 9 dangerous materials)　其他危险物质和物品（第 9 类危险货物）　系指运输期间具有其他类别所未包括的危险的物质。Miscellaneous dangerous substances and articles mean the materials in this class are materials and articles which, during transport, present a danger not covered by other classes.

Misfiring condition of an engine　发动机熄火状态　是其中的一个气缸缺乏燃油的喷射而导致的不正常现象（这引起了气缸中只有单纯的压缩或者膨胀）。Misfiring condition of an engine is the malfunction of one cylinder due to the absence of fuel injection (which results in a pure compression or expansion in the cylinder).

Missile magazine　导弹舱　供储存、输送或发射导弹用的舱室。

Mist　霭　系指轻雾或薄雾。

Miter wheel　正角斜齿轮

Mitering pump　限量泵

Mitigation　缓解　限制特定事件的任何负面效果。

Mix　混合

Mixed cycle　混合循环　等压和等容循环的组合循环。

Mixed flow pump　混流泵

Mixed oil　混合油

Mixed type's garbage　混合型垃圾　系指与不同处理或排放要求的其他有毒物质混合在一起的垃圾。

Mixing condensation　混合冷却

Mixtures containing crude oil　含原油的混合物

MKD (AIS) = AIS minimum keyboard and display　自动识别系统 (AIS) 最小化的键盘和显示器

Mobile auxiliary gas turbine　移动式辅燃气轮机　可以移动到各处工作的辅燃气轮机。

Mobile earth station　移动式地面站　系指在运行中或停在无特定地点期间意欲使用移动卫星服务的地面站。Mobile earth station means an earth station in the mobile satellite service intended to be used while in motion or during halts at unspecified points.

Mobile fire-fighting equipment　移动式消防设备

Mobile mooring system　移动式系泊系统　系指在海上系泊时间较短的工程船舶和浮式装置，诸如打捞船、起重船、半潜运输船、地质取芯船、潜水工作船、铺管船、钻井船（或驳船）、半潜式钻井平台、钻井辅助船等采用的辐射状（多点）锚泊定位系统。其主要特点是作业环境条件并不十分恶劣，一旦预报强风暴来临时可以撤离，即使是钻井船和半潜式钻井平台，遇到强风暴时，可提前脱开或收回隔水管，使其位移不受限制。移动式系泊系统一般采用悬链线（锚索）系统，且要求抛/起锚方便，因此通常采用大抓力拖曳埋置锚。在选择锚时，必须通过对各种环境条件下的系泊分析来确定锚的负荷。

Mobile offshore drilling unit (MODU) or units　海上移动式钻井平台 (MODU) 或平台　系指能够为勘探或开采诸如液态或气态碳氢化合物、硫或盐等海床下资源而从事钻井作业的海上建筑物。Mobile offshore drilling unit (MODU) or units is a vessel capable of engaging in drilling operations for the exploration for or exploitation of resources beneath the seabed such as liquid or gaseous hydrocarbons, sulphur or salt.

Mobile offshore drilling unit safety certificate　海上移动式钻井平台安全证书　系指对海上移动式钻井平台在按 1979 年海上移动式钻井平台构造和设备规则的规定进行检验后，或对 1991 年 5 月 1 日或以后建造的平台，经按 1989 年海上移动式钻井平台构造和规则进行检验后签发的证书。Mobile offshore drilling unit safety certificate means a certificate issued for mobile offshore drilling units after survey carried out in accordance with the provisions of the Code for the Construction and Equipment of mobile offshore drilling units, 1979, or, for the units constructed on or after 1 May 1991, the Code for the construction and equipment of drilling units, 1989.

Mobile offshore unit　海上移动平台　系指可根据需要从一个作业地点转移到另一个作业地点的海上平台,简称为平台。 Mobile offshore unit means a unit designed to be moved from one operating site to another as necessary, referred to as unit.

Mobile oil　滑油

Mobile packet data service(MPDS)　移动数据包服务

Mobile pump　可移式泵

MOCA　多功能模块化海洋工程组合体　一种采用"模块化 + 多功能 + 绿色能源"的新概念,由不同形态、不同功能的模块组合而成的海上油气资源勘探开发装备。MOCA 的辅助单体内置核动力电站驱动喷水推进器,可实现自航;而在组合体情况下其可同甲板连接,并可顶起甲板,为组合体提供浮力及动力定位能力。该组合体由 3 个单体——动力辅助单体、物资辅助单体和居住辅助单体组成,并通过不同的组合,实现风电安装、海上科考、海上铺管、海上度假等多种用途。组合体上的新颖能源装置——发电树,带有垂荡浮子—液压波浪发电装置、叶轮式洋流发电装置及太阳能电池板 3 套装置,能充分利用波浪、洋流和太阳能发电,清洁环保。

Mock　模拟的/虚假的/模仿的

Mode awareness　模式意识　基于综合桥楼系统(IBS)的各种运行的用途和知识。采用各种不同的运行模式时应遵守国际公司自动化方针规定的驾驶程序。 Mode awareness is based on the knowledge and purpose of various operation modes included in the IBS. Use of different operation modes should follow bridge procedures based on company automation policy.

Mode awareness　模式告知　在综合航行系统(INS)显示器或工作站的显示或标示的帮助下,航海人员对 INS 和其子系统的当前处于活动状态的控制、操作和显示模式的了解。 Mode awareness means the perception of the mariner regarding the currently active Modes of Control, Operation and Display of the INS including its subsystems, as supported by the presentations and indications at an INS display or workstation.

Mode of operating　操作模式　控制的模式,在此模式下动力定位系统可被操作,例如:(1)自动模式(自动船位和艏向控制);(2)独立的联合操纵杆模式(手动船位控制且具有可选择的自动或手动艏向控制);(3)手动模式(对每个推进器的螺距和速度、方位、启动和停止的单个控制)。

Model propellers　螺旋桨模型　将实际螺旋桨按比例缩小尺度而制成的供试验用的模型。

Model-ship correlation allowance　阻力换算修正值　视换算方法、船舶类型、大小、粗糙度以及测试条件等而定的,由船模阻力换算到实船阻力时所计及的修正值。

Modern shipbuilding mode　现代造船模式　以统筹优化理论为指导,应用成组技术原理,以中间产品为导向,按区域组织生产,壳、舾(装)、涂(装)作业在空间上分道,时间上有序,实现设计、生产、管理一体化,均衡、连续地总装造船。即由统筹优化的造船理念,面向生产的设计技术,均衡连续的作业流程,严密精细的工程管理,高效合理的生产组织等基本要素构成。造船总装化、管理精细化、信息集成化是建立现代造船模式的基本方向。

Modes of operation　作业模式(平台)　系指平台在作业点或迁移时操作或活动的条件或状态。平台的作业模式分为以下 4 种:(1)正常作业工况系指平台在作业点上作业或进行其他操作时承受与作业相适的设计限度内的组合环境载荷和作业载荷的状态;(2)自存工况系指平台承受最严重设计环境载荷时停止作业或其他操作,从而把抗环境能力提高到最大的状态;(3)迁移工况系指平台从一个地区迁移到另一个地区时的状态;(4)升降工况系指自升式平台在下放桩腿和升起主体结构或下降主体结构和拔起桩腿时的状态。 Modes of operation are conditions or manners in which a unit may operate or function while on location or in transit. The approved modes of operation of a unit are to include the following:(1)operating condition: a condition wherein a unit is on location for purposes of drilling or other similar operations, and combined environmental and operational loadings are within the appropriate design limits established for such operations;(2)survival condition: a condition during which a unit is subjected to the most severe environmental loadings for which the unit is designed and drilling or similar operations have been discontinued to maximize the capability of surviving such environmental loadings;(3)transit condition: all unit movements from one geographical location to another;(4)jacking condition: a condition during which a self-elevating unit lowers its legs and raises its hull structure or lowers its hull structure and raises its legs.

Modifications　改装(救生艇)　系指对某一经形式认可的救生艇释放和回收系统设计的更改,此类更改可能影响系统符合原认可要求或产品规定的使用条件。 Modifications are those changes to the design of an approved lifeboat release and retrieval system which may affect compliance with the original approval requirement or the prescribed conditions for use of the product.

Modified zig-zag maneuver test　变型 Z 形操纵试验　变换舵舷向角小于操舵角的 Z 形操纵试验。

MODU Code　MODU 规则　系指海上移动式钻井平台构造和设备规则。

Modular common rail fuel system(MCRS)　模块化共轨燃油系统　能够按容积精确地控制喷油量的系统。模块化共轨燃油系统使发动机在任何负载和速度下都能保持持续的高压喷油,高喷油压力可以保证燃油雾化良好,提高了燃油的燃烧效率,大幅降低了废气的排放量,使其满足 Tier II 的排放要求。

Modular unit　模块单元(可互换标准件)

Modulate　调节(调整)

Module　模块(组件,工作舱)

Moist　潮湿的

Moisture content　含水量　系指以代表性样品总湿重的百分比表示的该样品所含水、冰或其他液体的部分。 Moisture content means that portion of a representative sample consisting of water, ice or other liquid expressed as a percentage of the total wet mass of that sample.

Moisture eliminator　挡水板　空气沿其流动时能挡去水滴的一组曲折形薄板。

Moisture migration　水分迁移　系指货物由于振动和船舶运动而变得密实,使货物所含水分产生运动。水逐渐移位,这可能导致部分或全部货物出现流动状态。 Moisture migration means the movement of moisture contained in a cargo by settling and consolidation of the cargo due to vibration and ship's motion. Water is progressively displaced, which may result in some portions or all of the cargo developing a flow state.

Moisture removal equipment(water catcher)　去湿装置　在湿蒸汽区域工作的汽轮机中利用抽气、分离、加热等办法降低蒸汽湿度的装置。

Molasses/chemical carrier　糖蜜/化学品运输船

Molded base line　基线　系指基平面与中线面相交的直线,与横向垂直平面相交的直线。

Molded base line　型基线　系指通过船底板、下船体底板或沉箱底板上缘的水平线。 Molded base line is a horizontal line through the upper surface of the bottom shell, lower hull bottom shell or caisson bottom shell.

Molded breadth　型宽　系指船体型表面之间垂直于中线面的最大水平距离。

Molded case circuit breaker　塑料外壳式断路器　系指具有一个用模压绝缘材料制成的外壳作为断路器整体部件的断路器。

Molded depth　型深　系指在船的中横剖面处,自龙骨线沿垂直于基平面的方向量至舷弧线的距离。无特别注明时,通常系指量至上甲板弦弧线。

Molded depth line　舷弧基准线　通过中站处站线与舷弧之交点并平行于基线的直线。

Molded displacement　型排水量　任一水线下,船体型体积的排水量。

Molded displacement curve　型排水量曲线　船舶型排水量与吃水的关系曲线。

Molded draft　型吃水　(1)从型基线至勘定载重线之间的垂直距离;(2)自基平面沿其垂直方向量至任一水线的距离。对于具有龙骨设计斜度的船舶,则指自龙骨线或其延伸线沿垂直于基平面的方向量至水线的距离。无特殊注明时,通常尤指在中横剖面处,按以上所述量至设计水线的距离。

Molded hull surface　船体型表面　不包括附体在内的船体的外形的设计表面。对于金属船体系指通过船体骨架外缘的表面;对于木质、水泥、玻璃钢船体则指外板、甲板的外表面。

Molded lines　型线　系指船体型表面及有关附体型表面的外廓线、剖切线和投影线。

Molded volume(volume of molded form)　型排水体积　任一水线下,船舶裸船体所排开的水的体积。对于金属船体,不包括外板及附体的体积在内;对于木质、水泥、玻璃钢船体,则包括外板的体积在内,但不包括附体的体积。

Mollier diagram　(水蒸气)焓熵图

Moment asynchronous motor　力矩异步电动机　与一般鼠笼式异步电动机工作原理相同,只是在结构上稍有不同,它采用电阻率较高的导电材料(如黄铜)作为转子的导条和端环。力矩电动机允许长期低速运转,甚至堵转,发热相当严重,所以转子具有轴向通风,并外加鼓风机,以带走电动机的热量,其机械特性调节是通过调节其输入端电压。

Moment of inertia　惯性力矩

Moment of wind pressure　风压倾斜力矩　由风力作用使船舶倾斜的外力矩。其值为 0.001 与船舶受风面积、风压、计算风力作用力臂的乘积。

Moment to trim one centimeter　每厘米纵倾力矩　使船舶纵倾 1 cm 所需的力矩。

Momentum theory(actuator theory)　动量理论　将推进器看作是能突然增高所在处水流压力的鼓动器,但不涉及推进器的机构及如何使水的压力增高的一种理论。

Momentum thickness of boundary layer　边界层动

量厚度　表明由于边界层引起的动量传递损失的一个参数。对于平面流：$\theta = \int_0^\infty U/U_0(1 - U/U_0)\,dy$，式中，$U_0$—边界层外的流速；$U$—边界层内的流速。

Monitor　监控器(消防龙头,消防炮)
Monitor and control system　监控系统
Monitor display unit　监视器显示装置
Monitor log　监听记录
Monitored data　监测数据
Monitored equipment　受监测设备
Monitored machinery　受监测机械
Monitored object　监测对象
Monitored parameter　监控参数
Monitored ship　受监测船舶
Monitoring　监视/监控　系指定期查核仪表显示信息和周围环境以便发现异常现象的行为。Monitoring means an act of constantly checking information from instrument displays and environment in order to detect any irregularities.

Monitoring alarm　监测报警器
Monitoring and alarm systems　监视和报警系统　系指具有下列功能的系统：(1)自检和监视功能；(2)报警功能；(3)报警显示和报警应答；(4)监视和报警整定值修改功能；(5)报警闭锁功能。　Monitoring and alarm system means a system which have following functions: (1) Function for self-checking and monitoring; (2) Function for alarming; (3) Function for alarm indication and acknowledgement; (4) Function for monitoring and modifying set values of alarming; (5) Function for locking of alarms.

Monitoring and warning desk in engine room　机舱监测报警台

Monitoring box for control of communicating light　通信灯控制箱　安装在驾驶室内部标有各种通信灯工作情况。并可对其进行集中控制的装置。

Monitoring device　监测装置(监视装置)
Monitoring equipment　监控设备(检测设备)　系指用于评定原型压载水处理技术是否正常工作而安装的设备。Monitoring equipment refers to the equipment installed for assessment of the correct operation of the prototype ballast water treatment technology.

Monitoring instrument　监控仪器
Monitoring parameter　监控参数
Monitoring procedure　监测程序
Monitoring station　监测台(监控台除外)　是机械和设备的测量仪表、指示器、警报器集中在一起的地方。在这里可以获得它们的运转状态的必要信息，但是在提供船舶控制站上，涉及一个监测台的有关规范中的规定不能应用到有关的监测台。　Monitoring station (excluding control station) is a position where measuring instruments, indicators, alarms, etc. for the machinery and equipment are centralized and necessary information to grasp the operating condition of them can be obtained. Where, however, a monitoring station is provided with the ship in addition to a control station, the requirements of the relevant Rules relating to a monitoring station do not apply to the monitoring station concerned.

Monitoring system　监测系统(监听系统)　是一个通过检测不正确功能(测量变化量同规定值相比较)来观察设备是否正常工作的系统。Monitoring system is a system designed to observe the correct operation of the equipment by detecting incorrect functioning (measure of variables compared with specified value).

Monitoring system on ship's activity　船舶活动监控系统　主要依托全球定位系统(GPS)、无线电传输技术(GPRS/GSM)以及计算机网络技术、地理信息管理(GIS)技术，以定位数据和船舶数据为基础，以高性能计算机运算分析为主要实现模式。对船舶活动实施全天候监督监控，并实时在海监部门的终端监控电脑显示船舶动态位置，为海洋执法管理人员提供有效数据支持。该系统还可实现越界报警，以防止越界活动情况发生，并能识别船舶所属的企业。

Monitoring tests　监测试验　为评价接船后或者修船后实际航行所引起的性能变化的试验。

Monitoring workstation　监视工作站　可以连续检查设备和环境的工作站；当几名船员在桥楼上工作时，其用于在驾驶和操纵工作站处解除驾驶人员和/或由船长，备用驾驶人员和/或引航员执行操纵和咨询功能。Workstation from where equipment and environment can be checked constantly; when several crew are working on the bridge it serves for relieving the navigator at the navigating and manoeuvring workstation and/or for carrying out control and advisory functions by the master, back-up officer and/or pilot.

Monitoring　监视　系指定期查核仪表显示信息和周围环境，以便发现异常现象的行为。Act of constantly checking information from instrument displays and environment in order to detect any irregularities.

Mono-ammonium phosphate(MAP)　磷酸铵　无嗅，呈灰褐色颗粒状。能产生极多粉尘。吸湿。散装MAP 的 pH 值为4.5，含有水分时可能产生极强的腐蚀作用。该货物为不燃物或失火风险低。该货物吸湿，受潮会结块。该货物会使遮盖舱底污水井的粗麻布或帆布腐烂。长期连续载运该货物可能对结构有破坏作用。

Mono-ammonium phosphate(MAP) is odourless and comes in the form of brownish-grey granules. It can be very dusty. It is hygroscopic. Bulk MAP has a pH of 4.5 and in the presence of moisture content can be highly corrosive. This cargo is non-combustible or has a low fire-risk. This cargo is hygroscopic and will cake if wet. This cargo will decompose burlap or canvas cloth covering bilge wells. Continuous carriage of this cargo may have detrimental structural effects over a long period of time.

Mono-hull HSC 单体高速船 系指只有一个船体的高速船。 Mono-hull craft is a high speed craft with one hull.

Mono-hull ship 单体船 一种其单壳体可作为排水型的船或根据水力支承的情况为半滑行艇线型或滑行艇线型的船舶。 Mono-hull ship is a ship whose single hull may be of displacement form or of a semi-planning or planning form subject to some support by hydrodynamic lift.

Mono-wall(membrane tube wall, finned tube wall) 膜式水冷壁 由鳍片管或光管排构成,相邻管沿鳍片或借扁钢带纵向对焊连成一片的水冷壁。

Monoxide 一氧化物

Monoxide carbon 一氧化碳

Monthly mean temperature(MAMDAT) 月算术平均温度 系指相应月份的日统计平均温度的算术平均。 Monthly mean temperature(MAMDAT) is the average of the mean daily temperature for the month in question.

Moon pool 船井 又称月池。在半潜式钻井装置或钻井船的中部所开的钻井孔。

Mooring arrangement 系缆设备 将船舶停靠于码头、浮筒、船坞或其他船舶的设备。通常指系缆、系缆具以及系缆机械等。

Mooring boat 带缆艇 为大船系泊浮筒时系缆和解缆用的船艇。

Mooring capstan 系泊绞盘 用以收绞系缆绳的绞盘。

Mooring chain 系船链 用作系缆的链条。

Mooring cleat 带缆羊角 固定在甲板或舷墙上,供带缆用的羊角状系索栓。

Mooring components 系泊组件 系指一般种类的部件,诸如锚链、钢丝绳、合成纤维绳、水泥墩、绞车/绞盘、导缆器和锚等。 Mooring components mean those general class of components such as chain, steel wire rope, synthetic fiber rope, clump weight, buoy, winch/windlass, fairlead, and anchor.

Mooring cup 系缆穴 供小船系靠用的凹入船体外板的系索栓。

Mooring fittings(mooring appliances) 系缆具 配置在船舶甲板上,用以在带缆操作时系缚和引导缆绳的各种器具的统称。主要有带缆桩、拖缆桩、带缆羊角、导绳器、导向滚轮等。

Mooring line 系泊缆 用以将船舶紧固于码头、船坞、浮筒或相邻船的缆绳。其包括:艏缆(head line)、艉缆(stern line)、艏横缆(bow breast line)、艉横缆(stern breast line)、前艏倒缆(forward bow spring)、后艉倒缆(after stern spring)等。

Mooring machinery 系泊机械 收放系泊缆绳的机械。

Mooring pipe 导缆孔 装设在舷墙上的闭孔状导缆器。

Mooring ring 系索环 可供系结绳索用的眼环。

Mooring speed 系缆速度 绞缆筒在额定系缆拉力下的工作速度。

Mooring swivel 双链转环 抛八字锚时,用以连接两根锚链,以防止锚链互相绞缠的可转动的专用环。

Mooring system of FPSO 浮式生产储/卸油船的系泊系统 是影响浮式生产储/卸油船安全性的关键组成部分,它包括允许带有一定压力的液体通过并能360°旋转而不泄漏的部件系统。该系统的主要组成部件是产品分配器(production distribution unit, PDU)或称为旋转接头(swivel),使油气可以通过软管和旋转接头进行不间断地输送。在不同的海况条件和作业需求下,浮式生产储/卸油船的系泊系统也各不相同地应用在多种系泊系统中,以单点转塔系泊系统(single point mooring system, SPM)的应用最为广泛,它使系泊的船舶始终处于系泊力最小的状态。此外还有内转塔式系泊的浮式生产储/卸油船。

Mooring winch 系泊绞车 用以收绞系泊缆绳的绞车。

Morphine 吗啡 是通过化学手段从鸦片中衍生的,其纯净形式为白色晶体。它经常掺杂别的东西,其颜色有白色、奶油色或米色直至咖啡色。可作为无色液体针剂进行医疗注射。也可作为丸药和针剂进行商品化生产。以这种形式出现时,可能有淡淡的氨水味或臭鱼味。 Morphine is chemically derived from opium. In its pure form it consists, it is often adulterated and its colour may range from white, cream or beige to a dark coffee colour. It is also found in a medical injection form as a colourless liquid in ampoules. Both pills and ampoules may be commercially produced. In this form it may smell faintly of ammonia or rotting fish.

Morse signal lamp 通信灯 按编码符号,用可见

光与船外进行通信联络的灯具。

Mosquito screen 防蚊网 装于窗框上,用以防阻蚊蝇的细眼网。

Most significant IMO instruments 最重要的 IMO 文件 系指:(1)经修正的 1974 年国际海上人命安全公约(SOLAS 74);(2)经 1978 年议定书修正的 1973 年国际防止船舶造成海上污染公约(MARPOL 73/78);(3)1966年载重线公约(LL66);(4)经修正的 1978 年国际海员培训、发证和值班国际公约(STCW 78);(5)经修正的 1972 年国际海上避碰规则公约(COLREG 72);(6)1969 年国际吨位丈量公约(TONNAGE 69)。也需考虑联合国海洋法(ENCLOS)。 Most significant IMO instruments mean: (1) the International Convention for the Safety of Life at Sea, 1974 (SOLAS 74), as amended; (2) the International Convention for the Prevention of Pollution from ships, 1973 as modified by the Protocol of 1978 relating thereto(MARPOL 73/78), as amended; (3) the International Convention on Load Lines, 1966 (LL66); (4) the International Convention on Standards of Training, Certification and Watchkeeping for Seafarers, 1978 (STCW 78), as amended; (5) the Convention on the International Regulations for Preventing Collisions of Sea, 1972 (COLREG 72), as amended; and (6) the International Convention on Tonnage Measurement of ships, 1969, (Tonnage 69); Regard should also be given to the United Nations Convention on the Law of the Sea(UNCLOS).

Motion display modes 真运动显示模式 系指本船以其真运动移动的一种显示。 True motion means a display across which own ship moves with its own true motion.

Motion stability index 运动稳定性指数 表征船舶运动方向稳定性的特征方程的根: $A\sigma^2 + B\sigma^2 + C\sigma + D = 0$, $\psi = \psi_1 e_1^{-\sigma^2} + \psi_2 e_2^{-\sigma^2} + \psi_3 e_3^{-\sigma^2}$。

Motions of the moored unit 系泊平台运动 系指系泊平台在风力(定常)、海流力和波漂力的联合作用下将从无环境力作用时的初始位置偏移至新的平均位置,在此平均位置,系泊系统的复原力将平衡所施加的定常力。波浪将对于此新的平均位置引起系泊平台的波频和低频运动。 Motions of the moored unit means that under combined actions of steady wind, current and wave-drift forces, the moored unit will move from its initial equilibrium position where zero environmental force is acting on the unit, to a new mean position where the mooring system will have developed sufficient restoring force to balance the steady applied forces. Wave-induced oscillatory motions and low frequency motions of the unit take place about this new mean position.

Motor 电动机 是一种将机械功率转变为电功率的机器。

Motor alternator 电动交流发电机组

Motor armature 电动机电枢

Motor branch circuit 电动机分支电路 系指仅向一台或多台电动机及其控制器供电的电路。

Motor driven feed pump 电动给水泵 由电动机驱动的泵。

Motor lifeboat 机动救生艇 依靠动力机械推进的救生艇。

Motor oil 马达油

Motor sailer 机帆船 备有风帆装置或形似帆船的小型机动船舶。

Motor ship 内燃机船 以内燃机作为主机的船舶。包括柴油机船、燃气轮机船和汽油机船。

Motor vessel 机动船舶 系指由动力驱动的船,而非风力驱动的帆船。 Motor vessel means a power driven vessel which is not a sailing vessel.

Moulded baseline 基线 基线为通过平台底板上缘的一条水平线。 Moulded baseline is a horizontal line extending through the upper surface of the bottom plating.

Moulded breadth 型宽 系指:(1)船舶的最大宽度。对金属船壳的船舶是在船舯部量至两舷肋骨型线;对船壳为任何其他材料的船舶则是在船中部量至两舷船壳的外表面;(2)船舯在露天甲板下的最大型宽。 Moulded breadth is: (1) the maximum breadth of the ship measured amidships to the moulded line of the frame in a ship with a metal shell and to the outer surface of the hull in a ship with a shell of any other material; (2) the greatest moulded breadth, in m, measured amidships below the weather deck.

Moulded breadth of air-way system 坞体型宽 系指从坞体两侧肋骨外边缘量取的横截面宽度。

Moulded depth 型深 (1)系指从龙骨上缘量到船舷处上甲板下缘的垂直距离,对木质船舶和铁木混合结构的船舶,垂直距离是从龙骨口的下缘量起。如果船舶中央横剖面具有凹形,或装有加厚的龙骨翼板时,垂直距离是从船底平坦部分向内延伸与龙骨侧面相交的一点量起;(2)具有圆弧形舷边的船舶,型深是量到甲板型线和船舷外板型线相交之点,这些线的延伸是把该舷边看作是设计为角形的;(3)当上甲板为阶梯形甲板,并且其升高部分延伸超过决定型深的一点时,型深应量到此甲板较低部分的引出虚线,此虚线平行于甲板升高部分。 Moulded depth means: (1) the moulded depth is the vertical distance measured from the top of the keel to the underside of the upper deck at side. In wood and composite ships the distance is measured from the lower edge of the keel rabbet. Where the form at the lower part of the mid-ship section is of a hollow character, or where thick garboards are fitted, the distance is measured from the point where the line of the flat of the bottom continued inwards cuts the side of the keel; (2) in ships having rounded gunwales, the moulded depth shall be measured to the point of intersection of the moulded lines of the deck and side shell plating, the lines extending as though the gunwales were of angular design; (3) where the upper deck is stepped and the raised part of the deck extends over the point at which the moulded depth is to be determined, the moulded depth shall be measured to a line of reference extending from the lower part of the deck along a line parallel with the raised part.

Moulded depth 型深（平台） 系指：(1)对自升式平台和水面式平台而言，D 为平台长度中点处沿舷侧从基线量到干舷甲板梁上缘的垂直距离；(2)对柱稳式平台和坐底式平台而言，D 为平台长度中点处沿舷侧从基线量至下壳体最上层连续甲板梁上缘的垂直距离。 Moulded depth D(m) are: (1) for the purpose of self-elevating units and surface-type units, D is the vertical distance measured at the middle of the length L from moulded baseline to top of the deck beam at side on the freeboard deck; (2) for the purpose of column-stabilized units and submersible units, D is the vertical distance measured at the middle of the length L from moulded baseline to top of the deck beam at side on the uppermost continuous deck of lower hull.

Moulded depth(fishing vessel) 型深（渔船） (1)系指在船中处，从龙骨线量至工作甲板舷侧处横梁上缘的垂直距离；(2)对于具有圆弧形舷缘的船舶，船深应量至甲板型线延伸线与舷侧外板延伸线交点处；(3)工作甲板呈梯级状的渔船，且其甲板升高延伸线超过决定船深的那一点时，则船深应量至甲板较低部分的与升高部分相平行的延伸线。 Moulded depth: (1) is the vertical distance measured from the keel line to the top of the working deck beam at side; (2) In vessels having rounded gunwales, the moulded depth shall be measured to the point of intersection of the moulded lines of the deck and side shell plating, the lines extending as though the gunwale were of angular design; (3) Where the working deck is stepped and the raised part of the deck extends over the point at which the moulded depth is to be determined, the moulded depth shall be measured to a line of reference extending from the lower part of the deck along a line parallel with the raised part.

Moulded depth(SWATH) 型深（小水线面双体船） 系指船长 L 中点处（船中）截面由基线量至干舷甲板横梁上缘的垂直距离。 Moulded depth means the vertical distance measured at mid-craft from the base line to the top of the freeboard deck beam at side.

Moulded displacement $\triangle$ (t) 型排水量 系指在海水（密度 ρ = 1.025 t/m³）中吃水 T 时的排水量。 Moulded displacement means at displacement draught T, in sea water(density = 1.025 t/m³).

Moulded draught 型吃水 系指在舯横剖面上自型基线量到夏季载重水线的垂直距离。对于具有立龙骨的船舶型基线取自船底板上表面与立龙骨的交线。 Moulded draught T is the distance, in m, measured vertically on the midship transverse section, from the moulded base line to the summer load line. In the case of a ship with a solid bar keel, the moulded base line is to be taken at the intersection between the upper face of the bottom plating with the solid bar keel at the middle of length L.

Mouth 进口（进入管，狭窄部分）
Mouth bell 喇叭口
Mouth piece 泵吸口
Mouth piece 传话筒
Movable arm 可动臂
Movable decks 可移动甲板 通常是由腹板构件和顶部甲板组成的箱式结构，其他形式的构造等个别考虑。 Movable decks are generally to be constructed as pontoons comprising a web structure with top decking. Other forms of construction will be individually considered.

Movable hook component 活动钩体系 系指艇钩装置中与艇索的连接件直接接触的部分，其移动时使救生艇从艇索脱开。 Movable hook component is that part of the hook assembly in direct contact with the connection with the lifeboat falls which moves to enable release from falls.

Movable joint in discharging pipeline 排泥管活动接头 排泥管线上能上下、左右转动的接头。
Movable joint in suction pipeline 吸泥管活动接头 吸泥管的活动段与固定段之间活络连接的装置。
Moving blade(rotor blade, bucket) 动叶片 装配在轮盘式轮鼓上的叶片。
Moving-weight stabilizer 移动重力式减摇装置 利用移动重力形成减摇力矩的船舶减摇装置。
Mud breaker 碎泥刀 安装在自扬链斗式挖泥船的泥井里，形状如铣齿刀，用以捣碎泥块杂物，同时排除大石块的滚刀。
Mud collector 泥箱 （集泥器，污水过滤器）

Mud drum(water drum) 下锅筒 具有积聚并定期排放锅水中沉渣的功用,置于水管锅炉蒸发管束下部的容水部件。

Mud guard 挡泥板 设在斗桥槽两侧和泥阱周围,用以防止链斗运转时泥浆溅出的薄围壁板。

Mud lighter 泥驳 用以运输泥沙的驳船。

Mud tank 泥浆舱 贮存钻井用泥浆的舱。

Muffle 消声(闭式烤炉,玻璃灯罩)

Muffler 灭音器

Multi-stage flash evaporation 多级闪发 热海水经过若干串联的闪发室逐次闪发的闪发蒸发。

Multi-beam sounding system 多波束测深系统 一种同时发射数十个细窄波束以获得船舶正横方向数十个水深点的测深系统。它由窄波束回声测深仪和信号数字处理机两大部分组成。该系统可精确测出航行障碍物的位置、深度和范围以及海底地貌。并能绘出海底三维显视图,测深效率比一般单一波束回声测深仪要高。

Multi-bladed rudder 多叶舵 一根舵杆上装有两个或两个以上舵叶的舵。

Multi-casing steam turbine 多缸汽轮机 具有两个以上气缸的汽轮机。

Multi-cell hovercraft(multi skirt air-cushion vehicle) 多气室气垫船 以高于大气压的空气充入船底多个隔室以形成气垫的全垫升气垫船。

Multi-chamber buoyancy system 多室浮力系统 通过将气胀式救生衣分成两个或多个独立的隔室来提供浮力,以使其中一个隔室产生机械损坏时,其他隔室仍可工作和提供浮力,从而为已浸水的使用者提供帮助的系统。 Multi-chamber buoyancy system is the system that divides the buoyancy provided by an inflatable lifejacket into two or more separate compartments, such that if mechanical damage occurs to one, others can still operate and provide buoyancy so as to aid the user when immersed.

Multi-cylinder 多缸

Multi-decked ship 多甲板船 设有两层或两层以上甲板的船舶。

Multidimensional arrays 多维数组

Multi-drum winch 多卷筒绞车 由一台原动机带动一个钢丝绳卷筒和若干个电缆卷筒,而各卷筒可单独进行工作的一种绞车。

Multi-effect evaporation 多效蒸发 加热介质的热量被多次用于蒸发海水,即前一效应产生的二次蒸汽作为下一效应的加热蒸汽的蒸发方式。

Multi-engine geared drive 多机共轴齿轮传动 由两台或两台以上主机通过离合器和减速齿轮箱使功率合并由一根轴输出的传动方式。

Multi-fuel internal combustion engine 多种燃料内燃机 可燃用柴油、汽油、煤气、天然气及其他燃料等两种以上燃料的内燃机。

Multi-function display(MFD) 多功能显示(多功能显示器) 系指:(1)1个视频显示器,能同时或通过一系列页显示来自综合桥楼系统一个以上的初始信息;(2)单一视角显示单元,能同时或通过一系列可选页面显示源自一个以上单一功能的信息。 Multi-function display means:(1)a single visual display unit which can present, either simultaneously or through a series of selectable pages, information from more than one operation of an integrated bridge system;(2)a single visual display unit that can present, either simultaneously or through a series of selectable pages, information from more than a single function.

Multi-hulled ship 多体船 将两个或两个以上船体上部用强力构架联成一个整体的船。

Multi-modes transport consignee 多式联运收货人 系指有权从多式联运经营人接收货物的人。 Multi-modes transport consignee means the person entitled to receive the goods from the multi-modes transport operator.

Multi-modes transport consignor 多式联运托运人 系指与多式联运经营人签订多式联运合同的人。 Multi-modes transport consignor means the person who concludes the multi-modes transport contract with the multi-modes transport operator.

Multi-modes transport contract 多式联运合同 系指:(1)多式联运经营人凭以收取运费、负责履行或实现履行国际多式联运的合同;(2)以至少通过两种不同的运输方式运送货物的合同。 Multi-modes transport contract means:(1)a contract whereby a multimodal transport operator undertakes, against payment of freight, to perform or to procure the performance of international multimodal transport;(2)a single contract for the carriage of goods by at least two different modes of transport.

Multi-modes transport document 多式联运单证 系指证明多式联运合同的单证。该单证可以在适用法律的允许下,以电子数据交换信息取代,而且(1)以可转换方式签发;或者(2)表明记名收货人,以不可转换方式签发。 Multi-modes transport document means a document evidencing a multi-modes transport contract and which can be replaced by electronic data interchange messages insofar as permitted by applicable law and are:(1)issued in a negotiable form;(2)issued in a non-negotiable form indicating a named consignee.

Multi-modes transport operator 多式联运承运人

系指实际完成或承担完成此项运输或部分运输的人，不管他是否与多式联运经营人属于同一人。 Multi-modes transport operator means any person who concludes a multi-modes transport contract and assumes responsibility for the performance thereof as a carrier.

Multi-modes transport operator(MTO) 多式联运经营人 系指：(1)其本人或通过代其行事的他人订立多式联运合同的任何人，他是委托人，而不是发货人的代理人和参加多式联运的承运人的代理人或代表他们行事，他承担履行合同的责任；(2)签订一项多式联运合同并以承运人身份承担完成此项合同责任的任何人。

Multi-modes transport operator means:(1)any person who on his own behalf or through another person acting on his behalf concludes a multimodal transport contract and who acts as a principal, not as an agent or on behalf of the consignor of the carriers participating in the multimodal transport operations, and who assumes responsibility for the performance of the contract;(2)any person concludes a multi-modes transport contract and assumes responsibility for the performance thereof as a carrier.

Multi-orifice nozzle 多孔喷嘴

Multi-phase ejector 多相喷射器 利用蒸馏装置中产生的淡水加压后作为工作水，引射蒸发器中产生的蒸汽，其排出的介质为液-气双相混合物的喷射器。

Multi-phase synchronous generator 多相同步发电机 是一种其交流电路之设置使其在接线端上产生具有一定相位关系的两个或多个对称交变电动势的同步发电机。多相同步发电机通常有：产生两个彼此相对相移90°电角交变电动势的两相同步发电机和产生三个互相相对相移120°电角交变电动势的三相同步发电机。供船上使用的多相同步发电机一般为三相发电机，特殊情况下可以是两相发电机。

Multi-pivol type davit 跨步型艇架 放艇时，吊艇臂依次绕后，前支点倾倒，从而吊艇出舷的重力式艇架。

Multi-plate clutch(multi-disc clutch) 多盘式离合器 有三个或三个以上摩擦盘的离合器。

Multiple bucket ladder 复式斗桥 由主、副斗桥组成的斗桥。

Multiple irregularities in the section 横截面上多种缺陷 对于厚度$S \leqslant 10$ mm 或 $a \leqslant 10$ mm，可能需要规定某些参数。即理论上可能出现各种缺陷的总和。在这种情况下，对于几种等级所确定的数字，所有允许偏差值的总和应该是减少的。但是对于单个缺陷值 $\geqslant h$，例如对于单个孔隙，应不至于超过其相应允许值。见图 M-9。 For thickness $S \leqslant 10$ mm or $a \leqslant 10$ mm special parameters may be necessary, e. g. a theoretically possible accumulation of individual imperfections. In this case, the total of all permitted deviations for the determined figures of the several classification levels should be reduced. However, the value of a single imperfection $\geqslant h$; e. g. for a single pore, should not be exceeded. See Figure M-9.

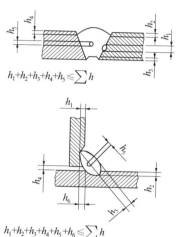

$h_1+h_2+h_3+h_4+h_5 \leqslant \sum h$

$h_1+h_2+h_3+h_4+h_5+h_6 \leqslant \sum h$

图 M-9 横截面上多种缺陷
Figure M-9 Multiple irregularities in the section

Multiple sheave block 多并滑车 由两个或两个以上相同滑轮并列组成的滑车。

Multiple valve 复式阀

Multiple-superimposed type frequency converter 多重叠加型变频器 是一种由多个低压单相电压型变频器(H桥功率单元)串联而成中压变频器的每一相的图表结构的变频器。

Multiplexed control system 多路控制系统

Multi-pressure stage 多压级

Multipurpose anchor winch 多用途起锚机 由起锚机和系泊绞车所组成的有起锚及系缆等功用的系船机械。

Multi-purpose chemical carrier 多用途化学品船

Multi-purpose controller(MPC) 多功能控制器

Multipurpose galley machine 厨用多功能机 具有多种功能的食品加工机械。

Multi-purpose research ship 多用途调查船 具有海洋调查研究、开发和运输等多用途的船舶。实验室为类似集装箱的活动单元结构式。根据调查科目可随时吊装上船，不必像综合调查船设置众多的实验室，提高了船的使用效率和经济性。

Multi-purpose vessel 多用途货船 其货舱结构被设计成适合于载运两种或两种以上货物的船舶。多用途货船可分为多用途杂货船和多用途散货船两种。多用途杂货船由杂货船演变而来,装载杂货可同时装载集装箱、重大件货物。多用途散货船由散货船演变而来,装载散货同时又装载集装箱。

Multi-reduction gear(multiple reduction gear) 多级减速齿轮箱 通过三对或三对以上齿轮的啮合作用,实现多次减速的减速齿轮箱。

Multi-shaft(rotor) gas turbine 多轴燃气轮机 具有多根独立的压气机涡轮轴的燃气轮机。

Multi-shafting arrangement 多轴布置

Multi-speed change(multi-change speed) 多级变速 系指轴系转速的多挡变换。

Multi-speed motor 多速电动机 是一种能以两种和两种以上的转速中任何一种转速运行的电动机。每种转速实际上均与电动机的负载无关。诸如具有两个电枢绕组的直流电动机,或具有其绕组能进行多种变极组合的感应电动机。

Multi-stages feed heating 多级预热给水 用主机中间级撇气及辅机排气来预热锅炉给水的多级预热形式。

Multi-step planning boat 多断级滑行艇 艇底具有两个或两个以上横向断级,滑行时艇底形成前后多个滑行面的滑行艇。

Multi-streamers seismic research vessel 多缆物探船 一种三维高性能深水物探船,又称深水油气地球物理3D地震勘探船。从事三维地震采集作业并能对三维地震作业资料现场处理,具有采集数据高速传输系统的专用物理勘探的工程船。其配有12~18根8 000 m地震采集电缆(相邻电缆间距100 m)和双震源共8排气枪阵列,在地震综合导航系统的综合控制下,高压空气气枪震源激发地震信号,由采集电缆接收地下底层反射地震信息并传至地震采集记录系统完成地震信息的记录。见图M-10。

图 M-10 多缆物探船
Figure M-10 Multi-streamers seismic research vessel

Municipal and commercial waste 城市和商业污水 系指包括城市污水、商业污水、医疗污水和其他特征的污水等的污水。

Munitions 军火 系指完整的装备(例如导弹、炮弹、鱼雷、爆破装置等,装有炸药、推进剂、烟火剂或起爆成分),用于进攻、防御、训练或非作战行动,包括含有炸药的武器系统部件。 Munitions are a complete device (e. g. missile, shell, mine, demolitions store, etc., charged with explosives, propellant, pyrotechnics, or initiating composition), for use in conjunction with offensive, defensive, training, or non-operational purposes, including those parts of the weapon systems containing explosives.

Munk moment 不稳定动力偶 三维细长体在理想流体中运动所产生的在前半体与后半体力量相等而作用力相反的一种使攻角增大,稳定性下降的纯力偶。

Mushroom anchor 菌型锚 长期系泊用的菌状或伞状锚。

Mushroom vent 蕈形通风筒(蕈形通风头)

Mushroom ventilator 蕈形通风筒 呈蕈状,并附有可开闭水密盖的通风筒。

Muster area 集合区域 平台上人员可以躲避意外情况直至登上救生艇的区域。

Muster list 应变部署表(平台) 系指详细说明通用报警系统的信号,以及这些警报发出时每人在各种作业模式下应采取的行动,指明他们应去的地方和要求他们履行的一般职责(如有)的表。应变部署表中应包括下列事项的职责:(1)平台水密门、防火门、阀、进/排气孔、泄水孔、舷窗、天窗、舱门和其他类似开口的关闭;(2)救生艇筏和其他救生设备的属具配备;(3)救生艇筏的准备和降落;(4)其他救生设备的一般准备工作;(5)来访者的集合;(6)通信设备的使用;(7)指派处理火灾的消防队的人员配备;(8)关于使用消防设备和装置的专门职责;(9)直升机甲板上的应急职责;(10)在碳氢化合物或硫化氢溢出失控时的专门职责,包括紧急关断。

Muster list is a list specified details of the general alarm system signals and also the action to be taken in all operating modes by every person when these alarms are sounded, indicating the location to which they should go and the general duties, if any, they would be expected to perform. The following duties should be included in the muster list: (1) closing of the watertight doors, fire doors, valves, vent inlets and outlets, scuppers, sidescuttles, skylights, portholes and other similar openings in the unit;(2) equipping of the survival craft and other life-saving appliances;(3) preparation and launching of survival craft;(4) general preparation of other life-saving appliances;(5) muster of visitors;(6) use of communication equipment; (7) manning of fire parties assigned to deal with fires; (8) special duties assigned in respect to the use of fire-fighting equipment and installations; (9) emergency duties on the helicopter deck; and (10) special duties assigned in the event of an uncontrolled escape of hydrocarbons or hydrogen sulphide, including emergency shutdown.

Muster list(emergency station bill) 应变部署表 明确列出每位船员在发生火灾、船体漏水及弃船时的岗位职责的图表。

Muster list and emergency instructions 应急部署表与应急须知 系指为船上人员配备的一份在紧急情况下必须遵循的、明确的须知。如为客船,这些须知应使用船旗国要求的一种或数种语言或英语写成。符合相关要求的应急部署表与应急须知在船上各显著部位显示,包括驾驶室、机舱和船员起居处所。同时应在乘客舱室内张贴配有适当文字的示意图,并在集合站及其他乘客处所的显著位置予以展示,以告知乘客:(1)他们的集合站;(2)他们在紧急情况下必须采取的重要行动;(3)救生衣的穿着方法。 Muster list and emergency instructions means a copy of clear instructions to be followed in the event of an emergency provided for every person on board. In the case of passenger ships these instructions shall be drawn up in the language or languages required by the ship's flag State and in the English language. Muster lists and emergency instructions complying with the relative requirements shall be exhibited in conspicuous places throughout the ship including the navigating bridge, engine-room and crew accommodation spaces. Illustrations and instructions in appropriate languages shall be posted in passenger cabins and be conspicuously displayed at muster stations and other passenger spaces to inform passengers of: (1) their muster station; (2) the essential actions they must take in an emergency; (3) the method of donning lifejackets.

Muster station 集合点 在紧急情况下撤离或弃船之前人员可以安全集合的地点。

Mutual independence between components 设备间的相互独立 系指设备的功能和动力源均不依赖除中央动力源以外的某一公用设备或系统。Mutual independence between components means that function of the components and their power supply is not dependent on some common component or system except for a central power supply.

MW 系指 MCR 或燃油燃烧装置 80% 的额定功率。 MW refers to the MCR or 80% of the power rating of the fuel oil combustion unit.

N

$N(m)$ 自基线至船体横剖面水平中和轴的垂直距离(m) N means the vertical distance (m) from the base line to the horizontal neutral axis of the hull transverse section.

NAAA = National Oceanic and Atmospheric Administration [美]国家海洋和大气管理局

Nacelle 机舱 系指风电系统中将机械能转换为电能的风能转换系统的组成部件。机舱中装备了传动系统和发电机。在一些风机中,变频器和变压器也设置在机舱内。

Naked hull resistance test 裸船模阻力试验 在水池中拖曳不带附体的船模,测出各种航速时阻力的试验。

Name 名称

NANDYMAX bulk carrier 灵便型散货船 系指其主尺度略小于巴拿马型散货船的,也可通过巴拿马运河的船舶。

Nanotechnology(NT) 纳米技术 是用单个原子、分子制造物质的科学技术,即在纳米尺度上研究物质的特性和相互作用的技术。其研究结构尺寸在 0.1~100 纳米范围内材料的性质和应用。纳米技术是一门应用科学,其目的在于研究于纳米规模时,物质和设备的设计方法、组成、特性以及应用,它是现代科学(混沌物理、量子力学、介观物理、分子生物学)和现代技术(计算机技术、微电子和扫描隧道显微镜技术、核分析技术)结合的产物,纳米科学技术又将引发一系列新的科学技术,例如:纳米物理学、纳米生物学、纳米化学、纳米电子学、纳米加工技术和纳米计量学等。纳米技术是许多如生物、物理、化学等科学领域在技术上的次级分类。

Nansen bottle 南森采水器 又称颠倒采水器。是一种可在预定水层采取水样并测温的金属圆筒。它的两端装有活门,在下放过程中,活门处于开启位置,达到预定深度后,从甲板打下使锤,采水器颠倒,活门关闭,完成采水样工作。

Nansen opening-and-closing net 南森开/闭式网 备有可控开闭网口装置的用于采集预定地点和预定水

层浮游生物的网具。

Naphtha 石脑油

NARA = National Archives and Record Administration ［美］国家档案与资料管理局

Narrow band direct printing equipment（NBDP） 窄带直接印字电报终端 是解决莫尔斯电报使用不便而研制的海上通信系统。它是在船舶上电台与岸上电台或船舶电台直接使用中频/高频波段，以遇险、安全及一般电传通信为目的的自动收发电报装置。

Narrow-tipped blade 窄稍叶 一般系指与同面积的椭圆叶相比，叶梢部分较窄的螺旋桨叶。

Narrow waters 狭窄水道 系指在其航向线任一侧都不允许船舶以相应航速航行相当于30min距离的水域。Waters with restricted freedom of course setting and where pilotage conventionally is the foremost navigational method.

Narrowband system "窄频"系统

National Association of Corrosion Engineers（NACS） ［美］腐蚀工程师协会

National Ballast Information Clearing-house（NBIC） 美国压载水情报交换所 系指海岸警卫队和由NISA管辖的史密斯学会环境保护研究中心共同经营的美国压载水情报研究机构。National Ballast Information Clearing-house（NBIC）means the National Ballast Information Clearing-house operated, by the Coast Guard and the Smithsonian Environmental Research Center as mandated under the National Invasive Species Act of 1996.

National environmental protection agency（NEPA） ［美］国家环保局

National History Protection Act（NHPA） ［美］国家历史保护法

National institute of standard and technology（NIST） ［美］国家标准与技术研究院 隶属于美国商业部的技术管理部门，前身是1901年建立的联邦政府的第一个物理科学实验室，位于马里兰州的Gaithersburg，是目前NIST总部所在地，占地234公顷，另一部分，位于科罗拉多州的Boulder，占地84公顷，还有两个设在大学的联合研究所，JILA在Colorado大学，CARB在Maryland大学。NIST的任务是为提高劳动生产率、促进贸易和改善生活质量、提高计量、标准和技术。NIST证明系指产品已经根据NIST SRM（Standard Reference Materials）测试，并符合相关测试要求。常见的NIST证明产品包括：计时器、标度砝码、转速计/流速计、声级计量器、电子万用表、体温计、时钟、压力计、风速计、Ph测试仪、测微计以及测光计等等。

National Invasive Species Act（NISA） ［美］有害入侵物种法 系指《1996年国家有害入侵物种法》，该法重新批准和修订了（美国）1990年防止和控制外来水生物污染法令（NANPCA）。

National maritime administration 国家海事主管机关 系指由登记国根据其立法设立的国家当局或主管机关，按照该立法，负责执行有关海上运输的国际协议，并负责适用有关在其管辖和控制下的船舶的规则和标准。National maritime administration means any State authority or agency which is established by the State of registration in accordance with its legislation and which, pursuant to that legislation, is responsible, for the implementation of international agreements concerning maritime transport and for the application of rules and standards concerning ships under its jurisdiction and control.

National Maritime Conference 国际海事会议 系指由德国联邦经济技术部主持的常规会议。会议内容是关于海上经济问题（航运、造船、航运工程技术等）。2000年至2009年已举行过6次会议。2005年首次为海上风电专门组织了研讨会。

National missile defense（NMD） ［美］国家导弹防御系统 冷战时期，美国为了取得对苏联的军事优势，于1983年提出"战略防御"计划（又称"星球大战"计划）。1993年，克林顿宣布"星球大战"时代结束，取而代之的是"NMD"（国家导弹防御系统）和"TMD"（Theater missile defense，战区导弹防御系统）。 国家导弹防御系统系指为保护美国本土免遭战略导弹攻击的武器系统。该系统由20世纪80年代的"战略防御计划"和20世纪90年代初的"对付有限打击全球防护系统"演变而来的。美国发展NMD和TMD将打破全球的战略平衡与稳定，阻碍核裁军的进程，破坏国际防扩散的努力，并可能引发新一轮的军备竞赛，所以受到世界上大多数国家的反对。

National pollution discharge elimination system（NPDES） ［美］污染控制排放系统

National Referral Center（NRC） ［美］全国咨询中心

National Research Council（NRC） ［美］全国科学研究委员会

National Security Agency（NSA） ［英］国家安全局 又称军事情报5处［Military Intelligence 5（MI 5）］成立于1909年，已有近90多年历史，原名为秘密勤务局（Secret Service Bureau），因第一次世界大战国内安全需要设立，1916年更名为MI 5，1931年又更名为安全局（Security Service）。迄今，该局与英国另一国外情报机构秘密情报局（Secret Intelligence Service，又称为MI 6）分别负责英国国内外情报及安全工作。即军情五处（MI

5），是世界上最具神秘色彩的谍报机构之一。1905 年由英国陆军大臣 R·B·霍尔丹实施的军队改革促使军事情报部门的成立。但是总参谋部为情报部门的归属问题却争论不休，争论的结果导致军情五处的成立，它起先归属于陆军部，后来由内政部接管。自成立以来，军情五处在对付颠覆和恐怖活动上立下累累战功，一度拯救英国于危亡之中。现主要从事反间谍、反渗透工作。

National shipping line 国家航运公司 某一国家的国家航运公司系指其总管理处及其实际控制设在该国，并为该国有关当局或依据该国法律所承认的使用船舶的运营商。牵涉两个或两个以上国家的合营事业所拥有和经营的航运公司，其权益的大部分为这些国家的公（或）私人利益所拥有，而其总管理处及其实际控制设在这些国家中之一国，则可被这些国家的有关当局承认为国家航运公司。 National shipping line means a national shipping line of any given country is a vessel-operating carrier which has its head office of management and his effective control in that country and is recognized as such by an appropriate authority of that country or under the law of that country. Lines belonging to and operated by a joint venture involving two or more countries and in whose equity the national interests, public and or private, or those countries have a substantial share and whose head office of management and whose effective control is in one of those countries can be recognized as a national line by the appropriate authorities of those countries.

National Television Standards Committee (NTSC) [美]国家电视标准委员会 NTSC 负责开发一套美国标准电视广播传输和接收协议。此外还有两套标准：逐行倒相（PAL）和顺序与存色彩电视系统（SECAM），用于世界上其他的国家。NTSC 标准从它们产生以来除了增加了色彩信号的新参数之外没有太大的变化。NTSC 信号是不能直接兼容于计算机系统的。

National tonnage 国家吨位

National waters 国内水域 系指船旗国政府规定的该国水域。 National waters mean the waters of a flag State, as specified by the Government of that State.

Natural circular frequency of heaving 垂荡固有圆频率 船舶作无阻尼自由垂荡用圆周运动表达时的角速度，$\omega_z = 2\pi/T_z$。

Natural circular frequency of pitching 纵摇固有圆频率 船舶作无阻尼自由纵摇用圆周运动表达时的角速度，$\omega_\Phi = 2\pi/T_\Phi$。

Natural circular frequency of roll 横摇固有圆频率 船舶作无阻尼自由横摇用圆周运动表达时的角速度，$\omega_\theta = 2\pi/T_\theta$。

Natural circulation cooling system 自然循环冷却系统

Natural disaster 自然灾害 系指地球上的自然变异，包括人类活动诱发的自然变异，无时无地不在发生，当这种变异给人类社会带来危害时，即构成自然灾害。因为它给人类的生产和生活带来了不同程度的损害，包括以劳动为媒介的人与自然之间，以及与之相关的人与人之间的关系。灾害都是起消极的或破坏的作用。所以说，自然灾害是人与自然矛盾的一种表现形式，具有自然和社会两重属性，是人类过去、现在、将来所面对的最严峻的挑战之一。

Natural draft(natural draught) 自然通风

Natural fiber 天然纤维 是自然界存在和生长的、具有纺织价值的纤维。全世界天然纤维的产量很大，并且在不断增加，是纺织工业材料的重要来源。

Natural frequency 固有频率（自然频率）

Natural frequency of heaving 垂荡固有频率 船舶作无阻尼自由垂荡时，单位时间内的垂荡循环次数，$f_z = 1/T_z$。

Natural frequency of pitching 纵摇固有频率 船舶作无阻尼自由纵摇时，单位时间内的纵摇循环次数，$f_\theta = 1/T_\theta$。

Natural frequency of roll 横摇固有频率 船舶作无阻尼自由横摇时，单位时间内的横摇循环次数，$f_\Phi = 1/T_\Phi$。

Natural gas 天然气 是一种碳氢化合物和少量的各种非碳氢化合物（例如 CO_2、氦、硫化氢和氮）的混合物。它以气态或液态的形式与储藏在地下储油层的原油混合在一起。 Natural gas is a mixture of hydrocarbon compounds and small amounts of various non-hydrocarbon (e.g., carbon dioxide, helium, hydrogen sulfide, and nitrogen) existing in gaseous phase or in solution with crude oil in natural underground reservoirs.

Natural gas(dry) 天然气（干燥） 系指在正常运转压力和温度下不会凝结，其主要成分为甲烷并含有一些乙烷和少量重烃（主要是丙烷和丁烷）的气体。天然气的成分随天然气的来源和气体加工过程有所不同。其典型的成分及所占体积（%）如表 N-1 所示。该气体可以作为压缩天然气或液化天然气进行储存和分配。

表 N-1　天然气
Table N-1　Natural gas

甲烷　Methane(C1)	94.0%
乙烷　Ethane(C2)	4.7%
丙烷　Propane(C3)	0.8%
丁烷　Butane(C4+)	0.2%
氮　Nitrogen	0.3%
气体密度　Density gas	0.73 kg/dm³
液体密度　Density liquid	0.45 kg/dm³
热值(低)　Calorific value(low)	49.5 MJ/kg
甲烷值　Methane number	83

Natural life of hull 船体自然寿命　又称物理寿命。系指船体从开始投入使用、由于受到各种环境条件的影响使船体产生耗损、破坏直至报废所经过的时间。船体自然寿命与平时经常性的维护保养和必要时的修理有关，即通过对船体的维修能延长其自然寿命。

Natural period of heaving 垂荡固有周期　船舶作无阻尼自由垂荡时，单位时间内的垂荡循环次数，$T_z = 1/f_z$。

Natural period of pitching 纵摇固有周期　船舶作无阻尼自由纵摇时，每一纵摇循环所经过的时间间隔，$T_\theta = 1/f_\theta$。

Natural period of roll 横摇固有周期　船舶作无阻尼自由横摇时，每一横摇循环所经的时间间隔，$T_\Phi = 1/f_\Phi$。

Natural person 自然人　是在自然状态下出生的人。这是一个法律概念。基于出生而为民事权利和义务主体的人。与法人相对。在中国和其他一些国家称为公民。但公民仅指具有一国国籍的自然人，而自然人还包括外国人和无国籍人。自然人是在自然状态之下而作为民事主体存在的人。享有权利并承担义务。自然人民事主体资格的法律特征：自然人主体资格具有广泛性。即任何人都要参加民事法律关系，不论其是否愿意，都要受到民事法律关系的调整。自然人主体资格的平等性。民法上的平等是机会平等，而不是实质平等。所有的人都有平等的民事权利，有平等的民事义务。

Natural resin 天然树脂　主要来源于植物渗(泌)出物的无定形半固体或固体有机物。特点是受热时变软，可熔化，在应力作用下有流动倾向，一般不溶于水，而能溶于醇、醚、酮及其他有机溶剂。天然树脂主要用作涂料，也可用于造纸、绝缘材料、胶黏剂、医药、香料等的生产过程；有些可作装饰工艺品的原料(如琥珀)；还有的如加拿大胶，其折光指数与普通玻璃相似，故作为显微镜等光学器材的透明胶黏剂。天然树脂可根据树脂组分、树脂形成的历史进行分类。

Natural resin paint 天然树脂漆

Natural rubber(NR) 天然橡胶　是由巴西三叶橡胶树分泌的天然胶乳，经凝固、加工而得到的弹性固体物，其主要成分是聚异戊二烯，含量在90%以上，此外还含有少量的蛋白质、类脂物和有机酸、糖类及灰分等。由于加工方法的不同以及橡胶制品的需要，天然橡胶产品可以分为天然生胶和浓缩天然胶乳两大类，前者如烟片胶、风干胶片、绉胶片、颗粒橡胶、橡胶粉及其他特种橡胶等，后者有离心浓缩天然胶乳、膏化浓缩的天然胶乳、蒸发浓缩天然胶乳及其他特种胶乳等。天然生胶主要用于制造各种轮胎、输送带、工业胶管、胶鞋等，浓缩天然胶乳则主要用于制造浸渍制品、海绵制品、输液胶管、铸模制品、地毯、胶粘剂与涂料等。

Natural stopping coasting 惯性尚航　停车后，任船舶依靠自身惯性运动继续尚动的停船方式。

Natural ventilation 自然通风　系指不需要动力进行的通风。通过空气管道和/或其他适当设计的开口进行空气流通。 Natural ventilation means ventilation that is not power generated. An air flow is supplied by air ducts and/or other adequately designed openings.

Natural vibration 自然振动(固有振动)

Natural-circulation boiler 自然循环锅炉　靠水和水汽混合物的比重差造成锅水循环的锅炉。

Nautical chart 海图　是地图的一种，是以表示海洋区域制图现象的一种地图。海图中数量最多的航海图，除其内容不同于陆图外，在表示方法上也有许多不同于陆图的地方，如：多采用墨卡托投影；没有固定的比例尺系列；深度起算面不用平均海面而用特定的深度基准面；分幅沿海岸或航线进行；在邻幅间还有重叠部分；有自己特有的编号方法；符号设计原则和制图综合原则也略有不同；为保证航行安全，航海图出版后要不间断地进行修正，始终保持现势性等。港湾图，主要显示港湾、锚地，图上详细表示沿岸地形、港湾设施、海底地貌、助航设备、航行障碍物等要素，比例尺通常大于 1∶10 万，供舰船进出港口、锚地，驻泊避风，在港湾布、扫水雷、障碍，组织合成军队登陆和抗登陆作战、训练、研究和实施港湾建设等使用。

Nautical chart or nautical publication 海图或航海出版物 系指专用的图或书，或支持这种图或书的经特殊编辑的数据库，由政府主管当局，经授权的水文局或其他相关的政府机构正式颁布，用于满足航海要求。
Nautical chart or nautical publication is a special-purpose map or book, or a specially compiled database from which such a map or book is derived, that is issued officially by or on the authority of a Government, authorized Hydrographic Office or other relevant government institution and is designed to meet the requirements of marine navigation.

Naval Authority 海军主管机关 由海军指定的负责提供有关采购和舰船支持法规的一个或多个机构。海军主管机关还可负责确定适当的标准、审查和认证。它可以是政府部门、法定当局、LR 或具有适当身份的独立组织。 Naval Authority means an authority or authorities nominated by the Owner responsible for providing regulation associated with procurement and support of the ship. The Naval Authority may also be responsible for identifying appropriate standards, auditing and certification. The Naval Authority could be a Government department, Statutory Authority, LR or an independent organization with appropriate standing.

Naval ship 军用船 各类军事用途舰船的统称。

Navigable waters 可航行水域 系指：(1) 1988 年 12 月 27 日总统行政法令（5928 号）中定义的美国领海和内海。(2) 具有与经修正的联邦水域防污染法令第 502 条中相同含义的水域。 Navigable waters means: (1) the territorial sea of the United States (as defined in Presidential Proclamation 5928 of December 27 1988) and the internal waters of the United States. (2) has the same meaning as in section 502 of the Federal Water Pollution Control Act, as amended.

Navigating and manoeuvring workstation 导航和操纵工作站 位于雷达控制台后方 350mm 的船舶主要工作站。它被设想为以坐姿/站立姿势工作，并具有最佳的可视性并集成了信息和操作设备。它旨在安全有效地操纵船舶，特别是在需要快速的操作程序时。
Navigating and manoeuvring workstation is main workstation for ship's command 350 mm behind the radar console. It is conceived for working in seated/standing position with optimum visibility and integrated presentation of information and operating equipment. It is designed to operate the ship safely and efficiently, in particular when a fast sequence of action is required.

Navigating radar 导航雷达 是用来探测水上目标的方位和距离的设备。它不受气候影响，可以全天候引导船舶进出港口、码头和海上安全航行。导航雷达按波段可分为 X 波段（探测信号波长 3cm）导航雷达和 S 波段（探测信号波长 10 cm）；按发射功率可分为中型（发射功率大于 25 kW）导航雷达和小型（发射功率小于 10 kW）导航雷达。见图 N-1。

图 N-1 导航雷达
Figure N-1 Navigating radar

Navigating system 导航系统 是确保海上运输和海洋工程船舶航行安全准确地引导船舶按预定航线迅速地到达目的地的系统。其由磁罗经（标准罗经、操舵罗经、备用标准罗经）、陀螺罗经［包括航向记录器、方位分罗经、航向分罗经（按需配置）］、自动操舵仪、舵角指示器、推进器转速指示器、回转速指示器、9GHz 雷达/带 ARPA、9GHz 雷达、3GHz 雷达/带 ARPA、电子海图显示器和信息系统、无线电导航（罗兰 C、台卡导航仪、欧米伽导航仪、卫星导航系统、全球定位系统）、测深仪系统、计程仪系统（水压式计程仪、电磁式计程仪、多普勒计程仪、声相计程仪）、驾驶室声响接收系统、白昼信号灯、自动识别系统、航行数据记录仪、新船远程识别和跟踪系统、船钟和风速风向仪等组成。

Navigation 导航（航行） 系指：(1) 计划船舶的航线以及确定船舶的位置和航向，并执行航向变化和航速改变的任务;(2)相对于水域和交通来决定,执行并维持航向和航速的所有相关任务。 Navigation means: (1) planning of the ship's route and determination of position and course of the ship, execution of course alterations and speed changes; (2) all tasks relevant for deciding, executing and maintaining course and speed in relation to waters and traffic.

Navigation cathode ray tube 导航分显示器 又称导航CRT。是利用阴极射线管将电信号转换成直观的图像或字符的电子仪器。导航CRT作为导航计算机的输入输出设备之一,它包括垂直定时系统、水平定时系统、存储器、字符发生器与多路转换器、电视同步电路发生器和计算机标志与中断逻辑电路等部分。它用来概括地显示导航系统的工作情况,显示的数据有经、纬度、时间,航向,航速以及船舶相对于测线的位置。也可作为连机的诊断工具,以八进制、十进制、浮点或ASCII码的形式显示存储器的内容。总之,导航CRT可以监视整个导航系统的工作情况。

Navigation commading telephone system 航行指挥电话系统 它是驾驶室与航行有关部位之间进行指挥通话使用的,是保证船舶安全航行的联络通信的系统。这种通信系统一般采用共电式电话或增音式声力电话。总机安装在驾驶室内,其容量根据需装用户数来确定。电话机安装在各航行有关部位,如机舱、应急发电机室、船长室、轮机长室、船首、船尾和消防控制站等处。

Navigation control data 航行控制数据 选择在任务站上为手动和自动控制船舶运动提供信息的任务。 Navigation control data means a task that provides information for the manual and automatic control of the ship's movement on a task station.

Navigation laboratory 航海实习室 安装有航海仪器和航海分显示器,供学员从事航海实习用的工作舱室。该舱室应靠近海图室和驾驶室。

Navigation light (NL) 航行灯 系指下列灯:(1) COLREG第21条定义的桅灯、舷灯、拖带灯、环照灯、闪光灯;(2) COLREG第23条对气垫船要求的环照黄色闪光灯;(3) COLREG第34(b)条要求的操纵号灯。光源包括灯、底座、安装位置和减少照明角度的装置。Navigation light (NL) means the following lights: (1) mast-head light, sidelights, stern-light, towing light, all-round light, flashing light as defined in Rule 21 of COLREG; (2) all-round flashing yellow light required for air-cushion vessels by Rule 23 of COLREG; and (3) manoeuvring light required by Rule 34 (b) of COLREG. The light source includes lamps, its housing, placing and means for delimiting the angle of lighting.

Navigation light controller (NLC) 航行灯控制器 系指能对航行灯进行操作控制的装置。Navigation light controller (NLC) means a device enabling operational control of a navigation light.

Navigation lights 号灯 根据海上避碰规则配置的,用以表明夜间船舶动态或作业状态的灯具。

Navigation radar 导航雷达 其工作原理:发射的电波以直线传播,当电波遇到物标时便会产生反波,若测出来回的电波传播时间,就测出了雷达至物标的距离(无线电波传播的速度不变)。用雷达测向常用的是最大信号测向法。在观测时一旦发现物标时,只要转动天线,使显示在荧光屏上回波信号最强,则这时天线所指方向就是物标方向角。其缺点是:因为在方向性图轴线方向偏离不大时,与最大辐射方向相比功率密度变化很小,不易区分,所以这种测向法较难获得高精确度的方位。利用雷达进行导航首要的工作是测定船位。导航雷达可测定距离与方位角,再根据海图上已知的陆标即可使用地文航海的类似方向进行定位并导航。所以导航雷达定位方法可用距离-方位法,两距法、两距离加方位等方法。

Navigation room 导航定位室 装有各种高精度无线电定位仪器和卫星导航定位仪器,进行导航定位的工作舱室。

Navigation season of the Saint Lawrence Seaway 圣劳伦斯航道航行期 系指圣劳伦斯航道开发公司和管理当局根据天气、结冰状态或船舶航行交通的需要,指定在年度中该航道可通航的时期。 Navigation season of the Saint Lawrence Seaway means the annual period designated by the corporation and the Manager, that is appropriate to weather and ice conditions or vessel traffic demands, during which the Seaway is open for navigation.

Navigation system 导航系统 在保证航行安全的前提下,按预定的航路到达预定的水域或地点的系统或设备。导航系统可分为地文导航、天文导航、无线电导航以及由罗经、计程仪等组成的船位推算导航系统。地文导航——利用地面目标(在海上则利用岛岸目标),并已知目标位置,测得方位、距离或夹角等来确定船位。天文导航——利用自然天体,已知天体的星历,测量其方位、高度,进而推算船位。无线电导航——利用无线电波,测得相对于已知导航台站的方位、距离、距离差等来确定船位。推算法导航——通过导航仪器提供的航速、航向,不断修正风流压的影响来推算船位。惯性导航——利用水平平台上的东西向及南北向的加速计,来不断确定船位。借助导航系统,可以准确地控制和掌握船舶自身的航向、航速运动状态所处地理位置和水域的情况,以避免碰撞、搁浅、触礁、迷航等事故。对于舰船来说,导航系统还担负着向现代化武器系统提供舰船各项运动参数的任务,离开了导航系统武器系统将无法发挥作用。

Navigation workstation 航行工作站 驾驶人员可以针对水流和交通情况,决定、实施和保持航向和航速的所有有关任务的工作站。航行工作站的仪表和控制器应使驾驶人员能从事以下工作:(1)分析交通状况;(2)监视船位、航向、航迹、航速、时间、螺旋桨转速和螺距、舵角、水深、转向角速率以及风速和风向;(3)改变航向和航速;(4)实现船内和船外通信;(5)发出和接收声信号;(6)控制航行灯;(7)监视和应答航行警报;(8)确认驾驶人员保持良好状态和值班警觉意识;及(9)记录航行数据。 Navigation workstation means a workstation at which the navigator may carry out all tasks relevant for deciding, executing and maintaining course and speed in relation to waters and traffic. The instrumentation and controls at the navigation workstation should allow the navigator to: (1) analyse the traffic situation; (2) monitor position, course, track, speed, time, propeller revolutions and pitch, rudder angle, depth of water, rate of turn, and wind speed and direction; (3) alter course and speed; (4) effect internal and external communications; (5) give and receive sound signals; (6) control navigational lights; (7) monitor and acknowledge navigational alarms; (8) confirm his well-being and watch-keeping awareness; and (9) record navigational data.

Navigational bridge visibility 驾驶室可视范围 系指船长不小于55m,1998年7月1日或以后建造的船舶,从其驾驶位置上所见的海面视域。并应满足下列要求:(1)从驾驶位置上所见的海面视域,在所有吃水、纵倾和甲板货状态下,自船首前方至任何一舷10°范围内均不应有超过两倍船长或500 m(取其小者)的遮挡。(2)在驾驶室外正横前方从驾驶室所见的海面视域内任何货物、起货装置或其他障碍物造成的盲视扇形区域的遮挡不应超过10°,盲视扇形区域的总弧度不应超过20°。在盲视区之间的可视扇形区域应至少为5°。但在上述的视区内每一单独盲视区均应不超过5°。(3)从驾驶位置上所见的水平视域应延伸为一个不小于22.5°的扇面,即从正前方至船舶任何一舷不小于22.5°的正横后方向。(4)从每一驾驶室翼台所见的水平视域应延伸一个不小于22.5°的扇面,即从船首另一侧至少45°经正前方,然后从正前方至180°至船舶相同一舷的正尾方向。(5)从主操舵位置所见的水平视域应延伸为一个从正前方至船舶每一舷至少60°的扇面。(6)船舷应从驾驶室翼桥上可见。(7)驾驶室甲板以上的驾驶室正前窗下部边缘高度应尽可能保持低位。任何情况下,该下部边缘均不得成为障碍而遮挡上述的前视视域。(8)驾驶室正前窗上部边缘应有一个水平前视范围,当船舶在大浪中纵摇时,应确保驾驶人员在驾驶位置上有一个自驾驶室甲板以上1 800 mm的视觉高度。主管机关如认为1 800 mm视觉高度不合理和不切实际,可允许降低该视觉高度,但不应少于1 600 mm。(9)驾驶室窗应满足下列要求:①为有助于避免反射,驾驶室正前窗应自垂直平面顶部向外倾斜,其角度不小于10°且不大于25°;②驾驶室窗之间的框架应保持最低数量,且不应设置在任何工作台的正前方;③不应设置偏振及着色玻璃窗;④不管天气状况如何,在任何时候至少两扇驾驶室正前窗应提供清晰的视域。此外根据驾驶室形状,其他一些的窗也应提供清晰的视域。 Navigational bridge visibility means a view of the sea surface from the conning position for ships of not less than 55 m in length, constructed on or after 1 July 1998, and shall meet the following requirements: (1) The view of the sea surface from the conning position shall not be obscured by more than two ship lengths, or 500 m, whichever is the less, forward of the bow to 10° on either side under all conditions of draught, trim and deck cargo; (2) No blind sector, caused by cargo, cargo gear or other obstructions outside of the wheelhouse forward of the beam which obstructs the view of the sea surface as seen from the conning position, shall exceed 10°. The total arc of blind sectors shall not exceed 20°. The clear sectors between blind sectors shall be at least 5°. However, in the view described in (1), each individual blind sector shall not exceed 5°; (3) The horizontal field of vision from the conning position shall extend over an arc of not less than 22.5°, that is from right ahead to not less than 22.5° abaft the beam on either side of the ship; (4) From each bridge wing the horizontal field of vision shall extend over an arc at least 22.5°, that is from at least 45° on the opposite bow through right ahead and then from right ahead to right astern through 180° on the same side of the ship; (5) From the main steering position, the horizontal field of vision shall extend over an arc from right ahead to at least 60° on each side of the ship; (6) The ship's side shall be visible from the bridge wing; (7) The height of the lower edge of the navigation bridge front windows above the bridge deck shall be kept as low as possible. In no case shall the lower edge present an obstruction to the forward view as described in this regulation; (8) The upper edge of the navigation bridge front windows shall allow a forward view of the horizon, for a person with a height of eye of 1 800 mm above the bridge deck at the conning position, when the ship is pitching in heavy seas. The Administration, if satisfied that a 1 800 mm height of eye is unreasonable and impractical, may allow reduction of the height of eye but not to less than 1 600 mm;(9) Windows shall meet the following requirements: ①To help avoid reflections, the bridge front windows shall be inclined from the vertical plane top out, at an angle of not less than 10° and not more than 25°. ② Framing between navigation bridge windows shall be kept to a minimum and not be installed immediately forward of any work station. ③Polarized and tinted windows shall not be fitted. ④A clear view through at least two of the navigation bridge front windows and, depending on the bridge configuration, an additional number of clear-view windows shall be provided at all times, regardless of weather conditions.

Navigational bridge 导航桥楼 系指船舶进行移动作业的地方。 Navigational bridge is the area of the bridge where transit operation is performed.

Navigational draft 外形吃水 系指包括在任何附体放射性下突出物在内的船舶最低点至水面的垂直距离。

Navigational repair/voyage repair 航修 对船舶营运期间影响航行的较小故障、海损或一般事故的紧急修理。为了尽量不影响船舶营运,常在船舶停港作业时间中进行,必要时由航修站或船厂派员随船进行抢修。

Navigation 导航 系指计划船舶的航线以及确定船舶的位置和艏向,并执行艏向变化和航速改变的任务。 Navigation is a process of planning, reading, and controlling the movement of a ship from one place to another.

Navigator 驾驶人员 系指导航、操作桥楼设备并操纵船舶的航行人员。 Navigator is a person navigating, operating bridge equipment and manoeuvring the ship.

NAVTEX NAVTEX接收机 是由IMO和IHO制定的国际海事无线电直通电报系统。它能自动接收广播信息,如航行、气象警报和24h值班的搜索营救报警信号。 NAVTEX is an international maritime radio telex system sponsored by IMO and IHO, which automatically receives the broadcast telex information such as navigational, meteorological warnings and search and rescue (SAR) alerts on a 24-hour watch basis.

NAVTEX 海事安全信息广播和接收系统 NAVTEX is a system for broadcast and reception of maritime safety information.

Navy 海军 舰船的操作者,海军也可以作为船东。 Navy is the operator of the ship. The Navy may also be the Owner.

Navy area theater ballistic missile defense (NATB-

MD) 海军区域战区弹道导弹防御 海军区域战区弹道导弹防御计划涉及装备标准导弹的宙斯盾巡洋舰和驱逐舰。将对宙斯盾武器系统进行改进,使其能用AN/SPY-1雷达探测、跟踪战区弹道导弹,并用SM-2 Block ⅣA导弹将其拦截。SM-2 Block ⅣA导弹由SM-2 Block Ⅳ导弹改进而来,增加了一个红外导引头,改进了战斗部、自动驾驶仪和引信,使其能拦截战区弹道导弹。这一计划包括宙斯盾探测战区弹道导弹能力的验证。目前还计划研制用户作战评估系统(UOES),以使早期参战单位能进行试验,并在国家紧急情况下提供有限的大气层内防御战术弹道导弹的能力。

Navy navigation satellite system 海军导航卫星系统 测定"子午卫星"的多普勒频移进行导航的系统。美国首先在1958年开始研究卫星导航,1964年研究成功提交美国海军使用。定名为"子午卫星系统",后来又称为海军导航卫星系统,于1967年公开为民用。海军导航卫星系统的主要组成部分有:海军导航卫星网、跟踪海军导航卫星并测定和预报卫星轨道的全套地面设备,接收多普勒信息及进行导航计算的船舶导航设备。

Navy ocean surveillance satellite (NOSS) 海军卫星监视系统

NC cutting machine 数控切割机 系指用数字程序驱动机床运动,随着机床运动时,随机附带的切割工具对物体进行切割。这种机电一体化的切割机就称之为数控切割机。

Near-coastal voyage 近岸航行(近岸航程,沿海航行) 系指一国主管机关所定义的在该国海岸附近的航行。 Near-coastal voyage means a voyage in the vicinity of the coast of a State as defined by the Administration of the State.

Nearest land 距最近陆地 系指距按照国际法划定领土所属领海的基线,但下述情况除外:就MARPOL附则Ⅴ而言,在澳大利亚东北海面"距最近陆地",系指距澳大利亚海岸下述各点的连线:自南纬11°00′东经142°08′的一点起,至南纬10°35′东经141°55′的一点;然后至南纬10°00′东经142°00′的一点,然后至南纬9°10′东经143°52′的一点;然后至南纬9°00′东经144°30′的一点;然后至南纬10°41′东经145°00′的一点;然后至南纬13°00′东经145°00′的一点;然后至南纬15°00′东经146°00′的一点;然后至南纬17°30′东经147°00′的一点;然后至南纬21°00′东经152°55′的一点;然后至南纬24°30′东经154°00′的一点;然后至澳大利亚海岸南纬24°42′东经153°15′的一点所画的一条连线。 The term "from the nearest land" means from the baseline from which the territorial sea of the territory in question is established in accordance with international law, except that, for the purposes of the Annex to MARPOL,"from the nearest land"off the north-eastern coast of Australia shall mean from a line drawn from a point on the coast of Australia in: latitude 11°00′S, longitude 142°08′E to a point in latitude 10°35′S, longitude 141°55′E, thence to a point latitude 10°00′S, longitude 142°00′E, thence to a point latitude 09°10′S, longitude 143°52′E, thence to a point latitude 09°00′S, longitude 144°30′E, thence to a point latitude 10°41′S, longitude 145°00′E, thence to a point latitude 13°00′S, longitude 145°00′E, thence to a point latitude 15°00′S, longitude 146°00′E, thence to a point latitude 17°30′S, longitude 147°00′E, thence to a point latitude 21°00′S, longitude 152°55′E, thence to a point latitude 24°30′S, longitude 154°00′E, thence to a point on the coast of Australia in latitude 24°42′S, longitude 153°15′E.

Neck bearing 下舵承 设置在舵杆穿出船体外板处的一般仅承受径向力的轴承。

Needle bearing 滚针轴承

Needle lift 针阀升程 喷油器中阀针的行程值。

Needle valve 针杆阀

Negative freight 负运价 系指航运企业不仅不从货主或代理商处收取货物运输费用,反而向货主或代理商支付费用以抢夺货源。负运价不等于没有钱赚。在负运价的情况下赚钱,一般船东会这样操作,在起始港口付一部分钱给货主,货物抵达目的港后,再从收货人哪里收取附加费和代理费,包括币值变动附加费、港口附加费、码头拥挤费等,其中代理费属于可收可不收的范畴。若收取的这些费用高于支付给货主的费用,就能够赚钱。当然,这首先要看目的港的客户是否接受。另一种方法,就是拼箱,将几个没有装满的集装箱里的货物拼到一个集装箱里,成本会降低很多,也可在一定程度上弥补负运价带来的损失。

Negative list 负面清单 是"负面清单管理模式"的简称。负面清单管理相当于投资领域的"黑名单",列明了企业不能投资的领域和产业。学术上的说法是,凡是针对外资的与国民待遇、最惠国待遇不符的管理措施,或业绩要求、高管要求等方面的管理限制措施,均以清单方式列明。在《中国(上海)自由贸易试验区总体方案》中,总体方案提出探索建立负面清单管理模式。上海自贸区负面清单按国民经济行业分类,列出18个门类,89个大类,419个中类,1069个小类,190条管理措施。约占试验区内1069个小经济行业分类的17.8%。对于未列入负面清单的外商投资一般项目,最快4天企业可以拿到营业执照、机构代码和税务登记等。

Negative vertical bending moment 负垂向弯矩 为中垂弯矩。 Negative vertical bending moment is a sagging moment.

Negative vibration isolation 消极隔振

Negotiated bidding 谈判招标 又叫议标,它是非公开的,是一种非竞争性的招标。这种招标由招标人物色几家客商直接进行全同谈判,谈判成功,交易达成。

Negotiating bank 议付银行 是根据开证行在议付信用证中的授权,买进受益人提交的汇票和单据的银行。议付系指由被授权议付的银行对汇票/单据付出对价。如果只审查单据而不支付对价并不构成议付。议付行是准备向受益人购买信用证下单据的银行,议付行可以是通知行或其他被指定的愿意议付该信用证的银行,一般是出口商所在地银行。

Negotiation 协商 是通过结构化地交换相关信息而改进有关共同观点或共同计划的过程,也即协商是协作双方为达成共识而减少不一致性或不确定性的过程。

Negotiation leadership 谈判领导人员 企业委派专门人员,或者是从有关人员中选择合适人选担任。

Negotiation quota 磋商限额 即对个别限额外的某些产品在原则上规定一定的配额数,如出口超过该额数,双方按一定程序进行磋商谋求解决,在双方未达成一致意见前,进口国可单方面实行进口限制。

Neighboring ship suction 邻船吸力 两船对遇或并列航行过分靠近时,由于其间水流加速、压力下降产生的易致互相吸引碰撞的互吸力。

NERR = National Estuarine Research Reserve [美]国家港湾搜索预备队

Nested 内嵌的

Net barter terms of trade(NBTT) 价格贸易条件 又称净贸易条件。其只考虑到进出口价格变化对一国进口能力产生的影响,它是以一国一定时期的出口价格指数与同期进口价格指数的比率来反映。具体计算方法为: N = (PX/PM) × 100% 其中:N—贸易条件;PX—出口商品价格指数;PM—进口商品价格指数。价格贸易条件大于1说明该国出口商品价格上升幅度超过进口商品价格的上升幅度,或出口商品价格下降幅度小于进口商品价格下降幅度(即出口商品价格相对于进口商品价格在上涨),该国的贸易条件改善了,有利于该国进口能力的提高。价格贸易条件小于1说明该国出口商品价格上升幅度低于进口商品价格的上升幅度,或出口商品价格下降幅度大于进口商品价格下降幅度(即出口商品价格相对于进口商品价格在下降),该国的贸易条件恶化了,进口能力受到削弱。

Net capacity cargo carrying/Capacity cargo deadweight 净载重吨

Net cost of averting a fatality(NCAF) 避免死亡/人的净成本 一种成本测算方式。即附加成本减去风

Net crane rate 净装卸率

Net efficiency 净效率

Net face plate thickness t_f (mm) 面板净厚度 t_f 系指普通扶强材或主要支撑构件面板的净厚，mm。t_f means a net face plate thickness, in mm, of ordinary stiffener or primary supporting member.

Net horse power 净功率

Net load 有效载荷(净载荷,净负载)

Net moment of inertia I_s 净惯性矩 I_s 系指：(1)普通扶强材或主要支撑构件(视具体情况而定,不计带板)对平行于板的中和轴的净惯性矩，cm^4；(2)扶强材或主要支撑构件连同宽度 s 船壳板对其与板材平行的中和轴的净惯性矩，cm^4。Net moment of inertia I_s：(1) I_s, in cm^4 of ordinary stiffener or primary supporting member, as the case may be, without attached plating, around its neutral axis parallel to the plating; (2) a net moment of inertia, in cm^4, of the stiffener or the primary supporting member, with attached shell plating of width s, about its neutral axis parallel to the plating.

Net polar moment of inertia I_p 净极惯性矩 I_p 系指普通扶强材或主要支撑构件与板材连接处的净极惯性矩(cm^4)。Net polar moment of inertia I_p means a net polar moment of inertia, in cm^4, of ordinary stiffener or primary supporting member, about its connection to plating.

Net present value (NPV) 净现值 也称投资的净现金流量,是财务管理上的一个概念。其处理方法是将投资的未来现金流量,按行业基准折现率或者其他设定的折现率全部折现成投资始日的价值。

Net pressure load 有效压力载荷

Net rack 轻便式物品架 安装在床铺旁舱壁上,供旅客存放轻便物品用的架子。

Net section modulus of attached plating Z 带板的净剖面模数 Z 系指普通扶强材或主要支撑构件连同宽度 b_p 带板的净剖面模数，cm^3。Net section modulus of attached plating Z means a net section modulus, in cm^3, of an ordinary stiffener or a primary supporting member with attached plating of width b_p.

Net sectional area A_s (cm^2) 净横剖面积 A_s 系指扶强材或主要支撑构件连同宽度 s 带板的净横剖面积(cm^2)。Net sectional area A_s means a net sectional area, in cm^2, of the stiffener or the primary supporting member, with attached plating of width s.

Net sectional moment of inertia I_w 净剖面惯性矩 I_w 系指普通扶强材或主要支撑构件与板材连接处的净剖面惯性矩，cm^4。Net sectional moment of inertia I_w means a net sectional moment of inertia, in cm^4, of ordinary stiffener or primary supporting member about its connection to plating.

Net shear sectional area A_{sh} (cm^2) 净剪切横剖面积 A_{sh} 系指扶强材或主要支撑构件净剪切横剖面积(cm^2)。Net shear sectional area A_{sh} means a net shear sectional area, in cm^2, of the stiffener or the primary supporting member.

Net shipping mass m_N 净装运质量 m_N 净装运质量 m_N 应包括制造厂商随船提供的所有永久性及散装的设备,但装运的材料除外,此质量以吨(t)或千克(kg)表示。Net shipping mass m_N shall include all permanent and loose equipment delivered with the craft by the manufacturer, but no shipping materials. The mass is expressed in tons, t, or kilograms, kg.

Net sounder (trawlink) 网位仪 安装在拖网上用以指示网口位置和开度的声学探测装置。

Net thickness 纯厚度(净厚度) 系指不包括腐蚀部分或其他附加部分的厚度。Net thickness is the thickness that does not include corrosion addition and voluntary addition.

Net thickness approach 净厚度方法 其原理是：(1)将新造船阶段的强度计算所用的厚度与营运阶段可接受的最小厚度直接联系起来；(2)使结构的腐蚀状态在船舶整个使用寿命周期能够被明确。The philosophy behind the net thickness approach is to: (1) provide a direct link between the thickness used for strength calculations during the new building stage and the minimum thickness accepted during the operational phase; (2) enable the status of the structure with respect to corrosion to be clearly ascertained throughout the life of the ship.

Net thickness of the plating attached t_p (mm) 带板的净厚度 t_p 系指普通扶强材或主要支撑构件带板的净厚度(mm)。Net thickness of the plating attached t_p means a net thickness, in mm, of the plating attached to an ordinary stiffener or a primary supporting member.

Net thickness offered $t_{net-offered}$ 净提供厚度 为总提供厚度减去 t_C 所得到的净厚度，mm。按下式计算：
$t_{net-offered} = t_{gros-offered} - t_C = t_{as-built} - t_{voluntary-addition} - t_C$。
Net thickness offered $t_{net-offered}$ is the net thickness, in mm, obtained by subtracting t_C from the gross thickness offered, as follows: $t_{net-offered} = t_{gross-offered} - t_C = t_{as-built} - t_{voluntary-addition} - t_C$.

Net thickness required $t_{\text{net-required}}$ 净要求厚度 为按规范满足所有结构强度要求所需的净厚度,mm,并四舍五入到最靠近的 0.5 mm。Net thickness required $t_{\text{net-required}}$ is the net thickness, in mm, as required by the Rules that satisfy all the structural strength requirements, rounded to the closest half millimeter.

Net ton 净吨

Net tonnage(NT) 净吨位 系指1969年国际船舶吨位丈量公约各项规定确定的船舶有效容积。即扣除不能用来载货或载客的处所后得到的船舶可营运容积。不能用来载货或载客的处所包括船员的生活起居处所、船舶机械和装置处所、航行设备处所、安全设备处所和压载处所等。根据我国《法定检验规则》,净吨位的计算,以丈量得到的各载货处所的总容积为基准,并考虑乘客定额以及船舶总吨位和船型尺度,用公式计算求得。向船舶征收的各种港口使用费,如船舶吨税、船舶港务费、引航费、码头费、系解缆费、船舶服务费等,一般以船舶净吨位作为计算的依据。

Net web thickness t_w(mm) 腹板净厚 t_w 系指普通扶强材或主要支撑构件的腹板净厚(mm)。Net web thickness t_w means a Net web thickness, in mm, of ordinary stiffener or primary supporting member.

Net weight 实物净重 系指:(1)除去直接接触商品的包装材料质量后,所表示出来的纯商品的质量;(2)货物本身的质量。按净重计系指以毛重扣除皮重的质量作为交货时计量数量和计价的基础。

Net width between inner walls 坞内净宽 浮船坞内坞墙之间或其上固定建筑物之间的最小距离。

Netsonde 网口探鱼仪 利用声波在水中的传播特性,将仪器探测部分安装在拖网上纲上,用来测量网深和探测网口上下或周围鱼群的仪器。

Network 网络 系指计算机之间进行数据传递和交换的通信网。Network means a communication net for data transfer and exchange between computers.

Network equipment 网络设备 是连接到网络中的物理实体。网络设备的种类繁多,且与日俱增。基本的网络设备有:计算机(无论其为个人电脑或服务器)、集线器、交换机、网桥、路由器、网关、网络接口卡(NIC)、无线接入点(WAP)、打印机和调制解调器。由于路由器、交换机、集线器等网络设备是由许多紧密的电子元件组成的,因此务必要将它们放置在干燥的地方,以防止潮湿引起电路短路。

Network equipment provider(NEP) 网络设备供应商

Network information center(NIC) 网络信息中心 系指为用户提供网络信息资源服务的网络技术管理机构。其主要职责是对网上资源进行管理和协调。

Network interface card(NIC) 网卡 是一个物理设备,类似于网关。通过它,网络中的任何设备都可以发送和接收数据帧。又称网络适配器。台式机一般采用内置网卡来连接网络。网卡是局域网中最基本的部件之一,它是连接计算机与网络的硬件设备。无论是双绞线连接、同轴电缆连接还是光纤连接,都必须借助于网卡才能实现数据的通信。它的主要技术参数为带宽、总线方式、电气接口方式等。它的基本功能为:从并行到串行的数据转换,数据包的装配和拆装,网络存取控制,数据缓存和网络信号。目前主要是8位和16位网卡。不同网络中,网络接口卡的名称是不同的。例如,以太网中称为以太网接口卡,令牌环网中称为令牌环网接口卡等。

Network operation system(NOS) 网络操作系统 是网络的心脏和灵魂,是向网络计算机提供服务的特殊的操作系统。它在计算机操作系统下工作,使计算机操作系统增加了操作所需的能力。

Network printer(NP) 网络打印机 系指通过打印服务器(内置或者外置)将打印机作为独立的设备接入局域网或者Internet,从而使打印机摆脱一直以来作为电脑外设的附属地位,使之成为网络中的独立成员,成为一个可与其并驾齐驱的网络节点和信息管理与输出终端,其他成员可以直接访问使用该打印机。网络打印机的硬件构成分为打印部分和网络部分,用户在选购时考虑性能不妨也从这两个方面出发,这两方面的性能共同决定了整机的性能。网络打印机一般具有管理和监视软件,通过管理软件可以从远程查看和干预打印任务,对打印机的配置参数进行设定,绝大部分的网络打印管理软件都是基于WEB方式的,简单快捷。

Network processor(NP) 网络处理器 根据国际网络处理器会议(Network Processors Conference)的定义表明,网络处理器是一种可编程器件,它特定的应用于通信领域的各种任务,比如数据包处理、协议分析、路由查找、声音/数据的汇聚、防火墙、QoS等。网络处理器件内部通常由若干个微码处理器和若干硬件协处理器组成,多个微码处理器在网络处理器内部 并行处理,通过预先编制的微码来控制处理流程。网络处理器(NP)是专门为处理数据包而设计的可编程处理器,能够直接完成网络数据处理的一般性任务。

Network protocol 网络协议 系指为计算机网络中进行数据交换而建立的规则、标准或约定的集合。例如,网络中一个微机用户和一个大型主机的操作人员进行通信,由于这两个数据终端所用字符集不同,因此操作人员所输入的命令彼此不认识。为了能进行通信,规定每个终端都要将各自字符集中的字符先变换为标准字符集的字符后,才进入网络传送,到达目的终端之后,再变换为该终端字符集的字符。当然,对于不相容终端,除了需变换字符集字符外,其他特性,如显示格式、行长、行数、屏幕滚动方式等也需作相应的变换。

Network software 网络软件 系指在计算机网络环境中,用于支持数据通信和各种网络活动的软件。连入计算机网络的系统,通常根据系统本身的特点、能力和服务对象,配置不同的网络应用系统。其目的是为了本机用户共享网中其他系统的资源,或是为了把本机系统的功能和资源提供给网中其他用户使用。为此,每个计算机网络都制订一套全网共同遵守的网络协议,并要求网中每个主机系统配置相应的协议软件,以确保网中不同系统之间能够可靠、有效地相互通信和合作。

Network terminal 网络终端 是一种专用于网络计算环境下的终端设备。与 PC 相比没有硬盘、软驱、光驱等存储设备。

Neutral angle of propeller blade 无推力转角 可调螺距螺旋桨叶由设计螺距到推力为零时叶柄转动的角度。

Neutral axis 中性轴(中和轴)

Neutral axis zone 中和轴区域 对散货船而言,只包括甲板区域和舱底区域之间的板材,如:(1) 舷侧外板;(2) 内壳板(如有)。 For bulk carriers, the neutral axis zone includes the plating only of the items between the deck zone and the bottom zone, as for example: (1) side shell plating; (2) inner hull plating, if any.

Neutral packing 中性包装 系指商品和内外均无标明生产国别、地名和厂商的名称,也不标明商标或牌号的包装。主要是为了适应国外市场的特殊要求,如转口销售,有可能你的买家不是最终的买家,只是一个中间商,所以要使用中性包装。或者为了打破某些进口国家的关税和非关税壁垒,并适应交易的特殊需要,它是出口国家厂商加强对外竞销和扩大出口的一种手段。

Neutral rudder angle 压舵角 舵叶对称面在船舶保持直航时的位置与零舵角的位置之间的夹角。

Neutral State 中立国 系指在发生武装冲突时,对交战的任何一方都不采取敌对行动的国家,分为战时中立国和永久中立国两种。中立国的权利和义务在海牙第 5 公约《中立国和人民在陆战中的权利和义务公约》和第 13 公约《关于中立国在海战中的权利和义务公约》中做出了规定。在冲突里中立属于一种主权受限,和不结盟是有区别的。许多国家在第二次世界大战时宣布中立。但二战中大部分中立国都被占领,二战末期全欧洲只剩爱尔兰、瑞典、瑞士(包含列支敦士登)、土耳其仍维持中立。后来有人质疑这几个国家当时的中立性,认为爱尔兰秘密参与同盟国,瑞典、瑞士和土耳其与纳粹德国有微妙的关系。

Neutron bomb 中子弹 是一种以高能中子辐射为重要杀伤力的低质量氢弹,其目的是杀伤敌方人员,对建筑物和设施破坏相对较少,带来的放射性污染较低,尽管从未曾在实战中使用过,但军事家仍将其称为战场上的"战神"。技术拥有国:中国、美国、俄罗斯、法国(中子弹:没有氢弹技术的免谈)。见图 N-2。

图 N-2 中子弹
Figure N-2 Neutron bomb

New concept FPSO 新概念浮式生产储/卸油船 为了解决环境污染问题,提高 FPSO 系统的环保性能,采用液化石油气(LPG)作为清洁能源的液化石油气-浮式生产储/卸油船(LPG-FPSO)和采用液化天然气(LNG)作为清洁能源的液化天然气-浮式生产储/卸油船(LNG-FPSO)等,是开发海上边际油气田及回收海上油田放空气的有效工具。此外还有将油气钻井设备并入浮式生产储/卸油船的浮式钻探、生产储/卸油船(FPSO)等新概念浮式生产储/卸油船。

New concept high speed Ro/Ro passenger ship with three hulls 新概念高速三体客滚船 一种采用三体型船体,以氢燃料电池为主要能源的新型高速客滚船。其首部形似高速动车的车头具有优良的流体动力性能,能有效减少水阻力和风阻力,防止波浪拍击。船体根据不同航速和航程的需求,可调整侧体相对位置形成有利兴波干扰,从而提高航速。同时该船采用新型贯穿全船的 3 个压载筒体来取代传统压载水舱,以 3 条海水可在舱内流动的大型管道"水流箱"来达到无压载水的目的并通过中部筒体的贯通供水,提高喷水推进效率。其还在合适的位置布置自动开合式水密横舱壁,增强船舶的抗沉性。该船型从根本上解决了三体船停靠难和实现了船舶"零压载"的问题。

New concept VLCC 新概念 VLCC 采用 V 型船体和创新的货舱设计,使零压载成为可能,采用 V 型船体后还减少了往返行程的潮湿表面,降低了方型系数,提高了船舶能效。该船型配备了两台双燃料低速主机,以液化天然气作为燃料,以轻柴油作为引燃燃料。而且该船型的辅助锅炉以回收货油的蒸发气（挥发性有机物）为燃料，为货油泵生产蒸发气。该船型与传统的 VLCC 相比，可使二氧化碳减排 54%，氮氧化物减排 80% 以上，硫化物和颗粒物质减排 95%，能耗减少 25%。新概念 VLCC 上层建筑结构甲板上设有两个符合国际海事组织（IMO）要求的 C 型压力罐，可存储 13 500 m^3 的液化天然气，足以推动该船航行 25 000 n mile（海里）。该船型还可采用双燃料中速主机和纯天然气主机。该船型具有三大优势：比传统油船更环保、具有在现有技术基础上创新与优化的可能性；与采用燃油为燃料的传统油船相比更加经济。

New container 新集装箱 系指在国际集装箱安全公约生效时或生效后开始制造的集装箱。 New container means a container the construction of which was commenced on or after the date of entry into force of the International Convention for safe container.

New Energy and Industrial Technology Development Organization（NEDO） ［日］新能源产业技术综合开发机构

New generation TLP 第三代张力腿平台

New installation 新设备（渔船） 系指在国际渔船安全公约生效之日或生效之日后全部安装于船上的设备。 New installation means an installation wholly installed on board a vessel on or after the date of entry into force of the International Convention for the safety of fishing vessels.

New installation 新装置 系指 2009 年香港国际安全与环境无害化拆船公约生效日以后船上安装的系统、设备、装置或船上的其他材料。 New installation means the installation of systems, equipment, insulation, or other material on a ship after the date on which Hong Kong International convention for the safe and environmentally sound cycling of ships, 2009 enters into force.

New lifeboat release and retrieval system 新救生艇释放和回收系统 系指按经 MSC.320(89) 决议修正的 LAS 规则的第Ⅳ章 4.4.7.6 认可的救生艇释放和回收系统。 New lifeboat release and retrieval system is a lifeboat release and retrieval system that has been approved in accordance with paragraph 4.4.7.6 of chapter Ⅳ of the LAS Code, as amended by resolution MSC.320(89).

New mandatory ship reporting system "In the Galapagos particularly sensitive sea area(GALREP)" 加拉帕戈斯特殊敏感海域（PASS）（GALREP）中新的强制性船舶报告系统

New Panamax 新巴拿马型船 系指尺度大于巴拿马型船和巴拿马加大型船，且符合巴拿马运河新船闸尺度和吃水限制，即船长 366 m、船宽 49 m、热带淡水（TFW）吃水 15.2 m 限制的所有船舶。 New Panamax mean all vessels with dimensions greater than Panamax or Panamax Plus that comply with the size and draft limitations of the new locks; namely, 366 m in length by 49 m in beam by 15.2 m TFW draft.

New passenger ship 新客船 系指：(1)在 2016 年 1 月 1 日或以后签订建造合同，或无建造合同但在 2016 年 1 月 1 日或以后安放龙骨或处于类似阶段的船舶；或 (2)在 2016 年 1 月 1 日以后 2 年或 2 年以上交船的船舶。 New passenger ship is a passenger ship: (1) for which the building contract is placed, or in the absence of a building contract, the keel of which is laid, or which is in a similar stage of construction, on or after 1 January 2016; or (2) the delivery of which is two years or more after 1 January 2016.

New ship 新船 系指：(1)在 MARPOL 附则Ⅳ生效之日或以后签订建造合同，或无建造合同但在 MARPOL 附则Ⅳ生效之日或以后安放龙骨或处于类似阶段的船舶。(2)在 MARPOL 附则Ⅳ生效之日后经过 3 年或 3 年以上交船的船舶。(3)①在 2009 年香港国际安全与环境无害化拆船公约生效日或以后签订建造合同的；②如无建造合同，在 MARPOL 公约生效日 6 个月或以后，已铺设龙骨或处于类似建造阶段的，③在 2009 年香港国际安全与环境无害化拆船公约生效日 30 个月或以后交船的船舶。(4)船级社规范生效之日及以后签订建造合同的新建船舶。(5)2013 年 1 月 1 日或以后签订建造合同。(6)如无建造合同，2013 年 7 月 1 日或以后安放龙骨或处于类似建造阶段。(7)2015 年 7 月 1 日或以后交付的船舶。 New ship means a ship: (1) for which the building contract is placed, or in the absence of a building contract, the keel of which is laid, or which is at a similar stage of construction, on or after the date of entry into force of this Annex. (2) the delivery of which is three years or more after the date of entry into force of this Annex. (3)①for which the building contract is placed on or after the entry into force of this Convention; or ②in the absence of a building contract, the keel of which is laid or which is at a similar stage of construction on or after six months after the entry into force of this Convention; or ③the delivery of which is on or after 30 months after the entry into force of Hong Kong International convention for the safe and environmentally sound cycling of ships, 2009. (4) a ship contracted for construction on or after the date of entry into force of an society Rules. (5) for which the building contract is placed on or after 1 January 2013. (6) in the absence of a building contract, the keel of which is laid or which is at a similar stage of construction on or after 1 July 2013. (7) the delivery of which is on or after 1 July 2015.

New special trade passenger ship 新特种业务客船 系指在特种业务客船协定生效之日或以后安放龙骨或处于相应建造阶段的特种业务客船, 或者在特种业务协议生效之日或以后第一次运送特种业务旅客的客船。 New special trade passenger ship means a special trade passenger ship the keel of which is laid, or which is at a similar stage of construction, on or after the date of entry into force of the special trade agreement, or a ship which carries special trade passengers for the first time on or after that date.

New technologies and treatments (ballast water discharge) 新技术和新方法(压载水排放) 系指如有新开发的技术和方法, 可以替代或与现有方法联合使用。这样的处理方法包括升温、过滤、紫外光消毒等经认可的其他方法。 New technologies and treatments mean that if new and emergent treatments and technologies prove viable, these may substitute for, or be used in conjunction with, current options. Such treatments could include thermal methods, filtration, disinfection including ultraviolet light, and other such means as approved.

New type propeller 新型推进器 包括导管螺旋桨、半浸式螺旋桨和吊舱式推进器。

New unit 新建平台 系指船级社规范生效之日及以后安放龙骨或处于相应建造阶段的平台。 New unit means a unit the keel of which is laid or which is at a similar stage of construction, on or after the date on which the Society Rules enter into force.

New vessel (fishing vessel) 新船(渔船) 系指下列在托列莫利诺斯国际渔船安全公约生效之日以后建造的渔船:(1)签订了建造或重大改建合同的渔船。(2)在托列莫利诺斯国际渔船安全公约生效之日以前签订建造或重大改建合同, 而交船期是在托列莫利诺斯国际渔船安全公约生效之日以后3年或3年以上的渔船;(3)无建造合同的渔船:①安放了龙骨的渔船;②某特定船舶结构同型的渔船;③分段装配已经开始, 至少使用了50 t 或整个结构材料总数的1%, 以少者为准。 New vessel is a vessel for which, on or after the date of entry into force of the Torremolinos international convention for the safety of fishing vessels: (1) the building or major conversion contract is placed; (2) the building or major conversion contract has been placed before the date of entry into force of the Torremolinos international convention for the safety of fishing vessels, and which is delivered three years or more after the date of such entry into force; (3) in the absence of a building contract: ①the keel is laid; or ②construction identifiable with a specific vessel begins; or ③assembly has commenced composing at least 50 t or 1% of the estimated mass of all structural material whichever is less.

New York Protocol of 1946 1946年纽约议定书 系指修正"国际鸦片公约""限制麻醉药品制造和管制麻醉药品运销公约""取缔非法贩运危险毒品公约"的议定书。 New York Protocol of 1946 means a protocol amending the International opium conventions, the Convention for limiting the manufacture and regulating the distribution of narcotic drugs and the Convention for the suppression of illicit traffic in dangerous drugs.

NEWBUILDCON 新标准造船合同 波罗的海国际航运公会(BINCO)为广大船东和造船厂推出新的造船合同。以代替单纯由造船厂提供的造船合同文本开展新船建造计划, 并能为船东和造船厂达成协议的坚实平台, 同时保证当事双方之间达成的最终协议包含任何造船合同都应有的关键条款。新标准造船合同共分为两部分:第一部分是框架图, 该框架图将造船合同内最主要的条款组成了34个表格, 并用简洁的文字进行表述;第二部分与我国通用造船合同相似。从内容上看新标准造船合同涵盖了一般造船合同所必需的所有内容和条款, 但其编排的次序、叙述的方式均有所改变, 并强化了环保概念, 增加了环保条款, 要求造船厂提供"绿色证书"及危险材料清单, 并明确"适用大、小造船厂及所有船型";此外在纠纷解决条款中, 该合同还加入了专家判定条款。从表面上看, 波罗的海国际航运公会(BINCO)在制订新标准合同时, 虽然在某些方面对造船厂做出了让步, 但更多仅对船东有利条款的存在使其离了公正性, 而这一切源于该合同起草时未能广泛征求并采纳广大造船厂的建议和意见。由于缺少造船厂的声音, 由波罗的海国际航运公会(BINCO)制定的该份标准合同在船舶、财务、生产、交船、法律以及杂项等6大方面均存在偏袒船东的条款, 其中不少属于"霸王条款", 因此, 该合同受到冷遇, 特别在亚洲, 进而导致"水土不服"。

Nicaragua canal 尼加拉瓜运河 运河线路从加勒比海侧的蓬塔戈尔达河, 沿杜乐河进入尼加拉瓜湖, 再到太平洋岸的布里托河口, 全长约276 km, 是巴拿马运河长度的3倍。航道规划底部宽度为230~520 m, 水深27.6 m, 航道的通过能力约9100艘/年, 可通行船舶最大吨位达40万吨, 计划5年内建成。40万吨的大船, 从委内瑞拉开往中国, 将节约两个月的航行时间;而从上海到巴尔的摩, 走尼加拉瓜运河航线要比苏伊士运河短4 000 km, 比绕过好望角短7 500 km。修建期间预计创造5万就业岗位,运营期间预计创造20万就业岗位。工程将耗资500亿美元,预定2019年竣工,2020年投入使用。该运河规划的长度将是巴拿马运河的3倍多。中国富商王靖拥有的HKND集团在香港成立尼加拉瓜大运河开发公司承建运河并获得100年经营权利。2014年12月22日,尼加拉瓜运河在尼加拉瓜里瓦斯市开工建设。尼加拉瓜运河完工后将联通大西洋和太平洋,贯通美国东海岸到亚洲、南美洲东岸到亚洲、美国西海岸到欧洲等繁忙、重要的海洋贸易航线。尼加拉瓜运河工程的构想由来已久,一旦建成可能成为全球航运的主航线。随着船舶大型化趋势加剧,中美洲唯一的一条跨洋运河巴拿马运河的通航能力已得严重不足。虽然进行了扩建,但目前运营的30万吨级超大型油船(VLCC)、1.8万箱集装箱船、40万吨级超大型矿砂船(VLOC)依然不能通过,而且,各国船舶常年排队通行巴拿马运河的状况也难有明显改善,这使得在中美洲开凿第二条运河成为迫切需要。尼加拉瓜运河的诞生将顺应船舶超大型化发展的趋势,为船东提供更多、更经济的选择。尼加拉瓜运河建成并开通,将为相关各方带来什么?尼加拉瓜运河建成并开通就惠及全球,特别对北美东海岸及加勒比海地区,如美国、墨西哥等国家的集装箱贸易以及石油、天然气、粮食等资源出口带来机遇,同时促进运河两岸国家的经济发展、自由贸易区建设、交通基础设施改善等。尼加拉瓜运河将成为国际主航线,除惠及周边地区和国家外,也对世界所有海运大国都有好处。由于几乎目前所有类型的大船都能通过该运河,因此,亚洲-美东、美西-欧洲、亚洲-南美洲及美东-美西的航线都可通过该运河。尼加拉瓜运河建成并开通将对三个国家的益处最大,那就是尼加拉瓜、美国和中国。从航运企业的角度来说,尼加拉瓜运河建成并开通则将大大降低其运输成本。首先,相关航线的航程缩短将降低运营成本及时间成本;其次,尼加拉瓜运河允许超大型船舶通行,使得航运企业的规模效益战略有了"用武之地";再次,尼加拉瓜运河建成并开通后,航运企业将免受在巴拿马运河排队一等多天的拥堵之苦;最后,尼加拉瓜运河的出现,将使得巴拿马运河不再拥有一家独大的"优势",竞争导致通行费用降低值得期望。对于尼加拉瓜运河来说,如果能够按照当前的设计实现投产,将与拓宽后的巴拿马运河形成较大竞争,特别是在大型集装箱船和液化气船运输方面。同时,尼加拉瓜运河还具备巴拿马运河没有的优势,即能够通过包括VLGC、好望角型散货船及VLOC等在内的全球原油及铁矿石的主力船型。世界上运河和交通要道的开通,一般都会使得对应的船型出现。如苏伊士运河对应的苏伊士型油船系指满载状况下吃水不超过17.5 m,可以通过苏伊士运河的最大油船。该船型以装载10万吨原油为其设计载重量,载重量一般不超过15万吨。巴拿马运河对应的巴拿马型船系指其最大尺寸为总长294.1 m,型宽32.3 m,吃水12 m的船舶。一般载重量在6.5万吨左右。好望角航线对应的好望角型散货船,其最大尺寸为总长280 m、型宽45 m、载重量为18万吨的船舶。尼加拉瓜运河同样也会出现尼加拉瓜型船,其肯定比现有的苏伊士型、巴拿马型、超巴拿马型、好望角型船舶都大。可以说,尼加拉瓜运河的建设为船型的未来发展创造了巨大的想象空间。对于集装箱船来说,近年来的大型化趋势明显,超大型集装箱船也为航运企业带来了看得见、摸得着的成本效益,目前中、韩等国的船企已经在为设计建造22 000 TEU集装箱船做准备。而富有企业也在为取得更高的运营效益推动集装箱船进一步大型化,推出22 000~24 000 TEU集装箱船。可以预见,尼加拉瓜运河开通将使这种庞然大物布置在美洲航线成为可能。而这又将反过来进一步增强航运企业订造更大集装箱船的愿望。此外,由于尼加拉瓜运河将满足目前最大油船、液化气船、散货船、矿砂船通过的条件,因此,相关的超大型船舶的需求还将加大。此外,尼加拉瓜运河开通将在一定程度上打破巴拿马运河在船舶大小及数量限制方面的瓶颈,使通过中美洲运河的亚洲-美东、美西-欧洲、亚洲-南美洲及美东-美西的航线在全球运输市场中占据更重要的位置。相关地区,特别是美东地区的港口应建设更深的码头,以迎接更大型船舶的到来;造船界则应认真分析运河开通后的船型需求趋势,以提前应对,研发更适应需求的船型,以免在未来市场陷入被动。

Niche area 不利区域 系指由于不同的水动力,受涂层系统磨损或破损影响,或施涂不当或未施涂,而可能更容易受生物污底影响的船上区域,例如通海阀箱、艉部推力器、螺旋桨轴、进水口格栅、干坞支撑条等。

Niche area mean areas on a ship that may be more susceptible to biofouling due to different hydrodynamic forces, susceptibility to coating system wear or damage, or being inadequately, or not, painted, e. g., sea chests, bow thrusters, propeller shafts, inlet gratings, dry-dock support strips, etc.

Nickel alloy steel 镍合金钢

Nickel aluminum bronze 镍铝青铜

Nickel base alloy 镍基合金 系指在650~1000 ℃高温下有较高的强度与一定的抗氧化腐蚀能力等综合性能的一类合金。按照主要性能又细分为镍基耐热合金,镍基耐蚀合金,镍基耐磨合金,镍基精密合金与镍基形状记忆合金等。高温合金按照基体的不同,分为:铁基高温合金,镍基高温合金与钴基高温合金。其中镍基高温合金简称镍基合金。

Nickel brass 镍黄铜(镍铜锌合金)

Nickel bronze　镍青铜

Nickel chrome steel　镍铬钢　俗称不锈钢。实际应用中，常将耐弱腐蚀介质腐蚀的钢称为不锈钢，而将耐化学介质腐蚀的钢称为耐酸钢。由于两者在化学成分上的差异，前者不一定耐化学介质腐蚀，而后者则一般均具有不锈性。不锈钢的耐蚀性取决于钢中所含的合金元素。铬是使不锈钢获得耐蚀性的基本元素。

Nickel- chrome steel (nickel chromium steel) (Ni Cr S)　镍-铬钢

Nickel content　镍含量

Nickel iron battery　镍铁蓄电池组

Nickel manganese steel　镍锰钢

Nickel plating　镀镍

Nickel silver　镍银合金

Nickel steel (NS，ns)　镍钢

Nickel steel tube　镍钢管

Nickelage　镀镍

Nickeline　锡基密封合金(铜镍锌合金)

Nickel-plated　镀镍的

Nicrite　镍铬耐热合金

Nippon Electric Company　日本电气公司　即日本电气股份有限公司(日文:日本电气株式会社)简称日本电气或日电或 NEC，是一家跨国信息技术公司，总部位于日本东京港区(Minato-Ku)。NEC 为商业企业、通信服务以及政府提供信息技术(IT)和网络产品。

NISA = National Invasive Species Act　[美]国家有害入侵物种法

Niskin sterile water sampler　尼斯金无菌采水器　系指尼斯金设计的预先杀菌的容量为1.3升的聚乙烯袋式采水器。

Nitrile-butadiene rubber　丁腈橡胶

Nitrogen generator　氮气发生器　系指能生成高纯度氮气，其中 CO_2 含量以体积计不超过5%的装置。 Nitrogen generator is a device which is capable of delivering high purity nitrogen with CO_2 content not exceeding 5% by volume.

Nitrogen generator system　氮气发生器系统　系指采用使压缩空气通过空心纤维半渗透膜或吸附材料分离空气与其组成气体的方式而获得惰性气体的系统。氮气发生器系统包括一个供气系统和任意数目的薄膜或吸附件，这些薄膜或吸附件所必须达到的额定容量应至少以体积表示的船舶的最大排气量的125%。 Nitrogen generator system means a system produced inert gas by separating air into its component gases by passing compressed air through a bundle of hollow fibers, semi-permeable membranes or absorber materials. A nitrogen generator consists of a feed air treatment system and any number of membrane or absorber modules in parallel necessary to meet the required capacity which is to be at least 125% of the maximum discharge capacity of the ship expressed as a volume.

Nitrogen Oxide (NO_x) emissions　氮氧化物 NO_x 排放　系指氮氧化物总排放量，按 NO_2 总加权排放量计算，并以2008 NO_x 技术规则所规定的相关试验循环和测量方法确定。 Nitrogen Oxide(NO_x) emissions means the total emission of nitrogen oxides, calculated as the total weighted emission of NO_2 and determined using the relevant test cycles and measurement methods as specified in the 2008 NO_x Code.

Nitrogen pumping system　液氮泵送系统

Nitrogen storage　液氮储存装置

NLS　有毒液体物质　NLS means Noxious Liquid Substances.

NLS certificate　NLS 证书　系指根据 MARPOL73/78 的规定所颁发的"国际防止散装运输有毒液体物质污染证书"。 NLS certificate means an international pollution prevention certificate for the carriage of noxious liquid substances in bulk issued under MARPOL73/78.

NLS tanker　有毒液体物质(NLS)液货船　系指经建造或改建用于散装运输有毒液体物质货物的船舶。包括 MARPOL 附则 I 定义的核准用于散装运输全部或部分有毒液体物质货物的船舶。 NLS tanker means a ship constructed or adapted to carry a cargo of Noxious Liquid Substances in bulk and includes an "oil tanker" as defined in Annex I of MARPOL when certified to carry a cargo or part cargo of Noxious Liquid Substances in bulk.

NLS = Noxious Liquid Substances　有毒液体物质.

No more favorable treatment (NMFT)　一视同仁的责任

NOAA = National Oceanic and Atmospheric Administration　[美]国家海洋和大气局

Node　节点　系指数据通信联络的内部连接点。 Node means a point of interconnection to a data communication link.

No-decompression diving　不减压潜水　通常在深度不大，水下作业时间不长的情况下，当水底工作结束后，潜水员可不经减压，直接上浮出水，而且不致发生减压病的潜水。

Nodular cast iron　球墨铸铁

Nodular graphite　球状石墨

Nodular iron casting　球墨铸铁件

Noise　噪声(杂声，杂波)　是发声体做无规则振

动时发出的声音。声音由物体振动引起,以波的形式在一定的介质(如固体、液体、气体)中进行传播,而通常所说的噪声污染系指人为造成的噪声。从生理学观点来看,凡是干扰人们休息、学习和工作的声音,即不需要的声音,统称为噪声。当噪声对人及周围环境造成不良影响时,就形成噪声污染。产业革命以来,各种机械设备的创造和使用,给人类带来了繁荣和进步,但同时也产生了越来越多而且越来越强的噪声。噪声污染对人、动物、仪器仪表以及建筑物均构成危害,其危害程度主要取决于噪声的频率、强度及暴露时间。

Noise attenuator 噪声衰减器

Noise control measures 噪声控制措施 系指降低噪声级的办法。常用的船上噪声控制措施包括:(1)结构噪声控制;(2)吸声;(3)隔声;(4)消声;(5)隔振;(6)阻尼敷层;(7)浮动地板;(8)机舱噪声控制;(9)通风管路系统低噪声设计等。

Noise cover 消声罩

Noise elimination 消声

Noise enclosure 隔声罩

Noise generator 噪声发生器

Noise identification 噪声识别

Noise level 噪声级 系指根据 ISO 2923(1996)测量所得的 A 加权声压级。 Noise level means the A-weighted sound pressure level measured in accordance with ISO 2923.

Noise measuring set 噪声测量仪

Noise ranging sonar 噪声测距声呐

Noise reduction coating 减噪涂层

Noise signature 噪声特征

Noise sound index(NSI) 噪声指数 系指评价气垫船噪声的烦恼程度的标准。

Noise source 噪声源

Noise source by fluid dynamics 流体动力噪声源

Noise source by machinery 机械噪声源

Noise source on board ship 船上噪声源 系指船上各种产生噪声的来源地。其包括(1)机械设备噪声源——主机以及发电机组、空调、风机、液压设备和带泵的设备等辅机;(2)流体动力噪声源——螺旋桨脉动压力、空泡、兴波、涡流等;(3)空气动力噪声源——发动机进排气、通风系统进排气和涡流等;(4)外部噪声源——主要是靠码头、货物装卸等噪声;(5)其他噪声源——如广播、鸣笛和乘客脚步声及开关门声等。

Noise statistics 噪声统计

Noise suppressor 消声器

Noise survey report 噪声检验报告 系指每艘船舶按船上噪声等级规则的规定在船上备有的一份噪声检验报告。 Noise survey report means a noise survey report made for each ship in accordance with the Code on Noise Levels on Board Ships.

Noise susceptibility 噪声感受性(噪声灵敏性)

Noise test 噪声试验

Noise trial 噪声特性测定试验

Noiseless 无噪声的(静的)

Noiseless running 无声运转

Noise-modulated 噪声调制的

Noisiness 噪度 系指中心频率为 1 000 Hz 的频带上,声级为 40 dB 的噪声,感觉噪度定义为 1 noy。

Nokia Corporation 诺基亚公司 系指总部位于芬兰埃斯波,主要从事移动通信产品生产的跨国公司。诺基亚公司成立于 1865 年,当时以造纸为主业,后来逐步向胶鞋、轮胎、电缆等领域扩展,最后发展成为一家手机制造商。自 1996 年以来,诺基亚公司连续 14 年占据市场份额第一。面对新操作系统的智能手机的崛起,诺基亚公司全球手机销量第一的地位在 2011 年第二季被苹果及三星双双超越。

No-load 空载(无负荷)

Nominal 额定的(名义上的,不同的,轻微的)

Nominal addendum 标称齿顶高

Nominal application rate 额定施放率 在泡沫灭火系统中系指单位面积的额定流速。Nominal application rate is the nominal flow rate per area.

Nominal diameter 公称直径

Nominal dimension 公称尺寸(名义尺寸)

Nominal dry film thickness(NDFT) 名义干膜厚度 NDFT is the nominal dry film thickness.

Nominal filling rate 额定填充率 系指在泡沫灭火系统中的额定泡沫生产量与被保护处所的面积之比,以 m/min 为单位。 Nominal filling rate is the ratio of nominal foam production to the area, i.e., expressed in m/min.

Nominal filling time 额定填充时间 系指在泡沫灭火系统中被保护处所的高度与额定充注率之比,以 min 为单位。 Nominal filling time is the ratio of the height of the protected space to the nominal filling rate, i.e., expressed in minutes.

Nominal foam expansion ratio 额定泡沫发泡倍数 系指在泡沫灭火系统中所生成泡沫的体积与产生这些泡沫的泡沫溶液体积之比。 Nominal foam expansion ratio is the ratio of the volume of foam to the volume of foam solution from which it was made.

Nominal foam production 额定泡沫生产量 系指在泡沫灭火系统中单位时间产生的泡沫量,即额定流速乘以额定泡沫发泡倍数,以 m³/min 为单位。 Nominal foam production is the volume of foam produced per time unit, i. e., nominal flow rate times nominal foam expansion ratio, expressed in m³/min.

Nominal horse power 名义功率

Nominal load 标称载荷(额定负载)

Nominal outside diameter 公称外径

Nominal pitch 标称螺距 变螺距螺旋桨叶面上有代表性的某一处螺距。径向变螺距螺旋桨常取半径 $0.7R$ 处螺距,符号写作 $P_{0.7R}$。

Nominal pitch ratio 标称螺距比

Nominal pressure 标定压力 系指最大许用工作压力。 Nominal pressure means the maximum permissible working pressure.

Nominal pressure for piping 管系的标定压力 系指管系中最大允许工作压力。标定压力应根据下列条件确定:①内压:内压 P_{nint} 应取下列较小者: $P_{nint} \leqslant P_{sth}/4$ 或 $P_{nint} \leqslant P_{lth}/2.5$。式中, P_{sth}——短期液压试验的爆破压力,MPa; P_{lth}——长期(> 100 000h)液压试验的爆破压力,MPa。②外压:外压 $P_{next} \leqslant P_{col}/3$ 式中, P_{col}——管子破损压力,MPa。 Nominal pressure for piping means the maximum permissible working pressure. The nominal pressure is to be determined from the following conditions: ①Internal pressure: for an internal pressure P_{nint}, the following is to be taken, whichever is smaller: $P_{nint} \leqslant P_{sth}/4$ or $P_{nint} \leqslant P_{lth}/2.5$; where: P_{sth}—short-term hydrostatic test failure pressure, in MPa; P_{lth}—long-term hydrostatic test failure pressure(100 000 h), in MPa. ②External pressure: For an external pressure: $P_{next} \leqslant P_{col}/3$; where: P_{col}—pipe collapse pressure, in MPa.

Nominal pressure of the boiler with super-heater 带过热器锅炉的公称压力 制造厂或使用人设定的过热器出口的最高蒸汽压力,是过热器安全阀门的调整标准压力。 Nominal pressure of the boiler with super-heater is the maximum steam pressure of super-heater outlet designed by the manufacturer or the owner, and is used the standard for setting the safety valve of super-heater.

Nominal rating 额定值(标准规格)

Nominal slip ratio 标称螺距比 系指螺旋桨滑距对螺距的比值。 $S_R = [P - (V_A/n)]/P = 1 - V_A/nP$,式中, V_A——进速; n——螺旋桨转速; P——面螺距或标称螺距。

Nominal speed 公称速度 系指在锚链抛入海中 82.5 m(3 节锚链长)不到达海底的状态下提升 55 m(2 节锚链长)时的平均速度。 Nominal speed is the average speed of recovery of 55 m(two lengths) of anchor chain cables when 82.5 m (three lengths) of the cables are submerged and freely suspended at commencement of lifting.

Nominal stress 公称应力(标称应力,名义应力) 是不包括结构不连续和焊道形状引起的应力集中的应力,在考虑位置处的剖面计算。 Nominal stress is global stress calculated in a sectional area, disregarding the local stress-raising effects of the structural discontinuities, weld bead shape, etc.

Nominal stress 名义应力 是在结构构件中考虑宏观几何影响,但不计由于结构不连续和焊缝存在引起的应力集中的应力。 Nominal stress is the stress in a structural component taking into account macro-geometric effects but disregarding the stress concentration due to structural discontinuities and to the presence of welds.

Nominal value 标称值(额定值)

Nominal wake 标称伴流 系指未受螺旋桨影响的轴向伴流。

Nominal wall thickness 公称壁厚

Nominal weight 标称质量

Nominal width 名义宽度

Nominal yield stress R_y (N/mm²) 材料名义屈服应力 系指材料的名义屈服应力。除另有规定者外,应取值为 $235/k$ N/mm²。 Nominal yield stress R_y means nominal yield stress of the material, to be taken equal to $235/k$ N/mm², unless otherwise specified.

Nomination 指定

Non gas free(N. G. F) 未除气 系指在测试当时未充分排出有毒和爆炸气体。 Non gas free(N. G. F) means not sufficiently free at the time if test from toxic and explosive gas.

Non- hazardous area 非危险区域 系指在满足一定条件的前提下视为无危险的区域,即,气体安全区域。 non- hazardous area means an area which is not considered to be hazardous, i. e. gas safe, provided certain conditions are being met.

Non watertight door 非水密门 无密性要求,仅供分隔之用的门。

Non-absorbent material 不吸收性材料

Non-absorbing 不吸收的

Non-actionable subsidy 不可申诉的补贴 系指普遍性实施的补贴和事实上并没有向某些特定企业提供的补贴。包括:(1)不属于特殊补贴的补贴,即属于普遍性的补贴;(2)扶植企业的科研活动、更高水平的教育或建立科研设施所提供的补贴,但属于工业科研项目的扶植不得超过其成本的75%或其竞争开发活动成本的50%;(3)扶植落后地区的经济补贴;(4)为适应新的环境保护要求,扶植改进现有设备所提供的补贴。

但这种补贴仅限于改造成本的20%。上述这些补贴不可诉诸争端解决,尽管如此,却要求缔约方将这类补贴情况提前、及时通知各缔约方,如果有异议,也须磋商解决。

Non-adiabatic 非隔热的

Non-adjacent cabin 相隔舱室 系指不与相邻舱室有共同船体结构(舱壁或者甲板)的舱室。

Non-agreement voluntary export quotas 非协定自动出口配额制

Non-automatic 非自动的

Non-axisymmetric buckling 非轴对称屈曲

Non-burning type steam reservoir 无须燃烧的贮汽器

Non-cavitating(sub-cavitation) 无空泡 水中不存在空泡的情况。但按无空泡情况设计的螺旋桨在一定的操作情况下也可能产生部分空泡。

Non-cavitating hydrofoil 无空泡水翼 不发生空泡现象的水翼。

Non-classed ship 非入级船舶 系指除入级船舶外的船舶。 Non-classed ship is any ship which is not classed.

Non-cohesive cargoes 非黏性货物 系指在干燥时不具有黏性的货物。如:(1)硝酸铵;(2)硝酸铵基化肥(A类、B类和无危害类);(3)硫酸铵;(4)无水硼砂;(5)硝酸钙化肥;(6)蓖麻籽;(7)磷酸二铵;(8)磷酸一铵;(9)氯化钾;(10)钾碱;(11)硝酸钾;(12)硫酸钾;(13)硝酸钠;(14)硝酸钠和硝酸钾的混合物;(15)过磷酸盐;(16)尿素。 Non-cohesive cargoes are the cargo which are non-cohesive when dry:(1)Ammonium nitrate;(2)Ammonium nitrate based fertilizes(type A, type B and non-hazardous);(3)Ammonium sulphate;(4)Borax, Anhydrous;(5)Calcium nitratertilizer;(6)Castor beans;(7)Diammonium phosphate;(8)Monoammonium phosphate;(9)Potassium chloride;(10)Podash;(11)Potassium nitrate;(12)Potassium sulphate;(13)Sodium nitrate;(14)Sodium nitrate and potassium nitrate mixture;(15)Superphosphate;(16)Urea.

Non-cohesive material 非黏性物质 系指在运输期间由于滑动作用而易于移动的干燥物质。 Non-cohesive material means dry materials that readily shift due to sliding during transport.

Non-combustible 不燃的

Non-combustible impurities 不可燃的杂质

Non-combustible material 不燃材料 系指:(1)某种材料加热到约750℃时,既不燃烧,也不发出足以造成自燃的易燃蒸发气体,此系通过规定的试验程序确定的,并取得主管机关的同意。除此以外的任何气体材料,均为可燃材料;(2)一般来说,仅由玻璃、混凝土、陶瓷制品、天然石料、砌石及常用金属和合金材料制成的产品可认为是不燃材料。其不经试验和认可即可使用于船上。 Non-combustible material means:(1)a material which neither burns nor gives off flammable vapours in sufficient quantity for self-ignition when heated to approximately 750 ℃, this being determined to the satisfaction of the Administration by an established test procedure. Any other material is a combustible material;(2)in general, products made only of glass, concrete, ceramic products, natural stone, masonry units, common metals and metal alloys are considered being non-combustible and they may be installed without testing and approval.

Non-condensable gas purger 空气分离器 用以驱除不凝性气体如氧、氢、氮、氯、水汽、油气和其他碳氢化合物等混合气体,提高制冷效应的压力容器。

Non-condension engine 无冷凝器蒸汽机

Non-conformity 不合格 系指:(1)不满足某项要求;(2)有客观证据表明的不满足规定要求的观察情况。 Non-conformity means:(1)non-fulfillment of a requirement;(2)an observed situation where objective evidence indicates the non-fulfilment of a specified requirement.

Non-conformity(NC) 不符合项 系指在审核中发现的,有物证证明的,诸如没有完成符合公司的安全管理体系、ISM规则以及其他相关国际公约和规则的事项。

Non-convention ship 非公约船舶 系指除公约船舶外的船舶。 Non-convention ship is any ship which is not a convention ship.

Non-corrosive metal 耐蚀金属

Non-destroyed test(NDT) 非破坏性试验(无损检测)

Non-destructive detection [non-destructive inspection(NDI), non-destructive test, non-destructive testing] 无损检测(无损探伤,非破坏性检验)

Non-destructive means(non-destructive method) 无损检测法

Non-destructive test standard 无损检验标准

Non-destructive testing(NDT) 无损探伤试验 系指利用目视检验、放射性探伤试验、超声波探伤试验、超声波试验、磁粉探伤试验、渗透试验和其他无损揭示焊缝缺陷和瑕疵的方法。 Non-destructive testing(NDT) means visual inspection, radiographic testing, ultrasonic testing, magnetic particle testing, penetrant testing and other non-destructive methods for revealing defects and irregularities.

Non-destructive testing equipment 无损检测设备（无损探伤设备）

Non-dimensional circulation 无量纲环量数　螺旋桨盘的单位圆周长度上环量对进速的比值。

Non-discrimination 非歧视原则　在世界贸易组织的管辖域内，各成员应公平、公正、平等地、一视同仁地对待其他成员的包括货物、服务、服务提供者或企业、知识产权所有者或持有者等在内的与贸易有关的主体和客体。非歧视原则要求各成员无论在给予优惠待遇方面，还是按规定实施贸易限制方面，都应对所有其他成员一视同仁，即"最惠国待遇"，不应在本国和外国的产品、服务或人员之间造成歧视，要给予他们"国民待遇"。非歧视原则是世界贸易组织各项协定、协议中最重要的原则。在世界贸易组织中，非歧视原则主要是通过最惠国待遇原则和国民待遇原则来实现的。

Non-displacement mode 非排水状态　系指当由非静水力全部或为主地支持船舶重力时船舶正常营运的状态。Non-displacement mode means the normal operational regime of a craft when Non-hydrostatic forces substantially or predominantly support the weight of the craft.

Non-effective superstructure 无效上层建筑　系指位于船中 $0.4L$ 区域以外或长度小于 $0.15L$ 或小于 12 m 的所有上层建筑。Non-effective superstructure are all superstructures being located beyond $0.4L$ amidships or having a length of less than $0.15L$ or less than 12 m.

Non-electrically controlled diesel engine 非电控柴油机　系指没有采用电子控制的柴油机。Non-electrically controlled diesel engine means diesel engine without electronic control system.

Non-elestic coupling 非弹性联轴节

Non-employee-developed technology 非职务技术成果　系指技术职务成果以外的技术成果，即完成技术职务成果的个人自行研究开发，并主要不利用本单位的物质和技术条件所完成的成果。

Non-essential equipment 非重要设备　系指短时间不运转不会对船舶推进和操舵有损害，也不会危及乘客、船员、货物、船舶以及机械安全的设备。Non-essential equipment is that whose temporary disconnection does not impair propulsion and steerability of the ship and does not endanger the safety of passenger, crew, cargo, ship and machinery.

Non-essential piping systems 非基本类管路系统　是为居住和娱乐条件所安装的，属于辅助类设备，包括以下设备：(1)加热系统；(2)空调系统；(3)舰船内卫生设备和淡水系统。Non-essential piping systems are those systems installed for conditions of habitability and recreation, are within the ancillary category and include the following: (1) Heating systems; (2) Air conditioning systems; (3) Domestic sanitary and fresh water systems.

Nonessential system 非重要系统　系指除关键系统和重要系统之外的系统。

Non-expansion steam engine 不膨胀式蒸汽机　蒸汽在气缸内利用压力推动活塞做功而不进行膨胀的往复式蒸汽机。

Non-expansive engine 无膨胀器的蒸汽机

Non-ferrous metals 有色金属　狭义的有色金属又称非铁金属，是铁、锰、铬以外的所有金属的统称。广义的有色金属还包括有色合金。有色合金是以一种有色金属为基体(通常大于50%)，加入一种或几种其他元素而构成的合金。有色金属通常系指除去铁(有时也除去锰和铬)和铁基合金以外的所有金属。有色金属可分为重金属(如铜、铅、锌)、轻金属(如铝、镁)、贵金属(如金、银、铂)及稀有金属(如钨、钼、锗、锂、镧、铀)。简史据考证，铜、铅、锡、汞、金、银和铁都是史前的金属。8000年以前，人类就知道使用铜，7000年前知道有铅，3000年前就在炼丹术中使用汞，2000年前已出现了炼锡技术。92种有色金属在17世纪前被认识和应用的有9种。

Non-fixed price 非固定价格　系指合同当事人在进行交易时，对于合同标的的具体价格不是采用固定规定的方法，而是采用只规定一个确定价格的方法或时间，或暂定一个价格，待日后根据情况予以规定。国际商品市场变化莫测，价格的巨涨暴跌屡见不鲜，为了减少风险，促成交易，提高合同的履约率，在合同价格的规定方面日益采取了一些变通做法，即运用非固定价格（即"活价"）。

Non-governmental organization(NGO) 非政府组织　系指一个不属于政府、不由国家建立的组织。虽然从定义上包含以营利为目的的企业，但该名词一般仅限于非商业化、合法的、与社会文化和环境相关的倡导群体。NGO通常是非营利组织，他们的基金至少有一部分来源于私人捐款。现在该名词的使用一般与联合国或由联合国指派的权威NGO相关。20世纪80年代以来，人们在各种场合越来越多地提及非政府组织（NGO）与非营利组织（NPO），把非政府组织与非营利组织看作在公共管理领域其作用日益重要的新兴组织形式。

Non-hazardous area 非危险区域　预期不可能大量存在爆炸性气体的环境，不要求对电气设备的结构、安装和使用采取专门措施的区域。Non-hazardous area means an area in which an explosive gas atmosphere is not expected to be present in quantities such as to require special precautions for the construction, installation and use of electrical apparatus.

Non-heat elimination equipment 非散热设备 系指在自由空气条件和试验用标准大气条件规定的大气压力(86 kPa～106 kPa)条件下,在温度稳定后测得的表面最热点温度与环境温度之差小于 5 K 的受试设备。

Non-indigenous aquatic nuisance prevention and control act(NANPCA) (美国)1990 年防止和控制外来水生物污染法令

Non-indigenous species 非本地物种 系指本地范围外的物种,不论其是由人类故意或意外传播的还是通过自然过程传播的。 Non-indigenous species mean any species outside its native range, whether transported intentionally or accidentally by humans or transported through natural processes.

Nonlinear indication 非线性缺陷显示 系指长度为不足厚度 3 倍的圆形或椭圆形的缺陷显示。 Nonlinear indication means an indication of circular or elliptical shape with a length less than three times the width.

Non-linear interference suppression 非线性干扰抑制 系指非线性连接结构在强辐射场的激励下产生并向外辐射的电磁干扰。非线性连接结构系指金属构件连接部位上电气特性是非线性的,也就是说,它的伏安特性是非线性的。如船上露天甲板上的船体结构,舾装设施以及金属管道、电缆、活动金属扶手、栏杆、链环、起艇架、回转起货机、钢丝绳、门、窗、舱口盖的铰链等。若处理不当,容易形成非线性连接结构。随着船舶航行产生振动,这些连接结构或构件,时而接触,时而断开,类似于电气开关接通或断开一样,在强电磁场中被激励而产生新的频率分量向外辐射。有的露天甲板船舶结构连接部位被腐蚀或氧化,对高频激励形成的干扰源进行检波,而产生很宽的个别频谱分量,通常称它为"锈蚀螺栓"效应。

Non-metallic inclusion 非金属夹杂物

Non-metallic material 非金属材料 由非金属元素或化合物构成的材料。自 19 世纪以来,随着生产和科学技术的进步,尤其是无机化学和有机化学工业的发展,人类以天然的矿物、植物、石油等为原料,制造和合成了许多新型非金属材料,如水泥、人造石墨、特种陶瓷、合成橡胶、合成树脂(塑料)、合成纤维等。这些非金属材料因具有各种优异的性能,为天然的非金属材料和某些金属材料所不及,从而在近代工业中的用途不断扩大,并迅速发展。

Non-oil tanker 非油船 系指除油船外的船舶。 Non-oil tanker means a ship other than oil tankers.

Non-open indication 关闭缺陷显示 系指去除磁粉后不能看见的缺陷显示或者使用透视探像剂的颜色对比检测不出的缺陷显示。 Non-open indication means an indication that is not visually detectable after removal of the magnetic particles or that cannot be detected by the use of contrast dye penetrant.

Non-open top ship 非敞口船 系指位于干舷甲板上露天部分的货舱口及其他开口设有风雨密舱盖的船舶。

Non-opening circle window(fixed circle window) 固定圆窗 装于舱壁或附装于门、天窗上,供透光之用,不能开启的具有一定密性的圆窗。

Non-opening side scuttle 固定舷窗 仅供透光用而不能开启的舷窗。

Non-opening square window(fixed square window) 固定矩形窗 装于舱壁或天窗盖上,能保持风雨密封,仅供透光,不能开启的非圆形窗的统称。

Non-preferential quota 非优惠性配额 系指在进口配额内的商品征收原来的进口税,超过进口配额的商品,征收惩罚性关税。

Non-profit organization(NPO) 非营利组织 就是不以营利为目的的组织结构。这个概念的产生晚于 NGO,大致出现于 20 世纪 80 年代的美国,之后兴盛于全球。

Non-propulsion 非机动推进 系指未设置用于航行目的的推进设备,或已设置的推进机械仅用于侧推、作业操作和拖航时辅助推进等目的。

Non-read rate(NRR) 拒识率 系指不能识别的条码符号数量与条码符号总数量的比值,即不同的条码应用系统对以上指标的要求不同。

Non-release or minimal release of ballast water 不排放或减少排放压载水 系指在更换压载水或其他处理方法不可行的情况下,压载水可以留存在液舱或货舱内。如这样也不可行时,按港口国应急策略,应仅排放少量必要的压载水。 Non-release or minimal release of ballast water means the case in which where ballast water exchange or other treatment options are not possible, ballast water may be retained in tanks or holds. Should this not be possible, the ship should only discharge the minimum essential amount of ballast water in accordance with port State' contingency strategies.

Non-retractable fin stabilizer 非收放型减摇鳍装置(不可收放式减摇鳍,固定式减摇鳍) 固定曝露在船体外的减摇鳍。外伸两舷的鳍板不能伸缩、折叠的减摇鳍装置。

Non-return 止回的(不倒转的)

Non-return device 止回装置

Non-return suction 止回吸入口 只能吸入，防止倒灌的单向流通的吸入口。

Non-return valve 止回阀

Non-reversible engine 不可逆转式发动机

Non-rust steel 不锈钢

Non-rusting 不锈的(防锈的)

Non-sailing boat 非帆艇 并非以风力作为其主要推进手段的艇，其 $A_S < 0.07 \times (m_{LDC})^{2/3}$。Non-sailing boat is the boat for which the primary means of propulsion is other than by wind power, having $A_S < 0.07 \times (m_{LDC})^{2/3}$.

Non-self priming pump 非自吸泵

Non-self propelled vessel 非机动船舶 系指一艘不装推进装置，或虽装有推进装置但在通过运河时不予使用的船舶，也被称为"死船拖带"(dead tow)。Non-self propelled vessel means a vessel which either does have installed means of propulsion, or has installed means of propulsion which does not function during transit. It is also referred to as dead tow.

Non-self-propelled unit 非自航平台 系指无推进装置的平台。Non-self-propelled unit means a unit which is not a self-propelled unit.

Non-solidifying substance 非固化物质 系指不是固化的有毒物质。Non-solidifying substance means a noxious liquid substance, which is not a solidifying substance.

Non-sparking fan 无火花风机(防爆风机) 如果一台风机不论在正常或异常情况下，均不会产生火花，则将其视为无火花风机。Non-sparking fan means a fan which is considered as non-sparking if in either normal or abnormal conditions, it is unlikely to product sparks.

Non-sufficient funds 空头支票 系指出票人在签发支票后，由于存款不足，支票持有人在向付款银行提示支票要求付款时，就会遭到拒付，这种支票称为空头支票。

Non-tariff barrier 非关税壁垒 又称非关税贸易壁垒，系指一国政府采取除关税以外的各种办法，对本国的对外贸易活动进行调节、管理和控制的一切政策与手段的总和，其目的是试图在一定程度上限制进口，以保护国内市场和国内产业的发展。非关税壁垒大致可以分为直接的和间接的两大类：前者是由海关直接对进口商品的数量、品种加以限制，其主要措施有：进口限额制、进口许可证制、"自动"出口限额制、出口许可证制等；后者系指进口国制定严格的条例和标准，间接地限制商品进口，如进口押金制、苛刻的技术标准和卫生检验规定等。

Non-transferable credit 不可转让信用证 系指受益人不能将信用证的可执行权利转让给他人的信用证。

Non-uniform rational B-spline (NURBS) 非均匀有理 B 样条

Non-vnlnerable 非薄脆性

Non-watertight door 非水密门 无密性要求，仅供分隔之用的门。

Noisy running 有杂声的运转

Normal ballast condition 正常压载工况 所指压载(无货物)工况为：(1)压载舱可为满载、部分压载或空舱；(2)任一货舱或可在海上装载压载水的货舱应为空舱；(3)螺旋桨应全浸没；和(4)船舶应尾倾且应不超过 $0.15L_{BP}$，在评估螺旋桨浸深和纵倾时，可用艏艉垂线处吃水。Normal ballast condition is a ballast (no cargo) condition where: (1) the ballast tanks may be full, partially full or empty. (2) any cargo hold or holds adapted for the carriage of water ballast at sea are to be empty; (3) the propeller is to be fully immersed; and (4) the trim is to be by the stern and is not to exceed $0.015L_{BP}$. In the assessment of the propeller immersion and trim, the draughts at the forward and after perpendiculars may be used.

Normal ballast draught 正常压载吃水 T_{bal-n} 正常压载吃水(m)系指装载手册中的正常压载工况下从船中型基线量起的出港吃水。Normal ballast draught T_{bal-n} in m is the draught at departure given for the normal ballast condition in the loading manual, measured from the moulded base line at amidships.

Normal conditions 正常状态 系指与航行操作有关的所有系统和设备在设计极限和环境条件范围内，如气候、交通，都不会造成值班人员超负荷工作。其包括日常操作、工况及在一定程度上要求采取预防性措施或行动，以继续运行的状态。Normal conditions mean that when all systems and equipment related to navigation operate within design limits, and environmental conditions such as weather and traffic do not cause excessive workload to the officer of the watch. Normal conditions include daily operations and working conditions, as well as situations which, to some degree, require precautionary measures or actions for a continued operation.

Normal crew working spaces 船员正常的工作处所 包括进行日常维护任务或对海上作业的机器进行就地控制的处所，并且包括特殊货物处所，如 SOLAS II 中定义的滚装处所和车辆处所，适用于运载车辆的所有类型的船舶。Includes spaces where routine maintenance tasks or local control of machinery operated at sea are undertaken and includes special category spaces like Ro-Ro spaces and vehicle spaces as defined in SOLAS II on all types of ship that carry vehicles.

Normal displacement 正常排水量 系指军用舰船标准排水量加上 50% 的燃料、滑油、备用锅炉水时的排水量。

Normal distribution 正态分布 有对称于平均值的钟形特征的曲线的一种常用理论分布规律。

Normal operating conditions 正常作业工况 系指：(1)平台在作业点上作业或进行其他操作时承受与作业相适应的设计限度内的组合环境载荷和作业载荷的状态。该平台可以浮在海面或坐底。对自升式平台而言，系指在规定的环境条件下，自升式平台满载并升到预定标高进行正常作业时的状态。(2)船上所有与桥楼主要功能有关的系统和设备都在设计限值内工作，且气候条件或交通并不会给操作人员增添过度的工作负担。 Normal operating conditions are: (1) a condition wherein a unit is on location to perform drilling or other related functions, and combined environmental and operational loading are within the appropriate design limits established for such operations. The unit may be either afloat or supported by the sea bed. For the purpose of self-elevating drilling unit, normal operating condition refers to a condition wherein a fully loaded unit is raised to a predetermined elevation above the sea surface for normal drilling operations. (2) when all shipboard systems and equipment related to primary bridge functions operate within design limits, and weather conditions or traffic, do not cause excessive operator workloads.

Normal operating conditions 正常运行条件（正常航行情况） 系指：(1)船舶在正常状态中借助自动驾驶仪或用任何自动控制系统在手动操作下，能以任何艏向，如为地效翼船运行则能以任何允许飞高安全营运的风浪条件；(2)船舶在任何首航向的情况下均能安全航行，不论手动操作、自动驾驶仪辅助操作或借助任何置于正常模式的自动控制系统操作。 Normal operating conditions mean: (1) a condition that the wind and sea conditions in which the craft can safely operate at any heading and if in ground effect, at any allowable altitude while operated manually with auto-pilot assistance or with any automatic control system in normal mode; (2) those in which the craft will safely cruise at any heading while manually operated, auto-pilot assisted operated or operated with any automatic control system in normal mode; and mooring arrangements and towing rope winches.

Normal operation of the main propulsion machinery 主推进机械正常操作 系指使用调速器和所有安全装置时在正常输出工况下的操作。 Normal operation of the main propulsion machinery is an operation at normal output condition, under using the governor and all safety devices.

Normal operational and habitable condition 正常操作和居住条件 系指船舶(平台)作为一个整体，其机器、设施，保证推进的设备和辅助设施、操舵能力、安全航行、消防和防浸水设施、内外通信和信号、脱险通道和应急救生艇绞车，以及设计要求的舒适居住条件，均处于正常工作和发挥效用的状态。 Normal operational and habitable condition is a condition under which the ship as a whole, the machinery, services, means and aids ensuring propulsion, ability to steer, safe navigation, fire and flooding safety, internal and external communication and signals, means of escape, and emergency boat winches, as well as the designed comfortable conditions of habitability are in working order and functioning normally. All electrical services necessary for maintaining the ship in normal operational and habitable conditions are essential services and services for habitability.

Normal perfomance test for coating 涂料的常规性能试验 系指包括以下性能指标的试验：(1)颜色及外观；(2)附着力；(3)黏度；(4)细度；(5)固体含量；(6)体积分数；(7)适用期；(8)贮存稳定性；(9)遮盖力；(10)可溶性盐含量测试；(11)干燥时间；(12)柔韧、弯曲性；(13)耐冲击性；(14)漆膜光泽；(15)漆膜色泽；(16)闪点；(17)铅笔硬度；(18)密度；(19)漆膜厚度；(20)流挂性；(21)耐磨性；(22)杯突试验；(23)对面漆适应性；(24)漆膜孔隙率测量；(25)老化评级；(26)不挥发物中金属锌含量；(27)挥发性有机物定性、定量；(28)防污漆锡总量；(29)防污漆 DDT 含量；(30)涂料中有害元素检测(铅、铬、镉、汞、砷)；(31)涂料成分分析；(32)涂层电化学性能测试；(33)涂层表面张力、表面能、接触角；(34)防污漆铜离子渗出率；(35)防污漆磨蚀率；(36)石油罐导静电涂料电阻率测定；(37)混凝土保护涂层黏结力；(38)抗氯离子渗透性测试(混凝土涂层)；(39)混凝土保护涂层耐碱性；(40)耐水性；(41)耐油性；(42)耐液体介质；(43)耐盐水；(44)耐温变性；(45)耐湿热；(46)耐酸、碱性；(47)抗气泡性；(48)耐浸泡性；(49)防锈漆阴极剥离试验；(50)防污漆与阴极保护相容性；(51)甲板漆防滑性测试(干态、湿态、油润态)；(52)甲板漆耐缆绳磨损测试；(53)工程项目现场涂层测试(涂层厚度、附着力等第三方检测)等。 Normal performance test for coating mean those performance tests included: (1) color and appearance; (2) adhesion; (3) viscosity; (4) fineness of grind; (5) non-volatile matter by weight; (6) non-volatile matter by volume; (7) pot life; (8) package stability;(9) maskage power;(10) determination of water-soluble salts;(11) drying time;(12) flexibility;(13) impact resistance;(14) film gloss;(15) film colour;(16) flash point;(17) hardness by pencil test;(18) density;(19) determination of film thickness;(20) determination of sagging of paint;(21) abrasion resistance;(22) cup test;(23) compatibility with top-coat;(24) porosity of paint films;(25) rating scheme of degradation of coats;(26) zine-content in non-volatile matter;(27) determination of VOC;(28) total tin centent in paint;(29) DDT content in paint;(30) harmful element content of coatings(Pb、Cr、Cd、Hg、As);(31) component analysis of coating;(32) electrochemical test of coat;(33) surface tension, contact angle, surface energy of coatings;(34) leaching of copper ion for antifouling paint;(35) polishing rate for antifouling paint;(36) electrical resistivity of antistatic coating in petroleum tanks;(37) adhesion of concrete paint;(38) leaching rate of chloride ion for concrete paint;(39) alkali resistance of concrete coating;(40) water resistance;(41) oil resistance;(42) liquid medium resistance;(43) salt water resistance;(44) temperature-alternate resistance;(45) heat and humidity resistance;(46) acid/alkali resistance;(47) resistance to blister for anticorrosive paint;(48) resistance to immersion test for anticrrosive paint;(49) cathodic disbondment resistance anticorrosive paint;(50) cathodic disbondment for antifouling paint;(51) determination of slip resistance of antislip deck paint;(52) abrasion test for deck paint; and (53) field test of project (DFT, adhesion surface as third parts), etc.

Normal pressure 标定压力 系指按有关要求的各种最大许用压力。 Normal pressure means the maximum permissible working pressure which should be determined in accordance with the relevant requirements.

Normal strength hull structural steel 普通强度船体结构钢 系指最小屈服强度为 235 N/mm² 的钢材。 Normal strength hull structural steel means the steel having a specified minimum yield stress of 235 N/mm².

Normalizing(N) 正火(N) 系指:(1)将轧制的钢材在 Ar_3 大于临界温度的奥氏体在结晶温度区域进行加热并放置于空气中冷却的热处理工艺。在该工艺中生产的轧制钢材的粒度的大小已被精细化,故其力学性能得以改进,(2)对钢材加热到转化温度,随后进行空气冷却,使得有可能实现晶粒细化和回火的过程。
(1) Normalizing(N) involves heating rolled steel above the critical temperature, Ar_3, and in the lower end of the austenite recrystallzation region followed by air cooling. The process improves the mechanical properties of as rolled steel by refining the grain size;(2)a process of heating steel above the transformation temperature followed by an air cooling to permit recrystallization and tempering.

Normalizing N condition 正火状态 N 系指一种在轧后的单独奥氏体化热处理,这种处理可以细化晶粒大小,改进钢材的力学性能。 Normalizing N condition refers to an additional heating cycle of rolled steel above the critical temperature, Ac3, and in the lower end of the austenite recrystallisation region followed by air cooling. The process improves the mechanical properties of as-rolled steel by refining the grain size.

Normalizing rolling NR 正火轧制 NR 该工序是使材料的最后变形在正火温度范围内实现,导致材料状态相当于正火状态的工序。因此,正火轧制工序可作为正火热处理的一种直接等效方法被接受。 Normalizing rolling NR, also known as controlled rolling, is a rolling procedure in which the final deformation is carried out in the normalizing temperature range, resulting in a material condition generally equivalent to that obtained by normalizing.

North America free trade agreement(NAFTA) 北美自由贸易协定 1989 年,美国和加拿大两国签署了《美加自由贸易协定》(North American Free Trade Agreement)。1991 年 2 月 5 日,美、加、墨三国总统同时宣布,三国政府代表从同年 6 月开始就一项三边自由贸易协定正式展开谈判。经过 14 个月的谈判,1992 年 8 月 12 日,美国、加拿大及墨西哥三国签署了一项三边自由贸易协定——北美自由贸易协定。

North Sea area 北海区域 系指北海本身,包括下列界限之内的海区:(1)北纬 62°以南和西经 4°以东的北海海域;(2)斯卡格拉克海峡,南至斯卡晏角以东北纬 57°44.8′;和(3)英吉利海峡及其西经 5°以东和北纬 48°30′以北的入口处。 North Sea area means the North Sea proper including seas therein with the boundary between:(1) the North Sea southwards of latitude 62° N and eastwards of longitude 4° W;(2) the Skagerrak, the southern limit of which is determined east of the Skaw by latitude 57° 44.8′ N; and(3) the English Channel and its approaches eastwards of longitude 5° W and northwards of latitude 48° 30′ N.

North up display 北朝上显示 系指采用电罗经输入(或等效)且北处于图上端的方位角稳定的显示。 North up display means an azimuth stabilized presentation which uses the gyro input(or equivalent) and north is uppermost on the presentation.

North West European waters 西北欧水域 包括北海及其入海口、爱尔兰海和其入海口、凯尔特海、英吉利海峡和其入海口，及紧靠爱尔兰西部的东北大西洋部分，该区域以下述各点连线为界：法国海岸上的北纬48°27′；北纬48°27′；西经6°25′；北纬49°52′；西经7°44′；北纬50°30′；西经12°；北纬56°30′；西经12°；北纬62°；西经3°；挪威海岸上的北纬62°；丹麦和瑞典海岸上的北纬57°44.8′。 North West European waters include the North Sea and its approaches, the Irish Sea and its approaches, the Celtic Sea, the English Channel and its approaches and part of the North East Atlantic immediately to the west of Ireland. The area is bounded by lines joining the following points: 48°27′N on the French coast; 48°27′N; 006°25′W; 49°52′N; 007°44′W; 50°30′N; 012°W; 56°30′N; 012°W; 62°N; 003°W; 62°N on the Norwegian coast; 57°44.8′N on the Danish and Swedish coasts.

Northern Refuse 北方保护区 系指在最北部的一点开始顺时针方向连接下列坐标点的恒向线所包围的区域：北纬45°45.0′，西经86°00.0′；粮船湾火山岛的西海岸，比弗岛的南海岸；北纬45°30.0′，西经85°30.0′；北纬45°30.0′，西经85°15.0′；北纬45°25.0′，西经85°15.0′；北纬45°25.0′，西经85°20.0′；北纬45°20.0′，西经85°20.0′；北纬45°20.0′，西经85°40.0′；北纬45°15.0′，西经85°40.0′；北纬45°15.0′，西经85°50.0′；北纬45°10.0′，西经85°50.0′；北纬45°10.0′，西经86°00.0′。Northern Refuse means the area enclosed by thumb line connecting the coordinates, beginning on the northernmost point and proceeding clockwise: 45°45.0′N, 086°00.0′W; Western shore of High Island, southern shore of Beaver Island; 45°30.0′N, 085°30.0′W; 45°30.0′N, 085°15.0′W; 45°25.0′N, 085°15.0′W; 45°25.0′N, 085°20.0′W; 45°20.0′N, 085°20.0′W; 45°20.0′N, 085°40.0′W; 45°15.0′N, 085°40.0′W; 45°15.0′N, 085°50.0′W; 45°10.0′N, 西经085°50.0′W; 45°10.0′N, 086°00.0′W.

Northwestern Polytechnical University (NPU) [中]西北工业大学 简称西工大，位于古都西安，是中华人民共和国工业和信息化部直属的一所以航空、航天、航海工程为特色，工、理为主，管、文、经、法协调发展的研究型、多科性、开放式全国重点大学，是国家"985工程""211工程"重点建设高校，入选"2011计划""111计划""卓越工程师教育培养计划"，是"卓越大学联盟""中俄工科大学联盟"成员，中管副部级建制，设有研究生院。学校最早可以追溯到1938年国立北洋工学院、国立北平大学工学院、国立东北大学工学院、私立焦作工学院在汉中组建的国立西北工学院；1946年，国立西北工学院迁至咸阳。1950年更名为西北工学院。1957年10月，西北工学院与西安航空学院合并组建西北工业大学。1960年，学校被国务院确定为全国重点大学；1970年2月，哈尔滨工程学院航空工程系整体并入，今日的西北工业大学蓬然成型。截至2014年2月，学校有全日制在校生26 408人，其中全日制博士研究生3 348人，全日制硕士研究生7 855人，本科生14 424人，外国留学生781人。

Norwegian ship-owners association and Norwegian shipbuilders association 挪威船东与船厂协会

Nose plate (leading edge plate) 舵叶导边板 系指舵叶前端处的舵板。

Nose-tail lime 头尾线 连接螺旋桨叶的叶切面两端的线。

Not fully developed sea 未充分发展风浪 在一定风速下，风区和风时不够长，风浪要素尚未达到充分发展值的风浪。

Notch 切口 系指焊接造成的构件不连续处。Notch means a discontinuity in a structural member caused by welding.

Notch stress 切口应力（缺口应力） 系指：(1)焊缝趾部的峰值应力。它考虑结构几何形状和焊缝的存在引起的应力集中。切口应力应采用热点应力乘以疲劳切口因子求得；(2)是在槽口处如焊缝的根部或开口的边缘的峰值应力。峰值应力考虑由于槽口存在的应力集中；(3)在焊接结构中，应力集中区系指诸如焊趾之类产生疲劳裂纹的位置，该处的总应力定义为缺口应力。Notch stress: (1) is defined as the peak stress at the weld toe taking into account stress concentrations due to the effects of structural geometry as well as the presence of welds. Notch stress is to be obtained by multiplying hot spot stress by fatigue notch factor; (2) a peak stress in a notch such as the root of a weld or the edge of a cut-out. This peak stress takes into account the stress concentrations due to the presence of notches; (3) the location where occurs a fatigue crack as the welding toe. The total stress at the location is defined as the notch stress.

Notched bar impact test 缺口冲击试验 是测定以焦耳(J)为单位的冲击功的试验。Notched bar impact test is a test determined the impact energy in joules (J).

Note 提醒项目 对检验而言，特指船级社或进行法定检验的机构，在证书或检查报告上注明的，提醒船舶在某时间以前该完成的检验项目，应满足的某种新的规则、公约要求，以及对证书上的某个特定事项的说明等。

Notebook computer (NB) 笔记本电脑 又称笔记型、手提或膝上电脑（Laptop computer/Laptop），是一种小型、可携带的个人电脑，通常重1~3kg。其发展趋势是体积越来越小，质量越来越轻，而功能却越来越强大。像

Netbook，也就是俗称的上网本，跟 PC 的主要区别在于其便携带方便。与台式计算机相比，它们是完全便携的，而且消耗的电能和产生的噪声都比较少。但是，它们的速度通常稍慢一点，而且对图形和声音的处理能力也比台式计算机稍逊一筹。此外，笔记本电脑的价格也比台式计算机昂贵。但是，它们之间的价格差距正在缩小——笔记本电脑价格的下降速度比台式计算机更快，而其实际销售量在 2005 年 5 月首次超过了台式计算机。

Notice of arbitration　仲裁通知

Notice of proof　举证通知书　是仲裁庭向当事人发出的一份通知，告知当事人举证责任的分配原则与要求，举证期限逾期提交证据、不提交证据以及举证不能的法律后果。

Notice of respondence to action　应诉通知书

Notification of crew and passengers　通知船员和乘客　系指将失火情况通知船员和乘客，以便安全撤离。为此，应装设一套通用应急报警系统和一套公共广播系统。Notification of crew and passengers is to notify crew and passengers of a fire for safe evacuation. For this purpose, a general emergency alarm system and a public address system shall be provided.

Notunder command light　操纵失控灯　用以表明船舶由主机故障、舵失灵、搁浅等原因而操纵失去控制的信号，而于船舶规定处所增设的两盏红环照灯。

Novel intelligent bridge system　新型智能船桥系统　该系统将雷达、电子海图显示与信息系统（ECDIS）、指挥舵轮等集成到一个工作站上，所有功能都采用统一的操作界面、调色板的转换和船桥上所有显示屏调光都可以通过中心的多功能显示器进行调节。

Novel life-saving appliance or arrangement　新颖救生设备或装置　系指具有 SOLAS 公约第Ⅲ章或规则之规定未能全部包括的新型特征，但达到等效的或更高的安全标准救生设备或装置。Novel life-saving appliance or arrangement is a life-saving appliance or arrangement which embodies new features not fully covered by the provisions of this chapter or the Code but which provides an equal or higher standard of safety.

Novel ship　新颖船舶

Novel systems or equipment　新型系统或设备　系指具有新功能，虽然未将 SOLAS 的条款全涵盖，但提供了一个至少是等效的安全标准的系统或设备。Novel systems or equipment are those systems or equipment which embody new features not fully covered by provisions of SOLAS V but which provide an at least equivalent standard of safety.

Novel/new technology or design　新技术/设计　新技术系指在指定的应用领域里没有文献记录可循的技术，也就是说，在该技术应用于实际操作时，没有文献为其是否能够满足特定的功能要求提供保证。这意味着新技术是已知领域里无迹可寻的技术或新环境下的成熟技术或新环境下无迹可寻的技术。

NO_x emission　氮氧化物排放　系指氮氧化物总排放量，按二氧化氮（NO_2）总加权排放量计算并以有关规定的相关试验循环和测量方法确定。

NO_x emission control（NEC NO_x）　氮氧化物排放控制　系指对柴油机的 NO_x 排放量的控制。 NO_x emission control (NEC NO_x) means control of NO_x emission from diesel engines.

NO_x Technical Code 1997　1997 年 NO_x 技术规则　系指经国际海事组织修正的 1997 年 MARPOL 缔约国大会决议 2 通过的船用柴油机氮氧化物排放控制技术规则，但这些修正案应按照 MARPOL 公约第 16 条的规定予以通过并生效。 NO_x Technical Code means the Technical Code on Control of Emission of Nitrogen Oxides from Marine Diesel Engines adopted by Resolution 2 of the 1997 MARPOL Conference, as amended by the Organization, provided that such amendments are adopted and brought into force in accordance with the provisions of article 16 of the present Convention.

NO_x Technical Code 2008　2008 年 NO_x 技术规则　"2008 年船用柴油机氮氧化物排放控制技术规则"的简称。2008 年 10 月 MEPC.177(58) 决议通过对"NO_x 技术规则"进行修订，成为新的"2008 NO_x 技术规则"（Amendments to the technical Code on control of emission of nitrogen oxides from marine diesel engines—NO_x Technical Code 2008），并于 2010 年 7 月 1 日起生效。

Noxious cargo　有毒货物

Noxious gas　有毒气体

Noxious liquid substance　有毒液体物质　系指：（1）在"国际散装化学品规则"第 17 章或 18 章中列入污染别栏，或现行 MEPC.2/通函规定的物质；或（2）根据"1973 年国际防止船舶造成污染公约 1978 年议定书"附则修正案第 6.3 条规定的暂定为 X、Y 或 Z 类的物质；（3）国际海上人命安全公约附录Ⅱ第 1(6) 条定义的有毒液体物质。 Noxious liquid substance means:（1）any substance indicated to the Pollution category column of chapter 17 or 18 of the International bulk chemical code, or the current MEPC;（2）circular or provisionally assessed under the provisions of regulation 6.3 of the amendments to the Annex of the Protocol of 1978 relative to the International Convention for the Prevention of Pollution from ships, 1973, as falling into category X, Y, or Z; (3) noxious liquid substances as defined in regulation 1(6) of Annex II of the Convention.

Noxious liquid substances/other substances 有毒液体物质/其他物质 系指以 OS(其他物质)形式被列入《国际散装化学品规则》第18章污染类别栏目中的物质,并经评定认为不能列入附则Ⅱ定义的 X,Y 或 Z 类物质之内,因为这些物质如从洗舱或排放压载作业中排入海,目前认为对海洋资源、人类健康、海上的休憩环境或其他合法的利用并无危害。排放仅含有被列为"其他物质"的物质的舱底水或压载水或其他残余物或混合物,不受该附则任何要求的约束。 Noxious liquid substances/other substances mean those substances indicated as OS (other substances) in the pollution category column of chapter 18 of the IBC Code which have been evaluated and found to fall outside category X, Y or Z as defined of the Annex Ⅱ because they are, at present, considered to present no harm to marine resources, human health, amenities or other legitimate uses of the sea when discharged into the sea from tank cleaning or de-ballasting operations. The discharge of bilge or ballast water or other residues or mixtures containing only substances referred to as "other substances" shall not be subject to any requirement of the Annex.

Noxious liquid substances/category X X 类有毒液体物质 这类有毒液体物质如从洗舱或排放压载作业中排放入海,将被认为会对海洋资源或人类健康产生重大危害,因而应严禁向海洋环境排放该类物质。 Noxious liquid substances/category X which, if discharged into the sea from tank cleaning or de-ballasting operations, are deemed to present a major hazard to either marine resources or human health and, therefore justify the prohibition of the discharge into the marine environment.

Noxious liquid substances/category Y Y 类有毒液体物质 这类有毒液体物质如从洗舱或排放压载作业中排放入海,将被认为会对海洋资源或人类健康产生危害,或对海上的休憩环境或其他合法利用造成损害,因而对排放入海的该类物质的质和量应采取严格的限制措施。 Noxious liquid substances/category Y which, if discharged into the sea from tank cleaning or de-ballasting operations, are deemed to present a hazard to either marine resources or human health or cause harm to amenities or other legitimate uses of the sea and therefore justify a limitation on the quality and quantity of the discharge into the marine environment.

Noxious liquid substances/category Z Z 类有毒液体物质 这类有毒液体物质如从洗舱或排放压载作业中排放入海,将被认为会对海洋资源或人类健康产生较小的危害,因而对排放入海的该类物质应采取较严格的限制措施。 Noxious liquid substances/category Z which, if discharged into the sea from tank cleaning or de-ballasting operations, are deemed to present a minor hazard to either marine resources or human health and therefore justify less stringent restrictions on the quality and quanty of the discharge into the marine environment.

Nozzle 导管 是一个包在螺旋桨外面的圆环形结构。 Nozzle is a circular structural casing enclosing the propeller.

Nozzle 喷嘴(消防水枪,熔嘴,喷水器) 系指由叶轮整流过的水喷出的那一部分。 Nozzle is the portion that injects the rectified water from the impeller.

Nozzle angle 喷口角(喷管倾角)
Nozzle area 喷嘴截面面积
Nozzle block 喷嘴块(喷嘴组)
Nozzle box 喷嘴室 把从蒸汽室来的蒸汽引向第一级喷嘴的空间。
Nozzle burner 燃油喷嘴
Nozzle exit momentum 喷出动量
Nozzle friction 喷管通道摩擦
Nozzle governing 喷嘴调节 用改变调节级喷嘴进气面积,即进气度的办法来改变蒸汽流量从而调节蒸汽轮机功率的方式。
Nozzle holder 喷嘴座
Nozzle hole 喷嘴孔
Nozzle loss 喷嘴损失
Nozzle propeller 导管推进器
Nozzle ring 喷嘴环
Nozzle spray 喷油雾
Nozzle throat area 喷嘴喉部截面面积
Nozzle tube 喷管
Nozzle wall 喷管壁
NPDES = National Pollution Discharge Elimination System [美]国家排污限制系统
NPS = National Park Service [美]国家公园管理局
NS1 ships NS1 类舰船 该类舰船包括用于调度飞机或设备的舰船和可以作为指挥中心的舰船。这类舰船设计用于在世界范围航行,并通常由 NS2 类舰船辅助。典型的 NS1 类舰船船长超过 140 m,并具备 10 000 t 或以上排水量,包括:航空母舰(aircraft carrier)、直升机和两栖作战舰船(helicopter carrier)和攻击型舰船(attack ship)。 NS1 ships cover those ships used for the deployment of aircraft or equipment and ships which may be used as centers of command. Designed for world-wide operation

and usually supported by ships from the NS2 category. Typically it will cover ships above 140 m in length with a deep displacement of 10 000 t or more. It will include aircraft carriers, helicopter carrier and attack ship.

NS2 ships　NS2 类舰船　该类舰船包括作为编队成员或作为独立单元来护卫 NS1 类舰船的舰船。它们有各种类型，可具备多重角色，包括防空、反潜、海上防卫、岸上辅助，其设计为世界范围航行。该类舰船船长为 70～140 m，排水量为 1 300～20 000 t。NS2 类舰船可为巡洋舰（cruisers）、护卫舰（frigates）、驱逐舰（destroyers）、两栖攻击舰（amphibious assault ship）、轻巡洋舰（corvettes）或类似舰船。　NS2 ships cover those ships used to defend NS1 ships as part of a task force or act as independent units. They may have a variety of sole or multiple roles including air defence, anti submarine, sea defence, shore support and will be designed for worldwide operation. Typically it will cover ships of length 70 m to 140 m with displacements of 1 300 t to 20 000 t. NS2 ships may be described as cruisers, frigates, destroyers, amphibious assault ship, corvettes or similar.

NS3 ships　NS3 类舰船　系指 NS1 类舰船和 NS2 类舰船中没有包含的战斗舰船。其排水量低于 1 500 t。它们可独立运行或作为编队成员，通常被设计和建造用于特定任务，例如扫雷（mine sweeping）、登陆（beach landings）、海岸防卫（coastal defense）或快速巡逻（fast patrol）等。可对其规定有限的服务区域，通常排水量≤1 500 t。NS3 类舰船可为扫雷艇（mine-sweeper）、登陆舰（landing craft）、高速巡逻艇（fast patrol craft）、近海巡逻艇（offshore patrol vessel）、快速打击艇（fast strike craft）、巡逻艇（patrol ship）、猎雷艇（minehunter）和布雷艇（minelayer）等。　Naval ships 3 cover those ships that have a front line role but are not covered by NS2 ships and NS1 ships. This category includes a variety of ships typically below 1 500 t displacement. They may operate independently or as part of a task force and are usually designed and constructed for specific roles such as mine sweeping, beach landings, coastal defense or fast patrol duties. A restricted service area may be specified. NS3 ships may be described as mine-sweeper, landing craft, fast patrol craft, offshore patrol vessel, fast strike craft, patrol ship, minehunter and minelayer etc.

NSA4 ships　NSA 类舰船　系指军用辅助船舶。该类舰船包括民用和军用辅助舰船。它们可具备多种功能，包括运送军事人员或其他人员、弹药、车辆、储备和燃料并给其他舰船输送这类物品。它们不具备明确的进攻性但可以具备一定的自卫能力。一般来说，这类舰船应符合船级社舰船入级规范第 3 篇第 17 章的任何有关要求，并应尽可能满足适于该类船型的国际公约要求。任何与适用的国际公约不符之处，均需得到海军或海军主管机关、或相关国家主管部门同意。由海军公告的设计和营运计划表示应被 LR 接受，其中应表示舰船用途，如所载设备、人员、储备和燃料。授予 NSA 标志的舰船在舰船机械级别中应获得※LMC、※LMC、［※］LMC、LMC 或 MCH 标志。对应按船级社钢质海船规范和规则适当章节的要求进行设计的气垫船（hovercraft）、扫雷艇（mine-sweepers）、登陆舰（landing ships），还应有专门的要求。NSA4 类舰船包括补给船（replenishment ship）、供油船（oil supply ship）、测量船（survey ship）、船坞登陆舰（landing ship dock）、两栖作战运输坞舰（amphibious transport dock）、滚装船（Ro-Ro ship）、运兵船（troop carrier）、车辆运输船（vehicle carrier）、登陆艇（landing craft）、气垫供应船（air cushioned support vehicle）和海上补给船（replenishment ship）等。　NSA ships mean the Naval auxiliary ships. They cover those auxiliary naval ships used for the support of civil and naval operations. They may have a variety of roles including the movement of military and other personnel, ammunition, vehicles, stores and fuels and the transfer of such to other naval ships. They do not have a defined offensive role but may have a limited self-defence capability. In general, the ships will comply with Society's Rules for Ships and any relevant requirements in Chapter 3.17 and satisfy as far as practicable the requirements of the International Conventions applicable to the ship type. Any deviation from the applicable International Convention requires agreement with the Navy or Naval Authority and where applicable the National Administration. A Design and Operating Scenario Statement declared by the Navy stating the role of the ship in terms of the carriage of equipment, personnel, stores and fuels is to be acceptable to the society. The assignment of the NSA ship type notation is dependent on the ship being in Machinery Naval Class with the ※LMC, ※LMC,［※］LMC, LMC or MCH notations. There are special requirements for hovercraft, mine-sweepers, landing ships, which are to be designed in accordance with the appropriate sections of relevant the society's Rules and Regulations. NS4 ships will include replenishment ship, oil supply ship, survey ship, landing ship dock, amphibious transport dock, Ro-Ro ship, troop carrier, vehicle carrier, landing craft, air cushioned support vehicle and replenishment ship etc.

n-th spectral moment　n 阶谱矩　谱密度的 n 阶原点矩。

Nuclear cargo safety certificate　核能货船安全证

书   对核能货船而言，系指由 SOLAS 公约缔约国政府授权的个人或组织，按经 1988 年议定书修订的 1974 年国际海上人命安全公约的规定签发的安全证书。 Nuclear cargo safety certificate means that for the purpose of nuclear cargo ship, means a Safety certificate issued by person or organization authorized under the authority of the Government of under the provisions of the International convention for the safety of life at sea, 1974, as modified by the Protocol of 1988 relating thereto.

Nuclear cargo ship safety certificate or nuclear passenger ship safety certificate   核能货船安全证书或核能客船安全证书   系指代替相应的货船安全证书或客船安全证书。应对每艘核动力的船舶签发 SOLAS 第Ⅷ章要求的证书。   Nuclear cargo ship safety certificate or nuclear passenger ship Safety Certificate means that a certificate used in place of the Cargo Ship Safety Certificate or Passenger ship safety certificate, as appropriate. Every nuclear powered ship shall be issued with the certificate required by SOLAS chapter Ⅷ.

Nuclear damage   核损害   系指由于核燃料或放射性产物或废料的放射性，或者由于核燃料或放射性产物或废料与有毒、易爆性及有害性相结合，而引起或产生的人身伤亡以及财产的灭失或损害。由上述原因而引起或产生的任何其他灭失、损害或费用，只要在所适用的国内法有规定的情况下，并在规定的范围内才可包括。   Nuclear damage means loss of life or personal injury and loss or damage to property which arises out of or results from the radioactive properties or a combination of radioactive properties with toxic, explosive or other hazardous properties of nuclear fuel or of radioactive products or waste, any other loss, damage or expertise so arising or resulting shall be included only if and to the extent that the applicable national law so provides.

Nuclear fuel   核燃料   系指能够通过自身的核裂变过程产生能量，并且被用于或意图用于核动力船舶的任何物质。   Nuclear fuel means any material which is capable of producing energy by a self-sustaining process of nuclear fission and which is used or intended for use in a nuclear ship.

Nuclear incident   核事故   系指造成核损害的任何事故或起源于相同的一系列事故。   Nuclear incident means any occurrence or series of occurrences having the same origin which causes nuclear damage.

Nuclear passenger ship safety certificate   核能客船安全证书   对核能客船而言，系指由 SOLAS 公约缔约国政府授权的个人或组织，按经 1988 年议定书修订的 1974 年国际海上人命安全公约的规定签发的供国际航行/短途国际航行用的证书。   For the purpose of nuclear passenger ship, nuclear passenger ship safety certificate means a Certificate for an international voyage/short international voyage Issued by person or organization authorized under the authority of the Government of under the provisions of the International convention for the safety of life at sea, 1974, as modified by the Protocol of 1988 relating thereto.

Nuclear power plant   核动力装置   系指使用或将使用核反应堆作为船舶推进或任何其他目的的动力源的任何动力装置。   Nuclear power plant means any power plant in which a nuclear reactor is, or is to be used as, the source of power, whether for propulsion of the ship or for any other purpose.

Nuclear reactor   核反应堆   又称为原子能反应堆或反应堆，是能维持可控自持链式核裂变反应，以实现核能利用的装置。核反应堆通过合理布置核燃料，使得在无须补加中子源的条件下能在其中发生自持链式核裂变过程。反应堆这一术语应覆盖裂变堆、聚变堆、裂变聚变混合堆，但一般情况下仅指裂变堆。技术拥有国：中国、美国、俄罗斯、英国、法国、印度（高温气冷反应堆技术）见图 N-3。

图 N-3   核反应堆
Figure N-3   Nuclear reactor

Nuclear Regulatory Commission (NRC)   ［美］核管理委员会

Nuclear ship   核动力船舶   系指配备有核动力设备的船舶。   Nuclear ship means any ship equipped with a nuclear power plant.

Nuclear suppliers group (NSG)   核供应国集团   成立于 1975 年，是一个由拥有核供应能力的国家组成的多国出口控制机制。该组织在国际防核扩散及核出口控制领域发挥重要作用。现有 40 个成员国（2004 年 5 月）。1975 年，加拿大、法国、联邦德国、日本、英国、美国和苏联等 7 个主要核出口国于在伦敦多次召开会议，讨论加强和完善核不扩散的政策和措施及敏感核材料和设备的出口控制等问题，并通过了《核转让准则》和《触发清单》，外界称其为"伦敦俱乐部"，又称"核供应国集团"。该集团的宗旨是通过加强核出口管制，防止敏感物项出口到未参加《不扩散核武器条约》的国家。

Nuclear weapon 核武器 利用核反应的光热辐射、冲击波和感生放射性造成杀伤和破坏作用，以及造成大面积放射性污染，阻止对方军事行动以达到战略目标的巨大杀伤力武器。主要包括裂变武器（第一代核武器，通常称为原子弹）和聚变武器（亦称为氢弹，分为两级及三级式）。亦有些还在武器内部放入具有感生放射的轻元素，以增大辐射强度扩大污染，或加强中子放射以杀伤人员（如中子弹）。核武器也叫核子武器或原子武器。1945 年 7 月 16 日，美国在新墨西哥州成功爆炸了世界上第一颗原子弹，它的成功标志着世界从此进入核武器时代。1952 年 11 月 1 日，美国在太平洋岛上又成功爆炸了世界上第一颗氢弹。

Nuclear weapon miniaturized technology 核武器小型化技术 技术拥有国：中国、美国、俄罗斯、法国、英国等。见图 N-4。

图 N-4 核武器小型化技术
Figure N-4 Nuclear weapon miniaturized technology

Nuclear-powered ship 核动力船 以核能作为推进动力的船舶。

Nuclear-powered submarine 核潜艇 系指以核反应堆为动力源的潜艇，全称核动力潜艇。它具有航行速度高、自给力大、攻击力强、续航时间长、能在水下长时间隐蔽活动等优点。核潜艇可分为战略核潜艇和攻击型核潜艇。核潜艇水下续航能力能达到20万海里，自持能力达 60～90 天。世界上第一艘核潜艇是美国的"鹦鹉螺"号，1954 年 1 月 24 日首次开始试航，它宣告了核动力潜艇的诞生。目前全世界公开宣称拥有核潜艇的国家有 6 个，分别为：美国、俄罗斯、英国、法国、中国、印度（印度的歼敌者号核潜艇在建）。其中美国和俄罗斯拥有核潜艇最多。核潜艇的出现和核战略导弹的运用，使潜艇发展进入一个新阶段。装有核战略导弹的核潜艇是一支水下威慑的核力量。见图 N-5。

图 N-5 核潜艇
Figure N-5 Nuclear-powered submarine

Nucleate boiling 核状沸腾 在传热温差低于 23～27 ℃时，水在传热面上因受热而形成一个个小气泡，逐渐扩大、上升，最后脱出水面的沸腾。

Nucleus of cavitation 空泡核 混于水中或附着于物体上，在略高于水汽压力时可由此产生空泡的气体或水汽的微粒。

Number of blades 叶数 系指螺旋桨的叶片数。

Number of input shaft 传动装置输入轴数 传动装置和主机连接端的轴的数量。

Number of maximum continuous revolutions 主机最大持续转数 系指主机在最大持续输出功率时的转数。

Number of maximum continuous revolutions 最大持续转速 额定功率状态下主机的转速。 Number of maximum continuous revolutions is the number of revolutions at the maximum continuous output.

Number of teeth 齿数

Numeric controlled milling machine for propeller 数控螺旋桨加工铣床 系指采用数控技术加工螺旋桨的铣床。

Numeric controlled plotter 数控绘图机 系指采用数控技术的绘图机。

Numerical value pool 数值水池 数值水池的基本原理是用计算机模拟流体流动，求解流体运动方程，模拟海洋结构物的运动和受力，用软件实现甚至超越物理水池的功能，从而设计出更优化的船型和深海平台。该技术是船舶与海洋工程装备研发由物理水池向虚拟仿真试验转变的最重要支撑技术，是国际水动力学发展的最新方向，也是船舶与海洋工程装备设计方法的革命性转变，具有极其重要的理论和工程意义。数值水池可以实现包括雷诺平均数值模拟、大涡模拟、直接数值模拟等模拟流体流动的三种类型。其不但可实现用于船舶试验的拖曳水池、空泡水洞、耐波性水池、操纵性水池和用于海洋工程的风浪流水池的功能，还可以揭示流体流动现象产生的原因，对流场的观察精细程度远远超过物理水池，并最终实现实际海洋结构物的模拟，结合海洋风浪气象预报，预测船舶沿着不同航线航行时的动力响应，从而帮助船东选择节能、安全、舒适的最佳航线。

NURBS surface 非均匀有理B样条曲面
Nut 螺母（螺帽）
Nut hexagon 六角螺母
Nut locking device 螺母的紧锁装置
Nut of propeller shaft（propeller nut） 推进器轴螺帽 推进器轴上用以固定推进器的螺帽。
Nut wrench 螺母扳手
Nutrients analyzer for seawater 海水营养盐分析器 系指自动分析海水中的硅酸盐、硝酸盐、亚硝酸盐和磷酸盐的仪器。
Nylon 尼龙（酰胺纤维）
Nylon rope 尼龙索

O

O_2 content meter 氧气含量表 一种通用的氧气含量测量装置。

Oar 桨 一种上部为圆杆，下部作板状，支于船侧的人力推进器。

Oar-propelled lifeboat 划桨救生艇 依靠划桨和风帆推进的救生艇。

Object 标的 是合同权利义务指向的对象。标的条款必须清楚地写明标的的名称、以使标的特定化，能够界定权利义务的量。

Objection 抗辩

Objection to execution 执行异议 系指执行过程中，案外人对执行标的提出书面异议的，人民法院应当自收到书面异议之日起15日内审查，理由成立的，裁定中止对该标的的执行，理由不成立的，裁定驳回。案外人、当事人对裁定不服，认为原判决、裁定错误的，依照审判监督程序办理；与原判决、裁定无关的，可以自裁定送达之日起15日内向人民法院提起诉讼。

Objective evidence 客观证据 系指根据观察、测量或试验得到的，且可证实的定性或定量的关于SMS要素的存在和实施情况的信息、记录或事实的陈述。Objective evidence means quantitative or qualitative information, records or statements of fact pertaining to safety or to the existence and implementation of a safety management system element, which is based on observation, measurement or test and which can be verified.

Objective evidence 物证

Objective reason 客观原因

Oblique model towing test 偏模直拖试验 在常规狭长水池中，使船模以规定的不同漂角和/或舵角按等速沿直线被拖曳，用以确定船舶运动位置导致的试验。

Oblique sea 斜浪 系指介于迎浪与横浪（或随浪与横浪）之间行进的波浪。 Oblique sea is waves propagating in a direction between head and beam sea（or following and beam sea）.

Observation 电离层观测 研究极地上空电离层的变化规律及对通信的干扰。

Observation 观察结果（观察项） 系指在安全管理审核期间所得出的，并能够被客观证据证实的实事陈述。 Observation means a statement of fact made during a safety management audit and substantiated by objective evidence.

Observation 观察项 系指在审核期间基于客观证据做出的事实陈述或建议，但不是不合格。 Observation means statements of fact or proposals made during an audit which are based on objective evidence but are not nonconformity.

Observation mast 瞭望桅 位于船前部，其上设有瞭望台，供登高瞭望用的桅。

Observation ship（oceanographic research ship） 海洋调查船 各种专门从事海洋科学调查研究用船的统称，包括海洋综合调查船和海洋专业调查船。用于海洋科学考察、研究、测量或勘探的船舶，通称海洋调查船。船上设有各种专门的仪器和设备，以及供科研人员在海上进行工作和生活的实验室、住舱和公用场所等必要的设施；并能在海洋进行多科目的海洋调查、研究工作。

Observed universe ship 宇宙观测船 专门从事大气上层气象观测，研究大气与海洋之间热交换，调查宇宙与海洋之间互相作用、相互干扰的船舶。

Observer's porthole 观察孔
Obsolete 废旧的
Obstacle 障碍 系指位于直升机甲板上供直升机移动区域内的任何物体或其变化，或延伸至一个为保护飞行中队的直升机所设的限定面之上的任何物体或其他部分。 Obstacle is any object, or part thereof, that is located on an area intended for the movement of a helicopter on a helideck or that extends above a defined surface intended to protect a helicopter in flight.

Obstacle clearing capability 越障高度 全垫升气垫船能超越地面障碍的高度。

Obstacle-free sector 无障碍区 系指一个复合的周界区,起始于直升机甲板上最终进场/起飞区域(FATO)边缘处的一个参照点并从该点展开,由两个部分组成:一个在直升机甲板以上,一个在直升机甲板以下,用于保证飞行安全。该区内仅允许存在规定的障碍。Obstacle-free sector is a complex surface originating at, and extending from, a reference point on the edge of the FATO of a helideck, comprised of two components, one above and one below the helideck for the purpose of flight safety within which only specified obstacles are permitted.

Obstructive sound 妨碍声 就是妨碍人们行动的声音,无论声音大小,只要它妨碍人们谈话交流、妨碍开会、妨碍打电话或手机通话、妨碍学习、妨碍睡眠等有损于人们的欲望、愿望的声音。

Occasion survey(OS) 临时检验 系指不属于各种定期检验的任何检验。按检验船舶的不同部分,该检验可以定义为船体、机械、锅炉、电气和自动控制及遥控系统等临时检验。船舶发生下列情况时,船东或其代理人应申请临时检验:(1)船名、船籍港、船旗和船东或经营人变更;(2)遭受影响入级的船舶及其设备的损坏;(3)港口国当局检查;(4)涉及入级的任何修理或改装或更换时;(5)检查的延期或建议。 Occasional survey is any survey which is not a periodical survey. The survey may be defined as an occasional survey of hull, machinery, boilers, electrical installations, automatic and remote control systems, etc., depending on the part of the ship concerned. An occasional survey is to be requested by the owner or his agent in any of the following cases: (1) change of the ship's name, port of registry, flag, the owner or operator; (2) damage which affects the class of the ship or its equipment; (3) port State control inspection; (4) any repair or alteration or conversion which affects class; (5) postponement of surveys or recommendations.

Occupational accidents 职业事故 与工作场所危险有关的事故(跌倒、滑倒、挤压等),而不是油气在压力下的危险。这些事故通常涉及单个人员。

Occurrence 事故 系指可能会降低安全程度的一种情况。 Occurrence is a condition involving a potential lowering of the level of safety.

Occurrence of a failure 直接发生的故障 系指直接引起其他的故障情况,应一起考虑类似的所有故障。 Occurrence of a failure means a failure which leads directly to further failures, all those failures are to be considered together.

Ocean areas 远洋航区(大洋航区) 是一个区域范围,这个区域的距离等于船舶以正常的航行速度至少航行 30 min 的路程,在这个范围内,任意方向的行程设置的自由度是不受限制的。 Ocean areas are areas in which the freedom of course setting in any direction for a distance equivalent to at least 30 minutes sailing with the navigating speed of the ship is not restricted.

Ocean bill of lading 海运提单 是托运人和承运人的合同、承运人收到货物的收据及物权凭证。详细解释为:是承运人收到货物后出具的货物收据,也是承运人所签署的运输契约的证明,提单还代表所载货物的所有权,是一种具有物权特性的凭证。运输单据的种类很多,包括海运提单(Ocean bill of lading)、海运单(Sea waybill)、航空运单(Air waybill)、铁路运单(Rail waybill)、货物承运收据(Cargo receipt)和多式联运单据(MTD)等。海运提单绝大多数情况下是货权的凭证(在一些交易中由于出现特殊情况,也会出现问题)。卖方(发货方)将货物交给承运人(船方)后,承运人向卖方开具一套提单。

Ocean bottom seismograph 海底地震仪 设置在海底用来进行天然地震及人工地震记录的仪器。由地震波记录装置、检波器及地震记录、安置仪器的密封压力舱、系留装置、回收装置等组成。

Ocean defense equipment 海上防务装备 主要系指各类海洋军事装备和海上执法装备。在海洋防务装备领域中,拥有从 2 000 t～10 000 t 海上公务执法船,提高了我国海上执法水平;从 1 000 t～10 000 t 级节能环保型海监、渔政、缉私等系列化海上执法船,加大了对区域内油气资源、远洋渔业资源及海上安全等的监管和保护力度。

Ocean development equipment 海洋开发装备 主要系指包括海洋油气资源、渔业资源、风能等在内的各类海洋资源勘探、开采、存储、加工等方面的装备。如 3 000 m 半潜式钻井平台"海洋石油 981"号、亚洲首艘新一代 12 缆物探船"海洋石油 720"号、全球首艘集钻井、水上工程、勘探功能于一体的 3000 m 深海勘察船"海洋石油 708"号,以及自升式钻井平台、半潜式钻井平台、钻井船、深水物探船、深水勘察船。

Ocean development ship 海洋开发船 系指专门从事海洋开发的船舶。

Ocean energy 海洋能 系指依附在海水中的可再

生能源。海洋通过各种物理过程接收、储存和散发能量,这些能量以潮汐、波浪、温度差、盐度梯度、海流等形式存在于海洋之中。其中,海洋温差发电是将海洋表面的温海水和深层(800~1000m)的冷海水之间约10 ℃~25 ℃的温度差作为能源转化成电能的发电系统。

Ocean engineering[1]　海洋工程(广义)　系指以开发、利用、保护、恢复海洋资源为目的,并且工程主体位于海岸线向海一侧的新建、改建、扩建工程。具体包括:围填海、海上堤坝工程、人工岛、海上和海底物资储藏设施、跨海桥梁、海底隧道工程、海底管道、海底电(光)缆工程、海洋矿藏资源勘探开发及其附属工程、海上潮汐电站、波浪电站、温差电站等海洋能源开发利用工程、大型海水养殖场、人工岛礁工程、盐田、海水淡化等海水综合利用工程、海上娱乐及运动、景观开发工程以及国家海洋主管部门向环境保护主管部门规定的其他海洋工程。具体系指:(1)应用海洋学,其他有关基础科学和技术学科开发利用海洋所形成的综合技术学科,包括海岸工程、近海工程和深海工程;(2)新建或改建的与海洋有关设施的总称。

Ocean engineering[2]　海洋工程(狭义)　海洋工程始于为海岸带开发的海岸工程。随着开采大陆架海域的石油与天然气,以及海洋资源开发和空间利用规模不断扩大,与之相适应的近海工程成为近30年来发展最迅速的工程之一。其主要标志是出现了钻探与开发石油(天然气)的海上平台,作业范围已由水深10 m以内的近岸水域扩展到了水深300m的大陆架水域。海底采矿由近岸浅海向较深的海域发展,现已能在水深1 000 m的海域钻井采油,在水深6 000 m的大洋进行钻探,在水深4 000 m的洋底采集锰结核。海洋潜水技术发展也很快。已能进行饱和潜水,载人潜水器下潜深度可达10 000 m以上,还出现了进行潜水作业的海洋机器人。这样,大陆架水域的近海工程(又称离岸工程)和深海水域的深海工程均已远远超出海岸工程的范围,所应用的基础科学和工程技术也超出了传统海岸工程的范畴,从而形成了新型的海洋工程。海洋工程的结构形式很多:常用的有重力式建筑物、透空式建筑物和浮式建筑物。重力式建筑物适用于海岸带及近岸浅海水域,如海堤、护岸、码头、防波堤、人工岛等,以土、石、混凝土等材料筑成斜坡式、直墙式或组合式的结构。透空式建筑物适用于软土地基的浅海,也可用于水深较大的水域,如高桩码头、岛式码头、浅海平台等。其中海上平台以钢材、钢筋、混凝土等建成,可以是固定式的,也可以是活动式的。浮式建筑物主要适用于水深较大的大陆架海域,如钻井船、船式平台、半潜式平台等,可以用作石油和天然气勘探开采平台、浮式储油船和炼油厂、浮式电站、浮式飞机场、浮式海水淡化装置、无人深潜器,用于

遥控海底采矿的生产系统等。

Ocean fishery research ship　海洋渔业调查船　专门从事水产资源、渔场、海洋环境以及渔具、渔法等调查研究的船舶。根据船的规模和具体任务,船上可设置鱼类实验室、海洋研究室、化学实验室、浮游生物及底栖生物实验室、渔捞研究室等。

Ocean going instrumentation ship　远洋综合测量船　系指专门从事海洋水文、气象、水声、地质、地球物理、生物等多项海洋科学调研用的远洋调查船。

Ocean going rescue & salvage ship　远洋救生打捞船　系指在远洋航行时,担负海上防险救助任务、搜救失事船舶及船员的船舶。

Ocean going scientific research ship　远洋科学考察船

Ocean going training ship　远洋训练船　系指用于远洋教学和实习的船舶。

Ocean rescue tug　远洋救助拖船　用于海难救助、远洋拖带海上钻井平台、起抛锚、潜水和打捞沉船等作业的拖船。

Ocean scientific investigation equipment　海洋科考装备　主要系指各类专门用于海洋资源、环境等科学调查和实验活动的装备。在海洋科考装备领域中,"向阳红"系列科考船、"远望"系列航天测量船、"雪龙"号极地科考船和"科学"号综合海洋科考船,为提高我国海洋科考能力做出了重要贡献。当前,正重点开展极地科考破冰船与远洋通信监测等产品的研发。为我国海洋科考提供更加先进可靠的装备,解决我国长期以来制约我国极地科学考察事业发展的瓶颈问题,提升我国在极地海洋区域开展综合考察的能力。

Ocean thermal energy conversion (OTEC)　海洋温差发电　主要是透过大海表层温度较高的温海水与深海底层温度较低的冷海水之间的温度差来进行发电、利用冷热之间的流动转换来产生电力。

Ocean thermal energy conversion system (OTECS)　海洋温差发电系统

Ocean towing　海上拖航　系指从指定避风港口间或沿航线安全锚地间的商业拖航作业,其中考虑气象条件。

Ocean transport equipment　海洋运输装备　主要系指包括豪华邮轮、液化天然气(LNG)船在内的各类海洋运输船舶。在海洋运输装备领域除散货船、油船、集装箱船三大主流船型的常规船舶外,还有LNG船、万箱级集装箱船、大型滚装船、超大型气体运输船(VLGC)、双燃料/LNG等新动力运输船、冰区加强/极地运输船舶等高技术船舶。

Oceangoing ship　远洋船舶　系指:(1)在美国政

府管辖下和在国际航线上营运的船舶;(2)在美国政府管辖下和准许其远洋航行的船舶;(3)在美国政府管辖下和准许其在距陆地3海里以外的沿海地区航行的船舶;(4)在美国政府管辖下和根据有关规定在任何时候在美国领海边界以外的公海上航行的船舶;(5)在除美国以外的国家管辖下营运的船舶。注:专门在北美的五大湖或与之相连的支流水域或专门在美国和加拿大内陆水域中航行的加拿大或美国的船舶不视为"远洋船舶"。Oceangoing ship means a ship that (1) is operated under the authority of the United States and engages in international voyages; (2) is operated under the authority of the United States and is certificated for ocean service; (3) is operated under the authority of the United States and is certificated for coastwise service beyond three miles from land; (4) is operated under the authority of the United States and operates at any time seaward of the outermost boundary of the territorial sea of the United States; (5) is operated under the authority of a country other than the United States. Note: A Canadian or U.S ship being operated exclusively on the Great Lakes of North America or their connecting and tributary waters, or exclusively on the internal waters of the United States and Canadian is not an "oceangoing" ship.

Ocean-going vessel 远洋船 适宜于在大洋和国际航线上航行的船舶。

Oceanic survey 大洋调查 对洋区和极地水域进行的海洋调查。它又分为多学科(海洋水文、气象、地质、化学和生物)综合性调查和单学科专题性调查。

Oceanographic research vessel for multi-purpose 海洋综合调查船 专门从事海洋水文、气象、水声、地质、地球物理、生物等多项海洋科学调研用的海洋调查船。

Oceanographic research vessel for single-purpose 海洋专业调查船 专门从事单项海洋科学调研用的海洋水文气象调查船、海洋地质调查船、海洋生物调查船等。

Oceanographic ship 海洋考察船 从事包括气象、水文、地理、海底地质、海洋矿物资源、海洋生物资源、海洋环境等海洋考察和调查的船舶。其能航行于世界各大洋,并能进行全天候的考察作业。海洋考察船装备有各种科学仪器、设备和先进的通信和交通工具,如卫星通信设备、直升机等。

Octane 辛烷

Octane value 辛烷值

Octave 倍频程 系指使用频率 f 与基准频率 f_0 之比等于 2 的 n 次方,即 $f/f_0 = 2$ 的 n 次方,则 f 称为 n 次倍频程。人耳听音的频率范围为 20 Hz 到 20 kHz,在声音信号频谱分析一般不需要对每个频率成分进行具体分析。为了方便起见,人们把20Hz 到 20kHz 的声频范围分为几个段落,每个频带成为一个频程。频程的划分采用恒定带宽比,即保持频带的上、下限之比为一常数。这样将一个倍频程划分为 3 个频程,称这种频程为 1/3 倍频程。

Oertz rudder 流线型舵轴舵 舵柱与舵叶连合成同一流线型剖面的舵。

Of bearing and distance 方位距离定位法 在已知点上测得测量船的方位和距离的定位方法。方位可以经纬仪测定,距离可以微波测距仪测定,也可以在船上用六分仪观测已知点的垂直角间接取得距离,但后者的定位精度较前者低。

Of operational limitations 操作限制清单 系指适用"1974 年国际海上人命安全公约"第 1 章的客船,列出对该船所有操作进行限制的文件。该清单中应包括对"1974 年国际海上人命安全公约"规则的任何免除、航区的限制、天气的限制、海况的限制、许用载荷的限制、纵倾限制、速度限制以及其他任何限制,不论这些限制是由主管机关强制规定或是在设计或建造阶段就已制订。

Of side intersection 侧方交会定位法 前方交会法与后方交会法相结合的一种定位法。在某一控制点上架设经纬仪观测船到另一控制点之水平角,以图解法或计算法确定船位。这种方法的定位精度比前方交会法低,比后方交会法高。

Off hire 停租 是租赁双方各自的原因或合同规定的相关事宜造成不能正常进行租赁活动。

Off-design performance (varying duty, off-design equilibrium running) 变工况 发动机组在非设计工况下工作的一切运行工况的统称。

Offer 报价 分为证券报价、投标报价、产品报价。证券报价是证券市场上交易者在某一时间内对某种证券报出的最高进价或最低出价。投标报价,是商家竞投某项目时愿意出的价格。产品报价系指卖方通过考虑自己产品的成本,利润,市场竞争力等因素,公开报出的可行的价格。报价太高,容易吓跑客户,报价太低,客户一看就知道你不是行家里手,不敢冒险与你做生意。出口商对出口货物的控制方面,在 FOB 价条件下,由于是进口商与承运人联系派船的,货物一旦装船,出口商即使想要在运输途中或目的地转卖货物,或采取其他补救措施,也会颇费一些周折。

Offer 发盘 (1)在国际贸易实务中,发盘也称报盘、发价、报价。法律上称之为"要约"。发盘可以是应对方询盘的要求发出,也可以是在没有询盘的情况下,直接向对方发出。发盘一般是由卖方发出的,但也可以由买方发出,业务称其为"递盘";(2)是一方当事人以缔结合同为目的,向对方当事人提出合同的条件,希望对方当事人接受的意思表示。

Office automation(OA)　办公自动化系统　系指利用现代通信技术、办公自动化设备和电子计算机系统或工作站来实现事务处理、信息管理和决策支持的综合自动化。实现办公自动化的系统是建立在计算机局域网基础上的一种分布式信息处理系统,所以又称办公信息系统。OA系统是一种人机系统,其核心设备是电子计算机系统或OA工作站。OA系统包括信息采集、信息加工、信息传输和信息存取等4个基本环节。

Officer　高级船员　系指经由国家的法规或规则评定,或在无任何相应的法规或规则时,由集体商议或按惯例确定的拥有高级船员头衔,而非船长的人员。Officer means a person other than a master ranked as an officer by national laws or regulations, or, in the absence of any relevant laws or regulations, by collective agreement or custom.

Officer of the Saint Lawrence Seaway　圣劳伦斯航道官员　系指由圣劳伦斯航道开发公司或管理当局雇用的负责某些操作或航道使用的人员。　Officer of the Saint Lawrence Seaway means a person employed by the Saint Lawrence Seaway Development Corporation or the Management to direct some phase of the operation or use of the Seaway.

Officer of the watch　值班驾驶员　系指:(1)当值的负责航行安全和桥楼操作的人员,直至被其他合格的驾驶人员替换;(2)正在操作驾驶室设备和操纵船舶的驾驶员。　Officer of the watch means:(1)a person responsible for the safety of navigation and bridge operations until relieved by another qualified officer;(2)person responsible for operating of bridge equipment and manoeuvring of the ship.

Off-line UPS unit　脱机型UPS装置　系指平时内部支电路发生异常或脱离事先设定好的界限值时,经过逆变器负载供电的装置。　Off-line UPS unit means an electrical power where under normal operation the output load is powered from the bypass line and only transferred to the inverter if the bypass supply fails or goes outside preset limits.

Offset　抵消

Offsets of rudder sections　舵剖面型值　流线型舵剖面轮廓线上各对应点距舵叶剖面中心线的距离。一般用舵宽与舵叶最大厚度的百分数表示。

Offshore access system(OAS)　海上通道系统　是一个悬浮在两个液压缸上可伸展的舷梯,通过获取相对于基础的船舶运动来稳定其自身的移动。

Offshore auxiliary vessel　海洋工程辅助船　该类船舶系指为海洋油气勘探开采提供配套服务的辅助工程船舶,主要包括三用工作船、平台供应船、油田守护船、破冰船、油田消防船、浮油回收船、多功能营救船、油田交通船等。见图O-1。

图 O-1　海洋工程辅助船
Figure O-1　Offshore auxiliary vessel

Offshore company　离岸公司　系指在离岸法区内依据其离岸公司法规范成立的有限责任公司或股份有限公司。当地政府对这类公司没有任何税收,只收取少量的年度管理费,同时,所有的国际大银行都承认这类公司,为其设立银行账号及财务运作提供方便。具有高度的保密性、减免税务负担、无外汇管制三大特点。著名的离岸管辖区有许多是前英属殖民地,如开曼群岛,英属维尔京群岛等,因此这些地区在很大基础上保留了英国的法律体系和司法制度。离岸公司与一般有限公司相比,主要区别在税收上。与通常使用的按营业额或利润征收税款的做法不同,离岸管辖区政府只向离岸公司征收年度管理费,除此之外,不再征收任何税款。

Offshore construction vessel(OCV)　海洋工程施工船

Offshore drilling unit　海上钻井平台　是主要用于钻探井的海上结构物。平台上装钻井、动力、通信、导航等设备,以及安全救生和人员生活设施,是海上油气勘探开发不可缺少的手段。主要分为移动式平台和固定式平台两大类。其中按结构又可分为:(1)移动式平台:坐底式平台、自升式平台、钻井船、半潜式平台、张力腿式平台、牵索塔式平台;(2)固定式平台:导管架式平台、混凝土重力式平台、深水顺应塔式平台。

Offshore engineering　近海工程　又称离岸工程。20世纪中叶以来发展很快。主要是在大陆架较浅水域的海上平台、人工岛等的建设工程,还有在大陆架较深水域的建设工程,如浮船式平台、移动式半潜平台(semi-submersible mobile unit)、自升式平台(jack-up platform)、石油和天然气勘探开采平台、浮式储油船、浮式炼油厂、浮式飞机场等建设工程。

Offshore fishing ship 近海渔船 系指在距离200 n mile以内作业（包括跨洋性近海作业），垂线间长30~50 m之间的中型渔船。

Offshore installation 海上安装

Offshore loading station 近海装卸站

Offshore operating vessel 海洋工程作业船 系指能独立从事海洋工程作业，为海洋油气勘探开发工程系统提供配套装备的工程作业和技术支持服务的工程船舶。目前主要有海洋资源调查船、地球物理勘探船、海洋工程起重船、导管架下水用的辅助船、海洋工程铺管/缆船、水下工程作业船、海洋工程综合检测船等。

Offshore structure 近海装置结构物 系指一种由海底支承或固定在海底上的浮式或非浮式结构物。其设计是以该结构物拟安装现场的基础情况以及长期环境条件为依据。海底与平台的连接物，可能是桩柱、直接支撑、系泊索或锚等。近海装置由下述一项或几项组成：（1）平台结构；（2）水下管路系统和立管；（3）近海设施——①机械、电力和管道系统；②立管和生产设备。

Offshore subsea construction vessel (OSCV) 水下施工船

Offshore supply ship 海上供应船 系指：（1）专为近海作业的海上设施、船舶供应物资和食品等补给的船舶；（2）系指既有近海供应船的特征，亦具有拖曳作业能力的船舶；（3）主要从事给海上平台运送物料、材料及设备的船舶；（4）设计为起居和桥楼上层建筑在船舶前部，用于在海上装卸货物的露天载货甲板在后部的船舶；（5）主要从事运送物品、设备和材料至近海设施上，并在船前部设计有居住处所和桥楼、在船后部有为海上货物操作的露天装货甲板的船舶。Offshore supply means: (1) a ship dedicated to supplying food, stores, etc. to installations and ships engaged in offshore operations; (2) a ship is capable of operating as offshore supply ships and of towing operations; (3) which is primarily engaged in the transport of stores, materials and equipment to offshore installations; (4) which is designed with accommodation and bridge erections in the forward part of the vessel and on exposed cargo deck in the after pare for the handling of cargo at sea; (5) a vessel which is engaged primarily in the transport of stores, materials and equipment to offshore installations and designed with accommodation and bridge erections in the forward part of the vessel and an exposed cargo deck in the after part for the handling of cargo at sea.

Offshore support vessel (OSV) 海洋工程支援船

Offshore survey 近海调查 对距离大陆200 n mile以内的海域进行的海洋调查。它又分为多学科（海洋水文、气象、物理、地质、化学和生物）综合性调查和单学科专题性调查。

Offshore tug/supply ship 近海供应拖船 系指既有近海供应船的特征，亦具有拖曳作业能力的拖船。Offshore tug (supply ship) means a ship is capable of operating as offshore supply ships and of towing operations.

Offshore wind energy converter (OWEC) 海上风电转换系统

Offshore wind farm 海上风电场 系指建立在开放海域上的风电场。多指水深10m左右的近海风电。与陆上风电场相比，海上风电场的优点主要是不占用土地资源，基本不受地形地貌影响，风速更高，风电机组单机容量更大（3~5兆瓦），年利用小时数更高。但是，海上风电场建设的技术难度也较大，建设成本一般是陆上风电场的2~3倍。其由若干风力涡轮机组成的，将风作为能源转动其转子而发电。一个典型的风电场包括塔架、机舱和转子。见图O-2。

图O-2 海上风电场
Figure O-2 Offshore wind farm

Offshore wind farm installation vessel 风电安装船

Offshore support vessels (OSV) 海洋石油支持船 系指用于海上油气勘探、开发、生产等各阶段期间从事作业支持的专用船舶的通称。即指进行与勘探、开发、生产、储运和油田废弃有关的供应、抛起锚、ROV/ROT支持作业、检测或准备（守护）服务的船舶。海洋石油支持船的法律定位系指执行与石油不直接相关、或不在认可当局所规定的沿海国规章制度范围的合同作业中的一项或多项服务功能的船舶。海洋石油支持船(OSV)是一艘特别地设计用于提供起抛锚服务和拖带半潜式平台及拖带装载有海洋石油平台和生产模块的船驳的起抛锚和近海供应船。该船也用作为油田生产平台的守护救助船并通常装备有用于对外消防、营救作业和浮油回收的设备。具有拖曳锚供应功能的船也用于为所有类型的平台的常规供应服务，输送除甲板货物之外的液态钻井液和固体散料等钻采用物资。海洋石油支持船(OSV)的主要功能及 CCS 船级附加标志,根据设备配置和船体结构等的不同,包括但不仅限于以下基本功能的一种或多种,并适合于恶劣海域(包括冰区)或热带海域操作:(1)海上钻采用物资,包括散货(干或液散货)载运供应(Offshore supply ship);(2)守护作业(应急响应与救助)(Stand-by ship);(3)拖曳功能,包括海上设施限位、连接支持作业(Offshore tug);(4)锚作,包括海上设施就位支持及工程施工支持等功能(Offshore tug);(5)提油支持功能(Offshore tug);(6)海上人员转运功能(Offshore supply ship 或 Traffic ship);(7)破冰支持功能(Icebreaking,ice class B1);(8)海上设施生产支持功能(酸化、压裂、固井或防砂支持)(Offshore supply ship);(9)对外消防灭火功能(Fire fighting ship 1,2 或 3);(10)海上设施拖曳锚作索具打捞或维修支持(Offshore tug);(11)潜水支持功能(Offshore tug);(12)浮油回收功能(取决于设备配置)(Oil recovery ship A or B);(13)海底管道勘测功能(Research ship)。

Ogival section(round-back section) 弓形切面 对螺旋桨而言,叶面是一直线,叶背是部分圆弧或抛物线,两端较尖,最大厚度在弦长中点的叶切面。

Oil 油类(加油,涂油) 对"73/78 防污公约"附则 Ⅰ 而言,油类系指任何形式的石油,包括原油、燃油、油泥、油渣和炼制品在内,以及该公约附则Ⅰ的附录Ⅰ所列"油类清单"中的物质。动植物油归入"有毒液体物质",而不属于本定义的油类。 For the purposes of the "MARPOL73/78 Annex Ⅰ", oil means petroleum in any form including crude oil, fuel oil, oil refuse and refined products(other than those petrochemicals which are subject to the peovisions of Annex Ⅱ of the MARPOL73/78 and without limiting the generality of the foregoing, includes the substances listed in appendix Ⅰ to Annex Ⅰ).

Oil absorption 吸油量

Oil barge 油驳 系指舱内装载原油或石油产品的驳船。 Oil barge is a barge carrying crude oil or oil products within holds.

Oil boom 围油栅 系指由浮体、拦油壁、围裙(Skirt)强力材等组成的,将溢出的油包围起来并引导至适宜处所,进行回收处理的工具。

Oil buffer 油缓冲器(油减振器)

Oil bunker 油舱(油柜)

Oil burner 油燃烧器 将燃料油和空气供入炉膛进行燃烧的设备。

Oil burning equipment 油燃烧设备

Oil burning installation 燃油装置 系指被用于为燃油锅炉输送燃油或被用于为内燃机输送加热燃油的设备,并包括用于处理油压超过 0.18 MPa 表压力的任何油泵、过滤器和加热器。

Oil burning unit 燃油单元

Oil can 油壶

Oil cargo residue 残货油 系指来自货舱舱和货油泵舱底处的任何固态、半固态、乳化态或液货油的残油,包括但不限于泄出、漏出和排出的油、油泥、黏结物、油渣、沉淀物、石蜡以及油的各种组成成分。术语"残货油"也称为"货油残油"。 Oil cargo residue means any residue of oil cargo whether in solid, semi-solid, emulsified, or liquid form from cargo tanks and cargo pump room bilges, including but not limited to, drainages, leakages, exhausted oil, muck, clingage, sludge, bottoms, paraffin(wax), and any constituent component of oil. The term "oil cargo residue" is also known as "cargo oil residue".

Oil channels 油槽(油路)

Oil clearance 油隙

Oil cloth(oil skin, oil cloth) 油布(防水布)

Oil collector 集油器

Oil compartment 油舱

Oil consumption 耗油量

Oil container 储油器

Oil containment boom 围油栏(油堰)

Oil contamination 油污

Oil content meter 油分计 系指能对水中所含油分进行检测并显示的一种仪表,通常包含在排油监控装置中。

Oil content meter(OCM) 油分浓度计

Oil cushion 用油缓冲(油垫)

Oil dash pot 油缓冲器

Oil discharge monitoring and control(ODMC) operational manual 排油监控(ODMC)操作手册 系指每艘设有排油监控系统的油船均应按主管机关批准的操作手册进行系统操作的须知。 Oil discharge monitoring and control(ODMC) operational manual means that every oil tanker fitted with an oil discharge monitoring and control system shall be provided with instructions as to the operation of the system in accordance with an operational manual approved by the Administration.

Oil discharge monitoring and control system 排油监控系统(排油监控装置) 对"73/78 防污公约"附则Ⅰ而言,排油监控系统系指监控含油压载水或其他油污水从货舱区域排入海中的系统,它包含下列项目:(1)一个油分计以测量排放物的油含量 ppm。油分计应按 MARPOL 附则Ⅰ的这些指南和技术条件附件中的规定获得认可并经证明考虑了载运货物的范围。(2)流速指示系统以测量排入海中的排放物的速度。(3)船速指

Oil discharge monitoring and control system(ODM)

装置以指出船舶速度(kn)。(4)船舶位置指示装置以指出船舶位置——经度和纬度。(5)取样系统以把排放物代表性样品传送至油分计。(6)舷外排放控制以终止舷外排放。(7)启动连锁以防止任何排放物向舷外排放，除非监控系统完全处于工作状态。(8)控制部分包括：①处理机，它接收排放物的含油量、排放物流速和船速信号，并将这些数值换算成每海里的排油量升和排油总量；②提供报警和向舷外排放控制提供命令信号的设备；③提供数据记录的记录设备；④展示目前操作数据的数据显示器；⑤在监控系统发生故障时使用的越控系统；⑥提供信号给启动连锁以防止在监控系统完全运作前排放任何排放物的设备。 For the purpose of the "MARPOL 73/78 Annex I", oil discharge monitoring and control system, referred to in these Guidelines and Specifications as:(1) monitoring system is a system which monitors the discharge into the sea of oily ballast or other oil-contaminated water from the cargo tank areas and comprises the items an oil content meter to measure the oil content of the effluent in ppm. The meter should be approved in accordance with the provisions contained in the annex to these Guidelines and Specifications and be certified to take into account the range of cargoes carried;(2) a flow rate indicating system to measure the rate of effluent being discharged into the sea;(3) a ship speed indicating device to give the ship's speed in knots;(4) a ship position indicating device to give the ship's position, latitude and longitude;(5) a sampling system to convey a representative sample of the effluent to the oil content meter;(6) an overboard discharge control to stop the overboard discharge;(7) a starting interlock to prevent the discharge overboard of any effluent unless the monitoring system is fully operational; and (8) a control section comprising:①a processor, which accepts signals of oil content in the effluent, the effluent flow rate and the ship's speed and computes these values into liters of oil discharged per nautical mile and the total quantity of oil discharged;②means to provide alarms and command signals to the overboard discharge control;③a recording device to provide a record of data;④a data display to exhibit the current operational data;⑤a manual override system to be used in the event of failure of the monitoring system; and ⑥means to provide signals to the starting interlock to prevent the discharge of any effluent before the monitoring system is fully operational.

Oil discharge monitoring and control system (ODM) 排油监控系统

Oil discharge monitoring equipment(ODME) 排油监测设备

Oil fuel unit

Oil discharge monitoring system(ODM) 排油监控系统　对排出的污水中的油含量进行监控的装置。

Oil discharge pipe 输油管

Oil distributing box 配油箱

Oil drip tray 滴油盘

Oil engine power plant(internal combustion engine power plant) 内燃机动力装置　以内燃机作为主机的船舶动力装置。有柴油机动力装置、汽油机动力装置和煤气机动力装置等。

Oil exhaust valve 排油阀

Oil feeder 加油器

Oil filled switch 油开关

Oil film 油膜

Oil filter 滤油器

Oil filtering equipment 滤油设备　系指15ppm舱底水分离器,可包括一个分离器、过滤器或凝聚过滤器的任意组合,也可是一个设计用于产生的排出物含油量不超过15ppm的单一装置。 Oil filtering equipment is a 15ppm bilge separator and may include any combination of a separator, filter or coalescer and also a single unit designed to produce an effluent with oil content not exceeding 15ppm.

Oil filtering system 滤油系统　系指具有包括滤油设备、管路、附件及仪表在内的一整套系统。

Oil fire 油类火灾

Oil fired boiler 燃油锅炉

Oil fog test 油雾法(油雾试验)

Oil fuel 燃油/油类燃料　系指船舶所载有并用作其推进和辅助机器的燃料的任何油类。 Oil fuel means any oil used as fuel in connection with the propulsion and auxiliary machinery of the ship in which such oil is carried.

Oil fuel burning system 燃油燃烧系统

Oil fuel capacity 燃油容量　系指允装率为98%时的舱容,m^3。 Oil fuel capacity means the volume of a tank in m^3, at 98% filling.

Oil fuel pump 燃油泵

Oil fuel pumping unit 燃油泵组

Oil fuel tank 燃油舱　系指装载燃油的液舱。但不包括在正常操作中不装燃油的液舱,如溢油舱。 Oil fuel tank means a tank in which oil fuel is carried, but excludes those tanks which would not contain oil fuel in normal operation, such as overflow tanks.

Oil fuel unit 燃油装置　系指:(1)准备为燃油锅炉输送燃油或准备为内燃机输送加热燃油的设备。并包括用于处理油类而压力超过0.18 N/mm²的任何压力油泵、过滤器和加热器;(2)包括为在压力超过0.18 N/

621

mm² 情况下为锅炉和主机(包括燃气轮机)准备和提供预热过或未预热燃料的任何设备。 Oil fuel unit is:(1) the equipment used for the preparation of oil fuel for delivery to an oil-fired boiler, or equipment used for the preparation for delivery of heated oil to an internal combustion engine, and includes any oil pressure pumps, filters and heaters dealing with oil at a pressure of more than 0.18 N/mm²; (2) oil fuel unit includes any equipment for the preparation of oil fuel and delivery of oil fuel, heated or not, to boilers and engines (including gas turbines) at a pressure of more than 0.18 N/mm².

Oil gage 油位表
Oil gage stick 量油杆
Oil gravity tank 重力油柜
Oil groove 油槽
Oil hardened steel 油淬钢
Oil hardening 油淬硬化
Oil head tank 重力油柜
Oil header 集油箱
Oil hole 油孔
Oil hose davit 油软管吊柱
Oil immersion "o" (electrical equipment) "o"油浸型(电气设备) 电气设备或电气设备的部件浸没在保护性液体中,以使得在该液体之上或在其外壳之外的爆炸性环境不可能被点燃的保护形式。 Oil immersion "o" (electrical equipment) means a type of protection in which the electrical apparatus or parts of the electrical apparatus are immersed in a protective liquid in such a way that an explosive atmosphere which may be above the liquid or outside the enclosure cannot be ignited.
Oil injection 喷油
Oil inlet valve 进油阀
Oil insulator 阻油器(防油器)
Oil less bearing 无油轴承
Oil level mark 油位标志
Oil like NLS 油性有毒液体物质
Oil limiter 限油器
Oil line 油管路
Oil loading terminal 装油港(装油码头)
Oil lubricating system 润滑油系统
Oil lubrication 油润滑
Oil measures cans 量油壶
Oil mist detector 油雾检测器
Oil mixture 油混合物 系指任何油类物质的混合体。 Oil mixture means a mixture with any oil content.
Oil mop machine 抹油机

Oil outlet 出油口(放油口)
Oil oven 燃油炉灶
Oil passageway 油管
Oil pier 油料码头
Oil pipe 油管
Oil pollution 油污(石油污染)
Oil pollution prevention index E 油污染指数 E 系指由一个表示油舱设计的防污染性能的无因次指数。该指数是三个泄油参数(零泄油概率、平均泄油和极端泄油)的函数。 Oil pollution prevention index E is a non-dimensional index expressing the oil pollution prevention performance of a tanker. It is a function of the three oil outflow parameters, " probability of zero oil outflow", " mean oil outflow" and "extreme oil outflow".
Oil pollution prevention system 防油污系统
Oil pressure 油压
Oil pressure differential controller 压差控制器 当制冷压缩机润滑油压力源之低压端即曲轴箱压力与高压端压力间压力差小于某一调定值时,能自动切断电动机电源的保护装置。
Oil pressure regulating valve 调压阀 液压调节系统中保持油压稳定的阀门。
Oil pressure test 油压试验
Oil pressure test for tightness 油压密封试验
Oil product 石油产品
Oil production 油气生产
Oil proof 耐油性(防油)
Oil pump 油泵
Oil purifier 分油器(净油器)
Oil quench (oil quenching) 油淬火
Oil receiver 储油器
Oil record book 油类记录簿 每艘 150 GT 及以上的油船以及 400 GT 及以上除油船以外的其他船舶,均应备有一份油类记录簿第 I 部分(机器处所作业),每艘 150 GT 及以上的油船,还应备有一份油类记录簿第 II 部分(货物/压载作业)。 Oil record book means that every oil tanker of 150 gross tonnage and above and every ship of 400 gross tonnage and above other than an oil tanker shall be provided with an oil record book, Part I (machinery space operations). Every oil tanker of 150 gross tonnage and above shall also be provided with an oil record book, Part II (cargo/ballast operations).

Oil record book, part II-cargo/ballast operations 油类记录簿第 II 部分(货物/压载的作业) 系指包括下列项目的记录簿:(1)货油的装载;(2)航行中货油的过驳;(3)货油的卸载;(4)货油舱的清洁压载舱的压载;

(5)货油舱的清洗(包括原油洗舱);(6)压载水的排放,但从专用压载舱排放者除外;(7)排放污油水舱的水;(8)污油水舱排放作业后,所使用的阀门或类似装置的关闭;(9)污油水舱排放作业后,为清洁压载舱与货油和扫舱管路隔离所需阀门的关闭;(10)残油的处理。 Oil record book, part Ⅱ-cargo/ballast operations is a book included followings:(1) loading of oil cargo;(2) internal transfer of oil cargo during voyage;(3) unloading of oil cargo;(4) ballasting of cargo tanks and dedicated clean ballast tanks;(5) cleaning of cargo tanks including crude oil washing;(6) discharge of ballast except from segregated ballast tanks;(7) discharge of water from slop tanks;(8) closing of all applicable valves or similar devices after slop tank discharge operations;(9) closing of valves necessary for isolation of dedicated clean ballast tanks from cargo and stripping lines after slop tank discharge operations; and(10) disposal of residues.

Oil record book, part Ⅰ-machinery space operations 油类记录簿第Ⅰ部分(机器处所的作业) 系指包括下列项目的记录簿:(1)燃油舱的压载或清洗;(2)燃油舱污压载水或洗舱水的排放;(3)油性残余物(油泥或其他残油)的收集和处理;(4)机器处所积存的舱底水向舷外排放或处理;(5)添加燃油或散装润滑油。 Oil record book, part Ⅰ-machinery space operations is a book included followings:(1) ballasting or cleaning of oil fuel tanks;(2) discharge of dirty ballast or cleaning water from oil fuel tanks;(3) collection and disposal of oil residues (sludge and other oil residues);(4) discharge overboard or disposal otherwise of bilge water which has accumulated in machinery spaces; and(5) bunkering of fuel or bulk lubricating oil.

Oil recovery equipment 浮油回收设备

Oil recovery ship 浮油回收船 系指建造成或改装成专门从事或兼用的水面浮油回收作业的钢质海船。浮油回收船可分为下列3类:(1)具有油回收设备和回收油贮存舱及排放设备的浮油回收船;(2)具有油回收设备,但是不具有回收油贮存舱及排放设备的浮油回收船;(3)具有油回收设备,回收闪点高于60℃浮油的浮油回收船。 Oil recovery ship means a steel ship constructed or altered for being specially or partially engaged in the recovery of oil floating on the sea. The oil recovery ships may be divided into three categories as follows:(1) Oil recovery ship with cargo tank;(2) Oil recovery ship without cargo tank; and(3) Oil recovery ship not suitable for products with a flash point of 60 ℃ and less.

Oil recovery ship not suitable for products with a flash point of 60 ℃ and less 具有油回收设备,回收闪点高于60℃浮油的浮油回收船 系指回收闪点(闭杯试验)超过60℃、雷特蒸气压力低于大气压力的水面浮油的钢质海船。

Oil recovery ship of category A A类浮油回收船 系指在可能受溢油源的失火和爆炸影响的区域进行作业的浮油回收船。 Oil recovery ship of category A is an oil recovery ship operating in the area which will possibly be affected by fire or explosion caused by an oil spill source.

Oil recovery ship of category B B类浮油回收船 系指在不在溢油源的失火和爆炸影响的区域进行作业的浮油回收船。 Oil recovery ship of category B is an oil recovery ship operating in the area which will not be affected by fire or explosion caused by an oil spill source.

Oil recovery ship with cargo tank 具有油回收设备和回收油贮存舱及排放设备的浮油回收船 系指设有回收油贮存舱及排放设备的,回收闪点(闭杯试验)不超过60℃、雷特蒸气压力低于大气压力的水面浮油的钢质海船。

Oil recovery ship without cargo tank 具有油回收设备,但是不具有回收油贮存舱及排放设备的浮油回收船 系指不设回收油贮存舱及排放设备的,回收闪点(闭杯试验)不超过60℃、雷特蒸气压力低于大气压力的水面浮油的钢质海船。

Oil refuse 油渣

Oil residue 残油(油渣) 系指:(1)残货油;和(2)在机器处所出现的,由泄出、漏泄和排出的油形成的固态、半固态、乳化态或液态的其他残油;(3)船舶正常操作过程中产生的残余废油产物,例如由主机或辅机的燃油或润滑油净化产生的残余废油产物,来自滤油设备的分离废油,滴油盘收集的废油,以及废弃的液压油和润滑油。 Oil cargo residue means:(1) oil cargo residue, and(2) other residue of oil whether in solid, semi-solid, emulsified, or liquid form, resulting from drainages, leakages, exhausted oil, and other similar occurrences from machinery spaces;(3) the residual waste oil products generated during the normal operation of a ship such as those resulting from the purification of fuel or lubricating oil for main or auxiliary machinery, separated waste oil from oil filtering equipment, waste oil collected in drip trays, and waste hydraulic and lubricating oils.

Oil residue(sludge) drain tanks 残油(油泥)泄放舱 这是2008年11月12日通过的 MEPC.1/Circ.642通函"处理机器处所含油废弃物系统2008 导则(含舱底水综合处理系统指南 IBTS)"中的定义。残油(油泥)泄放舱是:(1)用于收集分离器分离油泥和其他泄放的残油(油泥)的舱柜;(2)此舱柜无任何措施可对油泥进行处理及泄放;(3)此柜有一吸口接头仅用于连接油泥收集泵以将残油(油泥)排入残油(油泥)舱。 The residual oil (sludge) discharge tanks are defined in the 2008 Guide lines for waste containing systems (including guide lines for inteqrated bilge water treatment systems) adopted by MEPC. 1/Circ. 642 circular on 12 November 2008. Oil residue(sludge) drain tanks are:(1) tanks intended to receive separated sludge from purifiers and other oil residue (sludge) drains;(2) tanks without any means for disposal of and drains;(3) tanks with suction connection for a sludge collecting pump only capable of discharging to the oil residue (sludge) tank(s).

Oil residue(sludge) incineration systems 残油(油泥)焚烧系统 这是2008年11月12日通过的MEPC. 1/Circ.642通函"处理机器处所含油废弃物系统2008导则(含舱底水综合处理系统指南 IBTS)"中的定义。残油(油泥)焚烧系统是用于对船舶在海上航行中产生的残油(油泥)进行焚烧的系统。残油(油泥)焚烧系统可为:(1)备有合适的残油(油泥)处理系统的主蒸汽锅炉及辅蒸汽锅炉;(2)备有合适的残油(油泥)处理系统的热媒油系统加热炉;(3)备有合适的残油(油泥)处理系统并设计为用于焚烧油泥的焚烧炉;或(4)备有合适的残油(油泥)处理系统的惰性气体系统。残油(油泥)焚烧系统还应符合 MARPOL 附则Ⅵ第16条的规定。 The residual oil (sludge) discharge tanks are defined in the 2008 guide lines for waste containing systems (including guide lines for inteqrated bilge water treatment systems) adopted by MEPC. 1/Circ. 642 circular on 12 November 2008. Oil residue (sludge) incineration systems are systems proving incineration of oil residue (sludge) generated on board seagoing ships. Oil residue (sludge) incineration systems could be:(1) main and auxiliary steam boilers with appropriate oil residue (sludge) processing systems;(2) heaters of thermal fluid systems with appropriate oil residue (sludge) processing systems;(3) incinerators with appropriate oil residue (sludge) processing systems designed for sludge incineration; or (4) inert gas systems with appropriate oil residue (sludge) processing systems. (see MEPC. 1/Circ. 642). Oil residue (sludge) incineration systems shall conform to regulation 16 in MARPOL Annex Ⅵ.

Oil residue(sludge) service tank 残油(油泥)日用舱 这是2008年11月12日通过的EPC. 1/Circ.642通函"处理机器处所含油废弃物系统2008导则(含舱底水综合处理系统指南 IBTS)"中的规定,残油(油泥)日用舱是除残油(油泥)和其他废油用的残油(油泥)舱之外的另一个舱。该舱具有措施用以泄出水分(处理)并随后在残油(油泥)焚烧系统中处理残油(油泥)。该舱还应配有合适的如同"73/78 防污公约"附则Ⅰ第12.2.2条规定走向的排放接头。为改善燃烧及热值,应配有一个供应燃油的接头。 The residual oil (sludge) discharge tanks are defined in the 2008 guide lines for waste containing systems (including guide lines for inteqrated bilge water treatment systems) adopted by MEPC. 1/Circ. 642 circular on 12 November 2008. Oil residue(sludge) service tank should be provided with means for drainage of water (disposal) and subsequent disposal of the oil residue (sludge) in the oil residue(sludge) incineration system. The oil residue(sludge) service tank should be provided in addition to the oil residue(sludge) tank for oil residue (sludge) and other waste oils. It should be equipped with suitable drainage facilities terminating as provided for in regulation 12.2.2 of MARPOL Annex Ⅰ. With a view to improving combustibility and calorific value, a fuel oil supply connection should be provided (see MEPC. 1/Circ. 642 and MEPC. 1/Circ. 676).

Oil residue(sludge) tanks 残油(油泥)舱 这是2009年7月17日通过的 MEPC. 187(59)决议对"73/78防污公约"附则Ⅰ新增的定义,于2011年1月1日起生效。残油(油泥)舱系指储存残油(油泥)的舱柜,通过标准排放接头和其他任何认可的处理措施可从该舱直接处理油泥。 Residue oil (sludge) tank is a new definition of Annex Ⅰ to "MARPOL 73/78" adopted by MEPC. 187 (59) on 17 july 2009, with effect from 1 January 2011. Oil residue (sludge) tanks are the tanks which hold oil residue (sludge) directly from which oil residue(sludge) may be disposed through the standard discharge connection or any other approved means of disposal.

Oil residue(sludge) 残油(油泥) 系指船舶在正常营运过程中产生的残余废油产物。例如由主机或辅机的燃油或润滑油净化产生的残余废油产物,来自滤油设备的分离废油、滴油盆收集的废油以及废弃的液压油和润滑油。 Oil residue(sludge) means the residual waste oil products generated during the normal operation of a ship such as those resulting from the purification of fuel or lubricating oil for main or auxiliary machinery, separated waste oil from oil filtering equipment, waste oil collected in drip trays, and waste hydraulic and lubricating oil.

Oil resisting paint 耐油漆
Oil resistivity 耐油性
Oil ring bearing 油环轴承
Oil room 油料间

Oil scraper ring　刮油环
Oil seal　油封
Oil sealing gland　油封装置
Oil separator　油分离器(分油器)　装在制冷压缩机排出口与冷凝器之间,分离和收集排气中随带的润滑油,并使其返回制冷压缩机的曲轴箱或集中在集油器中的器具。
Oil separator space　分油器处所
Oil sight glass　油观察玻璃
Oil skimmer　浮油回收船　系指专门从事或兼用的水面浮油回收作业的钢质海船。
Oil skimmer　浮油撇除器
Oil skimmer(oil-catcher) and oily-water treatment ship　浮油回收船及油污水处理船　是为维护港口水域洁净而产生的一种环保船型。
Oil sorbent　油吸材
Oil sounding rod　量油杆
Oil spill source　溢油源　系指海面浮油的泄漏源头,例如来自油船、水下石油管道、海上钻井平台等。
Oil spray　油雾
Oil stern bearing　油润滑尾管轴承
Oil stone　油石
Oil storage barge　储油驳　系为生产平台不断产出的原油提供储存设施,并在穿梭油船因天气恶劣无法靠泊装油的情况下提供缓冲储存容积的驳船。
Oil storage facility　储油设施
Oil storage platform　储油平台　系指具有储油设施,可在海上为生产平台所生产的原油提供缓冲储存的平台。
Oil storage ship　储油船　系指为生产平台不断产出的原油提供储存设施,并在穿梭油船因天气恶劣无法靠泊装油的情况提供缓冲储存容积的油船。
Oil storage system　储油系统　系指为海上石油生产系统所生产的原油提供缓冲储存的系统。
Oil sump　油底壳(储油槽)　柴油机底部存放汇集从部件中流出来的滑油的机壳。
Oil supply　供油
Oil supply facility　供油设施
Oil supply ship/Fuel supply boat　供油船　系指专供油料的供应船舶。
Oil tank　油舱(油箱,油柜)
Oil tank hatch　油舱舱口　油舱上的舱口。
Oil tank ladder　油舱梯　设置在货油舱内的梯。
Oil tank paint　油舱漆
Oil tank vent　燃油舱透气管
Oil tank washing machine　油舱清洗机

Oil tanker　油船　(1)传统定义(仅就船舶的功能而言):油船系指建造为或改造为主要在其装货处所装运散装油类的船舶;(2)对《73/78防污公约》附则Ⅰ而言,油船系指建造为或改造为主要在其装货处所装运散装油类的船舶,并包括全部或部分装运散装货油的兼装船、《73/78防污公约》附则Ⅱ中所定义的任何"NLS液货船"和经修正的《1974 SOLAS 公约》第Ⅱ-1/3.20条中所定义的任何气体运输船。(3)FPSO和FSU不是油船,不得用于运输石油,除非经船旗国和相关沿海国根据航程予以明确同意,可在异常和罕见情况下将所采石油运至港口。　(1) For the purposes of its function, Oil tanker means a ship constructed or adapted primarily to carry oil in bulk in its cargo spaces;(2) For the "MARPOL73/78 Annex Ⅰ", Oil tanker means a ship constructed or adapted primarily to carry oil in bulk in its cargo spaces and includes combination carriers, any "NLS tanker" as defined in Annex Ⅱ of the MARPOL73/78 and any gas carrier as defined in regulation 3.20 of ch.Ⅱ-1 of SOLAS 74 (as amended), when carrying a cargo or part cargo of oil in bulk;(3) FPSOs and FSUs are not oil tankers and are not to be used for the transport of oil except that, with the specific agreement by the flag and relevant coastal States on a voyage basis, produced oil may be transported to port in abnormal and rare circumstances.

Oil tanker delivered after 1 June 1982　在1982年6月1日以后交船的油船　系指:(1)在1979年6月1日以后签订建造合同的油船;(2)无建造合同,在1980年1月1日以后安放龙骨或处于类似建造阶段的油船;(3)在1982年6月1日以后交船的油船;(4)经重大改装的油船:①在1979年6月1日以后签订改建合同;或②无改建合同,在1980年1月1日以后改建工程开工;或③在1982年6月1日以后改建工程完成。　Oil tanker delivered after 1 June 1982 means an oil tanker:(1) for which the building contract is placed after 1 June 1979; or (2) in the absence of a building contract, the keel of which is laid or which is at a similar stage of construction after 1 January 1980; or (3) the delivery of which is after 1 June 1982; or (4) which has undergone a major conversion:①for which the contract is placed after 1 June 1979; or ②in the absence of a contract, the construction work of which is begun after 1 January 1980; or ③ which is completed after 1 June 1982.

Oil tanker delivered before 6 July 1996　在1996年7月6日以前交船的油船　系指:(1)在1993年7月6日以前签订建造合同的油船;(2)无建造合同,在1994年1月6日以前安放龙骨或处于类似建造阶段的油船;

(3)在1996年7月6日以前交船的油船;(4)经重大改装的油船:①在1993年7月6日以前签订改建合同;或②无改建合同,在1994年1月6日以前改建工程开工;或③在1996年7月6日以前改建工程完成。 Oil tanker delivered before 6 July 1996 means an oil tanker:(1) for which the building contract is placed before 6 July 1993; or(2) in the absence of a building contract, the keel of which is laid or which is at a similar stage of construction before 6 January 1994; or(3) the delivery of which is before 6 July 1996; or(4) which has undergone a major conversion:①for which the contract is placed before 6 July 1993; or ② in the absence of a contract, the construction work of which is begun before 6 January 1994; or③which is completed before 6 July 1996.

Oil tanker delivered on or after 1 February 2002 在2002年2月1日或以后交船的油船 系指:(1)在1999年2月1日或以后签订建造合同的油船;(2)无建造合同,在1999年8月1日或以后安放龙骨或处于类似建造阶段的油船;(3)在2002年2月1日或以后交船的油船;(4)经重大改装的油船:①在1999年2月1日或以后签订改建合同;或②无改建合同,在1999年8月1日或以后改建工程开工;或③在2002年2月1日或以后改建工程完成。 Oil tanker delivered on or after 1 February 2002 means an oil tanker:(1) for which the building contract is placed on or after 1 February 1999; or(2) in the absence of a building contract, the keel of which is laid or which is at a similar stage of construction on or after 1 August 1999; or(3) the delivery of which is on or after 1 February 2002; or(4) which has undergone a major conversion:①for which the contract is placed on or after 1 February 1999; or ② in the absence of a contract, the construction work of which is begun on or after 1 August 1999; or ③which is completed on or after 1 February 2002.

Oil tanker delivered on or after 1 January 2010 在2010年1月1日或以后交船的油船 系指:(1)在2007年1月1日或以后签订建造合同的油船;(2)无建造合同,在2007年7月1日或以后安放龙骨或处于类似建造阶段的油船;(3)在2010年1月1日或以后交船的油船;(4)经重大改装的油船:①在2007年1月1日或以后签订改建合同;或 ②无改建合同,在2007年7月1日或以后改建工程开工;或③在2010年1月1日或以后改建工程完成。 Oil tanker delivered on or after 1 January 2010 means an oil tanker:(1) for which the building contract is placed on or after 1 January 2007; or(2) in the absence of a building contract, the keel of which is laid or which is at a similar stage of construction on or after 1 July 2007; or(3) the delivery of which is on or after 1 January 2010; or(4) which has undergone a major conversion:①for which the contract is placed on or after 1 January 2007; or ②in the absence of a contract, the construction work of which is begun on or after 1 July 2007; or ③which is completed on or after 1 January 2010.

Oil tanker delivered on or after 6 July 1996 在1996年7月6日或以后交船的油船 系指:(1)在1993年7月6日或以后签订建造合同的油船;(2)无建造合同,在1994年1月6日或以后安放龙骨或处于类似建造阶段的油船;(3)在1996年7月6日或以后交船的油船;(4)经重大改装的油船:①在1993年7月6日或以后签订改建合同;或②无改建合同,在1994年1月6日或以后改建工程开工;或 ③在1996年7月6日或以后改建工程完成。 Oil tanker delivered on or after 6 July 1996 means an oil tanker:(1) for which the building contract is placed on or after 6 July 1993; or(2) in the absence of a building contract, the keel of which is laid or which is at a similar stage of construction on or after 6 January 1994; or(3) the delivery of which is on or after 6 July 1996; or(4) which has undergone a major conversion:①for which the contract is placed on or after 6 July 1993; or ②in the absence of a contract, the construction work of which is begun on or after 6 January 1994; or ③which is completed on or after 6 July 1996.

Oil tanker delivered on or before 1 June 1982 在1982年6月1日或以前交船的油船 系指:(1)在1979年6月1日或以前签订建造合同的油船;(2)无建造合同,在1980年1月1日或以前安放龙骨或处于类似建造阶段的油船;(3)在1982年6月1日或以前交船的油船;(4)经重大改装的油船:①在1979年6月1日或以前签订改建合同;或②无改建合同;在1980年1月1日或以前改建工程开工;或③在1982年6月1日或以前改建工程完成。 Oil tanker delivered on or before 1 June 1982means an oil tanker:(1) for which the building contract is placed on or before 1 June 1979; or(2) in the absence of a building contract, the keel of which is laid or which is at a similar stage of construction on or before 1 January 1980; or (3) the delivery of which is on or before 1 June 1982; or(4) which has undergone a major conversion:①for which the contract is placed on or before 1 June 1979; or ②in the absence of a contract, the construction work of which is begun on or before 1 January 1980; or ③which is completed on or before 1 June 1982.

Oil tanker with double hull 双壳油船 系指具有满足规范规定间距要求的双壳,单甲板和小尺度舱口,载运原油或石油产品的船舶。 Oil tanker with double hull is a ship that distance between two hulls in compliance with the Rules, single deck and small-size hatches, carrying crude oil or oil products.

Oil temperature 油温

Oil tempering 油浴回火

Oil test 油密性试验(验油)

Oil tight 油密(油封)

Oil tightness 油密性 不透油的性能。

Oil tightness test 油密性试验

Oil transfer system 原油外输系统 是在 FPSO 上向外输送原油的系统。根据运输油船的靠泊方式,可分为串靠式、旁靠式、串旁联合式和 FPSO/FPSO 式以及 FPSO/单点浮筒终端式等。其一般由舱内吸口、液压遥控阀、货油外输泵、计量标定器、惰气系统、外输软管、软管滚筒、液压动力系统、仪表和管汇等组成。

Oil trap 集油器 收集制冷系统中分离出来的润滑油的容器。

Oil tray 油盘

Oil vapour 油气

Oil waste 油性废弃物 系指拟废弃的含油水。

Oil way 油路

Oil way chisel 油槽錾

Oil wiper 刮油器

Oil, gas and water processing areas 油、气、水处理区 系指油、气、水处理系统,包括计量分离设备所在的区域。 Oil, gas and water processing areas are those areas where oil, gas and water processing installations including metering and separating devices are located.

Oil/ballast tank 油/压载兼用舱 系指油船营运时的货油舱和在矿砂船或散装货船运航时作为压载舱(也包括用作空舱使用时)的油舱。 Oil/ballast tank is a tank which is used as a cargo oil tank when the ship is not in the dry cargo mode and which is used as a ballast tank or void space when the ship is in the dry cargo mode.

Oil/water interface detector 油水界面探测仪(油水界面探测器) (1)探测污油水舱中油水界面位置的装置;(2)一种仪器,能迅速而正确地测定污油水舱(或其他舱)中油/水的分界面。这种仪器的技术条件应按照 MEPC.5(XIII)号决议"油水界面探测器技术条件"。

Oil/water interface detector 油水界面探测仪

Oil-based mud 油基泥浆

Oil-burning auxiliary boiler 燃油辅助锅炉 是柴油机动力船舶上为供船上工作与生活用蒸汽和热水的装置。一般需要与给水泵、燃油泵及鼓风机等协同工作。

Oil-burning boiler(oil-firing boiler) 燃油锅炉 以燃油作为燃料的锅炉。

Oil-collecting agent 集油剂 系指由扩散压成的化学药剂所制成化合物,用来收集海面浮油。其在海上有效性可持续6 h 以上。

Oil-covered waters 浮油水域

Oiled 加油

Oil-electric drive 柴油机电力传动

Oilfield services 油田服务

Oil-free compressor(non-lubricated compressor) 无润滑压缩机 气体或传动部分无液体润滑剂的压缩机。

Oil-gas separation 油气分离 将从油井中采出的原油和伴生天然气分离,以便于后续分别对油和气进行工艺处理。

Oil-impervious coating 耐油涂层

Oiliness 油气(油质,含油)

Oiling 加油(加油法)

Oiling point 加油点(需要润滑处)

Oilkin 油布

Oil-like substances 类油物质

Oil-lubricated bearing 油润滑轴承

Oil-lubricated friction pair 油润滑摩擦副

Oilmeter 量油计

Oil-protected enclosure 防油外壳 系指其制成使可能积聚在其内部的油蒸发气或无压力的自由油不会妨碍所封装设备有效运行或引起损坏的外壳。

Oilrec = oil recovery 溢油回收

Oil-recovery ship(oil skimmer) 浮油回收船 (1)专门从事港口、海上油田等发生溢油、造成浮油大面积污染时进行浮油回收和消除污染工作的船舶。船上设有吸入设备将浮油和水吸入回收舱内,并设多级沉淀舱将油水分离;(2)装有浮油捕集器,用以收集和处理水面浮油的船。

Oil-serve-motor 伺服机 通过压力油的作用而动作的原动机。

Oil-tight 油密的

Oil-tight bulkhead 油密舱壁 在规定压力下能保持不透油的舱壁。

Oil-tight enclosure 油密外壳 系指制成使其可能存在于周围大气中的油蒸发气或无压力的自由油不能进入其内部的外壳。

Oil-tight floor 油密肋板 在规定压力下能保持不透油的肋板。

Oil-tight hatch cover 油舱盖 设置于油舱口,用

抗油材料作为填料,传动零件在启闭时,不致因碰击而产生火花的舱口盖。

Oil-tight joint 油密接头(油密接合)
Oil-tight riveting 油密铆
Oil-tight transverse bulkhead 油密横舱壁
Oiltightness 油密性 不透油的性能。
Oilways 油槽(油路)
Oily 油类 系指包括但不限于原油、燃油、油泥、油渣、残油、炼制品在内的固态、半固态、乳化态或液态的石油,且不限于以上所述的物质。包括 MARPOL73/78 附则 I 的附录 1 中所列的物质。"油类"不包括动物油和植物油,也不包括 MARPOL73/78 附则 II 中所列举的有毒液体物质。 Oily means petroleum whether in solid, semi-solid, emulsified, or liquid form, including but not limited to crude oil, fuel oil, sludge, oil refuse, oil residue, and refined products, and without limiting the generality of the foregoing, includes the substances listed in appendix 1 of MARPOL73/78. "oil" does not include animal and vegetable based oil or noxious liquid substances (NLS) designated under annex II of MARPOL73/78.

Oily ballast 含油压载水 利用货油舱作为压载水舱后所造成的含油压载水。

Oily ballast tank 含油压载舱

Oily bilge separator 舱底水油水分离器 在舱底水排出前将舱底水中的油分离出来的设备。

Oily bilge water 含油舱底水 对 MARPOL73/78 附则 I 而言,含油舱底水系指可能被由机器处所中的渗漏或维护工作产生的油污染的水。进入舱底水系统(包括舱底水阱、舱底水管系、内底或舱底水储存柜)的任何液体被视为含油舱底水。 For "MARPOL73/78" Annex I, Oily bilge water means water which may be contaminated by oil resulting from things such as leakage or maintenance work in machinery spaces. Any liquid entering the bilge system including bilge wells, bilge piping, tank top or bilge holding tanks is considered oily bilge water.

Oily bilge water holding tanks 含油舱底水储存柜 对 MARPOL73/78 附则 I 而言,含油舱底水储存柜系指在其排放、过驳或处理前收集含油舱底水的舱柜。 For "MARPOL73/78" Annex I, oily bilge water holding tanks are tanks collecting oily bilge water prior to its discharge, transfer or disposal.

Oily drains. 含油泄放水 系指诸如从使用油的设备中漏泄出来的以及从正常情况下可能含有油的设备中泄放出来的泄放水。 Oily drains means drains such as those resulting from the leakage of equipment used for oil and drains from equipment which under normal circumstances may contain oil.

Oily mixture 油性混合物 系指:(1)任何含油分的混合物。包括船底污水、油性生活污水、残油(油泥)、含油的压载水和货油舱的洗舱水。(2)以任何形式,任何油的成分组成的混合物。油性混合物包括但不限于:①舱底污水;②货舱污水(例如货油舱洗舱水、油性污水和油性残渣);③残油;④来自货油舱或燃油舱的含油的压载水。 Oily mixture means: (1) a mixture with any oil content, including bilge slops, oily wastes, oil residues (sludge), oily ballast water, and washings from cargo oil tanks; (2) a mixture, in any form, with any oil content. Oil mixture includes, but is not limited to:①slops from bilges;②slops from oil cargoes (such as cargo tank washings, oily waste, and oily refuse);③oil residue;④oily ballast water from cargo or fuel oil tanks.

Oily mud 油基泥浆 又称油基钻井液。其基本组成是油、水、有机黏土和油溶性化学处理剂。油基泥浆具有抗高温、抗盐钙侵蚀特点,有利于井壁稳定,润滑性好、对油气层损坏小,广泛应用于各类钻井平台。作为提高采收率、保障钻井安全,油基泥浆在钻采作业中起着重要作用。理论上,低毒矿物油油基泥浆基本上无毒,但不意味着一旦溢入海洋不会对海洋生态造成严重影响,或者可以排放到海里。即使在大陆上,矿物油油基泥浆也绝对不能排放。

Oily rags 油性抹布(含油碎布) 系指浸透油的抹布。 Oily rags means rags soaked with oil.

Oily slop tank 含油污水舱

Oily waste 含油废弃物 系指残油(油泥)和含舱底水。 Oily waste means oil residues (sludge) and oily bilge water.

Oily water 含油污水

Oily-water discharge monitoring system (ODMS) 排油监视系统

Oily water separating installation 油水分离装置

Oily water separator 油水分离器

Oily-water separating equipment 油水分离设备 通常系指设在油船上的用于从已被油污染的水中将油分分离出来的一套设备,包括具有沉淀、撇油等功能。

OLEO load OLEO 载荷 定义为引起阻尼器和轮胎的组合结束它们移动的载荷。OLEO 载荷通常不应用于确定在飞行甲板起落架的载荷。OLEO 载荷不总是反映飞机降落在舰船上强加的载荷。OLEO 载荷的比率可用于确定飞机起落架的载荷动态分布。 OLEO load is defined as the load which will cause the damper and tyre combination to reach the end of their travel. OLEO loads should not generally be used to determine loads from the un-

Oman area of the Arabian Sea

dercarriage on the flight deck. OLEO loads do not always reflect the loads that can be imposed by an aircraft landing on a ship. The ratios of OLEO loads may be used to determine the dynamic distribution of load from the undercarriage.

Oman area of the Arabian Sea 阿拉伯海的阿曼区域 系指包括下列坐标包围的海区:北纬22°30′.00;东经059°48′.00;北纬23°47′.27;东经060°35′.73;北纬22°40′.62,东经062°25′.29;北纬21°47′.40;东经063°22′.22;北纬20°30′.37;东经062°52′.41；北纬19°45′.90;东经062°25′.97;北纬18°49′.92;东经062°02′.94;北纬17°44′.36;东经061°05′.53;北纬16°43′.71;东经060°25′.62;北纬16°03′.90；北纬059°32′.24;东经058°15′.20；东经058°58′.52;北纬14°36′.93;东经058°10′.23;北纬14°18′.93;东经057°27′.03;北纬14°11′.53;东经056°53′.75;北纬13°53′.80;东经056°19′.24;北纬13°45′.86;东经055°54′.53;北纬14°27′.38;东经054°51′.42;北纬14°40′.10;东经054°27′.35;北纬14°46′.21;东经054°08′.56;北纬15°20′.74;东经053°38′.33;北纬15°48′.69;东经053°32′.07;北纬16°23′.02;东经053°14′.82;北纬16°39′.06;东经053°06′.52。 Oman area of the Arabian Sea means the sea area enclosed by the following coordinates: 22°30′.00 N;059°48′.00 E;23°47′.27 N;060°35′.73 E;22°40′.62 N;062°25′.29 E;21°47′.40 N;063°22′.22 E;20°30′.37 N;062°52′.41 E;19°45′.90 N;062°25′.97 E;18°49′.92 N;062°02′.94 E;17°44′.36 N;061°05′.53 E;16°43′.71 N;060°25′.62 E;16°03′.90 N;059°32′.24 E;15°15′.20 N;058°58′.52 E;14°36′.93 N;058°10′.23 E;14°18′.93 N;057°27′.03 E;14°11′.53 N;056°53′.75 E;13°53′.80 N;056°19′.24 E;13°45′.86 N;055°54′.53 E;14°27′.38 N;054°51′.42 E;14°40′.10 N;054°27′.35 E;14°46′.21 N;054°08′.56 E;15°20′.74 N;053°38′.33 E;15°48′.69 N;053°32′.07 E;16°23′.02 N;053°14′.82 E;16°39′.06 N;053°06′.52 E.

OMBO(one man bridge operation) 一人驾驶 系指桥楼和驾驶室的布置,以及航行设备和系统能适合于1人操纵的方式。 One man bridge operation means the arrangement and navigational equipment are suitable for one man control.

OMBO ship OMBO 船舶 系指一人桥楼操作的船舶。 OMBO ship means one man bridge operated ship.

Omega navigation system 奥米茄导航系统 奥米茄系统是英文 Omega 的译音。一种甚低频连续波相位双曲线无线电导航系统。由八个发射台组成;分别设置在挪威(A 台)、利比里亚(B 台)、美国夏威夷(C 台)、美国北达科他州(D 台)、法国留尼汪岛(E 台)、阿根廷(F 台)、澳大利亚(G 台)、日本(H 台)。

Omission 遗漏

Omnibearing propelling plant 全方位推进装置 系指可自由转动,能产生任何方向推力的装置。

Omni-directional thruster 全向推力器

On/off hire survey 租赁契约前后检验

On board B/L 已装船提单 系指轮船公司已将货物装上指定船舶后所签发的提单。

On deck B/L 舱面提单 系指货物装在船舶甲板上运输所签发的提单。故又称甲板提单。在这种提单中应注明"在舱面"字样。

On deck risk 舱面货物险 该附加险承保装载于舱面的货物被抛弃或海浪冲击落水所致的损失。一般来讲,保险人确定货物运输保险的责任范围和厘定保险费时,是以舱内装载运输为基础的。但有些货物因体积大或有毒性或有污染性或根据航运习惯必须装载于舱面,为对这类货物的损失提供保险保障,可以加保舱面货物险。加保该附加险后,保险人除了按基本险责任范围承担保险责任外,还要依舱面货物险对舱面货物被抛弃或风浪冲击落水的损失予以赔偿。由于舱面货物处于暴露状态,易受损害,所以保险人通常只是在"平安险"的基础上加保舱面货物险,以免责任过大。

On hand 手持 频繁使用或特别重要的设备/控制装置。 for equipment/control units used frequently or special important.

On sale 廉价出售/贱价抛售

Onboard conditions 船上条件 系指发动机:(1)安装在船上并与其驱动的实际设备相连接;(2)处于运行状态以实现设备的用途。 Onboard conditions means that an engine is:(1) installed on board and coupled with the actual equipment which is driven by the engine; and (2) under operation to perform the purpose of the equipment.

On-board diagnostic(OBD) 车载诊断系统 该系统随时监控发动机的运行状况和尾气后处理系统的工作状态。一旦发现有可能引起排放超标的情况,会马上发出警示。当系统出现故障时,故障灯(MIL)或检查发动机(Check engine)警告灯亮,同时 OBD 系统会将故障信息存入存储器,通过标准的诊断仪器和诊断接口可以以故障码的形式读取相关信息。根据故障码的提示,维修人员能迅速准确地确定故障的性质和部位。从20世纪80年代起,美、日、欧等各大汽车制造企业开始在其生产的电喷汽车上配备 OBD,初期的 OBD 没有自检功能。比 OBD 更先进的 OBD-Ⅱ在20世纪90年代中期产生,美国汽车工程师协会(SAE)制定了一套标准

Onboard local area net　船上局域网

Onboard monitoring manual(OMM)　船上监测手册　系指包括以下项目的手册:(1)用于评估废气滤清系统性能和洗涤水监测的传感器,及其运行、维护和校准要求;(2)废气排放测定和洗涤水监测位置以及必要的辅助服务的详细情况,例如样品传输线和样品处理装置以及相关的操作或维护要求;(3)采用的分析仪,其操作、维护和校准要求;(4)分析仪零位和跨度检查程序;(5)与监测系统正常工作或其证明符合的相关的其他信息或数据。　Onboard monitoring manual(OMM) means a manual included:(1) the sensors to be used in evaluating EGC system performance, and washwater monitoring, their service, maintenance and calibration requirements;(2) the positions from which exhaust emission measurements and washwater monitoring are to be taken together with details of any necessary ancillary services such as sample transfer lines and sample treatment units and any related service or maintenance requirement;(3) the analysers to be used, their service, maintenance, and calibration requirements;(4) analyzer zero and span check procedures; and (5) other information or data relevant to the correct functioning of the monitoring systems or its use in demonstrating compliance.

Onboard NO_x verification procedures　船上NO_x验证程序　系指规定的初次发证检验或换证、年度或中间检验对船上设备要求的程序,以证实符合有关规则的任何要求,亦即发动机发证申请方规定和主管机关认可的要求。　Onboard NO_x verification procedures mean a procedure, which may include an equipment requirement, to be used on board at initial certification survey or at the renewal, annual or intermediate surveys, as required, to verify compliance with any of the requirements of relevant Code, as specified by the applicant for engine certification and approved by the Administration.

Onboard training and drills　船上培训和演习　系指:(1)培训船员熟悉船舶的布置和可能需要其使用的任何灭火系统和设备的位置和操作;(2)紧急逃生呼吸装置的使用训练应视为船上培训的一部分;(3)对承担灭火职责的船员,应通过开展船上培训和演习对其履行职责的能力进行定期评估,以发现需要提高的方面,从而确保其灭火技能方面的适应性得以保持,并确保灭火组织处于操作就绪状态;(4)船上使用船舶灭火系统和设备的训练应按有关规定予以规划和实施;(5)应按有关规定进行消防演习并做记录。　Onboard training and drills are:(1) Crew members shall be trained to be familiar with the arrangements of the ship as well as the location and operation of any fire-fighting systems and appliances that they may be called upon to use.(2) Training in the use of the emergency escape breathing devices shall be considered as part of on-board training.(3) Performance of crew members assigned fire-fighting duties shall be periodically evaluated by conducting on-board training and drills to identify areas in need of improvement, to ensure competency in fire-fighting skills is maintained, and to ensure the operational readiness of the fire-fighting organization.(4) On-board training in the use of the ship's fire-extinguishing systems and appliances shall be planned and conducted in accordance with relative provisions.(5) Fire drills shall be conducted and recorded in accordance with relative provisions.

On-bottom stability of a unit　平台的坐底稳性　系指平台坐底或桩腿插入海床后,在浮力、重力和海床对平台的作用力的联合作用下,抵抗因环境载荷的作用而引起平台倾覆和整体滑移的能力。　On-bottom stability of a unit is the ability of that unit resting on the seabed or with legs penetrating the seabed to counteract the capsizing and sliding of the unit caused by the environmental forces, under combined actions of buoyancy, weight and seabed reaction.

Once-through flash evaporator　直流式闪发蒸发器　盐水经一次蒸发后其剩余部分全部排除的闪发蒸发器。

One cylinder engine　单缸发动机(试验用发动机)

One equipment concept　一个设备概念　通过综合SOLAS多个强制性设备的功能而被认定为一个设备类型的设备。　One equipment concept means the equipment which is recognized as one type of equipment by integrating the function of mandatory equipment of SOLAS of a plural number.

One hour rating　一小时功率　超过持续功率允许连续运行一小时的最大功率。通常为持续功率的110%。

One hour revolution　一小时转速　相应于一小时功率时的转速。

One man bridge operation(OMBO)　一人驾驶　系指仅1人在桥楼操纵船舶的方式。　One man bridge operation (OMBO) means one man operating a ship at bridge.

One-compartment subdivision　一舱制　系指船舶一舱破损进水后,仍能满足破舱水线不超过限界线要求的设计指标。

One-man bridge operated ship　一人驾驶船舶

One-plate clutch　单盘式离合器　一个摩擦盘的离

合器。

One-third octave band sound pressure level of propeller 1/3 倍频程螺旋桨频带声压级　系指带宽内螺旋桨噪声的声压级,中心频带为第 i 号倍频程频带声压级。

One-third octave pressure spectrum level 1/3 倍频程声压谱(密度)级　系指在第 i 号 1/3 倍频程中心频率 $\xi(i)$ 处的声压谱(密度)级。

One-way macerator system 单向碎渣机系统　是一种新型食物粉碎机。其由组合处理器和真空系统组成。用于卫生地收集和处理食品废物,能大量地处理肉骨头、鱼皮和海鲜贝壳以及糊状物、土豆和米饭。食品废物在厨房内被碎渣机粉碎和磨碎,然后用真空系统输送到专门的储柜中,以确保船舶在禁止排放区内运营时该系统仍能持续运转。

On-line information 在线信息

On-line system 联机系统

On-load release 有载脱开　系指当艇钩装置负载时,打开救生艇释放和回收系统的动作。On-load release is the action of opening the lifeboat release and retrieval system whilst there is load on the hook assemblies.

On-scent commander 现场指挥　指定在特定搜寻区域内对海面搜救工作进行协调的救助单位的指挥。On-scent commander means the commander of a rescue unit designed to co-ordinate surface search and rescue operations within a specified search area.

Onshore wind farm 陆上风电场　系指建立在陆上的风电场。

OOW = Officer of the navigational watch 值班驾驶人员　负责导航和桥楼操作安全的人员。OOW is a person responsible for the safety of navigation and bridge operations.

Open barge 敞舱驳　货舱上方全敞开的货驳。

Open cheque 空白支票　在支票出票时,对若干必要记载事项未记载,即完成签章并予以交付,而授权他人在其后进行补记,经补记后才使其有效成立的支票。

Open connected bucker chain 间隔斗链　前后两个斗链以链接板间隔安装的斗链。

Open cooling water system 开式冷却水系统　主、辅柴油机采用舷外水直接冷却的冷却系统。

Open cover 预约保险　是保险双方约定总的保险范围并签订预约保险合同的长期保险。预约保险合同内载明承保货物范围、保险责任范围、每批货物的最高保险金额、保险费率、保险费结算办法等。对于长期、大量投保的被保险人可采用这种保险。我国进口货物大都采用这种承保方式。国外预约保险合同往往列明价值条款、船舶条款,有时还附有费率表等。

Open cycle 开式循环　从大气中吸入空气,经压缩、燃烧、做功后再进入大气的燃气轮机循环。

Open cycle gas turbine plant 开式循环燃气轮机装置　按开式循环工作的燃气轮机动力装置。

Open deck spaces 开敞甲板处所　系指:(1)救生艇和救生筏登乘与降落站以外的开敞甲板处所和围蔽的游步甲板处所。如考虑将围蔽的游步甲板处所归入此类,其应无大的失火危险,即其内应只设有甲板家具。此外,此类处所还应通过固定开口进行自然通风。(2)露天处所(上层建筑和甲板室外面的处所)。(3)没有失火危险的开敞甲板处所、露天处所。 Open deck spaces are: (1) open deck spaces and enclosed promenades clear of lifeboat and liferaft embarkation and lowering stations. To be considered in this category, enclosed promenades shall have no significant fire risk, meaning that furnishings shall be restricted to deck furniture. In addition, such spaces shall be naturally ventilated by permanent openings; (2) air spaces (the space outside superstructures and deckhouses); (3) open deck spaces and air spaces having no fire risk.

Open deck 开敞甲板　系指两端开口或一端开口的,通过发布在侧壁或上部甲板的固定开口提供遍及整个甲板长度的充分有效的自然通风的甲板。 Open deck means a deck that is open on both ends, or is open on one end and equipped with adequate natural ventilation that is effective over the entire length of the deck through permanent openings distributed in the side panels or in the deck above.

Open decks(fishing vessel) 露天甲板(渔船)　系指:(1)露天甲板处所和围蔽游廊、渔获物初加工处所、洗鱼处所以及无火险的类似处所;(2)外面的露天处所。 Open decks are: (1) open deck spaces and enclosed promenades, spaces for processing fish in the raw state, fish washing spaces and similar spaces containing no fire risk; (2) the air spaces outside superstructures and deckhouses.

Open end 开口端(艏、艉端)

Open feed system 开路供电系统(开式给水系统)

Open forging 自由锻造/无型锻造　是利用冲击力或压力使金属在上下砧面间各个方向自由变形,不受任何限制而获得所需形状及尺寸和一定机械性能的锻件的一种加工方法,简称自由锻。自由锻造所用工具和设备简单,通用性好,成本低。同铸造毛坯相比,自由锻消除了缩孔、缩松、气孔等缺陷,使毛坯具有更高的力学性能。锻件形状简单,操作灵活。因此,它在重型机器及重要零件的制造上有特别重要的意义。

Open gauging device 开式液位测量装置 系指通过液货舱开口,将测量仪表放置于货物或其蒸发气之中,例如空挡液位测量孔的装置。 Open gauging device means a device which makes use of an opening in the tanks and may expose the gauger to the cargo or its vapour. An example of this is the ullage opening.

Open indication 开放缺陷显示 系指去除磁粉后也能看见的缺陷显示或者使用透视探像剂的颜色对比检出的缺陷显示。 Open indication means an indication visible after removal of the magnetic particles or that can be detected by the use of contrast dye penetrant.

Open port/open harbor 不冻港/开放港 系指冬季不会结冰,船舶能正常进出的港口,尤指高纬度地区(如俄罗斯、北欧、加拿大等)冬季不结冰的港口。我国的大连、旅顺、秦皇岛是我国北方终年不冻港。从地理学角度看,高纬度地区的不冻港主要受制于这样几个因素:暖性洋流流经港口海域;海水的盐度及港口附近有无河流注入,一般有河流注入的港口,海水盐度被河水冲淡,容易结冰;港口海域水体对太阳热能的储存能力。

Open RO-RO spaces 开敞滚装处所 系指下列滚装处所:(1)搭载的乘客能进入的;(2)①在其两端是敞开的 或 ②在一端有开口并在其侧板或甲板顶部或其上面分布有永久性开口,其总面积至少是处所侧壁总面积的10%;(3)开放船首、尾方向的两侧末端或其中的一个末端,拥有该处所面总面积10%以上的侧面积,从安装在甲板正方向的固定开口处或从上部,具有能提供遍及整个长度的充分有效的自然通风的滚装处所。 Open RO-RO spaces are those RO-RO spaces:(1)to which any passengers carried have access; and(2)which either;①are open at both ends; or ②have an opening at one end and are provided with permanent openings distributed in the side plating or deckhead or from above, having a total area of at least 10% of the total area of the space sides;(3)those RO-RO spaces that are either open at both ends or have an opening at one end, and are provided with adequate natural ventilation effective over their entire length through permanent openings distributed in the side plating or deck head or from above, having a total area of at least 10% of the total area of the space sides.

Open session 公开场合 是人与人发生交往和接触的地方,是公有公用之活动场所。即向公众开放的、允许公众自由出入的场所和场合,比如街道、田野、广场、集市、公园、娱乐和体育馆等场所。

Open socket 开式索节 具有叉头连接端的索节。

Open space 开敞处所(游艇) 对游艇而言,系指开敞的甲板空间。

Open space 开敞处所 系指(1)除围蔽处所和半围蔽处所之外的处所。(2)其内无滞流区域(通过风和自然对流使蒸发气迅速地弥散)在开敞空气位置中的处所。典型的空气速度应很少小于0.5m/s,且应经常大于2m/s。 Open space means:(1)these spaces other than enclosed spaces and semi-enclosed spaces. (2)a space in open air situation without stagnant areas where vapors are rapidly dispersed by wind and natural convection. Typical air velocities should rarely be less than 0.5 m/s and should frequently be above 2 m/s.

Open stokehold draft 开式炉舱通风 炉舱保持敞开的一种锅炉强力通风的形式。

Open superstructure 开敞上层建筑 系指非封闭的上层建筑。 Open superstructure is not enclosed superstructure.

Open top container 开顶集装箱 系指便于装运重型基础设施的集装箱。

Open type frahm tanks 开式减摇水舱 对称设置在船舶两舷,用以减少船舶横摇,其下部与舷外水接通,上部与大气接通的开式水舱。

Open vehicle space 开敞的车辆区域 系指下述情况:(1)车辆区域的外板无开口,并且前后端无舱壁,则该区域上部甲板的开口面积应满足下式:$a/A \geq 1/2$,式中,a—上部甲板开口的面积;A—车辆甲板的面积;(2)若车辆甲板两边外板有开口且其开口尽可能布置为横跨整个车辆区域的全长,则开口面积应满足下式:$[(a/A)+(5/3)(S_a/S_A)] \geq 1/2$,式中:$a$—上部甲板开口的面积,$A$—车辆甲板的面积,$S_a$—车辆区域一侧外板的开口面积,$S_A$—车辆区域一侧外板的侧面积;(3)开放船首尾方向的两侧末端或其中的一个末端,拥有该处所面总面积10%以上共计面积的侧面积,从安装在甲板正方向的固定开口处或从上部,具有能提供遍及整个长度的充分有效的自然通风的车辆处所。 Open vehicle space means the followings:(1)the bulkhead is not provided at the end of fore and after, and openings are not provided on the shell plating of vehicle area. In this case, the area of openings on the upper deck of considering area is to be complied with the followings:$a/A \geq 1/2$;Where:a—area of opening on the upper deck;A—area of vehicle deck;(2)When the openings are provided on the both side shell plating in vehicle area, the area of opening is comply with the following:$[(a/A)+(5/3)(S_a/S_A)] \geq 1/2$, where:$a$—area of opening on the upper deck,A—area of vehicle deck,

S_a—area of opening on one side in vehicle area, S_A—area of shell plating on one side in vehicle area; (3) those vehicle spaces either open at both ends, or have an opening at one end and are provided with adequate natural ventilation effective over their entire length through permanent openings distributed in the side plating or deckhead or from above, having a total area of at least 10% of the total area of the space sides.

Open water propeller efficiency 螺旋桨敞水效率 孤立的螺旋桨在开阔静水中或均匀进流中工作时推力功率对收到功率的比值 $\eta_0 = P_T/P_D = TV_A/2(nQ)$，式中，$P_T$—推力功率，$P_D$—收到功率，$T$—螺旋桨推力，$V_A$—进速，$n$—螺旋桨转速，$Q$—螺旋桨转矩。

Open wire 明线 系指平行架设在电线杆上的架空线路。它本身是导电裸线或带绝缘层的导线。虽然它的传输损耗低，但是由于易受天气和环境的影响，对外界噪声干扰比较敏感，已经逐渐被电缆取代。

Open yacht 敞开艇 系指从艇首至艇尾范围不具有风雨密的连续露天甲板的游艇。

Open-feed system 开式给水系统 凝水从冷凝器出来经热井再供入锅炉的给水系统。

Opening 孔（开度，打开状态）

Opening bank 开证行 系指接受开证申请人的要求和指示或根据其自身的需要，开立信用证的银行。开证行一般是进口人所在地银行。开证行是以自己的名义开立信用证下的义务负责的。虽然开证行同时受到开证申请书和信用证本身两个契约约束，但是根据UCP500第3条规定，开证行依信用证所承担的付款、承兑汇票或议付或履行信用证项下的其他义务的责任，不受开证行与申请人或申请人与受益人之间产生纠纷的约束。开证行在验单付款之后无权向受益人或其他前手追索。

Opening meeting 审核开始会议 系指由审核机构、被审核单位的相关人员参加的，旨在说明审核的目的、日程等的开始会议。

Opening moment 开启力矩

Opening of the valve 阀的开度

Opening pressure 开启压力

Opening ratio 开孔比值 系指除切口和扇形切口外的所有开孔面积的总和与舱壁总面积之比值。Opening ratio means the ratio of the sum of areas of openings, except slots and scallops, to the area of the bulkhead.

Opening side scuttle 活动舷窗 可启闭的舷窗

Opening square window 活动矩形窗 可启闭的矩形窗

Opening time (of mechanical switching device) (机械开关电器的) 断开时间 系指开关电器从断开操作开始瞬间到所有极的弧触头都分开瞬间为止的时间间隔。对断路器而言：(1) 直接导致的断路器，断开开始的瞬间系指电流增大到足以导致断路器动作开始的瞬间。(2) 利用任何辅助电源动作的断路器，断开开始的瞬间系指辅助电源施加于断开脱扣器的开始瞬间。注意：①断开开始的瞬间，即发出断开指令（例如施加脱扣电流等）的瞬间。②对于断路器，"断开时间"通常被称为"脱扣时间"。严格地说脱扣时间是表示断开操作开始的瞬间到断开命令变成不可逆的瞬间之间的时间间隔。

Opening (electrical equipment) 开口（电气设备） 系指任何小孔，或并非指定用于防止气体或蒸发气通过的门、窗或板。Opening (electrical equipment) means any aperture, or door, window or panel not designed to prevent the passage of gas or vapour.

Openings in freeboard deck 干舷甲板上的开口 参照经修正的ILLC（MSC.143(77)）决议第18(2)条，系指在干舷甲板上除货舱口、机舱开口、人孔和与甲板齐平的小舱口以外的，由封闭的上层建筑，或甲板室、或强度相当的和由风雨密的升降口来保护的开口。 Ref. ILLC, as amended [Resolution MSC.143(77) Reg. 18(2)], openings in freeboard deck mean openings other than hatchways, machinery space openings, manholes and flush scuttles, which protected by an enclosed superstructure or by a deckhouse or companionway of equivalent strength and weather-tightness.

Openings in superstructures 上层建筑上的开口 参照经修正的ILLC（MSC.143(77)）决议第18(2)条，系指在露天的上层建筑甲板上或在干舷甲板上的甲板室顶部，通往干舷甲板以下的处所或围蔽的上层建筑以内处所的，由有效的甲板室或升降口来保护的开口。Ref. ILLC, as amended [Resolution MSC.143(77) Reg. 18(2)], openings in superstructures mean openings in an exposed superstructure deck or in the top of a deckhouse on the freeboard deck which give access to a space below the freeboard deck or a space within an enclosed superstructure, which protected by an efficient deckhouse or companionway.

Openings in superstructures having height less than standard height 高度小于标准高度的上层建筑的开口 参照经修正的ILLC [MSC.143(77)] 决议第18(3)条，系指在高度小于标准高度的后升高甲板或上层建筑上的，如甲板室高度等于或大于标准后升高甲板高度，则甲板室顶部的设有可接受关闭装置的开口。但是，如果该甲板室的高度至少为一个标准上层建筑高度，则该开口不必用坚固的甲板室或升降口来保护，在高度小于标准上层建筑高度的甲板室上甲板室顶部的开口可以用类似的方法来处理。 Ref. ILLC, as amended [Resolution MSC. 143(77) Reg. 18(3)], openings in superstructures having height less than standard height mean openings in the top of a deckhouse on a raised quarterdeck or superstructure of less than standard height, having a height equal to or greater, than the standard quarterdeck height, which provided with an acceptable means of closing but need not be protected by an efficient deckhouse or companionway provided the height of the deckhouse is at least the height of the superstructure. Openings in the top of the deckhouse on a deckhouse of less than a standard superstructure height may be treated in a similar manner.

Openings normally closed at sea 通常在航行时关闭的开口 系指用以保证内部开口的水密完整性,且通常在航行时关闭的出入门和舱盖,并应在该处和驾驶室装设显示这些门或舱盖是开启或关闭的设施。每一个此类门或舱盖必须张贴一个告示牌,其大意是不能让它开着。这类门或舱盖的使用应经值班驾驶员批准。 Openings normally closed at sea means access doors and access hatch covers normally closed at sea, intended to ensure the watertight integrity of internal openings, are to be provided with means of indication locally and on the bridge showing whether these doors or hatch covers are open or closed. A notice is to be affixed to each such door or hatch cover to the effect that it is not to be left open. The use of such doors and hatch covers is to be authorized by the officer of the watch.

Openings permanently kept closed at sea 在海上保持永久关闭的开口 系指为保证内部开口的水密完整性,在海上保持永久关闭的其他关闭装置,应有一个告示牌张贴其上,其大意是必须保持关闭。用螺栓紧固盖子的人孔不必设该告示牌。 Openings permanently kept closed at sea means other closing appliances which are kept permanently closed at sea to ensure the watertight integrity of internal openings are to be provided with a notice which is to be affixed to each such closing appliance to the effect that it is to be kept closed. Manholes fitted with closely bolted covers need not be so marked.

Open-jet type cavitation tunnel 水柱式空泡试验水筒 测试段为自由射流的空泡试验水筒。

Open-riser stairs 透光梯步(透空梯步) 没有封板的梯步。

Open-top compartment 开敞舱室 系指每 $1m^3$ 净舱容积应至少具有 $0.3 m^3$ 直接通向大气的舱室。

Open-top container ship 敞口集装箱船 系指:(1)具有双层底、双壳、舷顶设抗扭箱、甲板开口大,载运集装箱或用双层底、具有抗扭箱或其他等效的单层壳舷侧结构代替,但货舱无舱口盖的船舶;(2)一种特殊设计的集装箱船,其 1 个或多个货舱没有设置舱口盖;(3)就"1969 年吨位丈量公约"的适用范围而言,敞口集装箱船系指设计用于载运集装箱且其结构像敞开的"U"形的船舶,货舱舱口的净开口总面积有不小于 66.7% 为"敞口"形式,具有双层底及其上面的上甲板上无舱口盖的高侧壁建筑,在型吃水以上无全通甲板。 Open-top container ship means:(1)a ship with double bottom and double side skin construction with torsion box girders fitted at top sides, large deck openings, carrying containers, or as alternative, single side skin construction with double bottom and torsion box girders or equivalent structure, but no hatch covers for holds;(2)a containership especially designed so that one or more of the cargo holds need not be fitted with hatch covers;(3) for the purpose of application of the 1969 Tonnage Measurement Convention, open-top containership means a ship which is designed for the carriage of containers and which is constructed like an open "U", with not less than 66.7% of the total cargo hatchway clear opening area to an "open-top" configuration, with a double and above the moulded draught.

Open-top ship 敞口船舶 系指位于干舷甲板上露天部分的舱口,无风雨密舱盖设备,其他舱口符合风雨密要求的船舶。

Open-type electric motor 开启型电机 系指具有通风开孔,以允许外部冷却空气在电机绕组上方或周围通过的电机。

Open-water propeller test 螺旋桨敞水试验 在未受扰动的水中进行的螺旋桨模型单独试验。

Open-water test(open-water test) boat 敞水试验箱 螺旋桨试验时用以安装螺旋桨测力计的水密箱。

Operating agreement 业务协定 系指"国际海事卫星组织业务协定"及其附件。 Operating agreement means the Operating Agreement on the International Maritime Satellite Organization(INMARSAT), including its Annex.

Operating and maintenance manual 操作和维护手册

Operating area 操纵区域 系指操纵舱室以及船的操纵舱室两侧和接近操纵舱室延伸到船侧的部分。 Operating area is the operating compartment and those parts of the craft on both sides of, and close to, the operating compartment which extend to the craft's side.

Operating book for oil skimmer 浮油回收船的操作手册 系指一本经批准的,包含浮油回收作业的准备及其在作业过程中需采取的安全措施的手册。其中包含:(1)设备与布置:①回收油贮存舱布置;②回收油输送系统;③可燃气体测量仪表;④其他有关设备。(2)操作准备:①检查船上所有设备,以确定哪些是符合有关规定的设备;②非永久性设备的安装和紧固;③某些管路的封堵;④装配空气管;⑤切断不符合有关规定的电气设备的电源;⑥关闭安全区域与气体危险区域之间的开口;⑦启动附加的通风设备;⑧将冷却水泵转换到低位海水吸入口;⑨设置禁止使用明火和非合格防爆电气设备等的标志牌。(3)浮油回收作业:①关于距溢油源安全距离及有关注意事项的指导性条文;②作业过程中,在开敞甲板上和可能聚集可燃气体的处所中进行可燃气体的探测,如在甲板上发现有可燃气体扩散时,则船舶应立即移开;③如可燃气体在围蔽处所中扩散,则应采清洗、通风、排空相邻的液舱等措施;④正压通风发生故障后须采取的应急措施;⑤防止回收油贮存舱溢油;⑥排出。(4)回收油贮存舱和管路的清洗和除气。

Operating coefficient 作业系数(起重设备) 系指考虑起重设备作业频次与载荷状态所给的裕度系数。

Operating compartment 操作舱室(操纵室) 系指进行驾驶和操纵船舶的围蔽区域。 Operating compartment means the enclosed area from which the navigation and control of the craft is exercised.

Operating conditions 作业工况 系指平台为进行钻井作业而在井位上时所处的工况,且其环境载荷与作业载荷的组合在为这种作业所确定的适当设计限度之内。根据情况,该平台可以是浮在海面或是支撑在海床上。 Operating conditions mean these conditions wherein a unit is on location for the purpose of conducting drilling operations, and combined environmental and operational loadings are within the appropriate design limits established for such operations. The unit may be either afloat or supported on the seabed, as applicable.

Operating conditions (fishing vessel) 营运条件(渔船) 系指经主管机关认可的各种营运条件和种类,包括以下各项:(1)满载燃料、备品、冰、渔具等开往渔场;(2)满载渔获物离开渔场;(3)满载渔获物和10%备品,燃料等到达基地港;(4)装载 20% 渔获物和 10% 备品,燃料等到达基地港。 Operating conditions means those conditions mean number and type of operating conditions shall be to the satisfaction of the Administration and shall include the following, as appropriate: (1) departure for the fishing grounds with full fuel, stores, ice, fishing gear, etc.; (2) departure from the fishing grounds with full catch; (3) arrival at home port with full catch and 10% stores, fuel, etc.; and (4) arrival at home port with 10% stores, fuel, etc. and a minimum catch, which shall normally be 20% of full catch but may be up to 40% provided the Administration is satisfied that operating patterns justify such a value.

Operating console 控制台

Operating current (of a over-current relay or release) (过电流继电器和脱扣器的)动作电流 系指当电流等于或大于此值时,继电器或脱扣器动作的电流值。

Operating draft (floating dock) 作业吃水 d_D (m) 系指浮船坞在平浮状态下,浮船坞承载达到其设计的举升能力船舶时的吃水。

Operating draught 营运吃水 系指反映实际部分装载状态或满载状态的任何营运吃水。 Operating draught means any operating draught reflecting actual partial or full load conditions.

Operating gear 操作装置(传动装置)

Operating handle 操纵杆(操纵手柄)

Operating hour 运转时数

Operating instruction 操作规程/操作指南(操作说明书,操作须知)

Operating instructions of survival craft 救生艇筏操作须知 系指在救生艇筏及其降落操纵器上或附近设置的告示或标志,其应包括:(1)有示意图说明操纵器的用途及此项设备的操作程序,并告知有关须知或注意事项;(2)在应急照明的条件下,容易看清;(3)使用符合 IMO 组织建议案的符号。 Operating instructions of survival craft means a posters or signs provided on or in the vicinity of survival craft and their launching controls and included: (1) illustrate the purpose of controls and the procedures for operating the appliance and give relevant instructions or warnings; (2) be easily seen under emergency lighting conditions; (3) use symbols in accordance with the recommendations of IMO.

Operating lever 操纵杆(工作杆)

Operating life 使用寿命

Operating limitations 操纵限制 系指高速船在控制、可操纵性和性能及其所遵循的运营程序方面所受的限制。 Operating limitations means the craft limitations in respect of handling, controllability and performance and the craft operational procedures within which the craft is to operate.

Operating load 工作载荷(操作负载,运行负载)

Operating maintenance 运行维修(日常维护)

Operating manual 操作手册

Operating manual for an unit　平台操作手册　系指一份可供所有人员随时使用的操作手册，作为在正常情况和预料到的紧急情况下安全操作平台的指南。该手册除了介绍平台总体情况外，还应包括所有对平台至关重要的操作程序及指导和限制条件，以确保作为入级依据的装载条件和环境条件在作业中不应被超过；该手册应简明扼要易懂。每本手册都应有目录表和索引并且有可互相参考的，在平台上能方便地查到的有关详细资料；操作手册所包括的内容至少应符合主管机关和/或 MODU 规则的有关规定。　Operating manual for an unit means a copy of operating manuals provided onboard and be readily available to all concerned as a guidance for the safe operation of the unit for both normal and envisaged emergency conditions provided. The manuals are, in addition to providing the necessary general information about the unit, to contain guidance and limitations on and procedures for the operations that are vital to the unit, ensuring that the loading and environmental conditions, on which the classification is based, will not be exceeded in service. The manuals are to be concise and be compiled in such a manner that they are easily understood. Each manual is to be provided with a contents list, an index and wherever possible cross-referenced to additional detailed information which is to be readily available onboard; the contents of the manual are at least to comply with relevant provisions of the Administration and MODU Code.

Operating mechanism　操作装置　系指操作者通过去激发活动钩体的打开和释放的设施。它包括操作手柄、连接件/软轴和静水压力连锁(如设有)。　Operating mechanism is the means by which the operator activates the opening, or release, of the movable hook component. It includes the operating handle, linkages/cables and hydrostatic interlock, if fitted.

Operating mode at first transitional regime　第一过渡阶段航行模式(地效翼船)　系指船由浮力、水动升力(作用于船体和侧浮舟上)及气垫升力(作用于主升力翼上)支撑。由于形成气垫的气动布局不同，动力气垫船和动力气垫地效翼船的气垫升力占总升力的比例大于动力增升地效翼船。船加速时会经过两个阻力峰：即气垫阻力峰和船体从排水状态过渡至滑行模式的阻力峰。在这种模式下，不同作用中心的水动力和气垫升力将改变船的纵倾直至其由飞行控制面(如水平安定面及水平舵)来平衡的航行模式。　Operation mode at first transitional regime means a mode at which the craft is supported by buoyancy and hydrodynamic lift (acting on hull and side buoys) and air cushion lift (acting on main lifting wing). The cushion lift is a larger portion of total lift force for DACC and DACWIG than PARWIG due to the aerodynamic configuration for forming the air cushion on these craft. As the craft accelerates, it will transit through two drag peaks: first, as the cushion passes through hump speed and later as the hull moves from displacement into planning mode. In this mode, the hydrodynamic and air cushion lift forces, acting through their respective centers of effort, will alter the trim unless balanced by adjustments to flying control surfaces, for example the tail stabilizer (elevators).

Operating mode at second transitional regime　第二过渡阶段航行模式(地效翼船)　系指在起飞速度时，船总重完全由动力气垫升力和气动升力支撑。船舶持续加速时可产生额外升力，从而使船升高至巡航高度，此时需调整主升力翼襟翼以维持平衡。在该模式下，维持主翼升力中心比船的重心略靠前是非常重要的，这样可使船降落或水上迫降时略有艉倾。　Operation mode at second transitional regime means a mode at which when the speed is take-off speed, the total WIG weight is supported by the dynamic air cushion lift and aerodynamic lift. As the craft continues to accelerate, an excess of lift is generated allowing the craft to rise to cruise elevation, and the main lifting wing flaps to be adjusted to level the craft out. In this mode, it is important that lift from the main wing acts slightly forward of the centre of gravity of the craft, to give tail down tendency when landing or ditching.

Operating panel　操作面板
Operating personnel　操作人员
Operating position　操作位置(作业位置)
Operating pressure　工作压力
Operating record　操作记录
Operating region　工作范围(航区)
Operating repair　日常维护检修
Operating rod　操作杆
Operating room　手术室(操作室)
Operating signal　操作信号
Operating site　作业场所
Operating state　操作状态(运行状态)
Operating station　操纵站　系指操纵室内设有必需的航行、操纵和通信设施的限制区域。在此区域执行航行、操纵、通信、指挥、下达舵令和瞭望观测等业务。　Operating station means a confined area of the operating compartment equipped with necessary means for navigation, manoeuvring and communication, and from where the functions of navigating, manoeuvring, communication, commanding, conning and lookout are carried out.

Operating stress 工作应力(操作应力)
Operating system 控制系统(操作系统)
Operating system(OS) 操作系统 是管理和控制计算机硬件与软件资源的计算机程序,是直接运行在"裸机"上的最基本的系统软件,任何其他软件都必须在操作系统的支持下才能运行。操作系统是用户和计算机的接口,同时也是计算机硬件和其他软件的接口。操作系统的功能包括管理计算机系统的硬件、软件及数据资源,控制程序运行,改善人机界面,为其他应用软件提供支持,让计算机系统所有资源最大限度地发挥作用,提供各种形式的用户界面,使用户有一个好的工作环境,为其他软件的开发提供必要的服务和相应的接口等。实际上,用户是不用接触操作系统的,操作系统管理着计算机硬件资源,同时按照应用程序的资源请求,分配资源,如:划分 CPU 时间,内存空间的开辟,调用打印机等。

Operating test 操作试验
Operating time 演算时间(工作时间,运行时间)
Operating values 操作值 系指柴油机参数,如发动机日志中所载的与 NO_x 排放量性能有关的气缸峰值压力、排气温度等。这些数据与载荷有关。 Operating values are engine data, like cylinder peak pressure, exhaust gas temperature, etc., from the engine log which are related to the NO_x emission performance. These data are load-dependent.

Operating valve 操纵阀
Operating zone 操作区(工作区)
Operation 运转(运行)
Operation characteristic 操作性能(工作性能)
Operation control center(OCC) 运行控制中心
Operation desk 操纵台(控制台)
Operation instruction 操作规程(操作说明书,操作须知)
Operation item 工艺项目
Operation manual 操作说明书(操作条令,操作手册)
Operation parameter 运行参数
Operation test 操作试验(运行试验,动作试验)
Operation time 演算时间(工作时间,运行时间)
Operation trim 营运纵倾
Operation waters 营运水域
Operational accuracy 操作精度(实际精度)
Operational bridge functions 作业桥楼功能 与船舶作业有关的功能。这些功能包括:操纵功能、甲板设备操作(用于起抛锚、溢油回收和货物装卸操作)、救助作业、监控内部安全系统、与桥楼上作业和遇险情况下的安全相关的外部通信以及内部通信、靠泊作业功能。 Functions related to ship handling in relation to the operation the vessel is engaged in. Such functions are: manoeuvring functions, deck equipment operation (for anchor handling, oil recovery and cargo transfer operations), rescue operation, monitoring of internal safety systems, external and internal communication related to safety in bridge operation and distress situations, docking functions.

Operational bridge 作业桥楼 海上作业工作站所在的地区。 Operational bridge is the area of the bridge where workstations for offshore operations are located.

Operational check 操作检查
Operational constraint 操作限制
Operational display area 操作显示区 用于以图表显示海图和雷达信息的显示区,不包括用户对话区。在海图显示器上,为海图显示区。在雷达信息显示器上,为包括雷达图像的区域。 Operational display area is an area of the display used to graphically present chart and radar information, excluding the user dialogue area. On the chart display this is the area of the chart presentation. On the radar display this is the area encompassing the radar image.

Operational element 操作元素
Operational envelope 营运包 营运包用运行速度、浪高、排水量、航区、温度、运动、飞行甲板操作和雷达等来定义舰船的业务,由船东提供。 The operational envelope defines the ship's service in terms of operational speeds, wave heights, displacements, service area, temperatures, motions, flight deck operations and radar, and is to be provided by the owner.

Operational instruction 操作说明
Operational level 操作级(人员) 系指与下列内容有关的责任级别:(1)在海船上担任负责航行或轮机值班的高级船员或被指定为周期无人机舱的值班轮机员或担任无线电操作员;(2)在相同责任范围内的管理级人员的指导下按照正规的程序对指定职责范围内所有职能的履行保持直接的控制。 Operational level means the level of responsibility associated with: (1) serving as officer in charge of a navigational or engineering watch or as designated duty engineer for periodically unmanned machinery spaces or as radio operator on board a ship; (2) maintaining direct control over the performance of all functions within the designated area of responsibility in accordance with proper procedures and under the direction of an individual serving in the management level for that area of responsibility.

Operational limit 营运范围

Operational limitation 操作限制

Operational loadings 作业载荷 系指在作业期间所受到的除环境载荷及部件自重以外的其他载荷。

Operational loads 营运载荷 (1)包括空船重力(钢材重力和舾装、机器和永久性的设备)、浮力载荷(船舶浮力)、可变载荷(货物、压载水、备品和消耗品、人员、临时设备)、其他载荷(拖船和停泊载荷、拖带载荷、锚泊和系泊载荷、起重设备载荷);(2)一般是静载荷。包括空船重力、浮力载荷、可变载荷和其他载荷。营运载荷系因船舶的运行和操纵而产生。 Operational loads include: (1) lightship weight (steel weight and outfit, machinery and permanent equipment); buoyancy loads (buoyancy of the ship); variable loads (cargo, ballast water, stores and consumables, personnel, temporary equipment); other loads (tug and berthing loads, towing loads, anchor and mooring loads; lifting appliance loads); (2) Operational loads are generally static loads. They are grouped into lightship weight, buoyancy loads, variable loads and other loads. The operational loads occur as a consequence of the operation and handling of the ship.

Operational manual for oil recovery ship 浮油回收船操作手册 系指一本经批准的并包含浮油回收作业的准备及其在作业过程中需采取安全措施的手册。其应包括:(1)设备与布置:①回收油储存舱布置;②回收油输送系统;③可燃气体测量仪表;④其他有关设备;(2)操作准备:①检查船上所有设备,以确定哪些是符合有关规定的设备;②非永久性设备的安装和紧固;③某些管路的封堵;④装配空气管;⑤切断不符合有关规定的电气设备的电源;⑥关闭安全区域与气体危险区域之间的开口;⑦启动附加的通风设备;⑧将冷却水泵转换到低海水吸入口;⑨设置禁止使用明火和非合格防爆电气设备等的标志牌;(3)浮油回收作业:①关于距溢油源安全距离及有关注意事项的指导性条文;②作业过程中,在开敞甲板上和可能聚积可燃气体的处所中进行可燃气体的探测,如在甲板上发现有可燃气体传播时,则船舶应立即移开;③如可燃气体在围板处所中传播,则应采取清洗、通风、排空相邻舱等措施;④正压通风发生故障后须采取的应急措施;⑤防止回收油储存舱溢油;⑥排出;(4)回收油储存舱和管路的清洗和除气。

Operational mode 操作模式(动力定位) 系指控制模式,在此模式下动力定位系统可被操作,例如:(1)自动模式(自动船位和船向控制);(2)独立的联合操纵杆模式(手动船位控制且具有可选择的自动或手动的舷向控制);(3)手动模式(对每个推进器的螺距和速度、方位,启动和停止的单个控制)。 Operational mode is the manner of control under which the DP-system may be operated, e. g. (1) automatic mode (automatic position and heading control); (2) independent joystick mode (manual position control with selectable automatic or manual heading control); (3) manual mode (individual control of pitch and speed, azimuth, start and stop of each thruster).

Operational modes 操纵模式 依据海域而定的操作模式。 Operational modes are modes of operation depending on the sea area.

Operational pressure 工作压力

Operational procedure 操作程序(运行程序)

Operational readiness of life-saving appliances 救生设备使用准备状态 系指在船舶离港以及航行期间,所有的救生设备均应处于工作状态,并随时可用。 Operational readiness of life-saving appliances means an order which before the ship leaves port and at all times during the voyage, all life-saving appliances shall be in working order and ready for immediate use.

Operational reliability 使用可靠性(工作可靠性)

Operational requirement 运用要求

Operational restriction 营运限制

Operational skill 操作技能

Operational speed 营运航速(地效翼船) 系指在地效状态中减少推进功率时的正常营运航速。 Operational speed is the normal operating speed at reduced level of propulsion power in ground effect mode.

Operational speed 营运速度 系指最大航速的90%。 Operational speed is 90% of maximum speed.

Operational status 操作状态

Operational test 操作试验

Operational testing 操作试验

Operational waste 作业废弃物 (1) MARPOL其他附则未涵盖的船上收集的产生于船舶正常维护或作业过程或用于货物存储和装卸的所有固体废弃物(包括泥浆)。作业废弃物也包括货舱和外部洗涤水中包含的清洁剂和添加剂。作业废弃物不包括灰水、船底水或气体对船舶作业至关重要的类似排出物。(2)所有与货物相关联的,除残油和有毒液体物质之货物残余物以外的废物、维修废物、货物残余物。"作业废弃物"包括船上焚烧炉和燃煤锅炉产生的烟灰和炉渣(即加热时不完全燃烧熔化在一起的物质),但不包括按废弃物处理塑料黏结物或按附则I废弃物处理的的油性抹布。 Operational waste means: (1) all solidwasters (including slurries) not covered by other Annexes to 73/78 MARPOL that are collected on board during normal maintenance or operations of a ship, or used for cargo stowage and handling. Operational wasters also include cleaning agents and additives contained in cargo hold and external wash water. Operational wasters do not include grey water, bilge water, or other similar discharges essential to the operation of a ship. (2) all cargo associated waste, maintenance waste, and cargo residues other than oil residues and NLS cargo residues. "Operational waste" includes ashes and clinkers (i.e. a mass of incombustible matter fused together by heat) from shipboard incinerators and coal burning boilers but does not include plastic clinkers, which are treated as waste, or oily rags, which are treated as waste.

Operational/functional modules 操作/功能模块
系指包含对航行系统的操作/功能要求的模块。Operational/functional modules mean modules comprising the operational/functional requirements for navigational systems.

Operator 操作者 系指:(1)船东或任何其他的组织或个人,诸如宣称负责船舶维护的船厂或裸船租赁方的管理者;(2)许多海上油田是由一个以上公司共同开发的。其中之一的业主被指定为操作者,其负责操作一大群设施和所有与其相连的流送管、管路和其他设施。Operator means: (1) the owner of the vessel or any other organization or person, such as the Manager, or the shipyard, or the bareboat charterer, who declares to be in charge of the maintenance of the ship; (2) Many offshore field are co-owned by more than one company. One of the owners is designated as the operator, with the responsibility to operate the host facility and all its associated flowlines, pipelines, and other facilities.

Operator 操作员 船舶定员中承担使用 DP 设备的任何成员,如动力定位操作员(DPO)、船长、值班轮机员、轮机长、电气技术员、张紧索操作员、无线电报务员。也可译为代表船东操作船舶的成员。

Operator 经营人 系指:(1)所有人或光船承租人,或经正式转让承担所有人或光船承租人的责任的其他任何自然人或法人;(2)经签发许可证的国家授权经营核动力船舶的人员,或者在由某一缔约国经营核动力船舶的情况下,即指该国。 Operator means: (1) the owner or bareboat charterer, or any other natural or juridical person to whom the responsibilities of the owner or bareboat charterer have been formally as signed; (2) the person authorized by the Licensing State to operate a nuclear ship, or where a Contracting State of the convention on the liability of operators of nuclear ships operates a nuclear ship, that State.

Operator of a transport terminal 运输港站经营人
系指在其业务过程中,在其控制下的某一区域内或在其有权出入或使用某一区域内,负责接管国际运输的货物,以便对这些货物从事或安排从事与运输有关的服务的人。但是,凡属于根据适用于货运的法律规则自身为承运人的人,不视为运输港站经营人。 Operator of a transport terminal means a person who, in the course of his business, undertakes to take in charge goods involved in international carriage in order to perform or to procure the performance of transport related services with respect to the goods in an area under his control or in respect of which he has a right of access or use. However, a person is not considered an operator whenever he is a carrier under applicable rules of law governing carriage.

Opiates and opioids 鸦片制剂和类鸦片 鸦片制剂是从罂粟中衍生的麻醉药品。鸦片是干的罂粟"汁",并含有吗啡和可待因。从吗啡中不难制造出海洛因,其纯净形状为白色粉末,效力是吗啡的两倍多。鸦片制剂在医疗中可作为镇痛剂、止咳和治疗腹泻药物使用。违法鸦片及其衍生物(吗啡和海洛因)的主要供应来源是东南亚缅甸、泰国和老挝的所谓"金三角"地区的罂粟田和西南亚阿富汗、巴基斯坦的"喀布尔"三角和"金月牙"地区的罂粟田。在东地中海一直到东南亚的其他地区内,也有少量的制造。根据以往查获的情况,最有可能的货源港口是曼谷、新加坡、滨城、巴生港、孟买、加尔各答、卡拉奇和哥达斯纳巴鲁。但是,出产地区内的其他大多数港口曾被毒贩所利用。吗啡和海洛因都是通过化学手段从鸦片中衍生。鸦片转换成吗啡,是个相对简单的化学加工过程,通常是在罂粟田附近的临时的实验室中进行。大约 10 kg 鸦片制造 1 kg 鸦片制剂,3 kg 鸦片制剂制造 1 kg 海洛因(即 30 kg 鸦片制造 1 kg 海洛因)。海洛因是通常使用的名称,表示一种含有二乙烯吗啡或二乙烯吗啡盐的制剂。它是经由吗啡碱的完全乙烯化衍生的半合成制品。鸦片制剂有多种形式:(1)生鸦片;(2)配好的鸦片;(3)鸦片渣;(4)医用鸦片;(5)吗啡;(6)药物海洛因;(7)合成剂;(8)半合成剂;(9)可待因。 Opiates and opioids mean that opiates are drugs derived from the opium poppy. Opium is the dried "milk" of the poppy and contains morphine and codeine. From morphine it is not difficult to product heroin which is, in its pure form, a white powder over twice as potent as morphine. Opiates have medical uses as painkillers, cough suppressants and anti-diarrhoea treatments. The main sources of supply for illicit opium and its derivatives, morphine and heroin, are the poppy field of the so-called "Golden Triangle" area of Burma, Thailand and Laos in

South East Asia and the "Kabul Triangle" area of Afghanistan, Pakistan and Iran in South West Asia. It is produced in similar quantities in other areas of the Eastern Mediterranean through to South East Asia. Most likely ports of origin, based on past seizures, are Bangkok, Singapore, Penang, Port Klang, Bombay, Calcutta, Karachi and Kota Kinabalu. However, most other ports within the area of production have been used by drug traffickers. Both morphine and heroin are chemically derived from opium. Opium is converted to morphine in a relatively simple chemical process that usually takes place in a makeshift laboratory near the poppy fields. It takes about 10 kg of opium to produce 1 kg of opium and 3kg of opium to produce 1 kg of heroin(i. e. 30 kg of opium to produce 1 kg of heroin). Heroin is a name commonly used to describe a preparation containing diacetyl morphine base or its salts. It is a semi-synthetic product derived from the complete acelylation of morphine base. Opiates may appear in various forms: (1) raw opium; (2) prepared opium; (3) opium dross; (4) medicinal opium; (5) morphine; (6) diamorphine(herion); (7) synteties; (8) semi-synteties; (9) codeine.

Opinion 意见 系指个人在某些问题刺激的情况下,所给予的的口头或书面陈述的反映。可以用于描述个人对事物的解释、期望与评估。

Opium dross 鸦片渣 这是在吸管中被抽吸后剩余的物质。由于燃烧不完全和挥发,可能会保留某些鸦片的特性,并含有相当数量的吗啡。被抽吸后呈烧焦状,鸦片的气味会长时间在空气中飘荡。 Opium dross is the substance remaining in the pipe after smoking. Due to incomplete combustion and volatilization, it can retain some characteristics of opium and contain a considerable amount of morphine. It will have a charred appearance and the smell of opium will linger in the air long after smoking.

Opposed pistons 对向活塞(对置活塞)

Opposed-cylinder internal combustion engine 对置气缸内燃机 两气缸中心线重合或平行,并对称地排列在曲轴中心线两边的内燃机。

Opposed-piston internal combustion engine 对置活塞内燃机 在一个共同的气缸内有两个相对运动的活塞,当两活塞相接近时,两者顶部与气缸壁共同组成燃烧室的内燃机。

Opposite flow condenser 逆流式冷凝器

Opposite pressure 反压力

Optical 光纤 系指传输光信号的有线信道,简称光纤。光纤是由华裔科学家高锟(Charles Kuen)发明的,他被认为是"光纤之父"。在1970年美国康宁(Corning)公司制造出了世界上第一根实用化的光纤,随着加工制造工艺的不断提高,光纤的衰减不断下降,世界各国干线传输网络主要是由光纤构成的。光纤中光信号的传输是基于全反射原理,光纤可以分为多模光纤(Multi-mode fiber/MMF)和单模光纤(Single mode fiber/SMF),多模光纤中光信号具有多种传播模式,而单模光纤中只有一种传播模式。光纤的信号光源可以有发光二极管(Light-emitted dioxide/LED)和激光。实际应用中使用的光波波长主要在1.31和1.55两个低损耗的波长窗口内,如Ethernet网中的1000Base-LX物理接口采用1.31波长的光信号。计算机局域网中也出现了850 nm波长的信号光源,如Ethernet网中的1000Base-SX物理口就采用这样的光源。LED光源光谱纯度低,不同波长的光信号在光纤中传播速度不同,因此随着距离的增加,光信号传播会发生色散,造成信号的失真,限制了光纤传输的距离,因此对于长距离的传输,每隔一段距离都需要对信号进行中继。单模光纤的色散要比多模光纤要小得多(在多模光纤中还存在模式色散),因而无中继传输距离更长,采用光谱纯度高的激光源传输时引起的色散则更小。见图O-3。

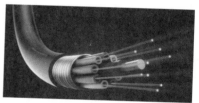

图O-3 光纤
Figure O-3 Optical

Optical acoustic underwater survey system 光纤水声探测系统 系指由高灵敏度的水下声信号传感器组成的系统,能将水声振动转换为光信号,通过光纤传到信号处理系统。与传统压电式水听器相比,这种光纤水听器灵敏度高出2~3倍,可探测到极微弱信号。而且,其耐腐蚀,损耗少,适合远距离传输。它既能"防守",例如可以在重要港口组成被动式警戒网络,对水下目标和水面目标同时进行监测,防范敌对势力利用水下蛙人、微型潜艇进行渗透;它也能"攻击",凭借体积小、质量轻的特点可以组成大型拖曳阵列,配备给水面反潜舰船、潜艇、反潜直升机等,实现对水下目标的精确探测。

Optical character recognition(OCR) 光学字符识别 系指电子设备(例如扫描仪或数码相机)检查纸上打印的字符,通过检测暗、亮的模式确定其形状,然后用字符识别方法将形状翻译成计算机文字的过程;即,针对印刷体字符,采用光学的方式将纸质文档中的文字转换成为黑白点阵的图像文件,并通过识别软件将图像中的文字转换成文本格式,供文字处理软件进一步编辑加工的技术。如何消除错误或利用辅助信息提高识别正确率,是 OCR 最重要的课题,ICR(Intelligent character recognition)的名词也因此而产生。衡量一个 OCR 系统性能好坏的主要指标有:拒识率、误识率、识别速度、用户界面的友好性、产品的稳定性,易用性及可行性等。

Optical fiber amplifier(OFA) 光纤放大器 系指运用于光纤通信线路中,实现信号放大的一种新型全光放大器。根据其在光纤线路中的位置和作用,一般分为中继放大、前置放大和功率放大 3 种。与传统的半导体激光放大器(SOA)相比,OFA 不需要经过光电转换、电光转换和信号再生等复杂过程,可直接对信号进行全光放大,具有很好的"透明性",特别适用于长途光通信的中继放大。可以说,OFA 为实现全光通信奠定了一项技术基础。

Optical fiber cable 光缆 (1)是为了满足光学、机械或环境的性能规范而制造的,它是利用置于包覆护套中的一根或多根光纤作为传输媒质并可以单独或成组使用的通信线缆组件。光缆主要是由光导纤维(细如头发的玻璃丝)和塑料保护套管及塑料外皮构成,光缆内没有金、银、铜铝等金属,一般无回收价值。(2)光缆是一定数量的光纤按照一定方式组成缆心,外包有护套,有的还包覆外护层,用以实现光信号传输的一种通信线路。即:由光纤(光传输载体)经过一定的工艺而形成的线缆。光缆的基本结构一般是由缆芯、加强钢丝、填充物和护套等几部分组成,另外根据需要还有防水层、缓冲层、绝缘金属导线等构件。

Optical pumping magnectometer 光泵磁力仪 又称光吸收磁力仪。利用气态原子受光刺激产生顺磁共振现象制成的磁力测流仪器。

Optimum diameter 最佳直径 设计螺旋桨时,在一定条件下效率最高的直径。

Optimum distribution of circulation 最佳环流分布 在一定的流场中螺旋桨能量损失最小的环流分布。

Optimum revolutions 最佳转速 设计螺旋桨时,在一定条件下效率最高的转速。

Option 续建选择权 系指在船东与造船厂之间签订的建造合同中,具有继续建造船舶与否的选择权。Option means that an option for building additional ships to the same approved plans is given in the contract for construction signed between the owner and the shipbuilder.

Oral hearing 开庭审理 系指人民法院在完成审判前的准备工作之后,在法院或其他适宜场所设置的法庭,对案件进行审理的过程。开庭审理是法院审判程序的核心阶段。审判程序可分为开庭审理前的准备、法庭审判(即开庭审理)、生效裁判执行等基本阶段。开庭审理的结果就是裁判(即判决或裁定),法院做出的裁判在满足生效的条件后即成为生效裁判,生效裁判进入执行程序予以执行。

Orally inflated PFD 吹气气胀式个人漂浮装置 用嘴吹气以产生浮力的个人漂浮装置。 Orally inflated PFD is the PFD inflated by mouth to produce buoyancy.

Orange peel grab 多腭抓斗 由 3 ~ 6 瓣斗体组成,形如橘瓣,适宜于抓石块的抓斗。

Orbital pipe welder 管子环缝自动焊机

Orbital radius 轨圆半径 波浪中水质点所作轨圆运动的半径。

Order B/L 指示提单 系指提单上的收货人拦填写"凭指定"或"凭某某人指定"字样。这种提单可经过背书转让,故其在国际贸易中广泛使用。

Order MTD 多式联运单据

Order to sign a statement of repentance 责令具结悔过

Ordered rudder angle 指令舵角 对操舵系统给定的舵角。

Ordinate station 站 系指在型线图中,沿基线将垂线间长或设计水线长分成若干间距的各点,及其在半宽水线图上沿中线面的相应投影点。

Ordure container 粪便柜

Ore carrier 矿砂船 系指:(1)通常为在整个货物长度区域内具有单甲板、双道纵舱壁和双层底的自行推进的船舶,主要用于仅在中央货舱内载运矿砂货物。(2)在装货处所具有单甲板、两道纵舱壁、双层底,仅仅中心舱用于运输矿砂的船舶。见图 O-4。 Ore carrier is:(1) a sea-going self-propelled ship which is constructed generally with single deck, two longitudinal bulkheads and a double bottom throughout the cargo length area and intended primarily to carry ore cargoes in the centre holds only.(2) a ship constructed with single deck, 2 longitudinal bulkheads and double bottom in cargo spaces, with only center holds used for carrying ore. See Fig. O-4.

Ore/bulk/oil carrier 矿/散/油船 系指具有双壳、单甲板、双层底、顶边舱和底边舱,载运油或散装干货(包括矿砂)的船舶。 Ore/bulk/oil is a carrier with double hull, single deck, double bottom, topside tanks and side or side and some centre holds used for carrying oil.

Organic carbon normalized distribution coefficient (Koc) 有机碳正态分布系数

Organic fiber material 有机纤维材料 系指使用

图 O-4　矿砂船
Figure O-4　Ore carrier

棉、麻等植物纤维及木质、竹质纤维制品等制成的吸声材料。但船舶舱室的吸声隔声处理禁止使用。因为它们不能防火、防蛀和防潮。

Organic light-emitting diode（OLED）　**有机发光二极管**　又称为有机电激光显示，由美籍华裔教授邓青云在实验室中发现，由此展开了对 OLED 的研究。OLED 显示技术具有自发光的特性，采用非常薄的有机材料涂层和玻璃基板，当有电流通过时，这些有机材料就会发光，而且 OLED 显示屏幕可视角度大，并且能够节省电能。OLED 显示技术具有自发光的特性，采用非常薄的有机材料涂层和玻璃基板，当有电流通过时，这些有机材料就会发光，而且 OLED 显示屏幕可视角度大，并且能够节省电能，从 2003 年开始这种显示设备在 MP3 播放器上得到了应用。

Organization　**组织**　系指国际海事组织。Organization means the International Maritime Organization.

Organization for economic cooperation and development（OECD）　**世界经济合作与发展组织（经合组织）**　经合组织帮助各国政府通过经济增长、金融稳固、贸易与投资、技术、创新、创业精神以及发展合作等方式促进繁荣，缓解贫困。它也正致力于帮助各国政府确保经济和社会的发展不以环境的急剧退化为代价。经合组织的其他目标包括创造人均就业机会、社会平等以及洁净高效的治理。经合组织秘书处设在巴黎，其 30 个成员国为澳大利亚、奥地利、比利时、加拿大、捷克共和国、丹麦、芬兰、法国、德国、希腊、匈牙利、冰岛、爱尔兰、意大利、日本、韩国、卢森堡、墨西哥、荷兰、新西兰、挪威、波兰、葡萄牙、斯洛伐克共和国、西班牙、瑞典、瑞士、土耳其、英国和美国。

Organization recognized by the competent Authority（ies）　**经主管机关认可的组织**　系指主管机关按照公约附则第 16.2 和第 16.3 条所指定的代表主管机关承担相关工作的组织。Organization recognized by the competent Authority（ies）means an organization designated by the Competent Authority（ies）in accordance with regulation 16.3 of the Annex to the Convention to undertake relevant tasks on behalf of the Competent Authority（ies）.

Orgeveryic light-emitting diode　**无机发光二极管**

Orientation directional sound wave　**定向声波**　是一种非杀伤性新概念武器。它能精确控制声音传播方向，可实现远距离喊话，警ň示。美国海军率先将其用于驱赶海盗和防范恐怖袭击。此外，它还发射强噪声，刺激听觉器官和中枢神经，使人丧失意识，甚至失去行动能力，因此，也被用于驱赶暴乱人群或城市维稳场合。定向声波武器还可为海上设施设立多道防御圈，例如，对进入 1 000 ~ 3 000 m 范围的不明船舶喊话和沟通，在 100 ~ 1 000 m 范围内警告，如果对方拒绝听从继续靠近，则可调节发射功率，在尽量不伤害对方人员的情况下，令其失去行动能力。据介绍，海上执法船也可利用这种设备，维护我国合法权益。

Orifice　**孔（喷嘴，喷管）**
Orifice meter　**孔板流量计**
Orifices　**孔口**
Original equipment manufacturer（OEMs）　**原型设备制造厂**
Original evidence　**原始证据**　就是直接来源于案件客观事实的证据，即人们通常所说的"第一手材料"。如合同的原件。
Original sound version（OSV）　**原声版本**
Oscilator test　**振动仪试验**　在风洞或船模试验水池内，用连接船模前后部的振动仪，强制船模振动，以测定其作用力和力矩的一种试验。
Oscillate　**振荡（摆动）**
Oscillating cam　**摇摆凸轮**
Oscillating element　**振荡环节**
Oscillating fin　**摇动水翼推进器（橹）**
Oscillating load　**振荡载荷**
Oscillating net machine　**刺网振网机**　利用振动原理抖落刺囊于网目中的渔获物的机械。
Oscillation　**振荡（摆动）**
Oscillation absorber　**减振器**
Oscillation damping　**减振**
Oscillation frequency　**振荡频率**
Oscillation period　**振荡周期**
Oscillation sensor　**振荡传感器**
Oscillation test　**振荡试验**
Oscillator　**振荡器（振子）**
Oscillator test　**振荡器试验（振动仪试验）**
Oscillatory　**振荡的（摇摆的）**
Oscillatory period　**振荡周期**

Oscillatory wave 振荡波
Oscillogram 波形图
Oscillograph 示波器
Oscillographic trace 示波器迹线
Osprey military transport aircraft "鱼鹰"运输机 可以搭载坦克、装甲车和火炮等武器，同时一次还能运送24名战斗人员，飞行速度又远高出一般的战斗直升机的3倍以上。"鱼鹰"相对于直升机的最大特点是航程远、无空中加油时航程为3 000 km左右，即具有"全球部署能力"，比如，"鱼鹰"从美国海岸飞往夏威夷仅需8小时，飞往太平洋中部岛屿也仅需1天多，而采用普通直升机则至少需要一周时间。

Other control stations 其他控制站 系指船上除机舱集控室、驾驶室控制站和就地控制站以外的控制站。Other control stations mean those control stations except centralized control station of engine room, bridge control station and local control station.

Other dry cargo holds 其他干货舱 除滚装货舱以外的所有用于装载非液体货物的处所及通往这些处所的围壁通道。Other dry cargo holds mean all spaces other than Ro-Ro cargo holds used for non-liquid cargo and trunks to such spaces.

Other finishes 其他表面涂料 系指作为可燃性甲板表面底材和薄板，适合对舱壁、内装及天花板的表面进行施工。不过，防止扶手及梯道踏板打滑或者是与此相同程度的小范围所使用的表面涂料不必遵循此规定。并且也不适用于家具。如果防振胶不是由构造部构成时便可以使用。Other finishes mean combustible flooring of deck covering and combustible veneers applied on surfaces of bulkheads, linings and ceilings. However, those surface materials used for handrails and non-skid strips of stairs or other surface materials used only for equally small areas of application may not be required to satisfy these requirements. And it does not apply to furniture. Vibration damping rubber may be used unless it forms a part of the structural integrity.

Other loads 其他载荷 系指拖船的停泊载荷、拖带载荷、锚泊和系泊载荷、起重设备载荷。Other loads include tug and berthing load/towing loads, anchor and mooring loads, lifting appliance loads.

Other machinery spaces 其他机器处所 系指：(1)电器设备间(自动电话交换室、空调管道处所)；(2)除A类处所以外的装有推进装置、锅炉、燃油装置、蒸汽机和内燃机、发电机和主要电动机械、加油站、冷藏室、防摇装置、通风机和空调机的处所，以及类似的处所和通往这些处所的围壁通道。Other machinery spaces are: (1) Electrical equipment rooms (auto-telephone exchange and air-conditioning duct spaces); (2) Spaces containing propulsion machinery, boilers, oil fuel units, steam and internal combustion engines, generators and major electrical machinery, oil filling stations, refrigerating, stabilizing, ventilation and air conditioning machinery, and similar spaces, and trunks to such spaces, excluding machinery spaces of category A.

Other machinery spaces and pump rooms 其他机器处所和泵舱 系指A类机器处所和泵舱以外的处所，包括推进机械、锅炉、蒸汽机和内燃机、发电机和主要电动机、泵、加油站、冷藏装置、防摇装置、通风机和空调装置，以及类似处所及通往这些处所的围壁通道。Other machinery spaces and pump rooms are those spaces, other than category A machinery spaces and cargo pump rooms, containing propulsion machinery, boilers, steam and internal combustion engines, generators and major electrical machinery, pumps, oil filling stations, refrigerating, stabilizing, ventilation and air-conditioning machinery, and similar spaces, and trunks to such spaces.

Other materials and products 其他材料和产品 系指诸如铸铁件制造的部件，如准许还有铜和铜合金制造的产品、铆钉、锚、锚链、起重机、桅、起重柱、吊杆、零件和钢丝绳索等。Other materials and products means such as parts made of iron castings, where allowed, products made of copper and copper alloys, rivets, anchors, chain cables, cranes, masts, derrick posts, derricks, accessories and wire ropes, etc.

Other product offsets 其他产品补偿 系指当所交易的设备本身不生产物质产品，或设备所生产的直接产品非对方所需或在国际市场上不好销时，可由双方根据需要和可能进行协商，用回购其他产品代替的补偿方式。

Other relevant evidence 其他有关证据

Other services 其他设备 系指为保持船舶正常的海上航行工作状态以及适于居住条件，除重要设备和最低舒适居住条件所需设备外，还需的设备，包括：(1)货物装卸和货物保管设备；(2)除适于居住条件设备外的日常生活服务设施；(3)除动力定位用推力器外的推力器。Other services mean those services such as the following, which are additional to those essential services and services necessary for minimum comfortable conditions of habitability, are considered necessary to maintain the ship in a normal sea-going operational and habitable condition: (1) cargo handling and cargo care equipment; (2) hotel services, other than those required for habitable conditions;

(3) thrusters, other than those for dynamic positioning.

Other spaces in which flammable liquids are stowed 储藏易燃液体的其他处所 系指：(1)油漆间；(2)存放易燃液体的储间(包括染料、药品等)；(3)实验室(室内存放易燃液体)。 Other spaces in which flammable liquids are stowed are: (1) paint lockers; (2) storerooms containing flammable liquids (including dyes, medicines, etc.); (3) laboratories (in which flammable liquids are stowed).

Other substances 其他物质 就 MARPOL 附则Ⅱ而言，以 OS(其他物质)形式列入《国际散装化学品规则》第18章污染类别栏目中的物质，并经评定认为不能列入附则Ⅱ第6.1条定义的 X、Y 或 Z 类物质之外，因为这些物质如从洗舱或排放压载作业中排放入海，目前认为对海洋资源、人类健康、海上休憩环境或其他合法利用并无危害。排放仅含有被列为"其他物质"的物质的舱底水或压载水或其他残余物或混合物，不受附则Ⅱ任何要求的限制。 For the purpose of the regulations of the AnnexⅡ, MARPOL, other substances indicated as OS(Other Substances) in the pollution category column of chapter 18 of the International Bulk Chemical Code which have been evaluated and found to fall outside Category X, Y or Z as defined in regulation 6.1 of AnnexⅡ because they are, at present, considered to present no harm to marine resources, human health, amenities or other legitimate uses of the sea when discharged into the sea from tank cleaning of deballasting operations. The discharge of bilge or ballast water or other residues or mixtures containing only substances referred to as "Other Substances" shall not be subject to any requirements of AnnexⅡ.

Out board(derrick) boom 舷外吊杆 定位双杆操作时，吊杆头端伸出舷外的一根吊杆。

Out rigger 舷外支架 从上平台伸出舷外的，对整个装置起稳定和增加浮力作用的构架。

Out rigger for crude burner 试油燃烧器外伸支架 安装在钻井装置上，外伸一定距离，其最外端装有试油燃烧器，并可随风向调整其方向的支架。

Outboard derrick boom 舷外吊杆 定位双杆操作时，吊杆头端伸出舷外的一根吊杆。

Outboard drive 舷外机传动装置

Outboard drive unit 舷外推进装置(舷外挂机组)

Outboard engine(out board motor, outboard motor) 舷外发动机(挂机) 悬挂在船尾部或船侧包括推进装置在内的内燃机。

Outboard motor 挂机 挂于船尾，由内燃机和舵组成的小型推进装置。

Outboard motor boat 舷外挂机艇

Outboard propulsion unit 舷外推进装置(舷外挂机组)

Outboard shafting 舷外轴承

Outboard valve 舷外阀

Outer continental shell(OCS) 外大陆架 对美国而言，系指大致从一个州的海岸线3海里向外到200海里的区域范围。在《联合国海洋法公约》谈判的过程中，一些国家希望把自己能够管到的大陆架扩展得越远越好。《联合国海洋法公约》规定，所有的沿海国都可以主张200海里宽的大陆架。对于一些拥有宽广大陆架的国家，还可以扩到200海里之外，但最远不得超过350海里或者2 500 m等深线以外100海里，这就是所谓的"外大陆架"。值得注意的是，并不是所有国家都能够实际获得200海里的大陆架，也并不是每个国家都能有"外大陆架"，只有满足一定自然条件和法律条件的国家，才有这样主张的资格。

Outer inspection 外部检查 系指用目测检查，查明部件是否有变形和其他缺陷，如裂纹或过度磨损和锈蚀等。

Outer propellers 外侧螺旋桨

Outer screw 外螺旋桨 船上装有两对边螺旋桨时，位于外侧的一对螺旋桨。

Outer shafting(outboard shafting) 外侧轴系 四轴推进船舶中，远离船中线面的一对轴系。

Outer sheath 外护套

Outer strake 外列板 系指两边均搭接在相邻列板外侧的列板。

Outfitting 舾装 是船舶制造工艺里的一种，泛指在各个生产阶段的安装工程，涵盖设备、管系、通风、电气、铁舾、内舾、武备等等各个方面。分为分段舾装、船坞(船台)舾装和码头舾装(把管子、阀门和其他大型设备及装潢材料装上船，该过程涉及的专业最多，是船舶建造过程中最乱也是最容易发生事故的过程)。

Outfitting quay 舾装码头 系指在船舶下水后进行船舶舾装工作的专用码头。

Outfitting shop 舾装车间 系指从事船舶舾装件生产、装配、调试及存储的场所。

Outflow temperature 排出温度

Outgassing 除气(去气)

Outgoing steamer 离港蒸汽机船

Outhaul whip 外牵索 将传送吊车自补给船牵回接收船的绳索。

Outlet 出口(排出，排出管)

Outlet channel 出水槽

Outlet end 输出端

Outlet pressure 出口压力

Outlet pressure gauge 出口压力表

Outlet valve 出口阀

Output 输出(输出量,效率)

Output power 输出功率

Output work 输出功

Outreach 舷外跨距　船舶正浮状态下吊杆偏出舷外时,吊钩中心线与舷边之间,在垂直于中线面方向上的最大水平距离,一般在最大船宽处量取。

Outreach of boat davit 吊艇架跨距　船舶正浮时,艇吊出后,吊艇索距装置该吊艇架的甲板处舷边之间的水平距离。

Outrigger 桅肩　在桅柱的上端向两端伸展,一般供设置吊杆装置的千斤座或千斤眼板的承载构件。

Outside admission 外进气

Outside calipers 外卡钳

Outside diameter 外径

Outside lap 外搭接(外搭接边,外余面)

Outside paint 外表漆

Outside water 舷外水

Outward turning 外旋　推船向前时,边螺旋桨下部背船中线面的旋向。

Ovality($\triangle_0$)　椭圆度　这是衡量管子横剖面非圆度的指标,其定义如下:$\triangle_0 = (D_{max} - D_{min})/(D_{max} + D_{min})$ 椭圆度会降低破坏压力,低的椭圆度可忽略不计的,但高的椭圆度能影响管子相互之间的焊接,因此椭圆度必须降低到最小。 This is one measure of pipe cross section out of roundness defined as follows: $\triangle_0 = (D_{max} - D_{min})/(D_{max} + D_{min})$ Ovality reduces the collapse pressure, and as such, low values are desirable. High ovality can impede the welding of pipes to each other, and consequently must be minimized.

Over bend 上弓段

Over current 过电流　系指超过额定电流的任何电流。

Overall accuracy 总精度

Overall coefficient of heat transfer 传热系数

Overall efficiency 总效率(综合效率)

Overall factor of merit 全面效能因数　船身效率与相对旋转效率乘积除以型排水体积表示的总阻力系数的商值。$F_M = C_{T\triangledown}/\eta_D \eta_R$,式中,$C_{T\triangledown}$——用型排水体积表示的总阻力系数,$\eta_D$——船身效率,$\eta_R$——相对旋转效率。

Overall integration 全盘一体化　这就是将区域内各成员国在所有经济部门加以一体化的形态,欧洲经济共同体和已经解体的经互会属于此类。

Overall length L_{OA}(m)　总长　对游艇而言,系指从艏柱最前端量至艉封板或艉柱后缘的距离,不包括任何其他突出物。

Overall length of floating dock 浮船坞总长　系指艇平台前端到艇平台后端包括各种伸出浮体外的构件在内的最大长度(如无伸出构件,即为浮体长)。

Overall mechanical efficiency 总机械效率

Overall size 外形尺寸

Overall survey 全面检查　系指用于报告船体结构整体状况和确定进行附加近观检查范围的检验。Overall survey is a survey intended to report on the overall condition of the hull structure and determine the extent of additional close-up surveys.

Overbalance 过平衡　系指舵压力中心位于舵杆轴线之前的状态。

Over-ballast operation for bridge crane and equipment carrier 大件运输船超压载作业　系指:(1)船舶为了在港内装卸成套大件或项目设备进行压载作业,使船舶吃水达到不超过作业吃水标志的主要状态;和(2)船舶在达到(1)压载状态下进行限定条件下的港内移船作业,如过桥等。

Overboard 舷外的

Overboard blow off valve 舷外排污阀

Overboard caisson 舷外排水管　将FPSO在海上进行油、水、气分离过程中产生大量的含油污水排出舷外的排水管。其通常布置在舷侧外板,高度从主甲板至水线以下,用支撑结构与舷侧外板相连。

Overboard discharge 舷侧排水口/排水舷口　自船内向舷外排水的出口。

Overboard discharge control 舷外排放控制　系指在报警条件下自动按程序停止排放物向舷外排放并在整个报警期间防止排放的装置。该装置可布置成关闭舷外阀或终止相关的泵(如合适)。 Overboard discharge control is a device which automatically initiates the sequence to stop the overboard discharge of the effluent in alarm conditions and prevents the discharge throughout the period the alarm condition prevails. The device may be arranged to close the overboard valves or to stop the relevant pumps, as appropriate.

Overboard discharge pipe 舷外排出管

Overboard discharge valve 舷外排出阀　位于舷内紧靠舷侧排水口安装的阀。

Overboard piping 排往舷外管路

Overboard scupper 舷外排水孔(舷外排水口)

Over-current discrimination 过电流选择性　系指两个或多个过电流保护电器之间的动作特性的配合。

在给定的范围内出现过电流时,指定在此范围动作的保护电器动作,而其他保护电器不动作。

Over-current protection 过电流保护(过载保护) 系指一种闭环期间在过电流时动作,中断流向被保护器件的电流,并保持此中断状态的效能。

Over-current protective co-ordination of over-current protective device 过电流保护电器的过电流保护协调 系指两个或多个过电流保护电器串联起来,用以保证过电流选择性保护和/或后备保护的实现。

Over-current relay or release 过电流继电器或脱扣器 系指当电器或脱扣器中的电流超过预定值时,使机械开关电器有延时或无延时地动作的继电器或脱扣器。注意:在某些情况下,预定值取决于电流的上升率。

Overfill 满溢

Overfill alarm 溢流报警器

Overflow 溢流(溢出)

Overflow alarm 溢流报警

Overflow blowdown 溢流排盐 利用溢流法从蒸发器液面排除盐水的方法。

Overflow device 泥舱溢流装置 由用于溢流的孔、管、槽、挡板和节流阀等组成,使泥舱中的水溢流至舷外的装置。

Overflow device 溢流装置

Overflow fuel oil tank 燃油溢油舱

Overflow lubricating oil tank 滑油溢油舱 供贮存加滑油时,由滑油舱通过溢流管溢出的滑油的舱室。

Overflow main pipe(overflow main) 溢流总管 有两个以上的油舱(柜)时,为汇集各舱溢流而设置的总管路。

Overflow method 溢流法 该方法是把深海海水从压载舱底泵入使压载水从舱顶连续不断地溢出,直到更换掉足够的压载水量以减少仍停留在舱中的目标生物数目。根据IMO的有关规定,在深海更换的压载水量应为压载水舱容积的3倍。该方法的优点是容易操作,且对船舶的强度、稳性和纵倾影响不大;缺点是双层底和艏艉尖舱难以彻底冲洗,可能引起超压损坏、泵和管系超负荷,而且不适合在寒冷的环境操作。

Overflow oil source 溢油源 系指海面浮油的泄漏源头,例如来自油船、水下石油管道、海上钻井平台等。

Overflow pipe 溢流管 在液舱注满时将液体溢流的管路。

Overflow protection 溢流保护

Overflow tap 溢流旋塞

Overflow valve 溢流阀

Overflow weir 溢流口(溢流挡板、溢流筒)

Overflowing weld metal 焊缝金属外溢

Overhang 悬伸部 泛指突出于设计水面以外的任何船体水上部分,尤指突出于设计水线两端位于艏艉垂线以外者。

Overhaul 检修 对营运期中的船舶按规定周期每隔2~3个小修的间隔期,并结合定期检验而进行的计划性修理。其目的是对船体结构、主辅机、锅炉、轴系以及其他设备等进行全面性的检查和修理,同时还需重点消除在小修中不能解决的重大缺陷,以保证船体强度的恢复和机械设备的安全运转。

Overhaul 大修(拆修,检修)

Overhaul period(working life, operating life, overhaul life, time between overhauls) 翻修期限 燃气轮机在第一次翻修前能正常使用的时间。

Overhauling inspection/Open-up inspection 拆检 系指将机械设备拆开进行的检查。

Overhead bucket dredger 高架链斗挖泥船 具有较高斗塔,由链斗挖出的泥沙倒入泥井后,通过长的溜泥槽或卸泥输送带送到岸边填泥区的链斗挖泥船。

Overhead line 架空输电线路 系指用绝缘子将输电导线固定在直立于地面的杆塔上以传输电能的输电线路。它由导线、架空电线、绝缘子串、杆塔、接地装置等组成。导线由导电良好的金属制成,有足够粗的截面(以保持适当的通流密度)和较大曲率半径(以减小电晕放电)。超高压输电则多采用分裂导线。架空电线(又称避雷线)设置于输电导线的上方,用于保护线路免遭雷击。重要的输电线路通常用两根架空电线。绝缘子串由单个悬式(或棒式)绝缘子串接而成,需满足绝缘强度和机械强度的要求。每串绝缘子个数由输电电压等级决定。杆塔多由钢材或钢筋混凝土制成,是架空输电线路的主要支撑结构。架空输电线路在设计时要考虑其受到的气温变化、强风暴侵袭、雷闪、雨淋、结冰、洪水、湿雾等各种自然条件的影响,还要考虑电磁环境干扰问题。架空输电线路所经路径还要有足够的地面宽度和净空走廊。

Overhead obstructions 顶部障碍物 系指通道上方包括扶强材在内的甲板和梯台结构。Overhead obstructions mean the deck or stringer structure including stiffeners above the means of access.

Overhead tank 重力柜

Overheat 过热

Overheated zone 过热区

Overheating 过热

Overheating of bearings 轴承过热

Overheating protection 过热保护

Overland common points(OCP)　内陆地区　系指根据美国运费率规定,以美国西部 9 个州为界,也就是以洛杉矶山脉为界,其以东地区,均为内陆地区的范围。

Overland transportation all risks　陆运一切险　其承保责任范围与海上运输货物保险条款中的"一切险"相似。保险公司负责赔偿被保险货物在运输途中遭受暴风、雷电、洪水、地震等自然灾害,或由于运输工具遭受碰撞、倾覆、出轨,或在驳运过程中因驳运工具遭受搁浅、触礁、沉没、碰撞,或由于遭受隧道坍塌、崖崩或失火、爆炸等意外事故所造成的全部或部分损失,负责保险货物在运输途中由于一般外来原因所造成的全部或部分损失。

Overland transportation risks　陆运险　包括货物在运输过程中因遭受自然灾害或意外的事故所造成的货物部分或全部损失。

Overlap　焊瘤　在焊接过程中,熔化金属流淌到焊缝之外未熔化的母材上所形成的金属流。焊缝表面存在焊瘤会影响美观,并易造成表面夹渣。

Overlap propeller　交叠螺旋桨　前后错开布置,轴线相互靠近,其盘面有部分重叠的两个螺旋桨。

Over-lap type hatch cover(lift-rolling hatch cover)　层叠型舱盖　舱盖板分层叠置的滚移式舱盖。

Overload　超载(过载)

Overload capacity　过载容量(超负荷能量)

Overload output　超负荷功率　主机在规定的时间内允许使用的最大功率。

Overload preventing device　防过载设备(过载保护装置)

Overload protection　过载保护(防过载)

Overload pull　过载拉力　系指起锚机必需的短时过载能力。Overload pull means the capability of the windlass necessary to withstand an overload pull for a short time.

Overload release　过载释放(过载释放装置)

Overload test　过载试验(超负载试验)

Overload trial　过载试验(过负荷试验)

Overload trip　过载保护器(过载自动脱扣器)

Overload valve　过载阀

Overloads　过载　系指电机承受下列电流或过转矩而不发生有害变形的能力:(1)直流发电机——50%过电流 15 s;(2)交流发电机——50%过电流 2 min;(3)直流电动机——50%过转矩 15 s;(4)多相异步电动机——60%过转矩 15 s;(5)多相凸极同步电动机——50%过转矩 15 s;(6)多相隐极同步电动机——35%过转矩 15 s;(7)多相异步结构(线绕转子)同步电动机——35%过转矩 15 s。Overloads mean the macnine capacity withstanded on test, the flowing overcurrent or excess torque without injury:(1) D. C. generators—50% overcurrent 15 s;(2) A. C. generators—50% overcurrent 2min;(3) D. C. motors—50% excess torque 15 s;(4) polyphase asynchronous motors—60% excess torque 15 s;(5) polyphase salient synchronous motors—50% excess torque 15 s;(6) polyphase non-salient synchronous motors—35% excess torque 15 s;(7) polyphase asynchronous construction synchronous motors(winding rotator)—35% excess torque 15 s.

Overlook　忽略(俯视,漏看,眺望)

Overpower protection　功率过载保护

Overpressure　超压　系指由于爆炸引起相对于大气压力产生变化的压力。既有正值也有负值。 Overpressure is a pressure relative to ambient pressure caused by a blast, both positive and negative.

Overpressure protection　超压保护

Overpressure relief valve(personal flotation device)　过压安全阀(个人漂浮装置)　在气胀式系统中可使用的,用以避免由于过压引起破坏可能性的阀。 Overpressure relief valve is the valve which may be used in an inflatable system to avoid the likelihood of destruction caused by overpressure.

Over-quench　过冷淬火

Override　越控　系指越过控制过程中的某一程序或某一安全保护功能,在短时间内强制机电设备继续运行以保证船舶安全的特殊控制措施。 Override means the special control measures for the skipping of a certain procedure or a certain safeguard action so as to effect compulsory operation to the machinery or electrical equipment for a short period to ensure the safety of the ship.

Override arrangements　越控设施　系指为了防止安全系统的部分或全部功能暂时停止而使用的装置,在设置该装置,要按下列要求进行:(1)当越控设施投入工作时,在机械和设备的有关控制站应有明确标识显示。(2)越控设施应能防止无意地投入工作。 Where arrangements for stopping temporarily the functions of safety system in part or in whole(hereinafter referred to as "override arrangements") are provided, the following requirements should be followed:(1) Visual indication is to be given at the relevant control stations of the machinery and equipment when an override is operated.(2) The override arrangements are to be such that inadvertent operation is prevented.

Override arrangements　过载布置　是对部分或全部暂时停止安全系统的功能而做出的布置。 Override arrangements are arrangements for stopping temporarily the functions of safety system in part or in whole.

Override button 越控按钮

Overriding authority 优先(超越)权限　ISM 规则要求公司在其 SMS 中赋予船长的,在关于船舶安全及防污染方面优先于其他事项的指挥权和处置权。

Overriding control system 越控系统

Overriding device 越控装置　系指不管发出任何报警信号,都无视该信号的使其继续保持从前状态的装置。Overriding device means a device to make keeping the current function of the equipment, though a set alarm signal in it would be taken place.

Overshoot angle 超越角　在 Z 形操纵试验中,从换操反舵瞬时的转舵角到最大转舵角之间的差值。

Overspeed 超速

Overspeed governor 超速调速器

Over-speed governor (emergency governor, emergency over-speed trip) 超速限制器　当柴油机超过规定的极限转速时,能使燃油减少或停止供油的装置。

Overspeed protection 超速保护

Over-speed protection device (emergency governor) 超速保护装置　当汽轮机达到极限速度时,发出信号,通过综合开关,如油遮断器使速关阀关闭,汽轮机紧急停车的装置。

Overspeed test 超速试验

Overspeed trial 超速试航

Overspeed trip 超速自动脱扣

Overspeed tripping device 超速脱扣装置

Overtaking sea 追浪　在左、右舷 165°~180°之间方向内与船遭遇而波速高于船速的波浪。

Overtemperature alarm 过热报警器

Owner 船东(货主、业主)　系指:(1)注册船东、船东、管理者或其他方面,他们负责保持船舶的航行能力,并特别关心保持船级的有关规定;(2)采购和支持通常由政府部门负责。在某些情况下,海军可从其他船东包租船舶进行营运,该情况下这些规则中定义的船东应与船级社规范达成一致;(3)各缔约国国家法律规定的所有人或承租人或受托人,如双方有协议,该承租人或受托人将承担对集装箱的维修和检验的责任;(4)对澳大利亚海事安全局海事指令而言,系指与用于对船舶进行装卸的岸上设备有关,并包括对用于装卸作业之设备具有所有权的人员。Owner mean: (1) the Registered Owner or the Disponent Owner or the Manager or any other party having the responsibility to keep the ship seaworthy, having particular regard to the provisions relating to the maintenance of class; (2) Generally, this will be the government department responsible for naval procurement and support. In certain circumstances, the Navy may operate ships chartered from other Owners, in which case the Owner as defined in these regulations is to be agreed with the society on a case by case basis; (3) The owner as provided for under the national law of the Contracting Party or the lessee or bailee, if an agreement between the parties provides for the exercise of the owner's responsibility for maintenance and examination of the container by such lessee or bailee; (4) For the purpose of the maritime order of the Australian maritime safety authority (AMSA), means a person that in relation to shore equipment used in loading or unloading a ship, includes the person having possession of the purpose of the loading or unloading operation.

Oxidizing substances (Class 5.1 dangerous materials) 氧化性物质(第 5.1 类危险货物)　系指虽然不一定可燃,但一般通过释放氧气而可能引起其他物质燃烧或促使其燃烧的物质。 Oxidizing substances mean the materials in this class are materials while in them-selves not necessarily combustible, may, generally by yielding oxygen, cause, or contribute to, the combustion of other material.

Oxygen 氧

Oxygen analysis and gas detection equipment 氧气分析和气体探测设备　系指在运输可能释放有毒或易燃气体或可能在货物处所中造成氧气耗尽的散装货物时,测量有毒或易燃气体或氧气浓度的适用仪器。Oxygen analysis and gas detection equipment means an appropriate instrument for measuring the concentration of gas or oxygen in the air detailed when transporting a bulk cargo which is liable to emit a toxic or flammable gas or cause oxygen depletion in the cargo space, such an instrument shall be to the satisfaction of the Administration.

Oxygen bottle 氧气瓶

Oxygen breathing apparatus 氧气呼吸器

Oxygen compressor 氧气压缩机　舰船制氧装置中,将氧气压缩至贮氧舱中所用的压缩机。

Oxygen content meter 氧量表

Oxygen content of condenser 冷凝水含氧量　溶解在冷凝水中的氧气量。

Oxygen corrosion 氧腐蚀

Oxygen cutting 氧气切割　简称气割,它具有设备简单、灵活方便、质量好等优点,它适用于切割厚度较大、尺寸较长的废钢,如:大块废钢板、铸钢件、废锅炉、废钢结构架等。对废汽车解体和旧船舶解体更能发挥其灵活方便的作用,它不受场地狭窄或物件大小的局限,可以在任何场合下进行作业。除使用气割加工炼钢炉料外,还可以在废钢中割取有使用价值的板、型、管等材料,供再生产使用。所以氧气切割是废钢铁加工的主要方法之一,目前在金属回收部门应用十分广泛。

Oxygen cutting equipment 氧气切割设备

Oxygen cutting machine 氧气切割机

Oxygen fuel gas welding(OFW) 气焊 利用可燃气体与助燃气体混合燃烧生成的火焰为热源,熔化焊件和焊接材料使之达到原子间结合的一种焊接方法。助燃气体主要为氧气,可燃气体主要采用乙炔、液化石油气等。所使用的焊接材料主要包括可燃气体、助燃气体、焊丝、气焊熔剂等。特点是设备简单不需用电。设备主要包括氧气瓶、乙炔瓶(如采用乙炔作为可燃气体)、减压器、焊枪、胶管等。由于所用储存气体的气瓶为压力容器、气体为易燃易爆气体,所以该方法是所有焊接方法中危险性最高的之一。氧气瓶的外面为蓝色,金属中金银材料最好但较贵且质量重,其次为铜,其氧化性较弱,铜的氢氧化物为蓝色。所以氧气瓶的颜色应该为蓝色。

Oxygen gouging 火焰氧气刨

Oxygen hose 氧气软管

Oxygen lance cutting 氧气枪切割

Oxygen mask 氧气面罩 提供了一个可以把呼吸需要的氧气从储罐中转入到人体肺部的方法。主要有医用氧气面罩、航空氧气面罩/航空乘客使用的氧气面罩以及焊工使用的面罩等种类,对于治疗疾病、保护乘客及飞行员安全起到了重要的作用。主要由塑料、有机硅、或橡胶制成。

Oxygen measuring equipment 氧气测量仪

Oxygen meter 氧气探测仪

Oxygen respirator 供氧呼吸器

Oxygenometer 测氧仪

Oxyhydrogen cutting 氢氧切割

Oxyhydrogen torch 氢氧吹管(氢氧割炬)

Oxyhydrogen welding 氢氧焊接

Oxyplane machine 氧刨机

Oxypropane cutting 氧丙烷气割

Ozone 臭氧

Ozone consumption 臭氧消耗量

Ozone depleter 臭氧消耗物质

Ozone depleting potential(ODP) 臭氧消耗趋势

Ozone generator 臭氧发生器 由臭氧发生管、延时继电器、排风扇等组成。通常为冷藏舱或冷藏库内防毒、杀菌、除异味和消毒之用,使空气中氧气变成臭氧的装置。

Ozone technology 臭氧技术

Ozone depleting substances 消耗臭氧物质 系指1987年消耗臭氧层物质蒙特利尔议定书第1条第4款中定义的并在该议定书附件A,B,C或E中所列的受控物质。在船上可能有的"消耗臭氧物质"包括但不限于下各项(表O-1):溴氯二氟甲烷、溴三氟甲烷、1,2-二溴化物-1,1,2,2-四氟乙烷、三氯氟甲烷、二氯二氟甲烷、1,1,2-三氯-1,2,2-三氟乙烷、1,2-二氯-1,1,2,2-四氟乙烷和氯五氟乙烷等。除2020年1月1日前允许含有氢化氯氟烃的新装置以外,所有船上应禁止使用含有消耗臭氧物质的新装置。

表 O-1 消耗臭氧物质
Table O-1 Ozone depleting substances

Halon1211	溴氯二氟甲烷
Halon1301	溴三氟甲烷
Halon2402	1,2-二溴化物-1,1,2,2-四氟乙烷(亦称 Halon114B2)
CFC-11	三氯氟甲烷
CFC-12	二氯二氟甲烷
CFC-113	1,1,2-三氯-1,2,2-三氟乙烷
CFC-114	1,2-二氯-1,1,2,2-四氟乙烷
CFC-115	氯五氟乙烷

Ozone-depleting substances record book 消耗臭氧物质记录簿 系指按其质量单位(kg)记录消耗臭氧物质,且在任何情况下及时记录下列内容的记录本:(1)含消耗臭氧物质的设备的全部或部分重新充注;(2)含消耗臭氧物质的设备的修理或维修;(3)消耗臭氧物质向大气的排放:①故意排放;②非故意排放;(4)消耗臭氧物质向陆基接收设备设施的排放;和(5)向船舶供应消耗臭氧物质。Ozone-depleting substances record book means a record bookrecorded in terms of mass(kg) of substance and completed without delay on each occasion, in respect of the following:(1)recharge, full or partial, of equipment containing ozone-depleting substances; (2) repair or maintenance of equipment containing ozone-depleting substances;(3)discharge of ozone-depleting substances to the atmosphere: ① deliberate; and ② non-deliberate; (4)discharge of ozone-depleting substances to land-based reception facilities; and(5)suppy of ozone-depleting substances to the ship.

P

P3 strategy shipping alliance P3 战略航运联盟
即在全球 20 大班轮公司中排名前三名的航运企业马士基、地中海航运和达飞于 2014 年 6 月 18 日联合发起成立的集装箱班轮联合组织。该联盟计划通过在英格兰和威尔士设立一级有限责任合伙制的网络中心，统一负责在亚洲-欧洲、跨大西洋和跨太平洋航线上集装箱班轮的运营事务。按照"P3 联盟"此前公布的计划，3 家企业将在三条航线上部署总计 260 万 TEU 的运力联合运营，其目的在于通过使用更大吨位的船舶，进一步减少主要航道的货船数量，其成员保有运力合计 642 万 TEU，占全球集装箱总运力的比重达到 38%，分别是现有 G6、CKYH 联盟的两倍和 3 倍。在此之前，美国联邦海事委员会在 2015 年 3 月决定批准"P3 联盟"在美国生效；欧洲委员会也在 6 月初宣布有条件通过"P3 联盟"，一旦中国政府最终放行，这一组织便可以开始正式运作。不过中国最终还是否决了该联盟的成立。其实"P3 联盟"刚刚宣布成立之时，全球航运市场高度关注，当时就有业内专家指出"P3 联盟"或涉嫌垄断。航运市场的集中垄断、议价操纵等行为，在全球市场中都被明确禁止和反对，"P3 联盟"一旦形成后可能对目前航运市场的公平竞争的现状产生负面影响。"P3 联盟"形成了交易方紧密型联盟，在亚洲-欧洲航线集装箱班轮运输服务市场可能具有排除、限制竞争效果，而参与集中的经营者不能证明该集中对竞争产生的有利影响明显大于不利影响或者符合社会公共利益。"P3 联盟"各方通过共同设立网络中心，在合作航线上的所有船舶进行日常管理，交易方仅保留对船舶的技术管理权。这一网络中心权力巨大，对未使用箱位的销售、停航等都可由其协调处理或直接决定。在费用分摊方面，"P3 联盟"通过规定包括租船费、燃油费、港口费、运河费等在内航次成本，将合作涉及的所有航线分为若干个结算组统一结算成本，并分摊，而不是像传统联盟那样独自核算。

PA 超声相控阵技术

Package(pkg) 包装（标准部件,成套设备）

Package switch(PS) 分组交换技术 也称包交换，是将用户传送的数据划分成一定的长度，每个部分叫作一个分组，通过传输分组的方式传输信息的一种技术。它是通过计算机和终端实现计算机与计算机之间的通信，在传输线路质量不高、网络技术手段还较单一的情况下，应运而生的一种交换技术。

Packaged auxiliary unit(auxiliary modules) 组装式辅机组 设备及其管路附件等组装成整体单元的成套设备。

Packaged boiler 总装锅炉 为便于运输及快速安装而将鼓风机、水泵、油泵和自动控制装置等成套辅助设备与锅炉本体在制造厂组装成的成套总体机组。

Packaged dangerous goods 包装危险货物 系指装于容器、移动式液货舱、货运集装箱或其他运载工具内的任何危险货物。此外还包括那些曾经运输过危险货物的空容器和空的移动式液货舱等。除非这样的容器和移动式液货舱已进行了清洁并被允许用于安全运输。 Packaged dangerous goods means any dangerous cargo contained in receptacle, portable tank, height container or vehicle. In addition, it includes an empty receptacle/portable tank which has previously been used for carriage of dangerous substance, unless such receptacle or tank has been cleaned and permits transport with safety.

Packaged form 包装形式 系指 IMDG 规则中对有害物质所规定的盛放形式。 Packaged form is defined as the forms of containment specified for harmful substances in the IMDG Code.

Packet processing 包处理

Packing 填料（垫料,包装）

Packing block 填料函（垫块,压垫盖）

Packing drawer 填料钩针

Packing gland 填料函（密封压盖,密封套）

Packing hook 填料钩

Packing knife 填料割刀

Packing list 装箱单 是发票的补充单据，它列明了信用证（或合同）中买卖双方约定的有关包装事宜的细节，便于国外买方在货物到达目的港时供海关检查和核对货物，通常可以将其有关内容加列在商业发票上，但是在信用证有明确要求时，就必须严格按信用证约定制作。装箱单的作用主要是补充发票内容，详细记载包装方式、包装材料、包装件数、货物规格、数量、质量等内容，便于进口商或海关等有关部门对货物的核准。装箱单(Packing list)：在中文"装箱单"上方的空白处填写当单人的中文名称地址，"装箱单"下方的英文可根据要求自行变换。

Packing material 用作填料的材料

Packing piece 填密片（衬垫）

Packing ring 密封圈

Packing screw 填料螺丝起

Packing spring 填料弹簧

Packing stick 塞填料的杆

Packing unit 密封元件

Packing washer 填密垫圈

Pad 衬垫(垫板,座板)

Pad eye 眼板(导向滑车)

Pad hook 带钩板

Padding method 隔绝法 对于液货舱的环境控制,系指将液体、气体或蒸发气充入液货舱及相关管系,使货物与空气隔绝并维持这种状态。 For the purpose of environmental control for cargo tanks, padding method means a method by filling the cargo tank and associated piping systems with a liquid, gas or vapour which separates the cargo from the air, and maintaining that condition.

Paddle 短桨 短柄的桨。用时双手持柄划水前进。

Paddle steamer, paddle vessel 明轮船 用安装在两舷或艏艉端的明轮推进的船舶。

Paddle wheel 明轮 装在船的两侧或船尾部,大部分露出水面,外形似车轮,沿轮周有若干叶蹼的一种推进器。

Pain threshold 痛阈 各种能引起疼痛的刺激,在其刺激强度非常微弱时,并不令人感到疼痛;当刺激达到一定强度时才感到疼痛。所谓"痛阈"系指引起疼痛的最低刺激量。痛阈降低会使人对疼痛的敏感性提高,正常的刺激,如体内生理范围的改变亦可出现痛感。痛阈的高低因人而异,且受多种因素影响,比如年龄、性别、性格、心理状态以及致痛刺激的性质等等。痛阈升高对于超出痛阈的刺激,也不能感受疼痛或反应延迟,对疾病不能早期发现、早期诊断,丧失早期治疗的机会。

Pain threshold sound pressure 痛阈声压 系指使人耳产生疼痛感觉时的声压(20Pa)。

Paint 涂料 系指涂于物体表面能形成具有保护、装饰或特殊功能涂膜的一类液态或固态材料的总称。 Paint means a general term for liquid or solid material coated on the surface to form a film having protection, decoration or special functions.

Paint store 油漆间 船上供贮存油漆和油漆工具的房间。

Painter 系艇索 配置在救生艇上,一根系于艇前端,带有套环和卸扣,主要用于艇落水后能和船留住,又能迅速解脱,另一根固定在艏柱上备用的专供系艇用的索具。

Painting dressing 涂装 系指将涂料涂覆于基体表面,形成具有保护、装饰或特殊功能涂层的过程。 Painting dressing means a process to coat on the surface to form a coating having protection, decoration or special function.

Painting dressing specifications 涂装说明书 系指明确表述下列技术要求的文件:(1)建造前钢材预处理质量要求,全船建造过程中各主要部位或区域的二次除锈质量要求以及适用标准;(2)全船各主要部位或区域的涂层配套要求(包括所选用的涂料品种、牌号、颜色、涂料适用的涂装条件、涂装度数和每道涂层的干膜厚度);(3)各部位涂层膜厚的分布要求;(4)所选用的每一种涂料的物理和化学性能参数。

Paired comparison 成对比较方法 以专家判断来定量估计人误概率。不要求专家做出任何定量分析,而只需要他们去比较一系列含有 HEPs 要求的成对任务,决定哪个最容易产生失误及其相对等级,最后通过对数变换获得 HEPs 的评估。

Pallet 集运架(托盘) 系指由两层相互连接的甲板组成的装载平台,其分别允许诸如叉臂、柄、梁或吊索等起吊设备进入。 Pallet means a load carrying platform having two interconnected decks separated to permit the entry of lifting equipment, such as fork arms, tines, bars or slings.

Pallet carrier 集装箱运输车 系指装载集装箱的大型运输车辆。

Palindrome 回文

PAM type single-pull hatch cover 套置型舱盖 舱盖板横剖面成梯形。开舱时各舱盖板可套合排列置放,以减小收藏位置的滚翻式舱盖。

Panama canal 巴拿马运河 位于中美洲横穿巴拿马地峡,总长82 km,最宽达304 m,最窄处也有152 m,水深在13~15 m。该运河直接连接太平洋和大西洋,是重要的世界航运通道之一。2007年巴拿马开始扩建工程。工程完成后最大允许通行长366 m,宽48.8 m,吃水15.3 m的大型船舶。为此,美国东海岸新泽西等港口以及巴西东北部地区的港口目前正在进行改建和扩建,以使接纳超大型货船。有分析认为,届时巴拿马运河的通航能力将比目前提高50%以上,集装箱船通过量的提高幅度更大。扩建前的巴拿马运河由于河道狭窄,每天仅能通过40艘船舶,同时有近100艘船舶在等待过河。扩建后,能够通过巴拿马运河的船舶范围扩大。能够航行阿芙拉型油船、12万吨左右的Mini-cap型散货船、1.3万TEU以下集装箱船和大型液化天然气船以及超大型气体运输船(VLGC),但阿芙拉型油船和Mini-cap型散货船都不是当前原油和干散货航运市场的主力船型。如果尼加拉瓜运河建成并开通,将对巴拿马运河产生很大的冲击,巴拿马运河的日渐式微将不可避免。

Panama Canal Authority (ACP) 巴拿马运河管理局 巴拿马政府授权管理运河的主管机构。

Panama canal tonnage 巴拿马运河吨位 凡通过

巴拿马运河的船舶，根据巴拿马运河当局规定的丈量规范所核定的登记吨位。

Panama Canal universal measurement system(PC/UMS) 巴拿马运河统一丈量系统(PC/UMS) 系指根据1969年通用丈量系统制定，在确定船舶总的容积时采用了前者的参数，以及由巴拿马运河管理局发布的一些附加变动的丈量系统。 Panama Canal universal measurement system(PC/UMS) means the system based on the universal measurement system, 1969, using its parameters for determining the total volume of a vessel with the additional variations established by the Panama Canal Authority.

Panamax 巴拿马型船舶 系指符合巴拿马运河现有船闸尺度和吃水限制，即船长294.1 m、船宽32.3 m、热带淡水(TFW)吃水12.04 m限制的所有船舶。 Panamax mean all vessels that comply with the size and draft limitations of the actual locks, namely, 294.1 m in length by 32.3 m in beam by 12.04 m TFW draft.

Panamax bulk carrier 巴拿马型散货船 系指船舶主尺度受巴拿马运河通航限制的散货船。一般按巴拿马运河允许通过的最大尺度进行设计。其载重量(DWT)60 000～75 000 t的散货船。

Panamax oil tanker 巴拿马型油船(PANAMAX) 系指载重量为60 000～75 000 t的油船。

Panamax plus 巴拿马加大型船 系指经授权其热带淡水(TFW)吃水可增至大于12.04 m，但小于或等于15.2 m，且准予通过巴拿马运河新船闸的所有巴拿马型船舶。 Panamax plus mean all Panamax vessels authorized for TFW drafts greater than 12.04 m up to 15.2 m and approved for transit of the new locks.

Panamax vessels 巴拿马型船 系指载重量(DWT)60 000～75 000 t的船舶。是一种专门设计的适合巴拿马运河船闸的大型船舶，这些船舶的船宽和吃水受到巴拿马运河船闸闸室的严格限制，越来越多的船舶在建造时精确的匹配巴拿马运河船闸的限制，以便在适应巴拿马运河的航道的前提下运送尽量多的货物。很多干散货，比如谷物，主要通过巴拿马级船舶运送。此类船舶的日益增加正成为巴拿马运河的一大难题。巴拿马级船舶，由于其非常精确地适合巴拿马运河，留出的余量很小，船舶必须受到非常精确的控制，过闸时间也可能因此得以延长，并且过闸时间限定在白天。因为一些大型船舶不能安全地通过盖拉德人工渠，运河只能以单向通过的方式来保证这些大型船舶的安全。

Panel 板格 对钢夹层板而言，系指主要结构件围的钢夹层板。

Panel of arbitrator 仲裁员名册

Panting 挠振 船舶在垂荡和纵摇时，艏、艉部壳板受浪击引起的凹凸脉动现象。

Panting arrangement 强胸结构 艏、艉端用以抵抗水冲击力减少局部振动的加强结构。

Panting beam 强胸横梁 强胸结构中，用以支撑舷侧纵桁的横构件。

Pantries containing no cooking appliances 没有烹调设备的配餐室 包含以下设备，并且具备这种设备的食堂不视为烹饪器具室。(1)每台最大功率为5 kW的自动咖啡机、吐司炉、洗碟机、微波炉、热水炉和类似设备。(2)每台最大功率为2 kW且表面温度不大于150 ℃的电热烹调板和食物保温电热板、烹饪器具室。 Pantries containing no cooking appliances may contain the following devices. However, a dining room containing such appliances is not to be regarded as apantry: (1) Coffee automats, toasters, dish washers, microwave ovens, water boilers and similar appliances each of them with a maximum power of 5 kW. (2) Electrically heated cooking plates and hot plates for keeping food warm each of them with a maximum power of 2 kW and a surface temperature not above 150 ℃.

Paper product 纸制品

Parachute drogue 漂流伞 用漂流测流法测量某一深度海流的装置。由浮子、漂流伞、重物等组成。

Parachute rocket 降落伞火箭 供船舶遇难时发射，能在高空中发出迅连于降落伞下的红光的一种信号烟火。

Paraffin oil 煤油

Paraffin test 煤油涂检法

Paraffin wax 固体石蜡

Parallel arrangement 并联装置(并联电路)

Parallel connection 并联

Parallel flow air register 平流式调风器 在喷油器头部处具有一次风轮，经风轮分流导出一股旋转的一次风，而掠过风轮的二次风则是一般平行气流的调风器。

Parallel jaw vice 平口虎钳

Parallel middle body 平行中体 设计水线下具有同样横剖面的船体水下部分。

Parallel middle-body length 平行中体长 平行中体前后端之间的水平距离。

Parallel operation 并联运行

Parallel operation test 并车运行试验

Parallel processing technology(PPROC) 并行处理技术 是40年来在微电子、印刷电路、高密度封装技术、高性能处理机、存储系统、外围设备、通信通道、语言开发、编译技术、操作系统、程序设计环境和应用问题等研究和工业发展的产物。并行计算机具有代表性的应用领域有：天气预报建模、VLSI 电路的计算机辅助设计、大型数据库管理、人工智能、犯罪控制和国防战略研究等，而且其应用范围还在不断地扩大。并行处理技术主要是以算法为核心，并行语言为描述，软硬件作为实现工具的相互联系而又相互制约的一种结构技术。

Parallel running test 多机并车试验 为检查多机传动共轴推进装置中主机并车工作可靠性、功率分配均匀性及减速运转性能所做的试验。

Parallel shift maneuver test 平移操纵试验 使船舶移动到与原航线平行的新航线上，用以确定实船对舵角低频率响应的一种操纵试验。

Parallel waterline 吃水平行段 平行于中线面的设计水线中部线段。

Parallel-axis gear (parallel-axis arrangement) 平行轴传动减速齿轮箱 各传动轴互相平行的减速齿轮箱。

Parallel-axis gear combined with right-angle gear 圆锥-平行轴减速齿轮箱 小型船舶中利用一对锥齿轮将主机功率传到齿轮减速器第一级小齿轮，因而主机轴线与轴系中线呈 V 字形状的传动方式时所用的减速齿轮箱。

Parallel-planetary gear 平行轴-行星减速齿轮箱 第一级采用平行轴齿轮，第二级则为行星齿轮结构的减速齿轮箱。

Parametric rolling 参数横摇 是大型集装箱船一个显见的问题，船体横摇运动与遭遇波浪频率之间形成共振增加横摇角。

Parametric rolling prediction 参数横摇周期

Paratrooper 伞兵 又称空降兵，主要是以空降到战场为作战方式。其特点是装备轻型化、高度机动化、兵员精锐化。一般独立建制为师级或旅级，直接隶属于军团一级或更高级别的指挥机构。

Paratrooper heavy equipment airdrop technology 伞兵重装空投技术 该技术拥有国：美国、俄罗斯、乌克兰、中国。见图 P-2。

Parcel post receipt 邮包收据 是邮包运输的主要单据，它既是邮局收到寄件人的邮包后所签发的凭证，也是收件人凭以提取邮件的凭证，当邮包发生损坏或丢失时，它还可以作为索赔和理赔的依据。但邮包收据不是物权凭证。

Parcel post transport 邮包运输 是一种较简便的运输方式。各国邮政部门之间订有协定或公约，各国的邮包可以相互传递，从而形成国际邮包运输网。由于国

图 P-2 伞兵重装空投技术
Figure P-2 Paratrooper heavy equipment airdrop technology

际邮包运输具有国际多式联运和"门到门"运输的性质，加之手续简便，费用也不高，故其成为国际贸易中普遍采用的运输方式之一。

Parentheses 括号

Parity 海盗行为 下列行为中的任何行为构成海盗行为：(1)私人船舶或私人飞机的船员、机组成员或乘客为私人目的，对下列对象从事的任何非法的暴力或扣留行为，或任何掠夺行为：①在公海上对另一艘船舶或飞机，或对另一艘船舶或飞机上的人或财物；②在任何国家管辖范围以外的地方对船舶、飞机、人、财物；(2)明知船舶或飞机成为海盗船舶或飞机的事实，而自愿参加其活动的任何行为；(3)教唆或故意便利(1)或(2)项所述行为的任何行为。 Parity consists of any of the following acts: (1) any illegal acts of violence or detention, or any act of depredation, committed for private ends by the crew or the passengers of a private ship or a private aircraft, and directed: ①on the high seas, against another ship or aircraft, or against persons or property on board such ship or aircraft; ②against a ship, aircraft, persons or property in a place outside the jurisdiction of any State; (2) any act of voluntary participation in the operation of a ship or of an aircraft with knowledge of facts making it a pirate ship or aircraft; (3) any act of inciting or of intentionally facilitating an act described in subparagraph(1)or(2).

Parking distance control system(PDC) 停车距离控制系统 该系统主要是协助驾驶者方便停车，尤其在都会区 PDC 是有其需要性，此套系统就是俗称的倒车雷达，PDC 系统通常会于车的后保险杠或前后保险杠设有雷达侦测器，用以侦测前后方的障碍物，此套系统主要是要协助驾驶者侦测前后方无法看到的障碍物，或停车时与其他车辆的距离，除了方便停车外更可以保护您的车身。

Part 部件 系指单个分系统、设备或模块。 Part is individual subsystem, equipment or module.

Part fixed price 部分固定价格 为了照顾双方的利益,可采用部分固定价格,部分非固定价格的做法,或者分批作价的办法,交货期近的价格在订约时固定下来,余者在交货前一定期限内进行作价。

Part flow system 分流系统 系指能有效提供有代表性的舷外排出物样本,以便在所有正常操作条件下进行视觉显示的系统。分流系统包括下列部件:(1)取样探头;(2)样水管系;(3)样水输送泵;(4)显示装置;(5)样水排放装置;以及根据样水管路的直径;和(6)冲洗装置。 Part flow system means a system which can effectively provide a repersentative sample of the overboard effluent for visual display under all normal operating conditions. Part flow system includes: the following: (1) sampling probes; (2) sample water piping system; (3) sample feed pump(s); (4) display arrangements; (5) sample discharge arrangements, and, subject to the diameter of the sample piping; (6) flushing arrangement.

Part load tuning 部分负荷调整 为执行 IMO 环保会关于"新船能效设计指数 EEDI"的要求,柴油机制造厂商采用了一种能使柴油机在部分负荷时降低燃油消耗的调整方法。

Partial award 部分裁决

Partial cavitation 部分空泡 物体,例如螺旋桨叶,一处或几处有时产生空泡的情况。

Partial discrimination 部分选择性保护 系指在有两个或两个以上过电流保护电器串联的电路中,当发生过电流故障时,只是在一定的短路电流范围内,才能做到最接近故障点的保护电器起保护作用,而不会导致其他保护电器动作的过电流选择性保护。 Partial discrimination is an overcurrent discrimination where, in the presence of two or moreover-current protective devices in series, the protective device closest to the fault effects the protection up to a given level of short-circuit current without causing the other protective devices to operate.

Partial integrations 部分综合 系指不包括航线监测和避碰任务的小范围综合。 Partial integrations mean the smaller integrations which are not covering the tasks "route monitoring" and "collision avoidance".

Partial load condition 部分负荷工况。

Partial load line 部分载重线 是空船吃水加上空船吃水与最深分舱载重线之间差值的 60%。 Partial load line is the light ship draught plus 60% of the difference between the light ship draught and deepest subdivision load line.

Partial load line draught d_P 部分载重吃水 d_P 系指空船吃水加上空船吃水与满载吃水之间差值的 60%。部分载重吃水 d_P 以 m 计。 Partial load line draught d_P is the light ship draught plus 60% of the difference between the light ship draught and the load line draught d_S. The partial load line draught d_P shall be measured in m.

Partial loading hold 部分装载舱 系指在货物处所内散装谷物未按经平舱的满载舱的要求装载或平舱的装载舱。部分装载舱的谷物表面也应平整。

Partial re-liquefaction system(PRS) 天然气再液化系统 系指将 LNG 船舱内自然蒸发气化的天然气再液化为 LNG 并送入舱内储存的设备。这套系统的特点是:无须使用液化必需的冷冻剂,因此船上不必另外配置动力设备和冷冻压缩机设备。该系统可直接将存储舱内蒸发气的天然气用作冷冻剂。

Partial shipment 分批装运 系指一笔成交的货物,分若干次装运。

Partial subdivision draught d_p 部分分舱吃水(部分分舱载重线) 系指轻载航行吃水加上轻载航行吃水与最深分舱吃水之差的 60%。 Partial subdivision draught is the light service draught plus 60% of the difference between the light service draught and the deepest subdivision draught.

Partial loss 部分损失 在海上运输途中,船舶、货物或其他财产遭遇共同危险,为了解除共同危险,有意采取合理的救难措施所直接造成的特殊牺牲和支付的特殊费用,称为共同海损。在船舶发生共同海损后,凡属共同海损范围内的牺牲和费用,均可通过共同海损情况,由有关获救受益方(即船方、货方和运费收入方)根据获救价值按比例分摊,然后再向各自的保险人索赔。共同海损分摊涉及的因素比较复杂,一般均由专门的海损理算机构进行理算(Adjustment)。

Partial-admission stage 部分进气级 只在一部分圆周上配置静叶,不是全周进气的汽轮机级。

Partially decked boat 部分甲板艇 其舷弧线区域至少 2/3 的水平投影设有甲板板、住舱、遮蔽或硬质舱口罩盖(其符合 ISO 12216 的水密性要求,且指定用于把水排出舷外)的艇。注意:该舷弧区域包括在距艇首 $L_H/3$ 范围之内的所有区域,以及从艇的周边(不包括艉板)向艇内 100mm 的区域。 Overpressure relief valve means the boat in which at least two-thirds of the horizontal projection of the sheerline area is equipped with decking, cabins, shelters or rigid covers which are watertight according to ISO 12216 and designed to shed water overboard. Note: The sheerline area includes the total area within $L_H/3$ from the bow and also the area 100 mm inboard from the periphery of the boat(excluding the transom).

Partially filled compartment 部分装载舱 系指散

装谷物未按有关方法装载的任何货物处所。Partially filled compartment refers to any cargo space where the bulk grain is not loaded in the relevant manner.

Partially immergence test 部分浸水试验 使螺旋桨模型部分露出水面,为检查其性能变化和吸气情况而进行的试验。

Partially underhung rudder (semi-balanced rudder) 半悬挂舵 舵的上半部支承于舵柱或挂舵臂的舵柱上,下半部呈悬挂状的舵。

Particle swarm optimization(PSO) 粒子群优化算法 又称粒子群算法、微粒群算法、或微粒群优化算法。是通过模拟鸟群觅食行为而发展起来的一种基于群体协作的随机搜索算法。通常认为它是群集智能(Swarm intelligence,SI)的一种。它可以被纳入多主体优化系统(Multiagent optimization system,MAOS),是由 Eberh 博士和 kennedy 博士发明。

Particular average 单独海损 与共同海损相对应,系指保险标的物在海上遭受承保范围内的风险所造成的部分灭失或损害,即指除共同海损以外的部分损失。这种损失只能由标的物所有人单独负担。单独海损的损失,由受损者自己承担,而共同海损的损失则由受益各方根据获救利益的大小按比例分摊。单独海损是由所承保的风险直接导致的船、货的损失,而共同海损是为解除或减轻风险,人为地有意识地采取合理措施造成的损失。单独海损只涉及损失方个人的利益,而共同海损是为船货各方的共同利益所受的损失。

Particularly sensitive sea area (PSSA) 特别敏感海区 系指由于其公认的生态学的、社会经济的或科学原因的重要性,并可能对国际海上活动损害具有脆弱性,通过国际海事组织的行动需要特殊保护的海区。Particularly sensitive sea area (PSSA) means an area that needs special protection through action by IMO because of its significance for recognized ecological or socioeconomic or scientific reasons and which may be vulnerable to damage by international maritime activities.

Particulars 船舶资料 系指包括下述内容的信息:(1)船名;(2)IMO 编号;(3)船旗国;(4)船籍港;(5)船东;(6)船级;(7)船级登记号;(8)总吨位;(9)载重量(公吨);(10)建造日期。 Particulars means a information included the following:(1) ship's name;(2) IMO number;(3) flag State;(4) port of registry;(5) owner;(6) classification Society;(7) class ID;(8) gross tonnage;(9) deadweight(metric tons);(10) date of build.

Partition bulkhead (screen bulkhead) 轻型舱壁 只起分隔舱室的作用,而不承担载荷的轻型舱壁。

Parts of billion (ppb) 十亿分之一 ppb means a parts per billion.

Parts per million (ppm) "百万分之一" (1)气体分析仪基于摩尔单位测量 ppm,假定每个总量摩尔的位置理想微摩尔($\mu mol/mol$),但使用 ppm 与 NO_x 技术规则中的单位保持一致;(2)每百万分水中的含油量(体积)。 ppm means:(1) parts per million. It is assumed that ppm is measured by gas analysers on a molar basis, assuming ideal micro-moles of substance per of total amount ($\mu mol/mol$), but ppm is used in order to be consistent with units in the NO_x Technical Code;(2) parts of oil per million parts of water by volume.

Party 缔约国 系指国际海员培训、发证和值班标准公约已对之生效的国家。 Party means a State for which the International Convention on standard of training, certification and watch-keeping for seafarers has entered into force.

Party autonomy 当事人意思自治 是仲裁制度的核心。当事人在仲裁中享有更多的自主权,当事人可以选择仲裁员、选择仲裁程序、合意选择开庭的地点、时间和开庭时使用的语言,甚至是最后解决纠纷应该适用的实体法律。

Pass valve 通海阀

Passage passageway 通道 船上供人员通行的过道。

Passage (ship) 通过(船舶) 系指为了以下目的,通过领海的航行:(1)穿过领海但不进入内水或停靠内水以外的泊船处或港口设施;(2)驶往或驶出内水或停靠这种泊船处或港口设施。通过应继续不停和迅速进行。通过包括停船和下锚在内,但以通过航行所附带发生的或由于不可抗力或遇难所必要的救助遇险或遭难的人员、船舶或飞机的目的为限。 Passage means navigation through the territorial sea for the purpose of:(1) traversing that sea without entering internal waters or calling at a roadstead or port facility outside internal waters; or(2) proceeding to or from internal waters or a call at such roadstead or port facility. Passage shall be continuous and expeditious. However, passage includes stopping and anchoring, but only in so far as the same are incidental to ordinary navigation or are rendered necessary by force majeure or distress or for the purpose of rendering assistance to persons, ships, or aircraft in danger or distress.

Passenger 乘客 系指除下列人员以外的每一个人:(1)船长和船员或在船上以任何职业从事或参加该船业务工作的其他人员;(2)1 岁以下的儿童。 Passenger means every person other than:(1) the master and the members of the crew or other persons employed or engaged in

any capacity on board a ship on the business of that ship;
(2) a child under one year of age.

Passenger/cargo ship 客货船 载运乘客超过 12 人，又能载货的船舶。

Passenger A ship A 类客船 系指满足下列条件的客船：在其规定的营运航线的任何地点出事，有很大把握能在以下三者中的最短时间内将船上所有乘客和船员救出：(1) 救生艇、筏内的人员因受冻以至伤亡的时间；(2) 与该航线所处的环境条件和地理特点相适应的时间；(3) 4h 以载客不超过 450 人的客船。

Passenger B ship B 类客船 系指 A 类客船以外的客船，这类船的机械和安全系统的设置应保证：一旦某一个舱发生破损且舱内的主要机械和安全系统失效，该船仍能保持安全航行的能力。

Passenger barge 客驳 专供载运旅客的驳船。

Passenger boat 小型客船 系指船长 20 m 以下的客船。Passenger boats is a passenger ships less than 20 m in length.

Passenger cargo vessel 客货船 以载运旅客为主，兼运一定数量货物的船。

Passenger ferry 旅客渡船 系指货物渡船兼用运送乘客定员在 12 人以上的船舶。Passenger ferry means the cargo ferry of carrying passengers not less than 12 persons in addition to vehicles.

Passenger load 乘客定额 系指船上所载运的最大乘客数。Passenger load is the maximum number of passengers on board.

Passenger semi-submersible craft 水下式观光半潜水艇 系指用于运送或搭载乘客，并能在水下观光旅游的自由自航的半潜水艇。其特点是可下潜到水下运行，但部分结构仍露出水面的机动船舶。Passenger semi-submersible craft is a craft self-propelled free semi-submersibles transporting or carrying passengers and capable of underwater sightseeing, with some portion of their structure above water surface.

Passenger ship 客船 系指：(1) 持有①按现行国际海上人命安全公约条款而颁发的安全证书，或者②乘客证书的船舶；(2) 运载乘客超过 12 人的船舶；(3) 主要用来载运乘客，并按公布的定期航班运行的船舶。船上所有的空间凡标明和认证作为旅客使用的都将被包括在船的总容积计算中。Passenger ship means: (1) a ship in respect of which there is in force either ①a safety certificate issued in accordance with the provisions of the International Convention for the Safety of Life at Sea for the time being in force or ②a passenger certificate; (2) a ship which carries more than twelve passengers; (3) a vessel that principally transports passengers and runs on fixed published schedules. All the spaces that have been identified and certified for the use or possible use of passengers are to be included in the total volume calculation of the vessel.

Passenger ship safety certificate 客船安全证书 系指经检验和检查并符合 1974 年 SOLAS 公约第Ⅱ-1 章、第Ⅱ-2 章、第Ⅲ章和第Ⅳ章要求以及其他有关要求之后，为每艘船舶签发的客船安全证书，应永久性附有一份客船安全证书的设备记录簿。Passenger ship safety certificate means a certificate called a Passenger ship safety certificate shall be issued after inspection and survey to a passenger ship which complies with the requirements of chapters Ⅱ-1, Ⅱ-2, Ⅲ and Ⅳ and any other relevant requirements of SOLAS 1974. A record of equipment for the passenger ship safety certificate shall be permanently attached.

Passenger spaces 乘客处所 (1) 拟供乘客使用的所有区域，且其包括以下各项：①乘客舱室；②公共处所（例如餐厅、诊疗室、起居室、阅览和游戏室、健身室、走廊、商店）；③开敞甲板休闲区域。(2) 那些提供乘客居住和使用的处所，不包括行李舱、贮藏室、食品舱和邮件舱。在所有情况下，容积和面积均应计及型线为止。Passenger spaces mean: (1) all areas intended for passenger use, and include the following: ①Passenger cabins; ②Public spaces (e.g. restaurants, hospital, lounges, reading and games rooms, gymnasiums, corridors, shops); ③Open deck recreation areas. (2) those spaces which are provided for the accommodation and use of passengers, excluding baggage, store, provision and mail rooms. In all cases volumes and areas are to be calculated to moulded lines.

Passenger submersible craft 水下观光潜水艇 系指用于运送或搭载乘客，并能在水下观光旅游的自由自航的潜水艇。Passenger submersible craft is a craft self-propelled free submersibles transporting or carrying passengers and capable of underwater sightseeing.

Passenger vessel 客船 系指主要用来载运乘客，并按公布的定期航班运行的船舶。船上所有的空间凡标明和认证作为旅客使用的都将被包括在船的总容积计算中。For the purpose of Panama Canal, passenger vessel means a vessel that principally transports passengers and runs on fixed published schedules. All the spaces that have been identified and certified for the use or possible use of passengers are to be included in the total volume calculation of the vessel.

Passenger/cargo ship 客货船 专门运送旅客及其行李、邮件和部分货物的船舶。客货船的机舱位于中部,为"中机型船"。客货船中设置多道水密纵横舱壁,且能起到防火的作用。大部分客货船采用双桨双舵,先进的客货船在船首还配有侧推器。

Passengers 乘客 是除了船员和乘员以外的每个人。Passengers are every person other than the crew and embarked personnel.

Passengers in transit 过境旅客 系指从国外一个国家乘船抵达某港,然后乘船或其他交通工具继续旅行去外国的旅客。Passengers in transit means a passenger who arrives by ship from a foreign country for the purpose of continuing his journey by ship or some other means of transport to a foreign country.

Passengers' accompanied baggage 旅客携带的行李 系指旅客携带在同一艘船上的,可能包括货币在内的财产,不论此财产是否为本人所有,只要不是按照某一运输合同或其他类似协议所运载的物品。Passengers' accompanied baggage means the property, which may include currency, carried for a passenger on the same ship as the passenger, whether in the personal possession or not, so long as it is not carried under a contract of carriage or other similar agreement.

Passing of satellite 卫星通过 对于观察者而言,卫星从地面上升起到降落至地平线以下的这段运行过程称为卫星通过。在卫星通过时,依次测量卫星的位置,从而计算出船位。因此卫星通过时间越长,获得的轨道数据越多。有利于提高定位精度。

Passing through 正在通过 系指正在通过某一船闸或通过船闸室任意一端引导墙所封闭的水域。Passing through means it is transiting through a lock or through the waters enclosed by the approach walls at either end of a lock chamber.

Passive anti-rolling tank 被动式减摇水舱 当船舶横摇时,两舷水舱的水在自然产生往复运动,利用左右水舱水体的重力差产生抵抗船舶横摇的力矩来达到船舶减摇目的的减摇水舱。

Passive constrained damping layer 被动式约束阻尼层

Passive electronically scanned array(PESA) 无源电子扫描阵列雷达

Passive failure 被动失效 对于设备的操作状态没有实时影响,且不能被监控回路检测出来,但在某些情况下可能导致系统失效。Passive failure has no immediate effect on the operating conditions of the installations and is not detected by the monitoring circuits but could lead, in certain conditions, to a failure of the system.

Passive quota arrangement 被动配额管理 系指由于进口国对某种商品的进口实行数量限制,并通过政府间多、双边贸易协定谈判,要求出口国控制出口数量,从而出口国对这类出口实施数量限制,称作被动配额管理。被动配额管理的出口商品主要是纺织品。设限国有美国、加拿大、欧盟国家和土耳其。根据《纺织品与服装协定》,自2005年1月1日起,纺织品被动配额全部取消,全球纺织品贸易实现一体化。

Passive risk control 被动风险控制 系指没有采取措施控制风险。

Passive system 被动系统 对计算机系统而言,系指要求手工输入数据的计算机系统。For the purpose of calculating system, passive system means a calculating system which requires manual data entry.

Passive tank stabilization system 被动水舱式减摇装置 由船舶本身摇摆导致舱内液位变化的水舱式减摇装置。

Past positions 先前位置 相同时间间隔的先前位置标示——被跟踪目标或报告目标及本船位置的坐标。用于显示先前位置的坐标可以是相对的或真实的。Past positions mean equally time-spaced past position marks of a tracked or reported target and own ship. The co-ordinates used to display past positions may be either relative or true.

Patch(wooden patch) 堵漏板 船舶破损时,用以堵拦漏洞的各种板件,有软边堵漏板、堵漏夹板、活页堵漏板等。

Patent clips 有节稳索夹头 能套夹住有节稳索上任一柱形金属节,以调整该索的工作长度和传递该索作用力的连接件。

Patent license 专利许可证

Patrol 监视指导船

Pay by installment 分期付款 系指在产品投产前,在卖方向买方提供出口许可复印本的情况下,买方用汇付方式,先向卖方交付部分货款或定金,其余货款按生产进度付款。

Payee 收款人/受款人 票据的主债权人。系指受领汇票所规定金额的人,通常是出口人或其指定的银行。

Payer 付款人 通常是进出口业务中的进口人或其指定的银行。

Paying bank 付款银行 系指信用证上被开证行指明国际贸易单证制作须履行付款责任的银行,一般为开证行本身或开证行的代理行。

Payload 有效载荷 系指:(1)可变钻井载荷和在浮箱和立柱内可消耗液体重力之和。(2)舰船为实现其操作要求而装载的设备和储备。Payload is:(1) defined as the combination of variable drilling load and the consumable liquids in the pontoons and columns. (2) the equipment and stores that are carried by the vessel for the purposes of fulfilling it operational requirements.

Payment 付款　对即期汇票,在持票人提示汇票时,付款人即应付款;对远期汇票,付款人经过承兑后,在汇票到期日付款。付款后,汇票上的一切债务即告终止。

Payment as per schedule 进度款　系指根据工程施工进度,按合同中的规定的比例交付的款项。

Payment by installments 分期付款　系指在产品投产前,在卖方向买方提供出口许可复印本的情况下,买方用汇付方式,先向卖方交付部分货款或定金,其余货款按生产进度付款。

Payment in advance 预付货款　系指进口人先将货款用汇付的方式交给出口人,出口人立即或在一定时间内发运货物。预付货款的做法主要是出口人对进口人不大信任,或是买卖的商品在国际市场上是抢手货,所以要预收货款作为担保。

Payment order/warrant 支付令　是人民法院依照民事诉讼法规定的督促程序,根据债权人的申请,向债务人发出的限期履行给付金钱或有价证券的法律文书。是人民法院根据债权人的申请,依法做出的督促债务人为一定给付义务的法律文书。这是处理债权债务关系明确的民事、经济纠纷的最好办法,但只能体现在债务人接到支付令之日起15日内,不向法院提出书面异议方可实现。债务人对债权债务关系没有异议,但对清偿能力、清偿期限、清偿方式提出不同意见的,不影响支付令的效力。若法院裁定终结督促程序,支付令自行失效,债权人可以提出诉讼。

PC N 极地航行　系指授予具有极地水域航行能力的船舶的附加标志;其中N为如下之一:1—全年在所有极地水域航行;2—全年在中等厚度的多年冰龄状况下航行;3—全年在第二年冰龄状况(可包括多年夹冰)下航行;4—全年在当年厚冰状况(可包括旧夹冰)下航行;5—全年在中等厚度的当年冰龄状况(可包括旧夹冰)下航行;6—夏季/秋季在中等厚度的当年冰龄状况(可包括旧夹冰)下航行;7—夏季/秋季在当年薄冰状况(可包括旧夹冰)下航行。

Peak frame 尖舱肋骨　艏、艉尖舱内的肋骨。

Peak short-circuit current 峰值短路电流　系指可达到的短路电流的最大可能瞬时值。Peak short-circuit current mean the maximum possible instantaneous value of the prospective(available) short-circuit current.

Peak sound pressure 峰值声压

Peak to peak sound pressure 峰到峰声压

Peanuts(in shell) 花生(带壳)　一种食用的棕黄色干果。含水量不定。粉尘极多。可能自燃。该货物为不燃物或失火风险低。　Peanuts(in shell) are an edible, tan coloured nut. It is variable moisture content. It is extremely dusty. It may heat spontaneously. This cargo is non-combustible or has a low fire-risk.

Pearl boat 海珍品采集船　带有轻型潜水装备,用以采集海参、鲍鱼、珍珠贝等珍贵海产品的渔船。

Peat moss 泥煤苔　是在泥沼、泥塘、沼泽、泥炭沼泽和沼地露天开采出来的。有各种类型,包括鲜类泥煤、莎草泥煤和草本泥煤。物理特性取决于有机物质、水和空气含量、植物的分解和分解程度。其范围可以包括植物残骸的高纤维黏合物(在天然状态下挤压时流出清水至略带颜色的水)以及完全分解的基本不定形物质(挤压时固体极少或无液体分离)等。晾干的泥煤通常密度低、压缩性高且含水量高,在去天然状态下饱和时,按照水的质量计,它能保留90%或更多水分。货物处所和相邻处所缺氧和二氧化碳增加。装载时粉尘存在爆炸的风险。在未经压缩的草泥表面走动或放下沉重机械时宜小心。按质量计含水量大于80%的泥煤苔宜仅用经特别装备或专门建造的船舶运载。粉尘可能刺眼、鼻和呼吸器官。　Peat moss is a surface mined from mires, bogs, fens, muskeg and swamps. Types include moss peat, sedge peat and grass peat. Physical properties depend on organic matter, water and air content, botanical decomposition and degree of decomposition. May range from a highly fibrous cohesive mass of plant remains which when squeezed in its natural state exudes clear to slightly coloured water, to a well decomposed, largely amorphous material with little or no separation of liquid from solids when squeezed. Typically air-dried peat has low density, high compressibility and high water content; in its natural state it can hold 90% or more of water by weight of water when saturated. Oxygen depletion and an increase in carbon dioxide in cargo and adjacent spaces. Risk of dust explosion when loading. Caution should be exercised when walking or landing heavy machinery on the surface of uncompressed Peat Moss. Peat Moss having a moisture content of more than 80% by weight should only be carried on specially fitted or constructed ships. Dust may cause eye, nose and respiratory irritation.

Pebbles(sea) 卵石　圆卵石。极易滚动。无特殊危害。该货物为不燃物或失火风险低。　Pebbles(sea) are round pebbles. It rolls very easily. No special hazards. This cargo is non-combustible or has a low fire-risk.

Pedestal fairlead 导向滚轮　装在甲板上并配有台座,用以引导缆绳通向缆绳卷车的滚轮导缆器。

Pedestal of bearing　轴承支座

Peep hole(periscope hole)　看火孔　用以观察炉膛内燃烧情况的小窗。

Peer-peer protocol(PPP)　端对端协议

Pellets(concentrates)　丸粒(精矿)　已制成丸粒的精矿。含水量最大为6%。无特殊危害。该货物为不燃物或失火风险低。 Pellets (concentrates) are concentrate ore which has been palletized. Moisture up to 6%. No special hazards. This cargo is non-combustible or has a low fire-risk.

Penalty　罚金/违约金　系指:(1)合同当事人一方未履行义务而向对方支付约定金额的罚金;(2)债权人或债务人完全不履行或不适当履行债务时,必须按约定给付他方的一定数额的金钱。

Pendent navigation light　吊升式号灯　使用时临时悬挂在规定位置上的号灯。

Pendent rigging of navigation lights　号灯吊升索具　供吊升号灯用的绳索及其卸扣、滑车等配件的总称。

Penetrating device for bunched cables　成束电缆贯穿装置　系指由一个四周为金属框架,其中装有多根非金属电缆密封材料所组成的器件。其嵌在甲板、舱壁或设备外壳的开口处,使电缆可由此穿过甲板、舱壁或进入设备而不会损伤其原来的防火或水密完整性。

Penetrating oil　渗透油

Penetration　渗透(熔深,贯穿件)

Penetration corrosion　渗透腐蚀

Penetration degree　渗透率

Penetration piece　通舱管件

Penetration test(PT)(dye-check, colour check)　着色探伤

Penetration testing　渗透试验(渗漏试验)　系指通过低表面张力液体的应用来证实舱室边界不存在连续渗漏的试验方法。

Perceived noiseness　感觉噪度　系指与人们主观判断噪声的"吵闹"程度成比例的质量。

Perfectly brittle　脆性破坏　系指一个构件失效后,不能在继续承受载荷。

Perfectly ductile　延性破坏　系指一个构件失效后,在某种程度上还能继续工作。

Performance capability rating　性能评价　系指利用有关的资料,核算出在标准的环境条件下,保持船舶的艏向及位置的时间的百分比。 Performance capability rating is calculated by the relative data and this rating indicates the percentage of time that a ship is capable of holding heading and position under a standard set of environmental conditions.

Performance check　性能检查　系指代表性的选择简短定性实验以确定综合桥楼系统的正确操作和主要功能。 Performance check is a representative selection of short qualitative tests, to confirm correct operation or essential functions of the integrated bridge system.

Performance criteria　性能标准　系指工程方面用来表述判断试验设计是否足够的可测量数值。 Performance criteria are measurable quantities stated in engineering terms to be used to judge the adequacy of trial designs.

Performance decision A　性能判据A　系指在试验过程中和试验以后,受试设备均应连续地进行预期的工作,无有关设备标准和制造厂商制订的技术条件规定的性能降低或者功能的丧失。

Performance decision B　性能判据B　系指在试验以后,受试设备能满意地工作,无有关设备标准和制造厂商制订的技术条件规定的性能降低或者功能的丧失。在试验过程中,允许有能自行恢复的功能或性能的降低或丧失存在,但不允许发生实际工作状态的改变和储存资料的变化。

Performance decision C　性能判据C　系指在试验过程中和试验以后,允许有有关设备标准和制造厂商制订的技术条件规定的功能或性能的暂时降低或丧失,但其功能能自行恢复,或者能以进行上述标准和技术条件规定的某种控制操作得以恢复。

Performance indicators　绩效指标　系指在IMO组织的任务声明的背景下和战略方向的基础上制定的指标,其旨在达到下列目的:(1)安全航行;(2)可靠航运;(3)合乎环保要求的航运;(4)有效航运;(5)可持续航运;(6)采用可行的最高标准;(7)文件实施;(8)能力建设。绩效指标包括:(1)加入公约;(2)生效;(3)实施和符合;(4)人命丧失;(5)船舶灭失;(6)保安失效;(7)海盗和武装抢劫;(8)船舶造成的水污染;(9)船舶造成的空气污染和CO_2排放;(10)环境道德;(11)港口国控制和扣船率;(12)港口国控制不合格率;(13)欺诈性证书;(14)技术援助的落实;(15)综合技术合作计划(ITCP)的可持续性;(16)周期时间;(17)IMO作用;(18)目标型标准;(19)其他联合国机构的工作;(20)航运效率——便利国际海上运输。 Performance indicators mean those indicators developed in the context of the IMO's mission statement and on the basis of the strategic directions, with the aim of achieving the following; (1) safe shipping; (2) secure shipping; (3) environment sound shipping; (4) efficient shipping; (5) sustainable shipping; (6) adoption of the highest practicable standards; (7) implementation of instruments;(8)capacity-building. Performance indicators include:(1)accessions to convention;(2)entry into force;(3)implementation and compliance;(4)lives lost;(5)ships lost;(6)security failures;(7)piracy and armed robbery;(8)ship-generated water pollution;(9)ship-generated air pollution and CO_2 emissions;(10)environment conscience;(11)PSC detention rate;(12)PSC non-compliance rate;(13)fraudulent certificates;(14)delivery of technical assistance;(15)sustainability of ITCP;(16)cycle time;(17)IMO's role;(18)goal-based standards;(19)work of other UN bodies;(20)efficiency of shipping-facilitation of international maritime traffic.

Performance parameter 性能参数

Performance record 性能记录

Performance reduce 性能降低 系指任何器件、设备和系统的工作性能与其正常性能相比,有非期望的偏高。

Performance requirement 性能要求

Performance shaping factor 行为影响因子 对人的行为产生积极或消极影响的因素。例如:检验、情景压力、个人动机、人机界面等因素。

Performance SLC 履约备用信用证 对履约责任进行担保。

Performance standard 性能标准

Performance standard for protective coatings (PSPC) 保护涂层 系指授予特定处所满足 IMO 有关保护涂层性能标准的船舶的附加标志,并其后缀可加一个或多个 B,C,D 和 V 标志,其含义如下:B—所有类型船舶专用海水压载舱处所施用的保护涂层;C—原油船货油舱处所施用的保护涂层;D—散货船双舷侧处所施用的保护涂层;V—散货船和油船的空舱处所施用的保护涂层。注意:B,C,D 和 V 可以单独也可以组合使用。

Performance standard for protective coatings for dedicated seawater ballast tanks in all types of ships and double-side skin spaces of bulk carrier 所有类型船舶专用海水压载舱和散货船双舷侧处所保护涂层性能标准

Performance standard for shipboard simplified voyage data records(S-VDR) 船载简化航行数据记录仪性能标准

Performance test 性能试验(运行试验) 系指为证实设备符合其设备标准(技术条件)所有性能要求的全面试验。

Performance test/operation test/function test 效用试验 对机器设备进行的证明其功能、效用的试验。

Performance testing 性能测试

Performance trial 性能试验

Performing carrier 履行承运人 系指承运人以外的,实际履行全部或部分运输的船舶所有人、承租人或经营人。 Performing carrier means a person other than the carrier, being the owner, charterer or operator of the ship, who actually perform the whole or a part of the carriage.

Perimeter lights 周界灯 系指可在降落区域或其上方从所有方向看见的绿色灯,勾画出降落和起飞区(TLOF)周界的灯。 Perimeter lights mean these lights delineated by green lights visible omnidirectionally from on or above the landing area.

Period 周期 系指声波振动一次循环所需要的时间。亦即声波传过一个波长的时间,或一个完整的周期波通过波线上某一点的时间。

Period of encounter 遭遇周期 相邻两波峰或波谷以不变的角度与船舶相遇时,经过船体上一定点的时间间隔。$T_E = 1/f_E$。

Periodical audit 定期审核 系指为确认形式认可和工厂认可证书保持持续有效而进行的审核。 Periodical audit means an audit for confirming continued compliance with the certificate of type approval A or the certificate of works approval.

Periodical survey 定期检验 是对与特定证书有关项目进行检验以确保其处于良好状态,并且适合营运船舶所要进行的营运业务。

Periodical survey on WIG craft 地效翼船的定期检验 系指在地效翼船安全证书签发周年日期前后 3 个月内进行的检验。 Periodical survey on WIG craft is a survey within three months before or after each anniversary date of the WIG Craft Safety Certificate.

Periodically unattended 周期性无人值班

Periodically unattended machinery space 周期无人值班机器处所 由于机器处所内广泛采用了具有各种自动功能(自动起停、自动检测、自动显示、自动报警等)的设备及管路、阀门和仪表,因此轮机员不需时刻位于机器处所进行值班,俗称无人机舱。

Periodically unattended machinery spaces 定期无人机舱 系指包括主推进装置和有关机械以及所有主供电电源设备的舱室,在各种运行情况下(包括操纵之在内),可以定时地而不需日夜值班。 Periodically unattended machinery spaces means those spaces containing main propulsion and associated machinery and all sources of main electrical supply which are not at all time manned under all operating conditions, including manoeuvring.

Peripheral 外围设备 系指执行计算机系统辅助工作的设备，如键盘、显示器、数据存储器、打印机、传感器、控制元件、电源等。 Peripheral means a device performing an auxiliary action in the system, e. g. keyboard, display, data storage device, printer, sensor, control unit, power source.

Peripheral jet hovercraft 周边射流气垫船 由位于船底四周的喷口喷射空气，形成气幕以封闭并维持气垫的全垫升气垫船。

Peripheral pump 旋涡泵 一种在泵运转时，有一股液体在泵的轴向截面中，沿纵向作螺旋状流动，并在螺旋流动过程中将能量传递给另一股低速液流而使之获得扬程的泵。按叶轮结构形式分为开式叶轮泵和闭式叶轮泵两种。

Perlite rock 珍珠岩 黏土状，多粉尘。淡灰色。无嗅。含水量 0.5% ~ 1%。无特殊危害。该货物为不燃物或失火风险低。 Perlite rock is clay-like and dusty. Light grey. It is odourless. Moisture: 0.5% to 1%. No special hazards. This cargo is non-combustible or has a low fire-risk.

Permanent awning 固定天幕 通常用硬质材料制成的固定设置的天幕。

Permanent continuous coaming 永久性连续围板 系指位于货油舱最尾端和甲板室前端舱壁间的适当的位置，并且高于甲板上 50 mm 的围板。 Permanent continuous coaming means a suitable place between the aft extreme end of cargo oil tanks and the front bulkheads of deckhouses and it is not to be made lower than 50 mm above the upper edge of shear strakes.

Permanent Court of Arbitration 常设仲裁法院 是 1899 年海牙《和平解决国际争端公约》的缔约国，根据公约第 20 ~ 29 条的规定，于 1900 年在荷兰海牙建立的。

Permanent hardness 永久硬度 水中由沸腾时不分解而能在锅炉金属壁面形成水垢的钙、镁盐以及其他非重碳酸钙、镁盐所构成的硬度。

Permanent label (LPG system) 永久性标记（LPG 系统） 固定在位置上的擦不掉的标记。 Permanent label is the label indelibly marked label secured in place.

Permanent mooring anchor 固定锚 供船舶长期停泊或使浮筒、浮标系留于水面的锚。通常有单爪锚、螺旋锚、菌型锚等。

Permanent mooring system 永久性系泊系统 系指用于在海上停留时间较长的浮式装置，诸如半潜式生产平台、Spar 平台、浮式生产储油船（FSO）、浮式生产储存和卸油装置（FPSO）等系泊定位系统。该类装置设计使用寿命在 10 年以上，而且在遇到强风暴时，一般不可能离开位置，更不可能快速回收水下立管，系泊系统的失效将造成严重后果。因此，应根据该系统设计规定的最恶劣环境条件所做的系泊分析来确定系留点（锚）的负荷以及系泊系统的形式和安装方式，所选择的锚应十分可靠，且需要更加充分地了解预定抛锚地点的底质条件海床形状。永久性系泊系统目前主要由两种形式，即悬链式和绷紧式。悬链式通常采用大抓力拖曳埋置锚，如同移动式系泊系统。绷紧式系泊系统则可采用桩锚、吸力桩和板锚等。

Permanent openings 永久开口 系指纵向倾斜 10°、横向倾斜 20°时，从外板垂直方向距救生筏末端偏离 2 m 的范围。 Permanent openings are to be arranged at the outside of the areas within the survival craft length plus 2 m from its ends in vertical direction on the shell plating under conditions of trim of up to 10° and heel of list of up to 20°.

Permanent-magnet generator 永磁发电机 系指磁通是由一个或几个永久磁铁提供的一种发电机。

Permeability 渗透率 系指在该处所内，假定被水浸占的容积与该处所总容积之比。假定破损处所的渗透率如表 P-1 所示： Permeability of a space means the ratio of the volume within that space which is assumed to be occupied by water to the total volume of that space. The permeabilities of spaces assumed to be damaged shall be as table P-1.

表 P-1 渗透率
Table P-1 Permeability

处所 Spaces	渗透率 Permeabilities
物料贮存处所 Appropriated to stores	0 ~ 0.95
起居处所 Occupied by accommodation	0 ~ 0.95
机器处所 Occupied by machinery	0.85

续表 P-1

处所　Spaces	渗透率　Permeabilities
留空处所　Voids	0.95
用于装载消耗液体处所　Intended for consumable liquids	0～0.95
用于装载其他液体处所　Intended for other liquids	0～0.95

Permeability for solid bulk cargo　固体散货渗透率　系指货物微粒、颗粒或任何较大料块间的可进水体积与散货总体积之比。 Permeability for solid bulk cargo means the ratio of the floodable volume between the particles, granules or any larger pieces of the cargo, to the gross volume of the bulk cargo.

Permeability in relation to a space　处所的渗透率　系指该处所可能被水浸占的体积与该处所总体积之比。 Permeability in relation to a space is the ratio of the volume within that space which is assumed to be occupied by water to the total volume of that space.

Permeability μ　渗透率 μ　某一处所的渗透率 μ 系指为该处所浸水容积与浸没容积之比。Permeability μ of a space is the proportion of the immersed volume of that space which can be occupied water.

Permeability μ **of a space**　某一处所的渗透率　系指该处所能被水浸占的浸水容积比例。SOLAS 公约第 II-I 章的分舱和破损稳性计算书中，每一普通舱室或某一舱室的一部分的渗透率应按表 P-2 规定取值：Permeability μ of a space is the proportion of the immersed volume of that space which can be occupied by water. For the purpose of the subdivision and damage stability calculations of Chapter II-I of SOLAS, the permeability of each general compartment or part of a compartment shall be as table P-2.

表 P-2　某一处所的渗透率
Table P-2　Permeability μ of a space

处所　Spaces	吃水 d_s 时的渗透率 Permeability at draught d_s	吃水 d_p 时的渗透率 Permeability at draught d_p	吃水 d_1 时的渗透率 Permeability at draught d_1
干货处所 Dry cargo spaces	0.70	0.80	0.95
集装箱处所 Container spaces	0.70	0.80	0.95
滚装处所 Ro-Ro spaces	0.90	0.90	0.95
液货 Cargo liquids	0.70	0.80	0.95

如经计算证实，渗透率可用其他数字。 Other figures for permeability may be used if substantiated by calculations.

Permissible clearance　允许间隙

Permissible design stress　许用设计应力

Permissible length　许可舱长　船舶各处主水密舱的最大许可长度。即舱长中点处的可浸长度和分舱因数的乘积。

Permissible payload　最大许用载荷　系指集装箱最大营运总质量或额定质量与空箱质量之差。Maximum permissible payload means the difference between maximum operating gross mass or rating and tare.

Permissible rope length of drum　卷筒容绳量　卷

筒上容许卷绕钢丝绳的长度。

Permissible stress 许用应力

Permit to operate high-speed craft 高速船营运许可证书 系指对符合 HSC 规则 1.2.2 至 1.2.7 和 1.8 要求的对高速船颁发的高速船营运许可证书。 Permit to operate high-speed craft means a certificate called a Permit to operate high-speed craft issued to a craft which complies with the requirements set out in paragraphs 1.2.2 to 1.2.7 of the 1994 HSC Code or the 2000 HSC Code, as appropriate.

Permit to operate WIG craft 地效翼(WIG)船营运许可证书 系指主管机关签发营运许可证书以证明其符合地效翼(WIG)船临时指南的要求。 Permit to operate WIG craft means that a permit to operate issued by the Administration to certify compliance with the provisions of the Interim Guidelines for WIG craft.

Permit to work 准许工作 主管机关凭此对某一系统执行确定的操作。这些准许是为一特定时间和特定工作进行认可和颁发的。在该时间内此工作得以完成。

Perpendicular 艉垂线 系指夏季载重水线上舵柱后端处水线的垂线。对于无舵柱的船舶,艉垂线为夏季载重线上舵杆中心线处水线的垂线。 After perpendicular is the perpendicular to the waterline at the after side of the rudder post on the summer load waterline. For ships without rudder post, the after perpendicular is the perpendicular to the waterline at the centre of the rudder stock on the summer load waterline.

Perpendiculars 艏、艉垂线 艏、艉垂线应取自长度 L 的前后两端。艉垂线应与计量长度的水线上的艏柱前缘相重合。 Forward and after perpendiculars shall be taken at the forward and after ends of the length L. The forward perpendicular shall coincide with the foreside of the stem on the waterline on which the length is measured.

Persistence tests 留存试验 系指通过确定相关条件下的半衰期的模拟测试系统对留存进行评估的试验。生物降解衰减试验可用于显示物质可随时降解。半衰期的确定应包括对相关化学品的评估。 Persistence tests mean those tests in which persistence should preferably be assessed in simulation test systems that determine the half-life under relevant conditions. Biodegradation screening tests may be used to show that the substances are readily biodegradable. The determination of the half-life should include assessment of relevant chemicals.

Persistent, bioaccumulation and toxicity (PBT) 留存、生物体内积累和产生毒性

Person 人员 系指船员和乘客。 Person means member of the crew and passengers.

Person 法人 系指:(1)个体户、商行、国营或民营公司、合伙人、联合体、州政府、市政府、委员会、州政府下属的机构或任何营利性的实体。(2)个体人、合伙人、商行、公司或集团,但不包括公务船的个体人或(3)个体户、公司、合伙人、有限责任公司、联合体、州政府、市政府、州政府的委员会或下属机构,或者联邦政府承认的任何印第安人部落政府。 Person means:(1)an individual, firm, public or private corporation, partnership, association. State municipality, commission, or any interstate body;(2)an individual, partnership, firm, corporation, or individual on board a public vessel;or(3)an individual, corporation, partnership, limited liability company, association, state, municipality, commission or political subdivision of a state, or any federally recognized Indian tribal government.

Person 人员 系指任何个人或合伙关系、或任何公共或私营机构,而不论其是否为法人,包括国家或其任何所属部门。 Person means any individual or partnership, or any public or private body whether corporate or not, including a State or any of its constituent subdivisions.

Person in charge 负责人员 对澳大利亚海事安全局海事指令而言,系指在澳大利亚某港口承担船舶装卸的除船长以外的某个人。 For the purpose of the maritime order of the Australian maritime safety authority(AMSA), Person in charge means a personal, other than master, undertaking to load or unload a ship at a port in Australian.

Personal digital assistant(PDA) 个人数字助手 又称掌上电脑,可以帮助人们完成在移动中工作,学习,娱乐等。按使用来分类,分为工业级 PDA 和消费品 PDA。工业级 PDA 主要应用在工业领域,常见的有条码扫描器、RFID 读写器、POS 机等都可以称作 PDA;消费品 PDA 包括的比较多,智能手机、平板电脑、手持的游戏机等。顾名思义就是辅助个人工作的数字工具,主要提供记事、通信录、名片交换及行程安排等功能。

Personal digital cellular(PDC) 个人数字蜂窝电话

Personal equipment(fire-fighter's outfit) 个人配备(消防员的装备) 包括:(1)防护服,其材料应能保护皮肤不受火焰的热辐射,并不受蒸汽的灼伤和烫伤,衣服的外表应是防水的;(2)由橡胶或其他不导电材料制成的消防靴;(3)一顶能对撞击提供有效防护的消防头盔;(4)一盏认可型的电安全灯(手提灯),其照明时间至少为 3h。油船配置的电安全灯及拟用于危险区域的电安全灯应为防爆型;(5)太平斧的手柄应具有高电压绝缘。 Personal equipment(fire-fighter's outfit)shall consist of the following:(1)protective clothing of material to protect the skin from the heat radiating from the fire and from burns and scalding by steam. The outer surface shall be water-resistant;(2) boots of rubber or other electrically non-conducting material;(3) rigid helmet providing effective protection against impact;(4) electric safety lamp(hand lantern) of an approved type with a minimum burning period of 3 h. Electric safety lamps on tankers and those intended to be used in hazardous areas shall be of an explosion-proof type;(5) axe with a handle provided with high-voltage insulation.

Personal flotation device(PFD) 个人漂浮装置 一种衣服或装置,在水中正确穿着和使用时,将为使用者提供规定数量的浮力,以增加其生存可能性。 Personal flotation device is the garment or device which, when correctly worn and used in water, will provide the user with a specific amount of buoyancy which will increase the likelihood of survival.

Personal jurisdiction 个人管辖权

Personal survival kit(PSK) 个人救生用具 个人救生用具的内容在表 P-3 中列出。 A sample of the personal survival kit(PSK) is listed in the table P-3.

表 P-3　个人救生用具
Table P-3　Personal survival kit

属具　Equipment	数量　Quantity
防护服 Clothing	
头盔(真空包装) Head protection(vacuum packed)	1 个
颈、面护具(真空包装) Neck and face protection(vacuum packed)	1 具
手保护——连指手套(真空包装) Hand protection—Mitts(vacuum packed)	1 双
手保护——手套(真空包装) Hand protection—Gloves(vacuum packed)	1 双
脚保护——短袜(真空包装) Foot protection—Socks(vacuum packed)	1 双
脚保护——靴子(真空包装) Foot protection—Boots(vacuum packed)	1 双
保温服(真空包装) Insulated suit(vacuum packed)	1 件
认可的救生服 Approved immersion suit	1 件
保暖内衣(真空包装) Thermal underwear(vacuum packed)	1 套
其他 Miscellaneous	

续表 P-3

属具 Equipment	数量 Quantity
手取暖器 Handwarners	1 个(240h)
太阳眼镜 Sunglasses	1 副
救生蜡烛 Survival candle	1 支
火柴 Matches	2 盒
哨子 Whistle	1 个
饮水杯 Drinking mug	1 个
袖珍折刀 Penknife	1 把
手册(极地救生) Handbook (Polar Survival)	1 本
用具包 Carrying bag	1 个

Personalized domain name service (PDNS) 个性化域名服务

Personnel alarm 人员报警 系指用来证实在机器处所单独值班的轮机员安全的报警。 Personnel alarm means an alarm to confirm the safety of the engineer on duty when alone in the machinery spaces.

Personnel hazard goal 1 防止人员危险目标 1 在使用或运输的材料可能产生对人员有害的蒸发气或烟气的情况下，舰船的布置应使那些蒸发气和烟气的影响降至最低；烟气的产生及其毒性的潜在性应符合 SOLAS 公约第 2 章 B 部分第 6 条的要求。 Where materials used or carried may develop vapors and smoke dangerous to personnel, the ship is to be arranged so as to minimise the effects from those vapors and smoke. The potential for smoke generation and its toxicity is to be in accordance with SOLAS Chapter 2, Part B, Regulation 6.

Personnel hazard goal 2 防止人员危害目标 2 在舰船人员可能被合理地指派灭火的情况下，应为每一名消防队员提供充足的防护装备。消防员装备应满足 SOLAS 公约第Ⅱ-2 章 C 部分第 10.10 条的要求。 Personnel hazard goal 2 means where personnel may be reasonably expected to fight a fire, there is to be adequate provision of protective equipment for each member of the firefighting party. The equipment is to be of a standard applicable for its intended application: fire-fighters' outfits are to be in accordance with SOLAS Chapter Ⅱ-2, Part C, Regulation 10.10.

Personnel hazard objective 人员危险目标 应采取所有可能的方法来防止火灾对人员造成的危害。 Personnel hazard objective means that all reasonable measures are to be taken to prevent hazards to personnel as a result of fire.

Personnel on board 船上人员(载员) 系指乘载于船上的所有人员，包括乘客、船员和特殊人员。 Personnel on board mean all personnel who are carried on board, including passengers, crew and special personnel.

Personnel protection goal 1 人员保护目标 1 撤

离人员应被保护，避免低温或曝露等环境的不利影响。通常，浸水服和防暴露服应按照 SOLAS 公约第Ⅲ章 B 部分 7.3 条要求提供。 Personnel protection goal 1 means that evacuated personnel are to be protected from the adverse effects of the environment such as hypothermia or exposure. In general, the provision of immersion suits and anti-exposure suits is to be consistent with SOLAS Chapter Ⅲ, Part B, Regulation 7.3.

Personnel protection goal 2 人员保护目标 2 救生艇筏应配备充足，保持在有关主管当局所规定的时间内，人员免于饥饿和脱水。 Personnel protection goal 2 means that survival craft are to be equipped with sufficient provisions to keep personnel free from starvation and dehydration for a period of time specified by the Naval Authority.

Personnel protection objective 人员保护目标 撤离人员应随时受到保护，直至其可从救生艇筏上获救。 Personnel protection objective means that evacuated personnel are to be kept protected until such time as they can be rescued from the survival craft.

Personnel safety 人员安全 系指所有参与油田作业人员的安全。

Petition for retrial 再审申请书

Petrol engine 汽油发动机（汽化器式发动机）

Petroleum 石油 系指包括所有的石油产品，例如：石油、仰光油、缅甸油和由石油、树脂、烟煤、片岩、页岩、泥煤和其他沥青物质中提炼出的油类，以及由石油和上述的任何油类制出的任何产品，例如：挥发油、煤油、汽油、燃料油、甲苯、粗石蜡等。石油可划分为：(1) A 级品——凡上述产品或未提及的任何其他产品，其闪点在 23 ℃ (73 ℉) 以下者；(2) B 级品——凡上述产品或未提及的任何其他产品，其闪点在 23 ℃ (73 ℉) 至 66 ℃ (150 ℉) 之间者；(3) C 级品——凡上述产品或未提及的任何其他产品，其闪点在 66 ℃ (150 ℉) 以上者。 Petroleum means all products such as: rock oil, Rangoon oil, Burmah oil, oil made form petroleum, rosin boghead, coal, schist, shale, peat(such as benzene, kerosene, gasoline, fuel oil, toluene, paraffin wax etc.). Petroleum is classified: (1) Grade A—those of the above mentioned products or any other not mentioned and having a flash point below 23 degrees Centigrade (73 degrees Fahrenheit); (2) Grade B—those of the above mentioned products or any other not mentioned and having a flash point between 23 degrees Centigrade (73 degrees Fahrenheit) and 66 degrees Centigrade (150 degrees Fahrenheit); (3) Grade C—those of the above mentioned products or any other not mentioned and having a flash point above 66 degrees Centigrade (150 degrees Fahrenheit).

Petroleum asphalt 石油沥青 系指石油原油经提炼出汽油、煤油、柴油及润滑油等石油产品后，再经处理而成的副产品。 Petroleum asphalt means co-product of petroleum crude oil extracted to petroleum product such as gasoline, coal oil, diesel oil and lubricating oil, and then treated.

Petroleum asphalt carrier 沥青船 系指专门从事载运散装熔化的石油沥青的无限航区或有限航区的船舶。 Petroleum asphalt carrier means a carrier specialized in carrying melted petroleum asphalt in bulk in unrestricted navigation area and restricted navigation area.

Petroleum coke (calcined or un-calcined) 石油焦炭（经煅烧或未经煅烧） 炼油产生的黑色细碎残留物，呈粉末和小块状。未经煅烧的石油焦炭如不按有关规定装载和运输，则易自热和自燃。该货物为不燃物或失火风险低。 Petroleum coke is a black, finely divided residue from petroleum refining in the form of powder and small pieces. Un-calcined petroleum coke is liable to heat and ignite spontaneously when not loaded and transported under the relevant provisions. This cargo is non-combustible or has a low fire-risk.

Petroleum gas 石油气

Petroleum liquefied gas 液化石油气

Petroleum (petroleum oil) 石油

Petty officer 专职船员 系指担负警戒职务或特殊责任，并按国家的法规或规则，或在无任何相应的法规或规则时，由集体商议或按惯例确定的归类于专职船员的普通船员。 Petty officer means a rating serving in a supervisory position or position of special responsibility who is classed as petty officer by national laws or regulations, or, in the absence of any relevant laws or regulations, by collective agreement or custom.

PFD with secondary donning 二次穿着的个人漂浮装置 需要做附加的穿戴或调节，以从正常穿着位置移至其起作用位置的个人漂浮装置。注：袋型装置为该类型 PFD 的例子，其通常要求这种附加的定位。 PFD with secondary donning is the PFD for which additional donning or adjustment that is needed to place the PFD in its functioning position from the position it is normally worn. Note: Pouch-type devices are examples of the type of PFDs which usually require such additional positioning.

pH value pH 值 表示水密度酸性或碱性的指示值，pH = 7 (中性), pH > 7 (碱性), pH < 7 (酸性)。

Phase alternating line (PAL) 逐行倒相 是电视广播中色彩调频的一种方法。

Phase ambiguity 相位多值性 无线电相位导航设备中,测量相位差的装置对于主、副台信号相位差的反映呈周期性的现象。即测量相位差装置的输出电压,只能反映出主、副信号相位差小于360°的部分(小于一个相位周期的小数部分)。当用这种设备实际测定点位时,就必须消除多值性。

Phase angle in ship motions 船舶运动相角 船舶摇荡等运动与波浪间的相位差。

Phase plane analysis 相平面分析法 用船舶转艏角加速度对转艏角速度的相平面图,分析Z形操纵试验数据的求解二阶非线性操纵运动方程的一种图解方法。

Phase plane plot 相平面图 用船舶转艏角加速度对转艏角速度,即一般用变量的导数对变量绘出的相轨曲线图。

Phase revolution 相位周值 当电磁波相位差变化2π(即一个相位周期)时,所对应的实际距离(或距离差)变化的长度。用于以长度为单位的距离(或距离差)或以相位周期为单位的距离差之间进行换算。

Phase shift 相位漂移 在无线电相位导航仪器中,由于电子元件参数和电磁波传播速度变化所引起的附加相位漂移。设立检查台(或检测台)实际测定相位漂移的大小,对测量结果加以改正,可保证定位精度。

Phase stability 相位稳定性 系指无线电相位导航系统的附加相位差的稳定程度。相位稳定性与导航仪器的质量有关,是影响测量相位差精度的重要因素之一。在仪器出厂和开始测量工作前,须对仪器进行相位稳定性检验。

Phased array radar 相控阵雷达 即有源电子扫描阵列雷达(active electronically scanned array, AESA)或无源电子扫描阵列雷达(passive electronically scanned array, PESA),系指一类通过改变天线表面阵列所发出波束的合成方式,来改变波束扫描方向的雷达。这种设计有别于机械扫描的雷达天线,可以减少或完全避免使用机械马达驱动雷达天线便可达到涵盖较大侦测范围的目的。目前使用的电子扫描方式包括改变频率或相位的方式,将合成的波束发射的方向加以变化。电子扫描的优点包含着扫描速率高,改变波束方向的速率快,对于目标信号测量的精确度高于机械扫描雷达,同时免去机械扫描雷达天线驱动装置可能发生的故障。相控阵雷达有相当密集的天线阵列,在传统雷达天线面的面积上可安装上千根相控阵天线,任何一根天线都可收发雷达波,而相邻的数根天线即具有一个雷达的功能。扫描时,选定其中一个区块(数根天线单元)或数个区块对单一目标或区域进行扫描,因此整个雷达可同时对许多目标或区域进行扫描或追踪,具有多个雷达的功能。由于一个雷达可同时针对不同方向进行扫描,再加之扫描方式为电子控制而不必由机械转动,因此资料更新率大大提高,机械扫描雷达因受限于机械转动频率因而资料更新周期为秒或10秒级,电子扫描雷达则为毫秒或微秒级。因而更适合对付高速机动目标。此外由于可发射窄波束,因而也可充当电子战天线使用,如电磁干扰甚至是构想中发射反相位雷达波来抵消探测电波等。见图P-3。

图P-3 相控阵雷达
Figure P-3 Phased array radar

Phased array radar technology 相控阵雷达技术 该技术拥有国:美国、以色列(Elta)、英法德合作(GT-DAR公司)、荷兰(TNO物理电子实验室)、瑞典(NGRA)、俄罗斯(Tikhomirov NIIP设计局)、中国(洛阳电子所)、印度(雷达开发实验室)。

Phenolic aldehyde glass fiber felt 酚醛玻纤毡

Phenolic foamed plastics 酚醛泡沫塑料

Phenomenon of drawing air 吸气现象 螺旋桨轴线在水面以下深度过小或有部分桨叶露出水面时发生的将空气吸入水中的现象。其影响是降低推力、转矩和效率。

Phosphate(defluorinated) 磷酸盐(脱氟) 颗粒状,类似细砂。干燥运。深灰色。不含水分。无特殊危害。该货物为不燃物或失火风险低。 It is granular, similar to fine sand and shipped dry. It is dark grey in colour. No moisture content. It has no special hazards. This cargo is non-combustible or has a low fire-risk.

Phosphate rock(calcined) 硝酸盐岩(经煅烧) 通常呈细磨岩屑或小球状。粉尘极多。无特殊危害。该货物为不燃物或失火风险低。该货物吸湿,受潮会结块。 It is usually in the form of fine ground rock or prills. It is extremely dusty and hygroscopic. It has no special hazards. This cargo is non-combustible or has a low fire-risk.

This cargo is hygroscopic and will cake if wet.

Phosphate rock (uncalcined) 硝酸盐岩(未经煅烧) 是磷和氧化合而成的一种矿石。视产地而定呈棕黄色至深灰色，干燥且多粉尘。含水量0%~2%。该货物视产地而定，可能具有流动性，但一旦密实后就不易移动。无特殊危害。该货物为不燃物或失火风险低。 Phosphate rock is an ore in which phosphorus and oxygen are chemically united. Depending on the source, it is tan to dark grey, dry and dusty. Moisture: 0% to 2%. Depending on its source this cargo may have flow characteristic, but once settled it is not liable to shift. It has no special hazards. This cargo is non-combustible or has a low fire-risk.

Phosphine 磷化氢 无色、易燃、剧毒并有腐烂鱼的气味。磷化氢作用于中枢神经系统和血液。磷化氢中毒的症状表现为胸部有压迫感、头痛、眩晕、全身虚弱、厌食和口渴难止。在 $2\ 000 \times 10^{-6}$ 浓度下几分钟就有生命危险，在 $400 \sim 600 \times 10^{-6}$ 浓度下有生命危险，0.3 ppm 是能忍受几小时而不出现症状的最大浓度。应不准长期暴露于该气体中。 Phosphine is colourless, flammable and highly toxic and has the odour of rotting fish. Phosphine acts on the central nervous system and the blood. The symptoms exhibited by phosphine poisoning are an oppressed feeling in the chest, headache, vertigo, general debility, loss of appetite and great thirst. Concentrations of $2\ 000 \times 10^{-6}$ for a few minutes and 400 to 600×10^{-6} are dangerous to life. 0.3 ppm is the maximum concentration tolerable for several hours without symptoms. No long-term exposures to this gas shall be permitted.

Photoelectric colorimeter 光电比色计 应用有色溶液对单色光的吸收程度来确定溶液浓度的仪器。由光学系统、滤光计、比色计、光电池(光电管)和显示部分组成。

Photo-electrically controlled cutting machine 光电切割机 系指利用光电技术进行切削物料的机器。

Photographic system 电子照相系统 系指以磁盘代替化学感光胶片的照相机，它通过光学镜头接收光像，再转投到电荷耦合摄像机上，转换为电信号后记录在小巧的方型磁盘上。因此，其结构与传统的利用感光胶片成像的照相机有很大的区别。主要部件包括镜头、电耦合器件、磁盘驱动机构、磁盘及快门等。快门采用电子控制，速度范围宽，便于控制。曝光等也都采用电子控制，所以可实现许多自动功能，并可进行连续拍摄，功能十分齐全优异。图像信息记录在磁盘上，保存和复制都很容易，但须经过配套的播映机和电视机才能显示出来。电子照相机的出现，使家庭的图像记录方式进入一个新的领域。

Photon computer 光子计算机 是一种由光信号进行数字运算、逻辑操作、信息存贮和处理的新型计算机。它由激光器、光学反射镜、透镜、滤波器等光学元件和设备构成，依靠激光束进入反射镜和透镜组成的阵列进行信息处理，以光子代替电子，光运算代替电运算。光的并行、高速，自然地决定了光子计算机的并行处理能力很强，具有超高运算速度。光子计算机还具有与人脑相似的容错性，系统中某一元件损坏或出错时，并不影响最终的计算结果。光子在光介质中传输所造成的信息畸变和失真极小，光传输、转换时能量消耗和散热量极低，对环境条件的要求比电子计算机低得多。随着现代光学与计算机技术、微电子技术相结合，在不久的将来，光子计算机将成为人类普遍的工具。

Phychological addiction 心理毒瘾 系指一种状态。在这种状态中，毒品催生一种满足感和反复消费毒品以产生快感或避免不适的冲动。在这种情况下，大脑逐渐对毒品产生依赖性，但人体可能没有这种依赖性。戒除症状不像人体毒瘾那样明显，但仍可能有易怒、突然发脾气、迷恋于再次服用。不合理的举动、感到受害的症状。 Phychological addiction means a condition in which the drug promotes a feeling of satisfaction and a drive to repeat the consumption of the drug in order to induce pleasure or avoid discomfort. In this case the mind develops a dependence on the drug although there may be no physical dependence. Withdrawal symptoms are not as in physical addiction but there may still be irritability, fits of anger, fixation on taking a further dosage, irrational behaviour, feelings of victimization, etc.

Physical addiction 人体毒瘾 系指当体内毒品数量明显减少时由人体失调所显示出的一种状态。这些失调形成一种脱瘾或戒除综合征，具有作为各种毒品特征的肉体和精神症状或征兆。就人体毒瘾而言，人体逐渐形成对毒品的渴望。当停止使用毒品时，戒除症发作，其中有些症状在身体上的表现形式为过多出汗，不断要喝水、搔痒、肌肉抽搐、易怒、腹泻、肌肉痉挛，极端的情况为昏迷和死亡。如果人体毒瘾发作，人体需要逐步增大毒品剂量，以达到同样程度的陶醉或"快感"。对剂量的增加察觉得越快，据说人体的耐受力就越大。 Physical addiction means a state that shows itself by physical disturbance when the amount of drug in the body is markedly reduced. The disturbances from a withdrawal or abstinence syndrome composed of somatic and mental symptoms and signs which are characteristic for each drug type. In the case of physical addiction the body develops a craving for the drug. Withdrawal symptoms occur when the drug is withheld and some of the symptoms are physically visible in the form of excessive sweating, constant desire for liquids, scratching, twithing of muscles, irritability, diarrhea, muscle spasm and in extreme cases, coma and death. Where physical addiction occurs the body requires progressively larger doses of the drug to achieve the same level of intoxication or "high". The quicker this increase this noticed the higher the body tolerance is said to be.

Physical and chemical analysis for lubricating oil 润滑油油品理化分析 系指：（1）测量并评定润滑油品质变化，如润滑油中氯化物含量，润滑油氧化、硫化及添加剂劣化等情况；（2）测定润滑油中杂质，如尘垢、燃油稀释及水分稀释等。（3）测定润滑油物理性能的变化，如黏度、总酸值、总碱值等。润滑油油品理化分析能够评定润滑油的状态，正确掌握润滑油换油期，在保证机械摩擦副良好润滑状态的情况下，尽可能地延长润滑油使用寿命。 Physical and chemical analysis for lubricating oil means: (1) Measuring and assessing the change of quality of lubricating oil, e.g. chloride content in lubricating oil, oxidation and sulphidation of lubricating oil and deterioration of additives; (2) Detecting the contaminants in lubricating oil, e.g. dust, diluted fuel and water; (3) Detecting the change of physical performance of lubricating oil, e.g. viscosity, total acid number and total base number. Assessing the lubricating oil condition to indicate proper intervals for replacing oil and extend the service life of lubricating oil so far as practicable with the machinery friction pairs being well lubricated.

Physical security 实地保安 系指保安管理的一部分，旨在为阻扰侵入保安防御的企图而布置的有关物理性障碍。 Physical security means the part of the security management, the purpose of which is to arrange physical obstruct to attempt of intruding the security defense.

Phytosanitary certificate 植物检疫证书

Pickling 酸洗（酸浸，酸蚀）

Pickling bath 酸洗池

Picture element V/Pixel 像元 （1）亦称像素或像元点。是组成数字化影像的最小单元。在遥感数据采集，如扫描成像时，它是传感器对地面景物进行扫描采样的最小单元；在数字图像处理中，它是对模拟影像进行扫描数字化时的采样点；（2）是显示设备中构成图像的最小单位，虽然像素非常的小，但是无数像素组织在一起才构成了一幅完整的图画。

Pictured Rocks National Lakeshore 彩岩石国家湖滨区 系指就在或邻近国家公园管理局管理的苏必利尔湖的地方，从沿着密歇根沿岸向东位于北纬46°26′21.3″，西经089°36′43.2″陆地的一点到北纬46°40′22.2″西经085°59′58.1″处。Pictured Rocks National Lakeshore means the site on or near Lake Superior, administered by the National Park Service, a point of land at from 46°26′21.3″N, 089°36′43.2″W, eastward along the Michigan, shoreline to 46°40′22.2″N, 085°59′58.1″W.

Pier 突堤码头 又称娱乐码头。娱乐码头为那些富裕的度假者提供了专用的散步场所。现在，这些码头则与廉价的娱乐活动密切相关。娱乐码头从海岸突出到海里，以铁条或木条制成的格子状构架为支撑。它有着木质人行道，并且路的尽头有亭子或其他建筑物。

Piexiglass/Methyl methacrylate (PMMA) 有机玻璃 是一个通俗的名称。该高分子透明材料的化学名称叫聚甲基丙烯酸甲酯，是由甲基丙烯酸甲酯聚合而成的高分子化合物。是一种开发较早的重要热塑性塑料。有机玻璃分为无色透明，有色透明，珠光，压花有机玻璃4种。有机玻璃俗称亚克力、中宣压克力、亚格力。有机玻璃具有较好的透明性、化学稳定性、力学性能和耐候性，易染色，易加工，外观优美等优点。有机玻璃又叫明胶玻璃、亚克力等。

Piezoelectric microphone 压电传声器

Pig iron 生铁 铸造生铁时铸成20 kg的铁锭，有28个等级。在任意堆放时，生铁约占表观体积的50%，无特殊危害。该货物为不燃物或失火风险低。 Foundry pig iron is cast in 28 grades into 20 kg pigs. In a random heap, pig iron occupies approximately 50% of the apparent volume. It has no special hazards. This cargo is non-combustible or has a low fire-risk.

Piggy-back microbiolegical sampler 复背式微生物采集器 采集海洋中微生物样品的一种厚壁橡皮球附在南森采水器上构成。

Pile 单桩 系指：（1）用大型液压锤打入海底的大直径的、长4~8 m的钢管；（2）直径约1 m的管桩，可打入海底牢牢固定基础结构。由焊接在机舱结构基点上的桩套管推进。

Pile driver piler 打桩船 用于水上打桩作业的船舶。船体为钢箱型结构，在甲板的端部装有打桩架，可前俯后仰以适应施打斜桩的需要。打桩船为非自航船，用推（拖）船牵引就位。打桩船广泛应用于桥梁、码头、水利工程施工。打桩船的生产过程包括：钢板切割，分段制作，船台阶段，设备调试阶段和试航阶段。打桩船配备打桩锤、桩架、附属设备。打桩船一般配有工作间或维修间，可以对简单的问题（例如焊接）进行修复。与国内其他打桩船相比，"海威951"打桩船有三大特点：（1）该船集打桩和吊装为一体，一次最大起吊质量达240 t可以做-27°~+25°的变幅；（2）可插打18.5°的斜桩，其桩基施工的仰俯角度在国内为最大；（3）95 m

高的桩架可进行3次变幅折叠，变幅后的桩架距水面高度仅有25.4 m，可以抵达长江中下游的任何施工地点。

Pile driving barge 打桩船 系指在甲板端上设有打桩设备，专为水上工程打桩用的船舶。Pile driving is a barge fitted with pile driving equipment at end or centre of deck, dedicated to pile driving in water.

Pile driving flame 打桩架 用以支承龙口和吊桩的架子。分为：(1)多能打桩架——能旋转俯仰，其龙口能伸缩和左右倾斜适合于在各种情况下作业的打桩架。(2)摆动式打桩架——龙口设在架的中部或稍偏高些，可左右摆动，用于打左右斜向支撑桩的打桩架。(3)俯仰式打桩架——可以前后俯仰，其俯仰角可调节的打桩架。(4)吊龙口打桩架——可由主装架把龙口吊起并前伸一定距离的打桩架。

Pile-supported platform 桩基平台/桩支承平台 系指：(1)利用桩腿作为支承结构的平台；(2)以打进海底的细长基础构件或桩柱支承的平台。

Pillar 支柱 系置于不由船壳或舱壁支撑甲板之间的垂直支撑结构。 Pillar means a vertical support placed between decks where the deck is unsupported by the shell or bulkhead.

Pilot 引水员 是船东雇用，协助船舶进出港口、狭窄水道等复杂水域的人员。因为船长不一定会经常进出停靠的港口、水道等，对该水域的水深、潮汐、水流等不太熟悉，为了航线安全，临时雇用熟悉当地的水域的引水员协助船长进出港口等。

Pilot boat 引航船/领航船 系指专门从事引水业务的船舶。 Pilot boat means a ship dedicated to pilot service.

Pilot exciter 副励磁机 系指为另一台励磁机的励磁提供全部或部分磁场电流的电源装置。

Pilot flag 引水员旗 悬挂在船上，表示引水员正在领航的旗。

Pilot fuel oil 引燃油 系指用来点燃气缸内的甲烷/空气混合物的燃油。

Pilot signals 引水信号

Pilot valve 油遮断器 液压保护系统中，在接到各种保护装置的动作信号后，发出信号，使速关阀关闭的综合开关。

Pilot vavle 操纵阀(导阀) 根据感受元件的指令使主阀动作的阀。

Pilot vessel/pilot boat 引水船 又称引航船，领港船。用于接送引航员的船舶。一般为小型专用交通艇。对于需去外海的引水船，排水量亦有达数百吨者，并设有引航员居住舱室及娱乐、休息处所。外国船舶进入领海后必须由本国引航员登船驾驶入港和出港。

Pilot wheel 导轮

Pilot-joint inspection boat 引航联检船 专用于接送引航员和港务监督、海关、检疫所、边防站联合检查人员的船舶。

Pilston type scavenging pump 活塞式扫气泵

Pin 销钉，插头，压住

Pin-holes 针孔

Pinion(pinion wheel) 小齿轮(副齿轮,传动齿轮) 系指齿数较少的齿轮。 Pinion is a gear with the less teeth.

Pipage 管路(管系,用管道输送)

Pipe 管(管系,导管)

Pipe accessories 管子附件

Pipe arrangement 管系布置

Pipe barge 运管驳船 系指为铺管船运载及供应铜管的驳船。

Pipe bend 管子弯头

Pipe bender 弯管机

Pipe bending shop 弯管车间

Pipe burying machine 埋管机

Pipe carrier 运管船 系指运送管承,管架,管座的船舶。

Pipe casing 管子套

Pipe chair 管承(管座)

Pipe clamp 管夹

Pipe coating 管面涂层

Pipe coupling 管件接头

Pipe cutter 管子割刀

Pipe cutting machine 截管机(管子切割机)

Pipe hanger 管吊架 支撑管子的架子。

Pipe head 管口

Pipe holder 管道支架(管夹,管箍)

Pipe layer 铺管船 系指设有铺管专用设备的船舶。 Pipe layer is a ship provided with special equipment for laying pipes.

Pipe laying 铺设管子

Pipe line 管系(导管)

Pipe passage 管路

Pipe pillar 管形支柱

Pipe rack 管架

Pipe racking machine 排管机

Pipe rail 管子栏杆

Pipe ramp 管滑道

Pipe recess 管道凹穴

Pipe stanchion 管支柱

Pipe steel 管子钢

Pipe stock and dies　管子丝攻扣（丝扳）
Pipe strap　管子卡箍
Pipe support　管子支架
Pipe system　管系
Pipe tap　管螺纹丝锥（管用攻丝锥）
Pipe testing　管子试压检验
Pipe tongs　管子钳
Pipe tracker sensor unit　海底管道探测跟踪单元
Pipe tunnel　管隧　系指在内底和外板之间沿船中舯舱线延伸的空舱，构成对舱底水管路、压载水管路和机舱至液舱的其他管路的保护处所。Pipe tunnel means a void space running in the mid-ships fore and aft lines between the inner bottom and shell plating forming a protective space for bilge, ballast and other lines extending from the engine room to the tanks.
Pipe tunnel plating　管隧板
Pipe turnbuckle　套筒松紧螺丝
Pipe vice (pipe wrench)　管子虎钳
Pipe welder (pipe welding machine)　焊管机
Pipe-burying barge　埋管驳船　系指借助于船载埋管机完成埋管作业的驳船。
Pipe-in-pipe system　套管系统　这是由两根同轴管组成的系统。内管通常在高温和高压情况下输送碳氢化合物，外管暴露于水中并指定承受外部压力。在这两根管子之间的环形套管是空的或填满绝缘材料。套管系统通常作热绝缘用，有时对内管实行机械保护。Pipe-in-pipe system is a system of two concentric pipes. The inner one carries hydrocarbons, usually at elevated pressure and temperature. The outer one is exposed to the water and is designed to carry external pressure. The annulus between the two pipes is either empty or contains insulation material. Pipe-in-pipe systems are usually used for thermal insulation and sometimes for mechanical protection of the inner tube.
Pipe-joint piece　管接头
Pipe-laying barge/pipeline layer　铺管船　系指设有铺管专用设备，如起重机、拖管绞车等，供在水域铺设石油或天然气输送管用的船舶。
Pipe-laying barge　铺管驳船　系指浅吃水、驳船型，可在近海、内河、沼泽地等区域进行管道铺设、架接等作业的驳船。
Pipelaying equipment　敷管设备
Pipe-laying ship (PLS)　铺管船　系指设有铺管专用设备，如起重机、拖管绞车等，供在水域铺设石油或天然气输送管用的船舶。
Pipe-laying system　铺管系统
Pipeline　管线（导管，管道安装）　系指在陆地或近海用来运输液体的管系。Pipeline is a piping used to transport fluids on land or offshore.
Pipeline burying machine　埋管机
Pipeline inspection gage (PIG)　管线检测计　系指通过管路推进去的，供检验及清洁管线用的装置。Pipeline inspection gage (PIG) is a device pulled through a pipeline for inspection as well as cleaning.
Pipeline welding　管路焊接
Pipes/piping systems　管子/管系（塑料管）　系指由塑料制成的管子、附件、管子接头、连接方法以及要求符合性能标准的任何内外衬、护层与涂层。Pipes/piping systems means those made of plastics and include the pipes, fittings, system joints, method of joining and any internal or external liners, coverings and coatings required to comply with the performance criteria.
Pipette　吸液管
Piping　管路（管系，管路布置，缩孔）　定义为包括下列部件：（1）管子；（2）法兰、螺栓连接件和其他管接头；（3）膨胀元件；（4）阀、包括液压和气动执行机构，及附件；（5）吊架及支座；（6）挠性软管；（7）泵罩。Piping is defined to include the following components: (1) pipes; (2) flanges and bolting and other pipe connections; (3) expansion elements; (4) valves, including hydraulic and pneumatic actuators, and fittings; (5) hangers and supports; (6) flexible hoses; (7) pump housings.
Piping and piping systems　管道和管系　（1）管道包括管子及其连接件、弹性软管、膨胀节、阀及阀控制系统，其他附件（滤器、液位计等）和泵壳。（2）管系包括管道以及所连设备。例如柜柜、压力容器、热交换器、泵和离心分离机，但不包括锅炉、涡轮机、内燃机和减速齿轮箱。Piping and piping systems are: (1) includes pipes and their connections, flexible hoses and expansion joints, valves and their actuating systems, other accessories (filters, level gauges, etc.) and pump casings; (2) piping systems include piping and all the interfacing equipment such as tanks, pressure vessels, heat exchangers, pumps and centrifugal purifiers, but do not include boilers, turbines, internal combustion engines and reduction gears.
Piping diagram (piping plan)　管系图（管路图）
Piping system　管路系统　（1）包括管子及附件，例如：膨胀接头、阀门、管路接头、管路支架、挠性软管等，以及直接与管件连接的设备，例如：泵、热交换器、压缩空气瓶、独立舱柜等。其不包括主/辅机，例如：柴油机、蒸汽机和燃气轮机、锅炉、减速齿轮箱等；（2）是管、阀门及管附件的总称。Piping system are: (1) including pipes and fittings such as expansion joints, valves, pipe

joints, support arrangements, flexible tube lengths etc., and components in direct connection with the piping such as pumps, heat exchangers, air receivers, independent tanks, etc. It does not include main and auxiliary machinery such as oil engines, steam and gas turbines, boilers, reduction gears, etc;(2) a general term of pipes, valves and pipe fittings.

Piping systems 管系(塑料) 包括管子、附件、系统的接头符合其性能标准所要求任何内外衬垫、护套和敷层。 Piping systems means those made of plastic(s) and include the pipes, fittings, system joints, method of joining and any internal or external liners, coverings and coatings required to comply with the performance criteria.

Piping test 管系试验

Piping(LPG system) 管路(LPG 系统) 刚性金属结构的管道。 Piping means the pipeline of rigid metallic construction.

Piracy 海盗行为 下列行为中的任何行为的构成海盗行为:(1)私人船舶或私人飞机的船员、机组成员或乘客为私人目的,对下列对象所从事的任何非法的暴力或扣留行为,或任何掠夺行为;①在公海上对另一艘船舶或飞机,或对另一艘船舶或飞机上的人或财物;②在任何国家管辖范围以外的地方对船舶、飞机、人或财物;(2)明知船舶或飞机成为海盗攻击的事实,而自愿参加其活动的任何行为;(3)教唆或故意便利(1)或(2)项所述的任何行为。 Piracy consists of any the following acts: (1) any illegal acts of violence or detention, or any act of depredation, committed for private ends by the crew or the passengers of a private ship or a private aircraft, and directed: ①on the high seas, against another ship or aircraft, or against persons or property on board such ship or aircraft; ②against a ship, aircraft, persons or property in a place outside the jurisdiction of any State; (2) any act of voluntary participation in the operation of a ship or of an aircraft with knowledge of facts making it a pirate ship or aircraft; (3) any act of inciting or of intentionally facilitating an act described in subparagraph(1) or (2).

Piston 活塞(柱塞) 直接承受气缸内燃气压力,在气缸内作往复运动的零件。

Piston boss 活塞套管

Piston clearance 活塞间隙

Piston compressor 活塞式压缩机

Piston cooling 活塞冷却

Piston cooling pump 活塞冷却泵(活塞冷却液泵) 抽送柴油机活塞冷却介质的泵。

Piston crown 活塞头

Piston displacement(swept volume) 气缸工作容积 气缸的横断面积与活塞行程的乘积。

Piston engine 活塞式发动机

Piston eye 活塞孔

Piston gudgeon pin 活塞销

Piston head 活塞头

Piston mean speed 活塞平均速度 活塞在气缸中移动速度的平均值。

Piston oiling 活塞加油法

Piston pin 活塞销 连接活塞与连杆的零件。

Piston pin hole 活塞销孔

Piston pump 活塞泵

Piston ring 活塞环 装在活塞上的开口环形零件。

Piston ring end clearance 活塞环对口间隙 活塞环在气缸内对口两端间的距离。

Piston ring groove 活塞环槽

Piston ring pliers 拆装活塞环的钳子

Piston ring side clearance 活塞环端面间隙 活塞环在槽内沿活塞高度方向的间隙。

Piston ring sticking 活塞环结胶 活塞环因活塞温度过高或其他原因,部分或全部胶结在槽内的现象。

Piston rod 活塞杆 连接活塞与十字头的杆形零件。

Piston scraping 拉缸 活塞和缸套的光洁面上产生和气缸中心线平行的高低不平条纹的现象。

Piston seizure 咬缸 活塞和缸套互相黏结或卡住的现象。

Piston skirt 活塞裙

Piston slide valve 圆气阀

Piston underside pump scavenging 活塞底泵扫气 利用活塞往复运动时,在其底部和气缸的特殊结构间压缩空气进行的扫气。

Piston valve 活塞式阀

Piston water cooler 活塞水冷却器 在水冷式活塞的闭式冷却水系统中,用舷外水冷却淡水的热交换器。

Pitch 螺距(螺旋桨) 一定半径处,母线上一点绕轴线一周沿轴向前进的距离。螺旋桨叶面为一等螺旋面者,此距离亦即螺旋桨的螺距。

Pitch 纵摇(俯仰角) 是船舶沿 y 轴的角运动。 Pitch is the ship's angular motion about the y axis.

Pitch angle 螺距角(桨叶安装角) 螺旋桨叶在一定半径处的面节线与垂直于轴线的平面间的角度。$\Phi = \mathrm{arctg}(P/2\pi r)$,式中,$P$—面螺距,$r$—叶切面所在半径。

Pitch angle of screw propeller 螺旋桨螺距角

Pitch angle 纵摇角

Pitch control link　螺距控制杆
Pitch control sensor　纵摇控制传感器
Pitch diameter　节圆直径
Pitch factor　螺距因数　螺旋桨实效螺距对面螺距的比值。
Pitch frequency　纵摇频率
Pitch indication transmitter　螺距指示发送器
Pitch indicator　螺距指示器
Pitch of propeller　螺旋桨螺距
Pitch period T_P　纵摇周期 T_P　由下式得出：$T_P = \sqrt{\dfrac{2\pi\lambda}{g}}$，式中，$\lambda = 0.6(1 + T_{LC}/T_S)L$。Pitch period T_P means a pitch period, in s, given by: $T_P = \sqrt{\dfrac{2\pi\lambda}{g}}$; where: $\lambda = 0.6(1 + T_{LC}/T_S)L$.
Pitch point　节点　在螺旋桨图中，有时在正视图上由轴中心向垂直于螺旋桨基准线方向量出局部螺距除以 2π 的值所得之点。
Pitch prill　沥青球　在煤焦化过程中用焦油制成。呈黑色，有特殊气味。挤压成特有的铅笔形状以方便运输。货物在 40 ℃ ~ 50 ℃ 下变软。熔点 105 ℃ ~ 107 ℃。受热熔化。可燃，燃烧时产生黑色浓烟。粉尘可能刺激皮肤和眼睛。该货物通常失火风险低。但是，货物粉末容易点燃并可能引起火灾和爆炸。装载或卸货期间宜特别注意防火。　Pitch prill is made from tar produced during the coking of coal. It is black with a distinctive odour. It is extruded into its characteristic pencil shape to make handling easier. Cargo softens between 40 ℃ to 50 ℃. Melting point: 105 ℃ to 107 ℃. Melts when heated. Combustible, burns with a dense black smoke. Dust may cause skin and eye irritation. Normally this cargo has a low fire-risk. However powder of the cargo is easy to ignite and may cause fire and explosion. Special care should be taken for preventing fire during loading or discharging.
Pitch ratio　螺距比　螺旋桨螺距对其直径的比值。
Pitch setting lever　螺距整定杆
Pitch template　螺距样板
Pitching　纵倾（纵摇）　船舶绕横轴所做的周期性角位移运动。
Pitching depth　艏纵倾深度
Pitching moment　纵摇力矩
Pitching motion　纵摇运动
Pitching stress　纵摇应力
Pitch-up　艉倾（上仰）
Pitting corrosion　麻点腐蚀　系指局部范围内材料的厚度点状/或小面积减少，且厚度减少量大于其周围材料的平均腐蚀厚度。麻点腐蚀密度见图 P-4。Pitting corrosion is defined as scattered corrosion spots/areas with local material reductions which are greater than the general corrosion in the surrounding area. The pitting intensity is shown in Figure P-4.

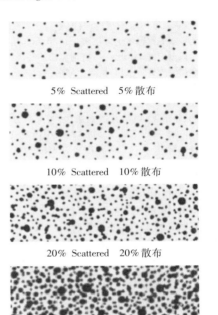

5% Scattered　5% 散布

10% Scattered　10% 散布

20% Scattered　20% 散布

30% Scattered　30% 散布

50% Scattered　50% 散布

图 P-4　麻点密度图
Figure P-4　Pitting intensity diagrams

Pivot　枢轴（旋转，滚动）
Pivot joint　轴连接
Pivot point　回转枢心　船舶中线面与垂直于该平面并通过回转轨迹曲率中心的垂直线的交点，即船上横向速度为零的一点。
Pivotal　中枢的/关键的
Pivoted duck(steering nozzle)　可转导管　可绕竖轴转动一定角度，对船兼起转向作用的导管推进器的导管。

Place an order with… 订货

Place of business 营业地点

Place of offshore company 离岸公司注册地 通常认为,在离岸地或离岸中心注册但不在当地经营,同时在登记、监督、信息披露、税务、管理和国际义务方面享有法律规定的特殊政策的商业组织通常被称为离岸公司。"离岸"的含义系指投资人的公司注册在离岸管辖区,但投资人不用亲临当地,其业务运作可在世界各地的任何地方直接开展。近年来,一些国家和地区纷纷以法律手段制定并培养出一些特别宽松的经济区域,一般称为离岸管辖区。目前,世界上知名的离岸公司注册地:英属维尔京群岛、开曼群岛、百慕大、文莱、塞舌尔、萨摩亚、毛里求斯、塞浦路斯等,其本身都是英国皇家属地,或英联邦国家,继承英国普通法,为大部分国家承认,而另外一些流行注册地,如新加坡、中国香港、英国、美国的特拉华州、卢森堡等国家和地区,本身就属于国家社会中主流的发达地区,为国际社会熟知。与通常按营业额或利润征收税的做法不同,离岸公司注册地政府只向离岸公司征收年度管理费。此外,几乎所有的离岸管辖区均明文规定,公司对股东资料、股权比例、收益状况等享有保密权利,可以不对外披露。目前,几乎所有的国际大银行都承认这类公司。它们可以在银行开设账号、财务运作上极其方便。

Place of performance of the contract 合同履行地

Place of refuge 庇护地/避难处 系指船舶遭遇恶劣气候条件时,能为其提供庇护的任何天然或人工的遮蔽地区。 Place of refuge is any naturally or artificially sheltered area which may be used as a shelter by a craft under conditions likely to endanger its safety.

Place of ship arrest 船舶扣押地

Place of the accident 事故发生地

Plaintiff 原告 在民事方面,以自己的名义提起诉讼,请求法院保护其权益,因而使诉讼成立的人。中国《行政诉讼法》规定,公民、法人或其他组织认为行政机关和行政机关工作人员的具体行政行为侵犯其合法权益,有权依照该法向人民法院提起诉讼成为原告。

Planar motion mechanism 平面运动机构 在快速性水池拖车上装设可作正弦形振动,振幅频率相同而相位差可以调整的两个支柱,使船模在水平面或垂向平面内产生规定的运动,以测定船舶位置导数、旋转导数、加速度导数等数据的试验设备。

Plane bulkhead 平面舱壁 由平舱壁板和骨架构成的舱壁。

Plane section of screw blade 叶平面切面 螺旋桨叶与一个垂直于其基准线的平面相交的面。

Plane sound wave 平面声波 系指波阵面为平面的声波。可以将各种远离声源的声波近似地看成平面波。排水量船舶的水螺旋桨和气垫船的空气螺旋桨发出的噪声、船舶内除机舱和辅机舱以外的舱室内的噪声,特别是从隔壁房间传来的噪声可按平面声波处理。

Planet carrier 行星架 用来支持行星轮的零件。

Planet gear(planetary gear) 行星轮(行星齿轮)

Planet gear pinion 行星小齿轮

Planetary hoist 行星式差动滑车组

Planetary reduction gear 行星式减速齿轮箱 外环内齿轮静止不动,与其啮合的行星轮绕中间的太阳轮公转的减速齿轮箱。高速轴为输入功率的太阳轮,低速轴为输出功率的行星轴,两者做同方向运动。

Planetary-parallel gear 行星-平行轴减速齿轮箱 第一级采用行星齿轮,第二级则为平行轴齿轮传动的减速齿轮箱。

Plank frictional resistance 平板摩擦阻力 平板在深水中,沿其平面方向作等速直线运动,水表面不产生波浪,也不产生涡流时所测得的阻力。

Plankton sampling 浮游生物取样 利用浮游生物网具或其他采集浮游生物的器具在水层中进行水平或垂直拖曳采集浮游生物样品。根据采集对象不同,有大、中、小三种类型网。如要进行微型浮游生物取样,则须用采水器采水后经沉淀或滤膜过滤取得。

Planned maintenance scheme(PMS) 计划保养系统(PMS) 系指船舶机械(包括电气设备),根据规范的有关要求和设备制造厂商说明书的规定,由船东制订一套详细的周期维修保养计划,通过该计划在船上的贯彻和实施,使船舶机械始终保持在良好的技术状态。对这种船舶机械采用周期性维修保养的计划管理,称为计划保养系统。 Planned maintenance scheme(PMS) means a detailed periodical maintenance plan for machinery(including electrical installations) onboard made by the owner in accordance with the relevant requirements of the Rules and the specifications of manufacturers. Through the implementation on board, the machinery will be always kept in good technical condition. Such planned management of periodical maintenance for machinery is called planned maintenance scheme.

Planned maintenance scheme survey on ship's machinery 船舶机械计划保养系统检验 系指一种由船东申请,经船级社批准的以船舶机械计划保养系统来替代船舶机械(包括电气设备)的循环检验和特别检验的制度。

Planned price 计划价格 系指国家各级行政部门有计划地制定的价格。是国民经济计划的一部分。在中国，由国家统一制定的价格为计划价格，根据统一领导、分级管理的原则，计划价格由中央行政部门和各级地方行政部门分别规定、调整和管理，可分中央计划价格和地方计划价格。

Planned working hours 计划工时 系指预先制定的，一小时所做正常工作量的劳动计量单位。工时管理主要是员工个人工作统计管理功能，生产管理人员可以详细地、逐项活动地查看生产和即时劳动力数据，特别是活动级劳动力信息，辅助生产管理人员利用从车间获得的效率数据，实时监控生产流程，并有提高生产率、控制劳动力成本方面制订科学的企业决策。工时的具体含义，根据情境的不同，可有几种情况：对于任务，系指完成任务所需的人员总数；对于工作分配，系指分配给资源的工时量；对于资源，系指为完成所有任务而分配给资源的总工时量。

Planning and documentation workstation 计划和文档工作站 计划航线以及记录船舶航行的所有情况的工作站。Workstation at which voyages are planned and where all facts of ship's operation are documented.

Planning boat (glider) 滑行艇 在高速航行时，仅部分艇体接触水面，其重力大部分靠直接作用于艇底的水动力所支持的艇。

Planning draft 滑行艇 系指利用高速运动的滑行面所产生的水升动力支承船重力的高速艇。其具有速度高和体积小的特点，因而无论在军事上或民用交通方面都具有相当的重要性。军用滑行艇有炮艇、鱼雷艇和导弹艇等，可承担巡逻、护卫、猎潜，甚至攻击大型舰艇的使命；民用滑行艇可按用途分为交通艇、游艇和运动竞赛艇等。从种类分，滑行艇可分为：无断级滑行艇(step-less planning draft，艇底无断级，滑行时整个艇底形成一个滑行面的滑行艇)、单断级滑行艇(single step planning draft，艇底具有一个横向断级，滑行时艇底形成前后两个滑行面的滑行艇)、多断级滑行艇(multi-step planning draft，艇底具有两个或两个以上横向断级，滑行时艇底形成前后多个滑行面的滑行艇)和三点式滑行艇(three-point planning draft，艇底形成"品"字形的三个滑行面的滑行艇)。从艇型分，滑行艇可分为：折角艇型(hard chine bull，具有较高的升阻比，所有横剖面线型在舭部具有明显折角的典型滑行艇型)、圆舭线型(round bilge hull，所有横剖面线型在舭部均为光顺曲线的艇型)、混合艇型(mixed hull，艏部横剖面线型采用圆舭型、艉部横剖面线型采用折角型的艇型)、深V艇型(deep "vee" hull，艉部横剖面具有较大的横向斜升角的折角艇型)、深Ω艇型(deep "Ω" hull，艉部横剖面采用"Ω"型，并且具有较大横向斜升角的折角艇型)、海撬艇型(sea sled，带有凹形底部的无断级的艇型)、双折角艇型

(double hard chine bull，横剖面线型两舷各具有双折角的艇型)、有多折角艇型(multiple hard chine hull，横剖面线型两舷各具有多折角的艇型)。

Planning mode 滑行状态 对地效翼船的操作模式而言，系指船舶在水面的稳定状态，此时船舶的重力主要由水动力支承。For the purpose of WIG craft operational modes, planning mode denotes the mode of steady state operation of a craft on water surface by which the craft's weight is supported mainly by hydro-dynamic forces.

Planning number 滑行数 滑行面总阻力与其升力的比值。$P_n = D_T/L_D = R_T/L_D$，式中，D_T 或 R_T—滑行面总阻力；L_D—滑行面动升力。

Plant 设备(装置，工厂，车间)

Plant barge 产品加工船 又称工厂船/浮动工厂。安装有成套的加工生产产品或半成品设备的船舶的总称。多为驳船型，非自航。停泊在原料生产地附近，就地取材，就地加工，完成的半成品、产品可由水路运送，以解决原料产地远离工厂造成的许多困难。也可用于发电和海水淡化等或对缺水、缺电地区进行供电、供水。对有些产品在陆上生产会影响城市环境的，改在海上生产，可解决污染问题。工厂船制造方便，造价较低、投产快、机动性强，修理时可拖到附近的港口船厂进行。近年来，许多国家大量设计制造各种各样的工厂船，如天然液化气生产船、发电船、尿素生产船、纸浆生产船、锯木船、海水淡化船等以及成套的生产设备用以出口，可减少在国外安装调试设备的困难。

Plant fiber 植物纤维 是广泛分布在种子植物中的一种厚壁组织。它的细胞细长，两端尖锐，具有较厚的次生壁，壁上常有单纹孔，成熟时一般没有活的原生质体。植物纤维在植物体中主要起机械支持作用。植物纤维与人类生活的关系极为密切。植物纤维环保材料是利用稻壳、稻草、麦秸秆、玉米秸秆、棉花秆、木屑、竹屑等农作物秸秆或其他植物的秆茎，先制成 10~200 目的植物纤维粉料，再用特殊工艺制成混合原料，然后模压或注塑成型的一种新型环保材料；天然成分可达 60%~90% 以上；可制成一次性餐饮容器具、可控降解容器、工艺品、日用品、建筑墙体材、工业包装等物品。

Plaster board 石膏板 是以建筑石膏为主要原料制成的一种材料。它是一种质量轻、强度较高、厚度较薄、加工方便以及隔音、绝热和防火等性能较好的建筑材料，是当前着重发展的新型轻质板材之一。石膏板已广泛用于住宅、办公楼、商店、旅馆和工业厂房等各种建筑物的内隔墙、墙体覆面板(代替墙面抹灰层)、天花板、吸音板、地面基层板和各种装饰板等。我国生产的石膏板主要有：纸面石膏板、装饰石膏板、石膏空心条板、纤维石膏板、石膏吸音板、定位点石膏板等。

Plastic 塑料 系指:(1)任何含有一种或多种高分子有机物合成的聚合物作为其主要成分的固体物质。在合成聚合物时通过加热或压力、或两者兼有之的办法将纤维素加入制成品而生成或成形。塑料系指由可分解的淀粉的化合物组成,并由 ①人工合成或 ②天然形成,且适用的"可分解的"塑料制品。天然形成的塑料,例如通常在海洋环境中出现的蟹壳或其他类型的贝壳,不属于塑料。塑料具有的材料性质,其范围从硬脆性到软弹性的。塑料通常在船上的各种用途包括但不限于:食品包装用品、个人卫生用品、包装用品(塑料薄膜、塑料瓶、塑料容器和塑料绳)、船舶构造(玻璃纤维和层压构件、板壁、管路和绝缘物、塑料地板、塑料地毯、纤维板、胶黏剂和电工部件及电子元件)、可处理的餐具和塑料杯(包括苯乙烯制品)、塑料包、塑料纸、塑料浮子、人工合成渔网、单股纤维的钓鱼绳、绑带、安全帽和人工合成的缆绳和缆索。(2)经增强或未经增强的热塑性材料或热固塑性材料两种,诸如聚氯乙烯 PVC 与纤维增强塑料 FRP。 Plastic means:(1) any garbage that is solid material, that contains as an essential ingredient one or more synthetic organic high polymers, and that is formed or shaped either during the manufacture of the polymers or polymers or during fabrication into a finished product by heat or pressure or both "Degradable" plastic, which are composed of combinations of degradable starches and are either: ①synthetically produced or ②naturally produced but harvested and adapted and adapted for use, are plastic. Naturally produced plastics such as crab-shells and other types of shells, which appear normally in the marine environment, are not plastics. Plastics possess material properties ranging from hard and brittle to soft and elastic, plastics are used for a variety of marine applications including, but not limited to: food wrappings, products for personal hygiene, packaging(vaporproof barriers, bottles, containers, and lines), ship construction(fiberglass and laminated structures, siding, insulation, flooring, carpets, fabrics, adhesives, and electrical and electronic components)、disposable eating utensils and cups(including styrene products)、bags, sheeting, floats, synthetic fishing line, strapping bands, hardhats, and synthetic ropes and lines;(2) both thermoplastic and thermosetting plastic materials with or without reinforcement, such as PVC and fibre reinforced plastics—FRP.

Plastic 塑料制品 系指含一种或多种高分子聚合物为关键成分的固体材料。其通过聚合形成(成型)或通过加热和/或加压形成成品,诸如聚氯乙烯——PVC 与纤维增强塑料——FRP。塑料的材料属性可为硬且脆,也可为软且有弹性。就 MARPOL 附则 V 而言,"一切塑料制品"系指包含或由任何形式塑料制品所组成的所有垃圾、包括合成缆绳、合成渔网、塑料垃圾袋和塑料制品的焊焚烧炉灰渣。 Plastic means a solid material which contains as an essential ingredient one of more high molecular mass polymers and which is formed(shaped) during either manufacture or the polymer or the fabrication into a finished product by heat and/or pressuire, such as PVC and fiber reinforced plastics——FRP. Plastics have material properties ranging from hard and brittle to soft and elastic. For the purpose of the Annex V to MARPOL, "all plastics" means all garbage that consists of or includes in any form, including synthetic ropes, synthetic fishing nets, plastic garbage bags and incinerator ashes from plastic products.

Plastic film 塑料薄膜 是以合成树脂为基本成分的高分子有机化合物成平面状可成卷的柔软包装材料的总称。

Plastic garbage bag 塑料垃圾袋

Plastics material 塑料 是一种有机物质,它可以是热固的或热塑的,且在其竣工状态可含有增强材料或添加剂。 Plastics material is regarded as an organic substance which may be thermosetting or thermoplastic and which, in its finished state, may contain reinforcements or additives.

Plate evaporator 板式蒸发器 加热元件为波形薄板的蒸发器。

Plate keel 平板龙骨 船底中线处的一列外板。

Plate packing plate fill 片式填料

Plate performance 单板性能

Plate structure 板结构 对极地航行的船舶而言,板结构系指与船体相连且承受冰载荷作用的加筋的板单元。这些要求适用于舷内范围的下列之中的小者:(1)毗邻平行的强肋骨或纵桁的腹板高度;或(2)与板结构相交的骨材构件高度的 2.5 倍。

Plate type contact freezer 接触式平板冻结装置 由多层可上、下、左、右移动的中间空的金属冷却板组成,货物与冷却板紧密接触,载冷剂在冷却板空间内流动吸取货物热量的冻结装置。

Plate type cooler 板式冷却器

Plate type heater 板式加热器

Plate vibration 板振动

Plate vibrator 平板振动器

Plate/paging acoustic filter 片式/百叶式消音器

Plated structures 板架结构 系指与船体相连,承受冰载荷的加强板单元。 Plated structures are those stiffened plate elements in contact with the hull and subject to ice loads.

Platform　平台　系指用于海上油气资源勘探与开发的移动式平台、固定式平台、顺应式结构的总称。由上部结构、设施与设备、支承结构等组成。

Platform decks　平台甲板　系指强力甲板以下，不计入船体总纵强度的不连续甲板。 Platform decks are the non-continuous decks below the strength deck, which are not considered to be effective decks for longitudinal strength.

Platform support vehicle（PSV）　平台供应船　系指主要给海上平台输送人员和物资的船舶。

Plat-forming　平颤运动　当气垫船、水翼船等高速艇在波长远小于其船长的小浪或涟波上航行时，由于航速快，从一个波峰跳到另一个波峰时船体对固定坐标的纵倾很少变化，而形成的基本上沿海平面一定高度略有周期性微小升降的运动。

Plausibility of data　数据的合理性　表示数据值在各自数据正常范围内的特征。 Plausibility of data is the quality representing, if data values are within the normal range for the respective type of data.

Playback equipment　播放设备（回放设备）　系指：（1）与记录介质和记录时所用的格式相兼容且用于恢复数据的设备，包括与原始数据设备相适应的显示硬件和软件。播放设备通常不安装在船上，就这些性能标准而言，播放设备也不被视为简化航行数据记录仪的一部分。（2）任何数据介质，其附带回放软件、操作说明书和为将商用流行的膝上型笔记本电脑连接至 VDR 所要求的任何专用部件。 Playback equipment means: (1) the equipment, compatible with the recording medium and the format used during recording, employed for recording the data. It includes also the display or presentation hardware and software that are appropriate to the original data source equipment. Playback equipment is not normally installed on a ship and is not regarded as part of S-VRD for the purpose of these performance standards. (2) any data medium with the playback software, the operational instructions and any special parts required for connecting a commercial off-the-shelf laptop computer to the VDR.

Playback software　回放软件　系指一份软件程序，其具有下载存储数据并回放其信息的能力。该软件应与商用流行的膝上型笔记本电脑所可用的操作系统相兼容，对于在 VDR 中采用的非标准或专有格式存储的数据，该软件应能将其转换为公开的行业标准格式。 Playback software means a copy of the software program to provide the capability to download the stored data and play back the information. The software should be compatible with an operating system available with commercial off-the-shelf laptop computers and where non-standard or proprietary formats are used for storing the data in the VDR, the software should convert the stored data into open industry standard formats.

Pleasure craft　游艇　系指专供游览的任何推进方式的船舶，而不是供付费旅客摆渡用的。 Pleasure craft means a vessel, however propelled, that is used exclusively for pleasure and that does not carry passengers who have paid a fare for passage.

Pleasure vessel　游览船　按 2004 年《商船（小型商船和引港船）航运规则》的定义，系指：（1）个人拥有的，并在使用时仅为船东或亲属或船东的朋友在自由航行或漂移状态下用作运动或游览的船。（2）公司拥有的，并在使用时仅为公司的雇员或官员、或其亲属或朋友在自由航行或漂移状态下用做运动或游览的船。（3）会员俱乐部会员或其代表拥有的，在使用时，会员俱乐部会员或其亲属或客人仅作运动或游览用的，并且除付入会员俱乐部安排作为日常使用的会员俱乐部基金外，使用该船不需付费的船舶。 Pleasure vessel means that as defined in the Merchant Shipping (Small Commercial Vessels and Pilot Boats) Regulations 2004, means: (1) a vessel which is owned by an individual, and at the time it is being used, the vessel is used only for the sport or pleasure of the owner or the immediate family or friends of the owner, and it is on a free voyage or excursion; (2) a vessel which is owned by a body corporate, and at the time it is being used, the vessel is used only for the sport or pleasure of employees or officers of the body corporate, or their immediate family or friends, and the vessel is on a free voyage or excursion; (3) a vessel which is owned by or on behalf of the members of a members' club, and which, at the time it is being used, is used only for the sport or pleasure of a member of that club, his immediate family or his guest, and for the use of which no payment is made other than a payment into the funds of the members club which funds are applied for the general use of the members club.

Pleasure yacht　游艇　系指供游览用的小艇。 Pleasure yacht is a small craft used for pleasuring.

Pledge　信约

PLEM　管线端部管汇　PLEM means a pipeline end manifold.

Plenum chamber hovercraft　增压室气垫船　以高于大气压的空气充入围裙包围的船底空腔内，形成气垫的全垫升气垫船。

Plotter　绘图仪　是一种自动化的绘图装置。在计算机指令的控制下，通过控制接口将计算机语言转换成绘图仪的控制信号,控制 X,Y 方向马达的转动和抬笔、落笔等动作,从而绘出图形、标记或字符。如先把要绘制的图形数字化,并记录在磁带上,然后在磁带控制器的控制下,可进行脱机绘图。

Plough in of WIG craft 地效翼船的埋艉 系指航行中的船舶因阻力持续增加而产生的一种被动的运动,通常与动力气垫局部破坏有关。 Plough in of WIG craft is an involuntary motion involving sustained increase in drag of a craft at speed, usually associated with partial collapse of an air cushion.

Plug 塞栓(插头,填充)

Plug hole 塞孔(船底放水孔)

Plummer block keep 中间轴压盖 中间轴承的上盖。

Plunger and bore of injection pump 喷油泵柱塞偶件 由柱塞和套筒零件组成的喷油泵主要精密偶件。

Plunger effective stroke 柱塞有效行程 至喷油开始到终了期间柱塞的行程。

Plunger pump 柱塞泵 柱塞呈细长形的一种往复泵,一般用于压力较高的场合。

Plunger type wave generator 冲箱式造波机 装置在船模试验池中,由机械或液压驱动的楔形箱体沿滑槽入水作垂向运动,并自动控制调节其振幅、速度、相位等,以产生不同人工波形的设备。

Plural 复数的/吸收

Plutonium 钚 系指由辐射性核燃料再加工提炼的材料,其为同位素的合成混合物。 Plutonium means the resultant mixture of isotopes of that material extracted from irradiated nuclear fuel from reprocessing.

Pneumatic clutch(electromagnetic clutch) 气胀型离合器 以压缩空气充入气胎进行操纵,利用圆柱面接合的离合器。

Pneumatic clutch(hydraulic clutch) 气动离合器 利用气压力实现离合的离合器。

Pneumatic dredger 气动挖泥船 装有空压机、气动泥泵和空气分配器等主要设备,利用压缩空气通过气动泥泵进行吸、排泥的吸扬挖泥船。

Pneumatic impact spanner 气动螺母扳手

Pneumatic machinery 气动机械(风动机械)

Pneumatic plant 气动装置

Pneumatic rube fire alarm system 空气管式火警烟雾信号系统

Pneumatic tank gauge 气动液位计

Pneumatic test 气压试验

Pneumatic valve 气动阀

Pneumatic wave generator 气动式造波机 在船模试验水池中,用鼓风机控制倒置钟形箱内的空气,以制造人工波浪的设备。

POE switching hub 网络有源交换器

Point(in wiring) 接点(敷线中) 拟用来连接灯具或把用电设备接至电源的固定敷线的任一端子。

Point corrosion 点腐蚀 系指分散的腐蚀点/区域,其腐蚀程度大于其周围区域的总体腐蚀程度。

Point of sales(POS) 销售点情报管理系统 是一种配有条码或 OCR 码技终端阅读器,有现金或易货额度出纳功能。其主要任务是对商品与媒体交易提供数据服务和管理功能,并进行非现金结算。POS 是一种多功能终端,把它安装在信用卡的特约商户和受理网点中与计算机联成网络,就能实现电子资金自动转账,它具有支持消费、预授权、余额查询和转账等功能,使用起来安全、快捷、可靠。大宗交易中基本经营情报难以获取,导入 POS 系统主要是解决零售业信息管理盲点。连锁分店管理信息系统中的重要组成部分。

Point sound source 点声源 系指声源的几何尺寸,比声波波长小得很多时,或测量点离声源相当远时,即离声源的距离比声源的尺寸大 5~10 倍以上,则可将该声源看成一个点。

Point to point protocol(PPP) 点对点通信协定 是为在同等单元之间传输数据包这样的简单链路设计的链路层协议。这种链路提供全双工操作,并按照顺序传递数据包。设计目的主要是用来通过拨号或专线方式建立点对点连接发送数据,使其成为各种主机、网桥和路由器之间简单连接的一种共通的解决方案。

Poisson distribution 泊松分布 适用于取样次数多、概率小统计平均值又与其方差相等的离散概率分布规律。

Poisson's ratio ν 泊松比 除另有规定者外,应考虑取值为 0.3。 Unless otherwise specified, a value of 0.3 is to be taken into account.

Polar class 极地级别 系指根据 IACS 统一要求授予船舶的等级,见表 P-4。 Polar class means the class assigned upon IACS United Requirements, see table P-4.

表 P-4　极地级别
Table P-4　Polar class

极地级别 Polar class	冰况描述 General description
PC1	在所有冰盖水域航行 Operation in all covered waters
PC2	在中等厚度的多年冰龄状况下航行 Operate in moderate multi-year ice condition
PC3	在第二年冰龄可包括多年夹冰的状况下航行 Operation in second-year ice which may include multi-year ice inclusions
PC4	在当年厚冰可包含旧夹冰的状况下航行 Operation in thick first-year ice which may include old ice inclusions
PC5	在中等厚度的当年冰龄可包含旧夹冰的状况下航行 Operate in medium first-year ice which may include old ice inclusions
PC6	在薄到中等厚度的当年冰龄可包含旧夹冰的状况下航行 Operate in thin to medium first-year ice which may include old ice inclusions
PC7	在薄的当年冰龄可包含旧夹冰的状况下航行 Operate in thin first first-year ice which may include old ice inclusions

Polar class ship　极地级别船舶　系指已授予极地级别的船舶。　Polar class ship means a ship for which a Polar Class has been assigned.

Polar diagram of stability　稳性极线图　在已知船舶各横倾角时的横稳心半径的情况下，所绘制的用以求取相应浮心、稳心位置、形状稳性臂、复原力臂及动稳性臂的曲线图。

Polar expedition laboratory　极地考察实验室　安装极地考察仪器和实验设备，专门从事极地海域的气象、水文、化学、大气高能物理、地质、生物和冰川与流水等各要素的观测和研究用的科学考察工作舱室。

Polar light and skylight observation room　极光夜光观测室　位于船舶高处，使用荧光计和全天候照相机，观测和记录极光、夜光出现的频率、形态和强度，并研究极光夜光各种现象与地球活动和太阳活动关系的工作舱室。

Polar orbiting satellite service　极轨道卫星业务　系指利用极轨道卫星接收和转发来自卫星应急无线电示位标(EPIRB)的遇险报警，并提供其位置的业务。　Polar orbiting satellite service means a service which is based on polar orbiting satellites which receive and relay distress alerts from satellite EPIRBs and which provides their position.

Polar region　极地　泛指地球南北两端纬度66.5′以上的区域，即南极和北极。

Polar ship　极地船舶　系指在极地航行的船舶。极地船舶划分为 A、B 和 C 三类，见表 P-5。

表 P-5　极地船舶分类
Table P-5　Category of polar ship

分类 Category	定义 Definition	
极地冰盖水域 A　Polar ice covered waters	在冰封达到或超过10%的极地水域航行的船舶 Ships that may operate in ice-covered waters with 10% or more of ice	极地级或等同 Polar class or equivalent

续表 P-5

分类 Category	定义 Definition	
极地开敞水域 B Polar open waters	在冰封少于10%,可导致船舶结构风险的水域航行的船舶 Ships that may operate in ice-covered waters with less than 10% ice, where it may pose a structural risk.	评估/冰区加强或其他缓解措施(例如破舱稳性) Assessment/ice-strengthening or other mitigating measures(such as damage stability).
极地无冰水域 C Polar ice free waters	在无冰或冰封少于10%,无结构风险的水域航行的船舶 Ships that may in water with zero to 10% ice cover, where it does not pose a structural risk.	无冰区加强 No ice-strengthening.

Polar waters 极地水域 包括北极和南极水域。Polar waters includes both Arctic and Antarctic waters.

Pole and line fishing boat 竿钓渔船 在甲板两舷设有钓鱼平台,以活、鲜诱饵,用竿钓方式捕鱼的钓渔船。

Polish 抛光(磨光,使发亮)
Polisher 抛光器
Polishing 抛光
Polishing compound 抛光剂
Polishing machine 抛光机
Polishing tool 抛光工具

Pollutant 污染物 系指 MARPOL 公约中附则 I 定义的油、油类混合物和燃料,附则 II 定义的有毒液体物质,以及附则 III 定义的散装运输的固体物质,也称有害物质。 Pollutant means the substances defined as oil, oily mixture and oil fuel in Annex I; noxious liquid substances in Annex II; and solids when carried in bulk, which are also identified as harmful substances in Annex III of the MARPOL Convention.

Pollution 污染损害(污染) 系指:(1)油类从船上的溢出或排放引起的污染在该船之外所造成的灭失或损害,不管发生于何处,但是,对环境的损害(不包括此种损害的利润损失)的赔偿,应限于已实际采取或将要采取的合理恢复措施的费用;(2)预防措施的费用及预防措施组成造成的新的灭失或损害。 Pollution means:(1)loss or damage outside the ship by contamination resulting from the escape or discharge of oil from the ship, wherever such escape or discharge may occur, provided that compensation for impairment of the environment other than loss of profit from such impairment shall be limited to costs of reasonable measures of reinstatement actually undertaken or to be undertaken;(2)the costs of preventive measures and further loss or damage caused by preventive measures.

Pollution 沾污 系指货物在运输途中受到其他物质的污染所造成的损失。

Pollution category 污染类别 系指按 73/78 防污公约附则 II 所确定的有关货物的污染类别。其中 Z 系指该货物的污染类别已被评定为 Z 类。OS 系指该货物已被评定并查明其污染类别不属于 X,Y 或 Z 类。 Pollution category means the pollution category assigned to each product under Annex II of MARPOL 73/78. In which the letter Z means the product was evaluated pollution categories Z and OS means the product was evaluated and found to fall outside categories X, Y, or Z.

Pollution damage 污染损害 系指:(1)由于船舶溢出或排放油类(不论这种溢出或排放发生在何处),在运油船舶本身以外因污染而产生的灭失或损失,但是对环境损害的赔偿,除这种损害所造成的盈利损失外,应限于已实际采取或行将采取的合理复原措施的费用;(2)预防措施的费用和因预防措施而造成的进一步灭失或损害。 Pollution damage means:(1)loss or damages caused outside the ship carrying oil by contamination resulting from the escape or discharge of oil from the ship, wherever such escape or discharge may occur provided that compensation for the impairment of the environment other than loss of profit from such impairment shall be limited to costs of reasonable measures of reinstatement actually undertaken or to be undertaken;(2)the costs of preventive measures and further loss for damages caused by preventive measures.

Pollution of harmful amuatic organisms and pathogens in ballast water 压载水有害水生物及病原体污染

Pollution of the marine environment 海洋环境的

污染 系指人类直接或间接把物质或能量引入海洋环境,其中包括河口湾,以致造成或可能造成损害生物资源和海洋生物,危害人类健康,妨碍包括捕鱼和海洋的其他正当用途在内的更重要活动,损害海水使用质量和减损环境优美等有害影响。 Pollution of the marine environment means the introduction by man, directly or indirectly, or substances or is likely to result in such deleterious effects as harm to living resources and marine life, hazards to, hindrance to marine activies, including fishing and other legitimate uses of the sea, impairment of quality for use of sea water and reduction of amenities.

Pollution prevention 防污染

Pollution prevention equipment 防污染设备 系指根据有关规定安装的防污染设备包括以下部分:(1)15 ppm 舱底水分离器;(2)15 ppm 舱底水报警装置;(3)自动关停装置。 Pollution prevention equipment means pollution prevention equipment installed in a ship in compliance with relevant regulation comprises:(1) 15 ppm bilge separator;(2) 15 ppm bilge alarm; and (3) automatic stopping device.

Polyamide(PA)/Nylon 尼龙/耐纶 是分子主链上含有重复酰胺基团—[NHCO]—的热塑性树脂总称。包括脂肪族 PA,脂肪—芳香族 PA 和芳香族 PA。其中,脂肪族 PA 品种多,产量大,应用广泛,其命名由合成单体具体的碳原子数而定。尼龙是美国杰出的科学家卡罗瑟斯(Carothers)及其领导下的一个科研小组研制出来的,是世界上出现的第一种合成纤维。尼龙的出现使纺织品的面貌焕然一新,它的合成是合成纤维工业的重大突破,同时也是高分子化学的一个重要里程碑。

Polychlorinated biphenyls(PCB) 多氯联苯(PCB) 系指联苯分子(两个苯环被一个碳-碳键连在一起)上的氢原子被可多至 10 个氯原子取代而形成的芳香族化合物。对于所有船舶,应禁止新装含有多氯联苯的材料。 Polychlorinated biphenyls(PCB) means aromatic compounds formed in such a manner that the hydrogen atoms on the biphenyl molecule(two benzene rings bonded together by a single carbon-carbon bond) may be replaced by up to ten chlorine atoms. For all ships, new installation of materials which contain Polychlorinated biphenyls shall be prohibited.

Polycyclic aromatic hydrocarbons carbons(PAHs) 多环芳烃

Polyester fiber 聚酯纤维 由有机二元酸和二元醇缩聚经纺丝所得的合成纤维。工业化大量生产的聚酯纤维是用聚对苯二甲酸乙二醇酯制成的,中国的商品名为涤纶。聚酯纤维是当前合成纤维的第一大品种,主要用于衣着和室内装饰。短纤维可以纯纺,也可与天然纤维及其他化学纤维混纺;长丝可加工制得弹力丝,也可制成轮胎帘子线、工业绳索、传动带、滤布及渔网等,还可用作电绝缘材料以及制备人造血管等。

Polyether foam plastics 聚醚乙烯泡沫塑料

Polymer damping materials 聚合物阻尼材料

Polymer-fuel 聚合燃料

Polymorphism 多态性

Polystyrene foam plastic board 聚苯乙烯泡沫塑料板

Poly-tropic efficiency(small-stage efficiency, infinitesimal – stage efficiency) 多变效率 当工作过程中压力和温度的变化是一个微量时,无限小级的绝热效率,压气机:$\eta_p > \eta_{ad}$,涡轮:$\eta_p < \eta_{ad}$。

Polyurethane foam 聚氨酯泡沫塑料 是异氰酸酯和羟基化合物经聚合发泡制成,按其硬度可分为软质和硬质两类,其中软质为主要品种。一般来说,它具有极佳的弹性、柔软性、伸长率和压缩强度;化学稳定性好,耐多种溶剂和油类;耐磨性优良,较天然海绵大 20 倍;还有优良的加工性、绝热性、黏合性等性能,是一种性能优良的缓冲材料,但价格较高。

Pontoon 趸船 是具有或没有自航能力,不配船员或配船员的浮动装置。趸船的主尺度比和普通海船的主尺度比有差别。趸船通常用来承受甲板载荷或载运工作设备(例如起重设备、打桩设备等),而无供载运货物的货舱。 Pontoon is unmanned or manned floating unit with or without self-propulsion. The ratios of the main dimensions of pontoon deviate from those usual for seagoing ship. Pontoon is designed to usually carry deck load or working equipment(e. g. lifting equipment, rams etc.) and have no holds for the carriage of cargo.

Pontoon 方驳 通常视为:(1)非自航的;(2)无船员的;(3)仅限装载甲板货;(4)方形系数等于或大于 0.9;(5)船宽/型深比大于 3.0;(6)除设有带垫料的盖、关闭的小人孔外,在甲板上没有舱口。 Pontoon is considered to be normally:(1) non self-propelled;(2) unmanned;(3) carrying only deck cargo;(4) having a block coefficient of 0.9 or greater;(5) having a breadth/depth ratio of greater than 3.0;(6) having no hatchways in the deck except small manholes closed with gasketed covers.

Pontoon 浮箱 系指位于两个坞墙之下的箱体结构。

Pontoon barge 箱形驳 系指方形的,甲板上装载不易受水侵蚀的货物的驳船。 Pontoon barge is a square barge carrying water-resistant cargoes on deck.

Pontoon boat 囤船/趸船 系指停泊水面,供船舶停靠,上下旅客、装卸货物等作码头用或作水上住宿、仓库、车间、泵站等用的非机动船舶。是具有或没有自航能力,不配船员或配船员的浮动装置。趸船的主尺度比和普通海船的主尺度比有差别。趸船通常用来承受甲板载荷或载运工作设备(例如起重设备、打桩设备等),而无供载运货物的货舱。

Pontoon deck 浮箱甲板 系指整体型或分离型浮船坞浮箱最上部的一层甲板,用于抬船、坞修或下水作业。

Pontoon bridge 舟桥 系指用船舶架设的浮桥,主要是用于抗洪、水利、交通等部门。见图 P-5。

图 P-5 舟桥
Figure P-5 Pontoon bridge

Pontoon deck 抬船甲板 用以铺设龙骨墩,抬举船舶的甲板。

Pontoon floating dock 浮箱式浮船坞 系指由连续坞墙与由几个浮箱组成的浮体连接而成的浮船坞。其浮箱可与坞墙永久链接或可分离。

Pontoon hatch cover 吊装式舱口盖/箱形舱盖 系指由金属或玻璃钢制成的箱形剖面舱盖板组成的拼装舱盖。

Pontoon wharf-boat 趸船 停泊水面,供船舶停靠,上下旅客、装卸货物等作码头用或作水上住宿、仓库、车间、泵站等用的非机动船舶。

Pontoon-less TLP 无浮箱张力腿平台

Pool fire 油槽失火 系指释放可燃液体和/或积聚在液体表面形成油槽的凝结气体,可燃蒸发气在积聚液体的表面上进行燃烧。 Pool fire is a release of a flammable liquid and/or condensed gas that accumulates on a surface forming a pool, where flammable vapors burn above the liquid surface of the accumulated liquid.

Poop 艉楼 系指:(1)自艉垂线向前延伸到艏垂线后某一位置的上层建筑。艉楼可以起始于艉垂线后的某一位置;(2)船最后端封闭上层建筑之下的处所。 Poop is:(1) a superstructure which extends from the after perpendicular forward to a point which is aft of the forward perpendicular. The poop may originate from a point aft of the aft perpendicular;(2) the space below an enclosed superstructure at the extreme aft end of a ship.

Poop deck 艉楼甲板 船舶后端遮蔽甲板之上的第一层甲板。 Poop deck is the first deck above the shelter deck at the aft end of a ship.

Pooping sea 淹艉浪 追艉浪峰涌上船尾部的大浪。常见于船长接近等于波长,特别是船速与波速也接近相等时。

POOR coating condition 差的涂层状况 系指在检验的区域中,有超过 20% 或更大范围的涂层普遍脱落,或有 10% 或更大范围的涂层产生硬质锈皮的状态。 POOR coating condition is a condition with general breakdown of coating over 20% or more of areas or hard scale at 10% or more of areas under consideration.

POOR condition 差的状况 对涂层而言,系指在检验的区域中,有超过 20% 或更大范围的涂层普遍脱落,或有 10% 或更大范围的涂层产生硬质锈皮。 For the purpose of coating, POOR condition means a condition with general breakdown of coating over 20% or more of areas or hard scale at 10% or more of areas under consideration.

Poor fit-up fillet welds 装配不良的角接焊缝 系指被连接构件之间的间隙过大或不适当。超过有关允许值的间隙可通过相对大的焊喉厚度来补偿。见图 P-6。 All excessive or inadequate gap between the components being joined. Gaps exceeding the relevant limit may be compensated by a correspondingly large throat thickness. See Figure P-6.

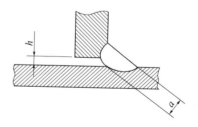

图 P-6 装配不良的角接焊缝
Figure P-6 Poor fit-up fillet welds

Poor restart 不良的再引弧

Popliteal height 提升高度 从脚凳到人的大腿下侧的垂直距离。 The vertical distance from the footrest to the underside of the thigh.

Populate 填充

Pore(pores) 气隙(气穴)

Pore of destination 目的港 系指最终卸货的港口。

Porosity 孔隙(孔隙度)

Porosity(gas pocket) 气孔 焊接时,熔池中的气泡在凝固时未能逸出而形成的空穴。由于气孔的存在,焊缝的有效截面减小,过大的气孔会降低焊缝的强度,破坏焊缝的致密性。

Porosity and pores 气孔和气穴 为了评估在有底漆的材料上进行角接焊缝时出现的孔穴(延伸到焊缝表面的条状孔除外),所用的极限值与集结孔穴的极限值有关。 For the evaluation of pores as they occur during the over-welding of shop primers in fillet welds(except for worm holes which extend to the surface), the limits imposed for clusters of pores shall be relevant.

Porpoise 海豚 海洋中哺乳动物,属鲸类。出没于欧洲沿海的中部或南部,一些河流下游也可发现。长成后长度可达2 m。

Porpoising 海豚式运动 高速艇在达到某一航速时,可能由于重心与水动力中心的位置配合不当,而使艇产生的由周期性纵摇和垂荡相耦合的犹如海豚逐浪式的运动。

Port 端口 系指处在外部电磁环境中的设备的特殊接口,通过它骚扰可被感受或发射出去。

Port² 港口 (1)系指:①装卸站(群),其组合成一个单位进行作业,并视作为港口;②选择视为港口的某一个港口的主管机关或其他机构;③港口主管机关特别指定作为港口的场地或设施;(2)任何通常用于船舶装载、卸载、维修和锚泊的港口、码头、近海装卸站、船厂和修船厂或锚地,或船舶可以停靠的任何其他地方。 Port means:(1)①a group of terminals,which combines to act as a unit and be considered a port;②a port authority or other organization that chooses to be considered a port;③a place or facility that has been specifically designated as a port by the port authority;(2)any port, terminal, offshore terminal, ship repair yard or roadstead which is normally used for the loading, unloading, repair and anchoring of ships, or any other place at which a ship can call.

Port administration 港口当局 系指船舶装货或卸货的港口所在国的有关当局。 Port administration means the appropriate authority of the country in the port of which the ship is loading or unloading.

Port charge/port dues 港口费 船舶进出港口和在港作业、停留所要支付的一切费用,又称港口使用费。港口费的费目繁多。不同国家,甚至一国的不同港口,或同一港口的不同码头,因为费用分别由提供不同劳务的机构征收,而收费规定又不同,所以费目名称不尽相同,计费办法也各异。

Port city 港口城市 系指船舶装卸作业所在的城市。 Port city is the city where the vessel loads or offloads.

Port duty generating set 停泊发电机组

Port duty generator 停泊发电机

Port engine 左舷主机(左主机) 系指布置在机舱左舷侧的主机。

Port facility 港口设施 系指由缔约国政府或由指定的当局确定的发生船/港界面活动的场所,其中包括锚地、等候靠泊区域和进港航道等区域。 Port facility is a location, as determined by the Contracting Government or by the Designated Authority, where the ship/port interface takes place. This includes areas such as anchorages, waiting berths and approaches from seaward, as appropriate.

Port facility security officer 港口设施保安员 系指被指定负责制定、实施、修订"港口设施保安计划"以及与船舶保安员和公司保安员进行联络的人员。 Port facility security officer means the person designated as responsible for the development, implementation, revision and maintenance of the port facility security plan and for liaison with the ship security officers and company security officers.

Port facility security plan 港口设施保安计划 系指为确保在港口设施范围内的港口设施和船舶、人员、货物、货物运输单元、船舶物料免受保安事件威胁的措施而制订的计划。 Port facility security plan means a plan developed to ensure the application of measures designed to protect the port facility and ships, persons, cargo, cargo transport units and ship's stores within the port facility from the risks of a security incident.

Port of disembarkation 离船港 系指根据国际偷渡公约的规定,该偷渡者被送交岸上适用的主管机构所在地的港口。 Port of disembarkation means the port at which the stowaway is delivered to the appropriate authority in accordance with the provisions of the International Convention relating to stowaways.

Port of embarkation 上船港 系指在船上被发现的偷渡者登船的港口或该港附近地点。 Port of embarkation means the port or place in the vicinity thereof at which a stowaway boards the ship on which he is found.

Port of register 登记港 系指船舶注册的港口。

Port of shipment 装运港 系指货物起始装运货物的港口。

Port or place of departure 出发港或地点 系指船舶已锚泊或系泊的任何港口或地点。Port or place of departure means any port or place in which a vessel is anchored or mooring.

Port or place of destination　目的港或地点　系指船舶驶往(以便锚泊或系泊)的任何港口或地点。 Port or place of destination means any port or place in which a vessel is bound to anchor or moor.

Port reduction gear　左机减速齿轮箱　具有两个以上推进器的船舶中,设置在左舷侧的减速齿轮箱。

Port security officer(PSO)　港口保安员

Port side　左舷　由船艉向船艏看,船的左侧即为左舷。

Port State　港口国　系指对包括船舶在北极冰覆盖水域内抵达的任何目的港行使管辖权的国家。 Port State means a State whose area of jurisdiction includes any destination port of a ship where such port lies within Arctic ice-covered waters.

Port State authority　港口国当局　系指港口国政府授权执行或实施有关国际航运管理措施的标准和规则的任何组织和机构。 Port State authority means any official or organization authorized by the Government of a port State to administer or enforce standards and regulations relevant to the implementation of natural and international shipping control measures.

Port state control(PSC)　港口国监督　是港口国当局对进入其港口的外国船舶进行的一种监督和检查。其目的是确认船员、船舶技术状态和操作符合适用的国际公约和相关法律的要求。港口国监督检查(PSC)的依据是 IMO 和国际劳工组织制定的国际公约及其修正案。PSC 检查项目:(1)船舶证书、文件和手册;(2)船体、机器和设备的状态;(3)机器、设备、仪器使用和操作要求;(4)船员配置、劳动和生活条件等。

Port state control survey　港口国监督检验　由港口当局进行的,针对入港船舶的检验。检验的依据除了国际公约以外,还有港口国当局的相关法律规定。

Port(-duty) pump(harbor pump)　停泊泵　专供船舶在停泊时使用的泵。

Portable compressor　可移式压缩机组　非固定安装并可用人力搬动的压缩机组。

Portable device　便携设备　系指具有便携性的小体积数字设备,且可以包括移动通信设备,例如个人数字蜂窝(PDC/personal digital cellular)电话、个人通信工具等。

Portable extinguisher　手提式灭火器

Portable fan　可移式风机　非固定安装,并可用人力搬动的风机。

Portable fire(fighting) unit　可移式消防泵机组　由独立动力机如汽油机、柴油机等驱动的,非固定安装并可用人力搬移的消防泵机组。

Portable fire extinguisher　可移式灭火装置(手提式灭火器)　可以随时移动的灭火装置。

Portable form applicator　手提式泡沫枪

Portable instrument for gas measurement　便携式气体测量设备　系指为液货船配备的,用于测量可燃蒸发气浓度的便携式仪表及足够的备件。 Portable instrument for gas measurement means a portable instrument for measuring flammable vapour concentrations, together with a sufficient set of spares equipped with for tankers.

Portable lavatory　可移动式厕所　即活动厕所。

Portable pump(mobile pump)　可移式泵　非固定安装并可用人力搬动的泵。

Portland cement　普通水泥　系指由淡水、砂子或其他符合要求的材料按一份水泥两份砂子的比例混合而成。 Portland cement is to be mixed with fresh water and sand or other satisfactory substances, in the proportion of about one part of cement to two of sand.

Portugal, Italy, Ireland, Greece and Spain (PIIGS)　欧猪五国　这是国际债券分析家和国际经济媒体对欧洲 5 个较弱经济体的贬称。对经济不景气、出现债务危机的葡萄牙(Portugal)、意大利(Italy)、爱尔兰(Ireland)、希腊(Greece)和西班牙(Spain),这 5 个欧洲国家因其英文国名首字母组合"PIIGS"类似英文单词"pigs"(猪),故名。特别指各国的主权债券市场,这些国家的公共赤字也都超过了 3%。

Position 1　位置1　参照经修正的国际载重线公约[MSC.143(77)决议第 13 条],位置 1 包括:(1)露天干舷甲板和后升高甲板;(2)位于沿艏柱前端至自龙骨顶部向上量至最小型深85%处水线,艏垂线起 0.25L_{LL} 前方的露天上层建筑甲板。 Ref. ILLC, as amended[Resolution MSC. 143 (77) Reg. 13], position 1 includes: (1) exposed freeboard and raised quarter decks; (2) exposed superstructure decks situated forward of 0.25 L_{LL} from the perpendicular, at the forward side of the stem, to the waterline at 85% of the least moulded depth measured from the top of the keel.

Position 2　位置2　参照经修正的国际载重线公约[MSC.143(77)决议第 13 条],位置 2 包括:(1)位于沿艏柱前端至自龙骨顶部向上量至最小型深85%处水线,艏垂线起 0.25L_{LL} 后方,且在干舷甲板以上至少一个标准上层建筑高度的露天上层建筑甲板;(2)位于沿艏柱前端至自龙骨顶部向上量至最小型深85%处水线,艏垂线起 0.25L_{LL} 前方,且在干舷甲板以上至少两个标准上层建筑高度的露天上层建筑甲板。 Ref. ILLC, as amended[Resolution MSC. 143(77) Reg. 13], position 2 includes: (1) exposed superstructure decks situated aft of $0.25 L_{LL}$ from the perpendicular, at the forward side of the stem, to the waterline at 85% of the least moulded depth measured from the top of the keel and located at least one standard height of superstructure above the freeboard deck; (2) exposed superstructure decks situated forward of $0.25 L_{LL}$ from the perpendicular, at the forward side of the stem, to the waterline at 85% of the least moulded depth measured from the top of the keel and located at least two standard heights of superstructure above the freeboard deck.

Position keeping 定位 系指在正常的控制系统的操作范围和环境条件下保持指定的位置。为此，应考虑环境动态效应(或风、浪、流)的主动补偿。

Position mooring system 定位系泊系统 系指通常设置在柱稳式平台或水面式平台上的一种定位系统。其一般采用辐射式悬链状系泊系统。该系统将由下列部件和设备组成：(1)锚或锚桩；(2)锚索，包括锚链、钢丝绳；(3)锚索附件，包括链扣、连接环、缆端嵌环、快速释放装置；(4)导向装置，包括导向弯管、导向滑轮及导向孔；(5)掣链/缆器；(6)锚机/绞车；(7)构成定位系泊系统的其他结构件或机械件。定位系泊系统应设计成在任一个锚索突然失效时，不会导致其他锚索相继失效。Position mooring system means a positing system fitted on bottom-stabilized or surface-type units normally, which is radial type catenary mooring systems. The systems will consist of: (1) Anchors or anchor piles; (2) Anchor lines, including chain cables and steel wires; (3) Anchor line fittings, including shackles, connecting links, wire rope terminations and quick-release devices; (4) Fairleads, including bent pipes, guide rollers and guide holes; (5) Chain/wire stoppers; (6) Winches/windlasses; (7) Other structural or mechanical items forming part of the position mooring system. The position mooring system is to be designed so that a sudden failure of any single anchor line will not cause progressive failure of remaining lines in the anchoring arrangement.

Position of exposed deck openings 露天甲板开口 其部位如下：位置 I——位于露天干舷甲板及艉楼甲板上和距艏垂直线 $0.25 L_f$ 处向前范围内的上层建筑露天甲板上；位置 II——位于距艏垂直线 $0.25 L_f$ 处后方的位置以及从干舷甲板上部处至少有一个标准船楼高度的上层建筑露天甲板上。位于距艏垂直线 $0.25 L_f$ 处前方的位置以及从干舷甲板上部处至少有两倍的标准船楼高度的层建筑露天甲板上。位于艏端 $0.25 L_f$ 后的上层建筑露天甲板上。 Position I——upon exposed freeboard and raised quarter decks, and upon exposed superstructure decks situated forward of a point located $0.25 L_f$ from the forward perpendicular. Position II——upon exposed superstructure decks situated abaft $0.25 L_f$ from the forward perpendicular and upon exposed superstructure decks situated forward of a point located $0.25 L_f$ from the forward perpendicular and located at least two standard heights of superstructure above the freeboard deck.

Position reference system 坐标参照系统 系指测量位置和艏向的系统。Position reference system means a system measuring the position and/or heading of the unit.

Position sensitivity 位置敏感性 当船舶在狭窄水道中偏离航道中线航行时，由于两侧压力不平衡而产生的使船舶向近岸一侧横移的特性。

Positional stability 位置稳定性 船舶在运动中受扰动偏离平衡状态，当扰动完全消除后船舶重心仍能循原有横向位置或深度沿直线方向运动的性能。

Positioning beacon 定位信标

Positive heave 正垂荡 为沿 z 轴正向的直线运动(向上)。 Positive heave is translation along positive z-axis (upwards).

Positive horizontal bending moment 正水平弯矩 为右舷受拉，左舷受压。 Positive horizontal bending moment is tension on the starboard side and compression on the port side.

Positive longitudinal acceleration 纵向正加速度 为沿 x 轴正向加速度(向前)。 Positive longitudinal acceleration is acceleration along positive x-axis(forward).

Positive pitch 正纵摇 为船首下降，船尾上升。 Positive pitch is bow down and stern up.

Positive roll 正横摇 为右舷下降，左舷上升。 Positive roll is starboard down and port side up.

Positive stability 正稳性 系指船舶在移去一横倾力矩后回复到其初始位置的能力。Positive stability is the ability of a ship to return to its original position after the removal of a heeling moment.

Positive surge 正纵荡 为沿 x 轴正向的直线运动(向前)。 Positive surge is translation along positive x-axis (forward).

Positive sway 正横荡 为沿 y 轴正向的直线运动(朝左舷)。 Positive sway is translation along positive y-axis(towards port side of vessel).

Positive transverse acceleration 横向正加速度 为沿 y 轴正向加速度(朝左舷)。 Positive transverse acceleration is acceleration along positive y-axis(towards portside of vessel).

Positive vertical acceleration 垂向正加速度 为沿

z 轴正向加速度（向上）。 Positive vertical acceleration is acceleration along positive z-axis (upwards).

Positive vertical bending moment 正垂向弯矩 为中拱弯矩。 Positive vertical bending moment is a hogging moment.

Positive yaw 正艏摇 为船首朝左舷旋转，船尾朝右舷旋转。 Positive yaw is bow rotating towards portside of vessel and stern towards starboard side.

Positive-action valve 直接作用阀

Positive-displacement pump 容积泵 利用其工作容积的变化，完成吸入和压出过程，以抽送液体的泵。

Postal parcel all risk 邮包一切险 其承保责任范围类似于海运货物一切险。

Postal parcel risk 邮包险 系指承保邮包通过海、陆、空三种运输工具在运输途中由于自然灾害、意外事故或外来原因所造成的包裹内物件的损失。邮包运输险的险别分为邮包险和邮包一切险。

Postal service 邮政服务

Postponed repairs 缓装工程 系指待主要工程结束后进行的工程。

Potable water treatment 饮水处理 淡化后的成品水在作为饮用水使用前所进行的杀菌和增添人体需要的某些矿物质元素等处理。

Potash 钾碱 呈褐色、粉红色或白色，制成颗粒状晶体产品。无嗅，吸湿。无特殊危害。该货物为不燃物或失火风险低。该货物吸湿，受潮会结块。 Brown, pink or white in colour, potash is produced in granular crystals. It is odourless and hygroscopic. No special hazards. This cargo is non-combustible or has a low fire-risk. This cargo is hygroscopic and will cake if wet.

Potassium chloride 氯化钾 褐色、粉红色或白色粉末。氯化钾制成颗粒状晶体产品。无嗅，溶于水，吸湿。该货物虽归入无危害类别，但潮湿时颗粒会造成严重的腐蚀。该货物为不燃物或失火风险低。该货物吸湿，受潮会结块。 It is brown, pink or white in colour, powder. Potassium Chloride is produced in granular crystals. It is odourless and is soluble in water and hygroscopic. Even though this cargo is classified as non-hazardous, it may cause heavy corrosion when wet. This cargo is non-combustible or has a low fire-risk. This cargo is hygroscopic and will cake if wet.

Potassium nitrate (UN 1486) 硝酸钾 透明、无色或白色结晶粉末或晶体。吸湿。潮湿时会氧化。与可燃物质的混合物容易点燃并可能猛烈燃烧。该货物吸湿，受潮会结块。 Potassium nitrate (UN 1486) is transparent, colourless or white crystalline powder or crystals and hygroscopic. Oxidises when wet. Mixtures with combustible materials are readily ignited and may burn fiercely. This cargo is hygroscopic and will cake if wet.

Potassium sulphate 硫酸钾 坚硬晶体或粉末。无色或白色。无特殊危害。该货物为不燃物或失火风险低。 It is hard crystals or powder. It is colourless or white. No special hazards. This cargo is non-combustible or has a low fire-risk.

Potential flow 势流 其流速 U 等于流势 ϕ 的梯度的无旋流。

Potential head 势头 用以表示单位质量流体所具有的势能的流体在基线以上的标高。

Potential loss of life (PLL) 潜在人员伤亡 作为每年死亡人数的预期值，是测算社会风险的一种简单方法。PLL 是一种航行整体，是通过后果和频率呈现的风险总和，是潜在的所有的可能发生的事故的总和。

Potential wake 势伴流 船舶前进时，由于势流使水在螺旋桨处产生的伴流。

Pour-in-place epoxy chocking compound 环氧树脂垫座浇注膏

Powder fire-extinguishing mediums 干粉灭火剂 以碳酸氢钠、碳酸氢钙为主要成分的粉末状灭火剂。

Power actuating system 动力转舵系统 系指提供动力以转动舵杆的液压设备，由一个或几个操舵装置动力设备，连同有关的管系、附件以及舵转动机构组成。各个动力转舵系统可共用某些机械部件，即舵柄、舵扇和舵杆，或共用与这些有同样用途的部件。 Power actuating system means the hydraulic equipment provided for supplying power to turn the rudder stock, comprising a steering gear power unit or units, together with the associated pipes and fittings, and a rudder actuator. The power actuating systems may share common mechanical components, i.e. tiller quadrant and rudder stock, or components serving the same purpose.

Power bilge pump 动力舱底泵 使用电动机、蒸汽机或其他动力机械驱动的舱底泵。 Power bilge pump means. The bilge pump is drived by motor, steam engine or other power machinery.

Power distribution 功率分支 发动机轴通过一个小齿轮同时与两个以上的大齿轮啮合而进行功率分配的传动方式。

Power distribution gear box 功率分支轴并列减速齿轮箱 采用功率分支的减速齿轮箱。

Power inverter 功率逆变器 是一种平均功率的流动方向在其内部是从直流电路到交流电路的变换器。

Power loading coefficient 功率载荷系数 用螺旋

桨收到功率表示其载荷的无因次系数。$C_P = P_D / [(1/2)\rho V_A^2 (\pi D^2/4)]$，式中，$P_D$—收到功率；$\rho$—水的质量密度；$V_A$—进速；$D$—螺旋桨直径。

Power management system(PMS) 功率管理系统　系指能使发电机随负荷的变动而启动和停止的系统。当没有足够的功率启动大功率的负荷时，应阻止大功率设备的启动，并按要求启动备用发电机，然后再启动所需的负载。功率管理系统应具有充足的冗余或适当的可靠性。Power management system means a system which is to be arranged to perform load-dependent starting and stop of generators. This system is to block starting of large consumers when there is not adequate running generator capacity, and to start up back-up generators as required, and hence to permit requested loads start to proceed. The power management system is to have adequate redundancy or reliability.

Power of Attorney 授权委托书　是委托他人代表自己行使自己的合法权益，委托人在行使权力时需出具委托人的法律文书。而委托人不得以任何理由反悔委托事项。被委托人如果做出违背国家法律的任何权益，委托人有权终止委托协议，在委托人的委托书上的合法权益内，被委托人行使的全部职责和责任都将由委托人承担，被委托人不承担任何法律责任。

Power output of charging device 充电功率　系指蓄电池组的标称电压乘最大充电电流值。Power output of charging device is to be calculated from the maximum charging current multiplied by the rated voltage of batteries.

Power plant 动力装置(电站)

Power plant concealment 动力装置隐蔽性　动力装置在正常工作时，所发出的声、光、电、热、磁、放射性、烟等物理现象运行隐蔽程度。

Power plant economy 动力装置经济性　动力装置在造价、折旧、燃料、管理、维修等各项费用开支和效率方面、综合经济性能。

Power plant heat rate 动力装置耗热率　主机发出单位千瓦·小时功，动力装置所消耗的热量。

Power plant maneuverability 动力装置操纵性　动力装置在规定的指令下，正倒车换向操纵的灵活性。

Power plant relative mass 动力装置相对质量　整个动力装置的质量与船舶满载排水量(军用舰船按正常排水量)的比值。

Power plant responsibility 动力装置机动性　动力装置在规定的指令下，启动、加速和加载的性能。

Power plant service reliability 动力装置可靠性　动力装置在预定的运行条件下长期正常工作的能力。

Power plant specific mass 动力装置单位质量　动力装置总质量与主机输出功率的比值。

Power plant thermal efficiency 动力装置的热效率　主机输出功率的热当量与单位时间内动力装置消耗燃料的总热量之比。

Power plant viability 动力装置生命力　动力装置遇到局部损坏后，仍能维持运行的能力。

Power pump 动力泵　由原动机驱动的泵。

Power rating 额定功率

Power refer to engine room unit area 机炉舱单位面积功率数　机舱或机炉舱单位面积的主机功率数。

Power refer to engine room unit length 机炉舱单位长度功率数　机舱或机炉舱单位长度的主机功率数。

Power refer to engine room unit volume 机炉舱单位容积功率数　机舱或机炉舱单位容积的主机功率数。

Power regulate system 功率调节系统

Power regulator 功率调节器

Power related unbalance(PRU) 单位功率的不平衡力矩　这是衡量柴油机(尤其是低速机)不平衡力矩大小的指标，根据该数据大小可粗略判断是否需采取减振措施。

Power reserve 功率储备

Power room 主发电机舱

Power source 动力源(电源)

Power station 电站

Power supply 供电(电源,动力源)

Power supply system 供电系统　是由电源系统和输配电系统组成的产生电能并供应和输送给用电设备的系统。

Power system 动力系统　动力装置中为主机服务的辅助机械设备、管路及附件的总称。

Power system for laboratory 实验室电源系统　向实验室仪器、设备供电的系统。由电源分电箱、稳压器、专用电源装置及其线路等组成。

Power take-off unit 动力输出传送装置

Power test 功率试验

Power tool chipping 动力工具处理　系指利用动力工具除去基底表面锈蚀和异物的工艺过程。Power tool chipping means a technological process to remove rest and obstructions from ground surface by power tool.

Power turbine(free turbine, free power turbine) 动力涡轮　其轴与燃气发生器转子无机械连接，所产生的动力全部作为有用功输出的涡轮。

Power unit 动力单元　是由液压泵、驱动电动机构成的总装件。Power unit is the assembly formed by hydraulic pump and its driving motor.

Power ventilation　机械通风

Power-driven vessel　机动船　系指用机器推进的任何船舶。　Power-driven vessel means any vessel propelled by machinery.

Power-operated system　动力操纵系统

Power-operated watertight door fault alarm　动力驱动的水密门故障报警　系指指示液压箱中液位低、气压低或液压蓄能器中储存的能量缺失，以及动力驱动的滑动水密门的电力供应中断的报警。　Power-operated watertight door fault alarm: alarms which indicate low level in hydraulic fluid reservoirs, low gas pressure or loss of stored energy in hydraulic accumulators, and loss of electrical power apply for power-operated sliding watertight doors.

Powers of an officer of the competent Authority when searching a vessel　主管当局的官员在搜查船舶时所具有的权力　根据各主管当局在本国法律范围内的职责，法律可允许主管当局的官员自由接触船舶的任何部位及船上货物。此外，他可以:(1)在装货之前给货物加上标记或促使货物加上标记;(2)对船上或任何处所或任何集装箱内装载的任何货物上锁、封存、加上标记或监护;(3)如果拒绝或没有钥匙，砸开任何处所或集装箱。主管当局的上述官员可有以下权限:(1)当为制止海上非法贩运而有必要登船或搜查船舶时，采取这些行动;(2)除国家法律另有规定外，逮捕任何违法者并可实施处罚或罚款。当主管当局的官员起诉时，根据国家法律的相应规定，可要求船长和其他责任方承担刑事责任。　Powers of an officer of the competent Authority when searching a vessel means that the law may permit subject to the powers of individual authorities within national law an officer of the competent Authority to have free access to every part of the ship and its cargo. Additionally he may: (1) mark, or cause to be marked, any goods before loading; (2) lock up, seal, mark or secure any goods carried in the ship, or in any place, or in any container; (3) break open any place or container which is locked if the keys are withheld or otherwise unavailable. Such officers of the competent Authorities may have authority to:(1) board or search ships when these actions are necessary to suppress illicit trafficking by sea; (2) arrest any offender and may impose sanctions or fines, and order arrest, unless otherwise laid down in the legislation of the country.

ppm display　ppm 显示器　是 15 ppm 舱底水报警装置的数字标示的显示器。　ppm display is a numerical scale display of the 15 ppm bilge alarm.

Pre-action system　预动作系统　系指将自动喷水器接至内部充满带压或不带压空气管路上的喷水系统。在该系统内，在喷水器同一布置区域范围内还装设附加的探测系统。探测系统动作后，将一个允许流入喷水管路的阀门打开，水流将从任何一个已开启的自动喷水器排出。　Pre-action system means a system employing automatic nozzles or sprinklers attached to a piping system containing air that may or may not be under pressure, with a supplemental detection system installed in the same area as the nozzles or sprinklers. Actuation of the detection system opens a valve that permits water to flow into the piping system and to be discharged from any nozzles or sprinklers that have operated.

Pre-alarm level　预报警水位　系指使货舱内传感器启动的较低一侧水位(一般为 0.5 m, 对于单货舱的货船为 0.3 m)。　Pre-alarm level means the lower level (0.5 m, single hold cargo ships: not less than 0.3 m)at which the sensor(s)in the cargo hold space will operate.

Precautionary principle　风险管理原则(预防原则)　(1)是低风险的情况可能存在，但无法达到零风险。就此而言，风险应通过确定每种情况中可接受的风险度来进行管理的原则;(2)是风险管理在做出假设和提出建议时纳入某种预防等级，以弥补信息的不确定、不可靠和不充分。因此，信息的缺乏或不确定应作为潜在风险的指示的原则。　Precautionary principle is: (1) a principle that low risk scenarios may exist, but zero risk is not obtainable, and as such risk should be managed by determining the acceptable level of risk to each instance;(2) a principle that risk assessments incorporate a level of precaution when making assumptions, and making recommendations, to account for uncertainty, unreliability, and inadequacy of information. The absence of, or uncertainty in, any information should therefore be considered an indicator of potential risk.

Precision long screw grinder　高精度长丝杆磨床

Pre-combustion　预燃

Pre-combustion chamber　预燃室　燃料先在气缸盖内的预热室中混合发火燃烧，然后沿通道进入气缸内继续燃烧的燃烧室。

Pre-compression　预压缩

Pre-cooler　预冷器

Pre-cooling　预冷　将热货冷却到接近其冷藏温度的过程。

Precursor　前体　系指加工可卡因或海洛因成品需要的一种化学物质，其分子呈现在成品的分子结构中。如果没有前体，不可能获得最终产品。在获得成品之前，必须拥有这种前体。　Precursor means a chemical substance which is needed for processing of a finishing product, either cocaine or heroin, its molecules will be present in the molecule of the finished product. If the precursor is not used, the final product cannot be obtained. Before obtaining the finished product it is necessary to have this precursor.

Pre-discharge alarms of fire extinguishing system 灭火剂施放前的报警　系指在任何滚装处所和通常有人工作或出入的其他被保护处所设置的施放灭火剂的听觉和视觉自动报警装置。该报警系统应在灭火剂施放前至少工作20s。 Pre-discharge alarms of fire extinguishing system means an automatic audible alarm system provided in any ro-ro spaces and other spaces for personnel normally work and access. The system is to be operated least 20s before the medium is released.

Prefabrication line for steel plates　钢板预处理

Preference tripping　优先脱扣　系指当任何一台发电机过载或类似工况时,为了确保向重要用户的供电而使不重要电路的保护装置自动分断的布置。Preference tripping is such an arrangement that the protective devices for unimportant circuits are opened automatically in order to ensure the power supply for vital services, when any one generator becomes overloaded or likely.

Preferential duties　特惠税　系指对从某个国家或地区进口的全部商品或部分商品,给予的特别优惠的低关税或免税待遇。但其不适用于从非优惠国或地区进口的商品。特惠税有的是互惠,有的是非互惠的。

Preferential quota　优惠性配额　系指在进口配额内的商品享受优惠关税,超过进口配额的商品征收原来的最惠国待遇税。

Preferential trade arrangements　优惠贸易安排　是经济一体化较低级和松散的一种形式,系指在实行优惠贸易安排的成员国间,通过协议或其他形式,对全部商品或部分商品规定特别的优惠关税。

Preferred noise criteria curves(PNCC)　较佳噪声标准曲线

Pre-focusing base bulb　预聚焦基座灯泡　系指其灯丝相对于其灯座为精确定位的发光灯泡。

Pre-heat(preheating)　预热

Preheat zone　预热区

Pre-heater　预热器

Preheating temperature　预热温度

Preheating time　预热时间

Pre-ignition　早燃　可燃混合气在点火之前的自然燃烧现象。

Preliminary approval/approval of general design　初步设计的初步批准　是审批机关发表声明说明提出的概念设计符合审批机关制定的规则、法规和/或适用标准的意图,尽管设计没有被充分开发的过程。

Preliminary document review　文件预审　系指对公司ISM体系文件的初步审核,以避免重大疏漏发生。

Preliminary examination/Pretrial　预审

Preliminary hearing　预备庭

Preliminary offsets　原始型值　取自设计图纸未经放样光顺的型值。

Preliminary system safety assessment(PSSA)　初步系统安全评估　是对建议系统结构的系统分析,其目的是显示较低层次的故障如何导致功能危险评估(FHA)所确认的功能危险。初步系统安全评估(PSSA)应为设计者提供系统所需的所有安全要求,并证明建议的结构能达到功能危险评估(FHA)所确认的安全目标。初步系统安全评估是一个互动的过程且在不同的开发阶段进行。在最低层次,初步系统安全评估确定与安全有关的硬件和软件提出要求。初步系统安全评估通常的形式为故障树分析(也可以使用从属图和马尔科夫分析法),它也应处理从共同原因考虑引起的安全问题。 Preliminary system safety assessment(PSSA)is a systematic analysis of the proposed system architecture. Its purpose is to show how failures at a lower hierarchical level can lead to the functional hazards identified in the FHA. The PSSA should provide the designer with all necessary safety requirements of the system and demonstrate that the proposed architecture can meet the safety objectives identified by the FHA. The PSSA is an interactive process and conducted at different development stages. At the lowest level, the PSSA determines the safety related design requirements of hardware and software. The PSSA usually takes the form of a Fault Tree Analysis(Dependence Diagram and Markov Analysis may also be used). It should also address safety issues arising from common cause considerations.

Preliminary table of offsets　原始型值表　由原始型值编制而成的型值表。

Preparation　配制品(配置品)　系指含有一种或多种活性物质(包括任何添加剂)的商业配方。配制品还包括船上产生的用于压载水管里的活性物质,以及为满足MARPOL公约要求在使用活性物质的压载水管理系统中形成的任何相关化学品。 Preparation means any commercial formulation containing one or more active substances including any additives. This term also includes any active substances generated onboard for purposes of ballast water management and any relevant chemicals formed in the ballast water management system that make use of active substances to comply with the MARPOL Convention.

Prepared opium　配好的鸦片　通过水萃取、过滤和蒸发等多种方法,对生鸦片处理后得到可以抽吸的制品。通常呈黑色易碎的块或剥下的皮,像生鸦片一样略有令人作呕的气味。 Prepared opium is produced by treating raw opium with various methods of water extraction, filtration and evaporation to obtain a product suitable for smoking. It usually appears as a black, brittle mass or parings and may smell faintly sickly like raw opium.

Pre-purchase survey 买船前检验

Prescribed 规定 系指由各国的法规或规则,或者由主管机关所做的规定。 Prescribed means prescribed by national laws or regulations or by the competent authority.

Prescribed person 指挥人员 对澳大利亚海事安全局海事指令而言,系指代表最接近的澳大利亚海事安全局检验办公室的验船师。 For the maritime order of the Australian Maritime Safety Authority (AMSA), prescribed person means the surveyor in charge of the nearest AMSA survey office.

Prescriptive requirements 规定性要求 系指:(1)SOLAS 公约的 B,C,D,E 或 G 部分规定的构造特性、限定的尺寸或消防安全系统。(2)在规则中对建造特性、限制范围、火灾安全装置的限定。 Prescriptive requirements means: (1) the construction characteristics, limiting dimensions, or fire safety systems specified in parts B, C, D, E or G of SOLAS. (2) the construction characteristics, limiting dimensions, or fire safety systems.

Presence of a witness 出庭作证

Present 陈述/到场

Presentation 提示 是持票人将汇票提交付款人要求承兑或付款的行为,是持票人要求取得票据权利的必要程序。提示又分付款提示和承兑提示。

Preservation 保全

Preservation 防腐(维护,保藏)

Preservation of boiler 锅炉的封存

Preservation of evidence 证据保全 系指法院在起诉前或在对证据进行调查前,依据申请人的申请或当事人的请求,以及依职权对可能灭失或今后难以取得的证据,予以固定和保存的行为。法院解决民事纠纷,认定案件事实是必不可少的环节,认定案件事实必须依靠证据。由于民事案件的事实是过去发生的事实,民事案件的起诉、审理直至判决需要一个过程,所需的证据可能会由于没有及时地收集而因人为或客观的原因灭失或难以取得,于是为了维护当事人的合法权益,为了法院公正审理民事案件,需要采取一定的措施,将可能灭失或以后难以取得的证据固定和保存下来,因此,有必要建立这样一套保全证据的制度。

Preservation of fruit (CF) 水果保鲜 载运水果货物的冷藏舱应保证其气密性以防止舱内货物相互生感染或带来不良影响(如某一品种的水果所吐出的气体,会促使另一品种的水果加快成熟等)并设有适当数量的温度计,以对货舱进行测温,以及设有永久性的 CO_2 成分指示设备,以达到水果保鲜的目的。

Preservation of maritime claim 海事请求保全 系指海事法院根据海事请求人的申请,为保障其海事请求的实现,对被请求人的财产所采取的强制措施。海事请求保全是民事诉讼中财产保全的一种特殊形态。

Preservation of property 财产保全 系指人民法院在诉讼开始后,或者在诉讼开始前,为保证将来判决的顺利执行,面对争议财产或与案件有关的财产,依法采取的各种强制性保护措施的总称。

Preservative 防腐剂(防腐的)

Preserve evidence 证据保全

Pre-shipment inspection certificate 装船前检验证书 即 PSI 检验证书。

Presiding arbitrator 首席仲裁员 系指由劳动争议仲裁委员会在业已组成的劳动争议仲裁庭组成人员中指定的,负责仲裁庭工作的专职或兼职仲裁员。

Press fit 压力配合

Press fitting 压配附件(固紧附件)

Press forging 压锻

Press quenching 模压淬火(加压淬火)

Press type 压紧型

Pressed up 满载 系指液舱完全装满而无因纵倾或不足透气引起的空隙。任何低于 100% 的装载,例如日常营运中的 98% 装载率都不能视为满载。 Pressed up means the condition in which the tank completely full with no voids caused by trim or inadequate venting. Anything less than 100% full, for example the 98% condition regarded as full for operational purposes, is not acceptable.

Pressure 压力

Pressure accumulation 蓄压器(压缩空气储存器)

Pressure alarm 压力警报器

Pressure amplification factor 压力放大系数

Pressure attenuation factor 压力衰减因子

Pressure calculation point 压力计算点 对于受非均布载荷的垂向板,取板的下缘。对于次要骨材,一般取其跨距中点,如骨材上压力为非线性分布时,设计压力取跨距中点压力与骨材两端压力平均值中之大者,对于主要骨材,取其承载区域的中点。

Pressure container 压力容器(受压容器) 系指承受外压或内压的受压容器及其附件。

Pressure control 压力控制

Pressure control head(PCH) 压力控制头
Pressure curve 压力曲线
Pressure detector 压力检波器 又称水听器,海洋检波器。接收弹性波在水中转播时所产生振动信号的一种敏感元件。
Pressure difference 压力差
Pressure distribution 压力分布
Pressure drop 压力降
Pressure element 压力元件
Pressure equalizer 均压器
Pressure gauge 压力表
Pressure gradient 压力梯度
Pressure head 压头 用以表示单位质量流体所具有的压能的单位面积压力能支持的流体高。$H = p/w$,式中,p—单位面积压力;w—单位体积流体的质量。
Pressure intensity 压力强度(压强)
Pressure lubrication 压力润滑
Pressure maintenance 压力维持
Pressure measuring set 声压测量仪 测量水中脉冲声或连续声信号声压值(或声压级)的仪器。
Pressure meter 压力表
Pressure point 压力中心(压力点)
Pressure propagation 压力传播
Pressure pump 加压泵(压力泵) 安装在甲板泵舱或甲板压缩机舱内,与深井泵串联运转,用于冷冻式液化气船上。冷冻液货经加压泵升压后再到加热器升温,再卸载到岸上常温压力容器中。
Pressure ratio 增压比 压气机排气绝对压力(一般系指总压力)与进气压力的比值。
Pressure reducing valve(pressure regulating valve) 减压阀(压力调节阀)
Pressure regulation system(LPG system) 压力调节系统(LPG系统) 具有一个或多个调节器,用来把该系统的高压减低至所要求的器具的标称压力的系统。Pressure regulation system means the system incorporating one or more regulators to reduce the high pressure of the system to the required nominal pressure of the appliances.
Pressure regulator 压力调节器
Pressure reinforcement layer 压力增强层
Pressure relief valve 安全阀
Pressure resistance 压阻力 流体经物体时,由流体法向力的合力所形成的阻力。
Pressure resistance strength test 耐压强度试验
Pressure response factor 压力响应因子
Pressure sensing device 压力传感器
Pressure sensitive adhesive 压敏胶黏剂

Pressure sensitive impedance glue 压敏阻尼胶
Pressure side 压力面
Pressure stage 压力级 由一列静叶和一列动叶组成的汽轮机级,分冲动式和反动式。
Pressure strainer 压力滤器
Pressure tank 压力液货舱 系指设计压力(表压)大于0.07MPa的液货舱。压力液货舱应为独立液货舱,对其结构设计应按照公认的对压力容器的设计标准。Pressure tank means a tank having a design pressure greater than 0.07 MPa gauge. A pressure tank shall be an independent tank and shall be of a configuration permitting the application of pressure-vessel design criteria according to recognized standards.
Pressure taper 压力表接头(测压点)
Pressure test 压力试验
Pressure testing 受压试验
Pressure tide gauge 压力式验潮仪 敷设于海底,通过静水压变化来测量潮位的仪器。
Pressure transducer 压力传感器 一种将所测得的压力转换为电量或机械量的敏感元件。压力传感器主要有应变式、振弦式、电位器式、压电晶体等类型。
Pressure tube 压力管
Pressure type mechanical oil atomizer(pressure oil atomizer) 机械压力雾化喷油器 依靠燃油泵的压力使燃油从喷孔以高速射流或旋流喷出时完成雾化的喷油器。
Pressure type wave meter 压力式测波仪 通过测量由波浪引起的海底或海中某一深度上的水动压力变化来记录波浪的仪器。
Pressure vacuum breaker 压力真空破坏器
Pressure vacuum breaking device 压力真空断开装置
Pressure vacuum relief 压力真空释放
Pressure vacuum relief valve 压力真空释放阀
Pressure vacuum valve 压力真空阀
Pressure ventilation 压力通风
Pressure vessel 压力容器 系指:(1)用于储存压力大于或小于环境压力而温度不限的液体的焊接或者无缝容器、液压或气动中的动力流体气缸也是压力容器;(2)贮存对其顶部形成超过一个大气压的气体或液体的容器,包括热交换器,但不接触火焰、燃气或高温气体。Pressure vessel means:(1) a welded or seamless used for the containment of fluids at a pressure above or below the ambient pressure and at any temperature. Fluid power cylinders in hydraulic or pneumatic plants also considered pressure vessel;(2) a vessel which contains gas or liquid, intended for the pressure exceeding the atmospheric pressure

at its top. It includes heat exchangers and does not contact with flame combustion gas of hot gas.

Pressure vessel for human occupancy　有人压力容器

Pressure volume diagram　压容图

Pressure water spraying system　压力喷水系统

Pressure(vessel)　设计压力(容器)　系指被制造厂商用于决定容器最小尺寸的压力。该压力不能小于最大工作压力, 由安全阀的压力设置限定, 按适用的规则说明。Design pressure is the pressure used by the manufacturer to determine the scantlings of the vessel. This pressure cannot be taken less than the maximum working pressure and is to be limited by the set pressure of the safety valve, as prescribed by the applicable Rules.

Pressure/Tensile Reinforcement　压力/抗拉增强层

Pressure/vacuum valve　压力/真空阀　是设计成在封闭容器内在设定极限值内保持压力和真空的装置。Pressure/vacuum valve is a device designed to maintain pressure and vacuum in a closed container within preset limits.

Pressurization "p"(electrical equipment)　"p"正压型(电气设备)　通过在外壳内维持高于外部大气压力的保护气体, 以防止外部可能为爆炸性的大气进入的技术。注意: IEC 60079-2 对采用此技术保护之电气设备的设计、制造和使用提供指南。Pressurization "p" (electrical equipment) means a technique of guarding against the ingress of the external atmosphere, which may be explosive, into an enclosure by maintaining a protective gas therein at a pressure above that of the external atmosphere. Note: IEC 600792-2 gives guidance on the design, construction and use of electrical apparatus protected by this technique.

Pressurized compartment　加压舱

Pressurized enclosure　加压外壳

Pressurizer unit　加压器(稳压器)

Prestress　预应力

Pretension　预张力　系指在无环境载荷的状态下, 通过绞车绞紧提供系泊缆的初始张力。

Pretreatment　预处理

Preventive measure　预防措施　系指事件发生后为防止或减轻污染损害而由任何人所采取的任何合理措施。Preventive measure means any reasonable measures taken by any person after an incident has occurred to prevent or minimize pollution damage.

Preventive risk control　预防性风险控制　系指通过风险控制措施降低事故的可能性。

Prewash　预洗　在"73/78 防污公约"附则 II 第13条"有毒液体物质残余物排放控制"中规定如下: 已被卸完 X 类物质的货舱, 在船舶离开卸货港口之前, 应予以预洗; 对 Y 或 Z 类物质, 如未按《程序和布置手册》的要求而进行了卸载, 在船舶离开卸货港口之前, 应予以预洗; 对 Y 类高黏度或固化物质, 应予以预洗。预洗应按"73/78 防污公约"附则 II 的附录 6 的要求进行。According the Regulation 13 of the "MARPOL73/78" Annex II, a tank from which a substance in category X has been unloaded shall be prewashed before the ship leaves the port of unloading; If the unloading of a substance of category Y or Z is not carried out in accordance with the Manual, a prewash shall be carried out before the ship leaves the port of unloading; For high-viscosity or solidifying substances in category Y, a prewash shall be carried out. The prewash shall be carried out according to the Appendix 6 of the Annex II.

Prewash procedures　预洗程序　系指使用在具体船上配置的洗舱设施和设备的特殊要求的程序。并包括下列内容:(1)拟使用的洗舱机位置;(2)污油水泵出程序;(3)洗舱机的循环次数(或时间);(4)热洗要求和(5)最低操作压力。Prewash procedures mean those procedures contained specific requirements for the use of the tank washing arrangements and equipment provided on the particular ship and include the following;(1) cleaning machine positions to be used;(2) stops pumping out procedure;(3) requirements for hot washing;(4) number of cycles of cleaning machine (or time); and (5) minimum operating pressure.

Pre-washing system　预洗系统

Price auction　价格拍卖　系指发行机构按价格及购买数额由高到低依次出售, 额满为止, 这是公募拍卖方式的具体拍卖方法之一。

Price bargain　讨价还价

Price difference　差价　系指同一种商品由于交易条件的不同而产生的价格上的差异。在国际贸易业务中, 影响商品差价的主要因素是所使用的贸易术语的不同而造成的。

Price quotations　报价

Price sealed auction　密封价格拍卖　又称招标式拍卖, 采用这种方法时, 先由拍卖人公布每批商品的具体情况和拍卖条件等, 最后由各买方在规定时间内, 将自己的出价密封递交拍卖人, 以供拍卖人进行审查比较, 决定将该批商品卖给哪一个竞拍者。这种方法不是公开竞买, 拍卖人有时要考虑除价格以外的其他因素。

Primarily to carry dry cargo in bulk　主要用于运输散装干货　系指设计主要用于装运散装干货和运输

以散装形式载运及装卸并占用船舶全部或部分处所的货物。Primarily to carry dry cargo in bulk means primarily designed to carry dry cargoes in bulk and to transport cargoes which are carried, and loaded or discharged, in bulk, and which occupy the ship's cargo spaces exclusively or predominantly.

Primary air 一次风 与燃料油首先接触,使其着火并维持稳定火焰的空气。

Primary air(main combustion air) 主燃空气 进入主燃区的空气。

Primary alarming point 主要报警点

Primary alarms 一级警报 系指某种状态需要立即引起注意以防出现紧急情况的警报。Primary alarms mean those alarms which indicate a condition that requires prompt attention to prevent an emergency condition.

Primary application structure 平台的主要构件 系指对平台结构整体完整性有重要作用的构件。Primary application structure is structural member essential to the overall integrity of the unit.

Primary application structure of column-stabilized units 柱稳式平台的主要构件 包括:(1)立柱、下壳体和上壳体、斜撑和水平撑杆等的外壳板(作为特殊构件者除外);(2)组成箱型或工字型支承结构的上壳体或上平台的甲板板、重型翼板和舱壁(作为特殊构件者除外);(3)作为交接点局部加强或使交接点处结构连续的舱壁、甲板或平台和骨架(作为特殊构件者除外);(4)悬臂式直升机甲板和救生艇平台;(5)重型底座和设备支撑,如钻台基座、起重机座、锚索导向装置及其支撑结构。Primary application structure of column-stabilized units includes:(1)shell plating of vertical columns, lower and upper hulls, and diagonal and horizontal braces except where the structure is considered as special application;(2)deck plating, heavy flanges and bulkheads within the upper hull or platform which form "box" or "I" type supporting structure except where the structure is considered as special application;(3)bulkheads, decks or flats and framing which provide local reinforcement or continuity of structure in way of intersections, except areas where the structure is considered as special application;(4)cantilevered helicopter decks and lifeboat platforms;(5)heavy substructures and equipment supports, e.g. drill floor substructure, crane pedestals, anchor line fairleads and their supporting structure.

Primary application structure of self-elevating units 自升式平台的主要构件 包括:(1)柱形桩腿的外板;(2)桁架式桩腿的全部骨材;(3)平台主体中组成箱型或工字型主支承结构的舱壁板、甲板板、舷侧板及底板;(4)升降机座的支撑结构;(5)桩靴或沉垫的外板,以及最初传递桩腿载荷的构件;(6)将主要集中载荷或均布载荷分散到结构中去的桩靴或沉垫支承结构的内部舱壁和骨架;(7)悬臂式直升机甲板和救生艇平台;(8)重型底座和设备支撑,如钻台基座、钻井悬臂梁和起重机座。Primary application structure of self-elevating units includes:(1)external plating of cylindrical legs;(2)plating of all components of lattice type legs;(3)combination of bulkhead, deck, side and bottom plating within the upper hull which form "box" or "I" type main supporting structure;(4)jack-house supporting structure;(5)external shell plating of footings or mats and structural members which receive initial transfer of loads from the legs;(6)internal bulkheads and framing of supporting structure of footings or mats which are designed to distribute major concentrated or uniform loads into the structure;(7)cantilevered helicopter decks and lifeboat platforms;(8)heavy substructures and equipment supports, e.g. drill floor substructure, drilling cantilevers and crane pedestals.

Primary application structure of submersible units 坐底式平台的主要构件 包括:(1)立柱、下壳体和上壳体、斜撑及抗滑桩等的外壳板(作为特殊构件者除外);(2)组成箱型或工字型支承结构的上壳体或上平台的甲板板、重型翼板和舱壁(作为特殊构件者除外);(3)作为交接点局部加强或使交接点处结构连续的舱壁、平台或甲板和骨架(作为特殊构件者除外);(4)悬臂式直升机甲板和救生艇甲板;(5)重型底座和设备支撑,如钻台基座、钻井悬臂梁和起重机座。Primary application structure of submersible units includes:(1)shell plating of vertical columns, lower and upper hulls, and diagonal braces and anti-slip spuds except where the structure is considered as special application;(2)deck plating, heavy flanges and bulkheads within the upper hull or platform which form "box" or "I" type supporting structure except where the structure is considered as special application;(3)bulkheads, flats or decks and framing which provide local reinforcement or continuity of structure in way of intersections, except areas where the structure is considered as special application;(4)cantilevered helicopter decks and lifeboat platforms;(5)heavy substructures and equipment supports, e.g. drill floor substructure, drilling cantilever and crane pedestals.

Primary barrier 主屏壁/主防壁 对散装运输液化气船而言,系指当货物围护系统会有两层周界时被用于装货的内层构件。

Primary bridge functions 桥楼主要功能 系指在

具体水域、交通和气象条件下,与确定、执行和保持船舶安全航向、航速和船位有关的功能。这些功能包括:航线计划功能、导航功能、避碰功能、操纵功能、靠泊作业功能、船内安全系统监测功能和与桥楼操作及遇险情况下的安全有关的船内通信和船外通信。 Primary bridge functions means functions related to determination, execution and maintenance of safe course, speed and position of the ship in relation to the waters, traffic and weather conditions, such functions are: route planning functions, navigation functions, collision avoidance functions, docking functions, monitoring of internal safety systems, external and internal communication related to safety in bridge operation and distress situations.

Primary container　主容器

Primary controls　主控制器　系指船舶在航行时用于安全操纵船舶所必需的所有控制设备,包括应急状态下所要求的控制设备。 Primary controls are all control equipment necessary for the safe operation of the craft when it is under way, including those required in an emergency situation.

Primary distribution system　一次配电系统　系指与发电机有电气连接的系统。Primary distribution system is a system having electrical connection with the generator.

Primary essential services　主要设备　系指那些需要保持连续工作的设备。主要设备列举如下:(1)操舵装置;(2)可调螺距螺旋桨的执行系统;(3)推进所必需的主辅柴油机和涡轮机的扫气鼓风机、燃油供给泵、燃油阀冷却泵、滑油泵和冷却水泵;(4)蒸汽装置或蒸汽透平船以及船上用于供给主要设备蒸汽的辅锅炉用的强制通风机、供水泵、水循环泵、凝水泵、燃汽装置;(5)配有滑油泵和冷却水泵的作为推进/操纵唯一手段的全回转推力器;(6)带滑油泵和冷却水泵的电力推进装置的电气设备;(7)发电机和供给上述设备的相关电源;(8)供给上述设备的液压泵;(9)重燃料油的黏度控制设备;(10)用于主要设备的控制、监测和安全装置/系统;(11)决定于推进所必需的主辅机电能的调速器;(12)柴油机和燃气轮机的启动设备。船上人员和旅客通常可接近和使用的那些部位的主照明系统也要考虑为(包括为)主要设备。 Primary essential services are those which need to be maintained in continuous operation. Examples of equipment for primary essential services are the following: (1) steering gear; (2) actuating systems of controllable pitch propellers; (3) scavenging air blowers, fuel oil supply pumps, fuel valve cooling pumps, lubricating oil pumps and cooling water pumps for main and auxiliary engines and turbines necessary for the propulsion; (4) forced draught fans, feed water pumps, water circulating pumps, condensate pumps, oil burning installations, for steam plants or steam turbines ship, and also for auxiliary boilers on ship where steam is used for equipment supplying primary essential services; (5) azimuth thrusters which are the sole means for propulsion/steering with lubricating oil pumps, cooling water pumps; (6) electrical equipment for electric propulsion plant with lubricating oil pumps and cooling water pumps; (7) electric generators and associated power sources supplying the above equipment; (8) hydraulic pumps supplying the above equipment; (9) viscosity control equipment for heavy fuel oil; (10) control, monitoring and safety devices/systems for equipment for primary essential services; (11) speed regulators dependent on electrical energy for main or auxiliary engines necessary for propulsion; (12) starting equipment of diesel engines and gas turbines. The main lighting system for those parts of the ship normally accessible to and used by personnel and passengers is also considered (included as) a primary essential service.

Primary filter　粗滤器

Primary frame spacing(m)　主要骨架间距　系指主要支撑构件间的距离,沿弦长在跨距中点量取。Primary frame spacing is defined as the distance between the primary supporting members, measured at mid-span along the chord.

Primary holes　主燃孔　供空气进入火焰筒主燃区的通孔。

Primary members　主要构件　在板架系统中是支撑次要构件和起重要作用的构件,例如:(1)船底结构-肋板、船底和内底强横构件和桁材;(2)甲板结构-甲板强横梁和纵桁;(3)舷侧结构-强肋骨和舷侧纵桁;(4)舱壁-舱壁竖桁和舱壁水平桁。舱壁或甲板之间须有额外支撑时,深腹板肋骨可用于支撑舱壁或甲板间的主要构件。 Primary members are those members supporting secondary members and will be the predominant members in grillage systems, e.g. (1) bottom structure-floors, bottom and inner bottom transverse and girders; (2) deck structure-deck transverses and girders; (3) side structure-side transverses and side stringers; (4) bulkheads-vertical webs and bulkhead stringers. Deep web frames are members supporting primary members accessible.

Primary refrigerant　制冷剂

Primary sample　初始样品　系指由放置于接收船燃油集管处的取样获取的在添加燃油的整个过程中收集的交付船舶的燃油的代表性样品。 Primary sample is the representative sample of the fuel delivered to the ship collected throughout the bunkering period obtained by the sampling equipment positioned at the bunker manifold of the receiving ship.

Primary stress 主应力(初应力)

Primary suit closure(immersion suit) 衣服的主要外罩(救生服) 为穿戴此救生服所用的外罩。 Primary suit closure is the closure used in the donning of the suit.

Primary support members 主要支撑构件 系指确保船壳和液舱界限例如双层底肋板和桁材、舷侧横向结构、甲板横梁以及舱壁水平桁和纵舱壁垂直桁材等的总体结构完整性的梁、桁材和水平桁类型的构件。 Primary support members means members of the beam, girder or stringer type which ensure the overall structural integrity of the hull envelope and tank boundaries, e. g. double bottom floors and girders, transverse side structure, deck transverses, bulkhead stringers and vertical webs on longitudinal bulkheads.

Primary surface preparation 一次表面处理

Primary insurance 基本险 系指不需附加在其他险别之下的,可以独立承保的险别,简单地说,能够独立投保的保险险种称为基本险。

Prime mover(primer mover) 原动机 用以直接驱动甲板机械的动力机械。如电动机、液压机、蒸汽机等。

Primer coat 底漆 系指车间底漆涂装后在船厂涂装的涂层系统的第一层涂层。 Primer coat is the first coat of the coating system applied in the shipyard after shop primer application.

Primer the fuel system 使燃油系统充油

Priming 汽水共腾 由于锅炉或蒸发器中水强烈蒸发而使水位涨起,汽空间骤减,大量水分随蒸汽逸出的现象。

Priming funnel 注入漏斗

Priming life 灌注寿命 系指两种原材料混合并搅拌,其从开始搅拌到固化的时间。

Priming right 灌注扶正 通过向位于浸水舱另一侧的舱室灌水,减少破损进水后船舶过大倾斜的措施。

Principal 委托人 (1)系指委托拍卖人拍卖物品或者财产权利的公民、法人或者其他组织。(2)是某项商务合同之债务人或执行者,又是向银行提出申请该保函之申请人。

Principal 主要的

Principal axes 主轴线

Principal dimensions 主要尺度

Principal procurator 主诉检察官 是其中涉及检察机关具有代表性的基本岗位及职能的方方面面,是优秀检察官原型与艺术再现相融合的结晶。

Principle coordinate planes 主坐标平面 系指固定于船舶上,用以确定型值及船舶各部分位置的直角坐标系统的坐标轴平面。其中 x-y 平面为基平面,x-z 平面为中线面,y-z 平面为中站面。

Principle dimensions 主尺度 表示船体外形大小的基本量度,即船长、船宽和船深。通常以垂线间长×型宽×型深表示。

Principle of competition 公平竞争原则 系指各个竞争者在同一市场条件下共同接受价值规律和优胜劣汰的作用与评判,并各自独立承担竞争的结果。

Principle of fairness and reasonableness 公平合理原则 系指处理事情公正符合情理的原则。

Principle of national treatment 国民待遇原则 是国际上关于外国人待遇的最重要的制度之一,其基本含义系指一国以对待本国国民之同样方式对待外国国民,即外国人与本国人享有同等的待遇。

Principle of transparency 透明度原则 系指缔约方所实施的与国家贸易有关的法令、条例、司法判决、行政决定,都必须公布,使各成员国及贸易商熟悉。一成员方政府与另一成员方政府所缔结的影响国家贸易的协定,也必须公布,以防止成员方之间不公平的贸易,从而造成对其他成员方的歧视。

Print 打印 通常系指把电脑或其他电子设备中的文字或图片等可见数据,通过打印机等输出在纸张等记录物上。

Printed circuit board(PCB) 印制电路板 又称印刷线路板,是重要的电子部件,是电子元器件的支撑体,是电子元器件电气连接的载体。由于它是采用电子印刷术制作的,故被称为"印刷"电路板。几乎会出现在每一种电子设备当中,如果在某样设备中有电子零件,那么它们也都是镶在大小各异的PCB上。除了固定各种小零件外,PCB的主要功能是提供上面的各项零件的相互电气连接。随着电子设备越来越复杂,需要的零件越来越多,PCB上头的线路与零件也越来越密集了。见图P-7。

图 P-7 印制电路板
Figure P-7 Printed circuit board(PCB)

Printer 打印机　是计算机最基本的输出设备之一。它将计算机的办理结果打印在纸上。

Printing current meter 印刷海流计　一种机械的转子式自记测流仪器。用于锚泊船舶或浮标上,能自动记录给定时间间隔内的平均流速和瞬时流向。

Prioritization/priority 优先顺序/优先性　根据严重程度、功能、顺序等的警报排序。Prioritization/priority means the ordering of alerts in terms of their severity, function, sequence, etc.

Priority pollutant 重点控制的污染物　系指联邦政府法规第 40 篇第 401.15 条中所列的有毒污染物。Priority pollutant means the list of toxic pollutants listed in 40 CFR 401.15.

Prismatic coefficient (longitudinal coefficient, cylindrical coefficient) 棱形系数　与基线相平行的任一水线下,型排水体积与其相对应的水线长、最大横剖面浸水面积的乘积之比。在无特别注明时,通常系指设计水线处omitted。对于民用船舶,各水线处棱形系数的计算,其中船长和型宽常用垂线间长和满载水线宽。

Private chatroom 私人聊天室

Private key 密码　系指经当事方同意为确保传输的真实性和完整性而采用的任何技术上适当方式,如一组数码和/或字母。Private key means any technically appropriate form, such as a combination of numbers any/or letters, which the parties may agree for securing the authenticity and integrity of a transmission.

Private-prosecuting case 自诉案件　在我国各级法院审理案件以起诉作为审判前提。如果没人向法院"告状",法院则不予审理。法院审理刑事案件,分公诉和自诉两种。自诉案件系指公诉案件的对称。系指被害人、或其法定代理人、近亲属为追究被告人的刑事责任,直接向司法机关提起诉讼,并由司法机关直接受理的刑事案件。

Privilege 优惠

Proactive tariff 保护关税　系指以保护本国工业或农业发展为目的而征收的关税。为达到保护的目的,保护关税的税率比较高。有时税率高达 100% 以上,等于禁止进口,成为禁止关税(prohibited duty)。保护关税又分为工业保护关税和农业保护关税。工业保护关税是以保护国内工业发展所征收的关税。工业保护关税原以保护国内幼稚工业为目的。但资本主义发展进入垄断时期后,发达国家为了垄断国内市场,往往对衰落的工业,甚至高度发达的工业进行保护,这种关税成为超保护关税。农业保护关税是为保护国内农业发展所征收的关税。第二次世界大战后,一些国家,如欧洲经济共同体等通过农业保护关税保护其农业的发展。

Probability 概率　系指出现特定事件的可能性。Probability means the likelihood of occurrence of a specific event.

Probability level 概率等级　系指可接受的概率范围。每小时的风险,且基于船舶预期平均营运时间。有 5 种概率等级:(1)极不可能的;(2)极少可能的;(3)很少可能的;(4)相当可能的;(5)经常的。Probability level means an acceptable probability range and should be established as the risk per hour in ground effect operation, based on the expected mean operating time for the craft. Five probability levels are distinguished: (1) extremely improbable; (2) extremely remote; (3) remote; (4) reasonably probable; and (5) frequent.

Probability of dangerous failure (PFD$_{avg}$) (平均要求)危险失效概率

Probability of ignition 引燃可能性　系指防止可燃材料或易燃液体被引燃。为达到这一目标,应满足以下功能要求。(1)应采取控制易燃液体渗漏的措施;(2)应采取限制易燃蒸发气聚集的措施;(3)应限制可燃物质的可燃性;(4)应限制着火源;(5)应将火源与可燃材料和易燃液体隔离开;(6)应将液货舱内的空气保持在不会发生爆炸的范围内。Probability of ignition is to prevent the ignition of combustible materials or flammable liquids. For this purpose, the following functional requirements shall be met: (1) means shall be provided to control leaks of flammable liquids; (2) means shall be provided to limit the accumulation of flammable vapours; (3) the ignitability of combustible materials shall be restricted; (4) ignition sources shall be restricted; (5) ignition sources shall be separated from combustible materials and flammable liquids; and (6) the atmosphere in cargo tanks shall be maintained out of the explosive range.

Probability of occurrences 事故的概率

Probability of zero oil outflow 零泄油概率　该数代表在发生碰撞或搁浅时无货油从油船漏出的概率。例如,如果该参数为 0.6,则可预期碰撞或搁浅事故 60% 不会发生泄油。Probability of zero oil outflow represents the probability that no cargo oil will escape from the tanker in case of collision or stranding. If, e.g., the parameter equals 0.6, in 60% of all collision or stranding accidents no oil outflow is expected to occur.

Probable error 概率误差　又称或然误差。在一组测量值中,测量误差在 ±r 之间的测量次数占总测量次数的 50%,则 r 值为概率误差。正态分布时,概率误差 $r = 0.6745\sigma$。

Probe 传感器(探测器,探头)

Probe refueling device 探头加油装置 能自动启闭和锁住的由补给船上的输油软管端部探头与接收船上的接收头组成的加油装置。

Probe unit 测试装置(检测器,探头)

Procedural order 程序令

Procedural risk control 程序风险控制 系指依靠操作人员按照规定程序控制风险。

Procedure document 程序文件 对 SMS 体系而言,系指保证船舶航行安全及环保的相关工作程序、规定,其中包括具体的操作指南、检查清单等。

Procedure execution 执行程序

Procedures and Arrangement Manual 程序和布置手册 根据"73/78 防污公约"附则Ⅱ第 14 条的要求,核准装运 X、Y 或 Z 类物质的每艘船应备有经主管机关批准的《程序和布置手册》。该手册应有符合附则Ⅱ附录 4 的标准格式。《程序和布置手册》的主要目的是为船舶高级船员确定实际安排和所有有关操作程序。即为符合附则Ⅱ的要求而必须遵守的货物操作、洗舱、污水处理及液货舱压载和排放。 According the Regulation 14 of the "MARPOL73/78" Annex Ⅱ Every ship certified to carry substances of category X, Y or Z shall have on board a Manual approved by the Administration. The Manual shall have a standard format in compliance with appendix 4 to Annex Ⅱ. The main purpose of the Manual is to identify for the ship's officers the physical arrangements and all the operational procedures with respect to cargo handling, tank cleaning, slops handling and cargo tank ballasting and de-ballasting which must be followed in order to comply with the requirement of the Annex Ⅱ.

Procedures and arrangements manual(P & A Manual) 程序和布置手册(P & A 手册) 系指每艘核准散装运输有毒液体物质的船舶均应持有一份经主管机关认可的程序和布置手册。 Procedures and arrangements manual(P & A Manual) means that every ship certified to carry noxious liquid substances in bulk shall have on board a Procedures and arrangements manual approved by the Administration.

Process 加工过程(进程)

Process annealing 程序退火

Process flow diagram 工艺流程图

Process of refrigeration 制冷过程

Processing deck 加工甲板 处理渔获物的甲板区。

Processing device 处理装置

Processing equipment 处理设备(加工设备,工艺设备)

Processing machinery 加工机械

Processing unit 加工处理单元

Proctor/Fagerberg test procedure 普氏/法氏试验程序 系指用于细粒和相对粗粒精矿或最大粒径为 5 mm 的类似物质的试验方法。该方法不适用于煤或其他多孔物质。 Proctor/Fagerberg test procedure means a test method for both fine and relatively coarse-grained ore concentrates or similar materials up to a top size of 5 mm. This method should not be used for coal or other porous materials.

Producing platform 生产平台 系指为油气生产提供一处水面以上的干环境,以利于油气处理装置正常运行以及操作人员从事正常的作业的平台。

Product 产品 系指:(1) 船上的机械、设备、材料和涂装的涂层。(2) 制造过程的成果。 Product means: (1) machinery, equipment, materials and applied coatings on board a ship. (2) a result of the manufacturing process.

Product carrier 成品油船 系指:(1) 从事除原油以外的油类运输的油船;(2) 用于装载成品油贸易的船舶。 Product carrier means: (1) an oil tanker engaged in the trade of carrying oil other than crude oil; (2) an oil tanker engaged in the trade of carrying oil other than crude oil.

Product certification 产品证书 包括下列可供选择的文件:(1) 船级社产品证书;(2) 工厂证书;(3) 试验报告。 Product certification involves the following alternative documents: (1) the Society product certificate; (2) works certificate; (3) test report.

Product craft 产品艇 系指除原型艇或首制艇以外的游艇。

Product design 产品设计 以规范、规则为依据,以目标规格为限制条件,对现有产品进行改进或创新设计。

Product identification 产品识别 对质量管理体系而言,系指制造厂商应建立和坚持一种制度,使产品在整个生产、交货和安装阶段都能在有关图纸,说明书或其他文件中进行识别。 For the purpose of quality management system, product identification means that the manufacturer is to establish and maintain a system for identification of the product to relevant drawings, specifications or other documents during all stages of production, delivery and installation.

Product inspection(anti-corrosion) 产品检验(防腐) 结构防腐中使用的原材料(如涂料、牺牲阳极)或设备(如外加电流装置)在其生产过程中或出厂时进行的检验。 Product inspection (anti-corrosion) means the inspection of raw material(e. g. paint, sacrificial anode) or equipment(e. g. external current device) used for structural anti-corrosion carried out during the process of production or before delivery.

Product of combustion 燃烧产物

Product offsets 产品补偿

Product oil 成品油　系指原油以外的任何其他油类。Product oil means any oil other crude oil.

Product performance 产品性能

Product planing 产品规划　即新产品的开发计划是企业中长期规划和年度经营方针目标的重要组成部分。

Product quality 产品质量

Product tankers 成品油船　系指:(1)从事除原油以外的油类运输的油船;(2)用于装载成品油贸易的船舶。

Production area 生产区　系指包括井口区及处理区的区域。Production area means an are included wellhead area and processing area.

Production distribution unit(PDU) 产品分配器

Production platform 生产平台/采油平台　系指设有油气处理装置,可在海上进行油气分离及处理的平台。

Production storage tanker/Oil production and storage vessel 生产储油船　系指用作油气处理及原油储存的船舶。

Production support ship 生产支持船　系指油田日常生产所需生产和生活资料的运输及生产设施维护的船舶。

Production test ship 生产测试船　系指进行海上石油早期生产及延长测试的船舶。

Production unit 生产平台　系指主要用于油气处理的平台。Production unit means a unit mainly used for oil and gas processing.

Products 货品　系指有毒液体物质及危险化学品的总称。Products are the collective term used to cover both noxious liquid substances and dangerous chemicals.

Products inspection 产品检验　系指通过设计评估、对最终产品和/或其制造过程中检查和试验,对产品所进行的符合性评价过程。Products inspection means the process of evaluating the compliance of the products with applicable requirements through design approval, examination and testing of the final products and/or during their manufacturing.

Products quotation 产品报价　系指卖方通过考虑自己产品的成本,利润,市场竞争力等因素,公开报出的可行的价格。

Professional export processing zone 专业性出口加工区　即在区内只准经营某种出口加工产品。

Profile 外廓线　系指中线面与船体型表面的交线。

Profile bending machine 肋骨冷弯机　系指采用冷加工方法使肋骨成型的机械。

Profile drag of blade element 叶元体外形阻力　螺旋桨叶元体的摩擦阻力与压力之和。

Profit 利润/津贴

Program control 程序控制　是一种预定的顺序期望值可更改的控制模式。Program control is a pattern of control that desired values can be changed in the predetermined schedule.

Programmable logic controller(PLC) 可编程控制器　是在大规模集成电路的基础上建立起来的集计算机和接口电路于一体的微型计算机。它具有算术运算功能,并有强大的逻辑功能,包括延时、计数、条件和步进顺序控制,能适用于单一的电动机控制直至最高级的自动化控制。

Programming language 编程语言　也称"计算机语言",其种类非常多,总的来说可以分成机器语言、汇编语言、高级语言3大类。电脑每做的一次动作,一个步骤,都是按照已经用计算机语言编好的程序来执行的,程序是计算机要执行的指令的集合,而程序全部都是用大众所掌握的语言来编写的。所以人们要控制计算机一定要通过计算机语言向计算机发出命令。目前通用的编程语言有两种形式:汇编语言和高级语言。

Progressive flooding 连续的进水　系指事先未假定破损的空间的附加进水,可能通过某些开口和管子发生这种附加进水。Progressive flooding is the additional flooding of spaces which were not previously assumed to be damaged. Such additional flooding may occur through some openings or pipes.

Progressive liberalization 逐步自由化　系指保留措施一经取消,将不能再恢复,从而防止新壁垒的产生。

Prohibited duty 禁止关税　有时税率高达100%以上,就是禁止关税。

Prohibited goods 禁运货物　系指:(1)IMDG规则规定禁止运输的任何货物;(2)未列入散装危险化学品规则、散装液化气体规则或固体散装货物规则的任何散装危险货物;(3)未列入《船舶适装证书》的危险货物;(4)未列入"危险货物清单"的危险货物;(5)不满足这些规范中对货物要求的任何货物。Prohibited goods means:(1)any goods which are specified by IMDG code as carriage prohibited;(2)bulk dangerous cargoes not listed in the code of dangerous chemical in bulk, the code of liquefied gases in bulk or solid bulk code;(3)dangerous cargoes that are listed on ship certificate of fitness;(4)dangerous cargoes that are listed in dangerous cargo manifest;(5)any goods which do not fulfill cargo requirements in these Rules.

Prohibited subsidy 禁止使用的补贴　又称禁止的

补贴。包括:(1)在法律上或事实上与出口履行相关的补贴,即出口补贴,并在协议中列出了具体的《出口补贴清单》。(2)其他由公共开支的项目。(3)国内含量补贴,即指前述及补贴只与使用国产货物相联系,而对进口货物不给补贴。

Projected angle of roll 横摇投影角 船体中线面绕固定纵坐标轴 x_0 转动时离垂向平面的角位移。

Projected area ratio 投射面比 螺旋桨投射面积对盘面积的比值。

Projected outline 投影轮廓 螺旋桨叶边缘在垂直于轴线的平面上的投影。

Projected sail area A_S 帆的投影面积 A_S 艇的帆投影面积 A_S 由在上风航行时安装于张帆杆、斜帆桁、斜帆撑杆或其他帆桁上所有帆的投影面积,加上在该艇航行时永久性地安装在适合带帆的桅上的最远处各前支索上的前三角帆投影面积的总和进行计算,但不包括重叠部分,纵帆前缘和后缘,且取直线。每根桅上的前三角帆的面积应按下式计算求得:$lJ/2$,式中 l 和 J 如图 P-8 所示,在桅的前边、前支索的后边、甲板边板处的甲板线之间进行测量。如桅之间的前支索未达甲板,则前三角帆的面积应按图 P-8 所示得出(P 和 E),但其只适用于带着可能装于有关支索上的帆时。在计算帆的投影面积时不包括帆桁的面积,但对翼桅除外。注意:翼桅的特点在于其横截面,其截面的后端平缓地过渡至帆,于是就提供了推动力。这种桅的横截面通常为椭圆形,很少为圆形或箱形。 Projected sail area of a vessel, A_S, is calculated as the sum of the projected profile areas of all sails that may be set when sailing to windward which are attached to booms, gaffs, sprits or other spars, plus the foretriangle area(s) to the outermost forestay(s) permanently attached during operation of the craft to that mast for which suitable sails are carried, without overlaps, luffs and leeches, taken as straight lines. The foretriangle area for each mast shall be the area given by $lJ/2$, where l and J are measurements between the forward side of the mast, the aft side of the forestay and the deckline at sidedeck, as shown in Figure P-8. Where forestays between masts do not reach the deck, the area of the foretriangle shall be taken as shown in Figure P-8 (P and E), but only if sails are carried that may be set on the stays concerned. The area of spars is not included in the calculation of the projected sail except for wing masts. Note: A wing mast is characterized by its cross-section which shows a smooth transition at the aft end into the sail, thus contributing to its driving force. Cross-sections of masts are usually elliptic; they are less often circular or box-shaped.

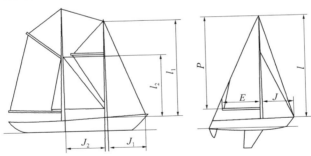

图 P-8 帆面积的测量
Figure P-8 Measurement of sail area

Projection area 投射面积 螺旋桨各叶投射轮廓内面积之和。

Promenade deck 游步甲板 客舱两侧或后端供人员在室外活动的一层甲板。

Promise 承诺 系指受要约人同意接受要约的全部条件以缔结合同的意思表示。承诺的法律效力在于一经承诺并送达于要约人,合同便宣告成立。

Promissory note 本票 就是由发票人签发的载有一定金额,承诺于指定到期日由自己无条件支付给收款人或执票人的票据。

Promissory notes 期票 系指到规定日期才能领取商品或货币的票据,定期支付货币的票据。一种信用凭证,由债务人签发的,载有一定金额,承诺在约定的期限由自己无条件地将所载金额支付给债权人的票据。债务人为出票人,债权人或持票人为受款人。期票一经发出,出票人就负有绝对付款的义务,如到期不予支付,受款人有权诉诸法庭。

Promotional freight rate 促进贸易性运费率 系

指为了促进运输有关国家的非常规出口货物而制订的运费率。Promotional freight rate means a rate instituted for promoting the carriage of non-traditional exports of the country concerned.

Prompt and thorough repair 及时和全面修理/立即彻底修理 系指在检验时完成使验船师满意的永久性修理。在这种情况下,可以不在船级上加任何附带条件。Prompt and thorough repair is a permanent repair completed at the time of survey to the satisfaction of the Surveyor, therein removing the need for the imposition of any associated condition of class or recommendation.

Prompt shipment 即刻装运

Proof load 验证负荷 集装箱系固设备在生产试验中使用的试验负荷。最小验证负荷按表 P-6 确定。

表 P-6 设计破断负荷和验证负荷
Table P-6 Design break load and proof load

	最小设计破断负荷(kN) SWL≤400	最小设计破断负荷(kN) SWL≥400	最小验证负荷(kN) SWL≤400	最小验证负荷(kN) SWL≥400
绑扎装置:				
钢丝绳	3×SWL			
杆(高强度钢)	2×SWL		1.5×SWL	
链(低碳钢)	3×SWL			
链(高强度钢)	2.5×SWL			
其他系固设备:	2×SWL	SWL+400	1.5×SWL	SWL+200

注意:(1)高强度钢的屈服应力应不小于 315 N/mm²。
(2)若不采用钢材而采用其他材料,则对其破断负荷和验证负荷,将另行考虑。
(3)SWL——安全工作负荷,kN。

Proof load, in relation to materials handling equipment 与物资装运设备有关的验证载荷 对澳大利亚海事安全局海事指令而言,系指根据澳大利亚海事安全局海事指令所确定的设备的验证载荷。For the purpose of the maritime order of the Australian maritime safety authority(AMSA), proof load, in relation to materials handling equipment means the proof load for that equipment determined in accordance with the order of the Australian maritime safety authority.

Proofs of documents identifying subject eligibility of the parties 当事人双方主体资格证明文件

Propagating buckle 屈曲的蔓延 如果外部压力高得足以使局部破坏存在蔓延的趋势,并在这一过程中压垮大剖面结构,则在外部压力下的破坏通常是局部的。Collapse under external pressure is usually local, if the external pressure is high enough the local collapse has the tendency to propagate and in the process flatten large sections of the structure.

Propane storage 支撑剂储存装置

Propellant 推进剂 喷射推进器所用的工作介质。如水、空气、水气混合物等。

Propeller 推进器(螺旋桨) 系指直接通过与水的作用来提供船舶推进力的装置,如螺旋桨、喷水器等,并包括为该装置传递机械能的任一设备,如轴承、齿轮传动装置等。按其加工成型方式分:(1)整体浇注成型的一体式螺旋桨(solid type propeller);(2)桨叶和桨毂分别加工成型后用螺栓组装成一体的组合式螺旋桨(build up type propeller)。按螺距可变与否分:固定螺距螺旋桨(fixed pitch propeller)和可变螺距螺旋桨(controllable pitch propeller)。见图 P-9。

图 P-9 推进器
Figure P-9 Propeller

Propeller arrangement 螺旋桨装置
Propeller assembly 推进器组合系统
Propeller boss cap fins 桨毂帽鳍 是固定在螺旋桨桨壳帽上的小鳍。其叶数与桨叶数相同，直径为螺旋桨直径的25%左右。其作用主要是消除毂涡，回收螺旋桨漩涡能量，从而起到节能和减振降噪的作用。见图P-10。

图 P-10 桨毂帽鳍
Figure P-10 Propeller boss cap fins

Propeller cap 毂帽 装在毂后，用以掩护艉轴螺帽并使水流顺畅离开螺旋桨的护罩。
Propeller design according to theory 螺旋桨理论设计 按照螺旋桨环流理论进行的螺旋桨设计。
Propeller design charts 螺旋桨设计图谱 根据定距螺旋桨系列模型敞水试验数据或理论计算结果绘制的供螺旋桨设计用的曲线图。
Propeller disc 螺旋桨盘 在螺旋桨平面内以螺旋桨中点为圆心，以螺旋桨半径为半径的圆。
Propeller effect on turning 螺旋桨偏转效应 由于螺旋桨转动而引起的船体偏转现象。
Propeller efficiency behind ship 螺旋桨船身效率 螺旋桨在船后或船模后工作时的推力功率对收到功率的比值。$\eta_B = P_T/P_D$ 或 $TV_A/2\pi nQ$，式中，P_T—推力功率；P_D—收到功率；T—螺旋桨推力；V_A—进速；n—螺旋桨转速；Q—螺旋桨转矩。T, V_A, n 和 Q 皆由船模自航试验量得。
Propeller machinery 推进机械（主机机组）
Propeller noise 螺旋桨噪声 （1）在螺旋桨及其附近由于空泡坍毁所产生的不规则尖锐爆裂和打击声。亦可由于机械作用产生的噪声。（2）螺旋桨叶片振动和螺旋桨空泡引起的噪声。
Propeller noise prediction 螺旋桨噪声预测
Propeller performance curve 螺旋桨性能曲线
Propeller plane 螺旋桨平面 通过螺旋桨中点，垂直于轴线的平面。
Propeller post 螺旋桨柱 用以支持螺旋桨轴的构件。是艉柱的一部分。
Propeller pump 轴流泵
Propeller race(slip stream) 螺旋桨尾流 螺旋桨后方受过螺旋桨扰动的水流。
Propeller racing 螺旋桨飞车 船舶在风浪中航行，部分螺旋桨露出水面，转速剧增，并伴有强烈振动的现象。
Propeller reference lines 螺旋桨基准线 垂直于螺旋桨轴线，在正视图和侧视图上作为垂向基准线的直线。
Propeller revolution 螺旋桨转速
Propeller shaft 螺旋桨轴/艉轴 系指轴系中安装推进器的轴。
Propeller shaft and stern tube survey 螺旋桨轴和艉轴管检验
Propeller shaft and tube survey 螺旋桨轴和艉轴管检验
Propeller shaft condition monitoring 螺旋桨轴状态监控
Propeller shaft king 1 or stern tube shaft 1 第一种推力轴或第一种艉轴管 系指采取经船级社形式认可的防海水腐蚀措施或使用获得认可的耐腐蚀性材料制成的轴。其余的都称第二种推力轴或第二种艉轴管。
Propeller shaft king 1 or stern tube shaft 1 is the shaft which is provided with effective measures against corrosion by sea water, or the shaft which is made of approved corrosion resistance material. The propeller shaft or stern tube shaft other than specified above is kind 2.
Propeller shaft line 推进轴系
Propeller shaft survey(PS) 轴检 对螺旋桨轴每5年进行一次的检查。轴检可分为：（1）正常轴检（ordinary PS）。抽轴、拆桨检查艉轴承的磨损情况，对艉轴的装桨部进行磁粉探伤（magnetic particle/flux test）以获得下一个5年的使用期；（2）部分轴检（partial PS）。通过检查记录，分析润滑油中的金属离子的含量等方法来确认艉轴系统的状态。在被确认满足要求时，船级社会根据轴系类型的不同，同意将进行正常轴检的时间推后1年~3年。
Propeller shafting 螺旋桨轴系
Propeller speed 螺旋桨转速
Propeller speed indicator 螺旋桨转速指示器
Propeller stern-tunnel(semi-duct) 螺旋桨艉隧 船尾装螺旋桨部分船底向上凹入以容纳较大直径螺旋桨的隧道。

Propeller strut propeller strut　轴支架　装在船体外部的艉轴支架。

Propeller-excited vibration　螺旋桨激振

Propeller-hull vortex　连船梢涡　由于物体某处水流作用,每当一螺旋桨叶接近某处时被引至船体继而离开的叶梢空泡。

Propeller-hull vortex(PHV)　连体空泡　系指由于伴流场不均匀所产生的空泡。连体空泡会导致加大艉部的剧烈振动,还会引起不必要的噪声问题。

Propeller-tail shaft assembly　螺旋桨-艉轴组合体

Propelling engine　推进发动机

Propelling machinery(propulsion machinery)　推进装置　动力装置中船舶推进用的动力机械及其直接有关配套设备,包括传动装置、轴系等的总称。

Propelling machinery space　推进机器处所(主机舱)

Propelling plant　推进装置

Propelling power space　推进机械处所　系指吨位丈量中,包括主机舱、主锅炉舱、轴隧,以及规则规定的其他与推进机械有关的各处所。

Propelling system　推进系统

Propelling unit　推进装置

Property　财产　系指带有所有权的、包括契约权在内的任何物品。 Property means anything that can be the subject of a right of ownership, including contractual rights.

Proposal　建议

Proposing Member Government　提出申请的成员国政府　系指一个向国际海事组织提交确定特别敏感海区及其相关保护措施申请的成员国政府。 Proposing Member Government means a Member Government (or Government) submitting an application for particularly sensitive sea area(PSSA) identification within its Associated Protective Measures to IMO.

Proprietors club　业主俱乐部　系指非会员俱乐部以外的任何俱乐部。 Proprietors club means any club that is not a members club.

Propulsion　推进(推进装置)

Propulsion and electrical operating station　推进和电气操纵部位

Propulsion auxiliary systems　辅助推进系统　包括推进系统正常操作所需的所有系统。其可包括或可能包括:(1)供油系统(包括日用油柜,但不包括燃油注入、输送和净化系统);(2)为主机、齿轮箱、轴系轴承、艉轴管等服务的润滑油系统;(3)供离合器、可调距螺旋桨、喷水装置倒车偏导板、启动系统等使用的液压油系统;(4)服务于推进系统任何部件的或用来冷却燃油管路、润滑油管路和液压油管路等的淡水冷却系统;(5)用来冷却推进系统任何部件的或上述任何系统的海水冷却系统;(6)加热系统(使用电、蒸汽或热流体);(7)启动系统(空气、电动、液压);(8)动力源(空气、电动、液压);(9)控制、监测和安全系统;(10)通风装置,如需要时(例如向原动机供给燃烧的空气或冷却空气)。 Propulsion auxiliary systems include all the systems that are necessary for the normal operation of a propulsion system. It includes or may include:(1)the fuel oil supply system from and including the service tanks. The fuel oil filling, transfer and purifying systems are not included;(2) the lubricating oil systems serving the engines, the gearbox, the shaftline bearings, the stern tube, etc. The lubricating oil filling, transfer and purifying systems are not included;(3) the hydraulic oil systems for operating clutches, controllable pitch propellers, waterjet reverse deflectors, starting systems, etc. ;(4)the fresh water cooling systems serving any component of the propulsion system or used for cooling the fuel oil circuits, the lubricating oil circuits, the hydraulic oil circuits, etc. ;(5)the sea water cooling systems used for cooling any component of the propulsion system or any of the aforementioned systems;(6) the heating systems (using electricity, steam or thermal fluids);(7) the starting systems (air, electrical, hydraulic);(8) the power supply (air, electrical, hydraulic);(9) the control, monitoring and safety systems;(10) the ventilation installation where necessary (e.g. to supply combustion air or cooling air to the primer movers).

Propulsion characteristic test　推进特性试验　测定船用发动机推进特性-功率与船速关系的试验。

Propulsion control room　推进控制舱

Propulsion devices　推进装置　系指直接提供推进力的装置。包括机器设备以及主要用来提供推进力的任何导管、桨叶、流体进口和喷嘴等。 Propulsion devices are those which directly provide the propulsive thrust and include machinery items and any associated ducts, vanes, scoops and nozzles, the primary function of which is to contribute to the propulsive thrust.

Propulsion diesel　推进柴油机(推进用柴油机)

Propulsion engine　主推进发动机(推进发动机)

Propulsion equipment　推进设备

Propulsion generator　推进发电机(主推进发电机)

Propulsion machinery　推进装置(推进机械)　系指为驱动推进器而提供机械能的装置,如柴油机、涡轮机、电动机等。

Propulsion machinery control position 主机控制站

Propulsion machinery output 推进机械输出功率
系指推进机械能连续发出的最大总输出功率。如机械输出功率受到技术措施的限制或船舶应遵照规则的限制,则推进机械输出功率应取受到限制的输出功率。

Propulsion machinery room 推进装置舱

Propulsion machinery space 推进机器处所(主机舱)

Propulsion motor 推进电机

Propulsion motor exciter 推进电动机励磁机

Propulsion motor room 推进电机舱

Propulsion of ships 船舶推进 研究驱动船舶前进的推进器结构和水动力性能,船体与推进器间的相互影响,推进器设计及其试验的一门学科。

Propulsion plant 推进装置

Propulsion power 推进动力 系指船舶登记证书或其他官方文件上出现的以千瓦计的功率。Propulsion power means the power in kilowatts which appears on the ship's Certificate of Registry or other official document.

Propulsion quality 推进性能

Propulsion shafting 推进轴系

Propulsion shafting centerline 推进装置轴系中心线

Propulsion system 推进系统(推进装置) 系指为船舶提供推力的系统。其包括:(1)原动机(包括驱动泵等);(2)传递扭矩的设备;(3)推进电机(如适用);(4)将扭矩转换成推力的设备;(5)操作所需的辅助系统;(6)控制、监测和安全系统。Propulsion system is a system that provides thrust to the ship. It includes:(1) the prime mover, including the integral equipment, driven pumps, etc.;(2) the equipment intended to transmit the torque;(3) the propulsion electric motor, where applicable;(4) the equipment intended to convert the torque into thrust;(5) the auxiliary systems necessary for operation;(6) the control, monitoring and safety systems.

Propulsion trial 推进性能试验

Propulsion unit 推进装置

Propulsion-steering unit 推进器-舵组合体

Propulsive arrangement 推进装置

Propulsive auxiliaries 推进辅机 动力装置中直接为主机和主机机组服务的辅机。

Propulsive coefficient 推进系数 船体有效功率对主机指示功率、制动功率或轴功率的比值。

Propulsive efficiency (quasi-propulsive coefficient) 推进效率 船的有效功率对螺旋桨收到功率的比值。
$\eta_R = P_E/P_D$,式中,P_E—有效功率;P_D—收到功率。

Propulsive engine 推进发动机(主机)

Propulsive performance 推进性能

Propulsive power 推进功率 供给推进系统(如空气螺旋桨、水螺旋桨、喷水推进器或喷气推进器等)的功率。

Propulsive system 推进系统

Propulsor 推进器(推进发动机)

Propulsor fitted to podded drive 吊舱驱动推进器

Prospective current(with respect to a switching device) 预期短路电流(针对开关电器) 系指当开关电器的每一极由一阻抗可以忽略不计的导体代替时,电路中可能流过的短路电流。Prospective current(with respect to a switching device) means the short-circuit current that would flow in the circuit if each pole of the device were replaced by a conductor of negligible impedance.

Protected against submersion enclosure 潜水型外壳 系指当其浸在水中时,其内部的设备能在规定的压力和时间条件下有效运行的外壳。

Protected areas 被保护区域 系指:(1)在被保护处所之内,要求由固定式水基的局部使用灭火系统予以保护的区域;(2)安装了固定式水基的局部使用灭火系统的机器处所。Protected areas are:(1) those areas within a protected space which are protected by a fixed water-based local application fire-fighting system;(2) a machinery space where a fixed water-based local application fire-fighting system is installed.

Protected hazard spaces 被保护危险处所 系指具有火灾危险的处所。Protected hazards spaces means the risk parts.

Protected space 被保护处所 系指:(1)安装了固定式水基的局部使用灭火系统的机器处所;(2)被保护危险部分所处的区域。Protected space is:(1) a machinery space where a fixed water-based local application fire-fighting system is installed;(2) the spaces which the protected hazards are located.

Protected waters service 保护水域航区 系指邻近砂岩、暗礁、防波堤或其他海岸特征的遮蔽水域中,以及岛屿间的遮蔽水域中的航区,例如:"Storebaelt Bridge 保护水域航区"。Protected waters service mean the service in sheltered water adjacent to sand banks, reefs, breakwaters or other coastal features, and in sheltered water between islands, e.g. "Protected Waters Service at Storebaelt Bridge".

Protecting covering 保护外套

Protecting jacket 防护套

Protection 防护

Protection & indemnity Clubs 保赔协会 系指国际保赔协会。其业务性质主要是承保船舶的船东责任保险，而他们的年结日期是每年的 2 月 20 日。

Protection "n" "n"防护型(无火花型) 适用于在正常运行中不可能点燃周围爆炸性气体环境，且不会出现可能导致点燃之故障的电气设备的保护形式。注：IEC 60079-15 规定了使用此防护方法之设备的结构特征和试验要求。 Protection "n" means a type of protection applied to electrical apparatus such that, in normal operation, it is not capable of igniting a surrounding explosive gas atmosphere and a fault capable of causing ignition is not likely to occur. Note: IEC 60078-15 specifies the constructional features and test requirement for apparatus using this method of protection.

Protection coatings in water ballast tanks (PC-WBT) 压载水舱的保护涂层

Protection device 保护装置(安全装置)

Protective bars (guard bar) 护栅 门窗上用以防护玻璃的金属条。

Protective casing 保护封闭罩壳 系指管状的封闭罩壳或其他非圆型的封闭管道。Protective casing means a closed cover in the form of a pipe or other closed ducts of non-circular shape.

Protective coating 防护涂层 (1)是施加的涂层系统，以防止结构件腐蚀，它不包括预制件底漆；(2)防止结构件免受腐蚀的最终涂层，不包括车间底漆；(3)是防护结构件锈蚀的最后一道涂层；(4)环氧树脂或等效物。其他涂层系统只要在应用中符合制造厂商的规定和适当的维护，可以考虑作为替代品予以接受。Protective coating is: (1) the coating system applied to protect the structure from corrosion. This excludes the prefabrication primer; (2) a final coating protecting the structure from corrosion; (3) a final coating protecting the structure from corrosion; (4) usually to be epoxy coating or equivalent. Other coating systems may be considered acceptable as alternatives provided that they are applied and maintained in compliance with the manufacturer's specifications.

Protective coatings for steel ships 钢质船舶的保护涂层 通常应为硬质涂层。其他涂料系统(如软涂层)，只要按制造厂商的规范应用并得以适当维修，可被视为可以接受的替代品。Protective coatings for steel ships should usually be hard coatings. Other coating systems (e.g. soft coating) may be considered as alternatives provided they are applied and properly maintained in compli-

ance with the manufacturer's specification.

Protective cover (personal flotation device) 保护罩(个人漂浮装置) 通常置于 PFD 功能元件之上，以防止其遭到物理损伤或被外部物体绊住的罩子。注意(1)：此保护罩可设计成具有附加的物理特性，也就是使此 PFD 可适合于在对象曝露于附加的危险(如严重的磨损、熔融金属飞溅、火等)中时使用。注意(2)：气胀式 PFD 的充气室是功能元件的一个例子。Protective cover is the cover that is normally in place over the functional elements of a PFD in order to protect them from physical damage, or snagging on external objects. Note(1): The protective cover may be designed to provide additional properties i.e. to make the PFDs suitable for use when the subject is exposed to additional hazards, e.g. significant abrasion, molten metal splash, flame and fire. Note(2): The inflatable chamber of an inflatable PFD is an example of a functional element.

Protective current density 保护电流密度 系指使被保护物体电位维持在保护电位范围内所需的极化电流密度。Protective current density means a necessary polarization current density for the potential of protected object to be maintained within the protection potential range.

Protective device 保护装置(安全装置) 系指能对整定值的偏离产生响应和发出信号，进而能防止设备产生损害的装置。Protective device means a device capable of making a response to and giving a signal for a deviation from a set value so as to prevent damage to the protected equipment.

Protective enclosure 防护外壳 系指其上所有可以与带电部件或旋转部件(光滑的旋转表面除外)直相通的开孔，在其上均采用结构部件或金属丝网、挡板、格栅、拉制的金属网或其他措施加以限制，以防止与危险部件偶然接触的外壳。这种外壳的开孔应做成不会让直径大于 1/2 英寸(in)的棒状物进入，但当裸露带电部件与防护物的距离大于 4 英寸(in)时，则开孔的形状可以做成不会让直径大于 3/4 英寸(in)的棒状物进入。

Protective equipment 防护设备

Protective film 保护膜

Protective function 保护功能

Protective gear 防护装置

Protective grating 防护格栅

Protective location of segregated ballast 专用压载舱的保护位置 系指在设计或布置专用压载舱时应按照"73/78 防污公约"附则Ⅰ的要求对其在舷侧及船底的投影面积达到一定的数值，以提供一种在船舶发生搁浅或碰撞时防止油类外流的保护措施。

Protective material　保护材料

Protective measure　保护措施

Protective packing　保护箱

Protective potential range　保护电位范围　系指使金属腐蚀速率达到预定保护要求的极化电位值的区间。 Protective potential range means an interval of polarization potential values for the metallic corrosive rate to meet the determined protective requirements.

Protective trade　保护贸易　系指国家运用权力，通过高额关税进口许可证、外汇管制等限制进口措施，来保护本国市场，防止外国商品竞争，同时对本国商品给予津贴。

Protective trade policy　保护贸易政策　系指国家广泛利用各种措施对进口和经营领域与范围进行限制，保护本国的产品和服务在本国市场上免受外国产品和服务的竞争，并对本国出口的产品和服务给予优待与补贴。国家对于贸易活动进行干预，限制外国商品、服务和有关要素参与本国市场竞争。

Protector　防护装置（护罩）

Protest for non-payment　拒付证书　系指由付款地的公证人对提交承兑或付款的支票、汇票付款人拒不承兑或拒付所出具的证书。

Protocal　议定书　系指缔约国对条约或协定的解释、补充、修改或延长有效期以及关于某些技术性问题所议定缔结的国际法律文件。

Prototype　样箱　系指按定型设计系列制成或准备制造的具有代表性的集装箱。 Prototype means a container representative of those manufactured or to be manufactured in a design type series.

Prototype　原型　系指：(1) 为评价设计的符合性而按设计形式系列制造的模型产品。(2) 这些制造的或按设计形式系列制造的有代表性的集装箱。 Prototype means: (1) a model product manufactured to the design which is to be evaluated for compliance with applicable requirements. (2) a container representative of those manufactured or to be manufactured in a design type series.

Prototype ballast water treatment technology　原型压载水处理技术　系指 2004 年国际船舶压载水和沉积物控制和管理公约的第 D-4 条规定的任何一压载水处理设备综合系统，其参与试验和评估是否符合或拆除超出 2004 年国际船舶压载水和沉积物控制和管理公约的第 D-2 条规定的性能标准的程序，包括处理设备、所有相关控制设备、监控设备和取样设备。原型压载水处理技术可以是一种机械、物理、化学或生物学处理过程，单独或组合采用不采用活性物质去除、无害化或避免吸入或排放压载水和沉积物中的有害生物和病原体。原型压载水处理技术可在压载水吸入或排放时，航行期间或在这些阶段的组合情况下进行工作。 Prototype ballast water treatment technology means any integrated system of ballast water treatment equipment as under regulation D-4 of the International Convention for the Control and Management of Ships' Ballast Water and Sediments, 2004, participating in a programme for testing and evaluation with the potential of meeting or exceeding the ballast water performance standard in regulation D-2 of the International Convention for the Control and Management of Ships' Ballast Water and Sediments, 2004, including treatment equipment, all associated control equipment, monitoring equipment and sampling facilities. A prototype ballast water treatment technology may be a mechanical, physical, chemical, or biological unit process, either singularly or in combination that may or may not use Active Substances that remove, render harmless, or avoid the uptake or discharge of Harmful Aquatic Organisms and Pathogens within ballast water and sediments. Prototype ballast water treatment technologies may operate at the uptake or discharge of ballast water, during the voyage or in any combination of these phases.

Prototype craft　原型艇　系指为评估设计的符合性而按设计制造的模型游艇。

Prototype engine　原型机（样机）

Prototype test　原型试验　系指为评价产品的设计，对产品原型包括其材料和部件所进行的试验和测试。原型试验可以采用破坏性试验。 Prototype test means testing and measurement of the prototype including its materials and components for evaluating the product design. The prototype test may be a destructive test.

Protrusion　突出物　系指延伸到一艘船的船体任何部分以外的物体，无论其是永久性或临时性的，但主锚除外。 Protrusion mean anything that extends beyond any portion of the hull of a vessel, whether it is permanent or temporary, except for the main anchors.

Proven technology　成熟技术　系指一定环境下的领域里有文献可循的技术。

Provider of sampling and testing services　抽样和测试服务提供方　系指以营利为基础，为输送到船上的船用燃油提供测试和抽样服务的评估燃油质量参数（包括硫含量）的公司。 Provider of sampling and testing services means a company that, on a commercial basis, provides testing and sampling services of bunker fuels delivered to ships for the purpose of assessing quality parameters of these fuels, including the sulphur content.

Provision for doubtful items　不可预见费　又称为

预备费，系指在工程投资概（估）算中，预留的为支付施工中可能发生的、比预期的更为不利的水文、天气、地质及其他社会、经济条件而需增加的费用，一般以总投资的某一百分数计。

Provision store 粮食库 船上供贮存粮食的仓库。

P_s（kN/m^2） 静水压力 p_s is the still water pressure, in kN/m^2.

PSC 港口国检查 系指港口国当局针对抵港的外国船舶实施的，以船员、船舶技术状况和操作要求为检查对象，以确保船舶和人命财产安全，防止海洋污染为宗旨的一种监督与检查。

Pseudo 伪 当一个位置测量系统作为另一个位置基准接口到 DP 控制系统时使用的前缀。

Pseudo code 伪代码

P_{SF}（kN/m^2） 进水工况下的静水压力。 P_{SF} is the still water pressure in flooded conditions(kN/m^2).

Psilocin/psilobycin 二甲-4-羟色胺/二甲-4-羟色胺磷酸 迷幻剂之一，是浅粉红色或黄色液体，有药丸或药片形式。没有气味。 Psilocin/psilobycin is one of hallucinogens, which is found as a pale pink or yellowliquid and in pill or tablet form. It is odourless.

PSPC 保护涂层 系指授予特定处所满足 IMO 有关保护涂层性能标准的船舶的附加标志。其后缀一个或多个 B、C、D 和 V 标志，其含义如下：B—授予各类型船舶专用海水压载舱处所施用的保护涂层；C—原油船货油舱处所施用的保护涂层；D—散货船双舷侧处所施用的保护涂层；V—散货船和油船的空舱处所施用的保护涂层。注：B,C,D 和 V 可以单独也可以组合使用。

PSV = platform supply vessel 平台供应船 A PSV is a vessel carrying out cargo operations.

Psychrometor 湿度计（干湿球）

Psychrometric chart 温湿图 根据一定大气压力下湿空气的物理性能参数，含热量、含湿量、温度、相对湿度、比容、水蒸气分压力绘制成的说明空气处理过程的曲线图。

Pub. L = Public Law 公共法

Public address system 广播系统 是以扬声器向船上人员正常到达的所有处所发布信息的设备。 Public address system is to be a loudspeaker installation enabling the broadcast of messages into all spaces where people on board are normally present.

Public address/general alarm(PA/GA) system 公共广播报警系统 系指用于在服务区内执行公共寻呼和紧急广播功能的系统。该系统主要由 3 部分组成：(1)主机部分，包括交换控制器、数字管理服务器、数字线路卡、冗余监控板、数字 PCM 连接板、数字交换接口板、基础光纤板、数字远端线路卡、数字音频记录器、数字 I/O 板、数字音频处理器、功率放大器、继电器板、扬声器监控单元、模块架；(2)设备，包括扬声器、话站、遥控单元及音量控制器；(3)电缆，包括主机与终端连接的电路及主机连接到其他系统的所有电缆。

Public affair ship 公务船 系指由政府部门拥有或经营，并仅用于政府执行公务的非商业性服务的船舶。 Public affair ship means a ship owned or operated by the government and used only for non-commercial service.

Public authorities 公共管理当局（公共当局） 系指：(1)某一国家负责实施和执行该国法律和规则的机构或官员，这些法律和规则系指与便利国际海上运输交通附则中的"标准"和"推荐做法"的任何有关的法律和规则；(2)一个国家负责应用和执行该国有关防止偷渡者进入和寻求成功解决偷渡案件责任分配指南任何方面法律和规则的机构或官员。 Public authorities means: (1) the agencies or officials in a State responsible for the application and enforcement of the laws and regulations of that State which relate to any aspect of the Standards and Recommended Practices contained in the Annex of the convention on facilitation of international maritime traffic; (2) the agencies or officials in a State responsible for the application and enforcement of the laws and regulations of that State which relate to any aspect of the Guidelines on the prevention of stowaway incidents and the allocation of responsibilities to seek the successful resolution of stowaway cases.

Public correspondence 公共信息通信 系指办事处和站点出于其为公众服务的原因，而必须接收并转送的电信。 Public correspondence means any telecommunication which the offices and stations must, by reason of their being at the disposal of the public, accept for transmission.

Public places 公开场合 就是向公众开放的、允许公众自由出入的场所和场合，比如街道、田野、广场、集市、公园、娱乐和体育馆等场所。

Public spaces 公共处所 系指：(1)供乘客使用的处所，包括酒吧、小卖部、吸烟室、主要座位区、娱乐室、餐厅、休息室、走廊、盥洗室和其他类似的处所，并可包括买品部；(2)起居处所中用作大厅、餐厅、休息室的部分以及类似的固定围蔽处所。 Public spaces are: (1) those spaces allocated for the passengers and include bars, refreshment kiosks, smoke rooms, main seating areas, lounges, dining rooms, recreation rooms, lobbies, lavatories and similar spaces, and may include sales shops; (2) those portions of the accommodation which are used for halls, dining rooms, lounges and similar permanently enclosed spaces.

Public switched telephone network(PSTN) 移动网与公用电话交换网 系指公用电话交换网。即人们日常生活中常用的电话网。众所周知,PSTN 是一种以模拟技术为基础的电路交换网络。

Public vessel 公务船 系指这样的一艘船舶:(1)政府拥有的,或裸船(干租)的形式)出租和使用的;(2)不从事商业性服务的。 Public vessel means a vessel that:(1) is owned or demise chartered, and operated by a government of any country;(2) is not engaged in commercial service.

Published TFW maximum draft 公布的热带淡水最大吃水(巴拿马运河) 系指考虑到夏通湖水位及由于受运河限制认为有必要的其他限制,并而由海事作业主任公布的在夏通湖水中浸没的最深点。 For the purpose of Panama Canal, published TFW maximum draft means the deepest point of immersion in Gatun Lake waters as promulgated by the Maritime Operations Director, taking into account the water level of Gatun Lake and other limitations deemed necessary because of restrictions in the Canal.

Pull out maneuver test(meander test) 回舵操纵试验 先操一指定舵角,待达到定常转艏角速度后回复到正舵,然后连续记录其瞬时转艏角速度直至定常状态,以测定船舶舵向不稳定回线环高的一种试验。

Pull-out torque 失步转矩 系指同步电动机在施以额定频率的额定电压及正常励磁情况下,在同步转速时产生的最大持续转矩。

Pulsating rate of revolution 转速波动率 内燃机在不变负荷运转时,在一定时间内测得的最大转速或最小转速和平均转速之差与平均转速的百分比。

Pulse 脉冲(脉动,半周期)

Pulse system supercharging 脉冲增压 利用几根较细的排气管,使整个气缸的高速排气冲击功能直接作用在涡轮上的废气涡轮增压方式。

Pulse width modulation(PWM) 脉冲宽度调制

Pulsed high magnetic field facility(PHMFF) 脉冲强磁场实验装置 这是我国"十一五"重大科技基础设施建设项目中的两个子项目之一,合肥物质科学研究院联合中国科学技术大学共建稳态强磁场实验装置,华中科技大学建设脉冲强磁场实验装置。目前,该装置已跻身世界"最佳"的脉冲场之列,并在电源设计和磁体技术方面取得的成就位列世界顶级。

Pumice 浮石 火山爆发形成的多孔岩石。灰白色。无特殊危害。该货物为不燃物或失火风险低。 Pumice is highly porous rock of volcanic origin. Greyish-white. No special hazards. This cargo is non-combustible or has a low fire-risk.

Pump 泵(泵送,抽吸)
Pump barrel(pump bucket) 泵的水缸
Pump bucket dredger 自扬/链斗式挖泥船
Pump capacity 泵的容量
Pump case 泵壳体
Pump compartment 泵舱
Pump cut-in pressure 泵接入压力
Pump diffuser 泵扩压器
Pump drive assembly 泵传动组件
Pump foundation 泵底座
Pump group 泵组
Pump house 泵站(泵舱)
Pump in 用泵吸入
Pump inlet 泵进口(泵吸入口)
Pump jet propeller 泵式喷水推进器
Pump jet propulsion 泵式喷水推进
Pump knock 泵敲击
Pump out 用泵排出
Pump output 泵功率
Pump priming system 泵引水系统 系指船舶系统离心泵启动时,引水用的专用真空系统。由真空泵、真空柜、管道和附件等组成。

Pump room 泵舱 系指:(1)位于装货区域内装有用于装卸压载水和燃油的泵及其辅助设备的处所;(2)位于货舱区域包含有用于装卸压载水和燃油或者船舶营运标志未涵盖的货物的泵和附近的舱。 Pump room is:(1) a space, located in the cargo area, containing pumps and their accessories for the handling of ballast and oil fuel;(2) a space, located in the cargo area, containing pumps and their accessories for the handling of ballast and fuel oil, or cargoes other than those covered by the service notation granted to the ship.

Pump room bulkhead 泵舱舱壁
Pump room sea valve(pump room sea-suction valve) 泵舱通海阀 为利用货油泵泵入压载用的海水,而在泵舱内设置的通海阀。
Pump suction chamber 泵吸入室
Pump system 泵系统
Pump unit 泵机组(泵装置)
Pumped up 用泵打进
Pumping 泵吸
Pumping and drainage plan 泵及泄水管系图
Pumping arrangement 泵吸装置(泵吸布置图)
Pumping main 泵送总管(排出总管)
Pumping performance 泵吸性能

Pumping system　泵吸系统(泵送系统)

Pump-room ventilator　油泵舱通风机　用以抽除油船货油泵舱中油气的防爆式通风机。

Punt pole　篙　在浅水中用人力撑船前进用的竿。

Purchaser supplied material　买方供应的材料　对质量管理体系而言,系指制造厂商应建立和维持的程序以控制买方供应的材料。　For quality management system, purchaser supplied material means that the manufacturer is to establish and maintain documented procedures for the control of purchaser supplied material.

Purchasing　采购　对质量管理体系而言,系指制造厂商应保证所采购的材料和所选用的服务符合规定的要求。　For the purpose of quality management system, purchasing means that the manufacturer is to ensure that purchased material and services conform to specified requirements.

Purchasing data　采购资料　对质量管理体系而言,系指每份采购文件都应清楚描述所订购的材料和服务的内容,尽可能包括下列各项:(1)型号,类别,等级或其他精确的特征;(2)名称或其他确切的特征以及说明书,图纸,加工要求,检查项目及其他有关数据。　For quality management system, purchasing data mean that each purchasing document should contain a clear description of the material or service ordered including as applicable, the following:(1) the type, class, grade, or other precise identification;(2) the title or other positive identification and applicable issue of specifications, drawings, process requirements, inspection instructions and other relevant data.

Purchasing power parity(PPP)　购买力平价法　是以国内商品价格与国家同种商品基准价格比率的加权平均值为购买力平价计算的方法。ICP 是由联合国统计局、世界银行等组织主持的一项旨在提供 GDP 及其组成部分的国际一致价格和物量的跨国比较体系。

Pure car carrier(PCC)　小轿车运输船　系指专门运输小轿车的船舶。

Pure car/truck carrier(PCTC)　轿车/卡车运输船　系指专门运输轿车和卡车的船舶。

Pure loss of stability　纯稳性丧失

Pure tone　纯音　系指最简单的声波。它是单一频率的声音。其具有一定频率和振幅的正(余)弦压力波。其传播的速度是由气体媒介质的体积弹性模量 k 及气体媒介质的密度 ρ 所决定的。也就是说,是气体的温度和气体的压力所决定的。

Purge　清除

Purging　除气　系指在液舱内用一般的空气替代惰性气体或碳氢化合物的过程,以便对液舱进行检查或修理。　Purging is the process of replacing the inert gas or hydrocarbon mixture in a tank with normal air for tank inspection or repair.

Purging　驱气　系指向已处于惰性状态的舱柜输入惰性气体以求:(1)进一步降低含氧量;(2)使烃气浓度降低到即使让空气引入舱内亦不致在舱内形成可燃的混合气体。　Purging means sending inert gas into an inerted tank to:(1) further reduce oxygen content;(2) reduce hydrocarbon gas content such that the entry of air will not create a flammable gas mixture.

Purging valve　放气阀

Purging(electrical equipment)　驱气(电气设备)　在对设备施加电压之前,使足够容积的保护气体通过正压外壳及其导管,以使任何爆炸性气体环境的浓度降至大大低于爆炸下限。　Purging(electrical equipment) means that passing of sufficient volume of protective gas through a pressurized enclosure and its ducts before the application of voltage to the apparatus to reduce any explosive gas atmosphere to a concentration well below the lower explosive limit.

Purification　过滤(净化)

Purified lubricating oil tank　滑油净油柜　贮存经净化处理后的清净润滑油的柜。

Purifier　分水机构　分离油中水分的机构。

Purifier　净化器(过滤器)

Purifier hot water tank　油分离机热水柜　贮存油分离机分离筒水封用热水的柜。

Purifier operating water tank　油分离机工作水柜　贮存油分离机工作用水的柜。

Purifing water device　净水装置

Purity　纯净度

Purpose built container space　集装箱区域　系指为了安全承载集装箱而设有导轨架的货物区域。　Purpose built container space is a cargo spaces fitted with cell guides for stowage securing of containers.

Purse seiner　围网渔船

Purse winch　围网括纲绞机　围网作业中用于绞收围网括纲的绞机。

Push and towing arrangement　推-拖设备　船上供推、拖其他船舶、浮物或供承受推-拖的所有专用结构、部件和机械的统称。

Push boat/Pusher　推船　系指艏部装有顶推装置,供顶推其他船用的船舶。通常在艉部还装有拖曳装置,既能顶推,又能拖曳。

Push due to surface slope　江面坡度推力　船舶在有坡度的江、河面顺水航行时,船体重力沿江、河面的分力。

Pushed beam 承推梁 设置于供顶推的驳船尾端,专用于传递推力和防撞的水平或竖向的构件。

Pusher 顶推船 系指专门用于顶推非自航货船的船舶。与拖船相比,顶推运输时驳船在前,推船在后,整个船队有较好的机动性,阻力减小,航速提高,不再需要驳船上的舵设备和操舵人员,从而降低了运输成本。

Pusher tug-barge combination 顶推船-驳船组合体 系指由一艘顶推船和一艘驳船组成具有下列特征的船队:(1)顶推船借助于一种除缆绳、链索或其他索具以外的铰接式机械连接装置,将其锁紧在驳船尾部的凹槽内;(2)铰接式连接装置允许顶推船和驳船之间有一个纵摇的自由度,即:顶推船/驳船能绕铰接连接装置的轴线(横向轴)转动,并可在组合体情况下,从顶推船上遥控和就地操纵连接装置相连或脱开。在整个营运航程中,顶推船与驳船始终保持所述连接状态;(3)铰接连接装置脱开后,顶推船和驳船可保持独立停泊或各自作业。见图 P-11。 Pusher tug-barge combination connected in articulated manner is a combination of a pusher tug and a barge with the following characteristics: (1) the pusher tug is secured in the barge notch by mechanical means of any type whatsoever other than just wire ropes, chains, lines or other tackles; (2) the articulated connector allows pitch between the pusher tug and the barge in one degree of freedom, i. e. rotation of the pusher tug/barge about the (transverse) axis of the connector. The combination is capable of being connected or disconnected remotely from the pusher tug or locally from the barge. (3) The pusher tug and the barge remain so connected throughout the voyage; when the articulated connector is disconnected, both the pusher tug and the barge may remain separately in mooring or operation. See Fig. P-11.

图 P-11 顶推船-驳船组合体
Figure P-11 Pusher tug-barge combination

Pusher/push-boat 推船 艏部装有顶推装置,供顶推其他船用的船舶。通常在艉部还装有拖曳装置,既能推顶,又能拖曳。

Push-in bitt 插入式带缆桩 将桩头插入甲板内,并与之联固的带缆桩。

Push-towing 顶推 推船或以拖带为主的拖船用船首顶着被推船舶或浮物被拖船或浮物一起行驶的一种拖带作业。

Push-towing steering rope 操纵绳 顶推系结部件中,起主缆作用,又可收放,以摆动推船,增加顶推船队回转性能的系缆。

P_w (kN/m^2) 波浪压力或动压力。 P_w is the wave pressure or dynamic pressures (in kN/m^2).

P_{wF} 进水工况下的波浪压力。 P_{wF} is the wave pressure in flooded conditions (kN/m^2).

Pylon 桨塔 全垫升气垫船上空气螺旋桨的 Z 形传动轴系中竖立于甲板上的部分及其外罩。有固定式及可摇头转向式两种。

Pyramidal waves 三角浪 常由于风向突然突变或不同方向风浪与涌叠加而形成的角锥形波浪。

Pyrite (containing copper and iron) 黄铁矿(含铜和铁) 二硫化铁,含铜和铁。含水量0%～7%。粉尘极多。无特殊危害。该货物为不燃物或失火风险低。 Iron disulphide, containing copper and iron. Moisture 0% to 7%. It is extremely dusty. No special hazards. This cargo is non-combustible or has a low fire-risk.

Pyrites, calcined (calcined pyrites) 黄铁矿,经过煅烧(煅烧黄铁矿) 呈粉尘至微粒状,是化学工业的副产品。各类金属硫化物在化学工业中用于生产硫酸,或加工提取金属元素——如铜、铅、锌等。该副产品有相当大的酸性,特别在水或潮湿的空气中,此时的pH 值为 1.3～2.1。潮湿时对钢有极强的腐蚀作用。吸入粉尘对人体有刺激性并有害。货物可能流态化。该货物为不燃物或失火风险低。 Dust to fines, Calcined Pyrites is the residual product from the chemical industry where all types of metal sulphides are either used for the production of sulphuric acid or are processed to recover the elemental metals-copper, lead, zinc, etc. The acidity of the residue can be considerable, in particular, in the presence of water or moist air, where pH values between 1.3 and 2.1 are frequently noted. Highly corrosive to steel when wet. Inhalation of dust is irritating and harmful. Cargo may liquefy. This cargo is non-combustible or has a low fire-risk.

Pyrophyllite 叶蜡石 天然水合的硅酸铝。白垩色,可能多粉尘。团块:75%;碎石:20%;颗粒:5%。无特殊危害。该货物为不燃物或失火风险低。 Pyrophyllite is a natural hydrous aluminum silicate. It is chalk-white in colour. It may be dusty. Lumps：75%，rubble：20%，fines：5%. It has no special hazards. This cargo is non-combustible or has a low fire-risk.

Pyrotechnical light 信号烟火 供船舶遇难时使用，能发生烟气、声光，作为求援、呼救等信号用的各种烟火的统称。

Pyrotechnical light chest 信号烟火箱 供船上存放信号烟火用的专用箱。

Pyrotechnics lockers 烟火室 其设计与建造应符合适用的小弹药库、弹药室或弹药箱要求。 Pyrotechnics lockers are to comply with the requirements for small magazines, magazine lockers or boxes as appropriate.

Q

QA = Quality Assurance 质量保证
QC = Quality Control 质量管理
QCP = Quality Control Plan 质量管理计划

QQ mailbox QQ邮箱 系指腾讯公司2002年推出，向用户提供安全、稳定、快速、便捷电子邮件服务的邮箱产品，已为超过1亿的邮箱用户提供免费和增值邮箱服务。QQ邮件服务以高速电信骨干网为强大后盾，拥有独立的境外邮件出口链路，免受境内、外网络瓶颈影响，全球传信。采用高容错性的内部服务器架构，确保任何故障都不影响用户的使用，随时随地稳定登录邮箱，收发邮件通畅无阻。QQ邮箱一些细节方面做得很好，例如可以显示发信进度，还有随着节日更换的Logo等等。每次QQ邮箱的更新都会给人耳目一新的感觉。QQ邮箱开始改变是在2006年。在2005年末QQ邮箱推出了3.0Beta版，获得内部测试资格后容量由5M直接升级到了1G,附件增加到20M。3.0Beta版的界面风格还是很漂亮的。

Q_{sw} 所考虑船体横剖面处的设计静水剪力。 Q_{sw} is the design still water shear force, in kN, at the hull transverse section considered.

Quadrant 舵扇 装在舵头上的带导槽或带齿的扇形舵柄。

Quadrant gear 扇形齿轮

Quadrantal davit 弧齿型吊艇架 摇动手柄，借助螺杆带动使吊艇臂沿下端的弧齿齿牙倒出，以吊放小艇出舷的摇倒式吊艇架。

Quadrature spectrum 正交谱 互谱中，表示两个随机函数中有90°相位差的组成分量间的相对能量大小的虚数部分。其表达式为 $Q_{1,2}(\omega) = (1/2\pi)\int_{-\infty}^{+\infty} R_{1,2}(\tau)\sin\omega\tau d\omega$。

Quadrennial through survey (QTS) 年度彻底检验 系指一年一度的肉眼检验，而检验是在许可的情况下尽量详细进行，以对所检验部分的安全程度得出可靠的结论。

Qualitative failure analysis 定性的故障分析 系指应根据单一故障基准而不是按同时发生的两个独立的故障进行的分析。 Qualitative failure analysis should be based on single failure criteria, not two independent failures occurring simultaneously.

Quality 质量 是表示标的的内在质量和外观形态的综合。

Quality of Service (QoS) 服务质量 系指一个网络能够利用各种基础技术，为指定的网络通信提供更好的服务能力，是网络的一种安全机制，是用来解决网络延迟和阻塞等问题的一种技术。在正常情况下，如果网络只用于特定的无时间限制的应用系统，并不需要QoS，比如Web应用，或E-mail设置等。但是对关键应用和多媒体应用就十分必要。当网络过载或拥塞时，QoS能确保重要业务量不受延迟或丢弃，同时保证网络的高效运行。在RFC 3644上有对QoS的说明。

Quality policy 质量方针 对质量管理体系而言，系指制造厂商应制定管理方针或质量目标并保证在工作的所有阶段中，都贯彻和坚持这些方针和目标。 For quality management system, quality policy means that the manufacturer is to define management policies and objectives or quality and ensure that these policies and objectives are implemented and maintained throughout all phases of the work.

Quality system documentation 质量体系文件 对质量管理体系而言，系指制造厂商应建立并执行质量管理制度，使之能保证材料或服务能与规定的要求相一致，包括有关的规定。 For quality management system, quality system documentation means that the manufacturer is to establish and maintain a documented quality system capable of ensuring that material or services conform to the specified requirements, including the relevant requirements.

Quantitative easing (QE) 量化宽松 系指中央银

行在实行零利率或近似零利率政策后,通过 购买国债等中长期债券,增加基础货币供应,向市场注入大量流动性的干预方式。其中,"量化"指的是扩大一定数量货币的发行;"宽松"则是减少银行储蓄必须投资的压力。当银行和金融机构的有价证券被央行收购,新发行的钱币便被成功地投入私有银行体系。与利率杠杆等传统工具不同,宽松被视为一种非常规的工具。比较央行在公开市场中对短期政府债券所进行的日常交易,量化宽松政策所涉及的政府债券,不仅金额要庞大得多,而且周期较长。一般来说,只有在利率等常规工具不再有效的情况下,货币当局才会采取这样极端的做法。

Quantity 数量 系指对货物的计量。

Quantized analysis 量化分析 就是将一些不具体、模糊的因素用具体的数据来表示,从而达到分析比较的目的。

Quantum 量子 是一个物理量的最小单位,光子、电子都是量子的一种。在微观世界中,量子有许多神奇的特性。在传统的计算机中,每个电子元件只能表示"1"或"0",即"开"或"关"两种状态,而量子则可以同时表示多种状态,这就意味着很少的量子比特就能实现大量经典比特才能完成的计算。"如果一台量子计算机的单次运算速度达到目前民用电脑CPU的级别,那么一台64位量子计算机的速度将是目前世界上最快的'天河二号'超级计算机(每秒33.86千万亿次)的545万亿倍"。量子信息技术利用光子传递的量子密码则能保证信息传输的"无条件安全"。光子非常脆弱,一经测量就遭受破坏,因此,只有通信双方利用光子这一特点生成的密码,就有了一种"魔力",只要第三方看一眼就会消失。量子信息技术在提高运算速度和信息安全方面都能突破现有信息技术的瓶颈。对量子的研究,在过去的100年已经成为现代科技的支柱,发展出了诸如激光、核磁共振、高温超导、巨磁阻等科学进展。可以期待量子研究在未来将会给人们带来更多的惊喜。也将给人们的生活带来巨大的改变。如:(1)"绝对安全"的通信方式和"只有你知我知"的密码保护。量子保密通信是唯一被严格证明无条件安全的通信方式,可以从根本上解决国防、金融、政府、商业等领域的信息安全问题,并将使互联网变得更加安全,例如用户的搜索、支付等私密信息将无法被复制窃取。未来不用再担心需要保密的信息在传输过程中被窃取和破译。(2)借助强大的运算能力改造传统行业。量子计算机能够大规模、多线程处理数据,所建立的数据模型精度将有指数级的提升。比如可以把某种病毒体内的各种成分分析得清清楚楚,可以制造针对性很强的药物,还可以在大量数据中分析筛选出特定性,例如通过罪犯身上的某个特点,利用量子计算机就能很快地锁定目标,不用在茫茫大海中筛选和搜寻。我国在量子通信研究和光量子计算机方面都处于世界领先地位,已经实现"弯道超车"。由我国自主研发的世界首颗"量子卫星"即将发射,将与地面上的光学实验室进行"互动"。结合将于2016年开通的通信保密干线"京沪干线",我国将有望初步构建广域量子通信体系。但是,量子计算机多项研究仍处于实验室阶段。目前面临的最大瓶颈是:科学家还没有找到足够抗干扰的系统来实现实用性的量子计算机。因此,量子信息技术还要加大投入,其研究任务重而道远。

Quantum bit (Qbit) 量子比特 从物理学的角度,人们习惯于根据量子态的特性称为量子比特(Qubit 或 Qbit)、纠缠比特(Ebit)、三重比特(Tribit)、多重比特(Multibit)和经典比特(Cbit)等等。这种方式让人眼花缭乱,并且对量子比特的描述要根据具体的物理特性来描述。

Quantum computer 量子计算机 是一种全新的基于量子理论的计算机,遵循量子力学规律进行高速数学和逻辑运算、存储及处理量子信息的物理装置。量子计算机的概念源于对可逆计算机的研究。量子计算机应用的是量子比特,可以同时处在多个状态,而不像传统计算机那样只能处于0或1的二进制状态。量子计算机主要运用在做量子系统的模拟、线性方程组量子算法以及单原子量子信息存储等科学领域。

Quantum information technology (QIT) 量子信息技术

Quantum memory 量子存储器 世界上首个存储单光子存储器在中国诞生。

Quantum of foreign trade 对外贸易量 系指以货币所表示的对外贸易值经常受到价格变动的影响,因而不能准确地反映一国对外贸易的实际规模,更不能使不同时期的对外贸易值直接比较。为了反映进出口贸易的实际规模,通常以贸易指数表示,其办法是按一定期的不变价格为标准来计算各个时期的贸易值,用进出口价格指数除进出口值,得出按不变价格计算的贸易值,便剔除了价格变动因素,就是贸易量。然后,以一定时期为基期的贸易量指数同各个时期的贸易量指数相比较,就可以得出比较准确反映贸易实际规模变动的贸易指数。贸易量原意是指用数量、质量、长度、面积、容积等计量单位表示的反映一定时期内贸易规模的指标。就一种商品来说,用计量单位表示是十分容易的。但是,由于参加贸易的商品种类繁多,计量单位的标准各不相同,价值有大有小,差别很大,无法统一衡量,用计量单位来统计对外贸易或国际贸易的规模是不现实的。因此,为了反映贸易实际规模的发展变化,只能剔除价格变动的影响,以一定时期的不变价格来计算贸易量,以达到不同时期的可比较性。

Quantum physics 量子物理学 是为描述远离人们的日常生活经验的抽象原子世界而创立的,但它对日常生活的影响无比巨大。没有量子力学作为工具,就不可能有化学、生物、医学以及其他每一个关键学科的引人入胜的进展。没有量子力学就没有全球经济可言,因为作为量子力学的产物的电子学革命将人们带入了计算机时代。同时,光子学的革命也将人们带入信息时代。量子物理的杰作改变了人们的世界,科学革命为这个世界带来了福音,也带来了潜在的威胁。中科大测出量子纠缠速度下限,标志中国在"绝对保密"的量子通信这个未来战略性领域继续领跑全球。

Quantum private communication technology 量子保密通信技术

Quarantine 检疫 是以法律为依据,由国家授权的特定机关对有关生物及其产品和其他相关商品实施科学检验鉴定与处理,以防止有害生物在国内蔓延和扩散。

Quarantine boat/ vessel 检疫艇/船 专用于接送检疫人员的艇/船。

Quarter rudder 边舵 在多舵的船上,设置在艉部两侧的舵。

Quartering sea 艉斜浪 在左、右舷105°～165°之间方向内与船舶遭遇的波浪。

Quartz 石英 结晶块。无特殊危害。该货物为不燃物或失火风险低。 Crystalline lumps. No special hazards. This cargo is non-combustible or has a low fire-risk.

Quartzite 石英岩 是一种坚实的颗粒状变质砂岩,含有石英。呈白色、红色、褐色或灰色,尺寸大小不一,有大的岩石也有卵石。可按半碎的状态或分级尺寸进行装运。无特殊危害。该货物粉尘的腐蚀性很强。该货物为不燃物或失火风险低。 Quartzite is a compact, granular, metamorphosed sandstone containing quartz. It is white, red, brown or grey in colour and its size varies from large rocks to pebbles. It may also be shipped in semi-crushed and graded sizes. No special hazards. Dust of this cargo is very abrasive. This cargo is non-combustible or has a low fire-risk.

Quasi-steady cavities 准定常空泡 附着于物体,其外形就时间上平均来说是稳定的空泡。

Quats level 配额水平 通常以协定缔结前一年的实际出口量,或以原协定最后一年的配额为基础进行协商,确定新协定第一年数额,然后再确定其他年的年增率。

Quay mooring 码头系泊

Quay/Wharf 码头 系指船舶停靠、人员上下、物资补给等的场所。

Quenching and tempering QT 淬火-回火 QT
(1)是一种热处理过程将钢材加热至超过 A_{c3} 的适宜温度,然后在适宜的冷却剂中进行冷却,使钢材中的微观结构硬化,接着进行回火,其过程是将钢材再次加热至不高于 A_{c1} 的适宜温度,以改善其微观结构而恢复其韧性。(2)(包括淬火)系指大于 A_{c3} 的适当的温度下加热后,为了硬化钢材的精细组织,利用适当的冷却剂进行冷却的热处理过程。淬火后加的回火系指改进钢材的精细组织,为恢复韧性,在小于 A_{c1} 的适当温度下进行再加热的热处理工艺。 Quenching and tempering QT is: (1) a heat treatment process in which steel is heated to an appropriate temperature above the A_{c3} and then cooled with an appropriate coolant for the purpose of hardening the microstructure, followed by tempering, a process in which the steel is re-heated to an appropriate temperature, not higher than the A_{c1} to restore the toughness properties by improving the microstructure; (2) quenching involves a heat treatment process in which steel is heated to an appropriate temperature above A_{c3}, and then cooling with an appropriate cooling for the purpose of hardening the microstructure. Tempering subsequent to quenching is a process in which the steel is reheated to an appropriate temperature not higher than the A_{c1} to restore toughness properties by improving the microstructure.

Questionnaires 问题单

Quick action valve 速动阀

Quick adjusting 快速调节

Quick backup 快速备份 就是将CF卡上所有的资料全部传输至硬盘中。

Quick closing emergency valve 速闭应急阀

Quick closing hatch cover 速闭舱盖 设有内外操作的联动把手,能迅速启闭的舱口盖。

Quick closing valve 速闭阀(快关阀)

Quick freezing 速冻(快速冷冻) 在规定的时间内将货品的温度迅速降到指定温度的过程。

Quick freezing capacity 速冻能力

Quick freezing tank 速冻舱 能将渔获物或其他食品迅速冻结成冻品的舱室。

Quick power increasing test 快速升负荷试验 检查船用主汽轮机的空负荷升到各规定工况的最短时间的试验。

Quick release valve 速放阀

Quick venting valve 高速透气阀

Quick-closing controlling gear for sliding door 水密滑门速闭装置 在应急时可远距离操纵,一般依靠动力传动的快速关闭水密滑门的装置。

Quick-closing hatchcover 速闭舱盖 设有内外操作的联动把手,能迅速启闭的舱口盖。

Quick-draining recess 快速泄水凹体 符合 ISO 11812 中"快速泄水艇舱和凹体"之所有要求的凹体。注意:按照其特性,对某一设计类别,艇舱可考虑为快速泄水型,但对更高的设计类别,也许不考虑为快速泄水型。 Protective cover is the recess fulfilling all the requirements of ISO 11812 for "quick-draining cockpits and recesses". Note: According to its characteristics, a cockpit may be considered to be quick-draining for one design category, but may be not for a higher one.

Quickly closing valve (emergency shut down valve) 速闭阀 安装在机舱内的燃油舱、滑油舱等可燃性液体的储存出口处,供火灾发生时紧急关闭这些舱柜内的阀门。速闭阀分为可以是气动的,也可以靠钢丝人工拉动两种。其控制场所要设置在机舱棚外。

Quick-reversing test 快速换向试验 检查船用主汽轮机在全负荷下快速换向的最短时间的试验。

Quiet submarine 安静型潜艇

Quill shaft (torque shaft) 联动轴 起一定挠性作用并将主机功率通过空心小齿轮传递出去的轴。

Quitrent 免疫税

Quota 配额/定额 系指:(1)一国政府在一定时期内对某些敏感商品的进口或出口进行数量或金额上的控制,其目的旨在调整国际收支和保护国内工农业生产,是非关税壁垒措施之一;(2)在合理的劳动组织和合理地使用材料和机械的条件下,预先规定完成单位合格产品所消耗的资源数量的标准,它反映一定时期的社会生产力水平的高低。对于每一个施工项目,都测算出用工量,包括基本工和其他用工。再加上这个项目的材料,包括基本用料和其他材料。对于用工的单价,是当地根据当时不同工种的劳动力价值规定的,材料的价值是根据前期的市场价格制定出来的预算价格。定额的特性,是由定额的性质决定的。

Quota arrangement 配额管理 是国家对外经济贸易活动管理的一种手段,国际间根据惯例和协定对一些商品在进出口数量上进行一定限制的管理。

Quotation offer 报价单 主要用于供应商给客户的报价,类似价格清单。如果是交期很长的物料或是进口的物料,交期(delivery time)和最小订单量(minimum quantity order/MQO)很重要,因为进口的物料需要报关,交期会很久,所以需要注明,交期一般用"周"(week)表示,最小订单量一般用"千个"(kpcs)表示,有特殊情况的需要备注说明,如:12 天系指工作日,需要 7 天报关等。其中,英文报价单和中文报价单略有区别。报价单的格式原则遵循于具体相关事件与法律法规条文。

Quote 报价 系指根据需求部门的要求,通过报价,比价,估价,议价,制作价格分析表提供具有竞争力的价格给业务部门。

Quote offer 证券报价 是证券市场上交易者在某一时间内对某种证券报出的最高进价或最低出价,报价代表了买卖双方所愿意出的最高价格,进价为买者愿买进某种证券所出的价格,出价为卖者愿卖出的价格。

Quotient 商数

Q_{wv} (kN) 所考虑船体横剖面处的波浪垂向剪力 Q_{wv} means the vertical wave shear force, in kN, at the hull transverse section considered.

R

r/min 每分钟转数 r/min is a revolutions per minute.

R1 1 类航区 系指授予离岸的距离不超过 200 海里(夏季/热带)或 100 海里(冬季)航行船舶的附加标志。 Service category 1 (R1) means a service restriction notation assigned the ship which navigates within 200 (summer/tropical) or 100 (winter) n mile off shore.

R2 2 类航区 系指授予离岸的距离不超过 20 海里(夏季/热带)或 10 海里(冬季)航行船舶的附加标志。 Service category 2 (R2) means a service restriction notation assigned the ship which navigates within 20 (summer/tropical) or 10 (winter) n mile off shore.

R3 3 类航区 系指授予遮蔽水域航行船舶的附加标志。 Service category 3 (R3) means a service restriction notation assigned the ship which navigates within sheltered waters.

Racing boat (runner boat) 赛艇 专供竞赛用的小艇。

Rack 齿条(网架,餐具架)

Rack and pinion jacking system 齿条齿轮升降装置

Rack and pinion steering gear 齿条齿轮操舵装置

Racking 扭变 由于水平力导致集装箱的端壁和侧壁的变形。

Radar[1] 测雨雷达 利用雨滴、云层、冰晶、雪花等对电磁波的散射作用来探测大气象中的降水现象的雷达。

Radar[2] 雷达 是 Radio detection and ranging("无线电探测与定位")的缩写。雷达的基本任务是探测感兴趣的目标,测定有关目标的距离、方向、速度等状态参数。雷达主要由天线、发射机、接收机(包括信号处理机)和显示器等部分组成。

Radar[3] 雷达(无线电方向和距离) 用来确定反射物体和发射装置的距离和方位的无线电系统。(Radio direction and ranging) A radio system means a system which allows the determination of distance and direction of reflecting objects and of transmitting device.

Radar anti-stealth technology 雷达反隐身技术 系指使雷达探测、跟踪、定位隐身目标而采用的技术。可通过采取扩展雷达的工作频段、改进雷达的探测性能、发展新技术体制雷达等途径,提高雷达的反隐身能力。

Radar beacon 雷达信标 通过产生雷达信号标明其位置和身份来应答雷达传送的航标。A navigation aid which responds to the radar transmission by generating a radar signal to identify its position and identity.

Radar detection false alarm 雷达探测故障报警 雷达故障报警概率表示杂波将穿过探测阈,并将在只有杂波存在时称为一个目标的概率。 Radar detection false alarm is the probability of a radar false alarm represents the probability that noise will cross the detection threshold and be called a target when only noise is present.

Radar mast 雷达桅 主要供装设雷达天线用的桅杆。

Radar platform 雷达平台 供安装雷达天线用的平台。

Radar plotting 雷达标绘 系指目标检测、跟踪、参数计算和信息显示的全过程。 Radar plotting is the whole process of target detection, tracking, calculating of parameters and display of information.

Radar room 雷达室 供安置雷达发射和接受设备并设置相应操作部位的房间。

Radar target 雷达目标 位置和运动由连续的雷达距离和方位测量确定的任何固定或移动目标。 Any object fixed or moving whose position and motion is determined by successive radar measurements of range and bearing.

Radar target enhancer 雷达目标增强器 一个电子雷达发射器,其输出功率为所收到的限幅外未经处理的雷达脉冲的放大形式。 Radar target enhancer means an electronic radar reflector, the output of which is an amplified version of the received radar pulse without any form of processing except limiting.

Radar technology 雷达技术 在国际上首次实现对车辆等典型人造目标的三维高分辨成像。该项技术在地理遥感和军事侦察领域有很好的应用前景。

Radar-based system 雷达系统 基于雷达信号测量的位置基准系统。该信号由被动应答器(如 Radascan、RADius)反射。

Radial boat davit 转出式吊艇柱 操作时,绕竖轴摆动,用以吊艇出舷的其上部通常形成弯钩的柱子。

Radial clearance 径向间隙

Radial deformation 径向变形

Radial distribution of pitch 径向螺距分布 螺旋桨叶面各半径处螺距不同者,其螺距沿半径方向的变化情况。

Radial type internal combustion engine 星型内燃机 气缸围绕一根曲轴排列成五列或五列以上的内燃机。

Radial-flow compressor(radial compressor) 辐流式压气机 由于在轴向交替布置的静叶和动叶叶列的作用使空气或其他气体压缩并基本上沿径向流动的压气机。

Radial-flow steam turbine 辐流式汽轮机 蒸汽在膨胀过程中是在与汽轮机轴心线相垂直的平面内作径向流动的汽轮机。分离心式和向心式两种。

Radial-flow turbine(radial-flow turbine) 辐流式涡轮 由于交替布置的静叶和动叶叶列的作用使高温燃气或其他气体膨胀并基本上沿径向流动的涡轮。

Radiative forcing(RF) 辐射强迫 是对地球-大气系统失去能量平衡的一种量度。它用静辐射通量表示,单位为 W/m^2,当它取正值时,使地球升温;反之则使其变冷。它是影响地球变暖和全球气候变化最重要的因素。

Radially varying pitch 径向变螺距 螺旋桨叶面各半径处的螺距不尽相同者。

Radiation damping 辐射阻尼 因辐射引起一个发射体系的运动的衰减,是谱线致宽的主要原因之一。经典电动力学理论把发射(或吸收)光的原子当作谐振子,辐射是由激发谐振子的振动产生的。由于辐射,谐振子受到阻尼力的作用,结果辐射出的电磁波的振幅不断衰减,这样就会得到具有一定宽度的谱线。也就是说,系指由于振动系统引起邻近质量的振动,使系统的能量逐渐向四周辐射出去,转变为波动能量的阻尼。

Radiation heat loss 散热损失 从蒸馏装置辐射

散失到外界的热量。

Radiation test（RT） 射线探伤

Radiative forcing（RF）capacity 辐射强迫值 定义为相对于工业化前(1750年)的差值,并以 W/m^2 为单位表述。

Radiator 散热器 供暖用的散发加热工质热量的器具。

Radiator cooling 散热水箱冷却 用风冷式散热水箱作为淡水冷却器的一种闭式冷却方式。

Radiator insulation 散热器隔振垫

Radio communication system 无线电通信系统 其主要任务是完成通信指挥、协同通信、情报警报通信。无线电通信是利用电磁波来传递信息完成通信的一种方式。其特点是:使用灵活、通信距离远、通信速度快,但结构复杂。它主要由发信设备、发信天线、收信天线、收信设备和终端设备等组成。(1)无线电发信设备,它主要有两个任务:一是产生一个具有一定功率的调频振荡;二是完成调制任务(调制就是用从消息转换来的原始电信号去控制等幅高频振荡的某一参数,如振幅、频率、相位的过程)。经过调制的高频振荡信号称为已调波。未经过调制的高频振荡信号称为载波,它起着运载消息的作用。常用的调制方法有:幅度调制,简称调幅;频率调制,简称调频;单边带调制等。无线电发信设备通常由振荡器、中间放大器、调制器、输出放大器等组成。(2)无线电收信设备,其作用:一是选择和放大信号,从众多电台发射出来的电磁波产生的感应电动势中选出需要的信号并加以放大,二是解频,从已调波中把原来代表消息的调制信号检测出来,三是还原,就是把解调出来的低频电依赖通过终端设备还原为原来所要传递的信息。例如把音频电信号通过耳机或喇叭还原为声音。(3)单边带通信:在通信中,被传送的消息包括在两个边带中。而且每一个边带都包含有完整的被传送的消息。因此,对不失真地传送消息来说,只要发送其中一个边带就可以了。这种只发送一个边带,而把载频和另一个边带完全抑制掉不发射的方式,叫单边带通信。(4)通信终端,通信就是传递信息的过程,虽然通信系统的主要设备是收发信机,但在实际进行通信时,双方只有收发信设备还不行,在发信端还必须还有某种设备把语言、文字、图像等原始信号变成电信号,输送给发信机发送出去。在接收端,再把收信机接收到的电信号经某种设备还原为原来的信号、文字、图像等信号。这两种位于发收信过程两端的设备,称为通信终端。无线电通信的终端设备,除简单的电键、话筒、耳机、喇叭外,目前还普遍使用电传打字机、快速收发终端机、超快速收发终端机、语言保密机和汉字终端等。(5)船用天线:天线是无线通信来发射或接收电磁波的。

Radio communication workstation 无线电通信工作站 用于控制遇险,安全和一般通信的外部通信工作站。Workstation for external communication distress, safety and general communication.

Radio detecting and ranging（RADAR） 雷达 利用超高频无线电测定目标方位及距离的探测装置。

Radio directions finding 无线电测向 用无线电测向仪测量无线电台的方位角。测向仪可固定在地面,也可以安装在船上。而被测方位的发射台可安装在船上,也可固定在地面。被测方位的发射台固定在地面作为导航台使用时,则称为无线电信标或无线电导向台。无线电测向的精度低,作用距离短,因此国外一般已不使用这种方法导航。

Radio duties 无线电职责 包括(如为适当时)根据"无线电规则""国际海上人命安全公约"及由主管机关自行决定的有关国际海事组织建议案中的值班和技术保养及修理。

Radio frequency identification（RFID） 射频识别 是一种非接触式的自动识别技术,它通过射频信号自动识别目标对象并获取相关数据,识别工作无须人工干预,作为条形码的无线版本,RFID技术具有条形码所不具备的防水、防磁、耐高温、使用寿命长、读取距离大、标签上数据可以加密、存储数据容量更大、存储信息更改自如等优点,其应用将给零售、物流等产业带来革命性变化。是一种通信技术,可通过无线电信号识别特定目标并读写相关数据,而无须识别系统与特定目标之间建立机械或光学接触。

Radio frequency modulator 射频调制器 简称调制器,被分为邻频调制器和隔频调制器,是前端电视机房的主要设备之一。邻频调制器,采用4路调制一体设计,将4台邻频调制器整合为一台,充分节省机房机柜空间,方便安装调试。酒店宾馆使用的4路调制器按专业级的固定频道邻频调制器设计,中频则采用声表滤波器处理,有效地保证了残留边带特性,使之适用于邻频传输系统。邻频调制器采用无源、高选择性单频道滤波器,有效地保证了调制器良好的寄生输出抑制性能。

Radio navigation 无线电导航 最基本任务是为船舶的航行安全提供其所在的地理位置,有的无线电导航仪还能提供航速与航向。它是一种有别于陀螺导航的导航方式。陀螺导航是自主式的,即它不依靠于任何外部信息,而无线电导航则必须依赖外部提供的无线电信号。无线电导航系统的工作与传播条件有关,它易受人为的或天然的无线电干扰。但其具有不受气象条件和时间的限制,启动时间极短,在工作距离上也可不受视距的限制,自动化程度较高,安装方便且占用空间少,耗电省、操作简单、定期精度高等。目前无线电导航仪中占主导地位的是罗兰 C 导航系统和 GPS 导航星全球定位系统。

Radio officer 电报员 系指持有一级或二级无线电报员证书或有按"无线电规则"规定颁发的水上移动业务无线电通信报务员一般证书的人,他在"国际海上人命安全公约"所要求的船上无线电台工作。 Radio officer means a person holding a first class or second class radiotelegraph operator's certificate or a radio-communication operator's general certificate for the maritime mobile service issued under provision of the Radio regulations, who is employed to the radiotelegraph station of a ship which is required to have such a station by the international convention for the safety of life at sea.

Radio officer 无线电报务主任 系指持有符合"无线电规则"规定的至少为水上活动业务无线电通信报务员一般证书,或一级或二级无线电报务员证书,并从事于符合有关规定的船舶无线电台工作的人员。 Radio officer means a person holding at least a radio-communication operator's general certificate for the maritime mobile service or a first or second class radiotelegraph operator's certificate complying with the Radio Regulations, who is employed in the radiotelegraph station of a vessel which is provided with such a station in compliance with relevant regulations.

Radio operator 无线电报务员 系指持有符合"无线电规则"规定的无线电报务员专用证书的人员。 Radio operator means a person holding a radiotelegraph operator's special certificate complying with the Radio Regulations.

Radio personnel 无线电人员 系指主管机关满意的,能胜任遇险和安全无线电通信的人员。这些人员应持有无线电规则中规定的适用的证书,在遇险时,应指定其中任何人员担负起无线电通信的主要责任。

Radio position fixing 无线电定位法 在陆地(或外层空间)建立若干个导航台;利用电磁波传播特性在测出其电气参数后,从而测定船舶相对于导航台的几何位置的方法。这种定位法广泛应用于海洋导航与定位。无线电定位系统根据位置线的形式不同有各种系统,如测向系统、测距系统(圆-圆)、测向-测距系统、双曲线系统、椭圆-双曲线系统、圆-双曲线系统等。

Radio regulations 无线电规则 系指任何时候实施的最新"国际电信公约"所附或视为其附件的"无线电规则"。 Radio regulations means the Radio regulations annexed to, or regarded as being annexed to, the most recent International Telecommunication Convention which is in force at any time.

Radio relay buoy 海上无线电中继浮筒

Radio room 报务室 供安置无线电通信设备,进行对外通信联络用的房间。

Radio waves laboratory 无线电波实验室 应用甚高频、高频和甚低频无线电发射机和接收机,通过接收回波信号强度,研究极地上空电介特性,为无线电通信提供资料的工作室。

Radio wind room 无线电测风室 又称测风雷达室,应用测风雷达,对高空风进行监测来获取高空风在三维空间变化的工作室。观测时可应用测风雷达独立监测,也可和无线电探空仪配合观测。

Radioactive material surface contaminated objects (SCO-1), non-fissile or fissile-excepted UN 2913 非裂变的或预计裂变的表面受到放射性物质污染的物体(SCO-1) SCO-1 的放射毒性低。包括非放射性物质的固态物体,其表面分布有放射性物质:(1)可接近表面的非固定污染按 300 cm^2(表面如小于 300 cm^2 则取其面积)平均值计,对 β 和 γ 辐射源及低毒性 α 辐射源不超过 4 Bq/cm^2,或对所有其他 α 辐射源不超过 0.4 Bq/cm^2;(2)可接近表面的固定污染按 300 cm^2(表面如小于 300 cm^2 则取其面积)平均值计,对 β 和 γ 辐射源及低毒性 α 辐射源不超过 4×10^4 Bq/cm^2,或对所有其他 α 辐射源不超 4×10^4 Bq/cm^2;(3)不可接近表面的非表面的固定污染按 300 cm^2(表面如小于 300 cm^2 则取其面积)平均值计,对 β 和 γ 辐射源及低毒性 α 辐射源不超过 4×10^4 Bq/cm^2,或对所有其他 α 辐射源不超 4×10^4 Bq/cm^2。低放射性。该货物为不燃物或失火风险低。 The radioactivity of SCO-1 is low. This schedule includes solid objects of non-radioactive material having a radioactive material distributed on its surfaces which: (1) the non-fixed contamination on the accessible surface, averaged over 300 cm^2 (or the area of the surface if less than 300 cm^2), does not exceed 4 Bq/cm^2 for beta and gamma emitters and low-toxicity alpha emitter, or 0.4 Bq/cm^2 for all other alpha emitters; (2) the fixed contamination on the accessible surface, averaged over 300 cm^2 (or the area of the surface if less than 300 cm^2), does not exceed $4\times10^4 Bq/cm^2$ for beta and gamma emitters and low-toxicity alpha emitters, or $4\times10^3 Bq/cm^2$ for all other alpha emitters; and (3) the non-fixed contamination plus the fixed contamination on the inaccessible surface, averaged over 300 cm^2 (or the area of the surface if less than 300 cm^2), does not exceed 4×10^4 Bq/cm^2 for beta and gamma emitters and low-toxicity alpha emitters, or 4×10^3 Bq/cm^2 for all other alpha emitters. It is low radioactivity. This cargo is non-combustible or has a low fire-risk.

Radioactive material, low specific activity (LSA-1) non-fissile or fissile-excepted (UN 2912) 低比活度(LSA-1)非裂变的或预计裂变的放射性物质　包括含有天然生成放射性核素(如铀、钍)的矿石以及这类矿石(包括金属、混合物、化合物)的天然或贫化铀和钍精矿。放射毒性低。有些物质可能有化学危害。该货物为不燃物或失火风险低。 This schedule includes ores containing naturally occurring radionuclides (e.g., uranium, thorium) and natural or depleted uranium and thorium concentrates of such ores, including metals, mixtures and compounds. It is low radiotoxicity. Some materials may possess chemical hazards. This cargo is non-combustible or has a low fire-risk.

Radioactive materials (Class 7 dangerous materials) 放射性物质(第7类危险货物)　系指所含放射性核素使托运货物的放射性浓度和总放射性均超过 IMDG 规则 2.7.7.2.1 至 2.7.7.2.6 所规定值的任何物质。 Radioactive materials mean the materials in this class are any materials containing radionuclides where both the activity concentration and the total activity in the consignment exceed the values specified in 2.7.7.2.1 to 2.7.7.2.6 of the IMDG Code.

Radioactive products or waste 放射性产物或废料　系指伴随核动力船舶使用核燃料产生的,通过中子辐照而具有放射性的任何物质,包括核燃料。 Radioactive products or waste means any material, including nuclear fuel, made radioactive by neutron irradiation incidental to the utilization of nuclear fuel in a nuclear ship.

Radioposition reference 无线电位置基准　通过大气传输无线电波的任一种位置基准,如 Microfix, Syledis。

Radiosonde detecting room 探空仪检测室　又称探空仪检定室。为确保探空仪具有一定的精度,在施放前利用仪器对其温度、湿度和气压感应器分别进行灵敏度和基点检查。并对气压和湿度感应器不合格者就地检查,绘制检定曲线的工作舱室。

Radiosotopic laboratory 放射性同位素实验室　用于放射性示踪和放射性污染样品(海区的大气样品、生物和地质样品等)的处理、分析和测定,研究其放射性污染水平和放射性核素分布和变化的工作舱室。该舱室在建造、使用和管理上,应按国家颁发的《放射防护规定》执行。

Radiotelegraph auto alarm 无线电报自动报警器　系指业经认可并能响应无线电报警信号的自动报警接收设备。 Radiotelegraph auto alarm means an approved automatic alarm receiving apparatus which responds to the radiotelegraph alarm signal.

Radiotelephone operator 无线电报务员/无线电话务员　系指持有按"无线电规则"规定颁发的适当的证书者。 Radiotelephone operator means a person holding an appropriate certificate issued under the provisions of the Radio regulations.

Radius of blade circle 叶元半径　平旋推进器各叶轴线所在圆周半径。

Radius of leading edge 导边半径　螺旋桨桨叶在某一半径处切面导边的曲率半径。常以叶切面厚度或螺旋桨半径的百分比表示。

Radius of propeller 螺旋桨半径　螺旋桨桨叶梢至其轴线的垂直距离。

Radius of trailing edge 随边半径　螺旋桨桨叶在某一半径处切面随边的曲率半径。常以叶切面厚度或螺旋桨半径的百分比表示。

Raft davit 吊筏架　用以承吊可吊救生筏的,通常为一带有支座的移动吊杆装置。

Raft/float 筏　用竹、木、浮囊等连成的简易水上运载、作业的工具。

Rag 碎布

Rail stanchion (handrail stanchion) 栏杆柱　支承扶栏和横栏的直立短柱。

Rail stanchion for feathering tread 可转栏杆柱　舷梯上能随梯架斜度改变而调整本身位置的栏杆柱。

Rail stanchion spacing 栏柱间距　相邻两栏杆柱之间的距离。

Rail stay 栏杆撑座　支撑栏杆柱的斜向撑杆。

Rail waybill 铁路运单　分为国际铁路联运和通往港澳的国内铁路运输,分别使用国际铁路货物联运单和承运货物收据。当通过国际铁路办理货物运输时,在发运站由承运人加盖日戳签发的运单叫"铁路运单"(Railway bill)。铁路运单是由铁路运输承运人签发的货运单据,是收、发货人同铁路之间的运输契约。

Railing 栏杆　装置在甲板、平台或顶棚边缘,由栏杆柱、扶栏和横栏或链索等组成的船上安全防护属具。

Railing ladder 舷外挂梯　临时挂在舷墙或栏杆处,供人员登、离船用的短梯。

Railing spacing 横栏间距　上下相邻两横栏之间的距离。

Railway transport 铁路运输　系指利用铁路进行出口货物运输的一种方式。铁路运输一般不受气候条件的影响,可保障全年的正常运输,而且运量大,速度快,有高度的连续性,运转过程中可能遭受的风险小。

Railway type horizontal berth 轨道平移式船台

Rain stanchion (handrail stanchion) 栏杆柱　支

承扶栏和横栏的直立短柱。

Rain-wing craft（aerofoil boat） 冲翼艇 利用安装在船体上的机翼贴近水面或地面飞行时所产生的表面效应升力支持艇重,能在水面航行或腾空低飞的艇。

Raised deck 升高甲板 系指下陷的上层建筑甲板,其下方没有甲板。Raised deck is the sunken superstructure deck below which no deck is provided.

Raised foredeck ship 艏升高甲板船 上甲板艏部作台阶形升高的船舶。

Raised helicopter port 升高的直升机港 位于高出陆地建筑物上的直升机港。

Raised manhole cover 升高人孔盖 用于周缘设有升高围板的开口上的人孔盖。

Raised quarter deck ship 艉升高甲板船 上甲板尾部作台阶形升高的船舶。

Raised quarterdeck 后升高甲板 是自艉垂线向前延伸的上层建筑。一般而言,其高度小于标准上层建筑高度,并有完整的前舱壁(非开启式舷窗装有带有效风暴盖,人孔盖用螺栓固定)。如果前舱壁因设有门和通道开口而不是完整的,则该上层建筑应视为艉楼。Raised quarterdeck is a superstructure which extends forward from the after perpendicular, generally has a height less than a normal superstructure, and has an intact front bulkhead (side scuttles of the non-opening type fitted with efficient deadlights and bolted man-hole covers). Where the forward bulkhead is not intact due to doors and access openings, the superstructure is then to be considered as a poop.

Rake 桨叶倾斜 系指桨叶母线叶梢与桨叶顶部与根部的延长线和螺旋桨轴线相交在同一点的与螺旋桨轴线正交的垂线间的水平距离,后倾认为是正值,而前倾则认为是负值。Rake is the horizontal distance between the line connecting the blade up to the blade root and the vertical line crossing the propeller axis in the same point where the prolongation of the first line crosses it, taken in correspondence of the blade up. Aft rakes are considered positive, fore rakes are considered negative.

Rake 纵斜 螺旋桨叶母线不垂直于轴线时,母线与螺旋桨平面间的距离。一般系指叶梢处距离。向后倾斜者为正。

Rake angle 倾斜角(纵斜角) 是叶片生成线上某一点的切线方向和同一点处垂直线之间任一点的夹角。如果叶片生成线直的,那么只有一个倾斜角;如果为曲线,那么将有无数个叶片倾斜角。Rake angle is the angle at any point between the tangent to the generating line of the blade at that point and a vertical line passing at the same point. If the blade generating line is straight, there is only one rake angle; if it is curved there are an infinite number of rake angles.

Rake of blade section 叶面纵斜 螺旋桨桨叶在某半径处切面的纵斜。$i = r\tan\theta$,式中,r—叶切面所在半径;θ—螺旋桨纵斜角。

Rake of keel（designed drag, keel drag） 龙骨设计斜度 系指龙骨呈直线时与设计水线的纵向不平行度。其值以龙骨线的延伸线与艏艉垂线的两交点距设计水线的高度差表示。

Rake ratio 纵斜比 螺旋桨纵斜对其直径的比值。

Raked bow 前倾型艏 艏柱侧影呈直线前倾或接近直线前倾形式的船首部。

Ramjet engine 冲压喷气式发动机

Ramp 坡道 包括艉坡道(stern ramp)和侧坡道(side ramp)两种。

Ramp door 舌门 装在艏门或艉门之内,翻开后可作为跳板,收起后可作为保证船舶密性的一道水密门。

Ramp-type pier 斜坡式码头 系指呈斜坡形的码头。斜坡延伸至水下,且适应大的潮差,但只适用于吃水浅的小型渡船。

Ram-type hydraulic steering gear 柱塞式液压舵机 采用柱塞式液压缸的液压舵机。

Random error 随机误差 系指测量系统在测量过程中,凡是不能精确估算的因素引起的随机性误差,即在同一条件下获得的观测值,其大小及符号是个随机变量。各种形式的噪声干扰引起的误差属于随机误差。

Random function 随机函数 随所含随机自变量变化的函数。

Random variable 随机变量 在固定条件下每个观测值带有偶然性,不能从其过去历程预定的变量。

Range light 后桅灯 设置在船舶后桅上的桅灯。

Range of repairs 修理范围 系指船舶修理工程中的全部项目。

Ranging system 测距系统 是一种在船上用测量其相对于已知地面站的距离来定位的系统。测量距离的手段有双程测距和单程测距系统两种,雷达定位属于前者,后者需在船上装有高精度频标,在原子钟的精度日益提高的情况下,到20世纪70年代末期以来已逐步使用单程测距系统定位。在船上测到两个地面站的距离,便可得到两条圆位置线,两者交点即为船位,故这种系统又称"圆-圆系统"(rang-rang system)。

Rapid shutdown of ventilation system 通风系统迅速关闭装置 系指分离具备能够一次性切断通风装置的阀门或者是与其切断速度相同的关闭装置。如果在能够密封的滚装货物区域设置了固定式气体灭火装置时,以便使其能够在该货物区域外实施密闭。 Rapid shutdown of ventilation system is to be provided with dampers which can be isolated by a single action or closing appliances capable of isolation at equivalent closing speed. And Ro-Ro spaces capable of being sealed are to be capable of being sealed from a location outside of such cargo spaces, if they are protected with a fixed gas fire extinguishing system.

Rasorite(anhydrous) 斜方硼砂(无水) 一种颗粒状黄白色物质,粉尘极少或无粉尘。有腐蚀性。吸湿。无特殊危害。该货物为不燃物或失火风险低。该货物吸湿,受潮会结块。 Rasorite (anhydrous) is a. granular, yellow-white crystalline material with little or no dust. It is abrasive and hygroscopic. It has no special hazards. This cargo is non-combustible or has a low fire-risk. This cargo is hygroscopic and will cake if wet.

Raster chart display system 光栅海图显示系统 系指一种航行信息系统。它用来自传感器的位置信息显示 RNC 来帮助航海人员计划航线和监控航线,如有要求,还可显示其他关于航行的信息。 Raster chart display system (RCDS) means a navigation information system displaying RNCs with positional information from navigation sensors to assist the mariner in route planning and route monitoring, and if required, display additional navigation-related information.

Raster navigational chart(RNC) 光栅航海图 系指由政府授权的航道测量机构发出或经其授权分发的纸质海图的复制品。 Raster navigational chart (RNC) means a facsimile of a paper chart originated by or distributed on the authority of, a government-authorized hydrographic office. RNC is used in these standards to mean either a single chart or a collection of charts.

Rat guard 防鼠板 船舶靠泊时,为隔阻老鼠防止其窜到船上而扣设在系缆上的薄挡板。

Rate 费率 系指缴纳费用的比率。如保险业的费率系指投保人向保险人交纳的金额与保险人承担赔偿金额的比率。也指电话在一个计费单元内收取的费用,如:国内长途 0.30 元/min 等。

Rate of exchange 汇兑率

Rate of pitch adjusting 调距速率 可调螺距螺旋桨调节螺距时引手柄转动的速度。

Rate of regulating speed in instantaneous 瞬时调速率 内燃机在持续功率运转工况下突卸或突增负荷后的瞬时增加或降低转速和负荷改变前的转速之差与持续转速的百分比。

Rate of regulating speed in stability 稳定调速率 内燃机在持续功率运转工况下突增或突卸负荷后的稳定转速和负荷改变前的转速之差与持续转速的百分比。

Rate of revolution 转速 单位时间内转动的周数。

Rate of turn 回转率 系指每个时间单位船首向的变化。 Rate of turn is change of bearing per time unit.

Rated capacity 额定容量(额定功率,额定能力)

Rated condition 额定工况

Rated current of fuse 断路器的额定电流 系指配有专门的过电流脱扣器的断路器,在制造厂商规定的环境空气温度下,所能连续承受的最大电流值,不会超过电流承受部件规定的温度限值。

Rated insulation voltage of fuse 断路器绝缘电压 系指电介质测试电压(通常大于 $2U_1$)和爬电距离所涉及的电压值。额定工作电压的最大值不得超过额定绝缘电压,即 $U_e \leqslant U_1$。

Rated load 缆绳端点负荷(或公称负荷) 在卷筒绕有单层绳索的条件下,绞车以公称速度收绳时,在卷筒出绳处测得的最大绳索拉力。

Rated operational voltage of fuse 断路器额定工作电压 系指断路器在正常(不间断的)情况下工作时的电压。

Rated output 额定功率(燃气轮机) 系指压气机进口处空气温度 15 ℃,绝对压力 0.1 MPa,冷却水进口温度 15 ℃(如适用时)的标准环境条件下,燃气轮机长期运转的最大有效功率。 Rated output is defined as the maximum effective power developed by gas turbines for continuous running under standard ambient conditions of 15 ℃ air temperature at compressor inlet, 0.1 MPa absolute pressure and, where applicable, a cooling water temperature 15 ℃.

Rated output(continuous output) 持续功率 系指主机在现行有关标准规定的环境条件下允许长时期持续运转的输出功率。

Rated power for engines driving electric generators 发电机的柴油机的额定功率 系指在基准的环境条件下,柴油机经过车间试验后设定的并在考虑了电网的超负荷后能发出的标称功率。 Rated power for engines driving electric generators is the nominal power, taken at the net of overload, at ambient reference conditions, which the engine is capable of delivering as set after the works trials.

Rated power(rated output) 额定功率 系指:(1)"73/78 防污公约"附则Ⅵ的 2008 NO_x 技术规则第 13 条和有关规则适用的船用柴油机的铭牌及技术案卷中载明的最大持续额定输出功率;(2)在基准环境条件下,在调速器允许的最大转速下柴油机经过车间试验后设定的能发出的最大功率(燃油停止功率)。(1) Rated power means the maximum continuous rated power output as specified on the nameplate and in the Technical File of the marine diesel engine to which regulation 13 of "MARPOL 73/78" Annex Ⅵ; (2) the maximum power at ambient reference conditions which the engine is capable of delivering as set after works trials (fuel stop power) at the maximum speed allowed by the governor.

Rated pressure 额定压力

Rated speed 额定转速 系指:(1)船用柴油机铭牌及技术案卷中载明的在额定功率输出时的每分钟的曲轴转数;(2)燃气轮机额定功率时的相应转速。 Rated speed is: (1) the crankshaft revolutions per minute at which the rated power occurs as specified on the nameplate and in the Technical File of the marine diesel engine; (2) is defined as the speed corresponding to the rated output of gas turbine.

Rated speed (nominal speed) 额定绳速 绞车在承受额定卷筒拉力时,能保持的最大平均绳速。

Rated stock torque 额定转舵扭矩 当船舶以最大航速前进时,舵机在规定最大舵角下,转动舵杆的最大扭矩。

Rated sustain current 额定持续电流 系指以有效值安培数或直流安培数表示的设计限值,开关或断路器在持续承载这种电流时将不超过可测温升限值。

Rated value 额定值(评定值)

Rated wind pressure level 计算风力作用力臂 船舶正浮时,受风面积中心到假定水动力中心的垂向距离。

Rated working hours 定额工时 系指根据一定时期的社会生产力水平的高低。对于每一个施工项目,测算出的用工量。

Rated working pressure 额定工作压力 系指拟运转的液压装置的最大工作压力。 Rated working pressure means the maximum service pressure at which a hydraulic device is intended to operate.

Rating 普通船员 系指船长或高级船员以外的船员。 Rating means a member of the ship's crew other than the master or an officer.

Rating 定额(功率)

Ratio of above water and submerged lateral areas 水上水下侧面积比 船体水上部分与水下部分在船体中线面上的投影面积之比。

Ratio of dependence on foreign trade 对外贸易依存度 又称对外贸易系数。以一国对外贸易额与该国 GNP 或 GDP 比率来表示,用以反映一国经济发展对对外贸易的依赖程度。对外贸易依存度一般通过进口系数和出口系数来反映,也可以用一国对外贸易总额与该国 GNP 或 GDP 计算出来的比率直接反映。一般说来,一国对外贸易依存度直接受经济发展水平及自然资源拥有的状况、对外贸易政策、国内市场容量等因素的影响。

Ratio of loading and unloading 货车装卸率

Rational ship design (RSD) 合理的船舶设计

Raukine cycle 完全膨胀循环 将理想循环的等温过程更换成等压过程的热循环是蒸汽动力装置的基本热力循环。

RAVE 阿尔法·文图斯研究项目 该项目包括阿尔法·文图斯海上风电试验场的附带生态研究。由德国联邦环境部(BMU)资助5 000万欧元。整个研究项目分为15个独立项目。

Raw opium 生鸦片 开始是一种浓稠的、深褐色或几乎是黑色的黏性物质,然后变得像干草那样坚硬,随着时间的推移,它成为像封蜡一样坚硬的褐色/黑色块,略微有些黏性,这取决于其年份。通常要注意确保其不会干透,因为如果变得坚硬易脆,其价值会大为降低。生的鸦片是不可以抽吸的,只有转换成配制好的鸦片才可以抽吸。生鸦片不大可能有识别标志。可用玻璃纸或聚乙烯包在防水纸内,防止生鸦片干透。曾发现有聚乙烯袋和玻璃纸袋放在罐中或用粗麻布或厚帆布包裹。生鸦片有一种类似干草的香甜、油性、刺鼻的气味。在一定距离时,该气味不会令人不快,但在靠近时或在没有通风的狭窄处所,则令人作呕。其包装的方法会减少通过气味被探测到的机会。 Raw opium starts as a thick, dark brown or almost black sticky substance, hardens to the consistency of liquorice and then; with time, to a hard brown/black slightly sticky mass like sealing wax, depending on its age. Care is usually taken to ensure that it does not dry out since it loses much of its value if it becomes hard and brittle. In its raw state opium cannot be smoked. It is smoked only after conversion to prepared opium. Raw opium is unlikely to have identification marks. It may be wrapped in cellophane or polythene inside waterproof paper in order to stop the raw opium drying out. Polythene or cellophane bags have been found inside tins or wrapped in sacking or sailcloth. Raw opium has a sweet, only, pungent aroma, reminiscent of hay. It is not an unpleasant smell from a distance, but is sickly and nauseous when close up or in a confined space without ventilation. Its method of packing is designed to reduce the chance of detection by smell.

Rayleigh distribution 瑞利分布 可由方差相等的两独立正态分布合成的,有不对称特征的分布规律。其概率密度函数表达式为 $P(x) = (x/\sigma^2) e^{-x^2/2s^2}, s \leqslant \infty$。

Reach 复向期 在 Z 形操纵试验中,从开始首次操舵的瞬时到船舶向右舷转首后返回原转艏的瞬时之间的时间间隔。

Reach 回转中心纵距 回转试验中,沿初始直线航线方向从开始转舵瞬时的船舶重心位置至定常回转中心的纵向距离。

Reactance 电抗 部分电路的正弦电流和同一频率电位差的电抗是该电流与电位差之间相角差的正弦乘以该有效电位差之比的乘积。此时,在所述的这部分电路内,不含有电源。对于交变电流之各分量的电路电机电抗是各不相同的。若 $e = E_{1\omega}\sin(\omega t + \alpha_1) + E_{2\omega}\sin(\omega t + \alpha_2) + \cdots$, $i = I_{1m}\sin(\omega t + \beta_1) + I_{2m}\sin(\omega t + \beta_2) + \cdots$,则对于不同分量的电抗 $X_1, X_2, \cdots$ 为 $X_1 = E_{1\omega}\sin(\alpha_1 - \beta_1)/I_{1m} = E_1\sin(\alpha_1 - \beta_1)/I_1$, $X_2 = E_{2\omega}\sin(\alpha_1 - \beta_2)/I_{2m} = E_2\sin(\alpha_2 - \beta_2)/I_2, \cdots$。

Reaction engine 反作用式发动机(喷气式发动机)

Reaction rudder(twisted rudder/asymmetrical rudder) 反应舵 又称不对称舵。(1)改变舵的前缘的形状,使其产生一个附加推力的舵。(2)舵叶的前部,以螺旋桨轴线为界,上下向左右舷相反方向扭曲,具有整流作用的舵。(3)通过扭曲舵叶剖面来适应螺旋桨的尾流使舵产生一个正推力的舵。除节能外,反应舵应有利于解决舵空泡。

Reaction type steam turbine 反动式汽轮机 蒸汽在静叶和动叶中都进行膨胀,使转子因其动叶在蒸汽的冲动力和反作用力的共同推动下转动做功的汽轮机。

Reactive search 反应性搜查 是对特定威胁或情报表明船上已放置了包裹或袋子做出反应而实施的搜查。这种搜查还可以在 2 级或 3 级保安或在威胁升级时用作防范措施。反应性搜查应遵循下列原则:(1)不允许船员搜查他们自己的区域,以防他们有可能参与毒品走私活动并将包裹或袋子藏匿在他们工作区域或个人的区域;(2)应根据具体的计划或安排进行搜查,搜查工作必须得到严格的控制;(3)对两人搜查组应给予特殊考虑,一人搜查"高处",一人搜查"低处",如果发现可疑物件,两人中一人留守,另一人报告发现的情况;(4)搜查人员应能识别可疑包裹或袋子;(5)应有一套标记或记录"干净"区域的系统;(6)为了防止搜查期间非法移动货物,应控制人员走动。如果这不适用,在搜查过的和未搜查的区域之间走动的人员应接受搜身;(7)搜查人员应与搜查控制者保持联系,或许可使用超高频/甚高频无线电;(8)搜查人员应备有关于发现可疑包裹或袋子如何处理的明确指南;(9)搜查人员应注意,走私人员可能会将包裹或袋子做成与周围情况相匹配的物件,如机舱工具箱。 Reactive search means a search which would be carried out in reaction to a specific threat or piece of intelligence indicating that a package or bag has been places on board. It can also be used as a precaution at level 2 or 3 or during times of heightened threat. A reactive search should comply with the following principles:(1) crew members should not be allowed to search their own areas in case they are involved in a drug smuggling operation and have concealed packages or bags in their own work or personal areas;(2) the search should be conducted according to a specific plan or schedule and must be carefully controlled;(3) special consideration should be given to search parties working in pairs with one searching "high" and one searching "low". If a suspicious object is found, one of the pair can remain on guard while the other reports the find;(4) searchers should be able to recognize a suspicious package or bag;(5) there should be a system for marking or recording "clean" areas;(6) to prevent the illicit movement of goods during a searching, the movement of persons should be controlled. Where this is not applicable, persons should be subject to search when transiting between searched and unsearched areas;(7) searchers should maintain contract with the search controllers, perhaps by UHF/VHF radio;(8) searchers should have clear guidance on what to do if a suspicious package or bag is found;(9) searchers should bear in mind that smugglers may try to match the package or bag to the background, such as a tool box in an engine room.

Reactor 电抗器 系指以将电抗引入电路为主要目的的一种器件。即电抗器是为把电抗引入电路,以用于诸如电动机启动、变压器并联以及控制电流的一种器件。

Reactor room 反应堆舱 核动力船舶上,供安置反应堆的舱室。

Reactor room air compressor 堆舱压缩机 核潜艇反应堆舱减压用的压缩机。此种压缩机一般要求无人操作。

Readable 可读性 在从操作人员正常眼高位置到地平线以上的水平扇区 225°和垂直扇区 90°以下至 60°之间范围内。Within a horizontal sector of 225° and vertical sector from 90° below to 60° above the horizon from the operators normal eye position.

Reader 读写器 一般认为是射频识别,即 RFID 的读写终端设备。它不但可以阅读射频标签,还可以读写数据,故称为读写器。若只能阅读,不能擦写,则称为读卡器或射频识别器。读写器应用非常广泛,主要应用于身份识别、货物识别、安全认证和数据收录等方面,具备安全、准确、快速、扩展、兼容性强等特点。读写器的天线是发射和接收射频载波信号的设备,它主要负责将读写器中的电流信号转换成射频载波信号并发送给电子标签,或者接收标签发送过来的射频载波信号并将其转化为电流信号,读写器的天线可以外置也可以内置,天线的设计对阅读器的工作性能来说非常重要,对于无源标签来说,其工作能量全部由阅读器的天线提供。

Readily accessible 易达性 在紧急状态下无须使用工具就可快速和安全地到达并进行有效使用的能力。
Readily accessible is capable of being reached quickly and safely for maintenance or effective use under emergency conditions without the use of tools.

Reagent 试剂 系指一种用于引起化学反应的制品,但如果能得到相同的化学反应,也可被另一种试剂取代。前提必须是这种类型的制品。试剂可能是一种制品,或是可以引起相同化学反应的另一种具有类似特性的制品:这一种可能被另一种取代。 Reagent means a product used to provoke a chemical reaction, but which is replaceable by another reagent if the same chemical reaction is obtained. The precursor must be a product of this type. The reagent may be one type of product, or another with similar properties which provokes the same chemical reaction: one may be substituted by another.

Real net weight 实际净重 系指从商品总质量中扣除内外包装的质量后的实际质量。

Real products of inertia 惯性积 船舶实体各质点的质量与该点分别至两坐标平面的垂直距离三者乘积的总和。

real radius of gyration 惯性半径(船舶实体) 船舶转动时假定其实体质量等效集中在一点上,从该点至转动轴线的距离。亦即等于船舶实体惯性矩对质量的比值的平方根 $\sqrt{\dfrac{I}{m}}$,式中,I—船舶实体惯性矩;m—船舶实体质量。

Real time video 实时视频

Rear stern tube 后艉轴管 多轴系船舶的侧轴上,艉轴管分三段组成时的后段部分。

Reasonable weather 适宜气候 适宜气候定义为风力等于或低于蒲氏风级6级,且:(1)包括在营运包络线中的海况,海况适中,足以保证舰船的甲板中有偶尔上浪或根本无上浪。(2)不削弱舰船营运效率的运动。
Reasonable weather is defined as wind strengths of force six or less on the Beaufort scale, associated with: (1) Sea states within the operational envelope which are sufficiently moderate to ensure that green water is taken on board at infrequent intervals only or not at all. (2) Motions such as do not impair the efficient operation of the ship.

Reasonably probable (Probability of occurences)
适当可能的(事件的概率) 对事故的概率而言,系指在一特定船上的总使用期间内,不可能经常发生的,但可能发生几次的(最坏 $10^{-5} \sim 10^{-3}$)。 For the purpose of probability of occurrences, reasonably probable is one which is unlikely to occur often but which may occur several times during the total operational life of a particular craft (between 10^{-5} and 10^{-3}).

Rebuilding 重建
Receipt 收据
Received for shipment B/L 备运提单 又称收讫待运提单。系指船公司已收到托运货物等待装运期间所签发的提单。

Receiver 接收人 系指:(1)实际接收卸于一当事国港口或码头的摊款货物的人员,但是,如果在接收时实际接收该货物的人系受任何缔约国管辖的另一人的代理人并且该代理人向有害有毒物质基金指明了该委托人,则该委托人应被视为接收人;(2)在当事国的并按该国的法律被视为卸于某一当事国港口或码头的摊款货物接收人的人员,但是根据此种国家法律所接收的摊款货物总量应与(1)项所接收的总量基本一致。 Receiver means either: (1) the person who physically receives contributing cargo discharged in the ports and terminals of a State Party, provided that if at receipt the person who physically receives the cargo acts as an agent for another who is subject to the jurisdiction of any State Party, then the principal shall be deemed to be receiver, if the agent discloses the principal to the HNS Fund, or (2) the person in the State Party who in accordance with the national law of that State Party is deemed to be the receiver of contributing cargo discharged in the ports and terminals of a State Party, provided that the total contributing cargo received according to such national law is substantially the same as that which would have been received under (1).

Receiver equipment for European satellite navigation system (GALILEO) 欧洲卫星导航系统(GALILEO)接收设备 系指包括欧洲卫星导航系统正确行使拟定功能所必需的所有部件和元件。欧洲卫星导航系统(GALILEO)接收设备至少包括下列设施:(1)能接收欧洲卫星导航系统(GALILEO)信号的天线;(2)欧洲卫星导航系统(GALILEO)接收机和处理器;(3)取用纬度/经度位置计算值的工具;(4)数据控制和接口;(5)位置显示以及如有要求,其他的输出格式。 Receiver equipment for European satellite navigation system (GALILEO) means those equipments which includes all the components and units necessary for the system properly to perform its intended functions. The Galileo equipment should include the following minimum facilities: (1) antenna capable of receiving Galileo signals; (2) Galileo receiver and processor; (3) means of accessing the computed latitude/longitude position; (4) data control and interface; and (5) position display and, if required, other forms of output.

Receiving bank 汇入行 又称解付行(Paying bank)是接受汇出行的委托解付款项的银行,汇入行通常是汇出行在收款人所在地的代理行。

Receiving centre 接收中心 无线电导航设备的船台测量距离差的接收点。一般与测深中心不重合,其位置主要受船台接收天线的架设位置,形状及船舶上层建筑物的二次辐射的影响。

Receiving inspection 进货检查 对质量管理体系而言,系指制造厂商应保证所有进厂的材料在没有检验或证实其符合规定要求以前,不能被使用或加工。在确定进货检查的数量和性质时,应考虑到供应商所实施的管理方法和他们所提供的书面质量证明。 For quality management system, receiving inspection means that the manufacturer is to ensure that all incoming material is not be used or processed until it has been inspected or otherwise verified as conforming to specified requirements. In establishing the amount and nature of receiving inspection, consideration is to be given to the control exercised by the supplier and documented evidence of quality conformance supplied.

Reception facilities 接收设备 (1)对"73/78 防污公约"附则Ⅰ而言,接收设备系指防污公约各缔约国应承担的义务,保证在其装油站、修理港以及船舶需要排放残油的其他港口提供足够的接收油船和其他船舶留存的残油和油性混合物的设备,以满足船舶使用的需要,而不对船舶造成不当延误。(2)对"73/78 防污公约"附则Ⅱ而言,接收设备系指防污公约各缔约国应承担的义务,为保证船舶使用其港口、装卸站或修理港的需要而提供如下接收设备:①船舶货物作业港、站应设有足够的设备,以接收船舶由于执行附则Ⅱ而留待处理的含有有毒液体物质的残余物和含有该有毒物质残余物的混合物,而不对相关船舶造成不当延误;②从事 NLS 船修理的船舶修理港,应备有足够设备,以接收到达该港的船舶所含有的有毒液体物质的残余物的混合物。(3)对"73/78 防污公约"附则Ⅳ而言,接收设备系指防污公约各缔约国应承担的义务,保证在其港口和近海装卸站提供足够的生活污水接收设备,以满足船舶使用的需要,而不对船舶造成延误。(1)For "MARPOL73/78" Annex Ⅰ, Reception facilities means that, the Government of each Party to the present Convention undertakes to ensure the provision at oil loading terminals, repair ports, and in other ports in which ships have oily residues to discharge, of facilities for the reception of such residues and oily mixtures as remain from oil tankers and other ships adequate to meet the needs of the ships using them without causing undue delay to ships. (2) For "MARPOL73/78" Annex Ⅱ, the government of each Party to the Convention undertakes to ensure the provision of reception facilities according to the needs of ships using its ports, terminals or repair ports as follows:①Ports and terminals involved in ships cargo handling shall have adequate facilities for the reception of residues and mixtures containing such residues of noxious liquid substances resulting from compliance with this Annex Ⅱ, without undue delay for the ships involved. ②Ship repair ports undertaking repairs to NLS tankers shall provided facilities adequate for the reception of residues and mixtures containing noxious liquid substances for ships calling at that port. (3) For "MARPOL73/78" Annex Ⅳ, reception facilities means that, the government of each Party to the Convention undertakes to ensure the provision of facilities at ports and terminals for the reception of sewage, without causing delay to ships, adequate to meet the needs of the ships using them.

Recess 凹体 可能积水的露天的任何容积。例如舾舱、围井、由舷墙或围槛围成的开敞容积或区域。注意:设置符合 ISO 12216 要求的关闭装置的住舱、遮蔽区域或储藏室不属于凹体。 Recess means any volumes open to the sky that may retain water. For example, cockpits, wells, open volumes or areas bounded by bulwarks or coamings. Note:Cabins, shelters or lockers provided with closures according to the requirements of ISO 12216 are not recesses.

Recess corrosion 凹槽腐蚀 系指扶强材与扶强材,扶强材与板材的焊缝连接处典型的局部厚度减少。

Recharging 再充电(再补气,再充灭火剂,再装填,再补料)

Recipient port 受体港 系指排放压载水的港口或地点。 Recipient port is the port or location where the ballast water is discharged.

Reciprocal credit 对开信用证 系指两张信用证的开证申请人互以对方为受益人而开立的信用证。对开信用证的特点是第一张信用证的受益人(出口人)和开证申请人(进口人)就是第二张信用证的开证申请人和受益人。

Reciprocating compressor 往复式压缩机

Reciprocating engine 往复式发动机(活塞式发动机)

Reciprocating internal combustion engine 往复式

内燃机 燃料在气缸内燃烧,产生热能,并通过活塞作往复运动,使热能转变为机械功的内燃机。

Reciprocating motion 往复运动

Reciprocating pump 往复泵

Reciprocating refrigeration compressor 活塞式制冷压缩机 由于活塞在汽缸内的往复运动,使气态制冷剂被吸入压缩和排出的制冷压缩机。

Reciprocation pump 往复泵 利用活塞或柱塞作往复运动时,缸内工作容积的变化,完成吸入和压出过程,以抽送液体的泵。

Recirculating 再循环

Recirculating arrangement 再循环装置

Recirculating pump 再循环泵

Recirculation flash evaporator 再循环闪发蒸发器 盐水大部分再循环使用的闪发蒸发器。

Reclamation craft 吹泥船 系指具有吸管、吸嘴等设备的挖泥船。 Reclamation craft is a dredger fitted with suction pipes, nozzles, etc.

Recognized classification societies 公认的船级社 系指国际船级社协会(IACS)成员单位的船级社,其在浮式结构物方面具有公认的和丰富的经验,并制订了采油活动中使用的装置的入级/认证的规范和程序。 Recognized classification societies mean those societies that are members of the International Association of Classification Societies (IACS), with recognized and relevant competence and experience in floating structures and with established rules and procedures for classification/certification of installations used in petroleum-related activities.

Recognized facility 认可的机构 系指由美国海岸警卫队提出的作为其认可机构的实验室或机构。 Recognized facility means any laboratory or facility listed by the Coast Guard as a recognized facility under this part.

Recognized security organization 经认可的保安组织 系指经授权开展SOLAS第Ⅺ-2章或ISPS规则A部分所要求的评估或批准,或签证活动,具有适当保安专长并具备适当船舶和港口操作方面知识的组织。如中国船级社。 Recognized security organization means the organization authorized to carry out assessment or approval or certification as required in chapter Ⅺ-2 of SOLAS Convention or part A of the ISPS Code, with appropriate security speciality and with appropriate knowledge in ship and port operations. China classification society is one of the recognized organizations.

Recognized standards 公认标准 系指为主管机关所接受的适用的国际标准或国内标准,或由某组织制订并保持的符合国际海事组织通过的各项标准并经过主管机关认可的标准。 Recognized standards means applicable international or national standards acceptable in the Administration or standards laid-down and maintained by an organization which complies with the standards adopted by the Organization and which is recognized by the Administration.

Recombination type transducer 复合型传感器

Recommendation 指定项目 对检验而言,特指船级社或进行法定检验的机构,在证书或检查报告上注明的对船舶提出的,有关检查实施、缺陷修理及整改的劝告或指示。

Recondition 热喷涂 系指一系列过程,在这些过程中,细微而分散的金属或非金属的涂层材料,以一种熔化或半熔化状态,沉积到一种经过制备的基体表面,形成某种喷涂沉积层。它是利用某种热源(如电弧、等离子喷涂或燃烧火焰等)将粉末状或丝状的金属或非金属材料加热到熔融或半熔融状态,然后借助火焰本身或压缩空气以一定速度喷射到预处理过的基体表面,沉积而形成具有各种功能的表面涂层的一种技术。利用由燃料气或电弧等提供的能量。

Recooling 回冷 冷却货物或冷冻货物在从陆上冷库运到船上的过程中,温度略有上升,在进入冷藏货舱、冷藏库后冷却到冷藏温度的过程。

Record book of engine parameters 发动机参数记录簿 系指发动机参数检查法使用的、记录可能影响发动机 NO_x 排放的所有参数变化包括构件和发动机的设定值的文件。 Record book of engine parameters is the document used in connection with the Engine Parameter Check method for recording all parameter changes, including components and engine settings, which may influence NO_x emission of the engine.

Record of file review(by lawyers) 阅卷笔录 系指办案人员或辩护律师在阅读案件卷宗时所作的记录。有保证诉讼活动顺利进行的作用,对查阅案卷宗、复查案件、核对证据、弄清事实、进行辩护、总结经验、提高办案效率和质量均有重要作用。阅卷笔录是律师在阅读案件卷宗的过程中,把案情事实和认定事实的证据、确定案件性质的依据以及对案件所作处理及其法律依据等加以分类和整理,并进行摘录的文字材料。阅卷笔录可以帮助律师从繁多的卷宗材料中理出头绪,深入了解案件实际情况,为下一步辩护词的制作和庭审做好准备。

Record of oil discharge monitoring and control system for the last ballast voyage 最近一次压载航行的排油监控系统记录 系指根据 MARPOL 73/78 附则 1 第15 条(4),(5),(6)和(7)的规定,每艘 150 总吨及以上的油船应装有一个记录器,以提供每海里排放升数和总排放量,或含油量和排放率。这种记录应能鉴别时间和日期,并应至少保存3年。 Record of oil discharge monitoring and control system for the last ballast voyage means that subject to provisions of paragraphs (4),(5),(6) and (7) of regulation 15 of Annex I of MARPOL 73/78, every oil tanker of 150 gross tonnage and above shall be fitted with an oil discharge monitoring and control system approved by the Administration. The system shall be fitted with a recording device to provide a continuous record of the discharge in litres per nautical mile and total quantity discharged, or the oil content and rate of discharge. This record shall be identifiable as to time and date and shall be kept for at least three years.

Recorder 录音机
Records 记录 对质量管理体系而言,系指制造厂商应制定并执行一种收集、使用和贮存质量记录的制度。质量记录的保存期限应予以书面规定,并应得到有关委员会的同意。 For the purpose of quality management system, records mean that the manufacturer is to develop and maintain a system for collection, use and storage of quality records. The period of retention of such records is to be established in writing and is to be subject to agreement by the Committee.

Records of navigational activities 航行活动记录 系指从事国际航线的船舶,在船上保存的有关航行活动和事件的记录本。这些活动和事件系对航行安全有重大影响,且其中的细节必定足以恢复关于该航次的1份完整的记录。

Recovery of execution 执行回转 又称再执行,系指在案件执行中或者执行完毕后,据以执行的法律文书被法院或其他有关机关撤销或者变更后,执行机关对已被执行的财产重新采取执行措施,恢复到执行程序开始时的状况的一种补救制度。执行回转制度是针对执行发生的错误而采取的一种补救措施,我国《民事诉讼法》第233条和《执行规定》第109条对此做出规定。

Recovery time 救助艇的回收时间 系指该艇被提升至某一位置,而使艇上的人员可从该处登上大船甲板所需的时间。回收时间包括在救助艇上做的回收准备工作所需的时间,诸如抛投和系住艉缆,连接救助艇与降落设备,以及提升救助艇的时间。回收时间不包括把降落设备降低至回收救助艇的位置所需的时间。 Recovery time is the time required to raise the boat to a position where persons on board can disembark to the deck of the ship. Recovery time includes the time required to make preparations for recovery on board the rescue boat such as passing and securing a painter, connecting the rescue boat to the launching appliance, and the time to raise the rescue boat. Recovery time does not include the time needed to lower the launching appliance into position to recover the rescue boat.

Rectification 补正
Rectifier 整流器 是一种平均功率的流动方向在其内部是从交流电路到直流电路的变换器。
Recurrent 复发的 对事故的概率而言,系指包括经常的和适当可能的这两者(概率)总范围的术语。 For the purpose of probability of occurrences, recurrent is a term embracing the total range of frequent and reasonably probable.

Recurrent(probability of occurences) 经常发生的(事件的概率) 是包含全部频繁的和适当可能的范围内的一个术语。 Recurrent is a term embracing the total range of frequent and reasonably probable.

Recycling 拆船 系指拆解和回收部件和材料以供再加工的活动。 Recycling means the activity of segregating and recovering components and materials for reprocessing.

Recycling company 拆船公司 系指拆船厂的拥有者或从拆船厂的拥有者处承担拆船活动的经营责任并在承担该责任的同时同意承担2009年香港国际安全与环境无害化拆船公约规定的所有职责和责任的任何其他组织或个人。 Recycling company means the owner of the ship recycling facility or any other organization or person who has assumed the responsibility for operation of the ship recycling activity from the owner of the ship recycling facility and who on assuming such responsibility has agreed to take over all duties and responsibilities imposed by the Hong Kong International Convention for the safe and environment sound recycling of ship, 2009.

Red alert 红色警戒 动力定位(DP)应急状态。位置和/或首向丧失已经发生或不可避免。

Red alert condition of DP 动力定位红色警戒状态 系指动力定位应急状态。位置丧失或位置丧失不可避免的状态。

Red all (round) light 红环照灯 显示红色光的环照灯。
Red flare 红火号 手持的点燃后能发出红色火焰的一种信号烟火。
Red globular shape (red ball) 红球 红色球体状或由两正交的红色圆片构成的号型。
Red Sea area 红海区域 系指红海本身,包括苏伊士湾和亚喀巴湾,南以拉斯西尼(北纬12°28.5′,东经43°19.6′)和胡森穆拉得(北纬12°40.4′,东经43°30.2′)之间的恒向线为界。 Red Sea area means the Red Sea

proper including the Gulfs of Suez and Aqaba bounded at the south by the rhumb line between Ras si Ane (12°28.5′N, 43°19.6′E)and Husn Murad (12°40.4′N, 43°30.2′E).

Red side light (port side light) 红舷灯 设置在船舶左舷侧，自艏至左舷112.5°内显示红色光的舷灯。

Red stern light 红艉灯 船舶通过苏伊士运河时专用的显示红色光的艉灯。

Red visual signal 红色的视觉信号 系指危险或需要立即采取行动的警告信号。例如：重要设备停止运转；水、油等温度或压力达到临界值；重要电路失电。Red visual signal means an indication warning of danger immediate action or a situation which required immediate action. For example: operation failure of essential equipment temperature or pressure of water/oil to a critical value; power failure of essential circuits.

Re-delivery of vessel 还船 系指租借合同到期，承租人将船舶还给船舶所有人。

Re-distillation 再蒸馏 多级闪发蒸馏装置中，前一级冷凝器内的蒸馏水在进入下一级冷凝器时有些蒸馏水再次转化成蒸汽的现象。

Reduced pressure 降压

Reduced propulsive efficiency 推进系数减值 船舶在风浪中由于螺旋桨的负荷增加、飞车或吸气以及船身效应的减小等影响，而使推进系统较静水中航行时下降的数值。

Reduced weld thickness (fillet weld) 缩减的焊缝厚度（角接焊缝） 明显缩减焊缝厚度的角接焊缝不视为缺陷，但条件是实际的焊缝厚度可通过较深的熔敷来进行补偿并因这样做而达到规定的尺寸。见图R-1。A fillet weld with an obviously smaller weld thickness should not be regarded as a defect provided that the actual weld thickness is compensated by a deeper penetration and by so doing achieving the specified size. See Figure R-1.

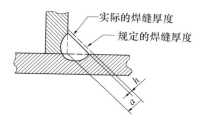

图 R-1 缩减的焊缝厚度
Figure R-1 Reduced weld thickness

Reducing factor 折减系数 将柔性构件的剖面积化为相当的刚性构件的剖面积时所乘的系数。

Reducing valve 减压阀

Reduction factor of circulation 环流减额因数 在一定半径处有限叶数对无限叶数螺旋桨叶元体的环量比值。

Reduction gear 减速齿轮箱（减速齿轮，减速装置，减速传动装置） (1)通过传动齿轮来实现减速并传递功率的齿轮传动组件；(2)由联轴器或耦合器与减速齿轮箱组成的仅起减速作用的传动设备。

Reduction gear for ahead and astern 倒顺车减速传动装置 通常是由离合器和减速齿轮箱组成的倒顺车传动设备。

Reduction gear lubricating oil gravity tank 减速齿轮箱滑油重力柜 在重力式滑油系统中，布置在一定高度依靠重力将滑油供给减速齿轮箱各润滑点的油柜。

Reduction gear lubricating oil settling tank 减速齿轮箱滑油澄清柜 沉淀减速齿轮箱滑油中杂质、水分的柜。

Reduction gear lubricating oil storage tank 减速齿轮箱滑油贮存柜 贮存减速齿轮箱用的新润滑油的柜。

Reduction gear lubricating oil tank 减速齿轮箱滑油循环柜 汇集减速齿轮箱滑油供再循环使用的柜。

Reduction gear of combined output of the engines 并车减速齿轮箱 将两台或多台发动机的功率集中到一个输出轴并拖动一个轴系的减速齿轮箱。

Reduction gear of combined output of the four-engines 四机轴并车减速齿轮箱 由四台发动机拖动一根轴系的减速齿轮箱。

Reduction gear of combined output of the triple-engines 三机轴并车减速齿轮箱 由三台发动机拖动一根轴系的减速齿轮箱。

Reduction gear of combined output of the twin-engines 双机轴并车减速齿轮箱 由两台发动机拖动一根轴系的减速齿轮箱。

Reduction gear room 减速器舱 供安置主机减速器的舱室。

Reduction ratio 减速比（传动比，减缩比） 当输入轴转速大于输出轴转速时两者的比值。

Reductor 减速器（减压器）

Reducing temperature device 降温装置

Redundancy 冗余 系指：(1)当发生单项失效时，单元或系统保持或恢复其功能的能力。它可通过例如设置多重单元、系统或其他实现同一功能的措施来实现；(2)完成一个所需功能，存在一个以上的方法。Redundancy means: (1)the ability of a component or system to maintain or restore its function, when a single failure has occurred. Redundancy can be achieved, for instance, by installation of multiple components, systems or alternative means of performing a function; (2) the existence of more than one means for performing a required function.

Redundancy 备份　系指为了完成一个特定的功能而以一种以上的方法保存。　Redundancy is the existence of more than one means for performing a required function.

Redundancy 冗余度(多余,冗余,重复)　是设备或系统在一次故障发生时保持或恢复其功能的能力。冗余度可通过诸如装设更多装置或替代措施以执行同一功能来获得。　Redundancy is the ability of a component or system to maintain or restore its function when one failure has occurred. Redundancy can be achieved for instance by installation of more units or alternative means for performing a function.

Redundancy design 冗余设计　系指当系统或设备发生故障时所采用的备用技术手段,依靠其即刻参与相应的功能而使动作状态持续运行,或者通过恢复原有功能而使运行状态得以延续。　Redundancy design means standby technical means to be used in case of system or equipment fault, by which a corresponding function can be taken over to continue an action or an original function are restored to continue the operation condition.

Redundancy rate 冗余率

Redundancy redundancies 冗余度

Redundancy test procedure 冗余度试验程序

Redundant 多余的(冗余的)

Redundant design 冗余设计

Redundant risk control 冗余风险控制　在风险控制失败时能够保持稳健状态。

Reel 暗礁

Reel 系缆绞车　卷起和收藏缆索的装置。

Reel barge 卷筒铺管驳船　系指将卷绕在大直径卷筒上的钢管从船上铺设到海底的驳船。

Reel lay/ Flex lay 卷筒铺管法　是为铺设复合管、脐带缆和电缆而开发出来的一种快速铺管方法,也称为柔性铺管法。

Reeling 卷放　系指管路的一种敷设方法。如果将一段长的管子预先绕到一个大直径的卷盘上,随后通过脱离卷轴的方法敷设在近海的场地上。　Reeling is a pipeline installation method where a long section of line is pre-wound onto a large diameter reel and subsequently installed at an offshore site by unspooling.

Re-engine 更机　系指更换主机。

Reentry 重定位　浮式钻井装置由于风浪或钻井作业故障等原因迫使其离开原井位,后来认为有必要而重新找到原井口继续进行钻井的一种定位作业。

Refer cargo carrier 冷藏货物运输船　简称冷藏船。专用于运输肉类、鱼类、水果等易腐烂变质货物的船舶。其船型与杂货船相似。此类船舶的货舱都具有良好的隔热功能,装备制冷设备。近年来,由于冷藏集装箱运输发展迅速,在一定程度上代替了冷藏船。但在远洋捕鱼船队中,冷藏船还发挥着重大的作用。

Reference period 参照周期　系指作为确定基本变量基础使用的时间周期。　Reference period means the period of time used as basic for determining the values of basic variables.

Reference target 参数目标　系指表示把关联的被跟踪静止目标作为地面稳定速度参照的符号。　Reference target means a symbol indicating that the associated tracked stationary target (e.g. a navigational mark) is used as a speed reference for the ground stabilization.

Reference temperature 基准温度　系指货物蒸发气压力与压力释放阀的设定压力值相当时的温度。　Reference temperature is the temperature at which the vapour pressure of the cargo corresponds to the set pressure of the pressure-relief valve.

Reference value A_{wr}　参照值 A_{wr}　系指以最初三年收集的售价为依据,并根据2010年世界范围内船用燃油的平均硫含量监测指南5~11节中确定的世界范围内船用残余燃油的平均硫含量。　Reference value A_{wr} means the value of the worldwide average sulphur content in residual fuels supplied for use on board ships, based on the first three years of data collected and as determined on the basis of paragraphs 5 to 11 of the 2010 Guidelines for monitoring the worldwide average sulphur content of fuel oils supplied for use on board ships.

Reference waterline WL_{ref}　基准水线 WL_{ref}　在满载备用状态时的水线。　Reference waterline is the waterline in the fully loaded ready-for-use condition.

Reference wave steepness 波陡参考值

Reflected sound wave 反射声波　系指当入射至两媒介质的界面时,一部分会经界面反射返回原来的介质中的声波。

Reflection survey 反射波勘探　利用地震反射波的特点进行的地震勘探,寻找石油和天然气的有利构造或储油圈团。

Reformates 重质油

Refraction survey 折射波勘探　利用地震折射波的特点进行的地震勘探,常用于浅层地震勘探和深部地层探测。

Refreshment kiosks 小卖部 系指供应点心和饮料并设有总功率不超过 5 kW 或以下和曝露加热表面温度不超过150℃的食品加热设备的非围蔽处所。 Refreshment kiosks are those spaces which are not enclosed, serving refreshments and containing food warming equipment having a total power of 5 kW or less and with an exposed heating surface temperature not above 150 ℃.

Refrigerant (refrigerant) 制冷剂 是一种用来传递和保持冷藏室内冷量的冷却介质。Refrigerant is a cooling medium which is used to transmit and maintain the cool in the refrigerated chamber.

Refrigerant condenser 制冷剂冷凝器

Refrigerant recovery equipment 制冷剂回收设备

Refrigerate 冷藏(制冷)

Refrigerated cabinet 冷藏柜

Refrigerated cargo (CRS) 货物冷藏 船上应配备的冷藏装置能保证在海水最高温度下维持冷藏货物的最低温度或范围,以达到货物在冷藏的条件下保证货品质量的目的。

Refrigerated cargo compartment/refrigerated cargo hold 冷藏货舱 设有冷藏设备,专供装载冷藏货物的货舱。

Refrigerated cargo installation 货物冷藏装置

Refrigerated cargo installation notation 冷藏装置附加标志

Refrigerated cargo ship 冷藏货物运输船 系指专门设计用于在货舱内载运冷藏货物的船舶。Refrigerated cargo ship means a ship designed exclusively for the carriage of refrigerated cargoes in holds.

Refrigerated chamber 冷藏室 是由制冷机组进行冷却的任何处所。一个冷藏室可能是有关的装货处所或船上任何的服务处所,例如厨房。 Refrigerated chamber is any space which is refrigerated by a refrigerating unit. A chamber may be a cargo space or any other ship service space, such as for instance the galley.

Refrigerated chamber (refrigerated store) 冷藏库 船上供冷藏食品用的仓库。

Refrigerated container 冷藏集装箱 系指带有自备冷藏系统的标准集装箱。该系统位于此集装箱的外形尺寸之内,且可由外部电源供电驱动。此冷藏系统可为"以夹子夹上去"或为整体式的制冷装置。 Refrigerated container means a standard container with a self-contained refrigeration system, located within the outer dimensions of the container, which can be driven by electrical power fed from an external power supply. The refrigeration system may be either a 'clip-on' or an integral type of cooling unit.

Refrigerated door 冷库门(隔热门)

Refrigerated hold 冷藏舱

Refrigerated seawater fish carrier 冷海水保鲜运输船 用制冷装置冷却海水,以冷海水作为冷媒,达到渔获物保鲜的运输船。

Refrigerated ship/cold storage ship 冷藏船 设有冷藏设备,专运冷藏货物的船舶。

Refrigerated store room 冷藏室(冷藏库)

Refrigerating 冷藏(制冷)

Refrigerating capacity 冷藏容量

Refrigerating compressor 制冷压缩机 用于制冷系统中的压缩机。

Refrigerating hatch cover 冷藏舱口盖

Refrigerating installations 冷藏设备(冷藏装置) 系指制冷装置、冷藏室的隔热装置和与冷藏室相关的其他装置。 Refrigerating installations means refrigerating machinery, insulation for refrigerated chambers and other related appliances in refrigerated chambers.

Refrigerating machine 制冷机

Refrigerating machinery (refrigerating plant) 制冷装置 系指组成制冷周期的系列制冷设备,包括压缩机、冷凝器、接收器、蒸发器、冷却器、管道系统和配套装置,制冷泵和压缩机的驱动电机、自动控制器和电气设备等。 Refrigerating machinery means a set of refrigerating units which compose refrigerating cycle, consisting of compressors, condensers, receivers, evaporators, coolers, piping and fittings, driving motors for the compressors and refrigerant pumps, automatic controllers, and electrical equipments.

Refrigerating system 制冷系统 由制冷压缩机、附属设备、阀件等通过管路连接而组成的用以制冷的系统。

Refrigerating system management plan 冷藏系统管理计划

Refrigerating unit 制冷机组(制冷设备) (1)是由一台或多台原动机,一台或多台压缩机,一台冷凝器以及全部必需的相关部件组成的冷却冷藏处所的一个独立的气-液系统。如制冷机组采用冷媒(盐水)时,该机组也应包括盐水冷却器(蒸发器)和盐水泵;(2)在制冷装置中例如像那些为了操作冷藏周期所需要的压缩机,马达、冷凝器、蒸发器、泵等机器。 Refrigerating unit:(1) includes one or more compressors driven by one or more prime movers, one condenser and all the associated ancillary equipment necessary to form an independent gas-liq uid system capable of cooling refrigerated chambers. When the installation includes a secondary refrigerant (brine), the refrigerating unit is also to include a brine cooler (evaporator) and a pump; (2) in general such machinery as compressors, motors, condensers, evaporators, pumps, etc. necessary to operate refrigerating cycles among the refrigerating machinery.

Refrigeration 冷藏（制冷）
Refrigeration cycle 制冷循环
Refrigeration equipment 制冷设备
Refrigeration machinery compartment 制冷机舱
Refrigeration medium pump 液泵 制冷系统中抽送液态制冷剂的泵。
Refrigeration system control (RSC) 冷藏系统控制 系指对制冷剂的臭氧消耗的控制。Refrigeration system control (RSC) means control of ozone consumption by coolant.
Refrigeration test 制冷试验
Refrigerator (domestic refrigerator) 冷藏机（冰箱） 系指：(1)箱体周围有绝热层并带有制冷装置的贮藏物品的冷藏箱。(2)俗称制冷压缩机。是利用制冷剂介质由液体变气体过程需吸热而起到制冷效果的装置。
Refrigerator room 冷冻机室 供安置冷冻机及其控制设备的房间。
Refrigerator space 冷藏舱（冷藏处所）
Refuel (refueling) 加燃料（加油）
Refueling and oil storage facility 加油和储油设施
Refund and rebates 回扣
Refuse 废物（垃圾）
Refuse disposal system 垃圾处理装置
Regenerative air heater 回转式空气预热器 借传热体或烟气和空气输送管的回转，使传热体与烟气和空气发生连续交替的接触，从而实现空气加热的空气预热器。
Regenerative cycle 回热循环 利用回热器将涡轮排气的部分余热用来加热进入燃烧室的空气的燃气轮机循环。
Regenerative gas turbine plant 回热循环燃气轮机装置 按回热循环工作的燃气轮机动力装置。
Regenerative steam engine 撤气式蒸汽机 从气缸间的容器中抽出部分已做过部分功的蒸汽来加热锅炉给水的蒸汽往复机。
Regenerative steam turbine 回热式汽轮机 从中间级中抽出部分蒸汽供加热锅炉给水的汽轮机。
Regenerative steam turbine plant 回热循环汽轮机装置 实现回热循环的汽轮机动力装置。

Regenerator 回热器 利用涡轮排气的余热以加热进入燃烧室或加热器前的空气或其他气体的一种热交换器。
Reginnal planning 区域规划 系指超越德国最小领土区域（自治区）的跨学科顶级规划，其责任和使命是提供可持续的空间开发，使空间开发的社会和经济要求与生态系统平衡，从而可长期在大面积领域内实施。
Region of icebergs guarded by the ice patrol 冰区巡逻所警戒冰山区 系指纽芬兰大浅湾附近冰山区的东南、南及西南界限。 Region of icebergs guarded by the ice patrol means the south-eastern, southern and south-western limits of the region of icebergs in the vicinity of the Grand Banks of Newfoundland.
Regional price in economy and trade block 区域性经济贸易集团内的价格 第二次世界大战后，成立了许多区域性的经济贸易集团。在这些经济贸易集团内部，形成了区域性经济贸易集团内价格，如西欧经济集团的共同农业政策中的共同价格。
Register breadth 登记宽度 按吨位丈量规则所规定的，在船舶登记以及在船舶有关证书中所载列的船舶宽度。
Register depth 登记深度 按吨位丈量规则所规定的，在船舶登记以及在船舶有关证书中所载列的船舶深度。
Register length 登记长度 按吨位丈量规则所规定的，在船舶登记以及在船舶有关证书中所载列的船舶长度。
Register of ships 船舶录 船级社对批准入级的船舶(授予入级符号和附加标志后)将其主要特征要素和细节汇编成册，供船舶有关方，如船厂、船东、保险商、货运方和租船方等提供信息的文件。
Register ton (vessel ton) 登记吨 吨位的单位，每登记吨等于 100 ft^3/2.83 m^3。
Registered length 登记船长
Registered letter 挂号信 挂号邮件必须到邮局营业窗口交寄，经邮局营业人员验视、封装、书写、邮资等，符合挂号邮件的要求，邮局办理挂号手续，并出具国内挂号信函收据；平常邮件不给出收据。挂号邮件在邮局内部处理和传递过程中都要逐件地进行登记；平常邮件不进行登记。挂号邮件在未收到时，可以拿收据到原来的邮局办理查询手续，进行查询；平常邮件不接受查询。挂号邮件属于邮局的责任发生丢失损毁时，按规定进行赔偿；平常邮件不负责赔偿。
Registered tonnage 注册吨位 ①衡量民用船容积或营运能力的一种数值。其演变过程很久，后来确定的丈量规则以 100 英尺³（或 2.83 m³）为 1 登记吨或丈量吨。源于 1854 年英国总丈量师莫逊（George Moorsom）所提出的莫逊法则。其后为各国普遍采用，仅在细节上略有差异。一艘船舶的登记吨数常称吨位。按国家颁布的船舶丈量规范来核定，载于发给船舶的吨位证书中，是对船舶征税、收费的依据。在国际上，一般互相承认其有效。总吨位表示船体可围蔽处所包含的总容积。但双层底、艏尖舱、艉尖舱等不用于营运部分的不计入；艏楼、桥楼、艉楼等上层建筑如无永久性围蔽装置，亦作为免除处所不计入总吨位内。净吨位是从总吨位中对若干按丈量规范认为非营运处所（如机舱、船员居住和活动处所、为操作和维护船舶所需备有的空间等）作为减除处所，将其容积由总吨位减去所得的数值。1969 年签订的《国际船舶吨位丈量公约》于 1982 年 7 月 18 日生效后，这种吨位丈量方法对国际航行的新建船舶（长 24 m 以下的船舶除外）已不再适用，对现有船舶在公约生效之日起在 12 年内仍可保留其原有的吨位，但现有船舶若经过改建或改装，其总吨位有实质上的变更或船舶所有人要求适用公约规定者除外。②按 1969 年《国际船舶吨位丈量公约》对一艘国际航行的民用船舶勘定的吨位。其计算总吨位（GT）的公式为：$GT = k_1 V$，式中，$k_1 = 0.2 + 0.02 \lg V$（或取公约附录 2 表中所示的数值），V 为船舶的围蔽处所总容积（m³），计算净吨位的公式为：$NT = K_2 V_2 (4d/3D)^2 + k_3 (N_1/0.1 N_2)$，其中 $(4d/3D)^2$ 应不大于 1.0，$K_2 V_2 (4d/3D)^2$ 应不小于 $0.25GT$，NT 应不小于 $0.30GT$；V_2 为多个载货处所总容积（m³），$K_2 = 0.2 + 0.02 \lg V_2$（或取公约附录 2 表中所示的数值），$k_3 = 1.26[(GT + 10\,000)/10\,000]$，$D$ 为船舯处的型吃水（m）；N_1 为不超过 8 个铺位的客舱中的旅客数；N_2 为其他旅客数；当 $N_1 + N_2$ 小于 18 时，N_1、N_2 均取作零。式中型吃水 d 应为 按现行《国际船舶载重线公约》勘定的夏季载重线（木材载重线除外）；对客船应为按现行《国际海上人命安全公约》或其他国际规定所勘定的最深分舱载重线；不适用上述国际公约的船舶应为按国家要求所勘定的夏季载重线或最大的许可吃水；对其他船取船舯处的型深 D 的 75% 作为型吃水。

Regression analysis 回归分析 对随机的分析数据，如其相关系数足够大时，用最小二乘法等确定其表示变量间线性关系的算式中有关系的统计分析方法。

Regular sea 规则波 通常系指由远处风浪传播而成的涌而不随风力而变，其传播方向、波高及周期较为稳定的二因次波列。其波面可用简单函数式表达。

Regulating arm 调节杆

Regulating equipment 调节设备

Regulating lever 调整杆

Regulating screw 调节螺钉

Regulating spring 调节弹簧

Regulating system 调节系统（控制系统）

Regulating valve 调节阀

Regulation 调节（规则，条例）

Regulation D-1 D-1 标准 这是"压载水置换标准"的代号，它是"2004 年船舶压载水和沉积物控制和管理国际公约"规定的，在该公约生效的前后一段时间内运行船舶必须执行的压载水置换程序及要求，以避免船舶压载水无序随意排放，当而后在船舶执行"压载水性能标准"（D-2 标准）后，此 D-1 标准即停止执行。

Regulation D-2 D-2 标准 这是"压载水性能标准"的代号，它是"2004 年船舶压载水和沉积物控制和管理国际公约"规定的压载水允许排放必须达到的压载水性能标准。为此，船上必须装设压载水处理装置对压载水进行处理。D-2 标准的要求是：(1) 每立方米中最小尺寸大于或等于 50 微米的可生存生物少于 10 个；(2) 每毫升中最小尺寸小于 50 微米但大于或等于 10 微米的可生存生物少于 10 个；(3) 排放的指示微生物不应超过如下规定的浓度：①每 100 毫升小于 1 cfu 的有毒霍乱弧菌（01 和 0139）（cfu = 菌落形成单位）或小于 1 cfu/1 克的浮游动物样品（湿重）；②每 100 毫升 250 cfu 的大肠杆菌；③每 100 毫升 100 cfu 的肠道球菌。 Requirements of D-2 standard: (1) less than 10 viable organisms per cubic meter greater than or equal to 50 micrometers in minimum dimension; (2) less than 10 viable organisms per millilitre less than 50 micrometers in minimum dimension and greater than or equal to 10 micrometers in minimum dimension; (3) discharge of the indicator microbes shall not exceed the specified concentrations described below: ① Toxicogenic Vibrio cholerae (01 and 0139) with less than 1 colony forming unit (cfu) per 100 milliliters or less than 1 cfu per 1 gram (wet weight) zooplankton samples; ② Escherichia coli less than 250 cfu per 100 millilitres; ③ Intestinal Enterococci less than 100 cfu per 100 milliliters.

Regulations 规则 系指现行公约所附的规则。Regulations means the Regulations annexed to the present Convention.

Regulative 调整的（调节的调整器）

Regulator 调节器（稳定器）

Reheat 再热

Reheat boiler 再热锅炉

Reheat cycle 再热循环

Reheat factor 重热效率 反映多级压气机或涡轮中前面级的损失对后面级工作过程影响的系数。对压气机是各级绝热压缩功之和与总绝热压缩功之比；对涡轮则是各级绝热膨胀功之和与总绝热膨胀功之比。

Reheat steam turbine 再热式汽轮机 蒸汽膨胀到一定压力后被全部抽出再次加热到一定的过热温度,然后返回汽轮机重新膨胀做功的汽轮机。

Reheat steam turbine plant 再热循环汽轮机装置 实现再热循环的汽轮机动力装置。

Reheat treatment 再热处理

Reheated steam engine 再热蒸汽机 蒸汽部分膨胀后从气缸间的容气器处被全部抽出,经再热器加热到一定的过热温度,再到下一气缸继续膨胀做功的蒸汽往复机。

Reheater 再热器 使汽轮机高压级抽气回到锅炉中再次加热,以提高蒸汽过热温度的装置。

Reheating cycle 再热循环 在蒸汽发动机中,抽出一部分做过部分功的蒸汽来加热到一定过热温度,然后返回蒸汽发动机继续膨胀做功的热力循环。

Reheating furnace 再热炉(重热炉)

Reimbursement trade 补偿贸易 又称产品返销,系指交易的一方在对方提供信用的基础上,进口设备技术,然后以该设备技术所生产的产品,分期抵付进口设备技术的价款及利息。

Reinforce 增强

Reinforced concrete ship 钢筋混凝土船 用钢筋混凝土作为船体结构材料的船舶。

Reinforced insulation 加强绝缘(增强绝缘)
(1)用于带电部件的一种单一绝缘系统,它对电击所提供的保护等级,在有关 IEC 标准规定的条件下与双重绝缘相等效。注意:术语"绝缘系统"并不是指绝缘必须为一均质层。它可以由数层组成,但各层不能如辅助绝缘或主绝缘那样单独试验;(2)附加在带电部件,提供一定程度保护以防止电击,等效于双重绝缘的一种单绝缘系统。基本绝缘并非指绝缘必须是单层的,它可由数层组成,但不能单独作为辅助绝缘或基本绝缘进行试验。
Reinforced insulation is:(1)the single insulation system applied to live parts which provides a degree of protection against electric shock equivalent to double insulation under the conditions specified in the relevant IEC standard. Note: The term " insulation system " does not imply that the insulation must be one homogeneous piece. It may comprise several layers which cannot be tested singly as supplementary or basic insulation;(2)the single insulation system applied to live parts, which provides a degree of protection against electric shock equivalent to double insulation. The term "insulation system" does not imply that the insulation must be one homogeneous piece. It may comprise several layers which cannot be tested singly as supplementary or basic insulation.

Reinsurance 再保险 保险人为了减轻自身承担的保险责任而将其不愿意承担或超过自己承保能力以外的部分保险责任转嫁给其他保险人或保险集团承保的行为,又称分保。因这种办理保险业务的方法有再一次进行保险的性质,故称再保险。

Rejection risk 拒赔险 系指保险公司对被保险货物因在港口被进口国有关当局拒绝进口或没收予以负责,并按照被拒绝进口或没收货物的保险价值赔偿。但投保这项保险时,投保人必须提供货物进口所需的进口许可证和限额等一切手续,这些手续都齐全,保险人才接受投保。假若被保险货物起运后,进口国宣布实行任何禁止或禁运,保险公司仅承担赔偿运到出口国或转口到其他目的地所增加的运费,但最多不超过该批货物的保险价值。应该注意的是:保险公司在下列原因造成的任何损失不负赔偿责任:(1)违反上述四种条件中的任何一款;(2)市价跌落;(3)被保险货物记载的错误,商标或标记的错误,贸易合同或其他文件发生的错误或遗漏;(4)违反内地国有关当局关于出口货物的有关规定;(5)在被保险货物起运前,进口国已宣布实行禁运或禁止。

Related interests 有关利益 系指直接受到海上事故影响或威胁的沿岸国的利益。例如:(1)在海岸、港口或河口处的活动,包括构成有关人员基本谋生手段的渔业活动;(2)有关地区的旅游景点;(3)沿岸居民的健康与有关地区的福利,包括保护海洋生物资源和野生物。

Related interests means the interests of a coastal State directly affected or threatened by the maritime casualty, such as:(1) maritime coastal, port or estuarine activities, including fisheries activities, constituting an essential means of livelihood of the persons concerned;(2) tourist attractions of the area concerned;(3) the health of the coastal population and the well-being of the area concerned, including conservation of living marine resources and wildlife.

Relative 亲属 系指兄弟、姊妹、祖辈和直系孙辈。
Relative are brother, sister, ancesto, or lineal descendant.

Relative bearing 舷角 艏向与物标方位线之间的夹角。以艏向为零度,向右称右舷××度,向左称左舷××度,最大为180°。

Relative bearing 相对方位 系指自本船参照位置的目标位置的方向,表示为自本船艏向的角位移。
Relative bearing means a direction of a radar's position from own ship's reference location expressed as an angular displacement from own ship's bearing.

Relative blade temperature 叶片相对温度 以叶片平均温度与冷却介质温度之差对燃气平均温度与冷

却介质温度之差的比值表示涡轮叶片冷却效果的特性参数。

Relative course　相对航向　系指相对于本船方向的目标运动方向(方位)。 Relative course means a direction of motion of a radar relative to own ship's direction (bearing).

Relative density　相对密度　系指一定条件下,货物的质量与等体积淡水的质量之比值。

Relative motion　相对运动　系指相对航向和相对速度的结合。 Relative motion means a combination of relative course and relative speed.

Relative motion display mode　相对运动显示模式　系指本船的位置保持固定且所有目标相对本船移动的一种显示。 Relative motion display mode means a display on which the position of own ship remains fixed, and all targets move relative to own ship.

Relative rotation efficiency　相对旋转效率　螺旋桨船后效率对其敞水效率的比值,用以表明船后不均匀流场对螺旋桨的影响。$\eta_R = \eta_B / \eta_0$,式中,η_B—螺旋桨船后效率;η_0—螺旋桨敞水效率。

Relative speed　相对速度　系指相对本船速度数据的目标速度。 Relative speed means a speed of a target relative to own ship's speed data.

Relative turning diameter　相对回转直径　定常回转直径与船长的比值。

Relative vector　相对矢量　系指相对本船运动的目标的预计运动。 Relative vector means a predicted movement of a target relative to own ship's motion.

Relay-contactor control　继电接触器控制　是一种由专用主令控制器操纵,通过接触器对电动机启动、停止、变速和换向,并具有过载和失压保护功能的控制方式。

Release (of a mechanical switching device)　(机械开关电器的)脱扣器　系指与机械开关电器相连的、用其来释放保持机构而使开关电器打开或闭合的部件。

Release gear　脱钩装置

Release lever　放松杆(释放杆)

Release pressure　释压

Release spring　安全弹簧

Release valve　释压阀

Releaser　释放器(排除器)

Releasing hook　联动脱钩(释放钩)

Relevant chemicals　相关化学品　系指压载水管理系统使用过程中或之后在压载水或接收环境中产生的转换或反应产物,其可能会对船舶安全、水环境和/或人体健康造成影响。 Relevant chemicals means transformation or reaction products that are produced during and after employment of the ballast water management system in the ballast water or in the receiving environment and that may be of concern to the ship's safety, aquatic environment and/or human health.

Relevant indication　关联显示　系指评价中要求的非连续的状态或模样奇异的缺陷显示。不管是什么样的尺寸,尺寸超过 1.5 mm 才可视为关联显示。 Relevant indication means an indication that is caused by a condition or type of discontinuity that requires evaluation. Only indications, which have any dimension greater than 1.5 mm, shall be considered relevant indication.

Relevant laws　有关法律　系指按照有关国际协议、规定须受管制事项的有关地区法律,而就某一有关营运者而言,系指其装置所在地区的此类法律。

Reliability　可靠性　系指系统或部件在一个规定的时间间隔内执行其自身任务而无故障的能力。 Reliability means the ability of system or a component to perform its required function without failure during a specified time interval.

Reliability analysis　可靠性分析　分析失效情况及原因、检查、维护和修理以及工作时间和故障时间的定量评估。

Reliability block diagram　可靠性方块图　系指表示功能块与分析等级之间关系的逻辑图。 Reliability block diagram is a logical figure showing the relations between the functional blocks and the analytic level.

Relief　安全阀

Relief device　安全释放装置

Relief outlet　放泄口

Relief valve　释放阀(安全阀,泄压阀)

Relieve　释放

Reliever　解脱器(减压装置)

Relieving　消除(解除)

Relight　再点火(重新启动)

Re-liquefaction　再液化

Re-liquefaction installation　再液化装置

Re-liquefaction of gas (LG)　气体再液化　液化气体运输船载运的液化气体在一定的条件下可能会蒸发。为货品和运输安全起见,将蒸发的气体进行再次液化的过程。

Re-liquefaction system　再液化系统　系指将运输过程中不断蒸发的液货产生的超压蒸发气重新冷凝液化的系统。以降低舱内的压力。

Re-liquefaction unit　气体再液化装置

Re-machine　再加工

Remainder 余数

Remittance 汇付 也称汇款,是付款方通过银行将应付款项汇交收款方的支付方式。属于商业信用,采用顺汇法。汇付的优点在于手续简便、费用低廉。汇付的缺点是风险大,资金负担不平衡。汇付是国际贸易中最简单的结汇方式,简称汇款。在发达国家之间,由于大量的贸易是跨国公司的内部交易,而且外贸企业在国外有可靠的贸易伙伴和销售网络,因此,汇付是主要的结算方式。在分期付款和延期付款的交易中,买方往往用汇付方式支付货款,但通常需辅以银行保函或备用信用证,所以又不是单纯的汇付方式了。开户银行账户的资料可向开户银行查询。

Remittance 汇款

Remitter 汇款人 即汇出款项的人,在进出口贸易中,汇款人通常是进口人。

Remitting bank 汇出行 系指受汇款人委托而汇出款项的银行。汇出行通常是汇款人所在地银行,其职责是按汇款人要求将款项汇给收款人。

Remitting bank 托收银行 系指接受委托人的委托,完成收款业务的银行。一般而言,托收行是债权人所在地的银行。

Remote 很少可能性 对事故的概率而言,系指不可能每艘船舶都发生,但在同一类型的许多船舶的总使用期间内,其中的少数船可能发生。 For the purpose of probability of occurrences, remote is one which is unlikely to occur to every craft but may occur to a few craft of a type over the total operational life of a number of craft of the same type.

Remote automatic depth control system 遥控自动深度控制系统 将定深器装配在海洋地震电缆上,由船上进行遥控操作,能使海洋地震电缆控制在规定深度作业,并能自动控制海洋地震电缆在水下 0~30 m 范围内上浮下潜的一种控制系统。

Remote control 遥控 系指:(1)由操作人员通过机械、电气、电子、气动、液压、电磁(无线电)及光学方式或其组合方式远距离对设备进行操作的控制;(2)通过一个电气的或其他连接件来控制一个远距离设备。 Remote control means:(1)control of a device by an operator from a distance through mechanical, telerctrifcal, electronic, pneumatic, hydraulic, electromagnetic (radio)or optical means or combination;(2)the control from a distance of apparatus by means of an electrical or other link.

Remote control device 遥控器 是一种用来遥控机械的装置。现代的遥控器,主要是由集成电路电板和用来产生不同信息的按钮所组成。而客车门遥控器是采用最新技术编码解码,以闪断的方式控制门泵电磁阀以达到开关自动门的目的。用于客车(大巴、中巴)遥控开、关车门,避免驾驶员每次都需要上车开门的烦恼。遥控器的发射部分的主要元件为红外发光二极管。它实际上是一只特殊的发光二极管,由于其内部材料不同于普通发光二极管,因而在其两端施加一定电压时,它便发出的是红外线而不是可见光。

Remote control device for main engine 主机遥控装置 为加强动力装置的自动化及集中控制,以消除操作差错,提高反应的灵敏程度,改善轮机人员的工作条件而设置的远距离控制装置。常用的有驾驶室遥控和集控室遥控。按其传动的方式有机械式、气动式、电动式、电-气或电-液组合式等。机械式:结构简单、工作可靠、无须能源,通常用滑轮钢索传动、软轴传动、链轮钢索传动及杠杆传动等。其缺点是操作费力,传动距离短(一般不超过 15 m),动作到位的精确度较差。气动式:结构比较简单、工作可靠、准确度和灵敏度较高,传动力矩大,能进行较远距离控制(可达 100 m 左右)。其缺点是装置质量大,需要气源供气,装置结构及管路布置复杂。电动式:传动距离不受限制,传动灵敏且精确度高,易实现多位控制、程序控制、连锁及自动化。其缺点是传动力矩小,元件数量多,维修比较复杂。液压式:具有较高的传动精度和灵敏度,传动力矩大,工作平稳可靠。其缺点是系统复杂,维修困难。电-气或电-液组合式:指令由电力输送,而执行部分用液力或气力实施,从原理上吸取了两者的优点,但缺点是需用两种能源,易出故障或失效。

Remote control equipment 遥控设备

Remote control from bridge(BRC) 驾驶室遥控 系指推进装置由驾驶室控制站控制,机器处所周期无人值班。 Remote control from bridge(BRC)means main propulsion machinery remotely controlled from bridge control space(BCS).

Remote control handle 遥控手柄

Remote control system 遥控系统 系指:(1)操作人员可以控制船舶的推力、推力方向及旋转力的一种半自动化控制系统;(2)从一个控制地点操作若干装置的所有必要设备组成的系统。在该控制地点操作者不能直接观察其动作的结果;(3)由操作人员通过机械、电气、电子、气动、液压、电磁(无线电)及光学方式或其组合方式远距离对设备进行操作的控制。 Remote control system is:(1)a semi-automatic control system, which enables the operator to give a defined thrust (force and direction)and a turning moment to the vessel;(2)remote control systems comprise all equipment necessary to operate units from a control position where the operator cannot directly observe the effect of his actions;(3)control of a device by an operator from a distance through mechanical, electrical, electronic, pneumatic, hydraulic, electromagnetic (radio) or optical means or combination thereof.

Remote control valve 遥控阀

Remote controlled Niskin sampler 尼斯金遥控采水器 由吊装在一个架上的多个采水瓶组成的装置。可用压力或从甲板上通过一根单芯电缆逐个控制。

Remote display unit 遥控显示装置

Remote method invocation (RMI) 远程方法调用

Remote operated vehicle(ROV) 遥控潜水器 即水下机器人,有时为了区别于地面和空中的 ROV(Remote control vehicle),又将海洋中所用的 ROV 称为 Remotely operated underwater vehicle。系统组成包括:动力推进器、遥控电子通信装置、黑白或彩色摄像头、摄像俯仰云台、用户外围传感器接口、实时在线显示单元、导航定位装置、自动舵手导航单元、辅助照明灯和凯夫拉零浮力拖缆等单元部件。功能多种多样,不同类型的 ROV 用于执行不同的任务,被广泛应用于军事、海岸警卫、海事、海关、核电、水电、海洋石油、渔业、海上救助、管线探测和海洋科学研究等各个领域。见图 R-2。

图 R-2 遥控潜水器
Figure R-2 Remote operated vehicle(ROV)

Remote operating panel 遥控操作面板

Remote operating position 遥控操作位置

Remote probability 很少可能的概率 系指故障情况在每艘船舶的整个使用期内不大可能发生,但考虑某种类型的若干船舶的整体使用期限时可能会发生几次(最坏 10^{-5})。 Remote probability means the failure conditions are unlikely to occur to each craft during its total life but may occur several times when considering the total operational life of a number of craft of a type (at worst 10^{-5}).

Remote reading 遥测读数

Remote release device of fall-safe type 故障防范型的远程关闭装置 系指以远程操作方式,脱开卡销或其他同等器具。同时,在其装置发生故障时,门可以自行关闭。 Remote release device of fall-safe type means the system which releases hooks or other equivalent device by remote operation, and that the door is automatically closed even in case of failure of the system.

Remote sensing 遥感

Remote shutdown 遥控切断

Remote switch 遥控开关

Remote terminal 远程终端

Remote(probability of occurences) 微小的(事件的概率) 是对于每一艘船舶都不大可能发生的,但是在一些同型船舶的整个运行生命期内有少数这种类型的船上会发生。 Remote is one which is unlikely to occur to every ship but may occur to a few ships of a type over the total operational life of a number of ships of the same type.

Remote-control system for main engine and contraollable pitch propeller 主机和可调螺距螺旋桨的遥控系统 系指从船舶的驾驶室对主机和可调螺距螺旋桨进行远距离操作,以控制船舶航速的系统。

Remote-controlled towing hook 遥控拖钩 能远距离控制迅速解脱拖缆的拖钩。一般在驾驶室进行遥控。

Remotely controlled closing means 遥控关闭装置

Remotely controlled stop device 遥控停止装置

Remotely controlled vehicle (ROV) 遥控深潜器 系指用于进行多种水下作业的潜水装置。其作业范围从在敷设管路时观察其着底情况到设备(如总管)等的操作情况。 Remotely controlled vehicle (ROV) is a submarine vehicles used to conduct many underwater operations ranging from observation of pipeline touchdown during installation to operation of equipment such as manifolds, etc.

Remotely operated actuator 遥控启动装置

Remotely operated valve 遥控操作阀

Remotely operated vehicle 遥控作业器

Remotely operated vehicle (ROV) 有缆遥控水下机器人 一种通过脐带与水面连接来获得动力,并由经验丰富的驾驶员进行操纵的水下机器人。由于配备脐带缆,使 ROV 变得轻便、简单可靠、作业能力强;但是脐带缆也限制机器人的活动范围,容易被不明物体刮碰缠绕。

Removing rust on an aerial platform 高空除锈作业车 系指高空除锈作业的机动工具。

Render 递交

Renew an offer 重新报价

Renewable energy act (REA) 可再生能源法案 2000 年以来该法案对可再生能源馈入公共电网进行了相关规定。以下是关于海上风电场电力馈入的具体规定:运营的前 12 年是 13 欧分每千瓦时。若该项目到 2015 年 12 月 31 日仍在运营中,其间电价涨到 15 欧分每千瓦时。运营 12 年后电价降低到 3.5 欧分每千瓦时。然而,若风电场距离海岸 12 海里以上,并且建址水深超过 20m,高初始电价将持续 12 年以上。

Renewal audit 更新审核 系指长期证书到期前 3 个月内必须进行的换证审核。

Renewal examination 重新检查 系指类似于符合性检查的检查。Renewal examination means a similar examination to the compliance examination.

Renewal survey 换证检验 对每艘 150GT 及以上的油船和 400GT 及以上的其他船舶而言,换证检验系指按主管机关规定的间隔期限进行的检验。但不得超过 5 年。但国际防止船舶造成污染公约附则 1 第 10.2.2, 10.5、10.6 或 10.7 条适用者除外。换证检验应确保其结构、设备、系统、附件、布置和材料完全符合上述附则的适用要求。 For every oil tanker of 150 gross tonnage and above, and every other ship of 400 gross tonnage and above, renewal survey means the survey proceeded at intervals specified by Administration, but not exceeding five years, except where regulation 10.2.2, 10.5, 10.6, or 10.7 of the MARPOL Annex Ⅰ is applicable. The renewal survey shall be such as to ensure that the structure, equipment, system, fitting, arrangements and material fully comply with the applicable requirements of the above Annex.

Renewal survey for the International Certificate of Inventory of Hazardous Materials 《国际有害物质清单证书》换证检验 系指验证香港国际安全与环境无害化拆船公约附则第 5 条所要求的有害物质清单 第Ⅰ部分符合 SRC 的要求的检验。检验合格后,换发有效期由主管机关规定,但最多不超过 5 年的 ICIHM。Renewal survey means which at intervals specified by the Administration, but not exceeding five years. This survey shall verify that Part Ⅰ of the Inventory of Hazardous Materials required by regulation 5 complies with the requirements of the Hong Kong International Convention for the Safe and Environmentally Sound Recycling of Ships, 2009.

Renewal survey on WIG craft 地效翼船的换证检验 系指按主管机关规定的时间间隔进行的检验,但不超过 5 年。Renewal survey on WIG craft is a survey at intervals specified by the Administration but not exceeding 5 years.

Renewal thickness $t_{renewal}$ 换新厚度 即最小许可厚度,mm。低于此值则应对结构构件换新:$t_{renewal} = t_{as-built} - t_c - t_{voluntary-addition}$。Renewal thickness is the minimum allowable thickness, in mm, below which renewal of structural members is to be carried out: $t_{renewal} = t_{as-built} - t_c - t_{voluntary-addition}$.

Repair completion 修理完工 系指计划的修理工程,按合同要求完成的状态。

Repair ship 修理船 具有必要的修理设施,专为船舶进行临时性抢修或小修用的船舶。

Repair work 修理工程 系指船体勘验、船舶机械设备拆检、修理、装配和调试等作业。

Repairable failure in the machinery 机械的可修复的故障 系指设计上预定能在船上修复的故障,并满足下列条件:(1)机械应设计布置成能在船上修理;(2)船上应备有可用以永久修复或临时修复该机械所必需的,超出规范对具体设备规定的备件或完整的备件组;(3)船上应备有进行修理工作的工具、维修说明书和其他必需的设备。Repairable failure in the machinery is a failure which is planned to be repaired on board and for which the following conditions are fulfilled:(1) the machinery is arranged and designed to allow for repair work at sea;(2) spare parts or complete spare units necessary for permanent or provisional repair are kept on board also to the extent that these spare parts exceed specific Rule requirements for the actual components;(3) tools, instruction material and other necessary facilities to perform the repair work are found on board.

Repairs with add cost 加账工程 系指超过合同中规定工程以外的工程。

Replenishment at sea (underway replenishment) 海上补给 在海上由补给船对舰船、渔船等进行定期或不定期的补充物资给养的作业。

Reply immediately 速复

Reply promptly 即复

Report of Planned Start of Ship Recycling(RPSSR) 拆船计划开工报告

Report prepared by the appraiser 鉴定报告

Report prepared by the expert 专家报告

Representative sample 代表性样品 系指具备与被取样品总量的平均特性相同的物理和化学特性的产品试样。Representative sample is a product specimen having its physical and chemical characteristics identical to the average characteristics of the total volume being sampled.

Representative space/tank 代表性处所/舱室 系指具有类似的形状和用途按照类似的防腐蚀系统的形式设置,可以反映别的处所/舱室状态的代表性处所/舱室。选择代表性处所/舱室时要考虑该船的运行及修理记录和被发现的结构薄弱地区或可疑地区。Representative space/tank is a space/tank which is expected to reflect the conditions of other spaces/tanks of similar type and service and with similar corrosion prevention systems. When selecting representative spaces/tanks, account is to be taken of the service and repair history on board and identifiable critical structural areas and/or suspect areas.

Representative tanks 代表性液舱 系指能反映类似型号、用途和具有类似防腐系统的其他液舱的液舱。在选择代表性液舱时，应考虑到其营运和修理史以及可视为相同的临界和/或可疑区域。Representative tanks are those which are expected to reflect the condition of other tanks of similar type and service and with similar corrosion protection systems. When selecting representative tanks account should be taken of the service and repair history onboard and identifiable critical and/or suspect areas.

Representative test sample 代表性物质 系指数量足够的样品，用以检测托运货物的理化特性是否达到规定的要求。 Representative test sample means a sample of sufficient quantity for the purpose of testing the physical and chemical properties of the consignment to meet specified requirements.

Representatives 代表（国际海事卫星组织） 就议定书成员国、总部国和签字国而言，系指赴国际海事卫星组织的代表，在具体情况下，系指代表团团长、副团长和顾问。Representatives in the case of Parties to the Protocol, the Headquarters Party and Signatories means representatives to INMARSAT and in each case means heads of delegations, alternates and advisers.

Requested materials for arbitration 仲裁申请材料 对简易程序（争议金额在人民币50万元以下）而言：(1)《仲裁申请书》一式共3份；(2)仲裁协议、合同文本及附件共3份；(3)支持仲裁请求的票据、信函、材料、及与本案有关的证明材料、证明文件一式共3份；(4)证据目录清单3份；(5)当事人双方主体资格证明文件。对一般程序（争议金额超过人民币50万元）提交材料内容与简易程序相同，除前第5项外，其他材料均应提供一式5份。

Required alert or indicator 要求的报警器或指示器 系指经修正的"1974年海上人命安全公约"（1974年SOLAS公约）、相关的规则["散装运输危险化学品船舶构造和设备规则"（BCH）、"潜水系统规则"、"国际消防安全系统规则"（FSS）、"国际气体运输船舶规则"、"2000年高速船规则"（2000 HSC）、"国际散装运输危险化学品船舶构造和设备规则"（IBC）、"国际散装运输液化气体船舶构造和设备规则"（IGC）、"国际海运危险货物规则"（IMDG）、"国际救生设备规则"（LSA）、"2009年近海移动钻井平台规则"（MODG）以及"核商船规则"]；经修正的"经1978年议定书修订的1973年国际防止船舶造成污染公约"（MAPOL73/78）,"托里莫利诺斯渔船安全公约 1993 年托里莫利诺斯议定书"[1993年托里莫利诺斯（SFV）议定书],"安全配员原则"；"惰性气体系统（IGS）指南"；"蒸发气释放控制系统（VEC）标准"；"驾驶室航行值班报警系统（BNWAS）的性能标准"，以及"经修订的综合航行系统（INS）的性能标准"所要求的报警器或指示器。 Required alert or indicator means an alert or indicator required by the International Convention for the Safety of life at sea,1974 (1974 SOLAS Convention), as amended, associated codes [BCH, Diving, FSS, Gas Carrier, 2000 HSC, IBC, IGC, IMDG, LSA, 2009 MODU, and Nuclear Merchant Ship Codes]; the International Convention for the prevention of Pollution from Ships, 1973, as modified by the Protocol of 1978 relating thereto (MARPOL, 73/78), as amended the Torremolinos Protocol of 1993 relating to the Torremolinos International Convention for the Safety of Fishing Vessel [1993 Torremolinos (SFV) Protocol]; the Principles of Safe Manning; the Guidelines for Inert Gas Systems (IGS); the Standards for Vapour Emission Control Systems (VEC); the Performance Standards for a Bridge Navigational Watch Alarm System (BNWAS); and the Revised Performance Standards for the Integrafed Navigation System (JNS). Any other alerts and indicators are referred to in this Code as non-required alerts or indicators.

Required EEDI 要求的能效设计指数 系指有关要求对特定尺度和船型所允许的获得的能效设计指数的最大值。 Required EEDI is the maximum value of attained EEDI that is allowed by relevant requirement for the specified ship type and size.

Required renewal thickness t_{ren} 要求的换新厚度

Required thickness at annual survey t_{annual} 年检中所要求的厚度 系指由建造厚度 $t_{asObuild}$ 减去船东额外余量 t_{own} 和总损耗允差 t_{was} 而得的厚度。

Requirement for lifesaving equipment 救生设备配置定额 按船舶航区、类型、主尺度对船上配置某种形式救生设备的数量所做的规定。

Re-registered 重新注册 系指船舶在一登记地区和所有权同时发生时进行的重新注册。Re-registered means re-registered on the occasion of a simultaneous change in the territory of registration and ownership of the vessel.

Rescue 救助 (1)人员被带到最终安全地点的过程。(这个定义不包括进行搜索和救助的管理能力，但涵盖了在紧急情况下定位和救助人员的能力)。(2)援救遇险人员，向其提供基本的医疗需求或其他需求，并把他们转移到安全处所的行为。 Rescue is: (1) a process by which personnel are taken to an ultimate point of safety. (This definition does not include the ability to conduct Search and Rescue but covers the ability to locate and rescue personnel in an emergency). (2) an operation to retrieve persons in distress, provide for their initial medical or other needs, and deliver them to a place of safety.

Rescue & salvage barge 救捞驳 系指协助救捞船进行海上救助、打捞作业用的非自航驳船。用于运送和储存救助、打捞、抢险工作所必需的各种器材、备品、淡水和燃料等。有良好的耐波性,能在恶劣气象下进行拖带作业。

Rescue bell 救生钟 配置在救生船上,能与失事潜艇的救生钟平台对口相接,专用于从事潜艇中营救艇员集体脱险出水的、钟型耐压容器。

Rescue boat 救助艇 系指:(1)任何一种易于推进、易于操纵,且能由少数船员既容易又能迅速使之下水,并适用于营救落水人员上船的小艇;(2)为救助遇险人员及集结救生艇筏而设计的艇。 Rescue boat is: (1) an easily propelled highly manoeuvrable boat capable of being easily and quickly launched by a small number of crew and adequate for rescuing a man overboard; (2) a boat designed to rescue persons in distress and to marshal survival craft.

Rescue co-ordination centre 救助协调中心 在搜救区域内负责推动各种搜救服务有效组织的和协调搜救工作指挥的单位。 Rescue co-ordination centre means a unit responsible for promoting efficient organization of search and rescue services and for co-ordination the conduct of search and rescue operations within a search and rescue region.

Rescue of personnel goal 1 人员救助目标1 每艘舰船应进行切实可行的设计,防止人员意外落入海中的危险:(1)抛绳器应满足 SOLAS 公约第Ⅲ章 B 部分第 18 条的要求;(2)通常,所有舰船上救生圈的配置应满足 SOLAS 公约第Ⅲ章 B 部分第 7.1 条的要求。关于救生圈的标记,所有救生圈都应标记舰船的识别号码。 Rescue of personnel goal 1 means that every ship is to be designed to prevent the risk of an accidental man overboard situation as far as is practicable: (1) Line throwing appliances are to be in accordance with SOLAS Part B, Chapter Ⅲ, Regulation 18; (2) In general, the provision of life buoys on all ships is to be in accordance with SOLAS Chapter Ⅲ, Part B, Regulation 7.1. With regard to the marking of life buoys, all life buoys are to be marked with the ship's identification number.

Rescue of personnel goal 2 人员救助目标2 每艘舰船都应有合适的配置以救助落水中人员,在舰船上:(1)救助艇降放和回收布置应符合 SOLAS 公约第Ⅲ章 B 部分第 17 条的要求;(2)救助艇的存放应符合 SOLAS 公约第Ⅲ章 B 部分第 14 条的要求。 Rescue of personnel goal 2 means that every ship is to be suitably equipped for the mass rescue of personnel from the water, on board: (1) Rescue boat launching and recovery arrangements are to be in accordance with SOLAS Part B, Chapter Ⅲ, Regulation 17. (2) Rescue boats are to be stowed in accordance with SOLAS, Part B, Chapter Ⅲ, Regulation 14.

Rescue of personnel goal 3 人员救助目标3 每艘舰船及其救生艇筏应安装设备以确保舰船及其救生艇、筏在必要时可有效就位和回收;(1)每艘舰船的两舷至少各配置一个雷达应答器。雷达应答器应定位安装,使其能用于任何救生艇,救生筏除外;(2)提供的烟火信号应满足 SOLAS 公约第Ⅲ章 B 部分第 6.3 条的要求。 Rescue of personnel goal 3 means that every ship and its survival craft is to be fitted with equipment to ensure that the ship and its survival craft can be efficiently located and retrieved as necessary: (1) Each ship is to be fitted with at least one radar transponder on both sides. The radar transponders are to be located such that they can be readily deployed on any survival craft, other than the life rafts. (2) The provision of flares is to be in accordance with SOLAS Chapter Ⅲ, Part B, Regulation 6.3.

Rescue of personnel objective 人员救助目标 每艘舰船及其救生设备应配备合适的装备以便于定位和将落水人员从水中救起。 Rescue of personnel objective means that every ship and its life saving equipment are to be suitably equipped to locate and rescue personnel from the water.

Rescue ship 救助船/救生船 系指担负海上防险救助任务、搜救失事船舶及船员的船舶。 Rescue ship is a ship engaged in rescue operations at sea for ships and crew.

Rescue sub-centre 救助分中心 在搜救区域的特定地区内为辅助救助协调中心而设置的附属于该中心的单位。 Rescue sub-centre means a unit subordinate to a rescue co-ordination centre established to complement the latter within a specified area within a search and rescue region.

Rescue tug 救助拖船 备有压缩空气、排水、潜水和消防等设备,供抢救搁浅、触礁和失去机动能力的船舶以及拖曳打捞浮筒等兼负水上救助任务的拖船。

Rescue unit 救助单位 由受过训练的人员组成并配有适于迅速执行搜救工作设备的单位。 Rescue unit means a unit composed of trained personnel and provided with equipment suitable for the expeditious conduct of search and rescue operations.

Rescue 救助 指定的船舶为安全起见,将遇险落水的本船人员,或者协助海上平台、驳船、生产模块/船舶或其他船舶将其遇险落水人员救起的作业。 An operation where a defined vessel is, either bringing own personnel being in distress in the water to safety, or is assisting an offshore platform, barge, production module/vessel or another ship in bringing their personnel being in distress in the water to safety.

Research ship 调查船 系指专用于海洋科学考察研究、测量勘探等的船舶。 Research ship means a ship specialized for marine research and study, survey, exploration, etc.

Reservation 保留(储备)

Reserve 储备(储量,储备品)

Reserve bunker 备用燃料舱(备用煤舱)

Reserve buoyancy 贮备浮力 系指设计水线以上船体水密部分的体积所提供的浮力,其值通常以设计排水体积的百分数表示。

Reserve capacity 储备容量(储备功率)

Reserve feed water 储备给水

Reserve generator 备用发电机

Reserve navigation light 备用号灯 为防备固定配置的号灯因故不能工作而配备的由蓄电池、燃油等供给能源的号灯。

Reserve power 储备功率 为了克服速度亏损所增加的主机功率。

Reserve protection 后备保护

Reserve thickness $t_{reserve}$ 预留厚度 即预计在一个2.5年的检验间隔期内可能发生的厚度减薄量,mm ($t_{reserve}=0.5$ mm)。 Reserve thickness is the thickness, in mm, to account for anticipated thickness diminution that may occur during a survey interval of 2.5 year ($t_{reserve}=0.5$ mm)。

Reserved space 备用舱(备用处所)

Reservoir 储器(蓄水池)

Residual 剩余的(残余的,留剩的)

Residual accident event 残余意外事件 海上设施设计时没有考虑到的意外事件,因而将会成为海上设施风险等级的一部分。

Residual fuel 残余燃油 系指:(1)输送到船上并用于燃烧的燃油,其在40℃时的运动黏度大于11.00 cst (mm²/s);(2)输送到船上并在船上使用的用于燃烧目的的燃油在50℃时其运动黏度大于或等于30.0 cst。 Residual fuel means:(1)the fuel oil for combustion purposes delivered to and used on board ships with a kinematic viscos-ity at 40℃ greater than 11.00 centistokes(mm²/s);(2)fuel oil for combustion purpose delivered to and used on board ships with a kinematic viscosity at 50℃ greater than or equal to 30.0 centistoke.

Residual fuel oil 残余燃料油

Residual fuel pump 渣油泵

Residual gas 残余废气 完成一个工作循环后,残留在气缸内并参与下一工作循环的废气。

Residual oil 残油(油渣)

Residual resistance 剩余阻力 从船模或实船的总阻力减去由摩擦阻力公式计算出来的相当平板摩擦阻力所得之值。

Residual resistance coefficient 剩余阻力系数 以剩余阻力对动压力与湿表面乘积的比值,表示剩余阻力特征的一个无量纲的参数。$C_R = R_R/[(1/2)\rho V^2 S]$,式中,$R_R$—总阻力;$V$—船速;$S$—湿面积;$\rho$—流体质量密度。

Residual risk 残余风险 风险处理后剩余的风险。

Residual wave height (H_{rw}) 剩余波高 取为超过运行区域20%的有义波高:$H_{rw}=0.90$ Hs (m)。 Residual wave height H_{rw} is to be taken as the significant wave height that has a 20 percent probability of being exceeded for the service area:$H_{rw}=0.90$ Hs (m)。

Residual wave period range (T_{rrange}) 剩余波浪周期范围 取为标准设计波浪周期:$T_{rrange}=T_{drange}$ (s)。 Residual wave period range T_{rrange} is to be taken as the standard design wave period range:$T_{rrange}=T_{drange}$ (s)。

Residual wave period T_{rw} 剩余波浪周期 取为标准设计波浪周期:$T_{dw}=Tz$ (s)。 Residual wave period T_{rw} is to be taken as the standard design wave period:$T_{dw}=Tz$ (s)。

Residue 残余物 系指任何需处理的有毒液体物质。 Residue means any noxious liquid substance which remains for disposal.

Residue/water mixture 残余物/水混合物 系指以任何目的加入水的残余物(例如清洗油舱、加压载水、舱底含油污水)。 Residue/water mixture means residue to which water has been added for any purpose (e.g. tank cleaning, ballasting, bilge slops).

Residues and mixtures containing NLSs(NLS residues) 含有毒液体物质的残余物和混合物(NLS 残余物) 系指:(1)因不满足收货方技术规格书要求而留在船上的任何 A,B,C 或 D 类有毒液体物质;(2)将有毒液体物质交付收货方后仍存在船上的任何部分的 A,B,C 或 D 类有毒液体物质,包括但不限于在舱底和污水阱内的糊状物,在液舱的残液和留在管路内的物质;或(3)黏有 A,B,C 或 D 类有毒液体物质的任何材料,包括但不限于舱底污液、压载水、软管滴水盘的液体和洗舱水。Residues and mixtures containing NLSs means:(1) any category A,B,C or D NLS cargo retained on the ship because it fails to meet consignee specifications;(2) any part of A, B, C or D NLS cargo remaining on the ship after the NLS is discharged to the consignee, including but not limited to puddles on the tank bottom and in sumps, clingage in the tanks, and substance remaining in the pipes; or (3) any material contaminated with A, B, C or D NLS cargo, including but not limited to bilge slops, ballast, hose drip pan contents, and tank wash water.

Resistance　阻力　作用于物体,与物体前进运动方向相反,阻止物体前进的力。

Resistance acoustic filter　抗性消音器　不是以吸声材料直接吸收声能为主达到消音的,而是应用阻抗失配的原理阻挡声音通过的消音器。其在结构上完全不同于阻性消音器,而是依靠管道截面的突变或旁ávr共振腔来改变声波传播的阻抗,从而产生声能的反射、干涉及共振吸声来降低消音器向外辐射的声源,达到消音的目的。

Resistance augment factor　阻力增额因数　螺旋桨推力对总阻力的比值。$1 + \alpha = T / R_T$,式中,T—螺旋桨推力;R_T—船的总阻力。

Resistance augment fraction　阻力增额分数　阻力增额对船的总阻力之比值。$\alpha = (T - R_T)/ R_T$,式中,T—螺旋桨推力;R_T—船的总阻力。

Resistance coefficient (drag coefficient)　阻力系数　以阻力对动压力与湿表面积乘积的比值。表示阻力特征的一个无量纲的参数。$C_R = R/[(1/2)\rho V^2 S]$ 或 $C_D = D/[(1/2)\rho U^2 A]$,式中,$R$ 或 D—阻力;S 或 A—湿面积或面积特征值;V—船速;U—流速;ρ—流体密度。

Resistance coefficient corresponding to K_T and K_Q　对应阻力系数　将船舶的阻力看作是螺旋桨的有效推力,对照推进系数 K_T 和转矩系数 K_Q 使用的相应船舶阻力系数 $K_R = R/(n^2 D^2)$,式中,R—螺旋桨所克服的舱壁阻力;ρ—水的密度;n—螺旋桨转速;D—螺旋桨直径。

Resistance coefficient for model-ship correlation allowance　阻力换算修正系数　以阻力换算修正值对动压力与湿表面积乘积的比值。表示阻力换算修正值特征的一个无量纲系数:$C_A = R_A /[(1/2)\rho V^2 S]$,式中,$R_A$—船模实船阻力换算修正值;$S$—湿面积;$V$—船速;$\rho$—水的密度。

Resistance of bilge eddy　舭涡阻力　船舶航行中消耗于产生舭部漩涡的能量所形成的阻力。

Resistance of hovering craft　气垫船的快速性　除空气阻力外还有水阻力。气垫船距水面的高度很小,围裙总难免与水面接触,特别是在波浪上,不可避免地有围裙阻力分量。气垫船在水面航行时,相当于水面上有一个矩形压力面在运动,水的表面自然要兴波,产生兴波阻力,兴波阻力随速度加大而增加,超过峰点后就随速度增大而减少,兴波阻力及其峰值阻力与气垫长宽比有关,还与单位长度气垫压力有关。此外,为了形成气垫向气垫内供气,而航行时带着这些气量会形成空气动量阻力。侧壁式气垫船的阻力包括:空气阻力、气垫兴波阻力、动量阻力(momentum resistance)、围裙浸湿阻力(wetting resistance)、围裙触地阻力(skirt over ground resistance)、侧壁阻力(sidewall resistance 包括侧壁摩擦阻力、侧壁兴波阻力、侧壁型阻力)、附体水阻力和波浪附加阻力等。全垫升气垫船的阻力包括:空气阻力、气垫兴波阻力、空气动量阻力、围裙浸湿阻力和波浪附加阻力等。

Resistance of ships (drag of ships)　船舶阻力　系指:(1)研究船舶航行时所受阻力,船体线型设计及其试验的一门学科;(2)流体作用于船体上,与船舶运动方向相反,阻止船舶运动的力。

Resistance of steering surface　操纵面阻力　由操纵船舶航行方向的操纵面,如舵与稳定翼等所产生的阻力。

Resistance of trailing stream (wake resistance)　尾流阻力　用测量尾流中动量的方法所求得的阻力。

Resisted rolling　有阻尼横摇　船舶在水中受阻尼作用时的横摇。

Resister　电阻器　系指其主要用途是将电阻引入某一电路的一种器件。电路中为实现操作、保护或控制而使用的电阻器通常是由若干个电阻元件组合而成。通常所供应的电阻器是由绝缘介质或埋置在绝缘介质内的金属线、金属带、金属铸件或碳化合物组成。绝缘介质可用来密封和支承此种电阻材料,例如瓷管型的电阻器或只在其支承点处提供绝缘;又例如安装在金属构架上的重负载型金属带或铸铁栅型电阻器。

Resistive and resistance compound acoustic filter　阻抗复合型消音器

Resistive sound eliminator　阻性消音器　系指一种将吸声材料固定在气流通过的管道周围或按一定方式排列在管道中,当声波进入消音器中,引起多孔材料中的空气和纤维振动,由于摩擦和黏滞阻力,使一部分声能转化为热能而耗散的消音器。

Resolution　分辨率　是业界衡量打印质量的一个重要标准。它本身表现了在每英寸的范围内喷墨打印机可打印的点数。单色打印时 DPI 值越高打印效果越好。而彩色打印时情况比较复杂。通常打印质量的好坏要受 DPI 值和色彩调和能力的双重影响。由于一般彩色喷墨打印机的黑白打印分辨率与彩色打印分辨率可能会有所不同,所以选购时一定要注意看制造厂告诉你的分辨率是哪一种分辨率,是否是最高分辨率。一般至少应选择在360DPI以上的喷墨打印机。

Resolutions 决议 系指:(1)自愿地由各船旗国作为其国家规则,一旦采用,对该国即具有强制性;(2)通常只涉及有限的技术领域,例如 IMO 第17届大会决议 A.695(17)"在406MHz运行的自由漂浮式卫星应急无线电示位标的性能标准"等;(3)决议是制订规则和公约的第一步;(4)船舶设计、建造中用得最多的决议为大会(Assembly, A)、海上安全委员会(Maritime Safety Committee; MSC)和海上环境保护委员会(Maritime Environment Protection Committee, MEPC)等的决议。

Resonance 共振(谐振,共鸣)
Resonance characteristic 谐振特性
Resonance check test 共振检查试验
Resonance curve 共振曲线(谐振曲线)
Resonance effect 共振现象(谐振现象)
Resonance frequency 共振频率(谐振频率)
Resonance inspection test 共振检查试验
Resonance level 谐振级(共振级)
Resonance sound absorption structure by sheet 薄板(膜)共振吸声结构 系指在薄板或薄膜的背面安装龙骨和垫块使背面留有一定的空间,形成共振的声学空腔。

Resonance zone 谐振区(共振区)
Resonant 共振的(谐振的)
Resonant(synchronous rolling) 谐摇 船舶在波浪中的遭遇周期与船舶的固有摇摆周期相等时的同步摇摆。
Resonant frequency 共振频率(谐振频率)
Resonant load 瞬时载荷
Resonant rolling 共振横摇(谐振横摇)
Resonant vibration 谐振(共振)
Resonator 谐振器(共鸣器)
Resorber 吸收器(吮吸器)
Resources 资源 系指区域内在海床及其以下原来位置的一切固体、液体或气体矿物资源,其中包括多金属结核。 Resources mean all solid, liquid or gaseous mineral resources in *situ* in the area at or beneath the sea bed, including poly-metallic nodules.

Respiratory sensitization 呼吸道过敏 系指符合以下情况的货物确定为呼吸道过敏剂:(1)如果证实该物质可导致人体呼吸道过敏症状;和/或(2)如果相关动物测试结果呈阳性;和/或(3)如果货物标为呼吸过敏剂,且无证据证明非呼吸过敏剂。 Respiratory sensitization means: A product is classified as a respiratory sensitizer: (1) if there is evidence in humans that the substance can induce specific respiratory hypersensitivity; and/or (2) where there are positive results from an appropriate animal test; and/or (3) where the product is identified as a skin sensitizer and there is no evidence to show that it is not a respiratory sensitizer.

Respondent 被申请人
Respondent/Defendant 仲裁被诉人
Responder 应答器 一种发射应答器。其中电子脉冲沿一根电缆发送询问。通常安装在 ROV 中,并沿 ROV 脐带发送询问。
Response amplitude operator 响应幅值算子 频率响应函数绝对值的平方值。 $\mid H(\omega) \mid^2 = S_{yy}(\omega)/S_{xx}(\omega)$。
Response times 响应时间 旨在反映警报声开始后撤离前的行为所花费的总时间。这包括信号感知能力和指令解析之类问题,个人反应时间和所有其他各种撤离前的行为。 Response times are intended to reflect the total time spent in pre-evacuation movement activities beginning with the sound of the alarm. This includes issues such as cue perception provision and interpretation of instructions, individual reaction times, and performance of all other miscellaneous pre-evacuation activities.

Responsibility and authority 职责和职权 对质量管理体系而言,系指在质量管理系统中的高级人员的职责和职权,应有明确的文件规定。 For the purpose of quality management system, responsibility and authority means that the responsibilities and authorities of senior personnel within the quality system are to be clearly documented.

Responsible person 负责人员(资格人员) 系指:(1)由雇主、该船的船长或该装置的所有者指定的人员。在这种情况下对指定职责负责并且具有足够的知识和经验以及适合承担该职责的人员;(2)具有足够的理论知识和实践经验并能对危险空气存在或随后将在该处所产生的可能性做出报告性评估通报的人员。 Responsible person means:(1) a person appointed by the employer, the master of the ship or the owner of the gear, as the case may be, to be responsible for the performance of a specific duty or duties and who has sufficient knowledge and experience and the requisite authority for the proper performance of the duty or duties;(2) a person authorized to permit entry into an enclosed space and having sufficient knowledge of the procedures to be followed.

Responsible person 责任人员(澳大利亚海事安全局海事指令) 系指与材料与物资装运设备有关的个人,他是有能力的和合格的,并(1)对以下负责:①该设备的制造厂商;②执行这些设备的入级或发证计划的船级社;③有能力的试验机构。根据澳大利亚海事安全局海事指令的要求进行各种试验和相关的全面检查,且颁发与该设备有关的试验的证书;(2)对以下负责:①如该设备为船上的设备,对船东或船厂;②如该设备为岸上的设备,对业主;③对执行该设备的入级或发证计划的船级社。根据澳大利亚海事安全局海事指令的要求,除了进行与试验相关的检查外,还应对该设备进行全面检查,或对不要求永久性标志安全工作载荷的起货机械确定其安全工作载荷。就设备的试验,与试验相关的全面检查和设备的发证而言,责任人员应为由船级社委派或授权且称职的合格人员或者是从事物资装运设备试验和发证的机构。 For the purpose of the maritime order of the Australian maritime safety authority (AMSA), responsible person means a person who, in relation to materials handling equipment, is competent and qualified and; (1) is responsible to: ①the manufacturer of that equipment; ②a classification society in pursuance of a scheme of classification or certification of such equipment; or ③a competent testing establishment. For carrying out any testing and associated thorough examination and issuing certificates of in respect of that equipment as required by the order of the Australian maritime safety authority; (2) is responsible to: ①the owner or master of the ship, where that equipment is the ship's equipment; or ②the owner, where that equipment is shore equipment; or ③ a classification society in pursuance of a scheme of classification or certification of such equipment. For carrying out thorough examinations, other than those associated with testing, of that equipment required by the maritime order of the Australian maritime safety authority or for determining the safe working load of cargo gear that is not required to be permanently marked with a SWL.

Restoring moment (righting moment) 复原力矩
船舶在外力作用下倾斜时,重力和浮力所形成的力矩。使船回复平衡位置的复原力矩为正值,反之为负值。

Restricted area 限制区域 系指船上那些一旦具有非法攻击行动目的人员登船即可能对船舶、人员等造成严重威胁的敏感区域(包括进入这些区域的通道),包括如果损坏或非法窥视,可能对船舶、船上人员、操作构成威胁的任何区域。限制区域可包括:(1)驾驶室;(2)机器控制室;(3)无线电/通信室;(4)A 类机器处所和控制站;(5)通风机及其控制室;(6)饮用水、泵、总管处所;(7)保安、监视设备和系统及其控制处所;(8)危险物质和货物、非随身行李存放处所;(9)货泵及其控制处所;(10)船员和人员舱室;(11)安全和应急设施存放处所;(12)电力控制/设备室;(13)照明控制室;(14)舵机舱;(15)货物处所;(16)甲板货储存区域;(17)船舶备件和重要维修设备场所;(18)船舶操作的紧急和备用设备处所;(19)紧急出口和疏散路线以及集合站。 Restricted area means the sensitive areas (including access to these areas) likely to cause serious threat to the ship, persons etc., by those purpose of illegal attack, including any areas where danger will happen to the ship, persons onboard the ship and operation if these ares are damaged or spied on. Restricted area may include: (1) maneuvering console; (2) machinery control room; (3) radio/ communication room; (4) type A machinery space and control station; (5) ventilating fan and its control room; (6) drinking water, pump, main pipe space; (7) security, surveillance equipment and system and their control space; (8) storage spaces of dangerous substance and cargoes, unaccompanied baggage; (9) cargo pumps and their control room; (10) crew and accommodation; (11) safety and emergency equipment storage; (12) power control / equipment room; (13) light control room; (14) tiller room; (15) cargo spaces; (16) storage area of deck cargoes; (17) storage spaces of ship's spare parts and spare equipment; (18) spaces of emergency ship operation and spare equipment; (19) emergency exit, evacuation way and assembly station.

Restricted condition 约束条件 系指与船舶保安有关的贸易航线区域、港口设施、船舶结构特点、货物、人员,包括船员和旅客特点而设定的条件。 Restricted condition means trading navigation areas, port facilities, construction of the ship, persons related to ship security, including the assumption of the characteristics of the crew and passengers.

Restricted gauging device 限制式液位测量装置
系指将此装置伸入液货舱内,使用时,允许少量货物蒸发气或液体逸入大气;不使用时,这种装置是完全封闭的,其设计应确保在打开这种装置时不致使舱内货物(气体或气雾)发生危险外溢。 Restricted gauging device means a device which penetrates the tank and which, when in use, permits a small quantity of cargo vapour or liquid to be exposed to the atmosphere. When not in use, the device is completely closed. The design shall ensure that no dangerous escape of tank contents (liquid or spray) can take place in opening the device.

Restricted service 有限航区 分为两类:(1)沿海或规定航行区域中航行的船舶;(2)在受保护水域或延伸保护水域中航行的船舶;(3)是 1 类航区、2 类航区和 3 类航区的统称。各类航区的航行限制如表 R-1 所示。 Restricted service is broken down into two broad categories

(1) ships operating coastal or specified operating areas;
(2) ships operating within protected or extended protected waters; (3) restricted service is a generic term of the service categories 1, 2 and 3 as shown in table R-1.

表 R-1 各类航区的航行限制
Table R-1 Limitation for navigation of service categories

类别 Category	航行限制 Limitation for navigation 距岸距离/海里 Distance to land /n mile
1 类航区 1 Service category	200(夏季/热带*)　100(冬季*) 200 (summer/tropical*)　100 (winter*)
2 类航区 2 Service category	20(夏季/热带*)　10(冬季*) 20 (summer/tropical*)　10 (winter*)
3 类航区 3 Service category	遮蔽水域 Sheltered waters**

注意：* 季节区按 1966 年国际载重线公约附则 Ⅱ 附录 Ⅰ 的规定。* Seasonal areas as specified in Annex II to the International Convention on Load Lines, 1966.

** 遮蔽水域包括海岸与岛屿、岛屿与岛屿围成的遮蔽条件较好、波浪较小的海域，且该海域内岛屿与岛屿之间、岛屿与海岸之间横跨距离不超过 10 n mile，或具有类似条件的水域。** Sheltered waters include the sea areas between an island and the shore and between islands with a distance of less than 10 n miles in between, which forms a comparatively good sheltered or similar condition with a little wave.

Restricted speed range　转速禁区

Restricted visibility　能见度不良　任何由于雾、霾、下雪、暴风雨、沙暴或任何其他类似原因而使能见度受到限制的情况。　Restricted visibility means any condition, in which visibility is restricted by fog, mist, falling snow, heavy rainstorms, sandstorms or any other similar causes.

Restricted water effect　狭窄航道效应　由于航道狭窄，水与船体相对速度加大，以及船波反射等使船舶航行状态改变，阻力增加等的现象。

Restricted water resistance increment　狭窄航道阻力增值　船舶在狭窄航道航行时，由于狭航道效应所增加的阻力。

Restricted zone　限制区

Resultant velocity of approaching flow　实效合速　对叶元体计入感生速度的水合速。

Retained sample　留存样本(留存样品)　系指根据"73/78 防污公约"附则 Ⅵ 第 18.8.1 条规定，从初始样品中提取交付给船舶的燃油代表性样本。　Retained sample is the representative sample in accordance with regulation 18.8.1 of Annex Ⅵ to MARPOL 73/78, of the fuel delivered to ship derived from the primary sample.

Retarding　阻滞(延迟,减速)

Retarding ignition　延迟点火

Retractable drive unit　可伸缩式推进装置

Retractable fin (retractable fin stabilizer)　收放式减摇鳍

Retractable fin stabilizer　伸缩式减摇鳍装置　外伸两舷的鳍板可缩入鳍箱内的减摇鳍装置。

Retractable hydrofoil　可收水翼　有利于低速航行、停靠码头或在浅水区航行。将结构形式设计成能向上翻起离开水面，或水翼各部件可以折叠或水翼支柱能缩入艇体的水翼。

Retractable steerable thruster　可转推力器　能旋转并可伸至船底以下的产生推力的器具。

Retractable water rudder　可升降水舵　气垫船狭窄航道或拥挤港口降速航行时，依靠水动力操纵航向的，不用时可升离水面的舵。

Retractable fender system　可伸缩防撞装置

Retreat during court session without permission　中途退庭

Retrial procedure　再审程序　一般系指申诉和审判监督程序，学理上称之为审判监督程序是法院对经过生效裁判的案件复核审理的法律程序，系指人民法院、人民检察院对于已经发生法律效力的判决和裁定，如果发现在认定事实或适用法律上确有错误，依法提出并进行重新审理的程序。民事再审程序，是我国民事诉讼中的一项重要制度,该程序强调无论在事实认定或法律适用上,只要有错误即应通过再审制度加以纠正,贯彻了我国有错必纠、有错必改、实事求是、司法公正的司法理念。

Retrieval 拯救 系指安全寻回幸存者。 Retrieval is the safe recovery of survivors.

Retro-reflective material 逆向反光材料 系指能使射入光束向相反方向反射的材料。Retro-reflective material is a material which reflects in the opposite direction a beam of light directed on it.

Retro-reflective material(immersion suit) 全反射材料(救生服) 把光束反射至其发出点处的材料。Retro-reflective material means the material that reflects light beams back to their point of origin.

Return flow oil atomizer 回油式喷油器 从喷油器中引回部分油量,以调节喷油量的可调式压力雾化喷油器。

Return satellite 返回式卫星 系指卫星发射入轨之后,就在太空执行任务,一般是不需要返回的,但有的卫星却需要回到地面,如侦察卫星获得的情报,科学实验卫星携带的实验品等。这就是返回式卫星。研制返回式卫星是卫星发展史上的一个重要突破。返回式卫星是作为观测地球的空间平台。返回式卫星所获取的各种对地观测信息资料,可以带回地面进行分析处理和详细研究。

Return satellite technology 返回式卫星技术 又称大气层再入技术。该技术拥有国:中国、美国、俄罗斯、英国和法国(没有这一条的不算真的有"洲际"导弹技术,不过关的导弹会在再入大气时直接销毁。奇怪的是,英国和法国从未展示过返回式卫星技术,但是仍然宣称拥有洲际导弹)。见图 R-3。

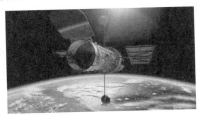

图 R-3 返回式卫星技术
Figure R-3 Return satellite technology

Return trip charter 来回程租船 系指来回一个航次的租船方式。

Returnable cargo unit 可重复使用的货物单元 系指货物包装在其中的包装物,拟将货物从托运人转运至收货人,并:(1)设计成作为单件货来转运;(2)不是冷藏集装箱或中间散装集装箱;(3)装有构成整体的起吊附件;(4)拟回收并重复使用,(5)可拆或不可拆的。Returnable cargo unit means a packaging, into which cargo is packed, intended to convey cargo from consignor to consignee and that:(1)is designed to be handled as a single unit;(2)is not a freight container or intermediate bulk container;(3)is fitted with integral lifting attachments;(4)is intended for return and subsequent re-use;(5)may or may not be collapsible.

Returned offsets(finished offsets) 完工的型值 经放样光顺修正后的型值。

Returned table of offsets 完工的型值表 由完工的型值编制而成的型值表。

Revenue tariff 财政关税 又称收入关税。系指以增加过境财政收入为目的的而征收的关税。对进口商品征收财政关税时,必须具备以下3个条件:(1)征税的进口货物必须是国内不能生产或无代用品的商品,以避免对国内市场形成冲击;(2)征税的进口货物在国内必须有大量消费;(3)关税税率必须适中,否则达不到增加财政收入的目的。

Reverberant field 混响声场 系指声音经过物体多次反射后到达受声点的反射形成的声场。

Reverberation measurement 混响测量 测量海面、海底或海中气泡等因素所产生的反向散射在接收机上的无规律叠加结果。常见的测量项目有混响时间及混响强度随时间、频率和距离的变化关系。

Reversal time 换向时间 从正车或倒车操纵开始至倒车或正车开始运转为止所需的时间。

Reverse frame 内底横骨 框架肋骨上缘,支承内底板的横向骨材。

Reverse osmosis methods 反渗透法 将海水加压到渗透压力以上,使海水中的溶剂即淡水通过半渗透性薄膜而析出的海水淡化方法。

Reversed flow 逆流 出现在漩涡或分离区的局部流体中,与主流方向相反的分流。

Reverser 反向机构 系指通过将从喷嘴喷出来的水流反向,从而将船向后推的机构。Reverser is the device to thrust the ship to go astern by reversing the flow direction of the water injected from the nozzle.

Reverser 换向器(倒转机构,换向开关)

Reversibility 可逆性

Reversible engine 可逆转式发动机

Reversible propeller 可反转螺旋桨 其桨叶可扭转较大角度,轴的转向不变但能产生向后拉力使船倒行的一种可调螺距螺旋桨。

Reversible pump 变向泵 能够变换进出口液流

方向的泵。

Reversing arrangement 换向装置

Reversing cam 换向凸轮

Reversing clutch 液压离合器 利用液压力实现离合的离合器。

Reversing device 换向装置

Reversing engine 可反转发动机

Reversing gas turbine（direct-reversing gas turbine） 倒顺车燃气轮机 动力涡轮的叶片设计成双层结构，因而可有正车和倒车两种转向的燃气轮机。

Reversing gear 换向机构 （可倒转齿轮）改变柴油机曲轴回转方向的机构。

Reversing rudder 反射舵 在螺旋桨不逆转情况下，可借操纵螺旋桨两侧的活动弧形片，将螺旋桨尾流导向一定方向或反向，使船舶转向或倒车的装置。

Reversing thermometer 颠倒温度表 一种测量海洋（或湖泊）表层以下某点水温的特殊玻璃水银温度表。安装在颠倒架上使用，通过把温度表颠倒的方法使水银柱断开，提取到船上读数。

Reversing valve 换向阀

Reversion time 换向时间 从换向操纵开始起到柴油机开始反向工作为止所需的时间。

Revision recommendation 修订建议案

Revocable letter of credit 可撤销信用证 系指在开证后，开证行无须事先征得受益人同意就有权修改其条款或者撤销的信用证。

Revolution direction 转向

Revolution indicator 转向指示器

Revolution perminute 每分钟转数

Revolution recorder 转速记录器 记录螺旋桨转速的设备。

Revolution recorder 转速记录器

Revolution speed 转速

Revolutive noise 旋转噪声 是由发动机的工作叶片在旋转中周期性地打击周围的空气质点，引起空气的压力波动而产生的噪声。

Revolving letter of credit 循环信用证 系指信用证被全部或部分使用后，其金额又恢复到原金额，可再次使用，直至达到规定的次数或规定的总金额为止。

Revolving scavenging pump 旋转式扫气泵

Reward 报酬 系指员工为组织工作而获得的所有各种他认为有价值的东西。不仅包括物质方面的，还包括精神、心理等方面的。

Reynolds number 雷诺数 表面仅考虑黏性力作用的水动力相似的一个无量纲参数。其通用表达式为

下：$R_n = VL/\nu$，式中，V—速度；L—长度；ν—流体运动黏性系数。对于螺旋桨：$R_n = UC/\nu$，式中，U—螺旋桨 0.7R 处叶切面合速；C—螺旋桨 0.7R 处叶切面弦长。

RGB color model（RGB） 三原色光模式 又称 RGB 颜色模型或红绿蓝颜色模型，是一种加色模型，将红（Red）、绿（Green）、蓝（Blue）三原色的色光以不同的比例相加，以产生多种多样的色光。RGB 颜色模型的主要目的是在电子系统中检测，表示和显示图像，比如电视和电脑，但是在传统摄影中也有应用。在电子时代之前，基于人类对颜色的感知，RGB 颜色模型已经有了坚实的理论支撑。RGB 是一种依赖于设备的颜色空间：不同设备对特定 RGB 值的检测和重现都不一样，因为颜色物质（荧光剂或者染料）和它们对红、绿和蓝色的单独响应水平随着制造厂商的不同而不同，甚至是同样的设备不同的时间也不同。

Rib collar 带状加强环 沿艇的四周装设的承受重载荷的，无论该艇是否使用总是预定充气的管状环。

Rib collar is the heavy duty tubular collar fitted around the periphery of the boat and always intended to be inflated whenever the boat is being used.

Rice steamer（steam rice cooker） 蒸汽饭锅 用蒸汽加热的煮饭锅。

Rich limit 富油极限 保持燃烧室稳定工作的空气燃料比的下限。

Ridding screw（turnbuckle） 松紧螺旋扣 由套筒和螺杆组成，当扭动套筒时，能使螺杆伸缩，以便张紧或放松绳索的索具配件。

Ride the wave 骑浪运动 小型船舶顺浪航行，当航速与波速相近时，船舶被波峰举起的危险航态。

Rig 钻机

Rigging fittings 索具配件 索具中连接、系固和夹扣绳索或滑车的金属附件。如索具钩、卸扣、套环、系索栓、眼板、眼环和松紧螺旋扣等。

Rigging line tower platform 拉索塔平台 是柔性固定平台。垂直的塔式桁架结构的下端同重力基座连接支承在海底，顶端的工作平台被托出水面之上，多根支索，一端系在塔架上，另一端连接锚，锚链和重块系留在海底，使塔架保持直立。

Rigging/ Riggers 索具 系指：（1）供吊卸或固定物件用的各种绳索、链条及其配件的总称。配件包括各种滑车及供连接、系固和夹扣绳索、链条用的金属件，诸如钩、卸扣、套环、索节、系索栓、眼板和松紧螺旋扣等。（2）包括各桅杆、桅桁（桅横杆）、帆桁（帆脚杆）、斜桁和所有索、链以及用来操作这些的用具的总称。系指为了实现物体挪移，系结在起重机械与被起重物体之间的受力工具，以及为了稳固空间结构的受力构件。

Right angle drive propulsion 直角传动推进

Right angle steering propeller 直角传动转向螺旋桨

Right hand propeller（right turning propeller） 右旋螺旋桨

Right of protection of the coastal State 沿海国的保护权　系指沿海国可在其领海内采取必要的步骤—防止非无害通过。 Right of protection of the coastal State means that the coastal State may take the necessary steps in its territorial sea to prevent passage which is not innocent.

Right of recourse 追索权　系指汇票受到拒付,持票人对于其前手(违书人、出票人)有请求其归还汇票钱数及费用的权利。

Right offer 现场报价　系指现场提供适当的产品;提供适当的价格;提供适当的服务。

Right ramp 直跳板　系指与船体中心线平行设置的跳板。其使用有一定限制。它可以在锯齿形码头、凹槽形的码头和设有专用带缆桩,水域较宽阔的码头停靠,对于一般沿岸码头则无法停靠装卸。

Right-hand revolving engine unit 右转机组　正车时,由功率输出端向自由端看,作顺时针方向旋转的主机组。

Right-hand turning 右旋　由船后向前看,螺旋桨推船向前时,为顺时针的旋向。

Righting level（restoring level） 复原力臂　船舶在外力作用下倾斜,其稳心又位于重心之上时,重心与浮心作用线的距离。

Righting lever GZ 扶正力臂 GZ　在横剖面上,浮心与重心之间的水平距离。注意:扶正力臂等于扶正力矩除以质量(kg)与重力加速度(9.806 m/s^2)的乘积,且以米表示。 Righting lever GZ is the distance in both the horizontal and transverse planes between the centre of buoyancy and the centre of gravity. Note: Righting lever is equal to the righting moment divided by the product of mass, in kilograms, and acceleration due to gravity (9.806 m/s^2) and is expressed in meters.

Righting moment RM 扶正力矩 RM　在静水中处于指定的横倾角时,由于艇的重心横向偏离艇体之浸水部分的浮心而形成的恢复力矩。注意(1):此扶正力矩随横倾角变化而变化,且通常与横摇角的关系采用图表的方式进行标绘。利用已知的艇体形状和重心的位置,通过计算机可精确地得出各扶正力矩。也可应用其他近似的方法。此扶正力矩实质上随艇体形状、重心位置、艇的质量和纵倾姿态的变化而变化。注意(2):扶正力矩以 N·m 表示。 Righting moment RM is that at a specific heel angle in calm water, the restoring moment generated by the transverse offset of the centre of gravity of the boat from the centre of buoyancy of the submerged part of the hull. Note(1): The righting moment varies with heel angle and is usually plotted graphically against heel angle. Righting moments are most accurately derived by computer from a knowledge of the hull shape and the location of the centre of gravity. Other more approximate methods are also available. The righting moment varies substantially with hull form, centre of gravity position, boat mass and trim attitude. Note(2): Righting moment is expressed in Newton meters.

Rigid buoyant material 刚性浮体　一般由硬质泡沫浮力块或水密箱体组成围设于刚性救生筏的周缘,既提供浮力,又保证稳性的物体。

Rigid connection PB combination-pusher 固定式联结顶推船驳船组合体—顶推船　系指由一艘顶推船和一艘驳船组成的船队。顶推船通过艏部机械装置锁紧在驳船尾部凹槽内,顶推船与驳船之间无相对运动,且在营运中始终保持联结状态。顶推船为组合体的组成部分。 Rigid connection PB combination-pusher means a combination of a pusher tug and a barge wherein the pusher tug is secured in the barge notch by mechanical means. There is no relative motion between the tug and the barge, resulting in the two vessels acting as a single unit in a seaway. The pusher tug is a component part of the combination.

Rigid connection PB combination-barge 固定式联结顶推船驳船组合体—驳船　系指由一艘顶推船和一艘驳船组成的船队。顶推船通过艏部机械装置锁紧在驳船尾部凹槽内,顶推船与驳船之间无相对运动,且在营运中始终保持联结状态。驳船为组合体的组成部分。 Rigid connection PB combination-barge means a combination of a pusher tug and a barge wherein the pusher tug is secured in the barge notch by mechanical means. There is no relative motion between the tug and the barge, resulting in the two vessels acting as a single unit in a seaway. The barge is a component part of the combination.

Rigid coupling（solid coupling） 刚性联轴器　轴系中无弹性和桡性,可以同时传递推力和扭矩的联轴器。

Rigid inflatable boat 刚性可充气艇　系指装有附在整个艇体上的,充气管道的艇。 Rigid inflatable boat means a vessel with inflatable tubes, attached to a solid hull. The tubes are inflated during normal craft operation.

Rigid lay 刚性铺管

Rigid liferaft 刚性救生筏　由刚性浮体的筏底组成的救生筏。

Rigid member 刚性构件　在抵抗总纵弯曲时不

致失稳的骨材及其带板。

Rigid polyvinyl chloride foam plastic board　硬质聚氯乙烯泡沫塑料板

Rigid shaft　刚性轴　其一次临界转速高于额定转速的汽轮机轴。

Rigid sides　固定舷侧　救生艇按有关规范规定必须具备的和艇体固定成一体的舷侧结构。

Rigidity sandwich plate　刚性夹心板

Rigidly connected pusher tug-barge combination　固定式顶推船—驳船组合体　系指由一艘顶推船和一艘驳船组成具有如下特征的船队：顶推船借助于一种除缆绳、链索或其他索具以外的机械连接装置，将其锁紧在驳船尾部的凹槽内，该连接装置使顶推船和驳船之间无任何相对运动。在整个营运航程中，顶推船与驳船始终保持所述连接状态。 Rigidly connected pusher tug-barge combination is a combination of a pusher tug and a barge with the following characteristics: The pusher tug is secured in the barge notch by mechanical means of any type whatsoever other than justwire ropes, chains, lines or other tackles, and there is no relative motion between the pusher tug and the barge. The pusher tug and the barge remain so connected throughout the voyage.

Ring dam　浓度环　分水机构中具有不同内径，可按油料浓度选用的一组环状零件。

Ring plate　眼环　带环的眼板。

Ring propeller (banded propeller)　带环螺旋桨（加环螺旋桨）　在叶梢或一定半径处有圆环连接各桨叶的螺旋桨。

Ring valve　环状阀　阀片呈环状的气阀。

Ringing　鸣振　系指冲击荷载引起的垂荡/纵摇/横摇运动。是对冲击荷载的响应，故系统阻尼的影响较小。

Ripple voltage　纹波电压　系指纹波电压的幅值为 U_{max} 与 U_{min} 之差值。 Ripple voltage is the amplitude of the ripple voltage represented by the difference between U_{max} and U_{min}.

Rippling　齿面点蚀　在接触应力的反复作用下，齿轮金属表面产生疲劳裂纹并逐渐扩展而导致表面金属微粒剥落的麻班状损坏。

Riser　立管　又称隔水管。系指将海底管路或油井连接到海面上设施的管系。立管系统由下而上组成为：下立管组（lower marine riser package, LMRG）、立管本体组件、伸缩立管、泥浆出口管及导流器组件。下立管组中的环形（万能）BOT、液压连接器、电—液阀组（双组）分别属于 BOT 组和控制系统。下立管组中属于立管系统的主要部件为桡性。立管本体组件主要有立管（高强度无缝或直缝钢管或轻合金管）、连接接头、压井/防喷管线（BOP 控制系统和多路传输线也附于其上）和浮力模块组成。伸缩立管在钻井设备下水安装后始终处于随着平台（船）升沉上下伸缩运动，密封设计采用上下两组，上组圆柱形密封采用两个半圆形，便于磨损后在工作状态下使用下密封而更换上密封，并用压缩空气推动，使磨损的耐磨橡胶密封件得以自动补偿。导流器主要用于当钻井遇浅层气时将钻井与上部井孔密封而将其中的钻具外径浅层气导出顺风向的船舷外点燃烧掉，以保护平台（船）及作业人员的安全。见图 R-4。

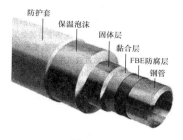

图 R-4　立管
Figure R-4　Riser

Riser　上升管　汽水混合物在其中作上升运动的水管。

Riser system　立管系统　系指由立管、立管支撑、所有附件和腐蚀防护系统组成的系统。

Riser system at sea　海上隔水管系统　系指为浮式钻井平台与井口之间提供流体流动通道的，为节流、压井和其他管线提供支承，为钻井工具进入井内提供导向，为防喷器的安装和回收提供服务的系统。其主要部件包括张力装置、导流装置、伸缩管节（含张力环）、联管器、隔水管管节、桡性/球型接头、隔水管底部组件（LM-RP）、节流压井和附属管线、浮力装置、隔水管安装设备及特种设备。

Riser tensioner　隔水器张力器

Riser turret　立管转塔

Riser turret mooring (RTM)　立管转塔系泊

Riser-less light well intervention (RLWI)　无立管修井

Risk　风险　系指在一定的时间内，由于系统行为的不确定性（主要系指发生了意料之外的事故）给人类带来危害的可能性的度量，即频率和后果严重性的组合。这种可能性既可采用频率（单位时间事件的发生率）表示，也可以采用概率来表示（具体环境下事件发生的概率）。由于系统的危险因素并加以排序，对排序后的危险因素依次确定减小发生频率和产生后果的措施。

Risk 风险　系指：(1)事故的频率(概率)和有关的危险等级；(2)对负面事件对人员、环境或物质财产带来的危害的表述。风险表示为事故发生的概率和后果；(3)后果的频率和严重性组合。　Risk means:(1) the frequency (probability) of occurrence and the associated level of hazard;(2) the expression of the danger that an undesired event represents to persons, to the environment or to material property. The risk is expressed by the probability and consequences of an accident;(3) the combination of the frequency and the severity of the consequence.

Risk acceptance 风险接受　接受某一风险的决策。

Risk analysis 风险分析　是研究风险、风险概率和风险造成后果的技术。

Risk assessment 风险评估　是客观分配具体事件的可能性和结果的逻辑过程。风险评估可定性或定量，如果以系统化和严格的方式完成，可以是有价值的决策辅助工具。　Risk assessment is a logical process for objectively assigning the likelihood and consequences of specific events. Risk assessments can be qualitative or quantitative, and can be a valuable decision aid if completed in a systematic and rigorous manner.

Risk assessment matrix 风险评估矩阵表　以定性的指标描述危险事件的发生频率和可能产生的后果，以定量的指标描述风险。

Risk assessment methods 风险评估方法　系指用于初步系统安全评估过程评估风险的方法。包括：(1)故障树分析(FTA)或从属图(DD)；(2)故障模式影响和界限分析(FMECA)；(3)故障模式和影响概要(FMES)；(4)区域危险分析(ZHA)。　Risk assessment methods mean a method used for risk assessment in the PSSA process:(1) fault tree analysis (FTA) or dependence diagrams (DD);(2) failure modes effect and criticality analysis (FMECA);(3) failure modes and effects summary (FMES);and (4) zonal hazard analysis (ZHA).

Risk avoidance 风险避免　不介入风险情况的决策或摆脱风险情况的行动。

Risk contribution tree(RCT) 风险分析树　系指构成分析树的所有故障树和事件的组合。Risk contribution tree (RCT) means the combination of all fault trees and event trees that constitute the (RCT).

Risk control 风险控制　实施风险管理决策的行动。

Risk control measure 风险控制措施　系指控制单一风险因素的方法。　Risk control measure means a means of controlling a single element of risk.

Risk control option(RCO) 风险控制选项/风险控制方案　系指风险控制措施的组合。　Risk control option (RCO) means a combination of risk control measures.

Risk evaluation 风险评价　将预计风险同给定风险准则进行比较以确定风险重要性的过程。

Risk evaluation criteria 风险评估准则　通常系指风险的可容忍度的标准。这些标准是可容忍风险的极限值。

Risk evaluation criteria 风险评估准则　系指用于评价风险的可接受性/可忍受度的准则。

Risk identification 风险识别　发现、列出风险因素并辨别其特征的过程。

Risk management 风险管理　协调涉及风险的行动以指导并管理组织。

Risk management system 风险管理系统　组织的管理系统中涉及管理风险的一系列因素。

Risk of boats 过驳险　系指投保货物过驳过程中，受到损坏或损失而获得赔偿的险种。

Risk perception 风险感知　基于一系列价值或利害关系、利益相关者看待风险的方式。

Risk reduction 风险降低　所采取的减少涉及风险的概率和/或负面效果的行动。

Risk transfer 风险转移　为了风险与另一方分担损失或利益。

Risk treatment 风险处理　选择或实施措施以改变风险的过程。

Risk-based design 基于风险的设计　系指设计的过程有风险评估的支持或设计依据来源于风险评估。也就是说，它是一种结构化和系统化的方法论，目的是通过运用风险分析和成本效益评估来保证安全性能和成本效果。

Risk-benefit analysis 风险效益分析

Risk-contribution tree 风险贡献树　风险贡献树分析作为一种图示表明不同的事故类别和子类别中风险分布的途径，采用这种混合可以找出每一类事故可能的原因，从而表示出每一类事故直接原因的逻辑组合。

River suction dredger 吸泥船

River water quick-settling unit 江水快速澄清装置　内河船舶使用江水作生活用水时，将江水中的泥沙快速沉淀的处理装置。

Riveted ship 铆接船　船体结构中主要用铆接方法联结的船舶。

RO/RO container ship 滚装/集装箱船　系指兼装集装箱的滚装船。

RO/RO passenger HSC 高速客滚船　系指载客且载小客车的高速船。　High speed RO-RO passenger craft is a high speed craft carrying passengers and cars.

RO/RO passenger ship 滚装客船(客滚船) 系指具有滚装处所或特种处所的客船。 RO/RO Passenger ship means a passenger ship with RO-RO spaces or special category spaces.

RO/RO ship 滚装船 系指具有多层甲板、双层底、能装载车辆或使用车辆装卸集装箱或托盘货的船舶。其结构适用于借助轮子以滚进和滚出的方式，不需要起重设备进行吊装或吊卸货物的、400GT 以上，可以载客 12 人以上的客、货滚装船以及专用于汽车和各类车辆运输的载车船称为滚装船。其上层建筑高大，货舱内设有多层纵通的车辆甲板，不设横舱壁，上下甲板之间设有可供车辆行驶的斜坡道，有的滚装船还设有升降机，用来代替斜坡道，车辆可通过坡道或升降机进入上下舱内。以车辆为运输单元的滚装船，船与岸都不需要设置起重设备，因此装卸效率高，实现了货物门对门的运输。但舱容利用率低。此外，滚装船设有调整纵倾和横倾的压载水舱。一般还设有减摇水舱。见图 R-5。

图 R-5　滚装船
Figure R-5　RO/RO ship

RO/RO space 滚装处所 系指通常不做任何分隔并延伸至船舶的大部分长度或整个长度的处所。该处所通常能以水平方向装载或卸下油箱内装有自用燃油和/或载有货物[以包装或散装形式载于铁路或公路车辆(包括公路或铁路油罐车)、拖车、集装箱、货盘、可拆container柜之内或之上，或类似装载装置或其他容器之内或之上的包装或散装货物]的机动车辆。 RO/RO space are spaces not normally subdivided in any way and normally extending to either a substantial length or the entire length of the ship in which motor vehicles with fuel in their tanks for their own propulsion and/or goods [packaged or in bulk, in or on rail or road cars, vehicles (including road or rail tankers), trailers, containers, pallets, demountable tanks or in or on similar stowage units or other receptacles] can be loaded and unloaded normally in a horizontal direction.

Road oil 铺路沥青

Road transport 公路运输 是一种现代化运输方式。它不仅可以直接运进或运出对外贸易货物，而且是车站、港口和机场集散进出口货物的重要手段。

Roaming 漫游 从字面上解释系指随意游玩，漫无目的地游走。对移动电话用户而言，漫游系指移动电话用户在离开开户地区或本国时，仍可以在其他一些地区或国家继续使用他们的移动电话的一种业务。

Robot[1] 机器人 是自动执行工作的机器装备。它既可以接受人类指挥，又可以运行预先编排的程序，也可以根据以人工智能技术制定的原则纲领行动。

Robot[2] 军用机器人 是一种用于军事领域的具有某种非人功能的自动机。机器人的国际名称"罗伯特"，原意是用人手制造的工人。该技术拥有国：美国、日本、英国、法国、中国。见图 R-6。

图 R-6　军用机器人
Figure R-6　Robot

Rock climbing wall 攀岩墙(豪华邮轮)

Rock crusher/ Rock breaking barge 碎石船 系指在甲板端部或中部开槽处设有一套碎石锤装置用以击碎水底岩石的船舶。

Rock drilling & blasting barge 钻孔爆破船

Rockwool 岩棉 又称岩石棉，是矿物棉的一种。岩棉是以天然岩石如玄武岩、辉长岩、白云石、铁矿石、铝矾土等为主要原料，经高温熔化、纤维化而制成的无机质纤维。

Rod (or truss) element 杆(或桁架)单元 是线单元，仅具有轴向刚度，且沿单元长度其横剖面面积不变。Rod (or truss) element is line element with axial stiffness only and constant cross-sectional area along the length of the element.

Roll 正横摇 是船舶沿 x 轴的角运动。Roll is the ship's angular motion about the x axis.

Roll landing machine gangplank rise down system 滚装船/渡船大门启闭跳板升降系统

Roll on/roll off system (Ro/Ro) 滚装装卸 用拖车、叉车等运载工具将货物直接带进带出船舶的装卸方法。

Roll on-roll off ship 滚装船　系指一种设计和制造成能装载车辆或使用车辆装卸集装箱或托盘货物的专用船舶。　Roll on-roll off ship is a ship specially designed and constructed for the carriage of vehicles, and cargo in pallet form or in containers, and loaded/unloaded by wheeled vehicles.

Roll period T_R (s)　横摇周期 T_R　由下式得出：$T_R = 2.3\ K_r/(GM)^{1/2}$，式中，$K_r$—所考虑装载工况的横摇回转半径(m)，当 K_r 未知时可用表 R-2 的值。GM—所考虑装载工况的稳性高度(m)，当 GM 未知时可用表 R-2 的值。T_R means a roll period, in s, are given by: $T_R = 2.3\ K_r/(GM)^{1/2}$; where K_r—Roll radius of gyration, in m, in the considered loading condition. When K_r is not known, the values indicated in Table R-2 may be assumed. GM—Metacentric height, in m, in the considered loading condition. When GM is not known, the values indicated in Table R-2 may be assumed.

表 R-2　K_r 值和 GM 值

装载工况	K_r	GM
满载工况（隔舱或均匀装载）	$0.35B$	$0.12B$
正常压载工况	$0.45B$	$0.33B$
重压载工况	$0.40B$	$0.25B$

Roll radius of gyration K_r (m)　横摇回转半径 k_r (m)　K_r means a roll radius of gyration, in m.

Roll single amplitude θ　横摇单幅值 θ　由下式得出：$\theta = [9,000\ (1.25 - 0.025\ T_R)\ f_p\ k_b]/(B + 75)\pi$；式中，$k_b$—系数，无舭龙骨的船舶，取 $k_b = 1.2$；有舭龙骨的船舶，取 $k_b = 1.0$。Roll single amplitude θ means a roll single amplitude, in deg; given by: $\theta = [9,000\ (1.25 - 0.025\ T_R)\ f_p\ k_b]/(B + 75)\pi$; where: k_b: Coefficient taken equal to: $k_b = 1.2$ for ships without bilge keel; $k_b = 1.0$ for ships with bilge keel.

Rollback　撤回

Rolled sections　轧制型材　系指包括球扁钢、不等边角钢和不等边不等厚角钢等的型材。Rolled sections mean those sections including bulb-flats, unequal angle bars, and unequal angle bars of unequal thickness, etc.

Roller fairlead　滚轮导缆器　设置在船舷，由数个独立滚轮并组成的列导缆器。有开式、闭式等类型。

Rolling　横摇　船舶绕纵轴所作周期性的角位移运动。

Rolling and hinged cradle davit　滚轮铰链型艇架　吊艇臂下端前面设一副滚轮，后面设铰链，放艇时滚轮沿直线斜轨滚出，同时后铰链转动，使吊艇臂外移和倾倒，从而吊艇出舷的重力式艇架。

Rolling and tipping hatch cover　翻滚式舱口盖　由多块舱口盖板组成，相邻两舱口盖板之间用链条连接。在舱口盖板至舱口端处，舱口盖板翻转至直立状态，排列存放于舱口盖板搁架上的舱口盖。

Rolling cradle type davit　滚动式艇架　吊艇臂下端设两副滚轮，放艇时前滚轮沿下斜的直线滑轨滚动，而后滚轮沿另一滑轨滚动，使吊艇臂外移并倾倒，从而吊艇出舷的重力式艇架。

Rolling hatch cover　滚动式舱盖　利用滚轮使舱盖启闭的机械装置。

Rolling tank stabilization system (anti-rolling tank stabilization system)　水舱式减摇装置　利用两舷水舱液位变化产生的减摇力矩以减少的船舶摇摆的装置。

Roll-moment of inertia　横摇惯性矩

Roll-on/roll-off ship (driven-on/driven-off ship)　滚装船　将带有滚车底盆的集装箱或装在托盘上的其他货物作为一个货物单元，用拖车或叉车带动直接开进开出船舱的船舶。

Rolltite cover drum (drum for rolltite cover)　舱盖板卷筒　滚卷式舱盖中，用以卷收舱盖板的专用卷筒。

Roll-up hatch cover (rolltite hatch cover)　滚卷式舱盖　舱盖沿纵向分为互相铰接的多块舱盖板，由专设的卷筒卷收，开启的机械舱盖。

Roofers flux　屋顶用柏油

Room sealed appliance (LPG system)　密封室的器具 (LPG 系统)　该器具具有一个燃烧系统，此系统进入燃烧室的空气和排出的燃烧产物都经过与封闭的燃烧室相连接的密闭管道。Room sealed appliance means the appliance having a combustion system in which incoming combustion air and outgoing products of combustion pass through sealed ductwork connected to the enclosed combustion chamber.

Rooms containing furniture and furnishings of restricted fire risk　设有限制失火危险的家具和设备的房间　系指配置具有危险性小的家具及其有备用品房间（船员室、公用室、办公室或是其他居住区域）：(1) 框架式家具，如书桌、衣橱、梳妆台、柜台或餐饮柜，其可使用完全认可的不燃材料制成。这些家具的表面可用厚度为 2 mm 的可燃性薄板做最后处理；(2) 独立式家具，如椅子、沙发或桌子，其骨架应用不燃材料制成；(3) 帷幔、窗帘以及其他悬挂的纺织品材料，其阻止火焰蔓延的性能不低于质量为 0.8 kg/m 的毛织品，且根据《耐火试验程序规则》确定；（4）地板覆盖物具有低播焰性；（5）舱壁、衬板及天花板的外露表面，应具有低焰传播性；（6）装有垫套的家具具有阻止着火和火焰蔓延的性能，且根据《耐火试验程序规则》确定；（7）床上用品具有阻止着火和火焰蔓延的性能，且根据《耐火试验程序规则》确定。 Rooms containing furniture and furnishings of restricted fire risk are those rooms containing furniture and furnishings of restricted fire risk (whether cabins, public spaces, offices or other types of accommodation) in which: (1) case furniture such as desks, wardrobes, dressing tables, bureaus, dressers, are constructed entirely of approved non-combustible materials, except that a combustible veneer not exceeding 2 mm may be used on the working surface of such articles; (2) free-standing furniture such as chairs, sofas, tables, are constructed with frames of non-combustible materials; (3) draperies, curtains and other suspended textile materials have qualities of resistance to the propagation of flame not inferior to those of wool having a mass of mass 0.8 kg/m, this being determined in accordance with the Fire Test Procedures Code; (4) floor coverings have low flame spread characteristics; (5) exposed surfaces of bulkheads, linings and ceilings have low flame-spread characteristics; (6) upholstered furniture has qualities of resistance to the ignition and propagation of flame, this being determined in accordance with the Fire Test Procedures Code; and (7) bedding components have qualities of resistance to the ignition and propagation of flame, this being determined in accordance with the Fire Test Procedures Code.

Root cavitation 叶根空泡 螺旋桨叶根处产生的空泡。

Root concavity 根部凹陷 见图 R-7。

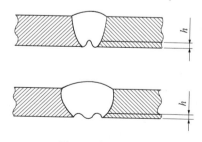

图 R-7 根部凹陷
Figure R-7 Root concavity

Root notch 根部缺口

Root thickness 根厚 螺旋桨叶根处略去桨叶与桨毂之间填角料后的切面最大厚度。

Root vortex 根涡 在螺旋桨叶根处泄出的尾涡。

Root's blower 罗茨鼓风机 转子截面两瓣呈"8"字形，或三瓣呈"品"字形的容积式风机。

Rope shackle 索具卸扣 主要供连接索具中的绳索、链条、滑车等用的卸扣。

Rope socket 索节 装配于钢丝绳末端并供连接用的一种索具配件。分为开式和闭式两种。

Rope stopper (nipper) 掣索器 能迅速掣固绳索的夹具。

Rope storage winch 储绳绞车 仅用于储存绳索的轻载功率的绞车。它与驱绳绞车联合使用时，要求两者同步。

Rope turnishing tools 缆绳工具 船上打绳结和插接缆绳使用的专用工具的统称。主要有木笔、铁笔、木槌、钢錾等。

Ropeless connection 无绳系结 顶推船队在编队时不靠绳索进行系结和传力的一种系结方式。

Ropeless linkage 无绳系结装置 推船和被推驳船之间，除绳索外的各种系结装置的统称。

Ro-Ro cargo holds 滚装货舱 系指 SOLAS 公约第Ⅱ-2/3.14 和 3.18 条定义的滚装货物处所和特种处所。 Ro-Ro cargo holds are Ro-Ro cargo spaces and special category spaces as defined in SOLAS, regulation Ⅱ-2/3.14 and 3.18.

Ro-Ro cargo ship 滚装货船 系指：(1)设计用于载运滚装运货单元的船舶。(2)具有多层甲板的设计用于载运空的小汽车和卡车的滚装货船。 Ro-Ro cargo ship means: (1) a ship designed for the carriage of roll-on/roll-off cargo transportation units. (2) a multi deck roll-on/roll-off cargo ship designed for the carriage or empty cars and trucks.

Ro-Ro cargo ship (volume carrier) 滚装货船(容积型船舶) 系指设计成载运运货单元的每米车道的载重量小于4吨的滚装货船。 Ro-Ro cargo ship (volume carrier) means a cargo ship, with a deadweight per lane-meter less than 4 tons/m, designed for the carriage of cargo transportation units.

Ro-Ro cargo ship (weight carrier) 滚装货船(质量型船舶) 系指设计成载运运货单元的每米车道的载重量不小于4t的滚装货船。 Ro-Ro cargo ship (weight carrier) means a cargo ship, with a deadweight per lane-meter of 4 tons/m or above, designed for the carriage of cargo transportation units.

Ro-Ro cargo spaces 滚装装货处所 系指通常设有以任何方式加以分割，并且延伸到大体上为整个船舶的长度，在此范围内货物(成套组装或散装货，在有轨机动车或汽车里或上面的货物、运输车辆,包括火车和公路油罐车、拖车、集装箱拖车、货盘、可卸油罐或在类似贮存容器里面或上面的货物,或其他容器),通常可以在水平方向上装卸的场所。 Ro-Ro cargo spaces are spaces not normally subdivided in any way and normally extending to either a substantial length or the entire length of the ship in which motor vehicles with fuel in their tanks for their own propulsion and/or goods (packaged or in bulk, in or on rail or road cars, vehicles, including road or rail tankers, trailers, containers, pallets, demountable tanks or in or on similar stowage units or other receptacles)can be loaded and unloaded normally in a horizontal direction.

Ro-Ro craft 滚装船 系指设有一个或以上滚装处所的船舶。 Ro-Ro craft is a craft fitted with one or more Ro-Ro spaces.

Ro-Ro passenger ship 客滚船 系指:(1)具有 SOLAS 公约第Ⅱ-2/3 条定义的滚装装货处所或特种处所的客船;(2)载运乘客超过 12 人的滚装船。 Ro-Ro passenger ship means:(1)a passenger ship with Ro-Ro cargo spaces or special category spaces as defined in regulation Ⅱ-2/3, SOLAS;(2)a Ro-Ro ship carrying more than 12 passengers.

Ro-Ro ship 滚装船 系指:(1)其构造适合于以滚进和滚出的方式,来装卸车辆和集装箱/或托盘化货物的船舶;(2)具有多层甲板、双层底、能装载车辆或使用车辆装卸集装箱或托盘货的船舶。 Ro-Ro ship means:(1)a ship constructed for loading and unloading vehicles and containers/or cargo in pallet form by rolling in and rolling out;(2)a ship with multi-tier decks and double bottom, carrying vehicles or cargo in pallet form or in containers and loaded/unloaded by wheeled vehicles.

Ro-Ro spaces 滚装处所 系指:(1)非正常分隔的并延伸至船舶大部分长度或整个长度的处所。该处所能以水平方向正常装卸油箱内备有自用燃油的机动车或者货物(包装或散装、用于公路或铁路装载的有车厢和无车厢车辆,包括公路或铁路油槽车、拖车、集装箱、货盘、可拆箱柜、类似装载装置或其他容器);(2)通常不予分隔并通常延伸至船舶的大部分长度或整个长度的处所,能以水平方向正常装卸油箱内备有自用燃料的机动车辆以及/或货物[在铁路或公路车辆、(包括公路或铁路有槽车的)汽车、拖车、集装箱、货盘、可拆槽罐之内或之上,或在类似装载单元或其他容器之内或之上的包装或散装货物]。 Ro-Ro spaces are:(1)spaces not normally subdivided in anyway and normally extending to either a substantial length or the entire length of the ship in which motor vehicles with fuel in their tanks for their own propulsion or goods (packaged or in bulk, in or on rail or road cars, vehicles, including road or rail tankers, trailers, containers, pallets, demountable tanks or in or on similar stowage units or other receptacles can be loaded and unloaded normally in a horizontal direction);(2)spaces not normally subdivided in any way and normally extending to either a substantial length or the entire length of the ship in which motor vehicles with fuel in their tanks for their own propulsion or goods (packaged or in bulk, in or on rail or road cars, vehicles (including road or rail tankers), trailers, containers, pallets, demountable tanks or in or on similar stowage units or other receptacles)can be loaded and unloaded normally in a horizontal direction.

Rotary compressor 回转式压缩机 压缩机轴作回转运动,无往复运动部件的一类容积式压缩机的统称。

Rotary force derivatives 水动力旋转导数 作用于船舶水动力分力对角速度的偏导数。如:$X_p = \partial X/\partial p$ 等。

Rotary moment derivatives 水动力矩旋转导数 作用于船舶水动分力矩对角速度的偏导数。如:$K_p = \partial K/\partial p$ 等。

Rotary oil burner (rotary mechanical atomizing oil burner) 旋转式燃烧器 将燃油注在高速旋转的部件上,使构成油膜而后甩出以完成雾化的燃烧器。

Rotary pump 回转泵 泵轴作回转运动,无往复运动部件的容积泵的统称。

Rotary retaining torque 止转转矩 系指电动机在施以额定频率的额定电压时,其转子在所有角度位置上静止时产生的最小转矩。

Rotary screw propeller 全回转式螺旋推进器 是一种能绕竖轴作 360°旋转的导管螺旋桨。因此可以产生任何方向的推力使船舶实现正航、倒航和斜航。船上设有两个全回转推进器时,可以使船舶平行横移和原地回转。全回转式螺旋推进器主要用于港作拖船、工程救护船等对操作性有特殊要求的船舶上。见图 R-8。

图 R-8 全回转式螺旋推进器
Figure R-8 Rotary screw propeller

Rotary silica-gel dehumidifier 转动式吸湿装置 利用吸附剂的吸湿和再生能力,将其装于筒内在吸湿和再生的空气回路之间旋转,以进行吸湿的装置。

Rotary vane 转片 转叶式液压舵机中,与转轴一起转动的叶片。

Rotary vane steering gear 转叶式舵机 采用旋转叶片式液压缸的液压舵机。

Rotary vane type strewing gear 转叶式舵机 是一种转舵角度可达70°的电动液压舵机。其适用于需要操纵性能好的工程船上,如耙吸式挖泥船、半潜重载船等。

Rotating arm test 旋臂试验 在旋臂水池中,强制船模做圆周运动以测定船舶运动旋转导数和位置导数等数据的试验。

Rotating cylinder rudder 转柱舵 在舵叶前装有主动旋转的垂向圆柱的舵。

Rotating head 旋转头

Rotating stall(propagating stall) 旋转失速 当压气机流量减少或在轴向上大大降低时,在环形通道中出现一个或几个沿着动叶旋转方向传播的不稳定工作状态。

Rotation reversal test(reversing test) 换向试验 检查可逆转内燃机换向操作或换向机构、倒顺离合器的可靠性和灵活性的试验。

Rotational inertia 转动惯量

Rotational velocity 旋转速度

Rotator 旋转器(转子)

Rotor 转子 系指由三片连接到风能转换系统的机舱上轮毂的叶片组成的涡轮机的转动机构。由动叶、轴和轮盘或轮鼓等组成。

Rotor bearing 转子轴承

Rotor star 星型叶轮 风能转换系统的组成部分,由3个叶轮和轮毂组成,简称叶轮。

Rough fiber glass 粗玻璃纤维

Rough wearing fiber 粗糙纤维

Roughness allowance 粗糙度补贴系数 由于粗糙度所增加的摩擦阻力系数。

Roughness height(roughness magnitude) 粗糙程度 表示物体表面粗糙的程度。

Roughness resistance 粗糙度阻力 船体表面的黏性阻力与同尺度的水力光滑面黏性阻力的差额。

Rough-sea resistance 汹涛阻力 船舶航行于大风浪中,较在静水中航行时所增加的阻力,包括风浪作用以及船身运动等所增加的阻力。

Rounded sheer strake 圆弧舷顶列板 呈圆弧形的舷顶列板。

Rounding errors 舍入误差

Rounding of edges 边缘圆度

Route monitoring 航线监测 系指针对预先计划的航线和水域,持续监测本船位置的航行任务。Route monitoring means the navigational task of continuous surveillance of own ships position in relation to the pre-planned route and the waters.

Route planning 航线计划 系指预先根据航行海区确定船舶的航向线、回转半径和航速。Pre-determination of course lines, turns radius and speed in relation to the waters to be navigated.

Router 路由器 是一种控制寻径的网络设备。它在互连网中从多条途径中搜索通信量最少的一条网络途径提供应用户通信。

Routes passing through regions of icebergs guarded by the Ice Patrol 穿越冰区巡逻所警戒冰山区的航线 系指:(1)在加拿大的北大西洋沿岸各港口(包括从北大西洋经坎索水道和卡伯特海峡抵达的内陆港口)和从北大西洋经直布罗陀海峡或直布罗陀海峡以北抵达的欧洲、亚洲或非洲各港口之间的航线(经过各类冰区最南限界的航线除外);(2)在纽芬兰累斯角西部的加拿大北大西洋沿岸各港口(包括从北大西洋经坎索水道卡伯特海峡抵达的内陆港口)和纽芬兰累斯角北部的加拿大北大西洋沿岸各港口之间经纽芬兰累斯角的航线;(3)在美国的大西洋和海湾沿岸各港口(包括从北大西洋经坎索水道卡伯特海峡抵达的内陆港口)和从北大西洋经直布罗陀海峡或直布罗陀海峡以北抵达的欧洲、亚洲或非洲各港口之间的航线(经过各类冰区最南限界的航线除外);(4)在美国的大西洋和海湾沿岸各港口(包括从北大西洋经坎索水道卡伯特海峡抵达的内陆港口)和纽芬兰累斯角北部的加拿大北大西洋沿岸各港口之间经纽芬兰累斯角的航线。Routes passing through regions of icebergs guarded by the Ice Patrol means:(1) routes between Atlantic coast ports of Canada (including inland ports approached from the North Atlantic through the Gut of Canso and Cabot Straits) and ports of Europe, Asia or Africa approached from the North Atlantic through or north of the Straits of Gibraltar (except routes which pass south of the extreme limits of ice of all types);(2) routes via Cape Race, Newfoundland between Atlantic coast ports of Canada (including inland ports approached from the North Atlantic through the Gut of Canso and Cabot Straits) west of Cape Race, Newfoundland and Atlantic coast ports of Canada north of Cape Race, Newfoundland;(3) routes between Atlantic and Gulf Coast ports of the United States of America (including inland ports approached from the North Atlantic through the Gut of Canso and Cabot Straits) and ports of Europe, Asia or Africa approached from the North Atlantic through or north of the Straits of Gibraltar (except routes which pass south of the extreme limits of ice of all types); (4) routes via Cape Race, Newfoundland between Atlantic and Gulf Coast ports of the United States of America (including inland ports approached from the North Atlantic through the Gut of Canso and Cabot Straits) and Atlantic Coast ports of Canada north of Cape Race, Newfoundland.

Routine steering gear 常规舵机 是船上的一种大型甲板机械。舵机的大小由外舾装按照船级社的规范决定，选型时主要考虑扭矩大小。

ROV 遥控操作的水下作业的潜水器

Row/stack 排/垛 集装箱堆/垛的横向位置标识。

Rubber bearing 橡胶轴承 以橡胶作为轴衬材料的轴承。

Rubber boat 橡皮艇 以橡胶布作为艇体基本材料的艇。

Rubber foam insulation material 橡塑发泡保温材料

Rubber isolation pad 橡胶隔振垫

Rudder[1] 舵 用以驾驶船舶的装置。普通型舵有垂直艉鳍、能从左舷35°移至右舷35°；舵的特征由面积、长宽比和形状表示。 Rudder is a device, usually of an aerofoil or flat section, that is used to steer a ship. A common type has a vertical fin at the stern and is able to move from 35°sport to 35° starboard; rudders are characterized by their area, aspect ratio, and shape.

Rudder[2] 平衡舵 舵杆轴线位于舵叶导边后面一定的距离。以使舵压力中心接近舵杆轴线而减少转舵舵扭矩的舵。

Rudder actuator 操舵油缸(转舵机构) 系指直接将液压转变为机械运动以转动舵的部件。Rudder actuator is the component which converts directly hydraulic pressure into mechanical action to move the rudder.

Rudder and steering gear 舵设备 舵及其支撑部件和操舵装置的总称。

Rudder angle[1] 舵角 舵叶绕舵杆轴线转动偏离正舵位置的角度。

Rudder angle[2] 舵角(方位推进器) 系指当配置方位推进器作为主推进装置时，推进器的平均舵角。Rudder angle mean thruster angle when main propulsion is azimuth thrusters.

Rudder area ratio 舵面积比 舵面积与船舶垂线间长和设计吃水的乘积之比值。

Rudder arm 舵臂 平板舵中连接舵板和下舵杆的加强筋。

Rudder balancing ratio 舵平衡比 舵杆轴线前的舵面积与整个舵面积的比值。

Rudder bearing (rudder bearer) 舵承 固定在船体上用以支承舵的轴承装置的统称。

Rudder blade 舵叶 产生舵压力的舵的主体部分。

Rudder brake 舵掣 系指使舵能在任何舵位刹住不动的摩擦制动器。

Rudder breath 舵宽 系指舵叶的导边至随边之间的距离。

Rudder carrier 上舵承 位于舵头处，用以承受舵的重力及其所受到的径向和轴向力的轴承。

Rudder coupling 舵杆接头 用以连接舵叶与舵杆的接头。通常有法兰接头、嵌接接头和插入式接头等。

Rudder effect 舵效 在操舵后产生的改变航向或使船舶升降的效应。

Rudder effectiveness test 舵效试验 船舶在指定航速沿直线航前进时，下令停车并操右满舵，当转艏角达1°时改操左满舵，以确定其舵效航速的一种试验。

Rudder frame 舵构架 舵叶中用以支撑舵板的骨架。

Rudder head 舵头 系指舵杆上部与舵机或舵柄连接的部分。

Rudder height 舵高 系指沿舵杆轴线方向，舵叶上缘至下缘的直线距离。

Rudder horn 挂舵臂 系指支承半悬挂舵的臂状构件。

Rudder movement 舵的行程 舵从一舷限位位置到另一舷限位位置的时间。

Rudder pintle 舵销 系指用以将舵连在舵柱或臂上的销轴或螺柱。一般制成锥状体按其部位和作用不同可以分别称为上舵销、下舵销等。

Rudder plate 舵板 系指构成舵叶的平板或曲面板。

Rudder post 舵柱 用以支持不平衡舵的柱子，是艉柱的一部分。

Rudder pressure 舵压力 船舶运动时，作用在舵叶上的流体动压力的合力。

Rudder rake 舵缘倾斜 舵叶导边缘与随边缘的倾斜。

Rudder section 舵剖面 与舵杆轴线垂直的舵叶剖面。

Rudder setting angle 舵侧斜角 舵安装时，舵杆轴线与船中线面所成的尖角。

Rudder spindle 舵轴 代替舵柱，供舵叶套在其上转动的固定轴。

Rudder stock 舵杆 用以连接舵叶和舵机或舵柄

的转动杆件。

Rudder stock liner 舵杆衬套 紧套在舵杆上,用以与舵承作相对运动起耐磨防蚀作用的衬套。

Rudder thickness ratio 舵厚度比 舵剖面的最大厚度与舵宽的比值。

Rudder torque 舵杆扭矩 舵压力对舵杆轴线的作用力矩。

Rudder yoke 横舵柄 横向地套在舵头上与舵杆形成十字的舵柄。

Rudder-stop on deck 甲板舵角限位器 设置在舵机舱平台甲板上,用以防止舵的转动角度超过允许值的器具。

Rule length 规范船长 L (m) 系指:(1)沿夏季载重线,由艏柱前缘量至舵柱后缘的距离,或如无舵柱,由艏柱前缘量至舵杆中心的距离。L 应不小于夏季载重水线总长的 96%,且不必大于夏季载重水线总长的 97%;(2)对无舵杆的船舶(如设有全向推力器的船舶)规范船长 L 应取等于夏季载重水线总长的 97%;(3)对设有特殊艏柱或艉柱装置的船舶规范船长 L 应按具体情况予以考虑。 Rule length L: (1) is the distance, in m, measured on the summer load waterline, from the forward side of the stem to the after side of the rudder post, or to the centre of the rudder stock where there is no rudder post. L is to be not less than 96% and need not exceed 97% of the extreme length on the summer load waterline; (2) in ships without rudder stock (e.g. ships fitted with azimuth thrusters), the rule length L is to be taken equal to 97% of the extreme length on the summer load waterline; (3) in ships with unusual stem or stern arrangements, the rule length L is considered on a case by case basis.

Rules[1] 规范 是船舶、海上平台及相关产品的设计、制造、检验及试验的依据。规范的修改通报与规范具有同等效用。制订规范的主要依据为:(1)IMO(国际海事组织)、ILO(国际劳工组织)、IACS(国际船级社协会)等所通过的有关公约、规则、决议、统一要求等适用部分;船级社采用国际海事组织标准、国际劳工组织标准,其目的是为了海上平台安全和防止污染,仅引用其技术条款;(2)有关理论和科研成果;(3)使用经验。就船级社办理船舶或海上平台入级的责任而言,规范规定了平台结构和重要机械的尺寸、所用材料的质量、结构和机械建造标准、检验和试验要求以及保持其良好状态的条件。 Rules are the basis for the design, manufacturing, surveys and tests of ship、units and related products. The amendments to the Rules are equally authentic as the Rules. The main input for formulation of the Rules is as follows: (1) the applicable parts of the relevant conventions, codes, resolutions adopted by IMO (International Maritime Organization) and ILO (International Labor Organization), and unified requirements by IACS (International Association of Classification Societies). The IMO and IIO standards adopted by the Society are for the safety of units and pollution prevention, introducing only the technical regulations thereof; (2) relevant scientific theories and research findings; (3) service experiences. For the purpose of a Society's responsibility for dealing with classification business, the Rules have stipulated the scantlings of unit's structures and essential machinery, the quality and structure of material employed, the standards of machinery manufacturing, requirements for surveys and tests as well as maintenance conditions.

Rules[2] 规范 系指:(1)船级社颁布的入级规范、专用规范、指南和计算软件的统称;(2)由船级社理事会所接受的作为入级依据的所有规则。该规则是船级社制订或采用有关的国际海事标准,对国际海事标准和国际海事标准适用范围之外的其他标准(如适用)的解析也包括在内。 Rules are: (1) the generic term of the classification rules, special rules, guidelines and computers softer released by a society; (2) to be understood as all those Rules requirements accepted by the Board as basis for classification, which are developed by the society or adopted from relevant international maritime standards, interpretations of international maritime standards and standards for ships outside the scope of international maritime standards as appropriate, are included.

Rules for the management, operation and financing of the north Atlantic Ice patrol 北大西洋冰区巡逻、运作和费用规则

Run 去流段 设计水线下,最大横剖面或平行中体以后的船体部分。

Rung 横档 系指垂直扶梯的梯级或垂直面上的梯级。 Rung means the step of a vertical ladder or step on the vertical surface.

Runner 导索 系指用于吊起或放下载荷的钢索。 Runner means a wire rope used for hoisting or lowering a load.

Runner 涡轮 耦合器中接受泵轮输出的工作液体的能量从而拖动从动轴的工作轮。

Running rigging 动索 承载时穿绕过滑车而运动的钢索。如货索等。

Russian Federation Security Central Office 俄联邦安全总局 是俄罗斯国家反间谍与情报侦察机构,是直接履行联邦安全局机构主要活动方向的,在职权范围内为俄联邦安全保障领域实施国家管理的,并协调有权

从事反间谍行动的各联邦执行权力机构活动的联邦执行权力机构。

Rust 生锈 系指货物在运输过程中发生锈损现象。

Rust risk 锈损险 投保平安险和水渍险的基础上加保此险，保险人负责赔偿承保的货物在运输过程中由于生锈而造成的损失。但生锈必须是在保险期内发生的，如原装船时就已生锈，保险公司不负责。此外，在海上保险实务中，保险人一般不就裸装的金属材料承保锈损险。

Rust 锈蚀

Rutile sand 金红石砂 褐色至黑色颗粒砂。有腐蚀性。干燥装运。颗粒多粉尘。无特殊危害。该货物为不燃物或失火风险低。 Rutile sand is a fine particled brown to black sand. It is abrasive and shipped dry. It may be dust. No special hazards. This cargo is non-combustible or has a low fire-risk.

S

S video port 独立视频信号端子 简称 S 端子，是一种将视频数据分成两个单独的信号（光亮度和色度）进行传送的模拟视频信号，不像合成视频信号（Composite video）是将所有信号打包成一个整体进行传送。

SA1 运行区域1 覆盖了世界范围运行的非限制船舶。运行区域1包括在所有其他运行区域内的操作。 Service area 1 covers ships having unrestricted world-wide operation. Service area 1 includes operation in all other service areas.

SA2 运行区域2 主要覆盖了在热带和温带地区运行的船舶。该运行区域包含了运行区域 SA1 所要求的海域操作。 Service area 2 is primarily intended to cover ships designed to operate in tropical and temperate regions. This service area excludes operating in sea areas for which a SA1 notation is required.

SA3 运行区域3 主要覆盖了在热带地区运行的船舶。该运行区域不包含运行区域 SA1 和 SA2 所要求的海域操作。 Service area 3 is primarily intended to cover ships designed to operate in tropical regions. This service area excludes operating in sea areas for which a SA1 or SA2 notation is required.

SA4 运行区域4 主要覆盖了在遮蔽水域运行的船舶。该运行区域不包含运行区域 SA1、SA2 或 SA3 所要求的海域操作。 Service area 4 covers ships designed to operate in sheltered water, This service area excludes operating in sea areas for which a SA1, SA2 or SA3 notation is required.

Sacrificial anode 牺牲阳极 系指依靠自身的腐蚀速率的增加，而使与之耦合的阴极获得保护的电极。船舶牺牲阳极可使用铝合金阳极和锌合金阳极，不推荐使用镁合金阳极。牺牲阳极应装有钢芯，钢芯应使阳极腐蚀后仍能保持住阳极。牺牲阳极应有足够的刚度，以避免与其基座发生共振。 Sacrificial anode means an electrode to protect its coupled cathode depending on increment of its own corrosive rate. The aluminum alloy anode or zinc alloy anode is to be used as ship's sacrificial anode, magnesium alloy anode is not to be recommended to use. The sacrificial anode is to be fitted with steel core, which is capable of maintaining the anode after it is deteriorated. The sacrificial anode is to have sufficient rigidity to prevent resonance with its backing.

Saddle 鞍座 悬吊在承载索上用以支承软管的鞍形座。

Saddle key 鞍形键（空底键）

Safe allowable stress 安全许用应力

Safe and environmentally friendly 安全和利于环保 系指船舶应具备足够的强度、完整性和稳性，使因结构失效（包括结构崩溃）而导致进水或丧失水密完整性所造成的船舶灭失或含有环境污染的风险减至最低。利于环保还包括船舶由可回收利用的环保材料建造。安全还包括用于安全通道、脱险、检查和妥善维护及便利安全操作的船舶结构、装置和布置。 Safe and environmentally friendly means the ship shall have adequate strength, integrity and stability to minimize the risk of loss the ship or pollution to the marine environment due to structural failure, including collapse, resulting in flooding or loss of watertight integrity. Environmentally friendly also includes the ship being constructed of materials for environmentally acceptable recycling. Safety also includes the ship's structures, fittings and arrangements providing for safe access, escape, inspection and proper maintenance and facilitating safe operation.

Safe area in the context of a casualty 事故中的安全区域 系指从可居住性角度而言，任何未进水或发生火灾的主竖区以外的区域。该区域可安全地容纳船上所有人员，使其不受生命或健康威胁，并向他们提供基本服务。 Safe area in the context of a casualty is, from the perspective of habitability, any area(s) which is not flooded or which is outside the main vertical zone(s) in which a fire has occurred such that it can safely accommodate all persons onboard to protect them from hazards to life or health and provide them with basic services.

Safe haven 安全避风港 系指能提供安全进入和保护免受风暴干扰的各种港口和避风锚地。Safe haven means a harbour or shelter of any kind which affords safe entry and protection from the force of weather.

Safe loading and unloading of cargos manual 安全装卸货物手册 系指为使船长在货物装卸作业过程中,防止船体结构中产生过大的应力,而配备的供船上货物安全装卸作业时使用的须知。该手册应至少包括下列内容:(1)船舶稳性资料;(2)加压载和减压载的速率和能力;(3)内底板上单位表面积的最大许用载荷;(4)每舱最大许用载荷;(5)有关船体结构强度的一般装卸须知,包括在装卸货物、压载作业及航行期间的最不利作业状态的任何限制;(6)任何特别的限制,例如主管机关或由其认可的组织所施加的最不利作业状态的任何限制(如适用);(7)如要求强度计算,在装卸货物及航行期间船体上的最大许用载荷和力矩。

Safe passage (safe route) 安全通道

Safe procedure 安全程序

Safe working limit 安全工作极限 一艘船舶或一家公司考虑到特定的设备故障和现场工地所强加的限制。为了动力定位(DP)装置安全工作所设定的环境极限。

Safe working load 安全工作载荷(安全工作负载) 系指:(1)许用的最大货物质量。在这个载荷以下能够安全地操作传送装置装卸和装卸货物的坡道,简写为 SWL,其单位是吨。(2)与物资装运设备某部件有关的载荷,责任人员认为该载荷是可能作用在部件上的最大载荷,以使该部件在正常工作时具有充分的安全裕度。(3)对巴拿马运河而言,安全工作载荷 SWL 应不超过其设计载荷的 80%。Safe working load is: (1) the maximum allowable mass of cargoes specified by the Rules with which the cargo gear and the cargo ramp can be safely operated. It is abbreviated to "SWL" and expressed in ton (t). (2) in relation to an article of materials handling equipment, the load that a responsible person considers is the maximum load that may be imposed on that article in order to allow an adequate margin of safety in the normal operation of the article. (3) for Panama Canal, safe working load should not exceed 80 percent of the design load.

Safe working load for detachable gear 可卸零部件的安全工作载荷 系指可卸零部件经设计和试验证明能承受的最大载荷。此最大载荷应不小于起重设备在安全工作载荷下,可卸零部件会受到的最大载荷。

Safe working load for hoisting apparatus 起重设备的安全工作载荷 系指经正确安装的起重设备在设计作业工况下证明其能吊运的最大静载荷。

Safe working load for union purchase 双杆操作安全负荷 吊杆装置以定位双杆操作方式装卸货物时的安全工作负荷。

Safe working load of a loose gear 传送装置的安全工作载荷 系指允许最大货物的质量。在这个载荷以下能够安全地操作传送装置,简写为"SWL",并且其单位是吨。对于吊货滑车,根据以下(1)或(2)分别定义其工作载荷:(1)单个有槽滑车的安全工作载荷系指被单个有槽滑车的头部装置所安全吊起的最大货物质量。该质量能够保证钢索安全地通过滑车的槽轮。(2)多个有槽滑车的安全工作载荷系指被多个有槽滑车的头部装置所安全吊起的最大货物质量。Safe working load of a loose gear is the maximum allowable mass of cargoes specified by the Rules with which the loose gear can be used safely. It is abbreviated to "SWL" and expressed in tons (t). For cargo blocks, the safe working load is defined according to (1) or (2) below: (1) the safe working load of a single sheave block is the maximum mass of cargoes that can be safely lifted by that block when it is suspended by its head fitting and the mass is secured to a wire rope passing round its sheave; (2) the safe working load of a multiple sheave block is the maximum mass of cargoes that may be applied to its head fitting of the block.

Safe working load, etc 安全工作载荷等 系指起重机系统允许的最小角度和其他限制条件、动臂起重机的安全工作载荷、最大回转半径和其他限制条件等。用来装卸货物其他机械被船级社认定必需的安全工作载荷和其他限制条件,装卸货物的坡道被船级社认定必需的安全工作载荷和其他限制条件。Safe working load, etc. are safe working load, allowable minimum angle and other restrictive conditions in case of the derrick systems, safe working load, maximum slewing radius and other restrictive conditions in case of the jib cranes, safe working load and other restrictive conditions deemed necessary by the society in case of other machinery used for the loading and unloading of cargo, and safe working load and other restrictive conditions deemed necessary by the Society in case of the cargo ramps.

Safe working pressure 安全工作压力

Safe working stress 安全工作应力

Safe-for-entry 进入安全 系指符合下列标准的处所:(1)空气中的氧气含量和易燃蒸发气的浓度在安全

限值以内；(2)空气中的任何有害物质在允许的浓度值以内；(3)与适任人员授权的工作相关的任何残余物或材料，在现有空气条件下按指示操作时不会产生不受控制的有毒物质释放或不安全的易燃蒸发气的浓度。 Safe-for-entry means a space that meets the following criteria: (1) the oxygen content of the atmosphere and the concentration of flammable vapours are within safe limits; (2) any toxic materials in the atmosphere are within permissible concentrations; and (3) any residues or materials associated with the work authorized by the competent person will not produce uncontrolled release of toxic materials or an unsafe concentration of flammable vapours under existing atmospheric conditions while maintained as directed.

Safe-for-hot work 热作安全 系指符合下列标准的处所：(1)具备安全、非爆炸的条件，包括除气条件，用于电弧或气焊设备、切割或气割设备或其他明火形式，以及进行加热、打磨或产生火花的作业；(2)符合进入安全要求；(3)热作后不会导致现有空气条件的改变；(4)为防止产生火焰或火焰扩张，所有相邻处所都已进行清洁或惰气化或充分处理。 Safe-for-hot work means a space that meets the following criteria: (1) a safe, non-explosive condition, including gas-free status, exists for the use of electric arc or gas welding equipment, cutting or burning equipment or other forms of naked flame, as well as heating, grinding, or spark generating operations; (2) Safe-for-entry requirements are met; (3) existing atmospheric conditions will not change as a result of the hot work; and (4) all adjacent spaces have been cleaned, or inerted, or treated sufficiently to prevent the start or spread of fire.

Safeguard 保护措施（安全装置，保护器）
Safety 安全 系指不存在威胁生命、肢体和健康的不可接受的风险水平（由于非故意行为）。
Safety alarm 安全报警器
Safety and limit protection (safety and limit control) 安全及极限保护 为保证锅炉安全运行所采取的包括过高压力、过高温度、过低水位、安全点火、熄火、低风压、过高油温、过低油温及失电压等各种保护措施的通称。
Safety and protective system 安全保护系统
Safety area[1] 安全区 直升机港中环绕进场和起飞区，除飞行导航实施外无其他任何障碍物的规定区域。其目的是减少偶然偏离进场和起飞区的直升机遭受损伤的危险。
Safety area[2] 安全区域 系指 0 类危险区域和 1 类危险区域以外的区域。 Safety area means an area other than Hazardous zone of Category 0 and Hazardous zone of Category 1.

Safety assessment 安全评估 系指对船舶功能和执行功能的系统设计的系统评估。它使用公认的方法来确认故障情况，确立安全目标和要求，并评估所实施的系统。 Safety assessment means a systematic evaluation of the craft functions and the design of systems performing these functions. It uses recognized methods to identify failure conditions, establish safety objectives and requirements and evaluate the implemented system.

Safety audit 安全性审查 不可能导致意外事故或导致财产损失和环境破坏的设备状态和操作程序。
Safety belt 安全带
Safety brake 安全闸
Safety centre 安全中心 系指处理紧急情况的控制站。安全系统的运作、控制和/或监测是安全中心的组成部分。 Safety centre is a control station dedicated to the management of emergency situations. Safety systems' operation, control and/or monitoring are an integral part of the safety centre.

Safety certificate for passenger ships 客船安全证书 对客船而言，系指由 SOLAS 公约缔约国政府授权的个人或组织，按经 1988 年议定书修订的 1974 年国际海上人命安全公约的规定对客船签发的供国际航线/短途国际航线用的安全证书。 For passenger ships, safety certificate for passenger ships means a safety certificate for a short international voyage issued by person or organization authorized under the authority of the Government of under the provisions of the International convention for the safety of life at sea, 1974, as modified by the Protocol of 1988 relating thereto.

Safety chain 安全链（备用链条）
Safety clearance 安全间隙
Safety coefficient 安全系数
Safety communication 安全通信 系指为保证整个装置和人员生命安全而设置的单向听觉、视觉警报和语音广播以及双向的听觉、视觉和语音通信。 Safety communication means one-way audible and visual alarm, voice broadcast, and two-way audible, visual and voice communication provided for ensuring the safety of the complete installation and the personnel.

Safety condition 安全状况
Safety cover 保险帽（护罩）
Safety critical 安全标准 包括风险因素，是对危险的必要防范。
Safety cut-out 安全切断
Safety deck 安全甲板 位于坞墙顶甲板下，用以限制进水,使船坞控制在一定沉深位置的连续水密甲板。

Safety device(protective system) 保护装置 当燃气轮机的转速、燃气温度、燃油压力等重要参数达到极限值或燃烧室点火失败熄火时发出信号,通过综合开关,如有遮断器使燃气轮机紧急停车的装置。

Safety device in venting systems 透气系统的安全装置 系指在透气系统中设有的防止火焰进入液货舱的装置。这些装置的设计、试验和安装位置应符合主管机关依据国际海事组织制定的指南所规定的要求。液位测量孔不得用于平衡压力。液位测量孔应装有自行关闭并密封的盖。在这些开口上不允许设置阻焰器和防火网。 Safety device in venting systems is a device provided in venting system to prevent the passage of flame into the cargo tanks. The design, testing and locating of these devices shall comply with the requirements established by the Administration based on the guidelines developed by IMO. Ullage openings shall not be used for pressure equalization. They shall be provided with self-closing and tightly sealing covers. Flame arresters and screens are not permitted in these openings.

Safety disk 安全膜片

Safety equipment survey 安全设备检验

Safety equivalency 安全等效 系指使用弹性条例,而运用规定来达到与规章要求等效的目的,即实现安全标准和防止污染。

Safety factor 安全因数(安全系数)

Safety falling velocity 安全降落速度 救生载具安全吊卸放落的速度范围。

Safety flap 安全瓣(安全挡板)

Safety goals 风险目标 通过具体的目标来度量油田设施作业的安全性。这些目标应有助于避免事故或抵御事故后果。

Safety hardener 安全固化剂(低毒固化剂)

Safety inserting depth 安全插入深度 自升式钻井平台或桩靴自升式钻井平台的桩腿插入海底硬泥(不包括淤泥厚度)中能保持稳定状态所要求的最小入土深度。

Safety integrity 安全完整性 系指安全相关系统在规定的时间内、规定的条件下,成功地实现所需安全功能的概率,即安全功能能够被有效执行的能力。

Safety integrity level(SIL) 安全完整性水平/等级 系指安全相关系统实现某种安全功能的能力大小。用于衡量安全完整性能力。其根据"平均要求危险失效"概率值分为4个离散的等级,即SIL.1-SIL.4。

Safety level 安全水准/安全等级 是一个表征船舶性能和对站着和坐着的人员的加速度—载荷结果严重性之间关系的数值,船舶性能以水平单幅加速度(g)和加速度速率(g/s)为代表。 Safety level is a numerical value characterizing the relationship between ship performance represented as horizontal single amplitude acceleration (g) and rate of acceleration (g/s) and the severity of acceleration-load effects on standing and sitting humans.

Safety management certificate(SMC) 安全管理证书 系指主管机关或其认可的组织为每艘船舶签发的安全管理证书。在签发安全管理证书前,主管机关或其认可的组织应验证该公司及其船上管理符合经认可的安全管理体系的要求。 Safety management certificate means a safety management certificate issued to every ship by the Administration or an organization recognized by the Administration. The Administration or an organization recognized by it shall, before issuing the safety management certificate, verify that the company and its shipboard management operate in accordance with the approved safety management system.

Safety management manual 安全管理手册 系指用来描述和实施安全管理体系SMS的文件,并包括下列文件:(1)SMS的程序文件和工作须知的清单;(2)公司的概况(包括组织机构图和分支机构);(3)船级社要求的关于SMS的其他文件。

Safety management objectives of the company 公司的安全管理目标 包括:(1)提供船舶营运的安全方法及安全工作环境;(2)对所有已标识的危害建立防范措施;(3)持续提高岸上及船上人员的安全管理技能,包括安全及环境保护方面的应急部署。 Safety management objectives of the company includes: (1) provide for safe practices in ship operation and a safe working environment; (2) assess all identified risks to its ships, personnel and the environment and establish appropriate safeguards; and (3) continuously improve safety management skills of personnel ashore and aboard ships, including preparing for emergencies related both to safety and environmental protection.

Safety management system(SMS) 安全管理体系 系指能使公司人员有效实施公司的安全及环境保护方针所建立并文件化的体系。 Safety management system means a structured and documented system enabling Company personnel to implement effectively the Company safety and environmental protection policy.

Safety margin 安全裕度 系指评估替代设计所使用的方法和假设不确定性所做的允许的调整,例如:性能标准的确定或用来评估危险后果的工程模型。

Safety margin means adjustments made to allow for uncertainties in the methods and assumptions used to evaluate an alternative design, e. g. in the determination of performance criteria or in the engineering models used to assess the consequences of a hazard.

Safety membrane 安全膜片
Safety net 安全网
Safety objective 安全目标 作业应关注的人员、环境及资产安全目标。
Safety operation 安全操作 处理本船上的紧急情况和遇险情况,或协助其他船舶和海上设施处理这种情况。Handling of emergency and distress situations on board own ship or assisting other vessels and offshore installations in such situations.
Safety pin 安全销(保险销)
Safety precaution 安全预防措施
Safety protective device 安全保护装置
Safety provision 安全措施
Safety related automatic functions 安全相关的自动功能 系指直接涉及船舶或人员危险的自动功能,如目标追踪。Safety related automatic functions means the automatic functions that directly impinge on hazards to ship or personnel, e. g. , target tracking.
Safety relief valve 安全阀
Safety requirement 安全要求 系指可予以确认并用以验证实施情况的详细陈述。Safety requirement means a statement in a specification that can be validated and against which an implementation can be verified.
Safety system 安全系统 是:(1)一个为了防止损害机械设备的自动操作的系统。当操作机械设备的时候如果产生严重影响机器功能的阻碍的时候,安全系统应该做出下列这些动作:①开启备用机械设备或装置;②降低机械或设备的输出;③切断燃料和电力的供应进而停止机械设备的运行。(2)当发生危及主推进装置、锅炉、电站以及其他重要机电设备的严重故障时,能使发生故障的机电设备,按下列 3 种类型自动产生保护性动作的系统:①类:立即停止运行,如主柴油机紧急停车,锅炉紧急停炉,紧急切断用电设备电源等,而且非经人工复位,该设备不应再投入运行;②类:暂时调节到可以勉强运行的状态,如降低功率或转速等;③类:启动和投入备用设备,以恢复正常的运行状态。Safety system is:(1)a system which operates automatically, in order to prevent damages to the machinery and equipment in case where serious impediments to functioning should occur on them during operation so that one of the following actions will take place:①starting of standby machinery or equipment; ②reduction of outputs of the machinery or equipment; ③shutting off the fuel or power supplies thereby stopping the machinery or equipment;(2)a systems which will operate automatically for safeguarding the machinery or electrical equipment in question in the following three modes of operation in case of serious faults endangering the main propulsion machinery, boilers, electric generating plants and other essential machinery or electrical equipment:①Mode(a):immediate shutdown. e. g. emergency stop of main engines, emergency cutoff of boiler fuel oil supply and emergency cutoff of electric power supply to consumers. And such machinery or equipment is not to be put into operation again if without the manual resetting. ②Mode(b):the operation of the machinery is temporarily adjusted to the prevailing conditions, e. g. , by reducing the output or rotation speed of the machinery;③Mode(c):the normal operating conditions are restored by starting of standby machinery.

Safety systems 安全系统(应急防护系统) 包括船舶操作所需的所有系统。其包括:(1)灭火系统;(2)舱底水系统;(3)通信系统;(4)航行灯;(5)救生设备;(6)预防导致失火或损坏情况的机械安全系统。Safety systems include all the systems that are necessary for the safety of the ship operation. They include:(1)fire fighting systems;(2)bilge system;(3)communication systems;(4)navigation lights;(5)life-saving appliances;(6)machinery safety systems which prevent of any situation leading to fire or catastrophic damage.

Safety tap 安全旋塞
Safety tread 防滑板
Safety valve 安全阀 装设在锅炉等压力容器上,其内部压力超过规定安全工作限值时即自动起跳排放蒸汽或其他工质的阀。
Safety valve operation test 安全阀调整试验 检查安全阀的开启和关闭的压力范围的试验。
Safety voltage 安全电压 (1)在各导体之间或在任一导体与地之间的不超过交流 50V 方均根值的电压,在电路中可用安全隔离变压器或带分离绕组的变流器与电源隔离;在各导体之间或任一导体与地之间的不超过直流 50V 的电压,在电路中其与更高电压的电路相隔离。(2)用一个安全隔离变压器将一个电路与电源相隔离,在导体间或任何导体对地之间为不超过 50V 交流均方根值的电压。(3)与较高电压电路相隔离的一个电路在各导体之间或任何导体对地之间为不超过 50V 的直流电压。Safety voltage means:(1)a voltage which does not exceed 50 V a. c. r. m. s. between conductors, or between any conductor and earth, in a circuit isolated from the

supply by means such as a safety isolating transformer, or convertor with separate windings. Voltage which does not exceed 50 V d.c. between conductors, or between any conductor and earth, in a circuit which is isolated from higher voltage circuits; (2) voltage which does not exceed 50 V a. c. r. m. s. between conductors, or between any conductor and earth, in a circuit isolated from the supply by means such as a safety isolating transformer; (3) voltage which does not exceed 50 V d.c. between conductors or between any conductor and earth in a circuit isolated from higher voltage circuits.

Safety workstation 安全工作站 监视显示和集中于船舶自身的安全系统的操作元素的工作站。Workstation at which monitoring displays and operation elements of systems serving

SAR = search and rescue 搜救

Safety-level approach 安全水平法 为针对 IMO 决策过程中基于风险的方法的结构性的应用。

Sag bend 触地段

Sagbend 悬链线下拐折点 系指在敷管过程中，靠近海底的那部分管子被弯曲成自然悬链线的一点。The section of pipe near the seabed that bends as the pipe acquires a natural catenary shape during installation.

Sagged 悬垂段

Sagging 中垂/下沉(下垂) 船体中部下垂，艏艉部上翘的弯曲状态。

Sail 帆 由帆布等材料制成，呈三角形、四边形或多边形，挂在桅上承受风力借以推船前进的蓬状物。

Sailer (sailing boat) 帆船 利用帆，依靠风力推进的船舶。

Sailing boat 帆艇 以风力作为主要推进方式的艇，当它迎风航行时，其以 m^2 表示的一次可张开的所有帆之总投影面积（帆的投影面积 A_S） $A_S \geq 0.07 \times (m_{LDC})^{2/3}$，$m_{LDC}$ 为艇的满载排水质量（kg）。Sailing boat is the boat for which the primary means of propulsion is by wind power, having a total profile area expressed in m^2, of all sails that may be set at one time when sailing closed hauled (Projected sail area A_S) of $A_S \geq 0.07 \times (m_{LDC})^{2/3}$, m_{LDC} is loaded displacement mass of boat.

Sailing vessel 帆船 系指任何驶帆的船舶，包括装有推进机器而不使用者。Sailing vessel means any vessel under sail provided that propelling machinery, if fitted, is not being used.

Sailmaker's tools 缝帆工具 船上为缝制帆布制作所使用的专用工具。

Saint Lawrence Seaway 圣劳伦斯航道 系指蒙特利尔港和伊利湖之间的深水航道，包括所有船闸、运河和连接水域，部分深水航道的邻近水域和其他已授权由公司或管理局进行经营、管理和控制的位于任何处所的运河和工程。Saint Lawrence Seaway means the deep waterway between the Port of Montreal and Lake Erie and includes all locks, canals and connecting and contiguous waters that are part of the deep waterway, and all other canal and works, wherever located, the management, administration and control of which have been entrusted to the Corporation or the Manager.

Sale by proxy 代销

Sale tax 销售税 是以特定消费品为课税对象所征收的一种税，属于流转税的范畴。我国现行消费税是 1994 年税制改革中新设的一种税种。

Sales contract 买卖合同 是一方转移标的物的所有权于另一方，另一方支付价款的合同。转移所有权的一方为出卖人或卖方，支付价款而取得所有权的一方为买受人或者买方。买卖是商品交换最普遍的形式，也是典型的有偿合同。根据合同法第 174 条、第 175 条的规定，法律对其他有偿合同的事项未作规定时，参照买卖合同的规定；互易等移转标的物所有权的合同，也参照买卖合同的规定。

Sales package 销售包装 又称内包装或小包装。是直接接触商品并随商品进入零售网点和消费者或用户直接见面的包装。适应商品市场竞争和满足多层次消费要求，不断向销售包装发出要求改进与创新的信息。这些信息不仅仅是针对销售包装的，其实也是泛指整个包装工业的；没有各个包装专业的同步发展与提高，销售包装就会处于"无米之炊"的境地。

Sales promotion 促销 就是营销者向消费者传递有关本企业及产品的各种信息，说服或吸引消费者购买其产品。促销以达到扩大销售的目的。企业所从事的这种市场营销活动叫作"促销"。

Salinity 盐度 (1) 1 公升水中以氯离子的毫克当量表示的氯盐含量；(2) 水溶液中溶解盐的含量。

Salinity gauge (salinometer) 盐度计

Salinity indicator 盐度指示器

SALM 单锚系泊 SALM means the single-anchor mooring.

SALM system with chain and rigid riser 带锚链与刚性立管的 SALM 系统 这种系统的锚腿由锚链和刚性立管组成，刚性立管与海底基座连接，锚链则与上部浮筒和刚性立管连接，两端均设有方向接头，刚性立管下部与海底油管连接，原油通过立管从上部旋转接头处输入浮动软管，送到油船上。

SALM system with chain leg 带锚链腿的 SALM 系统 系指其上部为浮筒，油船的首缆系在浮筒上，浮筒下部连接一根锚腿（锚链），其下端系在基座上，海底油管与基座上的一个旋转接头连接的系统。与 CALM 系统不同之处是：输油管不进入浮筒，而从旋转接头处用水下浮管将油输送到油船上。这种 SALM 系统用于水深 30～50 m 处。

SALM system with rigid riser 带刚性立管的 SALM 系统 系指具有较长的浮筒，其浮筒下部直接与一个刚性立管连接，而立管固定在海底基座上，海底油管附着在立管上连接到浮筒下部的旋转接头上，浮筒可旋转，内部有通道，可供工作人员出入，以便检查旋转接头，原油从旋转接头处通过漂浮的软管输送到油船上的系统。

Salt 盐/盐类 白色细精粒。含水量不定，最高可达5.5%。该货物极易溶解。如水进入货舱，船舶因盐的溶解（潮湿底层的形成和货物的移动）而存在失去稳性的危险。无特殊危害。该货物为不燃物或失火风险低。 Salt is a fine white grains and its moisture variable to 5.5%. This cargo is highly soluble. In the case of ingress of water into the holds, there is a risk to the loss of the stability of the ship through dissolution of this cargo (formation of a wet base and shifting of cargo). It has no special hazards. This cargo is non-combustible or has a low fire-risk.

Salt bath anneal 盐浴退火

Salt bath quenching 盐浴淬火

Salt cake 芒硝 非纯净的硫酸钠。呈白色。颗粒状，干燥装运。无特殊危害。该货物为不燃物或失火风险低。 Salt cake is a impure sodium sulphate. It is white in colour. Granular, shipped dry. It has no special hazards. This cargo is non-combustible or has a low fire-risk.

Salt content 盐含量（盐度）

Salt content of condenser 冷凝水含盐量 溶解在冷凝水中的氯化钠等盐的数量。

Salt deposit 盐沉淀（盐沉积物）

Salt fog test（saltwater ingestion test） 喷盐雾试验 在燃气轮机进口处喷入一定规格的盐雾以检查其主要零部件的材料及涂层的耐腐蚀性以及进气装置效用的长期台架负荷试验。

Salt rock 盐岩 白色。含水量0.02%。无特殊危害。该货物为不燃物或失火风险低。 It is white in colour. Moisture content 0.02%. It has no special hazards. This cargo is non-combustible or has a low fire-risk.

Salt spray test 喷盐水试验

Salt temperature deep cable winch 温盐深电缆绞车（STD 绞车） 为收放 STD 等测量仪器进行水文要素（水深、水温）和海水化学要素（盐度）观测用的绞车。

Salt water 盐水（海水）

Salt water ballast 海水压载

Salt water ballast space 海水压载处所

Salt water ingestion test 喷盐雾试验

Salt water pump 盐水泵

Salvage and rescue ship 打捞救生船 系指从事打捞和救生工作的船舶。

Salvage barge 打捞驳 系指设有压缩空气、排水、潜水和消防等打捞设备，用于打捞水下沉船、沉物为主兼负救生的驳船。

Salvage charge 救助费用 系指被保险标的遭遇保险责任范围以内的灾害事故时，由保险人和被保险人以外的第三者采取救助行为，对于此种救助行为，按照国际法规规定，获救方应向救助方支付相应的报酬，所支付的该项费用，被称为救助费用，它属于保险赔付范围。

Salvage pump 救助泵 打捞或救助作业用的泵。

Salvage ship 打捞船/救生船 系指设有打捞设备，用于打捞水下沉船、沉物的船舶。 Salvage ship is a ship provided with equipment for salvaging sunken ships or other objects.

Salvage tug（rescue tug） 救助拖船 系指以救助海损（难）船舶为主的拖船。航区有近海、远海。

Salvage vessel 救捞船 设有压缩空气、排水、潜水和消防等设备，以打捞任务为主兼负救生的机动船舶。

Salver 救助人 系指提供与救助作业直接相关的服务的任何人。 Salver means any person rendering services in direct connection with salvage operations.

Sampan 舢板 依靠桨、橹推进，作为短途运输用的无甲板的小艇。舢板结构简单，吃水浅，操作方便，可以进行海上救生、舷外作业和装载人员登岸等。一般备有1～6把桨的舢板称为小型舢板，备有8～16把桨的舢板称为中型舢板。

Sample 样品（标本，试验样品） 系指用于试验/检验的代表性的产品。样品的选取能在特性、特征、制造质量上代表或覆盖申请产品检验的产品或系列产品。 Sample means a representative product used for test/inspection. The selected sample is to be, in respect to performance, characteristics and manufacturing quality, capable of representing or covering the products or product series to be inspected.

Sample 试件 从一个试验单元量中为截取试样所选的制品，例如钢板和管材。

Sample extraction smoke detection systems 抽烟式探火系统

Sample of thermosetting resin 热固化树脂试样

Sample point（sewage treatment plant） 取样点（污水处理装置） 系指无须打开液舱、空舱或通风口，就可人工收集流入液体和排出液体之代表性试样的点。

Sample point means a point for manual collection of a representative sample of influent and effluent without opening tanks, voids or vents.

Sample shipment 试销

Sampling 取样（采样，脉冲调制）

Sampling facilities 取样设备（取样设施） 系指：（1）对原型压载水处理技术程序认可和监督导则要求的经处理或未经处理的压载水进行取样的装置；（2）安装用于取样的设备。Sampling Facilities refers to:（1）the means provided for sampling treated or untreated ballast water as needed in these Guidelines;（2）the equipment installed to take the sample.

Sampling methods 取样方法 系指以下列的任一种方法取得初始样品的方法:（1）手动阀门设置持续滴流取样;（2）按时间比例的自动取样;（3）按流量比例的自动取样。 Sampling methods are those methods obtained the primary sample by one of the following methods:（1）manual valve setting continuous drip sample; or（2）time-proportional automatic ampler;（3）flow-proportional automatic ampler.

Sampling point（ballast water piping） 取样点（压载水管系） 系指压载水管系中进行取样的位置。

Sampling point means that place in the ballast water piping where the sample is taken.

Sampling procedures 取样程序 对质量管理体系而言，系指制造厂商所采用的为证明一组产品是否合格的取样技术，其应用的步骤应与规定的要求相符或经验船师的同意。 For the purpose of quality management system, sampling procedures means where sampling techniques are used by the manufacturer to verify the acceptability of groups of products, the procedures adopted are to be in accordance with the specified requirements or are to be subject to agreement by the surveyors.

Sampling unit 取样器

Sand 砂 通常为细颗粒。砂包括：铸造用砂、钾长石砂、石英砂、硅砂、钠长石砂。有腐蚀性且多粉尘。吸入硅砂粉尘能导致呼吸系统疾病。硅砂颗粒易随空气流动和被吸入。工业用砂可能涂有树脂，如受热（55 ℃~60 ℃）会结块。该货物为不燃物或失火风险低。 Sand is usually fine particles and abrasive and dusty. Sands included in this schedule are: foundry sand, potassium feldspar sand, characteristics, silica sand. Inhalation of silica dust can result in respiratory disease. Silica particulates are easily transported by air and inhaled. Industrial sand may be coated with resin and will cake if exposed to heat（55 ℃ to 60 ℃）. This cargo is non-combustible or has a low fire-risk.

Sand（sand hole） 砂眼

Sand box 砂箱 贮存消防用黄沙的木质或金属箱。

Sand piling barge 打砂桩船 系指将空心桩打入水下软地层里，然后用大型喷砂泵将砂注入桩孔以形成砂桩的打桩船。

Sand pump 泥浆泵

Sand pump bucket dredger 自扬链斗挖泥船 由链斗挖取的泥砂，可直接依靠本船专用机械搅成泥浆，然后由泥泵扬出的链斗挖泥船。

Sand spreading barge 砂石撒铺船

Sand-filled apparatus "q"（electrical equipment） "q"充砂型（电气设备） 当其带电部件完全地埋入大量的粉末状材料中时，认为其为"充砂"型设备。注意：IEC 60079-5 规定了使用此防护方法之设备的结构特征和试验要求。 Sand-filled apparatus "q"（electrical equipment）means an apparatus, which is considered "sand-fitted" when all its live parts are entirely embedded in a mass of powdery material. Note: IEC 60078-5 specifies the constructional features and test requirement for apparatus using this method of protection.

Sandwich plate 夹心板 是由两层成型金属面板（或其他材料面板）和直接在面板中间发泡、熟化成型的高分子隔热内芯组成。这些夹芯板成品便于安装、轻质、高效。填充系统使用的也是闭泡分子结构，可以杜绝水汽的凝结。外层钢板的成型充分考虑了结构和强度要求，并兼顾美观，内面层成型为平板以适应各种需要。

Sanitary 卫生的

Sanitary and similar spaces 卫生间及类似处所 系指：（1）公共卫生设施、淋浴室、盆浴室、厕所等；（2）小洗衣间；（3）室内游泳场所；（4）起居处所内无烹调设备的单独配膳室。个人卫生设施应视为所在处所的一部分。 Sanitary and similar spaces are:（1）communal sanitary facilities, showers, baths, water closets, etc.（2）small laundry rooms.（3）indoor swimming pool area.（4）isolated pantries containing no cooking appliances in accommodation spaces. Private sanitary facilities shall be considered a portion of the space in which they are located.

Sanitary certificate 卫生证书

Sanitary discharger 卫生水排泄孔

Sanitary ejector 粪便喷射泵 使用压力水将粪便

从粪便柜中抽出喷射到船外的设备。

Sanitary fixtures 卫生设备 各种清洁卫生设备和用品的统称。

Sanitary inspection certificate 卫生检验证书 是对食用动物产品、食品检验后出具的证书。

Sanitary piping system（sewage piping） 粪便水排泄系统 将大小便器内的污水直接排至舷外或先排至粪便柜再利用喷射器或泵排至船外的系统。

Sanitary pressure tank 卫生水压力柜 利用柜内被压缩的空气压力,把来自舷外的水压送到船上卫生设备用水处的密封式水柜。

Sanitary pump 卫生泵(卫生水泵) 抽送卫生用水的泵。

Sanitary system 卫生水系统 供应卫生设备的舷外水冲洗系统。

Sanitary tank 粪便柜(卫生水柜) 专门收集全船粪便水的柜。

Sanitary treating arrangement 粪便水排出处理装置 对粪便水进行消毒处理的装置。

SAR services 救助服务 系指遇险监控、通信、协调和搜救职能的行为,包括通过利用公共和私人资源,提供医疗建议、基本医疗援助或医疗疏散。这些资源包括协作的飞机、船舶和其他浮动工具及设施。搜救服务还包括安排幸存者离开的援助船。 SAR services means the performance of distress monitoring, communication, co-ordination and search and rescue functions, including provision of medical advice, initial medical assistance, or medical evacuation, through the use of public and private resources including co-operating aircraft, vessels and other craft and installations. SAR services include making arrangements for disembarkation of survivors from assisting ships.

SART 搜救应答器 SART is search and rescue transponder.

Satellite cloud picture 卫星云图 气象卫星所获得的有关图像资料。

Satellite cloud picture receiving room 卫星云图接收室 接收各种气象卫星所发出的各种云图的工作室。因云图照片需及时冲洗,应附有专用暗室。

Satellite communication system 卫星通信系统 利用通信卫星和两个以上的地面站就构成卫星通信系统。其主要组成：(1)通信卫星——它是卫星通信中最关键的设备。其基本功能是接收来自地面站和船舶站的电波信号,经过变频、放大,再向地面发射。目前国际卫星通信中大多数采用静止卫星,又称同步卫星。它与地球自转同速、同向运动,这样的卫星相对地面是静止不动的。它的运动轨道就叫静止轨道或同步轨道。目前使用的国际全球通信系统,除了在地球各地可建立通信站之处,还必须在共视区内建立地面站。它们分别与大西洋、印度洋和太平洋上空的通信卫星构成全球性的卫星通信系统。(2)卫星地面站——进行卫星通信的地面设施。其任务是向卫星发射无线电信号,同时接收卫星转发下来无线电信号。卫星地面站按其机动性能方为固定站、半固定的移动站和移动站(包括船载站、车载站和机载站)三种;按电气性能分为标准站和非标准站两种。一个标准地面站包括：天线系统、发射系统、接收系统、终端系统和电源系统。

Satellite elevation 卫星仰角 卫星至观察者的连线与观察者当地水平面之间的夹角。卫星通过最接近点时的仰角为最大仰角。当卫星最大仰角大于70°时位置线的交角很小,所引起的定位误差较大,且航速误差和天线高度误差引起的定位误差也较大。如果最大仰角达到90°时,根本无法定位。当卫星最大仰角小于15°时获得的多普勒计数太少,且对流层的折射加强和电离层的厚度增大,故折射误差引起的定位误差增大。一般规定卫星通过的最大仰角在15°～70°范围之内,才能有效地观测。

Satellite fix modes 卫星定位模式 在卫星定位程序中给出了三种定位模式：二维、三维和VN模式,这三种定位模式适用于三种不同的工作情况。(1)二维模式可获得船舶的经度、纬度和接收机本地振荡频率的频漂。当使用好的速度传感器时(速度误差小于0.25 kn)二维模式适用于海上航行船舶的定位。其误差取决于速度误差和天线高度误差。(2)三维模式可获得精确的经度、纬度、天线高度和频漂。当船舶靠码头时综合多次卫星定位可得到精度在10 m以内的位置值,其精度随定位次数增加而提高。同时能精确测定船载天线的实际高度,三维模式还能精确标定无线电助航设备。(3)VN模式可获得经度、纬度、北向速度和频漂。当所使用的速度传感器的速度误差大于或等于0.25节(kn)时,VN模式适用于海上航行船舶的定位。

Satellite forecast 卫星预报 根据卫星地面站获得的最新轨道参数,推算出卫星在未来时刻空间位置的预报。在卫星导航设备中装入卫星预报程序就可实现卫星预报,这对于计划测量工作十分有用。当操作员提出预报请求时,该程序将打印出卫星预报表。表中给出卫星编号,卫星升起时间,任何选定经度、纬度上的卫星仰角,卫星过顶时间,卫星飞行方向等数据。

Satellite launching technologies 卫星发射技术 技术拥有国：中国、俄罗斯、美国、法国、日本、英国、印度、以色列、伊朗。见图S-2。

Satellite navigation 卫星导航 利用人造地球卫星发出的无线电信号,求出载体相对于卫星的位置,再

图 S-2 卫星发射技术
Figure S-2 Satellite launching technology

图 S-3 卫星导航技术
Figure S-3 Satellite navigation technology

根据地面站测出的卫星相对于地面的位置进行计算，从而确定载体在地球上的位置。卫星导航的优点是：全球性、全天候性，比天文定位精度高和自动化程度高。目前卫星导航的主要缺点是不能连续定位，一般需要间隔 1~2 h 才能获得一次准确位置，低纬区可长达 4 h。

Satellite navigation computer 卫星导航计算机 卫星导航系统所用的计算机包括地面设备和用户设备。前者是计算中心的计算机，它根据各跟踪站送来的多普勒数据经折射校正后用来计算卫星坐标，预报轨道参数，分析遥测数据，编制卫星上设备的工作程序，确定传输给卫星的指令和信息。后者是卫星导航接收机的计算机，它接收来自卫星的多普勒频移，双频道电离层折射校正，卫星轨道参数和精确的时标。接收来自各传感器的航向、航速以及包括天线高度、总天数和初始经纬度的输入常数。

Satellite navigation system 卫星导航系统 一种利用人造地球卫星为航行体进行导航定位的新技术。它具有全球、全天候、定位精度高以及安全、经济、可靠等优点。包括 1964 年第一个卫星导航系统，即 NNSS 子午仪卫星导航系统（第一代）；GPS 导航是全球定位系统（第二代）以及正在研制中的仪 CNNS 为代表的第三代卫星导航系统。卫星导航定位技术可用于一切需导航定位服务的军事和民用领域，范围包括海面上的各类舰船（艇）导航、陆地上的车辆导航、地图测绘等；空中的空中管制、飞机进场、着陆、武器的反射等。

Satellite navigation technology 卫星导航技术 系指采用导航卫星对地面、海洋、空中和空间用户进行导航定位的技术。导航卫星如同太空灯塔。卫星导航综合了传统导航的优点，实现了全球、全天候、高精度的导航定位。技术拥有国：中国（北斗）、美国（GPS）、俄罗斯（GLONASS）、欧盟（伽利略）。见图 S-3。

Satellite orbit parameter 卫星轨道参数 卫星定位，必须首先知道卫星在空间的准确位置。椭圆就是用来确定卫星飞行的轨道在空间的位置，以及卫星在该轨道上的位置。

Satellite parameters 卫星参数 卫星发射的两分钟电文有 157 个字。其中与导航有关的 25 个字是卫星参数，与导航无关的其余的字仍是保密的。25 个卫星参数是用来确定卫星轨道位置的，其中 17 个是固定参数，表示卫星的平均轨道，它在两次注入信息之间（12~16 h）保持常数。由于平均轨道是描述开普勒椭圆的，这对卫星实际飞行轨迹只是近似描述，因此在两分钟电文内还需要包括 8 个可变参数。

Satellite system 卫星系统 系指空间部分、控制空间部分的安排和访问空间部分的网络控制装置。 Satellite system means the space segment, the arrangements for controlling the space segment and the network control facilities controlling the access to the space segment.

Satellite update 卫星更新 把符合标准的卫星所确定的船位取代同一时刻的推算船位。并以卫星定出的船位为新的起点进行后面的推算。卫星定位的计算结果确定了推算船位和卫星更新位置之间的差值。每次定位计算完毕时绘出：沿航线更新值（卫星更新位置超过推算船位为止），正横航线更新值（卫星更新位置偏右为负），径向更新值，北向更新值（向南为负），东向更新值（向西为负）。

Satellite-launched ballistic missile（SLBM） 人造卫星发射的弹道导弹

Satisfactory ventilation 合格的通风 系指每小时换气次数不低于 12 次的通风。海上移动平台上的开敞处所被认为具有合格的通风。 Satisfactory ventilation means ventilation giving at least 12 air changes per hour. Open spaces in mobile offshore drilling units are considered to be satisfactorily ventilated.

Saturate 使饱和（浸透）

Saturated air 饱和空气

Saturated air content of water 水的饱和空气含量 水在饱和点所含的空气量。以百万分数计。

Saturated steam 饱和蒸汽 饱和温度下的蒸汽。

Saturated vapor 饱和蒸发气体

Saturation 饱和

Saturation diving 饱和潜水 是一种适用于大深度条件下,开展长时间作业的潜水方式。按照国际惯例,当潜水作业深度超过 120 m,时间超过 1 h,一般都采用饱和潜水方式。简单来说,在几十米的水下,人需要呼吸压缩后的普通空气。随着深度增加,人在水下呼吸普通空气时,其中的氮气在高压下易引发"氮麻醉",而且呼吸阻力也随之增大。这时人只能通过呼吸氦等惰性气体和氧的混合气来进行更深的潜水作业。由于惰性气体被吸入后溶入人体血液,这就决定了潜水员在水下作业时间越长,上浮减压的过程就越长,潜水作业效率将大大降低。如果不按规程进行减压,溶解在人体内的惰性气体将在潜水员的关节或身体组织中形成气泡,会造成严重的减压病,甚至危及生命。当潜水员在某一压力下连续停留 24 h 后,身体组织中溶解的惰性气体量就达到了最大限度。严格地说,潜水员这时吸入的惰性气体与呼出的量相等,处于动态平衡。这种潜水方式,就称为"饱和潜水"。作为唯一一种可使潜水员直接暴露于高压环境开展水下作业的潜水方式,饱和潜水已广泛应用于失事潜艇救援、海底施工作业、水下资源勘探、海洋科学考察等军事和民用领域。

图 S-4 饱和潜水系统
Figure S-4 Saturation diving system

Saturation diving system 饱和潜水系统 安装在潜水作业母船上进行饱和潜水作业所用的技术装备的总称。分下潜式加压舱系统(即深潜系统)、水下居住舱系统和闸式可潜器系统三种类型。其任务是为潜水员提供一个在高压环境下休息、睡觉、吃饭、卫生和外出潜水作业的条件,并在工作结束时对潜水员进行减压,使他们安全地返回大气环境中。该系统一般由甲板居住舱、潜水钟及其吊放架、潜水钟吊放系统、高压逃生舱、潜水钟主脐带缆、控制集装箱、环境控制单元和热水单元、气体供应、氦气回收系统和观察通信、高压气体储存容器、应急发动机、空气压缩机、工具箱以及工具下放平台等组成。可适用潜水深度一般为 200 ~ 300 m,最大可达 500 m。见图 S-4。

Saturation life support unit 饱和潜水的生命维护装置

Saturation point 饱和点

Saturation ratio 饱和度

Sauna 桑拿房 系指一种温度通常在 80 ℃ ~ 120 ℃ 之间的加温室,其热量由一种热表面提供(如电加热炉)。该加温室还可包括加热炉所在的处所和邻近的浴室。 Sauna is a hot room with temperatures normally varying between 80 ℃ ~ 120 ℃ where the heat is provided by a hot surface (e. g. by an electrically-heated oven). The hot room may also include the space where the oven is located and adjacent bathrooms.

Sawdust 锯屑 木质细颗粒。如不洁净、不干燥和有油污会自燃。易使货物处所缺氧。 Sawdust is a fine particle of wood. It is spontaneous combustion if not clean, dry and free from oil. It is lable to cause oxygen depletion within the cargo space.

SBG (See-Berüfsgenossenschaft) 德国海员工伤事故保险联合会 英文全称为"Professional Accident Insurance for Seaman"。

SBT 专用压载舱 SBT means segregated ballast tank.

Scab (scabs) 铸疤(瑕疵)

Scaffold staging 脚手架

Scale 瘢痕

Scale deposit 水垢

Scale effect 尺度效应(缩尺效应) 几何相似的物体间,由于大小不同而不能同时满足所有有关动力相似定律,而引起物体所受的力如重力、黏性力和表面张力或力矩系数以及流态等的差异。

Scale preventive 防水垢剂

Scale rate 生锈率(积垢率)

Scale ratio 尺度比/比例 实物与其模型尺度的比值。

Scale separator 氧化皮清除器(清污器)

Scale solvent 水垢的溶剂

Scaled 锅垢（刮垢,鳞状的）

Scaled a boiler 去除锅炉的水垢

Scalene scrapers 三角刮刀

Scallop 扇形孔 系指加强构件上的开孔，用以连续焊接的板缝。Scallop means a hole cut into a stiffening member to allow continuous welding of a plate seam.

Scanner 扫描仪 是利用光电技术和数字处理技术，以扫描方式将图形或图像信息转换为数字信号的装置。扫描仪通常被用于计算机外部仪器设备，通过捕获图像并将之转换成计算机可以显示、编辑、存储和输出的数字化输入设备。扫描仪对照片、文本页面、图纸、美术图画、照相底片、菲林软片，甚至纺织品、标牌面板、印制板样品等三维对象都可作为扫描对象，提取和将原始的线条、图形、文字、照片、平面实物转换成可以编辑及加入文件中的装置。扫描仪中属于计算机辅助设计（CAD）中的输入系统，通过计算机软件和计算机，输出设备（激光打印机、激光绘图机）接口，组成网印前计算机处理系统，而适用于办公自动化（OA），广泛应用在标牌面板、印制板、印刷行业等。

Scantling draught T_{sc} 结构吃水 是满足关于船舶尺度的强度要求下的最大设计吃水。结构吃水应不小于相应于核定干舷的吃水。Scantling draught is the maximum design draught, in metres, at which the strength requirements for the scantlings of the ship are met. The scantling draught is to be not less than that corresponding to the assigned freeboard.

Scantlings 尺度 系指结构构件的尺寸和厚度。Scantlings are those size and thickness of structural components.

Scarfing bracket 嵌接肘板 用在两个偏置结构项目之间的肘板。Scarfing bracket is a bracket used between two offset structural items.

SCART 欧洲标准的音视频信号

SCART on 启用欧洲标准的音视频信号输出

Scattering phenomenon 散射现象 系指声波辐射在表面很粗糙的障碍物上，这障碍物表面的起伏程度与声波波长相当，或者障碍物的大小与波长差不多，入射的声波就会向各个方向散射的现象。

Scavenge 扫气（换气，清扫）

Scavenger 扫气装置（换气管）

Scavenger relief device 扫气箱释放装置

Scavenging 扫气（换气，清除） 借助于进排气口或气门间的压力差，用新鲜可燃混合气或空气将燃烧产物扫出气缸的过程。

Scavenging air receiver 扫气箱 二冲程柴油机扫气口前的储气室间或箱型构件。

Scavenging air trunk 扫气总管

Scavenging blower 扫气鼓风机

Scavenging duct 扫气管

Scavenging fan 扫气风机

Scavenging period 扫气阶段

Scavenging port 扫气孔

Scavenging pressure 扫气压力 扫气口前的空气压力。

Scavenging pump 扫气泵（换气泵） 供二冲程柴油机完成扫除废气和充入新鲜空气之用的压气装置。

Scavenging ratio 扫气系数 每一个循环通过气缸的扫气室气量与进排气口或进排气口关闭后留在气缸内的新鲜空气气量之比。

SCC SO_x 排放控制区符合证书 SCC means the SECA compliance certification.

Scheduled airline transport 班机运输 系指采用具有固定开航时间、航线和停靠航站的飞机运输货物的方式。通常为客货混合型飞机，货舱容量较小，运价较贵，但由于航期固定，有利于客户安排鲜活商品或急需商品的运送。

Scheme A 方案 A 对船舶废气清洗系统试验和检验而言，系指通过台架试验和参数检查，证明废气清洗系统排放符合性的一种检验和发证方法。

Scheme B 方案 B 对船舶废气清洗系统试验和检验而言，系指通过对废气中 SO_x 排放极限连续监测，证明废气清洗系统排放符合性的一种检验方法。

Schiling rudder 西林舵 是一种由近似 JIS 翼型、舵上下端加制流板且随边处加装鱼尾构成的矩形舵。其展弦比通常为 1，其流体动力特征介于常规矩形舵和襟翼舵之间。舵叶自身的强度是该舵设计的关键。

Schneider propeller 直翼推进器

Schooner bar 恩古诺酒吧（豪华邮轮）

Schooner guy 吊杆间牵索 定位双杆操作吊杆装置中，连接舱口吊杆和舷外吊杆头端的绳索。

Schottel propulsion unit 悬挂式转向推进装置

Science citation index（SCI） 美国《科学引文索引》 于 1957 年由美国科学信息研究所在美国费城创办，是由美国科学信息研究所（ISI）1961 年创办出版的引文数据库。SCI（科学引文索引）、EI（工程索引）、ISTP（科技会议录索引）是世界著名的三大科技文献检索系统，是国际公认的进行科学统计与科学评价的主要检索工具，其中以 SCI 最为重要，创办人为尤金·加菲尔德（Eugene Garfield）。

Science-based principle 基于科学原则 系指风险管理基于使用科学方法和分析的最佳可用信息的原则。

Science-based principle is a principle that risk assessments are based on the best available information that has been collected and analysed using scientific methods.

Scleroscope 回跳硬度计

Scoop condenser（scoop circulation condenser） 自流式冷凝器　将船舶高速航行时形成的水动压头在专门设计的管道中转化为静压头，以克服冷凝器与冷却水管道的阻力而使冷却水自动流经冷却水管的冷凝器。

Scoop-circulating water system 直流式循环水系统　利用船舶在高速航行时迎面水流动压头，使冷却水流过主冷凝器的循环水系统。

Scope 放链长度　锚泊时，放出船外的锚链的长度，也有以水深的倍数表示的。

Scope of cases 受案范围　上海仲裁委员会国际航运仲裁院可以受理涉及海事海商、仓储物流、航运金融、港口建设等争议案件。具体包括：（1）海事海商——海上水陆货物运输、碰撞救助和打捞、船舶设计、建造、修理、租赁及买卖、船舶经营管理相关争议；（2）仓储物流——仓储、物流、公路运输、航空运输、货运代理；（3）航运金融——涉及航运业的保险、融资、抵押；（4）港口建设——水上、水下工程、港口作业、岸线作业；（5）其他事项——航运企业经营管理中产生的争议、船员劳务、旅客和行李运输、渔业生产以及当事人协议仲裁的其他事项。

Scoring 齿面胶合　当润滑失效时，齿面金属发生直接接触，而因摩擦系数骤增，局部温度过高，金属互相黏连，较软的金属齿面被较硬的金属齿面撕下一条条沟纹的破坏现象。

Scotch boiler 苏格兰锅炉（圆筒形锅炉）

Scouring device 消除吸附力装置（冲洗装置）　在沉垫升离海底或拔桩前，通过布设在其中的高压水管和喷嘴对海底泥面冲高压水，以消除或克服海底泥面对沉垫、桩靴或桩腿的吸附力而设的装置。

SCP SO_x 排放控制区符合计划　SCP means the SECA compliance plan.

Scrap brass 废黄铜（黄铜屑）

Scrap iron 铁屑（废铁）

Scrap metal 废金属　废铁或废钢所包括的黑色金属范围很广，主要供回收利用。无特殊危害。该货物为不燃物或失火风险低。但货物含有金属碎屑（易自燃的金属细碎旋屑）时除外。 Scrap iron or steel covers an enormous range of ferrous metals, principally intended for recycling. No special hazards. This cargo is non-combustible or has a low fire-risk except when cargo contains swarf (fine metal turnings liable to spontaneous combustion).

Scrap steel 废钢　既是钢铁厂生产过程中不成为产品的钢铁废料（如切边、切头等）以及使用后报废的设备、构件中的钢铁材料，成分为钢的叫废钢。钢铁工业主要的铁源为铁矿石。每生产 1 吨钢，大致需要各种原料（如铁矿石、煤炭、石灰石、耐火材料等）4～5 吨，能源折合标准煤（指发热值为 7 000 千卡/公斤的煤）0.7～1.0 吨。而利用废钢做原料直接投入炼钢炉进行冶炼，每吨废钢可再炼成近 1 吨钢，可以省去采矿、选矿、炼焦、炼铁等过程，显然可以节省大量自然资源和能源。

Scrap waste 碎片（碎屑,废料）

Scrape 刮（削,铲刮）

Scrape and touch-up 除锈补漆

Scraper 刮刀

Scraper ring 刮油胀圈（刮油环）

Scrapers and paint brushes 敲铲油漆工具　供船上敲铲铁锈和涂刷油漆、涂料的各种工具的统称。

Scratch 划痕（刮痕,划道）

Scratch wire brush 钢丝刷（金属丝刷）

Screen 挡光板　号灯座上，为限制号灯的光强而设置的板。

Screen 滤网（荧光屏,屏蔽）

Screen 屏幕/显示屏　系指根据一个或多个显示来提供直观信息的装置。 Screen means a device used for presenting visual information based on one or several displays.

Screen box 银幕匣　带有卷幕装置，用以收藏银幕的长形匣子。

Screen door 钢丝网门　在门框架上敷以钢丝网的门。

Screen/panel door 纱板两用门　上半截装窗纱，起纱门作用，亦可将附设的活络板拉起，盖住纱窗，起一般板门作用的门。

Screw 螺钉（旋入，螺旋桨）

Screw（suction cutter）dredge 绞吸式挖泥船　系指利用铰刀不断搅松水底砂土，然后吸取及输送泥浆的船舶。它可以挖起较硬而密实的土质。

Screw alley 推进器轴隧（艉轴隧,地轴弄）

Screw aperture 螺丝孔

Screw axis 螺旋桨轴（旋转轴）

Screw blade 螺旋桨桨叶

Screw block 千斤顶

Screw bolt 螺栓

Screw boss 螺旋桨毂

Screw brake 蜗杆闸

Screw clamp 螺旋夹

Screw clip 螺旋夹扣　利用活节螺栓旋紧的夹扣。

Screw compressor 螺杆压缩机　利用螺杆的挤压

作用,输送压缩空气的压缩机。
Screw die 螺丝绞扳(扳手)
Screw down 用螺钉钉住(用螺丝拧紧)
Screw down valve 螺旋阀
Screw driver 螺丝刀(起子)
Screw driving skylight controlling gear 螺杆式天窗传动装置 利用螺杆,带动套筒螺母启闭天窗盖的天窗传动装置。
Screw eye 螺丝眼
Screw feeder 螺旋送料机
Screw gauge 螺纹规
Screw house power 螺旋桨功率(马力)
Screw in 旋入(拧进)
Screw jack 螺旋起重器(螺旋千斤顶)
Screw joint 螺纹接合
Screw out 旋出(拧出)
Screw pitch gauge 螺距规
Screw plug 旋塞
Screw propeller (screw) 螺旋桨 一般装在船的尾部,有两个或较多的叶与毂相连,叶的向后一面为螺旋面或近似于螺旋面的一种船用推进器。
Screw propeller blade 螺旋桨叶 螺旋桨的叶状部分。
Screw propeller dynamometer 螺旋桨测力仪 测量螺旋桨模型推力、转矩和转速的设备。
Screw pump 螺杆泵 利用螺杆副旋转时,其啮合容积的变化,完成吸入和压出过程,以抽送液体的泵。
Screw rack 螺旋齿条
Screw rivet 螺旋铆钉
Screw shackle 螺旋销卸扣
Screw shaft 艉轴(螺旋桨轴)
Screw shaft conditions monitoring (SCM) 螺旋桨轴状态监控 对螺旋桨轴用润滑油进行各种测试分析,掌握轴承磨损状态,以确定润滑油的劣化状态。
Screw ship (propeller vessel) 螺旋桨船 系指用螺旋桨推进的船舶。
Screw slip stopper 螺旋滑钩制动器
Screw stay bolt 牵条螺栓
Screw steering gear 螺杆传动操舵装置(螺杆式舵机) 系指:(1)以左、右纹螺杆与拉杆等为传动部件的人力操舵装置;(2)采用螺杆、拉杆构件传动舵杆的舵机。
Screw stretcher 松紧螺旋扣
Screw tap 螺丝攻
Screw terminal 接线柱
Screw thread 螺纹
Screw type cable releaser 螺旋弃链器 用螺杆装置控制脱钩的弃链器。
Screw type chain cable compressor 螺旋止链器 用一根有左右旋螺纹的螺杆,调节一对夹链爪臂的开口,以夹住或松开锚链的止链器。
Screw type compressor 螺杆式压缩机
Screw type refrigerating compressor 螺杆式制冷压缩机 通过啮合螺杆的旋转作用来实现对气态制冷剂的吸入、压缩和排出的制冷压缩机。
Screw type ventilator 螺旋桨式风机(轴流通风机)
Screw up 旋紧(拧紧)
Screw vice 螺旋虎钳
Screw wrench 活络扳手
Screwed 用螺丝拧紧的(带螺纹的)
Screwed dowel 定位螺钉
Screwshaft 螺旋桨轴(艉轴)
Screwshaft condition monitoring system 螺旋桨轴状态监测系统
Screwshaft survey 螺旋桨轴检验
Scriber 画线针(画线器)
Scroll 涡形管(蜗壳)
Scrubber 净气器(滤尘器,洗涤器,洗涤塔)
Scuffing 拉缸(划伤,齿轮咬结)
Scull 橹 一种外形略似桨,下部切面略作半圆形,可左右摇荡并在改变摇向时将橹杆扭转,借以产生推力的人力推进器。
Scum 浮渣(泡沫,除去浮渣)
Scum dish 浮渣盆
Scum pan 浮渣盆(杂渣盆)
Scum vavle 上排污阀
Scumming 除去浮渣
Scupper 泄水孔/甲板排水孔 系指将甲板上的水直接或通过管路流出的任何开孔。 Scupper means any opening for carrying off water from a deck, either directly or through piping.
Scupper pipe 甲板泄水管
Scupper shoot 舷侧排水孔
Scupper shutter 舷樯排水口盖
Scupper system 排水口装置(甲板排水系统)
Scupper valve 舷侧排出阀(舷侧止回阀)
Scuttle 小开口 系指甲板或别处通入某一舱室的小开口,通常设有盖或罩、门。 Scuttle means a small opening in a deck or elsewhere, usually fitted with a cover or lid or a door for access to a compartment.
Scuttle butt 水桶(饮水保温桶)
SDME 速度和距离测量设备 SDME is a speed

SDME (BT) = speed and distance measuring device (bottom track) 速度和距离测量装置(相对于底部)

SDME (WT) = speed and distance measuring device (water track) 速度和距离测量装置(相对于水面)

Sea 海 系指除各国内部水域以外的所有水域。 Sea means all marine waters other than the internal waters of States.

Sea anchor (floating anchor) 浮锚 当船舶遇到风暴时,可抛出并悬浮于水中,从而使船首顶着风浪,以减轻船舶漂流的带有系缆锥状的帆布袋。

Sea area 海域 为世界海洋的小区域,具有汇集的统计波浪数据。 Sea area is small area of the world's oceans for which statistical wave data has been collected.

Sea area A1 A1 海区 系指至少由一个具有连续 DSC 报警能力的甚高频(VHF)海岸电台的无线电话所覆盖的区域。该区域可由各缔约国政府规定。 Sea area A1 means an area within the radiotelephone coverage of at least one VHF coast station in which continuous DSC alerting is available, as may be defined by a Contracting Government.

Sea area A2 A2 海区 系指除 A1 海区以外,至少由一个具有连续 DSC 报警能力的中频(MF)海岸电台的无线电话所覆盖的区域。该区域可由各缔约国政府规定。 Sea area A2 means an area, excluding sea area A1, within the radiotelephone coverage of at least one MF coast station in which continous DCS alerting is available, as may be defined by a Contracting Government.

Sea area A3 A3 海区 系指除 A1 和 A2 以外,由具有连续报警能力的 INMARSAT 对地静止卫星所覆盖的区域。 Sea area A3 means an area, excluding sea areas A1 and A2, within the coverage of an INMARSAT geostationary satellite in which continuous alerting is available.

Sea area A4 A4 海区 系指 A1 海区、A2 海区和 A3 海区以外的区域。 Sea area A4 means an area outside sea area A1, A2 and A3.

Sea bed gravimeter 海底重力仪 将重力仪密封沉到海底,由遥测遥控装置进行相对重力测量的仪器。由浮子、外壳、仪器、吊架及电缆组成。

Sea chest 海水阀箱/海底门 系指控制海水进入船内作业的操纵站。

Sea chest vibration 海水吸入箱振动

Sea cock 通海旋塞

Sea state 海况 表示对波浪情况的分类,通常由有义波高和波浪周期来定义,此外,还需要定义相应的波浪能量分布。标准的海况定义见表 S-1。 Sea state is an expression used to categories wave conditions and is normally defined by a significant wave height and wave period, a suitable wave energy distribution may also be defined. A list of standard sea state definitions is shown in Table S-1.

表 S-1 海况数据
Table S-1 Sea state data

海况级别 Sea State Number	有义波高/m Significant Wave Height/m		持续风速/kn(注意1) Sustained Wind Speed / Knots (See Note 1)		海况概率百分数 Percentage probability of sea state	常见波周期/s(注意3) Modal Wave Period /s(See Note 3)	
	范围 Range	均值 Mean	范围 Range	均值 Mean		范围(注意3) Range (See Note 3)	最大可能 Most probable
0~1	0~0.1	0.05	0~6	3	0.7	—	—
2	0.1~0.5	0.3	7~10	8.5	6.8	3.3~12.8	7.0
3	0.5~1.25	0.88	11~16	13.5	23.7	5.0~14.8	7.5
4	1.25~2.5	1.88	17~21	19	27.8	6.1~15.2	8.8
5	2.5~4	3.25	22~27	24.5	20.64	8.3~15.5	9.7
6	4~6	5	28~47	37.5	13.15	9.8~16.2	12.4
7	6~9	7.5	48~55	51.5	6.05	11.8~18.5	15.0
8	9~14	11.5	56~63	59.5	1.11	14.2~18.6	16.4
>8	>14	>14	>63	>63	0.05	18.0~23.7	20.0

注意(Notes):(1) 在充分发展的海域,周围的风保持在海面以上 19.5m 处。(Ambient wind sustained at 19.5 m above surface to generate fully-developed seas.)

(2) 利用公式 $V_2 = V_1 (H_2/19.5)^{(1/7)}$，转换至另一高度 H_2。[To convert to another altitude, H_2, apply $V_2 = V_1 (H_2/19.5)^{(1/7)}$]

(3) 对于给定波高范围的周期，其最小值为5%，最大值为95%。(Minimum is 5% and maximum is 95% for periods given wave height range.)

(4) 这里的波浪周期系指常见周期或峰值周期，T_p。对于充分发展的海域的跨零周期，T_z，可以通过表达式 $T_z = T_p/1.4$ 求得。(The wave period shown here is the modal or peak period, T_p. The zero crossing period, T_z, may be derived by the expression $T_z = T_p/1.4$ for fully developed seas.)

Sea connection 通海件（通海接头）

Sea connector 通海连接件

Sea corrosion resistance testing 耐海水腐蚀试验

Sea explorer 深海滑翔机 是一种新型无人水下侦察系统，又称无人水下航行器，主要由卫星控制其工作。水下工作时完全处于自主工作方式。其利用净浮力和姿态角调整获得推进力，能源消耗极小，只在调整净浮力和姿态角上消耗少量能源，具有效率高、续航力大（可达上千公里）。深海滑翔机的航行速度较慢，但其具有制造成本和维护费用低，可重复使用，可大量投放等特点，满足了长时间、大范围海洋探索的需要。近5、6年来，以能耗小、成本低、航程大、运动可控、部署便捷的混合推进技术为特征的新一代深海滑翔机成为国际研究新趋势。其具备独立在水下全天候工作的能力，在海洋科学、海洋军事等领域发挥重要作用。目前，世界海洋强国普遍把深海滑翔机先进研究成果应用到军事装备设计中，相关的技术和产品向中国提供。我国把深海滑翔机研制列入"十二五""863"计划海洋技术领域主持项目。项目的主要目标是开展深海滑翔机工程化技术研究，提高深海滑翔机系统的综合性能、可靠性和稳定性，解决深海滑翔机远程监控、海上应用以及观察数据处理等技术问题，使深海滑翔机达到实用化装备水平。见图 S-5。

图 S-5 深海滑翔机
Figure S-5 Sea explorer

Sea grasses 海草

Sea ice observation 海冰观测 船舶测冰主要是测流测冰。项目通常有：冰量、冰密集度、冰型、外貌特征、冰块大小、冰流方向和速度、冰区边线和冰厚。

Sea inlet valve 海水进水阀

Sea intake 海水进口（海水吸入口）

Sea magin 海况裕度 系指船舶设计时考虑到今后船体污脏、风浪等各种因素，为维持原有航速而在设计时须留有的一定的功率储备。

Sea opening 通海孔

Sea stabilization mode 海面稳定模式 系指航向信息系参照海面并采用回转仪或等效装置和水速记录仪输入数据作为参照的显示模式。Sea stabilization mode means a display mode in which speed and course information are referred to the sea, using gyro or equivalent and water speed log input as reference.

Sea state 海况 表示对波浪情况的分类，通常由有义波高和波浪周期来定义，此外，还需要定义相应的波浪能量分布。标准的海况定义见表 S-1。Sea state is an expression used to categories wave conditions and is normally defined by a significant wave height and wave period, a suitable wave energy distribution may also be defined. A list of standard sea state definitions is shown in Table S-1.

Sea state observation 海面状况观测 利用目测的方法观测在风力作用下的海面外貌特征。如波峰的形状、峰顶的破碎程度和浪花出现的多少。海面状况一般可分为10级。

Sea suction 海水吸入口

Sea suction valve 压载水吸入阀 装在船舷边，直接自舷外吸入压载用水的阀。

Sea surface meteorological observation 海面气象观测 对海面气象及有关水文要素进行的观测。观测项目通常有：海面能见度、云、天气现象、风、气温、湿度、气压、表层水温、海浪、海冰等。

Sea surface meteorological observation room 海面气象观察室 对海面气象各要素：能见度、云、天气现象、风、气温、湿度、气压等进行观测；并对观测资料进行分析的工作室。该室应视野开阔。

Sea temperature 海水温度

Sea test (sea trials) 海上试验（试航，航行试验）
系指在系泊试验完成后进行的试验，其应包括以下各项：(1) 验证主辅机，包括监测、报警和安全系统在实际使用条件下的正常运行；(2) 当其中一台重要辅机变得不能运行时，对推进能力的校核；(3) 在需要时，通过读取必需的读数来检测危险的振动。 Sea trials are those tests conducted after the trials at the moorings and including the following: (1) demonstration of the proper operation of the main and auxiliary machinery, including monitoring, alarm and safety systems, under realistic service conditions; (2) check of the propulsion capability when one of the essential auxiliaries becomes inoperative; (3) detection of dangerous vibrations by taking the necessary readings when required.

Sea thermometer 海水温度计

Sea trail 试航 系指船舶（包括小船，巨轮，潜艇）。通常是建造中的最后一个阶段，进坞大修后也有试航过程。在开阔水域试航，持续几小时甚至几天。通过试航可以测试船舶的机动性能及是否适航（Seaworthiness）。试航一般测量船舶的航速、操纵性、设备及其他安全特性。一般造船厂技术部门代表、管理及发证官员、船舶所有人代表要出席试航过程。成功的试航会使船舶获得相应证书，船舶所有人可以接船。

Sea valve 海底阀

Sea valve (sea suction valve) 通海阀 装于海水门上面的舷外水吸入阀。

Sea water 海水 天然的海洋水。

Sea water acoustics measurement 海水声学测量
对海水中的声速、海水质点振速、声压、声强、压缩量、声阻抗、声反射、声散射、声吸收、声衰减系数、声混响等进行的测量。

Sea water chemical laboratory 海水化学实验室
用于现场测定海水中化学各要素：氯度、盐度、溶解氧、pH 值、碱度、活化硅酸盐、活化磷酸盐、亚硝酸盐、硝酸盐、铵、各种微量元素及重金属元素等各化学组分的含量，并研究其分布状况和变化规律的工作室。

Sea water chemical sampling 海水化学取样 为分析海水化学成分，在现场采集水样的全部过程。根据分析海水中的常量组分、微量元素和有机成分以及放射性同位素等不同的要求，可选用有关类型的采水设备，以防止水样污染或增大水样体积，以利于富集和分析。

Sea water circulating pump 海水循环泵 蒸汽动力装置船舶中抽送冷凝器冷却用海水的泵。

Sea water cooling pipe 海水冷却管

Sea water cooling pump 海水冷却泵

Sea water corrosion 海水腐蚀（盐水腐蚀）

Sea water corrosion-resisting steel 耐海水腐蚀钢

Sea water density ρ 海水密度 取值为 1.025 t/m^3。Sea water density is taken equal to 1.025 t/m^3.

Sea water desalination 海水淡化 借助某种物理或化学方法将海水转化为淡水的过程。

Sea water desalting plant 海水淡化装置 用于海上淡化的全套设备、控制系统和连接管道、附件的总称。

Sea water evaporator 海水蒸发器 蒸发海水的设备。

Sea water filter (sea water strainer) 海水滤器 设于舷外水进入总管处的过滤器。

Sea water immersion test 浸海水试验

Sea water jacket space 海水腔

Sea water lubricated stern tube bearing 海水润滑艉管轴承

Sea water optical measurement 海水光学测量
利用光学仪器和设备对海水的透明度、水色、光照度等海水光学性质所进行的现场测定。

Sea water pressure tank 海水压力柜

Sea water pump 海水泵

Sea water service system 海水系统 船上供应生活洗涤用海水的系统。

Sea wave 海浪 主要指表层海水受外力影响而发生的起伏现象。根据波浪的特征分风浪、涌和混合浪等。

Sea wave observation 海浪观测 对风浪和涌浪的波面时空分布及其外貌特征进行的海面现场观测。观测项目通常有海面状况、波形、波向、周期和波高。

Sea waybill 海运单 又称海上运送单或海上货运单，正面内容与提单的基本一致，但是印有"不可转让"的字样。它是"承运人向托运人或其代理人表明货物已收妥待装的单据，是一种不可转让的单据，即不须以在目的港揭示该单据作为收货条件，不须持单据寄到，船主或其代理人可凭收货人收到的货到通知或其身份证明而向其交货"。

Seabed mining ship 海底矿物采集船

Seabed piping system 海底管道系统 系指用于输送石油、天然气或水等介质海底管道工程设施的所有组成部分，包括平管、立管、支撑构件、管道附件、防腐系统、配重层、泄漏检测系统、报警系统、应急关闭装置等。见图 S-6。

Seaborne machinery 船上设备（船上机械）

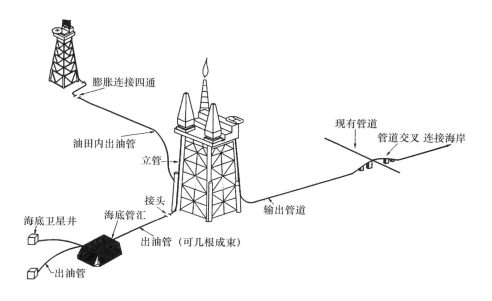

图 S-6　海底管道系统
Figure S-6　Seabed piping system

Seafarer　船员　系指船上以任何资格受雇、任职或工作的任何人。　Seafarer means any person who is employed or engaged or works in any capacity on board a ship.

Sea-fresh water corrosion　海-淡水交替腐蚀

Seagoing service　海上服务资历　系指与签发的证书或其他资格有关的船上服务。　Seagoing service means service on board a ship relevant to the issue of a certificate or other qualification.

Seagoing ship　远洋船　系指在公海航行的船舶，即沿海岸和岸到岸航行的船舶。　Seagoing ship is a ship navigating on the high seas, i. e. areas along coasts and from coast to coast.

Sea-going ship　海船　系指除了仅在内陆水域中或者遮蔽水域或港湾所适用的区域以内或与此两者紧邻的水域中航行的船舶以外的船舶。　Sea-going ship means a ship other than those which navigate exclusively in inland waters or in waters within, or closely adjacent to, sheltered waters or areas where port regulations supply.

Seagoing vessel　远洋船舶　系指在联邦法规登记第 7 篇建立的领海边界线外营运的商业用途的船舶。它不包括专门在内陆水域航行的船舶。　Seagoing vessel means a vessel in commercial service that operates beyond the boundary line established by 46 CFR part 7. It does not include a vessel that navigates exclusively on inland waters.

Seakeeping　耐波性　船舶在风浪中遭受由于外力干扰所产生的各种摇荡运动以及砰击上浪、失速、飞车和波浪弯矩等，仍能维持一定航速在水面安全航行的性能。

Seakeeping laboratory　耐波性试验池　能产生不同特征的规则波或不规则波，用于进行耐波性试验的船模试验池。

Seal　密封（填料函，图章）

Seal assembly　密封装置

Seal gland　气封　减少高压蒸汽从高压端经汽轮机转子和气缸之间的间隙向外泄漏或减少外部空气经转子和气缸之间的间隙泄漏进低压部分的装置。

Seal leakage　气封

Seal test　密封试验

Seal washer　密封垫圈

Seal welding　封焊

Sealant material　密封材料

Seale　水垢　由于锅水中析出的低溶解度的杂质黏附于锅筒、管壁、联箱表面而形成的沉积物。

Sealed　密封的

Sealed arrangement　密封装置

Sealed cabin　密封舱

Sealed container　密封集装箱（密封容器）

Sealed sweeper　密封清管器

Sealing　密封（焊封,堵塞）

Sealing and locking device　舱盖的密封压紧装置　保证舱盖水密要求和在风浪冲击下不移动,在舱盖四周和相邻盖板间设置的压紧装置。

Sealing arrangement　密封装置

Sealing device　密封装置

Sealing gland　密封压盖

Sealing plug　密封（塞堵,闷头）

Sealing ring　密封环

Sealing strip　密封条

Sealing surface　密封面

Sealing system　密封系统

Seam　边接缝　系指板长边的连接缝。

Seaming　缝（接缝）

Seamless　无缝的

Seamless copper tube　无缝铜管

Seamless pressure pipe　无缝压力管

Seamless steel tube　无缝钢管

Seamless pipe / Seamless tube　无缝管　是一种具有中空截面、周边没有接缝的长条钢材。无缝钢管是一种具有中空截面、周边没有接缝的圆形,方形,矩形钢材。无缝钢管是用钢锭或实心管坯经穿孔制成毛管,然后经热轧、冷轧或冷拨制成。无缝钢管具有中空截面,大量用作输送流体的管道,钢管与圆钢等实心钢材相比,在抗弯抗扭强度相同时,质量较轻,是一种经济截面钢材,广泛用于制造结构件和机械零件,如石油钻杆、汽车传动轴、自行车架以及建筑施工中用的钢脚手架等。

Seaplane　水上飞机　系指包括为能在水面操纵而设计的任何航空器。　Seaplane includes any aircraft designed to manoeuvre on the water.

Search and rescue centre　搜救区域　在规定的范围内提供搜救服务的区域。　Search and rescue centre means an area of defined dimensions within which search and rescue services are provided.

Search and rescue co-operation plan　搜救合作计划　系指适用 SOLAS 公约第 I 章,沿固定航线航行的每艘客船所备有的一份在紧急情况下与合适的搜救机构合作的计划。　Search and rescue co-operation plan means that passenger ships to which chapter I of SOLAS Convention applies shall have on board a plan for co-operation with appropriate search and rescue services in event of an emergency.

Search and rescue mode　搜救模式　系指进行搜救行动船舶操作的显示模式。　Search and rescue mode means a display mode for operations of a ship involved in search and rescue actions.

Search and rescue service　搜救服务　系指通过利用公共和私人资源（包括协作的飞机、船舶和其他浮动工具及设施）执行遇险监控、通信、协调和搜救职能,包括提供医疗建议、初步医疗救助或医务转移。　Search and rescue service is that the performance of distress monitoring, communication, co-ordination and search and rescue functions, including provision of medical advice, initial medical assistance, or medical evacuation, through the use of public and private resources including co-operating aircraft, ships, vessels and other craft and installations.

Search and rescue transponder (SART)　搜救应答器

Search engine (SE)　搜索引擎　系指根据一定的策略、运用特定的计算机程序从互联网上搜集信息,在对信息进行组织和处理后,为用户提供检索服务,将用户检索相关的信息展示给用户的系统。搜索引擎包括全文搜索引擎、目录搜索引擎、元搜索引擎、垂直搜索引擎、集合式搜索引擎、门户搜索引擎与免费链接列表等。代表性产品有 Google、Baidu、sogou、soso、360 等。百度和谷歌等是搜索引擎的代表。一个搜索引擎由搜索器、索引器、检索器和用户接口 4 个部分组成。搜索器的功能是在互联网中漫游,发现和搜集信息。

Sea-star TLP　海星张力腿平台　是一款用于边际油田或小区块的开发,承载能力小于 MOSES 的小型张力腿平台。

Seat　基座（阀座,座位）

Seat cushion　座垫

Seat of propeller shaft end　推进器轴端密封　推进器轴与推进器轴孔配合面间的密封填料。

Seating　基座（阀座）

Seating of boiler　锅炉座

Seatless bottom blow vavle　无座下排污阀

Seawater desalinator/ Sea water desalting plant　海水淡化装置　用于海水淡化的全套设备、控制系统和连接管道、附件的总称。一般采用真空式,海水的蒸发和蒸汽的冷凝是在具有一定真空度的工作容器内完成,既可降低水的沸点温度,减少结垢、便于清除;也有利于对动力装置冷却水中废热的利用,以提高装置的经济性。目前船用蒸馏海水淡化装置真空度都大于 80%。

Seawater filter　海水过滤装置

Sea-water pump　海水泵　抽送海水的泵。

Seaway station　航道站　系指圣劳伦斯航道开发公司或管理局经营的无线电站。　Seaway station means a radio station operated by the Saint Lawrence Seaway Development Corporation or the Manager.

Seaworthiness survey 耐航鉴定

Seaworthiness 适航性　船舶有足够的耐波性及在稳性、船体结构、各种设备、燃料、给养等方面符合航行法定要求，能保证安全航行的性能。

SECA SO_x 排放控制区　SECA means SO_x emission control area.

Secchi disc 透明度盘　用以测量海水透明度的直径为 30 cm，漆成白色的木质或金属圆盘。它在海水中的最大可见深度。定义为海水透明度。

Second barrier 次屏壁　系指用以防止一次屏壁失效时出现危险的技术措施，如：保护周围不受燃料柜泄漏影响的燃料柜的第二层外壳。　Second barrier means a technical measure which prevents the occurrence of a hazard if the first barrier fails, e.g., second housing of a tank protecting the surroundings from the effect of tank leaks.

Second deck 第二甲板　上甲板以下的第一层连续甲板。

Second engineer officer 大管轮　系指级别仅低于轮机长，并在轮机长不能工作时由其负责船舶的机械推进的轮机员。　Second engineer officer means the engineer officer next in rank to the chief engineer officer and upon of the incapacity of the chief engineer officer.

Second filter 细滤器

Second generation intact stability criteria 二代完整稳性参数横摇标准

Second generation intact stability criterion 第二代完整稳性标准　是在第一代完整稳性基础上，进一步引入稳性直接评估的概念，对瘫船稳性、纯稳性丧失、过度加速、参数横摇和骑浪或纵摇进行计算，为新颖船型、超长度比船舶的稳性提供个性化的稳性评估。

Second island chain 第二岛链　系指涵盖中国台湾东部海区外侧的那条弧形岛屿带，北起日本的南方岛屿（包括小笠原群岛、硫黄列岛），中接美国的马里亚纳群岛、关岛等岛屿，南接加罗林群岛组成的链形岛屿带。

Second overshoot angle 第二超越角　系指 Z 形操纵试验中紧接着第三次操舵增加的偏离值。Second overshoot angle is the additional heading deviation experienced in the zig-zag test following the third execute.

Second structure member 次要骨材　系指纵骨、横梁（横骨）等船体结构中的次要构件。

Secondary air 二次风（延燃空气）　（1）使油雾与空气良好混合并完全燃烧所需的空气的主体部分。（2）进入延易燃区的空气。

Secondary alarms 二级警报　系指所有其他警报。　Secondary alarms mean all other alarms.

Secondary application structural members of the unit 平台结构的次要构件　系指其失效不可能影响平台结构整体完整性的不重要的构件。　Secondary application structural members of the unit are structural members of minor importance, failure of which is unlikely to affect the overall integrity of the unit.

Secondary application structure of column-stabilized units 柱稳式平台的次要构件　包括：(1) 立柱、上壳体或上平台、下壳体、斜撑和水平撑杆等的内部结构，包括舱壁、扶强材、平台或甲板和桁材 (作为主要或特殊构件者除外)；(2) 上平台甲板或上壳体甲板板 (作为主要或特殊构件者除外)；(3) 某些长径比小的大直径立柱 (交接处除外)；(4) 直升机甲板和甲板室；(5) 救生艇甲板。Secondary application structure of column-stabilized units: (1) internal structure including bulkheads, stiffeners, flats or decks and girders in vertical columns, upper hull or platform, lower hulls, diagonal and horizontal bracing, which are not considered as primary or special application; (2) upper platform decks or decks of upper hull except areas where the structure is considered as primary or special application; (3) certain large-diameter vertical columns with low length-to-diameter ratios, except in intersections; (4) helicopter platforms and deckhouses; (5) lifeboat platforms.

Secondary application structure of self-elevating units 自升式平台的次要构件　包括：(1) 柱形桩腿内部骨架，包括隔壁和桁材；(2) 平台主体内部舱壁和骨架 (作为主要构件者除外)；(3) 桩靴或沉垫支承结构的内部舱壁和骨架 (作为主要或特殊构件者除外)；(4) 平台主体甲板板、舷侧板和底板 (作为主要构件者除外)；(5) 直升机甲板和甲板室；(6) 救生艇平台。Secondary application structure of self-elevating units: (1) internal framing, including bulkheads and girders, in cylindrical legs; (2) internal bulkheads and framing members of upper hull structure except where the structure is considered as primary structure; (3) internal bulkheads and framing members of footings or bottom mat supporting structures except where the structure is considered primary or special application; (4) deck, side and bottom plating of upper hull except where the structure is considered as primary application; (5) helicopter platforms and deckhouses; (6) lifeboat platforms.

Secondary application structure of submersible units 坐底式平台的次要构件　包括：(1) 立柱、上壳体或上平台甲板、下壳体等的内部结构，包括舱壁、扶强材、平台或甲板、和桁材（作为主要或特殊构件者除外）；(2) 上平台甲板或上壳体甲板板（作为主要或特殊构件者除外）；(3) 某些长径比小的大直径立柱（交接处除外）；(4) 直升机甲板和甲板室；(5) 救生艇甲板。 Secondary application structure of submersible units: (1) internal structure including bulkheads, stiffeners, flats or decks and girders in vertical columns, upper hull or platform decks and lower hulls, which are not considered as primary or special application; (2) upper platform decks or decks of upper hull except areas where the structure is considered as primary or special application; (3) certain large-diameter vertical columns with low length-to-diameter ratios, except in intersections; (4) helicopter platforms and deckhouses; (5) lifeboat platforms.

Secondary barrier 次屏壁 对散装运输液化气船而言，系指货物围护系统中被设计成能暂时容纳可能从主屏壁泄漏的液货的液密外层构件，同时也为了防止船体结构的温度会下降至不安全的程度。装运在常压（大气压）下温度低于 –10 ℃ 的货品需要设置次屏壁。

Secondary de-rusting 二次除锈 系指对涂有车间底漆（或其他涂层）的表面因热膨胀或机械原因引起底漆（或涂层）损伤而返锈的部位，再次进行表面处理的工艺过程。Secondary de-rusting means a technological process of surface retreatment for the surface with shop-coating (or other coating) where rust appears again due to printing paint (or coating) damage caused by heat influence or mechanical damage.

Secondary distribution system 二次配电系统 系指与发电机无电气连接的系统，例如用双绕组变压器加以隔离的系统。Secondary distribution system is a system having no electrical connection with the generator, e.g. a system isolated by a double-wound transformer.

Secondary essential services 次重要设备 系指那些不必连续工作的设备。次重要设备列举如下：(1) 锚机；(2) 侧推力器；(3) 燃油驳运泵和燃油处理装置；(4) 滑油驳运泵和滑油处理装置；(5) 重燃料油预热器；(6) 海水泵；(7) 启动空气和控制空气压缩机；(8) 舱底泵、压载泵和横倾泵；(9) 消防泵和其他灭火介质泵；(10) 机舱和锅炉间的通风机；(11) 保持危险货物处于安全状态所必须考虑的设施；(12) 航行灯、辅助设备和消防设备；(13) 船内安全通信设备；(14) 探火和报警设备；(15) 水密关闭器的电气设备；(16) 发电机和供给上述设备的动力；(17) 供给设施设备的液压泵；(18) 货物容量的控制、监测和安全系统；(19) 供次重要设备的控制、监测和安全装置/系统；(20) 环境控制处所的冷风系统。Secondary essential services are those services which need not necessarily be in continuous operation. Examples of equipment for secondary essential services are the following: (1) windlasses; (2) thrusters; (3) fuel oil transfer pumps and fuel oil treatment equipment; (4) lubrication oil transfer pumps and lubrication oil treatment equipment; (5) preheaters for heavy fuel oil; (6) sea water pumps; (7) starting air and control air compressors; (8) bilge, ballast and heeling pumps; (9) fire pumps and other fire-extinguishing medium pumps; (10) ventilation fans for engine and boiler rooms; (11) services considered necessary to maintain dangerous cargo in a safe condition; (12) navigation lights, aids and signals; (13) internal safety communication equipment; (14) fire detection and alarm systems; (15) electrical equipment for watertight closing appliances; (16) electric generators and associated power supplying the above equipment; (17) hydraulic pumps supplying the above mentioned equipment; (18) control, monitoring and safety for cargo containment systems; (19) control, monitoring and safety devices/systems for equipment for secondary essential services; (20) cooling system of environmentally controlled spaces.

Secondary flow 次级流 由于边界层形状和主流中存在的压力分布所导致的横向流动。

Secondary frame spacing s (m) 次要骨架间距 系指普通扶强材间的距离，沿弦长在跨距中点量取。Secondary frame spacing is defined as the distance between ordinary stiffeners in m, of ordinary stiffeners, measured at mid-span along the chord.

Secondary hardening 退火后二次硬化

Secondary holds 延燃孔 供空气进入火焰筒过渡区的通孔。

Secondary lifesaving equipment 辅助救生用具 泛指船上除救生载具以外，直接用于救助人员的器具。通常包括救生圈、救生衣、救生带、救生浮索和抛绳器具等。

Secondary means 辅助措施

Secondary members 次要构件 一般系指板的扶强构件，如肋骨、纵骨、横梁、舱壁扶强材、组合肋板的骨材等。Secondary members: the stiffeners of the plate are generally regarded as secondary members, i.e. frames, longitudinals, beams, bulkhead stiffeners and members of bracket floors, etc.

Secondary refrigerant 载冷剂 从冷源将热量转给正在蒸发的制冷剂的流体。

Secondary securing device 次紧锁装置 系指设在前甲板的小舱口上的紧锁装置，例如一个滑动螺栓、搭扣或松动配合的垫板，即使主紧锁装置松动或脱开时也能保持舱口盖就位。它应设在舱口盖铰链对面的一侧。Secondary securing device means an independent device fitted with at small hatches on the fore deck e. g. by means of a sliding bolt, a hasp or a backing bar of slack fit, which is capable of keeping the hatch cover in place, even in the event that the primary securing device became loosened or dislodged. It is to be fitted on the side opposite to the hatch cover hinges.

Secondary stress 次级应力

Secondary structural member 交叉构件/次要构件 系指下列构件：(1) 纵向舱壁的下部列板；(2) 露天甲板板（一般）；(3) 舷侧板。Secondary structural member means the following member: (1) lower strake of longitudinal bulkhead; (2) deck plating exposed to weather, in general; (3) side plating.

Secondary suit closure(immersion suit) 衣服的次要外罩（救生服） 穿戴者在水中可能启用的附加外罩。Secondary suit closure is the additional closure which may be operated by the wearer in the water.

Secondary surface treatment 二次表面处理 系指对涂有车间底漆（或其他涂层）的表面，因热影响或进行原因引起原涂层损伤而返锈的部位，再次进行表面处理的过程。

Secondhand goods 旧货/二手货

Secretariat 秘书处 系指设秘书长一人及若干职员，负责处理各种日常事务。此外，国联还有不少附属机构，其中较重要的是国际法庭和国际劳工组织。

Secretary-General 秘书长 系指国际海事组织秘书长。Secretary-General means the Secretary-General of the Organization.

Section 分隔 系指指示装置记录的火灾探测和手动呼叫点的集合。Section means group of fire detection and manually operated call points as reported in the indicating units.

Section 分区 对国际消防安全系统规则（FSS 规则）而言，系指指示装置标示的一组火警探测器和手动报警按钮。For the purpose of the International Code for fire safety systems (FSS Code), section means a group of the detectors and manually operated call points as reported in the indicating unit(s).

Section board 区配电板 用于控制来自配电板的电功率，且将其分配至其他区配电板、分配电箱或最后分路的开关装置和控制装置的组合件。Section board is a switchgear and control gear assembly for controlling the supply of electrical power from a switchboard and distributing it to other section boards, distribution boards or final sub-circuits.

Section identification capability 分区识别能力 对国际消防安全系统规则（FSS 规则）而言，系指系统具有识别已触发的探测器手动报警按钮所在分区的能力。For the purpose of the International Code for fire safety systems (FSS Code), section identification capability means a system with the capability of identifying the section in which a detector or manually operated call point has activated.

Section moduli at bottom $Z_{AB}(m^3)$ 船底处的剖面模数 $Z_{AB}(m^3)$ Z_{AB} means a section moduli, in m^3, at bottom.

Section moduli at deck $Z_{AD}(m^3)$ 甲板处的剖面模数 $Z_{AD}(m^3)$ Z_{AD} means a section moduli, in m^3, at deck.

Section modulus for bottom 船底剖面模数 系指中横剖面对船底基线而言的剖面模数。

Section modulus for deck 甲板剖面模数 系指中横剖面对强力甲板边线而言的剖面模数。

Section modulus of mid-ship section 舯横剖面模数 系指船体中部横剖面的最小剖面模数。

Section of pipe 管段

Section of screw blade 叶切面 螺旋桨叶与同轴线圆柱面相交的面。讲叶切面形状时一般系指展平的切面。

Sectional area curve, curve of sectional areas 横剖面面积曲线 表示沿船长各横剖面处设计水线以下横剖面面积的曲线。

Sectional header boiler ("Cock and Wileox" boiler) 分联箱横水管锅炉 用立置波形联箱连接横置在炉膛上方的水管，并与锅筒连通构成并列的循环回路的水管锅炉。

Sectional survey (fixed line observation) 断面调查 系指在调查海区布设若干有代表性的调查断面。于一定的时间内在调查断面测站上进行观测。

Sectoral integration 部门的一体化 系指区域内各成员国就一种或几种产业（或商品）形成的内部市场。

Secunty check instrument by teraHz wave 太赫兹安检仪 系指能接受人体发出的太赫兹波并处理转换，形成人体的二维太赫兹"快照"的仪器。当衣物中隐藏有物品时，吸收阻拦太赫兹波的正常轨迹，使"快照"中对应物品的位置和人体背景产生对比，从而探知物品的存在。

Securing 系固 系指为防止在整个航程中货物落水失灭所采取固定货物的预防措施。对于重型货物或异常外形尺寸货物，应采取适当的预防措施，以确保不发生船船结构损坏，并在整个航程中保持足够的稳性。在滚装船上货物单元和货物运输单元的装载和运输过程中，应特别注意这种船上货物单元和货物运输单元上的系固装置，以及系固点和捆索的强度。

Securing bolt 销紧螺栓（保险螺栓）

Securing device 系固设备（集装箱） 是所有用于系固和绑扎集装箱的固定式和便携式装置的总称。

Securing device 紧固装置 是一种通过防止门绕其铰链转动而使其保持关闭状态的装置。 Securing device is a device used to keep the door closed by preventing it from retating around its hinges.

Securing device on door 门的紧固装置 系指：(1)用于防止门绕绞链转动或绕附连于船上的附件旋转，使门保持关闭状态的一种装置；(2)使门保持关闭，防止其绕铰链转动的一种装置。 Securing device on door means: (1) a device used to keep the door closed by preventing it from rotating around its hinges or around pivoted attachments to the ship; (2) a device used to keep the door closed by preventing it from rotating around its hinges.

Securing manual 货物系固手册 系指除固体和液体散装船外的船舶,在船舶整个航程中,对货物、货物单元和货物运输单元进行装载、堆放和系固作业的指导性文件。

Security 安保 由于个人或群体的不理想行为为没有造成生命、健康、财产和环境风险。

Security auditor at sea 海上保安审核员 系指按规定要求，接受过正式培训的，船级社实施船舶保安计划审批和船上审核的人员。

Security equipment 保安设备 协作在保安计划规定的具有保安功能的设备。注意：(1)保安设备是船舶一个或多个位置安装、配备和设置的,预定用于阻止、监视、探测、观察潜在保安威胁的一个或多个装置或措施。从广义意义上说,保安设备可包括设备、装置、技术系统、标识、结构关闭装置等技术系统；(2)某些设备具有双重作用,诸如甲板照明用于正常的船舶操作,也发挥保安作用。

Security incident 保安事件 系指威胁船舶（包括海上移动式钻井平台和高速船）,或港口设施或任何船/港界面活动,或任何船到船活动有关保安的任何可疑行为或情况。 Security incident means any suspicious act or circumstance threatening the security of a ship, including a mobile offshore drilling unit and a high-speed craft, or of a port facility or of any ship/port interface or any ship-to-ship activity.

Security level 保安等级 系指企图造成保安事件或发生保安事件的风险级别的划分。 Security level means the qualification of the degree of risk that a security incident will be attempted or will occur.

Security level 1 保安等级1（正常,船舶和港口正常操作） 系指保持最低保安措施的等级。 Security level 1 means the level for which minimum appropriate protective security measures shall be maintained at all times.

Security level 2 保安等级2（加强） 系指由于保安事件危险性升高而应在一段时间内保持适当的附加保护性保安措施的等级。 Security level 2 means the level for which appropriate additional protective security measures shall be maintained for a period of time as a result of heightened risk of a security incident.

Security level 3 保安等级3（特殊） 系指当保安事件可能或即将发生(尽量可能尚无法确定具体目标)时应在一段时间内保持进一步的特殊保护性措施的等级。 Security level 3 means the level for which further specific protective security measures shall be maintained for a limited period of time when a security incident is probable or imminent, although it may not be possible to identify the specific target.

Security measures 保安措施（治安措施） 系指：(1)根据国际协议制定并执行的用以改进在船上、港口区域、设施以及国际供应链中移动货物的保安以发现和防止非法行为的措施；(2)为改善船上和港区治安以防止危害船上旅客和船员的非法行为而采取的国际上认可的措施。 Security measures means: (1) those measures developed and implemented in accordance with international agreements to improve security on board ships, in port areas, facilities and of goods moving in the international supply chain to detect and prevent unlawful acts; (2) the internationally agreed measures to improve security on board ships and port areas to prevent unlawful acts against passengers and crew on board ships.

Security non-conformity（SNC） 保安不符合项

Security threat 保安威胁 系指根据威胁情景和发生的可能性经过船舶保安评估加以确定的对船舶、人员或港口设施形成潜在威胁的状况。 Security threat means the status of potential threat determined to the ship, persons or port facilities in accordance with the threat situation and possibility.

Sedative drugs 镇静药 与酒一样有抑制神经的作用,产生相似的效果。镇静药有两种形式：巴比妥酸盐和甲喹酮。纯正的镇静药都是白色的粉末。这种物质有成百上千个商品名,其形式有药丸、药片和胶囊。在正常范围之外发现的所有显而易见的医用制剂都应引起怀疑。所有形式的商品通常都是没有气味的。

Sedative drugs mean those drug which depress the nervous system in the same way as alcohol, producing similar effects. They come in two forms: barbiturates and methaquqlone. In their pure form all are white powders. There are many hundreds of brand when the substance are found as pill, tablets and capsules. All discoveries of apparent medical preparations outside their normal context should be regarded as suspicious. All forms are normally odourless.

Sediment 沉淀（污垢,沉积物）

Sediment collector 沉淀槽（沉淀箱）

Sediment toxicity 沉积物毒性

Sedimentation and treatment unit 污水沉淀和处理装置

Sediments 沉淀物　系指船上压载水沉淀后产生的固体物质。 Sediments mean matter settled out of Ballast Water within a ship.

See-Berüfsgenossenehaft（SBG/professional accident insurance association for seamen） 德国海上职业联合会　对挂德国旗新船船的安全、事故预防和防污染进行法定检验的机构之一。总部设在汉堡的，代表德国"劳工和社会福利部"负责船舶安全、操作安全、海洋环境保护、港口情况控制、工作时的医疗保护和工伤事故预防等事项的机构。SBG颁布了"事故预防规则"（UVV）。其许多事项委托GL代理。

Seed cake (UN 2217)(with not more than 1.5% oil and not more than 11% moisture) 油含量不大于 1.5%,含水量不大于 11% 的种子饼　含油种子是用溶剂法萃取油或经机械压榨后所剩的残留物。本细目包括的谷物和谷物产品出自以下物质：烤制物质、大麦芽粒、甜菜、糠粒、酿酒用谷粒、柑橘渣粒、椰子、椰肉、玉米麸、棉籽、粕、麸粒、落花生粗粉、玉米糁、亚麻籽、玉米、含油粗粉、谷粕糠粒、皂脚饼粕、油饼、棕桐仁、花生、谷物糠粒、细糠粒、油菜籽、碎米、米糠、红花籽、含油种子粕、大豆、斯特拉瑟颗粒、葵花籽、烤制粗粉。以上物资可以渣、粗粉、饼、颗粒及粕形态装运。可能缓慢自热，如受潮或含有未氧化油的比例过高则可能自燃。易氧化，随后导致货物处所氧气减少。可能产生二氧化碳。 Seed cake (UN 2217) (with not more than 1.5% oil and not more than 11% moisture) are residue remaining after oil has been extracted by a solvent process from oil-bearing seeds. The cereals and cereal products included in this schedule are those derived from: Barley malt pellets; Beet; Bran pellets; Brewer's grain pellets; Citrus pulp pellets; Coconut; Copra; Corn gluten; Cotton seed; Expellers; Gluten pellets; Ground nuts, meal; Hominy chop; Linseed; Maize; Meal, oily; Mill feed pellets; Niger seed, expellers; Oil cake; Palm kernel; Peanuts; Pellets, cereal; Pollard pellets; Rape seed; Rice broken; Rice bran; Safflower seed; Seed expellers, oily; Soya bean; Strussa pellets; Sunflower seed; Toasted meals. The above may be shipped in the form of pulp, meals, cake, pellets and expellers. May self-heat slowly and, if wet or containing an excessive proportion of unoxidized oil, ignite spontaneously. It is liable to oxidize, causing subsequent reduction of oxygen in the cargo space. Carbon dioxide may be produced.

Seed cake, containing vegetable oil [UN 1386(a)] mechanically expelled seeds, containing more than 10% of oil or more than 20% of oil and moisture combined. 经机械压榨的种子,油含量大于 10% 以及油和含水量合计大于 20%,含植物油种子饼（a） 含油种子经机械压榨后的残留物。本细目包括的谷物和谷物产品出自以下物质：烤制物质、大麦芽粒、甜菜、糠粒、酿酒用谷粒、柑橘渣粒、椰子、椰肉、玉米麸、棉籽、粕、麸粒、落花生粗粉、玉米糁、亚麻籽、玉米、含油粗粉、谷粕糠粒、皂脚饼粕、油饼、棕桐仁、花生、谷物糠粒、细糠粒、油菜籽、碎米、米糠、红花籽、含油种子粕、大豆、斯特拉瑟颗粒、葵花籽、烤制粗粉。以上物资可以渣、粗粉、饼、颗粒及粕形态装运。可能缓慢自热，如受潮或含有未氧化油的比例过高则可能自燃。易氧化，随后导致货物处所氧气减少。可能产生二氧化碳。 Seed cake, containing vegetable oil UN 1386 (a) mechanically expelled seeds, containing more than 10% of oil or more than 20% of oil and moisture combined are residue remaining after oil has been expelled mechanically from oil-bearing seeds. Containing more than 10% of oil or more than 20% of oil and moisture combined. The cereals and cereal products included in this schedule are those derived from: Bakery materials; Barley malt pellets; Beet; Bran pellets; Brewer's grain pellets; Citrus pulp pellets; Coconut; Copra; Corn gluten; Cotton seed; Expellers; Gluten pellets; Ground nuts, meal; Hominy chop; Linseed; Maize; Meal, oily; Mill feed pellets; Niger seed, expellers; Oil cake; Palm kernel; Peanuts; Pellets, cereal; Pollard pellets; Rape seed; Rice broken; Rice bran; Safflower seed; Seed expellers, oily; Soya bean; Strussa pellets; Sunflower seed; Toasted meals. The above may be shipped in the form of pulp, meals, cake, pellets and expellers. May self-heat slowly and, if wet or containing an excessive proportion of unoxidized oil, ignite spontaneously. It is liable to oxidize, causing subsequent reduction of oxygen in the cargo space. Carbon dioxide may be produced.

Seepage 渗透（渗漏）

Segment 扇形体

Segment rack 扇形齿条

Segmented model test 分段拼模试验 用由若干分段拼成的船模所进行的振荡和约束船模的试验。

Segregated 分隔的（分开的）

Segregated ballast 分隔压载 系指舱内装压载水，这种舱与货油系统和燃油系统完全隔开，并永久地作为装载压载水或装载油类或有毒物质以外的压载或货物。Segregated ballast means the ballast water introduced into a tank which is completely separated from the cargo oil and fuel oil system and which is permanently allocated to the carriage of ballast or to the carriage of ballast or cargoes other than oil or noxious substances.

Segregated ballast 专用压载水 (1)①对"73/78防污公约"附则Ⅰ而言，专用压载水系指装入这样一个舱内的压载水，该舱与货物油及燃油系统完全隔绝并固定用于装载压载水，或固定用于装载"73/78防污公约"各附则中所指各种油类或有毒物质以外的货物；②"73/78防污染公约"附则Ⅰ还对"专用压载水"的定义做了统一解释如下：(a)可通过一可拆短管与一个货油泵相连接的方法，在紧急时排放专用压载水。在这种情况下，专用压载的连接管上应装有止回阀，以防止油进入专用压载舱。可拆短管应装在泵舱内明显的位置，其附近应显著放有限制其使用的永久性告示；(b)为延伸管路，当货物油或燃油管路穿过专用压载舱，以及专用压载管路穿过货物油或燃油舱时，滑动型连接器不应采用。(2)对"73/78 防污公约"附则Ⅱ而言：专用压载水系指装入一个舱内的压载水，该舱与货物及燃油系统完全隔绝并固定用于装载压载水，或固定用于装载"73/78防污公约"各附则中所定义的各种油类或有毒物质以外的压载水或货物。 Segregated ballast means:(1)①For "MARPOL73/78" Annex Ⅰ, segregated ballast means the ballast water introduced into a tank which is completely separated from the cargo oil and fuel oil system and which is permanently allocated to the carriage of ballast or to the carriage of ballast or cargoes other than oil or noxious liquid substances as variously defined in the Annexes of the MARPOL73/78;②Nevertheless, provision may be made for emergency discharge of the segregated ballast by means of a connection to a cargo pump through a portable spool piece.(a)In this case non-return valves should be fitted on the segregated ballast connections to prevent the passage of oil to the segregated ballast tanks. The portable spool piece should be mounted in a conspicuous position in the pump-room and a permanent notice restricting its use should be prominently displayed adjacent to it.(b) Sliding type couplings should not be used for expansion purposes where lines for cargo oil or fuel oil pass through tanks for segregated ballast and where lines for segregated ballast pass through cargo oil or fuel oil tanks.(2)For "MARPOL73/78" Annex Ⅱ, segregated ballast means ballast water introduced into a tank permanently allocated to the carriage of ballast or cargoes other than oil or noxious liquid substances as variously defined in the Annexes of the MARPOL73/78, and which is completely separated from the cargo and oil fuel system.

Segregated ballast arrangement 专用压载布置

Segregated ballast pump 专用压载泵

Segregated ballast tank(SBT) 专用压载水舱 系指：(1)用于装载专用压载水的压载水舱，该舱与货物及燃油系统完全隔绝并固定用于装载压载水，或固定用于装载"73/78 防污公约"各附则中所定义的各种油类或有毒物质以外的压载水或货物；(2)对载重量为20 000 t及以上的新原油船及载重量为30 000 t及以上的新成品油船要求设立的、专用于储存压载水而与货舱完全分立的舱。

Segregated ballast tank system 独立的压载舱系统 系指不在货油舱装载海水的压载舱系统。Segregated ballast tank system means that sea water ballast should not be carried in cargo oil tanks.

Segregated ballast tanks(SBT) 专用压载舱

Segregation 偏析 合金中各组成元素在结晶时分布不均匀的现象称为偏析。焊接熔池一次结晶过程中，由于冷却速度快，已凝固的焊缝金属中化学成分来不及扩散，造成分布不均，产生偏析。显微偏析系指发生在一个或几个晶粒之内，包括 枝晶偏析、晶间偏析、晶界偏析和胞状偏析。熔池一次结晶时，由于柱状晶体的不断长大和推移，会把杂质"赶"向熔池中心，使熔池中心的杂质含量比其他部位多，这种现象称为区域偏析。以后随着熔池的散热，结晶又重新开始，形成 周期性的结晶，伴随着出现结晶前沿 液体金属中杂质浓度的周期变动，产生周期性的偏析称为层状偏析。

Seiko Epson 精工爱普生（セイコーエプソン） 通常简称爱普生，是一家成立于1942 年，总部位于日本长野县的上市(TYO:6724)公司。公司主要生产喷墨打印机、激光打印机、点阵式打印机、扫描器和手表［显然其母公司精工(Seiko)品牌更加知名］、桌上型计算机、商务和家用投影机、大型家庭剧院电视、机器人以及工业自动化设备、POS 发票打印机和收银机、笔记型计算机、集成电路、LCD 元件和其他相关的电子设备。目前，在全世界各地有许多分公司。

Seine netter 围网渔船 系指采用一种带状网具从事围网作业的、围捕水域中、上层集群性鱼类的船舶。

Seismic air compressor room for seismic source 地

震震源空压机室　装有空气压缩机组、高压空气瓶和其他附属设备，为海洋地震气枪震源提供高压空气的舱室。

Seismic amplifier　**地震放大器**　将压力检波器接收来的微弱地震信号进行放大的装置。

Seismic cable winch　**地震电缆绞车**　为收放地震漂浮组合电缆进行海洋地质调查用的绞车。

Seismic detailed survey　**地震详查**　海上地震勘探比例尺为 $1:50\,000 \sim 1:100\,000$，网格密度为 2×2 km $\sim 2 \times 4$ km。在已知构造上查明构造的特点（范围、形态、目的层厚度、水下地层接触关系、高点位置、闭合度与相邻构造的关系、断层的分布及其大小等）提供最有利的含油气带，为钻准备井位的一种勘查。

Seismic playback apparatus　**地震回放仪**　又称地震模拟计算机。用于处理模拟磁带地震记录，提供解释资料的仪器。

Seismic profiler　**地震剖面仪**　地震勘探资料数据处理的硬件之一，作为电子计算机的专用输出设备，以显示时间剖面或深度剖面。

Seismic prospecting　**地震勘探**　是利用人工产生的地震波在弹性不同的地层内传播规律来勘测地下地质情况的方法。地震波在地下传播过程中，当地层岩石的弹性参数发生变化，从而引起地震波场发生变化，并发生反射、折射和透射现象，通过人工接收变化后的地震波，经数据处理、解释后即可反映出地下地质结构及其岩性，达到地质勘查的目的。地震勘探方法可分为反射波法、折射波法和透射波法等三大类，目前的地震勘探主要采用反射波法。

Seismic reconnaissance survey　**地震普查**　海上地震勘探比例尺为 $1:100\,000 \sim 1:200\,000$，网格密度为 4×8 km $\sim 5 \times 10$ km。目的在于进一步划分区域性的地质构造单元和寻找有利于含油、气的局部构造，为更详细的调查作业准备的一种调查。

Seismic survey ship　**海底探测船**

Seismic wave　**地震波**　由震源产生的弹性振荡波。

Seismoscope　**地震示波器**　显示地震波形的一种设备。

Seize　**卡住（咬住）**

Seizing　**滞住**（材料黏附在模子上，捆绑）

Selected location　**选定位置**

Selected target　**被选目标**　手动选择的用于在独立数据显示区内显示字母数字信息的目标。该目标以"被选目标"符号表示。 Selected target is a manually selected target for the display of detailed alphanumeric information in a separate data display area. The target is displayed by a "selected target" symbol.

Selection and approval of sub-contractors and suppliers　**选择、认可分包商和供应商**　对质量管理体系而言，系指制造厂商应建立和保存可以被接受的供应商的记录。这些厂商的选择和对他们实施管理的范围和形式与产品的型号和服务工作的内容以及这些厂商以往所显示出的能力和业绩相适应。应用文件建立新供应商的认可步骤并记录对供应商的评价（如已进行过）。这些记录都应保存，在验船师要求时，应予提供。 For the purpose of quality management system, selection and approval of sub-contractors and suppliers mean that the manufacturer is to establish and maintain records of acceptable suppliers and sub-contractors. The selection of such sources, and the type and extent of control exercised, is to be appropriate to the type of product or service and the suppliers' or sub-contractors' previously demonstrated capability and performance. Documented procedures for approval of new suppliers are to be established and records of vendor assessments (where carried out) are to be maintained and made available to the Surveyors upon request.

Selection of the coating system　**涂层系统的选择**

Selective catalytic reduction（SCR）　**选择性催化还原技术**　是目前国际上得到广泛应用的烟气脱硝技术。该技术系指在催化剂和氧气存在的条件下，在一定的范围（$320 \sim 420$ ℃）内，还原剂（如氨）有选择地将烟气中的氮氧化物还原成氮气和水，来减少氮氧化物排放的技术。因为整个反应具有选择性和需要催化剂，故被称为选择性催化还原技术。这种装置是用于处理柴油机排气的一套设备，用以减少排气中的 NO_x 含量；而且利用尿素的催化还原反应使柴油机（尤其是低速柴油机）废气中的氮氧化物降低的一套装置，该装置体积较庞大。

Selective corrosion　**选择腐蚀**

Selective quenching　**局部淬火**

Selective tempering　**局部回火**

Selective tripping　**选择性脱扣**　在具有串联连接的保护装置的电路中发生故障的情况下，为了维持对其余正常电路的供电，仅使最接近故障点的保护装置自动分断的布置。 Selective tripping means such an arrangement that only the protective device nearest to a fault point is opened automatically in order to maintain the power supply to the rest of sound circuits, in the event of a fault in the circuit having protective devices connected in series.

Selectivity　**选择性**（选择能力）

Selector　**选择器**（转换器，波段开关）

Selector vavle　**选择阀**

Self-propulsion point of model under tank conditions 试验池船模自航点 在定航速船模自航试验中，船模阻力包括阻力增额超过推力的数值等于船模与实船摩擦阻力系数差异影响时的状况。

Self stable integrity platform (SSIP) 自稳式一体化张力腿平台

Self unloading system 自卸系统 系指与散货转运有关的在船上的装置和设备的布置，其设计应在不依靠抓斗、管道或与起重机或吊杆装置有关的类似的可拆卸零部件的情况下有效进行卸载。 Self unloading system means, in relation to the handling of bulk cargoes, an arrangement of devices and equipment on board a ship, designed so that unloading is effected without recourse to the use of grabs, tubs or similar items of loose gear in conjunction with crane or derricks.

Self noise 自噪声 系指舰船、潜艇航行时产生的噪声。自噪声的主要来源是机械噪声、螺旋桨噪声和水动力噪声。

Self-acting PFD 自作用的个人漂浮装置 除气胀类型的检验和重新装备外，浮力由固有的设施（固有浮性材料）或由适当的通过在浸水时自动起作用的系统实现的设施（气胀式）提供的，当正确的穿戴时，不要求使用者进一步动作的个人漂浮装置。 Self-acting PFD means the PFD in which buoyancy is provided by permanent means (inherently buoyant material) or by suitable means (gas inflation) effected by a system which automatically activates upon immersion and which, except for the inspection and rearming of inflatable types, when correctly donned requires no further action by the user.

Self-acting plug 自泄艇底塞 一种不用旋放即可向外泄水又能防止倒流的底塞。

Self-adjusting bearing 自动调节轴承

Self-adjustment 自动调节

Self-aligning device 对中限制器 在收、放电缆时使电缆保持对准布缆机履压块 V 形槽的中心，防止电缆跳槽的装置。

Self-assessment 自我评定 系指提交方为评定其散货船和/或油船设计和建造规范是否符合国际散货船和油船目标型船舶建造标准的目标和功能要求。 Self-assessment means the submitter assesses its rules for the design and construction of bulk carriers and/or oil tankers for conformity with the goals and functional requirements as set out in the International goal-based ship construction standard for bulk and oil tanker.

Self-cleaning centrifuge 自动去污油分离机 系指能在机器连续工作的条件下，能自动定期消除分离筒内积聚的残渣的油分离机。

Self-cleaning lubricating oil filter 自动清洗滑油滤清器 当滤器进出口大时能自动消除滤器内杂质的滑油滤器。

Self-closing device 自闭装置

Self-closing vavle 自闭阀

Self-contained air conditioner 立柜式空气调节器 装于甲板上供个别舱室使用并带有制冷装置的立式空气调节器。

Self-contained air conditioning unit 独立式空调装置

Self-defense 自行辩护

Self-elevated marching cutter suction dredger 自升-步进式绞吸式挖泥船

Self-elevating drilling unit 自升式钻井平台 系指具有足够浮力运送该装置到达指定位置的船体的装置。在指定位置上，船体升高到海面上预先指定的高度，并支撑在它的桩腿上，而桩腿由海底支撑。这些装置的桩腿可以设计成插入海底，可以装有大型分段或桩靴，或可以附到底部沉垫上。 Self-elevating drilling unit means a unit having hull with sufficient buoyancy to transport the unit to the desired location. On location, the hull is raised to a predetermined elevation above the sea surface on its legs, which are supported by the sea bed. The legs of such units may be designed to penetrate the sea bed, may be fitted with enlarged sections or footings, or may be attached to a bottom mat.

Self-elevating platform 自升式工作平台 工作时水下部分着底，平台则沿桩腿升高到海面一定高度，完井后平台降到海面，桩腿升起，再依靠拖船或本身推进装置移至新的井位，适宜于在浅海进行钻井作业的钻井平台。按接地形式分为桩腿自升式、桩靴自升式和沉垫自升式钻井平台。

Self-elevating platform for civil engineering 水上施工平台 通常备有起重机、打桩机、转机、空压机、水泵、发电机等动力设备，可在水面移动并依靠升船的机构沿桩腿升降，供水域工程的施工用的自升式平台。

Self-elevating unit 直升式平台 系指具有活动桩腿，能将壳体升至海面以上并降回海里的平台。 Self-elevating unit is a unit with movable legs capable of raising its hull above the surface of the sea and lowering it back into the sea.

Self-emptying assemblage 吹填排泥装置 由泥泵将泥舱中泥沙经排泥管吹至填泥区的装置。

Self-excited oscillations (self-excited vibration) 自激振动

Self-igniting buoy light 自亮灯浮 配设在救生浮具上,在浮具投水后能自动发光,以使在夜间使人易于发觉的可浮灯具。

Self-investigating case 自侦案件

Self-localization and self-elevating barge 自定位自升式驳船 系指具有在某些特定天气条件下在施工地周围自行移动能力的驳船。但是,在往返港口时,还是需要拖船协助它们移动。

Self-locking gear 自动闭锁装置

Self-lubricating bearing 自润滑轴承

Self-oiling 自动加油

Self-priming 自吸(自驱动)

Self-priming capacity 自吸能力

Self-priming pump 自吸泵(自动引水泵) 具有不需补给引液的结构,能自行将吸入管路中的空气抽除干净,然后抽送液体的泵。

Self-priming type 自驱动型(自吸型)

Self-propelled bucket dredger 自航链斗挖泥船 系指用链斗挖取的泥沙,可通过溜泥槽装入本船的泥舱,并自航至深水区自动卸泥的链斗挖泥船。

Self-propelled grab hopper dredger 自航抓斗挖泥船 由抓斗抓取的泥沙,可装入本船的泥舱,并自航至深水区自动卸泥的抓斗挖泥船。

Self-propelled ship 机动船舶 系指在正常操作中,采用机械推进方式而无须其他船舶帮助的船舶。

Self-propelled unit 自航平台 系指不需要外界帮助而能自行迁航的平台。 Self-propelled unit means a unit designed for unassisted sea passages.

Self-propelled vessel 机动船 依靠本船主机产生的动力来推进的船舶。

Self-propulsion and self-elevating platform 自航自升式平台船 又称风力涡轮机安装船。系指能够利用自身的吊车吊装基础风力涡轮机,并依赖自身的动力将它们运到海上施工地的船舶。作业时,自航自升式平台船将平台升到距离海面安全的工作高度,然后使用吊车及各种船载索具和起重设备安装基础或风力涡轮机。

Self-recording apparatus 自动记录器

Self-recording unit 自动记录装置

Self-releasing 自动释放 系指存放于露天甲板上的救生筏及救生浮具的固结索具,在船舶遇难下沉时的自行解脱。

Self-stabilization of the craft 自稳 系指船舶依靠其自身的特性保证稳定性。Sel-stabilization of the craft is stabilization ensured solely by the craft's inherent characteristics.

Self-stand type 立式(柜式)

Self-sustaining speed 最低稳定转速 燃气轮机机组在启动装置脱开后的最低稳定转速。

Self-tilting rotary cradle 自摇式转盘滑道 系指可以360°转动、人力操纵的、船舶下水用的滑道。

Self-unloading barge 自卸泥驳 为对开泥驳,其左右舷门在泥舱端舱壁处采用铰链连接,以便在舱底打开时泥舱绕纵向轴转动。

Self-unloading bulk carrier 自卸散货船 系指一种设有自动连续卸货的自卸系统的散货船。分为适合于装载煤炭、矿石、石膏石、黄沙、盐块状散货的自卸船和适合于装载散货水泥、化肥、化工原料的粉状散货自卸船。

Self-ventilating type motor 自通风型电机 系指其风冷空气是靠与其组装在一起的设施来进行环流的电机。

Semi- automatic welding 半自动焊接 系指诸如金属弧焊接或药芯焊丝焊接时,仅焊丝自动供给,焊接则用手操作。 Semi- automatic welding is used to describe processes in which the weld is made manually by a welder holding a gun through which the electrode wire is continuously fed such as metal arc welding or flux-cored arc welding, etc.

Semi-submersible drilling unit 半潜式钻井装置 系指浮动式或坐底式从事钻井作业的稳柱式钻井平台。

Semi-automatic 半自动 这一术语是用来描述焊工手执连续送丝的焊炬,用人工进行焊接的工艺。Semi-automatic is used to describe processes in which the weld is made manually by a welder holding a gun through which the wire is continuously fed.

Semi-bulit-up crankshaft 半组合曲轴

Semi-closed cycle 半闭式循环 工质一部分自成密闭系统,另一部分则按半开式循环工作的燃气轮机循环。

Semi-closed feed water system 半闭式给水系统 凝水从冷凝器出来送到锅炉的过程中不与空气接触,但调节水柜为非密闭式的给水系统。

Semi-closed gas turbine plant 半闭式循环燃气轮机装置 按半闭式循环工作的燃气轮机动力装置。

Semi-conductor 半导体 系指常温下导电性能介于导体(Conductor)与绝缘体(Insulator)之间的材料。半导体在收音机、电视机以及测温上有着广泛的应用。如二极管就是采用半导体制作的器件。半导体系指一种导电性可受控制,范围可从绝缘体至导体之间的材料。无论从科技或是经济发展的角度来看,半导体的重要性都是非常巨大的。今日大部分的电子产品,如计算机、移动电话或是数字录音机当中的核心单元都与半导体有着极为密切的关联。常见的半导体材料有硅、锗、砷化镓等,而硅更是各种半导体材料中,在商业应用上最具有影响力的一种。

Semi-conductor material　半导体材料　是导电能力介于导体与绝缘体之间的物质。半导体材料是一类具有半导体性能、可用来制作半导体器件和集成电路的电子材料,其电导率在 $10(U-3) \sim 10(U-9)\Omega/cm$ 范围内。

Semi-conductor rectifier　半导体整流器　是由形成结的导体和半导体所组成的一种器件。这种结对于从两个方向上流经该结的电流呈现出不同的电阻。其结果是只能在一个方向上有有效电流流动。半导体整流堆是由一个或几个半导体整流元件所组成的一种单柱形结构。

Semi-conductor refrigerating plant　热电制冷装置　由半导体元件组成热电堆,通入电流后产生冷、热端的制冷装置。

Semi-conductor switching device　半导体开关电器　系指利用半导体的导电可控性接通和/或阻断电流的开关电器。在船舶电力配电系统中不使用"半导体开关电器"。

Semi-container ship　半集装箱船　只有一部分货舱作为集装箱专用舱,其余货舱仍与一般杂货船相同的集装箱船。

Semicountersunk head　半沉头

Semicountersunk rivet　半沉头铆钉

Semi-diesel engine　半柴油机(热球式发动机)

Semi-dock building berth　半坞式船台　系指临水一端设有闸门,使船台尾部不受水淹影响的一种纵向倾斜船台。建造船舶时,关闭闸门将水抽干,使整个船台成为干燥的工作场地。船舶建成后,开启闸门进行下水。半坞式船台可缩短船舶下水的滑程,降低船台首部高度,充分利用水域近旁的船台面积,便于船尾水下部分的施工和下水操作。适用于厂区受到限制、水位差或潮差变化较大的地区。

Semi-enclosed locations　半围蔽处所　系指自然通风条件由于存在顶板、风障和舱壁之类结构而与开敞甲板上者有显著差异,且布置为不会发生气体扩散的场所。 Semi-enclosed locations are locations where natural conditions of ventilation are notably different from those on open decks due to the presence of structures such as roofs, windbreaks and bulkheads and which are so arranged that dispersion of gas may not occur.

Semi-enclosed space　半围蔽处所　系指:(1)由于具有诸如顶板、风障和舱壁等结构,以致其自然通风条件与在开敞甲板上的处所有显著差异,且其布置又使气体不易扩散的处所。(2)由于有诸如甲板和/或舱壁限制,致使这些处所内的自然通风条件与露天甲板上有明显不同的处所。 Semi-enclosed space is:(1)a space limited by top plate, wind break and bulkhead, etc. in such a manner that the natural conditions of ventilation in the space are notably different from those obtained on weather deck and the gas is uneasily diffuse;(2)a space limited by decks and/or bulkheads in such a manner that the natural conditions of ventilation in the spaces are notably different from those obtained on open deck.

Semifinished item (semifinished material, semifinished product)　半成品

Semi-fixed coordinate system　半固定坐标系

Semi-floating gate transistor(SFGT)　半浮栅晶管体　作为一种新型的微电子基础器件,它的成功研制将有助于我国掌握集成电路的核心技术,从而在芯片设计与制造上逐渐获得更多话语权。

Semi-FPU　半潜式浮式生产装置

Semi-gross weight　半毛重

Semi-hermetic refrigerating compressor unit　半封闭式制冷压缩机组　电动机和制冷压缩机的曲轴箱联为一体无轴封闭装置的压缩机组。

Semihermetric refrigerating unit　半封闭制冷机

Semi-magnetic controller　半磁控制器　系指其基本功能中仅一部分是依靠由电磁操作的器件完成的一种电控制器。

Semi-mechanized gas-shielded metal arc welding (GMAW)　半自动气体保护金属电弧焊

Semi-mechanized welding processes　半自动焊工艺　如重力电弧焊(Gravity arc welding)、自动电阻焊(Auto-contact welding)、带垫板的单面焊(Single-side welding with backing)、无气体保护的药芯焊丝金属极电弧焊(Flux-cored wire metal-arc welding without shielding gas)等。

Semi-menbrane tank　半薄膜液货舱

Semi-protective casing　半防护外壳　系指通常在其上半部的所有开孔的防护方式如同"防护外壳"的情况一样,而其他开孔则让其敞开着的外壳。

Semi-refrigerated & semi-pressure gas tanker　半冷半压式液化气船　系指:(1)在较低的温度下将气体液化。一般将货舱温度控制在 $-45℃ \sim -104℃$,压力控制在 $0.5 \sim 0.7$ MPa 范围内。由于气体压力的降低,使得货舱的壁厚度可以减薄,从而减轻了货舱的质量。采用这种方式运输舱内要隔热绝缘,并且设置制冷装置,此类船舶适合运输大多数液化石油气和化学品气体。(2)在特定的温度(低温)及压力条件下装运货物,温度和压力的控制值根据货品决定。半冷半压式液化气船有专用船和多用途船两类，船的载货总容积大多在 15 000～30 000 m^3 范围。液货罐需包覆绝热物。船上设有再液化装置。

Semi-refrigerated & semi-pressurized LPG 半冷半压型液化石油气船

Semi-refrigerated gas tanker 半冷式液化气船 在较低的温度下将气体液化。一般将货舱温度控制在 -45 ℃～-104 ℃，压力控制在 0.5～0.7 MPa 范围内。由于气体压力的降低，使得货舱的壁厚度可以减薄，从而减轻了货舱的质量。采用这种方式运输舱内要隔热绝缘，并且设置制冷装置，此类船舶适合运输大多数液化石油气和化学品气体。

Semi-rigid foamed plastics 半硬质泡沫塑料 是泡沫塑料的一种，系指弹性模量 70～700 MPa 的泡沫塑料。泡沫塑料的特点：(1)容重很轻，可减轻包装质量，降低运输费用；(2)具有优良的冲击、振动能量的吸收性，用于缓冲防振包装能大大减少产品的破损；(3)对温、湿度的变化适应性强，能满足一般包装情况的要求；(4)吸水率低、吸湿性小，化学稳定性好，本身不会对内装物产生腐蚀，且对酸、碱等化学药品有较强的耐受性；(5)导热率低，可用于保温隔热包装，如冰激凌杯、快餐容器及保温鱼箱等；(6)成型加工方便，可以采用模压、挤出、注射等成型方法制成各种泡沫衬垫、泡沫块、片材等。容易进行二次成型加工，如泡沫板材经热成型可制成各种快餐容器等。另外，泡沫塑料块也可用黏合剂进行自身黏接或与其他材料黏接，制成各种缓冲衬垫等。

Semi-slewing ramp 半旋转跳板 系指以船中心位置点能向一侧旋转的跳板。其特点在于对码头无特殊要求，使船可适应任何港口，并能平行滚装货物。

Semi-submerged condition 半潜状态 系指举升甲板被水淹没，仅依赖上层建筑或甲板室或浮提提供储备浮力的状态。 Semi-submerged condition means any condition in which lifting cargo deck is submerged and reserve buoyancy is provided by the superstructure or deckhouse or pontoon only.

Semi-submerged small-water-plane area ship 小水线面半潜船 为改善在波浪上的航行性能，将上部船体通过若干剖面呈流线型的支柱支承在两个鱼雷状的潜体上，其水面附近处的水线面比较小的船舶。

Semi-submersible accommodation platform 半潜式生活平台

Semi-submersible barge 半潜驳

Semi-submersible drilling platform 半潜式钻井平台 是由坐底式钻井平台演变而来的一种移动式钻井装备。该类平台主要由水下浮体和水面上的平台，通过若干根立柱连接组成的平台。浮体为平台提供大部分浮力；立柱用于连接工作平台和浮体，支撑工作平台。工作平台是半潜式钻井平台海上工作的主要结构、装配有钻井设备、钻井器材、起吊设备、安全救生、人员生活设施，以及动力、通信、导航等设备。当平台工作时，水下浮体潜入水中一定深度，海面波浪对浮体的扰动较小，平台能在水面上保持稳定和平稳。半潜式钻井平台的作业水深最大可达 500 m。

Semi-submersible drilling rig 半潜式钻井平台 一种海上钻井装置。上部为工作甲板，下部为两个下船体，用支撑立柱连接。工作时下船体潜入水中，甲板处于水上安全高度，水线面积小，波浪影响小，稳定性好、自持力强、工作水深大。

Semi-submersible drilling unit 半潜式钻井装置 系指浮动式或坐底式从事钻井作业的稳柱式钻井平台。 Semi-submersible drilling unit means a column-stabilized unit designed for drilling operation, either afloat or supported by sea bed.

Semi-submersible floating foundation 半潜型浮式基础结构 系指由带压水板的三立柱半潜式平台和对称式系泊系统组成的，用于海上风机塔柱承坐的结构。

Semi-submersible heavy lift vessel 半潜船 在装卸作业或操作需要时能呈现半潜状态的船舶。

Semi-submersible maximum drilling water depth 半潜最大工作水深 半潜式钻井装置能进行半潜钻井作业的最大水深。

Semi-submersible minimum drilling water depth 半潜最小工作水深 半潜式钻井装置能进行半潜钻井作业的最小水深。

Semi-submersible platform 半潜式平台 (1)是漂浮作业的海上平台。通常由下船体、立柱和上平台组成，作业时下船体沉没水中一定深度处。(2)平台主体部分沉没于海面以下的钻井平台。它由平台甲板、立柱和下体（或沉箱）组成。

Semi-submersible self-propelled heavy-lift vessel (SSHLV) 半潜式自航工程船 又称半潜船。具有半潜、自航、重型甲板载货功能的工程船。主要用于装运超大件货物，既可以靠岸装载，也可以半下沉进行漂浮装载。与一般运输船相比，半潜功能是半潜式自航工程船具有的独特之处，其主要工作原理是首先让船舶下沉到要求的深度，待装载的货物漂移至船舶甲板后，再使船舶上浮将待装载的货物托起，货物固定在甲板上并航行运输至目的地，然后船舶再度下沉至要求的深度，将装载的货物浮移出船舶。其在海洋工程应用包括：(1)钻井平台和海洋模块运输；(2)海洋工程建造；(3)海上浮托法安装；(4)Spar 和张力腿平台深水浮体的运输和安装;(5)海上平台拆除;(6)其他应用,如作为作业船进行深水打桩、水下遥控机器人(ROV)支持、SCR安装、海底电缆铺设、深水用海上管子运输、铺设和深水临时维修基地(船坞功能)等。

Semi-submersible suction dredger(SSD) 半潜式绞吸式挖泥船

Semi-submersible transport ship 半潜运输船 系指在海上深水油气田开发中主要用于大型水面设施的运输的船舶。半潜运输船可以运输海洋平台船体、海上平台的上部组块或整体海洋平台。见图 S-7。

图 S-7 半潜运输船
Figure S-7 Semi-submersible transport ship

Semi-submersible vessel 半潜船 系指:(1)有较大开敞露天载货甲板,艏部或艉部有较高上层建筑或甲板室或浮箱,在装卸货物作业过程中呈现半潜状态的船舶;(2)在装卸作业或操作需要时能呈半潜状态的船舶。
Semi-submersible vessel means:(1) a ship, designed to load and unload cargoes in semi-submerged condition, with large open weather cargo deck and fitted with a forward or aft high superstructure or deckhouse or pontoon;(2) capable of being semi-submersible when needed during loading and unloading or operation.

Semi-superconductive homo-polar machine 半超导单极电机 是利用法拉第圆盘原理加上超导励磁绕组制成的电机。

Semi-synteties 半合成剂 通常呈无气味的药丸或针剂,如地劳迪德(dilaudid)、奥姆尼硼(omnipon)。Semi-synteties usually appear as odurless pills or amnipouls, for example dilaudid, omnipon.

Semi-taut 半张紧

Semi-watertight openings 半水密开口 系指配有半水密关闭方式的内部开口,其能承受一个固定水压头(相当于与破损后中间进水阶段平衡最高水线浸没水压头)。若它们位于破损后最终平衡水线以下,半水密开口可能导致连续进水。Semi-watertight openings means the internal openings fitted with semi-watertight means of closure, which are able to sustain a constant head of water corresponding to the immersion relevant to the highest waterline after damage at the equilibrium of the intermediate stages of flooding. Semi-watertight openings may lead to progressive flooding if they are located below the final equilibrium waterline after damage.

Semi-wet laboratory 干湿实验室 介于干室与湿室之间,主要用于采水测温、测定生产力、放射同位素、微生物、化学分析和综合实验用的工作室。

SENC = system electronic navigational chart 电子航行海图系统 由 ECDIS 转换 ENC 以供适当使用的数据库,通过适当的方式更新 ENC 以及由船员添加的其他数据。A database resulting from the transformation of the ENC by ECDIS for appropriate use, updates to the ENC by appropriate means and other data added by the mariner.

Send in person 派人送达

Senhouse slip 锚链脱钩 套在锚链末端处,即使在锚链受张力情况下,能通过敲击而迅速解脱的连接用钩。

Sensibility 灵敏度(敏感性)

Sensibility information and data 敏感性信息和资料 系指那些一旦落入他人之手,能协助其策划和攻击船舶保安体系的信息和资料。敏感性信息和资料可包括:(1)有关保安计划、安排的信息;(2)脆弱性评估报告/信息;(3)保安设备的规格和位置;(4)保安设备的局限性信息;(5)敏感性电文等。

Sensing element 敏感元件(灵敏元件)

Sensing probe 传感器(敏感元件)

Sensing technology 传感技术 是关于从自然信源获取信息,并对之进行处理(变换)和识别的一门多学科交叉的现代科学与工程技术,它涉及传感器(又称换能器)、信息处理和识别的规划设计、开发、制造/建造、测试、应用及评价改进等活动。

Sensitive 敏感的

Sensitiveness 灵敏度

Sensitivity 灵敏度(敏感性)

Sensitivity analysis 敏感性分析 系指对于给定的模式或计算方法,个别输入参数对于结果变化的影响分析。Sensitivity analysis means an analysis to determine the effect of changes in individual input parameters on the results of a given model or calculation method.

Sensitivity coefficient 灵敏系数 系指表征保护电器动作灵敏性的系数。对过电流保护电器而言,可表示为灵敏系数=金属性短路的最小短路电流计算值/保护电器动作额定值。注意:这里的"最小短路电流计算值"系指在单台最小发电机供电情况下,被保护电路末端短

Sensor 传感器 系指:(1)检测环境、柴油机工况及各种运行参数的器件。主要有曲轴转速传感器、压力传感器、温度传感器、湿度传感器、气体成分传感器、上止点位置传感器、流量传感器及振动传感器等,根据各类柴油机电控系统的需要选用;(2)简化航行数据记录仪外部的任何装置,简化航行数据记录仪与之连接并从它那里得到需记录的数据;(3)能够得知货舱或其他区域是否有水的信号启动装置;(4)一种助航装置(测量装置),可具有或不具有自身的显示器,相应进行处理、控制和自动操作系统或综合航行系统(INS)提供信息。 Sensor means:(1) the apparatus detecting environment conditions, working conditions and various operating parameters of the diesel engine. There are crankshaft speed sensor, pressure sensor, temperature sensor, humidity sensor, gas composition sensor, TDC position sensor, mass flow sensor and vibration sensor, which are selected according to the need of electronic control system of various diesel engines; (2) any unit external to the S-VDR, to which the S-VDR is connected and from which it obtains data to be recorded;(3) a unit fitted at the location being monitored that activates a signal to identify the presence of water at cargo hold or other locations;(4) a navigational aid (measuring device), with or without its own display, processing and control as appropriate, automatically providing information to operational systems or INS.

Sensor 水位探测器 系指装设在受监控场所的组件,它能触发信号表明该场所有水。 Sensor means a unit at the location being monitored that activates a signal to identify the presence of water at the location.

Sensor system (DPS) 传感器系统(动力定位系统) 系指测量船舶航向、船舶运动、风速风向的传感器。 Sensor system (DPS) means these sensors which measure at least vessel heading, vessel motion, wind speed and direction.

Sensor/source modules 传感器/信息源模块 系指包含传感器/信息源要求的模块。 Sensor/source modules mean modules comprising the sensor/source requirements.

Sentence/ finding of "guilty" 有罪判决

Sentinel vavle 信号阀

Separate 分隔 系指例如某一货物管系或货物透气系统不与另一货物管系或货物透气系统相连接。 Separate means that a cargo piping system or cargo vent system, for example, is not connected to another cargo piping or cargo vent system.

Separate discharge 独立排水孔

Separated berth 分隔统铺 常见于我国南方内河客船上,在相邻两铺位间用活动的铺间挡板隔开的互相连成一片的床铺。

Separated by a complete compartment or hold from 用整个舱室或货船舱隔离 系指垂直或水平隔离。如果甲板不防火和防液,则仅可接受纵向隔离,即用介于中间的整个舱室隔离。 Separated by a complete compartment or hold from means either a vertical or a horizontal separation. If the decks are not resistant to fire and liquid, then only a longitudinal separation, i.e. by an intervening complete compartment, is acceptable.

Separated cavity 分离空泡 完全脱离物体的空泡。

Separated flow 分离流 从固体面分离开来的流动。

Separated from 隔离 系指当甲板下积载时,装入不同的货舱中。只要中间隔舱甲板防火和防液,则垂直隔离,即在不同舱室积载可视为与此隔离等效。 Separated from means that in different holds when stowed under deck. Provided an intervening deck is resistant to fire and liquid, a vertical separation, i.e. in different compartments, may be accepted as equivalent to this segregation.

Separated longitudinally by an intervening complete compartment or hold from 用介于中间的整个舱室或货舱纵向隔离 系指如仅有垂向隔离,则不符合这一要求的隔离。 Separated longitudinally by an intervening complete compartment or hold from means vertical separation alone does not meet this requirement.

Separated sludge 分离油泥 系指燃滑油分离所产生的油泥。 Separated sludge is sludge resulting from purification of fuel and lubricating oil.

Separating plate 分离盘 油分离机中,用以将水分及杂质在离心力场作用下自油中分离的圆锥形盘状零件。

Separation 间隔(分隔,分离) 系指诸如一货物管系统或货物透气系统不与另一货物管系统或货物透气系统相连接。对于这种隔离,可以通过设计或操作方法予以实现。但在液货舱内不应采用操作方法来实现,而应采取下列方式之一的措施:(1)装设可拆短管或阀,并盲断管端;(2)装设两个串联的盲通两用法兰,并在这两个盲通法兰之间的管内设有能探测是否渗漏的装置。

Separator 分离器(隔板,脱模剂) 是把混合的物质分离成两种或两种以上不同的物质的机器。天然气出口处的压力控制阀通常是压力式调节阀或配套压力变送器、控制器、气源的气动薄膜调节阀等。出油阀通常为配套液位传感器、控制器、气源的气动薄膜调节阀或浮子液面调节器操纵的出油调节阀等。

Separator vessel 油水分离器容器
Sequence 顺序（使程序化，交替）
Sequence of control 控制程序
Sequence of instructions 指令顺序
Sequence of legs 桩腿顺序 自升式钻井平台桩腿的编号次序。
Sequential control 顺次控制 是一种以一个再决定的顺序自动开展控制的模式。 Sequential control is a pattern of control that can be carried out automatically in the redetermined sequence.
Sequential method 顺序法 系指用于装载压载水的压载水舱先抽空，然后用替换的压载水重新注满的过程，达到的置换率至少为压载水体积的95%。 Sequential method means a process by which a ballast tank intended for the carriage of ballast water is first emptied and then refilled with replacement ballast water to achieve at least a 95% volumetric exchange.
Sequential operation 程序操作
Sequential pumping out method（ballast water exchange） 排空注入方法（压载水排放） 系指按顺序将每个舱内压载水排空后立即再注满的压载水交换法。如可能，应采用扫除舱泵或喷射泵。 Sequential pumping out method is a ballast water exchange method in which each ballast tank, in a certain sequence, is pumping out and then fully refilled with water at once. Stopping pumps or ejectors should be used it possible.
Sequentiel couleur a memoire（SECAM） 存色彩电视系统 又称塞康制，SECAM是法文缩写，意为"按顺序传送彩色与存储"，是一个首先用在法国模拟彩色电视系统，系统化一个8 MHz宽的调制信号。
Serial communication 串行传输 即串行通信，系指使用一条数据线，将数据一位一位地依次传输，每一位数据占据一个固定的时间长度。其只需要少数几条线就可以在系统间交换信息，特别适用于计算机与计算机、计算机与外部设备之间的远距离通信。
Serial sister ships 系列姐妹船 系指按相同的入级批准图纸建造的船舶。包括未来船东最终行使选择权的特定可选的船舶。如系列船建造合同签订后1年之内行使续建选择权，则该可选建的船舶将被认为相同的系列姐妹船的一部分。
Series and parallel pump（series-and-parall elcconnection pump） 串并联泵 一种双级离心泵，每一级各有其吸入及排出接管，通过转换阀，可将这两级离心泵串联或并联工作的泵。

Series of ship-model 系列船模 有次序地改变船模中某些参数而构成的成套船模。
Series-turbo-charging 复式增压 系指机械增压和废气涡轮增压串联使用或交替使用两者联合的增压方式。
Series-wound motor 串励电动机 是一种磁场电路与电枢电路串联连接的换向器式电动机。该电动机轻载运行时的转速比满载运行时的转速高得多。
Serious accidents 严重事故
Serious corrosion 严重腐蚀
Serious damage 严重损坏
Serious fault 严重故障
Serious injury 严重伤害 系指人员受到的使其丧失能力的伤害，从受伤之日起7天内人员正常活动不能持续72 h以上。 Serious injury means an injury which is sustained by a person, resulting in incapacitation where the person is unable to function normally for more than 72 hours, commencing within seven days from the date when the injury was suffered.
Server 服务器 是一种高性能计算机，作为网络的节点，存储、处理网络上80%的数据、信息，因此也被称为网络的灵魂。
Service 服务 系指英国劳氏船级社为业主提供的服务。
Service area 运行区域 系指一组海域。运行区域规定了对船舶操作海域的限制。Service area refers to a collective group of sea areas. The service area specifies the limits of the ship's operational area.
Service business 服务业务 系指向AOL用户提供增值服务，包括整合的消息、音乐下载等。
Service category 1 1类航区 系指离岸不超过200海里（夏季/热带）或100海里（冬季）的航区。 Service category 1 means a service within 200 (summer/tropical) or 100 (winter) n mile off shore.
Service category 2 2类航区 系指离岸不超过20海里（夏季/热带）或10海里（冬季）的航区。 Service category 2 means a service within 20 (summer/tropical) or 10 (winter) n mile off shore.
Service category 3 3类航区 系指遮蔽水域。 Service category 3 is sheltered waters.
Service condition 营运状态（运转工况，使用条件）
Service generator 日用发电机
Service generator unit 日用发电机组
Service hour 使用时数
Service life 使用寿命（使用年限）

Service load 使用载荷　系指船体结构在设计工况下及其核准的营运气象限制规定的波高、风级条件下运行时,其某部分结构可能受到的最大载荷。

Service manual 使用说明书(使用条令,修理说明书)

Service power（continuous service output, normal output） 经济功率　主机在船舶经济航速下的长期运转功率。

Service pump 日用泵(通用泵,杂用泵)

Service reliability 使用可靠性

Service restriction 沿海航区营运限制　系指航行于距岸不超过 20 n mile 的水域,且船舶在其经营的航线上,满载并以其营运航速航行至庇护地:对客船不超过 4 h;对货船不超过 8 h。如上述某些水域的海况较为恶劣,则可视情况对上述距离提出更严格的要求。如船旗国主管机关或其所在营运区的海岸主管机关对该水域有特定距离的规定时,则应根据该主管机关的规定执行。

Service restriction notation 航区限制标志　表明船舶的入级仅限于在有关主管机关同意的航区或条件下营运的标志,例如保护水域航区。 Service restriction notation is a notation indicating that a ship has been classed on the understanding that it will be operated only in suitable areas or conditions which have been agreed by the Classification Committee, e. g. protected waters service.

Service ship 辅助船/工作船　是用于在另一艘船和设备之间接收和运送液货,或反向作业的船舶。Service ship is a ship which receives and transports liquid cargoes between a facility and another ship and vice versa.

Service short-circuit breaking capacity 运行短路分断能力　系指按规定的试验程序所规定的条件,包括断路器连续承载其额定电流能力的短路分断能力。

Service spaces 服务处所　系指:(1)设有加热食品的设备,但不带有曝露加热表面的烹调设备的配膳室、储存舱柜、卖品部、储藏室和封闭的行李间等围蔽处所。无烹调设备的此类处所可以有:①自动咖啡机、烤面包机、洗碗机、微波炉、开水壶以及类似用具,每一用具的最大功率为 5 kW;②电加热的烹饪盘以及食品保温加热盘,每一用具的最大功率为 2 kW,且表面温度不应超过 150 ℃;(2)用作厨房、设有烹调设备的配膳室、储物间、邮件及贵重物品室、储藏室、不属于机器处所组成部分的工作间,以及类似处所和通往这些处所的围蔽通道。 Service spaces are:(1) those enclosed spaces used for pantries containing food warming equipment but no cooking facilities with exposed heating surfaces, lockers, sales shops, store-rooms and enclosed baggage rooms. Such spaces containing no cooking appliances may contain: ①coffee automats, toasters, dish washers, microwave ovens, water boilers and similar appliances, each of them with a maximum power of 5 kW;②electrically heated cooking plates and hot plates for keeping food warm, each of them with a maximum power of 2 kW and a surface temperature not above 150 ℃;(2)those spaces used for galleys, pantries containing cooking appliances, lockers, mail and specie rooms, storerooms, workshops other than those forming part of the machinery spaces, and similar spaces and trunks to such spaces.

Service spaces（high risk） 具有较大失火危险的服务处所　系指厨房、设有烹调设备的配膳室、油漆间具有面积为 4 m² 及以上的小间和储藏室、不属于机器所组成部分的工作间和易燃液体储存处所。 Service spaces（high risk）mean galleys, pantries containing cooking appliances, paint lockers, lockers and store rooms having areas of 4 m² or more, spaces for the storage of flammable liquids and workshops other than those forming part of the machinery spaces.

Service spaces（low risk） 具有较小失火危险的服务处所　系指面积小于 4 m² 的小间和储藏室(储存易燃液体的处所除外)、干燥室、洗衣间和盥洗室。 Service spaces（low risk）mean lockers and store rooms not having provisions for the storage of flammable liquids and having areas less than 4 m² and drying rooms and laundries.

Service speed¹ 服务航速(常用航速)　运输船舶在平常海况下经常使用的航速。

Service speed² 营运航速　系指最大航速的 90%。 Service speed is 90% of maximum speed.

Service speed Froude number 服务航速的傅汝德数

Service telephone communication system 日用电话通信系统　它是为了有关舱室进行日常工作和生活上的联络通信而装设的。这种通信系统一般采用自动电话设备。自动电话设备的布置根据工作需要和使用要求而定。为了保证船长室、轮机长室电话在必要时能优先于其他用户进行呼叫和通话,船长室或轮机长室电话应为特殊用户。在船舶停靠码头时,自动电话系统能与岸上电话网相连。

Service trade 服务贸易　又称劳务贸易,系指国与国之间互相提供服务的经济交换活动。服务贸易有广义与狭义之分,狭义的服务贸易系指一国以提供直接服务活动形式满足另一国某种需要以取得报酬的活动。广义的服务贸易既包括有形的活动,也包括服务提供者与使用者在没有直接接触下交易的无形活动。服务贸易一般情况下都是指广义的。

Service water 生活用水

Serviceability 可维修性（使用可靠性，供给能力，操作性能）

Serviceability limit state 使用性极限状态 与正常使用有关的使用性极限状态，包括：(1)可能减少结构的工作寿命或影响构件的有效性或外形的局部损伤；(2)影响有效使用和构件外形或设备功能的不可接受的变形。 Serviceability limit state, which concerns the normal use, includes:(1) local damage which may reduce the working life of the structure or affect the efficiency or appearance of structural members;(2) un-acceptable deformations which affect the efficient use and appearance of structural members or the functioning of equipment.

Serviceable life 使用寿命（使用年限）

Services for habitability 居住便利用途 系指为船员和旅客提供最低快捷环境而使用的重要用途。维持居住便利用途的装置如下：(1)烹调；(2)取暖；(3)船内冷藏/冷冻装置；(4)机械通风；(5)卫生水和淡水；(6)向上述装置供电的发电机和相关电源。 Services for habitability are those services which need to be in operation for maintaining the vessel's minimum comfort conditions for the crew and passengers. Examples of equipment for maintaining conditions of habitability are as follows:(1) cooking;(2) heating;(3) domestic refrigeration;(4) mechanical ventilation;(5) sanitary and fresh water; and (6) electric generators and associated power sources supplying the above equipment.

Services necessary for minimum comfortable conditions of habitability 最低舒适居住条件所需设备 如下：(1)烹调设备；(2)供暖设备；(3)生活用冷藏装置；(4)机械通风设备；(5)盥洗用水和淡水。 Services such as the following are considered necessary for minimum comfortable conditions of habitability:(1) cooking;(2) heating;(3) domestic refrigeration;(4) mechanical ventilation;(5) sanitary and fresh water.

Services which are considered necessary for minimum comfortable conditions of habitability 满足最低宜居条件所需设备 如下：(1)烹调设备；(2)采暖设备；(3)生活用冷藏装置；(4)机械通风设备；(5)盥洗用水和淡水。 Services such as the following are considered necessary for minimum comfortable conditions of habitability:(1) cooking;(2) heating;(3) domestic refrigeration;(4) mechanical ventilation;(5) sanitary and fresh water.

Service-type manufacture "服务型"制造 系指将更多的服务理念融入生产当中。其主要包含两层意思：一是生产企业通过与客户密切合作，最大程度地满足客户所需；二是生产企业与上游企业展开合作，进而达到核心竞争力的高度协同。对于造船企业来说，一方面要根据船东的个性化需求进行定制生产和服务，另一方面还要与铸锻配套企业、钢铁企业加强协作，优化生产组织，降低物料损耗，进而达到快速提升工艺水平和增强自主创新能力的目的。

Servo actuator 伺服执行装置（伺服执行机构）

Servo cylinder 伺服缸（随动缸）

Servo element 伺服元件

Servo mechanism 伺服机构

Servo mechanism theory 伺服机构理论

Servo motor 伺服电动机

Servo regulating system 随动调节系统

Servo valve 伺服阀

Set 安装（用具，仪器，装置）

Set bolt 固定螺栓（止动螺栓，冲头销）

Set net pile hammer 定置网打桩机 在定置网作业中，向水底打固定网具定置桩头的机械。

Set stud 定位螺柱

Set top box (STB) 机顶盒 系指电视机顶上的盒子，是扩展电视机功能的一种家用电器。根据接收的信号种类，机顶盒分为模拟机顶盒和数字机顶盒。模拟机顶盒接收模拟信号，数字机顶盒接收数字信号。数字机顶盒是一种多媒体终端，有类似于家用电脑的硬件体系结构和专用的实时操作系统及应用软件。数字机顶盒是在有线电视网络状态下工作，有线电视网采用模拟传输，因此必须对数字信号进行调制和解调才能在模拟信道传输。调制解调器是系统关键的组成部分，在技术上类似电话调制解调器的原理，但采用了更高的调制方法，下行多采用 64QAM 或 256QAM，在 DVB-C (Digital Video Broadcast by Cable) 和 DAVIC 中采用 64 QAM 作为标准调制方法，以 Motorola 的 MC92305QAM 解调芯片为例，在 7 M 模拟带宽上采用 64 QAM 调制的数字信号速率可达 42 Mbit/s，上行采用两种方式，一种是采用电话线作为上行信道，另一种是采用双向 HFC 网的上行通道，采用 HFC 网时采用 QPSK 作为调制方案。

Set-back 翘度 螺旋桨叶切面的导边或随边处与面节线之间的距离。

Set-down 下沉运动 是一种低频响应，其引起张力腿的动张力也是慢变的张力。

Setting 设定（安置，调整，号料） 对"73/78 防污公约"附则Ⅵ的 2008 NO_x 技术规则而言，系指影响发动机 NO_x 排放性能的可调整部件的调整。 For the purposes of the MARPOL73/78 Annex Ⅵ Setting means adjustment of an adjustable feature influencing the NO_x emissions performance of an engine.

Setting pressure 设定压力
Setting value 整定值（调定值）
Settle 使稳定（解决）
Settlement 沉陷　自升式钻井平台在海底承载能力较小的海区，因波浪、潮流等作用导致的相对下陷。
Settlement of exchange 定期结汇　系指议付行根据向国外付款行索汇函往返需要的时间与付款行正常审单付款的时间，预先确定一个固定的结汇期限，期满时，无论是否收到货物，都会主动地将货款兑换成人民币存入出口方银行。定期结汇系指银行在收到出口企业提交的出口单证，经与信用证有关条款审核无误后，根据不同地区、不同索汇路线，以及即期或远期等具体情况，结合银行办理各项手续必需的合理的工作日，规定一定的结汇时间，到期由银行主动将外汇结付给出口企业的一种结汇方式。
Settler (settling tank) 沉淀器（沉淀柜）
Settling tank 沉淀舱　系指用作保存污染水，并随后与油分离的一个液舱。　Settling tank is a tank intended for the retention of polluted water and its subsequent separation from oil.
Seven-roller leveller 七辊校平机　是板材加工中常用的具有7根辊筒的设备。校平机的选型主要取决于被校带材的厚度、材质和要求。材料越厚所需结构刚性要越好，辊数越少，辊径越大，功率越大（幅宽一定），反之亦然。校平机设置有上压模和下压模，其中上压模固接在液压缸的推杆上，液压缸缸体固定在支撑架上，在上压模和下压模内各设置有独立的冷却水路，该冷却水路的出口和入口分别位于上压模或下压模的上方。校平机主要应用于矫正各种规格板材及剪切成块的板材。
Severe condition 恶劣条件
Severe damage to the environment 对环境造成的严重破坏　系指经受到影响的国家或船旗国评估的对环境造成极为有害影响的环境破坏。　Severe damage to the environment means damage to the environment which, as evaluated by the State(s) affected, or the flag State, as appropriate, produces a major deleterious effect upon the environment.
Severe ice condition 严重冰况
Severe icing 严重冰冻
Severe sliding wear particle 严重滑动磨粒
Severe storm conditions 强风暴（自存）工况　系指平台承受最严重设计环境载荷时停止作业或其他操作，从而把抗环境能力提高到最大的状态。平台可以浮起或坐底。Severe storm conditions mean these conditions wherein a unit may be subjected to the most severe environmental loading for which the unit is designed. Drilling operations are assumed to have been discontinued due to the severity of the environmental loading. The unit may be either afloat or supported on the seabed, as applicable.
Severe stress 危险应力
Sewage 生活污水（污水）　（1）对"73/78 防污公约"附则Ⅳ而言，生活污水系指下列这些排出物：①任何形式的厕所和小便池的排出物和其他废弃物；②医务室（药房、病房等）的面盆、洗澡盆和这些处所排水孔的排出物；③装有活畜禽货的处所的排出物；或 ④混有上述排出物的其他废水。（2）对 ISO5749-1:2004(E)而言，生活污水是从盥洗室，小便池和坐浴盆，包括添加剂、医疗区域（药房、病房等）和在这些区域的洗澡盆，浴桶和排水管，装有活畜禽货的处所和混有上述其他类型排出物的废水。（3）人体的粪便和从厕所和其他拟接收和留存人体粪便的容器来的废弃物。　（1）For "MARPOL73/78 Annex Ⅳ", Sewage means:①drainage and other wastes from any form of toilets and urinals;②drainage from medical premises (dispensary, sick bay, etc.) via wash basins, wash tubs and scuppers located in such premises;③drainage from spaces containing living animals; or ④other waste waters when mixed with the drainages defined above. (2) For ISO 15749-1:2004(E), Sewage is wastewater, from water closets, urinals and bidets, including additives;medical areas (pharmacy, hospital, etc) and from washing basins in those areas, bath tubs and water discharges;spaces housing living animals and other types of wastewater if mixed with contaminated water already mentioned. (3) Human body wastes and the wastes from toilets and other receptacles intended to receive or retain body waste.
Sewage comminuting and disinfecting system 生活污水粉碎和消毒系统　系指能对生活污水进行粉碎和消毒的一套系统，用于船舶在离最近陆地不到3海里时临时储存生活污水。
Sewage disposal system 生活污水处理系统
Sewage disposal unit 生活污水处理装置
Sewage disposal vessel 粪便处理船　为港内船舶卸运和处理粪便污水的船舶。
Sewage ejector 生活污水排射器
Sewage management plan 生活污水管理计划
Sewage management procedure 生活污水管理程序
Sewage pollution prevention certificate 防止生活污水污染证书
Sewage pump 粪便泵（污物泵，生活污水泵）　抽送粪便等污物的泵。
Sewage recovery vessel 垃圾回收船　系指专门用

于回收垃圾的船舶。

Sewage-removal & cleaning boat for hydropower station 水电站清污保洁船 系指确保水电站安全运转的环境保护的特种工程船舶。

Sewage system 生活污水系统 系指配备下列装置之一的系统;(1)生活污水处理装置,该装置经主管机关形式认可,并考虑到 IMO 组织制定的标准和试验方法;或(2)经主管机关认可的生活污水粉碎和消毒系统,该系统应配备令主管机关满意的各项设施,用于船舶在离最近陆地不到3海里的临时储存生活污水;或(3)集污舱,该集污舱的容量应参照船舶营运情况、船上人数和其他相关因素,能存放全部生活污水,并使主管机关满意。集污舱的构造应使主管机关满意,并应设有能指示其集存数量的目视装置。 Sewage system means a syatem equipped with one of the following means;(1)a sewage treatment plant which shall be of a type approval by the Adiministration taking into account the standards and test methods developed by the IMO;or (2)a sewage comminuting and disinfecting system approved by Administration. Such system shall be fitted with facilities to the satisfaction of the Adiministration, for temporary storage of sewage when the ship is less than 3 n miles from the nestest land;or (3)a holding tank of the capacity to the satisfaction of the Adimin-istration for the retention of all sewage, having regard to the operation of the ship, the number of persons on board and other relevant factors. The holding tank shall be constructed to the satisfaction of the Adiministration and shall have a means to indicate visually the amount of the contents.

Sewage tank 生活污水储存舱(污水柜,粪便柜)

Sewage treating ship 污水处理船 系指专门用于回收、处理污水的船舶。

Sewage treatment 生活污水处理 为使污水达到排水某一水体或再次使用的水质要求对其进行净化的过程。污水处理被广泛应用于建筑、农业、交通、能源、石化、环保、城市景观、医疗、餐饮等各个领域,也越来越多地走进寻常百姓的日常生活。

Sewage treatment plant 生活污水处理装置 系指能对生活污水进行无害化处理的一套设备,经处理后可达到排放标准。

Sewage treatment systems(STS) 污水处理系统

Sewerage treatment space 生活污水处理舱

Shackle 卸扣 开口端用栓或螺栓封闭,供连接绳索、环或链条的 U 形或 Ω 形金属连接件。主要有索具卸扣、起重卸扣、锚链卸扣、浮筒卸扣等。

Shaft 轴(竖井,通风井)

Shaft alley escape 轴隧应急出口

Shaft alley pump 轴隧排水泵

Shaft bearing 轴承

Shaft block 轴承座

Shaft bracket 轴支架(艉轴架,人字架)

Shaft bush 轴衬

Shaft coupling 联轴节

Shaft coupling bolt 轴法兰螺栓 轴法兰的连接螺栓。

Shaft disengaged teat 脱轴试验 脱开主机组与轴系的连接,对主机组进行的检查试验。

Shaft displacement 轴的位移

Shaft drive assembly 轴传动组件

Shaft driven generator 轴带发电机 是由主机驱动船用发电机,利用主机富裕功率达到节能的目的。在 20 世纪 70 年代初,中东石油危机后得到迅速发展。采用主机轴带发电,早在直流电制时代各国在中、小型船舶上就已应用。随着电力电子技术的发展,轴带发电装置作为节约运行费用及改善机舱管理运行条件的有效手段,以其特有的优越性引起了各国造船界和航运界的重视。

Shaft efficiency 轴效率

Shaft flange(coupling flange) 轴法兰 与轴成一体用以连接两段轴的轴端圆盘。

Shaft generator system 轴带发电机系统

Shaft generator-motor 轴带发电-电动机组

Shaft gland 轴封

Shaft horse power 轴功率(马力)

Shaft journal 轴颈

Shaft line 轴线

Shaft line efficiency 轴系效率

Shaft lining(shaft facing) 轴包覆 保护轴身的保护层。

Shaft locked test 锁轴试验 检查多桨推进装置在锁紧某一轴而用其余轴系推进时的工作可靠性的试验。

Shaft monitoring system 轴系状态监测系统 系指通过在艉管前后轴承的几个截面上安装位移传感器,实现对艉轴位置的监测,以便分析艉轴形态,并结合轴系静态校中和船舶轴系的操作状态,评估轴系的校中状况和轴系的工作性能的系统。

Shaft neck 轴颈

Shaft passage 轴隧(地轴弄)

Shaft pipe 艉轴管

Shaft power 轴功率 P_0 系指轮机入级时的最大推进轴功率(kW)。 Shaft power (P_0) is the maximum propulsion shaft power, in kW, for which the machinery is to be classed.

Shaft revolutions detector 轴转数传感器
Shaft seal 轴封
Shaft speed transducer 轴转速传感器
Shaft steady bearing foundation 艉轴支撑座
Shaft stool 轴支座
Shaft structure 轴结构
Shaft strut 轴架
Shaft torque limiter 轴扭矩限位器
Shaft trailed test 拖轴试验 检查多桨推进装置在某一轴系自由旋转而用其余轴系推进时的工作可靠性的试验。

Shaft tunnel / shaft tunnel 轴隧（轴隧端室） 系指：(1)轴隧两端的扩大部分；(2)自机舱至船尾，围罩轴系并可供人员通行的水密隧道。起保护轴系和便于船员对轴系进行检查和维修用的，连接机舱和船尾的水密通道。对于中机型船舶，轴隧较长，轴隧的高度约 2 m，宽度为 1.2 m～1.8 m。对于单桨船，轴隧的中心线偏向右舷，留有通道供船员走动。轴隧的末端在艉尖舱处设有应急围井通至露天甲板，作为逃生用。轴隧分为拱顶和平顶两种结构形式。

Shaft tunnel alley 艉轴隧
Shaft-driven alternator 轴带交流发电机
Shafting (shafting system) 轴系 船舶中，连接主机和推进器的整个传动系统。
Shafting arrangement 轴系布置 在全面考虑空间地位、主机机型以及如何有效利用螺旋桨推力和保证主机安装与工作可靠等各种因素的情况下，对船上整个轴系所做的合理布置。
Shafting brake 轴系制动器 用以限制轴系转动的设备。
Shafting connecting bolt 轴系连接螺栓
Shafting efficiency 轴系效率
Shafting foundation 轴系基座
Shafting hanger 轴系吊架
Shafting transmission efficiency 轴系效率（轴系传动效率） 联结主机组与推进器的轴系的传递效率。
Shafting vibration 轴系振动
Shafting-transmitted steering gear 操舵轴传动操舵装置 以操舵轴和伞齿轮或滚子链轮为传动部件的人力操舵装置。
Shake 振动（冲击，裂缝）
Shake table test 振动台试验
Shallow defect 浅层缺陷
Shallow hardening 浅表硬化（表面硬化，表面淬火）
Shallow profiler 浅层剖面仪 绘制和记录海底表层地质状态的仪器装置。

Shallow water 浅水 取决于DP船舶正在其中承担工作的水深。可能有必要做进一步考虑。
Shallow water effect 浅水效应 由于水浅，水对船体的相对速度加大，以及船波变为浅水波等影响，使船舶航行状态改变、阻力增加等现象。
Shallow water resistance increment 浅水增阻 船在浅水区航行，由于浅水效应所增加的阻力。
Shallow water towing tank 浅水船模试验水池 能模拟在浅水中航行的船模试验水池。
Shallow water wave 浅水波 水深小于波长的一半时，水质点沿近于椭圆形轨迹运动的波浪。
Shallowly submerged hydrofoil 浅浸式水翼 翼航时水翼浸深小于弦长，水翼升力随浸深的改变而急剧变化，利用水翼的浅浸效应保持艇的飞高及纵、横稳定性的水翼。
Shallow-water vessel 浅水船 适宜于浅水区航行的船舶。

Shanghai Ship Design And Research Institute (SDARI) 上海船舶研究设计院——中国民用船舶设计领域"领航者" 上海船舶研究设计院（SDARI）成立于1964年，是中国船舶工业集团有限公司旗下具有国际影响力的民用船舶设计单位之一。上海船舶研究设计院设计的各类海洋装备主要包括散货船、集装箱船、油船、化学品船、多用途船和滚装船等运输船舶以及救助船、打桩船、浮船坞、海洋工程辅助船、半潜船、综合勘察船、吊管架下水驳和深水起重铺管船、海洋平台等工程船舶。上海船舶研究设计院经过近 10 年的研究设计实践，在滚装船领域取得迅猛发展，船舶类型已由单一滚装船型扩展至大型客滚船（PCTC）领域，逐步形成在国内领先的设计能力。

Shanghai Shipbuilding Technology Research Institute (SSTRI) 上海船舶工艺研究所 作为中国船舶工业集团有限公司所属专门从事船舶、海洋工程以及各类钢结构建造工艺研究的综合性应用研究所，成立至今已走过 50 年的发展历程。伴随着中国船舶工业的发展壮大、转型升级，上海船舶工艺研究所以逐步发展成为拥有 6 个专业研究室，6 家全资产业子公司的经营科研型应用研究所。上海船舶工艺研究所专业范围覆盖舰船和海洋工程总体建造工艺。先进连接技术与自动化装备、非金属材料与防腐涂装技术、船舶应用软件开发、国际新规范研究与应对。钢结构制造监理及建设工程质量检测等。50 多年来，该所共取得科研成果 750 余项，其中国家（省部）级成果 200 余项，这些成果已经在船舶、海洋工程建造以及市政工程等各个领域得到应用，取得了显著的经济效益和社会效益。

Shapes　号型　悬升在船上用以表示船舶动态和作业状态的具有特定形状的有色几何体。

Share　股份/股票

Share certificate　股票　系指股份的证书。

Sharp-bottomed ship（sharp-built ship）　尖底船　舭部升高较大的船舶。

Shear wave　剪切波　系指传播方向与介质质点的振动方向垂直的波。又称横波，S 波。

Shearing force curve　剪力曲线　船体或构件弯曲时，其横剖面内的剪力沿船长或构件长度方向变化的曲线。

Shearing stress of fluid　流体切应力　黏性流体流动时，作用在流体内部或流体与固体面之间单位面积上的切向抵抗力。

Sheath　护套（阳极）

Sheave　滑轮　滑车中，具有绳槽供绳索穿绕的轮子。

Shedder plates　挡货板　系指用以提高槽型舱壁和甲板骨架构件稳性的斜板。　Shedder plates that are fitted to minimize pocketing of residual cargo in way of corrugated bulkheads.

Sheer　舷弧　甲板边线的纵向曲度。其值以各站处自舷弧基线沿垂向量至舷弧线的距离，亦即以各站处舷弧线与型深的高度差表示。当舷弧线高于舷弧基线时为正值，低于舷弧基准线时为负值。

Sheer at after perpendicular　艉舷弧　在艉垂线处的舷弧。

Sheer at deck center　脊弧　甲板中线的纵向曲度。

Sheer at forward perpendicular　艏舷弧　在艏垂线处的舷弧。

Sheer line　舷弧线　甲板边线在侧视方向的投影线。

Sheer plan　纵剖线图　系指型线图中，中线面及与其相平行的各纵向垂直平面与船体型表面的交线在侧视方向的投影图。

Sheer strake　舷顶列板　系指船侧外板的顶列板。　Sheer strake means the top strake of a ship's side shell plating.

Sheerline　甲板舷弧线　甲板与船体的交线，对于圆弧形甲板边缘，为正常的交线；或者如未设甲板或船体延伸至甲板以上（舷墙），则为船体的上缘线。　Sheerline is the intersection between deck and hull, for rounded deck edges the natural intersection, or, where no deck is fitted or the hull extends above the deck (bulwark), the upper edge of the craft's hull.

Sheet anchor　备用锚　供船舶备用的艏锚。

Sheet brass　薄黄铜板

Sheet cavitation（laminar cavitation）　片状空泡　螺旋桨叶导边附近水流脱离叶背形成的定常或准定常银白色薄片空泡。

Sheet cavity　片状空泡　当螺旋桨叶剖面的冲角较正常冲角大得很多时，由于叶背上产生很大的压力降便会产生片状空泡。片状空泡处于准定常稳定状态，对剥蚀无多大危险。当冲角较正常冲角小得很多时，在叶剖面的压力面上也会出现类似的现象，称为面空泡。这种空泡和弦长相比通常很小，试验表明它和吸力面产生的片状空泡不同，往往导致空泡剥蚀。

Sheet gasket　垫片（垫密片）

Sheet iron　薄铁板

Sheet lead　铅皮（铅板）

Sheet metal work　钣金工作

Sheet steel　薄钢板/钢皮　系指厚度不大于 3 mm 的钢板。常用的薄钢板厚度为 0.5～2 mm，分为板材和卷板供货。薄钢板钢号为 B0～B3 的冷扎或热扎钢板。对薄钢板的要求：表面平整、光滑，厚度匀称，允许有紧密的氧化铁薄膜，不得有裂痕、结疤等缺陷。工艺分为热轧薄钢板和冷轧薄钢板。主要用于汽车、电器设备、车辆、农机具、容器、钢制家具等。

Sheet zinc　薄锌板

Shelf coil air cooling type freezer　吹风搁架冻结装置　由通风设备和与货物直接接触的架状冷却盘管组成的冻结装置。

Shelf life of battery　蓄电池的贮存寿命　系指蓄电池在规定条件下的贮存期间，在此期间的最后仍能保持其规定的性能。　Shelf life of battery means the duration of storage under specified condition at the end of which a battery retains the ability to give a specified performance.

Shelf plate　顶板　舱壁凳顶的水平板材。　Shelf plate is a horizontal plate located on the top of a bulkhead stool.

Shell　车壳　滑车上，用以传递作用力，保护滑轮和防止绳索脱槽的外壳。

Shell（or bending plate）element　壳（或弯曲板）单元　是板单元，具有面内刚度和面外弯曲刚度，且厚度不变。　Shell（or bending plate）element is plate element with in-plane stiffness and out-of-plane bending stiffness with constant thickness.

Shell and tube condenser　壳管式冷凝器　圆筒状，两端焊有管板，管板上扩胀或焊有很多冷却管，制冷剂在管外冷凝成液体，冷却水在管内流动带走制冷剂冷凝热量的热交换器。

Shell envelope plating　外壳板　系指构成有效船体梁的外板。Shell envelope plating mean a shell plating forming the effective hull girder.

Shell expansion　外板展开图　系指在肋位上按肋骨的型线围长展开的外板布置图。因船体外形具有双曲度，按几何方法难以精确展开，故只能按肋位横向展开。一般船体是左右对称的，故仅画出其中的某一舷。图中表明外板及其骨架的布置，外板的尺寸和厚度、边接缝、端接缝的位置和连接方式，分段的划分线的位置、外板开口的位置和加强覆板的形状和尺度，以及在各有冰区加强时冰区带的范围等。

Shell plate (shell plating)　外板　构成船体底部、舭部及舷侧外壳的板。

Shell-and-tube type condenser　壳管式冷凝器

Shell-and-tube type cooler　壳管式冷却器

Shell-and-tube type evaporator　壳管式蒸发器

Shelter area　庇护区　平台上全体人员紧急情况下可以在一段时间内保持安全的区域。

Shelter deck ship　遮蔽甲板船　最上一层连续甲板为遮蔽甲板，该甲板及其甲板间舱壁上设有可供敞开或封闭使用的吨位开口，适于载轻货的船。

Sheltered water　遮蔽水域　风程等于或小于6海里的水域。Water where the fetch is 6 n mile or less than 6 n mile.

Sheltered water service restriction　遮蔽航区营运限制　系指航行于沿海航区内由海岸与岛屿，岛屿与岛屿围成的遮蔽条件较好、波浪较小的海域，在该海域内岛屿之间、岛屿与海岸之间距离不超过10海里；或离岸不超过10海里的水域，船舶满载并以其营运航速航行，航程不超过2 h，并限制在风力不超过7级(蒲氏风级)且自测波高不超过2.0 m的海况下航行。

Sheltered waters　遮蔽水域　系指：(1)海岸与岛屿、岛屿与岛屿围成的遮蔽条件较好、波浪较小的海域，且该海域内岛屿之间、岛屿与海岸之间横跨距离不超过10海里，或具有类似条件的水域；(2)风程等于或小于6海里的水域。 Sheltered waters mean: (1) the sea areas between an island and the shore and between islands with a distance of less than 10 n miles in between, which forms a comparatively good sheltered or similar condition with a little wave; (2) the waters where the fetch is 6 n miles or less.

Shield tunneling machine　盾构机　全名叫盾构隧道掘进机，是一种隧道掘进的专用工程机械。现代盾构掘进机集光、机、电、液、传感、信息技术于一体，具有开挖切削土体、输送土碴、拼装隧道衬砌、测量导向纠偏等功能，涉及地质、土木、机械、力学、液压、电气、控制、测量等多门学科技术，而且要按照不同的地质进行"量体裁衣"式的设计制造，可靠性要求极高。盾构掘进机已广泛用于地铁、铁路、公路、市政、水电等隧道工程。横断面宽10.122 m、高7.27 m、400吨的世界最大矩形盾构机在我国郑州下线。

Shielded thermocouple　保护套式热电偶

Shielding facilities　防护装置(屏蔽设备)

Shifting　移位

Shift　上挡转换键　系指电脑键盘上的键，通常印上向上箭头标记。普遍有两颗，多在Ctrl键之上；亦有一些键盘只设有一颗Shift键。常用于中英文转换。它为辅助控制键，可以和其他键一起使用。实际上，只要在执行上述重启命令时，按住"Shift"键，再按下"确定"按钮，一切都可省略，电脑会立刻重新进入Windows。在"资源管理器"中拖放以".exe"结尾的程序文件到其他文件夹时，会发现文件并没有被移走，而只是在目标文件夹中创建了一个快捷方式。

Shifting　位移(转换，船舶纵向移动)

Shifting pump　移注泵

Ship　船舶　系指：(1)在海洋环境中运行的任何类型的船舶，包括水翼船、气垫船、潜水船、水上的艇、筏和固定式或浮动式平台；(2)任何类型的由自身驱动或由他船拖带进行海上航行的船舶(包括浮动船艇)；(3)各种在海上环境下营运或操作的船舶，包括半潜船、浮动式船舶、浮动平台、自升式平台、浮式储油装置(FSU)和浮式生产、储存和卸油装置(FPSO)，包括被拆除船上设备的船舶或被拖的船舶；(4)为运输散装油类货物而建造或改装的任何类型的海船和海上航行器，但是，能够运输油类和其他货物的船舶，仅在其实际运输散装油类货物时，以及在此种运输之后的任何航行(已证明船上没有此种散装油类运输的残余物者除外期间，才应视作船舶；(5)任何种类的海船和海上航行器；(6)用于国际海上商务中运输货物、旅客或货物和旅客两者兼有的任何自航式海船。 Ship means: (1) a vessel of any type whatsoever operating in the marine environment and includes hydrofoil boats, air-cushion vehicles, submersibles, floating craft and fixed or floating platforms; (2) any sea-going vessel of any type whatsoever, including floating craft, whether self-propelled or towed by another vessel, making a sea voyage; (3) a vessel of any type whatsoever operating or having operated in the marine environment and includes submersibles, floating craft, floating platforms, self elevating platforms, Floating Storage Units (FSUs), and Floating Production Storage and Offloading Units (FPSOs), including a vessel stripped of equipment or being towed; (4) any sea-going vessel and seaborne craft of any type whatsoever constructed or adapting for the carriage of oil in bulk as cargo, provided that a ship capable of carrying oil and other cargoes shall be regarded as a ship only when it is actually carrying oil in bulk as cargo and during any voyage following such carriage unless it is proved that it has no residues of such carriage of oil in bulk aboard; (5) any seagoing vessel and seaborne craft, of any type whatsoever; (6) any self-propelled sea-going vessel used in international seaborne trade for the transport of goods, passengers, or both.

Ship 船舶 对 2009 年香港国际安全与环境无害化拆船公约而言,系指在海洋环境中正在服役或已在服役的任何类型的船舶,包括潜水船、浮式船艇、浮式平台、自升式平台、浮式储存装置和浮式生产储存和卸载装置,包括被拆除船上设备的船舶或被拖船舶。 For the Hong Kong International Convention for the Safe and Environmentally Sound Recycling of Ships, 2009, ship means a vessel of any type whatsoever operating or having operated in the martine environment and includes submersibles, floating craft, floating platforms, self elevating platforms, Floating Swtorage Units (FSUs), and Floating Production Storage and Offshore Units (FPSOs), including a vessel stripped of equipment or being towed.

ship breadth B (m) 船宽 B 系指在船舶的最宽处,由一舷的肋骨外缘量至另一舷的肋骨外缘之间的水平距离。 Breadth of ship B is the horizontal distance measured over the main frames at the widest part of the ship.

Ship breaking/Ship demolition/Ship recycling 拆船 系指在拆船厂内进行的旨在回收部件和材料再加工和再利用,并妥善处理有害物质和其他材料的船舶全部或部分拆除活动,包括与此相关的操作,如现场储存和处理部件和材料,但不包括在其他各拆船厂内进一步加工或处理。

Ship charter 船舶租赁

Ship constructed 在建船舶 系指安放龙骨或处于类似建造阶段的船舶。船舶改建为化学品液货船时,不论其建造日期为何时,开始建造的日期应视作化学品液货船的建造日期。该改建规定不适合用于符合下列全部条件的船舶改装:(1)1986 年 7 月 1 日前建造的船舶;(2)船舶已核准根据《装化学品规则》仅载运由该规则确定的只具有污染危害物质的货品。 Ship constructed means a ship the keel of which is laid or which is at a similar stage of construction. A ship converted to a chemical tanker, irrespective of the date of construction, shall be treated as a chemical tanker constructed on the date on which such conversion commenced. This conversion provision shall not apply to the modification of a ship, which complies with all of the following conditions: (1) the ship is constructed before 1 July 1986; (2) the ship is certified under the Bulk Chemical Code to carry only those products identified by the Code as substances with pollution hazards only.

Ship delivered after 31 December 1979 在 1979 年 12 月 31 日以后交船的船舶 系指:(1)在 1975 年 12 月 31 日以后签订建造合同的船舶。(2)无建造合同,在 1976 年 6 月 30 日以后安放龙骨或处于类似建造阶段的船舶。(3)在 1979 年 12 月 31 日以后交船的油船。(4)经重大改装的船舶:①在 1975 年 12 月 31 日以后签订改建合同;②无改建合同,在 1976 年 6 月 30 日以后改建工程开工;③在 1979 年 12 月 31 日以后改建工程完成。 Ship delivered after 31 December 1979 means a ship: (1) for which the building contract is placed after 31 December 1975; (2) in the absence of a building contract, the keel of which is laid or which is at a similar stage of construction after 30 June 1976; (3) the delivery of which is after 31 December 1979; (4) which has undergone a major conversion: ①for which the contract is placed after 31 December 1975; ②in the absence of a contract, the construction work of which is begun after 30 June 1976; ③which is completed after 31 December 1979.

Ship delivered on or after 1 August 2010 在 2010 年 8 月 1 日或以后交船的船舶 系指:(1)在 2007 年 8 月 1 日或以后签订建造合同的船舶。(2)无建造合同,在 2008 年 2 月 1 日或以后安放龙骨或处于类似建造阶段的船舶。(3)在 2010 年 8 月 1 日或以后交船的船舶。(4)经重大改装的船舶:①在 2007 年 8 月 1 日后签订改建合同;②无改建合同,在 2008 年 2 月 1 日后改建工程开工;③在 2010 年 8 月 1 日后改建工程完成。 Ship delivered on or after 1 August 2010 means a ship: (1) for which the building contract is placed on or after 1 August 2007. (2) in the absence of a building contract, the keels of which are laid or which are at a similar stage of construction on or after 1 February 2008. (3) the delivery of which is on or after 1 August 2010. (4) which have undergone a major conversion: ①for which the contract is placed after 1 August 2007; or ②in the absence of contract, the construction work of which is begun after 1 February 2008; or ③which is completed after 1 August 2010.

Ship delivered on or before 31 December 1979 在 1979 年 12 月 31 日或以前交船的船舶 系指:(1)在 1975 年 12 月 31 日或以前签订建造合同的船舶。(2)无建造合同,在 1976 年 6 月 30 日或以前安放龙骨或处于类似建造阶段的船舶。(3)在 1979 年 12 月 31 日或以前交船的船舶。(4)经重大改装的船舶:①在 1975 年 12

月 31 日或以前签订改建合同;②无改建合同,在 1976 年 6 月 30 日或以前改建工程开工;③在 1979 年 12 月 31 日或以前改建工程完成。 Ship delivered on or before 31 December 1979 means a ship:(1) for which the building contract is placed on or before 31 December 1975.(2) in the absence of a building contract, the keel of which is laid or which is at a similar stage of construction on or before 30 June 1976.(3) the delivery of which is on or before 31 December 1979.(4) which has undergone a major conversion:①for which the contract is placed on or before 31 December 1975; or ② in the absence of a contract, the construction work of which is begun on or before 30 June 1976; or ③which is completed on or before 31 December 1979.

Ship dimensions 船舶主尺度 包括:(1)主要用于船舶安全操纵和避让、离靠码头和通过桥梁的船舶最大尺度;(2)记录在船舶丈量证书上作为船舶管理重要资料的船舶登记尺度;(3)与船舶的主要航海性能有关的船型主尺度。

Ship earth station(SES) 船舶地面站

Ship electric propulsion system 船舶电力推进装置 一般系指采用电动机械带动螺旋桨来推动船舶运动的装置。电力推进船舶采用中高速柴油机,相比传统船舶使用的低速柴油机体积小、噪声低,同时柴油机在额定转速稳定运行,效率高、使用寿命长。由电机调速实现船舶速度与航向控制。

Ship emergency response service(SERS) 船舶应急响应服务 根据船东选择,预先建立船舶稳性与结构强度有关的电子数据库,当船舶遭遇海上碰撞、搁浅、溢油等事故后,提供快速的稳性、强度与溢油量评估,提出应急处理建议,协助船舶脱离危险。

Ship energy efficiency 船舶能效 IMO 主要围绕船舶能效设计指数(EEDI)和船舶能效计划(SEEMP)等相关导则,解决应用问题。MEPC 65 排除了包括平台、驳船和钻井装置等非机械方式推进的船舶以及破冰船适用 EEDI 的要求;完成了适用于 Phase 0 阶段的"2013 年恶劣海况下维持船舶操纵性最小推进功率导则"草案;通过了"用于 EEDI 的基准线计算导则"的修正案,内容涉及车辆滚装船、滚装货船、客滚船的 EEDI 计算方法及要求,通过了"2013 年具有非传统推进的游乐型客船的 EEDI 基准线计算导则",内容涉及专门用于采用非传统推进系统的游乐型客船的定义,基准线导则及计算的基准线、EEDI、基准线的折减率等。

Ship energy efficiency management plan 船舶能效管理计划 简称 SEEMP。

Ship energy efficiency management system(SEEMS) 船舶能效管理系统 该系统是一项针对 SEEMP 的集成设计,覆盖船型优化、航速优化、航线优化、资源管理和船队管理等方面。

Ship equipment 船舶设备 系指携带或安装在船上的物资装运设备。 Ship equipment means materials handling equipments carried or installed on board a ship.

Ship flooding valve 沉船阀 军用舰船仅在特殊情况下使用的,可将舷外水大量放入舱内的阀。

Ship form 船型 就法定要求而言,系指 SOLAS 公约所定义的船舶类型,就船级要求而言,系指船级社入级规范所给出的船型附加标志所对应的船舶类型。

Ship furniture 船用家具 供船上人员学习、工作和日常生活使用,与陆上家具相比,通常具有轻巧、不易滑倒并与船体相配合等船用特点的床、柜、椅、箱架等器具的统称。

Ship having an unusual large freeboard 非常规富裕干舷船舶 系指实际夏季干舷(相应于指定最大载重线的干舷)大于下列方法指定的最小夏季干舷与上层建筑标准高度(参考 1966 年国际载重线公约第 33 条)之和的船舶。即,此处的最小夏季干舷系指,从某船舶的干舷甲板向下取与上述上层建筑高度相等的距离,假定该处有甲板,将该设定甲板视作干舷甲板,按上述公约计算所得的干舷。 Ship having an unusual large freeboard is the ships having a actual summer freeboard (corresponding to the assigned load line) which is not less than the sum of minimum summer freeboard and standard height of superstructure (See Regulation 33 of "1966 International Convention on Load Line").

Ship hull vibration 船体振动

Ship insurance 船舶保险 是以各种类型船舶为保险标的,承保其在海上航行或者在港内停泊时遭到的因自然灾害和意外事故所造成的全部或部分损失及可能引起的责任赔偿。狭义的船舶保险系指船舶营运险,可分为基本险、附加险和特殊附加险三种。基本险包括全损险和一切险;附加险必须在投保全损险或一切险后才能投保。广义的船舶保险系指船舶所处状态而定的,包括船舶营运险、船舶建造险、船舶停航险、船舶修理险、拆船保险和集装箱保险等。

Ship loading calculation 船舶装载计算 系指为确定船舶的质量(即排水量)和重心位置,并利用静水力数据确定船舶的吃水和横倾的计算。在装载计算中一般不包括横倾计算。但对于某些船型,如设置边克令吊的船舶,必须进行横倾计算。

Ship machinery 船用设备(船用机械)

Ship magnetic bearing survey 船磁方位调查 选择平静的磁区,以船在 0°、225°、90°、315°、180°、45°、270°、135°,8 个方位通过无磁性浮标穿越测量、而对船

舶磁场进行相应的方位改正。

Ship maneuvering simulator 船舶操纵模拟器 用计算机控制的固定操纵台和活动操纵台,对船舶操纵和运动做实时模拟的装置。

Ship mechanical ventilation 船舶机械通风 由机械动力将舱外新鲜空气送入舱内或将舱内污浊空气抽出排入大气的通风方式。

Ship model experiment tank 船模试验水池 供进行船舶模型试验的水池。

Ship model towing tank 船模拖曳试验水池 可在其中拖曳船模进行试验的船模试验水池。

Ship mortgage act 船舶抵押法

Ship motions in waves 船舶摇荡运动 船舶由于风浪等外力作用所产生的纵摇、横摇、艏摇、纵荡、横荡、垂荡及其耦合运动。

Ship name flag 船名旗 由四面字母旗组成一串,用以表示船名的号旗。

Ship natural ventilation 船舶自然通风 利用外界空气—天然风对船舶的压力—正压或负压和舱内外温度差异所产生的重力作用,促使舱内空气获得更换的通风形式。

Ship navigating system at sea 船舶海上航行导航系统 系指能够在对海上气象条件进行分析计算的基础上,自动选择所消耗燃料最小的航线,从而达到减少温室气体排放目的的系统。

Ship noise measurement 船舶噪声测量

Ship of a similar type 相似类型船舶 系指不包括附加船体特征,例如鳍板的船型(以型线表示,例如型线侧视图和横剖型线图)和主要参数与基础船舶大部分相同的船舶。 Ship of a similar type Ship of a similar type means a ship of which hull form (expressed in the lines such as sheer plan and body plan) excluding additional hull features such as fins and of which principal particulars are largely identical to that of the base ship.

Ship oil pollution emergency plan (SOPEP) 油污应急计划 对150总吨以上的油船及400总吨以上的其他船舶要求制订的、包含油污事故报告程序、紧急联系当局及人员名单、为减少溢油及控制溢油事故应采取的行动说明、与当地政府进行协调的程序等文件。

Ship optimization 船舶优化 就EEDI而言,系指船舶为了满足某阶段EEDI参考线值的要求而采取的优化措施。如在第一阶段可采取的优化措施有:(1)配置轴带发电机,提高能源利用效率;(2)在螺旋桨后安装反转的小型螺旋桨,提高推进效率;(3)船底使用气泡润滑,降低船舶阻力;(4)优化线型,发掘船型快速性的潜能;(5)空船结构质量控制与优化,提升载重量。在第二阶段可采取的优化措施有:(1)安装废热回收装置,能够有效降低辅机功率消耗,EEDI降低10%左右;(2)使用天然气清洁燃料,由于天然气燃料碳含量较低,且热值较高,不仅能够降低C_F(碳转换系数),还可以降低油耗。因此,使用天然气能降低EEDI约20%~30%;(3)降速航行,因为航速与功率呈三次方关系,所以为得到较高航速需要付出更高的功率代价。因此,一定程度降低航速,不仅可以降低EEDI,而且能够有效节约燃料,提高船舶运营的经济性;(4)应用新型复合材料,降低空船质量;(5)利用风能和太阳能等清洁能源。

Ship performance monitoring system 船舶性能检测系统

Ship recycling 拆船 系指在拆船厂内进行的旨在回收部件和材料供再加工和再利用,并妥善处理有害物质和其他材料的船舶全部或部分拆解活动,包括与此相关的操作,如现场存放和处理部件和材料,但不包括其在其他拆船厂内的进一步加工或处置。 Ship recycling means the activity of complete or partial dismantling of a ship at a ship recycling facility in order to recover components and materials for reprocessing and re-use, whilst taking care of hazardous and other materials, and includes associated operations such as storage and treatment of components and materials on site, but do not include their further processing or disposal in separate facilities.

Ship recycling facility 拆船厂 系指用于拆船的特定区域,包括场地、船厂和设施。 Ship recycling facility means a defined area that is a site, yard or facility used for the recycling of ships.

Ship recycling facility(ies) plan (SRFP) 拆船厂计划 系指经缔约国授权的拆船厂制定的一份计划。该计划应由拆船公司的董事会或相应管理部门通过,并应参照IMO组织制定的指南,包括如下内容:(1)用于确保工人安全及保护人员健康和环境方针,包括制定各项目标以最大限度减少及在切实可行的范围内消除拆船对人员健康和环境造成的不利影响;(2)用于确保实施2009年香港国际安全与环境无害化拆船公约规定的要求,实现拆船公司方针中确定的目标和持续改进用于拆船作业的程序和标准的系统;(3)确定雇主和工人在进行拆船作业时的任务和职责;(4)为拆船厂安全和环境无害化作业提供相应信息和工人培训的程序;(5)应急部署和响应计划;(6)用于执行拆船的监控系统;(7)用于显示拆船如何进行的记录保持系统;(8)用于报告有关危害或可能危害工人安全、人员健康和环境的排放、排泄、事件和事故的系统;和(9)用于报告职业疾病、事故、伤害及气体对工人安全和人员健康的不利影响的系统。 Ship recycling facility(ies) plan means a plan developed by the ship recycling facility (ies) authorized by a Party. The plan shall be adopted by the board or the appropriate governing body of the recycling company, and shall include: (1) a policy ensuring workers' safety and the protection of human health and the environment, including the establishment of objectives that lead to the minimization and elimination to the practicable extent of the adverse effects on human health and the environment caused by ship recycling; (2) a system ensuring implementation of the requirement set out in the Hong Kong International Convention for the safe and environmentally sound recycling ship, 2009, to achieve the goals set out in the policy of the recycling company, and to continuously improve of the procedures and standards used in the ship recycling operations; (3) identification of roles and responsibilitilies for employers and workers when conducting ship recycling operations; (4) a programme for providing appropriate information and training of workers for the safe and environmentally sound operation of the ship recycling facility; (5) an emergency preparedness and response plan; (6) a system for monitoring the performance of ship recycling; (7) a record-keeping system showing how ship recycling is carried out; (8) a system for reporting discharges, emissions, incidents and accidents causing damage, or with the potential of causing damage, to workers' safety, human health and the environment; and (9) a system for reporting occupational diseases, accidents, injuries and other adverse effects on workers' safety and human health.

Ship recycling plan (SRP) 拆船计划 系指在任何拆船活动开始前,拆船厂参照IMO组织制定的指南制定的一份船舶特定的计划。拆船计划应:(1)参照船东提供的信息制定;(2)使用授权拆船厂的缔约国所接受的语音制定;如果使用的文字非英文、法文或西班牙文,该拆船计划应有其中一种文字的译文,但主管机关确信无此必要时除外;(3)包括建立、保持和监控进入安全和热作安全条件及如何对包括有害物质清单所列材料在内的材料类型和数量进行管理的有关信息;(4)根据2009年香港国际安全与环境无害化拆船公约正文第16.6条文的声明,经对拆船厂进行授权的主管当局明确批准或默认批准。该主管当局应按第24条在收到拆船计划的3个工作日内将其书面回执发送至拆船厂、船东和主管机关,此后:①如缔约国要求对拆船计划的明确批准,主管当局应将其对拆船计划的批准或拒绝的决定以书面通知发送到拆船厂、船东和主管机关;②如缔约国要求对拆船计划的默认批准,回执中应写明14天评审期的结束日期,主管当局应在此14天评审期内将其对拆船计划的任何书面反对通知拆船厂、船东或主管机关;如未通知此类书面反对,则拆船计划将视为予以批准;(5)一旦经按上述(4)的要求予以批准,提供给主管机关或任何经其指定的验船师或认可的组织供其检查时使用;(6)如使用于一个以上拆船厂,列出拟使用的拆船厂并说明每一经授权的拆船厂的具体拆船活动及出现的顺序。 Ship recycling plan means a plan developed by the ship recycling facility (ies) prior to any recycling of a ship, taking into accout the guidelines developed by the IMO. The ship recycling plan shall: (1) be developed taking into account information provided by the ship owner; (2) be developed in the language accepted by the Party authorizing the ship recycling facility, and if the language used is not English, French or Spanish, the ship recycling plan shall be translated into one of these languages, except the Adiministration is satisfied that the translation is not necessary; (3) include information concerning, inter alia, the establishment, maintenance, and mornitoring of safe-for-entry and safe-for-hot work conditions and how the type and amount of materials including those identified in the inventory of Hazardous material will be managed; (4) in accordance with the declaration deposited pursuant to Article 16.6 in the Hong Kong International Convention for the safe and environmentally sound recyding ship, 2009 be either explicitly or tacitly approved by the Compent Authority authorizing the ship recycling facility. The Competent Authority shall send written acknowledgement of receipt of the ship recycling plan to the ship recycling facility, ship owner and Adiministration within three working days of its recept in accordance with Regulation 24. Thereafter: ①where a Party requires explicit approval of the ship recycling plan, the Competent Authority shall send written notification of its decision to approve or deny tha ship recycling plan to the ship recycling facility, ship owner and Adiministration; ②where a Party requires tacit approval of the ship recycling plan, the acknowledgment of recept shall specify the end date of a 14-day review period. The Competent Authority shall notify any written objection to the ship recycling plan to the ship recycling facility, ship owner and Adiministration within 14-day review period. Where no such written objection has been notified, the ship recycling plan shall be deemed to be approved; (5) once approved in accordance with (4), be made available for inspection by the Adiministration, or any nominated surveyors or organization recognized by it; (6) where more than one ship recycling facility is ued, identify the ship recycling facilities to be used and specify the recycling activities and the order in which they occur at esch authorized ship recycling facility.

Ship relative motions 船舶相对运动 系指海面相对船舷的垂向振荡运动。其数值及正负符号由吃水 T_1 的水线面量取。 Ship relative motions are the vertical oscillating translations of the sea waterline on the ship side. They are measured, with their sign, from the waterline at draught T_1.

Ship repair and maintenance system (CWBT) 船舶维修保养体系(CWBT) 是将传统的船舶设备管理和国际上插卡式船舶设备管理相结合,形成集计划、管理、指导于一体的一种新颖、科学、实用的船舶设备管理模式,简称为 CWBT。它是由船舶(Chuanbo)、维修(Weixiu)、保养(Baoyang)、体系(Tixi)4 个词的汉语拼音按各词首字母排列而成。 Ship repair and maintenance system (CWBT) combines the traditional onboard equipment management with the internationally used card-inserted type management to form a new, scientific and practical mode of onboard equipment management, including planning, management and guidance. Its abbreviation CWBT consists of the initials of Chinese phonetic letters for the four Chinese words Chuanbo(ship), Weixiu(repair), Baoyang(maintenance) and Tixi(system).

Ship repair contract 船舶修理合同 系指船东与船厂签订的船舶修理合同。

Ship safety construction certificate 货船构造安全证书 系指对 500GT 及以上的货船,经检验满足《1974 年国际海上人命安全公约》第Ⅰ/10 条所述关于货船检验的要求,并符合关于灭火设备和防火控制图等要求以外的第Ⅱ-1 章和第Ⅱ-2 章适用的要求的货船,由主管机关及其认可的组织签发的货船构造安全证书。

Ship safety equipment certificate 货船设备安全证书 系指对 500GT 及以上的货船,经检验满足《1974 年国际海上人命安全公约》第Ⅱ-1 章和第Ⅱ-2 章的有关要求以及任何其他有关要求的货船,由主管机关及其认可的组织签发的货船设备安全证书。

Ship safety management system (SSAMS) 船舶安全管理体系 系指能使公司人员有效实施公司的安全及环境保护方针所建立的和文件化的体系。

Ship security 船舶保安 是当指定的船舶及其上人员、货物、设备和操作得到保护以防止非法行为和恐怖主义的危害并防止损失时所达到的状况。 Ship security means the status reached when the designed ship and personnel, cargoes, equipment and operations onboard the ship get protected to prevent from the illegal activities and acts of terrorism.

Ship security alarm system (SSAS) 船舶保安报警系统 系指 SOLAS XI-2 章第 5 条所要求的设备。

Ship security assessment (SSA) 船舶保安评估 是船舶保安计划制定和更新过程的基本组成部分,应至少包括下列要素:(1)确定现有保安措施、程序、操作;(2)确定并评价应予重点保护的船上关键操作;(3)确定船上关键操作可能受到的威胁及其发生的可能性,以确定并按优先顺序排定保安措施;和(4)确定基础措施、方针和程序中的弱点,包括人为因素。 Ship security assessment is the basic part to develop and renew ship security plan. It is at least to include the following elements:(1) determining the existing security measures, procedures and operations;(2) determining and evaluating key onboard operations to be wemphatically protected;(3) determining the possibility of key onboard operations to be threatened and the likelihood of happening, and establish the security arrangement according to the priority (4) determining the vulnerabilities in basic facilities, policy and procedures, including human elements.

Ship security identification 船舶保安威胁辨识 系指辨识海上保安威胁的存在并确定其特性(可包括实施保安破坏、制造保安事件的行为模式、逃避保安措施的方法,攻击目标等)的过程。 Ship security identification means process of identifying the existence of maritime security threat and determining its characteristics (including action modes of security tampering, security event, method of avoiding security incident, attacking object, etc.).

Ship security officer (SSO) 船舶保安员 系指由公司指定在船上负责船舶保安并对船长负责的人员,其职责包括实施和维护船舶保安计划以及与公司保安员和港口设施保安员进行联系。 Ship security officer means the person on board the ship, accountable to the master, designated by the Company as responsible for the security of the ship including implementation and maintenance of the ship security plan and liaison with the Company security officer and port facility security officers.

Ship security plan (SSP) 船舶保安计划 系指为确保在船上采取旨在保护人员、货物、货物运输单元、船舶物料及船舶免受保安事件威胁的措施而制订的计划。 Ship security plan means a plan developed to ensure the application of measures on board the ship designed to protect persons on board, cargo, cargo transport units, ship's stores or the ship from the risks of a security incident.

Ship security plan and associated records 船舶保安计划和相关记录 系指每艘船舶随船携带经主管机关批准的船舶保安计划。该计划应就 ISPS 规则 A 部分所定义的一个保安等级做出规定。船舶保安计划所涉及的以下活动的记录应至少按主管机关规定的最低期限保存在船上；（1）培训、演习和训练；（2）保安状况受到的威胁和保安事件；（3）保安事件受到破坏；（4）保安等级的改变；（5）与船舶直接保安状况（例如对船舶或对船舶所停留或曾经停留的港口设施的具体威胁）有关的通信；（6）保安活动的内部审核和评审；（7）对船舶保安评估的定期评估；（8）对船舶保安计划的定期评估；（9）对保安计划任何修正的实施；（10）船上保安设备的保养、校准和测试，包括对船舶保安警报系统的测试。 Ship security plan and associated records means that each ship shall carry on board a ship security plan approved by the Administration. The plan shall make provisions for the three security levels as defined in part A of the ISPS Code. Records of the following activities addressed in the ship security plan shall be kept on board for at least the minimum period specified by the Administration: (1) training, drills and exercises; (2) security threats and security incidents; (3) breaches of security; (4) changes in security level; (5) communications relating to the direct security of the ship such as specific threats to the ship or to port facilities the ship is, or has been under; (6) internal audits and reviews of security activities; (7) periodic review of the ship security assessment; (8) periodic review of the ship security plan; (9) implementation of any amendments to the plan; (10) maintenance, calibration and testing of any security equipment provided on board, including testing of the ship security alert system.

Ship security system 船舶保安体系 系指为验证ISPS规则符合性而被检查的船上实施的程序、文件和有关记录的体系。

Ship service power unit 船用电站

Ship sounding 船舶测深

Ship stabilizer (ship-stabilizing gear) 船舶减摇装置 利用重力、陀螺力或水动力等形成减摇力矩以减少船舶摇摆的装置。

Ship stores 船舶备品 系指船用物品和零配件的总称。

Ship structure access 船舶结构通道 系指船上用于全面检查、近观检查和厚度测量的通道。 Ship structure access is ship's means of access to carry out overall and close-up inspections and thickness measurements.

Ship structure access manual 船舶结构通道手册 系指包括每一处所下述资料的手册：（1）该处所出入通道图，并有相应的技术说明和尺寸；（2）每一处所内能进行全面检查的通道图，并有相应的技术说明和尺寸。图中应标示该处所内每一区域可从何处检查；（3）该处所内能进行近观检查的通道图，并有相应的技术说明和尺寸。图中应标示临界结构区域的位置，是否为永久通道或是便携式通道，以及每一区域可以从何处检查；（4）检查和维护保养所有出入通道和附属设备结构强度的须知，其中应考虑处所内任何腐蚀气体的影响；（5）当使用筏进行近观检查和厚度测量时，应有安全指导须知；（6）任何便携式通道安全安装和使用须知；（7）一份所有便携式通道的清单；（8）船上通道定期检查和维护保养的记录。船舶结构通道手册应经主管机关批准，一份最新版本的副本应保存在船上。 Ship structure access manual shall include the following for each space: (1) plans showing the means of access to the space, with appropriate technical specifications and dimensions; (2) plans showing the means of access within each space to enable an overall inspection to be carried out, with appropriate technical specifications and dimensions. The plans shall indicate from where each area in the space can be inspected; (3) plans showing the means of access within the space to enable close-up inspections to be carried out, with appropriate technical specifications and dimensions. The plans shall indicate the positions of critical structural areas, whether the means of access is permanent or portable and from where each area can be inspected; (4) instructions for inspecting and maintaining the structural strength of all means of access and means of attachment, taking into account any corrosive atmosphere that may be within the space; (5) instructions for safety guidance when rafting is used for close-up inspections and thickness measurements; (6) instructions for the rigging and use of any portable means of access in a safe manner; (7) an inventory of all portable means of access; (8) records of periodical inspections and maintenance of the ship's means of access. A ship structure access manual is to be approved by the Administration, an updated copy of which shall be kept on board.

Ship survival capability and location of cargo tanks 船舶残存能力和液货舱的位置 系指适用于散装运输"国际散装运输危险化学品船舶构造和设备规则（IBC规则）"危险化学品的船舶，应能承受在某种外力作用下船体遭受假定破损后浸水的正常影响。此外，为了保护船舶和环境，对某种类型船舶的液货舱应加以保护，以防其因船舶与例如码头或拖船接触后产生较小破损而引起渗漏，并且应采取保护措施以防其因船舶碰撞或触礁而引起破损，即把液货舱布置在船内距船体外板不小于规定的最小距离之处。对于所假定的破损以及液货舱与船体外板之间的距离，均取决于所装货品的危险程度。 Ship survival capability and location of cargo tanks means that ships, subject to the Construction and Equipment of Ships Carrying Dangerous Chemicals in Bulk (IBC) Code, shall survive the normal effects of flooding following assumed hull damage caused by some external force. In addition, to safeguard the ship and the environment, the cargo tanks of certain types of ships shall be protected from penetration in the case of minor damage to the ship resulting, for example, from contact with a jetty or tug, and given a measure of protection from damage in the case of collision or stranding, by locating them at specified minimum distances inboard from the ship's shell plating. Both the assumed damage and the proximity of the cargo tanks to the ship's shell shall be dependent upon the degree of hazard presented by the products to be carried.

Ship tank liquid sloshing 船舶液舱晃荡 系指液货船（VLCC，LNG，LPG）在波浪中航行时，舱内的液体产生的晃荡现象。常见的液舱晃荡现象有：驻波、行进波、水跃和组合波，以及漩涡飞溅等非线性现象。强烈的晃荡直接导致严重的海损污染事件。

Ship telegraph 机舱传令钟

Ship to ship (STS) activity 船到船活动 系指涉及物品或人员从一船到另一船转移的任何与港口设施无关的活动。 Ship to ship (STS) activity means any activity not related to a port facility that involves the transfer of goods or persons from one ship to another.

Ship type development 船型开发

Ship vapour connection 船舶蒸汽接头 系指船上固定的蒸汽收集系统与一套设备或另一艘船的收集系统之间的连接点。用于此目的的船上软管或装卸臂，均视为船舶蒸汽控制系统的一部分。 Ship vapour connection is the point of interface between the ship's fixed vapour collection system and the collection system of a facility or another ship. Hoses or loading arms on board, carried for the purpose of these Rules, are considered part of the vapour control system of the ship.

Ship ventilation 船舶通风 利用自然通风或机械通风方式将舱外新鲜空气输送到舱内工作场所、居住舱室或货舱内，以改善船员及旅客的生活、工作条件及待运货品的质量的技术。

Ship vibration 船舶总振动

Ship wave systems 船波 船舶由于航行而在水面形成的波系。有散波和横波。

Ship with capability of icebreaking ice 具有破冰能力船舶 系指对于具有航行冰区的加强要求，且航行于当年结冰水域，具有独立破冰能力的非破冰专用船舶。 Ship with capability of icebreaking ice means ships not specially designed for icebreaking duties and navigating in first-year ice conditions, complying with the requirements for ice strengthening and having independent icebreaking capability.

Ship with China registry 中国籍船舶 系指在中华人民共和国登记或将登记的船舶。

Ship with CNG fuel system 压缩天然气动力装置的船舶 系指配备压缩天然气系统，以压缩天然气作为主机燃料的船舶。

Ship with dual fuel system 双燃料动力装置船舶 系指配备双燃料系统，以天然气和柴油作为主机燃料的船舶。

Ship with LNG fuel system 液化天然气动力装置船舶 系指配备液化天然气系统，以液化天然气作为主机燃料的船舶。

Ship with LPG fuel system 液化石油气动力装置船舶 系指配备液化石油气系统，以液化石油气作为主机燃料的船舶。

Ship with non-China registry 外国籍船舶 系指非中国籍船舶。

Ship with several decks 多层甲板 系指干舷甲板之上有多层甲板的船舶，例如客船。 Ship with several decks means a ship having several decks above the bulkhead deck, such as passenger ships.

Ship year 船年 是基于劳氏船级社提供的船舶报告计算得出的，即将全年服役的每一艘船舶记为1个船年，而当年交付的新船则按服役的日期数计算船年，例如当年7月1日交付的新船记为0.5船年，报废的船舶按照同样的方法计算在役期间的船年，将统计的年份内每年的船年合计为总船年数，作为事故频率的计算基数。

Ship (the control of bio-fouling) 船舶（生物污底控制） 系指在水生环境中营运的任何类型的船舶，包括水翼艇、气垫船、潜水艇、水上艇筏、固定或浮动平台、浮式储存装置（FSU）以及浮式生产储存和卸货装置（FPSO）。 Ship means a vessel of any type whatsoever operating in the aquatic environment and includes hydrofoil boats, air-cushion vehicles, submersibles, floating craft, fixed or floating platforms, floating storage units (FSUs) and floating production storage and off-loading units (FPSOs).

Ship/port interface 船/港界面活动 系指当船舶受到涉及船舶与港口之间人员、物品移动或港口服务提供等行为直接和密切影响时所发生的互交活动。 Ship/port interface means the interactions that occur when a ship is directly and immediately affected by actions involving the movement of persons, goods or the provisions of port services to or from the ship.

Ship/vessel 船 航行或停泊于水域的运载、作战、作业的工具。是各种船、舰、艇、舢板、筏、以及水上浮动作业平台等的总称。

Ship's control signal device 船舶操纵信号设备 是供船舶驾驶员了解各种航行机械运行情况，得以正确指挥并进行操纵，提高船舶操纵性能，实现船舶正常安全航行的重要设备。

Ship's electric power system 船舶电力系统 是由船舶电源装置、配电装置、船舶电网和电力负荷按一定方式连接的整体，是船上电能产生、传输、分配和消耗等全部装置和网络的总称。

Ship's energy efficiency management certificate (SEEMC) 船舶能效管理证书 系指根据中国船级社"船舶能效管理认证规范"要求实施能效管理体系的船舶，经船舶现场审核，认为满足规范要求而签发的证书（自愿）。

Ship's equipment 船舶设备 系指除船舶备件以外的任何用于船上的可移动但不带消费性质的物品，包括诸如救生艇、救生用具、家具、船舶舾装及其他类似物品等附属物。 Ship's equipment means those articles, other than ship's spare parts, on board a ship for use thereon, which are removable but not of a consumable nature, including accessories such as lifeboat, life-saving device, furniture, ship's apparel and similar items.

Ship's form, hull form 船体型线 船体型表面的剖切线和投影线所确定的船体外型。

Ship's loading and machinery loading 船舶负载和机舱负荷 系指船体、甲板、航行和安全负载、推进和辅机负荷、机舱通风和辅助设备和船舶的一般负载所需的全部负荷组。

Ship's maintenance system 船舶维修保养体系 是将传统的船舶设备管理和国际上插卡式船舶设备管理相结合，形成集计划、管理、指导于一体的一种新颖、科学、实用的船舶设备管理模式。

Ship's meteorograph 船舶气象仪 主要用来测量风向、风速、温度、相对湿度、大气压力等气象参数的仪器。它通过键盘输入航向、航速值，经计算机处理后，能显示真风向、真风速。还具有20 min的平均相对风向、平均相对风速。

Ship's parameter 船舶要素 系指包括船舶主尺度、总吨位、载重吨、吃水、乘客人数等决定标准应用的主要船舶参数。

Ship's position keeping 船位保持 在控制系统正常的操作范围和环境条件下维持想要的船位。

Ship's primary movement 船舶的主要运动 系指船舶的纵向、横向和艏向回转运动。Ship's primary movement means the longitudinal directional, lateral directional and heading-rotational movement of the ship.

Ship's representative 船舶代表 系指船长或负责接收燃油和做记录的主管高级船员。 Ship's representative is the ship's master or officer in charge who is responsible for receiving bunkers and documentation.

Ship's significant alteration 船舶重大改装 系指现有船舶一个或多个重大特征实质性改变的修理、改造和改装。重大特征包括如下类别：(1)船舶主尺度，如新增导致船长改变的相同型宽和型深的船体平行中体、新增导致船长和/或型宽和/或型深改变的船体段；(2)船型，如货船改装成客船/客滚船、货船改装成油船、货船改装成化学品船/液化气船、货船改装成散货船、改变船型附加标志；(3)船舶分舱水平，如 $A/R_{改装后} < A/R_{改装前}$，$A/R_{改装后} < 1$；(4)船舶承载容量，增加货物载运容量、增加乘客人数；(5)乘客居住处所；(6)延长船舶营运寿命，如单壳油船改装成双壳油船、货舱和/或货物区域整体更换、更换或增加救生设备或装置、更换或增加乘客居住处所。

Ship's spare parts 船舶备品 系指船上所带的供船舶用于修理或替代的物品。 Ship's spare parts mean those articles of a repair or replacement nature for incorporation in the ship in which they are carried.

Ship's speed V_0 船速 V_0 系指总有效驱动功率全部传递到螺旋桨上时，船舶在静水中、夏季载重水线时的预期最大前进速度（kn）。 Ship's speed V_0 is expected maximum ahead speed of the ship in calm water (kn) at the summer load waterline, when the total available driving power is acting exclusively on the propeller.

Ship's stores 船用物料 系指船上使用的物品，包括消费物品，载于上卖给旅客和船员的物品、燃料和润滑油，但不包括船舶设备的备品。 Ship's stores are those goods for use in the ship, including consumable goods, goods carried for sale to passengers and crew members, fuel and lubricants, but excluding ship's equipment and ship's spare parts.

Ship's structure design 船舶结构设计 为：(1)具有内在冗余。船舶结构以分级方式工作，因而在该等级体系中处于较低级别的构件失效，不会立即导致该等级体系中较高级别构件失效；(2)将永久变形减至最低限度。可以接受局部板格或个别扶强板材的永久变形，但这不得影响到结构完整性、围护完整性或结构系统，或其他系统的性能；(3)将营运中发生破裂的事故减至最低限度，尤其在影响到结构完整性或围护完整性，影响到结构系统或其他系统的性能，或难以检查和修理的部位；(4)具有足够的结构冗余，以在结构意外受损（例如轻微碰撞导致任一舱室进水）时幸存。 Ship's structure is designed such that: (1) it has inherent redundancy. The ship's structure works in a hierarchical manner and, as such, failure of structural elements lower down in the hierarchy should not result in immediate consequential failure of elements higher up in the hierarchy; (2) permanent deformations are minimized. Permanent deformations of local panel or individual stiffened plate members may be acceptable provided that this does not affect the structural integrity, containment integrity or the performance of structural or other systems; (3) the incidence of in-service cracking is minimized, particularly locations which affect the structural integrity or containment integrity, affect the performance of structural or other systems or are difficult to inspect and repair; (4) it has adequate structural redundancy to survive in the event that the structure is accidentally damaged, for example, minor impact leading to flooding of any compartment.

Ship = borne wave recorder 随船浪高仪 安装在船上，利用激光、雷达、超声波等非接触方式或水压式，用以测量海上绝对波高的仪器。

Ship-based anti-missile system (SAMS) 舰载反导系统

Ship-based fire control system 舰载火力控制系统 系指舰船控制火炮、火炮群或导弹发射器瞄准和射击的整套设备。

Shipboard audit 船上审核 系指审核机构对船舶管理公司所管理船舶的审核。

Shipboard fittings 船用配件/船体舾装设备 系指正常系泊船舶用的系缆桩与缆柱、系缆器、立式滚轮、导缆孔和用于正常拖带船舶的类似部件。其他部件如绞盘、绞车等不包含在本节要求中。任何船舶配件与支撑结构的焊接、螺栓或其他紧固连接是船舶配件的部分，应满足该船舶配件所适用的任何工业标准。 Shipboard fittings mean bollards and bitts, fairleads, stand rollers and chocks used for the normal mooring of the ship and similar components used for the normal towing of the ship. Other components such as capstans, winches, etc. are not covered by this section. Any weld, bolt or other fastening connecting the shipboard fitting to the supporting hull structure is part of the shipboard fitting and subject to any industry standard applicable to such fitting.

Shipboard incineration 船上焚烧 系指将船舶正常作业时产生的废物或其他物质在船上进行焚烧。船上焚烧应只允许在船上焚烧炉中进行。应禁止下列物质在船上焚烧：(1) MARPOL 附则Ⅰ、Ⅱ或Ⅲ规定的货物残余物或有关的被污染的包装材料，(2)多氯联苯；(3) MARPOL 附则Ⅴ定义的含有超过微量重金属的垃圾；(4)含有卤素化合物的精炼石油产品；(5)不在船上产生的污泥和油渣；(6)废气清洗系统的残余物。禁止在船上焚烧聚氯乙烯，但在已颁发IMO型认可证书的船上的焚烧炉内焚烧除外。在船舶正常操作过程中产生的污泥和油渣的处所焚烧也可以在主、辅发电机和锅炉内进行，但在这种情况下，不能在码头、港口和河口内进行。 Shipboard incineration means the incineration of wastes or other matter on board a ship, if such wastes or other matter were generated during the normal operation of that ship. Shipboard incineration shall be allowed only in a shipboard dincinerator. Shipboard incineration of the following substances shall be prohibited: (1) residues of cargoes subject to Annex Ⅰ, Ⅱ or Ⅲ to MARPOL or related contaminated packing materials; (2) polychlorinated biphenyls (PCBs); (3) garbage, as defined by Annex Ⅴ to MARPOL, containing more traces of heavy metals; (4) relined petroleum product containing halogen compounds; (5) sewage sludge and slude oil either of which is not generated on board the ship; (6) exhaust gas cleaning system residues. Shipboard incineration of polyvinyl chlorides (PVCs) shall be prohibited, except in shipboard incinerators for which IMO Type Approval Certiftcates have been issued. Shipboard incineration of sewage sludge and sludge oil generated during normal operation of a ship may also take place in the main or auxiliary power plant or boilers, but those cases, shall not take place inside ports, harbours and estuaries.

Shipboard incinerator 船用焚烧炉 系指以焚烧为主要目的而设计的船上设备。 Shipboard incinerator means a shipboard facility designed for the primary purpose of incineration.

Shipboard installation 船载装置 系指安装在船上用于连接岸电的设备。一般包括插头/插座、岸电连接配电柜(板)、变压器、岸电接入控制屏(通常组合在主配电板上)、岸电电缆和电缆管理系统。

Shipboard machinery system 船上动力装置

Shipboard marine pollution emergency plan for noxious liquid substances 船上有毒液体物质污染应急计划 对每艘150GT及以上核准装载散装有毒液体物质的船舶而言，船上有毒液体物质污染应急计划系指以IMO 制定的"制订船上油类和/或有毒液体物质海洋污染应急计划的指南"为基础，并以船长和高级船员的工作语言书写的计划。该计划至少应包括：(1)根据IMO 编制的"制订船上油类和/或有毒液体物质海洋污染应急计划的指南"MARPOL 公约第8条和议定书Ⅰ要求的由船长或其他负责人员报告有毒液体物质污染事故所遵循的程序;(2)在发生有毒液体物质污染事故时应与之联系的当局或人员名单;(3)在事故发生后有船上人员为减少或控制排除有毒液体物质所立即采取的措施的详细说明书;(4)在处理污染时与政府和地方当局协调船上行动的程序和联络点。 For every oil tanker of 150 gross tonnage and above certified to carry noxioue liquid substances in bulk, shipboard marine pollution emergency plan for noxious liquid substances means a plan prepared based on "the Guidelines for the development of shipboard marine pollution emergency plans for oil and / or noxious liquid substances" developed by IMO, and written in the working language of the master and officers. The plan shall conssst at least of:(1)the procedure to be followed by the master or other persons having charge of the ship to report a noxious liquid substances pollution incident as required in Artcle 8 and Protocol I of the MARPOL Convention, based the Guidelines for the development of shipboard marine pollution emergency plans developed by IMO;(2)the list of authorities, or persons to be contacted in the event of a noxious liquid substances pollution incident;(3)a detailed description of the action taken immediately by persons on board to reduce or control the discharge of noxious liquid substances following the incident;(4)the procedures and point of contacted on the ship for co-ordinating shipboard action with national and local authorities in combating the pollution.

Shipboard marine pollution emergency plan (SMPEP) 船上海洋污染应急方案

Shipboard oil pollution emergency plan 船上油污染计划 系指按照 1978 年议定书修正的 1973 年防止船舶造成污染的国际公约(MARPOL73/78)的法则 I 第 26 条的规定进行编制的。该计划的目的是向船长和高级船员提供指导,说明当污染事故已发生或可能发生时将采取的步骤。该计划也包括 MEPC 第 54(32)号决议导则要求的全部信息和操作说明。附则包括本计划涉及的所有人员的姓名、电话号码等以及其他参考资料。船上油污染计划必须存放在船上方便利用的地方,并应使用英文,或如果是非英文,则使用该船船长和驾驶员的工作语言编制。该计划必须包括下列 6 项条款,第 7 项非强制性条款可根据船东的意见决定是否列入。(1)引言——该条款必须包括下列内容:①介绍的内容。该计划介绍的内容必须包括下列文字(对于在南极区域航行的船舶,该计划介绍的内容必须包括下列文字和解释:它们是符合对南极环境保护条约议定书的要求的)。②一般资料——船名、呼号、官方的编号、国际海事组织(IMO)授予的国际编号和基本特征。(2)序言——该条款应包括对该计划的目的和用途的解释以及表明船上计划与其他岸基计划之间的关系如何。(3)报告的要求。该计划的这一条款应包括与下列情况有关的信息:①报告的时间。每当事故涉及下列情况时均应进行报告:(a)由于船舶或其设备遭受到破坏或为保障船舶安全或救护海上人员而导致油类或油性混合物的排放;(b)船舶运营期间油类或油性混合物的排放超过允许的数量和油量瞬间排放率;(c)可能的排放。在确定是否排放时考虑的因素可包括但不限于,船舶的地理位置和最接近的陆地或其他航行的障碍物、气候、潮汐、海流、海况和交通的密度。在发生碰撞、搁浅、火灾、爆炸、结构失效、浸水或货物移动,或由于操舵装置、推进装置、发电系统或船上主要的导航设备的故障所引起的事故,船长应进行报告。②要求的信息。该计划的这一条款应包括负责接受和处理事故的主管机关的一份通知单。该通知单包括在最初通知单和附加通知单中提供的信息。最初通知单应尽可能多地提供信息,如适用时还应包括增补的信息。但是,该最初通知单不得延误紧急收集所有的信息。通知单要放在有提醒的地方。③联系人员:(a)该计划的本条款必须提供列出主管机关的联系人员、港口的联系人员和船舶利益的代表者名录的索引;(b)对于油类或油性混合物的实际排放或可能排放,该报告必须符合 MARPOL 议定书 I 中所规定的程序。报告应直接提交给最近港口;(c)对于南极区域,除符合上述(3)③(b)款的要求外,报告还应直接提交各个可能受到影响的南极站。(4)控制排放的步骤。该计划的本条款必须包括对提出下列方案的程序的讨论:①操作上的溅出:该计划必须提出消除溅出在甲板上的油类的步骤。该计划还应提供指导,以确保适当处理再生油类和净化材料:(a)管子泄漏:该计划必须对管子泄漏的处理提供特别指导;(b)液舱泄漏:该计划必须包括处理液舱溢流的步骤。必须提供替代的方法,例如驳运货油或燃油到空舱或淡水舱,或利用备有泵把过量的货油或燃油驳运到岸上;(c)船体泄漏:该计划必须提出由于怀疑船体泄漏引起溅出的应对步骤,包括指导采取措施,以降低液舱内的油类由于内部驳运或排至岸时产生的压头。如果不可辨别出现泄漏的具体液舱,则应提供判断泄漏部位的步骤。这些步骤必须综合考虑对船体梁的应力和船舶稳性作用的影响。②由于严重事故引起的溅出:对下列每一种严重事故,在本计划中应按这些条款各种核对用的清单或当出现特定严重事故时确保船长考虑所有相关因素的前提措施构成的单独条款来处理。核对用的清单,还必须识别出分派给各类人员的拟定的任务。查阅现有的防火控制图和花名册足以识别各类人员在下列情况时的职责:(a)搁浅;(b)火灾或爆炸;(c)碰撞;(d)船体失效;(e)严重横倾。③除上述所要求的核对用的清单和各类人员职责任务以外,本计划应包括——(a)确保人员和船舶安全的预防措施,评估该船的损坏程度和提出进一步采取的相应措施;(b)评估破舱稳性和总纵强度的资料与船级社联系以获取这些资料。本条款中的规定没有提出超过法律或规则所要求的条款以外进行破舱稳性图或计算的要求;(c)在结构大面积损坏的情况下应采取减轻负荷的措施。本计划应包括船与船之间驳运货油所采取措施方面的信息。本计划中可参考现有的公司制订的指南。在本计划中应持有这些公司在船与船之间的驳运货油作业程序的副本。如适用,本计划应提出与沿海的地方政府或港口所在地政府相协调的措施。(5)国家与地方的协调:①该计划的这一条款必须包括沿海的地方政府、当地政府或其他有关机构协助船长采取初步行动的信息。如沿海主管当局没组织协调时,则这些信息应包括协调船长对事故做出反应的指南。特殊区域的详细资料可作为本计划的附录;②对于南极区域,船东或驾驶员必须制订对在实施该船舶行动中可能出现的紧急情况采取敏捷和有效应答的行动计划;③为符合本条②款的要求,有关当局的机构可颁布一个命令,由在南极区域航行的该机构的公务船提供敏捷和有效的应答。(6)附录:①24h联络的信息和替代指定联络的方法。考虑到人员的变换、电话、电传和传真号码的变更,这些资料应经常更新。也应提供更理想的通信方法的明确指南。②下列清单,每一清单作为单独的附录:(a)负责接受和处理事故报告的沿海的当地政府官员或机构的清单;(b)定期地观察港口的官员或代理人的清单,当不可能这样做时,船舶一旦到达港口,船长必须立即索取有关当地报告程序的详细资料;(c)持股的所有投资者,例如船东和货主、保险商和银行的清单;(d)指定谁将负责通知上述的投资者,其中应列出优先获通知的投资者清单。③年度检查和变更的记录。(7)非强制性的规定。如果船东提出将本条款包括在本计划中时,则应包括下列类型的信息或可适用的其他信息:①各种图表。②应答设备或消除油类溢出组织方式。③公共事务的经验。④档案。⑤计划演练。⑥取得应答资格的人员。(8)各条款的索引。该计划必须按有关的要求编制。 Shipboard oil pollution emergency plan is written in accordance with the requirements of Regulation 26 of Annex 1 of the International Convention for the prevention of pollution from ships 1973, as modified by the protocol of 1978 relating thereto (MARPOL 73/78). The purpose of the plan is to provide guidance to the master and officers on board the ship with respect to the steps to be taken when a pollution incident has occurred or is likely to occur. The plan contains all information and operational instructions required by the guideline of Resolution MEPC, 54(32). The appendices contain names, telephone numbers, etc. of all contacts referenced in the plan, as well as other reference material. The shipboard oil pollution emergency plan must be available on board in English and in the working language of the master and the offices of the ship if it is not English. The plan must contain the following six sections. A seventh non-mandatory section may be included at the shipowner's discretion: (1) Introduction— The section must contain the following: ①introductory text. The introductory text of the plan must contain the following language (for ships operating in Antarctica, the introductory text of the plan must contain the following language and explain that they are in accordance with the Protocol on Environmental Protection to the Antarctic Treaty). ②general information—the ship's name, call sign, official number, International Maritime Organization (IMO) international number, and principal characteristics. (2) Preamble. The section must contain an explanation of the purpose and use of the plan and indicate how the shipboard plan relates to other shore-based plans. (3) Reporting requirements— this section of the plan must include information relating to the following: ①when to report: reports shall be made whenever an incident involves: (a) a discharge of oily mixture resulting from damage to the ship or its equipment, or for the purpose of securing the safety of a ship or saving life at sea; (b) a discharge of oily mixture during the operation of the ship in excess of the quantities or instantaneous rate permitted; (c) a probable discharge. Factors to be considered in determining whether a discharge is probable include, but are not limited to, ship location and proximity to land or other navigational hazards, such as, weather, tide, current, sea state, and traffic density. The master must make a report in cases of collision, grounding, fire, explosion, structural failure, flooding or cargo shifting, or an incident resulting in failure or breakdown of steering gear, propulsion, electrical generating system, or essential shipboard navigational aids. ②information required. This section of the plan must include a notification form that contains information to be provided the initial and follow-up notifications for Administration who accepts and takes care of the incident. The initial notification should include as much of the information on the form as possible, and iuclude supplemental information as appropriate. However, the initial notification must delay pending collection of all information. Copies of the form must be placed at the location(s) on the ship from which notification may be made. ③whom to contact: (a) this section of the plan must make reference to the appendices listing coastal state contacts, port contacts, and ship interest contacts; (b) for actual or probable discharges of oil, or oily mixtures the reports must comply with procedures described in MARPOL Protocol 1. The reports shall be directly submitted to the nearest port; (c) for Antarctica, in addition to compliance with Paragraph (3)③(b) of the section, reports shall also be directly submitted to any Antarctic station that may be affected. (4) steps to control a discharge. This section of the plan must contain a discharge procedures to address the following scenarios: ①operational spills. This plan must outline procedures for removal of oil spilled and contained on deck, the plan must also provide guidance to ensure proper disposal of recovered oil and clean-up materials; (a) pipe leakage, the plan must provide specific guidance for dealing with pipe leakage; (b) tank overflow. The plan must include procedures for dealing with tank overflows. It must provide alternatives such as transferring cargo or bunkers to empty or slack tanks, or readying pumps to transfer the excess ashore; (c) hull leakage. The plan must outline procedures for responding to spills due to suspected hull leakage, including guidance on measures to be taken to reduce the head of oil in the tank involved either by internal transfer or discharge ashore. Procedures to handling to situations where it is not possible to identify the specific tank from which leakage is occurring must also be provided. Procedures for dealing with suspected hull fractures must be included. These procedures must take into account the effect of corrective actions on hull stress and stability; ②spills resulting from casualties. Each of the casualties listed below must be treated in the plan as a separate section comprised of various checklists or other means which will ensure that the master considers all appropriate factors when addressing the specific casualty. These checklists must be tailored to the specific ship. In addition to the checklists, specific personnel assignments for anticipated tasks must be identified. Reference to existing fire control plans and muster lists is sufficient to identify personnel responsibilities in the following situations: (a) grounding; (b) fire or explosion; (c) collision; (d) hull failure; (e) excessive list. ③In addition to the checklist and personnel duty assignments required by paragraph above of this section, the plan must include: (a) priority actions to ensure the safety of personnel and the ship, assess the damage to the ship, and take appropriate further action; (b) information for making damage stability and longitudinal strength assessments, or contacting classification societies to acquire such information. Nothing in this section shall be construed as creating a requirement for damage stability plans or calculations beyond those required by law or regulation; (c) lightening procedures to be followed in cases of extensive structural damage. The plan must contain information on ship transfer of cargo. Reference may be made in the plan according to existing company guides, A copy of such company procedures for ship to ship transfer operations must be kept in the plan. The plan must address the coordination of this activity with the coastal or port state, as appropriate. (5) national and local coordination. ①This section of the plan must contain information to assist the master in initiating action by the coastal State, local government, or other involved Parties. This information must include guidance to assist the master with organizing a response to the incident, if a response not be organized by the shore authorities. Detailed information for specific areas may be included as appendices to the plan. ②for Antarctica, a vessel owner or operator must make a plan for prompt and effective response action to such emergencies as might arise in the performance of his vessel's activities; ③To comply with paragraph (b) of this section, an agency of the United States government may promulgate a directive providing for prompt and effective response by the agency's public vessels operating in Antarctica. (6) Appendices must include the following information: ①twenty-four hour contact information and alternates to the designated contacts. These details must be routinely updated to account for personnel changes and changes to telephone, telex, and telefacsimile numbers. Cleat guidance must also be provided regarding the preferred means of communication. ②The following lists, each identified as a separate appendix: (a) a list of agencies of officials and state administrations responsible for receiving and processing incident reports; (b) a list of agencies of officials in regularly visited ports. When this is not feasible, the master must obtain details concerning local reporting procedures upon arrival in port; (c) a list of all parties with a financial interest in the ship such as ship and cargo owners, insurers, and salvage interests; (d) a list which specifies who will be responsible for informing the parties listed and the priority in which they must be notified; ③A record of annual reviews and changes. (7) non-mandatory provisions. If this section is included by the shipowner, it should include the following types of information or any other information, that may be appropriate; ①diagrams; ②response equipment or oil spill removal organizations; ③public affairs practices; ④record-keeping; ⑤plan exercising; ⑥individuals qualified to respond. (8) index of sections. The plan must be organized as relative requirements.

Shipboard pH meter 船用 pH 计 供船上采样测定海水 pH 值的仪器。

Shipboard simplified voyage data records (S-VDR) 简化航行数据记录仪(S-VDR) 系指一个完整的系统。包括输入数据的接口、数据处理和编码装置、最终记录介质、电源及专用的备用电源。 Shipboard simplified voyage data records(S-VDR) means a complete system, including any items required to interface with the sources of input data, for processing and encoding the data, the final recording medium, the power supply and dedicated reserve power source.

Shipboard Technology Evaluation Program (STEP) 船上技术评估计划 系指美国海岸警卫队拟用于促进研究、开发和船上试验压载水管理系统的研究计划。船上技术评估计划的要求可在 http//www. uscg. mil/environmentaj _standards/下载。 Shipboard Technology Evaluation Program (STEP) means a Coast Guard research program intended to facilitate research, development, and shipboard testing of effective BWMS. STEP requirements are located at http//www. uscg. mil/environmentaj _standards/.

Shipboard testing 船上试验 系指为确认系统符合国际船舶压载水和沉积物控制和管理公约第 D-2 条规定的标准,按压载水管理系统认可导则附件第 2 部分的规定,在船上进行的压载水管理系统的全面测试。 Shipboard testing is a full-scale test of a complete BWMS carried out on board a ship according to Part 2 of the Annex to the Guidelines for approval of ballast water management systems, to confirm that the system meets the standards set by Regulation D-2 of the International Convention for the Control and Management of Ships' Ballast Water and Sediments, 2004.

Shipboard wave meter 船舷测波仪 在船舷下一定部位安装水压计和垂直加速度计,通过检测波浪引起的水压变化及船体升沉来记录波浪的仪器。

Shipborne barge 船载驳 系指一种独立非自航船舶,其设计和装备使在载重的情况下装载在船上。 Shipborne barge means an independent, non-self-propelled vessel, specifically designed to be lifted in a loaded condition and stowed on board a ship.

Shipbroker 船舶经纪人 系指在租船市场上,提供船舶和货源,互通情报的人员。通常在租船市场上,船舶经纪人与租船人、船东聚集在一起,进行无形贸易——租船交易,即转让船舶的使用权。

Shipbuilders Association of Japan 日本造船工业协会

Shipbuilding insurance 船舶建造险 亦称建造人风险保险,系指承保船舶在建造过程中发生的各种风险造成的物质损失、费用和责任的保险,属于船舶保险的一种,简称"船建险"。它承保造船厂的整个造船过程,从建造、试航直至交船,包括建造该船所需的材料、机械设备在船厂范围内装卸、运输、保管、安装以及船舶下水、进出船坞、停靠码头过程中发生保险事故造成的损失、责任和费用。它是以新造船舶为保险标的的保险活动,性质属于工程建设保险。由于船舶建造过程中有许多水上活动,如下水、试航等,所以也具备一些船舶保险的内容。

Shipbuilding shop 船体车间 系指进行船体分段切割、加工、装配等作业的场所。

Shipment as soon as possible 尽速装运

Shipment consolidation 集中运输 系指航空公司把若干单独发运的货物(每一货主货物要出具一份航空运单)组成一整批货物,用一份总运单,整批发运到目的地,由航空公司在那里的代理人收货,报送、分拨后交给实际收货人的运输方式。

Shipment consolidation policy 集中运输策略

Shipment note 托运单 俗称下货纸,是托运人根据贸易合同和信用证条款内容填制的,向承运人或其代理办理货物托运的单证。承运人根据托运单内容,并结合船舶的航线、挂靠港、船期和舱位等条件考虑,认为合适后,即接受托运。托运单是运货人和托运人之间对托运货物的合约,其记载有关托运人与送货人相互间的权利义务。运送人签收后,一份给托运人当收据,货物的责任从托运转至运送人,直到收货人收到货物为止。如发生托运人向运送业要求索赔时,托运单位必备的文件。运送业输入托运单上数据的正确与否,影响后续作业甚大。

Shipment 装货 系指向停靠在码头等待装载的船舶吊装货物的作业。

Ship-model 船模 将实船按比例缩小尺度而制成的供试验用的模型。

Ship-model correlation factor for propulsion efficiency 推进效率换算因数 估计的实船推进效率对由船模试验得出的推进效率的比值:$K_1 = \eta_{DS}/\eta_{Dm}$,式中,$\eta_{DS}$—实船的推进效率;$\eta_{Dm}$—船模的推进效率。

Ship-model resistance dynamometer 船模阻力仪 测量船模阻力的仪器。

Ship-model resistance test 船模阻力试验 测量船模在不同航速时阻力的试验。

Ship-model self-propulsion test 船模自航试验 由船模内动力机构驱动所装的螺旋桨推进船模,以预测实船的快速性和分析螺旋桨与船体间相互作用的试验。

Ship-owner 船东 系指登记注册为船舶拥有者的个人或公司,或无登记注册而拥有该船舶的个人或公司,或任何其他组织或个人,诸如管理者或光船承租人,其已从船舶拥有者处承担船舶营运的责任。但船舶系国家拥有并由在该国注册为船舶经营者的公司营运时,船东就是该公司。该定义也包括船舶出售或交付拆船厂之前的一个限定期内船舶的拥有者。 Ship-owner means the person or persons or company registered as the owner of the ship or, in the absence of registration, the person or persons or company owning the ship or any other organization or person such as the manager, or the bareboat charterer, who has assumed the responsibility for operation of the ship from the owner of the ship. However, in the case of a ship owned by a State and operated by a company which in that State is registered as the ship's operator, "owner" shall mean such company. This term also includes those who have ownership of the ship for a limited period pending its sale or handing over to a Ship Recycling Facility.

Shipowner 船舶所有人(船东) 系指:(1)拥有或经营船舶者,不论其是一个人、一个公司或其他法律实体,以及代表船舶所有人或经营者的任何人;(2)船舶所有人、承租人、经理人和经营人。 Shipowner means: (1) one who owns or operates a ship, whether a person, a corporation or other legal entity, and any person acting on behalf of the owner or operator; (2) the owner, charterer, manager and operator of the seagoing ship.

Shipper 托运人 系指:(1)为了运送对他有利益而与一个公会或航运公司订立或表示愿意订立合约性或其他协议的个人或实体;(2)由其本人或以其名义或代其与承运人订立海上货物运输合同的任何人,或是由其本人或以其名义或代其将货物实际交给与海上货物运输合同有关的承运人的任何人。 Shipper means: (1) a person or entity who has entered into, or who demonstrates an intention to enter into, a contractual aggrement or other arrangement with a conference or shipping line for the shipping of goods from which he has a beneficial interest; (2) any person by whom or in whose name or on whose behalf a contract of carriage of goods by sea has been concluded with a carrier, or any person by whom or in whose name or on whose behalf the goods are actually delivered to the carrier in relation to the contract of carriage by sea.

Shipper 装货人 系指承担装货作业的人员。

Shipper's export declaration 托运人出口报关清单

Shipper's guarantee 托运人保证书 系指托运人提供的银行保函。

Shipping advice 装运通知 是在采用租船运输大宗进出口货物的情况下,在合同中加以约定的条款。规定这个条款的目的在于明确买卖双方的责任,促使买卖双方相互配合,共同做好船货衔接工作。

Shipping bill 货物装船清单 系指托运人交付承运人的货物清单。

Shipping claim 装运索赔

Shipping insurance 航运保险 是以船舶及其附属品为保险标的的保险业务。狭义而言,航运保险又称水险,主要包括船舶险、货运险和保赔保险三大类。货运险针对船上所运输的各类货物,船舶险以各类船舶本身为保险标的,而保赔保险的定义和范畴目前在业界海存在一定的分歧,不过一般是用来指代船东保险协会承保的责任风险的保险。而从广义上讲,有不少学者认为,应该根据船舶所处的状态,将船舶营运险、船舶建造险、船舶停航险、船舶修理险、拆船保险等险种均纳入航运保险的范畴之中。

Shipping mark 运输标志 又称唛头,它通常是由一个简单的几何图形和一些英文字母、数字及简单的文字组成。运输标志的作用是使货物在运输过程中的每个环节便于识别,以免发生错装、错运、错转、错交和无法交付等情况;当由于某种原因发生票货分离时,也便于港航工作人员能很快地确认货物所有人。运输标志的内容繁简不一,由买卖双方根据商品特点和具体要求商定。

Shipping order(S/O) 装货单 是接受了托运人提出装货申请的船公司,签发给托运人的用以命令船长将承运的货物装船的单据。它既能用作装船的依据,又是货主用以向海关办理出口货物申报手续的主要单据之一,所以又叫关单。对于托运人来讲,它是办妥货物托运的证明。对船公司或其代理来讲是通知船方接受装运该批货物的指示文件。

Shipping permit 货物承运单

Shipping ton 货运吨

Shipping value 货运价格 系指货物运输的单价。

Shipping' organization 托运人组织 系指促进、代表或保护托运人的利益,并为其所代表的托运人的感觉到有关当局——如这些当局要这样做的话——承认其具有此种身份的会社或相等团体。 Shipping' organization means an association or equivalent body which promotes, represents and protects the interests of shippers and, if those authorities so desire, is recognized in that status by the appropriate authority or authorities of the country whose shippers it represents.

ShipRight notation ShipRight 标志 此标志表明已满足了英国劳氏船级社一个或多个的 ShipRight 程序。

将根据所应用的 ShipRight 程序是强制性的或自愿的,分别授予船级标志或描述性注解,即:(1)对于大型和结构复杂的船舶,有关船体设计和建造的这些程序,是为入级所强制,在此情况下,相应的 ShipRight 标志是一种船级标志,并在《船名录》第 4 栏中刊出。当这些程序是在自愿基础上采用时,相应的 ShipRight 标志是一种描述性注解,并在《船名录》第 6 栏中刊出;(2)其余的 ShipRight 程序对入级目的而言是自愿的,将授予描述性注解并在《船名录》第 6 栏中刊出。 ShipRight notation is a notation indicating that one or more of LR's ShipRight procedures have been satisfactorily followed. Class notations or descriptive notes will be assigned according to whether the ShipRight procedures are applied on a mandatory or voluntary basis, i. e. (1) the procedures relating to the design and construction of the hull are mandatory for the classification of large and structurally complex ships. In such cases, the associated ShipRight notation is assigned as a class notation and will be listed in Column 4 of the Register Book. When these procedures are applied on a voluntary basis, then the associated ShipRight notation is assigned as a descriptive note and will appear in Column 6 of the Register Book. (2) The remaining ShipRight procedures are voluntary for the purposes of classification, and are assigned as descriptive notes and will be listed in Column 6 of the Register Book.

Ships constructed　　建造的船舶　　系指安放龙骨或处于类似建造阶段的船舶。 Ships constructed means ships the keels of which are laid or which are at a similar stage of construction.

Ships general arrangement　　船舶总布置　　系指根据任务书的要求,对全船的舱室、上层建筑、通道以及各种主要设备、装置、系统等所做的全面统一的规划和布局。

Ships having an unusual large freeboard　　非常规富裕干舷船舶　　系指实际夏季干舷(相应于指定最大载重线的干舷)大于下列方法指定的最小夏季干舷与上层建筑标准高度(参考 1966 年国际载重线公约第 33 条)之和的船舶。 Ships having an unusual large freeboard are the ships having the actual summer freeboard (corresponding to the assigned load line) which is not less than the sum of minimum summer freeboard and standard height of superstructure (See Regulation 33. of "1966 International Convention on Load Line").

Ships length　　船长　　总长。 Ships length is the length over all.

Ships noise measurement　　船舶噪声测量　　对各种船舶类型的不同航速,在不同的航向由螺旋桨噪声、流体动力学噪声和机械噪声等所组成的总噪声的功率谱进行的测量。

Ships of novel design　　设计新颖的船舶　　系指:(1)那些具有非常规线型、尺度比例、航速和结构布置的船舶;(2)那些具有船级社规范所规定以外的非常规线型、尺度比例、航速和结构布置的船舶。 Ships of novel design mean:(1) those of unusual form, proportions, speed and structural arrangements;(2) those ships of unusual form, proportions, speed and structural arrangements outside those reflected in Rules of the Society.

Ships required to take part in the CANREP system　　要求加入加纳利群岛船舶强制报告系统的船舶　　系指载重量为 600 吨或以上,经过加纳利群岛,或者驶入或驶加纳利港口,或者在加纳利群岛的岛间航行,并载有下列货物的液货船:(1)在 15 ℃时密度大于 900 kg/m³ 的重原油;(2)在 15 ℃时密度大于 900 kg/m³ 在 50 ℃时动黏度大于 180 mm²/s 重燃油;(3)沥青,焦煤油及其乳状物。 Ships required to take part in the CANREP system means those tankers which are of 600 t deadweight and above, either transiting the Canary Islands or sailing to or from Canarian ports or involved in inter-island navigation, carrying the following:(1) heavy-grade crude oils with a density greater than 900 kg/m³ at 15 ℃;(2) heavy-grade crude oils with a density greater than 900 kg/m³ at 15 ℃ or kinematic viscosity greater than 180mm²/s at 50 ℃;(3) bitumen, coal tar and their emulsions.

Ships staff　　船上人员　　系指船员、船东和与运营管理有关人员的总称。

Ships' manning　　船舶配员　　系指各缔约国政府,各自对本国船舶保持实行或在必要时采取措施,以确保所有船舶从海上人命安全观点出发,配备足够数量和能胜任的船员。每艘船舶应备有一份由主管机关颁发的适当的最少安全配员证书或等效证明,作为所需的最少安全配员的凭证。

Shipshape fitness center & SPA　　水疗按摩中心(豪华邮轮)

Ship-shaped offshore structure　　近海船式结构物　　系指在几何尺寸和功能方面具有类似于油船形状的近海浮动结构物。 Ship-shaped offshore structure is a floating offshore structure that has a similar shape to a tanker-type ship in terms of geometry and functions.

Shipside doors　　舷门　　系指在船体舷侧开设的供引航员登离船用的舷门或供乘客登船或货物、车辆进出船舱的舷门。

Ship-side valve　　船舷阀　　系指安装在船底或船侧的阀门。 Ship-side valve means the valve attached to bottom or side of ship.

Ship-type drilling unit 船式钻井平台 系指配备推进机械的水面式钻井平台。 Ship-type drilling unit means a surface-type drilling unit with propulsion machinery.

Ship-type unit 船式平台 水面式平台之一,系指具有推进机械的水面式平台。 Ship-type unit is one of surface-type units, which is a surface-type unit having propulsion machinery.

Shock absorber 减振器(阻尼器,缓冲器)
Shock absorber rubber 减振器橡胶
Shock absorber support 减振器支架
Shock action 冲击作用
Shock attenuator 减振器
Shock piston type corer 振动活塞式取样管 用电振动器使取样管插入海底取样的装置。适于浅海和港湾的泥质。可进行砂质沉积区域的采样。
Shock reducer 缓冲器
Shock resistance 耐振强度
Shock resisting tool steel 抗振工具钢
Shock ring 减振环
Shock simulating machine 冲击模拟试验机(冲振模拟试验机)
Shock strength 抗振强度(抗冲击强度)
Shock stress 冲击应力
Shock test 冲击试验
Shock testing-machine 冲击试验装置
Shock tunnel 激波风洞 先利用激波压缩实验气体,再利用定常膨胀方法产生高超音速实验气流的风洞。它由一个激波管以及连于其后的喷管、实验段等风洞主要部件组成。我国的 JF12 激波风洞,试验时间 100 毫秒三倍于国外,是国际上最先进的高超声速风洞之一。
Shock wave 冲击波(激波)
Shock wave attenuation 冲击波衰减
Shock wave audio filter 声表滤波器 系指利用压电晶体制作的滤波元件。
Shock wave audio(SWA) Authorware 公司的声音文件 属于 MPEG 2 level 3 压缩格式,在低频下较 mp3 优秀。把后缀名直接改为 mp3 就可以播放。
Shock-free entry 水冲击进流 止于螺旋桨叶切面的流线在一定攻角小范围内可以顺易地连于切面轮廓以避免吸力峰的进流情况。
Shoe 桩靴 是一种箱式或八角形的底脚,与海床接触时,它能增加支腿底脚的面积,且在下面两方面起作用:(1)降低地面压强,加强抗击穿现象的安全系数;(2)由于底脚面积增大,穿入海床的程度降低。此外,设置桩靴的目的是在一定程度上穿透到海床中,避免支腿存在下冲现象,从而发生击穿现象。这是桩靴的一个非常重要的特点;(3)为增大桩腿的支承面积和减少桩腿在软海底的插入深度而设置在自升式钻井平台桩腿下部的可拆式的箱型结构。

Shoe self-elevating drilling platform 桩靴自升式钻井平台 在每根桩腿接触海底端设有一个桩靴,以增大桩端的承载面积,减少插入海底的深度的自升式钻井平台。

Shop primer 车间底漆 (1)是用于表面预加工之后和装配之前的一种薄涂层,作为构件装配期间的防腐蚀措施;(2)在制造过程中,构件表面处理后和装配之前使用的一种薄的防护保护涂层;(3)加工前涂在钢板表面的底漆,通常在自动化车间喷涂(在涂层系统第一道涂层之前)。 Shop primer is: (1) a thin coating applied after surface pretreatment and prior to fabrication as a protection against corrosion; (2) a thin coating applied after surface preparation and prior to fabrication as a protection against corrosion during fabrication; (3) the prefabrication primer coating applied to steel plates, often in automatic plants (and before the first coat of a coating system).

Shopping areas 购物廊(豪华邮轮)
Shore crane 岸上起重机 系指属岸上设备的起重机,并包括能从起重机室内进行操作或控制的各种设备,如:(1)永久性安装的辅助设备;(2)设计成与起重机一起使用的可拆设备。 Shore crane means a crane that is a shore equipment and includes any equipment operated or controlled from the cabin of that crane, such as: (1) permanently attached auxiliary equipment, or (2) detachable equipment designed for use with the crane.

Shore discharging pipeline 陆上排泥管 在排泥管线中,埋设或架设在陆地的排泥管。
Shore equipment 岸上设备 系指并非船上所属设备的物资装运设备,并包括浮式起重机。 Shore equipment means materials handling equipment that is not a ship equipment and includes a floating crane.
Shore hardness 回跳硬度(肖氏硬度)
Shore leave 登岸假期 系指船舶在港停留期间,在地理或时间许可范围内允许船员登岸,若有,可以由公共当局决定。 Shore leave means the permission for a crew member to be ashore during the ship's stay in port within such geographical or time limit, if any, as may be decided by the public authorities.

Shore power system with high-volt 高压岸电系统 系指允许装有特殊设备的船舶在码头停泊时接入码头的岸电电源,船舶可以从岸电系统获得所需的电力,而无须使用船上的动力设施,从而达到改善港口空气质量,绿色环保的目的。

Shore power technology　岸电技术　是优化电力系统的有效手段,通过高效率地利用港口的岸上电能,可节约燃油,减少二氧化碳、硫氧化物、氮氧化物等有毒气体的排放,减少船上发电机的损耗及机舱保养工作,减少船上发电机组工作时间,降低机舱振动噪声。

Shore sclercscope　肖氏回跳硬度计

Shore-based installation　岸基装置　系指安装在码头或港口,用于向船舶提供岸电的设备。一般包括高压配电箱、变压器、变频器(适用时)和码头岸电插座箱。

Shore-side personnel　岸上人员　系指:包括从事如下各种工作的人员:①编制散货运输单证;②交付散货供运输;③接收散货供运输;④装卸散货;⑤编制散货装载/积载图;⑥向/从船舶装载/卸载散货和实施适用规则和规范,或检验或检查是否符合适用规范和规则;⑦以主管当局确定的其他方式参与散货的装卸和运输;⑧但岸上人员不包括(a)ISPS 规则 A 部分 13.1 提及的公司保安人员和相应岸基人员;(b)ISPS 规则 A 部分 13.2 和 13.3 提及的船舶保安人员和船上人员;(c)ISPS 规则 A 部分 18.1 和 18.2 提及的港口设施保安人员、相应港口设施保安人员和承担具体保安职责的港口设施人员。Shore-side personnel:cover individuals such as those who:①prepare transport documents for bulk cargoes;②offer bulk cargoes for transport;③accept bulk cargoes for transport;④handle bulk cargoes;⑤prepare bulk cargoes' loading/stowage plans;⑥load/unload bulk cargoes into/from ships;and enforce or survey or inspect for compliance with applicable rules and regulations;⑦are otherwise involved in the handling and transport of bulk cargoes as determined by the competent authority;⑧however,shore-side personnel not cover:(a) the company security officer and appropriate shore-based personnel mentioned in section A/13.1 of the ISPS Code;(b) the ship security officer and the shipboard personnel mentioned in sections A/13.2 and A/13.3 of the ISPS Code;(c) the port facility security officer, the appropriate port facility security personnel and the port facility personnel having specific security duties mentioned in sections A/18.1 and A/18.2 of the ISPS Code.

Short blade　短叶片　一般系指径高比 $D/L \geqslant 10$ 的叶片。

Short blast　短声　系指历时约1s的笛声。 Short blast means a blast of about one second's duration.

Short bossing　短轴包架　舷侧艉舭穿过船底处,用以顺接该处船体表面并局部围封艉轴的轴包架。

Short certificate　短期证书　系指完成初次和/或换证审核后,在全期证书签发前,船级社所签发的不超过5个月的证书(DOC/SMC)。

Short circuit capacity　短路容量　系指电缆及其绝缘导体能耐受最大短路电流所产生的机械应力和热效应的能力。 Short circuit capacity means the capacity of cables and their conductors, which is to be capable of withstanding the mechanical and thermal effects of the maximum short-circuit.

Short circuit capacity of cable　电缆短路容量　系指当电力系统发生故障时,在系统的保护装置切除故障的短暂时间内,电缆将通过比额定电流大许多倍的短路电流。该短路电流可使导体的温度急剧上升,可能导致电缆的绝缘变形和断裂、护套熔化,以及电缆线芯熔断等故障。电缆短路容量取决于引起电缆绝缘的电气和物理性质发生显著变化的温度以及能够维持短路电流的持续时间。

Short deckhouse　短甲板室　系指非长甲板室的甲板室。 Short deckhouse means a short deckhouse, not a long deckhouse.

Short imperfection　短的缺陷　系指在任意的 100 mm长的焊缝中,一个或多个缺陷的总长度不大于 25 mm或对于长度小于 100 mm 的焊缝,缺陷的长度最大为 25% 焊缝长度。 Short imperfection is one or more imperfections of total length not greater than 25 mm in any 100 mm length of the weld or a maximum of 25% of the weld length for a weld shorter than 100 mm.

Short international voyage　短程国际航行　系指在航行中,船舶距离能够安全安置乘客和船员的港口或地点不超过200海里的国际航行。起航国最后停靠港至最终目的港之间距离与返航航程均应不超过600海里。最终目的港系指船舶开始返航回到起航国前的计划航次中的最后停靠港。 Short international voyage is an international voyage in the course of which a ship is not more than 200 n miles from a port or place in which the passengers and crew could be placed in safety. Neither the distance between the last port of call in the country in which the voyage begins and the final port of destination nor the return voyage shall exceed 600 n miles. The final port of destination is the last port of call in the scheduled voyage at which the ship commences its return voyage to the country in which the voyage begins.

Short message service(SMS)　短信息　简称短信、香港称短讯、台湾称简讯。短信息是手机服务的一种,由英文"简短的信息服务"翻译而来。在多数手机上都可使用。有时也称为信息、短信息、短信、文字信息。短信服务原先是为 GSM 系统手机所设计的,但现在几乎在任何手机系统上都能通用,如3G手机。

Short pennant 短缆 系指三角板或龙须缆/链与拖船拖缆连接的一段缆索。

Short range missile(SRM) 近程导弹 系指用于毁伤战役战术目标的导弹。其射程通常在1000 km以内,多属近程导弹。它主要用于打击敌方战役战术纵深内的核袭击兵器、集结的部队、坦克、飞机、舰船、雷达、指挥所、机场、港口、铁路枢纽和桥梁等目标。战术导弹种类繁多。有打击地面目标的地对地导弹、空对地导弹、舰对地导弹、反雷达导弹和反坦克导弹;打击水域目标的岸对舰导弹、空对舰导弹、舰对舰导弹、潜对舰导弹和反潜导弹;打击空中目标的地对空导弹、舰空导弹和空对空导弹等。这些导弹采用的动力装置有固体火箭发动机、液体火箭发动机和各种喷气发动机。战术导弹的弹头(战斗部)有普通装药弹头、核弹头和化学、生物战剂弹头等。20世纪50年代以后,常规战术导弹曾在多次局部战争中被大量使用,成为现代战争中的重要武器之一。

Short range yacht 短程游艇 系指500 GT以下的现有游艇或300GT以下,(1)限制在实际最大风级为蒲氏4级和(2)在距离安全避风港60海里范围内下营运的新艇。Short range yacht means an existing vessel under 500 GT or a new vessel under 300 GT, (1) restricted to operate in forecast or actual wind of a maximum Beaufort Force 4, and (2) within 60 nautical miles of a safe haven.

Short shank anchor 短杆锚 锚杆较一般为短的无杆锚。

Short superstructure 短上层建筑 不符合长上层建筑条件的上层建筑。

Short ton 短吨 等于0.907吨(t)或2 000磅(lb)。

Shortage 短量 系指货物在运输过程中发生质量短少。

Shortage risk 短量险 投保平安险和水渍险的基础上加保此险,保险人负责赔偿承保的货物因外包装破裂或散装货物发生数量损失和实际质量缺损的损失,但不包括正常运输途中的自然损耗。被保险人对于包装货物的短少,应当提供外包装发生破裂现象的证明;对于散装货物,则以装船质量和卸船质量之间的差额作为计算短量的依据。

Short-circuit 短路 在正常情况下电路中处于不同电压的两点或多点,通过一比较低的电阻或阻抗偶然或有意的连接。Short-circuit means that the accidental or intentional connection, by a relatively low resistance or impedance, of two or more points in a circuit, which are normally at different voltages.

Short-circuit breaking capacity 短路分断能力 系指在规定的条件下,包括开关电器出线端短路在内的分断能力。

Short-circuit current 短路电流 系指在电源不变情况下,由于故障或误操作引起短路而产生的过电流。Short-circuit current means an over-current resulting from a short circuit due to a fault or an incorrect connection of negligible impedance without any change of the supply.

Short-circuit making capacity 短路接通能力 系指在规定的条件下,包括开关电器出线端短路在内的接通能力。

Short-circuit release 短路脱扣器 系指用作短路保护的过电流脱扣器。

Short-crested irregular waves 短峰不规则波 系指由来自许多方向的一系列长峰规则波叠加而成的三因次波。

Short-crested waves 短峰浪 由不同方向、不同特征波列组成的三维不规则波系。

Short-delivery & non-delivery 短少和提货不着 系指货物在运输途中被遗失而未能运到目的地,或运抵目的地发现整件短少,未能交给收货人。

Short-distance transmission 短程传输 通常依靠电磁感应来实现。电磁感应电力传输主要以磁场为媒介射频来实现,其传输距离能达到10 m,但传输功率较小,多为汽车配件、助听器及人体植入仪器等电子设备供电。根据供电距离的不同,它可以分为短程、中程和远程传输3大类:(1)短程传输,短程传输距离上限是10 mm;(2)中程传输(Middle-distance transmission)可以通过电介,通过初级和次级线圈感应产生电流,可以隔着很多非金属材料进行传输。一般适用于小型便携式器与电子设备之间的传输。(3)远程传输(Long-distance transmission)主要利用微波或激光技术。微波传输是将电能转化为微波,让微波经自由空间传送到目标位置,在转化成直流电能,提供给负载。

Short-term credit 短期信贷 系指贷款期限在1年以下的信贷。

Short-tern DOC 短期DOC证书 是在长期证书被签发下来之前,由进行审核机构的分支机构签发的证书。

Short-time short-circuit current 短延时短路脱扣器 系指用于在短延时结束后动作的过电流脱扣器。

Short-time withstand current 短时耐受电流 系指在规定的使用和性能条件下,电路或在闭合位置上的开关电器在指定的短时间内所能承载的电流。

Shot(cable length) 链节 锚链的组成单元。按链节在整根锚链中所处的位置不同分别有锚端链节、中间链节、末端链节以及脱钩链节。

Shoulder　肩　系指进流段或去流段紧邻平行中体的部分。

Shoulder foam applicator　肩背式泡沫枪

Shoulder region　肩部区　是艏部区内主冰带区的一部分,从艏部区的后边界向前至船侧平直部分的前端界线之间的范围:(1)对冰级 1AS 和冰级 1A,此距离等于 0.04LR;(2)对冰级 1B 和冰级 1C,此距离等于 0.02LR。舷侧平直部分无明显的前端界线时,则肩部区域的前端边界取为:对冰级 1AS 和冰级 1A 取艏垂线后 0.32LR 处;对冰级 1B 和冰级 1C 取为 0.36LR 处。肩部区的范围,其后边界以前的区域,对冰级 1AS 和 1A 不少于 10 m。对冰级 1B 和 1C 不少于 4 m。 Shoulder region is a part of the main ice belt zone in the forward region and extends from the aft boundary of the forward region to forward of the forward borderline of the flat side of the hull by a distance of (1) 0.04LR for Ice Classes 1AS and 1A or (2) 0.02LR for Ice Classes 1B and 1C. Where no clear forward borderline of the flat side of the hull is discernible, the forward boundary of the shoulder region is to be taken 0.32LR aft of the forward perpendicular for Ice Classes 1AS and 1A or 0.36LR aft of the forward perpendicular for Ice Classes 1B and 1C. The extent of the shoulder region forward of its aft boundary is not to be taken as less than 10 m for Ice Classes 1AS and 1A or 4 m for Ice Classes 1B and 1C.

Shovel　消防铲　供消防用的专用铲子。

Shrink　缩套(收缩)

Shrink hole/shrinkage cavity　缩孔　系指铸件在冷凝过程中收缩而产生的孔洞,形状不规则,孔壁粗糙,一般位于铸件的热节处。

Shrinkage　收缩(收缩量)

Shrinkage fit　收缩配合

Shrinkage fitting　缸套装配

Shrinking　收缩(缩套,冷缩配合)

Shroud ratline　桅支索登攀梯(桅索具登攀梯)　在桅的两侧支索上,以一定间距设置的由横杆或绳索构成,供人员登攀用的梯步。

Shroud ring　围带　从顶部将若干动叶包箍成组的金属带。

Shrunk welded crankshaft　热套曲轴

Shunt generator　并励发电机　是一种直流发电机,其全部励磁通常来自一个由电阻较高的许多匝导线组成的绕组。对于自励发电机,这一绕组并联于电枢电路;对于他励发电机,则将其接至另一台发电机的负载侧或别的直流电源。

Shunt-wound motor　并励电动机　是一种其磁场电路或与电枢电路并联或与独立的励磁电压源并联的直流电动机。

Shunt-wound release　分励脱扣器　系指由电压激励的脱扣器。注意:电压源可与主电路电压无关。

Shut　关闭

Shut-down　关闭(停车)

Shut-down a engine　关车

Shut-down device　切断装置

Shut-off　关闭(切断,断路)

Shut-off device　切断装置

Shut-off valve(LPG system)　截止阀(LPG 系统)　使器具与气源相隔离的装置。 Shut-off valve is the device to isolate an appliance from the gas supply.

Shutter　百叶窗　系指舱棚或舱室门上的,非水密的透气窗。

Shutter　盖(挡板,百叶窗)

Shutter of sluice vavle　闸阀的闸板

Shuttle tanker　穿梭油船　具备在浮式生产储油船(FPSO)、浮产储油船(FSO)、采油平台、岸基之间进行原油驳运的能力,艏部装载系统可以在海上浮态和供给与接收双方都处于运动状态下实现原油的准确传输。船上装备 DP-2 动力定位系统,拥有可调螺距螺旋桨主推进器,首尾各配备伸缩式侧推进器和管隧式侧推器,具有自动导航,无人机舱等系统功能,并装备有减少挥发有机化合物(VOC)排放系统,有利于对环境污染的控制,成为符合环境低碳设计要求的绿色海洋工程船舶。

Side　舷(端,侧面)

Side anchoring winch　边锚绞车　在用锚定位的工程船上,设置在艏、艉两舷。用以收放左、右舷锚索,使船横移的绞车。

Side bend test on seam　焊缝侧弯试验　对焊缝而言,系指确定焊接接头在横断面上的韧性的试验。 For the seam, side bend test on seam means a test determined the ductility of the welded joint in the cross-sectional plane.

Side bunker　边燃油舱

Side buoys　侧浮舟(侧浮体)　系指主翼翼尖处产生浮力的结构。 Side buoys means the buoyant structures at the tip of the main wing.

Side construction　舷侧结构　系指从艉部肘板到上甲板间,船体舷侧的板架结构。舷侧结构主要承受水的压力、波浪冲击力,船体上甲板以上的上层建筑以及甲板设备和货物的质量,是保证船体横向强度的侧壁水密性的重要结构。舷侧结构也分横骨架式和纵骨架式两种。

Side discharge assemblage　边抛装置　使由泥泵

吸起的泥沙不输往泥舱而通过伸出舷外的悬臂桁架上的排泥管直接抛到近旁水流中的装置。

Side hopper barge 侧开泥驳 能打开两舷舣部活动门自动卸泥的泥驳。

Side inspection 现场检查 系指为确认经验证的文件所述的情况而对拆船厂进行的检查。 Side inspection means an inspection of the ship recycling facility confirming the condition described in the verified documentation.

Side keelson 旁内龙骨 单层船底结构中，中线面两侧的纵桁。

Side lights 舷灯 系指右舷的绿灯和左舷的红灯，各在112.5°的水平弧内显示的灯光。其装置要使灯光从船的正前方到各自一舷正横后22.5°内分别显示。长度小于20 m的船舶其舷灯可以合成一盏，装设于艏艉中心线上。 Sidelights means a green light on the starboard side and a red light on the port side each showing an unbroken light over an arc of the horizon of 112.5°and so fixed as to show the light from right ahead to 22.5°abaft the beam on its respective side. In a vessel less than 20 m in length the sidelights may be combined into one lantern carried on the fore and aft centerline of the vessel.

Side longitudinal 舷侧纵骨 舷侧外板上的纵骨。

Side paddle wheels 旁轮 装在船两侧的明轮。

Side plate of deck house 甲板室侧壁 甲板室两侧的围壁。

Side ports（side door） 舷门 开设于干舷甲板与最大载重水线之间船体外板上，供人员、货物、车辆等出入用的门。

Side projection blade 叶侧视投影 螺旋桨叶边缘在侧视图上的投影。

Side propeller 舷侧推进器

Side roller 防护滚轮 装在舷梯上，为防止梯身擦伤船体而设置的滚轮装置。

Side scan sonar 侧扫声呐 又称海底地貌探测仪或旁侧声呐。一种在水中向两侧发射声波，以探测障碍物或海底地貌的仪器设备。

Side screws 边螺旋桨 装在船两侧的螺旋桨。

Side screws in different rotations 错车 在同一时间内，船舶两侧螺旋桨开以不同车速或分别开正、倒车导致船首转向的操纵方式。

Side scuttle（side light） 舷窗 参照经修订的ILLC[MSC.143（77）决议第23（2）条，系指面积不超过0.16 m² 的圆形或椭圆形开口，面积超过0.16 m² 的圆形或椭圆形开口应作为窗处理。 Ref. ILLC, as amended [Resolution MSC.143（77）Reg.23（2）], side scuttles are round or oval openings with an area not exceeding 0.16 m². Round or oval openings having areas exceeding 0.16 m² are to be treated as windows.

Side seat 边座板 沿救生艇舷缘布设的供人员乘坐的纵向板材。

Side shell 舷侧外板 系指舣部甲板以上组成船体外壳板舷侧部分的船体外壳板。 Side shell means shell envelope plating forming the side portion of the shell envelope above the bilge plating.

Side slip 横移 由于推进器和舵等作用所引起的船舶横向移动。

Side stringer 舷侧纵桁 位于舷侧上的纵桁。

Side structure 舷侧结构 系指在底部结构与上层建筑舷侧处之间的外板及其肋骨和桁材的总称。 Side structure is defined as shell plating with stiffeners and girders between the bottom structure and the uppermost deck at side.

Side thruster 侧推器/侧推装置 一种轴线与船纵中平面垂直的小型螺旋桨。安装在船首的称为艏侧推器；安装在艉部的称为艉侧推器。其功能主要是提高大型船舶离靠码头的操纵性能。

Side trawler 舷拖网渔船

Side wings 侧翼 系指侧浮舟外的组合翼。 Side wings mean composite wings outside side buoys.

Side-board 餐具柜 存放各种餐具的专用柜子。

Side-opening bow door 边绞链式艏门 系指通过2个或多个位于近舷侧的绞链绕垂直轴向外旋转开启，或利用绞接于门上和船上的水平连杆作水平移动开启的艏门。 Side-opening bow door means a bow door opened either by rotating outwards about a vertical axis through two or more hinges located near the outboard edges or by horizontal translation by means of linking arms arranged with pivoted attachments to the door and the ship.

Side-opening door 侧开式舷门 系指：（1）通过位于两侧边缘的2个或以上的铰链绕垂直轴，或通过装有连接船体和门的枢轴装置的水平转换臂向外开启的门。通常侧开式舷门应成对布置。（2）舷门绕通过2个或以上位于门外侧的铰链的垂直轴向外转动，或采用在门和船上配枢轴的连杆作水平移动而开启。侧开式门要求成双设置。 Side-opening door means：（1）a door opened either by rotating outwards about a vertical axis through two or more hinges located near the outboard edges or by horizontal translation by means of linking arms arranged with pivoted attachments to the door and the ship. It is anticipated that side-opening bow doors are arranged in pairs. （2）opened either by rotating outwards about a vertical axis through two or more hinges located near the outboard edges or by horizontal translation by means of linking arms arranged with pivoted attachments to the door and the craft. It is anticipated that side-opening doors are arranged in pairs.

Side-rolling hatch cover 侧开式舱口盖 由一个或两个向一侧或两侧移动的舱盖组成的舱口盖。移动后将舱盖板水平放置在搁架上。此舱口盖需要大的甲板面积布置搁架，一般只适用于大型的船舶。

Sidewall air-cushion vehicle 侧壁气垫船 左右两侧有壁插入水中，仅在艏、艉两端允许空气逸出，只能在水面运行的气垫船。

Sideways maneuvering unit 横向操纵装置（侧推器，横向推力器）

Sight draft 即期汇票 即见票即付的汇票。包括载明即期付款、见票即付或提示付款以及未载明付款日的汇票。逾期后再经承兑或背书的汇票，对该种承兑人或背书人而言，应视为即期汇票。即期汇票一般以提示日为到期日，持票人持票到银行或其他委托付款人处，后者见票必须付款的一种汇票，这种汇票的持票人可以随时行使自己的票据权利，在此之前无须提前通知 付款人准备履行义务。即期汇票只可存入指定的受款人户口，因而与支票相比，遗失款项的风险较低。

Sight glass 观察窗

Sight hole 视察孔

Sight letter of credit 即期信用证 系指开证或付款行收到符合信用证条款的汇票和单据后，立即履行付款义务的信用证。

Sight opening 检视孔

Sighting port 观察窗（观察孔）

Sighting survey 目视检验

Sign a contract 签约

Sign contract 签订合同

Signal 信号 系指提供系统或设备状况信息的听觉指示。Signal is audible indication giving information about the condition of a system or equipment.

Signal control board in wheel house 驾驶室信号控制板（驾驶室信号控制台） 安装在驾驶室内，能示明号灯、助航仪电源、管制照明、各类报警、雾笛、测向仪、广播等工作情况，并能对其进行集中控制的板式装置。

Signal control desk in wheel house 闪光灯控制箱 安装在驾驶室内对闪光灯进行集中控制并标明闪光灯工作情况的装置。

Signal equipment 信号设备 船上供对外发出能被人的视角和听觉所感受的各种信号的设备。

Signal failure 单个故障 系指部件或系统出现的一个故障，考虑会造成下列影响中的一个或两个：(1)部件或系统的功能损失；(2)功能的退化达到了明显降低船舶、人员和环境的安全的程度。Signal failure means a failure appear in a part or a system that will result in following one or two affections: (1) function loss of a part or a system; (2) degradation of the function to the degree of lowering safety of vessels, crew or environment.

Signal flag (code flag) 号旗 供船舶通信及标识用的各种旗帜的统称。

Signal gun 信号枪 供船上发射信号弹用的枪支。

Signal mast (communication mast) 信号桅 专供装设和悬挂避碰、呼救和通信用信号设备的桅杆。

Signal smoke 信号烟雾 供船舶遇难时燃点，能发出橙黄色浓烟雾的一种信号烟火。

Signal source 信号源 系指VDR所连的任何外部传感器或装置，VDR从其获得将被记录的信号和数据。Signal source means any sensor or device external to the VDR, to which the VDR is connected and from which it obtains signals and data to be recorded.

Signal stay 信号张索 支张在两桅之间或桅与船体间，供悬挂号旗、号型等信号设备用的绳索。

Signal yard 桅横桁 固接在顶桅上部带有悬挂号旗及号型的滑车或眼环的横向杆件。

Signaling mirror 信号镜 配置在救生载具上，供登乘人员利用反光向施救人员联络的特制小镜。

Signatory 签字者(国际海事卫星组织) 系指业务协定对其生效的议定书成员国或议定书成员国所指定的某个实体。Signatory means either a Party to the protocol or an entity designated by a Party to the Protocol for which the operating agreement is in force.

Signature 签字

Significant alteration (floating dock) （浮船坞的）重大改装 系指现有浮船坞以增大主尺度和/或增加举升能力的改装或改建。

Significant amplitude of roll 三一横摇幅值 在不规则横摇记录中，按横摇幅值大小顺序排列取其中大端1/3区间的平均值。

Significant wave height 三一平均波高 将在不规则海面上定点或定时观测的波高数据，按大小序列，取其中大端的1/3区间的平均值，所标征的有代表性的波浪的高度。

Significant wave height 有义波高 系指1/3的最大波高的平均高度，其近似地相当于由有经验的观察者所估计的波高。某些波将为此高度的2倍。 Significant wave height is the mean height of the highest one-third of the waves, which approximately corresponds to the wave height estimated by an experienced observer. Some waves will be double this height.

Signpost 路标/指示牌

SIL = safety integrity level 安全完整性等级 见表 S-2。

表 S-2 安全完整性水平/等级

安全完整性等级 SIL	1	2	3	4
执行设计功能时要求的平均要求危险失效概率 PFD_{avg}	$\geqslant 10^2 \sim 10^1$	$\geqslant 10^3 \sim 10^2$	$\geqslant 10^4 \sim 10^3$	$\geqslant 10^5 \sim 10^4$

Silence 消音 手动停止听觉信号。Silence means that manual stopping of an audible signal.

Silence boiler 废气锅炉消声器

Silencer 消音器/消声器 是抑制发动机废气排放所产生的噪声，普遍安装在排气管的艉端部分，成为现代车辆不可或缺的组件。

Silent power source installation 低噪声电源装置 水声调查船用的特种电源装置。是蓄电池组成低噪声发电机组。低噪声发电机组是具有减振、隔声设施的柴油发电机组。它传至海洋的噪声低于海洋环境的噪声。

Silent ventilator 低噪声风机 采用特殊措施将噪声级控制在低水平的风机。

Silicate cotton wool 炉渣棉

Silicomanganese (low carbon) 硅锰合金（低碳）（具有已知危险特性或已知会释放气体）（硅含量 25% 以上） 硅锰合金是一种极重的货物，有灰色氧化表层的银色金属物质。遇水可能释放氢，是一种可能在空气中形成爆炸性混合物的易燃气体，在类似的情况下并可能产生磷化氢和砷化氢，两者均为剧毒气体。该货物易减少货物处所的氧气含量。该货物为不燃物或失火风险低。Silicomanganese (low carbon) (with known hazard profile or known to evolve gases) (with silicon content of 25% or more) is an extremely heavy cargo, it is silvery metallic material with a grey oxide coating. In contact with water it may evolve hydrogen, a flammable gas that may form explosive mixtures with air and may, under similar conditions produce phosphine and arsine, which are highly toxic gases. This cargo is liable to reduce oxygen content in a cargo space. This cargo is non-combustible or has a low fire-risk.

Silver alloy 银合金（银焊焊料）

Silver brazing alloy 银焊焊料

Similar refuse 类似的废弃物

Similar stage of construction 类似建造阶段 系指在此阶段：(1) 可辨认出某一具体船舶建造开始；(2) 该船业已开始的装配量至少为 50 t，或为全部结构材料估算质量的 1%，取较小者。Similar stage of construction means the stage at which: (1) the specific ship's beginning of construction is identifiable; (2) assembly of that ship has commenced comprising at least 50 t or one percent of the estimated mass of all structural material, whichever is less.

SIMOPS 模拟操作

Simple apparatus 简单设备 系指电器元件符合电路本质安全性能的电气元件或结构简单的元件组合。下列设备认为是简单设备：(1) 无源元件，如开关、接线盒、电阻和简单半导体装置；(2) 具有明确参数的储能元件，如电容器或电感器，当确定系统综合的安全性时，应考虑其量值；(3) 产生能量的能源，如热电偶、光电池，它们产生的电压、电流和功率不大于 1.5 V, 100 mA 和 25 mW，这些能源的电容或电感应考虑。

Simple catenary riser (SCR) 简单悬链式立管

Simple cycle 简单循环 工质依次通过压气机、燃烧室和涡轮的简单形式的燃气轮机循环。

Simple cycle gas turbine plant 简单循环燃气轮机装置 按简单循环工作的燃气轮机动力装置。

Simple harmonic vibration 简谐振动 系指其运动规律是用时间为变量的正（余）弦曲线来描写的，最简单、最基本的周期直线振动。

Simple harmonic wave 简谐波 系指简谐振动在空间传递时形成的波动。其波函数为正弦或余弦函数形式。简谐波是最简单的机械波，其他机械波可以看成是数个不同简谐波合成的。在波的传播方向上振动状态完全相同的两个质点间的最短距离称为波长，用 λ 表示。波速为波长和频率的乘积（$v = \lambda f$），表示波在的传播速度。空间等相位各点连接成的曲面称波面，波所到达的前沿各点连接成的曲面必定是等相面，称波前面或波阵面。常根据波面的形状把波动分为平面波、球面波和柱面波等，它们的波面依次为平面、球面和圆柱面。

Simple measuring equipment 简单测量设备 系指诸如直尺、卷尺、焊角规、千分尺等量具。Simple measuring equipment means rulers, measuring tapes, weld gauges，micrometers，etc.

Simple operator action 操作员简单动作 通过不超过 2 个硬键或软键动作完成的程序，不包括任何必要的光标移动或所有数码激活语言。 Simple operator action means a procedure achieved by no more than two hardkey or soft-key actions, excluding any necessary cursor movements, or voice actuation using programmed codes

Simple search 简单检索 系指单一因素的检索。

Simple support 简支 系指扶强材端部削斜，或扶强材仅连接至板上的端部连接方式。 Simple support means an end attachment manner in which the stiffener end are sniped or the stiffeners are connected to plating only.

Simple toilet 活动厕所 悬挂于小船的船尾外缘，便于收放的轻便而又简易的厕所。

Simplex 单工制（单缸的）

Simplex DP 无冗余的 DP 控制系统

Simplex nozzle 单油路喷油嘴 只有一个油路和喷油孔的喷油嘴。

Simplex pump 单缸泵

Simplex rudder 舵轴舵 套在舵轴上的流线型平衡舵。

Simplified voyage data recorder 简化航行数据记录仪（S-VDR） 系指一个完整的系统，包括输入数据的接口，数据处理和编码装置，最终记录介质，电源及专用的备用电源。Simplified voyage data recorder means a complete system, including any items required to interface with the sources of input data, for processing and encoding the data, the final recording medium, the power supply and dedicated reserve power source.

Simulated diving 模拟潜水 模拟水下状态，在加压舱内人工创造的高气压环境条件或同时注水的情况下，所进行的类似潜水的活动。

Simulated ventilation pipeline 模拟通风管路

Simulation filter 模拟滤器

Simultaneous disengaging gear 联动脱钩装置 在救生艇放落水面后，能保证前后艇吊钩同时解脱的装置。

Simultaneous flooding 同时浸水

Simultaneous ignition 同时点火

Sing-around velocimeter 环鸣声速仪 利用环鸣法测量声波在水中传播速度的仪器。

Singing 唱音/谐鸣 系指：(1)由于螺旋桨表面有刻痕或其他局部缺陷可能导致发生空泡的临界转速降低而引起螺旋桨共振产生的噪声。(2)螺旋桨由于所受水动力不稳定引起叶边振动所发出的断续鸣声。

Singing vibration （螺旋桨）鸣音振动

Single anchor leg mooring（SALM）system 单锚腿系泊系统 系指用一根锚腿伸入海底固定于海底锚固基础上的系统。其形式有：(1)带锚链腿的 SALM 系统；(2)带锚链与刚性立管的 SALM 系统；(3)带刚性立管的 SALM 系统。

Single anchoring leg storage system（SALS） 单锚腿储存系统

Single armed propeller strut 单臂轴支架 只有一根支撑的艉轴架。

Single atom iridiumcatalyst 单原子铱催化剂 系指在氧化铝或其他载体上浸渍活性金属铱并经过特殊处理制得的催化剂。根据不同的使用要求，催化剂的活性金属铱含量通常为 0.3% ~ 30%。属于肼单元推进剂用自发性催化剂，一般具有多次冷启动性能。用于航天卫星姿态控制发动机中单组元推进剂肼的催化分解。中国世界首创成单原子铱催化剂，可用于卫星推进剂，能够降低金属用量，提供催化效率，节约催化剂成本。

Single bottom 单底 没有内底的单层船底。

Single bottom ship 单底船 没有内底的船舶。

Single bottom structure 单底结构 系指在舭部角上拐点以下的外板及其肋骨和桁材的总称。 Single bottom structure is defined as shell plating with stiffeners and girders below the upper turn of bilge.

Single buoy mooring（SBM） 浮筒单点系泊系统

Single buoy storage（SBS）system 单浮筒储存系统 系指专用于系泊生产储油卸油船的系统。在该系统中，储油船通过一个刚性臂（轭架）与浮筒连接。这种形式避免了由于浮筒与储油船各自运动状态不一致而造成的连接钢丝绳负荷的突然增加，而且，在恶劣的海况下，不会发生浮筒与储油船碰撞的现象。

Single buoy system 浮筒单点系统

Single casing riser 单屏立管

Single collar thrust bearing 单环推力轴承

Single Convention on Narcotic Drugs 麻醉品单一公约 系指 1961 年纽约的麻醉品单一公约及其 1972 年修正案。 Single Convention on Narcotic Drugs means the Single Convention on narcotic drugs, and its amending Protocol of 1972.

Single decked ship 单甲板船 只有一层甲板的船舶。

Single end windlass 单侧起锚机 只具有一个锚链轮的卧式起锚机。

Single expansion engine 单胀式蒸汽机

Single factorial terms of trade 单因素贸易条件 是在价格贸易条件的基础上，考虑出口商品劳动生产率的变化对该国贸易利益的影响。计算方法为：$S = (P_X/P_M) \times Z_X$ 其中：S——单因素贸易条件；P_X——出口商品价格指数；P_M——进口商品价格指数；Z_X——出口商品劳动生产率指数。其结果大于 1 时，表明该国贸易条件得到改善；小于 1 时，则表明该国的贸易条件恶化。

Single failure　单个故障（单一故障）　系指：(1) 部件或系统出现的一个故障，可能会造成下列影响中的一个或两个：①部件或系统的功能损失；②功能的退化达到了明显降低船舶、人员或环境的安全的程度；(2) 装置内的 1 个以上的配件造成故障结果的情况，应一起考虑由结果造成的所有故障。 Single failure means: (1) a failure appearing in a part or a system that will result in following one or two affections: ①function loss of a part or a system; ②degradation of the function to the degree of lowering safety of vessels, crew or environment; (2) a failure which results in failure of more than one component in a system (common cause failure), all the resulting failures are to be considered together.

Single fluke anchor　单爪锚　系指只有一个锚爪的有杆锚。

Single gas fuel engine　单一气体燃料发动机　系指只能依靠气体燃料运转且不能转换到依靠燃油运转的发动机。 Single gas fuel engine means a power generating engine capable of operating on gas-only, and not able to switch over to oil fuel operation.

Single harmonic content　单次谐波　某次谐波的有效均方根值与基波均方根值之比，用百分数表示。 Single harmonic content is a ratio of the effective r.m.s. value of that harmonic to the r.m.s. value of the fundamental, expressed in percent.

Single hull tanker　单壳油船　系指因不满足 MARPOL 附则Ⅰ中第 13F 条款规定的，其将要在满 25 年时被淘汰的，装运原油或石油产品的单壳、单底或双底的油船。

Single lens reflex (SLR)　单镜头反光相机　简称单反相机。

Single operator action　操作员单一动作　通过不超过 1 个硬键或软键动作（不需移动光标）或使用编程代码启动语音来完成的程序。 Single operator action means a procedure achieved by no more than one hard-key or soft-key action, excluding any necessary cursor movements, or by voice actuation using programmed codes.

Single pitch amplitude Φ　纵摇单幅值 Φ　由下式得出：$\Phi = f_p (960/L)(V/C_B)^{1/4}$。 Single pitch amplitude Φ means a single pitch amplitude, in deg, given by: $\Phi = f_p(960/L)(V/C_B)^{1/4}$.

Single plate clutch　单片离合器

Single plate rudder　平板舵　舵叶由木质或金属平板制成的舵。

Single point mooring (SPM)　单点系泊　系指系泊和转运装置，其在海底管道和系泊船舶（海上浮式装置、油船等）之间提供一种联系，需要时可供输送流体货物用，船舶可系固在上面，且在环境载荷作用下，所系船舶能绕系泊点转动。 Single point mooring means a mooring and transferring arrangements providing a connection between submarine pipelines and mooring ships (offshore floating units, oil tankers, etc.) and if necessary, it can transfer liquid cargo. Ships may be moored to such arrangements and moored ships can turn around the mooring point under environmental loading.

Single point mooring system (SPM)　单点转塔系泊系统

Single quotes　单引号

Single reduction gear　单级减速装置

Single rope grab　单索抓斗　用一根绳索和一个滚筒控制启闭、抛落和提升的抓斗。

Single row ball bearing　单排滚珠轴承

Single shaft turbine　定轴涡轮　所产生的动力一部分用来驱动压气机，其余部分则作为有用功输出的单轴气轮机中的涡轮。

Single side skin bulk carrier　单壳散货船　系指其一个或以上的货舱的边界是舷侧外板或有 2 道水密边界，而其中之一是舷侧外板，且其间距小于 1 000 mm 但大于 760 mm。（水密边界之间的距离，是垂直于舷侧外板量取的）的散货船。 Single side skin bulk carrier is a bulk carrier in which one or more cargo holds are bound by the side shell only or by two watertight boundaries, one of which is the side shell, which is less than 1 000 mm but more than 760 mm apart. The distance between the watertight boundaries is to be measured perpendicular to the side shell.

Single skin member　单壳构件　单壳构件所指的构件是由理想化的梁组成腹板，其顶折边由附板构成底折边由面板构成。 Single skin member is defined as a structural member where the idealized beam comprises a web, with a top flange formed by attached plating and a bottom flange formed by face plating.

Single stage compression refrigerating system　单级压缩制冷系统　制冷剂只经制冷压缩机一次压缩即进入冷凝器放热液化的制冷系统。

Single stage compressor　单级压缩机

Single strut shaft bracket　单臂轴支架

Single thread　单螺纹

Single trawler　单拖网渔船　单船进行拖网作业的拖网渔船。

Single unit floating dock　整体式浮船坞　坞墙与浮体做成整体的浮船坞。

Single well oil production system（SWOPS）　单井式石油生产系统　系指采用水下井口及动力定位生产储油船（兼作运输油将原油运回港口）进行海上石油早期生产与延长测试的海上石油生产测试系统。

Single-buoy mooring　单浮筒系泊　系指借助单个浮筒进行的系泊。 Single-buoy mooring is a mooring connected by a single buoy.

Single-casing steam turbine　单缸汽轮机　只有一个气缸的汽轮机。

Single-effect evaporation　单效蒸发　加热介质的热量在蒸发器内仅被利用一次，即蒸发器产生的二次蒸汽其汽化热由外部冷源吸收，不再用于蒸发海水的蒸发方式。

Single-flow steam turbine　单流式汽轮机　进入低压缸的蒸汽只向一个方向流动的汽轮机。

Single-hydrofoil craft　单水翼艇　只在艉部装有水翼的水翼艇。

Single-layer board　单层板

Single-machine traction　单机拖动　是一种由母线供电无电源变换装置按逻辑信号方式进行控制的不调速或有级调速的电力拖动方式。

Single-pass boiler　单烟道锅炉　具有单侧排烟烟道的锅炉。

Single-phase circuit　单相电路　系指由单一交变电动势供电的电路。单相电路通常经由两根导线供电。这两根电线中的电流从电源向外算起相位差180°或半周。

Single-phase synchronous generator　单相同步发电机　是一种在其接线端上产生单一交变电动势的发电机。这种发电机输出以两倍频率脉动的电功率。

Single-pull hatch cover　滚翻式舱盖　沿舱口纵向分为若干互相牵连并装有滚轮的舱盖板，开舱时滚移、翻转排列并收藏于舱口一端，关舱时则反向移行，依次就位的机械舱盖。

Single-purchase winch（single-drum winch）　单卷筒绞车　具有一个卷筒的绞车。

Single-reduction gear（single-reduction unit）　单级减速齿轮箱　通过一对齿轮的啮合作用实现减速的减速齿轮箱。

Single-shaft（rotor）gas turbine　单轴燃气轮机　压气机和定轴涡轮共同连接在一根轴上的燃气轮机。

Single-stage flash evaporation　单级闪发　热海水在闪发室内仅经过一次闪发的闪发蒸发。

Single-step planning boat　单断级滑行艇　艇底具有一个横向断级，滑行时艇底形成前后两个滑行面的滑行艇。

Sink　汇　流体对称地向所有方向汇集的一点。

Sinkage experiment　沉浮试验　浮船坞在水上进行的单独下沉和起浮的试验。

Sinusoidal maneuver test　正弦形操纵试验　使舵角按正弦形变化操纵，以确定船舶对舵角的高频率响应的一种模型操纵试验。

Sinusoidal propeller　直翼推进器

Sinusoidal wave　正弦波　波面轮廓呈正弦的简谐波。其表达式为 $\xi = \xi_A \sin(k_x - \omega_i)$。

Siren　号笛（报警器）

Sister block　双滑车　由两个绳索绕向相反且共旋转轴互成90°的单滑车组成的多饼滑车。

Sister ship　姊妹船　系指与系列中的第一艘船具有相同主尺度、总布置、舱容图和结构设计的船舶。Sister ship is a ship having the same main dimensions, general arrangement, capacity plan and structural design as those of the first ship in a series, and built in the same shipyard.

Site inspection　现场检查　系指为确认经验证的文件所述的情况而对拆船厂进行的检查。 Site inspection means an inspection of the ship recycling facility, confirming the condition described by the verified documentation.

Site witness　现场见证　系指根据认可的检查和试验计划或与之同等程度的要求现场参与预定检验项目，以核查是否满足检验要求。

Situation awareness　状况告知　系指航海人员根据状况做出及时反应的要求，对提供的航行和技术信息的了解，对其含义的理解以及对其近期状况的预测。状况告知包括模式告知。 Situation awareness is these mariner's perception of the navigational and technical information provided, the comprehension of their meaning and the projection of their status in the near future.

Situational awareness　状态意识　是船员对INS工作站通过的航行和技术信息的感知，对其含义的理解以及对其近期状况的推测。 Situational awareness is the mariner's perception of the navigational and technical information provided at the INS workstation, the comprehension of their meaning and the projection of their status in the near future.

Six Fathom Scarp Mid-Lake Special Protection Area　六斜坡湖心特别保护区　系指在其最北部的一点开始顺时针方向连接下列坐标点的恒向线所包围的区域：北纬 44°55.0′, 西经 082°33.0′; 北纬 44°47.0′, 西经 082°18.0′; 北纬 44°39.0′, 西经 082°12.0′; 北纬 44°27.0′, 西经 082°13.0′; 北纬 44°27.0′, 西经 082°20.0′; 北纬 44°17.0′, 西经 082°25.0′; 北纬 44°17.0′, 西经 082°30.0′; 北纬 44°28.0′, 西经 082°40.0′; 北纬 44°51.0′, 西经 082°44.0′; 北纬 44°53.0′, 西经 082°44.0′; 北纬 44°54.0′, 西经 082°40.0′。Six Fathom Scarp Mid-Lake Special Protection Area mean the area enclosed by rhumb lines connecting the following co-ordinates, beginning on the northernmost point and proceeding clockwise: 44°55.0′N, 082°33.0′W; 44°47.0′N, 082°18.0′W; 44°39.0′N, 082°12.0′W; 44°27.0′N, 082°13.0′W; 44°27.0′N, 082°20.0′W; 44°17.0′N, 082°25.0′W; 44°17.0′N, 082°30.0′W; 44°28.0′N, 082°40.0′W; 44°51.0′N, 082°44.0′W; 44°53.0′N, 082°44.0′W; 44°54.0′N, 082°40.0′W.

Skate　防撞滑架　设置在救生艇外缘前、后处，卸艇时用以承受与船体的碰撞以及滑越船体突出结构，当艇落水后又可和艇体脱开的架子。

Skeg　侧向鳍　系指作为翼的一部分或附于翼上以减小吸入性气动阻力或增加静态或动态气垫的有效性的一种垂直或倾斜的翼板或一种三维结构。Skeg is a vertical or inclined profiled plate or a volumetric construction, which forms part of or is attached to a wing for the purpose of decreasing the inductive aerodynamic resistance or increasing the effectiveness of static or dynamic air cushions.

Skew　侧斜　在螺旋桨正视图上，叶面参考线偏离母线的情况。可在正视图上用叶梢处侧斜角或叶端至螺旋桨基准线垂直距离表示。

Skew angle　侧斜角　系指从螺旋桨中心引出的射线和叶片中心线切线，和通过叶顶的从螺旋桨中心引出的射线构成的角度。Skew angle is the angle between a ray starting at the centre of the propeller axis and tangent to the blade midchord line and a ray also starting at the centre of the propeller axis and passing at the blade tip.

Skew angle of propeller　螺旋桨的侧斜角　系指在桨叶的投影图上，从桨叶的叶梢到桨毂内径中心的连线与桨毂内径的中心向桨叶的半宽弦线的轨迹线做一切线，这 2 根线的夹角。大侧斜螺旋桨的侧斜角大于 25°；而小侧斜螺旋桨的侧斜角小于等于 25°。

Skewed propellers　侧斜螺旋桨　系指该螺旋桨叶片的侧斜角不为零。Skewed propellers are propellers whose blades have a skew angle other than 0.

Skew-induced rake　侧斜轴向位移　螺旋桨某半径处叶面参考线由于侧斜所造成的轴向位移。在母线后者为正号。

Skew-induced rake　侧斜轴向位移

Skid rail　滑道

Skin diving　裸潜　在水中不使用专门的水下呼吸器。只使用简单的通气管、面具、脚蹼或保暖潜水服的潜水活动。

Skin raft　皮筏　用树枝编成框架，下面缚上几只充气的橡皮轮胎或兽皮囊而成的简易水上运载工具。

Skin sensitization　皮肤过敏　（1）符合以下情况的货品确定为会引起皮肤过敏：①如果证明相当数量的人员在皮肤接触该货品后发生过敏；②有相关动物检验的阳性结果。（2）如果采用皮肤过敏的辅助测试，则不低于 30% 的动物反应可认定为阳性。如果采用非辅助性测试，则不低于 15% 的动物反应可认定为阳性。（3）当从鼠耳膨胀测试（MEST）或局部淋巴结化验（LLNA）中取得阳性结果，即足以证明该货品将导致皮肤过敏。Skin sensitization means: (1) a product is classified as a skin sensitizer: ①if there is evidence in humans that the substance can induce sensitization by skin contact in a substantial number of persons; ② where there are positive results from an appropriate animal test. (2) When an adjuvant type test method for skin sensitization is used, a response of at least 30% of the animals is considered as positive. For a non-adjuvant test method a response of at least 15% of the animals is considered positive. (3) When a positive result is obtained from the Mouse Ear Swelling Test (MEST) or the Local Lymph Node Assay (LLNA), this may be sufficient to classify the product as a skin sensitizer.

Skirt (flexible skirt)　围裙　连接于气垫船刚性气道出口处，由柔性材料制成各种形式的用以维持气垫的裙形装置。围裙的出现，使气垫船从实验研究段向工程阶段飞跃。使用围裙的优点：垫升功率大大减少；越障能力明显提高使之真正具有两栖性，其快速性、操纵性以及耐波性有所改善等。围裙对于气垫船相当于轮胎对于汽车的作用。围裙发展经历如下阶段：（1）初期采用的围裙是以延长喷口的形式出现的。这种形式既有形成周边射流气幕的优点，又有保证柔性封闭提高两栖性的特点，但其阻力性能是越峰性能欠佳，围裙成形的稳定性也不好。稍后，在此基础上将围裙的上部扩大或扩压室，以期保证气幕射流稳定。这样围裙的横截面与火腿相似，称为火腿形围裙。火腿形围裙由于扩压室存在，不会产生皱折或外翻，但围裙的索链连接容易损坏。后来在围裙外层和内层装有较大的隔片，隔片与围裙内外层用尼龙线缝制连接。这种连接比索链连接有所改进，但尼龙线连接强度较差。（2）目前普遍使用的是囊指围裙(bag fingers skirt) 这种围裙由囊和手指两部分组成。囊的底部开有囊孔，气流通过囊孔进入手指形成气垫。手指系指形围裙，连接在囊的下面，各手指互不连接，能单独活动，这样把囊与运行中容易磨损的部件分离开来。手指容易磨损，磨损后可以更换手指，而不必更换大囊。囊指型围裙最大的优点是结构简单、制造方便，安装维修容易，使用寿命长，越障能力和耐波性均较好。(3) 锥形围裙：有的气垫船艏围裙的高度较大，船艉围裙的较低，这样形成了一个锥度，称为锥形围裙。它有一定的抬头角度，有助于提高波浪性。(4) 低囊压比围裙：早期设计的囊指围裙，囊压比一般均为 1.5～1.6 左右，垫升功率消耗太大。现在为了节省垫升功率，囊压比一般取 1.15～1.2，提高了经济性。(5) 响应围裙(responsive skirt) 增加围裙囊的宽度，使围裙囊的外凸肩大大增加（即围裙伸出刚性船体部分大大增加）使之不易缩进，同时使囊的随波起伏响应增大，因而称为响应围裙。它可以大大减少船的运动幅值和运动加速度。为了保证全垫升气垫船的稳定性，船底设有纵向和横向稳定围裙。稳定围裙有十字形和 T 字形两种，又为保证全垫升气垫船尾部围裙不产生库水现象，因此在围裙囊下面设有锥囊、锥筒的小囊。侧壁式气垫船围裙比较简单，围裙有深手指和囊指形，船艏围裙有双囊形和三囊形。内河侧壁式气垫船也有采用滑板式艏封板用以防止积砂。实践证明，尼龙织物外涂天然橡胶或人工合成橡胶是最为理想的围裙材料，选择材料时应当考虑抗拉强度、抗撕裂强度、剥离强度（即抗脱层能力）、尼龙织物的柔软性和抗老化能力以及低温特性特。

Skirt lift 提裙装置 为了控制纵倾、横倾，能在垫升航态时把部分围裙提升及放下，以调节逸气面积及方向的装置。

Skylight 天窗 系指设有或不设有窗玻璃的甲板开口，供机舱、住舱等通风用。 Skylight is a deck opening fitted with or without a window pane and serving as a ventilator for engine room, quarter, etc.

Skylight coaming 天窗框架 沿天窗开口周缘凸出顶棚的框架。

Skylight flap 天窗折合盖

Skylight gear 天窗传动装置（天窗启闭机构）

Skylight grating 天窗格栅

Skylight hood (skylight lid, skylight cover) 天窗盖 用铰链联结于天窗框架上，装有防护栅或配有钢丝玻璃的盖子。

Skylight quadrant 天窗扇形支撑齿条

Skylight quadrant (skylight controlling gear) 天窗传动装置 启闭天窗盖的机构。

Slack meter 电缆松紧指示器 设在布缆控制室内，能自动显示瞬时布缆松紧度的仪器。

Slacken a bolt 旋松螺栓

Slag marks 夹渣 焊接后残留在焊缝的熔渣。由于夹渣的存在，焊缝的有效截面减小，过大的夹渣也会降低焊缝的强度和致密性。

Slag wool 矿渣棉

Slamming 砰击 船舶在垂荡和纵摇时，船体与波浪间产生猛烈局部冲击的现象。多发生在艏部。

Slant leg jack-up drilling unit 倾斜桩腿自升式钻井平台 系指斜置桩腿使诸桩腿在海底的间距增大，以改善平台稳定性的自升式钻井平台。

Slanted plates 挡货板 用以提高槽形舱壁和骨架构件结构稳定性的斜板。

S-Lay S 式敷管法 系指管子以垂直（或接近垂直）方向离开敷管船并其悬浮段形成 S 形的一种管路安放方法。 S-Lay is a pipeline installation method where the pipe leaves the installation vessel in a vertical (or nearly vertical) orientation, and the suspended section takes an "S" shape.

Sled 拖运器 系指锚泊海底的结构。它作为流通管的开始点，并通过跨接管连接到油井和总管。 Sled is an anchored seabed structure that acts as the starting point for a flowline and connects to a well or manifold with a jumper tube.

Sleeping AIS target 静止自动识别系统（AIS）目标 指示在特定位置，且配备了 AIS 的船舶的存在和方向的目标。该目标以"静止目标"符号显示，被激活前不显示附加信息。 Sleeping AIS target is a target indicating the presence and orientation of a vessel equipped with AIS in a certain location. The target is displayed by a sleeping target symbol. No additional information is presented until activated.

Sleeping Bear Dunes National Lakeshore 睡熊沙丘国家湖滨区 系指就在或邻近国家公园管理局管理的苏必利尔湖的地方，其包括北神灵岛、南神灵岛和密歇根湖岸线，从沿岸线向东位于北纬 44°42′45.1″西经 086°12′18.1″北部陆地的一点到北纬 44°57′12.0″，西经 085°48′12.8″处。 Sleeping Bear Dunes National Lakeshore means the site on or near Lake Michigan, administered by the National Park Service, that includes North Maniton Island, South Manitou Island and the Michigan shoreline from a point of land at 44°42′45.1″N, 086°12′18.1″W, north and eastward along the shoreline to 44°57′12.0″N, 085°48′12.8″W.

Sleeve 衬套（套筒）

Sleeve bearing 套筒轴承

Sleeve coupling (box coupling) 套筒联轴器 通常用来连接艉轴与推进器轴的圆筒形刚性联轴器。

Sleeve joint 套筒接合

Sleeve nut 套筒螺母

Sleeve weld 套筒焊接

Slender body theory 细长体理论 用假设船宽和吃水对船长的比值都很小的一次方或更高次方的参数来展开船舶运动的速度势，并按如自由表面和船壁等边界条件求解的一种理论。

Slenderness ratio 吊杆长细比 吊杆长度与吊杆中央断面的惯性半径之比。

Slew type floating crane 转机式起重船 设有装在转盘上能旋转的起重机的起重船。

Slewing angle 吊杆偏角 吊杆向舷外偏转时，其轴线与船体中线面之间的夹角。

Slewing ramp 旋转跳板 系指可以向两舷旋转的跳板。它是由斜跳板发展而来的。它克服了斜跳板只能使船一侧靠岸的缺点更增强了船舶的使用性能，但其结构复杂，造价亦高。

Slewing winch 回转吊杆绞车 用以回转和保持吊杆位置的绞车。

Slide shaft 滑动轴

Slide valve 滑阀

Slide valve diagram 滑阀图

Sliding 滑动（滑移）

Sliding area 滑动面

Sliding bearing 滑动轴承

Sliding block 滑块（滑动龙骨墩） 沿着装置在机架上的导板上下滑动的部件。

Sliding door (sluice door) 水密滑门（滑移门） 可沿紧贴舱壁的导轨移动以进行开闭的水密门。

Sliding expansion joint 滑动膨胀接头

Sliding foot 滑动底脚

Sliding friction 滑动摩擦

Sliding friction force 滑动摩擦力

Sliding key 滑键

Sliding pressure starting 滑参数启动 汽轮机进气压力和温度在启动过程中逐渐增高的启动方式。

Sliding type 滑动式

Sliding type expansion joint 滑动式膨缩头

Sliding window 扯窗 采用上下或左右移动的开闭形式，用人力提拉、定位或扯移的窗。

Slim-line task 钢丝作业 系指用于回收遗失在井下的工具、井径切割、下放和回收井塞、设置和移除钢丝可回收阀和记忆测井的作业方式。

Sling bridge pier 吊桥式码头 系指通过吊桥与船舶连接的码头。这种码头对潮差和风浪的适应性强，但初投资和维修费用大。

Sling hook 吊艇环 位于吊艇索端或浮动滑车下用以直接承吊小艇的套环。

Slings 吊具 供装卸货物用的各种辅助吊装工具的统称。

Slip 滑脱（滑距） (1)螺旋桨每转进程小于其螺距的现象；(2)螺旋桨螺距与其每转进程之差。

Slip 滑差 系指感应电机中，其同步转速与运行转速之差。可以用下述方法表示：(1)同步转速的百分数；(2)同步转速的十进制小数；(3)直接以每分钟转数表示。

Slip 转差率 耦合器泵轮与涡轮的转速差对泵轮转速的比值。$S = (n_1 - n_2)/n_1$；式中，n_1—泵轮转速；n_2—涡轮转速。

Slip expansion joint 滑套式膨胀接头

Slip hook 速脱钩 固结在拖网渔船或围网渔船尾部或舷侧，用以钩住或迅速脱出网具的钢索，以便完成预定操作的特制的活络铁钩。

Slip joint 滑动接合（伸缩式连接）

Slip speed 滑速 螺旋桨转速与螺距的乘积与其进速之差。

Slip stopper 链扣止链器 系指将链条的一端固定在甲板上，另一端连有脱钩式夹扣，以止住锚链的止链器。

Slip type 滑动型

Slip type coupling 滑动型联轴节 系指液压联轴节、电磁联轴节或等效者。 Slip type coupling signifies hydraulic coupling, electro-magnetic coupling or the equivalent.

Slip-on buckle arrestor 滑动式屈曲限制器 系指一个沿着管路滑动的环。其用于对管路进行局部加强以防止破坏。定期沿管路安放，它用来阻止新出现的屈曲蔓延。 Slip-on buckle arrestor is a ring that slips over the pipeline, used to locally strengthen the pipeline against collapse. Placed periodically along the line, it is used to stop an incoming propagating buckle.

Slip-on joint 滑套接头

Slipper guide 滑块导板

Slipper 滑块

Slipway 滑道 系指船舶下水时，从陆上滑落到水中的通道。

Slit tube 有缝管

Slitless 无缝的

Slop 含油污水（废油，污油）

Slop chute 污水斜槽

Slop piping 废液管系

Slop tank　污油水舱（废油舱）　系指：（1）专用于收集舱柜排出物、洗舱水和其他油性混合物的舱柜；（2）计划以货油舱清洗后，舣部的存储和装载货油为目的所设置的油舱，在矿砂船或散装货船航行时装载闪点低于 60 ℃ 油料的油舱；（3）油船用以从洗舱之后的货油舱收集油水混合物的液舱。　Slop tank means：（1）a tank specifically designated for the collection of tank drainings, tank washings and other oily mixtures；（2）a tank which is provided mainly for the carriage of tank washings and cargo oil and which is designed to be capable of loading oil whose flash point does not exceed 60 ℃ when the ship is in the dry cargo mode；（3）a tank in an oil tanker which is used to collect the oil and water mixtures from cargo tanks after tank washing.

Slop tank arrangements　污油水舱装置　系指包括污油水舱在内，连同附属的污油水泵、管路、阀门、仪表等一整套装置。

Slope　斜度（斜率，坡度）

Slope variated slipway　斜坡滑道　系指具有一定坡度的、供船舶下水用的滑道。

Sloping vehicle ramps　车辆坡道

Slopping bulkhead　斜舱壁　不与基平面垂直的舱壁。

Sloshing impact load　晃荡冲击载荷

Slot　槽（切口，箱位）

Slot　井口槽　用于安装钻井、井架以及为便于钻井平台退出井位而设在钻井甲板端部的凹形开槽。

Slot jack-up drilling unit　槽口自升式钻井平台　系指其钻柱从钻台下通过甲板直达海底进行钻井，将甲板结构做成凹形槽口的自升式钻井平台。

Slotted head screw　槽头螺钉

Slotted nut　有槽螺母

Slow adhead　慢速顺车

Slow astern　慢速倒车

Slow combustion　缓慢燃烧

Slow cooling　缓冷

Slow infrared transfer　低速红外线　系指其传道输送速度在每秒 115.2 kbits 的红外线。

Fast infrared transfer　高速红外线　系指传道输送速度在每秒 1 mbits 或 4 mbits 的红外线。

Slow speed diesel　低速柴油机

Slow-speed steam turbine（low-speed steam turbine）　低速级组　为提高低速经济性而设置的，在低速时投入运行，高速时空转的汽轮机级组。

Sludge　油渣（油泥，污泥）　锅水中不溶性杂质的沉淀物。

Sludge collecting pumps　油泥收集泵　系指能用于从任何残油（油泥）产生设备或排入残油（油泥）舱以外的舱柜抽吸并仅排入残油（油泥）舱的泵。　Sludge collecting pumps are pumps capable of taking suction from any oil residue (sludge) producing equipment or tank, other than an oil residue (sludge) tank(s), and discharging only to oil residue (sludge) tank(s).

Sludge collecting tank　油泥舱（残渣收集器，油渣舱）

Sludge content　含渣量

Sludge hole　出泥孔（放油渣孔）

Sludge oil（waste）　残油　系指来自燃油或润滑油分离器的油泥，主机或辅机的废弃润滑油，或舱底水分离器、油过滤装置或滴油盘的废油。残油的成分为：（1）75% 重燃油的残油；（2）5% 废润滑油（3）20% 乳化油。　Sludge oil (waste) means sludge from the fuel oil or lubricating oil separators, waste lubricating oil from main or auxiliary machinery, or waste oil from bilge water separators, oil filtering equipment or drip trays. Sludge oil consisting of：（1）75% sludge oil from HFO；（2）5% waste lubricating oil（3）20% emulsified water.

Sludge pipe　油泥管

Sludge pump　油渣泵（油泥泵）

Sludgeless oil　无渣油

Sluice door　闸门

Sluice vavle　闸阀

Slushing oil　防锈油（滑脂）

Small diameter line　小直径管路　这是用于油船在卸货完成时将所有货油泵及货油管路泄空的一条管路，通常连接到扫舱装置，并排往岸上，连接于货油汇集管阀门的向舷外的一侧。新油船的小直径管路的横剖面面积不应超过船上主卸货管路直径的 10%。

Small hatch cover　小舱盖　系指装用于尺寸较小的舱盖上，舱盖板多为整块，用填料和压紧器保持密性的舱口盖。

Small magazines　小弹药库　是开放地设在上甲板外的舱室。受形状和尺寸的限制，这种弹药库不可进入，库内物资只能从外面处理。有些军火需要安全地进行长期储存，也有些军火需要时刻处于待用状态，这种弹药库就是为了存放这两种军火而专门设计和建造的。　Small magazines are compartments opening off the upper deck which are restricted by shape and size, which does not permit walk-in and where the contents are handled from outside. Small magazines are to be specifically designed and constructed for the safe permanent or ready use stowage of munitions defined in the armaments requirement.

Small oil fuel tank 小型燃油舱　系指单个最大容积小于 30 m³ 的燃油舱。　Small oil fuel tank is an oil fuel tank with a maximum individual capacity no greater than 30 m³.

Small openings 小开口　系指:(1)从甲板或别处通入某一舱室的小开口,通常设有盖或罩,或门;(2)在一个横剖面内强力甲板或船底上尺寸小于上述规定的开口。不必从船体梁横剖面面积内扣除。假如: $\sum b_s$ ≤0.06($B- \sum b$),式中, $\sum b_s$ —所考虑横剖面内强力甲板或船底上小开口的总宽度,m; $\sum b$ —所考虑横剖面内大开口的总宽度,m;小开口的总宽度 $\sum b_s$ 如不符合以上标准,仅从船底横剖面面积中扣除超出宽度的部分。　Smaller openings mean:(1) a small opening in a deck or elsewhere, usually fitted with a cover or lid or a door for access to a compartment;(2) one transverse section those above of in the strength deck or bottom area do not need to be deducted from the sectional areas included in the hull girder transverse sections, provided that: $\sum b_s$ ≤ 0.06 × (B- $\sum b$), where: $\sum b_s$ —total breadth of small openings, in m, in the strength deck or bottom area at the transverse section considered; $\sum b$ —total breadth of large openings, in m, at the transverse section considered. Where the total breadth of small openings $\sum b_s$ does not fulfill the above criteria, only the excess of breadth is to be deducted from the sectional areas included in the hull girder transverse sections.

Small profit and quick-returns (SPQR)　薄利多销

Small size TLP 小型张力腿平台　系指主要用于柔性开发系统,即湿树模式的张力腿平台。

Small vessel 小船　系指载重线长度小于 24 m 或吨位小于 150 t 的,1968 年 7 月 21 日前安放龙骨或处于类似建造阶段的船舶。　Small vessel means a vessel of less than 24 meters in load line length, or a vessel of less than 150 tons, where the keel of that vessel was laid, or where the vessel was at a similar stage of construction, before 21st July 1968.

Small water-plane area twin hull (SWATH)　非高速小水线面双体船　系指其最大航速限定为 V < 30(kn),且在设计中不考虑船体产生的水动力及其特性的小水线面双体船。式中,V—船舶处于最大营运质量状态,以核定的最大持续推进功率,在静水中航行能达到的速度。

Small waterplane area twin hull craft (SWATH)　小水线面双体船　系指:(1)为改善耐波性和减少兴波阻力,将双体船的片体在水线处缩小形成狭长流线型截面的双体船。其主船体由连接桥结构连接左右两个片体组成。每一片体包括上船体,支柱体和下潜体;(2)具有小水线面面积,且片体水下部分呈鱼雷状的一种特殊船型的双体船。　Small waterplane area twin hull craft means:(1) double hull craft with narrow linear section of the hulls at waterline in order to improve sea-keeping ability and reduce wave-making resistance. The main body is composed of two semi-hulls connected by the left and right cross-decks, either of which includes upper hull, vertical struts and lower hulls;(2) a special type of catamaran with small waterplane area, and with underwater portions of hulls being formed in shape of torpedo.

Small waterplane area twin hull (HSC)　高速小水线面双体船　系指其最大航速满足 $V \geq 3.7 \nabla^{0.1667}$ (m/s)的小水线面双体船。式中,V—船舶处于最大营运重量状态,以核定的最大持续推进功率,在静水中航行能达到的速度; ∇ —设计水线对应的体积,m³。

Small water-plane area twin-hull ship 小水线面双体船　属高科技、高附加值、高性能船舶。不论是在航行状态,还是停泊状态,该类船舶的耐波性都明显优于其他常规船舶。因其晕船率低、出航率高、高海况时失速率低、快速性好,小水线面双体船被誉为"不晕船的船""全船候船"。此外,这种船舶还具有操纵灵活,可原地回转、甲板宽敞、作业空间大、安全裕度高、水下声辐射小等特点。目前,小水线面双体船被广泛应用于声呐搜索、海洋调查、海上交通、海洋工程作业、缉私、监督与救助、旅游等方面。小水线面双体船同时具有较明显的缺点,摩擦阻力大、船体结构复杂、质量比相同排水量的单体船大,增加了结构设计难度,吃水变化敏感,设计、建造、使用过程中,要对船的质量及其分布加权控制,必须有压载调整补偿系统。见图 S-8。

Smartphone 智能手机　是一种运算能力及功能比传统手机更强的手机。

Smaze 烟霾　系指烟和霾混合而成的空气污染物,样子类似于烟雾,但是湿度稍小。　Smaze is a mixture of smoke and haze similar to smog in appearance, but it is less damp in consistency.

Smith correction 波浪浮力修正　由于考虑到波浪中水分子运动所产生的惯性力的影响,而对按静水压力计算的浮力所作的修正。

Smith Mcibtyre mud sampler 史密斯-麦金太尔取泥器　又称弹簧采泥器。用于大型底栖生物调查的采集器。

图 S-8 小水线面双体船
Figure S-8 Small water-plane area twin-hull ship

SMM 汉堡海事展览会
Smog 烟雾 系指烟和雾混合而成的又黑又浓的空气污染物。
Smoke accumulator 聚烟器（使用于抽烟式的探火系统）
Smoke alarm 烟气报警器
Smoke box 烟箱
Smoke box door 烟箱门
Smoke cover 防烟套
Smoke deflector 导烟板
Smoke detecting device 感烟检测器（测烟器）
Smoke detecting plant 感烟探测装置
Smoke detecting system 感烟探测系统
Smoke detection appliance 感烟探测设备
Smoke detection system 烟气探测系统
Smoke detector 感烟探测器（烟气探测器）
Smoke extraction system 抽烟系统
Smoke flue 烟道（烟囱）
Smoke funnel 烟囱
Smoke hatch 通烟口
Smoke indicator 烟色指示器
Smoke mask 防烟面罩
Smoke mask (fire mask) 防烟面具 供消防人员穿戴的面具。
Smoke observation device 烟气观测装置
Smoke pipe 排烟管（烟管）
Smoke pipe fire alarm system 烟管火警系统
Smoke signal 烟雾信号（发烟信号）
Smoke stack 烟囱
Smoke stack hood 烟囱帽
Smoke-free 无烟的
Smokeless coal 无烟煤
Smoke-tight integrity 防烟完整性
Smoke-tight or capable of preventing the passage of smoke 烟密或能防止烟气通过 系指用不燃材料或阻燃材料制成的分隔能阻止烟气通过。 Smoke-tight or capable of preventing the passage of smoke means that a division made of non-combustible or fire-restricting materials is capable of preventing the passage of smoke.
Smoking room 吸烟室
Smooth 光滑的（弄平滑）
Smooth surface 光滑面
Smooth water service area 静水区域（平水区域） 系指湖水、河川和港内的水域。 Smooth water service area means water area within lakes, rivers and harbours.
Smothery station ventilation 灭火站室通风 为使灭火站内温度不致太高并排除可能从钢瓶内泄漏的氮气或二氧化碳等惰性气体而进行的自然或机械通风。
Snake wave generator 蛇形造波机 在试验水池中，用大量连成一行的各单元造波机，可各自调整幅值、频率和相位以制造复杂人工波形的设备。
Snaking 纵向破裂 系指由于热膨胀而引起海底管路的屈曲。 Snaking means a seafloor line buckling of a pipeline due to thermal expansion.
Snapper (grab sampler) 表层采样器 海洋底部表层采样用的装置。主要部件为一对"铁钳"式抓斗组成。
Snatch block 开口滑车 具有便于脱卸绳索的有开口的滑车。
Snifting valve 吸气阀
Snubber 减振器（缓冲器,消声器）
Snubber fitting 减振接头
Social deviance event 社会异常事件 主要指一些偶发事件阻碍合同的履行。如罢工、骚乱等。这些事件既不是自然事件，也不是政府行为，而是人为的行为，但对于合同当事人来说，在订约时是不可预见的，因此也可以成为不可抗力的事件。
Social risk 社会风险 系指整个群体（船员、港口雇员或整个社会）所蒙受的平均风险，这里的风险一般指死亡风险。它用于描述事故发生概率与事故造成的人员受伤或死亡人数的相互关系。社会风险主要用 F-N 曲线来衡量。
Society 船级社 系指办理船舶入级的船级社。 Society means the Classification Society with which the ship is classed.
Society product certificate 船级社产品证书 由验船师签署的文件，其阐明：(1) 该产品符合规范要求；(2) 在检定该产品时已进行过试验；(3) 已对该发证产品进行取样试验；(4) 有验船师在场时或者按特定的协议对该产品进行过试验。 Society product certificate means

a document signed by a surveyor stating: (1) conformity of the certified product with Rule requirements. (2) that tests are carried out on the certified product itself; (3) that tests are made on samples taken from the certified product itself; (4) that tests are performed in presence of the surveyor or in accordance with special agreements.

Socio-technical approach to human reliability assessment 人因可靠性评估的社会技术方法 属于专家判断方法。没有对人误行为作出显示的处理，而是使用影响图表达多层次 PSFs 与目标任务间的关系。

Socket 插座 系指安装在插座引出线端上，可容纳一个连接插头的器件。

Socket joint 活节接合

Socket spanner 套筒扳手

Socket type joint 插座接合（套管接合）

Soda ash (dense and light) 苏打灰 （密质和轻质）粉末状，由白色无嗅细粒和粉尘组成。用盐和石烧制而成。溶于水。苏打灰碰到油就会毁坏。无特殊危害。该货物为不燃物或失火风险低。Soda ash is in powdery, composed of white, odourless grains and dust. It is made by the combustion of salt and limestone. It is soluble in water. Soda ash is ruined in contact with oil. It has no special hazards. This cargo is non-combustible or has a low fire-risk.

Sodium nitrate (UN 1498) 硝酸钠 无色、透明、无嗅晶体。吸湿并溶于水。虽不可燃，但其与可燃物质的混合物容易点燃并快速猛烈燃烧。该货物吸湿，受潮会结块。Sodium nitrate is a colourless, transparent, odourless crystals. It is hygroscopic and soluble in water. Although it is non-combustible, mixtures with combustible material are readily ignited and may burn fiercely. This cargo is hygroscopic and will cake if wet.

Sodium nitrate and potassium nitrate mixture (UN 1499) 硝酸钠和硝酸钾混合物 一种吸湿的混合物，溶于水。虽不可燃，但其与可燃物质的混合物容易点燃并快速猛烈燃烧。该货物吸湿，受潮会结块。Sodium nitrate and potassium nitrate mixture (UN 1499) is a hygroscopic mixture, soluble in water. Although it is non-combustible, mixtures with combustible material are readily ignited and may burn fiercely. This cargo is hygroscopic and will cake if wet.

Soft chine 圆舭部 呈圆角的舭。

Soft coating 软涂层 系指用以原油产品或羊毛脂类物质为基料的不干或半干性涂层。Soft coating means non-dried or semi-dried coating with the crude oil product or wool grease substance as base material.

Soft steel/low carbon steel 低碳钢 为碳含量低于 0.25% 的碳素钢，因其强度低、硬度低而软，故又称软钢。它包括大部分普通碳素结构钢和一部分优质碳素结构钢，大多不经热处理用于工程结构件，有的经渗碳和其他热处理用于要求耐磨的机械零件。低碳钢退火组织为铁素体和少量珠光体，其强度和硬度较低，塑性和韧性较好。因此，其冷成形性良好，可采用卷边、折弯、冲压等方法进行冷成形。这种钢还具有良好的焊接性。含碳量从 0.10% 至 0.30%。低碳钢易于接受各种加工如锻造、焊接和切削，常用于制造链条、铆钉、螺栓、轴等。

Soft water 软水

Soft wood 软木

Soften (softening) 软化（漏气，软水剂）

Softener 软化剂

Soft Ether 虚拟网卡 就是模拟以太网卡的工作顺序，可以模拟 HUB 功能使用 tunnel 特性，实现 VPN 的功能。

Soft-tank 软舱 在 Spar 平台结构中，系指提供压载的结构部分。又称压载舱。

Software 软件 系指：(1) 与计算机系统操作有关的程序、数据和文件，如应用程序（包括自检程序）、操作系统程序等；(2) 一个程序和计算机系统操作的文档。Software means: (1) programs, data and documentation associated with the operation of a computer system such as, application programs (including self-check program), operational programm, etc.; (2) the program, procedures and associated documentation pertaining to the operation of the computer system.

Software registry 软件注册

Sogou "搜狗" 系指一项非常强大的搜索技术和先进的 p2p 下载技术。

Sogou search engine "搜狗"搜索引擎 是搜狐公司于 2004 年 8 月 3 日推出的全球首个第三代互动式中文搜索引擎，域名为 www.sogou.com。"搜狗"以搜索技术为核心，致力于中文互联网信息的深度挖掘，帮助中国上亿网民加快信息获取速度，为用户创造价值。

Soil pipe 污水管（粪便管）

Solar cell 太阳能电池

Solar energy 太阳能

Solar energy ship 太阳能船舶 一种采用太阳能、柴电动力混合能源技术的"绿色"船舶。船舶配备全电脑控制太阳能收集器，可高效收集太阳能。其通过 3 种模式为其系统提供动力：纯太阳能、推进蓄电池组模式、纯柴油发电机组模式及混合模式。其中纯太阳能、推进蓄电池组模式是把太阳能转换成电能，贮存在推进蓄池组中,再通过输电和交直流转换设备供电,最终实现电力推进,此时船舶完全利用太阳能航行,全船处于静音状态,航速为 0~4 节(kn)。纯柴油发电机组模式是由两套柴油发电机组向动力系统供电实现电力推进,此时该船航速为 6~8 节。混合模式是太阳能、推进蓄电池组与 1 套柴油发电机组同时使用,通过变电、并网、输电设备等最终实现电力推进,此时其航速为 4~6 节。该船装有可跟踪太阳的可变太阳翼,用于接收太阳能并将其转换为电能。翼的两面都能接收太阳能,其有 3 种操作模式:手动、半自动和全自动。当选用全自动模式时,其方位将根据风力、风向和变化的阳光自动优化。此外,还有由太阳能、风能和燃油提供动力的船舶。

Solar energy sightseeing boat 太阳能游览船

Solar enery electric propulsion 太阳能电力推进 系指利用装在船(艇)顶上的太阳能电池,把太阳能直接转变成电能,驱动推进电动机的电力推进方式。

Solar reduction gear 恒星式减速齿轮箱 中间太阳轮静止不动,而输入功率的行星架和输出功率的外环内齿轮作同方向转动的减速齿轮箱。

Solarium 日光浴室(豪华邮轮)

SOLAS A pack SOLAS A 型封堵器 系指满足 IMO《国际救生设备规则》要求的救生筏应急封堵器。SOLAS A pack means a liferaft emergency pack complying with the requirements of the IMO International Life-Saving Appliances Code.

SOLAS Convention 国际海事人命安全公约 国际海事人命安全公约是以相继颁布的形式,被普遍认为是涉及商船安全的所有国际协定中最为重要的公约。其第一版是为了响应"泰坦尼克"号灾难 1914 年被国际社会正式实施的;第二版于 1929 年颁布;第三版于 1948 年颁布;第四版 1960 年颁布。1960 年版公约是国际海事组织在其创立后的首要任务。SOLAS 已经过连续多年的增补修订。 SOLAS Convention (International Convention for the safety of life at sea) in its successive forms is generally regarded as the most important of all international treaties concerning the safety of merchant ships. The first version was internationally adopted in 1914, in response to the Titanic disaster, the second in 1929, the third in 1948, and the fourth in 1960. The 1960 Convention was the first major task for IMO after the Organization's creation. SOLAS was continually updated by amendments through the years.

Sole arbitrator 独任仲裁员

Sole license 排他许可证

Sole piece 舵柱底骨 系指舵柱底部连接螺旋桨柱和舵柱或支承下舵销的杆材。

Solenoid vavle 电磁阀

Solid bulk cargo 固体散货 系指除液体或气体以外,由粒子、颗粒或较大块状物质组成的任何物质,成分通常一致,并直接装入船舶的货物处所而无须任何中间围护形式。 Solid bulk cargo means any material, other than liquid or gas, consisting of a combination of particles, granules or any larger pieces of material, generally uniform in composition, which is loaded directly into the cargo spaces of a ship without any intermediate form of containment.

Solid bulk cargo carrier 固体散货船 专门从事运输大宗固体散货货物,如煤炭、矿石、谷物、水泥、饲料等的船舶。这类船舶多为单甲板尾机型船。根据运输货物的种类和船舶结构形式的不同,此类船舶又可分为通用型散货船、专用型散货船和自卸式散货船。此类船舶一般有固定航线,在国际海上货物运输中占较大的比例。

Solid cargo/oil hold 固体货物/油、货兼用舱 系指用矿砂或散装货船进行运航时,作为固体货物荷载舱或油船航行时作为货油舱使用的处所。 Solid cargo/oil hold is a compartment which is used as a solid cargo stowing hold when the ship is in the dry cargo mode and which is used as a cargo oil tank when the ship is not in the dry cargo mode.

Solid flange coupling 整体法兰联轴节

Solid floor 主肋板 用板材制成的肋板。

Solid forging 实锻(整体锻件)

Solid inclusions (other than copper and tungsten) 固体夹杂物(铜、钨除外) 固体夹杂物包括氧化物。如果在横截面上出现几种夹杂物 $h_1, h_2, h_3, \cdots$,则其总和为 $\sum h = h_1 + h_2 + h_3 + \cdots$。 Solid inclusions include oxide inclusions. if several inclusions $h_1, h_2, h_3, \cdots$ are present in a section, the sum is $\sum h = h_1 + h_2 + h_3 + \cdots$。

Solid injection diesel 无气喷射式柴油机

Solid ink-jet printer 固体喷墨打印机

Solid matter 固体物质

Solid propeller 整体螺旋桨 系指(包括桨毂和叶片)构成一体的螺旋桨。Solid propeller is a propeller (including hub and blades) cast in one piece.

Solid shaft 实心轴 整个轴段上无轴向通孔的轴。

Solid state 固态

Solid state disk (SSD) 固态硬盘 是一种基于永久性存储器,如内存,或永久性存储器、同步动态随机存取储存器等。

Solid waste 固态废弃物(固态废物) 其成分为:(1)50% 食物废弃物;(2)50% 垃圾(包括约 30% 纸;约 40% 硬纸板;约 10% 破布;约 20% 塑料);(3)混合物的

湿度可达50%和不燃固态物质可达7%。 Solid waste consisting of: (1) 50% food waste; (2) 50% rubbish containing (including approx. 30% paper; 40% cardboard; approx. 10% rags; approx. 20% plastic); (3) the mixture will have up to 50% moisture and 7% incombustible solids.

Solid waste disposal act [美国]浓污水处理法令

Solidifying NLS 凝固的有毒液体物质 系指具有下列熔点的A、B或C类有毒液体物质: (1)熔点高于0℃但低于15℃, 在卸货时的温度高出其熔点不足5℃; (2)熔点等于或大于15℃, 在卸货时的温度高出其熔点不足10℃。 Solidifying NLS means category A、B or C NLS that has a melting point: (1) greater than 0 ℃ but less than 15 ℃, that the temperature is less than above its melting point at the time it is unloaded; or (2) 15 ℃ or greater and a temperature, that the temperature is less than 10 ℃ above its melting point at the time it is unloaded.

Solidifying substance 固化物质 系指有毒液体物质,即: (1)物质的熔点低于15℃, 处于卸载时熔点以上不到5℃的温度; (2)物质的熔点等于或高于15℃, 处于卸载时熔点以上不到10℃的温度。 Solidifying substance means a noxious liquid substance which: (1) in the case of a substance with a melting point of less than 15 ℃, is at a temperature of less than 5 ℃ above its melting point at the time of unloading; or (2) in the case of a substance with a melting point of equal to or greater than 15 ℃, is at a temperature of less than 10 ℃ above its melting point at the time of unloading.

Solidity of blade elements 叶元体实度 螺旋桨一定半径处叶元体的弦长对周向间距的比值。即其间隙比的倒数。

Solids-water partition coefficient K_d 固体-水分布系数

Solution 解决方案 就是针对某些已经体现出的, 或者可以预期的问题、不足、缺陷、需求等等, 所提出的一个解决整体问题的方案(建议书、计划表), 同时能够确保加以有效的执行。解决方案不局限于解决本次问题, 它应该避免相关问题的出现, 警示相关的人员, 并且能够做到经验的传承积累。这时, 可以参考戴尔的磁盘保护解决方案, 这是企业需要进行快速备份大量数据时, 提供数据保护的第一道防线, 可以根据企业的需求, 恢复单个文件, 或者现有服务器的整个映像, 在减轻主数据丢失对组织影响的同时, 还降低管理数据保护的成本。

Solution 溶液(解决,解释)

Solution time 溶解时间

Solution treated condition 固熔热处理状态

Solution treatment 固溶热处理 系指在将合金加热到一定温度, 保持充分的时间使一种或几种溶质融入其固溶体并进行急速冷却, 制止其析出提高其硬度的处理。 Solution treatment is the heating and holding an alloy at a temperature at which one (or more) constituent enters into solid solution, then cooling the alloy rapidly to prevent the constituent from precipitating and to improve the hardness.

Solvent 溶剂 系指包括在化学配方中的一种化学物质。为了引起反应来溶解和消除杂质, 需要这种物质, 使制品更容易处理。 Solvent means a chemical substance which is included in the formula, its presence is required in order to cause a reaction when dissolving and eliminating impurities, thus marking the product easier to handle.

Sonar 声呐 其全称为声音导航和测距。是一种利用声波在水下的传播特性, 通过电声转换和信息处理, 完成水下探测和通信任务的电子设备。声呐技术出现至今已有100多年的历史。世界上第一部声呐仪是一种被动式的聆听装置, 主要用来侦察冰山。到20世纪中期, 不但有回声探测设备(主动式声呐), 还有靠受听舰船发出的噪声进行探测的被动式声呐, 有装在船壳上的大型声呐, 也有可以拖曳的拖曳声呐, 有直升机用的吊放式声呐, 也有固定翼飞机用的声呐浮标。其中, 主动式声呐由简单的回声探测仪演变而来, 能主动发射声波, 然后收测目标回波进行判断, 适用于探测冰山、暗礁、沉船、水深、鱼群、水雷和关闭发动机的隐蔽潜艇等; 被动式声呐则由简单的水听器演变而来, 能够收听目标发出的噪声, 判断出该目标的位置和特性。

Sonar control room 声呐室 供安置声呐发射和接受设备并设置相应操作部位的房间。

Sonar transducer room 声呐舱 供安置声呐升降装置的舱。

Sons of the American Revolution (SAR) 美国革命之子组织

Soot 煤烟灰(结炭)

Soot blower 吹灰器 用蒸汽或压缩空气的高速射流去除受热面上积灰的装置。

Soot collector 烟灰收集器

Sooting 结炭

Sophisticated sonar system 复杂声呐系统

SOSO "腾讯"搜索网站 是"腾讯"公司主要的业务单元之一。网站于2006年3月正式发布并开始运营。主要为网民提供实用便捷的搜索服务, 同时承担"腾讯"全部搜索业务, 是"腾讯"整体在线生活战略中重要的组成部分之一。SOSO目前主要包括网页搜索、综合搜索、图片搜索、音乐搜索、论坛搜索、搜吧等16项产品,通过互联网信息的及时获取和主动呈现,为广大用户提供实用和便利的搜索服务。用户既可以使用网页、音乐、图片等搜索功能寻找海量的内容信息,也可以通过搜吧、论坛等产品表达和交流思想。2013 年 9 月 16 日,"腾讯"公司宣布以 4.48 亿美元战略入股搜狗,并将旗下的搜索和 QQ 输入法并入"搜狗"现有的业务中,"腾讯"将持有新"搜狗"36.5% 的股份。

Sound 声音(声测深,坚固的)

Sound absorbing 吸声 是将吸声材料(或吸声结构)衬贴或悬挂在舱室内部,当声波通过吸声材料或射到吸声材料表面时,依靠材料的吸声作用,声能减少或转换为其他能量的过程称为吸声。

Sound absorbing 吸声的

Sound absorbing paint 吸声涂料(吸声漆)

Sound absorbing structure 吸声结构

Sound-absorbing 新型吸音

Sound absorption coefficient 吸声系数 是表示吸声材料或吸声结构的吸声能力的参数,用 α 表示,其公式为: $\alpha = E_a/E_i = (E_i - E_r)/E_i = 1 - r$ 式中,r ——反射系数;$r = E_r/E_i$,它是按照吸音材料进行分类的。说明不同材料有不同吸音质量分贝(dB),是声压级大小的单位(声音的大小)。人们使用吸声系数频率特性曲线描述材料在不同频率上的吸声性能。按照 ISO 标准和国家标准,吸声测试报告中吸声系数的频率范围是 100 ~ 5kHz。将 100 ~ 5 kHz 的吸声系数取平均得到的数值是平均吸声系数,平均吸声系数反映了材料总体的吸声性能。一般认为 NRC 小于 0.2 的材料是反射材料,NRC 大于等于 0.2 的材料才被认为是吸声材料。当需要吸收大量声能降低室内混响及噪声时,常常需要使用高吸声系数的材料。

Sound absorption material 吸声材料 系指对入射声能有吸收作用的材料。是具有较强的吸收声能、减低噪声性能的材料。借自身的多孔性、薄膜作用或共振作用而对入射声能具有吸收作用的材料,超声学检查设备的元件之一。吸声材料要与周围的传声介质的声特性阻抗匹配,使声能无反射地进入吸声材料,并使入射声能绝大部分被吸收。选用吸声材料,首先应从吸声特性方面来确定合乎要求的材料,同时还要结合防火、防潮、防蛀、强度、外观、建筑内部装修等要求,综合考虑进行选择。

Sound absorptive amount 吸收量 系指吸声处理面积与其吸声系数的乘积,单位为平方米,m^2。

Sound alarm system 声报警系统(音响报警系统)

Sound and optics test 声光试验

Sound barrier 隔声屏障 是应用建筑学的产物,主要用于室外,是用来遮挡声源和接收者之间直达声的设施,特性是声音的频率增加一倍,屏障的降噪量增加 3 dB,屏障的高度加倍,其减噪量增加 6 dB。

Sound damping device 消声装置

Sound energy density 声能密度 系指在声场中由声扰动使单位体积的介质中储存了的声能量(往复动能 + 形变势能)。这单位体积中的能量称为声能密度。

Sound field 声场 系指有声波存在的弹性媒质所占有的空间。声场可以无限大,也可以指某局部的空间。媒质可以是气体、液体和固体,环境声学涉及的媒质主要是大气。声场又可以分为自由声场(free field)和混响声场(reverberant field)、扩散声场(diffuse sound field)等。把室内的直达声和混响声都计算在内时,在离声源中心距离为 r 的某点,声压平方的平均值 p ,ρ_0 是空气的密度;c 是空气中的声速,Q_θ 是声源在 θ 方向辐射的指向性因数,即该点的声强与声功率相等的无指向性声源在相同距离上的声强之比;$R = S\gamma/(1 - \gamma)$ 称为房间常数;S 是室内总表面积;γ 是室内诸表面的平均吸声系数。

Sound flow-based acoustic filter 声流式消音器

Sound insulating material 隔声材料 系指把空气中传播的噪声隔断、隔绝、分离的一种材料、构件或结构。隔声材料材质的要求是密实无孔隙或缝隙;有较大的质量。隔声材料或构件,会因使用场合不同,测试方法不同而得出不同的隔声效果。隔声材料可使透射声能衰减到入射声能的 $10^{-3} \sim 10^{-4}$ 或更小。为方便表达,其隔声量用分贝的计算方法表示。

Sound insulation 隔声 用构件将噪声源和接收者分开,隔离空气噪声的传播,从而降低噪声污染程度。采用适当的隔声设施,能降低噪声级 20 ~ 50 dB。这些设施包括隔墙、隔声罩、隔声幕和隔声屏障等。

Sound insulation measure 隔声措施 系指将声源和接收者进行隔离的噪声控制措施。主要包括:(1)声源隔声设计:采用隔声罩结构,用隔声结构把噪声源封闭起来,使噪声局限在一个小的空间内;(2)对接受者的隔声设计:采用隔声间,如控制室,把需要安静的场所用隔声结构围蔽起来,使外面的噪声传不进去;(3)噪声传递途径的隔声设计:采用隔声屏结构,在噪声源与受噪声干扰的位置之间设立隔声屏障,隔挡噪声向接受位置传播。

Sound insulation panels 隔声壁板 系指用来阻挡声源与受声点之间直达声的屏障板。

Sound intensity 声强 又称能流密度。系指在声场中某处在单位时间内通过单位波阵面上的声能。

Sound interval 频程 又称频带。两个声或其他信号的频率间的距离,是频率的相对尺度。以高频与低频的率比的对数来表示。此对数通常以2为底,单位称为倍频,可以以10为底的对数表示,此时单位为10倍程。

Sound level meter 声级计 是根据国际标准和国家标准而制造的,按照一定的频率计权和时间计权测量声压级的仪器。它是声学测量中最常用,最基本的仪器,适用于船舶机舱内的机器噪声测量、各舱室的噪声测量和船舶甲板上空的环境噪声测量。

Sound power 声功率 系指声源在单位时间内辐射的总能量。单位为瓦,W。

Sound power level 声功率级 声功率与参考声功率之比的10为底的对数乘以10。单位为分贝,dB。声功率级的计算公式为:$L_w = 10 \lg (w/w_0)$ 式中,L_w—声功率级,dB;w—声功率,W;w_0—参考声功率;W。空气中参考声功率为1×10^{-12} W。

Sound pressure 声压 就是大气压受到声波扰动后产生的变化,即为大气压强的余压,单位为帕[斯卡],Pa。它相当于在大气压强上的叠加一个声波扰动引起的压强变化。由于声压的测量比较容易实现,通过声压的测量也可以间接求得质点速度等其他物理量,所以声学中常用这个物理量来描述声波。

Sound pressure level 声压级 系指声音或噪声的声压级,单位 dB,由下式给出:$L_p = 10 \lg (P/P_0)$ 其中:P—声压,P_a;P_0—基准声压($= 20\mu P_a$)。A计权声压级L_{PA},单位 dB,它可以用 IEC 651 对声级计测量所规定的A计权频率求得。

Sound proof house 隔声屋 是一种隔声装置。它与隔声罩一样,用作隔声的固定装置。它由各种复合隔声结构(如轻质多层复合隔声壁、隔声门、隔声窗、浮动地板等)所组成,具有良好的隔声性能的,且独立的房间,固定在机舱地板上或固定在机舱的舱壁上的壳体结构装置。

Sound ray 声射线 系指自声源出发表示声波传播方向和传播途径的带有箭头的线。

Sound reduction index 隔声量 系指墙或其他构件一侧的入射声功率与另一侧声功率级之差。单位为分贝;dB。隔声量计算公式为:$R = 10 \lg (1/\tau)$ 式中,R—隔声量,dB;τ—透射系数。

Sound rocket 音响火箭 供船舶遇难时发射,能在高空中发出强烈爆炸声的一种火箭式信号烟火。

Sound signal appliances 音响信号器具 能发出声响作为船舶避碰信号用的各种器具的总称。

Sound signal shell 音响榴弹 供船舶遇难时点燃,能发出强烈爆炸声的一种榴弹式信号烟火。

Sound source 声源 系指发出声音的振动物体。

声音是由物体的振动产生的。一切发声的物体都在振动。物理学中,把正在发声的物体叫声源。如:正在振动的声带、正在振动的音叉、敲响的鼓等都是声源。但是声源是不能脱离其周围的弹性介质的,空间中同样的物体,同样的振动状态,如果脱离了弹性介质,那么就不能产生声波了,这时的振动着的物体不是声源。它可以是固体、液体和气体。

Sound velocimeter 声速仪 是用来测量介质中声音传播速度的一种仪器。在海水中,声速是随着温度、盐度和压力的不同而变化的。测量不同海区、不同深度的声速。不仅为声呐提供确切的声速值以修正其作用距离,有效地发挥声呐的作用;而且还可以积累各海区的声速资料,为海洋考察、海洋开发和声呐作用距离预报提供依据。该仪器的基本工作原理是测量声波通过两定点间的传播时间。

Sound velocity 声速 系指声振动在介质中的传播速度(波速度)或等位相面(波阵面)传播的速度(相速度)。单位为米每秒,m/s。

Sound wave 声波 系指弹性媒质中传播的压力、应力、质点位移、质点速度等的变化或几种变化的组合。

Sound wave diffraction 声波的衍射/绕射 系指声波在传播过程中遇到障碍物或孔洞时,如果波长大于障碍物尺寸,则声波可以绕过障碍物边缘传播的现象。

Sound eliminator 消声器 是一种能阻止声音传播而又允许气体流通的装置,是控制空气噪声通过管道向外传播的有效措施。船舶采用的消声器有:阻性消声器、抗性消声器、阻抗复合式消声器、微穿孔吸声结构消声器。小孔喷注消声器和多孔扩散消声器等。

Sound eliminator by multiport diffusion 多孔扩散消声器 系指利用多层金属丝网(其有效出流面积大于排气管口的横截面积)的细小孔隙,将中、高压气流过滤成数个小气流、降低排空压力和流速从而降低声辐射的消声器。

Sound eliminator by small jet spray 小孔喷注消声器

Sound eliminator with sound absorbent perforated structure 微穿孔板吸声结构消声器 系指由微穿孔板组合而成的消声器。其既有阻性又有抗性消声性能。其特点是:高速气流通过时,仍然有较好的消声效果。

Sound-absorbed bricks with resonance cavities 共振腔吸声砖

Sound-absorbing foam 吸音泡沫塑料

Sound-absorbing panels 吸音板 系指板状的具有吸声减噪作用的材料,主要应用于影剧院、音乐厅、博物馆、展览馆、图书馆、审讯室、画廊、拍卖行、体育馆、报告厅、多功能厅、酒店大堂、医院、商场、学校、琴房、会议室、演播室、录音室、KTV 包房、酒吧、工业厂房、机房、家庭降噪等对声学环境要求较高及高档装修的场所。具有吸声、环保、阻燃、隔热、保温、防潮、防霉变、易除尘、易切割、可拼花、施工简便、稳定性好、抗冲击能力好、独立性好、性价比高等优点。

Sound-absorbing porous material　多孔吸声材料　系指能把入射的声能通过材料空隙的空气和材料纤维的振荡把声能转变成热能吸收掉的材料。

Sounded（sounding）　测深（液舱测量）

Sounder　声信号仪器（测深锤、测深仪）

Sounding arrangement　测深装置

Sounding device　测深器（回声测深仪）

Sounding hole　测深孔

Sounding installation　测深装置

Sounding lead line　测深绳（水砣绳）

Sounding of alarm　声响报警

Sounding pipe　测量管　供测深尺测量水舱和油舱液位深度而专设的管件。

Sounding rod　测深杆

Sounding room　测深室　装有各种测深仪器，进行水深测量的工作室。

Sound-proof door　隔声门　敷有消声材料，能起隔声作用的门。

Sound-proof room　隔声室

Sound-proof wall　隔声墙　声波在媒体中传递，也就是能量在传递，动量在分散，声波，比如在空气中传递，它的声强与频率的平方根、振幅的平方成正比。单看与频率的关系，频率越高声强越大，当一个物体在空气中运动时，其速度与声波的速度相等时，由多普勒效应可知，频率可以说是无穷大，这时声波的能量都集中在一个点上，从理论上讲能量是无穷大的，在物体运动达到与声速同带时，好像遇到了一面无法超越的墙，也就是说理论上物体运动速度不可能超过这面墙，故叫隔声墙。

Source　信息源　系指产生的数据或信息的装置（如：海图数据库），作为综合航行系统（INS）的组成部分自动向综合航行系统（INS）提供信息。Source means a device, or location of generated data or information（e.g. chart database）, which is part of the INS automatically providing information to INS.

Source　源　流体对称地向所有方向流出的一点。

Source cabin　源舱室　作为激励源的机械设备所在的舱室。

Source of category 1 release　1 类释放源　系指在正常的工况下会有释放的释放源。Source of category 1 release means a source of release available under normal conditions.

Source of category 2 release　2 类释放源　系指在正常的工况下不大可能出现释放，即使释放也只持续很短时间的释放源。Source of category 2 release means a source of release unlikely available under normal conditions and if available, only for a short time.

Source of continuous release　连续释放源　系指连续或近似连续释放的释放源。Source of continuous release means a source of continuous or nearly continuous release.

Source of energy　能源

Source of ignition　火源

Source of release　释放源　系指气体燃料系统内可燃气体、蒸发气体或液体可能释放出能形成爆炸性气体环境的部位或地点。如任何阀门、可拆卸式管接头、管垫圈、压缩机或泵密封装置等。Source of release means a point or location from which a gas, vapour, mist or liquid may be released into the atmosphere so that an explosive atmosphere may be formed, for example, any valve, detachable pipe joint, pipe packing, compressor or pump seal in the gas fuel system.

Source of release（electrical equipment）　释放源（电气设备）　气体、蒸汽、雾或液体可通过其释放至周围环境，以在正常运行条件下可形成爆炸性气体环境的点或部位，例如在液货管系中的阀和法兰。Source of release（electrical equipment）means a point or location from which a gas, vapour, mist or liquid may be released into the atmosphere so that an explosive atmosphere may be formed under normal operating conditions, for example, valves and flanges in cargo piping systems.

Source of trouble　故障源

Source program　源程序　系指未经编译的，按照一定的程序设计语言规范书写的，人类可读的文本文件。通常由高级语言编写。源程序可以是以书籍或者磁带或者其他载体的形式出现，但最为常用的格式是文本文件，这种典型格式的目的是为了编译出计算机可执行的程序。将人类可读的程序代码文本翻译成为计算机可以执行的二进制指令，这种过程叫作编译，由各种编译器来完成。一般用高级语言编写的程序称为"源程序"。

Southern South African waters　南非的南部区域　系指下述坐标范围内的海域：南纬 31° 14′，东经 17° 50′；南纬 31° 30′，东经 17° 12′；南纬 32° 00′，东经 16° 52′；南纬 34° 06′，东经 16° 52′；南纬 34° 06′，东经 17° 24′；南纬 36° 58′，东经 20° 54′；南纬 36° 00′，东经 22° 30′；南纬 35° 14′，东经 22° 54′；南纬 34° 30′，东经 26°

00′;南纬33°48′,东经27°25′;南纬33°27′,东经27°12′。 Southern South African waters means the sea area enclosed by the following co-ordinates: 31°14′S;17°50′E;31°30′S;17°12′E;32°00′S;16°52′E;34°06′S;17°24′E;36°58′S;20°54′E;36°00′S;22°30′E;35°14′S;22°54′E;34°30′S;26°00′E;33°48′S;27°25′E;33°27′S;27°12′E.

SO_x emission control (SEC SO_x) 硫化物(SO_x)排放控制 系指对船上所用的所有燃油硫含量的控制。 SO_x emission control (SEC SO_x) means control of sulphur content of all oil fuel used on board.

SO_x emission control area (SECA) 硫化物(SO_x)排放控制区 系指要求对船舶SO_x排放采取特殊强制措施以防止、减少和控制SO_x造成大气污染以及随之对陆地和海洋区域造成不利影响的区域。 SO_x emission control area means an area where the adoption of special mandatory measures for SO_x emissions from ships is required to prevent, reduce and control air pollution from SO_x and its attendant adverse impacts on land and sea areas.

Space[1] 处所 系指船上的永久或临时三维结构或隔间,诸如,但不限于,液货舱或干货舱;泵舱或机舱;储藏柜;含有易燃或燃烧用液体、气体、或固体的罐柜;其他舱室;爬行处所;隧道(例如轴隧);或进入通道。处所中的气体环境是其限界内的整体容积。 Space means a permanent or temporary three-dimensional structure or compartment on a ship such as, but not limited to, cargo tanks or holds; pump or engine rooms; storage lockers; tanks containing flammable or combustible liquids, gases, or solids; other rooms; crawl spaces; tunnels (i. e. shaft alleys) ; or access ways. The atmosphere within a space is the entire volume within its bounds.

Space[2] 处所(舱室) 系指:(1)包括货物舱和油箱的各个独立的舱室;(2)货舱,液舱,与货舱相连的空隔舱及管隧,包括甲板及船体外部各个独立的区域;(3)分离的舱室,例如货舱和液舱。 Space is: (1) a separate compartment including holds and tanks; (2) separated compartments including holds, tanks, cofferdams and void spaces bounding cargo holds, decks and the outer hull; (3) separate compartments such as holds, tanks.

Space aircraft 空天飞机 集飞机、搭载器、航天器等多种功效与一身,并直接加速进入地球轨道,既能在大气层内进行高倍声速飞行,又能进入环绕地球太空轨道的飞行器,将是21世纪控制空间、争夺制天权的关键武器装备之一。

Space between shielded rooms 屏蔽间处所 对散装运输液化气船而言,系指不论是其全部还是部分由绝热材料或其他材料所填充的主屏壁和次屏壁之间的处所。

Space debris 太空碎片 系指包括废油箱、火箭外壳、失效的卫星以及太空爆炸的碎片,将对人类的空间活动构成了日益严重的威胁。

Space occupied by machinery 机器处所

Space segment 空间段 系指卫星以及跟踪、遥测、指令、控制、监测和辅助卫星运行所需的有关设施和设备。 Space segment means the satellites, and the tracking, telemetry, command, control, monitoring and related facilities and equipment required to support the operation of these satellites.

Space sound absorber 空间吸声体 系指一种把吸声材料或吸声结构分散悬挂在室内或大厅内,离壁面有一定距离的空间中的,用以降低室内噪声或改善室内音质的吸声构件。

Spacer (packing piece/root piece) 隔叶片 装在相邻叶片之间保证叶片节距及通道宽度用的具有叶根形状的隔离件。

Spaces 处所 系指:(1)独立的舱室,包括货舱、液舱、邻接货舱、甲板和外壳板的隔离舱和空舱。(2)分隔的舱室,诸如液舱、泵舱和空舱。对储油平台的原油/压载兼用舱,当发现显著腐蚀时,也应视为压载舱。(3)船体和上层建筑包括金属板桅房的所有独立间隔。整体油舱被认为是独立处所。 Spaces are: (1) independent compartments including holds and tanks, adjacency holds, cofferdam between deck and outer shell, and voids. (2) Separate compartments such as tanks, pump rooms and voids. A tank which is used for both the storage of crude oil and salt water ballast on board oil storage units will be treated as a ballast tank when substantial corrosion has been found in that tank. (3) All separate compartments within the hull and superstructure including plated masts. Integral tanks are considered to be independent spaces.

Spaces where penetration of oil products is possible 油可以渗入的区域 系指油(燃料油、润滑油、液压油、热介质油)所经过的所有机器(净化器、泵、油舱)及管系(阀门、法兰、筛网过滤器、流量仪表等)附近运转及整修作业时,泄漏、飞溅的燃料油或燃料油蒸发气飞至热绝缘材的可能性。但不适合机械室内的热绝缘管。在此区域没有打孔的绝热金属板或者连接部分,使用蒸发气保护玻璃纤维布密封遮蔽。 Spaces where penetration of oil products is possible means the spaces located in the vicinity of all types of equipment (purifiers, pumps and tanks) and pipe fittings (valves, flanges, strainers, flow-meters, etc.) handling oils (fuel oil, lubricating oil, hydraulic oil and thermal oil) with possible involvement of oils or oil vapours leaked or splashed during operation or in maintenance work to reach out thermal insulation. However, these do not apply to thermal insulation of pipes in machinery spaces. The fire insulation in such spaces can be covered by metal sheets (not perforated) or by vapour-proof glass cloth accurately sealed at the joint.

Spaceship 宇宙飞船 又称载人飞船,是一种运送航天员,货物到达太空并安全返回的一次性使用的航天器。它能基本保证航天员在太空短期生活并进行一定的工作。它的运行时间一般是几天到半个月,一般乘2~3名航天员。截至目前只有美国、俄罗斯和中国具备建造和回收宇宙飞船的技术。

Spacing 间距
Spacing of frame 骨材间距 S (m) 对次要骨材取次要骨材间距,对主要骨材取其承载面积的平均宽度。
Spacing of longitudinals 纵骨的间距 系指两相邻纵骨理论线之间的距离。
Spacing of tubes 管距
Spade of jam 钳口开度
Span 跨距 构件两支点间的距离。
Span 翼展 翼形体两端间垂直于运动方向的距离;对悬臂翼则指翼端至根部的距离。
Span bearing 千斤眼板座 用以支承千斤眼板,并使该眼板在座中左右转动的座子。
Span block 千斤滑车 用于千斤索具中的滑车。
Span chain 千斤链 在千斤索具中,为使千斤索可改变其长度而配置的长环链。
Span eye 千斤眼板 (1)安装于起重柱顶部或桅肩等处供系挂千斤索具用。(2)装于吊杆头端处,专供连接千斤索具用的眼板。
Span of a primary supporting member with end brackets 具有端部肘板的主要构件的跨距 取两端肘板上的1/2 主要支撑构件高度处之间的长度。 Span (m) of a primary supporting member with end brackets is taken between points where the depth of the bracket is equal to half the depth of the primary supporting member.
Span of a primary supporting member without end bracket 无端部肘板的主要构件的跨距 取两支撑之间的构件长度。 Span (m) of a primary supporting member without end bracket is to be taken as the length of the member between supports.
Span of frame 骨材跨距 l (m) 当骨材端部不设置肘板时,跨距点取在骨材端部。当骨材端部设置肘板时,跨距点可取在肘板长度之半处。主要骨材支撑处的结构如能有效地防止该骨材在该处转动或位移,则认为支撑处可取作该主要骨材的端部。
Span of ordinary stiffener or primary supporting member l (m) 普通扶强材或主要支撑构件的跨距 l (m) 系指普通扶强材或主要支撑构件的跨距,视具体情况沿弦长量取。 l means a span, in m, of ordinary stiffener or primary supporting member, as the case may be, measured along the chord.
Span rigging (spab tackle) 千斤索具 系指悬吊杆头端并用以调整吊杆仰角或工作位置的系索索具。一般包括滑车千斤索以及卸扣、眼板、套环、千斤链等连接零件。
Span winch (hanger winch) 千斤索绞车 用以调节吊杆幅度和支持吊杆上负荷的绞车。
Span wire (topping lift) 千斤索 在千斤索具中,用以系挂吊杆和承载吊杆负荷,位于吊杆头端至千斤座之间并包括曳引分支的整根钢索。
Span-guy derrick 千斤-牵索吊杆装置 由千斤索和牵索相互贯通的两组索具操纵吊杆的吊杆装置。
Spanner 扳手(扳紧器)
Span-wire rig 跨索索具 在补给船与接收船之间,只有一条绳索连接的传送干货的索具。
Spar 立柱式平台(深水浮筒平台) 系指将甲板设置在用系泊缆定位的垂直圆筒形的船体上的一种平台。 Spar is a type of platform in which the deck is on a vertical cylindrical vessel held in place with mooring lines.
Spar platform 筒形平台 一种具有大直径、深吃水的圆柱体结构的、造价相对较低的第一代圆筒式海上平台。通常它由平台上体、平台主体和系泊系统组成。并在平台底部设置固定压载舱,以降低重心高度,使其低于浮心,因此稳定性极高。由于采用深吃水结构,波浪对平台主体底部的冲力通常很小,而且垂向运动、横摇和纵摇的固有周期一般在20 s以上,高于海浪特征周期,因此在波浪上的垂向运动性能较好。筒形平台被认为是一种特别适用于深海的海上平台,其工作水深范围从350~3 000 m。
Spar/DDCVs 立柱式平台/深海箱式生产平台
Spare 多余的(空间的,节省的,备件)
Spare cylinder 备用储气瓶
Spare equipment 备用设备
Spare gears 备件(备用装置)
Spare list 备件明细表
Spare oil tank 备用油柜
Spare part list 备件、备品表
Spare parts 备件/备品 是一个通用词,所有与

设备有关的零件都可以用作备品备件。按词面来说就是备用的物品和备用的零件。无论在维修还是制造，还是在各个领域，都需要提前准备一些物品和零件，这些提前准备并会在不久的将来使用上的物品和零配件在文字上就定为"备品"。备品系指未完成的加工品，如待精铣的金属零件、待蘸火的滚珠等。机械设备中备品就是为易损件准备的备用件，以防止因为易损件损坏、失效等原因使得生产停顿。备品是一个广义的含义，对不同的行业，只是所备用的"物质"不一样而已。

Spare parts compartment　备件舱
Spare parts kit　备品零件箱
Spare parts stowage　备件储藏室
Spare parts supply　备件供应
Spare pump　备用泵
Spare radio room　备用报务室　系指作为应急使用的报务室。
Spare room　备用室
Spare towline or emergency towing　备用拖缆和应急拖缆　系指主拖缆发生故障后，用于代替主拖缆或临时稳定被拖物的拖缆。
Spares　备件（备用的，可分让的）
Spares for accommodation outfit　舱室备品　为补偿或修复损耗的舱室设备而储存在船上的各种设备、器具和材料的统称。
Spark　火花（打火花）
Spark advance　点火提前
Spark arrest　火花避雷器（火花制止器，火花罩）
Spark arrester　火星熄灭器（火花避雷器，火花制止网，火花熄灭器）　防止火星进出引起失火的器具。
Spark catcher　火花防止器
Spark distributor　火花分配器
Spark extinguisher　火花熄灭器
Spark gap　火花隙（火花放电器，避雷器）
Spark gap distance　火花间隙
Spark ignition　火花点火
Spark lighter　火花点火器
Spark plug　火花塞
Spark quench　火花猝熄
Spark retarding　火花延迟
Spark source[1]　电火花源
Spark source[2]　电火花振源　利用电容贮存电能，通过置于水中的电极放电，在产生电火花的同时造成电极周围的海水振动的振源。
Spark test　火花鉴别（火花试验）
Speaking pipe　通话管
Speaking terminal　传声口
Speaking trumpet　传声器
Special additional risk　特别附加险　是以导致货损的某些政府行为风险作为承保对象的，它不包括在一切险范围，不论被保险人投保何种基本险，要想获取保险人对政府行为等政治风险的保险保障，必须与保险人特别约定，经保险人特别同意。否则，保险人对此不承担保险责任。我国保险公司开办的特别附加险现有6种。

Special administrative region (SAR)　特别行政区　简称特区，系指根据宪法规定，在中华人民共和国行政区域范围内设立的、享有特殊法律地位、实行资本主义制度和资本主义生活方式的地方行政区域。特别行政区是我国为以和平方式解决历史遗留下来的香港问题和澳门问题而设立的特殊的地方行政区域。特别行政区的建立构成了我国单一制的一大特色，是马克思主义国家学说在我国具体情况下的创造性运用。特别行政区的成立，有助于维持香港与澳门的繁荣和稳定。此外，中国大陆也有经国务院审批的特别行政区，如卧龙特别行政区。

Special application structure of column-stabilized units　柱稳式平台的特殊构件　包括：(1) 立柱与上平台甲板和上、下壳体交接部分的外壳板；(2) 组成箱型或工字型支承结构且承受主要集中载荷的上壳体或平台的甲板、重型翼板、外壳板和舱壁；(3) 撑杆的结点；(4) 主要结构构件交接处承受集中载荷的外部肘板、部分舱壁、平台和骨架；(5) 立柱、上平台甲板及上壳体或下壳体连接处提供适当对齐和足够载荷传递的"贯穿"构件。

Special application structure of column-stabilized units include: (1) shell plating in way of the intersections of vertical columns with platform decks and upper and lower hulls; (2) portions of deck plating, heavy flanges, shell boundaries and bulkheads within the upper hull or platform which form "box" or "I" type supporting structure and which receive major concentrated loads; (3) major intersections of bracing members; (4) external brackets, portions of bulkheads, flats and frames which are designed to receive concentrated loads at intersections of major structural members; (5) "through" material used at connections of vertical columns, upper platform decks and upper or lower hulls which are designed to provide proper alignment and adequate load transfer.

Special application structure of self-elevating units　自升式平台的特殊构件　包括：(1) 与沉垫或桩靴相连接部分的桩腿垂直结构；(2) 含有新颖构造桁架式桩腿结构中的连接部位，包括使用的铸钢件。 Special application structure of self-elevating units include: (1) vertical leg structures in way of intersections with individual footings

Special application structure of submersible units
坐底式平台的特殊构件　包括:(1)大直径立柱与上平台甲板或上壳体、下壳体交接部分的外壳板;(2)组成箱型或工字型支架结构且承受主要集中载荷的上壳体或平台的甲板板、重型翼板、外壳板和舱壁;(3)在重要结构构件交接处承受集中载荷的部分舱壁、平台和骨架;(4)立柱、上平台甲板及上壳体或下壳体连接处提供适当对齐和足够载荷传递的"贯穿"构件。 Special application structure of submersible units include:(1) shell plating in way of the intersections of large-diameter vertical columns with platform decks or upper and lower hulls;(2) deck plating, heavy flanges, shell boundaries and bulkheads within the upper hull or platform which form "box" or "I" type supporting structure and which receive major concentrated loads;(3) portions of bulkheads, flats and framing which are designed to receive concentrated loads at intersections of major structural members;(4) "through" material used at connections of vertical columns, upper platform decks and upper or lower hulls which are designed to provide proper alignment and adequate load transfer.

Special area　特殊海区　系指:(1)这样的有关海域,在该海域中,由于其海洋学的和生态学的情况以及其运输的特殊性质等方面公认的技术原因,需要采取防止海洋油污染的特殊强制办法,以防止生活污水污染海洋;(2)这是"73/78防污公约"几个附则中所划定的区域,这个特殊区域系指这样的一个海域,在该海域中,由于其海洋学的和生态学的情况以及其交通的特殊性质等方面公认的技术原因,需要采取特殊的强制办法以防止油类物质、有毒液体物质、垃圾等污染海洋。①就 MARPOL73/78 附则Ⅰ而言,有下列一些特殊区域:地中海区域、波罗的海区域、黑海区域、红海区域、海湾区域、亚丁湾区域、南极区域、西北欧水域、阿拉伯海的阿曼区域、南非南部水域;②就 MARPOL73/78 附则Ⅱ而言,特殊区域为:南极区域。③就 MARPOL73/78 附则Ⅳ而言,特殊区域为地中海区域、波罗的海区域、黑海区域、红海区域、海湾区域、北海区域、南极区域和大加勒比海区域。④就 MARPOL73/78 附则Ⅴ而言,有下列一些特殊区域:地中海区域、波罗的海区域、黑海区域、红海区域、海湾区域、北海区域、南极区域、大加勒比海区域。⑤就 MARPOL73/78 附则Ⅵ而言,排放控制区相当于特殊区域,有下列一些特殊区域:NO_x 排放控制区—北美区域。SO_x 排放控制区—波罗的海区域、北海区域、北美区域。 Special area means:(1) a sea area where for recognized technical reasons in relation to its oceanographical and ecological condition and to the particular character of its traffic, the adoption of special mandatory methods for the prevention of sea pollution by sewage is required;(2) a sea area defined in Annexes to "MARPOL73/78" where for recognized technical reasons in relation to its oceanographical and ecological condition and to the particularcharacter of its traffic, the adoption of special mandatory methods for the prevention of sea pollution by oil, NLS, bargage are required:①For "MARPOL73/78" Annex Ⅰ, the special area are defined as follows:the Mediterranean Sea area, the Baltic Sea area, the Black Sea area, the Red Sea area, the Gulfs area, the Gulfs of Aden area, the Antarctic area, the North West European waters, the Oman area of the Arabian Sea, the Southern South African waters;②For "MARPOL73/78" Annex Ⅱ, the special areas are defined as follows:the Antarctic area;③For "MARPOL73/78" Annex Ⅳ, the special area are defined as follows:the Mediterranean Sea area, the Baltic Sea area, the Black Sea area, the Red Sea area, the Gulfs area, the North Sea area, the Antarctic area, the Wider Caribbean region;④For "MARPOL73/78" Annex Ⅴ, the special areas are the Mediterranean Sea area, the Baltic Sea area, the Black Sea area, the Red Sea area, the Gulfs area, the North Sea area, the Antarctic area and the Wider Caribbean Region;⑤For "MARPOL73/78" Annex Ⅵ, the special area are "emission control area" and defined as follows: NO_x emission control area: North American emission control Area. SO_x emission control area: the Baltic Sea area, the North Sea area, the North American emission control area.

Special ballast tank　专用压载水舱

Special carriage control conditions　特殊运输控制条件　系指为避免危险反应而采取的特殊措施,包括:(1)抑制—加入化合物(通常为有机化合物)来延缓或阻止某些不良化学反应如腐蚀、氧化或聚合;(2)稳定—加入某种物质(稳定剂)来避免化合物、混合物或溶剂改变形态或化学特性。这种稳定剂可延缓反应速率、保持化学成分平衡、防止氧化、保持颜色和其他成分的乳化状态或防止胶状颗粒受到冲击;(3)惰化—在液舱的膨胀余位内加入气体(通常是氮气),以防止可燃性货物/气体混合物的产生;(4)温度控制—保持货物温度在一定范围内,以避免有害反应或者保持液体的低黏度,方便泵系工作;(5)衬垫和通风—仅适用于特别情况下的特殊货品。 Special carriage control conditions refer to specific measures that need to be taken in order to prevent a hazardous reaction. They include:(1) inhibition—the addition of a compound (usually organic) that retards or stops an undesired chemical reaction such as corrosion, oxidation or polymerization; (2) stabilization—the addition of a substance (stabilizer) that tends to keep a compound, mixture or solution from changing its form or chemical nature. Such stabilizers may retard a reaction rate, preserve a chemical equilibrium, act as antioxidants, keep pigments and other components in emulsion form or prevent the particles in colloidal suspension from precipitating; (3) inertion—the addition of a gas (usually nitrogen) in the ullage space of a tank that prevents the formation of a flammable cargo/air mixture; (4) temperature control—the maintenance of a specific temperature range for the cargo in order to prevent a hazardous reaction or to keep the viscosity low enough to allow the product to be pumped; (5) padding and venting—only applies to specific products identified on a case by case basis.

Special category spaces 特种处所 系指:(1)用以载运在油箱内备有自用燃料的机动车辆的围壁处所,包括以载运货车的处所,此处所能让上述车辆驾驶进出,并有乘客进出通道;(2)舱壁甲板以上或以下能让车辆驾驶进出,并有乘客可以进入通道的围蔽处所。如车辆的总净空高度不超过10m,特种处所可设置一层及以上的甲板。 Special category spaces are: (1) those enclosed spaces intended for the carriage of motor vehicles with fuel in their tanks for their own propulsion, into and from which such vehicles can be driven and to which passengers have access for embarking and disembarking, including spaces intended for the carriage of cargo vehicles; (2) those enclosed vehicle spaces above and below the bulkhead deck, into and from which vehicles can be driven and to which passengers have access. Special category spaces may be accommodated on more than one deck provided that the total overall clear height for vehicles does not exceed 10 m.

Special closing appliances 专用关闭装置 在承受压力或密性方面有专门要求的关闭装置。

Special consideration or specially considered (in connection with close-up surveys and thickness measurements) 特殊考虑或(与近观检验和厚度测量相关的)专门考虑 系指应给予足够的近观检查和厚度测量,以确认在涂层之下结构的实际平均状况。 Special consideration or specially considered (in connection with close-up surveys and thickness measurements) means sufficient close-up inspection and thickness measurements are to be taken to confirm the actual average condition of the structure under the coating.

Special court 专门法院 是我国统一审判体系—法院体系中的一个组成部分,它和地方各级法院共同行使国家的审判权。其包括军事法院、海事法院、铁路运输法院和森林法院。

Special drawing right (SDR) 特别提货权 系指国际货币基金的记账单位。 Special drawing right (SDR) means the unit of account as defined by the International Monetary Fund.

Special duties notation 特种任务标志 表明船舶的设计、改造或布置适于执行船型标志和货种标志所包含以外的特种任务的标志,例如调查研究。具有特种任务标志的船舶并不因此而妨碍其执行适合于该船的其他任务。 Special duties notation is a notation indicating that the ship has been designed, modified or arranged for special duties other than those implied by the type and cargo notations, e.g. research. Ships with special duties notations are not thereby prevented from performing any other duties for which they may be suitable.

Special economic zones 经济特区 1979年4月邓小平首次提出要开办"出口特区",后于1980年3月,"出口特区"改名为"经济特区",并在深圳加以实施。按其实质,经济特区也是世界自由港区的主要形式之一。以减免关税等优惠措施为手段,通过创造良好的投资环境,鼓励外商投资,引进先进技术和科学管理方法,以达到促进特区所在国经济技术发展的目的。经济特区实行特殊的经济政策、灵活的经济措施和特殊的经济管理体制,并坚持以外向型经济为发展目标。1979年7月,中共中央、国务院同意在广东省的深圳、珠海、汕头三市和福建省的厦门市试办出口特区。1980年5月,中共中央和国务院决定将深圳、珠海、汕头和厦门这四个出口特区改称为经济特区。同年8月26日,第五届全国人民代表大会常务委员会第15次会议批准《广东省经济特区条例》。截至目前中国大陆地区共有7个经济特区。中国经济特区诞生于20世纪70年代末、80年代初,成长于90年代。经济特区的设置标志中国改革开放进一步发展。1992年中国加快改革开放后经济特区模式移到国家级新区,上海浦东等国家级新区新的特区扩大改革等发展起来,成为中国新一轮改革重要标志。1992年中国另一个改革高地国家级新区诞生,标志着中国新一轮改革起航。

Special extraneous risks 特殊外来风险 系指由于政治、军事、国家法令、政策及行政措施等特殊外来原因所造成的风险称为特殊外来风险。例如战争、罢工、取不到货、拒绝收货等。

Special freight rate 特别运费率 系指除促进贸易性运费率外,可由有关当局各方商定的优惠运费率。 Special freight rate means a preferential freight rate, other than a promotional freight rate, which may be negotiated between the parties concerned.

Special laboratory 专用实验室 又称专业实验室。安装专用仪器及实验设备，仅能从事单项目或单学科的试验、观测、记录和分析用的科学研究实验室。

Special machinery survey 轮机特殊检验

Special members of main structure 主体结构的特殊构件(半潜式平台) 系指重要构件连接处应力集中的构件，它的破坏将危及整个平台的安全，如，立柱和沉垫及上平台相交处的节点，水平桁撑和斜撑的节点以及某些重要构件的节点等。

Special operation condition 特殊作业工况 对起重设备而言，系指起重设备设计时所考虑的主要工况超过起重作业实际工况，包括：(1)船舶横倾和/或纵倾大于标准作业工况规定；(2)作业于无遮蔽的海域；(3)起重设备工作时的风速超过 20 m/s，相应风压超过 250 Pa；(4)起重时，起重荷重不是处于静止状态；(5)起重荷重的运动受到外力的制约；(6)起重作业的性质，即作业的频次和动载特性与有关规定的因素载荷不一致。

Special painting dressing 特殊涂装 系指适用于成品油船与化学品船的液货舱内部的涂装。特种船舶的压载水舱内涂装可以参照应用。 Special painting dressing means exclusively the painting dressing in the interior of cargo tank onboard the oil product carriers and chemical carriers, the painting dressing of the interior of ballast tank of special ships may be carried out by reference to it.

Special permit 特别许可证 系指根据有关规定经过事先申请而为特殊用途发给的许可证。 Special permit means permission granted specifically on application in advance and accordance with relevant requirement.

Special personnel 特殊人员 系指乘客或船员或1岁以下儿童以外与船舶特殊用途有关的或在船上进行特殊工作而乘载于船上的所有人员。特殊人员数量作为参数出现时，包括船上所载的乘客数量应不超过12名。特殊人员被认为具有良好的身体，对船舶布置有相当的了解并在离港前受过安全程序及船上安全设备操作训练，包括：(1)船上从事科研、非商业考察和调研的科学家、技术人员和考察人员；(2)为开发适合海上专门职业的航海技能而参加培训和实际航海经验的人员。此类培训应符合经主管机关批准的培训计划；(3)在不从事捕捞的加工船上从事捕鱼、鲸及其他海洋生物资源的人员；(4)在打捞船上的打捞人员，在布缆船上的布缆人员，在地震测量船上的地震测量人员，在潜水支撑船上的潜水人员，在铺管船上的铺管人员以及在起重船上的起重机操作人员；(5)主管机关认为可以归入此类等(1)~(4)所述相类似的其他人员。 Special personnel means all persons who are not passengers or members of the crew or children of under one year of age and who are carried on board in connection with the special purpose of that ship or because of special work being carried out aboard that ship. Wherever in this Code the number of special personnel appears as a parameter, it should include the number of passengers carried on board which may not exceed 12. Special personnel are expected to be able bodied with a fair knowledge of the layout of the ship and to have received some training in safety procedures and the handling of the ship's safety equipment before leaving port and include the following: (1) scientists, technicians and expeditionaries on ships engaged in research, non-commercial expeditions and survey; (2) personnel engaging in training and practical marine experience to development seafaring skill suitable for a professional career at sea. Such training should be in accordance with a training programme approved by the Administration; (3) personnel who process the catch of fish, whales or other living resources of the sea on factory ships not engaged in catching; (4) salvage personnel on salvage ships, cable-laying personnel on cable-laying ships, seismic personnel on seismic survey ships, diving personnel on diving support ships, pipe-laying personnel on pipe layer and crane operating personnel on floating cranes; (5) other personnel similar to those referred to in (1) ~ (4) who, in the opinion of the Administration, may be referred to this group.

Special processes 特殊生产过程 对质量管理体系而言，系指那些对于产品在随后检查和试验中都无法证实其有效性的工序，除有关规定外，还应该根据成文的步骤进行持续监控。 For quality management system, special processes mean those processes where effectiveness cannot be verified by subsequent inspection and test of the product are to be subjected to continuous monitoring in accordance with documented procedures in addition to the relevant requirements.

Special purpose ship safety certificate 特殊用途船舶安全证书 系指经按特殊用途船舶安全规则前言第7节规定的各种 SOLAS 证书外，还应签发一份特殊用途船舶安全证书。该证书的有效期限和有效性应取决于1974年 SOLAS 公约中关于货船的有关规定。如果为一艘小于 500GT 特殊用途船舶签发证书，则该证书应注明按1.2的规定可接受的放宽范围。 Special purpose ship safety certificate means a certificate issued for a special purpose ship, in addition to SOLAS certificates as specified in Paragraph 7 of the Preamble of the Code of safety for special purpose ships. The duration and validity of the certificate should be governed by the respective provisions for cargo ships in SOLAS 1974. If a certificate is issued for a special purpose ship of less than 500 gross tonnage, this certificate should indicate what extent relaxations in accordance with Paragraph 1.2 can be accepted.

Special purpose ship(SPS) 特殊用途船舶 系指因船舶功能的需要而载有 12 名以上特殊工作人员的机械自航船舶。特种用途船舶包括:(1)从事研究、考察和调研的船舶;(2)从事海员训练船舶;(3)不从事捕捞的海象和鱼类加工船;(4)不从事捕捞的其他海洋资源加工船;(5)主管机关认为可以归入此类的,其设计特点和作业模式类似于(1)~(4)的其他船舶。 Special purpose ship means a mechanically self-propelled ship which, by reason of its function, carries on board more than 12 special personnel including passengers. Special purpose ships to which this Code applies include the following types: (1) ships engaged in research, expeditions and survey; (2) ships for training of marine personnel; (3) whale and fish factory ships not engaged in catching; (4) ships processing other living resources of the sea, not engaged in catching; (5) other ships with design features and modes of operation similar to ships referred to in (1) ~ (4) which in the opinion of the Administration may be referred to this group.

Special research ship 专业调查船 用于对海洋进行单学科为主的海洋调查研究的船舶。如从事海洋音响调查的水声调查船和专门为渔捞进行调查的渔业调查船,以及用木制成或玻璃钢制成的非磁性船舶进行磁性、电离层等海洋调查的船舶。

Special rules 专门规范 系指只有专门内容,且与入级规范配套使用的规定。 Special rules are such provisions that have special content and will be used in combination with the classification rules.

Special spaces 特种处所 系指可供乘客出入的围蔽滚装处所。如果车辆的总净高度合计不超过 10 m,特种处所可设在一个以上甲板上。 Special spaces are those enclosed vehicle spaces above and below the bulkhead deck, into and from which vehicles can be driven and to which passengers have access. Special category spaces may be accommodated on more than one deck provided that the total overall clear height for vehicles does not exceed 10 m.

Special structural member 特种构件 系指下列构件:(1)与强力甲板相接的舷顶列板;(2)强力甲板的甲板边板;(3)与纵向舱壁相接的甲板板;(4)舭列板;(5)纵向连续的舱口围板。 Special structural member means the following members: (1) sheer strake at strength deck; (2) stringer plate in strength deck; (3) deck strake at longitudinal bulkhead; (4) bilge strake; (5) continuous longitudinal hatch coamings.

Special structural members of the unit 平台的特殊构件 系指在关键载荷传递点和应力集中处的主要构件。例如:(1)自升式平台的特殊构件,包括:①与沉垫或桩靴相连接部分的桩腿垂直结构;②含有新颖构造桁架式桩腿结构中的连接部位,包括使用的铸钢件。(2)柱稳式平台的特殊构件:①立柱与上平台甲板和上、下壳体交接部分的外壳板;②组成箱型或工字型支承结构且承受主要集中载荷的上壳体或平台的甲板板、重型翼板、外壳板和舱壁;③撑竿的结点;④主要结构构件交接处承受集中载荷的外部肘板、部分舱壁、平台和骨架;⑤立柱、上平台甲板及上壳体或下壳体连接处提供适当对齐和足够载荷传递的"贯穿"构件。(3)坐底式平台的特殊构件:①大直径立柱与上平台甲板或上壳体、下壳体交接部分的外壳板;②组成箱型或工字型支承结构且承受主要集中载荷的上壳体或平台的甲板板、重型翼板、外壳板和舱壁;③在重要结构构件交接处承受集中载荷的部分舱壁、平台和骨架;④立柱、上平台甲板及上壳体或下壳体连接处提供适当对齐和足够载荷传递的"贯穿"构件。 Special structural members of the unit mean those portions of primary structural members which are in way of critical load transfer points, stress concentrations, such as: (1) special application structure of self-elevating units:①vertical leg structures in way of intersections with individual footings or with the mat structure;②intersection of lattice type leg structures which incorporate novel construction, including the use of steel castings. (2) special application structure of column-stabilized units:①shell plating in way of the intersections of vertical columns with platform decks and upper and lower hulls;②portions of deck plating, heavy flanges, shell boundaries and bulkheads within the upper hull or platform which form "box" or "I" type supporting structure which receive major concentrated loads;③major intersections of bracing members;④external brackets, portions of bulkheads, flats and frames which are designed to receive concentrated loads at intersections of major structural members;⑤"through" material used at connections of vertical columns, upper platform decks and upper or lower hulls which are designed to provide proper alignment and adequate load transfer. (3) special application structure of submersible units:①shell plating in way of the intersections of large-diameter vertical columns with platform decks or upper and lower hulls;②deck plating, heavy flanges, shell boundaries and bulkheads within the upper hull or platform which form "box";"I" type supporting structure which receive major concentrated loads;③portions of bulkheads, flats and framing which are designed to receive concentrated loads at intersections of major structural members;④" through" material used at connections of vertical columns, upper platform decks and upper or lower hulls which are designed to provide proper alignment and adequate load transfer.

Special survey of boiler　锅炉的特殊检验

Special survey(SS)　特别检验　又称船舶换证检验。在5年间隔期内进行的,对船舶的全面检验。检验时要求进坞、测厚等。特别检验也是对船级证书进行更新的检验。

Special trade　特种业务　系指在以下指定区域内,通过海上国际航线,运送大量特种业务旅客:南面以南纬20°线为界,从非洲东岸至马达加斯加西岸,再沿马达加斯加西岸和北岸至东经50°,然后沿东经50°子午线至南纬10°,再沿恒向线至南纬8°东经75°一点,然后沿恒向线至南纬11°东经120°一点,再沿南纬11°线至东经141°03′;东面以东经141°03′子午线为界,从南纬11°至新几内亚南岸,再沿新几内亚南岸、西北岸至东经141°03′一点沿恒向线至人民答那峨东北海岸北纬10°一点,再沿莱特、萨马和吕宋三岛西岸至吕宋岛的苏尔港,然后至苏尔港沿恒向线至香港;北面以亚洲南海岸为界,自香港至苏伊士;西面 以非洲东岸为界,自苏伊士至南纬20°一点。　Special trade means the conveyance of large numbers of special trade passengers by sea on international voyages within the area special below:on the south bounded by the parallel of latitude 20°S from the east coast of Africa to the west coast of Madagascar, thence the west and north coast of Madagascar to longitude 50°E, thence the meridian of longitude 50°E to latitude 10°S, thence the rhumb line to the point latitude 8°S, longitude 75°E, thence the rhumb line to the point latitude 12°S, longitude 120°E, thence the parallel of latitude 11°S to longitude 141°03′E;on the east bounded by the meridian of longitude 141°03′E from latitude 11°S to the south coast of New Guinea, thence the south, west and north coast of New Guinea to the point longitude 141°03′E, thence the rhumb line from the north coast of New Guinea at the point 141°03′E to the point latitude 10°N, at the north-east coast of Mindanao, thence the west coasts of the island of Leyte, Samar and Luzon to the Port of Sual (Luzon Island), thence the rhumb line from the Port of Sual to Hong Kong;on the north bounded by the south coast of Asia from Hong Kong to Suez;on the west bounded by the east coast of Africa from Suez to the point latitude 20°S.

Special trade passenger　特种业务旅客　系指特种业务中在露天甲板、上甲板和/或甲板间载运的旅客,这些处所的旅客均超过8人。　Special trade passenger means a passenger carried in special trades in spaces on the weather deck, upper deck and/or between decks which accommodate more than eight passengers.

Special trade passenger ship　特种业务客船　系指运送大量特种业务旅客的机动客船。　Special trade passenger means a mechanically propelled passenger ship which carries large numbers of special trade passengers.

Special trade passenger ship safety certificate　特种业务客船安全证书(1971)　系指根据"1971年特种业务客船舱室要求议定书"的规定颁发的特种业务客船格式的安全证书。　Special trade passenger ship safety certificate means a safety certificate for special trade passenger ships issued under the provisions of the special trade passenger ships agreement.

Special trade passenger ship space certificate　特种业务客船舱室证书(1973)　系指根据1973年特种业务客船舱室要求的议定书规定颁发的特种业务客船舱室证书。　Special trade passenger ship space certificate means a certificate issued under the provisions of the Protocol on space requirements for special trade passenger ship, 1973.

Special trade passenger ship space certificate　特种业务客船舱室证书(1974)　系指根据"1974年特种业务客船协定"的规定,签发的特种业务客船舱室证书。　Special trade passenger ship space certificate means a certificate issued under the provisions of the special trade passenger ships agreement, 1974.

Special trade passenger ships　特种业务客船　根据"1971年特种业务客船协定"的规定,主管机关或其认可的组织为特种业务客船签发的安全证书。或根据"1973特种业务客船舱室要求议定书"的规定,颁发的特种业务客船舱室证书。

Special trade system　专门贸易体系　又称特殊贸易体系。系指贸易国家进行对外贸易统计所采用的统计制度之一。它以货物经过结关作为统计进出口的依据。

Special vessel　特种船舶　系指主要用作摆渡、挖泥和捕鱼等用途的船舶。Special vessel means a ship intended primarily for ferry service, for dredging, for fishing, etc.

Special welding processes　特殊焊接工艺　如:(1)螺柱焊;(2)闪光对接焊;(3)摩擦焊;(4)激光焊;(5)堆焊;(6)环状盘管焊缝的环绕焊;(7)机器人焊等。　Special welding processes mean:(1) stud welding;

(2) flash butt welding; (3) friction welding; (4) laser-beam welding; (5) build-up welding; (6) orbital welding of circumferential pipe welds; (7) robot welding etc.

Specialized handling system 专用的转运系统 系指使用专用的物资装运设备进行货物装卸的一种方法。它通过吊索、吊盘、吊槽、集运架或类似的起货机械来替代手动转运货物,并包括回收装置/传送带或通过扭曲锁绑扎或捆扎,或类似设施将多个单元连在一起进行转运。该方法不必按有关澳大利亚海事安全局要求进行单独的试验和标记。 Specialized handling system means a method of loading or unloading cargo that employs specialized material handling equipment designed to dispense with manual handling of cargo by sling, tray, tub, pallet or similar cargo gear, and includes reclaiming devices, conveyor belts, and the handling of multiple units connected together by twistlocks, banding or strapping, or similar appliances, that have not been individually tested and marked in accordance with the order of the Australian Maritime Safety Authority.

Specially short tanks 长度特别短的货舱 系指低于邻接货舱的长度0.3倍以下的货舱。Specially short tanks are holds having a length equal to or less than 30% of the length of adjacent holds.

Specially suitable compartment 特别适合装载谷物的舱 系指至少由2块垂直或倾斜的纵向谷物隔壁构成的货物处所。该隔壁与舱口边纵桁相一致,或位于能限制谷物任何横向移动作用的位置。如果纵隔壁是倾斜的,则隔壁相对于水平面的倾斜角应不小于30°。Specially suitable compartment refers to a cargo space which is constructed with at least two vertical or slopping, longitudinal, grain-tight divisions which are coincident with the hatch side girders or are so positioned as to limit the effect of any transverse shift of grain. If slopping, the divisions are to have an inclination of not less than 30° to the horizontal.

Special-purpose ship 专用船 系指由于其功能原因而在船上搭载有包括乘客在内的12名以上专业人员,例如从事调查、训练和钻井等活动的船舶,以及鱼品加工船。Special-purpose ship means a ship which by reason of its function carries on board more than 12 special personnel including passengers, e.g. ships engaged in research, training and drilling as well as fish factory ships.

Specific absorption rate (SAR) 特定吸收率

Specific automation equipment 特殊自动控制装置 是对以下第1种特殊自动控制装置、第2种特殊自动控制装置和第3种特殊自动控制装置的总称:(1)第1种特殊自动控制装置—对遥控控制的压载舱排放压载水的布局的自动操纵系统,对散装液体货物的遥控控制装卸系统、机械传动的打开或关闭设备,主发动机的自动记录设备,遥控控制的系泊设备布局和控制点的空调布局的自动或遥控的控制设备;(2)第2种特殊自动控制装置—除了上述第1种特殊自动控制装置以外的—燃料油供应布局,冷藏集装箱的中央监控设备,货物软管操作的绞盘,自动甲板清洗布局,码头遥控控制系泊设备的布局,电力操控的绳梯缠绕设备和紧急拖曳索绞盘的自动或遥控的控制设备。(3)第3种特殊自动控制装置—除了上述第(2)种特殊自动控制装置以外的—对机械、机械中央控制系统、在驾驶室外面的主发动机和操舵装置的遥控控制布局,货物存储舱底的高水平的警报装置,独立的系泊设备和拖曳索绞盘的自动或遥控的控制设备。 Specific automation equipment is a general term for Class 1 specific automation equipment, Class 2 specific automation equipment and Class 3 specific automation equipment detailed below:(1) Class 1 specific automation equipment: automatic and remote controlled ballasting / de-ballasting arrangement, automatic steering system, remote controlled handling system for liquid cargo in bulk, power-driven opening and closing devices, automatic recording devices for main engine, remote-controlled mooring arrangements and air-conditioning arrangements for control stations; (2) Class 2 specific automation equipment: in addition to these in (1): automatic or remote equipment for remote-controlled fuel oil filling arrangements, centralized monitoring device for refrigerating containers, cargo hose handling winches, automatic deck washing arrangements, remote-controlled mooring arrangements at ship-sides, power-operated pilot ladder winding appliances and emergency towing rope winches; (3) Class 3 specific automation equipment: in addition to these in (2): automatic and remote control equipment centralized monitoring systems for machinery, centralized control systems for machinery, remote control arrangements for main engines and steering gear at the outside of the navigating bridge, high-level alarm device for cargo hold bilge, independent automatic or remote-controlled mooring arrangements and towing rope winches.

Specific combustion intensity (volumetric heat-release rate) 燃烧室热容强度 在单位时间、单位容积或单位面积内,燃烧室所产生的热量与进口空气总压之比。

Specific density 比密度

Specific design 具体设计 是初步设计的发展,开发。具体设计符合通用分析关于已确定的风险控制方案的结果和审批机关的要求。具体设计是在审批机关发表的声明的基础上开发出来的。

Specific duties 从量税 是以商品的质量、数量、容量、长度和面积等计量单位为标准计征的关税。从量税的计算公式如下：从量税额 = 商品数量 × 每单位从量税。各国征收从量税,大部分是以商品的质量为单位来征收,但各国对应纳税的进口商品质量计算的方法各有不同。一般有以下3种:(1)毛重(gross weight)法。又称总质量法,即包括商品内外包装在内的总质量计征税额;(2)半毛重(semi-gross weight)法。又称半总质量法,即对商品总质量扣除外包装后的质量计征其税额。这种办法可分为两种:①法定半毛重法。即从商品总毛重中扣除外包装的法定质量后,再计征其税额;②实际半毛重法。即从商品总毛重中扣除外包装的实际质量后,计征其税额;(3)净重(net weight)法。又称纯质量法。即在商品总质量中扣除内外包装的质量后,再计征其税额。这种办法又有2种:①法定净重法(legal net weight)。即从商品总质量中扣除内外包装的法定质量后,再计征其税额;②实际净重(real net weight)法。即从商品总质量中扣除内外包装的实际质量后,再计征其税额。

Specific elongation 延伸率

Specific environment condition 规定的环境条件 系指规定的风速、水流和浪高,在这种环境条件下船舶能进行预期的操作,抗冰载荷可不予考虑。 Specific environment condition are the specified wind speed, sea current and wave height under which the vessel is designed to carry out its intended operations. Ice-resistant load may not be considered.

Specific flow of persons 规定的人流 系指每单位时间及所涉及通道的每单位有效宽度通过脱险通道出口的一点的逃生人员的数量。 Specific flow of persons is the number of evacuating persons past a point in the exit route per unit time per unit of effective width of the route involved.

Specific fuel consumption 燃油消耗率

Specific gravity 重力密度

Specific heat capacity 比热容 又称比热容量,简称比热(specific heat),是单位质量物质的热容量,即使单位质量物体改变单位温度时的吸收或释放的内能。比热容是表示物质热性质的物理量。通常用符号 c 表示。比热容的单位是复合单位。最初是在18世纪,苏格兰的物理学家兼化学家 J·布莱克发现质量相同的不同物质,上升到相同温度所需的热量不同,而提出了比热容的概念。水的比热容较大,在工农业生产和日常生活中有广泛的应用。对于气候的变化有显著的影响。在同样受热或冷却的情况下,水的温度变化小一些,水的这个特征对气候影响很大。

Specific heat consumption 单位耗热量 加热量与蒸馏水产量之比。

Specific heat load of condenser 冷凝器单位热负荷 单位时间内,冷凝器每平方米冷却面积传递的热量。

Specific humidity 比湿(湿度比) 系指一团由干空气和水汽组成的湿空气中的水汽质量与湿空气的总质量之比。若湿空气与外界无质量交换,且无相变,则比湿保持不变。以 g/g 或 g/kg 为单位,通常大气中比湿都小于 40g/kg。

Specific jurisdiction 专门管辖

Specific load 单位载荷(比载)

Specific lubricating oil consumption 滑油消耗率 单位时间内主机或辅机输出功率的滑油消耗量。

Specific operating limits 规定的作业范围 系指规定的允许船位偏离某一设定点的范围。Specific operating limits are those specified allowable position deviations from a set point.

Specific output 比出量(单位输出)

Specific person 特殊人员 系指特种用途船上除乘客或船员或1岁以下儿童以外,与船舶的特殊用途有关的或在船上进行特殊工作而乘载于船上的人员。

Specific power 比效率 燃气轮机输出功率与通过压气机的总空气流量的比值。

Specific power consumption 单位耗电量 (海水淡化装置)总耗电量与淡水产量之比。

Specific requirement 具体要求(专门要求,特定要求)

Specific residual elongation 规定残余伸长率

Specific responsibility of companies 公司的具体责任 系指公司应确保船长在任何时候船上有资料可供缔约国政府正式授权的官员使用,使其能确定:(1)谁负责指派船员或当前以任何职能身份在船上受雇或工作的其他人员;(2)谁负责决定船舶的使用;(3)如果船舶按租船合同的条款使用,则谁租船合同的各方。 Specific responsibility of companies means that a company shall ensure that the master has information available on board, at all times, through which officers duly authorized by a Contracting Government can establish:(1)who is responsible for appointing the members of the crew or other persons currently employed or engaged on board the ship in any capacity on the business of that ship;(2)who is responsible for deciding the employment of the ship;(3)in cases where the ship is employed under the terms of charter party(ies), who are the parties to such charter party(ies).

Specific route 特定航线 系指船舶专门从事于2

个或几个规定的港口之间的航行。

Specific steam consumption 单位耗汽量 在以蒸汽为加热介质的蒸馏装置中蒸馏水产量与加热蒸汽量之比。

Specific test 特殊试验

Specific transformer 特种变压器 系指用于特殊用途的变压器。如整流变压器和对焊变压器等。应具有满足特定用途的性能。

Specific ultimate strength 比极限强度

Specific volume 比容 系指单位质量的物质所占有的容积,用符号"V"表示。其数值是密度的倒数。即,$V=1/\rho$。比容在大气科学中应用很多,可以运用到位涡的定义等。赤道地区由于温度很高,盐度很低,因而表面海水的密度很小,约 1.023;亚热带海区盐度虽然很高,但温度也很高,所以密度仍然不大;极地海区由于温度很低,所以密度最大。在垂直方向上,海水的结构总是稳定的,密度向下递增,在海洋上层密度垂直梯度较大,约从 1 500 m 开始,密度的垂直梯度很小,在深层,密度几乎不随深度而变化。水的汽化热为 40.8 千焦/摩尔,相当于 2 260 千焦/千克。

Specific weight 质量密度

Specification 说明书(一览表)

Specification of reserve parts tools and accessories 备件、工具及附件明细表

Specifications 技术条件(设计任务书,分类)

Specifications for materials 材料规格

Specified coastal service 特定的沿海航区 系指沿海航区,其地理范围将在《船名录》中注明,且出海距离通常不超过 21 海里,但船籍国主管机关或其所在营运区的海岸主管机关对"沿海航区"有其他的特定距离规定者除外,例如"印度尼西亚沿海航区"。 Specified coastal service mean the service along a coast, the geographical limits of which will be indicated in the Register Book, and for a distance out to sea generally not exceeding 21 nautical miles, unless some other distance is specified for "coastal service" by the Administration with which the ship is registered, or by the Administration of the coast off which it is operating, as applicable, e.g. "Indonesian coastal service".

Specified design life 规定的设计寿命 系指船舶假定处于营运和/或环境条件和/或腐蚀环境下的名义期限,用于选择相应的船舶设计参数。但船舶的实际使用寿命期限可能更长或更短,这取决于船舶整个寿命周期内的实际营运条件以及维护情况。 Specified design life is the nominal period that the ship is assumed to be exposed to operating and/or environmental conditions and/or the corrosive environment and is used for selecting appropriate ship design parameters. However, the ship's actual service life may be longer or shorter depending on the actual operating conditions and maintenance of the ship throughout its life cycle.

Specified elongation 规定伸长率

Specified level 规定水平

Specified method 规定的方法

Specified minimum tensile strength 标定抗拉强度下限值

Specified minimum yield stress 标定的最小屈服应力(标定的最小屈服强度)

Specified operating and environmental conditions 规定的营运和环境条件 系指对船舶在其整个寿命期限内预定航行的区域限定,涉及港内、水道和海上的货物和压载作业工况(包括中间工况)。 Specified operating and environmental conditions are defined as the intended operating area for the ship throughout its life and cover the conditions, including intermediate conditions, arising from cargo and ballast operations in port, waterways and at sea.

Specified operating or service areas 规定的航行和服务区域 系指可在 2 个或多个港口或地理特征地点之间航行,或者在规定的地理区域,例如在"红海区","Piraeus ~ Thessaloniki 以及在爱琴海内的岛屿间"航行。 Specified operating or service areas may be service between two or more ports or other areas of geographical features, or service within a defined geographical area such as "Red Sea Service", "Piraeus to Thessaloniki and Islands within the Aegean Sea".

Specified rated load 额定载荷(额定负载)

Specified requirement 规定要求

Specified route service[1] 特定航线航区 系指 2 个或多个港口之间或其他特定地理区之间航区,须在"船名录"中注明,例如:(1)"伦敦至鹿特丹的航区";(2)"伦敦、鹿特丹及汉堡间的航区";(3)"太平洋热带区航区";(4)"大湖及圣劳伦斯至蒙斯港航区";(5)"红海、东地中海及黑海航区"。 Specified route service means the service between two or more ports or other areas of geographical features which will be indicated in the Register Book, e.g.: (1) "London to Rotterdam service"; (2) "London, Rotterdam and Hamburg service"; (3) "Pacific Tropical Zone service"; (4) "Great Lakes and St. Lawrence to Pt. du Monts service"; (5) "Red Sea, Eastern Mediterranean and Black Sea service".

Specified route service[2] 特定航线营运 系指船舶专门从事于 2 个或几个港口之间的营运。 Specified route service means the service between two or more designated ports.

Specified value 规定值
Specify 指定（详细说明）
Specimen configuration 试样外形
Specimen test 试样试验
Speckle 斑点（斑纹，加斑点）
Spectral analysis 频谱分析　系指分析噪声能量在各个频率上的分布特性。
Spectral density 谱密度　用以表示波浪、船舶摇荡运动等在每单位频带宽度内组成分量平均能量的一种频率的函数。
Spectral measurement 频谱测量　系指在频域内测量信号的频率分量，以获得信号的多种参数和信号所通过的网络的参数。频谱测量虽属电子测量范围，但它除了对电信号进行分析研究以外，还可以借助各种传感器或转换器对各种非电量信号（水声、振动、生物、医学、各种随机过程和瞬态过程如爆炸、导弹发射、水声混响、舰船噪声和鱼雷噪声等）进行分析研究，从而改进其设计。它是在噪声控制工程中必做的一项工作，它提供了噪声源或声场的噪声特性与频率的关系，为进一步的噪声控制提供依据。
Spectrogram 频谱图　是以中心频率为横坐标；以各频率成分对应的强度（声压级或声强级等）为纵坐标的，而做出声压级-频率的关系曲线图。
Speech intelligibility index 语言清晰度指数　系指一个正常的语言信号（如音节、单词、句子等）能被听懂的百分数。
Speed 速率（速度，转速）
Speed (m/s) of persons along the escape route 沿着脱险通道的人员流速　沿着脱险通道的人员流速取决于人员的特定流量和脱险通道的类型。 Speed (m/s) of persons along the escape route depends on the specific flow of persons and on the type of escape facility.
Speed adjustment device 转速调整器
Speed and distance measuring equipment (SDME) 速度和测量设备
Speed and powering of ship 船舶快速性　表征船舶在静水中直线航行速度的性能。
Speed characteristic test 速度特性试验　内燃机在油量控制机构不变的情况下测定随转速而变化的各个主要参数的试验。
Speed correction method 航速修正法　系指根据英国船舶研究协会提出的运用海流对航速进行修正的方法。
Speed corrector 速度误差校正器

Speed governing 调速
Speed governor 调速器　当转速偏离给定值时，能发出信号，通过调节系统使转速恢复到给定值的机构。
Speed limiter (speed limiting device) 限速器
Speed loss in waves 风浪失速　推进动力装置功率调定后，船舶在风浪中较在静水航行中航速的降低值。
Speed mdicator (speed meter) 速度计（转速表）
Speed of advance 进速　螺旋桨相对于水沿轴线前进的速度。
Speed of ignition 点火速度
Speed of ship V　船舶的速度 V　系指干净船底以满载吃水状态航行于平稳海面时在最大持续功率下可获得的船舶的设计速度（kn）。 Speed of ship V is the designed speed in knots obtainable with clean bottom at calm sea and at the designed summer load line with the engine running at maximum continuous rating.
Speed over ground (SOG) 对地航速（SOG）　系指船上测量的相对于大地的船速。 Speed over ground (SOG) means a speed of the ship relative to the earth, measured on board of the ship.
Speed reduction 主动减速　船舶在风浪中航行，为了减少风浪对船舶的不利影响，主动调低主机功率，使航速比静水中下降的数值。
Speed reduction 减速
Speed reduction gear 减速机构（减速齿轮）
Speed regulating valve 调速阀
Speed sensor 速度传感器
Speed signal 速率信号
Speed test (speed trial) 航速试验（速率试验，测试试航）
Speed through water (STW) 对水速度　相对于水面的船速。 Speed through water is the speed of the ship relative to the water surface.
Speed V (kn) 航速　系指最大营运前进速度，kn。是船舶按其设计在海上以夏季载重线吃水和相应于最大持续功率（MCR）下的螺旋桨转速（r/min）能保持的航速。 Speed means a maximum ahead service speed. It means the greatest speed which the ship is designed to maintain in service at her deepest sea-going draught at the maximum propeller RPM and corresponding engine MCR (Maximum Continuous Rating).
Speed variation test 变速试验
Speedboat 快艇　是小型高速船舶的总称。主要用途为执勤、水上救援、娱乐、体育等。快艇也根据其用途分为巡逻艇、救生艇、游艇、赛艇等。
Speke anchor 斯贝克锚 一种改良型的霍尔锚。通过增加锚臂的长度，同时用锚冠板增强了锚的稳定性，使锚爪更容易入土。
Spherical bearing 球形轴承（球面轴承）
Spherical elbow union 球面弯头活管接
Spherical hinge 球铰链
Spherical joint 活节接合（球形接合）
Spherical linkage 球形铰链
Spherical sound wave 球面声波 系指在各向同性的均匀媒质中，从一个表面同步胀缩的点声源发出的声波。其波阵面为一个以点声源为球心的球面。
Spherical vavle 球形阀
Spherical wear particle 球状磨损粒子
Spheroidal graphite cast iron 球墨铸铁 是通过球化和孕育处理得到球状石墨，有效地提高了铸铁的力学性能，特别是提高了塑性和韧性，从而得到比碳钢还高的强度。球墨铸铁是20世纪50年代发展起来的一种高强度铸铁材料，其综合性能接近于钢，正是基于其优异的性能，已成功地用于铸造一些受力复杂、强度、韧性、耐磨性要求较高的零件。球墨铸铁已迅速发展为仅次于灰铸铁的、应用十分广泛的铸铁材料。所谓"以铁代钢"，主要指球墨铸铁。
Spheroidal graphite iron casting 球墨铸铁件
Spheroidize annealing/ Spheroidizing 球化退火 是使钢中碳化物球化而进行的退火工艺。将钢加热到Ac1以上20～30℃，保温一段时间，然后缓慢冷却，得到在铁素体基体上均匀分布的球状或颗粒状碳化物的组织。
Spigot and faucet joint[1] 套筒接合（插管接合）
Spigot and socket joint[2] 套筒接合（窝接，插承接合）
Spigot joint 套筒接合
Spike type box grab 齿式爪箱器
Spill nozzle 回油式喷油嘴 具有回油通道使部分燃油可以再循环的喷油嘴。
Spill retainment flat 溢出物挡板
Spill valve 溢流阀
Spillage 溢油量
Spilling oil 溢油
Spilling oil source 溢油源
Spilt hydrofoil 分裂式水翼 由中间分开并且两舷对称的两部分构成的水翼。
Spin magnetic resonance（SMR） 磁共振 系指自旋磁共振现象。其意义上较广，包含：(1)核磁共振（Nuclear magnetic resonance/NMR）；(2)电子顺磁共振（Electron paramagnetic resonance/EPR）或称电子自旋共振（Electron spin resonance/ESR）；(3)用于医学检查的主要是磁共振成像（Magnetic resonance imaging/MRI）。比如，电流通过一根导线，会在导线周围形成磁场，当电磁波频率与机体振动频率一致时，会产生共振。磁共振成像技术由于其无辐射、分辨率高等优点被广泛地应用于临床医学与医学研究。一些先进的设备制造厂商与研究人员一起，不断优化磁共振扫描仪的性能、开发新的组件。例如：德国西门子公司的1.5T超导磁共振扫描仪具有神经成像组件、血管成像组件、心脏成像组件、体部成像组件、肿瘤程序组件、骨关节及儿童成像组件等。其具有高分辨率、磁场均匀、扫描速度快、噪声相对较小、多方位成像等优点。磁共振催生"隔空充电"早在100多年前，科学家就发现电力可以转化为电磁波在空气中传播。当时有科学家制作了两个铜线圈，让其中一个连接电源，负责发出能量；另一个连接灯泡，负责接收能量；两个线圈中间没有任何电线连接。由于这两个线圈拥有相同的电磁波振动频率，电能最终成功地在两个线圈之间进行传输。
Spindle 轴（芯轴，杆）
Spindle gland 芯轴填料函压盖
Spindle nose 轴头
Spindle oil 锭子油
Spiral 螺旋（螺旋，螺旋的）
Spiral angle 螺旋角
Spiral curve 螺线
Spiral propellers 螺环推进器 桨叶为部分环形，叶面螺距由前向后逐渐增大的螺旋推进器。
Spiral strakes 螺旋侧板 系指在Spar平台壳体外侧设置的用来抑制Spar平台涡激运动的结构物。
Spirit thermometer 酒精温度计
Splash 飞溅
Splash lubrication 飞溅润滑
Splash plate 防溅挡板 设置在蒸发液面或闪发孔口上方防止汽水混合物溅入冷凝器的挡板。
Splashproof 防溅的
Splashproof type 防溅式
Splashtight door 溅密门
Splatter 溅泼
Spline 花链（齿条）
Splined shaft 花键轴
Splinter 片（裂片，木片，劈开，扯裂）
Split 裂缝（裂纹，裂口，劈开）
Split bolt 开口螺栓（开尾螺栓，带开尾销螺栓）
Split cotter（split pin） 开尾销（开口销）
Split hopper barge 对开式泥驳 系指整个主船体

可从纵中剖面处打开而达到卸泥目的的驳船。 Split hopper barge is a barge that entire main hull is to be opened along longitudinal centerline for unloading.

Split hopper barge or split hopper dredger 对开泥驳/对开挖泥船 系指由沿船体纵向分开的2个独立的半体组成,2个半体通过设在泥舱前后端壁旁的甲板铰链和液压装置相互连接的泥驳或挖泥船。 Split hopper barge or split hopper dredger means a barge or dredger composed of two independent half hulls separating longitudinally along the hull, with the half hulls being connected by deck hinges and hydraulic installation at the fore and aft transverse bulkheads of the hopper.

Split key 开尾销
Split pin extractor 起开尾销器
Split pin hole 开尾销孔
Split type dump hopper barge 对开泥驳 由左右2个对称的半船体在艏艉端铰接而成,可自动开底进行卸泥的泥驳。
Splite 拆
Split-shaft gas turbine 分轴燃气轮机 燃气发生器和动力涡轮分别配置在两根彼此独立的轴上的燃气轮机。
Spoil hoppers（mud hold） 泥舱 挖泥船或驳船上,用以装载泥沙的舱。
Spoke of a wheel 轮轴
Spoke sheave 辐轮
Spontaneous combustion 自燃
Spontaneous ignition temperature（SIT） 自燃温度
Spool 线轴（短管,四通）
Spooling gear 排缆装置 使缆绳依次存放在绞车卷筒上的导向排列装置。
Sporadic dumping 偶然性倾销 系指某一商品的生产厂商为避免存货的过量积压,于短期内向海外市场大量低价销售该商品。这种倾销方式是偶然发生的,一般无占领国外市场、排挤竞争者之目的,而且因为持续的时间较短,不至于打乱进口国的市场秩序、损害其工业。因此,国际社会一般对这种偶发性倾销通常不采取反倾销措施。
Sport universal vehicle（SDV） 运动型多功能车
Spot 点（焊点,斑点）
Spot annealing 局部退火
Spout 壶嘴（管）
Spray 喷雾（油雾）
Spray angle 喷雾角 燃油自喷油嘴喷口喷出所形成的雾化锥角。

Spray film evaporator 卧管淋膜蒸发器 水喷淋在卧管换热面上形成薄膜蒸发的蒸发器。
Spray hood（personal flotation device） 防溅罩（个人漂浮装置） 为了减少或去除波浪飞溅,或防止水进入使用者的通气孔,从而增加使用者在恶劣海况下生存的机会,在使用者通气孔前面带有或放置的罩子。 Spray hood is the cover brought or placed in front of the airways of a user in order to reduce or eliminate the splashing of water from waves or the like onto the airways and thereby to promote the survival of the user in rough water conditions.
Spray paint 喷漆 系指使用喷涂工具进行涂装的作业方式。
Spray pump 喷雾泵
Spray resistance 飞溅阻力 高速船航行时,使水飞溅所消耗的能量所形成的阻力。
Spray test 喷雾试验（雾化试验）
Spray tightness 风雨密 在风雨中能不使雨水透入舱室内的密封性。
Spray water 喷淋水（雾状水）
Sprayer 喷雾器
Spraying 喷射（喷涂）
Spraying nozzle 水雾喷嘴
Spray-tight door 风雨密门 符合风雨密要求的门。
Spraytightness 风雨密 在风雨中能不使雨水透入舱室的密性。
Spread mooring 分布式系泊
Spread mooring drilling ship 辐射系泊定位钻井船 系指依靠在船体四周呈辐射状分布的多根系泊缆进行定位的钻井船。
Spread spectrum wireless data communication 扩频无线数据通信
Spreader 集装箱吊具 吊装集装箱的专用吊具。
Spring 弹簧（斜缆,移船缆）
Spring balance 弹簧秤
Spring balanced hatch cover 弹簧平衡舱盖 利用弹簧弹力平衡舱盖自重,以减少翻启用力的舱口盖。
Spring base 弹簧座
Spring buffer 弹簧缓冲器
Spring constant 弹簧常数
Spring cotter 弹簧销
Spring coupling 弹簧联轴节
Spring damper 弹簧减振器
Spring handle 弹簧把手 装有弹簧可借弹力自动扣锁关闭的把手。
Spring line 倒缆 自船艏部向后或自船艉部向前

引出的系缆。前者称前倒缆,后者称后倒缆。
Spring lock 弹簧锁(弹簧锁扣)
Spring range 弹力范围
Spring retainer 弹簧保险圈
Spring seat 弹簧座
Spring sheet holder 金属片弹簧夹
Spring shock absorber 弹簧减震器
Spring steel 弹簧钢 主要是有良好的弹性,又由于它是在动载荷环境条件下工作的,所以对制造弹簧的材质最主要的应有高的屈服强度;在承受重载荷时不引起塑性变形;应有高的疲劳强度,在载荷反复作用下具有长的使用寿命;并有足够的韧性和塑性,以防在冲击力作用下突然脆断。
Spring stiffness 弹簧刚度(弹性刚度)
Spring surge 弹簧颤动
Spring suspension 弹簧悬置
Spring tab 弹簧调整片
Spring towing hook 弹簧拖钩 装有缓冲弹簧,能承受挂拖时冲击载荷的拖钩。
Spring unit 弹簧组件
Spring washer 弹簧垫圈
Spring wire 弹簧钢丝
Springing 弹振 系指简谐强迫振动。其本质上是船舶在波浪中的一种振动行为,与船舶在波浪中的相对运动有关。当船体梁的垂向二节点固有频率与波浪遭遇频率接近时,船体梁将发生共振。弹振对船体梁结构强度产生不利影响,其主要影响船体梁的疲劳强度。
Springingness 弹性(有弹力)
Spring-loaded governor 弹簧式调速器
Spring-loaded pressure 弹簧式压力计
Spring-loaded safety vavle 弹簧式安全阀
Sprinkler 洒水器(喷水器)
Sprinkler fire-extinguishing system 喷水灭火系统
Sprinkler pump 喷水泵
Sprinkler system 喷水器系统(喷淋系统)
Sprocket 链轮(链轮铣刀)
Sprocket wheel (chain sprocket) 链条卷条(链条卷轮) 带动或固定传动循环链条的专设链轮或组合链轮。
Spud 定位桩 安装在工程船上,以电动绞车或液压机械控制其起落,用以固定船位的桩。
Spud gantry (spud carriage) 定位桩架 用以支持定位桩的构架。
Spud-can 桩靴 是自升式海洋平台重要的组成部分。是为增大桩腿的支承面积和减少桩腿在软海底

的插入深度而设置在自升式钻井平台桩腿下部的可拆式的箱型结构。它起着支撑平台的作用。桩靴结构的可靠性对于保证平台稳定具有决定性的作用。在桩靴设计过程中,须考虑的载荷主要是从桩腿传递下来的垂直轴向力和土壤对桩靴底部的支撑反力。
Spud-can jack-up drilling unit 沉垫型自升式钻井平台
Spur-gear 正齿轮
Spur-wheel 正齿轮
Square file 方锉
Square groove 方槽
Square head bolt 方头螺栓
Square netter 敷网渔船 利用撑杆和提放网设备,把网具事先敷设水中,用诱集方法使鱼入网而进行捕捞作业的渔船。
Square nut 方形螺母
Square spanner 方形扳手
Square steel 方钢 系指用方坯热轧出来的一种方形材料,或者由圆钢经过冷拔工艺,拔出来的方形材料。方钢理论质量计算公式:边长×边长× 0.00785 = kg/m。方钢是实心的棒材。区别于方管,空心的,属于管材。类似的棒材还有圆钢,六角钢,八角钢。
Square thread 方形螺纹
Squid angling boat / Squid tackle vessel 鱿鱼钓船 使用鱿鱼钓具,专门捕捞鱿鱼的钓渔船。
Squid angling machine 自动鱿鱼钓机 具有引诱鱿鱼捕食、拉鱼上钩,自动连续起放钓线等多种功能的机械。
Squirrel-cage winding 鼠笼绕组 是一种通常不绝缘的,其导体沿电机周边均匀分布,且采用连续端环连接的一种永久短路的绕组。主要用于感应电机。
Stabilisation 稳定(稳定化,减摇)
Stabilising 稳定化处理 系指进行低温加热通常250℃,然后慢慢冷却,消除内部应力的热处理。Stabilising is the relief of residual internal stresses by heating to a predetermined temperature, usually 250 ℃ in the region, then cooling slowly.
Stability[1] 安定(稳定的)
Stability[2] 稳性 系指船舶或海洋结构物在规定时间及环境条件下抗御倾覆(防翻船)的能力。这种能力是由环境条件(造成翻船的各种外力,如风、浪、流、破损时进水等)及船舶自身具有的扶正能力两方面因素所决定的,而环境条件和船的自身扶正能力是多种多样的,随机的。近代稳性定义为船舶在风浪的作用下保持其不翻(或防止翻船)的概率。
Stability at large angles (of inclination) 大倾角稳

性　船舶作大角度(一般为 10°以上)横向倾斜的稳性。

Stability criteria　稳性标准

Stability criterion number　稳性标准数　船舶倾覆力矩与规定的倾覆力矩的比值，或倾覆力臂与规定的倾覆力臂的比值，是船舶稳性的一个考核指标。

Stability curves　稳性曲线

Stability in damaged condition (flooded stability)　破损稳性　系指船舶因破损而船体内进水后的稳性。

Stability information　稳性资料　系指向船长提供的经主管机关同意的必要资料，以使船长能用迅速而简便的方法获得有关各种营运状态下船舶稳性的正常指导。应将一份稳性资料的副本提供给主管机关。这些资料应包括：(1)确证符合有关完整和破损稳性要求的最小营运初稳性高度(GM)对吃水的曲线图或表格，也可选择相应的最大许用重心垂向位置(KG)对吃水的曲线图或表格，或与这些曲线图等效的其他资料；(2)有关横贯进水装置的操作说明；(3)破损后维持要求的完整稳性和稳性所必需的所有其他数据和辅助措施。稳性资料应表明在营运纵倾范围超过 ±0.5% L_s 的情况下各种纵倾的影响。对必须满足 SOLAS 公约 B-1 部分稳性要求的船舶，稳性资料按有关分舱指数的计算确定，方式如下：d_s、d_p 和 d_l 三种吃水的最小要求 GM(或最大的许用重心垂向位置 KG)等于残存因数(S_i)所用相应装载情况的 GM(或 KG 值)。对中间吃水，所用的值应通过线性内插法求得，仅用于最深分舱吃水和部分分舱吃水之间以及部分载重量和轻载航行吃水之间的 GM 值。还应考虑完整稳性标准，即按这两种标准为每个吃水保留最小要求 GM 值中的最大者，或最大的许用 KG 值中的最小值。如果分舱指数按不同纵倾计算，可用同样方式确立若干要求的 GM 曲线。当最小营运初稳性高度(GM)对吃水的曲线图或表格不适用时，船长应确保营运工况不偏离经研究采用的装载工况，或通过计算验证符合该装载工况的稳性标准。 Stability information means information supplied to the master, with such information satisfactory to the Administration which is necessary to enable him by rapid and simple processes to obtain accurate guidance on the stability control of the ship under varying conditions of service. A copy of the stability information shall be furnished to the Administration. The information should include: (1) curves or tables of minimum operational metacentric height (GM) versus draught which assures compliance with the relevant intact and damage stability requirements, alternatively corresponding curves or tables of the maximum allowable vertical centre of gravity (KG) versus draught, or with the equivalents of either of these curves; (2) instructions concerning the operation of cross-flooding arrangements; (3) all other data and aids which might be necessary to maintain the required intact stability and stability after damage. The stability information shall show the influence of various trims in cases where the operational trim range exceeds ±0.5% of L_s. For ships which have to fulfil the stability requirements of Part B-1 of SOLAS, information referred to in Paragraph 2 are determined from considerations related to the subdivision index, in the following manner: minimum required GM (or maximum permissible vertical position of centre of gravity KG) for the three draughts d_s, d_p and d_l are equal to the GM (or KG values) of corresponding loading cases used for the calculation of survival factor S_i. For intermediate draughts, values to be used shall be obtained by linear interpolation applied to the GM value only between the deepest subdivision draught and the partial subdivision draught and between the partial load line and the light service draught respectively. Intact stability criteria will also be taken into account by retaining for each draft the maximum among minimum required GM values or the minimum of maximum permissible KG values for both criteria. If the subdivision index is calculated for different trims, several required GM curves will be established in the same way. When curves or tables of minimum operational metacentric height (GM) versus draught are not appropriate, the master should ensure that the operating condition does not deviate from a studied loading condition, or verify by calculation that the stability criteria are satisfied for this loading condition.

Stability instrument　稳性仪　系指安装在特定的船舶上，用以确定在任何作业装载工况下满足稳性手册中对该船舶规定的稳性要求。稳性仪由硬件和软件组成。 Stability instrument is an instrument installed on board a particular ship by means of which it can be ascertained that stability requirements specified for the ship in the Stability Booklet are met in any operational loading condition. A stability instrument comprises hardware and software.

Stability standard　稳性标准　为保证船舶航行安全，按船舶类型、航区自然条件和船舶营运等情况所制定的有关稳性的标准。

Stabilization control system　稳定控制系统　系指用于稳定船舶状态的主要参数：横倾、纵倾、航向、高度及控制船舶运动：横摇、纵摇、艏摇、升沉的一种系统。但不包括与船舶安全营运无关的那些设备，诸如减少船舶运动或垫航控制的系统。稳定控制系统的主要部件可包括如下：(1)执行装置，诸如舵、水翼、襟翼、围裙、风扇、喷水器、可回转和可控制的螺旋桨、传输液体的泵等;(2)驱动执行装置的动力机械;(3)搜集和处理数据并做出判断,发出指令的稳定设备,诸如传感器、逻辑处理器和自动安全控制器等。 Stabilization control system is a system intended to stabilize the main parameters of the craft's attitude: heeling, trim, course and height and control of the craft's motions: roll, pitch, yaw and heave. This term excludes devices not associated with the safe operation of the craft, e. g. motion-reduction or ride-control systems. The main elements of a stabilization control system may include the following: (1) devices such as rudders, foils, flaps, skirts, fans, water jets, tilting and steerable propellers, pumps for moving fluids; (2) power drives, actuating stabilization devices; (3) stabilization equipment for accumulating and processing data for making decisions and giving commands such as sensors, logic processors and automatic safety control.

Stabilization device 稳定装置 系指依靠稳定控制系统中的装置产生的力来控制船舶位置的一种装置。

Stabilization device means a device as enumerated in stabilization control system with the aid of which forces for controlling the craft's position are generated.

Stabilization modes 稳定模式 (1)地面稳定:航速和航向信息系统参照地面并采用地面轨迹输入数据或 EPFS 作为参照的显示模式;(2)海面稳定:航速和航向信息系统参照海面并采用回转仪或等效装置和水速记录仪输入数据作为参照的显示模式。 Stabilization modes means: (1) ground stabilization: display mode in which speed and course information are referred to the ground, using ground track input data, or EPFS as reference; (2) sea stabilization: display mode in which speed and course information are referred to the sea, using gyro or equivalent and water speed log input as reference.

Stabilization system 稳定系统 系指用于稳定船舶航行姿态的主要参数:横摇、飞行纵倾、纵摇、艏向和飞高及控制船舶运动:横摇、纵摇、艏摇和升沉的一种系统。该术语不包括与船舶安全营运无关的装置。稳定系统的主要部件可包括如下各项:(1)舵、翼、襟翼、围裙、风扇、可回转和可控制的螺旋桨、传输液体的泵之类装置;(2)启动稳定装置的动力驱动装置;(3)积累和处理数据以做出决定和发出指令的稳定设备,诸如传感器、逻辑处理器和自动安全控制装置等。 Stabilization system is a system intended to stabilize the main parameters of the craft's attitude: roll, flight trim, pitch, heading and altitude and control of the craft's motions: roll, pitch, yaw and heave. This term excludes devices not associated with the safe operation of the craft. The main elements of a stabilization system may include the following: (1) devices such as rudders, foils, flaps, skirts, fans, tilting and steerable propellers, pumps for moving fluids; (2) power drives actuating stabilization devices; (3) stabilization equipment for accumulating and processing data for making decisions and giving commands, such as sensors, logic processors and automatic safety control.

Stabilization trial 减摇试验

Stabilizator(stabilizer, stabiliser) 稳定器(减摇装置,稳定剂)

Stabilized gyrocompass 平台罗经 把传统的平台与罗经结合在一起,应用电磁控制技术代替传统的摆式控制方式,从而使性能得到提高。平台罗经除能提供较高精度的方位信息外,还能提供其他系统如雷达、武器等所必需的姿态信息—纵横摇角。新型平台罗经一般都具有短期惯导功能,加之其价格比高精度的惯导系统低得多,因此平台罗经正成为各种船舶最基本的主流陀螺导航装备。平台罗经一般由主仪器、修正系统(计算机)、稳定回路和同步复示系统等部分组成。

Stabilized shunt generator 稳定并励发电机 这种发电机除增加一个串励磁场绕组外,与并励发电机相同。此串励绕组提供的励磁占适当比例,因而不需要采用均压线就可实现满意的并联运行。此型发电机的电压调节应与并励发电机的要求相一致。

Stabilized shunt motor 稳定并励电动机 是一种为了防止转速升高或随着负载增加使转速稍有降低而增加了一个弱串励绕组的并励电动机。

Stabilizer compartment 减摇装置舱(减摇装置室)

Stabilizer control gear 减摇控制设备 测量船舶横摇摆幅参数将其信号放大并控制执行机构动作的设备。

Stabilizer room 减摇装置舱

Stabilizing device 稳定装置(减摇装置)

Stabilizing equipment 稳定设备(减摇装置,消摆装置)

Stabilizing fin 减摇鳍(稳定翼,舷侧可控鳍) 外伸于两舷。控制攻角以产生不同水动力矩的鳍板。

Stabilizing tank(anti-rolling tank) 减摇水舱 为减缓船的横摇而设置的专用水舱。

Stable 稳定的

Stable equilibrium 稳定平衡

Stable operation condition 稳定运转状态

Stable running 稳定运行

Stable state 稳态

Stable vibration 稳定振动

Stack 烟囱(排气管,堆装)

Stack factor 堆叠系数 系指流至形成堆叠的集装箱的实际热量,与可能流至(假定其所有表面被完全暴露于货舱温度中的)相同的集装箱的热量之比。 Stack factor means the ratio of the actual heat flowing into the containers forming the stack, to the heat which would flow into the same containers if all their surfaces are completely exposed to the hold temperature.

Stack of containers 集装箱堆(集装箱块) 系指由紧固件垂向相连的 N 个集装箱。 Stack of containers consists of "N" containers connected vertically by securing devices.

Staff member 职员(国际海事卫星组织) 系指总干事和由国际海事卫星组织所雇用的全日制工作的并受人事条例规定制约的任何人员。 Staff member means the Director General and any person employed full time by INMARSAT and subject to its staff regulations.

Stage efficiency 级效率 压气机或涡轮一个级的绝热效率。压气机:$\eta_p > \eta_s > \eta_{ad}$;涡轮:$\eta_p < \eta_s < \eta_{ad}$;式中,$\eta_p$—整台压气机或涡轮的多变效率;$\eta_{ad}$—整台压气机或涡轮的绝热效率。

Stage for liferaft 救生筏架 用以存放救生筏,通常又作为滑放救生筏的斜轨的专门架子。

Stage heating 中间加热

Stage matching 级间配合 在给定环面尺寸和形状的多级轴流式压气机中,为避免当非设计工况时,密度比的变化引起各级轴向速度分布的变化,使高压级或低压级中首先出现失速阻塞现象,而在设计时考虑的各级工作之间的协调与配合。

Stage of speed 速度级

Stagnation pressure 驻点压力 速度为零的流线分岔点上的压力。

Stainless 不锈的

Stainless clad 不锈包层钢

Stainless steel 不锈钢

Stainless steel grinding dust 不锈钢研磨粉 褐色团块,含水量 1~3%,可能产生粉尘,无特殊危害,该货为不燃物或失火风险低。 Stainless steel grinding dust is a brown lump, moisture content is 1% to 3%. It may give off dust. It has no special hazards. This cargo is non-combustible or has a low fire-risk.

Stair landing 扶梯平台 装于扶梯中部,供缓步的平台。

Stair soffit board 防尘板 紧贴在舱室楼梯步下面,为防止灰尘落到楼梯下的板。

Stairway and passageway arrangement 梯道布置 为保证人、物在各层甲板及其上、下之间能安全和顺利的通行,而对全船扶梯、通道所做的全面、统一的规划和布置。

Stairways 梯道(通道) 系指内部梯道、升降机、完全围蔽的紧急脱险围阱、自动扶梯(完全设在机器处所内者除外),以及通往上述梯道的环围。对此,仅在一层甲板设有环围的梯道应视为未用防火门与其隔开的处所的一部分。 Stairways mean interior stairways, lifts, totally enclosed emergency escape trunks, and escalators (other than those wholly contained within the machinery spaces) and enclosures thereto. In this connection, a stairway which is enclosed only at one level shall be regarded as part of the space from which it is not separated by a fire door.

Stakeholder 利益相关方 能够影响风险、被风险所影响、或认为自己被风险所影响的任何个人、团体或组织。

Stale B/L 过期提单 (1)系指提单晚于货物到达目的港。(2)是向银行交单时间超过提单签发日期 21 天,这种滞后交到银行的提单。银行有权拒收。

Stall 失速 由于冲角过大或冲击波的作用而造成的压气机叶片表面上气流附面层显著分离因而叶片性能急剧下降的现象。

Stall margin 喘振边界

Stalling rudder angle 临界舵角 舵升力开始停止增加时的舵角。

Stamp forging 压锻(落锻,型锻)

Stamp out 冲出(模压,灭火,冲孔)

Stamping 冲击(锤击,压碎)

Stamping forming 冲压成形

Stand by air compressor 备用空气压缩机 做好启动准备,随时可以投入使用的空气压缩机。

Stand pipe 竖管 直接与锅炉及压力容器相连的下列部件:(1)管及喷嘴;(2)以将阀门连接到锅炉及压力容器为目的的,由直接与锅炉及压力容器相连的管和法兰构成的贯通件;(3)人孔、清洁孔或检查孔等的安装环。 Stand pipe means the following attached directly to body of boilers and pressure vessels:(1)pipe and nozzle;(2)penetration piece consisting of flange and pipe attached directly to boilers and pressure vessels for the purpose of fitting the valve attached directly to body of boilers and pressure vessels;and (3) rings for installation of manhole, mud holes and peep holes.

Standard 标准 系指船舶适用的国际公约和规则、船旗国主管机关要求和船级社规范、指南。

Standard acceleration of gravity g_0　标准重力加速度　Standard acceleration of gravity g_0 = 9.81 m/s²。

Standard chartered bank　渣打银行　又称标准渣打银行、标准银行,是一家总部在伦敦的英国银行。它的业务遍及许多国家,尤其是在亚洲和非洲,在英国的客户却非常少,2004 年其利润的 30% 来自香港地区。渣打银行的母公司渣打集团有限公司则于伦敦证券交易所及香港交易所上市,亦是伦敦金融时报 100 指数成分股之一。渣打银行是一家历史悠久的英国银行,在维多利亚女皇的特许(即"渣打"这个字的英文原义)下于 1853 年建立。是获得皇家特许而设立的,专门经营东方业务。1858 年在上海设立分行。1859 年在香港设立分行,当时第一任总经理为英国人麦加利。渣打银行的总公司渣打集团有限公司除了在伦敦证券交易所、香港证券交易所上市外,还在印度的孟买证券交易所以及印度国家证券交易所上市,是伦敦金融时报 100 指数成分股之一,是英国伦敦家喻户晓的著名银行,在全球拥有 1 470 多家分支机构,遍布全世界 56 个国家和地区,总资产超过 800 亿人民币。

Standard container　标准集装箱　系指按船级社的"集装箱认证设计图"或其他符合 ISO 1492/2 要求的经认可的"集装箱认证设计图"制造的 40 ft 相当单位(FEU)的标准生产集装箱。集装箱可为标称型或"高立方"("high-cube")型。 Standard container means a forty-foot equivalent unit (FEU) standard production container constructed in compliance with the Society's Container Certification Scheme, or another recognised Container Certification Scheme in accordance with ISO 1492/2 requirements. The container may be the normal or "high-cube" type.

Standard cubic foot of gas (scf)　天然气的标准立方英尺　天然气的体积是在相对于一个大气压和 60 ℉时,用标准立方英尺(scf)来量度的;而米制单位是在相对于 1 bar(10^5 Pa)和 0 ℃时用标准立方米(m^3)来量度的。 Gas is measured in standard cubic feet, corresponding to the volume at one atmosphere and 60 ℉. Metric measurements are in normal cubic meter (N − m^3), corresponding to the volume at one bar and 0 ℃.

Standard deviation　标准差　方差的平方根。其表达式为: $\sigma = (D\xi)^{1/2}$。

Standard deviation of wave period Tdsd　波浪周期的标准偏差　在运行区域所有的海况下,跨零周期的标准偏差 Tdsd = Tsd (s)。 Standard deviation of the zero crossing periods for the service area Tdsd = Tsd (s).

Standard discharge connections　标准排放接头　系指用于使接收设备与船上的排放管路相连接的接头。排放接头法兰的标准尺寸见表 S-2。 Standard discharge connections are those connections used for connecting reception facilities with the ship's discharge pipeline. Standard dimensions of flanges for discharge connections are shown in Table S-2.

表 S-2　标准排放接头尺寸
Table S-2　Standard discharge connections

规格　Description	尺寸　Dimension
外径　Outside diameter	210 mm
内径　Inner diameter	按照管子的外径　According to pipe's outside diameter
螺栓圈直径　Bolt circle diameter	170 mm
法兰槽口　Slots in flange	直径 18 mm 的孔 4 个,等距离分布在上述直径的螺栓圈上,开槽口至法兰外沿,槽口宽 18 mm 4 holes, 18 mm in diameter equidistantly placed on a bolt circle of the above diameter, slotted to the flange periphery. The slot width is 18 mm
法兰厚度　Flange thickness	100 mm
螺栓和螺帽:数量、直径 Bolts and nuts:quantity and diameter	4 个,每个直径 16mm,长度适当。 4, each of 16 mm in diameter and of suitable length

续表 S-2

规格 Description	尺寸 Dimension
法兰应设计为能接受最大内径不大于 100 mm 的管子,以钢或其他等效材料制成,表面平整,连同一个适当的垫圈,应能承受 600 kPa 的工作压力	The flange is designed to accept pipes up to a maximum internal diameter of 100 mm and shall be of steel or other equivalent material having a flat face. This flange, together with a suitable gasket, shall be suitable for a service pressure of 600 kPa

Standard displacement 标准排水量 系指军用舰船空载排水量加上全额的人员、淡水、粮食、燃料、滑油、弹药、供应品、给水以及锅炉、冷凝器内所保持一定水位的水,但不包括燃料、滑油和备用锅炉水时的排水量。

Standard display 标准显示 当海图首次在 ECDIS 上显示时应显示的信息级别。为制定航线或监控航线而提供的信息级别可被航海者需要由其进行修改。在进行航线计划和航线监督时至少应用的显示模式。标准显示包括:(1)基本显示;(2)干燥线;(3)浮标、灯标其他助航设备和固定结构;(4)航道、海峡等边界;(5)可视图项和雷达上的显示图项;(6)禁航区和限航区;(7)海图比例边界;(8)注意事项的显示;(9)船舶航线划定系统和导航航线;(10)群岛海上航路。 Standard display means the level of information that should be shown when a chart is first displayed on ECDIS. The level of the information it provides for route planning or route monitoring may be modified by the marine according to the mariner's needs. Standard display is the display intended to be used as a minimum during route planning and route monitoring. The chart content is listed as following:(1) display base;(2) drying line;(3) buoys, beacons, other aids to navigation and fixed structures;(4) boundaries of fairways, channels, etc.;(5) visual and radar conspicuous features;(6) prohibited and restricted areas;(7) chart scale boundaries;(8) indication of cautionary notes;(9) ships' routeing system and ferry routes;(10) archipelagic sea lanes.

Standard fire test 标准耐火试验 系指:(1)将需要试验的舱壁或甲板试样置于试验炉内,加温到大致相当于下列标准时间—温度曲线的一种试验,试样暴露表面面积应不小于 4.65 m²,其高度(或甲板长度)不小于 2.44 m,试样应尽可能与所设计的构件近似,并在相当位置包括至少有关接头。标准时间—温度曲线应是连接下列各点的一条光滑曲线:自开始至满 5 min 时,538 ℃;自开始至满 15 min 时,704 ℃;自开始至满 30 min 时,843 ℃;自开始至满 60 min 时,927 ℃。(2)将相关舱壁或甲板的试样置于试验炉内,根据《耐火试验规则》规定的试验方法加温到大致相当于标准时间—温度曲线的一种试验。 Standard fire test is:(1) one in which specimens of the relevant bulkheads or decks are exposed in a test furnace to temperatures corresponding approximately to the standard time-temperature curve. The specimen shall have an exposed surface of not less than 4.65 m² and a height (or length of deck) of 2.44 m resembling as closely as possible the intended construction and including where appropriate at least one joint. The standard time-temperature curve is defined by a smooth curve drawn through the following points: at the first s min—538 ℃; at the first 10 min—704 ℃; at the first 30 min—843 ℃; at the first 60 min—927 ℃;(2) a test in which specimens of the relevant bulkheads or decks are exposed in a test furnace to temperatures corresponding approximately to the standard time-temperature curve in accordance with the test method specified in the Fire Test Procedures Code.

Standard for vapor emission control system 货物蒸发气体排放控制系统标准

Standard fresh water 标准淡水 温度为 15 ℃时,密度为 101.87 kg·s²/m²,运动黏度系数为 $1.139\,02 \times 10^{-6}$ m²/s,含盐度为零的水。

Standard height of superstructure 上层建筑标准高度 上层建筑的标准高度定义在表 S-3。 Standard height of superstructure is defined in following table S-3.

表 S-3　上层建筑的标准高度
Table S-3　The standard height of superstructure

干舷船长(m) Load line length L_{LL}	标准高度(m) Standard height h_S,	
	后升高甲板 Raised quarter deck	其他所有上层建筑 All other superstructures
$L_{LL} \leq 30$	0.90	1.80
$30 < L_{LL} < 75$	$0.9 + 0.00667(L_{LL} - 30)$	1.80
$75 \leq L_{LL} < 125$	$1.2 + 0.012(L_{LL} - 75)$	$1.8 + 0.01(L_{LL} - 75)$
$L_{LL} > 125$	1.80	2.30

Standard of competence　适任标准　系指按照"海员培训、发证和值班标准国际公约"所列的国际公认标准并结合所规定的知识理解和所显示的技能标准或水平，为正规地履行船上的有关职能应达到的熟练程度。Standard of competence means the level of proficiency to be achieved for the proper performance of function on board a ship in accordance with the international agreed criteria as set forth herein and incorporating prescribed standards or levels of knowledge, understanding and demonstrated skill.

Standard operation condition　标准作业工况　对起重设备而言，系指在确定安全工作载荷时所处的作业工况，包括:(1)起重设备工作时，船舶处于横倾5°、纵倾2°;(2)在港内作业;(3)起重设备工作时，风速不超过20 m/s，相应风压不超过250 Pa;(4)起重负荷的运动不受外力的制约;(5)起重作业的性质，即作业的频次和动载荷特性与有关规定的因素载荷相一致。

Standard specifications for shipboard incinerators　船用焚烧炉标准技术条件(船用焚烧炉标准技术规范)

Standard power　标准功率

Standard radar reflector　标准雷达反射器　安装在海平面以上3.5 m且在X波段的有效反射面积为10 m²的基准反射器。Standard radar reflector is a reference reflector mounted 3.5 m above sea level with 10 m² effective reflecting area at X-band.

Standard rate of revolution　标准转速

Standard reference materials (SRM)　标准基准材料

Standard Rock Protection Area　标准岩石保护区　系指以位于北纬47°10′57″，西经087°13′34″的标准岩石灯塔为圆心的半径6英里范围内的区域。Standard Rock Protection Area means the area within a 6 mile radius from Standard Shoal, at 47°10′57″, N, 087°13′34″W.

Standard subsidiary Charpy V notch test specimen　标准辅助V形缺口试样

Standard subsidiary specimen　标准辅助试样

Standard subsidiary test specimen　标准辅助试样

Standard test　标准试验

Standard test bar　标准试件

Standard test specimen　标准试样

Standard time-temperature curve　标准时间-温度曲线　系指由下式定义的时间—温度曲线: $T = 345 \lg(8t + 1) + 20$, 式中, T—为平均炉温(℃); t—为时间(min)。Standard time-temperature curve means the time-temperature curve defined by the formula: $T = 345 \lg(8t + 1) + 20$, where: T is the average furnace temperature (℃), t is the time (min).

Standard type single-pull hatchboard　竖置型舱盖　舱口开启后，舱盖板竖直排列置放的滚翻式舱盖。

Standard variance　标准差

Standard water temperature　标准水温

Standardization　标准化(标定)

Standardize　作为标准(统一,使合标准)

Standardized symbol　标准化符号

Standards　标准　系指认可的标准,诸如英国标准(BS)、欧洲共同体认可的标准化用的欧洲标准(EB)、国际电工委员会标准(IEC)和国际标准化组织标准(ISO), 还应包括其经修正或替换的任何标准。Standards mean those standards such as BS (British Standard), EN (European Standard accepted by the European Committee for Standardization, CEN), IEC (International Electro-technical Commission) and ISO (International Organization for

Standardization）identified in the Code，including any standards which amend or replace them.

Standby condition 待机工况 系指钻进停止，钻杆取出排放立根盒，无钩载，隔水管与井口连接，船舶可以通过调节艏向装置抵抗环境载荷的工况。

Standby fan 备用风机

Standby generating set 备用发电机组

Standby generator 备用发电机

Standby letter of credit（SBLC） 备用信用证 又称担保信用证，系指不以清偿商品交易的价款为目的，而以贷款融资或担保债务偿还为目的所开立的信用证。开证行保证在开证申请人未能履行其应履行的义务时，受益人只要备具备用信用证的规定向开证行开具汇票，并随附开证申请人未履行义务的声明或证明文件，即可得到开证行的偿付。备用信用证只适用《跟单信用证统一惯例》（500号）的部分条款，为UCP600部分条款。备用信用证最早流行于美国，因美国法律不允许银行开立保函，故银行采用备用信用证来代替保函，后来其逐渐发展成为国际性合同提供履约担保的信用工具。

Stand-by machine 备用机械

Standby machinery 备用设备（备用机械）

Standby pump 备用泵 做好启动准备，随时可以投入使用的泵。一般与值班泵轮换使用。

Standby redundancies 备用冗余

Standby set 备用机组

Stand-by ship 守护船 系指设有救助及医疗设备，为钻井平台执行看守、值班及协助抛锚起锚等作业的辅助船舶。

Stand-by vessel 警卫船 对风电场而言，系指负责监控可能撞击风电场的船舶和/或在禁区内周边巡航的船舶。警卫船必须全天候不停地巡逻，确保风电场项目区域的安全。

Standing rigging 静索 承载时不穿过滑车或穿过滑轮而无运动的钢索，如稳索、桅支索等。

Star board side 右舷 按船尾向船首的视向，船的右侧称为右舷。

Star carrier 行星齿轮架

Star reduction gear 行星式减速齿轮箱 行星架静止不动，因此行星轮只有自转不做公转，而中间输入功率的太阳轮和输出功率的外环内齿轮作反方向转动的减速齿轮箱。

Starboard engine 右舷主机 系指布置在机舱右舷侧的主机。

Starboard reduction gear 右机减速齿轮箱 具有2个以上推进器的船舶中，设置在右舷侧的减速齿轮箱。

Starring up pump 启动用泵 供启动时使用的泵。如向锅炉注水的泵、启动时用的滑油泵等。

Starter 启动器（启动装置，启动机） 系指用来使电动机从静止状态加速到正常转速或使电动机停止转动的一种电控制器。为使电动机在2个旋转方向上都能启动而设计的启动器，具有附加的逆转功能，因而应称为控制器。

Starter without contacts 无触点启动器（软启动器） 是采用晶集管替代接触器所组成的启动器。当晶集管采用全导通控制方式时为直接启动器，当晶集管的辅助组件采用微处理器后，能连续检测电动机电流和发热状态而实现某种控制作用时，又称为软启动器，甚至做到软停止功能。

Starting 启动 燃气轮机由静止状态进入运行状态的过程。

Starting（performance）test 启动性能试验 为检查内燃机的启动可靠性、启动条件和启动程序完善性所进行的试验。

Starting air bottle 启动空气瓶

Starting air compressor 启动用空气压缩机 产生柴油机启动压缩空气的压缩机。

Starting air consumption 启动耗气量 内燃机每次正常启动所需的空气量。

Starting air distributor 启动空气分配器 根据启动时的需要，将压缩空气按次序送入个气缸气动阀的装置。

Starting air manifold 启动空气总管

Starting air pressure 启动空气压力

Starting air system 启动空气系统 主、辅柴油机启动用的压缩空气系统。由空气压缩机、空气瓶、管路及附件等组成。

Starting air tank（starting air bottle） 启动空气瓶 柴油机启动用的压缩空气储存容器。

Starting arrangement 启动器（启动装置）

Starting button 启动按钮

Starting characteristics 启动特性

Starting device 启动装置 将燃气轮机从静止状态带动到启动转速的装置。

Starting function 启动功能

Starting interlock 启动连锁 系指当国际防止船舶造成污染公约（MARPOL 73/78）要求使用监控系统时在监控系统充分运行前防止排放阀打开或防止其他等效装置进行运行的设备。Starting interlock is a facility which prevents the initiation of the opening of the discharge valve or the operation of other equivalent arrangements before the monitoring system is fully operational when using the monitoring system required by the MARPOL 73/78.

Starting performance test　启动性能试验

Starting sequence　启动程序

Starting system　启动系统

Starting test　启动试验

Starting time　启动时间　内燃机由启动操作开始到自行连续稳定运转所需要的时间。

Starting torque　启动转矩

Starting value　初值（起始值）

Starting valve　启动阀　供给启动空气的阀门。

State of registration　登记国　系指船舶在其船舶登记簿上登记的国家。State of registration means the State in whose Register of Ships a ship has entered.

State of the ship's registry　船舶登记国　就登记的船舶而言，系指船舶进行登记的国家，就未登记的船舶而言，系指船旗国。State of the ship's registry means the State of registration of the ship in relation to registered ships, and the State whose flag the ship is flying in relation to unregistered ships.

Statement　陈述

Statement of appointing arbitrator　指定仲裁员声明

Statement of completion　完工声明　系指拆船厂签发的确认拆船已经按2009年香港国际安全与环境无害化拆船公约完工的声明。Statement of completion means a confirmatory statement issued by the ship recycling facility that the ship recycling has been completed in accordance with Hong Kong International convention for the safe and environmentally sound cycling of ships, 2009.

Statement of completion of ship recycling (SCSR)　拆船完工声明

Static air cushion　静力气垫　系指：(1) 当船面上静止或缓慢地移动时，由喷口、螺旋桨或风扇直接对准机翼或船的其他基面与水面之间区域而产生的足以支撑船总质量的高压区；(2) 从推进发动机和船体和/或翼下的其他发动机集中空气的产生的高压区。气垫船或侧壁式气垫船（SES）在船周边形成气封以在船底下获取气垫。装有产生静力气垫的船舶推进器的地效翼船，通常会在零速或低速时关闭主翼的襟翼以维持静力气垫。这是俄罗斯动力气垫船（DACC）以及中国和俄罗斯动力气垫地效翼船（DACWIG）的机理。 Static air cushion means: (1) a high-pressure region originating from air jets, propellers or fans that are directed between an airfoil or other base plane of the craft and water surface when the craft stays still or moves slowly over surface and is sufficient to support the craft's total weight; (2) a high-pressure region generated by directing air from the propulsion engine or other engine underneath the craft's body and/or wings. An ACV or SES has seals right around the periphery for containing the air cushion under the base plane of the craft. A WIG with bow thrusters designed to create a static air cushion will normally close the main-wing flap at zero or low speed to assist retaining the static air cushion. This is the basis for the DACC in Russia and the DACWIG in China and Russia.

Static balancing test　静平衡试验

Static converter (changer)　静止变换器　是一种采用静止整流器件，诸如半导体整流器或可控整流器（晶闸管）、晶体管、电子管和磁放大器将交流功率转变为直流功率或将直流功率转变为交流功率的变换器。

Static delivery head　泵排出的静压头

Static equilibrium　静平衡

Static load stresses　静载应力　系指仅由静载荷引起的应力，其载荷包括平台作业重力载荷和处于漂浮或坐底状态时的自身重力以及相对应的浮力和/或底部反力，所对应的工况称为静载工况。 Static load stresses means stresses due to static loadings only, where the static loads include operational gravity loadings and weight of the unit, with the unit afloat or resting on the sea bed in calm water, and counterforce corresponding buoyancy and/or bottom. The corresponding condition is called static loading condition.

Static pressure regulator　静压调节器　自动调节送风管中静压的设备。

Station chain　台链　无线电导航系统岸台配置的形式，常用的形式有星形和链式配置两种。

Stationary auxiliary gas turbine　固定式辅燃气轮机　固定在一定位置工作的辅燃气轮机。

Stationary blade　固定叶片（定子隔板）

Stationary blade of turbine　汽轮机的固定叶片

Stationary vane　定叶　转叶式液压舵机中，固定在液压缸体上不转动的叶片。

Statistical analyzer of noise　噪声统计分析仪　是用来测量噪声级的统计分布，并直接指示累积百分声级 L_N 的一种噪声测量仪器。

Statistical energy analysis　统计能量分析　系指使用能量作为独立的动力学变量来解决复杂系统的固体结构和流体声场之间的耦合动力学问题的一种分析方法。

Statistical model of noise　噪声统计模型

Status lights　状态灯　系指对平台上存在可能危及直升机或其他乘员的情况发出警告的灯。 Status lights are these lights installed to provide warning that a condition exists on the unit which may be hazardous for the helicopter or its occupants.

Status of damaged security 破坏保安状态 系指由于非法行为对船舶、财产、设施以及人员造成的后果。
Status of damaged security means the consequence caused by an illegal act to the ship, properties, facilities and persons.

Statutory and certification survey 法定发证检验 系指按照主管机关的规定进行检验,以确认规定的要求已被满足。检验完成时,船级社将按照授权的范围签发有关的法定噪声和提供/签署有关的检验文件。

Statutory survey 法定检验 又称条约检验。是船舶登记国(又称船旗国政府)根据其政府法令,遵照国际公约及规则对国际航行和国内航行的入级和非入级船舶进行的一种强制性检验。除船旗国政府的法令、法规所规定的检验外,船舶航区所及航道和港口的有关规则所规定的检验,也可归入法定检验范围。经法定检验合格后,将由船旗国政府授权的海事管理部门签发法定证书。证书的有效期一般为5年。检验的依据是由联合国下属的国际海事组织(IMO)和国际劳工组织(ILO)以及国际电信联盟(ITU)颁布的国际公约及其修正案。包括海上安全(marine safety)、海洋污染(marine pollution)、责任与补偿(liability and compensation)、其他(other subject)、根据SOLAS公约强制适用的规则(codes and instruments made mandatory under/SOLAS)、根据MAPOL公约强制适用的规则(codes and instruments made mandatory under/MAPOL)、协议(protocol)、根据国际劳工组织强制适用的规则(codes and instruments made mandatory under/ILO)等方面。对于挂"方便旗"(如巴拿马、利比里亚旗等)的船舶主要应满足上述IMO等的公约和规则;对于挂"非方便旗"(如美、英、德、挪威旗等)的船舶还需满足相应各船旗国的有关法规(如USCG、DOT、SBG、NMD等)的要求。法定检验机关可以是船旗国政府、也可以是船旗国政府授权的船级社或其他海事服务人员。法定检验的种类有:(1)初次检验;(2)年度检验;(3)定期检验;(4)中间检验;(5)换证检验;(6)附加检验;(7)船底外部检验等。颁发的证书有:(1)货船安全结构证书;(2)货船安全设备证书;(3)货船无线电报安全证书;(4)船舶最低安全配员证书;(5)免除证书;(6)国际吨位证书;(7)国际防止油污染证书;(8)国际防止生活污水污染证书;(9)国际船舶载重线证书;(10)国际防止散装运输有毒液体物质污染证书;(11)国际散装运输液化气体适装证书;(12)危险品适装证书;(13)散装固体物质适装证书;(14)特殊用途船舶安全证书;(15)船舶起重设备检验和试验证书;(16)钢索检验和试验证书等。

Stay(shroud) 桅支索 牵张桅杆,增强其承载能力的钢索或牵条。

Stay in a dock 住坞 系指停放在船坞内,以便对船舶进行检查和修理。

Stay light(forward anchor light) 前锚灯 设置在船舶艏部的锚灯。

Stayed derrick post 支索起重柱 柱身全部或部分由一根或多根支索所支持的起重柱。

Stayed flat surfaces 有支撑的平壁板(锅炉) 系指用焊装牵条或焊装牵条和弯边系固的平壁板。Stayed flat surfaces mean the flat plates supported by welded-in stays or by welded-in stays and flange connection.

Stayed mast 支索桅 除在甲板处紧固外,在其上端还有支索牵张的桅。

STCW Convention 国际海员培训、发证和值班规则 于1978年修订的国际海员培训、发证和值班规则设定了海上商船船长、高级船员和值班人员的资格认定标准。STCW于1978年在伦敦举办的国际海事组织会议正式通过,并于1984年生效。该公约于1995年经较大修订。 STCW Convention is the International Convention on standards of Training, Certification and Watch-keeping for seafarers (STCW), 1978, as amended, sets qualification standards for masters, officers and watch personnel on seagoing merchant ships. STCW was adopted in 1978 by conference at the International Maritime Organization (IMO) in London, and entered into force in 1984. The Convention was significantly amended in 1995.

STD/SV system 温盐深声速测量系统 用于测量河水的温度、盐度(或电导率)和流速流向随深度变化的装置。

Steady flow 定常流 系指与波浪无关的水质点单向运动。

Steady high magnetic field 稳态强磁场 强磁场与极端低温、超高压一样,被列为现代科学实验最重要的极端条件之一,为物理、化学、材料和生物等学科研究提供了新途径,对于发现和认识新现象、揭示新规律具有重要作用。强磁场下的核磁共振,又是生命科学、医学、脑科学研究的必要工具。例如,在农业、人类健康密切相关的生物大分子研究中,相当多的样品只能是液态的,只有在强磁场条件下才能研究。强磁场可分为稳态强磁场和脉冲强磁场两大类,其对应的发生装置又分为稳态强磁场装置和脉冲强磁场装置。

Steady high magnetic field facility(SHMFF) 稳态强磁场实验装置 是我国"十一五"重大科技基础设施建设项目之一。稳态强磁场将建设成20~40特斯拉稳态混合磁体、高功率水冷磁体和超导磁体等9台稳态磁体,及相关配套设施。建设地点为安徽省合肥科学岛,由中科院合肥物质科学研究院与中国科学技术大学共同承建。SHMFF 可应用于同步加速器及核磁共振装置等。见图 S-9。

图 S-9 稳态强磁场实验装置
Figure S-9 Steady high magnetic field facility

Steady state tracking 稳定状态跟踪 系指在稳定运动时开始跟踪一个目标:(1)捕获过程完成后;(2)未操纵目标或本船;(3)无目标交换或干扰。 Steady state tracking means a tracking of a target, proceeding at steady motion:(1) after completion of the acquisition process;(2) without a manoeuvre of target or own ship;(3) without target swap or any disturbance.

Steady time 稳定时间 从负荷或转速突然变化时起到转速稳定时止所需的时间。

Steam 蒸汽(蒸汽的)

Steam accumulator 储气器

Steam anchor capstan 蒸汽起锚绞盘 蒸汽驱动的起锚绞盘。

Steam and water purity test (thermal chemical test) 热化学试验 对新投运或大修后的锅炉,在规定的运行方式下所做的鉴定和调整汽、水品质的试验。

Steam annealing 蒸汽退火

Steam bleeder 抽汽装置

Steam blowing 蒸汽吹洗

Steam boiler 蒸汽锅炉

Steam cargo winch 蒸汽起货机 蒸汽驱动的起货绞车。

Steam casing 汽套

Steam chamber of boiler 锅炉汽包

Steam chest 蒸汽室 在速关阀之后,调节阀之前,为均匀配气而设置的汽室。

Steam circuit 蒸汽循环路线

Steam condenser 凝汽器

Steam condensing zone 蒸汽凝结区域 表面式冷凝器中使汽轮机排汽和其他方面排来的蒸汽的绝大部分冷凝为水的区域。

Steam condition 蒸汽参数 表示蒸汽物理状态如压力、温度、焓、熵等的数值指标。

Steam consumption 蒸汽耗量

Steam consumption rate 汽耗率 以蒸汽为工质的主、辅机每单位功率、单位时间的蒸汽消耗量。

Steam cushion 汽垫

Steam distribution device 配汽机构 根据负荷变化的要求,通过改变调节阀开度以调节进汽量的机构。

Steam dome 汽室

Steam drum 上锅筒(汽包,汽鼓) 置于水管锅炉顶部,具有造成锅炉水循环并分离蒸汽中水分作用的集汽、容水部件。

Steam ejector 蒸汽(喷射)抽气泵

Steam ejector gas freeing system 蒸汽喷射油气抽除装置 利用蒸汽喷射器,通过各油舱的专用管子将油气抽出的装置。

Steam electric propulsion 蒸汽-电力推进

Steam engine power plant 蒸汽机动力装置 以蒸汽机作为主机的动力装置。

Steam extinguishing 蒸汽灭火

Steam flow path 蒸汽通道

Steam gage 蒸汽压力表

Steam generator 蒸汽发生器 系指用来产生蒸汽的热交换器以及相连的管系的统称。除非另有说明,通常在规范中锅炉的要求也适用于蒸汽发生器。 Steam generator is a heat exchanger and associated piping used for generating steam. In general in these rules, the requirements for boilers are also applicable for steam generators, unless otherwise indicated.

Steam hauling machine 蒸汽绞车

Steam heater 蒸汽加热器

Steam heating system 蒸汽供热系统 采用蒸汽作为加热工质的舱室取暖系统。

Steam hoist 蒸汽绞车

Steam horn(steam syren) 汽笛

Steam inlet pressure gauge 蒸汽进口压力表

Steam jacket 汽套

Steam jet 蒸汽喷射(喷汽口)

Steam jet air-ejector pump 蒸汽喷射空气抽除泵

Steam lap 进汽余面

Steam oil atomizer 蒸汽雾化喷油器 利用压力蒸汽高速射流雾化燃油的喷油器。

Steam output 蒸汽出量

Steam pipe 蒸汽管

Steam piping 蒸汽管系

Steam power plant 蒸汽动力装置 主机以蒸汽为工质的船舶动力装置。有汽轮机动力装置和蒸汽轮机动力装置等。

Steam pressure 蒸汽压力

Steam pressure measuring equipment 蒸汽压力测量设备

Steam pressure pipe 蒸汽受压管

Steam pump 蒸汽泵 用蒸汽直接驱动泵活塞轴抽送液体的泵。

Steam purity 蒸汽纯度 通常以蒸汽中杂质含量多寡表示的蒸汽纯洁度。

Steam quality 蒸汽品质 蒸汽中携带水分和盐分的程度。

Steam raising 升汽

Steam rate（specific steam consumption） 汽耗率 以增强为工质的主、辅机每单位功率、单位时间的增强消耗量。

Steam re-heater 蒸汽加热式再热器（蒸汽再热器） 在再热循环汽轮机装置中用锅炉新蒸汽加热在汽轮机内做过部分功的蒸汽的热交换器。

Steam seal 汽封

Steam separator 汽水分离器 将水分从蒸汽中分离出来的器具。

Steam sink 蒸汽洗池 用蒸汽加热水温,以供洗菜、洗碗或消毒餐具,也可供煮菜或煮汤的池。

Steam siren 蒸汽笛

Steam slide vavle lose motion 蒸汽泵滑阀的空动

Steam smothering 蒸汽灭火

Steam smothering arrangement 蒸汽窒息灭火装置

Steam smothering system 蒸汽灭火系统（蒸汽窒息灭火系统） 采用低压饱和蒸汽作为灭火剂的灭火系统。

Steam space 蒸汽腔

Steam steering gear 蒸汽舵机 蒸汽驱动的舵机。

Steam storage 蒸汽贮存器

Steam superheater 蒸汽过热器

Steam supply 供汽（送汽）

Steam temperature automatic regulating system 蒸汽温度自动调节装置 利用减温器降低过热和再热蒸汽温度或从烟气侧改变过热器的热负荷从而自动调节锅炉出口蒸汽温度的装置。

Steam test 蒸汽试验

Steam test for leak tightness 蒸汽密封性试验 用额定压力的蒸汽检查各接口处如焊缝、法兰接口等在热态下密封性的试验。

Steam tight 汽密

Steam tightness 汽密性

Steam towing winch 蒸汽拖缆机 蒸汽驱动的拖缆机。

Steam trap 蒸汽阻汽器（阻汽器）

Steam trial 蒸汽动力装置试验

Steam turbine 汽轮机（蒸汽轮机） 水蒸气在主要是由静叶片和动叶片组成的通流部分中连续膨胀,使其部分热能转换成动能,最后由转子输出机械功的一种叶片式动力机械。

Steam turbine gas freeing system 汽轮机油气抽除装置 利用小型汽轮抽风机,将油舱内油气驱除的装置。

Steam turbine power plant 汽轮机动力装置 以汽轮机作为主机的动力装置。

Steam turbine room 汽轮机舱

Steam turbine single-up test 汽轮机单缸试验 为检查主汽轮机组在应急情况下单独使用高压缸或低压缸带动轴系时的性能所做的试验。

Steam turbine-driven alternator 蒸汽轮机驱动交流发电机组

Steam whistle 汽笛

Steam windlass 蒸汽起锚机 利用蒸汽驱动的起锚机。

Steam-heated oven 蒸汽加热炉

Steam-heated steam generator 蒸汽加热的蒸汽发生器

Steaming 蒸发

Steam-rate test（specific steam consumption test） 汽耗率试验 为测定主、辅汽轮机在各种负荷下的汽耗率所做的试验。

Steam-tight 汽密的

Steam-tight joint 汽密接头（汽密接合）

Steel 钢

Steel alloy 合金钢 系指钢中除含硅和锰作为合金元素或脱氧元素外,还含有其他合金元素（如铬、镍、钼、钒、钛、铜、钨、铝、钴、铌、锆和其他元素等）,有的还含有某些非金属元素（如硼、氮等）的钢。合金钢中由于含有不同种类和数量的合金元素,并采取适当的工艺措施,便可分别具有较高的强度、韧性、淬透性、耐磨性、耐蚀性、耐低温性、耐热性、热强性、红硬性等特殊性能。

Steel cargo pipe 钢质液货管

Steel casting/ Cast steel 铸钢/铸钢件 （1）是在凝固过程中不经历共晶转变,用于生产铸件的铁基合金,为铸造合金的一种。铸钢是以铁、碳为主要元素的合金,碳含量0～2%。铸钢分为铸造碳钢、铸造低合金钢和铸造特种钢3类。水玻璃吹二氧化碳工艺由于水玻璃加入量高,溃散性差,旧砂不能再生,已面临被淘汰;碱性酚醛树脂自硬砂无磷、无硫、工艺性能好,推广应用前景好;酯硬化水玻璃自硬砂无毒、无味,旧砂可再生回用,现已在铁道部、冶金部、通用机械广泛应用,生产出几公斤到几百吨的铸钢件;(2)用铸钢制作的零件。与铸铁性能相似,但比铸铁强度好。铸钢件造型工艺的基本原则:质量要求高的面或主要加工面应放在下面;大平面应放在下面;薄壁部分应放在下面;厚大部分应放在上面;应尽量减少砂芯的数量;应尽量采用平直的分型面。铸钢件在铸造过程中易出现气泡、角度定位不准确等缺点,在长期使用中就有可能出现机壳断裂的现象。铸钢件的产量仅次于铸铁,约占铸件总产量的15%。铸钢件主要应用于强度、塑性和韧性要求更高的机械零件。

Steel catenary riser(SCR) 钢悬链式立管

Steel chrome 铬钢 具有较高的抗氧化性和耐蚀性。为了不同的使用环境,常加入其他元素如钼、钒、钨、钛、铌、硼等元素。常用的应用领域和钢种有12CrMo(高、中压蒸汽导管)、20CrMo(叶片)、35CrMo(650 ℃以下长期使用的零件)、1Cr5Mo(650 ℃以下再热器、550 ℃以下侵蚀性强的石油化工设备)、1Cr12Mo(450 ℃以下叶片)。

Steel forged 锻钢

Steel forging 钢锻件(钢锻造)

Steel grating 钢格栅

Steel hawser 钢缆 用作缆绳的钢丝绳。

Steel or other equivalent material 钢或其他等效材料 系指本身或由于所设置隔热物,经过标准耐火试验规定的适用爆火时间后,在结构性和完整性上与钢具有等效性能的任何不燃材料(例如设有适当隔热材料的铝合金)。 Steel or other equivalent material means any non-combustible material which, by itself or due to insulation provided, has structural and integrity properties equivalent to steel at the end of the applicable exposure to the standard fire test (e.g. aluminium alloy with appropriate insulation).

Steel sheet 薄钢板 薄钢板系指厚度不大于3 mm的钢板。常用的薄钢板厚度为0.5~2 mm,分为板材和卷板供货。薄钢板钢号为B0-B3 的冷扎或热扎钢板。对薄钢板的要求:表面平整、光滑,厚度匀称,允许有紧密的氧化铁薄膜,不得有裂痕、结疤等缺陷。工艺分为热轧薄钢板和冷轧薄钢板。主要用于汽车、电器设备、车辆、农机具、容器、钢制家具等。

Steel tube 钢管

Steels tough at subzero temperatures 低温韧性钢

Steep S-riser 陡S形立管

Steep wave riser 陡波立管

Steering auxiliary systems 辅助操舵系统 包括推进系统正常操作所需的所有系统。其包括或可包括:(1)淡水冷却系统;(2)海水冷却系统;(3)动力供应(气、电力、液力);(4)控制、监测和安全系统。 Steering auxiliary systems include all the systems that are necessary for the normal operation of a steering system. It includes or may include: (1) the fresh water cooling systems; (2) the sea water cooling systems; (3) the power supply (air, electrical, hydraulic); (4) the control, monitoring and safety systems.

Steering chain 操舵链 操舵用的操舵链条。

Steering control system 操舵系统 从驾驶室操纵舵机的系统。一般有机械、液压和电力等类型。

Steering gear 舵机/操舵装置 是保证船舶安全航行、作业并使船舶具有良好操纵性的重要设备。其作用是使船舶按照操纵的要求开展船舶的航行方向。舵机按传动形式可分为人力舵机、电动舵机和液压舵机。

Steering gear alarm 操舵装置报警 系指指示操舵装置系统故障或其他异常状况的报警,例如过载报警、断相报警、无电压报警和液压油柜低液位报警。 Steering gear alarm means an alarm which indicates a malfunction or other abnormal condition of the steering gear, e.g. overload alarm, phase failure alarm, no-voltage alarm and hydraulic oil tank low-level alarm.

Steering gear control system 操舵装置控制系统 系指用于将舵令由驾驶室传至舵机装置动力设备之间的一系列设备。操舵装置控制系统由发送器、接收器、液压控制泵及电动机、电动机控制器、管路和电缆组成。 Steering gear control system is equipment by which orders are transmitted from the navigating bridge to the steering gear power units. Steering gear control systems comprise transmitters, receivers, hydraulic control pumps and their associated motors, motor controllers, piping and cables.

Steering gear for active rudder 主动舵舵机 用以转动主动舵的舵机,此舵机转动舵叶偏离零位最大可达90°。

Steering gear power actuating system 操舵装置动力传动系统 系指为转动舵杆而提供动力的液压设备,其包括动力装置或机组及与其相关的液压管系和附件以及操舵机构。但如设有2台或以上的液压传动系统,则舵柄、舵扇等机械部件可共用。 Steering gear power actuating system is the hydraulic equipment provided for supplying power to turn the rudder stock, comprising a power unit or units, together with the associated hydraulic pipes and fittings, and a rudder actuator. The power actuating system may share common mechanical components, i.e., tiller, quadrant, etc.

Steering gear power unit 操舵装置动力设备
(1)如为电动操舵装置,系指电动机及有关的电气设备;(2)如为电动液压操舵装置,系指电动机及有关的电气设备和与之相连接的泵;(3)如为其他液压操舵装置,系指驱动机及与之相连接的泵。 Steering gear power unit is:(1) in the case of electric steering gear, an electric motor and its associated electrical equipment;(2) in the case of electrohydraulic steering gear, an electric motor and its associated electrical equipment and connected pump;(3) in the case of other hydraulic steering gear, a driving engine and connected pump.

Steering nozzle with rudder 带舵可转导管 后端连有舵叶的可转导管。

Steering rod 操舵拉杆 舵链传动操舵装置中,直接传递动力用的拉杆。

Steering shafting 操舵轴 操舵轴传动操舵装置中用以传递扭矩的传动轴。

Steering stability 操舵稳定性 即拖船保持航向的能力。 Steering stability, i.e. stable course maintaining capacity of the tug.

Steering stand 操舵台 设有操舵轮或操舵手柄以及转换开关、传动机构、航海仪表等供驾驶人员使用的操纵台。

Steering system 操舵系统 系指控制船舶艏向的系统。其包括:(1)动力执行系统;(2)将扭矩传输到操舵装置的设备;(3)操舵装置(例如:舵、可转动的推力器、喷水操舵装置等)。 Steering system is a system that controls the heading of the ship. It includes:(1) the power actuating system;(2) the equipment intended to transmit the torque to the steering device;(3) the steering device (e.g. rudder, rotatable thruster, water-jet steering deflector, etc.).

Steering wheel 操舵轮 专供驾驶人员转舵用的手轮。

Steering wire 操舵索 系指操舵用的绳索。

Step-back relay 跳返继电器 系指当电动机电枢电流或线电流升高时动作,以限制电动机的峰值电流的继电器。此外,当这种高电流的起因被排除时,跳返继电器可以动作,以取消这种限制。

Stepped gear ratio (stage gear ratio) 级传动比 一对相啮合的齿轮中,其主动齿轮转速与被动齿轮转速的比值。

Sterilization cabine 餐具消毒柜 利用蒸汽、高温消毒餐具的柜子。

Sterilizer 消毒器(消毒器)
Sterilizing 消毒
Sterilizing device 杀菌设备(自封存装置)
Sterilizing facilities 消毒设备
Stern 艉 系指船舶的艉部。
Stern anchor 艉锚 设在船艉部的专用锚。
Stern anchoring winch 艉锚绞车 在用锚定位的工程船上,设置在艉端,用以收放艉锚索使船身保持定向位移的绞车。
Stern barrel (stern roll) 艉部滚筒 为使起放网时减少绳索与船体的磨损,而设置于渔船上甲板艉端,横跨船宽的长滚筒。
Stern boss 艉柱轴毂
Stern bush 艉轴衬套
Stern bush clearance 艉轴衬套间隙
Stern door 艉门 供车辆出入或进行某种专业操作之用的设在船艉端的门。
Stern fueling rig 艉加油索具 向位于补给船艉后的接收船供油的软管系统。
Stern gear 呆木装置(艉轴管装置,艉轴传动装置)
Stern gland 艉轴填料箱压盖
Stern line 艉缆 系指船舶带缆时,从船艉部向后引出的缆绳。
Stern ramp 艉滑道 在拖网渔船或捕鲸母船的尾部,为使网具或鲸鱼能全部从海面拖曳到甲板上而做成的弧形曲面或斜平面的滑道。
Stern ramp trawler 艉滑道拖网渔船 船艉部设有起放拖网用的滑道的拖网渔船。
Stern shaft 艉轴
Stern shaft seal 轴封 一种摩擦密封或填料函,用以防止压缩机或其他流体输送设备轴与轴承之间的液体泄漏。轴封是防止泵轴与壳体处泄漏而设置的密封装置。常用的轴封形式有填料密封、机械密封和动力密封。往复泵的轴封通常是填料密封。当输送不允许泄漏介质时,可采用隔膜式往复泵。旋转式泵(含叶片式泵、转子泵等)的轴封主要有填料密封、机械密封和动力密封。
Stern sheave 艉滑轮 安装在布缆船的甲板艉端,在布缆时供导引电缆的滑轮。
Stern thruster 艉推力器
Stern tube 艉轴管 轴穿过该管达到螺旋桨,用作轴系的后轴承,可用水或油润滑。 Stern tube is a tube through which the shaft passes to the propeller and acts as after bearings for the shafting and may be water or oil lubri-

cated.

Stern tube arrangement 艉轴管装置

Stern tube bearing 艉轴管轴承 艉轴管中，支承艉轴或推进器轴的轴承。

Stern tube bearing strip 艉轴管轴承衬条

Stern tube bulkhead 艉轴管舱壁

Stern tube bushing 艉轴衬套

Stern tube enclosure 艉管密封装置

Stern tube gland 艉轴管填料函压盖 艉轴管填料函中，用以压紧填料的零件。

Stern tube lubricating oil 艉轴管滑油

Stern tube lubricating oil pump 艉轴管滑油泵 抽送润滑艉轴管轴承用油的泵。

Stern tube nut 艉轴管螺母 安装艉轴管用的圆螺母。

Stern tube sealing oil pump 艉轴管油封用油泵

Stern tube stuffing 艉轴管料填函 设在艉轴管前端的密封装置。

Stern tube-retaining strip 艉袖管轴承止动嵌条

Sternlight 艉灯 系指安装在尽可能接近船艉的白灯，在135°的水平弧内显示不间断的灯光。其装置要使灯光从船的正后方到左、右舷67.5°内显示。 Sternlight means a white light placed as nearly as practicable at the stern showing an unbroken light over an arc of the horizon of 135° and so fixed as to show the light 67.5° from right aft on each side of the vessel.

Stiff shafting 刚性轴系 临界转速大于标定转速的轴系。

Stimulant drugs 兴奋药物 主要的兴奋剂有安非他明盐和硫酸盐、维洛沙素、甲基苯异丙基苄胺、氯芬他明、芬坎法明、甲非他明、亚甲基二氯苯丙胺（MDA）、帕吗啉、苯二甲吗啉、苯丁胺、哌苯甲醇和普罗啉坦。合法制造的安非他明制品，含有硫酸盐或磷酸盐吸收的麻醉药。在不同国家的市场上销售的有药片、胶囊、糖浆或配剂。除哌苯甲醇是白色晶体外，其他兴奋剂的纯净形状都是白色粉末。有成百上千个商品名。通常都是药片、药丸或胶囊，但偶尔也有注射用的针剂。这些全都是兴奋药物，但芬坎法明已经解除管制，凭处方可以购买。按照制造厂商的图表是可以识别每一种胶囊和药丸。关于药丸或药片的直径、颜色、形状和标记的信息，可以从船舶上用无线电报发送到下一个停泊港口，以便进行初步的辨认。违禁品的颜色各不相同，从白色粉末或灰白色粉末到黄色粉末或褐色粉末，取决于类型及杂质和掺杂物的量。它们通常是潮湿的，由于存在溶剂的残留物，因此具有难闻的特殊气味。也有小的胶质胶囊和药片。在正常范围之外发现的所有显而易见的医用制剂都应引起怀疑。通常情况下，所有兴奋药物都是没有气味的。纯净的安非他明可能有氨水味或"鱼腥"味。 Stimulant drugs means that the main stimulants are amphetamine salts and sulphate, phenmetrazine, benzphetamine, chlorphenamine, fencamfamine, mephentamine, methylendioxyamphetamine (MDA), pemoline, phendimetrazine, phentermine, pipradol and prolintane. Amphetamine products, legally manufactured, contain the drug in the form of the sulphate or phosphate salt. They are marketed in different countries as tablets, capsules, syrups or elixirs. In pure form all are white powders except pipradol which is found as white crystals. There are many hundreds of brand names. They are usually found in pill or tablet form or as capsules, but occasionally in ampoules for injection. All are stimulant drugs, but fencamfamine has decontrolled to prescription availability. Identification of individual pill and capsules is possible by consulting manufacturers' charts. Information such as the diameter of pill or tablet, colour, shape and markings can be radioed from the ship to the next port of call to obtain a tentative identification. Illicit products vary in colour from a white or off-white powder to yellow or brown depending on the type and amount of impurities and adulterants. They are often damp, with a characteristic unpleasant odour due to the presence of solvent residues. They can be found as small gelatin capsules and as tablets. All discoveries of apparently medical preparations outside their normal context should be regarded as suspicious. All are normally odourless. Pure forms of amphetamine may smell faintly ammoniac or "fishy".

Stimulation 增产作业 系指采用压裂、补充射孔、洗井和控制出砂量等措施扩大油井的出油通道，提高油井的渗透率，从而提高油井产量的方法。

Stimulation vessel 增产作业船 系指用于井下对油层进行酸化、压裂和防砂等处理作业任务，以维持或提高油气产量的船舶。其结构形式为船形。

Stinger 托管架 系指长得像吊杆的刚性或铰接式的结构物。其支撑来自铺管船一定距离的管路。 Stinger is a long, boom-like structure, rigid or articulated, that supports the pipeline for a certain distance as it comes a lay-vessel.

Stirling engine 外燃机（外燃式往复机，斯特林发动机）

Stock 生物群体

Stock anchor 有杆锚 系指在锚杆上部装有垂直于锚爪平面的锚横杆，其锚爪与锚杆通常为整体的锚。

Stock holders 股东 系指股票的持有人。

Stockless anchor 无杆锚 系指锚爪与锚杆可相对转动的无横杆的锚。
Stokehold escape 锅炉舱安全出口
Stokehold floor 锅炉舱花钢板
Stokehold ladder 锅炉舱梯
Stokehold skylight 锅炉舱天窗
Stone catcher 沉石箱 装设于吸泥管线上，用以沉积泥浆中的铁块、石头等杂物，以免进入泥浆的箱子。
Stone chippings 石屑 无特殊危害。该货物为不燃物或失火风险低。 Stone clippings have no special hazards. This cargo is non-combustible or has a low fire-risk.
Stone dumper 抛石驳 系指运载石块到指定地区后利用横倾或其他方法自动抛石块于水底的船舶。它通过压载水舱水量的调节，改变船舶重心与浮心的横向位置，使船舶横倾，自动卸下装于甲板上的石块，然后压载水舱的阀门自行排水使船扶正。压载水舱的进排水阀门可由拖船遥控。操作时，船上不宜留人，以免落水。该船结构简单、卸石迅速，用于海港筑堤、桥墩等工程。还有一种翻斗式抛石船，在甲板上装有 6 对用液压缸顶升的大翻斗，沿船体中剖面处开一长条槽，石块自槽内卸下，抛石准确，每一翻斗可吊到码头上预装石块，可缩短装船时间，提高效率。
Stop gear 停止装置（制动装置）
Stop posts for towline 拖缆限位器 设置于拖船两舷，用以限制拖缆摆动幅度，以防损坏在其前部的船体结构和舱面设备的部件。
Stop rod（stop bar） 制动条 用以防止轴衬材料沿周向移动的铜条。
Stop stopping valve 截止阀
Stop vavle 停气阀
Stopper 限位器 系指限制舵扇或舵柄在任何一舷运动的装置。限位器以及与船壳的底座结构应坚固，以在舵杆设计屈服扭矩下不超过所用材料的屈服点。 Stopper means a device limites a motion of quadrants or tillers on either side. The stoppers and their foundations connected to the ship's hull are to be of strong construction so that the yield point of the applied materials is not exceeded at the design yield moment of the rudder stock.
Stopper 制动器（制链器，塞子）
Stopper bolt 制动螺栓
Storage container 储藏容器 对氨冷冻装置而言，储藏容器系指储藏补充用气体的容器。 For refrigerating machinery using ammonia as refrigerant, storage container means used for storing gas for replenishment.
Storage unit 储油平台 系指用于储油作业的平台。 Storage unit means a unit used for oil storage.
Store-rooms, workshops, pantries, etc. 储藏室，工作间，配膳室等 系指：(1) 不属于主厨房的配膳室；(2) 主洗衣间；(3) 大干燥间（甲板面积大于 4 m²）；(4) 杂物间；(5). 邮件和行李室；(6) 垃圾间；(7) 工作间（不是机器处所、厨房等的一部分）；(8) 面积大于 4 m² 的储藏间和储物间，存放易燃液体的处所除外。 Store-rooms, workshops, pantries, etc. are: (1) main pantries not annexed to galleys; (2) main laundry; (3) large drying rooms (having a deck area of more than 4 m²); (4) miscellaneous stores; (5) mail and baggage rooms; (6) garbage rooms; (7) workshops (not part of machinery spaces, galleys, etc.); (8) lockers and store-rooms having areas greater than 4 m², other than those spaces that have provisions for the storage of flammable liquids.
Stores and supplies 船舶物料 系指：(1) 用于船舶的保养、维护、安全、操作或航行的材料；(2) 用于船舶的乘客或船员安全或舒适的材料，包括船舶乘客或船员的伙食。
Storm covers 风暴盖 系指在便于使用的情况下装设在窗的外侧的盖子，可以是铰链式或拆卸式。 Storm covers are fitted to the outside of windows, where accessible, and may be hinged or portable.
Storm hook 风暴钩 为防止船舶航行时桌、椅等活动家具的移动，专装于桌、椅底板下，使之与甲板固联的拉钩。
Storm oil 镇浪油 救生艇或船上配备的，供在风暴情况下散布于海面，利用其表面张力缓和附近海浪，以减轻对艇的颠簸和碎浪拍击的植物油或动物油。
Storm oil bag 布油袋 为使镇浪油尽量匀薄地散布于海面而使用的一种可扣附于浮锚上的盛镇浪油的小布口袋。
Storm rails（grab rail） 风暴扶手 装设在甲板室围壁、走廊或通道中，供人员在恶劣海况时扶靠用的扶手。
Storm scupper 风浪排水孔
Storm valve 防浪阀
Storn line 艉缆 船舶带缆时从船尾部向后引出的系缆。
Stowage 堆装 系指在甲板上和甲板下装运货物、货物单元和货物运输单元的方式。
Stowage factor 积载因素 系指装载每吨货物所占用的平均舱容，以 m³/t 计。积载因素一般包括货物间或货与货舱之间所不能利用的亏损舱容。货物的轻重和包装方式是影响积载因素的主要因素。通常将各类货物的积载因素列成表格（积载因素表）供查用。如船舶舱容系数(即全船货舱总容积与总载重量之比)大于货物的积载因素,表示货物按预定的载货量装船,且船上舱容有剩余。反之,未装到预定的载货量,而船上舱容已被占尽,即载货量未充分利用。如船中所载货物有两种不同的积载因素,设大的为 b,小的为 a,全船载货容积为 V^3,净载货量为 c 吨,取 x 为应装的大积载因素货物,y 为应装的小积载因素货物,则 $y = (cb - V)/(b - a)$ 吨,$x = (V - ca)/(b - a)$ 吨。现代杂货船的发展趋势是逐渐提高舱容系数,以装载积载因素较大的轻泡货物。

Stowage ratio 装载比例 系指深度冷冻货物相对于香蕉或冷藏货物的比例。除特别规定外,认为此装载比例为 50% 深度冷冻货物和 50% 冷藏货物。 Stowage ratio means the proportion of deep frozen cargo in relation to banana or chilled cargoes. Unless specifically stated, the stowage ratio will be deemed to be 50% deep-frozen and 50% chilled cargo.

Stowage survey 积载鉴定

Stowaway 偷渡者 对防止偷渡者进入和寻求成功解决偷渡案件责任分配指南而言,系指:(1)未经船长或船东或任何其他负责人员同意。藏匿于船上或之后装载于船上的货物中,在船舶离港前在船上被发现或在到达港卸货时在货物中被发现并由船长向有关当局作为偷渡者报告的人员;(2)在任何港口或该港附近地点,未经船舶所有人或船长,或掌管船舶的任何其他人员的同意而潜入船内,并在该船驶离上述港口或地点后仍留在船上的人。 For guidelines on the prevention of stowaway incidents and the allocation of responsibilities to seek the successful resolution of stowaway cases, stowaway means: (1) a person who is secreted on a ship, or in cargo which is subsequently loaded on the ship, without the consent of the shipowner or the master or any other responsible person, and who is detected on board the ship after it has departed from a port, or in the cargo while unloading it in the port of arrival, and is reported as a stowaway by the master to the appropriate authorities; (2) a person who, at any port or place in the vicinity thereof, secretes himself in a ship without the consent of the ship-owner or the master or any other person in charge of the ship and who is on board after the ship has left that port or place.

Straight B/L 记名提单 系指提货单上的收货人栏内填明特定收货人名称,只能由该特定收货人提货,由于这种提单不能在流通,故其在国际贸易中很少采用。

Straight blade (parallel sided blade) 直叶片 叶型和安装角沿叶高不变的叶片。

Straight compound gas turbine 顺排式燃气轮机 低压压气机与低压涡轮相连接,高压压气机与高压涡轮轴相连接的燃气轮机。

Straight run 直馏汽油(直馏油)

Straight run residue 直馏渣油

Straight shackle 直形卸扣 普通的 U 形卸扣。

Straight tube acoustic filter 直管式消音器

Straight-boom sheath (screw davit, Columbus type davit) 倒杆型吊艇架 摇动手柄,使伸缩机构的螺杆和套筒伸长,从而倒出吊艇杆以吊艇出舷的摇倒式艇架。

Straightening vane (straightener blade, guide vane) 整流叶片 其出口气流一般是轴向的、位于轴流式压气机末级动叶后的第二列出口导叶。

Straighter 矫直机 是对金属型材、棒材、管材、线材等进行矫直的设备。矫直机通过矫直辊对棒材等进行挤压使其改变直线度。一般有 2 排矫直辊,数量不等。也有两辊矫直机,依靠两辊(中间内凹,双曲线辊)的角度变化对不同直径的材料进行矫直。主要类型有压力矫直机、平衡滚矫直机、鞋滚矫直机、旋转反弯矫直机等。

Straight-through telephone system 直通电话系统 它是 2 个重要故障部位的专用电话,保证快速通话的联络通信。这种通信系统一般采用共电式电话或声力电话。

Straight-through-flow combustion chamber 顺流式燃烧室 火焰筒外侧的空气流向与火焰筒中燃气流向一致的燃烧室。

Strainer 过滤器(滤网,拉紧装置)

Strainer type separator of oily bilge water 滤网式舱底水油分离器 利用滤网对含油的舱底水进行过滤,并将水中的油吸附在滤网上的油水分离器。为了提高油水分离的效果,装置顶部的加热管用来提高油污水的温度,降低油的黏度。

Strainer-plate type separator of oily bilge water 网板组合式舱底水油分离器 由多块网板组合而成的舱底水油分离器。

Straining screw 松紧螺旋扣

Strait of Gibraltar 直布罗陀海峡 是位于西班牙最南部和非洲西北部之间(西经 5°36′,北纬 35°57′)、沟通地中海与大西洋的海峡。直布罗陀海峡长 58 km(36 哩),最窄处在西班牙的马罗基(Marroqui)角和摩洛哥的西雷斯(Cires)角之间,宽仅 13 km(8 英里)。海峡平均深度 375 m,面积 5.5 km²,属于亚热带地中海气候。

Strapdown-type stabilized gyrocompass 捷联式平台罗经 其最大特点是无框架结构,因而价格十分低廉。它采用捷联式陀螺仪,即将陀螺仪直接固联在载体上,直接测量载体相对于惯性空间的转动角速率,并将测量信息送给微处理器,实时计算出载体相对于惯性空间的姿态运动。捷联系统的实现需要动态范围较大的惯性敏感器和计算速度很高的计算机。计算机技术的飞速发展,也促进捷联惯性系统的迅猛发展和广泛应用。在微机迅速发展的同时,适合捷联应用的新型陀螺也应运而生,如捷联式动力调谐陀螺仪和全固态光学陀螺仪。"常固态"系指光学陀螺仪内部既无流体又无旋转部件,仪表各部分质量之间无相对运动,全部固化在一起。光学陀螺仪按光的传播路线类型不同分为两类。光在环形激光谐振腔中传播的一类叫环形激光陀螺仪;光在圆形光纤围传播的另一类叫光纤陀螺仪。光学陀螺由于没有机械活动部件,对机电型陀螺的许多误差源不敏感。它比机电型陀螺更能承受环境的振动与冲击,它的测量范围大、可靠性高、结构简单、坚固、成本低,且能直接数字输出,因而是联捷惯导的理想器件。

Strategic homeland intervention, enforcement and logistics division(SHIELD) 神盾局 全称为"国土战略防御攻击与后勤保障局"。

Streamline rudder 流线型舵 舵叶剖面呈流线型的复板舵。

Strength deck 强力甲板 通常系指最高一层的连续甲板。具有等效结构的其他甲板也可视为强力甲板。若上甲板是阶梯形的,如舰船带有后甲板的情况,则强力甲板是阶梯形的。 Strength deck is normally the uppermost continuous deck. Other decks may be considered as the strength deck provided that such decks are structurally effective. Where the upper deck is stepped, as in the case of vessels with a quarter deck, the strength deck is stepped.

Strength of boom 吊杆强度 吊杆在仅承受轴向压力时和在承受轴向压力的同时又承受附加弯矩时的稳定性能力。

Striker(fuse) 撞击器(熔断器) 系指熔断体的机械部件。当熔断器动作时释放所需的能量,以促使其他部件或指示器动作,或者提供互锁。

Strikes risk 罢工险 是承保因罢工者、被迫停工工人、参加工潮、暴动和民众斗争的人员,采取行动造成保险货物的损失。对于任何人的恶意行为造成的损失也负责。对上述的各种行动和行为所引起的共同海损的牺牲、分摊和救助费用也由保险公司赔偿。罢工险负责的损失都必须是直接损失,对于间接损失是不负责任的。例如,因为罢工劳动力不足,或者无法使用劳动力对堆存在码头的货物,遇到大雨无法采取覆盖防雨布的措施而遭淋湿受损;因为罢工,没有劳动力对冷冻机添加燃料或使动力中断冷冻机停机,而使冷冻货物遭受到化冻变质的损失等。此外,由于罢工引起的费用损失,如港口工人罢工无法在原定港口卸货,改到另外一个港口卸货引起的增加运输费用,均属于间接损失,不予负责。伦敦协会货物险条款中专门有一条罢工险除外条款,明确对罢工、被迫停工、工潮、暴动或民变等造成保险货物的损失不予负责。但如加保了罢工险则对上述罢工险除外的责任予以负责,并应将罢工险除外条款打上删除印章。

Striking price under International Commodity Agreement 国际商品协定下的协定价格 通常采用最低价格和最高价格等办法来稳定食品价格。当有关商品价格降到最低价格以下时,就减少出口;或用缓冲基金收购商品;当市场价格超过最高价格时,则扩大出口或抛售缓冲存货。

String board 封板 舱室梯上,置于梯步踏面下及侧面的垂直板。

String literal 字符串

Stringer 梯架 船梯上用以支承梯步的主构梁。

Stringer 梯台 系指:(1)扶梯的架构,(2)设置于船侧外板、横舱壁和纵舱壁等的扶强水平板结构。对于形成双壳船侧区域具有宽度小于 5 m 的压载舱,如水平板结构提供船侧外板或纵舱壁上扶强材或大于特设肋板宽度为大于 600 mm 的连续通道,其水平板结构可视为梯台和纵方向永久出入通道。为提供梯台上的安全通行和横腹板的安全出入,用于永久出入通道的设于纵桁板的开口应具有栏杆或箱形盖子。 Stringer means: (1) the frame of a ladder; (2) the stiffened horizontal plating structure fitted on the side shell, such as transverse bulkheads and/or longitudinal bulkheads in the space. For ballast tanks of less than 5 m width forming double side spaces, the horizontal plating structure is credited as a stringer and a longitudinal permanent means of access, if it provides a continuous passage of 600 mm or more in width past frames or stiffeners on the side shell or longitudinal bulkhead. Openings in stringer plating utilized as permanent means of access shall be arranged with guard rails or grid covers to provide safe passage on the stringer or safe access to each transverse web.

Stripe coating 预涂 系指对关键区域边缘、焊缝、不易喷涂区域等位置的预先涂刷,以保证良好的涂料附着力和恰当的涂层厚度。 Stripe coating is planting of edges/welds/hard to reach areas, etc., to ensure good paint adhesion and proper paint thickness in critical areas.

Stripping 扫舱

Stripping device 扫舱装置 油船卸油过程的最后阶段,需将泵吸系统无法抽吸的剩余货油留得越少越好,而采用较小排量的泵并通过较小直径的管路将这些剩余货油尽可能多地排出，这一套泵吸装置称为扫舱装置。

Stripping efficiency 扫舱效率

Stripping pipe lines 清舱管路 设在货油舱底部剩油聚集最深处，专门用以吸净货油舱剩油的较小管径的专用管路。

Stripping pump 扫舱泵

Stripping quantity 扫舱量

Stripping system 扫舱系统（残油扫舱系统）

Stroke 行程（冲程，打击） 活塞运行在上下两止点间的距离。

Stroke adjusting shaft 行程调整轴

Stroke adjustment 行程调整

Stroke bore 行程缸径比

Stroke-bore ratio 行程缸径比 行程与缸径之比。

Structural accessibility 结构可达性 系指船舶的设计和建造，应提供足够的进入所有空间处所和内部结构的通道设施，以能对结构进行全面检查、近观检查和厚度测量。Structural accessibility means that the ship is to be designed and constructed to provide adequate means of access to all spaces and internal structures to enable overall and close-up inspections and thickness measurements。

Structural integrity 结构完整性 系指防止由于热量造成的强度降低而使船舶结构部分或全部破坏。为此，船舶结构中使用的材料应确保结构完整性不会由于失火而被削弱。Structural integrity is to prevent partial or whole collapse of the ship structures due to strength deterioration by heat. For this purpose, materials used in the ship's structure shall ensure that the structural integrity is not degraded due to fire.

Structural integrity goal 结构完整性目标 舰船中结构所用材料应确保不因火灾而降低结构完整性：火灾中结构完整性的布置应符合SOLAS公约第Ⅱ-2章C部分第11条的要求。Materials used in the construction of the ship's structure are to ensure that structural integrity is not degraded due to fire; the arrangements for structural integrity following a fire are to be in accordance with SOLAS Chapter Ⅱ-2, Part C, Regulation 11.

Structural integrity objective 结构完整性目标 在火灾发生后应保持足够的结构完整性，以避免因受热而在力产生变化的情况下，舰船结构整体或部分倒塌。Sufficient structural integrity is to be maintained following a fire so as to prevent the whole or partial collapse of the ship's structures due to strength deterioration by heat.

Structural noise control 结构噪声控制

Structural noise form equipments 设备的结构噪声 系指船舶固体构件（如机器基座、船体结构等）在机械设备激励下产生的振动。通常以速度级或者加速度级表示结构噪声的大小。

Structural strength 结构强度 系指结构或结构构件承受作用和提供足够刚度的能力。Structural strength is the capacity of a structure or structural component to withstand actions and provide adequate rigidity.

Structural sweat 船体结构结露 货舱内的空气露点温度高于甲板和船壳板的温度时，船体结构表面产生的凝水现象。

Structural testing 结构试验 系指以确认液舱密闭性和确认的结构是否合适为目的而进行的水压试验。如因实际困难导致无法进行水压试验［如实际上无法适用液舱顶部要求的水压头（water head）压力时］，可以水压-气压试验（hydro-pneumatic test）代替。水压-气压试验时，试验状态应尽可能接近液舱的实际载重状态。Structural testing is a hydrostatic test carried out to demonstrate the tightness of the tanks and the structural adequacy of the design. Where practical limitations prevail and hydrostatic testing is not feasible (for example when it is difficult, in practice, to apply the required water head pressure at the top of the tank) hydro-pneumatic testing may be carried out instead. When a hydro-pneumatic testing is performed, the conditions should simulate, as far as practicable, the actual loading of the tank.

Structure repair manual (SRM) 结构修理手册

Strut bearin 艉轴架轴承 艉轴架中，用以支承推进器轴的轴承。

Stub 存根

Stud 螺柱（链环横档，双头螺栓）

Stud bolt 双头螺栓

Stud chain 有档锚链 用有档锚链环连接而成的锚链。

Stud fixed manhole cover 螺栓固定人孔盖 盖板依靠螺栓和人孔围板或座板紧固和保持密性的人孔盖。

Stud tube water wall 螺栓管水冷炉壁

Stud welding 螺柱焊

Stud welding gun 螺柱焊枪

Student visa 学生签证 系指拥有学生资格的留学凭证。

Studless chain 无档锚链 用无档锚链环连接而成的锚链。

Study on coping with image crisis 形象危机应对报告

Stuffing box[1] 舵杆填料函 设置在舵杆穿出船体处，通常与舵承和在一起的水密封装置。

Stuffing box[2]　填料函
Stuffing box bulkhead　艉轴孔舱壁（有填料函舱壁）
Stuffing box casing　填料函外壳
Stuffing box gland　填料函压盖
Stuffing box recess　填料函凳
Stuffing tube　填料函
Sub-bottom prospecting　浅地层剖面测量　用剖面仪对海底进行连续测量作业，以获取海底浅地层结构剖面。
Sub-centimetre wave　亚厘米波
Sub-control station　辅助控制站　是一个除了主推进装置的局部控制站外能够控制主推进装置的控制站，它被安装在船桥上有主控台的船舶的机房内。Sub-control station is such a control station at which the main propulsion machinery is capable of being controlled, except for local control station for the main propulsion machinery, which is provided in the machinery room of the ship provided with a main control station on bridge.
Subcooled liquid　过冷液体
Sub-cooling　局部冷却（再冷却）
Subcritical HAZ　亚临界热影响区
Subcritical annealing temperature　亚临界退火温度
Subcritical rotor　亚临界转子
Subdivision and stability information　分舱和稳性资料　系指适用73/78防污染公约附则Ⅰ第25条的每艘油船，均应按认可的格式备有为确保符合本条各项规定所必需的有关货油装载和分配的资料以及关于船舶符合该条所确定的破舱稳性标准能力的资料。Subdivision and stability information means that every oil tanker to which Regulation 25 of Annex I of MARPOL 73/78 applies shall be provided in an approved form with information relative to loading and distribution of cargo necessary to ensure compliance with the provisions of this regulation and data on the ability of the ship to comply with damage stability criteria as determined by this regulation.
Subject matter　标的　是合同当事人双方权利和义务所共同指向的对象。它是合同成立的必要条件，是一切合同的必备条款。系指经济合同当事人双方权利和义务共同指向的对象，如货物、劳务、工程项目等。它是合同成立的必要条件，是一切合同的必备条款。标的的种类总体上包括财产和行为，其中财产又包括物和财产权利，具体表现为动产、不动产、债权、物权、无形体财产权等；行为又包括作为、不作为等。标的和标的物并不是永远共存的。一个合同必须有标的，而不一定有标的物。在保险的种类中，公众责任保险，它是一种无形财产保险，没有实际标的物。举例说明，在提供劳务的合同中，标的是当事人之间的劳务服务的给付行为。而在劳务合同中，就没有标的物。
Subject matter jurisdiction　主题管辖权
Submarine escape　潜艇艇员水下脱险　潜艇失事抢修无效时，艇员利用各种方法脱险出水的活动过程。
Submarine mass escape　潜艇艇员集体脱险　艇员依靠救生钟或深潜救生艇等救生设备集体离艇出水的潜艇艇员水下脱险。
Submarine propeller noise　潜艇螺旋桨噪声　系指螺旋桨在水下推进时产生的噪声。
Submarine rescue　潜艇救生　系指使用救生船援救失事潜艇的活动。
Submarine rocket（SubRoc）　反潜火箭
Submarine-launched ballistic missile（SLBM）　潜射弹道导弹　技术拥有国：中国、美国、俄罗斯、英国、法国。见图S-10。

图S-10　潜射弹道导弹
Figure S-10　Submarine-launched ballistic missile

Submerge　下潜（浸水）
Submerge interval of floating dock　下沉时间
（1）浮船坞从工作吃水至最大沉深吃水所需的时间；
（2）钻井装置坐底作业时，其沉垫或下浮体从漂浮状态下沉至海底并达到作业状态所需要的时间。
Submerged　水下的（下潜的，沉没的）
Submerged cargo pump　潜液泵
Submerged dredge pump　潜水泥泵　装设在耙头架或铰刀架上，深潜水中进行吸排泥的水密封闭式泥泵。
Submerged fish-pump　潜水鱼泵　使用时泵体浸没在水中的鱼泵。
Submerged turret loading（STL）　水下装载转塔
Submerged turret production（STP）　水下生产转塔

Submerged-tube evaporator 浸管式蒸发器 加热管组浸没在水中工作的蒸发器。

Submersible carrier 潜水器母船 系指以潜水器作为调查工具的船舶。

Submersible compression chamber 下潜式加压舱 从船上吊放入水中工作的加压舱。

Submersible drilling platform 坐底式钻井平台 系指仅在坐底情况下从事钻井作业的稳柱式钻井平台。

Submersible drilling unit 坐底式钻井平台 系指仅在坐底情况下从事钻井作业的稳柱式钻井平台。Submersible drilling unit means a column-stabilized unit designed for drilling operation solely when supported by sea bed.

Submersible drilling unit/ bottom-supported drilling unit 坐底式钻井平台 系指坐落于海底时钻井,起浮后可拖航至另一地点作业的移动式钻井平台。

Submersible lifeboat 可潜救生艇 主要配置在油船或海上石油钻井平台装置上,能供火灾时搭载人员后潜入水下驶离着火海区的救生艇。

Submersible motor 潜水电机 可长期潜在水下正常运转的电机。电机的结构为全封闭充油式,电机内部压力高于外部的水压。

Submersible navigation light 潜水式号灯 能耐受海水腐蚀和承受一定压力的,具有较高水密要求和电气绝缘性能的号灯。

Submersible pump 潜水泵 与电动机组合成一体浸没在水中工作的泵。在船舶上主要作应急排水之用。

Submission of tender 投标 它系指投标人应招标人的约请,依照招标的要供和条件,正在规定的时间内向招标人递价,争取中标的行为。

Submitter 提交方 系指申请 IMO 组织验证其散货船和/或油船设计和建造规范符合国际散货船和油船目标型船舶建造标准的任何主管机关或认可组织。Submitter means any Administration or recognized organization that requests the IMO organization to verify that its ship design and construction rules for bulk carriers and/or oil tankers conform to the International goal-based ship construction standard for bulk and oil tanker.

Submitter/design team 发布/设计小组 系指一个行政部门认可的,由船东、建造者或设计者成立的设计小组。当有其他的设计和布置要求的便利的评估时,该小组可能还要有一位在船舶设计、消防安全和/或操作方面有必要知识和经验的专家。其他成员可能来自验船师、船舶经营者、安全工程师、设备制造商、人为因素专家、船舶建造师和海洋工程师。

Sub-organization 分支机构 系指在公司控制下并在相同安全管理体系(SMS)覆盖下,作为公司一部分的办事处。

Subsea production system 水下生产系统 系指由水下井口、水下基盘及管汇、水下储油中心、输油中间站等整套水下生产设备及海底管道组成的海上油气生产系统。

Subses manifold 海底集油管汇

Subses tieback 海底回接系统

Subsidy 补贴 系指政府或任何公共机构对企业提供的财政捐助和政府对收入或价格的支持。其范围包括:(1)政府直接转让资金,即赠予、贷款、资产注入;潜在的直接转让资金或债务,即贷款担保;(2)政府财政收入的放弃或不收缴;(3)政府提供货物或服务,或购买货物;(4)政府向基金机构拨款,或委托、指令私人机构履行前述(1)~(3)的职能;(5)构成1994年关贸总协定第16条含义的任何形式的收入或价格支持。

Subsonic compressor 亚声速压气机 各级动叶进口气流速度为亚声速的压气机。

Substance 物质(实质,内容)

Substances liable to spontaneous combustion (Class 4.2 dangerous materials) 易自燃物质(第4.2类危险货物) 系指遇空气后不需提供能量即能自热的物质,但自燃性物质除外。 Substances liable to spontaneous combustion mean the materials in this class are materials, other than pyrophoric materials, which, in contact with air without energy supply, are liable to self-heating.

Substances prohibited to incinerate on shipboard 船上禁止焚烧的物质 系指下列物质:(1)1973年国际防止船舶造成污染公约(MARPOL)的附则Ⅰ、Ⅱ或Ⅲ规定的货物残余物或有关的被污染的包装材料;(2)多氯联苯(PCB);(3)上述公约附则Ⅴ定义的含有超过微量重金属的垃圾;(4)含有卤素化合物的精炼的石油产品;(5)不在船上产生的污泥和油渣;(6)废气滤清系统的残余物。也应禁止在船上焚烧聚氯乙烯(PVC),但在已颁发IMO形式认可证书的船上焚烧炉除外。 Substances prohibited to incinerate on shipboard means the following substances:(1)residues of cargoes subject to Annex Ⅰ、Ⅱ or Ⅲ of MARPOL or related contaminated packing materials;(2)polychlorinated biphenyls (PCBs);(3)garbage, as defined by Annex Ⅴ of MARPOL, containing more than traces of heavy metals;(4)refined petroleum products containing halogen compounds;(5)sewage sludge and sludge oil either of which are not generated on board the ship;(6)exhaust gas cleaning system residues. Shipboard incineration of polyvinyl chlorides (PVCs) shall be prohibited, except shipboard incinerator for which an IMO Type Approval Certificates have been issued.

Substances which, in contact with water, emit flammable gases (Class 4.3 dangerous materials) 遇水散发易燃气体的物质(第4.3类危险货物) 系指与水发生相互作用后易自燃或产生的易燃气体达到危险数量的物质。 Substances which, in contact with water, emit flammable gases mean the materials in this class are solids which, by interaction with water, are liable to become spontaneously flammable or to give off flammable gases in dangerous quantities.

Substantial 有实质的(本质的,重大的)

Substantial corrosion 显著腐蚀 系指依据测量厚度评定腐蚀类型,腐蚀的程度虽然未超过船级社规定的许用极限但是超过了许用极限的75%的腐蚀状态。对依据国际船级社协会(IACS)的共同构造规范建造的船舶来说,许用极限系指腐蚀类型的评估按照测厚的结果所测的厚度在 $t_{net}+0.5$ mm 和 t_{net} 之间的腐蚀状态。 Substantial corrosion is an extent of corrosion such that assessment of corrosion pattern indicates the wastage in excess of 75% of allowable margins, but within acceptable limits. For vessels built under the IACS Common Structural Rules, substantial corrosion is an extent of corrosion such that the assessment of the corrosion pattern indicates a gauged (or measured) thickness between $t_{net}+0.5$ mm and t_{net}.

Substantial modification of a marine diesel engine 船用柴油机的实质性改变 对"73/78 防污公约"附则Ⅵ的 2008 NO_x 技术规则而言,系指:(1)对安装在 2000 年 1 月 1 日或以后建造的船上的发动机而言,实质性改变系指:可能造成发动机超出第 13 条规定的适用排放极限的发动机的改装。如果技术档案中所指的不改变排放性能的常规发动机部件部分更换,不论是一部分还是多部分部件被替换,均不视为实质性改变。(2)对安装在 2000 年 1 月 1 日以前建造的船上的发动机而言,实质性改变系指对发动机做了增加由 6.3 条所述的简单测试方法确定的其现有排放特性,使超出 6.3.11 所述的允许值的任何改装。这些改变包括,但不限于其运转或技术参数(例如:改变凸轮轴、燃油喷射系统、空气系统、燃烧室构造,或发动机定时校准)的改变。就附则Ⅵ第 13.2 条的适用范围而言,按第 13.7.1.1 条的核准认可方法的安装或按第 13.7.1.2 条的证书,均不视为实质性改变。 For "MARPOL73/78" Annex Ⅵ, substantial modification of a marine diesel engine means: (1) for engines installed on ships constructed on or after 1 January 2000, substantial modification means any modification made to an engine that could potentially cause the engine to exceed the applicable emission limit set out in Regulation 13. Routine replacement of engine components by parts specified in the Technical File that do not alter emission characteristics shall not be considered a "substantial modification" regardless of whether one part or many parts are replaced; (2) for engines installed on ships constructed before 1 January 2000, substantial modification means any modification made to an engine which increases its existing emission characteristics established by the Simplified Measurement method as described in 6.3 in excess of the allowances set out in 6.3.11. These changes include, but are not limited to, changes in its operations or in its technical parameters (e.g. changing camshafts, fuel injection systems, air systems, combustion chamber configuration, or timing calibration of the engine). The installation of a certified Approved Method pursuant to Regulation 13.7.1.1 or certification pursuant to Regulation 13.7.1.2 is not considered to be a substantial modification for the purpose of the application of Regulation 13.2 of the Annex Ⅵ.

Substantially interesting State 有重大利益关系的国家 系指某一国家:(1)为海难或海上事故所涉及的船舶的船旗国;(2)海难或海上事故所涉及的沿海国;(3)海难对该国的环境(包括其按国际法得到承认的水域和领土的环境)造成严重或很大破坏;(4)海难或海上事故的后果对该国或其拥有管辖权的人工岛、设施或结构造成或预示造成严重破坏;(5)由于海难的发生,造成该国国民死亡或严重受伤;(6)具有进行过安全调查的国家认为对调查有用的重要信息;(7)出于某些其他原因,确定海上安全调查国认为重要的利益。 Substantially interesting State means a State: (1) it is the flag State of a ship involved in a marine casually or marine incident; (2) which is the coastal State involved in a marine casualty or marine incident; (3) whose environment was severely or significantly damaged by a marine casualty (including the environment of its waters and territories recognized under international law); (4) where the consequences of a marine casualty or marine incident caused, or threatened, serious harm to that State or to artificial islands, installations, or structures over which it is entitled to exercise jurisdiction; (5) where, as a result of a marine casualty, nationals of that State lost their lives or received serious injuries; (6) that has important information at its disposal that the marine safety investigating State(s) consider useful to the investigation; (7) that for some other reason establishes an interest that is considered significant by the marine safety investigating State(s).

Substantive defense 实体答辩

Substantially similar in relation to use of an Active Substance or preparation　在使用活性物质或制剂方面的实质相似　系指活性物质或制剂的应用方法及注入压载水管理系统(BWMS)的注入点与获得认可的系统没有明显不同。 Substantially similar in relation to use of an Active Substance or preparation means the method of application and point of injection of the Active Substance or Preparation to the BWMS are not significantly different to that in the system granted approval.

Substar　星下点　指卫星到地心的连线与地面的交点。卫星运行是星下点的轨迹称为卫星的子轨迹。

Subsurface buoy system（undersea buoy system）　水下浮标系统　又称潜标系统。是系泊在海面下的浮标系统。它通常与声释放器配合使用，可按指令回收。回收时发出声指令，释放机构使其自行脱钩弃锚，浮标上浮水面，发出信号以供寻迹。

Subsystem　子系统　系指具有相似共振形式的模态群。

SubVue　反潜雷达

Suck　吸入（抽吸，表面浅洼型缩孔）

Suck-in　吸取

Sucking pump　抽入泵

Suck-out　吸出（抽出）

Suction　抽吸（吸入吸口）

Suction back　吸力面

Suction caisson　吸力沉箱　是一种吸力埋置锚。具有较小的高度与直径比，穿透底质的深度比吸力桩小得多。吸力沉箱在水中的重力构成了它的大部分垂向支持能力。其承受垂向负荷的能力主要是其自身重力加上一些表面摩擦力和内部的吸力,承受水平负荷的能力则取决于裙板的穿透深度和受到剪切的各层底质之间的摩擦力。

Suction dredger(SD)　吸扬挖泥船/直吸式挖泥船　系指：(1) 在用松土装置把水底泥沙搅成泥浆的同时，利用泥泵通过吸嘴和泥管吸进泥浆，经排泥管，扬至泥舱或填泥区的挖泥船。包括耙吸、绞吸、冲吸和气动式挖泥船等；(2) 利用泥泵产生的真空和离心作用，从吸泥头经吸管进泥浆，再通过排泥管输送至卸泥区的船舶。它机构简单，但仅能吸松软土砂。

Suction embedded plate anchor(ESPLA)　吸力埋置式板锚

Suction filter　吸入过滤器

Suction flap　吸气瓣（吸泥舌门）

Suction follower　吸力跟踪器　如同吸力桩，一端封闭，另一端开槽插入板锚的装置。

Suction force　吸力

Suction gear　吸泥装置

Suction head　吸头/泵吸入压头　以法兰连接于吸泥管的端部，用以吸入泥浆的装置。

Suction hopper dredger　耙吸式挖泥船　系指具有良好的船体线型和推进设备，并有连接于吸泥管的波浪补偿器，能在汹涌波涛海面上进行挖泥施工的船舶。它自备泥舱，能在航行时挖泥装舱，并自行运泥抛卸。

Suction lift　泵吸高

Suction line liquid accumulator（suction trap）　回气桶　专供分离和收集随盘管或蒸发器带来的液滴的容器。

Suction manifold　吸入总管

Suction passage　吸入道（吸口）

Suction piece　吸入管接头

Suction pile　吸力桩　又称负压桩。适用于深水系泊系统，主要用于黏土型底质，也可用于细砂或颗粒层，能承受很高的系泊索的水平和垂直载荷。吸力桩必须根据底质情况专门设计，其施工复杂，安装费用高，安装和拆卸潜水泵均需要潜水员协助，通常用于永久性系泊系统。吸力桩一般为钢质圆筒形筒体结构，底部敞开，顶部是封闭的。安装时，首先把吸力桩下降到海底，靠其自重使筒体入底质。然后，不断地抽去吸力桩内的水，使筒体内部压力下降。内外压力差产生的垂直向下的压力作用在筒顶部，使筒体不断地被压入土中，直至筒体内的水全部抽光，紧贴底质为止。拔桩时，则按相反的程序，将水注入筒内，将筒体顶出底质。

Suction pipe　吸泥管/吸入管　系指泥泵吸口前的一段输泥管。

Suction pipeline　吸入管路

Suction plate dredge（SPD）　吸盘式挖泥船

Suction pressure　吸气压力　压缩机吸入口处的压力。

Suction pressure regulator　吸气压力调节阀　自动控制制冷压缩机吸气压力使之保持稳定的阀。

Suction pump　抽吸泵

Suction resistance　吸入阻力

Suction screen　吸口滤网

Suction side　吸力面

Suction surface　吸力面

Suction valve　吸入阀

Suction ventilater　吸风机

Suction well　吸水井

Suction wells in cargo tanks　货油舱吸井　货油舱底部向下凹入的一个类似井的布置，用于安置货油吸管端部的吸入滤网，以使油船卸油时所留残油尽可能少。

Suction-stabilized boundary layer　吸力稳定边

界层

Sue and labour expenses 施救费用 亦称营救费用，系指被保险货物在遭遇承保的灾害事故时，被保险人或其代理人、雇用人为避免、减少损失采取各种抢救、防护措施时所支付的合理费用。

Suez canal 苏伊士运河 位于埃及东北部，是连接大西洋和印度洋的著名国际海运通道，承担着全世界14%的海运贸易。亚欧之间除石油外的一般货物海运，80%经过苏伊士运河。2014年，埃及政府计划在已有145年历史的苏伊士运河沿线修建一条新的运河。这个投资额将近90亿美元的项目旨在进一步繁荣苏伊士运河沿岸贸易。该项目有长达35 km的区域需要进行"干料挖掘"，另有37 km需要拆建和加深，这意味着总长度为163 km的现有苏伊士运河将被拓宽。这将让每日通过运河的船舶数量从25艘增加到90艘以上。目前，该项目的资金问题已通过众筹方式解决。

Suez Canal tonnage 苏伊士运河吨位 系指按苏伊士运河规则规定据以征收船舶各项税金和费用的吨位。是根据1873年君士坦丁堡国际委员会所订丈量规则求得的净吨位，并正式记载于各国主管机关签发的专用吨位证书之中。在计算征收税金时，还应考虑到在上述证书发出以后净吨位的任何变化。Suez Canal tonnage means a tonnage on which fill dues and charges to be paid by vessels, as specified in the Suez Canal regulations, are assessed, is the net tonnage resulting from the system of measurement laid down by the International Commission held at Constantinople in 1873, and duly entered, on the special certificates issued by the competent authorities in each country. In assessing the dues, any alteration of net tonnage subsequent to the delivery of the above mentioned certificates is taken into account.

Suez canal waters 苏伊士运河水域 系指运河本身，以及与运河本身、赛得港和苏伊士港相邻的运河管理局管辖范围内的水域。运河本身的长度，对于从塞得港湾进入的船舶而言，自西支流（west branch）Km 3.710处算起，或对于从东进口航道进入的船舶而言，自东支流（east branch）Km 1.333处算起，直至苏伊士港HM.3处为止，包括大苦湖（Great Bitter lake）的两个航道和所有运河支流在内。至于运河的宽度，当堤岸高出水面时，以该堤岸为界线；当堤岸没入水中时，运河的宽度以通过水下堤岸与相应于计及船舶航行下沉量的最大允许吃水平面交点的垂线为界线。包括出入的航道，有塞得港东西进口航道，通往运河进口处的苏伊士港东航道。Suez canal waters mean the canal proper, and the waters within the Canal Authority concession adjacent to the Canal proper, port Said harbor and port Suez. The canal proper: as to its length, is reckoned to run from Km 3.710 west branch for vessels entering form port Said harbor and from Km 1.333 east branch for vessels entering through the east Approach channel to Hm.3 at Suez, including the two channels of the G.B.L and all Canal by-passes. As to its width, the Canal is bounded by two banks when they are immersed in water. If the banks are submerged, the width of the Canal is limited to the perpendiculars at the point of intersection of the submarine bank with the horizontal plane corresponding to the maximum draught authorized including squat access channels there to port Said eastern and western entrance channels, Suez entrance channel which includes the port of Suez eastern channel leading to the Canal entrance.

Suezmax tankers 苏伊士型油船 系指满载状况下可以通过苏伊士运河的最大油船，即吃水不超过58英尺。

Sugar 糖 视糖的类型而定，可能为褐色或白色细粒，含水量极低，约0～0.05%。由于糖溶于水，进水后可能会随船舶的运动而在货舱内部形成气囊。由此产生的危害与可流态化货物构成的危害相似。应该认识到水如进入货舱，船舶的稳性因糖的溶解（液态底层的形成和货物的移动）而有风险。Depending on types, sugar may be either brown or white granules, with very low moisture content to the order of 0% to 0.05%. As sugar dissolves in water, ingress of water may result in the creation of air pockets in the body of the cargo with the ship's motion. The hazards are then similar to the hazards presented by cargoes which may liquefy. In case of ingress of water into the holds, the risk to the stability of the ship through dissolution of sugar (formation of a liquid base and shifting of cargo), should be recognized. This cargo is highly soluble.

Suit system 救生服系统 救生服以及与该救生服一起使用之任何其他制品的组合。Suit system is the combination of a suit and any other products which are used in conjunction with it.

Suitable repair condition 适修状态 系指船舶停靠码头，处于待修状态。

Suite 套（套间）

Suite of equipment 成套设备

Sulfidation attack (sulfur corrosion) 硫化腐蚀

Sulfuric acid 硫酸

Sullage 淤泥（污水，垃圾）

Sulphate of potash and magnesium 硫酸钾和硫酸镁 颗粒状褐色物质。水溶液几乎为中性。视制造厂商工艺而定，可能稍有气味。熔点72 ℃。含水量0.02%。无特殊危害。该货物易溶解。该货物为不燃物或失火风险低。 Sulphate of potash and magnesium is granular light brown material. Solution in water is almost neutral. It may have a slight odour, depending on the process of manufacturer. Its melting point is 72 °C and moisture content is 0.02%. It has no special hazards. This cargo is highly soluble. This cargo is non-combustible or has a low fire-risk.

Sulphur（formed, solid） 硫黄（成形、固体） 酸性气体加工或炼油作业提取的一种副产品。经成形过程将硫黄从溶态转为特定固体形态（如球粒、颗粒、丸粒、锭或薄片），呈鲜黄色，无嗅。本细目不适用于碾碎、块状和粒硫黄或未经上述成形过程的酸性气体加工或炼油作业的副产品。该货物为不燃物或失火风险低。货物如遇火，可能产生有害气体。该货物按有关规定装卸和装运时，对人体组织或船舶无腐蚀或粉尘危害。 Sulphur（formed, solid）is a co-product recovered from sour gas processing or oil refinery operations that has been subjected to a forming process that converts sulphur from a molten state into specific solid shapes（e.g., prills, granules, pellets, pastilles or flakes）; it is bright yellow in colour; odourless. This schedule is not applicable to crushed, lump and coarse-grained sulphur（Sulphur UN 1350）, or to co-products from sour gas processing or oil refinery operations not subjected to the above-described forming process. This cargo is non-combustible or has a low fire risk. If involved in a fire, cargo may generate harmful gases. When handled and shipped in accordance with the provisions of the schedule, this cargo poses no corrosion or dust hazards for human tissue or vessels.

Sulphur dioxide 二氧化硫 潮湿时有腐蚀性，吸入有毒，造成灼伤，对眼睛和呼吸系统有刺激。 Sulphur dioxide is corrosive when wet, toxic if inhaled, causes burns, and is an irritant to the eyes and respiratory system.

Sulphur elimination 脱硫（除硫）

Sulphur oxides(SO_x) 硫氧化物

Sulphur（UN 1350）（crushed lump and coarse grained） 硫黄（碎块或粗粒） 火山地区存在的一种游离矿物质。呈黄色，易碎，不溶于水，但受热易熔化。硫黄是在潮湿状态下装载。特别是在装载和卸载期间以及卸货后和扫舱时，易燃和粉尘爆炸。该货物可能容易着火。该货物为不燃物或失火风险低。 Sulphur（UN 1350）（crushed lump and coarse grained）is a mineral substance found free in volcanic countries. It is yellow in colour, brittle, insoluble in water, but readily fusible by heat. Sulphur is loaded in a damp or wet condition. It is inflammable and is easy to cause dust explosion especially during loading and unloading and after discharge and cleaning. This cargo may ignite readily. This cargo is non-combustible or has a low fire-risk.

Summary procedure 简易程序 中国 1996 年修订后的刑事诉讼法，为提高诉讼效率，增设了简易程序。但从目前司法实践来看，简易程序的适用在各地不是很平衡，大部分法院没能很好地适用简易程序，适用率达到 30% 的不是很多，未能有效地发挥程序分流的功能，立法意图也没能得到很好的实现。

Sump 集存槽（沉淀柜，污水井）

Sump tank 油底壳

Sun gear 太阳轮 行星齿轮传动装置中位于中央的可做定轴转动或固定不转的齿轮。

Sunk manhole cover 埋置人孔盖 周缘设有下陷围板的开口上使用的，不凸出板件表面的人孔盖。

Super atomic bomb 氢弹 也被称作热核弹，是利用原子弹爆炸的能量点燃氢的同位素氘等轻原子核的聚变反应瞬时释放出巨大能量的核武器。它的爆炸过程大致是裂变—聚变—裂变。它的特点是借助热核反应产生的大量中子轰击 U-238，使 U-238 发生裂变反应。这种氢铀弹的威力非常大，放射性尘埃特别多，所以是一种"肮脏"的氢弹。1953 年 8 月 14 日，苏联总理马林科夫宣布苏联已不再垄断氢弹的生产了。人类所制造破坏力最大的爆炸装置为苏联于 1961 年试爆的"沙皇氢弹"（代号"伊凡"），其原有设计拥有一亿吨 TNT 当量，但基于种种考虑，其实际制造当量约为 5 000 万吨。技术拥有国：中国、美国、俄罗斯、法国。见图S-12。

图 S-12 氢弹
Figure S-12 Super atomic bomb

Super computer 超级计算机 系指技术能力，特别是计算速度为世界顶尖的电子计算机。它的体系设计和运作机制都与人们日常使用的个人电脑有很大区别。超级计算机能够执行一般个人电脑无法处理的大资料量与高速运算的电脑，其基本组成组件与个人电脑的概念无太大差异，但规格与性能则强大许多，是一种超大型电子计算机。具有很强的计算和处理数据的能力，主要特点表现为高速度和大容量，配有多种外部和外围设备及丰富的、高功能的软件系统。现有的超级计算机运算速度大都可以达到每秒一太（Trillion,万亿）次以上。超级计算机是计算机中功能最强、运算速度最快、存储容量最大的一类计算机，多用于国家高科技领域和尖端技术研究，是一个国家科研实力的体现，它对国家安全、经济和社会发展具有举足轻重的意义。是国家科技发展水平和综合国力的重要标志。

Super heavy torpedo 超级重型鱼雷 系指鱼雷中体型和质量较大、航程较远、战斗部威力较大的鱼雷，可载于水面舰艇或潜艇，主要用于攻击敌方水面舰船和潜艇等目标。重型鱼雷长 5~8 m，直径通常为 533 mm，大的 550 mm 甚至达到 650 mm，总重 1 000~2 000 kg 不等，航速为 20~60 kn，最大航深 350~914 m，最大航程一般为 15~46 km，战斗部装药通常为 120~260 kg，动力装置有电动力和热动力两种。重型鱼雷的制导方式为有线制导加主动或被动声自导，而轻型鱼雷一般无须线导，只有主/被动声自导。这是因为重型鱼雷航程较远，先用线导把鱼雷导向目标附近，最后转换成主/被动声自导。如果没有线导，鱼雷声自导不可能捕获远距离目标；而没有主/被动声自导，鱼雷的命中精度不高。这与远程反舰导弹需要中段惯性制导加末段主动雷达寻的道理是一样的。鱼雷线导控制系统由导线、放线器和信号传输设备组成。导线一般为直径小于 1.2 mm，芯线直径小于 0.4 mm 的特制导线，具有较强的拉力和抗腐蚀能力。鱼雷发射后，射击控制系统通过导线传输指令，控制鱼雷的航向、航速、航深和姿态；鱼雷则通过导线向发射舰船连续传回自身的工作状态、位置、运动姿态，以及目标的方位、距离和干扰情况等信息。射击控制系统根据目标和鱼雷的运动参数，经处理后，向鱼雷发出指令，把鱼雷导向目标。当鱼雷进入声自导作用距离时，启动自导系统，先以低速被动声导进行搜索，发现目标后转入自动跟踪、识别，在一定时候转入主动声自导对目标精确定位和攻击。此时，被动声自导和线导都处于监控状态，一旦鱼雷受到干扰或未命中目标，则自动转为线导，重新搜索目标再次攻击。冷战结束后，未来的海战主要在浅水海域进行，对原来适合于深海作战的重型鱼雷将进行改进，提高浅水攻击性能。由于浅水中声环境比较复杂，对主/被动声自导系统的要求更高，人工智能将引入声自导系统，发展新的智能鱼雷；线导的导线将采用光纤，战斗部采用定向爆破装药，可减少药量而爆破威力不减。由于反舰导弹的发展，重型远程鱼雷将主要用于反潜、反水面舰的作用将下降。该技术拥有国：俄罗斯（暴风雪）、伊朗（鲸鱼）、中国（空泡1型）、美国。见图 S-13。

图 S-13　超级重型鱼雷
Figure S-13　Super heavy torpedo

Super large bore engine 超大缸径发动机

Super long stroke engine 超长冲程发动机

Super micro computer（SMC） 超微电脑

Super sea trial 超级海上试验 系指不再分别进行建造厂商海上试验和海军接收海上试验，而是把两次海上试验合并为一次的海上试验。超级海上试验与传统海上试验相比，不仅节省燃油和人力，也节省了时间。

Super ventilated hydrofoil 全通气水翼 整个水翼的吸力面都发生通气现象，并与大气相连通的水翼。

Super-cavitation 超空泡 是一种奇特的物理现象，随着水下物体运动速度的加快，其上所承受的水压反而会减小，一旦水压减小到一定程度时，与水下物体接触的水就会气化，形成空泡。通过一定的技术手段使"气泡"把整个航行物包裹起来，形成一种"气体外衣"，就可使物体始终航行在自己制造的超空泡内部，从而最大限度地避免水的黏性阻力，实现高速航行。

Super-cavitation technologies 超空泡技术

Super-cavitation torpedo "超空泡"鱼雷 系指其艏部装有空泡发生器，航行时能产生空泡，从而最大限度地降低鱼雷遇到的水的阻力，实现高速航行的鱼雷。"超空泡"鱼雷的最高速度超过音速，敌方很难拦截。

Supercharged boiler（pressure fired boiler） 增压锅炉 用锅炉排烟驱动废气涡轮，再由与后者同轴的压气机向炉膛供入空气以建立增压燃烧，构成联合循环运行的锅炉。

Supercharged diesel 增压式柴油机

Supercharged diesel engine 增压柴油机 大气经过增压器提高到一定压力后进入气缸的柴油机。

Supercharged steam generator 增压蒸汽发生器

Supercharger 增压器 提高内燃机进气压力的装置。

Supercharging 增压 为增加进入气缸的可燃混合气或空气量而采取的提高内燃机进气压力的措施。

Supercharging（supercharge） 增压（增压作用）

Supercharging compressor 增压压气机

Supercharging pressure 增压压力 在增压器出口处新鲜空气的压力。

Supercharging pump 增压泵

Superconducting magnetohydrodynamic fluid propulsor 超导磁流体推进器　系指用吸入导流管和喷出导流管替代螺旋桨,在强大的电磁作用下,海水高速进入吸入导流管,经加速后由喷出导流管射出,推动潜艇前进的推进器。

Superconducting magnetohydrodynamic fluid submarine 超导磁流体潜艇　系指利用超导磁流体推进器产生的推力前进的潜艇。其特点是:(1)由于没有螺旋桨拍打水流,轴承、齿轮系统的简化减少摩擦,使潜艇航行时的噪声降低至极低,几乎可实现"零噪声";(2)该潜艇可以潜得更深,且极为灵活,可在水中"跳舞"。其艇身外壳由新型高强度玻璃钢制成,抗压力是普通潜艇的3倍。一艘超导磁流体潜艇配备6个以上的磁流体推进器。它们之间是相互独立的,任意改变其中某几个推进器的电流方向和强度,即可改变潜艇的航行状态,实现快速左转、右转、上浮或下沉,比传统的潜艇灵活得多;(3)高速旋转的螺旋桨推进系统受到机械材料强度的限制,因此传统潜艇的速度也受到限制。相比之下超导磁流体推进器的磁体、电极都是相对静止的固定装置,可以通过增强电压提供超大输出功率,从而提高潜艇的航行速度;(4)超导磁流体潜艇的推进系统和电池舱位于艇身两侧,且充分利用舱室空间,布局灵活,不像传统潜艇那样将庞大的螺旋桨推进系统置于艉部。动力系统大幅瘦身,这意味着火力系统可以扩容,搭载更多威力强大的鱼雷和潜射导弹,使潜艇的水下攻击力更强。

Superconducting propulsion system 超导电力推进装置

Superconduction electrical machine 超导电机　系指绕组由超导体(超导线)制成的电机。

Superconductive material 超导材料(超导体)　系指具有超导电性的材料。其主要特性有:(1)完全导电性。在某一温度以下,其电阻降为零,而在此温度以上才具有电阻,此温度称为临界温度。电阻为零的状态称为超导态,具有电阻的状态称为正常态;(2)完全抗磁性。超导体处在磁场中时,将磁力线排除在外,即超导体内部 $B=0$ 。完全抗磁性又称迈斯纳效应。除临界温度外,超导体还有2个临界参数,即临界磁场强度和轮机电流密度。在临界温度以下,当超导体所处磁场强度超过某一值时,超导体就从超导态变为正常态,此磁场强度值称为轮机磁场强度;在临界温度和临界磁场强度以下,当超导体内通过的电流密度超过某一值时,超导体也将从超导态变为正常态(此现象称为失超),此电流密度值称为临界电流密度。

Superconductivity 超导电性　系指某些金属、合金或化合物在深低温下电阻消失的现象。

Superheat 过热

Superheated steam 过热蒸汽　温度高于同一压力下饱和温度的蒸汽。

Superheater 过热器　将饱和蒸汽加热成过热蒸汽的装置。

Superheater element 过热器元件

Superheater temperature alarm 蒸汽过热器超温报警

Superheater tube 过热器管

Superheater unit 过热器组

Superheating 过热的

Superheating surface 过热面

Superheating temperature 过热温度

Superhigh pressure boiler 超高压锅炉

Superintendent 机务代表　系指在修船过程中,船方主持轮机、电气等专业的负责人。

Superior Shoal Protection Area 苏必利尔湖(威斯康星州)沙滩保护区　系指以位于北纬48°03.2′,西经087°06.3′处的苏必利尔沙滩为圆心的半径6英里范围内的区域。Superior Shoal Protection Area means the area which a 6 mile radius from the center of Superior Shoal, at 48°03.2′N, 087°06.3′W.

Superphosphate 过磷酸盐　灰白色。含水量0~7%。无特殊危害。该货物为不燃物或失火风险低。该货物吸湿,受潮会结块。 It is greyish-white in colour. Moisture:0% to 7%. It has no special hazards. This cargo is non-combustible or has a low fire-risk. This cargo is hygroscopic and will cake if wet.

Superphosphate (triple, granular) 过磷酸盐(三重,颗粒)　呈颗粒状,深灰色,视产地而定会有很多粉尘。无特殊危害。该货物为不燃物或失火风险低。该货物吸湿,受潮会结块。Granular in form, is dark grey colour and, depending on its source, can be dusty. It has no special hazards. This cargo is non-combustible or has a low fire-risk. This cargo is hygroscopic and will cake if wet.

Superpressure 超压

Supersaturated 过饱和的

Supersaturated air 过饱和空气

Supersaturated vapour 过饱和蒸发气

Supersaturating 过饱和

Supersonic compressor 超声速压气机　各级动叶进口气流速度为超声速的压气机。

Supersonic anti-ship missile 超音速反舰导弹　是一种危险的武器,由法国航空航天公司和联邦德国MBB公司联合研制的新一代反舰导弹。超音速反舰导弹技术拥有国:中国、法国、德国、俄罗斯、印度(与俄罗斯合作）。见图 S-14。

图 S-14　超音速反舰导弹
Figure S-14　Supersonic anti-ship missile

Superstructure　上层建筑　（1）对强度评估而言，上层建筑被定义为在强力甲板上的甲板建筑，从舰船的一舷延伸至另一舷，或其侧壁板距船壳板向内小于船宽 B 的 4%；（2）甲板结构，不包括位于干舷甲板以上的烟囱，在干舷甲板或以上。Superstructure means: (1) for strength assessment a superstructure is defined as a decked structure on the strength deck, extending from side to side of the vessel, or with its side plating being less than 4% of the breadth B, inboard of the shell plating; (2) decked structure, not including funnels, which is on or above the freeboard deck.

Superstructure effectiveness　上层建筑有效度　系指上层建筑参与总纵弯曲的程度。

Supervision　监督　系指有系统及持续的调查或记录变数或过程。

Supplementary cock　辅助旋塞（备用旋塞）

Supplementary insulation　辅助绝缘　系指除基本绝缘以外，为在基本绝缘损坏时防止电击而提供保护所用的单独绝缘。Supplementary insulation means the independent insulation applied in addition to basic insulation in order to provide protection against electric shock in the event of a failure of basic insulation.

Supplementary repairs with add cost　追加工程　系指超出修船合同修理工程以外增加的工程。

Supplementary water　锅炉补加水

Suppletion vavle　附加阀

Supplier　供应商　系指提供产品的公司，其可以是制造厂商、贸易商或代理商。Supplier means a company which provides products, which may be a manufacturer, a trader or an agency.

Supplier credit　卖方信贷　是出口国为了支持本国机电产品、成套设备、对外工程承包等资本性货物和服务的出口，由出口国银行给予出口商的中长期融资便利。

Supplier's representative　供应商代表　系指负责交付燃油和做记录的加油船上的人员，或在从岸上直接向船舶添加燃油时，负责交付和做记录的人员。Supplier's representative is the individual from the bunker tanker who is responsible for the delivery and documentation or, in the case of deliveries direct from the shore to the ship, the person who is responsible for the delivery and documentation.

Supplier's declaration of conformity (SDoC)　供应商符合声明　编写供应商符合声明的目的是确保所提供的相关材料声明符合"材料声明"的要求，并标识负责实体。只要产品在船上，其供应商符合声明应保持有效。编写供应商符合声明的供应商应建立公司政策（company policy，可使用认可的质量管理体系）。关于供应商生产或销售的产品中化学物质管理的公司政策应涵盖：在文件中清晰描述涉及产品中化学物质管理的规则和要求，文件应予以保存和维护；以及在获取部件和产品的原材料时，应在评估后选择供应商，且应获取其提供的化学物质信息。供应商符合声明应包括：（1）唯一标识号（unique identification number）；（2）签发方的名称和联系地址；（3）符合声明主题的标识（例如，名称、形式、型号和/或其他相关补充信息）；（4）符合声明；（5）签发日期和地点；（6）代表签发方的经授权人员的签名（或等效的批准标记）、姓名和职务。

Supply chain　供应链　系指涉及从原材料至成品的材料和货物的供应和采购的系列实体。Supply chain means the series of entities involved in the supply and purchase of materials and goods, from raw materials to final product.

Supply pump　供给泵

Supply ventilation　进气通风

Supply vessels　供应船　是布置有艏部上层建筑和一个用作载货的艉部宽敞露天甲板的单甲板船。

Supply/anchor-handling tug　拖曳/锚作/供应船　系指为海上石油钻井平台服务为主的多用途拖船。航区有近海、远海。

Support level　支持级（人员）　系指在操作级和管理级人员的指导下，与在海船上履行指派的任务、职责或责任有关的责任级别。Support level means the level of responsibility associated with performing assigned tasks, duties or responsibilities on board a ship under the direction of an individual serving the operational or management level.

Support system for masters　船长决策支持系统

Supporting device　支撑装置　是一种将门的外载

荷或门内载荷传递至紧固装置、然后再从紧固装置传递至船体结构的装置,或通过除紧固装置之外的其他装置,诸如铰链、限位器或其他固定装置,能将门的载荷传递至船体结构的装置。　　Supporting device is a device used to transmit external or internal loads from the door to a securing device and from the securing device to the ship's structure, or a device other than a securing device, such as a hinge, a stopper or other fixed device, that transmits loads from the door to the ship's structure.

Supporting evidence　支持证据

Suppressed area　被删除区域　操作人员设定的不进行目标捕获的区域。　　Suppressed area is an area set up by the operator within which targets are not acquired.

Suppression　抑制(扑灭,消除)

Supreme People's Court　最高人民法院　成立于1949年10月22日,依照中华人民共和国《宪法》规定,中华人民共和国最高人民法院是中华人民共和国最高审判机关,负责审理各类案件,制定司法解释,监督地方各级人民法院和专门人民法院的审判工作,并依照法律确定的职责范围,管理全国法院的司法行政工作。中华人民共和国最高人民法院由全国人民代表大会选举产生的最高人民法院院长、副院长、审判员组成,对全国人民代表大会及其常务委员会负责。

Supreme People's Procuratorate　最高人民检察院　成立于1954年,是最高国家检察机关,与国务院一样同属中央国家机关序列。其主要任务是领导地方各级人民检察院和专门人民检察院依法履行法律监督职能,保证国家法律的统一和正确实施。最高人民检察院对全国人民代表大会和全国人民代表大会常务委员会负责,地方各级人民检察院对产生它的国家权力机关和上级人民检察院负责。人民检察院作为国家法律监督机关,依照法律规定代表国家独立行使检察权,不受任何行政机关、社会团体和个人的干涉。

Surface air cooler　表面式空气冷却器　由多排金属翅片管组成的使空气在流经管外表面时得到散热冷却和去湿的空气冷却器。

Surface blow　上排污　在锅筒水汽蒸发表面下泄放锅水的过程。

Surface blow valve　上排污阀

Surface blow-off cock　上排污旋塞

Surface boiling　表面沸腾

Surface condenser　表面式凝汽器

Surface condition　表面状态

Surface cooler　表面式冷却器

Surface effect craft　表面效应船(水面效应船)　系指一种借助浸在水中的永久性硬结构完全或部分保持气垫的高速船舶。　　Surface effect craft is a high speed craft with air cushion being wholly or partially.

Surface-effect ship(SES)　表面效应船(水面效应船)　系指一种借助永久浸在水中的硬结构可全部或部分保持气垫的气垫船,如双体气垫船、侧壁气垫船。　　Surface-effect ship(SES) is an air-cushion vehicle whose cushion is totally or partially retained by permanently immersed hard structures, such as air cushion catamaran, side wall hovercraft.

Surface friction　表面摩擦

Surface gradient sampler　表层梯度采水器　利用聚苯乙烯泡沫浮子保持浮在水面上的一种泵吸式供采集近表层梯度变化水样的采水器。

Surface mark　表面痕迹

Surface of fracture　表面张力

Surface peening　表面清理　系指基底表面经二次除锈后,涂装前除去油污、盐分、水、垃圾和杂物等清理工作的总称。　　Surface peening means a general term for peening work to remove sludge oil, salt, water, garbage and sundries prior to painting dressing after the secondary derusting of ground surface.

Surface piercing hydrofoil　割划式水翼　翼航时水翼割划水面,利用水翼浸水面积增减引起的升力变化而保持艇的飞高和纵、横稳定性的水翼。有V型、弧形、梯形等多种形式。

Surface pore　表面孔隙

Surface pretreatment　表面预处理　系指建造前对钢板或型材以物理方法或化学方法除去表面氧化皮、铁锈和异物并涂装车间底漆的工艺过程。　　Surface pretreatment means a technological process to remove scale, rust and obstructions from steel plates or sections by mechanical or chemical method and dress shop-coating prior to construction.

Surface treatment　表面处理　系指为改善涂层与基底间的结合力和防腐效果,在涂装之前用物理方法或化学方法处理基底表面,以达到符合涂装要求的措施。　　Surface treatment means a measure to treat the ground surface to meet the painting dressing requirement by mechanical or chemical method prior to painting dressing in order to improve the bonding force between coating and ground and the anticorrosion effect.

Surface ventilation　表面通风　系指在货物表面以上对货物处所进行的通风。　　Surface ventilation means ventilation of the space above the cargo.

Surface wave　表面波　当固体介质表面受到交替变化的表面张力作用时,质点作相应的纵横向复合振动;此时,质点振动所引起的波动传播只在固体介质表面进行,故称表面波。表面波是横波的一个特例。

Surface-type drilling unit 水面式钻井平台 系指具有单体或多体结构排水型船体的、在漂浮状态下从事钻井作业的平台。 Surface-type drilling unit means a unit with displacement hull of single or multiple hull construction designed for drilling operation in the floating condition.

Surface-type unit 水面式平台 系指具有单个或多个船形或驳船形排水型船构造的、在漂浮状态下作业的平台。可分为船式平台和驳船式平台:(1)船式平台:系指具有推进机械的水面式平台;(2)驳船式平台:系指无推进机械的水面式平台。 Surface-type unit means a unit having a ship-type or barge-type displacement hull or hulls for operation in the floating condition:(1)a ship-type unit is a surface-type unit having propulsion machinery;(2) a barge-type unit is a surface-type unit having no propulsion machinery.

Surge(surging) 喘振 压气机的排气压力和空气流量以特定的频率进行的脉动变化。

Surge limit(surge line) 喘振边界 由压气机特性曲线图上所示的在各种转速下各喘振开始点连接而成的边界线。

Surge margin 喘振裕度 用压气机特性曲线上工作线偏离喘振线的程度来表示压气机工作稳定性的指标。

Surge tank 调节水柜 凝水—给水系统中用于均衡变负荷时水量的柜子。

Surge-preventing system 防喘装置 为改善压气机的启动性能,扩大其稳定工作范围,防止放气阀、放气带和可转静叶发生喘振的装置。

Surging characteristics 喘振特性

Surging condition 喘振工况 压气机发生喘振时的工况。

Surging(surge) 纵荡(喘振)

Surplus air 过剩空气

Surrounding pressure 环境压力

Survey 检验 系指为了完成船级的授予或保存,或者在主管机关委任的权限范围内,验船师所参加的工作。 Survey means an intervention by the Surveyor for assignment or maintenance of Class, or interventions by the Surveyor within the limits of the tasks delegated by the Administrations.

Survey lines 测线 又称航线,是船舶的航行测量路线。一般系指船舶的计划航线。为进行海洋地质勘探和海道测量,出测前根据测量任务布设好测线,并把测线标绘在海图上,舰船沿测线进行航行测量。

Survey methods which the Surveyor is directly inrolved 验船师直接相关的检查方法 系指:(1)巡查—对船舶建造技能的相关程序、活动及相关文件,是否符合船级及政府要求,进行独立的不预定的基本检查的行为;(2)讨论—为了追踪和识别,及确认是否符合船级及政府代表要求进行固定的持续性的文件讨论的行为;(3)到场—为了检查符合检查条件,对必要的范围进行协商的检查及按照试验方案或其他相同的措施进行预定检查的参观。 Survey methods which the Surveyor is directly involved in:(1) patrol—the act of checking on an independent and unscheduled basis that the applicable processes, activities and associated documentation of the shipbuilding functions are to conform to classification and statutory requirements;(2) review—the act of examining documents in order to determine traceability, identification and to confirm that processes continue to conform to classification and statutory requirements;(3) witness—is the attendance at scheduled inspections in accordance with the agreed Inspection and Test Plans or equivalent to the extent necessary to check compliance with the survey requirements.

Survey of acoustic absorption in the ocean 海洋声吸收测量 对海洋中各种电解质引起的声吸收及海水中气泡、悬浮粒子、浮游生物等引起的声吸收作所做的全面调查。测量各种声的吸收系数与频率、深度的关系。

Survey of PMS for machinery means a system 船舶机械计划保养系统检验 系指一种由船东申请,经船级社批准的以船舶机械计划保养系统来替代船舶机械(包括电气设备)的循环检验或特别检验的制度,称为船舶机械计划保养系统检验。 Survey of PMS for machinery means a system means a system which is applied for by the owner and approved by a society to substitute the PMS for machinery for the system of continuous or special machinery (including electrical installations) survey.

Survey of the radio station 无线电台的检验(平台) 系指下列检验:(1)无线电台投入使用前,由发证的主管机关或其授权的代表检验;(2)当平台转移到处于另一沿海国的行政管辖下时,该国或其授权代表检验;(3)MODU 规则证书的每一周年日前 3 个月或后 3 个月内,由主管机关和/或沿海国的官员或各自授权代表进行的定期检验。 Survey of the radio station is the survey as following:(1) carried out by the Administration which issues the license or its authorized representative before the radio station is put into service;(2) when the unit is moved and comes under the administrative control of another coastal State, a survey may be carried out by that State or its authorized representative; (3) within three months before or after the anniversary date of the MODU Code certificate, a periodical survey carried out by an officer of the Administration and/or the coastal State or their respective authorized representative.

Survey points 测点 船舶在测线上实施测量的点。利用地震法进行海洋地质勘探时又称炮点，测点上应同时获得测量数据和定位数据。测点的间距既可按距离来确定，也可按时间来确定。每根测线上的测点应按先后次序进行编号。在测量时要记录测点号、测量时间、测量数据和导航定位数据等。

Survey report 噪声检验报告 系指对每艘船舶按"船上噪声等级规则"的规定进行噪声检测后，由主管机关或其认可的组织签发的检验报告。

Survey to be held 待检验 系指检验等待进行，直到完成。 Survey to be held is interpreted to mean that the survey is being held until it is completed.

Survey work 检验工作

Surveying ship 水道测量船

Surveyor 验船师 系指代表船级社执行与入级有关的任务和检验职责的技术雇员。 Surveyor means technical staff acting on behalf of the Society to perform tasks in relation to classification and survey duties.

Survivability 生存性 系指船舶遭受袭击后在某种程度上维持运转的概率。生存性主要包括以下2个方面：(1)易受攻击性，威胁对象捕获和击中舰船，并在舰船上爆炸的概率；(2)易损性，在威胁对象袭击成功的情况下，船舶能够生存下来且在某种程度上维持运转的概率。生存性通常是综合了易受攻击性和易损性两方面因素的计算结果。易恢复性十分重要，因为它能对整艘舰船的易损性产生系统性的重大影响。易恢复性可以定义为船舶在遭受攻击后恢复运行水平的能力。 Survivability is defined as the probability that a ship can remain operational to some degree following an attack. Survivability is divided into two main aspects: (1) susceptibility, the probability of a threat acquiring, reaching and detonating on a ship; (2) vulnerability, the probability that a ship will be able to survive a successful attack and operate at a certain level. Survivability is normally calculated as the product of susceptibility and vulnerability. Recoverability is an important aspect as it has a significant influence on the vulnerability of the overall ship as a system. It can be defined as a measure of the ability of the ship to reach a particular level of operation higher than that immediately following the hit.

Survival condition[1] 生存工况 系指隔水管与海底井口脱开，船可以通过调节艏向装置抵抗环境载荷的工况。

Survival condition[2] 自存工况 系指极端环境条件下，自升式平台不能继续作业，但可通过调整可变载荷或放弃部分载荷以及其他措施以达到某种较为安全的状态。 Survival condition refers to a condition wherein a unit has to suspend drilling operation under extreme environmental conditions and need to adjust variable loads, reject some loads or take other measures for the purpose of keeping the unit in a safe condition.

Survival craft[1] 救生艇筏 系指：(1)弃船时，用以容纳船上人员的艇筏，包括救生艇、救生筏以及在此情况下，适于防护和保护人员的经认可的其他任何艇筏；(2)从弃船时候起能维持遇险人员生命的艇筏。 Survival craft mean: (1) those craft provided for accommodating the persons on board in the event of abandonment of the vessel and includes lifeboat, life-craft, and any other craft approved as suitable for the protection and preservation of persons in such circumstances; (2) a craft capable of sustaining the lives of persons in distress from the time of abandoning the craft.

Survival craft[2] 救生载具 救生设备中用以在水域承载人员使其不致浸泡水中的器具。一般系指救生艇和救生筏。

Suspect areas 可疑区域 系指船体结构内的结构性恶化可能性容易增加的位置，可能包括：(1)对钢质船体，为腐蚀和/或疲劳破裂的区域；(2)对铝合金船体，为疲劳破裂的区域和两种金属连接的附近区域；(3)对合成材料船体，为受到冲击损坏的区域；(4)对高速船舶，为易受砰击的底部结构前部区域。 Suspect areas are locations within the hull structure vulnerable to increased likelihood of structural deterioration and may include: (1) for steel hulls, areas of corrosion and/or fatigue cracking; (2) for aluminium alloy hulls, areas of fatigue cracking and areas in the vicinity of bimetallic connections; (3) for composite hulls, areas subject to impact damage; (4) for high speed ships, areas of the bottom structure forward prone to slamming damage.

Suspender 吊货架 由吊环绳索、吊钩和连接框架或撑杆组成的吊具。

Suspension height 支悬高度 吊杆根部叉头横销中心至千斤索具悬挂处的千斤板板中心的垂直距离。

Suspension height-boom length ratio 悬高杆长比 支悬高度与吊杆长度的比值。

Suspension hook (lifting hook) 吊艇钩 设置在救生艇上，供艇被吊卸用的吊钩装置。

SWA filter 声表滤波器 系指利用压电晶体制作

的滤波元件。

Swanneck ventilator 鹅颈通风筒 通风口朝下，筒体弯成鹅颈状的通风筒。

Swash bulkhead 防荡舱壁 系指液舱中设计成能防止液体晃动的非水密部件。 Swash bulkhead is a non-watertight partition in a tank designed for avoiding sloshing.

Swath-high speed craft 小水线面双体高速船 系指具有小水线面面积，而且片体水下部分呈鱼雷状的一种特殊船型的双体高速船。 Swath-high speed craft is a special type of catamaran high speed craft with small water-plane area and with underwater portions of hulls being formed in shape of torpedo.

Sweat and heating risk 受潮受热险 投保平安险和水渍险的基础上加保此险。保险人负责赔偿承保的货物因气温突然变化或由于船上通风设备失灵致使船舱内水汽凝结、受潮或受热所造成的损失。

Sweep boat 清扫船 系指清扫水面杂物，保护港口水面卫生的船舶。在自航船上设置收集水面上漂浮杂物的装置和垃圾舱或储放箱。按收集方式可分为导入法、铺集法和捞取法。船体为方驳型，有单体和双体之分。小型的排水量不到20吨，大型的多在100吨以上，舱容达几十立方米。也有采用集装箱存放、吊卸垃圾的。为提高其利用率现已发展为多用途港作船，兼任水面清扫、污油回收及港口拖曳等作业。见图S-15。

图 S-15 清扫船
Figure S-15 Sweep boat

Sweep winch 扫海绞车 用于扫海测量船上，作为收放和拖曳扫海具及绳索进行扫测海底沉碍物的绞车。

Sweeper 扫海船 又称"扫海艇"。用以查明航道或水域中在一定深度范围内有无障碍物并确定其位置的船舶。也可用于搜查沉落海底的物件。其作业方法是在船艉部拖一扫海具。扫海具由沉锤、浮子、竖标和钢索等组成。分水上部分和水下部分。按探查要求，预定水下部分的深度，一旦碰到障碍物，水面部分竖标翻倒，即可查明障碍物位置。又按搜查宽度要求，可单船作业，也可双船作业，类似单拖渔船和对拖渔船。扫海船作业水域深度不大，离岸也较近，多属小型船。

Sweeping 扫海具 由船舶拖曳进行扫海测量的工具。

Sweeping ship 扫海测量船 也称扫测船。用于探测水下障碍物的船舶。船上装有海具、侧声呐和其他扫测仪器设备以便进行扫测。

Sweeping trains 扫海趟 又称扫海带，扫海具一次扫过的区域。

Swell compensator 波浪补偿器 由液压缸、蓄压器和其他仪表配件等组成。用以在有风浪的水面上作业时，使耙头紧贴泥面的装置。

Swept volume 扫海体积 系指船宽乘以吃水乘以航程。 Swept volume means ship breadth multiplied by draft and multiplied by distance traveled.

Sweptback hydrofoil 后掠式水翼 平面形状具有后掠角的水翼。

Swimming organism sampling 浮泳生物取样 对具有一定游泳能力的海洋生物，在预定的时间、范围和水层，利用标准网具进行拖曳以及利用其他渔具进行试捕，来获取样品的方法。

Swinging boom derrick 摆动吊杆式 又分为摆动吊杆式（swinging boom system）、联动式（union purchase system）和平衡式（counterweight system）三大类。

Swinging boom system 单杆操作 系指摆动一根单杆来装卸货物的操作方法。

Swinging derrick 摆动吊杆装置 不配置专用牵索绞车的普通吊杆装置。

Swinging pneumatic conveyor 可吊式气力输送机 系指卸料时吊入舱内，直接将舱内的物料输送到岸上的货舱内的机械。

Swinging system 转向系统 系指用于控制船舶运动方向的系统，包括舵和操舵装置等。Swinging system is a system used for controlling movement direction of a ship, such as ruder and steering gear, etc.

Swirl combustion chamber 涡流室燃烧室 利用涡流式的形状和通道在压缩过程中使空气造成有规则的涡流与喷入该室的燃料进行较完善的混合，再进入主燃烧室以达到较完全燃烧的燃烧室。

Switch fabric 交换矩阵 是背板式交换机上的硬件结构，用于在各个线路板卡之面实现高速的点到点连接。

Switch gear 配电装置 系指用来接受和分配电能，并对电网进行保护的装置。船用配电装置种类，按其用途分为：(1)主配电板—用来控制和监视主发电机的工作，并将主发电机产生的电能，通过主电网或直接给用电设备配电；(2)应急配电板—用来控制和监测应急发电机的工作,并将应急发电机产生的电能,通过应急电网或直接给用电设备配电;(3)蓄电池充放电板——用来控制和监测充电发电机或充电整流器,对蓄电池组机械充放电工作,并通过低压电网或直接给用电设备配电;(4)交流配电板——当船舶采用直流电制时,由于多数通信导航设备仍需要交流电源,所以需装设交流变流机组。交流配电板用来控制和监测交流变流机组,并给用电设备配电;(5)岸电箱——当船舶停靠码头时,将岸上电源接至船上,通过主配电板(或应急配电板)给用电设备配电;(6)区配电板——介于主配电板或应急配电板与分电箱(又称为分配电板或分配电箱)之间,用以向分电箱和最后支路供电的配电板;(7)分电箱——用以向成组的最后支路供电,并装有保护装置。按其使用目的和性质通常又可分为电力分配箱(又称为分配电板)、专门分电箱(又称为分配电板)、无线电分配电板(又称为无线电源板)、助航通信分配电板和专用设备分配电板(如冷藏集装箱电源板);(8)电工试验板——接有全船各种电源和必要的检测仪表,专供船上检修和校验各种用电设备的配电板。

Switch gear and control gear assembly 开关设备和控制设备组件 系指一个或多个开关电器,连同控制、测量、信号、保护和调节设备等的组合,由制造厂商负责加上所有电气和机械的内部连接件和结构件组装完成的组件。 Switch gear and control gear assembly is one or more switch gear together with control, measurement, signal, protection and adjustment, etc. and the assembly assembled by the internal connection, and structural members of all the electrical and machinery installations fitted by the manufacturers.

Switchboard 配电板 用于控制由电源所产生的电功率,且将其分配至用电设备的开关装置和控制装置的组合装置。 Switchboard is a switch gear and control gear assembly for the control of power generated by a source of electrical power and its distribution to electrical consumers.

Switchboard and screen 配电板与配电屏 系指发电机与配电板接受来自发电装置的电能并将其直接或间接地分配给所有由该发电装置供电的设备。分配电板基本上是发电机与配电板的一部分(用汇流条馈线接至发电机与配电板,由于方便和经济的原因,远距离布置),其功用是向船上某一区域的照明、加热及电力电路分配电能。配电屏接受来自配电板或分配电板的电能而后分配给用电装置或其他配电屏或配电箱。配电箱是一个被封闭在金属箱内的配电屏。

Switching device 开关电器 系指用于接通或分断一个或几个电路中电流的电器。注意:一个开关电器

可以完成一个或两个操作。

Swivel 旋转接头 是将流体介质从静止的管道输入到旋转或往复运动的设备中,再从旋转接头中排出的连接密封装置,简称"旋转接头"。

Swivel joint 旋转接头(水龙头接头)

Swivel shackle 转环卸扣 转环与卸扣合为一个整体的卸扣。

Symmetrical cable 对称电缆 是由若干对叫做芯线的双导线放在一根保护套内制成的,为了减小每对导线之间的干扰,每一对导线都做成扭绞形状,称为双绞线,同一根电缆中的各对线之间也按照一定的规律扭绞在一起,在电信网中,通常一根对称电缆中有 25 对双绞线,对称电缆的芯线直径在 0.4～1.4 mm,损耗比较大,但是性能比较稳定。对称电缆在有线电话网中广泛应用于用户接入电路,每个用户电话都是通过一对双绞线连接到电话交换机,通常采用的是 22～26 号线规的双绞线。双绞线在计算机局域网中也得到了广泛的应用,Ethernet 中使用的超五类线就是由 4 对双绞线组成的。

Symmetrical short-circuit current 对称短路电流 系指预期(可达到的)短路电流交流对称分量的均方根值。如有直流分量应不计算在内。 Symmetrical short-circuit current means the r. m. s. value of symmetrical component of a prospective (available) short-circuit current, it is to be excluded, if D. C. component comes into existence.

Synchro steering gear 同步舵机

Synchro-asynchro steering gear 同异步舵机

Synchronous current commutator 同步变流机 是一种把电动机和发电机两种作用结合在一个电枢绕组内,且由一个磁场激励的变流机。通常用于将交流功率转变为直流功率。

Synchronous generator 同步发电机 是一种将机械功率转变为电功率的同步交流电机。

Synchronous motor[1] 同步电机 系指正常运行时的平均转速与其所接系统的频率精确地成正比的一种电机。

Synchronous-motor[2] 同步电动机 是一种将电功率转变为机械功率的同步电机。

Syncrolift 升船机 又称"举船机"。利用机械装置升降船舶以克服航道上集中水位落差的通航建筑物。由承船厢、支承导向结构、驱动装置、事故装置等组成。

Synteties 合成剂 通常是以药丸或针剂形式出现。没有气味的药丸往往是白色的,但也可能有不同的颜色,例如,陪替丁。 Synteties normally appear in pill or ampoule form. Those pills are odourless and are often white but may vary in colour, for example pethidine.

Synthetic fishing net 合成渔网

Synthetic Rope 合成缆绳

Syntheticr or designer drugs 合成药或化合致幻药 系指通过母体物质的化学改性而衍生的非法药品，后者有时相当于药理化合物。这一类包括 MDMA—3-4 亚甲基双氧苯丙胺（MDMA），以摇头丸著称，是一种滥用的物质，可归类于所谓的化合致幻药。于 1910 年由曼尼斯和雅克布逊合成，并于 1914 年由麦克实验室在德国作为减食欲药注册了专利，但没有在市场上销售。直到 20 世纪 70 年代和 80 年代才再次被使用，这一次是用于药物治疗试验，并在 1985 年证明对动物有毒害神经的作用，被归入限用的物质。为娱乐用途在秘密实验室制造，并以名为 MDMA 的销售在欧洲和美国引起了"锐舞"运动，其特点是在快节奏的聚会上，在饮料中掺入氨基酸和咖啡因以达到兴奋的效果。 Sytheticr or designer drugs means a illicit drugs derived from chemical modification of matrix substances, the latter sometimes corresponding to pharmacological compounds. This category includes MADA (Ecstasy). 3.4 methylene-dioxymethanphetamine (MADA), popularly known as "Ecstasy", is a substance of abuse belonging to the group of so-called designer drugs. It was synthesized in 1910 by Nanis and Jacobson and patented by Merck Laboratories in Germany in 1914 as an anorexic drug, but not marketed. Not until the 1970s and 1980s was it used again, this time for drug treatment testing, and in 1985 it was shown to have a neurotoxic effect on animals and classified as a restricted substance. It is made in clandestine laboratories for recreational use, and in the form known as MDMA, it has given rise in Europe and the United States to the "rave" movement, which is characterized by high-tempo parties at which the drinks are mixed with amino-acids and caffeine to achieve a stimulant effect.

Synthetics 合成纤维 是化学纤维的一种，是用合成高分子化合物做原料而制得的化学纤维的统称。它以小分子的有机化合物为原料，经加聚反应或缩聚反应合成的线型有机高分子化合物，如聚丙烯腈、聚酯、聚酰胺等。从纤维的分类可以看出它属于化学纤维的一个类别。

Synthetics 合成药 迷幻剂之一，外形为粉末、粗制药丸，或无色液体。没有气味。 Synthetics are one of hallucinogens, which are found in powder, crude pill or tablet form, or as colourless liquids. It is odorless.

Syren 号笛（报警器，注射器）

Syringe for lubrication 滑油注射器

System 系统

System alerts 系统警报 系指与设备故障或丢失（系统故障）相关的警报。 System alerts mean these alerts related to equipment failure or loss (system failures).

System calibration 系统修正 在综合导航系统中为提供精确的导航能力，而对声呐/罗经的校正进行定期检查。也就是通过一些方法来取得下列修正值；ABIA（方位角偏差）、TADJ（热敏电阻速度偏差）或 VADJ（声速仪的速度偏差）。取得修正值的办法有；测场法、卫星校正法、自动卫星校正法和无线电助航设备校正法。

System drawing 系统图

System electronic navigation chart (SENC) 系统电子航海图 （1）选择一个数据库，以制造厂商内部 ECDIS 格式由整个 ENC 内容及其各次重新的无损转换而成。该数据库由 ECDIS 用于显示生成的海图和其他导航功能，等同于一张最新的纸质海图。SENC 还可以包含航海人员增加的信息和其他来源的信息。（2）由适用的 ECDIS 将电子航行海图（ENC）转换而得到的数据库，可通过适当的设施和由海员补充其他数据来更新 ENC。

System electronic navigation chart (SENC) mean: (1) a database, in the manufacture's internal ECDIS formal, resulting from the lossless transformation of the entire ENC contents and its update. It is this database that is accessed by ECDIS for the display generation and other navigational functions, and is equivalent to an up-to-date paper chart. The SENC may also contain information added by the mariner and information from other sources; (2) a database resulting from the transformation of the ENC by ECDIS for appropriate use, ENC can be updated by appropriate means and other data added by the mariner.

System failure 系统故障 系指推进系统、操舵系统或发电站（包括它们的辅助系统）的主动或从动部件的任何故障，如管子或电缆这类部件也应考虑，但仅需考虑单个故障。 System failure means any failure of an active or passive component of a propulsion system, steering system or power generation plant, including their auxiliary systems. Components such as pipes or electric cables are also to be considered. Only single failure needs to be considered.

System integration 系统集成 是由单一指定的一方来管理，根据确定的程序来执行，并且在设计师和船东/操作者之间达成共识。 System integration is to be managed by a single designated party, and is to be carried out in accordance with a defined procedure, agreed between the designer and the owner/operator.

System integrator 系统综合者 系指负责确保综合航行系统（INS）符合综合航行系统性能标准要求的组织。 System integrator means the organization responsible for ensuring that the INS complies with the requirements of the performance standard for integrated navigation system.

System interaction objective 系统间干涉目标 防火方法或系统引起的火灾相关或非火灾相关危险的可能性应尽可能地保持在最低水平。系统的设计应尽可能地确保系统工作时不致使其他系统的性能受到非疏忽性降低。 The possibility of fire protection measures or systems causing fire related, or non-fire related hazards are to be kept to a level that is as low as is reasonably practicable. Systems are to be designed, as far as is practicable, to ensure that the operation of that system will not inadvertently degrade the performance of any other system.

System of pipes 管系

System position 系统位置 系指综合航行系统（INS）内通过至少2个定位传感器计算出的位置。System position means a position calculated in the INS out of at least two positioning sensors.

System raster navigational chart database（SRNC） 系统光栅航海图数据库 系指 RCDS 转换 RNC 以通过适当方式对 RNC 进行更新而产生的数据库。 System raster navigational chart database (SRNC) means a database resulting from the transformation of the RNC by the RCDS to include updates to the RNC by appropriate means.

System safety assessment（SSA） 系统安全评估 是对实际系统所做的系统评估，用以证明实际达到功能危险评估（FHA）所确认的安全目标和初步系统安全评估（PSSA）推导出的安全要求。系统安全评估通常是基于系统安全评估故障树分析。 System safety assessment (SSA) is a systematic assessment of the actual system to demonstrate that safety objectives from the FHA and derived safety requirements from the PSSA are actually met. The SSA is usually based on the preliminary system safety assessment (PSSA) fault tree analysis.

System software 系统软件 系指控制和协调计算机及外部设备，支持应用软件开发和运行的系统，是无须用户干预的各种程序的集合，主要功能是调度，监控和维护计算机系统；负责管理计算机系统中各种独立的硬件，使得它们可以协调工作。系统软件使得计算机使用者和其他软件将计算机当作一个整体而不需要顾及到底层每个硬件是如何工作的。

Systematic error 系统误差 系指测量系统所固有的一种误差。即在同一条件下获得的观测值，其大小及符号保持不变或者有规律地变化，可以通过大量的实验或理论计算确定这类误差的数值，对测量结果进行修正。

T

T valve 三通阀（T型阀）

Table for recording 记录桌 供记录用的、桌面呈倾斜状的小桌。

Table of offsets 型值表 系指由型值组成表列形式的表或表册。

Table vice 管钳（管式虎钳）

Tablet personal computer（Tablet PC） 平板电脑 也叫平板计算机，是一种小型、方便携带的个人电脑，以触摸屏作为基本的输入设备。它拥有的触摸屏（也称为数位板技术）允许用户通过触控笔或数字笔来进行作业而不是传统的键盘或鼠标。用户可以通过内建的手写识别、屏幕上的软键盘、语音识别或者一个真正的键盘（如果该机型配备的话）实现输入。平板电脑由比尔·盖茨提出，应支持来自高通骁龙处理器，Intel、AMD和ARM的芯片架构，平板电脑分为ARM架构（代表产品为ipad和安卓平板电脑）与X86架构（代表产品为Surface Pro）后者X86架构平板电脑一般采用intel处理器及Windows操作系统，具有完整的电脑及平板功能，支持exe程序。平板电脑的发展伴随着通信技术大发展日新月异，作为一项新兴技术，CDMA、CDMA2000 正迅速风靡全球并已占据18%的无线市场。截至2012年，全球CDMA2000用户已超过2.56亿，遍布70个国家的156家运营商已经商用3G CDMA 业务。

Tachograph 转速表
Tachometer 转速计（流量表）
Tachometer indicator 转速表指示器
Tachometer transmitter 转速计传感器
Tachometry 转速测定法
Tachoscope 手提转速计
Tackle 船用索具 包括各种桅杆、桅桁（桅横杆）、帆桁（帆脚杆）、斜桁和所有索、链以及用来操作这些的用具的总的术语。系指为了实现物体挪移系结在起重机械与被起重物体之间的受力工具，以及为了稳固空间结构的受力构件。

Tackle purchase 滑车组 滑车与绳索配合在一起工作的整个滑车系统。通常分单滑车组、复滑车组和联合滑车组等类型。

Taconite pellets 铁燧岩丸粒 矿石。灰色圆形钢丸。含水量2%。无特殊危害。该货物为不燃物或失火风险低。 Taconite pellets are ore, grey, round steel pellets. Moisture content is 2%. It has no special hazards. This cargo is non-combustible or has a low fire-risk.

Tactical diameter 回转圈直径/战术直径　系指船舶自发出操舵令位置到舯向改变离初始航向 180°在船中点量取的距离,其沿垂直于船舶初始航线的方向量得。Tactical diameter is the distance travelled by the midship point of a ship from the position at which the rudder order is given to the position at which the heading has changed 180° from the original course. It is measured in a direction perpendicular to the original heading of the ship.

Tactical Missile Defense(TMD)　战术导弹防御

Tail journal　艉轴颈

Tail shaft　艉轴（螺旋桨轴）　系指连接推进器的轴。

Tail shaft passage　艉轴隧（轴隧）

Tail shaft survey　艉轴检验

Tail-stabilizer anchor　艉翼式锚

Tailswing　抓斗机尾部半径　抓斗机棚后端至抓斗机转轴中心线间的距离。

Taint of odor　串味　系指货物受到其他异味物品的影响而引起串味导致的损失。

Taint of odor risk　串味险　投保平安险和水渍险的基础上加保此险。保险人负责赔偿承保的食用物品（如食品、粮食、茶叶、中药材和香料）、化妆品原料等因受其他物品的影响而引起的串味损失。该险主要承保被保险货物在运输过程中配载不当而受其他物品影响,引起的串味损失。如茶叶、香料与皮张、樟脑等堆放在一起产生异味,不能使用。

Take off/landing mode　起飞/着陆状态　对地效翼船操作模式而言,系指从滑行状态演变至地效状态的暂时状态,反之亦然。For WIG crafts' operational modes, take off/landing mode denotes the transient mode from the planning mode to the ground effect mode and vice versa.

Take-home boiler　应急航行锅炉　主锅炉发生故障时,船舶应急航行时使用的锅炉。

Take-home engine unit　应急航行机组　主推进装置发生故障时,船舶应急航行时使用的机组。

Taken in charge　接管　系指货物已提交给多式联运经营人运送并由其接受。Taken in charge mans that the goods have been banded over and accepted for carriage by the multimodes transport operator.

Take-off and landing area(TLOF)　直升机降落和起离区　直升机可着降和起离的承载区域。

Take-off mode for carrier-borne aircraft　航母舰载机起飞方式　(1)蒸汽弹射起飞——美国大多数航母舰载机采用这种方式。目前世界上仅有美国有能力制造大功率弹射器;(2)滑跃起飞——为避免蒸汽弹射起飞的缺点,一些国家采用此种方式航母跑道尽头有一定的仰角,比如我国"辽宁号"航母;(3)垂直/短距离起飞——以英国"海鹞"战斗机和美国"鱼鹰"运输机为代表。

Take-off speed　起飞速度　系指在该速度下,船从水面抬升以进入地效飞行模式。在起飞后,船体完全脱离水面,阻力突降,从而迅速加速至巡航速度。Take-off speed means a speed at which, the craft will lift off from the water surface to reach true flying mode in ground effect. After take-off, a daylight clearance will exist under the hull bottom, and the resistance will suddenly drop, allowing swift acceleration to cruising speed.

Talc　滑石　是一种极为柔软、发白、绿色或浅灰色天然水合硅酸镁。有一种独特的细腻或滑溜感。无特殊危害。该货物为不燃物或失火风险低。Talc is an extremely soft, whitish, green or greyish natural hydrated magnesium silicate. It has a characteristic soapy or greasy feel. It has no special hazards. This cargo is non-combustible or has a low fire-risk.

Tandem articulated gear (tandem gear)　串联式减速齿轮箱　各级齿轮沿轴向顺序排列的减速齿轮箱。

Tandem compound turbines　多缸单轴涡轮机

Tandem engine　串列式发动机

Tandem gear　单式串联减速齿轮箱　功率不分支的两级串联式减速齿轮箱。

Tandem piston　串联活塞

Tandem propellers　串列螺旋桨　在同一轴上安装前后两个或更多的转向相同的螺旋桨。

Tandem system　串列式水翼系统　前后两个水翼尺度相近,水翼面积大致均匀分布于艇的重心前后的水翼系统。

Tangential flow fan　切流式风机　具有径向叶片,其出口处平均空气流动方向切于叶轮外周的风机。

Tank　舱柜　系指为船舶永久结构所形成并设计为装运散装液体的围蔽处所。Tank means an enclosed space which is formed by the permanent structure of a ship and which is designed for the carriage of liquid in bulk.

Tank barge　液货驳　系指无自航设施的液货船。

Tank bulkhead　液舱舱壁　系指用于装载液货、压载水或液体燃料的液舱的边界舱壁。Tank bulkhead is a boundary bulkhead in tank for liquid cargo, ballast or bunker.

Tank capacity　液舱容积　液舱扣除舱内骨架所占空间后的容积。

Tank capacity 舱容 当船艇静止在基准水线 WL_{ref} 上时,各舱柜的有用的净容积。 Tank capacity is the net usable volume of the tank(s) for the craft at rest at the reference waterline, WL_{ref}.

Tank cleaning installations 货油舱洗舱设备 货油船洗舱设备中为完成洗舱任务所设置的洗舱海水加热器、洗舱喷水器、洗舱泵等设备。

Tank cleaning machines 洗舱机

Tank cleaning opening 洗舱孔/洗舱开口 系指油船甲板上,供洗舱用的开孔。

Tank cleaning procedure 洗舱程序

Tank cleaning pump 洗舱泵

Tank cleaning seawater heater (butter worth heater) 洗舱海水加热器 为提高洗舱效率,加热洗舱用的海水的专用加热器。

Tank cleaning system 货油舱洗舱系统 为油船换装不同油类或入坞修理前进行洗舱所设置的全套管路及货油舱洗舱设备。

Tank container 液货集装箱 系指适用于装运酒类、油类或液态化学物品的集装箱。

Tank deck 坦克甲板 登陆舰艇上停放坦克用的甲板。

Tank finished automobile full autonomy develop 完全自主坦克整车研制 该技术拥有国:德国、美国、中国、俄罗斯、英国、法国、意大利。见图 T-1。

图 T-1 完全自主坦克整车研制
Figure T-1 Tank finished automobile full autonomy develop

Tank gas-freeing installations 货油舱油气驱除装置 为油舱卸空后,驱除货油舱内油气和空气的混合物而设的专用装置。有抽吸和吹送两种形式。

Tank gauge transmitter 液货舱液面计传感器

Tank gun 坦克炮 是现代坦克的主要武器。坦克主要在近距离作战,坦克炮在 1 500～2 500 m 距离上的射效高,使用可靠,用来歼灭和压制敌人的坦克、装甲车,消灭敌人的有生力量和摧毁敌人的火器与防御工事。坦克炮是由小口径地面炮演变而来的。现代坦克炮是一种初速高、身管长的加农炮。它的主要诸元有口径、穿甲弹的初速、全装药杀伤爆破榴弹和减装药杀伤爆破榴弹的初速、破甲弹的初速、发射速度、高低射界、方向射界、炮弹质量和弹药基数等。该技术拥有国:德国(莱茵金属)、英国(皇家兵工厂)、美国(通用动力)、法国、瑞士(瑞士军械公司)、俄罗斯、中国(北方工业)、以色列(军事工业公司)、印度、乌克兰/美国(MIAI)、日本90、韩国(KIAI 使用莱茵公司许可证生产的 120 滑膛炮,莱茵莱茵,新时代的克鲁伯)。见图 T-2。

图 T-2 坦克炮
Figure T-2 Tank gun

Tank jet cleaning machine 洗舱器 接在高压水软管上,吊入油舱内。依靠高压热舷外水推动旋转喷嘴,在各位置上冲洗油舱的一种器具。

Tank laser pressing device 坦克激光压制装置 该技术拥有国:中国(99 式)。

Tank level gauge 舱柜液位计

Tank level indicator system 液舱液位指示系统

Tank level sensor 货舱液位传感器

Tank pumping system 液货舱泵吸系统

Tank room 燃料柜舱室 系指包括所有燃料柜接头与所有燃料柜阀门在燃料柜周围的气密处所。Tank room means the gastight space surrounding the bunker tank, containing all tank connections and all tank valves.

Tank ship 液货船 系指有动力或风帆推进的液货船。

Tank steaming-out piping system 蒸汽熏舱系统 供给蒸汽以熏蒸附在舱壁及结构上的油垢,使其软化剥落的系统。一般往往借用油舱内装设的蒸汽灭火系统进行熏舱。

Tank test 水池试验 系指模型拖曳试验、模型自航试验和螺旋桨模型敞水试验。如果数字试验在造船厂和船东同意的具有文件证明的条件下进行,数字试验可等同于模型试验予以接受。Tank test means model towing tests, model self-propulsion tests and model propeller open water tests. Numerical tests may be accepted as equivalent to model tests if they are preformed under documented conditions agreed by the ship-owner and ship-builder.

Tank top 舱顶 系指构成货油舱底的水平板材。 Tank top means a horizontal plating forming the bottom of a cargo tank.

Tank ventilation system 液舱通风系统

Tank vessel 液货船 系指建造或改建的专门在船舱中装载散装液货的船舶。

Tankage 箱装肥料（或饲料） 屠宰场加工车间的动物有机物质的下脚，已晒干。粉尘极多，易于自热并可能易于点燃。可能传染病菌。 Tankage is the dried sweeping of animal matter from slaughterhouse floors. It is very dusty. It is subject to spontaneous heating and possible ignition. It is possibly infectious.

Tank-cleaning vessel 洗舱船 采用混有蒸汽的海水或用温海水冲洗等方法，专为油船清洗货油舱用的船舶。

Tanker 液货船 （1）是建造或改建用来散装运载易燃液货的船舶。如：化学品船、矿石、散货、油兼装船、矿石、油、易燃液体物质液货船、液化气体运输船、浮油回收船、油船等；（2）其构造主要适用于载运散装液体货物的货船；（3）运输散装易燃液体的任何船舶。液货船应符合《1974/1978 国际海上人命安全公约》（SOLAS74/78）的标准，并必须在属于国际船级协会（IACS）会员之一的，经认可的船级社根据运输的易燃液体（石油）入级，而且仍然受该船级社监督；（4）MARPOL 公约附则Ⅰ第 1 条中定义的油船或附则Ⅱ第 1 条中定义的化学品船；（5）大部分货舱用于载运散装液体货物的船舶及临时载运油类的船舶；（6）其构造主要适用于载运散装液体货物的货船；（7）专门建造用来散装运输液化石油产品、液化化学品、液化食用油和液舱（该液舱既是液体的一部分又是船上所有货物的运载体）、装载液化可燃气船舶。 Tanker is: (1) a cargo ship constructed or adapted for the carriage in bulk of liquid cargoes of an inflammable nature such as: chemical tanker, combination carrier/OBO, combination carrier/OOC, flammable liquid substance tanker, liquefied gas carrier, oil recovery ship, oil tanker, etc. ; (2) a cargo ship which is constructed primarily to carry liquid cargoes in bulk; (3) any vessel that transport bulk inflammable liquids. She shall comply with standards of SOLAS74/78 and must be classified in one of the recognized classification societies belonging to the IACS to carry inflammable liquids (petroleum) and still under its supervision;

(4) an oil tanker as defined in Regulation 1 of Annex I or a chemical tanker as defined in Regulation 1 of Annex II of the present Convention; (5) a ship in which the greater part of the cargo space is constructed or adapted for the carriage of liquid cargoes in bulk and which is not, for the time being, carrying a cargo other than oil in that part of its cargo space; (6) a cargo ship which is constructed primarily to carry liquid cargoes in bulk; (7) any vessel specifically constructed for carrying bulk cargoes of liquid petroleum products, liquid chemicals, liquid edible oils and liquefied gases in tanks which form both an integral part and the total cargo carrying portion of that vessel.

Tanker airplane 空中加油机 主要是使用在军事用途上，其利用机身内部装载的燃料，在空中为其他具有接收设备的飞机补充燃料，使得许多军用飞机的航程不再受到起飞时其最大燃料装载量的限制。该技术拥有国：美国（KC-767）、英国（L-1011）、俄罗斯（Ⅱ-78）、中国（HY-6）、法国（KG-135FR）。见图 T-3。

图 T-3 空中加油机
Figure T-3 Tanker airplane

Tanker piping system 油船系统 油船上为完成货油装卸，保证油船安全和保持油舱清洁的系统。包括：货油装卸、货油舱透气、清舱、洗舱油气驱除、蒸汽薰舱、货油加热、烟气防爆、甲板洒水系统等。

Tanks for oil residues (sludge) 残油（油泥）舱 系指接收 MARPOL 附则Ⅰ要求不能以其他方式处理的残油（油泥）诸如由于净化燃油、各种润滑油和机器处所中的漏油所产生的残油油舱。 Tanks for oil residues (sludge) means a tank receives the oil residues (sludge) which cannot be dealt with otherwise in accordance with the requirements of the Annex I, MARPOL, such as those resulting from the purification of fuel and lubricating oils leakages in the machinery spaces.

Tanks with smooth walls 有平滑壁的舱 系指具有平坦舱壁的舱，并应包括"油类/散货/矿砂船"的主舱壁。这类舱可建有小深度的垂直框架。垂直波形舱壁可视作平滑壁。 Tanks with smooth walls mean those tanks with smooth bulkheads, and including the main cargo tanks of oil/bulk/ore carriers which may be constructed with vertical framing of a small depth. Vertically corrugated bulkheads are considered smooth walls.

Tanks, voids and auxiliary machinery spaces having little or no fire risk 极少或无失火危险的液舱、空舱及辅机处所 系指：(1) 构成船体结构部分的水舱；(2) 空舱及隔离舱；(3) 不设置具有压力润滑系统的机器的辅机处所，且在该处所内禁止储存可燃物品。例如：①通风机和空调机室；②锚机室；③舵机室；④减摇设备室；⑤电力推进电动机室；⑥设有分区配电板和除浸油式电力变压器 (10 kVA 以上) 以外的纯电器设备舱室；⑦轴隧和管隧；⑧泵及制冷处所 (不输送或使用易燃液体)；(4) 为上述处所服务的围蔽围阱；(5) 其他围蔽围阱。如管道和电缆围阱。 Tanks, voids and auxiliary machinery spaces having little or no fire risk are: (1) water tanks forming part of the ship's structure; (2) voids and cofferdams; (3) auxiliary machinery space which do not contain machinery having a pressure lubrication system and where storage of combustibles is prohibited, such as: ①ventilation and air-conditioning rooms; ②windlass room; ③steering gear room; ④stabilizer equipment room; ⑤electrical propulsion motor room; ⑥rooms containing section switchboards and purely electrical equipment other than oil-filled electrical transformers (above 10 kVA); ⑦shaft alleys and pipe tunnels; ⑧spaces for pumps and refrigeration machinery (not handling or using flammable liquids); (4) closed trunks serving the spaces listed above; (5) other closed trunks such as pipe and cable trunks.

Tap bolt 带头螺栓

Tap gas-thread 螺丝攻

Tape recorders 磁带录音机 系指发明于 1929 年或 1930 年的记录和重放声音的装置。

Tapered liner 楔形垫片

Tapering 锥形

Tapering spindle 锥形轴

Tapioca 木薯淀粉 粉末和颗粒的干燥、多粉尘混合物。可能自热并使货物处所缺氧。该货物为不燃物或失火风险低。 Tapioca is a dry, dusty mixture of powder and granules. It may heat spontaneously with oxygen depletion in the cargo space. This cargo is non-combustible or has a low fire.

Tanker airplane risk 空中加油机

Tapper top 螺母丝锥 (机用丝锥)

Taps 夹层

Tare 空箱质量 系指包括固定附属装置在内的空集装箱的质量。 Tare means the mass of the empty container, including permanently affixed ancillary equipment.

Tare weight 皮重 系指集装箱空载的质量，包括装置的永久性设备。 Tare weight means the weight of the empty container including permanently affixed ancillary equipment.

Target language 目标语言 是一种利用特定语言输入目标语言的方法。该方法在电子装置中设置输入模块、执行模块、数据库模块和显示模块；输入模块与执行模块连接，执行模块与数据库模块和显示模块相连；执行特定语言输入目标语言是先设定目标语言，然后触发查询模式，在使用者以不同于目标语言的特定语言输入欲查询字词后，会查询出所有查询字词所对应目标语言的目标字词组，并将查询所得的目标字词组显示出来，最后由使用者由显示的目标字词组中的选定一个目标字词进行输入。

Target price 指标价格 是欧盟市场内部以生产效率最低而价格最高的内地中心市场的价格为准则而制订的价格。这种价格一般比世界市场的价格高。为了维护这种价格水平，还确定了干预价格，一旦中心市场的实际市场价格跌到干预价格水平，有关机构便使从市场上购进货物，以防止价格继续下跌。

Target species 目标物种 系指缔约国确定的满足特定标准而表明可能损害或破坏环境、人体健康、财产或资源的物种，其定义是针对特定港口、国家或生物地理区域。 Target species are those species identified by a Party that meet specific criteria indicating that they may impair or damage the environment, human health, property or resources and are defined for a specific port, State or biogeographic region.

Target swap 目标交换 系指被跟踪目标的雷达数据与另一个被跟踪目标或非跟踪雷达回波错误地联系在一起的情况。 Target swap means a situation in which the incoming radar data for a tracked target becomes incorrectly associated with another tracked target or a non-tracked radar echo.

Target tracking (TT) 目标跟踪 (TT) 系指为建立目标运动而观察雷达目标位置变化的计算机程序。该目标为被跟踪目标。 Target tracking (TT) means a computer process of observing the sequential changes in the position of a radar target in order to establish its motion. Such a target is a tracked target.

Target-tracking radar (TTR) 目标跟踪雷达

Target useful life 目标使用寿命 为防腐蚀保护方法或使用的耐腐蚀材料的设计耐用期的目标值，以年计。 Target useful life is the target value, in year, of the durability for which the means of corrosion protection or utilization of corrosion resistance material is designed.

Target's predicted motion 目标的预计运动 系指基于雷达上目标距离和方位的由先前测量确定的现在运动以线性外推法对目标的未来航向和速度的预测。

Target's predicted motion means a prediction of a target's future course and speed based on linear extrapolation from its present motion as determined by past measurement of its range and bearing on the radar.

Tariff 关税 是海关依法对进出国境或关境的货物、物品征收的一种税。是主权国度或单独关税区海关对收支国境或关境的货品或物品征收的流转税,它有权对不切合国度法令划定的收支口物品不予放行。

Tariff barrier 关税壁垒 系指用征收高额进口税和各种进口附加税的办法,以限制和阻止外国商品进口的一种手段。贸易壁垒可以提高进口商品的成本从而削弱其竞争能力,起到保护国内生产和国内市场的作用。它还是在贸易谈判中迫使对方妥协让步的重要手段。世界贸易组织对其极力反对,并通过谈判将其大幅削减。

Tariff protection principle 关税保护原则 系指仅允许"以关税作为保护手段",原则上不允许其他一切非关税措施,现在实行的数量限制等非关税措施都要逐步转化为关税保护。然后,成员国之间在互惠互利的基础上进行关税减让谈判,逐步降低关税。

Tariff quota 关税配额 系指在某些商品进口数量或金额达到规定的额度后,继续进口便需提高关税的配额管理。关税配额不绝对限制商品的进口总量,而是在一定时期内对一定数量的进口商品,给予低税、减税或免税的待遇,对超过此配额的进口商品,则征收较高的关税或附加税和罚款。自动出口配额系指出口国家或地区在进口国家的要求或压力下,"自动"规定某一时期内(一般为3年)某些商品对该国出口的限制额。

Tariff reduction and exemption principle 关税减免原则 是减征关税和免征关税的简称,是海关全部或部分免除应税货品纳税人的关税给付义务的一种行政措施。

Tarpaulin 防水盖布 覆盖在拼装舱盖的舱盖板上,包蔽舱口以保持密性用的防水布。

Task analysis 任务分析 系指对比夏天需要和操作人员能力的一系列方法,通常为了改善人的行为,比如采用减少失误的方法。

Task station 任务站 系指具有专用控制装置,能提供显示和操作任何航行任务的多功能显示器。任务站是工作站的一部分。Task station means a multifunction display with dedicated controls providing the possibility to display and operate any navigational tasks. A task station is part of a workstation.

Taut 张紧式
Taut mooring 张紧式系泊
Taut wire 张紧索 用一根张紧的钢缆垂直于某个海床上的重物或水平于附近的某个外部固定物体上的位置基准。

Taut-wire measuring gear 拉线测速装置 装有转速计和自动同步机,能在控制室显示布缆的速度和里程以便控制布缆作业的装置。

Tax of price differences 差价税 又称差额税。是按国内市场和国际市场的价格差额对进口商品征收的关税。当某种本国生产的产品国内价格高于同类的进口商品时,为了削弱进口商品的竞争能力,保护国内生产和国内市场,按国内价格与进口价格之间的差额征收的关税。由于差价税是随着国内外价格差额的变动而变动的,因此它是一种滑动关税(siding duty)。对于征收差价税的商品,有的规定按价格差额征收,有的规定在征收一般关税以外另行征收,这种差价税实际上属于进口附加税。例如,欧盟对冻牛肉进口,首先征收20%一般进口税,然后根据每周进口价格与欧盟的内部价格变动情况征收变动不定的差价税。

Tayloe's diameter constant 直径系数 一种表示螺旋桨转速、直径与进速间关系的系数。

Tayloe's propeller coefficient based on delivered power 收到功率系数 一种表示螺旋桨转速、收到功率与进速之间关系的系数。$B_p = nP_D^{0.5}/V_A^{2.5}$,式中,$n$—螺旋桨转速,以 r/mim 计;$P_D$—收到功率;$V_A$—进速,以 kn 计。

Tayloe's propeller coefficient based on thrust power (Bu coefficient) 推力功率系数 一种表示螺旋桨转速、推力功率与进速之间关系的系数。$B_D = nP_T^{0.5}/V_A^{2.5}$,式中,$n$—螺旋桨转速,以 r/mim 计;$P_T$—推力功率;$V_A$—进速,以 kn 计。

Taylor quotient Γ 泰勒系数 Γ 泰勒系数被定义为 $\Gamma = \dfrac{V_m}{\sqrt{L_{WL}}}$,式中,$V_m$—适当的航速(m/s);$L_{WL}$—水线长(m)。Taylor quotient is defined as $\Gamma = \dfrac{V_m}{\sqrt{L_{WL}}}$; where: V_m is the appropriate ship speed in m/s. L_{WL} is waterline length in m.

Taylor wake factor 伴流因数 螺旋桨进速对船速的比值。$1 - \omega = V_A/V$,式中,V—船速;V_A—螺旋桨转速。

**Taylor wake fraction determined from thrust iden-

tity　等推力伴流分数　螺旋桨分别在敞水中和船模后进行试验时，在推力系数相同条件下得出的伴流分数。

Taylor wake fraction determined from torque identity　等转矩伴流分数　螺旋桨分别在敞水中和船模后进行试验时，在转矩系数相同条件下得出的伴流分数。

Taylor wake friction　伴流分数　船速与螺旋桨进速之差对船速的比值。$\omega = (V - V_A)/V$，式中，V—船速；V_A—螺旋桨进速。

TBCC　涡轮基组合动力

t_c（**mm**）　t_c**腐蚀增量（mm）**　t_c is a corrosion addition, in mm.

TCPA = the time to closest point of approach　预期相遇最近点时间

Tean factor verification　精益要素验证

Tean ideal　精益概念

Tean management　精益管理

Tean product design　精益产品设计　系指在最短的时间内按照规范、规则的要求，将产品目标说明中所描述的要素用工程语言具体直观表达出来，设计出刚好满足用户要求，成本最低的产品的设计。

Tean product development planning　精益产品开发规划　利用精益理论和机会漏斗对各种产品开发机会进行认真筛选，最终形成既能真正满足顾客要求，又能符合企业发展方向的产品开发项目和市场推介时机。

Tea-set rack　茶具架　固定在舱壁上，供安放水瓶和茶杯用的架子。

Technical barriers to trade（TBT）　技术性贸易壁垒　又称"技术性贸易措施"或"技术壁垒"，是以国家或地区的技术法规、协议、标准和认证体系（合格评定程序）等形式出现，涉及的内容广泛，涵盖科学技术、卫生、检疫、安全、环保、产品质量和认证等诸多技术性指标体系，运用于国际贸易当中，呈现出灵活多变、名目繁多的规定。

Technical code on control of emmision of hydrogen oxide from marine diesel engines　船用发动机氮氧化物（NO_x）排放控制技术规则

Technical Committee（TC）　技术委员会

Technical Committee for Anti-corrosion Standardization（TCAS）　[美]国防腐蚀标准化技术委员会

Technical condition　技术状况

Technical consulting contract　技术咨询合同　系指当事人一方为另一方就特定技术项目提供可行性论证、技术预测、专题技术调查、分析评价报告所订立的合同。

Technical contract　技术合同　是当事人就技术开发、转让、咨询或者服务订立的确立相互之间权利和义务的合同。技术合同的标的与技术有密切联系，不同类型的技术合同有不同的技术内容。技术合同履行环节多，履行期限长。技术合同的法律调整具有多样性。当事人一方具有特定性，通常应当是具有一定专业知识或技能的技术人员。技术合同是双向、有偿合同。

Technical Cooperation Committee（TC）　技术合作委员会（技合会）

Technical data sheet　技术规格书（涂层）　为涂料生产厂商的产品规格书。包含与涂料及其涂装有关的详细技术性说明和资料。 Technical data sheet is manufacturers' product data sheet for paint which contains detailed technical instruction and information relevant to the coating and its application.

Technical defect　技术缺陷　系指船体结构或其机械、设备或附件的部分，或运行的缺陷或故障。

Technical delivery system　机械输送系统　系指采用链式输送机集料，螺旋输送机提升送料的输送系统。

Technical design（detail design）　技术设计（详细设计）　系指根据船东和船厂签订的建造合同的技术附件和合同设计进行更深入的设计工作，来确定各系统的技术形态。技术设计分为三个阶段：（1）完成送船级社涉及船舶安全方面的送审图纸；（2）完成船东要求审查的、主要涉及使用要求方面的图纸资料；（3）详细设计所要求的设计船厂如何施工的图纸资料。详细设计实际上是包括技术设计在内的一部分施工设计的内容。技术设计完成后，整个船舶的技术形态已完全确定，解决了造一艘什么样的船的问题。

Technical file　技术案卷　系指含有可能影响发动机 NO_x 排放的参数，包括发动机构件和设定值的所有详细资料的记录。 Technical file is a record containing all details of parameters, including components and settings of an engine, which may influence the NO_x emission of the engine.

Technical group　技术组　对 2009 年香港国际安全与环境无害化拆船公约而言，系指由来自缔约国、IMO 组织成员、联合国及其专门机构、与 IMO 组织有协议的政府间组织已经在 IMO 组织中具有咨询地位的非政府组织的代表组成的小组。该小组最好应包括机构和实验室的代表。他们应在物质对环境的归趋和影响、毒理影响、海洋生物学、人类健康、经济分析、风险管理、造船、国际航运、职业健康和安全方面具有专业知识或具有客观评审某提议的技术优点所必需的其他领域的专业知识。 For Hong Kong International convention for the safe and environmentally sound cycling of ships, 2009, technical group means a group which comprise representatives of the Parties, Members of IMO Organization, the United Nations and its Specialized Agencies, intergovernmental organizations having agreements with IMO Organization, and non-governmental organizations in consultative status with IMO Organization, which should preferably include representatives of institutions and laboratories with expertise in environmental fate and effects of substances, toxicological effects, marine biology, human health, economic analysis, risk management, shipbuilding, international shipping, occupational health and safety or other fields of expertise necessary to objectively review the technical merits of a proposal.

Technical life of hull　船体技术寿命　系指船体从开始投入使用直至因技术落后而被淘汰所经历的时间。

Technical service contract　技术服务合同　系指当事人一方以技术知识为另一方解决特定技术问题所订立的合同。技术服务合同中包括技术培训合同和技术中介合同。技术培训合同系指当事人一方委托另一方对指定的专业技术人员进行特定项目的技术指导和专业训练所订立的合同。技术服务合同可由委托人直接与被委托人协商约定后签订,也可以由委托人通过中介与被委托人协商签订,不管采用什么方式,必须符合技术服务合同的特征,并不得遗漏其主要条款。

Technical management department (TMD)　技术管理部

Technical/pneumatic delivery system　机械/气力混合输送系统　系指:(1)一种采用气流床集料,螺旋输送机提升送料的输送系统。这种系统适合于岸上有相应的接料系统;(2)一种采用链式输送机集料,螺旋输送机提升,气力送料的,可直接通过管道向料舱送料,而不需要在岸上设置与船上螺旋输送机出口相连接的输送系统。

Technician　技术人员　系指熟悉生产技术、产品性能和技术发展动向的技术员、工程师或总工程师,在谈判中可负责对产品性能、技术质量标准、产品验收、技术服务等问题的谈判,也可与商务人员密切配合,为价格决策作技术参谋。

Technique A　A 工艺　如同 M 工艺,只是使用具有大热量输入率(相对于被焊接的厚度有较大的焊道尺寸)的程序。该程序可用于厚度为 20 mm,用 4 道或 4 道以下焊成的焊缝;或厚度为 35 mm,用 8 道或 8 道以下焊成的焊缝。Technique A— as for M but using a procedure with a high heat input rate (large bead size relative to thickness welded). This would apply to welds made by four or less runs in 20 mm thickness, or eight or less runs in 35 mm.

Technique for human error rate prediction　人为失误率预测技术　系指涵盖了任务分析、人为失误识别、人为失误模型和人为失误量化的综合性方法,主要用在人的可靠性分析(HRA)中,评估与某些因素有关的人为误差引起的系统变化,并采用减少系统内发生失误可能的措施,以提高整个系统的安全水平。

Technique M　M 工艺　系指使用自动多道焊工艺的焊丝-焊剂/垫背组合或焊丝-气体/垫背组合(可能是附加充填材料)的工艺。 Technique M means the technique for wire-flux or wire-gas in combination with backing material (and maybe supplementary filler materials) used with an automatic multi-run technique.

Technique S　S 工艺　系指使用半自动多道焊工艺的焊丝-气体/垫背组合的工艺。Technique S means the technique for wire-gas/backing combinations used with semi-automatic multi-run technique.

Technique T　T 工艺　系指使用自动双面单道焊工艺的焊丝-焊剂或焊丝-气体组合的工艺。 Technique T means the technique for wire-flux or wire-gas combinations used with an automatic two-run technique.

Techno super-liner (TSL)　高技术超级班轮　为迎接 21 世纪高速海上运输需要的研究项目,其目标是提供航速 50 节、装载 1 000 t、航程 500 海里的高耐波性海上定期运输班轮。采用大型高速水翼小水线面船、带水翼的双体水垫船。日本注重水翼双体船方案,俄罗斯和美国则采用大型地效翼船和洲际喷气客机方案。

Technological achievements　技术成果　系指利用科学技术知识、信息和经验做出的产品、工艺、材料及其改进等技术方案。

Technological achievements file　技术成果文件　系指专利申请书、科学技术奖励申报书、科技成果登记书的确认技术成果完成者身份和授予荣誉的证书和文件。

Technology development contract　技术开发合同　系指当事人之间就新技术、新工艺和新工艺的新材料及其系统的研究开发所订立的合同。其包括委托开发合同和合作开发合同。其客体是尚不存在的有待开发的技术成果,其风险由当事人共同承担。

Technology entertainment design (TED)　科技、娱乐、设计

Technology transfer　技术转让　又称技术转移。技术转移系指技术在国家、地区、行业内部或之间以及技术自身系统内输入与输出的活动过程。技术转移包括技术成果、信息、能力的转让、移植、产业化、引进、交流和推广普及等。技术转让和引进主要形式有以下几种:采用成套设备引进和转让、合作生产、补偿贸易及合资经营四种形式。技术转让是技术贸易的一种主要类型,技术市场上的技术转让系指技术成果由一方转让给

另一方的经营方式。技术转让系指拥有技术的当事人一方将现有技术有偿转让给他人的行为。

Technology transfer contract 技术转让合同 系指当事人双方就现有特定技术权益的转让所订立的明确相互权利义务关系的协议。具体地说,系指以专利申请权转让、专利权转让、技术秘密转让和专利实施许可为目的,明确相互权利义务关系的协议。

Tee T 形材(三通管,T 形接头)
Tee fitting T 形三通管接头
Tee saddle joint 三通管接头
Teeth 齿
Teeth spacing 齿间距
Telecommunication (德国)邮政与电信管理局 总部设在汉堡的、德国邮政部下属的、负责对无线电设备进行认可和检查的机构。GPS、无线电测向定位设备须由 BSH 和 BAPT 认可。无线电系统图、天线布置图和无线电站设备的安装布置图等应送 BAPT 认可。

Tele-fax 传真
Telegraph gong and indicator for controllable pitch propeller 调距螺旋桨传令钟和指示器 系指在设有调距螺旋桨的船上,控制调距螺旋桨的角度以实现对船的航速、前进和后退的控制的设备。调距螺旋桨传令钟用于驾驶室与机舱之间传递控制命令;调距螺旋桨的舵角指示器用于驾驶室与机舱之间传递执行的舵角循环。

Telegraphic money 电汇
Telegraphic transfer(T/T) 电汇 是汇出行应汇款人的申请,拍发押电报或电传给在另一个国家的分行或代理行(即汇入行)指示解付一定金额给收款人的一种汇款方式。

Telemetering STD profiling system 遥测温盐深剖面仪 用电缆把水下探头测得的温盐深信号传送到水上的装置,能实时地显示和记录海水温度、盐度随深度变化的仪器。

Telemetry 遥测技术 在 FPSO/FSU 和穿梭油船之间的一种 UHF 通信系统。沿着货油传输过程由一系列连锁的制动装置构成,建立一条绿色线路以允许进行货油传输,绿色线路中断会导致货油传输自动关闭。

Telescope 潜望镜 系指从海面下伸出海面或从低洼坑道伸出地面,用以窥探海面或地面上活动的装置。其构造与普通地上望远镜相同,唯另加两个反射镜,使物光经两次反射而折向眼中。潜望镜常用于潜水艇、坑道和坦克内用以观察敌情。处于水下航行状态的潜艇观察海平面和空中情况的唯一手段便是借助潜望镜。而多数潜艇均安装有两部潜望镜,其中一部攻击潜望镜和一部观察潜望镜。见图 T-4。

Telescopic joint 伸缩节 也称管道伸缩节、膨胀

图 T-4 潜望镜
Figure T-4 Telescope

节、补偿器,伸缩器。伸缩节分为波纹伸缩节、套筒伸缩节、方形自然补偿伸缩节等几大类型,其中以波纹伸缩节较为常用;套管伸缩节由能够作轴向相对运动的内外套管组成。伸缩节主要用于补偿管道因温度变化而产生的伸缩变形,也用于管道因安装调整等需要的长度补偿。

Telescopic pipe(telescopic tube) 伸缩管(套管)
Telescople mast 伸缩桅 为可降低其高度而做成上段可缩下的桅杆。
Telethermometer 遥测温度计
Television(TV) 电视机
Television photograph cable winch 电视摄影电缆绞车 为收放摄影或电视装置及电缆进行海洋底质和海洋生物调查用的绞车。
Telex 电传
Temperature 温度
Temperature alarm 过热报警器
Temperature and time combination 时间-温度关系
Temperature coefficient 温度系数
Temperature composating maximum relay 温度补偿过载继电器 系指在电流超过某一预定值的任意电流值时动作,且其动作基本上与环境温度无关的器件。
Temperature conductivity indicator 温度-电导率测量仪 用于现场测量温度和电导率,并可确定盐度的仪器。
Temperature control 温度控制
Temperature detector 检温计
Temperature distribution 温度分布
Temperature entrope diagram 温熵图(T-S 图)
Temperature field 温度场
Temperature gauge 温度计
Temperature gradient 温度梯度,温度差

Temperature humidity and wind gradiometer 海面温度、湿度、风梯度仪 探测海面以上几米内的空气温度、湿度和风速等气象要素随高度变化的仪器。

Temperature monitoring system 温度监控系统

Temperature regulating valve 温度调节阀

Temperature rise 温升

Temperature sensing device 温度传感器（温度敏感器，温度变送器）

Temperature sensing device for cargo pump 液货输送泵温度探测装置 安装在泵壳、传动轴等处的温度检测装置。

Temperature sensitivity 温度灵敏度

Temperature sensor 温度传感器

Temperature strain 温差应变

Temperature stress 温差应力

Temperature switch 温度开关（调温开关）

Temperature transducer（temperature transmitter） 温度传感器 一种将所测的温度转换为电量或机械量的敏感元件。主要有热敏电阻、铂电阻、铜电阻、石英晶体、复合式温度传感器等类型。

Temperature traverse quality（TTQ） 温度分布品质 判明燃烧室出口燃气温度分布均匀性并反映掺混区中热燃气与冷空气混合效果的指标。$\tau = (T_{max} - T_3)/(T_3 - T_2)$；式中，$T_{max}$——最高出口温度；$T_3$——平均出口温度；$T_2$——平均进口温度。

Temperature warping stress 温度弯翘应力

Temperature-controlled 温度控制的

Temperature-resistant 耐温的（耐热的）

Template 模板 置于海底，在桩腿打入海底之前为了提供精确的着落位置的板。

Template platform 基盘式平台 系指导管架和预先安装并固定于海底的水下基盘连接成一体的平台。

Temporary anchoring equipment 临时锚泊设备 包括锚、锚链、锚机等，只供迁移、移船和在港口以及遮蔽水域中锚泊使用。整套设备是按平台受到中等程度的环境载荷作用时仍可稳住平台的原则要求而设置的。 Temporary anchoring equipment including anchors, cables, winches, etc. intended to be used while not in a working condition but during transit and location move and for anchoring in harbors or in sheltered areas. The complete set of equipment is to be designed to hold a unit in position when exposed to moderate environmental loads.

Temporary external supply 临时外来电源供电 系指平台上的设备由岸电或其他外来电源供电。Temporary external supply means supply of electricity is from a source on shore or elsewhere for arrangements of the unit.

Temporary hardness 暂时硬度 水中由加热时将分解成易于排除的难溶性沉淀析出物的重碳酸钙、镁盐所构成的硬度。

Tencent Holdings Limited 腾讯控股有限公司 简称"腾讯"，是一家民营IT企业，成立于1998年11月，由马化腾、张志东、许晨晔、陈一丹、曾李青创办，总部在中国广东深圳，是中国最大的互联网综合服务提供商之一。"腾讯"公司主要产品有IM软件、网络游戏、门户网站以及相关增值产品。

Tencent instant chat tools（QQ） "腾讯"即时聊天工具 是腾讯公司开发的一款基于Internet的即时通信（IM）软件。腾讯QQ支持在线聊天、视频通话、点对点断点续传文件、共享文件、网络硬盘、自定义面板、QQ邮箱等多种功能，并可与多种通信终端相连。其标志是一只戴着红色围巾的小企鹅。QQ覆盖Microsoft Windows、OS X、Android、iOS、Windows phone等多种主流平台。

Tender bidding 投标 系指投标人应招标人的约请，依照投标的要求和条件，在规定的时间内向招标人递价，争取中标的行为。

Tender bond standby LC 投标备用信用证 系指开证行应开证申请人的请求对受益人开立的承诺某项义务的凭证。根据《ISP98》规定，备用信用证有以下几种：预付款备用信用证、履约备用信用证、投标备用信用证和进出口备用信用证。除以上几种之外，还可以根据需要由开证行开出融资备用信用证、借款备用信用证和保险备用信用证等。投标备用信用证是开证申请人（投标人）要求开证行为其保证，保证投标人在开标前不能中途撤标或片面修改标书内容，并保证投标人中标后签约和履约的信用证。

Tender offer 投标报价 系指承包商采取投标方式承揽工程项目时，计算和确定承包该工程的投标总价格。投标报价的第二步是最重要的中心环节，即是根据现有定额做概（预）算，据编好的工程量清单上价，这里一定要讲究对号入座、量体裁衣。投标报价技巧的作用体现在可以使实力较强的投标单位取得满意的投标成果；使实力一般的投标单位争得投标报价的主动地位；当报价出现某些失误时，可以得到某些弥补。投标报价是一个相当复杂的过程，过去那种单一严谨的计价方式受到了市场经济的强烈冲击，首当其冲的是取费项目和取费标准。

Tensile compression rigidity（TCR） 拉压刚度

Tensile strength 抗拉强度

Tensile strength（Ultimate stress） 拉伸强度（极限应力） 系指在单轴向拉伸试验时测得的最大应力。对于韧性材料，其取决于缩颈现象的出现。 Tensile strength（Ultimate stress）is the maximum engineering stress recorded in a un-axial tension test. For a ductile material, it corresponds to the onset of necking.

Tensile test　拉伸试验

Tensile test on round tensile specimens　圆棒型拉伸试样的焊缝拉伸试验　系指确定焊缝金属的抗拉强度、屈服强度或 0.2% 规定非比例伸长应力、断面收缩率和伸长率。对于高温钢，如有必要，还应测定高温下的 0.2% 规定非比例伸长应力。　Tensile test on round tensile specimens means determination of the tensile strength, yield strength or 0.2% proof stress, reduction in area and elongation of the weld metal. Where necessary in the case of high-temperature steels, the 0.2% proof stress at elevated temperatures shall also be established.

Tensile test on seam　焊缝拉伸试验　对焊缝而言，系指确定焊缝的抗拉强度、断裂的位置及形式，以及如适用时，系指垂直于焊缝的试样伸长率的试验。　For the seam, tensile test on seam means determination of the tensile strength of a seam, position, type of fracture and, where appropriate, the elongation of specimens located at right angle to the seam.

Tensile tests on cruciform-tensile specimens　十字型拉伸试样的焊缝拉伸试验　系指确定焊缝金属的拉伸剪切强度的试验。　Tensile tests on cruciform-tensile specimens mean those tests determine the tensile shear strength of the weld metal.

Tension leg platform（TLP）　张力腿平台　系指通过垂直的系缆连接到海底的垂直系泊的浮动式平台。该平台通过张紧缆绳或张力腿将浮式半潜式平台结构系于海底，适用于 150～2 000 m 水深海域。其结构由平台主体、浮体、张力腿、上部设施模块、顶张力口立管、悬链式立管和锚桩基础所构成。整个平台通过由钢管组成的张力腿与固定于海底的锚桩相连，船体浮力使得张力腿始终处于张紧状态，从而使平台保持垂直方向稳定。

Tensioner　张力器/张紧器　系指敷管船上对悬浮管路施加张紧力的装置。　Tensioner is a device on a lay-vessel for applying tension to the suspended pipeline.

Tentative price　暂定价格　系指在合同中先订立一个初步价格，作为开立信用证初步付款的依据，待双方确定具体的价格后再进行最后清算，多退少补。

Tentative rules　暂行规范　是适用于新领域的规范。为了得到对规范所反映的意见，在一定的时间内船级社保留对其做出调整的权利。　Tentative rules are Rules applying to new fields to which the Society reserves the right to make adjustments during a period in order to obtain the intention reflected in the Rules.

TeraHertz wave　太赫兹波　系指频率在 0.1 THz 到 10 THz 范围的电磁波，波长大概在 0.03 mm 到 3 mm 范围，介于微波与红外之间，能够穿透衣服，人体自身会产生并向外辐射的电磁波。THz 波（太赫兹波）或成为 THz 射线（太赫兹射线）是从 20 世纪 80 年代中后期，才被正式命名的，在此以前科学家们将统称为远红外射线。

Terminal　码头　系指存放收到的来自水上运输的有害有毒物质的任何场地，包括由管道或其他设备与此种场地相连的任何离岸设施。　Terminal means any site for the storage of hazardous and noxious substances received from waterborne transportation, including any facility situated off-shore and linked by pipeline or otherwise to such site.

Terminal　装卸站　系指：(1) 位于美国或由美国管辖的可航行水域的用于或拟用于作为有害物质传输或其他装卸的港口或设施或近岸结构物；(2) 用于或拟用于作为有害物质传输或其他装卸的港口或设施的岸上设施或近岸结构物。就 MARPOL73/78 附则 V 而言，商业性的渔业码头设施、游艇码头设施和采矿及采油工业的岸上基地可作为装卸站，因为通常这些码头设施为船舶提供停靠和其他服务，包括垃圾的处理。Terminal means: (1) an onshore facility or an offshore structure located in the navigable waters of the United States or subject to the jurisdiction of the United States and used, or intended to be used, as a port or facility for the transfer or other handling of a harmful substance; (2) an onshore facility or an offshore structure, or intended to be used, as a port or facility for the transfer or other handling of a harmful substance, for Annex V of MARPOL 73/78, the commercial fishing facilities, recreational boating facilities, and mineral and oil industry shorebases to be terminals, since these facilities normally provide wharfage and other services, including garbage handling, for ships.

Terminal high-altitude area defense(THAAD)　末段高空区域防御系统　即"萨德"反导系统。该系统是美国全球导弹防御系统的一个子系统。由美国航空航天制造商洛克希德·马丁公司承担主要的研发和生产，是一种可车载机动部署的反导系统。也是战区导弹防御（TMD）体系的地基高层防御部分，它比低层系统拦截的高度高，距离远，既能在大气层内又能在大气层外撞击杀伤拦截。具备在大气层内外拦截来袭的短程、中程和远程洲际弹道导弹的独特能力。一套"萨德"反导系统通常由指挥中心、上部地面 X 波段雷达、6 部 8 联装反射装置和 48 枚拦截弹组成，其拦截高度介于大气层内 40 km 以上至大气层外 150 km 之内，射程可达 200 km，可以击中中超音速 8 倍以上速度发射弹道导弹。而现有的导弹防御系统通常都是在距离地面 10～15 km 左右的末段低空进行拦截。"萨德"反导系统的最大亮点在于它的 X 波段雷达。"萨德"反导系统的核心装备 AN/TPY-2 雷达探测距离最远可达 2 000 km,其分辨率非常高,可以完成探测、搜索、追踪、目标识别等多功能任务。THAAD 雷达成功跟踪了拦截弹和靶弹,标志着在接下来的飞行试验中作为主控雷达迈出了重要的一步。THAAD 计划由美国国防部弹道导弹防御局(BMDO)管理,陆军防空与导弹防御计划执行办公室和陆军 THAAD 项目办公室执行。THAAD 拦截弹发射 2 min 后,动能杀伤器在稠密大气层与目标偏差几百米,丢失了目标。其设计要求是为美国陆军提供区域弹道导弹防御能力,拦截中近程弹道导弹,并对洲际导弹具有有限的拦截能力,由于威力超过著名的"爱国者",因此被外界称为"反导之王"。该系统最突出的特色是所有装备均能塞进 C-130 战术运输机的货舱里,从而方便全球机动部署。理论上,其具备在 200 km 外打掉高速弹道导弹的能力。

Terminal installation 油站 系指任何可以接收来自水路的油类、贮存散装油类的基地,包括位于近岸并与该基地连接的设备在内。 Terminal installation means any site for the storage of oil in bulk which is capable of receiving oil from waterborne transportation, including any facility situated off shore and linked to such site.

Terminal reheat air conditioning system 末端再加热空气调节系统 部分舱外新鲜空气和部分舱内循环空气两者混合经集中式空气调节器处理到较低温度后,由通风机输送到通风管路,并经过各舱室出风口处的加热器进行加热,根据舱内温度调节器的控制,自动或手动调节所需温度再送入舱内的空气调节系统。

Terminal representative 码头代表 系指船舶装卸货物的码头或其他设施使用方指定的人员。其负责码头或设施对特定船舶进行作业。 Terminal representative means a person appointed by the terminal or other facility, where the ship is loading or unloading, who has responsibility for operations conducted by that terminal or facility with regard to the particular ship.

Terminal temperature difference 传热端差 冷凝器喉部进口蒸汽温度与冷却水出口温度之差。

Terminal vapour connection 终端蒸汽接头 系指终端的蒸汽收集系统与蒸汽收集软管或装卸臂的连接点。 Terminal vapour connection is that point at which the terminal vapour collection system is connected to a vapour collection hose or arm.

Terms of payment 支付条件 系指订立合同时,合同的一方向另一方转移支付费用的条款。

Terms of reference 审理范围书

Terms of trade 贸易条件 系指一个国家出口商品价格与进口商品价格之比。应该注意的是:这里的贸易条件是所有出口商品价格与所有进口商品价格之间的比率。贸易条件是用来衡量一定时期内一个国家出口相对于进口的盈利能力和贸易利益的指标,反映该国的对外贸易状况,一般以贸易条件指数报收,在双边贸易中尤为重要。采用的贸易条件有 4 种不同形式:价格贸易条件、收入贸易条件、单要素贸易条件和双要素贸易条件。它们从不同的角度衡量一国的贸易所得。其中价格贸易条件最有意义,也最容易根据现有数据进行计算。这样,一方面需要计算进、出口商品的平均价格,同时还要考虑的基期的选择问题,因而,现实中的贸易条件计算一般采用出口商品价格指数与进口商品价格指数直接进行比较的做法。

Territorial seas[1] 领海 系指沿着与深海直接接壤的那部分海岸,有直接管辖权的法定水域轮廓线测得的海区,该轮廓线标志着内海与外海的界线。该线离海岸线距离为 3/12 海里。 Territorial seas means the belt of the seas measured from the line of ordinary low water along that portion of the coast which is in direct contact with the open sea and line marking the seaward limit of inland waters, and extending seaward a distance of 3/12 n miles.

Territorial seas[2] 领海(美国) 系指沿着直接接壤的那部分海岸,有直接管辖权的法定水域轮廓线测得的海区。该轮廓线标志着内海和外海的界线。该线离海岸线的距离为 3 海里。 Territorial seas means the belt of the seas measured from the line of ordinary low water along that portion of the coast which is in direct contact with the open sea and the line marking the seaward limit of inland waters, and extending seaward a distance of 3 n miles.

Tertiary holds 掺混孔 供空气进入火焰筒掺混区的通孔。

Test 试验

Test at constant speed of revolution 定进速试验 保持进速不变,改变转速的螺旋桨模型试验。

Test basin with $x-y$ plot carriage 坐标定位拖车试验池 在桥式主拖车上,装设可作横向移动的副拖车,使船模可同时调整纵向和横向分速度循指定轨迹拖航的一种船模试验水池。

Test batch 试验批 供试验用的一批制品。其由下列制品组成:(1)相同的金属元素制成并源自同一炉的;(2)形状相同和尺寸相同(厚度相同的板材和带材)的;(3)成型工艺相同的;(4)交货状态相同的;(5)热处理相同的。

Test cock 试验旋塞

Test condition 试验条件(试验工况)

Test data 试验数据

Test desk 试验台

Test document 试验文件

Test equipment(test facilities) 试验设备

Test expiry date 试验生效日期 系指根据国际海上人命安全公约指定试验程序可做试验用的和后续认可任何产品的最后的日期。 Test expiry date means the last date on which the given test procedure may be used to test and subsequently approving any product under the Convention.

Test expiry date 试验无效日期(耐火试验) 系指给定的试验程序可试验的最终日期,之后的任何产品的认可必须按 SOLAS 公约规定执行。 Test expiry date means the last date on which the given test procedure may be used to test and subsequently approving any product under the SOLAS Convention.

Test fixture 试验夹具(试验台)

Test fluid"A" 试验液体 A 是一种用于舱底水分离器形式认可试验的,符合 ISO8217 的残余船用燃油,型号 RGM35(密度在 15 ℃下不小于 980 kg/m³)。 Test fluid "A" means a marine residual fuel oil in accordance with ISO 8217, type RMG 35 (density at 15℃ not less than 980 kg/m³), which is used for type approval test related to 15 ppm bilge separators.

Test fluid"C" 试验液体 C 是一种用于舱底水分离器形式认可试验的油和淡水的乳化混合液,混合比例为 1 kg 该液体由以下成分组成:(1)947.8 g 淡水;(2)25.0 g 试验液体 A;(3)25.0 g 试验液体 B;(4)0.5 g 干型表面活性剂(十二烷基磺酸钠盐);(5)1.7 g 氧化铁,其粒度分布状况为 90% 小于 10 μm,其余的最大粒度为 100 μm。 Test fluid "C" means a mixture of an oil-in-fresh water emulsion, in the ratio whereby 1 kg of the mixture consists of: (1)947.8 g of fresh water; (2)25.0 g of test fluid "A"; (3)25.0 g of test fluid "B"; (4)0.5 g surfactant (sodium salt of dodecylbenzene sulfonic acid) in the dry form; (5)1.7 g "iron oxides" with a particle size distribution of which 90% is less than 10 microns, the remainder having a maximum particle size of 100 microns), which is used for type approval test related to bilge separtors.

Test for combustion control 锅炉调整试验 为得到锅炉在承受各种负荷时的最佳燃烧条件而对燃油集及烟、风系统所进行的配合调整试验。

Test procedure guide on liquid tank and lightness test 液舱及密性试验程序导则 是对 SOLAS 公约第 Ⅱ-1/11 条(每艘船舶所有双层底处所、内壳、艏尖舱、液舱均要进行水压试验)的简化,如果选择代表性舱室进行水压试验,姊妹船的后续船可减免结构试验或一定情况下可以用气压或液压-气动试验替代水压试验等。

Test report 试验报告 由制造厂商签署的文件。其阐明:(1)该产品符合规范要求;(2)在现有产品上进行抽样试验。 Test report means a document signed by the manufacturer stating: (1) conformity with Rule requirements; (2) that tests are carried out on samples from the current production.

Test section 试验段 从试件上截取并供制备一个或一个以上的试样用的某一段材料。如板中的板条。

Test specimen 试样 从试验段中截取的一段,其可以是机加工的也可以是非机加工的,具有规定尺寸并用作相应的试验。

Test speed V 测试速度 系指不小于相应 85% 主机最大输出功率时船速的 90% 的速度。 Test speed V is a speed of at least 90% of the ship's speed corresponding to 85% of the maximum engine output.

Testimony of witness/Affidavit 证人证言 是证据的一种,是证人就自己所知道的案件情况向法院或侦查机关所作的陈述。因为各种诉讼案件都是社会上发生的,案件一经发生,往往就会被人所感知,因此借证人的证言来查清案件事实为古今中外的法律所重视,也是各国民事诉讼中运用最广泛的一种证据形式。

Testing ashore 陆上试验 系指对安装上船前的生活污水处理装置的试验,例如工厂试验。 Testing ashore is a testing carried out on a sewage treatment plant prior to installation e. g. testing in the factory.

Testing establishment 试验机构 对澳大利亚海事安全局海事指令而言,系指为试验和检验起货机械而设置的机构。 For maritime order of the Australian Maritime Safety Authority (AMSA), testing establishment means an establishment equipped for the testing and examination of cargo gear.

Testing on board 船上试验 系指对已安装上船的生活污水处理装置进行形式认可的试验。 Testing on board is a testing, for the purpose of type approval, carried out on a sewage treatment plant that has been installed on a ship.

Tethered submersible 系缆可潜器 通过一根有时兼作电缆用的缆索与水面保障船相连接,并可在水下一定范围内自行推进,适于深度小于 500 m 的水区作业的潜水船。

TEU 标准集装箱 系指国际集装箱计量单位(20ft 换算值),其尺寸为 20 英尺×8 英尺×8.5 英尺。 TEU means the International measure standard for a container (20-foot equivalent unit), which is 20 ft×8 ft×8.5 ft.

t_f（mm） t_f 普通扶强材或主要支撑构件的面板净厚，（mm）。 t_f is the net face plate thickness (mm) of ordinary stiffener or primary supporting member.

The Construction and Equipment of Ships Carrying Dangerous Chemicals in Bulk（IBC Code） 散装运输国际散装运输危险化学品船舶构造和设备规则（IBC 规则）

The Inter-American Drug Abuse Control Commission（CICAD） 美洲国家间毒品滥用管制委员会

The International Code for Fire Safety Systems 国际消防安全系统规则 即"Fire Safety Systems Code"（"消防安全系统规则"），系国际海事组织的海上安全委员会 MSC.98（73）决议通过。The "Code" was adopted by the Maritime Safety Committee of the IMO by resolution MSC.98（73）.

The United National Framework Convention on Climate Change（UNFCCC） 联合国关于气候变化框架公约

Theater / War zone 战区 系指大规模战争中根据不同战情而划分的作战范围内。为实现战略计划、执行战略任务而划分的作战区域。也泛指进行战争的区域。它一般是根据战略意图和军事、政治、经济、地理等条件在战前或战争爆发后确定的，并可在战争中根据情况变化适时调整或建立新的战区。战区的范围是发展的。随着军事技术和海军、空军的发展，作战空间不断扩大，战区已由陆地扩大到海域、空域。

Theater ballistic missile defense（TBMD） 战区弹道导弹防御 由美国防部掌管的核心弹道导弹防御计划，是美国战区弹道导弹防御（TBMD）体系的基石，主要包括爱国者 PAC-3、水师区域 TBMD、THAAD（战区高空区域防御）和水师全战区防御等 4 大系统。

Theater missile defense（TMD） 战区导弹防御 TMD 计划是美国总统克林顿于 1993 年提出的，其前提是认为冷战后"战区弹道导弹"在第三世界国家中迅速扩散，并已成为美国前沿部队及海外盟友面临的主要威胁。美国认为，所有威胁不到美国本土的弹道导弹，都属于"战区弹道导弹"，只有能够打到美国本土的弹道导弹，才是"战略弹道导弹"。因此，TMD 是相对于防御"战略弹道导弹"的"国家导弹防御系统"（NMD）而言的。TMD 与 NMD 共同构成了美国"弹道导弹防御"（BMD）构想的两大内容，其开发工作由美国国防部弹道导弹防御局具体负责。"战区"系指"美国本土以外，由一个联合司令部和专门司令部管辖的地区"。因此，战区导弹防御系统是"用于保护美国本土以外一个战区免遭近程、中程或远程弹道导弹攻击的武器系统"。美国军方对于战区导弹的防卫有三种主要策略：（1）在来袭导弹发射前侦察到并将其摧毁；（2）在来袭导弹发射升空时将其摧毁；（3）在来袭导弹飞行途中或重回大气层时予以拦截摧毁。TMD 的设想由低层防御和高层防御两部分组成。低层防御设想包括"爱国者-3"（PAC-3）、"扩大的中程防空系统"（MEADS）、"海军区域防御"（NAD）系统。高层防御系统包括陆军"战区高空区域防御"（THAAD）系统、"海军战区防御体系"（NTW）、空军"助推段防御"（BPI）。其中，"爱国者-3""海军区域防御"系统、"陆军"战区高空区域防御"系统""海军战区防御体系"构成 TMD 的核心和重点开发项目。

Theft 偷窃 一般系指暗中的窃取，不包括公开的劫夺。

Theft, pilferage and non-delivery risk（TPND） 偷窃、提货不着险 投保平安险和水渍险的基础上加保此险，保险人负责赔偿对被保险货物因被偷窃，以及被保险货物运抵目的地后整件未交的损失。但是，被保险人对于偷窃行为所致的货物损失，必须在提货后 10 天内申请检验，而对于整件提货不着，被保险人必须取得责任方的有关证明文件，保险人才予赔偿。

Theoretic weight 理论质量 对于一些按固定规格生产和买卖的商品，只要其质量一致，每件质量大体是相同的，所以一般可以从其件数推算出总质量，这就是理论质量。

Theoretical air 理论空气量 理论上满足燃料完全燃烧所必需的空气量。

Thermal balance 热平衡

Thermal balance test 热平衡试验

Thermal conductivity 导热系数 K 是材料绝热材料最重要的性能指标之一。对于均质、各向同性的物体，在稳态一维热流情况下，导热系数指在稳定传热条件下，1 m 厚的材料，两侧表面的温差为 1 度（K,℃）在 1 s 内，通过 1 m^2 面积传递的热量，用 K 表示，单位为瓦/米·度（$W/m·K$）。

Thermal expansion 热膨胀

Thermal inkjet technology 热感应式喷墨技术 是利用一个薄膜电阻器，在墨水喷出区中将小于 0.5% 的墨水加热，形成一个气泡。这个气泡以极快的速度（小于 10 μs）扩展开来，使墨滴从喷嘴中喷出。

Thermal oil and hot water units 热油加热器和热水加热器 无论是使用燃油还是使用废气来加热均包括在内。Thermal oil and hot water units include the units using either fuel oil or exhaust gas for heating.

Thermal oil boiler 热油锅炉

Thermal oil heater 热油加热器 系指用火焰或燃气加热的加热器。Thermal oil heater means a heater heated by flame or combustion gas.

Thermal oil heating system 热油加热系统

Thermal oil installation 热油设备 热油通过火焰、燃气或其他高温气体加热后再用作货油或燃油加热或用作产生辅助用蒸汽或热水的加热源的设备。Thermal oil installation is the arrangement in which thermal oil is heated and circulated for the purpose of heating cargo or fuel oil for production of steam and hot water for auxiliary purpose.

Thermal oil pipeline 热油管路

Thermal oil units 热油加热器 系指由燃油、废气或电加热有机液体(热油)的一种加热装置,其中被加热有机液体的温度始终低于该液体在大气压下的沸点温度。

Thermal protective aid 保温用具 系指采用低导热率的防水材料制成的袋子或衣服。Thermal protective aid is a bag or suit made of waterproof material with low thermal conductance.

Means of access 通道设施 系指能够确保安全进入船舶结构的设施,以便对结构进行全面检查、近观检查和厚度测量。通道设施应在船舶结构通道手册中予以说明。Means of access are mean capable of ensuring safe access to the structures, in order to proceed overall and close-up inspection to ship structures and thickness measurements. The means of access are to be described in a Ship Structure Access Manual.

Thermal protective lifejackets (TP-lifejackets) 保温救生服 系指具有保温能力的救生衣。它适用于受过训练的人员,而对于未经训练的乘客是不大适用的。Thermal protective lifejackets (TP-lifejackets) mean the immersion suit offered buoyancy and thermal protection. They are intended for use by trained personnel and are less suitable for use by untrained passengers.

Thermal resistance 热阻(R) 系指热量在热路径上遇到的阻力,反映介质或介质之间传热能力的大小,表明了 1 W 热量所引起的温升大小,单位为($m^2 \cdot K/W$)。

Thermal transmittance 传热系数 系指在稳定传热条件下,围护结构两侧空气温差为 1 度(K,℃),1 h 内通过 1 m^2 面积传递的热量,单位是瓦/(平方米·度)($W/m^2 \cdot K$)。

Thermal treatment 热处理

Thermoelectric cooling unit 热电制冷装置(温差电制冷装置)

Thermoelectric couple 热电偶

Thermoelectric pyrometer 热电高温计

Thermoelectric refrigerating unit 热电制冷装置

Thermo-electric type air conditioner (semi-conductor air conditioner) 热电式空气调节器 带有热电制冷装置的空气调节器。

Thermo-mechanical rolling (thermo-mechanical controlled processing)(TM)(TMCP) 温度-形变轧制[温度-形变控制工艺 TM(TMCP)] 作为控制轧制的一种,不但轧制温度,而且也对轧压出量进行严格控制,一般大部分轧制在奥氏体再结晶区域进行,在 Ar_3 临界温度区域[包括奥氏体和铁素体(Ferrite)的双向(dual-phase)复合区域]结束轧制的轧制工艺。温度-形变轧制结束后加速冷却,直接淬火(direct-quenching)或回火处理应另行获取有关船级社的认可后方可适用。This is a procedure which involves the strict control of both the steel temperature and the rolling reduction. Generally a high proportion of the rolling reduction is carried out close to the Ar_3 temperature and may involve the rolling in the dual phase temperature region. Unlike controlled rolled (normalized rolling) the properties conferred by TM (TNCP), cannot be reproduced by subsequent normalizing or other heat treatment. The use of accelerated cooling on completion of TM-rolling may also be accepted subject to the special approval of the relevant Society.

Thermo-mechanical rolling TM 热机轧制 这是一种必须严格控制钢材温度和轧制减薄量的方法。通常在接近 Ac_3 温度下实施高比例的轧制减薄量,并可在两相温度区域内进行轧制。不同于控制轧制法(正火轧制)热机轧制具有的性能不可能由随后的正火或其他热处理方法再产生。Thermo-mechanical rolling TM is a procedure which involves the strict control of both the steel temperature and the rolling reduction. Generally a high proportion of the rolling reduction is carried out close to Ac_3 temperature and may involve the rolling in the dual phase temperature region. Unlike controlled rolling (normalized rolling) the properties conferred by TM cannot be reproduced by subsequent normalizing or other heat treatment.

Thermo-mechanically controlled rolling TM 热机械控制轧制 TM 是在严格控制温度及厚度缩减的条件下,使轧制在近于及可能小于铁素体完全形成的规定温度时进行的工序。其结果是在微观结构及力学性能上均非正火轧制材料所能及。Thermo-mechanically controlled rolling TM is a procedure which involves the strict control of both the steel temperature and the rolling reduction.

Thermometry 测温法

Thermoplastic plastic 热塑性塑料 系指具有加热软化、冷却硬化特性的塑料。人们日常生活中使用的大部分塑料属于这个范畴。加热时变软以至流动,冷却变硬,这种过程是可逆的,可以反复进行。聚乙烯、聚丙烯、聚氯乙烯、聚苯乙烯、聚甲醛、聚碳酸酯、聚酰胺、丙烯酸类塑料、其他聚烯烃及其共聚物、聚苯醚、氯化聚醚等都是热塑性塑料。热塑性塑料中树脂分子链都是线型或带支链的结构,分子链之间无化学键产生,加热时软化流动。冷却变硬的过程是物理变化。

Thermo-regulator 温度调节器

Thermos flask rack 热水瓶架 固定在舱壁上,供安放水瓶用的架子。

Thermosetting 热固的(热硬的)

Thermosetting material 热固性材料

Thermosetting plastic 热固性塑料 以热固性树脂为主要成分,配合以各种必要的添加剂通过交联固化过程形成制品的塑料。在制造或成型过程的前期为液态,固化后即不溶不熔,也不能再次热熔或软化。常见的热固性塑料有酚醛塑料、环氧塑料、氨基塑料、不饱和聚酯、醇酸塑料等。热固性塑料与热塑性塑料共同构成合成塑料中的两大组成体系。热固性塑料又分甲醛交联型和其他交联型两种类型。

Thermosetting resin 热固性树脂

Thermostat 温度控制器 与执行机构配套,对气体、蒸汽或液体进行双位式调节的温度升降而使电路闭合或断开的控制器。

Thermostat(thermo) 恒温器(恒温箱)

Thermostat-controlled 恒温器控制的

Thermostatic bimetal 恒温双金属

Thermostatic expansion valve 热力膨胀阀 带有温包并能感受出口气体过热度变化而自动节流调节制冷剂供液量的阀。

Thermostatic valve 恒温阀

Thermostatically controlled valve 恒温控制阀

Thermostatic-throttle valve 恒温节流阀

Thermotolerant coliforms 耐热大肠杆菌 系指在44.5 ℃情况下,48 h 内可从乳糖产生气体的大肠杆菌群。有时称这些微生物为"粪便大肠杆菌";但由于并非所有这些微生物都来自粪便,所以现在认为"耐热大肠杆菌"一词更为合适。Thermotolerant coliforms mean the group of coliform bacteria which produce gas from lactose in 48 h at 44.5 ℃. These organisms are sometimes referred to as "faecal coliforms"; however, the term "thermotolerant coliforms" is now accepted as more appropriate, since not all of these organisms are of faecal origin.

Thick-film hybrid integrated circuit 厚膜混成集成电路 系指由单独的半导体设备和被动元件,集成到衬底或线路板所形成的小型化电路。

Thickness for voluntary addition $t_{\text{voluntary-addition}}$ 自愿增加厚度 为除 t_c 以外,船东自愿增加的额外腐蚀磨损余量,mm。Thickness for voluntary addition $t_{\text{voluntary-addition}}$ is the thickness, in mm, voluntarily added as the owner's extra margin for corrosion wastage in addition to t_c.

Thickness measurement 测厚 系指测量物体的厚度。由于现代化的要求,很多时候都需要运用到测量物体的厚度以达到实际的要求。这样,测量厚度就在慢慢地发展起来。

Thickness of blade section 叶切面厚度 螺旋桨的叶切面垂直于其面节线方向的最大厚度。

Thickness of pipe 管壁厚度

Thickness of shaft flange 轴法兰厚度 轴法兰圆盘的边厚。

Thickness of top plate 舱盖板顶板厚度 每块舱盖板最上面板件的厚度。

Thimble 套环 带有绳槽,供绳索末端环绕扎结,以防止绳、缆过度弯曲和磨损,并连接他物的圆形或心形金属圆环。

Thin film evaporation 薄膜蒸发 借助某种方法使海水在蒸发传热面上形成极薄水膜而受热转化为蒸汽的相态变化。

Thin-film integrated circuit 薄膜集成电路

Thinner 稀释剂

Third island chain 第三岛链 系指以美国的夏威夷群岛为中心,北起阿留申群岛、南到大洋洲的一些群岛组成的链形岛屿带。

Third-country shipping line 第三国航运公司 系指在两国之间使用船舶经营货运而并非该两国国家航运公司的运输商。Third-country shipping line means a vessel-operating carrier in its operations between two countries, which is not a national shipping line.

Thorough examination 全面检查 系指按有关要求进行的仔细的目视检查。Thorough examination means a detailed visual examination in accordance with relative requirement.

Thread 梯步 是船梯上供人员踏登的部件。

Thread 螺纹

Thread cutting screw 攻丝螺丝

Thread damage 滑丝

Thread gauge 螺纹规

Thread joint 螺纹接头

Thread plug 螺纹塞

Threaded tube 有螺纹的管子

Threading tool 绞螺纹的工具

Three cylinder platform 三柱体式平台 系指由

三根桩腿支持的平台。

Three dimensional management and control system at sea 海洋立体管控系统 系指以卫星通信、无线电、岸基有线等各类无线、有线通信手段为基础，以分布在各类执法平台上的指挥控制设备为核心，将卫星、巡逻机、无人机、直升机、各类公务船、执法人员、岸基指挥所、车辆、水下监测节点等分布在岸、海、空、水下的海洋执法平台、节点及其执法装备、人员有机组合在一起，发挥1+1>2的作用，并与民用信息系统相融合，实现信息互通、数据互享、优势互补、协同行动使我国的海上综合执法与管控能力倍增的系统。该系统通过开展态势综合分析、海上情ること研判、执法方案辅助拟制及执法过程推演、执法命令下达和反馈、海上船舶实时监控及识别、海上水文气象信息监测等方面的研究，能够实现多渠道信息获取、多手段信息传输、多样化信息展示、多途径接警处理和服务，执法过程实时监控和记录等目的。

Three dimensional print（TDP） 3D 打印 即快速成型技术的一种。它是一种以数字模型文件为基础，运用粉末状金属或塑料等可黏合材料，通过逐层打印的方式来构造物体的技术。3D打印通常是采用数字技术材料打印机来实现的。常在模具制造、工业设计等领域被用于制造模型，后逐渐用于一些产品的直接制造，已经有使用这种技术打印而成的零部件。该技术在珠宝、鞋类、工业设计、建筑、工程和施工（AEC）、汽车、航空航天、牙科和医疗产业、教育、地理信息系统、土木工程、枪支以及其他领域都有所应用。目前，我国能够生产优于美国的激光成型钛合金构件，成为目前世界上唯一掌握激光成型钛合金大型主承力构件制造且付诸实施的国家。

Three dimesional underway replenishment 海上立体补给舰 主要采用横向、纵向和垂向方式，用于向航母战斗编队、舰船供应正常执勤所需的燃油、航空燃油、弹药、食品、备件等补给品，是专门用来在战斗中帮助编队的舰船，其特殊设计允许它装设战舰级的远端维修系统，并且减少所有辅助维修系统的能量需求，因此被广泛地在作战任务中使用。

Three layers exchange technique 三层交换技术 就是二层交换技术+三层转发技术。

Three legged foundation 三脚桩基础 系指突出在海洋表面上的钢管结构物。基础有三条桩腿，每一根桩腿与桩套管连接，在与打入海底的锚桩灌浆固定。三脚桩基础的优点是穿过波浪区的面积与单桩一样小，因为是一根单独的钢管，如同一架相机三脚架在海底展开，因此提供巨大的稳定性，以防弯矩。此外，锚桩到基础中心距离很长，有能力抵抗由风机和海浪引起的十分强烈的垂直振动力和弯矩。特别是因为具有单独的钢管，有可能如计算单桩一样计算因海浪引起的负载。三脚桩基础通常用于工作水深非常深（25～50 m）的海上油气行业。

Three point planning boat 三点式滑行艇 艇底形成品字形的3个滑行面的滑行艇。

Three-compartment subdivision 三舱制 系指船舶相邻三舱破损进水后，仍能满足破舱水线不超过限界线要求的设计指标。

Three-dimensional（3D） 三维图形 在计算机里显示3D图形，就是说在平面里显示三维图形。不像现实世界里，真实的三维空间，有真实的距离空间。计算机里只是看起来很像真实世界，因此在计算机显示的3D图形，就是让人眼看上就像真的一样。

Three-island ship 三岛式船 上甲板上设有较短艏楼、桥楼和艉楼的船舶。

Three-layer packet-switch technology 三层分组交换技术

Three-layer plywood 三夹板 又叫多层板和三合板，层数不同叫法不同，其优劣主要看原料。现在家装中使用的主要是饰面三夹板，即在工厂中已经将非常薄的实木饰面贴在三夹板上。饰面三夹板使用方便，价格也便宜。

Three-phase circuit 三相电路 系指由相位相差1/3周（120°）的交流电动势供电的一种组合电路。实际上这种相位可以偏离规定相角几度。

Three-phase short-circuit current 三相短路电流 系指当三相同时短接在一起时，就产生三相短路状态，此时在每相中所产生的短路电流。它是一个复杂的时间函数。该短路电流由对称短路电流（交流分量）和直流分量组成。 Three-phase short-circuit current means short-circuit current occured in each phase, when all three phase are simulaneously shorted together, while three-phase short-circuit current condion occurs. The resulting current is a complex time dependent function occurring in each phase. The current contains both A. C. and D. C. components typically.

Three-piece type floating dock 三段式浮船坞 分艏、舯、艉三段固定连接而成的浮船坞。

Three-roller bender 三芯辊床 系指具有三根滚轴的辊床。

Three-way cock 三通旋塞

Three-way valve 三通阀（三路阀）

Threshold level 阈值水平 系指均质材料中的浓度值。 Threshold level means a concentration value in homogeneous material.

Threshold price 门槛价格 即从"指标价格"中扣

除有关货物从进口港运到内地中心市场所付一切开支的余额。这种价格是差价税估价的基础。

Threshold value 极限值 系指调查中的化学品浓度极限,在极限内可以假定符合 2001 年船舶防污底系统公约的相关要求。 Threshold value means the concentration limit of the chemical under investigation below which compliance with the relevant provisions of the International Convention on the control of harmful anti-fouling system on ships, 2001, may be assumed.

Throttle 节流(节流阀,操纵阀)
Throttle control 节流控制
Throttle dowm 节流(关气门)
Throttle gate 节流门
Throttle governing 节流调节 用改变节流阀开度的办法来改变蒸汽的有效热降和流量,从而改变蒸汽轮机功率的调节方式。
Throttle link 节流连杆
Throttle process 节流过程
Throttle valve 节流阀
Throttled 塞住
Throttling 节流
Throttling process 节流过程
Throttling valve 节流阀
Through B/L 联运提单 系指经过海运和其他运输方式组成联合运输时,由第一程承运人所签发的包括全程运输的提单。但签发联运提单的承运人,一般只承担他负责运输的一段航程内的货损责任。
Through bolt 贯穿螺栓
Through bracket 贯通肘板 贯穿壁板连接骨材使之保持连续的肘板。
Through drive 直接传动
Through holt 贯穿螺栓 贯穿汽缸体、机架、机座等,使之连成刚性整体的螺栓。
Through scavenging 直流扫气
Through stay 长牵条,贯穿牵条
Throw 曲柄臂(冲程,投掷)
Throw into dear 使齿轮啮合
Throw of out gear 使齿轮脱开
Throw out of action 使停止动作
Throwing line 抛射绳 船舶在进行施救作业时,用专门抛绳器具定向抛射至所需处,作为引缆用的细绳索。
Throw-over claw 换向爪
Thrust 推力
Thrust bearing (thrust block) 推力轴承(止推轴承) 用以承受推进器所产生的推力及拉力的轴承。

Thrust bearing foundation (thrust block seat) 推力轴承底座 推力轴承的下支承体。
Thrust block 推力块
Thrust block keep 推力轴承盖 推力轴承的上盖。
Thrust coefficient 推力系数 表示螺旋桨推力的无因次系数。$K_T = T/\rho n^2 D^4$,式中,T—螺旋桨推力;ρ—水的质量密度;n—螺旋桨桨速;D—螺旋桨直径。
Thrust collar 推力环 推力轴上承受推进器产生的推力的环。
Thrust deduction 推力减额 就螺旋桨而言,相当于阻力增额的推力的减少值。
Thrust deduction factor 推力减额因数 船舶的总阻力对螺旋桨推力的比值。$1 - i = R_T / T$,式中,T—螺旋桨推力;R_T—船的总阻力。
Thrust deduction fraction 推力减额分数 推力减额对螺旋桨推力的比值。$i = (T - R_T)/T$,式中,T—螺旋桨推力;V_A—进速;n—螺旋桨转速;Q—螺旋桨转矩。
Thrust face 推力面
Thrust indicator 推力指示器
Thrust journal 推力轴颈
Thrust loading coefficient 推力载荷系数 用螺旋桨推力表示其载荷的无因次系数。$C_{Th} = T /[(1/2) \rho V_A^2 (\pi D^2/4)]$,式中,$T$—螺旋桨推力;$\rho$—水的质量密度;$V_A$—进速;$D$—螺旋桨直径。
Thrust meter 推力计 在模型试验或实船试验中用以测量螺旋桨推力的设备。
Thrust of duct 导管推力 导管推进器中导管所产生的推力。
Thrust of ideal propeller 理想推进器推力 根据动量理论理想推进器所能发出的推力。
Thrust of impeller 叶轮推力 导管推进器中叶轮所产生的推力。
Thrust of propeller 螺旋桨推力
Thrust pad (thrust shoe) 推力块 推力轴承中的承压块。
Thrust power 推力功率 螺旋桨推力与进速的乘积。推力与进速可在实船或船模上测得。
Thrust pressure 止推压力
Thrust ratio of ducted propeller 推力比 导管推进器的叶轮推力对总推力的比值。
Thrust shaft (thrust block shaft) 推力轴 轴系中用以使推力轴承传递给船体的单独的轴段。
Thrust shaft bracket 节能助推轴支架 一种通过将其两只机翼绕支臂转动一定的角度,使机翼剖面与来

流形成一个合适的攻角,从而水流在螺旋桨盘面上产生一个与螺旋桨旋向相反的预旋流,并使轴支架支臂所围的区域内的水流获得加速,和进入螺旋桨盘面的流速较为均匀,从而改善螺旋桨的进流,提高螺旋桨的推进效率的轴支架,节能轴支架有时还能起到降低阻力的作用。

Thrust surface　推力表面
Thrust transducer　推力传感器
Thrust unit　推力器
Thrust washer　止推垫圈
Thruster　推力器(助推器)　系指装在可旋转的喷管内或在船舶中的特殊横向通道内的螺旋桨或喷水推进器。推力器可用于推进、机动和操舵或者它们的任何组合功能。在固定喷管内的推进螺旋桨不能认为是推力器。Thruster is a propeller or a water-jet installed in a revolving nozzle or in a special transverse tunnel in the ship. A thruster may be intended for propulsion, manoeuvring and steering or any combination thereof. Propulsion propellers in fixed nozzles are not considered thrusters.

Thruster(DP system)　推进器(动力定位系统)　系指管隧推进器、全回转推进器、固定或可调螺距螺旋桨推进器。其驱动方式可分为电动机、柴油机或液压传动。Thruster (DP system) are pipe tunnel thrusters, azimuth thrusters, thrusters for fixed or variable pitch blades, driven by electric motors, diesel engines, or hydraulically.

Thruster assisted position mooring systems　推力器辅助定位系泊系统　系指通常设置在柱稳式平台或水面式平台上的一种定位系统。该系统由下列部件和设备组成:(1)锚或锚桩;(2)锚索,包括锚链、钢丝绳;(3)锚索附件,包括卸扣、连接环、缆端嵌件、快速释放装置;(4)导向装置,包括导向弯管、导向滑轮及导向孔;(5)掣链／缆器;(6)锚机／绞车;(7)构成定位系泊系统的其他结构件或机械件;(8)推力器及其原动机、传动轴和机构、电气设备;(9)推力器的控制、报警和安全系统。Thruster assisted position mooring systems means a positing system fitted on bottom-stabilized or surface-type units normally. The systems will consist of:(1) anchors or anchor piles;(2) anchor lines, including chain cables and steel wires;(3) anchor line fittings, including shackles, connecting links, wire rope terminations and quick-release devices;(4) fairleads, including bent pipes, guide rollers and guide holes;(5) chain/wire stoppers;(6) winches/windlasses;(7) other structural or mechanical items forming part of the position mooring system;(8) thruster and its prime mover, gearing, shafting, electrical installations;(9) control, alarm and safety systems for thrusters.

Thruster system(dynamic positioning systems)　推力器系统(动力定位系统)　系指用于动力定位的推进器及其控制系统。推进器系统包括:(1)具有驱动装置和必要的附属系统,(包括管路)的推进器;(2)有动力定位系统的控制的主推进器和舵;(3)推力器电子控制设备;(4)手动推进器控制器;(5)相关的电缆和电缆布线。Thruster system (dynamic positioning systems) means all thrusters and their control units used for dynamic positioning, including:(1) thrusters with drive units and necessary auxiliary systems including piping;(2) main propellers and rudders if these are under the control of the DP-system;(3) thruster control electronics;(4) manual thruster controls;(5) associated cabling and cable routing.

Thruster units　推进器装置　系指以下所有装置:(1)推力器、推力器的传动装置及控制推力器的速度、颠簸及方位的推力器控制硬件;(2)控制动态位置是所使用的主推进装置及其他的推进装置。Thruster units comprise the following:(1) thruster, power transmission gears driving thruster, thruster control hardware for control of thruster speed, pitch and heading;(2) main propellers and other propulsion units when these are included in dynamic poisitioning control mode.

Thruster-assisted mooring　推力器辅助系泊　系指用推力器或主推进装置(如果是如此设计)来辅助船舶锚泊的系统。Thruster-assisted mooring' is the use of thrusters and main propulsion, if so designed, to supplement the ship's anchoring system.

Thumb screw　翼形螺钉

Thunder Bay National Marine Sanctuary　桑德湾国家海上避难所　系指就在或邻近美国国家海洋和大气管理局指定的休伦湖的地方。其由沿阿尔皮纳县的北部和南部边界之间的一般高位基准点,抄近路穿过海流的入口并向着湖的方向从沿纬度线至西经83°的这些点延伸而形成近似矩形的区域作为边界。该边界的坐标是:北纬45°12′25.5″,西经083°23′18.6″;北纬45°12′25.5″,西经083°00′00″;北纬44°51′30.5″,西经083°00′00″;北纬44°51′30.5″,西经083°19′17.3″。Thunder Bay National Marine Sanctuary means the site on or near Lake Huron designated by the National Oceanic and Atmospheric Administration as the boundary that forms an approximately the ordinary high water mark between the northern and southern boundaries of Alpena County, cutting across the mouths of rivers and streams and lakeward from those points along latitude lines to longitude 83 degrees west. The coordinates of the boundary are:45°12′25.5″ N, 083°23′18.6″W;45°12′25.5″ N, 083°00′00″W;44°51′30.5″ N,

083°00′00″W;44°51′30.5″N,083°19′17.3″W.

Thwart 横坐板 沿救生艇舷缘布设的供人员乘坐的横向板材。

Thwartship thruster 侧推器

Ticket service/Tickets 票务服务

Tickets issuance 出票 系指出票人签发票据并将其交付给收款人的票据行为，出票为基本票据行为或主票据行为。出票人签发支票后，在规定的提示期间内，不得撤销对付款人代表其付款的委托，否则就会影响持票人的票据权利，也使票据失去信用。

Tidal flat survey 滩涂调查 对高潮淹没，低潮出露的软底潮间带所进行的科学调查。调查内容有潮间带水文、化学、底质、地貌、生物和环境污染等。

Tide table 潮汐表 潮汐预报表的简称。它预报沿海某些地点在未来一定时期的每天潮汐情况。在航运方面，有些水道和港湾须在高潮前后才能航行和进出港；在军事方面，有时为了选择有利的登陆地点和时间，就必须考虑和掌握潮汐的情况；在生产方面，沿海的渔业、水产养殖业、农业、盐业、资源开发、港口工程建设、测量、环境保护和潮汐发电等，都要掌握潮汐变化的规律。潮汐表就是为这些方面服务的。

Tie back 回接

Tie plate 甲板牵条 为加强木甲板而在其下设置的金属板条。

Tie rod 贯穿螺栓（系紧杆，拉条）

Tier[1] 层 系指甲板室范围的量度。一个甲板室层包括甲板和外部围壁。通常，第一层系指位于干舷甲板上的层。 Tier is defined as a measure of the extent of a deck house. A deck house tier consists of a deck and external bulkheads. In general, the first tier is the tier situated on the freeboard deck.

Tier[2] 层数 集装箱堆垛的垂向位置的标识。堆垛最下层位置的集装箱为第一层。

Tier Ⅲ-Verification of conformity (Goal-based ship construction standard) 第三层-符合性验证 系指：(1)主管机关按 SOLAS 第 Ⅺ-1/1 条规定所认可的组织的散货船和油船设计和建造规范，或者按 SOLAS 第 Ⅱ-1/3-1 条使用且与认可组织规范等效的主管机关的国家规范，应根据 IMO 组织制订的"散货船和油船目标型船舶建造标准符合性验证指南"验证其第一层的目标和第二层的功能要求。符合性验证的最终决定应由 IMO 组织的海上安全委员会做出，该委员会并应将该决定通知所有缔约国政府；(2)术语"验证"（单词 verify 及其任何变体）系指已将上述散货船和油船设计和建造规范与散货船和油船目标型船舶建造标准进行比较，并已查明与散货船和油船目标型船舶建造标准的目标和功能要求相符或一致；(3)主管机关或认可组织的散货船和油船设计和建造规范一经验证为符合散货船和油船目标型船舶建造标准，此符合性在规范修改后应仍视为有效。但对规范修改的验证应未出现不一致之处。除海上安全委员会另行决定外，经过符合性验证而采用的任何规范修改应适用于在规范修改生效日或以后签订建造合同的船舶。 Tier Ⅲ Verification of conformity means:(1) the rules for the design and construction of bulk carriers and oil tankers of an organization which is recognized by an Administration in accordance with the provisions of SOLAS Regulation Ⅺ-1/1, or national rules of an Administration used as an equivalent to the rules or a recognized organization according to SOLAS Regulation Ⅱ-1/3-1, shall be verified as conforming to the Tier Ⅰ goals and Tier-Ⅱ functional requirements, based on the guidelines developed by the IMO organization. The final decision on verification of conformity shall be taken by the Maritime Safety Committee of the IMO organization which shall inform all Contracting Governments of the decision;(2) the term "verification" (and any variation of the word "verify") means that the rules for the design and construction for bulk carriers and oil tankers as described above have been compared to the standards and have been found to be in conformity with or are consistent with the goals and functional requirements as set out in the standards; (3) once the rules for the design and construction of bulk carriers and oil tankers of an Administration or recognized organization have been verified as being in conformity with the standards, this conformity shall be considered to remain in effect for rule changes, provided that no verification of rule changes has resulted in a non-conformity. Unless the Maritime safety Committee decides otherwise, any rule changes introduced as a result of verification of conformity shall apply to ships for which the building contract is placed on or after the date on which the rule change enters into force.

Tier Ⅰ-Goals (Goal-based ship construction standard) 第一层—目标 系指船舶应按规定的设计寿命设计和建造为安全和利于环保的船舶，如在规定的营运和环境条件下营运和维护得当，可在其整个寿命期限内保持安全并利于环保。 Tier Ⅰ-Goals means ships shall be designed and constructed for a specified design life to be safe and environmentally friendly, when properly operated and maintained under the specified operating and environmental conditions, in intact and specified damage conditions, throughout their life.

Tier Ⅱ-Functional requirements (Goal based ship construction standard) 第二层—功能要求（适用于无

限航区散货船和油船） 系指：(1) 设计：①设计寿命—规定的设计寿命应不小于25年。②环境条件—船舶应按北大西洋环境条件和相关的长期海况散布图进行设计。③结构强度—（a）总体设计：船舶构件的设计应与处所的用途相容并确保一定的结构连续性。船舶构件的设计应便利所有预定货物的装卸，以避免由装卸设备造成损坏，这种损坏会危及结构安全；（b）变形和失效模式：结构强度评估应考虑过大偏移和失效模式，包括但不限于屈曲、屈服和疲劳；（c）极限强度：船舶应设计成具有足够的极限强度。极限强度计算应包括船体梁的极限能力以及相关的板材和扶强材的极限强度，并基于功能要求(1)②的环境条件按总纵弯矩极限验算；（d）安全裕度：船舶应设计成具有合适的安全裕度—（d-1）在完整工况下以净尺寸可承受按船舶设计预计寿命的环境条件和相应的装载工况，这应包括均匀满载和隔舱满载、部分装载、多港口和压载航行，以及压载水管理工况下的装载和卸载操作中的偶尔超量/超载，视对所定船级的适用范围而定；（d-2）适合于所有设计参数，其计算涉及一定的不确定性，其中包括载荷、结构模型、疲劳、腐蚀、材料瑕疵、建造工艺差错、屈曲、剩余强度和极限强度。④疲劳寿命—设计疲劳寿命应不小于船舶的设计寿命，并应以功能要求(1)②的环境条件为基础；⑤剩余强度—船舶应设计成具有足够的强度，以承受碰撞、搁浅或进水之类规定破损工况下的波浪载荷和内部载荷。剩余强度计算应考虑到船体梁的进行储备能力，包括永久性变形和后屈曲特性。对此，应在合理可行的范围内尽量对可预见的实际情景进行调查；⑥腐蚀防护—（a）涂层寿命：涂层应按制造厂商有关表面处理、涂层选择、涂装和维护的规定进行涂装和维护。如要求进行涂装，应规定设计涂层寿命。实际涂层寿命可比设计涂层寿命长或短，这取决于船舶的实际工况和维护情况。涂层应按舱室的预定用途、材料和其他防腐系统（如阴极保护或其他替代措施）的应用来选择。（b）腐蚀增量：腐蚀增量应加在净尺寸上，并应适合于规定的设计寿命。腐蚀增量应基于所接触的水、货物或腐蚀性大气之类腐蚀介质或所受到的机械磨损，以及结构是否有防腐蚀系数（如涂层、阴极保护或其他替代措施）保护来确定。设计腐蚀率（毫米/年）应根据由使用经验和/或加速模型试验确定的统计资料进行评估。实际腐蚀率可能比设计腐蚀率大或小，这取决于船舶的实际工况和维护情况；⑦结构冗余—船舶的设计和构造应具有冗余，使任一加强构件的局部损坏（如局部永久变形、破裂或焊缝损坏）不会导致整个加筋板格立即随之崩溃；⑧水密和风雨密完整性—船舶应按船舶预定用途设计成具有合适水密和风雨密完整性，且船体开口的系固装置具有适当强度和冗余度；⑨人为因素考虑—船舶的结构和装置应使用工效学原理进行设计和布置，以确保操作、检查和维护时的安全。这些考虑应包括，但不限于梯道、垂直梯、跳板、走道和用作检查通道的站立平台、工作环境、检查和维护以及操作便利；⑩设计透明度—船舶设计的过程应可靠、受控和透明，并按确认新建船舶安全所需的程度予以开放，并应充分考虑到知识产权。随时可供使用的文件和资料应包括主要目标型参数和所有可能限制船舶营运的相关设计参数。(2) 建造：①建造质量程序—船舶应按受控的和透明的质量生产标准进行建造，并充分考虑知识产权。船舶建造质量程序应包括，但不限于材料规格、制造、对中、装配、连接和焊接工艺、表面处理和涂装。②建造检验—应结合船型和设计，制订船舶建造简单的检验计划。检验计划应含有一套要求，其中包括规定建造检验的程度和范围，并确定在检验过程中需要特别注意的区域，以确保建造符合强制性船舶建造标准。(3) 营运期间考虑：①检验和维护—船舶的设计和建造应便利检验和维护，特别要避免形成过于狭窄的处所，以便进行适当的检验和维护活动。应确定在船舶整个寿命期限内，检验过程中需要特别注意的区域，这特别应包括在选择船舶设计参数时假定的所有必要的营运期间检验和维护。②结构的通达—船舶的设计、建造和设备应能提供到达所有内部构件的适当检验通道，以便利全面检查、近观检查以及测厚。(4) 回收利用考虑：船舶的设计和建造应使用可回收利用的环保材料，且不危及船舶安全和不降低操作效率。 Tier Ⅱ—Functional requirements (applicable to bulk carriers and oil tankers in unrestricted navigation) mean: (1) Design: ①design life: the specified design life shall not be less than 25 years; ②environmental conditions: ships shall be designed in accordance with North Atlantic environmental conditions and relevant long-term sea state scatter diagrams; ③structural strength: (a) general design: the ship's structural members shall be of a design that is compatible with the purpose of the space and ensures a degree of structural continuity. The structural members of ships shall be designed to facilitate load/discharge for all contemplated cargoes to avoid damage by loading/discharging equipment, which may compromise the safety of the structure; (b) deformation and failure modes: the structural strength shall be assessed against excessive deflection and failure modes, including but not limited to bucking, yielding and fatigue; (c) ultimate strength: ships shall be designed to have adequate ultimate strength. Ultimate strength calculations shall include ultimate hull girder capacity and related ultimate strength of plates and stiffeners, and be verified for a longitudinal bending moment based on the environmental conditions in functional requirement (1)②;(d) safety margins: ships shall be designed with suitable safety margins: (d-1) to withstand, at net scantlings, in the intact condition, the environmental conditions anticipated for the ship's design life and the loading conditions appropriate for them, which shall include full homogeneous and alternate loads, partial loads, multi-port and ballast voyage, and ballast management condition loads and occasional overruns/overloads during loading/unloading operations, as applicable to the class designation; (d-2) appropriate for all design parameters whose calculation involves a degree of uncertainty, including loads, structural modeling, fatigue, corrosion, material imperfections, construction workmanship errors, buckling, residual and ultimate strength; ④fatigue life: the design fatigue life shall be not less than the ship's design life and shall be based on the environmental conditions in functional requirement (1)②; ⑤residual strength: ships shall be designed to have sufficient strength to withstand the wave and internal loads in specified damaged conditions such as collision, grounding or flooding. Residual strength calculations shall take into account the ultimate reserve capacity of the hull girder, including permanent deformation and post-buckling behaviour. Actual foreseeable scenarios shall be investigated in this regard as far as is reasonably practicable; ⑥protection against corrosion: (a) coating life: coatings shall be applied and maintained in accordance with manufacturer's specifications concerning surface preparation, coating selection, application and maintenance. Where coating is required to be applied, the design coating life shall be specified. The actual coating life may be longer or shorter than the design coating life, depending on the actual conditions and maintenance of the ship. Coatings shall be selected as a function of the intended use of the compartment, materials and application of other corrosion prevention systems, e.g. cathodic protection or other alternatives; (b) corrosion addition: the corrosion addition shall be added to the net scantling and shall be adequate for the specified design life. The corrosion addition shall be determined on the basis of exposure to corrosive agents such as water, cargo or corrosive atmosphere, or mechanical wear, and whether the structure is protected by corrosion systems, e.g. coating, cathodic protection or by alternative means. The design corrosion rates (mm/year) shall be evaluated in accordance with statistical information established from service experiences and/or accelerated model tests. The actual corrosion rate may be greater or smaller than the design corrosion rate, depending on the actual conditions and maintenance of the ship; ⑦structural redundancy: ships shall be of redundant design and construction so that localized damage (such as local permanent deformation, cracking or weld failure) of any stiffening structural member will not lead to immediate consequential collapse of the complete stiffened panel; ⑧watertight and weathertight integrity: ships shall be designed to have adequate watertight and weathertight integrity for the intended service of the ship and adequate strength and redundancy of the associated securing devices of hull openings; ⑨human element considerations: ship's structures and fittings shall be designed and arranged using ergonomic principles to ensure safety during operations, inspection and maintenance. These considerations shall include, but not limited to, stairs, vertical ladders, ramps, walkways and standing platforms used for means of access, the work environment, inspection and maintenance and the facilitation of operation; ⑩design transparency: ships shall be designed under a reliable, controlled and transparent process made accessible to the extent necessary to confirm the safety of the new as-built ship, with due consideration to intellectual property rights. Readily available documentation shall include the main goal-based parameters and all relevant design parameters that may limit the operation of the ship; (2) Construction: ①construction quality procedures: ships shall be build in accordance with controlled and transparent quality production standards with due regard to intellectual property rights. The ship construction quality procedures shall include, but not be limited to, specifications for material, manufacturing, alignment, assembling, joining and welding procedures, surface preparation and coating; ②construction survey: a survey plan shall be developed for the construction phase of the ship, taking into account the ship type and design. The survey plan shall contain a set of requirements, specifying the extent and scope of the construction survey(s) and identifying areas that need special attention during the survey(s), to ensure compliance of construction with mandatory ship construction standards; (3) in-service considerations: ①survey and maintenance: ships shall be designed and constructed to facilitate case of survey and maintenance, in particular avoiding the creation of spaces too confined to allow for adequate survey and maintenance activities. Areas shall be identified that need special attention during surveys throughout the ship's life. In particular, this shall include all necessary in-services survey and maintenance that was assumed when selecting ship design parameters; ②structural accessibility: the ship shall be designed, constructed and equipped to provide adequate means of access to all internal structures to facilitate overall and close-up inspections and thickness measurements;(4) Recycling considerations;ship shall be designed and constructed of materials for environmentally acceptable recycling without compromising the safety and operational efficiency of the ship.

Tight fit　紧密配合

Tight joint　密封接合

Tightly　上紧

Tightness device　密封装置

Tightness test　密性试验

Tightness　密性　系指不透过流体的性能。

Tightning device　紧固装置

Tight-strong seam　强固紧密接缝（强固紧密焊缝）

Tilde　波浪号

Tiller　舵柄　装在舵头上用以转动舵的臂状构件。

Tiller tie-bar　舵柄连杆　连接两个或两个以上舵柄用的杆件。

Timber　木材　系指木料或锯材、斜木、杆材、纸浆原材和所有其他散装或包装成形的木材，但不包括木质纸浆或类似货物。　Timber means sawn wood or lumber, cants, logs, poles, pulpwood and all other types of timber in loose or packaged forms. The term does not include wood pulp or similar cargo.

Timber carrier　运木船　系指专运原木和木材，配备系固设备的普通船舶。　Timber carrier is a general ship dedicated log or timber carriers, provided with securing equipment.

Timber deck cargo　木材甲板货　系指在干舷甲板或上层建筑甲板上露天部分装载的木材，但不包括纸浆或类似货物。　Timber deck cargo means a cargo of timber carried on an uncovered part of a freeboard or superstructure deck. The term does not include wood pulp or similar cargo.

Timber load line　木材载重线　系指：(1) 符合国际载重线公约中有关船舶结构的某些条件的勘划在船上的一条特殊载重线，并且在使用此载重线时其货物应按 1991 年装载木材甲板货船舶安全操作规则[A.715(17)决议]规定的条件堆装和紧固；(2) 木材甲板货可以认为是给船舶一定的附加浮力和增加抗御海浪的能力。为此，对运载木材甲板货的船舶，可以允许按照有关规定的计算减少干舷，并根据有关规定在船舷勘划标志。但是，为允许和使用上述载运木材的特殊干舷，木材甲板货应符合有关规定的某些条件，并且船舶本身也

应符合有关规定中列出的有关船舶构造的某些条件。

Timber load line means:(1) a special load line assigned to ships complying with certain conditions related to their construction set out in the International Convention on Load Lines and used when the cargo complies with these stowage and securing conditions of the Code of Safe Practice for Ships Carrying Timber Deck Cargoes, 1991 [Resolution A. 715 (17)];(2) timber deck cargo may be regarded as giving a ship a certain additional buoyancy and a greater degree of protection against the sea. For that reason, ships carrying a timber deck cargo may be granted a reduction of freeboard calculated according to the relative provisions and marked on the ship's side in accordance with the relative provisions. However, in order that such special freeboard may be granted and used, the timber deck cargo shall comply with certain conditions, and the ship itself shall also comply with certain conditions relating to its construction.

Timber shop　木工车间　系指从事木作加工、装配、存储等的场所。

Time bill　远期汇票　系指在一定期限或特定日期付款的汇票。远期汇票的付款时间，有以下几种规定办法：(1) 见票后若干天付款（At XX days after sight）；(2) 出票后若干天付款（At XX days after date）；(3) 提单签发日后若干天付款（At XX days after date of Bill of Lading）；(4) 指定日期付款（Fixed date）。

Time charter　定期租船　是船舶所有人将船舶出租给承租人，供其使用一定时间的租船运输。承租人也可以将此租船作为班轮或程租船使用。

Time charter　期租　系指按约定期限的租船方式。

Time charter on trip basis(TCT)　航次期租　系指以完成一个航次运输为目的，按完成航次所花的时间，按约定的租金率计算租金的一种租船方式。

Time charter party　定期租约

Time domain computation　时域计算

Time for a complete cycle　Z 形操纵周期　在 Z 形操纵试验中，从开始操舵的瞬时到向右舷和随后向左舷转舵完成一个循环所经过的时间间隔。其无量纲值为：TV/L。

Time letter of credit　远期信用证　系指开证行或付款行收到信用证的单据时，在规定期限内履行付款义务的信用证，是银行（即开证行）依照进口商（即开证申请人）的要求和指示，对出口商（即受益人）发出的、授权出口商签发以银行或进口商为付款人的远期汇票，保证在交来符合信用证条款规定的汇票和单据时，必定承兑，等到汇票到期时履行付款义务的保证文件。

Time limit to launch 降落时限 系指船上所有救生载具从开始释放起到全部降落到水面操作完毕所允许的最长时间限度。

Time of arrival 抵港时间 系指船舶刚到达某一港口停留的时间,而不论是锚泊或停靠码头。Time of arrival means time when a ship, first comes to rest, whether at anchor or at a dock, in a port.

Time of rudder movement 转舵时间 当船舶以最大航速前进时,舵从一舷规定的角度转至另一舷规定的角度所需要的最长时间。

Time of shipment 装运期 是卖方装运货物的期限。

Time required for ballasting of floating 上浮时间 钻井装置的沉垫或下浮体从海底上浮至漂浮正常状态时所需的时间。

Time required for ballasting of immersing 下沉时间 (1)钻井装置坐底作业时其沉垫或下浮体从漂浮状态下沉至海底并达到作业状态所需的时间。(2)浮船坞从工作吃水下沉至最大沉深吃水所需的时间。

Time sheet 装卸时间表 系指装卸作业的安排表。

Time to check yaw 纠偏期 在Z形操纵试验中,从换操反舵瞬时到达最大转艏角瞬时之间的时间间隔。其对于向右舷偏移的无量纲值为 $B = t_a V/L$。

Time window 时间窗口 系指允许实施特定种类检验的时间跨度。

Timing 正时 通常以上行下止点为基准,用曲轴回转角表示的进排气门或进排气口及喷油阀、超动阀等开启和关闭的相对时间。

Tip cavitation 叶梢空泡 螺旋桨叶梢处由于梢涡产生的形状如丝带卷的空泡。

Tip losses 叶梢损失 螺旋桨叶近梢部分环流的降低。

Tip vortex 梢涡 在螺旋桨叶梢处泄出的尾涡。

Tip-nub ratio 轮毂比 轴流式压气机或涡动叶片的轮毂直径,即内径与叶尖直径,即外径的比值。

Tipping 艉落 船舶纵向下水过程中,当重心离开滑道末端的力矩仍大于浮力对滑道末端力矩的瞬间所发生的艉部下落的现象。

T-max tanker T型油船

To ditch 迫降 系指航空器被迫在水面上降落。To ditch means a case in which an aircraft makes a forced landing on water.

TOFD 衍射时差技术

Toggle 切换

Toilet equipment 盥洗设备 各种供洗澡、洗脸、洗脚等用的设备。

Tolerable risk 可容忍的风险 根据ISPS规则目标和海上安全方针,已降至公司可接受程度的风险。Tolerable risk means the risk reduced to the extent acceptable to the company in accordance with the objective and maritime security policy of ISPS Code.

Tolerance range 容许范围 系指加到极限值上的数值范围,表明由于公认的分析误差的原因,可以接受测得的超过极限值的浓度范围,而且这样不会影响符合的设定。Tolerance range means the numerical range added to the threshold value indicating the range where detected concentrations above the threshold value are acceptable due to recognized analytical inaccuracy and thus do not compromise the assumption of compliance.

Toluene 甲苯 高度易燃(闪点4 ℃),遇空气能形成爆炸性混合物,有刺激性,严重危及健康,对生殖器官有毒性。Toluene is highly flammable (flash point of 4 ℃), can from an explosive mixture with air, it is irritant and can cause acute health risk, it is toxic to reproductive organs.

Tonal sound 有调声 系指含有易于听觉的单音调声音。

Tonnage 吨位/总吨 在民用船的内部容积,以登记吨计。

Tonnage breadth 量吨宽度 在吨位丈量中,为计算量吨甲板下位于量吨甲板或其分段长度各等分点处的舱内横剖面积,于其量吨深度的各等分点上,沿水平方向按规则规定所量取的舱内宽度。

Tonnage certificate (certificate of admeasurement) 吨位证书 主管部门对按照吨位丈量规范已测定总吨位和净吨位的船舶所颁发的吨位丈量证书。

Tonnage deck 量吨甲板 在吨位丈量中,按规则规定作为基准的甲板。

Tonnage depth 量吨深度 在吨位丈量中,为计算量吨甲板下位于量吨甲板或其分段长度各等分点处的舱内横剖面积,在各该处按规则规定所量取的舱深。

Tonnage length 量吨长度 在吨位丈量中,为计算量吨甲板下全部容积,按规则规定所量取的量吨甲板长度。

Tonnage mark 吨位标志 具有两组登记吨位的两层以上全通甲板船舶,按吨位丈量规范要求的式样和位置勘绘在其两舷,视浸水位置,用以确定该船在该航次中按何组登记吨位计算的标志。

Tonnage measurement 吨位丈量 为核定船舶总吨位和净吨位所进行的丈量和计算。

Tonnage measurement of ships 船舶吨位丈量

系指核定船舶总吨位和净吨位。

Tonnage of hatchways 舱口吨位 上甲板以上为舱口围板所围封的吨位。

Tooth 齿

Tooth bar 齿条（齿杆）

Tooth face 齿面

Tooth gear 齿轮（齿轮传动机构）

Tooth outline 齿轮外形

Tooth pitch 齿节（齿距）

Tooth surface stress 齿面应力

Toothed quadrant 齿形扇弧

Top dead center 上死点

Top dead center (upper dead center) 上止点 活塞在气缸内与曲轴中心线距离达最大值的位置。

Top deck 坞墙顶甲板 位于坞墙顶，承受纵向强度的连续甲板。

Top drive-drilling system (TDS) 顶部驱动钻井系统

Top end 顶点

Top end bearing 十字头轴承

Top mast 顶桅 系指桅肩以上，自桅柱向上延伸或固接在桅柱上或安装在门形桅横向构杆上的，主要用以装设号灯、天线和桅横桁等的桅上部结构。

Top plate[1] 舵叶顶板 舵叶顶面的封板。

Top plate[2] 顶板（钢夹层板） 钢夹层板中底板另一侧的钢板。

Top speed 最大速度

Top tensioned riser (TTR) 顶张式立管/刚性立管 是一种采油立管。

Top-down 自下而上法 系指基于各国海运业燃油消耗量统计值计算海运温室气体排放量的方法。按下式计算：$E = FC_1 \times EF_1$ 式中，E 为海运温室气体分类的温室气体排放量；FC_1 为不同类型燃料的消耗量；EF_1 为不同类型燃料海运温室气体分类的排放因子。

Topmast light (truck flashing light) 桅顶灯 设置在船舶桅杆顶部的灯具。

Topmast shroud 顶桅支索 牵张顶桅用的绳索。

Topper 千斤调整索 千斤索具中，在吊杆无载荷时用以调节吊杆位置的长环链。

Topping bracket 千斤座 由千斤眼板和千斤眼板座两主要零件所组成，装于起重柱顶部、桅肩或其他船体杆件上供悬系千斤索具的座子。

Topping winch 千斤索卷车 在吊杆无载荷时，调节吊杆仰角，用以收放千斤索的卷车。

Topping-off 加装作业 系指把液货从一艘辅助船输送至另一艘装载船，使装货船可在吃水较深的状态下装载。Topping-off is the operation of transfer of liquid cargo from a service ship to another ship in order to load the receiving ship at a deeper draft.

Topping-up pump 增压泵

Topside wing tank/ Topside tank 顶边舱/顶舱 系指散货船上，位于货舱顶部两侧角隅处的舱。

Torque coefficient 转矩系数 表示螺旋桨转矩的无因次系数。$K_Q = Q/\rho n^2 D^5$，式中，Q—螺旋桨转矩；ρ—水的质量密度；n—螺旋桨桨速；D—螺旋桨直径。

Torque margin 转矩裕度 系指推进系统可以承受大于额定转矩时，且不会使电动机与发电机失步的转矩增量。

Torque meter 转矩计 在模型试验或实船试验中用以测量螺旋桨转矩的设备。

Torque of propeller 螺旋桨转矩 螺旋桨所受到的或为克服水阻力所需的转矩。

Torsion dynamometer 扭力测功器

Torsion failure 扭转损坏

Torsion stress 扭应力

Torsional frequency 扭振频率

Torsional vibration 扭转振动 系指船体横剖面绕其纵轴扭转的振动。

Torsional vibration absorber 减扭振器

Torsional vibration check device 扭振检查器

Torsional vibration damper 扭振减振器/扭振阻尼器

Torsional vibration measurement 扭振测量

Torsional vibration of shafting 轴系扭转振动 轴系相对于自身角位移的振动。

Torsional vibration stresses 扭转振动应力 系指由有关的谐波阶数的合成所对应的交变扭矩得出的应力。Torsional vibration stresses are the stresses resulting from the alternating torque corresponding to the synthesis of the harmonic orders concerned.

Total assembly time t_A 总集合时间 系指最长的个人集合时间。Total assembly time t_A is the maximum individual assembly time.

Total cost for repairs 工程总价 系指修船工程的总费用。

Total deadweight (gross deadweight) 载重量 系指民用船舶所允许装载的可变载荷的最大值，其值等于满载排水量与空船质量之差，包括燃料、淡水、食物、船员、旅客、货物。

Total discrimination 完全选择性保护 系指在有2个或2个以上过电流保护电器串联的电路中，当发生过电流故障时，只是最接近故障点的保护电器起保护作用,而不会导致其他保护电器动作的过电流选择性保护。 Total discrimination is an over-current discrimination where, in the presence of two or more over-current protection devices in series, the protective device on the load side effects the protection without causing the other protective devices to operate.

Total displacement 总排水量 任一情况下,型排水量加上附体排水量。对于金属船体尚包括外板排水量在内。

Total efficiency 总效率

Total elevated load of a self-elevating unit 自升式平台总提升载荷 系指:(1)轻载排水量,但不包括桩腿和桩腿箱的质量;(2)所有船用和钻井用的设备及其管系;(3)液体可变载荷;(4)固体可变载荷;(5)合成的钻井载荷的总和。 Total elevated load of a self-elevating unit is the combination of:(1) the lightweight, but excluding the weight of the legs and spud cans;(2)all shipboard and drilling equipment and associated piping;(3) liquid variables;(4) solid variables;(5) combined drilling loads.

Total engine output P 主机总输出功率 P 是推进机械能在系统赖以分级的最大扭矩对应的螺旋桨转速下,可以连续传输给推进系统的最大输出功率。如果机械的输出功率因技术方法或适用于舰船的任何规则而受限,则 P 应取受限的输出功率。 Total engine output P is the maximum output the propulsion machinery can continuously deliver to the propulsion system with the propeller(s) operating at the revolutions per minute at the maximum torque for which the system is to be classed. If the output of the machinery is restricted by technical means or by any Regulations applicable to the ship, P, shall be taken as the restricted output.

Total gear ratio (total velocity ratio) 总传动比 多级传动齿轮中,第一级主动齿轮转速与末级被动齿轮转速的比值。

Total harmonic distortion (THD) 总谐波畸变 去掉基波之后,余下的谐波均方根值与基波均方根值之比,用百分数表示。 Total harmonic distortion means a ratio of the r.m.s. value of the residue, after elimination of the fundamental, to the r.m.s. value of the fundamental, expressed in percent.

Total head 总压头 速头、压头与势头之和。$H = U^2/2g + p/w + Z$,式中,U —流速;g —重力加速度;w —单位体积流体的质量;p —单位面积压力;Z —势头。

Total hydraulic head 总水压头

Total installing power 总装机功率

Total jack-up loads 最大升船能力 自升式钻井平台的各桩腿提升能力的总和。

Total length of leg 桩腿总长 包括桩靴或沉垫在内的从桩腿最顶端至其最底端的总长度。

Total load 总载荷(总负载)

Total loss 全部损失 系指被保险货物的全部遭受损失,全损有实际全损(actual total loss)和推定全损(constructive total loss)之分。

Total loss only 全损险 系指投保货物全部遭受损失的险种。

Total overall clear height for vehicles 车辆用整体高度 系指形成一水平分区的甲板和甲板肋骨之间距离的总和。 Total overall clear height for vehicles is the sum of distances between deck and web frames of the decks forming one horizontal zone.

Total power of ACV 气垫船总功率 系指气垫船的垫升功率与推进功率之和。

Total pressure 总压力 静压力和动压力之和。总压力 $= (p + wZ) + (1/2)\rho U^2$;式中,$p$ —单位面积压力;w —单位体积流体的质量;Z —势头;ρ —流体质量密度;U —流速。

Total rake 总纵斜 系指螺旋桨纵斜与侧斜轴向位移之和。一般系指叶梢或叶端至螺旋桨平面间的距离。

Total resistance 总阻力 通常系指船舶在静水中所受的摩擦阻力、黏压阻力、兴波阻力之和。可在船模试验或实船拖曳试验时拖索上测得。

Total resistance coefficient 总阻力系数 以总阻力对动压力与湿表面乘积的比值,表示总阻力特征的一个无量纲的参数。$C_T = R_T / [(1/2)\rho V^2 S]$,式中,$R_T$ —总阻力;V —船速;S —湿面积或其他面积特征量如 $\nabla^{2/3}$ 等;ρ —液体质量密度。

Total resistance of model 船模总阻力 船模试验拖索上所测得的阻力。

Total resistance of ship 实船总阻力 由船模阻力经过分析而换算而得的实船的总阻力或指实船拖曳试验时,拖索上测得的阻力。

Total salinity 总含盐度 以滤过的水溶液在110 ℃下的干馏残渣量表示的水中所有溶盐的总和。

Total static head 总静压头

Total thrust of ducted propeller 导管推进器总推力 导管推进器的叶轮推力与导管推力之和。

Total variable drilling load (VDL) 总可变钻井载荷 系指下列 4 种工况下的载荷(轻载迁移、深吃水迁移、作业和生存工况)。 Total variable drilling loads (VDL) are defined as four conditions (light transit, deep draft transit, operational and survival conditions).

Total weight of lifeboat 救生艇总重 救生艇载足额定乘员和规定属具时的全部质量。

Totally enclosed and fan-cooled motor 全封闭扇冷型电机 系指采用预期组装在一起,但处在封闭部件外部的一个或多个风扇进行外部冷却的全封闭型电机。

Totally enclosed bridge 全围蔽的桥楼 设有围蔽桥楼翼台的桥楼,意味着围蔽的桥楼翼台是驾驶室的组成部分。Totally enclosed bridge is a bridge without open bridge wings, meaning that bridge wings form an integral part of an enclosed wheelhouse.

Totally enclosed motor 全封闭型电机 系指封闭的能阻止机壳内、外部之间空气交换,但此封闭尚不足以称之为气密型电机。

Totally enclosed space 全围蔽处所 系指无通道且无通风的处所。 Totally enclosed space is a space which has no means of access and no ventilation.

Touchdown and lift-off area (TLOF) 降落和起飞区(TLOF) 系指一个动态承载区域,直升机可在该区域降落或起飞。对于直升机甲板,假设最终进场/起飞区域(FATO)和降落和起飞区(TLOF)重合。Touchdown and lift-off area (TLOF) is a dynamic load-bearing area on which a helicopter may touch down or lift off. For a helideck it is presumed that the FATO and the TLOF will be coincidental.

Touch-down point(TDP) 触地点

Touch-down zone(TDZ) 触地区

Tourist visa 旅游签证 是签证的一种。一般是为了方便游客而开发旅游资源而设立的一种快速签证方式,相应的受限制也大。一般来说,有效期和停留期都较短,且只能够用来从事旅游相关活动。

Tourist/passenger submersibles 观光潜水器 系指能够载运乘客在水下运行、浮出水面并保持水面航行的自航载器。也是一种可搭载未经任何特殊训练的游客邀游海底的高科技旅游设施。

Towage 拖航 系指整个拖航作业过程,包括从始发港接拖、拖带航行到目的港交船的全过程。

Towage/tugboat charge/towage dues 拖船费 系指在港口租用拖船的费用。

Towed 拖带 系指船舶被拖或推着通过该水域。Towed means pushed or pulled through the aters.

Towed objects 被拖物 通常系指非机动船,如驳船、起重船、打桩船、挖泥船、打捞船、布设船、趸船和浮船坞及水上设施,如浮式装置、水上建筑、移动式平台和其他水上建筑物,以及机械推进装置损坏丧失推进

能力的机动船舶,但不包括应急拖带和救助拖带的船舶。

Towed STD profiling system 拖曳式温盐深剖面仪 能在船舶航行中用铠装电缆拖曳水下装置测量海水温度、盐度(或电导率)等要素随深度变化的仪器。

Tower 塔架 系指用螺栓将2根或2根以上的钢结构件连接而成的高耸钢结构物。不定型产品,对不同的工程情况可以根据其要求由设计人员设计塔架架构。塔架的用途很广泛,主要用于工厂烟囱的支撑、大型建筑物支撑、水塔塔架、监控工程、通信工程及海上风电场等其他特殊用途。主要结构及材料:三柱钢管(法兰连接)四柱角钢(连接板)。

Tower platform 塔式平台 系指以格栅形混凝土沉箱(兼作储油罐)作基础,向上伸出2~4根空心混凝土柱支撑上部结构的大型重力式平台。

Towing 拖曳 可定义为接受另一艘船的动力辅助或给另一艘船提供动力辅助。Towing can be defined as either receiving motive assistance from, or rendering it to, another vessel.

Towing apparatus 拖带用具 除拖船外,其他船舶拖带他船或被他船拖带时用以紧固拖缆的用具。

Towing arch(towing gallow) 拖曳弓架 拖船上专用于联络拖钩或拖曳滑车,借以传递拖力于船体结构的弓形构件。

Towing beam 拖缆承梁 横跨拖船艉部甲板,支立于两舷,用以支撑拖缆,以防拖缆摆动时损伤人员和舱面设备的拱梁。

Towing block 拖曳滑车(曳纲束锁) (1)拖船上用以引导拖缆绞车上的拖缆使转向船艉并承受挂拖拖力的专用滑车;(2)安装在单拖网渔船艉部或舷侧,在拖曳网具过程中,用以锁住2根曳钢于一点的一个带有滑轮的锁状装置。

Towing carriage 引船小车 设置在浮船坞上,用以牵引被修理船舶进出坞用的拖车。

Towing chock 拖缆孔 供拖缆通过的导缆孔或转动导缆孔。

Towing depth distance recorder 拖曳航行深度距离记录仪 用于航行中测定浮游生物采集器的深度和拖曳航行距离的仪器,亦可用于其他拖曳式仪器中。

Towing draft 平台拖航吃水 钻井平台在拖航状态时的吃水。

Towing embedded anchor 拖曳埋置锚 系指以其抓力与自重之比,即抓重比(或抓力系数)表征抓持特性的锚。随着锚重的增加,抓重比将会下降。拖曳埋置锚一般只能承受水平力,不考虑承受垂直力。因此,与之相连的锚索应有足够的长度,并尽可能使得锚索在锚处

与海底相切。根据用途和抓持能力,拖曳埋置锚可分为4类:(1)船用锚;(2)一般性海洋工程用锚;(3)较高性能海洋工程用锚;(4)高性能海洋工程用锚。表 T-1 为拖曳埋置锚的分级及在泥或砂中的抓持特性。

表 T-1　质量为 10 t 的各种拖曳埋置锚抓持特性
Table T-1　Holding characteristic of towing embedded anchors of 10 t weight

锚的级别	抓力系数	锚的名称	抓持特性
A	33～55	史蒂芙帕瑞锚、史蒂芙夏克锚、布鲁斯锚、FFTS 锚	具有极大的穿透底质的能力
B	17～25	布鲁斯铸钢锚、布鲁斯 TS 锚、霍克锚	锚柄如弯曲的胳臂,有利于穿透底质
C	14～26	史蒂汶锚、史蒂芙菲克斯锚、史蒂芙莫特锚、阔鳍型德尔泰锚	具有开敞的锚冠以及相对较短的锚柄和稳定杆,锚爪转动的枢轴接近重心处
D	8～15	丹福斯锚、轻量型锚、斯达托锚、穆尔法斯特/近海钻井Ⅱ型锚、博斯锚	在锚爪的后部设置较长的稳定杆兼作锚爪转动的枢轴,并具有相对较长的锚柄
E	8～11	AC-14 锚、斯托克司锚、Snugstow 锚、Weld-hold 锚	具有很短的稳定器(锚冠部分)锚爪转动的枢轴在后部,并具有相对较短横截面为矩形的锚柄
F	4～6	美国海军无杆锚、Beyers 锚、Union 锚、斯贝克锚	锚柄截面为矩形,无稳定杆,锚的稳定依靠锚冠
G	<6	单爪锚、海军锚、Dredger 锚、Mooring 锚	锚爪面积小,稳定杆设在锚柄端部

Towing equipment　拖曳设备　系指拖船和被拖物上专为拖曳作业而设置的设备。包括拖船上的拖缆机、拖钩、拖索拱架、拖缆滚筒、拖缆孔(导缆孔)、缆绳架、拖销、鲨鱼钳等以及被拖物上设置的拖力点(拖力眼板或拖桩)、拖缆孔(导缆孔)等。

Towing force in self-propulsion test(friction correction for self-propulsion test)　自航试验拖力　在定转速船模自航试验中,为抵消船模与实船摩擦阻力系数的差异以达到相当于实船自航情况所加于船模的拖力。

Towing force of ship　船舶拖曳力　一船拖带另一船或别物航行时的拉力。

Towing gear　拖曳索具　系指拖船和被拖物上专为拖曳作业而使用的索具。包括主拖索、备用拖索、龙须缆(链)、短缆、三角眼板、拖曳环、卸扣和应急拖缆等。

Towing gear of hatchcover　舱盖曳行装置　机械舱盖中,拖曳舱盖板使其移动就位的装置。

Towing hook　拖钩　船上用于扣住拖缆并能迅速解脱的专用器具。

Towing hook platform　拖钩台　位于拖钩下方,用以支持拖钩自重并便于拖钩摆动的小平台。

Towing length　拖带长度　系指从拖船船尾量至最后一艘被拖物后端的水平距离。

Towing light　拖带灯　系指具有与艉灯系统特性的黄灯。Towing light means a yellow light having the same characteristics as the stern light.

Towing master　拖航船长　系指拖带航行的管理负责人。拖航作业时,拖船的船长可以指定为拖航船长。

Towing operations　拖带作业　包括一艘或多艘能够协助海上平台,驳船和生产模块/船舶从一个位置移动到另一个位置或保持其规定位置的海工作业船的作业方式。Towing operations is an operation including one or more offshore service vessels capable to assist offshore platforms, barges and production modules/vessels in moving from one position to another, or in keeping their defined position.

Towing pennant(towing line)　拖缆　系指用于有效拖带船舶的一根长绳。Towing pennant means a long rope which is used to tow a ship effectly.

Towing point　拖力点　系住被拖物上专门用于连接拖缆或龙须缆/链的设施,包括拖力眼板或拖桩。

Towing post　拖桩　与船体主要结构牢固联结,供套扣拖缆的立柱。

Towing rope（towline,towrope） 拖缆 拖船上专用于挂拖其他船舶或浮物的缆绳。

Towing speed 拖曳航速 船或船模被拖曳或船舶拖曳它物时的航速。

Towing tackle 曳行索具 机械舱盖中,拖曳舱盖板用的索具。

Towing winch 拖缆绞车（拖缆机） 卷收并存放拖缆或推拖船队其他系船用缆专用的绞车。

Towknee 拖柱 设置于推船或分节驳艏端,专供传递推力的柱架结构件。

Tow-line 纤 在内河用人力或畜力曳船的绳索。

Towling 主拖缆 系指用于拖船与被拖物的拖带连接缆。

Toxic and corrosive substances 有毒和腐蚀性物质 系指符合 IMO"国际海上危险物质规则（IMDG 规则）"修正案规定的物质。Toxic and corrosive substances are those which are listed in the IMO "International Maritime Dangerous Goods Code (IMDG Code)", as amended.

Toxic substances (Class 6.1 dangerous materials) 有毒物质（第6.1类危险货物） 系指如吞入或吸入,或接触皮肤会造成死亡或严重伤害,或危及人类健康的物质。Toxic substances mean the materials liable either to cause death or serious injury or to harm human health if swallowed or inhaled, or by skin contact.

Toxic to mammals by prolonged exposure 长期接触对哺乳动物的毒性 系指:(1)如果某货物符合下列任一标准,则标为长期接触毒性类型:已知或怀疑将导致癌症、诱导突变、影响后代、神经中毒、损伤免疫系统或其他非致命剂量但造成特殊器官不可逆性中毒及其他相关影响。(2)这类影响可通过 GESAMP 危险品档案或其他已知途径获取。 Toxic to mammals by prolonged exposure means:(1) a product is classified as toxic by prolonged exposure if it meets any of the following criteria: it is known to be, or suspected of being a carcinogen, mutagen, reprotoxic, neurotoxic, immunotoxic or exposure below the lethal dose causing specific organ oriented systemic toxicity (TOST) or other related effects. (2) Such effects may be identified from the GESAMP Hazard Profile of the product or other recognized sources of such information.

Toxicity tests 毒性试验 系指采用急性和/或慢性生态毒性数据（最好包括敏感生命阶段）对毒性标准急需评估的试验。Toxicity tests mean those tests used for the assessment of the toxicity criterion by means of acute and/or chronic ecotoxicity data, ideally covering the sensitive life stages.

t_p（mm） 普通扶强材或主要支撑构件带板的净厚(mm) t_p is the net thickness, in mm, of the plating attached to an ordinary stiffener or a primary supporting member.

T_P（s） 纵摇周期 由下式得出:$T_P = \sqrt{\frac{2\pi\lambda}{g}}$,式中,$\lambda = 0.6(1 + T_{LC}/T_S)L$。Pitch period, in s, is given by: $T_P = \sqrt{\frac{2\pi\lambda}{g}}$, where: $\lambda = 0.6(1 + T_{LC}/T_S)L$.

T_R（s） 横摇周期 由下式得出:$T_R = 2.3 k_r/(GM)^{1/2}$,式中,k_r—所考虑装载工况的横摇回转半径,m,当 k_r 未知时可用表 T-2 中的值;GM—所考虑装载工况的稳性高度,m,当 GM 未知时可用表 T-2 的值;横摇单幅值,由下式得出:$T_B = [9000(1.25 - 0.025 T_R)f_p, k_b]/(B+75)\pi$,式中,$k_b$—系数,无舭龙骨的船舶 $k_b = 1.2$,有舭龙骨的船舶 $k_b = 1.0$。 Roll period, in s, is given by: $T_R = 2.3 k_r/(GM)^{1/2}$, where: k_r—roll radius of gyration, in m, in the considered loading condition. When k_r is not known, the values indicated in Table 3-4 may be assumed; GM—metacentric height, in m, in the considered loading condition. When GM is not known, the values indicated in Table 3-4 may be assumed. Roll single amplitude, in deg, is given by: $= [9000(1.25 - 0.025 T_R)f_p, k_b]/(B+75)\pi$, where: k_b—coefficient taken equal to: $k_b = 1.2$ for ships without bilge keel; $k_b = 1.0$ for ships with bilge keel.

表 T-2　k_r 值和 GM 值
Table T-2　k_r value and GM value

装载工况	k_r	GM
满载工况 （隔舱或均匀装载）	$0.35B$	$0.12B$
正常压载工况	$0.45B$	$0.33B$
重压载工况	$0.40B$	$0.25B$

Traces of oil 油迹 在水面上或水面下能用肉眼看得到的油类痕迹。

Track 航迹 系指被跟踪的对地轨迹。 Track means a path to be followed over ground.

Track (path) 航迹 船舶航行时,其重心运动的轨迹。

Track adjuster 履带张紧器

Track angle 轨迹角 潜水船舶在垂向平面内航行时,从水平面到其重心处瞬时航速矢量间的垂向夹角。顺时针为正。

Track control 航迹控制 系指沿航迹对船舶的控制。Track control means control of the ship movement along a track.

Track plotting system 航迹绘图仪 简称航迹仪。是现代航海中常用的一种设备。其功能是在船舶航行过程中,根据组合导航系统提供的舱位信息,陀螺罗经提供的航向信息,计程仪提供的航速信息,在海图上自动实时地绘制出本船航行的航迹。目前航迹仪已发展为电子显示型和硬拷贝型2种,从结构到功能都发生了极大的改观,已从单一化绘制航迹发展为多功能绘图系统。电子显示式航迹仪包括:CRT 显示式电子海图系统和 X-Y 光电投影式航迹点显示设备。前者利用计算机图形显示技术,将预先数字化后存入内存的海图资料显示在屏幕上,系统通过数字接口接收外部导航定位信息,把航迹线及各种航海标记叠加在海图上。该系统配有键盘输入设备,操作人员可根据键盘命令对海图显示进行修正,也可输入若干个转向点,做出计划航线,作为实际航线的参考。

Track reach[1] 停船轨迹(停船迹程) 系指船舶执行全速倒车指令直到船停止在水中时船舯点走过的距离。Track reach is the distance along the path described by the mid-ship point of a ship measured from the position at which an order for full astern is given to the position at which the ship stops in the water.

Track reach[2] 航迹航程 系指从发出全速倒车试验指令时起至船舶停在水中为止,船舯点沿其轨迹量取的距离。Track reach is the distance along the path described by the mid-ship point of a ship measured from the position at which an order for full astern is given to the position at which the ship stops in the water.

Track recorder 航迹记录仪 是用来绘制本船航迹的终端设备。它以计算机为中心,通过标准数字接口与导航系统(或定位系统)或其他设备进行数字通信,从导航系统送来的(实时)数字信号经过微机处理,根据不同的海图基准纬度及其比例值输出绘图笔在海图上移动的纵、横坐标增量,并将其转化为 X 方向和 Y 方向的执行机构动作,使绘图笔在海图上自动完成海图作业。

Track way type davit 滑架型艇架 放艇时,吊艇臂依靠下端两副滚轮沿轨架滑出并倾倒,从而吊艇出舷的重力式艇架。

Tracking 跟踪 系指观察运动中的目标位置相继变化的过程。Tracking is a process of observing the sequential changes in the position of a target, to establish its motion.

Tracking radar 跟踪雷达 系指能连续跟踪一个目标并测量目标坐标的雷达。它还能提供目标的运动轨迹。跟踪雷达一般由跟踪距离支路、方位角跟踪支路和仰角跟踪支路组成。它们各自完成对目标的距离、方位和仰角的自动跟踪,并连续测量目标的距离、方位和仰角。相干脉冲多普勒跟踪雷达还具有多普勒频率跟踪的能力,并能测量目标的径向速度。

Tracking satellite ship (tracking-surveying ship) 卫星跟踪船 又称靶场测量船,在海上用于跟踪和测量飞行体轨迹的船舶。是一种在海上自由移动的特殊的"地面跟踪测量试验基地",是发展宇宙空间科学的活动的多用途的科学试验船舶。船上设有专门设备对卫星进行跟踪测量工作。见图 T-5。

图 T-5 卫星跟踪船
Figure T-5 Tracking satellite ship
(tracking-surveying ship)

Traction winch 驱绳绞车(牵引绞车) 带有1个或2个摩擦卷筒的单层卷绕收放绳索的绞车,要与储绳绞车联合使用。

Tractor pusher tug at sea 海上顶推船组 系指由一艘推船(与常规拖船相似)和一艘驳船(艉部有凹槽)组成,有插销连接装置把两船连接一起实施推航的船组。

Trade liberalization 贸易自由化 系指一国对外国商品和服务的进口所采取的限制逐步减少,为进口商品和服务提供贸易优惠待遇的过程或结果。无论是以往的关贸总协定,还是现在的世贸组织,都是以贸易自由化为宗旨。理论基础来源于亚当·斯密和大卫·李嘉图的比较优势论。根据以上分析,人们其所以没有采取现有国际贸易的某一理论和政策,而提出和运用了产业组织理论中"可竞争市场"的概念,意在从中国是一个发展中和转型中的大国的实际出发,坚持经济市场化和贸易自由化的方向,博采各家之长,特别是其适用于我方的部分,将其加以本土化的改造,来解释中国的国际贸易,借以构造国际贸易的中国模式。

Trade mark 商标

Trade quota 贸易配额

Trademark 定牌 在国际贸易中,买方要求卖方在出售的商品或包装上标明买方指定的商标或牌名的做法。一般对于国外大量的长期的稳定的订货，可以接受买方指定的商标。在我国出口业务中，我方同意使用定牌，是为了利用买主的经营能力和他们的企业信誉或品牌声誉，以提高商品售价和扩大销售数量。

Traffic in transit 过境运输 系指人员、行李、货物和运输工具通过一个或几个过境国家领土的过境，而这种通过不论是否需要转运、入舱、分卸或改变运输方式，都不过是以内陆国领土为起点或终点的旅运全程的一部分。Traffic in transit means transit of persons, baggage, goods and means of transport across the territory of one or more transit States, when the passage across such territory, with or without trans-shipment, warehousing, breaking bulk or change in the mode of transport, is only a portion of a complete journey which begins or terminates within the territory of the land-locked State.

Traffic management 流量管理 是基于网络的流量现状和流量管控策略，对数据流进行识别分类，并实施流量控制、优化和对关键 IT 应用进行保障的相关技术。

Traffic separation scheme（TSS） 分道航行系统 为不同航向设置了独立的航道区域。主要服务繁忙航道，以减少因船舶航道拥挤造成碰撞的风险。分道航行的规划与高速公路相似。每个航向一个区，中间设有禁止通行的航道。阿尔法·文图斯海上风电试验场位于两个分道航行系统的中间。

Traffic ship 交通船 系指不属客运业务范围，用以运送人员的船舶。Traffic ship means a ship used for transporting personnel, but not as passenger transport service.

Trail speed 试航航速 新建船舶在适当平静的深水中，主机在额定工况下，测试所得的航速。

Trailing edge 随边 螺旋桨叶片的随边系叶片导边的另一方。Trailing edge of a propeller blade is the edge of the blade opposite the leading edge.

Trailing edge bar 舵叶艉材 舵叶后端的加强构件。

Trailing suction hopper dredger（TSHD） 耙吸式挖泥船 系指具有耙头等挖泥设备的挖泥船。是一种边走边挖，且挖泥、装泥和卸泥等全部工作都由自身来完成的挖泥船。耙吸式挖泥船用疏浚装舱法开挖航道，满舱后，驶向倾倒区，在倾倒区内把船舱的疏浚土卸入水中，然后返回挖槽，重复上一轮的工作。

Trailing vortex 尾涡 按螺旋桨环流理论，就所在位置而言，遗留于螺旋桨后水中的螺旋线形涡的部分。

Trails 轨迹 通过目标雷达回波形式显示航线的轨迹可为真实的或相对的。Trails are tracks displayed by the radar echoes of targets in the form of an afterglow. Trails may be true or relative.

Train ferry 火车渡船 专运列车的渡船。

Train/RO-RO passenger ship 铁路车辆客滚船 系指载运乘客超过 12 人，又能载运铁路车辆的客滚船。见图 T-6。

图 T-6 铁路车辆客滚船
Figure T-6 Train/RO-RO passenger ship

Training 培训 对质量管理体系而言，系指制造厂商应遵循一套招聘员工和培训的方针，以便提供每一工种所需技艺的充分劳动力。对于从事生产管理，特殊加工检查和试验或质量管理等维护活动的人员是否具有相应经验或经过适应培训，应保持适宜的纪录。For the of quality management system, training means that the manufacturer is to follow a policy for recruitment and training which provides an adequate labour force with such skills required for each type of work operation. Appropriate records are to be maintained to demonstrate that all personnel performing process control, special processes inspection and test or quality system maintenance activities have appropriate experience or training.

Training manual on life-saving appliances 救生设备培训手册 系指在每一个船员餐厅和娱乐室，或每一个船员舱室内配备的供救生设备培训用的手册。培训手册可分为若干分册，应包括关于船上所配备的救生设备和最佳救生方法的须知和资料，并应用易懂的措辞写成；如有可能应配以图解说明。这些资料的任何部分都可以用视听的教材形式提供。下列各项应予详细解析：(1) 救生衣、救生服和抗暴服的穿着法（按适用者）；(2) 在指定地点集合；(3) 救生艇筏和救助艇的登乘、降落和离开，包括（如适用）海上撤离系统的使用；(4) 在救生艇筏内降落的方法；(5) 从降落设备脱开；(6) 降落区域内防护方法与防护设备的用法（如适用）；(7) 降落区域的照明；(8) 所有救生属具的用法；(9) 所有探测装备的用法；(10) 用图解说明无线电救生设备的用法；(11) 海锚的用法；(12) 发动机及其附件的用法；(13) 救生艇筏和救助艇的回收,包括存放和系固;(14)暴露的危害性和穿用保暖衣服的必要性;(15)为救生而使用救生艇筏设备的最佳方法;(16)拯救的方法,包括直升机救助装置(吊绳、吊篮、吊担架)、连裤救生圈、海岸救生工具和船舶抛绳设备的用法;(17)应变部署表与应变须知所列出的所有其他措施;(18)救生设备应急修理须知。培训手册应以船上工作语言撰写。Training manual on life-saving appliances means a manual provided in each crew's mess room and recreation room or cabin for life-saving appliances training. The training manual, which may comprise several volumes, shall contain instructions and information, in easily understood terms illustrated wherever possible, on the life-saving appliances provided in the ship and on the best methods of survival. Any part of such information may be provided in the form of audio-visual aids in lieu of the manual. The following shall be explained in detail: (1) donning of lifejackets, immersion suits and anti-exposure suits, as appropriate; (2) muster at the assigned stations; (3) boarding, launching, and clearing the survival craft and rescue boats, including, where applicable, use of marine evacuation systems; (4) method of launching within the survival craft; (5) release from launching appliances; (6) methods and use of devices for protection in launching areas, where appropriate; (7) illumination in launching areas; (8) use of all survival equipment; (9) use of all detection equipment; (10) with the assistance of illustrations, the use of radio lifesaving appliances; (11) use of drogues; (12) use of the engine and accessories; (13) recovery of survival craft and rescue boats including stowage and securing; (14) hazards of exposure and the need for warm clothing; (15) best use of the survival craft facilities in order to survive; (16) methods of retrieval, including the use of helicopter rescue gear (slings, baskets, stretchers), breeches-buoy and shore life-saving apparatus and ship's line-throwing apparatus; (17) all other functions contained in the muster list and emergency instructions; (18) instructions for emergency repair of the life-saving appliances. The training manual shall be written in the working language of the ship.

Training manuals 培训手册(训练手册) 系指详细解析以下内容的小册子:(1)有关烟气危害、电气危险、易燃液体和船上类似常见危险的一般消防安全操作和预防措施;(2)关于灭火行动和灭火程序的一般须知,包括报告火灾及使用手动报警按钮的程序;(3)船舶各种报警的含义;(4)灭火系统和设备的操作和使用;(5)防火门的操作和使用;(6)挡火闸和挡烟闸的操作和使用;(7)脱险通道和设备。同时,应在每一船员餐厅和娱乐室或在每一船员居住室内配备一本培训手册。培训手册应用船舶的工作语言写成。培训手册可分为若干分册,应包括上述所要求的须知和资料,并应用易懂的措辞写成,如有可能应配以图解说明。这些资料的任何部分可以用视听教材形式提供,用以代替手册。Training manuals mean manuals explained the following in detail: (1) general fire safety practice and precautions related to the dangers of smoking, electrical hazards, flammable liquids and similar common shipboard hazards; (2) general instructions on fire-fighting activities and fire-fighting procedures including procedures for notification of a fire and use of manually operated call points; (3) meanings of the ship's alarms; (4) operation and use of fire-fighting systems and appliances; (5) operation and use of fire doors; (6) operation and use of fire and smoke dampers; (7) escape systems and appliances. And a training manual shall be provided in each crew's mess room and recreation room or in each crew's cabin. The training manual shall be written in the working language of the ship. The training manual, which may comprise several volumes, shall contain the instructions and information required above in easily understood terms and illustrated wherever possible. Any part of such information may be provided in the form of audio-visual aides in lieu of the manual.

Training programme 培训计划 系指针对船舶操作所有方面的指导和实践经验所确定的课程,其与主管国家海事机构提供的基本安全培训相类似。 Training programme means a defined course of instruction and practical experience in all aspects of ship operations, similar to the basic safety training as offered by the maritime institutions in the country of the Administration.

Training research ship 实习调查船 用于对学员进行海洋科学的教学、实习及科学研究的船舶。船上设有海洋水文、海洋化学、海洋物理、海洋生物、海洋地质、地貌、海洋气象和航海等实验室及各种型号绞车。并有供学员上课的教室和其他生活设备。见图T-7。

Training ship 训练船 系指用于海上教学和实习的船舶。 Training ship means a ship used for teaching and practice at sea.

Trans shipment 转船 如货物没有直达船或一时无合适的船舶运输,而需通过中途港转运的称为转船。

Trans shipment port 转船港 系指实现货物中途换装其他船舶的港口。

Trans-Atlantic Trade Partnership Agreement (TTIP) 跨大西洋贸易和伙伴关系协定 是美国与欧盟正在进行谈判磋商的一项贸易协定。目标是建立一个覆盖8亿人口的自由贸易区。

图 T-7　实习调查船
Figure T-7　Training research ship

Trans-boundary movement　跨境运输　系指对废弃物的海上运输,即从一个国家管辖的区域到达或通过另一个国家管辖的区域,或者到达或通过没有任何国家管辖的区域,但此种运输至少应涉及 2 个国家。 Trans-boundary movement means maritime transport of wastes from an area under the national jurisdiction of one country to or through an area under the national jurisdiction of another country, or to or through an area not under the national jurisdiction of any country, provided at least two countries are involved by the movement.

Transboundary movement of waste　废弃物越境转移　系指将废弃物从一国的管辖区域运抵或运经他国的管辖区域,或运抵或运经无任何国家管辖的区域,但运输中至少涉及 2 个国家。 Transboundary movement of waste means any shipment of wastes from an area under the national jurisdiction of one country to or through an area under the national jurisdiction of another country, or to or through an area not under the national jurisdiction of any country, provided at least two countries are involved in the movement.

Transcoder(X CDR)　变码器　系指完成 GSM 规范规定的话音编码。用于压缩来自交换机的语音信号,以便更有效的使用 GSM 频谱和陆地接口。

Transfer　正横矩　回转试验中,从初始直线航线至转艏角 90°瞬时,船舶重心自初始直线航线向回转圈中心一侧横移的距离。

Transfer arrangements　登离船装置　系指能使引航员从船舶任一舷安全登船和离船的装置。在所有船舶上,当从海平面至登船处或离船处的距离超过 9 m,并欲将舷梯或引航员机械升降器或其他同样安全方便使用的装置与引航员软梯一起供引航员登船或离船使用时,则应在每舷均应这种设备,除非该设备能够转移以供任一舷使用。船舶应设置下列任一装置,以供安全方便地登船或离船:(1)引航员软梯,所需爬高不小于 1.5 m,离水面高度不超过 9 m,其位置和系固应做到:①避开任何可能的船舶排水孔;②在平行船体长度范围内,并尽实际可能在船艏一半船长范围内;③每级踏板稳固地紧靠船舷,如结构特性,例如护舷材妨碍上述规定的实施,应做出使主管机关满意的特别布置,以确保人员能安全登船和离船;④引航员软梯的单一长度能从登船处或离船处抵达水面,并充分考虑所有装载工况和船舶纵倾及 15°的不利横倾;安全加固点、卸扣和系索的强度应至少与扶手索相同;(2)当从水面至登船处的距离超过 9 m 时,与引航员软梯相连的舷梯,或其他同样安全与方便的装置。舷梯应导向船艉设置。在使用时,舷梯的下端应稳固地紧靠在平行船体长度范围内的船舷,并应尽可能在船艏一半船长范围内,且避开所有的排水孔;(3)引航员机械升降器,其位置应在平行船体长度范围内,并应尽可能在船艏一半船长范围内,且避开所有的排水孔。 Transfer arrangements means arrangements provided to enable the pilot to embark and disembark safely on either side of the ship. In all ships where the distance from sea level to the point of access to, or egress from, the ship exceeds 9 m, and when it is intended to embark and disembark pilots by means of the accommodation ladder, or by means of mechanical pilot hoists or other equally safe and convenient means in conjunction with a pilot ladder, the ship shall carry such equipment on each side, unless the equipment is capable of being transferred for use on either side. Safe and convenient access to, and egress from, the ship shall be provided by either: (1) a pilot ladder requiring a climb of not less than 1.5 m and not more than 9 m above the surface of the water so positioned and secured that:①it is clear of any possible discharges from the ship;②it is within the parallel body length of the ship and, as far as is practicable, within the mid-ship half length of the ship;③each step rests firmly against the ship's side, where constructional features, such as rubbing bands, would prevent the implementation of this provision, special arrangements shall, to the satisfaction of the Administration, be made to ensure that persons are able to embark and disembark safely;④the single length of pilot ladder is capable of reaching the water from the point of access to, or egress from, the ship and due allowance is made for all conditions of loading and trim of the ship, and for an adverse list of 15°; the securing strong point, shackles and securing ropes shall be at least as strong as the side ropes; (2) an accommodation ladder in conjunction with the pilot ladder, or other equally safe and convenient means, whenever the distance from the surface of the water to the point of access to the ship is more than 9 m. The accommodation ladder shall be sited leading aft. When in use, the lower end of the accommodation ladder shall rest firmly against the ship's side within the parallel body length of the ship and, as far as is practicable, within the mid-ship half length and clear of all discharges;(3) a mechanical pilot hoist so located that it is within the parallel body length of the ship and, as far as is practicable, within the mid-ship half length of the ship and clear of all discharges.

Transfer function　传递函数　系统的输入值转变为输出值的一种运算子。即系统输出量的拉普拉斯变换对输入量的拉普拉斯变换的比值。

Transfer platform　输油平台　系指供停靠穿梭油船并将原油输送到穿梭油船上运走的固定式平台。

Transfer pump　驳运泵（调运泵，输送泵）

Transfer valve　输送阀

Transferable credit　可转让信用证　系指信用证的受益人可以要求授权付款、承担延期付款的责任、承兑或议付的银行，或当信用证是自由议付时，可以要求信用证中特别授权的转让银行，将信用证全部或部分转让给一个或数个受益人的信用证。遵照 UCP 第 44 条规定，出口商可要求进口商开具可转让信用证，可转让信用证的受益人(Beneficiary)作为转让人(Transferer)，通过银行称转让银行(Transferable bank)将信用证金额(Amount)的全部或部分，一次转让给出口商所在地或异地口岸的分支机构，或给异地各货源的供应商，即第二受益人(Second beneficiary)，由第二受益人按规定的产品，在规定的时间内分批装船，制单结汇。

Transformer for power and lighting　电力和照明变压器　系指用于电力系统的输出、变电和配电的变压器。如动力变压器、照明变压器、安全灯变压器和隔离变压器等。通常可制成单相或三相，要求具有较高的效率及合适的短路电压。

Transformer oil　变压器油

Transformer substation　变电站　系指使用变压器改变供电电压的设施。变电站由开关设备、变压器、控制中心以及测量和安全设备组成。

Transient condition (transient operation, transient running)　过渡工况　发动机组随外界负荷需求情况的变化从一工况转变到另一工况的过渡过程。

Transist satellite　子午卫星　又称海军导航卫星，是用于导航的一种人造卫星，其轨道倾角约 60°；即大致在通过地球南北极的椭圆形轨道上运转。星体直径约 50 cm，有四块向卫星供电的太阳能电池板，卫星总重在 55~80 kg，姿态稳定杆长 30 m，其端点有一个 105 kg 的小球用来稳定卫星。

Transit　通过　系指使用整个航道或部分航道，不论是上航还是下航。　Transit means to use the Seaway, or a part of it, either up-bound or down-bound.

Transit condition　迁移工况　系指平台从一个地区迁移到另一个地区时的状态。可分为油田内迁移工况和远洋迁移工况。　Transit condition means all unit movements from one geographical location to another. Transit condition is divided into field transit condition and ocean transit condition.

Transit passage　过境通行　系指按照有关规定，专为公海或专属经济区的一部分和公海或专属经济区的另一部分之间的海峡，为继续不停迅速过境的目的而行使的航行和飞越自由。但是，对继续不停和迅速过境的要求，并不排除在一个海峡沿岸国入境条件的限制下，为驶入、驶离该国或自该国返回的目的而通过海峡。　Transit passage means the exercise in accordance with relevant requirement of the freedom of navigation and over-flight solely for the purpose of continuous and expeditious transit of the strait between one port of the high seas or an exclusive economic zone and another part of the high seas of an exclusive economic zone. However, the requirement of continuous and expeditious transit does not preclude passage through the strait for the purpose of entering, leaving or returning from a State bordering the strait, subject to the conditions of entry to that State.

Transit State　过境国　系指位于内陆国与海洋之间以及通过其领土进行过境运输的国家，不论其是否具有海岸。　Transit State means a State, with or without a sea-coast, situated between a land-locked State and the sea through whose territory traffic in transit passes.

Transit tax　过境税　又称通过税。系指一国对于通过其关境的外国货物所征收的关税。目前，大多数国家在外国商品通过其领土时，不再征收过境税，只是征收少量的准许费、印花费、登记费等。

Transit trade　过境贸易　系指别国出口货物通过本国国境，未经加工改制，在基本保持原状条件下运往另一国的贸易活动。过境贸易有两种类型：一种称之为直接过境贸易，如在海运情况下，外国货物到港后，在海关监管下，从一个港口通过国内航线装运到另一个口岸，而后离境；有时，不需卸货转船。直接过境完全是为了转运而通过某国国境，承办过境的国家由此可获得与转运相关的费用。另一种为间接过境，系指外国货物到港后，先存入海关保税仓库，未经加工改制后，从海关保税仓库提出，运出国境的活动。

Transit visa　过境签证　公民取得前往国家（地区）的入境签证后搭乘交通工具时，途径第三国家（地区）的签证称为过境签证。

Transit zone 过境区 是沿海国家为了便利邻国的进出口货运,开辟某些海港、河港或边境城市作为货物过境区,过境区对过境货物简化通关手续,免征关税或只征小额的过境费用。

Transition minimized differential signaling(TMDS) 最小化传输差分信号 系指通过逻辑算法将原始信号资料转换成10位,前8为资料由原始信号经运算后获得,第9位指示运算的方式,第10位用来对应直流平衡等。

Transition period 过渡阶段 回转试验中,船舶所处转舵阶段与定常阶段间的中间阶段。

Transition temperature 输送温度 系指能使抗断裂韧度显著降低的临界温度。Transition temperature is the temperature below which fracture toughness does a significant reduction.

Transitional flow 过渡流 处于层流和湍流之间的一种不稳定流态。

Transitional mode 过渡状态 对地效翼船操作模式而言,系指从排水状态演变至滑行状态的一个中间短暂状态,反之亦然。 For operational modes of WIG craft, transitional mode denotes the transient mode from the displacement mode to the step-taxi mode and vice versa.

Transitional sea 过渡风浪 系指当风作用在开阔海面,随风力作用时间的增加,波高及周期亦随之增大的风浪。

Transitions 转换处 系指在出口系统中,通道的类型(例如从走廊到楼梯)或尺寸有变化,或通道合并或分开的地点。 Transitions are those points in the egress system where the type (e. g. from a corridor to a stairway) or dimension of a routes changes or where routes merge or ramify.

Transitions of escape routes 脱险通道转移点 系指出口系统中那些通道尺寸形式(例如从走廊转入梯道)改变或合并或分叉的部位。 Transitions of escape routes are those points in the egress system where the type (e. g. from a corridor to a stairway) or dimension of a route changes or where routes merge or ramify.

Transit-oriented development (TOD) patterns 公共交通导向发展模式

Translation type rolling hatchboard 滚移式舱盖 舱盖板下方设置滚轮,可平行滚移启闭的机械舱盖。

Translator 翻译人员 由熟悉外语和有关知识,善于与人密切配合、工作积极、纪律性强的人员担任。

Transmission 传输 系指一个或一个以上的文电通过电子传输作为一个发送单元共同向外传递,其中包括标题和结尾数据。 Transmission means one or more messages electronically sent together as one unit of dispatch which includes heading and terminating data.

Transmission capacity 传递能力 用主机功率和转速的比值来表示的选择减速齿轮箱的指标。传递能力 $=N/n$,式中,N—主机额定输出功率;n—主机额定输出转速。

Transmission dead time 倒车传令时滞 从发出倒车命令瞬时到实际开始执行倒车操纵的瞬时之间所经过的时间间隔。

Transmission device 传动装置 系指将主机发出的动力,经过轴系传给推进器的装置。它包括2部分:轴段及后传动装置。(1)轴段各部分如下:①艉轴:艉轴的一端安装螺旋桨且在水中工作,另一端经过艉管(或艉轴管)进入船体内部与中间轴连接。②中间轴:轴系中的连接轴段,根据轴系的长短可以有一段或多段,彼此之间用联轴节相联结。③推力轴:艉轴上的螺旋桨产生很大的推力,通过推力轴承传给船体,推力轴承中的轴段称为推力轴。在各轴段上根据需要设有以下装置:①联轴节:各轴段之间靠联轴节来联结,用以传递转动力矩和轴向推力。②轴承:轴承的作用是支承轴段。在船体外艉部水中支承艉轴的轴承称为支架轴承,在船体内支承中间轴的称为支点轴承,承受螺旋桨产生轴向推力使船体运动的轴承称为推力轴承。③密封装置:艉轴从船体外穿入船内及中间轴穿过水密舱壁处都要设置密封装置,称为隔舱填料函。④刹轴装置:在轴系不工作时,为了减少磨损或便于对轴系及主机进行检修,一般在轴系中设有刹轴装置。⑤转轴装置:在日常检试和维护中通过转轴装置可慢速转动轴系,有时也可通过转轴装置转动主机的曲轴。(2)后传动装置。舰船动力装置中,在主机和轴段之间还设置了离合器、减速齿轮箱及弹性联轴节等部件,统称为后传动装置。离合器的主要作用是根据船舶运动的需要使主机和轴系能及时准确地分离或结合。它对于船舶运动的机动性有很大的影响,是轴系中最重要的部件之一。离合器按其传递扭矩的方式的不同分为三类:摩擦副式、液力式和电磁式。摩擦副式离合器,它利用摩擦副之间的摩擦力传递力矩。它通常和回行减速器及弹性联轴节组合成一个整体。这种装置操作简便但传递的力矩较小。摩擦副式离合器按照摩擦副之间的紧压力来源不同可分为:①机械力(人力)。通过杠杆和弹簧组成的传动机构将摩擦副紧或脱开这种方式传动的力矩较小(约为3 000 kg-m);②液压力。利用压力油推动压板使摩擦副离合,由于油的压强较高,且液压活塞的面积可做得较大,因此压紧力可相当大,可达8 500 kg-m;③气压力。利用压缩空气压力使摩擦副离合,摩擦副可以是片状、圆锥形和圆柱形的,可传递力矩85 000 kg-m;④电磁力。利用磁性吸力控制摩擦片的离合,这种方式传递的力矩较小,但易于遥控。液力离合器:它的主要元件是两个工作轮——泵轮(主机轮)及涡轮。泵轮和涡轮之间形成环形工作腔,且被众多径向的直叶片隔成许多小的环形流道。液力离合器处于结合状态时,工作腔中充满了滑油,主动轴带动泵轮转动时,它就像离心泵一样,赋予泵轮内的滑油以动能。这股具有动能的滑油通过环形流道冲向涡轮,并在涡轮的环形流道内将动能转变为使涡轮转动的机械能,带动从动轴转动。液力离合器具有以下特点:①由于它靠液体传递动力,在泵轮和涡轮之间没有机械联系。即在发动机与轴系之间是弹性连接从而对扭振振动具有良好的隔绝作用,这一特性使液力离合器常被用来作为动力传动系统中的弹性连接元件。②改变液力离合器的充油度,可以控制传动扭矩的大小,可在发动机转速不变的情况下改变轴系的转速;③当螺旋桨被卡住时可以对发动机起保护作用。且仍能保持很大的扭矩用于克服障碍;④工作中泵轮与涡轮之间存在滑差,其传动效率降低;⑤泵轮与涡轮之间充油与放油的时间较长,机动性较差;⑥尺寸、质量比较大。电磁离合器:它有2个转子分别与发动机及轴系连接,利用电磁感应原理实现离合。其特点是与液力离合器相似,但其滑差较大,制动力矩小,因此实现紧急倒车的功能较差。除了以上3种离合器以外,在联合动力装置中还必须采用自动同步离合器(s.s.s离合器),其主要功能是:①COADG 组合方式中,在巡航状态下使巡航机组与加速机组完全脱离;②在转入全速航行时,当加速机组启动后,转速达到一定值时自动并入;③从全速转入巡航状态时,当加速机组转速降到一定值时自动脱开。在 CODAG 组合方案中,在柴油机输出轴上装有自动同步离合器,其作用是从柴油机驱动的巡航状态转入到燃气轮机驱动的合速状态的过程中,当燃气轮机转速超越柴油机转速时,能防止柴油机超速而自动脱开。减速齿轮箱,在高性能船中高速柴油机和燃气轮机等发动机的转速比螺旋桨允许的工作转速要高许多,因此减速齿轮箱就成为后传动装置中不可缺少的组成部件。其基本功能是降低转速,在动力装置中减速齿轮箱还有其他方面的作用:①调整与改善轴系的布置,如 V 型传动,另外减速器输入轴与输出轴上下、左右安排,也可使发动机与轴系的中心线在不同方向进行调整;②采用并车齿轮箱(2个输入端,1个输出端)可实现发动机同轴并车,增大轴的输出功率,或采用1个输入端,两个输出端实现功率分支;③回行齿轮箱可在发动机转向不变的情况下,使轴系实现正反转运行;④在 CODOG 组合方式的柴-燃联合动力装置中使用双齿轮箱以便柴油机在不同工况下实现并车,并能发出其全部功率。弹性联轴节,它的功能是使发动机与轴系实现弹性连接,从而允许两者的连接有较大的偏差,同时弹性联轴节具有良好的阻尼减振特性,可以起缓冲作用。主要由内轮和外轮两部分组成,由轮轴(花键轴)与主动轴相连,外轮(侧轮)与从动轴相连,在内外轮之间径向布置若干组金属簧片,其一端与外轮元件固定,另一端嵌入花键轴的槽中,利用金属簧片的自由支撑作用,达到内外部件间的扭矩传递。在空间中充满滑油,工作时,簧片间的相对滑动与滑油从侧板间间隙被挤入相邻的油腔的运动形成了高阻尼,削弱了扭振。在传递力矩时,金属簧片的挠曲变形使联轴节具有较好的弹性,起到缓冲使用。除了金属簧片以外,还有采用橡胶作为吸收振动的弹性元件。橡胶的挠性和弹性比金属簧片好,传动无噪声,能吸收扭振能量,但橡胶易老化,但与油类接触,工作时产生的热量不易散出。

Transmission efficiency 船用传动装置效率 传动装置输出功率与输入功率的比值。

Transmission gear size 传动装置外形尺寸 传动装置不包括附属设备的长、宽、高的最大尺寸。

Transmission mechanism 传动装置

Transmission ratio 传动比

Transmission shaft 传动轴(中间轴)

Transmission sound wave 透射声波 系指当入射至两媒介质的界面时,一部分将进入另一种介质中的声波。

Transmission system 传动系统(传递系统)

Transmitting gear 传送齿轮

Transom 艉封板/方艉端面 系指船舶艉端的结构布置和形状。 Transom means structural arrangement and forms of the aft end of the ship.

Transom beam B_T 艉板宽度 B_T 在甲板舷弧线上或甲板舷弧线以下的艉板处的艇体的最大宽度,不包括延伸部分、把手和附件。[注意(1):如将防溅条作为艇纵材或滑行面的一部分,则在测量艉板宽度时可将其包括在内。注意(2):对具有圆弧形或钢管艉柱的艇,或者艉板宽度小于该艇最大宽度之半的艇,其艉板宽度 B_T 为艇体的后 1/4 长度向前至艉艏的在甲板舷弧线上或甲板舷弧线以下的最宽处的宽度]。 Transom beam B_T is the maximum width of the hull at the transom at or below the sheerline, excluding extensions, handles and fittings. Note 1: Where spray rails act as chines or part of the planning surface, they are included in the transom beam measurement. Note 2: For craft with a rounded or pointed stern or with a transom beam of less than half the maximum beam of the craft, the transom beam, B_T, is the widest beam at or below the sheerline at the aft quarter length of the hull forward of the stern.

Transom stern 方型艉 艉端呈平面或曲面，且与中线面成直角或近似直角，船体型线的纵剖线和个各水线则凸出终止于该面的船艉部。

Transonic compressor 跨声速压气机 各级动叶进口气流相对速度沿部分叶高是超声速的轴流式压气机。

Trans-Pacific Partnership Agreement（TTP） 跨太平洋伙伴关系协定 美国推动的一项名为"跨太平洋伙伴关系协定"的贸易协议，主张在9个签约国之间互免进口关税。

Transpacific Trade And Investment Partnership（TTIP） 跨太平洋贸易和投资伙伴关系协定

Transponder 发射应答器 在海床上的装置，响应从船上HPR发出的声响询问并给出船舶相对位置。

Transport contract 运输合同 系承运人将旅客或者货物从起运地点运输到约定地点，旅客、托运人或者收货人支付票款或者运输费用的合同。收货人的主要权利是：承运人将货物运到指定地点后，持凭证领取货物的权利；在发现货物短少或灭失时，有请求承运人赔偿的权利。对于联合运输过程中的货物灭失或毁损的赔偿责任以及赔偿数额，首先适用法律的特别规定或国际公约的规定；发生损害的运输区段不能确定的，由多式联运经营人负赔偿责任，承运人之间的内部责任依约定或法定分配。

Transport document 运输单据 系指证明船舶所有人与发货人之间的运输合同的文件，如海上运货单、提单或多式联运单据。Transport document means a document evidencing a contract of carriage between a shipowner and a consignor, such as a sea waybill, a bill of lading or a multimodal transport document.

Transport efficiency 输送效率（运输效率）

Transport equipment 运输设备 系指某种具有永久性特征的设备。该设备把货物结合或集合为单一的单元进行运输。其包括集装箱、中间散装集装箱、可重复使用货物单元和船载驳，但不包括起货机械。Transport equipment means equipment of a permanent character that facilitates the transport of a combination or aggregation of cargo as a single unit and includes containers, intermediate bulk containers, returnable cargo units and shipborne but does not include cargo lift equpment.

Transport package 运输包装 系指以满足运输储存要求、保护产品为主要目的的包装。它具有保障产品的安全，方便储运装卸，加速交接、点验等作用。

Transport related service 与运输有关的服务 系指诸如堆存、舱储、装/卸货、积载、平舱、隔垫和绑扎等的服务。Transport related service means such service: warehousing, loading, unloading, stowage, trimming, dunnaging and lashing.

Transport ship 运输船 各种专门担负运输任务的船舶的统称。包括客船、客货船、货船、渡船、驳船等。

Transportable moisture limit（TML）of a cargo which may liquefy 可流态化物质的适运水分极限（TML） 系指对于使用不符合固体散货安全操作规则第7.3.2节特别规定的船舶因数的货物而言，视为安全的最大含水量。该极限值用主管当局认可的使用程序确定。Transportable moisture limit（TML）of a cargo which may liquefy means the maximum moisture content of the cargo which is considered safe for carriage in ships not complying with the special provisions of subsection 7.3.2 on BC Code It is determined by the test procedures, approved by a competent authority.

Transshipment B/L 转船提单 系指从装运港装货的船舶，不直接驶往目的港，而需中途换装另外船舶所签发的提单。在这种提单上要注明"转船"字样。

Transverse bulkhead 横舱壁 系指沿船舶宽度的方向设置的舱壁。

Transverse bulkhead complete（甲板强横梁） 完整的横向舱壁 系指包括纵桁系统及相邻构件的构件。Transverse bulkhead complete means structural members including girder system and adjacent structural members.

Transverse bulkhead lower part 横舱壁下部 系指包括梁系及相邻结构构件。Transverse bulkhead lower part means structural members including girder system and adjacent structural members.

Transverse center of buoyancy 浮心横向坐标 船舶浮心至中线面的距离。

Transverse center of floatation 漂心横向坐标 船舶漂心至中线面的距离。

Transverse center of gravity 重心横向坐标 船舶重心至中线面的距离。

Transverse hold frame 主肋骨 系指从防撞舱壁至艉尖舱之间包括机器处所在内的，最下层甲板下的肋骨。

Transverse metacenter 横稳心 船舶横向倾斜时的稳心。

Transverse metacenter height 横稳心高 横稳心在重心之上的高度。

Transverse metacenter radius 横稳心半径 船舶横向倾斜时的稳心半径。

Transverse propeller（transverse thrust unit） 横向推力器

Transverse ring 横向环肋 系指在双层底肋板、

竖桁和甲板横向桁材处,船体横截面内所有横向材料。

Transverse ring means all transverse material appearing in a cross-section of the ship's hull, in way of a double bottom floor, vertical web and deck transverse girder.

Transverse section 横剖面 (1)系指包括所有的纵向构件,例如船壳板、甲板纵骨及大梁、舷侧、船底、内底及漏斗形舷侧外板、纵舱壁及最上面的舷顶边舱的船底板。对于横框架结构船舶,横剖面包括临近框架结构及其横坡面上的终端连接。(2)系指包括在横剖面上的甲板,船侧外板,船底板,内底板,纵舱壁上的纵骨和纵桁等总纵附件。横向结构的船舶包括沿着横剖面交叉的纵骨及其端部的肘板。 Transverse section includes: (1) all longitudinal members such as plating, longitudinals and girders at the deck, sides, bottom, inner bottom and hopper side plating, longitudinal bulkheads and bottom plating in top wing tanks. For transversely framed vessels, a transverse section includes adjacent frames and their end connections in way of transverse sections. (2) all longitudinal members such as plating, longitudinals and girders at the deck, sides, bottom, inner bottom and longitudinal bulkhead. For transversely framed vessels, a transverse section includes adjacent frames and their end construction in way of transverse sections.

Transverse section of a bulk carrier 散货船的横剖面 系指包括所有的纵向构件,例如船壳板、甲板、舷侧、底部、内底、顶边舱、纵舱壁、顶边舱底等的纵桁和纵骨。 Transverse section of a bulk carrier includes all longitudinal members such as plating, longitudinals and girders at the deck, sides, bottom, inner bottom, hopper sides, longitudinal bulkheads and bottom in top wing tanks.

Transverse section of a double skin bulk carrier 双壳散货船的横剖面 系指包括所有的纵向构件,例如船壳板、甲板、舷侧、底部、内舷侧、底边舱和顶边舱的纵骨和纵桁。 Transverse section of a double skin bulk carrier includes all longitudinal members such as plating, longitudinals and girders at the deck, sides, bottom, inner sides, hopper sides and top wing inner sides.

Transverse space of two legs 桩腿横距 三条、四条桩腿等的自升式钻井平台垂直于设计拖航方向两桩腿中心线之间的距离。

Transverse stability 横稳性 船舶横向倾斜时的稳性。

Transverse strength 横向强度 船体结构承受相应载荷的能力。

Transverse system of framing 横骨架式 横向骨架较密、纵向骨架较疏的骨架形式。

Transverse thruster 侧推器(侧向推力器,横向推力器) 系指发出侧向推力,以用于机动目的的侧向推力器。 Transverse thruster is an athwartship thruster developing a thrust in a transverse direction for manoeuvring purposes.

Transverse vibration 横向振动

Transverse wave 横波 系指:(1)质点的振动方向与声波传播的方向相互垂直的形式。这仅在固体介质中发生,如机器振动传递至船体结构的噪声就是横波。(2)船舶航行时,自艏、艉端向后扩散的且与船航行方向垂直的波系。

Transverse web frame 横向强框架 系指连接纵向构件的主要横向桁材。 Transverse web frame mean those primary transverse girders which join the ships longitudinal structure.

Transversely framed ship 横骨架船 船体结构中横向骨材比纵向骨材较密的船舶。

Transversely framed side structures 横骨架式舷侧结构 系指由水平桁架的横向肋骨组成的舷侧结构。 Transversely framed side structures means the structures built with transverse frames possibly supported by horizontal side girders.

Trap 陷阱

Travel expenses 差旅费

Travel time 行进时间(移动时间) 系指船上所有人员从听到通知并向集合站移动直至到达登乘站的时间。 Travel time is defined as the time it takes for all persons on board to move from where they are upon notification to the assembly stations and then on to the embarkation stations.

Traveling crane 行车 包括集装箱吊(container crane)、单梁吊(trolley hoist, monorail crane)、天花板行车(overhead traveling crane)和电动卷扬机(electric hoist)四类。

Trawl drum 拖网卷筒机 由卷筒和原动机组成。用来绞收并卷绕拖网网衣的机械。

Trawl hoisting gantry 吊网门形架 渔船上,为便于起网操作而设置的带有各种导向滑车的门形构架。

Trawl winch 拖网绞钢机 由摩擦鼓轮、绳索卷筒和动力源组成,用来绞收拖网曳钢和起吊渔获物的机械。包括串联式绞钢机、并联式绞钢机、分列式绞钢机和卧式绞钢机等。

Trawler/trawler boat 拖网渔船 是用拖网来捕捞水域底层及中下层鱼虾等水产物的渔船。拖网渔船上的拖网,是利用甲板上的绞车来收网的。拖网渔船捕捞到的渔获物,储藏在船上的冷库中。拖网捕鱼是一种效果好、适用范围广的捕鱼方法。

Tray 盘 系指物资装运设备的一个部件,其设计用于反复转送货物。该盘具有附件以供起吊或其他转送之用,但并不包括运输设备或集运架。 Tray means an article of materials handling equipment designed for repeated use in conveying cargo. It has attachments by which it may be hoisted or otherwise conveyed, but it does not include transport equipment or a pallet.

Tray tower 盘式塔

Tread 踏板 系指斜梯的梯级或垂直通道开口的踏板。 Tread means the step of an inclined ladder or step for the vertical access opening.

Treated sewage 经处理的污水 系指符合经修正的"联邦水域防污染法令"和"公共法"(Pub. L)106.554的第ⅩⅣ篇"某些阿拉斯加大型邮轮营运",以及根据上述两者之一颁布之规则的所有适用的废液排放限制标准和处理要求的污水。 Treated sewage means sewage meeting all applicable effluent limitation standards and processing requirements of the Federal Water Pollution Control Law 106-554 "Certain Alaskan Cruise Ship Operations" and regulations promulgated under either.

Treatment rated capacity(TRC) 额定处理功率 系指压载水管理系统进行形式认可的最大连续功率(m^3/h)。其表述了为满足国际船舶压载水和沉积物控制和管理公约 D-2 标准的要求,压载水管理系统每单位时间能处理压载水的数量。 Treatment rated capacity (TRC) is the maximum continuous capacity expressed in cubic metres per hour for which the BWMS is type approved. It states the amount of ballast water that can be treated per unit time by the BWMS to meet the standard in Regulation D-2 of the international Convention for the Control and Management of Ships' Ballast Water and Sediments, 2004.

Treatment(the control of ships' bio-fouling) 处理(船舶生物污底控制) 系指可能使用机械的、物理的、化学的或生物的方法来消除或抑制入侵或潜在入侵水生物种在船上结垢或繁殖的过程。 Treatment means a process which may use a mechanical, physical, chemical or biological method to remove or render sterile, invasive or potentially invasive aquatic species fouling a ship.

Treaty on Antarctica 南极条约 1959年12月1日由阿根廷、澳大利亚、比利时、智利、法兰西共和国、日本、新西兰、挪威、南非联邦、苏维埃社会主义共和国联盟(1991年苏联解体后分裂出15个国家:东斯拉夫三国、波罗的海三国、中亚五国、外高加索三国、摩尔多瓦)、大不列颠及北爱尔兰联合王国和美利坚合众国政府等国建立,承认南极洲永远继续专用于和平目的和不成为国际纠纷的场所或对象,是符合全人类的利益的;确认在南极洲进行科学调查方面的国际合作对科学知识有重大的贡献。旨在约束各国在南极洲这块地球上唯一一块没有常住人口的大陆上的活动,确保各国对南极洲的尊重。该条约中规定,南极洲系指南纬60°以南的所有地区,包括冰覆盖层,总面积约5 200万平方千米。条约的主要内容是:南极洲仅用于和平目的,促进在南极洲地区进行科学考察的自由,促进科学考察中的国际合作,禁止在南极地区进行一切具有军事性质的活动及核爆炸和处理放射物,冻结目前领土所有权的主张,促进国际在科学方面的合作。

Treaty on the Non-Proliferation of Nuclear Weapons(NPT) 不扩散核武器条约 又称"防止核扩散条约"或"核不扩散条约",制定目的是禁止核扩散,具体是禁止非核国拥有核武器、核国向非核国转让核武器的国际法律条约。该条约在1969年6月12日于联合国大会上通过。

Trebling bit(Tribit) 三重比特

Tree 海底采油树 系指由管路和安装在采油井顶部的,用来调节进出采油井液体流量的阀组成的装置。 Tree is the assemblage of tubes and valves mounted at the top of a well used to regulate the flows in and out of the well.

Trencher 开沟机 用于管子或控制线埋设的水下运载工具。

Trial manoeuvre 试航操纵 为导航和避碰目的,通过显示所有被跟踪目标和AIS目标预计未来状况作为本船模拟操纵的结果,协助操作员进行计划操纵的设备。 Trial manoeuvre is facility used to assist the operator to perform a proposed manoeuvre for navigation and collision avoidance purposes, by displaying the predicted future status of all tracked and AIS targets as a result of own ship's simulated manoeuvres.

Trials at the moorings 系泊试验 是验证以下各项的试验:(1)机器的满意运行;(2)提供命令的快速和容易响应性能;(3)各种装置的保护,对以下各项:①机械部件的保护;②人员的安全保护;(4)清洗、检查和维护的方便性。如认为上述各项并不满意,仍需要修正或更新的话,则船级社保留在这些修正或更换后随所有或部分项目重做系泊试验的权利。 Trials at the moorings are to demonstrate the following: (1) satisfactory operation of the machinery;(2) quick and easy response to operational commands;(3) protection of the various installations, as regards:①the protection of mechanical parts;②the safeguards for personnel;(4) accessibility for cleaning, inspection and maintenance. Where the above features are not deemed satis-

Triangular plate 三角眼板 开有三个孔供连接起货索具用的三角形金属板。

Triatic stay（jumper stay） 桅间张索 支张在两桅上端之间的绳索。

Tribble skid way 立根输道 将单根钻杆接成立根并把它们不断输送到钻井平台和堆场的一整套联合装置。

Trichloroethylene evaporator 三氯乙烯蒸发器 以蒸发三氯乙烯（C_2HCl_3）液体来清洗滑油冷却器内污油的设备。

Tricolour stern light 三色艉舷灯（三色舷灯） 设置在船舶规定处，并在水平面上 360°内，及自艏向至左舷 112.5°为红光，至右舷 112.5°为绿光，自左、右舷 112.5°至 180°内为白光，显示红、绿、白三色不间断光的灯具。

Trim 纵倾 系指船艏吃水与船艉吃水之差，吃水分别在前端点和后端点量取，不计及龙骨斜度。Trim is the difference between the draught forward and the draught aft, where the draughts are measured at the forward and aft terminals respectively, disregarding any rake of keel.

Trim（ming）angle 纵倾角 船舶纵倾时的水线平面与正浮时的水线平面之间的夹角。

Trim by the bow 艏倾 船舶相对于正浮状态，艏吃水增大的纵向倾斜。

Trim by the stern 艉倾 船舶相对于正浮状态，艉吃水增大的纵向倾斜。

Trim of WIG craft 地效翼船的纵倾 在排水或滑行等其他状态下，系指艏吃水和艉吃水之间的差异。Trim of WIG craft means that when applied to the displacement or other modes up to and including planning, means the difference between forward and aft draughts.

Trim stability and longitudinal strength calculation 纵向稳性及纵向强度计算

Trim/Longitudinal inclination 纵倾 船舶自正浮位置向船艏或船艉倾斜的一种浮态。船舶纵倾后，船体中纵剖面仍垂直于水面，舯剖面与铅垂平面相交的角度，也即船舶在正浮时的水线面与纵倾后水线面相交的角度，成为纵倾角。

Trimming 平舱 系指对货物处所内的部分或全部货物进行的任何平整工作。平整中可利用装货喷管或滑槽，也可以利用移动机械、设备或人工进行。Trimming means any levelling of a cargo within a cargo space, either partial or total, by means of loading spouts or chutes, portable machinery, equipment or manual labour.

Trimming moment 纵倾力矩 使船舶产生纵向倾斜的外力矩。

Trimming pump 纵倾平衡泵（纵倾平衡系统泵） 抽送舱柜中的水或油类以调节船舶纵倾平衡的泵。

Trimming tank 纵倾压载水舱 为调整起重船工作时产生的纵倾，设置在远离起重机另一端甲板下面的专用压载水舱。

Trinitrotoluene（TNT） 三硝基甲苯 又称黄色炸药，是一种烈性炸药，由 J·威尔勃兰德发明，呈黄色粉末或鱼鳞片状，难溶于水，可用于水下爆破。由于威力大，常用来做爆药。爆炸后呈负氧平衡，产生有毒气体。性质稳定，不易爆炸，即使直接被子弹击中也不会引爆。需要雷管进行引爆。

Triple expansion engine 三胀式蒸汽机 蒸汽依次在高压、中压及低压缸内进行三次膨胀，将蒸汽热能转换成机械功的蒸汽往复机。

Triple-drum boiler（triple-header boiler） 三锅筒锅炉 具有 3 个锅筒的水管锅炉。

Triple-purpose carrier/oil bulk and pre carrier 三用散货船 在构成一个周期的各阶段中，可分别载运石油、粮食或煤、矿砂的散货船。

Triple-shaft reduction gear 三轴并列减速齿轮箱 由汽轮机高、中、低压气缸拖动 1 个轴系的减速齿轮箱。

Triplex DP 三重 DP 三重 DP 控制系统。能用投票决定并从 3 个系统中去掉 1 个有缺陷的系统。

Tripod 三脚式桩基 海上风能转换系统几种可能的桩基中的一种。

Tripod mast 三角桅 由 1 根竖立的主杆和两根斜杆所构成的桅杆。

Tripping bracket 防倾肘板 系指用以加强承受压缩力构件的肘板，以抵抗扭力。 Tripping bracket means a bracket used to strengthen a structural member under compression, against torsional forces.

Tripping bracket of foundation 基座肘板 基座桁侧面的肘板。

Tripping device 脱扣装置（释放装置）

Trochoidal wave 坦谷波 波峰陡、波谷平坦，其一切水质点绕固定中心的轨圆按定长角速度旋转的几何规则波。其波形任一质点的坐标为 $\chi = \lambda/2\pi + \gamma \sin\theta$，$z = \cos\theta$。

Trojan 木马 是一种计算机病毒。它是目前比较流行的病毒文件，它通过将自身伪装吸引用户下载执行，向施种木马者提供打开被种者电脑的门户，使施种者可以任意毁坏、窃取被种者的文件,甚至远程操控被种者的电脑。使你的机器远程被黑客控制,经常是为拒绝服务攻击(DOS)。

Trolling boat (troller) 拖钓渔船 在船尾或船舷拖曳若干条鱼饵钓,以钩吊上层鱼类,如马鲛鱼、金枪鱼等的钓渔船。

Tropical cyclone 热带气旋 系指国际气象组织所属各国家气象服务机构的通用术语。根据地理位置也可以使用飓风、台风、气旋、强热带风暴等词。 Tropical cyclone is the general term used by national meteorological services of the meteorological organization. The term hurricane, typhoon, cyclone, severe tropical storm, etc. may also be used, depending on the geographical location.

Tropical freeboard 热带干舷 是从夏季干舷中减去夏季吃水的1/48。

Tropical fresh water(TFW) 热带淡水 系指夏通湖的热带淡水,其密度在29.4 ℃(85 °F)时为0.995 4 g/cm³。[注意:大船在进入淡水后通常会增加7.5~10 cm(3~4 in)艏倾]。 Tropical fresh water(TFW) mean the tropical fresh water of Gatun Lake, density 0.995 4 tons/m³ at 29.4 ℃. (Note: Transition to fresh water frequently alters the trim of large vessels 7.5 to 10 centimeters by the head).

Tropospheric refraction correction 对流层折射校正 从地面到离地面20 km处的低空大气层称为对流层。当电波穿过对流层时将产生折射,由此引起的定位误差约0.02 n mile(40 m),对流层的折射影响取决于该层内的大气结构和卫星仰角,因此不能由设备来测定和消除,对于高精度的接收机可采用数字的方法进行对流层折射校正,根据要求也可采用近似公式。

Truch 桅冠 位于顶桅顶端,设有悬挂号旗或支张无线电天线用的滑轮、桅顶灯、避雷针等的半球状结构。

True bearing 真方位 系指自本船基准位置或其他目标位置的目标方向,以自真北的角位移表示。 True bearing means direction of a target from own ship's reference location or from another target's position expressed as an angular displacement from true north.

True course 真航向 系指相对地面或海面的目标运动方向,以真北的角位移表示。 True course means a direction of motion relative to ground or to sea, of a target expressed as an angular displacement from true north.

True motion 真运动 系指本船以其真运动移动的一种显示。 True motion means a display across which own ship moves with its own true motion.

True projection of blade section 叶切面实投影

螺旋桨图中,叶切面在相应半径的水平线上的投影。

True speed 真速度 系指相对地面或海面的目标的速度。 True speed means a speed of a target relative to ground, or to sea.

True vector 真矢量 系指代表目标预计真运动的矢量,显示参照地面的航向和航速。 True vector means a vector representing the predicted true motion of a target, showing course and speed with reference to the ground.

Trunk 围壁通道 是类似于甲板室的甲板建筑物。但不设有下甲板。 Trunk is a decked structure similar to a deckhouse, but not provided with a lower deck.

Trunk 围井 为供人员上下、设备、货物的升降或通风照明等目的设置的竖向围壁通道。

Trunk facilities 通道设施 系指门、人孔和诸如服务处所和管隧之间的类似开口。

Trunk opening 通道开口 系指货物处所和管隧之间的人孔。

Trunk piston type diesel 筒形活塞柴油机

Trunk piston type diesel engine 筒型活塞柴油机 活塞直接与连杆相连的柴油机。

Trunk ventilator 围壁通风筒

Truss spar 桁架柱体式平台/桁架型立柱式平台 第二代立柱平台。是在传统立柱基础上缩短了圆柱的高度,取而代之的是增加桁架和重块,并在桁架结构上增加一些阻尼板以进一步改善立柱平台的垂荡性能。

Trust certificate 信托证书 是证明其持有人已将股份及表决权转让与表决权受托人并不受益人地位的法律凭证,该证书的持有人有权要求公司对表决权受托人支付的股息、红利等。

Trust of propeller 螺旋桨推力 螺旋桨轴向发出的力。

Tsunami (seismic wave) 海啸波 由于海底地震或火山爆发而在海上形成的破坏性极大的海浪。

Tube 油管 是在钻探完成后将原油和天然气从油气层运输到地表的管道,它用以承受开采过程中产生的压力。油管的外径一般为60.3 ~ 114.3 mm。

Tube arrangement 管道布置

Tube bender 弯管机

Tube bending machine 弯管机

Tube brush 管刷

Tube bundle (tube nest) 管束 由一定数量的冷却水管按规定方式排列在一起的组合体。

Tube cleaner 清管器

Tube drawing bench 排管机

Tube expander 扩管器(管辘,辘)

Tube joint 管接头(管节点)

Tube packing 管子填料函垫装 利用某种填料、金属环和压紧套，使冷却水管紧固在管板孔中的一种方法。

Tube plate（tube sheet） 管板 表面式冷凝器中固定冷却水管两端的板。

Tube shaft（stern shaft） 艉轴（艉管轴） 轴系中，从舱内伸出舷外与推进器轴联结的轴。

Tube stopper 管塞

Tube support plate 支承隔板 在表面式冷凝器的蒸气空间中用来支撑冷却水管的板。

Tube welding 管子焊

Tube wrench 管子钳

Tube-annular combustion chamber（can-annular combustion chamber, cannular combustion chamber） 环管燃烧室 在内外壳之间的环腔内配置的若干独立管状火焰筒组成的燃烧室。

Tubeshaft 艉管轴

Tubeshaft survey 艉管轴检验

Tubing 管路（管系）

Tubing coupling 管件连接接头

Tubing joint 油管接头

Tubular 管的（管形的，由管构成的）

Tubular boiler 水管锅炉

Tubular column 管柱

Tubular combustion chamber 管型燃烧室 一般沿周向均匀布置的由管型外壳及火焰筒组成的燃烧室。

Tubular cooler 管式冷却器

Tubular electrode 管状焊条

Tubular heater 管式加热器

Tubular joint 管状接头

Tubular member 管状构件

Tubular pillar 管型支柱 用管型材料制成的支柱。

Tug 拖船 系指设有拖曳设备，专用于在水上拖曳船舶或其他浮体的船舶。Tug is a ship fitted with towing equipment, dedicated to towing ships or other floating objects on water.

Tug 港作拖船 系指在港口及其附近作业的拖船。

Tug boat / tug / towboat 拖船 系指自身有动力装置，设有拖曳设备，专用于在水上拖曳船舶或其他物体的船舶。

Tug master 拖船船长 系指拖航作业时拖船的船长。

Tug, anchor handling and supply vessel 三用拖船 系指拖曳钻井平台转移井位，协助就位和起、抛锚作业，以及向钻井平台与生产平台供应钻管、泥浆、水泥、燃油、钻井用水等生产物品和食品、淡水等生活补给品的船舶。

Tug/offshore supply/fire fighting ship 多用途拖船 系指具有供应、消防等3种以上功能的拖船。Tug/offshore supply/fire fighting ship is a multi-purpose tug with more than three functions such as supply, fire safety etc.

Tuggage 挂拖 拖船通过拖钩和拖缆挂连船后与被拖的船舶或浮物一起行驶的一种拖带作业方式。

Tumble home 内倾 系指设计水线以上船长中部舷侧表面向内倾斜。其值以最大横剖面上自横梁上缘线与龙骨外缘线之交点至中线面的水平距离与设计水线半宽值之差表示。

Tuna long-line ship 金枪鱼延绳钓船 系指利用延绳钓的方法钓获金枪鱼的渔船。

Tungsten gas arc welding（TIG） 钨极惰性气体保护焊 系指用钨棒作为电极，用氩或氦作为保护气体的焊接工艺。电弧熔化母材形成接头，必要时还可加入填充焊丝。钨极惰性气体保护焊的特点是电弧稳定，输入能量易于控制。因此多用于焊接尺寸精度要求较高、材料易于过热脆化和在空气中易于氧化的工件。

Tungsten inclusion 钨夹杂物

Tuning factor for heaving 垂荡调谐因数 垂荡固有周期对波浪遭遇周期的比值。$\Lambda_Z = T_Z / T_E = \omega_E / \omega_Z$。

Tuning factor for pitching 纵摇调谐因数 纵摇固有周期对波浪遭遇周期的比值。$\Lambda_\theta = T_\theta / T_E = \omega_E / \omega_\theta$。

Tuning factor for roll 横摇调谐因数 横摇固有周期对波浪遭遇周期的比值。$\Lambda_\Phi = T_\Phi / T_E = \omega_E / \omega_\Phi$。

Tunnel bearing（tunnel shaft bearing） 中间轴承 支承中间轴旋转运动的轴承。

Tunnel escape 轴隧逃生口/轴隧应急出口 系指设在轴隧端室邻近艉尖舱舱壁处，直通露天甲板的应急出口。

Tunnel flat 轴隧平台

Tunnel head line 隧道顶线 系指隧道型艉船舶，通过其轴系中心线并垂直于基平面的平面与船体型表面底部的交线。

Tunnel shaft 艉轴

Tunnel stern 隧道型艉 为提高推进效率，增大螺旋桨直径，将螺旋桨所在处船体底部线型局部设计成隧道状拱形曲面，并向前延伸与相邻的船体型表面平顺连接的船艉部。

Tunnel stool 中间轴承座 中间轴承承受负荷的底座。

Tunnel thruster 导管推力器（管隧推力器）

Tunnel trunk 轴隧通道

Tunnel type air cooling freezer 隧道吹风冻结装置　冻结舱内一侧布置有密集的冷却盘管和通风设备，货物在另一侧隧道内自动缓慢连续地进出的冻结装置。

Tunnel way 地轴弄走道

Turbidimeter 浊度计　利用光透射或散射（或两者兼用）方法测量海水浊度的仪器。

Turbine 涡轮机（透平机）　使高温燃气或其他气体在其叶片通流部分中膨胀并将部分热量变为机械功输出的叶轮式动力机械。

Turbine driven 透平传动

Turbine driven feed pump 汽轮给水泵　由汽轮机驱动的给水泵。

Turbine dynamo 涡轮直流发电机

Turbine electric generator 透平发电机

Turbine foundation 涡轮机底座

Turbine oil 涡轮机油

Turbine oil storage tank 涡轮机滑油贮存柜　贮存废气涡轮增压器或辅汽轮机用的润滑油的柜。

Turbine performance characteristic diagram (curve) 涡轮特性曲线　表示涡轮各特性参数之间关系的曲线。

Turbine plenum engine 涡轮增压发动机　依靠涡轮增压器来加大发动机进气量的发动机。涡轮增压器利用发动机排出的废气作为动力来推动增压器中的涡轮，涡轮转动的同时带动增压器中的压气机的叶轮，叶轮压缩通过滤清器管道送来的新鲜空气，再送入气缸。当发动机转速加快时，高速的废气推动涡轮提速，空气压缩程度就加大，进气量大幅度增加，气缸内燃料燃烧更充分，动力输出也就更高。见图 T-8。

图 T-8　涡轮增压发动机
Figure T-8　Turbine plenum engine

Turbine plenum technology 涡轮增压技术　增加气缸中压缩量的技术。在引入涡轮增压技术前，柴油机气缸中的空气靠自然吸气来完成。但自然吸气存在空气密度低，燃料燃烧不充分的严重缺陷。引入涡轮增压技术后，涡轮增压器可以大幅度提高柴油机的进气密度，使燃料燃烧更充分，大幅度地增加柴油机输出功率，同时可以减少有害气体的排放量。

Turbine room 涡轮机舱

Turbine steam seal system 气封系统　用蒸汽防止汽轮机转子和气缸之间漏气或空气漏入的设备、管路及附件。

Turbine washing test 清试洗验　为求得消除压力机叶片积垢的最佳清洗参数和条件，在燃气轮机低速运转下所做的喷入清洗介质的试验。

Turbine-driven oil pump lubricating oil tank 汽轮货油泵滑油柜　油船中，贮存供汽轮机货油泵使用的润滑油的柜。

Turbine-driven pump 涡轮机驱动泵

Turbo fuel 燃气轮机燃料

Turbo-alternator 涡轮交流发电机组

Turbo-blower 汽轮鼓风机（透平鼓风机）　汽轮机驱动的汽轮鼓风机组。

Turbo-charged diesel 涡轮增压柴油机

Turbo-charger 废气涡轮增压器

Turbo-charger lubricating pump 增压器滑油泵　将润滑油抽送至涡轮增压器轴承所用的泵。

Turbo-charging 涡轮增压

Turbo-compressor 涡轮压缩机

Turbo-electric propulsion 涡轮机电力推进

Turbo-feed pump 涡轮给水泵

Turbo-generator 涡轮发电机组

Turbo-generator room 涡轮发电机舱

Turbo-machinery（turbo-maching） 叶轮机械　诸如压气机和涡轮等以流体为工质，以叶片为进行能量转换的主要工作元件的工作机械或动力机械。

Turbo-pump 涡轮泵

Turbo-supercharger 废气涡轮增压器

Turbo-supercharger matching test 增压器配机试验　检查增压器与柴油机配合性能是否良好的试验。

Turbo-supercharging 废气涡轮增压　用排气驱动涡轮增压器进行的增压。

Turbulence 湍流（紊流，湍流度）

Turbulence detector 湍流探测器　探测边界层是否属于湍流流态的设备。

Turbulence Reynolds stress 湍流雷诺应力（紊流雷诺应力）

Turbulence stimulater 激流装置/湍流装置　船模

在试验水池中进行拖曳试验时,安装在船模首部,促使船模边界层变成湍流的装置。

Turbulent boundary layer 湍流边界层(紊流边界层)

Turbulent flow 湍流 速度和方向都出现明显紊乱变动的流态。

Turbulent flow air register 旋流式调风器 一种带有旋流器使燃烧用空气形成旋转气流喷入炉膛的调气器。

Turn 转动(转数,扭曲)

Turn buckle 旋转张紧器

Turn counting dial 转动计数度盘

Turn down ratio 燃烧室调节比 燃烧室的最大燃烧量与最小燃烧量的比值。

Turn table 网台 围网或拖网渔船上,供盘放网具和起放网操作用的平台。

Turnbuckle 松紧螺丝扣(花篮螺丝,伸缩螺杆)一种形式的张紧装置,用于连接绑扎杆和绑扎点。

Turning 转向 系指船舶在规定的风况和海况下以最大营运航速航行时,船舶航向的变化。Turning is the change of direction of a craft's track at its normal maximum operating speed in specified wind and sea conditions.

Turning ability 回转能力 是衡量船舶使用满舵回转的能力。结果是"艏向改变90°时的最小纵距"和由"艏向改变180°时的最小纵距"定义的"战术直径",对最终回转直径的分析还有其他益处。Turning ability is the measure of the ability to turn the ship using hard-over rudder. The result being a minimum "advance at 90° change of heading" and "tactical diameter" defined by the "transfer at 180° change of heading", analysis of the final turning diameter is of additional interest.

Turning ability index 回转性指数 船舶回转时用以表示定常回转角速度与对应舵角关系的一种指数。

Turning circle 回转圆 船舶以不变的舵角回转时,其重心的运动轨迹。

Turning circle manoeuvre 回转性能试验 系指在偏航率几乎为零时,分别操左、右35°舵角或该测试速度下允许的最大舵角时的操纵性。 Turning circle manoeuvre is the manoeuvre to be performed to both starboard and port with 35° rudder angle or the maximum rudder angle permissible at the test speed, following a steady approach with zero yaw rate.

Turning circle test 回转试验 用全航速和最大舵角及15°舵角,分别向左、右舷进行回转操纵达540°转艏角回转圈的试验。亦宜另加中速和低速的试验。

Turning device 回转装置

Turning force 旋转力

Turning gear(barring gear) 盘车装置(盘车机,回转装置) 盘车时使汽轮机转子间断或连续地低速转动的装置。

Turning gear test 盘车试验 用盘车装置对主汽轮机转子进行盘车操作,以初步检查主汽轮机装配质量以及盘车装置工作可靠性的试验。

Turning lag index 应舵指数 用以表示船舶操舵时对舵角响应的时间的滞后的一种指数。

Turning moment 回转力矩

Turning out gear 摇倒机构 摇倒式艇架中,供摇转而使吊艇臂或吊艇杆倾倒的机构,一般有伸缩螺杆和螺杆、弧形齿牙等形式。

Turning performance 回转性能

Turning period 回转周期 回转试验中,船舶从初始直线航向转360°时所经的时间间隔。

Turning quality 回转性 船舶应能绕瞬时回转中心做圆弧运动的性能。

Turning speed 转速

Turning test from zero-speed 零速回转试验 操最大舵角,从零速度用半速正车分别向左、右舷执行回转操纵达180°转艏角为止的试验。

Turning upper platform 可转上平台 能旋转的舷梯上平台。

Turret mooring 转塔单点系泊

Turret mooring system(TMS) 转塔式系泊系统

TV remote control device 电视机遥控器

t_w(**mm**) 普通扶强材或主要支撑构件的腹板净厚(**mm**) t_w means the net web thickness (mm) of a ordinary stiffener or a primary supporting member.

Tween deck 中间甲板 其在此处所用英语"tween deck"是"between deck"的缩写,其位于上甲板和货油舱的舱顶之间。 Tween deck is an abbreviation of between decks, placed between the upper deck and the tank top in the cargo tanks.

Tween-deck bulkhead 甲板间舱壁 两层甲板之间的舱壁。

Tween-deck frame 甲板间肋骨 两层甲板之间的肋骨。

Tween-deck tonnage 甲板间吨位 总吨位中位于量吨甲板与最上层全通甲板之间的吨位。

Twelve-nautical-mile zone 12 海里区域 系指沿海国家领水区域。即离岸12海里(22.2 km)的沿海区域。

Twenty-foot equivalent unit(TEU) 标准集装箱 系指一只集装箱的国际度量标准(相当于20ft单元),

$20'×8'×8.5'$。 Twenty-foot equivalent unit (TEU) means an international measure standard for a container (20-foot equivalent unit) which is $20'×8'×8.5'$.

Twin pontoon column stabilized semi-submersible drilling unit 双下浮垫柱稳半潜式钻井平台 系指为简化结构及改善拖航性能,将下部结构设计成2个前后向的箱形或其他形状浮体的柱稳、半潜式钻井平台。

Twin pump 双联泵

Twin rope grab 双索抓斗 用2根绳索和2个滚筒控制启闭、抛落和提升的抓斗。

Twin span derrick (twin tackles derrick) 双千斤索吊杆装置 无牵索索具,仅有2组左、右分开的千斤索具操纵吊杆的吊杆装置。

Twin-single pump 双联单作用泵

Twisted blade 扭叶片 安装角沿高变化的叶片,分等截面扭叶片和变截面扭叶片两种。

Twistlock 扭锁 安装在角件中的设备,用于系固上层的集装箱,可承受拉力、压力和剪切力。

Twist-pair 双绞线 是综合布线工程中最常用的一种传输介质。双绞线采用了一对互相绝缘的金属导线互相绞合的方式来抵御一部分外界电磁波干扰。把两根绝缘的铜导线按一定密度互相绞在一起,可以降低信号干扰的程度,每一根导线在传输中辐射的电波会被另一根线上发出的电波抵消。任何材质的绝缘导线绞合在一起都可以叫作双绞线,同一电缆内可以是一对或一对以上双绞线,一般由两根22～26号单铜导线相互缠绕而成,也有使用多根细小铜丝制成单根绝缘线的(这与集肤效应有关)。实际使用时,双绞线是由多对双绞线一起包在一个绝缘电缆套管里,典型的双绞线有一对的,有四对的,也有更多对双绞线放在一个电缆套管里,这些人们称之为双绞线电缆。

Two stage bidding 两段招标 系指无限竞争招标和有限竞争招标的综合方式,先用公开招标,再用选择性招标,分两段进行,一般适用于技术复杂的大型招标。

Two-compartment subdivision 两舱制 系指船舶相邻两舱破损进水后,仍能满足破舱水线不超过限界线要求的设计指标。

Two-cycle engine 两冲程发动机

Two-cylinder turbine 双缸式涡轮机

Two-drum boiler (double-header boiler) 双锅筒锅炉 具有2个锅筒的水管锅炉。

Two-element air ejector 双组空气抽逐器

Two-nodded vibration 双节点振动

Two-pass condenser 两路式冷凝器

Two-phase jet propulsion 双态喷射推进 用水或空气或燃气作为推进剂的喷射推进。

Two-screw pump 双螺杆泵

Two-stage air ejector 双级空气抽逐器

Two-stage supercharging 两级增压 两只增压器串联应用的增压方式。

Two-stage turbocharged technology 两级涡轮增压技术 以往的柴油机采用一级涡轮增压技术,其产生的压力相对较低,使压缩进气缸的空气量有限,气缸中的燃料不能充分燃烧,这不仅浪费了大量燃料,也使柴油机的废气排放量偏高。通过使用两级涡轮增压器来增大压力,增加气缸中的空气量,解决一级涡轮增压器压力不足的问题。两级涡轮增压器系统是由两个大小不同的涡轮增压器串联组成。其工作原理是:利用发动机工作产生的废气的能量,驱动体积较小的、增压度高的涡轮增压器(第一级),然后再驱动体积大的、增压度较低的涡轮增压器(第二级)。低压比涡轮增压器的压缩机将周围的空气压缩,然后经过一个直接相连的冷却器,将压缩后的空气传送到高压比涡轮增压器的压缩机中,在此之前被压缩过的空气再次被压缩,再经过一个空气冷却器,传送到发动机气缸。经过两次压缩,可以大大增加进入气缸的空气量,从而使发动机中的燃料燃烧更充分,大大提高发动机的输出功率和功率密度,并大幅减少有毒气体的排放量。该技术能大幅提高柴油机的效率、输出功率和功率密度能够提升10个百分点,同时降低燃油消耗和二氧化碳排放量。

Two-strake internal combustion engine (two-cycle internal combustion engine) 两冲程内燃机 活塞往复一次完成一个工作循环的内燃机。

Two-stroke double acting type 两冲程双作用式

Two-stroke engine 两冲程发动机

Two-way cock 双通旋塞

Two-way hydraulic dynamometer 双向水力测功器 一种大功率双向负载测功设备。可用于动力装置的功率测量和民用原动机包括风电齿轮箱等负载装置的测试,可进行低速大功率、双向、复杂工况的测试。

Two-way pump 双通阀

Type 形式 就一个救生艇释放和回收系统的设计而言,系指给定安全工作负荷、品牌和型号下一个完全相同的救生艇释放和回收系统(因此对构造材料、设计布置或尺寸的任何更改即为对型号的更改)。 In relation to the design of a lifeboat release and retrieval system, type means an identical lifeboat release and retrieval system of given safe working load, make and modle (thus any change to the materials of construction, design arrangement or dimensions constitutes a change of type).

Type "A" freeboard ships A型干舷船舶

Type "A" or type "B" freeboard ship A型干舷船

舶和 B 型干舷船舶

Type "B" freeboard ship B 型干舷船舶

Type "A" ship "A"型船舶 参照经修正的国际载重线公约 MSC.143(77)决议 第 27.1 条,系指:①专为装运散装液体货物而设计;②其露天甲板具有高度完整性,仅设有通向货舱的小型出入口,并以钢质或等效材料的水密填料盖关闭;③载货时,货舱具有低渗透率。对 A 型船舶应根据经修正的 1966 年国际载重线公约中所述的要求核定干舷。 Ref. ILLC, as amended [Resolution MSC.143(77) Reg. 27.1], Type A ship is one which:①is designed to carry only liquid cargoes in bulk;② has a high integrity of the exposed deck with only small access openings to cargo compartments, closed by watertight gasketed covers of steel or equivalent material;③ has low permeability of loaded cargo compartments. A Type A ship is to be assigned a freeboard following the requirements reported in the International Load Line Convention 1966, as amended.

Type "B + " ship "B + "型船舶 就 1966 年国际载重线公约而言,系指因舱盖设置而增加干舷的货船。 For the International Convention on Load Line, 1966, Type "B + " ship is a cargo ship with increased freeboard on account of hatch cover arrangement.

Type "B-100", "B-60"ship "B-100","B-60"型船舶 就 1966 年国际载重线公约而言,系指因其抗损的能力而减小干舷的 "B" 型船舶。 For the International Convention on Load Line, 1966, Type "B-100" or "B-60" ship is a cargo ship of type "B" with reduced freeboard on account of their ability to survive damage.

Type 1 ship 1 型船舶 对于船舶残存能力和液货舱位置而言,系指用于运输散装运输国际散装运输危险化学品船舶构造和设备规则(IBC 规则)第 17 章中对环境或安全有非常严重危险的货品的化学品船,需用最有效的预防措施消除其漏逸。因而 1 型船舶是用于运输具有重大危险性货品的化学品船。1 型船舶应能经受住最严重的破损标准,其液货舱应位于舷内离外板具有重大规定距离之处。 For ship survival capability and location of cargo tanks, a type 1 ship is a chemical tanker intended to transport products specified in Chapter 17 of the construction and equipment of ships carrying dangerous chemicals in bulk (IBC) Code, products with very severe environmental and safety hazards which require maximum preventive measures to preclude an escape of such cargo. Thus, a Type 1 ship is a chemical tanker intended for the transportation of products considered to present the greatest overall hazard. Accordingly, a Type 1 ship shall survive the most severe standard of damage and its cargo tanks shall be located at the maximum prescribed distance inboard from the shell plating.

Type 2 ship 2 型船舶 对于船舶残存能力和液货舱位置而言,系指用于运输散装运输国际散装运输危险化学品船舶构造和设备规则(IBC 规则)第 17 章中对环境或安全有相当严重危险的货品的化学品船,需用有效的预防措施消除其漏逸。2 型船舶是用于运输危险性相继减少的货品的化学品船。 For ship survival capability and location of cargo tanks, a Type 2 ship is a chemical tanker intended to transport products specified in Chapter 17 of the construction and equipment of ships carrying dangerous chemicals in bulk (IBC) Code, products with appreciably severe environmental and safety hazards which require significant preventive measures to preclude an escape of such cargo. A Type 2 ship is intended to transport products of progressively lesser hazards.

Type 3 ship 3 型船舶 对于船舶残存能力和液货舱位置而言,系指用于运输散装运输国际散装运输危险化学品船舶构造和设备规则(IBC 规则)第 17 章中对环境或安全有足够严重危险的货品的化学品船,需用中等程度的围护以增加其在破损条件下的残存能力。3 型船舶是用于运输危险性相继减少的货品的化学品船。 For ship survival capability and location of cargo tanks, a Type 3 ship is a chemical tanker intended to transport products specified in Chapter 17 of the construction and equipment of ships carrying dangerous chemicals in bulk (IBC) Code, products with sufficiently severe environmental and safety hazards which require a moderate degree of containment to increase survival capability in a damaged condition. A Type 3 ship is intended to transport products of progressively lesser hazards.

Type A independent tank A 型独立液货舱

Type A public space(measurements of noise) A 型公共处所 系指噪声通常较高的航行时持久有人的围蔽舱室或娱乐处所、迪斯科舞厅。 Type A public space(Measurements of noise) are those closed rooms normally manned at sea or recreational spaces where noise is generally high (discotheques).

Type A rigging system A 型索具系统 系指在支柱顶部左舷和右舷具备 2 个支索滑车的索具系统,所以这些支索滑车可以作为顶部提升装置。 Type A rigging system is rigging system having two guy tackles on port and starboard sides of the top of the post so that these guy tackles may also serve as topping lifts.

Type A rudder A 型舵 系指有上、下舵销的舵。

Type A rudder means a rudder with upper and bottom pintles.

Type A WIG craft　A 类地效翼船　系指:(1)经核准只能在地效区内营运的船舶;(2)无地面效应,不能飞行的船。　Type A WIG craft means:(1) a craft which is certified for operation only in ground effect;(2) a craft not capable of operation without the ground effect.

Type approval　形式认可　系指:(1)船级社通过产品的设计认可和产品制造管理体系审核,以确认申请认可的制造厂商具备持续生产符合船级社规范要求的产品的能力的评定过程。根据产品制造管理体系的证实程度,分为形式认可 A 和 B 两级,其中申请形式认可 B 的制造厂商应具有申请认可产品的生产和测试能力,并具有有效的质量控制制度;申请形式认可 A 的制造厂商除具备形式认可 B 的条件外,还应建立并保持一个至少符合 ISO 9000 标准的质量保证体系,使其产品的质量保持持续稳定;(2)船级社认为它是连续生产的代表,证明一个产品、一组产品或一个系统满足规范的认可过程。　Type approval means:(1) the evaluation process whereby the requesting manufacturer's ability to produce consistent products in compliance with Society's rule is confirmed by a Society through design approval of products and audit of manufacturing management system. Depending on the validation of the manufacturing management system, the type approval is to be classified as A or B. A manufacturer requesting type approval B is to be capable of producing and testing the products to be approved and have an effective quality control system. A manufacturer requesting type approval A is to be qualified for type approval B and in addition, to establish and maintain a quality assurance system complying at least with ISO 9000 so as to meet the specified level of product quality consistently;(2) a approval process for verifying compliance with the Rules of a product, a group of products or a system, and considered by the Society as representative of continuous production.

Type approval A　形式认可 A　系指除满足形式认可 B 的要求外,制造厂商应建立和实施一个至少符合 ISO 9000 标准或等效标准的资料管理体系,并具有船级社批准的按规范要求进行检验和试验程序。　Type approval A means that in addition to complying with the requirements for type approval B, the manufacturer is to establish and implement a quality management system complying at least with ISO 9000 or another equivalent standard and has inspection and test procedures approved by a Society according to its rule.

Type approval B　形式认可 B　系指制造厂商具备适宜的产品生产和测试设备,并建立有效的质量控制制度。　Type approval B means the manufacturer has appropriate production and test equipment and has established an effective quality control system.

Type approval of products　产品形式认可　由产品的设计认可和制造评估两部分组成。Type approval of products consist of design approval and manufacturing assessment.

Type B independent tank　B 型独立液货舱

Type B public spaces (Measurements of noise)　B 型公共处所　系指航行时噪声可能稍高的持续有人的封闭舱室,如餐厅、酒吧、电影院、赌场、休息室。　Type B public spaces (Measurements of noise) are those closed rooms permanently manned at sea where noise may be moderately high, such as restaurants, bars, cinemas, casinos, lounges.

Type B rigging system　B 型索具系统　系指左舷和右舷的支索末端与顶部提升装置末端有一个三角板连接装置的索具系统,所以顶部提升装置的张力可以吸收支索的松弛。　Type B rigging system is a rigging system having a deltaplate connecting the end of topping lift and ends of port and starboard side guy ropes so that the tension of topping lift may absorb the slackening of guy ropes.

Type B rudder　B 型舵　系指有下舵承和下舵销的舵。　Type B rudder means a rudder with the neck bearing and the bottom pintle.

Type B ship　B 型船舶　(1)参照经修正的国际载重线公约[MSC.143(77)决议第27.5条],系指凡未列入[3.19.1条]关于 A 型船舶规定的所有船舶应视为 B 型船舶。对 B 型船舶应根据经修正的1966年国际载重线公约中所述的要求核定干舷;(2)设有未列入关于 A 型船舶规定的船舶应认为是 B 型船舶。B 型船舶按1966年国际载重线公约及修正案要求勘划干舷;(3)就1966年国际载重线公约而言,系指"A"型船舶以外的,具有钢质水密舱口盖的船舶。　(1) Ref. ILLC, as amended (Resolution MSC. 143 (77) Reg. 27.5), all ships which do not come within the provisions regarding Type A ships stated in [3.19.1] are to be considered as Type B ships. A Type B ship is to be assigned a freeboard following the requirements reported in the International Load Line Convention 1966, as amended;(2) All ships which do not come within the provisions regarding Type A ships are to be considered as Type B ships. A Type B ship is to be assigned a freeboard following the requirements reported in the International Load Line Convention 1966, as amended;(3) For the International Convention on Load Line, 1966, A Type B ship is a cargo ship other than Type A, with steel weathertight hatch covers.

Type B WIG craft B类地效翼船 系指:(1)经核准可临时把营运飞高提高至地效作用范围以外的有限高度,但不超过表面以上150 m的船舶。(2)能在地效应区外作短时且增加高度有限的,以越过某一船舶、障碍物飞行或具有其他用途的船舶。此类"高飞"的最大高度应小于国际民航组织(ICAO)规定的最小安全飞高。 Type B WIG craft means:(1) a craft which is certified to temporarily increase its altitude to a limited height outside the influence of ground effect but not exceeding 150 m above the surface. (2) a craft capable to increase its altitude limited in time and magnitude outside influence of the ground effect in order to over fly a ship, an obstacle or for other purpose. The maximal height of such an "over flight" should be less than the minimal safe altitude of an aircraft prescribed by ICAO.

Type B-100 ship B-100型船舶 参照经修正的国际载重线公约[MSC.143(77)决议第27.10条],B-100型船舶是船长超过100 m的任何B型船舶,并在根据经修正的1966年国际载重线公约中适用要求核定干舷时,取自表的干舷值的减小值应不大于相应船长在表"B"和"A"上所列数值之差的100%。 Ref. ILLC, as amended [Resolution MSC.143(77) Reg. 27.10], a Type B-100 ship is any Type B ship of over 100 meters in length which, according to applicable requirements of in the International Load Line Convention 1966, as amended, is assigned with a value of tabular freeboard which can be reduced up to 100 percent of the difference between the "B" and "A" tabular values for the appropriate ship lengths.

Type B-60 ship B-60型船舶 参照经修正的国际载重线公约[MSC.143(77)决议第27.9条],B-60型船舶是船长超过100 m的任何B型船舶,并在根据经修正的1966年国际载重线公约中适用要求核定干舷时,取自表的干舷值的减小值应不大于相应船长在表"B"和"A"上所列数值之差的60%。 Ref. ILLC, as amended [Resolution MSC.143(77) Reg. 27.9], a Type B-60 ship is any Type B ship of over 100 meters in length which, according to applicable requirements of in the International Load Line Convention 1966, as amended, is assigned with a value of tabular freeboard which can be reduced up to 60% of the difference between the "B" and "A" tabular values for the appropriate ship lengths.

Type C public spaces (Measurements of noise) C型公共处所 系指航行时要求环境噪声较低的持续有人的围蔽舱室,如:演讲厅、图书馆、剧院。 Type C public spaces are those closed rooms permanently manned at sea requiring relatively low background noise, such as lecture rooms, libraries, theatres.

Type C rigging system C型索具系统 系指具备一个连接板用来连接两侧(或单侧)支索末端并沿着起重机支柱来引导顶部提升装置,所以支索的松弛可以被顶部提升装置的张力所吸收。 Type C rigging system is a rigging systems having a connecting block connecting the end of guy rope(s) of both sides (or of one side) and the topping lift led along the derrick post so that the slackening of guy rope(s) may be absorbed by the topping lift.

Type C rudder C型舵 系指在下舵承以下处无舵承的舵。 Type C rudder means a rudder having no bearing below the neck bearing.

Type C WIG craft C类地效翼船 系指:(1)经核准可在地效作用范围以外且超过表面以上150 m营运的船舶;(2)能从地面起飞,在超出ICAO规定的最小安全飞高巡航的船舶。 Type C WIG craft means:(1) a craft which is certified for operation outside of ground effect and exceeding 150 m above the surface;(2) a craft capable to take-off from the ground and cruise at an altitude that exceeds the minimal safety altitude of an aircraft prescribed by ICAO.

Type D public spaces D型公共处所 系指不要求环境噪声很低的航行时间隙时使用的围蔽舱室或通道,如:大厅、门廊、商店、走廊、梯道、运动房、健身房。 Type D public spaces are those closed rooms intermittently used at sea or passages which do not require very low background noise, such as halls, atriums, shops, corridors, staircases, sport, rooms, gymnasiums.

Type D rudder D型舵 系指有下舵承和固定下舵销的"航海者"型舵。 Type D rudder means a mariner type rudder with neck bearing and pintle, of which lower end is fixed.

Type E rudder E型舵 系指有双舵销,且下舵销为固定的"航海者"型舵。 Type E rudder means a mariner type rudder with two pintles, of which lower ends are fixed.

Type I marine sanitation device I型船用卫生装置 系指根据《美国联邦水域防污染法令》§159.123和§159.125中所述的试验条件所产生的排放水中所含的大肠杆菌数不超过1 000个/100 mL和无可见漂浮固体物的装置。 Type I marine sanitation device means a device that, under the test conditions described in §159.123 and §159.125, produces an effluent having a fecal coliform bacteria count not greater than 1,000 per 100 milliliters and no visible floating solids.

Type Ⅰ ships 第Ⅰ类船舶 系指符合下述规定之一的船舶:(1)甲板大开口船舶,应考虑大开口处船体梁垂向弯曲和水平弯曲以及扭转和横向载荷所产生的合成应力;(2)可能非均匀装载的船舶,即货物和/或压载可以是不均匀分布。船长 120 m 以下的船舶,如设计考虑了货物和压载的不均匀分布,则属于第Ⅱ类船舶;(3)化学品船和气体运输船。 Type Ⅰ ships means those ships complying with one of the following provisions:(1)for ships with large deck openings, combined stresses caused by vertical bending and horizontal bending as well as torsional loading and transverse loading of hull beam at large openings are to be considered;(2)for ships with non possible even loading, i.e., cargo and/or ballast can be unevenly distributed. For ships of less than 120 m in length, where uneven distribution of cargo and ballast is considered in design, they belong to Type Ⅱ ships;(3)Ships carrying chemicals and gases.

Type Ⅱ marine sanitation device Ⅱ型船用卫生装置 系指根据《美国联邦水域防污染法令》§159.126 和 §159.126a 中所述的试验条件所产生的排放水中所含的大肠杆菌数不超过 200 个/100 mL,且漂浮的固体物不超过 150 mg/L 的装置。 Type Ⅱ marine sanitation device means a device that, under the test conditions described in §159.126 and §159.126a, produces an effluent having a fecal coliform bacteria count not greater than 200 per 100 milliliters and suspended solids not greater than 150 milligrams per liter.

Type Ⅱ ships 第Ⅱ类船舶 系指符合下述规定之一的船舶为第Ⅱ类船舶:(1)其布置使得货物和压载分布的变化可能性很小的船舶,以及在定期航线和以固定贸易方式营运的船舶,对于这类船舶的装载手册应给予足够重视;(2)除第Ⅰ类以外的船舶。 Type Ⅱ ships mean those ship complying with the following one of the provisions:(1)for ships with little change in arrangement of cargo and ballast, and ship for regular line and in service for fixed trade, special attention is to be paid to Loading Manual;(2)ships other than those belong to Type Ⅰ.

Type Ⅲ marine sanitation device Ⅲ型船用卫生装置 系指设计成防止处理过或未处理过的污水或污水中引出的废液排至舷外的装置。 Type Ⅲ marine sanitation device means a device that is designed to prevent the overboard discharge of treated or untreated sewage or any waste derived from sewage.

Type notation 船型标志 表明船舶的布置和构造符合该型船舶特定规范要求的标志,例如挖泥船。 Type notation is a notation indicating that the ship has been arranged and constructed in compliance with particular Rules intended to apply to that type of ship, e.g. dredger.

Type of case furniture 箱柜式家具 系指书桌、衣柜、梳妆台。 Type of case furniture means those desks, wardrobes, dressing tables and dressers.

Type of container 集装箱的形式 系指经主管机关认可的设计形式。 Type of container means the design type approved by the Administration.

Type of defect 缺陷的种类 由渗透检测检出的缺陷的种类如:(1)裂纹—认定为裂纹的;(2)圆形状缺陷—除裂纹以外的缺陷,其长度系小于宽度的 3 倍;(3)船上缺陷—除裂纹以外的缺陷,其长度大于宽度的 3 倍;(4)连续缺陷—多个圆形状缺陷或船上缺陷几乎在同一船上连续存在(缺陷相互显示形状的长度为小于 2 mm),认定为是连续缺陷的. 而且,连续缺陷的长度应系各个缺陷的长度和相互距离之和。 Defects detected by the liquid penetrant test are divided into following types:(1)cracks—the defects regarded as cracks;(2)circular defects—the defects other than cracks, in which the length is less than 3 times the width;(3)liner defect—the defects other than crack, in which the length equal to or greater than 3 times the width;(4)aligned defect—aligned defects consisting of two or more liner or circular defects which are almost aligned and the spacings between them do not exceed 2 mm. The length of an aligned defect is to be equal to the sum of the lengths of all individual defects and all spacings between them.

Type of internal combustion engine 柴油机类别 根据下列指标定义:(1)缸径;(2)冲程;(3)喷油方式(直接或间接喷射);(4)燃油种类(液体、双燃料、气体);(5)工作循环(4 冲程、2 冲程);(6)扫气系统(自然吸气、增压);(7)额定转速和/或平均有效压力下的单缸额定功率;(8)增压系统(脉冲系统、定压系统);(9)吸气冷却系统(有或没有中间冷却器,级数)(10)气缸排列方式(直列、V 型)。 Type of internal combustion engine is defined by:(1)the bore;(2)the stroke;(3)the method of injection (direct or indirect injection);(4)the kind of fuel (liquid, dual-fuel, gaseous);(5)the working cycle (four-stroke, two-stroke);(6)the gas exchange (naturally aspirated or suoercharged);(7)the rated power per cylinder at rated speed and/or mean effective pressure;(8)the method of pressure charging (pulsating system, constant pressure system);(9)the charging air cooling system (with or without intercooling, number of stages);(10)cylinder arragement (in-line, vee).

Type of protection 防爆形式　系指为防止电器设备引起周围爆炸性气体环境点燃而采取的特殊措施。

Type survey certificate for yacht 游艇形式检验证书　系指通过原型艇或/和艏制艇的全面检验或试验,证实其代表一个游艇形式混合规定要求的文件。

Type test (testing) 形式试验　系指:(1)按规定的试验方法对产品样品,包括其材料和部件所进行的试验,以确认其符合指定标准或技术规范的全部要求。形式试验可以采用破坏性试验。(2)是一个公正的、提供独立的第三方验证的过程,它表明某一项机械或设备已令人满意地经受了功能性的形式试验。　Type test means:(1) testing of the sample including its materials and components by a specified method for confirming compliance with all requirements of designated standard(s) or technical specifications. The type test may be a destructive test. (2) an impartial process that provides independent third-party verification that an item of machinery or equipment has satisfactorily undergone a functional type test.

Type-2G liquefied gas carrier 2G 型液化气船　系指装运对环境危害小于 IG 型液化气船货品的船舶。其要求采取相当严格的保护措施,以防止货物漏逸。

Type-2PG liquefied gas carrier 2PG 型液化气船　系指船长为 150 m 或以下载运相当危险液化气货品的船舶,它要求采取相当严格的保护措施,以防止货物漏逸,且货物是装载在独立的 C 型舱内(该舱释放阀最大调定值至少为 0.7 MPa,设计温度为 -55 ℃以上)。

Type-3G liquefied gas carrier 3G 型液化气船　系指装运对环境危害较小的货品,如氮和制冷气体等的液化气船。其要求采取中等的保护措施,以防止货物漏逸。

Type-A independent tank A 型独立液货舱　用于全冷冻液化气船装运不低于 -55 ℃的货品,主要是平板结构的液货舱。

Type-B independent tank B 型独立液货舱　专用于液化天然气船的液货舱。B 型独立舱只需要局部次屏壁,其有球形(MOSS)和菱形(SPB)两种。

Type-C independent tank C 型独立液货舱　系指常用于全压式液化天然气船(圆筒形或球形压力容器)及半冷半压式液化气船(圆筒形式或双联圆筒形)的,不要求次屏壁的液货舱。

Type-IG liquefied gas carrier IG 型液化气船　系指装运对环境危害最大的货品,如氯、溴丙烷、二氧化硫和环氧乙炔等的液化气船。其要求采取最严格的保护措施,以防止货物漏逸。

Type-series container 定型系列集装箱　系指按照批准的定型设计制造的任何集装箱。Type-series container means any container manufactured in accordance with the approved design type.

Typhoon 台风

U

U bolt U 形螺栓

U pipe (U tube) U 形管

U-pipe manometer U 形管压力计

U tube tank U 型管减水舱

U tube type Frahm tanks U 形减摇水舱　对称设置在船舶两舷,相互连通,但与舷外水隔绝,用以减少船舶横摇的封闭式水舱。

U disk/Universal serial bus (USB) 跟读盘/U 盘　也叫 USB 闪存盘,是一种小型的硬盘。

U.S inspected ships 美国检验的船舶　系指要求按美国联邦政府法规第 46 篇第 2.01-7 条的规定检验和发证的船舶。　U.S inspected ships mean ships required to be inspected and certificated under 46 CFR2.01-7.

U.S International Trade Commission (USITC) [美]国际贸易委员会

U.S.-Dominican Republic-Central America Free Trade Agreement (DR-CAFTA) 中美洲自由贸易协定　是美国与中美洲 5 国(尼加拉瓜、洪都拉斯、萨尔瓦多、危地马拉、哥斯达黎加)以及加勒比地区的多米尼加共和国在 2005 年 10 月 20 日签署了关于七国间全面贸易的协议。美国建立此一自由贸易协定的目的是建立一个类似北美自由贸易协定(NAFTA)的自由贸易区,以消除美国与以上国家之间的贸易关税,该自由贸易协定原本应于 2006 年 1 月 1 日正式生效,因哥斯达黎加未能核准美国肉品检查机制而使正式生效日期往后延至同年 3 月 1 日生效。

U-bend U 形弯管

Ullage 舱空量　系指液舱空余空间所表示的数量。　Ullage is the quantity represented by the unoccupied space in a tank.

Ullage 膨胀容积　液舱、液柜或液货桶中为适应液体因温度升高发生膨胀等情况而留出的空间容积。

Ultimate bending moment of ship hull 船体极限弯矩 中横剖面中距离中和轴远的纵向强力构件上的应力达到材料的屈服极限时,整个剖面相应地所能承受的弯矩值。

Ultimate capacity(ultimate output) 最大功率

Ultimate factor of safety 极限安全因数(限安全系数)

Ultimate limit state(LUS) 最终极限状态 系指最大承受载荷能力的极限状态,或在一定情况下,最大适用应变或变形的极限状态;包括:(1)剖面、构件或连接由于断裂或过度变形达到最大抵抗能力;(2)整个或部分结构失稳。 Ultimate limit state, which corresponds to the maximum load-carrying capacity, or in some cases, the maximum applicable strain or deformation, includes:(1) attainment of the maximum resistance capacity of sections, members or connections by rupture or excessive deformations;(2) instability of the whole structure or part of it.

Ultimate load 极限载荷

Ultimate minimum tensile strength R_m(N/mm²) 材料最小极限抗拉强度 系指材料的最小极限抗拉强度。 Ultimate minimum tensile strength R_m means ultimate minimum tensile strength of the material.

Ultimate point(s) of safety 最终安全地点 要在设计中明确,并且可以是其他的另一艘舰船、飞机或干的陆地。 Ultimate point(s) of safety is/are to be declared in the Design Disclosure, and can, amongst other things, be another vessel, aircraft or dry land.

Ultimate short-circuit breaking capacity 极限短路分断能力 系指按规定的试验程序所规定的条件,不包括断路器连续承载其额定电流能力的短路分断能力。

Ultimate strain 极限应变

Ultimate strength 极限强度(最大强度)

Ultimate stress 极限应力

Ultimate tensile strength(UTS) 极限抗拉强度(抗拉强度)

Ultimate tension 拉力极限

Ultra high frequency(UHF) 特高频 系指波长范围为1 m~1 dm,频率为300~3 000 MHz的无线电波,常用于移动通信和广播电视领域。无线电射频根据频率和波长的不同,可以划分为不同的波段,其中微波频段的波长范围为1 mm~1 mm,频率范围为300 MHz~300 GHz。特高频只能以地面空间波的形式进行有效传播。特高频主要用于短途通信,可以用小而短的天线作收发,适合移动通信。由于特高频频率超过300 MHz,地面(土壤或海水)造成的衰减随频率增加迅速加大,地表面波在较短的距离上就已衰减掉,因而只有高出地面的

直射波存在,也就是说,特高频只能以地面空间波的形式进行有效传播。

Ultra high molecular weight polyethylene fiber 超高分子量聚乙烯纤维 是最新的制造防弹衣材料,就是凯幅拉的升级版。该技术拥有国:荷兰、美国、日本、中国。见图 U-1。

图 U-1 超高分子量聚乙烯纤维
Figure U-1 Ultra high molecular weight polyethylene fiber

Ultra large crude carrier(ULCC) 超大型油船 系指载重量(DWT)300 000 t以上的油船。

Ultra large ore carrier(ULOC) 超大型矿砂船

Ultra Panama maximum ship 超巴拿马型船舶 系指载重量(DWT)75 000 t以上的船舶。

Ultra-light aerogel 超轻气凝胶 系指浙江大学高分子系高超教授的课题组制备出了一种0.16 mg/mm³的超轻气凝胶,刷新了目前世界上最轻材料的记录。

Ultrapure Water(UPW) 美国超纯水协会

Ultrasonic method 超声波法

Ultrasonic inspection 超声波检查 是利用超声产生的波在人体内传播时,通过示波屏显示体内各种器官和组织对超声的反射和减弱规律来诊断疾病的一种方法。超声波具有良好的方向性,当在人体内传播过程中,遇到密度不同的组织和器官,即有反射、折射和吸收等现象产生。根据示波屏上显示的回波的距离、弱强和多少,以及衰减是否明显,可以显示体内某些脏器的活动功能,并能确切地鉴别出组织器官是否含有液体或气体,或为实质性组织。

Ultrasonic inspection equipment 超声波检查设备 系指进行超声波检查时所采用的仪器、设备。

Ultrasonic leakage test 超声波查漏试验 将超声波发射装置放在被测试的舱室、货舱内,用超声波接收探头在舱室的门、窗、舱盖以及舱口密封条处探测有无间隙、泄漏的方法。如发现漏点,可在耳机里听到报警声。这种方法可取代冲水试验检测货舱舱盖的密性。

Ultrasonic test(UT) 超声波探伤(超声波检测)

Ultrasonic testing 超声波试验(渗漏试验) 系指

通过超声波来证实接缝密性的试验方法。

Ultrasound 超声 系指频率大于 20 000 Hz 的声音。

Ultrathin keyboard 超薄键盘

Ultraviolet radiation method 紫外线照射法

Ultraviolet ray(UVR) 紫外线 来自太阳辐射的一部分，它由紫外光谱区的三个不同波段组成，从短波的紫外线 C 到长波的紫外线 A。

UMA ship UMA 船 系指一种其周期性无人值班机器处所的操作系统符合有关规定的并已经注册的船舶。 UMA ship is the ship whose operating system for periodically unattended machinery spaces comply with the relative requirements and is registered.

Umbilical 脐带 在潜水母船与潜水钟之间、ROV 和类似装置（也包括潜水员和潜水钟之间的潜水员脐带）之间进行生命支持和通信系统的连接载体。

UN GHS = United National Globally Harmonized System for Classfication and Labeling of Chemicals(UN GHS) 联合国化学品全球标记和协调制度

UN/EDIFACT 联合国行政、商业、运输电子数据交换规则 UN/EDIFACT means the United Nations Rule for electronic data interchange for administration, commerce and transport.

Unassisted craft 非受援船 系指受援船以外的任何地效翼客船。 Unassisted craft is any passenger WIG craft other than an assisted craft.

Unattend 不值班（无人值班）

Unattended 无人管理的（无人值班的）

Unattended appliance(LPG system) 无须照看器具（LPG 系统） 其功能无须操作者经常照看，且可自动地周期性接通和断开的器具。注意：无须照看器具的例子为水加热器、冰箱或舱室加热器。不认为燃气炉、烘箱和燃气灯为无须照看器具。 Unattended appliance means the appliance intended to function without the constant attention of an operator and which may cycle on and off automatically. Note: Examples of unattended appliances are water heaters, refrigerators and cabin heaters. Stoves, ovens and gas lamps are not considered to be unattended appliances.

Unattended engine room 无人机舱

Unattended machinery operation 无人值班机械操作 是以下（1）~（7）所列的机械和设备的一个操作，在一个预先设定的期间没有指定专人对这些设备进行操作和监视。（1）主推进机械（除了在电力推进船舶中的推进发电装置）；（2）调距螺旋桨；（3）蒸汽发电装置；（4）发电机组（包括在电力推进船舶中的推进发电装置）；（5）在（1）~（7）中所列的机械和设备的辅助机械；（6）燃料油系统；（7）舱底系统。 Unattended machinery operation is an operation of machinery and equipment specified as following (1) to (7) without watchkeeping personnel with the specific duty of the operation and surveillance during a predetermined period. (1) main propulsion machinery (propulsion generating set in electric propulsion ships are excluded); (2) controllable pitch propeller; (3) steam generating set; (4) electric generating set (propulsion generating set in electric propulsion ship are included); (5) Auxiliary machinery associated with machinery and equipment listed in (1) to (7); (6) fuel oil systems; (7) bilge systems.

Unattended machinery space 无人值班机器处所

Unauthorized broadcasting 未经许可的广播 系指船舶或设施违反国际规则在公海上播送旨在使公众收听或收看的无线电广播或电视播放，但遇难呼叫的播送除外。 Unauthorized broadcasting means the transmission of sound radio or television broadcasts from a ship or installation on the high seas intended for reception by the general public contrary to international regulations, but excluding the transmission of distress calls.

Unbalanced rudder 不平衡舵 整个舵叶都在舵杆后面的舵。

Uncertainty analysis 不确定性分析 研究关键问题中变量的不确定性，在这些问题中知识库通过观察值和模型呈现出来。也就是说，不确定性分析旨在通过相关变量不确定性的量化，为关键问题提出分析方法。当某项研究涉及统计模型时，不确定性分析和敏感性分析用来检查研究的稳健性。

Uncertainty factor 不确定系数 系指泄漏事故损失的不确定性。

Uncertainty phase 不明阶段 对船舶及船上人员的安全处于不明的情况。 Uncertainty phase means a situation wherein uncertainty exists as to the safety of a vessel and the persons on board.

Unclean B/L 不清洁提单 系指船公司在提单上对货物表面状况或包装有不良或存在缺陷等批注的提单。在结算中，银行通常不接受这样的提单。

Unconfirmed letter of credit 不保兑信用证 系指开证银行开出的信用证没有经另一家银行保兑。当开证银行资信好和成交金额不大时，一般使用这种不保兑的信用证。

Unconstrained damped plate 自由阻尼层板

Undamped free vibration 无阻尼自由振动 系指一个振动的物体不受任何阻尼力的影响，在其回复力作用下所做的振动。

Under deck operating 甲板下作业 系指在以船壳板、隔壁和顶盖围闭处所内或其上的作业。

Underscores 下画线

Under water completion 水下完井 海底钻井作业完成后,把井口系统安装在海底泥线上的作业。

Undercut 咬边 由于焊接参数选择不当,或操作工艺不正确,使焊缝边缘留下的凹陷。咬边会减小母材的工作截面,并可能在咬边处造成应力集中,因此要求平顺地过渡。见图 U-2。

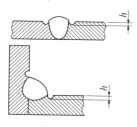

图 U-2 咬边
Figure U-2 Undercut

Under-deck tonnage 甲板下吨位 总吨位中位于量吨甲板以下的吨位。

Underhung rudder 悬挂舵 仅在船体内部设有支承点,而舵叶悬挂在船体外面的舵。

Under-keel clearance parameter 富裕水深参数 船舶吃水与自船底到河底的距离的比值。

Undersea construction boat 水下建设船 系指专门用于水下生产系统和立管等水下设施的安装和维护作业的船舶。

Undertake 承包

Undertaking 承接

Undervoltage protection 欠电压保护 系指在电压降低或失电时断开电路以防止电动机自动重新启动。

Undervoltage protection means that when operating on the reduction or failure of voltage, it is going to cause and maintain the interruption of power in the circuit until the motor is deliberately restarted.

Undervoltage release 欠电压释放 系指在电压降低或失电时断开电路,当电压恢复时电动机可重新自行启动,但应避免过大的电压降落或过大的冲击电流。当电压在额定电压的 85% 以上时,保护电器应允许电动机启动。当电压低于额定电压的 20% 左右,且在额定频率下,保护电器应断开电路,同时在需要时应有一定的延时。 Undervoltage release means when operating on the reduction or failure of voltage, but so arranged that the motor restarts automatically and without excessive starting current on restoration of voltage. The protective devices are to allow the motor to start when the voltage is above 85% of the rated voltage, and are without failure to intervene when the voltage is lower than approximately 20% of the rated voltage, at rated frequency, and with time delay when necessary.

Undervoltage/low voltage protection release 欠电压/低电压保护 系指一种保护器件在电压降低或发生故障时动作,中断主电路中电力,并保持此中断状态的效能。

Underwater absorption meter 水下吸收仪 测量水中光吸收系数的仪器。由点光源和放置在离光源适当距离上且垂直于光束的辐照度计所组成。

Underwater accommodation system 水下居住舱系统 是在深潜系统基础上增加一个水下居住舱组成大规模的饱和潜水系统。可分三班作业,其中:两个班在甲板加压舱中休息,一个班以水下居住舱为基地外出作业,然后进行轮换适用于大型水下工程施工和水下科学研究。

Underwater acoustic research ship 水声调查船 用于对海洋声速场、声传播、吸收、混响、散射、起伏、噪声、水声反射等水声物理学和应用方面考察研究的船舶。船上设有海洋水文、海洋物理、海洋化学实验室;并有无声或低噪声电站以及有关的水声放射和接收设备。

Underwater acoustic system(underwater communication system) 水下通信装置 利用声波在海水中传播信息以达到通信联络目的的设备。

Underwater acoustic transducer 水声换能器 简称换能器。用来在水中进行声-电或电-声转换的器件。将电信号转换成声信号向水中发射的叫发射换能器。将接收到的水中声信号转换成电信号的叫接收换能器,又称水听器。

Underwater camera 水下照相机 用于水下摄影的装置。它适合水中摄影的镜头及水密外壳,大多数采用短焦距和较小幅度的胶片。见图 U-3。

Underwater discharge outlet 水下排放口 根据"MARPOL73/78"附则 II 的规定,凡核准载运 X、Y 或 Z 类物质的船舶,应开设一个或几个水下排放口。

Underwater drilling and blasting ship 炸礁船 装有多台可移动的钻岩机,在清理航道时,为爆破礁石进行钻孔、装药和引爆用的船舶。

Underwater irradiance meter 水下辐照度仪 测量水中光辐照度的仪器。由收集器、光能检测器和显示器构成。

Underwater observation boat 水下观测船

Underwater plasma cutting machine 等离子水下

图 U-3　水下照相机
Figure U-3　Underwater camera

切割机　系指在水下作业时采用的等离子切割机。其配合不同的工作气体可以切割各种氧气切割难以切割的金属，特别是对于有色金属（不锈钢、铝、铜、钛、镍）切割效果更佳；其主要优点在于切割厚度不大的金属的时候，等离子切割速度快，尤其在切割普通碳素钢薄板时，速度可达氧切割法的 5~6 倍，切割面光洁、热变形小、几乎没有热影响区。

Underwater scattering meter　水下散射仪　测量海水介质对光的散射特性的仪器。由产生低发散度光束的光源、检测器及有关光学系统组成。

Underwater ship，submarine　潜水船　可潜没在水面以下航行的船舶。

Underwater sonar monitoring system　水下声呐监听系统　是在海底敷设电缆，通过水下听声器来搜集潜艇发出的声纹、磁场数据，继而通过复杂的运算和数据集成，探测水下目标的行动轨迹。

Underwater sound integrated measuring set　水声综合测量仪　用以测量水声换能器和水声设备的电声学参数的综合测量设备。

Underwater sound positioning system（USPS）　水声定位系统　系指利用声波在水中的传播提供定位功能的系统。水声定位系统有：长基线（LBL）、短基线（SBL）、超短基线（USBL）和长/超短基线（LUSBL）四种形式。

Underwater sound projector　水声发射器　把电信号转换为水中传播的声信号的换能器。

Underwater television camera　水下电视摄像机　水下进行电视摄像的装置。

Underwater transmissometer　水下透射率仪　用于测量海水透射率的仪器。由光源、产生平行光束及接收光的光学系统、光检测器及显示记录装置组成。

Underway　在航　系指船舶不是处于锚泊或绑到岸上或搁浅的状态。　Underway means that a vessel is not at anchor，or made fast to the shore，or aground.

Underwriter Laboratories Inc.（UL）　[美]保险商实验室　又称 UL 安全试验所，是美国最有权威的，也是世界上从事安全试验和鉴定的较大的民间机构。它是一个独立的、非营利的、为公共安全做试验的专业机构。它采用科学的测试方法来研究确定各种材料、装置、产品、设备、建筑等对生命、财产有无危害和危害的程度；确定、编写、发行相应的标准和有助于减少及防止造成生命、财产受到损失的资料，同时开展实情调研业务。UL 始建于 1894 年，初始阶段 UL 主要靠防火保险部门提供资金维持运作，直到 1916 年，UL 才完全自立。经过近百年的发展，UL 已成为具有世界知名度的认证机构，其自身具有一整套严密的组织管理体制、标准开发和产品认证程序。UL 由一个有安全专家、政府官员、消费者、教育界、公用事业、保险业及标准部门的代表组成的理事会管理，日常工作由总裁、副总裁处理。UL 在美国本土现有 5 个实验室，总部设在芝加哥北部的 Northbrook 镇，同时在台湾和香港分别设立了相应的实验室。2003 年 1 月 13 日 UL 在中国和中国权威的检验认证机构——中国检验认证（集团）有限公司注册，成立的全球第一家合资子公司成立 UL-CCIC，现总部设在上海，并在上海、广州、苏州、北京和重庆均设有分公司，苏州和广州为公司的主要测试基地。UL 是一家产品安全测试和认证机构，她也是美国产品安全标准的创始者。在一个多世纪的岁月里，UL 已对成百上千种产品和部件进行了相关的安全标准测试并按照国际标准评估其管理系统。

Underwriter survey　保险检验/保险鉴定

Unexposed zones　非曝露区（非露天区域）　系指与船舷距离大于 $0.04B$ 处设置的上层建筑或甲板室的边界。　Unexposed zones are the boundaries of deckhouses set in from the ship's side at a distance greater than $0.04B$.

UNFCCC　联合国气候变化框架公约

UNFCCC Copenhagen Climate Conference　联合国关于气候变化框架公约哥本哈根气候大会

Unfired pressure vessel　非受火压力容器　系指：(1)受火压力容器以外的任何压力容器；(2)任何不与燃烧气体或者燃烧器的火焰接触的容器或有过热危险的热压力容器。　Unfired pressure vessel is：(1) any pressure vessel other than fired pressure vessel；(2) any pressure vessel which is not exposed to fire from burners or combustion gases or otherwise heated pressure vessel with a risk of overheating.

Un-flattened fully loading hold　未经平舱的满载舱　系指在货物处所的舱口范围内散装谷物装满到可能最大程度，在舱口范围以外可处于其自然休止角位置的以下两类舱：(1)具有主管机关认为合理的装置或结构布置，如添注管道、开孔甲板等；并且在计算时考虑谷物自由流入甲板下形成的几何空挡，计算结果应满足稳性标准，并经主管机关批准；（2）符合下列要求的专用舱：每一货物处所至少有两道垂直的或与水平面倾斜角至少30°的倾斜的、纵向的、谷密的隔壁，该隔壁可以与舱口边缘桁重合，也可在其他能有效限止谷物横向移动的位置。专用舱内的两端可准许免于平舱。

Unified combatant command 联合作战司令部
各国媒体谈及的"联合作战司令部"基本上是按照美军的联合作战司令部的模式而言，在美军，联合作战司令部有着明确的概念。根据1986年《戈德华特-尼科尔斯法》，"联合作战司令部"系指"负有范围广泛、连续性强的任务，并由两个及两个以上军种的部队组成的军事司令部"。目前，美军共设有9个联合作战司令部。其中，包括6个按地理划分的战区司令部；3个按功能划分的专业职能性司令部。6个战区司令部包括美国北方司令部、中央司令部、欧洲司令部、太平洋司令部、南方司令部和非洲司令部。3个职能司令部分别为美国特种作战司令部（为美国陆海空军以及陆战队提供特别作战支援）、美国战略司令部（运用战略威胁力量以及协调太空武器运用）和美国运输司令部（为地区司令部提供全球运输支援）。目前，美国采取的是作战指挥系统和行政/后勤指挥系统"分权"的指挥体制。两个指挥系统的职能各有侧重。作战指挥系统是总统——国防部长——联合作战司令部。该系统的主要职能是拟定作战计划、对部队实施作战指挥、负责军事咨询和对军事行动进行战略指导以及组织联合训练等。而行政/后勤指挥系统是总统——国防部长——军种部长。该系统的主要职能是负责军种的编制、武器装备采购和保养、人事管理、军种部队的技术训练和战术训练，以及对本军种提供后勤支援等。也就是说，作战司令部除履行作战职能外，对其他"养兵"的军事问题一概不管。平时联合司令部没有对部队的管理权，战时组建联合部队（由各军种提供部队）时，联合司令部在国防部长的授权下，才能对军种部提供的部队行使"作战指挥权、作战控制权、战术控制权和支援权"。简单而言，美军的养兵和用兵体系是分开的；养兵就像办企业，需要有各种预算、行政管理，越细致越好。而打仗用兵就要指挥系统越减越好，效率越高越好；而对稍纵即逝的战机，既怕做错决策，但更怕晚决策和不决策。世界新军事变革以来，战争加速向以信息为主导，以网络为支撑的空天地海一体化联合作战演进，呈现出高强度、高速度、非线性、大变化、快节奏等鲜明特点。为此，世界主要国家军队均寻求建立以现代化信息系统为支撑的"扁平化"指挥体制，这也是美国的联合作战司令部为代表的联合指挥机制所具备的优点。

Unified interpretations（UI） 统一解释 系指针对在执行IMO公约或建议的要求中所产生的问题而通过的决定。这些决定包括对IMO公约或建议的统一解释，公约中的这些问题是要求主管机关满意，或者是含糊的。统一解释涉及：（1）化学品规则（IMO Chemical Code，CC IMO）；（2）国际海上避碰规则（IMO COLREG）；（3）耐火试验程序（Fire Test Procedure，FTP）；（4）散装液化气（Liquefied gasses in bulk，GC）；（5）高速船规则（High Speed Craft Code，HSC）；（6）载重线公约（Load Line Convention，LL）；（7）防污染公约（MARPOL Convention，MPC）；（8）载客潜水艇（Passenger Submersible Craft，PASSUB）；（9）海上人命安全公约（SOLAS，SC）；（10）吨位丈量（Tonnage Measurement，TM）。

Unified requirements（UR） 统一要求 系指IACS对船舶的统一要求。是与船级社的特定规范和实践直接相关的决定。经船级社主管机关批准，在该船级社的规范中应载入此统一要求。统一要求是最低要求。每一船级社可以制定更详细的要求。统一要求涉及：A部分——系泊和锚泊；D部分——移动式近海钻井装置；E部分——电气；F部分——防火；G部分——气体运输船；K部分——推进器；L部分——分舱、稳性、载重线；M部分——机械装置；N部分——航行；P部分——管路和压力容器；S部分——船舶强度；W部分——材料和焊接；Z部分——检验和证书。

Uniflow 单向流动

Uniflow scavenging 直流扫气 由气缸一端的进气口向气缸另一端的排气口或排气门进行扫气的形式。

Uniflow steam engine 单流蒸汽机（单流式蒸汽机） 蒸汽在各气缸中作单向流动并进行一次膨胀做功的蒸汽往复机。

Uniform Customs and Practice for Commercial Documentary Credits（UCP500） 跟单信用证统一惯例
是调整信用证的法律关系的规定，属于国际惯例。虽然我国并未加入该惯例，但是中国银行率先在其所开立的信用证中标注"依据UCP500开立"，随后所有银行进行效仿，所以实践中我国银行仍受UCP500的约束，发生纠纷，UCP500在我国的仲裁机构及法院均有约束力。国际商会为明确信用证有关当事人的权利、责任、付款的定义和术语，减少因解释不同而引起有关当事人之间的争议和纠纷，调和各有关当事人之间的矛盾，于1930年拟订一套《商业跟单信用证统一惯例》，并于1933年正式公布。

Uniform flow 均匀流 所有速度矢量是平行且等的流动。

Uniform National Discharge Standard for Vessels of the Armed Forces 美国军用舰船统一排放标准

Uniform Rules for Collection, ICC Publication No. 322 商业单据托收统一规则 国际商会为统一托收业

务的做法,减少托收业务各有关当事人可能产生的矛盾和纠纷,曾于 1958 年草拟《商业单据托收统一规则》。(1995 年再次修订,称为《托收统一规则》。国际商会第 522 号出版物(简称《URC522》),1996 年 1 月 1 日实施。《托收统一规则》自公布实施以来,被各国银行所采用,已成为托收业务的国际惯例。需要注意的是,该规则本身不是法律,因而对一般当事人没有约束力。只有在有关当事人事先约定的条件下,才受该惯例的约束。

Uninspected vessel 无须检验的船舶 系指根据美国联邦政府法规第 46 篇第 1 章的规定不要求进行检验的船舶。Uninspected vessel means any vessel that is not required to be inspected under 48 CFR Chapter 1.

Uninterruptible power system (UPS) 不间断供电电源系统(不间断电源) 系指长期连续工作的电源系统。在船舶主电源失电的事件中向负载连续提供电源的装置。由整流电源、变流器、大功率切换开关、控制电路、储能装置(如蓄电池)等组成。

Union 接合(管节)

Union joint 联管节接头

Union purchase system (burtoning system) 定位双杆操作 为缩短钩吊周期,使两根单杆位置固定,两吊杆的吊货索连在一个吊钩上,利用分别收放吊货索装卸货物的操作方法。

Unit 平台 系指设计成浮动式或坐底式的海上移动结构物或船舶。 Unit means a mobile offshore structure or vessel, designed for operation afloat or supported on sea bed.

Unit cargo ship 单元货物运输船 运输单元货物,如集装箱、车辆、原木等货物的船舶。例如集装箱船(container vessel)、车辆运输船(vehicle carrier)、集装箱/原木运输船(container/lumber carrier)、集装箱/车辆运输船(container/vehicle carrier)、拖车运输船(trailer carrier)、冷藏货物/车辆运输船(refrigerated cargo/vehicle carrier)、车辆渡船(vehicle ferry)等。

Unit resting on the seabed 坐底式平台 系指由下壳体和数根立柱支承海面以上的上壳体构成的平台,它只适合于浅水作业,作业时下壳体坐落在海底上,并由立柱支承上壳体上的全部重力。 Unit resting on the seabed means a unit with its upper hull above sea surface supported by lower hull and some columns, only suitable for operations in shallow waters. Its lower hull is on the sea bed during operation and the upper hull is supported solely by columns.

Unit test/test batch 试验单元量/试验批 交付的材料或制品中试验结果所涵盖的部分。由一定数量的、由同一炉冶炼、且形状和尺寸相同的制品或同一轧制长度的材料(钢板或带钢),或单个制品(大型锻件或铸件)。

Unit/batch inspection 单件/单批检验 系指由船级社验船师为签发产品证书,对产品逐件或逐批进行的检验。 Unit/batch inspection means the unit-by-unit or batch-by-batch inspection of products by the Surveyor for the purpose of issuing a products certificate.

United Kingdom ship 联合王国(英国)船舶 系指 1995 年英国商船运输条例第 85(2)节中定义的船舶。 United Kingdom ship means a vessel as defined in Section 85(2) of the Merchant Shipping Act 1995 (c.21).

United Nations (UN) number 联合国编号 系指"联合国危险品运输专家委员会"所提出的建议案中所示的每一种产品的编号。

United Nations Convention against illicit traffic in narcotic drugs and psychotropic substances 联合国禁止非法贩运麻醉药品和精神药品公约 系指 1988 年维也纳联合国禁止非法贩运麻醉药品和精神药品公约。1988 年 12 月 20 日签订的维也纳公约于 1990 年 12 月 11 日在国际上生效,截止到 2006 年 1 月,该公约有 179 个当事国,包括 57 个签字国。 United Nations Convention against illicit traffic in narcotic drugs and psychotropic substances means the United Nations Convention against illicit traffic in narcotic drugs and psychotropic substances, 1988, Vienna. The Vienna Convention of 20 December 1988 came into international force on 11 December 1990, by January 2006 it had 179 States Parties, including 57 signatories.

United Nations Framework Convention on Climate Change (UNFCCC) 《联合国气候变化框架公约》 简称《框架公约》,是一个国际公约,是联合国政府间谈判委员会就气候变化问题达成的公约,于 1992 年 5 月在纽约联合国总部通过,1992 年 6 月在巴西里约热内卢召开的有世界各国政府首脑参加的联合国环境与发展会议期间开放签署。1994 年 3 月 21 日,该公约生效。《联合国气候变化框架公约》是世界上第一个为全面控制二氧化碳等温室气体排放,以应对全球气候变暖给人类经济和社会带来不利影响的国际公约,也是国际社会在对付全球气候变化问题上进行国际合作的一个基本框架。

United Nations office on drugs and crime (UNODC) 联合国毒品与犯罪署

United state of coast guard (USCG) 美国海岸警卫队 是美国政府授权对船舶执行法定检验的主要主管机关。1915 年成立,隶属于财政部。1967 年起属运输部,2001 年改属国土安全部。根据条例规定,海岸警备队属于军事部门,在战时听从海军指挥。海岸警备队总部设在华盛顿(哥伦比亚特区),下设两个司令部(即大西洋区及其海防司令部和太平洋区及其海防司令部）以及 10 个地区司令部，每一个地区管辖附近几个港口和船舶检验处。对拟进入美国可航水域的外国船舶，除必须遵守各种国际公约和规则外，还应符合美国海岸警卫队、海事管理局、联邦海事委员会等机构的有关规定。这些规定载于美国联邦政府法规（CFR）的有关篇章中。如 USCG 对美国船和进入美国可航水域的外国船的防污染要求载于 CFR 第 33 篇第 1 章的第 0 分章"防污染"中。该分章的第 151 节适用于"装运油类、有毒液体物质、垃圾、城市和生活污水和压载水的船舶"。该分章的第 159 节为"船用卫生装置"。

United States 美国 系指：(1) 各州、哥伦比亚特区、关岛、美属萨摩亚、维尔京群岛、波多黎各联邦、北马里兰群岛联邦和任何美国行使主权的其他领地；(2) 包括各个州、哥伦比亚特区、波多黎各州、维尔京群岛、关岛、美属萨摩亚、运河地区和太平洋群岛托管领地。United States means: (1) the States, the District of Columbia, Guam, American Samoa, the Virgin Island, the Commonwealth of Puerto Rico, the Commonwealth of the Northern Mariana Islands, and any other territory of possession over which the United States exercises sovereignty; (2) includes the States, the District of Columbia, the Commonwealth of Puerto Rico, the Virgin Islands, Guam, American Samos, the Canal Zone, and the Trust Territory of the Pacific Islands.

United States Department of Commerce Maritime Administration ［美］运输部海事局

United States Department of Defense (DOD/DoD) 美国国防部 是属于美国军队的部门。它的中心是五角大楼。国防部的领导是美国国防部长。按照美国法律，部长须为文官。美国国防部成立于 1947 年 9 月 18 日，前身为美国战争部，总部设于五角大楼。美国国防部的中心是五角大楼，它由六个实体部门组成，分别是国防部部长办公室、军事部门、参谋长联席会议、统一作战指挥、国防机构和国防部门现场活动。美国国防部当前体系由国防部长办公厅、参谋长联席会议、3 个军种部、10 个联合作战司令部、国防部所属 16 个局和 6 个专业机构组成。

United States inspected ships 美国检验的船舶 系指要求按美国联邦法规第 46 篇第 2.01-7 条的规定检验和发证的船舶。

Unitized cargo handling system 成组装卸 用吊货网兜或货板，成组装卸货物的方式。

Universal chock 转动导缆孔 配合自动缆绳卷车使用的能按缆绳方向转动的导缆孔。

Universal code council (UCC) 统一编码委员会

Universal coordinated time (UTC) 国际坐标时

Universal coupling 万向联轴节（万向接头）

Universal joint 万向接头

Universal joint ball 万向节球

Universal joint knuckle 万向接头关节

Universal laboratory 通用实验室

Universal product code (UPC) 统一产品代码 是美国统一代码委员会制定的一种商品用条码，主要用于美国和加拿大地区，在从美国进口的商品上可以看到。

Universal screw spanner 万向螺丝扳手

Universal screw wrench 通用扳手（活络扳手）

Universal time coordinated (UTC) 国际协调时间 又称世界标准时间、世界统一时间或世界协调时间，1884 年在美国华盛顿召开的国际经度会议上，各国为了克服时间上的混乱，根据经度将全球划分为 24 个时区，各时区以国际协调时间，加减数字表示。它是最主要的世界时间标准，其以原子时的秒长为基础，在时刻上尽量接近于格林尼治平时。中国大陆采用 ISO 8601:2000 的国家标准 GB/T 7408-2005《数据元和交换格式 信息交换 日期和时间表示法》中亦称之为协调世界时。中国台湾采用 CNS 7648 的《资料元及交换格式 - 资信交换 - 日期及时间的表示法》（与 ISO 8601 类似）称之为世界协调时间。美国幅员辽阔，横跨多个地理时区，在实际操作中执行分时区制。美国及其属地共有 9 个时区，涵盖美国本土、阿拉斯加州、夏威夷州以及波多黎各、美属萨摩亚、关岛的地区。其中，美国本土有 4 个时区，自东向西分别是北美东部时区、北美中部时区、北美山区时区和太平洋时区。实行分时区制的一般都是国土面积较大的国家，如俄罗斯、加拿大、美国、巴西、印度尼西亚、澳大利亚、哈萨克斯坦、墨西哥等。也有一些横跨两个以上时区的经度的国家采用统一时区制。如日本、印度、伊朗、阿根廷、沙特阿拉伯等。如果一个国家人口分布没有明显的两级化或多极化，那么完全可以采用中间地区所处的标准时，没有硬分成多个时区的必要。例如欧洲大陆使用 UTC + 1 时间的区域，西起西班牙、东到挪威，跨 3 个时区依然统一时间，显然是为了方便做出的选择。中国国土虽然横跨经度很大，但绝大多数人口集中于中间两个时区的范围，中东部地区人口分布相对均匀，采用统一的北京时间，对于生产调配、交通运输显然是必要的。而美国、加拿大、澳大利亚等国家人口聚集区集中于国土两侧，首都又正好位于其中一侧，故采用分时区制。

UNIX system UNIX 操作系统 是一个强大的多用户、多任务操作系统，支持多种处理器架构，最早由 KenThompson、DennisRitchie 和 DouglasMcIlroy 于 1969 年在 AT&T 的贝尔实验室开发。

Unload barge 卸砂驳

Unload test(free-running test) 空负荷试验 用蒸汽冲动汽轮机转子使之达到额定转速,并在空负荷下进一步检查其装配质量的试验。分正车空负荷试验和倒车空负荷试验。

Unloading 卸载 系指把船上的货物转运至船外的某一地点,并包括相关的作业,诸如解开货物和从紧固装置上移开夹子和销子。 Unloading means to convey cargo located on board a ship to a location outside the ship, and includes associated operation such as unlashing of cargo and removing clamps and pins from securing devices.

Unmanned engine room 无人机舱

Unmanned under-water vehicle(UUV) 无人水下潜器 是一种能够依靠自身携带能源和自身的自治能力管理自己,完成各种被赋予的使命的无人运载器。见图 U-4。

图 U-4 无人水下潜器
Figure U-4 Unmanned under-water vehicle(UUV)

Unpleasant sound 不愉快声 系指使人难受、感到突然发生的声音,或给人以厌恶的声音。如锯木头的声音,小孩的啼哭声。

Unprotected openings 没有保护的开口 系指未设有至少风雨密关闭方式的开口,如果位于复原力臂曲线正值范围内,和如果位于破损后水线以下(在任何进水阶段),其可能导致连续进水。 Unprotected openings are openings which are not fitted with at least weathertight means of closure. Unprotected openings may lead to progressive flooding if they are situated within the range of the positive righting lever curve or if they are located below the waterline after damage(at any stage of flooding).

Un-regularity orbital transfer technology for ballistic missile 弹道导弹无规律性变轨技术 该技术拥有国:美国、俄罗斯、中国。见图 U-5。

Unresisted rolling 无阻尼横摇 船舶在水中不受阻尼作用时的横摇。

图 U-5 弹道导弹无规律性变轨技术
Figure U-5 Un-regularity orbital transfer technology for ballistic missile

Unrestricted service 无限航区 系指:(1)从事国际航行的船舶;(2)船舶在无限制水域航行。 Unrestricted service means:(1)a ship engaged on international voyages;(2)the areas for navigation are not limited.

unstable motion 不稳定运动 在所受扰动完全消除后,不能再恢复到新的平衡状态的船舶运动。

Unstable or hysteresis loop height 不稳定回线环高 在用螺线操纵试验数据绘制的特征曲线上,当零转艏角速度时,其左、右两舷转艏角速度间的距离。

Unstable or hysteresis loop width 不稳定回线环宽 在用螺线操纵试验数据绘制的特征曲线上,当零转艏角速度时,其左、右两舷舵角值间的距离。

Unstayed derrick post 无支索起重柱 系指无支索支张仅由甲板支承的起重柱。

Unsteady degree of revolution 转速不稳定度 内燃机在不变负荷运转时,在一定时间内测得的最大转速、最小转速之差与平均转速之百分比。

Unsupported span 无支承跨距(无支撑跨距) 系指在两根支承桁材之间的加强筋实际长度,或者包括端部连接(肘板)在内的加强筋长度。 Unsupported span is the true length of the stiffeners between two supporting girders or else their length including end attachments(brackets).

Unsupported span of corrugated bulkhead elements 槽形舱壁单元的无支承跨距 系指其在船底或甲板之间的长度,在垂直桁之间或水平桁之间的长度。如槽形舱壁构件连接至刚度较低的箱形构件,除非另有计算证明,否则这些箱形的深度应包括在跨距内。 Unsupported span of corrugated bulkhead elements is their length between bottom or deck and their length between vertical or horizontal girders. Where corrugated bulkhead elements are connected to box type elements of comparatively low rigidity, their depth is to be included into the span unless otherwise proved by calculations.

Unsymmetrical flooding　不对称进水　系指船内左右两侧不对称的进水。

Untreated sewage　未经处理的生活污水　系指未经已获形式认可的生活污水处理装置处理，或未经粉碎和消毒的生活污水。　Untreated sewage means sewage that has not been treated by a type approved sewage treatment plant, or that has not been comminuted and disinfected.

Upheaval buckling　剧变屈曲　系指由于热膨胀使一段偏离海沟的埋置管路的垂直屈曲。Upheaval buckling means the vertical buckling of a section of buried pipeline out of its trench due to thermal expansion.

Upper between deck　上层中间甲板　系指露天甲板下面的甲板，或船舶具有舷侧开口时为上甲板以下的甲板。Upper between deck means the deck below the weather deck or, in ships with side openings, the deck below the upper deck.

Upper bunk bed test　上床铺试验　系指火源布置在一个上床铺上，并点燃放在枕块中前部（朝向门外）的引燃物的试验。Upper bunk bed test means a test in which fire is arranged in one upper bunk bed with the igniter located at the front (towards door) centerline of the pillow.

Upper cargo purchase block (derrick head cargo block)　上吊货滑车　吊货滑车组中位于吊杆头端处的滑车。

Upper dead center　上死点

Upper deck　上甲板　系指最高一层露天全通甲板，在露天部分上的一切开口，设有永久性的水密关闭装置。而且在该甲板下面船旁两侧的一切开口，也有永久性的水密关闭装置。如船舶具有阶梯型上甲板，则最低的露天甲板线与平行于甲板较高部分的延伸线作为上甲板。Upper deck is the uppermost complete deck, exposed to weather and sea, which has permanent means of weathertight closing of all openings in the weather part thereof, and below which all openings in the sides of the ship are fitted with permanent means of watertight closing. In a ship having a stepped upper deck, the lowest line of the exposed deck and the continuation of that line parallel to the upper part of the deck is taken as the upper deck.

Upper explosive limit (UEL)　爆炸上限 (UEL)　在空气中易燃气体、蒸发气或雾的某一浓度，在此浓度以上，将不会形成爆炸性气体环境。Upper explosive limit (UEL) means a concentration of flammable gas, vapour or mist in air, above which an explosive gas atmosphere will not be formed.

Upper flammable limit (UFL)　燃烧上限

Upper hull　上壳体　系指柱稳式或坐底式平台的上部平台结构。Upper hull is the upper platform structure of a column-stabilized or submersible unit.

Upper hull (SWATH)　上船体（小水线面双体船）　对小水线面双体船而言，系指包括主甲板及以下至支柱体以上的结构。For SWATH, upper hull means the structure from below of main deck to above of vertical struts.

Upper ice waterline (LIWL)　高位冰水线　系指船舶在冰区航行预定水线最高点的包络线。该水线可为折线。Upper ice waterline is to be the envelope of the highest points of the waterline at which the ship is intended to operate in ice. The line may be a broken line.

Upper lubricator package (ULP)　顶部防喷系统总成

Upper pintle (top pintle)　上舵销　系指靠近舵叶顶端，通常装有锁紧装置的舵销。

Upper platform　上平台　半潜式钻井装置中，位于立柱之上起钻井平台作用的结构体。

Upper platform　舷梯上平台　装于舷梯出口处的舷梯平台。

Upper rail (main rail)　扶栏　系指设在栏杆柱顶部的一根供人员扶靠用的杆件。

Upper support of bucket ladder　斗桥上支承　固定在斗桥上端，安装在斗塔前端斜撑轴承内，能使斗桥绕之起落的轴和轴套的总称。

Upper tumbler　上导轮　安装在斗塔顶部平台上的专用承座上，由动力驱动。用以带动斗链连续运转的正四边形或正五边形带轴主动导轮。

Upper-air observation room　高空气象观察室　对探空仪进行监视、基准测定，以及对其发出的无线电信号进行接收和分析的工作舱室。

Uppermost continuous deck　上层连续甲板　系指船体的最高一层全通甲板。Uppermost continuous deck is the uppermost deck which extends from the stem to the stern.

Upright pump　立式泵

Upright shaft　立轴

Upright ship condition　船舶迎浪状态　在此状态下，船舶遭遇波浪时产生在 x-z 平面内的船舶运动，即纵荡、垂荡和纵摇。Upright ship condition means a condition in which the ship encounters waves which produce ship motions in the x-z plane, i.e. surge, heave and pitch.

UPS unit　UPS 装置　系指电力变换装置、开关或蓄电池的组合，当输入电源丧失时能用于对连续负载供电的电源装置。UPS means a source of electrical power with converters, switches and batteries, constituting for maintaining continuity of load power in case of input power failure.

Upstream industry　上游产业　系指处在整个产业链的开始端，包括重要资源和原材料的采掘、供应业以及零部件制造和生产的行业，这一行业决定着其他行业的发展速度。

Up-take ventilator　排风帽　供排出或抽出空气用的通风帽。

Urea　尿素　白色颗粒状、无嗅货品。含水量小于 1%。吸湿。无特殊危害。该货物为不燃物或失火风险低。该货物吸湿，受潮会结块。含有水分的尿素（纯或不纯）可能损坏油漆表面或腐蚀钢材。 Urea is a white, granular, and odourless commodity. Moisture content is less than 1%. It is hygroscopic. It has no special hazards. This cargo is non-combustible or has a low fire-risk. This cargo is hygroscopic and will cake if wet. Urea(either pure or impure) may, in the presence of moisture, damage paintwork or corrode steel.

Urea-formaldehyde foam　脲醛泡沫塑料

Usability　易用性/可用性　系指在一年的时间中确保系统不失效的时间比率。是衡量其优劣的重要质量指标，办公自动化（OA）作为协同办公的工具，时刻处理着人机交互，其易用性尤为重要。

USB local area network　有线网卡

Useful deadweight(net capacity)　净载重量　载重量中允许装载货物与旅客，包括行李及携带物品在内的最大值。

User configured presentation　用户设定的显示　用户为手头特定工作设定的显示器图像。图像显示可包括雷达和/或海图信息，以及其他航行与船舶相关数据。 User configured presentation is a display presentation configured by the user for a specific task at hand. The presentation may include radar and/or chart information, in combination with other navigation or ship related data.

User dialogue area　用户对话区　显示区域，包括数据域和/或菜单，菜单主要以字母、数字形式交互显示，输入或选择操作参数、数据和命令。 User dialogue area is an area of the display consisting of data fields and/or menus that is allocated to the interactive presentation and entry or selection of operational parameters, data and commands mainly in alphanumeric form.

User input device(UID)　用户输入设备　例如：键盘、舵柄、操纵杆、舵轮、按钮等。Example: keyboard, tiller, joystick, helm, pushbutton, etc.

User interface　用户界面　也称使用者界面。是系统与用户之间进行交互和信息交换的媒介，它实现信息的内部形式与人类可以经受形式之间的转换。用户界面是介于用户与邮件而设计彼此之间交互沟通相关的软件。

User selected presentation　用户选择的显示　系指由用户确定的用于手头特定任务的辅助显示。该显示可包括雷达和/或海图信息，以及其他与航行或船舶相关的数据。 User selected presentation means an auxiliary presentation configured by the user for a special task at hand. The presentation may include radar and/or chart information, in combination with other navigation or ship related data.

U-shaped section　U 型剖面　底部近似弧形而舷侧明显地趋于垂直，如同 U 形的横剖面。

UST　超汽轮机装置　是一种新的汽轮机装置。其采用了再加热蒸汽法，提高热能的有效利用率。除采用高压和低压汽轮机之外，UST 采用了一台中压汽轮机。从锅炉出来的蒸汽驱动高压汽轮机运行后，排出的蒸汽又回到锅炉进行加热，以便驱动中压汽轮机和低压汽轮机。与传统的蒸汽轮机相比，UST 减少了热能的消耗，从而降低了油耗。

UTC　协调世界时　UTC is the universal time co-ordinated.

Utility patent　实用新型专利　系指对产品的形状、构造或者其结合所提出的适于实用的新的技术方案。实用新型专利所指的产品形状，系指产品所具有的、可以从外部观察到的确定的空间形状。无确定形状的产品，如气态、液态、粉末状、颗粒状的物质或材料，其形状不能作为实用新型产品的形状特征。实用新型专利所指的产品构造系指产品的各个组成部分的安排、组织和相互关系。

Utility program　实用程序　系指系统软件中的一组常用程序。它向操作系统、应用软件和用户提供经常需要的功能，如编辑程序、连接装配程序、文卷操作程序和调试程序等。文卷操作程序的主要功能有显示、转储、修改、编辑等。显示系指输出文卷的目录、内容、属性或图形；转储系把信息从一种存储介质复制到相同或不同的另一种存储介质上，如从磁盘转储到磁带。修改系指改变文卷的内容和属性。调试程序用以排除程序中的错误，主要提供追踪、监控等功能。利用调试程序可以方便地发现程序的错误并加以排除。

U-type internal combustion engine　U 型内燃机　两列立式内燃机并列，两根曲轴用齿轮连接到一根功率输出轴上的内燃机。

V

V block　V形气缸体
V engine　V形发动机
V groove pulley　三角槽皮带轮
V notch impact test　V型缺口冲击试验
V notch specimen　V形缺口试样
Vacuum air pump　真空抽气泵
Vacuum diode　真空二极管
Vacuum feed　真空补水(冷凝器)
Vacuum fishpump　气力吸鱼泵　利用鼓风机的能量抽吸渔获物的泵。
Vacuum gage　真空压力计
Vacuum meter　真空表
Vacuum pump air-removal system　抽气真空泵组　为保持冷凝器内真空度,用真空泵将其中不凝结、不溶解的气体抽出排入大气的成套设备。
Vacuum space of condenser　冷凝器真空空间
Vacuum system of ventilation　抽吸通风系统
Vacuum test[1]　真空盒法(真空试验)
Vacuum test[2]　真空试验　用其不抽真空的方法检查焊缝质量的试验。其原理与气密试验相仿。
Vacuum treatment　真空处理
Vacuum trip device　低真空保护装置　当冷凝器真空下降到极限值时发出信号,通过综合开关,如油遮断器使速关阀关闭而使汽轮机紧急停车的装置。
Vacuum valve　真空阀(电子管)
Vacuum-pumping testing　抽真空试验(渗漏试验)　系指将一个盒子置于接缝之上并通过在盒内创建真空以发现任何泄漏的带有渗漏指示方法且适用于填角焊或对接焊的试验方法。
Valid arbitration clause　有效的仲裁条款　系指包括三方面的意思,即意思的表示、请求仲裁的意思和仲裁事项及选定的仲裁机构。例如"因本合同引起的或与本合同有关的任何争议,双方同意请上海仲裁委员会按其现行有效的仲裁规则进行仲裁,仲裁裁决是终局的,对双方均有约束力。"
Valid period　有效期限
Valid statutory certificates　有效的法定证书　系指有关国际公约的法定检验证书应始终保持有效,且应在所规定的时间伸缩范围内进行这些公约中所规定的检验。Valid statutory certificates means that the statutory certificates of the applicable international conventions are to be valid at all times, and the surveys prescribed in the conventions are to be carried out within the time windows prescribed.
Validate　验证
Validity time of a claim　索赔时效　系指法律规定的被保险人和受益人享有的向保险公司提出赔偿或给付保险金权利的期间。
Valise for liferaft　气胀筏容具　用以存放气胀救生筏,投掷入水后又不会妨碍救生筏翻转和自动充气的筒或包裹。
Valuation　估价/计价
Value of foreign trade　对外贸易额　又称对外贸易值,它是由一国或地区一定时期进口总额与出口总额构成的,是反映一国对外贸易规模的重要指标之一。一般都用本国货币表示,也有用国际上通用的货币表示的。联合国编制和发表的世界各国对外贸易额的资料,是以美元表示的。
Valve　气阀　控制压缩机吸气或排气过程的组件。
Valve actuating device　阀门传动装置　驱使配气机构工作的传动装置。
Valve adapter　电子管适配器(阀的接头)
Valve attached directly to boilers and pressure vessels　直接连接在锅炉及压力容器上的阀门　系指使用螺栓、法兰、竖管或隔片与锅炉、压力容器相连的阀门,包括螺纹截止阀。 Valve attached directly to boilers and pressure vessels is the valve attached to body of boilers or pressure vessels by stud bolts, flange, stand pipe or distance piece, and includes the screw down check valve.
Valve box　阀箱
Valve bronze　阀用青铜
Valve cap　阀盖
Valve case　阀体
Valve chamber　阀室
Valve chest　阀箱(阀壳)
Valve clearance　气门间隙　冷态柴油机在挺杆滚轮与凸轮圆柱部分相接触时气孔口杆或其气门盖面与相应的摇臂或其延伸件端面间的间隙。
Valve cock　旋塞
Valve control　阀门控制
Valve disk　阀盘
Valve gear　阀装置
Valve gear (tappet gear)　配气机构　保证按规定的要求进行进气或排气的机构。

Valve handle 阀手柄
Valve hole 阀孔
Valve lap 阀接触面
Valve lift (valve travel) 阀升程(气门升程) 进气门与排气门的行程值之差。
Valve operation 阀门操作
Valve operator adapter 阀门操纵转接器
Valve position indicator 阀位指示器
Valve remote control mechanism 阀门遥控机构 阀门的远距离操纵机构。
Valve remote emergency shut-down mechanism 阀门遥控速关机构 紧急时,远距离切断如燃油柜等出口阀的紧急关闭机构。
Valve reseater 阀座磨合器
Valve rod 阀杆
Valve seal 阀封(阀座)
Valve seat carbon deposit 气门座积碳 气门座上附着燃烧生成物的现象。
Valve seize 阀咬住
Valve spigot 阀插口
Valve stem 阀杆
Valve stem sticking 气门杆结胶 由于滑油分裂成碳和其他物质,积于气门杆和导管之间导致气门杆运行困难或胶住的现象。
Valve washer 阀座垫圈
Valve-regulated sealed batteries 阀控式密封电池 系指在正常情况下单体电池封闭,但具有当内部压力超过预定值时允许液体逸出装置的电池。单体电池一般不能添加电解液。 Valve-regulated sealed batteries are batteries whose cells are closed under normal conditions but which have an arrangement which allows the escape of gas if the internal pressure exceeds a predetermined value. The cells cannot normally receive addition to the electrolyte.
Vam hydraulic steering gear 往复式液压舵机 是电动液压舵机的一种。其推舵机构按其动作方式属于往复式的舵机。其多采用的有柱塞式油缸的拨叉式推舵机构和采用活塞式油缸的摆缸式推舵机构。
Vanadium attack 钒灰腐蚀 当锅炉用含钒成分较高的燃料燃烧时,由于烟气中低熔点的钒灰黏附于锅炉过热器管壁而引起的金属腐蚀。
Vanadium content 燃油中钒含量
Vanadium ore 矾矿石 粉尘可能有毒。该货物为不燃物或失火风险低。 Dust may be toxic. This cargo is non-combustible or has a low fire-risk.
Vane 风向标(叶片,桨叶)
Vane cascade 叶栅

Vane hydraulic steering gear 转叶式液压舵机 是电动液压舵机。其推舵机构按其动作方式属于回转式舵机中的一种。
Vane pitch 叶片节距
Vane pump 叶片泵(叶轮泵,滑片泵)
Vane type electric-hydraulic gear 转叶式电动液压传动装置
Vane type pump 叶轮泵
Vane wheel 螺扇轮(叶轮) 叶轮作折扇形,有柄连接于水面以上的驱动轴。各叶依次进出水面的一种螺旋推进器。
Vaned wheel propulsion 翼轮推进
Vapor 二次蒸汽 海水受蒸汽加热蒸发所产生的蒸汽。
Vapor compression distillation 压气式蒸馏 利用蒸发器内产生的蒸汽经压缩提高其压力和温度作为蒸发器的加热蒸汽而后冷凝为蒸馏水的蒸馏方法。
Vapor compression distillation plant 压气式蒸馏装置 按蒸汽压缩式蒸馏方法进行海水淡化的全套装置。
Vapor compressor 蒸汽压缩机 压气式蒸馏装置中用于压缩蒸发蒸汽的机械。
Vapor gravitative separation 汽水自然分离 蒸汽自蒸发液面上升时其中夹带的盐水滴因自身的重力自然沉降而使蒸汽净化的现象。
Vapor scrubber 洗汽器 备有淡水槽或喷淋蒸馏水的装置,对经过粗分离的蒸汽进行洗涤的净化设备。
Vapor volume load 容积负荷 蒸发蒸汽量即体积流量与蒸发器内容汽空间容积的比值。
Vaporization heat 汽化热
Vaporizing nozzle 蒸发式喷油嘴 燃油以低压喷入置于主燃区中的蒸发管形成油雾后再自蒸发管喷出并与主燃空气逆方向混合以获得良好雾化的喷油嘴。
Vaporizing uil range 气化油灶 使燃油加热气化并加压后进行燃烧的燃油炉灶。
Vaportor 蒸馏器
Vapour balancing 蒸发气平衡 系指通过蒸发气收集系统将装货船由于装载货物而排出的货舱蒸发气,送回到供货设备的货舱内或容器内。 Vapour balancing is the transfer of vapour displaced by incoming cargo from the tank of a ship receiving cargo into a tank of a facility delivering cargo via a vapour collection system.
Vapour collection system 蒸发气收集系统 系指用于收集液货船货舱内散发出的蒸发气,并将其输送至蒸发气处理系统的管路和软管装置。 Vapour collection system is an arrangement of piping and hoses used to collect

vapour emitted from a ship's cargo tank and to transport the vapour to a vapour processing unit.

Vapour control system(VCS) 蒸发气控制系统

Vapour control system-transfer(VCS-T) 蒸发气控制系统—中转

Vapour emission collection system 蒸发气排放收集系统

Vapour emission control 蒸发气排放控制

Vapour emission control system 蒸发气排放控制系统

Vapour line 蒸发气管路

Vapour lock 气封

Vapour locking 气塞现象

Vapour manifold 蒸发气总管

Vapour phase 气相

Vapour point 蒸发点

Vapour pressure 水汽压力 相应于水温度的饱和水汽压力。

Vapour pressure 蒸发气压力 系指在规定温度下用帕斯卡(帕)表示的在液体上面的饱和蒸发气的平衡压力。 Vapour pressure is the equilibrium pressure of the saturated vapour above a liquid expressed in Pascals (Pa) at a specified temperature.

Vapour processing unit 蒸发气处理系统(蒸发气处理单元) 系指对从液货船收集来的蒸发气进行回收、消除或分散的蒸发气控制系统的一部分。 Vapour processing unit is that component of a vapour control system that recovers, eliminates or disperse or vapour collected from a ship.

Vapour recovery system 蒸发气回收系统

Vapour refrigeration cycle 蒸发气制冷循环

Vapour reliquified system 蒸发气再液化系统

Vapour treatment system 蒸发气处理系统

Vapourization(vapourize) 蒸发(气化)

Vapourizer 蒸发器(汽化器)

Vapourous cavity 含汽空泡 在低于水汽压力下,突然产生的几乎仅含有水汽的空泡。

Vapourous cavity 含汽空泡

Vapour-proof 防气的

Vapour-tight 气密的

Variable capacity pump(variable delivery pump, variable displacement pump, variable output pump, variable volume pump) 变量泵 本身具有能够调节其出口流量的机构的泵。

Variable drilling load 钻井可变载荷 系指在主甲板上的载荷和立柱上的载荷之和。如果出现吊钩、钻管和立管张紧器载荷,则这部分也作为钻井可变载荷的一部分。 Variable drilling load is defined as the combination of loads located above the main deck and loads in the columns. The hook, rotary and riser tensioner loads, if occurring, are part of the variable drilling loads.

Variable exhaust closing(VEC) 可变排气关闭

Variable flow hydraulic pump 流量可调液压泵

Variable force 可变力(交变力)

Variable gear 变速齿轮

Variable injection timing(VIT) 可变喷油定时

Variable loads(varying load) 可变载荷 系指货物、压载水、备品和消耗品、人员和临时设备。 Variable loads are cargo, ballast water, stores and consumables, personnel, temporary equipment.

Variable pitch propeller 可调螺距螺旋桨(变螺距螺旋桨)

Variable pitch reversing propeller 可调螺距可逆转螺旋桨

Variable speed pump 变速泵

Variable stroke pump 变程泵

Variable turbine area(VTA) 涡轮增压技术 使柴油机处在任何负荷和速度运行时,都能根据燃油的喷入量自动、持续、精确地匹配压缩空气的进入量,解决了传统涡轮增压器只能在事先设定的发动机负荷点实现最大效率的问题。这大大提高了燃料燃烧的效率,节约了燃油,大幅削减了碳氢化合物、二氧化碳、烟煤的排放量。

Variable value 变值

Variable-angle nozzle 可旋转导叶

Variable-gain amplifier(VGA) 可变增益放大器 是一种通过改变电路某一参量对放大器增益进行调节的放大器,广泛应用于无线通信、医疗设备、助听器、磁盘驱动等领域。

Variance 方差 随机变量 ξ 与其期望值 $M\xi$ 之差的平方的统计平均值。其表达式为 $D\xi = \sigma^2 = M(\xi - M\xi)^2$。

Varying acceleration 变加速度

Varying duty 变工况

Varying pitch 变螺距 螺旋桨叶面沿半径方向或圆周方向各点螺距有变化者。

Varying pitch propeller 径向变螺距螺旋桨

V-drive V形传动 轴系中线布置呈V字形的传动方式。

Vector modes 矢量模式 系指:(1)真矢量:代表目标预计真运动的矢量,显示参照地面的航向和航速;(2)相对矢量:相对本船运动的目标的预计运动。

Vector modes means: (1) for the true vector: vector representing the predicted true motion of a target, showing course and speed with reference to the ground; (2) for the relative vector: predicted movement of a target relative to own ship's motion.

Vee belt 三角皮带

Vee slot V形槽

Vee-groove pulley 三角皮带盘

Vegetable chamber (vegetable cold storage) 蔬菜库

Vegetable oil tanker 植物油运输船 系指载运植物油的船舶。Vegetable oil tanker is a tanker carrying vegetable oil.

Vehicle area 车辆区域 系指设置了车辆门的车辆装载区。Vehicle area means the vehicle loading area providing the vehicle door.

Vehicle carrier 汽车运输船 系指专门运输成品小型轿车为主,兼运商务车或大型车辆的纯汽车运输船。

Vehicle deck 车辆甲板 系指车辆通过的甲板或车辆区域内的车辆装载甲板。Vehicle deck means the deck providing passageway of vehicles or vehicle loading deck provided in vehicle area.

Vehicle hold 汽车舱 供装载汽车用的舱。

Vehicle ramp 车辆跳板 供车辆登、离船用的跳板。

Vehicle spaces 车辆处所 系指:(1)拟用于装载油箱内备有自用燃料的机动车辆的货物处所;(2)滚装区域之外有汽车的区域。Vehicle spaces are:(1) cargo spaces intended for carriage of motor vehicles with fuel in their tanks;(2) those cargo spaces other than ro/ro spaces.

Vehicle-carried station 车载电台

Velocity azimuth display(VAD) 速度方位显示器

Velocity decrement 速度亏损 船舶航行时,由于各种原因而引起的航速减小。

Velocity head 速头 用于表示单位质量流体所具有的动能的高度值。$h_a = U^2/2g$,式中:U——流速;g——重力加速度。

Velocity hyper-sonic shock tunnel 倍音速激波风洞 该装置拥有国:中国。见图V-1。

Velocity indicator 速度指示器

Velocity level 速度级 系指速度与参考速度之比的以10为底的对数乘以20。单位为分贝。

Velocity meter(velometer) 速度计(测速器)

Velocity of evaporation 蒸发速度

Velocity of piston 活塞运动速度

图 V-1 倍音速激波风洞
Figure V-1 Velocity hyper-sonic shock tunnel

Velocity of rotation 回转速度

Velocity of undisturbed flow 未受扰动流速 远离物体,未受其影响处的流速。

Velocity potential 速度势 描述无旋运动的流体时,可导出流体任一点上的速度的函数(流速U等于速度势Φ的梯度)。

Velocity ratio 速度比 汽轮机级的圆周速度与级中气流的理想速度或静叶或喷嘴的出口速度之比。

Velocity stage 速度级 在较小的速度比下工作的,一个叶轮上有两列动叶的汽轮机级。

Vendee 买受人 系指以最高应价购得拍卖标的物的竞买人。

Veneer 胶合板 是由木段旋切成单板或由木方刨切成薄木,再用胶黏剂胶合而成的三层或多层的板状材料,通常用奇数层单板,并使相邻层单板的纤维方向互相垂直胶合而成。胶合板是家具常用材料之一,是一种人造板。一组单板通常按相邻层木纹方向互相垂直组坯胶合而成,通常其表板和内层板对称地配置在中心层或板芯的两侧。用涂胶后的单板按木纹方向纵横交错配成的板坯,在加热或不加热的条件下压制而成。层数一般为奇数,少数也有偶数。纵横方向的物理、力学性质差异较小。常用的有三合板、五合板等。胶合板能提高木材利用率,是节约木材的一个主要途径。亦可供飞机、船舶、火车、汽车、建筑和包装箱等作用材。通常的长宽规格是:1 220×2 440 mm,而厚度规格则一般有:3 mm、5 mm、9 mm、12 mm、15 mm 和 18 mm 等。主要树种有:山樟、柳桉、杨木、桉木等。胶合板的主要产地:印度尼西亚和马来西亚。

Venetian ventilator 百叶式通风筒

Vent condenser 逸气冷凝器 在压气式蒸馏装置中,使冷凝器中伴随空气逸出的蒸发冷凝的设备。

Vent gutter 通风道

Vent hole 通风孔(排气孔)

Vent line 通风管路逸气系统

Vent line　通气管
Vent main　透气总管
Vent pipe　通风管(透气管)
Vent piping system　透气管系
Vent system　换气系统　可分为独立式和集中式两种。独立式系指各舱的排气管是相互独立的;而集中式则是把多个货舱的排气管集中到一个排气桅上。
Vent valve　通风阀
Vent vapor heat loss　逸气热损失　蒸馏装置冷凝器逸气带走的热量。
Ventage　通气口(通风系统,排泄)
Vented batteries　通风式蓄电池　系指电解液能更换且在充电时会自由释放气体的蓄电池。通风式蓄电池的结构应能承受船舶的运动以及它所处的大气环境(盐雾、油等)。 Vented batteries are those in which the electrolyte can be replaced and freely releases gas during periods of charge and overcharge. Vented batteries are to be constructed to withstand the movement of the ship and the atmosphere(salt mist, oil, etc.) to which they may be exposed.
Vented disc　通风盘式
Ventilated container　通风集装箱　系指适用于装运新鲜的蔬菜、水果等怕闷热货物的集装箱。
Ventilating arrangement　通风布置
Ventilating cowl　通风帽　配置在通风筒围板上,借以连通大气的通风筒部件。
Ventilating hood　通风罩
Ventilating pipe　通气管
Ventilating pipe division lock　防火风闸　通风管穿过防火舱壁时,为防止火焰通过风管蔓延而设置的隔断装置。
Ventilating piping(ventilating system)　通风管系　供桥楼内部的工作处所和生活处所,以及机舱、货舱等舱室通风用的管系。通风方法主要采用自然通风、机械通风和空调系统。
Ventilation[1]　充气　在水中形成并保持空气泡的过程。包括连续从水面吸入空气的"自然充气"和用辅助装置连续输入空气的"强制充气"。
Ventilation[2]　通风　系指从货物处所外向内交换空气,使处所内积聚的易燃气体或蒸发气降低到爆炸下限以下,或使处所内有毒气体、蒸发气或粉尘含量维持在安全水平。 Ventilation means exchange of air from outside to inside the cargo space to remove any build-up of flammable gases or vapours to a scale point below the lower explosive limit(LEL), or for toxic gases, vapours or dust to a level to maintain a safe atmosphere in a cargo space.
Ventilation and temperature control system　通风和温度控制系统
Ventilation appliance　通风设备
Ventilation cover　通风盖
Ventilation device　通风设备
Ventilation ducts　通风管道　系指由钢或等效材料制成的,但长度一般不超过 2 m,且有效横截面积不超过 0.02 m^2 的短导管。如符合下列条件,则不必使用不燃材料:(1)导管用低播焰性材料制成的;(2)对于在2010年7月1日或以后建造的船舶,导管应由耐热的不燃材料制成,其内外表面用低播焰性的薄膜覆盖,并且在各种情况下,对所使用厚度,导管表面区域的卡路里值不超过 45 MJ/m^2。(3)导管只用在通风装置的末端;(4)导管位于沿该导管自"A"或"B"级分隔(包括连续"B"级天花板)上的开口量起不小于 600 mm 的位置。 Ventilation ducts mean short ducts of steel or equivalent material. However, generally not exceeding 2 m in length and with a free cross-sectional area not exceeding 0.02 m^2. The ventilation ducts do not have to be non-combustible materials, subject to the following conditions:(1)the ducts are made of any material which has low flame-spread characteristics;(2) on ships constructed on or after 1 July 2010, the ducts shall be made of heat resisting non-combustible material, which may be faced internally and externally with membranes having low flame-spread characteristics and, in each case, a calorific value not exceeding 45 MJ/m^2 of their surface area for the thickness used;(3)the ducts are only used at the end of the ventilation device;(4)the ducts are situated not less than 600 mm, measured along the duct, from an opening in an "A"or"B"class division including continuous "B"class ceiling.
Ventilation fan　通风机　压力较低的风机,一般供通风换气用。
Ventilation fence　通风栅
Ventilation fitting　通风属具　设置在露天甲板上的通风口处,借以吸入或排出空气的舱面属具配件的统称。
Ventilation index　通风指数
Ventilation method　通风法　对于液货舱的环境控制,系指强制通风或自然通风。 For environmental control for cargo tanks, ventilation method means forced or natural ventilation.
Ventilation of machinery spaces　机器处所的通风　系指在正常情况下,机器处所应有充分的通风,以防止油气聚集。 Ventilation of machinery spaces means that a ventilation of machinery spaces shall be sufficient under normal conditions to prevent accumulation of oil vapour.

Ventilation opening 通风开口
Ventilation piping system 通风管系
Ventilation plan 通风系统图
Ventilation plant 通风装置
Ventilation procedures 通风程序 系指使用在具体船上配置的液货舱通风系统或设备的特定要求的程序,并包括下列内容:(1)拟使用的通风系统位置;(2)风机的最小流量或速度;(3)对货物管路、泵、过滤器等通风的程序;(4)确保完工后液货舱干燥的程序。 Ventilation procedures mean those procedures contains specific requirements for the use of the cargo tank ventilation system, or equipment, fitted on the particular ship and shall include the following:(1) ventilation positions to be used;(2) minimum flow or speed of fan;(3) procedures for ventilating cargo pipeline, pumps, filters, etc.;(4) procedures for ensuring that tanks are dry on completion.
Ventilation system 通风系统(集装箱) 系指采用机械风机向货舱供给空气和/或从货舱抽取空气的强制通风装置。 Ventilation system means a forced ventilation arrangement using mechanical fans to supply and/or extract air from the hold space.
Ventilation tubing 通风管系
Ventilation umbrella 通风帽
Ventilator 通风筒(通风机,通风管道) 装设在露天甲板上,供船舱通风用的筒状器具。
Ventilator coaming 通风道围板
Ventilator cowl 喇叭式风斗(通风帽)
Ventilator hood 通风斗罩
Ventilator opening 通风筒开口
Ventilator pipe 通风筒
Ventilator trunk 通风筒围板(通风管道) 位于通风筒根部,焊固在甲板上作为连接通风帽的筒形围板。
Venting arrangement 透气装置
Venting pipe 透气管
Venting system 通气管系统(透气系统)
Venting valve 透气阀
Ventube 帆布通风筒
Venture capital (VC) 风险投资 是由职业金融专家投入到新兴的、迅速发展的、有巨大竞争潜力的企业中去的一种权益资本。风险投资以一定方式吸收机构和个人的资金,投向于那些不具备上市资格的中小企业和新兴企业,尤其是高新技术企业,无须风险企业的资产抵押担保。它多以股份的开式参与投资,其目的就是为了帮助所投资的企业尽快成熟,从而使资本增值。风险投资通过转让股权而收回资金,继续投向其他风险企业。
Venturimeter 文丘里流量计(喉管流量计)
Verification 验证 系指:(1)通过检查和提供客观证据确认已满足规定的要求。验证是对一项资产在其寿命期内各个阶段系统和独立的检查,以确认该资产是否符合和/或继续符合其技术规范的某些或全部要求。(2)通过提供客观证据,对适用的审核准则得到满足的认定。客观证据可以通过观察、测量、试验或其他手段获得。
Verification (and any variation of the word "verify") 验证(单词 verify 及其任何变体) 系指已将散货船和/或油船设计和建造规范与国际散货船和油船目标型船舶建造标准进行比较,并以查明与国际散货船和油船目标型船舶建造标准的目标和功能要求相符或一致。 Verification (and any variation of the word "verify") means the rules for the design and construction of bulk carriers and oil tankers have been compared to the International goal-based ship construction standard for bulk and oil tanker and have been found to be in conformity with or are consistent with the goals and functional requirements as set out in the International goal-based ship construction standard for bulk and oil tanker.
Verification audit or audit 验证审核或审核 系指对提交方的规范、自我评定和支持性文件进行评估以确认信息的有效性和可靠性这一过程。审核的目的是根据对所做的工作的抽查来评定所提交的规范对国际散货船和油船目标型船舶建造标准的符合性。 Verification audit or audit means the process of evaluating the submitter's rules, self-assessment and supporting documentation to ascertain the validity and reliability of information. The purpose of the audit is to assess the conformity of the submitted rules with the International goal-based ship construction standard for bulk and oil tanker, based on work done on a sampling basis.
Verification of purchased material and services 采购材料及服务的验证 对质量管理体系而言,系指制造厂商应保证能赋予验船师以权力,使他们能在所需购材料或服务的原始供应点或接收站验证这些材料和服务是否符合规定的要求。验船师的验证并不能解除制造厂商提供可接受材料的责任,也不能保证今后不发生拒收。 For quality management system, verification of purchased material and services mean that the manufacturer is to ensure that the Surveyors are afforded the right to verify at source or upon receipt that purchased material and services conform to specified requirements. Verification by the Surveyors shall not relieve the manufacturer of his responsibility

to provide acceptable material nor is it to preclude subsequent rejection.

Verifier 验证方 系指按照能效设计指数自愿验证临时指南进行自愿 EEDI 验证的组织，包括主管机关、船级社和其他组织，其拥有进行能效设计指数验证所必需的技术专家。 Verifier means an organization which conducts the voluntary EEDI verification in accordance with the interim guidelines for voluntary verification of the energy efficiency design index, including Administrations, classification Societies and other organizations which possess technical expertise necessary for conducting the EEDI verification.

Vermiculite 蛭石 一种云母类矿石。灰色。平均含水量 6～10%。可能产生粉尘。无特殊危害。该货物为不燃物或失火风险低。 Vermiculite is a mineral of the mica group. It is grey. Average moisture: 6% to 10%. It may give off dust. It has no special hazards. This cargo is non-combustible or has a low fire-risk.

Vermiculite board 蛭石板 是一种用蛭石制成的隔热材料。

Vermiculite brick 蛭石砖 以膨胀蛭石为原料、加黏结剂制成的具有正规形状的隔热制品。有水泥膨胀蛭石制品、水玻璃膨胀蛭石制品和沥青膨胀蛭石制品等。

Vertical accommodation 垂直融通 系指上下年度内组与组之间，项与项之间的留用额。

Vertical axis 垂直轴

Vertical axis propeller 竖轴推进器

Vertical boiler 立式锅炉

Vertical boiler with horizontal water tubes 立式横水管锅炉 外形立置，水和汽水混合物在与水平成一定倾斜角度的钢管内流动，烟气在管外流动的锅炉。

Vertical boiler with upright water tubes 直烟管立式锅炉 外形立置，水和汽水混合物在垂直布置的钢管内流动，烟气在管外流动的锅炉。

Vertical boring machine 立式镗床 系指镗刀与工件垂直的镗床。

Vertical bow 直立型艏 艏柱的侧投影与艏垂线重合或接近重合，呈直线形式的船首部。

Vertical center of gravity 重心垂向坐标 船舶重心至基平面的距离。

Vertical centers of buoyancy 浮心垂向坐标 船舶浮心至基平面的距离。

Vertical compressor 立式压缩机 汽缸中心线与水平面垂直的压缩机。

Vertical contract audit 垂直的合同审核 是国际船级社协会(IACS)通过对某一特定合同过程的审核，以评定其质量体系实施的正确程度。IACS QSCS(质量体系发证程序)审核组负责进行这些审核。 Vertical contract audit is an IACS audit which assesses the correctness of application of the quality system through audit of the process for a specific contract. The IACS QSCS (Quality System Certification Scheme) audit team is responsible for carrying out these audits.

Vertical control surfaces 垂直艉翼 艉翼中用以提供垫升航态时的航向稳定性和水平面内操纵性的部分，包括垂直安定面和空气方向舵。

Vertical distance from the base line to the horizontal neutral axis of the hull transverse section N (m) 自基线至船体横剖面水平中和轴的垂直距离 N(m) Vertical distance from the base line to the horizontal neutral axis of the hull transverse section N means a vertical distance, in m, from the base line to the horizontal neutral axis of the hull transverse section.

Vertical distance of the highest point of the tank from the baseline Z_{top} 自液舱最高点至基线的垂直距离 Z_{top} 对压载货舱而言，Z_{top} 系指自舱口围板顶部至基线的垂直距离。 For ballast holds, Z_{top} means vertical distance, in m, of the highest point of the tank from the baseline.

Vertical flow type 垂直吹风式 冷风从冷风机出风口送入冷藏舱底部的木格栅，沿着格栅隙缝吹出的冷风经过货堆后通过回风管或直接通过冷风机室顶端的回风棚回入冷风机的吹风方式。

Vertical girder 竖桁 支承舱壁扶强材或槽型舱壁板的竖向构件。

Vertical integration 垂直一体化 它是由经济发展水平不同的国家形成的经济一体化组织。

Vertical internal combustion engine 立式内燃机 气缸中心线垂直于基座安装面的内燃机。

Vertical ladder 垂直扶梯(直梯) 系指倾斜角为大于 70° 小于 90° 的扶梯，横向倾斜角应小于 2°。 Vertical ladder means a ladder of which the inclined angle is 70° and over up to 90°. A vertical ladder shall not be skewed by more than 2°.

Vertical lay 垂直铺管

Vertical lay system (VLS) 垂直铺管系统

Vertical net-hauler (vertical seine-selective machine) 立式滑轮起网机 V 形槽轮竖立安装的围网起网机。

Vertical position of propeller 螺旋桨垂向位置 螺旋桨中点在船基平面以上的距离。

Vertical positions of longitudinal metacenter 纵稳

心垂向坐标　纵稳心至基平面的距离。

Vertical positions of transverse metacenter　横稳心垂向坐标　横稳心至基平面的距离。通常系指初横稳心至基平面的距离。

Vertical prismatic coefficient　垂向棱形系数　与基线相平行的任一水线下，型排水体积与其相对应的水线面面积、平均型吃水的乘积之比。在无特别注明时，通常系指设计水线处者。对于民用船舶，各水线处垂向棱形系数之计算，其中船长和型宽常用垂线间长和满载水线宽。

Vertical projection fish finder　垂直探鱼仪　利用回声原理，探测渔船底下方鱼群的仪器。

Vertical pump　立式泵　系指回转泵的轴为立轴，或往复泵的缸体为立式的泵。

Vertical return tube boiler("Cochran" boiler)　立式横烟管锅炉　在炉膛上部设置布有横烟管的水空间，顶部大多为半球形壳体构成的蒸汽空间，整体外形呈立式筒体的锅炉。

Vertical search engine　垂直搜索引擎　是针对某一个行业的专业搜索引擎，是搜索引擎的细分和延伸，是对网页库中的某类专门的信息进行一次整合，定向分字段抽取出需要的数据进行处理后再以某种形式返回给用户。垂直搜索是相对通用搜索引擎的信息量大、查询不准确、深度不够等提出来的新的搜索引擎服务模式，通过针对某一特定领域、某一特定人群或某一特定需求提供的有一定价值的信息和相关服务。其特点就是"专、精、深"，且具有行业色彩，相比通用搜索引擎的海量信息无序化，垂直搜索引擎则显得更加专注、具体和深入。能否提供全面权威的行业信息，能否拥有行业资源是垂直搜索引擎发展的门槛。

Vertical shaft propeller　竖轴推进器

Vertical sliding door　直摇式水密滑门　门板沿垂直方向升降开闭的水密滑门。

Vertical spindle　立轴(垂直心轴)

Vertical stabilizing fin　垂直安定面　垂直艉翼中用以提供垫升航态时的航向稳定性而不能转动的部分。

Vertical stretched tank(VST) design　垂直拉伸增加舱容(VST)设计　对LNG船而言，系指采用液舱垂直拉伸增加舱容的设计。采用VST设计在球罐型液舱的建造中也不需要增加新的设施或对原有的建造设备进行大量的改造。

Vertical struts(SWATH)　支柱体(小水线面双体船)　系指上船体以下至下潜体以上在设计水线面结构附近的狭长垂直结构，其截面呈扁薄，舯艏艉端为流线型。支柱体有多种形式，根据每一片体所拥有的支柱体数量，分别称为单支柱体或双支柱体。Vertical struts mean long and narrow vertical streamlined structural members that extends from the lower hulls to the haunch or cross-deck at design waterline, whose section is flat and thin. There are various forms of vertical struts, which may be one or two struts associated with each lower hull.

Vertical struts breadth　支柱体宽度　对小水线面双体船而言，系指单个支柱体的最大宽度。

Vertical struts depth under waterline　水线下支柱体深度　对小水线面双体船而言，系指位于船中处从水线面量至下潜体与支柱体壳板交线处的垂直距离。

Vertical struts waterline length L_S　支柱体水线长度　对小水线面双体船而言，系指船舶静浮于水面时，沿设计水线处量得支柱体的最大长度。对于前后独立设置的支柱体，取一个前后同一方向上的各支柱体长度之和。

Vertical tube evaporator　竖管式蒸发器　加热管组的竖立直管的蒸发器。

Vertical type induction unit　立式诱导器　装于甲板上，冷风或热风从顶部或侧面出风棚吹出的诱导器。

Vertical vibration　垂直振动/垂向振动　系指船体在纵中剖面上所产生的垂直弯曲振动。

Vertical warping winch(capstan)　立式绞缆机　又称绞盘。用来驾收或放出缆索的动力机械。在船舶靠泊过程中使用。立式绞缆机的卷筒中心线为垂直位置。

Vertical wave bending moment M_{WV} **(kN-m)**　波浪垂向弯矩 M_{WV}　系指所考虑船体横剖面处的波浪垂向弯矩：(1)中拱工况：$M_{WV} = M_{WV,H}$；(2)中垂工况：$M_{WV} = M_{WV,S}$。M_{WV} means vertical wave bending moment, in kN-m, at the hull transverse section considered: (1) $M_{WV} = M_{WV,H}$ in hogging conditions; (2) $M_{WV} = M_{WV,S}$ in sagging conditions.

Vertical wave shear force Q_{WV} **(kN)**　波浪垂向剪力(Q_{WV})　系指所考虑船体横剖面处的波浪垂向剪力。Q_{WV} means vertical wave shear force, in kN, at the hull transverse section considered.

Vertical web　垂直隔板　舵构架中与舵杆轴线平行的构件。

Vertical-down welding　垂直下行焊

Very high frequency(VHF)　甚高频　系指频带由30~300 MHz的无线电电波。比甚高频无线电频率低的是高频(HF)，比甚高频无线电的频率高的是特高频(UHF)。甚高频全向信标(VHF Omni-directional Range, VOR)，是一种用于航空的无线电导航系统。甚高频多数是用作电台及电视台广播，同时又是航空和航海的沟通频道。VOR发射机发送的信号有两个：一个是相位固定的基准信号；另一个信号的相位随着围绕信标台的圆周角度是连续变化的，也就是说各个角度发射的信号的相位都是不同的。见图 V-2。

图 V-2　甚高频
Figure V-2　Very high frequency(VHF)

Very high holding power anchors(VHHP)　超大抓力锚　亦即按船级社材料规范适用要求证明为抓力高于普通锚抓力的锚，不需在海底作预调整或在海底做特殊布置。超大抓力锚如用作艏锚，每只锚质量应不小于相应普通无杆锚质量的 50%。超大抓力锚的质量一般应小于或等于 1 500 kg。　Very high holding power anchors(VHHP) mean anchors for which a holding power higher than that of ordinary anchors has been proved according to the applicable requirements of the Society's Rules for Materials, do not require prior adjustment or special placement on the sea bottom. Where VHHP anchors are used as bower anchors, the mass of each anchor is to be not less than 50%, of that required for ordinary stockless anchors. The mass of VHHP anchors is to be, in general, less than or equal to 1 500 kg.

Very large crude carrier(VLCC)　大型油船　系指载重量(DWT)200 000 t 以上的油船。见图 V-3。

图 V-3　大型油船
Figure V-3　Very large crude carrier(VLCC)

Very large floating ular-ring base(VLFUB)　深远海环形超大型浮式基地

Very serious marine casualty　非常严重的海难　系指海难涉及船舶全损人员或死亡或对环境造成严重破坏。　Very serious marine casualty means a marine casualty involving the total loss of the ship or death or severe damage to the environment.

Very small aperture terminal(VSAT)　甚小口径终端　是一种微型卫星通信终端，它使用卫星通信领域的最新研究成果来让用户接入稳定的卫星，通过卫星的网络连线，提供宽带上网的服务。

Vessel　船舶　系指：(1)作为或能作为一种水上运输工具的各类水上船艇，包括非排水型船、地效翼船和水上飞机；(2)各种水上运输工具或其他用于或能用于作为水上运输工具的人造的机械装置。　Vessel means: (1) every description of water craft, including non-displacement craft, WIG craft and seaplanes, used or capable of being used as a means of transportation on water; (2) every description of watercraft or other artifical contrivance used, or capable of being used, as a means of transportation on water.

Vessel constrained by her draught　限制吃水的船舶　限制由于吃水与可航水域的水深和宽度的关系致使其偏离所驶航向的能力严重受到限制的机动船舶。　Vessel constrained by her draught means a power-driven vessel which, because of her draught in relation to the available depth and width of navigable water, is severely restricted in her ability to deviate from the course.

Vessel engaged in fishing　从事捕鱼的船　系指使用网具、绳钩、拖网或其他使其操纵性能受到限制的渔具捕鱼的任何船舶，但不包括使用曳绳钩或其他并不使其操纵性能受到限制的渔具捕鱼的船舶。　Vessel engaged in fishing means any vessel fishing with nets, lines, trawls or other fishing apparatus which restrict manoeuvrability of the vessel, but does not include a vessel fishing with trolling lines or other fishing apparatus which do not restrict manoeuvrability of the vessel.

Vessel monitoring system　船舶监控系统　系指由一个或多个缔约国政府建立的用于监控悬挂其国旗的船舶动态的系统。船舶监控系统也可以通过船舶搜集建立该系统的缔约国政府规定的信息。　Vessel monitoring system means a system established by a Contracting Government or a group of Contracting Governments to monitor the movements of the ships entitled to fly its or their flag. A vessel monitoring system may also collect information from the ships specified by the Contracting Government(s) which has established it.

Vessel not under command　失控的船舶　系指在某些例外的环境下，不能按规范的要求进行操纵，因此未能避开另一艘船舶航道的船舶。　Vessel not under command means a vessel which through some exceptional circumstance is unable to manoeuvre as required by these rules and is therefore unable to keep out of the way of another vessel.

Vessel operating manual(VOM)　船舶操纵手册　系指包括有关实现和使用自动和综合桥楼综合系统的公司方针的手册。　Vessel operating manual(VOM) is a manual which incorporates the Company policy for implementing and using automation and the Integrated Bridge System.

Vessel restricted in her ability to manoeuvre　操纵能力受到限制的船舶　系指由于工作性质，使其按国际海上避碰规则要求进行操纵，因而不能给他船让路的船舶。操纵能力受到限制的船舶应包括，但不限于下列船舶：(1)从事敷设、维修、打捞助航标志、海底电缆或管道的船舶；(2)从事疏浚、测量或水下作业的船舶；(3)在航中从事补给或转运人员、食品或货物的船舶；(4)从事发放或回收航空器的船舶；(5)从事清除水雷作业的船舶；(6)从事拖带作业的船舶，而该项拖带作业使拖船及其被拖船偏离所驶航向的能力严重受到限制者。　Vessel restricted in her ability to manoeuvre means a vessel which from the nature of her work is restricted in her ability to manoeuvre as required by the International Regulations for preventing collision at sea and is therefore unable to keep out of the way of another vessel. Vessels restricted in their ability to manoeuvre shall include but not be limited to: (1) a vessel engaged in laying, servicing or picking up a navigation mark, submarine cable or pipeline; (2) a vessel engaged in dredging, surveying or underwater operations; (3) a vessel engaged in replenishment or transferring persons, provisions or cargo while underway; (4) a vessel engaged in the launching or recovery of aircraft; (5) a vessel engaged in mine clearance operations; (6) a vessel engaged in a towing operation such as severely vestricting the towing vessel and her tow in their ability to deviate from their course.

Vessel with class notation DP-1　有 DP-1 附加标志的船舶　系指可在规定的环境条件下，自动保持船舶的位置和艏向，同时还应设有独立的集中手动船位控制和自动艏向控制的船舶。　Vessel with class notation DP-1 means a vessel which can keep the position and heading of the vessel under the specified environment conditions. And at the same time, independent, concentrated manual control of vessels position and automatic heading control is to be fitted.

Vessel with class notation DP-2　有 DP-2 级附加标志的船舶　系指可在出现单个故障(不包括一个舱室或几个舱室的损失)后，可在规定的环境条件下，在规定的作业范围内自动保持船舶的位置和艏向的船舶。　Vessel with class notations DP-2 means a vessel which can automatically keep the position and heading of the vessels when single failure(excluding loss of a cabin or cabins) occurs under the specified environment condition ad in specified operating field.

Vessel with class notations DP-3　有 DP-3 附加标志的船舶　系指可在在出现任一故障(包括由于失火或进水造成一个舱室的完全损失)后，可在规定的环境条件下，在规定的作业范围内自动保持船舶的位置和艏向的船舶。　Vessel with class notations DP-3 means a vessel which can automatically keep the position and heading of the vessel when any failure(including entirely loss of a cabin caused by fire or flood) occurs under the specified environment condition and in specified operating field.

Vessel　船舶　系指用于或能被用来作为美国水域上的运输设施的各种类型的水运工具或其他人造的机械装置。　Vessel includes every description of watercraft or other artificial contrivance used, or capable of being used, as a means of transportation on the waters of the United States.

Vessels and aircraft　船舶和飞机　系指任何类型的海、空运载工具。并包括不论是否以自力推进的气垫船和浮艇。　Vessels and aircraft mean waterborne or airborne craft of any type whatsoever. The expression includes air cushioned craft and floating craft, whether self-propelled or not.

Vest-type PFD　背心型个人漂浮装置　穿在使用者上部躯干上像背心一样的个人漂浮装置。　Vest-type PFD is the PFD covering the upper trunk of the user like a vest.

Veterinary inspection certificate　兽医检验证书　是证明出口动物产品或食品经过检疫合格的证件。适用于冻畜肉、冻禽、禽畜罐头、冻兔、皮张、毛类、绒类、猪鬃、肠衣等出口商品。

VHF omni-directional range(VOR)　甚高频全向信标　是一种用于航空的无线电导航系统。

Viable organisms　活性有机物　系指存活的有机物及其任何生命阶段。　Viable organisms are organisms and any life stages thereof that are living.

Vibrate　振动　描述机械系统运动或位置的量值相当于某一平均值或大或小交替地随时间变化的现象。

Vibrated and radiated noise　振动辐射噪声

Vibration　总体振动　系指凡在船舶绝大部分均能感受到的振动。

Vibration ability　振动能力
Vibration absorber　减振器
Vibration absorbing coating　吸振涂层
Vibration analysis　振动计算(振动分析)
Vibration at fundamental frequency　基频振动
Vibration control system　减振装置(控振装置)
Vibration damper　阻尼器
Vibration damping　减振(振动阻尼)
Vibration damping covering　减振覆层
Vibration deadener　减振器
Vibration detector　振动传感器
Vibration durability test　振动耐久试验(耐振试验)
Vibration elimination　抑制振动(消振)
Vibration frequency　振动频率
Vibration hazard　振动损害
Vibration in phase　同相振动
Vibration indicator　振动指示器
Vibration intensity　振动强度
Vibration isolating pad stiffness　隔振垫刚度
Vibration isolating rubber　隔振橡胶
Vibration isolating spring　隔振弹簧

Vibration isolation[1]　隔振　系指为减小或控制机器设备或结构振动传递,在振源与安装结构之间插入弹性元件或隔振装置(隔振器),降低或减小振源输入结构振动能量,从而减小结构噪声的产生与传播的措施。

Vibration isolation[2]　阻振　就是采用阻尼减振方法的简称,即用附加的子系统连接于需要减振的结构或系统以消耗振动能量,从而达到控制振动水平的目的。

Vibration isolation mass　阻振质量　一般为金属块,其剖面有正方形、矩形或圆形。可以焊在基座腹板或振动传播途径的船体结构上。

Vibration isolation method　阻振方法　是采用阻尼减振方法的简称,即用附加的子系统连接于需要减振的结构或系统以消耗振动能量,从而达到控制振动水平的目的。阻尼减振技术能降低结构或系统在共振频率附近的动响应和宽带随机激励下响应的均方根值,以及消除由于自激振动而出现的动不稳定现象。阻尼减振有两种方式,一类是非材料阻尼,如各种成型的阻尼器,另一类是材料阻尼,如各种黏弹性阻尼材料以及复合材料等。目前黏贴在结构上的自由阻尼敷层和约束阻尼敷层应用很广泛。前者利用拉伸变形来消耗振动能量,后者则利用剪切变形来消耗振动能量。尤其是多层约束阻尼层,往往较之前种方法更为有效。如美国 F-4 战斗机的武器发射装置的中央腹板由于宽带激励下的多模态共振而迅速破坏。粘贴了多层约束阻尼层后,由于在其工作温度条件下的多个模态上都提供了一定的损耗因子,解决了这种振动疲劳造成的破坏问题。

Vibration isolator　隔振器/隔振体　系指连接设备和基础,用以减少和消除由设备传递到基础的,振动力和由基础传递到设备振动的弹性元件。

Vibration level　振动级　通过采用 ISO 6954 的两种版本中任一版本的值:(1)如采用 ISO 6954:1984,此振动级定义为:在 5~100 Hz 的频率范围内,具有最大重复工况的有代表性的稳态振动期间,甲板结构振动的单一振幅峰值(mm/s)。对于 5 Hz 以下的频率,这一对振动级的要求采纳(与 5 Hz 时的加速度相对应的)恒定的加速度曲线。(2)如采用 ISO 6954:2000,此振动级定义为:在 1~80 Hz 的频率范围内,稳态运行期间振动的总的频率加权有效值。通常,应优先采用 ISO 6954:2000 标准,如采用 ISO 6954:2000 有实际的困难,且船东与建造方之间已就此达成协议,则可采用 ISO 6954:1984。 Vibration level means the value defined by the application of either of the two versions of the ISO 6954 standard: (1) Where ISO 6954:1984 is applied, for frequency range 5 Hz to 100 Hz, the vibration level is defined as the single amplitude peak value of deck structure vibration during a period of steady state vibration, representative of maximum repetitive behaviour, in mm/s, over the frequency range 5 Hz to 100 Hz. For frequencies below 5 Hz, the requirements for vibration levels follow constant acceleration curves corresponding to the acceleration at 5 Hz. (2) Where ISO 6954:2000 is applied, for frequency rage 1 Hz to 80 Hz, the vibration level is defined as the overall frequency weighted r. m. s. value of vibration during a period of steady-state operation over. Grenerally speaking, ISO 6954:2000 will be given priority, ISO 6954:1984 can be adopted if there is difficulty in application of ISO 6954:2000, and agreement has been achived between the shipowner and the constractor.

Vibration load　振动载荷
Vibration meter　测振仪
Vibration monitoring　振动监测
Vibration node　振动节点
Vibration of lower order　低阶振动
Vibration of membrane　薄膜振动
Vibration of string　弦振动
Vibration pad　隔振垫　系指把具有一定弹性的软材料,如橡胶、软木、毛毡、海绵、泡沫塑料等,制成各种垫子形状的隔振元件。其中使用最普遍的是橡胶隔振垫,尤其适合小型机器设备的隔振。
Vibration post 防振柱
Vibration recording equipment 振动记录仪
Vibration sensor 振动传感器
Vibration signal 振动信息
Vibration signature 振动特征
Vibration strength 抗振强度(振动强度)
Vibration stress 振动应力
Vibration suitability test 振动适应性试验
Vibration table 振动台
Vibration test 振动试验
Vibration testing machine 抗振试验机
Vibration theory 振动理论
Vibration transducer 振动传感器
Vibration trial 振动特性测定
Vibration waveform 振动波形
Vibrational period 振动周期
Vibration-isolating element 隔振元件
Vibration-proof 耐振的
Vibration-proof rubber 抗振橡胶
Vibration-proof stay 防振支柱
Vibration-proof structure 防振结构(减振结构)
Vibrator 振动器(振子,断续器)
Vibratory fatigue 振动疲劳
Vibratory force 振动力
Vibratory loading 振动载荷
Vibratory response 振动响应
Vibratory stress 振动应力
Vibratory torque 振动扭矩
Vibrobench 振动台
Vibrograph 示振仪(振动记录仪)
Vibrometer 振动计
Vibrorecord 振动记录
Vibroscope 振动仪
Vibroshock 减振器(缓冲器)
Vibrostand 振动试验台
Vice 老虎钳(钳紧,缺陷)
Vice bench 钳工台(钳工桌)
Vicker's diamond hardness(VDH) 维氏钻石硬度
Vicker's hardness number(VHN) 维氏硬度值
Vicker's hardness test 维氏硬度试验
Vicker's pyramid hardness 维氏角锥硬度
Vickers hardness 维氏硬度
Vicker's hardness tester 氏硬度计 维氏硬度计
Victual waste 废弃食品 系指任何已变质或未变质的食品的废弃物。Victual waste means any spoiled or unspoiled food waste.
Video arcade 游艺室(豪华邮轮)
Video call 视频通话 又称视频电话,分为沿IP线路以及沿普通电话线路两种方式。视频通话通常系指基于互联网和移动互联网(3G互联网)端,通过手机之间实时传送人的语音和图像(用户的半身像、照片、物品等)的一种通信方式。
Video cards 视频卡 又称视频采集卡,按照其用途可以分为广播级视频采集卡,专业级视频采集卡,民用级视频采集卡。
Video cassette recorder(VCR) 盒式磁带录像机 系指使用空白录像带并加载录像机进行影像的录制及存储的监控系统设备。它带有的磁带用来录制电视广播节目的声音及视频留作以后播放。许多VCR有自己的调谐器(用于电视节目接收)和可程序定时器(用于自动在某个时间录制特定频道的节目)。
Video graphics array port(VGAP) 视频图形阵列端子 也叫D-Sub接口。VGA接口是一种D型接口,上面共有15针,分成3排,每排5个。VGA接口是显卡上应用最为广泛的接口类型,绝大多数的显卡都带有此种接口。迷你音响或者家庭影院拥有VGA接口就可以方便地与计算机的显示器连接,用计算机的显示器显示图像。
Video graphics array(VGA) 视频图形阵列 实际上有几种不同分辨率。最常见的是640×480(每像素4比特,16种颜色可选)。还有320×200(每像素8比特,256种颜色可选)和720×400的文本模式。
Video phone 视频电话 是利用电话线路实时传送人的语音和图像(用户的半身像、照片、物品等)的一种通信方式。如果说普通电话是"顺风耳"的话,视频电话就既是"顺风耳",又是"千里眼"了。视频电话分为沿IP线路以及沿普通电话线路两种方式。
Video port 视频端口 是家庭影院用于与显示设备(比如电视)连接的接口,通过这些端口,可以将电影等图像在电视等设备上播放。此外,部分家庭影音套装还带有视频刻录功能,因此需要通过视频端子做输入。视频端子有不同类型,购买家庭影院套装时尽量挑接口齐全的产品,尤其是最常见的接口,这样可以更方便地与各种设备连接。目前最基本的视频端子是复合视频端子(也叫AV端子)、S端子;另外常见的还有色差端子、VGA端子、DVI端子、HDMI端子。
Video signal 视频信号 系指电视信号、静止图像信号和可视电视图像信号。对于视频信号可支持3种制式:NTSC、PAL、SECAM。视频信号在重放过程中,由于磁头旋转不均匀和磁带运行速度不稳定,以及磁带伸缩等因素,会使重放的视频信号产生抖动,即时间轴发生变动，产生了时基误差，这种影响表现在亮度信号是同步信号周期性中晃动，而表现在色度信号上是副载波频率和相位的变化，并引起图像色调失真。作为 S-Video 的进阶产品色差输出将 S-Video 传输的色度信号 C 分解为色差 C_r 和 C_b，这样就避免了两路色差混合解码并再次分离的过程，也保持了色度通道的最大带宽，只需要经过反矩阵解码电路就可以还原为 RGB 三原色信号而成像，这就最大限度地缩短了视频源到显示器成像之间的视频信号通道，避免了因烦琐的传输过程所带来的图像失真，所以色差输出的接口方式是目前各种视频输出接口中最好的一种。视频信号与脉冲波形类似，也适合采用直流耦合。

Video tape recorder 录像机 系指供记录电视图像及伴音信号的机器。通常系指磁带录像机。

View 观点

Vigilance system 值班安全系统 系指显示当班驾驶员在驾驶室正常值班的系统。该系统应不影响驾驶功能的实施。 Vigilance system means a system indicating that an alert officer of the navigational watch is present on the bridge. The system is not to cause undue interference with the performance of bridge fuctions.

Viking crown lounge 维京皇冠酒廊（豪华邮轮）

Virdual simulation system 虚拟仿真系统 由计算机生成的一个实时三维环境的系统。利用其在概念设计中可以利用触摸屏来选择或修改模型产品的造型、色彩、装修风格等。在渲染和生成十分逼真的三维模型时通过各种专用的传感交互设备与虚拟物体进行交互操作，让用户看到的是全彩色景象、听到的是虚拟环境中的音响、感觉（皮肤、嗅觉、振动等）到的是虚拟环境反馈的作用力，从而使使用者产生一种身临其境的感觉。与传统设计相比，能大大减少投放市场的风险性、也为决策者寻找商机，判断概念设计能否进一步开发生产，提供更好的依据，缩短了设计周期，保证产品开发一次性成功。

Virtual metacentric height 修正后初稳心高 经过自由液面修正和其他修正后的初稳心高。

Virtual pitch 实效螺距 螺旋桨不发生推力时的每转进程。

Virtual private network(VPN) 虚拟专用网络

Virtual slip ratio 实效滑距比 螺旋桨实效螺距与每转进程之差对实效螺距的比值。$S_V = [P_V - (V_A/R)]/P_V = 1 - V_A/n P_V$，式中：$V_A$——进速；$n$——螺旋桨转速；$P_V$——实现效螺距。

Visa 签证

Visa procedures 签证手续

Viscoelastic material 黏弹性材料 是一种兼有黏性流体特性又具有弹性固体特性的高分子聚合物材料。

Viscosimeter 黏度计

Viscosity coefficient 黏性系数

Viscosity controlling system for MGO 船用低硫油(MGO)黏度控制系统 一种通过控制船用低硫油温度进而改变其黏度，从而使其符合现有船用柴油机使用要求的装置。该装置由制冷系统、冷却水系统和油循环系统组成。其特点是采用了对油箱进行冷却的专利技术、制冷系统采用了不会对臭氧层起破坏作用的冷却剂，冷却水系统采用的是闭式循环，选用 30% 的乙二醇环保型水溶剂作为介质，可以最低限度降低热损失。该系统还采用模糊控制技术可以实时监视设备压力、温度和流量从而使压缩机、水泵和三通阀处于最佳工作状态。

Viscosity damping 黏性阻尼

Viscosity test 黏性试验

Viscous flow (viscous fluid) 黏性流（黏性流体） 流体内部和流体与固体面之间均具有切应力作用的实际流体的流动。

Viscous friction 黏性摩擦

Viscous pressure resistance 黏压阻力 黏性流体经物体面上时，由边界层加厚与边界层分离以及感生涡流所产生的法向分力的积分所形成的阻力。

Viscous resistance 黏性阻力 系指船舶航行时，由于水有黏性，在船体表面应力场变化所产生的阻力。

Visible trade 有形贸易 系指买卖那些看得见、摸得着的物质性商品的活动，也称其为货物贸易。有形贸易是"无形贸易"的对称，系指商品的进出口贸易。由于商品是可以看得见的有形实物，故商品的进出口被称为有形进出口，即有形贸易。有形商品的种类繁多，按照联合国《国际贸易商品标准分类》进行分类。

Visitor visa 访问签证

Visitors 来访者 系指不定期派到平台上的人员。Visitors are personnel not regularly assigned to the unit.

Visor bow door 罩壳式艏门 系指靠纵向布置的升降臂，通过 2 个或多个位于近门顶的门主要结构上的绞链绕一水平轴向上和向外开启的艏门。 Visor bow door means a bow door opened by rotating upwards and outwards about a horizontal axis through two or more hinges located near the top of the door and connected to the primary structure of the door by longitudinally arranged lifting arms.

Visor doors 上翻铰链式门 艏门绕过 2 个或 2 个以上铰链的水平轴向上和向外翻而开启，该铰链位于靠近门的上缘并通过纵向设置的吊臂与门的主要结构相连接。 Visor doors are opened by rotating upwards and outwards about a horizontal axis through two or more hinges located near the top of the door and connected to the primary structure of the door by longitudinally arranged lifting arms.

VISS 船舶虚拟仿真数据平台

Visual aids 视觉辅助设备　系指在平台上设置的一个风向标，以尽可能地表明整个直升机甲板上的实际风况。在需要直升机夜间操作的平台上须装备发光的风向标，并在直升机甲板上设有的如下标志：(1)用0.3 m宽的白线连续勘划的周界；(2)平台的名称须用与背景有较大反差的颜色标在直升机甲板上，且位于有障碍的一边，其字母高度不小于1.2 m；(3)位于直升机甲板中心的目标环，采用黄颜色涂制，其内径等于0.5LD，环的宽度为1 m；(4)白色"H"位于降落区域中心，水平方向位于无障碍扇形区域中线上，"H"的字母应为3 m高，1.8 m宽，其笔画宽度为0.4 m；(5)直升机甲板的无障碍扇形区域标志应表明无障碍区域的起始点、扇形区域的限制方向和甲板的设计尺寸(LD 或 RD 的数值)。

Visual aids are the wind direction indicators fitted on the unit to show the actual wind condition for the whole helicopter deck so far as possible. Where the helicopter is to operate at night, the unit is to be fitted with a luminescent wind direction indicator and fitted with marks on helicopter deck, as follows: (1) the perimeter with a continuous white line of 0.3 m width; (2) the name of the unit is to be marked with letters having a height not less than 1.2 m in a contrasting color on the deck at the side with obstructions; (3) the aiming circle at the center of the deck is to be 1 m wide with internal diameter of 0.5 LD, painted in yellow; (4) the letter "H" is to be painted at the center of the landing area and horizontally at centerline of the obstruction-free sector in 0.4 m wide white lines forming a letter 3 m high and 1.8 m wide; (5) the obstruction-free sector is to be marked to indicate the starting point and restricted direction(s) of the sector and the design size of the deck (LD or RD value).

Visual alarm 视觉报警(灯光报警,视觉报警器)
Visual alarm unit 光报警单元(视觉报警单元)
Visual check 目视检查(外观检查)
Visual display unit (VDU) 视频显示器
Visual indication 视觉显示　系指按通灯光或在所处场所各种明暗情况下人眼看得见的其他装置。
Visual indication means indication by activation of a light or other device that is visible to the human eyes in all levels light or dark at the location where it is situated.

Visual inspection 目视检查/外观检验　系指利用光学(放大)设施对焊缝的表面和背面进行彻底检验，以检查其外部特征。应检查下列特征：(1)完整性；(2)尺寸精确度；(3)符合规定的焊缝形状；(4)无不允许的外部缺陷。Visual inspection means a complete inspection on the surfaces and back sides of the welds, with the aid of optical (magnifying) appliances where necessary, to check their external characteristics. The following characteristics shall be checked: (1) completeness; (2) dimensional accuracy; (3) compliance with the specified weld space; (4) absence from inadmissible external defects.

Visual signal 视觉信号　视觉信号的颜色应符合表 V-1 的规定。The color code for visual signal is to comply with the requirements of Table V-1.

表 V-1　视觉信号的颜色表
Table V-1　Color Code of Visual Signal

颜色 color	含义 Meaning	说明 Explanation	应用举例 Example
红 red	危险或报警 Danger or alarm	危险或需要立即采取行动的警告 Warning of danger or a situation which requires immediate action	重要设备停止运转；水、油等温度或压力达到临界值；重要电路失电 Operation failure of essential equipment; temperature or pressure of water, oil, etc. to a critical value; power failure of essential circuits
黄 yellow	注意 Caution	状态的改变或即将改变 Change or impending change of conditions	温度或压力值异常但未达到临界值 Temperature or pressure is abnormal but not to a critical value

续表 V-1

颜色 color	含义 Meaning	说明 Explanation	应用举例 Example
绿 green	安全（正常运转或正常工作状态）Safety (normal operating or working conditions)	安全状态指示 Indication of a safe situation	机械正常运转；液体正常循环；温度、压力和电流等在限定值内 Normal operation of machinery; normal circulation of liquids; temperature pressure, current, etc. within the limited value
蓝 blue	指导/信息（根据需要给予特定的含义）Instruction/information (specific meaning assigned as needed in the case considered)	可赋予上述红、黄、绿三色未涉及的特定含义 Blue may be given a meaning which is not covered by the three above colors: red, yellow and green	电动机准备启动；空载发电机准备合闸；停转电动机加热电路接通 Motors begin to start; unloading generators begin to switch on; heating circuit of stopping motor is connected
白 white	无具体含义 No specific meaning assigned	任何含义，可在认为红、黄、绿三色不适用时 Any meaning, to be used if red, yellow or green is considered not applicable	对地绝缘指示；同步指示；电话呼叫；自动控制的设备 Earthing insulation indication; synchroscope; telephone calling; equipment by automatic control

Vlasov's curves 符拉索夫曲线 在型线图的各站横剖面上，以吃水为纵坐标，相应的横剖面面积之半，以及该面积对中线面和基平面的净矩为横坐标所绘制的三组积分曲线。

Voice chat 语音聊天 是一个人工智能的聊天工具。

Voice frequency(VF) 声频(音频)

Voice messages 语音信息

Voice over internet protocol(Voip) 网络语音电话业务 简称 IP 电话，系指在 Internet 上通过 TCP/IP 协议实时传送语音信息的应用。最初的 IP 电话是个人计算机与个人计算机之间的通话，通话双方为自己的计算机安装声卡及相关软件，并且约定好时间同时上网进行通话，现在通常称为网上语音聊天。随后发展到通过网关把因特网与传统电话网联系起来，实现从普通电话机到普通电话机的 IP 电话。IP 电话从形式上可分为 4 种：PC—PC、电话—PC、PC—电话、电话—电话。现在人们所说的真正意义上的 IP 电话，也是最具有商业价值的 IP 电话是通过 Internet 实现从普通电话到普通电话之间的通话。过去 IP 电话主要应用在大型公司的内联网内，技术人员可以复用同一个网络提供数据及语音服务，除了简化管理，更可提高生产力。

Voice pipe 通话管

Voice tube 传话管

Void space 空舱处所（留空处所） 系指：(1)舱壁甲板以下的围蔽处所，位于油船货物区域或散货船货物长度区域以内和前部，不包括：①专用海水压载舱；②载货处所；③任何物质的存储处所（如：燃油、淡水、备品）；④任何安装机械设备的处所（如：货油泵、压载泵、艏推进器）；⑤人员正常使用的处所；⑥船长 160 m 及以上散货船的双舷侧处所，这些处所应符合 MSC.215(82)决议通过的所有类型的船舶专用海水压载舱和散货船双舷侧处所涂层性能标准；(2)货物区域内的液货舱外部的围蔽处所，但不包括液货处所、压载舱、燃油舱、货泵舱、泵舱和人员正常使用的任何处所。 Void space is:(1) an enclosed space below the bulkhead deck, within and forward of, the cargo area of oil tankers or the cargo length area of bulk carriers, excluding:①a dedicated seawater ballast tank;②a space for the carriage of cargo;③a space for the storage of any substance (e. g. oil fuel, fresh water, provisions);④a space for the installation of any machinery(e. g. cargo pump, ballast pump, bow thruster);⑤any space in normal use by personnel;⑥a double-side skin space of bulk carriers of 150 m in length and upwards which shall comply with the performance standard for dedicated seawater ballast tanks in all types of ships and double-side skin spaces of bulk carriers adopted by resolution MSC. 215 (82); (2) an enclosed space in the cargo area external to a cargo tank, other than a hold space, ballast space, oil fuel tank, cargo pump-room, pump-room, or any space in normal use by personnel.

Voids 露底

Voith schncider ship 平旋推进器船 用平旋推进器推进的船。

Voith-scheider propeller(cycloidal propeller) 直叶推进器 又称竖轴推进器。转盘绕垂直轴转动，可以调整直叶为不同的角度发出不同方向的推力的推进器。此类推进器多用于港口作业船舶或对操纵性有特殊要求的船上。

Voith-Schneider propeller(epicyloidal propeller) 外摆线推进器 叶的轨迹是一外摆线，推进器盘面每转一周各叶也自转一周的平旋推进器。

Volatile content 易挥发物含量

Volatile matter 易挥发物质

Volatile organic compounds(VOC) 挥发性有机化合物系指液货船所载的液货产生的蒸发气体。

Volatilies 挥发物

Volatility 挥发性(不稳定,易变状态)

Vold 空舱 系指舱壁围蔽的留空处所。 Vold is an enclosed empty space in a ship.

Voltage of standard system 标准系统电压 系指:(1)对于交流系统—110 V,115 V,120 V,220 V,230 V和240 V。(2)对于直流系统—12 V,24 V,32 V和120 V。

Volocity stage 速度级

Voltage tolerance 电压偏差 在正常工作状态下偏离正常使用电压的最大值,但不包括瞬态和周期性的电压变化。注意:电压偏差是稳定偏差,并包括在电缆和电压调整器特性中的电压降,也包括由环境条件引起的变化。

Voltage transient 电压瞬变 电压的突然变化(不包括尖峰脉冲)。该电压在扰动开始之后,超出额定电压偏差限值之外,且在规定的恢复时间之内(时间范围为秒)返回并保持在这些限值之内。

Voltage transient recovery time 电压瞬变恢复时间 从电压超出额定偏差到电压恢复并保持在额定偏差限值内所经过的时间。

Voltage unbalance tolerance 电压不平衡偏差 电压最高的相与电压最低的相之间的电压差。

Volume of the craft V 艇的容积 V 小艇的容积 $V(m^3)$ 按下式求得: $V = V_H + V_S$,式中, V_H 为艇体容积, m^3; V_S 为上层建筑容积, m^3。该艇的容积应通过公认的造船学方法或按近似估算求得。 Volume of the craft means the volume, V, in cubic meters, of a small craft is given by the following formula: $V = V_H + V_S$, where V_H is the volume of the hull, in cubic meters; V_S is the volume of the superstructure, in cubic meters. The volume of the craft shall be established either by accepted naval architectural methods or by an approximate assessment.

Volume of the hull V_H 艇体容积 V_H 艇体容积应使用近似方法,按下式确定: $V_H = 0.15 L_H (B_0 D_0 + B_{20} D_{20} + B_{40} D_{40} + B_{60} D_{60} + B_{80} D_{80} + B_{100} D_{100})$。见图 V-4。 Volume of the hull V_H is the volume determined by using the approximate method, the volume of the hull shall be determined as follows(Figure V-4): $V_H = 0.15 L_H (B_0 D_0 + B_{20} D_{20} + B_{40} D_{40} + B_{60} D_{60} + B_{80} D_{80} + B_{100} D_{100})$.

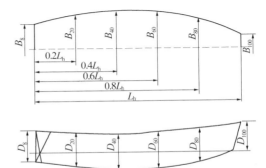

图 V-4 容积测定

Figure V-4 Volume measurement

Volume of the moulded displacement ▽ 型排水体积 系指以型吃水算得的型排水体积,不包括轴包架。 Volume of the moulded displacement ▽ means volume of displacement excluding bossings, taken by the moulded draught.

Volume of the superstructure V_S 上层建筑容积 V_S 系指甲板舷弧线/舷侧处甲板线以上的上层建筑每一部分容积之和。仅在一侧有开口的任何处所在计算时均应包括在内。在这种情况下,开口系指所涉及的面积不大于10%者。小于 0.05 m³ 的容积可忽略不计。 Volume of the superstructure, V_S, shall be the sum of the volume for each part of the superstructure above the sheerline/deck at the side. Any space that is open at no more than one side shall be incorporated in the calculation. Open in this sense means that no more than 10% of the area may be covered. Volumes of less than 0.05 m³ shall be omitted.

Volume RO-RO carrier 容积型滚装货船 系指

每米车道载重量小于 4 t 的滚装船。

Volume-dependent total cost function [$f(v)$] 容量与总成本依赖关系函数[$f(v)$] 可以看出单位溢油成本随着每吨溢油的美元价格的下降而下降。

Voluntary export quotas 自动出口配额制 又称自动限制出口,也是一种限制进口的手段。所谓自动出口配额制是出口国家或地区在进口国的要求或压力下,"自动"规定某一时期内(一般为 3～5 年)某些商品对该国的出口限制,在限定的配额内自行控制出口,超过配额即禁止出口。它是在二战后出现的非关税壁垒措施,出口限制实际上是进口配额制的变种,同样起到了限制商品进口的作用。它的重要特点就是带有明显的强制性。"自动"出口限制往往是出口国在面临进口国采取报复性贸易措施的威胁时被迫做出的一种选择。自动出口配额制带有明显的强制性。进口国家往往以商品大量进口使其有关工业部门受到严重损害,造成所谓"市场混乱"为理由,要求有关国家的出口实行"有秩序的增长",自动限制商品出口,否则就单方面强制限制进口。"自动"出口配额制与绝对进口制在形式上略有不同。绝对进口制是由进口国直接控制进口配额来限制商品的进口,而"自动"出口配额制则是由出口国直接控制这些配额对指定进口国家的出口。但就进口国来说,"自动"出口配额制和绝对配额制一样,都起到了限制进口的作用。

Voluntary thickness addition $t_{\text{voluntary-addition}}$ 自愿增加厚度 即在 t_c 的基础上,船东自愿增加的额外腐蚀余量,mm。 Voluntary thickness addition is the thickness in mm, voluntarily added as the Owner's extra margin for corrosion wastage in addition to t_c.

Volute 蜗壳

Volute pump 螺旋泵(蜗壳离心泵)

Volute spring 锥形弹簧

Vortex cavitation 涡空泡 出现在叶梢和毂部的空泡,分别称为梢涡空泡和毂涡空泡。这种空泡对螺旋桨剥蚀无直接影响,但梢涡空泡在导管内表面崩溃时会导致导管的剥蚀。对于大方形系数船舶,如果伴流场很不均匀,桨船之间可能产生连体空泡。产生连体空泡主要是船尾线型不太理想,而不是螺旋桨本身设计问题。连体空泡会导致冲突尾部的剧烈振动,还会引起不必要的噪声问题。

Vortex chamber (swirl chamber) 旋流器 安装在喷油嘴周围使进入火焰筒的空气产生旋转运动以获得稳定和完全燃烧的叶片式配气机构。

Vortex induced vibration (VIV) 涡激振动

Vortex line 涡线 作为环流流速场载体或核心的线。

Vortex noise 涡流噪声 当高速气流通过旋转叶片时,由于流场不均匀性,负载流量与转速变化使流入叶栅气流的入射角往往不能正好维持叶型的冲角为零,因而产生气流的附面层脱离以及气流的旋转而脱离所产生的噪声。

Vortex pump 旋涡泵

Vortex-induced motion (VIM) 涡激运动 是海洋中的流引起的运动响应,定常流和交变流都会引起涡激运动。

Voyage[1] 航程 系指:(1)船舶预定从专属经济区外的某一港口或地点进入五大湖区域或乔治·华盛顿桥以北的哈德逊河,包括中间停靠在专属经济区范围内某一港口或地点的运输路线。(2)驶往任何美国港口或地点的任何运输路线。 Voyage means: (1) any transit by a vessel destined for the Great Lakes or the Hudson River, north of the George Washington Bridge, from a port or place outside of the EEZ, including intermediate stops at a port or place within the EEZ. (2) any transit by a vessel destined for any United States port or place.

Voyage[2] 航次(航行) 通常系指从一个港口出发至下一个港口出发的时间段,包括偏移。 Voyage general means the period between departure from a port to the departure from the next port. Voyage includes an excursion.

Voyage charter[1] 程租 系指按航次计算的租船方式。

Voyage charter[2] 定程租船 又称程租船,是海洋运输的一种方式,系指租船人向船东租赁船舶按租船合同规定的航程进行货物运输。这是一种以航程为基础的租船方式,船舶所有人按双方事先议定的运价与条件向租船人提供船舶全部或部分舱位,在指定的港口之间进行一个或多个航次运输,以完成指定货物运输业务。

Voyage charter party 定程租约

Voyage data recorder (VDR) 航行数据记录仪(VDR) 系指一个完整的系统,包括与输入信号源相连所需的任何项目、数据处理和编码装置、最终记录介质、回放设备、电源和专用的备用电源。见图 V-5。 Voyage data recorder (VDR) means a complete system, including any items required to interface with the sources of input signals, their processing and encoding, the final recording medium, the playback equipment, the power supply and dedicated reserve power source. As can be seen in Fig. V-5.

Voyage data recorder system-certificate of compliance 航行数据记录仪系统符合证书 系指一份进行过年度性能试验的,包括所有传感器的航行数据记录仪系统的符号证书。该试验应由认可的试验或检修机构进行,以验证所记录数据的精度、持续时间和可恢复性。

图 V-5 航行数据记录仪(VDR)
Figure V-5 Voyage data recorder(VDR)

另外还应进行试验和核查,以确定所有防护外罩和辅助定位装置的适用性。船上应保留一份由试验机构颁发的载明符合日期和适用性能标准的符号证书的副本。

Voyage data recorder system-certificate of compliance means a certificate of compliance for voyage data recorder system(including all sensors) subjected to an annual performance test. The test shall be conducted by an approved testing or servicing facility to verify the accuracy, duration and recoverability of the recorded data. In addition, tests and inspections shall be conducted to determine the serviceability of all protective enclosures and devices fitted to aid location. A copy of the certificate of compliance issued by the testing facility, stating the date of compliance and the applicable performance standards, shall be retained on board the ship.

Voyage planning workstation 航程计划工作站 系指驾驶员可以进行下列工作而不影响船舶实际航行的工作站:(1)检查和更新海图和其他有关文件;(2)第一次航程设计成一系列航程基准点、航向、航速和转向;(3)计算航程中各点的估计到达时间;(4)确定并绘制船舶位置(图)。 Voyage planning workstation means a workstation at which the navigator may carry out the following tasks without affecting the actual navigation of the vessel:(1)examine and update charts and other relevant documentation;(2)plan a voyage as a series of waypoints, courses, speeds and turns;(3)calculate an estimated time of arrival at various points on the voyage;(4)determine and plot the ship's position.

Voyage planning 航程规划 收集与预计航程相关的信息;在适用的海图中绘制航线和回转半径;危险区域的指示,现有船舶的航线和报告系统,船舶交通服务,涉及海洋环境保护考虑和安全航速的区域。 Voyage planning in cludes the following task:gathering information relevant to the contemplated voyage; the plotting of course lines and turn radii of the intended voyage in appropriate charts: indication of areas of danger, existing ships' routing and reporting systems, vessel traffic services, areas involving marine environmental protection considerations and safe speed.

Voyage plan 航程计划 由船桥楼团队开发和使用的综合泊位指南,用于确定最佳航道,识别沿线的危害,并确定桥楼管理团队的管理以确保船舶的安全航行。 Voyage plan means a comprehensive, berth to berth guide, developed and used by the vessel's bridge team to determine the most favourable passage, to identify hazards along the track, and to make out the bridge team management to ensure the vessel's safe passage.

Voyage repair 航修 对船舶营运期间影响航行的较小故障、海损或一般事故的紧急修理。为了尽量不影响船舶营运,常在船舶停港作业时间中进行,必要时由航修站或船厂派员随船进行抢修。

Voyage replenishment ship 综合补给舰 主要用于向航母战斗编队、舰船供应正常执勤所需的燃油、航空燃油、弹药、食品、备件等补给品,是专门用来在战斗中帮助编队中的舰船,其特殊设计允许其设战舰级的远端维修系统,并且减少所有辅助维修系统的能量需求,因此被广泛地在任务中使用。

V-shaped section V型剖面 底部相对尖瘦而舷侧明显呈直线外倾,如同V字形的横剖面。

VSSP = Vessel Specific Sampling Plan 船舶专项取样计划

V-type compressor(V-form compressor) V型压缩机 汽缸中心线倾斜于水平面,并是V形排列的压缩机。

V-type derrick post V形起重柱 由两根分别向两舷倾斜的柱子所组成的起重柱。通常专指供装设相邻两舱共用的重型吊杆。

V-type internal combustion engine V型内燃机 两列气缸夹角成小于180°的V形,共用一根曲轴的内燃机。

Vulnerability 易损性 舰艇对军事威胁手段的耐受性可以用易损性这一术语来表述,易损性系指舰艇遭受特定威胁手段打击时丧失运行能力的概率。 The resistance of a vessel to loadings from military threats can be described by the term vulnerability which is the probability that once hit by a specified threat a vessel will lose capability.

W

Wadden sea national park 瓦登海国家公园 系指德国北海沿岸的一个面积辽阔的自然保护区,以浅滩著称。这里的水域遍布浅滩,也因此成为世界著名的海滩漫步区。瓦登海受严格的自然保护条例制约,相关利用也受到严格限制。从法律上讲,该区域由3个国家公园组成:石荷州浅滩海国家公园、卡萨克森州浅滩海国家公园和汉堡浅滩海国家公园。自2009年以来,联合国教科文组织将该国家公园认定为世界自然遗产。

Wading rod(current pole) 测流标杆 用漂流测流法观测海流用的随流飘动的直立木杆或竹竿。上端露出水面。用测流绳自船上放出。

Waigaoqiao Ship-building Company 外高桥造船公司——后来居上的第一船厂 上海外高桥造船公司成立于1999年,是中国船舶工业集团有限公司旗下的上市公司——中国船舶工业股份有限公司的全资子公司。公司全资拥有上海外高桥造船海洋工程有限公司,控股上海江南长兴重工有限责任公司、上海外高桥海洋工程设计有限公司、上海中船船用锅炉有限公司、中船圣汇装备有限公司,参股上海江南长兴造船有限责任公司。外高桥造船公司累计承建并交付的17万吨级和20万吨级散货船占全球好望角型散货船船队比重的11.3%,30万吨级超大型油船(VLCC)累计交付量占全球VLCC船队的8.3%;同时,公司在建世界最大18 000 TEU级集装箱船和83 000 m³大型液化气运输船(VLGC)。在海洋工程业务领域,公司先后承建并交付了15万吨级、17万吨级、30万吨级海上浮式生产储油船(FPSO),第六代的3 000 m深水半潜式钻井平台"海洋石油981"号,正在建造的海洋工程产品有JU2000E型/Cj48型自升式直接平台和PX-121系列海洋平台供应船;标志着该公司在自升式钻探平台领域已经形成系列化生产能力。自2005年起,该公司造船总量和经济效益连续多年位居国内造船企业前列。2011年,公司完工交船38艘,成为中国第一家年造船完工总量突破800万载重吨大关的船厂,被誉为"中国第一船厂"。

Wake(wake velocity) 伴流 船舶航行时使船艉附近水分子产生的速度。具有相对于螺旋桨的轴向、周向和径向三种分量;通常系指螺旋桨盘面毂外部分的平均或等效轴向伴流。

Wake equalizing duct(inflow compensative nozzle) 补偿导管 位于船艉部螺旋桨前中剖面两侧,偏置于桨轴上方,具有机翼形剖面的半圆形或长L形的导流装置。补偿导管一方面加速螺旋桨上部进流,使桨盘面进流更加均匀,提高了推进效率,另一方面减少船艉部的流动分离,降低了形状阻力,同时可调整螺旋桨进流的预旋程度,减少艉流中旋转能力的损失。如果补偿导管的安装角度适当,还可产生附加推力。另外补偿导管可以与其他节能装置,如补偿导管+舵球、补偿导管+桨壳帽鳍等一起配合使用。

Wake simulation 伴流模拟 在空泡试验水筒测试段上游装置格栅、水管循环、船模等为控制流场而进行的伴流分布的模拟。

Wake stream homing torpedo 尾流自导鱼雷 尾流自导鱼雷是利用尾流自导技术来制导的鱼雷。它是一种自导鱼雷,属非声自导鱼雷一类。尾流自导鱼雷是利用敌方舰船航行时产生的尾流,来进行跟踪,它是依靠测定敌舰艇航行时在水中形成的尾流来判定目标。尾流包括声尾流、湍流尾流、气泡尾流、磁性尾流、光尾流和热尾流等。声尾流系指舰艇开过之后,海水由于惯性运动,从而产生噪声;而气泡尾流系指舰艇螺旋桨产生的气泡,这些小气泡可能在海水中滞留很长时间。目前的尾流自导技术主要是依靠探测舰艇的声尾流和气泡尾流。该技术拥有国:俄罗斯(65型)、美国(Mk45F)、中国(鱼6/7/820)、意大利(A184Mod3)、法国(F17Mod2)、瑞典(TP61)。见图W-1。

图 W-1 尾流自导鱼雷
Figure W-1 Wake stream homing torpedo

Wake-adapted propeller 适应伴流螺旋桨 对螺旋桨叶各半径处螺距的选择适应所在半径的周向伴流,以使船与螺旋桨系统能量损失最小的螺旋桨。

Wall bottom 坞墙底 浮箱式浮船坞上,与浮箱分界处的坞墙底面。

Wall socker 壁式插座

Wall temperature 壁温

Wall thickness eccentricity δ_0 壁厚偏心率 系指主要用于测量无缝钢管壁厚变化的百分数,其定义如下：$\delta_0 = (t_{max} - t_{min})/(t_{max} + t_{min})$。大的偏心率是不确定的，因为其妨碍焊接和能对一些力学性能产生不利的影响。 Wall thickness eccentricity (δ_0) is one measure of wall thickness variations present primarily in seamless pipe and defined as follows: $\delta_0 = (t_{max} - t_{min})/(t_{max} + t_{min})$. Large eccentricity is undesirable as it impedes welding and can adversely affect on some mechanical properties.

Wall thickness t　壁厚　对于每种标准管径都有几种标准壁厚。例如对于管径 $D = 9.625$ 英寸，可提供 20 种标准壁厚，范围从 0.125 m ~ 1.000 m，对于无缝钢管选择一种标准壁厚是很重要的，因为其规定在管子轧制厂采用的一套内部的制造工具。相反，埋弧焊（UOE）管非标准值的壁厚是能易于调节的。　For every standard pipe diameter, there are several standard wall thicknesses. For example, for $D = 9.625$ in, 20 standard thicknesses are available, ranging from 0.125 to 1.000 m. Selecting one of the standard thicknesses is important for seamless pipe, as they define a set of internal tools available at the pipe mill for their manufacture. By contrast, UOE pipe thicknesses other than the standard values can be easily accommodated.

Wall tube insulator　管壁绝热
Wall ventilator[1]　舱室壁通风筒
Wall ventilator[2]　流吸式排风帽　具有筒体收缩段，使空气局部增速，形成低压以吸出舱内空气的排风帽。
War risk　战争险　又称战乱险，包括战争险和内乱险，系指外国投资者在东道国的投资因当地发生战争等军事行动或内乱，而导致损失的风险。
Warm(**warmer**)　加温器（加温，加热）
Warm-up valve　暖机阀
Warming-up(**warm-up**)　暖机（加热，加温）　汽轮机启动时，为减少气缸与转子的热膨胀差及热应力的影响，一面进行低速转动，一面送入少量蒸汽使汽轮机各部件均匀受热的过程。
Warming-up and starting test　暖机启动试验　确定主、辅汽轮机合理的暖机确定程序的试验。
Warming-up time　暖机时间
Warning　警告　系指（1）无须驾驶台工作人员立即关注或行动的情况。显示警告是为预防起见，让驾驶员注意到情况发生变化，虽然这些情况尚无紧迫危险，但如不采取行动也会有变成危险的可能性；（2）要求驾驶团队立即注意但不必立即采取行动的状况。警告是出于预防的原因使驾驶团队意识到状况的变化不是即刻危险，但可能成为危险。　Warning means: (1) a condition requiring no-immediate attention or action by the bridge team. Warnings are presented for precautionary reasons to make the bridge team aware of changed conditions which are not immediately hazardous, but may become so, if no action is taken; (2) the condition requiring immediate attention, but no immediate action by the bridge team. Warnings are presented for precautionary reasons to make the bridge team aware of changed conditions which are not immediately hazardous, but may become so if no action is taken.

Warning device　报警装置（警告装置）
Warning mark　警告性标志　又称危险货物包装标志。凡在运输包装内装有爆炸品、易燃物品、有毒物品、腐蚀物品、氧化剂和放射性物质等危险货物时，都必须在运输包装上标明用于各种危险品的标志，以示警告，便于装卸、运输和保管人员按货物特性采取相应的防护措施，以保护物资和人身的安全。
Warping block　绞滩滑车（绞滩绳）　（1）设置于内河船首部两舷，专供导引绞滩缆的固定导向滑车；（2）内河船上通过急流险滩时放出，供卷收以牵曳本船上滩的绳索。
Warping end(**warping head**, **warping drum**, **gypsy**)　绞缆筒　安装在绞车或起锚机等轴端，用以绞引而不储存缆索的筒体。
Warping rope　绞滩缆　内河船舶上水过急流险滩时放出，供卷收以牵曳本船上滩的缆索。
Warping winch　绞缆机　用以收绞系泊缆绳的机械。
Warsaw-Oxford Rules　华沙-牛津规则　是国际法协会专门为解释 CIF 合同而制定的规则。
Warship　军舰　系指属于一国海军，并且有识别其国籍的外部标志，由该国政府正式委任并在海军名册上登记的军官指挥，并且配备服从正规海军纪律的船员的船舶。　Warship means any ship belonging to the naval force of a State and bearing the external marks distinguishing warships of its nationality, under the command of an officer duly commissioned by the Government of such State and whose name appears in the Navy List, and manned by a crew who are under regular naval discipline.
Wash bulkhead　制荡舱壁　是液舱内具有开口的舱壁或部分舱壁。　Wash bulkhead is a perforated or partial bulkhead in a tank.
Wash down　洗刷
Wash water pump　冲洗水泵
Washdeck gear　甲板冲洗装置
Wash-deck gear　甲板清洗工具　供清洗甲板用的板刷、拖畚、水桶等用具的统称。

Washdeck hose 冲洗甲板软管
Washdeck pump 甲板冲洗泵
Washdeck system 甲板冲洗系统
Washer lock 止推垫圈(保险垫片)
Washer of nut 螺母垫圈
Washer snap 开口垫圈
Washing arrangement 洗舱装置 液货舱进行清洗用的一套装置,包括泵、管路、喷射设备、仪表及控制设备等。
Washing equipment 清洗装置 消除压气机和涡轮通道内,特别是叶片上附着的沉积物及减少腐蚀的装置。
Washing heater 洗舱加热器
Washing out boiler 洗刷锅炉
Washing water 洗涤水(洗涤用水) 船上供生活上洗涤用的水。
Washings 洗涤剂(涂料)
Waste carrier 废物运输船
Waste cotton 废纱头
Waste crushing-disposal methods 垃圾粉碎处理法 系指用粉碎机将垃圾粉碎,使其粒度达到排放要求后再排放入海。长期航行在禁止投弃垃圾海域的船舶应在舷侧装一个储存柜。粉碎后的垃圾储存在里面,当船舶航行至非管制海域时再排放。
Waste directly jettison methods 垃圾直接投弃法 系指当船舶在管制海域时,先将垃圾保存在船上,待船舶航行在非限制海域时,再将垃圾直接排放入海。此法的缺点是储存的垃圾会腐烂而产生臭味等不卫生问题。
Waste disposer 废物清除器
Waste gas flue 废气道
Waste gas suoerheater 废气过热器
Waste gas supercharger 废气增压器
Waste heat 废热
Waste heat boiler 废热锅炉
Waste heat flue 废热道
Waste heat utilization 废热利用
Waste heating 废气加热
Waste incineration treatment methods 垃圾焚烧处理法 系指将可燃烧的垃圾(固体垃圾和液体垃圾)送入焚烧装置内焚烧掉。垃圾焚烧装置是焚烧船上垃圾的防污染设备。主要由焚烧炉、燃烧器、风机、废油柜、燃油泵及电气控制箱等组成。
Waste oil 废油
Waste oil treating ship 污油处理船
Waste pipe 排泄管(废气管,污水管)

Waste product 废品
Waste receptacle 废物箱(废物库)
Waste steam economizers 废气经济器 系指强制循环废气热交换器,不能直接提供蒸汽,而要用燃油锅炉作为蒸汽贮存器来提供蒸汽的称为废气经济器。Waste steam economizers are forced-circulation exhaust gas heat exchangers not capable of directly supplying steam, and requiring fired boilers as steam storage for supplying steam.
Waste steam turbine 废气透平
Waste tank 废液舱
Waste-heat recovery(waste-heat utilization) 废热利用 通常系指柴油机排气和冷却主机的淡水的余热利用。
Wastes 垃圾 对固体散货安全操作规则而言,系指含有该规则中第 4.1、4.2、3、5.1、6.1、8 或 9 类物质的规定所适用的一种或多种组分或受其沾染,而且除倾倒、焚烧或其他处置外无直接用途的一些固体物质。For the purpose of BC Code, wastes are solid materials containing or contaminated with one or more constituents which are subject to the provisions of the BC Code applicable to materials of Classes 4.1, 4.2, 3, 5.1, 6.1, 8 or 9 and for which no direct use is envisaged but which are carried for dumping, incineration or other methods or disposal.
Wastes or other matter 废物及其他物质 系指任何种类、任何形状或任何式样的废弃材料和物质。Wastes or other matter means material and substance of any kind, form or description.
Watch alarm 值班报警 系指当驾驶员的值班功能丧失(缺席、缺乏警觉、对另一报警没有反应等)时,从驾驶室传至船长和后备驾驶员的报警。 Watch alarm means alarm that is transferred from the bridge to the master and the back-up navigator in case of any officer of the watch deficiency(absence, lack of alertness, no response to the other alarm/warning, etc.).
Watch engineer room 值班轮机员室
Watchdog 监视器 系指定期监测软件和硬件运行情况的系统。 Watchdog means a system which monitors the software and hardware well running at regular intervals.
Watchkeeping cabin 机舱值班室
Water ballast capacity 压载水舱容量
Water ballast double bottom tank 双层底压载水舱
Water ballast main 压载水总管
Water ballast space 压载水舱
Water ballast suction 压载水吸管
Water ballast tank(WBT) 压载水舱 船舶空载

时，用来盛放压载物的舱室的总称。压载物是使船舶稳定便于操作的重物，通常为海水。作此用途的舱室包括双层底的深舱和翼舱。在不装压载物时，根据水箱和船舶种类的具体情况，这些舱室可盛放淡水、燃油甚至装载货物。

Water ballast tank vent 压载水舱通气管

Water based coating 水性涂料 系指用水作溶剂或者作分散介质的涂料。其主要包括水溶性涂料、水稀释性涂料、水分散性涂料(乳胶涂料)等三种。水溶性涂料(水性重防腐涂料)又分为水性丙烯酸涂料、水性环氧涂料、水性无机硅酸锌涂料等。水性涂料主要应用于船体的机舱内部、上层建筑内部、舾装件的空气瓶、箱柜等。

Water boiler 沸水器 用蒸汽、燃气或电等热源煮沸饮水的器具。

Water box 水室 表面式冷凝器中冷却水管进口和出口端对冷却水起均匀分配和集流作用的，由封头和管板形成的空间。

Water circulation test (water cycle test) 水循环试验 检查锅炉在承受各种负荷时水循环可靠性的试验。

Water cooled compressor 水冷式压缩机 汽缸或冷却器均用水冷却的压缩机。

Water cooled motor 水冷电动机

Water-cooled refrigerated containers 水冷式冷藏集装箱

Water cooling system 冷却水系统

Water depth above keel blocks 龙骨墩以上吃水 从中龙骨墩的上表面至水面的垂直距离。

Water depth h (m) 水深 h (m) h 为从海底到平均低水位海平面的垂直距离加上天文潮和风暴潮的潮高。Water depth h as used herein is the vertical distance from the seabed to the mean low water level plus the height of astronomical and storm tides.

Water exhaust valve 排水阀

Water exit 出水

Water feeding system 给水系统

Water filter 滤水器

Water filter tank 水过滤柜 滤去食用水中杂质的过滤柜。

Water fire extinguishing system 水灭火系统 直接采用舷外水进行灭火的系统。

Water fog applicator 水雾施放装置(水雾枪)

Water freeing arrangement 排水布置

Water funnel 水漏斗(水筒)

Water gland 水封压盖

Water glass 水玻璃

Water head 水头(水位差)

Water heating 水暖

Water ingress detection pre-alarm 进水探测预报警 系指当水位达到散货船或单舱货船的货舱或其他处所的低位时所发出的报警。Water ingress detection pre-alarm means an alarm given when the water level reaches a lower level in cargo holes or other spaces on a bulk carrier or single hold cargo ship.

Water inlet manifold 进水总管

Water inlet valve 进水阀

Water intake 进水口

Water intake duct 水进流管道 系指将水从进水口处吸进到叶轮进口处的那一部分。Water intake duct is the portion that leads the water drawn from the water intake to the impeller inlet.

Water jacket 水套(水衣)

Water jacket space 水腔(水套)

Water jet 喷水(喷水推进器)

Water jet propeller 喷水推进器 是一种新型的特种动力装置，与常见的螺旋桨推进方式不同，喷水推进的推力是通过推进水泵喷出的水流的反作用力来获得的，并通过操纵舵及倒舵设备分配和改变喷流的方向来实现船舶的操纵。在滑行艇、穿浪艇、水翼艇、气垫船等中、高速船舶上得到了应用。见图 W-2。

图 W-2 喷水推进器
Figure W-2 Water jet propeller

Water jet propulsion 喷水推进 是一种新型的推进方式，与常见的螺旋桨推进方式不同，喷水推进的推力是通过推进水泵喷出的水流的反作用力来获得的，并通过操纵舵及倒舵设备分配和改变喷流的方向来实现船舶的操纵。

Water jet propulsion arrangement 喷水推进器(喷水推进装置)

Water jet propulsion system 喷水推进系统

Water jet pump 喷水泵

Water jet ship 喷水推进船 系指利用喷水推进装置产生推力的船舶。

Water jet test 冲水试验(喷水试验)

Water jet unit 喷水推进装置

Water leakage alarm 漏水报警器
Water leakage detection system 水渗漏探测系统
Water leakage detector 漏水显示器
Water level alarm 水位报警器
Water level control 水位控制
Water level detector 水位探测器/水位传感器 系指能够探测到货舱或其他区域进水后,能鸣笛报警的装置。由传感器和指示器构成。 Water level detector means a system comprising sensors and indication devices that detect and warn of water ingress in cargo holds and other spaces.
Water level indicator 水位指示器
Water locks 水闸
Water lubricated pump 水润滑泵
Water lubricated stern-tube 水润滑艉管
Water main 水总管
Water mist 细水雾
Water mist nozzle 细水雾喷嘴
Water mist system 细水雾系统
Water monitor system 水灭火系统(水炮系统)
Water pressure main 水压总管
Water pressure tank (elevated tank) 压力水柜 借柜内被压缩的空气压力,把水压送到船上各用水处的封闭式水柜。有淡水压力柜、卫生水压力柜等。
Water pressure test 水压试验
Water pressure test for strength 水压强度试验
Water pressure test for tightness 水压紧密性试验
Water pressure tester 水压试验装置
Water pressure testing (water test) 水压试验
Water purification system 水净化系统
Water quality monitoring ship 水质监测船 系指设置专用仪器仪表,对重要水域的水质及生态指标实时进行监测和数据分析,掌握其变化规律,寻觅污染源头,向管理部门及时提供决策依据的船舶。
Water quality monitoring system 水质监测系统 又称水质多要素监测装置。测量海水温度、盐度、电导率、深度、溶解氧、pH 值、浊度、氧化还原电位等多种要素的装置。
Water reactive substances 遇水反应物质 这些物质分为以下 3 类,见表 W-1。 These are classified into three groups as table W-1.

表 W-1 遇水反应物质
Table W-1 Water reactive substances

遇水反应指数(WRI) Water reactive index (WRI)	定 义 Definition
2	接触水后,产生有毒、易燃或腐蚀性气体或气雾的化学品。 Any chemical which, in contact with water, may produce a toxic, flammable or corrosive gas or aerosol.
1	接触水后,发热或产生无毒、不可燃、无腐蚀性气体的化学品。 Any chemical which, in contact with water, may generate heat or produce a non-toxic, non-flammable or non-corrosive gas.
0	接触水后,不产生上述 1 类或 2 类反应的化学品。 Any chemical which, in contact with water, would not undergo a reaction to justify a value of 1 or 2.

Water regulating valve 水量调节阀 根据制冷压缩机冷凝压力自动调节冷凝器水流量,并使冷凝压力保持稳定的阀。
Water sampling and temperature reading room 采水测温室 用于放置采水器并进行现场读温和量取水样的工作室。该室位于上甲板,靠近水文绞车、水文室和化学室。
Water screen system 水幕系统 军用船舶和消防船上,用以产生水幕,作降温、灭火和隔火防护用或在进行原子、化学防护时作冲洗甲板用的系统。
Water service system 供水系统
Water siphone 虹吸

Water soluble salts limit 水溶性盐含量

Water spray fire extinguisher 水雾灭火器

Water spray holder 喷水器夹环

Water spray nozzle 水雾喷枪

Water spray system 水雾系统　对消防船而言，系指能保证水线以上的船体，包括上层建筑、甲板室及水炮座和其他消防设备的外部垂直面得到保护的系统。

For fire-fighting ship, water spray system means the system which is capable of ensuring a protection of all outside vertical areas of hull above waterline, including superstructures, deckhouses, seatings for water monitors and other fire-fighting equipment.

Water spray system 喷淋系统　安装在液化气船货舱的上甲板曝露部分、居住区前壁、货泵舱等区域，以冷却、防火、保护船员为目的，而非为灭火之用的装置。

Water spray system 水雾系统（水幕系统）

Water spray/jet type nozzle 水雾/喷水两用枪

Water spraying system 喷水灭火系统

Water sprinking system 淋水系统

Water stabilizing pocket 平衡水袋　装设在气胀救生筏底部四周，用以增加筏的稳定性的水袋。

Water supply 供水（给水）

Water tank 水柜（水舱）

Water tanker 运水船/供水船　系指载运淡水的船舶。Water tanker is a ship carrying fresh water.

Water tap 水龙头

Water temperature observation 水温测量　利用表面温度表、颠倒温度计或其他测温仪器对海水温度进行的现场观测。

Water temperature observation 水温观测

Water tender（water ship） 供水船　专供淡水的供应船。

Water tightness 水密性　不会透水的性能。

Water to steam ratio 水汽比率

Water treatment 水处理（软水处理）

Water trough 水槽

Water tube boiler 水管锅炉

Water tunnel 水漏斗（水筒）

Water vapour 水蒸发气

Water wall（water screen/ cooling wall） 水幕/水冷壁　置于炉膛中接受辐射热的由管排组成的水冷炉墙或水冷隔墙。

Water-and-oil separator 油水分离器　分离压缩介质中的液滴的容器。

Water-augmented air-jet propulsion 加水喷水推进　通过喷雾嘴将水加入空气中，增加推进剂质量以提高推进效能的一种双态喷射推进。

Water-based extinguishing medium 水基灭火剂　含有或不含能增强灭火能力的活性物质的淡水或海水。

Water-based extinguishing medium is fresh water or sea water with or without additives mixed to enhance fire-extinguishing capability.

Water-based fire-extinguishing systems 水基灭火系统　使用水基灭火剂进行灭火的灭火系统。

Water-cooled 水冷的

Water-cooled internal diesel engine 水冷内燃机　用水冷却的内燃机。

Water-entry angle 入水角　系指在救生艇最初入水时，降落滑道与水平面形成的角度。Water-entry angle is the angle between the horizontal and the launch rail of the lifeboat when it first enters the water.

Watering 注水

Watering can 喷水壶

Water-ingress detection main alarm 进水探测主报警　系指当水位达到散货船或单舱货船的货舱或其他处所的主报警水位时所发出的报警。 Water-ingress detection main alarm means an alarm given when the water level reaches the main alarm level in cargo holds or other spaces on a bulk carrier or single hold cargo ship.

Water-jet 喷水推进器　系指在管状箱体（或导管）中装有叶轮的设备。箱体的形状能保证叶轮产生喷水，其强度足以提供一个正的推力。喷水推进器可具有改变喷水方向的设施，以提供操纵船舶的功能。 Water-jet is equipment constituted by a tubular casing（or duct）enclosing an impeller. The shape of the casing is such as to enable the impeller to produce a water-jet of such intensity as to give a positive thrust. Water-jets may have means for deviating the jet of water in order to provide a steering function.

Water-jet dredger（WID） 喷水疏浚船　简称冲吸式挖泥船。系指配备大流量低扬程水泵，通过船上设置的管路系统，在经由水下高压喷嘴以设定的角度冲刷沉积于河床的泥沙，使其悬浮于底部的水流之中，并在水流的作用下形成异重流，使其在附近高程较低的地方再次沉积下来，即是借助流体扰动的原理达到冲吸效果的挖泥船。

Water-jet propulsion 喷水推进　是利用推进泵喷出水流的反作用力来推动船舶前进，并通过操舵倒航设备分配和改变喷流方向来实现船舶操纵。喷水推进的优点：(1)推进泵叶轮运行平稳，水下噪声小；(2)推进泵叶轮在高速范围内较螺旋桨有更好的抗空泡性能，从而能有更高的推进效率；(3)推进泵比螺旋桨更适用于重载荷以及限制直径的场合；(4)喷水推进适应变工况的能力极强,在工况多变的船舶上能充分利用主机功率,延长主机寿命;(5)具有优异的操纵性和动力定位性能;(6)推进泵叶片在管道中不易受损,可靠性好。喷水推进的缺点:(1) 由于增加了管道中水的质量,加大了船舶的排水量;(2)在水草或杂物较多的水域,进口容易出现堵塞现象而影响航速;(3)推进泵叶轮拆换比螺旋桨复杂。

Water-jet propulsion systems and azimuth or rotatable thrusters 喷水推进系统和全回转推力器 系指包括叶轮、主轴、水进流管道、喷嘴、导向机构和反向机构等部件的系统,其从进流管道中吸入水并通过喷嘴将增加速度的水排出以产生推力。 Water-jet propulsion systems and azimuth or rotatable thrusters means a system including impeller, main shaft, water intake duct, nozzle, deflecter and reverser that receives water through an inlet duct and discharges it through a nozzle at increased velocity to produce propulsive thrust.

Waterline beginning 水线前段 水线平行段以前的设计水线前部线段。

Waterline breadth 水线宽 与基平面相平行的任一水线平面处,船体型表面之间垂直于中线面的最大水平距离。

Waterline breadth (SWATH) 水线宽(小水线面双体船) 系指船舶静浮于水面时,沿设计水线量得支柱体的最大型宽之和。 Waterline breadth means the maximum moulded breadth measured from the design waterline with no lift or propulsion machinery.

Waterline breadth B_{WL} 水线宽 B_{WL} 通常为设计吃水处测得的最大型宽。 Waterline breadth B_{WL} is generally the greatest moulded breadth, in m, measured at the design draught.

Waterline coefficient (waterplane coefficient) 水线面系数 与基平面相平行的任一水线面的面积与其相应的水线长、水线宽的乘积之比。在无特别注明时,通常系指设计水线处的。对于民用船舶,各水线处水线面系数之计算,其中船长和型宽常用垂线间长和满载水线宽。

Waterline d_B 水线 就 MARPOL 而言,系指自船舯处的型基线至相应于 30% 船深 D_S 的水线之间的垂直距离,m。 For MARPOL, waterline d_B is the vertical distance, in m, from the moulded baseline at mid-length to the waterline corresponding to 30% of the depth D_S.

Waterline ending 水线后段 水线平行段以后的设计水线后部线段。

Waterline length 水线长 与基平面相平行的任一水线平面与船体型表面首尾端交点间的水平距离。

Waterline length L_{WL} 水线长度 L_{WL}(小艇) 在指定条件下的水线长度 L_{WL} 应取在垂直于艇中线面的两垂向平面之间平行于基准水线和艇的中心线的距离,其一个平面通过艏柱与水平面的最前交点,另一平面通过艇体与水平面的最后交点。 Waterline length L_{WL} for a designated condition shall be measured parallel to the reference waterline and craft centerline as the distance between two vertical planes, perpendicular to the center-plane of the craft, one plane passing through the foremost intersection of the stem with the flotation plane and the other through the aftermost intersection of the hull and the flotation plane.

Waterline plane coefficient C_{wp} 水线面系数 对小水线面双体船而言,系指按下式算得的系数:
$$C_{wp} = \sum A_{wl} / \sum l_{s1} B_{wl1}$$
式中:A_{wl}—沿设计水线截得的一个片体中的第一个支柱体水线面积,m^2;l_{s1}—沿设计水线量得的一个片体中的第一个支柱体的最大长度,m;B_{wl1}—沿设计水线量得的一个片体中的第一个支柱体的最大型宽,m;若在一个片体中设置了前后独立的支柱体,即 ≥2,则应将 A_{wl} 和 l_{s1} 与 B_{wl1} 所围的矩形面积按公式计入一个片体所含的总量。

Waterline WL 水线 WL 在型线图的侧剖面或横剖面上以直线显现,但其真实形状在半宽图中显现的漂浮平面与艇体的交线。 Waterline WL is the intersection between the flotation plane and the hull which appears as a straight line in either the sheer plan or the body plan, but in its true form in the half-breadth plan.

Water-lubricated bearing 水润滑轴承 以水作为润滑剂的轴承。

Water-plane 水线面 由水线围封的平面。

Water-proof glass 防水玻璃

Water-proof test 防水试验

Water-proof type 防水式

Water-protected enclosure 防水外壳 系指其制成使可能出现在该外壳内的潮气或漏水不会妨碍设备有效运行的外壳。对于电动机或发电机机壳,在轴的周围可能出现的漏泄可认为是允许的,只要能防止其进入油并设有使电动机或发电机壳自动泄水的措施即可。

Water-resistant 防水的

Waters of the Alexander archipelago 亚历山大群岛水域 系指靠近或在阿拉斯加东南部,在美国管辖下的下列水域:(1)自北纬 58°11′44″ 西经 136°39′25″ 的一点(靠近"开普斯宾塞灯塔")起,然后向东南方向,从测量太平洋和"迪克森入海口"的领海宽度用的基线向外延伸 3 n mile 划一根线,但该线与下列 5 对点连成的短程线(测地线)相交的水域除外:北纬 58°05′17″ 西经

136°33′49″与北纬 58°11′41″西经 136°39′25″(Cross 海峡);北纬 56°09′40″西经 134°40′00″与北纬 55°49′15″西经 134°17′40″(Chatham 海峡);北纬 55°49′15″西经 134°17′40″与北纬 55°50′30″西经 133°54′15″(Sumner 海峡);北纬 54°41′30″西经 132°01′00″与北纬 54°51′30″西经 131°20′45″(Clarence 海峡),北纬 54°51′30″西经 131°20′45″与北纬 54°46′15″西经 130°52′00″(Revillagigedo 海峡);(2)上述(1)款中每一这样的短程线(测地线)的一部分,都位于从这 5 对点所在的亚历山大群岛水域的外限线测量领海宽度用的基线向外延伸 3 n mile 的水域以外。

Waters of the United States 美国水域 系指联邦政府法规第 33 篇 2.38 中定义的美国管辖的水域。包括美国可航行水域。对于联邦政府法规第 33 篇 §151 的 C 分节和 D 分节,可航行水域包括根据 1988 年 12 月 27 日总统行政法令第 5928 号指定的海岸线向海外延伸 12 海里的领海。 Waters of the United States means waters subject to the jurisdiction of the United States as defined in CFR 33 Part 151. Subparts C and D, the navigable waters include the territorial sea as extended to 12 nautical miles from the baseline, pursuant to Presidential Proclamation No. 5928 of December 27, 1988.

Water-softening plant (de-mineralizing plant) 给水软化器 去除给水中引起污垢的镁、钠杂质等的设备。

Watertight 水密 系指:(1)构件尺寸和布置在完整和破损工况中可能产生的水头下,能防止水从任何方向进入。在破损工况中,水头应考虑在平衡时,包括进水的中间阶段中最差的状态;(2)如果一个边界或关闭装置,其设计能在周围结构的设计压头下,防止水从任何方向通过,就可认为是水密的;(3)防止水在周边结构设计水压作用下从任何方向进入的结构。Watertight means:(1)having scantlings and arrangements capable of preventing the passage of water in any direction under the head of water likely to occur in intact and damaged conditions. In the damaged condition, the head of water is to be considered in the worst situation at equilibrium, including intermediate stages of flooding;(2)a boundary or closing appliance is considered watertight, if it is capable of preventing the passage of water in either direction under a head of water for which the surrounding structure is designed;(3)a capability of preventing the passage of water through structure in any direction under the head of water for which the surrounding structure is designed.

Watertight aircase 空气箱 救生艇内用以提供内部浮力的水密箱。

Watertight bulkhead 水密舱壁 在规定压力下能保持不透水的舱壁。

Watertight door 水密门 符合水密要求的各种门的统称。根据门板厚度和楔形把手的多少,一般有重型水密门和轻型水密门之分。

Watertight enclosure 水密外壳 系指制成使其从直径不小于 1 英寸(in)的软管喷出的流水在 35 英尺(ft)的水头下从距离 10 ft 处的任意方向喷射在其上,历时 15 min 而无泄漏的外壳。该软管的喷嘴应具有均匀的 1 in 的内径。

Watertight floor 水密肋板 在规定压力下能保持不透水的肋板。

Watertight in relation to a structure 与某一结构有关的水密 系指在完整或破损情况中可能产生的水压下能够防止水从任何方向穿过该结构。Watertight in relation to a structure means the capability of preventing the passage of water through the structure in any direction under the head of water likely to occur in the intact or damaged condition.

Watertight lamp 水密灯具 系指当曝露在除浸水以外的任何条件下时,其结构能防止水进入(包括有接线装置、接线和灯泡的)外罩的任何部分中的封闭式航行灯。

Watertight of unit 平台的水密 系指防止水在周边结构设计水压作用下从任何方向进入浮力结构。Watertight of unit means that capability of preventing the passage of water through structure in any direction under the head of water for which the surrounding structure is designed.

Watertight openings 水密开口 系指配有水密关闭方式的内部开口,其能承受一个固定水压头,相当于该开口最低边缘和舱壁/干舷甲板间距离的水压头。水密开口不会导致连续进水。 Watertight openings are those internal openings fitted with watertight means of closure, which are able to sustain a constant head of water corresponding to the distance between the lowest edge of this opening and the bulkhead/freeboard deck. Watertight openings do not lead progressive flooding.

Watertight recess 水密凹体 符合 ISO 11812 中"水密尾舱和凹体"之所有要求的凹体。注意:此术语仅指与水密性和门槛高度有关的要求,而与泄水的要求无关。 Watertight recess is the recess fulfilling all the requirements of ISO 11812, for "watertight cockpits and recesses". NOTE: This term only implies requirements in respect of watertightness and sill heights, but not those for drainage.

Watertight sliding door 水密滑动门

Watertight subdivision 水密分舱 系指利用水密舱壁将船体分成若干水密舱室,以满足船舶抗沉性要求一种设计措施。

Watertight test 水密试验 往水舱、油舱内灌水来检验其水密性的一种试验。但可有条件地用气密试验所代替。

Watertightness 水密性 不透水的性能。

Water-tube boiler 水管锅炉 水和汽水混合物在管内流动,烟气在管外流过,受热面由水管构成的锅炉。

Waukegan Special Protection Areakegan Specia 沃基根(美国伊利诺州东北部城市)特别保护区 系指由从最北的一点开始并顺时针方向连接下列坐标的恒向线所包围的区域:北纬42°24.3′,西经087°29.3′;北纬42°13.0′,西经087°25.1′;北纬42°12.2′,西经087°29.1′;北纬42°18.1′,西经087°33.1′;北纬42°24.1′,西经087°32.0′。Waukegan Special Protection Area means the area enclosed by rhumb lines connecting the following coordinates, beginning on the northernmost point and proceeding clockwise; 42°24.3′N, 087°29.3′W; 42°13.0′N, 087°25.1′W; 42°12.2′N, 087°29.1′W; 42°18.1′N, 087°33.1′W; 42°24.1′N, 087°32.0′W.

Wave 浪 对设计环境而言,系指对每一个特定的设计项目均应规定有义波高与周期范围。For the design environmental, the significant wave height and period range is to be defined for each relevant design case.

Wave age 波龄 波速与风速的比值。

Wave bending moment 波浪弯矩 船舶在波浪上运动时,相对于静水状态所增加的弯矩。

Wave circular frequency 波浪圆频率 波浪质点轨圆运动的角速度。

Wave coefficient C_{wv} 波浪系数 系指根据规范船长确定的系数,见表W-2。Wave coefficient C_{wv} is a coefficient determined according to rule length, as can be seen in table W-2.

表 W-2 波浪系数 C_{wv}
Table W-2 Wave coefficient C_{wv}

规范船长	C_{wv}
$150 \leq L < 300$	$10.75 - [(300 - L)/100]^{2/3}$
$300 \leq L < 350$	10.75
$350 \leq L < 500$	$10.75 - [(L - 350)/150]^{3/2}$

Wave crest 波峰 波形剖面的最高点,或泛指波形突出部分。

Wave damper 消波器 布置在船模试验水池池壁,以消除波浪反射作用的设备。

Wave detector of effective values 有效值检波器

Wave direction angle 波向角 波浪传播方向与x_0轴或其他参考方向间的水平夹角。

Wave energy 波能 波系中包括动能和位能在内并与波幅的平方成比例的能量。

Wave frequency 波浪频率 系指在单位时间内通过固定点的波浪的数量。Wave frequency means the number of waves passing a fixed point in unit time.

Wave frequency 波浪频率 单位时间内,波浪周期性地产生峰或谷的次数。

Wave frequency(WF) 波频 系指波浪的频率。

Wave front[1] 波前 系指声波传播处于最前沿的波阵面。

Wave front[2] 波阵面 系指声波在传播过程中,在某一时刻从波源向外发出的波所到达的空间内,所有相位相同的媒介质点所形成的面,又称为等相面,简称"波面"。

Wave generator 造波机 在船模试验水池中,制造波浪的设备。有冲箱式、摇板式和压缩空气式等。

Wave height 波高 相邻波峰与波谷间的垂向距离。

Wave height recorder 浪高仪 测量和记录波高的仪器。

Wave impact load 波浪冲击载荷 船舶在波浪上运动时,由于波浪的冲击而作用在船体上的载荷。

Wave length 波长 系指声波在一个周期中传播的距离。亦即同一波线上两个相邻的周期差为2π的质点之间的距离,就是一个完整波的长度。

Wave loads 波浪载荷 系指:(1)由于海水压力和船舶运动引起,并假定与诱导波浪有相同周期的载荷;(2)由波浪产生的、作用在结构物或部件上的载荷。Wave loads are:(1)those due to wave pressures and ship motions, which can be assumed to have the same wave encounter period;(2)the action on structure or component arising from waves.

Wave loads calculation system(WALCS) 船海浮体载荷计算软件 系指船舶与海洋工程浮体水动力与波浪载荷计算专用软件。WALCS是适用于多种类型船舶与海洋结构物,如油船、散货船、浮式生产储油船,半潜式平台等的运动与波浪载荷计算的三维水动力软件。适用范围包括无限水深和有限水深,并可考虑航速的影响。其主要功能包括:浮体附加质量和波浪辐射阻尼计算、锚泊系统恢复力的计算、浮体上的压力分布、浮体剖面载荷计算、水动力分析与通用结构分析有限元软件的接口等。该链接能解决目前船舶与海洋工程水动力计算领域遇到的大部分工程问题。

Wave number 波数 其值等于单位距离上出现的波浪个数的 2π 倍数的一种表示波形的参数。其表达式为 $K = 2\pi/\lambda = \omega_\mathrm{w}^2/g$。

Wave observation laboratory 波浪观察室 用于波浪观察,研究其变化规律及产生的机制、分析波谱,并进行波浪预报的工作舱室。

Wave of translation (solitary wave) 浅水移动波 水深很浅时,由于水质点的垂向运动受到水底的限制,其轨迹运动保持近于水平状态,波形是一完全在水面上的波峰,波速只与水深及波高有关而与波长无关的波。

Wave parameter C 波浪参数 C 见表 W-3。

表 W-3 波浪参数 C
Table W-3 Wave parameter C

对 90 m $\leq L <$ 300 m 的船舶 (for 90 m $\leq L <$ 300 m)	$C = 10.75 - \{(300-L)/100\}^{1.5}$
对 300 m $\leq L \leq$ 390 m 的船舶 (for 300 m $\leq L \leq$ 390 m)	$C = 10.75$

Wave pattern resistance 波型阻力 用测量波型的方法所求得的兴波阻力部分。

Wave period 波浪周期 除非另有说明,应采取跨零波周期,即在平均值处向上跨过零点之间的平均时间间隔。References to wave period are to be taken as the zero crossing wave period, i.e. the average time interval between upward crossings of the mean value, unless otherwise stated.

Wave pierce craft 穿浪双体船 系指具有大宽长比和小水线面面积的一种特殊船型的双体高速船。Wave pierce craft is a special type of catamaran high speed craft with large aspect ratio and small waterplane area.

Wave pressure in flooded conditions p_WF (kN/m²) 进水工况下的波浪压力 p_WF p_WF means wave pressure, in kN/m², in flooded conditions.

Wave pressure or dynamic pressures p_W (kN/m²) 波浪压力或动压力 p_W p_W means wave pressure or dynamic pressures, in kN/m².

Wave probe 水池浪高仪 船模试验中,用以测定池中浪高的仪器。一般有电导式、电容式、超声波式等。

Wave pumping 波浪泵气效应 气垫船在垫升航态时,由于波浪的作用而引起气垫体积变化或气垫压力变化的效应。

Wave slope 波倾 与波峰线正交的垂向剖面上的波浪表面最大倾斜角。

Wave spectrum 波浪谱

Wave steepness 波陡 波高与波长的比值。

Wave trough 波谷 波形剖面的最低点,或泛指波形下凹部分。

Wave velocity 波速 波形前进的速度。

Wave wake 船波伴流 船前进时,在螺旋桨处由于船波运动使水分子产生的向前或向后的伴流。

Wave-breaking resistance 破波阻力 由于船首有波破碎,其重力能量耗散为黏性能量的一部分兴波阻力。

Wave-making resistance (wave-forming resistance) 兴波阻力 系指船舶航行时在船体周围及后方产生不断向外传播的波浪,使船体表面应力分布改变而引起的阻力。

Wavemeter winch 波浪仪绞车 作为收、放重力式测波仪、船舷测波仪等仪器,进行海浪观测工作的绞车。

Waves 波浪 系指由风和周期小于 25s 的波浪在海面上引起水的运动。Waves are the movement on the sea surface generated by wind and wave periods of less than 25 seconds.

Waves in ship direction 顺浪 包含随浪、追浪、淹艇浪等与船舶同方向前进的波浪。

Waypoint mode 航路点方式 是确定测线的一种方法。根据海区的特点,布设的测线是由若干个航路点(转向点)连接而成,也可由互不连续的若干折线段组成。

Weak coupling 弱耦合 系指耦合损耗因子在数值上明显小于每个子系统的内损耗因子的耦合子系统的连接。

Weak limit 贫油极限 保持燃烧室稳定工作的空气燃料比的上限。

Weakly non-linear model 弱非线性模型

Wear allowance 磨损许可量

Wear and tear 磨损(消耗)

Wear clearance 磨损间隙

Wear clearance limit 磨损极限间隙

Wear condition 磨损状况

Wear control line 磨损控制线

Wear down 磨耗

Wear gauge 磨损量规

Wear hardness 抗磨硬度(抗磨力)

Wear intensity 磨耗程度

Wear particle 磨损颗粒(磨粒)

Weather awning 风雨篷 为保护乘员免受曝露引起的伤害,并用以收集雨水而配备在救生艇上的篷罩。

Weather clothes 围帘 张设在船舷边,供挡风、雨雪用的布帘。

Weather deck 露天甲板 系指:(1)在上方且至少有两侧完全曝露于露天的甲板;(2)曝露于自然环境的甲板或一段甲板,其所有的舱口和开口均有风雨密关闭设施;(3)部分或全部露天的可能部分或全部为旅客利用的最上层全通甲板。(4)通常系指遭受海水和露天载荷的最低连续甲板。其应在设计初期阶段由船级社和建造者联合定义。 Weather deck is:(1)a deck which is completely exposed to the weather from above and from at least two sides;(2)a deck or section of deck exposed to the elements which has means of closing weathertight, all hatches and openings;(3)the uppermost continuous deck fully or partially exposed to weather which may be wholly or partially used by passengers;(4)generally the lowest continuous deck exposed to sea and weather loads. It is to be defined at the early stages of design in conjunction with LR and the Builder.

Weather forecasting room 天气预报室 又称天气分析室,对所获取的各种天气情报分析整理,并进行天气会商,公布本海区天气预报的工作舱室。

Weather ladder 露天梯 设置于敞露部位的甲板梯。

Weather operating 露天作业 系指在露天甲板或在顶盖完全移开的处所内的作业。

Weather protection 露天保护 系指在进行焊接施工的区域应能足以防止气候的影响,例如风雨、潮湿和寒冷等,而且必要时应进行预热。 Weather protection means the area to be welding shall be adequately protected against climatic, influences, such as wind, damp and cold and shall be preheated where necessary.

Weather protection(welding at low temperatures) 露天保护(低温焊接) 系指进行焊接施工的区域,特别是外场作业,应有能防风雨、潮湿和寒冷的遮蔽。在进行气体保护电弧焊的场所,应特别注意穿堂风。在恶劣的气候条件下露天作业时,建议用加热的方法在施焊前将拟焊接的边缘烘干。 Weather protection, welding at low temperatures means that the area in which welding work is performed-particularly outside is to be sheltered from wind, damp and cold. Where gas-shielded arc welding is carried out, special attention is to be paid to ensure adequate protection against draughts. When working in the open under unfavourable weather conditions, it is advisable always to dry welding edges by heating.

Weather protection for welding at low temperatures 在寒冷条件下焊接的露天保护 系指在进行焊接施工的区域,特别在室外应防风雨、潮湿和寒冷的影响。如进行气体保护电弧焊时,则应特别注意充分保护以防止穿堂风。当在不利的气候条件下露天进行施工时,建议通过加热烘干焊接边缘。当环境温度低于+5℃时,应采取额外措施,例如构件的围蔽、扩大初始加热和预热,特别在相对较低热量输入(单位长度焊缝的能量输入)焊接时,例如熔敷度较薄的角接焊缝或焊接厚壁的构件过程中热量快速耗散时。当环境温度低于-10℃时应尽可能不进行焊接。 Weather protection for welding at low temperatures means the area in which welding work is performed is to be sheltered from wind, damp and cold, particularly if out of doors. Where gas-shielded arc welding is carried out, special attention is to be paid to ensure adequate protection against draughts. When working in the open under unfavourable weather conditions it is advisable to dry welding edges by heating. At ambient temperatures below +5℃, additional measures shall be taken, suck as shielding of components, extensive preliminary heating and preheating, especially when welding with a relatively low heat input (energy input per unit length of weld), e.g. when laying down thin fillet welds or in the case of rapid heat dissipation, e.g. when welding thick-walled components. Wherever possible, no welding should be performed at ambient temperatures below -10℃.

Weather restriction Ⅰ 营运气象Ⅰ 系指地效翼船营运气象Ⅰ,即限制航行区域有义波高不超过3.0 m,且风力不超过7级(蒲氏风级)。

Weather restriction Ⅱ 营运气象Ⅱ 系指地效翼船营运气象Ⅱ,即限制航行区域有义波高不超过2.0 m,且风力不超过6级(蒲氏风级)。

Weather restriction Ⅲ 营运气象Ⅲ 系指地效翼船营运气象Ⅲ,即限制航行区域有义波高不超过1.0 m,且风力不超过5级(蒲氏风级)。

Weather restriction Ⅳ 营运气象Ⅳ 系指地效翼船营运气象Ⅳ,即限制航行区域有义波高不超过0.5 m,且风力不超过4级(蒲氏风级)。

Weather ship 天气船(气象船) 用于海洋气象进行长期定点观测的船舶。也可供飞机和其他船舶定位,提供航线的气象情报,并发出危险天气预报。

Weather-tight 风雨密 系指：(1)在临界设计工况规定范围内的任何风力和波浪状况下，水不会渗入船内；(2)在任何海况下，水都不会渗透进入船内；(3)如果一个边界或关闭装置，其设计能在任何海况下防止水进入船内，就可认为是风雨密的。 Weather-tight means: (1) that water will not penetrate into the craft in any wind and wave conditions up to those specified as critical design conditions; (2) that in any sea conditions water will not penetrate into the ship; (3) a boundary or closing appliance is considered weather-tight if it is capable of preventing the passage of water into the ship in any sea conditions.

Weathertight of unit 平台的风雨密 系指在任何海况下水都不能渗入平台。 Weathertight of unit means that in any sea conditions water will not penetrate into the unit.

Weathertight openings 风雨密开口 系指配有风雨密关闭方式的开口，其不能承受一个固定水压头，但在稳性正值范围可间隙地浸没。如位于破损后水线下（在任何进水阶段）风雨密开口可能连续进水。 Weathertight openings are those openings fitted with weathertight means of closure, which are not able to sustain a constant head of water, but they can be intermittently immersed within the positive range of stability. Weathertight openings may lead to progressive flooding if they are located below the waterline after damage (at any stage of flooding).

Weave maneuver test 小舵角 Z 形操纵试验 试图变换几种 ±5°以下小角度，以决定能应舵的不稳定回线环宽的 Z 形操纵试验。

Web 腹板 系指与板面垂直相接的加强构件的一部分。 Web means a section of a stiffening member attached perpendicular to the plated surface.

Web frame side transverse 强肋骨 用于局部加强或支承舷侧纵骨的，加大尺寸的肋骨。

Web height h_w (mm) 腹板高度 h_w 系指普通扶强材或主要支撑构件的腹板高度（mm）。 Web height h_w means a web height, in mm, of ordinary stiffeners or primary supporting members.

Web of foundation girder 基座腹板 基座桁的腹板。

Weber number 韦伯数 考虑在表面张力与惯性力作用下的流体动力相似的一个无量纲参数。一般表达式为 $W_n = U^2 L/R$ $R = \sigma/\rho$，式中：U—流速；L—长度；σ—单位长度表面张力；ρ—流体质量密度；R—运动毛细系数。

Webpage cannot be found 无法找到网页

Wechat "微信" 是腾讯公司于 2011 年 1 月 21 日推出的一款手机通信软件。其支持通过手机网络发送语音短信、视频、图片和文字，支持视频聊天，还能根据地理位置找到附近的人，以及通过朋友圈分享自己的生活，带给朋友们全新的移动沟通体验。支持 iOS、Android、塞班、WP 等多种平台手机，可以显示简体中文、繁体中文、英文、泰语、印尼语、越南语、葡萄牙语等 7 种界面。

Wedge 固桩块 用来充填桩腿与船体之间的空隙，使之牢固卡死在一起的楔形固定块。

Wedge (hatch wedge) 封舱楔 置于舱口四周，用以压紧封舱塞条的硬木楔板。

Wedge bracket 固桩架 固定桩腿和安装升降机构用的结构物。

Wedge clip 楔形夹扣 利用闩块压紧门框或舱盖围板上的楔块的夹具。

Wedge-shaped section 楔形切面 导边较薄，随边较厚的切面。

Weeding boat 除草船 系指用以清除各类影响生态及航运安全的恶性水草，如水葫芦、青苔等水生植物的船舶。

Weight curve 重力曲线 表示船舶在某一状态下，全部力量沿船长分布状况的曲线。

Weight displacement of naked hull 裸船体排水质量 实船或船模的裸船体水线以下部分所排开的质量：$W = \omega \nabla$，式中：W—裸船体排水质量的通用符号；ω—单位体积水的质量；∇—裸船体排水体积。

Weight list 质量单 是标示该物品内部实质意义的凭证。标签主要有非净重、净重、质量差异等参数。

Weight RO-RO carrier 载重型滚装货船 系指每米车道运载质量等于 4 t 或以上的滚装船。

Weighted sound reduction index 计权隔声量 R_w 表征建筑构件空气声隔声性能的单值评价量，用 R_w 表示，单位 dB。即表示墙、门或地板（在实验室内）整体隔声性能的一个单一数值，以分贝计。不同材料的隔声量频率特性曲线很不相同，为了使用单一指标比较不同材料及构造的隔声性能，人们使用计权隔声量 R_w。R_w 是将墙体材料的 100～3 150 Hz 频段内的 16 个倍频程（或 120 Hz 到 2 000 Hz 频段范围内的 5 个倍频程）隔声量与一组基准值按一定的方法进行比较后得出的。基准值是根据人体对不同频率声音感觉的灵敏度所制定的，标准评价曲线符合人耳低频不敏感的听觉特性。具体评价方法可参见国标 GBJ121-88"建筑隔声评价标准"。

Weighted swivel 压重转环 具有压重作用的转环。

Weighting network 计权网络

Weighting sound level 计权声级

Weighting 计权 系指有用信号功率与无用的噪

声功率之比。通常以分贝计量，因为功率是电流和电压的函数，所以信噪比也可以用电压值来计算，即信号电平与噪声电平之比值，只是计算公式稍有不同。

Weld crack 焊接裂缝 在船舶或平台建造过程中，并在焊接应力及其致脆因素共同作用下，焊接接头中局部地区的金属原子结合力遭到破坏而形成新界面产生的裂缝。

Weld metal(WM) 焊缝金属

Weld splatter 焊渣飞溅

Welded pipe(continuous) 焊接管(连续焊成的) 系指通过连续焊工艺形成一条纵向焊缝的钢板加工而成的钢管。这种钢管的主要类型分为埋弧焊(UOE 或 SAW)钢管和电渣焊(ERW)钢管两种。埋弧焊钢管直径为 16 英寸(in)或稍厚，而电渣焊(ERW)钢管的直径范围从 2.375~24 英寸(in)。 Welded pipe(continuous) is a pipe that formed from plate with one longitudinal seam produced by a continuous welding process. The two main categories are UOE(or submerged-arc welding, SAW) pipe and electric resistance welded(ERW) pipe. UOE pipe diameter is 16 in or higher while REW pipe diameters range from 2.375 to 24 in.

Welded ship 焊接船 船体结构中主要用电焊方法连接的船舶。

Welding apertures 避焊孔 系指为便于横向构件定位后进行对接焊缝或角接焊缝的焊接(后续的)而开设的孔。 Welding apertures means a hole used for the (later) execution of butt or fillet weld following the positioning of transverse members.

Welding distortion 焊接变形 因焊件的不同而表现为翘起、角变形、弯曲变形、波浪变形的焊接缺陷。

Welding proceduce specifications(WPS) 焊接程序大纲

Welding procedure qualification record(PQR) 工艺评定报告书(PQR) 系指根据(WPQT)中规定的要求焊接经认可的试验材料时，对于适用的实际焊接特性的记录，无损检测和力学性能试验的结果报告。 Welding procedure qualification record(PQR) is a record of the actual parameters employed during welding of the qualification test piece according to the requirement of WPQT, and resulting from the non-destructive inspection and mechanical testing.

Welding procedure qualification record(WPQR) 焊接工艺合格记录 在焊接合格试件时选用的实际参数的记录和无损探伤试验及力学试验的结果。 Welding procedure qualification record(WPQR) is record of the actual parameters employed during welding of the qualification test piece, and resulting from the non-destructive testing and mechanical testing.

Welding procedure qualification test(WTQT) 焊接工艺合格试验 为证实按具体的工艺规程进行的焊接确实符合规定的要求而进行的试验。当要求进行焊接工艺合格试验时，该试验应在与实际生产相适应的环境下进行，并符合施焊前规定的最低要求。合格试验应有验船师见证。 Welding procedure qualification test (WTQT) is a test carried out in order to demonstrate that a weld made according to a specific procedure specification meets the given requirements. When welding procedure qualification is required, the tests must be performed in the environment applicable to the actual production and meet the specified minimum requirements prior to commencing the production welding. The qualification test is to be witnessed by the Surveyor.

Welding procedure qualification tests(WPQT) 焊接程序认可试验(WPQT) 系指为了确认按规定的焊接工艺评定进行的焊缝是否能满足所要求的条件而进行的试验。 Welding procedure qualification tests (WPQT) is a test carried out in order to demonstrate that a weld made according to a specific welding procedure specification meets the given requirements.

Welding procedure specification(WPS) 焊接工艺评定(WPS) 系指详细记载有适用于特定焊缝的焊接特性、焊接工艺和材料等规程。 Welding procedure specification(WPS) is a specification of materials, detailed methods, welding parameters etc, to be applied in the welding of a particular joint.

Welding procedure tests in the user's works 在用户工厂进行的焊接程序试验 系指开始焊接装配前在船级社的监督下按有关要求对在车间的条件下不同应用场合规定的范围在用户工厂进行的焊接程序试验。工地条件(防风雨、焊接设备、操作工具、焊工、制造容差等)和任何预期的极端冷成形作业和适用的材料和/或焊缝的热处理应包括在焊接程序试验内。 Welding procedure tests in the user's works mean those tests carried out under the Society's supervision in user's works before starting the fabrication work according to the relevant requirements for the different areas of application under workshop conditions. Workplace conditions(weather protection, welding equipment, operating jigs, welders, production allowances etc.) and any intended extreme cold-forming operations as well as heat treatments of the materials and/or the welds where applicable shall form an integral part of the welding procedure tests.

Welding procedures specifications(WPS) 焊接工艺规程 对特定焊接接头采用的材料、焊接方法、操作和各种参数的规定。经认可的焊接工艺规程至少应包括下述与焊接操作有关的资料：(1)材料、标准、等级和处理；(2)标称的厚度/直径范围(尺寸)；(3)焊接工艺；(4)接头/坡口设计；(5)焊接位置和方向；(6)焊接材料、商标、焊条/焊丝直径、保护气体、焊剂和认可等级；(7)焊接顺序(焊道/焊层的次数和顺序)；(8)焊接参数：电压、电流、极性和焊接速度；(9)预热和层间温度；(10)焊后热处理。 Welding procedures specifications (WPS) mean specifications of materials, detailed methods, practices and parameters employed in the welding of a particular joint. Welding procedures specifications subjected to approval are to contain as a minimum the following information as relevant for the welding operation: (1) material, standard, grade and modification; (2) nominal thickness/diameter range(dimensions); (3) welding process; (4) joint/groove design; (5) welding position and direction; (6) welding consumables, trade name, electrode/wire diameter, shielding gas, flux and recognized classification; (7) welding sequence(number and order of passes / layers); (8) welding parameters: voltage, current, polarity and welding speed; (9) preheat and interpass temperature; (10) post weld heat treatment.

Welding processes 焊接工艺 系指：(1)手工焊接工艺：垂直下行焊、深熔焊、带垫板的单面焊等；(2)半自动焊工艺：重力电弧焊或自动电阻焊、带垫板的单面焊、无气体保护的药芯焊丝金属极电弧焊等；(3)全自动焊：埋弧焊、熔弧和药芯焊丝金属极电弧焊、多极埋弧焊、单面焊、填角接焊和双面填角接焊、带或不带易熔焊丝导向喷嘴电极的电渣焊、气体保护金属极电弧焊、电气焊等；(4)特殊焊接工艺或特殊用途：螺柱焊、闪光对接焊、摩擦焊、激光焊、堆焊、环状盘管焊缝的环绕焊、机器人焊等。 Welding processes means: (1) manual welding processes: vertical-down welding, deep penetration welding, single-side welding with backing etc.; (2) semi-mechanized welding processes: gravity arc or automatic resistance welding, single-side welding with backing, flux-cored wire metal-arc welding without shielding gas, etc.; (3) fully-mechanized welding processes: submerged arc welding, fusarc and flux-cored wire arc welding, multiple-electrode submerged arc welding, single-side welding, filler and double-fillet welding, electro-slap welding with and without fusible wire guide nozzle electrode(s), gas-shielded metal-arc welding, electro-gas welding, etc. (4) special welding processes or special applications: stud welding, flash butt welding, friction welding, laser-beam welding, build-up welding, orbital welding of circumferential pipe welds, robot welding, etc.

Welding shop 焊接车间 系指造船厂或焊接施工单位，其厂房和生产管理设施可视为一独立单位。分部和分承包商一般视为其必须符合有关要求的"独立"单位。特别是每个焊接车间必须配备固定的内部焊接监督人员。在焊接车间施工的外单位也认可为独立单位。 Welding shop means the welding production plant which, due to its space and organizational facilities, can be regarded as an independent unit. Branches and subcontractors shall generally be regarded as "independent" facilities which have to meet the relative requirements. In particular, each welding shop must have its own permanent in-house welding supervisory staff available. Outside companies working in welding shops may be approved as independent companies.

Well 井 是曝露于露天甲板上水能续集起来的区域。井视为由甲板结构的两个或多个边界围成的甲板区域。 Well is any area on the deck exposed to the weather, where water may be entrapped. Wells are considered to be deck areas bounded on two or more sides by deck structures.

Well 开槽 系指为布置挖泥设备，在挖泥船船体上的凹进部分或阱。 Well is a recess or well in the hull of dredgers for arranging the dredging gear.

Well area 井口区 系指钻井架、钻井设备或采油树和分配管汇所在区域。 Well area means an area where derrick, drilling plant or Christmas tree and distributing piping are located.

Well cementing ship 固井船 系指设有水泥储存舱、水泥搅拌机、水泥泵 以及供安装管具、填料用的机械设备，带有足够的水泥与淡水，为钻井作业进行固井的船舶。

Well control system 井控系统 系指能容纳井眼中的流体，使钻井液泵入井眼，在可控状态下使井流安全地泄出，并使浮动平台的隔水管能迅速与海底防喷器脱开的系统。该系统主要由防喷系统、压井和井流系统、导流系统以及控制和监控系统组成。

Well intervention 修井

Well intervention vessel 修井船 系指执行湿树井的修井等井干涉作业的船舶。其包括轻型修井船（无立管修井）和重型修井船（有立管修井），轻型修井船为船形，重型修井船为半潜式结构。

Well-deck barge/ Semi-hold barge 半舱驳 介于甲板驳与敞舱驳之间的一种中间形式的驳船。载货甲板呈井型,低于上甲板。在井型甲板边缘设有高于上甲板的挡货围板。半舱驳既保留了甲板驳装卸货物方便的特点,又使货物重心降低而改善了稳性,也便于驾驶人员瞭望。但与甲板驳一样不宜运载要求防水、防湿的货物。

Wellhead platform 井口平台 系指甲板上设有采油井口,可在海上进行油气采集的平台。

Western Basin 西海盆 系指霹雳岛正南方向线为界的西伊利湖的一部分。Western Basin means that portion of Lake Erie west by a line due south from Point Pelee.

Wet air pump 湿空气泵

Wet buckle 湿屈曲 这是导致管路局部破坏的一种屈曲形式,其结果依次是使管壁产生裂缝随后水灌入管路。 Wet buckle is a buckle that leads to localized collapse of a pipeline, which results in fracturing of the pipe wall, allowing water to flood the pipeline.

Wet bulb thermometer 湿球温度计

Wet compression chamber 湿加压舱 舱内能注水的加压舱。

Wet deck(SWATH) 湿甲板(小水线面双体船) 系指连接桥结构的最下暴露表面结构。Wet deck means the lower most plated surface of the cross structure.

Wet insulation 湿式保温

Wet laboratory 湿式实验室 适应湿度较高的海洋调查仪器和设备安装与使用,并有完善的供排水系统的工作室。该工作室可供各学科共用或专用。如对化学样品、生物样品和地质样品等分析用。

Wet lubrication 湿式润滑

Wet pipe system 湿管系统 是一个喷水灭火系统,上面布有一定数量的自动喷水器,而系统内充满水,且与供水系统相连接,当喷水器被火灾的热量激发打开后压力水立刻进入该系统。 Wet pipe system is: a sprinkler system employing automatic sprinklers attached to a piping system.

Wet preservation 湿保养 用碱度较高的软水或脱氧水灌满锅炉(包括过热器与经济器),借碱液与金属作用形成的氧化物保护膜来防止锅炉在较长停用期中发生腐蚀的措施。

Wet provision store 湿食品库

Wet pump room 湿泵舱 浮船坞下沉时,舱内允许进水的泵舱。

Wet steam 湿蒸汽

Wet submarine launch for nuclear bomb 潜射核弹湿发射 即先用专用的储水容器将水注满有导弹的发射筒,然后点燃导弹的火箭发动机,从而在燃料箱底与尾段壳体所衔接的空间里形成"钟形气腔",以缓冲发射时产生的气体动力,使导弹从没有专门排烟道的发射筒打出去。印度的 R-15 潜射导弹就采用这种发射方式。这种发射方式早已为俄罗斯和美国海军所抛弃。

Wet suit(immersion suit) 湿式服(救生服) 设计成在穿戴者落水时允许水进入和存在的衣服。 Wet suit is the garment designed to permit the entry and exit of water upon immersion.

Wet sump lubrication 湿底润滑 系指柴油机利用曲柄箱兼作油池,箱内保持一定油位的润滑形式。

Wet tree 湿式采油树 系指直接安装在海底的水下井口上的采油树。

Wet vacuum pump 湿空气泵

Wet well 活鱼舱 渔业船舶上,带有充气、换水装置,用以装载活鱼的舱室。

Wet-and-dry bulb thermometers 干湿泡温度计

Wetted surface 湿面积 通常系指物体浸没在水中部分的表面积。

Whale boat 捕鲸船 是进行捕鲸作业的专用渔船,船艏设捕鲸炮台,炮台上装有捕鲸炮,有效射程为 50~60 m。由于鲸资源严重衰退,近年来世界上已很少建造捕鲸船。

Whale factory ship 捕鲸母船 专门在渔场为捕鲸船就地加工鲸鱼和担负后勤支援任务的船舶。

Whale line buffer 鲸绳缓冲器 安装在捕鲸船上,连接于捕鲸绳末端的缓冲装置。

Whaling gun 捕鲸炮 设置在捕鲸船上,供猎捕鲸、鲨等海中巨型动物的专用炮。

Whaling harpoon 捕鲸铦 头部设有倒钩和二氧化碳爆破弹,尾部系以长绳,击中目标后可以将猎获物回收的与捕鲸炮配合使用的专用弹头。

Whaling ship 捕鲸船 使用捕鲸炮、弹头,专门猎鲸鱼的渔船。

Whaling winch 捕鲸绞机 能绞放捕鲸绳和绞拉鲸鱼靠近船侧的专用机械。

Wharfage 码头费 系指使用码头的费用。

What-if analysis technique 设问-处理分析技术 是危险辨别会议中使用的危险辨别技术。参加会议的主要有:协调人、记录员和经过认真挑选的有经验的研究此主题的人员。通常每组 7 人到 10 人。也是一种危险及处理措施的识别方法。它由一组专家通过不断提出"如果出现什么情况,该怎样处理的问题"的方式找出与某一(船舶)功能/系统有关的危险、其后果、安全性及可能的降低风险的措施。

Wheel 绳长计数器 又称滑轮计数器。用以测量放出或收回钢丝绳长度的器具。主要由滑轮,齿轮组、指针等组成。

Wheel casing　轮箱(齿轮箱)

Wheel efficiency　轮周效率　系指汽轮机级的有用功与级的理想可用能量之比。

Wheel gear noise　齿轮啮合噪声　系指燃气轮机的齿轮噪声。

Wheel handle　操舵手柄　装在操舵轮或操舵台上供操舵用的手柄。

Wheel spanner　开阀用扳手

Wheel tooth　轮齿

Wheel type suction dredger　斗轮式挖泥船　系指配备斗轮装置的挖泥船。

Wheelhouse　驾驶室　系指：(1)由负责船舶安全航行值班船员使用的操纵部位；(2)桥楼的围蔽部分。 Wheelhouse means: (1) the control position occupied by the officer of the watch who is responsible for the safe navigation of the vessel; (2) an enclosed area of the bridge.

Wheeling　回转(旋转)

Wheel-over-line　航路线　系指与新航线平行的线。船舶进行之字形航行，以消除任何偏离新航线的影响，同时考虑船舶保持必需的回转速率所要求的距离。Wheel-over-line means a line parallel to the new course line where the ship has to initiate a curved track to eliminate the effect of any offset with respect to the new course, taking into consideration the distance required for the ship to build up the necessary turn rate.

Wheel-over-point　航路点　系指船舶进行之字形航行的点。同时考虑船舶保持必需的回转速率所要求的距离。Wheel-over-point means a point where the ship has to initiate a curved track, taking into consideration the distance required for the ship to build up the necessary turn rate.

Wheels　轮系　装在彼此相邻的轴上并传递回转运动的一群互相啮合的齿轮。

When, where the vessel is ready　WWR 条件　系指船舶在何时、何地备妥就在何时、何地还船。

Whipping　颤振/击振　其本质上是船舶在波浪中的一种振动行为，与船舶在波浪中的相对运动有关。当船体梁受到瞬间剧烈的抨击作用时，船体梁会发生的瞬间高频振动。颤振对船体梁结构强度产生不利影响，其主要影响船体梁的极限强度。

Whirl　旋转(旋涡)

Whirling vibration　回旋振动(急旋振动)

Whistle　号笛(号角)　系指能够发出规定笛声并符合国际海上避碰规则附录Ⅲ所载规格的任何声响信号器具。 Whistle means any sound signalling appliance capable of producing the prescribed blasts and which complies with the specifications in Annex Ⅲ to the International regulations for preventing collision at sea.

Whistle(personal flotation device)　哨笛(个人漂浮装置)　可协助判断使用者位置，用嘴吹时发出可听得见声音的一种装置。 Whistle is the device which, when blown by mouth, produces an audible sound which can aid in the location of the user.

White all = round light　白环照灯　显示白色光的环照灯。

White diamond shape　白菱号型　正倒 2 个白色的圆锥体合用一个底部所构成的号型。

White noise statistics estimation　白噪声统计估计

White quartz　白石英　该货物为不燃物或失火风险低。硅石含量为 99.6%。无特殊危害。 This cargo is non-combustible or has a low fire-risk. It has 99.6% silica content. It has no special hazards.

White stern light　白艉灯　显示白色光的艉灯。

White visual signal　白色的视觉信号　系指无任何含义的，可在认为红、黄、绿三色不适用时应用的指示。例如：对地绝缘指示，同步指示灯，电话呼叫，自动控制的设备。 White visual signal means that an indication no specific meaning has been assigned, it can be used if, yellow and green is not applicable. For example, indications for earthing indication, synchroscope, telephone calling, equipment by automatic control.

Whole tine grab　全齿抓斗　左右两腭由厚钢齿做成。适宜挖掘硬质土和卵石的抓斗。

Wide blade　宽叶　高速舰艇为避免或减轻空泡影响所用宽度很大、厚度较小的螺旋桨叶。

Wideband　宽频　系指宽带网，也被称为"宽频网络"。是与传统互联网传播方式(窄带)所对立的一个概念。一般系指以能够实现视频点播的传输速率(512kb/s)为分界，将 512kb/s 及其以下的接入称为"窄带"，之上的接入方式则归类于"宽带"。宽频大多指新一代的高速传输服务，让用户接驳互联网或与互联网有关的服务时，速度可远远高于传统拨号上网服务所能支援的速度。现时国际上并无一致的"宽频"定义，一般而言，"宽频"一般系指速度由每秒数 100 倍千比特(kilobits per second 或 kb/s)至每秒数兆比特(Megabits per second 或 Mb/s)的互联网接驳服务(1 兆比特等于 1 000 倍千比特)。有线宽频机、ATM(异步传输模式)、Ethernet(以太网)、ADSL(非对称数码用户线路)及其他不同类型的 DSL(数码用户线路)则是一般用来提供宽频服务的技术。

Wideband system　"宽频"系统

Wider Caribbean region　大加勒比海区域　系指

墨西哥湾和加勒比海本身,包括其中的海湾或海区以及由以下边界组成的大西洋的一部分,在北纬30°自佛罗里达向东至西经77°30′然后连一条恒向线至北纬20°与西经59°的交叉点,然后再连一条恒向线至北纬7°20′与西经50°的交叉点,然后再连一条恒向线沿西南方向至法属圭亚那的东部边界。 Wider Caribbean region means the Gulf of Mexico and Caribbean Sea proper including the bays and seas therein and that portion of the Attantic Ocean within the boundary constituted by the 30°N parallel from Florida eastward to 77°30′W meridian, thence a ehunb line to the intersection of 20°N parallel and 59°W meridian, thence a ehunb line to the intersection of 7°20′N and 50°W meridian, thence a rhumb line drawn southwesterly to the eastern boundary of French Guiana.

Wide-tipped blade 宽梢叶 一般系指与同面积的椭圆叶相比,叶梢部分较宽的螺旋桨桨叶。

Width of the plating attached b_p(mm) 带板的宽度 b_p 系指扶强材或主要支撑构件带板的宽度(mm),用于屈服强度校核。Width of the plating attached b_p means a width, in m, of the plating attached to the stiffener or the primary supporting member, for the yielding check.

Wifi 无线网卡 是一种可以将个人电脑、手持设备(如pad、手机)等终端以无线方式互相连接的技术,事实上它是一个高频无线电信号。无线保真是一个无线网络通信技术的品牌,由Wi-Fi联盟所有。目的是改善基于IEEE 802.11标准的无线网络产品之间的互通性。有人把使用IEEE 802.11系列协议的局域网就称为无线保真。甚至把无线保真等同于无线网际网路(Wi-Fi是WLAN的重要组成部分)同理有Hi-Fi。2014年8月,网络安全研究人员鲁本·圣马尔塔表示,如果飞机WiFi开放度过高,可能会被黑客用于劫机。

WIG craft 地效翼船 系指:(1)具有多种航态的船舶。这种船在营运状态中,通过地效作用在水面或其他各种表面以上飞行,与该表面无连续的接触且在空中主要通过拟使用地效作用的一个翼(几个翼)、船体或各部件产生的气动升力予以支承;(2)是多模式海船,能以某一模式运行,即完全支撑于水或其他表面上方的空中,因此与这些表面无固定的接触,仅通过使用气动升力,即地面效应。地面效应是由于船机翼、船体或其他部件的下表面,与水面或其他表面之间的相互干涉,在船向前运动过程中产生的。在这些表面上运行的最大高度等于地效翼船总宽的100%。见图W-3。 WIG craft is: (1) a multimodal craft which, in its main operational mode, flies by using ground effect above the water or some other surface, without constant contact with such a surface, and supported in the air, mainly, by an aerodynamic lift generated on a wing(wings), hull, or their parts, which are intended to utilize the ground effect action;(2) a multimodal marine craft capable of operation in a mode where the craft is supported wholly in the air above water or some other surface, and not having constant contact with such surface, through the use of an aerodynamic lifting force known as ground effect. Ground Effect is generated during forward movement of the craft by interaction between the lower surfaces of the crafts wing(wings), hull, or their respective parts, and the water or other surface being traversed up to a maximum height above such surface equal to 100% of the overall width of the WIG craft. It can be seen in Figure W-3.

图 W-3 地效翼船
Figure W-3 WIG craft

WIG or Ekranoplan 地效翼船或地效飞机 通用名称,系指无特种增升措施的地效翼船。 WIG or Ekranoplan is a generic name, means a craft without special lift enhancement features.

Wildcard 通配卡

Wildcat[1] 锚链轮 与抛锚相啮合在一起抛锚链时传递动力的星形轮体。

Wildcat[2] 探井 系指经过地球物理勘探,证实有希望的地质构造,为了探明地下情况,寻找油、气田而钻探的井。

Winamp 数字媒体播放的先驱 由Nullsoft公司在1997年开发的软件。创始人Justin Frankel,该软件支持MP3,MP2,MOD,S3M,MTM,ULT,XM,IT,669,CD-Audio,Line-In等格式,至今已经从1.0版本升级到5.57版本。Winamp以其声音效果、播放列表和媒体库的功能而出名。2013年11月21日,Nullsoft公司宣布,于2013年12月20日正式关闭这款拥有15年历史的产品。2014年1月2日,"美国在线"公司旗下TechCrunch网站称,在线广播电台公司Radionomy从AOL公司手中收购Winamp和Shoutcast音乐服务,双方有望1月3日完成收购谈判,这款经典播放软件才免遭一劫。

Winch 绞车 有一个或数个不平置卷筒的绞引缆索的机械。

Winch platform 起货机平台 装设起货机的

Winch quick-release device 拖缆机快速释放装置
系指通过一个适当的装置或采用缆索终端不固定于的卷筒上的方法使缆索能够从拖缆机卷筒解脱的装置。

Winch quick-release device is a device that the unhooking of the rope from the winch drum is to be enabled by means of a suitable device or by using a rope whose terminal is not fixed to the drum.

Wind 风 对设计环境而言,系指用以确定风力的设计风速。通常可取海平面之上 10 m 处的一小时平均值。应考虑风速随海平面以上高度的改变而变化。风速梯度可计算如下: $V_H = V_{10}^{0.125}$,式中: V_H——海平面之上高度为 H 处的风速; V_{10}——海平面之上高度为 10m 处 1 小时平均风速。 For the design environment, design wind speed for wind force determination can normally be taken as the one-hour mean value referenced to 10 m above sea level. Account is to be taken of the variation of wind speed with height above sea level. The wind velocity gradient can be calculated from the following: $V_H = V_{10}^{0.125}$, where: V_H = wind velocity at H height above sea level; V_{10} = 1 hour mean wind speed referenced to 10 m above sea level.

Wind and water strake 干湿交变列板 系指轻、重载水线之间的舷侧外板。由于船舶的纵倾,这些列板在船长范围内可能变化。 Wind and water strake are the strakes of a ship's side shell plating between the ballast and the deepest load waterline. Due to vessel's trim, the strakes may vary over the length of the vessel.

Wind catcher 招风斗 可从开启的舷窗孔中伸出舷外,供引风用的工具。

Wind duration 风时 稳定状态的风在水面上吹过的持续时间。

Wind energy converter(WEC)/Wind turbine generator(WTG) 风能转换系统 即风力发电机。一种能将动能转换成电能的发电装置。一个风能转换系统一般由一个塔架、电机外壳和叶轮组成。风机安装在海上需要专门的基础结构上。

Wind force 风力 系指风的力量,即在任意风级上的某一定数,如 5 级或 7 级;也就是从风得到的机械力,气概与魄力;据测定,一台 55 kW 的风力发电机组,当风速每秒为 9.5 m 时,机组的输出功率为 55 kW;当风速每秒 8 m 时,功率为 38 kW;风速每秒为 6 m 时,只有 16 kW;而风速为每秒 5 m 时,仅为 9.5 kW。

Wind loadings 风载荷 系指作用于构件上的风力 F,其应按下式计算,并应确定合力作用点的垂直高度: $F = C_h C_s SP(kN)$,式中,P——风压,kPa;S——平台在正浮或倾斜状态时,受风构件的正投影面积,m²;C_h——受风构件的高度系数,其值可根据构件高度(构件型心到设计水面的垂直距离),由表 W-4 选取;C_s——受风构件形状系数,其值可根据构件形状由表 W-5 选取,也可根据风洞试验确定。

表 W-4 高度系数 C_h
Table W-4 Height coefficient C_h

海平面以上高度 h/m Height above sea level	高度系数 C_h Height coefficient
0 ~ 15.3	1.00
15.3 ~ 30.5	1.10
30.5 ~ 46.0	1.20
46.0 ~ 61.0	1.30
61.0 ~ 76.0	1.37
76.0 ~ 91.5	1.43
91.5 ~ 106.5	1.48
106.5 ~ 122.0	1.52
122.0 ~ 137.0	1.56
137.0 ~ 152.5	1.60
152.5 ~ 167.5	1.63
167.5 ~ 183.0	1.67
183.0 ~ 198.0	1.70
198.0 ~ 213.5	1.72
213.5 ~ 228.5	1.75
228.5 ~ 244.0	1.77
244.0 ~ 256.0	1.79
256 以上	1.80

表 W-5 形状系数 C_s
Table W-5 Shape Coefficient C_s

构 件 形 状 Component Shape	形状系数 Shape coefficient C_s
球 形	0.4
圆柱形	0.5
大的平面(船体、甲板室、平滑的甲板下表面)	1.0

续表 W-5

构 件 形 状 Component Shape	形状系数 Shape coefficient C_s
甲板室群或类似结构	1.1
钢索	1.2
井架	1.25
甲板下暴露的梁和桁材	1.3
独立的结构(起重机、梁等)	1.5

Wind power generation　风电　风电系指风能发电或者风力发电。属于可再生能源，清洁能源。风力发电是风能利用的重要形式，风能是可再生、无污染、能量大、前景广的能源。风电技术装备是风电产业的重要组成部分，也是风电产业发展的基础和保障。由于低风速风区分布广泛，需要风机通过不同的配置，满足特殊环境的使用要求。如满足长江流域的高雷暴和高湿度天气、中南地区的冰冻现象、西南地区的高原环境要求等。风能作为一种清洁的可再生能源，越来越受到世界各国的重视。其蕴量巨大，全球的风能约为 2.74×10^9 MW，其中可利用的风能为 2×10^7 MW，比地球上可开发利用的水能总量还要大10倍。见图W-4。

图 W-4　风电
Figure W-4　Wind power generation

Wind propulsion　风力推进　利用风力使船前进的船舶推进方式。

Wind tunnel test　风洞试验　模型安置在风洞中，用测力天平测定流体对船模各种姿态时的作用力和力矩等数据的试验。

Wind turbine installation vessel(WTIV)　自升式风电安装船　采用电力推进及动力定位技术，具有风力涡轮机运输安装功能的船舶。其甲板上安装有起重能力达800 t的吊机，用于安装风力涡轮机，采用柴电推进系统，满足根据不同的工作模式，灵活对电力进行优化

配置的要求，同时能为该船带来极佳的操纵性和可靠性，并减少燃料消耗量和废气排放量。

Wind velocity　风速　在水面规定的高度上风的前进速度。

Windage area A_{LV}　受风面积 A_{LV}　当艇以相应的装载状态正浮时，水线以上的艇体、上层建筑、甲板室和帆桁的侧向投影面积。注意(1)：在恶劣气候中航行时可能安装的小艇天篷和挡风板，例如艇舱防浪板、小艇舱口罩棚应予包括。注意(2)：受风面积以 m^2 表示。Windage area A_{LV} means the projected profile area of hull, superstructures, deckhouses and spars above the waterline at the appropriate loading condition, the boat being upright. Note(1): Canopies and screens are likely to be erected when sailing in bad weather, such as, e. g. cockpit dodgers, pram hoods. Note(2): Windage area is expressed in square meters.

Windfinding radar　测风雷达　用于探测高空不同大气层的水平风向、风速、气压、温度和湿度要素。为天气预报、飞行安全、武器试验提供气象资料。

Windjammer café　帆船咖啡厅(豪华邮轮)

Windlass(anchor windlass)　锚机(起锚机)　系指收放锚链用的机器。 Windlass means a machine for lifting and lowering the anchor chain.

Windlass overload pull　锚机过载负荷　系指锚机上瞬间所产生的提升最大负荷，应不小于工作负荷的1.5倍。 Windlass overload pull is the necessary temporary overload capacity of the windlass, and to be not less than 1.5 times the working load.

Windlass working load　锚机工作负载　系指由锚链的公称直径和等级决定的上卷锚和锚链时对锚链轮造成的切向拉力。 Windlass working load, derived from the nominal diameter and the grade of anchor chain cables, is the tensile force exerted upon the cable lifter in the tangential direction when the anchor and anchor chain cable are being hoisted.

Window　窗　(1)通常为矩形开口，根据公认的国家或国际标准在每个角隅具有相对于窗尺寸的圆弧、面积超过 $0.16 \ m^2$ 的圆形或椭圆形开口；(2)适合安装在船上的，不论其形状的各种船用窗。 Window is: (1) rectangular opening generally, having a radius at each corner relative to the window size in accordance with recognised national or international standards, and round or oval openings with an area exceeding $0.16 \ m^2$; (2) a ship's window, being any window, regardless of shape, suitable for installation aboard ships.

Window operated by turning handle(handle turn-

ing window）　手摇窗　可摇动手柄，带动机构，使之升降启闭的活动矩形窗。

Window type air conditioner　窗式空气调节器　装于舱壁上供个别舱室使用并带有制冷装置的空气调节器。

Window with counter weight　重力平衡窗　利用其自重，以便于启闭和定位的活动矩形窗。

Window with crank handle　摇窗　设有螺杆或扇形齿轮等传动装置，用人力或电力摇旋而使其开闭的窗。

Window with spring balance　弹簧平衡窗　利用弹簧弹力平衡窗的自重，以便于启闭和定位的活动矩形窗。

Windows phone　微软手机操作系统　它将微软旗下的 Xbox Live 游戏、Zune 音乐与独特的视频体验整合至手机中。2010 年 10 月 11 日晚上 9 点 30 分，微软公司正式发布了智能手机操作系统 Windows phone，微软将其使用接口套用了一种称为 Metro 的设计语言，并将微软以及其他的第三方的软件整合到了操作系统中。2011 年 2 月，诺基亚与微软达成全球战略同盟并深度合作共同研发。2012 年 3 月 21 日，Windows Phone 7.5 登陆中国。6 月 21 日，微软正式发布最新手机操作系统 Windows phone 8，Windows phone 8 采用与 Windows 7 相同的内核。2014 年，微软发布 Windows phone 8.1 系统，发布时提到 Windows phone 8.1 可以向下兼容，让使用 Windows phone 8 手机的用户也可以升级到 Windows phone 8.1。

Wine wave　风浪　系指当地风所引起且直到观测时仍处于风力作用下的海浪。按浪高从 0～9 划分为 10 级。

Wing　翼　系指气翼或其他在飞行时产生气动升力支承船舶重力的表面，可包括机身。　Wing denotes an air foil or other air lift generating surface to support the weight of the craft in flight and may include the fuselage.

Wing blade　边舵叶　多叶舵中，位于两侧的舵叶。

Wing in ground craft　地效翼船　系指重力由机翼利用其与贴近水表面或其他表面之间的地面表面效应所产生气动升力支持的船舶。　Wing in ground craft is a craft supported by using ground effect above the water or some other surface, without constant contact with such a surface and supported in the air, mainly, by an aerodynamic lift generated on a wing(wings) which are intended to utilize the ground effect action.

Wing in ground craft A　A 类地效翼船　系指只能在地面效应区内飞行的地效翼船。　Wing in ground craft A means a wing in ground craft certified for operation only in ground effect.

Wing in ground craft B　B 类地效翼船　系指能在地面效应区以外瞬时增加飞行高度并飞行一段有效距离的地效翼船。　Wing in ground craft B means a wing in ground craft certified to temporarily increase its altitude to a limited height outside the influence of ground effect but not exceeding a certain distance.

Wing in-ground effect craft　掠海地效翼船　是介于飞机、舰船和气垫船之间的一种新型高速飞行器。与普通飞机不同的是，地效飞行器主要在地面效应区范围内飞行，也就是贴近地面、水面飞行，而飞机主要在地面效应区外飞行；与气垫船不同的是，气垫船靠自身动力产生气垫，而地效飞行器靠地面效应产生气垫。在 1980 年，苏联实验了喷射推进的 Ekranoplan，就是这样的飞机，极限飞行质量可以达 1 000 吨，还在冷战时期的隐秘情况下测试，被美国称为"里海怪物"。大部分地效翼船都被设计为在水面上运作，因为水面比地面平滑和少障碍物，不单危险度较少，而且在其不运作时，还可以利用水面浮力来承受船体重力。

Wing nut　翼形螺母

Wing shaft　侧轴系　不在船中轴线上的轴。

Wing tank　边舱　系指：(1)以内壳纵舱壁和舷侧外板为界面的处所；(2)与船壳板相连的任何舱柜。　Wing tank is: (1) the space bounded by the inner hull longitudinal bulkhead and side shell; (2) any tank adjacent to the side shell plating.

Wing wall / Dock wall　坞墙　系指位于浮船坞两舷侧浮箱甲板之上的纵向连续箱形结构。其外板结构和内壁板结构也分别称为外坞墙和内坞墙。用以承受纵向强度的墙式主要构件。

Wing-in ground(WIG) craft　地效翼(WIG)船　系指一种多航态的船舶，其主要运行方式为利用地效作用贴近地面飞行。　Wing-in-ground(WIG) craft means a multimodal craft which, in its main operational mode, flies in close proximity to the surface by utilizing surface-effect action.

Wing-in-ground craft safety certificate　地效翼(WIG)船安全证书　系指对经初次检验或换证完成后，符合地效翼(WIG)船临时指南要求的地效翼(WIG)船签发的安全证书。　Wing-in-ground craft safety certificate means a WIG craft safety certificate issued after completion of an initial or renewal survey to a craft, which complies with the provisions of the Interim Guidelines for a WIG craft.

Wingwall　坞墙　位于浮船坞舷侧，用以承受纵向强度的墙式主要构件。

Wingwall passage　坞墙通道　从抬船甲板通至舷

外的坞墙出入口。

Winning party 胜诉方

Wiper ring 刮油环

Wiper tank 污油柜

Wire angle indicator 倾角器 在进行海洋观测时用来测量钢丝绳的倾角的器具。

Wire clip 紧索夹 简易的结扎钢索端头或连接卸扣用的索具配件。

Wire reel(hawser reel) 缆绳卷车 供收藏缆绳用的卷绳车。

Wire rope 钢索 系指索具中的钢丝绳。

Wire rope block 钢索滑车 穿绕钢丝绳的滑车。

Wire rope shackles / SHACKLES 船用索具卸克 又称卸克。连接各种绳索、链条、滑车等用的可拆卸的金属环形件。由本体和横销组成。

Wire rope steering gear 舵索传动操舵装置 以操舵索为传动部件的人力操舵装置。

Wire rope tensioner 钢丝绳张紧器

Wire sling 柃索 一般在其端部附有抓钩环或自成环形，供钩系被吊物用的短索或短链。

Wired headset 有线耳麦

Wired information channels 有线信道 系指传输媒介为明线有线信道示意图、对称电缆、同轴电缆、光缆及波导等一类能够看得见的媒介。有线信道是现代通信网中最常用的信道之一。如对称电缆（又称电话电缆）广泛应用于（市内）近程传输。

Wire-drag survey 扫海测量 在一定海区内进行面的探测，以查明该区是否存在航行障碍物的工作。

Wireless access point(WAP) 无线接入点 它是一个无线网络的接入点，用于无线网络的无线交换机，也是无线网络的核心，主要在媒体存取控制层 MAC 中扮演无线工作站及有线局域网络的桥梁。无线路由器主要有路由交换接入一体设备和纯接入点设备。一体设备执行接入和路由工作，纯接入设备只负责无线客户端的接入；纯接入设备通常作为无线网络扩展使用，与其他 AP 或者主 AP 连接，以扩大无线覆盖范围。无线接入点主要用于宽带家庭、大楼内部以及园区内部，典型距离覆盖几十米至上百米，目前主要技术为 802.11 系列。

Wireless application protocol(WAP) 无线应用协议 是一个使移动用户使用无线设备（例如移动电话）随时使用互联网的信息和服务的开放的规范。 WAP 的主要意图是使得袖珍无线终端设备能够获得类似网页浏览器的功能，因此其功能上有限。WAP1.X 规定无线设备访问的页面是用 WML（一种 XML 方言）语言编写的，但是 WAP2.0 将 XHTML-MP 作为主要内容格式。

Wireless charging 无线充电 又称作感应充电、非接触式感应充电，是利用近场感应，也就是电感耦合，由供电设备（充电器）将能量传送至用电的装置。该装置使用接收到的能量对电池充电，并同时供其本身运作之用。由于充电器与用电装置之间以电感耦合传送能量，两者之间不用电线连接，因此充电器及用电的装置都可以做到无导电接点外露。主流的无线充电标准有4种：Qi 标准、Power matters alliance(PMA)标准、Alliance for wireless power(A4WP)标准和 iNPOFi 技术。

Wireless charging by magnetic resonance 磁共振催生"隔空充电" 早在100多年前，科学家就发现电力可以转化为电磁波在空气中传播。当时有科学家制作了两个铜线圈，让其中一个连接电源，负责发出能量；另一个连接灯泡，负责接收能量；两个线圈中间没有任何电线连接。由于这两个线圈拥有相同的电磁波振动频率，电能最终成功地在两个线圈之间进行传输。通常，科技公司提到的无线充电技术系指可以将手机放在充电垫上充电，不需要充电线。例如：苹果、华为等品牌的手机，通过两线圈相互接触感应，可以进行电能转换。真正的无线充电技术，类似于"隔空充电"，像 WiFi 信号一样，通过空气等媒体来传输电能，其能量传输效率可以保持在70%左右，且都采用智能化控制，能自动识别负载是否完成充电，比不能识别是否充电完成的传统充电器更加环保，电能利用率更高。但是，传输距离和辐射问题，也限制着无线充电技术的推广应用。无线电波的弥散、吸收与衰减，被认为是无线充电的难点。电磁波在自由空间传输能量的过程中，会向四方散发，不易集中，定向性差，能量在无线传输过程中，各地各样的干扰会造成能量传输损耗。特别是微波漫射在空间，能量衰减更快。此外，使用无线充电的产品工作频率设定的范围是 50~60 Hz，理论上电磁辐射与普通小家电相当。但检测发现，无线充电的电磁辐射相较于无线通信要高得多。还有专家表示，产生辐射的不仅仅是无线充电器，电脑、手机、WiFi/电视、微波炉等同样有辐射，因而存在一个叠加辐射的问题。因此，应当在提高无线充电传输能力的同时，降低电磁辐射水平，将其下降到一个安全许可的范围内。这也许是无线充电技术发展应用的一大瓶颈。对汽车无线充电系统的检测还显示，磁感应传输电能会产生轻微辐射。在人体各器官中，心肺受到的影响较为明显。需作为重点防护对象。不过，由于车身的屏蔽作用以及电磁参数随距离的急剧衰减，车外的电磁安全指标明显优于车内。因此，就个体而言，最好不要置于无线充电的中心磁场。

Wireless charging technology 无线充电技术 就是利用磁共振、电磁感应等原理，实现电能在空气中传播的技术。

Wireless data communication 无线数据通信

Wireless data communication system 无线数据通信系统

Wireless fidelity (Wi-Fi) 无线上网技术 是一种可以将个人电脑、手持设备(如 pad、手机)等终端以无线方式互相连接的技术,是基于 IEEE 802.11 标准的无线局域网。事实上它是一个高频无线电信号。无线保真是一个无线网络通信技术的品牌,由 Wi-Fi 联盟所持有。目的是改善基于 IEEE 802.11 标准的无线网络产品之间的互通性。有人把使用 IEEE 802.11 系列协议的局域网就称为无线保真。甚至把无线保真等同于无线网际网路(Wi-Fi 是 WLAN 的重要组成部分)。关于"Wi-Fi"这个缩写词的发音,根据英文标准韦伯斯特词典的读音注释,标准发音为/wai.fai/因为 Wi-Fi 这个单词是两个单词组成的,所以书写形式最好为 Wi-Fi,这样也就不存在所谓专家所说的读音问题,同理有 Hi-Fi (/hai.fai/)。2014 年 8 月,网络安全研究人员鲁本·圣马尔塔表示,如果飞机 WiFi 开放度过高,可能会被黑客用于劫机。

Wireless information channels 无线信道 对于无线电波而言,它从发送端传送到接收端,其间并没有一个有形的连接,它的传播路径也有可能不只一条,但是,为了形象地描述发送端与接收端之间的工作,一般想像两者之间有一个看不见的道路衔接,把这条衔接通路称为信道。信道具有一定的频率带宽,正如公路有一定的宽度一样。

Wireless local area net access point 无线局域网接近点

Wireless local area network (WLAN) 无线局域网 也称 WLAN,WLAN 是利用无线技术在空中传输数据、话音和视频信号。作为传统布线网络的一种替代方案或延伸,无线局域网把个人从办公桌边解放了出来,使他们可以随时随地获取信息,提高了员工的办公效率。此外,WLAN 还有其他一些优点。它能够方便地实施联网技术,因为 WLAN 可以便捷、迅速地接纳新加入的雇员,而不必对网络的用户管理配置进行过多的变动。WLAN 还可以在有线网络布线困难的地方比较容易实施,使用 WLAN 方案,则不必再实施打孔敷线作业,因而不会对建筑设施造成任何损害。目前该系列包含以下 4 种规范:802.11、802.11$_a$、802.11$_b$ 以及 802.11$_g$。所有这 4 种协议都采用以太网协议和载波监听多路访问/冲突避免技术(CSMA/CA,替代了 CSMA/CD)来实现信道共享。

Wireless microphone 无线传声器

Wireless headset 无线"耳麦" 系指采用无线接收技术,由发射端和接收装置组成的耳麦。无线"耳麦"是一种数码随身听产品的产物。无线"耳麦"分为 3 个部分:(1)发声源;(2)接收器;(3)耳机。其主要功能是将手机或接收器传送来的信号转化为声音再传到人的耳朵里。按照无线"耳麦"的信号传送方式,可以分为:(1)"蓝牙"无线耳机;(2)红外线无线耳机;(3)2.4 G 无线耳机。红外耳机为早期蓝牙、2.4 G 无线技术未成熟时,无线耳机产品主要采用的传输技术,自从 2000 年后,"蓝牙"技术和 2.4 G 无线技术的成熟,在市场上红外线无线产品逐渐被"蓝牙"与 2.4 G 无线产品所替代。功能特点:具有无线传输、无线监护、无线语音聊天等功能;带有专业的学习方案,可学习各种家电遥控器(TV、VCR、DVB、DVD 等)的功能,也可用作数字机顶盒遥控;使用无线耳机,可在电视静音的情况下,自由欣赏精彩节目;将发射机置于老人、婴儿、病人等需要照顾的人身旁,使用接收机可收听到被照顾者的声音,方便轻松地护理等。

Wireless-operator's table 报务桌 专供安置收发报机供报务员工作的桌子。

Wiring chip 接线片 系指借助机械压力或焊接而使导电线连接于其上的一种导线连接器件。

With average/with particular average 水渍险 又称"单独海损险",英文原意系指单独海损负责赔偿,海洋运输货物保险的主要险别之一。这里的"海损"是自然灾害及意外事故,导致货物被水淹没,引起货物的损失。保险人对下列各项损失和费用,不负赔偿责任:(1)被保险人的故意行为或过失所造成的损失;(2)属于发货人所引起的损失;(3)在保险责任开始前,被保险货物已存在的品质不良或数量短差所造成的损失;(4)被保险货物的自然损耗、本质缺陷、特性以及市价跌落、运输延迟所引起的损失或费用。

Withdraw a piston 吊缸(拆卸活塞)

Withdraw an offer 撤回报价

Withdraw halfway form the hearing 中途退庭

Withdrawal 回避 系指被选定或者指定的仲裁员有下列情形之一的,应自行应仲裁委员会的要求回避:(1)是本案当事人或者代理人的近亲属;(2)与本案有利害关系;(3)与本案当事人、代理人有其他关系,可能影响公正仲裁的;(4)私自会见当事人、代理人;或者接受当事人、代理人请客送礼的。

Withdrawl 要约的撤回 系指要约人在发出要约后,在其尚未到达受要约人之前,或在要约到达受要约人的同时,以适当的方式将该项要约取消,使之失去作用。

Within reach 可及的范围 系指操作人员能接使用装置的距离:(1)对于操纵台站立定位,该距离相对于前方最大为 800 mm,在两侧为 1 400 mm;(2)对与操纵台距离 350 mm 的座位,该距离最大为 1 000 mm,对频繁使用的设备,该距离最大为 800 mm,这些设备应在易于接近的范围之内。 Within reach means within a distance the operator can reach and use a control unit: (1) from a standing position at a console this distance is regarded to be maximum 800 mm in forward direction and 1 400 mm sideways; (2) from a seated position, at a distance of 350 mm from a console, this distance is regarded to be maximum 1 000 mm, and maximum 800 mm for frequently used equipment, which is to be within easy reach.

Within the reach from standing position 在站位的伸手可及范围 操作人员可以到达的距离,并且可以在控制台旁边的站立位置使用用户输入设备(UID)。这个距离应该是前向 800 mm 和侧向 1 400 mm。 The distance within which the operator can reach and can use an UID from a standing position next to a console. This distance shall be 800 mm in forward direction and 1 400 mm sideways.

Within the reach of the officer of the watch in sitting position 坐在座位上的值班人员可及的范围 在值班人员伸手可及的范围内的操作和控制装置应在坐着的值班人员周围半径 1 000 mm 以内。 Operation and control units within the reach of the officer of the watch shall be within a radius of 1 000 mm around the seated officer.

Withstand load 支持负载 系指锚链轮制动器应能承受的锚链上最大静负载。 Withstand load means the maximum static load applied to chain cables which the windlass brake can withstand.

Witness 证人 系指目击者或见证人。

WMO 世界气象组织 WMO means the World Meteorological Organization.

Wood chip carrier 碎木运输船 系指专门设计和建造用于运输碎木的船舶。 Wood chip carrier is a ship designed and constructed for the carriage of wood chip.

Wood deck 木甲板 铺设木板的甲板。

Wood hatch board 木质盖板 由整块木板制成或数块木板拼合成,两端加扁钢框紧,并设有供起吊舱盖拉环的舱盖板。

Wood hatch covers 木质舱盖 由木质舱盖板组成的拼装舱盖。

Wood pellets 木球团 呈浅金黄色至深褐色,非常坚硬,轻易不被压碎。木球团的典型比密度为 1 100 ~ 1 700 kg/m³,散货密度为 600 ~ 750 kg/m³。木球团由木材加工过程中产生的锯屑、刨花和树皮等其他废木制成。除有规定者外,木球团通常无添加剂和黏合剂混入。原材料经碎裂、干燥和挤压而成木球团。原材料的压缩倍率约为 3.5,木球团成品的含水量一般为 4 ~ 8%。木球团在地区供暖和发电中用作燃料,且可用作装有火炉和壁炉之类小空间供暖的燃料。木球团由于其吸收性特点,也可用作动物的垫料。这类木球团的含水量一般为 8 ~ 10%。所装运的货物可能易于氧化,导致货物处所和相邻处所缺氧及一氧化碳和二氧化碳增加。吸入水分会膨胀。木球团如含水量高于 15%,在一段时间后可能发酵,导致产生窒息和易燃并可能引起自燃气体。木球团的装卸可能产生粉尘。粉尘浓度高时有爆炸风险。 Wood pellets are light blond to chocolate brown in colour, very hard and cannot be easily squashed. Wood pellets have a typical specific density between 1 100 to 1 700 kg/m³ and a bulk density of 600 kg/m³ to 750 kg/m³. Wood pellets are made of sawdust, planer shavings and other wood waste such as bark coming out of the lumber manufacturing processes. Normally there are no additives or binders blended into the pellet, unless specified. The raw material is fragmented, dried and extruded into pellet form. The raw material is compressed approximately 3.5 times and the finished wood pellets typically have a moisture content of 4 to 8%. Wood pellets are used as a fuel in district heating and electrical power generation as well as a fuel for small space heaters such as stoves and fireplaces. Wood pellets are also used as animal bedding due to the absorption characteristics. Such wood pellets typically have a moisture content of 8 to 10%. Shipments may be subject to oxidation, leading to depletion of oxygen and increase of carbon monoxide and carbon dioxide in cargo and communicating spaces. Wood pellet may swell if exposed to moisture. Wood pellets may ferment over time if moisture content is over 15%, leading to generation of asphyxiating and flammable gases which may cause spontaneous combustion. Handling of wood pellets may cause dust to develop. There is risk of explosion at high dust concentration.

Wood pulp pellets 木浆球团 球团呈褐色,非常坚硬,轻易不被压碎。质量轻,大小如个半软木瓶塞。球团由木屑压缩而成。该货物有化学危害。某些装运货物可能易于氧化,导致货物处所和相邻处所缺氧及一氧化碳和二氧化碳增加。该货物含水量为 15% 或以上时,失火风险低。含水量减少,则失火风险增大。 Wood pulp pellets are brown in colour very hard and cannot be easily squashed. They are light and are about half the size of a bottle cork. The pellets are made of compacted woodchips. This cargo possesses a chemical hazard. Some shipments may be subject to oxidation, leading to depletion of oxygen and increase of carbon dioxide in cargo and adjacent spaces. With moisture content of 15% or more, this cargo has a low fire-risk. As the moisture content decreases, the fire risk increases.

Wood-cement board 水泥木丝板 是用选定种类的晾干木料刨成细长木丝,经化学浸渍稳定处理后,木丝表面浸有水泥浆再加压成水泥木丝板。在20世纪40年代开始在欧洲广泛应用,目前已成为国际上应用范围很广的建筑材料。它实用性广、性能优异,有着耐腐、耐热、耐蚁蚀、易加工、与水泥、石灰、石膏配合性好、绿色环保等多种优点。

Woodchips 木片 机械削制的天然木材,大小如名片。该物质有化学危害。某些装运货物可能易于氧化,导致货物处所和相邻处所缺氧及二氧化碳增加。该货物含水量为15%或以上时,失火风险低。含水量减少,则失火风险增大。木片干燥时,能轻易地被外部火源点燃,易燃并能通过摩擦点燃。可能在不到48 h内出现氧气耗尽的情况。 Woodchips are natural timber mechanically chipped into the approximate size of a business card. This material possesses a chemical hazard. Some shipments may be subject to oxidation, leading to depletion of oxygen and increase of carbon dioxide in cargo and adjacent spaces. With moisture content of 15% or more, this cargo has a low fire-risk. As the moisture content decreases, the fire risk increases. When the woodchips are dry, they can be easily ignited by external sources, are readily combustible and can be ignited by friction. A condition with complete depletion of oxygen may be present in less than 48 hours.

Wooden plug 堵漏木栓 船舶破损时,用以堵塞小洞的木塞子。

Wooden vessel 木船 以木材作为船体结构主要材料的船舶。

Wooden wedge 堵漏楔 用以垫塞支撑柱两端和船体结构之间的空隙或堵塞船体裂缝的木楔。

Wood-wool board 木丝板 是用选定种类的晾干木料刨成细长木丝,经化学浸渍稳定处理后,木丝表面浸有水泥浆再加压成水泥木丝板,简称为木丝板。又称万利板。木丝板是纤维吸声材料中的一种有相当孔结构的硬质板,具有吸声、隔热、防潮、防火、防长菌、防虫害和防结露等特点。木丝板具有强度和刚度较高,吸声构造简单,安装方便、价格低廉等特点。最近20~30年由于无机纤维生产发展很快,如玻璃棉、矿棉板及岩棉板等吸声制品,不仅吸声系数高、质量轻、防火性能好、品种规格多,质量日益提高,而且价格也逐渐下降,在吸声应用中几乎替代了木丝板。

WordPress 个人博客系统 是一种使用PHP语言开发的博客平台,用户可以在支持PHP和MySQL数据库的服务器上架设属于自己的网站。也可以把WordPress当作一个内容管理系统(CMS)来使用。WordPress是一款个人博客系统,并逐步演化成一款内容管理系统软件,它是使用PHP语言和MySQL数据库开发的。用户可以在支持PHP和MySQL数据库的服务器上使用自己的博客。WordPress有许多第三方开发的免费模板,安装方式简单易用。不过要做一个自己的模板,则需要有一定的专业知识。比如至少要懂的标准通用标记语言下的一个应用HTML代码、CSS、PHP等相关知识。WordPress官方支持中文版,同时有爱好者开发的第三方中文语言包,如wopus中文语言包。WordPress拥有成千上万个各式插件和不计其数的主题模板样式。

Work hours/Hours of labor 工时 系指一小时所做正常工作量的劳动计量单位。工时管理主要是员工个人工作统计管理功能,生产管理人员可以详细地、逐项活动地查看生产和即时劳动力数据,特别是活动级劳动力信息,辅助生产管理人员利用从车间获得的效率数据,实时监控生产流程,并在提高生产率,控制劳动力成本方面制定科学的企业决策。工时的具体含义,根据情景的不同,可有几种情况:对于任务,系指完成任务所需的人员总数;对于工作分配,系指分配给资源的工时量;对于资源,系指为完成所有任务而分配给资源的总工时量。

Work instruction 施工说明书 对质量管理体系而言,系指制造厂商应编制和执行明确和完整详细的施工说明书。明确说明这些规定要求的贯彻和在设计研制和制造中执行。如果没有这种说明书,在上述工作中可能会造成不良影响。 For quality management system, work instruction means that the manufacturer is to establish and maintain clear and complete written work instructions that prescribe the communication of specified requirements and the performance of work in design, development and manufacture which would be adversely affected by lack of such instructions.

Work raft (paint raft) 工作筏 配置在船上,供船员划登浅滩,救捞落水人、物以及对水面附近船体进行检修和油漆等用的非救生专用的小型浮筏。

Work ratio 功比 燃气轮机机组输出功率与总膨胀功之比。

Work ship 工作船 系指水上工作的船舶。包括自升式平台、电缆铺设船和浮吊等。

Work boat 工作艇 系指非运动或非引航船的游艇供商业用的小船。 Workboat means a small vessel for commercial use other than a sport boat or a dedicated pilot boat.

Work-done factor 减功系数 在多级轴流压气机设计中，用调整做功量的方法考虑环面附面层的存在和发展的影响时所采取的修正系数。

Worker 工人 系指任何根据雇佣关系从事定期或临时工作的人，包括合同工。 Worker means any person who performs work, either regularly or temporarily, in the context of an employment relationship including contractor personnel.

Workforce churn 员工流失

Working accidents 操作事故 与非油气压力下的危险相关的其他危险造成的事故（如跌倒、挤压等），通常涉及单个人员。

Working cloth locker 污衣柜 专供船员存放油污工作服的衣柜。

Working deck (fishing vessel) 工作甲板（渔船）系指用于捕捞作业的最深作业水线以上的最低一层连续甲板。若船舶设有 2 层或多层连续甲板时，主管机关可允许将位于最深作业水线上面的较低一层甲板作为工作甲板。 Working deck is generally the lowest complete deck above the deepest operating waterline from which fishing is undertaken. In vessels fitted with two or more complete decks, the Administration may accept a lower deck as a working deck provided that that deck is situated above the deepest operating waterline.

Working draft 工作吃水 浮船坞在抬举船舶进行正常坞修工作时，在带有油水的状态下，由基线量至水面的垂直距离。

Working environment (immersion suit) 工作环境（救生服） 救生服系统的穿戴者能从事正常工作的环境。 Working environment means the environment in which the wearer of a suit system would engage in normal work.

Working freeboard 工作干舷 浮船坞处于正常坞修工作时，抬船甲板最低点离水面的垂直距离。

Working liquid 工作液体 系指用于船舶机械操作的任何油或油类物质。 Working liquid means any oil or oily substance used for the operation of the ship's machinery.

Working load 工作负载 系指在锚链轮链条出口处测得的拉力。 Working load means the tension measured at wildcats.

Working radius of grab machine 抓斗机工作半径 抓斗中心线至抓斗机转轴中心线间的距离。

Working range of derrick 吊杆工作范围 吊杆装置安全操作的工作范围。包括：舷外跨距、吊杆偏角、吊杆仰角以及吊钩所及的区域等的限制，对于重型吊杆装置还包含有相应的安全工作负荷和船舶横倾角等。一般在船舶的平面图表示之。

Working ship 工程船 专门从事水上工程的船舶。包括海洋开发船、挖泥船、起重船、打桩船、布设船、打捞救生船、浮船坞等。

Working ship type 工程船船型 见表 W-6。

表 W-6 工程船船型
Table W-6 Working ship type

1. 航道及港口服务船型		浮船坞	Floating Dock
航标船	Buoy Tender	沉箱浮坞	Caisson Dock
航标巡检船	Buoy Tour & Inspection Ship	修理船	Repair Ship
航标灯船	Beacon Light Boat	半潜驳	Semi-submersible Barge
破冰船	Ice Breaker	布缆船	Cable Layer
水底整平船	Knife Holder Ship	碎石船	Rock Breaking Barger
水道测量船	Surveying Ship	砂石撒铺船	Sand Spreading Barge
扫海船	Sweeper	卸砂船	Unload Barge
绞滩船	Floating Winch Station for Warping	海底矿物采集船	Seabed Mining Ship
垃圾船	Garbage Boat	采矿船	Mining Dredger
清扫船	Sweep Boat	采金船	Gold Dredger
消防船	Fire Boat	自升式工作平台	Self-elevating Platform

续表 W-6

检疫船	Quarantine Vessel	半潜式工作平台	Semi-submersible Platform
医院船	Hospital Ship	软体排铺设船	Mat-base Laying Ship
浮油回收船	Oil Skimmer	电焊工作船	Floating Welder
起重船	Floating Crane	**4. 挖泥船船型**	
港作拖船	Tug	耙吸式挖泥船	Trailing Suction Hopper Dredger
顶推船	Pusher	绞吸式挖泥船	Cutter Suction Dredger
供应船	Supply Vessel	斗轮式挖泥船	Bucket Wheel Suction dredger
领航船	Pilot Boat	吸盘式挖泥船	Dustpan Suction Dredger
污油处理船	Waste Oil Treating Ship	冲吸式挖泥船	Jetting Suction Dredger
污水船	Sewage Treating Ship	吸泥船	River Suction Dredger
钢扒船	Dozer Dredger	链斗式挖泥船	Bucket Dredger
2. 救助、打捞、潜水工作船船型		自扬、链斗式挖泥船	Pump Bucket Dredger
救生打捞船	Life-salvage Ship	抓斗式挖泥船	Grab Dredger
打捞船(驳)	Salvage Ship (Barge)	铲斗式挖泥船(正铲)	Dipper Dredger
救捞驳	Rescue & Salvage Barge	铲斗式挖泥船(反铲)	Back Hoe Dredger
救助船	Rescue Ship	拉铲式挖泥船	Dragline Dredger
救助拖船	Salvage Tug	喷射泵挖泥船	Jet Ejector Dredger
潜水工作船	Diving Ship	气力提升挖泥船	Air Lift Dredger
潜水工作驳	Diving Boat	吹泥船	Barge Unloading Dredger
潜水训练船	Diver Training Vessel	搅动挖泥船	Agitation Dredger
深潜器母船	Bathyscaph Support Vessel	开底泥驳	Hopper Dump Barge
水下观测船	Underwater Observation Boat	对开(开体)泥驳	Split Hopper Barge
3. 水域施工船型和平台		抛石驳	Stone Dumper
打桩船	Floating Pile Driver	**5. 其他工程船船型和浮体**	
打砂桩船	Sand Piling Barge	发电船	Generating Ship
压桩船	Floating Pile Presse	产品加工船	Plant Barge
深层软地基固化船	Deep Mixing Ship	海水淡化船	Desalinating Ship
打夯船	Compacting Hammer Barge	抛锚艇	Anchor Handling Boat
混凝土搅拌船	Floating Concrete Mixer	储油驳	Oil Storage Barge
钻探船	Drilling Ship	近海装卸站	Offshore Loading Station
钻孔爆破船	Rock Drilling & Blasting Barge	海上观测浮筒	Marine Research Buoy
甲板驳	Deck Cargo Barge	海上无线电中继浮筒	Radio Relay Buoy

Working spaces 工作处所 系指未包括在危险区域和机器处所之内,设有与钻井作业相关的设备和装置的开敞或围蔽处所。 Working spaces are those open or enclosed spaces containing equipment and process associated with drilling operations, which are not included in hazardous areas and machinery spaces.

Workover barge 修井驳船 系指设有修井用各种设备的小型驳船。

Workover platform 修井平台 系指专用于对油井施行井下作业使油井恢复或增加产量的平台。其结构形式和悬臂自升式钻井平台相似。

Work-over supporting barge 修井支持驳船 系指用于海洋工程油气田开发和修井项目的支持作业的非自航船舶。其作业水深一般为 30～60 m 同时在船上为海洋工程项目的作业人员能提供食宿服务。

Work-over unit 修井平台 系指主要用于修井作业的平台。 Work-over unit means a unit mainly used for work-over.

Works approval 工厂认可 系指船级社通过制造厂商的资料审查、认可试验和产品制造过程的审核,对产品制造厂商的产品生产条件和能力予以确认的评定过程。 Works approval means the evaluation process whereby the manufacturer's production conditions and ability is confirmed by a Society through document review, approval testing and verification of manufacturing process.

Works certificate 工厂证书 由制造厂商签署的文件,其阐明:(1)该产品符合规范要求;(2)试验在检定该产品时已进行过;(3)已对该发证产品进行取样试验;(4)该产品试验由有资格的部门见证和签署。 Works certificate means a document signed by the manufacturer stating:(1) conformity with Rule requirements;(2) that tests are carried out on the certified product itself;(3) that tests are made on samples taken from the certified product itself;(4) that tests are witnessed and signed by a qualified department.

Workshop 机修间 供修理各种机械设备用的房间。

Workstation 工作站 系指:(1)执行构成某一特定活动的一项或数项任务的部位;(2)与完成任务的所有与工作相关项目的组合,包括具有全部装置、设备和部件的控制台。 Workstation is:(1) a position at which one or several tasks constituting a particular activity are carried out;(2) the combination of all job-related items, including the console with all devices, equipment and the furniture, to fulfil certain tasks.

Workstation for communication 通信工作站 系指供船上操作人员操作和控制遇险与安全通信(GMDSS)和船内通信用设备的工作场所。 Workstation for communication is a work place for operation and control of equipment for distress and safety communication (GMDSS), and shipboard communication of ship opera-tions.

Workstation for primary bridge functions 桥楼主要功能工作站 当执行导航、航线监视、交通监视和操纵功能时,驾驶人员具有指挥视野和能监视船舶安全状态的工作场所。 Workstation for primary bridge functions meas a workplace with commanding view used by navigators when carrying out navigation, route monitoring, traffic surveillance and manoeuvring functions, and which enables monitoring of the safety state of the ship.

Workstation for safety operations 安全操作工作站 系指专门用于组织和控制内部应急和遇险作业,以及易于获取与船舶安全状态有关信息的工作场所。 Workstation for safety operations means a workplace dedicated organization and control of internal emergency and distress operations, and which provides easy access to information related to the safety state of the ship.

World Bank Group(WBG) 世界银行集团 简称世界银行,是由国际复兴开发银行、国际开发协会、国际金融公司、多边投资担保机构和国际投资争端解决中心 5 个成员机构组成;成立于 1945 年。1946 年 6 月开始营业。凡是参加世界银行的国家必须首先是国际货币基金组织的会员国。世界银行总部设在美国首都华盛顿,有员工 10 000 多人,分布在全世界 120 多个办事处。狭义的"世界银行"仅指国际复兴开发银行(IBRD)和国际开发协会(IDA)。按惯例,世界银行集团最高领导人由美国人担任,为期 5 年。2016 年 1 月 12 日世界银行集团行长金墉宣布任命杨少林为世界银行集团首席行政官,若阿金·莱维为世界银行集团首席财务官。首席行政官兼常务副行长是世行集团新设的一个职位,旨在将机构战略、预算与计划及信息技术等多项职能集中起来。

World closed market price 世界封闭市场价格 是一种因为买卖双方在一定的约束条件下形成的价格,由于非经济因素干预而不能反映国际间实际供求关系。

World conference on free trade area 世界自由贸易区大会

World Customs Organization(WCO) 世界海关组织

World free market price 世界自由市场价格

World Trade Organization(WTO) 世界贸易组织 简称世贸组织,1994 年 4 月 15 日,在摩洛哥的马拉喀什市举行的关贸总协定乌拉圭回合部长会议决定成立更具全球性的世界贸易组织,以取代成立于 1947 年的关贸总协定,1995 年 1 月 1 日正式开始运作,负责管理世界经济和贸易秩序,总部设在瑞士日内瓦蒙湖畔。世贸组织是一个独立于联合国的永久性国际组织,具有法人地位,在调解成员争端方面具有更高的权威性。与国际货币基金组织、世界银行一起被称为世界经济发展的三大支柱。1999 年 11 月 15 日,中国和美国签署关于中国加入世界贸易组织的双边协议。2001 年 11 月 10 日,中国被批准加入世界贸易组织。2001 年 12 月 11 日,中国正式成为其第 143 个成员。

World Wide Web 国际互联网/Web 开发技术
又称万维网,现已成为网民们查询网络、获取信息最为流行的手段。作为 20 世纪人类最伟大的发明之一,蔚然成为当今世界的发展潮流。它是随着 Internet 的普及使用而发展起来的一门技术,该项发明极大地方便了人们对 Internet 上资源的组织和访问。Web 是一种典型的分布式应用结构。Web 应用中的每一次信息交换都要涉及客户端和服务端。因此,Web 开发技术大体上也可以被分为客户端技术和服务端技术两大类。

World Court 国际法庭

World's number-one economy entity 世界头号经济体

Worldwide radio navigation system (WWRNS) 全球无线电导航系统 是在全球范围内利用无线电技术对飞机、船舶或其他运动载体进行导航和定位的系统。无线电导航技术的基本要素是测角和测距,因此可以组成测角—测角、测距—测距、测角—测距、测距差(双曲线)等多种形式的系统。

Worm and gear 蜗轮蜗杆

Worm feeder 螺旋送料机

Worm gear 蜗杆齿轮(蜗轮)

Worm gear ratio 蜗轮传动比

Worm holds 条形孔

Worm propeller 螺杆推进器 装于船的底部,在水中可借助水动力推船前进。登陆后也可使船爬行的螺杆状推进装置。

Worm rack 蜗杆螺纹齿条

Worm shaft 蜗杆

Worm thread 蜗杆螺纹

Worm type propeller 螺杆推进器

Worm wheel 蜗轮

Worm zine 腐蚀锌板

Worm-geared steering gear 双蜗杆式舵机 采用双蜗杆及内齿圈带动舵杆的舵机。

Worn 磨损

Worn-in journal 磨合轴颈

Worst case consequence 最坏情况后果 危险事故导致的健康、环境与安全的最坏后果。如果发生,所有重要的防御措施一定已经失效。

Worst case failure 最恶劣状况故障 动力定位(DP)系统最恶劣状况故障。该故障已作为设计基础,并由 FMEA 认可。通常与若干可能同时失效的推力器和发电机相关,并用于结果分析。

Worst intended conditions 预定的最不利情况 系指:(1)不需要特殊的引航技巧,船舶应能维持安全航行。但是,在所有相对于风和海况的艏向情况下的操纵,可认为不可能。对于在非排水状态下具有较高性能的船型,其性能和加速度也应在船舶处于预定最不利情况下及操纵时的排水模式下确定;(2)船舶证书中规定的该船从事预定营运的环境条件。应考虑诸如允许的最大风力、有义波高(包括波长和浪向的不利组合)、最低气温、能见度、安全操作水深等最差条件参数,以及主管机关认为在该地区营运的这种类型的高速船所需要的其他参数。 Worst intended conditions are:(1) those in which it shall be possible to maintain safe cruise without exceptional piloting skill. However, operations at all headings relative to the wind and sea may not be possible. For type of craft having a higher performance standard in non-displacement mode, the performance and accelerations shall also be established at displacement mode during operation in the worst intended condition;(2) the specified environmental conditions within which the intentional operation of the craft is provided for in the certification of the craft. This shall take into account parameters such as the worst conditions of wind force allowable, significant wave height (including unfavorable combinations of length and direction of waves), minimum air temperature, visibility and depth of water for safe operation and such other parameters as the Administration may require in considering the type of craft in the area of operation.

Worthing pump 双缸蒸汽泵

Wound-rotor motor 绕线转子式感应电动机 是一种感应电动机,其次级电路系由多相绕组或线圈所组成。该绕组或线圈的接线端以短路或通过适当电路予以闭合。当设有集流环或滑环时,也称之为滑环式感应电动机。

Wreck 海损 系指船舶和货物等在海上运输中遭遇自然灾害、意外事故或其他特殊情况,为了解除共同危险而采取合理措施所引起的特殊损失(即牺牲)和合理的额外费用。海上运输中,由于自然灾害或意外事故引起的船舶或货物的任何损失,如船舶因触礁、搁浅、碰撞、沉没、火灾、风灾、爆炸等造成船舶或货物的物质损失及费用损失等,均属海损。

Wreck repair 海损修理

Wreck salvation 防险救生 为了保障安全,预防船舶在航行或作业中发生险情,以及在一旦发生事故时及时地给予援救的一系列专业活动的总称。

Wrench 扳手

Wrench four way 四用扳手

Wrench socket 扳手套筒

Wrenching 扳紧

Written award 裁决书 系指一份包括下列内容的文件:仲裁请求、争议事实、裁决理由、裁决结果、仲裁费用和承担、裁决的日期和地点。当事人协议不写明争议事实和裁决理由的,以及按照双方当事人和解协议的内容做出裁决的,可以不写明争议事实和裁决理由。

Written counterclaim 反要求书 是一份写明具体的反要求事项、理由以及所依据的事实,并附上相关证据的文件。

Written defense 答辩书 是一份写明答辩的事实、理由并附上相关证据的文件。

Written form 书面形式 系指合同书、信件和数据电文(包括电报、电传、传真、电子数据交换和电子邮件)等可以有形地表现所载内容的形式。

WS = Workstation 工作站 系指执行构成某一特定活动的一项或多项任务,且能为安全和有效地执行任务提供所需信息、系统和设备,并开展桥楼团队合作的工作场所。 A workstation means a workplace at which one or several tasks constituting a particular activity are carried out, designed, arranged and located as required to provide the information, systems and equipment required for safe and efficient performance of dedicated tasks and bridge team co-operations.

WSOG 油井特殊操作指南

W-type internal combustion engine W 型内燃机 (1)两台 V 型内燃机排列成 W 型,两根曲轴用齿轮连接到一根功率输出轴上的内燃机;(2)三列气缸共用一根曲轴,其剖面呈▽形的内燃机。

X

X carrier X 船 系指除普通干货船以外的装运固体货物的机动船舶,按其装运的货物名称授予船型附加符号,其中 X 由具体货物名称替代。典型的船型包括:如水泥运输船,木材制品运输船,碎木运输船,甲板货船;冷藏货船,牲畜运输船,烟灰运输船和散糖运输船等。 X carriers mean self-propelled ships carrying solid cargo, other than general dry cargo ships. Type notations are to be assigned according to names of product carried and X to be substitide by such names. Typical type notations are: cement carrier, forest product carrier, wood chip carrier, deck cargo ship, refrigerated cargo ship, livestock carrier, fly ash carrier, and sugar carrier, etc.

x directions accelerations a_x x 向加速度 a_x 系指任何一点的 x 向加速度(m/s^2)。由下式得出:$a_x = C_{xG} g \sin\Phi + C_{xs} a_{surge} + C_{xp} a_{pitch\,x}$,式中:$a_{surge}$——纵荡引起的纵向加速度。由下式得出:$a_{surge} = 0.2 a_0 g$;$a_{pitch\,x}$——纵摇引起的纵向加速度($m/s^2$)。由下式得出:$a_{pitch\,x} = \Phi(180/\pi)(2\pi/T_P)^2 R$。 x directions accelerations a_x means that at any point, accelerations, in m/s^2, along x directions, given by $a_x = C_{xG} g \sin\Phi + C_{xs} a_{surge} + C_{xp} a_{pitch\,x}$, where: a_{surge}—The longitudinal acceleration due to surge, in m/s^2, is given by $a_{surge} = 0.2 a_0 g$, $a_{pitch\,x}$—Longitudinal acceleration due to pitch, in m/s^2; $a_{pitch\,x} = \Phi(180/\pi)(2\pi/T_P)^2 R$.

X-band X 波段 根据 IEEE 521-2002 标准,X 波段系指频率在 8~12 GHz 的无线电波波段,在电磁波谱中属于微波。而在某些场合中,X 波段的频率范围则为 7~11.2 GHz。通俗而言,X 波段中的 X 即英语中的"extended",表示"扩展的"调幅广播。

X-band radar(XBR) X 波段雷达 系指对火控、目标跟踪雷达的统称,其波长在 3 cm 以下。XBR 用于弹道导弹防御、测试、演习、训练,并协同观测比如太空碎片、航天飞机等的运动。

X-band super radar(detect 6 000 km) X 波段超级雷达(探测 6 000 km) 该技术拥有国:中国、美国、俄罗斯。见图 X-1。

图 X-1　X 波段超级雷达
Figure X-1　X-band super

X-barge X 驳船 系指专门运载特殊货物的驳船。X 代表特殊货物。

X-bow X 型船艏

Xerox 施乐 是美国施乐公司一个著名商标和品

牌。施乐公司于1906年成立于美国康涅狄格州费尔菲尔德县。作为商标，施乐只用来标识施乐公司的各种产品和服务。施乐商标放在名词的前面起到形容的作用：如施乐复印机、施乐打印机等。

X-ray radio-graphite inspection equipment X光检查设备

X-type internal combustion engine X型内燃机 四列气缸围绕一根曲轴交叉排列，其剖面呈X形的内燃机。

Y

Y alloy 高强度耐热铝合金（Y合金）
Y axis Y轴（Y轴线，纵坐标轴）
Y bend Y形弯头
y directions accelerations a_y y向加速度 a_y 系指任何一点的y方向加速度。由下式得出：$a_y = C_{yG} g \sin\theta + C_{yS} a_{sway} + C_{yR} a_{roll\,y}$，式中：$a_{sway}$—横荡引起的横向加速度。由下式得出：$a_{sway} = 0.3 a_0 g$；$a_{roll\,y}$—横摇引起的横向加速度。由下式得出：$a_{roll\,y} = \theta(\pi/180)(2\pi/T_R)^2 R$。y directions accelerations a_y means that at any point, accelerations, in m/s², along y directions, given by $a_y = C_{yG} g \sin\theta + C_{yS} a_{sway} + C_{yR} a_{roll\,y}$, where: a_{sway}—Transverse acceleration due to sway, in m/s², is given by $a_{sway} = 0.3 a_0 g$; $a_{roll\,y}$—Transverse acceleration due to roll, is given by $a_{roll\,y} = \theta(\pi/180)(2\pi/T_R)^2 R$.

Y joint 叉形接头（Y形接头，三通接头）
Y junction 三角分线杆（Y形交叉，Y形接头）
Y node Y型节点
Y piece Y形接头
Y pipe（**y tube**） 三通管（Y形接管）
Yacht（**pleasure craft**） 游艇 一种供游览、休闲、娱乐等用的非营业性机动艇，包括以整船租赁形式从事上述活动的船舶。为了减轻船艇的质量，游艇的船体一般采用高强度铝合金或玻璃钢等其他轻材料。游艇按动力分为无动力艇、机动和风帆三种。其中帆艇（帆船）又可分为无动力辅助帆艇和辅助动力帆艇。而机动游艇的发动机一般采用高速柴油机，备有风帆的游艇，有的还装舷外挂机。按尺寸分，国际标准游艇的规格是以英尺计算的：36英尺（ft）以上为小型游艇，36～60英尺（ft）为中型游艇，60英尺（ft）以上为大型豪华游艇。按材质分，有木质艇、玻璃钢艇、凯夫拉纤维增强的复合材料艇、铝质艇和钢质艇。按功能分，有运动型游艇、休闲型游艇和商务游艇等。

Yahoo! Inc.（**Yahoo!**） 雅虎 是美国著名的互联网门户网站，也是20世纪末互联网奇迹的创造者之一。其服务包括搜索引擎、电邮、新闻等，业务遍及24个国家和地区，为全球超过5亿的独立用户提供多元化的网络服务。同时也是一家全球性的因特网通信、商贸及媒体公司。

Yarrow boiler 三锅筒水管锅炉
Yaw 艏摇 系指：（1）船舶艏向变化而船舶航向无变化；（2）是船舶沿z轴的角运动。 Yaw is: (1) change of direction of a craft's heading without a change in the craft's track. (2) the ship's angular motion along the z axis.

Yaw angle 偏航角 船舶受扰动后从其原艏向至偏离原位的新艏向之间的水平夹角。顺时针为正。

Yaw checking ability 偏航纠正能力 是衡量某种回转状态下船舶对反向舵的响应，例如在标准Z形操纵时反向舵消除之前船舶向超越角。 Yaw checking ability of the ship is a measure of the response to counter-rudder applied in a certain state of turning, such as the heading overshoot reached before the yawing tendency has been cancelled by the counter-rudder in a standard zig-zag manoeuvre.

Yawing 艏摇 船舶在水平面内绕垂向轴所做的周期性角位移运动。

Yellow alert 黄色警戒 降级的动力定位（DP）状态，DP船舶对此具有一个事先部署好的响应，准备应对与DP红色警戒有关的风险。

Yellow alert condition of DP 动力定位黄色警戒状态 系指动力定位降级状态。通常，这是冗余动力定位设备中一个或多个项目已失效的状态，正在超越安全工作极限或可能艏向或位置的偏移。

Yellow stern light 黄艉灯 船舶拖带时专用的显示黄色光的艉灯。

Yellow visual signal 黄色的视觉信号 系指状态的改变或即将改变的指示。例如：温度或压力值异常，但未达到临界值的信号。 Yellow visual signal means an indication of change or impending change of conditions. For example: temperature or pressure is abnormal but not to a critical value.

Yi-Chang sampan 宜昌舢板 起源于我国宜昌地

区,配置在内河船上供工作或救生用的平底、雪橇形艏、方艉式舢板。

Yield failure 屈服损坏
Yield load 屈服载荷
Yield per net 网次产量　每起一次网的渔获量。
Yield phenomenon 屈服现象
Yield point 屈服点(流动点)
Yield point elongation 屈服点伸长
Yield point strain 屈服点应变
Yield strength(YS) 屈服强度
Yield stress(strength) σ_0 屈服应力(强度)　这是在 0.5% 比例伸长的条件下,管子材料单轴向拉伸或压缩试验时记录的应力。本书最初使用规范的应力定义为 σ_0,其代表在 0.2% 比例伸长的条件下的应力,对于硬度相当大的材料,这两个定义导致的差异仅为几个百分数。 Yield stress(strength) σ_0 is the stress recorded in a uni-axial tensile or compressive test on the pipe material at a strain of 0.5%. This book primarily uses the more conventional yield stress definition σ_0, that represents the stress at a strain offset of 0.2%, for materials with healthy hardening, the two definitions tend to differ by a few percent.
Yield to tensile ratio 屈强比
Yielding 屈服性的(流动性的,可压缩性的)
Yielding point 屈服点
Yoke 系泊刚臂

Yoke for accommodation ladder 舷梯吊架　吊梯装置中,用以使吊梯链索分列于舷梯梯架上的吊具。
Yoke-type PFD 轭架型个人漂浮装置　穿戴在人的颈部周围,通过腰部扎带固定的个人漂浮装置。Yoke-type PFD is the PFD worn around the neck and secured by a waist strap.
Young's modulus E(N/mm^2) 杨氏模量　应取为:(1)对普通钢材 $E = 2.06 \times 10^5$ N/mm^2;(2)对不锈钢 $E = 1.95 \times 10^5$ N/mm^2;(3)对铝合金 $E = 7.0 \times 10^4$ N/mm^2。 Young's modulus is to be taken equal to:(1) $E = 2.06 \times 10^5$ N/mm^2, for steels in general;(2) $E = 1.95 \times 10^5$ N/mm^2, for stainless steels;(3) $E = 7.0 \times 10^4$ N/mm^2, for aluminium alloys.
YUV 色差　是欧洲电视系统所采用的一种颜色编码方法(属于 PAL),是 PAL 和 SECAM 模拟彩色电视制式采用的颜色空间。在现代彩色电视系统中,通常采用三管彩色摄影机或彩色 CCD 摄影机进行取像,然后把取得的彩色图像信号经分色、分别放大校正后得到 RGB,再经过矩阵变换电路得到亮度信号 Y 和两个色差信号 B - Y(即 U)、R - Y(即 V),最后发送端将亮度和色差三个信号分别进行编码,用同一信道发送出去。这种色彩的表示方法就是所谓的 YUV 色彩空间表示。采用 YUV 色彩空间的重要性是它的亮度信号 Y 和色度信号 U、V 是分离的。

Z

Z(cm^3) 普通扶强材或主要支撑构件连同宽度 b_p 带板的净剖面模数　Z is the net section modulus (cm^3) of an ordinary stiffener or a primary supporting member with attached plating of width b_p.
z directions accelerations a_z z 向加速度 a_z　系指任何一点的 z 向加速度,m/s^2。由下式得出:$a_z = C_{zH} a_{heave} + C_{zR} a_{roll\ z} + C_{zp} a_{pitch\ z}$,式中:$a_{heave}$—垂荡引起的垂向加速度,m/s^2。由下式得出:$a_{heave} = a_0 g$;$a_{roll-z}$—横摇引起的垂向加速度,m/s^2。由下式得出:$a_{roll-z} = \theta(\pi/180)(2\pi/T_R)^2 y$;$a_{pitch\ z}$—纵摇引起的垂向加速度,m/s^2。由下式得出:$a_{pitch\ z} = \Phi(\pi/180)(2\pi/T_P)^2 |(x - 0.45 L)|$,式中:$|(x - 0.45 L)|$ 应取不小于 $0.2 L$。$R = z - \min(D/4 + T_{LC}/2, D/2)$。 z directions accelerations a_z means that at any point accelerations, in m/s^2, along z directions, given by $a_z = C_{zH} a_{heave} + C_{zR} a_{roll\ z} + C_{zp} a_{pitch\ z}$, where: a_{heave}—the vertical acceleration due to heave, in m/s^2, is given by $a_{heave} = a_0 g$; $a_{roll\ z}$—vertical acceleration due to roll, in m/s^2, is given by $a_{roll\ z} = \theta(\pi/180)(2\pi/T_R)^2 y$; $a_{pitch\ z}$—vertical acceleration due to pitch, in m/s^2, is given by $a_{pitch\ z} = \theta(\pi/180)(2\pi/T_R)^2 y$, where:$(x - 0.45L)$ is to be taken not less than $0.2L$; $R = z - \min(D/4 + T_{LC}/2, D/2)$.
Z_{AB}(m^3) Z_{AB} 船底处的剖面模数,m^3。 Z_{AB} is section moduli(in m^3), at bottom.
Z_{AD}(m^3) Z_{AD} 甲板处的剖面模数,m^3。 Z_{AD} is section moduli(in m^3), at deck.
Z-drive Z 形传动　轴系中线布置呈 Z 字形的传动方式。
Zero emission port 零排放港区　系指在该港区内,船舶无论是否满足机器处所舱底水的排放要求,其机舱舱底水不得在此港区内排放。
Zero man engine room 无人机舱

Zero speed maneuverability 零速操纵性 船舶零航速漂泊在海上时,保持船位所具备的能力。

Zero-lift line 零升力线 流体流向物体,产生升力为零的流线。

Zig-zag maneuver test Z 形操纵试验 系指船舶进行一定量的两交替操舵,每次操舵时船航向应达到明显偏离原航向。 Zig-zag maneuver test is the manoeuvre where a known amount of helm is applied alternately to either side when a known heading deviation from the original heading is reached.

Zimuth propulsion arrangements 全方位推进装置 系指包括 Z 型推进装置、导管(Duck)推进装置、REX–推进装置和舵桨装置等的装置。 Zimuth propulsion arrangements mean arrangements including Z propulsion arrangement, Duck propulsion arrangement, REX propulsion arrangement and steering oar arrangement, etc.

Zinc annunciator for protection 防腐锌阳极

Zinc anode cathodic protection system 锌阳极阴极防护系统

Zinc anodes 锌阳极 系指以锌制成的阳极。

Zinc ashes(UN 1435) 锌灰 遇水分或水易释放氢气(一种易燃气体)和有毒气体。该货物为不燃物或失火风险低。 In contact with moisture or water, it is liable to give off hydrogen, a flammable gas, and toxic gases. This cargo is non-combustible or has a low fire-risk.

Zinc coating(zinc coating layer) 镀锌层(镀锌)

Zinc galvanizing 镀锌

Zirconsand 锆石砂 通常为细粉,白色至黄色,从钛铁砂中提取,腐蚀性很强。可能多粉尘。干燥装运。无特殊危害。该货物为不燃物或失火风险低。 Zirconsand is usually fine white to yellow, very abrasive extracted from ilmenite sand. It may be dusty, is highly corrosive and shipped dry. It has no special hazards. This cargo is non-combustible, or a low fire-risk.

Zonal hazard analysis(ZHA) 区域危险分析(ZHA) 其目的是确定从安装(隔离、分离、保护等)和操作(保养工作等)设计产生的潜在风险区域。 Zonal hazard analysis(ZHA) means an analysis whose objective is to identify potential areas of risk arising from the design of the installation (segregation, separation, protection, etc.) and the operation (maintenance tasks, etc.).

Zone 隔离带 由船舶内的一组舱室组成,隔离带内部应设有必要的独立系统,以提供一个不含任何核、生、化与放射性污染的无毒区域。 Zone is a smaller group of compartments within the citadel with some or all of the independent systems necessary to provide a toxic free area that is free from any CBRN hazard.

Zone 0 0 区 对危险区域而言,系指爆炸性气体环境连续或长期存在的区域。

Zone 1 1 区 对危险区域而言,系指在正常运行时,可能出现爆炸性气体环境的区域。

Zone 2 2 区 对危险区域而言,系指在正常运行时,不可能出现爆炸性气体环境,即使出现,也是偶尔发生并且仅是短时间存在的区域。

Zone address identification capability 区域编址识别功能 系指一个具有单个识别探测器的系统。

Zone reheat air conditioning system 区域再加热空气调节系统 部分舱外新鲜空气和部分舱内循环空气两者混合经集中式空气调节器处理达到较低温度后,由通风机输送到通风管路,并经过各舱室出风口处的加热器进行加热,根据温度调节器的控制,自动或手动调节各区域所需温度,然后再送入各舱室内的空气调节系统。

Z-propulsion Z 向推进系统

ZTOP 自液舱最高点至基线的垂直距离 对压载货舱,ZTOP 为自舱口围板顶部至基线的垂直距离。 ZTOP means the vertical distance, in m, of the highest point of the tank from the baseline. For ballast holds, ZTOP is the vertical distance, in m, of the top of the hatch coaming from the baseline.

Zurich, Switzerland, chamber of commerce court of arbitration 瑞士苏黎世商会仲裁院

λ(m) 波浪长度(m) λ is the wave length, in m.

σ_x 船体梁正应力 σ_x is the hull girder normal stress, in N/mm^2.

Φ 纵摇单幅值(°) 由下式得出:$\Phi = f_p(960/L)(V/C_B)^{1/4}$。 Φ is the single pitch amplitude, in deg, is given by:$\Phi = f_p(960/L)(V/C_B)^{1/4}$.

以数字起首的词条

100 000-ton class aircraft carrier 10 万吨级航空母舰(拥有数量/制造能力) 系指拥有全通甲板,可搭载并起降固定翼飞机或直升机、排水量超过 100 000 t 的作战舰艇。其中以搭载直升机或短距垂直起降战机为

主的舰船被称为"准航空母舰"或直升机航空母舰。该技术拥有国：美国（11 艘/有制造能力）、印度（3 艘/无）英国（2/有）、法国（2/有）、巴西（2/无）俄罗斯（1/有）、中国（1/有）。见图 1-1。

图 1-1 10 万吨级航空母舰
Figure 1-1 100 000 ton class aircraft carrier

1/10 highest waves 十一保证率波高 系指将在不规则海面上定点或定时观测的波高数据，按大小序列，取其中大端的 1/10 区间的最低值，即有 1/10（从 ξ_{max} 到 $\xi_{1/10}$）的累积概率代表海上典型波浪上限的波高。

1/100 highest waves wave amplitude 百一保证率波高 系指将在不规则海面上定点或定时观测的波高数据，按大小序列，取其中大端的 1/100 区间的平均值，即有 1/100（从 ξ_{max} 到 $\xi_{1/100}$）的累积概率代表海上最大波浪的波高。

10°/10° zig-zag test 10°/10°Z 形操纵试验 系指紧接着使船舶偏离初始航向达 10°，按以下程序交替向两舷操 10°舵角：(1) 当船舶进入偏航率为零的稳定状态后，向右舷或左舷操 10°舵角（第 1 次操舵）；(2) 当船舶航向改变 10°时，反向操舵至左舷或右舷 10°舵角（第 2 次操舵）；(3) 当舵转至左舷/右舷后，船舶将以减低的回转速率继续沿初始航线方向回转。然后，船舶响应舵而转向左舷/右舷。当船舶航偏离初始航向左舷/右舷 10°时，再向右舷/左舷操 10°舵角（第 3 次操舵）。10°/10° zig-zag test is performed by turning the rudder alternately by 10° to either side following a heading deviation of 10° from the original heading in accordance with the following procedure: (1) after a steady approach with zero yaw rate, the rudder is put over to 10° to starboard or port (first execute); (2) when the heading has changed to 10° off the original heading, the rudder is reversed to 10° to port or starboard (second execute); (3) after the rudder has been turned to port/starboard, the ship will continue turning in the original direction with decreasing turning rate. In response to the rudder, the ship should then turn to port/starboard. When the ship has reached a heading of 10° to port/starboard of the original course, the rudder is again reversed to 10° to starboard/port (third execute).

20°/20° zig-zag test 20°/20°Z 形操纵试验 系指紧接着使船舶偏离初始航向达 20°，按以下程序交替向两舷操 20°舵角：(1) 当船舶进入偏航率为零的稳定状态后，向右舷或左舷操 20°舵角（第 1 次操舵）；(2) 当船舶航向改变 20°时，反向操舵至左舷或右舷 20°舵角（第 2 次操舵）；(3) 当舵转至左舷/右舷后，船舶将以减低的回转速率继续沿初始航线方向回转。然后，船舶响应舵而转向左舷/右舷。当船舶航偏离初始航向左舷/右舷 20°时，再向右舷/左舷操 20°舵角（第 3 次操舵）。The 20°/20° zig-zag test is performed by turning the rudder alternately by 20° to either side following a heading deviation of 10° from the original heading in accordance with the following procedure: (1) after a steady approach with zero yaw rate, the rudder is put over to 20° to starboard or port (first execute); (2) when the heading has changed to 20° off the original heading, the rudder is reversed to 20° to port or starboard (second execute); (3) after the rudder has been turned to port/starboard, the ship will continue turning in the original direction with decreasing turning rate. In response to the rudder, the ship should then turn to port/starboard. When the ship has reached a heading of 20° to port/starboard of the original course, the rudder is again reversed to 20° to starboard/port (third execute).

3D seismic research ship 三维地震物探船 其好比是海洋工程"联合舰队"中的"侦察船"，海洋物探船对海底地质条件的勘探，是海上油气开发的第一步。其主要任务是利用地震波探测技术采集水下 1 500 ~ 3 000 m 的地层信息，勘探油气资源。该类船舶一般配置 2 ~ 8 根地震采集电缆。国内新一代物探船工作水深可达 3 000 m，拖带 12 根 8 000 m 长的地震采集电缆、8 排双震源的气枪阵列，且平均每天勘探面积可达 60 ~ 120 km²。当作业时，尾部的 12 根电缆在水中呈扇形分布，它将高压空气通过气枪阵生成地震波，传播到河底深部的地层中，当地震波碰到岩层界面时，就会产生反射波，再传回物探船的接收装置并被记录下来，通过计算机处理后获得地震反射剖面，然后通过对地震反射剖面进行分析和解释后，编制出海洋油气田藏构造的关键路线图。

附录1 设计图纸上的缩写符号
Annex 1 Abbreviations and symbols on design drawing

An oil purif = an oil purifier 油分水机
An oil serv tk = an oil service tank 油日用柜
An oil sett tk = an oil settling tank 油沉淀柜
An oil stor tk = an oil storage tank 油储存柜
An oil trans pump = an oil transfer pump 油传输泵
ACC auto comb cont = automatic combustion control equipment 自动燃烧控制设备
Accomm ladder winch = accommodation ladder winch 舷梯绞车
ADE = above deck equipment 甲板上设备
AHD MANOEUV VALVE = ahead manoeuvring valve 正车操纵阀
AHD TURB = ahead turbine 前进透平
Atm condr = atmospheric condenser 大气冷凝器
Air clr = air cooler 空气冷却器
Air clr chem clean pump = air cooler chemical cleaning pump 空冷器化学清洗泵
Air clr clean w tk = air cooler cleaning water tank 空冷器清洁水舱
Air comp cfw clr = air compressor cooling fresh water cooler 空压机冷却淡水冷却器
Air comp cfw pump = air compressor cooling fresh water pump 空压机冷却淡水泵
Air comp csw pump = air compressor cooling seawater pump 空压机冷却海水泵
Air cond fan = air conditioning fan 空调机扇
Air cond hot circ pump = air conditioning hot water circulating pump 空调热水循环泵
Air cond ref c w pump = air conditioning refrigerating cooling water pump 空调制冷用冷却水泵
Air cond ref comp = air conditioning refrigerating compressor 空调制冷器压缩机
Air htr = air heater 空气加热器
Airplane warn LT = airplane warning light 对飞机警告灯
Alkal batt = alkaline battery 碱性电池
Alm log = alarm logger 报警记录仪
Anal monit = analogue monitor 模拟监视仪
Anchor LT = anchor light 锚灯
Ant chang = antenna changer 天线转换器
Ant multi = antenna multi-coupler 天线多路耦合器

AST MANOEUV VALVE = astern manoeuvring valve 倒车操纵阀
AST TURB. = stern turbine 倒车透平
AT = aftpeak tank 艉尖舱
Atmos condr circ pump = atmospheric condenser circulating pump 空调凝水器循环泵
Atmos drain tk = atmospheric drain tank 大气放泄舱
Auto alarm = auto alarm 自动警报
Auto keyer = auto keyer 自动调制器
Auto moor winch = auto mooring winch 自动绞缆机
Auto tent winch = auto tension winch 自动张力绞车
Aux air comp = auxiliary air compressor 辅空气压缩机
Aux air ejector = auxiliary air ejector 辅空气抽出器
Aux air reserv = auxiliary air reservoir 辅空气瓶
Aux blr = auxiliary boiler 辅锅炉
Aux cfw pump = auxiliary cooling fresh water pump 辅助冷却淡水泵
Aux circ pump = auxiliary circulating pump 辅助循环泵
Aux condr = auxiliary condenser 辅冷凝器
Aux csw pump = auxiliary cooling seawater pump 辅助冷却海水泵
Aux condr pump = auxiliary condensate pump 辅助凝水泵
Aux fd(w) pump = auxiliary feed water pump 辅助疏水泵
Aux gen = auxiliary generator 辅发电机
Aux GEN = auxiliary generator 辅发电机
Aux gen d eng = auxiliary generator diesel engine 辅发电机内燃机
Aux gen eng F.O. tk = auxiliary generator engine fuel oil tank 辅机燃油柜
Aux mach cfw pump = auxiliary machinery cooling fresh water pump 辅助机械冷却淡水泵
Aux mach csw pump = auxiliary machinery cooling seawater pump 辅助机械冷却海水泵
Aux ry = auxiliary relay 辅延时
Ballast = ballaster 镇流器
Ballast pump = ballast pump 压载泵
Ballast strip pump = ballast stripping pump 压载扫舱泵
BC receiv ant = broadcast receiving antenna 广播接收天线
BDE = blow deck equipment 甲板下设备

Berth LT = berth light   停泊灯
BFV = butterfly valve   蝶形止回阀
BG = bottom girder   底桁
Bi-color LT = bi-color light   双色灯
Bilge and ballast pump = bilge and ballast pump   舱底及压载泵
Bilge pump = bilge pump   舱底泵
Blr comp inject pump = boiler compound injection pump   锅炉复合物喷射泵
Blr comp tk = boiler compound tank   锅炉复合舱
Blr F. O. htr = boiler fuel oil heater   锅炉燃油加热器
Blr F. O. sett tk = boiler fuel oil settling tank   锅炉燃油沉淀柜
Blr ignit O. tk = boiler ignition oil tank   锅炉点火用油柜
Blr test pump = boiler test pump   锅炉试验泵
Blr w circ pump = boiler water circulating pump   锅炉水循环泵
Bnr = burner   燃烧器
Boat dk LT = boat deck light   艇甲板灯
Boat winch = boat winch   救生艇绞车
Bow thruster = bow thruster   艏侧推器
BP = bilge pump   舱底水泵
BP = bottom plating   舱底板
Br cont stand = bridge control stand   驾驶室控制台
Bread slicer = bread slicer   面包切片机
BT = ballast tank   压载舱
BT = bilge tank   污水舱
Bulbous bow war LT = bulbous bow warning light   球鼻艏警示灯
BV = ball valve   球阀
C．O．collect pump = cargo oil collecting pump   货物油收集泵
C．O．strip pump = cargo oil stripping pump   货油扫舱泵
C（B）oil trans pump = C（B）oil transfer pump   C（B）油传输泵
C（B）oil pump = C（B）oil pump   C（B）油分水机
C（B）oil serv tk = C（B）oil service tank   C（B）油日用柜
C（B）oil sttl tk = C（B）oil settling tank   C（B）油沉淀柜
C．O．p condr conds pump = cargo oil pump condenser condensate pump   货油泵冷凝器凝水泵
C．O．p condr circ pump = cargo oil pump condenser circulating pump   货油泵冷凝器循环泵
Camshaft charge l o clr = camshaft charge lubricating oil cooler   凸轮滑油冷却器
Camshaft L．O．tk = camshaft lubrication oil tank   凸轮轴滑油柜

Camshaft L. O. pump = camshaft lubricating oil pump   凸轮轴滑油泵
Capstan = capstan   绞盘
Cargo hold vent fan = cargo hold ventilating fan   货舱排气扇
Cargo LT = cargo light   货梯灯
Cargo pump = cargo pump   货油泵
Cargo ref c w pump = cargo refrigerating cooling water pump   货物制冷用冷却水泵
Cargo ref comp = cargo refrigerating compressor   货物制冷器压缩机
Cargo hold alarm = cargo hold alarm   冷藏货舱警报
Cargo winch = cargo winch   起货绞车
Cascade tk = cascade tank   叶栅舱
Caustic soda solute inject pump = caustic soda solution injection pump   苛性钠溶液喷射泵
CCP = chemical cleaning pump   化学清洗泵
Ceil LT = celling light   电池灯
CFD = cofferdam   隔离舱
Cfw pump = cooling fresh water pump   冷却淡水泵
CH = cargo hold   货舱
Chart t LT = chart table light   海图桌上灯
Chem. Clean pump = chemical cleaning pump   化学清洗泵
Circ pump = circulating pump   循环泵
CL = chain locker   锚链舱
Cland exh fan = gland exhaust fan   密封排气扇
Cland st condr = gland steam condenser   密封蒸汽冷凝器
Clarif = clarifier   分油机
Clean ballast pump = clean ballast pump   清洁压载泵
CO_2 mtr = CO_2 meter   二氧化碳表
Coffee urn = coffee urn   咖啡壶
Cold start F. O. burn pump = cold start fuel oil burning pump   冷启动燃油燃烧泵
Cold start F. O. htr = cold start fuel oil heater   冷启动燃油加热器
Cold start fd（w）pump = cold start feed water pump   冷启动给水泵
Cold w fountain = cold water fountain   冷水喷水机
Comb fd（w）htr = combined feed（water）heater   混合给水加热器
Comp gauge = compound gauge   复合计
Composite blr = composite boiler   补偿锅炉
Comput log = computer logger   计算记录仪
Condr pump = condenser pump   凝水泵
Cont air comp = control air compressor   控制空气压缩机

Cont air dehyd = control air dehydrator 控制用空气脱水器
Cont air reserve = control air reservoir 控制空气瓶
Cont console = control console 控制台
Contlr = controller 控制器
Contn ref cfw pump = container refrigerating cooling fresh water pump 集装箱制冷用冷却淡水泵
Contn ref csw pump = container refrigerating cooling seawater pump 集装箱制冷用冷却海水泵
Cool f w（stor）tk = cooling fresh water（storage）tank 冷却淡水储存舱
Cool f w exp tk = cooling fresh water expansion tank 冷却淡水膨胀舱
Cool f w oil sep tk = cooling fresh water oil separating tank 冷却淡水分离舱
Cool fw clr = cooling fresh water cooler 冷却淡水冷却器
COT = cargo oil tank 货油舱
CP = cargo pump 货油泵
Cpp L.O. grav tk = controllable pitch propeller lubricating oil gravity tank 可变螺距螺旋桨滑油重力柜
Cpp L.O. stor tk = controllable pitch propeller lubricating oil storage tank 可变螺距螺旋桨滑油储存柜
Cpp L.O. sump tk = controllable pitch propeller lubricating oil sump tank 可变螺距螺旋桨滑油油底壳沉淀柜
Cpp L.O. clr = controllable pitch propeller lubricating oil cooler 可调螺距螺旋桨滑油冷却器
Cpp L.O. pump = controllable pitch propeller lubricating oil pump 可调螺距螺旋桨滑油泵
Crosshead L.O. pump = crosshead lubricating oil pump 十字头滑油泵
Csw pump = cooling seawater pump 冷却海水泵
CWP = cooling water pump 冷却水泵
Cyl O. serv pump = cylinder oil service pump 气缸油总用泵
Cyl O. stor tk = cylinder oil storage tank 气缸油储存柜
Cyl O. trans pump = cylinder oil transfer pump 气缸油输送泵
Cyl O. measure tk = cylinder measuring tank 气缸油测量柜
D.O. serv tk = diesel oil service tank 柴油日用柜
D.O. sett tk = diesel oil settling tank 柴油沉淀柜
D.O. stor tk = diesel oil storage tank 柴油储存柜
D.O. trans pump = diesel oil transfer pump 柴油传输泵
D.C.M = direct current motor 直流电动机
D.O. purifier = diesel oil purifier 柴油分水机
Daily tk = daily tank 日用柜

Dange cargo flash LT = dangerous cargo flashing light 危险货物运输闪光灯
Data log = data logger 数据记录仪
Daylight sig LT = daylight signaling light 昼间信号灯
DBT = double bottom tank 双层底舱
Dc M = direct current motor 直流电动机
DD = upper deck 上甲板
Deaerat fd（w）htr = deaerating feed（water）heater 脱气机给水加热器
Deck crane = deck crane 甲板吊
Deck mach hyd pump = deck machinery hydraulic pump 甲板机械液压泵
Deep draft LT = deep draught vessel light 深吃水船灯
Desk LT = desk light 台灯
Desuphtr = desuperheater 减温器
DH = deck house 甲板室
Dial sw = dial switch 拨盘开关
Dial thmtr = dial thermometer 刻度温度计
Diesel gen（d/c）= diesel generator 内燃发电机
Diesel GEN = diesel generator 内燃发电机
Dim = dimmer 调光器
Dipole ant = dipole antenna 偶极子天线
Dish washer = dish washer 洗碗机
Dispens LT = dispensary light 诊疗灯
Disposer = disposer 除霜机
Dist plant = distilling plant 蒸馏设备
Dist plant brine pump = distilling plant brine pump 蒸馏系统卤水泵
Dist plant circ pump = distilling plant circulating pump 蒸馏设备循环泵
Dist plant comp inject pump = distilling plant compound injection pump 蒸馏设备复合物喷射泵
Dist plant eject pump = distilling plant ejector pump 蒸馏设备抽吸泵
Distil plant = distilling plant 蒸发单元
Distil pump = distillate pump 蒸馏泵
Distil w tk = distilled water tank 蒸发淡水舱
Distl plant comp tk = distilling plant compound tank 蒸发单元复合舱
Double ant = double antenna 双重天线
Double evap blr = double evaporation boiler 双蒸发锅炉
Dough mixer = dough mixer 和面机
Drain clr = drain cooler 放泄冷却器
Drain collect tk = drain collecting tank 放泄收集舱
Drain pump = drain pump 疏水泵
Drill mach = drilling machine 钻床

Drink w hyd tk = drinking water hydrophore tank 饮用水采样舱

Drink w press tk = drinking water pressure tank 饮用水压力舱

Drink w pump = drinking water pump 饮用水泵

Dry tumb = drying tumbler 干燥滚筒

DSC = digital selective call 数字选择呼叫

DSC = digital selective calling 数字选择呼叫

DT = deep tank 深舱

DTE = data terminal equipment 数据终端

DWP = drinking water pump 饮用水泵

EGC = enhanced group calling 增强群呼

EL.（earth l）= earth lamp 底线灯

Elect rice blr = electric rice boiler 电饭锅

Elect welder = electric welder 电焊机

ELT = emergency locator transmitter 应急定位发射器（飞机用）

EME = externally mounted equipment 外接设备

Emegr LT = emergency light 应急灯

Emerg air reserv = emergency air reservoir 应急空气瓶

Emerg fire pump = emergency fire pump 应急消防泵

Emerg gen = emergency generator 应急发电机

Emerg GEN = emergency generator 应急发电机

Emerg gen d eng = emergency generator diesel engine 应急发电机内燃机

Emerg gen eng F．O．tk = emergency generator engine fuel oil tank 应急发电机燃油柜

Eng cont rm unit clr = engine control room unit cooler 机舱冷却单元

Eng rm vent fan = engine room ventilating fan 机舱排气扇

Engr alm = engineer's alarm 轮机员传唤警报

Engr ext alm = engineer's extension alarm 轮机员传唤延长警报

EPIRB = emergency position indication radio beacon 应急无线电示位标

EPIRB = emergency position indicating radio beacon 应急无线电示位标

ER = engine room 机舱

Exh g eco turbo gen（exh g eco t/g）= exhaust gas economizer turbo generator 废气节能器涡轮发电机

Exh g eco turbo GEN（exh g eco t/g）= exhaust gas economizer turbo generator 废气节能器涡轮发电机

Exh gas blr = exhaust gas boiler 废气锅炉

Exh gas eco = exhaust gas economizer 废气节能器

Exh gas eco fd（w）pump = exhaust gas economizer feed（water）pump 废气锅炉给水泵

Exh vent fan = exhaust ventilating fan 排风扇

Ext desuphtr = external desuperheater 外部减温器

F．O．addit tk = fuel oil additive tank 燃油添加柜

F．O．buff tk = fuel oil buffer tank 燃油缓冲柜

F．O．drain tk = fuel oil drain tank 燃油放泄柜

F．O．grav tk = fuel il gravity tank 燃油重力柜

F．O．purif = fuel oil purifier 燃油分水机

F．O．sludge tk = fuel oil sludge tank 燃油污油柜

F．O．tk = fuel oil tank 燃油舱

F d fan = forced draught fan 强迫抽风扇

F v cfw pump = fuel valve cooling fresh water pump 燃油阀冷却淡水泵

F v cool f w clr = fuel valve cooling fresh water cooler 燃油阀冷却淡水冷却器

F v cool f w tk = fuel valve cooling fresh water tank 燃油阀冷却淡水舱

F v cool O．clr = fuel valve cooling oil cooler 燃油阀冷却油冷却器

F v cool O．tk = fuel valve cooling oil tank 燃油阀冷却油柜

F w eject pump = fresh water generator ejector pump 海水淡化装置抽吸泵

F w gen = fresh water generator 海水淡化装置

F w hyd tk = fresh water hydrophore tank 淡水采样舱

F w press tk = fresh water pressure tank 淡水压力舱

F w pump = fresh water pump 淡水泵

F．O．addit pump = fuel oil additive pump 燃油添加泵

F．O．serv．pump = fuel oil service pump 燃油服务泵

F．O．sup．pump = fuel oil supply pump 燃油供应泵

F．O．trans pump = fuel oil transfer pump 燃油传输泵

F．O．boost pump = fuel oil booster pump 燃油增压泵

F．O．burn．pump = fuel oil burning pump 燃油燃烧泵

F．V．C．O．pump = fuel valve cooling oil pump 燃油阀冷却油泵

Fax = facsimile 传真机

Fd（w）htr = feed（water）heater 给水加热器

Fd（w）pump = feed（water）pump 给水泵

FD = forecastle deck 艏楼甲板

Fd w auto cont system = feed water automatic control system 给水自动控制系统

Fd w reg = feed water regulator 给水调节器

Fd（w）overflow tk = feed（water）overflow tank 给水溢流舱

Fd（w）tk = feed（water）tank 给水舱

Fdw filter tk = feed water filter tank 给水过滤舱

Ferrous sulfate inject pump = ferrous sulfate injection pump

硫酸铁喷射泵
Filter L. O. pump = filter lubricating oil pump　滤器滑油泵
Fire and ballast pump = fire and ballast pump　消防及压载泵
Fire and g s pump = fire and general service pump　消防及通用泵
Fire pump = fire pump　消防泵
Fish LT = fishing light　渔灯
Fl. LT. (F. L) = fluorescent light　荧光灯
Flame proof bhd LT = flame-proof bulkhead light　耐火壁灯
Flame proof cen LT = flame-proof ceiling light　耐火电池灯
Flood LT = flood light　强光灯
Flw mtr = flow meter　流量计
FM = frame　肋骨
FOP = fuel oil pump　燃油泵
FOT = fuel tank　燃油舱
FP = fire pump　消防泵
Fry pan = frying pan　煎锅
Fryer = fryer　油炸锅
FT = forepeak tank　艏尖舱
Fwd stern tube seal O. pump = forward stern tube sealing oil pump　艉轴前油封泵
Fwd stern tube seal O. clr = forward stern tube sealing oil cooler　前端艉轴密封油冷却器
FWP = fresh water pump　淡水泵
FWT = fresh water tank　淡水舱
G s pump = general service pump　通用泵
Gangway LT = gangway light　舷梯灯
Gauge bd = gauge board　测量板
Gear d eng = geared diesel engine　齿轮传动内燃机
Gen eng cfw clr = generator engine cooling fresh water cooler　发电机主机冷却淡水冷却器
Gen eng cfw pump = generator engine cooling fresh water pump　发电机主机冷却淡水泵
Gen eng csw pump = generator engine cooling seawater pump　发电机主机冷却海水泵
Gen eng l o sett tk = generator engine lubrication oil settling tank　发电机主机滑油沉淀柜
Gen eng l o stor tk = generator engine lubrication oil storage tank　发电机主机滑油储存柜
Gen eng l o sump tk = generator engine lubrication oil sump tank　发电机主机滑油底壳沉淀柜
Gen eng L. O. prim pump = generator engine lubricating oil priming pump　发电机主机滑油泵

Gen eng L. O. pump = generator engine lubricating oil pump　发电机主机滑油泵
Germicidal LT = germicidal light　杀菌灯
GOC = general operator's certificate　普通操作员证书
Gov. amp. Hyd. pump = governor amplifier hydraulic pump　调速机放大器液压泵
Graph pnl = graphic panel　绘图板
Grinder = grinder　磨光机
Guy winch = guy winch　稳索绞车
GV = gate valve　闸阀
Ham slicer = ham slicer　火腿切片机
Hamburg cust LT = Hamburg custom light　汉堡海关灯
Hand lamp = hand lamp　手灯
Harbour radio tel = harbour radio telephone　港内用无线电话
Heavy F. O. trans pump = heavy fuel oil transfer pump　重燃油输送泵
Heavy F. O. serv tk = heavy fuel oil service tank　重油日用柜
Heavy F. O. purifier = heavy fuel oil purifier　重油分水机
Heavy O. sett tk = heavy oil settling tank　重油沉淀柜
Heel pump = heeling pump　倾斜泵
HF = high frequency　高频
High press fd (w) htr = high pressure feed (water) heater　高压给水加热器
Hot plate = hot plate　热板
Hot w circ pump = hot water circulating pump　热水循环泵
HP = hopper plate　底边舱顶板
HP TURB. = high pressure turbine　高压涡轮机
HT = hopper tank　底边舱
Huge vessel LT = Huge vessel flashing light　巨轮闪光灯
Hydrazine inject pump = hydrazine injecting pump　联氨喷射泵
Hydro extractor = hydro extractor　抽湿机
I . G. gen cfw pump = inert gas generator cooling fresh water pump　惰性气体发生器冷却淡水泵
I . G. gen csw pump = inert gas generator cooling sea water pump　惰性气体发生器冷却海水泵
I M = induction motor　异步电动机
IB = inner bottom　内底
Ice cream freez = ice cream freezer　雪糕机
Ice maker = ice making machine　制冰机
ID = intermediate deck　中间甲板
Ig cont system = inert gas control system　惰性气体控制系统
Ig dk seal pump = inert gas deck seal pump　惰性气体甲板

密封泵
Ig fan = inert gas fan　惰性气体扇
Ig scrub = inert gas scrubber　惰性气体涤气机
Ig scrub sw pump = inert gas scrubber sea water pump　惰性气体海水泵
Ig w separ = inert gas water separator　惰性气体水分离器
IH = inner hull　内壳
IME = internally mounted equipment　内接设备
Incand LT. (I. L) = incandescent light　白炽灯
INMARSAT = International Maritime Satellite Communication System　国际海事卫星系统
INMARSAT = International Mobile Satellite Organization　国际移动卫星协调组织
Insp tk = inspection tank　检查用舱
IP TURB = intermediate pressure turbine　中压涡轮机
Iron = iron　熨铁
ITU = International Telecommunication Union　国际电信联盟
Jacket and piston cfw pump = jacket and piston cooling fresh water pump　缸套冷却海水泵
Jacket cfw clr = jacket cooling fresh water cooler　缸套冷却淡水冷却器
Jacket cfw pump = jacket cooling fresh water pump　缸套冷却淡水泵
Jb = junction box　接线盒
Keros tk = kerosene tank　煤油柜
L. O. clr circ pump = lubricating oil cooler circulating pump　滑油冷却器循环泵
L. O. drain tk = lubricating oil drain tank　滑油放泄柜
L. O. grav tk = lubricating oil gravity tank　滑油重力柜
L. O. reserve tk = lubricating oil reserve tank　滑油备用柜
L. O. residue tk = lubricating oil residue tank　滑油残渣柜
L. O. sett tk = lubricating oil settling tank　滑油沉淀柜
L. O. sludge tk = lubricating oil sludge tank　滑油渣油柜
L. O. stor tk = lubricating oil storage tank　滑油储存柜
L. O. sump tk = lubricating oil sump tank　滑油底壳沉淀柜
L. O. tk = lubricating oil tank　滑油柜
L. O. dirty tk = lubricating oil dirty tank　滑油污油柜
L p s g condr = low pressure steam generator condenser　低压蒸汽发生器用冷凝器
L p s g drain clr = low pressure steam generator drain cooler　低压蒸汽发生器放泄冷却器
L p stm gen = low pressure steam generator　低压蒸汽发生器

L. O. clr = lubricating oil cooler　滑油冷却器
L. O. prim. pump = lubricating oil priming pump　滑油启动泵
L. O. pump = lubricating oil pump　滑油泵
L. O. trans. pump = lubricating oil transfer pump　滑油输送泵
Lapp mach = lapping machine　包装机
Lath mach = lathing machine　板条机
LB = longitudinal bulkhead　纵舱壁
Lead acid batt. = lead acid battery　铅蓄电池
Lev gauge = level gauge　水位计
Life boat radio = portable life boat radio equipment　救生艇用便携式无线电设备
Light oil tk = light oil tank　轻油柜
Loop ant = loop antenna　环状天线
LOP = lubricating oil pump　滑油泵
LOT = lubrication oil tank　滑油舱
Low press fd (w) htr = low pressure feed (water) heater　低压给水加热器
Lp fd (w) htr drain pump = low pressure feed (water) heater drain pump　低压给水加热器疏水泵
Lp stm gen aux fd (w) pump = low pressure steam generator auxiliary feed (water) pump　低压蒸汽发生器辅助给水泵
Lp stm gen fd (w) pump = low pressure steam generator feed (water) pump　低压蒸汽发生器给水泵
LP TURB. = low pressure turbine　低压透平
LS = low stool　下凳座
LUT = local user terminal　本地用户终端
Main air comp = main air compressor　主空气压缩机
Main air ejector = main air ejector　主空气抽出器
Main air reserv = main air reservoir　主空气瓶
Main blr = main boiler　主锅炉
Main cfw pump = main cooling fresh water pump　主冷却淡水泵
Main circ pump = main circulating pump　主循环泵
Main condr = main condenser　主冷凝器
Main conds pump = main condensate pump　主凝水泵
Main csw pump = main cooling seawater pump　主冷却海水泵
MAIN D. Eng/main eng = main diesel engine　主内燃机
Main eng aux blwr = main engine auxiliary blower　主机抽风机
Main eng driven GEN = main engine driven generator　轴带发电机
Main eng driven gen = main engine driven generator　主机

驱动发电机

Main eng F. O. htr = main engine fuel oil heater　主机燃油加热器

Main eng protect sys = main engine protection system　主机保护系统

Main eng remo cont sys = main engine remote control system　主机遥控系统

Main fd（w）pump = main feed（water）pump　主给水泵

MAIN GAS TURB. = main gas turbine　主燃气涡轮机

Main gen = main generator　主发电机

Main GEN = main generator　主发电机

Main gen d eng = main generator diesel engine　主发电机内燃机

Main gen turb = main generator turbine　主发电机涡轮机

Main receiv = main receiver　主接收机

MAIN STM TURE. / MAIN TURE. = main steam turbine　主蒸汽涡轮机

Main transmt = main transmitter　主变压器

Manoeuv LT = manoeuvring light　操纵灯

Mast head LT = mast head light　大桅顶灯

MCC = mission control center　COSPAS-SARSAT 的任务控制中心

MD = main deck　主甲板

Meat slicer = meal slicer　肉切片机

Merc LT.（M. L）= mercury light　汞灯

MF = medium frequency　中频

Micro wave oven = micro wave oven　微波炉

Mill mach = milling machine　铣床

Mirror LT = mirror light　镜前灯

Miscall crane = miscellaneous crane　杂用吊

MMSI = maritime mobile service identify number　海事移动通信识别码

Moor winch = mooring winch　绞缆机

Morse sig LT = Morse signal light　莫斯灯

MSI = maritime safety information　海事安全信息

Mult. d. eng = multiple diesel engine　多用途内燃机

Nuc LT = not under command light　操纵不灵灯

Natr LT = natrium light　钠灯

Nav LT = navigation light　航行灯

Nav LT = navigation light indicator　航行灯指示器

NBDF = narrow band direct printing telegraph equipment　窄带直接打字电报设备

NBDP = narrow band direct printing telegraphy　窄带直接打印电报

NNSS = navy navigational satellite system　美国海军导航卫星系统

Observ tk = observation tank　观察舱

Odoriz inject pump = odorizing injecting pump　除味剂喷射泵

Oil bilge separ = oil bilge separator　油水分离器

Oper O. sunp tk = operating oil sump tank　操作油沉淀舱

Oper O. sup tk = operating oil supply tank　操作油供给舱

Oper O. tk = operating oil tank　操作油舱

Oper, O. pump = operating oil pump　操作油泵

Oven = baking oven　烘烤炉

Ovhd crane = overhead crene　天车（高架起重机）

Parabo ant = parabolic antenna　抛物面反射天线

PC = passenger cabin　客舱

Pend LT = pendant light　吊灯

Perm w ballast pump = permanent water ballast pump　永久水压载泵

PGV = pressure gauge valve　压力表阀

Piston cfw clr = piston cooling fresh water cooler　活塞冷却淡水冷却器

Piston cfw pump = piston cooling fresh water pump　活塞冷却淡水泵

PLB = personel locator beacon　人员定位标（陆上用）

Portable LT = portable light　移动灯

Pot w pump = potable water pump　移动式水泵

Potato peeler = potato peeler　土豆剥皮机

Prim blr fd（w）pump = primary boiler feed（water）pump　初级锅炉给水泵

Project = flood lighting projector　强光投影灯

Prop warn LT = propeller warning light　螺旋桨警示信号灯

Prov hand crane = provision handling crane　伙食吊

Prov ref alm = provision refrigerating chamber alarm　伙食冷藏库警报

Prov ref comp = provision refrigerating compressor　伙食制冷器压缩机

Prov. ref c w pump = provision refrigerator cooling fresh water pump　伙食冷藏用冷却淡水泵

PRV = pressure reducing valve　减压阀

Pub add = public addressor　公共广播

Purif F. O. htr = purifier fuel oil heater　分水机燃油加热器

Purif hot w tk = purifier hot water tank　分水机热水舱

Purif L. O. htr = purifier lubricating oil heater　分水机滑油加热器

Purif oper w tk = purifier operating water tank　分水机操作水舱

Purif. L. O. pump = purifier lubricating oil pump　分水机

滑油泵
Purifd F.O. tk = purified fuel oil tank　清净燃油柜
Purifd L.O. tk = purified lubricating oil tank　清净滑油柜
Quarant LT = quarantine light　检疫灯
Radar buoy = radar buoy　雷达浮标
Radio = broadcast radio receiver　无线电广播接收机
Radio buoy = radio buoy　无线电浮标
Radio teleg and tel equip = radio telegraph and telephone equipment　无线电报、电话设备
RCC = rescue co-ordination center　救助协调中心
Recap = receptacle　插座
Recap with sw = receptacle with switch　带开关的插座
Reduct gear L.O. grav tk = reduction gear lubricating oil gravity tank　减速装置滑油重力柜
Reduct gear L.O. sett tk = reduction gear lubricating oil settling tank　减速装置滑油沉淀柜
Reduct gear L.O. stor tk = reduction gear lubricating oil storage tank　减速装置滑油储存柜
Reduct gear L.O. sump tk = reduction gear lubricating oil sump tank　减速装置滑油油底壳沉淀柜
Reduct gear L.O. clr = reduction gear lubricating oil cooler　减速齿轮滑油冷却器
Reduct gear = reduction gear　减速齿轮
Reduct. gear L.O. pump = reduction gear lubricating oil pump　减速齿轮滑油泵
Ref brine pump = refrigerating brine pump　制冷卤水泵
Ref chamb LT = ref. chamber light indicator　冷藏室灯光指示器
Ref conin aln = refrigerating container alarm　冷藏集装箱警报
Refrigerator = electric refrigerator　电冰箱
Reliq plant cfw pump = reliquefaction plant cooling fresh water pump　再液化装置冷却淡水泵
Reliq plant csw pump = reliquefaction plant cooling seawater pump　再液化装置冷却海水泵
Reserve LT = reserve light　备用灯
Rest manoeuv LT = restricted manoeuvring light　操纵受限灯
Resv receiv = reserve receiver　备用接收机
Resv transmt = reserve transmitter　备用变压器
Rice washer = rice washer　淘米机
RL = red pilot lamp　红色导航灯
ROC = restricted operator's certificate　限用操作员证书
Rock. arm L.O. pump = rocker arm lubricating oil pump　摇臂滑油泵
RR = radio regulation　（国际电信联盟颁发的）无线电规则
Rud ang ind = rudder angle indicator　舵角指示器
RV = relief valve　溢流阀
S w hyd tk = sea water hydrophore tank　海水采样舱
S w press tk = sea water pressure tank　海水压力舱
S w serv pump = sea water service pump　海水总用泵
Safety link = safety link　安全环
Sal ind = salinity indicator　盐度指示器
Sanit pump = sanitary pump　卫生泵
SAR Convention = international convention on maritime search and rescue　国际海事搜救公约
Search LT = search light　搜寻灯
Self suppt ant = self support antenna　自支撑天线
Sewage pump = sewage pump　粪便泵
Sewage treat unit = sewage treatment unit　生活污水处理装置
Shaft gen (s/g) = shaft driven generator　轴带发电机
Shaper = shaping machine　刨床
Ship abandon alm = ship abandon alarm　弃船警报
Ship serv air comp = ship service air compressor　船舶杂用空气压缩机
Ship serv air reserve = ship service air reservoir　船舶杂用空气瓶
Side LT = side light　侧灯
Side thruster = side thruster　侧推器
Sig bell = signal bell　信号铃
SLP = sludge pump　污泥泵
Sludge pump = sludge pump　污泥泵
Smoke detect = smoke detector　烟探测器
Sod LT = sodium light　钠灯
Soil pump = soil pump　泥浆泵
Soot blwr = soot blower　吹灰器
Sound p tel = sound powered telephone　声力电话
Soup boiler = soup boiler　煲汤机
SP = side plating　舷侧板
Spray pump = spray pump　喷射泵
SSB transcv = SSB transceiver　SSB 收发两用机
SSV = sea suction valve　通海阀
ST = sewage holding tank　生活污水舱
ST = slop tank　污油舱
ST = stringer　纵向加强材
St Lawrence anchor LT = St Lawrence anchor light　圣·劳伦斯航道锚灯
St Lawrence dk LT = St Lawrence deck light　圣·劳伦斯航道甲板灯
St Lawrence range LT = St Lawrence range light　圣·劳伦

斯航道范围灯
St Lawrence sig LT = St Lawrence sea way signal light 圣·劳伦斯航道信号灯
Start air comp = starting air compressor 启动空气压缩机
Steer aim = steering gear alarm 舵机警报
Steer LT = steering light 操舵灯
Stereo = stereophonic gramophone 立体声录音机
Stern gear = stern gear 舵机
Stern LT = stern light 艉灯
Stern thruster = stern thruster 艉侧推器
Stern tube L.O. clr = stern tube lubricating oil cooler 艉轴管滑油冷却器
Stern tube L.O. grav tk = stern tube lubricating oil gravity tank 艉轴管滑油重力柜
Stern tube L.O. pump = stern tube lubricating oil pump 艉轴管滑油泵
Stm sep drum = steam separation drum 蒸汽分离管
STR = steering gear room 舵机舱
Strip pump = stripping pump 扫舱泵
Suez search LT = Suez canal searchlight 苏伊士运河搜寻灯
Suez sig LT = Suez canal signaling light 苏伊士运河信号灯
Sup vent fan = supply ventilating fan 送风扇
Suphtr = superheater 过热器
SV = safety valve 安全阀
Sw = switch 开关
SWP = sea water pump 海水泵
Sy L = synchronizing lamp 同步灯
Syn scope = synchroscope 同步指示器
Sync condr = synchronous condenser 同步电容
Talk bk sys = talkbeck system 复述系统
TB = transverse bulkhead 横舱壁
Thyr m = thyristor motor 硅整流马达
Timer = timer 计时器/定时器
Tk clean htr = tank cleaning heater 洗舱加热器
Tk clean pump = tank cleaning pump 舱室清洗泵
Vacuum pump = vacuum pump 真空泵
Toaster = toaster 烤面包机
Tofu mach = tofu machine 豆腐机
Topping winch = 顶部绞车
Tow LT = towing light 拖带灯
Trans = transformer 变压器
TS = tweendeck space 甲板间舱
TT = topsider tank 顶边舱
Turbo charg. L.O. pump = turbo changer lubricating oil pump 增压器滑油泵
Turbo charg clean w tk = turbo charger cleaning water tank 增压器清洁水舱
Turbo charg L.O. grav tk = turbo charger lubricating oil gravity tank 增压器滑油重力柜
Turbo charg L.O. stor tk = turbo charger lubricating oil storage tank 增压器滑油储存柜
Turbo charg L.O. sump tk = turbo charger lubricating oil sump tank 增压器滑油油底壳沉淀柜
Turbo charge L.O. clr = turbo charge lubricating oil cooler 增压器滑油冷却器
Turbo gen (t/g) = turbo generator 涡轮发电机
Turbo GEN = turbo generator 涡轮发电机
Turbo gen condr circ pump (t/g condr circ pump) = turbo generator condenser circulating pump 涡轮发电机冷凝循环泵
Turbo gen condr conds pump (t/g gen condr conds pump) = turbo generator condenser condensate pump 涡轮发电机冷凝器冷凝泵
Turbo gen. L.O. pump (t/g l o pump) = turbo generator lubricating oil pump 涡轮发电机滑油泵
Turn. gear = turning gear 盘车机
TV ant = TV antenna 电视机天线
TV = television receiver 电视接收机
TW = transverse web 强横肋骨
Unit clr = unit cooler 单元制冷器
Univ cook mach = universal cooking machine 多用炊事机
Univ mach = universal machine 多用途机床
US = upper stool 上凳座
VHF = very high frequency 甚高频
VHFtel = VHF radio telephone equipment 甚高频无线电设备
VIR = video tape recorder 磁带式录像记录机
Wall LT = wall light 壁灯
Wash mach = washing machine 洗衣机
Waste oil incin = waste oil incinerator 废油焚烧炉
Water boiler = water boiler 热水机
Water tk = water tank 水舱
Whip ant = whip antenna 鞭状天线
Windlass = windlass 锚机
Wire ant = wire antenna 金属天线
WL = white pilot lamp 白色导向灯
Work shop unit clr = work shop unit cooler 机工间冷却单元
Wrong op alm = wrong operation alarm 误操作警报
WT = wing tank 边舱
YACLant = YACL antenna YACL天线

附录 2　国际公约和规则专属名词解释
Appendix 2　Interpretation of exclusive terms in international conventions and regulations

2.1　海上安全
Table 2.1　Maritime safety

1	International convention for the safety of life at sea, 1974 (SOLAS 1974)	1974 年国际海上人命安全公约(SOLAS 1974)
2	International convention on load lines, 1966 (LL 1966)	1966 年国际载重线公约(LL 1966)
3	Special trade passenger ships agreement, 1971 (STP 1971)	1971 年特种业务客船协议(STP 1971)
4	International regulations for preventing collisions at sea, 1972 (COLREC 1972)	1972 年国际海上避碰规则(COLREC 1972)
5	International convention for safe containers, 1972 (CSC 1972)	1972 年国际安全集装箱公约(CSC 1972)
6	Convention on the international maritime satellite organization, 1976 (INMARSAT 1976)	1976 年国际海事卫星组织公约(INMARSAT 1976)
7	The Torremolinos international convention for the safety of fishing vessels, 1977 (SFV 1977)	1977 年托雷莫利诺斯国际渔船安全公约(SFV 1977)
8	International convention on standards of training, certification and watchkeeping for seafarers, 1978 (STCW 1978)	1978 年海员培训、发证和值班标准国际公约(STCW 1978)
9	International convention on maritime search and rescue, 1979 (SRC 1979)	1979 年国际海上搜寻与救助公约(SRC 1979)
10	International convention on standards of training, certification and watchkeeping for fishing vessel personnel, 1995 (STCW-F 1995)	1995 年渔船人员培训、发证和值班标准国际公约(STCW-F 1995)

2.2　海洋污染
Table 2.2　Marine pollution

1	International convention for prevention of pollution of the sea by oil, 1954 (OILPOL 1954)	1954 年国际防止海上油污染公约(OILPOL 1954)
2	International convention relating to intervention on the high seas in cases of pollution casualties, 1969 (INTERVENTION 1969)	1969 年国际干预公海油污染事件公约(INTERVENTION 1969)

续表 2.2

3	Convention on the prevention of marine pollution by dumping of wastes and other matter, 1972 (LDC 1972)	1972 年防止倾倒废弃物及其他物质污染海洋公约 (LDC 1972)
4	International convention for the prevention of pollution from ships, 1973, as modified by protocol of 1978 relating thereto (MARPOL 73/78)	经 1978 年议定书修正的 1973 年国际防止船舶造成污染公约 (MARPOL 73/78)
5	International convention on oil pollution preparedness, response and co-operation, 1990 (OPRC 1990)	1990 年国际油污染准备、响应和合作公约 (OPRC 1990)
6	Protocol on preparedness, response and co-operation to pollution incidents by hazardous and noxious substances, 2000 (HNS-OPRC 2000)	2000 年由危险和有毒物质引起的污染事件的准备、响应和合作的议定书 (HNS-OPRC 2000)
7	International convention on control of harmful anti-fouling system on ships, 2001 (AFS 2001)	2001 年国际控制船舶有害防污底系统公约 (AFS 2001)
8	International convention for the control and management of ships' ballast water and sediments, 2004 (BWM 2004)	2004 年国际船舶压载水及沉淀物控制和管理公约 (BWM 2004)
9	International convention for the safe and environmentally sound recycling of ships	国际安全与环境无害化拆船公约

2.3 责任与补偿
Table 2.3　Liability and compensation

1	International convention on civil liability for oil pollution damage, 1969 (CLC 1969)	1969 年国际油污损害民事责任公约 (CLC 1969)
2	International convention on the establishment of an international fund for compensation for oil pollution damage, 1971 (FUND 1971)	1971 年设立国际油污损害赔偿基金公约 (FUND 1971)
3	Athens convention relating to the carriage of passengers and their luggage by sea, 1974 (PAL 1974)	1974 年海上载运旅客及其行李的雅典公约 (PAL 1974)
4	Convention relating to civil liability in field of maritime carriage of nuclear material, 1971 (NUCLEAR 1971)	1971 年海上核材料运输民事责任公约 (NUCLEAR 1971)
5	Convention on limitation of liability for maritime claims, 1976 (LLMC 1976)	1976 年海上索赔责任限制公约 (LLMC 1976)
6	International convention on liability and compensation for damage in connection with the carriage of hazardous and noxious substances by sea, 1996 (HNS 1996)	1996 年与危险和有毒物质载运有关损害的责任和赔偿的国际公约 (HNS 1996)
7	International convention on civil liability for bunker oil pollution damage, 2001 (BUNKERS 2001)	2001 年国际燃油污染损害民事责任公约 (BUNKERS 2001)

2.4 其他
Table 2.4 Other subjects

1	Convention on facilitation of international maritime traffic, 1965 (FAL 1965)	1965 年国际便利海上运输公约(FAL 1965)
2	International convention on tonnage measurement of ships, 1969 (TONNAGE 1969)	1969 年国际船舶吨位丈量公约(TONNAGE 1969)
3	Convention for the suppression of unlawful acts against the safety of marine navigation, 1988 (SUA 1988)	1988 年制止危及海上航行安全的非法行为公约(SUA 1988)
4	Protocol for the suppression of unlawful acts against the safety of fixed platforms located on the continental shelf, 1988 (SUAPROT 1988)	1988 年制止危及位于大陆架上的固定平台安全的非法行为议定书(SUAPROT 1988)
5	International convention on salvage, 1989 (SALVAGE 1989)	1989 年国际打捞公约(SALVAGE 1989)

2.5 作为 SOLAS 之一部分的强制性规则和决议
Table 2.5 Codes and Resolutions made mandatory as a part of SOLAS

对于已作为 SOLAS 之一部分的规则,具有与公约相同的性质和特点,迄今作为 SOLAS 一部分的规则有 17 项,它们的缩写、名称和相对应的 SOLAS 章号见表 2.5(a)。

表 2.5(a) 作为 SOLAS 之一部分的强制性规则
Table 2.5(a) Codes made mandatory as a part of SOLAS

缩 写	名 称	对应的 SOLAS 章号
IS Code	国际海事组织文件中所有类型船舶的完整稳性规则 Code on Intact Stability for All Types of Ships Covered by IMO Instruments	Ⅱ-1
2008 IS Code	2008 年国际完整稳性规则 (该规则的 A 部分具有强制性,B 部分为推荐性,2010 年 7 月 1 日生效,替代 IS Code) International Code on Intact Stability, 2008	Ⅱ-1
Noise Levels	船上噪声级规则 (该规则以 MSC.337(91)决议通过,将于 2014 年 7 月 1 日生效,替代 A.468(Ⅻ)决议"船上噪声级规则") Code on Noise Levels on Board Ships	Ⅱ-1
FTP Code	国际实施耐火试验程序规则 International Code for Application of Fire Test Procedures	Ⅱ-2
FSS Code	国际消防安全系统规则 International Code for Fire Safety Systems	Ⅱ-2
LSA Code	国际救生设备规则 International Life-Saving Appliances Code	Ⅲ

续表 2.5(a)

缩　写	名　称	对应的SOLAS章号
Grain Code	国际安全散装运输谷物规则 International Code for the Safe Carriage of Grain in Bulk	Ⅵ
IMSBC Code	国际海运固体散装货物规则 ［该规则2011年1月1日生效,替代"固体散装货物安全操作规则"(Code of Safe Practice for Solid Bulk Cargoes,BC Code)］ International Maritime Solid Bulk Cargoes Code	Ⅵ、Ⅶ
IMDG Code	国际海运危险货物规则 International Maritime Dangerous Goods Code	Ⅶ
IBC Code	国际散装运输危险化学品船舶构造和设备规则"MSC.4(48)" International Code for the Construction and Equipment of Ships Carrying Dangerous Chemicals in Bulk	Ⅶ
IGC Code	国际散装运输液化气船舶构造和设备规则"MSC.5(48)" International Code for the Construction and Equipment of Ships Carrying Liquefied Gases in Bulk	Ⅶ
INF Code	国际船舶安全运输密封装辐射性核燃料,钚和强放射性废料规则 International Code for the Safe Carriage of Packaged Irradiated Nuclear Fuel, Plutonium and High-Level Radioactive Wastes in Flasks on board Ships	Ⅶ
ISM Code	国际安全管理规则 International Safety Management Code	Ⅸ
1994 HSC Code	1994年国际高速船安全规则 International Code of Safety for High-Speed Craft, 1994	Ⅹ
2000 HSC Code	2000年国际高速船安全规则 International Code of Safety for High-Speed Craft, 2000	Ⅹ
ISPS Code	国际船舶和港口设施保安规则 International Ship and Port Facility Security Code	Ⅺ
2011 ESP Code	2011年国际散货船和油船检验期间的强化检查方案规则 ［该规则以A.1049(27)决议通过,将于2014年1月1日生效,替代A.744(18)决议］ International Code on the Enhanced Programme of Inspections During Surveys of Bulk Carriers and Oil Tankers, 2011	Ⅺ-1

　　同样,对于已作为SOLAS之一部分的决议,具有与公约相同的性质和特点,迄今作为SOLAS一部分的决议有5项,它们的决议号、名称和相对应的SOLAS章号见表2.5(b)。

表 2.5(b) 作为 SOLAS 之一部分的强制性决议
Table 2.5(b) Resolutions made mandatory as a part of SOLAS

缩 写	名 称	对应的 SOLAS 章号
MSC.215(82) 决议	所有类型船舶专用海水压载舱和散货船双舷侧处所保护涂层性能标准(简称"压载舱 PSPC") Performance standard for protective coatings for dedicated seawater ballast tanks in all types of ships and double-side skin spaces of bulk carriers	II-1
MSC.287(87) 决议	国际散货船和油船目标型船舶建造标准(即 GBS) International Goal-Based Ship Construction Standards for Bulk Carriers and Oil Tankers	II-1
MSC.288(87) 决议	原油船货油舱保护涂层性能标准(简称"货油舱 PSPC") Performance standard for protective coatings for cargo oil tanks of crude oil tanker	II-1
MSC.289(87) 决议	原油船货油舱防腐蚀替代措施的性能标准(简称"货油舱防腐蚀替代措施") Performance standard for alternative means of corrosion protection for cargo oil tanks of crude oil tankers	II-1
A.744(18) 决议	散货船和油船检验期间的强化检查方案指南 (至 2014 年 1 月 1 日,该决议将由以 A.1049(27)决议通过的 2011 ESP Code 所替代) Guidelines on the enhanced programme of inspections during surveys of bulk carriers and oil tankers	XI-1

迄今作为 SOLAS 一部分的 17 项规则和 5 项决议(这一规则和决议数还会增加)与 SOLAS 构成了一个完整的系列文件。

表 2.6 MAPOL 公约各附则
Table 2.6 Annex to MAPOL Convention

1	Annex I -Regulations for the prevention by oil	附则 I——防止油类污染规则	附则 I 于 1983 年 10 月 2 日生效并在"防污公约"缔约国之间替代早期生效并经 1962 年和 1969 年修正的"1954 年国际防止海上油污公约"。附则 I 的修正案已由 MEPC 通过并已生效。 Annex I entered into force on 2 October 1983 and, as between the Parties to MARPOL73/78, supersedes the International Convention for the prevention of pollution of the sea by oil, 1954, as amended in 1962 and 1969, which was then in force. A number of amendments to Annex I have been adopted by the MEPC and have entered into force.
2	Annex II -Regulations for the control of pollution by noxious liquid substances in bulk	附则 II——控制散装有毒液体物质污染规则	为便于该附则的实施,原版本在 1985 年经 MEPC.16(22)决议的修正,这次修正涉及泵吸、管路和控制要求。MEPC 在其 22 届会议上根据 1978 年议定书第 11 条还决定,"各缔约国应自 1987 年 4 月 6 日起受经修正的'73/78 防污公约'附则 II 各项规定的约束"。 To facilitate implementation of the Annex, the original text underwent amendments in 1985, by resolution MEPC,16(22), in respect of pumping, piping and control requirements. At its twenty-second session, the MEPC also decided that, in accordance with article of the 1978 Protocol, "Parties shall be bound by the provisions of Annex II of MARPOL 73/78 as amended from 6 April 1987"

续表 2.6

3	Annex Ⅲ-Regulations for the prevention of pollution by harmful substances carried by sea in packaged form	附则Ⅲ——防止海运包装有害物质污染规则	附则Ⅲ于1992年7月1日生效。但早在该生效日期前,MEPC与海上安全委员会(MSC)即一致同意将该附则通过IMDG规则予以实施。IMDG规则包括由MSC编制的涉及海洋污染的修正案(修正案25-89),这些修正案自1991年1月1日起实施。Annex Ⅲ entered into force on 1 July 1992. However, long before this entry into force date, the MEPC, with the concurrence of the Maritime Safety Committee (MSC), agreed that the Annex should be implemented through the IMDG Code. The IMDG Code had amendments covering marine pollution prepared by the MSC(Amendment 25-89).
4	Annex Ⅳ-Regulations for the prevention of sewage pollution from ships	附则Ⅳ——防止船舶生活污水污染规则	附则Ⅳ于2003年9月27日起生效。其后的修正案已由MEPC通过并已生效。Annex Ⅳ entered into force on 27 September 2003. Subsequent amendments have been adopted by the MEPC and have entered into force.
5	Annex Ⅴ-Regulations for the prevention of garbage pollution from ships	附则Ⅴ——防止船舶垃圾污染规则	附则Ⅴ于1988年12月31日生效。其后的修正案已由MEPC通过并已生效。Annex Ⅴ entered into force on 31 December 1988. Subsequent amendments have been adopted by the MEPC and have entered into force.
6	Annex Ⅵ-Regulations for the prevention of air pollution from ships	附则Ⅵ——防止船舶造成空气污染规则	附则Ⅵ由"防污公约"缔约国国际会议于1997年9月通过,是"经1978年议定书修订的1973年国际防止船舶造成污染公约"的附件。附则Ⅵ于2005年5月19日生效。Annex Ⅵ is appended to the Protocol of 1997 to amend the International Convention for the Prevention of Pollution from ships,1973, as modified by the Protocol of 1978 relating thereto, which was adopted by the International Convention of Parties to the MARPOL Convention in September 1997. Annex Ⅵ entered into force on 19 May 2005.

表 2.7 议定书
Table 2.7 Protocol

1	1988 protocol "74 SOLAS PROT(HSSC)88"	该议定书导入新的证书制度,使SOLAS的关联证书的检查时期、有效期保持一致,俗称协调体系。这样一来,所有货船的条约证书有效期为5年,客船为1年
2	Harmonized system of survey and certification (HSSC)	检验和发证协调体系

表 2.8 要求船上具备的证书和文件
Table 2.8 Certificates and documents required to be carried on board ships

	(1) 适用所参照公约的所有船舶	名称	参照
1	International tonnage measurement certification	国际吨位证书	1969 国际吨位丈量公约
2	International load line certificate	国际载重线证书	1966 年国际载重线公约及 1988 年议定书
3	International load line exemption certificate	国际载重线免除证书	1966 年国际载重线公约及 1988 年议定书
4	Coating technical file	涂层技术文件	SOLAS 公约;MSC.215(82)决议
5	Construction drawings	建造图纸	SOLAS 公约;MSC/Circ.1135 通函
6	Ship construction drawings	船舶建造案卷	SOLAS 公约;MSC/Circ.1343 通函
7	Intact stability booklet	完整稳性手册	SOLAS 公约;国际载重线公约
8	Damage control plans and booklet	破损控制图和手册	SOLAS 公约;MSC.1/Circ.1245 通函
9	Minimum safe manning document	最低安全配员文件	SOLAS 公约
10	Fire safety training manual	消防安全培训手册	SOLAS 公约
11	Fire control plan/booklet	防火控制图/小册子	SOLAS 公约
12	Onboard training and drills record	船上培训和演习记录	SOLAS 公约
13	Fire safety operational booklet	消防安全操作手册	SOLAS 公约
14	Maintenance plans	维护保养计划	SOLAS 公约
15	Training manual	培训手册	SOLAS 公约
16	Nautical charts and mautical publications	海图和航海出版物	SOLAS 公约
17	"International Code of Signals and a cope of Volmme III of IMASAR Manual"	"国际信号规则"和 IAMSAR 手册第Ⅲ卷	SOLAS 公约
18	Records of navigational activities	航行活动的记录	SOLAS 公约
19	Manoeuvring booklet	操纵手册	SOLAS 公约

续表 2.8

20	Certificates for masters, officers or ratings	船长、高级船员或普通船员等级证书	1978 年 STCW 规则
21	Records of hours of rest	休息时间记录	STCW 规则;"1996 年海员工时和船舶配员公约";"IMO/ILO 海员船上工作安排表制定和船员工作或休息时间记录格式指南"
22	International oil pollution prevention certification	国际防止油污染证书	MARPOL 公约
23	Oil record book	油类记录簿	MARPOL 公约
24	Shipboard oil pollution emergency plan	船上油污染应急计划	MARPOL 公约
25	International sewage pollution prevention certification	国际防止生活污水污染证书	MARPOL 公约
26	Garbage management plan	垃圾管理计划	MARPOL 公约
27	Garbage record book	垃圾记录簿	MARPOL 公约
28	Voyage data recorder system-certificate of compliance	航行数据记录仪系统符合证书	SOLAS 公约
29	Cargo securing manual	货物系固手册	SOLAS 公约;MSC.1/Circ.1353 通函
30	Document of compliance	符合证明	SOLAS 公约;ISM 规则
31	Safety management certificate	安全管理证书	SOLAS 公约;ISM 规则
32	International Ship Security Certificate (ISSC) or Ihterim International Ship Security Certificate	国际船舶保安证书或临时国际船舶保安证书	SOLAS 公约;ISPS 规则
33	Ship security plan and associated records	国际船舶保安计划和相关记录	SOLAS 公约;ISPS 规则
34	Continuous synopsis record (CSR)	连续概要记录	SOLAS 公约
35	International anti-fouling system certificate	国际防污底系统证书	AFS 公约
36	Declaration on anti-fouling system	防污底系统声明	AFS 公约
37	International air pollution prevention certificate	国际防止空气污染证书	MARPOL 公约
38	Ozone depleting substances record book	消耗臭氧物质记录簿	MARPOL 公约
39	Fuel oil changeover procedure and log-book (record of fuel changeover)	燃油转换程序和航海日志(燃油转换记录)	MARPOL 公约
40	Manufacturer's operating manual for incinerators	制造厂商的焚烧炉操作手册	MARPOL 公约
41	Bunker delivery note and representative sample	燃油交付单和代表样品	MARPOL 公约

续表 2.8

42	Technical file	技术案卷	NO_x 技术规则
43	Record book of engine parameters	发动机技术参数记录簿	NO_x 技术规则
44	Exemption certificate *	免除证书 *	SOLAS 公约
45	LRIT conformance test report	LRIT 符合性试验报告	SOLAS 公约；MSC.1/Circ 1307 通函
	(2)除上述(1)中的证书外,客船还应持有下列文件		
46	Passenger ship safety certificate	客船安全证书	
47	Special trade passenger ship safety certificate; Special trade passenger ship space certificate	特种业务客船安全证书,特种业务客船舱室证书	1971 年特种业务客船协定,1973 年特种业务客船舱室要求议定书
48	Search and rescue cooperation plan	搜救合作计划	SOLAS 公约
49	List of operational limitations	操作限制清单	SOLAS 公约
50	Decision support system for marsters	船长决策支持系统	SOLAS 公约
	(3)除上述(1)中的证书外,货船还应持有下列文件		
51	Cargo ship safety construction certification	货船构造安全证书	1974 年 SOLAS 公约；1988 年 SOLAS 公约议定书
52	Cargo ship safety equipment certification	货船设备安全证书	1974 年 SOLAS 公约；1988 年 SOLAS 公约议定书
53	Cargo ship safety radio certification	货船无线电安全证书	1974 年 SOLAS 公约；1988 年 SOLAS 公约议定书
54	Cargo ship safety certification	货船安全证书	1988 年 SOLAS 公约议定书
55	Document of authorization for the carriage of grain	谷物载运的批准文件	1974 年 SOLAS 公约；国际散装谷物安全载运规则
56	Ceritificate of insurance or other financial security in respect of civil liability for oil pollution damage (for each ship carrying more than 2 000 tons of oil in bulk as cargo)	关于油污损害民事赔偿责任的保险或其他财产保证的证书(对每艘载运 2 000 t 以上散装货油的船舶)	1969 年 CLC 公约
57	Ceritificate of insurance or other financial security in respect of civil liability for bunker oil pollution damage (for each ship of greater than 1 000 gross tonnage)	关于燃油油污损害民事赔偿责任的保险或其他财产保证的证书(对每艘大于 1 000 GT 的船舶)	2001 年燃油公约
58	Ceritificate of insurance or other financial security in respect of civil liability for oil pollution damage (for each ship carrying more than 2 000 tons of oil in bulk as cargo)	关于油污损害民事赔偿责任的保险或其他财产保证的证书(对每艘载运 2 000 t 以上散装货油的船舶)	1992 年 CLC 公约

续表 2.8

59	Enhanced survey report file	加强检验报告案卷	1974 年 SOLAS 公约
60	Record of oil discharge monitoring and control system for the last ballast voyage	最近一次压载航行的排油监控系统记录	MARPOL 公约
61	Oil discharge monitoring and control (ODMC) operational manual	排油监控操作手册	MARPOL 公约
62	Cargo information	货物资料	1974 年 SOLAS 公约, MSC/Circ. 665 通函
63	Ship construrure access manual	船舶结构通道手册	1974 年 SOLAS 公约
64	Bulk carrier booklet	散货船手册	1974 年 SOLAS 公约；散货船安全装卸操作规则（BLU 规则）
65	Crude oil washing operation and equipment manual (COW manual)	原油洗舱操作与设备手册（COW 手册）	MARPOL 公约
66	Condition assessment scheme (CAS) statement of compliance. CAS final report and review record	状态评估计划（CAS）符合证明；CAS 最终报告和审核记录	MARPOL 公约；MEPC 决议
67	Subdivision and stability information	分舱和稳性资料	MARPOL 公约
68	VOC management plan	VOC 管理计划	MARPOL 公约
	（4）除上述(1)和(3)中的证书外,适用时,任何散装运输有毒液体化学品物质的船舶,还应持有下列文件		
69	International pollution prevention certificate for the carriage of noxious liquid substances in bulk (NLS certificate)	国际防止散装运输有毒液体化学品物质污染证书（NLS 证书）	MARPOL 公约
70	Cargo record book	货物记录簿	MARPOL 公约
71	Procedures and arrangement manual (P&A manual)	程序和布置手册（P&A 手册）	MARPOL 公约
72	Shipboard marine pollution emergency plan for noxious liquid substances	船上有毒液体物质污染海洋应急计划	MARPOL 公约
	（5）除上述(1)和(3)中的证书外,适用时,任何化学品液货船,还应持有下列文件		
73	Certificate of fitness for the carriage of dangerous chemicals in bulk	散装运输危险化学品适装证书	BCH 规则
74	International certificate of fitness for the carriage of dangerous chemicals in bulk	国际散装运输危险化学品适装证书	IBC 规则

续表 2.8

	（6）除上述（1）和（3）中的证书外,适用时,任何气体运输船,还应持有下列文件		
75	Certificate of fitness for the carriage of liquefied cases in bulk	散装运输液化气体适装证书	GC 规则
76	International certificate of fitness for the carriage of liquefied cases in bulk	国际散装运输液化气体适装证书	IGC 规则
	（7）除上述（1）、（2）或（3）中的证书外,适用时,高速船,还应持有下列文件		
77	High-speed craft safety certificate	高速船安全证书	1974 年 SOLAS 公约;HSC 规则
78	Permit to operate high-speed craft	高速船营运许可证书	HSC 规则
	（8）除上述（1）、（2）或（3）中的证书外,适用时,任何载运危险货物的船舶,还应持有下列文件		
79	Document of compliance with the special requirement for ships carrying gangerous goods	对装运危险货物船舶特殊要求的符合证明	1974 年 SOLAS 公约
	（9）除上述（1）,（2）或（3）中的证书外,适用时,任何载运包装危险货物的船舶,还应持有下列文件		
80	Dangerous goods manfest or stowage plan	危险货物舱单或积载图	1974 年 SOLAS 公约;MARPOL 公约
	（10）除上述（1）,（2）或（3）中的证书外,适用时,载运 INF 货物的船舶,还应持有下列文件		
81	International certificate of fitness for the carriage of INF cargo	国际装运 INF 货物适装证书	1974 年 SOLAS 公约;INF 规则
	（11）除上述（1）,（2）或（3）中的证书外,适用时,任何核能船舶,还应持有下列文件		
82	Nuclear cargo ship safety certificate or nuclear passenger ship safety certificate, in place of the cargo ship safety certificate as appropriate	核能货船安全证书或核能客船安全证书,代替货船安全证书或客船安全证书	1974 年 SOLAS 公约
	其他非强制性的噪声和文件		

续表2.8

83	Special purpose ships	特种用途船舶	
	Special purpose ships safety certificate	特种用途船舶安全证书	2008年SPS规则;1974年SOLAS公约;1988年议定书
84	Offshore support vessel	近海供应船	
	Offshore support vessel document of compliance	近海供应船符合证明	MSC.235(82)决议
85	Certificate of fitness for offshore support vessel	近海供应船适装证书	A673(16)决议;MARPOL公约
86	Diving system	潜水系统	
	Diving system safety certificate	潜水系统安全证书	A.536(13)决议
87	Dynamically supported craft	动力支承船舶	
	Dynamically supported craft construction and equipment certificate	动力支承船舶构造和设备证书	A.373(X)决议
88	Mobile offshore drilling units	海上移动式钻井平台	
	Mobile offshore drilling units safety certificate	海上移动式钻井平台安全证书	A.414(XI)决议等;2009年MODU规则
89	Wing-in-ground(WIG)craft	地效翼船	
	Wing-in-ground(WIG)craft safety certificate	地效翼船安全证书	MSC/Circ.1054通函
90	Permit to operate WIG craft	地效翼船营运许可证书	MSC/Circ.1054通函
91	Noise levels	噪声等级	
	Noise survey report	噪声检验报告	A.468(XII)决议

* **Exemption certification 免除证书** 其种类很多,各船旗国对免除SOLAS公约中的某种可免除的设备的前提条件也各不相同。较常见的免除证书是每次干货船货舱中的固定式灭火装置(fixed fire fighting equipment, FFFE)。

附录3 不适用 IBC 规则的货物清单
Annex 3　List of products to which the IBC Code does not apply

序号	货物名称 Product name	污染类别 Pollution Category
1	Acetone　丙酮	Z
2	Alcoholic beverages, n.o.s.　含酒精饮料, n.o.s.	Z
3	Apple juice　苹果汁	OS
4	n-Butyl alcohol　正-丁醇	Z
5	sec-Butyl alcohol　仲-丁醇	Z
6	Calcium nitrate solution（50% or less）　硝酸钙溶液（50%或以下）	Z
7	Clay slurry　黏土泥浆	OS
8	Coal slurry　煤泥浆	OS
9	Diethylene glycol　二甘醇	Z
10	Ethyl alcohol　乙醇	Z
11	Ethylene carbonate　碳酸乙烯酯	Z
12	Glucose solution　葡萄糖溶液	OS
13	Glycerine　甘油	Z
14	Hexamethylenetetramine solutions　乌洛托品六胺溶液	Z
15	Hexylene glycol　己二醇	Z
16	Hydrogenated starch hydrolysate　氢化淀粉水解液	OS
17	Isopropyl alcohol　异丙醇	Z
18	Kaolin slurry　高岭土浆	OS
19	Lecithin　卵磷脂	OS
20	Magnesium hydroxide slurry　氢氧化镁浆	Z
21	Maltitol solution　麦芽醇溶液	OS
22	N-Methylglucamine solution（70% or less）　葡甲胺溶剂（70%或以下）	Z
23	Methyl propyl ketone　甲基丙基甲酮	Z
24	Molasses　糖蜜	OS
25	Noxious liquid,（11）n.o.s.（trade name…. contains…）Cat. Z　有毒液体,（11）n.o.s.（商品名…, 包含…）Cat. Z	Z

续表

序号	货物名称 Product name	污染类别 Pollution Category
26	Non-noxious liquid, (12) n.o.s. (trade name…, contains…) Cat. OS 非有毒液体, (12) n.o.s. (商品名……, 包含……) Cat. OS	OS
27	Orange juice (concentrated) 橘子汁(浓缩的)	OS
28	Orange juice (not concentrated) 橘子汁(非浓缩的)	OS
29	Polyaluminium chloride solution 聚氯化多铝溶液	Z
30	Polyglycerin, sodium salt solution (containing less than 3% sodium hydroxide) 甘油聚合物, 钠盐溶液(含3%以下氢氧化钠)	Z
31	Potassium formate solution 甲酸盐溶剂	Z
32	Propylene carbonate 碳酸丙烯	Z
33	Propylene glycol 丙二醇	Z
34	Sodium acetate solutions 乙酸钠溶液	Z
35	Sodium sulphate solution 硫酸钠溶液	Z
36	Sorbitol solution 山梨(糖)醇溶液	OS
37	Sulphonated polyacrylate solution 磺酸酯聚丙烯酸酯溶液	Z
38	Tetraethyl silicate monomer/oligomer (20% in ethanol) 四乙基硅酸单体/低聚体(20%乙醇溶剂)	Z
39	Triethylene glycol 三乙二醇	Z
40	Vegetable protein solution (hydrolysed) 植物蛋白溶液(水解)	OS
41	Water 水	OS

附录4 危险化学品清单
Annex 4 Dangerous chemicals list

Acetic acid 乙酸
Acetic anhydride 醋酐
Acetochlor 乙草胺
Acetone cyanohydrin 丙酮氰醇
Acetonitrile 乙腈
Acetonitrile (Low purity grade) 乙腈（低纯度）
Acid oil mixture from soyabean, corn (maize) and sunflower oil refining 从大豆、玉米及精炼向日葵油提取的酸性油混合物
Acrylamide solution (50% or less) 丙烯酰胺溶液（50%或以下）
Acrylic acid 丙烯酸
Acrylonitrile 丙烯腈
Acrylonitrile-Styrene copolymer dispersion in polyether polyol 聚醚多元醇分散体中的丙烯腈-苯乙烯共聚物
Adiponitrile 己二腈
Alachlor technical (90% or more) 甲草胺工艺（90%或以上）
Alcohol (C9-C11) poly (2.5-9) ethoxylate 乙醇(C9-C11)聚(2.5-9)乙氧基化物
Alcohol (C6-C17) (secondary) poly(3-6) ethoxylates 乙醇(C6-C17)（仲）聚(3-6)乙氧基化物
Alcohol (C6-C17) (secondary) poly(7-12) ethoxylates 乙醇(C6-C17)（仲）聚(7-12)乙氧基化物
Alcohol (C12-C16) poly (1-6) ethoxylates 乙醇(C12-C16)聚(1-6)乙氧基化物
Alcohol (C12-C16) poly (20+) ethoxylates 乙醇(C12-C16)聚(20+)乙氧基化物
Alcohol (C12-C16) poly (7-19) ethoxylates 乙醇(C12-C16)聚(7-19)乙氧基化物
Alcohols (C13+) 乙醇(C13+)
Alcohols (C8-C11), primary, linear and essentially linear 乙醇(C8-C11)，直链和主要直链的
Alcohols (C12-C13), primary, linear and essentially linear 乙醇(C12-C13)，直链和主要直链的
Alcohols (C14-C18), primary, linear and essentially linear 乙醇(C14-C18)，直链和主要直链的
Alkanes (C6-C9) 烷烃(C6-C9)
Iso-and cyclo-alkanes (C10-C11) 异烷烃与环烷(C10-C11)
Iso-and cyclo-alkanes (C12+) 异烷烃与环烷(C12+)
n-Alkanes (C10+) n-烷烃(C10+)
Alkaryl polyethers (C9-C20) 烷基聚醚(C9-C20)
Alkenoic acid, polyhydroxy ester borated 烷烯基酸,多羟基硼酸酯
Alkenyl (C11+) amide 烯基(C11+)胺
Alkenyl (C16-C20) succinic anhydride 烯基(C16-C20)琥珀酸酐
Alkyl acrylate-vinylpyridine copolymer in toluene 甲苯中烷基丙烯酸酯-乙烯基吡啶共聚物
Alkylaryl phosphate mixtures (more than 40% diphenyl tolyl phosphate, less than 0.02% ortho-isomers) 烷芳基磷酸酯混合物(二苯甲基磷酸酯40%以上,邻位异构物0.02%以下)
Alkylated (C4-C9) hindered phenols 烷化(C4-C9)受阻酚
Alkylbenzene, alkylindane, alkylindene mixture (each C12-C17) 烷基苯,烷基二氢茚,烷基茚混合物(各C12-C17)
Alkyl benzene distillation bottoms 烷基苯蒸馏物
Alkylbenzene mixtures (containing at least 50% of toluene) 烷基苯混合物(包含至少50%甲苯)
Alkyl (C3-C4) benzenes 烷基(C3-C4)苯
Alkyl (C5-C8) benzenes 烷基(C5-C8)苯
Alkyl(C9+) benzenes 烷基(C9+)苯
Alkyl (C11-C17) benzene sulphonic acid 烷基(C11-C17)苯磺酸
Alkylbenzene sulphonic acid, sodium salt solution 烷基苯磺酸,钠盐溶液
Alkyl (C12+) dimethylamine 烷基(C12+)二甲胺
Alkyl dithiocarbamate (C19-C35) 烷基二硫代氨基甲酸盐(C19-C35)
Alkyldithiothiadiazole (C6-C24) 烷基二硫代噻二唑(C6-C24)
Alkyl ester copolymer (C4-C20) 烷基酯共聚物(C4-C20)
Alkyl (C8-C10)/(C12-C14):(40% or less/60% or more) polyglucoside solution (55% or less) 烷基(C8-C10)/

（C12-C14）（40% 或以下/60% 或以上）聚葡糖苷溶液（55% 或以下）

Alkyl（C8-C10）/（C12-C14）：（60% or more/40% or less）polyglucosidesolution（55% or less） 烷基（C8-C10）/（C12-C14）（60% 或以上/40% 或以下）聚葡糖苷溶液（55% 或以下）

Alkyl（C7-C9）nitrates 硝酸烷基（C7-C9）酯

Alkyl（C7-C11）phenol poly（4-12）ethoxylate 聚（4-12）乙氧化烷基（C7-C11）酚

Alkyl（C8-C40）phenol sulphide 烷基（C8-C40）酚硫化物

Alkyl（C8-C9）phenylamine in aromatic solvents 烷基（C8-C9）芳香剂中的苯胺

Alkyl（C9-C15）phenyl propoxylate 烷基（C9-C15）苯基丙氧基化物

Alkyl（C8-C10）/（C12-C14）：（50%/50%）polyglucoside solution（55% or less） 烷基（C8-C10）/（C12-C14）：（50%/50%）聚葡糖苷溶液（55% 或以下）

Alkyl（C12-C14）polyglucoside solution（55% or less） 烷基（C12-C14）聚葡糖苷溶液（55% 或以下）

Alkyl（C8-C10）polyglucoside solution（65% or less） 烷基（C8-C10）聚葡糖苷溶液（65% 或以下）

Alkyl（C10-C20, saturated and unsaturated）phosphate 烷基（C10-C20, 饱和及不饱和）亚磷酸盐

Alkyl（C18+）toluenes 烷基（C18+）甲苯

Alkyltoluenesulphonic acid, calcium salts 烷基甲基苯磺酸, 钙盐

Alkyl sulphonic acid ester of phenol 酚的烷基磺酸酯

Allyl alcohol 丙烯醇

Allyl chloride 丙烯基氯

Aluminium sulphate solution 硫酸铝溶液

2-(2-Aminoethoxy)ethanol 2-(2-氨基乙氧基)乙醇

Aminoethyldiethanolamine/Aminoethylethanolamine solution 氨基乙荃二乙醇胺/氨基乙荃乙醇胺溶液

Aminoethyl ethanolamine 氨乙基乙醇胺

N-Aminoethylpiperazine N-氨乙基哌嗪

2-Amino-2-methyl-1-propanol 2-氨基-2-甲基-1-丙醇

Ammonia aqueous（28% or less） 氨水（28% 或以下）

Ammonium hydrogen phosphate solution 磷酸氢二铵溶液

Ammonium lignosulphonate solutions 铵磺化盐溶液

Ammonium nitrate solution（93% or less） 硝酸铵溶液（93% 或以下）

Ammonium polyphosphate solution 多磷酸铵溶液

Ammonium sulphate solution 硫酸铵溶液

Ammonium sulphide solution（45% or less） 硫化铵溶液（45% 或以下）

Ammonium thiosulphate solution（60% or less） 硫代硫酸铵溶液（60% 或以下）

Amyl acetate（all isomers） 乙酸戊酯（所有异构体）

n-Amyl alcohol 正戊醇

Amyl alcohol, primary 伯戊醇

sec-Amyl alcohol 仲戊醇

tert-Amyl alcohol 叔戊醇

tert-Amyl methyl ether 叔戊醇甲基醚

Aniline 苯胺

Aryl polyolefins（C11-C50） 芳基聚烯烃（C11-C50）

Aviation alkylates（C8 paraffins and iso-paraffins BPT 95～120 ℃） 航空烷基化汽油（C8 烷属烃及异构烷烃沸点 95～120 ℃）

Barium long chain（C11-C50）alkaryl sulphonate 钡长链（C11-C15）烷芳基磺酸酯

Benzene and mixtures having 10% benzene or more（i） 苯和含 10% 或以上苯的混合物（i）

Benzene sulphonyl chloride 苯磺酰氯

Benzenetricarboxylic acid, trioctyl ester 苯三酸, 三辛酯

Benzyl acetate 乙酸苄酯

Benzyl alcohol 苄甲醇

Benzyl chloride 苄基氯

Brake fluid base mix：Poly(2-8)alkylene（C2-C3）glycols/Polyalkylene（C2-C10）glycols monoalkyl（C1-C4）ethers and their borate esters 制动液混合物：聚(2-8)亚烃基（C2-C3）乙二醇/聚亚烃基（C2-C10）糖醇单烷基（C1-C4）乙醚及其硼酸盐

Bromochloromethane 溴氯甲烷

Butene oligomer 丁烯低聚物

Butyl acetate（all isomers） 乙酸丁酯（所有异构体）

Butyl acrylate（all isomers） 正丙烯酸丁酯（所有异构体）

tert-Butyl alcohol 叔丁醇

Butylamine（all isomers） 丁胺（所有异构体）

Butylbenzene（all isomers） 丁苯（所有异构体）

Butyl benzyl phthalate 邻苯二甲酸丁苄酯

Butyl butyrate（all isomers） 丁酸丁酯（所有异构体）

Butyl/Decyl/Cetyl/Eicosyl methacrylate mixture 乙基/癸基/十六烷基/二十烷基异丁烯酸混合物

Butylene glycol 丁二醇

1,2-Butylene oxide 1,2-环氧乙烷

n-Butylether 正丁醚

Butyl methacrylate 甲基丙烯酸丁酯

n-Butylpropionate 丙酸正丁酯

Butyraldehyde（all isomers） 丁醛（所有异构体）

Butyric acid 丁酸

gamma-Butyrolactone　γ-丁内酯
Calcium alkyl（C10-C28）salicylate　烷基（C10-C28）水杨酸钙
Calcium carbonate slurry　碳酸钙结晶浆液
Calcium hydroxide slurry　氢氧化钙浆液次氯酸钙溶液（15% 或以下）
Calcium hypochlorite solution（15% or less）　次氯酸钙溶液（15% 或以下）
Calcium hypochlorite solution（more than 15%）　次氯酸钙溶液（15% 以上）
Calcium lignosulphonate solutions　木素磺酸钙溶液
Calcium long-chain alkaryl sulphonate（C11-C50）　长链烷基磺酸钙（C11-C50）
Calcium long-chain alkyl（C5-C10）phenate　长链烷基（C5-C10）酚盐钙
Calcium long-chain alkyl（C11-C40）phenate　长链烷基（C11-C40）酚盐钙
Calcium long-chain alkyl phenate sulphide（C8-C40）　长链烷基酚盐硫化物钙（C8-C40）
Calcium long-chain alkyl salicylate（C13+）　长链烷基水杨酸钙（C13+）
Calcium long-chain alkyl（C18-C28）salicylate　长链烷基水杨酸钙（C18-C28）
Calcium nitrate/Magnesium nitrate/Potassium chloride solution　硝酸钙/硝酸镁/氯化钾溶液
epsilon-Caprolactam（molten or aqueous solutions）　ε-己内酰胺
Carbolic oil　酚油
Carbon disulphide　二硫化碳
Carbon tetrachloride　四氯化碳
Cashew nut shell oil（untreated）　漆树果壳油（未处理）
Castor oil　蓖麻油
Cetyl/Eicosyl methacrylate mixture　甲基丙烯酸十六～二十甘酯
Chlorinated paraffins（C10-C13）　氯化石蜡（C10-C13）
Chlorinated paraffins（C14-C17）（with 50% chlorine or more，and less than 1% C13 or shorter chains）　氯化石蜡（C14-C17）（含50%的氯或以上，及小于1%的C13 或短链）
Chloroacetic acid（80% or less）　氯乙酸（80%或以下）
Chlorobenzene　氯苯
Chloroform　氯仿
Chlorohydrin（crude）　（粗）氯乙醇
4-Chloro-2-methylphenoxyacetic acid，dimethylamine salt solution　4-氯-2-甲基苯氧基酸，二甲铵盐溶液
o—Chloronitrobenzene　邻-氯硝基苯

1-（4-Chlorophenyl）-4，4-dimethyl-pentan-3-one 2-or 3-Chloropropionic acid
1-(4-氯苯基)-4,4-二甲基-戊-3-单2-或 3-氯丙酸
Chlorosulphonic acid　氯磺酸
m-Chlorotoluene　间-氯甲苯
o-Chlorotoluene　邻-氯甲苯
p-Chlorotoluene　对-氯甲苯
Chlorotoluenes（mixed isomers）　氯甲苯(混有异构体)
Choline chloride solution　胆碱盐酸盐溶液
Citric acid（70% or less）　柠檬酸（70%或以下）
Coal tar　煤焦油
Coal tar naphtha solvent　煤焦油石脑油溶剂
Coal tar pitch（molten）　煤焦油沥青（熔化的）
Cocoa butter　可可油
Coconut oil　椰子油
Coconut oil fatty acid　椰子油脂肪酸
Coconut oil fatty acid methyl ester　椰子油脂肪酸甲酯
Copper salt of long chain（C17+）alkanoic acid　长链（C17+）烷基铜盐
Corn Oil　玉米油
Cotton seed oil　棉籽油
Creosote（coal tar）　杂酚油（煤焦油）
Cresols（all isomers）　甲酚（所有异构体）
Cresylic acid，dephenolized　甲酚基酸，脱酚
Cresylic acid，sodium salt solution　甲酚基酸，钠盐溶液
Crotonaldehyde　巴豆醛
1,5,9-Cyclododecatriene　1,5,9-环十二碳三烯
Cycloheptane　环庚烷
Cyclohexane　环己烷
Cyclohexanol　环己醇
Cyclohexanone　环己酮
Cyclohexanone，Cyclohexanol mixture　环己酮,环己醇混合物
Cyclohexyl acetate　乙酸环己酯
Cyclohexylamine　环己胺
1,3-Cyclopentadiene dimer（molten）　1,3-环戊二烯二聚物(熔化的)
Cyclopentane　环戊烷
Cyclopentene　环戊烯
P-Cymene　对-散花烃
Decahydronaphthalene　十氢化萘
Decanoic acid　癸酸
Decene　癸烯
Decyl/Dodecyl/Tetradecyl alcohol mixture　癸醇/十二（烷）基醇/十四（烷）基醇混合物
Decyl acrylate　丙烯酸癸酯

Decyl alcohol (all isomers)　癸醇(所有异构体)
Decyloxytetrahydrothiophene dioxide　癸基氧四氢噻吩
Diacetone alcohol　双丙酮醇
Dialkyl (C8-C9) diphenylamines　二烃基(C8-C9)二苯胺
Dialkyl (C7-C13) phthalates　二烃基(C7-C13)邻苯二甲酸酯
Dibromomethane　二溴甲烷
Dibutylamine　二丁胺
Dibutyl hydrogen phosphonate　二丁基磷酸氢盐
2,6-Di-tert-butylphenol　2,6-二-叔-丁基苯酚
Dibutyl phthalate　邻苯二甲酸二丁酯
Dichlorobenzene (all isomers)　二氯(代)苯(所有异构体)
3,4-Dichloro-1-butene　3,4-二氯-1-丁烯
1,1-Dichloroethane　1,1-二氯乙烷
Dichloroethyl ether　二氯乙醚
1,6-Dichlorohexane　1,6-二氯己烷
2,2-Dichloroisopropyl ether　2,2-二氯异丙醚
Dichloromethane　二氯甲烷
2,4-Dichlorophenol　2,4-二氯酚
2,4-Dichlorophenoxyacetic acid, diethanolamine salt solution　2,4-二氯苯氧基乙酸,二乙醇胺盐溶液
2,4-Dichlorophenoxyacetic acid, dimethylamine salt solution (70% or less)　2,4-二氯苯氧基乙酸,二甲胺盐溶液(70%或以下)
2,4-Dichlorophenoxyacetic acid, triisopropanolamine salt solution　2,4-二氯苯氧基乙酸,三异丙醇胺盐溶液
1,1-Dichloropropane　1,1-二氯丙烷
1,2-Dichloropropane　1,2-二氯丙烷
1,3-Dichloropropene　1,3-二氯丙烯
Dichloropropene/Dichloropropane mixtures　二氯丙烯/二氯丙烷混合物
2,2-Dichloropropionic acid　2,2-二氯丙酸
Diethanolamine　二乙醇胺
Diethylamine　二乙胺
Diethylaminoethanol　二乙胺基乙醇
2,6-Diethylaniline　2,6-二乙基胺
Diethylbenzene　二乙苯
Diethylene glycol dibutyl ether　二甘醇二丁醚
Diethylene glycol diethyl ether　二甘醇二乙醚
Diethylene glycol phthalate　邻苯二甲酸二甘醇酯
Diethylenetriamine　二乙撑三胺
Diethylenetriaminepentaacetic acid, pentasodium salt solution　二亚乙基三胺五乙酸,五钠盐溶液
Diethyl ether　二乙醚
Di-(2-ethylhexyl) adipate　二(2-乙基己基)己二酸酯

Di-(2-ethylhexyl) phosphoric acid　二(2-乙基己基)磷酸
Diethyl phthalate　邻苯二甲酸二乙酯
Diethyl sulphate　硫酸二乙酯
Diglycidyl ether of bisphenol A　双酚A的二环氧甘油醚
Diglycidyl ether of bisphenol F　双酚F的二环氧甘油醚
Diheptyl phthalate　二庚基邻苯二甲酸酯
Di-n-hexyl adipate　二-正-己基己二酸酯
Dihexyl phthalate　己二基邻苯二甲酸酯
Diisobutylamine　二异丁胺
Diisobutylene　二异丁烯
Diisobutyl ketone　二异丁基酮
Diisobutyl phthalate　邻苯二甲酸二异丁酯
Diisononyl adipate　己二酸二异壬酯
Diisooctyl phthalate　邻苯二甲酸二异辛酯
Diisopropanolamine　二异丙醇胺
Diisopropylamine　二异丙胺
Diisopropylbenzene (all isomers)　二异丙苯(所有异构体)
Diisopropylnaphthalene　二异丙基萘
N,N-Dimethylacetamide　N,N-二甲基乙酰胺
N,N-Dimethylacetamide solution (40% or less)　N,N-二甲基乙酰胺溶液(40%或以下)
Dimethyl adipate　二甲基己二酸酯
Dimethylamine solution (45% or less)　二甲胺溶液(45%或以下)
Dimethylamine solution (greater than 45% but not greater than 55%)　二甲胺溶液(45%以上但不超过55%)
Dimethylamine solution (greater than 55% but not greater than 65%)　二甲胺溶液(55%以上但不超过65%)
N,N-Dimethylcyclohexylamine　N,N-二甲基环己胺
Dimethyl disulphide　二甲基二硫化物
N,N-Dimethyldodecylamine　N,N-二甲基十二烷胺
Dimethylethanolamine　二甲基乙醇胺
Dimethylformamide　二甲基甲酰胺
Dimethyl glutarate　二甲基戊二酸
Dimethyl hydrogen phosphate　二甲基亚磷酸氢盐
Dimethyl octanoic acid　二甲基辛酸
Dimethyl phthalate　邻苯二甲酸二甲酯
Dimethylpolysiloxane　二甲基聚硅氧烷
2,2-Dimethylpropane-1,3-diol (molten or solution)　2,2-二甲基丙烷-1,3-二醇(熔融或溶液)
Dimethyl succinate　二甲基琥珀酸酯
Dinitrotoluene (molten)　二硝基甲苯(熔融)
Dinonyl phthalate　邻苯二甲酸二壬酯
Dioctyl phthalate　邻苯二甲酸二辛酯
1,4-Dioxane　1,4-二恶烷

Dipentene 二聚戊烯
Diphenyl 联苯
Diphenylamine (molten) 二苯胺(熔融)
Diphenylamine, reaction product with 2,2,4-Trimethylpentene 二苯胺与2,2,4-三甲基戊烯的反应物
Diphenylamines, alkylated 烷基二苯胺
Diphenyl/Diphenyl ether mixtures 联苯/二苯醚混合物
Diphenyl ether 二苯醚
Diphenyl ether/Diphenyl phenyl ether mixture 二苯醚/二苯基二苯醚混合物
Diphenylmethane diisocyanate 二苯甲烷二异氰酸酯
Diphenylol propane-epichlorohydrin resins 二苯丙烷—表氯醇树脂
Di-n-propylamine 二正丙胺
Dipropylene glycol 二丙基二醇
Resin oil, distilled 馏出的松香油
Dithiocarbamate ester (C7-C35) 二硫代氨基甲酸盐酯(C7-C35)
Ditridecyl adipate 双十三烷基己二酸酯
Ditridecyl phthalate 邻苯二甲酸(二)十三烷基酯
Diundecyl phthalate 双十一基甲邻苯二甲酸酯
Dodecane (all isomers) 十二烷(所有异构体)
tert-Dodecanethiol 叔十二烷硫醇
Dodecene (all isomers) 十二(碳)烯(所有异构体)
Dodecyl alcohol 十二(烷)醇
Dodecylamine/Tetradecylamine mixture 十二烷胺/十四(烷)胺混合物
Dodecylbenzene 十二烷基苯
Dodecyl diphenyl ether disulphonate solution 十二(烷)基联苯醚二磺酸酯溶液
Dodecyl hydroxypropyl sulphide 十二烷基羟基丙基硫化物
Dodecyl methacrylate 甲基丙烯酸十二酯
Dodecyl/Octadecyl methacrylate (mixture) 十二烷基/十八烷基异丁烯酸盐(混合物)
Dodecyl/Pentadecyl methacrylate mixture 甲基丙烯酸十二~十五酯混合物
Dodecyl phenol 十二烷基苯酚
Dodecyl Xylene 十二烷基二甲苯
Drilling brines (containing zinc salts) 钻井盐水(含有锌盐)
Drilling brines, including: calcium bromide solution, calcium chloride solution and sodium chloride solution 钻井盐水,包括溴化钙溶液,氯化钙溶液和氯化钠溶液
Epichlorohydrin 表氯醇
Ethanolamine 乙醇胺

2-Ethoxyethyl acetate 2-乙氧基醋酸乙酯
Ethoxylated long chain (C16+) alkyloxyalkylamine 长链(C16+)乙氧基化烷基烷氧基胺
Ethyl acetate 乙酸乙酯
Ethyl acetoacetate 乙酰乙酸乙酯
Ethyl acrylate 丙烯酸乙酯
Ethylamine 乙胺
Ethylamine solutions (72% or less) 乙胺溶液(72%或以下)
Ethyl amyl ketone 乙基戊基甲酮
Ethylbenzene 乙苯
Ethyl tert-butyl ether 乙基-叔丁基醚
Ethyl butyrate 丁酸乙酯
Ethylcyclohexane 乙基环己烷
N-Ethylcyclohexylamine N-乙基环己胺
SEthyl dipropylthiocarbamate S-乙基二丙硫代氨基酸酯
Ethylene chlorohydrin 乙撑氯醇
Ethylene cyanohydrin 乙撑氰醇
Ethylenediamine 乙二胺
Ethylenediaminetetraacetic acid, tetrasodium salt solution 乙二胺四乙酸,四钠盐溶液
Ethylene dibromide 二溴化乙烯
Ethylene dichloride 二氯化乙烯
Ethylene glycol 乙二醇
Ethylene glycol acetate 乙二醇醋酸酯
Ethylene glycol butyl ether acetate 乙二醇丁醚酯
Ethylene glycol diacetate 乙二醇二乙酸酯
Ethylene glycol methyl ether acetate 乙二醇甲基醚乙酸酯
Ethylene glycol monoalkyl ethers 乙二醇单烷基醚
Ethylene glycol phenyl ether 乙二醇苯基醚
Ethylene glycol phenyl ether/Diethylene glycol phenyl ether mixture 乙二醇苯基醚/二基乙二醇苯基醚混合物
Ethylene oxide/Propylene oxide mixture with an ethylene oxide content of not more than 30% by mass 环氧乙烷/环氧丙烷混合物(其中环氧乙烷按质量计含量不超过30%)
Ethylene-Vinyl acetate copolymer (emulsion) 乙烯-醋酸乙烯共聚物(乳剂)
Ethyl-3-ethoxypropionate 乙基-3-乙氧基丙酸酯
2-Ethylhexanoic acid 2-乙基己酸
2-Ethylhexyl acrylate 丙烯酸2-乙基己酯
2-Ethylhexylamine 2-乙基己胺
2-Ethyl-2-(hydroxymethyl)propane-1,3-diol (C8-C10) ester 2-乙基-2-(羟甲基)丙烷-1,3-二醇,C8-C10酯

Ethylidene norbornene 乙叉降冰片烯
Ethyl methacrylate 甲基丙烯酸乙酯
N-Ethylmethylallylamine N-甲基乙基丙烯
Ethyl propionate 丙酸乙酯
2-Ethyl-3-propylacrolein 2-乙基-3-丙基丙烯醛
Ethyl toluene 乙基甲苯
Fatty acid（saturated C13+） 脂肪酸（饱和 C13+）
Fatty acid methyl esters（m） 脂肪酸甲酯（m）
Fatty acids, 12+ 脂肪酸（12+）
Fatty acids, C8-C10 脂肪酸（C8-C10）
Fatty acids, essentially linear（C6-C18）2-ethylhexyl ester. 脂肪酸,基本为线形,C6-C18,2-乙基己基酯
Fatty acids, C16+ 脂肪酸（C16+）
Ferric chloride solutions 氯化铁溶液
Ferric nitrate/Nitric acid solution 硝酸铁/硝酸溶液
Fish oil 鱼油
Fluorosilicic acid（20-30%）in water solution 氟硅酸水溶液（20~30%）
Formaldehyde solutions（45% or less） 甲醛溶液（45%或以下）
Formamide 甲酰胺
Formic acid 甲酸
Furfural 糠醛
Furfuryl alcohol 糠醇
Glucitol/glycerol blend propoxylated（containing less than 10% amines） 葡萄糖醇/甘油混合丙氧基酯（含胺少于10%）
Glutaraldehyde solutions（50% or less） 戊二醛溶液（50%或以下）
Glycerol monooleate 甘油单油酸酯
Glycerol propoxylated 甘油丙氧基酯
Glycerol, propoxylated and ethoxylated 甘油丙氧基酯及乙氧基酯
Glycerol/sucrose blend propoxylated and ethoxylated 甘油/蔗糖混合丙氧基酯及乙氧基酯
Glyceryl triacetate 甘油三乙酸酯
Glycidyl ester of C10 trialkylacetic acid C10 三烷基醋酸缩水甘油酯
Glycine, sodium salt solution 甘氨酸,氯化钠溶液
Glycolic acid solution（70% or less） 乙醇酸溶液（70%或以下）
Glyoxal solution（40% or less） 乙二醛溶液（40%或以下）
Glyoxylic acid solution（50% or less） 二羟基乙酸溶液（50%或以下）
Glyphosate solution（not containing surfactant） 乙酸苄酯溶液（不含表面活性剂）
Groundnut oil 花生油
Heptane（all isomers） 庚烷（所有异构体）
n-Heptanoic acid n-庚酸
Heptanol（all isomers）（d） 庚醇（所有异构体）（d）
Heptene（all isomers） 庚烯（所有异构体）
Heptyl acetate 醋酸庚酯
1-Hexadecylnaphthalene / 1,4-bis（hexadecyl）naphthalene mixture 1-十六烷基萘/1,4-双（十六烷基）萘混合物
Hexamethylenediamine adipate（50% in water） 乙撑二胺己二酸酯（50%在水中）
Hexamethylenediamine（molten） 己撑二胺（溶化）
Hexamethylenediamine solution 乙撑二胺溶液
Hexamethylene diisocyanate 己二异氰酸酯
Hexamethylene glycol 己二醇
Hexamethyleneimine 六甲撑亚胺
Hexane（all isomers） 己烷（所有异构体）
1,6-Hexanediol, distillation overheads 1,6-己二醇,蒸馏塔顶馏分
Hexanoic acid 己酸
Hexanol 己醇
Hexene（all isomers） 己烯（所有异构体）
Hexyl acetate 醋酸己酯
Hydrochloric acid 盐酸
Hydrogen peroxide solutions（over 60% but not over 70% by mass） 过氧化氢溶液（浓度占60%以上,但不超过70%）
Hydrogen peroxide solutions（over 8% but not over 60% by mass） 过氧化氢溶液（体积占8%以上,但不超过60%）
2-Hydroxyethyl acrylate 2-羟乙基丙烯酸酯
N-(Hydroxyethyl)ethylenediaminetriacetic acid, trisodium salt solution 正-(羟乙基)乙二胺三乙酸,三钠盐溶液
2-Hydroxy-4-(methylthio)butanoic acid 2-羟基-4-(甲硫基)丁酸
Illipe oil 雾冰草脂
Isoamyl alcohol 异戊醇
Isobutyl alcohol 异丁醇
Isobutyl formate 甲酸异丁酯
Isobutyl methacrylate 异丁烯酸酯
Isophorone 异佛尔酮
Isophoronediamine 异佛尔酮二胺
Isophorone diisocyanate 异佛尔酮二异氰酸酯
Isoprene 异戊二烯
Isopropanolamine 异丙醇胺
Isopropyl acetate 乙酸异丙酯

Isopropylamine 异丙胺
sopropylamine (70% or less) solution 异丙胺(70%或以下)溶液
Isopropylcyclohexane 异丙基环己烷
Isopropyl ether 异丙醚
Lactic acid 乳酸
Lactonitrile solution (80% or less) 乳腈溶液(80%或以下)
Lard 猪脂
Latex, ammonia (1% or less)-inhibited 乳胶,氨(1%或以下),抑制的
Latex: Carboxylated styrene-Butadiene copolymer; Styrene-Butadiene rubber 乳胶;羧基化苯乙烯-丁烯共聚物;苯乙烯-丁二烯共聚物
Lauric acid 月桂酸
Ligninsulphonic acid, sodium salt solution 木质素磺酸,钠盐溶液
Linseed oil 亚麻籽油
Liquid chemical wastes 液态化学废料
Long-chain alkaryl polyether (C11-C20) 长链烷芳基聚醚(C11-C20)
Long-chain alkaryl sulphonic acid (C16-C60) 长链烷芳基磺酸(C16-C60)
Long-chain alkylphenate/Phenol sulphide mixture 长链烷基酚盐/苯酚硫黄混合物
L-Lysine solution (60% or less) L-赖氨酸(60%或以下)
Magnesium chloride solution 氯化镁溶液
Magnesium long-chain alkaryl sulphonate (C11-C50) 长链烷芳基磺酸镁(C11-C50)
Magnesium long-chain alkyl salicylate (C11+) 长链烷基水杨酸镁(C11+)
Maleic anhydride 顺丁烯二酐
Mango kernel oil 芒果核油
Mercaptobenzothiazol, sodium salt solution 巯基苯并噻唑钠盐溶液
Mesityl oxide 异亚丙基丙酮氧化物
Metam sodium solution 变位钠溶液
Methacrylic acid-alkoxypoly (alkylene oxide) methacrylate copolymer, sodium salt aqueous solution (45% or less) 甲基丙烯酸-烷氧基聚(氧化烯)甲基丙烯酯共聚物,钠盐水溶液(45%或以下)
Methacrylic acid 甲基丙烯酸
Methacrylic resin in ethylene dichloride 二氯化乙烯中的甲基丙烯酸
Methacrylonitrile 甲基丙烯腈

3-Methoxy-1-butanol 3-甲氧(基)-1-丁醇
3-Methoxybutyl acetate 3-甲氧丁基乙酸盐
N-(2-Methoxy-1-methyl ethyl)-2-ethyl-6-methyl chloroacet-anilide N-(2-甲氧基-1-甲基乙基)-2-乙基-6-甲基乙酰氯苯胺
Methyl acetate 乙酸甲酯
Methyl acetoacetate 乙酰乙酸甲酯
Methyl acrylate 丙烯酸甲酯
Methyl alcohol 甲醇
Methylamine solutions (42% or less) 甲胺溶液(42%或以下)
Methylamyl acetate 乙酸甲基戊酯
Methylamyl alcohol 甲戊醇
Methyl amyl ketone 甲-戊基(甲)酮
Methylbutenol 甲基丁醇
Methyl tert-butyl ether 甲基叔丁基醚
Methyl butyl ketone 甲基丁基酮
Methylbutynol 甲基丁炔醇
Methyl butyrate 丁酸甲酯
Methylcyclohexane 甲基环己烷
Methylcyclopentadiene dimmer 甲基环戊二烯二聚物
Methylcyclopentadienyl manganese tricarbonyl 甲基环戊三锰羧
Methyl diethanolamine 甲基二羟乙基胺
2-Methyl-6-ethyl aniline 2-甲基-6-乙基苯胺
Methyl ethyl ketone 甲基乙基酮
2-Methyl-5-ethyl pyridine 2-甲基-5-乙基吡啶
Methyl formate 甲酸甲酯
2-Methyl-2-hydroxy-3-butyne 2-甲基-2-羟基-3-丁炔
Methyl isobutyl ketone 甲基异丁基酮
Methyl methacrylate 甲基丙烯酸甲脂
3-Methyl-3-methoxybutanol 3-甲基-3-甲氧基丁醇
Methyl naphthalene (molten) 甲基萘(熔融)
2-Methyl-1,3-propanediol 2-甲基-1,3-丙二醇
2-Methylpyridine 2-甲基吡啶
3-Methylpyridine 3-甲基吡啶
4-Methylpyridine 4-甲基吡啶
N-Methyl-2-pyrrolidone N-甲基-2-吡咯烷
Methyl salicylate 水杨酸甲酯
alpha-Methylstyrene α-甲基苯乙烯
3-(methylthio)propionaldehyde 3-(甲硫基)丙醛
Molybdenum polysulfide long chain alkyl dithiocarbamide complex 聚硫化钼长链烷基二硫化异硫氰酯络合物
Morpholine 吗啉
Motor fuel anti-knock compounds (containing lead alkyls) 内燃机燃料抗爆化合物

Myrcene 月桂烯
Naphthalene (molten) 萘(熔融)
Naphthalenesulphonic acid-Formaldehyde copolymer, sodium salt solution 萘磺酸-甲醛共聚物,钠盐溶液
Neodecanoic acid 新癸酸
Nitrating acid (mixture of sulphuric and nitric acids) 硝化酸(硫酸和硝酸混合物)
Nitric acid (70% and over) 硝酸(70%及以上)
Nitric acid (less than 70%) 硝酸(低于70%)
Nitrilotriacetic acid, trisodium salt solution 次氮基三乙酸,三钠盐溶液
Nitrobenzene 硝基苯
Nitroethane 硝基乙烷
Nitroethane (80%)/Nitropropane (20%) 硝基乙烷(80%)/硝基丙烷(20%)
Nitroethane, 1-Nitropropane (each 15% or more) mixture 硝基乙烷,1-硝基丙烷(各占15%或以上)混合物
o-% Nitrophenol (molten) 邻-硝基苯酚(熔化的)
1-or 2-Nitropropane 1-或2-硝基丙烷
Nitropropane (60%)/Nitroethane (40%) mixture 硝基丙烷(60%)/硝基乙烷(40%)混合物
o-or p-Nitrotoluenes 邻-或对-硝基甲苯
Nonane (all isomers) 壬烷(所有异构体)
Nonanoic acid (all isomers) 壬酸(所有异构体)
Nonene (all isomers) 壬烯(所有异构体)
Nonyl alcohol (all isomers) 壬醇(所有异构体)
Nonyl methacrylate monomer 壬基异丁烯酸单体
Nonylphenol 壬基苯酚
Nonylphenol poly(4+) ethoxylate 壬基聚酚(4+)乙氧醚
Noxious liquid, NF, (1) n.o.s. (trade name…, contains…)ST1, Cat. X 有害液体,NF,(1)n.o.s.(商品名……,包含……)ST1, Cat. X
Noxious liquid, F, (2) n.o.s. (trade name…, contains…)ST1, Cat. X 有害液体,F,(2)n.o.s.(商品名……,包含……)ST1, Cat. X
Noxious liquid, NF, (3) n.o.s. (trade name…, contains…)ST2, Cat. X 有害液体,NF,(3)n.o.s.(商品名……,包含……)ST2, Cat. X
Noxious liquid, F, (4) n.o.s. (trade name…, contains…)ST2, Cat. X 有害液体,F,(4)n.o.s.(商品名……,包含……)ST2, Cat. X
Noxious liquid, NF, (5) n.o.s. (trade name…, contains…)ST2, Cat. Y 有害液体,NF,(5)n.o.s.(商品名……,包含……)ST2, Cat. Y
Noxious liquid, F, (6) n.o.s. (trade name…, contains…)ST2, Cat. Y 有害液体,F,(6)n.o.s.(商品名……,包含……)ST2, Cat. Y
Noxious liquid, NF, (7) n.o.s. (trade name…, contains…)ST3, Cat. Y 有害液体,NF,(7)n.o.s.(商品名……,包含……)ST3, Cat. Y
Noxious liquid, F, (8) n.o.s. (trade name…, contains…)ST3, Cat. Y 有害液体,F,(8)n.o.s.(商品名……,包含……)ST3, Cat. Y
Noxious liquid, NF, (9) n.o.s. (trade name…, contains…)ST3, Cat. Z 有害液体,NF,(9)n.o.s.(商品名……,包含……)ST3, Cat. Z
Noxious liquid, F, (10) n.o.s. (trade name…, contains…)ST3, Cat. Z 有害液体,F,(10)n.o.s.(商品名……,包含……)ST3, Cat. Z
Octane (all isomers) (正)辛烷(所有异构体)
Octanoic acid (all isomers) 辛酸(所有异构体)
Octanol (all isomers) 辛醇(所有异构体)
Octene (all isomers) 辛烯(所有异构体)
n-Octyl acetate 乙酸正辛酯
Octyl aldehydes 辛醛
Octyl decyl adipate 辛基癸基己二酸酯
Olefin-Alkyl ester copolymer (molecular weight 2000+) 烯烃-烷基酯共聚物(分子重2000+)
Olefin mixtures (C5-C7) 烯烃混合物(C5-C7)
Olefin mixtures (C5-C15) 烯烃混合物(C5-C15)
Olefins (C13+, all isomers) 烯(C13+,所有异构体)
alpha-Olefins (C6-C18) mixtures α-烯烃(C6-C18)混合物
Oleic acid 油酸
Oleum 发烟硫酸
Oleylamine 油酰胺
Olive oil 橄榄油
Oxygenated aliphatic hydrocarbon mixture 氧化脂肪族烃混合物
Palm acid oil 棕榈酸油
Palm fatty acid distillate 棕榈脂肪酸蒸馏物
Palm kernel acid oil 棕榈仁酸油
Palm kernel oil 棕榈仁油
Palm kernel olein 棕榈仁油脂
Palm kernel stearin 棕榈仁酸甘油酯
Palm mid fraction 棕榈中间馏出物
Palm oil 棕榈油
Non-edible industrial grade palm oil 非食用工业棕榈油
Palm oil fatty acid methyl ester 棕榈油脂肪酸甲酯
Palm olein 棕榈油精
Palm stearin 棕榈硬脂精

Paraffin wax 石蜡
Paraldehyde 仲醛
Paraldehyde-ammonia reaction product 仲醛-氨反应产物
Pentachloroethane 五氯乙烷
1,3-Pentadiene 1,3-戊二烯
Pentaethylenehexamine 五亚乙基六甲胺
Pentane（all isomers） 戊烷（所有异构体）
Pentanoic acid 戊酸
n-Pentanoic acid（64%）/2-Methyl butyric acid（36%）mixture 正戊酸（64%）/2-甲基丁酸（36%）混合物
Pentene（all isomers） 戊烯（所有异构体）
n-Pentyl propionate 正戊基丙酸
Perchloroethylene 全氯乙烯
Petrolatum 矿脂
Phenol 苯酚
1-Phenyl-1-xylyl ethane 1-苯基-1-二甲苯基乙烷
Phosphate esters, alkyl（C12-C14）amine 烷基（C12-C14）胺磷酸酯
Phosphoric acid 磷酸
Phosphorus, yellow or white 磷，黄的或白的
Phthalic anhydride（molten） 酞酐（熔融）
alpha-Pinene α-蒎烯
beta-Pinene β-蒎烯
Pine oil 松油
Polyacrylic acid solution（40% or less） 聚丙烯酸溶液（40%或以下）
Polyalkyl（C18-C22）acrylate in xylene 二甲苯中聚烷（C18-C22）丙烯酸盐（或酯）
Polyalkylalkenaminesuccinimide, molybdenum oxysulphide 二甲苯中聚烷烯烃琥珀酰亚胺，氧(代)硫化钼
Poly(2-8) alkylene glycol monoalkyl（C1-C6）ether 聚(2-8)烷基二醇单烷基(C1-C6)醚
Poly(2-8) alkylene glycol monoalkyl（C1-C6）ether acetate 聚(2-8)烷基二醇单烷基(C1-C6)醋酸醚
Polyalkyl（C10-C20）methacrylate 二甲苯中聚烷（C10-C20）异丁烯酸盐(或酯)
Polyalkyl（C10-C18）methacrylate/ethylene-propylene copolymer mixture 二甲苯中聚烷（C10-C18）异丁烯酸盐/乙烯-丙烯共聚物混合物
Polybutene 聚丁烯
Polybutenyl succinimide 聚丁烯琥珀酰亚胺
Poly(2+)cyclic aromatics 聚(2+)环芳香物
Polyether（molecular weight 1350+） 聚醚（分子量1350+）
Polyethylene glycol 聚乙二醇
Polyethylene glycol dimethyl ether 聚乙二醇二甲醚
Polyethylene polyamines 聚乙烯聚胺

Polyethylene polyamines（more than 50% C5-C20 paraffin oil） 聚乙烯聚胺(C5-C20 液状石蜡超过50%)
Polyferric sulphate solution 聚硫酸铁溶液
Poly（iminoethylene）-graft-N-poly（ethyleneoxy）solution（90% or less） 聚（亚氨基乙烯）-移植-N-聚（氧化乙烯）溶液（90%或以下）
Polyisobutenamine in aliphatic（C10-C14）solvent 脂族中聚异丁烯胺（C10-C14）溶剂
Polyisobutenyl anhydride adduct 聚丁烯基酐加合物
Poly(4+) isobutylene 聚(4+)异丁烯
Polymethylene polyphenyl isocyanate 聚亚甲基聚苯异氰酸酯
Polyolefin（molecular weight 300+） 聚烯烃（分子质量300+）
Polyolefin amide alkeneamine（C17+） 聚烯酰胺烯胺（C17+）
Polyolefin amide alkeneamine borate（C28-C250） 聚烯酰胺烯胺硼酸盐（C28-C250）
Polyolefin amide alkeneamine polyo 聚烯酰多羟基胺烯胺
Polyolefinamine（C28-C250） 聚烯烃胺（C28-C250）
Polyolefinamine in alkyl（C2-C4）benzenes 烷基（C2-C4）苯中聚烯烃胺
Polyolefinamine in aromatic solvent 芳香溶剂中聚烯烃胺
Polyolefin aminoester salts（molecular weight 2000+） 聚烯烃氨基酯盐（分子量2000+）
Polyolefin anhydride 聚烯酐
Polyolefin ester（C28-C250） 聚烯酯（C28-C250）
Polyolefin phenolic amine（C28-C250） 聚烯苯酚胺（C28-C250）
Polyolefin phosphorosulphide, barium derivative（C28-C250） 聚烯烃偶磷硫化钡衍生物（C28-C250）
Poly(5+) propylene 聚(5+)丙烯
Polypropylene glycol 聚丙二醇
Polysiloxane 聚硅氧烷
Potassium chloride solution 氯化钾溶液
Potassium hydroxide solution 氢氧化钾溶液
Potassium oleate 油酸钾
Potassium thiosulphate（50% or less） 硫代硫酸盐钾（50%或以下）
n-Propanolamine 正丙醇胺
beta-Propiolactone β-丙内酯
Propionaldehyde 丙醛
Propionic acid 丙酸
Propionic anhydride 丙酸酐
Propionitrile 丙腈
n-Propyl acetate 正乙酸丙脂

n-Propyl alcohol 正丙醇
n-Propylamine 正丙胺
Propylbenzene (all isomers) 丙苯(所有异构体)
Propylene glycol methyl ether acetate 丙二醇甲基醚乙酸盐
Propylene glycol monoalkyl ether 丙二醇单烷基醚
Propylene glycol phenyl ether 丙二醇苯基醚
Propylene oxide 氧化丙烯
Propylene tetramer 丙烯四聚物
Propylene trimer 丙烯三聚物
Pyridine 吡啶
Pyrolysis gasoline (containing benzene) 裂解汽油(含苯)
Rapeseed oil 菜籽油
Rapeseed oil (containing less than 4% free fatty acid) 菜籽油(含少于4%的脂肪酸)
Rapeseed oil fatty acid methyl esters 菜籽油脂肪酸甲醚
Rice bran oil 米糠油
Rosin 松香
Safflower oil 红花油
Shea butter 牛油果油
Sodium alkyl (C14-C17) sulphonates (60%-65% solution) 烷基钠(C14-C17)磺酸盐(60%~65%溶液)
Sodium aluminosilicate slurr 硅铝酸钠生料
Sodium benzoate 苯甲酸钠
Sodium borohydride (15% or less)/Sodium hydroxide solution 氢硼化钠(15%或以下)/氢氧化钠溶液
Sodium carbonate solution 碳酸钠溶液
Sodium chlorate solution (50% or less) 氯化钠溶液(50%或以下)
Sodium dichromate solution (70% or less) 重铬酸钠溶液(70%或以下)
Sodium hydrogen sulphide (6% or less)/ Sodium carbonate (3 or less) solution 氢亚硫酸钠(6%或以下)/碳酸钠(3%或以下)溶液
Sodium hydrogen sulphite solution (45% or less) 亚硫酸氢钠溶液(45%或以下)
Sodium hydrosulphide/Ammonium sulphide solution 氢硫化钠/硫化胺溶液
Sodium hydrosulphide solution (45% or less) 氢硫化钠溶液(45%或以下)
Sodium hydroxide solution 氢氧化钠溶液
Sodium hypochlorite solution (15% or less) 次氯酸钠溶液(15%或以下)
Sodium nitrite solution 亚硝酸钠溶液
Sodium petroleum sulphonate 石油磺酸钠
Sodium poly (4+) acrylate solutions 聚(4+)丙烯酸钠溶液
Sodium silicate solution 硅酸钠溶液
Sodium sulphide solution (15% or less) 硫化钠溶液(15%或以下)
Sodium sulphite solution (25% or less) 亚硫酸钠溶液(25%或以下)
Sodium thiocyanate solution (56% or less) 硫氰酸钠溶液(56%或以下)
Soyabean oil 豆油
Styrene monomer 苯乙烯单体
Sulphohydrocarbon (C3-C88) 硫氢化碳(C3-C88)
Sulpholane 磺基烷
Sulphur (molten) 硫(熔化的)
Sulphuric acid 硫酸
Sulphuric acid, spent 废硫酸
Sulphurized fat (C14-C20) 硫化脂肪(C14-C20)
Sulphurized polyolefinamide alkene (C28-C250) amine 硫化聚烯烃酰胺烯烃(C28-C250)胺
Sunflower seed oil 向日葵籽油
Tall oil, crude 粗妥尔油
Tall oil, distilled 精制妥尔油
Tall oil fatty acid (resin acids less than 20%) 妥尔油脂肪酸(树脂酸小于20%)
Tall oil pitch 妥尔油沥青
Tallow (动物)脂
Tallow fatty acid (动物)脂肪酸
Tetrachloroethane 四氯乙烷
Tetraethylene glycol 四甘醇
Tetraethylene pentamine 四乙撑五胺
Tetrahydrofuran 四氢呋喃
Tetrahydronaphthalene 四氢化萘
Tetramethylbenzene (all isomers) 四甲苯(所有异构体)
Titanium dioxide slurry 二氧化钛生料
Toluene 甲苯
Toluenediamine 甲苯二胺
Toluene diisocyanate 甲苯二异氰酸酯
o-Toluidine 邻甲苯胺
Tributyl phosphate 磷酸三丁酯
1,2,3-Trichlorobenzene (molten) 1,2,3-三氯苯(熔化的)
1,2,4-Trichlorobenzene 1,2,4-三氯苯
1,1,1-Trichloroethane 1,1,1-三氯乙烷
1,1,2-Trichloroethane 1,1,2-三氯乙烷
Trichloroethylene 三氯乙烯
1,2,3-Trichloropropane 1,2,3-三氯丙烷
1,1,2-Trichloro-1,2,2-Trifluoroethane 1,1,2-三氯-1,2,

2-氟烷

Tricresyl phosphate (containing 1% or more ortho-isomer)　磷酸三甲苯酯（含有1%或以上的正异构体）

Tricresyl phosphate (containing less than 1% ortho-isomer)　磷酸三甲苯酯（含有1%以下的原异构体）

Tridecane　十三(碳)烷

Tridecanoic acid　十三(烷)酸

Tridecyl acetate　十三烷基乙酸酯

Triethanolamine　三乙醇胺

Triethylamine　三乙胺

Triethylbenzene　三乙基苯

Triethylenetetramine　三乙撑四胺

Triethyl phosphate　磷酸三乙酯

Triethyl phosphate　亚磷酸三乙酯

Triisopropanolamine　三异丙醇胺

Triisopropylated phenyl phosphates　三异丙基苯磺酰磷酸盐

Trimethylacetic acid　三甲基乙酸

Trimethylamine solution (30% or less)　三甲胺溶液（30%或以下）

Trimethylbenzene (all isomers)　三甲苯(所有异构体)

Trimethylol propane propoxylated　三羟甲基丙烷聚乙氧酸酯

2,2,4-Trimethyl-1,3-pentanediol diisobutyrate　2,2,4-三甲基-1,3-戊乙醇 三异丁酸酯

2,2,4-Trimethyl-1,3-pentanediol-1-isobutyrate　2,2,4-三甲基-1,3-戊乙醇-1-异丁酸酯

1,3,5-Trioxane　1,3,5-三戊烷

Tripropylene glycol　三聚丙烯二醇

Trixylyl phosphate　e磷酸（三）二甲苯酯

Tung oil　桐油

Turpentine　松节油

Undecanoic acid　十一烷酸

1-Undecene　1-十一碳烯

Undecyl alcohol　十一醇

Urea/Ammonium nitrate solution　尿素/硝酸铵溶液

Urea/Ammonium nitrate solution (containing less than 1% free ammonia)　尿素/硝酸铵溶液（含1%以下游离氨）

Urea/Ammonium phosphate solution　尿素/磷酸铵溶液

Urea solution　尿素溶液

Valeraldehyde (all isomers)　戊醛（所有异构体）

Vegetable acid oil (m)　植物酸油(m)

Vegetable fatty acid distillates (m)　植物脂肪酸馏出物(m)

Vinyl acetate　乙酸乙烯

Vinyl ethyl ether　乙烯基乙基醚

Vinylidene chloride　二氯乙烯

Vinyl neodecanoate　新癸酸乙烯酯

Vinyltoluene　乙烯基甲苯

Wax　蜡

White spirit, low (15%-20%) aromatic　白节油,低于(15%~20%)芳香物

Xylenes　二甲苯

Xylenes/ethylbenzene (10% or more) mixture　二甲苯/乙苯(10%或以上)混合物

Xylenol　二甲苯酚

Zinc alkaryl dithiophosphate (C7-C16)　烷基锌二硫代磷酸盐(C7-C16)

Zinc alkenyl carboxamide　烷基锌甲酰胺

Zinc alkyl dithiophosphate (C3-C14)　烷基锌二硫代磷酸盐(C3-C14)

附录5　国际组织及其规则、标准中与船舶有关缩写词英汉对照

Annex 5　English-Chinese abbreviations concerning ships on International Marine Organization and their Codes

在涉及船舶合同、设计与建造的各种文件中，经常会遇到有关国际组织(特别是国际海事组织及其所属和相关组织)，各主要法定检验机构与船级社，国际海事组织公约、规则等及其中专用术语的缩写词。为便于正确理解这些缩写词的含义，按英文字母顺序选编了这些缩写词的"英汉对照"。

A

A = Assembly　大会
ABK = Addressed and binary broadcast acknowledgement　寻址二进制广播确认
ABM = Addressed binary and safety related message　寻址二进制与安全有关的信息
ABS = American Bureau of Shipping　美国船级社
AC = Alternating current　交流(电)
ACA = AIS regional channel assignment massage　AIS区域信道分配信息
ACC = Air cushion cavity + planing mono-hull craft + air cushion layer　气垫船 = 滑行单体船 + 气垫层
ACEP = American College of Emergency Physician　美国急救医师学会
ACM = Association for Computing Machine　(美国)计算机协会
ACON = Rolls-Royce's alarm and monitoring system　罗尔斯-罗伊斯报警和监视系统
ACOPS = Advisory Committee on Protection of the Sea　海洋保护咨询委员会
ACP = Autoridad del Canal de Panama (Panama Canal Authority)　巴拿马运河管理局
ACS = Channel management information source　信道管理信息源
ACS = Asia Classification Society　亚洲船级社协会
ACV = Air cushion vehicle　气垫船
ADCAP = Advanced capability　推进能力
ADCS = Automatic data collection system　自动数据收集系统
ADP = Automatic data processing　自动数据处理

AEA = American Electronics Association　美国电子协会
AFA = Amt für Arbeitsschutz　(德国)劳动保护局
AFNOR = Association Francaise de Normalisation　法国标准化协会
AFRAMAX = AFRAMAX oil tanker　阿芙拉型油船
AFS 2001 = International Convention on the Control of Harmful Anti-fouling Systems on Ships, 2001　2001年国际控制船舶有害防污底系统公约
AFS = Anti-fouling systems　防污底系统
Aft. = Afterward　在船尾(艉部的、后面的、向后)
AIR = Average individual risk　平均个体风险率
AIR = AIS inquiry request　AIS询问请求
AIS = (vessel's) Automatic identification system　(船舶)自动识别系统
AIS = Automated Information System　自动化信息系统
AISC = American Institute for Steel Constructure / American Steel Construction Association　美国钢结构协会
AISI = American Iron and Steel Institute　美国钢铁学会
AIS-SART = (vesse's) AIS search and rescue transponder　(船舶)自动识别系统搜救应答器
AIT = Auto ignition temperature　自燃温度
ALARP = As low as reasonably practicable　即ALARP原则，意为尽可能降低风险又能够实现
ALT = Articulated Loading Tower System　铰接式装卸塔系统
AMP = Alternative marine power　替代船用动力
AMPD = Average most probable discharge　平均可能最大排放(美国防污染要求)
AMS GROUP = Amendments group　修订小组
AMS = Alternate Management System　替代管理系统(USCG对压载水管理系统所许可的)
AMSA = Australian Maritime Safety Authority　澳大利亚海事安全局
AMVER = Automated merchant vessel reporting　商船自动报告(系统)
ANSI-American National Standard Institute　美国国家标准学会

1025

ANTS = Automatic navigation and track-keeping system 自动航行和航迹保持系统

AOC = All outside cabins 所有的外舱室

APEC = Asia Pacific Economic Cooperation 亚太经济合作组织

APHIS = Animal and Plant Health Inspection Service (of U. S. Department of Agriculture) （美国农业部）动植物检疫服务中心

API = American Petroleum Institute 美国石油学会

APJ = Absolute probability judgment 绝对概率判断方法

APLUS = Automatic pallet loading/unloading system 自动货盘装卸系统

AR = Aspect ratio 展弦比

ARCS = Admiralty raster chart services 海军光栅海图服务

ARE = Arab Republic of Egypt 阿拉伯埃及共和国

ARPA = Automatic radar plotting aids 自动雷达标绘仪

ARREST 1999 = International Convention on the Arrest of Ships 1999 1999年国际扣船公约

ARS = African Regional Standard 非洲地区标准

ARSO = African Regional Standardization Organization 非洲地区标准化组织

AS = Annual survey 年度检验

AS = Arab Standard 阿拉伯标准

ASCA = American Steel Construction Association 美国钢结构协会

ASEF = Asian Shipbuilding Experts' Forum 亚洲造船技术论坛

ASEP = Accident sequence evaluation program 事故顺序评估法

ASGD = Alarm signal generator device 报警信号发生器

ASIC = Application-specific integrated circuit 专用集成电路

ASLS = Automatic side loading pallet handling system 自动舷侧装货盘操作系统

ASME = American Society of Mechanical Engineers 美国机械工程师学会

ASMO = Arab Standardization Metrological organization 阿拉伯标准化与计量组织

ASNE = American Society of Naval Engineers 美国造船工程师学会

ASTM = American Society for Testing and Materials 美国试验与材料学会

ATA = Automatic tracking aid 自动跟踪仪

ATS = Annual through survey 年度全面检验

AVR = Automatic voltage regulator 自动稳压器

AWES = Association of West European Shipbuilders 西欧造船商协会

AWS = American Welding Society 美国焊接协会

AWT = Advanced Wastewater Treatment 先进的污水处理

B

babr = bar absolute 绝对压力

BAPT = Bundesamt für Post und Telekommunikation (Federal Office for Post and Telecommunication) （德国）邮政与电信管理局

barg = bar gauge (over-pressure) 棒规（过压）

BBD = Barrier block diagram 屏障方框图

bbls = barrels 桶

BBM = Broadcast binary message 广播二进制信息

BC Code = Code of safe practice for solid bulk cargoes 固体散装货物安全操作规则

BCE = Before the Christian Era 在基督教时代之前

BCH Code = Code for the Construction and Equipment of Ships Carrying Dangerous Chemicals in Bulk 散装运输危险化学品船舶构造与设备规则

BD = Blow down 放空

BDS = (Shipborne) BeiDou Satellite (Navigation System) （船载）北斗卫星（导航系统）

BER = Bit error rate 比特误码率

BFETS = Blast and fire engineering for topside system 顶边系统爆炸冲击与消防工程

BIIT = Built-in integrity tests 内置式完整性测试

BIMCO = Baltic & International Maritime Council 波罗的海国际海运公会

BIPAR = International Association on Producers of Insurance & Reinsurance 国际保险和再保险经纪人协会

BLEVE = Boiling liquid expanding vapour explosion 沸腾液体膨胀式蒸汽爆炸

BLG = Sub-Committee on Bulk-Liquids and Gases 散装液体和气体分委会

BLU Code = Code of practice for the safe loading and unloading of bulk carriers 散货船装卸安全操作规则

BMEP = Brake mean effective pressure 平均制动有效压力

BMP = Best management practices 最佳管理经验

BNWAS = Bridge navigational watch alarm system 桥楼航行值班报警系统

BOA = Beam overall （船舶）总宽

BOD = Biochemical oxygen demand 生化需氧量（生活污水排出物指标）

BOD5 = 5-day biochemical oxygen demand 5天生化需氧

量(生活污水排出物指标)
BOP = Blowout preventer　井喷防止器
BORA = Barrier and operational risk analysis　屏障与营运风险分析
BP = Between perpendicular　垂线间
BS = British standards　英国标准
BSH = Bundesamt für See Schiffahrt und Hydrographic (Federal Maritime and Hydrographic Agency)　(德国)海事与水文管理局
BSI = British Standards Institute　英国标准学会
BTU = British Thermal Unit　英国热量单位
BUNKERS 2001 = International Convention on Civil Liability for Bunker Oil Pollution Damage, 2001　2001年国际燃油舱油污损害民事责任公约
BV = Bureau Veritas　法国船级社
BW = Ballast water　压载水
BWDS = Ballast Water Discharge Standard　压载水排放标准
BWDS = Ballast Water Discharge Systems　压载水排放系统
BWM 2004 = International Convention for the Control and Management of Ships' Ballast Water and Sediments, 2004　2004年国际船舶压载水及沉积物控制和管理公约
BWMP = Ballast water management plan　压载水管理计划
BWMS = Ballast water management systems　压载水管理系统
BWMS = Bridge watch monitoring system　桥楼值班监测系统
BWRG = Review group on ballast water treatment technologies　压载水处理技术审核组
BWWG = Ballast Water Working Group　压载水工作组

C

C/L = Chain locker　锚链舱
C = Council　理事会
CA = Controlled atmosphere (systems)　受控大气(系统)
CAA = Collision avoidance aids　避碰导航装置
CABA = Computer aided business administrate　计算机辅助经营管理
CAC = Crew Accommodation comfort　船员起居舒适
CAD = Computer aided design　计算机辅助设计
CAE = Computer aided engineering　计算机辅助工程
CAI = Container Aid International　国际集装箱推广协会
CALM = Catenary anchor leg mooring system　悬链锚腿系泊系统
CANSI = China Association of the National Shipbuilding Industry　中国船舶工业行业协会
CAP = Condition assessment program　状态评估计划
CAPP = Computer aided process planning　计算机辅助工艺过程设计
CAS = China Association for Standardization　中国标准化协会
CAS = Condition assessment scheme　状态评估计划
CAT = Catamaran　双体船
CATB = CAT with bulbous bow　配球鼻艏的双体船
CBA = Cost benefic analysis　成本效益分析
CBD = Convention on biological diversity　生物多样性公约
CBI = Classification Bureau of Indonesia　印度尼西亚船级社
CBT = Clean ballast tanks　清洁压载舱
CC = Customer code　客户代码
CC = IMO Chemical Code　IMO化学品规则(IACS统一解释
CCA = Cause-consequence analysis　因果分析
CCC = Sub-Committee on Carriage of Cargoes and Containers (IMO)　货物和集装箱运输分委员会
CCF = Common cause failure　共因失效
CCIR = International Radio-Communication Consultative Committee　国际无线电通信咨询委员会
CCITT = Consultative Committee for International Telegraphy and Telephony　国际电报和电话咨询委员会
CCRP = Consistent common reference point　统一共同基准点
CCRS = Consistent common reference system　统一共同基准系统
CCS = China Classification Society　中国船级社
CCS = Combat control system　作战指挥系统
CCSDS = Consultative Committee for Space Data Systems　航天数据系统咨询委员会
CCTV = Closed Circuit Television　闭路电视
CD = Clear design　清洁设计
CD = Committee draft　(国际标准的)委员会文件
CDC = Centers for Disease Control　疾病控制中心
CDG = Carriage of dangerous good　危险品运输
CDM = Clean development mechanism　无污染发展机制
CDMA = Code Division Multiple Access　码分多址[船载北斗卫星导航系统]
CDV = Committee draft for vote　(国际标准的)供投票的委员会文件
CE = Council of Europe　欧洲委员会
CEFIC = European Chemical Industry Council　欧洲化学工业委员会

CEN = Comite Europeen de Nomalisation (European Committee for Standardization) 欧洲标准化委员会
CENELEC = Comite Europeen de Nomalisation Electrotechnique (European Committee for Electrotechnical Standardization) 欧洲电工标准化委员会
CEO = Chief executive officer 首席执行官(执行总裁)
CES = Coast earth station （卫星导航)沿岸地面站
CES = Cognitive environment simulator 认知环境模拟
CESA = Community of European Shipyards Association (previous known as the Association of European Shipbuilders and Shiprepairers) 欧洲造船协会共同体(前称为欧洲船厂和修船厂协会)
CET = Continuous event tree 连续事故树
CFD = Computational fluid dynamics 计算流体力学
CFO = Cargo fuel oil 货物燃油
CFR = Code of Federal Regulations （美国)联邦政府法规
CGCS = China Geodetic Coordinate System 中国大地坐标系[船载北斗卫星导航系统]
CIE = International Commission of Illumination 国际照明委员会
CIMS = Calculator integrated manufacturing system 计算机集成制造系统
CIN = Craft identification number 艇的识别号
Circ. = Circulars (International Maritime Organization) 通函(国际海事组织文件)
CIRM = International Radio-Maritime Committee 国际无线电海事委员会
CISPR = International Special Committee on Radio-Interference 国际无线电干扰特别委员会
CIT = Critical incident tecnique 危险事故咨询
CLC 1969 = International Convention on civil liability for oil-pollution damage,1969 1969年国际油污损害民事责任公约
CLC PROT 1992 = Protocol to the International convention on civil liability for oil pollution damage, 1969 (1992) 1969年国际油污损害民事责任公约1992年议定书
CLIA = Cruise Lines International Association 国际(定期)旅游船协会
CLIA = Cruise Line International Association 国际邮轮协会
CM = Construction monitoring 建造监测
CMI = Committee Maritime International 国际海事委员会
CMS = Continuous machinery survey 循环机械检验
CMS = Convention on the Conservation of Migratory Species of Wild Animals 保护野生动物的迁徙物种公约
CN = Classification Notes 规范修改通报

CNG = Compressed natural gas 压缩天然气
CNIS = China National Institute of Standardization 中国标准化研究院
COC = Certificate of compliance 符合证书
COD = Chemical oxygen demand 化学需氧量(生活污水排出物指标)
CODAG = Combined diesel and gas turbine 柴油机和燃气轮机组合
CODAM = Corrosion and damage database 腐蚀破损数据库
CODED = Combined diesel-electric and diesel 柴油机-电力和柴油机组合
CODLAG = Combined diesel-electric and gas 柴油机-电力和燃气轮机组合
CODOC = Combined Diesel engines and/or Gas turbines (power plant) 柴油机和/或燃气轮联合(动力)装置
CoG = Center of gravity 重心
COG = Course over ground 对地航向
COGAS = Combined gas turbine and steam 燃气轮机和蒸汽组合
COGES = Combined steam turbine and gas turbine 燃气轮机和蒸汽轮机组合
COLREG 1972 = Convention on the International Regulations for Preventing Collisions at Sea,1972 1972年国际海上避碰规则公约
COLREG = Interpretations of the COLREG 国际海上避碰规则的解释(IACS统一解释)
COLREG = International Regulations for the Prevention of Collisions at sea 国际海上避碰规则
COMAH = Control of major accident hazard 重大事故隐患监控
COMSAR = Sub-Committee on Radio Communications and Search and Rescue 无线电通信与搜救分委会
COMSAT = Communication Satellite Corporation （美国)通信卫星公司
COPANT = Pan American Technical Standards Committee 泛美技术标准委员会
COSAG = Combined steam and gas turbine 蒸汽轮机和燃气轮机组合
COT = Crude oil tank 原油舱
COTP = Captain of port (USCG) （美国海岸警卫队)港区司令官
COW(COWS) = Crude oil washing (system) 原油洗舱(系统)
CPA = Closest point of approach 最接近的目标点
CPI = Permanent International & European Commission of

Industrial Gases & Calcium Carbide 工业瓦斯和电石国际及欧洲常设委员会
CPP = Controllable pitch propeller 可调螺距螺旋桨
CPP = Cargo piping protected 货油管保护
CPPS = Permanent Commission for the South Pacific 南太平洋常设委员会
CPS = Coating performance standard 涂层性能标准
CPU = Central processing unit 中央处理设备
CRCH = Carriage of refrigerated containers in holds 货舱内装运冷藏集装箱
CREAM = Cognitive reliability and error analysis method 认知可靠性与认知错误分析方法
CRS = Coast radio station 海岸无线电台
CRS = Croatian Register of Shipping 克罗地亚船级社
CSA = Canadian Standards Association 加拿大标准协会
CSC 1972 = International Convention for Safe Containers, 1972 1972年国际安全集装箱公约
CSC = Container securing certificate 集装箱系固证书
CSE = Concept safety evaluation 概念安全性评价
CSM = Cargo securing manual 货物系固手册
CSMART = Center for simulator marine training 海洋模拟器训练中心
CSO = Company security officer 公司保安官员
CSQA = China Classification Society Quality Assurance LTD 中国船级社质量论证公司
CSQS = China shipbuilding quality standard 中国造船质量标准
CSR(CSO) = Continuous service rating (of engine) (主机)连续服务功率
CSR = Common structural rules (IACS)共同(结构)规范
CSR = Continuous synopsis record 连续概要记录
CSS Code = Code of safe practice for cargo stowage and securing 货物堆装和系固安全操作规则
CSSN = China Standards Service Net 中国标准服务网
CSTDMA = Carrier-sense time division multiple access(techniques) 载波侦听时间分割的多路存取(技术)
CTF = Coating technical file 涂装技术文件
CTU = Cargo transport units 货物运输单元
CVSSA = Cruise Vessel Security and Safety Act 邮轮保安和安全法案
CWA = Clean water act (美国)清洁水法
CWT = Cooling water tank 冷却水舱

D

DACC = Dynamic air cushion craft 动力气垫船
DACWIG = Dynamic air cushion (PAR)WIG 动力气垫型地效翼船
DAD = Decision action diagrams 决策行为图
DAE = Design accidental events 设计意外事件
DAL = Design accidental loads 设计意外载荷
DAL = Determinate accidental load 确定的事故载荷
DAP = Detail assembly procedure 详细组合要令
DASR = Document of authorization to undertake ship recycling 拆船厂授权书
DBMS = Data base management system 数据库管理系统
DC = Direct current 直流电
DCBT = Dedicated clean ballast tanks 专用清洁压载舱
DCE = Dangerous cargo endorsements 危险货物背书
DE = Sub-Committee on Ship Design & Equipment 船舶设计和设备分委会
DES = Discrete event simulation 离散事件仿真
DET = Discrete event tree 离散事件树
DETAM = Dynamic event tree analysis method 动态事件树分析法
DFD = Dual-fuel diesel 双燃料柴油机
DFDE = Dual-fuel diesel engine (-electric propulsion system) 双燃料柴油机(-电气推进系统)
DFEP = Dual-fuel electric propulsion 双燃料电力推进
DFGT = Dual-fuel gas turbine 双燃料燃气轮机
DFM = Dual-fuel machine 双燃料机械
DFO = Domestic fuel oil 生活用燃油
DFT = Dry film thickness 干膜厚度
DFU = Defined situations of hazard and accident 危险与事故状态定义
DG = Dangerous goods 危险品
DGAC = Dangerous Goods Advisory Council 危险品咨询理事会
DGLONASS = Differential Global Navigation Satellite Systems 差分全球导航卫星系统
DGPS = Differential global positioning system 差分全球定位系统
DHI = Deutsches Hydrographisches Institut (German hydrology institute) 德国水文研究所
DHS = Department of Homeland Security (美国)国土安全部
DHSV = Down home safety valve 井下安全阀
DIN = Deutsche Industrie Normen (German industrial standard) 德国工业标准
DIP = Dust-ignition-proof 防粉尘点燃
DIS = Draft information system 吃水信息系统
DIS = Draft international standard 国际标准草案
DLA = Dynamic loading analysis 动态载荷分析

DMA = Danish Maritime Authority　丹麦海事局
DMC = Distress message controller　遇险信息控制器
DMG = Distress message generator　遇险信息发生器
DMLC = Declaration maritime labour compliance　海事劳工符合声明
DN = Nominal diameter　公称的直径
DNV-GL = Det Norske Veritas-Germanischer Lloyd　挪威-德国船级社
DO = Diesel oil　柴油
DOC = Document of compliance　符合声明(拆船公约"有害物质清单编制指南"要求提供)
DOL = Direct on line　航线指引
DOS = Declaration of security　保安公告
DOT = Department of Transportation　(美国)运输部
DOT = Department of Trade　(英国)贸易部
DP(DPS) = Dynamic positioning system　动力定位(系统)
DSC Code = Code of safety for dynamically supported craft　动力支承艇安全规则
DSC = Direct self-control　直接自控制
DSC = Digital selective calling　数字选择性呼叫
DSC = Sub-Committee on The Carriage of Dangerous Goods Solid Cargoes and Containers　危险品、固体货物和集装箱运输分委会
DTC = Direct torque control　直接力矩控制
DTI = Department of Trade and Industry　(英国)贸易工业部
DW = Drill water　钻井用水
DWT = Deadweight tons　载重吨
DYLAM = Dynamic logic analysis methodology　动态逻辑分析分类法

E

E&P Forum = Oil Industry International Exploration and Production Forum　石油工业国际勘探和生产论坛
E&P Forum = Previous name of organization, now called OGP　勘探与生产论坛(组织原名称,现为OGP)
E&P = Exploration and production　勘探与生产
EAL = Environmentally Acceptable Lubricants　环保(可接受的)润滑油[美国环保署(EPA)船舶通用许可(Final 2013 VGP)要求]
EC = Electronic chart　电子海图
EC = The European Community　欧洲共同体
ECA = Emission control area　排放控制区
ECC = Engine control cabinet　发动机控制台
ECDB = Electronic chart data base　电子海图数据库

ECDE = Electronic chart display equipment　电子海图显示设备
ECDIS = Electronic chart display and information systems　电子海图显示和信息系统
ECE = Economic Commission for Europe　欧洲经济委员会
ECMC = European Container Manufacturers' Committee　欧洲集装箱制造厂商委员会
ECOR = Engineering Committee on Oceanic Resources　海洋资源工程委员会
ECR = Engine control room　发动机控制室
ECS = Electronic chart system　电子海图系统
EDCS = Electronic data collection system　电子数据收集系统
EEBD = Emergency escape breathing device　紧急逃生呼吸装置
EEC = The European Economic Community　欧洲经济共同体
EEDI = Energy efficiency design index　能效设计指数
EEEC = Commission of the European (Economic) Communities　欧洲(经济)共同体委员会
EEOI = Energy efficiency operational index　能效营运指数
EER = Escape, evacuation and rescue　逃生、撤离和救援
EES = Emergency Equipment Signs　应急设备标志(IMO大会决议 A.1116(30)脱险通道标志和设备位置标记 2017年12月5日)
EESLR = Risk due to explosion escalation by small leaks　因小泄漏而使爆炸升级所产生的风险
EET = Expanded event tree　扩展事件树
EEZ = The exclusive economic zone　专属经济区
EFRSLR = Risk due to fire escalation by small leaks　因小泄漏而使火灾升级所产生的风险
EGC Code = Code for Existing Ships Carrying Liquefied Gases in Bulk　现有的散装运输液化气船舶规则
EGC = Enhanced group call　强化群呼
EGC = Exhaust gas cleaning　废气清洗
EGCS = Exhaust gas cleaning systems　废气清洗系统
EGR = Exhaust Gas Recirculation　废气再循环
EIA = Electronic Industries Association　(美国)电子工业协会
EIA = Environmental impact assessment　环境影响评价
EIAPP = Engine international air pollution prevention certificates　发动机国际防止空气污染证书
EIF = Environmental impact factor　环境影响因素
ELSBM = Exposed location single buoy mooring system　旷海单浮筒系泊系统
ELT = Emergency location transmitter　应急位置发信机

EMBA = Executive master of business administration 高级管理人员工商管理硕士课程
EMC = Equipment monitoring control 监视控制设备
EMC = Electromagnetic compatibility 电磁兼容性
EMK = Emergency medical kit/bag 全套应急医疗用具
EmS = Emergency procedures for ships carrying dangerous goods 运输危险货物船舶的应急程序
EMS = Environment management system 环境管理体系
EMSA = European Maritime Safety Authority 欧洲海事安全局
EN = Equipment number 舾装数
EN = European Committee for Standardization 欧洲标准化委员会
EN = European Standards 欧洲标准
ENC = Electronic navigational chart or electronic nautical chart 电子航行海图
EP = Environmental protection 环境保护
EP = European Parliament 欧洲议会
EPA = Electronic plotting aid 电子标绘仪
EPA = Environmental position indicating radio beacon 指示环境位置无线电信标
EPA = Environmental Protection Agency （美国）环境保护署（环保署）
EPFS = Electronic position-fixing system 电子定位系统
EPIRB = Emergency position indicating radio beacon 应急无线电示位标
EQDC = Emergency quick dissociator 应急快速分离器
ER = Engine room 机舱
ERS = Emergency response service 应急响应服务
ES = Enhanced scantlings 增强的构件尺寸
ES = Environmental safety 环境安全
ESA = European Space Agency 欧洲空间机构
ESD = Emergency shut down 应急切断
ESD = Event sequence diagram 事件序列图
ESI = Environmental ship index 船舶环境指数
ESM = Electronic supervisory measures 电子监测装置
ESM = Environmental sound management 环境无害化管理
ESN = Enhanced survivability notation 加强残存能力标志
ESP Code 2011 = International Code on Enhanced Programme of Inspections during Surveys of Bulk Carriers and Oil tankers, 2011 2011年散货船和油船检验期间的强化检查方案国际规则
ESP = Enhanced survey program 强化检验方案
ESPH = Evaluation of safety and pollution hazards of chemicals 化学品安全和污染危害评估
ETA = Emergency towing arrangements 应急拖带装置
ETA = Estimated time of arrival 预计到达时间
ETA = European Tug owners Association 欧洲拖轮船东协会
ETA = Event tree analysis 事件树分析
ETAS = Emergency technical assistance service 应急技术支持服务
ETB = Emergency Towing Book 应急拖带手册
ETD = Embedded temperature detectors 埋入式温度检测器
ETM-A = EGC system technical manual for scheme A 方案A的EGC系统技术手册
ETM-B = EGC system technical manual for scheme B 方案B的EGC系统技术手册
ETS = European telecommunication standards 欧洲电信标准
ETSI = European Telecommunication Standards Institute 欧洲电信标准学会
ETV = Environmental Technology Verification (Program) (BWMS) 环境技术验证（方案）（压载水管理系统）
EU = European Union 欧洲联盟
EUROMOT = European Association of Internal Combustion Engine Manufacturers (previously known as the Association of European Manufacturers of Internal Combustion Engines) 欧洲内燃机制造厂商协会（前称为欧洲内燃机制造厂商协会）
EUROSPACE = European Industrial Space Study Group 欧洲工业空间研究组
EUT = Equipment under test 被试设备
EVTMS = Enhanced vessel traffic management system 增强的船舶通航管理系统
Ex = Explosion (protected) 爆炸（保护型）
EX-proof = Explosion proof equipment 防爆设备

F

FAA = The Federal Aviation Adiministration 美国联邦航空署
FACAT = Foil-assisted catamaran = catamaran (CAT) + hydrofoil (HYC) 水翼双体船
FAL 1965 = Convention on Facilitation of International Maritime Traffic, 1965 1965年国际便利海上运输公约
FAL = Facilitation Committee 便利运输委员会（便运会）
FAO = Food & Agricultural Organization of the United Nations 联合国粮农组织
FAR = Fatal accident rate 死亡事故率

FBB = Fleet broadband 高速宽带网
FBI = Federal Bureau of Investigation 联邦调查局
FC = Fruit carrier 水果运输船
FCC = Fully Cellular Containership 全箱格导轨式集装箱船(苏伊士运河航行规则 2015 年 8 月版)
FDA = Fatigue design assessment 疲劳设计评估
FDIS = Final draft international standard 最终国际标准草案
FE = Finite element 有限元
FEA = Finite element analysis 有限元分析
FEC = Forward error correction 向前纠错
FEM = Finite element method 有限元法
FES = Fire-fighting equipment signs 防火设备标志(IMO 大会决议 A.1116(30)脱险通道标志和设备位置标记 2017 年 12 月 5 日)
FFA = Fire fighting appliances 灭火装置
FFA = Fuctional failure analysis 功能失效分析
FFD = Free form deformation 几何造型技术
Fi-Fi = Fire fighting 消防,灭火
FL = Filling limit 充装极限(IMO 的 IGF Code 中采用)
FL = Fatigue life 疲劳寿命
FLACS = Flame accelerator software 促燃爆炸仿真软件
FMC = Federal Marine Commission (美国)联邦海事委员会
FMEA = Failure mode and effect analysis 故障模式和影响分析
FMECA = Failure mode, effect and criticality analysis 故障模式、影响和危害程度分析
FMS = Fleet management system 船队管理系统
FMSN = Fleet management system network 船队管理系统网络
FO = Fuel oil 燃油
FOE = Friends of the Earth International 国际地球之友
FOI = Floating offshore installation 浮式近海装置
FONASBA = Federation of National Association of Ship Brokers and Agents 国家船舶经纪人和代理商联合会
FOV = Field of vision 视野
FP = Fixed pitch propeller 定距螺旋桨
FP = Sub-committee on Fire Protection 防火分委会
FPD = Fall prevention devices 防跌落装置(救生艇钩)
FPP = Fixed pitch propeller 定螺距螺旋桨(定距桨)
FPPY = Fatalities per platform year 每平台年的死亡率
FPS = Floating production system 浮式生产系统
FPSO = Floating production, storage and offloading facilities 浮式生产储(存和卸)油船(装置)
FPU = Floating production unit 浮式生产装置
FR = Federal register (美国)联邦法规登记
FRC = Fast rescue craft 快速救生艇
Frl = Froude number based on waterline length, $V/(gL)^{0.5}$ 基于水线长的傅汝德数
Fr_L = Froude number 傅汝德数
FRP = Fibre reinforced plastics 玻璃纤维增强塑料
FSA = Formal safety assessment 综合安全评估
FSC = Flag State control 船旗国控制
FSI = Sub-committee on flag State Implementation 船旗国管理分委会
FSICR = Finnish-Swedish ice-class rules 芬兰-瑞典冰级规范
FSS Code = International code for fire safety systems 国际消防安全系统规则
FSS = Future ship safety 未来船舶安全
FSU = Floating storage units 浮式储油装置
FTA = Failure tree analysis 故障树分析
FTP = Fuel tank protection 燃油舱保护
FTP Code = International Code for application of fire test procedures 国际实施耐火试验程序规则
FTP = Fire test procedures 耐火试验程序
FUND 1971 = International Convention on the establishment of an international fund for compensation for oil pollution damage, 1971 1971 年设立国际油污损害赔偿基金公约
FUND PROT 1992 = Protocol to the International convention on the establishment of an international fund for compensation for oil pollution damage, 1971(1992) 1971 年设立国际油污损害赔偿基金公约 1992 年议定书
FW = Fresh water 淡水
FWBLAFFS = Fixed water-based local application fire-fighting system 固定式水基局部使用灭火系统
Fwd. = Forward 在船首(艏部的、前面的、向前)
FY = Fiscal year 财政年度

G

GA(P) = General arrangement(plan) 总布置(平面图)
GAS = Grounding avoidance system 避免搁浅系统
GATT = General Agreement on Tariffs & Trade 关税及贸易总协定(关贸总协定)
GBS = Goal-based (new ship construction) standards 目标型(新船建造)标准
GBS = Gravity base structure 重力基座结构
GC = Ships carrying liquefied gases in bulk (IACS 统一解释)散装液化气船
GC Code = Code for the Construction and Equipment of Ships Carrying Liquefied Gases in Bulk 散装运输液化气船舶

构造与设备规则
GCU = Gas combustion unit   燃气燃烧装置（气体燃烧单元）
GEMS = Generic error modeling system   普通错误模型系统
GEMS = Global environmental monitoring system   全球环境监察系统
GEO = Geosynchronous Earth Orbit (satellites)   地球同步轨道（卫星）[船载北斗卫星导航系统]
GESAMP = Joint Group of Experts on the Scientific Aspects of Marine Pollution   海洋污染科学问题联合专家组
GESAMP-BWWG = Joint Group of Experts on the Scientific Aspects of Marine Pollution-Ballast Water Working Group   海洋污染科学问题联合专家组—压载水工作组
GEZ = Ground effect zone   地效区
GF = Interpretations of the IGF Code   国际使用燃气或其他低闪点燃料船舶安全规则的解释(IACS统一解释)
GGA = Global positioning system fix data   全球定位系统定位数据
GHG = Green house gas   温室气体
GI = Greenpeace International   国际绿色和平组织
GIE = Group of Independent Experts   独立专家组
GIR = Group individual risk   群体－个体风险
GIS = Geographical information system   地理信息系统
GISIS = Global integrated shipping information system   全球综合航运信息系统
GLONASS = Global navigation satellite systems   全球导航卫星系统
GM = Metacentric height   稳心高度
GMDSS = Global maritime distress and safety system   全球海上遇险与安全系统
GNSS satellite fault detection   全球导航卫星故障检测
GNSS = Global navigation satellite system   全球导航卫星系统
GoM = Gulf of Mexico   墨西哥湾
GOST = Russia Standard   独联体（俄罗斯）标准
GPR = Green passport for recycling   绿色拆船护照
GPS = Global positioning system   全球定位系统
GR = Group risk   群体风险
Grain Code = International Code for the Safe Carriage of Grain in Bulk   国际安全散装运输谷物规则
GRP = Glass-fibre reinforced plastic   玻璃纤维增强塑料
GRT = Gross registered tons   注册总吨位
GSK = Group survival kit   团体救生工具
GSM = Global system for mobile   全球移动电话系统
GSM = Group special mobile   特殊移动电话群组
GST = Global telecommunication system   全球无线电通信系统
GWC = Grey water control   灰水控制

H

HACCP = Hazard analysis and critical control point   危害分析和关键点控制
HANDYMAX = HANDYMAX oil tanker   灵便型油船
HARDER = Harmonization of rules and design rational   规范与合理设计的协调
HAZID = Hazard identification   危险识别
HAZOP = Hazard and operability study   危险与可操作性研究
HC = Hydrocarbon   碳氢化合物,烃
HCLIP = Hydrocarbon leak and inventory project   油气泄漏与库存项目
HCM = Hull condition monitoring   船体状态监测
HCR = Human cognitive reliability   人的认知可靠性分析方法
HCS = Hatch cover strength   舱口盖加强
HCS = Heading control system   （船舶）艏向控制系统
HCSR = Harmonized common structural rules   (IACS)协调版共同(结构)规范
HDG = Heading   船艏向
HDTWPL = Heading, true, waypoint location   真艏向,航路点位置
HEA = Human error analysis   人为错误分析
HEAP = Human element analysing process (Formal safety assessment)   人的因素分析程序(综合安全评估)
HEART = Human error assessment and reduction technique   人为错误评估和减少技术
HEI = Human error identification   人为错误识别
HELMEPA = Hellenic marine environmental protection association   希腊海洋环境保护协会
HEP = Human error probability   人为错误概率
HES = Health, environment and safety   健康、环境与安全
HF = High frequency   高频
HFO = Heavy fuel oil   重燃油
HFTG = Human factors task group   人类因素任务组
HGO = Heavy grade oil   重质油
HIPPS = High integrity pressure system   高完整性压力保护系统
HKMD = Hong Kong Maritime Department   （中国）香港海事处
HLP = Hull life cycle programme   船体生命周期程序
HME = Harmful to the marine environment(substance)   对

海洋环境有害物质

HMI = Human machine interface　人机界面

HMIS = Hazardous material information system　危险材料信息系统

HMSI = Non-maintenance significant-iterms　非重点维护项目

HNS Convention 1996 = International Convention on Liability and Compensation for Damage in Connection with the Carriage of Hazardous and Noxious Substances by Sea, 1996　1996年与有害和有毒物质载运有关的破损的责任和赔偿的国际公约

HNS Protocol = Protocol on Preparedness, Response and Co-operation to Pollution Incidents by Hazardous and Noxious Substances, 2000　2000年由有害和有毒物质引起的污染事件的准备、响应和合作的议定书

HOE = Human and organization error　人和组织错误

HOF = Human and organizational factors　人为与组织因素

HONG KONG SRC 2009 = Hong Kong International Convention for the Safe and Environmentally Sound Recycling of Ships, 2009　2009年香港国际安全与环境无害化拆船公约

HPAV = High-performance marine vehicle　高性能船舶

HPMS = Hull planned maintenance scheme　船体计划维修制

HR = Human reliability　人因可靠性

HRA = Human reliability analysis　人因可靠性分析

HRMS = Human reliability management system　人因可靠性管理系统

HRS = Hellenic register of shipping　希腊船级社

HS = Hazardous substances　有害物质

HSC = High-speed craft code　（IACS统一解释）高速艇规则

HSC Code = International Code of Safety for High-Speed Craft　国际高速艇安全规则

HSE = Health and safety executive　（英国）健康与安全管理局

HSS = Hull surveillance systems　船体监视系统

HSSC = Harmonized system of survey and certification　检验和发证协调体系

HT = High temperature　高温

HTA = Hicrarchical task analysis　分层任务分析方法

HTSM = High temperature superconductor motor　高温超导同步电动机

HTW = Sub-Committee on Human Element, Training and Watchkeeping (IMO)　人为因素、培训和值班分委会

HVAC = Heating, ventilation and air conditioning (system)　取暖、通风和空调(系统)

HVSC = High voltage shore connection　高压岸电连接

HVSCS = High voltage shore connection systems　高压岸电连接系统

HW = Hardware　硬件

HW = High water　高潮

HYC = Hydrofoil catamaran　水翼双体船

HYF = Hydrofoil　水翼

HYSWAC = Hydrofoil small water plane area twin hull craft = small water plane area twin hull craft + hydrofoil　水翼型小水线面双体船（小水线面双体船+水翼）

I

I/O = Input/output　输入/输出

IACS = International Association of Classification Society　国际船级社协会

IADC = International Association of Drilling Contractors　国际钻井承包商协会

IAEA = International Atomic Energy Agency　国际原子能机构

IAIN = International Association of Institutes of Navigation　国际航海学会协会

IALA = International Association of Maritime Aids to Navigation and Lighthouse Authorities　国际灯塔协会/国际航道标志协会

IAMSAR manual = International aeronautical and maritime search and rescue manual　国际航空与航海搜救手册

IAMU = International Association of Maritime University　国际海事大学协会

IAPH = International Association of Ports and Harbors　国际港口协会

IAPP = International air pollution prevention (certificate)　国际防止空气污染(证书)

IAPRI = International Association of Packing Research Institutes　包装研究学会国际协会

IATA = International Air Transport Association　国际航空运输协会

IBA = International Bar Association　国际律师协会

IBC Code = International code for the construction and equipment of ships carrying dangerous chemicals in bulk　国际散装运输危险化学品船舶构造与设备规则(国际散化规则)

IBC = Intermediate bulk container　中间散装容器

IBIA = International Bunker Industry Association　国际船舶燃料工业协会

IBRD = International Bank for Reconstruction & Development 国际复兴开发银行
IBS = Integrated bridge system 综合桥楼系统
IBTA = International Bulk Terminal Association 国际散货码头协会
IBTS = Integrated bilge water treatment systems 综合舱底水处理系统
ICAO = International Civil Aviation Organization 国际民航组织
ICC = Integrated computer control 综合计算机控制
ICC = International Chamber of Commerce 国际商会
ICCL = Cruise Lines International Association ［previously known as the International Comcil of Cruise Lines(ICCL)］ 国际邮轮公司协会(前称为国际邮轮公司理事会)
ICCPP = Impressed current cathodic, protection 外加电流阴极保护
ICEM = Inter-Governmental Committee for European Migration 政府间欧洲移民委员会
ICFTU = International Confederation of Free Trade Unions 国际自由工会联合会
ICHCA = International Cargo Handling Coordination Association 国际货物装卸协调协会
ICIHM = International Certificate on Inventory of Hazardous Materials 国际有害物质清单证书(拆船公约"有害物质清单编制指南"要求提供)
ICMA = International Christian Maritime Association 国际基督徒海事协会
ICOMIA = International Council of Marine Industry Associations 国际船舶工业协会理事会
ICON DP = Rolls-Royce's dynamic positioning system 罗尔斯-罗伊斯动力定位系统
ICS = International Chamber of Shipping Limited 国际航运公会
ICS = International Chamber of shipping 国际箱式运输协会
ICS = International Code of Signals 国际信号代码
IDA = International Development Association 国际开发协会
IDC = International LRIT Data Center 国际(远程识别与跟踪)数据中心
IDE = International LRIT data exchange 国际(远程识别与跟踪)数据交换
IEC = International Electrotechnical Committee 国际电工委员会
IEE = International Energy Efficiency(Certificate) 国际能效(证书)

IEEC = International energy efficiency certificate 国际能效证书
IEEE = International of Electrical and Electronic Engineers 国际电气和电子工程师协会
IEP = Integrated electric propulsion(system) 综合电力推进(系统)
IES = Illuminating Engineering Society (美国)照明工程协会
IEV = International electrotechnical vocabulary 国际电工词汇
IFAC = International Federation of Automatic Control 国际自动控制联合会
IFAN = International Federation of Standard Users 国际标准用户联盟
IFAW = International Fund for Animal Welfare 国际爱护动物基金会
IFC = International Finance Corporation 国际金融公司
IFC = Republic of Singapore Navy Information Fusion Center 新加坡海军情报中心
IFEP = Integrated full electric propulsion system 综合全电力推进系统
IFP = Integrated fire protection (system) 综合防火(系统)
IFSMA = International Federation of Shipmasters' Associations 国际船长协会联合会
IGC Code = International code for the construction and equipment of ships carrying liquefied gases in bulk 国际散装运输液化气船舶构造与设备规则
IGC = Inert gas system 惰性气体系统
IGCT = Integrated gate commutated thyristors 集成门极换流晶闸管
IGF Code = International code on safety for natural gas-fueled ships 国际天然气燃料船舶安全规则
IGS = Inert gas system 惰性气体系统
IGSO = Inclined Geosynchronous Satellite Orbit 倾斜地球同步轨道(卫星)［船载北斗卫星导航系统］
IGU = International Gas Union 国际天然气联合会
IHM = Inventory of hazardous materials 有害物质清单
IHMA = International Harbour Masters' Association 国际港务长协会
IHO = International Hydrographic Organization 国际航道测量组织
IICL = Institute of International Container Lessors 国际集装箱出租商协会
IIDM = Ibero-Amerika Institute of Maritime Law 伊比利亚-美洲海事法学会

Ⅲ Code = IMO Instruments Implementation Code　IMO 文件实施规则

Ⅲ = Sub-Committee on Implementation of IMO Instruments　IMO 文件实施分委员会

IIR = International Institute of Refrigeration　国际制冷学会

IIWG = Inter-Industry Working Group　工业间工作组

IJS = Independent Joystick system　独立操纵杆系统

ILA = International Law Association　国际法律协会

ILAMA = International Lifesaving Appliance Manufacturers' Association　国际救生设备制造商协会

ILC = International Labor Convention　国际劳工公约

ILLC = International Convention on Load Lines　国际载重线公约

ILO = International Labour Organization　国际劳工组织

IMarEST = Institute of Marine Engineering, Science and Technology［previously known as the Institute of Marine Engineers(IME)］　船舶工程、科学和技术学会(前称为船舶工程师协会)

IMB = International Maritime Bureau　国际海事局

IMC = International Maritime Committee　国际海事委员会

IMCA = International Maritime Contractors Association　国际海事承包商协会

IMCO CONVENTION = Convention on the Inter-Governmental Maritime Consultative Organization　政府间海事协商组织公约

IMCO = Inter-Governmental Maritime Consultative Organization　政府间海事协商组织

IMDG Code = International Maritime Dangerous Goods Code　国际海运危险货物规则

IMEC = International Marine Equipment Committee　国际船用设备委员会

IMF = International Monetary Fund　国际货币基金组织

IMHA = International Maritime Health Association　国际海员健康协会

IMLA = International Maritime Lecturers Association　国际海事教员协会

IMLI = International Maritime Law Institute　国际海事法律学院(IMO 决议 A.1129(30)世界海事大学和国际海事法律学院学生访问 IMO 总部,2017 年 12 月 6 日通过)

IMO Convention = Convention on the International Maritime Organization　国际海事组织公约

IMOSAR manual = IMO search and rescue manual　国际海事组织搜索与营救手册

IMO = International Maritime Organization　国际海事组织

IMPA = International Maritime Pilot's Association　国际海上引航员协会

IMR = Inspection maintenance and repair　检查维护和修理

IMRF = International Maritime Rescue Federation (previously known as the International lifeboat Federation)　国际海上救助联合会(前称为国际救生艇联合会)

IMS = Integrated management system　综合管理体系

IMSBC Code = International maritime solid bulk cargoes code　国际海运固体散装货物规则

IMSO C 1976 = Convention on the International Maritime Satellite Organization, 1976　1976 年国际海事卫星组织公约

IMSO = International Mobile Satellite Organization　国际移动卫星组织

INEC = International Marine Equipment Committee　国际船用设备委员会

INF Code = International code for the safe carriage of packaged irradiated nuclear fuel, plutonium and high-level radioactive wastes in flasks on board ships　国际船舶安全装运密封装辐射性核燃料、钚和强放射性废料规则

INF = Irradiated nuclear fuel　核乏燃料

INMARSAT 1976 = Convention on the International Maritime Satellite Organization, 1976　1976 年国际海事卫星组织公约

INMARSAT OPERATING AGREEMENT = Operating agreement on the international maritime satellite organization　国际海事卫星组织使用协定

INMARSAT = International Maritime Satellite Organization　国际海事卫星组织

INS = Integrated navigation system　综合航行系统

INSA = International Shipowners Association　国际船东协会

INTERCARGO = International Association of Dry Cargo Shipowner　国际干货船东协会

INTERFERRY = International Association of Ferry　国际渡船协会

Intermanager = International Ship Managers Association(previously known as ISMA)　国际船舶管理人协会(前称为 ISMA)

INTERTANKO = International Association of Independent Tanker Owners　国际独立油船东协会

INTERVENTION 1969 = International convention relating to intervention on the high seas in cases of oil pollution casualties, 1969　1969 年国际干预公海油污事件公约

INTERVENTION PROT 1973 = Protocol relating to intervention on the high seas in cases of marine pollution by substances other than oil, 1973　1973 年干预公海非油类物

质污染议定书
IOI = International Ocean Institute　国际海洋学会
IOPC Fund = International Oil Pollution Compensation Fund　国际油污赔偿基金
IOPP Certificate = International Oil Pollution Prevention Certificate　国际防油污证书
IP = Institute of Petroleum　（英国）石油学会（标准）
IP = Intellectual property　知识产权
IPCEA = Insulated Power Cable Engineer Association　（美国）绝缘电力电缆工程师协会
IPIECA = International Petroleum Industry Environment Association　国际石油工业环境保护
IPOTF = International Tanker Owners Pollution Federation LTD (previously known as the International ship Suppliers Association)　国际油船船东防污染联合会
IPPIC = International Paint and Printing Ink Council (Control of Harmful Anti-Fouling Systems on Ships, AFS)　国际油漆和印刷油墨理事会（用于控制船舶有害的防污底系统）
IPS = Integrated Propulsion System　综合推进系统
IPTA = International Parcel Tankers Association　国际零担液货船协会
IR = Individual risk　个体风险
IRCC = Integral refrigerated container carrier　整体冷藏集装箱船
IRCS = Integrated radio communication system　综合无线电通信系统
IRCS = Shipboard integrated radio-communication system　船载组合无线电通信设备
IRNSS = Indian Regional Navigation Satellite System　印度区域导航卫星系统［船载北斗卫星导航系统中涉及的区域性导航系统］
fIRPA = Individual risk per annum　每年个体风险
IRRC = International ready for recycling certificate　国际适合拆船证书
IRS = Indian Register of Shipping　印度船级社
IRU = International Road Transport Union　国际道路运输联盟
IS Code 2008 = International Code on Intact Stability, 2008　2008年国际完整稳性规则
IS Code = Code on intact stability for all types of ships covered by IMO instruments (intact stability code)　国际海事组织文件中所有类型船舶的完整稳性规则（完整稳性规则）
IS = Intermediate survey　中间检验
ISA = Instrument Society of American　美国仪表协会
ISA = Instrumentation, System and Automatic Society　（国际）测量、系统和自动化协会
ISAF = International Sailing Federation　国际帆船联合会
ISC = International Chamber of Shipping　国际航运公会
ISCO = International Spill Control Organization　国际溢漏控制组织
ISCOS = Inter-Governmental Standing Committee on Shipping　政府间航运常设委员会
ISF = International Shipping Federation Limited　国际航运联合会
ISGOTT = International safety guide for oil tankers & terminals　国际油船与码头安全指南
ISM Code = International safety management code　国际安全管理规则
ISM = International safety management　国际安全管理
ISMA = International Ship Managers Association (Intermanaer)　国际船舶管理者协会
ISMA = International Superphosphate Manufacturers' Association Limited　国际过磷酸盐制造厂商协会
ISM Code = International management code for the safe operation of ships and for pollution prevention　国际船舶安全营运和防污染管理规则
ISMS = Information security management systems (ship construction file)　信息保安管理体系（船舶建造档案）
ISO = International Organization for Standardization　国际标准化组织
ISPP Certificate = International Sewage Pollution Prevention Certificate　国际防止生活污水证书
ISPS = International Ship and Port Facility Security Certificate　国际船舶和港口设施保安证书
ISPS Code = International ship and port facility security code　国际船舶和港口设施保安规则
ISSA = International Ship Suppliers and Services Association (previously known as International Ship Suppliers Association)　国际船舶供应商和服务协会（前称为国际船舶供应商协会）
ISSC = International Ship and Offshore Structure Congress　国际船舶与海洋工程结构大会
ISSC = International Ship Security Certificate　国际船舶保安证书
ISSETA = Inflatable Safety & Survival Equipment Trade Association　气胀式安全与救生设备贸易协会
ISU = International Salvage Union　国际海难救助联合会
ITB = Integrated Tug-Barge combination　推-驳组合体（巴拿马运河作业部 N-1-2018 航运通告）

ITCP = Integrated technical co-operation programme 综合技术合作计划

ITF = International Transport Workers' Federation ［Transferred to ITF from the International Confederation of free Trade Union(ICFTU,1961)］ 国际运输工人联盟［1961年自国际自由工会联合会(ICFTU)转至(ITF)］

ITRF = International Terrestrial Reference Frame (System) 国际地球参考框架(系统)［船载北斗卫星导航系统］

ITTC = International Towing Tank Conference 国际船模拖曳水池会议

ITU = International Telecommunication Union 国际电信联盟

IUA = International Underwriters Associations 国际保险商协会

IUCN = International Union for Conservation of Nature ［previously known as the World Conservation Union(IUCN)］ 国际自然保护联盟(前称为世界自然保护联盟)

IUMI = International Union of Marine Insurance 国际海上保险联盟

IWA = International workshop agreement 国际专题协议

IWS = In-water survey 水中检验

J

JBP = Joint Bulker Project （IACS)散货船联合项目组

JIP = Joint Industry Project 联合工业项目

JIS = Japanese industrial standards 日本工业标准

JMS = Japan marine standards 日本船舶标准

JMSA = Japan Marine Standards Association 日本船舶标准协会

JSA = Japanese Standards Association 日本标准协会

JTC 1 = Joint Technical Committee for Information Technology (ISO/IEC 第1联合技术委员会)信息技术联合技术委员会

JTP = Joint Tanker Project(IACS) 油船联合项目组

JWG = Joint MSC/MEPC Working Group MSC/MEPC(海上安全委员会/海上环境保护委员会)联合工作组

K

KPI = Key performance indicators 主要绩效指标

KRS = Korean Register of Shipping 韩国船级社

KS = Korean standards 韩国标准

L

LA = Link analysis 连接分析

LAN = Local area network 局域网

LAS = League of Arab States 阿拉伯国家联盟

LASA = Latin American Ship Owners' Association 拉丁美洲船东协会

Lat = Latitude 纬度

LC = Level control 水平控制

LC 1972 = Convention on the Prevention of Marine Pollution by Dumping of Wastes & Other Matters, 1972 1972年防止倾倒废弃物及其他物质污染海洋公约

LCAC = Landing craft, air cushion 气垫登陆艇

LCC = Large crude oil carrier 大型(原)油船

LCC = Life cycle cost 生命周期成本

LDC 1972 = Convention on the prevention of marine pollution by dumping of Wastes & Other Matters, 1972 1972年防止倾倒废弃物及其他物质污染海洋公约

LEG = Legal Committee 法律委员会(法委会)

LEL = Lower explosion limit (lower flammable limit) 爆炸下限(燃烧下限)

LFL = Lower flammability level 可燃性下限

LHNS Guidelines = Guidelines for the transport and handling of limited amounts of hazardous and noxious liquid substances in bulk on offshore support vessels 近海供应船散装有限数量有害和有毒液体物质的运输和装卸指南

LITP = Land-based information technology platform 陆基信息技术平台

LL = International Load Lines Certificate 国际载重线证书

LL 1966 = International convention on Load Lines,1966 1966年国际载重线公约

LL PROT 1988 = Protocol of 1988 relating to the International Convention on Load Lines,1966 1966年国际载重线公约1988年议定书

LL = Loading limit 加注极限(IMO 的 IGF Code 中采用)

LLL = Low location lighting 低位照明

LLMC1976 = Convention on limitation of liability for maritime claims,1976 1976年海事索赔责任限制公约

LNG = Liquefied natural gas 液化天然气

LO = Lubrication oil 润滑油

LOA = Length overall 总长

Long = Longitude 经度

LPG = Liquefied petroleum gas 液化石油气

LR = Lloyd's Register of Shipping (英国)劳氏船级社

LR = Long range 远程

LR1 = Long-range reply with destination for function request "A" A 类功能请求的远距离答复

LR2 = Long-range reply for function request "B, C, E and F" B, C, E 和 F 类请求的答复

LR3 = Long-range reply for function request "I, O, P, U and W" I, O, P, U 和 W 类请求的答复

LRAD = Long range acoustic devices 远程声学设备
LRF = Long range function 远距离功能
LRI = Long range interrogation 远距离询问
LRIT = Long-range identification and tracking（for ships）（船舶）远程识别与跟踪
LRQA = Lloyd's Register Quality Assurance Limited （英国）劳氏质量认证有限公司
LSA Code = International life-saving appliance code 国际救生设备规则
LSA = Life-saving appliance 救生设备
LSS = Life-Saving Systems and appliances signs 救生系统和设备标志［IMO 大会决议 A.1116（30）脱险通道标志和设备位置标记 2017 年 12 月 5 日］
LSW = Loudspeaker watch receiver 带扬声器的值班接收机
LT = Low temperature 低温
LTE = Low temperature environments 低温环境
LUT = Local user terminal 本地用户终端
LVSC = Low voltage shore connection（Systems） 低压岸电连接（系统）
LW = Low water 低潮

M

MAC = Medium access control 介质访问控制
MALS = Mitsubishi air lubrication system 三菱空气润滑系统
MARPOL 1973 = International convention for the prevention of pollution from ships,1973 1973 年国际防止船舶造成污染公约
MARPOL 73/78 = International convention for the prevention of pollution from ships,1973, as modified by the protocol of 1978 relating thereto 经 1978 年议定书修正的 1973 年国际防止船舶造成污染公约
MARPOL = Marine Pollution Convention 海洋污染公约
MARSEC = Marine security 海上安全
MARVS = Maximum Allowable Relief Valve Setting 最大容许的溢流阀设定值（IMO 的 IGF Code 中采用）
MAS = Mandatory annual survey 强制年度检验
MAS = Maritime assistance services 海事援助服务
MAWP = Maximum allowable working pressure 最大容许的工作压力（系指系统部件或气罐,IMO 的 IGF Code 中采用）
MBMs = Market-based measures 市场机制措施（温室气体减排）
MCAP = Condition assessment program of machinery 机械状态评估程序

MCB = Miniature circuit breaker 小型断电保护器回路
MCC = Motor control centre 电动机控制中心
MCC = Mission control center 执行控制中心
MCCB = Moulded case circuit breaker 断电保护器回路模板
MCM = Machinery condition monitoring 机械状态监测
MCR = Maximum continuous rating（of engine） （主机）最大持续功率
MD =（Hong Kong）Marine Department （中国）香港海事处
MD = Material document 材料声明（拆船公约"有害物质清单编制指南"要求由设备、部件和材料供应商提供）
MDCS = Mobile data collection system 移动数据收集系统
MDO = Marine diesel oil 船用柴油
Meg = Message 报文
MEIC = Most easily ignited concentration 最易点燃浓度
MEO = Medium-Earth Orbit（satellites） 中地球轨道（卫星）［船载北斗卫星导航系统］
MEPC = Maritime Environment Protection Committee 海上环境保护委员会（环保会）
MERSAR Manual = Merchant ship search and rescue manual 商船搜寻与救助手册
MES = Means of Escape Signs 脱险设施标志［IMO 大会决议 A.1116（30）脱险通道标志和设备位置标记 2017 年 12 月 5 日］
MES = Marine evacuation system 海上疏散系统
MF = Medium frequency 中频
MFAG = Medical first aid guide 急救指南
MFAG = Medical first aid guide for use in accidents involving dangerous goods 涉及危险货物的事故中采用的急救指南
MGO = Marine gas oil 船用轻质柴油
MGPS = Marine growth prevention system 海生物生长预防系统
MHB = Materials hazardous only in bulk 仅在散装时危险的材料（如散装运输会有危险,要求采取特殊措施的物质）
MHD = Magneto hydrodynamic drive 磁流体推进
MHI = Mitsubishi Heavy Industries 三菱重工
MIC = Minimum ignition current 最小点燃电流
MIE = Minimum ignition energy 最小点燃能量
MIL = Military specifications and standards （美国）军用规格和标准
MIL-SPEC = Military specifications （美国）军用规格
MIL-STD = Military standards （美国）军用标准
MIRA = Environment risk analysis（Miljørettet riskoanal-

yse) 环境风险分析
MIS = Manufacture information system 生产信息管理系统
MKD = Minimum keyboard and display 最小键盘和显示
MLC 2006 = Maritime Labour Convention, 2006 2006 年海事劳工公约
MLC = Marine Labour Convention 海事劳工公约
MLC = Maritime Labour Certificate 海事劳工证书
MM = Marine Marchande (法国)海运局
MMI = Man machine interface 人机界面
MMRTC = Mediterranean Maritime Research and Training Center 地中海海事研究和培训中心
MMSI = Maritime mobile service identity 海上移动业务识别码
MNOK = Million Norwegian kroner 百万挪威克朗
MOB = Man over board 人员落水
MOB = Mobile offshore base 移动式海上基地
MODU Code 2009 = Code for the Construction and Equipment of Mobile Offshore Drilling Units, 2009 2009 年海上移动式钻井平台构造和设备规则
MODU Code = Code for the construction and equipment of mobile offshore drilling units 移动式近海钻井装置构造和设备规则
MODU = Mobile offshore drilling unit 移动式海上钻井平台
MONALISA = Motorways and electronic navigation by intelligence at sea 高速公路和海上情报电子导航
MOU = Memorandum of Understanding 谅解备忘录
MOU = Mobile offshore units 移动式近海装置
MP = Main (propulsion) power 主(推进)动力
MPC = MARPOL Convention (IACS 统一解释)国际防止船舶造成污染公约
MPMS = Machinery planned maintenance scheme 机械计划维修制
MRCC = Maritime Rescue Co-ordination Center 海事救援协调中心
MROPC = Maritime regulations for the operation of the Panama Canal 巴拿马运河航行海事规则
MRP = Material resource plan 物料资源规划
MRU = Motion reference unit 运动基准装置
MSA China = Maritime Safety Administration China 中国海事局
MSC = Maritime Safety Committee 海上安全委员会(海安会)
MSCHOA = Maritime Security Center Hom of Africa 非洲海角海事保安中心
MSD = Marine sanitation device 海上卫生设备

MSDS = Material safety data sheets 材料安全数据单
MSF = Module support frame 模块支持构架
MSG = Maintenance steering group 维护控制组
MSI = Maritime safety information 海上安全信息
MSI = Maintenance significant group 重点维护项目
MSML = Maritime safety markup language 海上安全标记语言
MSP3 = 3rd Maritime safety package (欧洲联盟)第三套海上安全一揽子措施
MSRC = Marine Simulation and Resource Center 海洋模拟和资源中心
MTBF = Mean time between failure 平均失效间隔时间
MTIS = Marine Terminal Information System 海上码头信息系统
MTTF = Mean time to failure 平均失效时间
MWARC = Mobile world administrative radio conference 世界水上无线电行政会议

N

N/A = Not applicable 不适用
NACE = National Anti-Corrosion Engineer Institution 美国防腐蚀工程师协会
NANPCA = Non-indigenous Aquatic Nuisance Prevention and Control act of 1990 1990 年外来水生滋扰预防和控制法
NANPCA = The Non-indigenous Aquatic Nuisance Prevention and Control act of 1990 (美国)1990 年防止和控制外来水生物污染法令
NARA = National Archives and Records Administration (美国)国家档案与资料管理局
NASA = National Aeronautics and Space Adiministration 美国航空航天局
NATO = North Atlantic Treaty Organization 北大西洋公约组织
NAV = Sub-Committee on Safety of Navigation 航行安全分委会
NAVAREA = Navigational area 航行海区
NAVSEA = Naval Sea Systems Command (美国)海军海上系统指挥部
NAVTEX = Navigation telex 航行电传(航行气象告警电传接收机)
NBDP = Narrow band direct printing 窄带直接印字报
NBFM = Narrow band frequency modulation 窄带频率调制
NBL = Navigational bridge layout 驾驶桥楼布置
NBLES = Navigational bridge layout and equipment/systems

驾驶桥楼布置及设备/系统
NCS = Network coordination station 网络协调站
NCSR = Sub-Committee on Navigation, Communications and Search and Rescue （IMO）航行、通信与搜救分委会
NDFT = Nominal dry film thickness 名义干膜厚度
NDT = Non-destructive test 无损检测
NEC = NO_x emission control 氮氧化物排放控制
NEC = National electrical code （美国）国家电气法规
NECA = NO_x emission control area 氮氧化物排放控制区
NEMA = National Electrical Manufacturers Association （美国）国家电气制造商协会
NF = Normes Francaise 法国标准
NFPA = National Fire Protection Association 美国国家防火协会
NGHP carriers = Natural gas hydrate pellets carriers 天然气水合物运输船
NGHP = Natural gas hydrate pellets 天然气水合物
NIBS = Navigational integrated bridge system 航行的综合桥楼系统
NIOSH = National Institute for Occupational Safety and Health, UAS 美国国家职业安全和健康协会
NIOSH = National Institute Manufactures Association, USA 美国国家电气设备制造厂商协会
NISA = National Invasive Species Act of 1996 （美国）1996年国家有害入侵物种法
NISO = National Information Standards Organization （美国）国家信息标准组织
NK = Nippon Kaiji Kyokai 日本海事协会
NLS Certificate = International Pollution Prevention Certificate for Carriage of Noxious Liquid Substances in Bulk 国际防止散装运输有毒液体物质污染证书
NLS = Noxious liquid substances 有毒液体物质
NMD = Norwegian Maritime Directorate 挪威海事管理局
NMEA = National Marine Electronics Association （美国）国家航海电子协会
NOAA = National Oceanic and Atmospheric Administration （美国）国家海洋和大气管理局
Noise Levels = Code on noise levels on board ships 船上噪声级规则
NOPSA = National Offshore Petroleum Directorate 国家海上石油安全局（澳大利亚）
NORSOK = Norwegian Offshore standardization organization (Norsk Sokkels Konkurranseposisjon) 挪威石油标准化组织
NOS = National ocean survey 国家海洋调查
NOS = Network operation system 网络操作系统

NO_x = Nitrogen oxides 氮氧化物
NP = New work item proposal （国际标准的）新工作项目建议
NPD = Norwegian Petroleum Directorate 挪威石油管理局
NPDES = National pollution discharge elimination system （美国）国家污染物排放消减系统
NPV = Net present value 净现值
NPX = New Panamax (container ship) 新巴拿马型（集装箱船）
NR = Non-return (valve) 不归零制
NRT = Net registered tonnage 净登记吨
NS = Norwegian standard 挪威标准
NSC = Norwegian Ship Control 挪威船舶检验机构
NSI = Netherlands Shipping Inspection 荷兰航运检验局
NSM = National safety management (Code) （美国）国内安全管理（规则）
NSR = Northern sea route 北海航线
NSSN = National resource of global standards （美国）国家全球标准资源库
NSTC = Naval Ship Technical Committee (LR) 舰船技术委员会（英国劳氏船级社）
NTC 2008 = NO_x technical code 2008 2008年氮氧化物技术规则
NTNU = Norwegian University of Science and Technology 挪威科技大学
NTO = Man, technology and organization 人、技术和组织
NTS = Norwegian Technology Standards Institution 挪威技术标准协会
NTSC = (China) National Time Service Centre （中国）国家授时中心）[船载北斗卫星导航系统]
NTT = Number of TEUs transported TEU运输数
NTVRP = Non Tank Vessel Response Plans 非油船响应计划（美国防污染）
NUC = Not under control 失控
NUCLEAR 1971 = Convention relating to civil liability in the field of maritime carriage of nuclear material, 1971 1971年国际海上核材料运输民事责任公约

O

OAPEC = Organization of Arab Petroleum Exporting Countries 阿拉伯石油输出国组织
OAS = Organization of American States 美洲国家组织
OBO Carrier = Ore / bulk / oil carrier 矿（砂）/散（货）/油船
OCC = Operations coordination center 操纵协调中心
OCIMF = Oil Companies International Marine Forum 石油

公司国际海事论坛

OCM = Oil content meter 油分浓度计

OCSV = Offshore construction support vessels 近海构筑物供应船

OCTI = Central Office for International Rail Transport 国际铁路运输中央事务局

ODAS = Ocean data acquisition systems 海洋数据获得系统

OD-Box = Oil distribution box 燃油分配箱

ODM = Oil discharge monitoring and control systems 排油监控系统

ODME = Oil discharge monitoring equipment 排油监测设备

ODP = Ozone depletion potential 消耗臭氧层潜能值

ODS = Ozone depletion substance 消耗臭氧物质

OECD = Organization for Economic Cooperation and Development 经济合作与发展组织

OEM = Original equipment manufacturer 设备最初制造厂商

OER = Operational experience review 操作经验回顾

OGP = International Association of Oil and Gas Producers [previously known as the Oil Industry International Exploration and Production Forum (E&F Forum)] 国际石油天然气生产商联合会(前称为石油行业国际勘探和生产论坛)

OHSAS = Occupational health safety administration system 职业健康安全管理体系

OILPOL 1954 = International Convention for the prevention of pollution of the sea by oil, 1954 1954年国际防止海上油污染公约

OIML = International Organization of Legal Metrology 国际法制计量组织

OLF = The Norwegian Oil Industry Association (Oljeindustriens Landsforening) 挪威石油工业协会

OMBS = One man bridge operation 一人桥楼

OMM = Onboard monitoring manual 船上监测手册

OOW = Officer on watch 值班驾驶员

OP = Operations Department 作业部(巴拿马运河管理局所属机构)

OPEC = Organization of Petroleum Exporting Countries 石油输出国组织

OPRC 1990 = International Convention on oil pollution preparedness, response and co-operation, 1990 1990年国际油污准备、响应和合作公约

OPRC-HNS Protocol = Protocol on Preparedness, Response and Co-operation to Pollution Incidents by Hazardous and Noxious Substances, 2000 2000年由有害和有毒物质引起的污染事件的准备、响应和合作的议定书

OPRC-HNS = Oil pollution preparedness, response and co-operation and hazardous and noxious substances 油污准备、响应和合作,以及危险和有毒物质

OR = Overall risk 总风险

OREDA = Offshore reliability data handbook 离岸可靠性数据手册

ORIL = Open reversible inflatable liferaft 可逆充气敞开式救生筏

OS = Open Service 公开服务(船载北斗卫星导航系统)

OS = Offshore Standards 海洋工程标准

OSD = Owner ship data 本船数据

OSHA = Occupational safety and health act (美国)职业安全与健康条例

OSHMS = Occupational safety and health management system 职业安全健康管理体系

OSS = Offshore Service Specifications 近海设施技术条件

OSV Code = Code of safe practice for the carriage of cargoes and persons by offshore supply vessels 近海供应船货物和人员载运安全操作规则

OSV = Offshore supply vessel 近海供应船

OTS = Operational condition safety (operasjonell tilstand sikkerher) 安全运行条件

OU = Offshore Unit 近海装置

OVID = Offshore Vessel Inspection Database 海上船舶检查数据库

P

P&I = International Group of P&I Associations 国际保赔协会集团(P&I协会)

P&ID = Piping and instrumentation drawing 管路与仪表流程图

PAC = Passenger accommodation comfort 乘客起居舒适

PACSCAT = Partial air cushion supported catamaran = air cushion vehicle (ACV) + catamaran (CAT) 部分气垫支承的双体船(气垫装置+双体船)

PAL 1974 = Athens Convention Relating to the Carriage of Passengers & Their Luggage by Sea, 1974 1974年海上载运旅客及其行李雅典公约

PAL PROT 1976 = Protocol of the Athens Convention Relating to the Carriage of Passengers & Their Luggage by Sea, 1974 (1976) 1974年海上载运旅客及其行李雅典公约1976年议定书

PANAMAX = PANAMAX oil tanker 巴拿马型油船

PAR = Power assisted ram wing 动力增升型机翼

PAS = Pan American Standard　泛美标准
PAS = Publicly available specification　公众可获得的技术条件(IEC 的出版物)
PASSUB = Passenger submersible craft　(IACS 统一解释)载客潜水艇
PC/UMS = Panama Canal Universal Measurement System　巴拿马运河统一丈量系统
PC = Paired comparison　成对比较方法
PCAC = Passenger and crew accommodation comfort　乘客和船员起居舒适
PCAT = Planning catamaran　双体滑行艇
PCR = Performance capability rating（class notations）　作业能力等级(船级标志)
PCSO = Panama Canal security officer　巴拿马运河保安员
PCSOPEP = Panama Canal ship oil pollution emergency plan　巴拿马运河船舶油污应急计划
PCWBT = Protection coatings in water ballast tanks　压载水舱的保护涂层
PDOP = Position dilution of precision　位置精度损失
PDS = Petroleum geo-services　ASA 挪威油田服务集团
PE = Protective earth　接地线(保护线)
PER = Packet error rate　分组误码率
PES = Programmable electronic systems　可编程电子系统
PETRAD = Program for petroleum management and administration　石油经营与管理计划
PFEER = Prevention of fire and explosion and emergency response　火灾爆炸预防与应急响应
PFP = Passive fire protection　被动消防
PFSO = Port facility security officer　港口设施保安官员
PHA = Preliminary hazard analysis　初步危险分析
PI = Presentation interface　表示接口
PIANC = Permanent International Association of Navigation Congresses　航运会议国际常设协会
PIMS = Planned inspection and maintenance system　计划检查与维修体系
PL = Protective location of segregated ballast tanks　专用压载舱的保护位置
PLATO = Software for dynamic event tree analysis　动态事件树分析软件
PLL = Potential loss of life　潜在人员伤亡
PLS = Progressive limit state　累进极限状态
PM = Planned machinery maintenance scheme　计划的机械维修制度
PM = Preventive maintenance　定期维护
PMA = Panama Maritime Authority　巴拿马海事局
PMA = Permanent means of access　永久性通道设施

PMA = Phased mission analysis　阶段性任务分析
PMM = Permanent Magnet synchronous Motors　永磁同步电动机
PMMS = Preventive machinery maintenance scheme　预防的机械维修制度
PMS = Position monitoring system　位置监测系统
PMS = Power management system　功率管理系统
PNIC = Pleasure Navigation International Joint Committee　游览航行国际联合委员会
POB = Personnel on board　在船人员总数
POL = Polar orbit satellite system　极轨道卫星系统
Polar Code = International Code for Ships Operating in Polar Waters　国际极地水域营运船舶规则(简称"极地规则")
PORTS = physical oceanographic real-time system　物理海洋学实时系统
POST = Deutsche Bundepost　(德国)联邦电信局
POT = Protection of fuel and lubrication oil tanks　燃油/滑油保护
PPE = Pollution prevention equipment　防污染设备
PPM = Parts per million　百万分之……
PPR = Sub-Committee on Pollution Prevention and Response (IMO)　防污染及应对分委员会
PRA = Preliminary risk analysis　初步风险分析
PRA = Probability risk analysis (assessment)　概率风险分析(评估)
PRC = Piracy Reporting Center　盗版举报中心
PrHA = Primary hazard analysis　初步的危险分析
PRS = Polish Register of Shipping　波兰船级社
PRS = Position reference system　定位参照系统
PS = Port side　左舷
PSA = Petroleum Safety Assessment (Norway)　挪威石油安全管理局
PSA = Probabilistic safety assessment　概率安全评估方法
PSC = Port State control　港口国管理
PSD = Process shut down　过程切断
PSF = Performance shaping factor　行为影响因子
PSK = Personal survival kit　个人救生工具
PSP = Primary surface preparation　一次表面处理
PSPC = Performance standard for protective coatings　保护涂层性能标准
PSSA = Particularly sensitive sea area　特别敏感海区
PSV = Platform Supply Vessels　平台供应船
PT = Project team　项目组
PTI = Power take in　输入功
PTO = Power take off　提取功率

PTSA = The port and tanker safety 港口与油船安全
PVC = Polyvinyl chloride 聚氯乙烯
PWM = Pulse width modulation 脉宽调制
PWOM = Polar Water Operational Manual 极地水域营运手册

Q

QA = Quality assurance 质量保证
QAP = Quality assurance plan 质量保证计划
QC = Quality control 质量控制
QM = Quality management 质量管理
QMS = Quality management system 质量管理体系
QP = Quarters platform 居住平台
QRA = Quantified risk assessment 定量风险评估
QSCS = Quality system certification scheme 质量体系认证规划
QZSS = Quasi-Zenith Satellite System （日本）准天顶卫星系统[船载北斗卫星导航系统中涉及的区域性导航系统]

R

R&D = Research and development 研发
RABL = Risk assessment of buoyancy loss 浮力损失风险评估
RAC = Risk acceptance criteria 风险接受准则
RACON = Radar signal amplification 雷达信号放大
RAE = Residual accidental events 剩余意外事件
RAIM = Receiver autonomous integrity monitoring 接收机自动完整性监测
RAMS = Reliability, availability, maintainability, safety 可靠性、可用性、可维护性和安全
RAS = Replenishment at sea 海上航行补给
RCC = Refrigerated cargo carrier 冷藏货船
RCC = Rescue Coordination Center 搜救协调中心
RCCC = Refrigerated cargo container carrier 冷藏货集装箱船
RCDS = Raster chart display system 光栅海图显示系统
RCM = Reliability centred maintenance 以可靠性为中心的维修
RCN = Rule change notice 规范修改通报
RE = Resgistro Espagriol 西班牙船级社
REBLT = Refrigerated edible bulk liquid tanker 冷藏散装可食用液货船
REDS = Removable external data source 可移动的外部数据源
RELIQ = Re-liquefaction 再液化装置
RES = Residual strength 剩余强度

RFC = Refrigeration fish carrier 冷藏鱼货运输船
RFID = Radio frequency identification 射频识别
RIF = Risk influence factor 风险影响因素
RINA = Registro Italiano Navale 意大利船级社
RINA = Royal Institution of Naval Architects （英国）皇家造船师学会
RLV = Reference line value （船舶的 EEDI）基准线值
RMC = Recommended minimum specific GNSS data 建议的最小 GNSS 数据
RMC = Refrigeration machinery certificate 冷藏机械证书
RNC = Raster navigational chart 光栅航海图
RNE = Romania Naval of Shipping 罗马尼亚船级社
RNNS = Risk level (project) [risiko niva norsk sokkel] 风险等级
RO Code = Code for Recognized Organizations 被认可组织规则
RO = Recognized organization 被认可的组织
ROCC = Regional oil combating center 地区抗油污中心
RoHS = On restriction of the use of certain hazardous substances in electrical and electronic equipment （欧盟）关于在电气和电子设备中限制使用某些有害物质的指令
ROPME = Regional Organization for the Protection of the Marine Environment 海洋环境保护地区组织（由 8 个海湾地区国家组成的海洋环境保护组织）
ROV = Remote operated vehicle 遥控深潜器
RP = Recommended Practices 推荐规程
RPM = Revolutions per minute 每分钟转数
RPSSR = Report of planned start of ship recycling 拆船计划开工报告
RR = Radio regulations 无线电规则
RRM = Risk reducing measure 风险降低措施
RS = Russian Maritime Register of Shipping 俄罗斯船级社
RSA = ROPME sea area 海洋环境保护地区组织海区
RSC = Refrigeration system control 冷藏系统控制
RSM = Reynolds stress model 雷诺应力模型
RSO = Recognized security organization 认可的保安组织
RTCM = Radio Technical Commission for Marine Service 航海无线电技术委员会
RTD = Reference test device (a suitable size standard reference lifejacket) 基准测试装置，试验样衣（适当尺寸标准基准救生衣）
RTF = Run to failure 运行至失效
RW = Reduced weight anchors 剩余锚重量
Rx = Receive/receiver 接收/接收机

S

S(T)D = Starboard　右舷
S. hp = Shaft horsepower　轴功率(马力)
SAC = Sectional area curve　横剖面曲线
SAC = Standardization Administration of China　中国标准化管理委员会
SAFOP = Safety and operability study　安全性与可操作性研究
SAJ = Shipbuilders' Association of Japan　日本造船协会
SALM = Single anchor leg mooring system　单锚腿系泊系统
SALS = Single anchoring leg storage system　单锚腿贮存系统
SALS = Semi-automatic side loading pallet handling system　半自动舷侧装货盘操作系统
SALVAGE 1989 = International Convention on Salvage, 1989　1989年国际打捞公约
SAM = System action management　系统行为管理
SAMI = Security Association for the Maritime Industry　安全海运业协会
SAR 1979 = International Convention on Maritime Search & Rescue, 1979　1979年国际海上搜寻与救助公约
SAR = Search and rescue　搜救(搜寻与救助)
SARSAT = Search and Rescue Satellite-aided Tracking　搜救卫星辅助跟踪
SARSAT = Search and rescue satellite　搜救卫星(萨赛特)
SART = Search and rescue radar transponder　搜救(雷达)应答器
SAS = Special annual survey　特别年度检查
SBG = See-Berufsgenossenschaft (Professional Accident Insurance Association for Seamen)　(德国)海上职业联合会
SBM = Single buoy mooring　单浮筒系泊
SBS = Single buoy storage system　单浮筒贮储存系统
SBT = Segregated ballast tanks　专用压载舱
SBV = Stand by vessel　守备船
SC = Suez Canal　苏伊士运河
SC = Cargo Ship Safety Construction Certificate　(货船)结构安全证书
SC = SOLAS (IACS统一解释)　国际海上人命安全公约
SC = Sub-Committee　分技术委员会
SCA = Suez Canal Authority　苏伊士运河管理局
SCBA = Self contained breathing apparatus　自持式呼吸器具
SCF = Ship construction file　船舶建造档案
SCGT = Suez Canal gross tonnage　苏伊士运河总吨位
SCM = Screw shaft condition monitoring　艉轴状态监测
SCNT = Suez Canal net tonnage　苏伊士运河净吨位
SCR systems = Selective catalytic reduction systems　选择性催化还原法系统
SCR = Selective Catalytic Reduction　选择性催化还原
SCR = Safety case regulations　安全案例规范
SCR = Suez Canal regulation　苏伊士运河规则
SCS = Supply chain security　供应链保安
SCSR = Statement of completion of ship recycling　拆船完工声明
SCT = Suez Canal tonnage　苏伊士运河吨位
SCT = Scheduled on-condition task　定期有条件的任务
SCYTM = Suez Canal Vessel Traffic Management System　苏伊士运河船舶交通管理系统(苏伊士运河航行规则2015年8月版)
SDA = Structural design assessment　结构设计评估
SDC = Sub-Committee on Ship Design and Construction　(IMO)船舶设计与建造分委员
SDME = (Marine) speed and distance measuring equipment　(海上)速度和距离测量设备
SDN = Software defined network　软件定义网络
SDoC = Suppliers document of compliance　供应商符合声明(拆船公约"有害物质清单编制指南"要求提供)
SDR = Special drawing rights　特别提款权
SE = Cargo ship safety equipment certificate　(货船)设备安全证书
SEA(HSS) = Ship event analysis (hull surveillance systems)　船舶事件分析(船体监视系统)
SEC = Ship security　船舶保安
SEC = SO_x emission control　硫氧化物排放控制
SECA = SO_x emission control area　硫氧化物排放控制区
SECC = SO_x emission compliance certificate　SO_x排放符合证书
SEEMP = Ship energy efficiency management plan　船舶能效管理计划
SEMS = Safety and environmental management system　安全和环境管理体系
SENC = System electronic nautical chart　电子海图系统
SEPC = SO_x emission compliance plan　SO_x排放符合计划
SERS = Ship emergency response service　船舶应急响应服务
SES = Ship earth station　船舶地面站
SES = Standards Engineers Society　(美国)标准工程师

协会
SES = Surface effect ship 表面效应船
SEZ = Surface effect zone 表面效应区
SFA = Spectral fatigue analysis 疲劳谱分析
SFC = Specific fuel consumption 燃油消耗量参数(柴油机的 EIAPP 证书中要求)
SFT = Scheduled function task 定期功能检测
SFV 1977 = Torremolinos International Convention for the Safety of Fishing Vessels, 1977 1977 年托雷莫利诺斯国际渔船安全公约
SH = Safe hull 安全船体
SHARP = Systemic human action reliability procedure 系统的人的行为可靠性分析框架
SHCM = Safe hull construction monitoring 安全船体建造监测
SH-DLA = Safe hull-dynamic loading approach 安全船体动载荷方法
SHEMS = (Occupational) safety and health management system 职业安全和健康管理体系
SHERPA = Systemic human error reduction and prediction approach 减少和预测人为错误的系统方法
SIGTTO = Society of International Gas Tanker and Terminal Operators 国际气体船和终端经营者协会
SIL = Safety integrity level 安全完整性水平
SILENV = Ships oriented innovative solutions to reduce noise and vibration 船舶定向创新解决
SIRE = Ship inspection report program 船舶检查报告计划
SIS = Shanghai Institute of Standardization 上海标准化研究院
SIS = Swedish Standards Institute 瑞典标准学会
SIT = Spontaneous ignition temperature 自燃温度
SITP = Shipboard information technology platform 船上信息技术平台
SJA = Safe job analysis 安全工作分析
SLF = Sub-committee on Stability & Load Lines and Fishing Vessels Safety 稳性与载重线和渔船安全分委会
SLIM-MAUD = Success likelihood index method using multi-attribute utility decomposition 采用多重分解的成功可能性指标法
SLR = Risk due to small leaks 小泄漏引起的风险
SLS = SOLAS 公约的缩写(在 IMO 通函中采用)
SLSDC = (U.S.) Saint Lawrence Seaway Development Corporation (美国)圣劳伦斯航道开发公司
SLSMC = (Canadian) Saint Lawrence Seaway Management Corporation (加拿大)圣劳伦斯航道管理局

SMC = Safety management certificate 安全管理证书
SMCP = Standard marine communication phrases 标准海上通信术语
SMPEP = Shipboard marine pollution emergency plan 船上海洋污染应急计划
SMS = Safety management system 安全管理系统
SN = Safety of navigation equipment 航行安全设备
SNAME = Society of Naval Architects & Marine Engineers (美国)舰船工程师协会
SOC = Statement of compliance 符合证明
SOG = Speed over ground 对地速度
SOH = Scheduled overhaul 定期大修
SOLAS 1974 = International Convention for the safety of life at sea, 1974 1974 年国际海上人命安全公约
SOLAS PROT 1978 = Protocol of 1978 relating to the International Convention for the Safety of life at Sea, 1974 1974 年国际海上人命安全公约 1978 年议定书
SOLAS = International Convention for Safety of Life at Sea 国际海上人命安全公约
SOPEP = Shipboard oil pollution emergency plan 船上油污应急计划
S-O-R = Simulation-organism-response 刺激-机体-反应
SO_x = Sulphur oxides 硫氧化物
SPACE STP 1973 = Protocol on space requirements for special trade passenger ships, 1973 1973 年特种业务客船舱室要求议定书
Spec = Specification 技术条件
SPI = Ship/port Interface 船舶/港口界面
SPM = Single point mooring (system) 单点系泊(系统)
SPOT System = Speed, position and track system 航速、船位、航迹测定系统
SPREP = Secretariat of the Pacific Region Environmental Program 太平洋地区环境计划署秘书处
SPS Code = Code of safety for special purpose ships 特殊用途船安全规则
SPS = Standard position service 标准定位服务
SQAP = Software quality assurance plan 软件质量保证计划
SR = Cargo ship safety radio certificate 货船无线电安全证书
SRF = Ship recycling facility 拆船厂
SRFP = Ship recycling facility(ies) plan 拆船厂计划
SRNC = System raster navigational chart database 光栅航海图数据库系统
SRP = Submarine recycling program 潜艇循环利用计划
SRP = Scheduled replacement 定期替换

SRP = Ship recycling plan　拆船计划
SRR = Search and rescue region　搜救区域
S-R-R = Skill-based, rule-based, knowledge based　基于技能-规则-知识的行为模型
SRS = Ship reporting system　船舶报告制度
SS = Stainless steel　不锈钢
SS = Special survey　特别检验
SSA = Ship security assessment　船舶保安评估
SSA = Singapore Shipping Association　新加坡船东（航运）协会
SSAS = Ship security alert system　船舶保安警戒系统
SSB = Single sided band　单边带
SSD = Ship static data　船舶静态数据
SSD = Slow speed diesel　低速柴油机
SSE = Sub-Committee on Ship Systems and Equipment（IMO）船舶系统与设备分委员
SSIV = Sub-sea isolation valve　水下隔离阀
SSO = Ship security officer　船舶保安官员
SSP = Secondary surface preparation　二次表面处理
SSP = Ship security plan　船舶保安计划
SSPC = Society for Protective Coatings　保护涂层协会
SSR = Seafarer safety representative　海员安全代表
SSRS = Ship security report system　船舶保安报告系统
SSV = Safety regulation for seagoing ships　（德国）海船安全法规
ST = Shuttle tanker　穿梭油船
ST = Steam turbine　蒸汽轮机
STAHR = Sociotechnical approach to human reliability assessment　人因可靠性评估的社会技术方法
STCW 1978 = International Convention on Standards of Training, Certification & Watch-keeping for eafarers, 1978　1978年国际海员培训、发证和值班标准公约
STCW Code = Seafarers' training, certification and watch-keeping code　海员培训、发证和值班规则
STCW-F 1995 = International Convention on Standards of Training, Certification and Watch-keeping for ishing vessel Personnel, 1995　1995年渔船人员培训、发证和值班标准国际公约
Std. = Standard　标准
STEP = Shipboard Technology Evaluation Program　船上技术评估方案（步骤）（USCG对压载水管理系统所要求的）
STL = Submerged turret loading（arrangement）　水下旋塔装载（装置）
STP 1971 = Special Trade Passenger Ships Agreement, 1971　1971年特种业务客船协定

STS = Ship to ship（trans-shipment operation）　船对船（载驳作业）
STW = Sub-Committee on Standards of Training and Watch-keeping　培训和值班标准分委会
SUA 1988 = Convention for the Suppression of Unlawful act Against the Safety of Maritime Navigation, 1988　1988年制止危及海上航行安全的非法行为公约
SUA PROT 1988 = Protocol for the suppression of unlawful act against the safety of fixed platforms located on the continental shelf, 1988　1988年制止危及位于大陆架固定式平台安全的非法行为议定书
Sub Roc = Submarine rocket　反潜火箭
SUEZMAX = SUEZMAX oil tanker　苏伊士型油船
SUS = Stainless steel　不锈钢
S-VDR = Simplified voyage data recorder　简易航行数据记录仪
SW = Software or seawater　软件/海水
SWATH = Small water plane area twin hull ship　小水线面双体船
SWL = Safety working load　安全工作载荷

T

TA = Technical agreement　技术协议
TA = Type approval（certification）　形式认可（证书）
TB = Technical background　技术背景
TBT = Tributyl tins　三丁基锡
TBTO = Tributyl tins oxide　三丁基氧化锡
TC = Technical Committee　技术委员会
TC = Technical Co-operation Committee（IMO）技术合作委员会（技合会）
TCM = Main steam turbine condition monitoring　主汽轮机状态监测
TCM = Convention on International Combined Transport of Goods　国际货物运输公约
TCM = Tail-shaft condition monitoring　艉轴状态监测
TCPA = Time to closest point of approach　至最接近目标点的时间
TCS = Track control system　（船舶）航线控制系统
TDC Code 2011 = Code of safe practice for ships carrying timber deck cargoes, 2011　2011年载运木材甲板货船舶安全操作规则
TDMA = Time division multiple access（techniques）　时间分割的多路存取（技术）
TDRSS = Tracking and Data Relay Satellite System　跟踪和数据中继卫星系统
TDS = Technical data sheet（of coating）　（涂料）技术规

格书
TEU = Twenty-foot equivalent unit 标准(集装)箱
TEWI = Total equivalent warming impact 总当量温热影响
TFW = Tropical fresh water 热带淡水
TH = Thruster 推力器
THD = Total harmonic distortion 总谐波畸变
THD = Transmitting heading device 艏向发送装置
THERP = Technique for human error rate prediction 人为错误率的预测技术
TIG = Tungsten inert gas (welding) 氩弧焊
Timber Code = Code of safe practice for ships carrying timber deck cargos, 1991 1991年载运木材甲板货船舶安全操作规则
TLA = Timeline analysis 时间线分析
TLP = Tension leg platform 张力腿平台
TM = Tonnage measurement 吨位丈量
TMAS Tele-medicine assistance service 远程医疗辅助服务
TMHD = Transmitting magnetic heading device 磁罗经艏向发送装置
TMON = Tail shaft monitoring 艉轴监控装置
TMS = Taut mooring system 张紧式系泊系统
TMSA = Tanker management and self assessment 油船管理和自我评估
TONNAGE 1969 = International Convention on tonnage measurement of ships, 1969 1969年国际船舶吨位丈量公约
TP Lifejacket = Thermal protective lifejacket 保温救生衣
TP = Thermal protective 保温
TPC = Tunnel planning catamaran 隧道式滑行双体船
TR = Technical report 技术报告(IEC的出版物)
TR = Temporary refuge 临时避难所
TR = Turkish Register of Shipping 土耳其船级社
TRA = Total risk analysis 总风险分析
TRC = Time reliability correlation 时间可靠性关系
TS = Technical specification 技术条件(IEC的出版物)
TSA = Tabular scenatio analysis 表格式任务分析
TSA = Transportation security administration 运输安全管理
TSCE = Total-ship computing environment 美军全舰计算环境
TSCF = Tanker Structure Co-operative Forum 油船结构合作论坛
TSPP 1978 = International Conference on Tanker Safety and Pollution Prevention, 1978 1978年油船安全与防污染国际会议

TSS = Total suspended solids 悬浮固体总量(生活污水排出物指标)
TSS = Traffic separation scheme 分道航行制
TST = Technical condition safety (teknisk Sikkerhets Tilstand) 技术安全状态
TTA = Total TEU allowance 总的TEU容许数
TTS = Technical condition safety (teknisk Tilstand Sikkerhet) 安全技术条件
TTS = Total suspended solids 悬浮固体总量(生活污水排出物指标)
TVC = Torsional vibration calculation 扭转振动计算
TV-date = Torsional vibration data 扭转振动数据
Tx = Transmit/transmitter 发信/发信机

U

UCI = Unique consignment identifier 独特货运识别符
UEL = Upper explosive limit (upper flammable Limit) 爆炸上限(燃烧上限)
UFL = Upper flammability limit 可燃性上限
UFTAA = Universal Federation of Travel Agents Association 万国旅行社协会联合会
UI = Unified interpretation(IACS) 统一解释
UIC = Union international railway and containers 国际铁路与联盟集装箱
UKOOA = UK Offshore Association 英国海洋工程协会
UL = Underwriters Laboratories Inc. (美国)保险商试验所
ULCC = Ultra large crude oil carrier 特大型(原)油船(载重量30万吨以上)
ULCS = Ultra large container ship 超大型集装箱船
UMS = Universal Measurement System (巴拿马运河)统一丈量系统
UMS = Unattended machinery space(s) 无人值班机器处所
UN/ECE = United Nations Economic Commission for Europe 联合国欧洲经济委员会
UN = United Nations 联合国
UNCLOS = United Nations Convention on the law of the sea 联合国海洋法公约
UNCTAD = United Nations Conference on Trade & Development 联合国贸易和发展会议
UNDP = United Nations Development Programmer 联合国开发计划署
UNEP = United Nations Environment Programmer 联合国环境计划署
UNESCO = United Nations Educational, Scientific & Cultur-

al Organization 联合国教科文组织
UNFCCC = United Nations Framework Convention on Climate Change 联合国气候变化框架公约
UNIDO = United Nations Industrial Development Organization 联合国工业发展组织
UPS = Underwater production system 水下生产系统
UPS = Uninterrupted power system 不间断电源系统
UPU = Universal Postal Union 万国邮政联盟
UR = Unified requirements （IACS）统一要求
USCG = United States Coast Guard 美国海岸警卫队
UTC = Universal time coordinate 世界协调时

V

VAC = Volt-alternating current 交流电压
VBW = Dual ground/water speed 双对地/对水速度
VCA = Vertical contract audit 垂直合同审核
VCG = Vertical center of gravity 重心垂向位置
VCS = Vapor control systems 蒸发气控制系统
VDC = Volt-direct current 直流电压
VDL = VHF data link VHF 数据链
VDM = VHF data-link massage VHF 数据链路信息
VDM = Serial output message containing VDL information 包含 VDL 信息的串行输出报文
VDO = VHF data-link own-vessel massage VHF 数据链本船信息
VDR = Voyage data recorder 航行数据记录仪
VDU = Visual display unit 视频显示器
VEC = Valued ecological compounds 重要的生态组成部分
VEC = Vapour emission control 蒸发气控制
VEC-L = Vapour emission control, Lightering 蒸发气控制，含驳运状态
VGP = Vessel general permit （美国国家污染物排放消减）船舶通用许可
VHF = Very high frequency 甚高频
VIMSAS = Voluntary IMO member State audit scheme IMO 成员国自愿审核机制
VIP = Vacuum insulated panel 真空绝热板
VLC = Vertical launch system 垂直发射系统
VLCC = Very large crude oil carrier 超大型(原)油船(载重量17.5万吨以上，30万吨以下)
VLFS = Very large floating structure 超大型浮式结构
VLGC = Very large (fully refrigerated liquefied) gas garrier 超大型(全冷式)液化气运输船
VLOC = Very large ore carrier 超大型矿砂船
VLOO = Very large ore/oil carrier 超大型矿砂/石油运输船
VOC = Volatile organic compounds 挥发性有机化合物
VOHMA = Vessel Operators Hazardous Materials Association, INC 船舶经营人危险物质法人团体
VOS = Voluntary observing ships 任意的观察船
VPQ = Vessel particulars questionnaire 船舶概况问卷 (OCIMF)
VRP = Vessel response plan 船舶响应计划
VRU = Vertical reference unit 垂直基准装置
VSD = Voyage static data 航次静态数据
VSP = Variable system parameter 可变的系统参数（苏伊士运河航行规则2015年8月版）
VSP = Vessel sanitation program 船舶卫生计划
VSWR = Voltage standing wave ration 电压驻波比
VTG = Course over ground and ground speed 对地航向和对地速度
VTM = Vessel traffic management 船舶通航管理
VTMS = Vessel traffic management system 船舶通航管理系统
VTS = Vessel traffic service (centers) 船舶通航服务(中心)
VTS = Vessel traffic service (personnel) 船舶通航服务(人员)

W

WAAS = Wide area augmentation system 广域增量系统
WAN = Wide area network 广域网
WARC = World Administration Radio Conference 世界无线电行政会议
WB = Water ballast 压载水舱
WCL = World Confederation of Labour 世界劳工联合会
WCO = World Customs Organization 世界海关组织
WD = Working-group document （国际标准)工作组文件
WG = Working group （国际标准)工作组
WHO = World Health Organization 世界卫生组织
WiFi = Wireless fidelity 无线保真技术
WIG Craft = Wing-in-ground (craft) 地效翼船(艇)
WIPO = World Intellectual Property Organization 世界知识产权组织
WMARC = World mobile administration radio conference 世界水上无线电行政会议
WMO = World Meteorological Organization 世界气象组织
WMU = World Maritime University 世界海事大学
WNTI = World Nuclear Transport Institute 世界核运输协会
WOAD = Worldwide offshore accident database 全球离岸

工程事故数据库
WSD = Work sequence diagram　建造综合工程图
WSSN = World standards service network　世界标准服务网
WT/TT = Walk-through and talk = through　实践/讨论
WT = Watertight　水密
WTO = World Trade Organization　世界贸易组织
WW1 = The First World War (1914~1918)　第一次世界大战
WW2 = The Second World War (1939~1945)　第二次世界大战
WWF = World Wide Fund for Nature　世界自然基金
WWNWS = World wide navigational warning system　全球航行警告系统

Y

Y/D = Star/delta(starter)　星形/三角形(启动装置)

中 文 索 引

A

阿尔波斯特群岛的国家湖滨区 41
阿尔法·文图斯研究项目 720
阿芙拉型船舶 17
阿芙拉型油船 17
阿拉伯海的阿曼区域 629
阿拉伯海湾石油公司 401
阿拉斯加的适用水域 42
阿斯卡尼亚海洋重力仪 50
埃及环境保护法规(第 4 号,1994年) 271
霭 574
安保 777
安德森电源产品 35
安定(稳定的) 846
安静型潜艇 713
安静型先进推进系统 15
安全 757
安全瓣(安全挡板) 758
安全保护系统 757
安全保护装置 759
安全报警器 757
安全避风港 756
安全标准 757
安全操作 759
安全操作工作站 984
安全插入深度 758
安全程序 756
安全措施(防护检修) 538
安全措施 759
安全带 757
安全弹簧 732
安全等效 758
安全电压 759
安全阀 691

安全阀 732
安全阀 759
安全阀 759
安全阀调整试验 759
安全工作极限 756
安全工作压力 756
安全工作应力 756
安全工作载荷(安全工作负载) 756
安全工作载荷等 756
安全工作站 760
安全固化剂(低毒固化剂) 758
安全管理手册 758
安全管理体系(SMS)的基本要求 366
安全管理体系 540
安全管理体系 758
安全管理证书 758
安全和环境无害化拆船指南 400
安全和利于环保 755
安全及极限保护 757
安全甲板 757
安全间隙 757
安全降落速度 758
安全口盖(逃口盖) 294
安全链(备用链条) 757
安全膜片 758
安全膜片 759
安全目标 759
安全评估 757
安全切断 757
安全区 757
安全区域 757
安全设备检验 758
安全释放装置 732
安全水平法 760

安全水准/安全等级 758
安全通道 756
安全通信 757
安全完整性 758
安全完整性等级 816
安全完整性水平/等级 758
安全网 759
安全系数 53
安全系数 757
安全系统(应急防护系统) 759
安全系统 759
安全相关的自动功能 759
安全销(保险销) 759
安全性审查 757
安全许用应力 755
安全旋塞 759
安全要求 759
安全因数(安全系数) 758
安全与应急准备措施的有效性分析 270
安全与应急准备的功能要求 366
安全预防措施 759
安全裕度 758
安全闸 757
安全中心 757
安全装卸货物手册 756
安全状况 757
安装(用具,仪器,装置) 789
安装 461
安装稳性 461
安卓软件 35
氨基甲酸泡沫橡胶 109
鞍形键(空底键) 755
鞍座 755
岸电技术 811
岸电箱 490

岸基装置　811
岸距参数　70
岸上起重机　810
岸上人员　811
岸上设备　810
按键　489
暗礁　727
暗室　208
凹槽　323
凹槽腐蚀　723
凹度(空心度)　430
凹痕　222
凹体　723
凹形转子　320
奥米茄导航系统　629
澳大利亚海事安全局　57
澳大利亚货舱梯　57

B

巴拿马加大型船　652
巴拿马型船　652
巴拿马型船舶　652
巴拿马型散货船　652
巴拿马型油船(PANAMAX)　652
巴拿马运河　651
巴拿马运河吨位　651
巴拿马运河管理局　651
巴拿马运河航行海事规则　549
巴拿马运河统一丈量系统(PC/UMS)　652
巴塞罗那世界移动通信大会　71
巴氏货油舱清洗开口　103
罢工险　863
白环照灯　973
白菱号型　973
白色的视觉信号　973
白石英　973
白艉灯　973
白云石　249
白噪声统计估计　973
白昼　212

白棕缆　541
百度搜索引擎　67
百叶窗　813
百叶式通风筒　943
百一保证率波高　990
百一平均波高　560
摆动吊杆式　877
摆动吊杆装置　877
摆动蹼　319
摆动双杆操作　252
败诉方　524
扳紧　986
扳手(扳紧器)　833
扳手　986
扳手套筒　986
扳罾网　502
班机运输　766
班轮公会或公会　509
班轮条件下的船上交货　349
班轮运费　509
班轮运输　509
斑点(斑纹,加斑点)　843
瘢痕　765
板格　652
板架结构　676
板结构　676
板式加热器　676
板式冷却器　676
板式蒸发器　676
板振动　676
钣金工作　793
办公自动化系统　618
办事处　501
半闭式给水系统　782
半闭式循环　782
半闭式循环燃气轮机装置　782
半薄膜液货舱　783
半舱驳　971
半柴油机(热球式发动机)　783
半超导单极电机　785
半沉头　783

半沉头铆钉　783
半成品　783
半齿抓斗　403
半磁控制器　783
半导体　782
半导体材料　783
半导体开关电器　783
半导体整流器　783
半防护外壳　783
半封闭式制冷压缩机组　783
半封闭制冷机　783
半浮栅晶管体　783
半高底纵桁　403
半固定坐标系　783
半合成剂　785
半集装箱船　783
半进流角　403
半宽水线图　403
半冷半压式液化气船　783
半冷半压型液化石油气船　784
半冷式液化气船　784
半梁　403
半毛重　783
半潜驳　784
半潜船　784
半潜船　785
半潜式浮式生产装置　783
半潜式绞吸式挖泥船　785
半潜式平台　784
半潜式生活平台　784
半潜式自航工程船　784
半潜式钻井平台　155
半潜式钻井平台　784
半潜式钻井平台　784
半潜式钻井装置　782
半潜式钻井装置　784
半潜型浮式基础结构　784
半潜运输船　785
半潜状态　784
半潜最大工作水深　784
半潜最小工作水深　784

半水密开口　785
半围蔽处所　783
半围蔽处所　783
半坞式船台　783
半悬挂舵　655
半旋转跳板　784
半硬质泡沫塑料　784
半张紧　785
半自动　782
半自动焊工艺　783
半自动焊接　782
半自动气体保护金属电弧焊　783
半自由空间声场　403
半自由声场　403
半组合曲轴　782
伴流　958
伴流分数　886
伴流模拟　958
伴流因数　885
邦戎曲线　90
帮助　316
绑拖　28
绑扎点　492
绑扎桥　492
绑扎装置　492
包处理　650
包机运输　133
包坞　90
包销　306
包运合同　179
包装（标准部件，成套设备）　650
包装舱容　67
包装货物舱容　67
包装破裂险　95
包装危险货物　650
包装形式　650
包装形式的有害物质　407
饱和　765
饱和点　765
饱和度　765
饱和空气　764

饱和潜水　765
饱和潜水的生命维护装置　765
饱和潜水系统　765
饱和蒸发气体　765
饱和蒸汽　764
保安不符合项　777
保安措施（治安措施）　777
保安等级1（正常，船舶和港口正常操作）　777
保安等级2（加强）　777
保安等级3（特殊）　777
保安等级　777
保安设备　777
保安声明　216
保安事件　777
保安威胁　777
保持性能　488
保兑信用证　171
保函　399
保护（维护）　538
保护/维护　538
保护材料　705
保护措施（安全装置，保护器）　757
保护措施　705
保护电流密度　704
保护电位范围　705
保护封闭罩壳　704
保护功能　704
保护关税　696
保护贸易　705
保护贸易政策　705
保护膜　704
保护水域航区　703
保护套式热电偶　794
保护涂层　660
保护涂层　706
保护外套　703
保护箱　705
保护罩（个人漂浮装置）　704
保护装置（安全装置）　704

保护装置（安全装置）　704
保护装置　758
保留（储备）　738
保赔协会　704
保全　690
保热的（保温的，贮热的）　418
保税仓库　90
保税陈列场　90
保税工厂　90
保税货棚　90
保税区　90
保温材料/绝缘材料　462
保温层　462
保温集装箱　361
保温救生服　894
保温用具　894
保险　462
保险备用信用证　463
保险单　462
保险检验/保险鉴定　933
保险帽（护罩）　757
保险凭证　462
保修　399
保修期　399
保障的风险　399
堡垒（密闭区）　138
报酬　744
报关　201
报价　617
报价　692
报价　713
报价单　713
报警　25
报警闭锁功能　365
报警传送系统　26
报警功能　365
报警器（报警装置，报警显示屏）　26
报警设施　26
报警系统　26
报警显示和报警应答　365

索引

1053

报警指示器　26
报警装置(警告装置)　959
报警装置　26
报警装置　26
报务室　716
报务桌　979
爆燃　235
爆炸　309
爆炸品船舶　310
爆炸上限(UEL)　938
爆炸物　309
爆炸下限(LEL)　526
爆炸下限　496
爆炸限值　310
爆炸性气体环境　309
爆炸压力释放　309
北朝上显示　607
北大西洋冰区巡逻、运作和费用规则　754
北斗卫星导航系统　80
北方保护区　608
北海和波罗的海研究站　352
北海区域　607
北极冰覆盖区域　46
北极水域　47
北美驯鹿岛(卡里布岛)和西南浅滩保护区　117
北美自由贸易协定　607
贝氏计权　74
贝叶斯网络　74
备　67
备份　727
备件(备用的,可分让的)　834
备件(备用装置)　833
备件/备品　833
备件、备品表　833
备件、工具及附件明细表　842
备件舱　834
备件储藏室　834
备件供应　834
备件明细表　833

备品零件箱　834
备用报务室　834
备用泵　834
备用泵　853
备用舱(备用处所)　738
备用储气瓶　833
备用发电机　738
备用发电机　853
备用发电机组　853
备用风机　853
备用号灯　738
备用回收系统　67
备用机械　853
备用机组　853
备用空气压缩机　849
备用控制系统　66
备用锚　793
备用燃料舱(备用煤舱)　738
备用冗余　853
备用设备(备用机械)　853
备用设备　833
备用室　834
备用拖缆和应急拖缆　834
备用系统　67
备用信用证　853
备用油柜　833
备运提单　722
背景噪声　66
背景噪声级　66
背空泡　66
背散射扫描　66
背书　283
背心型个人漂浮装置　949
背压　66
背压调节器　66
背压式汽轮机　66
背压试验　66
背压脱扣器　66
倍频程　617
倍音速激波风洞　943
被保护处所　703

被保护区域　703
被保护危险处所　703
被动齿轮　258
被动风险控制　657
被动配额管理　657
被动失效　657
被动式减摇水舱　657
被动式约束阻尼层　657
被动水舱式减摇装置　657
被动系统　657
被激活的自动识别系统(AIS)目标　8
被删除区域　874
被申请人　740
被审核者　56
被涂油的　392
被拖物　906
被选目标　780
本票　699
泵(泵送,抽吸)　707
泵舱　707
泵舱　707
泵舱舱壁　707
泵舱通海阀　707
泵传动组件　707
泵的容量　707
泵的水缸　707
泵底座　707
泵功率　707
泵机组(泵装置)　707
泵及泄水管系图　707
泵接入压力　707
泵进口(泵吸入口)　707
泵壳体　707
泵扩压器　707
泵轮(叶轮,工作轮,钻子)　446
泵排出的静压头　854
泵敲击　707
泵式喷水推进　707
泵式喷水推进器　707
泵送总管(排出总管)　707

泵吸 707
泵吸高 868
泵吸口 580
泵吸入室 707
泵吸系统(泵送系统) 708
泵吸性能 707
泵吸装置(泵吸布置图) 707
泵系统 707
泵引水系统 707
泵站(泵舱) 707
泵组 707
比出量(单位输出) 841
比极限强度 842
比密度 840
比热容 841
比容 842
比湿(湿度比) 841
比特 84
比效率 841
笔记本电脑 608
舭部 81
舭部板 82
舭部半径 82
舭部切点 82
舭部升高 213
舭部升高角β 213
舭部吸水阀 82
舭列板 82
舭龙骨 82
舭涡阻力 739
舭斜剖线 81
舭肘板 81
闭海或半闭海 281
闭路电视 126
闭式测量装置 147
闭式给水系统 147
闭式滚装处所 148
闭式冷却水系统 147
闭式炉舱通风 148
闭式索节 148
闭式循环 147

闭式循环燃气轮机装置 147
闭锁 86
闭锁装置 520
庇护地/避难处 674
庇护区 794
蓖麻籽或蓖麻粉或蓖麻油渣或蓖麻片 119
壁厚 959
壁厚偏心率 958
壁式插座 958
壁温 958
避焊孔 970
避航区 47
避免死亡/人的净成本 592
避碰 155
避碰功能 155
边舱 977
边舱撑材 196
边舵 712
边舵叶 977
边绞链式艏门 814
边接缝 773
边界层 92
边界层动量厚度 576
边界层分离 92
边界层厚度 92
边界层能量厚度 284
边界层排挤厚度 244
边界条 92
边螺旋桨 814
边锚绞车 813
边抛装置 813
边燃油舱 813
边缘打磨 268
边缘腐蚀 268
边缘未对准 573
边缘应力 268
边缘圆度 752
边座板 814
编程语言 698
编译器 164

编组(分组) 399
扁锉 342
扁钢 342
变程泵 942
变电站 913
变工况 617
变工况 942
变加速度 942
变量泵 942
变量提升 429
变流机室 161
变螺距 942
变码器 912
变频电源 359
变坡滑道 523
变速泵 942
变速齿轮 942
变速齿轮舱 374
变速电动机 132
变速试验 843
变向泵 743
变形载荷 219
变型Z形操纵试验 576
变压器油 913
变值 942
便携设备 684
便携式气体测量设备 684
辩论 47
标称伴流 601
标称齿顶高 600
标称螺距 601
标称螺距比 601
标称螺距比 601
标称载荷(额定负载) 601
标称值(额定值) 601
标称质量 601
标的 614
标的 865
标定的最小屈服应力(标定的最小屈服强度) 842
标定抗拉强度下限值 842

标定压力　601
标定压力　607
标示　451
标准　849
标准　852
标准差　850
标准差　852
标准淡水　851
标准辅助 V 形缺口试样　852
标准辅助试样　852
标准辅助试样　852
标准功率　852
标准化（标定）　852
标准化符号　852
标准基准材料　852
标准集装箱　850
标准集装箱　892
标准集装箱　923
标准雷达反射器　852
标准耐火试验　851
标准排放接头　850
标准排水量　851
标准时间-温度曲线　852
标准试件　852
标准试验　852
标准试样　852
标准水温　852
标准系统电压　955
标准显示　851
标准岩石保护区　852
标准重力加速度　850
标准转速　852
标准作业工况　852
表层采样器　825
表层梯度采水器　874
表层温度表　98
表灯　373
表观波长　41
表观波高　41
表观波浪周期　42
表观横摇幅值　41

表观进速系数　41
表观螺距比　41
表面波　874
表面处理　874
表面沸腾　874
表面痕迹　874
表面孔隙　874
表面摩擦　874
表面清理　874
表面式空气冷却器　874
表面式冷却器　874
表面式凝汽器　874
表面通风　874
表面效应船（水面效应船）　874
表面效应船（水面效应船）　874
表面预处理　874
表面张力　874
表面张力波　109
表面状态　874
表式示号器　373
表压　541
表压力　373
濒危物种法　282
冰带　442
冰带区　442
冰带艏部区域　353
冰带艉部区域　17
冰带舯部区域　569
冰点（凝固点）　358
冰覆盖水域　444
冰级吃水　443
冰季　443
冰晶石　198
冰块扭矩　444
冰区加强　444
冰区加强的冰级　144
冰区驾驶员　443
冰区满载水线　443
冰区轻载水线　443
冰区巡逻服务　443
冰区巡逻所警戒冰山区　729

冰山和浮冰观察室　444
冰铜　187
冰雪负荷　443
柄（把手，搬运，处理）　404
并车　156
并车传动装置　156
并车减速齿轮箱　726
并车运行试验　652
并励电动机　813
并励发电机　813
并联　652
并联舵　147
并联运行　652
并联装置（并联电路）　652
并坞　247
并行处理技术　652
波长　966
波陡　967
波陡参考值　727
波峰　966
波幅　41
波高　966
波谷　967
波浪　967
波浪泵气效应　967
波浪补偿器　877
波浪参数 C　967
波浪长度(m)　989
波浪冲击载荷　966
波浪垂向剪力（Q_{WV}）　947
波浪垂向弯矩 M_{WV}　947
波浪浮力修正　824
波浪观察室　967
波浪号　902
波浪频率　966
波浪频率　966
波浪谱　967
波浪水平弯矩 M_{WH}　431
波浪弯矩　966
波浪系数　966
波浪压力或动压力 p_W　967

波浪压力或动压力　709
波浪仪绞车　967
波浪圆频率　966
波浪载荷　966
波浪中的侧向载荷　493
波浪中平均功率增值　560
波浪中平均推力增值　560
波浪中平均转矩增值　561
波浪中平均转速增值　449
波浪中平均阻力增值　560
波浪中阻力平均增值　560
波浪周期　967
波浪周期的标准偏差　850
波龄　966
波罗的海区域　70
波能　966
波频　966
波前　966
波倾　967
波数　967
波速　967
波向角　966
波形舱盖　189
波形图　643
波型阻力　967
波阵面　966
波阻峰值　437
波阻谷点　429
玻璃钢　322
玻璃棉　388
玻璃棉毡　387
玻璃水位计　388
玻璃纤维（玻璃丝）　322
玻璃纤维/玻璃棉/纤维玻璃　322
玻璃纤维　322
玻璃纤维　387
玻璃纤维增强塑料（玻璃钢）　322
剥蚀坑　294
剥蚀强度（侵蚀烈度，冲刷强度）　294
剥蚀损伤（剥蚀损坏）　294

播放设备（回放设备）　677
伯尔尼碳循环模型　81
驳船　504
驳船　71
驳船费　504
驳船式平台　72
驳运泵（调运泵，输送泵）　913
驳运作业　505
泊松比　678
泊松分布　678
泊位装货租船合同　81
薄板（膜）共振吸声结构　740
薄钢板/钢皮　793
薄钢板　858
薄黄铜板　793
薄利多销　824
薄膜（涂膜，效片）　323
薄膜集成电路　895
薄膜式液货舱　564
薄膜振动　950
薄膜蒸发　895
薄片滤器　324
薄铁板　793
薄锌板　793
补板　155
补偿导管　958
补偿金　162
补偿贸易　163
补偿贸易　731
补充裁决　11
补给泵　539
补给阀　539
补给水（补加水）　539
补给水　539
补给油柜　323
补贴　866
补正　725
补重系统　162
捕获（捉）　8
捕获/激活区　8
捕获的雷达目标　8

捕鲸船　972
捕鲸船　972
捕鲸绞机　972
捕鲸母船　972
捕鲸炮　972
捕鲸铦　972
捕捞设备　337
捕捞渔船　120
哺乳动物的毒性资料　209
哺乳动物急性中毒　10
不保兑信用证　931
不冻港/开放港　632
不对称进水　938
不符合项　602
不公开场合　148
不挂钩/解耦　217
不规则波　484
不规则工作状态　484
不含硝酸盐的化肥　322
不合格　602
不合理排放控制策略　484
不计名提单　76
不间断供电电源系统（不间断电源）　935
不兼容　449
不减压潜水　599
不可撤销信用证　484
不可抗力　350
不可逆转式发动机　484
不可逆转式发动机　605
不可燃的杂质　602
不可燃性　449
不可申诉的补贴　601
不可压缩流体　449
不可预见费　705
不可转让信用证　605
不扩散核武器条约　918
不利区域　598
不良的再引弧　682
不论到港顺序　355
不明阶段　931

不凝性气体(惰性气体) 353
不排放或减少排放压载水 604
不膨胀式蒸汽机 603
不平衡舵 931
不清洁提单 931
不确定系数 931
不确定性分析 931
不燃材料 449
不燃材料 602
不燃的 602
不同的运输方式 237
不完全的熔敷坡口 449
不完全燃烧 446
不稳定动力偶 583
不稳定回线环高 937
不稳定回线环宽 937
不稳定运动 937
不吸收的 601
不吸收性材料 601
不相容物质 449
不锈包层钢 849
不锈的(防锈的) 605
不锈的 849
不锈钢 605
不锈钢 849
不锈钢研磨粉 849
不愉快声 245
不愉快声 937
不允许载荷重新分布的屈曲能力(方法2) 98
不正常工作状态 2
不值班(无人值班) 931
布标船 208
布墩长 500
布尔运算 90
布缆船 104
布缆浮筒 104
布设船 494
布新尼斯克数 92
布油袋 861
步桥 368

钚 678
部分裁决 654
部分分舱吃水(部分分舱载重线) 654
部分负荷调整 654
部分负荷工况。 654
部分固定价格 654
部分甲板艇 654
部分进气级 654
部分浸水试验 655
部分空泡 654
部分损失 654
部分选择性保护 654
部分载重吃水 d_{P} 654
部分载重线 654
部分装载舱 654
部分装载舱 654
部分综合 654
部件 164
部件 653
部门的一体化 776

C

材料 552
材料安全数据单 555
材料的最低设计温度 571
材料规格 842
材料名义屈服应力 601
材料声明 553
材料声明标识号 553
材料系数 553
材料最小极限抗拉强度 930
财产 702
财产保全 690
财务人员 325
财政关税 743
裁决书 986
裁决书草案 255
采购 708
采购、设计、建造和安装调试 291
采购材料及服务的验证 945

采购资料 708
采金船 389
采矿船 256
采水测温室 962
采用挖深 268
采油船/钻井船 258
采油树 137
彩色鱼探仪 155
彩岩石国家湖滨区 669
参数横摇 653
参数横摇第二层薄弱性标准 501
参数横摇第一层薄弱性标准 501
参数横摇周期 653
参数目标 727
参照值 A_{wr} 727
参照周期 727
餐具 201
餐具柜 814
餐具消毒柜 859
残货油 620
残留物排放记录仪 243
残油(油泥) 624
残油(油泥)舱 624
残油(油泥)舱 883
残油(油泥)焚烧系统 624
残油(油泥)日用舱 624
残油(油泥)泄放舱 623
残油(油渣) 623
残油(油渣) 738
残油 823
残余废气 738
残余风险 738
残余燃料油 738
残余燃油 738
残余物/水混合物 738
残余物 738
残余意外事件 738
舱壁 101
舱壁 99
舱壁板 101
舱壁凳座 102

舱壁阀 102
舱壁隔填料函 102
舱壁甲板 101
舱壁甲板 101
舱壁结构 101
舱壁结构 102
舱壁龛 102
舱底报警 81
舱底泵 82
舱底泵数 82
舱底范围 81
舱底和压载水系统 81
舱底结构 91
舱底木铺板 91
舱底排水设备 83
舱底排水装置 83
舱底区域 91
舱底水 82
舱底水泵送装置 82
舱底水高位报警器 82
舱底水管系 82
舱底水过滤箱 82
舱底水和压载水的油水分离器 81
舱底水监视装置 82
舱底水水位探测与报警系统 82
舱底水水位监测装置 82
舱底水吸入阀 82
舱底水吸入口滤网 82
舱底水系统 82
舱底水液位报警装置 82
舱底水油水分离器 628
舱底水油水分离器 82
舱底水预处理柜 82
舱底水支管 81
舱底水直通支管 240
舱底水总管 534
舱底水总管 82
舱底水综合处理系统 464
舱顶 883
舱盖板 408
舱盖板长度 497

舱盖板顶板厚度 895
舱盖板卷筒 749
舱盖板宽度 95
舱盖板主梁深度 223
舱盖的密封压紧装置 773
舱盖分块长度 498
舱盖滚轮 408
舱盖滚轮 408
舱盖宽度 95
舱盖启闭装置 408
舱盖起升装置 408
舱盖深度 223
舱盖填料 408
舱盖拖曳装置 527
舱盖压紧装置 250
舱盖曳行装置 907
舱柜 881
舱柜液位计 882
舱空量 929
舱口 408
舱口驳 491
舱口吊杆 447
舱口端横梁 408
舱口吨位 904
舱口盖 408
舱口盖绞车 408
舱口梁 408
舱口梁承座 408
舱口梁销 76
舱口梯 408
舱口围板 149
舱口围板 408
舱口围罩 161
舱口纵桁 408
舱面货物险 629
舱面提单 629
舱面属具 215
舱内载运冷藏集装箱 118
舱棚 119
舱棚出入口盖 162
舱容 110

舱容 882
舱容图 110
舱室 162
舱室 7
舱室备品 834
舱室壁通风筒 959
舱室布置 467
舱室长度 $l_H(m)$ 498
舱室绝缘 101
舱室空气噪声 23
舱室门窗 104
舱室排水系统 236
舱室设备 6
舱室梯 104
舱室梯 162
舱室通风 512
舱室通风机 104
舱室噪声级规则 152
舱室属具 104
舱室总容积 396
舱室组 397
操舵拉杆 859
操舵链 858
操舵轮 859
操舵手柄 973
操舵索 859
操舵台 859
操舵位置 172
操舵稳定性 859
操舵系统 858
操舵系统 859
操舵油缸(转舵机构) 753
操舵轴 859
操舵轴传动操舵装置 792
操舵装置报警 858
操舵装置动力传动系统 858
操舵装置动力设备 859
操舵装置控制系统 858
操纵 541
操纵阀(导阀) 670
操纵阀 637

操纵杆(操纵手柄)　635
操纵杆(工作杆)　635
操纵号灯　541
操纵机构　541
操纵灵活性　404
操纵面(地效翼船)　540
操纵面阻力　739
操纵模式　638
操纵能力受到限制的船舶　949
操纵区域　634
操纵设备(航运设备)　404
操纵绳　709
操纵失控灯　609
操纵室　182
操纵台(控制台)　637
操纵限制　635
操纵性　540
操纵性　540
操纵性特征曲线　540
操纵站　636
操作/功能模块　639
操作舱室(操纵室)　635
操作程序(运行程序)　638
操作杆　636
操作规程(操作说明书,操作须知)　637
操作规程/操作指南(操作说明书,操作须知)　635
操作和维护手册　634
操作级(人员)　637
操作记录　636
操作技能　638
操作检查　637
操作精度(实际精度)　637
操作面板　636
操作模式(动力定位)　638
操作模式　575
操作区(工作区)　637
操作人员　636
操作事故　982
操作试验　638

操作试验(运行试验,动作试验)　637
操作试验　637
操作试验　638
操作手册　635
操作说明　637
操作说明书(操作条令,操作手册)　637
操作位置(作业位置)　636
操作系统　243
操作系统　637
操作显示区　637
操作限制　637
操作限制　638
操作限制清单　512
操作限制清单　617
操作信号　636
操作性能(工作性能)　637
操作元素　637
操作员　639
操作员单一动作　818
操作员简单动作　817
操作者　639
操作值　637
操作装置(传动装置)　635
操作装置　636
操作状态(运行状态)　636
操作状态　638
槽(切口,箱位)　823
槽口自升式钻井平台　823
槽轮　395
槽头螺钉　823
槽形舱壁　189
槽形舱壁单元的无支承跨距　937
槽形轨道　395
侧壁气垫船　815
侧方交会定位法　617
侧浮舟(侧浮体)　813
侧开泥驳　814
侧开式舱口盖　815
侧开式艉门　814

侧漂　256
侧扫声呐　814
侧推器(侧推装置)　493
侧推器(侧向推力器,横向推力器)　917
侧推器(横向推力器)　493
侧推器/侧推装置　814
侧推器　53
侧推器　899
侧推推力　493
侧推装置舱　493
侧向鳍　820
侧向推力器　493
侧向推力器Z形操纵试验　493
侧向推力器回转试验　493
侧斜　820
侧斜角　820
侧斜螺旋桨　820
侧斜轴向位移　820
侧斜轴向位移　820
侧移式舱盖　282
侧移式水密滑门　431
侧翼　814
侧轴系　977
测点　876
测风雷达　976
测厚　895
测距系统　718
测力传感器　350
测力仪摩擦修正曲线　360
测力仪校验设备　266
测量(记录,调节)　565
测量方法　373
测量管　831
测量液位口(量油口)　373
测量装置　373
测量装置　562
测流标杆　958
测漏器(探漏器)　495
测氢试验　441
测深(液舱测量)　831

测深杆 831

测深孔 831

测深器（回声测深仪） 831

测深绳（水砣绳） 831

测深室 831

测深装置 831

测深装置 831

测试段 562

测试速度 892

测试装置（检测器，探头） 697

测温法 894

测温式失火自动报警器 25

测隙器（量隙规，塞尺） 320

测线 875

测向系统 242

测烟式失火自动报警器 58

测氧仪 649

测雨雷达 714

测振仪 950

层 899

层次 423

层叠型舱盖 647

层交换技术 494

层流 490

层流底层 490

层数 899

层压胶木轴承 490

叉头横销 90

叉形接头（Y形接头，三通接头） 987

插入/篡改 480

插入式带缆桩 709

插入损失 460

插座 826

插座接合（套管接合） 826

插座引出线端 494

茶杯架 198

茶具架 886

差的涂层状况 682

差的状况 682

差动式减速齿轮箱 237

差分全球定位系统 236

差价 692

差价税 885

差旅费 917

差排式燃气轮机 196

差异化顺应式系泊系统 237

拆 845

拆除 243

拆船 725

拆船 795

拆船 797

拆船厂 797

拆船厂计划 797

拆船厂批准指南 400

拆船公司 725

拆船计划 798

拆船计划编制指南 401

拆船计划开工报告 735

拆船设施应做好准备进行处理的船上可见材料 553

拆船完工声明 854

拆检 646

拆装活塞环的钳子 672

柴-燃联合装置 156

柴油 237

柴油-液化天然气双燃料发动机 237

柴油澄清柜 237

柴油分离机 237

柴油机 237

柴油机-发电机电力推进 212

柴油机-燃气轮机船 237

柴油机参数记录簿 285

柴油机齿轮传动 374

柴油机船 237

柴油机的重大改装 538

柴油机电控系统 275

柴油机电力传动 627

柴油机动力装置 237

柴油机发电+燃气轮机动力推进装置 203

柴油机功率裕度 285

柴油机国际防止空气污染证书 285

柴油机滑油状态监控 237

柴油机类别 928

柴油机类型 287

柴油机推进 203

柴油机与燃气轮机联合动力装置 151

柴油机直接传动 237

柴油日用柜 237

柴油压缩机 237

柴油压缩机组 237

掺铒光纤放大器 293

掺混空气 239

掺混孔 891

掺混区 239

产地证明书 130

产品 697

产品报价 698

产品补偿 698

产品分配器 698

产品规划 698

产品几何技术规范 386

产品加工船 675

产品检验（防腐） 697

产品检验 698

产品录 512

产品设计 697

产品识别 697

产品艇 697

产品形式认可 926

产品性能 698

产品证书 697

产品制造管理体系 543

产品制造管理体系审核 56

产品制造过程审核 56

产品质量 698

铲斗 239

铲斗柄 239

铲斗吊杆 239

1061

铲斗机　239
铲斗式挖泥船(反铲)　66
铲斗式挖泥船　239
铲斗式挖泥船　66
铲斗式挖泥船　98
铲斗支架　239
颤振/击振　973
颤振　347
颤振计算(抖振计算)　347
颤振频率　347
颤振破坏　347
颤振试验　347
长柄火铲　333
长冲程柴油机　521
长的缺陷　521
长度吃水比　497
长度特别短的货舱　840
长吨　522
长峰规则波　522
长峰浪　522
长甲板室　521
长甲板室及短甲板室　521
长颈叶根　521
长宽比　497
长期 DOC 证书　363
长期记录介质　523
长期接触对哺乳动物的毒性　908
长期信贷　523
长期性倾销　523
长期演进　521
长牵条,贯穿牵条　897
长桥楼　520
长桥楼船　520
长上层建筑　521
长上层建筑及短上层建筑　521
长深比　497
长石块　320
长艏楼　521
长艏楼船　521
长艉楼　521
长艉楼船　521

长叶片　520
长轴　538
长轴包架　520
常穿服　173
常规舵机　753
常规潜水　185
常规潜艇　185
常规停船　185
常规拖船　385
常规污染物　185
常规鱼雷潜艇　185
常见有害物质在船舶上的分布　375
常设仲裁法院　661
敞舱驳　631
敞开艇　633
敞口船舶　634
敞口集装箱船　634
敞口集装箱船的干舷甲板　357
敞水试验箱　634
唱音/谐鸣　817
抄网　239
超巴拿马型船舶　930
超薄键盘　931
超长冲程发动机　871
超大缸径发动机　871
超大型矿砂船　930
超大型油船　930
超大抓力锚　948
超导材料(超导体)　872
超导磁流体潜艇　872
超导磁流体推进器　872
超导电机　872
超导电力推进装置　872
超导电性　872
超负荷功率　647
超高分子量聚乙烯纤维　930
超高压锅炉　872
超高液位　423
超高音速武器　14
超级海上试验　871

超级计算机　870
超级重型鱼雷　871
超空泡　871
超空泡技术　871
超汽轮机装置　939
超轻气凝胶　930
超声　931
超声波查漏试验　930
超声波法　930
超声波检查　930
超声波检查设备　930
超声波试验(渗漏试验)　930
超声波探伤(超声波检测)　930
超声速压气机　872
超声相控阵技术　650
超速　648
超速保护　648
超速保护装置　648
超速调速器　648
超速试航　648
超速试验　648
超速脱扣装置　648
超速限制器　648
超速自动脱扣　648
超微电脑　871
超细玻璃棉　313
超压　647
超压　872
超压保护　647
超音速反舰导弹　872
超越角　648
超载(过载)　647
潮湿的　576
潮汐表　899
车间底漆　810
车壳　793
车客渡船　109
车辆处所　943
车辆甲板　943
车辆坡道　823
车辆区域　943

车辆跳板 943
车辆用整体高度 905
车辆运输船 109
车用汽油 59
车载电脑 109
车载电台 943
车载网络 480
车载诊断系统 629
扯窗 822
掣链钩 236
掣锚链条 35
掣索器 750
撤回 749
撤回报价 979
撤离 300
撤离程序 301
撤离分析 300
撤离目标1(救生和救助) 300
撤离目标2(救生和救助) 300
撤离目标3(救生和救助) 300
撤离目标4(救生和救助) 301
撤离目标5(救生和救助) 301
撤离时间 301
撤离站和外部脱险通道 301
撤气调节阀 313
撤气管系 86
撤气式蒸汽机 729
撤销案件 243
撤销报价 107
尘密外壳 263
沉船阀 796
沉垫 552
沉垫长 498
沉垫裙板 552
沉垫型深 224
沉垫型自升式钻井平台 846
沉垫自升式钻井平台 552
沉淀(污垢,沉积物) 778
沉淀舱 790
沉淀槽(沉淀箱) 778
沉淀器(沉淀柜) 790

沉淀物 778
沉浮试验 819
沉积物毒性 778
沉石箱 861
沉陷 790
沉箱浮坞 105
陈述/到场 690
陈述 854
衬垫(垫板,座板) 651
衬套(套筒) 821
衬条 509
撑材 196
撑杆 93
成本/收益评估 190
成本/效益分析 189
成本加运费(……指定目的港) 189
成本加运费 189
成对比较方法 651
成对齿轮 374
成对转动 374
成品油 698
成品油船 697
成品油船 698
成人 14
成熟技术 705
成束电缆 105
成束电缆贯穿装置 659
成束敷设 102
成套设备 869
成像 446
成员国政府 564
成组装卸 936
承包 932
承兑 4
承兑交单 249
承接 932
承诺 699
承诺信 500
承推梁 709
承运货物收据 114

承运人 118
承载比 77
城市和商业污水 583
乘客 655
乘客 657
乘客处所 656
乘客定额 656
乘员 276
乘员定额(CL) 194
程序操作 787
程序风险控制 697
程序和布置手册(P&A 手册) 697
程序和布置手册 697
程序控制 698
程序令 697
程序退火 697
程序文件 697
程租 956
澄清油 139
澄油箱(燃油沉淀柜) 362
吃水 d(平台) 256
吃水(内河船)$d(m)$ 255
吃水(小水线面双体船) 255
吃水 254
吃水 255
吃水 T(艇) 256
吃水标志 255
吃水平行段 653
吃水型深比 255
持续改进原则 178
持续功率 178
持续功率 719
持续通风 179
持续有人操作的中心控制站 179
持续转速 178
持有人 429
持证人员 130
尺度 766
尺度比/比例 765
尺度效应(缩尺效应) 765

1063

齿 888
齿 904
齿动传动的手摇绞盘 374
齿动传动式柴油机 374
齿动水密门 374
齿间距 888
齿节（齿距） 904
齿轮（齿轮传动机构） 904
齿轮 374
齿轮泵 374
齿轮泵 374
齿轮比 374
齿轮齿条起重器 486
齿轮传动柴油机 374
齿轮传动式汽轮机 374
齿轮传动涡轮发电机组 374
齿轮传动涡轮机组 374
齿轮负荷系数 513
齿轮换向器 374
齿轮回转泵 374
齿轮减速装置（减速器） 374
齿轮面 374
齿轮啮合噪声 973
齿轮平面 374
齿轮外形 904
齿轮箱 374
齿轮箱体 374
齿轮轴 374
齿轮装置（传动装置,啮合） 374
齿轮装置 374
齿轮组（齿轮系） 374
齿面 341
齿面 904
齿面点蚀 746
齿面胶合 767
齿面应力 904
齿扇式舵机 374
齿式爪箱器 844
齿数 613
齿条（齿杆） 904
齿条（网架,餐具架） 713

齿条 374
齿条齿轮操舵装置 713
齿条齿轮升降装置 713
齿形联轴器 374
齿形扇弧 904
翅片管式散热器 325
翅片式的（有安定面的,装鳍的） 325
充电功率 687
充分发展的风浪 363
充分发展风浪 364
充满 323
充气 373
充气 944
充气袋囊 455
充气登乘滑梯 455
充气二极管 370
充气救生衣 455
充气效率 132
冲程长度 498
冲出（模压,灭火,冲孔） 849
冲动［冲击,冲量（自动化）］ 447
冲动式汽轮机 447
冲动式涡轮机 447
冲动式叶片 447
冲灰器 50
冲灰旋塞 50
冲击（锤击,压碎） 849
冲击 446
冲击波（激波） 810
冲击波衰减 810
冲击模拟试验机（冲振模拟试验机） 810
冲击试验 446
冲击试验 446
冲击试验 810
冲击试验机 446
冲击试验装置 810
冲击应力 810
冲击作用 810
冲量试验 447

冲水检验 431
冲水试验（喷水试验） 961
冲水试验 431
冲吸式挖泥船/吸盘式挖泥船 263
冲洗泵 347
冲洗阀 347
冲洗甲板软管 960
冲洗设备 347
冲洗水 347
冲洗水泵 959
冲箱式造波机 678
冲压成形 849
冲压喷气式发动机 718
冲翼艇 26
冲翼艇 718
抽风（通风）机 452
抽气泵 20
抽气阀 313
抽气口 313
抽气冷却器（喷射抽气冷凝器组） 20
抽气量 86
抽气喷射器（抽气器） 20
抽气器 20
抽气式汽轮机 86
抽气真空泵组 940
抽汽装置 856
抽取试验（抽提试验） 313
抽取显示器 313
抽入泵 868
抽水泵 313
抽吸（吸入吸口） 868
抽吸 313
抽吸泵 313
抽吸泵 868
抽吸式 313
抽吸通风系统 940
抽吸式火警报警系统 346
抽烟式探火系统 761
抽烟系统 825
抽样和测试服务提供方 705

抽真空试验(渗漏试验) 940
臭氧 649
臭氧发生器 649
臭氧技术 649
臭氧消耗量 649
臭氧消耗趋势 649
臭氧消耗物质 649
出发港或地点 683
出灰管 50
出灰装置 50
出口(排出,排出管) 644
出口(输出) 310
出口(太平门,排气管) 309
出口补贴 311
出口端 309
出口阀 645
出口加工区 310
出口奖励证制 129
出口配额 311
出口商品 310
出口商品价格指数 310
出口税 310
出口税 311
出口信贷 310
出口信贷国家担保制 310
出口信用保险 310
出口许可证 310
出口压力 645
出口压力表 645
出泥孔(放油渣孔) 823
出票 899
出入舱口 303
出入舱口盖 4
出水 961
出水槽 644
出水楔形体积 276
出庭作证 690
出油口(放油口) 622
出租 495
初步设计的初步批准 689
初步系统安全评估 689

初次检验(MARPOL) 457
初次检验(平台) 457
初次检验 457
初次入级检验 457
初次酸洗 336
初始舵角 457
初始回转/航向改变能力 458
初始人员密度 457
初始审核 457
初始事件 458
初始样品 694
初稳心 457
初稳心半径 457
初稳心高 457
初稳性 457
初稳性高度 GM(m) 564
初值(起始值) 854
初转期 458
除草船 969
除垢 224
除气(去气) 644
除气 372
除气 708
除气装置 371
除氢(脱氢) 440
除去浮渣 768
除外责任 305
除锈补漆 767
厨房 368
厨房机械 368
厨房设备 367
厨用多功能机 582
厨用烘箱 367
储备(储量,储备品) 738
储备给水 738
储备功率 738
储备容量(储备功率) 738
储藏容器 861
储藏室,工作间,配膳室等 861
储藏易燃液体的其他处所 644
储存柜主阀 537

储氢室 421
储气罐 372
储气器 856
储器(蓄水池) 738
储绳绞车 750
储液器 439
储油驳 625
储油船 625
储油平台 625
储油平台 861
储油器 620
储油器 622
储油设施 625
储油系泊装油装置 518
储油系统 625
处理(船舶生物污底控制) 918
处理、贮存、发送 404
处理器分享 385
处理设备(加工设备,工艺设备) 697
处理装置 697
处所(舱室) 832
处所 832
处所 832
处所的渗透率 662
触底吸力 91
触地点 906
触地段 760
触地区 906
触发电弧 46
触发事件 458
穿管系数 256
穿浪双体船 967
穿梭油船 813
穿越冰区巡逻所警戒冰山区的航线 752
传递函数 913
传递能力 914
传动比 258
传动比 915
传动链 374

传动系统（传递系统） 915
传动循环链 258
传动轴（中间轴） 915
传动装置 914
传动装置 915
传动装置输入轴数 613
传动装置外形尺寸 915
传感技术 785
传感器（敏感元件） 785
传感器(探测器,探头) 696
传感器/信息源模块 786
传感器 786
传感器系统（动力定位系统） 786
传话管 954
传话筒 580
传热端差 891
传热系数 419
传热系数 645
传热系数 894
传声口 834
传声器 834
传输 914
传送齿轮 915
传送吊车 426
传送装置 524
传送装置的安全工作载荷 756
传统张力腿平台/第一代张力腿平台 185
传真 888
传真机 319
船/港界面活动 801
船 801
船波 801
船波伴流 967
船舶（生物污底控制） 801
船舶 794
船舶 795
船舶 948
船舶 949
船舶安全管理体系 799
船舶保安 799

船舶保安报警系统 799
船舶保安计划 799
船舶保安计划和相关记录 799
船舶保安评估 799
船舶保安体系 800
船舶保安威胁辨识 799
船舶保安员 799
船舶保险 796
船舶备品 800
船舶备品 802
船舶不符合要求的明显理由 146
船舶残存能力和液货舱的位置 800
船舶舱室的空气噪声（船上噪声） 21
船舶操纵模拟器 797
船舶操纵手册 949
船舶操纵信号设备 802
船舶测深 800
船舶代表 802
船舶的尺度限制和吃水限制（巴拿马运河） 507
船舶的管理 404
船舶的建造 173
船舶的速度 V 843
船舶的维护 538
船舶的主要运动 802
船舶登记国 854
船舶抵押法 797
船舶地面站 796
船舶电力推进装置 796
船舶电力系统 802
船舶动力装置 547
船舶吨位丈量 903
船舶多通道立体播放软件 560
船舶辅机 544
船舶负载和机舱负荷 802
船舶高效率航海软件 269
船舶海上航行导航系统 797
船舶和飞机 949
船舶横浪状态 448

船舶活动监控系统 577
船舶机械计划保养系统检验 674
船舶机械计划保养系统检验 875
船舶机械通风 797
船舶监控系统 948
船舶检查指南 400
船舶减摇装置 800
船舶建造日期 212
船舶建造险 807
船舶结构监控 149
船舶结构设计 802
船舶结构通道 800
船舶结构通道手册 800
船舶经纪人 807
船舶空气调节 543
船舶扣押地 674
船舶快速性 843
船舶冷藏 547
船舶录 729
船舶绿色护照 393
船舶耐波性指标 451
船舶内部电气信号装置 467
船舶内舱室的噪声 469
船舶能效 796
船舶能效管理计划 796
船舶能效管理计划符合声明 164
船舶能效管理系统 796
船舶能效管理证书 802
船舶泥浆系统 511
船舶配员 809
船舶气象仪 36
船舶气象仪 802
船舶设备 796
船舶设备 802
船舶失电 85
船舶使用费 242
船舶所有人(船东) 808
船舶通道 5
船舶通风 801
船舶涂料海洋环境曝露试验 311
船舶推进 703

船舶拖曳力 907
船舶维修保养体系(CWBT) 799
船舶维修保养体系 802
船舶物料 861
船舶系统 548
船舶相对运动 799
船舶性能检测系统 797
船舶修理/修正补充合同 30
船舶修理合同 799
船舶虚拟仿真数据平台 953
船舶压载水管理计划编制指南 400
船舶摇荡运动 797
船舶要素 802
船舶液舱晃荡 801
船舶应急响应服务 796
船舶迎浪状态 938
船舶优化 797
船舶远程识别和跟踪系统 523
船舶远程识别和跟踪信息 523
船舶远程识别和跟踪用户 528
船舶运动加速度引起的动载荷和液舱晃动压力 264
船舶运动确定式处理法 235
船舶运动相角 667
船舶噪声测量 797
船舶噪声测量 809
船舶蒸汽接头 801
船舶制冷装置 547
船舶中压岸电系统 29
船舶重大改装 802
船舶重大改装合同日期 212
船舶主尺度 796
船舶专项取样计划 957
船舶转级检验 145
船舶装载计算 796
船舶资料 655
船舶自然通风 797
船舶总布置 809
船舶总体电磁兼容 274
船舶总振动 801

船舶综合电力系统 465
船舶租赁 795
船舶阻力 739
船长(地效翼船) 497
船长(非高速小水线面双体船) 496
船长(高速小水线面双体船) 496
船长(内河船)$L(m)$ 496
船长(渔船) 500
船长 496
船长 551
船长 809
船长、高级船员和普通船员等级证书 130
船长对船舶安全和保安的决定权 552
船长和船宽 496
船长决策支持系统 214
船长决策支持系统 873
船长室 109
船长中点 569
船磁方位调查 796
船到船活动 801
船底(小水线面双体船) 192
船底板/舵叶底板 91
船底处的剖面模数 $Z_{AB}(m^3)$ 776
船底防污染公约(AFS公约) 17
船底防污系统声明 17
船底肋骨 91
船底剖面模数 776
船底塞 91
船底外板 91
船底斜升线 346
船底纵骨 91
船东(货主、业主) 648
船东 808
船端部 282
船海浮体载荷计算软件 966
船级 142
船级 144
船级社 144

船级社 825
船级社产品证书 825
船级条件 143
船级条件 170
船级证书 143
船价鉴定 43
船井 578
船具(船用装置,船用齿轮组) 545
船具 41
船具数(舾装数) 292
船宽(特种用途船舶) 94
船宽(小水线面双体船) 94
船宽 93
船宽 94
船宽 B(内河船) 94
船宽 B(渔船) 94
船宽 B 795
船宽 B 94
船宽 B_s 94
船况鉴定 170
船龄 18
船名旗 797
船模 807
船模试验水池 797
船模拖曳试验水池 797
船模自航试验 807
船模总阻力 905
船模总阻力系数 152
船模阻力试验 807
船模阻力仪 807
船年 801
船旗国 339
船旗国履约评估内部标准 469
船旗国履约评估外部标准 312
船旗国主管机关 14
船上 NO_x 验证程序 630
船上电气设备所用防爆结构的类型 489
船上动力装置 803
船上焚烧 803
船上海洋污染应急方案 804

船上混合 86
船上技术评估计划 807
船上监测手册 630
船上禁止焚烧的物质 866
船上局域网 630
船上培训和演习 630
船上人员（载员） 665
船上人员 809
船上设备（船上机械） 771
船上审核 803
船上试验 807
船上试验 892
船上天然气燃料发动机装置安全暂行指南 467
船上条件 629
船上油污染计划 804
船上有毒液体物质污染应急计划 803
船上噪声源 600
船身效率 435
船式平台 810
船式钻井平台 810
船首和船尾处所 17
船首基准高度 H_b 92
船首向上显示 416
船首向上显示 92
船速 V_0 802
船体 435
船体车间 807
船体船级证书 435
船体分段吊装 51
船体号 436
船体横剖面对其垂向中和轴的惯性矩 (m^4) 486
船体横剖面对其水平中和轴的惯性矩 (m^4) 486
船体极限弯矩 930
船体技术寿命 887
船体监控系统 436
船体建造监控计划 181
船体建造监控计划 435

船体结构 436
船体结构结露 864
船体结构噪声 436
船体经济寿命 268
船体空缺面积 47
船体梁极限弯矩 M_U 436
船体梁载荷 435
船体梁正应力 989
船体排水航行模式（地效翼船） 436
船体强度 436
船体设备 435
船体寿命 436
船体型表面 576
船体型线 802
船体循环检验 178
船体循环检验系统 179
船体以上高度 421
船体振动 796
船体振动阻尼 436
船体支撑结构 435
船体自然寿命 587
船位保持 802
船位推算法 212
船艉 A 型吊架及拖缆绞车 55
船坞登陆舰 247
船舷测波仪 807
船舷阀 809
船舷冷却 435
船型 796
船型标志 928
船型开发 801
船型系数 153
船用 pH 计 807
船用泵 547
船用材料（耐海洋环境材料） 546
船用柴油 560
船用柴油机 544
船用柴油机的实质性改变 867
船用柴油机节能环保技术 284
船用产品 547

船用产品检验 547
船用产品证书 547
船用传动装置 548
船用传动装置效率 915
船用担架 350
船用低硫油（MGO）黏度控制系统 952
船用低硫油 568
船用电磁计程仪 276
船用电站 800
船用发动机氮氧化物（NO_x）排放控制技术规则 886
船用分光光度计 87
船用焚烧炉 803
船用焚烧炉标准技术条件（船用焚烧炉标准技术规范） 852
船用风扇 549
船用辅机 435
船用锅炉 544
船用合金 543
船用机械 546
船用计程仪 546
船用家具 796
船用减速齿轮箱 547
船用空调 543
船用炉灶 368
船用铝合金 543
船用锚 544
船用配件/船体舾装设备 803
船用配件 545
船用汽轮机 548
船用燃气轮机 545
船用燃气轮机机组 545
船用设备（船用机械） 796
船用设备（海用设备） 545
船用设备证书 545
船用设施 545
船用索具 880
船用索具卸扣 978
船用卫生装置 548
船用物料 802

船用小五金 104
船用压缩机 544
船用油分离机 544
船用有线对讲机 548
船用轴流泵喷水推进装置 63
船用装载仪 546
船用装置 546
船用装置证书 546
船用子母钟 547
船员/乘员 193
船员 772
船员安全通道的可接受的装置 3
船员不法行为 72
船员步桥 125
船员处所 194
船员起居舱室 193
船员室 194
船员物品 194
船员正常的工作处所 605
船载驳 807
船载简化航行数据记录仪性能标准 660
船载装置 803
船舯（渔船） 31
船舯 31
船舯部 128
船舯部 569
船舯部分 569
船舯吃水 569
船舯处最大冰级吃水 557
船舯最大吃水 557
喘振 875
喘振边界 849
喘振边界 875
喘振工况 875
喘振特性 875
喘振裕度 875
串并联泵 787
串励电动机 787
串联活塞 881
串联式减速齿轮箱 881

串列滑车 523
串列螺旋桨 881
串列式发动机 881
串列式水翼系统 881
串味 881
串味险 881
串行传输 787
窗 976
窗盖 213
窗框 388
窗帘匣 199
窗式空气调节器 977
窗座 338
床梯 81
吹风搁架冻结装置 793
吹风冷却 19
吹灰器 828
吹泥船 72
吹泥船 72
吹泥船 724
吹气气胀式个人漂浮装置 641
吹填排泥装置 781
吹熄极限 87
炊具 185
垂荡补偿器 419
垂荡调谐因数 921
垂荡幅值 33
垂荡固有频率 586
垂荡固有圆频率 586
垂荡固有周期 587
垂钓 36
垂线间长 L_{pp}（m） 497
垂向棱形系数 947
垂向正加速度 685
垂直安定面 947
垂直吹风式 946
垂直的合同审核 946
垂直扶梯（直梯） 946
垂直隔板 947
垂直加强筋的 A 型连接 171
垂直加强筋的 B 型连接 171

垂直拉伸增加舱容（VST）设计 947
垂直铺管 946
垂直铺管系统 946
垂直融通 946
垂直搜索引擎 947
垂直探鱼仪 947
垂直通路顺应式立管 164
垂直艉翼 946
垂直下行焊 947
垂直一体化 946
垂直振动/垂向振动 947
垂直轴 946
锤锻 403
纯厚度（净厚度） 593
纯净度 708
纯稳性丧失 708
纯稳性丧失第二层薄弱性标准 501
纯稳性丧失第一层薄弱性标准 501
纯音 708
磁场 533
磁场均衡器（补偿磁铁系统） 533
磁场强度 533
磁带机 533
磁带记录 533
磁带录音机 884
磁的 532
磁方位角 532
磁放大器 532
磁粉 533
磁粉检测（磁粉检查） 533
磁粉检测（磁粉检验） 533
磁粉检测（磁粉探伤） 533
磁粉检测 533
磁粉检测法（磁粉检测机,磁通量） 532
磁粉检测法 533
磁粉检测设备（磁粉探伤设备） 533

索引

1069

磁粉检测仪（磁力检测仪） 533
磁粉探伤 533
磁感应 533
磁共振 844
磁共振催生"隔空充电" 978
磁老化 532
磁力传动（带电磁离合器的传动装置） 533
磁力检测（磁力检验） 532
磁力检验（磁力检测法） 533
磁力勘探 533
磁力控制器 532
磁力裂纹检测仪 533
磁力裂纹探测（磁粉检测） 533
磁力启动器控制 180
磁力仪绞车 533
磁力仪探头 533
磁流体 533
磁流体推进核潜艇技术 534
磁罗经 532
磁盘 243
磁盘驱动 243
磁铁（磁性） 532
磁铁 533
磁通量（磁性焊剂,磁性熔剂） 533
磁吸引（磁引力） 532
磁性材料 533
磁性钢 532
磁性钢 533
磁性焊剂电弧焊 533
磁性焊剂气体保护电弧焊 533
磁性合金 532
磁性检测（磁性探伤） 532
磁性检测 533
磁性金属 533
磁性自动操舵装置 532
磁轴 532
次级流 775
次级应力 776
次紧锁装置 775
次屏壁 774

次屏壁 775
次声 456
次要构件 775
次要骨材 774
次要骨架间距 775
次重要设备 775
刺网 387
刺网绞盘 387
刺网起网机 387
刺网渔船 387
刺网振网机 642
从价税 14
从量税 841
从事捕鱼的船 948
粗玻璃纤维 752
粗糙程度 752
粗糙度补贴系数 752
粗糙度阻力 752
粗糙度阻力系数 152
粗糙纤维 752
粗滤器 335
粗滤器 694
促进贸易性运费率 699
促销 760
催化剂 120
催化剂块 120
脆性破坏 659
淬火-回火 QT 712
淬硬钢（硬化钢） 405
淬硬区（硬化区,硬化层） 405
淬硬性试验 405
存根 864
存色彩电视系统 787
磋商限额 592
错车 814
错误 294
错误代码 294

D

答辩 219
答辩书 986

打夯船 161
打捞驳 761
打捞船/救生船 761
打捞浮筒 103
打捞救生船 761
打扫干净 97
打砂桩船 762
打印 695
打印机 696
打桩船 344
打桩船 669
打桩船 670
打桩架 670
大舱盖 491
大舱口散货船 101
大侧斜螺旋桨 425
大侧斜螺旋桨和超大侧斜螺旋桨 427
大侧斜叶（大侧斜桨叶） 426
大齿轮 374
大多数 539
大副 134
大功率的 424
大管轮 774
大河谷国家纪念碑 393
大湖 392
大湖型散货船 393
大加勒比海区域 973
大件运输船 420
大件运输船 96
大件运输船超压载作业 313
大件运输船超压载作业 645
大开口 491
大括号 198
大理石碎粒 543
大量燃油 492
大楼并发症 99
大螺距螺旋桨 424
大麻 107
大麻植物 422
大面调查 339

索引

1070

大气波道　301
大气冷凝器　54
大气泄水柜　54
大倾角稳性　846
大容量采水器　492
大堂(豪华邮轮)　535
大写的(字母)　109
大型登陆舰　491
大型电动机　491
大型军用运输机　492
大型汽车运输船　491
大型污底　531
大型邮轮　197
大型油船　948
大修(拆修,检修)　646
大修　539
大洋调查　617
大圆航线　191
大主应力　539
大抓力锚　423
大抓力锚　506
呆木装置(艉轴管装置,艉轴传动装置)　859
代表(国际海事卫星组织)　736
代表性处所/舱室　735
代表性物质　736
代表性样品　735
代表性液舱　736
代理　18
代理　18
代理费　18
代理协议　18
代理银行　188
代销　760
代用材料　539
带板　70
带板的净厚度 t_p　593
带板的净剖面模数 Z　593
带板的宽度 b_p　974
带电的　512
带舵可转导管　859

带附体阻力试验　449
带刚性立管的 SALM 系统　761
带钩板　651
带冠叶片　464
带过热器锅炉的公称压力　601
带环螺旋桨(加环螺旋桨)　746
带键轴锥体　489
带缆船/起锚拖船　35
带缆艇　578
带缆羊角　578
带缆桩　84
带锚链腿的 SALM 系统　761
带锚链与刚性立管的 SALM 系统　760
带棉绒的棉籽(含水量不大于9%,油含量不大于20.5%)　509
带气化外输装置的液化天然气储存和再液化装置　513
带倾横摇　448
带绳扣滑车　86
带头螺栓　884
带指示器的电子方位线　293
带状加强环　744
待机工况　853
待检验　876
丹福氏锚　206
单板性能　676
单臂轴支架　817
单臂轴支架　819
单侧起锚机　817
单层板　819
单次谐波　818
单底　817
单底船　817
单底结构　817
单点系泊　818
单点系泊的浮筒　125
单点转塔系泊式浮式生产储/卸油船(FPSO)　354
单点转塔系泊系统　818
单独海损　655

单独显示　452
单断级滑行艇　819
单浮筒储存系统　817
单浮筒系泊　819
单杆操作　877
单缸泵　817
单缸发动机(试验用发动机)　630
单缸汽轮机　819
单个故障(单一故障)　818
单个故障　815
单个警报　452
单工制(单缸的)　817
单环推力轴承　817
单机拖动　819
单级减速齿轮箱　819
单级减速装置　818
单级闪发　819
单级压缩机　818
单级压缩制冷系统　818
单甲板船　817
单件/单批检验　935
单井式石油生产系统　819
单镜头反光相机　818
单卷筒绞车　819
单壳构件　818
单壳散货船　818
单壳油船　818
单流式汽轮机　819
单流蒸汽机(单流式蒸汽机)　934
单螺纹　819
单锚腿储存系统　817
单锚腿系泊系统　817
单锚系泊　760
单排滚珠轴承　818
单排直立式发动机　459
单盘式离合器　630
单片离合器　818
单屏立管　817
单式串联减速齿轮箱　881
单手柄控制器　487
单水翼艇　819

单索抓斗 818
单体船 578
单体高速船 578
单艇体吃水 T_C 107
单拖网渔船 819
单位功率的不平衡力矩 687
单位耗电量 841
单位耗汽量 842
单位耗热量 841
单位货运量的 CO_2 排放量 149
单位载荷(比载) 841
单舷侧结构的散货船 100
单相电路 819
单相同步发电机 819
单向流动 934
单向碎渣机系统 631
单效蒸发 819
单烟道锅炉 819
单一气体燃料发动机 818
单因素贸易条件 817
单引号 818
单油路喷油嘴 817
单元货物运输船 935
单原子铱催化剂 817
单胀式蒸汽机 817
单证 249
单轴燃气轮机 819
单爪锚 818
单桩 669
单桩腿负荷 486
淡水 360
淡水泵 360
淡水驳运泵 360
淡水舱(淡水柜) 供贮存淡水的舱。 360
淡水干舷 357
淡水管路 360
淡水冷却泵 360
淡水冷却器 360
淡水滤器 360
淡水压力柜 360

淡水雨淋 360
淡水雨淋险 360
弹道导弹 70
弹道导弹无规律性变轨技术 937
弹簧(斜缆,移船缆) 845
弹簧把手 845
弹簧保险圈 846
弹簧颤动 846
弹簧常数 845
弹簧秤 845
弹簧垫圈 846
弹簧调整片 846
弹簧刚度(弹性刚度) 846
弹簧钢 846
弹簧钢丝 846
弹簧缓冲器 845
弹簧减振器 845
弹簧减震器 846
弹簧联轴节 845
弹簧平衡舱盖 845
弹簧平衡窗 977
弹簧式安全阀 846
弹簧式调速器 846
弹簧式压力计 846
弹簧锁(弹簧锁扣) 846
弹簧拖钩 846
弹簧外露式安全阀 311
弹簧销 845
弹簧悬置 846
弹簧组件 846
弹簧座 845
弹簧座 846
弹力范围 846
弹性(有弹力) 846
弹性(恢复性,伸缩性) 272
弹性安装 342
弹性变形 271
弹性波 272
弹性的 271
弹性夹心板 272
弹性联轴节 271

弹性联轴器/弹性联轴节 342
弹性限度 506
弹性元件 272
弹性振动 272
弹药舱/弹药库 531
弹药舱通风 33
弹药室 532
弹药箱 531
弹药转运间 33
弹振 846
氮气发生器 599
氮气发生器系统 599
氮氧化物 NO_x 排放 599
氮氧化物排放 609
当量(等价,等值,同意义的) 292
当事人双方主体资格证明文件 700
当事人意思自治 655
当值轮机员 287
挡板(导流板,消声器) 67
挡板喷嘴 341
挡光板 767
挡火条 333
挡货板 793
挡货板 821
挡浪板 95
挡泥板 581
挡水板 576
档案(国际海事卫星组织) 46
导板 196
导边 494
导边半径 717
导标-角定位法 568
导弹舱 574
导弹护卫舰 400
导弹驱逐舰 400
导风轮 452
导管 261
导管 610
导管架[井架,(蒸发气)套,(水)套] 486

导管螺旋桨 261
导管螺旋桨 262
导管推进器 262
导管推进器 610
导管推进器总推力 905
导管推力 897
导管推力器(管隧推力器) 921
导航(航行) 588
导航 591
导航测风系统 187
导航定位室 589
导航分显示器 589
导航和操纵工作站 588
导航雷达 588
导航雷达 589
导航桥楼 591
导航系统 588
导航系统 589
导航与绘图系统 464
导缆滚轮 137
导缆孔 578
导缆器 317
导缆钳 136
导链滚轮 131
导链滚轮 490
导流挡板 400
导流管 546
导流器 400
导流罩舵(整流帽舵) 99
导轮 670
导热系数 K 893
导入管 456
导索 754
导体 170
导向阀 400
导向滚轮 658
导向滚柱 216
导向滑车(导块) 400
导向滑车 494
导向机构 219
导向叶片 400

导向支架 400
导向轴承(辅助轴承) 400
导向装置 400
导烟板 825
导叶(导向叶片) 400
导翼(导流叶片) 400
岛屿 485
倒臂型艇架 529
倒车传令时滞 914
倒车导板 66
倒车舵 341
倒车功率 53
倒车功率 53
倒车级 53
倒车排气室喷雾器 53
倒车燃气轮机 53
倒车时滞 306
倒车透平 53
倒车性能试验 53
倒杆型吊艇架 862
倒角 268
倒缆 845
倒顺车 19
倒顺车减速传动装置 726
倒顺车燃气轮机 744
到达工况 48
到港压载 48
到航线终点距离 245
德国风能研究所 236
德国海上职业联合会 778
德国海员工伤事故保险联合会 765
德国联邦海事与水文局 319
灯 490
灯标船 503
灯船 505
灯光传令钟 504
灯光传令钟 505
灯具(航行灯) 490
灯具贮藏室 490
灯泡 490

灯塔费 504
灯座 490
登岸假期 810
登乘 276
登乘甲板 276
登乘时间和降落时间 276
登乘时间及下水时间 276
登乘梯 276
登乘站 276
登记长度 729
登记船长 729
登记吨 729
登记港 683
登记国 854
登记宽度 729
登记深度 729
登离船装置 912
等待时间 492
等电位连接 485
等离子水下切割机 932
等螺距 173
等容循环 173
等势面 292
等势线 292
等体积倾斜 293
等体积倾斜水线 293
等体积倾斜轴线 293
等推力伴流分数 886
等推力法 562
等温退火 485
等温线跟踪仪 485
等响曲线 292
等效/弯曲刚度 292
等效材料 293
等效电动机 293
等效发电机 293
等效固定式气体灭火系统 293
等效静载荷系数 293
等效连续声压级 292
等效喷水器系统 293
等效设计波 292

等效设施　293
等效水基灭火系统　293
等效细水雾灭火系统　293
等效压力水雾灭火系统　293
等效证明文件　293
等压循环　173
等压增压　173
等噪度曲线　292
等值系数　152
等转矩伴流分数　886
等转矩法　562
低比活度（LSA-1）非裂变的或预计裂变的放射性物质　717
低播焰性　525
低潮高地　527
低负荷调整　527
低滑油压保护装置　525
低阶振动　950
低硫燃油　526
低硫燃油冷却系统　186
低硫油（MGO）黏度控制系统　183
低螺距螺旋桨　525
低密度材料　525
低黏度物质　527
低频　525
低频声　525
低频响应　526
低熔点合金（易熔合金）　300
低水位报警器（低水位信号）　526
低水位报警器　526
低水位吸入阀　526
低速操纵性　525
低速柴油机　525
低速柴油机　823
低速风筒　525
低速红外线　823
低速级组　823
低速推进　525
低碳　524
低碳钢　525
低碳钢　826

低位冰区水线　526
低位海水门　525
低温报警　526
低温腐蚀　526
低温韧性钢　858
低温退火　526
低温液货　526
低压　525
低压CO_2（灭火）系统　525
低压泵送系统　525
低压侧　525
低压端（LPG系统）　525
低压给水预热器　525
低压空调系统　525
低压喷焊器　458
低压气体管路　525
低压涡轮　527
低压系统　527
低压循环贮液器　525
低压压气机　527
低压压缩机　525
低压蒸汽发生器　525
低压蒸汽发生器给水泵　525
低压主汇流排　527
低噪声电路设计　525
低噪声电源装置　816
低噪声电子设计　527
低噪声风机　816
低噪声烧嘴　525
低噪声推进器　525
低噪声优化设计　527
低真空保护装置　940
滴油盘　621
滴油灶　258
滴状冷凝　259
抵港时间　903
抵消　618
底板（钢夹层板）　72
底边舱　92
底边舱斜板　431
底部吹泄阀（排污阀）　87

底部吹泄阀　91
底部防喷系统总成　526
底部修井装置总成　526
底舱　527
底脚螺栓　429
底开门　91
底栖生物　81
底栖生物拖网　81
底栖生物拖网　81
底栖拖网绞车　81
底栖拖网装置　81
底漆　695
底特律河国际野生动物保护区　235
底拖钓船　91
底拖网　91
底延绳钓　91
底纵桁　91
地　267
地　396
地磁测量船　386
地磁场电磁海流计　386
地基拦截弹　397
地理不利国　386
地面稳定模式　397
地面效应　396
地面效应作用区（地面效应区）　397
地球物理勘探　386
地球物理勘探船　386
地球站　267
地热流仪　387
地图/航线　543
地效区（GEZ）内飞行模式　348
地效翼（WIG）船　977
地效翼（WIG）船安全证书　977
地效翼（WIG）船营运许可证书　663
地效翼船　396
地效翼船　974
地效翼船　977

地效翼船的初次检验　458
地效翼船的定期检验　660
地效翼船的附加检验　12
地效翼船的换证检验　735
地效翼船的埋艏　678
地效翼船的纵倾　919
地效翼船分类　144
地效翼船或地效飞机　974
地效状态　396
地震波　780
地震电缆绞车　780
地震放大器　780
地震回放仪　780
地震勘探　780
地震剖面仪　780
地震普查　780
地震示波器　780
地震详查　780
地震震源空压机室　779
地质调查船　386
地质绞车　386
地质勘查　386
地质样品舱　386
地中海海区　564
地轴弄走道　922
递交　734
第1类消防船　330
第1类油船　120
第2类消防船　330
第2类油船　120
第3类消防船　331
第3类油船　121
第4.1类易燃固体　139
第4.2类易自燃物质　139
第4.3类 遇水产生可燃气体的物质　139
第5.1类具有氧化性的物质（氧化剂）　140
第6.1类有毒物质　140
第6.2类 感染性物质　140
第7层负载均衡　494

第7类放射性物质　140
第8类腐蚀性物质　140
第9类其他危险物质和物品　140
第Ⅰ类船舶　928
第Ⅱ类船舶　928
第Ⅲ类船舶　125
第Ⅳ类船舶　121
第二层—功能要求（适用于无限航区散货船和油船）　899
第二超越角　774
第二代完整稳性标准　774
第二岛链　774
第二过渡阶段航行模式（地效翼船）　636
第二甲板　774
第三层-符合性验证　899
第三代张力腿平台　596
第三岛链　895
第三国航运公司　895
第四代战机　354
第一层—目标　899
第一超越角　335
第一岛链　335
第一过渡阶段航行模式（地效翼船）　636
第一种推力轴或第一种艉轴管　701
缔约国　180
缔约国　655
颠倒温度表　744
点（焊点,斑点）　845
点对点通信协定　678
点腐蚀　678
点火（发火,起爆）　445
点火（燃烧）　335
点火　445
点火泵　335
点火次序　335
点火定时　445
点火定时器　445
点火器（发火器,点火剂）　445

点火器　339
点火试验　335
点火试验　445
点火试验　48
点火速度　843
点火提前　834
点火装置　445
点火装置　504
点声源　678
电报费　104
电报员　716
电茶壶　273
电炒锅　272
电传　888
电磁点火　533
电磁吊车（电磁吸盘式起重机）　532
电磁调速异步电动机　276
电磁发射　274
电磁阀　274
电磁阀　532
电磁阀　533
电磁阀　827
电磁辐射　274
电磁辐射　274
电磁辐射吸收率　2
电磁干扰　274
电磁焊　533
电磁火花塞　533
电磁兼容性　274
电磁离合器　274
电磁离合器　532
电磁连轴节　533
电磁喷水推进　440
电磁骚扰　274
电磁式脱扣器　533
电磁探头　533
电磁性噪声　274
电磁闸　532
电导率传感器　170
电点火　272

电点火器　272
电动泵　272
电动舵机　273
电动发动机组　273
电动辅机　272
电动给水泵　579
电动机　579
电动机电枢　579
电动机分支电路　579
电动减摇鳍装置　273
电动交流发电机组　579
电动起货机　272
电动起锚机　273
电动起锚绞盘　273
电动拖缆机　273
电动液压舱口盖　439
电动液压操舵装置（电动液压舵机）　273
电法勘探　273
电工车间　273
电工间　273
电焊工作船　345
电焊锚链　273
电荷耦合元件　132
电汇　888
电汇　888
电火花点火器　272
电火花源　834
电火花振源　834
电极式盐度计　273
电加热带　272
电解法船舶压载水管理系统　273
电抗　721
电抗器　721
电控柴油机　272
电控制器　272
电缆舱　105
电缆短路容量　811
电缆敷设船　105
电缆管理系统　105
电缆埋设机　104

电缆松紧指示器　821
电离层观测　614
电离层观察室　481
电离层折射校正　481
电力操舵系统　273
电力传动装置　272
电力和照明变压器　913
电力推进　272
电力推进操纵站（台）　173
电力推进船　273
电力推进控制装置　180
电力推进系统　272
电力推进系统　273
电力推进装置　272
电力网　273
电流放大器　198
电脑鼠标　167
电暖　272
电暖器　272
电暖器　273
电耦合（电磁联轴节）　272
电气和电子工程师协会　445
电气和电子设备的电磁兼容　274
电气两用饭锅　273
电气器具分支电路　272
电气设备（动力定位系统）　273
电气设备　273
电气设备工作的环境条件　290
电器用途　273
电取暖器　272
电热器　272
电热融霜　272
电热元件　272
电热元件　273
电容（电容量）　108
电容传声器　169
电容器　108
电渗析法　273
电视机　888
电视机遥控器　923
电视摄影电缆绞车　888

电视游乐器　368
电梯　502
电压不平衡偏差　955
电压偏差　955
电压瞬变　955
电压瞬变恢复时间　955
电压周期性变化　202
电源传导干扰抑制　467
电灶　272
电站　687
电子储藏　275
电子定位系统　291
电子定位系统　275
电子二极管　274
电子反压制　274
电子管适配器（阀的接头）　940
电子海图显示和信息系统　267
电子海图显示和信息系统　274
电子海图显示和信息系统标准显示器（ECDIS标准显示器）　268
电子海图显示和信息系统显示库　268
电子航行海图　275
电子航行海图系统　785
电子计算机室　167
电子监控系统（数据）　275
电子监听　275
电子监听装置　275
电子控制器　272
电子控制系统　275
电子滤波器　275
电子签证　275
电子扫描阵列雷达　275
电子商务　267
电子式测温计　274
电子束离子阱　274
电子数据交换　269
电子学　275
电子压制　275
电子邮件　275
电子邮件系统　275

电子照相技术　275
电子照相系统　668
电子支援　275
电阻焊接管　272
电阻器　739
垫舱物料　262
垫衬材料　372
垫块　323
垫料框　372
垫片（垫密片）　793
垫圈（垫片）　372
垫圈（垫片,垫密片,填料）　372
垫升　434
垫升动力限制　502
垫升风机　502
垫升功率　502
垫升航态　200
垫升和/或滑行模式（地效翼船）　200
垫升装置　502
淀粉样前体蛋白　34
吊臂叉头　47
吊舱驱动推进器　703
吊舱式推进装置　64
吊车　192
吊床　403
吊筏架　717
吊杆　224
吊杆抱合箍　90
吊杆变断面系数　316
吊杆叉头　224
吊杆长度　90
吊杆长细比　822
吊杆初挠度　457
吊杆附加弯矩　43
吊杆工作范围　982
吊杆箍　224
吊杆间牵索　766
吊杆零部件　90
吊杆偏角　822
吊杆偏心距/偏心半径　267

吊杆起重机　224
吊杆强度　863
吊杆式　224
吊杆台　224
吊杆托架　224
吊杆仰角　90
吊杆转枢　390
吊杆装卸　224
吊杆装置　224
吊杆装置　224
吊杆座　390
吊杆座滑车　420
吊杆座滑车眼板　420
吊缸（拆卸活塞）　979
吊钩　430
吊钩净高　417
吊钩梁　171
吊钩下交货的船上交货　349
吊钩卸扣　430
吊钩转环　112
吊钩转环　117
吊钩装置　112
吊货短链　111
吊货钩　112
吊货架　876
吊货索　115
吊货索间夹角　36
吊货索具　114
吊货索压重　190
吊货网兜　113
吊货卸扣　115
吊货眼板　114
吊架　107
吊具　822
吊梁（提升梁）　503
吊锚杆　35
吊锚索具　120
吊桥式码头　822
吊升式号灯　659
吊索垫片　515
吊索环　515

吊梯装置　490
吊艇臂　212
吊艇杆　87
吊艇钩　876
吊艇环　822
吊艇架　87
吊艇架额定负荷　515
吊艇架跨距　645
吊艇架座架　212
吊艇索　87
吊艇装置　88
吊网门形架　917
吊柱　212
吊装式舱口盖/箱形舱盖　682
吊装装卸　502
吊装装卸　503
钓获量　36
钓鱼船　508
调拨价格　28
调查船　738
调车机构试验　390
调风器　22
调和油料　86
调节（调整）　576
调节（规则,条例）　730
调节弹簧　730
调节阀（调整阀）　13
调节阀　390
调节阀　730
调节杆　390
调节杆　730
调节级　183
调节螺钉　730
调节螺丝　390
调节器（稳定器）　730
调节器传动装置（调节装置,调速装置）　390
调节设备　730
调节水柜　875
调节系统（控制系统）　730
调节系统　390

调解 168
调解愿望 168
调距螺旋桨传令钟和指示器 888
调距速度 719
调配 304
调速 843
调速电动机 13
调速电动机的基本转速 73
调速阀 843
调速器(调节阀,平衡器,稳定器) 390
调速器 390
调速器 843
调速器试验(突卸负荷试验) 390
调速试验 390
调速性能试验 390
调速性能试验 390
调停者 13
调压阀 622
调整(控制,操纵) 390
调整的(调节的调整器) 730
调整杆 730
调整螺母 519
调质钢 405
叠合板 165
叠合船模 252
碟状阀 243
蝶形阀 103
蝶形螺母 266
丁腈橡胶 599
顶板(钢夹层板) 904
顶板 793
顶边舱/顶舱 904
顶部防喷系统总成 938
顶部驱动钻井系统 904
顶部障碍物 646
顶点 904
顶篷 191
顶式诱导器 126
顶推 709
顶推船 709

顶推船-驳船组合体 709
顶桅 904
顶桅支索 904
顶张式立管/刚性立管 904
订货 674
定常流 855
定程租船 956
定程租约 956
定刺网 35
定点连续观测 35
定额(功率) 720
定额工时 720
定航速船模自航试验 288
定机变幅式起重船 529
定机式起重船 339
定价 339
定进速试验 891
定螺距螺旋桨 338
定牌 909
定蹼 338
定期检验 660
定期结汇 790
定期审核 660
定期无人机舱 660
定期租船 172
定期租船 902
定期租约 902
定深器 223
定深扫海测量 53
定时限过电流继电器或脱扣器 219
定位(寻位) 519
定位 685
定位螺钉 339
定位螺钉 768
定位螺柱 789
定位器(漏电防护装置) 519
定位倾斜仪 216
定位双杆操作 935
定位系泊系统 685
定位系泊系统的设计工况 225

定位销 519
定位信标 685
定位信号 519
定位桩 846
定位桩架 846
定向红外干扰技术 241
定向声波 642
定型系列集装箱 929
定性的故障分析 710
定叶 854
定置网打桩机 789
定轴涡轮 818
定柱沉垫式钻井平台 91
定转速船模自航试验 177
锭子油 844
东南亚国家联盟 52
董事会 87
动横倾角 265
动滑车 347
动力泵 687
动力舱底泵 686
动力操纵系统 688
动力单元 687
动力定位 254
动力定位 264
动力定位不间断供电电源系统 254
动力定位传感器 254
动力定位船舶 254
动力定位船舶 265
动力定位船舶 265
动力定位的控制器 265
动力定位的推进器系统 265
动力定位红色警戒状态 725
动力定位黄色警戒状态 987
动力定位控制系统 253
动力定位控制系统 254
动力定位控制系统 264
动力定位控制站 253
动力定位控制站 264
动力定位绿色警戒状态 394

动力定位设备等级　140
动力定位式钻井装置　265
动力定位推力器　254
动力定位系统 1　254
动力定位系统 2　254
动力定位系统　254
动力定位系统　265
动力定位系统 3　254
动力定位系统操作人员　254
动力定位系统的主要操作模式　536
动力定位装置　254
动力定位钻井船　265
动力定位钻井船　266
动力放大系数　264
动力工具处理　687
动力号笛　563
动力黏度　489
动力黏性系数　152
动力气垫　263
动力气垫船或地面效应器　203
动力气垫型地效翼船　203
动力牵索吊杆装置　562
动力驱动的水密门故障报警　688
动力输出传送装置　687
动力涡轮　687
动力系统　687
动力系统事故　253
动力相似　265
动力源(电源)　687
动力载荷　264
动力支承船舶构造和设备证书　265
动力转舵系统　686
动力装置(电站)　687
动力装置操纵性　687
动力装置单位质量　687
动力装置的热效率　687
动力装置耗热率　687
动力装置机动性　687
动力装置经济性　687

动力装置可靠性　687
动力装置生命力　687
动力装置相对质量　687
动力装置隐蔽性　687
动量理论　576
动摩擦　360
动平衡(惯性平衡)　454
动倾覆角　265
动索　754
动态频谱分析仪　264
动态稳定控制系统　265
动态应力变量　265
动弯矩　265
动稳性　265
动稳性臂　265
动稳性曲线　199
动稳性曲线　265
动物/植物油船　37
动物尸体　37
动物纤维　37
动压力　265
动压力　265
动叶片　580
动载荷　263
动载系数(起重设备)　264
动载系数 KV　264
动植物栖息保护区域　346
冻结能力　358
冻结器(间)(冻结设备)　358
冻结试验(抗冻性试验)　358
冻结速度　358
冻凝(冷藏)　358
冻融试验(冻结解冻试验)　358
冻住(水塞)　358
兜带救生圈　95
斗链　97
斗链节距　98
斗链张紧装置　98
斗链转速　98
斗轮式挖泥船　201
斗轮式挖泥船　973

斗轮式挖泥船　98
斗桥　98
斗桥吊架　98
斗桥起落装置　98
斗桥倾角　36
斗桥上支承　938
斗桥下支承　527
斗塔　128
陡 S 形立管　858
陡波立管　858
逗号　158
毒品　259
毒气防护　372
毒性试验　908
独家代理　306
独立泵　450
独立弹药库　450
独立的联合操纵杆　450
独立的联合操纵杆控制系统　450
独立的压载舱系统　779
独立架设燃油柜　356
独立开口　450
独立开口　485
独立排水孔　786
独立式惰性气体发生器　454
独立式空调装置　781
独立视频信号端子　755
独立通风系统　451
独立性(电气系统)　450
独立性(蒸汽控制系统)　450
独立性风险分析　450
独立液货舱　450
独立液压系统　450
独立装置　450
独任仲裁员　827
独占许可证　306
读写器　721
堵漏板　657
堵漏盒　127
堵漏螺丝杆　430
堵漏木栓　981

堵漏席　155
堵漏楔　981
堵漏用具　345
堵塞系数　86
渡船　320
镀镍　599
镀镍　599
镀镍的　599
镀锌(电镀)　368
镀锌　989
镀锌层(镀锌)　989
镀锌钢　368
端板　282
端部弧坑　282
端部接头　282
端到端测试　282
端到端方案　282
端对端协议　659
端盖　282
端盖垫料　282
端口　683
端面　282
端翼螺旋桨　324
短程传输　812
短程国际航行　811
短程游艇　812
短的缺陷　811
短吨　812
短峰不规则波　812
短峰浪　812
短杆锚　812
短甲板室　811
短桨　651
短缆　812
短梁　118
短量　812
短量险　812
短路　812
短路电流　812
短路电流直流分量　203
短路分断能力　812

短路接通能力　812
短路容量　811
短路脱扣器　812
短期DOC证书　812
短期信贷　812
短期证书　811
短上层建筑　812
短少和提货不着　812
短声　811
短时耐受电流　812
短纤维玻璃棉　524
短信息　811
短延时短路脱扣器　812
短叶片　811
短轴　573
短轴包架　811
断裂韧性　354
断路器　95
断路器的断开定额(额定断开电流)　94
断路器的额定电流　719
断路器额定工作电压　719
断路器绝缘电压　719
断面调查　776
煅烧氧化铝　30
锻钢　352
锻钢　858
锻工车间　352
锻件　352
锻接　334
堆舱压缩机　721
堆叠系数　849
堆装　861
对…有异议　132
对背信用证　66
对称电缆　878
对称短路电流　878
对船舶的武装抢劫　47
对地航速(SOG)　843
对地航向(COG)　191
对地航向　153

对海洋环境有害并与排放有关的(货物残余物)　407
对环境造成的严重破坏　790
对机舱传令钟(对机舱车钟)　285
对接缝　103
对开泥驳/对开挖泥船　845
对开泥驳　845
对开式泥驳　844
对开信用证　723
对流层折射校正　920
对水航向(CTW)　191
对水速度　843
对拖网渔船　102
对外加工装配业务　313
对外贸易　351
对外贸易地理方向　241
对外贸易额　940
对外贸易货物结构　165
对外贸易量　711
对外贸易依存度　351
对外贸易依存度　720
对外贸易政策　351
对向活塞(对置活塞)　640
对应阻力系数　739
对置活塞内燃机　640
对置气缸内燃机　640
对中限制器　781
对仲裁员回避的书面请求　132
对转螺旋桨　151
对转螺旋桨　179
兑换券　154
吨位/总吨　903
吨位标志　903
吨位丈量　903
吨位证书　903
趸船　681
趸船　682
囤船/趸船　682
盾构机　794
多变效率　681
多并滑车　582

多波束测深系统 581
多层甲板船 801
多断级滑行艇 583
多腭抓斗 641
多缸 581
多缸单轴涡轮机 881
多缸汽轮机 581
多功能控制器 582
多功能模块化海洋工程组合体 575
多功能显示(多功能显示器) 581
多桁架立柱式平台 126
多环芳烃 681
多机并车试验 653
多机共轴齿轮传动 581
多级变速 583
多级减速齿轮箱 583
多级闪发 581
多级预热给水 583
多甲板船 581
多卷筒绞车 581
多孔扩散消声器 830
多孔喷嘴 582
多孔吸声材料 831
多缆物探船 583
多路控制系统 582
多氯联苯(PCB) 681
多盘式离合器 582
多普勒海流计 251
多普勒计数 251
多普勒声呐 251
多气室气垫船 581
多式联运 157
多式联运承运人 581
多式联运单据 641
多式联运单据 157
多式联运单证 157
多式联运单证 581
多式联运合同 581
多式联运经营人 157
多式联运经营人 582

多式联运收货人 581
多式联运托运人 581
多室浮力系统 581
多艘无人水下航行器联合控制系统 156
多速电动机 583
多态性 681
多体船 581
多维数组 581
多相喷射器 582
多相同步发电机 582
多效蒸发 581
多压级 582
多叶舵 581
多用途调查船 582
多用途化学品船 582
多用途货船 583
多用途起锚机 582
多用途拖船 921
多余的（空间的,节省的,备件） 833
多余的(冗余的) 727
多支承舵 428
多种燃料内燃机 581
多种作业渔船 156
多重叠加型变频器 582
多轴并列式汽轮机 396
多轴布置 583
多轴燃气轮机 583
舵 422
舵 753
舵板 753
舵臂 753
舵柄 902
舵柄连杆 902
舵侧斜角 753
舵掣 753
舵承 753
舵的行程 753
舵杆 753
舵杆衬套 754

舵杆吊环 503
舵杆接头 753
舵杆扭矩 754
舵杆填料函 864
舵高 753
舵工 422
舵构架 753
舵厚度比 754
舵机/操舵装置 858
舵角(方位推进器) 753
舵角 753
舵角传令钟 422
舵角指示器 422
舵宽 753
舵链传动操舵装置 131
舵链导轮 494
舵链卷筒 131
舵面积 47
舵面积比 753
舵钮 400
舵平衡比 753
舵剖面 753
舵剖面型值 618
舵球鳍 190
舵扇 710
舵设备 873
舵实效展弦比 269
舵索传动操舵装置 978
舵头 753
舵销 753
舵效 753
舵效试验 753
舵压力 753
舵压力中心 127
舵叶 753
舵叶导边板 608
舵叶顶板 904
舵叶舵角限位器 406
舵叶艏材 910
舵缘倾斜 753
舵展弦比 51

舵轴　753
舵轴舵　817
舵柱　753
惰化法　454
惰轮(空转轮)　444
惰轮　444
惰性保护气体　454
惰性波纹立管　494
惰性的(不活泼的)　453
惰性化　454
惰性化　454
惰性气体　447
惰性气体　453
惰性气体保护　453
惰性气体发生器　453
惰性气体防爆系统　454
惰性气体鼓风机　454
惰性气体灭火及防火系统　453
惰性气体系统　445
惰性气体系统　453
惰性气体窒息灭火系统　453
惰性气体装置　453
惰性气体总管　453
惰性状态　454
惰转时间　444

E

俄联邦安全总局　754
鹅颈管接头(S形管接头,鹅颈连接法,S形弯管连接法)　390
鹅颈通风筒　877
额定持续电流　720
额定处理功率　918
额定的(名义上的,不同的,轻微的)　600
额定工况　719
额定工作压力　720
额定功率(燃气轮机)　719
额定功率　687
额定功率　719
额定泡沫发泡倍数　600

额定泡沫生产量　600
额定容量(额定功率,额定能力)　719
额定绳速　720
额定施放率　600
额定填充率　600
额定填充时间　600
额定压力　720
额定载荷(额定负载)　842
额定值(标准规格)　601
额定值(评定值)　720
额定转舵扭矩　720
额定转速　720
额外的、合理的实际开支　313
额外压载水　11
厄克曼海流计　271
轭架型个人漂浮装置　988
恶劣条件　790
恩古诺酒吧(豪华邮轮)　766
恩氏黏度　288
儿童救生衣　501
耳塞(耳罩)　266
饵料舱　67
二次表面处理　776
二次除锈　775
二次穿着的个人漂浮装置　666
二次风(延燃空气)　774
二次配电系统　775
二次蒸汽　941
二代完整稳性参数横摇标准　774
二级串联减速齿轮箱　261
二级减速齿轮箱　253
二级警报　774
二极管　239
二甲-4-羟色胺/二甲-4-羟色胺磷酸　706
二进制代码　83
二进制计数器　83
二氯甲烷　568
二审程序　42
二氧化硫　870

二氧化锰　540
二氧化碳捕捉和储运　109
二氧化碳灭火系统　149
二氧化碳排放强度指标　270
二氧化碳施放报警器　149
二氧化碳压缩机　109
二氧化碳运输船　149

F

发布/设计小组　866
发电船/电站船　386
发电机　386
发电机的柴油机的额定功率　719
发电机组(生成集)　386
发电机组　272
发电机组　386
发电机组航行试验　386
发电机组系泊试验　386
发动机(引擎)　285
发动机参数记录簿　724
发动机的启动器　286
发动机底座　286
发动机故障　287
发动机机架　285
发动机机座　285
发动机气缸　285
发动机启动器　286
发动机特性　286
发动机熄火状态　574
发动机噪声　285
发动机转速表　285
发动机座　285
发光二极管　504
发光效率　530
发挥想像/头脑风暴　93
发火次序　445
发火提前　445
发火滞后　445
发火滞后时间　445
发货人　172
发明人　57

发明专利　57

发盘　617

发票　481

发热量　419

发散衰减　246

发射应答器　916

发现　325

发证检验　131

乏气孔　308

罚金/违约金　659

阀插口　941

阀的开度　633

阀封(阀座)　941

阀盖　940

阀杆　941

阀杆　941

阀接触面　941

阀孔　941

阀控式密封电池　941

阀门操纵转接器　941

阀门操作　941

阀门传动装置　940

阀门控制　940

阀门遥控机构　941

阀门遥控速关机构　941

阀盘　940

阀升程(气门升程)　941

阀室　940

阀手柄　941

阀体　940

阀位指示器　941

阀箱(阀壳)　940

阀箱　940

阀咬住　941

阀用青铜　940

阀装置　940

阀座垫圈　941

阀座磨合器　941

筏　717

法定发证检验　855

法定检验　855

法定净重　496

法定质量　496

法兰(凸缘,缘板,折边)　341

法兰撑开器(拆开法兰用)　341

法兰接合　341

法兰连接　341

法兰连接交流发电机　341

法兰连接螺栓　341

法兰式接头　341

法兰应力(凸缘应力)　341

法兰圆角　341

法兰轴承　341

法令　8

法律人员　164

法人　496

法人　663

法院　191

法院院长　191

帆　760

帆布救火水龙　331

帆布具　107

帆布天幕　107

帆布通风筒　108

帆布通风筒　108

帆布通风筒　945

帆船　760

帆船　760

帆船咖啡厅(豪华邮轮)　976

帆的投影面积 A_s　699

帆缆舱　408

帆缆用具　88

帆艇　760

翻滚式舱口盖　749

翻梯装置　490

翻修期限　646

翻译人员　914

矾矿石　941

钒灰腐蚀　941

钒铁合金　322

反补贴税　41

反铲式挖泥船　239

反弹道导弹　38

反导拦截弹　40

反动度　221

反动式汽轮机　721

反横矩　489

反舰弹道导弹　15

反舰导弹　40

反平衡　417

反潜火箭　865

反潜雷达　868

反倾销法　39

反倾销税　38

反射波勘探　727

反射舵　744

反射声波　727

反渗透法　743

反时限过电流继电器或脱扣器　480

反坦克导弹　41

反卫星技术　40

反向机构　743

反向旋转螺旋桨　190

反向转动轴承　190

反斜杆　66

反压力　640

反要求书　986

反应堆舱　721

反应舵　721

反应螺旋桨　190

反应推进器　179

反应性搜查　721

反转试验　66

反作用式发动机(喷气式发动机)　721

返回式卫星　743

返回式卫星技术　743

泛加勒比海区域　117

方案A　766

方案B　766

方案报价设计/概念设计　168

方便旗　339

方驳　681
方槽　846
方差　942
方锉　846
方法(规律,秩序)　565
方法　43
方钢　846
方龙骨　71
方式(态度)　541
方头螺栓　846
方位距离定位法　617
方位推进系统　64
方位指示器　64
方艉浸宽　446
方艉浸深　446
方向、姿态和飞高控制系统　242
方向控制系统　241
方向控制装置/系统　236
方向谱密度　241
方向稳定性　242
方形扳手　846
方形螺母　846
方形螺纹　846
方形系数 C_B(驳船)　86
方形系数 C_b　86
方形系数曲线　199
方型艉　916
芳烃油类(不包括植物油)　48
防爆(安全)阀　310
防爆　310
防爆门　310
防爆式风机　310
防爆释放装置　310
防爆填料函　39
防爆外壳　310
防爆外壳　310
防爆形式　929
防冰装置　40
防波堤　94
防尘板　849
防尘外壳　263

防冲刷装置　40
防喘装置　875
防磁钢　40
防荡舱壁　877
防滴外壳　258
防冻阀　360
防冻系统　39
防毒面具　371
防腐(维护,保藏)　690
防腐剂(防腐的)　690
防腐蚀　38
防腐系统/防腐措施　188
防腐系统　38
防腐锌阳极　989
防过载设备(过载保护装置)　647
防护　704
防护格栅　704
防护滚轮　814
防护设备　704
防护套　704
防护涂层　704
防护外壳　704
防护装置(屏蔽设备)　794
防护装置(护罩)　705
防护装置　704
防滑板　759
防火　331
防火舱壁(耐火舱壁)　335
防火的(耐火的,不燃的)　335
防火风闸　327
防火风闸　205
防火风闸　944
防火服　326
防火控制图　326
防火门　328
防火门的吸持和释放系统　328
防火面罩　331
防火目标1　331
防火目标2　331
防火目标3　331
防火目标4　332

防火目标5　332
防火漆　332
防火漆　332
防火器材　332
防火墙　335
防火区　334
防火设备(消防设备)　332
防火设备(消防设备)　332
防火设备　331
防火设备　332
防火网　340
防火系统　332
防火险规则　326
防火闸　331
防火装置　335
防溅挡板　844
防溅的　844
防溅式　844
防溅外壳　41
防溅罩(个人漂浮装置)　845
防空导弹　20
防空火炮　38
防空识别区　20
防浪阀　861
防漏(避漏)　495
防漏　494
防漏系统(漏电防护装置)　495
防沫　39
防喷管　529
防喷器　87
防喷器　90
防喷器凹形区　65
防喷器叉车　90
防气的　942
防倾肘板　919
防热的(耐热的)　419
防鼠板　719
防水玻璃　964
防水的　964
防水盖布　885
防水垢剂　765

防水式 964
防水试验 964
防水外壳 964
防蚊网 579
防污底化合物和系统 39
防污底涂层系统 39
防污底系统 39
防污公约 附则 III "防止海运包装有害物质污染规则" 550
防污公约 附则 IV "防止船舶生活污水污染规则" 550
防污公约 附则 V "防止船舶垃圾污染规则" 550
防污公约 附则 I "防止油类污染规则" 550
防污公约 附则 II "控制散装有毒液体物质污染规则" 550
防污公约 附则 VI "防止船舶造成空气污染规则" 550
防污染 681
防污染设备 681
防污涂层 39
防污油漆 39
防险救生 985
防锈剂柜 38
防锈油（滑脂） 823
防烟面具 825
防烟面罩 825
防烟套 825
防烟完整性 825
防摇装置 40
防油外壳 627
防油污系统 622
防振结构（减振结构） 951
防振支柱 951
防振柱 951
防振装置 41
防止船舶垃圾污染检验指南 400
防止和制止从事国际海上运输船舶走私毒品、精神药物和前体化学品指南 401

防止热膨胀隔板 419
防止人员危害目标2 665
防止人员危险目标1 665
防止生活污水污染证书 790
防止偷渡者进入和寻求成功解决偷渡案件责任分配指南 401
防撞舱壁 155
防撞滑架 820
防撞门钩 250
防撞碰垫 320
防撞装置 40
妨碍声 615
房船 433
访问签证 952
放大器 33
放大因数 534
放电处理法 273
放链长度 767
放气 87
放气阀 22
放气阀 708
放弃潜水 2
放热（热量发射） 418
放热铬铁白金 321
放射性产物或废料 717
放射性同位素实验室 717
放射性物质（第7类危险货物） 717
放松杆（释放杆） 732
放艇安全索 501
　放艇联动装置 212
放泄口 732
飞机式水翼系统 25
飞机用大功率发电机及供电系统 23
飞机状态 24
飞剪型艏 147
飞溅 844
飞溅润滑 844
飞溅阻力 845
飞轮 348

飞轮式调速器（蒸汽机用） 348
飞球调速器 67
飞行甲板 342
飞行状态 348
飞行纵倾 342
飞行纵倾角 342
飞行纵倾速度 342
非本地物种 604
非薄脆性 605
非常的（格外的） 313
非常规富裕干舷船舶 796
非常规富裕干舷船舶 809
非常严重的海难 948
非敞口船 604
非弹性联轴节 603
非电控柴油机 603
非对称数字用户环路 53
非法行为 445
非帆艇 605
非高速小水线面双体船 824
非隔热的 602
非公约船舶 602
非固定价格 603
非固化物质 605
非关税壁垒 605
非机动船舶 605
非机动推进 604
非基本类管路系统 603
非金属材料 604
非金属夹杂物 604
非均匀有理B样条 605
非均匀有理B样条曲面 614
非裂变的或预计裂变的表面受到放射性物质污染的物体(SCO-1) 716
非黏性货物 602
非黏性物质 602
非排水状态 603
非破坏性试验(无损检测) 602
非曝露区(非露天区域) 933
非歧视原则 603
非入级船舶 602

非散热设备 604
非收放型减摇鳍装置（不可收放式减摇鳍,固定式减摇鳍） 604
非受火压力容器 933
非受援船 931
非水密门 601
非水密门 605
非危险区域 601
非危险区域 603
非线性干扰抑制 604
非线性缺陷显示 604
非协定自动出口配额制 602
非营利组织 604
非优惠性配额 604
非油船 604
非政府组织 603
非直接测量装置 452
非职务技术成果 603
非重要设备 603
非重要系统 603
非重要用途辅助锅炉 60
非轴对称屈曲 602
非自动的 602
非自航平台 605
非自吸泵 605
肥大船型 363
废钢 767
废黄铜（黄铜屑） 767
废金属 767
废旧的 614
废品 960
废气（SO$_x$）滤清系统 307
废气（SO$_x$）滤清系统技术手册 271
废气（锅炉）调节阀（废气转换阀） 307
废气（排气,排出） 307
废气道 960
废气分析 307
废气锅炉 307
废气锅炉 307
废气锅炉消声器 816

废气锅炉烟气调节阀 308
废气过热器 960
废气加热 960
废气加热经济器 307
废气经济器 960
废气滤清系统（EGCS）残余物 270
废气滤清系统 307
废气滤清系统 307
废气轮机（废气透平） 307
废气轮机（废气透平） 308
废气燃油交替式锅炉 29
废气燃油组合式锅炉 164
废气热交换器 307
废气透平 960
废气透平增压器 308
废气涡轮增压 922
废气涡轮增压器 307
废气涡轮增压器 307
废气涡轮增压器 308
废气涡轮增压器 922
废气涡轮增压器 922
废气涡轮增压器的滑油循环柜 308
废气涡轮增压器的滑油重力柜 308
废气涡轮增压器滑油冷却器 529
废气再循环系统 307
废气增压器 960
废弃食品 951
废弃物越境转移 912
废热 960
废热道 960
废热锅炉 307
废热锅炉 960
废热回收装置 307
废热利用 960
废热利用 960
废纱头 960
废物（垃圾） 729
废物及其他物质 960
废物清除器 960

废物箱（废物库） 960
废物运输船 960
废氧化铁或废海绵铁 484
废液舱 960
废液管系 822
废油 960
沸点 89
沸点升高 89
沸水器 258
沸水器 961
沸腾钢 270
沸腾蒸发 89
费尔索夫图谱 335
费率 719
费用/效益分析（成本效益分析） 189
费用 309
费用先付 132
分辨率 739
分布式系泊 845
分舱和稳性资料 865
分舱因数 316
分层（叠层） 490
分叉屈曲 81
分道航行系统 910
分段拼模试验 779
分段组装 86
分断时间 94
分隔 776
分隔 786
分隔的（分开的） 779
分隔结构 126
分隔统铺 786
分隔压载 779
分节驳 464
分离空泡 786
分离流 786
分离盘 786
分离器（隔板,脱模剂） 786
分离器（抽出器,脱模剂） 313
分离筒 92

分离油泥　786
分励脱扣器　813
分联箱横水管锅炉　776
分链器　131
分裂式水翼　844
分流器柔性接头　247
分流系统　654
分米波　214
分配电板　246
分配站　246
分配轴(柴油机)　541
分批装运　654
分期偿还时间　33
分期付款　657
分期付款　658
分区　776
分区识别能力　776
分散式制冷装置　214
分数　354
分水机构　708
分摊国政府　180
分析证书　34
分油器(净油器)　622
分油器处所　625
分杂机构　139
分支电路　93
分支机构　866
分轴燃气轮机　845
分组交换技术　650
芬兰-瑞典冰级规则　325
酚醛玻纤毡　667
酚醛泡沫塑料　667
焚烧炉　448
焚烧炉灰渣　448
焚烧炉装置　448
粉碎机　160
粪便泵(污物泵,生活污水泵)　790
粪便处理船　790
粪便柜(卫生水柜)　763
粪便柜　641
粪便喷射泵　762

粪便水　319
粪便水排出处理装置　763
粪便水排泄系统　763
粪便中的大肠杆菌　319
粪便贮容器　306
丰满型船首　87
风　975
风暴扶手　861
风暴盖　212
风暴盖　861
风暴钩　861
风暴压载舱　367
风程/风区长度　322
风电　976
风电安装船　619
风洞试验　976
风帆渔船　337
风机噪声　318
风浪　977
风浪排水孔　861
风浪失速　843
风冷内燃机　24
风冷式冷藏集装箱　24
风冷式压缩机　19
风力　975
风力推进　976
风量倍数　19
风能转换系统　975
风扇　318
风扇特性电阻器　318
风扇推进燃气轮机　318
风时　975
风速　976
风险　746
风险　747
风险避免　747
风险处理　747
风险分析　747
风险分析树　747
风险感知　747
风险贡献树　747

风险管理　747
风险管理系统　747
风险管理原则(预防原则)　688
风险降低　747
风险接受　747
风险控制　747
风险控制措施　747
风险控制选项/风险控制方案　747
风险目标　758
风险评估　747
风险评估方法　747
风险评估矩阵表　747
风险评估准则　747
风险评估准则　747
风险评价　747
风险识别　747
风险投资　945
风险效益分析　747
风险转移　747
风向标(叶片,桨叶)　941
风压倾覆力臂　501
风压倾斜力矩　576
风雨密　845
风雨密　845
风雨密　969
风雨密开口　969
风雨密门　845
风雨篷　968
风载荷　975
封板　863
封闭车辆区域　148
封闭处所　281
封闭的上层建筑　282
封闭集装箱　147
封闭式粪便水系统　281
封闭式空泡试验水筒　148
封闭式耐火救生艇　332
封闭式耐火救生艇　335
封闭式伸缩舷梯　281
封闭式外通风型电机　281
封闭式液位测量装置　147

封闭式自通风型电机 281
封舱塞条 73
封舱楔 969
封舱楔耳 408
封舱压条 520
封舱装置 408
封底泥驳 276
封焊 772
封口法兰 86
封头 259
峰到峰声压 658
峰值短路电流 658
峰值声压 658
蜂巢形柱体式平台/多柱型立柱式平台 126
蜂窝电话 126
蜂窝夹心板 430
蜂窝式消音器 430
缝（接缝） 773
缝帆工具 760
敷管设备 671
敷网渔船 846
弗劳霍夫风能及能源系统技术研究所 355
扶栏 938
扶强材或主要支撑构件净剪切横剖面积（cm^2）。 50
扶强材或主要支撑构件连同宽度 s 船壳板对其与板材平行的中和轴的净惯性矩（cm^4）。 485
扶强材或主要支撑构件连同宽度 s 带板的净横剖面积（cm^2）。 50
扶手(栏杆,把柄) 403
扶手支架(栏杆支架) 403
扶手支柱 404
扶手支座 102
扶梯平台 849
扶正力臂 GZ 745
扶正力矩 RM 745
服务 787
服务处所 788

服务航速（常用航速） 788
服务航速的傅汝德数 788
服务贸易 788
服务贸易总协定 375
服务器 787
服务业务 787
服务质量 710
氟利昂(制冷剂) 359
氟利昂制冷 359
氟橡胶 347
浮标系统绞车 102
浮船坞 344
浮船坞 344
浮船坞长 $L_D(m)$ 497
浮船坞的空坞质量 504
浮船坞飞桥 347
浮船坞举力 503
浮船坞挠度 219
浮船坞内通道 459
浮船坞总长 645
浮动地板 344
浮动滑车 343
浮动件 344
浮动结构物 345
浮阀（浮子阀） 343
浮筏 102
浮管 344
浮浆 102
浮力 350
浮力曲线 102
浮力载荷 102
浮锚 769
浮球阀 67
浮石 707
浮式储存-再液化船 361
浮式储油和卸油船 361
浮式储油装置 361
浮式生产、储油和卸油船 354
浮式生产储/卸油船的系泊系统 578
浮式生产储油/卸油船 344

浮式生产储油船 343
浮式生产储油装置 345
浮式生产系统 345
浮式生产装置 354
浮式液化天然气 343
浮式液化天然气生产储卸装置 344
浮式液化天然气装置 344
浮式再液化气储存装置 345
浮式钻井生产储油船 344
浮式钻井装置 344
浮式钻井装置 344
浮胎 102
浮态 343
浮体 343
浮体长 497
浮筒单点系泊系统 817
浮筒单点系统 817
浮筒绞车 132
浮筒损 102
浮托工程 345
浮箱 681
浮箱甲板 682
浮箱式浮船坞 682
浮心 127
浮心垂向坐标 946
浮心垂向坐标曲线 199
浮心轨迹 520
浮心横向坐标 916
浮心曲线 199
浮心纵向坐标曲线 199
浮性 102
浮性器材 346
浮延绳钓 344
浮泳生物取样 877
浮油回收船 623
浮油回收船 625
浮油回收船 627
浮油回收船操作手册 638
浮油回收船的操作手册 634
浮油回收船及油污水处理船 625

浮油回收设备　623
浮油撇除器　625
浮油水域　627
浮游生物光诱捕装置　504
浮游生物连续采集器　178
浮游生物取样　674
浮渣(泡沫,除去浮渣)　768
浮渣盆(杂渣盆)　768
浮渣盆　768
浮子开关(浮动开关)　345
符合船舶载运危险货物特别要求的文件　248
符合声明　216
符合证明　248
符拉索夫曲线　954
幅度差　33
辐流式汽轮机　714
辐流式涡轮　714
辐流式压气机　714
辐轮　845
辐射强迫　714
辐射强迫值　715
辐射系泊定位钻井船　845
辐射性核燃料(INF)规则　454
辐射性核燃料(INF)规则　472
辐射性核燃料　484
辐射阻尼　714
福斯特锚式拖网　352
辅泵　62
辅柴油机　61
辅给水管路　61
辅钩起重量　61
辅锅炉　60
辅机(辅助机械)　61
辅机(辅助机械)　61
辅机/辅助机械　61
辅机舱　61
辅机舱　61
辅机舱　62
辅机处所　61
辅机排气预热给水　62

辅空压机　60
辅冷凝器　60
辅冷凝器循环泵　60
辅汽轮机　62
辅汽轮机转移负荷试验　62
辅燃气轮机　61
辅推力轴承　62
辅蒸汽阀　62
辅蒸汽管系　62
辅助操舵系统　858
辅助操舵装置　62
辅助船/工作船　788
辅助措施　775
辅助斗桥　60
辅助发电机　61
辅助工具(辅助设备)　62
辅助锅炉舱　60
辅助救生用具　775
辅助救生用具属件　292
辅助绝缘　873
辅助控制站　61
辅助控制站　865
辅助控制装置　60
辅助设备　60
辅助输出装置　62
辅助通风管路　61
辅助推进/操纵装置　62
辅助推进器　62
辅助推进系统　62
辅助推进系统　702
辅助推进装置(辅助推进机械)　62
辅助系统　62
辅助旋塞(备用旋塞)　873
辅助装置(附加设备)　61
辅助装置(附加设备)　61
腐蚀(侵蚀)　294
腐蚀　188
腐蚀皮肤　189
腐蚀品(第8类危险货物)　189
腐蚀试验　189
腐蚀锌板　985

腐蚀增量　188
腐蚀值　294
付款　658
付款交单　249
付款人　657
付款银行　657
负垂向弯矩　592
负荷范围　514
负荷分配系数 K_y　514
负荷计算　513
负荷面积　513
负荷特性试验　513
负荷系数　516
负面清单　592
负排气余面　573
负数(数号,减)　573
负运价　592
负载　515
负载的气缸润滑　513
负载平衡　513
负载系数　518
负责人员(资格人员)　740
负责人员　663
附加标志　12
附加阀　873
附加工程　12
附加惯性积　11
附加惯性矩　11
附加检验　12
附加桥楼功能　11
附加审核　11
附加险　12
附加值　11
附加质量　11
附加阻力　12
附件(塑料管)　337
附件　54
附具　5
附录　42
附体　42
附体尺度效应增值　42

索引

1089

附体阻力　42
附体阻力系数　152
附涡　92
附则　37
附属生态研究　7
附属实验室　54
附属空泡　54
附着力　13
附着生物调查　54
复背式微生物采集器　669
复发的　725
复合材料　165
复合钢　138
复合式增压柴油机　165
复合视频端子　165
复合型传感器　724
复合型高性能船　438
复合型破乳剂　164
复合型阻尼金属板材　157
复合岩棉板（耐火）　332
复合装甲　164
复合阻尼结构　165
复励电动机　165
复励发电机　165
复式斗桥　582
复式阀　582
复式增压　787
复数的/吸收　678
复向期　720
复印机属模拟方式　187
复原力臂　745
复原力矩　741
复杂声呐系统　828
复杂循环　164
复杂循环燃气轮机装置　164
复制件　187
复制品　263
副励磁机　670
傅汝德数 F_n　360
傅氏伴流分数　361
傅氏伴流因数　361

富士通株式会社　362
富油极限　744
富裕水深参数　932
腹板　969
腹板高度 h_w　969
腹板净厚 t_w　594
覆板　252
覆板舵　252
覆盖层或覆盖结构　138

G

伽利略定位系统　367
改进措施　187
改良型吊杆装置　563
改向试验　132
改正措施请求书　187
改装（救生艇）　575
改装工程　185
盖（挡板,百叶窗）　813
盖斯林格联轴节　374
概率　696
概率等级　696
概率误差　696
概率误差圆　138
概念开发　168
干保养　260
干泵舱　260
干船坞　259
干底润滑　260
干粉灭火剂　686
干粉灭火系统　260
干管系统　260
干货船　259
干货物残余物管理计划　259
干加压舱　259
干膜厚度　236
干湿交变列板　975
干湿泡温度计　972
干湿实验室　785
干实验室　260
干式采油树　260

干式服（救生服）　260
干式离合器　259
干式蒸发器　260
干舷　356
干舷 F　357
干舷船长　357
干舷船长 L_f　498
干舷船长 L_{LL}　514
干舷船宽 B_f　94
干舷甲板　356
干舷甲板上的开口　633
干舷艏垂线　352
干舷艉垂线　17
干燥（去湿）　437
干燥法　260
干燥器　256
干燥室（干燥装置,炉）　489
甘蔗纤维板　67
杆（或桁架）单元　748
竿钓渔船　680
感觉噪度　659
感生拱度　452
感生速度　452
感生阻力　452
感温式自动失火报警器　59
感温探测器　417
感烟检测器（测烟器）　825
感烟式自动失火报警器　59
感烟探测器（烟气探测器）　825
感烟探测设备　825
感烟探测系统　825
感烟探测装置　825
感应电动机　452
感应电机　452
感应发电机　452
感应空调系统　452
感应熔炼炉　452
感应式盐度计　453
感应无线数据通信　210
感应子式同步发电机　453
刚性浮体　745

刚性构件　745
刚性固定　173
刚性夹心板　746
刚性救生筏　745
刚性可充气艇　745
刚性联轴器　745
刚性铺管　745
刚性轴　746
刚性轴系　860
钢　857
钢扒船　253
钢板预处理　689
钢材的最低设计温度　527
钢船涂层状况　150
钢锻件（钢锻造）　858
钢格栅　858
钢管　858
钢焊接接头的缺陷　446
钢或其他等效材料　858
钢筋混凝土船　731
钢缆　858
钢丝绳张紧器　978
钢丝刷（金属丝刷）　767
钢丝网门　767
钢丝网水泥船　321
钢丝作业　822
钢索　978
钢索滑车　978
钢悬链式立管　858
钢质船舶的保护涂层　704
钢质液货管　857
缸径　202
缸套冷却泵　486
缸套水泵　486
缸套装配　813
港口　404
港口　683
港口保安员　684
港口城市　683
港口船舶岸电供电技术　393
港口当局　683

港口费　683
港口国　684
港口国当局　684
港口国当局的应急策略　42
港口国监督　684
港口国监督检验　684
港口国检查　706
港口设备保安员　683
港口设施　683
港口设施保安计划　683
港口水域内作业的起重船　344
港区司令官（COTP）　109
港湾调查　405
港湾调查船　405
港务船　405
港作拖船　405
港作拖船　921
高倍泡沫灭火系统　423
高超声速飞行器　442
高超声速滑翔飞行器　442
高超声速可重复使用飞行器　434
高超音速　442
高超音速武器　428
高次谐波　424
高次谐波　426
高带宽数字内容保护技术　425
高弹性橡胶联轴节　423
高低水位　426
高低水位报警器　423
高低压控制器　423
高电压　425
高电阻合金　424
高度小于标准高度的上层建筑的开口　633
高发热值　425
高高水位　426
高隔声壁板　425
高硅氧玻璃棉　427
高恢复系数热电偶　424
高级船员　618
高级语言　423

高技术超级班轮　887
高架链斗挖泥船　646
高架门座起重机　368
高架索　423
高架索传送系统　426
高阶波　426
高阶振动　424
高阶振动　426
高精度长丝杆磨床　688
高空除锈作业车　734
高空气象观测　16
高空气象观察室　938
高流速损伤法　205
高锰钢　423
高密度固体散货　426
高黏度的B类有毒液体物质　425
高黏度的C类有毒液体物质　425
高黏度燃油　425
高黏度物质　428
高黏度有毒液体物质　425
高膨胀泡沫（高倍泡沫）　423
高频　423
高频声　423
高频响应/和频响应　423
高频振动　423
高强度（淬火加回火）细化晶粒结构钢　426
高强度船体结构钢　426
高强度船体结构钢　426
高强度低合金钢　425
高强度钢系数　426
高强度耐热铝合金（Y合金）　987
高强度铸铁　426
高清　426
高清晰度电视　423
高清晰度多媒体接口　423
高水位报警器（高水位报警）　425
高斯　373
高速柴油机　425
高速船　427
高速船　434

高速船安全证书 428
高速船模试验水池 425
高速船营运许可证书 663
高速航行补给 425
高速红外线 318
高速红外线 823
高速货船 112
高速机 428
高速客滚船 747
高速空调系统 425
高速空泡实验室 425
高速内燃机 425
高速双体船 252
高速通风 425
高速透气阀 712
高速小水线面双体船 824
高速游艇 425
高速诱导空气调节系统 428
高位冰区水线 938
高位海水门 424
高温报警器 425
高温作业 433
高效焊条及药芯焊丝 423
高效木材阻燃剂 270
高谐波项 426
高锌黄铜(硬黄铜,优质黄铜) 423
高性能船舶 424
高性能海洋工程用锚 427
高压安全阀 424
高压岸电系统 810
高压泵送系统 424
高压侧 424
高压端(供气压力端)(LPG 系统) 424
高压缸 424
高压给水预热器 427
高压管路(高压管系) 427
高压海水(HPSW)系统 424
高压绝缘栅双极晶体管 462
高压空调系统 424
高压空气系统 424

高压钠灯 424
高压气体燃料供应系统 424
高压燃油管 424
高压容器 424
高压水泵(挖泥船) 427
高压水喷射清洗 424
高压水清洁 424
高压水松土装置 427
高压逃生舱 427
高压涡轮 427
高压系统 428
高压压气机 427
高压压缩机 424
高压油泵 424
高压油泵 424
高压油管(喷油管) 361
高压增压 424
高压直流输电 428
高压指示表 424
高应力 424
高真空焊接烟尘治理设备 428
篙 708
锆石砂 989
搁置检验 494
割划式水翼 874
格林尼治标准时间 394
格网方式 395
格栅(格子,百叶栅) 395
格栅(格子板) 391
格栅(炉排) 391
格栅 391
格栅舱口(栅形舱盖) 391
格栅舱口(栅形舱盖) 391
格栅甲板 391
格栅框架 391
格栅平台(通道,走廊) 367
格栅铺板 395
格子板(地格板,踏脚格栅) 350
格子盖(格栅盖) 391
格子线 395
隔板 237

隔板气封 237
隔板套 118
隔舱通风闸阀 102
隔绝法 651
隔离 786
隔离舱/隔离空舱 153
隔离舱、空舱等 153
隔离带 989
隔离阀(隔流阀,截止阀) 485
隔热(热绝缘) 418
隔热材料 418
隔热层衬板 462
隔热的(绝热的) 418
隔热设施 462
隔热套(外罩,延迟) 490
隔热罩 462
隔声 829
隔声壁板 829
隔声材料 829
隔声措施 829
隔声间 485
隔声间 7
隔声量 830
隔声门 831
隔声屏障 829
隔声墙 831
隔声室 831
隔声屋 830
隔声罩 600
隔声罩 7
隔声罩 8
隔水导管 170
隔水器张力器 746
隔叶片 832
隔栅(围墙,雷达警戒网) 320
隔振 950
隔振弹簧 950
隔振垫 950
隔振垫刚度 950
隔振器/隔振体 950
隔振橡胶 950

隔振元件　951
个人博客系统　981
个人防护设备　452
个人风险　452
个人管辖权　664
个人集合时间　452
个人救生用具　664
个人配备(消防员的装备)　663
个人漂浮装置　664
个人数字蜂窝电话　663
个人数字助手　663
个人移动时间　452
个性化域名服务　665
各缸工作均匀性试验　202
各态历经性　293
铬钢　858
铬铁白金　321
铬铁矿石　137
铬丸　137
给水(供电,供给)　320
给水(进料,进刀)　320
给水　320
给水倍率　320
给水泵　320
给水处理　320
给水阀　320
给水方法　561
给水分析　320
给水供应　320
给水管　320
给水加热器　320
给水滤器　320
给水喷射泵　320
给水器(进料器)　320
给水软化器　965
给水设备　320
给水系统　961
给水预热器　320
给水止回阀　320
给水自动调节装置　58
根部凹陷　750

根部缺口　750
根厚　750
根涡　750
跟单汇票　249
跟单信用证　249
跟单信用证统一惯例　934
跟读盘/U 盘　929
跟踪　909
跟踪雷达　909
更机　727
更新审核　735
工厂认可　984
工厂证书　984
工程(工程的)　287
工程船　982
工程船船型　982
工程风险控制　288
工程机械　288
工程计算　288
工程可靠性　288
工程日志　288
工程设计　288
工程师(机械师,轮机员)　287
工程塑料　288
工程图　288
工程总价　904
工程总验收单　374
工控机　453
工人　982
工时　981
工业(产业,勤勉)　453
工业标准　453
工业标准化实践　453
工业毛毡　453
工艺流程图　697
工艺评定报告书(PQR)　970
工艺项目　637
工作场所　263
工作吃水　982
工作处所　983
工作船　981

工作筏　981
工作范围(航区)　636
工作负载　982
工作干舷　982
工作环境(救生服)　982
工作甲板(渔船)　982
工作救生衣　194
工作艇　487
工作艇　981
工作压力　636
工作压力　638
工作液体　982
工作应力(操作应力)　637
工作硬化系数　406
工作载荷(操作负载,运行负载)　635
工作站　984
工作站　986
弓形切面　620
弓形梯步　199
公布的热带淡水最大吃水(巴拿马运河)　707
公称壁厚　601
公称尺寸(名义尺寸)　600
公称速度　601
公称外径　601
公称应力(标称应力,名义应力)　601
公称直径　600
公共处所　706
公共法　706
公共故障模式　160
公共管理当局(公共当局)　706
公共广播报警系统　706
公共交通导向发展模式　914
公共信息通信　706
公海　424
公会承认的货物　389
公开场合　632
公开场合　706
公开招标　163

索引

1093

公量　170
公路运输　748
公平地　318
公平合理原则　695
公平竞争原则　695
公认标准　724
公认的船级社　724
公司(ISM 规则)　162
公司(ISPS 规则)　162
公司　187
公司保安员　162
公司的安全管理目标　758
公司的具体责任　841
公司能效管理计划　162
公司旗　433
公司旗杆　433
公司审核　162
公司审核员　162
公司现场审核　162
公务船　706
公务船　707
公约　184
公约船舶　185
功比　981
功率储备　687
功率调节器　687
功率调节系统　687
功率分支　686
功率分支轴并列减速齿轮箱　686
功率估计载荷因数　513
功率管理系统　687
功率过载保护　647
功率逆变器　686
功率试验　687
功率载荷系数　686
功率综合控制缸套冷却　465
功能　364
功能度　366
功能方块图　366
功能可靠性试验　366
功能弱化状态　219

功能试验　366
功能试验　366
功能顺序图　366
功能危险评估　366
功能性要求　366
攻角　36
攻丝螺丝　895
供电(电源,动力源)　687
供电船　385
供电连续性　177
供电系统　687
供给泵　873
供煤船　149
供气量　108
供气水翼　20
供汽(送汽)　857
供水(给水)　963
供水船　963
供水系统　962
供水压力　320
供体港　250
供氧呼吸器　649
供应船　873
供应链　873
供应商　873
供应商代表　873
供应商符合声明　873
供油　362
供油　625
供油船　625
供油单元　362
供油设施　625
供油提前角　458
拱度　106
拱度比　106
拱度修正因数　106
拱线　106
共定子电动发电机　160
共轨管(蓄能器)　160
共轨系统(共管系统)　160
共积谱　151

共同但有区别的责任　160
共同海损　375
共同结构规范　161
共同市场　160
共同原因　160
共同装载舱　160
共享文件　519
共因失效　160
共用天线系统　160
共用蓄能器(共轨管)　160
共振(谐振,共鸣)　740
共振的(谐振的)　740
共振横摇(谐振横摇)　740
共振检查试验　740
共振检查试验　740
共振频率(谐振频率)　740
共振频率(谐振频率)　740
共振腔吸声砖　830
共振曲线(谐振曲线)　740
共振现象(谐振现象)　740
沟槽(开槽,压槽)　395
沟槽腐蚀　395
钩吊周期　430
钩损　430
钩损险　430
构件长度　498
购买力平价法　708
购物廊(豪华邮轮)　810
估计功率(估计马力)　298
估价/计价　940
估价　298
估价工时　298
毂长　434
毂径比　434
毂帽　701
毂涡　434
毂涡空泡　434
毂直径　434
古柯膏　151
古柯叶　151
谷歌搜索引擎　390

谷物 390
谷物法 187
谷物装运的批准文件 248
股东 860
股份/股票 793
股票 793
骨材间距 S（m） 833
骨材跨距 l（m） 833
骨架 355
骨架层 467
骨料疏浚船 18
鼓动器 10
鼓风机（通风机、增压器、压气机） 85
鼓风机平台 318
鼓风机室 318
鼓风机透平 87
鼓风炉（高炉） 86
鼓轮式布缆机 259
鼓形控制器 259
固定（安装，确定） 338
固定（上紧,握紧） 318
固定（锁紧） 338
固定部件（不可卸零部件） 338
固定导流片 338
固定或浮动平台 338
固定记录介质 339
固定价格 339
固定桨叶 338
固定桨叶螺旋桨 338
固定矩形窗 604
固定栏杆 339
固定螺栓（止动螺栓，冲头销） 789
固定螺栓 429
固定锚 661
固定灭火系统 338
固定式采油平台 339
固定式低倍泡沫灭火系统 338
固定式顶推船—驳船组合体 746
固定式惰性气体灭火系统 338

固定式二氧化碳灭火系统 338
固定式辅燃气轮机 854
固定式高倍泡沫灭火系统 338
固定式甲板泡沫（灭火）系统 338
固定式甲板泡沫灭火系统 338
固定式局部使用灭火系统 338
固定式局部使用灭火系统启动报警 338
固定式局部水基灭火装置 338
固定式联结顶推船驳船组合体—驳船 745
固定式联结顶推船驳船组合体—顶推船 745
固定式灭火系统 338
固定式灭火装置 338
固定式泡沫灭火系统 338
固定式泡沫炮系统 338
固定式平台 338
固定式气溶胶灭火系统 338
固定式气体灭火系统 338
固定式生产系统 339
固定式水基局部使用灭火系统 339
固定式探火和失火报警系统 339
固定式碳氢化合物气体探测系统 338
固定式细水雾灭火系统 339
固定式消防系统 338
固定式压力水雾和细水雾灭火系统 339
固定式压力水雾灭火系统 339
固定式氧气分析仪 338
固定式应急消防泵 338
固定式钻井平台 338
固定艏鳍（固定船首减纵摇鳍） 338
固定天幕 661
固定网 338
固定舷侧 746
固定舷窗 604
固定相移 338

固定压载 338
固定叶片（定子隔板） 854
固定叶片 338
固定翼飞机 339
固定鱼泵 338
固定圆窗 604
固定支承的 339
固定坐标系 338
固化剂 405
固化物质 828
固井船 971
固溶热处理 828
固熔热处理状态 828
固态 827
固态废弃物（固态废物） 827
固态硬盘 827
固体-水分布系数 828
固体货物/油、货兼用舱 827
固体货物专用舱 306
固体夹杂物（铜、钨除外） 827
固体喷墨打印机 827
固体散货 101
固体散货 827
固体散货船 827
固体散货渗透率 662
固体散装货物规则 152
固体散装危险货物 207
固体石蜡 652
固体树脂 406
固体物质 827
固艇索具 87
固有动稳定性 456
固有风险控制 456
固有浮性材料 457
固有频率（自然频率） 586
固桩架 969
固桩块 969
故障 316
故障 317
故障安全 316
故障安全原则 316

故障标准　317
故障承受性　319
故障防范型的远程关闭装置　734
故障分析　316
故障模式和影响概要　317
故障模式与分析　348
故障情况　317
故障树分析　319
故障树分析　319
故障探测器（探伤器）　319
故障严重性　317
故障影响（故障后果）　317
故障源　831
顾问　15
顾问　174
雇主　280
刮（削,铲刮）　767
刮刀　767
刮油环　625
刮油环　978
刮油器　627
刮油胀圈（刮油环）　767
挂舵臂　753
挂号信　729
挂机　644
挂式痰盂　404
挂拖　921
关闭（切断,断路）　813
关闭（停车）　813
关闭　813
关闭缺陷显示　604
关闭设备　435
关车　813
关键部位　488
关键操作　488
关键区域　194
关键人为因素　195
关键图（概略原理图,解说图）　488
关键位置　195
关键系统　195
关键原则　488

关联目标　52
关联设备　52
关联显示　732
关税　200
关税　201
关税　885
关税保护原则　885
关税壁垒　885
关税减免原则　885
关税配额　201
关税配额　885
关税同盟　201
关税与贸易总协定　374
关于油污损害民事赔偿责任的保险和其他财务保证的证书　130
观察窗（观察孔）　815
观察窗　815
观察结果（观察项）　614
观察孔　614
观察项　614
观点　952
观光潜水器　906
管（管系,导管）　670
管板　921
管壁厚度　895
管壁绝热　959
管承（管座）　670
管道凹穴　670
管道布置　920
管道和管系　671
管道支架(管夹,管箍)　670
管的（管形的,由管构成的）　921
管吊架　670
管段　776
管滑道　670
管汇甲板(集管中心甲板)　541
管夹　670
管架　670
管件接头　670
管件连接接头　921
管接头（管节点）　920

管接头　671
管距　833
管口　670
管理　51
管理费　14
管理和操作　540
管理或处理　540
管理级（人员）　540
管理检查　540
管理者代表　540
管路（管系）　921
管路（LPG系统）　672
管路（管系,管路布置,缩孔）　671
管路（管系,用管道输送）　670
管路　670
管路焊接　671
管路系统　671
管螺纹丝锥(管用攻丝锥)　671
管面涂层　670
管内装卸率　519
管钳（管式虎钳）　880
管塞　921
管式加热器　921
管式冷却器　921
管束　920
管刷　920
管隧　671
管隧板　671
管系（导管）　670
管系（管路,总管）　508
管系（塑料）　672
管系　671
管系　880
管系布置　670
管系的标定压力　601
管系的等级　141
管系的耐火性　328
管系的设计温度　232
管系的设计压力　230
管系试验　672
管系图(管路图)　671

管线(导管,管道安装) 671
管线端部管汇 677
管线检测计 671
管形支柱 670
管型燃烧室 921
管型支柱 921
管支柱 670
管柱 921
管状构件 921
管状焊条 921
管状接头 921
管子/管系(塑料管) 671
管子附件 670
管子钢 670
管子割刀 670
管子焊 921
管子虎钳 671
管子环缝自动焊机 641
管子卡箍 671
管子栏杆 670
管子钳 671
管子钳 921
管子试压检验 671
管子丝攻扣(丝扳) 671
管子套 670
管子填料函垫装 921
管子弯头 670
管子支架 671
贯穿螺栓(系紧杆,拉条) 899
贯穿螺栓 897
贯穿螺栓 897
贯通肘板 897
惯常住所 403
惯性(惰性,惯量) 454
惯性半径(船舶实体) 722
惯性导航系统 454
惯性回转试验 454
惯性积 722
惯性力 454
惯性力矩 576
惯性式平台罗经 454

惯性淌航 587
惯性系统 454
惯性轴 454
惯性坐标系 454
盥洗设备 903
灌注泵 323
灌注扶正 695
灌注寿命 695
光报警单元(视觉报警单元) 953
光泵磁力仪 641
光船租船 71
光船租赁 71
光船租赁合同 71
光达技术 503
光电比色计 668
光电切割机 668
光弧 504
光滑的(弄平滑) 825
光滑面 825
光缆 641
光票 146
光票信用证 146
光强 504
光驱 161
光通量 504
光纤 640
光纤放大器 641
光纤水声探测系统 640
光纤水听器探测系统 235
光学字符识别 640
光源 504
光栅海图显示系统 719
光栅航海图 719
光子计算机 668
广播系统 706
广播桌 97
广船国际股份有限公司——海洋装备制造企业上市的"先驱" 399
广温性物种 300
广盐性物种 300
规(表,计,轨距) 367

规定 690
规定残余伸长率 841
规定的方法 842
规定的航行和服务区域 842
规定的环境条件 841
规定的人流 841
规定的设计寿命 842
规定的营运和环境条件 842
规定的作业范围 841
规定伸长率 842
规定水平 842
规定性要求 690
规定要求 842
规定值 843
规范 754
规范 754
规范船长 L(m) 754
规则 152
规则 730
规则波 730
硅锰合金(低碳) 816
硅石氧化铝 30
硅石氧化铝丸粒 30
硅铁(硅含量30%以上,但小于90%,包括砖形块) 321
硅铁合金(硅钢) 321
硅铁铝粉末 30
轨道平移式船台 717
轨迹 910
轨迹角 908
轨圆半径 641
柜床 80
滚动摩擦 360
滚动式舱盖 749
滚动式艇架 749
滚翻式舱盖 819
滚卷式舱盖 749
滚轮导缆器 749
滚轮铰链型艇架 749
滚移式舱盖 914
滚针轴承 592

滚柱导缆器 318
滚装/集装箱船 747
滚装处所 748
滚装处所 751
滚装船/渡船大门启闭跳板升降系统 748
滚装船 748
滚装船 749
滚装船 749
滚装船 751
滚装船 751
滚装货舱 750
滚装货船(容积型船舶) 750
滚装货船(质量型船舶) 750
滚装货船 750
滚装客船(客滚船) 748
滚装装货处所 750
滚装装卸 748
锅垢(刮垢,鳞状的) 766
锅炉(蒸汽发生器) 88
锅炉 89
锅炉半自动控制 89
锅炉报警器 88
锅炉补加水 873
锅炉舱 332
锅炉舱 89
锅炉舱安全出口 861
锅炉舱花钢板 861
锅炉舱棚(安装围栏的舱口) 322
锅炉舱棚 89
锅炉舱平台 332
锅炉舱梯 861
锅炉舱天窗 861
锅炉超负荷蒸发量 89
锅炉程序控制 89
锅炉的传热面积 419
锅炉的封存 690
锅炉的特殊检验 839
锅炉底座 89
锅炉点火泵 89
锅炉点火油柜 89

锅炉调整试验 892
锅炉额定蒸发量 89
锅炉二次鼓风机 89
锅炉附件 292
锅炉给水 88
锅炉给水泵 88
锅炉给水复合泵 88
锅炉鼓风机 88
锅炉和热油加热器检验 88
锅炉机动性试验 89
锅炉基座 88
锅炉及压力容器的附件 54
锅炉及压力容器的隔片 245
锅炉检验 89
锅炉炉膛 89
锅炉排污管路 88
锅炉汽包 856
锅炉前部 89
锅炉全自动控制 89
锅炉燃烧器 88
锅炉燃油泵 89
锅炉燃油澄清柜 89
锅炉燃油柜 89
锅炉燃油加热器 89
锅炉燃油日用柜 89
锅炉燃油系统 89
锅炉热工试验 88
锅炉热交换表面 89
锅炉热平衡 89
锅炉热效率 88
锅炉水舱 320
锅炉水强制循环泵 88
锅炉水循环 89
锅炉水循环系统 89
锅炉药剂柜 88
锅炉引风机 89
锅炉止回阀 88
锅炉重油 366
锅炉装置 89
锅炉自动控制装置 57
锅炉最大蒸发量 89

锅炉座 773
锅内处理 469
锅内设备 259
锅水 89
锅水浓度 89
锅水硬度 89
国别配额 190
国际比较项目 472
国际标准化组织 485
国际材料试验协会 442
国际茶叶委员会 479
国际传输中心 479
国际船舶保安证书 478
国际船级社协会 470
国际电工委员会 445
国际电信联盟-通信组 485
国际电信联盟-无线电报 485
国际吨位 479
国际吨位证书 479
国际多式联运 476
国际多式联运 476
国际多式联运单据 476
国际法庭 985
国际法院 472
国际反倾销法 469
国际防止船舶造成污染公约 550
国际防止船舶造成污染公约 551
国际防止散装运输有毒液体物质污染证书(NLS证书) 477
国际防止生活污水污染证书 477
国际防止油类污染证书 476
国际防止油污染证书 481
国际分工 473
国际服务贸易 477
国际复兴开发银行 470
国际公约 472
国际谷物规则 473
国际惯例 477
国际海上避碰规则 155
国际海事标准 476
国际海事承包商协会 474

国际海事会议 585
国际海事活动 474
国际海事人命安全公约 827
国际海事卫星系统 459
国际海事卫星组织 459
国际海事卫星组织 475
国际海事卫星组织的公务活动(空间段) 459
国际海事组织(IMO)压载水管理指南 475
国际海事组织 475
国际海事组织指南 475
国际海洋法法庭 479
国际海员培训、发证和值班规则 855
国际海运燃料排放 394
国际海运危险货物输规则(IMDG规则) 474
国际航空和海上搜救手册 469
国际航行(国际航线) 479
国际航运安全及防止污染管理规则 474
国际互联网/Web开发技术 985
国际货币基金组织 476
国际货物保险 471
国际货物买卖合同 179
国际货协运单 471
国际金融公司 473
国际竞争性招标 470
国际救生设备(LSA)规则 474
国际开发协会 473
国际控制有害船底防污系统证书 469
国际劳工组织 473
国际麻醉药品管制委员会 476
国际贸易 479
国际贸易地理方向 479
国际贸易货物结构 165
国际贸易术语解释通则 477
国际贸易委员会 479
国际民航组织 442

国际奈伏斯泰业务(国际NAVTEX) 476
国际能效符合证明 472
国际能效证书 473
国际散货船和油船目标型船舶建造标准 389
国际散装化学品规则 470
国际散装运输危险化学品的适装证书 471
国际散装运输液化气体的适装证书 471
国际商会 472
国际商会仲裁院 472
国际商品协定 472
国际商品协定下的协定价格 863
国际商事仲裁 472
国际商事仲裁协会 472
国际商务谈判 471
国际生产价格 477
国际市场价格 476
国际适合拆船证书 477
国际条约 479
国际铁路联运 477
国际通岸接头 478
国际投资纠纷解决中心 471
国际网络处理器会议 476
国际物品编码协会 469
国际消防安全系统规则(FSS规则) 472
国际消防安全系统规则 893
国际协调时间 936
国际新闻社 476
国际信号规则 472
国际信号旗 472
国际刑警组织 473
国际刑事法院 473
国际羊毛局 479
国际移动电话漫游 477
国际移动卫星组织 476
国际有害物质清单证书 472
国际运输 479

国际运输密封装辐射性核燃料货物的适装证书 471
国际运输委员会 479
国际载重线公约 444
国际载重线免除证书 474
国际载重线证书 475
国际装运INF货物适装证书 130
国际坐标时 936
国家吨位 586
国家海事主管机关 585
国家航运公司 586
国民待遇原则 695
国内生产总值 395
国内水域 586
国内铁路运输 250
过饱和 872
过饱和的 872
过饱和空气 872
过饱和蒸发气 872
过驳 505
过驳险 747
过大的不等边臂长的角接焊缝 304
过大的根部增强高 303
过大的焊缝厚度(角接焊缝) 304
过大的焊缝增强高(过大的增强高) 303
过大的凸出 303
过低水位报警 525
过电流 645
过电流保护(过载保护) 646
过电流保护电器的过电流保护协调 646
过电流继电器或脱扣器 646
过电流选择性 645
过度腐蚀 303
过渡舱 11
过渡舱 289
过渡风浪 914
过渡工况 913
过渡阶段 914

索引

1099

过渡流　914
过渡区　468
过渡状态　914
过境国　913
过境旅客　657
过境贸易　913
过境签证　913
过境区　914
过境税　913
过境通行　913
过境运输　910
过冷淬火　647
过冷度　168
过冷液体　865
过量空气系数　303
过量空气系数　303
过磷酸盐(三重,颗粒)　872
过磷酸盐　872
过滤(净化)　708
过滤法　323
过滤器(滤网,拉紧装置)　862
过滤系统　323
过滤元件　323
过平衡　645
过期提单　849
过热　646
过热　646
过热　872
过热保护　646
过热报警器　648
过热报警器　888
过热的　872
过热面　872
过热器　872
过热器管　872
过热器元件　872
过热器组　872
过热区　646
过热温度　872
过热蒸汽　872
过剩空气　875

过时的接口　222
过响声　303
过压安全阀(个人漂浮装置)　647
过载　647
过载保护(防过载)　647
过载保护器(过载自动脱扣器)　647
过载布置　647
过载阀　647
过载继电器　558
过载拉力　647
过载容量(超负荷能量)　647
过载试验(超负载试验)　647
过载试验(过负荷试验)　647
过载释放(过载释放装置)　647

H

海　769
海-淡水交替腐蚀　772
海岸带调查　150
海岸地面站　149
海岸工程　150
海岸警卫队　149
海岸值守单位　149
海冰观测　770
海草　770
海船　772
海盗行为　653
海盗行为　672
海道测量　441
海底采油树　918
海底地震仪　615
海底阀　771
海底管道探测跟踪单元　671
海底管道系统　771
海底回接系统　866
海底集油管汇　866
海底计程仪　91
海底矿物采集船　771
海底探测船　780
海底重力仪　769

海关　200
海关过境单证　201
海关过境手续　201
海关合作理事会税则目录　201
海关税则　201
海关艇　201
海军　591
海军导航卫星系统　591
海军锚　14
海军区域战区弹道导弹防御　591
海军系数　14
海军裕度　14
海军主管机关　588
海况　769
海况　770
海况持续时间　263
海况裕度　770
海浪　771
海浪观测　771
海里　569
海流观测　198
海流计　198
海面气象观测　770
海面气象观察室　770
海面搜寻协调船　186
海面温度、湿度、风梯度仪　889
海面稳定模式　770
海面状况观测　770
海难　544
海上安全调查　547
海上安全调查报告　547
海上安全调查当局　547
海上安全调查国家　548
海上安全记录　548
海上安全委员会　549
海上安全信息　550
海上安装　619
海上保安审核员　777
海上补给　735
海上哺乳动物和鸟类观察室　546
海上撤离系统　545

海上的(船用的,海运业)　543
海上顶推船组　909
海上防务装备　615
海上风电场　619
海上风电转换系统　619
海上服务资历　772
海上隔水管系统　746
海上供应船　619
海上环境保护委员会　549
海上货物运输合同　118
海上获救须知　462
海上立体补给舰　896
海上猎奇(豪华邮轮)　15
海上事故　545
海上事故　549
海上试验(试航,航行试验)　771
海上通道系统　618
海上拖航　616
海上无线电中继浮筒　716
海上移动平台　575
海上移动式钻井平台(MODU)或平台　574
海上移动式钻井平台安全证书　574
海上意外相遇规则　5
海上运输　118
海上钻井平台　618
海生物生长预防系统　545
海生物污底　354
海生物污底　545
海事安全信息广播和接收系统　591
海事保安审核员　550
海事赔偿责任限制制度　506
海事请求　549
海事请求保全　690
海事总验船师　134
海事作业部(巴拿马运河)　549
海水　548
海水　771
海水泵　771

海水泵　773
海水淡化　771
海水淡化船　245
海水淡化装置　771
海水淡化装置　773
海水阀(通海阀)　489
海水阀箱/海底门　769
海水腐蚀(盐水腐蚀)　771
海水光学测量　771
海水过滤装置　773
海水化学取样　771
海水化学实验室　771
海水进口(海水吸入口)　770
海水进水阀　770
海水冷却泵　771
海水冷却管　771
海水冷却控制系统　186
海水滤器　771
海水密度　771
海水腔　771
海水溶解氧现场测定仪　460
海水润滑艉管轴承　771
海水声学测量　771
海水温度　770
海水温度计　771
海水吸入口　770
海水吸入箱振动　769
海水系统　771
海水循环泵　771
海水压力柜　771
海水压载　761
海水压载处所　761
海水营养盐分析器　614
海水蒸发器　771
海损　405
海损　62
海损　985
海损事故　543
海损修理　985
海损载荷　205
海图　587

海图或航海出版物　588
海图室　132
海图桌　131
海豚　683
海豚式运动　683
海湾　73
海湾阿拉伯国家合作委员会　401
海湾区域　401
海峡船　132
海啸波　920
海星张力腿平台　773
海洋初级生产力测定　547
海洋磁力测量室　546
海洋大气腐蚀　544
海洋底栖生物实验室　544
海洋地球物理勘探　545
海洋地震测量室　548
海洋地震电缆　548
海洋地震电缆深度控制系统　548
海洋地震勘探　548
海洋地质地貌实验室　545
海洋地质调查船　545
海洋地质取样　545
海洋地质实验室　386
海洋调查船　614
海洋浮游生物实验室　547
海洋工程(广义)　616
海洋工程(狭义)　616
海洋工程辅助船　618
海洋工程施工船　618
海洋工程支援船　619
海洋工程作业船　619
海洋光学实验室　546
海洋化学实验室　544
海洋环境　544
海洋环境保护　544
海洋环境保护委员会　544
海洋环境的污染　680
海洋环境调查　545
海洋环境噪声测量　30
海洋技术协会　548

海洋开发船　615
海洋开发装备　615
海洋考察船　617
海洋科考装备　616
海洋矿藏资源调查　546
海洋立体管控系统　896
海洋能　615
海洋气象实验室　546
海洋气象学　568
海洋生物调查船　544
海洋生物生理生态实验室　546
海洋生物实验室　544
海洋生物资源调查　546
海洋声吸收测量　875
海洋石油支持船　619
海洋水产资源调查　545
海洋水声实验室　548
海洋水文观测　545
海洋水文化学实验室　545
海洋水文气象调查船　545
海洋水文实验室　545
海洋微生物取样　546
海洋微生物实验室　546
海洋温差发电　616
海洋温差发电系统　616
海洋污染调查　547
海洋污染监测实验室　547
海洋污染物　547
海洋物理实验室　546
海洋物探实验室　545
海洋渔业调查船　616
海洋运输装备　616
海洋质子磁力仪　547
海洋质子梯度仪　547
海洋重力测量室　545
海洋重力仪　545
海洋专业调查船　617
海洋综合调查船　617
海涌　397
海用微波测距定位系统　546
海域　769

海运单　771
海运货物集中运输　18
海运提单　615
海运提单　83
海珍品采集船　658
氦氧压缩机　421
氦氧混合气系统　422
氦氧潜水　421
含气空泡　372
含气空泡　372
含汽空泡　942
含汽空泡　942
含水量　576
含油舱底水　628
含油舱底水储存柜　628
含油废弃物　628
含油污水（废油，污油）　822
含油污水　628
含油污水舱　628
含油泄放水　628
含油压载舱　628
含油压载水　628
含有毒液体物质的残余物和混合物（NLS残余物）　738
含原油的混合物　574
含渣量　823
焓（热含量）　417
焓　289
焓降　417
焓熵图（$i\text{-}s$图）　289
汉堡海事展览会　825
焊缝侧弯试验　813
焊缝金属　970
焊缝金属外溢　646
焊缝拉伸试验　890
焊工的认可　44
焊管机　671
焊接变形　970
焊接车间　971
焊接程序大纲　970
焊接程序认可试验（WPQT）　970

焊接船　970
焊接工艺　971
焊接工艺规程　971
焊接工艺合格记录　970
焊接工艺合格试验　970
焊接工艺评定（WPS）　970
焊接管（连续焊成的）　970
焊接裂缝　970
焊瘤　647
焊枪（焊接钳）　401
焊渣飞溅　970
航标舱　503
航标船　102
航标灯　74
航标灯船　74
航标工作船　205
航标起重机　503
航标巡检船　102
航程　956
航程规划　957
航程计划　957
航程计划工作站　957
航次（航行）　956
航次期租　902
航道测量船　441
航道站　773
航海实习室　589
航迹　908
航迹　908
航迹航程　909
航迹绘图仪　909
航迹记录仪　909
航迹控制　909
航空发动机(空气发动机)　20
航空发动机　16
航空火力控制系统　63
航空母舰　24
航空母舰战斗群　24
航空汽油　63
航空器　24
航空特快专递　311

航空邮包收据 22
航空运单 25
航空运输 22
航空运输货物保险 23
航空运输险 23
航空运输一切险 22
航路点 973
航路点方式 967
航路线 973
航母舰载机降落方式 491
航母舰载机起飞方式 881
航区限制标志 788
航速 843
航速角 36
航速试验（速率试验，测速试航） 843
航速修正法 485
航速修正法 843
航线计划 752
航线监测 752
航向 190
航向-角定位法 568
航向保持能力 191
航向向上显示 191
航行灯 589
航行灯控制器 589
航行高度 557
航行工作站 590
航行活动记录 725
航行距离 245
航行控制数据 589
航行期间 263
航行数据记录仪（VDR） 956
航行数据记录仪系统符合证书 956
航行途中 281
航行指挥电话 159
航行指挥电话系统 589
航修 591
航修 957
航运保险 808

好望角航线 108
好望角型船 109
好望角型船舶 109
好望角型散货船 109
号灯 589
号灯吊升索具 659
号笛（报警器） 819
号笛（报警器,注射器） 879
号笛（号角） 973
号锣 389
号旗 815
号型 793
号钟 80
耗热量 417
耗油量 620
合成剂 878
合成缆绳 879
合成视频信号 165
合成纤维 879
合成纤维缆 322
合成药 879
合成药或化合致幻药 879
合成渔网 878
合法权益 496
合格 171
合格安全型 131
合格安全型设备 131
合格的通风 764
合金 28
合金钢 857
合理的船舶设计 720
合拢 293
合同 161
合同 179
合同检查 179
合同履行地 674
合同设计 179
和解 31
和解协议 31
河口调查 298
荷兰型材 429

核电池 54
核动力船 613
核动力船舶 612
核动力装置 612
核反应堆 612
核供应国集团 612
核能货船安全证书 611
核能货船安全证书或核能客船安全证书 612
核能客船安全证书 612
核潜艇 613
核燃料 612
核事故 612
核损害 612
核武器 613
核武器小型化技术 613
核状沸腾 613
核准值 131
盒式磁带录像机 951
赫兹 422
黑海区域 85
黑菱号型 84
黑球 85
黑色金属钻屑、削屑、旋屑或切屑（呈自然状态） 321
黑色鱼篮 84
黑双锥号型 251
黑水 85
黑匣子 84
黑柱号型 84
黑锥号型 84
很少可能的概率 734
很少可能性 733
恒定转矩电阻器 173
恒湿器 441
恒速电动机 173
恒温阀 895
恒温恒湿室 173
恒温节流阀 895
恒温控制阀 895
恒温器（恒温箱） 895

1103

恒温器控制的 895
恒温双金属 895
恒星式减速齿轮箱 827
恒压膨胀阀(自动膨胀阀) 58
恒张力绞车 173
桁材 387
桁架桅 105
桁架柱体式平台/桁架型立柱式平台 920
横波 917
横舱壁 916
横舱壁下部 916
横档 754
横舵柄 754
横骨架船 917
横骨架式 917
横骨架式舷侧结构 917
横截面上多种缺陷 582
横栏 468
横栏间距 717
横缆 95
横浪 76
横流扫气 195
横漂 256
横剖面 195
横剖面 917
横剖面面积及其净力矩曲线 199
横剖面面积曲线 776
横剖面上的纵向构件 522
横剖线 88
横剖线图 88
横倾 512
横倾角 36
横倾角 420
横倾力矩 421
横倾平衡泵 512
横倾平衡舱 292
横倾平衡水泵 421
横倾平衡系统 420
横倾水舱 421
横甩 97

横稳心 916
横稳心半径 916
横稳心垂向坐标 947
横稳心垂向坐标曲线 199
横稳心高 916
横稳性 917
横向操纵装置(侧推器,横向推力器) 815
横向传送 28
横向的(水平的) 431
横向环肋 916
横向甲板/横跨甲板 195
横向力 493
横向强度 917
横向强框架 917
横向推力器 916
横向振动(侧向振动) 493
横向振动 917
横向正加速度 685
横摇 749
横摇单幅值 θ 749
横摇调谐因数 921
横摇幅值 34
横摇固有频率 586
横摇固有圆频率 586
横摇固有周期 587
横摇惯性矩 749
横摇回转半径(m) 489
横摇回转半径 k_r(m) 749
横摇角 36
横摇投影角 699
横摇突倾 530
横摇周期 908
横摇周期 T_R 749
横移 814
横张索 212
横坐板 899
红海区域 725
红环照灯 725
红火号 725
红球 725

红色的视觉信号 726
红色警戒 725
红外反隐身技术 456
红外辐射测温仪 456
红外线传输 456
红外线通信灯 456
红舭灯 726
红舷灯 726
虹吸 962
后备保护 67
后备保护 738
后备驾驶人员 66
后部界限 17
后端壁 17
后端点(艉端) 17
后端点 18
后端区域结构 17
后果 172
后果评价 172
后肩 17
后冷凝器 17
后冷凝室 17
后冷却器 18
后冷式柴油机 18
后联轴节 66
后掠式水翼 877
后锚灯 17
后燃 17
后升高甲板 718
后体 17
后退措施 561
后桅灯 718
后桅杆 17
后艉轴管 722
后踵 18
厚膜混成集成电路 895
呼号 106
呼叫 106
呼吸充电 132
呼吸道过敏 740
呼吸阀(通气阀) 95

呼吸阀　95
呼吸器　95
忽略(俯视,漏看,眺望)　647
狐齿型吊艇架　710
弧坑　193
壶嘴(管)　845
互补性氧化金属半导体芯片　164
互感器(仪表用)　462
互换性　305
互惠贸易　317
互联网＋　480
互谱　196
互相关函数　196
互助索　99
护航　295
护航作业　295
护目镜　314
护目罩　314
护索环　115
护套(阳极)　793
护套空间(隔层空间)　486
护舷材　320
护栅　704
沪东中华——海上超级"LNG 船的摇篮"　434
花钢板　134
花格窗　493
花键轴　844
花链(齿条)　844
花生(带壳)　658
华沙-牛津规则　959
滑参数启动　822
滑差　822
滑车　86
滑车接头　86
滑车组　880
滑道　820
滑道　822
滑动(滑移)　822
滑动底脚　822
滑动接合(伸缩式连接)　822

滑动面　822
滑动摩擦　822
滑动摩擦力　822
滑动膨胀接头　822
滑动式　822
滑动式膨缩头　822
滑动式屈曲限制器　822
滑动型　822
滑动型联轴节　822
滑动轴　822
滑动轴承　822
滑阀　822
滑阀图　822
滑架型艇架　909
滑键　822
滑块(滑动龙骨墩)　822
滑块　822
滑块导板　822
滑轮　793
滑石　881
滑丝　895
滑速　822
滑套接头　822
滑套式膨胀接头　822
滑脱(滑距)　822
滑行数　675
滑行艇　675
滑行艇　675
滑行状态　675
滑油　575
滑油泵　529
滑油舱(柜)　528
滑油舱　529
滑油澄清柜　529
滑油磁性滤清器　533
滑油低压自动停车装置　529
滑油定期净化　528
滑油分离机　528
滑油分离加热器　529
滑油分析　528
滑油分析记录表　528

滑油分析资料　528
滑油分油器　529
滑油管系　528
滑油加热器　529
滑油净油柜　708
滑油老化　18
滑油冷却器　528
滑油连续净化　528
滑油滤清器　529
滑油输送泵　529
滑油温度　529
滑油稀释　529
滑油系统　529
滑油消耗量记录表　528
滑油消耗率　528
滑油消耗率　841
滑油循环柜　529
滑油压力　529
滑油溢油舱　646
滑油油渣柜　529
滑油重力柜　529
滑油贮存柜　529
滑油注入总管　529
滑油注射器　879
滑油状态　528
滑油状态监控系统　528
滑脂润滑法　392
滑脂油　392
化学不完全燃烧热损失　418
化学剂贮存舱　133
化学凝聚剂　133
化学品/成品油船　134
化学品/油液货船　133
化学品　133
化学品驳　133
化学品船　133
化学品液货船　133
化学注入　134
划痕(刮痕、划道)　767
划桨救生艇　614
画线针(画线器)　768

还船 726
还盘 190
环保材料 291
环保工作船 289
环保工作船 291
环保会 564
环保燃料技术 362
环保设备 291
环保挖泥船 291
环保型船 268
环槽绝缘体 395
环管燃烧室 921
环境 289
环境安全 291
环境保护 289
环境保护 291
环境保护附加标志 291
环境标准 291
环境毒瘾 289
环境符合性记录 290
环境工程师 290
环境管理体系 289
环境监测船 291
环境监控装置 291
环境控制 290
环境控制单元 289
环境控制系统 290
环境力 290
环境试验 291
环境条件 289
环境条件 290
环境温度 289
环境污染 289
环境无害化方法 291
环境效应 290
环境压力 875
环境严重性因素 291
环境异常 289
环境因素 290
环境影响评估 290
环境载荷（风、浪、流作用） 290

环境载荷 290
环境载荷条件 290
环境振动 291
环境转移条件 291
环境资源 291
环量 138
环流 138
环流减额因数 726
环鸣声速仪 817
环向增强层 430
环形防喷器 37
环形燃烧室 37
环形通气孔 37
环形推力轴承 154
环眼螺栓卸扣 314
环氧底漆（环氧涂料） 291
环氧富锌底漆 292
环氧基体系 292
环氧胶 291
环氧沥青防锈涂料 292
环氧膜 291
环氧泥子 291
环氧黏合剂/环氧胶黏剂 291
环氧泡沫塑料 291
环氧树脂 291
环氧树脂 291
环氧树脂 299
环氧树脂垫座浇注膏 686
环氧树脂裂片甲板涂料 292
环氧树脂黏合的 292
环氧树脂涂层 291
环氧树脂涂料 291
环氧树脂涂料 292
环氧值 291
环氧值 292
环照灯 28
环状阀 746
环状天线测向器 523
缓S形立管 494
缓冲器（冲击能吸收器） 446
缓冲器 810

缓冲系数 99
缓解 574
缓冷 823
缓慢燃烧 823
缓装工程 686
换舵艏向角 306
换气次数 19
换气气缸 304
换气系统 944
换向阀 744
换向机构 744
换向器（倒转机构，换向开关） 743
换向时间 743
换向时间 744
换向试验 752
换向凸轮 744
换向爪 897
换向装置 744
换向装置 744
换新厚度 735
换证检验 735
皇家赌场（豪华邮轮） 119
黄色的视觉信号 987
黄色警戒 987
黄色闪光灯 28
黄铁矿（含铜和铁） 709
黄铁矿，经过煅烧（煅烧黄铁矿） 709
黄铜/青铜 93
黄艇灯 987
晃荡冲击载荷 823
灰尘 263
灰口铸铁件（灰铁铸件，生铁铸件） 394
灰水 392
灰水 394
灰水储存舱 395
灰水处理系统 395
灰水的控制 392
灰水控制 395
灰水控制 402

灰铸铁　394
灰铸铁试样　394
挥发物　955
挥发性(不稳定,易变状态)　955
挥发性有机化合物　955
回避　979
回车　118
回舵操纵试验　707
回放软件　677
回归分析　730
回火　341
回接　899
回扣　729
回冷　724
回流扫气　524
回路　523
回气桶　868
回热器　729
回热式汽轮机　729
回热循环　729
回热循环汽轮机装置　729
回热循环燃气轮机装置　729
回声测深仪　224
回声测深仪　268
回声测深仪　268
回跳硬度(肖氏硬度)　810
回跳硬度计　767
回文　651
回旋振动(急旋振动)　973
回油式喷油器　743
回油式喷油嘴　844
回转(旋转)　973
回转泵　751
回转初速　43
回转吊杆绞车　822
回转横倾角　420
回转力矩　923
回转率　719
回转能力　923
回转圈直径/战术直径　881
回转式空气预热器　729

回转式压缩机　751
回转试验　923
回转枢心　673
回转速度　943
回转性　923
回转性能　923
回转性能试验　923
回转性指数　923
回转圆　923
回转中心　127
回转中心纵距　721
回转周期　923
回转装置　923
汇　819
汇出行　733
汇兑率　719
汇付　733
汇款　733
汇款人　733
汇率　304
汇票　311
汇票　83
汇入行　723
会议中心(豪华邮轮)　170
绘图机　255
绘图室　256
绘图仪　677
惠普　422
惠普公司　422
混合　574
混合骨架船　156
混合骨架式　156
混合光纤同轴网络　437
混合结构船　165
混合冷却　574
混合推进系统　438
混合型个人漂浮装置　438
混合型垃圾　574
混合循环　260
混合循环　574
混合硬盘　438

混合油　574
混流泵　574
混凝土搅拌船　343
混凝土重力式平台　168
混砂系统　86
混响测量　743
混响声场　743
混杂、沾污险　468
豁免　446
豁免与紧急行动原则　306
活动厕所　817
活动钩体系　580
活动矩形窗　633
活动栏杆　154
活动栏杆柱　235
活动式海底生物呼吸测量器　356
活动踏脚板　350
活动梯步　319
活动舷窗　633
活动仪器的接地装置　267
活动桌面板　349
活节接合(球形接合)　844
活节接合　826
活节连接器　487
活络扳手　768
活塞(柱塞)　672
活塞泵　672
活塞底泵扫气　672
活塞杆　672
活塞环　672
活塞环槽　672
活塞环端面间隙　672
活塞环对口间隙　672
活塞环结胶　672
活塞加油法　672
活塞间隙　672
活塞孔　672
活塞冷却　672
活塞冷却泵(活塞冷却液泵)　672
活塞平均速度　672
活塞裙　672

活塞式发动机　672
活塞式阀　672
活塞式扫气泵　670
活塞式压缩机　672
活塞式制冷压缩机　724
活塞水冷却器　672
活塞套管　672
活塞头　672
活塞头　672
活塞销（杆销，十字头销，轴头销）　400
活塞销　672
活塞销　672
活塞销孔　672
活塞运动速度　943
活鲜鱼运输船　118
活性物质　9
活性物质和配制品的数据集　210
活性有机物　949
活鱼舱　972
活鱼运输船　512
火车渡船　910
火的抑制（火灾抑制）　333
火管锅炉　334
火管锅炉　335
火和燃烧产物的抑制目标　326
火和燃烧产物抑制目标1　325
火花（打火花）　834
火花避雷器（火花制止器，火花罩）　834
火花猝熄　834
火花点火　834
火花点火器　834
火花防止器　834
火花分配器　834
火花间隙　834
火花鉴别（火花试验）　834
火花塞　834
火花熄灭器　834
火花隙（火花放电器，避雷器）　834

火花延迟　834
火警报警　334
火警锣（火警钟）　331
火警探测　327
火警信号　325
火警信号　333
火警指示板　331
火警指示灯　331
火警装置　325
火炬平台　341
火炬塔　341
火控雷达　334
火力控制系统　327
火炮基座　401
火炮平台　401
火势控制　326
火星熄灭器（火花避雷器，火花制止网，火花熄灭器）　834
火焰（火舌，燃烧）　339
火焰故障　339
火焰监控装置（LPG系统）　340
火焰炉（反射炉）　339
火焰钎焊（气焊）　339
火焰清净炬　339
火焰清理（火焰除锈）　339
火焰试验（焰色试验）　339
火焰速度　340
火焰探测器　339
火焰探测器　340
火焰筒　340
火焰筒气膜冷却　340
火焰退火　339
火焰弯管　339
火焰温度（着火温度）　340
火焰氧气刨　649
火焰制止器（防焰器）　339
火源　333
火源　831
火灾报警器　325
火灾的限制　177
火灾探测目标1　327

火灾探测目标2　328
火灾探测目标　328
火灾探测器（火警探测器）　328
火灾预防目标　332
货驳　358
货舱　112
货舱　429
货舱舱壁　112
货舱处所　429
货舱空气干燥　112
货舱口　112
货舱排水舷口　429
货舱梯　112
货舱通风　112
货舱围井　112
货舱液位传感器　882
货车装卸率　720
货船　111
货船　116
货船安全证书　115
货船安全证书　129
货船构造安全证书　115
货船构造安全证书　799
货船设备安全证书　115
货船设备安全证书　799
货船无线电安全证书　116
货钩牵索　116
货交承运人（指定地点）　355
货款/信贷　193
货品　698
货品的危险性　416
货物　110
货物　389
货物保险条款　462
货物报警　110
货物残余物　114
货物操作　114
货物长度区域　113
货物承运单　808
货物处所　116
货物处所　116

货物渡船　111
货物渡船　358
货物附属装置　111
货物记录簿　114
货物结露　117
货物控制室　111
货物控制站　111
货物冷藏　728
货物冷藏装置　728
货物贸易　389
货物坡道　114
货物区域　117
货物区域或货物长度区域　110
货物渗透率　114
货物提升设备　113
货物围护系统　111
货物系固手册　115
货物系固手册　777
货物运输　118
货物蒸发气体排放控制系统标准　851
货物装船清单　808
货物装卸设备　112
货物资料　112
货油/压载兼用舱　156
货油　114
货油泵　110
货油泵舱　114
货油泵舱管系　114
货油舱　114
货油舱　117
货油舱舱壁　117
货油舱舱壁隔离阀　101
货油舱管系　114
货油舱呼吸阀　114
货油舱气压指示器　114
货油舱清舱系统　114
货油舱区域　117
货油舱透气管系　117
货油舱吸井　868
货油舱洗舱设备　882

货油舱洗舱系统　882
货油舱油气驱除装置　882
货油阀　114
货油加热系统　114
货油清舱泵　114
货油软管　114
货油装卸系统　117
货油总管　534
货运吨　808
货运合同　110
货运价格　808
货种标志　113
获得的船舶能效设计指数　54
霍尔锚　403
霍尔木兹海峡　431

J

机舱　285,286,288,530,584
机舱布置　286,530
机舱舱壁(机器处所舱壁)　531
机舱舱壁　286
机舱舱口　286
机舱操纵台　286
机舱储藏室　286
机舱传令钟(车钟)　287
机舱传令钟(机舱车钟)　286
机舱传令钟(机舱传令指示器)　286
机舱传令钟　801
机舱传令钟发送器　287
机舱传令钟接收器(车钟接收器)　287
机舱底层地板　286
机舱辅机　288
机舱集控室　285
机舱集控室　286
机舱集控台　286
机舱集控站(室)　128
机舱集控站(室)　128
机舱记录　530
机舱记录台　286

机舱监测报警台　577
机舱检测、报警和控制系统　235
机舱结构　286
机舱开口　285
机舱控制室　286
机舱口　285
机舱棚(机舱口围阱)　530
机舱棚　286
机舱平台　288
机舱平台　286
机舱平台　286
机舱日志　286
机舱日志　286
机舱梯　286
机舱天窗　286
机舱通风机　286
机舱通风机间　286
机舱通风机间　286
机舱图　286
机舱行车(机舱起重吊车)　286
机舱应急舱底水阀　286
机舱应急舱底水管　286
机舱油柜　286
机舱噪声　286
机舱噪声级/机舱噪声电平　286
机舱值班室　960
机舱自动化　286
机舱自动化　530
机场旅客处理系统　25
机带泵　54
机电维护设备　562
机顶盒　789
机动船　688
机动船　782
机动船舶　579
机动船舶　782
机动救生艇　579
机帆船　579
机帆渔船　337
机工(机械师)　531
机加工(刨尽)　531

机加工工厂 531	机械操纵货油阀 563	机械式探温计 562
机加工精度 531	机械槽型 530	机械式探温计 562
机加工开裂 531	机械测深 563	机械试车 531
机加工面 530	机械除垢 562	机械手 562
机架(发动机机架) 285	机械处理(机械加工) 563	机械手 563
机匠 285	机械传动门 563	机械输送系统 886
机库甲板 404	机械的(机械能) 562	机械损伤 562
机炉舱 285	机械的可修复的故障 735	机械损失 563
机炉舱单位长度功率数 687	机械定时器 563	机械特性 562
机炉舱单位面积功率数 687	机械锻造 530	机械填垫复合物 530
机炉舱单位容积功率数 687	机械堆装设施 563	机械通风 563
机炉舱格栅 285	机械方法 563	机械通风 688
机炉舱通风 285	机械工程 288	机械通风系统 563
机铆 530	机械工程师 562	机械推进救生艇 563
机器报警 530	机械构件(零件) 530	机械推进装置 563
机器操纵室 530	机械化 563	机械弯管 562
机器操作 285	机械画(工程画) 562	机械效率 562
机器处所 832	机械混合物 563	机械性保护 563
机器处所(船舶及近海装置) 531	机械计划维修安排 530	机械压力雾化喷油器 691
机器处所(机舱) 286	机械加工后状态 530	机械遥控操纵 563
机器处所 530	机械加工裕度 530	机械噪声 530
机器处所的通风 944	机械加煤机 563	机械噪声源 600
机器处所和主厨房 531	机械夹杂物 562	机械增压 287
机器处所集中控制 560	机械开关的阀 563	机械振动 563
机器处所周期无人值班 531	机械开关电器 563	机械制图 562
机器处所周期性无人值班 57	机械空气循环系统 562	机械制造 530
机器控制站 285	机械控制执行结构 562	机械质量 531
机器人 748	机械连接(接线夹) 562	机械装置 530
机器效率 285	机械联动装置 563	机械装置 530
机损 530	机械零件 530	机械状态监测 530
机体 285	机械零位 563	机械自动控制系统 562
机务代表 872	机械摩擦副 530	机修间 984
机械(机器) 530	机械磨损 563	机翼切面 16
机械(机械装置,机械原理) 563	机械抛光(法) 563	机油柜 285
机械(装置)[细节(部件)] 562	机械喷射柴油机 563	机载空中预警和控制系统 23
机械/气力混合输送系统 887	机械清洗 562	机罩(支架) 430
机械 530	机械设备 562	机座(机架) 288
机械不完全燃烧热损失 418	机械设备底座 530	机座(发动机底座) 285
机械舱盖 562	机械升降机构 562	机座 285
机械操舵系统 563	机械式传令钟(机械车钟) 563	机座 285
机械操舵装置 563	机械式示振仪 563	机座 285

机座　530
机座纵桁　286
积分平均声级计　465
积分声级计　465
积极隔振　10
积载鉴定　862
积载因素　861
基本板格　275
基本结构图　174
基本绝缘　73
基本速率接口　73
基本显示　244
基本险　695
基础结构　354
基底　396
基地地球站　72
基地港　72
基地气象台　72
基料　72
基尼系数　387
基盘式平台　889
基频　366
基频振动　950
基平面　72
基线(渔船)　73
基线　576
基线　579
基线　73
基因技术　386
基于安全方针和目标的绩效　268
基于风险的设计　747
基于计算机的系统　168
基于科学原则　766
基于网络的,组合导航系统　444
基站　72
基站控制器　72
基站收/发台　72
基准标志(标距标记)　373
基准传送站　72
基准面　212
基准水线 WL_{ref}　727

基准温度　727
基座(阀座)　773
基座(阀座,座位)　773
基座　354
基座腹板　969
基座隔板　93
基座桁材　387
基座面板　316
基座肘板　919
绩效指标　659
激波风洞　810
激光打印机　492
激光电能传输　492
激光动态检测仪　492
激光绘图机　492
激光技术　492
激光扫描技术　492
激光束热处理　419
激光系统　492
激流装置/湍流装置　922
激振载荷　447
及时和全面修理/立即彻底修理　700
级差活塞　237
级传动比　859
级间冷却　480
级间密封　480
级间配合　849
级间水封　480
级效率　849
极不可能的概率　314
极地　679
极地船舶　679
极地航行　658
极地级别　678
极地级别船舶　679
极地考察实验室　679
极地水域　680
极端泄油参数　314
极端载荷　313
极光观测　57

极光夜光观测室　679
极轨道卫星业务　679
极区船　46
极少或无失火危险的液舱、空舱及辅机处所　884
极少可能的概率　314
极限安全因数(限安全系数)　930
极限报警　506
极限波浪周期范围 T_{xrange}　314
极限短路分断能力　930
极限风暴持续时间　263
极限抗拉强度(抗拉强度)　930
极限强度(最大强度)　930
极限应变　930
极限应力　930
极限载荷　506
极限载荷　930
极限值　897
极限重心垂向坐标曲线　199
极限转向角　507
极限状态　506
极值　314
即复　735
即刻可用性　461
即刻装运　700
即期付款交单　250
即期汇票　815
即期信用证　815
即时通信　461
即时通信软件　461
急件传递　20
急性皮肤中毒　10
急性期蛋白　11
急性释放　10
急性水毒性　10
急性吸入中毒　10
集成电路　464
集成监测系统　464
集成门极换流晶闸管　464
集成系统(计算机)　465
集存槽(沉淀柜,污水井)　870

集合点　584
集合区域　583
集合式搜索引擎　51
集合运输包装　129
集合站（登乘站）　51
集结的孔穴（集结孔隙）　149
集散式布风器　21
集输系统　373
集污舱　429
集线器　434
集油剂　627
集油器　620
集油器　627
集油箱　622
集运架（托盘）　651
集中操纵货油装卸系统　128
集中控制站　128
集中式空气调节器　127
集中式空气调节系统　127
集中式氧乙炔焊接设备　128
集中运输　807
集中运输策略　807
集装箱　174
集装箱绑架　492
集装箱插头点　176
集装箱船　176
集装箱船　176
集装箱导轨架/格栅　126
集装箱的形式　928
集装箱电源　176
集装箱吊具　845
集装箱堆（集装箱块）　849
集装箱堆垛　176
集装箱堆码　73
集装箱架　176
集装箱角附件　176
集装箱块　176
集装箱区域　708
集装箱式实验室　177
集装箱提单　176
集装箱系固件/集装箱紧固设备　176
集装箱系固系统　176
集装箱系固装置　176
集装箱运输　176
集装箱运输车　651
集装装卸　177
几何平均数　386
几何相似　386
几何相似的船模　387
挤塑式聚苯乙烯泡沫塑料板　314
脊弧　793
计（表，测量，满载吃水）　373
计程仪　520
计程仪舱　520
计划保养系统（PMS）　674
计划工时　675
计划和文档工作站　675
计划价格　674
计量（校准）　373
计量泵　565
计量吃水　562
计量船长　562
计量方形系数　561
计量设备（调节设备）　565
计量系统　373
计权　969
计权隔声量 R_w　969
计权声级　969
计权网络　969
计算长度（标准距离）　373
计算的人流　105
计算风力作用力臂　720
计算风速 V_w　105
计算机（自动控制）系统的分类　143
计算机　166
计算机 CPU 设计制造技术　225
计算机病毒　167
计算机存储　167
计算机辅助分析与计算　167
计算机辅助工程　166

计算机辅助工艺过程设计　167
计算机辅助计算　167
计算机辅助计算系统　167
计算机辅助计算与分析　167
计算机辅助教育　167
计算机辅助经营管理　166
计算机辅助设计　166
计算机辅助设计计算　166
计算机辅助制造　167
计算机集成制造系统　167
计算机技术　167
计算机图形学　167
计算机网络　167
计算机网络　167
计算机系统（动力定位系统）　167
计算机系统　167
计算机系统的重大改装　539
计算软件　167
计算型深　223
记录　725
记录桌　880
记名提单　862
技术案卷　886
技术成果　887
技术成果文件　887
技术服务合同　887
技术管理部　887
技术规格（技术要求）　288
技术规格书(涂层)　886
技术合同　886
技术合作委员会（技合会）　886
技术开发合同　887
技术缺陷　886
技术人员　887
技术设计（详细设计）　886
技术条件（设计任务书，分类）　842
技术委员会　886
技术性贸易壁垒　886
技术转让　887
技术转让合同　888

技术状况　886
技术咨询合同　886
技术组　886
继电接触器控制　732
寄售　172
寄售协议　18
加冰孔　443
加冰孔盖　443
加大链环　288
加工处理单元　697
加工龟裂　531
加工过程(进程)　697
加工机械　697
加工甲板　697
加厚无缝钢管　420
加拉帕戈斯特殊敏感海域(PASS)(GALREP)中新的强制性船舶报告系统　596
加拿大-欧盟自由贸易协定　106
加气喷水推进　23
加强检验报告卷宗　288
加强检验程序　288
加强绝缘(增强绝缘)　731
加燃料(加油)　729
加热(供暖)　419
加热(加温)　419
加热管　419
加热炉　419
加热盘管　419
加热喷灯　419
加热器(热源,加热工)　419
加热器部件　419
加热区通　419
加热设备(加热器)　419
加热时间　419
加热水倍率　419
加热弯曲试验　419
加热液舱　419
加热硬化(加热淬火)　418
加热元件　419
加热元件　419

加热蒸汽　419
加热周期　419
加热装置(暖气装置)　419
加热装置　419
加湿器　437
加水喷水推进　963
加速导管推进器　3
加速度参数 a_0　3
加速度级　3
加速段　3
加速机组　90
加速冷却(AcC)　3
加速燃气轮机　90
加速燃气轮机　90
加速性试验　3
加温器(加温,加热)　959
加稳定剂的鱼粉(鱼渣)(经抗氧处理)　337
加压泵(压力泵)　691
加压舱　166
加压舱　692
加压器(稳压器)　692
加压外壳　692
加油(加油法)　627
加油　627
加油点(需要润滑处)　627
加油和储油设施　729
加油器　621
加油设备　102
加账工程　735
加装作业　904
夹层　884
夹带盐水　289
夹管钳　395
夹壳联轴器　138
夹扣(夹具,桨柄)　395
夹扣　147
夹扣型　395
夹头　395
夹心板　762
夹渣　821

佳能公司　107
家居生活污水(生活废弃物)　250
家具充电　132
家用电脑　430
家用投影机　250
甲板　214
甲板安全索具和安全索　216
甲板边线　215
甲板驳　191
甲板驳　215
甲板部高级船员　216
甲板操纵隔离阀　216
甲板冲洗泵　216
甲板冲洗泵　960
甲板冲洗系统　216
甲板冲洗系统　960
甲板冲洗装置　959
甲板处的剖面模数 $Z_{AD}(m^3)$　776
甲板窗　215
甲板导缆孔　215
甲板舵角限位器　754
甲板货船　215
甲板货油管系　114
甲板机械　215
甲板机械下的木质垫块(机械座垫块)　530
甲板间舱壁　923
甲板间吨位　923
甲板间肋骨　923
甲板结构　215
甲板结构　216
甲板居住舱　215
甲板居住舱　6
甲板排水槽(排水管,斜槽)　391
甲板排水阀　216
甲板排水管　216
甲板排水及疏水系统　215
甲板剖面模数　776
甲板牵条　899
甲板强横梁　216
甲板清洗工具　959

1113

甲板区域 216
甲板洒水系统 216
甲板上浪 394
甲板室 215
甲板室 216
甲板室侧壁 814
甲板室甲板 215
甲板室围壁 215
甲板水沟 401
甲板水舷外排水口 215
甲板梯 215
甲板艇 215
甲板下吨位 932
甲板下作业 932
甲板舷弧线 793
甲板线 215
甲板泄水管 768
甲板用具 215
甲板中线 215
甲板纵骨 215
甲板纵桁 215
甲苯 903
甲级防火门 1
甲烷 565
甲烷值 565
钾碱 686
价格贸易条件 592
价格拍卖 692
价款/成本 189
价值检验证书 486
驾驶甲板/桥楼甲板 96
驾驶人员 591
驾驶室 973
驾驶室对驾驶室的通信 96
驾驶室航行值班报警系统 96
驾驶室集控台 95
驾驶室可视范围 590
驾驶室控制站 96
驾驶室信号控制板(驾驶室信号控制台) 815
驾驶室遥控 733

驾驶室遥控 95
驾驶台或驾驶位 172
驾驶信息显示 172
驾驶指挥位置 172
架空横梁 195
架空输电线路 646
架空纵梁 522
假底 318
假定的泄油量 442
尖舭部 405
尖舱肋骨 658
尖底船 793
尖嘴锤 540
坚持/依附 13
间断工作制 243
间断构件 466
间断纵向构件 466
间隔(分隔,分离) 786
间隔斗链 631
间接 452
间接辐射强迫(负强迫) 451
间接冷却 451
间接冷却式空气冷却器 451
间接冷却系统 451
间距 833
间隙(余隙,双体船的船体间的距离) 368
间歇性或掠夺性倾销 468
肩 813
肩背式泡沫枪 813
肩部区 813
监测报警器 577
监测程序 577
监测对象 577
监测试验 577
监测数据 577
监测台(监控台除外) 577
监测系统(监听系统) 577
监测装置(监视装置) 577
监督 873
监护者 400

监控参数 577
监控参数 577
监控器(消防龙头,消防炮) 577
监控设备(检测设备) 577
监控系统 577
监控系统的控制部分 183
监控仪器 577
监视/监控 577
监视 577
监视工作站 577
监视和报警系统 577
监视和报警整定值修改功能 365
监视器 960
监视器显示装置 577
监视指导船 657
监听记录 577
兼容性 162
兼装船 156
检测仪 461
检查(检测,验证) 302
检查 460
检查、维护和修复 461
检查和试验情况 460
检查清单 133
检查设备 461
检漏器(漏电检测仪) 494
检视孔 815
检温计 888
检修 646
检验 460
检验 875
检验工作 876
检验合格证书 130
检验和发证指南 400
检验强化制度 288
检疫 712
检疫艇/船 712
剪力曲线 793
剪切波 793
减除处所 217
减功系数 982

减价拍卖 263
减扭振器 904
减轻孔 504
减热器 418
减热蒸汽 234
减速 843
减速比(传动比,减缩比) 726
减速齿轮舱(齿轮传动舱) 374
减速齿轮箱(减速齿轮,减速装置,减速传动装置) 726
减速齿轮箱滑油澄清柜 726
减速齿轮箱滑油循环柜 726
减速齿轮箱滑油重力柜 726
减速齿轮箱滑油贮存柜 726
减速导管推进器 214
减速段 214
减速机构(减速齿轮) 843
减速器(减压器) 726
减速器舱 726
减温器 234
减压船模试验水池 222
减压阀(压力调节阀) 691
减压阀 726
减摇泵 40
减摇控制设备 848
减摇鳍(稳定翼,舷侧可控鳍) 848
减摇鳍 324
减摇鳍控制器 324
减摇鳍收放装置 324
减摇鳍转鳍机构 324
减摇曲线 199
减摇试验 848
减摇水舱 40
减摇水舱 848
减摇装置 40
减摇装置舱(减摇装置室) 848
减摇装置舱 848
减噪涂层 600
减账工程 107
减振(振动阻尼) 950

减振 642
减振覆层 950
减振环 810
减振接头 825
减振器(缓冲器,消声器) 825
减振器(平衡块减振器) 66
减振器(阻尼器,缓冲器) 810
减振器(缓冲器) 951
减振器 41
减振器 642
减振器 810
减振器 950
减振器 950
减振器橡胶 810
减振器支架 810
减振装置(控振装置) 950
减振装置(阻尼器) 41
减阻降耗技术 255
简单测量设备 816
简单检索 817
简单设备 816
简单悬链式立管 816
简单循环 816
简单循环燃气轮机装置 816
简化航行数据记录仪(S-VDR) 807
简化航行数据记录仪(S-VDR) 817
简谐波 816
简谐振动 816
简易程序 870
简支 817
碱度 27
碱洗 27
建设工程合同 180
建议 702
建造 173
建造标准 315
建造厂 99
建造的船舶 809
建造的高速船 192

建造的散货船 101
建造合同的日期 211
建造厚度 50
建造签约日 211
建造日期 210
建造说明书 174
建造裕度 99
建造者 174
建筑研究所 99
健康/安全和环境 417
健康证书 417
舰船总吨位 GT 396
舰员(舰艇) 193
舰员起居室 194
舰载垂直发射系统 118
舰载反导系统 803
舰载火力控制系统 803
舰载机 119
渐近稳定运动 53
渐开线凸轮 481
溅密门 844
溅泼 844
鉴定报告 735
鉴定检验 43
鉴定检验 444
鉴定人 43
键槽 488
键槽 488
键接 488
键孔 489
键盘(开关板,用键盘输入) 489
键配合 488
江面坡度推力 708
江面坡度阻力 255
江水快速澄清装置 747
桨 614
桨叉 196
桨毂 90
桨毂导流帽 434
桨毂帽鳍 701
桨毂密封 434

桨前扇形整流鳍 352
桨塔 709
桨叶倾斜 718
桨叶展开面积 309
降落和起飞区(TLOF) 906
降落滑道长度 493
降落滑道角度 493
降落伞火箭 652
降落设备或装置 493
降落时限 903
降落条件 493
降落装置 493
降膜蒸发器 318
降速电动机组 259
降温装置 726
降压 726
降阻剂 360
交变流 29
交叉构件/次要构件 776
交叉构件 475
交叉构件 480
交叉梁系(格架,格栅,板架) 395
交叉许可证 195
交船日期 212
交叠螺旋桨 647
交付 221
交换阀 304
交换矩阵 877
交换器(转换装置) 305
交换性 466
交货 221
交货不到险 317
交接船试验 221
交流磁力启动器 1
交流电 1
交流高压岸电系统 425
交流高压电气装置 425
交替推进系统 29
交通船 161
交通船 910
交直流两用电动机 3

浇铸轴承 508
胶合板 943
胶凝作用 374
焦炭 154
焦屑 154
角板(封槽板) 401
角阀 36
角加速度分量 36
角件 187
角配件 187
角频率 36
角速度 37
角速度分量 37
角柱 187
绞车 974
绞刀架起落装置 429
绞缆机 959
绞缆筒 959
绞螺纹的工具 895
绞盘 109
绞滩船 345
绞滩滑车(绞滩绳) 959
绞滩滑车 6
绞滩缆 959
绞吸式挖泥船 201
绞吸式挖泥船 767
铰刀 201
铰刀吊架 201
铰刀驱动装置 201
铰刀轴 201
铰翻型舱盖 428
铰接连接顶推船驳船组合体—推船 49
铰接式浮动塔 48
铰接式连接顶推船驳船组合体—驳船 48
铰接式载油平台 28
铰接式装卸塔系统 49
铰接柱 48
铰链式舱盖 428
铰链式人孔盖 428

铰链式艇架 428
铰链型艇架 428
矫直机 862
脚底止回阀 350
脚手架 765
搅动挖泥船 18
轿车/卡车运输船 708
较大失火危险处所 414
较高性能海洋工程用锚 426
较佳噪声标准曲线 689
阶梯式码头 490
阶梯式水翼 490
接触器 180
接触式平板冻结装置 676
接地 267
接地参考平面 397
接地系数 267
接地装置 397
接点(敷线中) 678
接管 881
接合(管节) 935
接合(塑料管) 487
接合 285
接合件(接缝器,非标短截管子) 487
接合器(适配器,转接器) 11
接近结构方式 4
接口 467
接口技术 467
接力泵船 343
接入齿轮传动装置 374
接收机 270
接收人 722
接收设备 723
接收中心 723
接受/验收 4
接受准则 4
接头(接合,接缝,关节) 487
接线片 979
接线柱 768
节 489

节点　599
节点　673
节流(关气门)　897
节流(节流阀,操纵阀)　897
节流　897
节流调节　897
节流阀　897
节流阀　897
节流过程　897
节流过程　897
节流控制　897
节流连杆　897
节流门　897
节能推进系统　284
节能助推轴支架　897
节省的全部时间　27
节圆直径　673
洁净　145
结构吃水　766
结构可达性　864
结构强度　864
结构试验　864
结构完整目标　864
结构完整性　864
结构完整性目标　864
结构修理手册　864
结构噪声　174
结构噪声控制　864
结果/后果　269
结霜　360
结炭　828
捷联式平台罗经　862
截管机(管子切割机)　670
截止阀(LPG系统)　813
截止阀　861
解决方案　828
解脱器(减压装置)　732
解析表达法　34
解析模型　34
解约日　107
介质　563

界面波　469
借款备用信用证　90
借款合同　518
金红石砂　755
金枪鱼延绳钓船　921
金属护套　564
金属极混合气体保护电弧焊　371
金属极气体保护电弧　371
金属硫化物精矿　564
金属卤化物灯　564
金属片弹簧夹　846
金属丝网　373
金属温度　564
金砖国家　93
金砖银行　95
襟舵角　341
襟叶舵　341
襟翼　341
襟翼舵　423
仅在散装时有危害的物质(MHB)　555
紧凑型半潜式平台　161
紧固(紧固件)　318
紧固件(接合件,扣闩)　318
紧固装置　902
紧固装置　318
紧固装置　777
紧急报警　277
紧急断电　278
紧急关闭　295
紧急阶段　278
紧急逃生呼吸装置　277
紧急停车装置　279
紧急停船　193
紧急停止装置/紧急关断装置　279
紧急无线电示位标　278
紧急状态(应急状态)　277
紧密配合　902
紧密性试验(泄漏试验,漏电试验)　495
紧索夹　978

紧压(机器等)　486
尽速装运　807
尽早实施2009年香港国际安全与环境无害化拆船公约中的技术标准　266
进场和起飞区　318
进出口关税及其他税　447
进出坞/上、下排　289
进度款　658
进风帽　253
进货检查　723
进距　14
进口(进入管,狭窄部分)　580
进口　463
进口阀箱　463
进流　455
进流段　289
进流段长　497
进气度　219
进气阀　21
进气阀　463
进气管(吸入管)　463
进气管　456
进气过滤器　22
进气机匣　459
进气口　459
进气门　459
进气提前　463
进气通风　873
进气系统　463
进气装置　21
进汽余面　856
进入　289
进入　456
进入安全　756
进入货舱和其他处所的通道　561
进入货物区域内各处所的通道　4
进入油船货物区域中处所的通道　4
进水　345
进水　456

进水阀 961

进水工况下的波浪压力 p_{WF} 967

进水工况下的波浪压力。 709

进水工况下的静水压力。 706

进水计算 345

进水角 253

进水角 289

进水角 345

进水角曲线 253

进水口 961

进水探测预报警 961

进水探测主报警 963

进水总管 961

进速 843

进速比 14

进速系数 14

进坞平面图 248

进油阀 622

近岸航行(近岸航程,沿海航行) 591

近岸吸力 71

近程导弹 812

近观检查 148

近观检验 148

近海船式结构物 809

近海调查 619

近海工程 618

近海供应船适装证书 129

近海供应拖船 619

近海海区 150

近海航区内作业的起重船 344

近海航区营运限制 393

近海渔船 619

近海装卸站 619

近海装置结构物 619

近距离加油索具 148

浸管式蒸发器 866

浸海水试验 771

浸水 clo 值(救生服) 446

浸水阀 345

浸水概率 345

浸水管 345

浸体长 500

浸体宽 94

禁区 350

禁用工况 457

禁运货物 698

禁止关税 698

禁止使用的补贴 698

京都议定书 489

经常的概率 360

经常发生的(事件的概率) 725

经处理的污水 918

经典比特 142

经典柱体式平台/传统型立柱式平台 142

经机械压榨的种子,油含量大于10%以及油和含水量合计大于20%,含植物油种子饼(a) 778

经纪人佣金 97

经纪人佣金条款 97

经济工况试验 361

经济功率 788

经济级组 268

经济器 268

经济特区 836

经济体 268

经济同盟 268

经济一体化 268

经耐火试验的贯穿装置 335

经平舱的满载舱 342

经认可的保安组织 724

经验 309

经营 103

经营人 639

经主管机关认可的组织 642

晶间腐蚀倾向试验 466

晶体二极管 198

精对中 325

精工爱普生(セイコーエプソン) 779

精加工(精机加工) 325

精矿 168

精矿 569

精确位置指示器 7

精神药物公约 185

精益产品概念开发 168

精益产品开发规划 886

精益产品设计 886

精益概念 886

精益管理 886

精益生产方式 495

精益要素验证 886

鲸炮台 408

鲸绳缓冲器 972

井 971

井架移动机构 224

井控系统 971

井口 258

井口槽 823

井口平台 972

井口区 971

井口升降甲板 503

井斜方位 241

井斜方位角 64

警报 26

警报发布 27

警报管理 27

警报管理 95

警报历史清单 27

警告 959

警告对人员有紧迫危害的报警 27

警告性标志 959

警戒阶段 27

警卫船 853

径高比 561

径向变螺距 714

径向变螺距螺旋桨 942

径向变形 714

径向间隙 714

径向螺距分布 714

净吨 594

净吨位 594

净功率 593
净惯性矩 I_s 593
净横剖面积 A_s 593
净厚度方法 593
净化器（过滤器） 708
净极惯性矩 I_p 593
净剪切横剖面积 A_{sh} 593
净空吃水 H_a 20
净空高度 417
净剖面惯性矩 I_w 593
净气器（滤尘器,洗涤器,洗涤塔） 768
净水装置 708
净提供厚度 593
净现值 593
净效率 593
净要求厚度 594
净载重吨 592
净载重量 939
净装卸率 593
净装运质量 m_N 593
竞买人／竞买者 81
竞争 163
静电绘图机 275
静力气垫 854
静摩擦 360
静平衡 854
静平衡试验 854
静水力曲线 244
静水力曲线 441
静水区域(平水区域) 825
静水压力 706
静水压试验 441
静水中的侧向载荷 493
静索 853
静稳性 564
静稳性臂 501
静压调节器 854
静压平衡装载（HBL）操作手册 441
静压式舱容计 441

静叶环 338
静叶片 338
静载应力 854
静止变换器 854
静止角 36
静止自动识别系统（AIS）目标 821
镜面负荷 301
纠缠比特 289
纠偏期 903
纠正措施 187
酒精温度计 844
旧货／二手货 776
救火水龙管 431
救火水龙喷嘴 331
救火旋塞（救火龙头） 326
救火钟 326
救捞驳 737
救捞船 761
救生打捞船 502
救生带 501
救生凳 501
救生筏 502
救生筏底面积 345
救生筏干舷 357
救生筏架 849
救生筏属具 292
救生服 446
救生服 502
救生服系统 869
救生浮 501
救生浮具 102
救生快艇 425
救生圈 501
救生设备(LSA)规则 502
救生设备 502
救生设备标志 550
救生设备布置图 502
救生设备培训手册 910
救生设备配置定额 736
救生设备使用准备状态 638

救生绳（舷沿救生索） 501
救生艇 501
救生艇 502
救生艇长 497
救生艇乘员定额 119
救生艇发动机 502
救生艇筏 876
救生艇筏操作须知 635
救生艇干舷 357
救生艇宽 95
救生艇立方容积 198
救生艇释放和回收系统 502
救生艇型深 223
救生艇用机器 87
救生艇属具 292
救生艇总重 906
救生衣 502
救生衣柜 502
救生衣架 502
救生用通信 161
救生载具 876
救生钟 737
救生属具 292
救助 736
救助 738
救助泵 761
救助船／救助船 737
救助单位 737
救助费用 761
救助分中心 737
救助服务 763
救助人 761
救助艇 737
救助艇的回收时间 725
救助拖船 737
救助拖船 761
救助协调中心 737
就地 447
就地控制 518
就地控制机构 518
就地控制站 518

居住便利用途 789
居住舱室 6
居住舱位置 6
居住处所 6
居住甲板/起居甲板 6
居住平台 7
居住塔楼 7
局部淬火 780
局部动应力幅值 518
局部腐蚀 518
局部腐蚀 519
局部故障后果 519
局部过大的熔深 518
局部回火 780
局部冷却(再冷却) 865
局部螺距 519
局部摩擦阻力 519
局部强度 519
局部退火 845
局部应力分量 519
局部载荷 519
局部振动 519
局部支撑构件 519
局域网 518
举升甲板 276
举升甲板 503
举证通知书 609
拒付 243
拒付证书 705
拒绝服务/DOS 攻击 222
拒识率 318
拒识率 604
拒收险 731
具体设计 840
具体要求(专门要求,特定要求) 841
具有端部肘板的主要构件的跨距 833
具有缓慢的火焰扩散的特性 408
具有较大失火危险的服务处所 788

具有较大失火危险的起居处所 6
具有较小失火危险的服务处所 788
具有较小失火危险的起居处所 7
具有了证书的 130
具有破冰能力 444
具有破冰能力船舶 801
具有油回收设备,但是不具有回油贮存舱及排放设备的浮油回收船 623
具有油回收设备,回收闪点高于60℃浮油的浮油回收船 623
具有油回收设备和回收油贮存舱及排放设备的浮油回收船 623
具有中等失火危险的辅机处所、货物处所、货油舱和其他油舱以及其他类似处所 61
具有中等失火危险的起居处所 7
剧变屈曲 938
距离定位法 568
距最近陆地 352
距最近陆地 591
锯屑 765
聚氨酯泡沫塑料 681
聚苯乙烯泡沫塑料板 681
聚芳酰胺纤维 45
聚合燃料 681
聚合物阻尼材料 681
聚磷酸铵 33
聚醚乙烯泡沫塑料 681
聚烟器(使用于抽烟式的探火系统) 825
聚酯纤维 681
卷放 727
卷筒 259
卷筒额定拉力 259
卷筒铺管驳船 727
卷筒铺管法 727
卷筒容绳量 662
决议 740
绝对传递率 2

绝对概率判断方法 2
绝对配额 2
绝对最大船舶速度 2
绝热系数 13
绝热系统 13
绝缘材料(隔热材料) 490
绝缘体 462
绝缘系统 462
军火 583
军舰 959
军用船 588
军用机器人 748
均压器 691
均匀沉浮方式 302
均匀流 934
均匀装载工况 430
均质材料 430
均质处理(均质化,均质化处理) 430
均质器(均化器) 430
菌型锚 583

K

卡口式快速接头 74
卡扣配合型连接器接头 74
卡住(咬住) 780
开舱 94
开槽 395
开槽 971
开槽焊缝(坡口焊缝) 395
开槽机(开沟机) 395
开槽机(开沟机) 395
开槽凿 395
开敞舱室 634
开敞处所(游艇) 632
开敞处所 632
开敞的车辆区域 632
开敞滚装处所 632
开敞甲板 631
开敞甲板处所 631
开敞上层建筑 632

开敞水域 147
开船旗 87
开底泥驳 431
开底泥舱 430
开顶集装箱 632
开动杆 285
开阀用扳手 973
开放缺陷显示 632
开工 159
开沟机 918
开关(按钮,扳手) 488
开关电器 878
开关设备和控制设备组件 878
开孔比值 633
开口(电气设备) 633
开口垫圈 960
开口端(艏、艉端) 631
开口滑车 825
开口螺栓（开尾螺栓,带开尾销螺栓） 844
开口销(扁销,销住) 351
开口销卸扣 352
开路供电系统(开式给水系统) 631
开坡口 395
开坡口焊(开槽焊) 395
开启力矩 633
开启型电机 634
开启压力 633
开氏温标(绝对温度) 488
开氏温标(绝对温度数) 488
开式给水系统 633
开式减摇水舱 632
开式冷却水系统 631
开式炉舱通风 632
开式索节 632
开式循环 631
开式循环燃气轮机装置 631
开式液位测量装置 632
开庭审理 641
开尾销（开口销） 844

开尾销(锚杆销) 352
开尾销 845
开尾销孔 845
开证申请人 42
开证行 633
勘定干舷 51
勘探 309
勘验工程 461
看火孔 659
康柏电脑公司 162
抗辩 614
抗沉性 462
抗腐能力 294
抗滑移稳性 38
抗挤压层 39
抗拉强度 889
抗磨层 41
抗磨硬度(抗磨力) 968
抗曝露服 39
抗倾覆稳性 40
抗热抗潮试验 417
抗湿性 437
抗谐鸣边 41
抗性消音器 739
抗摇设备(减横摇装置) 40
抗振工具钢 810
抗振强度（抗冲击强度） 810
抗振强度(振动强度) 951
抗振试验机 951
抗振橡胶 951
靠泊 247
靠泊作业 248
靠泊作业工作站/进坞工作站 248
靠舷索 355
苛性腐蚀 27
苛性钠运输船 125
科技、娱乐、设计 887
颗粒度 390
颗粒状炉渣 391
颗粒状轮胎橡胶 391
壳(或弯曲板)单元 793

壳管式冷凝器 793
壳管式冷凝器 794
壳管式冷却器 794
壳管式蒸发器 794
壳架等级电流 354
壳体振动(船体振动) 436
壳舾涂一体化 464
可比较的特征、货物和营运 162
可编程控制器 698
可编辑字段 269
可变桨叶角螺旋桨 13
可变力(交变力) 942
可变排气关闭 942
可变喷油定时 942
可变载荷 942
可变增益放大器 942
可剥离防护涂料 221
可拆联轴器 341
可拆卸零部件 524
可撤销信用证 744
可达性 5
可待因 152
可倒桅杆 154
可吊放救生筏 212
可吊放救生筏 236
可吊式气力输送机 877
可调变速电动机 13
可调螺距可逆转螺旋桨 942
可调螺距螺旋桨(变螺距螺旋桨) 942
可调螺距螺旋桨 13
可调螺距螺旋桨 184
可调面积小水线面双体船 47
可调排量泵 13
可动臂 580
可读性 721
可锻铸铁（韧性铸铁） 540
可锻铸铁 262
可锻铸铁件(马铁铸件,韧性铸件) 540
可反转发动机 744

可反转螺旋桨　743
可分离系泊　243
可分离转塔　243
可航行水域　588
可互换的　466
可及的范围　979
可接近(用于敷线的方法)　5
可接近(用于设备)　5
可接受的风险　4
可浸长度　345
可浸长度曲线　345
可卡因　151
可靠性　732
可靠性方块图　732
可靠性分析　732
可控被动式减摇水舱　184
可控被动水舱式减摇装置　184
可控浮筒　180
可控减摇鳍(主动式减摇鳍)　9
可控减摇鳍(主动式减摇鳍)　9
可控鳍(主动式鳍)　9
可控舵(减纵摇)鳍　9
可控水翼　184
可控约束阻尼层　184
可控制脱开　184
可控主动式减摇水舱　184
可流态化物质的适运水分极限(TML)　916
可流态货物　117
可能对健康造成危害的材料　555
可逆性　743
可逆转式发动机　743
可潜救生艇　866
可燃冰　158
可燃冰调查船　158
可燃气　340
可燃气体检测与报警系统　235
可燃气体探测报警系统　157
可燃气体探测器　158
可燃气体探测仪　158
可燃性(易燃性)　445

可燃性材料　158
可燃性极限　340
可燃液体　158
可热处理的　419
可容忍的风险　903
可申诉的补贴　8
可伸缩防撞装置　742
可伸缩式推进装置　742
可升降舵　503
可升降水舵　742
可升降转轮　49
可收水翼　742
可手动关闭　80
可听声　56
可维护保养能力　538
可维修性(使用可靠性,供给能力,操作性能)　789
可卸零部件　234
可卸零部件的安全工作载荷　756
可信度　170
可行性　319
可旋转导叶　942
可移动甲板　580
可移动式厕所　684
可移式泵　575
可移式泵　684
可移式风机　684
可移式灭火装置(手提式灭火器)　684
可移式消防泵机组　684
可移式压缩机组　684
可疑区域　876
可用度　62
可用配置　171
可用性　62
可再生能源法案　734
可重复使用的货物单元　743
可逐一识别　452
可转导管　673
可转静叶　13
可转栏杆柱　717

可转让信用证　913
可转上平台　486
可转上平台　923
可转推力器　742
克拉克溶解有机物采水器　139
克努森移液管　489
客驶　656
客舱　104
客舱阳台　104
客船　656
客船　656
客船安全证书　656
客船安全证书　757
客船船长决策支持系统　214
客船的舱壁甲板　101
客船上失水报警信号系统　325
客服中心(豪华邮轮)　400
客观原因　614
客观证据　614
客滚船　751
客户　147
客货船　656
客货船　656
客货船　656
客货船的载重线标志　514
客机反导系统　40
空白支票　631
空舱　955
空舱处所(留空处所)　954
空舱费　212
空车运行　355
空船排水量(平台)　505
空船排水量　503
空船排水量　505
空船排水量 LW　504
空船质量　505
空船质量测定　506
空船状态(空船质量)　505
空调机露点　24
空调室　24
空调系统(空调器)　19

空调装置(空调器)　19
空调装置(用)泵　19
空调装置　19
空负荷试验　937
空钩绳索　504
空间段　832
空间吸声体　832
空军前进引导员　352
空泡　126
空泡标准　126
空泡剥蚀　126
空泡核　613
空泡开始　126
空泡临界速度　195
空泡临界压力　195
空泡螺旋桨　126
空泡试验水筒　126
空泡数　126
空泡现象　126
空泡消失　234
空泡压力　126
空泡重溶装置　22
空气泵　22
空气舱柜　22
空气调节阀　19
空气调节装置　19
空气动力管(压力通风总管)　318
空气动力性噪声　20
空气动力噪声　16
空气阀　23
空气反应物质　22
空气方向舵　16
空气分离器　602
空气分配器　20
空气更新设备　22
空气管　21
空气管关闭装置　21
空气管式火警烟雾信号系统　678
空气管头　21
空气过滤器　20
空气加热器　21

空气兼溢流管　23
空气净化设备　22
空气净化系统　22
空气冷却器(冷风机)　19
空气冷却区　19
空气冷却式柴油机(风冷式柴油机)　19
空气滤器　22
空气滤清器(空气初滤器)　22
空气滤清消音器　20
空气滤清装置(空气过滤设备)　21
空气螺旋桨　22
空气螺旋桨船　25
空气泡　20
空气喷射式柴油机　21
空气瓶　19
空气瓶　22
空气瓶罐　19
空气启动阀　22
空气潜水　20
空气枪　21
空气枪激发控制器　21
空气枪振源　21
空气试验(渗漏试验)　19
空气室燃烧室　23
空气涡轮泵　23
空气污染指数　21
空气雾化喷油器　22
空气雾化喷油嘴　23
空气箱　965
空气压缩机　19
空气预热器　22
空气噪声　54
空气质量指数　22
空气轴承　19
空气阻力　22
空腔　126
空燃比(空气燃料比)　24
空射运载火箭/弹道导弹技术　24
空射运载火箭　24
空射运载火箭系统　24

空天飞机　16
空天飞机　832
空头支票　605
空投　24
空坞吃水　504
空坞排水量　503
空箱质量　884
空心板　429
空心活塞　429
空心螺钉　429
空心叶片　429
空心轴　429
空载(无负荷)　600
空载排水量　503
空载速度(惰速)　445
空中加油机　883
空中加油机　884
空中交通工具内部的噪声　469
空中预警机　23
空中预警与地面整合系统　23
空转　444
空转功率　444
空转轮(惰轮,中介齿轮)　444
孔(开度,打开状态)　633
孔(喷嘴,喷管)　642
孔板流量计　642
孔口　642
孔隙(孔隙度)　683
控告　446
控告人　446
控制　184
控制变压器　183
控制程序　787
控制船体受损程度方法　181
控制电路　180
控制阀　184
控制面　183
控制面板　182
控制器(动力定位系统)　184
控制器和测量系统(动力定位系统)　184

索引

1123

控制设备　180

控制设备　183

控制台　635

控制系统（操作系统）　637

控制系统　183

控制系统故障报警　183

控制轧制［正火轧制，CR（NR）］　184

控制站（平台）　183

控制站　183

控制装置/控制器　183

扣紧螺母　395

扣留　235

扣押　48

跨　74

跨步型艇架　582

跨大西洋贸易和伙伴关系协定　911

跨接　90

跨接管　487

跨境运输　912

跨距　833

跨区切换　404

跨声速压气机　916

跨索具　833

跨太平洋伙伴关系协定　916

跨太平洋贸易和投资伙伴关系协定　916

块截面积　86

块状装载工况　86

快递服务　190

快速备份　712

快速调节　712

快速攻击艇　318

快速换向试验　713

快速升负荷试验　712

快速泄水凹体　713

快速易损耗区域　519

快速自动往复传送系统　318

快艇　318

快艇　843

宽带 F 船站　342

宽度吃水比　94

宽频　973

宽频带频谱分析仪　70

宽频带噪声　97

宽稍叶　974

宽叶　973

旷海单浮筒系泊系统　311

矿/散/油船　641

矿砂船　641

矿物　570

矿物溶剂油　570

矿物油　570

矿渣棉　570

矿渣棉　821

框架集装箱　355

框架肋板　93

亏损　524

亏损舱容　97

馈电系统（供给系统）　320

馈线　320

扩大的保护水域航区　312

扩大危险区　311

扩管器（管辘，辘）　920

扩孔钻　309

扩口试验　309

扩流毂锥　179

扩频无线数据通信　845

扩散声场　237

扩压器　237

扩压因子　237

扩展条款　312

扩张器（扩管器,辘）　309

扩张型喷嘴　309

括号　653

L

垃圾(73/78 防污公约)　368

垃圾　960

垃圾处理系统　369

垃圾处理装置　369

垃圾处理装置　729

垃圾船　369

垃圾焚烧处理法　960

垃圾粉碎处理法　960

垃圾管理计划　369

垃圾管理计划　540

垃圾滑槽　369

垃圾滑槽　369

垃圾回收船　790

垃圾记录簿　369

垃圾污染　369

垃圾直接投弃法　960

拉铲式挖泥船　255

拉长石　490

拉丁美洲自由贸易协会　493

拉缸（划伤,齿轮咬结）　768

拉缸　672

拉筋　490

拉紧螺栓油压机　438

拉力极限　930

拉伸强度（极限应力）　889

拉伸试验　890

拉索（顺应式）塔　402

拉索塔平台　744

拉线测速装置　885

拉压刚度　889

喇叭口　580

喇叭式风斗（通风帽）　945

来访者　952

来回程租船　743

来源不明的物种　198

拦截弹　466

拦鱼池　336

拦柱间距　717

拦阻索　219

栏杆　717

栏杆插座　216

栏杆撑座　717

栏杆高　421

栏杆柱　403

栏杆柱　717

栏杆柱　717
蓝色的视觉信号　87
缆绳测力计　105
缆绳测力计　266
缆绳端点负荷(或公称负荷)　719
缆绳工具　750
缆绳缓冲器　102
缆绳计数器　190
缆绳卷车　978
缆索　408
浪　966
浪高仪　966
捞网　256
劳动生产率　489
劳动生产率指数　489
劳动仲裁申请书　42
劳工提供国　490
劳务补偿　490
老虎钳(钳紧,缺陷)　951
老化的结构　18
老鹰板　44
雷达(无线电方向和距离)　714
雷达　714
雷达　715
雷达标绘　714
雷达反隐身技术　714
雷达技术　714
雷达目标　714
雷达目标的捕获　8
雷达目标增强器　714
雷达平台　714
雷达室　714
雷达探测故障报警　714
雷达枪　714
雷达系统　714
雷达信标　714
雷诺数　744
肋板　345
肋骨　354
肋骨间距 a　354
肋骨冷弯机　354

肋骨冷弯机　698
肋片管（翅片管）　324
类似的废弃物　816
类似建造阶段　816
类油物质　627
累积谱　198
棱形系数　696
棱形系数曲线　199
冷备用系统　154
冷藏(制冷)　728
冷藏(制冷)　728
冷藏(制冷)　729
冷藏舱（冷藏处所）　729
冷藏舱　728
冷藏舱口盖　462
冷藏舱口盖　728
冷藏处所关闭报警　148
冷藏船　728
冷藏柜　728
冷藏货舱　728
冷藏货物运输船　727
冷藏货物运输船　728
冷藏机（冰箱）　729
冷藏集装箱　728
冷藏库　728
冷藏门　462
冷藏容量　728
冷藏设备(冷藏装置)　728
冷藏室（冷藏库）　728
冷藏室　728
冷藏系统管理计划　728
冷藏系统控制　729
冷藏装置附加标志　728
冷成形　154
冷吹运行　154
冷冻舱　358
冷冻机室　729
冷冻剂　358
冷冻艉拖网渔船　358
冷冻装置　358
冷海水保鲜运输船　728

冷库门（隔热门）　728
冷淋除垢　186
冷凝舱试验　169
冷凝器　169
冷凝器补偿器　169
冷凝器单位热负荷　841
冷凝器弹簧支座　169
冷凝器防摇滑动支座　169
冷凝器封头　169
冷凝器喉部　169
冷凝器壳体　169
冷凝器冷却面积　169
冷凝器流程　169
冷凝器密封性试验　169
冷凝器热负荷　169
冷凝器特性曲线　169
冷凝器通道　169
冷凝器真空补给水　313
冷凝器真空度　169
冷凝器真空空间　940
冷凝式汽轮机　169
冷凝收缩　358
冷凝水含盐量　761
冷凝水含氧量　648
冷凝压力　169
冷却倍率　138
冷却风扇　186
冷却货物　134
冷却水倍率　186
冷却水管　169
冷却水温度　186
冷却水系统　186
冷却水系统　961
冷却水压力　186
冷却叶片　185
冷却罩壳　186
冷却装置　186
冷热水管路　432
冷塞　154
冷态启动　154
冷态应急启动试验　277

冷态作业　186
冷饮用水机　258
厘米波　127
离岸公司　618
离岸公司注册地　674
离船港　683
离港工况　222
离港压载　222
离港蒸汽机船　644
离合　495
离合器　149
离合装置　285
离心泵　129
离心式风机　129
离心式压缩机　129
离心式制冷压缩机　129
理舱费在内的船上交货　349
理货(货物)服务处所　115
理论空气量　893
理论质量　893
理赔　138
理想流体　444
理想推进器推力　897
理想推进器效率　270
力和力矩传感器　350
力矩异步电动机　576
力学(机械学)　562
力学性能试验　562
立案流程　323
立根输道　919
立管　746
立管系统　746
立管转塔　746
立管转塔系泊　746
立柜式空气调节器　781
立式(柜式)　782
立式泵　938
立式泵　947
立式锅炉　946
立式横水管锅炉　946
立式横烟管锅炉　947

立式滑轮起网机　946
立式绞缆机　947
立式内燃机　946
立式镗床　946
立式压缩机　946
立式诱导器　947
立轴(垂直心轴)　947
立轴　938
立柱(稳柱)　155
立柱　156
立柱浮箱　155
立柱式平台(深水浮筒平台)　833
立柱式平台/深海箱式生产平台　833
立柱式平台　198
励磁机　305
利润/津贴　698
利息备付率　466
利益方　467
利益相关方　849
沥青玻璃纤维毡　51
沥青船　666
沥青矿棉毡　84
沥青球　673
沥青溶液　51
沥青型阻尼材料　51
沥青运输船　51
例外　302
例外情况　302
粒子群优化算法　655
连船梢涡　702
连动喷油器　374
连杆　171
连杆机构(联动装置)　509
连接(接合)　487
连接插头　171
连接传动装置　509
连接点　171
连接管　487
连接锚链(连接链环)　235
连接桥结构(小水线面双体船)

　196
连接卸扣　171
连锁(连锁装置)　467
连锁安全系统　468
连锁阀　467
连锁架(互联机构)　468
连锁开关　468
连锁气体阀　251
连锁气体阀　468
连锁装置　467
连体空泡　702
连同性　171
连续"B"级天花板或衬板　177
连续的进水　698
连续地层剖面仪　178
连续斗链　147
连续概要记录　178
连续工作制　179
连续构件　178
连续检验　178
连续进料　177
连续去污油分离机　178
连续缺陷显示　27
连续释放源　831
连续凸形液舱盖　179
连续油管　154
连续有人控制站　179
连续值班　179
联动脱钩(释放钩)　732
联动脱钩装置　817
联动轴　713
联动装置　467
联管节接头　935
联合国编号　935
　联合国毒品与犯罪署　935
联合国工业开发署　453
联合国关于气候变化框架公约　893
联合国关于气候变化框架公约哥哈根气候大会　933
联合国化学品全球标记和协调制度

931
联合国禁止非法贩运麻醉药品和精神药品公约 935
联合国气候变化框架公约 933
联合国危险品运输专家委员会 160
联合国行政、商业、运输电子数据交换规则 931
联合式压缩机 165
联合王国(英国)船舶 935
联合循环 156
联合运输单据 157
联合装置 156
联合作用的可能性 487
联合作战司令部 934
联机交换型UPS装置 508
联机系统 631
联结 509
联锁保护 467
联通管 195
联箱(集管,磁头,镦锻机) 416
联焰管 466
联运提单 897
联轴节 190
联轴节 791
联轴器 190
廉价出售/贱价抛售 629
镰刀型艇架 193
链斗 97
链斗式挖泥船 97
链斗式挖泥船 98
链斗挖泥船 98
链端卸扣 282
链接和(连杆连接) 509
链节 812
链径 131
链扣止链器 822
链轮(链轮铣刀) 846
链轮式天窗传动装置 131
链条滑车(链条绞辘,神仙葫芦) 563

链条卷条(链条卷轮) 846
良好海况区域 81
良好状况 389
梁单元 74
梁拱 106
梁拱线 106
梁肘板 74
粮食库 706
两舱制 924
两冲程发动机 924
两冲程发动机 924
两冲程内燃机 924
两冲程双作用式 924
两段招标 924
两级涡轮增压技术 924
两级增压 924
两截门 252
两路式冷凝器 924
两年度检验 81
两栖船 33
两栖调查船 33
两栖攻击舰/直升机登陆运输舰 33
两栖全地形战车 28
两栖挖泥船 33
两栖战舰 33
两栖状态 33
两用散货船 261
两爪锚 251
亮度 529
量吨长度 903
量吨甲板 903
量吨宽度 903
量吨深度 903
量化分析 711
量化宽松 710
量隙规(塞尺) 320
量油杆 622
量油杆 625
量油壶 622
量油计 627

量子 711
量子保密通信技术 712
量子比特 711
量子存储器 711
量子计算机 711
量子物理学 712
量子信息技术 711
聊天室/私人聊天室 133
瞭望(监视) 523
瞭望台 196
瞭望桅 614
裂缝(裂纹,裂口的,劈开) 844
裂化瓦斯油 192
裂纹 192
邻船吸力 592
邻近区域 13
邻频调制器 13
临界舵角 849
临界故障 195
临界结构区域 195
临界可靠性 195
临界空泡数 194
临界区域 194
临界设计工况 194
临界转矩 195
临时检验 615
临时锚泊设备 889
临时外来电源供电 889
临时证书 467
淋水系统 963
磷化氢 668
磷酸铵 577
磷酸二铵 236
磷酸盐(脱氟) 667
磷铁合金(包括砖块) 321
磷铁矿 484
灵便型船舶 404
灵便型散货船 584
灵便型油船(HANDYMAX) 404
灵敏度(敏感性) 785
灵敏度(敏感性) 785

灵敏度 785
灵敏系数 785
柃索 978
零排放港区 988
零升力攻角 36
零升力线 989
零速操纵性 989
零速回转试验 923
零泄油概率 696
领海(美国) 891
领海 891
领空 22
溜泥槽 388
溜泥槽起落装置 429
溜泥槽倾角 36
留存、生物体内积累和产生毒性 663
留存试验 663
留存样本(留存样品) 742
流(流通,流程) 346
流(通量,焊剂) 347
流 198
流程图(流图) 346
流刺网 344
流道结垢 346
流动时间 346
流动水分点 346
流动噪声 346
流动状态 346
流量管理 910
流量计(流量表) 346
流量可调液压泵 942
流量系数 346
流入液体(Q_i) 455
流水孔 255
流水孔 506
流送管 346
流速 199
流速 346
流态 346
流态临界速度 195

流体(液体,介质) 346
流体动力噪声源 600
流体切应力 793
流吸式排风帽 959
流线段 346
流线型舵 863
流线型舵轴舵 617
硫化腐蚀 869
硫化氢 440
硫化物(SO_x)排放控制 832
硫化物(SO_x)排放控制区 832
硫黄(成形、固体) 870
硫黄(碎块或粗粒) 870
硫酸 869
硫酸铵 33
硫酸钾 686
硫酸钾和硫酸镁 869
硫氧化物 870
馏分油 245
馏分油 245
六分仪后方交会定位法 568
六角扳手 422
六角螺母 422
六角螺母 614
六斜坡湖心特别保护区 820
龙骨 488
龙骨吃水 488
龙骨墩以上吃水 961
龙骨扶手 81
龙骨设计斜度 718
龙骨水平宽度 403
龙骨线 488
龙骨翼板 369
龙骨折角线 489
龙口 403
龙门刨床 252
龙门型吊架 368
龙须缆(链) 96
龙须缆/链顶点 96
垄断价格 13
漏出 494

漏风试验 494
漏气 371
漏气试验 371
漏水报警器 962
漏水显示器 962
漏泄(漏损,漏泄电阻) 494
漏液探测系统 495
撸 768
炉(炉膛,反应堆,熔炼) 366
炉板 119
炉箅摇动器 391
炉底(炉舱工作台) 335
炉排面积 391
炉墙 366
炉水废气预热器 307
炉水循环加速器 441
炉膛容积热负荷 418
炉膛烟气 346
炉条 326
炉渣棉 816
炉栅 391
卤代烃 403
卤化物灭火系统 403
陆基试验 491
陆上地面站 490
陆上风电场 631
陆上建筑物内部噪声 469
陆上交通工具内部的噪声 469
陆上排泥管 810
陆上试验 892
陆运险 647
陆运一切险 647
录像机 952
录音机 725
路标/指示牌 816
路由器 752
滤波器 323
滤水器 961
滤网(荧光屏,屏蔽) 767
滤网式舱底水油分离器 862
滤油器 621

滤油设备　621
滤油系统　621
露底　955
露点　236
露天保护(低温焊接)　968
露天保护　968
露天甲板(渔船)　631
露天甲板　968
露天甲板开口　685
露天甲板上的空气管头　21
露天梯　968
露天作业　968
旅客渡船　656
旅客携带的行李　657
旅游船　197
旅游船　306
旅游签证　906
铝箔　30
铝箔纸　30
铝合金　30
铝合金艇　30
铝熔炼副产品或铝再熔炼副产品　30
铝铁镍锰高级青铜　426
履带式布缆机　508
履带张紧器　908
履行承运人　660
履约备用信用证　660
律师　493
律师所代理人　55
绿环照灯　393
绿色船舶　393
绿色船舶规范　394
绿色的视觉信号　394
绿色环保技术　393
绿色建造　393
绿色设计　393
绿色营运　393
绿色照明　393
绿色状态　394
绿舷灯　394

绿舷灯　394
氯度滴定箱　136
氯化技术　136
氯化钾　686
氯甲烷制冷系统　568
卵石　658
掠海地效翼船　977
轮齿　374
轮齿　973
轮毂　434
轮毂比　903
轮毂高度　434
轮鼓　259
轮机班　285
轮机部　285
轮机部门　287
轮机长　134
轮机船级证书　530
轮机工程师(轮机员,船舶机械工程师)　544
轮机人员工作间　287
轮机日记　285
轮机入级附加标志　530
轮机特殊检验　837
轮机循环检验　178
轮机验船师　287
轮机员　287
轮机员报警系统　287
轮机员报警装置　288
轮机员呼叫铃（轮机员信号铃）　287
轮机员呼叫铃按钮　287
轮机员呼叫装置（机舱警报）　288
轮机员居住舱室　287
轮机员室　287
轮壳比　434
轮盘　242
轮胎粗碎块　149
轮系　973
轮箱（齿轮箱）　973
轮周效率　973

轮轴　845
罗茨鼓风机　750
罗经甲板　162
罗经自动操舵仪　60
罗兰C导航系统(LORAN)　521
罗兰C导航系统　524
罗兰D导航系统　524
罗兰导航系统　524
罗密欧与朱丽叶餐厅(豪华邮轮)　268
罗亚尔岛国家公园　485
螺钉(旋入,螺旋桨)　767
螺杆泵　768
螺杆传动操舵装置(螺杆式舵机)　768
螺杆式天窗传动装置　768
螺杆式压缩机　768
螺杆式制冷压缩机　768
螺杆艇　551
螺杆推进器　985
螺杆推进器　985
螺杆压缩机　767
螺环推进器　844
螺距(螺旋桨)　672
螺距比　673
螺距规　768
螺距角(桨叶安装角)　672
螺距控制杆　673
螺距样板　673
螺距因数　673
螺距整定杆　673
螺距指示发送器　673
螺距指示器　673
螺母(螺帽)　614
螺母扳手　614
螺母的紧锁装置　614
螺母垫圈　960
螺母丝锥(机用丝锥)　884
螺扇轮(叶轮)　941
螺栓　767
螺栓固定人孔盖　864

螺栓管水冷炉壁　864
螺丝刀（起子）　768
螺丝攻　768
螺丝攻　884
螺丝绞扳（扳手）　768
螺丝孔　767
螺丝眼　768
螺纹　768
螺纹　895
螺纹规　768
螺纹规　895
螺纹接合　768
螺纹接头　895
螺纹塞　895
螺线　844
螺线操纵试验　237
螺旋（螺线，螺旋的）　844
螺旋泵（蜗壳离心泵）　956
螺旋侧板　844
螺旋齿条　768
螺旋阀　768
螺旋虎钳　768
螺旋滑钩制动器　768
螺旋夹　767
螺旋夹扣　767
螺旋桨　768
螺旋桨-艉轴组合体　702
螺旋桨半径　717
螺旋桨测力仪　768
螺旋桨敞水试验　634
螺旋桨敞水效率　633
螺旋桨船　768
螺旋桨船后试验　80
螺旋桨船身效率　701
螺旋桨垂向位置　946
螺旋桨的侧斜角　820
螺旋桨的危险区域　282
螺旋桨飞车　701
螺旋桨功率（马力）　768
螺旋桨毂　767
螺旋桨横向位置　493

螺旋桨环流理论　138
螺旋桨基准线　701
螺旋桨激振　702
螺旋桨桨叶　767
螺旋桨桨叶最大倾斜角　558
螺旋桨浸深　446
螺旋桨理论设计　701
螺旋桨理想效率　444
螺旋桨螺距　673
螺旋桨螺距角　672
螺旋桨模型　575
螺旋桨盘　701
螺旋桨偏转效应　701
螺旋桨平面　701
螺旋桨设计图谱　701
螺旋桨式风机（轴流通风机）　768
螺旋桨特性曲线　132
螺旋桨推力　897
螺旋桨推力　920
螺旋桨尾流　701
螺旋桨艉隧　701
螺旋桨性能曲线　701
螺旋桨叶　768
螺旋桨叶元体理论　85
螺旋桨噪声　701
螺旋桨噪声频带声压级　70
螺旋桨噪声预测　701
螺旋桨直径　236
螺旋桨中点　129
螺旋桨轴（艉轴）　768
螺旋桨轴（旋转轴）　767
螺旋桨轴/艉轴　701
螺旋桨轴和艉轴管检验　701
螺旋桨轴和艉轴管检验　701
螺旋桨轴检验　768
螺旋桨轴系　701
螺旋桨轴状态监测系统　768
螺旋桨轴状态监控　701
螺旋桨轴状态监控　768
螺旋桨柱　701
螺旋桨转矩　904

螺旋桨转速　701
螺旋桨转速　701
螺旋桨转速指示器　701
螺旋桨装置　701
螺旋桨纵向位置　523
螺旋角　844
螺旋铆钉　768
螺旋面　421
螺旋弃链器　768
螺旋起重器（螺旋千斤顶）　768
螺旋起重器（螺旋千斤顶,螺旋夹）　486
螺旋式液压舵机　421
螺旋送料机　768
螺旋送料机　985
螺旋线　421
螺旋销卸扣　768
螺旋止链器　768
螺柱（链环横档,双头螺栓）　864
螺柱焊　864
螺柱焊枪　864
裸船模阻力试验　584
裸船体　71
裸船体排水质量　551
裸船体排水质量　969
裸船体阻力　71
裸机　71
裸潜　820
落锤撕裂试验　259
落地导向滑车　215
落地签证　491
落地式动力滑车　397
落地镗床　345
落管船　318

M

麻点腐蚀　673
麻醉品单一公约　817
马达油　579
马氏大抓力锚　426
码头　712

码头　890
码头代表　891
码头费　972
码头捐　247
码头收据　247
码头系泊　712
呜啡　578
埋管驳船　671
埋管船　103
埋管机　670
埋管机　671
埋弧焊　422
埋入式测温计　276
埋头螺钉　429
埋置人孔盖　870
霾(阴霾、灰霾)　416
买船前检验　690
买方供应的材料　708
买方信贷　103
买卖合同　760
买受人　943
买主　71
麦角酸二乙胺(LSD)　530
卖方信贷　873
卖主　71
脉冲(脉动,半周期)　707
脉冲宽度调制　707
脉冲强磁场实验装置　707
脉冲噪声　447
脉冲增压　707
满的(充分的,肥满线型)　362
满舵舵角　406
满溢　646
满载　690
满载吃水　217
满载吃水　363
满载吃水　363
满载吃水　516
满载吃水 d　513
满载航速　515
满载排水量 △(t)　515

满载排水量(地效翼船)　363
满载排水量(小水线面双体船)　363
满载排水量　363
满载排水量　363
满载排水量△　363
满载设计吃水 T_{full}　363
满载水线(内河船)　515
满载水线　515
满载水线长　515
满载状态　363
满载钻井作业负荷　257
满足最低宜居条件所需设备　789
慢跑道(豪华邮轮)　487
慢速倒车　823
慢速顺车　823
慢性释放　137
慢性水毒性　137
漫游　748
芒硝　761
盲板　86
盲区　86
猫道机　125
毛边/毛刺　341
毛细常数　109
毛毡　320
毛重　396
锚　34
锚　34
锚臂　34
锚臂长度　498
锚泊机械　34
锚泊设备　34
锚床　34
锚灯　35
锚浮标　34
锚杆　35
锚冠　34
锚横杆　35
锚环　35
锚机(起锚机)　976

锚机工作负载　976
锚机过载负荷　976
锚架　35
锚具　397
锚缆　104
锚缆　35
锚链　131
锚链　34
锚链舱　131
锚链管　131
锚链环　131
锚链轮　974
锚链破断试验载荷　95
锚链筒　408
锚链筒盖　98
锚链脱钩　785
锚链张紧器/链条张紧器　131
锚链转环　35
锚头　35
锚卸扣　35
锚穴　35
锚抓力　429
锚抓重比　35
锚爪　35
锚爪尖　35
锚爪袭角　36
锚爪折角　347
铆接船　747
贸易差额　67
贸易配额　909
贸易条件　891
贸易自由化　909
没有保护的开口　937
没有烹调设备的配餐室　652
眉毛板　314
煤　149
煤舱盖　149
煤泥　149
煤气发生炉　372
煤气机　370
煤烟灰(结炭)　828

煤油　488
煤油　488
煤油　652
煤油渗透试验　488
煤油涂检法　652
酶斯卡灵　564
每分钟转数　713
每分钟转数　744
每厘米吃水吨数曲线　199
每厘米纵倾力矩　576
每厘米纵倾力矩曲线　199
每月最低温度值（记录至今）　564
每月最高温度值（记录至今）　564
每转进程　14
美国　936
美国《科学引文索引》　766
美国标准化协会　30
美国材料与试验协会　53
美国超纯水协会　930
美国对外贸易定义　30
美国革命之子组织　828
美国国防部　936
美国国防部弹道导弹防御局　70
美国国防部高级研究计划局　208
美国国防情报局　219
美国海岸警卫队　935
美国海洋能源管理局　88
美国环境保护委员会/环境保护局　291
美国检验的船舶　929
美国检验的船舶　936
美国军用舰船统一排放标准　934
美国农业部动植物检疫服务中心　41
美国石油学会　31
美国水域　965
美国压载水情报交换所　585
美国仪表协会　485
美国仲裁协会　30
美军情报监视与侦察系统　159
美洲国家间毒品滥用管制委员会

　893
门板　250
门的紧固装置　777
门的净宽　147
门的锁紧装置　520
门顶缓冲器　250
门定位磁座　533
门槛　250
门槛价格　896
门框　250
门形桅杆　389
门形柱　389
锰钢　321
锰钢　540
锰含量　540
锰黄铜（高强度黄铜）　425
锰黄铜　540
锰矿砂（石）　540
锰青铜　540
锰铁（合金）　321
锰铁黄铜　540
迷宫式密封　490
迷宫式消音器　490
迷宫式压盖（曲径式压盖）　490
迷宫式音箱　490
迷幻剂　403
迷你好望角型散货船　570
米歇尔式轴承　568
秘书长　776
秘书处　776
密度　222
密封（焊封，堵塞）　773
密封（塞堵，闷头）　773
密封（填料函，图章）　772
密封材料　772
密封舱　773
密封的　494
密封的　772
密封垫圈　772
密封环　773
密封集装箱（密封容器）　773

密封集装箱　281
密封价格拍卖　692
密封接合　902
密封面　773
密封清管器　773
密封圈　650
密封试验　772
密封室的器具（LPG系统）　749
密封条　773
密封系统　773
密封压盖　773
密封元件　651
密封装置　772
密封装置　772
密封装置　773
密封装置　773
密封装置　902
密码　696
密性　902
密性试验　902
蜜尔沃基湖心特殊保护区　569
免除处所　305
免除处所　306
免除证书　306
免费　355
免费链接列表　355
免税　307
免疫税　713
免责　501
面（叶面，压力面）　315
面板　316
面板净厚度 t_f　593
面板宽度 b_f　316
面积速度值（AV）　47
面节线　316
面空泡　315
面螺距　316
面罩　316
瞄准点　19
灭火/消防　330
灭火　328

灭火剂(灭火介质) 330
灭火剂 313
灭火剂 330
灭火剂 334
灭火剂容器 329
灭火剂施放报警 329
灭火剂施放前的报警 689
灭火目标 2 330
灭火目标 1 329
灭火目标 330
灭火泡沫发生器 360
灭火器(熄灭器) 313
灭火器 329
灭火器 333
灭火枪 331
灭火沙箱 333
灭火系统(消防系统) 330
灭火系统(消防系统) 331
灭火系统 334
灭火系统效能试验 322
灭火站室通风 825
灭火装置(消防设备) 329
灭火装置(消防设备) 329
灭火装置 329
灭火装置 330
灭火装置室 335
灭音器 581
民用船舶 138
敏感的 785
敏感性分析 785
敏感性信息和资料 785
敏感元件(灵敏元件) 785
名称 584
名义干膜厚度 600
名义功率 601
名义宽度 601
名义应力 601
名誉主席 430
明轮 651
明轮船 651
明线 633

鸣振 746
模板 889
模块(组件,工作舱) 576
模块单元(可互换标准件) 576
模块化共轨燃油系统 576
模拟操作 816
模拟的/虚假的/模仿的 575
模拟机顶盒 34
模拟滤波器 817
模拟潜水 817
模拟通风管路 817
模拟信号 34
模式告知 575
模式意识 575
模数转换器 34
模压淬火(加压淬火) 690
模压外壳式断路器 237
膜传热系数 323
膜厚分布 323
膜式水冷壁 578
膜式压缩机 237
膜状沸腾 323
膜状冷凝 323
摩擦伴流 360
摩擦充电 132
摩擦副 360
摩擦滚筒 360
摩擦力 360
摩擦轮传动装置 360
摩擦式离合器 360
摩擦损失 360
摩擦头 360
摩擦系数 360
摩擦修正因数 360
摩擦硬化 360
摩擦阻力 360
摩擦阻力系数 152
摩擦阻尼 360
磨(磨光,磨碎) 395
磨耗 968
磨耗程度 968

磨合轴颈 985
磨痕(磨削裂纹) 395
磨孔 395
磨料(磨剂) 395
磨料 2
磨料嵌入物 2
磨轮保护镜片 395
磨石 395
磨碎机(磨床,砂轮机,磨工) 395
磨损(消耗) 967
磨损 985
磨损极限间隙 967
磨损间隙 967
磨损颗粒(磨粒) 968
磨损控制线 967
磨损量规 968
磨损试验 395
磨损许可量 967
磨损状况 967
磨削(打磨,研磨法) 395
抹油机 622
末端盖板 282
末端后果 282
末端弧坑裂纹 282
末端链环 282
末端再加热空气调节系统 891
末段高空区域防御系统 890
末叶片 324
沫状空泡 348
某一处所的渗透率 662
母线 386
母子式渔船 337
木材 902
木材甲板货 902
木材载重线 902
木材制品运输船 352
木船 981
木筏 520
木工车间 902
木工间 118
木工用具 118

索引

1133

木甲板　980

木浆球团　980

木马　919

木片　981

木球团　980

木薯淀粉　884

木丝板　981

木炭　132

木质舱盖　980

木质舱盖板　980

目标的预计运动　885

目标跟踪(TT)　884

目标跟踪雷达　884

目标交换　884

目标使用寿命　884

目标物种　884

目标型船舶建造　373

目标型新船建造标准　389

目标语言　884

目的地海关　201

目的港　683

目的港或地点　684

目的港支付运费　359

目录索引　242

目视检查(外观检查)　953

目视检查/外观检验　953

目视检验　815

苜蓿　27

N

纳米技术　584

耐波性　772

耐波性试验池　772

耐波性指标　194

耐腐蚀钢　188

耐腐蚀钢的标准　188

耐海水腐蚀钢　771

耐海水腐蚀试验　770

耐海洋大气腐蚀钢　544

耐航鉴定　774

耐火材料　332

耐火的(防火的)　340

耐火电缆　332

耐火分隔　332

耐火分隔　335

耐火合金　335

耐火救生绳　331

耐火聚酯树脂　332

耐火绝热材料　332

耐火绝缘　331

耐火绝缘　332

耐火门　332

耐火木材　332

耐火黏土/耐火泥　326

耐火黏土　132

耐火漆　332

耐火漆布　332

耐火试验　334

耐火试验程序　334

耐火完整性　331

耐火性　328

耐火性　332

耐火砖　326

耐火砖　332

耐磨机　395

耐燃性　340

耐热材料　418

耐热大肠杆菌　895

耐热的(抗热的,难熔的)　418

耐热钢　418

耐热合金　418

耐热合金钢　418

耐热漆　419

耐热试验(加热试验)　419

耐热性(耐热度)　418

耐热铸铁　418

耐蚀金属　602

耐蚀青铜　438

耐温的(耐热的)　889

耐压强度试验　691

耐油漆　624

耐油涂层　627

耐油性(防油)　622

耐油性　624

耐振的　951

耐振强度　810

南非的南部区域　831

南非国际贸易委员会　479

南极　38

南极调查船　38

南极区域(水域)　38

南极条约　918

南森采水器　584

南森开/闭式网　584

挠曲增大因数　534

挠性密封装置　342

挠性软管　342

挠性软管组件　342

挠性轴　342

挠性轴系　342

挠振　652

内(外)抗拉铠装层　459

内爆　446

内部浮力　456

内部检验　469

内部绝热液货舱　469

内部开口　469

内部空气泡沫系统　460

内部连接管　466

内部审核　468

内侧　459

内侧轴系　459

内场布线　455

内衬层　509

内齿轮　37

内搭接(内搭接边,内余面)　460

内底　459

内底板　459

内底边板　543

内底横骨　743

内底纵骨　459

内河船　459

内河推船　458

内甲板 469
内径 460
内径 469
内径规(塞规) 460
内径千分卡 460
内聚力 154
内壳 459
内控 469
内列板 459
内龙骨 488
内陆地区 647
内陆国 491
内螺纹 460
内螺纹 469
内螺纹管接头配件(阴螺纹管接头配件) 320
内螺旋桨 459
内牵索 456
内嵌的 592
内倾 921
内燃机 469
内燃机船 579
内燃机动力装置 621
内燃式燃气轮机装置 469
内损耗因子 205
内通道 467
内外搭接式 447
内效率 469
内斜轴系 247
内旋 481
内旋螺旋桨 481
内压护套层 469
内圆角(角焊缝,焊脚,填角料) 323
内在缺陷 457
内转塔 469
能发生自续分解的化肥 322
能耗强度指标 283
能见度不良 742
能谱 284
能效 283

能效标杆 284
能效方针 284
能效管理计划 283
能效管理体系 283
能效基准 284
能效目标 283
能效设计指数 283
能效数据 283
能效因素 283
能效营运指数 283
能效营运指数 284
能效指标 283
能引起额外失火危险的货物 113
能源 831
尼加拉瓜运河 597
尼龙(酰胺纤维) 614
尼龙/耐纶 681
尼龙索 614
尼斯金无菌采水器 599
尼斯金遥控采水器 734
泥泵 256
泥驳 431
泥驳 581
泥舱 845
泥舱容量曲线 199
泥舱溢流装置 646
泥管闸阀 373
泥浆泵 762
泥浆舱 581
泥浆净化平台 126
泥井 259
泥煤苔 658
泥门 262
泥门启闭装置 262
泥箱 580
逆火 341
逆流 743
逆流式冷凝器 640
逆螺线操纵试验 80
逆向反光材料 743
年度彻底检验 710

年度检验 37
年度审核 37
年度审核日期 37
年度死亡率 37
年度修理 37
年检中所要求的厚度 736
年盈利能力 37
黏弹性材料 952
黏度计 952
黏土 145
黏性流(黏性流体) 952
黏性摩擦 952
黏性试验 952
黏性物质 154
黏性系数 952
黏性阻力 952
黏性阻力系数 153
黏性阻尼 952
黏压阻力 952
黏压阻力系数 153
尿素 939
脲醛泡沫塑料 939
啮合 374
啮合 487
镍-铬钢 599
镍钢 599
镍钢管 599
镍铬钢 599
镍铬耐热合金 599
镍含量 599
镍合金钢 598
镍黄铜(镍铜锌合金) 598
镍基合金 598
镍铝青铜 598
镍锰钢 599
镍青铜 599
镍铁合金 321
镍铁蓄电池组 599
镍银合金 599
凝固的有毒液体物质 828
凝胶 374

凝聚剂 149
凝汽器 856
凝水-给水系统 169
凝水泵 169
凝水除油器 222
凝水观察柜 168
凝水过冷 169
凝水再循环管路 169
牛油杯(滑油杯) 392
牛油杯 392
牛油枪 392
牛脂填料 392
扭变 713
扭力测功器 904
扭锁 924
扭叶片 924
扭应力 904
扭振测量 904
扭振检查器 904
扭振减振器/扭振阻尼器 904
扭振频率 904
扭转损坏 904
扭转振动 904
扭转振动应力 904
农用船 19
浓度比 168
浓度环 746
暖菜柜 432
暖机(加热,加温) 959
暖机阀 959
暖机启动试验 959
暖机时间 959
暖机蒸汽管系 287
挪威船东与船厂协会 608
诺基亚公司 600

O

欧盟法令的编号 299
欧盟环境署 299
欧洲标准的音视频信号 766
欧洲船用设备指令 563

欧洲毒品和毒瘾监管中心 300
欧洲法院 299
欧洲风能协会 302
欧洲复兴开发银行 299
欧洲联盟 300
欧洲煤钢联营 299
欧洲投资银行 299
欧洲卫星导航系统(GALILEO)接收设备 722
欧洲卫星导航系统 367
欧洲委员会 299
欧洲物品编码协会 299
欧洲议会 300
欧洲原子能联营 299
欧洲自由贸易联盟 299
欧猪五国 684
偶发破损事故 6
偶极子 253
偶然事件 302
偶然性倾销 845
耦合 190
耦合器效率 190
耦合损耗因子 190

P

爬坡坡度 556
耙臂 255
耙头 255
耙头架 255
耙头架起落装置 429
耙吸式挖泥船 868
耙吸式挖泥船 910
拍卖 56
拍卖人 56
拍照手机 106
排/垛 753
排出温度 644
排出物 270
排除(免除) 358
排放(辐射,发射,放射) 280
排放 242

排放到专用接收装置 243
排放管 243
排放汇集管 242
排放控制 280
排放控制区 280
排风帽 939
排管机 670
排管机 920
排灌船 485
排灌运输船 485
排空 281
排空法 307
排空注入方法(压载水排放) 787
排缆装置 845
排泥管浮筒 256
排泥管活动接头 580
排气(乏气) 308
排气 308
排气冲程 308
排气阀 21
排气风扇(抽风机) 307
排气管(排出管) 308
排气管(排出管) 309
排气管(排气道) 308
排气管 307
排气管路 307
排气管系 308
排气机匣 309
排气口 308
排气门(排气阀) 308
排气门 307
排气歧管(排气阀箱) 307
排气器(抽风机) 308
排气室 307
排气室 307
排气通风 308
排气通风筒(抽风机) 308
排气凸轮 307
排气温度 307
排气温度 308
排气蜗壳 307

排气系统　308
排气线(示功图)　307
排气循环　307
排气压力(排出压力)　308
排气压力　242
排气再循环技术　307
排气装置　308
排气总管(废气集合器)　307
排气总管　308
排水布置　961
排水阀　961
排水管(逸气管,油舱的油气溢出管)　358
排水航态　436
排水航行机组　435
排水口　358
排水口装置(甲板排水系统)　768
排水量(船舶)　243
排水量(艇)　243
排水量　244
排水量数值图　441
排水量艇　244
排水模式　244
排水体积 V_D　244
排水舷口　358
排水型船　244
排水装置　561
排水装置　561
排水状态　244
排他许可证　827
排往舷外管路　645
排位(集装箱)　74
排污　86
排污阀　87
排泄管(废气管,污水管)　960
排烟(废气)　307
排烟管(烟管)　825
排烟热损失　418
排盐泵　87
排盐量　97
排盐喷射器　96

排盐热损失　97
排油阀　621
排油监测设备　621
排油监控(ODMC)操作手册　620
排油监控系统(排油监控装置)　620
排油监控系统　621
排油监控系统　621
排油监视系统　628
派人送达　785
攀岩墙(豪华邮轮)　748
盘　918
盘车(转车)　486
盘车　72
盘车电动机　72
盘车机　486
盘车机　486
盘车连锁装置　72
盘车试验　923
盘车装置(盘车机,回转装置)　923
盘管船　154
盘管冷却　154
盘面积　242
盘式联轴节　243
盘式塔　918
盘式消音器　243
盘线机　507
盘型高弹性离合器　23
旁底桁　91
旁路排气　103
旁轮　814
旁内龙骨　814
旁通调节　103
旁通阀　103
抛(喷)射处理　86
抛光(磨光,使发亮)　680
抛光　680
抛光工具　680
抛光机　680
抛光剂　680
抛光器　680

抛落式救生艇　355
抛锚艇　35
抛弃服　2
抛射绳　897
抛绳火箭　509
抛绳火箭筒　509
抛绳器　509
抛绳器具　502
抛绳枪　502
抛绳装置　509
抛石驳　861
炮钢　401
炮铜(锡锌青铜)　401
炮铜　401
泡沫　348
泡沫保温材料　349
泡沫玻璃　349
泡沫材料　349
泡沫发生器　349
泡沫发生设备　349
泡沫发生支管　349
泡沫浮力块　349
泡沫混合率　349
泡沫混合器　349
泡沫尖劈材料　348
泡沫灭火剂　348
泡沫灭火喷淋系统　349
泡沫灭火器　348
泡沫灭火器　348
泡沫灭火器　349
泡沫灭火器　360
泡沫灭火器　360
泡沫灭火系统　348
泡沫灭火系统　349
泡沫浓缩剂(泡沫灭火剂原液)　348
泡沫浓缩剂柜(泡沫液柜)　348
泡沫炮　349
泡沫喷枪　348
泡沫容器　348
泡沫溶液　349

泡沫生成液　349
泡沫水泥　349
泡沫塑料　349
泡沫橡胶　349
泡沫橡胶　349
泡沫液循环泵　168
泡沫原液泵　168
泡状空泡　97
培训　910
培训计划　911
培训手册（训练手册）　911
赔偿保证书　501
配备动力定位系统船舶的指南　401
配电板　878
配电板与配电屏　878
配电方式　246
配电中心　246
配电装置　877
配额/定额　713
配额管理　713
配额水平　712
配光曲线　504
配好的鸦片　689
配合公差　337
配气机构　940
配汽机构　856
配油箱　621
配载图　114
配制品（配置品）　689
配置数据　171
喷出（排出）　271
喷出动量　610
喷吹玻璃棉　87
喷管　270
喷管　487
喷管　610
喷管壁　610
喷管通道摩擦　260
喷火　487
喷口角（喷管倾角）　610

喷淋水（雾状水）　845
喷淋系统　963
喷流（射流）　487
喷墨打印机　458
喷漆　845
喷气燃料类　487
喷气式飞机　487
喷气推进　371
喷气推进　372
喷气推进船　24
喷洒灭火器　333
喷洒灭火系统　333
喷射（喷涂）　845
喷射(注射,注入,铸入)　458
喷射泵(喷油泵,高压油泵)　458
喷射泵（射流泵,抽气泵）　271
喷射泵　458
喷射处理和粗糙度　86
喷射定时　458
喷射管路　458
喷射气体柴油机推进系统　371
喷射器（排泄器）　269
喷射器（喷嘴喷油器）　458
喷射设备　458
喷射式冷却　487
喷射式汽化器　487
喷射提前　458
喷射推进　487
喷射推进器　487
喷射挖泥船　487
喷射压力　458
喷射滞后　458
喷射装置　458
喷水（喷水推进器）　961
喷水泵　846
喷水泵　961
喷水壶　963
喷水灭火系统　846
喷水灭火系统　963
喷水器（中心注管）　354
喷水器夹环　963

喷水器系统（喷淋系统）　846
喷水疏浚船　963
喷水推进　441
喷水推进　961
喷水推进　963
喷水推进船　441
喷水推进船　961
喷水推进器(喷水推进装置)　961
喷水推进器　961
喷水推进器　963
喷水推进系统　961
喷水推进系统和全回转推力器　964
喷水推进型船用燃气轮机　549
喷水推进装置　439
喷水推进装置　439
喷水推进装置　961
喷丸除锈/喷丸处理（喷砂处理）　395
喷丸设备（抛丸设备,喷丸处理）　395
喷雾（油雾）　845
喷雾泵　845
喷雾角　845
喷雾器　845
喷雾试验（雾化试验）　845
喷雾试验　349
喷雾系统　349
喷盐水试验　761
喷盐雾试验　761
喷盐雾试验　761
喷油　622
喷油泵柱塞偶件　678
喷油器（油头）　361
喷油器冷却器　458
喷油器头　458
喷油压力　458
喷油嘴　361
喷油嘴　458
喷油嘴　610
喷油嘴冷却泵　361

1138

喷注耗散型消音器　21
喷柱/喷水(射流,喷射器)　486
喷嘴(喷射器,喷吸器,弹射器)　271
喷嘴(消防水枪,熔嘴,喷水器)　610
喷嘴　458
喷嘴调节　610
喷嘴喉部截面面积　610
喷嘴环　610
喷嘴截面面积　610
喷嘴孔　610
喷嘴块(喷嘴组)　610
喷嘴失去作用试验　242
喷嘴室　610
喷嘴损失　610
喷嘴座　610
抨击　342
砰击　821
烹饪炊具室　185
膨胀(扩大,展开)　309
膨胀(扩张,胀开)　309
膨胀比　309
膨胀补偿　163
膨胀冲程　309
膨胀阀　309
膨胀阀　309
膨胀裂纹(膨胀开裂)　309
膨胀容积　929
膨胀温度计　309
膨胀箱(柜)　309
膨胀性(延伸性)　309
膨胀因数(膨胀系数)　309
膨胀珍珠岩　309
膨胀装置　309
碰垫　320
碰损　139
碰损、破碎险　139
批量生产　551
批量试验　73
批准　44

皮带张紧器　80
皮筏　820
皮肤过敏　820
皮重　884
疲劳极限状态　318
疲劳模式和结果分析　317
片(裂片,木片,劈开,扯裂)　844
片阀(舌阀,瓣阀,止回阀)　341
片式/百叶式消音器　676
片式填料　676
片状灰铸铁件　394
片状空泡　793
片状空泡　793
偏航角　987
偏航纠正能力　987
偏模直拖试验　614
偏位式减速齿轮箱　267
偏析　779
偏心半径　267
偏心冲击力　267
偏心度　267
偏心杆(蒸汽机)　267
偏心杆滑块　267
偏心滚轮　267
偏心滑轮　267
偏心环(偏心滑轮)　267
偏心环　267
偏心角　267
偏心距(偏心率,偏心度)　267
偏心轮(偏心的,离心的)　267
偏心轮　267
偏心轮机构　267
偏心起升装置　267
偏心运动　267
漂角　256
漂流瓶　256
漂流伞　652
漂心　127
漂心横向坐标　916
漂心纵向坐标曲线　199
飘尘　347

票汇　222
票务服务　899
票证　83
撇缆绳　419
拼装舱盖　503
贫油极限　967
频程　829
频繁的(事件的概率)　360
频率(频度)　359
频率分析器/谐波分析器　359
频率偏差　359
频率瞬变　359
频率瞬变恢复时间　359
频率选择表面　359
频率选择材料　359
频谱　359
频谱测量　843
频谱分析　843
频谱分析仪　359
频谱图　843
品质(等级,度)　390
品质检验证书　461
平安险　355
平板电脑　880
平板舵　818
平板龙骨　676
平板摩擦阻力　674
平板摩擦阻力系数　152
平板式玻璃水位表　342
平板叶　342
平板振动器　676
平舱　919
平舱的满载舱　323
平舱费在内的船上交货　349
平颤运动　677
平底船　342
平衡　292
平衡泵　292
平衡窗　190
平衡调制器　67

平衡舵 67
平衡舵 753
平衡阀 292
平衡阀 67
平衡滑阀 292
平衡滑阀 67
平衡活塞 262
平衡减压阀 67
平衡孔 67
平衡块减振器(液压弹簧式减振器) 439
平衡膨胀连接 292
平衡器(平衡装置,稳定器) 67
平衡式通风系统 67
平衡水袋 963
平衡装置 292
平滑阀 342
平甲板船 347
平接式 347
平静水域营运限制 106
平均吃水 560
平均个体风险率 62
平均横摇幅值 560
平均摩擦阻力 560
平均皮重 63
平均偏移 560
平均声压 62
平均速度 560
平均无故障时间 561
平均舷弧 62
平均泄油参数 560
平均修理所需时间 560
平均叶宽 560
平均叶宽比 560
平均有效压力 560
平均运费指数 62
平均张力 560
平均指示压力 560
平口虎钳 652
平流式调风器 652
平面 342

平面舱壁 674
平面分段流水线 51
平面声波 674
平面运动机构 674
平台 677
平台 935
平台操作手册 636
平台长 497
平台长度 $L(m)$ 499
平台的沉浮稳性 224
平台的风雨密 969
平台的破舱稳性 205
平台的水密 965
平台的特殊构件 838
平台的完整稳性 463
平台的舾装设备 292
平台的主要构件 693
平台的坐底稳性 630
平台供应船 677
平台供应船 706
平台甲板 677
平台结构的次要构件 774
平台结构构件分类 120
平台罗经 848
平台面积 47
平台设计载荷 228
平台升降负荷 514
平台升降海况 486
平台升降速度 486
平台拖航吃水 906
平台型深 223
平台重大改装 538
平行中体 652
平行中体长 652
平行轴-行星减速齿轮箱 653
平行轴传动减速齿轮箱 653
平旋推进器 202
平旋推进器船 955
平旋推进器节距 269
平旋推进器偏心点 267
平旋推进器叶角速 37

平移操纵试验 653
平置人孔盖(齐平人孔盖) 347
评估 301
评价体系 301
苹果操作系统 42
苹果平板电脑 484
苹果手机 484
屏蔽间处所 832
屏幕/显示屏 767
坡道 718
坡口半径 395
坡口焊接接头 395
坡口加工 395
坡口角度 395
坡口深度 395
迫降 246
迫降 903
破冰船 442
破冰船 444
破冰拖船 444
破冰型艏 444
破波阻力 967
破舱控制图和手册 204
破舱水线 345
破舱稳性 205
破舱稳性 205
破断载荷 95
破坏 154
破坏保安状况 154
破坏保安状态 855
破裂探测方法 354
破乳剂 222
破碎 97
破损控制甲板 204
破损控制手册 204
破损控制图和破损控制手册 204
破损稳性 847
铺管驳船 671
铺管船 670
铺管船 671
铺管船 671

铺管系统 671
铺间挡板 247
铺龙骨日期 212
铺路沥青 748
铺设管子 670
铺设塔 494
蒲氏风级 77
蒲式耳 103
普遍腐蚀 312
普遍腐蚀 375
普遍优惠制 385
普氏/法氏试验程序 697
普通船员 720
普通扶强材或主要支撑构件带板的净厚(mm) 908
普通扶强材或主要支撑构件的腹板高度(mm) 437
普通扶强材或主要支撑构件的腹板净厚(mm) 923
普通扶强材或主要支撑构件的跨距 l (m) 833
普通扶强材或主要支撑构件连同宽度 b_p 带板的净剖面模数 988
普通扶强材或主要支撑构件与板材连接处的净极惯性矩(cm^4) 481
普通扶强材或主要支撑构件与板材连接处的净剖面惯性矩(cm^4) 486
普通干货船 375
普通货船 375
普通链环 160
普通强度船体结构钢 436
普通强度船体结构钢 607
普通水泥 684
谱密度 843
蹼板 343
曝露区(露天区域) 311
曝露在货物中的材料 554

Q

七辊校平机 790
栖息地 403

期票 699
期望值 555
期租 902
其他表面涂料 643
其他材料和产品 643
其他产品补偿 643
其他干货舱 643
其他机器处所 643
其他机器处所和泵舱 643
其他控制站 643
其他设备 643
其他危险物质和物品(第9类危险货物) 574
其他物质 644
其他有关证据 643
其他载荷 643
脐带 931
骑浪和横甩第二层薄弱性标准 501
骑浪和横甩第一层薄弱性标准 501
骑浪运动 744
旗杆 339
旗钩 339
旗索 339
旗索滑车 86
旗箱 339
鳍板(减摇鳍,翼片,散热片) 323
鳍轴 325
气-水分离器 25
气-液换热器 512
气道(烟气通路) 371
气道控制系统 261
气垫 19
气垫长 199
气垫船 19
气垫船 20
气垫船 24
气垫船 433
气垫船的快速性 739
气垫船总功率 905

气垫宽 199
气垫面积 199
气垫推力 200
气垫系统 199
气动阀 21
气动阀 678
气动机械(风动机械) 678
气动离合器 678
气动螺母扳手 678
气动式造波机 678
气动挖泥船 678
气动液位计 678
气动装置 678
气阀 940
气封 371
气封 772
气封 772
气封 942
气封抽气器 387
气封冷凝器 387
气封平衡器 387
气封系统 922
气封压力调节器 387
气缸/机匣 119
气缸盖 202
气缸盖罩 202
气缸工作容积 672
气缸间隙 202
气缸控制单元 202
气缸润滑系统 202
气缸体 202
气缸油计量柜 202
气缸油贮存柜 202
气缸总容积 202
气割 339
气割 370
气割炬(气焊炬) 370
气罐(空气瓶) 19
气罐 371
气罐处所 371
气罐柜(LPG系统) 202

气罐箱(LPG 系统) 202
气焊 649
气焊焊剂 372
气焊炬喷嘴 372
气焊喷口 371
气焊设备 372
气焊条 372
气化油灶 941
气孔 371
气孔 370
气孔 683
气孔和气穴 683
气冷淬火 372
气冷式发动机 19
气力提升挖泥船 21
气力吸鱼泵 940
气流噪声 24
气门杆结胶 941
气门间隙 940
气门座积碳 941
气密/气密的 373
气密 372
气密的 372
气密的 942
气密缝 373
气密门 373
气密试验 22
气密性 372
气密性 373
气密性试验 372
气幕 19
气幕法 22
气凝胶 16
气刨 371
气泡 97
气泡船 25
气泡技术 97
气瓶室 369
气溶胶 16
气塞 21
气塞现象 942

气态的 372
气态燃料 372
气态碳氢化合物 372
气体(煤气,毒气) 369
气体安全处所 372
气体安全区域 373
气体保护电弧焊 369
气体保护电弧焊 373
气体保护焊 373
气体保护重力焊设备 373
气体爆震 371
气体常数 370
气体发生器 371
气体方程 370
气体防护 372
气体腐蚀 370
气体含量 370
气体激光器 371
气体夹杂物 371
气体检漏器(气体分析仪,气体检定器) 370
气体净化 370
气体净化 372
气体净化装置 372
气体侵蚀 371
气体驱逐装置 370
气体燃料技术 371
气体燃料总阀 552
气体渗氮淬火回火钢 371
气体渗氮合金钢 371
气体渗碳(充气渗碳) 369
气体探测报警 370
气体探测设备 370
气体探测与报警系统 370
气体退火 369
气体危险处所 370
气体危险处所 371
气体危险区域 370
气体危险区域 372
气体温度 372
气体系统 372

气体压力 372
气体压力调节阀 372
气体运输驳船 369
气体运输船 369
气体再液化 732
气体再液化装置 732
气隙(气穴) 683
气隙 21
气相 942
气象观测船 565
气象火箭 565
气象火箭备品室 565
气象火箭室 565
气象雷达 565
气象雷达观察室 565
气象情报室 565
气旋分离法 202
气压焊(加压气焊) 372
气压试验 678
气焰铜焊(气焰硬焊) 369
气闸 24
气闸室 25
气胀筏容具 940
气胀救生筏 455
气胀密封垫 455
气胀式救生筏 455
气胀式离合器 678
气胀式设备/充气式设备 455
弃船 2
弃船条款 2
弃船系统 2
弃船演习 2
弃锚器 105
汽车舱 943
汽车渡船 59
汽车运输船 59
汽车运输船 943
汽笛 308
汽笛 431
汽笛 856
汽笛 857

汽垫 856
汽封 857
汽缸套 202
汽缸油泵 202
汽耗率 856
汽耗率 857
汽耗率试验 857
汽化(蒸发) 301
汽化 372
汽化的(蒸发的) 302
汽化点 301
汽化沸腾 302
汽化热 941
汽轮给水泵 922
汽轮鼓风机(透平鼓风机) 922
汽轮货油泵滑油柜 922
汽轮机(蒸汽轮机) 857
汽轮机舱 857
汽轮机单缸试验 857
汽轮机的固定叶片 854
汽轮机动力装置 857
汽轮机油气抽除装置 857
汽密 857
汽密的 857
汽密接头（汽密接合） 857
汽密性 857
汽室 856
汽水分离器 857
汽水共腾 695
汽水自然分离 941
汽套 856
汽套 856
汽艇/摩托艇 493
汽油(瓦斯油) 371
汽油 372
汽油舱 372
汽油调和料类 372
汽油发动机（汽化器式发动机） 666
汽油机 372
汽油机船 373

汽油类 373
汽油箱 373
启动 853
启动按钮 853
启动程序 854
启动阀 854
启动功能 853
启动耗气量 853
启动空气分配器 853
启动空气瓶 853
启动空气瓶 853
启动空气系统 853
启动空气压力 853
启动空气总管 853
启动连锁 853
启动器（启动装置） 853
启动器(启动装置,启动机) 853
启动时间 854
启动试验 854
启动特性 853
启动系统 854
启动性能试验 853
启动性能试验 854
启动用泵 853
启动用空气压缩机 853
启动转矩 854
启动装置 853
启用欧洲标准的音视频信号输出 766
契约许可证 180
起、抛锚,拖带,供应 19
起、抛锚 19
起飞/着陆状态 881
起飞速度 881
起货滑车 110
起货机 117
起货机平台 974
起货机械 111
起货设备 112
起货设备检验 111
起货设备利用系数 111

起货索具 224
起货装置 111
起居、服务处所 6
起居处所 6
起开尾销器 845
起落绞车 429
起锚船 34
起锚机 35
起锚平均速度 62
起锚深度 419
起锚系缆机 35
起锚系缆绞盘 34
起抛锚/拖带/供应三用工作船 35
起泡 86
起泡剂(泡沫发生器) 360
起升联动杆 503
起艇机 88
起因 125
起重船/浮吊 343
起重船 192
起重船 343
起重船的船长 497
起重机/甲板吊 215
起重机 192
起重机或吊机 192
起重机械 429
起重机装卸 112
起重铺管船 503
起重设备 429
起重设备的安全工作载荷 756
起重生活平台 502
起重桅 224
起重柱 224
千斤-牵索吊杆装置 833
千斤调整索 904
千斤顶 767
千斤滑车 833
千斤链 833
千斤索 833
千斤索绞车 833
千斤索具 833

千斤索卷车　904
千斤眼板　833
千斤眼板座　833
千斤座　904
迁移工况　913
牵索　401
牵索滑车　401
牵索索具　402
牵索眼板　401
牵索柱　402
牵条螺栓　768
铅矿砂　494
铅皮（铅板）　793
签订合同　815
签订建造合同　180
签发许可证的国家　501
签署临时DOC的审核　56
签约　815
签证　129
签证　952
签证手续　952
签字　815
签字者（国际海事卫星组织）　815
前端壁　360
前端点（舱端）　353
前方交会定位法　568
前机舱　352
前肩　353
前联箱　360
前锚灯　855
前倾型艏　718
前体　350
前体　688
前桅灯　352
前桅杆　351
前踵　351
钳工台（钳工桌）　951
钳口开度　833
潜射弹道导弹　865
潜射核弹干发射　260
潜射核弹湿发射　972

潜射交互防御与进攻武器系统　444
潜水泵　866
潜水船　933
潜水电机　866
潜水工作驳/船　246
潜水工作船　247
潜水工作间　247
潜水母船　260
潜水泥泵　865
潜水器材舱　247
潜水器母船　866
潜水式号灯　866
潜水梯　246
潜水系统　247
潜水系统安全证书　247
潜水型外壳　703
潜水训练船　246
潜水用的中压空气系统　25
潜水鱼泵　865
潜水支持船　247
潜水钟　247
潜水钟　80
潜水钟吊放系统　247
潜水钟吊放系统　80
潜水钟控制系统　80
潜水钟主脐带　80
潜水作业　247
潜水作业母船　247
潜艇救生　865
潜艇螺旋桨噪声　865
潜艇艇员单人脱险　452
潜艇艇员集体脱险　865
潜艇艇员水下脱险　865
潜艇艇员脱险呼吸器　452
潜望镜　888
潜液泵　865
潜在人员伤亡　686
浅表硬化（表面硬化，表面淬火）　792
浅层剖面仪　792

浅层缺陷　792
浅吃水/空载吃水　504
浅吃水肥大型散货船　526
浅地层剖面测量　865
浅浸式水翼　792
浅水　792
浅水波　792
浅水船　792
浅水船模试验水池　792
浅水底吸力　91
浅水临界航速　195
浅水效应　792
浅水移动波　967
浅水增阻　792
欠电压/低电压保护　932
欠电压保护　932
欠电压释放　932
嵌接肘板　766
嵌入滑轮　99
嵌入式电力推进控制系统　460
嵌入式减速齿轮箱　467
枪(喷射器，注射器，焊枪)　401
强度计算用型深D_s　223
强放射性废料　426
强风暴(自存)工况　790
强固紧密接缝（强固紧密焊缝）　902
强肋骨　969
强力鼓风机　350
强力甲板　863
强力甲板的有效剖面积　270
强力可卡因　192
强力通风　350
强力通风　350
强力通风通道　350
强力循环锅炉　350
强迫振动　350
强胸横梁　652
强胸结构　652
强制横摇　350
强制空气循环　350

强制频率(强制振动频率)　350
强制扰动运动　350
强制通风　350
强制稳定　350
强制许可证　166
强制循环　350
强制循环泵　350
强制循环废气热交换器　350
强制循环锅炉　350
强制循环式润滑系统　350
强制预洗　540
抢修　336
敲铲油漆工具　767
敲缸　489
桥挡护板　131
桥楼　95
桥楼　96
桥楼驾驶值班报警系统　87
桥楼警报管理　70
桥楼警报管理系统　70
桥楼控制设备　96
桥楼上的集中监测控制站　128
桥楼上的主控制站　534
桥楼系统　96
桥楼翼台　96
桥楼主要功能　693
桥楼主要功能工作站　984
桥式起重机　96
翘度　789
切断装置　813
切断装置　813
切割机　201
切换　903
切口　608
切口应力(缺口应力)　608
切流式风机　881
切线器　104
侵蚀腐蚀　294
亲属　731
轻便式物品架　593
轻度冰况区域航行　443

轻吨　503
轻谷物　504
轻微后果　573
轻型舱壁　655
轻型吊杆　503
轻型舷窗　504
轻型修井作业　504
轻油　504
轻运行裕度　504
轻载航行吃水　504
氢弹　870
氢化氯氟烃　440
氢气瓶　440
氢气压缩机　440
氢氧吹管(氢氧割炬)　649
氢氧焊接　649
氢氧切割　649
倾倒　262
倾覆力臂　109
倾覆力矩　109
倾角器　978
倾斜/横倾　420
倾斜沉浮方式　420
倾斜角(纵斜角)　718
倾斜力矩　449
倾斜试验　449
倾斜试验重物证书　131
倾斜仪　147
倾斜桩腿自升式钻井平台　821
清舱管路　864
清除　708
清除垃圾　369
清楚的/清晰的　309
清管器　920
清洁剂　146
清洁水储存柜　146
清洁提单　145
清洁添加剂　146
清洁系数　146
清洁泄放水　146
清洁压载　145

清洁压载水　145
清洁压载水舱　146
清洁用具　146
清扫船　346
清扫船　877
清试洗验　922
清洗口　103
清洗周期　146
清洗装置　960
情报中心　156
请求人　138
求交算法　480
球鼻艏　99
球扁钢　99
球阀　67
球朊型剖面　99
球化退火　844
球铰链　844
球面声波　844
球面弯头活管接　844
球墨铸铁　599
球墨铸铁　844
球墨铸铁件　599
球墨铸铁件　844
球形阀　389
球形阀　844
球形铰链　844
球形蜗杆　389
球形止推轴承　67
球形轴承(球面轴承)　844
球轴颈　67
球状磨损粒子　844
球状石墨　599
区配电板　776
区域　47
区域编址识别功能　989
区域规划　729
区域内活动　10
区域危险分析(ZHA)　989
区域性经济贸易集团内的价格　729

区域再加热空气调节系统　989
曲柄　193
曲柄臂（冲程,投掷）　897
曲柄臂　193
曲柄箱透气管路　193
曲柄销　193
曲径密封（迷宫式密封）　490
曲径式密封圈　490
曲径式气封　490
曲径填料箱　490
曲轴　193
曲轴臂距差　193
曲轴箱　192
曲轴箱爆炸　193
曲轴箱防爆门　193
曲轴箱扫气　193
曲轴箱透气装置　193
驱气（电气设备）　708
驱气　708
驱绳绞车（牵引绞车）　909
屈服点（流动点）　988
屈服点　988
屈服点伸长　988
屈服点应变　988
屈服强度　988
屈服损坏　988
屈服现象　988
屈服性的(流动性的,可压缩性的)　988
屈服应力(强度)　988
屈服载荷　988
屈强比　988
屈曲　98
屈曲的蔓延　700
屈曲或破坏的局部化　519
屈曲强度　98
屈曲失效模式　99
取缔非法贩运危险毒品公约　184
取键工具　488
取暖、风和空调系统　419
取样（采样,脉冲调制）　762

取样程序　762
取样点（压载水管系）　762
取样点（污水处理装置）　762
取样方法　762
取样器　762
取样设备（取样设施）　762
去除锅炉的水垢　766
去流段　754
去流段长　498
去氢处理　441
去湿装置　576
全（周）进气透平　362
全包运费　359
全部大写的　27
全部损失　905
全齿抓斗　973
全充气螺旋桨　364
全船燃料消耗率　27
全船通风　436
全磁控制器　364
全电压启动器　28
全垫升气垫船　19
全垫升气垫船　33
全垫升气垫船　434
全反射材料（救生服）　743
全方位推进装置　629
全方位推进装置　64
全方位推进装置　989
全封闭扇冷型电机　906
全封闭式制冷压缩机　422
全封闭式制冷压缩机组　422
全封闭型电机　906
全高清　363
全工况燃气轮机　536
全功率　363
全功率试航（全功率试车）　363
全滑行模式或非排水模式　734
全回转导管推力器　64
全回转港作拖船　261
全回转起重船　364
全回转式螺旋推进器　751

全回转推进器（Z推,L推）　64
全回转推进系统　28
全回转推力器　64
全回转拖船　27
全集装箱船（巴拿马运河）　363
全甲板艇　364
全浸深水翼　364
全浸式水翼　364
全空泡　364
全空泡螺旋桨　364
全冷式液化气船　364
全面腐蚀/平均腐蚀　375
全面检查　164
全面检查　645
全面检查　895
全面效能因数　645
全盘一体化　645
全球变暖潜能值　389
全球定位系统　388
全球海上遇险和安全系统（GMDSS）标识　388
全球海事环境大会　389
全球海事宽带通信网络　97
全球配额　388
全球气候变暖　389
全球统一标识系统　34
全球卫星导航系统/俄罗斯格洛纳斯卫星导航系统　388
全球卫星定位系统（GPS）卫星导航仪　390
全球无线电导航系统　388
全球无线电导航系统　985
全球移动通信系统　389
全燃联合装置　156
全冗余　363
全上层建筑　363
全速　363
全速倒车　363
全速倒车试验　363
全速倒车停船惯性试验　363
全速试航　363

全速运转状态(全负荷运转状态) 363
全速正车 362
全速正车试验 362
全损赔偿 355
全损险 905
全天候飞机 28
全通气水翼 871
全围蔽处所 906
全围蔽的桥楼 906
全文索引 363
全向推进器 28
全向推力器(方位推力器,可转向推力器) 64
全向推力器 629
全压[满压(锅炉)] 363
全压式液化气船 364
全自动焊接工艺 364
缺口冲击试验 608
缺陷(故障,损伤,断层) 319
缺陷 218
缺陷的种类 928
缺陷种类 218
确定 235
确定的事故载荷 239
确立应急准备 298
确认 171
确认书 171
确认性审核 171
群岛 46
群岛国 46
群岛海道通过 46
群呼 149

R

燃点 340
燃料(燃油) 361
燃料爆震 361
燃料舱(燃油舱) 362
燃料舱 102
燃料舱 361

燃料的成分 361
燃料电池(燃料舱,燃料箱) 361
燃料电池 361
燃料电池电力推进装置 361
燃料电池推进系统 361
燃料电池组 361
燃料供应泵 362
燃料柜舱室 882
燃料气体(气体燃料) 361
燃料树脂 362
燃料税(碳排放税) 102
燃料条款 102
燃料消耗量 361
燃料消耗量试验 361
燃气发生炉 372
燃气发生器 371
燃气分离器 372
燃气高温保护装置 423
燃气功率 371
燃气鼓风机 370
燃气管 371
燃气混合物 371
燃气加热器 371
燃气轮机 372
燃气轮机舱 372
燃气轮机船 372
燃气轮机动力装置 372
燃气轮机动力装置 373
燃气轮机工作线 372
燃气轮机交流发电机组 372
燃气轮机交流发电机组 372
燃气轮机驱动交流发电机组 372
燃气轮机燃料 922
燃气轮机组装体 372
燃气密度 370
燃气喷射压缩机 65
燃气燃烧装置 370
燃气室 370
燃气收集器 370
燃气系统控制室 370
燃烧 158

燃烧 158
燃烧 455
燃烧臂 103
燃烧产物 698
燃烧范围 158
燃烧热 418
燃烧上限 938
燃烧室 158
燃烧室 158
燃烧室 326
燃烧室调节比 923
燃烧室热容强度 840
燃烧室外壳 158
燃烧效率 158
燃烧噪声 158
燃烧自动调节装置 57
燃油/油类燃料 621
燃油 361
燃油泵 362
燃油泵 362
燃油泵 621
燃油泵组 621
燃油驳运泵 362
燃油舱 362
燃油舱 621
燃油舱保护 362
燃油舱保护 362
燃油舱透气管 625
燃油沉淀 362
燃油沉淀柜 362
燃油传输作业 362
燃油单元 620
燃油的使用限制 507
燃油阀 362
燃油阀 362
燃油分析 361
燃油分油器 362
燃油辅助锅炉 627
燃油管 362
燃油管 362
燃油管理程序 362

燃油管路　361
燃油管系　362
燃油柜取出舱口　362
燃油锅炉　334
燃油锅炉　621
燃油锅炉　627
燃油控制器　361
燃油炉灶　622
燃油滤器　362
燃油喷射　361
燃油喷射　458
燃油喷射泵　361
燃油喷雾阀　362
燃油喷嘴　610
燃油燃烧系统　621
燃油燃烧装置　362
燃油日用柜　362
燃油容量　621
燃油深舱　362
燃油雾化　361
燃油系统　362
燃油消耗率　841
燃油泄油柜　362
燃油泄油柜　362
燃油溢油舱　646
燃油油渣柜　362
燃油增压泵　362
燃油质量设定　362
燃油中钒含量　941
燃油贮存柜　362
燃油转换操作　361
燃油装载　361
燃油装置　620
燃油装置　621
扰动力/激振力　305
扰动力矩/(激动力矩)　305
绕线转子式感应电动机　985
热备用(动力定位系统)　433
热备用系统　433
热变形温度　417
热处理　894

热处理钢　419
热处理炉　419
热处理炉　419
热处理温度　419
热处理状态　419
热传导　419
热传感器　418
热带淡水　920
热带干舷　920
热带气旋　920
热当量　293
热点　432
热点应力　432
热电高温计　894
热电偶　894
热电式空气调节器　894
热电制冷装置(温差电制冷装置)　894
热电制冷装置　783
热电制冷装置　894
热镀锌(热电镀)　432
热辐射　418
热腐蚀　417
热腐蚀　432
热腐蚀试验　3
热感应式喷墨技术　893
热功当量　562
热固的(热硬的)　895
热固化(热塑化,热硫化)　417
热固化树脂试样　762
热固树胶　432
热固性材料　895
热固性树脂　895
热固性塑料　895
热耗率　417
热化学试验　856
热货　432
热机　418
热机械控制轧制 TM　894
热机轧制　894
热加工　419

热交换　304
热交换　418
热交换器　418
热浸镀锌法　432
热浸镀锌钢丝　432
热浸法　432
热浸涂镀　432
热井　433
热绝缘试验　418
热空气加热(热空气供暖)　432
热力工程　418
热力膨胀阀　895
热量单位　419
热量发生器　418
热裂纹/热裂　417
热流量　418
热敏的　419
热敏涂料　419
热敏性(热灵敏度)　418
热挠曲温度　417
热能　418
热凝树脂　432
热排气融霜　432
热喷涂　724
热喷丸　432
热膨胀　893
热疲劳　418
热平衡　417
热平衡　893
热平衡计算　417
热平衡试验　417
热平衡试验　893
热气焊　432
热气通道　432
热球柴油机　433
热容量　417
热时效　417
热输入　418
热水单元　433
热水供暖(供热设备)　433
热水供暖系统　433

热水锅炉　433
热水加热器　433
热水瓶架　895
热水套　433
热水循环泵　419
热水循环泵　433
热水循环管路　433
热水循环回流管　433
热塑性塑料　894
热态启动　433
热态应急启动试验　278
热态作业　433
热套　419
热套曲轴　813
热弯　432
热线圈　417
热线图　417
热效率　417
热性能　418
热循环　417
热压条件　418
热盐水融霜　432
热应变(救生服)　419
热影响区　417
热油淬火　432
热油管路　894
热油锅炉　893
热油加热器　893
热油加热器　894
热油加热器和热水加热器　893
热油加热系统　894
热油设备　894
热源　418
热运行试验　418
热轧无缝钢管　432
热轧无缝钢管　432
热值　419
热值　419
热阻(R)　894
热作　433
热作安全　757

人工(机械)通风　49
人工操舵工作站　542
人工磁场电磁海流计　274
人工岛　49
人工调节　403
人工复原(人工重调)　403
人工海草　49
人工去污油分离机　404
人工神经网络　49
人工生物角膜　49
人工时效硬化处理　18
人机工程学(人类工程学)　293
人机界面　428
人机界面　437
人孔　540
人孔盖　541
人力操舵装置　404
人力操纵的(人工驱动的,手动的)　404
人力船　541
人力机械推进救生艇　542
人力抛投　493
人力起锚机　404
人力起锚绞盘　403
人力液压操舵装置　404
人体毒瘾　668
人体曝露情景　437
人为可靠性　437
人为失误　437
人为失误概率　437
人为失误恢复　437
人为失误率预测技术　887
人为失误评估和减小技术　437
人为因素　436
人为因素　437
人因可靠性分析　437
人因可靠性评估的社会技术方法　826
人员　663
人员　663
人员安全　666

人员保护目标1　665
人员保护目标2　666
人员保护目标　666
人员报警　665
人员救助目标1　737
人员救助目标2　737
人员救助目标3　737
人员救助目标　737
人员落水模式　541
人员危险目标　665
人造卫星发射的弹道导弹　764
人造橡胶　49
人字架　1
人字桅　84
认可(集装箱)　44
认可/验收(材料)　44
认可　44
认可　44
认可保持　538
认可的机构　724
认可方法　44
认可方法案卷　44
认可生效日期　44
认证　131
任何一点的 x 方向加速度(m/s^2)　63
任何一点的 y 方向加速度(m/s^2)　64
任何一点的 z 方向加速度(m/s^2)　64
任何一种水声位置基准　434
任务分析　885
任务站　885
韧性　262
日本电气公司　599
日本工业型材标准　487
日本国际商事仲裁协会　472
日本造船工业协会　807
日常维护检修　636
日光浴室(豪华邮轮)　827
日用泵(通用泵,杂用泵)　788

日用电话通信系统　788
日用发电机　787
日用发电机组　787
日用燃油柜　362
日用油柜　203
冗余　726
冗余度(多余,冗余,重复)　727
冗余度　727
冗余度试验程序　727
冗余风险控制　727
冗余率　727
冗余设计　727
冗余设计　727
容错　319
容积泵　686
容积负荷　941
容积曲线　108
容积图　108
容积型滚装货船　955
容量与总成本依赖关系函数[f(v)]　956
容许范围　903
溶剂　828
溶解时间　828
溶液(解决,解释)　828
熔断器　367
熔断器的指示器　451
熔断体　367
熔敷线　367
熔化极气体保护焊(金属极气体保护电弧焊)　371
熔化热　418
熔体　367
熔盐水箱　97
融霜　219
融霜贮液器　219
融资备用信用证　325
柔性构件　342
柔性管　342
柔性铺管　342
柔性铺管船　342

柔性塔　164
柔性塔架　164
柔性制造系统　342
乳化(乳状)　281
乳化油　281
入级　142
入级船舶　141
入级船舶　145
入级规范　144
入级活动　143
入级检验　140
入级检验　145
入境签证　289
入口(开始,进口,入港手续)　289
入门费　253
入侵水生物种　480
入水角　963
入水楔形体积　446
软舱　826
软管(LPG系统)　432
软管(挠性管,消防水带)　431
软管车(水龙带小车)　431
软管加长节　431
软管架(消防水龙管架)　431
软管接头　431
软管连接阀　431
软管连头　431
软化(漏气,软水剂)　826
软化剂　826
软件　826
软件注册　826
软木　826
软驱　346
软水　826
软涂层　826
瑞典斯德哥尔摩商会仲裁院　45
瑞利分布　720
瑞士苏黎世商会仲裁院　989
润滑(润滑法)　529
润滑　528
润滑　528

润滑故障　529
润滑环　529
润滑剂(浸渍润滑剂)　528
润滑系统　529
润滑性(含油性)　529
润滑性质　529
润滑油(滑油)　528
润滑油(滑油)　528
润滑油管　529
润滑油和调和油料　529
润滑油密性试验　529
润滑油试验　529
润滑油系统　622
润滑油油品理化分析　669
润滑脂　529
润磨油　395
弱非线性模型　967
弱耦合　967

S

撒手铜　51
洒水器(喷水器)　846
塞孔(船底放水孔)　678
塞栓(插头,填充)　678
塞填料的杆　651
塞住　897
赛艇　713
三舱制　896
三层分组交换技术　896
三层交换技术　896
三岛式船　896
三点式滑行艇　896
三段式浮船坞　896
三锅筒锅炉　919
三锅筒水管锅炉　987
三机轴并车减速齿轮箱　726
三夹板　896
三角槽皮带轮　940
三角分线杆(Y形交叉,Y形接头)　987
三角刮刀　766

三角浪 709
三角皮带 943
三角皮带盘 943
三角桅 919
三角眼板 919
三脚式桩基 919
三脚桩基础 896
三氯乙烯蒸发器 919
三色艉舷灯（三色舷灯） 919
三通阀（三路阀） 896
三通阀(T 型阀) 880
三通管(Y 形接管) 987
三通管接头 888
三通旋塞 896
三维波浪载荷计算软件系统 162
三维地震物探船 990
三维图形 896
三相电路 896
三相短路电流 896
三硝基甲苯 919
三芯辊床 896
三一横摇幅值 815
三一平均波高 815
三用散货船 919
三用拖船 921
三原色光模式 744
三胀式蒸汽机 919
三重 DP 919
三重比特 918
三轴并列减速齿轮箱 919
三柱体式平台 895
伞兵 653
伞兵人车一体式空投技术 464
伞兵重装空投技术 653
散波 247
散波角 36
散货/包装 390
散货/硫酸运输船 101
散货 99
散货船 99
散货船舱壁和双层底强度标准 100
散货船的船长 497
散货船的横剖面 917
散货船宽度 94
散货船压载舱 68
散货集装箱 99
散热的 418
散热量 417
散热器 715
散热器隔振垫 715
散热设备 417
散热水箱冷却 715
散热损失 418
散热损失 714
散热温度 309
散热值 309
散射现象 766
散装舱容(谷物舱容) 390
散装干货物残余物 101
散装谷物积载设备 390
散装谷物运输船 101
散装化学品规则 101
散装货船协调附加标志 BC-A 74
散装货船协调附加标志 BC-B 74
散装货船协调附加标志 BC-C 74
散装货物密度 101
散装泥浆舱 101
散装水泥舱 101
散装水泥运输船 127
散装危险化学品规则 151
散装危险化学品运输船 207
散装危险货物 207
散装液化气体规则 151
散装运输国际散装运输危险化学品船舶构造和设备规则(IBC 规则) 893
散装运输危险化学品适装证书 129
散装运输液化气体适装证书 130
桑德湾国家海上避难所 898
桑拿房 765
扫舱 863
扫舱泵 864
扫舱量 864
扫舱系统（残油扫舱系统） 864
扫舱效率 864
扫舱装置 863
扫海测量 978
扫海测量船 877
扫海船 877
扫海绞车 877
扫海具 877
扫海趟 877
扫海体积 877
扫掠式喷射 86
扫描仪 766
扫气（换气,清除） 766
扫气（换气,清扫） 766
扫气泵（换气泵） 766
扫气泵扫气 87
扫气风机 766
扫气鼓风机 766
扫气管 766
扫气阶段 766
扫气孔 766
扫气系数 766
扫气箱 766
扫气箱释放装置 766
扫气压力 766
扫气装置（换气管） 766
扫气总管 766
扫尾工程 325
色表 155
色差 988
色差端子 164
杀菌设备(自封存装置) 859
纱板两用门 767
砂 762
砂砾(颗粒,砂粒,金属屑) 395
砂石撒铺船 762
砂箱 762
砂眼 762

砂桩打桩船 345
山字钩 252
舢板 761
闪点 315
闪点 341
闪点 341
闪点测定器 341
闪点试验 341
闪发室 341
闪发原料油 341
闪发蒸发 341
闪光(通信)灯控制箱 341
闪光 341
闪光灯 341
闪光灯控制箱 815
扇面形刻度盘 318
扇形齿轮 374
扇形齿轮 710
扇形齿条 779
扇形孔 766
扇形块推力轴承 431
扇形体 778
扇形叶 318
商标 909
商船 159
商品 48
商品倾销 262
商品税 160
商数 713
商务签证 103
商务人员 103
商业本票 159
商业单据托收统一规则 934
商业发票 159
商业汇票 159
商业拖航 159
商用货船 564
熵 289
上部甲板至下端的距离 245
上层建筑 873
上层建筑标准高度 851

上层建筑长度 498
上层建筑高度 H 421
上层建筑或其他建筑物的高度 421
上层建筑容积 V_s 955
上层建筑上的开口 633
上层建筑有效 873
上层连续甲板 938
上层中间甲板 938
上船港 683
上船体(小水线面双体船) 938
上床铺试验 938
上挡转换键 794
上导轮 938
上导轮中心高 127
上吊货滑车 938
上舵承 753
上舵销 938
上翻铰链式门 952
上浮时间 903
上纲起锚联合绞机 156
上弓段 645
上锅筒(汽包,汽鼓) 856
上海船舶工艺研究所 792
上海船舶研究设计院——中国民用船舶设计领域"领航者" 792
上甲板 938
上紧 902
上壳体 938
上浪 216
上排污 874
上排污阀 768
上排污阀 874
上排污旋塞 874
上平台 938
上千斤滑车 224
上千斤滑车 551
上升管 746
上死点 904
上死点 938
上游产业 939

上止点 904
尚好的涂层状况 317
梢厚 85
梢涡 903
梢隙 85
哨笛(个人漂浮装置) 973
赊销 193
舌门 718
蛇形造波机 825
舍入误差 752
设备(装置,工厂,车间) 675
设备(装置,仪表) 42
设备 316
设备的独立工作 450
设备的结构噪声 864
设备的空气噪声 21
设备等级 292
设备互换性 292
设备间的相互独立 584
设备零件明细表 292
设定(安置,调整,号料) 789
设定压力 790
设计 225
设计波高 H_{dw} 234
设计波浪周期 T_{dw} 234
设计波浪周期范围 234
设计吃水(小水线面双体船) 227
设计吃水 227
设计吃水 234
设计初稳性高度 228
设计电压 233
设计工况 226
设计航速 232
设计静水剪力 Q_{sw} 232
设计静水弯矩 M_{sw} 232
设计举升能力 228
设计类别 D("遮蔽水域"中航行的类别) 226
设计类别 226
设计类别 A("远洋"航行的类别) 226

1152

设计类别 B("近海"航行的类别) 226
设计类别 C("沿海"航行的类别) 226
设计流速　227
设计螺距　234
设计排水量　234
设计破断载荷　225
设计认可　225
设计任务书　232
设计使用寿命　231
设计水线　234
设计水线　234
设计水线长 L_{WL}　500
设计水线长　234
设计水线宽　234
设计条件（温度）　227
设计拖力 T　228
设计危险　228
设计温度（容器）　232
设计温度　232
设计温度 t_D　232
设计新颖的船舶　809
设计压力（压力容器）　231
设计压力（管道和管系）　230
设计压力（管系和零部件）　230
设计压力（锅炉）　230
设计压力（零部件）　229
设计压力（容器）　692
设计压力（设备）　229
设计压力（稳性）　229
设计压力（制冷剂）　231
设计意外事件　225
设计意外载荷　225
设计应力（起重设备）　232
设计营运吃水　229
设计裕度　229
设计载荷（kN/m²）　228
设计载荷(钻井船)　228
设计载荷组合　228
设计者（机构）　234

设计值　233
设计装载工况　228
设计状态　231
设计纵倾　233
设问-处理分析技术　972
设有独立液货舱/整体液货舱的石油沥青船（独立液货舱/整体液货舱）　51
设有限制失火危险的家具和设备的房间　749
社会风险　825
社会异常事件　825
射流　487
射频调制器　715
射频识别　715
射线探伤　715
涉及使用可能对健康造成危害的材料　481
涉及有害物质的事故　448
涉外案件　351
申请方　42
申请人　42
伸手可及区域　404
伸缩杆　309
伸缩管（套管）　888
伸缩接头（膨胀接头）　309
伸缩接头　154
伸缩节　888
伸缩式减摇鳍装置　742
伸缩桅　888
伸张轮廓　309
伸张面比　309
伸张面积（展开面积）　309
砷化氢　48
深舱　218
深舱舱壁　218
深层软地基固化船　217
深吃水多用途柱体式平台　217
深弹舱　223
深弹滚架舱　223
深海/远洋渔船　217

深海工程　218
深海滑翔机　770
深浸式水翼　218
深井泵　218
深潜救生艇　217
深潜救生艇　217
深潜器　217
深潜器母船　73
深潜系统　217
深水不倒翁平台　218
深水底栖拖网绞车　217
深水航行　217
深水勘察船　218
深水锚　217
深水抛锚绞车　217
深水抛锚装置　217
深水抛锚装置　218
深温计绞车　224
深远海环形超大型浮式基地　948
神盾舰　15
神盾局　863
审查　480
审核/审计　56
审核　302
审核程序　56
审核计划　56
审核检查表　56
审核结束会议　148
审核开始会议　633
审核员　56
审理　417
审理范围书　891
审批程序　44
审批工作　44
审批机关　44
审批矩阵图　44
审批小组　44
甚高频　947
甚高频全向信标　949
甚小口径终端　948
渗漏/漏泄(逸流,漏出量)　495

渗漏试验（泄漏试验） 494
渗漏险 495
渗透（渗漏） 778
渗透(熔深,贯穿件) 659
渗透腐蚀 659
渗透率 659
渗透率 661
渗透率 μ 662
渗透试验（渗漏试验） 659
渗透油 659
升船机 878
升浮时间 344
升高的直升机港 718
升高甲板 718
升高人孔盖 718
升降工况 486
升降和锁紧系统 486
升降机 276
升降机报警 276
升降机构室 276
升降节距 486
升降口 155
升降口 162
升降设备 503
升降通道 429
升力 502
升力定律 502
升力面理论 503
升力涡 503
升力系数 152
升力线理论 503
升汽 857
升推联动系统 464
升阻比 255
升阻比 502
生产测试船 698
生产储油船 698
生产管理 543
生产过程 543
生产过程中的检查 459
生产机器和部件 453

生产平台/采油平台 698
生产平台 697
生产平台 698
生产区 698
生产信息管理系统 542
生产支持船 698
生存工况 876
生存性 876
生活舱室 512
生活平台 7
生活区 512
生活区 6
生活污水（污水） 790
生活污水储存舱（污水柜,粪便柜） 791
生活污水处理 203
生活污水处理 791
生活污水处理舱 791
生活污水处理系统 790
生活污水处理装置 203
生活污水处理装置 790
生活污水处理装置 791
生活污水粉碎和消毒系统 790
生活污水管理程序 790
生活污水管理计划 790
生活污水排射器 790
生活污水系统 791
生活用水 789
生活用水系统 250
生活质量指数 501
生火（锅炉工） 334
生火工具 335
生命支持系统 501
生铁 669
生物柴油 83
生物地理区域 84
生物调查船 440
生物计算机 83
生物降解功能 83
生物群 84
生物群体 860

生物燃油 84
生物燃油混合物 84
生物体内积累测试 83
生物污底（污垢） 83
生物污垢管理计划 84
生物研究所 84
生物样品舱 84
生物药效率（生物放大作用,生物集结） 83
生锈 755
生锈率（积垢率） 765
生鸦片 720
声报警系统（音响报警系统） 829
声表滤波器 810
声表滤波器 876
声波 830
声波的衍射/绕射 830
声场 829
声反射测量 8
声功率 830
声功率级 830
声光试验 829
声级计 830
声流式消音器 829
声呐 828
声呐舱 828
声呐电缆绞车 105
声呐室 828
声能密度 829
声频（音频） 954
声强 829
声散射测量 8
声射线 830
声释放器 8
声速 830
声速仪 830
声响报警 831
声响报警器 7
声信号仪器（测深锤,测深仪） 831
声学海流计 7

声讯服务　56
声压　830
声压测量仪　691
声压级　830
声音(声测深,坚固的)　829
声音充电　132
声应答器　8
声源　830
声阻抗　56
牲畜船　125
胜诉方　978
绳长计数器　972
绳扣　80
绳索滑车　322
绳网　259
圣劳伦斯航道　760
圣劳伦斯航道官员　618
圣劳伦斯航道航行期　589
剩余波高　738
剩余波浪周期　738
剩余波浪周期范围　738
剩余的(残余的,残留的)　738
剩余水线面面积　270
剩余阻力　738
剩余阻力系数　738
失步转矩　707
失电(动力定位系统)　85
失火　325
失火报警　325
失火报警系统　325
失火报警系统板　325
失火探测系统(探火系统)　327
失火探测系统　328
失火探测装置(探火装置)　327
失火危险(易燃性)　331
失控的船舶　948
失能调整生命年/质量调整生命年　242
失速　849
失误产生条件　294
失效模式　317

失踪的AIS目标　524
失踪的被跟踪的目标　524
失踪目标　524
施放装置(喷射器,喷射枪)　43
施工设计(生产设计)　235
施工说明书　981
施工装备　173
施救费用　869
施救浮索　102
施乐　986
湿保养　972
湿泵舱　972
湿底润滑　972
湿度计(干湿球)　706
湿度控制器　437
湿度器　441
湿管系统　972
湿加压舱　972
湿甲板(小水线面双体船)　972
湿空气泵　972
湿空气泵　972
湿面积　972
湿面积系数　153
湿球温度计　972
湿屈曲　972
湿食品库　972
湿式保温　972
湿式采油树　972
湿式离合器　437
湿式润滑　972
湿式实验室　972
湿式服(救生服)　972
湿蒸汽　972
十一保证率波高　990
十一平均波高　560
十亿分之一　655
十字缆桩　195
十字头　196
十字头式柴油机　195
十字头轴承　904
十字型拉伸试样的焊缝拉伸试验

　890
石膏　402
石膏板　675
石灰(生)　506
石灰石　506
石棉　50
石墨烯　391
石脑油　585
石屑　861
石英　712
石英岩　712
石油　666
石油　666
石油产品　622
石油焦炭(经煅烧或未经煅烧)　666
石油沥青　666
石油气　666
时间-温度关系　888
时间窗口　903
时域计算　902
识别速度　444
实测厚度　373
实船总阻力　905
实船总阻力系数　153
实地保安　669
实动工时　10
实锻(整体锻件)　827
实际承运人　10
实际舵角　10
实际价格　10
实际净重　722
实际皮重　10
实际全损　10
实际循环　10
实盘　335
实施检验　446
实时视频　722
实体答辩　867
实物净重　594
实习调查船　911

实效伴流　270
实效舵角　269
实效攻角　269
实效拱度　269
实效合速　742
实效滑距比　952
实效螺距　952
实心轴　827
实压力空泡数　126
实验室　489
实验室电话系统　490
实验室电源系统　687
实验室信号插座箱　490
实验室仪器系统　489
实用程序　939
实用新型专利　939
食品废弃物　350
食宿费　87
食用水　258
食用水臭氧消毒器　258
食用油　185
史密斯-麦金太尔取泥器　824
矢量模式　942
使……确信　539
使……成为(向去,做成)　539
使……明确(弄清楚)　539
使饱和(浸透)　764
使齿轮啮合　897
使齿轮脱开　897
使滚装船门固定的液压夹扣　438
使汽化　372
使燃油系统充油　695
使停止动作　897
使稳定（解决）　790
使用可靠性(工作可靠性)　638
使用可靠性　788
使用时数　787
使用寿命（使用年限）　787
使用寿命（使用年限）　789
使用寿命　635
使用说明书（使用条令,修理说明

书）　788
使用性极限状态　789
使用载荷　788
氏硬度计　维氏硬度计　951
示波器　643
示波器迹线　643
示功阀　451
示功器转动装置　451
示功图　451
示功图　451
示振仪(振动记录仪)　951
世界"封闭市场价格"　472
世界"自由市场价格"　473
世界封闭市场价格　984
世界海关组织　984
世界经济合作与发展组织(经合组
织)　642
世界贸易组织　984
世界气象组织　980
世界头号经济体　985
世界银行集团　984
世界自由贸易区大会　984
世界自由市场价格　984
事故　447
事故　5
事故　6
事故　615
事故场景　5
事故的概率　696
事故发生地　674
事故极限状态　5
事故类型　5
事故原因　5
事故中的安全区域　755
事故状态/事故工况　5
事件树分析　302
事态加剧　294
事态加剧因素　294
势伴流　686
势流　686
势头　686

试航　771
试航操纵　918
试航航速　910
试剂　722
试件　761
试水位旋塞　367
试销　762
试验　891
试验报告　892
试验池船模自航点　781
试验单元量/试验批　935
试验段　892
试验机构　892
试验夹具(试验台)　892
试验临界雷诺数　195
试验批　891
试验设备　892
试验生效日期　892
试验数据　892
试验台　892
试验条件(试验工况)　891
试验文件　892
试验无效日期(耐火试验)　892
试验旋塞　891
试验液体 A　892
试验液体 C　892
试样　892
试样试验　843
试样外形　843
试油燃烧器　196
试油燃烧器外伸支架　644
视察孔　815
视觉报警(灯光报警,视觉报警器)
　953
视觉辅助设备　953
视觉显示　953
视觉信号　953
视频电话　951
视频端口　951
视频卡　951
视频通话　951

视频图形阵列 951
视频图形阵列端子 951
视频显示器 953
视频信号 951
视听资料 56
视野 323
适当的最新纸质海图卷(APC) 44
适当可能的(事件的概率) 722
适航性 774
适任标准 852
适任机构(耐火试验) 163
适任人员[1] 163
适修状态 869
适宜气候 722
适应伴流螺旋桨 958
适装证书 129
释放 732
释放阀(安全阀,泄压阀) 732
释放阀最大的调定值 551
释放器(排除器) 732
释放源(电气设备) 831
释放源 831
释压 732
释压阀 732
收到功率 221
收到功率系数 885
收放绞车 1
收放式减摇鳍 742
收费航行或游览 356
收货人 172
收据 722
收款人/受款人 657
收缩(收缩量) 813
收缩(缩套,冷缩配合) 813
收缩配合 813
收妥结汇 7
收鲜船 103
手柄 395
手操纵的 403
手持 629
手动报警按钮 541

手动报警器 541
手动泵 403
手动操作(人工操作) 403
手动倒顺车机 403
手动调节器(手动调节,手动控制) 403
手动阀 542
手动火警器 403
手动火灾报警按钮 542
手动机械装置 404
手动空气压缩机 542
手动控制 541
手动控制方式(动力定位系统) 541
手动控制器 541
手动控制装置 541
手动离合器 542
手动气胀式个人漂浮装置 542
手动启闭门 404
手动启动(手动启动) 542
手动释放装置(手动投掷装置) 403
手动停车装置 403
手动越控功能 542
手动转轴机 404
手工(焊接) 541
手工焊接 542
手工焊接工艺 542
手工金属电弧焊 541
手工弯管 541
手拉葫芦 404
手轮式天窗传动装置 404
手轮式天窗传动装置 404
手旗 403
手钳 404
手术室(操作室) 636
手提式灭火器 403
手提式灭火器 684
手提式泡沫枪 684
手提转速计 880
手摇窗 977

手摇救火泵 403
守护船 853
首席信息官/首席资讯官 134
首席仲裁员 690
首制艇 494
艏 92
艏、艉部 350
艏、艉垂线 352
艏、艉垂线 663
艏部 351
艏部侧向推力器 493
艏部区(冰区加强) 92
艏部区的范围 353
艏侧推器(船首横向推力器) 92
艏沉 240
艏沉深度 258
艏吃水 256
艏吃水 350
艏垂线 353
艏端 351
艏端盖板 351
艏端区域结构 351
艏舵 92
艏封板 92
艏封门 353
艏干舷 F_F 358
艏横向推力器 53
艏滑轮架 92
艏尖舱 352
艏尖甲板 351
艏尖区域(艏尖舱) 351
艏缆 92
艏楼 351
艏楼甲板 351
艏螺旋桨 92
艏落 259
艏落高度 259
艏锚 92
艏锚绞车 416
艏门 92
艏倾 919

艏升高甲板船 718
艏推力器(艏转向装置) 92
艏舷弧 793
艏向 416
艏向和/或航迹控制系统 416
艏向角 416
艏向控制系统 416
艏向敏感性 416
艏斜浪 92
艏摇 987
艏摇 987
艏摇幅值 34
艏摇角 36
艏摇频率响应函数 359
艏与艉 92
艏踵区1 351
艏踵区2 351
艏肘板 95
艏纵倾深度 673
受案范围 767
受潮受热险 877
受风面积 A_{LV} 976
受火压力容器 334
受监测船舶 577
受监测机械 577
受监测设备 577
受让人 52
受热面 419
受热面蒸发率 301
受热受潮 419
受试设备 292
受体港 723
受压试验 691
受益人 81
受援船 52
受载日 494
授标 63
授权委托书 687
兽医检验证书 949
书架 90
书面审理 249

书面形式 986
书信电报 500
枢摩擦(轴颈摩擦) 487
枢轴(旋转,滚动) 673
疏水泵 255
输出(输出量,效率) 645
输出端 644
输出功 645
输出功率 645
输出线 310
输入功率 459
输入输出 459
输送阀 913
输送机 185
输送温度 914
输送效率(运输效率) 916
输油管 621
输油平台 913
蔬菜库 943
鼠笼绕组 846
鼠笼式感应电动机 453
竖管 849
竖管式蒸发器 947
竖桁 946
竖立式调查船 342
竖型舱盖 852
竖轴推进器 946
竖轴推进器 947
数据操作语言 209
数据处理 491
数据传输 210
数据的合理性 677
数据定义语言 209
数据检索 209
数据库 210
数据库管理系统 210
数据库模式定义语言 209
数据宿 209
数据通信 208
数据通信设备 208
数据线传输 209

数据源 209
数据终端 209
数据字段 209
数控绘图机 613
数控技术 167
数控螺旋桨加工铣床 613
数控切割机 591
数量 711
数量检验证书 461
数码单镜反光相机 238
数码静态相机 238
数码照相机 237
数位光碟播放机 263
数学型线 555
数值水池 613
数字地震仪 238
数字电视 238
数字化器绘图机 239
数字化视频光盘 238
数字机顶盒 239
数字录音机 238
数字媒体播放的先驱 974
数字式电子战系统 238
数字图像 238
数字图像处理 238
数字卫星广播系统标准 263
数字显示工作组 238
数字选择性呼叫 238
数字指示器 238
衰减器 55
衰减曲线 199
闩式弃链器 250
双板舱壁 252
双饼滑车组 401
双层壁燃油管系 362
双层底 251
双层底舱 251
双层底舱 91
双层底结构 251
双层底压载水舱 960
双船级船舶 251

双底船　251
双腭抓斗　138
双方同意　103
双风管空气调节系统　261
双杆操作安全负荷　756
双杆吊货钩　551
双缸式涡轮机　924
双缸蒸汽泵　985
双管板　252
双锅筒锅炉　924
双滑车　819
双机轴并车减速齿轮箱　726
双级空气抽逐器　924
双级压缩制冷系统　252
双甲板船　251
双绞线　924
双节点振动　924
双井架钻机　261
双卷筒绞车　252
双壳构件　252
双壳散货船　100
双壳散货船的横剖面　917
双壳散货船压载舱　68
双壳液舱　252
双壳油船　252
双壳油船　626
双联泵　924
双联单作用泵　924
双链转环　578
双流式汽轮机　252
双流双胀蒸汽机　196
双路循环锅炉　252
双螺杆泵　924
双盘式离合器　252
双频道卫星接收机　260
双屏立管　252
双屏立管　260
双千斤索吊杆装置　924
双曲线导航定位系统　441
双曲线系统　442
双燃料柴油机　261

双燃料动力装置船舶　801
双燃料发动机　261
双燃料发动机舱室　261
双色舷灯　81
双水翼艇　252
双索抓斗　924
双态喷射推进　924
双体船　120
双体调查船　120
双体高速船　120
双体近岸快速攻击艇　318
双体气垫船　434
双体水翼船　440
双通阀　924
双通旋塞　924
双头螺栓　864
双蜗杆式舵机　985
双务合同　81
双下浮垫柱稳半潜式钻井平台　924
双舷侧　253
双舷侧结构的散货船　100
双舷侧普通干货船　375
双舷侧散货船　252
双相不锈钢　262
双向水力测功器　924
双烟道锅炉　252
双因素贸易条件　251
双油路喷油嘴　262
双胀式蒸汽机　165
双支承舵　251
双重 DP　262
双重船级船舶　260
双重绝缘　252
双轴并列减速齿轮箱　253
双柱带缆桩　90
双爪锚　252
双锥筒推进器　469
双组空气抽逐器　924
双作业钻机　261
水玻璃　961

水舱式减摇装置　749
水槽　963
水产品运输船　336
水产品运输船　45
水池浪高仪　967
水池试验　882
水冲击进流　810
水处理(软水处理)　963
水弹性　440
水道测量船　876
水的饱和空气含量　764
水电加热器　272
水电站清污保洁船　791
水动力分量　440
水动力分量系数　153
水动力节能装置　440
水动力进角　440
水动力矩分量　440
水动力矩分量系数　153
水动力矩线加速度导数　3
水动力矩旋转导数　751
水动力线加速度导数　3
水动力性噪声　440
水动力旋转导数　751
水动力学　440
水分迁移　576
水封压盖　961
水垢　765
水垢　772
水垢的溶剂　766
水管锅炉　921
水管锅炉　963
水管锅炉　966
水柜(水舱)　963
水果保鲜　690
水过滤柜　961
水基灭火剂　963
水基灭火系统　963
水加热柜(热水柜)　433
水进流管道　961
水净化系统　962

水冷的 963
水冷电动机 961
水冷内燃机 963
水冷式冷藏集装箱 961
水冷式压缩机 961
水力半径 439
水力光滑 440
水力清理 438
水力式挖泥船 438
水力制动 438
水量调节阀 962
水疗按摩中心(豪华邮轮) 809
水龙头 963
水漏斗(水筒) 961
水漏斗(水筒) 963
水路测量 440
水路测量船 440
水密 965
水密凹体 965
水密舱壁 965
水密灯具 965
水密分舱 966
水密滑动门 966
水密滑门(滑移门) 822
水密滑门手动装置 404
水密滑门速闭装置 712
水密开口 965
水密肋板 965
水密门 965
水密门就地控制模式 518
水密试验 966
水密外壳 965
水密性 963
水密性 966
水面清扫船 346
水面式平台 875
水面式钻井平台 875
水灭火系统(水炮系统) 962
水灭火系统 961
水幕/水冷壁 963
水幕系统 962

水泥 127
水泥舱 127
水泥船 168
水泥搅拌舱 127
水泥木丝板 981
水泥熔块 127
水暖 961
水盘 258
水平安定面 431
水平吹风式 431
水平舵转舵机构(减摇鳍传动机构) 324
水平隔板 431
水平管 501
水平桁 431
水平铺管 431
水平融通 431
水平探鱼仪 431
水平位移补偿装置 431
水平艉翼 431
水平一体化 431
水平振动 431
水平轴 431
水平肘板 431
水平纵倾 501
水汽比率 963
水汽压力 942
水腔(水套) 961
水溶性盐含量 963
水润滑泵 962
水润滑艉管 962
水润滑轴承 964
水色计 155
水上飞机 773
水上排泥管 344
水上施工平台 781
水上水下侧面积比 720
水深 h(m) 961
水深 224
水深测量 562
水渗漏探测系统 962

水生物物种特遣队 45
水生植物 45
水声调查船 932
水声定位系统 933
水声发射器 933
水声换能器 932
水声学电缆绞车 441
水声综合测量仪 933
水室 961
水套(水衣) 961
水听器 441
水桶(饮水保温桶) 768
水头(水位差) 961
水位报警器 962
水位控制 962
水位探测器/水位传感器 962
水位探测器 786
水位指示器 962
水位自控系统 58
水温测量 963
水温观测 963
水文吊杆 441
水文调查船 441
水文服务 441
水文绞车 441
水雾/喷水两用枪 963
水雾灭火器 963
水雾灭火系统 339
水雾喷枪 963
水雾喷嘴 845
水雾施放装置(水雾枪) 961
水雾系统(水幕系统) 963
水雾系统 963
水下侧面积系数 152
水下的(下潜,沉没的) 865
水下电视摄像机 933
水下浮标系统 868
水下辐照度仪 932
水下观测船 932
水下观光潜水艇 656
水下检验 481

水下建设船　932
水下居住舱系统　932
水下目标纯方位角跟踪　77
水下排放口　932
水下清洗　481
水下散射仪　933
水下生产系统　866
水下生产转塔　865
水下声呐监听系统　933
水下施工船　619
水下式观光半潜水艇　656
水下通信装置　932
水下透射率仪　933
水下完井　932
水下吸收仪　932
水下照相机　932
水下装载转塔　865
水线　964
水线 WL　964
水线长　964
水线长度 L_{WL}（小艇）　964
水线后段　964
水线宽 B_{WL}　94
水线宽（小水线面双体船）　964
水线宽　964
水线宽 B_{WL}　964
水线宽度 B_{WL}　74
水线面　964
水线面处船长 Lw　496
水线面面积曲线　199
水线面系数　964
水线面系数　964
水线面系数曲线　199
水线前段　964
水线下支柱体深度　947
水线下最大船宽　556
水性涂料　961
水循环喷射器　138
水循环试验　961
水压紧密性试验　962
水压密封(水封)　439

水压密封(水封)　439
水压强度试验　962
水压试验(液压试验)　439
水压试验(液压试验)　439
水压试验　962
水压试验　962
水压试验装置　962
水压总管　962
水翼　349
水翼　440
水翼颤振　347
水翼船　440
水翼双体船　120
水跃　439
水闸　962
水蒸发气　963
水质监测船　962
水质监测系统　962
水中空气含量　19
水柱式空泡试验水筒　634
水渍险　979
水总管　962
睡熊沙丘国家湖滨区　821
顺次控制　787
顺浪　967
顺流式燃烧室　862
顺排式燃气轮机　862
顺序（使程序化,交替）　787
顺序法　787
顺应式结构/顺应式平台　164
顺应式平台　164
顺应式生产系统　164
顺应式桩承塔　164
瞬时调速率　719
瞬时继电器或脱扣器　462
瞬时声压　462
瞬时载荷　740
说明书（一览表）　842
司令官　159
私人聊天室　696
斯贝克锚　844

四冲程发动机　354
四冲程三胀式蒸汽机　354
四冲程循环　354
四冲程循环周期　354
四机轴并车减速齿轮箱　726
四路联箱　354
四用扳手　986
四爪锚　391
伺服电动机　789
伺服阀　789
伺服缸（随动缸）　789
伺服机　627
伺服机构　789
伺服机构理论　789
伺服元件　789
伺服执行装置（伺服执行机构）　789
松紧螺丝扣（花篮螺丝,伸缩螺杆）　923
松紧螺旋扣　744
松紧螺旋扣　768
松紧螺旋扣　862
送风头　21
搜救　760
搜救 AIS 应答器　25
搜救服务　773
搜救合作计划　773
搜救模式　773
搜救区域　773
搜救应答器　763
搜救应答器　773
搜索、预防和减少海上碎片公约　560
搜索引擎　773
苏必利尔湖(威斯康星州)沙滩保护区　872
苏打灰　826
苏格兰锅炉（圆筒形锅炉）　767
苏伊士型油船　869
苏伊士运河　869
苏伊士运河吨位　869

苏伊士运河水域 869
诉讼 512
诉讼时效 506
速闭舱盖 712
速闭舱盖 713
速闭阀（快关阀） 712
速闭阀 713
速闭应急阀 712
速动阀 712
速冻（快速冷冻） 712
速冻舱 712
速冻能力 712
速度比 943
速度传感器 843
速度方位显示器 943
速度和测量设备 843
速度和距离测量设备 768
速度和距离测量装置（相对于底部） 769
速度和距离测量装置（相对于水面） 769
速度级 849
速度级 943
速度级 955
速度计（转速表） 843
速度计（测速器） 943
速度禁区 350
速度亏损 943
速度势 943
速度特性试验 843
速度误差校正器 843
速度指示器 943
速放阀 712
速复 735
速率（速度,转速） 843
速率信号 843
速遣费 243
速遣日数 243
速头 943
速脱钩 822

塑料 676
塑料 676
塑料薄膜 676
塑料垃圾袋 676
塑料外壳式断路器 576
塑料制品 676
溯河产卵物种 34
酸洗(酸浸,酸蚀) 669
酸洗 7
酸洗池 669
酸液储存装置 7
算术平均值 63
随边 910
随边半径 717
随船浪高仪 803
随订单付款 119
随动调节系统 789
随机变量 718
随机函数 718
随机误差 718
随浪 350
碎布 717
碎木运输船 980
碎泥刀 580
碎片（碎屑,废料） 767
碎石船 748
隧道吹风冻结装置 922
隧道顶线 921
隧道型艉 921
损害 203
损害修井 458
损伤(结构) 203
损伤半径 205
损伤鉴定 205
损伤控制舱壁 204
损失浮力 524
损失浮力法 524
损失水线面面积 524
梭形切面 500
缩减的焊缝厚度（角接焊缝） 726
缩孔 813

缩套（收缩） 813
所得税 449
所考虑船体横剖面处的波浪垂向剪力 713
所考虑船体横剖面处的设计静水剪力。 710
所适用的国内法 42
所有船舶 27
所有类型船舶专用海水压载舱和散货船双舷侧处所保护涂层性能标准 660
所有油性混合物 27
所载散装干货的密度 ρ_c 222
所载液体的密度 ρ_L 222
索节 750
索具 744
索具钩 430
索具链 131
索具配件 744
索具卸扣 750
索赔 138
索赔 138
索赔清单 138
索赔时效 940
索赔书 138
锁定控制面稳定性 184
锁定装置(锁紧装置) 520
锁紧垫圈 520
锁紧螺钉(止动螺钉) 519
锁紧螺母(定位用附件,护板) 488
锁紧螺母(防松螺母) 486
锁紧装置(舵) 520
锁紧装置 520
锁口件 323
锁轴试验 791

T

塔架 906
塔式平台 906
踏板 918
踏步 350

踏脚格栅　350
台风　929
台卡导航系统　214
台链　854
台式计算机　234
抬船甲板　682
抬船试验　248
太赫兹安检仪　776
太赫兹波　890
太空碎片　832
太平门(安全门,火警应急出口)　328
太阳轮　870
太阳能　826
太阳能船舶　826
太阳能电池　826
太阳能电力推进　827
太阳能供应技术研究所　462
太阳能游览船　827
钛铁矿砂　445
钛铁矿土　445
泰勒系数 $\varGamma$　885
坍毁压力　154
摊款货物　180
摊销　33
滩涂调查　899
瘫船状态(平台)　212
瘫船状态　213
谈判领导人员　592
谈判招标　592
坦谷波　919
坦克激光压制装置　882
坦克甲板　882
坦克炮　882
探测　235
探测和报警　235
探测雷达　235
探测器(瞄准装置,选择器,寻线机)　325
探测器　235
探火报警　327

探火系统报警　327
探火与失火报警系统　327
探火与失火报警装置　327
探火装置(失火探测器)　327
探井　974
探空仪　20
探空仪检测室　717
探伤(故障检验)　319
探头加油装置　697
探针　320
碳当量　109
碳化硅　109
碳环式气封　109
碳氢化合物　440
碳氢气体　440
镗床　90
糖　869
糖蜜/化学品运输船　576
逃生/脱险(泄泄,应急出口)　294
逃生救生设备(应急逃生设备)　294
逃生目标(脱险)　294
逃生目标1(脱险)　294
逃生目标2(脱险)　294
逃生目标3(脱险)　294
逃生目标4(脱险)　294
逃生目标5(脱险)　294
逃生目标6(脱险)　294
逃生通道(脱险通道)　295
逃生通道　295
讨价还价　692
套(套间)　869
套管　119
套管　119
套管保护系统　486
套管头　119
套管系统　671
套环　895
套筒扳手　826
套筒焊接　822
套筒接合(插管接合)　844

套筒接合(窝接,插承接合)　844
套筒接合　822
套筒接合　844
套筒联轴器　822
套筒螺母　822
套筒松紧螺丝　671
套筒轴承　821
套置型舱盖　651
特别附加险　834
特别检验　839
特别敏感海区　655
特别适合装载谷物的舱　840
特别提货权　836
特别行政区　834
特别许可证　837
特别运费率　836
特定的沿海航区　842
特定航线　841
特定航线航区　842
特定航线营运　842
特定吸收率　840
特高频　930
特惠税　689
特殊海区　835
特殊焊接工艺　839
特殊考虑或(与近观检验和厚度测量相关的)专门考虑　836
特殊人员　837
特殊人员　841
特殊生产过程　837
特殊试验　842
特殊条件的区域　47
特殊涂装　837
特殊外来风险　836
特殊用途船舶　838
特殊用途船舶安全证书　837
特殊用途船舶装运爆炸品的储存仓库　532
特殊运输控制条件　835
特殊自动控制装置　840
特殊作业工况　837

特性比　132
特许证　132
特种变压器　842
特种处所　836
特种处所　838
特种船舶　839
特种构件　838
特种任务标志　836
特种业务　839
特种业务客船　839
特种业务客船　839
特种业务客船安全证书（1971）　839
特种业务客船舱室证书（1973）　839
特种业务客船舱室证书（1974）　839
特种业务旅客　839
腾讯控股有限公司　889
梯步　895
梯步间距　480
梯步深度　223
梯道（通道）　849
梯道　188
梯道布置　849
梯架　863
梯宽　95
梯台　863
梯斜度　448
锑矿石和残留物　40
提出申请的成员国政府　702
提单　83
提货单　221
提交方　866
提前排气　307
提裙装置　821
提升负荷/起升载荷　503
提升高度　682
提升环（个人漂浮装置）　503
提升机/水平舵　276
提示　690

提醒（当心）　125
提醒项目　608
体温过低（救生服）　442
替代的管理系统（AMS）　29
替代方法　29
替代设计和布置　29
替代通信系统　29
替代系统（涂层）　29
天车补偿器　196
天窗　821
天窗传动装置（天窗启闭机构）　821
天窗传动装置　821
天窗盖　821
天窗格栅　821
天窗框架　821
天窗扇形支撑齿条　821
天窗折合盖　821
天井　54
天幕　63
天幕钩　63
天幕桁架　63
天幕帘　63
天幕系索　63
天幕压条　63
天幕张索　63
天幕贮藏室　63
天幕柱　63
天气船（气象船）　968
天气预报室　968
天桥　171
天然菱镁矿石　532
天然气（干燥）　586
天然气　586
天然气储存系统　513
天然气的标准立方英尺　850
天然气再液化系统　654
天然树脂　587
天然树脂漆　587
天然纤维　586
天然橡胶　587

天线单元　38
天线杆　38
天线高度　38
填充　682
填料（垫料,包装）　650
填料弹簧　650
填料割刀　650
填料钩　650
填料钩针　650
填料函（垫块,压垫盖）　650
填料函（密封压盖,密封套）　650
填料函（压盖,密封装置）　387
填料函　387
填料函　865
填料函　865
填料函龛　865
填料函外壳　865
填料函压盖　865
填料螺丝起　650
填料箱凸缘　341
填料泄漏　495
填密垫圈　651
填密片（衬垫）　650
填塞（加注）　323
条形孔　985
条形码/条码　71
条形码打印机　71
跳板　368
跳返继电器　859
贴现　243
铁磁金属　321
铁合金　321
铁矿石　484
铁矿石丸　484
铁类金属（黑色金属）　321
铁梨木　506
铁梨木轴承　506
铁路车辆客滚船　910
铁路机动弹道导弹技术　70
铁路运单　717
铁路运输　717

铁谱分析　321
铁素体受压钢管　320
铁燧岩丸粒　880
铁屑(废铁)　767
铁质材料　321
听觉显示　56
听阈　56
听阈声压　56
停泊(用)泵　405
停泊泵　684
停泊发电机　405
停泊发电机　683
停泊发电机组　683
停泊和应急发电机间　404
停车　285
停车距离控制系统　653
停船冲程　417
停船轨迹(停船迹程)　909
停船横矩　493
停电期　85
停缸试验　201
停气阀　861
停增压器试验　201
停止装置(制动装置)　861
停租　617
艇　88
艇长　500
艇的容积 V　955
艇的识别号(CIN)　192
艇底塞　88
艇舵　88
艇帆　88
艇钩锁定装置　430
艇钩装置　430
艇甲板　87
艇宽　94
艇体长度 L_H　498
艇体宽度 B_H　74
艇体容积 V_H　955
艇桅　88
艇罩　87

艇舯型深 $D_{LWL/2}$　569
艇装置　87
艇座　87
通舱管件　659
通常在航行时关闭的开口　634
通道　4
通道　655
通道开口　920
通道设施　561
通道设施　894
通道设施　920
通风　944
通风布置　944
通风程序　945
通风道　943
通风道围板　945
通风斗罩　945
通风阀　944
通风法　944
通风盖　944
通风管(透气管)　944
通风管道　944
通风管路系统低噪声设计　525
通风管路逸气系统　943
通风管系　944
通风管系　945
通风管系　945
通风和温度控制系统　944
通风机(打风机)　318
通风机　944
通风机舱(通风机室)　318
通风集装箱　944
通风开口　945
通风孔(排气孔)　943
通风量　22
通风帽　944
通风帽　945
通风盘式　944
通风棚　21
通风设备　944

通风设备　944
通风式蓄电池　944
通风筒(通风机,通风管道)　945
通风筒　945
通风筒开口　945
通风筒围板(通风管道)　945
通风系统(集装箱)　945
通风系统图　945
通风系统迅速关闭装置　718
通风栅　944
通风罩　944
通风指数　944
通风属具　944
通风装置　945
通过(船舶)　655
通过　913
通海的(自由进水的)　355
通海阀　655
通海阀　771
通海件(通海接头)　770
通海孔　770
通海连接件　770
通海旋塞　769
通函　138
通焊孔　147
通话管　834
通话管　954
通孔尺寸　147
通量密度　347
通流部分　346
通流面积(流截面,流通截面)　346
通流能力　346
通配卡　974
通气管　944
通气管　944
通气管系统(透气系统)　945
通气口(通风系统,排泄)　944
通气桅(杆)　372
通索索具　103
通信单元　161
通信灯　578

通信灯控制箱 577
通信分支电路 161
通信工程 161
通信工作站 984
通信技术 161
通信录 12
通信探照灯 161
通信系统(动力定位系统) 161
通信系统的可用性 62
通烟口 825
通用扳手(活络扳手) 936
通用航空 375
通用机械 385
通用机械区 385
通用紧急报警 375
通用模型 386
通用实验室 936
通用应急报警系统 375
通知船员和乘客 609
通知银行 15
同步变流机 878
同步电动机 878
同步电机 878
同步舵机 878
同步发电机 878
同时点火 817
同时浸水 817
同相振动 950
同向谱 189
同心式减速齿轮箱 129
同异步舵机 878
同轴电缆 151
铜合金 186
铜合金管件 186
铜夹杂物 187
铜粒 186
统计能量分析 854
统计平均的每日温度 560
统计平均每日最高温度 560
统计平均值 561
统一编码委员会 936

统一产品代码 936
统一共同基准点 172
统一共同基准系统 172
统一解释 934
统一要求 934
筒形活塞柴油机 920
筒形锚链舱 202
筒形平台 833
筒型活塞柴油机 920
痛阈 651
痛阈声压 651
偷渡未遂者 55
偷渡者 862
偷窃 893
偷窃、提货不着险 893
头尾线 608
头罩(盖,罩形艉楼) 430
投保 103
投标 866
投标 889
投标报价 889
投标备用信用证 81
投标备用信用证 889
投弃式测温计 309
投射面比 699
投射面积 699
投影轮廓 699
透光尺寸 146
透光梯步(透空梯步) 634
透明度盘 774
透明度原则 695
透明性原则／一致性原则 172
透平传动 922
透平发电机 922
透气阀 945
透气管 945
透气管系 944
透气孔 21
透气系统的安全装置 758
透气装置 945
透气总管 944

透射声波 915
凸轮间隙 106
凸轮止链器 212
凸轮轴 106
凸轮轴滑油泵 106
凸轮轴滑油循环柜 106
凸轮轴张紧器 106
凸轮轴转角 104
凸形甲板 185
凸形转子 539
突出物 705
突堤码头 669
突然离去(离岸,划线) 539
突卸负荷试验 515
图书馆(豪华邮轮) 501
图像分辨率 445
图形终端 391
涂层标准 151
涂层合格预试验 151
涂层技术规格书（涂层技术条件）
　　151
涂层技术文件 151
涂层类型 151
涂层系统的选择 780
涂料 150
涂料 651
涂料的常规性能试验 606
涂料的海洋环境模拟加速试验 49
涂油器(注油器) 529
涂装 651
涂装说明书 651
湍流(紊流,湍流度) 922
湍流 923
湍流边界层(紊流边界层) 923
湍流雷诺应力（紊流雷诺应力）
　　922
湍流探测器 922
团体救生用具 397
推-驳组合体(巴拿马运河) 49
推-舱组合体(巴拿马运河) 49
推-拖设备 708

推船 708
推船 709
推定全损 174
推进(推进装置) 702
推进柴油机(推进用柴油机) 702
推进电动机励磁机 703
推进电机 703
推进电机舱 703
推进动力 703
推进发电机(主推进发电机) 702
推进发动机(主机) 703
推进发动机 702
推进辅机 703
推进功率 703
推进和电气操纵部位 702
推进机器处所(主机舱) 702
推进机器处所(主机舱) 703
推进机械(主机机组) 701
推进机械处所 702
推进机械输出功率 703
推进剂 700
推进控制舱 702
推进器(动力定位系统) 898
推进器(螺旋桨) 700
推进器(推进发动机) 703
推进器-舵组合体 703
推进器手动控制器 542
推进器轴端密封 773
推进器轴螺帽 614
推进器轴隧（艉轴隧，地轴弄） 767
推进器装置 898
推进器组合系统 701
推进设备 702
推进特性试验 702
推进系数 703
推进系数减值 726
推进系统(推进装置) 703
推进系统 702
推进系统 703
推进效率 703

推进效率换算因数 807
推进性能 703
推进性能 703
推进性能试验 703
推进轴系 701
推进轴系 703
推进装置(推进机械) 702
推进装置 702
推进装置 702
推进装置 702
推进装置 702
推进装置 703
推进装置 703
推进装置舱 703
推进装置轴系中心线 703
推力 897
推力比 897
推力表面 898
推力传感器 898
推力功率 897
推力功率系数 885
推力环 897
推力计 897
推力减额 897
推力减额分数 897
推力减额因数 897
推力块 897
推力块 897
推力面 897
推力器(助推器) 898
推力器 898
推力器辅助定位系泊系统 898
推力器辅助系泊 898
推力器系统(动力定位系统) 898
推力器自动控制(动力定位系统) 59
推力系数 897
推力载荷系数 897
推力指示器 897
推力轴 897

推力轴承(止推轴承) 897
推力轴承底座 897
推力轴承盖 897
推力轴颈 897
推压机构 196
退火后二次硬化 775
退火色斑 37
退货 389
退役 216
托管 540
托管架 860
托收 155
托收银行 733
托运单 807
托运货物 172
托运人 808
托运人保证书 808
托运人出口报关清单 808
托运人组织 808
拖船 921
拖船 921
拖船船长 498
拖船船长 921
拖船费 906
拖船和驳船组合体 465
拖带 906
拖带长度 907
拖带灯 907
拖带用具 906
拖带作业 907
拖底扫海测量 19
拖钓渔船 920
拖钩 907
拖钩快速释放装置 430
拖钩台 907
拖航 906
拖航船长 907
拖缆 907
拖缆 908
拖缆承梁 906
拖缆机快速释放装置 975

索引

1167

拖缆绞车(拖缆机) 908
拖缆孔 906
拖缆限位器 861
拖力点 907
拖网 255
拖网绞钢机 917
拖网卷筒机 917
拖网渔船 917
拖曳/锚作/供应船 873
拖曳 906
拖曳弓架 906
拖曳航速 908
拖曳航行深度距离记录仪 906
拖曳滑车(曳纲束锁) 906
拖曳埋置锚 906
拖曳设备 907
拖曳式温盐深剖面仪 906
拖曳索具 907
拖运器 821
拖轴试验 792
拖柱 908
拖桩 907
脱钩装置 732
脱机型 UPS 装置 618
脱开齿轮传动机构 374
脱扣装置(释放装置) 919
脱硫(除硫) 870
脱氢热处理 419
脱险口(逃生出口) 294
脱险通道 561
脱险通道的净宽 147
脱险通道的有效宽度 270
脱险通道人员密度 222
脱险通道图 561
脱险通道转移点 914
脱险围井(救生通道) 294
脱氧处理生物技术 222
脱氧剂(镇静剂) 489
脱氧器 214
脱氧作用(镇静作用) 489
脱轴试验 791

陀螺导航系统 402
陀螺罗经/电罗经 402
陀螺罗经室 402
陀螺式减摇装置 402
陀螺稳定平台 402
椭圆度 645
椭圆型艉 276
椭圆叶 276

W

挖泥船 256
挖泥船生产力 256
挖泥船作业限定海域 R1 256
挖泥船作业限定海域 R2 256
挖泥船作业限定海域 R3 256
瓦登海国家公园 958
外摆线齿轮装置 291
外摆线推进器 955
外板 794
外板展开图 174
外板展开图 794
外表漆 645
外部安全操作 313
外部安全相关信息 313
外部检查 644
外部检验 313
外部开口 313
外部审核 312
外侧螺旋桨 644
外侧轴系 644
外插法 313
外齿轮 313
外搭接(外搭接边,外余面) 645
外大陆架 644
外附的(外来的,无关的) 313
外高桥造船公司——后来居上的第一船厂 958
外观设计专利 229
外国籍船舶 801
外护套 644
外环境噪声 312

外汇分红 304
外汇倾销 305
外加电流阴极保护 312
外进气 645
外径 312
外径 645
外卡钳 645
外壳 282
外壳板 794
外壳防护等级 220
外廓线 698
外来风险 313
外列板 644
外螺纹管接头配件(阳螺纹管接头配件) 539
外螺纹接合器 539
外螺纹螺丝钉(阳螺纹螺丝钉) 539
外螺旋桨 644
外贸保税区 351
外牵索 644
外倾(外飘) 341
外燃机(外燃式往复机,斯特林发动机) 860
外燃式燃气轮机装置 312
外套 486
外特性曲线 312
外通道 312
外通风型电机 313
外推(外插) 313
外围设备 661
外形吃水 591
外形尺寸 312
外形尺寸 645
外旋 645
外转塔 313
弯管(弯头) 272
弯管车间 670
弯管机 670
弯管机 920
弯管机 920

弯管接头　276
弯管连接头　272
弯管通风筒(鹅颈式通风筒)　272
弯管用心轴　540
弯接头　272
弯矩曲线　80
弯曲比率　80
弯曲波　80
弯头套管　272
弯头套环/弯头联管节　272
弯头消音器　272
丸粒(精矿)　659
完车(本机使用完毕)　325
完工的型值　743
完工的型值表　743
完工声明　854
完工文件(竣工图)　325
完全经济一体化　164
完全满载　362
完全膨胀循环　720
完全退火　362
完全选择性保护　904
完全自主坦克整车研制　882
完税后交货到指定目的港　221
完整的横向舱壁(甲板强横梁)　916
完整的横向环状框架　164
完整的液舱　164
完整稳性　463
完整稳性手册　463
完整性　466
烷基化燃料　27
万能机械手(通用机械手)　385
万向接头　936
万向接头关节　936
万向节球　936
万向联轴节(万向接头)　936
万向螺丝扳手　936
网板架　368
网板组合式舱底水油分离器　862
网次产量　988

网关　373
网际协议　480
网卡　594
网口探鱼仪　594
网络　594
网络操作系统　594
网络处理器　594
网络打印机　594
网络日志　86
网络软件　595
网络设备　594
网络设备供应商　594
网络协议　594
网络信息中心　594
网络有源交换器　678
网络语音电话业务　954
网络终端　595
网台　923
网位仪　593
往复泵　724
往复泵　724
往复式发动机(活塞式发动机)　723
往复式内燃机　723
往复式压缩机　723
往复式液压舵机　941
往复运动　724
望鱼台　196
危害级别　409
危害周围环境　289
危难、紧迫和安全警报　246
危险　409
危险操作　416
危险程度和安全标记　219
危险处所　208
危险的控制　181
危险废物　414
危险和可操作性分析　409
危险化学品　207
危险化学品或有毒化学品　409
危险货物(巴拿马运河)　207

危险货物　207
危险货物　207
危险货物舱单或配载图　207
危险货物事故　207
危险结果　413
危险目标　208
危险气体　413
危险气体处所　370
危险气体环境　206
危险区域(电气设备)　410
危险区域(平台)　410
危险区域　208
危险区域　410
危险区域分级　410
危险区域和处所(浮油回收作业)　415
危险区域或处所　415
危险区域中的0类危险区(平台)　412
危险区域中的1类危险区(平台)　412
危险区域中的2类危险区(平台)　412
危险确定/危险识别　409
危险识别　416
危险通报　206
危险涂料　207
危险应力　790
危险影响　414
危险与事故状态定义　219
危险状态　409
微处理器　128
微穿孔板吸声结构消声器　830
微观结构试验　569
微软手机操作系统　977
微纤维玻璃棉　388
微小的(事件的概率)　734
微型污底　568
韦伯数　969
违约　218
违约金　325

违约责任　501

围蔽处所(游艇)　282

围蔽和敞开的上层建筑　281

围壁通道　920

围壁通风筒　920

围带　813

围井　920

围帘　968

围裙　820

围网　137

围网括钢绞机　708

围网引钢绞机　255

围网渔船　708

围网渔船　779

围堰　153

围油栏(油堰)　620

围油栅　620

桅　551

桅灯　551

桅灯　552

桅顶灯　904

桅顶通信灯　161

桅箍座板　551

桅冠　920

桅横桁　815

桅间张索　919

桅肩　645

桅设备　551

桅室　551

桅索具　551

桅梯　551

桅支索　855

桅支索登攀梯(桅索具登攀梯)　813

桅柱　526

维持船级的检验　140

维度　239

维护　538

维护保养、装配与拆卸,维修拆装　538

维护保养计划　538

维护保养记录　538

维护保养设备　538

维护保养手册　462

维护保养手册　538

维护保养说明书(维修说明书)　538

维护措施　538

维护等级　538

维护方案　538

维护管理计划　538

维护检查(定期检验)　538

维护控制设备　538

维京皇冠酒廊(豪华邮轮)　952

维氏角锥硬度　951

维氏硬度　951

维氏硬度试验　951

维氏硬度值　951

维氏钻石硬度　951

维修程序(维修工作)　538

维修程序(维修工作,日常例行维护保养)　538

维修费用　538

维修工作　538

维修工作量　538

维修设备　538

维修通道　4

维修中生成的垃圾　538

维修准备时间　538

卫生泵(卫生水泵)　763

卫生的　762

卫生间及类似处所　762

卫生检验证书　763

卫生设备　763

卫生水排泄孔　762

卫生水系统　763

卫生水压力柜　763

卫生证书　762

卫星参数　764

卫星导航　763

卫星导航计算机　764

卫星导航技术　764

卫星导航系统　764

卫星定位模式　763

卫星发射技术　763

卫星跟踪船　909

卫星更新　764

卫星轨道参数　764

卫星通过　657

卫星通信系统　763

卫星系统　764

卫星系统的覆盖区域　191

卫星仰角　763

卫星有效定位　270

卫星预报　763

卫星云图　763

卫星云图接收室　763

未充分发展风浪　608

未除气　601

未经处理的生活污水　938

未经平舱的满载舱　933

未经许可的广播　931

未平舱的满载舱　323

未熔合　225

未熔合　490

未受扰动流速　943

未完全焊透　449

伪　706

伪代码　706

位移(转换,船舶纵向移动)　794

位移法　244

位置1　684

位置2　684

位置敏感性　685

位置丧失　524

位置稳定性　685

位置线　508

位置线交角　36

尾流离心作用　269

尾流自导鱼雷　958

尾流阻力　739

尾涡　910

委托　67

委托人　695
委员会　160
艉　859
艉板宽度 B_T　915
艉部　17
艉部滚筒　859
艉部区　17
艉吃水　17
艉垂线（AP）　17
艉垂线（AP）　17
艉垂线　663
艉灯　860
艉端　17
艉封板/方艉端面　915
艉封门　17
艉浮　502
艉干舷 F_A　358
艉管密封装置　860
艉管轴　921
艉管轴检验　921
艉滑道　859
艉滑道拖网渔船　859
艉滑轮　859
艉机布置　17
艉机型船　285
艉加油索具　859
艉尖舱　17
艉尖舱壁　17
艉尖舱舱壁　17
艉尖区　18
艉缆　859
艉缆　861
艉楼　682
艉楼甲板　682
艉落　903
艉锚　859
艉锚绞车　859
艉门　859
艉旗杆　289
艉鳍　214
艉倾（上仰）　673

艉倾　919
艉伸部　190
艉升高甲板船　718
艉推力器　859
艉舷弧　793
艉斜浪　712
艉袖管轴承止动嵌条　860
艉翼　16
艉翼式锚　881
艉轴（螺旋桨轴）　768
艉轴（螺旋桨轴）　881
艉轴（艉管轴）　921
艉轴　859
艉轴　921
艉轴衬套　859
艉轴衬套　860
艉轴衬套间隙　859
艉轴管　791
艉轴管　859
艉轴管舱壁　860
艉轴管衬套　459
艉轴管滑油　860
艉轴管滑油泵　860
艉轴管料填函　860
艉轴管螺母　859
艉轴管填料函压盖　860
艉轴管油封用油泵　860
艉轴管轴承　860
艉轴管轴承衬条　860
艉轴管装置　860
艉轴架轴承　864
艉轴检验　881
艉轴颈　881
艉轴孔舱壁（有填料函舱壁）　865
艉轴隧（轴隧）　881
艉轴隧　792
艉轴填料箱压盖　859
艉轴支撑座　792
艉柱底骨　827
艉柱轴毂　859
温差应变　889

温差应力　889
温度　888
温度-电导率测量仪　888
温度-形变轧制〔温度-形变控制工艺TM（TMCP）〕　894
温度补偿过载继电器　888
温度场　888
温度传感器（温度敏感器，温度变送器）　889
温度传感器　889
温度传感器　889
温度调节阀　889
温度调节器　895
温度分布　888
温度分布品质　889
温度计　888
温度监控系统　889
温度开关（调温开关）　889
温度控制　888
温度控制的　889
温度控制器　895
温度灵敏度　889
温度梯度,温度差　888
温度弯翘应力　889
温度系数　888
温度自控系统　59
温熵图（T-S 图）　888
温升　889
温湿图　706
温室气体　394
温盐深电缆绞车（STD 绞车）　761
温盐深声速测量系统　855
文件（文件化）　249
文件及其修改的控制　249
文件评审　248
文件预审　689
文件终审　324
文丘里流量计（喉管流量计）　945
纹波电压　746
吻合效应　154
稳定（稳定化,减摇）　846

稳定并励电动机 848
稳定并励发电机 848
稳定的 848
稳定调速率 719
稳定化处理 846
稳定控制系统 847
稳定模式 848
稳定平衡 848
稳定器(减摇装置,稳定剂) 848
稳定设备(减摇装置,消摆装置) 848
稳定时间 856
稳定系统 848
稳定运行 848
稳定运转状态 848
稳定振动 849
稳定装置(减摇装置) 848
稳定装置 848
稳定状态跟踪 856
稳态 848
稳态强磁场 855
稳态强磁场实验装置 855
稳心 564
稳心半径 564
稳心高 564
稳心高度 457
稳心曲线 520
稳心图 564
稳性 846
稳性标准 847
稳性标准 847
稳性标准数 847
稳性极线图 679
稳性计算 105
稳性控制资料 204
稳性曲线 847
稳性十字曲线 195
稳性消失角 36
稳性消失角 φ_V 36
稳性仪 847
稳性资料 847

稳焰器 237
问题单 712
涡激运动 956
涡激振动 956
涡空泡 956
涡流室燃烧室 877
涡流噪声 956
涡轮 754
涡轮泵 922
涡轮传动装置 374
涡轮发电机舱 922
涡轮发电机组 922
涡轮给水泵 922
涡轮后废气温度 307
涡轮机(透平机) 922
涡轮机舱 922
涡轮机底座 922
涡轮机电力推进 922
涡轮机滑油贮存柜 922
涡轮机驱动泵 922
涡轮机油 922
涡轮基组合动力 886
涡轮交流发电机组 922
涡轮前排气温度 307
涡轮特性曲线 922
涡轮压缩机 922
涡轮增压 922
涡轮增压柴油机 922
涡轮增压发动机 922
涡轮增压技术 922
涡轮增压技术 942
涡轮直流发电机 922
涡线 956
涡形管(蜗壳) 768
蜗杆 283
蜗杆 985
蜗杆齿轮(蜗轮) 985
蜗杆螺纹 985
蜗杆螺纹齿条 985
蜗杆闸 767
蜗壳 956

蜗轮 985
蜗轮传动比 985
蜗轮蜗杆 985
沃基根(美国伊利诺伊州东北部城市)特别保护区 966
卧管淋膜蒸发器 845
卧龛 77
卧式泵 431
卧式发动机 431
卧式绞缆机 431
卧式壳管蒸发器 431
卧式内燃机 431
卧式烟管锅炉 431
卧室 81
污底(弄脏) 353
污底阻力 354
污滑油舱 529
污滑油柜 529
污染类别 680
污染损害(污染) 680
污染损害 680
污染物 680
污水舱 82
污水舱 83
污水沉淀和处理装置 778
污水处理 82
污水处理船 791
污水处理系统 791
污水管(粪便管) 826
污水井 83
污水疏水管系 255
污水斜槽 822
污压载水 242
污液相容性 162
污衣柜 982
污油舱 362
污油舱 362
污油处理船 960
污油柜 978
污油水舱(废油舱) 823
污油水舱装置 823

钨极惰性气体保护焊　921
钨夹杂物　921
屋顶用柏油　749
无触点启动器(软启动器)　853
无档锚链　864
无端部肘板的主要构件的跨距　833
无法找到网页　969
无缝的　773
无缝的　822
无缝钢管　773
无缝管　773
无缝铜管　773
无缝压力管　773
无浮箱张力腿平台　682
无杆锚　861
无规聚丙烯　53
无害通过(船舶)　459
无火花风机(防爆风机)　605
无机发光二极管　642
无机纤维材料　459
无脊椎动物　480
无键轴锥体　489
无空泡　602
无空泡水翼　602
无缆直治水下机器人　59
无冷凝器蒸汽机　602
无立管修井　746
无量纲环量数　603
无膨胀器的蒸汽机　603
无气喷射发动机　24
无气喷射式柴油机　24
无气喷射式柴油机　827
无人管理的(无人值班的)　931
无人机舱　931
无人机舱　937
无人机舱　988
无人水下潜器　937
无人值班机器处所　531
无人值班机器处所　931
无人值班机械操作　931

无冗余的 DP 控制系统　817
无润滑压缩机　627
无声运转　600
无绳系结　750
无绳系结装置　750
无失火危险或失火危险较小的辅机处所　61
无刷励磁机　97
无刷同步电机　97
无损检测(无损探伤,非破坏性检验)　602
无损检测法　602
无损检测设备(无损探伤设备)　603
无损检验标准　602
无损探伤试验　602
无填料函泵　387
无推力转角　595
无限航区　937
无限展弦比　455
无线"耳麦"　979
无线充电　978
无线充电技术　978
无线传声器　979
无线电报务员/无线电话务员　717
无线电报务员　716
无线电报务主任　716
无线电报自动报警器　717
无线电波实验室　716
无线电测风室　716
无线电测向　715
无线电导航　715
无线电定位法　716
无线电规则　716
无线电人员　716
无线电台的检验(平台)　875
无线电通信工作站　715
无线电通信系统　715
无线电位置基准　717
无线电职责　715
无线接入点　978

无线局域网　979
无线局域网接近点　979
无线上网技术　979
无线数据通信　979
无线数据通信系统　979
无线网卡　974
无线信道　979
无线应用协议　978
无效上层建筑　603
无须检验的船舶　935
无须燃烧的贮汽器　602
无须照看器具(LPG 系统)　931
无旋流　485
无烟的　825
无烟煤　825
无油轴承　622
无源电子扫描阵列雷达　657
无噪声的(静的)　600
无渣油　823
无障碍区　615
无支承跨距(无支撑跨距)　937
无支索起重柱　937
无支索桅　38
无阻尼横摇　937
无阻尼自由振动　931
无座下排污阀　773
五大湖区　392
五夹板　338
五维空间　337
坞(排)修　259
坞墩肘板　248
坞宽 B_D(m)　94
坞龙骨　248
坞内检验(平台)　260
坞内检验　248
坞内净宽　594
坞墙　977
坞墙　977
坞墙底　958
坞墙顶甲板　904
坞墙通道　977

坞深 D_D(m) 223
坞体型宽 579
坞修期 260
武备布置 48
物料资源规划 554
物流 520
物证 614
物质（实质，内容） 866
物资损失安全 553
物资装运设备 555
误差范围 294
误差椭圆 294
误识率 318
误食急性中毒 11
雾 349
雾笛 349
雾笛和信号控制装置 237
雾化器 54
雾角 349

X

西北欧水域 607
西海盆 972
西林舵 766
西欧造船家协会 52
吸出（抽出） 868
吸风机 868
吸口滤网 868
吸力 868
吸力沉箱 868
吸力跟踪器 868
吸力埋置式板锚 868
吸力面 868
吸力面 868
吸力稳定边界层 868
吸力桩 868
吸泥船 747
吸泥管/吸入管 868
吸泥管活动接头 580
吸泥装置 868

吸盘式挖泥船 868
吸气瓣（吸泥舌门） 868
吸气阀 825
吸气现象 667
吸气压力 868
吸气压力调节阀 868
吸取 868
吸热量 417
吸入（抽吸，表面浅注型缩孔） 868
吸入冲程 463
吸入道（吸口） 868
吸入阀 868
吸入管接头 868
吸入管路 868
吸入过滤器 868
吸入总管 868
吸入阻力 868
吸声 829
吸声材料 829
吸声的 829
吸声结构 829
吸声量 2
吸声墙砖 2
吸声涂料（吸声漆） 829
吸声系数 829
吸湿的 441
吸收量 829
吸收器（吮吸器） 740
吸收式制冷机组 2
吸水井 868
吸头/泵吸入压头 868
吸烟室 825
吸扬挖泥船/直吸式挖泥船 868
吸液管 671
吸音板 830
吸音泡沫塑料 830
吸油口 114
吸油口加热盘管 114
吸油量 620
吸振涂层 950

牺牲阳极 755
息税前利润 266
稀释的 239
稀释法 239
稀释剂 895
稀释物（Q_d） 239
舾装 644
舾装车间 644
舾装码头 644
锡基密封合金（铜镍锌合金） 599
熄火保护 339
熄火保护装置 339
膝上电脑 491
习惯皮重 200
洗舱泵（油船洗舱泵） 103
洗舱泵 882
洗舱程序 882
洗舱船 883
洗舱海水加热器 882
洗舱机 882
洗舱加热器 960
洗舱孔/洗舱开口 882
洗舱器 882
洗舱装置 960
洗涤剂（涂料） 960
洗涤设备 493
洗涤水（洗涤用水） 960
洗汽器 941
洗刷 959
洗刷锅炉 960
洗碗水 243
洗消室 217
洗消站 146
系泊定位式钻井装置 35
系泊刚臂 988
系泊机械 578
系泊绞车 578
系泊绞盘 578
系泊缆 578
系泊平台运动 579
系泊设备 35

系泊试验　918
系泊组件　578
系船链　578
系船设备　35
系固　776
系固设备（集装箱）　777
系缆绞车　727
系缆具　578
系缆可潜器　892
系缆设备　578
系缆速度　578
系缆穴　578
系列船模　787
系列姐妹船　787
系索环　578
系索栓（系索耳，羊角）　147
系艇索　651
系统　879
系统安全评估　880
系统电子航海图　879
系统故障　879
系统光栅航海图数据库　880
系统集成　879
系统间干涉目标　880
系统警报　879
系统软件　880
系统图　879
系统位置　880
系统误差　880
系统修正　879
系统综合者　879
系柱拖力　90
细长体理论　822
细锉　325
细滤器　325
细滤器　774
细水雾　962
细水雾喷嘴　962
细水雾系统　962
狭窄航道效应　742
狭窄航道阻力增值　742

狭窄水道　585
下半轴承　91
下层部分货舱和压载舱　527
下层中间甲板　526
下沉时间　865
下沉时间　903
下沉运动　789
下床铺试验　526
下导轮　527
下吊货滑车　526
下舵承　592
下舵杆　527
下舵销　526
下浮体　526
下浮体底与上甲板间距　421
下锅筒　581
下海产卵物种　120
下画线　932
下甲板　526
下降管　253
下壳体　526
下排污　91
下潜（浸水）　865
下潜式加压舱　866
下潜体(小水线面双体船)　526
下潜体长度　526
下潜体宽(小水线面双体船)　94
下潜体宽度　526
下水　493
下水驳　493
下水驳　493
下水曲线　493
下水重力　493
下死点　526
下游产业　253
下止点　91
下座板　526
夏比V型缺口冲击试验　132
先进舰炮系统　14
先进水面导弹系统　15
先前位置　657

纤　908
纤维板　322
纤维玻璃棉　388
纤维材料　322
纤维索　322
纤维增强材料　322
纤维增强塑料　322
纤维增强塑料艇　322
鲜活鱼运输船　360
弦长　137
弦振动　950
舷（端,侧面）　813
舷侧阀(通海阀)　436
舷侧结构　813
舷侧结构　814
舷侧排出阀（舷侧止回阀）　768
舷侧排水孔　768
舷侧排水口/排水舷口　645
舷侧推进器　341
舷侧推进器　814
舷侧外板　814
舷侧纵骨　814
舷侧纵桁　814
舷窗　814
舷灯　814
舷顶列板　793
舷弧　793
舷弧基准线　576
舷弧线　793
舷角　731
舷门　809
舷门　814
舷墙　102
舷墙顶舷材　102
舷樯排水口盖　768
舷桥　368
舷梯　368
舷梯　6
舷梯吊架　988
舷梯扶索　540
舷梯绞车　6

舷梯平台　368
舷梯上平台　938
舷梯下平台　526
舷梯中间平台　468
舷拖网渔船　814
舷外的　645
舷外吊杆　644
舷外吊杆　644
舷外发动机(挂机)　644
舷外阀　644
舷外挂机艇　644
舷外挂梯　717
舷外机　285
舷外机传动装置　644
舷外跨距　645
舷外排出阀　645
舷外排出管　645
舷外排放控制　645
舷外排水管　645
舷外排水孔(舷外排水口)　645
舷外排污阀　645
舷外水　351
舷外水　645
舷外推进装置(舷外挂机组)　644
舷外推进装置(舷外挂机组)　644
舷外支架　644
舷外轴承　644
舷缘　401
显色性　155
显示　245
显示　451
显示方向　245
显示分辨率　245
显示库　244
显示模式　244
显示器　244
显示器控制台　245
显示系统　245
显示系统　451
显示液　451
显著腐蚀　867

现场　447
现场报价　745
现场萃取采水器　460
现场检查　814
现场检查　819
现场见证　819
现场浇注的环氧树脂定位垫　292
现场审核　460
现场水色计　460
现场盐度计　460
现场指挥　631
现代造船模式　575
现行标准　308
现有船舶(渔船)　308
现有船舶　308
现有的救生艇释放和回收系统　308
现有发动机　308
现有集装箱　308
现有客船　308
现有平台　308
现有散货船　308
现有设备　308
现有特种业务客船　308
限额的融通　6
限界线　543
限量泵　574
限速器　843
限弯器　80
限位器(限制器)　507
限位器　861
限油器　622
限制吃水的船舶　948
限制处所　171
限制麻醉药品制造和管制麻醉药品运销公约　184
限制区　742
限制区域　741
限制式液位测量装置　741
线(管路,航线)　508
线加速度分量　508

线膨胀　508
线膨胀系数　508
线速度分量　509
线形缺陷显示　508
线性速度(LV)　509
线轴(短管,四通)　845
陷阱　917
相当平板　293
相当砂径粗糙度　293
相对GPS　390
相对方位　731
相对航向　732
相对回转直径　732
相对密度　732
相对矢量　732
相对速度　732
相对旋转效率　732
相对运动　732
相对运动显示模式　732
相隔舱室　602
相关保护措施　52
相关的管系　52
相关函数　187
相关化学品　732
相关人为因素　481
相控阵雷达　667
相控阵雷达技术　667
相邻舱室　13
相邻处所　13
相平面分析法　667
相平面图　667
相似类型船舶　797
相位多值性　667
相位漂移　667
相位稳定性　667
相位周值　667
相应的行动　43
相应速度　188
相遇最近点/相遇最近点时间　192
香槟酒吧(豪华邮轮)　132
香港国际金融中心　473

箱柜式家具　928
箱式冷却器　93
箱形驳　681
箱形龙骨　261
箱形龙骨　93
箱型横向结构　93
箱型中底桁　262
箱装肥料(或饲料)　883
响度　524
响度级　524
响应幅值算子　740
响应时间　740
向下进水　253
向下进水点　253
向心式涡轮　129
像元　669
橡胶隔振垫　753
橡胶和塑料绝缘碎料　137
橡胶轴承　753
橡皮艇　753
橡塑发泡保温材料　753
消波器　966
消除(解除)　732
消除吸附力装置(冲洗装置)　767
消毒　859
消毒检验证书　243
消毒器(消毒器)　859
消毒设备　859
消防　332
消防安全操作工程分析　287
消防安全操作手册　333
消防安全目标　332
消防安全培训手册　333
消防安全系统　333
消防泵　332
消防泵　333
消防铲　813
消防船(救火船)　334
消防船　326
消防船　334
消防阀(消火栓)　334

消防斧　334
消防功能要求　331
消防管　333
消防规范　332
消防火钩　331
消防救生部署表　326
消防控制站　327
消防龙头　438
消防炮　330
消防设备(灭火器具)　326
消防设备(灭火器具)　330
消防设备(灭火装置)　329
消防设备(灭火装置)　330
消防设备布置(消防装置)　332
消防设施　332
消防栓(消火栓)　331
消防水带　331
消防水带箱　331
消防水桶　326
消防水总管系统　331
消防毯　326
消防通道　331
消防拖船　331
消防巡逻　331
消防巡逻制度　331
消防演习　328
消防用具/消防设备　330
消防员装备(消防员装具)　334
消防员装备　330
消防员装备　334
消防员装备储存箱　331
消防站　332
消防总管　331
消防总管　331
消费型数码相机　168
消耗臭氧物质　649
消耗臭氧物质记录簿　649
消火栓(消防龙头)　438
消极隔振　592
消声(闭式烤炉,玻璃灯罩)　581
消声　600

消声器　600
消声器　8
消声器　830
消声罩　600
消声装置　829
消涡鳍　434
消音　816
消音器/消声器　816
硝酸铵　32
硝酸铵基化肥(UN 2067)　31
硝酸铵基化肥(UN 2071)　32
硝酸铵基化肥(无危害)　32
硝酸钙　105
硝酸钙化肥　105
硝酸钾　686
硝酸镁　532
硝酸钠　826
硝酸钠和硝酸钾混合物　826
硝酸铅　494
硝酸盐岩(经煅烧)　667
硝酸盐岩(未经煅烧)　668
销钉,插头,压住　670
销紧螺栓(保险螺栓)　777
销售包装　760
销售点情报管理系统　678
销售税　760
小舱盖　823
小齿轮(副齿轮,传动齿轮)　670
小船　824
小弹药库　823
小舵角Z形操纵试验　969
小轿车运输船　708
小开口　768
小开口　824
小孔喷注消声器　830
小块漂浮冰况区域航行　443
小卖部　728
小水线面半潜船　784
小水线面双体船　824
小水线面双体船　824
小水线面双体高速船　877

小型货船　110
小型客船　656
小型联箱式锅炉　66
小型燃油舱　824
小型文本文件　185
小型张力腿平台　824
小修　199
小直径管路　823
肖氏回跳硬度计　811
校准　118
效率（高效）　270
效用试验　270
效用试验　660
楔形垫片　884
楔形夹扣　969
楔形切面　969
协调世界时　939
协定自动出口配额制　19
协方差　191
协会货物保险条款　462
协商　592
协议配额　18
协议书　18
斜舱壁　823
斜度（斜率,坡度）　823
斜方硼砂（无水）　719
斜交轴传动减速齿轮箱　480
斜角甲板　36
斜浪　614
斜肋骨　107
斜坡滑道　823
斜坡式码头　718
斜剖线　236
斜梯　448
斜梯的梯段　342
斜跳板　448
斜置型舱盖　530
谐波（谐波的,谐函数）　407
谐波激励（谐波励磁）　407
谐波消除　407
谐和（谐波,调整）　407

谐和运动（简谐运动）　407
谐频（谐波频率）　407
谐摇　740
谐振（共振）　740
谐振动（谐波振动）　407
谐振级（共振级）　740
谐振器（共鸣器）　740
谐振区（共振区）　740
谐振特性　740
泄漏　495
泄漏痕迹　495
泄漏检查（漏电检查）　494
泄漏探测器　495
泄水管路　255
泄水柜　255
泄水孔/甲板排水孔　768
泄水冷却器　255
卸货　242
卸货泊位　243
卸扣　791
卸砂驳　937
卸载　937
蟹行　192
心理毒瘾　668
芯材　187
芯片　136
芯轴（型芯,紧轴）　540
芯轴填料函压盖　844
辛烷　617
辛烷值　617
锌灰　989
锌阳极　989
锌阳极阴极防护系统　989
新巴拿马型船　596
新标准造船合同　597
新船（渔船）　597
新船　596
新船能效设计指数　269
新船能效设计指数　283
新概念 VLCC　595
新概念浮式生产储/卸油船　595

新概念高速三体客滚船　595
新集装箱　596
新技术/设计　609
新技术和新方法（压载水排放）　597
新建平台　597
新救生艇释放和回收系统　596
新客船　596
新设备（渔船）　596
新特种业务客船　597
新型推进器　597
新型吸音　829
新型系统或设备　609
新型智能船桥系统　609
新颖船舶　609
新颖救生设备或装置　609
新造船舶初始入级日期　212
新造船入级检验　145
新蒸汽　512
新装置　596
信标　74
信道　455
信函　188
信号　815
信号灯杆　137
信号阀　786
信号镜　815
信号枪　815
信号设备　815
信号桅　815
信号烟火　710
信号烟火箱　710
信号烟雾　815
信号源　815
信号张索　815
信托证书　920
信息　455
信息采集　455
信息处理系统　456
信息传输　456
信息存取　455

信息服务平台 456
信息服务业务 456
信息和通信技术 455
信息技术 456
信息加工 456
信息网络系统 456
信息源 831
信息载体 208
信息终端公司 456
信用证 500
信元 126
信约 677
兴波阻力 967
兴波阻力系数 153
兴奋药物 860
星下点 868
星型内燃机 714
星型叶轮 752
行车 917
行程（冲程,打击） 864
行程调整 864
行程调整轴 864
行程缸径比 864
行程缸径比 864
行话/专业术语 486
行进时间（移动时间） 917
行李 529
行李舱 529
行李丢失或损坏 524
行式打印机 508
行为管理者 80
行为影响因子 660
行星-平行轴减速齿轮箱 674
行星齿轮 291
行星齿轮架 853
行星架 674
行星减速齿轮装置 291
行星轮（行星齿轮） 674
行星式差动滑车组 674
行星式减速齿轮箱 674
行星式减速齿轮箱 853

行星小齿轮 674
形式 924
形式认可 926
形式认可 A 926
形式认可 B 926
形式试验 929
形象危机应对报告 864
形状稳性臂 501
形状稳性臂曲线 195
形状稳性力臂曲线 352
形状系数 352
形状效应 352
形状因数 352
形状阻力 352
型吃水 576
型吃水 580
型基线 576
型宽（中部） 94
型宽 576
型宽 579
型排水量 576
型排水量 580
型排水量曲线 576
型排水体积 576
型排水体积 955
型排水体积曲线 199
型深（平台） 580
型深 D（小水线面双体船） 224
型深（内河船）D 223
型深（小水线面双体船） 580
型深（游艇） 224
型深（渔船） 580
型深 222
型深 576
型深 579
型线 576
型线图 509
型值表 880
性能标准 659
性能标准 660
性能参数 660

性能测试 660
性能记录 660
性能检查 659
性能降低 660
性能判据 A 659
性能判据 B 659
性能判据 C 659
性能评价 659
性能曲线图表 108
性能试验（运行试验） 660
性能试验 660
性能要求 660
汹涛阻力 752
雄性的（阳性的） 539
修订建议案 744
修井 971
修井驳船 984
修井船 971
修井隔水管系统 480
修井平台 984
修井平台 984
修井支持驳船 984
修理船 735
修理范围 718
修理工程 735
修理完工 735
修正后初稳心高 952
修正水层 187
锈蚀 755
锈损险 755
溴化锂 512
溴化锂-水吸收式制冷装置 512
溴化锂吸收式制冷装置 512
虚拟仿真系统 952
虚拟网卡 826
虚拟专用网络 952
需氧和厌氧情况下的环境后果和影响资料 209
需用系数 222
许可舱长 662
许可方 501

许可证　501

许可证贸易　501

许可证协议　501

许用设计应力　662

许用应力　663

续航力(持续时间)　283

续建选择权　641

蓄电池舱通风　73

蓄电池舱通风机　73

蓄电池的贮存寿命　793

蓄电池放电率　242

蓄电池容量　73

蓄电池室　73

蓄能器　7

蓄压器(压缩空气储存器)　690

悬臂梁　107

悬臂起重机设备　224

悬臂自升式钻井平台　107

悬垂段　760

悬高杆长比　876

悬挂舵　932

悬挂式动力滑车　551

悬挂式转向推进装置　766

悬链锚腿系泊系统　125

悬链式系泊　125

悬链线立管/柔性立管　125

悬链线锚腿系泊　106

悬链线下拐折点　760

悬锚舞厅(豪华邮轮)　34

悬伸部　646

旋臂试验　752

旋出(拧出)　768

旋紧(拧紧)　768

旋流器　956

旋流式调风器　923

旋入(拧进)　768

旋塞(开关,龙头,套管)　319

旋塞　768

旋塞　940

旋松螺栓　821

旋筒推进器　342

旋涡泵　661

旋涡泵　956

旋转(旋涡)　973

旋转接头(水龙头接头)　878

旋转接头　878

旋转力　923

旋转器(转子)　752

旋转失速　752

旋转式燃烧器　751

旋转式扫气泵　744

旋转速度　752

旋转跳板　822

旋转头　752

旋转噪声　744

旋转张紧器　923

选定位置　780

选频阻尼材料　359

选择、认可分包商和供应商　780

选择阀　780

选择腐蚀　780

选择器(转换器,波段开关)　780

选择税　29

选择性(选择能力)　780

选择性催化还原技术　780

选择性脱扣　780

学生签证　864

雪鲤鱼平台　42

熏蒸　364

熏蒸证书　364

巡航导弹　197

巡航功率　197

巡航航速　198

巡航机组　197

巡航级组　198

巡航汽轮机　198

巡航燃气轮机　197

巡回检测装置　210

巡检　209

巡洋舰　197

巡洋舰型艉部　197

询盘/询价　289

循环倍率　138

循环滑油舱　138

循环水倍率　138

循环水槽　138

循环水系统　138

循环信用证　744

循环载荷　202

训练船　911

蕈形通风筒(蕈形通风头)　583

蕈形通风筒　583

Y

压差控制器　622

压电传声器　669

压锻(落锻,型锻)　849

压锻　690

压舵角　595

压盖衬套　387

压盖密封(封闭装置)　387

压痕　447

压痕试验　131

压紧器　250

压紧楔　250

压紧型　690

压紧装置　395

压力/抗拉增强层　692

压力/真空阀　692

压力　690

压力表(液位压力计)　541

压力表　691

压力表　691

压力表接头(测压点)　691

压力差　691

压力传播　691

压力传感器　691

压力传感器　691

压力调节器　691

压力调节系统(LPG系统)　691

压力放大系数　690

压力分布　691

压力管　691

压力级 691
压力计 541
压力计的(用压力计量的) 541
压力计算点 690
压力检波器 691
压力降 691
压力警报器 690
压力铠装层 468
压力控制 690
压力控制头 691
压力滤器 691
压力面 691
压力配合 690
压力喷水系统 692
压力强度(压强) 691
压力曲线 691
压力容器(受压容器) 690
压力容器 691
压力容器的半球形封头 422
压力容器的平封头 342
压力润滑 691
压力润滑法 350
压力式测波仪 691
压力式验潮仪 691
压力试验 691
压力衰减因子 690
压力水柜 962
压力梯度 691
压力通风 691
压力维持 691
压力响应因子 691
压力液货舱 691
压力元件 691
压力增强层 691
压力真空表 541
压力真空断开装置 691
压力真空阀 691
压力真空破坏器 691
压力真空释放 691
压力真空释放阀 691
压力中心(压力点) 691

压敏胶黏剂 691
压敏阻尼胶 691
压配附件(固紧附件) 690
压气机 166
压气机喘振边界测定 166
压气机特性曲线 166
压气机涡轮共同工作线 166
压气式蒸馏 941
压气式蒸馏装置 941
压容图 692
压入配合 350
压缩比 166
压缩机滑油柜 19
压缩空气 165
压缩空气填角焊试验(渗漏试验) 165
压缩空气装置 21
压缩容积 147
压缩天然气(CNG)运输船 149
压缩天然气 166
压缩天然气动力装置的船舶 801
压缩压力 166
压头 691
压载 67
压载泵 67
压载舱换水 360
压载操纵系统 67
压载吃水 67
压载航速 70
压载兼用舱 70
压载水 68
压载水舱 68
压载水舱 70
压载水舱 960
压载水舱 960
压载水舱的保护涂层 704
压载水舱容量 960
压载水舱通海阀 68
压载水舱通气管 961
压载水处理 70
压载水处理设备 70

压载水处理系统 69
压载水更换标准(D-1) 68
压载水更换标准(D-2) 69
压载水工作小组 70
压载水公约 69
压载水管理 69
压载水管理计划 69
压载水管理系统 69
压载水管理指南和制定压载水管理计划指南 400
压载水管系 67
压载水交换 68
压载水交换率 304
压载水排放 68
压载水容量 68
压载水吸管 960
压载水吸入阀 770
压载水系统 67
压载水线 70
压载水有害水生物及病原体污染 680
压载水支管 67
压载水置换 68
压载水总管 67
压载水总管 960
压重转环 969
压桩船 344
压阻力 691
押汇 249
鸦片渣 640
鸦片制剂和类鸦片 639
鸭式水翼系统 107
哑终端 262
雅虎 987
亚丁湾区域 401
亚厘米波 865
亚历山大群岛水域 964
亚临界热影响区 865
亚临界退火温度 865
亚临界转子 865
亚声速压气机 866

亚太经济与合作组织 50
亚洲基础设施投资银行 50
亚洲开发银行 50
亚洲相互协作与信任措施会议 170
烟-水管锅炉 335
烟囱(排气管,堆装) 849
烟囱 825
烟囱 825
烟囱的调节风门 366
烟囱帽 825
烟道(烟囱) 825
烟道(烟管,烟囱) 346
烟道 346
烟道 366
烟道 371
烟斗式通风帽 192
烟管 366
烟管火警系统 825
烟灰收集器 828
烟灰运输船 347
烟火室 710
烟火探测(探火) 327
烟霾 824
烟密或能防止烟气通过 825
烟气报警器 825
烟气调节挡板 103
烟气分析器 369
烟气隔离阀 346
烟气鼓风机 453
烟气观测装置 825
烟气加热式再热器 372
烟气检测器(检烟器) 346
烟气灭火系统 346
烟气探测系统 825
烟气洗涤器 346
烟色指示器 825
烟台中集来福士海洋工程有限公司 137
烟雾 825
烟雾信号(发烟信号) 825

烟箱 326
烟箱 825
烟箱架 326
烟箱门 825
烟罩(排气罩) 371
淹艉浪 682
延迟点火 742
延期付款 219
延期支付 221
延燃孔 775
延伸(扩展) 311
延伸报警 311
延伸报警功能 311
延伸率 841
延伸式张力腿平台 312
延绳钓 521
延绳钓起线机 521
延绳钓渔船 521
延性破坏 659
延展性材料 262
严重冰冻 790
严重冰况 790
严重冰况区域航行 443
严重不合格 539
严重腐蚀 787
严重故障 787
严重滑动磨粒 790
严重伤害 787
严重事故 787
严重损坏 787
岩棉 748
沿岸国主管机关 14
沿岸渔船 150
沿海船 150
沿海国/沿岸国家 150
沿海国的保护权 745
沿海国的大陆架 177
沿海航区内作业的起重船 343
沿海航区营运限制 150
沿海航区营运限制 788
沿海区域 150

沿航线距离 245
沿周平均伴流 62
沿着脱险通道的人员流速 843
研阀 395
盐/盐类 761
盐沉淀(盐沉积物) 761
盐度 760
盐度计 760
盐度指示器 760
盐含量(盐度) 761
盐水(海水) 761
盐水 96
盐水泵 761
盐水泵 97
盐水阀 97
盐水膨胀箱 96
盐水膨胀箱 97
盐酸 440
盐岩 761
盐浴淬火 761
盐浴退火 761
檐板 199
衍射时差技术 903
眼板(导向滑车) 651
眼板(系缆钮,吊环板) 314
眼环 746
演练 256
演算时间(工作时间,运行时间) 637
演算时间(工作时间,运行时间) 637
演习 256
验残检验证书 461
验船师 548
验船师 876
验船师直接相关的检查方法 875
验水阀 373
验证(单词 verify 及其任何变体) 945
验证 940
验证 945

验证方 946
验证负荷 700
验证审核或审核 945
阳端接头 539
阳极 37
阳极 37
阳模 539
阳凸面 539
杨氏模量 988
氧 648
氧丙烷气割 649
氧腐蚀 648
氧化铝 29
氧化镁(烧僵) 532
氧化镁(未熟化) 532
氧化皮 326
氧化皮 339
氧化皮清除器（清污器） 765
氧化性物质（第5.1类危险货物） 648
氧量表 648
氧刨机 649
氧气测量仪 649
氧气分析和气体探测设备 648
氧气含量表 614
氧气呼吸器 648
氧气面罩 649
氧气瓶 648
氧气枪切割 649
氧气切割 648
氧气切割机 649
氧气切割设备 649
氧气软管 649
氧气探测仪 649
氧气压缩机 648
样品（标本,试验样品） 761
样箱 705
邀请发盘 481
摇摆凸轮 642
摇板式造波机 341
摇窗 977

摇倒机构 923
摇倒式艇架 529
摇动水翼推进器（橹） 642
遥测读数 734
遥测技术 888
遥测温度计 888
遥测温盐深剖面仪 888
遥感 734
遥控 733
遥控操作的水下作业的潜水器 753
遥控操作阀 734
遥控操作面板 734
遥控操作位置 734
遥控阀 734
遥控关闭装置 734
遥控开关 734
遥控启动装置 734
遥控器 733
遥控潜水器 734
遥控切断 734
遥控设备 733
遥控深潜器 734
遥控手柄 733
遥控停止装置 734
遥控拖钩 734
遥控系统 733
遥控显示装置 734
遥控自动深度控制系统 733
遥控自主水下机器人 59
遥控作业器 734
咬边 932
咬缸 672
药剂除垢 133
药物海洛因 237
要求的报警器或指示器 736
要求的换新厚度 736
要求的能效设计指数 736
要求加入加纳利群岛船舶强制报告系统的船舶 809
要素 316

要素验证 316
要约的撤回 979
椰肉(干燥) 187
业务标准数 194
业务流程自动化 103
业务协定 634
业主俱乐部 702
叶背 66
叶柄 85
叶侧的投影界限 506
叶侧视投影 814
叶长 497
叶端 282
叶根 85
叶根空泡 750
叶根圆角空泡 323
叶冠 464
叶厚比 85
叶间隙 368
叶蜡石 709
叶轮 446
叶轮 85
叶轮泵 446
叶轮泵 941
叶轮机械 922
叶轮面平均流速 560
叶轮推力 897
叶面 316
叶面比 85
叶面参考线 85
叶片 85
叶片泵(叶轮泵,滑片泵) 941
叶片节距 941
叶片侵蚀 85
叶片失效载荷 F_{ex} 85
叶片式通风机 318
叶片相对温度 731
叶片展开面积 85
叶片振动 85
叶平面切面 674
叶切面 776

叶切面表观投影　41
叶切面厚度　895
叶切面实投影　920
叶切面纵斜　718
叶梢　85
叶梢空泡　903
叶梢损失　903
叶数　613
叶元半径　717
叶元体　85
叶元体间隙比　368
叶元体进角　14
叶元体实度　828
叶元体外形阻力　698
叶元体效率　270
叶栅　941
叶栅作用　119
曳纲　102
曳行索具　908
液-气试验（静水压气动试验）　441
液泵　729
液泵供液　511
液舱　512
液舱舱壁　881
液舱高度　421
液舱及密性试验程序导则　892
液舱容积　881
液舱通风系统　883
液舱液位指示系统　882
液氮　511
液氮泵送系统　599
液氮储存装置　599
液化　512
液化气　509
液化气船　509
液化气船货舱结构　116
液化气体船（1G/2G/2PG/3G 型）
　510
液化气体货物　509
液化石油气（LPG）　511
液化石油气　527

液化石油气　666
液化石油气船　511
液化石油气动力装置船舶　801
液化石油气燃油系统　528
液化石油气作燃料动力游艇　528
液化天然气（LNG）运输船　512
液化天然气　511
液化天然气-浮式生产储油/卸油船
　513
液化天然气船　511
液化天然气的生产船　344
液化天然气动力装置船舶　801
液化天然气燃料推进船　513
液化天然气蒸发气体　512
液化天然气装载装置　512
液化乙烯船　512
液化易燃气体运输船　511
液货泵　511
液货驳　881
液货舱　117
液货舱泵吸系统　882
液货舱处所　511
液货舱的环境控制　180
液货舱气室　369
液货舱清洗系统　117
液货舱压力释放阀　117
液货舱液面计传感器　882
液货舱罩　191
液货船　883
液货船　511
液货船　882
液货船　883
液货船扫舱程序　117
液货船应急拖带装置　280
液货集装箱　882
液货输送泵温度探测装置　889
液货压缩机　111
液货蒸发器　111
液力变矩器　440
液力测功器　438
液力传动　438

液力吊杆装置　438
液力顶推铰链　439
液力减振器　439
液力减震器　438
液力铰链　439
液力离合器　438
液力联轴节　347
液力联轴节　438
液力耦合器　438
液力推动　439
液力作用阀　439
液式灭火器　347
液体　511
液体传动调速器　438
液体分离器　345
液体化学品船　511
液体化学品废弃物　511
液体类别　124
液体喷墨打印机　512
液体燃料　511
液体散货船　511
液体渗碳（充液渗碳）　511
液体输送系统　347
液体物质　512
液体压力　439
液位报警装置　511
液位测量系统　501
液位指示器　501
液位指示系统　501
液压爆炸试验（液压胀裂试验）
　438
液压泵　439
液压泵站　439
液压部件　438
液压操舵器　439
液压操舵系统　439
液压操纵隔离阀　439
液压操纵货油阀　439
液压操作系统　439
液压沉箱吊　438
液压处理　439

液压传动 439
液压传动门 438
液压船架小车 438
液压调节 439
液压调节器 439
液压动力(水力,液力) 439
液压动力舱 439
液压动力操纵系统 439
液压动力系统(液压动力单元) 439
液压舵机 439
液压发送器 439
液压阀 439
液压防碰设备 438
液压功率发送器 439
液压管路 438
液压管路 439
液压管路 439
液压机构 439
液压机械 439
液压机械手 439
液压件车间 438
液压绞盘 438
液压控制执行机构 438
液压离合器 744
液压连接 439
液压流体 439
液压马达 439
液压气缸单元 439
液压起货机 438
液压起锚机 439
液压起锚绞盘 438
液压起重机 438
液压设备 439
液压设备 439
液压升降机 439
液压升降机构 438
液压升降结构 438
液压锁(液压封闭,液阻塞) 439
液压拖钩 439
液压拖缆机 439

液压弯管 438
液压系统 439
液压蓄能器 438
液压遥控传动装置 439
液压油 439
液压油补给柜 439
液压油出口压力表 439
液压油缸 438
液压油循环柜 439
液压油压力泵 439
液压油贮存柜 439
液压折弯机 438
液压执行器 438
液压抓斗 439
液压装配 438
液压装载机 439
液压装置(液压设备) 438
液压装置 438
一般附加险 374
一般检验 385
一般禁止数量限制原则 276
一般外来风险 375
一般无线电通信 385
一般性能海洋工程用锚 385
一般许可证 385
一般用途外壳 385
一裁终局 324
一舱制 630
一次表面处理 695
一次风 693
一次配电系统 694
一个设备概念 630
一个项目的测量厚度 373
一级警报 693
一阶差分滤波器 335
一切险/保全险 27
一人驾驶 629
一人驾驶 630
一人驾驶船舶 630
一视同仁的责任 599
一小时功率 630

一小时转速 630
一氧化碳 578
一氧化物 578
衣服的次要外罩(救生服) 776
衣服的主要外罩(救生服) 695
医疗废弃物 563
医疗设备 563
医用鸦片 563
医院船 432
依赖性风险控制 222
仪表放大器 462
仪器 462
仪器检修舱 462
宜昌舢板 987
移驳绞车 72
移动设备应用程序 43
移动式地面站 574
移动式辅燃气轮机 574
移动式系泊系统 574
移动式消防设备 574
移动数据包服务 575
移动网与公用电话交换网 707
移动重力式减摇装置 580
移位 794
移位 243
移坞装置 247
移注泵 794
遗漏 629
乙烯运输船 299
已定义危险与事故状态 219
已检验的船舶 460
已装船提单 629
以色列情报和特殊使命局 462
以数字起首的词条 989
以太网 299
议定书 705
议付银行 592
议价 71
异步传输 53
异步传输模式 53
异步电机 53

异常的作用力　2
异常控制流　302
异径接管(异径接头)　449
抑制(扑灭,消除)　874
抑制目标　177
抑制振动(消振)　950
抑制装置　218
译电工作室　152
译员　480
易爆性/可燃性极限/范围　310
易达性　722
易挥发物含量　955
易挥发物质　955
易货贸易　72
易燃　341
易燃材料　340
易燃材料　455
易燃的　331
易燃的气体或蒸发气　340
易燃的雾　340
易燃固体(第4.1类危险货物)　340
易燃货品　455
易燃货物　340
易燃货物　455
易燃气体　455
易燃性　455
易燃性气体探测器　455
易燃压缩气　455
易燃液体(可燃液体)　340
易燃液体　455
易燃液体货品　455
易燃油　340
易熔材料　418
易受攻击区域　267
易损性　957
易用性/可用性　939
易于接近　267
易于阅读　267
易于执行　285
易自燃物质(第4.2类危险货物)

866
逸出阀　295
逸出蒸汽　295
逸流　495
逸气管(放气管)　295
逸气冷凝器　943
逸气热损失　944
意见　640
意外事故　5
意外事件　5
意外事件的作用力　5
意外效应　5
意外泄油性能　6
意外载荷　6
意向书　466
意向书　501
溢出物挡板　844
溢流(溢出)　646
溢流保护　646
溢流报警　646
溢流报警器　646
溢流阀　646
溢流阀　844
溢流法　346
溢流法　646
溢流管　646
溢流管路　345
溢流口(溢流挡板、溢流筒)　646
溢流排盐　646
溢流旋塞　646
溢流装置　646
溢流总管　646
溢油　844
溢油回收　627
溢油量　844
溢油源　625
溢油源　646
溢油源　844
翼　16
翼　977
翼板　341

翼轮推进　941
翼形螺钉　898
翼形螺母　977
翼展　833
因果分析　125
因素载荷　316
阴极保护　125
阴极保护设计计算书　105
阴极保护系统　125
阴极保护用防腐材料　189
阴螺纹　320
阴模(下模)　320
音频声　56
音响火箭　830
音响榴弹　830
音响信号器具　830
银焊焊料　816
银合金(银焊焊料)　816
银幕匣　767
银行保函　71
银行本票　119
银行付款信用证　193
银行汇票　71
引出线端　494
引船小车　906
引航船/领航船　670
引航联检船　670
引航员机械升降装置　563
引火源　445
引进方/受让方　501
引力精密测量　325
引燃可能性　696
引燃物　445
引燃油　670
引人入胜的　162
引入(进港,进刀,参加)　289
引水船　670
引水信号　670
引水员　670
引水员旗　670
饮水处理　686

饮用水泵　258
隐蔽处所或不能接近的处所　168
隐蔽工程　422
印第安纳沙丘国家湖滨区　451
印度尼西亚人民银行　71
印刷海流计　696
印制电路板　695
应变部署表(平台)　583
应变部署表　584
应变程序　278
应变片接线板　373
应答　7
应答器　740
应舵指数　923
应付利息　467
应急(事变,紧急关头,非常时期)　277
应急泵　278
应急部署表与应急须知　584
应急舱底排水系统　277
应急舱底水吸口　277
应急舱口/逃生舱口　294
应急操舵索具　279
应急操舵装置　277
应急操纵性　278
应急柴油发电机柴油柜　278
应急超速保护器　278
应急出口(擒纵机构)　295
应急淡水柜　278
应急灯(个人漂浮装置)　278
应急电源(应急动力源,应急供电)　278
应急电源　279
应急电站　278
应急舵　276
应急发电机　278
应急发电机室　278
应急发电机组　278
应急发电系统　278
应急发动机(备用发动机)　277
应急阀　280

应急风口　280
应急鼓风机　277
应急关闭和脱开1　295
应急关闭和脱开2　295
应急关断/事故停车　279
应急关断系统　279
应急规划　177
应急航行锅炉　881
应急航行机组　881
应急呼叫　277
应急救生艇　278
应急空气鼓风机　277
应急空气瓶组　277
应急灭火装置　278
应急排水系统　277
应急配电板　279
应急切断系统　277
应急切断装置　277
应急设备　277
应急逃生通道　278
应急停机(应急脱扣)　280
应急停止　279
应急通道　277
应急通道目标　277
应急通孔/脱险口(安全口)　295
应急拖带装置　279
应急脱开　277
应急脱开程序　269
应急脱扣装置(应急断开设备,应急跳闸装置)　280
应急脱险(紧急逃生)　277
应急围井/应急出口(救生口)　295
应急响应服务　278
应急消防泵　276
应急消防泵　278
应急消防梯　327
应急泄氨器　277
应急压载系统　277
应急用空气压缩机(应急空压机)　277
应急预案　231

应急装置　280
应急状况　279
应急状态关闭门模式　250
应急准备　278
应急准备分析　278
应急准备组织　278
应力调和函数　407
应诉通知书　609
应用程序　43
应用程序界面　43
应用软件　43
应用软件　43
英国标准协会　97
英国船舶研究协会　97
英国海外公民护照　97
英国劳氏船级社集团　397
英国劳氏船级社认为有效的制造商证书　542
英国劳氏船级社证书　528
英国伦敦国际仲裁院　520
英吉利海峡　288
英特尔公司　466
英制标准螺丝　288
迎浪　416
荧光灯　347
荧光高压汞灯　424
荧光计　347
荧光检验(荧光检测)　347
荧光探伤　347
萤石　347
营业地点　674
营业税　103
营运包　637
营运吃水　635
营运范围　638
营运航速(地效翼船)　638
营运航速　788
营运期间倾斜试验　460
营运气象Ⅰ　968
营运气象Ⅱ　968
营运气象Ⅲ　968

营运气象Ⅳ 968
营运水域 637
营运速度 638
营运条件(渔船) 635
营运限制 638
营运载荷 638
营运状态（运转工况,使用条件） 787
营运纵倾 637
硬舱 405
硬淬 405
硬度 407
硬度标准(硬度计) 406
硬度测量 406
硬度测试器 406
硬度测算 406
硬度极限 406
硬度试验 406
硬度试验仪(回跳硬度计,硬度试验机) 406
硬度数 406
硬度指数 405
硬化(凝固,淬火) 405
硬件 406
硬角 405
硬木 406
硬盘驱动器 405
硬盘消音器 405
硬硼酸钙石 154
硬涂层 405
硬质聚氯乙烯泡沫塑料板 746
佣金 160
佣金代理 160
永磁发电机 661
永磁发动机 532
永久开口 661
永久性标记(LPG系统) 661
永久性连续围板 661
永久性系泊系统 661
永久硬度 661
永远漂浮 30

勇气、力量、忠诚、能力 190
用泵打进 707
用泵排出 707
用泵吸入 707
用刚性肘板固定 171
用户对话区 939
用户界面 939
用户设定的显示 939
用户输入设备 939
用户选择的显示 939
用键固定(接通) 488
用介于中间的整个舱室或货舱纵向隔离 786
用螺钉钉住（用螺丝拧紧） 768
用螺丝拧的（带螺纹的） 768
用柔性肘板固定 171
用油缓冲(油垫) 620
用于预洗的最小水量 572
用整个舱室或货船舱隔离 786
用作填料的材料 650
优惠 696
优惠贸易安排 689
优惠性配额 689
优先(超越)权限 648
优先靠码头 356
优先顺序/优先性 696
优先脱扣 689
由主机驱动 535
邮包收据 653
邮包险 686
邮包一切险 686
邮包运输 653
邮件 534
邮件舱 534
邮政服务 686
油/压载兼用舱 627
油、气、水处理区 627
油-气分离器 21
油泵 622
油泵舱通风 114
油泵舱通风机 708

油驳 620
油布(防水布) 620
油布 627
油舱(油柜) 620
油舱(油箱,油柜) 625
油舱 620
油舱舱口 625
油舱盖 627
油舱集中换气系统 168
油舱漆 625
油舱清洗机 625
油舱梯 625
油舱压载水油分离装置 67
油槽(油路) 620
油槽(油路) 628
油槽 622
油槽失火 682
油槽錾 627
油船 625
油船系统 883
油淬钢 622
油淬火 622
油淬硬化 622
油底壳(储油槽) 625
油底壳 870
油分计 620
油分离机 129
油分离机工作水柜 708
油分离机热水柜 708
油分离器(分油器) 625
油分浓度计 620
油封 625
油封装置 625
油观察玻璃 625
油管 622
油管 622
油管 920
油管接头 921
油管路 622
油含量不大于1.5%,含水量不大于11%的种子饼 778

1188

油壶 620
油环轴承 624
油缓冲器(油减振器) 620
油缓冲器 620
油混合物 622
油迹 908
油基泥浆 627
油基泥浆 628
油井特殊操作指南 986
油开关 621
油可以渗入的区域 832
油孔 622
油类(加油,涂油) 620
油类 628
油类火灾 621
油类记录簿 622
油类记录簿第Ⅰ部分(机器处所的作业) 623
油类记录簿第Ⅱ部分(货油/压载的作业) 622
油量瞬间排放率 461
油料间 624
油料码头 622
油路 627
油密(油封) 627
油密舱壁 627
油密的 627
油密横舱壁 628
油密接头(油密接合) 628
油密肋板 627
油密铆 628
油密外壳 627
油密性 627
油密性 628
油密性试验(验油) 627
油密性试验 627
油膜 621
油泥舱(残渣收集器,油渣舱) 823
油泥管 823
油泥收集泵 823
油盘 627

油漆间 651
油气(油质,含油) 627
油气 627
油气分离 627
油气两用双燃料内燃机 260
油气生产 622
油区终端平台 323
油燃烧器 620
油燃烧设备 620
油软管吊柱 622
油润滑 622
油润滑摩擦副 627
油润滑尾管轴承 625
油润滑轴承 627
油石 625
油水分离器 628
油水分离器 963
油水分离器容器 787
油水分离设备 628
油水分离装置 628
油水供给船 362
油水界面探测仪(油水界面探测器) 627
油水界面探测仪 627
油田服务 627
油位标志 622
油位表 622
油位指示器 361
油温 627
油污(石油污染) 622
油污 620
油污染指数 E 622
油污水收集和处理系统 83
油污应急计划 797
油雾 625
油雾法(油雾试验) 621
油雾检测器 622
油吸材 625
油隙 620
油性废弃物 627
油性混合物 628

油性抹布(含油碎布) 628
油性有毒液体物质 622
油压 622
油压密封试验 622
油压试验 622
油浴回火 627
油渣(油泥,污泥) 823
油渣 623
油渣泵(油泥泵) 823
油站 891
油遮断器 670
油脂(涂油,润滑) 392
油脂杯(牛油杯) 392
油脂密封 392
油脂填料 392
鱿鱼钓船 846
游步甲板 699
游览船 677
游艇 677
游艇 677
游艇 987
游艇形式检验证书 929
游艺室(豪华邮轮) 951
游泳池(豪华邮轮) 536
有 DP-1 附加标志的船舶 949
有 DP-2 级附加标志的船舶 949
有 DP-3 附加标志的船舶 949
有槽螺母 823
有挡焊接锚链 273
有档锚链 864
有调声 903
有毒和腐蚀性物质 908
有毒货物 609
有毒气体 609
有毒气体和蒸发气 406
有毒气体探测装置 370
有毒物质(第 6.1 类危险货物) 908
有毒液体物质(NLS)液货船 599
有毒液体物质/其他物质 610
有毒液体物质 406

有毒液体物质 599
有毒液体物质 599
有毒液体物质 609
有毒有害气体排放控制 414
有缝管 822
有杆锚 860
有关当局 44
有关法律 732
有关利益 731
有害缺陷 406
有害水生物 406
有害水生物和病原体 406
有害损伤 406
有害物质/有毒物质 406
有害物质/有害材料 414
有害物质的控制 181
有害物质清单 480
有害物质清单应列的最少项目 572
有害有毒物质基金 428
有机玻璃 669
有机发光二极管 642
有机碳正态分布系数 641
有机纤维材料 641
有节稳索夹头 657
有缆遥控水下机器人 734
有螺纹的管子 895
有平滑壁的舱 883
有人处所 541
有人机舱 288
有人驾驶的(有人操纵的,有人管理的) 541
有人压力容器 692
有人值班机器处所 541
有色金属 603
有实质的（本质的,重大的） 867
有限的叶数影响 270
有限航区 741
有线耳麦 978
有线数据通信 104
有线网卡 939

有线信道 978
有效波陡最大值 492
有效波倾 270
有效的带板面积 269
有效的法定证书 940
有效的仲裁条款 940
有效功率(有效马力) 269
有效功率 269
有效接地系统 269
有效扣板 269
有效期限 940
有效热效率 93
有效上层建筑 270
有效声压 270
有效推力 270
有效脱离船舶 269
有效效率 269
有效卸货板 270
有效性 62
有效性原则 270
有效压力载荷 593
有效压缩比 269
有效载荷(净载荷,净负载) 593
有效载荷 657
有效值检波器 966
有效纵舱壁 269
有形贸易 952
有义波高 815
有源电子扫描阵/自动电子扫描相控阵 9
有源滤波器 9
有源相控阵雷达 9
有杂声的运转 605
有载脱开 631
有支撑的平壁板（锅炉） 855
有重大利益关系的国家 867
有阻尼横摇 739
有罪判决 786
右舷 853
右舷主机 853

右旋 745
右旋螺旋桨 745
右转机组 745
诱导比 453
诱导器 453
诱鱼灯船 336
淤泥（污水,垃圾） 869
余数 733
余弦波 189
鱼（散装） 336
鱼舱 336
鱼舱盖 336
鱼舱进水 345
鱼类加工船 336
鱼类加工船 336
鱼类实验室 444
鱼鳞式 147
鱼品加工 336
鱼品加工间 336
鱼探仪 336
鱼箱 336
渔船 336
渔船 337
渔船 337
渔获量 337
渔获物 336
渔获物处理 404
渔具 337
渔具舱 337
渔捞甲板 337
渔业补给船 336
渔业船 336
渔业船舶 336
渔业调查船 336
渔业辅助船 336
渔业辅助船 336
渔业基地船 336
渔业监督船 336
渔业监督船 336
渔业监视指导船 336
渔业检查船 336

渔业救助船 336
渔业实习船 336
渔业训练/调查船 337
渔业指导船 336
渔政船 336
与船舶建造有关的建造阶段 173
与概率水平对应的系数 354
与概率水平对应的系数 f_p 152
与海难相关的材料破损 552
与货物相关联的废弃物 110
与某一结构有关的水密 965
与设计温度有关的外部结构 313
与物资装运设备有关的验证载荷 700
与运输有关的服务 916
宇宙飞船 833
宇宙观测船 614
宇宙射线观测 189
宇宙射线观测室 189
雨淋系统 222
雨雪扫除器 147
语言翻译 491
语言清晰度指数 843
语音聊天 954
语音信息 954
预报警水位 688
预备庭 689
预处理 692
预定的最不利情况 985
预动作系统 688
预防措施 692
预防性风险控制 692
预付货款 658
预付款 14
预估费用/暂定价 298
预计到达时间 298
预警报 15
预警和控制系统飞机 23
预警机 266
预警雷达 266
预聚焦基座灯泡 689

预冷 688
预冷器 688
预留成本 189
预留厚度 738
预期短路电流（针对开关电器） 703
预期相遇的最近点 192
预期相遇最近点时间 886
预燃 688
预燃室 688
预热 689
预热器 352
预热器 689
预热区 689
预热时间 689
预热温度 689
预审 689
预算 99
预涂 863
预洗 692
预洗程序 692
预洗系统 692
预压缩 688
预应力 692
预约保险 631
预张力 692
预支信用证 38
域名服务 250
阈值水平 896
遇水反应物质 79
遇水反应物质 962
遇水散发易燃气体的物质（第4.3类危险货物） 867
遇险阶段 245
遇险情况 246
员工流失 982
原动机 695
原告 674
原声版本 642
原始型值 689
原始型值表 689

原始修理工程单 512
原始证据 642
原位 460
原型 705
原型机(样机) 705
原型设备制造厂 642
原型试验 705
原型艇 705
原型压载水处理技术 705
原油/成品油油船 197
原油 196
原油舱区 196
原油储存区 196
原油外输系统 627
原油洗舱 191
原油洗舱 197
原油洗舱操作与设备手册(COW手册) 196
原油洗舱设备 196
原油洗舱系统 197
原油洗舱装置 196
原油油船 196
原则 46
原子光谱分析 54
圆-双曲线系统 137
圆棒型拉伸试样的焊缝拉伸试验 890
圆舭部 826
圆弧舷顶列板 752
圆盘离合器 242
圆气阀 672
圆筒形浮式钻井生产储油船 202
圆筒形钻井平台 202
圆形卸扣 92
圆柱形通风帽 202
圆柱形月池 202
圆锥-平行轴减速齿轮箱 653
源 831
源舱室 831
源程序 831
远程方法调用 734

远程终端　734

远离　63

远期汇票　902

远期信用证　902

远洋船　617

远洋船　772

远洋船舶　616

远洋船舶　772

远洋航区(大洋航区)　615

远洋救生打捞船　616

远洋救助拖船　616

远洋科学考察船　616

远洋训练船　616

远洋综合测量船　616

约定皮重　166

约束模操纵性试验　109

约束条件　741

约束阻尼层　173

约束阻尼层结构　173

约束阻尼结构 黏弹性阻尼结构　173

月算术平均温度　578

月牙切面　193

钥匙箱　488

阅卷笔录　724

越控　647

越控按钮　648

越控设施　647

越控系统　648

越控装置　648

越障高度　615

云状空泡　148

允许的最大通过吃水(巴拿马运河)　556

允许的最小角度　28

允许间隙　662

允许载荷重新分布的屈曲能力(方法1)　98

允准的最大吃水(巴拿马运河)　556

运动毛细常数　152

运动黏性系数　152

运动稳定性指数　579

运动相似　489

运动型多功能车　845

运动坐标系　88

运费、保险费付至指定目的地　118

运费付至……　118

运费付至指定目的地　118

运费预付　359

运管驳船　670

运管船　670

运煤船　155

运木船　902

运肉船　560

运牲畜船　512

运输包装　916

运输标志　808

运输船　916

运输单据　916

运输港站经营人　639

运输工具　561

运输合同　179

运输合同　916

运输设备　916

运水船/供水船　963

运水果船　361

运算器　47

运行参数　637

运行短路分断能力　788

运行控制中心　637

运行区域1　755

运行区域2　755

运行区域3　755

运行区域4　755

运行区域　787

运行维修(日常维护)　635

运用要求　638

运载火箭　118

运载效率系数　515

运转(运行)　637

运转时数　635

Z

杂质　447

灾害　120

灾害性危险品船舶　413

灾难性影响(灾难性后果)　120

载驳船　72

载荷　513

载荷曲线　513

载荷系数　513

载货量　111

载货区域　110

载冷剂　775

载冷剂泵　186

载热体(传热介质)　417

载人航天　541

载运能力　108

载运危险货物船舶特殊要求的符合证明　249

载重吨位 DW　214

载重量(载货量,承载能力)　513

载重量　214

载重量　904

载重量鉴定　214

载重物车辆　420

载重系数　214

载重线　513

载重线　518

载重线标尺　214

载重线标志　518

载重线长 $L_L(m)$　514

载重线吃水　514

载重线方形系数 C_{bL}　514

载重型滚装货船　969

再保险　731

再充电(再补气,再充灭火剂,再装填,再补料)　723

再点火(重新启动)　732

再加工　732

再热　730

再热处理　731

再热锅炉 730
再热炉(重热炉) 731
再热器 731
再热式汽轮机 731
再热循环 730
再热循环 731
再热循环汽轮机装置 731
再热蒸汽机 731
再审案件 119
再审程序 742
再审申请书 666
再生通风机 22
再循环 724
再循环泵 724
再循环闪发蒸发器 724
再循环装置 724
再液化 732
再液化系统 732
再液化装置 732
再蒸馏 726
在 1979 年 12 月 31 日或以前交船的船舶 795
在 1979 年 12 月 31 日以后交船的船舶 795
在 1982 年 6 月 1 日或以前交船的油船 626
在 1982 年 6 月 1 日以后交船的油船 625
在 1996 年 7 月 6 日或以后交船的油船 626
在 1996 年 7 月 6 日以前交船的油船 625
在 2002 年 2 月 1 日或以后交船的油船 626
在 2010 年 1 月 1 日或以后交船的油船 626
在 2010 年 8 月 1 日或以后交船的船舶 795
在地面效应区外飞行(像飞机一样)模式 347
在海上保持永久关闭的开口 634

在海上使用的门 250
在寒冷条件下焊接的露天保护 968
在航 933
在建船舶 795
在近海海域作业 256
在使用活性物质或制剂方面的实质相似 868
在线服务公司 31
在线信息 631
在沿海海域作业 256
在用户工厂进行的焊接程序试验 970
在用配置 171
在站位的伸手可及范围 980
在遮蔽海域作业 256
暂定价格 890
暂时硬度 889
暂行规范 890
遭遇角 282
遭遇浪向 36
遭遇频率 359
遭遇圆频率 138
遭遇周期 660
早期生产系统 266
早燃 689
造波机 966
造成(补充) 539
造船厂评估 51
造气涂料焊条(气体保护型焊条) 373
造水机 360
噪度 600
噪声(杂声,杂波) 599
噪声测距声呐 600
噪声测量仪 600
噪声调制 600
噪声发生器 600
噪声感受性(噪声灵敏性) 600
噪声级 600
噪声检验报告 600

噪声检验报告 876
噪声控制措施 600
噪声控制指标分配 246
噪声识别 600
噪声试验 600
噪声衰减器 600
噪声特性测定试验 600
噪声特征 600
噪声统计 600
噪声统计分析仪 854
噪声统计模型 854
噪声源 600
噪声指数 600
责令具结悔过 641
责任人员(澳大利亚海事安全局海事指令) 740
责任人员[2] 163
增产作业 449
增产作业 860
增产作业船 449
增产作业船 860
增加裕度 399
增加质量法 11
增价拍卖 14
增强 731
增强残存能力 288
增强的船舶交通管理系统 302
增强群呼系统 288
增速比 449
增速齿轮 449
增速齿轮箱 449
增稳 57
增压(增压作用) 871
增压 871
增压泵 871
增压泵 90
增压泵 904
增压比 691
增压柴油机 871
增压锅炉 871
增压器 871

增压器滑油泵 922
增压器配机试验 922
增压式柴油机 871
增压室气垫船 677
增压涡轮 166
增压压力 871
增压压气机 871
增压蒸汽发生器 871
增氧装置 449
增值税 449
渣打银行 850
渣油泵 738
轧后(AR)状态 51
轧制 50
轧制型材 749
轧制状态 50
闸(闭锁装置,自动跟踪) 519
闸刀式制链器 72
闸刀止链器 501
闸阀 373
闸阀 823
闸阀的闸板 813
闸门 823
闸门室(水闸) 519
闸式可潜器系统 519
栅栏门 493
栅形阀 395
栅形膨胀阀 395
炸礁船 932
炸药 310
窄带直接印字电报终端 585
窄馏分油 417
窄稍叶 585
债权人会议 193
沾污 680
毡垫片 320
毡垫圈 320
毡填料 320
毡纤维 320
斩波器 137
展开轮廓 236

展开面积 236
展开盘面比 236
展弦比 51
战斗机 323
战区 893
战区弹道导弹防御 893
战区导弹防御 893
战术导弹防御 881
战争险 959
站 641
张紧式 885
张紧式系泊 885
张紧索 885
张力器/张紧器 890
张力腿平台 890
胀管式 219
障碍 614
障碍限制区(LOS) 507
招标 480
招标、投标买卖 480
招风斗 975
招商局集团有限公司 135
照度 445
照明 445
照明分支电路 505
照明引出线端 494
照相温度表 323
罩壳 490
罩壳式艉门 952
遮蔽航区内作业的起重船 344
遮蔽航区营运限制 794
遮蔽甲板船 794
遮蔽水域 794
遮蔽水域 794
折板式结构 428
折板式消音器 428
折边 341
折边 489
折边的(用法兰连接的,带凸缘的) 341
折叠式舱口盖 349

折叠收放式减摇鳍 349
折叠型舱盖 428
折减系数 726
折角线 489
折扣 243
折流板式 219
折流板式塔 67
折曲式 487
折射波勘探 727
蔗渣膳食纤维 67
蔗渣纤维P/VC复合材料 67
针阀升程 592
针杆阀 592
针孔 670
侦查阶段 480
侦查终结 168
珍珠岩 661
真方位 920
真航向 920
真空泵 301
真空表 940
真空补水(冷凝器) 940
真空抽气泵 940
真空处理 940
真空度 501
真空二极管 940
真空阀(电子管) 940
真空盒法(真空试验) 940
真空试验 940
真空压力计 940
真矢量 920
真速度 920
真运动 920
真运动显示模式 579
振荡(摆动) 642
振荡(摆动) 642
振荡波 643
振荡传感器 642
振荡的(摇摆的) 642
振荡环节 642
振荡频率 642

振荡器(振子) 642
振荡器试验(振动仪试验) 642
振荡试验 642
振荡载荷 642
振荡周期 642
振荡周期 642
振动(冲击,裂缝) 792
振动 949
振动波形 951
振动传感器 950
振动传感器 951
振动传感器 951
振动辐射噪声 949
振动活塞式取样管 810
振动级 950
振动计 951
振动计算(振动分析) 950
振动记录 951
振动记录仪 951
振动监测 950
振动节点 950
振动理论 951
振动力 951
振动耐久试验(耐振试验) 950
振动能力 950
振动扭矩 951
振动疲劳 951
振动频率 950
振动器(振子,断续器) 951
振动强度 950
振动试验 951
振动试验台 951
振动适应性试验 951
振动损害 950
振动台 951
振动台 951
振动台试验 792
振动特性测定 951
振动特征 951
振动响应 951
振动信息 951

振动仪 951
振动仪试验 642
振动应力 951
振动应力 951
振动载荷 950
振动载荷 951
振动指示器 950
振动周期 951
震源 284
镇静钢/脱氧钢 489
镇静药 777
镇浪油 861
争议 245
征询原、被告最后意见 174
蒸-燃联合装置 157
蒸发(气化) 942
蒸发 301
蒸发 857
蒸发点 942
蒸发管束 386
蒸发盘管 302
蒸发气处理系统(蒸发气处理单元) 942
蒸发气处理系统 942
蒸发气管路 942
蒸发气回收系统 942
蒸发气控制系统 942
蒸发气控制系统—中转 942
蒸发气排放控制 942
蒸发气排放控制系统 942
蒸发气排放收集系统 942
蒸发气平衡 941
蒸发气收集系统 941
蒸发气套的泄水阀 486
蒸发气体推进系统 89
蒸发气压力 301
蒸发气压力 942
蒸发气再液化系统 942
蒸发气制冷循环 942
蒸发气总管 942
蒸发器(汽化器) 942

蒸发器 302
蒸发器舱 302
蒸发热 302
蒸发式喷油嘴 941
蒸发速度 943
蒸发压力 301
蒸发压力调节阀 66
蒸馏法 245
蒸馏器 245
蒸馏器 941
蒸馏设备 245
蒸馏水 245
蒸馏水带热损失 245
蒸馏水柜 245
蒸馏水检验柜 168
蒸馏水检验柜 245
蒸馏装置(制淡水装置) 360
蒸馏装置 245
蒸汽(蒸汽的) 856
蒸汽(喷射)抽气泵 856
蒸汽-电力推进 856
蒸汽泵 857
蒸汽泵滑阀的空动 857
蒸汽参数 856
蒸汽出量 856
蒸汽吹洗 856
蒸汽纯度 857
蒸汽带水 119
蒸汽笛 857
蒸汽动力装置 857
蒸汽动力装置试验 857
蒸汽舵机 857
蒸汽发生器 856
蒸汽饭锅 744
蒸汽供热系统 856
蒸汽管 856
蒸汽管系 856
蒸汽锅(水壶) 488
蒸汽锅炉 856
蒸汽过热器 857
蒸汽过热器超温报警 872

1195

蒸汽耗量　856
蒸汽机-乏汽轮机联合装置　157
蒸汽机动力装置　856
蒸汽机货船　359
蒸汽加热的蒸汽发生器　857
蒸汽加热炉　857
蒸汽加热器　856
蒸汽加热式再热器（蒸汽再热器）　857
蒸汽绞车　856
蒸汽绞车　856
蒸汽进口压力表　856
蒸汽轮机驱动交流发电机组　857
蒸汽轮机与燃气轮机联合动力推进　154
蒸汽密封性试验　857
蒸汽灭火　856
蒸汽灭火　857
蒸汽灭火系统（蒸汽窒息灭火系统）　857
蒸汽凝结区域　856
蒸汽泡沫灭火系统　16
蒸汽喷射（喷汽口）　856
蒸汽喷射空气抽除泵　856
蒸汽喷射油气抽除装置　856
蒸汽品质　857
蒸汽起货机　856
蒸汽起锚机　857
蒸汽起锚绞盘　856
蒸汽腔　857
蒸汽试验　857
蒸汽室　856
蒸汽受压管　857
蒸汽通道　856
蒸汽退火　856
蒸汽拖缆机　857
蒸汽温度自动调节装置　857
蒸汽雾化喷油器　856
蒸汽洗池　857
蒸汽熏舱系统　882
蒸汽循环管路　856

蒸汽压力　857
蒸汽压力表　856
蒸汽压力测量设备　857
蒸汽压缩机　941
蒸汽窒息灭火装置　857
蒸汽贮存器　857
蒸汽阻汽器（阻汽器）　857
拯救　743
整笔运费租船　530
整定值（调定值）　790
整个系统配置　171
整流隔板　440
整流器　725
整流叶片　862
整流罩　317
整数　463
整体弹药库　463
整体法兰联轴节　827
整体腐蚀　388
整体螺旋桨　827
整体式浮船坞　105
整体式浮船坞　819
整体液货舱　464
整叶桨（定距螺旋桨）　339
正摆线推进器　489
正铲挖泥船　98
正常操作和居住条件　606
正常排水量　606
正常压载吃水 T_{bal-n}　605
正常压载工况　605
正常运行条件（正常航行情况）　606
正常状态　605
正常作业工况　606
正车级　19
正车透平　19
正齿轮　846
正齿轮　846
正垂荡　419
正垂荡　685
正垂向弯矩　686

正点观测　35
正浮　344
正横荡　685
正横航线距离　245
正横矩　912
正横摇　685
正横摇　748
正火（N）　607
正火轧制 NR　607
正火状态 N　607
正交力　195
正交谱　710
正角斜齿轮　574
正螺线试验　241
正平衡　417
正时　903
正式合同　352
正式授权的官员　262
正艏摇　686
正水平弯矩　685
正态分布　606
正稳性　685
正弦波　819
正弦形操纵试验　819
正在通过　657
正纵荡　685
正纵摇　685
证件　248
证据/凭证　302
证据保全　690
证据保全　302
证据保全　690
证据保全申请书　42
证据目录清单　302
证券报价　713
证人　980
证人证言　892
证书　129
证书　130
证书签发机关　485
政府间气候变化专门委员会　467

政府行为　390
支撑剂储存装置　700
支撑器　191
支撑装置　873
支承隔板　921
支持负载　980
支持级(人员)　873
支持证据　874
支付令　658
支付条件　891
支票　134
支索起重柱　855
支索桅　855
支线船　320
支悬高度　876
支柱　670
支柱体(小水线面双体船)　947
支柱体宽度　947
支柱体水线长度 Ls　947
知道时间/觉察期　63
知识产权　466
知识产权庭　466
执行　285
执行程序　697
执行逮捕　306
执行舵角　306
执行和解　168
执行回转　725
执行机构　10
执行时间　492
执行庭　306
执行异议　614
执行员　306
执行中止　243
执行终结　168
直布罗陀海峡　862
直达提单　240
直读式海流计　240
直方图　428
直管式消音器　862
直角传动推进　745

直角传动转向螺旋桨　745
直接补贴　241
直接产品补偿　240
直接传动　897
直接传动式汽轮机　241
直接发生的故障　615
直接辐射强迫(正强迫)　240
直接过境贸易　241
直接还原铁(B)　240
直接还原铁(C)　241
直接还原砖状块铁(A)(热铸)　240
直接计算　336
直接接触冷凝　241
直接冷却系统　240
直接连接　529
直接连接在锅炉及压力容器上的阀门　940
直接贸易　241
直接喷射燃烧室　241
直接式风冷系统　240
直接印字电报　242
直接蒸发　240
直接蒸发式空气冷却器　240
直接装注油管　240
直接作用阀　686
直径　236
直径系数　885
直立型艏　946
直流电　203
直流电动传令钟　203
直流阀　240
直流换向电机　203
直流均压机　203
直流扫气　897
直流扫气　934
直流式闪发蒸发器　630
直流式循环水系统　767
直馏汽油(直馏油)　862
直馏渣油　862
直升机港　421

直升机甲板(平台)　421
直升机甲板　421
直升机降落和起离区　881
直升机降落甲板　421
直升机设施　421
直升机运输服(救生服)　421
直升式平台　781
直跳板　745
直通电话系统　862
直形卸扣　862
直烟管立式锅炉　946
直摇式水密滑门　947
直叶片　862
直叶推进器　955
直翼推进器　766
直翼推进器　819
值班安全系统　952
值班报警　960
值班泵　263
值班驾驶人员　631
值班驾驶员　618
值班轮机员室　960
职能　364
职业事故　615
职员(国际海事卫星组织)　849
职责和职权　740
植物检疫证书　669
植物纤维　675
植物油运输船　943
止回的(不倒转的)　604
止回阀(背压阀)　66
止回阀　605
止回吸入口　605
止回装置　604
止裂器　192
止屈器　98
止推垫圈(保险垫片)　960
止推垫圈　898
止推压力　897
止转转矩　751
只读光盘　126

纸牌室(豪华邮轮) 110
纸制品 652
指标价格 884
指导性意见 400
指定(详细说明) 843
指定 601
指定安全地点 234
指定保税区 234
指定辩护 43
指定当局 234
指定地点的边境交货 221
指定地点的工厂交货 302
指定地点的货交承运人 355
指定目的港的成本加保险费加运费 190
指定目的港的船上交货 221
指定目的港的船上交货 355
指定目的港的装运港船上交货 355
指定人员 234
指定特殊区域和确定特别敏感海区导则("导则") 400
指定危险区 234
指定项目 724
指定一个特别敏感海区 444
指定仲裁员声明 854
指定装运港的船边交货 355
指挥和控制目标(消防控制) 159
指挥和控制目标(救生和人员撤离) 159
指挥和控制目标1(救生和人员撤离) 158
指挥和控制目标1(消防控制) 159
指挥和控制目标2(救生和撤离) 159
指挥和控制目标2(消防控制) 159
指挥人员 690
指挥视野 159
指挥仪室 334

指令舵角 641
指令顺序 787
指南 400
指示表刻度盘 451
指示灯(信号灯) 451
指示电路 451
指示阀 451
指示功 451
指示功率(额定功率) 451
指示功率 451
指示面板 451
指示面板 451
指示器(显示器,计量表) 451
指示器(指示设备) 451
指示器 451
指示热效率 451
指示式回声测深仪 451
指示提单 641
指示图 451
指示图 451
指示效率 451
指示压力 451
指示仪表 451
指示元件 451
指示值 451
制单结汇 250
制荡舱壁 959
制动功率 93
制动力矩 429
制动螺栓 861
制动器(制链器,塞子) 861
制动条 861
制动载荷 429
制冷(冻结,凝固,冻融) 358
制冷过程 697
制冷机 728
制冷机舱 729
制冷机组(制冷设备) 728
制冷剂 694
制冷剂 728
制冷剂回收设备 728

制冷剂冷凝器 728
制冷设备 729
制冷试验 729
制冷系统 728
制冷循环 729
制冷压缩机 728
制冷装置 728
制链器 131
制氢室 440
制造(组成) 539
制造厂商(船用卫生装置) 542
制造厂商(救生艇) 542
制造厂商 542
制造厂商颁发的根据"材料质量方案"的证书 543
制造厂商证明 543
制造厂商证书 542
制止(禁止) 457
质量(船舶实体) 551
质量 710
质量保证 710
质量单 969
质量方针 710
质量管理 710
质量管理计划 710
质量检验证书 461
质量密度 842
质量体系文件 710
质移 551
致船东的备忘录 564
致命事故发生率 318
窒息灭火系统 333
窒息灭火装置室 333
蛭石 946
蛭石板 946
蛭石砖 946
智库 93
智能手机 824
智能网络处理器 466
智能网络服务 466
智能型柴油机 466

滞后（绝热装置，绝缘层） 490
滞期费 222
滞期期限 222
滞期条款 222
滞燃 340
滞燃材料 340
滞住（材料黏附在模子上，捆绑） 780
置信限 171
中船澄西远航船舶（广州）越秀公司——华南海上改装修理"桥头堡" 134
中船黄埔文冲船舶有限公司 198
中垂/下沉（下垂） 760
中等冰况区域航行 443
中底桁 91
中断 85
中舵 127
中高速柴油机 569
中拱 428
中国-东盟自贸区 135
中国船舶工业系统工程研究院/海洋电子科技有限公司——海洋防务装备的"推动者" 136
中国船舶建造标准合同 135
中国船级社 134
中国对外经济贸易仲裁委员会 135
中国工程物理研究院 134
中国国际海运集装箱（集团）股份有限公司 135
中国国际金融有限公司 135
中国国际贸易促进委员会 134
中国海事仲裁委员会 135
中国籍船舶 801
中国联合通信有限公司 135
中国社会科学院 136
中国水域 135
中国网络通信有限公司 135
中国物品编码中心 48
中国移动通信集团公司 135

中和轴区域 595
中后机型船 18
中华人民共和国仲裁法 45
中机型船 31
中级人民法院 468
中间裁决 468
中间舱 128
中间的（中级的） 468
中间舵叶 127
中间盖板 466
中间机匣 468
中间加热 849
中间甲板 923
中间检验 468
中间肋骨 468
中间冷凝器 466
中间冷却（中冷） 466
中间冷却空气发动机 466
中间冷却器（内冷器） 466
中间散装集装箱 468
中间压力 468
中间支承 468
中间轴 468
中间轴 468
中间轴承 468
中间轴承 921
中间轴承座 468
中间轴承座 921
中间轴压盖 678
中介轴 312
中空（空心的，凹槽） 429
中空结构 429
中空填棉板 430
中括号 93
中立国 595
中立国船（战时） 355
中美洲自由贸易协定 929
中内龙骨 127
中频 568
中频弯管 564
中剖面系数曲线 199

中枢的/关键的 673
中速柴油机 564
中途退庭 742
中途退庭 979
中艉轴管 468
中线螺旋桨 129
中线面 127
中心锚泊钻井船 127
中心系泊定位钻井船 127
中性包装 595
中性轴（中和轴） 595
中压压缩机 468
中央处理机时间 128
中央处理器 128
中央井 129
中央警报管理 106
中央警报管理 127
中央警报管理-人机界面 106
中央警报管理-人机界面 128
中央警报管理系统 106
中央控制室 128
中央露天舞台（豪华邮轮） 129
中站 569
中站面 569
中止船级 141
中轴系 127
中子弹 595
中纵舱壁 127
中纵舱壁 522
中纵桁材 127
终端蒸汽接头 891
舯部区（冰区加强） 569
舯部区 569
舯垂线 569
舯干舷 F_M 358
舯横剖面 569
舯横剖面模数 776
舯横剖面系数 569
舯剖面 31
仲裁 45
仲裁被诉人 740

仲裁标的 45
仲裁裁决 63
仲裁程序/仲裁流程 45
仲裁代理人 45
仲裁法 45
仲裁费用表 45
仲裁规则 46
仲裁机构 45
仲裁申请材料 736
仲裁申请书 42
仲裁申诉人 138
仲裁声明书 46
仲裁时效 506
仲裁条款 45
仲裁庭 45
仲裁通知 609
仲裁委员会 45
仲裁委员会的渔业争议解决中心 336
仲裁委员会物流争议解决中心 520
仲裁文件 45
仲裁协议 45
仲裁员 46
仲裁员办理案件的特殊报酬 46
仲裁员名册 652
重柴油 420
重铲 405
重锤张力装置 190
重大改建 538
重大改装 28
重大后果(重大影响) 539
重大事故 538
重大修理 539
重点检查 488
重点控制的污染物 696
重吊杆座 420
重吊起货机 420
重定位 727
重复测量 263
重谷物 420

重级别油/重质油 420
重建 722
重晶石舱 72
重力(万有引力) 391
重力 391
重力波 391
重力测量 392
重力柜 646
重力活塞式取样管 392
重力加速度 3
重力加速度 391
重力勘探 391
重力勘探 392
重力路线测量 392
重力锚 391
重力密度 841
重力平衡舱盖 391
重力平衡窗 977
重力曲线 969
重力取样管 391
重力燃油柜 361
重力燃油柜 362
重力润滑系统 392
重力式测波仪 343
重力式滑油系统 391
重力式基础 391
重力式平台 392
重力式润滑系统 392
重力式艇架 392
重力输送直卸式散货船 391
重力水柜 391
重力稳性臂 501
重力循环 391
重力液舱 391
重力液货舱 392
重力仪绞车 391
重力油柜 622
重力油柜 622
重力载荷 392
重燃油 420
重热效率 730

重心 127
重心垂向坐标 946
重心横向坐标 916
重新报价 734
重新检查 735
重新注册 736
重型吊杆 420
重型舷窗 420
重型修井作业 420
重压载工况 419
重要标示 447
重要部件 295
重要舱室(舰船) 194
重要的管路系统 296
重要电话系统 447
重要辅机 295
重要辅助锅炉 295
重要管路和电缆线路 195
重要机械 296
重要设备(船舶) 298
重要设备(船舶或移动式近海装置) 298
重要设备(舰船) 296
重要设备 295
重要系统 447
重要用途 297
重要用途的所有其他锅炉(辅锅炉、废气锅炉、经济器和蒸汽加热蒸汽发生器) 60
重要用途的压力容器 296
重要用途辅锅炉 60
重要用途供电 296
重油 420
重质油 727
舟桥 682
周 202
周边射流气垫船 661
周界灯 660
周年日期 37
周期 660
周期无人值班机器处所 660

周期性无人值班　660
周日连续观测　246
周围压力　30
周向变螺距　138
周向感生速度　138
周向进流因数　138
周向速度　138
轴（竖井，通风井）　791
轴（芯轴，杆）　844
轴包板　91
轴包覆　791
轴包套/轴包架　91
轴衬　791
轴承　508
轴承　791
轴承衬　509
轴承高温报警装置　425
轴承过热　646
轴承间隙　77
轴承跨距　77
轴承烧损　76
轴承支座　659
轴承座　791
轴传动组件　791
轴带发电-电动机组　791
轴带发电机　791
轴带发电机系统　791
轴带给水泵　537
轴带交流发电机　792
轴的位移　791
轴法兰　791
轴法兰厚度　895
轴法兰螺栓　791
轴封　791
轴封　792
轴封　859
轴封泄漏　387
轴封泄漏　387
轴功率（马力）　791
轴功率 P_b　791
轴架　792

轴检　701
轴结构　792
轴颈（枢轴，航海日记）　487
轴颈　791
轴颈　791
轴颈填料（轴颈油封）　487
轴颈圆角　323
轴颈轴承　487
轴连接　673
轴流泵　63
轴流泵　701
轴流式风机　63
轴流式汽轮机　63
轴流式涡轮　63
轴流式压气机　63
轴扭矩限位器　792
轴隧（地轴弄）　791
轴隧（轴隧端室）　792
轴隧排水泵　791
轴隧平台　921
轴隧逃生口/轴隧应急出口　921
轴隧通道　922
轴隧应急出口　791
轴套　93
轴头　844
轴系（轴线）　508
轴系　792
轴系布置　792
轴系吊架　792
轴系横向振动　493
轴系基座　792
轴系连接螺栓　792
轴系临界转速　195
轴系内倾角　185
轴系扭转振动　904
轴系倾角　36
轴系外斜角　246
轴系效率（轴系传动效率）　792
轴系效率　791
轴系效率　792
轴系振动　792

轴系制动器　792
轴系中线　508
轴系状态监测系统　791
轴系纵向振动　63
轴线　791
轴线上叶厚　85
轴向感生速度　63
轴向进流因数　63
轴向速度　63
轴向位移保护装置　63
轴向振动　63
轴向柱塞泵　63
轴效率　791
轴支架（艉轴架，人字架）　791
轴支架　702
轴支座　792
轴转数传感器　792
轴转速传感器　792
肘板（管弯头，顶推架）　489
肘板　93
肘板长度 l_b(m)　497
肘板趾端　93
宙斯盾　23
宙斯盾战斗系统　15
竹筏　70
逐步自由化　698
逐个产品认可　119
逐件试验　452
逐位计算　84
逐项检查　235
逐行倒相　666
主/辅机集中操纵台　128
主安全功能　536
主报警水位　534
主泵　536
主冰带区　536
主舱底排水系统　534
主操舵部位　537
主操舵装置　537
主操舵装置　555
主操纵盘（主驾驶盘，主齿轮）　537

主操作板 536
主尺度 535
主尺度 695
主传动装置 537
主船体 536
主电源(平台) 537
主电源 537
主电站(主发电站) 535
主动齿轮 258
主动舵 9
主动舵舵机 858
主动风险控制 9
主动减速 843
主动减摇鳍(可控的减摇鳍) 8
主动配额管理 9
主动失效 9
主动式减摇鳍 8
主动式减摇鳍装置 9
主动式减摇水舱 9
主动式灭火系统 9
主动水舱式减摇装置 8
主动系统 9
主动转向装置 8
主斗桥 534
主耳轴装置 537
主发电机舱 687
主阀［总阀,(汽轮机)速关阀］ 537
主阀 537
主隔板 536
主给水管路 535
主钩起重量 536
主管当局/主管机关 13
主管当局 163
主管当局的官员在搜查船舶时所具有的权力 688
主管工程师 287
主管机关(集装箱) 14
主管机关 44
主管机关公认的实验室 490
主管人员 14

主管人员[3] 163
主锅炉 534
主海水进水孔 536
主海水排水孔 536
主桁 536
主汇流排 534
主机(主要机械) 536
主机 535
主机舱 535
主机舱警铃系统 535
主机操纵室 535
主机功率 286
主机和可调螺距螺旋桨的遥控系统 734
主机活塞杆填料函污滑油柜 538
主机机座 535
主机基座 535
主机控制台(主机操纵台) 535
主机控制台(主机操纵台) 536
主机控制站 703
主机驱动发电机组 535
主机燃油加热器 535
主机试车 535
主机输出功率 285
主机输出功率 535
主机特征数(机号) 530
主机遥控系统 535
主机遥控装置 733
主机转速表 535
主机转速测量系统 535
主机转速禁区 72
主机总输出功率 P 905
主机最大持续转数 613
主机最小功率 572
主监视报警装置机架 552
主节流阀 537
主截止阀(LPG系统) 537
主绝缘 73
主空气瓶 534
主空压机 534
主控开关 552

主控台(主指挥所) 534
主控台 534
主控制板 534
主控制阀 535
主控制器 694
主控制系统 535
主控制站 534
主缆 536
主肋板 827
主肋骨 535
主肋骨 916
主冷凝器 534
主冷凝器循环泵 534
主令开关直接控制 552
主螺栓(中心主轴) 489
主配电板(平台) 537
主配电板 537
主屏壁/主防壁 693
主汽轮机状态监测 537
主汽轮机组 537
主汽轮机组 537
主启动按钮 552
主启动阀 537
主燃空气 693
主燃孔 694
主燃气轮机 535
主燃气轮机组 535
主燃烧区 534
主任审核员 494
主容器 694
主竖区 537
主诉检察官 695
主题标签 408
主题管辖权 865
主体结构的特殊构件(半潜式平台) 837
主体结构的主要构件(半潜式平台) 536
主推进电动机配电板 536
主推进发动机(推进发动机) 702
主推进发动机和涡轮机 536

主推进机械处所　536
主推进机械正常操作　606
主推进系统　536
主推进装置(主推进机械)　536
主推进装置　536
主推力器　536
主拖缆　908
主循环水泵　534
主压载水舱通海阀　534
主要报警点　693
主要不符合项　539
主要部件　296
主要尺度　695
主要船级标志　534
主要的　695
主要的滑油管路　536
主要附件　534
主要功能(关键功能)　296
主要构件　694
主要骨材　537
主要骨架间距　694
主要结构(钢夹层板)　537
主要开口　536
主要零件　539
主要缺陷　539
主要设备　535
主要设备　535
主要设备　694
主要信息(关键信息)　296
主要营运　296
主要用于运输散装干货　692
主要支撑构件　695
主液舱透气阀　537
主应力(初应力)　695
主振动　537
主蒸汽阀　537
主蒸汽管　537
主蒸汽管系　537
主指挥站　534
主轴　537
主轴承　534

主轴承螺栓　534
主轴颈　536
主轴线　695
主转子　536
主坐标平面　695
煮炉　89
属具间　374
助理检察官　52
助理轮机员　52
助理审判员　52
住宿船　344
住坞　855
贮备浮力　738
贮备排水量　244
贮备系数　152
贮存舱　7
贮氢室　440
贮液器　511
注册登记　125
注册吨位　729
注入(喷射)　458
注入阀　323
注入管　323
注入兼测量管　323
注入漏斗　695
注入滤网　323
注水　963
注油头　392
驻点压力　849
柱面声波　202
柱塞泵　678
柱塞式液压舵机　718
柱塞有效行程　678
柱稳、半潜式生产平台　155
柱稳式平台(半潜平台)　156
柱稳式平台的次要构件　774
柱稳式平台的特殊构件　834
柱稳式平台的主要构件　693
柱稳式钻井平台　156
柱靴　350
柱靴柱稳半潜式钻井平台　350

铸疤(瑕疵)　765
铸钢/铸钢件　857
铸钢锚链　119
铸铁　320
铸造车间　119
铸造车间　354
铸造缺陷　119
抓斗　390
抓斗机　390
抓斗机工作半径　982
抓斗机尾部半径　881
抓斗式挖泥船　344
抓斗式挖泥船　390
抓斗挖泥船　390
抓斗装卸　112
抓斗装卸结构加强　390
抓缆钩　391
爪形离合器(爪盘联轴节)　486
专家(国际海事卫星组织)　309
专家报告　735
专家系统　309
专利代理人　501
专利实施　309
专利许可证　657
专门法院　836
专门管辖　841
专门规范　838
专门贸易体系　839
专业调查船　838
专业性出口加工区　698
专用船　840
专用的备用电源　217
专用的转运系统　840
专用关闭装置　836
专用滚装船　217
专用化学品船　306
专用集成电路　43
专用清洁压载泵　145
专用清洁压载舱操作手册　217
专用实验室　837
专用污液舱　217

专用压载泵　779
专用压载布置　779
专用压载舱　306
专用压载舱　765
专用压载舱　779
专用压载舱的保护位置　704
专用压载水　779
专用压载水舱　779
专用压载水舱　835
专用压载系统　452
专职船员　666
专属管辖　306
专属经济区　306
转差率　822
转出式吊艇柱　714
转船　911
转船港　911
转船提单　916
转动（转数,扭曲）　923
转动导缆孔　936
转动惯量　752
转动计数度盘　923
转动式吸湿装置　751
转舵阶段　540
转舵时间　903
转环卸扣　878
转换处　914
转机式起重船　822
转矩计　904
转矩系数　904
转矩裕度　904
转口贸易　289
转口贸易　468
转片　752
转鳍机构　325
转让　52
转艏性　191
转艏指数　191
转艏滞后　190
转枢座　390
转速　719

转速　744
转速　923
转速表　880
转速表指示器　880
转速波动率　707
转速不稳定度　937
转速测定法　880
转速调整器　843
转速计（流量表）　880
转速计传感器　880
转速记录器　744
转速记录器　744
转速禁区　742
转速修正　187
转塔单点系泊　923
转塔式系泊系统　923
转向　744
转向　923
转向闪光灯　341
转向系统　877
转向指示器　744
转叶式电动液压传动装置　941
转叶式舵机　752
转叶式舵机　752
转叶式液压舵机　941
转义符号　295
转柱舵　752
转子　752
转子轴承　752
桩基平台/桩支承平台　670
桩孔　496
桩腿　496
桩腿　496
桩腿标尺　550
桩腿沉箱　105
桩腿横距　917
桩腿入土深度　497
桩腿顺序　787
桩腿直径　236
桩腿自升式钻井平台　486
桩腿纵距　523

桩腿总长　905
桩靴　105
桩靴　810
桩靴　846
桩靴自升式钻井平台　350
桩靴自升式钻井平台　810
装船前检验证书　690
装货　807
装货单　808
装货甲板　111
装货区域　110
装货人　808
装货时间　518
装料斗　320
装配（配合）　337
装配不良的角接焊缝　682
装配工（钳工）　337
装配公差（配合公差）　337
装配间隙　337
装设防腐材料　509
装填物（装填器,熔接料,填充丝,浇铸设备）　323
装箱单　650
装卸发动机舱口（机舱顶可拆盖）　286
装卸发动机舱口（机舱顶可拆盖）　286
装卸率　112
装卸能力　112
装卸设备（搬运设备）　404
装卸设备　111
装卸时间　493
装卸时间表　903
装卸站　890
装油港（装油码头）　622
装油软管吊杆　431
装有选择性催化还原系统发动机　286
装运港　683
装运期　903
装运索赔　808

装运通知　808
装载　515
装载比例　862
装载工况　513
装载计算机系统　515
装载量　108
装载手册(货船)　517
装载手册　516
装载仪　516
装载仪 D　515
装载仪 G　515
装载仪 I　515
装载仪 S　515
装载仪系统　515
装载指导资料　516
装载状态　516
装置(机构、齿轮、滑车、器具、开动、啮合)　373
装置(装配、供给设备)　337
装置/设备　461
状况告知　819
状态(施工)检验(防腐)　169
状态灯　854
状态监控　169
状态监控设备　169
状态监控系统　170
状态评估计划(CAS)符合证明(CAS最终报告和审核记录)　169
状态意识　819
撞击器(熔断器)　863
追加工程　873
追究刑事责任　480
追浪　648
追索权　745
锥形　884
锥形弹簧　956
锥形高弹性离合器　23
锥形离合器　170
锥形轴　884
准定常空泡　712
准许工作　663

浊度计　922
着火(点燃)　445
着火点(燃点)　455
着火点　331
着火点　445
着陆速度范围　491
着色探伤　659
资料室　209
资料通信　209
资源　740
子午卫星　913
子系统　868
姊妹船　819
紫外线　931
紫外线照射法　931
自闭阀　781
自闭式防火门　328
自闭装置　781
自存工况　876
自带行李　104
自定位自升式驳船　782
自动　57
自动闭锁装置　782
自动边墩　57
自动舱底水排出装置　81
自动操舵控制　60
自动操舵仪　57
自动出口配额制　956
自动挡火闸　58
自动点火系统　58
自动电话　59
自动调节　781
自动调节轴承　781
自动舵　58
自动发音教学机　57
自动防火风闸　327
自动隔断装置　58
自动功率管理系统(动力定位系统)　58
自动关闭系统　58
自动柜员机　59

自动焊接　59
自动滑油注油器　58
自动化系统　59
自动记录器　782
自动记录装置　782
自动加油　782
自动监测装置　59
自动控制　57
自动控制阀　59
自动控制方式(动力定位系统)　58
自动控制功能　58
自动控制门　59
自动控制系统　58
自动雷达标绘仪　48
自动雷达标绘仪　58
自动喷水灭火系统　58
自动喷水灭火系统　59
自动喷水灭火系统　59
自动喷水器、探火及失火报警系统　59
自动喷水器系统　59
自动气胀式个人漂浮装置　59
自动启动器　59
自动启动装置　59
自动清洗滑油滤清器　781
自动去污油分离机　781
自动洒水器　59
自动识别系统(AIS)目标　25
自动识别系统(AIS)最小化的键盘和显示器　574
自动识别系统　25
自动识别系统目标　444
自动识别系统目标的激活　8
自动释放　782
自动送钻杆辊道　58
自动拖缆机　59
自动温度调节装置　59
自动系泊绞车　58
自动卸载　58
自动鱿鱼钓机　846
自动张紧器　59

1205

自浮记录介质 343
自航链斗挖泥船 782
自航平台 782
自航前阻力试验 52
自航试验拖力 907
自航抓斗挖泥船 782
自航自升式平台船 782
自基线至船体横剖面水平中和轴的垂直距离（m） 584
自基线至船体横剖面水平中和轴的垂直距离 N（m） 946
自激振动 781
自检和监视功能 366
自亮灯浮 782
自流式冷凝器 767
自耦变压器式启动器 60
自驱动型（自吸型） 782
自然人 587
自然通风 366
自然通风 586
自然通风 587
自然循环锅炉 587
自然循环冷却系统 586
自然灾害 586
自然振动（固有振动） 587
自燃的 845
自燃温度 845
自润滑轴承 782
自上而下法 92
自升-步进式绞吸式挖泥船 781
自升式风电安装船 976
自升式工作平台 781
自升式平台的次要构件 774
自升式平台的特殊构件 834
自升式平台的主要构件 693
自升式平台主体结构 436
自升式平台总提升载荷 905
自升式钻井平台 486
自升式钻井平台 781
自首 170
自诉案件 696

自通风型电机 782
自位式推力轴承 13
自稳 782
自稳式一体化张力腿平台 781
自稳性 456
自我评定 781
自吸（自驱动） 782
自吸泵（自动引水泵） 782
自吸能力 782
自下而上法 904
自相关函数 57
自泄艇底塞 781
自卸货系统 112
自卸泥驳 782
自卸散货船 782
自卸系统 781
自行辩护 581
自行浮起式救生艇筏 343
自扬/链斗式挖泥船 707
自扬链斗式挖泥船 762
自摇式转盘滑道 782
自液舱最高点至基线的垂直距离 989
自液舱最高点至基线的垂直距离 Z_{top} 946
自由边界区 355
自由变形方法 358
自由导向滑车 97
自由锻造/无型锻造 631
自由港 355
自由航速 355
自由横摇 355
自由降落的核准高度 358
自由降落加速度 358
自由降落下水 358
自由流孔 355
自由贸易 356
自由贸易区 356
自由贸易政策 356
自由抛落式救生艇 358
自由飘浮下水 343

自由扰动运动 355
自由声场 355
自由涡 356
自由心证制度 248
自由液面 356
自由液面减摇水舱 347
自由液面修正 356
自由液面影响 356
自由站立式组合立管 356
自由张紧索 356
自由振动 356
自由自航模操纵性试验 355
自由阻尼层 355
自由阻尼层板 931
自愿增加厚度 895
自愿增加厚度 956
自噪声 781
自侦案件 782
自治 60
自重载荷（起重设备） 214
自主配额 59
自作用的个人漂浮装置 781
字符串 863
纵舱壁 522
纵荡（喘振） 875
纵骨 522
纵骨的间距 833
纵骨架式 523
纵骨架式船 523
纵骨架式舷侧结构 523
纵桁 387
纵距 14
纵剖线 103
纵剖线图 793
纵倾（纵摇） 673
纵倾 522
纵倾 919
纵倾 919
纵倾调整 132
纵倾角 919
纵倾力矩 919

纵倾平衡泵（纵倾平衡系统泵） 919
纵倾压载水舱 919
纵稳心 522
纵稳心半径 522
纵稳心垂向坐标 946
纵稳心高 522
纵稳性 523
纵向波 523
纵向传送 523
纵向船体梁结构构件 522
纵向浮心（自艉垂线距离） 522
纵向隔壁 522
纵向联杆 523
纵向漂心 522
纵向破裂 825
纵向强力构件 523
纵向稳性及纵向强度计算 919
纵向下水船架 522
纵向振动 523
纵向正加速度 685
纵向重心（自艉垂线距离） 522
纵斜 718
纵斜比 718
纵摇（俯仰角） 672
纵摇单幅值 Φ 818
纵摇单幅值(°) 989
纵摇调谐因数 921
纵摇幅值 33
纵摇固有频率 586
纵摇固有圆频率 586
纵摇固有周期 587
纵摇角 36
纵摇角 672
纵摇控制传感器 673
纵摇力矩 673
纵摇频率 673
纵摇频率响应函数 359
纵摇应力 673
纵摇运动 673
纵摇周期 908

纵摇周期 T_p 673
纵轴 522
总部国（国际海事卫星组织） 417
总长 498
总长 645
总长 L_{oa} 500
总传动比 905
总代理 374
总吨 396
总吨位 396
总管（主管） 536
总管（主要管路） 536
总管/集管（分配阀箱） 541
总含盐度 905
总机械效率 645
总集合时间 904
总精度 645
总静压头 905
总可变钻井载荷 905
总贸易体系 385
总排水量 905
总排水体积曲线 199
总数 27
总水压头 905
总提供厚度 396
总体 289
总体安装方案 461
总体动态应力分量（主要应力） 388
总体设计 375
总体振动 949
总限额 18
总线（总管,主母线,主要的） 534
总效率（综合效率） 645
总效率 905
总谐波畸变 905
总压力 905
总压头 905
总要求厚度 396
总用泵 385
总灾难转移费用 395

总载荷（总负载） 905
总质量 27
总装锅炉 650
总装机功率 905
总装配 375
总装式燃烧器 401
总装图 375
总装运质量 m_G 396
总纵强度 523
总纵弯曲 522
总纵斜 905
总阻力 905
总阻力系数 905
综合安全评估 352
综合安全评估 361
综合补给舰 957
综合船用推进系统 464
综合导航系统 459
综合调查船 165
综合航行系统 465
综合检索 465
综合桥楼系统 464
综合实验室 165
综合水下作战系统 466
综合卫星导航系统 465
综合性出口加工区 165
综合性原则 165
综合业务数字网 465
走廊 188
走廊和楼梯的净宽 147
走廊试验 188
租船订舱 133
租船合同 132
租船运输 133
租船者 133
租赁合同 495
租赁契约前后检验 629
租期 133
阻碍 428
阻火的(滞燃的, 滞火剂) 332
阻抗复合式消音器 164

阻抗复合型消音器　739
阻力　739
阻力峰　437
阻力峰速度　437
阻力换算修正系数　739
阻力换算修正值　575
阻力系数　739
阻力增额　57
阻力增额分数　739
阻力增额因数　739
阻流阀（阻断阀）　86
阻尼材料　205
阻尼敷层　205
阻尼结构　205
阻尼力　205
阻尼力矩　205
阻尼挠性联轴节　342
阻尼器（缓冲器，减震器）　2
阻尼器　950
阻尼绕组　205
阻尼涂料　205
阻尼约束板　205
阻尼振动　205
阻燃材料　335
阻燃剂　340
阻燃型树脂　332
阻塞　137
阻塞比　86
阻塞边界　137
阻塞条件　198
阻塞效应　86
阻塞修正　86
阻性消音器　739
阻油器（防油器）　622
阻振　950
阻振方法　950
阻振质量　950
阻滞（延迟，减速）　742
组合　18
组合报警　397
组合导航系统　156

组合阀　157
组合活塞　156
组合警报　18
组合空气枪　21
组合立管塔　438
组合气枪控制器　21
组合启动器　157
组合曲轴　99
组合式浮船坞　464
组合式屈曲限制器　463
组合体长度 L_c　497
组合系统　465
组合应力　157
组合支柱　99
组限额　397
组织　642
组装螺旋桨　99
组装式辅机组　650
组装式曲轴　51
钻杆　257
钻机　744
钻井凹槽　258
钻井驳船　257
钻井吃水　257
钻井船　258
钻井定位　257
钻井工况　257
钻井供应船/钻井辅助船　258
钻井甲板　257
钻井可变载荷　942
钻井立管/钻井隔水导管　258
钻井平台　257
钻井平台　257
钻井平台　258
钻井区　257
钻井水舱　258
钻井系统　258
钻井液（泥浆）处理区　257
钻井装置　257
钻孔爆破船　256
钻孔爆破船　748

钻孔生物调查　257
钻模　170
钻探管架（杂志）　531
钻柱补偿器　257
最大安全速度(地效翼船)　558
最大爆发压力　556
最大长度（艇）L_{max}　557
最大长度　313
最大沉浮吃水　559
最大沉浮吃水 d_{max}（m）　558
最大沉深　557
最大沉深　559
最大沉深干舷　559
最大吃水（冰区航行）　557
最大吃水（艇）T_{max}　557
最大吃水　557
最大持续船速　559
最大持续功率　556
最大持续输出功率　556
最大持续转速　613
最大船长　557
最大船宽（巴拿马运河）　556
最大船宽　555
最大倒车速度　556
最大服务航速　558
最大复原力臂角　36
最大工作水深　557
最大工作外压　559
最大工作压力　560
最大功率　930
最大航速　559
最大横距　559
最大横剖面　558
最大横剖面系数　559
最大回转半径　558
最大货物密度　555
最大间隙　556
最大宽度（艇）B_{max}　556
最大宽度　313
最大连续功率　560
最大连续载荷　556

1208

最大排泥高度　557
最大排泥距离　556
最大排水量　557
最大偏移　558
最大起升高度　557
最大容量（最大功率，最大生产率）　556
最大设计吃水 T_{sc}　556
最大设计满载吃水　234
最大设计满载载重线　234
最大升船高度　557
最大升船能力　905
最大收放绳速度　557
最大输送率　559
最大速度　904
最大挖宽　557
最大挖深　557
最大稳定噪声级　559
最大型深 D_{max}　556
最大许用载荷　662
最大许用重心高度　555
最大叶宽　559
最大叶宽比　556
最大营运傅汝德数　558
最大营运前进航速　555
最大营运质量（t）　558
最大营运质量（t）　558
最大营运总质量或额定质量　558
最大允许传送率　555
最大允许启动时间（巴拿马运河）　555
最大允许响应时间　555
最大允许载货质量　558
最大张力　559
最大蒸汽输出　559
最大正常地效速度　558
最大转舵角　406
最大纵距　555
最大总长　557
最大总宽　559
最低安全配员文件　573

最低合理可行　50
最低合理可行范围　50
最低启动压力试验　573
最低屈服强度　526
最低燃烧下限　526
最低日统计的平均温度　527
最低舒适居住条件　570
最低舒适居住条件所需设备　789
最低稳定工作转速　573
最低稳定工作转速试验　573
最低稳定转速　782
最低稳定转速试验　527
最低月统计的平均温度　527
最恶劣状况故障　985
最高表面温度　559
最高人民法院　874
最高人民检察院　874
最后分路　324
最坏情况后果　985
最佳环流分布　641
最佳直径　641
最佳转速　641
最接近点　148
最近一次压载航行的排油监控系统记录　724
最轻载航行状态　505
最深分舱水线（分舱载重线）　218
最深分舱载重线　218
最深作业水线（渔船）　218
最小吃水（冰区航行）　571
最小吃水 T_{min}　571
最小尺寸　571
最小船舶操纵船速　573
最小船首高度　570
最小方差更新　573
最小工作水深　571
最小化　570
最小化传输差分信号　914
最小间隙　570
最小破断强度　570
最小全速前进船速（巴拿马运河）

572
最小设计压载吃水　571
最小绳速　193
最小收、放绳速度　572
最小舯吃水　571
最小挖深　571
最小压载吃水　570
最小正常地效速度　572
最严重冰况　313
最严重冰况区域航行　443
最终安全地点　930
最终裁决/最后裁决书　324
最终极限状态　930
最终记录介质　324
最终检查　324
最终进场/起飞区域（FATO）　324
最终试验　325
最终状态　324
最重要的 IMO 文件　579
左机减速齿轮箱　684
左螺纹　496
左舷　684
左舷主机（左主机）　683
左向螺母　496
左旋　496
左旋发动机　496
左旋螺旋桨　496
左转机组　496
作动筒张力器　439
作为标准（统一，使合标准）　852
作业场所　636
作业吃水 d_D（m）　635
作业废弃物　638
作业工况　635
作业模式（平台）　575
作业桥楼　637
作业桥楼功能　637
作业系数（起重设备）　635
作业载荷　638
作用力　8
作用效应　8

坐标参照系统　685
坐标测量机　186
坐标定位拖车试验池　891
坐底式平台　935
坐底式平台的次要构件　774
坐底式平台的特殊构件　835
坐底式平台的主要构件　693
坐底式钻井平台　866
坐底式钻井平台　866
坐底式钻井平台　866
坐底作业船底加强　91
坐坞强度　248
坐在座位上的值班人员可及的范围　980
座板　88
座垫　773

数字及其他

（DASR）拆船厂授权书　248
（NEC NO_x）　氮氧化物排放控制　609
（NOSS）　591
（Tayloe's design charts）B-δ 式设计图谱　103
（爆炸品）独立库房（独立式仓库）　450
（爆炸品）箱形库房（箱式仓库）　531
（爆炸品）整体库房（整体式仓库）　463
（爆炸性气体环境的）点火温度　445
（被动）减摇水舱　347
（德国）海事局　102
（德国）邮政与电信管理局　888
（对干扰的）抗扰度　467
（多极开关电器的）燃弧时间　46
（浮船坞）轻载排水量　503
（浮船坞）重大改装　815
（管端的）弯筒结合　319
（锅炉）坝（矮墙）　326
（锅炉）生火　334
（锅炉的）前管板　360
（锅炉炉门）防火筛（挡火屏）　333
（过电流或过载继电器或脱扣器的）电流整定值　199
（过电流或过载继电器或脱扣器的）电流整定值范围　199
（过电流继电器和脱扣器的）动作电流　635
（机械开关电器的）断开时间　633
（机械开关电器的）脱扣器　732
（机械式）断路器　137
（开关电器的）接通能力　539
（开关电器的或熔断器的）分断电流　95
（螺旋桨）鸣音振动　817
（螺旋桨）锁定试验　338
（美国）1990 年防止和控制外来水生物污染法令　604
（美国）国土安全部　236
（平均要求）危险失效概率　314
（平均要求）危险失效概率　696
（燃油）喷射角（喷射持续角）　458
（示功图）膨胀曲线　309
（水蒸气）焓熵图　576
（危害的）清单分析　133
（一极或熔断器的）燃弧时间　46
（蒸汽机）冲压缸　468
（蒸汽机）排气余面　307
（蒸汽机）旁通阀　468
（蒸汽轮机）低压缸　527
0 类危险区（平台）　411
0 类危险区域（浮油回收船）　414
0 类危险区域　409
0 区　989
1/3 倍频程螺旋桨频带声压级　631
1/3 倍频程声压谱（密度）级　631
10°/10°Z 形操纵试验　990
10 万吨级航空母舰（拥有数量/制造能力）　989
12 海里区域　923

1946 年纽约议定书　597
1973/1978 年国际防止船舶造成污染公约，经常修正，并由国际海事组织出版。　472
1974/1978 国际海上人命安全公约，经常修正，并由国际海事组织出版　472
1983 年国际运输公约法　479
1994 年高速船规则　427
1997 年 NO_x 技术规则　609
1-1 类船舶　120
1-2 类船舶　120
1-3 类船舶　120
1-D 号燃料油　362
1 号燃料油（煤油）　362
1 级动力定位系统　264
1 级锰青铜　390
1 级压力容器（PV-1）　139
1 类航区　713
1 类航区　787
1 类释放源　831
1 类危险区（平台）　411
1 类危险区域（浮油回收船）　414
1 类危险区域　409
1 区　989
1 型船舶　925
2000 年高速船规则　427
2004 年船舶压载水和沉积物控制和管理国际公约　472
2008 年 NO_x 技术规则　609
2009 年香港国际安全与环境无害化拆船公约　430
20°/20°Z 形操纵试验　990
20 国集团　397
2G 型液化气船　929
2PG 型液化气船　929
2-D 号燃料油　362
2 号燃料油　362
2 级动力定位系统　264
2 级镍锰青铜　390
2 级压力容器（PV-2）　139

2类航区　713
2类航区　787
2类释放源　831
2类危险区（平台）　411
2类危险区域　410
2区　989
2型船舶　925
3D打印　896
3G型液化气船　929
3级动力定位系统　264
3级镍锰青铜　390
3级压力容器（PV-3）　139
3类航区　713
3类航区　787
3型船舶　925
40英尺集装箱　352
4号燃料油　362
4级锰铝青铜　390
5号燃料油　362
6号燃料油　362
73/78防污公约　472
［德国］海上试验风场与基础设施公司　387
［美］陆军工程兵部队　48
［美］保险商实验室　933
［美］超微电脑股份有限公司　14
［美］腐蚀工程师协会　585
［美］国防腐蚀标准化技术委员会　886
［美］国际贸易委员会　929
［美］国际商业机器公司　471
［美］国家标准与技术研究院　585
［美］国家档案与资料管理局　585
［美］国家导弹防御系统　585
［美］国家电视标准委员会　586
［美］国家港湾搜索预备队　592
［美］国家公园管理局　610
［美］国家海洋和大气管理局　584
［美］国家海洋和大气局　599
［美］国家环保局　585
［美］国家历史保护法　585

［美］国家排污限制系统　610
［美］国家有害入侵物种法　599
［美］海岸带管理法　150
［美］核管理委员会　612
［美］科学信息研究所　462
［美］联邦航空管理局　319
［美］联邦航空委员会　319
［美］联邦环境保护局　291
［美］联邦贸易委员会　319
［美］联邦食品、药品及化妆品法　319
［美］联邦政府法规　152
［美］联邦政府行政法规汇编/美国联邦政府法规　131
［美］联邦咨询委员会　319
［美］全国科学研究委员会　585
［美］全国咨询中心　585
［美］数字设备公司　238
［美］污染控制排放系统　585
［美］运输部海事局　936
［美］政治军事局　102
［美国］联邦政府水污染控制法令　320
［美国］浓污水处理法令　828
［美国］有害入侵物种法　585
［日］新能源产业技术综合开发机构　596
［英］纺织品协定　97
［英］国家安全局　585
［英］政府通信信息部　390
［中］对外经济贸易仲裁委员会　351
［中］西北工业大学　608
A/B级标准耐火试验　1
A1海区　769
A2海区　769
A3海区　769
A4海区　769
AMD公司的加速并行处理技术　3
ATM无线接入通信系统　54
Authorware公司的声音文件　810

A-0级防火分隔　140
A-15级防火分隔　140
A-30级防火分隔　140
A-60标准　1
A-60级防火分隔　140
A-60级耐火分隔　1
A标准船站　2
A工艺　887
A级防火分隔　140
A级分隔　1
A级易燃气体　1
A类地效翼船　926
A类地效翼船　977
A类断路器　137
A类浮油回收船　623
A类机器处所　531
A类机械场所　121
A类警报　121
A类客船　121
A类客船　656
A类培训　121
A类游艇　122
A声级　1
A型吊架　1
A型独立液舱　55
A型独立液货舱　925
A型独立液货舱　929
A型舵　925
A型干舷船舶　924
A型干舷船舶和B型干舷船舶　924
A型公共处所　925
A型艉柱　56
A型索具系统　925
A组货物　397
B&W型给水调节器　66
B1*冰级标志　65
B1冰级标志　65
B2冰级标志　65
B3冰级标志　66
Bauschinger效应　73

BC-A 型散货船　74
BC-B 型散货船　74
BC-C 型散货船　74
b_f　81
b_p　93
B_{WL}线　103
B-0 级防火分隔　140
B-100 型船舶　927
B-15 级防火分隔　140
B-60 型船舶　927
B 标准船站（卫星通信）　65
B 级防火门　65
B 级分隔　65
B 级易燃气体　65
B 类地效翼船　927
B 类地效翼船　977
B 类断路器　137
B 类浮油回收船　623
B 类警报　122
B 类客船　122
B 类客船　656
B 类培训　122
B 类游艇　123
B 声级　65
B 型船舶　926
B 型独立液货舱　926
B 型独立液货舱　929
B 型舵　926
B 型干舷船舶　925
B 型公共处所　926
B 型球形独立液舱　97
B 型索具系统　926
B 组货物　397
C　104
CANREP 系统　107
CA 区域　104
CDT 绞车　126
CF 卡　161
CKYH 联盟　138
Clo 值（救生服）　147
CMA 船　149

C_P-C_{Th} 式设计图谱　192
C 标准船站（卫星通信）　104
C 级分隔　104
C 级易燃气体　104
C 类地效翼船　927
C 类警报　123
C 类培训　123
C 类游艇　123
C 盘　104
C 区域（螺旋桨）　104
C 声级　104
C 型独立液货舱　929
C 型舵　927
C 型公共处所　927
C 型索具系统　927
C 型圆柱形的独立液舱　198
C 组货物　397
D_F 最小型深　495
DMT　247
DOC 证书　248
DP 近乎失去　254
DP-1 船舶　254
DP-2 船舶　254
DP-3 船舶　254
DP 备份　66
DP 不希望事故　254
DP 船舶　254
DP 控制位置　253
DP 入级附加标志　253
DP 停工期　254
DP 危险观察　254
DVI 端子/数字视频接口　238
D-1 标准　730
D-2 标准　730
D 或 D 值　203
D 级可燃液体　203
D 类游艇　123
D 盘　203
D 声级　203
D 型舵　927
D 型公共处所　927

EAN 码　299
EGC 记录簿　270
EIAPP 证书　271
ETM"方案 B"　299
ETM"方案 A"　299
E 级可燃液体　266
E 型舵　927
FMEA 报告　348
FSS 规则　361
FTP 规则　361
F-N 曲线　348
F 标准船站（卫星通信）　315
F 级第Ⅰ法　566
F 级第Ⅱ法　567
F 级第Ⅲ法　567
F 级分隔　314
G6 联盟　367
GM　389
HCS-5 型船舶运动阻截系统　437
H 级标准耐火试验　403
H 级分隔　402
H 型内燃机　434
IAPP 证书　442
IAPP 证书　469
IC 法（货船）　565
IC 法，IIC 法和 IIIC 法　565
IC 法　565
IG 型液化气船　929
IHS 设计　464
IIC 法（货船）　566
IIC 法　566
IIC 法和 IIIC 法　566
IIIC 法（货船）　567
IIIC 法　567
IMDG 规则　446
IMO 海事安全委员会　446
IMO 航海安全委员会　446
IMO 要求　446
IMSBC 规则　447
INF 1 级船舶　139
INF 2 级船舶　139

INF 3 级船舶　139

INF 货物　454

ISO 货运集装箱　485

ISPS 规则　485

IS 规则　485

JP-1（煤油）喷气燃料　487

JP-3 喷气燃料　487

JP-4 喷气燃料　487

JP-5（煤油,重质）喷气燃料　487

J 式敷管法　487

J 形管路　487

K-J 式设计图谱　489

LPG 系统　528

LPG 运输船　528

LRIT 信息　528

LR 形式认可　528

LR 形式认可体系　528

LUF 系统　529

L_{WL} 线　530

L 声级　489

L 型吊架　489

MAFI 拖车　531

MARPOL 油船　551

MDAT 的月算术平均值　540

MODU 规则　576

M 标准船站（卫星通信）　530

M 工艺　887

n 阶谱矩　611

NAVTEX 接收机　591

NLS 证书　599

NS1 类舰船　610

NS2 类舰船　611

NS3 类舰船　611

NSA 类舰船　611

OLEO 载荷　628

OMBO 船舶　629

P3 战略航运联盟　650

pH 值　666

ppm 显示器　688

QQ 邮箱　710

ShipRight 标志　808

SOLAS A 型封堵器　827

SO_x 排放控制区符合计划　767

SO_x 排放控制区　774

SO_x 排放控制区符合证书　766

S 工艺　887

S 式敷管法　821

t_c 腐蚀增量（mm）　886

t_f　893

T 工艺　887

T 形材（三通管,T 形接头）　888

T 形三通管接头　888

T 型油船　903

UMA 船　931

UNIX 操作系统　936

UPS 装置　938

U 形管　929

U 形管压力计　929

U 形减摇水舱　929

U 形螺栓　929

U 形弯管　929

U 型管减摇水舱　929

U 型内燃机　939

U 型剖面　939

V 形槽　943

V 形传动　942

V 形发动机　940

V 形气缸体　940

V 形起重柱　957

V 形缺口试样　940

V 型内燃机　957

V 型剖面　957

V 型缺口冲击试验　940

V 型压缩机　957

WiFi 充电　132

WWR 条件　973

W 型内燃机　986

X 波段　986

X 波段超级雷达（探测 6 000 km）　986

X 波段雷达　986

X 驳船　986

X 船　986

X 光检查设备　987

X 类有毒液体物质　124

X 类有毒液体物质　610

x 向加速度 a_x　986

X 型船舶　986

X 型内燃机　987

Y 类有毒液体物质　124

Y 类有毒液体物质　610

y 向加速度 a_y　987

Y 形接头　987

Y 形弯头　987

Y 型节点　987

Y 轴（Y 轴线,纵坐标轴）　987

Z_{AB}　988

Z_{AD}　988

Z 类有毒液体物质　124

Z 类有毒液体物质　610

z 向加速度 a_z　988

Z 向推进系统　989

Z 形操纵试验　989

Z 形操纵周期　902

Z 形传动　988

"1973/1978 国际防止船舶造成污染公约"　551

"A"型船舶　925

"B+"型船舶　925

"B-100","B-60"型船舶　925

"d"隔爆型（电气设备）　340

"e"增安型（电气设备）　449

"i"本质安全电路　480

"m"密封型（电气设备）　281

"n"防护型（无火花型）　704

"n"防护型电气设备　273

"o"油浸型（电气设备）　622

"p"正压型（电气设备）　692

"q"充砂型（电气设备）　762

"百万分之一"　655

"贝来"给水调节器　67

"博客"　86

"超空泡"鱼雷　871

索引

"冲撞式"破冰法 155
"耳麦" 417
"服务型"制造 789
"国际安全管理（ISM）规则" 477
"国际船舶和港口设施保安（ISPS）规则" 478
"国际气体运输船规则（IGC规则）" 473
"国际散装危险化学品规则（IBC规则）" 471
"宽频"系统 973
"蓝牙"传输 87
"蓝牙"耳机 87
"蓝牙"技术 87
"连续式"破冰法 178
"耐火试验程序规则" 334
"室" 433
"搜狗" 826

"搜狗"搜索引擎 826
"蒜靶"软件 259
"腾讯"即时聊天工具 889
"腾讯"搜索网站 828
"微信" 969
"鹰击18" 266
"鱼鹰"运输机 643
"月女神"主动声呐系统 50
"云" 148
"云计算" 148
"窄频"系统 585
《国际鸦片公约》 477
《国际有害物质清单证书》初次检验 457
《国际有害物质清单证书》附加检验 12
《国际有害物质清单证书》换证检验 735

《国际有害物质清单证书》最终检验 324
《加快基础设施规划法案》 456
《控制危险废料越境转移及其处置巴塞尔公约》 73
《联合国气候变化框架公约》 935
《消防安全系统规则》 333
Ⅰ级绿色船舶 394
Ⅰ类船舶 123
Ⅰ型船用卫生装置 927
Ⅱ级绿色船舶 394
Ⅱ类船舶 121
Ⅱ类船舶 124
Ⅱ型船用卫生装置 928
Ⅲ级绿色船舶 394
Ⅲ型船用卫生装置 928
▽型内燃机 222

后　　记

《英汉船舶及海洋工程技术大辞典》(第 2 版)由船舶及海洋工程领域内众多知名专家在第 1 版的基础上,历时四年精心修订编辑而成,现在终于正式与读者见面了。在出版过程中,哈尔滨工程大学出版社领导非常关注和重视辞典的审校和编辑工作,多次与主编交换意见,并指定专人负责落实和解决有关问题。值此出版之际,向哈尔滨工程大学出版社表示诚挚的谢意。

另外,特别需要强调的是,本辞典的编委会学术阵容强大,各成员单位的大力支持是辞典顺利出版的基础保障。编者谨向中国船舶及海洋工程设计研究院、中国造船工程学会、上海交通大学、上海市船舶与海洋工程学会、上海研途船舶海事技术有限公司、武昌船舶重工集团有限公司、哈尔滨工程大学、江苏科技大学、中国海洋大学、天海融合防务装备技术股份有限公司等全体编委会成员单位表示诚挚的谢意。

船舶及海洋工程涉及的专业范围广,发展日新月异,加之编者学识水平限制,辞典在收录的准确性和及时性方面恐有不完善之处,期望广大读者提出宝贵意见,以便再版时改进。

编　者

2021 年 3 月于上海